W. Tang G. Eisenbrand

Chinese Drugs of Plant Origin

Chemistry, Pharmacology, and Use in Traditional and Modern Medicine

With 41 Figures

Springer-Verlag
Berlin Heidelberg New York
London Paris Tokyo
Hong Kong Barcelona
Budapest

Professor Dr. WEICI TANG
Professor Dr. GERHARD EISENBRAND

Lebensmittelchemie und Umwelttoxikologie,
Universität Kaiserslautern,
Erwin-Schrödinger-Straße,
D-6750 Kaiserslautern

ISBN-13: 978-3-642-73741-1 e-ISBN-13: 978-3-642-73739-8
DOI: 10.1007/978-3-642-73739-8

Softcover reprint of the hardcover 1st edition 1992

27/3145-543210 Printed on acid-free paper

Preface

Traditional Chinese medicine has been used for thousands of years by a large population. It is currently still serving many of the health needs of the Chinese people; and still enjoying their confidence it is practised in China in parallel with modern Western medical treatment. In addition to scientific organisations dedicated to modern Western medicine, e. g. the Chinese Academy of Medical Sciences and various medical schools, a series of parallel institutions have been established in China to promote traditional Chinese medicine, such as the Academy of Traditional Chinese Medicine and training institutions. Almost all hospitals in China have a department of traditional medicine. Furthermore, a large number of scientific journals are dedicated to traditional Chinese medicine, covering both experimental and clinical investigations.

Medicinal materials constitute a key topic in the treatment of disease according to traditional Chinese medicine. The Chinese Pharmacopoeia (1985 edition) is therefore divided into two separate volumes, Volume I containing traditional Chinese medicinal materials and preparations and Volume II containing pharmaceutics of Western medicine.

The oldest Chinese review of medicinal materials, Shennong Bencao Jing (100–200 A. D.), covered 365 herbal drugs. The classic compilation in this field, Bencao Gangmu (Compendium of Materia Medica), was published in 1578 by Li Shi-zhen and recorded as many as 1898 crude drugs of plant, animal and mineral origin.

About 50 years ago modern chemical and pharmacological methods were first used to investigate traditional medicinal materials in China. With the growth in our knowledge about chemistry, biochemistry, physiology, and pharmaceutics and the progress made in scientific instrumentation, there have been an increasing number of scientific reports characterizing the biological activities of the chemical constituents of Chinese medicinal materials. The majority of these papers have been in Chinese, scattered in a series of journals not easily available outside China.

The present book provides information on recent advances and perspectives for future research into Chinese medicinal materials, including a large number of reports from Chinese journals. It is hoped that this information may be of value for the development of new drugs and may stimulate further investigations. Since traditional Chinese medicine has its own theoretical system that is rather different from modern pharmacological science, we hope that this book may serve as a bridge between traditional Chinese medicine and modern Western medicine.

Our descriptions focus on the chemical constituents and the most important biological activities. Toxicological aspects, especially those relating to mutagenic and carcinogenic activities, are also given consideration.

Within this limited survey it was not possible to cover all the Chinese medicinal materials used in traditional and folk medicine. Altogether, more than 500 species and subspecies from about 130 genera and about 3000 chemical constituents have been described. Most subjects represent official drugs of plant origin selected from the Chinese Pharmacopoeia, Vol. I (1985). For completeness, the total list of official plant species in the Chinese Pharmacopoeia (1985) and in the appendix to it have been included. On the other hand, some unofficial plants with biologically active ingredients have also been considered. Examples are *Anisodus tanguticus* with anticholinergic alkaloids and drugs of potential use in the treatment of cancer, such as *Camptotheca acuminata, Cephalotaxus* spp., *Rabdosia* spp., and *Tripterygium wilfordii.*

Different plant species within the same genus are generally treated as one item. Exceptions are the genera *Artemisia, Panax,* and *Sophora.*

Some items from Chinese folk medicine have not been included in this edition because of limited space. Examples are *Ganoderma lucidum, Gossypium herbaceum, Huperzia serrata,* and *Zanthoxylum* species. They might be included in a later edition.

Most medicinal materials contain a large variety of known or still unknown compounds. Traditional Chinese medicine prefers modalities characterized by a combination of numerous individual materials, sometimes up to a hundred or more. It is obvious that the interaction between individual materials might be of considerable relevance for the biological effectiveness of these combinations. Studies on the chemical constituents and pharmacological activities of combined preparations have been very scarce up to now and an appropriate discussion of the effects of such combinations is therefore not possible at the present time.

Kaiserslautern, January 1992

W. Tang
G. Eisenbrand

Contents

Acanthopanax senticosus (Rupr. et Maxim.) Harms

1

1.1 Introduction

Ciwujia, Radix Acanthopanacis senticosi, is the dry root and rootstock of *Acanthopanax (Eleutherococcus) senticosus* (Rupr. et Maxim.) Harms (Araliaceae), which is collected in spring and fall. It is listed officially in the Chinese Pharmacopoeia and commonly known as "Siberian ginseng". It belongs to the same plant family as *Panax ginseng*. In addition, two galenic preparations of *A. senticosus* are also included in the Chinese Pharmacopoeia:

- Ciwujia Jingao, Extractum Acanthopanacis senticosi, prepared by extraction of the powdered root of *A. senticosus* with 75% ethanol and concentration of the extract
- Ciwujia Pian, Tabellae Acanthopanacis senticosi, prepared from the extract

The roots and rootstock of *A. senticosus* and its preparations have been used as a tonic in Chinese traditional medicine for a long time.

1.2 Chemical Constituents

From the roots and stems of *A. senticosus* collected in China, isofraxidin (*1-1*), sesamin (*1-2*), β-sitosterol (*1-3*), friedelin (*1-4*), and several polysaccharides have been isolated in addition to eleutherosides A, B (*1-8*), B_1 (*1-9*), C, D, E, I, K, L, and M [1]. The eleutherosides I, K, L, and M have also been isolated from the leaves of *A. senticosus* [2].

Isofraxidin is a derivative of coumarin, the lactone of coumarinic acid; sesamin is a lignan derivative; and β-sitosterol, a widely distributed plant sterol, has a stigmastane (*1-5*) carbon skeleton, whereas friedelin belongs to triterpenes derived from D:A-friedooleanane (*1-6*).

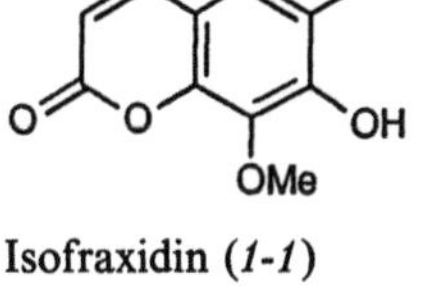

Isofraxidin (*1-1*)

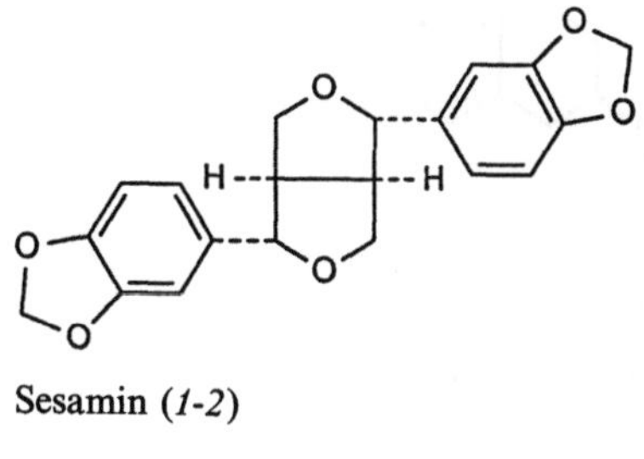

Sesamin (*1-2*)

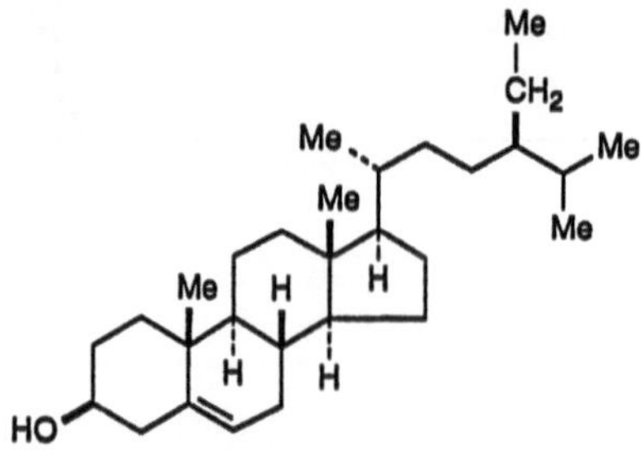

β-Sitosterol (*1-3*)

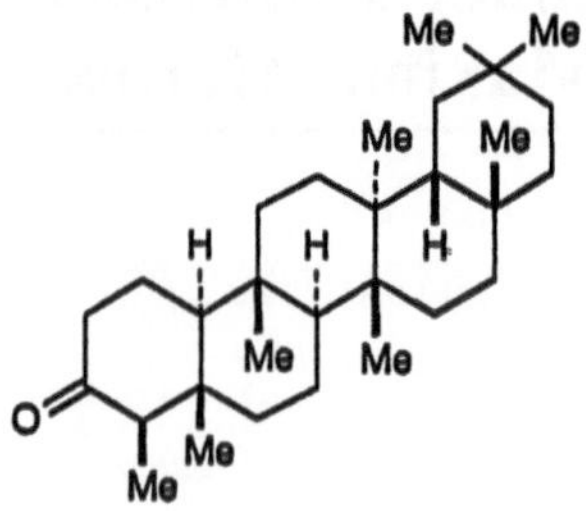

Friedelin (*1-4*)

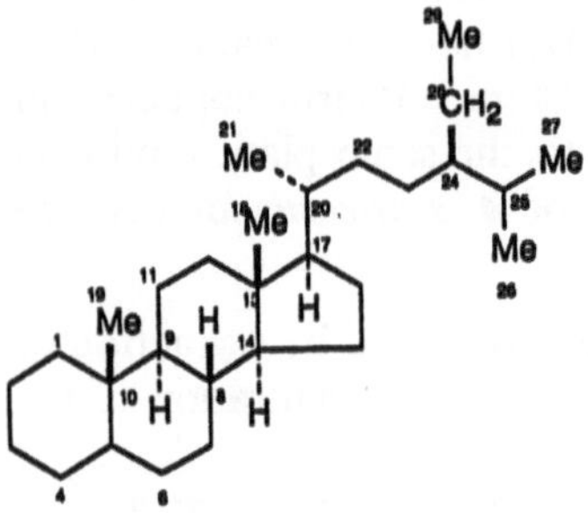

Stigmastane (*1-5*)

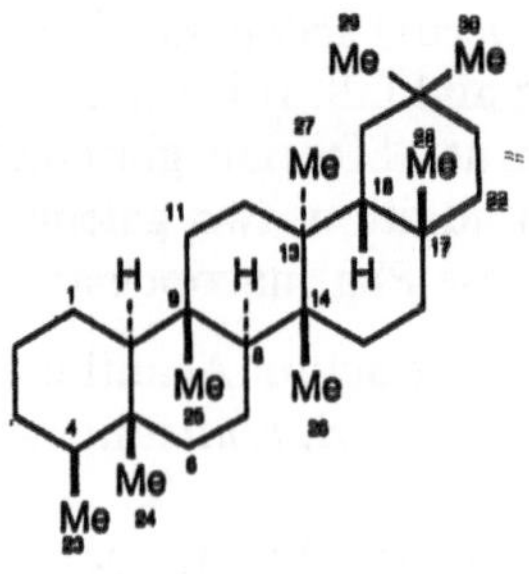

D:A-Friedooleanane (*1-6*)

The eleutherosides are glycosides with different aglycones. Thus, eleutheroside A (*1-7*) is a steroid glycoside with β-sitosterol as the aglycone; eutherosides I (*1-13*), K (*1-14*), L (*1-15*), and M (*1-16*) are triterpene glycosides with oleanolic acid as the aglycone; and eleutherosides D (*1-11*) and E (*1-12*) are epimeric syringaresinol diglycosides. The other eleutherosides are glycosides with simple aglycones. The most simple eleutheroside is eleutheroside C (*1-10*), which is ethyl α-D-galactopyranoside. Eleutheroside B is identical to syringin.

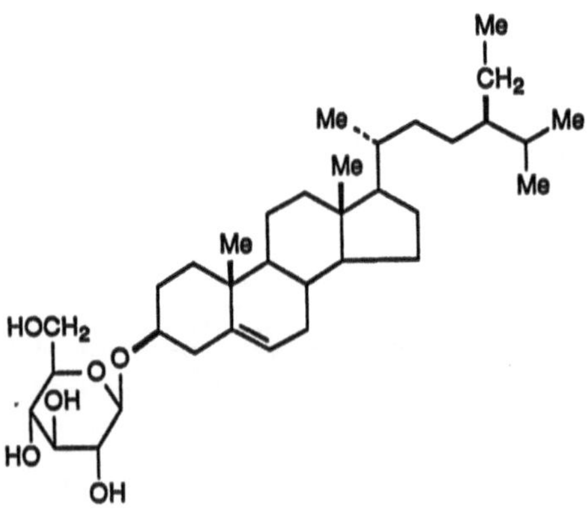

Eleutheroside A (*1-7*)

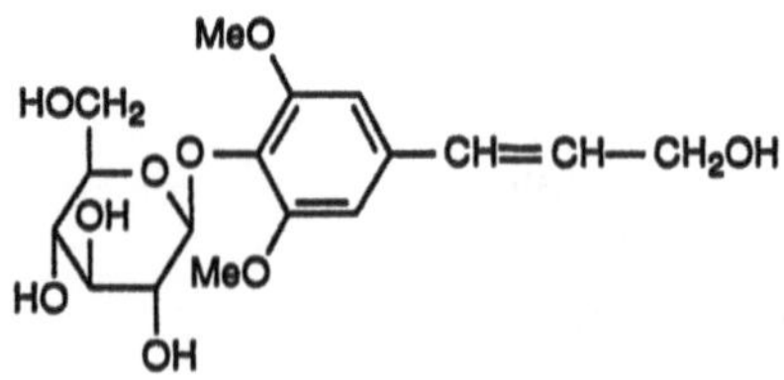

Eleutheroside B (*1-8*)

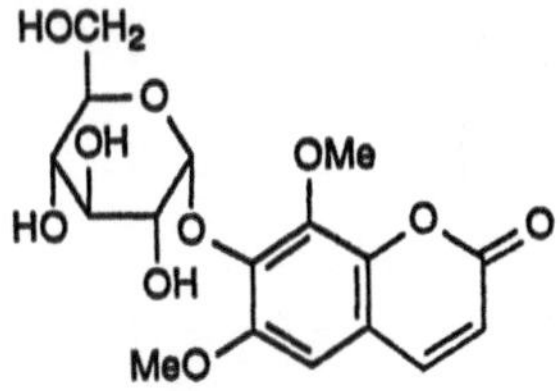

Eleutheroside B_1 (*1-9*)

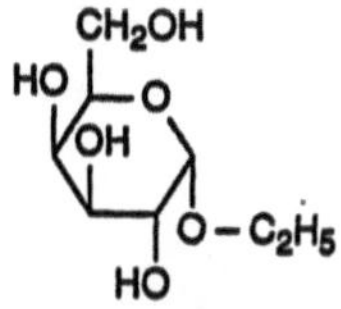

Eleutheroside C (*1-10*)

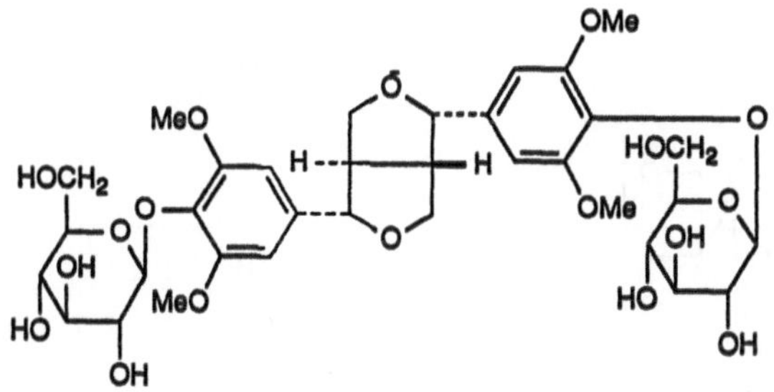

Eleutheroside D (*1-11*)

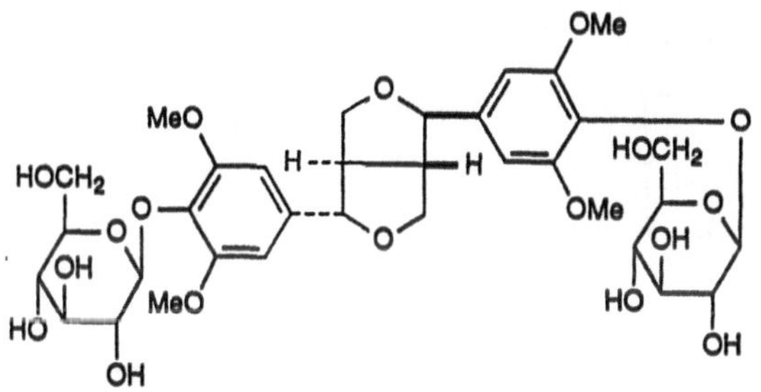

Eleutheroside E (*1-12*)

Recently, the isolation and structure determination of a series of minor saponins named ciwujianosides A_1, A_2, A_3, A_4, B, C_1, C_2, C_3, C_4, D_1, D_2, D_3, and E from the leaves of *A. senticosus* have been further reported. Like the eleutherosides, ciwujianosides A_1 (*1-17*), C_3 (*1-18*), C_4 (*1-19*) and D_1 (*1-20*) have oleanolic acid as aglycone. Ciwujianosides A_2 (*1-21*), B (*1-22*), C_1 (*1-23*), C_2 (*1-24*), D_2 (*1-25*), and E (*1-26*) have been found to possess 30-norolean-12,20(29)-dien-28-oic acid as aglycone, whereas the aglycone for ciwujianosides A_3 (*1-27*), A_4 (*1-28*), and D_3 (*1-29*) has been found to be mesembryanthemoidigenic acid [28, 29].

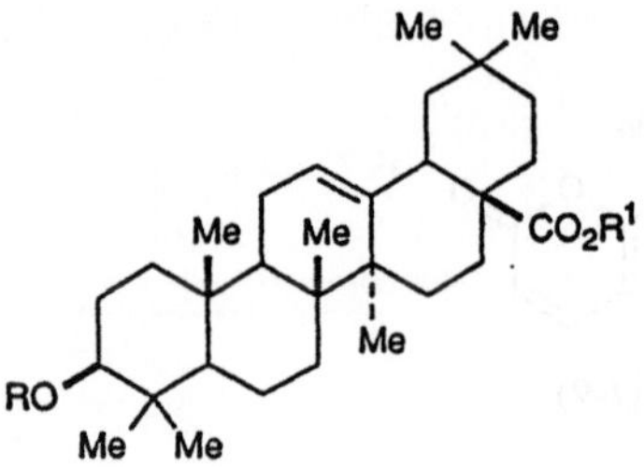

	R	R1
Eleutheroside I (*1-13*)	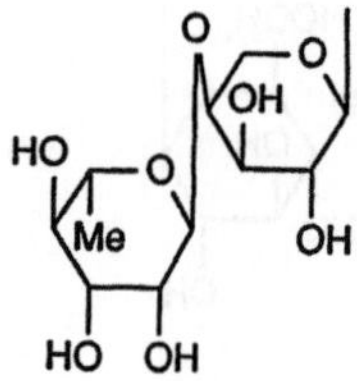	H
Eleutheroside K (*1-14*)	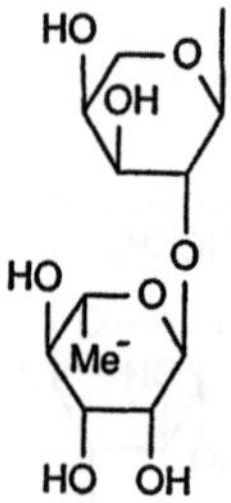	H
Eleutheroside L (*1-15*)	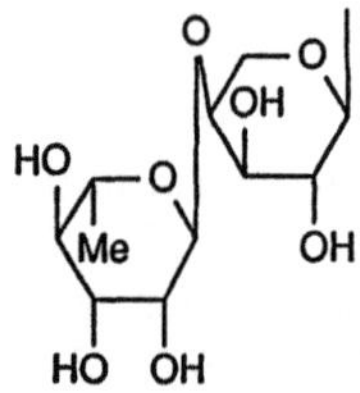	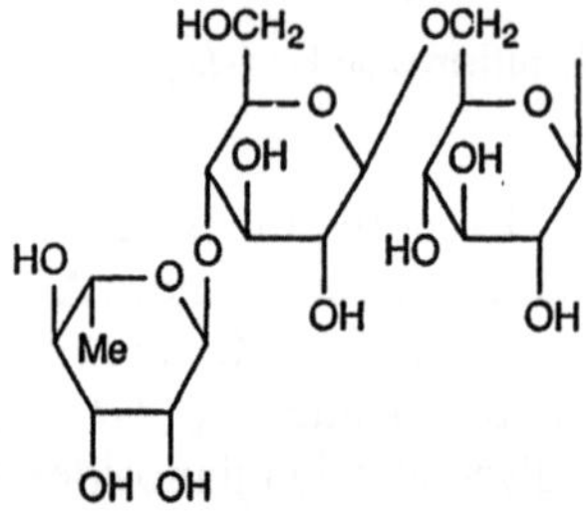
Eleutheroside M (*1-16*)	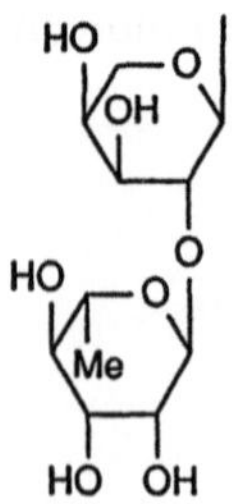	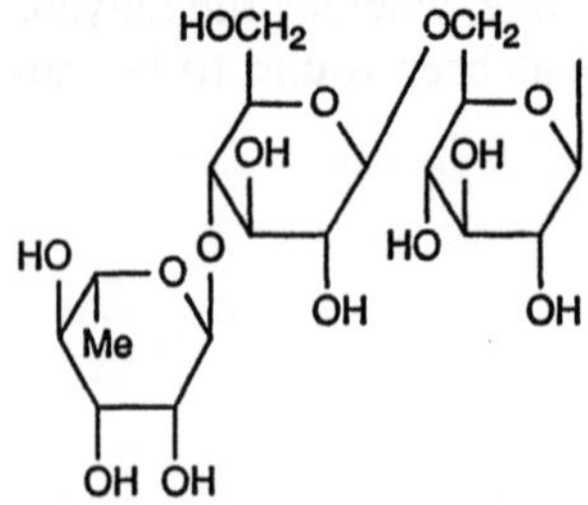

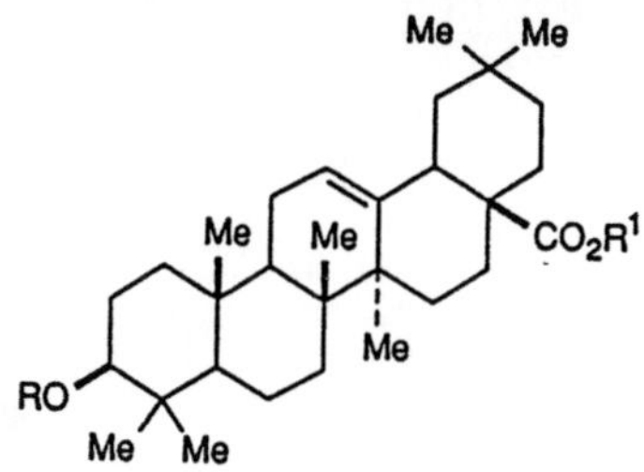

R R¹

Ciwujianoside A_1 (*1-17*)

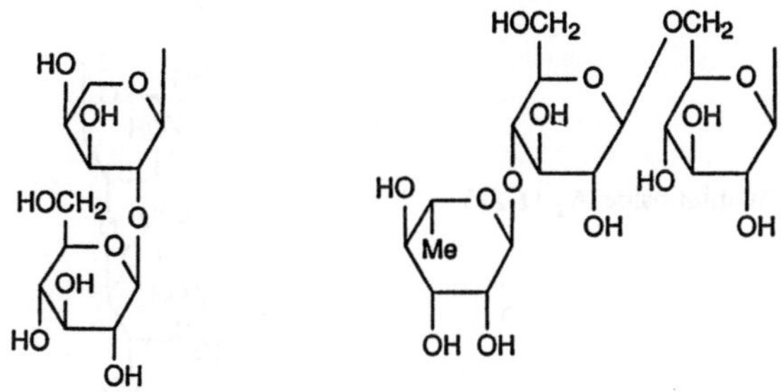

Ciwujianoside C_3 (*1-18*)

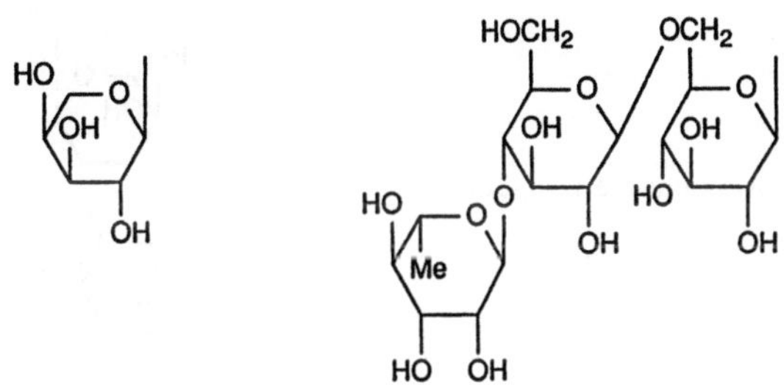

Ciwujianoside C_4 (*1-19*)

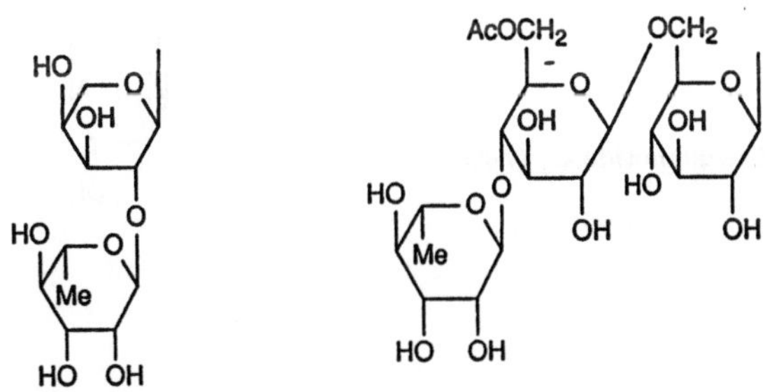

Ciwujianoside D_1 (*1-20*)

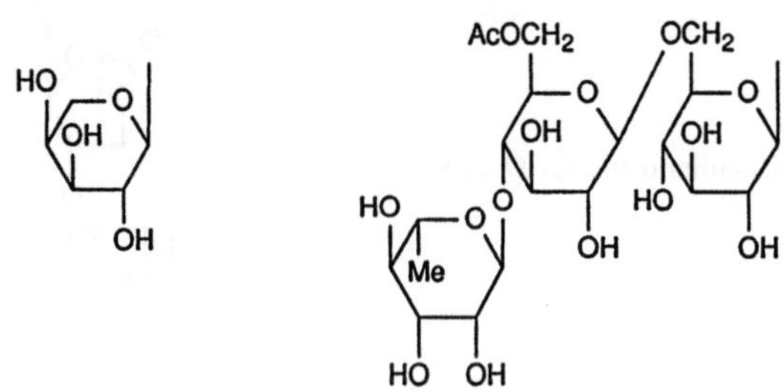

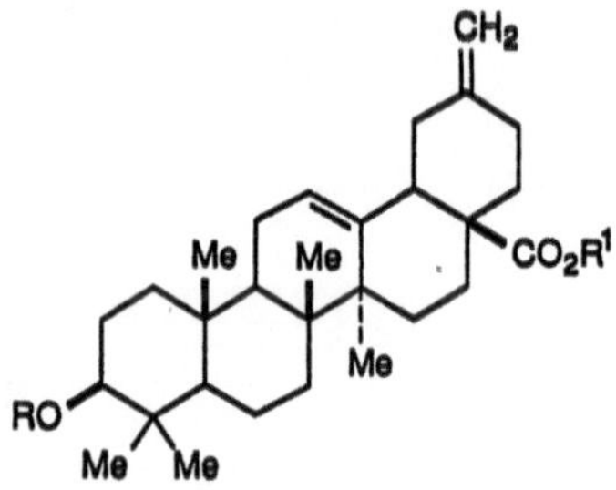

R | R[1]

Ciwujianoside A_2 (*1-21*)

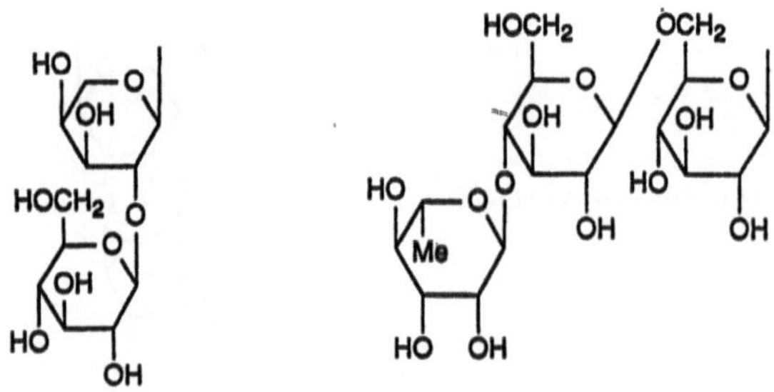

Ciwujianoside B (*1-22*)

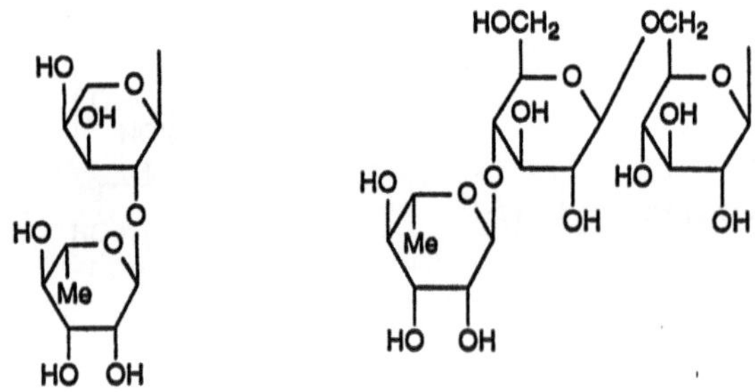

Ciwujianoside C_1 (*1-23*)

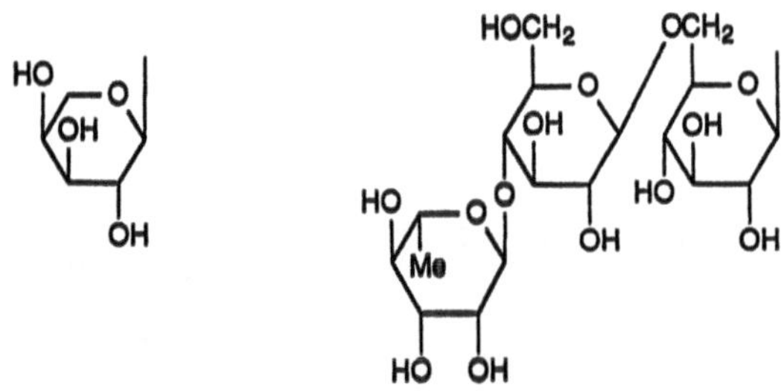

Ciwujianoside C_2 (*1-24*)

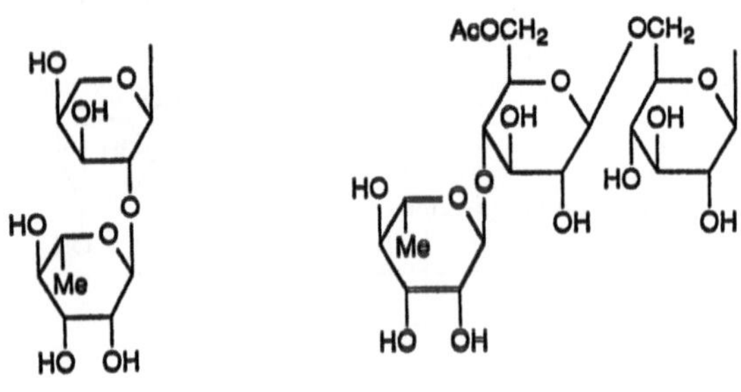

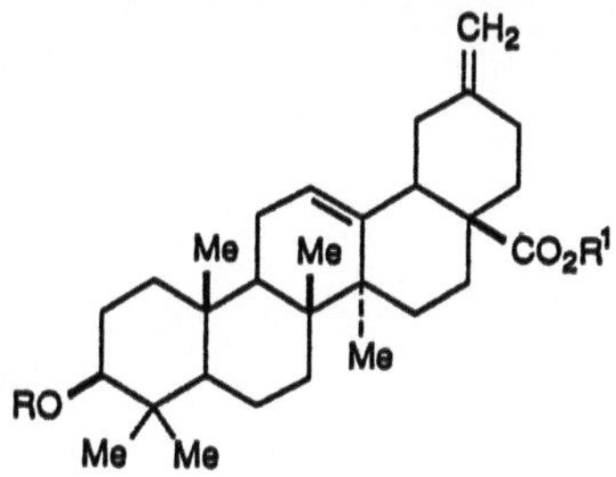

	R	R^1
Ciwujianoside D_2 (*1-25*)	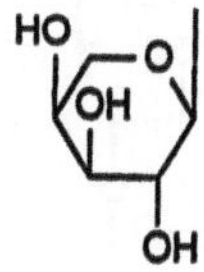	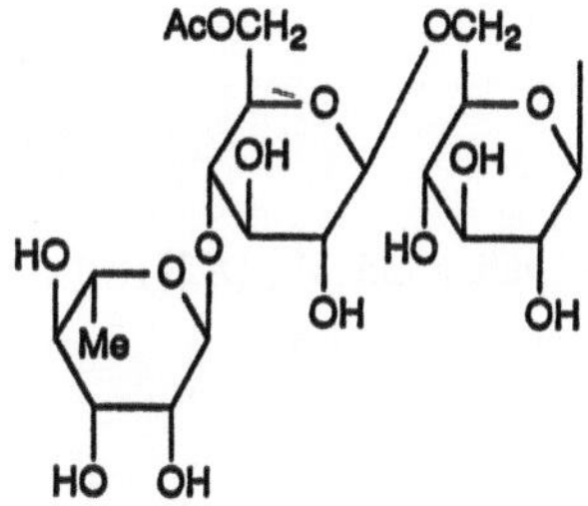
Ciwujianoside E (*1-26*)	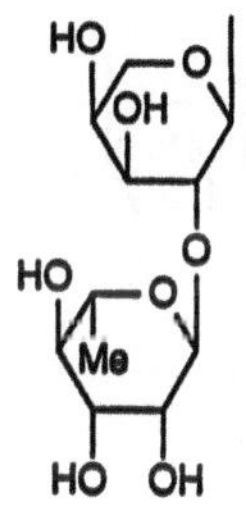	H

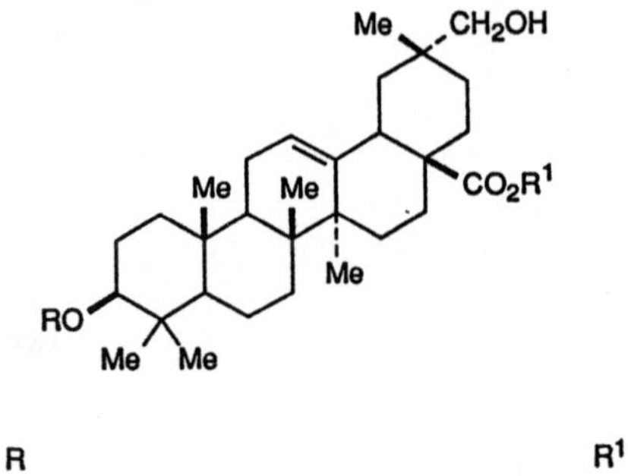

Ciwujianoside A_3 (*1-27*)

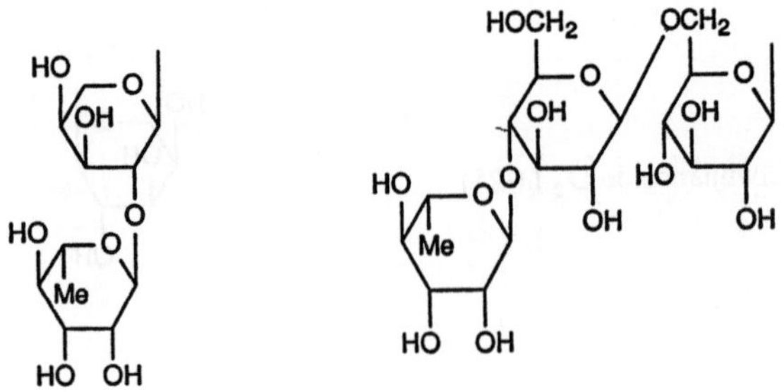

Ciwujianoside A_4 (*1-28*)

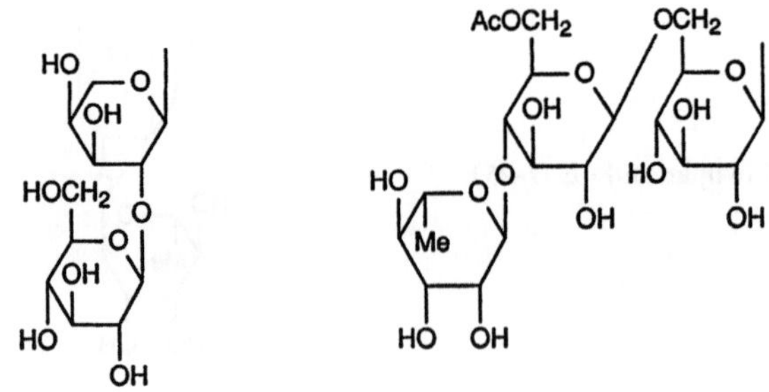

Ciwujianoside D_3 (*1-29*)

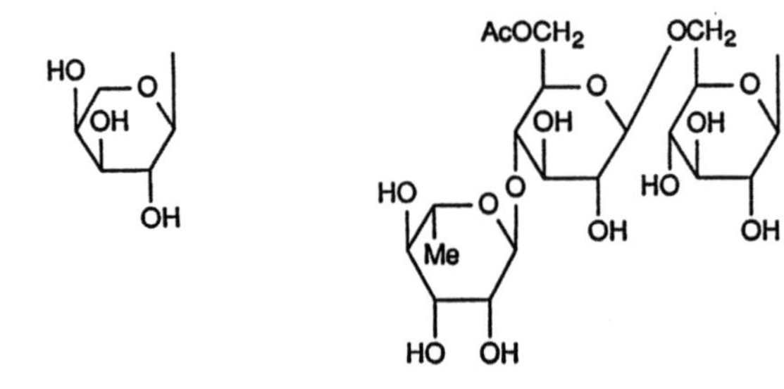

Furthermore, the isolation of 3,4-dihydroxybenzoic acid [3] and a number of glycans referred to as eleutherans A, B,C, D, E, F, and G [4] have also been reported.

The amount of syringin, isofraxidin, and total flavones were highest in the root, rhizome, and stem. Syringin and isofraxidin were not found in the leaf and fruit, whereas a large amount of total flavone was present in the leaf. The amounts of the above mentioned constituents in the root, rhizome, and stem were highest during May and October and lowest in July. The amounts were also different in *A. senticosus* collected from various geological regions [5]. In the root and rhizome of *A. senticosus* a syringin yield of 0.03% was obtained [6]. The syringin content in powdered roots or rhizomes decreased to 50% of the original value after 12 months storage and could no longer be detected after 3 years of storage under regular conditions [7].

1.3 Pharmacology

The water extract of *A. senticosus* prevented stress-induced decreases in rectal temperature and body and grip tonus and accelerated recovery from decreases in body and grip tonus in acutely stressed mice. These effects were attributed to syringin and syringaresinol-di-*O*-glucoside [8]. The water extract and syringaresinol also protected mice from stress-induced decreases in sex behaviour. They had no effect, however, on stress-induced increases in tyrosine hydroxylase activity in adrenal gland and hypothalamic regions and on corticosterone contents in adrenal gland and serum [9].

The extract of *A. senticosus* at a single i.p. dose of 40–320 mg/kg or at a dosage of 80–320 mg/kg within 4–5 days produced a sedative effect, resulting in increased sleep duration. It was also shown to produce an inhibition of hexobarbital metabolism in vitro, supporting enzyme inhibition rather than enzyme induction as a mechanism for its actions [10].

The alcohol extract of *A. senticosus* inhibited protein synthesis in cell-free rat liver microsomal systems to a greater extent than in polyribosomal systems. This effect was found to be concentration dependent. The water extract had less inhibitory activity and the pure glycosides, eleutherosides B and D, inhibited protein synthesis 10–20 times more than did the alcohol extract [11]; however, i.p. administration of the alcohol extract to rats stimulated protein biosynthesis in the pancreas, liver, and adrenal glands. The results are consistent with the observation that high levels of eleutheroside accumulate in adrenal glands [12].

Intraperitoneal administration of an extract containing mainly eleutherosides B and D to mice at daily dose of 18 mg/animal for 1 week increased the cytostatic activity of natural killer cells by about 200%. It appeared that the eleutherosides stimulated macrophagal T cell and possibly B cell mediated immunity [13]. A recent study on the immunomodulatory activity of the ethanol extract of *A. senticosus* administered orally to healthy volunteers for 4 weeks, showed a drastic increase in the absolute number of immune competent cells, especially T lymphocytes. No side effects were observed within 6 months [14].

Treatment of rats with an extract of *A. senticosus* for 14 days before γ-irradiation accelerated the restoration of blood nucleic acid levels to normal, delayed the nadir in blood leukocyte count for 1–3 days, and increased leukocyte count on days 10–30 after radiation compared to untreated, irradiated controls. The extract thus appeared to promote recovery from radiation effects rather than to protect against them [15]. Use of an aqueous extract of *A. senticosus* in combination with either cytarabine or N^6-(Δ^2-isopentenyl)adenosine had additive antiproliferative effects on L 1210 leukemia cells in vitro [16].

A crude polysaccharide component, PES, was obtained in 0.5% yield by treatment of a hot ethanol extract of powdered roots of *A. senticosus* with acetone. Polysaccharide components PES-A and PES-B were separated by chromatography of crude PES on DEAE-Sephadex A-25 and elution with water and 0.1 and 0.25 N NaCl solutions. PES-A and PES-B were recovered in 0.1 N and 0.25 N NaCl fractions, respectively, and further purified on DEAE-cellulose DE-32 to a final yield of 0.01% and 0.001% of the root, respectively. Gel filtration on Sephadex G-150 and G-200 showed molecular weights of 7000 for PES-A and of 76 000 for PES-B. Both PES-A and PES-B contained glucose, galactose, and arabinose. The molar ratios of glucose:galactose:arabinose were 33:2:1 for PES-A and 2:9:18 for PES-B [17]. The crude polysaccharide PES and the separated and purified components PES-A and PES-B were effective immunostimulating agents. They potentiated the antibody response against sheep red blood cells and stimulated phagocytosis by peritoneal macrophages of mice. They were also found to decrease toxic effects of thioacetamide and phytohemagglutinin in mice and to enhance resistance to X-ray irradiation [18]. Intraperitoneal administration into mice of PES at a dosage of 125 mg/kg for 5 days simultaneously with 0.2 mg bovine serum albumin (BSA) per animal markedly increased the serum levels of anti-BSA IgG and total anti-BSA antibodies but not the serum level of total IgG indicating that PES stimulates the immune activity of mice against invading foreign substances [19]. In vitro the polysaccharides caused a five- to tenfold increase in interferon titer in S 801 and S 7811 leukemic cell cultures [20].

In addition, a homogeneous glucan with a mean molecular weight of 150 000 and homogeneous heteroxylan with a mean molecular weight of 30 000 were isolated from an alkaline aqueous extract of *A. senticosus* by DEAE-Sepharose CL-6B and Sephacryl S-400 column chromatography. The crude polysaccharide mixture and the heteroxylan stimulated phagocytosis in in vitro and in vivo tests [21].

Furthermore, the glycans eleutheran A–G exerted a marked hypoglycemic effect in normal and in alloxan-induced hyperglycemic mice [4]. 3,4-Dihydroxybenzoic acid and its ethylester inhibited rat platelet aggregation [3].

1.4 *Acanthopanax gracilistylus*

Wujiapi, Cortex Acanthopanacis, is another item derived from the *Acanthopanax* plant and listed officially in the Chinese Pharmacopoeia. It is the dry root bark of *A. gracilistylus* W. W. Smith. The roots are collected in summer and fall, and the rootbark is peeled off and dried. It is used as an antirheumatic, antiedemic, and tonic preparation.

From the root bark of *A. gracilistylus*, sesamin, β-sitosterol, syringin, β-sitosterolglucoside, eleutheroside B_1, kaurenoic acid (*1-30*) 16-α-hydroxy-kauran-18-oic acid (*1-31*), and stearic acid have been isolated and identified [22, 23].

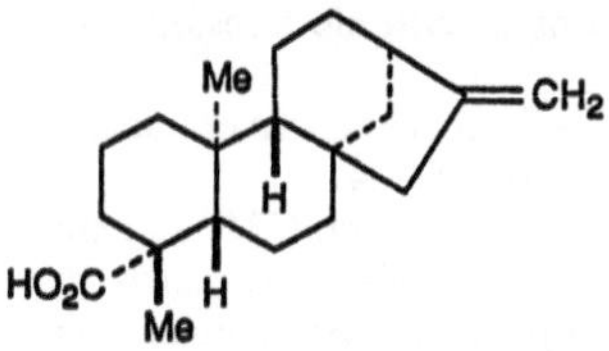

Kaurenoic acid (*1-30*)

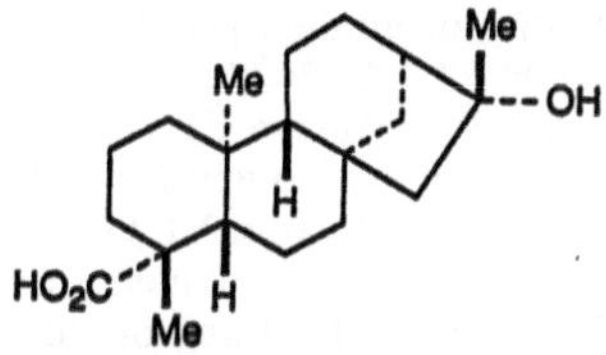

16-α-Hydroxy-kauran-18-oic acid (*1-31*)

The total glycoside fraction isolated from *A. gracilistylus* var. *pubescens* was administered i.v. to rabbits with acute myocardial ischemia produced by coronary artery occlusion. A significant decrease in heart rate and blood pressure was seen. The lactic acid concentration and creatine kinase activity were also significantly decreased. The total ST segment elevation within 8 h, the number of total pathologic Q waves, and the infarct size determined by precordial electrocardiogram mapping were markedly reduced [24].

The pharmacokinetics of eleutheroside B, one of the major active principles of *A. senticosus* has been studied. Tritiated eleutheroside B (5 mg/kg) was administered to rats i.p. Maximal levels of radioactivity were observed in blood 15 min after treatment. Urinary excretion of radioactivity reached 35%, 55%, and 90% of the administered dose at 2, 4, and 48 h, respectively. Only 2.5%–3.0% of the administered dose was excreted in the feces [25, 26]. Eleutheroside B is strongly bound to blood serum globulins and albumins and to a much lesser extent to lipids [27].

References [1]

1. Shih CL (1981) Study on chemical constituents in *Acanthopanax senticosus* Harms. Chin Pharm Bull 16:53
2. Frolova GM, Ovodov YS (1971) Triterpenoid glycosides of *Eleutherococcus senticosus* leaves. II. Structure of eleutherosides I, K, L and M. Khim Prir Soedin 618–622 (CA 76:59965 r)
3. Yun-Choi HS, Kim SO, Lee JR (1986) Platelet anti-aggregating plant materials. Korean J Pharmacogn 17:161 (CA 105:222802 p)
4. Hikino H, Takahashi M, Otake K, Konno C (1986) Isolation and hypoglycemic activity of eleutherans A, B, C, D, E, F, and G, glycans of *Eleutherococcus senticosus* roots. J Nat Prod 49:293–297
5. Xu ZB, Tong WJ, Yang G (1984) Assay of active constituents in different parts of manyprickle acanthopanax (*Acanthopanax senticosus*). Chin Trad Herb Drugs 15:224–226
6. Suu WJ, Sha ZI (1986) Determination of syringin in *Acanthopanax senticosus* by HPLC. Bull Chin Mat Med 11:234–235
7. Xu ZB, Wang MY (1984) Content variation of some chemical constituents of Ci Wu Jia (*Acanthopanax senticosus*) during storage. Chin Trad Herb Drugs 15:15–17
8. Takasugi N, Moriguchi T, Fuwa TS, Sanada S, Ida Y, Shoji J, Saito H (1985) Effect of *Eleutherococcus senticosus* and its components on rectal temperature, body and grip tones, motor coordination and exploratory and spontaneous movements in acute stressed mice. Shoyakugaku Zasshi 39:232–237 (CA 104:102464 n)
9. Nishiyama N, Kamegaya T, Iwai A, Saito H, Sanada S, Ida Y, Shoji J (1985) Effect of *Eleutherococcus senticosus* and its components on sex- and learning-behaviors and tyrosine

[1] Some of the works cited in this and in many subsequent reference lists are also summarized in *Chemical Abstracts*. In each case the appropriate citation is given in parentheses at the end of the reference.

hydroxylase activities of adrenal gland and hypothalamic regions in chronic stressed mice. Shoyakugaku Zasshi 39:238–242 (CA 104:102465p)

10. Medon PJ, Ferguson PW, Watson CF (1984) Effects of *Eleutherococcus senticosus* extracts on hexobarbital metabolism in vivo and in vitro. J Ethnopharmacol 10:235–241
11. Todorov IN, Sizova ST, Mitrokhin YI, German AV, Dardymov IV, Barenboim GM, Shulman ML (1984) Study of the pharmacokinetics and mechanism of action of *Eleutherococcus* glycosides. VII. Effect of the extract and of individual glycosides on protein biosyntheses. Khim Farm Zh 18:920–924
12. Todorov IN, Sizova ST, Kosaganova NY, Mitrokhin YI, German AV, Mitrofanova MA (1984) Pharmacokinetics and mechanism of action of glycosides of eleutherococci. Effect of an extract on the metabolism and biosynthesis of protein in several organs and tissues of rats. Khim-Farm Zh 18:529–533
13. Barenboim GM, Sterlina AG, Bebyakova NV, Ribokas AA, Fuks BB (1986) Investigation of the pharmacokinetics and mechanism of action of *Eleutherococcus* glycosides. VIII. Investigation of natural killer activation by the *Eleutherococcus* extract Khim Farm Zh 20:914–917
14. Bohn B, Nebe CT, Birr C (1987) Durchflußzytometrische Untersuchungen auf immunmodulatorische Wirkungen von *Eleutherococcus senticosus*-Extrakt. Arzneim-Forsch 37:1193–1196
15. Minkova M, Pantev T, Topalova S, Tenchova V (1982) Peripheral blood changes in *Eleutherococcus* pretreated mice exposed to acute gamma radiation. Radiobiol Radiother 23:675–678
16. Hacker B, Medon PJ (1984) Cytotoxic effects of *Eleutherococcus senticosus* aqueous extracts in combination with N^6-(Δ^2-isopentenyl)adenosine and 1-β-D-arabinofuranosylcytosine against L 1210 leukemia cells. J Pharm Sci 73:270–272
17. Xu RS, Feng SC, Fang ZY, Ye CQ, Zhai SK, Shen ML (1983) Polysaccharide components of the roots of *Acanthopanax senticosus* (Rupr. et Maxim.) Harms. Kexue Tongbao 28:185–187
18. Xu RS, Feng SC, Fan ZY, Ye CJ, Zhai SK, Shen ML (1982) Immunopotentiating polysaccharides of *Acanthopanax senticosus* (Rupr. et Maxim.) Harms. In: Wang Y (ed) Chem Nat Prod, Proc Sino-Am Symp 1980, Sci Press, Beijing, pp 271–274
19. Zhu CA, Tu GR, Shen ML (1982) Effect of polysaccharide from *Acanthopanax senticosus* on mouse serum type-specific antibodies. Chin Pharm Bull 17:178
20. Yang JC, Xu HZ, Liu JS (1984) Interferon induction by *Acanthopanax senticosus* polysaccharide and by sodium carboxymethyl starch in S 801 and S 7811 cell culture. Chin J Microbiol Immunol 4:329–330
21. Fang JN, Proksch A, Wagner H (1985) Immunologically active polysaccharides of *Acanthopanax senticosus*. Phytochemistry 24:2619–2622
22. Song XH, Xu GJ, Jin RL, Xu LS (1983) Studies on the identification of the Chinese drugs Wu Jia Pi. J Nanjing Coll Pharm 15–24
23. Xiang RD, Xu RS (1983) Studies on chemical constituents of the root bark of *Acanthopanax gracilistylus* W. W. Smith. Acta Bot Sin 25:356–362
24. Yang SG, Yan YF, Ye YW (1987) Effects of the total glycoside of *Acanthopanax gracilistylus* var. *pubescens* on myocardial infarct size in rabbits with occluded coronary artery. Bull Hunan Med Coll 12:23–27
25. Bezdetko GN, German AV, Shevchenko VP, Mitrokhin YI, Myasoedov NF, Dardymov IV, Todorov IN, Barenboim GM (1981) Study of the pharmacokinetics and action mechanism of *Eleutherococcus* glycosides. I. Incorporation of tritium into eleutherosides B, kinetics of its accumulation and excretion from the body of animal. Khim Farm Zh 15:28–33
26. German AV, Bezdetko GN, Mitrokhin YI, Chivkov GN, Shevchenko VP, Barenboim GM, Dardymov IV, Myasoedov NF, Todorov IN (1982) Study of the pharmacokinetics and mechanism of *Eleutherococcus* glycosides. II. Distribution of eleutheroside in organs and subcellular fraction. Khim Farm Zh 16:26–30
27. Bezdetko GN, German AV, Khasina EI, Shevchenko VP, Dardymov IV, Myasoedov NF, Barenboim GM, Todorov IN (1982) Study of the pharmacokinetics and mechanism of action of *Eleutherococcus* glycosides. V. Metabolism and the kinetics of binding with blood serum components. Khim Farm Zh 16:528–531
28. Shao CJ, Kasai R, Xu JD, Tanaka O (1988) Saponins from leaves of *Acanthopanax senticosus* Harms., Ciwujia: structures of ciwujianosides B, C_1, C_2, C_3, C_4, D_1, D_2, and E. Chem Pharm Bull 36:601–608
29. Shao CJ, Kasai R, Xu JD, Tanaka O (1989) Saponins from leaves of *Acanthopanax senticosus* Harms., Ciwujia: II. Structure of ciwujianoside A_1, A_2, A_3, A_4, and D_3. Chem Pharm Bull 37:42–45

Achyranthes bidentata Bl. 2

2.1 Introduction

Niuxi, Radix Achyranthis bidentatae is the dry root of *Achyranthes bidentata* Bl. (Amaranthaceae). The roots are dug and collected during winter when the above ground part of the plant has withered. It is officially listed in the Chinese Pharmacopoeia and used as a tonic.

The roots of *A. aspera* L., which is unofficial, have also been used in Chinese traditional medicine and folk medicine.

2.2 Chemical Constituents

Besides oleanolic acid (*2-1*) from *A. bidentata* and *A. aspera* [1], some insect molting substances were also isolated from the roots of *Achyranthes* species [2]. In a study on the biologically active compounds two insect molting hormones were isolated from *A. fauriei* and were identified as ecdysterone (*β*-ecdysone) (*2-2*) [3, 4] and inokosterone (*2-3*) [3–6]. Ecdysterone and inokosterone were also found in the root of *A. bidentata* [7], whereas ecdysterone was again obtained from the roots of *A. aspera* [2, 7, 8].

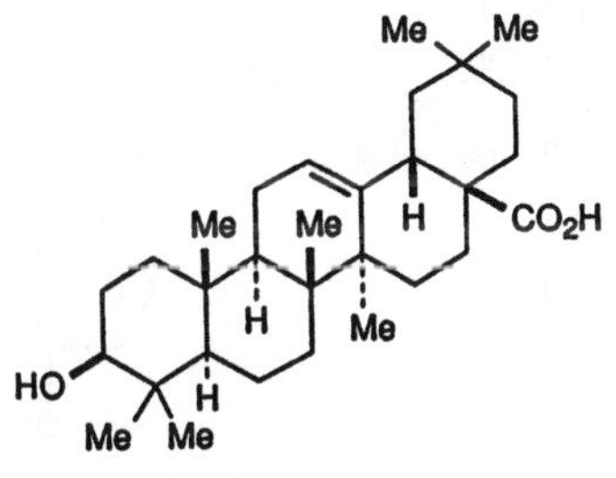

Oleanolic acid (*2-1*)

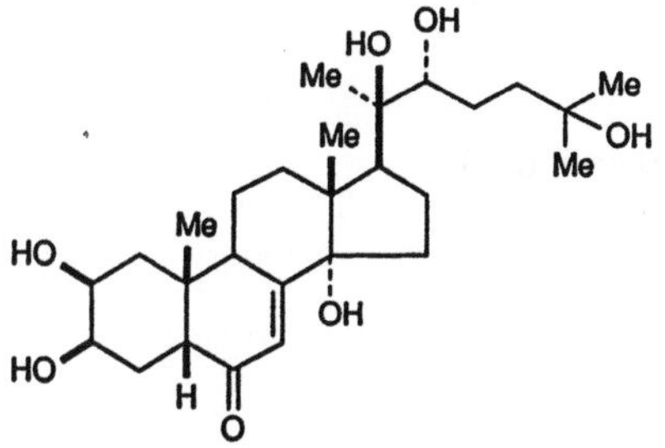

Ecdysterone (*β*-Ecdysone) (*2-2*)

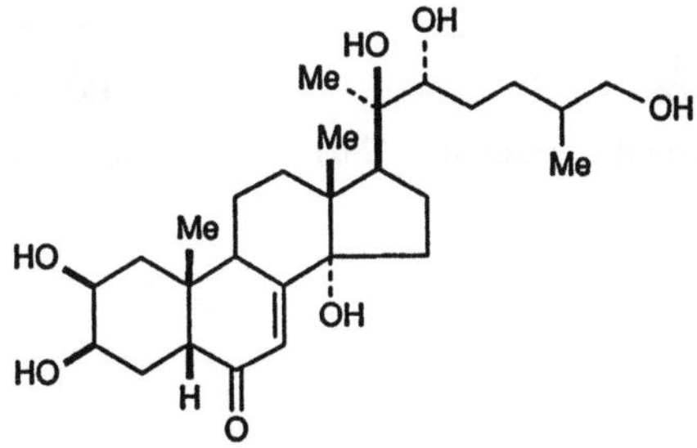

Inokosterone (*2-3*)

Oleanolic acid is a triterpene compound derived from oleanane (*2-4*) and is widely distributed in herbal medicine. The insect molting substances ecdysterone and inokosterone are of steroid nature with a cholestane (*2-5*) skeleton.

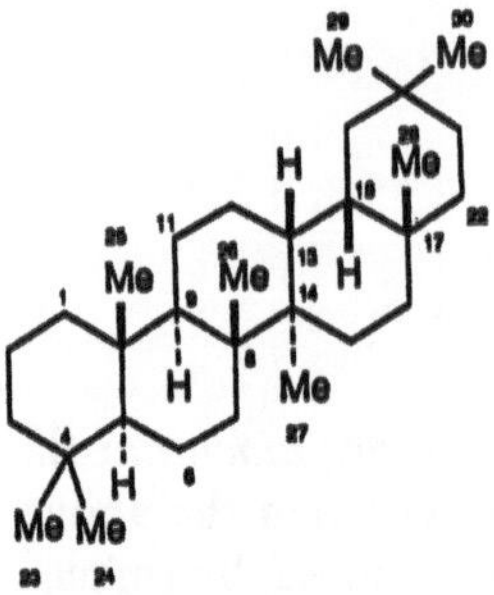

Oleanane (*2-4*)

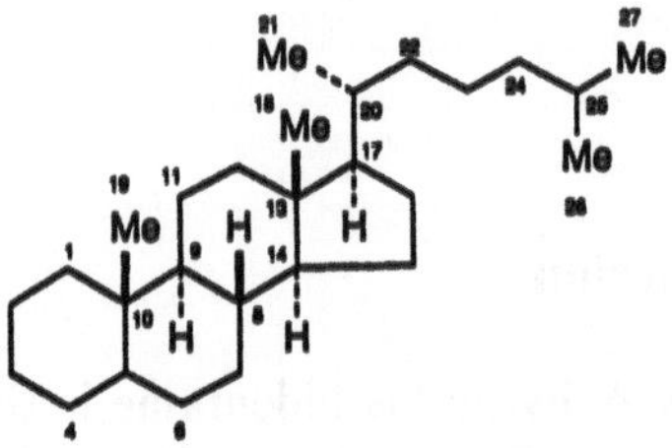

Cholestane (*2-5*)

Two saponins, achyranthes saponin A (*2-6*) and B (*2-7*), were isolated from seeds of *A. aspera* [9], and two saponins, achyranthes saponin C (*2-8*) and D (*2-9*), were isolated from the unripe fruits of *A. aspera* [10]. After saponification of the four saponins only oleanolic acid was determined as a sapogenin. The structures of the four saponins were also determined.

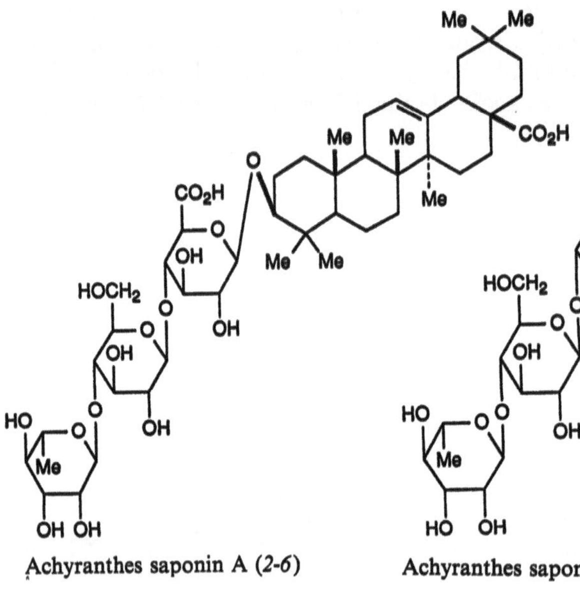

Achyranthes saponin A (*2-6*)

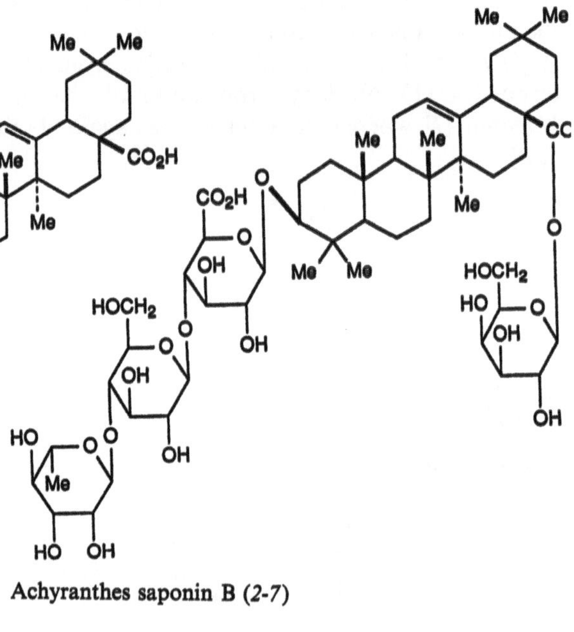

Achyranthes saponin B (*2-7*)

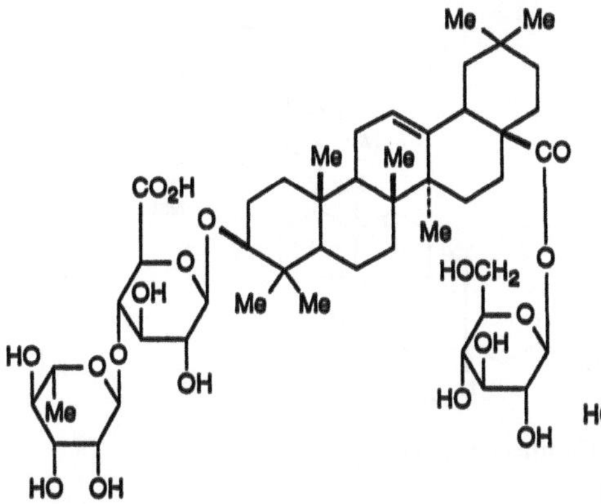

Achyranthes saponin C (*2-8*)

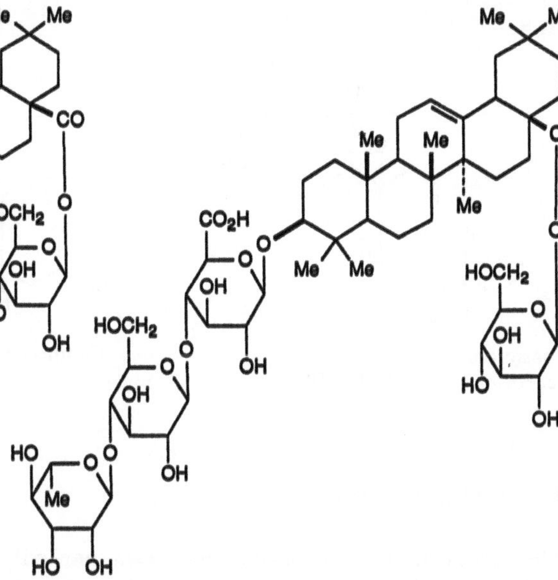

Achyranthes saponin D (*2-9*)

From the fresh root of *A. longifolia*, used in folk medicine, four compounds were isolated. They were identified as oleanolic acid, oleanolic acid glucuronide, ecdysterone, and ursolic acid on the basis of spectroscopic analyses [11].

Another Chinese traditional medicine of the family Amaranthaceae, officially listed in the Chinese Pharmacopoiea and used similarly to *Achyranthes* root, is Chuanniuxi, Radix Cyathulae, the dry root of *Cyathula officinalis*. It must be collected during fall and winter. *C. capitata*, an unofficial *Cyathula* plant was also used in folk medicine in China. The genus *Cyathula* is known to contain insect molting hormones [12]. From the root of *C. capitata* amarasterone A (*2-10*) and B (*2-11*) [13], capitasterone (*2-12*) [14], cyasterone (*2-13*) [15], isocyasterone, the 25-epimer of cyasterone [16], and sengosterone (*2-14*) [17] were isolated and structurally determined.

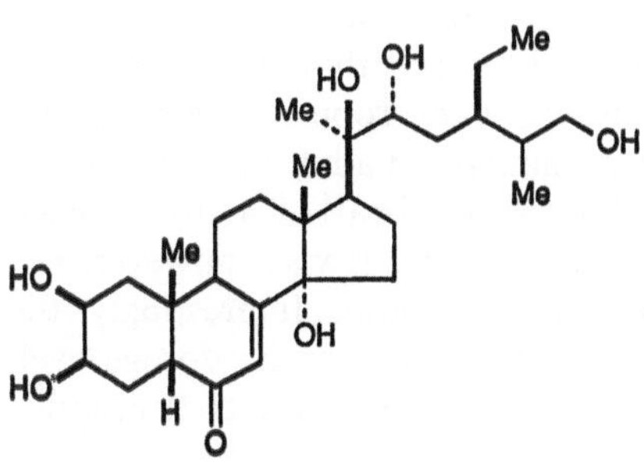

Amarasterone A (*2-10*)

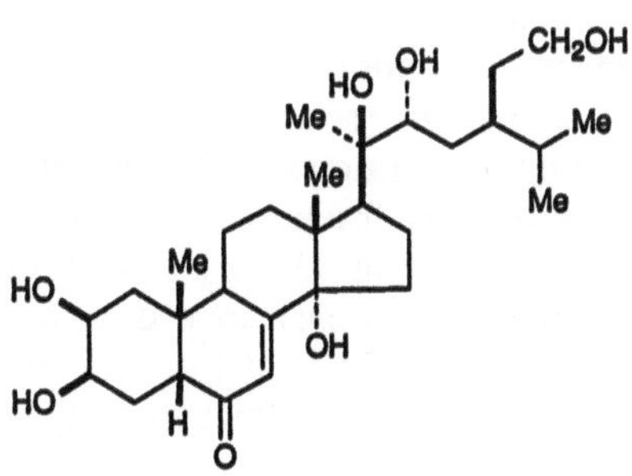

Amarasterone B (*2-11*)

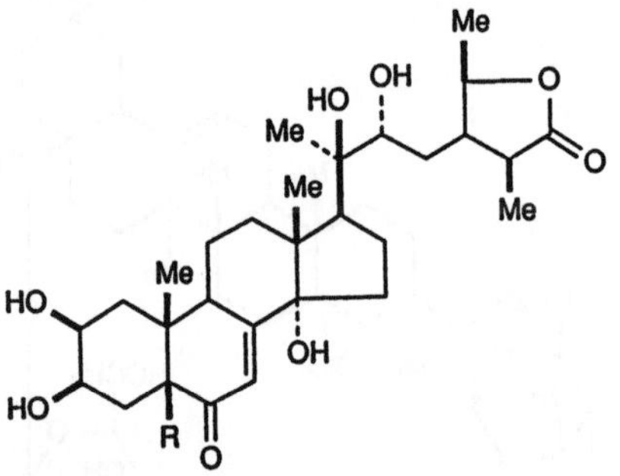

Cyasterone (*2-13*): R=H
Sengosterone (*2-14*): R=OH

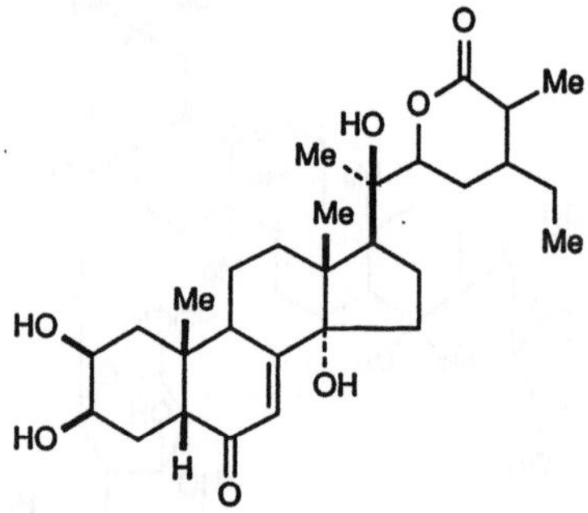

Capitasterone (*2-12*)

2.3 Pharmacology

The decoction of the root of *A. bidentata* markedly increased blood flow in rat hind limbs and had a vasodilatory effect. It also reduced the inflammation in rat paws that was induced by egg white. When administered to mice the decoction exhibited an analgesic action. In rabbits, i.v. injection of the aqueous extract elicited a prompt reduction in blood pressure. The hypotensive effect was reversible [18].

The mixture of saponins isolated from the seeds of *A. aspera* increased the force of contraction of hearts isolated from frog, guinea pig, and rabbit. At lower doses, the stimulating effect could be blocked by pronethalol and partly by mepyramine. At higher saponin doses, the effect was not blocked by pronethalol. The saponins also increased the tonus of the hypodynamic heart and the force of contraction of failing papillary muscle [19]. Perfusion of isolated rat heart with adrenaline or the saponins obtained from *A. aspera* increased the activity of phosphorylase A but had no effect on total phosphorylase activity [20].

The saponins isolated from the fresh root of *A. longifolia* by extraction with butanol had a contraceptive action and induced early abortion in pregnant mice [21]. The benzene extract fraction of the plant *A. aspera* showed a 100% abortive activity in the rabbit at a single dose of 50 mg/kg. It was reported to be neither estrogenic nor antiestrogenic nor androgenic in mice. Abortion was apparently not due to a deficiency in prolactin, growth hormone, or pituitary gonadotropins. The drug was nontoxic and nonteratogenic [21]. A contraceptive activity in adult female rats was also observed with a fraction of a butanol extract of the aerial part of *A. aspera* when administered orally at 75 mg/kg on days 1–5 post coitum. It was, however, not observed in hamsters even at a dose of more than 300 mg/kg. Interestingly, the butanol extract exhibited potent estrogenic activity at the contraceptive dosage level. A significant uterotropic effect was apparent even at only 5% of the contraceptive dose [22].

The pharmacology of ecdysterone and inokosterone in vertebrates was also investigated. Ecdysterone and inokosterone, applied at doses of 0.1–10 mg/kg i.p. or 1–100 mg/kg orally, did not affect respiratory, circulatory, or autonomic nervous systems or blood sugar levels. It had no antiinflammatory or muscle relaxant effects nor did it accelerate wound healing in rabbits, rats, or guinea pigs [23]. No anabolic, androgenic, or antiandrogenic effects of ecdysterone and inokosterone were observed in rats [23–25], indicating that these insect molting substances of plant origin apparently have no pharmacological effects in mammals. The only observed biological effect of ecdysterone and inokosterone was the suppression of hyperglycemia induced by glucagon in rats [23–25]. Oral administration of a mixture of ecdysterone and inokosterone at a daily dose to rats of 0.2–2 g/kg for 35 days did not produce any toxic effects [23].

Oleanolic acid was effective in the prevention of experimental liver damage induced by carbon tetrachloride (CCl_4) in rats. Treatment with oleanolic acid markedly reduced the elevation of serum glutamic-pyruvic transaminase (GPT) and liver triglyceride levels in rats intoxicated with CCl_4. The degeneration and necrosis of liver cells induced by CCl_4 were significantly diminished with oleanolic acid treatment. Moreover, the glycogen content in the liver cells of the treated rats was increased, and the damaged mitochondrial and endoplasmic structures of liver cells were restored [26].

A number of esters, amides, or mixed amides of oleanolic acid were synthesized and tested for antiulcer activity. 3-Hemisuccinato-oleanolic acid morpholinide, 3-hemisuccinato-oleanolic acid isopropylamide, and the mixed amide from oleanolic acid and succinic acid were the most active compounds in this series and were more effective than the known antiulcer agent carbenoxolone [27].

In addition, oleanolic acid inhibited the activation of Epstein-Barr virus induced by the tumor promotor 12-*O*-tetradecanoylphorbol 13-acetate (TPA) and the tumor promoting activity of TPA in mice. The inhibitory activity of oleanolic acid on tumor promotion by TPA was comparable to that of the known tumor promotion inhibitor retinoic acid [28].

References

1. Khastgir HN, Gupta PS (1958) Oleanolic acid from *Achyranthes aspera*. J Indian Chem Soc 35:529–530
2. Ogawa S, Nishimoto N, Okamoto N, Takemoto T (1971) Constituents of *Achyrantes* radix. VIII. Insect-molting substances in *Achyranthes* genus. 2. Yakugaku Zasshi 91:916–920
3. Takemoto T, Ogawa S, Nishimoto N (1967) Isolation of the molting hormones of insects from achyranthis radix. Yakugaku Zasshi 87:325–327
4. Takemoto T, Ogawa S, Nishimoto N (1967) Constituents of achyranthis radix. II. Isolation of insect molting hormones. Yakugaku Zasshi 87:1469–1473
5. Takemoto T, Ogawa S, Nishimoto N (1967) Constituents of achyranthis radix. III. Structure of inokosterone. Yakugaku Zasshi 87:1474–1477
6. Takemoto T, Hikino Y, Arihara S, Hikino H, Ogawa S, Nishimoto N (1968) Absolute configuration of inokosterone, an insect-moulting substance from *Achyranthes fauriei*. Tetrahedron Lett 2475–2478
7. Takemoto T, Ogawa S, Nishimoto N, Hirayama H, Taniguchi S (1968) Constituents of achyranthis radix. VII. The insect-molting substances in *Achyranthes* and *Cyathula* genera. Yakugaku Zasshi 88:1293–1297
8. Ikan R, Ravid U, Trosset D, Shulman E (1971) Ecdysterone: an insect molting hormone from *Achyranthes aspera*. Experientia 27:504–505
9. Hariharan V, Rangaswami S (1970) Structure of saponins A and B from the seeds of *Achyranthes aspera*. Phytochemistry 9:409–414
10. Seshadri V, Batta AK, Rangaswami S (1981) Structure of two new saponins from *Achyranthes aspera*. Indian J Chem [B] 20:773–775

11. Wu NJ, Zhang GQ (1982) Study on chemical constituents of Tu Niu Xi (*Achyranthes longifolia* Makino). Chin Trad Herb Drugs 13:437–438
12. Hikino H, Takemoto T (1972) Arthropod molting hormones from plants, *Achyranthes* and *Cyathula*. Naturwissenschaften 59:91–98
13. Takemoto T, Nomoto K, Hikino H (1968) Structure of amarasterone A and B, novel C_{29} insect-moulting substances from *Cyathula capitata*. Tetrahedron Lett 4953–4956
14. Takemoto T, Nomoto K, Hikino Y, Hikino H (1968) Structure of capitasterone, a novel C_{29} insect-moulting substance from *Cyathula capitata*. Tetrahedron Lett 4929–4932
15. Takemoto T, Hikino Y, Hikino H (1967) Structure of cyasterone, a novel C_{29} insect-moulting substance from *Cyathula capitata*. Tetrahedron Lett 3191–3194
16. Hikino H, Hikino Y (1970) Arthropod molting hormones. In: Herz W, Grisebach H, Scott AI (eds) Fortschritte der Chemie organischer Naturstoffe, vol 28. Springer, Berlin Heidelberg New York, pp 256–312
17. Hikino H, Nomoto K, Takemoto T (1969) Structure of sengesterone, a novel C_{29} insect-moulting substance from *Cyathula capitata*. Tetrahedron Lett 1417–1420
18. Sun SP, Li XH, Sun SG (1985) Pharmacological studies on *Achyranthes bidentata*. Henan Trad Chin Med 39–40
19. Gupta SS, Bhagwat AW, Ram AK (1972) Cardiac stimulant activity of the saponin of *Achyranthes aspera*. Indian J Med Res 60:462–471
20. Ram AK, Bhagwat AW, Gupta SS (1971) Effect of the saponin of *Achyranthes aspera* on the phosphorylase activity of rat heart. Indian J Physiol Pharmacol 15:107–110
21. Pakrashi A, Bhattacharya N (1977) Abortifacient principle of *Achyranthes aspera* Linn. Indian J Exp Biol 15:856–858
22. Wadhwa V, Singh MM, Gupta DN, Singh C, Kamboj VP (1986) Contraceptive and hormonal properties of *Achyranthes aspera* in rats and hamsters. Planta Med 52:231–233
23. Ogawa S, Nishimoto N, Matsuda H (1974) Pharmacology of ecdysones in vertebrates. In: Burdette WJ (ed) Invertebrate Endocrinology and Hormonal Heterophylly. Springer, Berlin Heidelberg New York, pp 341–344
24. Matsuda H, Kawabara T, Yamamoto Y (1970) Pharmacological studies of insect metamorphosing steroids from *Achyranthes* radix. Nippon Yakubutsugaku Zasshi 66:551–563 (CA 76:30743f)
25. Masuoka M, Orita S, Shino A, Matsuzawa T, Nakayama R (1970) Pharmacological studies of insect metamorphosing hormones. Ponasterone A, ecdysterone and inokosterone in the rats. Jpn J Pharmacol 20:142–156
26. Ma XH, Zhao YC, Yin L, Han DW, Ji CX (1982) Studies on the effect of oleanolic acid on experimental liver injury. Acta Pharm Sin 17:93–97
27. Wrzeciono U, Malecki I, Budzianowski J, Kierylowicz H, Zaprutko L, Beimeik E, Kostepska H (1985) Nitrogenous triterpene derivatives. X. Hemisuccinates of some derivatives of oleanolic acid and their antiulcer effects. Pharmazie 40:542–544
28. Tokuda H, Ohigashi H, Koshimizu K, Ito Y (1986) Inhibitory effects of ursolic and oleanolic acid on skin tumor promotion by 12-O-tetradecanoylphorbol 13-acetate. Cancer Lett 33:279–285

Aconitum spp. 3

3.1 Introduction

The aconite root is one of the most important and common drugs in Chinese traditional medicine and folk medicine. *Aconitum carmichaeli* Debx. and *A. kusnezoffii* Reichb. (Ranunculaceae) are now officially listed in the Chinese pharmacopoeia, which contains the following items regarding *Aconitum*.

- Chuanwu, Radix Aconiti, is the dry root of *A. carmichaeli* collected from June to August.
- Zhichuanwu, Radix Aconiti Preparata, is the root of *A. carmichaeli* prepared by soaking in water and then cooking in water for 4–6 h or steaming for 6–8 h.
- Fuzi, Radix Aconiti Lateralis Preparata, is the lateral root of *A. carmichaeli* prepared by different methods.
- Caowu, Radix Aconiti kusnezoffii, is the dry root of *A. kusnezoffii* collected in fall when the aerial part of the plant has withered.
- Zhicaowu, Radix Aconiti kusnezoffii Preparata, is the root of *A. kusnezoffii* prepared as described for Radix Aconiti Preparata.
- Caowuye, Folium Aconiti kusnezoffii, are the leaves of *A. kusnezoffii* collected in summer before the plant flowers.

Besides the above mentioned items, the following aconite species are included in the appendix of the pharmacopeia: *A. balfourii* Stapf (roots), *A. szechenyianum* Gay. (roots, leaves), *A. naviculare* Stapf (whole plant), *A. tanguticum* (Maxim.) Stapf (whole plant), and *A. kusnezoffii* (sprouts).

The aconite roots are very toxic and are used as analgesic and anesthetic agents in the treatment of neuralgic and rheumatic affections. The processed roots are less toxic because the alkaloid content is in part decomposed during the preparation process. The lateral roots are widely used as a cardiotonic and to improve blood circulation.

There are 167 species of *Aconitum* found in China, 44 of which have been used in medicine [1].

3.2 Chemical Constituents

The aconite plants are known to contain a number of C_{19} and C_{20} diterpene alkaloids which can be generally divided into two classes. The basic structures of the first class have a four ring system in common, derived from kaurane (*3-1*) with carbon atoms C-19 and C-20 connected to an amine thus yielding a cyclic amine designated as 7,20-cycloveatchane (*3-2*). An example of this class is the alkaloid songorine (*3-3*).

The second class contains a skeleton with a seven-membered ring, carbon atoms C-17 and C-19 being connected to an amine to form a heterocycle. The majority of aconite alkaloids have this ring system. The most important basic structure is aconitane (*3-4*), and the most important representative is aconitine (*3-5*).

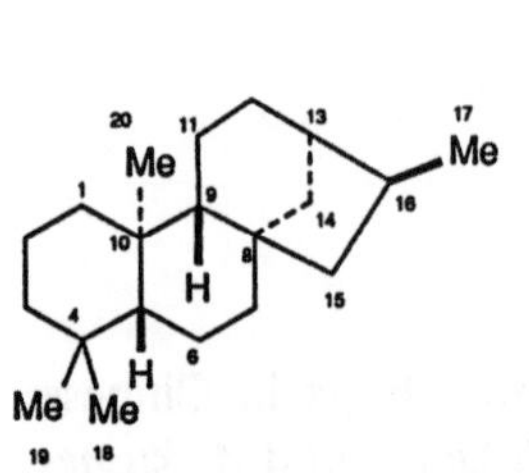

Kaurane (*3-1*)

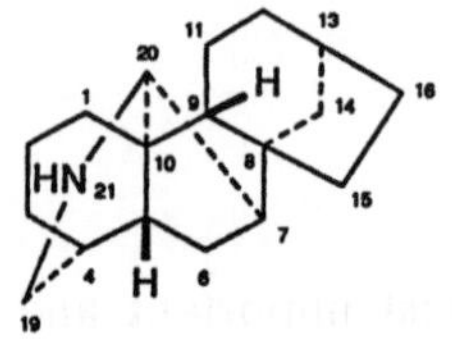

7,20-Cycloveatchane (*3-2*)

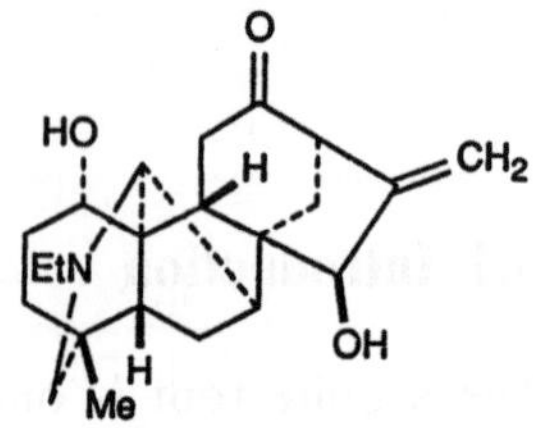

Songorine (*3-3*)

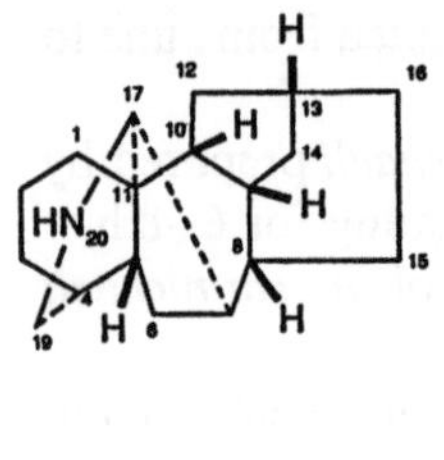

Aconitane (*3-4*)

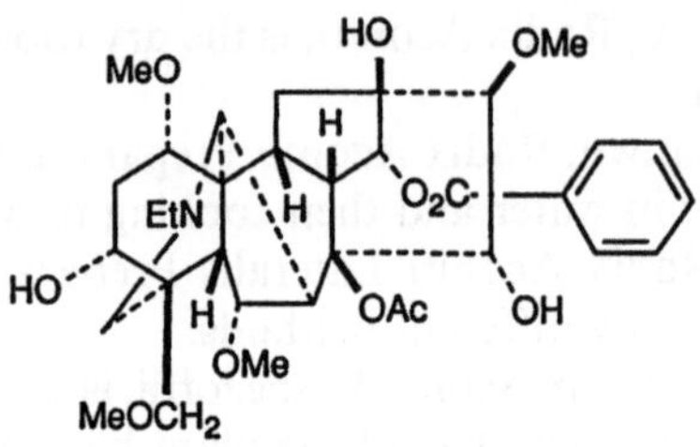

Aconitine (*3-5*)

3.2.1 Diterpene Alkaloids of *Aconitum carmichaeli*

Chen et al. were the first to study the chemical constituents of the roots and lateral roots of *A. carmichaeli* and to isolate six alkaloids. Four of them were identified as aconitine, mesaconitine, hypaconitine, and talatisamine. One of the other two, provisionally named Chuan-wu base A, was proved later to be identical with isotalatisidine [2]. Aconitine, mesaconitine, and hypaconitine are the main alkaloids present in *A. carmichaeli*.

Aconitine is the diester of the aminoalcohol aconine (*3-6*) and can be easily hydrolyzed to form benzoylaconine (*3-7*) by loss of the acetyl group and to subsequently form aconine by elimination of the benzoyl group. Since benzoylaconine and aconine are less toxic than aconitine, the toxicity of aconite roots decreases with increasing storage time or by processing.

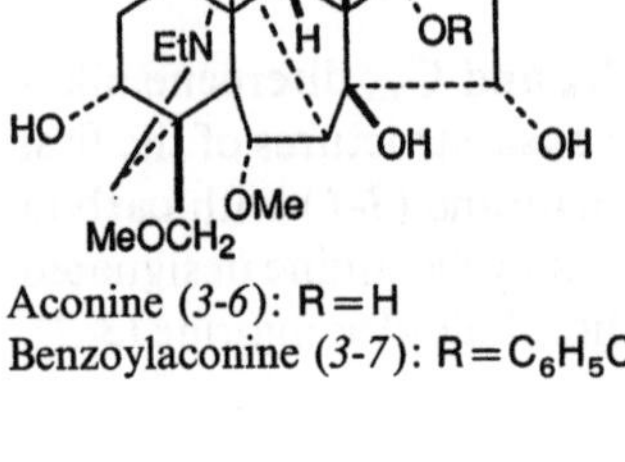

Aconine (*3-6*): R=H
Benzoylaconine (*3-7*): $R=C_6H_5CO$

By mild alkaline hydrolysis of aconitine with potassium carbonate in methanol at room temperature, considerable amounts of 8-*O*-methylaconine, together with smaller quantities of desbenzoyl-pyroaconitine (*3-8*) and 16-*epi*-desbenzoyl-pyroaconitine (*3-9*), were obtained along with aconine. Better yields of desbenzoyl-pyroaconitine and its 16-epimer were obtained when the hydrolysis was carried out by heating aconitine in ethanol with potassium carbonate [3].

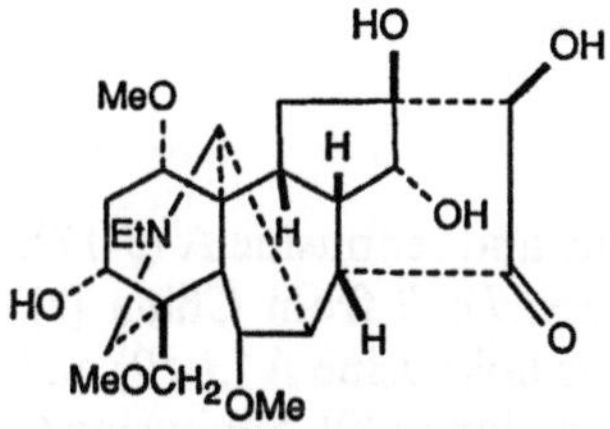

Desbenzoylpyroaconitine (*3-8*)

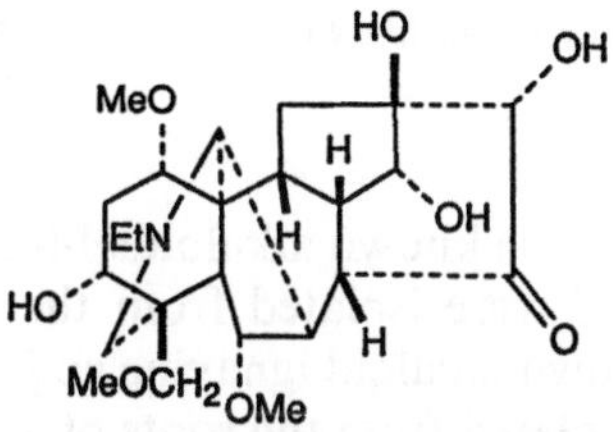

16-*epi*-Desbenzoylpyroaconitine (*3-9*)

Mesaconitine (*3-10*) is the *N*-methyl homologue of aconitine, whereas hypaconitine (*3-11*) represents the 3-dehydroxy analogue of mesaconitine.

Talatisamine (*3-12*), another aminoalcohol, differs from aconine by lacking three hydroxyl groups and one methoxyl group. The 1-demethyl derivative of talatisamine is isotalatisidine (*3-13*). The alkaloids were isolated by extraction of powdered aconite root with organic solvents followed by chromatography on an aluminum oxide column. Yields of hypaconitine, aconitine, and mesaconitine from *A. carmichaeli* were about 0.05%, 0.01%, and 0.006%, respectively [4]. The alkaloid content of aconite roots shows great geographic [5] and seasonal [6] variation. It increases from fall to spring and decreases significantly in summer.

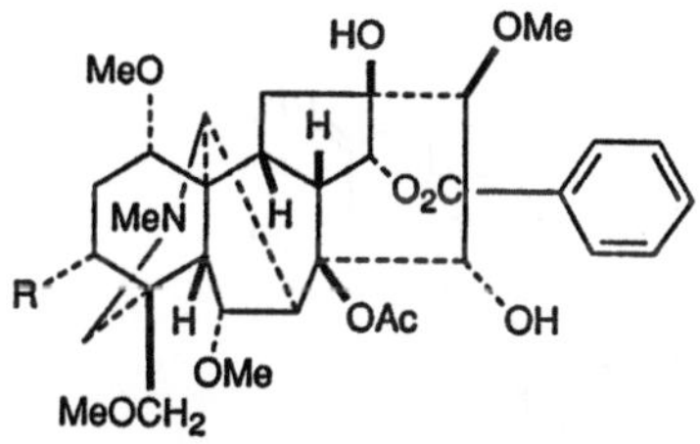

Mesaconitine (*3-10*): R=OH
Hypaconitine (*3-11*): R=H

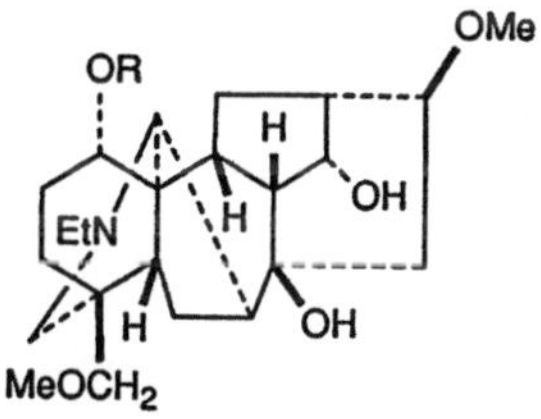

Talatisamine (*3-12*): R=CH_3
Isotalatisidine (*3-13*): R=H

A number of other alkaloids were later isolated and determined by different groups. Isolation of carmichaeline (*3-14*), which proved to be identical with karacoline from the root of *A. carmichaeli* [4]; the known alkaloids neoline (*3-15*) and songorine (*3-3*); and a new alkaloid fuziline (*3-16*), from the roots and lateral roots of *A. carmichaeli* [5–7], were reported. The structure of fuziline was ascertained by spectral and single crystal X-ray analyses. Neoline and fuziline are both aminoalcohols.

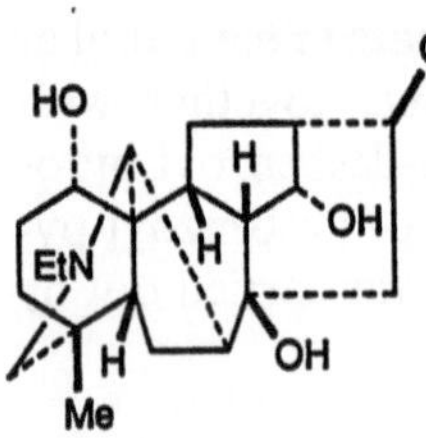

Carmichaeline (karacoline) (*3-14*)

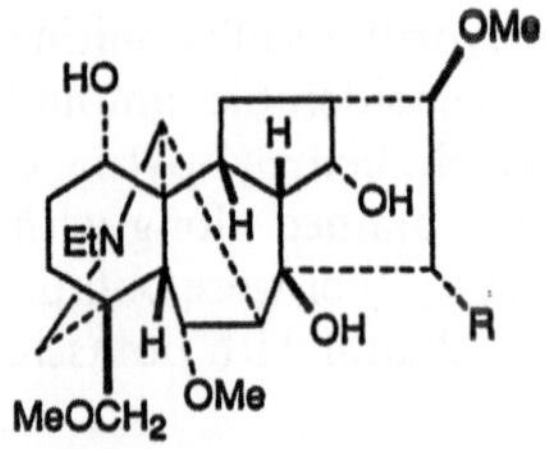

Neoline (*3-15*): R=H
Fuziline (*3-16*): R=OH

Furthermore, the known alkaloids 14-acetyltalatisamine and senbusine A (*3-17*), B (*3-18*), and C were isolated from the roots of *A. carmichaeli* from China [8], whereas the known alkaloid ignavine and the new alkaloids hokbusine A (*3-19*) and B (*3-20*) were isolated from the roots of *A. carmichaeli* from Japan [9]. Senbusine C was shown later to be identical with fuziline. Ignavine (*3-21*) is an alkaloid of the hetisan type (*3-22*) and hokbusine B an alkaloid of the aconitine type without a substituent on the nitrogen atom.

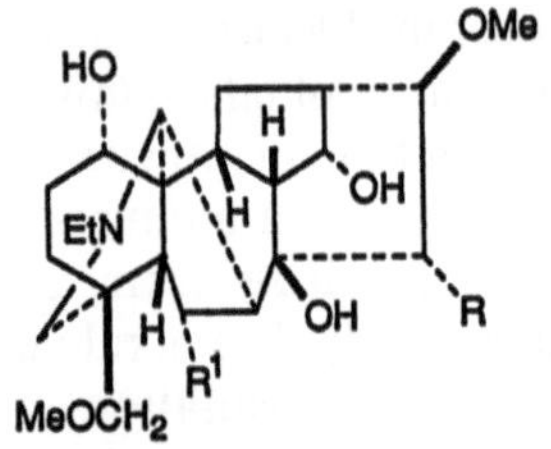

Senbusine A (*3-17*): R=H, R[1]=OH
Senbusine B (*3-18*): R=OH, R[1]=H

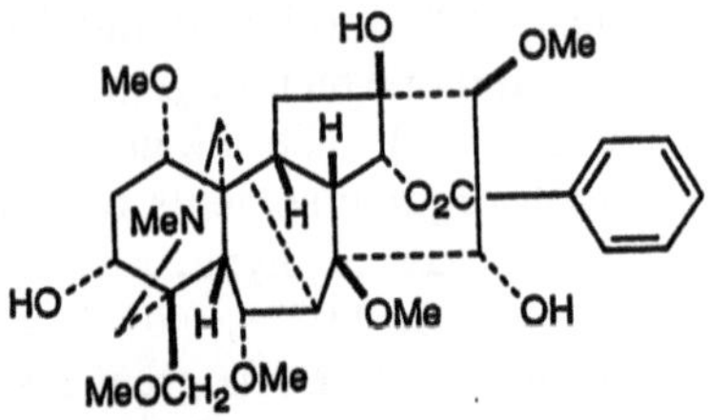

Hokbusine A (*3-19*)

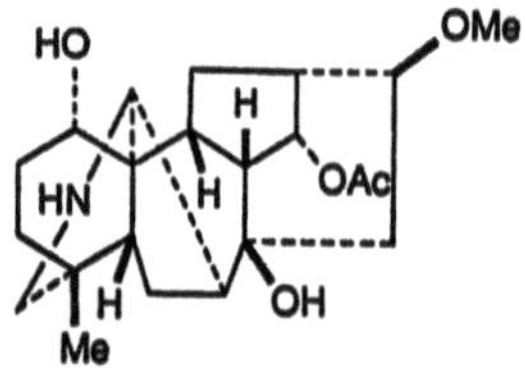

Hokbusine B (*3-20*)

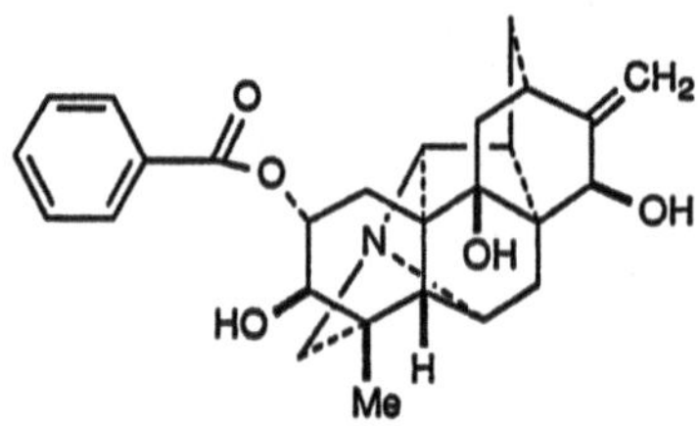

Ignavine (*3-21*)

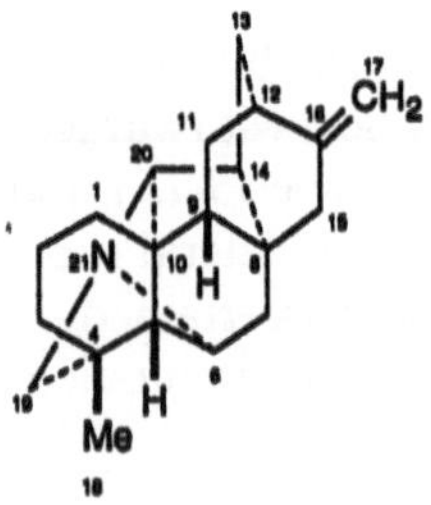

Hetisan (*3-22*)

The known alkaloid isodelphinine (*3-23*) [10] and the lipoalkaloids lipoaconitine, lipohypaconitine, lipomesaconitine, and lipodeoxyaconitine, in which the acetyl functions attached to the C-8 hydroxyl group of the corresponding parent alkaloids are replaced by fatty acid residues, were found in the processed roots of *A. carmichaeli* [11–13].

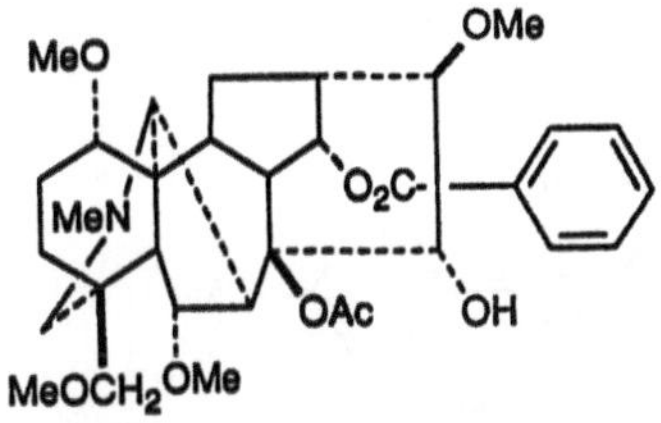

Isodelphinine (*3-23*)

3.2.2 Diterpene alkaloids of *Aconitum kusnezoffii*

From the second officially listed aconite plant, *A. kusnezoffii*, five alkaloids were isolated. Four of them were identified as 3-deoxyaconitine [1], hypaconitine, aconitine, and mesaconitine on the basis of their physical and chemical properties. The fifth alkaloid was a novel compound with the same skeleton as mesaconitine but with one more hydroxyl group at position C-10. It was named beiwutine (*3-24*) [14]. Beiwutine was later also detected in the lateral root of *A. carmichaeli* [15].

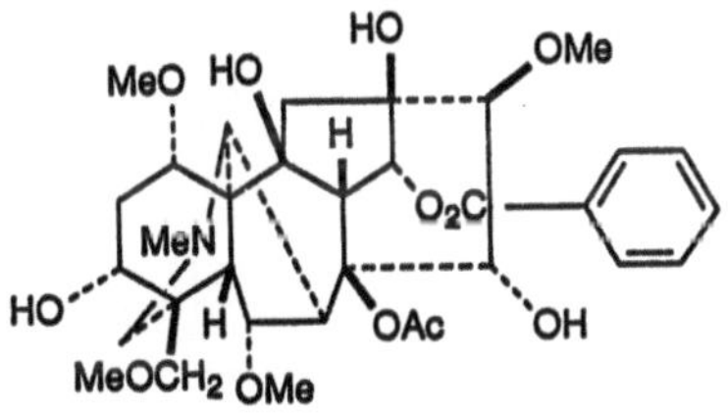

Beiwutine (*3-24*)

The amounts of the main alkaloids aconitine, mesaconitine, and hypaconitine in roots of *A. kusnezoffii* obtained from various sources were determined by high performance liquid chromatography. The roots contained about 0.03%–0.06% aconitine, 0.1%–0.6% mesaconitine and 0.02%–0.2% hypaconitine, depending on the source [16].

3.2.3 Diterpene Alkaloids of Unofficial *Aconitum* Species Used in Chinese Folk Medicine

More recently, a number of unofficial *Aconitum* species used in Chinese folk medicine were studied chemically and pharmacologically. A series of novel alkaloids were isolated and characterized.

3.2.3.1 *Aconitum hemsleyanum*

Yunaconitine (*3-25*), an aconine diester with a *p*-anisoyl group at position 14, was isolated from *A. hemsleyanum*. The structure was elucidated on the basis of spectral data and by chemical degradations. On hydrolysis with 5% methanolic KOH solution yunaconitine gave the corresponding aminoalcohol pseudaconine, acetic acid, and *p*-anisic acid [17]. Zhang et al. [18] isolated from *A. hemsleyanum* three alkaloids, guayewuanine A (*3-26*) and B. Guayewuanine B was found to be identical with yunaconitine; the structure of guayewuanine A was also elucidated.

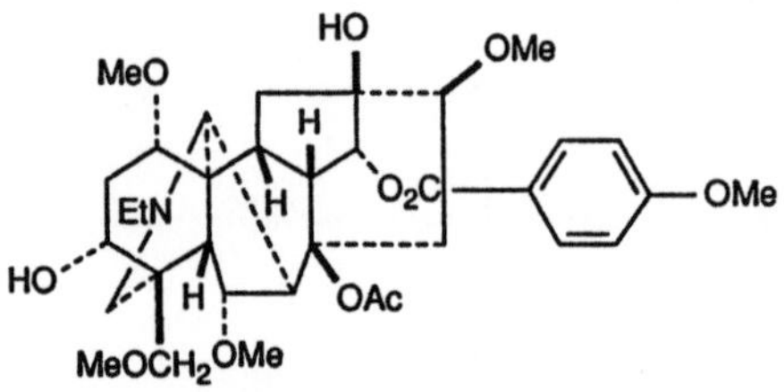

Yunaconitine (guayewuanine B) (*3-25*)

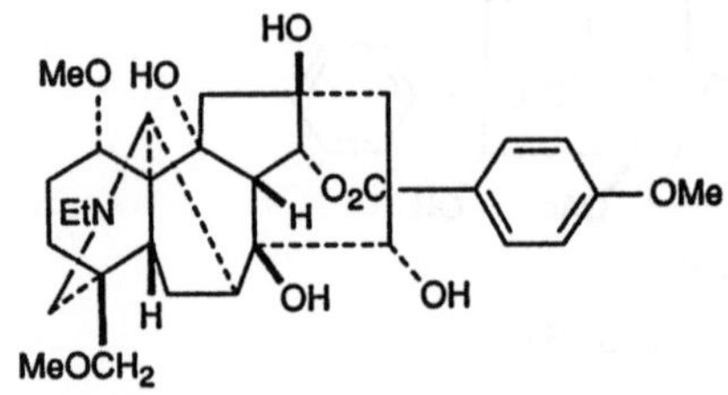

Guayewuanine A (*3-26*)

3.2.3.2 *Aconitum delavyi*

A new alkaloid, delavaconitine (*3-27*), was isolated and characterized from *A. delavyi*; in addition, yunaconitine was found [1].

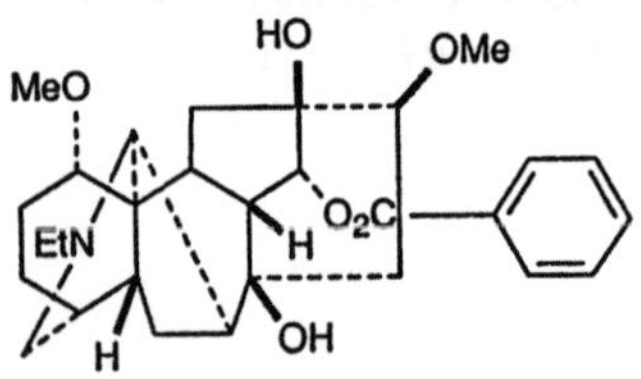

Delavaconitine (*3-27*)

3.2.3.3 *Aconitum nagarum* var. *lasiandrum*.

Four alkaloids were isolated from *A. nagarum* var. *lasiandrum*. Three were bullatine B, C, and G. Structure determination revealed that bullatine B and G were identical with neoline and songorine, respectively, whereas bullatine C was characterized as 14-acetylneoline. The location of the acetylated hydroxyl group at C-14 of neoline in bullatine C was ascertained by ^{1}H NMR spectroscopy [19]. Furthermore, denudatine (*3-28*), songoramine (*3-29*), virescenine (*3-30*), and flavaconitine (*3-31*) were also isolated and identified [20]. Denudatine possesses a 7,20-cycloatidane (*3-32*) skeleton that is related to 7,20-cycloveatchane (*3-2*).

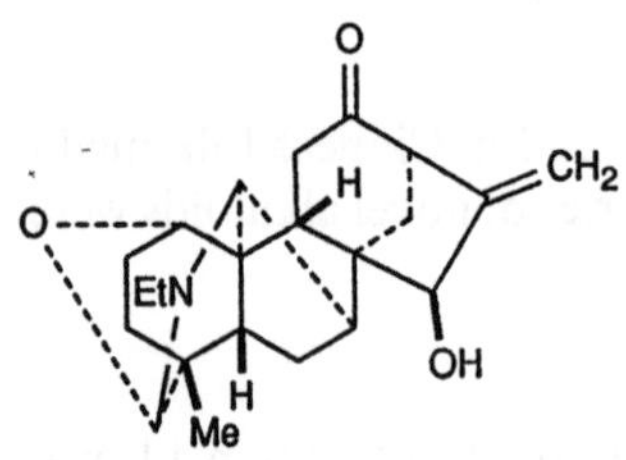

Songoramine (*3-29*)

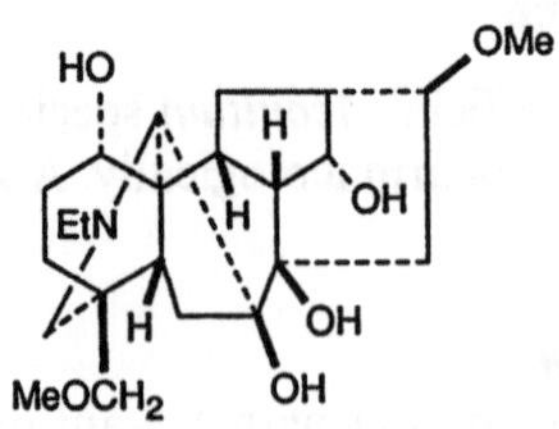

Virescenine (*3-30*)

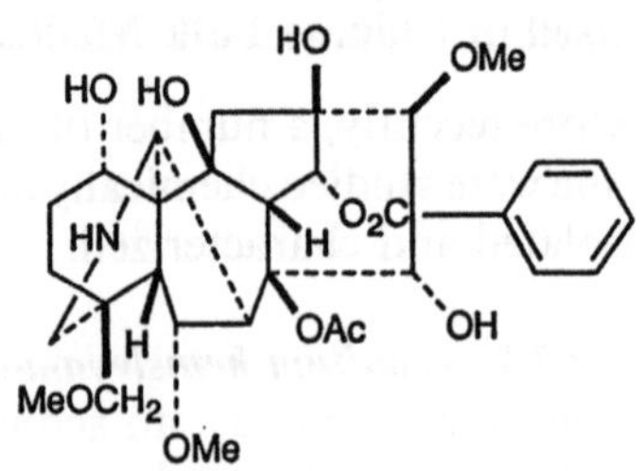

Flavaconitine (*3-31*)

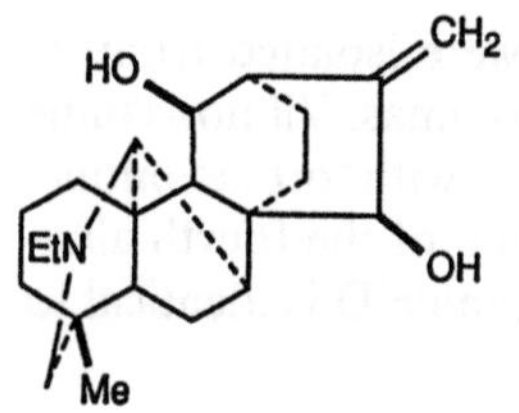

Denudatine (*3-28*)

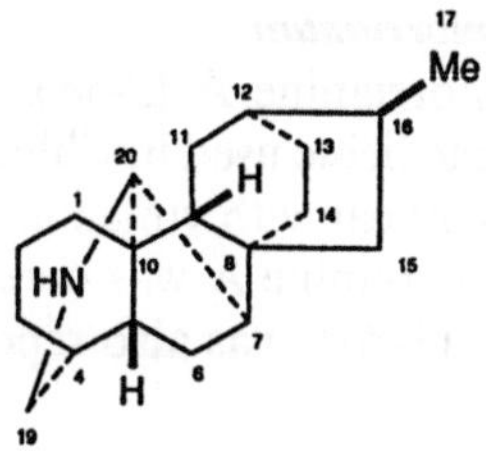

7,20-Cycloatidane (*3-32*)

3.2.3.4 *Aconitum nagarum* var. *heterotrichum f. dielsianum*

Mody et al. reported the isolation of two known alkaloids, aconitine and deoxyaconitine, and of a new alkaloid nagarine (*3-33*), from *A. nagarum* var. *heterotrichum*. They are used in Chinese folk medicine for treatment of rheumatic and neuralgic disorders [21]. The structure of nagarine was determined on the base of its ^{13}C NMR spectrum and was confirmed by a partial synthesis from delphisine (*3-34*) via pyroneoline (*3-35*). The structure of nagarine is unusual because it is the only C_{19} diterpene alkaloid in which the hydroxyl group at position C-15 exists in the β configuration.

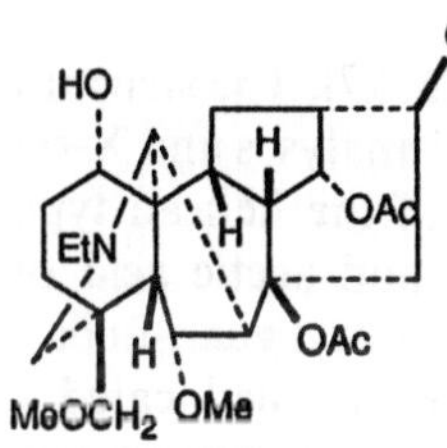

Delphisine (*3-34*)

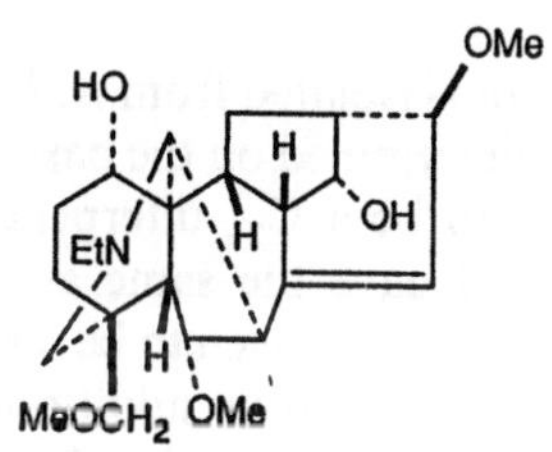

Pyroneoline (*3-35*)

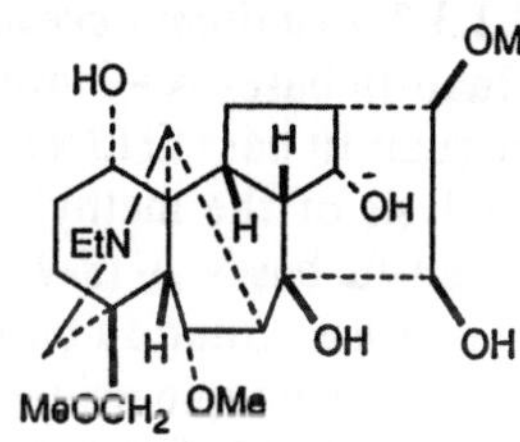

Nagarine (*3-33*)

Note that 10-hydroxymesaconitine, isolated from *A. nagarum* var. *lasiandrum*, has also been called nagarine [1].

3.2.3.5 *Aconitum sinomontanum*

Ranaconitine (*3-36*) and lappaconitine (*3-37*) are two known alkaloids isolated from *A. sinomontanum*. Both alkaloids are characterized by a methoxy group at C-14 and a hydroxyl group at C-4 that is esterified with *N*-acetylanthranilic acid [22, 23].

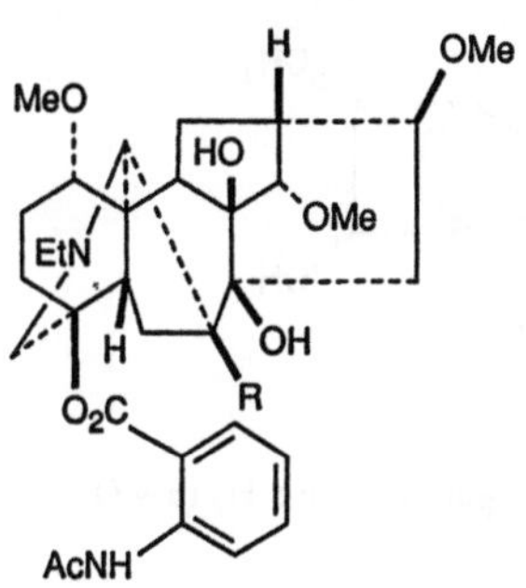

Ranaconitine (*3-36*): R=OH
Lappaconitine (*3-37*): R=H

3.2.3.6 *Aconitum vilmorrianum*

Three alkaloids, vilmorrianine A (*3-38*), B, and C (*3-39*), were isolated from *A. vilmorrianum*, a folk medicine used in China for treatment of traumas. Vilmorrianine B was shown to be identical with yunaconitine, vilmorrianine C with foresaconitine. The structure of vilmorrianine A was determined [24]; isolation of the fourth alkaloid, vilmorrianine D (*3-40*), was also reported [25]. Vilmorrianine D is identical to sachaconitine.

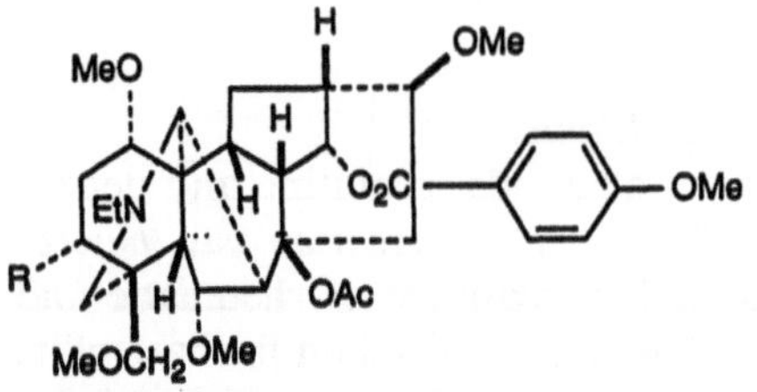

Vilmorrianine A (*3-38*): R=OH
Vilmorrianine C (*3-39*): R=H
(Foresaconitine)

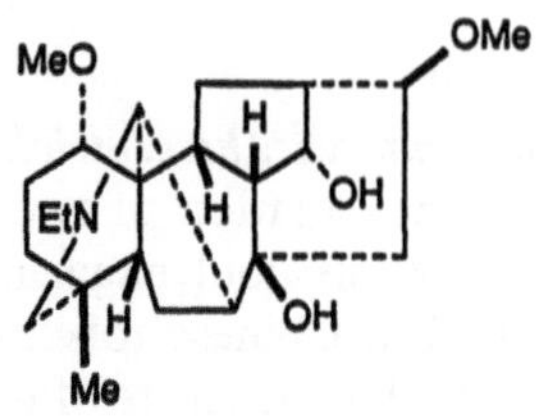

Vilmorrianine D (*3-40*)
(Sachaconitine)

3.2.3.7 *Aconitum koreanum*

Guan-fu bases A–G are alkaloids isolated from *A. koreanum* [26, 27]. The structure of guan-fu base G (*3-41*) was determined on the basis of spectral analyses and X-ray analysis of the methyl iodide to be a C_{20} diterpene alkaloid of the hetisan type. Guan-fu bases A (*3-42*) and G gave the same aminoalcohol and acetic acid on hydrolysis. Guan-fu base G was then proven to be identical with acetyl guan-fu base A. Spectroscopic evidence revealed that an additional acetyl group was located at position C-12. Guan-fu base G thus represents the first example of a hetisan type alkaloid with a C-14 tertiary alcohol and a missing oxygen group at C-15 [1]. Recently, alkaloid guan-fu base H (*3-43*) [28], 2-isobutyryl-14-hydroxyhitisine (*3-44*), 2-acetyl-14-hydroxyhitisine (*3-45*), and isoatisine (*3-46*) [29] were obtained from *A. koreanum*. Guan-fu base H was identified as atisinium chloride; 2-isobutyryl-14-hydroxyhetisine was designated as guan-fun base Z [30]; and the 2-acetyl analogues as guan-fu base Y [31]. *A. koreanum* has been used in folk medicine as an expectorant and as an analgesic.

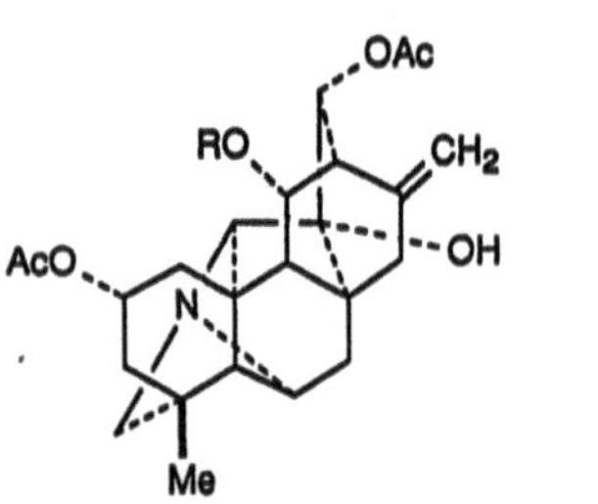

Guan-fu base A (*3-42*): R=H
Guan-fu base G (*3-41*): R=Ac

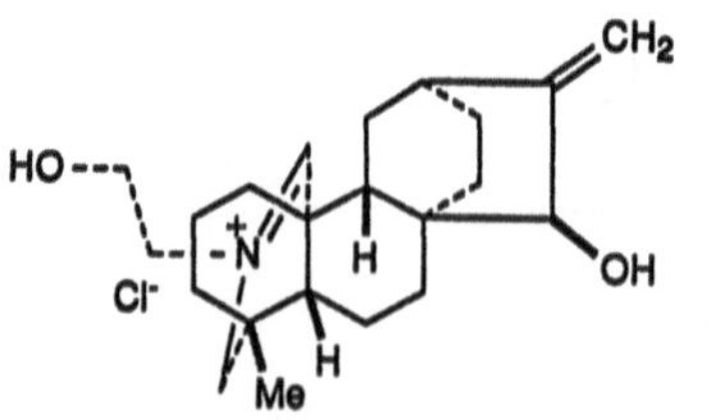

Atisinium chloride (guan-fu base H) (*3-43*)

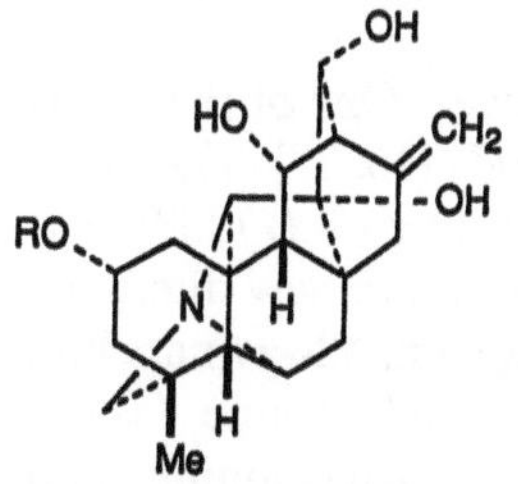

Guan-fu base Z (*3-44*): R=$(CH_3)_2$–CH–CO
Guan-fu base Y (*3-45*): R=Ac

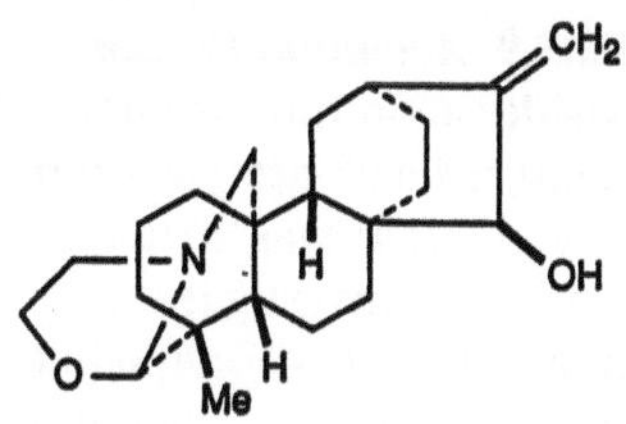

Isoatisine (*3-46*)

3.2.3.8 *Aconitum finetianum*

A. finetianum is an *Aconitum* species native to China and used in folk medicine to treat acute dysentery and enteritis. It induces relaxation of smooth muscle. From the roots of *A. finetianum*, delsoline (*3-47*), lycoctonine (*3-48*), avadharidine (*3-49*), and two as yet unknown alkaloids were isolated. In avadharidine the hydroxyl group at position C-4 is esterified with *N*-aminodioxobutylanthranilic acid [32].

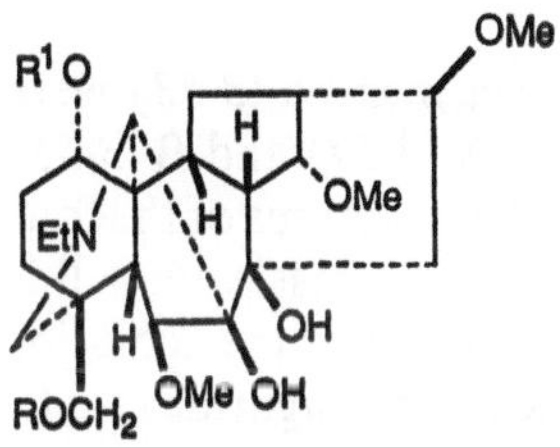

Delsoline (*3-47*): R=CH_3, R^1=H
Lycoctonine (*3-48*): R=H, R^1=CH_3

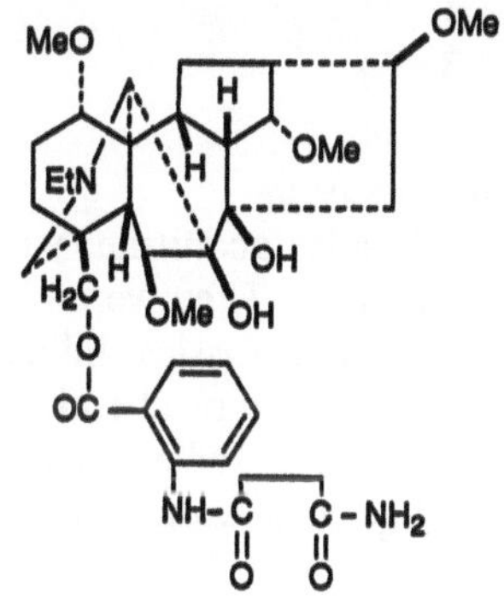

Avadharidine (*3-49*)

Jiang et al. [33–35] reported the isolation of eight diterpene alkaloids from *A. finetianum*. Five of them were the known alkaloids avadharidine, lycoctonine, ranaconitine, lappaconitine, and *N*-deacetyllappaconitine. One of the three new alkaloids, finaconitine (*3-50*), was assigned the structure of 10-β-hydroxyranaconitine on the basis of spectral data. The other two were determined to be *N*-deacetylranaconitine and *N*-deacetylfinaconitine.

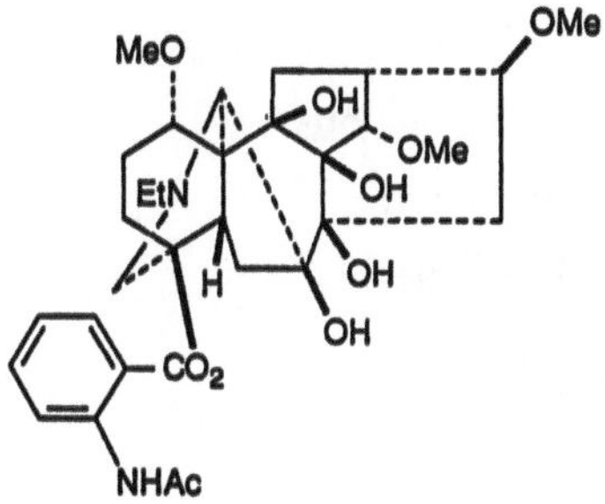

Finaconitine (*3-50*)

3.2.3.9 *Aconitum flavum*

Besides aconitine two new alkaloids were isolated from *A. flavum*. One of them was identified as 3-acetylaconitine by chemical and spectral methods [36, 37]. The other new alkaloid was structurally elucidated and named flavaconitine (*3-31*) [38]. Recently, a reinvestigation of *A. flavum* resulted in the isolation of five new diterpene alkaloids, dehydronapelline, 12-acetyl-luciduscuiline, 1-*epi*-napellin, 12-*epi*-napellin, and 1-demethylhypaconitine, along with napelline (*3-51*), luciduscuiline (*3-52*), aconitine, 3-acetylaconitine, deoxyaconitine, flavaconitine, benzoylaconine, and neoline [39].

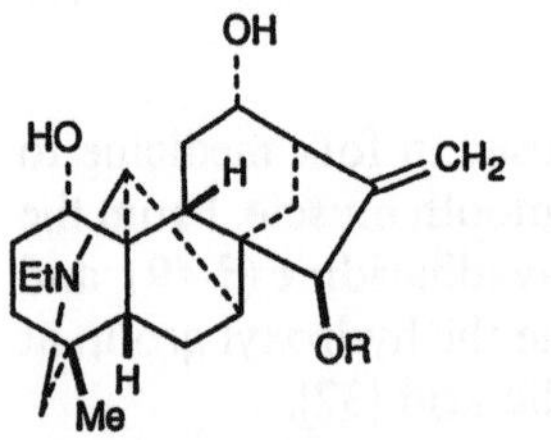

Napelline (*3-51*): R=H
Luciduscuiline (*3-52*): R=Ac

3.2.3.10 *Aconitum crassicaule*

Four new alkaloids besides aconitine, yunaconitine, and chasmanine (*3-53*) were isolated from *A. crassicaule*. They were named crassicauline A (*3-54*) and B (*3-55*) [40–42], crassicaulisine, and crassicaulidine (*3-56*) [43, 44]. The structures of the four new alkaloids were determined by chemical and spectral analyses. Crassicauline B is an alkaloid of the hetisan type, whereas the other three are alkaloids of the aconitine type. Crassicaulisine has been found to be structurally identical with nagarine (*3-33*).

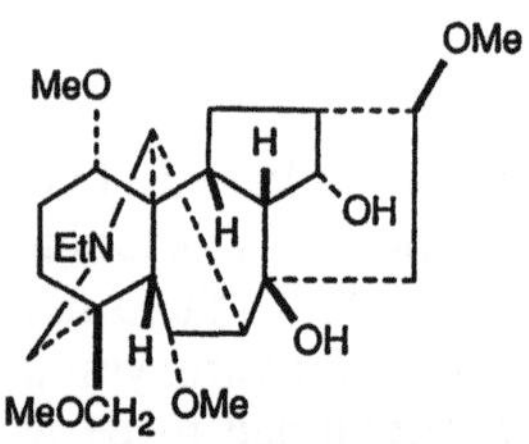

Chasmanine (*3-53*)

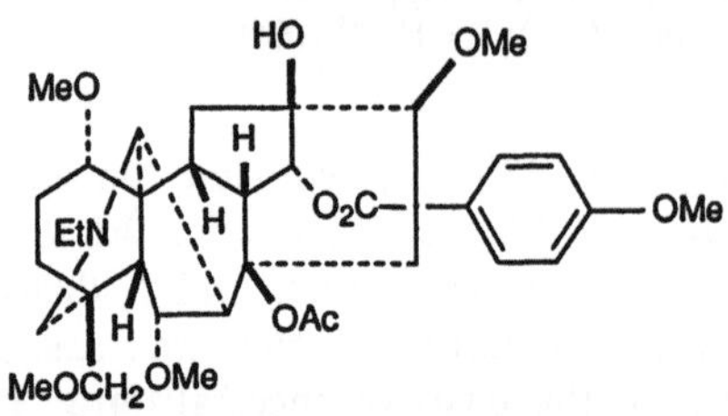

Crassicauline A (*3-54*)

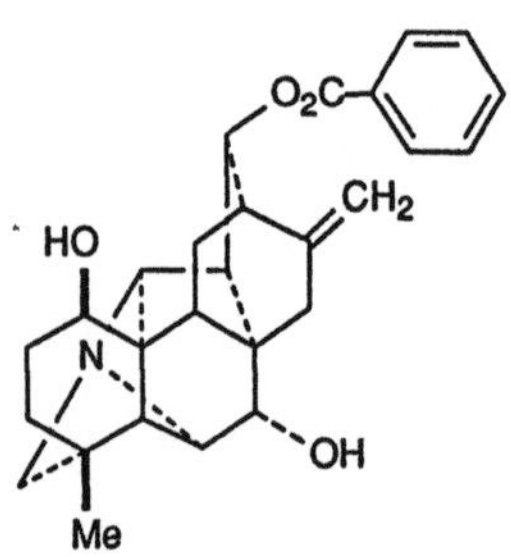

Crassicauline B (*3-55*)

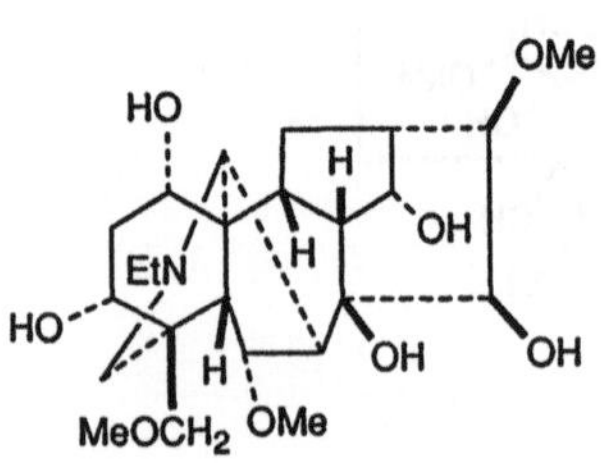

Crassicaulidine (*3-56*)

3.2.3.11 ***Aconitum franchetii***

Six alkaloids, indaconitine (*3-57*), chasmaconitine (*3-58*), chasmanine, talatisamine, ludaconitine (*3-59*) [45], and franchetine (*3-60*) [46], were isolated from *A. franchetii* and their structures elucidated. Ludaconitine and franchetine are two new alkaloids. Franchetine was the first example of a C_{19} diterpene alkaloid with a dihydropyrane ring.

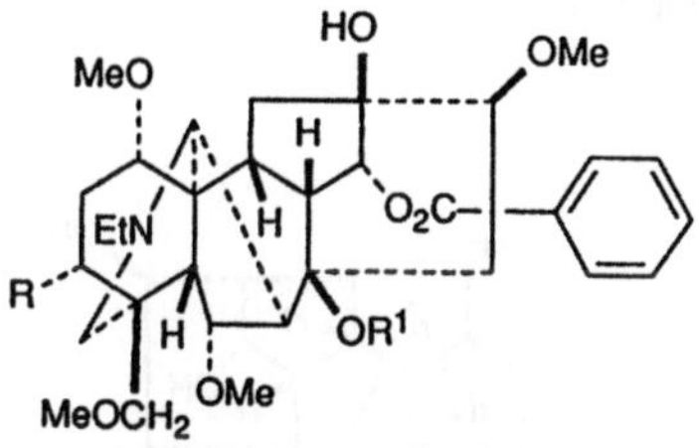

Indaconitine (*3-57*): R=OH, R^1=Ac
Chasmaconitine (*3-58*): R=H, R^1=Ac
Ludaconitine (*3-59*): R=OH, R^1=H

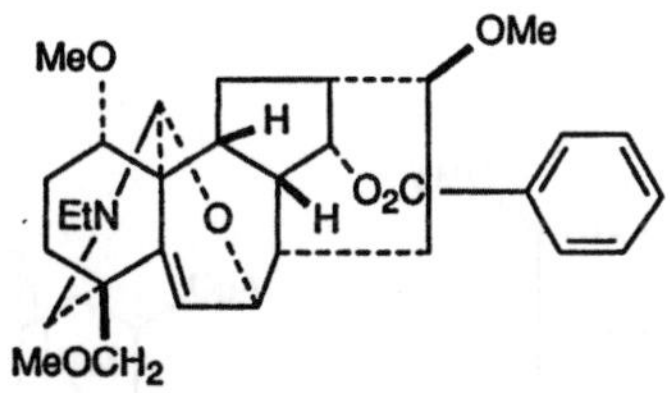

Franchetine (*3-60*)

3.2.3.12 ***Aconitum stapfianum* var. *pubipes***

Two alkaloids were isolated from *A. stapfianum* var. *pubipes*. They were identified by spectral analyses as aconosine (*3-61*) and its 14-acetyl derivative, dolaconine [46, 47].

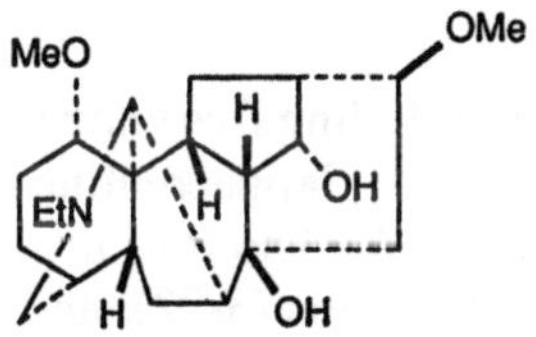

Aconosine (*3-61*)

3.2.3.13 ***Aconitum kongboense***

The diterpene alkaloid vilmorrianinc A was isolated from *A. kongboensc* [48].

3.2.3.14 ***Aconitum episcopale***

Five new alkaloids, episcopalisine (*3-62*), episcopalisinine (*3-63*), episcopalitine (*3-64*), episcopalidine (*3-65*) [49, 50], and scopaline (*3-66*) [25], were isolated from *A. episcopale* and their structures were elucidated by spectral methods. Episcopalidine is an alkaloid of the hetisan type, whereas the other four are C_{18} diterpene alkaloids of the aconitine type, lacking the substituent at position C-4. On consideration of structural features it appeared justified to classify the C_{18} diterpene alkaloids as a separate group and not as a 4-nor subgroup of the C_{19} diterpene alkaloids. In these C_{18} alkaloids the C-3 and, in most cases, the C-6 positions are not oxygenated.

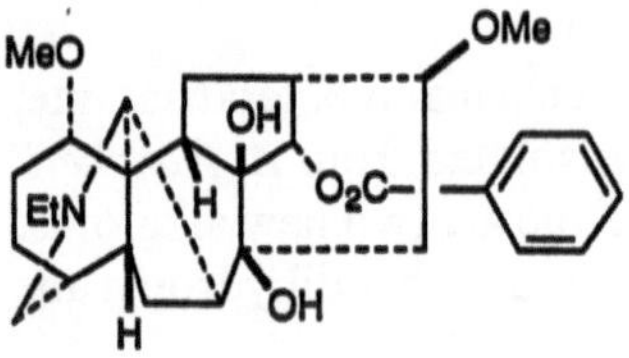

Episcopalisine (*3-62*)

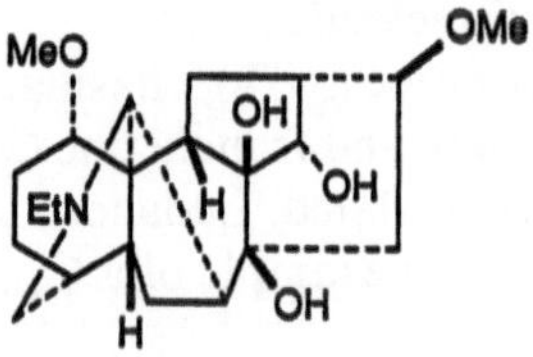

Episcopalisinine (*3-63*)

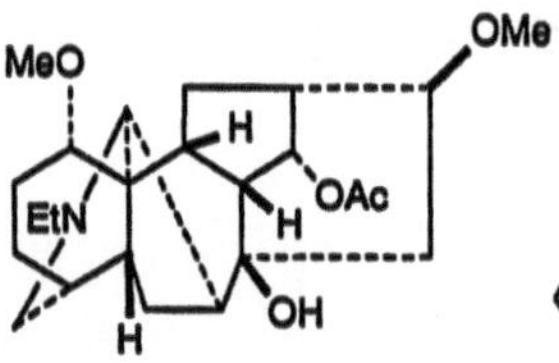

Episcopalitine (*3-64*)

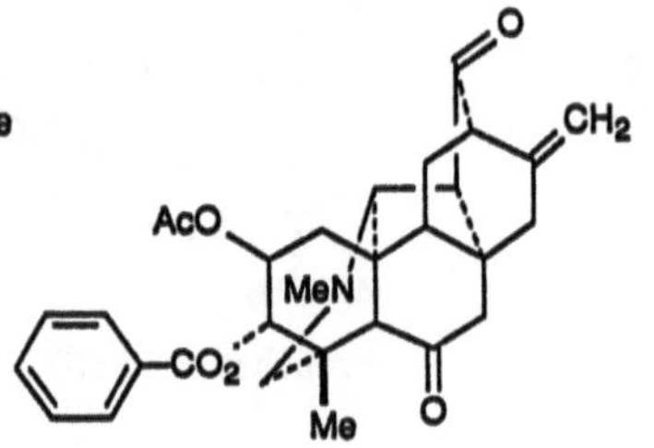

Episcopalidine (*3-65*)

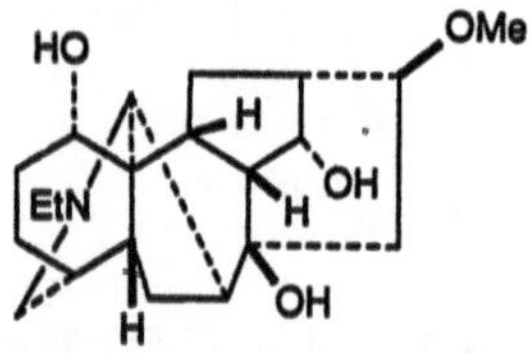

Scopaline (*3-66*)

3.2.3.15 *Aconitum teipeicum*

Four known diterpene alkaloids were isolated from *A. teipeicum* and identified as yunaconitine, neoline, talatisamine, and chasmanine [51].

3.2.3.16 *Aconitum barbatum* var. *puberulum*

Nine alkaloids were isolated from *A. barbatum* var. *puberulum* including five known and four new alkaloids. The five known alkaloids were ranaconitine, lappaconitine, septentriodine (*3-67*), septentrionine (*3-68*), and lycaconitine (*3-69*). The four new alkaloids were puberanine (*3-70*), puberanidine (*3-71*), puberaconitine (*3-72*), and puberaconitidine (*3-73*). These were characterized by physiochemical and spectral analysis [52, 53].

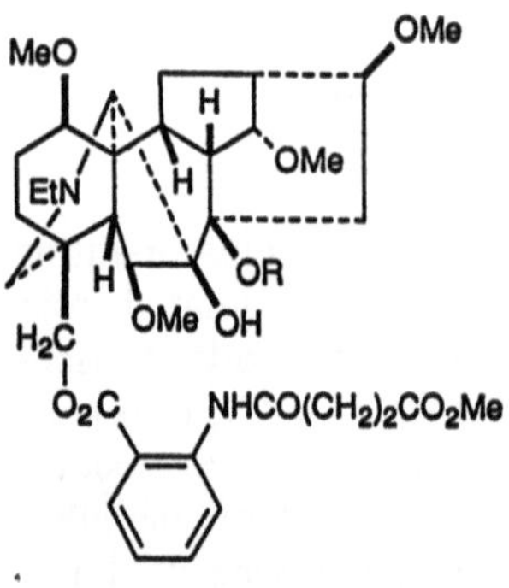

Septentriodine (*3-67*): R=H
Septentrionine (*3-68*): $R=CH_3$

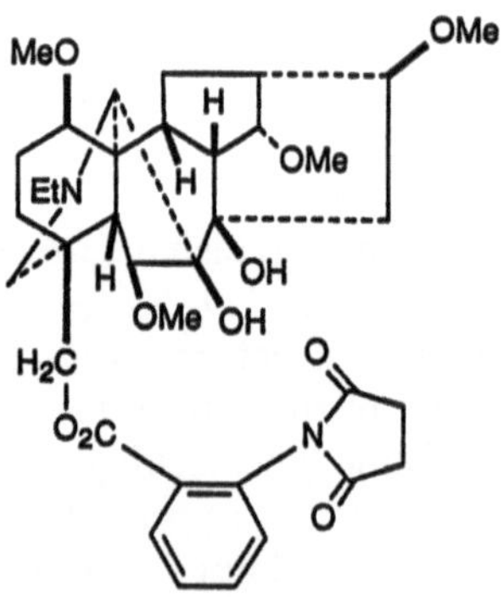

Lycaconitine (*3-69*)

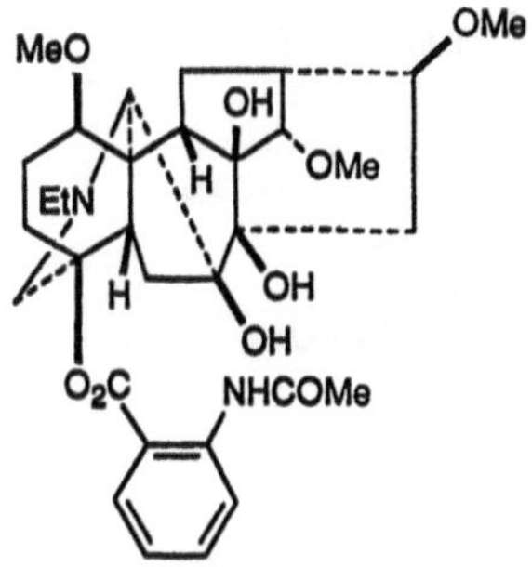

Puberanine (*3-70*)

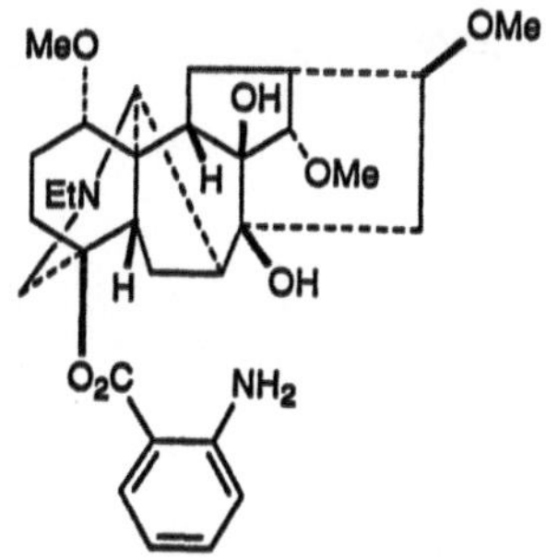

Puberanidine (*3-71*)

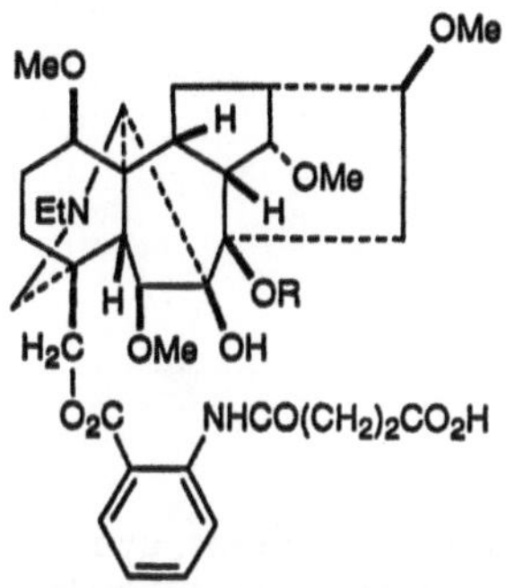

Puberaconitine (*3-72*): R=H
Puberaconitidine (*3-73*): $R=CH_3$

3.2.3.17 *Aconitum pendulum*

Three known alkaloids, hypaconitine, 3-acetylaconitine, and aconitine, and a new alkaloid, penduline (*3-74*), were isolated from *A. pendulum*. Penduline was identified as 3,13-dideoxyaconitine [54].

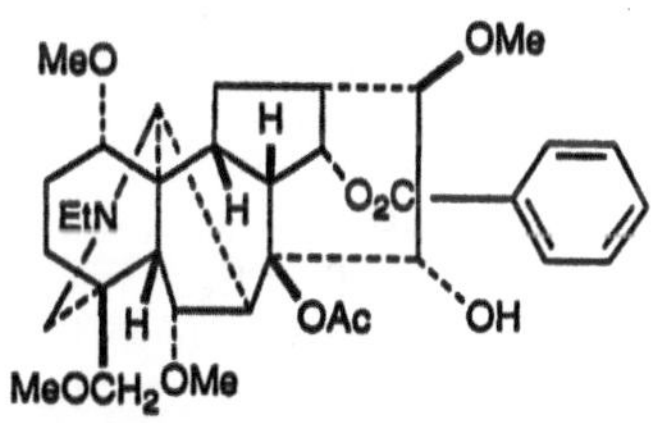

Penduline (*3-74*)

3.2.3.18 *Aconitum forestii*

Thus far, 10 alkaloids have been isolated from *A. forestii*. The main alkaloid foresaconitine (*3-39*), a new alkaloid of the aconitine type, was isolated and structurally identified. It was independently isolated from *A. vilmorrianum* and named vilmorrianine C [55]. Besides foresaconitine six other alkaloids were isolated, including five known alkaloids, crassicauline A, yunaconitine, chasmaconitine, aconosine, and cammaconine (*3-75*), and a new alkaloid, liwaconitine (*3-76*), a C_{19} diterpene alkaloid of the aconitine type containing two p-anisoyl moieties [56, 57]. Further studies revealed three new alkaloids, forestine (*3-77*), foresticine (*3-78*) [58], and 8-deacetylyunaconitine [59], which were structurally characterized by spectroscopy.

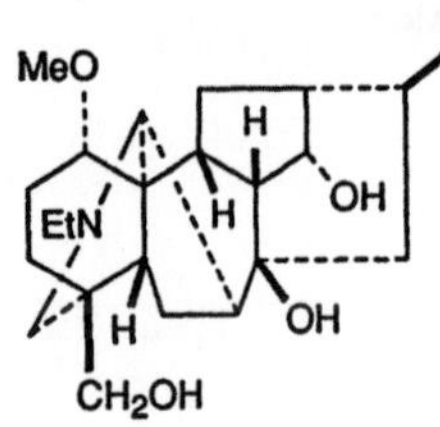

Cammaconine (*3-75*)

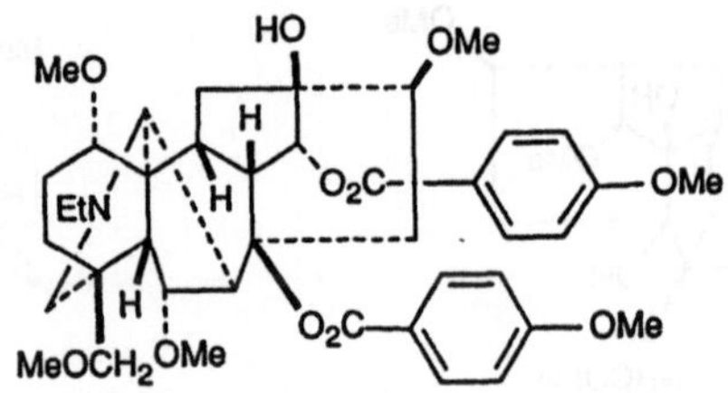

Liwaconitine (*3-76*)

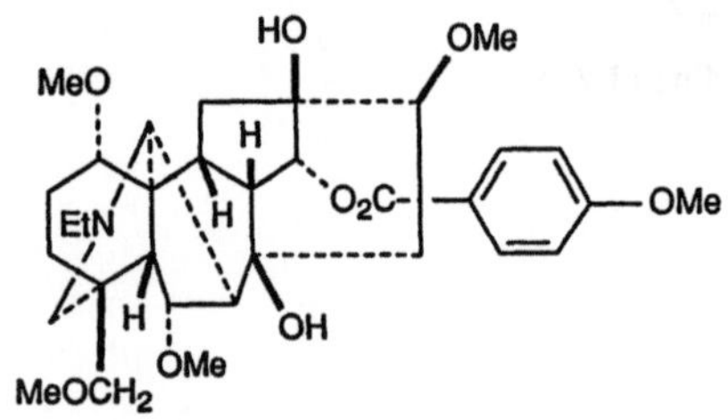

Forestine (*3-77*)

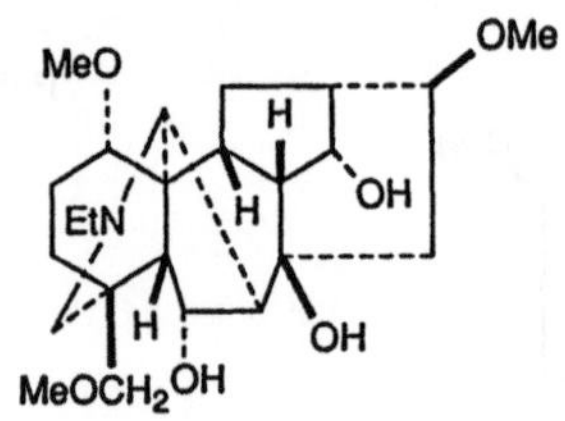

Foresticine (*3-78*)

3.2.3.19 *Aconitum gymnandrum*

Atisinium chloride [60], atisine (*3-79*), talatisamine, and two new alkaloids, gymnaconitine (*3-80*) and its 1-methyl analog methylgymnaconitine (*3-81*) [61], were isolated from *A. gymnandrum*. Gymnaconitine is the first *Aconitum* alkaloid to contain 3,4-*O*-dimethyl-caffeic acid ester.

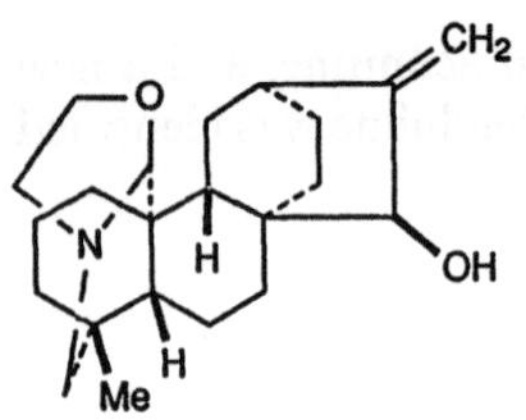

Atisine (*3-79*)

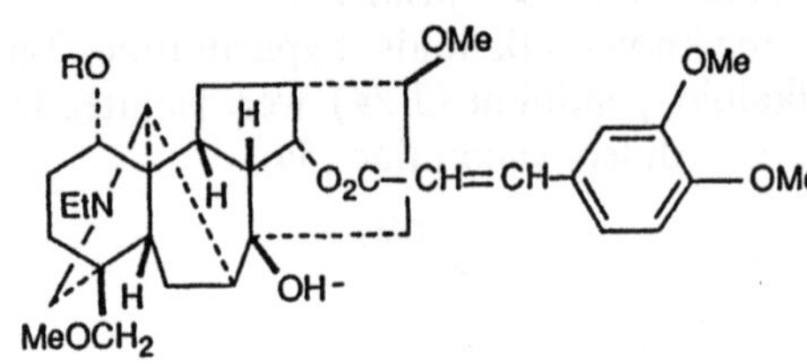

Gymnaconitine (*3-80*): R=H
Methylgymnaconitine (*3-81*): R=CH_3

3.2.3.20 *Aconitum karakolicum*

Aconitine, deoxyaconitine, neoline, and songorine were isolated from *A. karakolicum*. The content of aconitine in the root of *A. karakolicum* was 0.49%, a large amount compared with other *Aconitum* species [62].

3.2.3.21 *Aconitum pseudogeniculatum*

From the roots of *A. pseudogeniculatum* six diterpene alkaloids were isolated and structurally identified: denudatine, chasmanine, talatisamine, yunaconitine, crassicauline A, and vilmorrianine C. Yunaconitine is the major alkaloid [63].

3.2.3.22 *Aconitum polyschistum*

Three new diterpene alkaloids, polyschistine A (*3-82*), B (*3-83*), and C (*3-84*), were isolated from *A. polyschistum*. The structures of these three new alkaloids were determined on the basis of ^1H and ^{13}C NMR spectra [64]. Polyschistine A possesses an ethoxy group at position C-8 and polyschistine C has a secondary amine group.

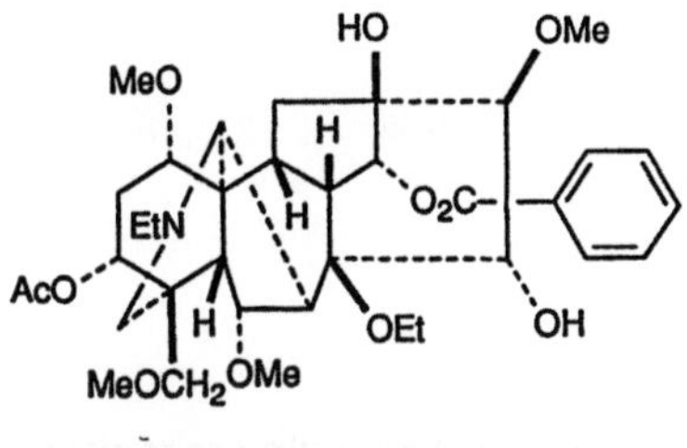

Polyschistine A (*3-82*)

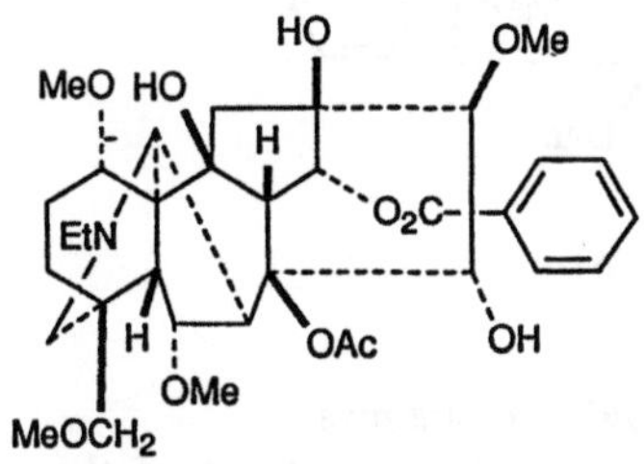

Polyschistine B (*3-83*)

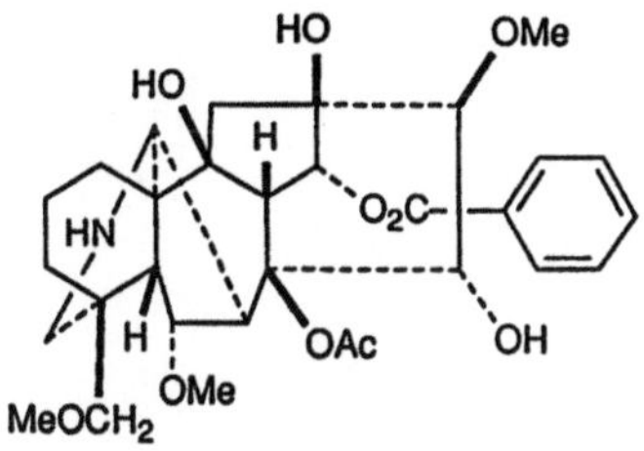

Polyschistine C (*3-84*)

3.2.3.23 *Aconitum longtounense*

The known alkaloids chasmanine, yunaconitine, and crassicauline A and the new alkaloids longtouconitine B [65] and 8-acetyl-14-benzoylchasmanine [66] were isolated from *A. longtounense*. Longtouconitine B was proved to be identical with forestine (*3-78*).

3.2.3.24 *Aconitum geniculatum*

Seven diterpene alkaloids were isolated from *A. geniculatum*. Six of them were identified as yunaconitine, crassicauline A, vilmorrianine C, talatisamine, chasmanine, and 8-deacetylyunaconitine. A new alkaloid, geniconitine (*3-85*), was structurally elucidated [67].

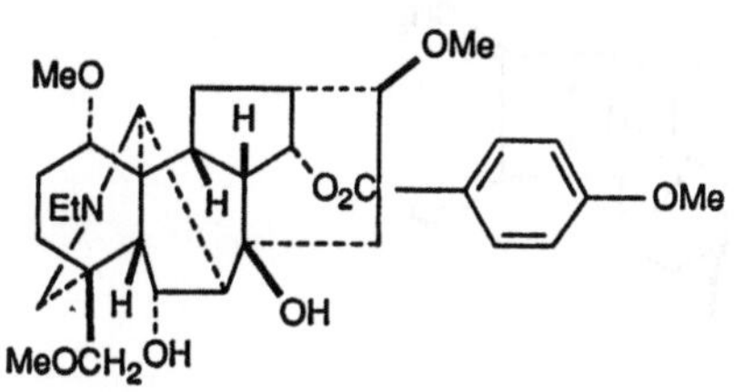

Geniconitine (*3-85*)

3.2.3.25 *Aconitum duclouxii*

A chasmaconitine analog with a 12-β-hydroxyl group, duclouxine (*3-86*), was isolated from the root of *A. duclouxii* together with aconitine [68].

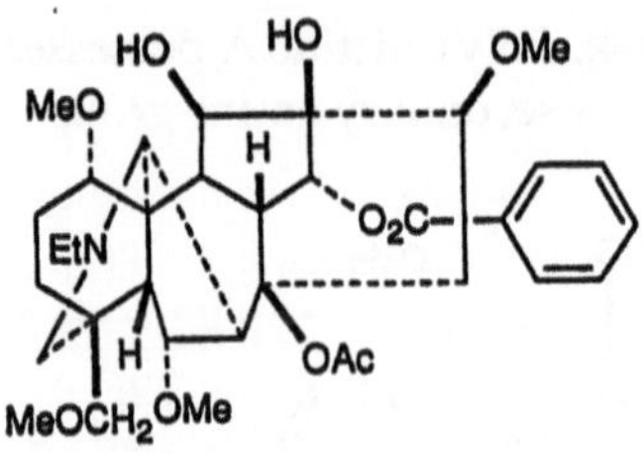

Duclouxine (*3-86*)

3.2.3.26 *Aconitum chinense*

From the cold methanol extract of the root of *A. chinense*, benzoylmesaconitine, neoline, ajaconine (*3-97*), ignavine, and fuziline (*3-16*) were isolated in addition to aconitine, mesaconitine, and hypaconitine [69].

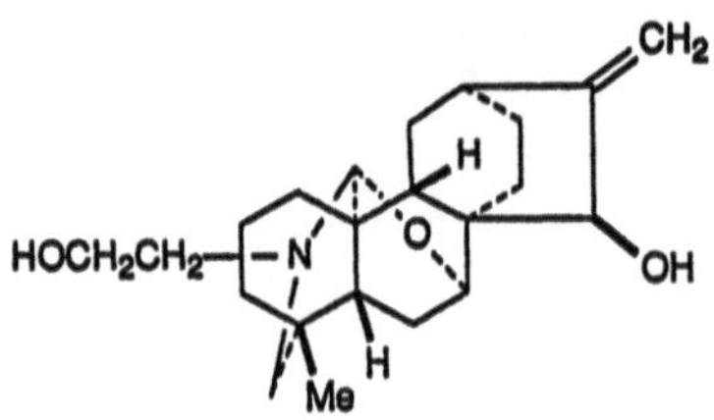

Ajaconine (*3-87*)

3.2.3.27 *Aconitum jinyangense*

A new alkaloid, jynosine, was isolated from the root of *A. jinyangense* together with denudatine and 14-acetyl-neoline. Jynosine was identified as 15-acetyldenudatine [70].

3.2.3.28 *Aconitum pseudohuiliense*

Denudatine (*3-28*) is the main alkaloid present in the root of *A. pseudohuiliense*. In addition, three new alkaloids related to denudatine were isolated and designated as lepenine (*3-88*), lepedine (*3-89*), and lepetine (*3-90*) [71].

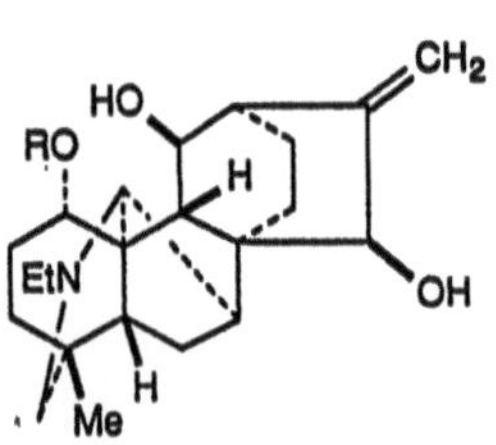

Lepenine (*3-88*): R = H
Lepedine (*3-89*): R = CH_3

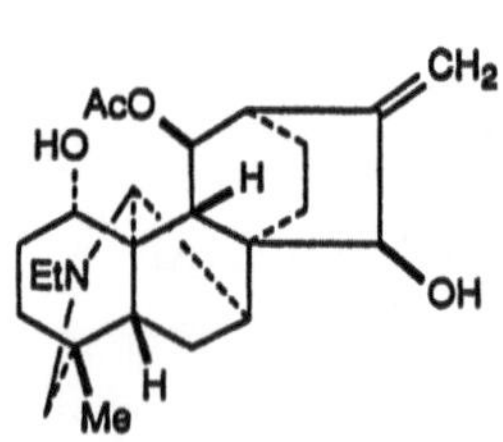

Lepetine (*3-90*)

3.2.3.29 *Aconitum scaposum* var. *vaginatum*

Three new diterpene alkaloids vaginatine (*3-91*), vaginaline (*3-92*), and vaginadine (*3-93*) were isolated from the root of *Aconitum scaposum* var. *vaginatum*. They are polyhydroxyaconitane derivatives. Whereas vaginatine is a tetrol, vaginalin is a 14-oxo derivative and vaginadine a 6,14-dioxo analogue [72].

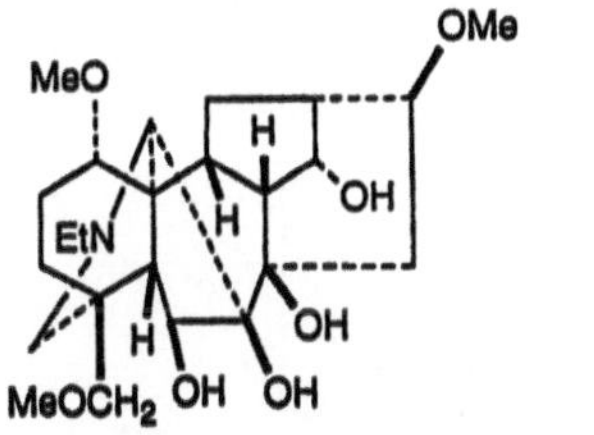

Vaginatine (*3-91*)

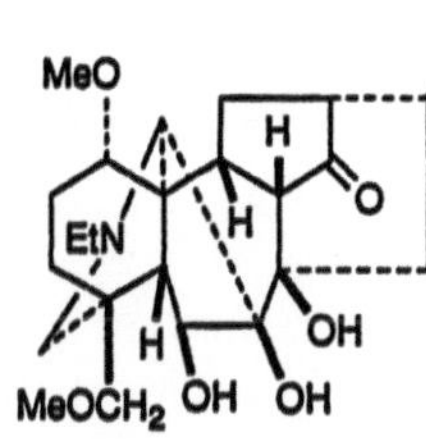

Vaginaline (*3-92*)

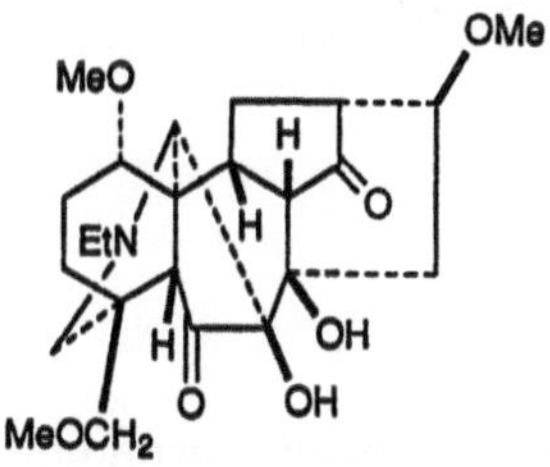

Vaginadine (*3-93*)

3.2.3.30 *Aconitum tanguticum*

Tanwusine (*3-94*), a new alkaloid with a hetisan skeleton, was isolated, together with atisine, heteratisine (*3-95*) and its benzoyl derivative, from the whole plant of *A. tanguticum* [73].

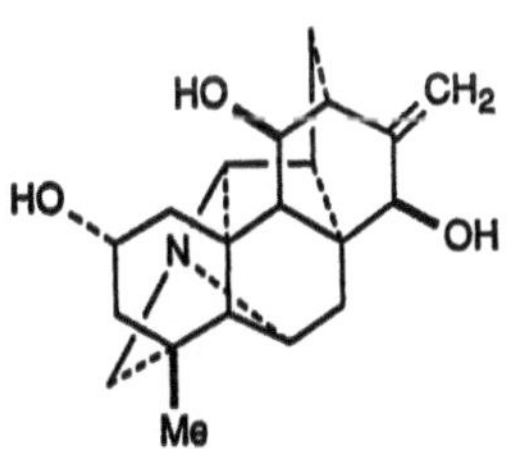

Tanwusine (*3-94*)

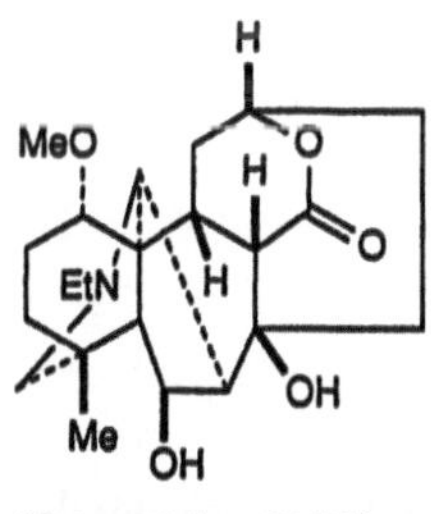

Heteratisine (*3-95*)

3.2.4 Chemical Constituents Other than Diterpene Alkaloids

Besides the diterpene alkaloids some other compounds, with or without nitrogen, were isolated from aconite roots and chemically and pharmacologically investigated.

From the roots of *A. carmichaeli*, coryneine chloride (*3-96*) [74] and salsolinol (*3-97*) [75] were discovered. From the roots of *A. japonicum*, used for a long time as heart stimulants, diuretics, and analgesic agents, higenamine (*3-98*) and yokonoside (*3-99*) [76, 77], were isolated and identified.

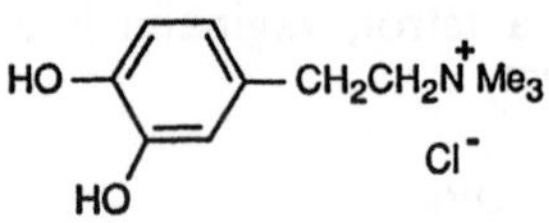

Coryneine chloride (*3-96*)

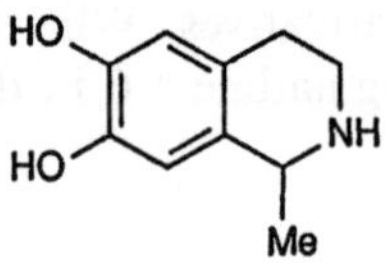

Salsolinol (*3-97*)

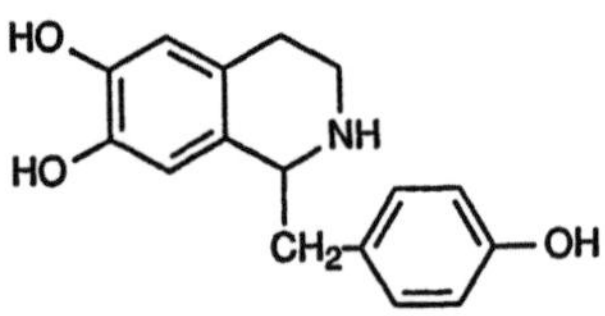

Higenamine (*3-98*)

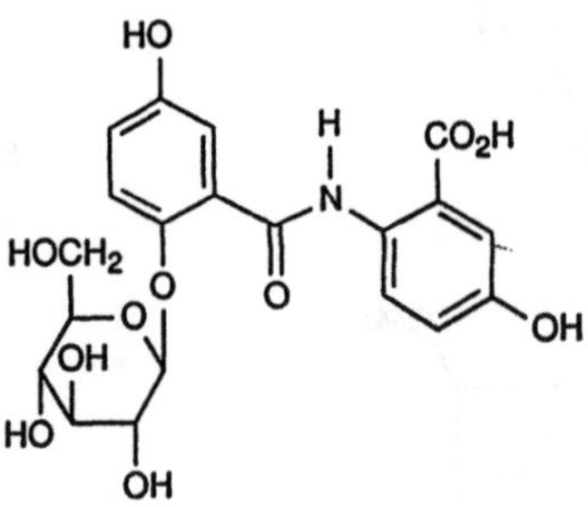

Yokonoside (*3-99*)

Recently, the isolation and determination of certain glycans, namely, aconitans A, B, C, and D, from the roots of *A. carmichaeli* were reported by Konno et al. [78]. Gel chromatography of aconitans A, B, C, and D over Sephacryl S-200 or S-300 gave molecular weights of approximately 8.2×10^3, 2.1×10^5, 4.3×10^3, and 4.2×10^4, respectively. The glycans contain a neutral sugar, an acidic sugar (aconitan C), and a peptide moiety in different molar ratios. The neutral sugar ccomponents were glucose for aconitan A and rhamnose, arabinose, mannose, galactose, and glucose, with a molar ratio of 0.6:0.1:2.3:1.0:1.2 for aconitan B and 1.1:1.0:0.2:1.0:0.8 for aconitan C. Aconitan D contains rhamnose, arabinose, galactose, and glucose with a molar ratio of 0.4:0.6:1.0:0.3. Aconitan A is composed of α-(1→6)-linked D-glucopyranose molecules with three branching points at O-3 [79].

3.3 Pharmacology

3.3.1 Toxicity

After i.v. injection into mice, the acute LD_{50} values of aconitine, mesaconitine, beiwutine, hypaconitine, 3-acetylaconitine, and deoxyaconitine were 0.22, 0.27, 0.42, 0.50, 1.01, and 1.90 mg/kg, respectively [80]. In subacute toxicity tests, rats treated with aconite root extract and mesaconitine at daily doses of 1.1 g/kg and 1.3 mg/kg, respectively, died within 3–6 days. In mice, aconite root extract at a daily dose of 800 mg/kg caused a decrease in the number of erythrocytes and in the serum levels of total protein and albumin [81]. In subchronic toxicity tests, a decrease in body weight was observed in rats treated with mesaconitine at a daily dose of 0.4 mg/kg. Glutamic-oxalacetic transaminase and lactate dehydrogenase levels also decreased in animals treated with raw or processed aconite roots at daily doses of 0.08–0.32

and 5–20 g/kg, respectively. Alkaline phosphatase was elevated in mice but lowered in rats after treatment with raw aconite root or mesaconitine. Pathological examination showed a slight focal cell infiltration in the liver of some mice treated with raw aconite root and mesaconitine. No pathologic change was observed in mice treated with processed aconite root at a daily dose of 1 g/kg [81].

Both respiratory depression in rabbits and heart fibrillation in guinea pigs caused by aconitine were antagonized by i.v. infusion of calcium chloride. Atropine counteracted the antagonistic effect of calcium choride on respiratory depression caused by large doses of aconitine [82]. Hydrocortisone was effective in treatment of *A. brachypodum* poisoning in rabbits [83].

Toxic effects of aconitine are mainly seen in the nervous system, effecting first excitation and then inhibition of the vagus and sensory nerves.

Aconitine also acts directly on cardiac muscle. Symptoms of intoxication include systemic paralysis, nausea, and vomiting, followed by dizziness, palpitation, intolerance of cold, irritability, delirium, hypotension, arrhythmia, shock, and coma. The common abnormal electrocardiographic signs include arrhythmia, ventricular tremor, atrioventricular block, and myocardial damage [84].

The toxicity of aconitine and nine analogues was tested in mice. High toxicity appeared to be associated with the presence of both an acetyl and a benzoyl group [85]. With respect to the toxicity of the roots of eight *Aconitum* species, that of those characterized by a lack of diester alkaloids and containing mainly C_{20} aminoalcohols or monoesterified C_{19} diterpene alkaloids, is comparatively low. In mice, i.v. LD_{50} values were 1600–3400 mg/kg. By contrast, the toxicity of those species containing diester bases, with an acetoxy residue at C-8 and benzoyloxy or anisoyloxy residues at C-14, is very high. The *Aconitum* species that contain mainly an aminoalcohol of C_{19} diterpene alkaloids display intermediate toxicity with LD_{50} values of 210–260 mg/kg [86].

Aconitine exhibits a noncompetitive, inhibitory effect on pig heart aconitase in vitro. This suggests a possible molecular basis for the toxic and pharmacologic actions produced in experimental animals by aconitine [87].

3.3.2 Arrhythmic Effects

Aconitine, mesaconitine, beiwutine, hypaconitine, 3-acetylaconitine, and deoxyaconitine administered i.v. to anesthetized rats induced arrhythmia. The potency of arrhythmia induction was of the order: beiwutine > mesaconitine > aconitine > 3-acetylaconitine > hypaconitine > deoxyaconitine [80]. Aconitine caused cardiac arrhythmia in mice. It can be used in experimental arrhythmia models to screen for antiarrhythmic drugs. The arrhythmia caused by aconitine was counteracted by i.v. administration of calcium [88]. Thus, when aconitine-induced arrhythmia is used to screen for antiarrhythmic effects of crude extracts, e.g., herbal drugs, interference from calcium that might be present in the extracts should be considered. Beiwutine appeared to be superior to aconitine in producing experimental arrhythmia for drug screening, since beiwutine administered i.v. to anesthetized rats caused arrhythmia but had little effect on blood pressure [80].

The arrhythmic activity of aconite alkaloids can be correlated with the presence of a benzoyl ester group; such compounds have been shown to be effective in inducing arrhythmia [85].

In anesthetized open-chest dogs aconitine, at doses that did not produce cardiac arrhythmia, did not affect prostaglandin E (PGE) and $PGF_{2\alpha}$ efflux into coronary sinus blood but increased PGE and $PGF_{2\alpha}$ efflux at doses that produced arrhythmia. The increased $PGF_{2\alpha}$ efflux in cardiac arrhythmia was not affected by antiarrhythmics. Thus, prostaglandins appear to be released in cardiac arrhythmia and the increased PGE efflux during arrhythmia may be a protective mechanism against sympathetic influences [89].

3.3.3 Analgesic Effects

Three widely occurring alkaloids in aconite roots, mesaconitine, aconitine, and hypaconitine, showed analgesic activity in mice [90]. An investigation on the contribution of central monoamines and the opiate receptor to mesaconitine-induced analgesia showed that the analgesic action of intracerebral mesaconitine was dose dependent, indicating that its activity is elicited through the central nervous system. Levallorphan did not affect the analgesic activity of mesaconitine, suggesting that its activity is not mediated via the opiate receptor. Thus, the analgesic activity of mesaconitine is closely related to responses involving the central catecholaminergic system, particularly the noradrenergic system [91].

Mesaconitine-induced antinociception could be significantly potentiated by cyclic adenosine monophosphate (cAMP). The phosphodiesterase inhibitor theophylline also significantly potentiated mesaconitine-induced antinociception. Furthermore, mesaconitine-induced antinociception was markedly increased by isoproterenol, a β-adrenoceptor agonist, and reduced by propranolol, a β-adrenoceptor antagonist. Apparently, the antinociceptive action of mesaconitine is potentiated through cAMP and occurs via stimulation of the central β-adrenergic system [92].

Lappaconitine [93], *N*-deacetyllappaconitine, *N*-deacetylranaconitine, and *N*-deacetylfinaconitine [94] all exhibited a strong analgesic activity in animal experiments. The median analgesic dose (ED_{50}) of lappaconitine in mice after i.p. administration was found to be 3.5 mg/kg. An anesthesia test on rabbit cornea revealed that the surface anesthetic potency of lappaconitine was eight times stronger than that of cocaine. Local anesthetic effects on sciatic nerve block in mice were five times those of cocaine; in the intracutaneous wheal test in the guina pig lappaconitine was found to be about equally active [94].

It is interesting to note that aconitine could be used as an agent to induce a writhing syndrome in mice, thus providing a suitable model system for assessing aspirin-like analgesic activity. The writhing appeared quickly and was of greater frequency and longer duration than that caused by other agents. Orally administered, nonnarcotic analgesics antagonized the aconitine-induced writhing more selectively than did narcotic analgesics. Potencies of some nonnarcotic analgesics tested decreased in the following order: acetylsalicylic acid, phenylbutazone, amidopyrine, phenacetin, sodium salicylate [95].

3.3.4 Antiinflammatory Activity

Aconite alkaloids, obtained from the extract of *A. carmichaeli* roots, inhibited the increased vascular permeability induced by acetic acid in mouse peritoneal cavity and that induced by histamine in rat intradermal sites; they also inhibited hind paw edema formation induced by carrageenin in rats and mice [96]. In rats, aconitine, mesaconitine, hypaconitine, benzoylaconitine, benzoylmesaconitine, and benzoylhypaconitine prevented inflammation induced by injection of 0.1 ml 1% carrageenin solution. Inflammation could be prevented by oral administration of aconitine at a dose of 0.1 mg/kg 30 min prior to the injection of carrageenin [97].

The mechanism of mesaconitine antiinflammatory activity was studied in experimental animals. Mesaconitine inhibited carrageenin-induced hind paw edema in sham operated mice and adrenalectomized mice. Hind paw edema, produced by subplantar injection of histamine, serotonin, and PGE_1, was suppressed by mesaconitine, indicating that mesaconitine elicits antiinflammatory activity at the early, exudative stage of inflammations; however, mesaconitine did not affect biosynthesis of the prostaglandins. Mesaconitine produced dose dependent antiinflammatory effects on hind paw edema and inhibited the increase in vascular permeability mediated by acetic acid and agar when administered intracerebrally at doses that induced analgesic activity. Thus, the antiinflammatory activity of mesaconitine appears to involve the central nervous system [98].

3-Acetylaconitine had antiinflammatory effects on vascular permeability, edema, and granuloma formation in mice and in rats [99]. Ignavine had antiinflammatory effects on acetic acid-induced writhing and carrageenin paw edema tests at oral doses of 50–100 mg/kg but did not produce undesirable effects such as sedation, motor incoordination, muscle relaxation, hypothermia, ulceration, or antidiuresis [100]. A series of aconine esters of fatty acids were prepared and tested for antiinflammatory and analgesic activities [101].

3.3.5 Other Pharmacological Activities

Yang et al. reported on the antihistamine effects of aconite alkaloids. The total alkaloids of *A. kusnezoffii* had antihistamine and antiacetylcholine effects on isolated guinea pig ileum; aconitine had the same action. Their antihistamine potency was much weaker than that of promethazine. The contraction of guinea pig ileum induced by egg albumin was antagonized by total alkaloids. The total alkaloids lowered the blood pressure in cats, and at higher doses the decrement became more significant and persistent. The i.p. LD_{50} of the total alkaloid was 98 μg/kg in mice and 0.14 mg/kg in guinea pigs [102].

Mesaconitine significantly stimulated incorporation of 5-[^{3}H] orotic acid into liver nuclear RNA 16 h after its administration. Among the aconite alkaloids mesaconitine exhibited the strongest activity. Stimulation could be inhibited by actinomycin D. Mesaconitine also increased incorporation of 5-[^{3}H]orotic acid into ribosomal RNA in mouse liver. When mesaconitine was added to an RNA polymerase preparation obtained from rat liver, the increase in enzyme activity was weak. RNA polymerase from the liver of rats previously treated with mesaconitine elicited increased incorporation of [^{3}H] cytidine monophosphate (CMP) into RNA compared with that from livers of untreated rats. Thus, aconite alkaloids, particularly mesaconitine, accelerate liver RNA synthesis mainly by increasing RNA polymerase activity [103]. Mesaconitine, aconitine, and hypaconitine accelerate protein synthesis in mouse liver. Aconitine, given orally to mice at a dose of 0.5 mg/kg, increased the uptake of [^{3}H] leucine into liver protein by 35% [104].

Intraperitoneal delsoline, at doses of 1.25–2.5 mg/kg in dogs or 10 mg/kg in rats, had a hypotensive effect. In conscious rats rendered hypertensive by ligation of the left renal artery, 75 mg/kg delsoline significantly lowered systolic blood pressure and heart rate and enhanced respiration amplitude [105].

Interestingly, guan-fu base A showed an antiarrhythmic activity in various animals. When given at a dose of 20–30 mg/kg i.v., it decreased the incidence of ventricular fibrillation from $CaCl_2$ in rats and from ouabain in anesthetized guinea pigs. The ventricular fibrillation threshold to electric stimulation of cats was elevated after i.v. administration of 2–8 mg/kg guan-fu base A [106]. Treatment of canine Purkinje fibers in vitro with guan-fu base A at 100 µg/ml for 20 min decreased the amplitude and V_{max} of the action potential and lengthened both its duration at 100% repolarization and the effective refractory period, suggesting that guan-fu base A has electrophysiological properties similar to antiarrhythmic drugs [107].

Nonditerpene alkaloid constituents such as higenamine had a significant cardiovascular effect. In anesthetized dogs, an aqueous solution of an alcohol extract of aconite root applied i.v. at a dose of 2 g/kg had no effect on heart rate or blood pressure, whereas higenamine given i.v. at a dose of 1–4 µg/kg increased heart rate and coronary artery circulation. In conscious dogs, aconite root extract caused hypertension and tachycardia. These effects were decreased by pretreatment with reserpine. Comparison of the effects of α- and β-blockers on the responses to higenamine and aconite root suggests that higenamine is a β-agonist, whereas aconite root has both α- and β-adrenergic agonist activities [108, 109]. In an in vitro study, contraction of isolated toad heart was enhanced by higenamine at a concentration of 10^{-8} g/ml [109]. Salsolinol, isolated from the lateral root of *A. carmichaeli*, also has cardiac and hypertensive activities [75].

The glycans aconitan A, B, C, and D, isolated by Konno et al. [78] from the root of *A. carmichaeli*, were found to possess a hypoglycemic activity when injected i.p. into normal mice. The plasma glucose levels of the experimental animals decreased in a dose dependent manner 7 h after administration. Aconitans B and D were more active than the other two aconitans and, 24 h after administration, still had remarkable effects. The main glycan aconitan A exhibited a hypoglycemic action 7 h after i.p. injection into alloxan-induced hyperglycemic mice.

References

1. Zhu YL, Zhu RH (1982) Studies on the diterpene alkaloids of the Chinese drug, *Aconitum* spp. Heterocycles 17:607–614
2. Chen Y, Chu YL, Chu JH (1965) The alkaloids of the Chinese drug, *Aconitum* species. IX. Alkaloids from Chuan-wu and Fu-tze, *Aconitum carmichaeli*. Acta Pharm Sin 12:435–439
3. Katz A, Rudin H (1984) Aconitum. VII. Mild alkaline hydrolysis of aconitine. Helv Chim Acta 67:2017–2022
4. Iwasa J, Naruto S (1966) Alkaloids from *Aconitum carmichaeli*. Yakugaku Zasshi 86:585–590
5. Chen SY, Liu YQ, Wang JC (1982) Alkaloidal constituents of cultivated Chuan Wu of Yunnan (China). Acta Bot Yunnan 4:73–75
6. Hikino H, Shiota S, Takahashi M (1983) Seasonal dynamics of the accumulation of aconitine alkaloids in *Aconitum carmichaeli* roots. Shoyakugaku Zasshi 37:68–72
7. Pelletier SW, Mody NV, Varughese KI (1982) Fuziline, a new alkaloid from the Chinese drug "Fuzi" (*Aconitum carmichaeli* Debx.). Heterocycles 18:47–49

8. Konno C, Shirasako M, Hikino H (1982) Pharmaceutical studies on *Aconitum* roots. IX. Structure of senbusine A, B and C, diterpene alkaloids of *Aconitum carmichaeli* roots from China. J Nat Prod 45:128–133
9. Hikino H, Kuroiwa Y, Konno C (1983) Pharmaceutical studies on *Aconitum* roots from Japan. J Nat Prod 46:178–182
10. Zhang DH, Li HY, Sang WL (1982) Studies on constituents of Fu Zi (*Aconitum carmichaeli*), a traditional Chinese medicine. II. Chemical constituents of a processed *A. carmichaeli*, Bai Fu Pian. Chin Trad Herb Drugs 13:481–484
11. Kitagawa I, Chen ZL, Yoshihara M, Yoshikawa M (1984) Chemical studies on crude drug processing. II. Aconiti tuber (1). On the constituents of "Chuan Wu", the dried tuber of *Aconitum carmichaeli* Debx. Yakugaku Zasshi 104:848–857
12. Kitagawa I, Chen ZL, Yoshihara M, Yoshimura Y (1984) Chemical studies on crude drug processing. III. Aconiti tuber (2). On the constituents of "Pao-fuzi", the processed tuber of *Aconitum carmichaeli* Debx. and biological activities of lipoalkaloids. Yakugaku Zasshi 104:858–866
13. Kitagawa I, Chen ZL, Yoshihara M, Yoshikawa M (1984) Chemical studies on crude drug processing. IV. Aconiti tuber (3). Quantitative determination of aconitine alkaloids in aconiti tuber by high performance liquid chromatography. Yakugaku Zasshi 104:867–872
14. Wang YG, Zhu YL, Zhu RH (1980) Alkaloids of the Chinese drugs, *Aconitum* spp. XIII. Alkaloids from Pei Cao Wu, *Aconitum kusnezoffii*. Acta Pharm Sin 15:526–531
15. Wang YZ, Han GY (1985) Alkaloids isolated from Jiangyou Fuzi (*Aconitum carmichaeli*). Acta Pharm Sin 20:71–73
16. Wang MZ, Li BL, Gao FY (1983) High-performance liquid chromatographic determination of the main alkaloids in Wu Tou (aconite). Acta Pharm Sin 18:689–694
17. Chen SY (1979) Structure of yunaconitine. Acta Pharm Sin 37:15–20
18. Zhang HQ, Zhu YL, Zhu RH (1982) Studies on the alkaloidal constituents of the root of *Aconitum hemsleyanum* Pritz. Acta Bot Sin 24:259–263
19. Wang HC, Zhu DZ, Zhao ZY, Zhu RH (1980) Studies on the Chinese drug, *Aconitum* spp. XII. Alkaloids from Xuan-Wu, *Aconitum nagarum* Stapf var. *lasiandrum* W.T. Wang and their structures. Acta Chim Sin 38:475–480
20. Chen SY, Li SH, Hao XJ (1986) The diterpenoid alkaloids of *Aconitum nagarum* var. *lasiandrum* and its chemotaxonomic significance. Acta Bot Sin 28:86–90
21. Mody NV, Pelletier SW, Chen SY (1982) The structure and partial synthesis of nagarine, a novel alkaloid from the Chinese drug *Aconitum nagarum* var. *heterotrichum* f. *dielsianum* W.T. Wang. Heterocycles 91–94
22. Chen SY, Liu YQ, Yang TR (1980) Chemical constituents of *Aconitum sinomontanum* Nakai. Acta Bot Yunnan 2:473–475
23. Wei BY, Kung HW, Chao CY, Wang HC, Chu JH (1981) Studies on Chinese *Aconitum* species. XVIII. Alkaloids in *Aconitum sinomontanum* Nakai. I. Chin Trad Herb Drugs 6:26–28
24. Yang CR, Hao XJ, Wang DZ, Zhou J (1981) Alkaloids of *Aconitum vilmorrianum* Kom. I. The structures of vilmorrianine A and C. Acta Chim Sin 39:147–152
25. Yang CR, Wang DZ, Wu DG, Hao XJ, Zhou J (1981) Carbon-13 NMR spectroscopic studies of several new diterpenoid alkaloids from *Aconitum* species. Acta Chim Sin 39:445–452
26. Liu JH, Wang HC, Kao YL, Chu JH (1981) Studies on *Aconitum* species in China. XVI. New alkaloids in *Aconitum coreanum*. 2. Chin Trad Herb Drugs 12:1–2
27. Kao HC, Yo FH, Chu JH (1966) The alkaloids of Chinese drugs from *Aconitum*. X. New alkaloids from Guan-Bai-Fu-Tsu, *Aconitum koreanum*. Acta Pharm Sin 13:186–194
28. Liu JH, Chen XF (1985) Structure of guanfu base H. J Nanjing Coll Pharm 16:58–60
29. Chen DC, Yan WM, Chu KD, Liang HR, Huang ZG (1986) Pharmacognosy and chemical constituents of Guanbaifu. J Beijing Coll Trad Chin Med 9:30–34
30. Reinecke MG, Watson WH, Chen DC, Yan WM (1986) A 2-D NMR structure determination of Guan-fu base Z, a new diterpene alkaloid from the Chinese herb Guan-Bai-Fu-Tzu (*Aconitum koreanum*). Heterocycles 24:49–61
31. Reinecke MG, Minter DE, Chen DC, Yan WM (1986) Alkaloids of the Chinese herb Guan-Bai-Fu (*Aconitum koreanum*); Guan-Fu base Y and A. Tetrahedron 42:6621–6626
32. Chen BR, Yang YF, Tian RM, Zhang CP, Xiao YZ, Liu MX (1981) Alkalaoids of *Aconitum finetianum* Hand-Mazz. I. Isolation and identification of alkaloids from *Aconitum finetianum* Hand-Mazz. Acta Pharm Sin 16:70–72

33. Jiang SH, Zhu YL, Zhu RH (1981) Studies on chemical constituents of *Aconitum finetianum*. Chin Pharm Bull 16:55
34. Jiang SH, Zhu YL, Zhao ZY, Zhu RH (1982) Studies on *Aconitum finetianum* Hand-Mazz. II. Chin Trad Herb Drugs 13:5
35. Jiang SH, Zhu YL, Zhu RH (1982) Studies on the Chinese drug, *Aconitum* spp. XX. Alkaloids from *Aconitum finetianum* Hand-Mazz. Acta Pharm Sin 17:283–288
36. Chang XR, Wang HC, Lu LM, Zhu YL, Zhu RH (1981) Studies on Chinese drug, *Aconitum* spp. XVII. Alkaloids from *Aconitum flavum* Hand-Mazz. Acta Pharm Sin 16:474–476
37. Chen DL, Liu YQ, Chu TT (1982) Alkaloids of *Aconitum flavum*. In: Wang Y (ed) Chem Nat Prod Sino-Am Symp, 1980, Science Press, Beijing, pp 238–239
38. Liu YQ, Chang GT (1982) Chemical constituents of *Aconitum flavum*. Chin Pharm Bull 17:243–244
39. Chen ZG, Lao AN, Wang HC, Hong SH (1987) Studies on the active principles from *Aconitum flavum* Hand-Mazz. The structures of five new diterpenoid alkaloids. Heterocycles 26:1455–1460
40. Wang FP, Fang CC (1981) Chemical study on alkaloids from *Aconitum crassicaule* W.I. Wang. Chin Pharm Bull 16:49
41. Wang FP, Fang QC (1981) Alkaloids from roots of *Aconitum crassicaule*. Planta Med 42:375–379
42. Wang FP, Fang QC (1982) Chemical study on alkaloids from the root of *Aconitum crassicaule*. II. Chin Pharm Bull 17:300–301
43. Wang FP, Fang QC (1983) Structures of crassicaulisine and crassicaulidine, two new diterpenoid alkaloids. Planta Med 47:39–42
44. Wang FP, Liang XT (1985) Structures of crassicauline B and crassicaulidine. Planta Med 51:443–444
45. Chen DH, Song WL (1982) Alkaloids from roots of *Aconitum franchetii*. Chin Trad Herb Drugs 13:8–12
46. Chen DH, Song WL (1983) Structure of franchetine. Acta Chim Sin 41:843–847
47. Luo SD, Chen WX (1981) Alkaloids from the roots of *Aconitum staphianum* Hand-Mazz. var. *pubipes* Wang. Acta Chim Sin 39:808–810
48. Wang FP, Fang QC (1982) Chemical studies on alkaloids from the roots of *Aconitum kongboense*. I. Chin Pharm Bull 17:395–396
49. Wang FP, Fang QC (1982) Chemical studies on alkaloids from the roots of *Aconitum episcopale*. I. Chin Pharm Bull 17:300
50. Wang FP, Fang QC (1983) Studies on alkaloids of *Aconitum episcopale* Le'vl. Acta Pharm Sin 18:514–521
51. Wang FP, Fang QC (1982) Study on the alkaloidal components from *Aconitum teipeicum* native to China. Acta Bot Sin 24:591–592
52. Yu DQ (1982) Alkaloids of *Aconitum barbatum* var. *puberulum*. Chin Pharm Bull 17:301
53. Yu DQ, Das BC (1983) Alkaloids of *Aconitum barbatum*. Planta Med 49:85–89
54. Liu LM, Wang HC, Zhu YL (1983) Studies on Chinese *Aconitum* spp. XIX. Alkaloids of *Aconitum pendulum* and their chemical structure. Acta Pharm Sin 18:39–44
55. Chen WS, Breitnaier E (1981) Foresaconitine, the main alkaloid from the roots of *Aconitum forestii* Stapf. Chem Ber 114:394–397
56. Wang CH, Chen DH, Song WL (1983) Chemical constituents in *Aconitum forrestii* Diels. Chin Trad Herb Drugs 14:5–7
57. Wang CH, Chen DH, Song WL (1983) Liwaconitine, a new diterpenoid alkaloid from *Aconitum forrestii*. Planta Med 48:55
58. Pelletier SW, Ying CS, Joshi BS, Desai HK (1984) The structures of forestine and foresticine, two new C_{19}-diterpenoid alkaloids from *Aconitum forrestii* Stapf. J Nat Prod 47:474–477
59. Chen SY, Liu YQ (1984) A new diterpenoid alkaloid 8-deacetylyunaconitine from the roots of *Aconitum forrestii*. Acta Bot Yunnan 6:338–340
60. Wu FG, Zhu ZQ (1983) Study on atisinium chloride from *Aconitum gymnandrum* Maxim. J Lanzhou Univ [Nat Sci] 19:183
61. Jiang SH, Guo SH, Zhou BN, Wang SX, Yi FS, Ji LJ (1986) Alkaloids from *Aconitum gymnandrum* Maxim. I. Acta Pharm Sin 21:279–284
62. Song WL, Chen DH, Wang LW, Xiago PG (1984) Chemical constituents and resource utilization of Duo Gen Wu Tou (*Aconitum karakolicum*). Chin Trad Herb Drugs 15:5–7

63. Chen DH, Song WL (1984) Alkaloidal constituents of the root of *Aconitum pseudogeniculatum*. Acta Bot Sin 26:82–84
64. Wang HC, Lao AN, Fujimoto Y, Tatsuno T (1985) The structure of polyschistine A, B, and C: three new diterpenoid alkaloids from *Aconitum polyschistum* Hand-Mazz. Heterocycles 23:803–807
65. Luo SD, Liu MM, Wu SB, Gun YH, Chen WX (1985) Alkaloid constituents from *Aconitum longtounense* T.L. Ming. Acta Chim Sin 43:577–580
66. Luo SD, Ruecker G (1986) Secondary alkaloids from *Aconitum longtounense*. Planta Med 412–413
67. Hao XJ, Chen SY, Zhou J (1985) Geniconitine, a new diterpenoid alkaloid from roots of *Aconitum geniculatum* Fletcher. Acta Bot Sin 27:504–509
68. Wang CY, Chen JB, Zhu YL, Zhu RH (1984) Alkaloids of *Aconitum duclouxii* Levl. and their structures. Acta Pharm Sin 19:445–449
69. Yan WM (1986) Chemical study on *Aconitum chinenese* cultivated in Japan. J Beijing Coll Trad Chin Med 9:34–36
70. Chen DH, Song WL (1981) Study on alkaloids of *Aconitum jinyangense* W.T. Wang. Acta Pharm Sin 16:748–751
71. Song WL, Li HY, Chen DH (1987) New diterpenoid alkaloids from *Aconitum pseudohuiliense*. Proc Chin Acad Med Sci Peking Union Med Coll 2:48–50
72. Jiang QP, Sung WL (1986) The diterpenoid alkaloids from *Aconitum scaposum* var. *vaginatum*. Heterocycles 24:877–879
73. Chen DH, Song WL (1985) Alkaloids of tangut monkshood (*Aconitum tanguticum*). Chin Trad Herb Drugs 16:338–342
74. Konno C, Shirasaka M, Hikino H (1979) Pharmaceutical studies on *Aconitum* roots. II. Cardioactive principle of *Aconitum carmichaeli* roots. Planta Med 35:150–155
75. Chen DH, Liang XT (1982) Studies on the constituents of lateral root of *Aconitum carmichaeli* Debx. (Fu Zi), a traditional Chinese medicine. I. Isolation and structural determination of salsolinol. Acta Pharm Sin 17:792–794
76. Kosuge T, Yokota M (1976) Studies on the cardiac principle of aconite root. Chem Pharm Bull 24:176–178
77. Kosuge T, Yokota M, Nagasawa M (1978) Studies on cardiac principle in aconite roots. I. Isolation and structural determination of higenamine. Yakugaku Zasshi 98:1370–1375
78. Konno C, Murayama M, Sugiyama K, Arai M, Murakami M, Takahashi M, Hikino H (1985) Isolation and hypoglycemic activity of aconitans A, B, C and D, glycans of *Aconitum carmichaeli* roots. Planta Med 51:160–161
79. Tomoda M, Shimada K, Konno C, Murakami M, Hikino H (1986) Validity of the oriental medicines. XCVIII. Antidiabetic drugs. 11. Structure of aconitan A, a hypoglycemic glycan of *Aconitum carmichaeli* roots. Carbohydr Res 147:160–164
80. Dong YL, Chen WZ, Ding GS (1981) Comparison of arrhythmic effects of aconitine and its 5 analogs. Acta Pharmacol Sin 2:173–176
81. Hikino H, Takahashi M, Konno C, Hashimoto I, Namiki T (1983) Pharmacological study of *Aconitum* roots. XV. Subacute and subchronic toxicity of *Aconitum* extracts and mesaconitine. Shoyakugaku Zasshi 37:1–9
82. Liu SF, Yang YZ (1983) Antiaconitine effects of calcium chloride and atropine. Acta Pharmacol Sin 4:110–113
83. Zhang HS, Kuang HM, Wang BX, Dai L (1982) Treatment of poisoning of *Aconitum brachypodum* in rabbits by hydrocortisone. Chin Pharm Bull 17:248–249
84. Yang QH (1985) Poisoning of *Aconitum* species and its prevention and treatment. Chin J Integr Trad Western Med 5:511–512
85. Zhou YP, Liu WH, Zeng GY, Chen DH, Li HY, Song WL (1984) Toxicity of aconitine and its analogs and their effects on cardiac contractive function. Acta Pharm Sin 19:641–646
86. Xiao PG, Wang LW, Tong YY (1983) Studies of the medicinal plants of the family Ranunculaceae in China. VIII. Correlation among the root morphology, phylogeny, main constituent and toxicity of 27 Chinese aconites. Chin J Pharm Anal 3:276–280
87. Hernanz A, Silio F (1983) Inhibition by aconitine of aconitase of pig heart. Comp Biochem Physiol [C] 76:335–338
88. Gao TL, Liao FL, Wang YS, Zhuang H (1981) Experimental arrhythmic models in mice and the factors affecting them. Chin J Cardiol 9:223–227

89. Mest HJ, Taube C, Foerster W (1981) Influence of aconitine and ouabain induced cardiac arrhythmias on the prostaglandin efflux in the canine coronary sinus blood and the influence of antiarrhythmic drugs. In: Foerster W (ed) Prostaglandins Thromboxanes. Pergamon, Oxford, pp 103–105
90. Hikino H, Ito T, Yamada C, Sato H, Konno C, Ohizumi Y (1979) Validity of oriental medicine. X. Pharmaceutical studies on *Aconitum* roots. VII. J. Pharmacobiodyn 2:78–83
91. Murayama M, Ito T, Konno C, Hikino H (1984) Mechanism of analgesic action of mesaconitine. I. Relationship between analgesic effect and central monoamines or opiate receptors. Eur J Pharmacol 101:29–36
92. Murayama M, Hikino H (1985) Validity of oriental medicine. LXXVI. Pharmaceutical studies on *Aconitum* roots. 20. Effect of cyclic AMP on mesaconitine induced analgesia in mice. Eur J Pharmacol 108:19–23
93. Tang XC, Zhu MY, Feng J, Wang YE (1983) Studies on pharmacologic effects of lappaconitine hydrobromide. Acta Pharm Sin 18:579–584
94. Jiang SH, Zhu YL, Zhao ZY, Zhu RH (1983) Studies on the Chinese *Aconitum* species. XXI. Study on *Aconitum finetianum* Hand-Mazz. Acta Pharm Sin 18:440–445
95. Bhalla TN, Bhargava KP (1980) Aconitine induced writhing as a method for assessing aspirin-like analgesic activity. J Pharmacol Methods 3:9–14
96. Hikino H, Konno C, Takata H, Yamada Y, Chizuko O, Ohizumi Y, Sugio K, Fujimura H (1980) Antiinflammatory principles of *Aconitum* roots. J Pharmacobiodyn 3:514–525
97. Murayama K (1981) Aconitine derivatives as inflammation inhibitors. Jpn Kokai Tokkyo Koho JP 81,120,620 (CA 96:46256y)
98. Hikino H, Takata H, Fujiwara M, Konno C, Ohuchi K (1982) Mechanism of inhibitory action of mesaconitine in acute inflammations. Eur J Pharmacol 82:65–71
99. Tang XC, Lin ZG, Cai W, Chen NA, Shen L (1984) Antiinflammatory effect of 3-acetyl-aconitine. Acta Pharmacol Sin 5:85–89
100. Saito H, Ueyama T, Naka N, Yagi J, Okamoto T (1982) Pharmacological studies of ignavine, an aconitum alkaloid. Chem Pharm Bull 30:1844–1850
101. Taiho Pharmaceutical (1983) Preparation of antiinflammatory, analgesic aconine alkaloids. Jpn Kokai Tokkyo Koho JP 58 52, 273(8352,273) (CA 99:76850s)
102. Yang YZ, Liu SF (1980) Antihistaminic effects of total alkaloids of *Aconitum kusnezoffii* Reichb. Acta Pharmacol Sin 1:131–133
103. Murayama M, Hikino H (1984) Stimulating actions on ribonucleic acid biosynthesis of aconitines, diterpenic alkaloids of *Aconitum* roots. J Ethnopharmacol 12:25–33
104. Murayama K (1982) Accelerators of protein formation in the liver. Jpn Kokai Tokkyo Koho JP 82,58,627 (CA 97:28606d)
105. Lian QS, Feng GH (1985) Hypotensive effect of delsoline. Acta Pharmacol Sin 6:37–40
106. Chen WZ, Dong YL, Zhang YF, Ding GS (1983) Antiarrhythmic effects of guanfu base A. Acta Pharmacol Sin 4:247–250
107. Zhang L, Gu PK, Zhao WB, Chen YL, Zhang JX, Jin ZJ, Chen WZ (1986) Effect of guanfu base A on action potential of canine Purkinje fibers. Acta Pharmacol Sin 7:234–236
108. Huang NH, Zhou YP, Liu WH, Fan LL, Tseng KY (1980) Comparison of cardiovascular effects of aconite root and higenamine in dogs. Acta Pharmacol Sin 1:34–39
109. Chon YP, Fan LL, Chang LY, Tseng KY (1978) Studies on the pharmacological effect of aconite root. I. Effect of higenamine on cardiovascular system. Natl Med J China 58:664–669

Acorus gramineus Soland. 4

4.1 Introduction

Shichangpu, Rhizoma Acori graminei, is the dry rootstock of *Acorus gramineus* Soland. (Araceae) collected in fall and winter. It is listed officially in the Chinese Pharmacopoeia and used as a digestant, an expectorant, and as a stimulant against digestive disorders, diarrhea, and epilepsy.

In the appendix of the Chinese Pharmacopoeia *A. calamus* L. is included.

4.2 Chemical Constituents

The rootstock of *A. gramineus* contains an essential oil from which α-asarone (*4-1*), β-asarone (*4-2*), γ-asarone (*4-3*) [1], and a compound named bisasaricin (*4-4*) [2] were isolated and identified. Irradiation of α-asarone in anhydrous ethanol with UV light for 5 h also yielded bisasaricin [2].

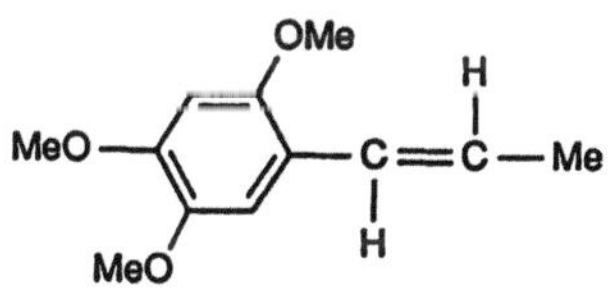

α-Asarone (*4-1*)

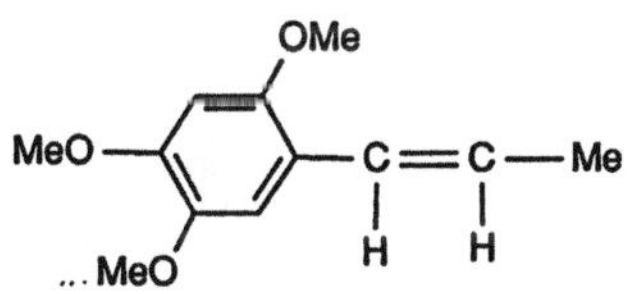

β-Asarone (*4-2*)

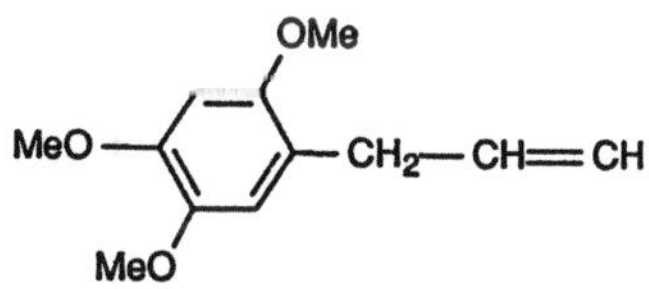

γ-Asarone (*4-3*)

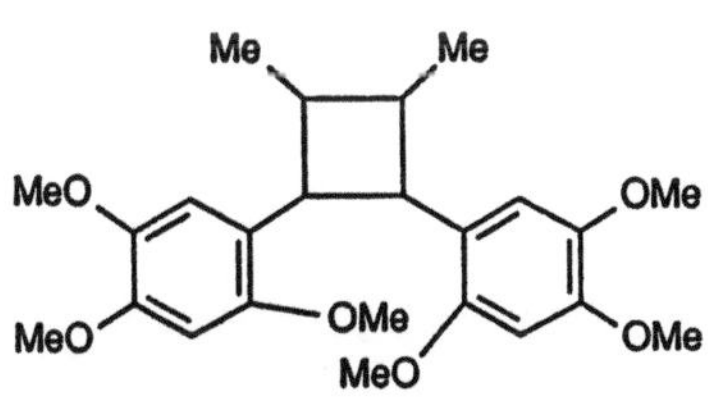

Bisasaricin (*4-4*)

4.3 Pharmacology

α-Asarone, β-asarone, and γ-asarone isolated from *A. gramineus* have been tested for spasmolytic activity. All three compounds had a spasmolytic effect on isolated guinea pig trachea and ileum contractions induced by acetylcholine, histamine, serotonin, and barium chloride. Among these three compounds, α-asarone was the most effective [1].

Bisasaricin had hypolipemic activity [2].

A long-term carcinogenicity study of the essential oil from *A. calamus* of Indian origin containing about 80% β-asarone bas been carried out in rats and resulted in induction of tumors in the duodenal region after oral administration [3]. The carcinogenic effect of the essential oil of *A. calamus* in experimental animals was ascribed to β-asarone [4]. A mutagenic activity of β-asarone in tests with *Salmonella typhimurium* has only been observed in strain TA 100 with S9 mixture [5]. In addition, β-asarone also showed a very strong effect on the induction of structural chromosome aberrations in human lymphocytes in vitro after metabolic activation [6].

References

1. Liu GQ, Sun JN, He ZZ, Jiang Y (1983) Spasmolytic effects of active principles of the essential oil of *Acorus gramineus*. Acta Pharmacol Sin 4:95–97
2. Yuan YH, Wang CW, Zhou XY (1982) Study on the hypolipemic principles on Shi Chang Pu (*Acorus gramineus* Soland.). Chin Trad Herb Drugs 13:387–388
3. Taylor JM, Jones WI, Hagan EC, Gross MA, Davis DA, Cook EL (1967) Toxicity of oil of calamus (Jammu variety). Toxicol Appl Pharmacol 10:405
4. Habermann RT (1971) Report of the Food and Drug Administration. Project P-155-70
5. Goggelmann W, Schimmer O (1983) Mutagenicity testing of β-asarone and commercial calamus drugs with *Salmonella typhimurium*. Mutat Res 121:191–194
6. Abel G (1987) Chromosomenschädigende Wirkung von β-Asaron in menschlichen Lymphocyten. Planta Med 251–253

Agrimonia pilosa Ledeb. 5

5.1 Introduction

Xianhecao, Herba Agrimoniae, is the dry above ground part of *Agrimonia pilosa* Ledeb. (Rosaceae) harvested in summer and fall when the plants flourish. It is listed officially in the Chinese Pharmacopoeia and used as a hemostatic, antimalarial, and antidysenteric agent.

5.2 Chemical Constituents

A. pilosa was found to contain a number of phenolic compounds. Thus, five agrimols A (*5-1*), B (*5-2*), C (*5-3*), D (*5-4*), and E (*5-5*) [1, 2] and agrimoniin, potentillin, pedunculagin [3], luteolin-7-glucoside, apigenin-7-glucoside, quercetin, ellagic acid, caffeic acid, and gallic acid [4] were detected.

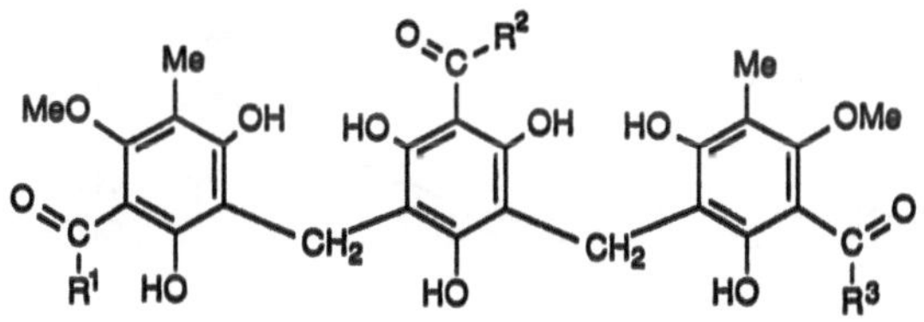

	R^1	R^2	R^3
Agrimol A (*5-1*)	$-CH(CH_3)-CH_3$	$-CH(CH_3)-CH_2-CH_3$	$-CH(CH_3)-CH_3$
Agrimol B (*5-2*)	$-CH_2-CH_2-CH_3$	$-CH(CH_3)-CH_2-CH_3$	$-CH_2-CH_2-CH_3$
Agrimol C (*5-3*)	$-CH_2-CH_2-CH_3$	$-CH_2-CH_2-CH_3$	$-CH_2-CH_2-CH_3$
Agrimol D (*5-4*)	$-CH(CH_3)-CH_3$	$-CH(CH_3)-CH_2-CH_3$	$-CH_3$
Agrimol E (*5-5*)	$-CH_3$	$-CH(CH_3)-CH_2-CH_3$	$-CH_3$

The structures of agrimols A, B, C, D, and E were elucidated by spectral analyses and chemical reactions. They are closely related compounds. In addition, the agrimols were synthesized by formaldehyde condensation of the corresponding phenolic moieties [5].

Potentillin (*5-6*) is a novel α-glucosyl ellagitannin and agrimoniin (*5-7*) is the corresponding dimer. They were isolated together with the previously known compound pedunculagin (*5-8*).

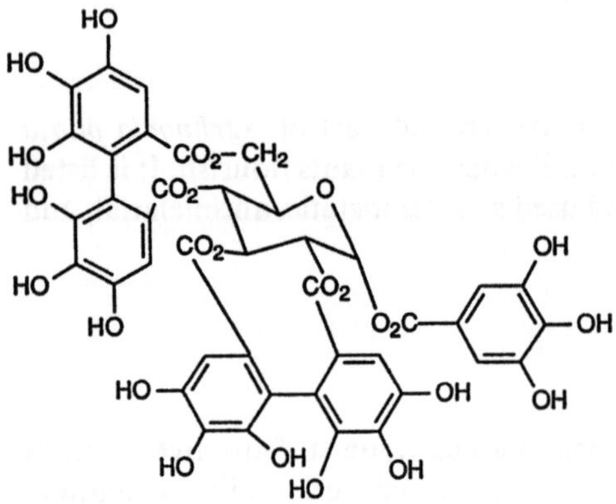

Potentillin (*5-6*)

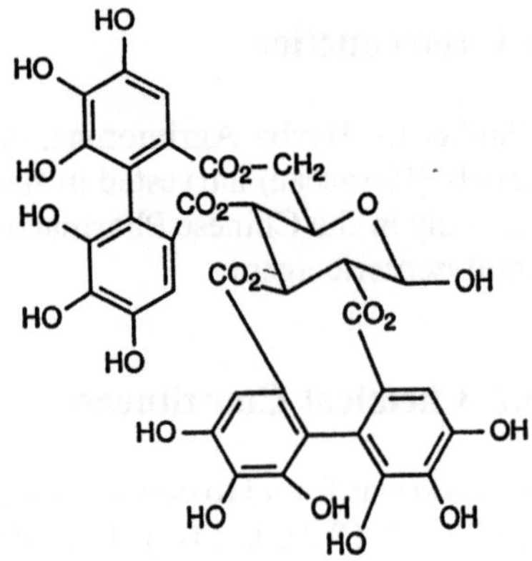

Pedunculagin (*5-8*)

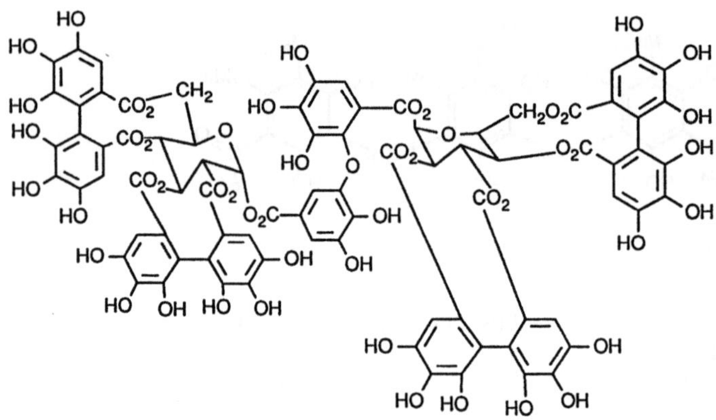

Agrimoniin (*5-7*)

Agrimonolide (*5-9*) was isolated from the rhizome of *A. pilosa*. Its structure was elucidated by chemical synthesis [6].

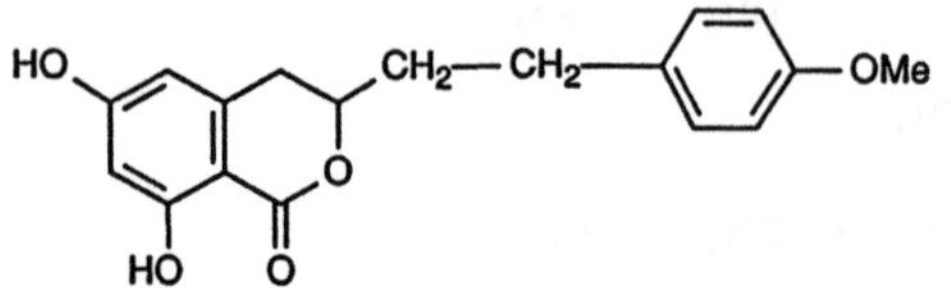

Agrimonolide (*5-9*)

From the sprout of *A. pilosa*, the phenolic compound agrimophol (*5-10*) was isolated and its structure determined [7]. The agrimophol content was 0.8% [8].

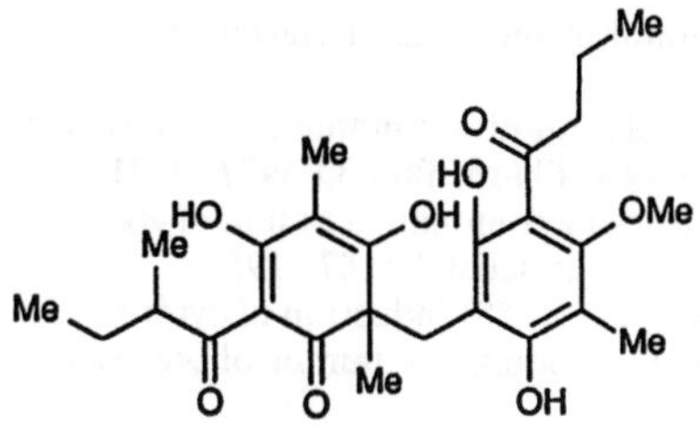

Agrimophol (*5-10*)

5.3 Pharmacology

Six compounds isolated from *A. pilosa*, luteolin-7-glucoside, apigenin-7-glucoside, ellagic acid, caffeic acid, quercetin, and gallic acid, were active against bacillary dysentery. Mixtures of luteolin-7-glucoside with ellagic acid, apigenin-7-glucoside with ellagic acid, apigenin-7-glucoside with caffeic acid, and luteolin-7-glucoside and apigenin-7-glucoside with quercetin were more active than the individual compounds [4].

The antitumor activities of various extracts from the roots of *A. pilosa* were also studied. Each extract was given to mice in a single i.p. dose 4 days before their i.p. inoculation with mouse mammary carcinoma MM2 cells. Nonsugar fractions of median polarity containing agrimoniin had antitumor activity. Agrimoniin itself had antitumor activity when given as a pre- or posttreatment. A single dose of 10–30 mg/kg prolonged the life span of mice bearing MM2 tumors or yielded cures when given i.v. or orally prior to or after tumor inoculation. Agrimoniin also inhibited the growth of MH-134 and Meth-A solid tumors in mice. It was strongly cytotoxic to MM2 cells in vitro, but the activity was almost completely abolished by the addition of fetal calf serum to the culture. Intraperitoneal injection of agrimoniin increased the number of peripheral white blood cells and the proportion of monocytes. The antitumor activity of agrimoniin appears to be due to its enhancement of the immune response [9–11].

References

1. Cheng CL, Chu TY, Wang HC, Huang PS, Chin GW (1978) Studies on the active principles of Shianhotsao. II. Structures of agrimol A, B, D and E. Acta Chim Sin 36:35–41
2. Shian Ho Tasao Working Group (1974) Active components of Shian Ho Tsao. Structure and synthesis of agrimol C. Kexue Tongbao 19:479–480
3. Okuda T, Yoshida T, Kuwahara M, Memon MU, Shingu T (1982) Agrimoniin and potentillin, an ellagitannin dimer and monomer having an α-glucose core. J Chem Soc Chem Commun 163–164
4. Su GS, Su SW, Zhu TR (1984) Studies on bacteriostatic components from *Agrimonia pilosa* Ledeb. J Shenyang Coll Pharm 1:44–50
5. Li LC, Cheng YP, Yu PL, Li Y, Kai YC, Wang TS, Chen IS (1978) Studies on the active principles of Shianhotsao. III. Syntheses of agrimol A, B, D and E. Acta Chim Sin 36:43–48

6. Yamato M, Hashigaki K (1976) Synthesis of *dl*-agrimonolide (constituent of the rhizome of *Agrimonia pilosa* Ledeb.) Chem Pharm Bull 24:200–203
7. Shenyang College of Pharmacy (1977) Elucidation of the structure of agrimophol. Acta Chim Sin 35:87–96
8. Sha SY (1980) TLC scanning determination of agrimophol in the buds of *Agrimonia pilosa*. Chin Pharm Bull 15:6–7
9. Miyamoto K, Koshiura R, Ikeya Y, Taguchi H (1985) Isolation of agrimoniin, an antitumor constituent, from the roots of *Agrimonia pilosa* Ledeb. Chem Pharm Bull 33:3977–3981
10. Miyamoto K, Kishi N, Koshiura R (1987) Antitumor effect of agrimoniin, a tannin of *Agrimonia pilosa* Ledeb., on transplantable rodent tumors. Jpn J Pharmacol 43:187–195
11. Miyamoto K, Kishi N, Murayama T, Furukawa T, Koshiura R (1988) Induction of cytotoxicity of peritoneal exudate cells by agrimoniin, a novel immunomodulatory tannin of *Agrimonia pilosa* Ledeb. Cancer Immunol Immunother 27:59–62

Ailanthus altissima (Mill.) Swingle

6

6.1 Introduction

Chunpi, Cortex Ailanthi, is the dry root bark or stem bark of *Ailanthus altissima* (Mill.) Swingle (Simaroubaceae). It can be peeled off throughout the year. This officially listed herbal medicine is used as an astringent, antidiarrheic, and hemostatic agent.

6.2 Chemical Constituents

The major constituents of the bark of *A. altissima* are bitter compounds of quassinoid nature. Isolation of a number of quassinoids was reported: amarolide (*6-2*), amarolide 11-acetate (*6-3*) [1, 2], alianthone (*6-4*) [3–6], glaucarubinone (*6-5*) [7, 8], 13(21)-dehydroglaucarubinone (*6-6*) [7, 9], 13(21)-dehydroglaucarubolone (*6-7*) [7, 10, 11], chaparrolide (*6-8*) [7, 12], chaparrinone (*6-9*) [7, 13, 14], shinjulactone A (*6-10*) [15], shinjulactone B (*6-11*) [16], shinjulactone C (*6-12*) [7, 17], shinjulactones D (*6-13*) and E (*6-14*) [18], shinjulactone F (*6-15*) [19, 20], shinjulactones G (*6-16*) and H (*6-17*) [21], shinjulactones I (*6-18*), J (*6-19*), K (*6-20*) [20], shinjulactone L (*6-21*) [22], shinjulactones M (*6-22*) and N (*6-23*) [23], and shinjudilactone (*6-24*) [7, 24]. These compounds are quite related and structurally derived from picrasane (*6-1*), except for shinjulactone B and shinjudilactone.

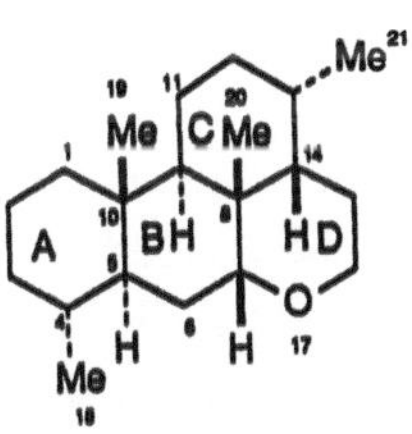

Picrasane (*6-1*)

Shinjulactone B is a C_{19} quassinoid compound with a quite original structure determined by X-ray diffraction [25]. It may be derived biogenetically from a C_{20} quassinoid via a 1,2-dioxo derivative by oxidative bond cleavage between C-1 and C-2, decarboxylation, contraction of ring B, and formation of the α,β-unsaturated γ-lactone [25].

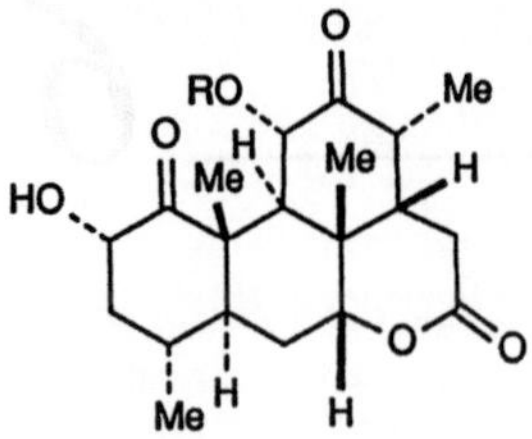

Amarolide (*6-2*): R=H
Amarolide 11-acetate (*6-3*): R=Ac

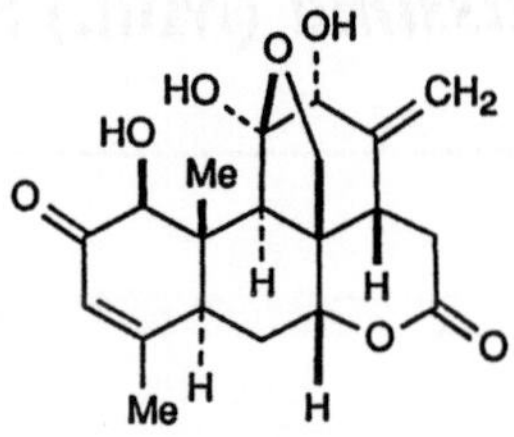

Ailanthone (*6-4*)

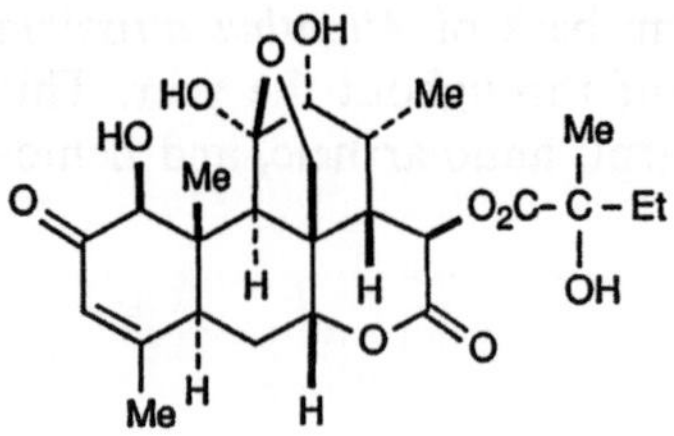

Glaucarubinone (*6-5*)

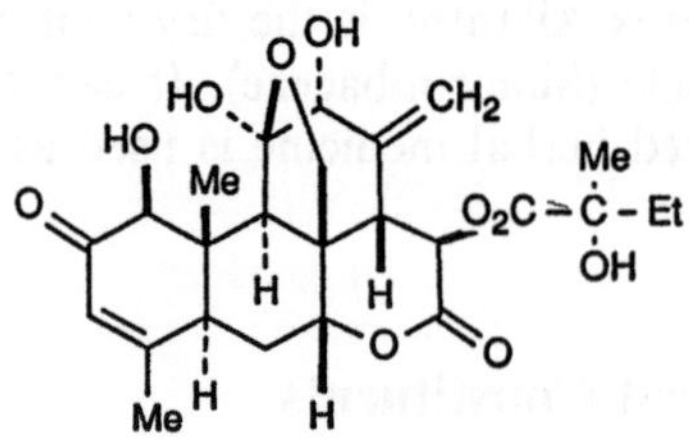

13(21)-Dehydroglau-
carubinone (*6-6*)

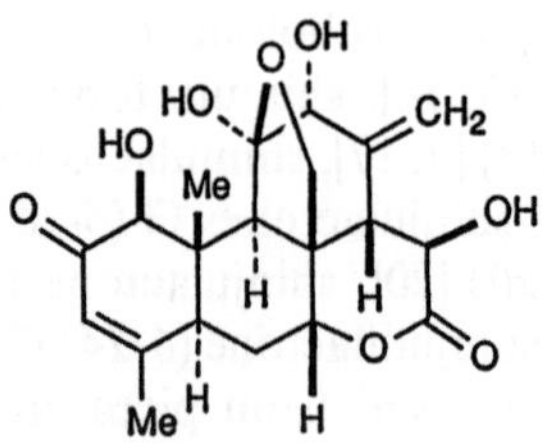

13(21)-Dehydroglau-
carubolone (*6-7*)

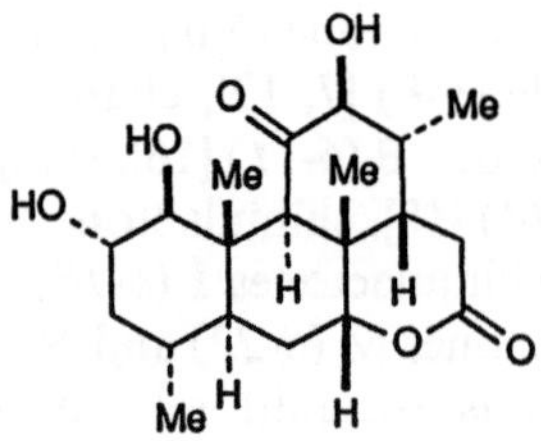

Chaparrolide (*6-8*)

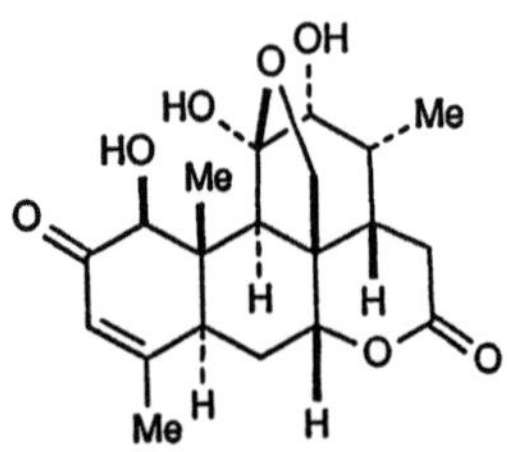

Chaparrinone (*6-9*)

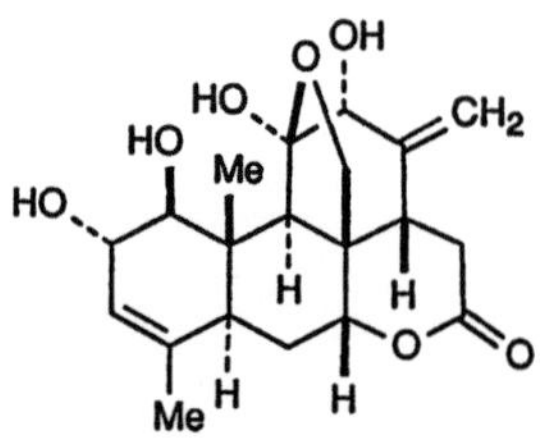

Shinjulactone A (*6-10*)
(2-Dihydroailanthone)

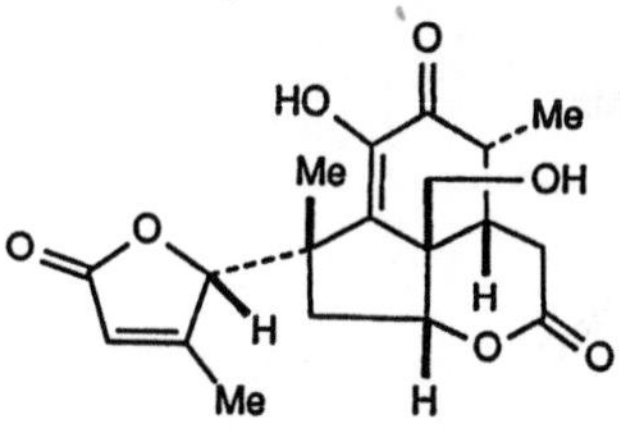

Shinjulactone B (*6-11*)

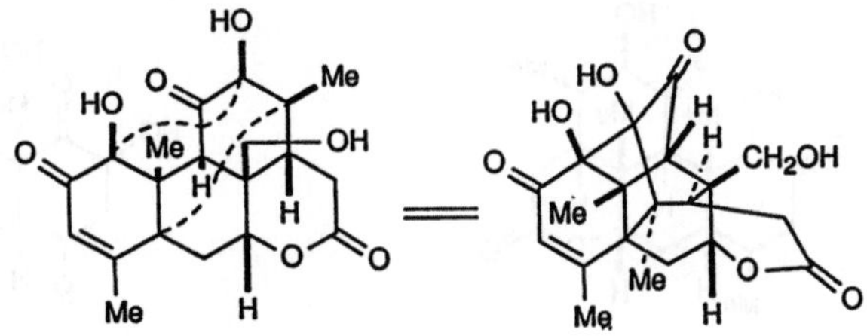

Shinjulactone C (*6-12*)

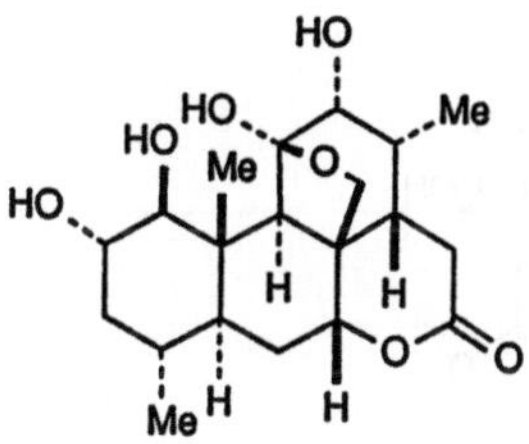

Shinjulactone D (*6-13*)

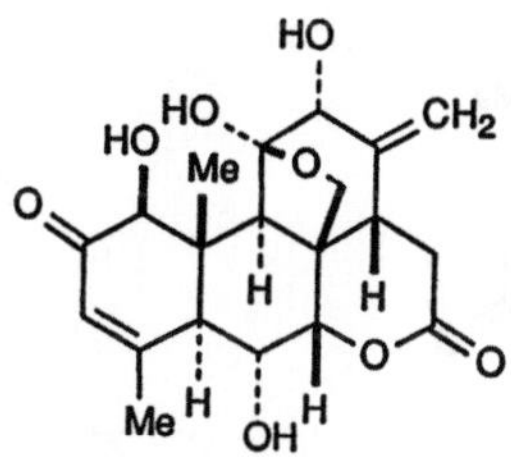

Shinjulactone E (*6-14*)

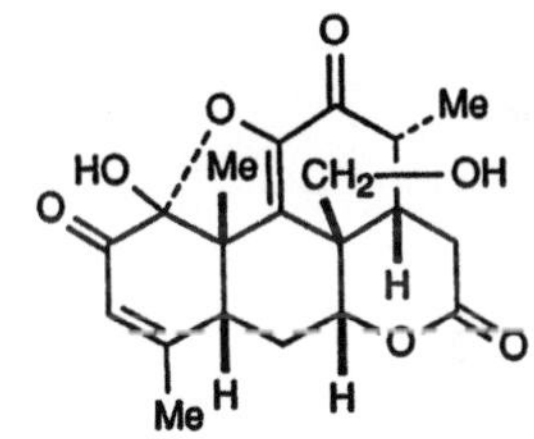

Shinjulactone F (*6-15*)

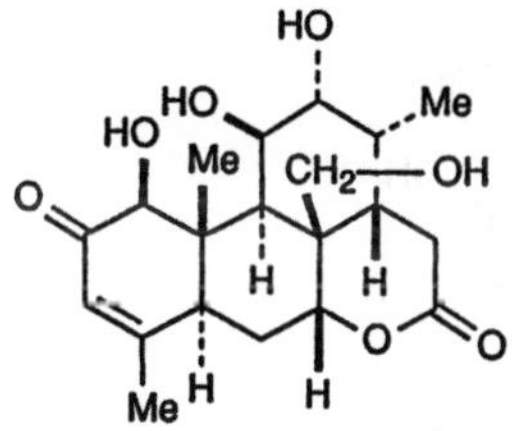

Shinjulactone G (*6-16*)

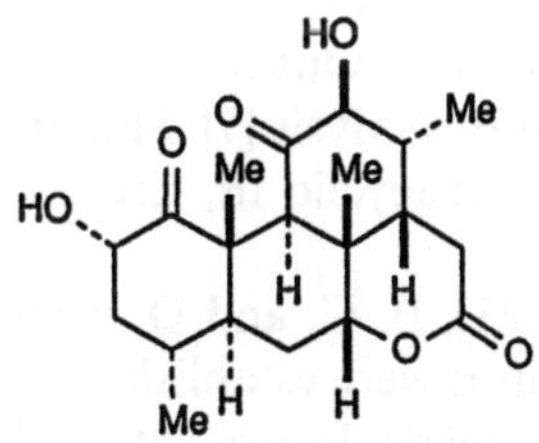

Shinjulactone H (*6-17*)

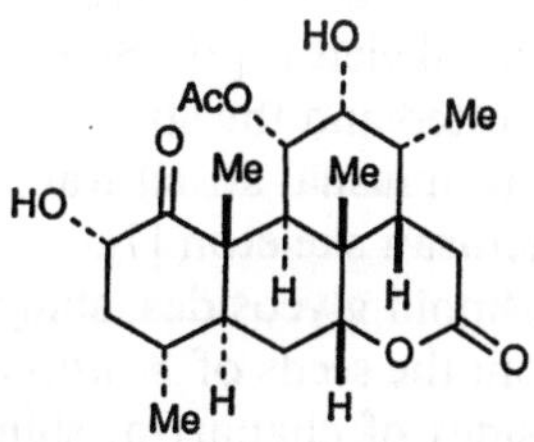

Shinjulactone I (*6-18*)

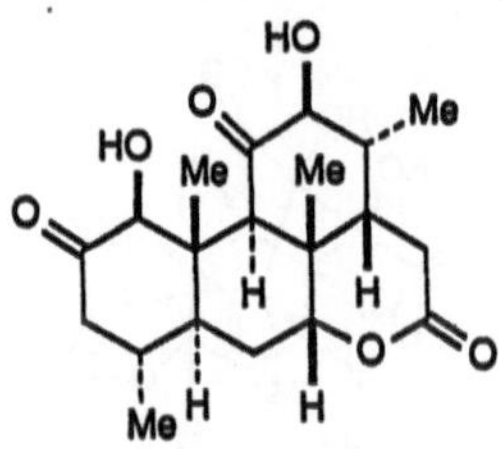

Shinjulactone J (*6-19*)

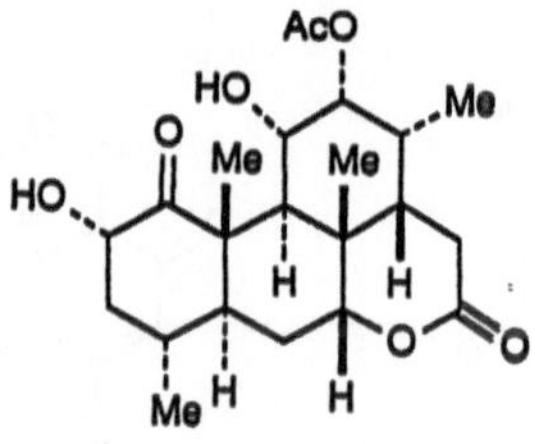

Shinjulactone K (*6-20*)

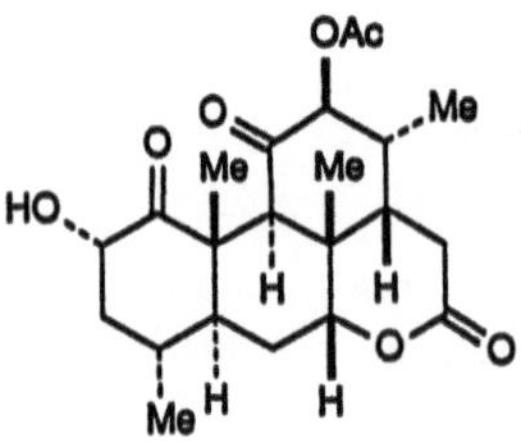

Shinjulactone L (*6-21*)

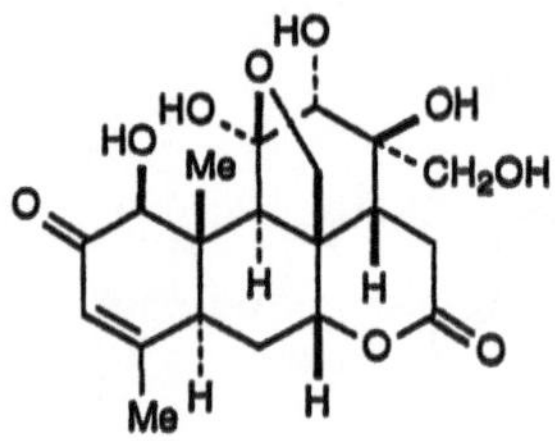

Shinjulactone M (*6-22*)

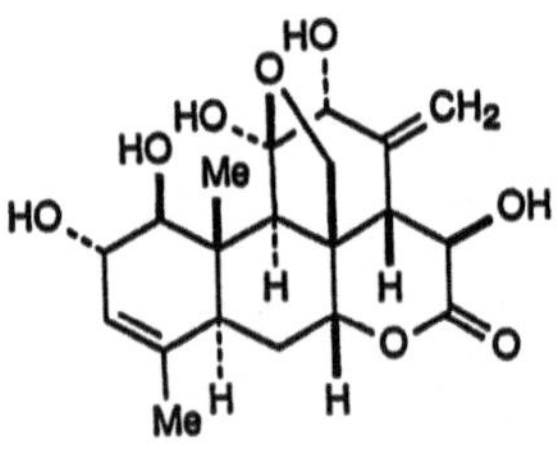

Shinjulactone N (*6-23*)

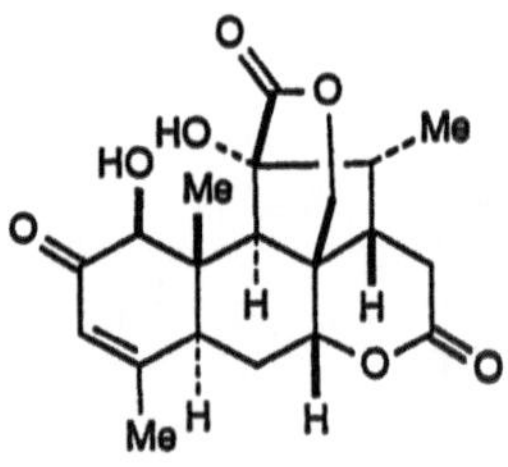

Shinjudilactone (*6-24*)

Shinjudilactone also does not represent the normal picrasan skeleton, but rather a migrated picrasane skeleton [7]. Biogenetic formation of shinjudilactone from ailanthone was suggested and the structure confirmed by X-ray analysis [7]. Shinjulactone C also has an unusual structural feature, namely, a hexacyclic 1α, 12α: 5α, 13α-dicyclo-9βH-picrasan skeleton [7].

Four bitter quassinoid glycosides, shinjuglycosides A (*6-25*), B, C, and D, were recently isolated from the seeds of *A. altissima*. Their structures were established as 2-β-D-glucopyranosides of chaparrin, shinjulactone A, amarolide 11-acetate, and amarolide, respectively [26]. The structure of shinjuglycoside A is shown as an example.

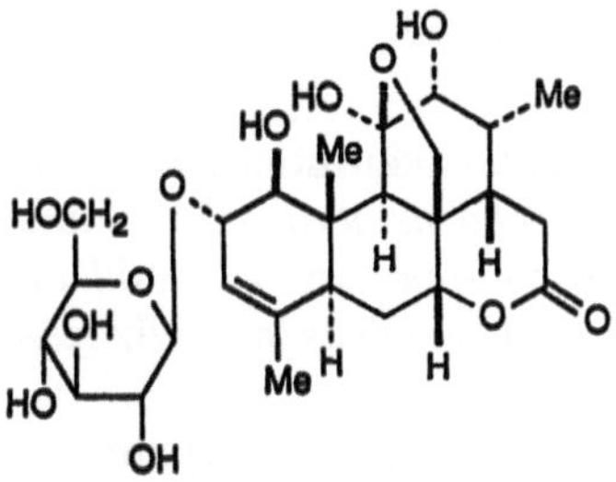

Shinjuglycoside A (*6-25*)
(Chaparrin-2-*O*-β-D-glucopyranoside)

In addition to the bitter quassinoids, a number of indole alkaloids were isolated from *A. altissima*. They were canthin-6-one (*6-26*), 1-methoxycanthin-6-one [27, 28], canthin-6-one-3-oxide (*6-27*) [27], 6-methoxy-β-carboline-1-carboxylic acid methyl ester [29], 1-acetyl-4-methoxy-β-carboline, 1-(2′-hydroxyethyl)-4-methoxy-β-carboline, 1-(1′,2′-dihydroxyethyl)-4-methoxy-β-carboline (*6-29*) [30, 31], 1-methoxy-canthin-6-one-3-oxide [30], 1-hydroxy-canthin-6-one (*6-28*) [30], 1-(2-hydroxy-1-methoxy)ethyl-4-methoxy-β-carboline, 5-hydroxy-methylcanthin-6-one, β-carboline-1-propionic acid, 1-carbamoyl-β-carboline, and 1-carbomethoxy-β-carboline [32]. Canthin-6-one, 1-methoxycanthin-6-one, and 4-methoxy-1-vinyl-β-carboline were isolated from the leaves of *A. altissima* [33].

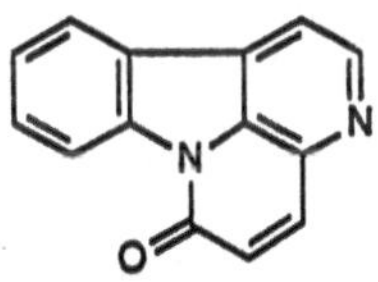

Canthin-6-one (*6-26*)

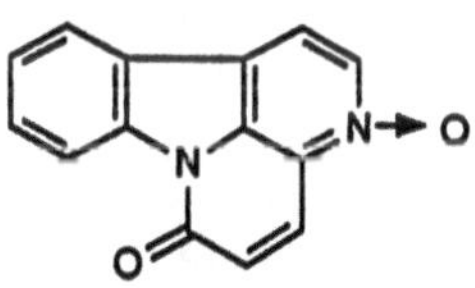

Canthin-6-one-3-oxide (*6-27*)

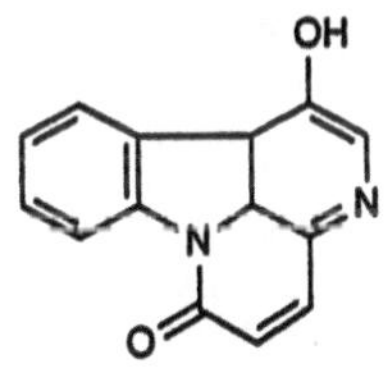

1-Hydroxy-canthin-6-one (*6-28*)

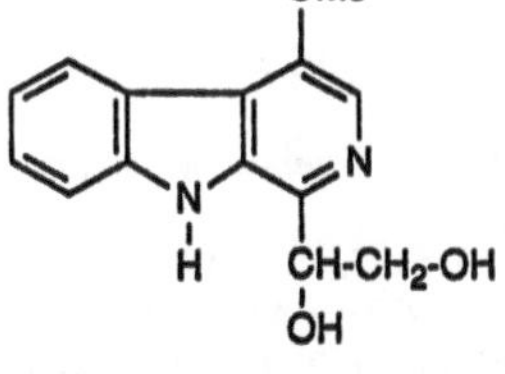

1-(1′,2′-Dihydroxyethyl)-4-methoxy-β-carboline (*6-29*)

The biosynthetic relationship between indole alkaloids of *A. altissima* was studied by administration of the precursor, [^{14}C] tryptophan, to cell cultures of *A. altissima*. The biosynthetic sequence of the alkaloids produced was found to be as follows: tryptophan → β-carboline-1-propionic acid → 4,5-dihydrocanthin-6-one → canthin-6-one → 1-hydroxycanthin-6-one → 1-methoxy-canthin-6-one → 1-methoxy-canthin-6-one-3-oxide. [^{14}C] tryptamine was ineffective as a precursor [34].

6.3 Pharmacology

Among the above mentioned quassinoids, glaucarubinone and ailanthone have been shown to be amebicidal in vitro against the parasite *Entamoeba histolytica* [35]. Some quassinoids markedly inhibited the growth of chloroquine-resistant *Plasmodium falciparum*. Glaucarubinone gave complete inhibition at 0.006 μg/ml, whereas chaparrinone had little effect even at 0.01 μg/ml [36].

Antimalarial activities of ailanthone and some related quassinoids against *P. berghei* have also been reported [37].

References

1. Casinovi CG, Bellavita V, Grandolini G, Ceccherelli P (1965) Occurrence of bitter substances related to quassin in *Ailanthus glandulosa*. Tetrahedron Lett 2273–2279
2. Stöcklin W, Stefanovic M, Geissman TA (1970) Amarolide, isolation from *Castela nicholsoni* Hook and revision of its structure. Tetrahedron Lett 2399–2402
3. Polonsky J, Fourrey JL (1964) Constituants des graines d'*Ailanthus altissima* Swingle. Structure de l'ailanthone. Tetrahedron Lett 3983–3990
4. Gaudemer A, Fourrey JL, Polonsky J (1967) Etude de la methylation par le diazométhane des hydroxyles de composés amers des simarubacées. Bull Soc Chim Fr 1676–1985
5. Casinovi CG, Ceccherelli P, Grandolini B, Bellavita V (1964) On the structure of ailanthone. Tetrahedron Lett 3991–3997
6. Naora H, Furuno T, Ishibashi M, Tsuyuki T, Takahashi T, Itai A, Iitaka Y, Polonsky J (1982) On the structure of ailanthone, a bitter principle from *Ailanthus altissima*. Chem Lett 661–662
7. Ishibashi M, Tsuyuki T, Murae T, Hirota H, Takahashi T, Itai A, Iitaka Y (1983) Constituents of the root bark of *Ailanthus altissima* Swingle. Isolation and X-ray crystal structures of shinjudilactone and shinjulactone C and conversion of ailanthone into shinjudilactone. Bull Chem Soc Jpn 56:3683–3693
8. Gaudemer A, Polonsky J (1965) Structure de la glaucarubinone, nouveau principe amer isole de *Simaruba glauca*. Phytochemistry 4:149–153
9. Polonsky J, Varon Z, Jacquemin H, Pettit GR (1978) The isolation and structure of 13,18-dehydroxyglaucarubinone, a new antineoplastic quassinoid from *Simaruba amara*. Experientia 34:1122–1123
10. Moron J, Rondest J, Polonsky J (1966) Sur la biosynthèse des constituants amers des simarubacées. Experientia 22:511–512
11. Kupchan SM, Lacadie JA (1975) Dehydroailanthinone, a new antileukemic quassinoid from *Pierreodendron kerstingii*. J Org Chem 40:654–656
12. Mitchell RE, Stöcklin W, Stefanovic M, Geissman TA (1971) Chaparrolide and castelanolide, new bitter principles from *Castela nicholsoni*. Phytochemistry 10:411–417
13. Polonsky J, Bourquignon-Zylber N (1965) Etude des constituants de *Hannoa klaineana* (simarubacee): chaparrinone et klaineanone. Bull Soc Chim Fr 2793–2799
14. Polonsky J, Baskevitch Z, Gottlieb HE, Hagaman EW, Wenkert E (1975) Carbon-13 nuclear magnetic resonance spectral analysis of quassinoid bitter principles. J Org Chem 40:2499–2504
15. Naora H, Ishibashi M, Furuno T, Tsuyuki T, Murae T, Hirota H, Takahashi T, Itai A, Iitaka Y (1983) Structure of bitter principles in *Ailanthus altissima*. Structure of shinjulactone A and revised structure of ailanthone. Bull Chem Soc Jpn 56:3694–3698
16. Furuno T, Naora H, Murae T, Hirota M, Tsuyuki T, Takahashi T, Itai A, Iitaka Y, Matsushita K (1981) Structure of shinjulactone B, a new bitter principle from *Ailanthus altissima*. Chem Lett 1797–1798
17. Ishibashi M, Murae T, Hirota H, Tsuyuki T, Takahashi T, Itai A, Iitaka Y (1982) Shinjulactone C, a new quassinoid with a 1α, 12α:5α, 13α-dicyclo-9βH-picrasane skeleton from *Ailanthus altissima* Swingle. Tetrahedron Lett 23:1205–1206
18. Ishibashi M, Furuno T, Tsuyuki T, Takahashi T, Matsushita K (1983) Structures of shinjulactones D and E, new bitter principles of *Ailanthus altissima* Swingle. Chem Pharm Bull 31:2179–2182

19. Ishibashi M, Yoshimura S, Tsuyuki T, Takahashi T, Itai A, Iitaka Y, Matsushita K (1984) Shinjulactone F, a new bitter principle with a 5βH-picrasane skeleton from *Ailanthus altissima* Swingle. Chem Lett 555–556
20. Ishibashi M, Yoshimura S, Tsuyuki T, Takahashi T, Itai A, Iitaka Y (1984) Sturcture determination of bitter principles of *Ailanthus altissima.* Structures of shinjulactones, F, I, J and K. Bull Chem Soc Jpn 57:2885–2892
21. Ishibashi M, Yoshimura S, Tsuyuki T, Takahashi T, Matsushita K (1984) Shinjulactones G and H, new bitter principles of *Ailanthus altissima* Swingle. Bull Chem Soc Jpn 57:2013–2014
22. Ishibashi M, Tsuyuki T, Takahashi T (1985) Structure determination of a new bitter principle, shinjulactone L, from *Ailanthus altissima.* Bull Chem Soc Jpn 57:2723–2724
23. Niimi Y, Tsuyuki T, Takahashi T, Matsushita K (1986) Structure determination of shinjulactones M and N, new bitter principles from *Ailanthus altissima* Swingle. Bull Chem Soc Jpn 59:1638–1640
24. Ishibashi M, Murae T, Hirota H, Naora H, Tsuyuki T, Takahashi T, Itai A, Iitaka Y (1981) Shinjudilactone, a new bitter principle from *Ailanthus altissima* Swingle. Chem Lett 1597–1598
25. Polonsky J (1985) Quassinoid bitter principles II. In: Herz W, Grisebach H, Kirby GW, Tamm CH (eds) Progress in the Chemistry of Organic Natural Products. Springer, Berlin Heidelberg New York, p 224
26. Yoshimura S, Ishibashi M, Tsuyuki T, Takahashi T, Matsushita K (1984) Constituents of seeds of *Ailanthus altissima* Swingle. Isolation and structures of shinjuglycosides A, B, C and D. Bull Chem Soc Jpn 57:2496–2501
27. Ohmoto T, Tanaka R, Nikaido T (1976) Studies on the constituents of *Ailanthus altissima* Swingle. On the alkaloidal constituents. Chem Pharm Bull 24:1532–1536
28. Szendrei K, Korbely T, Kreuzien H, Reisch J, Novak I (1977) β-Carboline alkaloids and coumarins from the root bark of the tree of heaven (*Ailanthus altissima* (Mill.) Swingle, Simaroubaceae). Herba Hung 16:15–21
29. Varga E, Szendrei K, Reisch J, Maroti G (1980) Indole alkaloids of *Ailanthus altissima.* Planta Med 40:337–339
30. Ohmoto T, Koike K, Sakamoto Y (1981) Studies on the constituents of *Ailanthus altissima* Swingle. II. Alkaloidal constituents. Chem Pharm Bull 29:390–395
31. Varga E, Szendrei K, Reisch J, Maroti G (1981) Indole alkaloids of *Ailanthus altissima.* II. Fitoterapia 52:183–186
32. Ohmoto T, Koike K (1984) Studies on the constituent of *Ailanthus altissima* Swingle. III. The alkaloidal constituents. Chem Pharm Bull 32:170–173
33. Souleles C, Waigh R (1984) Indole alkaloids of *Ailanthus altissima.* J Nat Prod 47:741
34. Crespi-Perellino N, Guicciardi A, Malyszko G, Minghetti A (1986) Biosynthetic relationship between indole alkaloids produced by cell cultures of *Ailanthus altissima.* J Nat Prod 49:814–822
35. Gillin FD, Reiner DS (1982) In vitro activity of certain quassinoid antitumor agents against *Entamoeba histolytica.* Arch Invest Med [Suppl 3] 13:43
36. Trager W, Polonsky J (1981) Antimalarial activity of quassinoids against chloroquine-resistant *Plasmodium falciparum* in vitro. Am J Trop Med Hyg 30:531–537
37. Bray DH, Boordman P, O'Neill MJ, Chan KL, Phillipson JD, Warhurst DC, Suffness M (1987) Plants as a source of antimalarial drugs. 5. Activities of *Ailanthus altissima* stem constituents and some related quassinoids. Phytother Res 1:22–24

Akebia quinata (Thunb.) Decne. 7

7.1 Introduction

Yuzhizi, Fructus Akebiae, is the dry ripe fruits of *Akebia quinata* (Thunb.) Decne., *A. trifoliata* (Thunb.) Koidz., or *A. trifoliata* (Thunb.) Koidz. var. *australis* (Diels) Rehd. (Lardizabalaceae) collected in summer and fall when the fruits have yellowed. It is listed officially in the Chinese Pharmacopoeia and is used in Chinese traditional medicine as an analgesic, antiphlogistic, and diuretic.

7.2 Chemical Constituents

A. quinata is rich in triterpene saponins that are present not only in fruits and seeds, but also in stems. A number of triterpene saponins have been isolated, mainly, with hederagenin (*7-1*) as the sapogenin, but also some with oleanolic acid as the aglycon. Thus, from the seeds of *A. quinata*, saponins A–G [1, 2]; from the stems, saponins referred to as akeboside St_{b-f}, St_h, St_j, and St_k [3, 4]; from the fresh pericarps, pericarp saponins A–H, J1–J3, and K [5]; and arjunolic acid, norarjunolic acid, and their glycosides [6] were isolated. The structures of saponins A–G (*7-2–7-8*); akebosides St_h (*7-9*), St_j (*7-10*), and St_k (*7-11*); pericarp saponins A, B–G (*7-12–7-16*), J1 (*7-17*), and J2 (*7-18*), and arjunolic acid (*7-19*), norarjunolic acid (*7-20*), and their glycosides (*7-21, 7-22*) were elucidated. Among them saponin A and pericarp saponin A; saponin C and pericarp saponin F; and akeboside St_h and pericarp saponin K were identical. Akeboside St_j and pericarp saponins B and E have an oleanolic acid aglycon, whereas the other saponins all have hederagenin as the sapogenin. The saponins are listed in Table *7-1*.

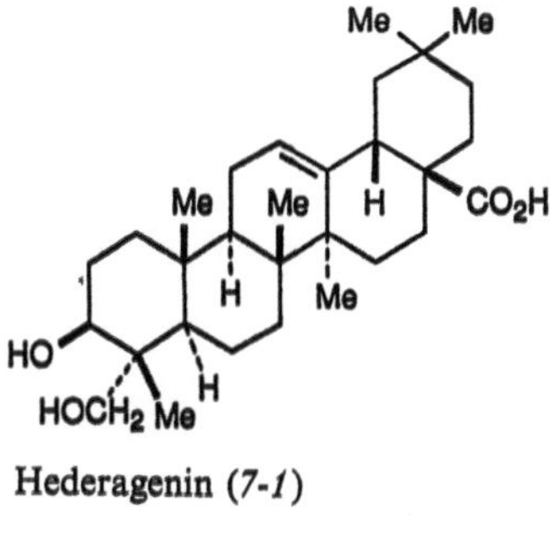

Hederagenin (*7-1*)

Table 7.1. Saponin components of *Akebia quinata*

Compound	Structure	Ref.
Akebia saponin A (Pericarp saponin A) Hederagenin 3-*O*-α-L-arabinopyranoside (*7-2*)	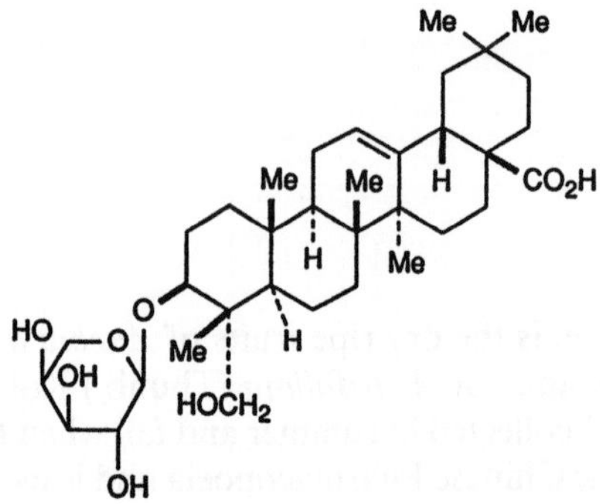	[1, 5]
Akebia saponin B Hederagenin 3-*O*-*β*-D-xylopyranosyl-(1→2)-α-L-arabinopyranoside (*7-3*)	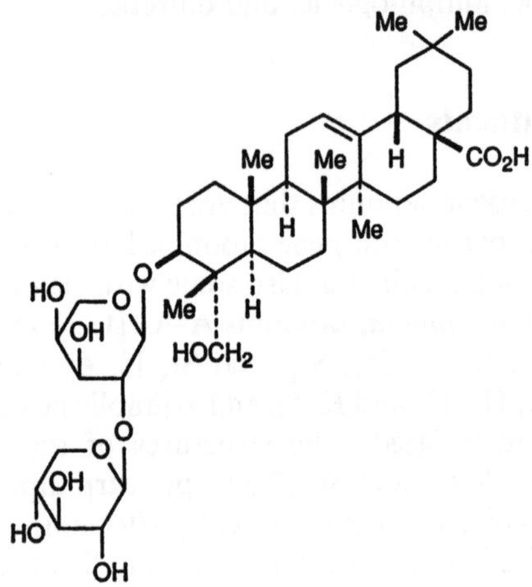	[1]
Akebia saponin C (Pericarp saponin F) Hederagenin 3-*O*-*β*-D-glucopyranosyl-(1→2)-α-L-arabinopyranoside (*7-4*)	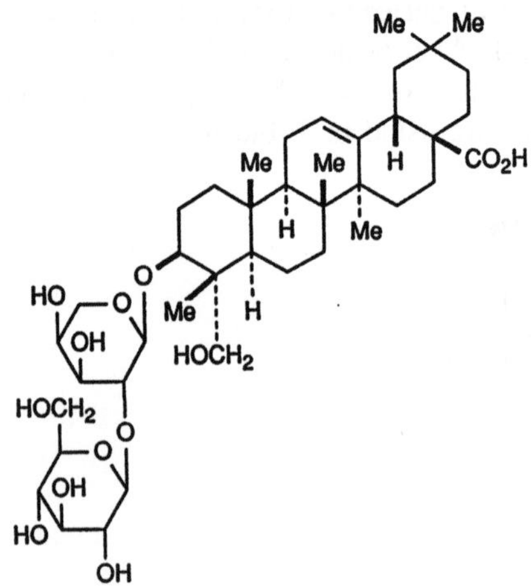	[1, 5]

Table 7.1. (continued)

Compound	Structure	Ref.
Akebia saponin D Hederagenin 3-*O*-α-L-arabinopyranoside 28-*O*-β-D-glucopyranosyl-(1→6)-β-D-glucopyranosyl ester (*7-5*)	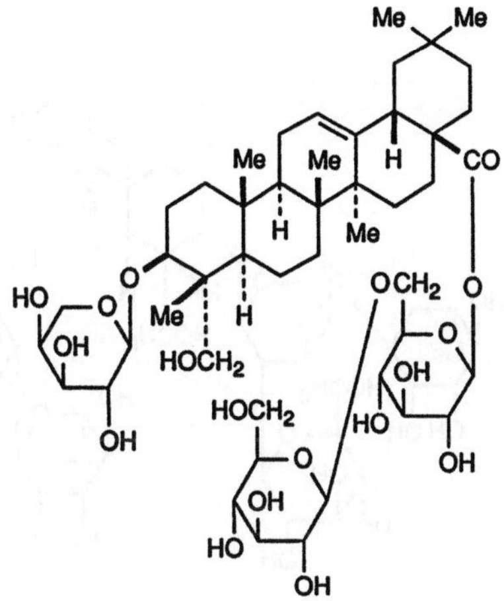	[2]
Akebia saponin E Hederagenin 3-*O*-β-D-xylopyranosyl-(1→2)-α-L-arabinopyranoside 28-*O*-β-D-glucopyrosyl-(1→6)-β-D-glucopyranosyl ester (*7-6*)	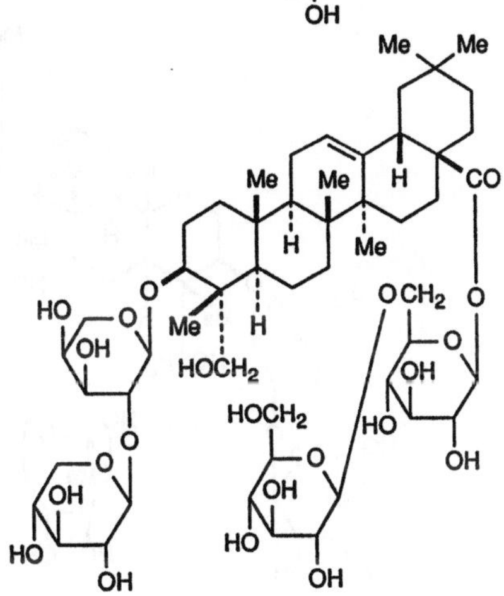	[2]
Akebia saponin F Hederagenin 3-*O*-β-D-glucopyranosyl-(1→2)-α-L-arabinopyranoside 28-*O*-β-D-glucopyranosyl-(1→6)-β-D-glucopyranosyl ester (*7-7*)	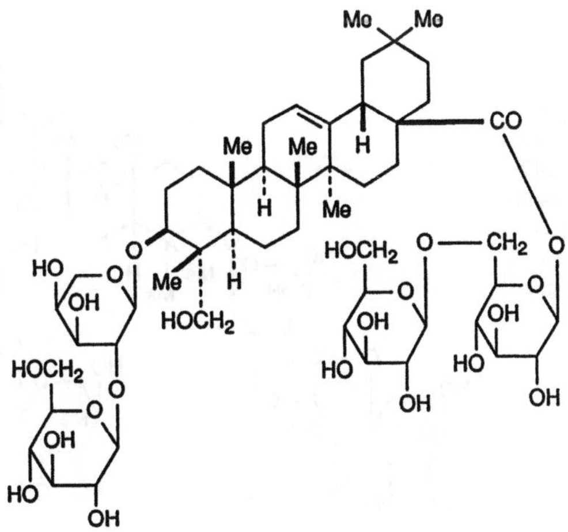	[2]

Table 7.1. (continued)

Compound	Structure	Ref.
Akebia saponin G Hederagenin 3-*O*-β-D-glucopyranosyl-(1→2)-[α-L-rhamnopyranosyl-(1→4)]-α-L-arabinopyranoside 28-*O*-α-L-rhamnopyranosyl-(1→4)-β-D-glucopyranosyl-(1→6)-β-D-glycopyranosyl ester (*7-8*)	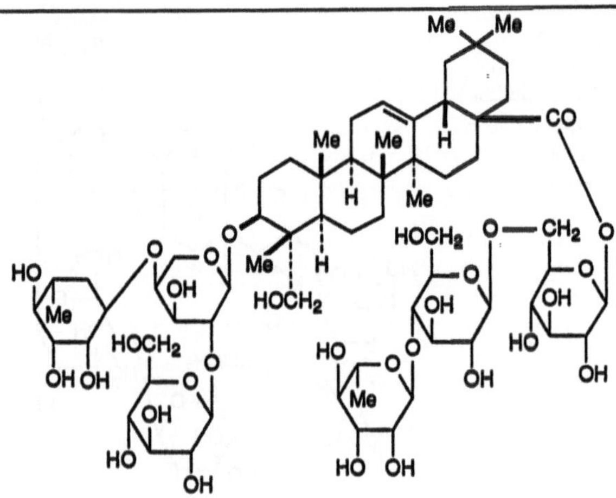	[2]
Akeboside St_h (Pericarp saponin K) Hederagenin 3-*O*-α-L-rhamnopyranosyl-(1→2)-arabinopyranoside 28-*O*-α-L-rhamnopyranosyl-(1→4)-β-D-glucopyranosyl-(1→6)-β-D-glucopyranosyl ester (*7-9*)	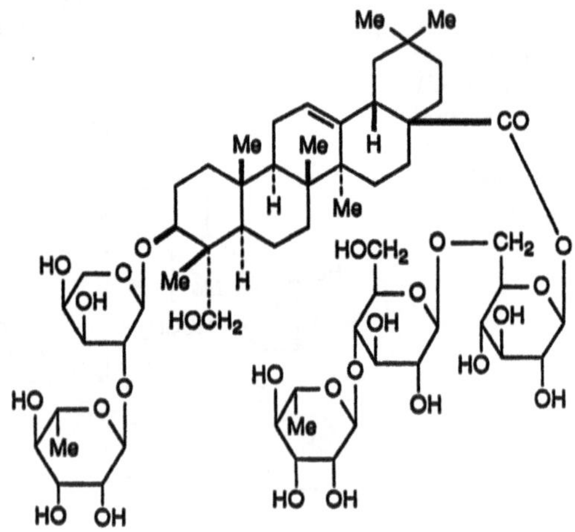	[3–5]
Akeboside St_j Oleanolic acid 3-*O*-α-L-rhamnopyranosyl-(1→6)-β-D-glucopyranosyl-(1→2)-α-L-arabinopyranoside 28-*O*-α-L-rhamnopyranosyl-(1→4)-β-D-glucopyranosyl-(1→6)-β-D-glucopyranosyl ester (*7-10*)	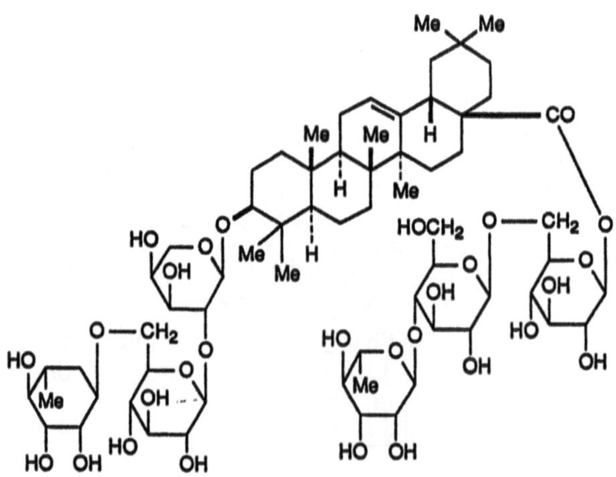	[3, 4]

Table 7.1. (continued)

Compound	Structure	Ref.
Akeboside St_k Hederagenin 3-*O*-α-L-rhamnopyranosyl-(1→6)-β-D-glucopyranosyl-(1→2)-α-L-arabinopyranoside28-*O*-α-L-rhamnopyranosyl-(1→4)-β-D-glucopyranosyl-(1→6)-β-D-glucopyranosyl ester (*7-11*)	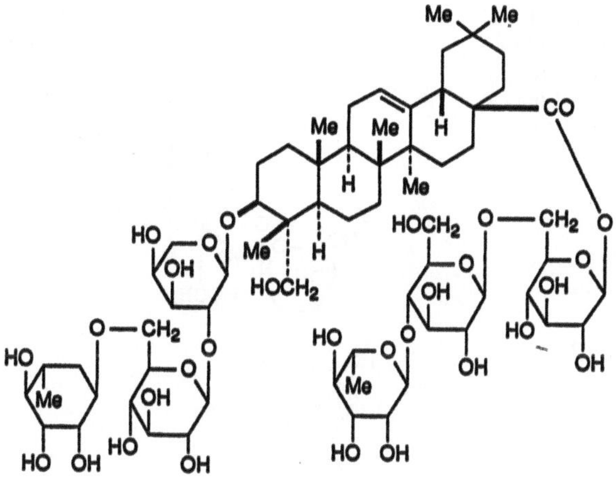	[3, 4]
Pericarp saponin B Oleanolic acid 3-*O*-α-L-rhamnopyranosyl-(1→2)-α-L-arabinopyranoside (*7-12*)	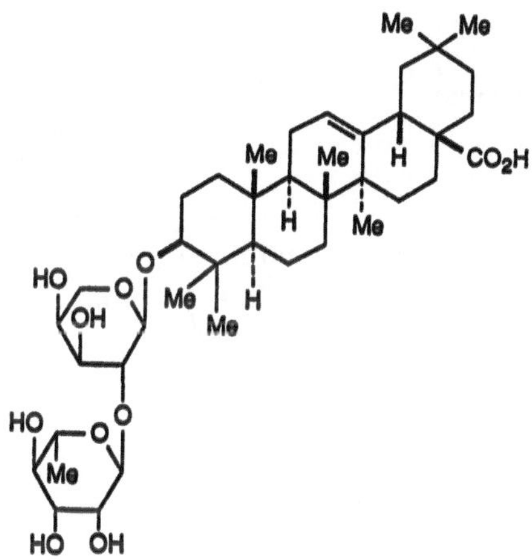	[5]
Pericarp saponin C Hederagenin 3-*O*-β-D-xylopyranosyl-(1→3)-α-L-arabinopyranoside (*7-13*)	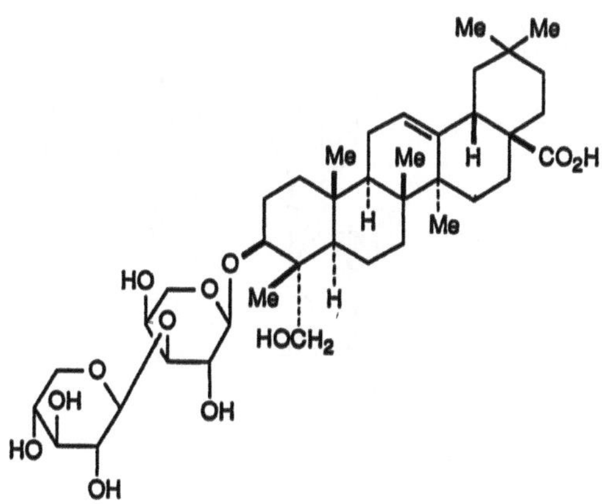	[5]

Table 7.1. (continued)

Compound	Structure	Ref.
Pericarp saponin D Hederagenin 3-*O*-α-L-rhamnopyranosyl-(1→2)-α-L-arabinopyranoside (*7-14*)	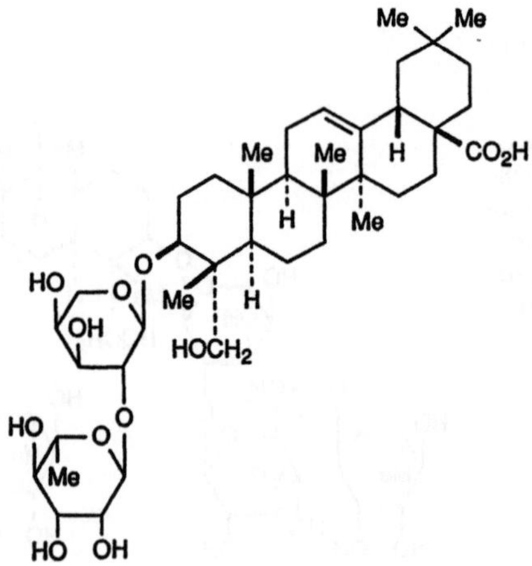	[5]
Pericarp saponin E Oleanolic acid 3-*O*-β-D-glucopyranosyl-(1→2)-α-L-arabinopyranoside (*7-15*)	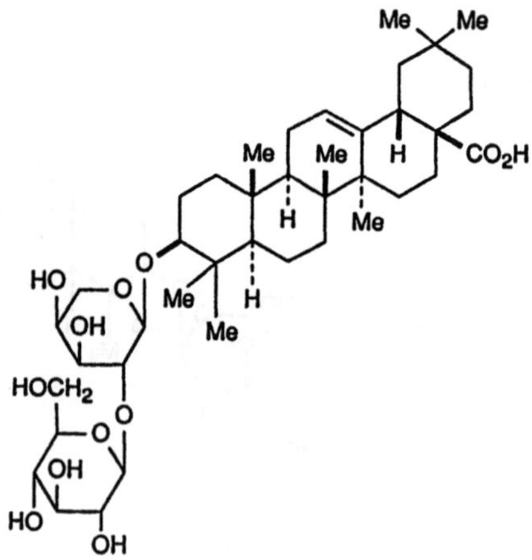	[5]
Pericarp saponin G Hederagenin 3-*O*-β-D-xylopyranosyl-(1→3)-α-L-rhamnopyranosyl-(1→2)-α-L-arabinopyranoside (*7-16*)	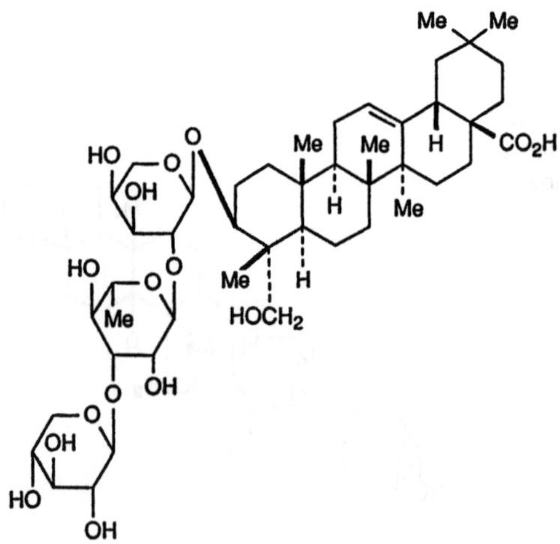	[5]

Table 7.1. (continued)

Compound	Structure	Ref.
Pericarp saponin J_2 Hederagenin 3-*O*-α-L-arabinopyranoside 28-*O*-α-L-rhamnopyranosyl-(1→4)-β-D-glucopyranosyl-(1→6)-β-D-glucopyranosyl ester (*7-17*)	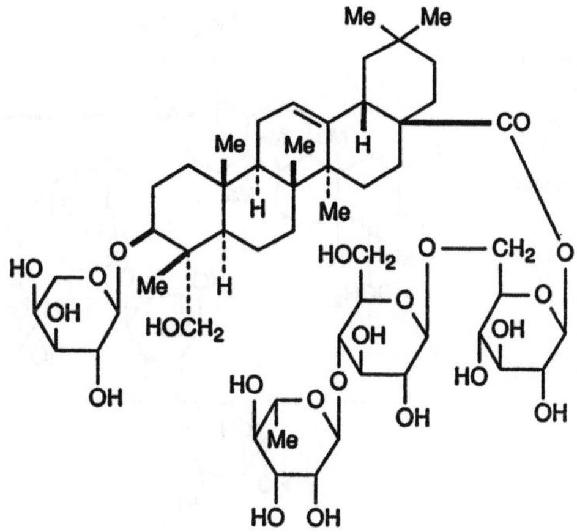	[5]
Pericarp saponin J_3 Oleanolic acid 3-*O*-α-L-rhamnopyranosyl-(1→2)-α-L-arabinopyranoside 28-*O*-α-L-rhamnopyranosyl-(1→4)-β-D-glucopyranosyl-(1→6)-β-D-glucopyranosyl ester (*7-18*)	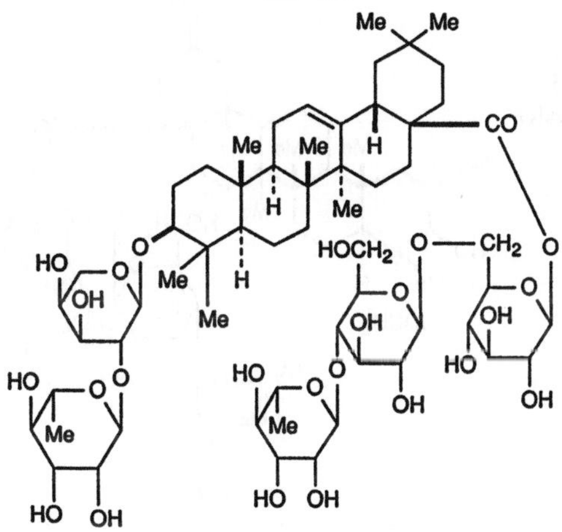	[5]
Arjunolic acid (*7-19*)	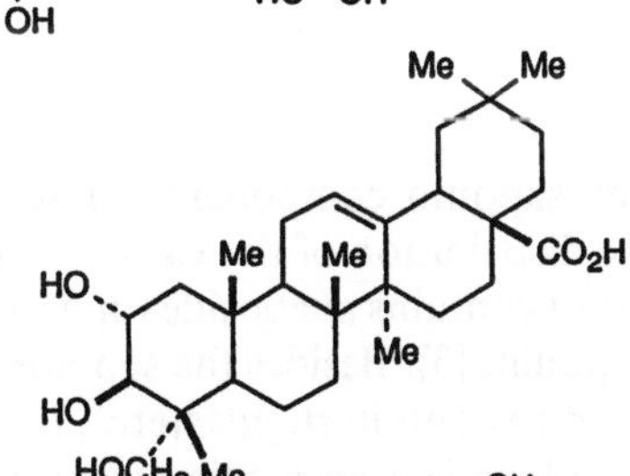	[6]
Norarjunolic acid (*7-20*)	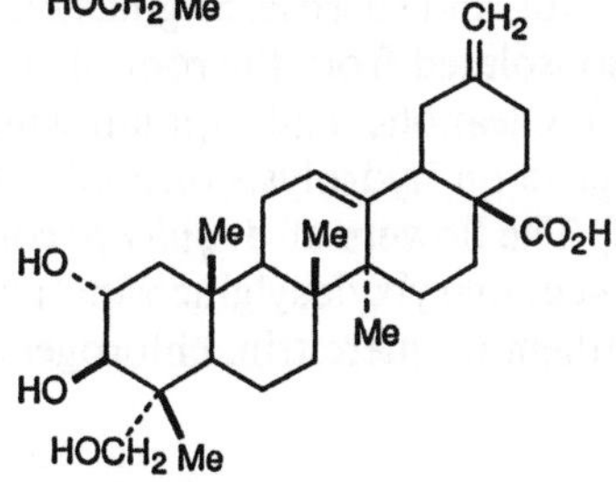	[6]

Table 7.1. (continued)

Compound	Structure	Ref.
Arjunolic acid 28-*O*-β-D-xylopyranosyl-(1→3)-α-L-rhamnopyranosyl-(1→4)-β-D-glucopyranosyl-(1→6)-β-D-glucopyranosyl ester (*7-21*)	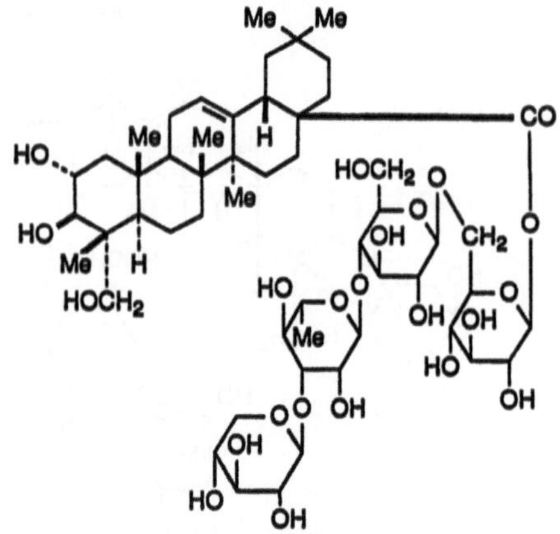	[6]
Norarjunolic acid 28-*O*-α-L-rhamnopyranosyl-(1→4)-β-D-glucopyranosyl-(1→6)-β-D-glucopyranosyl ester (*7-22*)	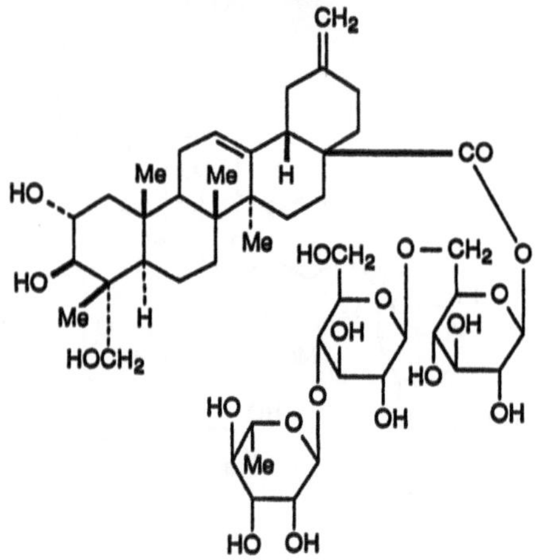	[6]

A comparison of the saponin components of seed, leaf, stem, and root of *A. quinata* showed that the distribution of the various saponins is different in different parts of the plant. The root contains akebosides and saponins with sugar residues but no 3-*O*-glucosides of saponins [3]. Besides the saponins described above, the stem of *A. quinata* was also found to contain stigmasterol, β-sitosterol, betulin, β-sitosterol-β-D-glucoside, myoinositol, and sucrose. Stigmasterol, β-sitosterol, and β-sitosterol-β-D-glucoside were also isolated from the root of *A. quinata* and from the root and stem of *A. trifoliata* [7]. Oleanolic acid saponin which was isolated from the dried root of *A. trifoliata*, gave on hydrolysis oleanolic acid, glucose, and rhamnose in equimolar amounts [8]. The flowers of *A. quinata* contained cyanidin 3-*p*-coumarylglucoside, cyanidin 3-*p*-coumarylxylosylglucoside, and an aroylglycoside of cyanidin in addition to chrysanthemin, quercitrin, chlorogenic acid, and caffeic acid [9].

7.3 Pharmacology

The aqueous and the methanol-ethereal extracts of *Akebia* stem, as well as oleanolic acid and hederagenin inhibited carrageenin-induced paw edema in rats and inhibited capillary permeability in mice. The extracts had diuretic and uricosuric activity in mice and rats; the aqueous extract caused central depression, sedation, and hypothermia. It had antipyretic and weak analgesic actions and stimulated intestinal motility and ileal contractility [10, 11].

References

1. Higuchi R, Miyahara K, Kawasaki T (1972) Seed saponins of *Akebia quinata* Decne. I. Hederagenin 3-*O*-glycosides. Chem Pharm Bull 20:1935–1939
2. Higuchi R, Kawasaki T (1972) Seed saponins of *Akebia quinata* Decne. II. Hederagenin 3,28-*O*-bisglycosides. Chem Pharm Bull 20:2143–2149
3. Kumekawa Y, Itokawa H, Fujita M (1974) The study on the constituents of *Clematis* and *Akebia* sp. III. The study on the structures of akebosides isolated from the stem of *Akebia quinata* Decne. Chem Pharm Bull 22:2294–2300
4. Fujita M, Itokawa H, Kumekawa Y (1974) Constituents of *Clematis* and *Akebia* subspecies. II. Saponins isolated from the stem of *Akebia quinata*. I. Yakugaku Zasshi 94:194–198
5. Higuchi R, Kawasaki T (1976) Pericarp saponins of *Akebia quinata* Decne. I. Glycosides of hederagenin and oleanolic acid. Chem Pharm Bull 24:1021–1032
6. Higuchi R, Kawasaki T (1976) Pericarp saponins of *Akebia quinata* Decne. II. Arjunolic and norarjunolic acids and their glycosides. Chem Pharm Bull 24:1314–1323
7. Fujita M, Itokawa H, Kumekawa Y (1974) Constituents of *Clematis* and *Akebia* subspecies. I. Distribution of triterpenes and other components. Yakugaku Zasshi 94:189–193
8. Sawai M, Nakamura S, Kitami F, Takezaki T, Taruya M (1972) Oleanolic acid glycoside from *Akebia*. Japanese patent no. 7229,964 (CA 77:156332k)
9. Ishikura N, Nagamizo N (1976) Anthocyanins from *Akebia* and *Stauntonia*. Phytochemistry 15:442–443
10. Tsen TT (1973) Pharmacological studies on the components of *Akebia longeracemosa*, especially on the chemical and pharmacological properties of *Akebia* saponins. Shikoku Igaku Zasshi 29:65–83
11. Yamahara J, Takagi Y, Sawada T, Fujimura H, Shirakawa K, Yoshikawa M, Kitagawa I (1979) Effects of crude drugs on congestive edema. Chem Pharm Bull 27:1464–1468

Alangium chinense (Lour.) Harms

8

8.1 Introduction

Alangium chinense (Lour.) Harms. (Alangiaceae) is a medicinal herb found in China. Its leaves, stems, and roots, especially the fibrous roots, are used in folk medicine as an analgesic, antirheumatic, and muscle relaxant. *A. chinense* is not officially listed in the Chinese Pharmacopoeia. Some other *Alangium* species are also used in folk medicine for the same indications. They are *A. platanifolium, A. salviifolium, A. kurzii, A. handelii*, and *A. chinense* var. *panciflorum*. Among them, *A. chinense* and *A. platanifolium* are the species most often used in folk medicine.

8.2 Chemical Constituents

A. chinense contains anabasine (*8-1*) as the major alkaloid component and active principle [1, 2].

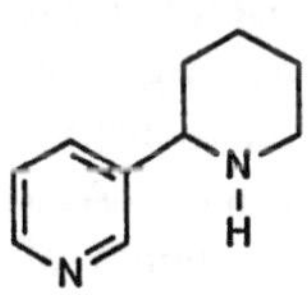

Anabasine (*8-1*)

A study of the distribution and amount of total alkaloids and anabasine in different plant parts showed that the amount of anabasine was highest in the fibrous root, followed by the rootlet, thick root, and leaves. The total alkaloid content in fibrous root, rootlet, and thick root was 0.16%, 0.04%, and 0.016%, respectively [1]. Thus, the amount of anabasine in the root of *A. chinense* is influenced not only by geographic factors, but also by the diameter of the root [2].

From the branches of *A. salviifolium* four alkaloids were isolated and identified: venoterpine (*8-2*), ankorine (*8-3*), cephaeline (*8-4*), and psychotrine (*8-5*) [3]. Anabasine was also detected in the root of *A. salviifolium* [2]. In addition, anabasine was found in *A. kurzii* [4], *A. chinense* var. *panciflorum* [2], *A. handelii*, and *A. platanifolium*, whereas ankorine was isolated from *A. kurzii* and venotropine from *A. chinense, A. handelii*, and *A. platanifolium* [4].

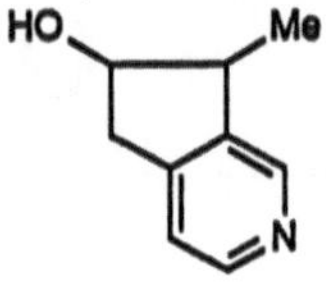

Venoterpine (*8-2*)

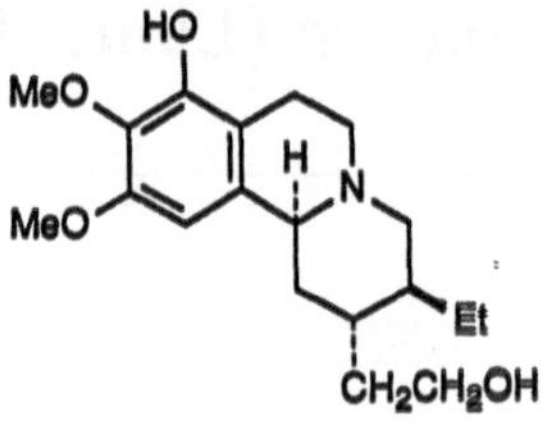

Ankorine (*8-3*)

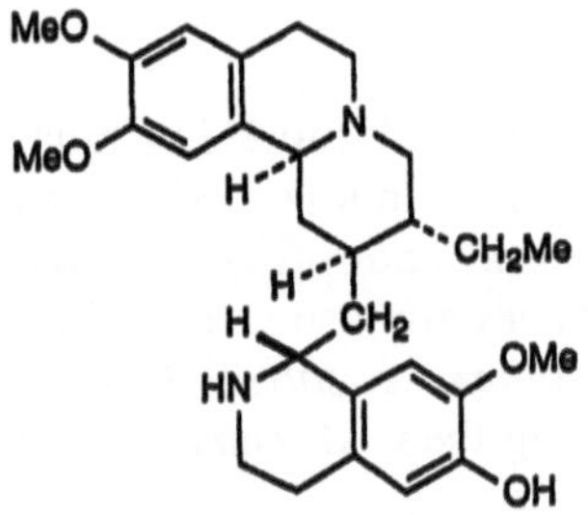

Cephaeline (*8-4*)

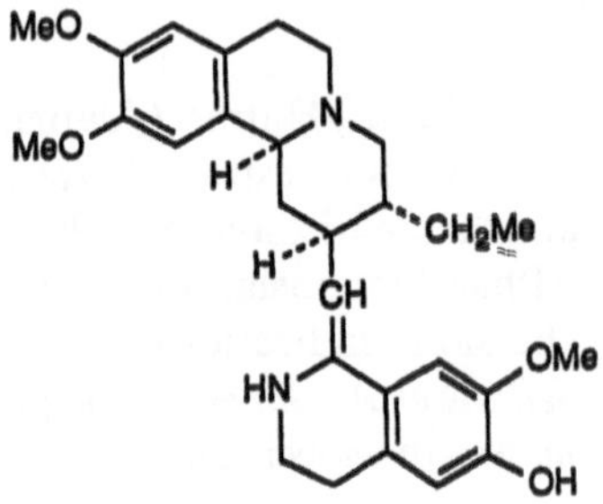

Psychotrine (*8-5*)

8.3 Pharmacology

Anabasine exerted a significant neuromuscular blocking effect on isolated rat diaphragm preparations. This action could be partly antagonized by neostigmine. On rat denervated diaphragms, anabasine inhibited the muscular response to acetylcholine. Anabasine had a depolarizing effect on isolated frog sartorius muscle. At a concentration of 0.83 μg/ml, it first increased and then decreased the frequency of end-plate potentials but did not alter the nerve terminal potential [5]. Intravenous injection of anabasine caused muscle relaxation in rabbits which lasted more than 1 h. Smooth muscle was only slightly relaxed [6]. In a clinical study, i.v. infusion of anabasine at single doses of 0.25–0.4 mg/kg caused muscle relaxation in more than 300 patients. In some patients, this effect lasted longer than 180 min [6].

The LD_{50} values of anabasine in mice were 14.7 mg/kg after i.p. and 4.3 mg/kg after i.v. administration. In rabbits, slight morphological changes were observed in the heart, liver, lung, and kidney, after i.v. injection of anabasine [6]. In vitro metabolism of anabasine by liver homogenates of rat, rabbit, or guinea pig resulted in two diastereoisomeric anabasine-*N*-oxides [7].

References

1. Shi DW, Wang ZW, Bi ZQ, Zeng XP, Shi DY (1983) Distribution and content of alkaloids in *Alangium chinense* (Lour.) Harms. Chin Trad Herb Drugs 14:165–166
2. Guo HS, Ying K, Xu HL, Du YX, Wang XM (1982) Distribution of anabasine and its contents of Alangiaceae plants in China. Chin Pharm Bull 17:390–391
3. Chen MJ, Hou LL, Zhu H (1980) Isolation and identification of alkaloids from *Alangium salviifolium* (Linn.f) Wangerin. Acta Bot Sin 22:257–259

4. Hou LL, Chen MQ, Zhu H (1981) Alkaloids in some species of *Alangium*. Chin Trad Herb Drugs 12:352–353
5. Yang QZ, Shu HD, Lin LR (1981) Blocking effect of anabasine on the neuromuscular junction. Acta Pharmacol Sin 2:84–88
6. Chang ZQ (1981) Study on *Alangium chinense* (Lour.) Harms, an herbal muscle relaxant. Bull Chin Mat Med 6:34–36
7. Beckett AH, Sheikh AH (1973) In vitro metabolic N-oxidation of the minor tobacco alkaloids, (−)-methyl-anabasine and (−)-anabasine to yield a hydroxyamine and a nitrone in lung and liver homogenates. J Pharm Pharmacol [Suppl] 25:171

Albizia julibrissin Durazz. 9

9.1 Introduction

Hehuanpi, Cortex Albiziae, is the dry stem bark of *Albizia julibrissin* Durazz. (Fabaceae) collected in summer or fall. The Chinese Pharmacopoeia requires for this official herbal medicine a qualitative determination of the saponin content by a foam test and by hemolytic activity on rabbit erythrocytes in physiological saline. The stem bark of *A. julibrissin* is used as a sedative and for treatment of trauma.

Hehuanhua, Flos Albiziae, is the dry inflorescence of *A. julibrissin* collected in summer when the flower blooms. It is also officially listed in the Chinese pharmacopoeia and used as a sedative.

9.2 Chemical Constituents

The stem bark of *A. julibrissin* showed a positive reaction when tested for saponins [1]. Among the sapogenins of triterpene type machaerinic acid methly ester (*9-1*), acacic acid lactone (*9-2*) [2], acacigenin B (*9-3*), machaerinic acid lactone (*9-4*) [3], and 16-deoxyacacigenin B (*9-5*) [4] were isolated and identified.

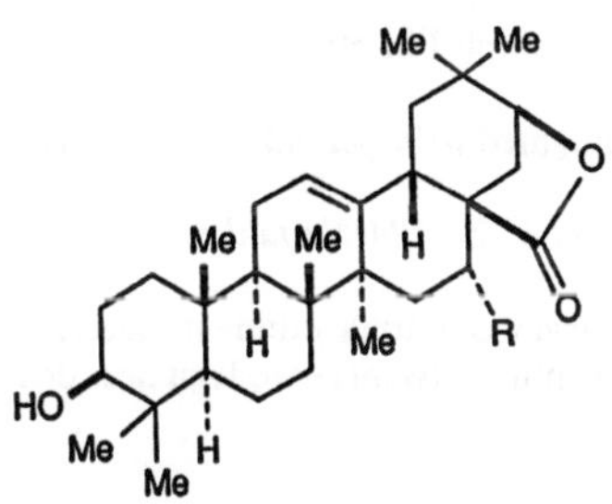

Machaerinic acid lactone (*9-4*): R=H
Acacic acid lactone (*9-2*): R=OH

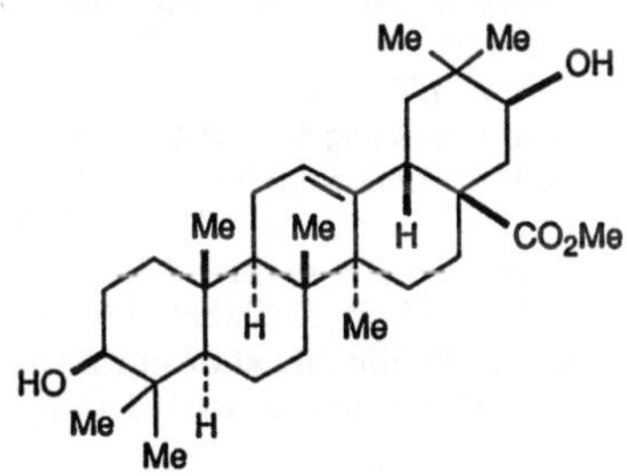

Machaerinic acid methyl ester (*9-1*)

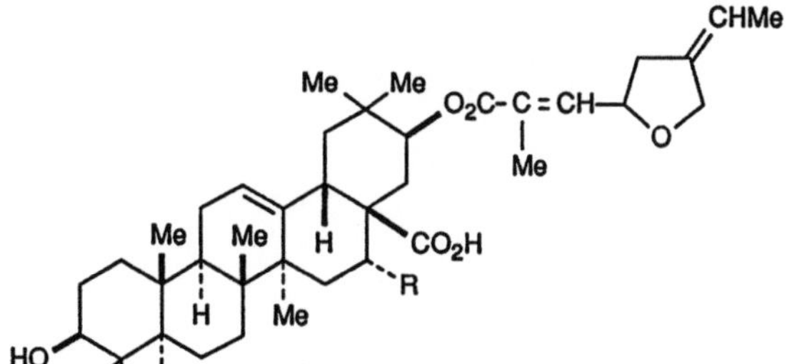

Acacigenin B (*9-3*): R=OH
16-Deoxyacacigenin B (*9-5*): R=H

In addition to the sapogenins, α-spinasteryl glucoside and 7,3′,4′-trihydroxyflavone were found in the stem bark of *A. julibrissin* [5]. From the lower of *A. julibrissin* cyanidin-3-glucoside (*9-6*) was identified [6].

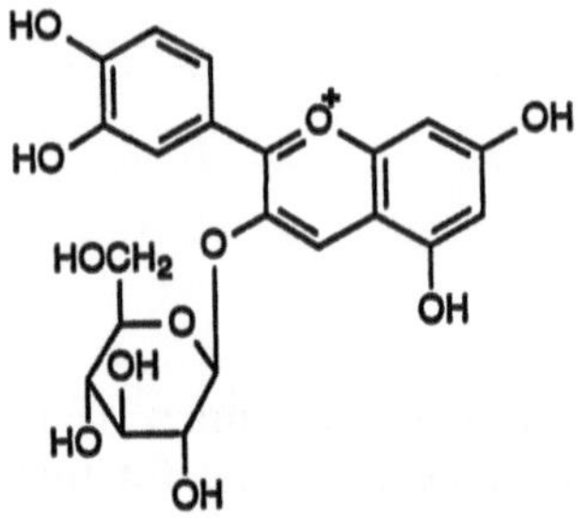

Cyanidin-3-β-D-glucopyranoside (*9-6*)

9.3 Pharmacology

The saponin fraction of the stem bark of *A. julibrissin*, from which machaerinic acid methyl ester and acacid acid lactone were isolated, had a strong uterotonic activity [2].

References

1. Wang CS (1982) Chemical identification of some traditional Chinese drugs containing saponins. Bull Chin Mat Med 7:13–14
2. You YH, Woo WS, Choi JS, Kang SS (1982) Isolation of a new sapogenin from *Albizzia julibrissin*. Arch Pharmacal Res 5:33–38 (CA 98:122814n)
3. Kang SS, Woo WS (1983) Sapogenins from *Albizzia julibrissin*. Arch Pharmacal Res 6:25–28 (CA 99:102292h)
4. Woo WS, Kang SS (1984) Isolation of a new monoterpene conjugated triterpenoid from the stem bark of *Albizzia julibrissin*. J Nat Prod 47:547–549
5. Chamsuksai P, Choi JS, Woo WS (1981) 3′,4′,7-Trihydroxyflavone in *Albizzia julibrissin*. Arch Pharmacal Res 4:1219–1231 (CA 96:196564m)
6. Ishikura N, Ito S, Shibata M (1978) Paper chromatographic survey of anthocyanins in Leguminosae. III. Identification and distribution of pattern of anthocyanins in twenty-two legumes. Bot Mag 91:25–30 (CA 89:103708d)

Alisma orientalis (Sam.) Juzep. 10

10.1 Introduction

Zexie, Rhizoma Alismatis, is the dry rhizome of *Alisma orientalis* (Sam.) Juzep. (Alismataceae). This official herbal drug is used in traditional Chinese medicine as a diuretic in the treatment of oliguresis and edema. It is also used to treat hyperlipidemia.

10.2 Chemical Constituents

Six triterpenes were isolated from the rhizome of *A. orientalis*: alisol A (*10-1*), alisol A monoacetate (*10-2*) [1–3], alisol B (*10-3*), alisol B monoacetate (*10-4*) [1, 2, 4], alisol C monoacetate (*10-5*) [4], and *epi*-arisol A (*10-6*) [1]. All have a dammarane (*10-7*) as a structural feature. The structure of alisol A was established by X-ray crystallography of crystalline alisol A triacetate; the structure of other alisols was determined by chemical and spectral analysis using alisol as a reference.

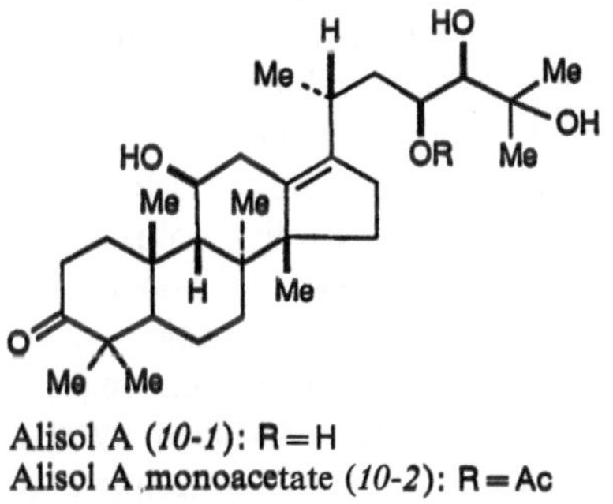

Alisol A (*10-1*): R=H
Alisol A monoacetate (*10-2*): R=Ac

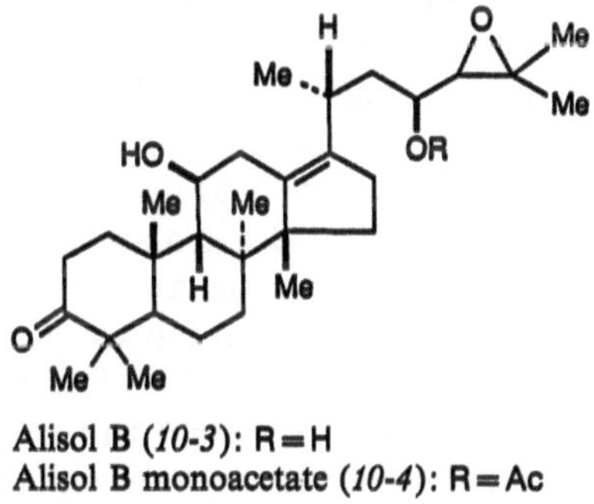

Alisol B (*10-3*): R=H
Alisol B monoacetate (*10-4*): R=Ac

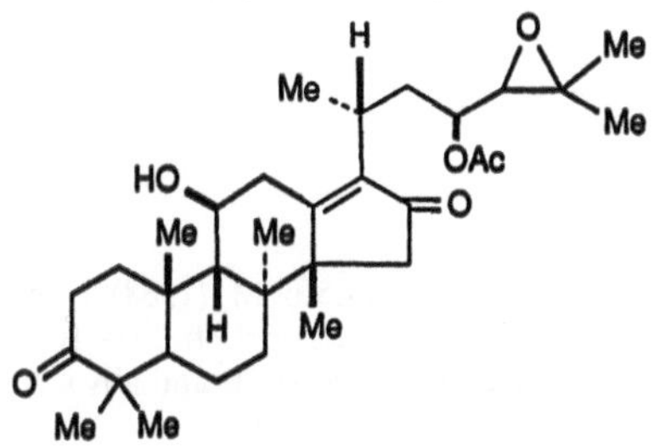

Alisol C monoacetate (*10-5*)

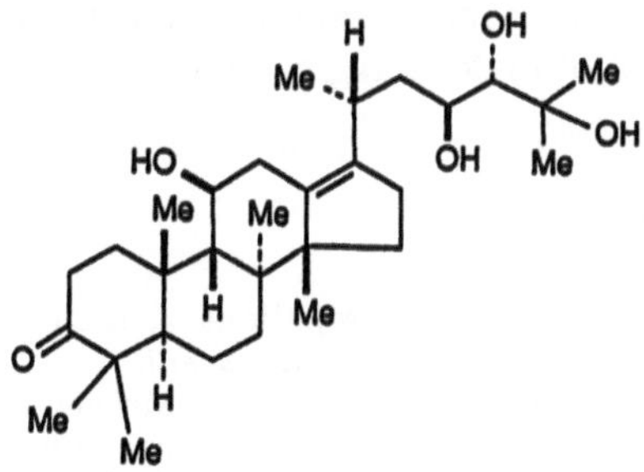

epi-Alisol A (*10-6*)

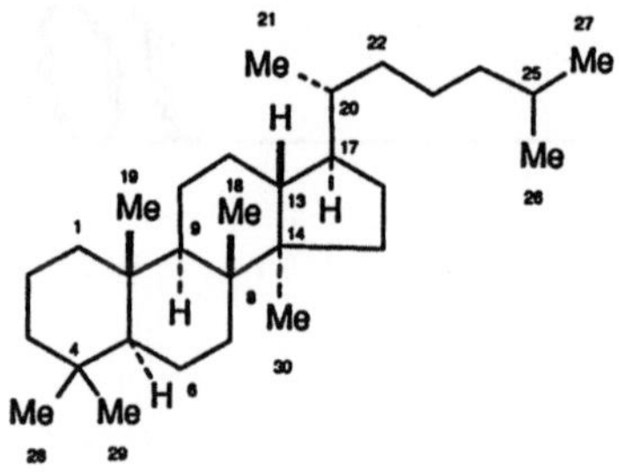

Dammarane (*10-7*)

Besides the triterpenes, two sesquiterpenes, alismol (*10-8*) and alismoxide (*10-9*), were isolated from the rhizome of *A. orientalis* and their structure determined [5].

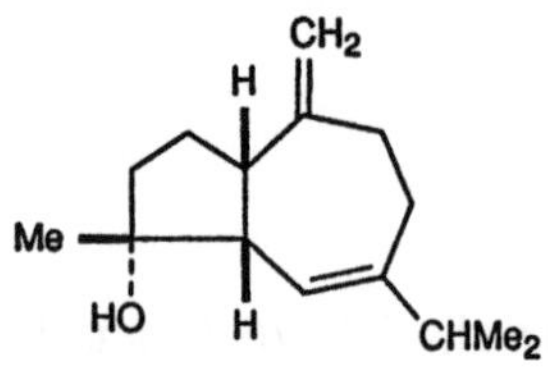

Alismol (*10-8*)

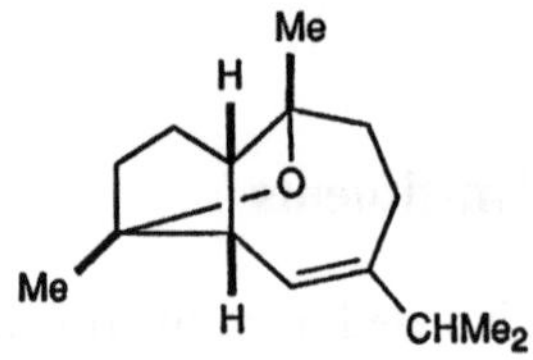

Alismoxide (*10-9*)

10.3 Pharmacology

Alisols A and B and their monoacetates as well as alisol C monoacetate showed significant activity against hypercholesterolemia in rats [6]. In a test of diuretic activity in saline-loaded mice and rats, none of the alisols produced changes in urinary volume or Na^+ excretion in mice at a dose of 100 mg/kg s.c.; however, in rats alisols A and B, given orally at doses of 30 mg/kg, produced a significant increase in Na^+ excretion. Alisol B has also been reported to increase urinary volume [7]. Furthermore, alisol A monoacetate, alisol B monoacetate, and alisol C monoacetate protected mice against carbon tetrachloride-induced liver damage, as indicated by serum GPT and triglyceride levels. Alisol C monoacetate was the most effective [8].

An acetone extract of *Alisma* rhizome inhibited contractions induced by angiotensin I in rabbit aortic strips. Alismol was found to be the active principle and the activity was dose dependent [9].

Sublingual administration of an extract of *A. orientalis* to mice resulted in strong inhibition of platelet aggregation [10].

References

1. Murata T, Shinohara M, Hirata T, Kamiya K, Nishikawa M, Miyamoto M (1968) New triterpenes of *Alisma plantago-aquatica* L. var. *orientale* Samuels. Tetrahedron Lett 103–108
2. Murata T, Shinohara M, Hirata T, Miyamoto M (1968) The structures of alisol B and alisol A monoacetate – occurrence of a facile acyl migration. Tetrahedron Lett 849–854
3. Murata T, Miyamoto M (1970) Biologically active triterpenes of alismatis rhizoma. II. Structures of alisol A and alisol A monoacetate. Chem Pharm Bull 18:1354–1361

4. Murata T, Shinohara M, Miyamoto M (1970) Biological-active triterpenes of alismatis rhizoma. IV. Structures of alisol B, alisol B monoacetate and alisol C monoacetate. Reactions of the α-hydroxy epoxide of the alisol B derivatives. Chem Pharm Bull 18:1369–1384
5. Oshima Y, Iwakawa T, Hikino H (1983) Sesquiterpenoids. LVIII. Alismol and alismoxide, sesquiterpenoids of *Alisma* rhizoma. Phytochemistry 22:183–185
6. Murata T, Imai Y, Hirata T, Miyamoto M (1970) Biological-active triterpenes of alismatis rhizoma. Chem Pharm Bull 18:1347–1353
7. Hikino H, Iwakawa T, Oshima Y, Nishikawa K, Murata T (1982) Efficacy of oriental drugs. XXXIV. Diuretic principles of *Alisma plantago-aquatica* var. *orientale* rhizomes. Shoyakugaku Zasshi 36:150–153 (CA 98:27656d)
8. Chang IM, Kim YS, Yun HS, Kim SO (1982) Liver-protective activities of alisol compounds against carbon tetrachloride intoxication. Korean J Pharmacogn 13:112–125 (CA 98:172944a)
9. Yamahara J, Matsuda H, Murakami H, Fujimura H (1986) The active principle of alismatis rhizoma which inhibits contractile responses in aorta. Chem Pharm Bull 34:4422–4424
10. Le ZS, Liu XF, Sun YL, Liu JZ, Yao SZ, Li YL, Chen XY, Fang LG (1985) Preliminary study on the inhibitory effect of 53 Chinese herbal drugs on platelet aggregation. Bull Chin Mat Med 10:44–45

Allium sativum L.

11

11.1 Introduction

Allium sativum L. (Liliaceae), garlic, is a well known spice and has been used worldwide as a folk medicine for treatment of various infectious diseases; prevention of coronary thrombosis, atherosclerosis, and stroke; and for treatment of hyperlipidemia and vascular disorders. It is included in the appendix of the Chinese Pharmacopoeia.

The Chinese Pharmacopoeia lists two additional items from *Allium* species.

- Jiucaizi, Semen Allii tuberosi, is the dry ripe seed of *A. tuberosum* Rottl. collected in fall when the seeds are ripe. It is used for treatment of polyuria, impotence, and lumbago.
- Xiebai, Bulbus Allii macrostemi, is the dry bulb of *A. macrostemon* Bge. collected in summer and fall. It is used as an antiasthmatic and antidiarrheic drug.

11.2 Chemical Constituents

11.2.1 Chemical Constituents of *Allium sativum*

Garlic is known to contain a number of organic sulfur compounds including the odoriferous substance allicin (*11-1*), which was isolated by extracting garlic with ethanol at room temperature [1]. The structure of allicin was determined as *S*-allyl 2-propene sulfinothioic acid ester [1, 2].

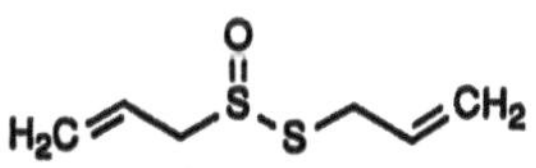

Allicin (*11-1*)

Later, the isolation of a sulfur containing amino acid derivative, *S*-allylcysteine *S*-oxide (alliin, *11-2*), was reported [2]. Alliin is an odorless solid with a pungent taste that is converted by the enzyme allinase into allicin (Fig. 11-1).

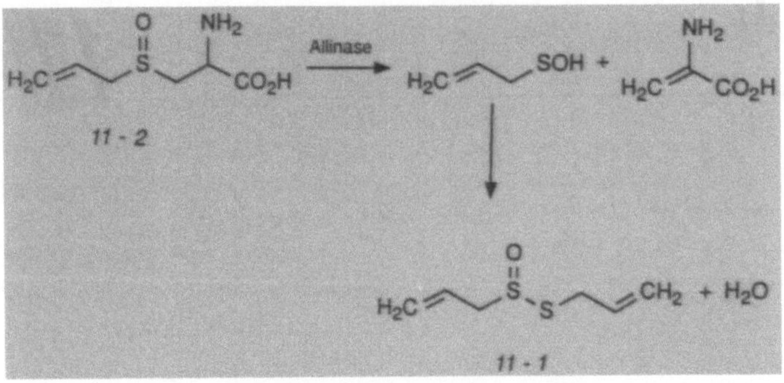

Fig. 11.1. Conversion of alliin to allicin by allinase

In subsequent investigations it was found that the cysteine sulfoxide fraction of garlic consists of 85% alliin along with 2% *S*-propylcysteine sulfoxide (*11-3*) and 13% *S*-methylcysteine sulfoxide [3].

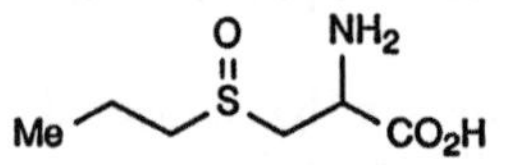

S-propylcysteine sulfoxide
(*11-3*)

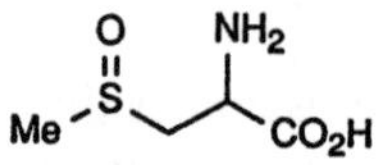

S-methylcysteine sulfoxide
(*11-4*)

Allinase activity on the *S*-substituted cysteine sulfoxide fraction from garlic extract yields allyl methanesulfinothioic acid ester and other symmetrical or asymmetrical sulfinothioic acid esters with methyl, propyl, and allyl substituents [3]. The presence of *S*-allyl-L-cysteine (*11-5*), which may be a precursor of alliin, has also been reported [4].

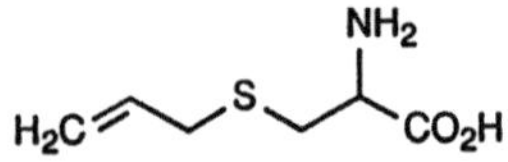

S-allylcysteine (*11-5*)

Additional volatile components of garlic extract were identified as allyl alcohol, methyl allyl disulfide, diallyl disulfide (*11-6*), dimethyl trisulfide, allyl methyl trisulfide, diallyl trisulfide (*11-7*) and sulfur dioxide [5].

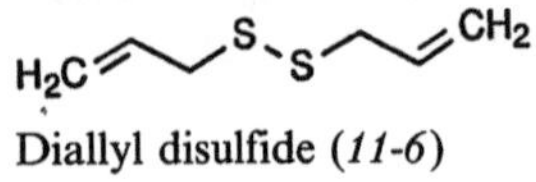

Diallyl disulfide (*11-6*)

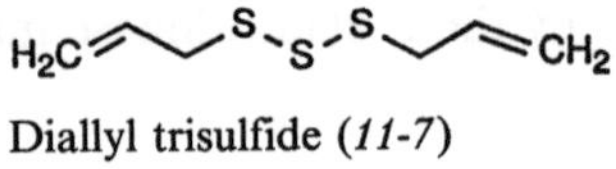

Diallyl trisulfide (*11-7*)

Allicin decomposed nearly completely at 20 °C within 20 h, giving diallyl disulfide as the major product and diallyl trisulfide, diallyl sulfide, sulfur oxide, and trace amounts of two 1,2-dithiin compounds: 3-vinyl-6H-1,2-dithiin (*11-8*) and 3-vinyl-4H-1,2-dithiin (*11-9*) [5].

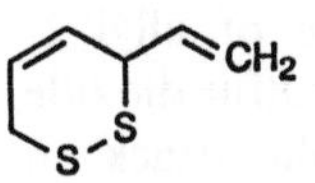

3-Vinyl-6H-1,2-dithiin (*11-8*)

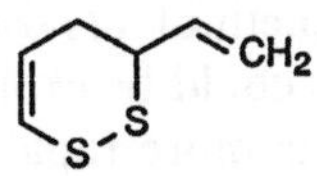

3-Vinyl-4H-1,2-dithiin (*11-9*)

Other cyclic sulfur compounds separated from the benzene fraction of the steam volatile oils from garlic are the trithiolane derivatives *cis-* and *trans*-3,5-diethyl-1,2,4-trithiolane (*11-10, 11-11*) and *cis-* and *trans*-3-methyl-5-ethyl-1,2,4-trithiolane (*11-12, 11-13*) [6].

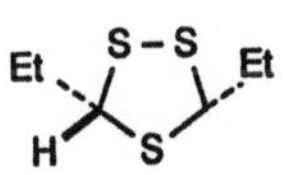

cis-3,5-Diethyl-1,2,4-trithiolane (*11-10*)

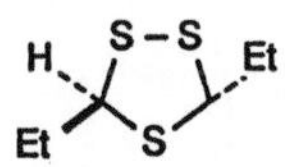

trans-3,5-Diethyl-1,2,4-trithiolane (*11-11*)

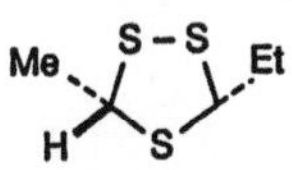

cis-3-Methyl-5-ethyl-1,2,4-trithiolane (*11-12*)

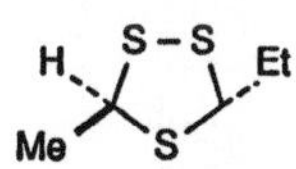

trans-3-Methyl-5-ethyl-1,2,4-trithiolane (*11-13*)

In garlic, 2-vinyl-1,3-dithiin (*11-14*) [7, 8] and allyl 1,5-hexadienyltrisulfide (*11-15*) [7] were also detected.

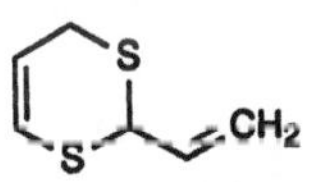

2-Vinyl-1,3-dithiin (*11-14*)

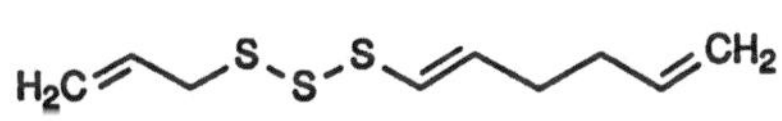

Allyl 1,5-hexadienyltrisulfide (*11-15*)

E and Z isomers of 4,5,9-trithiododeca-1,6,11-triene-9-oxide (ajoene) (*11-16, 11-17*) were isolated recently from garlic oil. Ajoene can be readily synthesized by decomposing allicin in acetone and water [8].

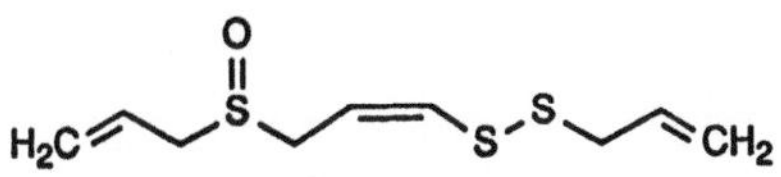

(Z)-Ajoene (*11-16*)

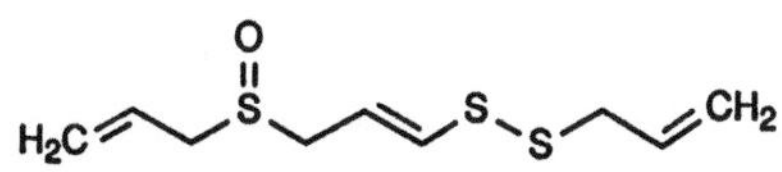

(E)-Ajoene (*11-17*)

Block et al. postulated that all of the sulfur containing products isolated from garlic are derived from allicin. Beta-elimination of allicin should yield 2-propenesulfenic acid (*11-18*) and thioacrolein (*11-19*). The latter compound is reported to dimerize to 3-vinyl-4H-1,2-dithiin and 2-vinyl-4H-1,3-dithiin [9–11]. *S*-allylthiolation of allicin should give a sulfonium ion (*11-20*), which could undergo β-elimination to a cation (*11-21*). Subsequent γ-addition of 2-propenesulfenic acid gives (E,Z)-ajoene. Hydrolysis of the sulfonium ion yields allyl alcohol (*11-22*) and diallyl trisulfide. Hydrolysis of allicin should give 2-propenesulfinic acid (*11-23*); β,γ-unsaturated sulfinic acids are known to readily lose sulfur dioxide. Diallyl disulfide could arise via attack of 2-propenethiol (*11-24*) on allicin. The absence of allyl 2-propenethiosulfonate (*11-25*) could be explained if the rate of loss of sulfur dioxide from 2-propenesulfinic acid is more rapid than its rate of nucleophilic attack on allicin. The mechanisms for formation of the respective sulfur containing products from allicin is summarized in Fig. 11-2 [8].

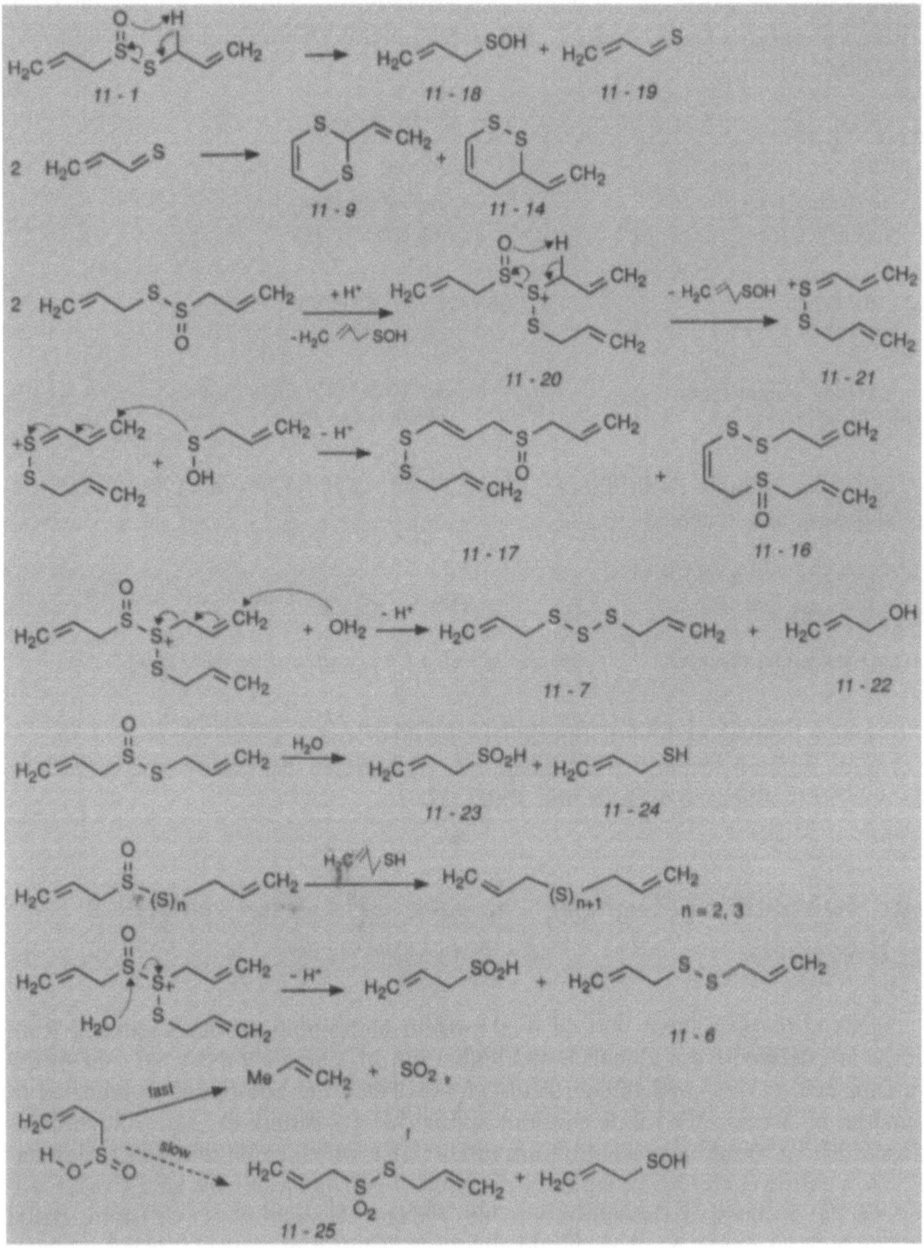

Fig. 11.2. Mechanisms for formation of sulfur-containing products from allicin

In addition, some sulfur containing acidic peptides such as γ-glutamyl-*S*-methylcysteine and its sulfoxide derivative, γ-glutamyl-*S*-(2-carboxy-propyl)-cysteinyl-glycine (*11-26*); γ-glutamyl-*S*-allylcysteine (*11-27*); and γ-glutamyl-*S*-propylcysteine as well as γ-glutamylphenylalanine without sulfur were found in garlic [12].

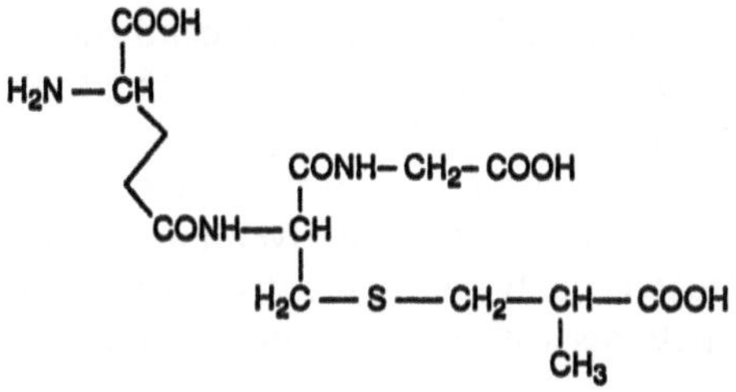

γ-Glutamyl-*S*-(2-carboxy-propyl)-cysteinyl-glycine (*11-26*)

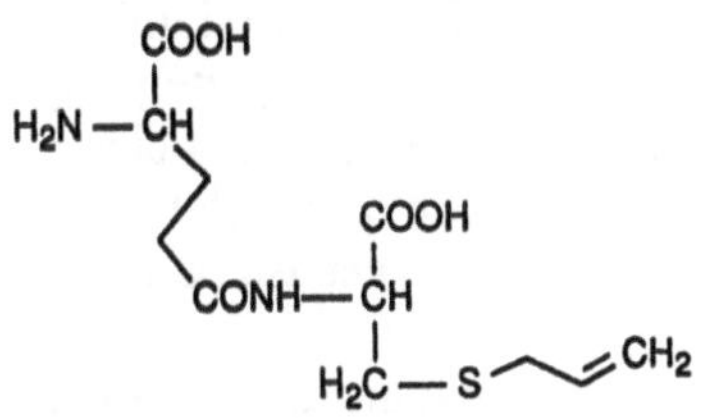

γ-Glutamyl-*S*-allyl-cysteine (*11-27*)

11.2.2 Chemical Constituents of *Allium tuberosum*

Seven sulfides: dimethyl sulfide, diallyl sulfide, methyl allyl disulfide, dimethyl trisulfide, diallyl disulfide, methyl allyl trisulfide and dimethyl tetrasulfide were identified together with 2 ketones, 18 alcohols, and 2 esters in the oil obtained from extraction of the steam distillate of *A. tuberosum*. Dimethyl disulfide and dimethyl trisulfide were the main volatile components [13].

11.3 Pharmacology

Garlic or its constituents exhibit various biological activities such as antibacterial, antifungal, antiviral, antitumor and antidiabetic effects. The greatest interest, however, has been focused on their anticholesterolemic and antithrombotic activities. Applied orally, garlic and its ethanol extract reduced plasma and liver cholesterol levels in male rats fed a diet containing 1% cholesterol. In particular, very low density lipoprotein and low density lipoprotein cholesterol fractions were reduced. The hypocholesterolemic principle of garlic was found in the ethanol extract and was stable when autoclaved at 120 °C for 1 h [14]. Oral administration of an aqueous extract of garlic to hypercholesterolemic patients for 2 months significantly reduced cholesterol levels. Withdrawal of treatment increased the cholesterol levels again [15].

Garlic administered at a daily dose of 5 g for up to 3 weeks to healthy humans with normal blood cholesterol level decreased serum triglycerides and decreased slightly total serum cholesterol [16]. Treatment at a daily dose of 10 g for 2 months significantly decreased blood cholesterol levels [17].

Experimental studies showed that cholesterol-induced hyperlipidemia in rats could be controlled by garlic feeding. Garlic treatment did not alter concentrations of circulating thyroid hormones or thyroidal uptake of radioiodine, suggesting that the hypolipidemic effect of garlic is not mediated by the thyroid [18].

Rabbits with cholesterol-induced atherosclerosis had lower levels of tissue and serum cholesterol and less marked changes in their aortas when given an alcohol extract of garlic [19]. Garlic significantly increased the fibrinolytic activity of blood from healthy individuals [16, 20] and from patients with old or acute myocardial infarction [20].

Allicin [21], methyl allyl trisulfide [22], ajoene [8], diallyl trisulfide, 2-vinyl-1,3-dithiin, and allyl 1,5-hexadienyl trisulfide [7] all inhibited platelet aggregation. Aggregation of human platelets induced by ADP or arachidonic acid was also inhibited when incubated with garlic. Thromboxane B_2 synthesis was almost completely suppressed and a new metabolite, identified as 10-hydroxy-11,12-epoxy-5,8,14-eicosatrienoic acid, was formed under these conditions. Thus, inhibition of platelet aggregation was related to alterations in the cyclooxygenase and lipoxygenase pathways [23, 24].

When a garlic extract containing 8 μM allicin was administered intragastrically to rats, a maximum of 0.4 μM in the intestine and 2.4 μM in the cecum was deteced after 4 and 6 h, respectively. A 50%–60% reduction in microflora was observed in the intestine 4 h after administration, but no such change was observed in the cecum even after 6 h [25].

Both allicin and a garlic extract had antifungal activity towards clinical isolates of *Candida albicans*. The activity was inhibited by L-cysteine or dithioerythritol [26]. Ajoene also had strong antifungal activity. The growth of both *Aspergillus niger* and *C. albicans* was inhibited by ajoene at <20 µg/ml [27]. Furthermore, garlic extract containing allicin, diallyl disulfide, and diallyl trisulfide possessed antiviral activity in vitro against influenza B virus and herpes simplex type 1 virus but not against Coxsackie B virus [28].

Ehrlich ascites tumor cells mixed with 2.8 mM allicin and incubated at 37 °C for 1 h were not lethal to mice when injected i.p. Likewise, cells from a mouse spontaneous mammary tumor, a mouse 20-methylcholanthrene-induced sarcoma, or Yoshida sarcoma cells were not tumorigenic when treated with >2.8 mM allicin before injection into mice [29]. Intratumoral injection of allicin into mice inoculated with sarcoma-180 tumor cells 24 h prior to allicin treatment resulted in a marked inhibition of tumor growth [30].

Addition of an aqueous garlic extract to cultures of *Salmonella typhimurium* strains decreased the number of (His^+) mutants induced by peroxides and especially γ-irradiation. It did not decrease the mutagenicity of adriamycin, mitomycin, sodium azide, 2-nitrofluorene, 1,2-epoxy-3,3,3-trichloropropane, or methyl-nitronitrosoguanidine. Thus, garlic extract has antimutagenic activity only against factors whose genotoxic activity is the result of radical formation and not against agents causing DNA damage by other mechanisms [31].

The first stage of tumor promotion caused by 12-*O*-tetradecanoyl-phorbol-13-acetate (TPA) in mouse skin carcinogenesis has been found to be suppressed by treatment with garlic extract. The garlic extract seems to be effective in inhibiting the initial events caused by TPA [32].

Oral administration of allicin to alloxan-induced diabetic rabbits produced dose-dependent hypoglycemia [33]. It also significantly improved glucose tolerance and increased serum insulin effect and liver glycogen synthesis [34].

The pharmacokinetics of allicin in rabbits by intravenous or rectal administration has been reported. The half-lives after intravenous or rectal applications were 2.3 and 1.5 h, respectively. Allicin was distributed in the lung and heart, with a minimal amount in the liver, and no accumulation was noted following rectal administration at a daily dose of 5 mg/kg for 5 days [35].

References

1. Cavallito CJ, Buck JS, Suter DM (1944) Allium, the antibacterial principle of *Allium sativum*. II. Determination of the chemical structure. J Am Chem Soc 66:1952–1954
2. Stoll A, Seebeck E (1948) Über Alliin, die genuine Muttersubstanz des Knoblauchöls. 1. Mitteilung über *Allium*-Substanzen. Helv Chim Acta 31:189–210
3. Freeman GG, Whenham RJ (1975) A survey of volatile components of some *Allium* species in terms of *S*-alk(en)yl-L-cysteine sulphoxides present as flavour precursors. J Sci Food Agric 26:1869–1886
4. Suzuki T, Sugii M, Kakimoto T, Tsuboi N (1961) Isolation of (−)-*S*-allyl-L-cysteine from garlic. Chem Pharm Bull 9:251–252
5. Brodnitz MH, Pascale JV, van Derslice L (1971) Flavor components of garlic extract. J Agric Food Chem 19:273–275
6. Kameoka H, Demizu Y, Iwase Y, Miyazawa M (1978) Structure of cyclic sulfur compounds from *Allium* genus. Tennen Yuki Kagobutsu Toronkai Koen Yoshihu, 21st, 199–205 (CA 90:69093 h)
7. Apitz-Castro R, Cabrera S, Cruz MR, Ledezma E, Jain MK (1983) Effects of garlic extract and of three pure components isolated from it on human platelet aggregation, arachidonate metabolism, release reaction and platelet ultrastructure. Thromb Res 32:155–169
8. Block E, Ahmad S, Catalfamo JL, Jain MK, Apitz-Castro R (1986) Antithrombotic organosulfur compounds from garlic: structural, mechanistic and synthetic studies. J Am Chem Soc 108:7045–7055
9. Bock H, Mohmand S, Hirabayshi T, Semkow A (1982) Gasphasen-Reaktionen. XXX. Thioacrolein: Das stabilste C_3H_4S-Isomere und seine PE-spektroskopischer Nachweis in der Gasphase. Chem Ber 115:1339–1348
10. Vedejs E, Eberlein TH, Varie DL (1982) Dienophilic thioaldehydes. J Am Chem Soc 104:1445–1447
11. Beslin P (1983) A facile synthesis of two thioacrolein dimers. A new entry to a flavor component in *Asparagus*. J Heterocycl Chem 20:1753–1754
12. Suzuki T, Sugii M, Kakimoto T (1961) New γ-glutamyl peptides in garlic. Chem Pharm Bull 9:77–78
13. Iida H, Hashimoto S, Miyazawa M, Kameoka H (1983) Volatile flavor components of nira (*Allium tuberosum* Rottl.). J Food Sci 48:660–661
14. Chi MS (1982) Effects of garlic products on lipid metabolism in cholesterol-fed rats. Proc Soc Exp Biol Med 171:174–178
15. Augusti KT (1977) Hypocholesterolemic effect of garlic, *Allium sativum* Linn. Indian J Exp Biol 15:489–490
16. Jain RC (1977) Effect of garlic on serum lipids, coagulability and fibrinolytic activity of blood. Am J Clin Nutr 30:1380–1381
17. Bhushan S, Sharma SP, Singh SP, Agrawal S, Indrayan A, Seth P (1979) Effect of garlic on normal blood cholesterol level. Indian J Physiol Pharmacol 23:211–214

18. Chaudhuri BN, Mukherjee SK, Mongia SS, Chakravaty SK (1984) Hypolipidemic effect of garlic and thyroid function. Biomed Biochim Acta 43:1041–1043
19. Jain RC (1978) Effect of alcoholic extract of garlic in atherosclerosis. Am J Clin Nutr 31:1982–1983
20. Bordia AK, Joshi HK, Sanadhya YK, Bhu N (1977) Effect of essential oil of garlic on serum fibrinolytic activity in patients with coronary artery disease. Atherosclerosis 28:155–159
21. Mohammad SF, Woodward SC (1986) Characterization of a potent inhibitor of platelet aggregation and release reaction isolated from *Allium sativum* (garlic). Thromb Res 44:793–806
22. Boullin DJ (1981) Garlic as a platelet inhibitor. Lancet 1 (8223):776–777
23. Block E, Iyer R, Grisoni S, Saha C, Belman S, Lossing FP (1988) Lipoxygenase inhibitors from the essential oil of garlic. Markovnikov addition of the allyldithio radical to olefins. J Am Chem Soc 110:7813–7827
24. Makheja AN, Vanderhoek JY, Bryant RW, Bailey JM (1980) Altered arachidonic acid metabolism in platelets inhibition by onion or garlic extracts. Adv Prostaglandin Thromboxane Res 6:309–312
25. Shashikanth KN, Basappa SC, Murthy VS (1985) Allicin concentration in the gut of rats and its influence on the microflora. J Food Sci Technol 22:440–442
26. Barone FE, Tansey MR (1977) Isolation, purification, identification, synthesis and kinetics of activity of the anticandidal component of *Allium sativum* and a hypothesis for its mode of action. Mycologia 69:793–825
27. Yoshida S, Kasuga S, Hayashi N, Ushiroguchi T, Matsuura H, Nakagawa S (1987) Antifungal activity of ajoene derived from garlic. Appl Environ Microbiol 53:615–617
28. Tsai Y, Cole LL, Davis LE, Lockwood SJ, Simmons V, Wild GC (1985) Antiviral properties of garlic: in vitro effects on influenza B, herpes simplex and Coxsackie viruses. Planta Med 51:460–461
29. Nakata T (1973) Effect of fresh garlic extract on tumor growth. Nippon Eiseigaku Zasshi 27:538–543 (CA 79:111680x)
30. Cheng HH, Tung TC (1981) Effect of allithiamine on sarcoma-180 tumor growth in mice. Taiwan I Hsueh Hui Isa Chih 80:385–393
31. De Martin R, Knasmueller S, Weniger P (1986) Antimutagen effect of garlic extract in the Ames assay. Oesterreichisches Forschungszentrum, Seibersdorf (Ber) OEFZS-4354 (CA 104:223865c)
32. Nishino H, Iwashima A, Itakura Y, Matsuura H, Fuwa T (1989) Antitumor-promoting activity of garlic extracts. Oncology 46:277–280
33. Augusti KT (1975) Studies on the effect of allicin (diallyl disulfide oxide) on alloxan diabetes. Experientia 31:1263–1265
34. Mathew PT, Augusti KT (1973) Effect of allicin (diallyl disulfide oxide) on alloxan diabetes. I. Hypoglycemic action and enhancement of serum insulin effect and glycogen synthesis. Indian J Biochem Biophys 10:209–212
35. Wang ZH, Hong XK, Qian WJ, Hu CX, Wang YR (1988) Serum concentration, tissue distribution and bioavailability of allicin in rabbits by GC analysis. Chin Trad Herb Drugs 19:540–541

Alpinia spp. 12

12.1 Introduction

Four *Alpinia* species are listed officially in the Chinese Pharmacopoeia:

- Hongdoukou, Fructus Galangae, is the dry ripe fruit of *Alpinia galanga* Willd. (Zingiberaceae) collected in fall when the fruits have turned red. It is used in the treatment of various gastric disorders, emesis, and diarrhea.
- Caodoukou, Semen Alpiniae katsumadai, is the dry, nearly ripe seed of *A. katsumadai* Hayata collected in summer and fall. The Chinese Pharmacopoeia requires a quantitative determination of the content of essential oil in the seed. The content should not be less than 1% (ml/g). It is used as an antiemetic and for treatment of stomach disorders.
- Gaoliangjiang, Rhizoma Alpiniae officinarum, is the dry root stock of *A. officinarum* Hance collected in late summer and early fall. It is used for treatment of dyspepsia, gastralgia, and emesis.
- Yizhi, Fructus Alpiniae oxyphyllae, is the dry ripe fruit of *A. oxyphylla* Miq. collected in summer and fall. It is used for treatment of diarrhea, polyuria, and gastralgia. A quantitative determination of the essential oil content in the seed is required by the Chinese Pharmacopoeia and should not be less than 1% (ml/g).

12.2 Chemical Constituents

12.2.1 Chemical Constituents of *Alpinia galanga*

A number of terpene and nonterpene components were isolated from the essential oil of the seeds of *A. galanga* and identified as 1′-acetoxychavicol acetate (*12-1*), 1′-acetoxyeugenol acetate (*12-2*), caryophyllene oxide, caryophyllenol I (*12-3*) and its 5-epimer caryophyllenol II, pentadecane, heptadec-7-ene, and methyl esters of some fatty acids [1].

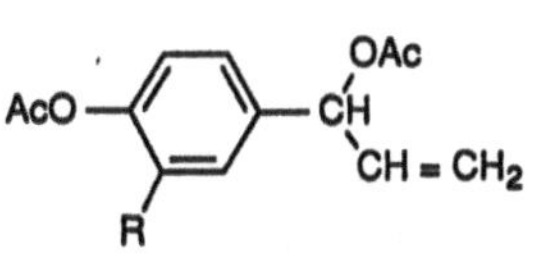

1′-Acetoxychavicol acetate (*12-1*): R=H
1′-Acetoxyeugenol acetate (*12-2*): R=OCH_3

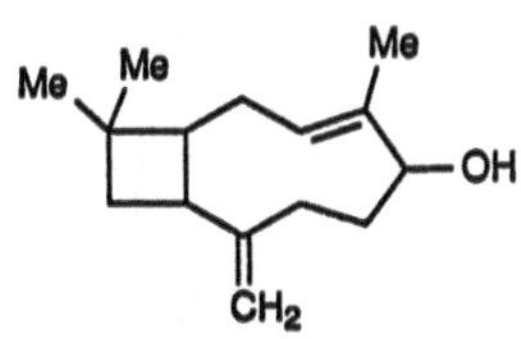

Caryophyllenol (*12-3*)

The components of the essential oil from the rhizome of *A. galanga* were identified as borneol, bornyl acetate, camphene, cineole, *p*-cymene, geranyl acetate, limonene, linalool, α-pinene, β-pinene, sabinene, γ-terpinene, α-terpineol, terpinolene [2, 3], 3-carene, citronellol, α-fenchene, α-fenchol, geraniol, geranial, isoborneol, *p*-menth-2-en-1-ol, myrcene, β-ocimene, α-phellandrene, β-phellandrene, sabinene hydrate, α-terpinene, α-thujene, β-thujone [2], β-bisabolene, carveol, β-caryophyllene, caryophyllene oxide, chavicol, chavicol acetate, citronellyl acetate, α-copaene, *p*-cymenol, β-farnesene, α-humulene, methyleugenol, neryl acetate, and santalene [3]. Depending on the origin, the constituents of *A. galanga* have a different essential oil composition compared to that of the rhizome. Thus, the major component in essential oil from the fresh or dry rhizome of *A. galanga* from Malaysia is β-farnesene [3], whereas those from Europe were α-pinene, β-pinene, limonene, and cineole [2].

Most of the terpenes present in the essential oil are mono- and sesquiterpenes with various carbon skeletons. Among the monoterpenes, myrcene (*12-4*) and β-ocimene (*12-5*) are aliphatic hydrocarbons; citronellol (*12-6*), geraniol (*12-7*), and geranial (*12-8*) are oxygenated aliphatic hydrocarbons.

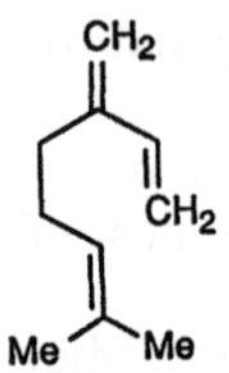

Myrcene
(*12-4*)

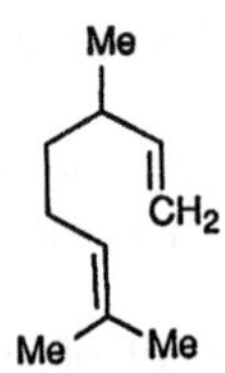

β-Ocimene
(*12-5*)

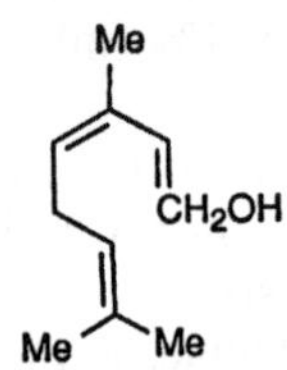

Citronellol
(*12-6*)

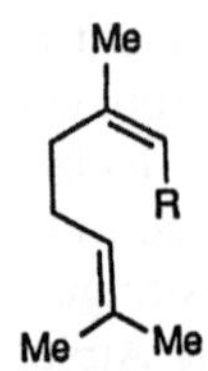

Geraniol (*12-7*): R = CH_2OH
Geranial (*12-8*): R = COH

α-Terpinene (*12-9*), γ-terpinene (*12-10*), terpinolene (*12-11*), limonene (*12-12*), α-phellandrene (*12-13*), and β-phellandrene (*12-14*) are common monocyclic monoterpenes; *p*-menth-2-en-1-ol (*12-15*) is representative for the oxygenated monocyclic monoterpenes.

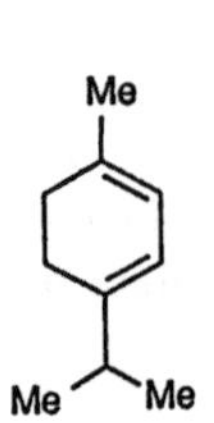

α-Terpinene (*12-9*)

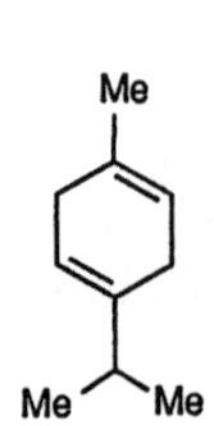

γ-Terpinene (*12-10*)

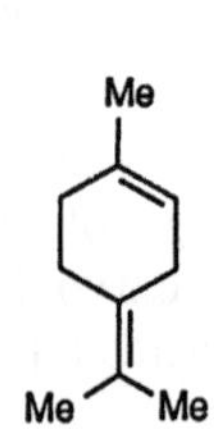

Terpinolene (*12-11*)

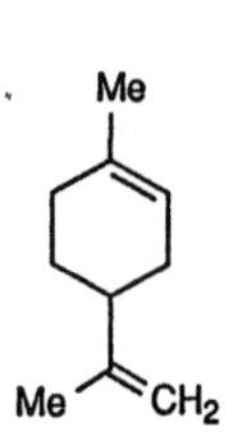

Limonene (*12-12*)

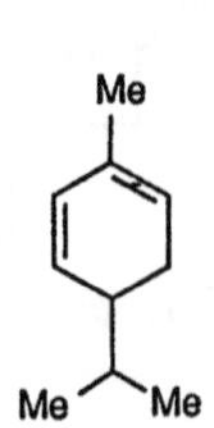

α-Phellandrene (*12-13*)

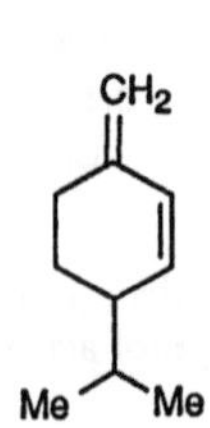

β-Phellandrene (*12-14*)

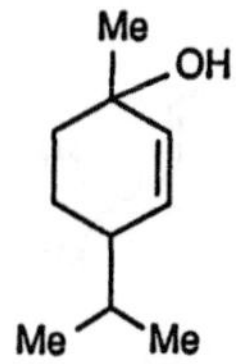

p-Menth-2-en-1-ol (*12-15*)

α-Thujene (*12-16*), sabinene (*12-17*), 3-carene (*12-18*), camphene (*12-19*), α-pinene (*12-20*), β-pinene (*12-21*), and α-fenchene (*12-22*) are bicyclic monoterpenes, whereas β-thujone (*12-23*), borneol (*12-24*), isoborneol (*12-25*), and α-fenchol (*12-26*) are oxygenated bicyclic monoterpenes.

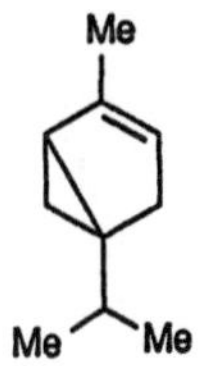

α-Thujene (*12-16*)

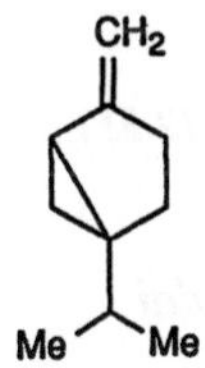

Sabinene (*12-17*)

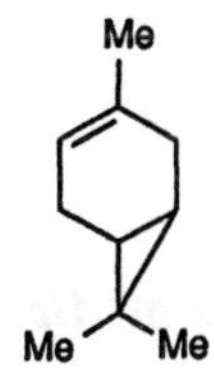

3-Carene (*12-18*)

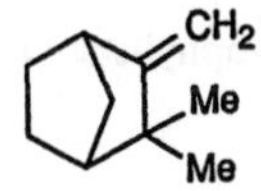

Camphene (*12-19*)

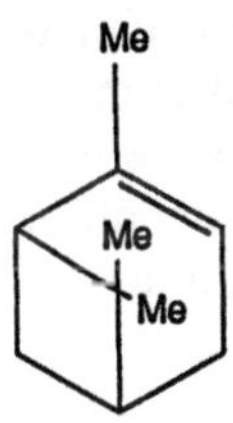

α-Pinene (*12-20*)

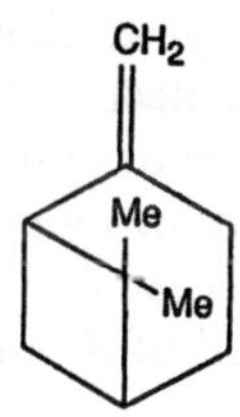

β-Pinene (*12-21*)

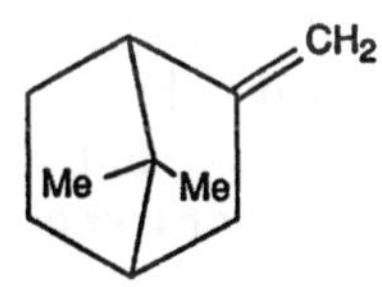

α-Fenchene (*12-22*)

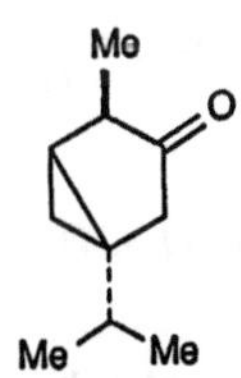

β-Thujone (*12-23*)

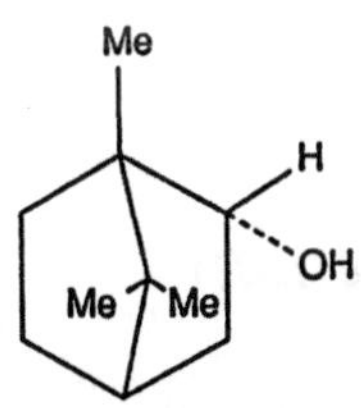

Borneol (*12-24*)

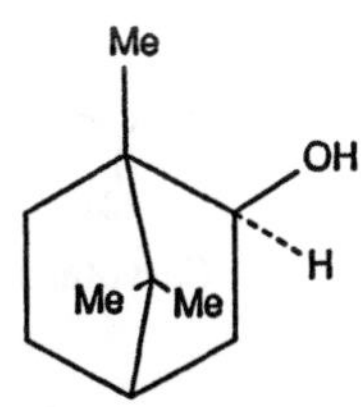

Isoborneol (*12-25*)

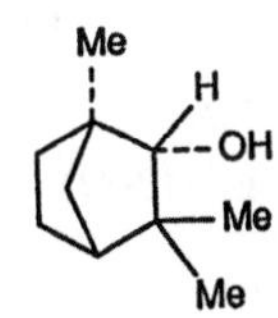

α-Fenchol (*12-26*)

Among the sesquiterpenes, β-farnesene (*12-27*) is alipathic; β-bisabolene (*12-28*) and α-humulene (*12-29*) are monocyclic; and β-caryophyllene (*12-30*) and α-copaene (*12-31*) are bicyclic and tricyclic sesquiterpenes, respectively.

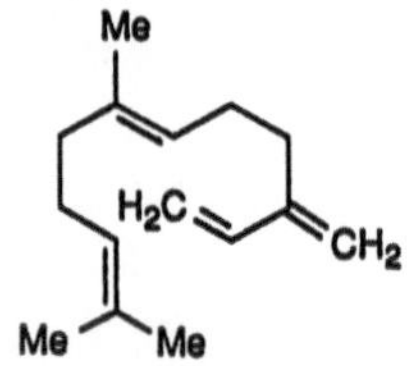

β-Farnesene (*12-27*)

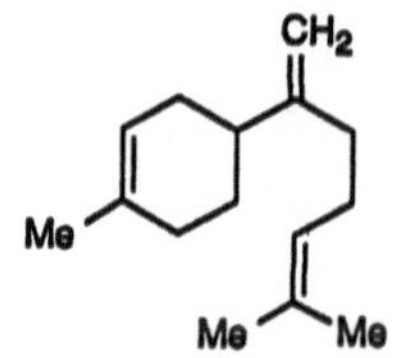

β-Bisabolene (*12-28*)

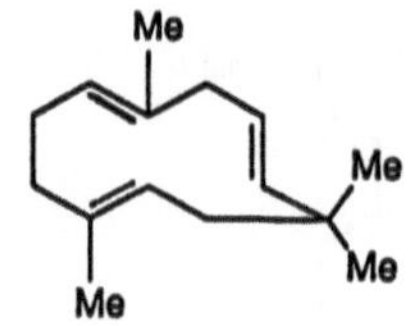

α-Humulene (*12-29*)

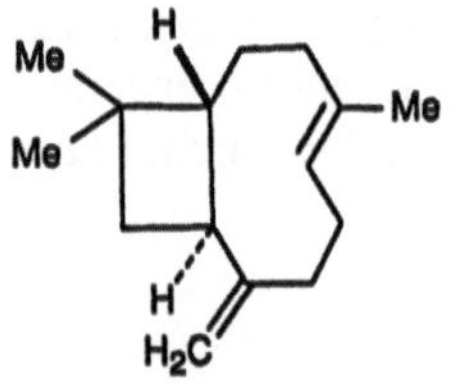

β-Caryophyllene (*12-30*)

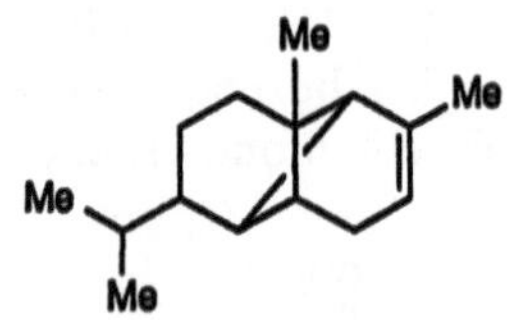

α-Copaene (*12-31*)

12.2.2 Chemical Constituents of *Alpinia katsumadai*

Saiki et al. reported that steam distillation of the seeds of *A. katsumadai* gave a 1.5% yield of an essential oil from which three main components were identified: cineole, α-humulene, and farnesol (*12-32*). Eight other compounds, identified as linalool, camphor, carvotanacetone (*12-33*), bornyl acetate, geranyl acetate, methyl cinnamate, and nerolidol (*12-34*) were also detected. Some other unidentified components, detected using combined gas chromatography-mass spectrometry, were presumed to be either sesquiterpene alcohols or monoterpenes [4]. These results differ from those of Lawrence et al. [5] who reported that camphor and bornyl acetate were the major components of the essential oil of *A. katsumadai*.

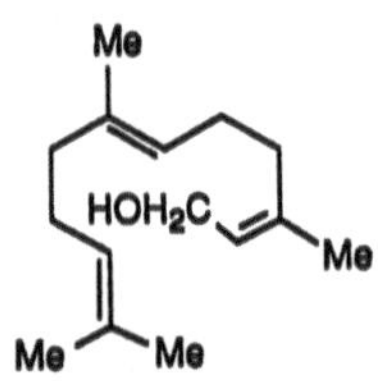

Farnesol (*12-32*)

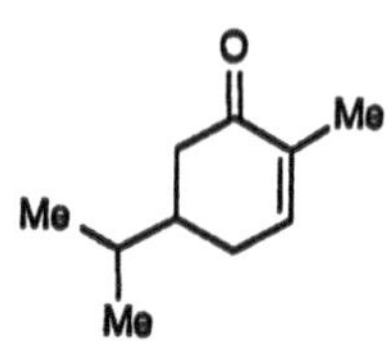

Carvotanacetone (*12-33*)

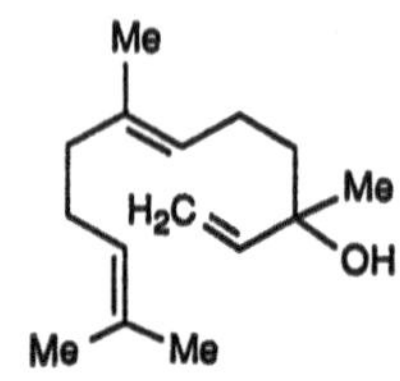

Nerolidol (*12-34*)

In addition, a number of new and known diarylheptanoids: 1,7-diphenyl-5-hydroxy-6-hepten-3-one (*12-35*); 1,7-diphenyl-3,5-dihydroxy-1-heptene; 1,7-diphenyl-5-hydroxy-1-heptene; 1,7-diphenyl-5-hydroxy-4,6-heptadien-3-one (*12-36*); 3,5-dihydroxy-1,7-diphenylheptane; and 1,7-diphenyl-4,6-heptadien-3-one; the flavanones alpinetin (*12-37*) and pinocembrin (*12-38*); and the chalcone cardamonin (*12-39*) were isolated from the seeds of *A. katsumadai*, together with farnesol and cinnamaldehyde [6].

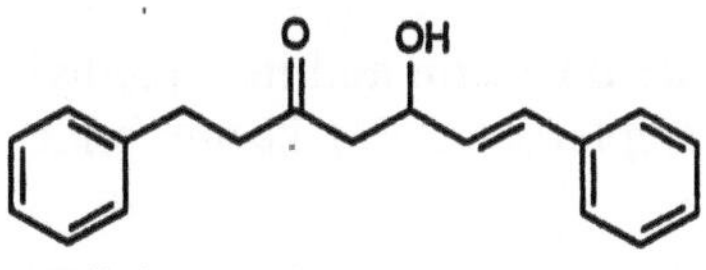

1,7-diphenyl-5-hydroxy-
6-hepten-3-one (*12-35*)

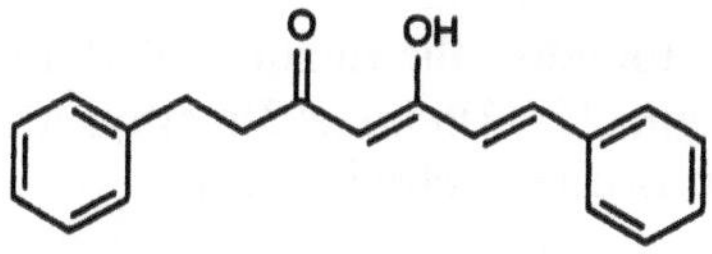

1,7-diphenyl-5-hydroxy-
4,6-heptadien-3-one (*12-36*)

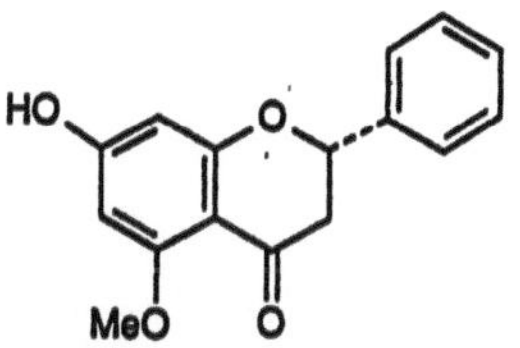

Alpinetin (*12-37*)

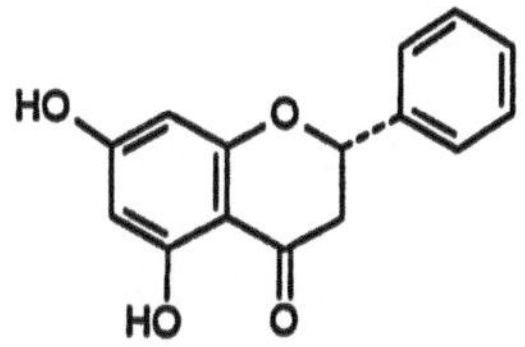

Pinocembrin (*12-38*)

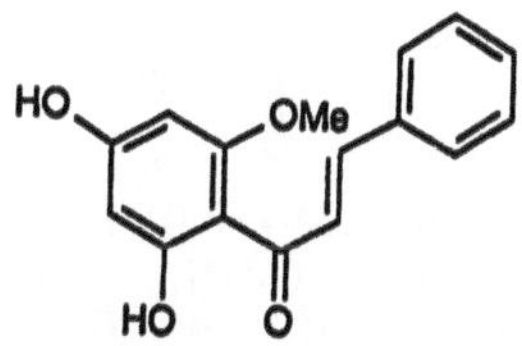

Cardamonin (*12-39*)

12.2.3 Chemical Constituents of *Alpinia officinarum*

A number of newly identified diarylheptanoids such as 5-hydroxy-7-(4-hydroxy-3-methoxyphenyl)-1-phenyl-heptan-3-one (*12-40*) [7]; 1,7-diphenylhept-4-en-3-one; 7-(4-hydroxy-3-methoxyphenyl)-1-phenylhept-4-en-3-one [8]; 7-(4-hydroxy-3-methoxyphenyl)-1-phenyl-heptan-3,5-dione; 7-(4-hydroxyphenyl)-1-phenyl-5-hydroxy-heptan-3-one [9]; 7-(4-hydroxyphenyl)-1-phenyl-heptan-3-one; 5-methoxy-1,7-diphenyl-heptan-3-one; and 5-methoxy-7-(4-hydroxyphenyl)-1-phenyl-heptan-3-one [10] were isolated from the rhizome of *A. officinarum*. Known diarylheptanoids were also found, e.g., 1,7-*bis*(4-hydroxy-3-methoxyphenyl)-5-hydroxy-heptan-3-one (*12-41*) [10].

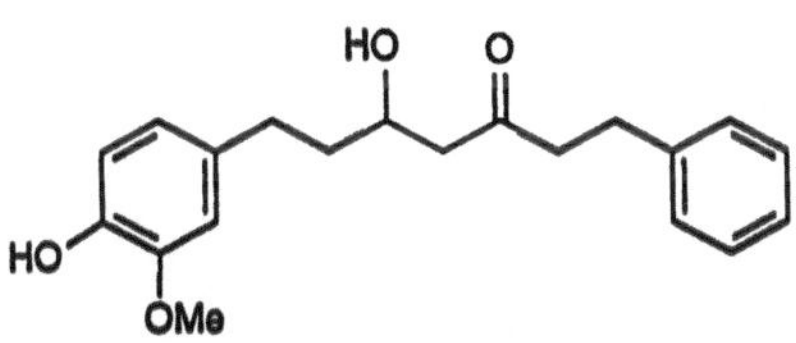

5-Hydroxy-7-(4-hydroxy-3-methoxyphenyl)-
1-phenyl-heptan-3-one (*12-40*)

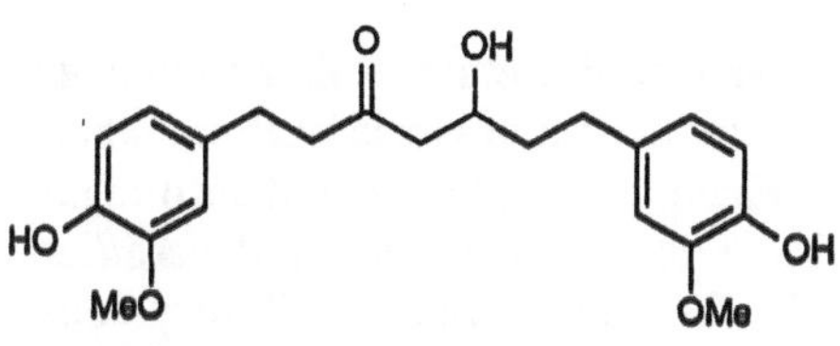

1,7-*bis*-(4-Hydroxy-3-methoxy-
phenyl)-5-hydroxy-heptan-3-one
(*12-41*)

Flavones found in the rhizome of *A. officinarum* are quercetin and its 3-methyl ether; galangin (*12-42*) and its 3-methyl ether; and kaempferol (*12-43*), kaempferide (*12-44*), and isorhamnetin [11, 12].

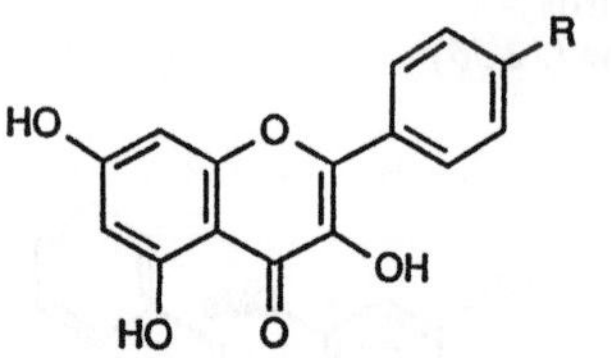

Galangin (*12-42*): R=H
Kaempferol (*12-43*): R=OH
Kaempferide (*12-44*): R=OCH_3

The rhizome of *A. officinarum* also contains a small amount of essential oil, the major component of which is cineole [13].

12.2.4 Chemical Constituents of *Alpinia oxyphylla*

Two diarylheptanoids, yakuchinone A (*12-45*) [14] and yakuchinone B (*12-46*) [15], and a sesquiterpene, nootkatol (*12-47*) [16], were isolated from the fruits of *A. oxyphylla.*

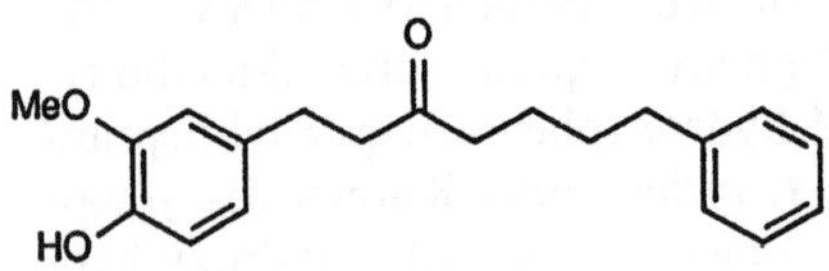

Yakuchinone A (*12-45*)

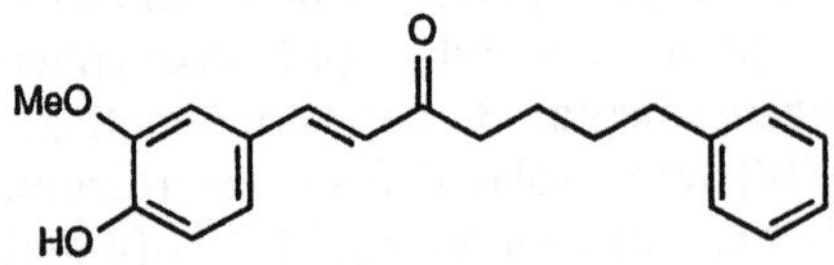

Yakuchinone B (*12-46*)

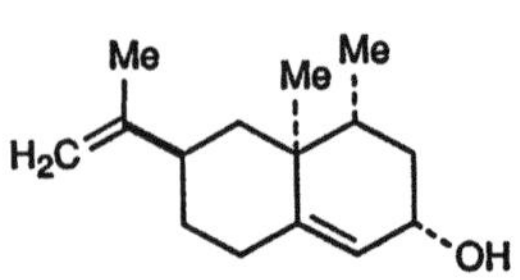

Nootkatol (*12-47*)

12.3 Pharmacology

In rats, 1′-acetoxychavicol acetate and 1-acetoxyeugenol acetate, isolated from *A. galanga*, had significantly inhibited gastric ulcers induced by pyloric ligation, when given i.p. at a dose of 1 – 10 mg/kg [1]. 1′-acetoxychavicol acetate also had antifungal activity against *Trichophyton mentagrophytes, T. concentricum, T. rubrum, Aspergillus niger, Penicillium expansum,* and *Rhizopus stolonifer*. The minimal inhibitory concentration of 1′-acetoxychavicol acetate for these dermatophytes ranged from 50 to 250 μg/ml.

The essential oil obtained by steam distillation of the rhizome of *A. galanga* showed bacteriostatic activity against *Mycobacterium tuberculosis* at a concentration of 25 μg/ml. The LD_{50} of this essential oil in guinea pigs by i.p. administration was 0.68 ml/kg [17].

The new diarylheptanoids isolated from *A. officinarum* were found to exhibit an inhibitory activity on prostaglandin biosynthesis [9], whereas yakuchinone A from *A. oxyphylla* showed a cardiotonic activity [18]. Yakuchinone A, at concentrations of 1–100 μg/ml, exhibited a dose dependent inotropic effect on guinea pig left atria [18]. The sesquiterpene nootkatol, from *A. oxyphylla*, acted as a calcium antagonist [16].

References

1. Mitsui S, Kobayashi S, Nagahori H, Ogiso A (1976) Constituents from seeds of *Alpinia galanga* Willd. and their antiulcer activities. Chem Pharm Bull 24:2377–2382
2. Scheffer JJC, Gani A, Baerheim SA (1981) Analysis of essential oils by combined liquid-solid and gas-liquid chromatography. V. Monoterpene in essential oil of *Alpinia galanga* (L.) Willd. Sci Pharm 49:337–346
3. De Pooter HL, Omar MN, Coolsaet BA, Shamp NM (1985) The essential oil of greater galanga (*Alpinia galanga*) from Malaysia. Phytochemistry 24:93–96
4. Saiki Y, Ishikawa Y, Uchida M, Fukushima S (1978) Essential oil from Chinese drug "caodoukou", the seeds of *Alpinia katsumadai*. Phytochemistry 17:808–809
5. Lawrence BM, Hogg JW, Terhune SJ, Pichitakul N (1972) Terpenoids of two *Amomum* species from Thailand. Phytochemistry 11:1534–1535
6. Kuroyanagi M, Noro T, Fukishima S, Aiyama R, Ikuta A, Itokawa H, Morita M (1983) Studies on the constituents of the seeds of *Alpinia katsumadai* Hayata. Chem Pharm Bull 31:1544–1550
7. Inoue T, Shinbori T, Fujioka M, Hashimoto K, Masada Y (1978) Studies on the pungent principle of *Alpinia officinarum* Hance. Yakugaku Zasshi 98:1255–1257
8. Itokawa H, Morita M, Mihashi S (1981) Two new diarylheptanoids from *Alpinia officinarum* Hance. Chem Pharm Bull 29:2383–2385
9. Kiuchi F, Shibuya M, Sankawa U (1982) Inhibitors of prostaglandin biosynthesis from *Alpinia officinarum*. Chem Pharm Bull 30:2279–2282
10. Itokawa H, Morita H, Midorikawa I, Aiyama R, Morita M (1985) Diarylheptanoids from the rhizome of *Alpinia officinarum* Hance. Chem Pharm Bull 33:4889–4893
11. Bleier W, Chirikdjian JJ (1972) Flavonoids of rhizoma galangae. Planta Med 145–151
12. Tunmann P, Tkotz H (1972) Flavonols and sterol glucosides in the root of *Alpinia officinarum*. Z Naturforsch [B] 27:323–324
13. Lin QS (1977) Chemistry of the Constituents in Chinese Medicinal Herbs. Scientific Press, Beijing, p 575
14. Itokawa H, Aiyama R, Ikuta A (1981) A pungent diarylheptanoid from *Alpinia oxyphylla*. Phytochemistry 20:769–771
15. Itokawa H, Aiyama R, Ikuta A (1982) A pungent principle from *Alpinia oxyphylla*. Phytochemistry 21:241–243
16. Shoji N, Umeyama A, Asakawa Y, Takemoto T, Nomoto K, Ohizumi Y (1984) Structural determination of nootkatol, a new sesquiterpene isolated from *Alpinia oxyphylla* Miquel possessing calcium-anatagonistic activity. J Pharm Sci 73:843–844
17. Chopra IC, Khajuria BN, Chopra CL (1957) Antibacterial properties of volatile principles from *Alpinia galanga* and *Acorus calamus*. Antibiot Chemother 7:378–383
18. Shoji N, Umeyama A, Takemoto T, Ohizumi Y (1984) Isolation of a cardiotonic principle from *Alpinia oxyphylla*. Planta Med 50:186–187

Amomum spp. 13

13.1 Introduction

Three items with *Amomum* species are officially listed in the Chinese Pharmacopoeia.

- Doukou, Fructus Amomi rotundus, is the dry ripe fruit of *Amomum kravanh* Pirre ex Gagnep. or *A. compactum* Soland ex Maton (Zingiberaceae). It is mainly used for treatment of stomach disorders.
- Caoguo, Fructus Tsaoko, is the dry ripe fruit of *A. tsao-ko* Crevost et Lemaire collected in fall when the fruits have ripened. It is used for stomach disorders, as a mucolytic agent, and as an antimalarial drug.
- Sharen, Fructus Amomi, is the dry ripe fruit of *A. villosum* Lour. or *A. longiligulare* T.L. Wu collected in summer or fall. It is used for gastrointestinal disorders.

13.2 Chemical Constituents

From the essential oil of the fruits and leaves of *A. kravanh* and *A. compactum*, cineole (*13-1*), α-pinene, β-pinene, comphene, limonene, *p*-cymene, terpinene, α-terpineol, and α-humulene were isolated and identified. Cineole is the major component and accounts for 60%–80% [1].

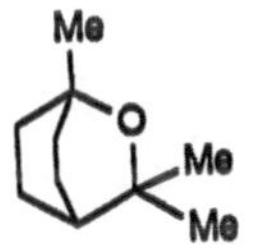

Cineole (*13-1*)

The essential oil of *A. villosum* and *A. longiligulare* contains bornyl acetate (*13-2*) and camphor (*13-3*) as major components. In addition, nerolidol and linalool were also found [2, 3].

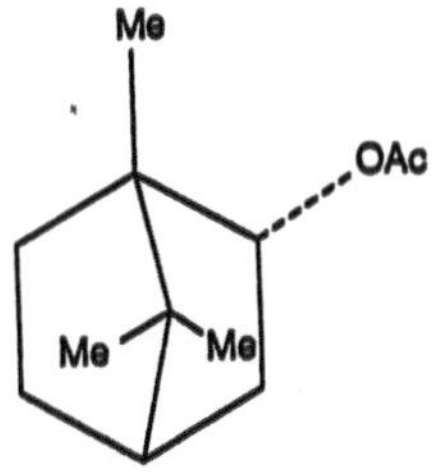

Bornyl acetate (*13-2*)

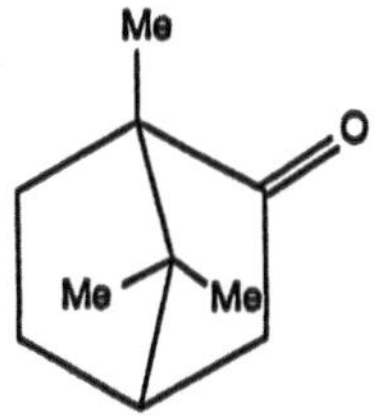

Camphor (*13-3*)

Recently, Lu et al. have performed morphological and chemical studies on 14 *Amomum* species found in China. It was shown that the plants could be divided into three groups, according to the major components in their seed essential oils and the morphological characteristics of their fruits. *A. chinense, A. longiligulare, A. thyrsoideum, A. villosum,* and *A. villosum* var. *xanthioides* have bornyl acetate and camphor as the major components in their seeds; *A. austrosinense, A. compactum, A. kravanh, A. subulatum,* and *A. tsao-ko* have cineole as the major component. Nerolidol or farnesol is the major component of the seed of *A. auranticum, A. dealbatum, A. maximum,* and *A. sericeum* [4].

13.3 Pharmacology

In a clinical study, patients with peptic ulcers were treated with the seeds of *A. villosum.* A marked curative effect was reported [5].

In rats given cineole at a dose of 800 mg/kg intragastrically, 1,8-dihydroxy-10-carboxy-*p*-menthane, 2-hydroxycineole, and 3-hydroxy-cineole were found to be the main metabolites. Cineole given as an aerosol to rats induced the cytochrome P450 system of the liver, but not of the lung [6].

References

1. Yu JG, Fang HJ, Li JT (1982) Essential oil of fruits and leaves of *Amomum kravanh* and *A. compactum.* Chin Trad Herb Drugs 13:4–7
2. Yang ZQ, Zhang J, Zhang BL, Qin XL (1985) Investigations on the quality of Sharen (*Amomum villosum* Lour.). Chin J Pharm Anal 5:351–358
3. Gu MX, Yang Y (1985) Gas chromatographic determination of bornyl acetate and camphor in essential oil of *Amomum* species. Chin J Pharm Anal 5:359–360
4. Lu BY, Zhu LF, Wu TL (1986) Correlation between the major components in the seed essential oil of Chinese cardamons and the morphology of their fruits. Guihaia 6:131–139
5. Huang ZY, Lin BH, Sun HX, Wu ZP, Wu QD, Huang YC, Gu CF, Rao DY, Huang ZL, Lin YQ, Chen ZM (1984) Therapeutic effect of *Alpinia japonica* on peptic ulcers. Fujian Med J 6:24–25
6. Madyastha KM, Chadha A (1986) Metabolism of 1,8-cineole in rat: its effects on liver and lung microsomal cytochrome P-450 systems. Bull Environ Contam Toxicol 37:759–766

Andrographis paniculata (Burm. f.) Nees

14

14.1 Introduction

Chuanxinlian, Herba Andrographitis, is the dry aerial part of *Andrographis paniculata* (Burm. f.) Nees (Acanthaceae) collected in early fall when the stems and leaves are growing exuberantly. This official herbal medicine is used as an antiinflammatory and antipyretic drug for treatment of cold, fever, laryngitis, and diarrhea.

Chuanxinlian Pian, Tabellae Andrographitis, is the tablet produced from the extract of *A. paniculata* and used for the same indications as the whole plant.

14.2 Chemical Constituents

The main constituent of *A. paniculata* is the diterpene lactone andrographolide. The bitter principle, a colorless, neutral crystalline substance, was first isolated by Boorsma from different parts of *A. paniculata* and named andrographide. Gorter proved that it is structurally a lactone and changed its name to andrographolide (*14-1*) [1]. Since then, this bitter principle has been the subject of a number of chemical investigations. More than 50 years later the exact structure of andrographolide was confirmed by spectral and chemical studies [2, 3]. It is a diterpene containing a γ-lactone ring connected to a decalin ring system via an unsaturated C_2 moiety. The crystal structure of andrographolide was determined by Smith et al. [4] and Fujita et al. [5].

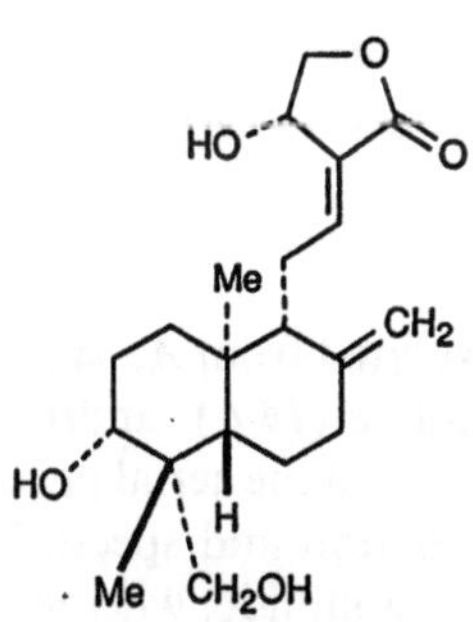

Andrographolide (*14-1*)

Andrographolide can be isolated from the aerial part of the plant by extraction with alcohol or with alkaline solutions. Hydrolysis of andrographolide under cleavage of the lactone ring yields salts of andrographolic acid (*14-2*) which can be reconverted into andrographolide by acidification.

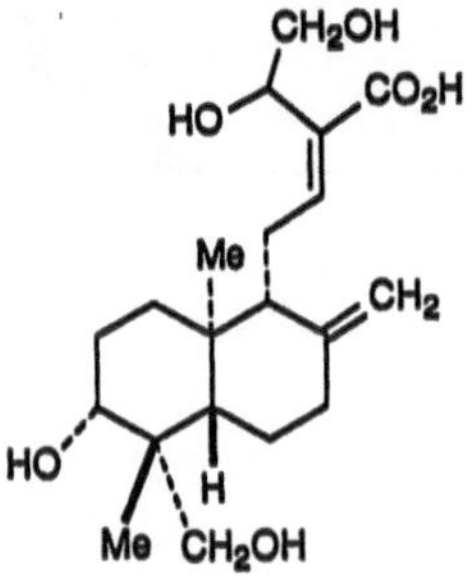

Andrographolic acid (*14-2*)

The second diterpene isolated from *A. paniculata* was the minor nonbitter constituent neoandrographolide (*14-3*), which was first described by Kleipool [6]. The structure of neoandrographolide was described by Chan et al. [7] as a diterpene glucoside. Balmain and Connolly reported the isolation and identification of three minor diterpene constituents from *A. paniculata*. They are structurally closely related to andrographolide and have been referred to as deoxy-didehydroandrographolide (*14-4*), deoxy-oxoandrographolide (*14-5*), and deoxyandrographolide [8].

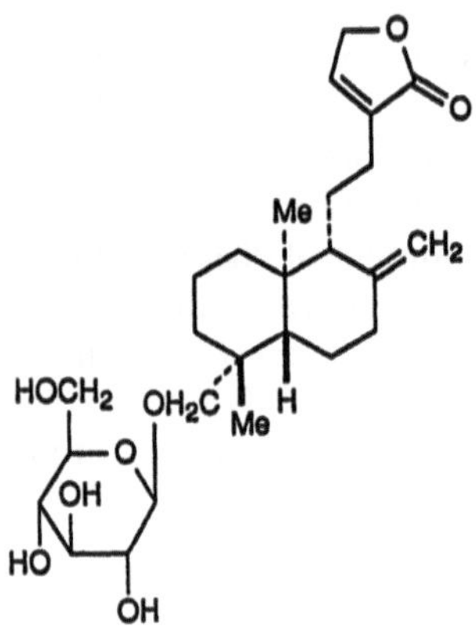

Neoandrographolide (*14-3*)

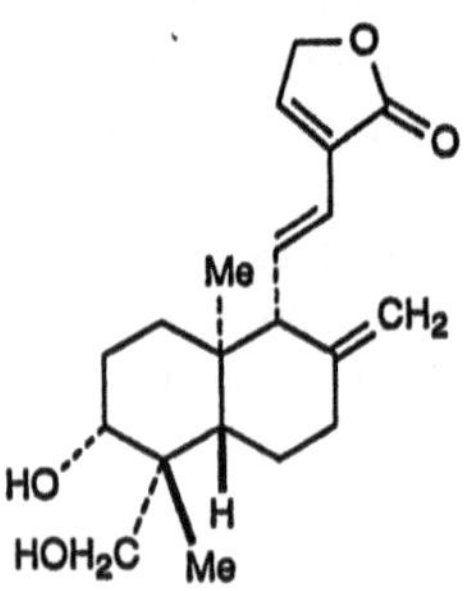

Deoxy-didehydroandrographolide (*14-4*)

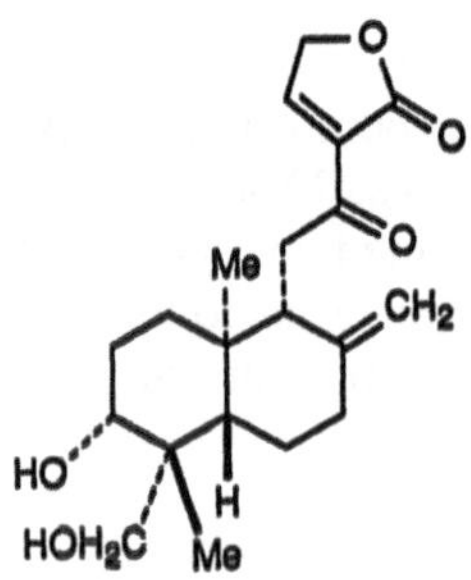

Deoxy-oxo-andrographolide (*14-5*)

Recently, another series of diterpene constituents has been isolated from *A. paniculata* and structurally elucidated. Thus, dideoxy-andrographolide (*14-6*), andrographiside (*14-7*), and its 14-deoxy analogue (*14-8*) were isolated from the aerial part of *A. paniculata* and structurally determined by chemical correlation and spectral analysis. Dideoxyandrographolide is the aglucone of neoandrographolide, whereas andrographiside is the glucoside of andrographolide [9–11]. Together with another new diterpene lactone, deoxy-methoxyandrographolide (*14-9*), dideoxyandrographolide and deoxyandrographiside were also reported by Fujita et al. and named andrograpanin and andropanoside, respectively [5].

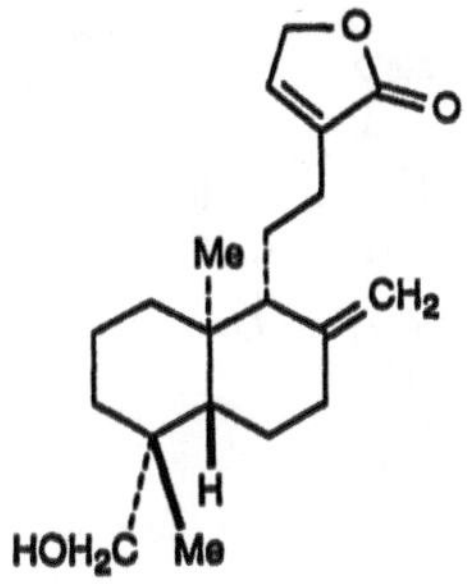

Dideoxy-andrographolide (*14-6*)
(Andrograpanin)

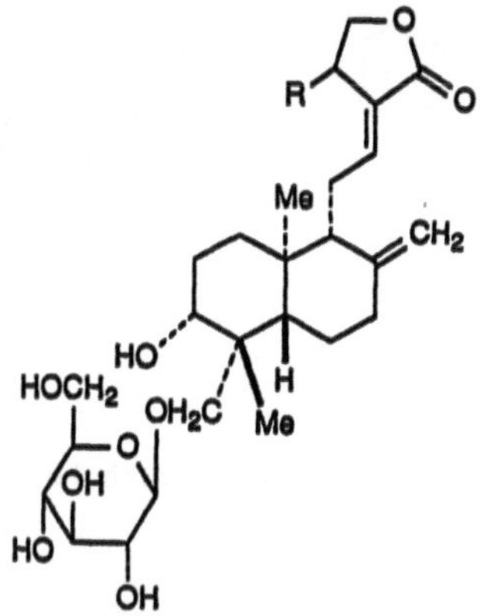

Andrographiside (*14-7*): R=OH
Deoxyandrographoside (*14-8*): R=H
(Ninandrographolide [12], Andropanoside)

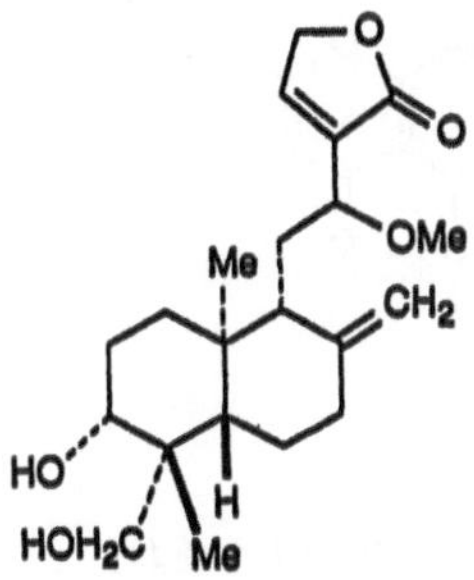

Deoxy-methoxy-andrographolide (*14-9*)

The amount of andrographolide in leaves and stems of *A. paniculata* were quantitatively determined by UV spectrophotometry after separation on silica gel plates. The stem contained 0.1%–0.4% and the leaf 2.6% andrographolide as peak values when collected in October. Regional variation in the andrographolide content was also observed [13]. More important for the andrographolide content is the harvest season. The leaves contain more than 2% andrographolide before the plant blooms; afterwards the content decreases to less than 0.5%.

As a water soluble andrographolide derivative, the sodium bisulfite adduct was synthesized for medical use as an antipyretic agent. The structure was described on the basis of spectral data [14, 15].

Besides diterpene lactones, flavone derivatives such as oroxylin and wogonin were isolated from the leaves of *A. paniculata* [16]. The isolation of a new flavanone glycoside, andrographidine A (*14-10*), and flavone glycosides, andrographidine B, C, D, E, and F (*14-11*–*14-15*), having uncommon *O*-substitution pattern from the root of *A. paniculata* has also been reported [17].

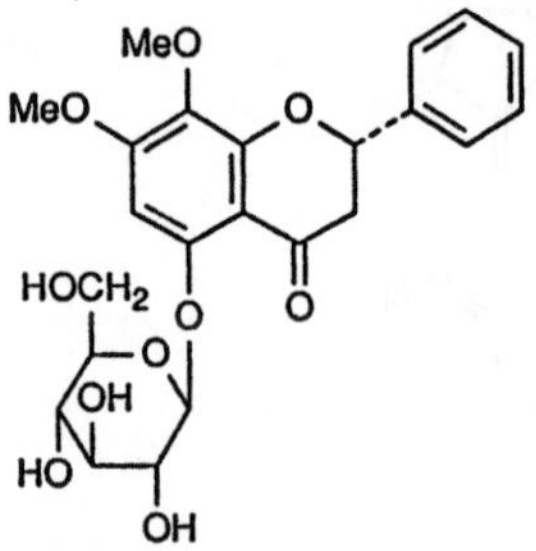

Andrographidine A (*14-10*)

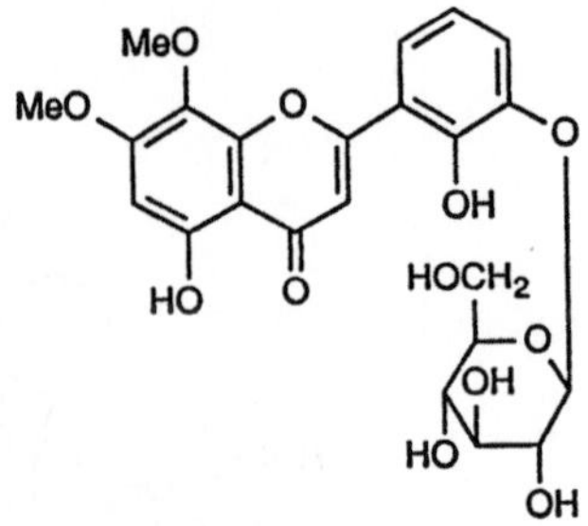

Andrographidine B (*14-11*)

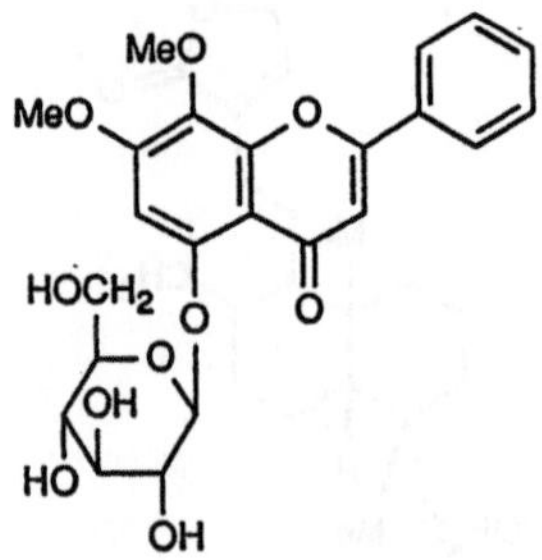

Andrographidine C (*14-12*)

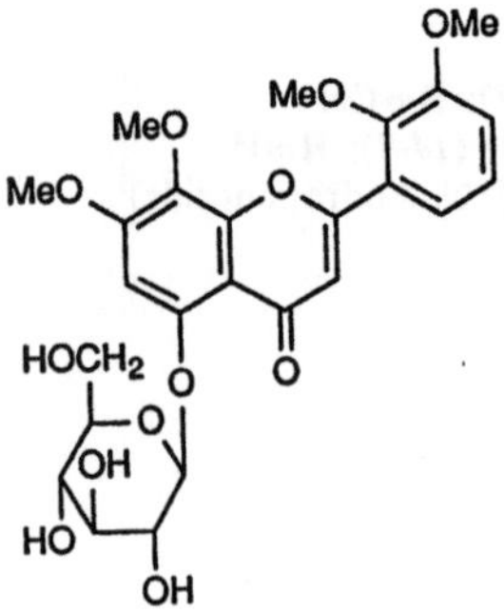

Andrographidine D (*14-13*)

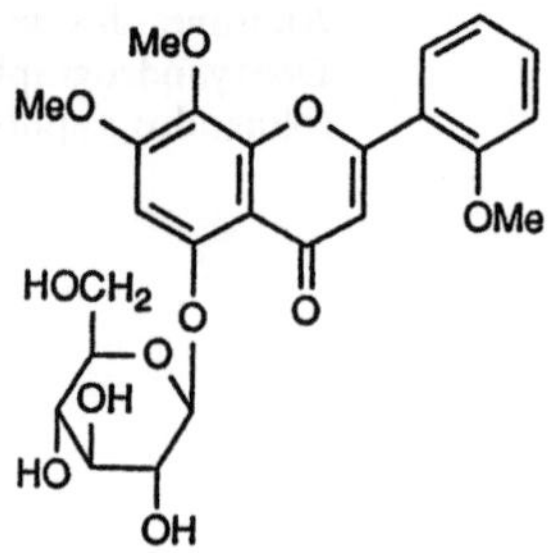

Andrographidine E (*14-14*)

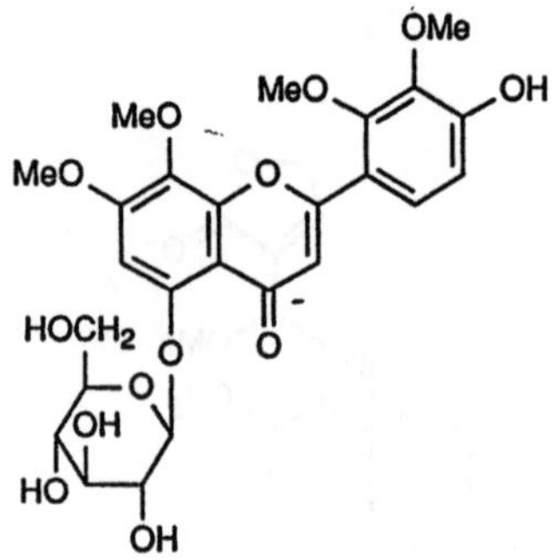

Andrographidine F (*14-15*)

Tissue cultures derived from hypocotyl and stem of *A. paniculata* have been shown to produce sesquiterpene lactones and apparently lack the ability of the whole plant to synthesize diterpenes. Three sesquiterpene lactones based on bisabolene were isolated and structurally elucidated. They are paniculide A (*14-16*), B (*14-17*), and C (*14-18*). The contents of paniculide A, B, and C in culture were 3–5, 20–25, and 8–12 mg/l, respectively. Andrographolide or related substances could not be detected in tissue culture [18]. The absolute configuration of paniculide B has been determined by X-ray crystallography of its *bis-p*-bromobenzoate [19, 20].

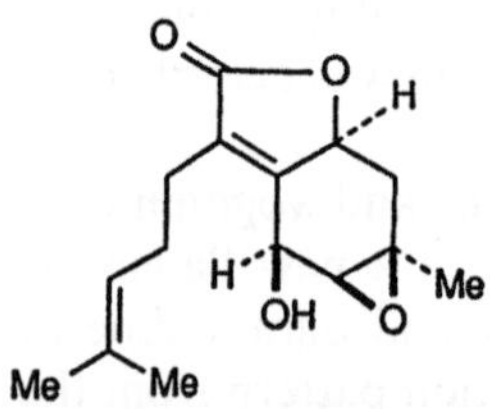

Paniculide A (*14-16*)

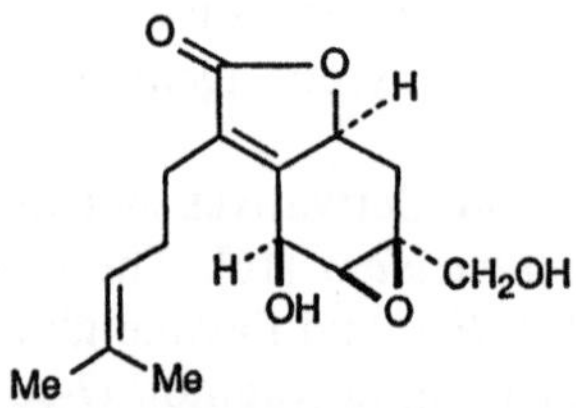

Paniculide B (*14-17*)

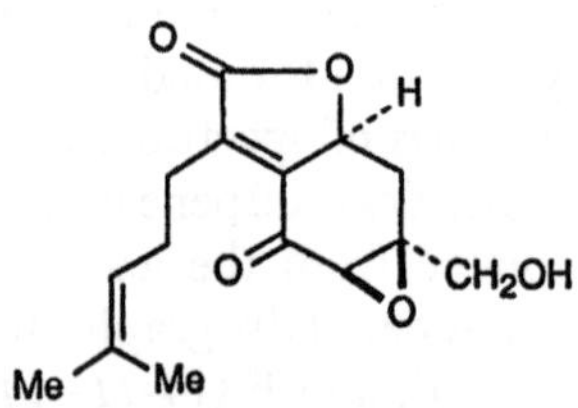

Paniculide C (*14-18*)

In addition, two acidic polysaccharides, PA and PB, were recently isolated from the pectin of *A. paniculata*. The acidic polysaccharide PA contains galactose, arabinose, and rhamnose in a ratio of 5:2.8:1 and 25.6% galacturonic acid. Partial acid hydrolysis and Smith degradation of PA yields β-(1→3)-D-galactan. The acidic polysaccharide, PB, contains galactose, arabinose, and rhamnose in a ratio of 3.4:1.7:1. Partial acid hydrolysis of PB yields α-(1→4)-D-galacturonan [21].

14.3 Pharmacology

Andrographolide and the related compounds, deoxyandrographolide, deoxydidehydroandrographolide, and neoandrographolide, were investigated for their pharmacological properties in experimental animals including mice, rats, and rabbits. They all showed varying degrees of antipyretic and antiinflammatory activity in animals with fever induced by 2,4-dinitrophenol or endotoxin, with edema caused by egg white, or with inflammation caused by croton oil. The antipyretic and antiinflammatory effects of andrographolide and related compounds, however, were lower than those of corticosteroids and of conventionally used nonsteroidal drugs. The pharmacological effect was highest with deoxydidehydroandrographolide, followed by deoxyandrographolide, neoandrographolide, and andrographolide. The minimal lethal dose of these compounds by oral administration was greater than 20 g/kg. The antiinflammatory effect of all four compounds disappeared in adrenalectomized animals, indicating a possible involvement of the pituitary and adrenal systems. Administration of the four compounds did not significantly affect inflammatory hyperplasia and migration of leukocytes into the inflammatory focus. The antiinflammatory mechanism was speculated to be different from that of conventional drugs [22].

Neither the extracts of leaves and stems of *A. paniculata* by oral administration nor andrographolide by s.c. or oral administration had an effect on blood sugar levels of normal or diabetic rats [23].

Pretreatment of dogs with leaves of *A. paniculata* at a dose of 500 mg/kg or with andrographolide at a dose of 5 mg/kg prevented the increase in serum levels of glutamic-oxalacetic transaminase (GOT) and GPT and the decrease in the liver levels of these enzymes when induced by oral administration of CCl_4. Simultaneous treatment did not show any effect. An inhibitory effect on formation of trichloromethyl free radical from CCl_4 is a likely explanation [24]. Pretreatment also caused a decrease of CCl_4-induced hepatic microsomal lipid peroxidation but only with a single dose and not with long-term administration of leaf extract or andrographolide. The protective action on CCl_4-induced hepatotoxicity of leaf extract was more significant than that of andrographolide [25]. NADPH-induced hepatic microsomal lipid peroxidation could not be significantly changed by oral administration of leaf extract or andrographolide [25].

After i.v. or intragastric administration to mice [^{3}H]andrographolide is rapidly absorbed and eliminated consistent with an open two compartment kinetic model. The $t_{\frac{1}{2}\alpha}$ and $t_{\frac{1}{2}\beta}$ values of i.v. administered [^{3}H]andrographolide were 0.06 and 4.7 h, respectively, whereas the corresponding values following intragastric application were 0.15 and 1.07 h, respectively. Intragastric [^{3}H]andrographolide was rapidly absorbed. It rapidly distributed to other organs, especially gallbladder, kidney, ovary, and lung. Radioactivity was low in spleen, heart, and brain. Approximately 90% of the intragastrically administered [^{3}H]andrographolide was excreted in urine and feces at 24 h and 94% after 48 h. The calculated biological availability of intragastric [^{3}H]andrographolide was only 44%. In the 48 h urine and liver extracts, only 10.8% and 10.7%, respectively, of the ^{3}H activity was accounted for by [^{3}H]andrographolide; the remaining ^{3}H activity was present as metabolites, indicating that [^{3}H]andrographolide was rapidly metabolized and excreted mainly as metabolites [26].

After i.v. injection of andrographolide sodium bisulfite at a dose of 100 mg/kg, no drug was detectable in blood 15, 30, or 60 min after application; however, at a dose of 250 mg/kg it was detectable in blood and urine. The concentration in the blood was 123–132 μg/ml at 30 min and 47–56 μg/ml at 60 min after injection. Concentrations of andrographolide sodium bisulfide in the urine 30 min, 60 min, and 6 h after injection were 17–37, 17–24, and 1–3 mg/ml, respectively. Two fractions were observed on chromatograms of collected urine samples [27].

References

1. Gorter K (1911) The bitter constituent of *Andrographis paniculata* Nees. Rec Trav Chim 30:151–160
2. Chan WR, Willis C, Cava MP, Stein RP (1963) Stereochemistry of andrographolide. Chem Ind 495
3. Cava MP, Chan WR, Stein RP, Willis CR (1965) Andrographolide, further transformations and stereochemical evidence; the structure of isoandrographolide. Tetrahedron 21:2617–2632
4. Smith AB, Toder BH, Carroll PJ, Donohue J (1982) Andrographolide: an X-ray crystallographic analysis. J Crystallogr Spectrosc Res 12:309–319
5. Fujita T, Fujitani R, Takeda Y, Takaishi Y, Yamada T, Kido M, Miura I (1984) On the diterpenoids of *Andrographis paniculata* X-ray crystallographic analysis of andrographolide and structure determination of new minor diterpenoids. Chem Pharm Bull 32:2117–2125
6. Kleipool RJC (1952) Constituents of *Andrographis paniculata*. Nature 169:33–34
7. Chan WR, Taylor DR, Willis CR, Bodden RL (1971) The structure and stereochemistry of neoandrographolide, a diterpene glucoside from *Andrographis paniculata*. Tetrahedron 27:5081–5091
8. Balmain A, Connolly JD (1973) Minor diterpenoid constituents of *Andrographis paniculata* Nees. J Chem Soc Perkin Trans I, 1247–1251
9. Hu CQ, Zhao BN (1981) Studies on the new diterpenoides of Chuan Xin Lian (*Andrographis paniculata*). Chin Trad Herb Drugs 12:531
10. Hu CQ, Zhao BN, Chou PN (1982) Isolation and structure of two new diterpenoid glucosides from *Andrographis paniculata* Nees. Acta Pharm Sin 17:435–440
11. Chen W, Liang XT (1982) Deoxyandrographolide-19-D-glucoside from the leaves of *Andrographis paniculata*. Planta Med 45:245–246
12. Meng ZM (1982) Isolation and identification of a new glucoside bicyclic diterpenoic lactone from *Andrographis paniculata* Nees. J Nanjing Coll Pharm 25–30
13. Zhu PY, Peng NB, Jiang WJ (1984) TLC-UV spectrophotometric determination of andrographolide in the leaves and stems of *Andrographis paniculata*. Chin J Pharm Anal 4:34–36
14. Qin GW, Wang PT, Chen CH, Lu HL (1981) Studies on structure of andrographolide sodium bisulfite. Chin Trad Herb Drugs 12:1–5
15. Meng ZM (1981) Studies on the structure of the adduct of andrographolide with sodium sulfite. Acta Pharm Sin 16:571–575
16. Zhu PY, Liu GQ (1984) Separation and identification of flavonoids in *Andrographis paniculata* Nees leaves. Chin Trad Herb Drugs 15:375–376
17. Kuroyanagi M, Sato M, Ueno A, Nishi K (1987) Flavonoids from *Andrographis paniculata*. Chem Pharm Bull 35:4429–4435
18. Allison AJ, Butcher DN, Connolly JD, Overton KH (1968) Paniculides A, B, and C, bisabolenoid lactones from tissue cultures of *Andrographis paniculata*. Chem Commun 1493

19. Anastasis P, Freer I, Gilmore C, Mackie H, Overton K, Swanson S (1982) Cyclisation of farnesyl pyrophosphate to γ-bisabolene in tissue cultures of *Andrographis paniculata*. J Chem Soc Chem Commun 268–270
20. Anastasis P, Freer I, Gilmore C, Mackie H, Overton K, Picken D, Swanson S (1984) Biosynthesis of γ-bisabolene in tissue culture of *Andrographis paniculata*. Can J Chem 62:2079–2088
21. Li ZP (1987) Studies on the poylsaccharides of *Andrographis paniculata* (II). Purification and characterization of pectins. J Northeast Norm Univ [Nat Sci] 35–39
22. Deng WL, Nie RJ, Liu JY (1982) Comparison of pharmacological effect of four andrographolides. Chin Pharm Bull 17:195–198
23. Ahmed M, Talukder SA (1977) Studies on the hypoglycemic activity of kalmegh (*Andrographis paniculata* Nees.) on the blood sugar level of rats. Bangladesh Pharm J 6:21–24 (CA:88, 315y)
24. Choudhury BR, Poddar MK (1984) Andrographolide and kalmegh (*Andrographis paniculata*) extract: effect on rat liver and serum transaminases. IRCS Med Sci 12:466–467
25. Choudhury BR, Poddar MK (1984) Andrographolide and kalmegh (*Andrographis paniculata*) extract: in vivo and in vitro effect in hepatic lipid peroxidation. Methods Find Exp Clin Pharmacol 6:481–485
26. Zheng ZY, Wan YD, He GX, Yi MG (1982) Pharmacokinetic study of ^{3}H-andrographolide. Chin Trad Herb Drugs 13:417–420
27. Lu HL, Chao HL, Chang SL, Chang CL, Ku YM, Shih YK (1981) Study on the metabolism of andrographolide derivative. Chin Trad Herb Drugs 12:6

Anemarrhena asphodeloides Bge. 15

15.1 Introduction

Zhimu, Rhizoma Anemarrhenae, is the dry rootstock of *Anemarrhena asphodeloides* Bge. (Liliaceae) dug in spring and fall. It has been used for a long time in traditional Chinese medicine as an antipyretic, sedative, and diuretic agent.

15.2 Chemical Constituents

The main constituents of the rhizome of *A. asphodeloides* are saponins and sapogenins of steroid nature, especially sarsasapogenin and its glycosides. The sapogenins isolated and reported are sarsasapongenin (*15-1*) [1, 2], and markogenin (*15-2*) [3, 4]. They are derived from spirostane (*15-3*). Sarsasapogenin is the major sapogenin found in the rhizome.

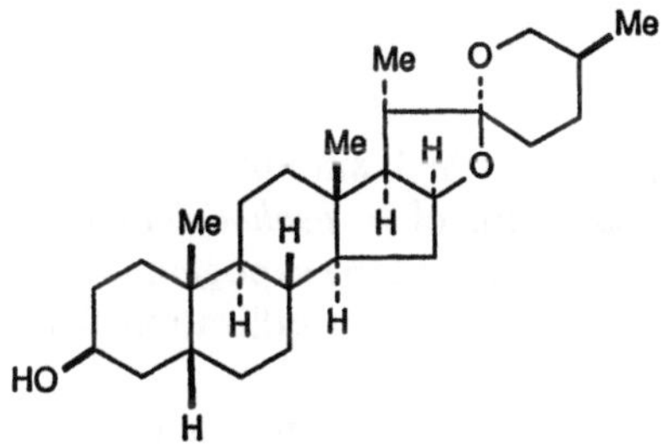

Sarsasapogenin (*15-1*)

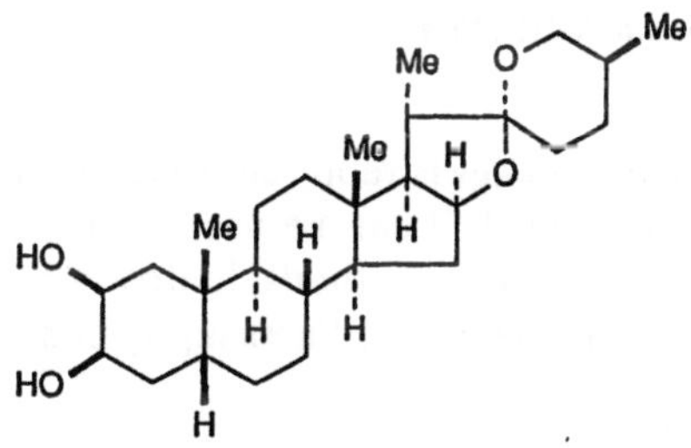

Markogenin (*15-2*)

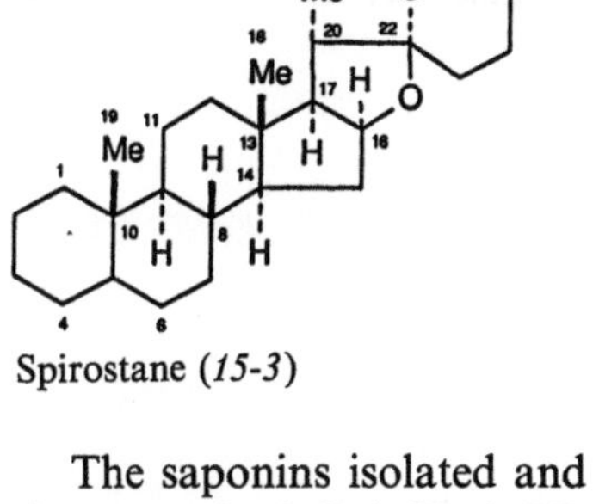

Spirostane (*15-3*)

The saponins isolated and reported are referred to as timosaponins, including timosaponin A-I, A-II, A-III, A-IV, B-I, and B-II [4, 5]. Timosaponin A-III (*15-4*) was obtained as a homogeneous substance and structurally determined [5]. Timosaponin A-III, after completely hydrolysis with 2N HCl, yielded equimolar amounts of sarsasapogenin, D-glucose, and D-galactose. Partial hydrolysis yielded sarsasapogenin and a prosapogenin. The latter was identical with timosaponin A-I which on further hydrolysis gave sarsasapogenin and D-galactose. From the hydrolysate, *O*-β-D-glucopyranosyl-(1→2)-D-galactopyranose, timobiose, was isolated. Timosaponin A-I could be synthesized by reaction of sarsasapogenin with tetra-*O*-acetyl-β-D-galactopyranosyl bromide. Thus, the structure of timosaponin A-III was prove to be sarsasapogenin *O*-β-D-glucopyranosyl-(1→2)-β-D-galactopyranoside. Timosaponin B-I is a sarsasapogenin trihexoside containing one glucose more than timosaponin A-III.

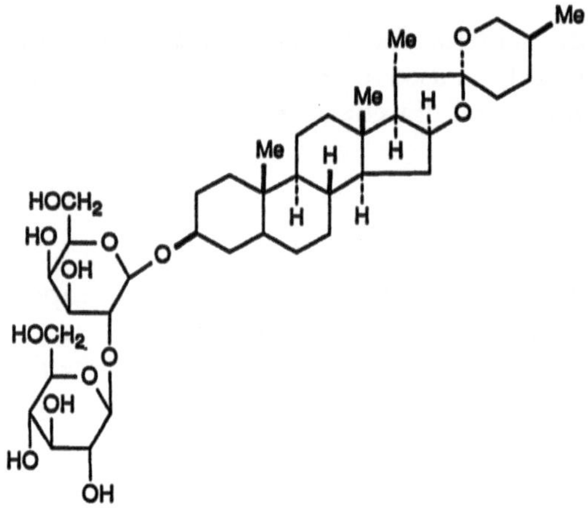

Timosaponin A-III (*15-4*)

A. asphodeloides is representative of plants containing a high amount of sarsasapogenin for which it can be used as a source [6]. Saponins of *A. asphodeloides* are extracted by routine methods, hydrolyzed with HCl/EtOH, and the sapogenins then extracted with benzene. Sarsasapogenin can be purified by recrystallization from methanol.

Besides the saponins and sapogenins, some norlignans with inhibitory activity on cAMP phosphodiesterase, such as hinokiresinol (*15-5*) and oxy-hinokiresinol, were isolated from the rootstock of *A. asphodeloides* by extraction with chloroform and structurally identified on the basis of spectroscopic data [7]. 2,6,4′-Trihydroxy-4-methoxy-benzophenone and *p*-hydroxy-phenyl-crotonic acid have also been detected [7].

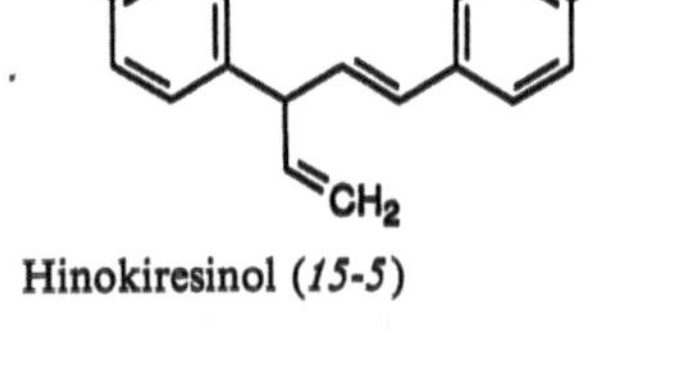

Hinokiresinol (*15-5*)

Recently, Takahashi et al. have reported the isolation of four glycans from the rootstock of *A. asphodeloides:* anemarans A, B, C, and D [8].

From the aerial part of *A. asphodeloides*, two xanthone C-glucosides, mangiferin (*15-6*) and isomangiferin (*15-7*), were isolated [9, 10] and structurally elucidated [10, 11].

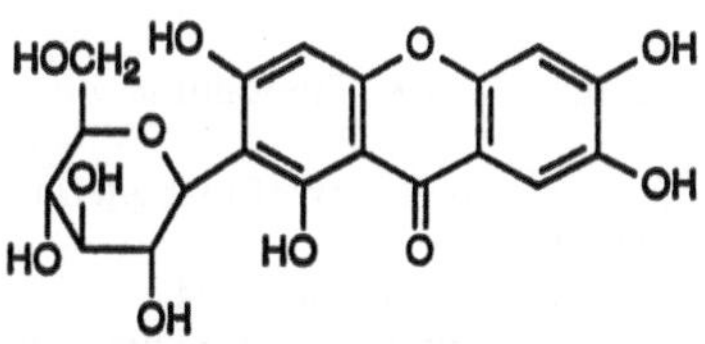

Mangiferin (*15-6*)

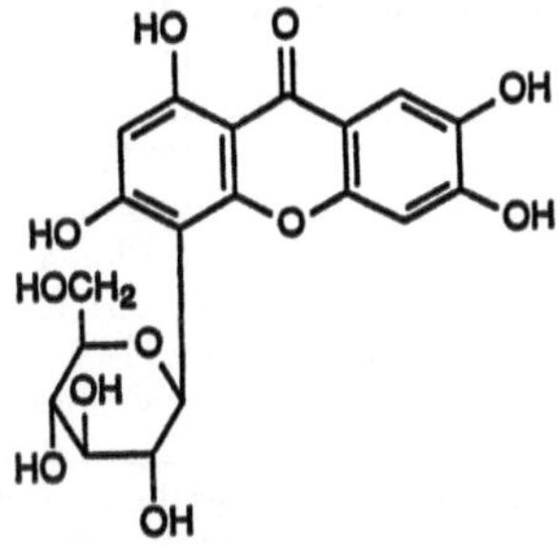

Isomangiferin (*15-7*)

Mangiferin was also isolated from the rootstock; contents varied with the collection period and were lowest in March (0.12%) and highest in April (1.26%) [12].

15.3 Pharmacology

The saponin fraction isolated from the rootstock of *A. asphodeloides* and its hydrolysis product, sarsasapogenin, as well as its hemisuccinyl derivative all showed potent inhibitory action on Na^+/K^+ ATPase and decreased oxygen uptake in thyroxine-treated liver. The inhibitory action of the hemisuccinyl derivative was even more potent than that of ouabain [13]. Sarsasapogenin also inhibited the Na^+/K^+ ATPase of human red blood cells in vitro. The inhibitory effect developed slowly and could be enhanced by external sodium ions and antagonized by external rubidium ions. The inhibitory activity of sarsasapogenin on the ATPase may be related to its antipyretic action [14].

The norlignans hinokiresinol and its oxy- and tetrahydro-derivative all showed significant inhibitory activity on cAMP phosphodiesterase in vitro. Moreover, hinokiresinol prolonged the duration of hexobarbital-induced sleep in mice when given at a dose of 25–100 mg/kg [7].

The four glycans, anemarans A, B, C, and D, showed a significant hypoglycemic effect in normal and in alloxan-induced hyperglycemic mice [8].

References

1. Takeda K, Okanishi T, Shimaoka A (1953) Steroidal components of domestic plants. Constituents of the rhizome of *Anemarrhena asphodeloides*. I. J Pharm Soc Jpn 73:29–31
2. Wu CH (1960) Steroid sapogenins in some Chinese medicinal plants. II. Acta Pharm Sin 8:66–69

3. Takeda K, Okanishi T, Shimaoka A (1954) Steroidal components of domestic plants. VII. Constituents of *Anemarrhena asphodeloides* 2. J Pharm Soc Jpn 74:1371–1373
4. Kawasaki T, Yamauchi T, Itakura N (1963) Saponins of *Anemarrhena* rhizomes. I. Yakugaku Zasshi 83:892–896
5. Kawasaki T, Yamauchi T (1963) Saponins of timo (*Anemarrhena* rhizome). II. Structure of timosaponin A-III. Chem Pharm Bull 11:1221–1224
6. Han GR (1983) Improved method for extraction of sarsasapogenin from *Anemarrhena asphodeloides* Bunge. Bull Chin Mat Med 10:133
7. Nikaido T, Ohmoto T, Noguchi H, Kinoshita T, Saitoh H, Sankawa U (1981) Inhibitors of cyclic AMP phosphodiesterase in medicinal plants. Planta Med 43:18–23
8. Takahashi M, Konno C, Hikino H (1985) Validity of the oriental medicines. LXXXVI. Antidiabetes drugs. 7. Isolation and hypoglycemic activity of anemarans A, B, C and D, glycans of *Anemarrhena asphodeloides* rhizomes. Planta Med 51:100–102
9. Morita N, Shimizu M, Fukuta M (1965) The medicinal resources. XXIV. Chimonin in *Anemarrhena* Root. Yakugaku Zasshi 85:374–375
10. Aritomi M, Kawasaki T (1969) A new xanthone C-glucoside, position isomer of mangiferin, from *Anemarrhena asphodeloides* Bunge. Tetrahedron Lett 941–944
11. Bhatia VK, Seshadri TR (1968) Synthesis of mangiferin. Tetrahedron Lett 1741–1742
12. Hong YF, Han GY, Zhou ZL, Wu MS (1985) Preliminary study of the variation of mangiferin contents in common anemarrhena (*Anemarrhena asphodeloides*) collected in different periods. Chin Trad Herb Drugs 16:175–177
13. Chen RQ, Yu ZY, Zhang XY, Zheng JJ, Koo TC (1982) Zhi mu (*Anemarrhena asphodeloides* Bge.) sapogenin is a powerful inhibitor of sodium, potassium ATPase. Acta Biochim Biophys Sin 14:159–164
14. Lu SL, Chen RQ, Gu TJ (1983) Inhibitory effect of sarsasapogenin on the sodium-potassium pump in human erythrocytes. Acta Acad Med Primae Shanghai 10:327–334

Anemone raddeana Regel

16

16.1 Introduction

Anemone raddeana Regel (Ranunculaceae) is a herbal plant native to China, the rhizome of which is used in folk medicine for treatment of rheumatism and phlebitis. It is included in the appendix of the Chinese Pharmacopoeia.

16.2 Chemical Constituents

A number of saponins, raddeanins A [1–4], B [1, 2, 5], C [1, 2, 6], D [1, 2 7], E, and F [8, 9], were isolated from the rhizome of *A. raddeana* and structurally determined. All have oleanolic acid (*2-1*) as a sapogenin. Thus, raddeanin A (*16-1*) is oleanolic acid 3-*O*-α-L-rhamnopyranosyl-(1→2)-β-D-glucopyranosyl-(1→2)-α-L-arabinopyranoside; raddeanin B is oleanolic acid 3-*O*-α-L-rhamnopyranosyl-(1→4)-α-L-arabinopyranoside and is identical with eleutheroside I (*1-13*); raddeanin C (*16-2*) is oleanolic acid 3-*O*-α-L-rhamnopyranosyl-(1→4)-α-L-arabinopyranoside 28-*O*-α-L-rhamnopyranosyl-(1→2)-β-D-glucopyranosylester; raddeanin D (*16-3*) is oleanolic acid 3-*O*-α-L-rhamnopyranosyl-(1→2)-β-D-glucopyranosyl-(1→2)-α-L-arabinopyranoside 28-*O*-α-L-rhamnopyranosyl-(1→4)-β-D-glucopyranosylester; raddeanin E (*16-4*) is oleanolic acid 3-*O*-β-D-glucopyranosyl-(1→2)-α-L-rhamnopyranosyl-(1→4)-α-L-arabinopyranoside; and raddeanin F (*16-5*) is oleanolic acid 3-*O*-β-D-glucopyranosyl-(1→2)-α-L-arabinopyranoside 28-*O*-α-L-rhamnopyranosyl-(1→4)-β-D-glucopyranosylester.

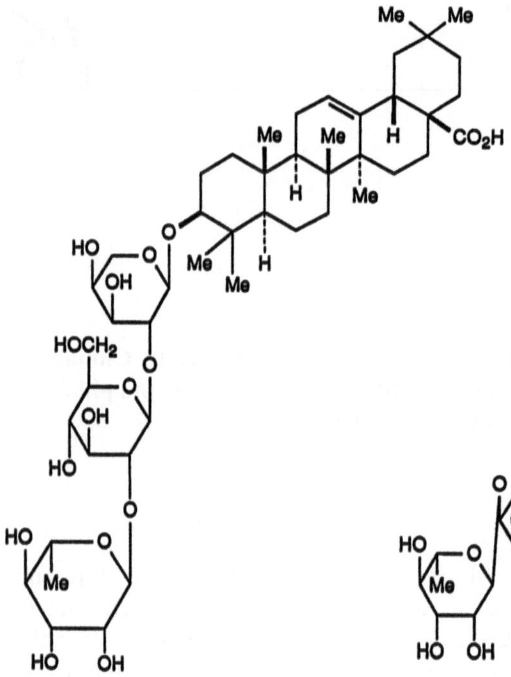

Raddeanin A (*16-1*)

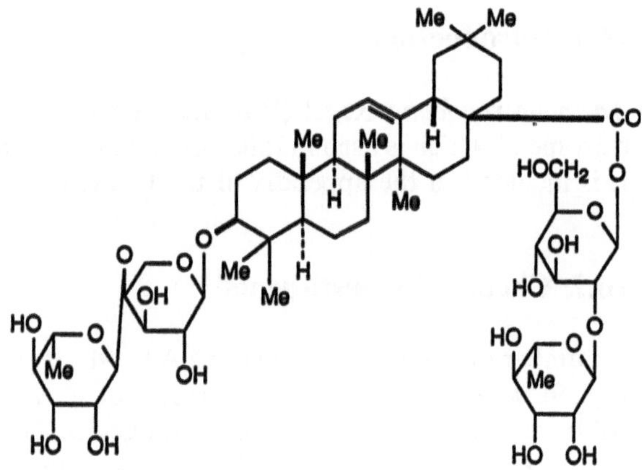

Radeanin C (*16-2*)

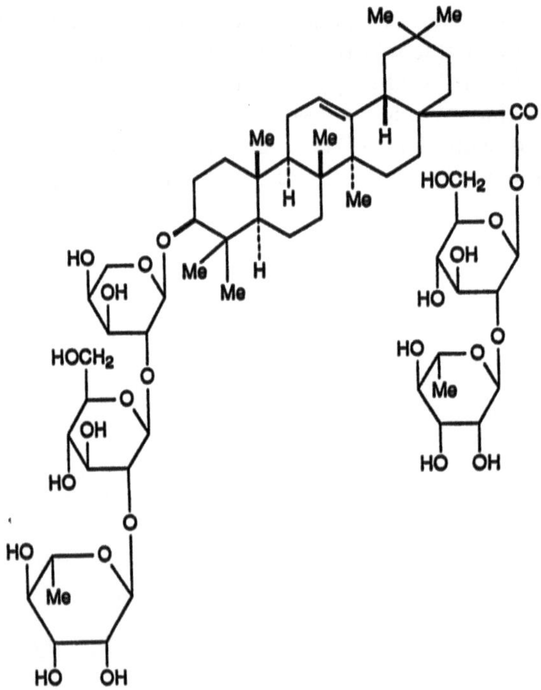

Raddeanin D (*16-3*)

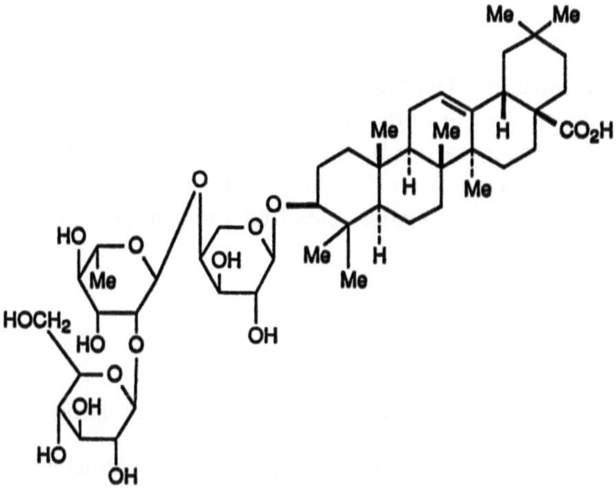

Raddeanin E (*16-4*)

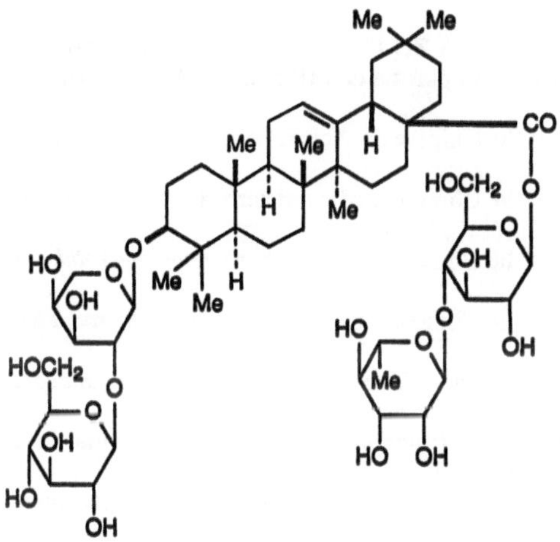

Raddeanin F (*16-5*)

In addition to the raddeanins, diosgenin, oleanolic acid, oleanolic acid 3-*O*-α-L-arabinopyranoside [10], and a lactone glucoside, ranunculin (*16-6*) [11], were isolated and identified.

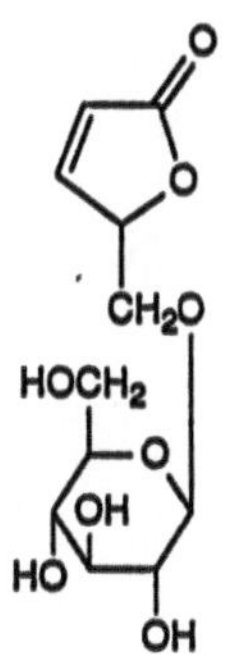

Ranunculin (*16-6*)

16.3 Pharmacology

DNA, RNA, and protein synthesis in sarcoma 180 and ascitic hepatoma cell cultures were inhibited by 30 μg/ml of raddeanin A (anemodeanin A). The growth of ascitic hepatoma cells was suppressed at this concentration by 81%. The ID_{50} of raddeanin A on DNA synthesis in ascitic hepatoma cells after 48 h was 21 μg/ml. Raddeanin A, given i.p. at a dose of 10 mg/kg for 5 days to mice, elevated the plasma cAMP levels by 61% [1, 4].

References

1. Wu FE, Zhu ZQ (1983) Study on the chemical composition of the traditional Chinese medicine, *Anemone raddeana* Regel. J Lanzhou Univ [Nat Sci] 19:188
2. Wu FE, Zhu ZQ (1984) Studies on the chemical constituents of the Chinese medicinal herb *Anemone raddeana* Regel. Kexue Tongbao [Foreign Lang] 29:1644–1645
3. Wu FE, Zhu ZQ (1984) Chemical constituents of the Chinese medicinal herb *Anemone raddeana* Regel. Acta Chim Sin 42:253–258
4. Liu LS, Xiao XH, Zhang LD, Zheng RL, Wu FE, Zhu ZQ (1985) Effects of anemodeanin A on DNA, RNA and protein of tumor cells in vitro and plasma cAMP in mice. Acta Pharmacol Sin 6:192–194
5. Wu FE, Zhu ZQ (1985) Chemical constituents of the Chinese medicinal herb *Anemone raddeana* Regel. III. Chem J Chin Univ 6:36–40
6. Wu FE, Zhu ZQ (1984) Chemical constituents of the Chinese medicinal herb *Anemone raddeana* Regel. IV. Acta Chim Sin 42:1266–1270
7. Wu FE, Zhu ZQ (1985) Chemical constituents of the Chinese medicinal herb *Anemone raddeana* Regel. V. Acta Chim Sin 43:82–86
8. Wu FE, Zhu ZQ (1985) Chemical constituents of the Chinese medicinal herb *Anemone raddeana* Regel. VI. Acta Chim Sin 43:692–697
9. Wu FE, Zhu ZQ (1984) Chemical composition of the Chinese medicine *Anemone raddeana* Regel. J Lanzhour Univ [Nat Sci] 20:164
10. Wu FE, Zhu ZQ (1983) Studies on the chemical constituents of the Chinese medicinal herb *Anemone raddeana* Regel. Chem J Chin Univ 4:595–599
11. Liu DY (1983) Isolation and identification of ranunculin from Liang Tou Jian (*Anemone raddeana*). Chin Trad Herb Drugs 14:532–533

Angelica spp. 17

17.1 Introduction

Angelica is a plant genus often used in traditional Chinese medicine. Officially listed in the Chinese Pharmacopoeia are three items:

- Baizi, Radix Angelicae dahuricae, is the dry root of *Angelica dahurica* (Fisch. ex Hoffm.) Benth. et Hook f. or *A. dahurica* var. *formosana* (Boiss.) Shan et Yuan (Apiaceae) collected between summer and fall when the leaves have turned yellow. It is used as an antipyretic and analgesic for cold, headache, and toothache.
- Danggui, Radix Angelicae sinensis, is the dry root of *A. sinensis* (Oliv.) Diels collected in late summer. It has been used as an analgesic in treatment of menstrual disorders, menorrhalgia, and rheumatalgia.
- Duhuo, Radix Angelicae pubescentis, is the dry root of *A. pubescens* Maxim. f. *biserrata* Shan et Yuan collected in late fall or early spring. It has been used as an analgesic and antirheumatic agent.

17.2 Chemical Constituents

17.2.1 Chemical Constituents of *Angelica dahurica* and *Angelica dahurica* var. *formosana*

The roots of *A. dahurica* and *A. dahurica* var. *formosana* are known to contain a number of coumarin and furocoumarin derivatives, listed in Table *17-1*. The furocoumarin byakangelicin (*17-13*) was also isolated from the fruits of *A. dahurica* var. *formosana*.

Among the furocoumarins, furo[3,2-*g*] [1] benzopyran-7-one is known as psoralen (*17-1*) and furo[2,3-*h*]-1-benzopyran-2-one as angelicin (*17-2*).

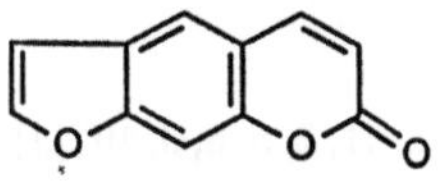

Psoralen (*17-1*)

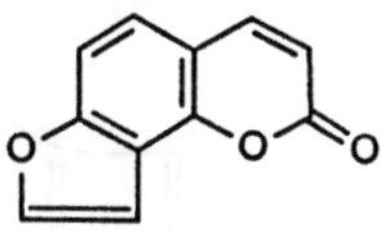

Angelicin (*17-2*)

Table 17.1. Chemical constituents of *Angelica dahurica* and *Angelica dahurica* var. *formosana*

Compound	Structure	Reference
Coumarin derivatives		
Coumarin (*17-3*)	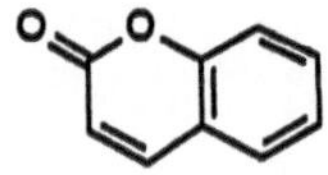	[8]
Scopoletin (*17-4*)	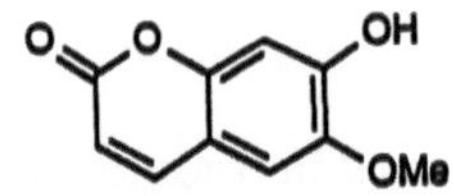	[5]
7-Demethyl-suberosin (*17-5*)	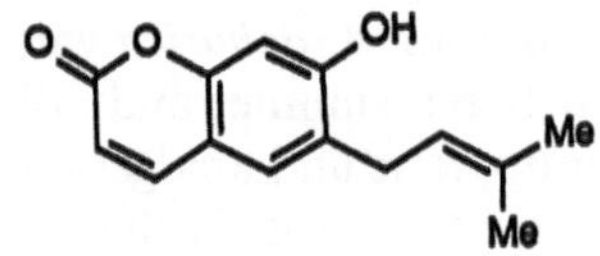	[7, 9]
6-Methoxy-7-(3-methyl-2-butenyloxy)coumarin (*17-6*)	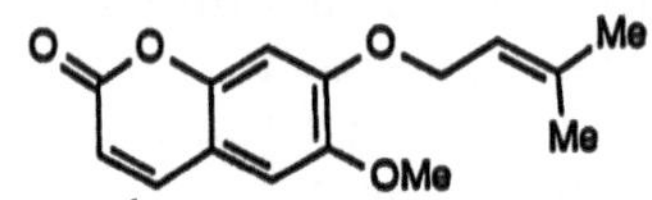	[9]
Cedrelopsin (*17-7*)	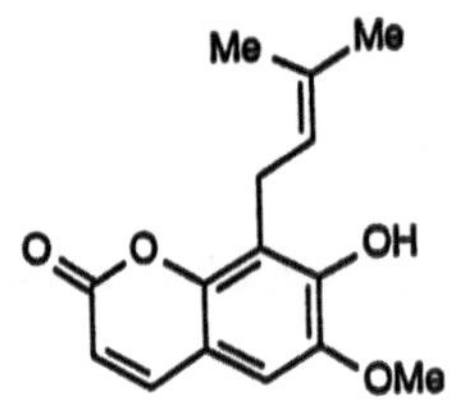	[9]
Furocoumarin derivatives		
Bergapten (*17-8*)	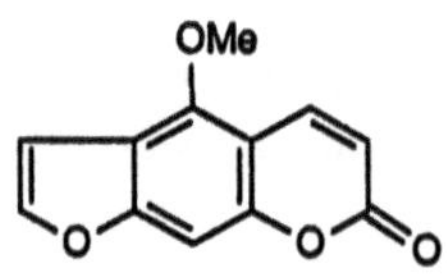	[2, 8, 12]
Imperatorin (*17-9*)	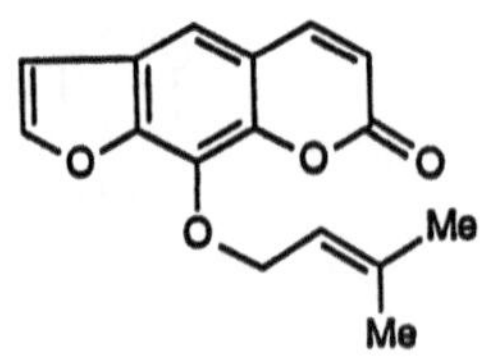	[1, 2, 5, 6, 8, 10, 11]

Table 17.1. (continued)

Compound	Structure	Reference
Isoimperatorin (*17-10*)	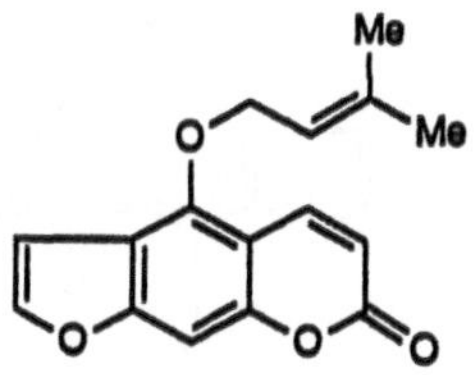	[1, 2, 6, 8, 10, 11]
Alloisoimperatorin (*17-11*)	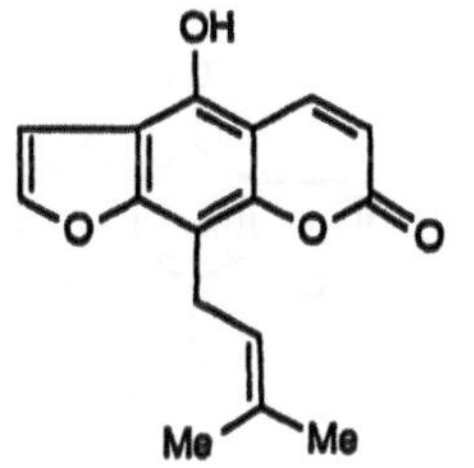	[5, 8]
Phellopterin (*17-12*)	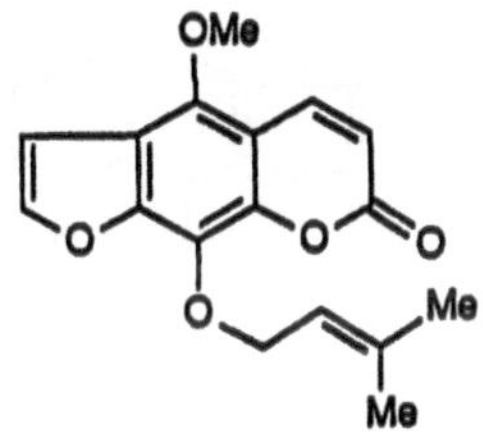	[1, 2, 5, 9]
Oxypeucedanin (*17-13*)	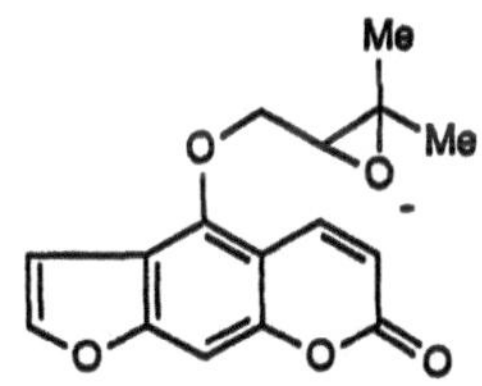	[1, 2, 5, 6, 8]
Oxypeucedanin hydrate (*17-14*)	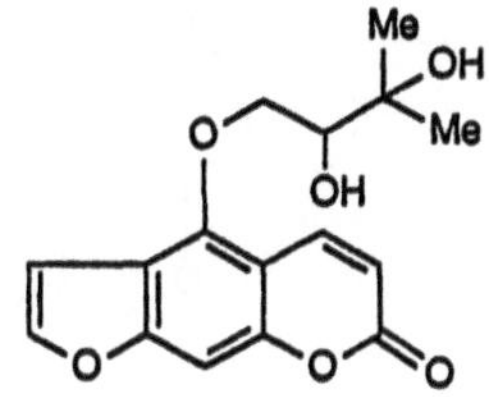	[2, 6, 8]

Table 17.1. (continued)

Compound	Structure	Reference
Byakangelicin (*17-15*)	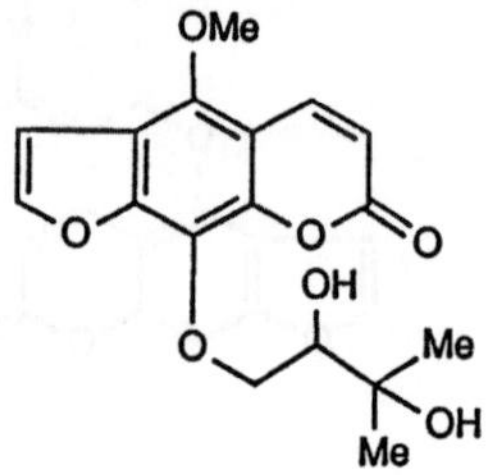	[1, 3, 5, 6, 8]
Byakangelicol (*17-16*)	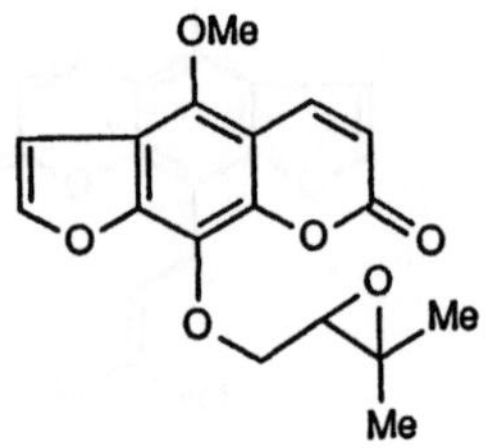	[1, 4, 5]
Anhydrobyakangelicin (*17-17*)	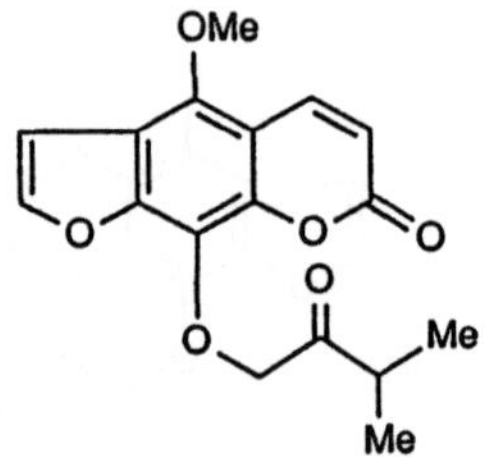	[5, 9]
tert-O-Methylbyakangelicin (*17-18*)	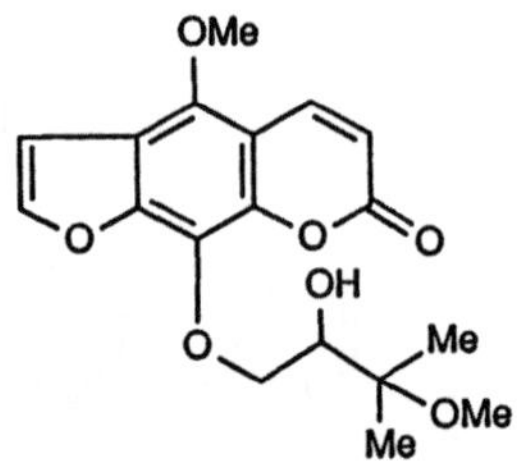	[9]
sec-O-Acetylbyakangelicin (*17-19*)	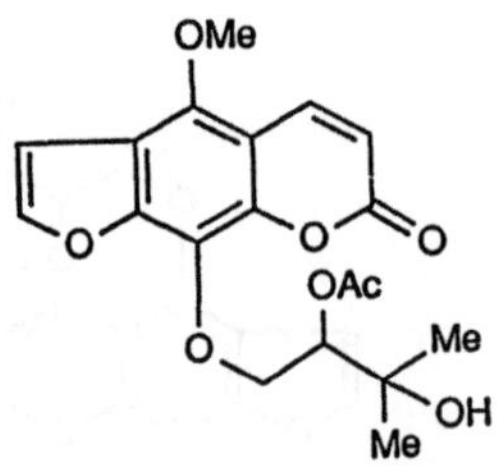	[9]

Table 17.1. (continued)

Compound	Structure	Reference
Senbyakangelicol (*17-20*)	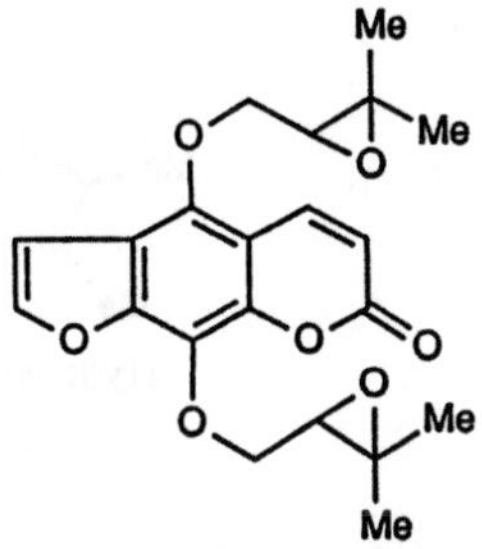	[7]
5-(2-Hydroxy-3-methoxy-3-methyl-butoxy)psoralen (*17-21*)	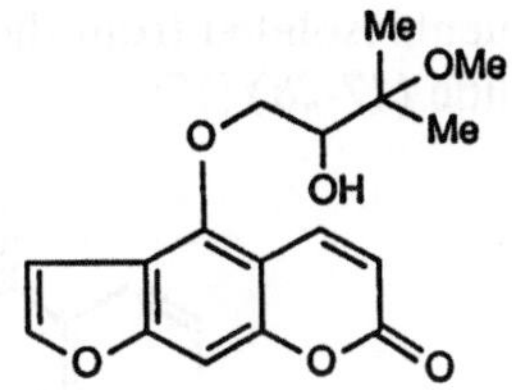	[9]
Marmesin (*17-22*)	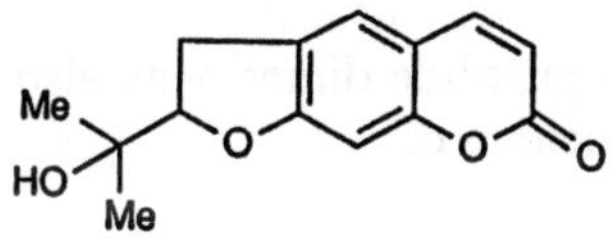	[5]
Xanthotoxin (*17-23*)	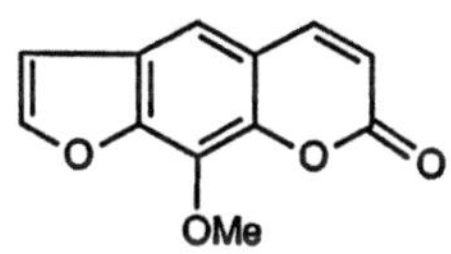	[5]
Neobyakangelicol (*17-24*)	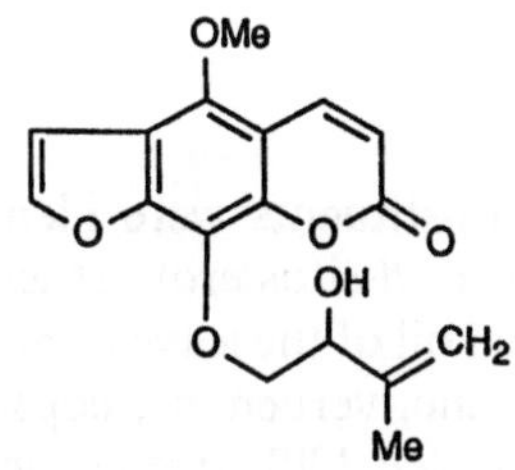	[5, 9]

Besides the coumarin and furocoumarin derivatives, stigmasterol [13], sitosterol [8] and some lactones such as β-angelica lactone (*17-25*) [12], 2-hydroxy-3,4-dimethyl-2-buten-4-olide (*17-26*) [14], γ-nonalactone, and γ-decalactone [15] were isolated and identified.

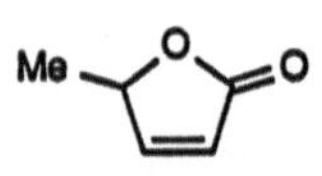

β-Angelica lactone (*17-25*)

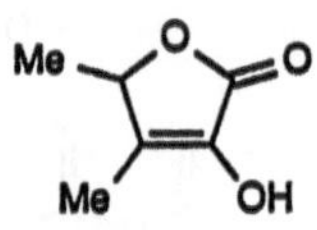

2-Hydroxy-3,4-dimethyl-2-buten-4-olide (*17-26*)

17.2.2 Chemical Constituents of *Angelica sinensis*

The most important constituents isolated from the roots of *A. sinensis* were ferulic acid (*17-27*) [16] and ligustilide (*17-28*) [17].

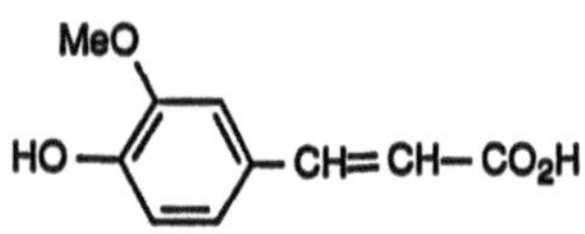

Ferulic acid (*17-27*)

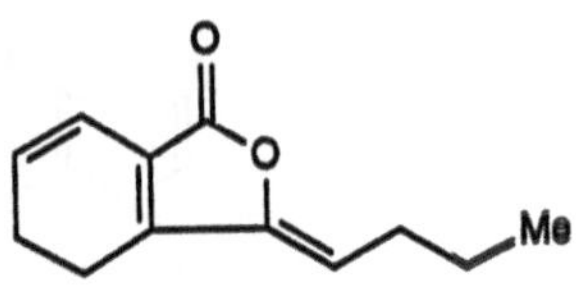

Ligustilide (*17-28*)

Angelicide (*17-29*) [18], a ligustilide dimer, was also isolated from the roots of *A. sinensis* and structurally elucidated.

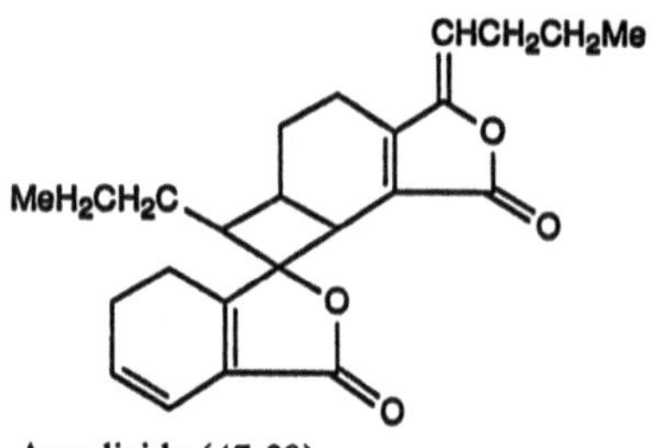

Angelicide (*17-29*)

Additional nonvolatile constituents were identified as brefeldin A (*17-30*), 6-methoxy-7-hydroxy-coumarin, β-sitosterol, stigmasterol, and β-sitosterol-D-glucoside [19]. From the essential oil of the leaves a number of terpene compounds such as α-pinene, myrcene, limonene, verbenone, copaene, β-farnesene, β-selinene, and β- and α-cadinene were detected [20]. Amino acids [21] and trace elements [22] present in the root of *A. sinensis* have also been studied.

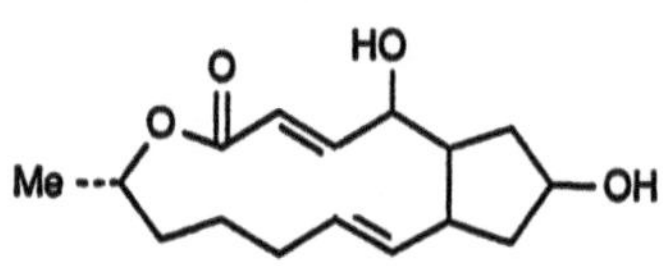

Brefeldin A (*17-30*)

17.2.3 Chemical Constituents of *Angelica pubescens*

A. pubescens contains a number of coumarin and furocoumarin derivatives, some of which were also found in *A. dahurica*. Hata and Kozawa have reported the isolation of osthol (*17-31*), bergapten, glabralactone (*17-32*), and angelol, from the root of *A. pubescens* [23]. Kozawa et al. have also reported isolation of psoralen, xanthotoxin, isopimpinellin (*17-33*), byakangelicin, coumurrayin (*17-34*), 7-methoxy-8-senecioyl-coumarin, scopoletin, and two prenylcoumarins structurally elucidated as 8-(3-hydroxyisovaleroyl)-5,7-dimethoxycoumarin and 5-isopentenyloxy-7-methoxy-8-senecioylcoumarin (*17-35*) [24].

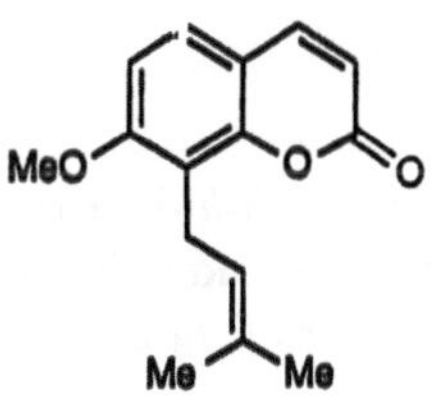

Osthol (*17-31*)

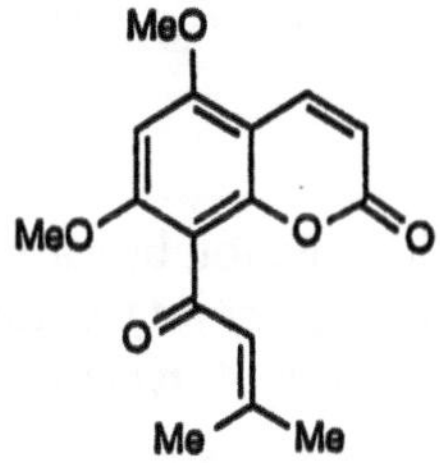

Glabralactone (*17-32*)

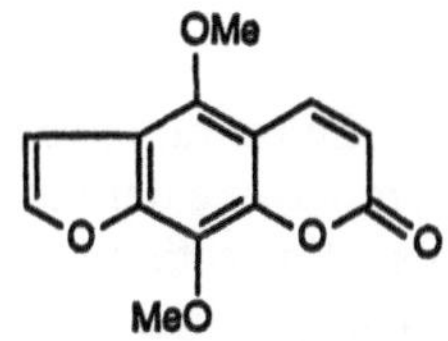

Isopimpinellin (*17-33*)

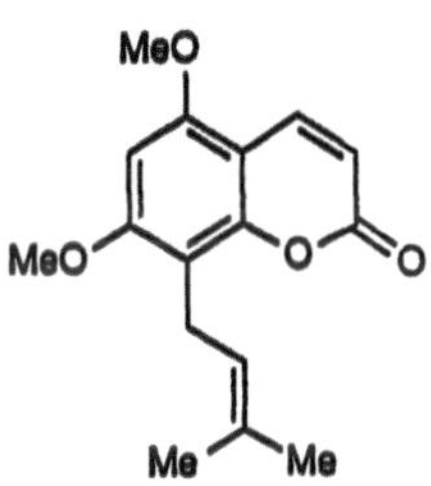

Coumurrayin (*17-34*)

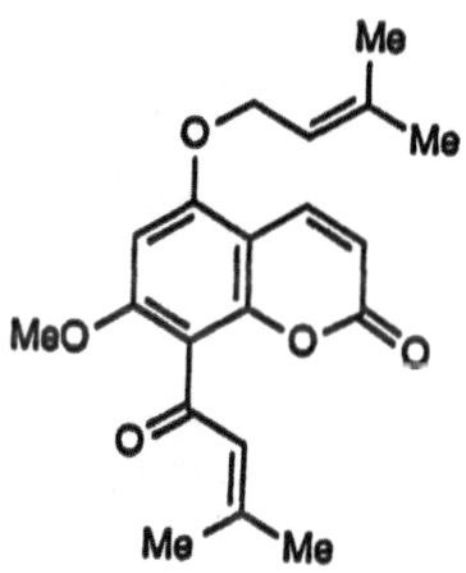

5-Isopentenyloxy-7-methoxy-8-senecioyl-coumarin (*17-35*)

The structure of angelol (angelol A) was the focus of several studies [23, 25, 26]. Baba et al. reported the isolation of some new angelol type prenylcoumarins, angelols B-H (*17-37*–*17-43*), together with previously known angelol A (*17-36*) [26]. The structures of these compounds were determined on the basis of spectral and chemical evidence. Angelol A and B; angelol C and F; and angelol D, G, and H are stereoisomers.

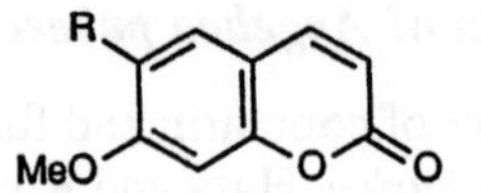

Angelol A (*17-36*), B (*17-37*): $R{=}HOC(CH_3)_2{-}CH(O_2C{-}C(CH_3){=}CH{-}CH_3){-}CH(OH){-}$

Angelol C (*17-38*), F (*17-41*): $R{=}HOC(CH_3)_2{-}CH(O_2C{-}CH(CH_3){=}CH_2{-}CH_3){-}CH(OH){-}$

Angelol D (*17-39*), G (*17-42*) H (*17-43*): $R{=}HOC(CH_3)_2{-}CH(OH)CH{-}O_2C{-}C(CH_3){=}CH{-}CH_3$

Angelol E (*17-40*): $R{=}HOC(CH_3)_2{-}CH(O_2C{-}CH_2{-}CH(CH_3)_2){-}CH{-}(OH){-}$

Chen et al. have isolated osthol, osthenol, bergapten, and the furo-[2,3-*h*]-benzopyran-2-one derivatives columbianadin (*17-44*), columbianetin (*17-45*), and columbianetin acetate (*17-46*) from the root of *A. pubescens* and *A. pubescens* f. *biserrata* [27, 28].

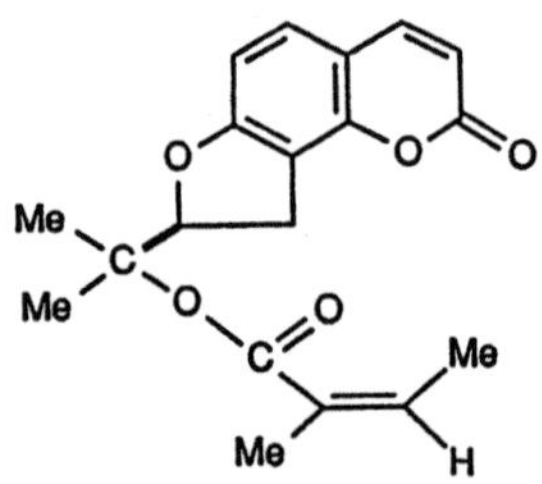

Columbianadin (*17-44*)

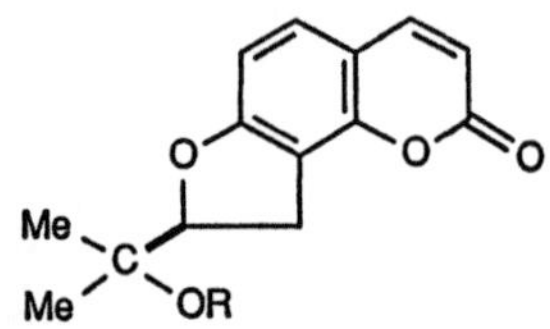

Columbianetin (*17-45*): R=H
Columbianetin acetate (*17-46*): R=Ac

From the fruits of *A. pubescens*, nine coumarins were isolated and identified as osthol, byakangelicin, byakangelicol, oxypeucedanin, umbelliferone, umbelliprenin (*17-47*), imperatorin, neobyakangelicol, and *sec-O*-acetylbyakangelicin (*17-19*) [29].

Umbelliprenin (*17-47*)

17.2.4 Chemical Constituents of Other *Angelica* Species Used in Chinese Folk Medicine

17.2.4.1 Chemical Constituents of *Angelica apaensis*

From the root of *A. apaensis* the furocoumarins oxypeucedanin, isoimperatorin, oxypeucedanin hydrate, byakangelicol, byakangelicin [30], pabulenol (*17-48*), and a new compound, apaensin (*17-49*) [31], were isolated. The structure of apaensin was determined on the basis of spectral analyses.

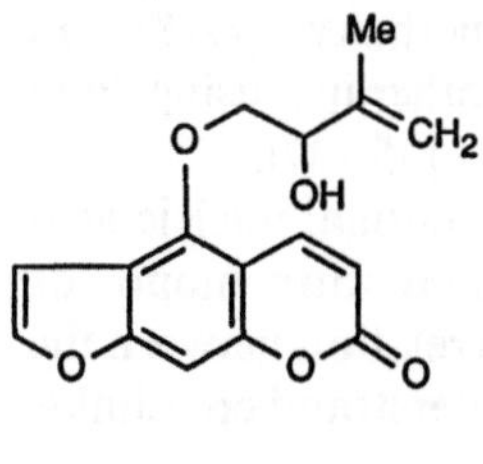

Pabulenol (*17-48*)

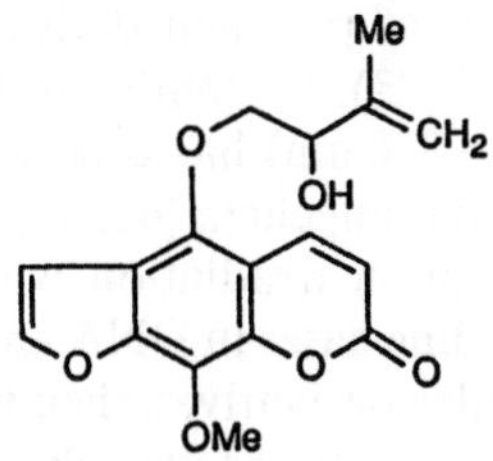

Apaensin (*17-49*)

17.2.4.2 Chemical Constituents of *Angelica omeiensis*

The furocoumarins isoimperatorin, oxypeucedanin, oxypeucedanin hydrate, byakangelicin, and coumarin were isolated from the root of *A. omeiensis* [32].

17.2.4.3 Chemical Constituents of *Angelica edulis*

From the root of *A. edulis* two new coumarins, edulisin I (*17-50*) and edulisin II (*17-51*), were isolated and their structure established by chemical studies and spectral analyses [33].

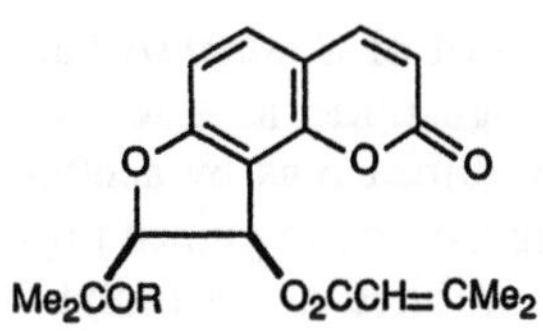

Edulisin I (*17-50*): R = –COCH=CH–C_6H_4–OH

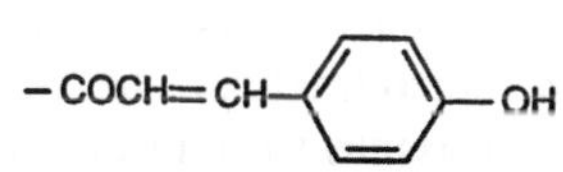

Edulisin II (*17-51*): R = –COCH=CMe_2

17.3 Pharmacology

17.3.1 Pharmacology of *A. dahurica* and its Constituents

Experiments to determine the effects of coumarins on the actions of adrenaline, ACTH, and insulin in fat cells isolated from rats showed that the furocoumarins oxypeucedanin, bergapten, xanthotoxin, imperatorin, and phellopterin activated adrenaline-induced lipolysis. Oxypeucedanin hydrate, imperatorin, and phellopterin also activated ACTH-induced lipolysis, whereas the furocoumarins byakangelicin, neobyakangelicol and isopimpinellin strongly inhibited insulin-stimulated lipogenesis. Therefore, the root of *A. dahurica* activates lipolytic hormones and selectively inhibits antilipolytic hormones [34].

Oral administration of psoralen derivatives, especially 5-methoxy- (*17-8*) and 8-methoxypsoralen (*17-23*), in combination with photochemotherapy using long wave UV radiation (>320 nm) has been used to treat psoriasis [35–37].

Psoralen [38, 39] and angelicin [40, 41] derivatives are bifunctional nucleic acid photoreagents which upon irradiation with long wave UV light form mono- or diadducts with pyrimidine bases in DNA and RNA. A direct correlation between the phototoxicity of psoralen derivatives, their induction of DNA interstrand crosslinks, and inhibition of DNA synthesis was observed [42].

Mutagenic [43, 44] and carcinogenic [45] activities of psoralen derivatives after photoactivation were reported. Evidence of the carcinogenic risk to humans of psoralen derivatives used in photochemotherapy was also described [46].

17.3.2 Pharmacology of *A. sinensis* and its Constituents

The aqueous extract of the root of *A. sinensis* and its ingredient ferulic acid inhibited rat platelet aggregation and serotonin release in vivo and in vitro [47]. These properties may explain the therapeutic effect of the extract of *A. sinensis* on cerebrovascular disorders. Moreover, both the alcohol extract and sodium ferulate, given i.v. to rats, had antiarrhythmic activity [48]. *A. sinensis* and ferulic acid potentiated the phagocytic activity of macrophages when given to mice [49].

Another major constituent, ligustilide, isolated from the root of *A. sinensis* had remarkable antiasthmatic and spasmolytic activity in vivo. Ligustilide, at a dose of 0.14 ml/kg i.p., inhibited the asthmatic reaction induced in guinea pigs by acetylcholine and histamine. Intravenous injection of 0.08 ml/kg liguistilide to guinea pigs inhibited histamine-induced reactions. In vitro, ligustilide exhibited a spasmolytic effect on isolated guinea pig trachea contracted by acetylcholine, histamine, or $BaCl_2$ and produced relaxation of the uncontracted trachea. The cAMP and cGMP content of guinea pig pulmonary and intestinal tissues were not altered, however, by treatment with ligustilide [50].

A number of homologues of ligustilide, including vinylphthalide, propenylphthalide, isobutenylphthalide, and butylphthalide, were synthesized and their biological activity was investigated. All of the compounds showed marked and comparable relaxing effects on animal tracheal smooth muscle, indicating that the phthalide moiety is the principal antiasthmatic component of phthalide derivatives of *A. sinensis*. The onset of action of these phthalides was fastest for phthalide and ethenylphthalide and decreased with increasing chain length of the derivatives; however, the effect of butylphthalide was faster than that of butenylphthalide. The relaxing effect of phthalides on smooth muscle was not associated with the β-adrenergic receptor and was not affected by stimulating the release of mediators from the adrenergic system. Phthalides showed potent antagonistic effects on acetylcholine- and histamine-induced tracheal muscle constriction, but their actions were not associated with histaminergic or cholinergic receptors. Phthalide had rapid and potent action against $BaCl_2$ on tracheal smooth muscle, suggesting that the relaxant effect of phthalide is due to its direct action on tracheal smooth muscle [51].

17.3.3 Pharmacology of *A. pubescens* and its Constituents

A. pubescens had antiinflammatory and analgesic effects on rat hind paw edema induced by carregeenan and on mouse writhing induced by acetic acid. The active principle was separated and identified as osthol [52].

References

1. Hata K, Kozawa M, Yen KY (1963) Pharmacognostical studies on umbelliferous plants. XIX. Chinese drug byakushi. 4. Coumarins of the roots of *Angelica dahurica* var. *dahurica* and *A. dahurica* var. *pai-chi*. Yakugaku Zasshi 83:606–610
2. Hata K, Kozawa M, Yen KY, Kimura Y (1963) Pharmacognostical studies on umbelliferous plants. XX. Chinese drug byakushi. 5. Coumarins of the roots of *Angelica formosana* and *A. anomala*. Yakugaki Zasshi 83:611–614
3. Yen KY (1967) Chemical constituents of umbelliferous plants of Taiwan. XI. Coumarins of the fruits of *Angelica dahurica* var. *formosana*. J Taiwan Pharm Assoc 19:38–40 (CA 73:32242)
4. Yoshida N, Murayama M, Murai H, Suminokura T, Ozaki M (1971) Byakangelicol from *Angelica*. Japanese patent no. 71 12,775 (CA 76:40411 q)
5. Saiki Y, Morinaga K, Okegawa O, Sakai S, Amaya Y, Ueno A, Fukushima S (1971) Coumarins of the roots of *Angelica dahurica*. Yakugaku Zasshi 91:1313–1316
6. Shlyunko EK, Shagova LI, Tikhomirova LI, Nadezhina TP (1977) Coumarins from *Angelica dahurica* roots. Khim Prir Soedin 280 (CA 87:114614e)
7. Fujiwara H, Yokoi T, Tani S, Saiki Y, Kato A (1980) Studies on constituents of Angelica dahuricae radix. I. On a new furocoumarin derivative. Yakugaku Zasshi 100:1258–1261
8. Zang HQ, Yuan CQ, Chen GY, Ding YM, Chen SQ, Deng YQ (1980) Study on the chemical constituents of the root of *Angelica dahurica* var. *formosana*. Chin Pharm Bull 15:2–4
9. Kozawa M, Baba K, Okuda K, Fukumoto T, Hata K (1981) Studies on chemical components of "Bai Zhi" (Suppl. 1). On coumarins from "Japanese Bai Zhi". Shoyakugaku Zasshi 35: 90–95
10. Lu LG, Cai YC (1982) Thin-layer chromatographic separation and UV spectrophotometric determination of imperatorin and isoimperatorin in Baizhi preparations. Chin J Pharm Anal 2:348–350
11. Zhu ZY (1982) Quality survey of the branched Bai-zhi (Angelica dahurica formosana). Chin J Pharm Anal 2:221–224
12. Ding YM, Zhang HQ (1981) Determination of coumarins in *Angelica dahurica* var. *formosana* by thin-layer chromatography and UV spectrophotometry. Chin Pharm Bull 16:16–17
13. Yen KY (1969) Chemical constituents study of umbelliferous plants in Taiwan. XIII. Steroid of the roots of *Angelica dahurica* var. *formosana*. J Taiwan Pharm Assoc 21:10–12 (CA 75: 95345 b)
14. Baba K, Matsuyama Y, Fukumoto M, Kozawa M (1985) 2-Hydroxy-3,4-dimethyl-2-buten-4-olide as a flavoring component of Japanese Bai Zhi. Planta Med 51:64–66
15. Tani S, Fujiwara H, Kato A (1984) Studies on constituents of *Angelica dahurica*. II. Identification of γ-nonalactone and γ-decalactone by GC and GC/MS as a part of the odor components. J Nat Prod 47:734
16. Lu RM, Ho LI, Lo SY (1980) Determination of ferulic acid in Danggui (*Angelica sinensis*). Chin Trad Herb Drugs 11:385–388
17. Lu RM, He LY, Fang HJ, Zhang XQ (1980) Thin layer chromatography and densitometry of liguistilide in Umbelliferae plants. Acta Pharm Sin 15:371–374
18. Chen YZ, Zhang HD, Chen NY, Zhao TZ, Wang MT (1984) Analysis of the ingredients of *Angelica sinensis* – determination of the structure of angelicide. Kexue Tongbao [Foreign Lang] 29:560–562
19. Chen YZ, Zhang HD (1984) Analysis of the chemical ingredients of *Angelica sinensis*. Chem J Chin Univ 5:515–520
20. Chen YZ, Li HQ, Chen NY, Ma XY (1985) Combined capillary gas chromatography-mass spectrometry for analysis of the essential oils of *Angelica sinensis* (Oliv) Diels leaves from Min country, Kansu Province, China. J Lanzhou Univ [Nat Sci] 21:130–132

21. Chen YZ, Zhang HD, Cai WT (1983) Study on the analysis of chemical constituents of *Angelica sinensis*. III. Determination of amino acids in Min-Gui. J Lanzhour Univ [Nat Sci] 19:194–195
22. Chen YZ, Zhang HD, Cai WT (1983) Study on the analysis of chemical constituents of *Angelica sinensis*. II. Determination of minor elements in Min-Gui. J Lanzhou Univ [Nat Sci] 19:192–193
23. Hata K, Kozawa M (1965) The constitution of angelol, a new coumarin isolated from the root of *Angelica pubescens* Maxim (Umbelliferae). Tetrahedron Lett 4557–4562
24. Kozawa M, Baba K, Matsuyama Y, Hata K (1980) Studies on coumarins from the root of *Angelica pubescens* Maxim. III. Structures of various coumarins including angelin, a new prenylcoumarin. Chem Pharm Bull 28:1782–1787
25. Hata K, Kozawa M (1968) Coumarins from the root of *Angelica pubescens*. II. Absolute configuration of angelol. Yakugaku Zasshi 88:293–298
26. Baba K, Matsuyama Y, Kozawa M (1982) Studies on coumarins from the root of *Angelica pubescens* Maxim. IV. Structures of angelol-type prenylcoumarins. Chem Pharm Bull 30:2025–2035
27. Chen SQ, Zhang HQ, Yuan CQ, Deng YQ, Chen GY (1982) Studies on coumarins from Chuan Du Hou (root of *Angelica pubescens* f. *biserrata*) and Zhe Du Huo (root of *A. pubescens*). Chin Pharm Bull 17:120
28. Chen SQ, Zhang HQ, Yuan CQ, Deng YQ, Chen QY (1982) Studies on the coumarins from the traditional Chinese medicines Chuan Du Huo (root of *Angelica pubescens* Maxim. f. *biserrata* Shan et Yuan) and Zhe Du Huo (root of *A. pubescens* Maxim.). Acta Pharm Sin 17:392–394
29. Hata K, Nishino T, Hirai Y, Wada Y, Kozawa M (1981) On coumarins from the fruits of *Angelica pubescens* Maxim. Yakugaku Zasshi 101:67–71
30. Sun HD, Lin ZW, Niu FT (1978) Study of Chinese drugs from Umbelliferae. I. Chemical constituents of the roots of *Angelica apaensis* Shan et Yuan. *Heracleum rapula* Fr., and *Heracleum scabridum* Fr. Acta Bot Sin 20:244–254
31. Sun HD, Lin ZW, Niu FD, Ding JK (1981) Studies on the Chinese drugs of Umbelliferae. IV. Structure of apaensin. Acta Bot Yunnan 3:279–281
32. Zhou AL, Du FM (1982) Isolation of coumarins from the root of *Angelica omeiensis*. Chin Trad Herb Drugs 13:6–8
33. Kawasaki C, Okuyama T, Shibata S, Iitaka Y (1984) Studies on coumarins of a Chinese drug "Qian Hu". VI. Coumarins of *Angelica edulis*. Planta Med 50:492–496
34. Kimura Y, Ohminami H, Arichi H, Okuda H, Baba K, Kozawa M, Arichi S (1982) Effects of various coumarins from roots of *Angelica dahurica* on actions of adrenaline, ACTH and insulin in fat cells. Planta Med 45:183–187
35. Parrish JA, Fitzpatrick TB, Tanenbaum L (1974) Photochemotherapy of psoriasis with oral methoxsalen and long wave ultraviolet light. N Engl J Med 291:1207–1211
36. Wolff K, Hönigsmann H, Gschnait F (1975) Photochemotherapy bei Psoriasis. Klinische Erfahrungen bei 152 Patienten. Dtsch Med Wochenschr 100:2471–2477
37. Wolff K, Hönigsmann H (1981) Clinical aspects of photochemotherapy. Pharmacol Ther 12:381–418
38. Gia O, Palu G, Palumbo M, Antonello C, Magno SM (1987) Photoreaction of psoralen derivatives with structurally organized DNA. Photochem Photobiol 45:87–92
39. Pathak MA (1984) Mechanism of psoralen photosensitization reactions. Natl Cancer Inst Monogr 66:41–46
40. Guiotto A, Rodighiero P, Manzini P, Pastorini G, Bordin F, Baccichetti F, Carlassare F, Vedaldi D, dall'Acqua F (1984) 6-Methylangelicins: a new series of potential photochemotherapeutic agents for the treatment of psoriasis. J Med Chem 27:959–966
41. Dall'Acqua F, Vedaldi D, Caffieri S, Guiotto A, Bordin F, Rodighiero G (1984) Chemical basis of the photosensitizing activity of angelicins. Natl Cancer Inst Monogr 66:55–60
42. Nielsen PE, Bohr V (1983) Phototoxic effects of four psoralens on L1210 cells. The correlation with DNA interstrand crosslinking. Photochem Photobiol 38:653–657
43. Babudri N, Pani B, Venturini S, Monti-Bragadin C (1986) Mutagenic activity of four furocoumarins: angelicin, 4,5'-dimethylangelicin, psoralen and 8-methylpsoralen on V-79 Chinese hamster cells. J Environ Pathol Toxicol Oncol 7:123–129
44. Averbeck D, Papadopoulo D, Quinto I (1984) Mutagenic effects of psoralens in yeast and V-79 Chinese hamster cells. Natl Cancer Inst Monogr 66:127–136
45. Mullen MP, Pathak MA, West JD, Harrist TJ, dall'Acqua F (1984) Carcinogenic effects of monofunctional and bifunctional furocoumarins. Natl Cancer Inst Monogr 66:205–210

46. Stern RS (1984) Carcinogenic risk of psoralen plus ultraviolet radiation therapy: evidence in humans. Natl Cancer Inst Monogr 66:211–216
47. Yin ZZ, Zhang LY, Xu LN (1980) Effect of Dang-Gui (*Angelica sinensis*) and its ingredient ferulic acid on rat platelet aggregation and release of 5-HT. Acta Pharm Sin 15:321–326
48. Cha L, Chien CC, Lu FH (1981) Antiarrhythmic effect of *Angelica sinensis* root, tetrandrine and *Sophora flavescens* root. Chin Pharm Bull 16:53–54
49. Xu LN, Ouyang R, Yin ZZ, Zhang LY, Li LX (1981) Effect of Dang Gui (*Angelica sinensis*) and its constituent ferulic acid on phagocytosis in mice. Acta Pharm Sin 16:411–414
50. Tao JY, Ruan YP, Mei QB, Liu S, Tian QL, Chen YZ, Zhang HD, Duon ZX (1984) Studies on the antiasthmatic effect of ligustilide of Dang Gui, *Angelica sinensis* (Oliv.) Diels. Acta Pharm Sin 19:561–565
51. Liu ZG, Song YT, Wang WL, Chen WQ, Wang XY (1982) Antiasthmatic effect of n-butenyl phthalide from Dang Gui and some other synthetic derivatives of phthalide. Chin Trad Herb Drugs 13:17–21
52. Kosuge T, Yokota M, Sugiyama K, Yamamoto T, Mure T, Yamazawa H (1985) Studies on bioactive substances in crude drugs used for arthritic diseases in traditional Chinese medicine. II. Isolation and identification of an anti-inflammatory and analgesic principle from the root of *Angelica pubescens* Maxim. Chem Pharm Bull 33:5351–5354

Anisodus tanguticus (Max.) Pasch 18

18.1 Introduction

Anisodus tanguticus (Max.) Pasch. (*Scopolia tangutica* Maxim.) (Solanaceae) is a medicinal plant growing in the Qinghai province of China and used in folk medicine as a spasmolytic and antiasthmatic agent.

The active component of the plant, anisodamine hydrobromide, and a form for injection are listed officially in the Chinese Pharmacopoeia Vol. II. They are used as anticholinergic agents in the treatment of gastrointestinal spasms and acute microcirculatory disorders, and as antidotes against poisoning with organic phosphorus compounds.

18.2 Chemical Constituents

From *A. tanguticus* the tropane alkaloids anisodamine [1], anisodine [2], hyoscyamine, scopolamine, tropine [3], apoatropine (*18-1*) [3, 4], 3α-(4,4,4-trichloro-2-phenylbutyryloxy)tropane (*18-2*) [4] and the nontropane alkaloid cuscohygrine (*18-3*) [3] were isolated and identified.

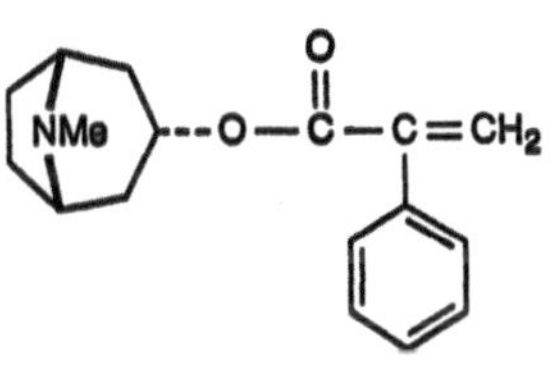

Apoatropine (*18-1*)

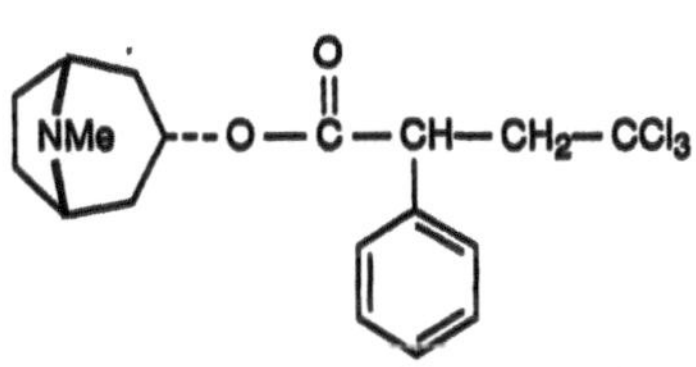

3α-(4,4,4-Trichloro-2-phenylbutyryloxy)-tropane (*18-2*)

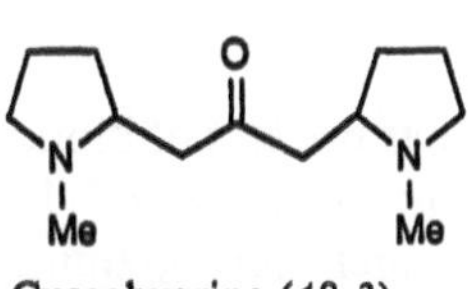

Cuscohygrine (*18-3*)

Anisodamine (*18-4*) and anisodine (*18-5*) are two new alkaloids first isolated from *A. tanguticus*. The structure and absolute configuration of anisodamine were determined on the basis of spectral data and by chemical correlation [1].

Anisodamine hydrobromide has two different melting points (159°–163 °C and 176°–178 °C) representing two crystal forms with different IR spectra which are interconvertible. The crystal form with the higher melting point was found to be more stable. No difference in the pharmacologic action and toxicity between the two forms were observed [5].

The structure of anisodine was determined from the hydrolysis products and spectral analyses [2]. The absolute configuration of (−)-anisodine and its side chain (−)-anisodinic acid were determined by chemical correlation [6].

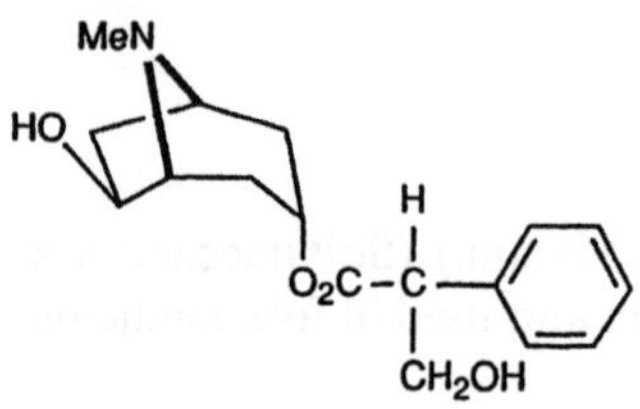

Anisodamine (*18-4*)

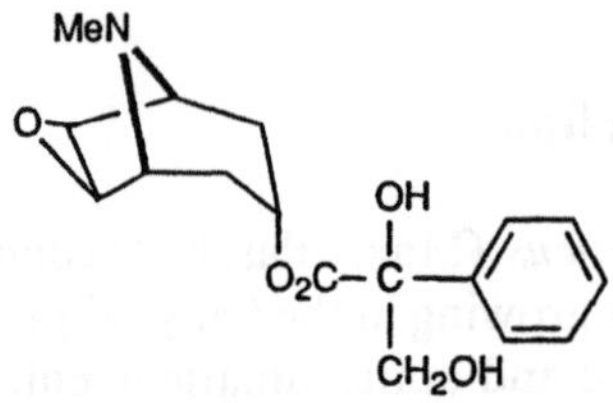

Anisodine (*18-5*)

The contents of hyoscyamine and anisodamine in *A. tanguticus* are dependent on the development of the plant. A peak content of anisodamine was found in 7–10 year old plants. In addition, the anisodamine content was greater in the upper parts than in the roots, whereas the hyoscyamine content was higher in the roots than in the upper parts of the plant. Regional and seasonal differences in hyoscyamine and anisodamine contents were also reported [7].

Anisodamine could be synthesized by acetylation of 6β-hydroxytropinone, followed by reduction with Raney nickel, acylation with α-(acetoxymethyl)-benzeneacetyl chloride, and finally hydrolysis of the diacetate [8, 9].

18.3 Pharmacology

Anisodamine was effectively used in treatment of shock. This property was observed in various kinds of experimental shock given to rabbits. Anisodamine significantly alleviated the progress of shock and increased the survival rate of the animals. The therapeutic effect was much better than that of other commonly used vasoactive drugs such as noradrenaline, dopamine, and aramine [10].

Anisodamine had a better therapeutic effect, as indicated by parameters of blood pressure and mortality, than noradrenaline in rabbits in hemorrhagic shock [11]. Nonetheless, without simultaneous and adequate volume replacement anisodamine produced some harmful effects on rabbits in shock. The mechanism of its antishock effect probably lies partly in dilating spastic arterioles and improving the microcirculation [12].

Cats under hemorrhagic shock treated with anisodamine showed significantly lower cathepsin D activity and amino nitrogen concentration in plasma than untreated shocked animals. Plasma myocardial depressant factor activity was significantly increased in these controls. Correspondingly, anisodamine reduced plasma accumulation of myocardial depressant factor [13], indicating that the antishock effect of anisodamine was probably partly due to a lysosome stabilizing action [14]. In hemorrhage shocked rats, anisodamine increased arterial blood pressure following retransfusion of blood and prolonged shock compensation time. Cathepsin D activity in plasma and liver lysosomes was also markedly decreased by treatment with anisodamine [15].

Anisodamine significantly increased the survival time of traumatically shocked rats but had no effect on cathepsin D activity; however, plasma myocardial depressant factor accumulation was decreased by anisodamine, indicating that prevention of the formation of this factor may be one of the protective mechanisms of the drug [16].

The antishock effect of anisodamine was studied in rabbits shocked by experimental superior mesenteric artery occlusion after i.v. and direct intraenteral infusion. Shock was reversed in a number of treated animals, the mortality rate decreased, and blood pressure increased. Ventral vagotomy did not prevent shock induced by superior mesenteric artery occlusion. Thus, the antishock effect of anisodamine may be due to the prevention of production of a shock inducing intestinal factor and not via blockade of the ventral vagal cholinergic system. The elevation of serum acid phosphatase observed in rabbits with superior mesenteric artery occlusion did not occur in anisodamine pretreated animals, indicating prevention of release of lysosomal enzymes induced by intestinal hypoxia [17, 18].

In *Escherichia coli* endotoxin shocked dogs, abnormalities in pulmonary circulation and function were markedly improved and the average survival time was prolonged by injection of anisodamine after the endotoxin injection [19]. The hemodynamic parameters in endotoxin shocked dogs were normalized or improved following i.v. administration of anisodamine; these improvements included restoration of systemic arterial pressure and total systemic peripheral resistance as well as increases in cardiac output [20, 21], indicating that the effects of anisodamine are not confined to vasodilatation and may be relatively complex.

In rabbits, infused i.v. anisodamine antagonized the decrease in arterial blood pressure and heart rate observed during induction of shock by *Escherichia coli* endotoxin. A slight decrease in renal sympathetic nerve discharge was observed in anisodamine treated rabbits under endotoxin shock in contrast to an increase in untreated shocked rabbits. No such effects were observed in endotoxin shocked rabbits when a bolus injection of anisodamine was given intracerebroventricularly. Anisodamine given by i.v. infusion also decreased renal sympathetic nerve discharge in normal rabbits but did not affect blood pressure or heart rate [22].

Anisodamine induced the release of serotonin from platelets in correlation with a protective effect against endotoxic shock in dogs [23]. Intravenous injection of endotoxin induced a significant increase in plasma levels of cAMP and cGMP in dogs. Anisodamine reduced the plasma levels of cGMP in control dogs and apparently prevented the rise in plasma levels of cGMP in shocked dogs whose survival times were prolonged [24, 25].

Plasma levels of 6-keto-$PGF_{1\alpha}$, a stable metabolite of prostacyclin, were markedly elevated in dogs administered *Escherichia coli* endotoxin; blood pressure decreased simultaneously. Intravenous injection of anisodamine countered these effects [26–28].

Anisodamine injected i.v. into anesthetized dogs at doses higher than those causing cholinergic blockade induced a dose dependent decrease in arterial blood pressure and heart rate [29]. In rats, anisodamine lowered the mean arterial pressure, cardiac output, and contractility of the left ventricle. After coronary ligation, anisodamine also reduced the heart rate. It apparently reduced oxygen consumption by the heart and thus may protect the heart during insufficient blood oxygen supply [30]. The beneficial effect of anisodamine in dogs after ligation of the coronary artery was also proven by electrocardiography. Elevation of the ST segment of the electrocardiogram in experimental acute myocardial ischemia was markedly decreased by i.v. administration of anisodamine [31]. Intraperitoneal administration of anisodamine to rats immediately after ligation of the coronary artery decreased the ischemic damage as evidenced by an electron microscopic study. Ultrastructural changes in cell nuclei, mitochondria, sarcoplasmic reticulum, and capillary vessels following ischemic insult were markedly decreased by anisodamine treatment [32, 33].

Anisodamine also increased the arteriolar diameter and red cell flow velocity in the skeletal muscle of the anesthetized rat. Arteriolar flow was three to four times greater than under control conditions. Mean arterial blood pressure was reduced immediately by 15%–20% for 60 min before normalization [34].

Intraperitoneal injection of anisodamine into rats for 21 days increased the cAMP concentration and cAMP/cGMP ratio in liver, kidney, and plasma. In mice, i.p. injection of anisodamine for 28 days increased bone marrow colony-forming cells and nucleated cells in the bone marrow. Microcirculation of the bone marrow was also increased by anisodamine as detected by capillary hyperproliferation; however, in vitro addition of anisodamine to marrow cell cultures did not increase the colony-forming cells, suggesting that anisodamine increases bone marrow function by promoting microcirculation [35].

Anisodamine did not show obvious effects on pulmonary microcirculation kinetics and pulmonary water and protein transport in sheep with chronic lung fistulas. In sheep with pulmonary injury induced by air embolization, anisodamine treatment at the early stage reduced water exudation from pulmonary capillary vessels apparently by reducing pressure in the vessels [36].

The phase transition temperature of dipalmitoyl phosphatidylcholine liposomes was decreased in the presence of the tropane alkaloids anisodamine, anisodine, scopolamine, and atropine in a concentration dependent fashion, reflecting a concentration dependent increase in the fluidity of phospholipid liposomes. These observations offer an explanation of the molecular mechanism of drug action on biological membranes [37, 38]. Anisodamine displayed the highest activity followed by atropine, scopolamine, and finally anisodine [39]. The interaction of anisodamine with dipalmitoyl phosphatidylcholine and dipalmitoyl phosphatidic acid liposomes was interpreted as a trigger mechanism [40]. Raman spectroscopic studies showed that the interaction of anisodamine with the neutral phospholipid membrane is essentially a polar-polar interaction [41]. The increase in membrane fluidity appears to be a result of the interaction of anisodamine mainly with the polar head of the phospholipid, affecting the hydrophobic region of the bilayer as detected by electron spin resonance [42, 43] and fluorescence polarization studies [43].

Electroencephalographic studies in conscious rabbits after intracerebroventricular injection of anisodamine, atropine, anisodine, and scopolamine suggested that anisodine and scopolamine are mainly central nervous system depressants, whereas anisodamine and atropine are stimulants. Intravenous injection of the four compounds into cats caused high electroencephalographic synchronization and blocked the electroencephalographic arousal response. In mice, the mean intracerebroventricular convulsant doses of anisodamine and atropine were not affected by a number of neurotransmitter antagonists and ligands, except for diazepam which led to an increase, suggesting that central stimulation by these two drugs is GABAergic in nature [44]. The electroencephalographic responses to anisodine and scopolamine were antagonized by physostigmine or arecoline. The effects of anisodine appear to be similar to, but weaker than those of scopolamine [45, 46].

Both the synthesis and release of acetylcholine by rat brain cortex slices stimulated by high K^+ medium were enhanced by incubation with anisodine [47]. In addition, the release of acetylcholine from cat sensorimotor cortex was increased by anisodine [48]. Anisodine inhibited learning and the consolidation and retrieval of short-term memory in rats without affecting long-term memory. The action of anisodine could be blocked by the cholinergic agonist arecoline [49]. The effects of anisodine on learning and memory appeared to be specific. The results support the view that normal function of the cholinergic and especially of the muscarinic system is necessary for memory formation in mammalian brain [50].

In experimental studies, anisodamine was as potent as atropine sulfate in spasmolytic activity. In vitro, it antagonized the contraction of intestine and bladder smooth muscle induced by acetylcholine. In vivo, anisodamine reduced intestinal tension to the same extent as atropine. Anisodamine, however, had fewer side effects on salivary glands, pupils, and the central nervous system; its central action was only $^1/_6-^1/_{20}$ as potent as atropine [51, 52].

Anisodamine inhibited histamine-induced bronchial smooth muscle contractions in guinea pigs [53]. In chicken, i.v. injection of anisodamine did not cause any change in tracheal ciliary movement, but did so in animals treated with cigarette smoke. In young rats the change in serous tracheal exudation under cold stress was found to be inhibited by anisodamine. Acetylcholine content in trachea increased under cold stress without affecting that in brain. Apparently, the expectorant effect of anisodamine was mainly due to its peripheral action [54]. Phagocytosis by macrophages obtained by bronchoalveolar lavage from dogs with oleic acid-induced lung injury was increased by anisodamine treatment. The surface of the macrophages from anisodamine treated animals was smooth and the cells contained many lipid particles. The number of macrophages and polymorphonuclear leukocytes decreased after anisodamine treatment [55]. This effect may result from an inhibition of oleic acid-induced leukocyte accumulation in the microvasculature [56]. Anisodamine and anisodine promoted in vivo phagocytosis of i.v. injected colloidal ^{198}Au by mouse liver and spleen [57].

The LD_{50} of anisodamine in mice was 350–430 mg/kg by i.p. and 123 mg/kg by i.v. administration. The minimal lethal dose by oral administration of anisodamine was 1600 mg/kg [51]. A pharmacokinetic study showed that the highest concentration of anisodamine, 15 min after i.v. injection into rats, was observed in the kidney. At 30 min, however, anisodamine concentration was highest in the pancreas followed by lung, heart, kidney, and spleen; the lowest concentration was observed in brain and plasma. Anisodamine had no significant accumulation effect in tissues. During the 24 h period after injection of anisodamine, urinary excretion was 39% in rats and 42%–49% in patients [58]. The biphasic half-lives of anisodamine injected i.v. into rats were 13 and 70 min [59].

A series of quaternary ammonium salts and N-oxides of anisodamine and anisodine was synthesized and tested for their anticholinergic activity. *N*-methyl-anisodaminium bromide, prepared by treating anisodamine with methyl bromide, had the most anticholinergic activity [60].

References

1. Chinese Academy of Medical Science (1976) Chemical studies on anisodamine. Acta Chim Sin 34:39–44
2. Hsieh CH, Wang L, Liu YL, Shang TM, Hsieh FC, Ke TL (1975) Structure of anisodine. Kexue Tongbao 20:52–53
3. He LY (1982) TLC separation and densitometric determination of tropine alkaloids. Chin Trad Herb Drugs 13:13–16
4. Yang JS, Chen YW, Feng HC (1981) Studies on trace amounts of alkaloids in *Anisodus tanguticus* Pascher. Chin Trad Herb Drugs 12:4–7
5. Liu YL, Xie FZ (1979) Study on the alkaloids of *Physochlaina physaloides* G. Don. Acta Pharm Sin 14:497–501
6. Xie JX, Yang JH, Zhao YX, Zhang CZ (1983) Absolute configuration of (−)-anisodine (a new ganglion blocking agent) and (−)-anisodinic acid. Sci Sin [B] [Engl] 26:931–935
7. Wang ZB, Wu XQ (1979) Variation of the contents of two alkaloids in *Anisodus tanguticus*. Acta Bot Sin 21:85–87
8. Xie JX, Liu CX, Jia XX, Zhou J (1975) Total synthesis of anisodamine. Kexue Tongbao 20:197–198
9. Xie JX, Zhou J, Jia XX, Liu CX, Xu HQ, Fang AS, Wang JZ, Xia BY (1980) Studies on the total synthesis of anisodamine. Acta Pharm Sin 15:403–409
10. Su JY, Wu LL, Tang CS (1983) Experimental study in rabbits of the antishock effect of anisodamine (654-2), and its mechanism of action. Resuscitation 10:173–184
11. Peking Medical College (1977) A comparison of the effectiveness of anisodamine and noradrenaline on hemorrhagic shock in rabbits. Chin Med J [Engl] 3:63–68
12. Peking Medical College (1978) A comparison of the effectiveness of anisodamine and norepinephrine on hemorrhagic shock in rabbits. Cardiology 63:160–167
13. Su JY, Hock CE, Lefer AM (1984) Beneficial effect of anisodamine in hemorrhagic shock. Naunyn-Schmiedebergs Arch Pharmacol 325:360–365
14. Su JY (1984) Mechanism of the antishock effect of anisodamine. Ziran Zazhi 7:638–639
15. Zhu Y, Su YJ (1985) Protective effect of anisodamine on hepatic lysosomes in hemorrhage-shocked rats. J Beijing Med Coll 17:161–164
16. Hock CE, Su JY, Lefer AM (1983) Salutary effects of anisodamine in murine traumatic shock. Circ Shock 11:211–216
17. Tang CS, Wu LL, Su JY (1981) Mechanism of antishock function of anisodamine. Protective effect on intestinal shock. Kexue Tongbao 26:1402–1404
18. Wu LL, Su JY (1982) Comparison of antishock effect of 654-2 (anisodamine) to dopamine and aramine. Natl Med J China 62:678–680
19. Sun RY, Guo HY, Zhou CF, Yan YZ (1982) Effect of anisodamine on pulmonary circulation and function of dogs in shock. Natl Med J China 62:458–460
20. Guo HY, Sun RY (1982) Effects of intravenously injected anisodamine (654) on hemodynamics in normal anesthetized and endotoxin-shocked dogs. Kexue Tongbao 27:320
21. Guo HY, Sun RY (1983) Effects of intravenous administration of anisodamine (654-2) on hemodynamics of nonshocked and endotoxin-shocked dogs. Kexue Tongbao [Foreign Lang] 28:142
22. Wang PP, Shi LB (1985) Effect of anisodamine on renal sympathetic nerve discharge during endotoxin shock in anesthetized rabbits. Acta Acad Med Primae Shanghai 12:309–311
23. Xu SH, Chen L, Su MC, Chen HC (1984) Significance of 5-HT in endotoxic shock and possible effect of anisodamine on 5-HT. Acta Acad Med Sin 6:142–144
24. Xiao DM, Wang ZG, Chen HC (1983) Variation in plasma levels of cAMP and cGMP in dogs with endotoxin shock and effects of anisodamine on these variations. Acta Acad Med Sin 5:202–204

25. Ye SJ (1984) Change of plasma levels of cyclic nucleotides (cAMP, cGMP) in septic shock. Kexue Tongbao [Foreign Lang] 29:1135–1136
26. Xiao DM, Wang ZG, Chen HC (1983) Changes in plasma levels of prostaglandin E and $F_{2\alpha}$ in *E. coli* endotoxin-shocked dogs and children with epidemic meningococcal meningitis. Acta Acad Med Sin 5:186–189
27. Xiao DM, Wang XL, Chen HC (1985) Changes of plasma 6-keto-prostaglandin $F_{1\alpha}$ levels during canine endotoxin shock. Acta Acad Med Sin 7:50–52
28. Xiao DM, Chen HC (1985) Changes in the plasma levels of PGE, PGF_2 and 6-keto-PGF_1 in canine endotoxic shock. Chin Med J [Engl] 98:501–506
29. Shi JC, Miao ZH, Yang MG, Yu ZJ, Zhang SL, Yang XY (1986) Negative effects of anisodamine on cardiovascular function in anesthetized dogs. Tianjin Med J 14:135–138
30. Han QD, Hao HQ, Wang X (1985) Effects of anisodamine on the hemodynamics of rats with experimental myocardial infarction. Xian Yixueyuan Xuebao 6:12–15
31. Shi JC, Miao ZH, Zhang SL, Zhou XY, Yu ZL (1984) The protective effect of anisodamine (654-2) on the acute ischemic myocardium in anesthetized dogs. Tianjin Med J 12:551–553
32. Han QD, Xi YP (1985) Effects of anisodamine on the ultrastructural changes of ischemic myocardium. J Beijing Med Coll 17:101–103
33. Yao T, Xiao YF, Sun XY, Tong HH, Yuan M, Xu ZY, Dai ZY (1984) Effects of anisodamine on cardiovascular activities in endotoxin-shocked dogs. Chin Med J [Engl] 97:871–876
34. Xiu RJ, DeLano FA, Zweifach BW (1985) Influence of anisodamine on microhemodynamics in skeletal muscle preparations. In: Chang HM (ed) Adv Chin Med Mater Res Int Symp 1984. World Science, Singapore, pp 545–552
35. Lu SF, Yu JL, Li YT, Zhao WJ, Zheng SX, Zou DB, Liu SJ, Yu SM, Sheng ZL (1986) Mechanism of anisodamine in treatment of aplastic anemia. Chin J Intern Med 25:10–12
36. Culver PL, Dodak P, Staub NC (1985) Effect of anisodamine on pulmonary microcirculation during experimental lung injury. Natl Med J China 65:389–392
37. Wang SM, Huang F, Fu YZ, Zhang ZL, Dong RJ, Hu CQ (1983) Differential scanning thermoanalysis of effects of tropane alkaloids on fluidity of phospholipid liposomes. Kexue Tongbao 28:60–62
38. Huang F, Wang SM, Hu CC (1983) The disordering effect of hyoscyamine drugs on phospholipid membranes. Biochim Biophys Acta 736:220–225
39. Wang SM, Huang F, Fu YZ, Zhang ZL, Dong RJ, Hu CQ (1983) The effect of hyoscyamine drugs on the fluidity of phospholipid model membranes assessed by a differential scanning calorimetric method. Kexue Tongbao [Foreign Lang] 28:693–696
40. Wang SM, Huang F, Fu YZ, Zhang ZL, Dong RJ, Hu CQ (1984) The interaction of anisodamine with phospholipid membranes. Kexue Tongbao [Foreign Lang] 29:529–532
41. Sun YT, Wang SM (1986) Raman spectroscopic study on the interaction of anisodamine and phospholipid membrane. Kexue Tongbao [Foreign Lang] 31:413–418
42. Huang F, Chen JW, Chao BC, Wang ZF (1984) ESR studies of the effect of anisodamine on the fluidity of dipalmitoylphosphatidic acid liposomes. Bopuxue Zazhi 1:481–485
43. Huang F, Chen JW, Fu YG, Wu YW, Wang ZF, Zhang ZY (1985) ESR and fluorescence polarization studies of the effect of anisodamine on the fluidity of phospholipid liposomes. Kexue Tongbao [Foreign Lang] 30:539–543
44. Peng JZ, Jin LR, Chen XY, Chen ZX (1983) Central effects of anisodamine, atropine, anisodine and scopolamine after intraventricular injection. Acta Pharmacol Sin 4:81–87
45. Peng JZ, Chen ZX, Chen XY (1982) Effects of anisodine of EEG and behavior of conscious cats. Acta Pharmacol Sin 3:78–81
46. Bian CF, Duan SM (1981) Experimental observation of the pharmacological effect of anisodine on central nervous system. Acta Pharm Sin 16:801–805
47. Niu XY, Ren ZH (1981) Effect of anisodine on synthesis and release of acetylcholine by rat cerebral cortex in vitro. Acta Pharm Sin 16:545–547
48. Huang JH, Chen ZX, Chen XY (1983) Effects of anisodine alone and in combination with chlorpromazine on spontaneous acetylcholine release from sensory-motor cortex of cats. Acta Pharmacol Sin 4:1–4
49. Han YF, Chen XY (1983) Effects of anisodine and other cholinergic drugs on learning and memory in spatial discrimination by mice. Acta Pharmacol Sin 4:220–225
50. Han YF, Chen XY (1984) Effects of anisodine and other cholinergic drugs on the conditioned response of the hippocampal θ-rhythm in rabbits. Acta Pharmacol Sin 5:166–170

51. Institute of Materia Medica, Chinese Academy of Medical Sciences (1973) Pharmacologic effects of anisodamine. Natl Med J China 269–273
52. Department of Pharmacology, Chinese Academy of Medical Sciences (1975) Pharmacological effects of anisodamine. Natl Med J China, 133–138
53. Chen XY, Wang ZQ, Yan YZ (1982) Studies on reactivity of bronchial smooth muscle. I. Effects of Yiqi Huoxue (YH) and anisodamine on bronchial smooth muscle. Acta Acad Med Sin 4:57–59
54. Yu SR, Ma JW (1982) Expectorant effect of anisodamine (654). Acta Acad Med Sin 4:258–260
55. Yu XY, Luo ZY, You JL, Luo H (1985) Effect of anisodamine (compound 654-2) on the cells from bronchoalveolar lavage in experimental respiratory distress syndrome. Bull Hunan Med Coll 10:13–18
56. Chen XY, Yan YZ, Xi PX, Zhang H (1985) Protective effect of anisodamine (654-2) on oleic acid induced lung injury. Acta Acad Med Sin 7:97–101
57. Wu LS, Yang GD, Jiang ML, Dong C (1984) Effects of henbane drugs on the phagocytosis of colloidal gold-198 by liver and spleen in mice. Acta Pharmacol Sin 5:140–142
58. Institute of Materia Medica, Chinese Academy of Medical Sciences (1973) Absorption, distribution and excretion of anisodamine. Natl Med J China 274–278
59. Yue TL, Wang GF, Song ZY (1979) Metabolism of ^{3}H-scopolamine and ^{3}H-anisodamine. Acta Pharm Sin 14:208–217
60. Xie JX, Yang JH, Zhang CZ (1981) Synthesis of quaternary ammonium salts and *N*-oxides of anisodamine and anisodine. Acta Pharm Sin 16:762–766

Ardisia japonica (Thunb.) Bl. 19

19.1 Introduction

Ardisia japonica (Thunb.) Bl. (Myrsinaceae) is a medicinal herb used in Chinese folk medicine. The whole plant is used as a bacteriostatic, tuberculostatic, hemostatic, and antiasthmatic drug in the treatment of tuberculosis, chronic asthma, and other diseases of the respiratory tract.

19.2 Chemical Constituents

Two new resorcinol derivatives, ardisinols I (*19-1*) and II (*19-2*), were isolated from a tuberculostatic fraction obtained from the water insoluble portion of an alcohol extract of *A. japonica* [1]. The structures of the ardisinols were determined as tridecenylresorcinol derivatives [2–4].

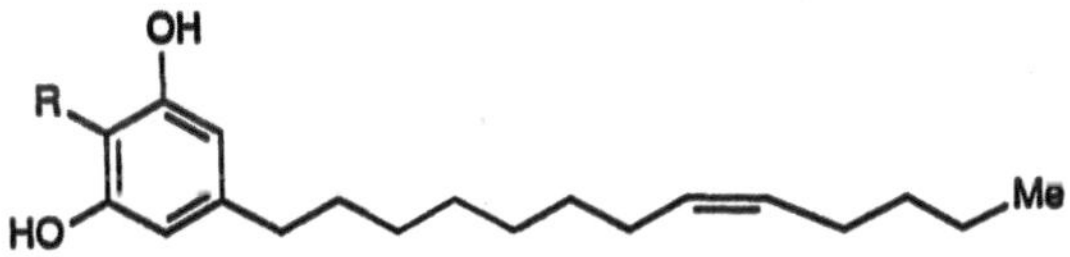

Ardisinol I (*19-1*): R=CH_3
Ardisinol II (*19-2*): R=H

Besides ardisinols, methylcardol (*19-3*), embelin (*19-4*) [1], 5-methoxy-3-(*cis*-pentadec-10-enyl)-1,4-benzoquinone (*19-5*) [5], bergenin (*19-6*), ilexol (*19-7*), and quercetin were also isolated from *A. japonica* [1]. Methylcardol is a homologue of ardisinol I with a C_{15} side chain, embelin is a 1,4-benzoquinone derivative with a saturated side chain.

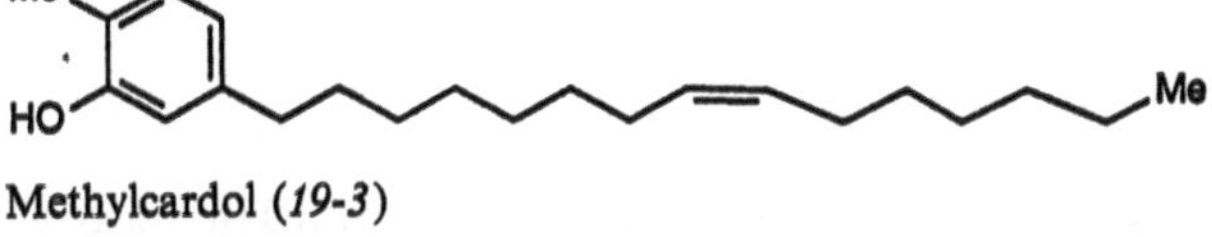

Methylcardol (*19-3*)

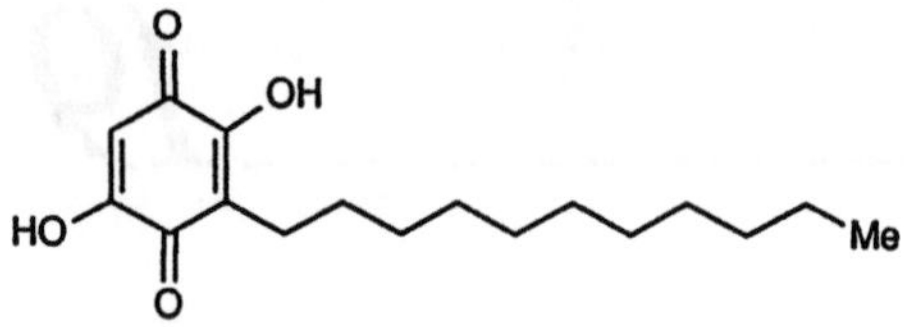

Embelin (*19-4*)

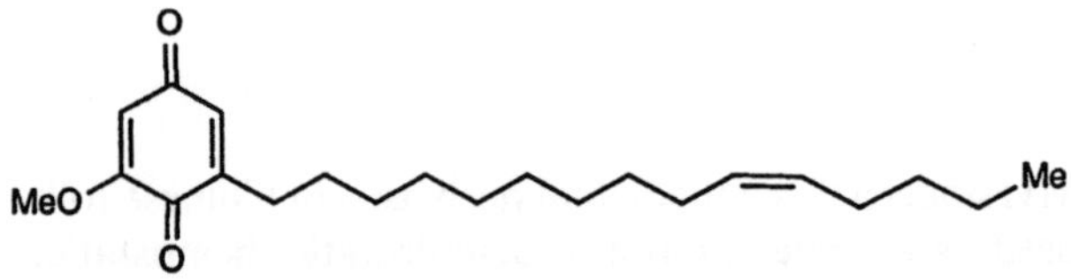

5-Methoxy-3-(*cis*-10-pentadecenyl)-
1,4-benzoquinone (*19-5*)

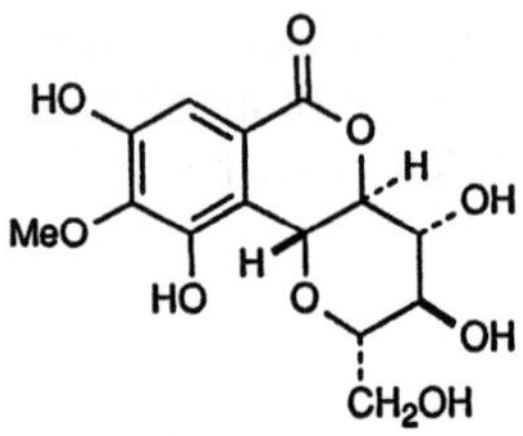

Bergenin (*19-6*)

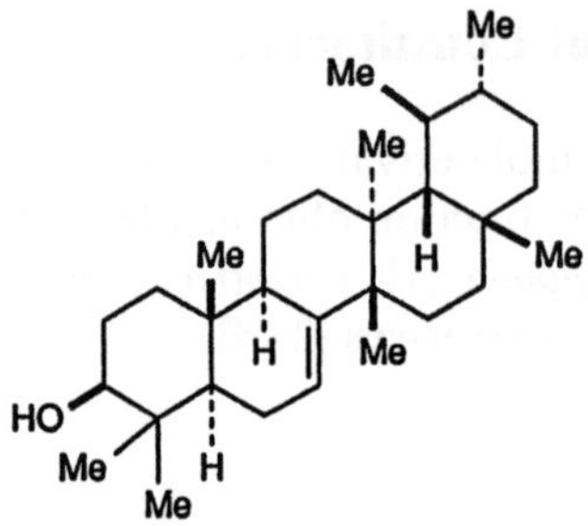

Ilexol (*19-7*)

19.3 Pharmacology

Ardisinol I and ardisinol II were both found to inhibit the growth of *Mycobacterium tuberculosis* in vitro. In clinical tests, the compounds were effective in 164 of 201 patients [5]. Bergenin is the active antitussive agent in *A. japonica* [1].

The 1,4-benzoquinones such as 5-methoxy-3-(*cis*-pentadec-10-enyl)-1,4-benzoquinone are inhibitors of arachidonate 5-lipoxygenase; this may account for their reported effectiveness in the treatment of asthma and inflammation [5].

19.4 Other *Ardisia* Species

19.4.1 *Ardisia hortorum*

A. hortorum is another *Ardisia* species used as an antipyretic and antiinflammatory agent against laryngitis. From the leaves and stems bergenin was isolated [6, 7].

19.4.2 *Ardisia alyxiaefolia*

A triterpene saponin which inhibited IgG production in rabbits was isolated from *A. alyxiaefolia.* After hydrolysis two sapogenins were obtained and identified as cyclamiretin A (*19-8*) and D (*19-9*) on the basis of their melting points, spectroscopic data, and the properties of their derivatives. Cyclamiretin A was shown to be the protosapogenin of the triterpene saponin [8].

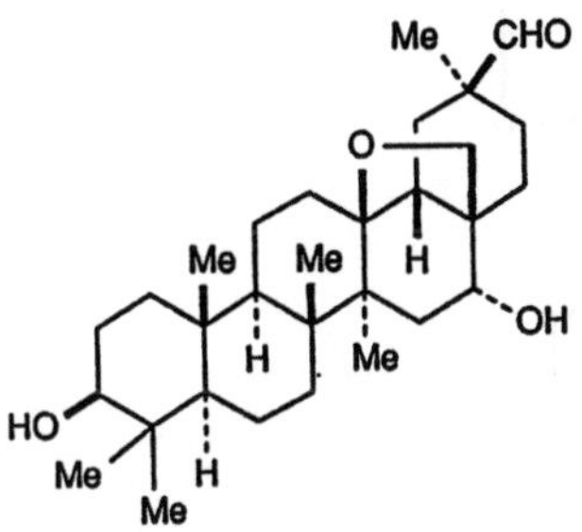

Cyclamiretin A (*19-8*)

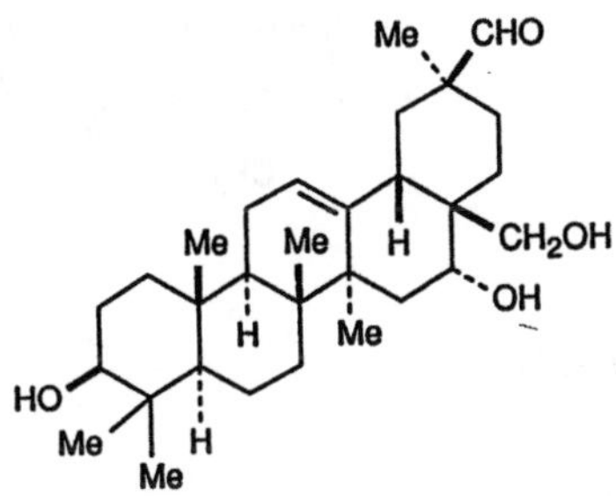

Cyclamiretin D (*19-9*)

19.4.3 *Ardisia faberi*

A. faberi is a folk medicine used for treatment of inflammation and coughing. A saponin with a new sapogenin named ardisiogenin (*19-10*) has been isolated. Ardisiogenin was structurally determined as 3β,22β-dihydroxy-olean-13,28-epoxy-30-al [9].

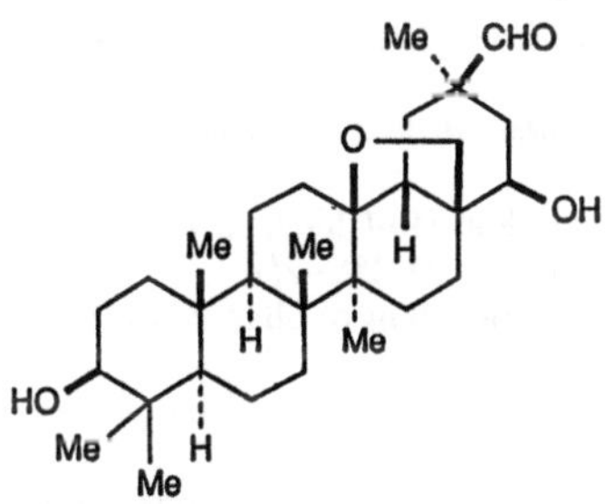

Ardisiogenin (*19-10*)

19.4.4 *Ardisia crenata*

From the root of *A. crenata* cyclamiretin A and its 3-*O*-α-L-arabinopyranoside [10], bergenin, friedelin, β-sitosterol, and rapanone [11] have been isolated and identified. Rapanone is the tridecyl homologue of embelin. In addition, the isolation of a cyclic depsipeptide FR 900359 (*19-11*) has also been reported [12]. Of particular significance of this cyclic depsipeptide are the novel amino acids *N*-methyldehydroalanine and *N,O*-dimethylthreonine [13]. This depsipeptide has been found to be active in inhibition of platelet aggregation. A dose-related hypotension in anesthetized rats caused by this compound was also observed [13]. Moreover, it showed cytotoxic activity in cultured rat fibroblasts and myelocytic leukemia cells [13].

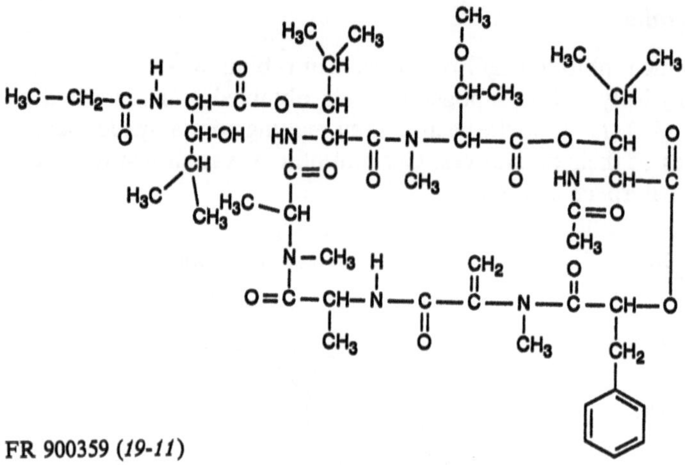

FR 900359 (*19-11*)

References

1. Huang BH, Chen WS, Hu Y, Qin CB, Zhang WJ, Liang BL (1981) Study on the chemical constituents of *Ardisia japonica* (Hornsted) Blume, an antitubercular Chinese herb. Acta pharm Sin 16:27–30
2. Liang BL, Yang ZX (1979) Structural studies in ardisin of *Ardisia japonica*. Kexue Tongbao 24:910–912
3. Hu Y, Chen WS, Huang BH, Liu LZ, Xu RS (1979) Structures of two new antitubercular compounds from *Ardisia japonica* Blume. Kexue Tongbao 24:207–209
4. Hu Y, Chen WS, Huang BH, Liu LZ, Xu RS (1981) Structure of antitubercular compounds from *Ardisia japonica* (Thunb.) Blume. Acta Chim Sin 39:153–158
5. Otsuka Pharmaceutical (1985) Extraction of 1,4-benzoquinones from plants. Jpn Kokai Tokkyo Koho JP 60 75, 442 (85 75,442) (CA 103:166145w)
6. Chu JH (1947) Constituents of the Chinese drug, Kai-ho-chien, *Ardisia hortorum*. Sci Rec 2:77–79
7. Hung SH, Chu JH (1957) The constituents of the Chinese drug, Kai-Ho-Chien, *Ardisia hortorum*. II. Identification of ardisic acid B as bergenin. Acta Chim Sin 23:255–258
8. Li GY, Li HY, Xu WK, Meng LS (1981) Chemical study of the sapogenins of Shao Nian Hong (*Ardisia alyxiaefolia* Tsiang). Acta Pharm Sin 16:554–556
9. Jiang ZH, Jia XS, Xu JW (1988) Chemical studies of ardisioside from faber ardisia (*Ardisia faberi*). Chin Trad Herb Drugs 19:530–532
10. Guan XT, Wang MT, Gong YM, Zhao TZ, Hong SH (1987) Sapogenin and secondary glycosides of coral ardisia (*Ardisia crenata*). Chin Trad Herb Drugs 18:338–341
11. Ni MY, Han L (1988) Studies on the chemical constituents of *Ardisia crenata* Sims. Bull Chin Mat Med 13:737–738
12. Miyamae A, Fujioka M, Koda S, Morimoto Y (1986) Molecular structure and absolute configuration of FR 900359, a novel cyclic depsipeptide from *Ardisia crenata* Sims (Myrsinaceae). Pept Chem 24:135–138
13. Fujioka M, Koda S, Morimoto Y, Biemann K (1988) Structure of FR 900359, a cyclic depsipeptide from *Ardisia crenata* Sims. J Org Chem 53:2820–2825

Areca catechu L. 20

20.1 Introduction

Binglang, Semen Arecae, is the dry mature seed of *Areca catechu* L. (Palmaceae) collected between late spring and early fall. It is recommended in traditional Chinese medicine as an anthelminthic, antimalarial, antidysenteric, and digestant for treatment of ascaris, tapeworms, fasciolopsiasis, malaria, dysentery, and dyspepsia.

Dafupi, Pericarpium Arecae, is the dry pericarp of the mature or immature fruits and is mainly used as a diuretic for treatment of edema.

20.2 Chemical Constituents

The areca nut is well known to contain a number of closely related alkaloids with a pyridine ring. The major alkaloid arecoline (*20-1*) was first reported about 100 years ago [1–3]. Later, arecaidine (*20-2*) [2–4], guvacine (*20-3*) [4–6], guvacoline (*20-4*) [7], arecolidine (*20-5*) [8], and isoguvacine (*20-6*) [6, 9] were isolated and structurally determined.

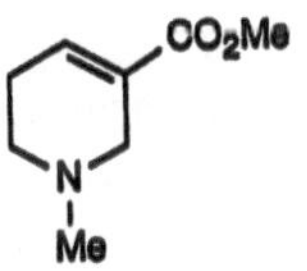

Arecoline (*20-1*)

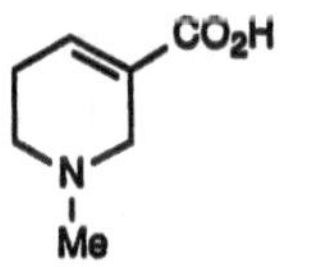

Arecaine (Arecaidine) (*20-2*)

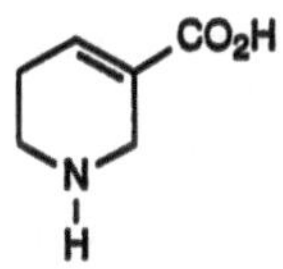

Guvacine (*20-3*)

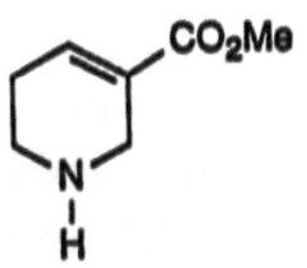

Guvacoline (*20-4*)

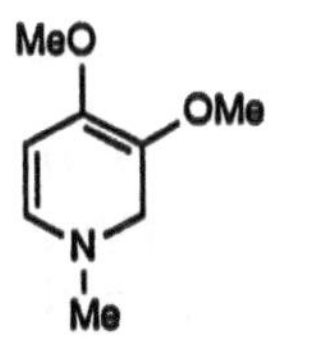

Arecolidine (*20-5*)

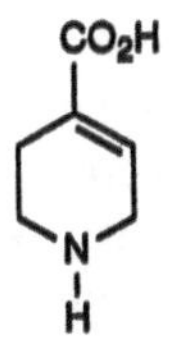

Isoguvacine (*20-6*)

The total alkaloid content was 0.15%–0.67%, in which 0.07%–0.50% is arecoline. The other five alkaloids are minor components. In addition to alkaloids, tannins, sitosterol, carbohydrate, saponins, and carotene were detected in areca nut [10].

Arecaine and its methyl ester arecoline could be synthesized from aliphatic compounds either by cyclization or directly from pyridine derivatives [11–19].

20.3 Pharmacology

Alkaloids, especially the most abundant one arecoline, are mainly responsible for the biological significance of the areca nut.

Arecoline hydrobromide was effective in treating tapeworm in geese [20] and dogs [21]. It exerts its anthelminthic action by causing the muscles of cestodes to relax and by causing the host to purge so that detached worms are removed [22]. Oral administration of 40 mg arecoline hydrobromide per dog daily for 5 days completely controlled tapeworm [23]; however, it had a low efficiency against ascaris in dogs [21]. Synthetic arecoline was recommended as an equivalent substitute for the natural compound [24]. In ducks, arecoline hydrobromide had a lethal dose of 7–8 mg/kg; the therapeutic dose was 0.5–1.0 mg/kg [25]. Arecoline is an acaricidal but not an insecticidal agent [26].

Most investigations on the biological activities of arecoline have focused on its neuropharmacologic properties. The central effects of arecoline were mydriasis, induction of behavioral abnormalities, impairment of conditioned reactions, and pseudoanalgesic properties. In mice, arecoline decreased motility, exploratory activity, and enhancement of motility induced by caffeine and amphetamine. These effects were suppressed by atropine but not by methylatropine [27].

Arecoline was similar to pilocarpin in increasing the acetylcholine and hydroxyindoleacetic acid levels in rat brain; however, the effect caused by arecoline could be blocked by atropine [28].

Administration of arecoline at a dose of 10 mg/kg, i.p. into rats age 10–30 days or more than 300 days caused tremors and an increase in cerebral acetylcholine content which was restricted to the telencephalon. In 5-day-old animals the same dose of arecoline caused neither tremors nor changes in acetylcholine content [29]. The intensity of the tremor was dose related, reached its maximum at 2–5 min after arecoline administration, and disappeared within 30 min. At a dose of 0.05 mg/kg, arecoline selectively increased local cerebral glucose utilization in the hippocampus and medial raphe, whereas higher doses produced more generalized metabolic enhancement. The selective increase in local cerebral glucose utilization by a low dose of arecoline in the hippocampus presumably is due to a specific action of arecoline [30].

Arecoline also stimulated the superior cervical ganglion of cats following intraarterial injection into the ganglion [31]. Since the effects of arecoline on the central nervous system are inhibited by atropine, they are presumably mediated by muscarinic receptors [32, 33]. Arecoline increased cGMP levels of rat corpus striatum slices in vitro with an EC_{50} value of 22 μM. The cGMP response corresponded directly to muscarinic receptor occupancy [34]. Inhibition of the Na^+/K^+ ATPase activity in rat brain cerebral cortex, striatum, thalamus, hippocampus, and medulla was time dependent following i.p. administration of arecoline at a dose of 5 mg/kg. The greatest decrease of Na^+/K^+ ATPase activity was detected in those brain structures rich in cholinergic innervation such as the striatum and hippocampus [35].

The cardiovascular effects of arecoline were studied in anesthetized dogs. After i.v. bolus administration of arecoline at dose of 0.01–100 μg/kg, mean arterial blood pressure was immediately reduced within 1 min after each dose but returned to the baseline value by 5 min after doses of 0.01–10 μg/kg, and by 20 min after a 30 μg/kg dose. Blood pressure was still depressed 20 min after a dose of 100 μg/kg. Cardiac output was significantly reduced only after the 30 or 100 µg/kg doses. In contrast to the bolus injection, i.v. infusion of arecoline at 0.3 or 1.0 µg/kg/min did not cause any changes in blood pressure, heart rate, or cardiac output. An infusion of 3.0 or 10 µg/kg/min produced dose dependent decreases in mean arterial blood pressure and heart rate [36].

Intraperitoneal injection of an aqueous extract of areca nut increased glutathione content in liver, kidney, and muscle of mice, whereas arecoline decreased it in almost all tissues [37]. Both the aqueous extract of areca nut and arecoline, given i.p. to mice, increased hepatic DNA and RNA content and stimulated hepatic DNA and RNA synthesis [38, 39].

The urine of rats that had received arecoline or arecaine was found to contain *N*-acetyl-*S*-(3-carboxyl-2-methyl-piperid-4-yl)-cysteine (*20-7*). Furthermore, a glucuronic acid conjugate was also found along with the *N*-oxide metabolite of arecoline [40, 41].

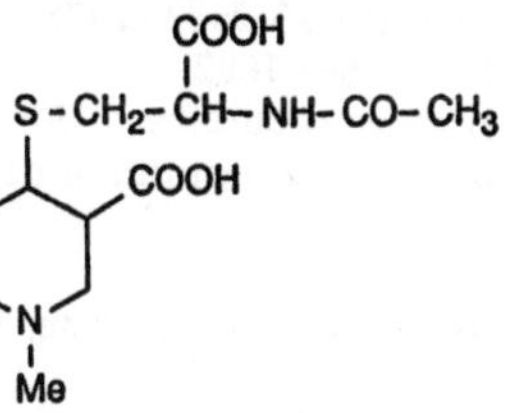

N-acetyl-*S*-(3-carboxyl-1-methyl-piperid-4-yl)-cysteine (*20-7*)

Arecaine and the related alkaloid guvacine inhibited uptake of GABA in brain slices when introduced directly into the brain [42–44]. Arecaine produced behavioral changes in mice, but only at a dose as large as 10–50 mg/kg, probably because arecaine and guvacine do not easily cross the blood-brain barrier [43] due to their carboxyl groups.

Only a few studies have been reported on the effects of arecoline in humans. Results suggested that small doses affect the human central nervous system in a manner similar to animals [45].

Areca nut is widely used as a chew in India and other countries. It is often consumed in fresh or processed form together with the betel leaf of *Piper betle* (Piperaceae) and a little lime, and with or without tobacco. Epidemiological studies have revealed a possible correlation between betel quid chewing and oral cancer [46]. Khanolkar [47] reviewed the distribution of oral cancer and chewing habits and concluded that habitual chewing of areca nut and betel leaf had no etiological role in the development of mouth cancer but that the inclusion of tobacco in the quid made it carcinogenic. In Malaysia, oral cancer is common in the Indian community but is rare among Malays. Indians and Malays both chew betel quid the only difference being that Indians add tobacco to their quid, whereas Malays usually do not [48].

In general, the International Agency for Research on Cancer concluded: There is sufficient evidence that the habit of chewing betel quid containing tobacco is carcinogenic to humans, but the evidence for carcinogenicity of chewing betel quid without tobacco to humans is inadequate; evidence for the carcinogenicity to experimental animals of either areca nut with and without tobacco or the aqueous extract is limited [49].

Recently, some new polyphenolic compounds designated as NPF-86IA, NPF-86IB, NPF-86IIA, and NPF 86IIB with inhibitory activity on 5′-nucleotidase from snake venom and rat liver membrane were isolated from the *Areca* nut. The average molecular masses of these four compounds have been estimated at 5620, 5000, 29400, and 8610, respectively [50]. In addition, these compounds have been found to display significant therapeutic activity by i.p. administration against i.p. implanted Ehrlich ascites carcinoma, but were not effective against L 1210 cells [51].

References

1. Jahns E (1888) Über die Alkaloide der Arecanuß. Chem Ber 21:3404–3409
2. Jahns E (1890) Über die Alkaloide der Arecanuß. 2. Mitteilung. Chem Ber 23:2972–2978
3. Freudenberg K (1981) Über die Alkaloide der Betelnuß. Chem Ber 51:1668–1682
4. Hess K, Leibbrandt F (1918) Synthese von *N*-Methyl-tetrahydropyridin-carbonsäure. I. Eine neue Bildungsweise des Arecaidins und des Arecolins. Zur Aufklärung der Konstitution des Guvacins und des Arecains. Chem Ber 51:806–820
5. Freudenberg K (1918) Über das Guvacin. Chem Ber 51:976–982
6. Winterstein E, Weinhagen A (1918) Nicotinic acid derivatives. II. Guvacine and isoguvacine. Z Physiol Chem 104:48–53
7. Hess K (1918) Über den Guvacin-methylester (Guvacolin) und sein natürliches Vorkommen. Chem Ber 51:1004–1006
8. Emde H (1915) Arecolidine. Apoth Ztg 30:240–241
9. Jahns E (1891) Über die Alkaloide der Arecanuß. 3. Mitteilung. Chem Ber 24:2615–2617
10. Raghavan V, Baruah HK (1958) Arecanut: India's popular masticatory = history, chemistry and utilization. Econ Bot 12:315–325
11. Arjungi KN (1976) Areca nut. Arzneimittelforschung 26:951–956
12. Wohl A, Johnson A (1907) Über Arecaidin und Arecolin. Chem Ber 40:4712–4719
13. Wohl A, Grosse E (1907) Über eine tertiäre Triacetalbase und den freien Arecaidin-aldehyd. Chem Ber 40:4719–4722
14. Dankova TF, Sidorova EA, Preobrazhenskii NA (1941) Synthesis of the alkaloid arecoline, the active constituent of betel nuts. J Gen Chem USSR 11:934–938 (CA 37:381)
15. Dobrowsky A (1952) Über eine einfache Arecolinsynthese. Monatsh 83:443–447
16. Knox LH (1950) Arecoline, US 2,506,458 (CA 45:671 c)
17. Preobrazhenskii NA, Malkov KM, Maurit ME, Vorobev MA, Vlasov AS (1957) Synthesis of alkaloid arecoline and its homologs. Zh Obshch Khim 27:3162–3170
18. Kinoshita N, Hamana M, Kawasaki T (1963) Reduction of substituted pyridinium salts with $NaBH_4$. I. Reduction of 1-methyl-3- and 4-methoxycarbonylpyridinium iodide with sodium borohydride. Yakugaku Zasshi 83:115–120
19. Kinoshita N, Hamana M, Kawasaki T (1962) Reduction of 1-methyl-3- and -4-methoxycarbonylpyridinium iodide with sodium borohydride. Chem Pharm Bull 10:753–755
20. Kisielweska K (1957) Treatment of helminthiasis in geese with arecoline hydrobromide. Wiad Parazytol 3:477 (CA 53:19131 f)
21. Ross IC (1924) The possible use of arecoline hydrobromide as an anthelmintic. J Comp Pathol Ther 37:246–259
22. Batham EJ (1946) Testing arecoline hydrobromide as an anthelminthic for hydatid worms in dogs. Parasitology 37:185–191
23. Stepanyan SG, Agaronyan AM, Chubaryan FA, Zakharyan VA, Pkhrikyan LV, Vardanyan AV (1977) New method for the dehelminthization of dogs with small doses of arecoline. Izv S Kh Nauk 20:85–90 (CA 87:127374 b)
24. Feoktistov PI (1953) Synthetic arecoline as anthelminthic in hymenolepiasis of geese. Veterinariya 30(8):31–32 (CA 48:7258 g)

25. Tsarev SG (1953) Antihelmintic action of synthetic arecoline. Veterinariya 30(8):30–31 (CA 48:7193e)
26. Bigg DCH, Purvis SR (1976) Muscarinic agonists provide a new class of acaricide. Nature 262:220–222
27. Nieschulz O (1967) Pharmacology of the active principle of betel. I. Central effects of arecoline. Arzneimittelforschung 17:1292–1297
28. Haubrich DR, Reid WD (1972) Effects of pilocarpine or arecoline administration on acetylcholine levels and serotonin turnover in rat brain. J Pharmacol Exp Ther 181:19–27
29. Westermann K, Fischer HD, Oelssner W (1970) Tremor and brain acetylcholine content of infantile and senile rats following arecoline. Acta Biol Med Ger 25:855–861
30. Soncrant T, Holloway HW, Rapoport SI (1985) Arecoline-induced elevations of regional cerebral metabolism in the conscious rat. Brain Res 347:205–216
31. Tripathi ON (1983) Arecoline induced nicotinic and muscarinic stimulation of the superior cervical ganglion of cat. Biomed Biochim Acta 42:275–282
32. Meltzer LT, Rosecrans JA (1981) Discriminative stimulus properties of arecoline: a new approach for studying central muscarinic receptors. Psychopharmacology (Berlin) 75:383–387
33. McKinney M, Richelson E (1984) The coupling of the neuronal muscarinic receptors to responses. Annu Rev Pharmacol Toxicol 24:121–126
34. Hanley MR, Iversen LL (1977) Muscarinic acetylcholine receptors and cyclic GMP in rat brain. Br J Pharmacol 59:503P–504P
35. Stojanovic T, Mrsulja BB (1985) Brain sodium-potassium activated adenosine triphosphatase: effect of arecoline in vivo. Jugosl Physiol Pharmacol Acta 21:21–25 (CA 103:206225f)
36. Beil ME, Goodman FR, Shlevin HH, Smith EF, III (1986) Evaluation of the cardiovascular effects of arecoline in the anesthetized dog. Drug Dev Res 9:203–212
37. Shivapurkar NM, Bhide SV (1978) Effect of betel nut constituents on sulfhydryl metabolism. Indian J Pharmacol 10:257–264
38. Shivapurkar NM, Bhide SV (1979) Effect of betel nut constituents on nucleic acid metabolism. Indian J Exp Biol 17:1141–1144
39. Shivapurkar NM, Bhide SV, Ranadive KJ (1978) Biochemical studies on betel nut constituents. Indian J Pharmacol 10:191–200
40. Boyland E, Nery R (1969) Metabolism of arecoline and arecaidine. Biochem J 112:33P
41. Boyland E, Nery R (1969) Mercapturic acid formation during the metabolism of arecoline and arecaidine in the rat. Biochem J 113:123–130
42. Johnston GAR, Krogsgaard-Larsen P, Stephanson A (1975) Betel nut constituents as inhibitors of γ-aminobutyric acid uptake. Nature 258:627–628
43. Lodge D, Johnston GAR, Curtis DR, Brand SJ (1977) Effects of the areca nut constituents arecaidine and guvacine on the action of GABA in the cat central nervous system. Brain Res 136:513–522
44. Krogsgaard-Larsen P (1980) Inhibitors of the GABA uptake systems. Mol Cell Biochem 31:105–121
45. Christie JE, Shering A, Ferguson J, Glen AIM (1981) Physostigmine and arecoline: effects of intravenous infusions in Alzheimer presenile dementia. Br J Psychiatry 138:46–50
46. Suraiya JN (1973) Medicine in ancient India with special reference to cancer. Indian J Cancer 10:391–402
47. Khanolkar VR (1944) Oral cancer in Bombay, India. A review of 1,000 consecutive cases. Cancer Res 4:313–319
48. Ahluwalia HS, Duguid JB (1966) Malignant tumors in Malaya. Br J Cancer 20:12–15
49. International Agency for Research on Cancer (1985) Betel-quid and areca-nut chewing. IARC Monogr Eval Carcinog Risk Chem Hum 37:141–200
50. Uchino K, Matsuo T, Iwamoto M, Tonosaki Y, Fukuchi A (1988) New 5′-nucleotidase inhibitors, NPF-86IA, NPF-86IB, NPF-86IIA, and NPF-86IIB from *Areca catechu*. Part 1. Isolation and biological properties. Planta Med 54:419–422
51. Iwamoto M, Matsuo T, Uchino K, Tonosaki Y, Fukuchi A (1988) New 5′-nucleotidase inhibitors NPF-86IA, NPF-86IB, NPF-86IIA, and NPF-86IIB from *Areca catechu*. Part 2. Antitumor effects. Planta Med 54:422–425

Aristolochia spp. 21

21.1 Introduction

Some *Aristolochia* species have found wide use in traditional Chinese medicine and folk medicine. The Chinese Pharmacopoeia lists the following items:

- Madouling, Fructus Aristolochiae, is the dry ripe fruit of *Aristolochia contorta* Bge. or *A. debilis* Sieb. et Zucc. (Aristolochiaceae) collected in fall when the fruits have become yellow. It is used in the treatment of respiratory diseases as an antitussive and antiasthmatic.
- Tianxianteng, Herba Aristolochiae, is the dry aerial part of *A. contorta* or *A. debilis* harvested in fall and used as a diuretic against edema and as an antirheumatic.
- Guangfangji, Radix Aristolochiae fangchi, is the dry root of *A. fangchi* Y. C. Wu ex L. D. Chou et S. M. Hwang collected in fall and winter. It is used as an antirheumatic and diuretic.
- Guanmutong, Caulis Aristolochiae manshuriensis, is the dry vine of *A. manshuriensis* Kom. harvested in fall and winter. It is used as a diuretic and antiphlogistic for treatment of edema and rheumatic pain.

Besides the four official species, a number of other *Aristolochia* species used in traditional Chinese medicine and folk medicine have been investigated chemically and pharmacologically.

21.2 Chemical Constituents

The chemical constituents of *Aristolochia* species [1] can be divided into three chemical groups: aristolochic acid derivatives, alkaloids, and sesquiterpenes. The aristolochic acids are derived from the phenanthrene system and bear a carboxyl function and a nitro substituent. When the nitro group is replaced by an amino group, the carboxyl group forms a lactam ring, giving a number of aristololactams [2]. The structures of aristolochic acid I (aristolochic acid A, *21-1*) and aristololactam (aristololactam I, *21-2*) are given below. One of the alkaloids isolated from some *Aristolochia* species is magnoflorine (*21-3*), which is structurally and phylogenetically related to aristolochic acid derivatives. The sesquiterpenes are mainly constituents of the volatile oil.

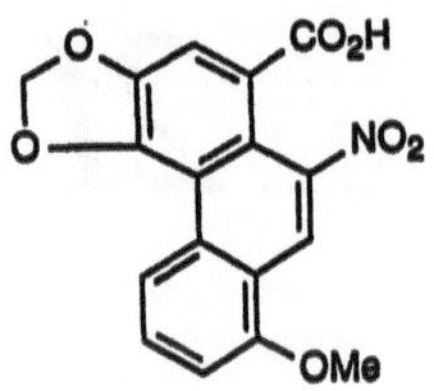

Aristolochic acid I
(*21-1*)

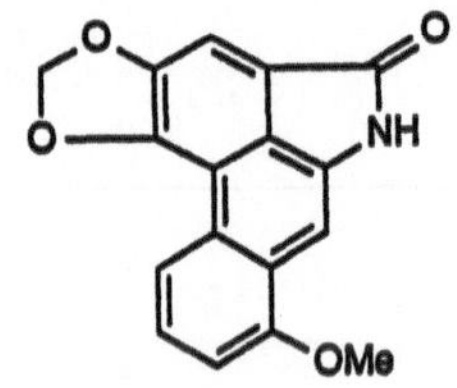

Aristololactam
(*21-2*)

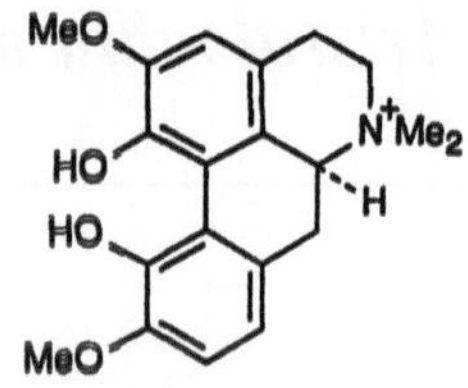

Magnoflorine
(*21-3*)

21.2.1 Chemical Constituents of *Aristolochia debilis*

The ripe fruit and aerial part of *A. debilis* are officially recognized by the Chinese Pharmacopoeia as mentioned above, but the roots are also widely used in traditional Chinese medicine as an antihypertensive agent.

21.2.1.1 Aristolochic Acid Derivatives

Tseng and Ku isolated aristolochic acid A [3] and debilic acid (*21-4*) [4, 5] from the roots of *A. debilis*. The yield of the main constituent aristolochic acid A, obtained from various sources, was 0.1%–0.6% as determined by a polarographic method [6].

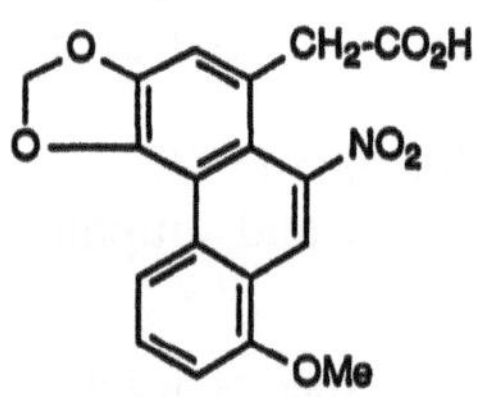

Debilic acid (*21-4*)

The isolation of 9-hydroxy-(*21-5*) and 9-methoxyaristolochic acid I (*21-6*) and aristolochic acid C (aristolochic acid IIIa, *21-7*) was also reported [7].

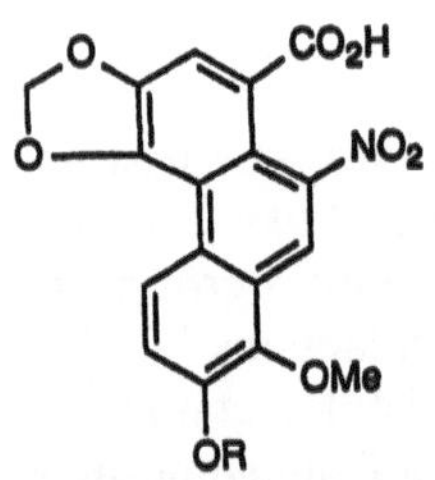

9-Hydroxyaristolochic acid I (*21-5*): R=H
9-Methoxyaristolochic acid I (*21-6*): R=CH_3

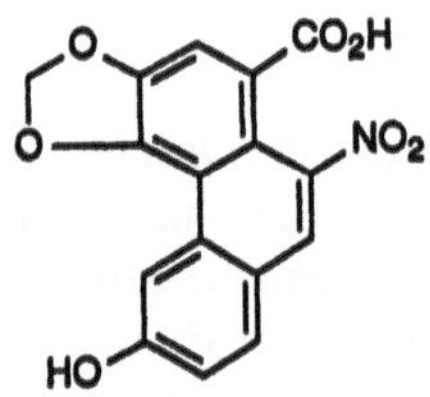

Aristolochic acid C (*21-7*)

Moreover, aristolochic acid II (*21-8*), aristolochic acid methyl ester, aristolochic acid IV (*21-9*), and aristolochic acid IVa (aristolochic acid D, *21-10*) were also isolated from the roots of *A. debilis* [8].

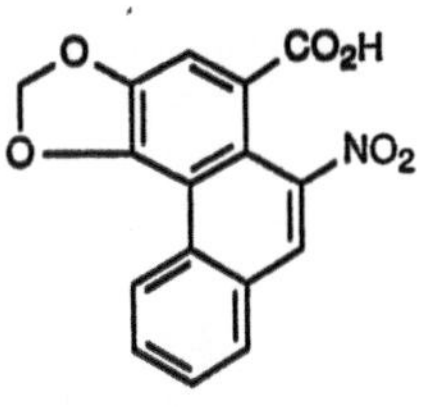

Aristolochic acid II
(*21-8*)

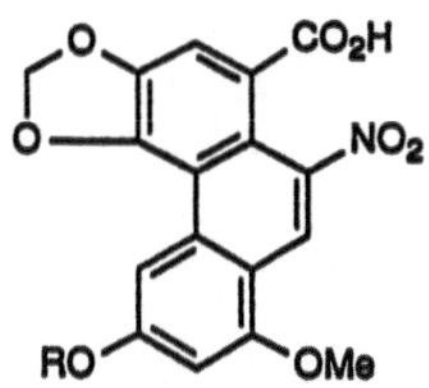

Aristolochic acid IV (*21-9*): R = CH_3
Aristolochic acid IVa (*21-10*): R = H

21.2.1.2 Alkaloidal Constituents

Magnoflorine (*21-3*), a quaternary ammonium base with hypotensive activity [9, 10], and cyclanoline (*21-11*) [11] were isolated from the roots of *A. debilis*. Another alkaloid, tetrandrine [12], and allantoin were also found in the roots of *A. debilis* [4].

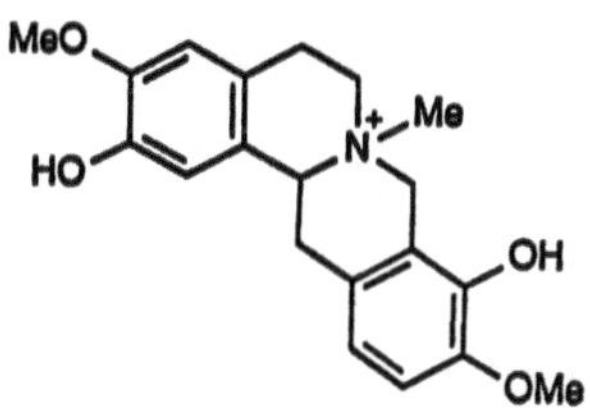

Cyclanoline (*21-11*)

21.2.1.3 Sesquiterpenes

Among the sesquiterpenes aristolone (*21-12*) was isolated from *A. debilis* and structurally elucidated [13, 14]. From 3 kg of dried *A. debilis* roots 12.5 g of aristolone was obtained.

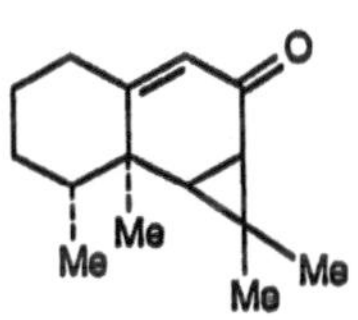

Aristolone (*21-12*)

The isolation of 9-aristolene (*21-13*), 1(10)-aristolene (*21-14*) and debilone (*21-15*) was also reported [15]. The fresh root of *A. debilis* contained a saturated sesquiterpene, 3-oxoishwarane (*21-16*), in addition to aristolone and its isomer 1(10)-aristolen-9-one (*21-17*) [16]. The latter was first isolated from the root of *Nardostachys chinensis* (Valerianaceae), used in folk medicine as a sedative [17]. 9-Aristolene, 1(10)-aristolene, and their related compounds all possess a 1H-cyclopropa[*a*]naphthalene skeleton.

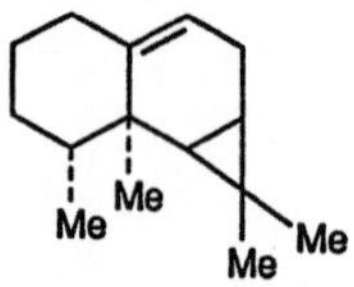

9-Aristolene (*21-13*)

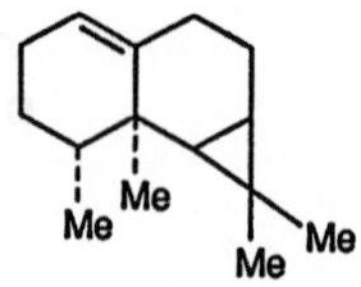

1(10)-Aristolene (*21-14*)

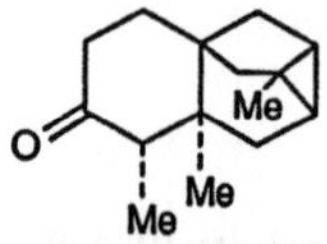

3-Oxoishwarane (*21-16*)

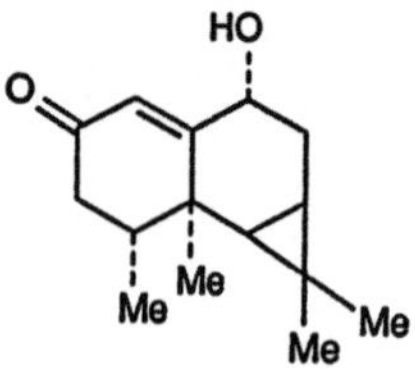

Debilone (*21-15*)

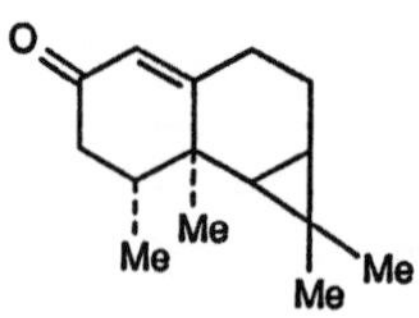

1(10)-Aristolenone (*21-17*)

21.2.2 Chemical constituents of *Aristolochia manshuriensis*

Aristolochic acid I methyl ester, aristolochic acid IV methyl ester, and aristolochic acid IVa were isolated from the stems of *A. manshuriensis* [18]. In addition a new glycoside, aristolochic acid D 6-*O*-*β*-D-glucopyranoside, was also isolated and named aristoloside or aristolochin (*21-18*) [18–20].

Magnoflorine was also detected in *A. manshuriensis* [20].

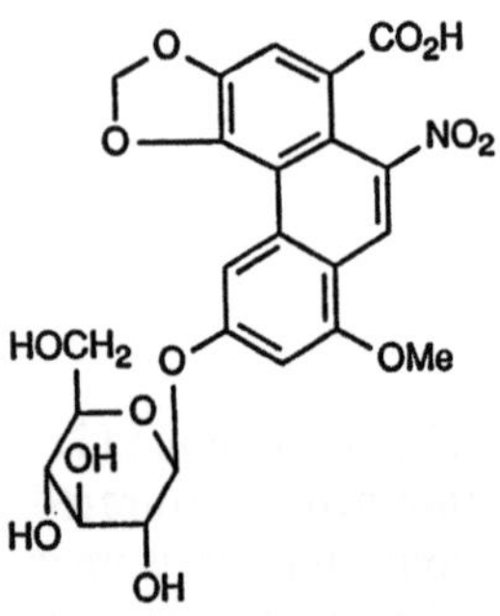

Aristoloside (Aristolochin) (*21-18*)

21.2.3 Chemical Constituents of *Aristolochia fangchi*

Thin layer chromatography (TLC) study of the constituents in the root of *A. fangchi* revealed that the ethanol extract contained aristolochic acid A, aristolochic acid D, aristolochic acid IV methyl ester, moupinamide, magnoflorine, *p*-coumaric acid, *N*-(*p*-hydroxyphenethyl)-*p*-coumaramide (*21-19*), syringic acid, palmitic acid, and β-sitosterol. The yields of aristolochic acids and magnoflorine in the root of *A. fangchi* were approximately 0.02% and 0.9%, respectively [21].

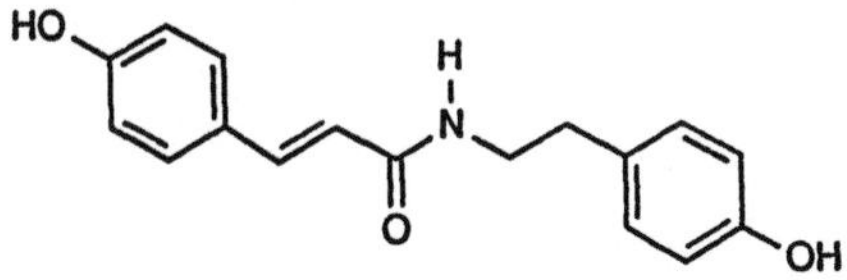

N-(*p*-Hydroxyhenethyl)-*p*-coumaramide (*21-19*)

21.2.4 Chemical Constituents of *Aristolochia contorta*

From the root of *A. contorta*, aristolochic acid A and a new compound aristolochic acid E (*21-20*), were isolated together with allantoin, daucosterol, magnoflorine, and β-sitosterol [22, 23].

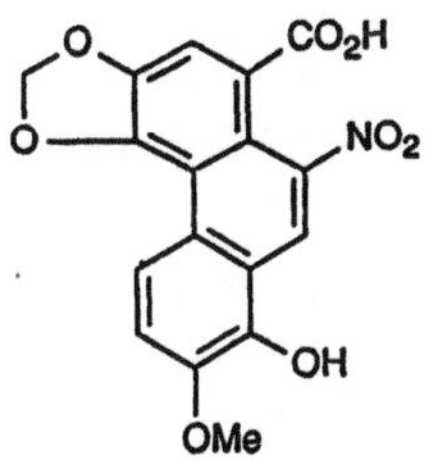

Aristolochic acid E (*21-20*)

21.2.5 Chemical Constituents of Unofficial *Aristolochia* Species Used in Chinese Medicine

21.2.5.1 *Aristolochia mollissima*

From the aerial part of *A. mollissima*, aristolochic acid A [24], aristololactam, aristolochic acid D [25, 26], and the sesquiterpene aristolactone (*21-21*) [27, 28] were isolated and identified. Recently, a new sesquiterpene lactone, mollislactone (*21-22*), was isolated and structurally elucidated [29].

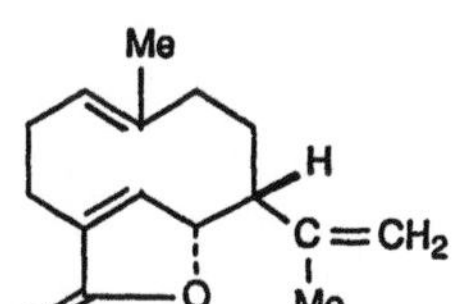

Aristolactone (*21-21*)

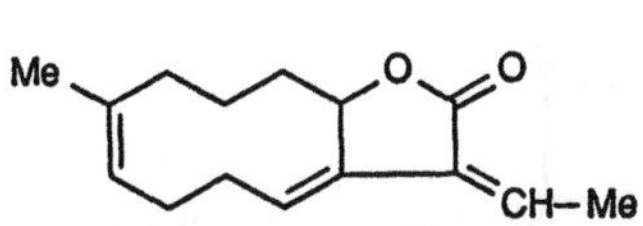

Mollislactone (*21-22*)

21.2.5.2 *Aristolochia kwangsiensis*

From the bulbs of *A. kwangsiensis*, aristolochic acid A was isolated with a yield of 0.06%, based on fresh weight [30, 31]. In addition, aristolochic acid IV methyl ester, aristolinic acid methyl ester (*21-23*), the denitro analogue of aristolochic acid IV methyl ester, allantoin, and β-sitosterol were found [32, 33].

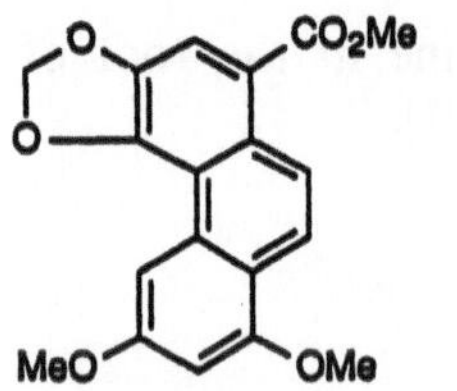

Aristolinic acid methyl ester (*21-23*)

21.2.5.3 *Aristolochia tuberosa*

From the roots of *A. tuberosa*, aristolochic acid A, aristolochic acid C, and a number of lactams and their glycosides were isolated [34]. The lactam constituents were all identified as analogues of aristololactam, such as 8-demethoxy-10-hydroxy- (*21-24*), 8-demethoxy-10-hydroxy-*N*-*β*-D-glucopyranosyl- (*21-25*), 8-demethoxy-10-acetoxy-*N*-*β*-D-glucopyranosyl- (*21-26*), and 10-methoxy-*N*-*β*-D-glucopyranosyl-aristololactam (*21-27*) [35].

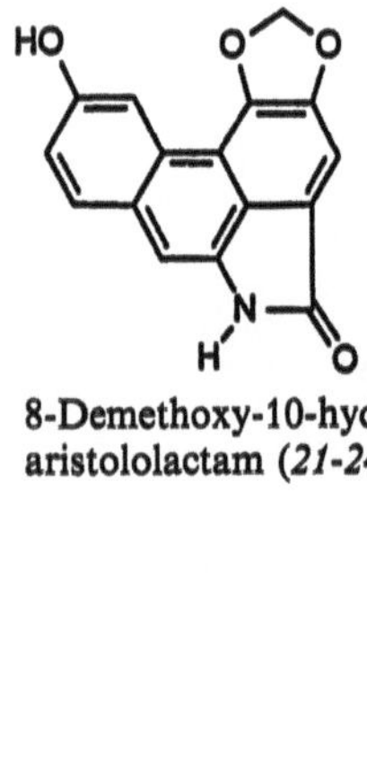

8-Demethoxy-10-hydroxy-aristololactam (*21-24*)

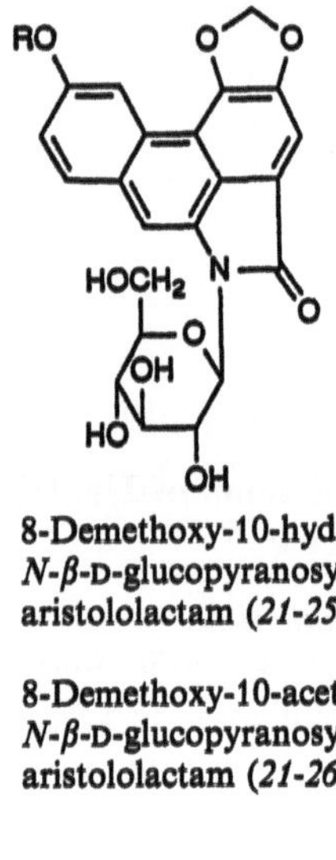

8-Demethoxy-10-hydroxy-*N*-*β*-D-glucopyranosyl-aristololactam (*21-25*): R=H

8-Demethoxy-10-acetoxy-*N*-*β*-D-glucopyranosyl-aristololactam (*21-26*): R=Ac

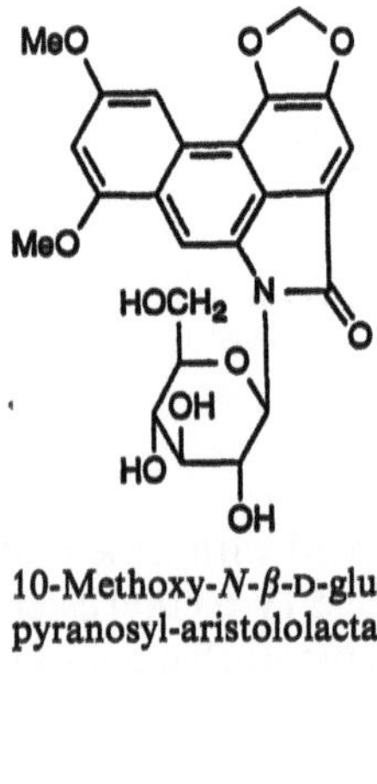

10-Methoxy-*N*-*β*-D-gluco-pyranosyl-aristololactam (*21-27*)

In addition to the aristolochic acid derivatives two 4,5-dioxoaporphine alkaloids, referred to as tuberosinone (*21-28*) and its *N*-*β*-D-glucoside, were isolated and their structures determined [36, 37].

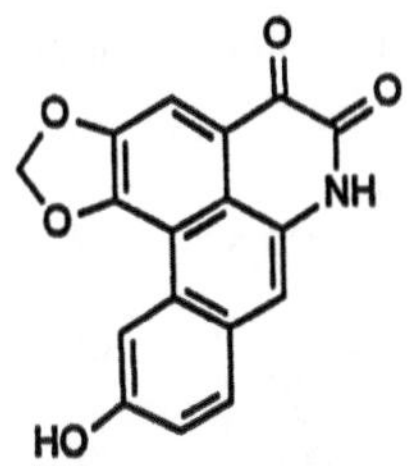

Tuberosinone (*21-28*)

21.2.5.4 ***Aristolochia tagala***

Six chemical constituents were isolated from the roots of *A. tagala*, four of them were identified as aristolochic acid A, aristolochic acid C, 9-hydroxyaristolochic acid I, and allantoin [38].

21.2.5.5 ***Aristolochia championii***

From the root of *A. championii* five chemical constituents were obtained and identified as aristolochic acid A, aristolinic acid methyl ester, aristolochic acid IV methyl ester, allantoin, and *β*-sitosterol [39, 40].

21.2.5.6 ***Aristolochia heterophylla***

From the roots of *A. heterophylla* aristolochic acid A, aristolochic acid D, magnoflorine, allantoin, and *β*-sitosterol were isolated and identified [41, 42].

21.2.5.7 ***Aristolochia moupinensis***

Tian et al. reported the isolation of aristolochic acid A, magnoflorine, allantoin, and *β*-sitosterol from the roots of *A. moupinensis* [42]. Xu and Sun reported the isolation of 13 constituents of the roots and stems. In addition to the four constituents reported by Tian, syringic acid, *p*-coumaric acid, aristolochic acid IV and its methyl ester, aristolochic acid D, palmitic acid, aristolochic acid II, *N*-(*p*-hydroxyphenethyl)-*p*-coumaramide, and moupinamide (*21-29*), have been isolated. The structure of moupinamide was confirmed from spectral data and chemical synthesis [21, 43]. Contents of aristolochic acids and magnoflorine in *A. moupinensis* were 0.05% and 1.1%, respectively, similar to those in *A. fangchi* [21].

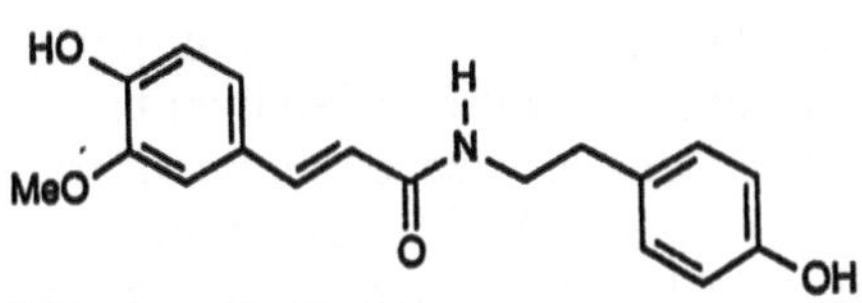

Moupinamide (*21-29*)

21.2.5.8 *Aristolochia versicolar*

Twelve constituents were obtained from the roots of *A. versicolar*. The identified compounds were: aristolochic acid A and its methyl ester, β-sitosterol and its D-glucoside, stigmastane-3,6-dione, stigmast-4-en-3,6-dione, and stearic acid [44]. Three new sesquiterpene lactones were also determined: versicolactone B (*21-30*), versicolactone C (*21-31*) [45], and yindailactone B (*21-32*) [46].

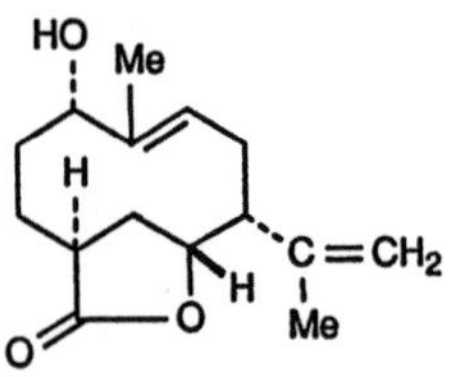

Versicolactone B (*21-30*)

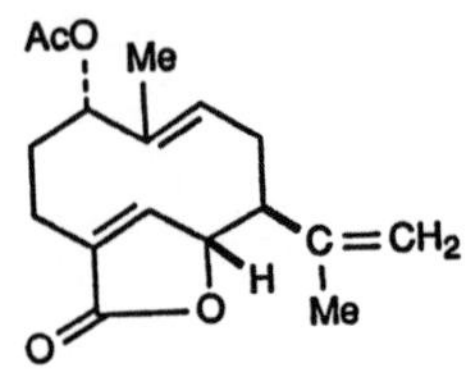

Versicolactone C (*21-31*)

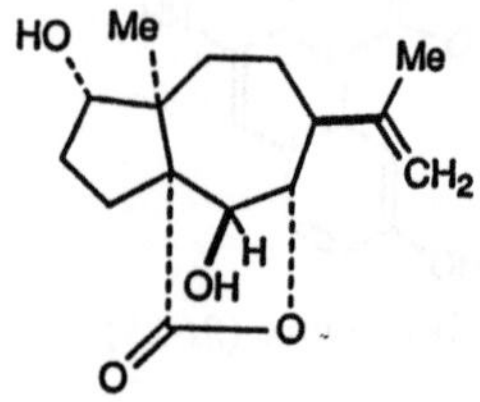

Yindailactone B (*21-32*)

21.2.5.9 *Aristolochia cinabaria*

Tuberosinone, its *N*-β-D-glucopyranoside, and aristolochic acid A were detected in wild and cultivated *A. cinabaria*. The yield of tuberosinone was 0.4% in both wild and cultivated *A. cinabaria*; yields of aristolochic acid A were 1.1% and 1.3% in wild and cultivated plants, respectively [47].

21.2.5.10 *Aristolochia austrozechuanica*

In the ligroin extract of *A. austrozechuanica* magnoflorine and β-sitosterol were detected by TLC [48].

21.3 Pharmacology

The biological activities of aristolochic acid A have been extensively studied. A number of gram-positive bacteria including *Staphylococcus, Streptococcus, Diplococcus, Bacillus, Sarcina*, and *Mycobacterium* are inhibited by aristolochic acid A at a concentration of 50–200 μg/ml. The concentration of aristolochic acid A needed to inhibit gram-negative bacteria and fungi were higher than 200 μg/ml [49].

Mice infected with *Staphylococcus aureus, Diplococcus pneumoniae*, or *Streptococcus pyogenes* were found to be protected from disease by i.p. administration of aristolochic acid A at a dose of 50 μg/kg. The phagocytic activity of peritoneal macrophages of treated mice was markedly stimulated [50]. Schunack et al. [51] could not prove an antibiotic activity of aristolochic acid A and aristolochic acid II against *Staphylococcus aureus* G-511 and *Escherichia coli* B. Pretreatment of pneumococcus infected mice with both aristolochic acids in a solution of sodium bicarbonate significantly increased the survival rate, but a similar effect to a lesser degree was also noted in control animals pretreated only with sodium bicarbonate solution. A correlation between the dose and effect of aristolochic acids could not be demonstrated.

Some results of studies on the immunostimulating and antitumor activities of aristolochic acid A contradicted each other. A remarkable prolongation of the survival time of mice bearing ascitic sarcoma-37 tumors treated with aristolochic acid A by i.p. administration at a daily dose of 1.25–5 mg/kg for 5 days was reported. Growth of mouse sarcoma-37 cells were found to be completely inhibited by incubation with aristolochic acid A. Treatment of mice with aristolochic acid A at a daily i.p. dose of 2.5–5 mg/kg for 3 days after s.c. implantation of sarcoma-37 cells resulted in 40%–50% inhibition of tumor growth [50].

Aristolochic acid A, administered orally, reduced the number of tumors induced by methylcholanthrene in mice. The antitumor effect by oral administration was better than by parenteral injection [52]. Sex differences in both toxicity and antitumor activity of aristolochic acid A in mice with Ehrlich ascites carcinomas were also noted: Acute toxicity of aristolochic acid A was higher in males, whereas females were more affected by chronic administration. At a dose below the ED_{50} (1.15 mg/kg), aristolochic acid A had higher antitumor activity in males than in females; at higher doses the reverse was observed [53].

Aristolochic acid analogues and derivatives such as α-nitro-stilbene, β-nitrostyrene, and 1-(2-nitro-2-phenylvinyl)-naphthalene were synthesized and their cytotoxicity and antitumor activity were studied. The substituted nitrovinyl function of the latter compound was postulated to participate in a Michael type reaction with cellular nucleophilic groups [54], a reaction that was regarded to be relevant for cytotoxic activity.

Aristolochic acid A increased oxygen consumption in a dose dependent manner in liver cells and splenocytes of mice [55]. The metabolic activity of guinea pig peritoneal macrophages and human leukocytes was also enhanced by aristolochic acid A, as shown by measuring oxygen consumption [56].

Aristolochic acid A and aristolochic acid II exhibited a stimulation of lucigenin-enhanced, opsonized zymosan-induced neutrophil chemoluminescence as a sensitive assay for immunostimulating activity [57]. In a leukocyte adherence inhibition test, an activity of aristolochic acid A could also be demonstrated; however, it was weaker than that of prednisolone [58]. Following the administration of aristolochic acid A to guinea pigs immunized with Q fever antigen, the antigen-induced decrease in bone marrow lymphocyte count was restored to normal levels much faster than was observed in untreated immunized controls. Following Q fever immunization in rabbits, the antibody titer did not differ between treated and control animals, but a significantly higher antibody titer was observed in immunized rabbits treated with aristolochic acid A and prednisolone [59].

In contrast to the results described above, Xing et al. [60] reported that aristolochic acid A did not prolong the survival time of tumor bearing mice, enhance the immune function of the mouse reticuloendothelial system, or the phagocytic activity of mouse peritoneal macrophages.

Studies on acute toxicity showed LD_{50} values of 14.3 mg/kg, i.p., and 48 mg/kg, orally, in mice [61].

No apparent abnormalities in various organs were observed in rats after i.p. injection of ≤4 mg/kg daily for 7 days [61].

Mutagenic and carcinogenic activities of aristolochic acids were extensively studied. Aristolochic acid A was proven to be a direct mutagen in *Salmonella typhimurium* strains TA 1537 and TA 100. The presence of S9 mix had only a minor enhancing effect on the number of induced revertants. Aristolochic acid A had no mutagenic effect on TA 1535, TA 1538, or TA 98 with or without S9 mix [62]. Aristolochic acid II had almost equal mutagenic potency [63].

Metabolites of aristolochic acid A and aristolochic acid II formed by rat liver were isolated and identified. Under anaerobic conditions the major metabolites of both aristolochic acids were the corresponding aristololactams, whereas under aerobic conditions aristolochic acid II was not metabolized at a measurable rate. The only metabolite found for aristolochic acid A was the O-demethylated product aristolochic acid Ia. Mutagenic activities of the three metabolites were tested in *S. typhimurium* strains TA 1537 and TA 100. The aristololactams were mutagenic in both strains when a metabolizing system was present. Aristolochic acid A, aristolochic acid II, and the corresponding aristololactams may exert their mutagenicity via a common reactive intermediate, probably the hydroxylamine. Aristolochic acid Ia was only very weakly mutagenic [64].

The mutagenic activity of aristolochic acid A was also tested in the granuloma pouch assay, which detects gene mutations induced in subcutaneous granuloma tissue of rats. After direct exposure of the target tissue, aristolochic acid A induced high frequencies of mutations at a relatively low cytotoxic level. Aristolochic acid A was more potent than methyl-nitro-nitrosoguanidine at equimolar doses. After oral administration of aristolochic acid A to rats dose dependent mutagenic activity was registered, whereas a very weak and inconsistent mutagenic effect was seen after systemic application of methyl-nitro-nitrosoguanidine. This suggested that aristolochic acid A was not detoxified efficiently after oral administration and could exert its mutagenic activity in extrahepatic tissues; whereas methyl-nitro-nitrosoguanidine was detoxified to a large extent [65].

The carcinogenic activity of aristolochic acid A has been demonstrated in experimental animals. Male and female rats treated orally with aristolochic acid A at daily doses of 0.1, 1.0, or 10.0 mg/kg developed a high incidence of tumors, dependent on dose and time. Rats treated for 3 months with 1.0 or 10.0 mg/kg aristolochic acid A developed severe papillomatosis of the forestomach with occasional signs of malignancy. Without further treatment, the rats developed squamous cell carcinomas in the forestomach 3–6 months later and formation of metastases. At the same time, there was anaplasia of the tubular epithelium; adenomas appeared in the renal cortex. The transitional epithelium of the renal pelvis and the urinary bladder showed hyperplasia, papillomas, or carcinomas [66]. Aristolochic acid A (10 mg/kg) caused extensive necrosis of the squamous epithelium of the forestomach followed by regeneration and hyperplasia, papilloma formation, and ultimately invasive squamous cell carcinoma [67]. Treatment at the lowest dose did not result in tumor development in the first 6 months, but papillomas or squamous cell carcinomas occurred in the forestomach after 12 and 16 months. In addition, hyperplasia of the transitional epithelium of the renal pelvis was found while the renal cortex and the urinary bladder remained normal [66].

The growth of i.p. inoculated Ehrlich ascites carcinoma cells in mice and the incorporation of [^{3}H] thymidine into Ehrlich ascites carcinoma DNA were increased >20% and 24%, respectively, by i.p. injection of aristolochic acid A. The hepatoma inducing efficiency of dimethyl yellow, administered i.p. to mice, was increased from 39% to 92% by simultaneous administration of aristolochic acid A, whereas the skin cancer inducing effect of benz[*a*]pyrene was not enhanced by aristolochic acid A [61].

Aristolochic acid A exhibited a contraceptive activity in female mice. Given orally it showed significant antiimplantation and early pregnancy interrupting effects which were not observed in rats. Neither estrogenic nor antiestrogenic actions were observed. Treatment with exogenous progesterone failed to prevent the pregnancy interrupting action. In addition, intraamniotic injection of aristolochic acid A into midterm pregnant dogs and rats led to termination of pregnancy. No significant changes were found in blood chemistry, hepatic or renal functions, or in the morphology of the main viscera when aristolochic acid A was administered intraamniotically to dogs at a single dose of 0.2–2 mg [68].

Magnoflorine is a hypotensive principle. In anesthetized cats, i.v. injection of 2 mg/kg magnoflorine produced a prompt and significant fall in blood pressure. Oral administration at a dose of 20–40 mg/kg also resulted in hypotension. The acute LD_{50} of magnoflorine by i.v. injection in mice was 20 mg/kg. Oral administration with a tenfold higher dose daily for 4 weeks neither elicited any toxic symptoms nor retarded growth [10].

References

1. Ding LS, Lou FC (1983) Chemical constituents of *Aristolochia* plants. Chin Trad Herb Drugs 14:424–432
2. Mix DB, Guinandeau H, Shamma M (1982) The aristolochic acids and aristolactams. J Nat Prod 45:657–666
3. Tseng KF, Ku YT (1958) Chemical studies on *Aristolochia* species. II. Relation between isoaristolochic acid and aristolochic acid. Acta Pharm Sin 6:174–177
4. Tseng KF, Ku YT (1958) Chemistry of *Aristolochia*. I. Constituents of *Aristolochia debilis*. Acta Pharm Sin 6:33–38
5. Ku YT, Tseng KF (1957) Structure of debilic acid. Kexue Tongbao 761–762
6. Zhang XQ, Xu LX (1982) Analysis of active principles in Chinese herbal drugs. IV. Determination of aristolochic acid in *Aristolochia debilis* Sieb. et Zucc. Chin J Pharm Anal 2:72–75
7. Chen ZL, Huang BS, Zhu DY, Yin ML (1981) Studies on the active principles of *Aristolochia debilis*. II. 7-Hydroxy-aristolochic acid A and 7-methoxy-aristolochic acid A. Acta Chim Sin 39:237–242
8. Huang BS, Wu LJ, Yin ML, Chen ZL (1985) Active principles of *Aristolochia debilis*. III. The isolation and identification of five derivatives of aristolochic acid. Chin Trad Herb Drugs 16:482–484
9. Tomita M, Kura S (1957) Ingredients of aristolochiaceous plants. I. Ingredients of *Aristolochia kaempferi* and *A. debilis*. Yakugaku Zasshi 77:812–814
10. Chang CC, Wang CK, Li CC, Shao IT, Pei YC, Chiang MY, Li T, Hsu TC (1964) Pharmacological studies on magnoflorine, a hypotensive principle from tu qing mu xiang. Acta Pharm Sin 11:42–49
11. Tomita M, Fukagawa K (1962) Ingredients of aristolochiaceous plants. V. Ingredients of *Aristolochia debilis*. Yakugaku Zasshi 82:1673–1675
12. Rücher G, Mayer R (1985) Bisbenzylisoquinoline alkaloids from roots of *Aristolochia debilis*. Planta Med 51:183–184
13. Furukawa S, Soma N (1961) Structure of aristolone, a constituent of *Aristolochia debilis*. I. Yakugaku Zasshi 81:559–565
14. Furukawa S, Soma N (1961) Structure of aristolone, a constituent of *Aristolochia debilis*. II. Yakugaku Zasshi 81:565–570
15. Krepinsky J, Jommi G, Samek Z, Sorm F (1970) Terpenes CCIV. Sesquiterpene constituents of *Aristolochia debilis*. Structure of debilone. Collect Czech Chem Commun 35:745–748
16. Nishida R, Kumazawa Z (1973) 3-Oxoishwarane, a new sesquiterpenic ketone from *Aristolochia debilis*. Agric Biol Chem 37:341–344
17. Rücker G (1968) 1(10)-Aristolenon-(2) and 1,8,9,10-Tetradehydroaristolanone-(2), neue Sesquiterpene-ketone vom Aristolan-Typ. Liebigs Ann Chem 717:221–224

18. Nakanishi T, Iwasaki K, Nasu M, Miura I, Yoneda K (1982) Aristoloside, an aristolochic acid derivative from stems of *Aristolochia manshuriensis*. Phytochemistry 21:1759–1762
19. Otsuka Pharmaceutical (1983) Aristolochin and its ester. Jpn Kokai Tokkyo JP. 58 152,897 (83 152,897) (CA 100:91346q)
20. Ding LS, Lou FC, Cao MC, Wang YF, He CG (1986) Chemical constituents of Guanmutong (Mutong Madouling) (*Aristolochia manshuriensis*). Chin Trad Herb Drugs 17:347–348
21. Xu LZ, Sun NJ (1984) Preliminary comparison of chemical constituents between *Aristolochia fangchi* and *A. moupinensis*. Bull Chin Mat Med 9:206–207
22. Lou FC, Ding LS, Li LL, Wu MY (1986) Chemical constituents of Bei Madouling (*Aristolochia contorta*). Chin Trad Herb Drugs 17:390–391
23. Lou FC, Ding LS, Wu MY, Li LL (1986) Chemical studies on *Aristolochia contorta* Bunge. II. The structure of aristolochic acid E. Acta Pharm Sin 21:702–705
24. Ding LS, Ho MS, Lou FC (1980) Studies on the chemical constituents of *Aristolochia mollissima*. II. Chin Trad Herb Drugs 11:487
25. He LX, Xue HZ, Xu YX, Weng J (1984) Studies on the constituents of *Aristolochia mollissima* Hance. Acta Bot Sin 26:527–531
26. He MS, Fang H, Lin AP (1982) Studies on *Aristolochia mollissima*. III. Chin Trad Herb Drugs 13:534–535
27. Ding LS, Lou FC (1980) Studies on the chemical constituents of *Aristolochia mollissima*. I. Chin Trad Herb Drugs 11:484–486
28. Hua ZQ, Xu XJ, Lou FJ, Ding LS (1983) Molecular and crystal structure of the compound $C_{15}H_{20}O_2$ from *Aristolochia mollissima* Hance. Jiegou Huaxue 2:197–200
29. Lou FC, Ding LS, Wu MY (1984) Studies on chemical constituents of *Aristolochia mollissima* Hance. IV. Determination of chemical structure of mollislactone. Acta Pharm Sin 18:684–688
30. Li SY, Yao Q (1981) Isolation and identification of aristolochic acid from *Aristolochia kwangsiensis*. Chin Pharm Bull 16:16–17
31. Li SY, Yao C (1981) Isolation and identification of the chemical constituents in *Aristolochia kwangsiensis* Chun et How. Chin Trad Herb Drugs 12:25
32. Chou FH, Liang PY, Chu SC, Wen C (1981) Chemical constituents in *Aristolochia kwangsiensis*. Chin Pharm Bull 16:56–57
33. Zhou FX, Liang PY, Qu CJ, Wen J (1981) Studies on chemical constituents of *Aristolochia kwangsiensis* Chun et How ex CF Liang. Acta Pharm Sin 16:638–640
34. Zhang XQ, Xu LX (1982) Assay of aristolochic acid in Zhu Sha Lian (*Aristolochia tuberosa*). Chin Trad Herb Drugs 13:448–451
35. Zhu DY, Jiang FX, Xu RS, Qiu YP, Chen XZ, Yu DJ (1981) Studies on the chemical constituents of Zhu Sha Lian (*Aristolochia tuberosa*). II. Structural elucidation of the lactam and its glycoside. Chin Trad Herb Drugs 12:529–530
36. Zhu DY, Wang BD, Huang BS, Xu RS, Qiu YP, Chen XZ (1982) Two new 4,5-dioxoaporphine alkaloids isolated from *Aristolochia tuberosa*. Heterocycles [Spec Issue] 17:345–347
37. Zhu DY, Wang BD, Huang BS, Xu RS, Qiu YP, Chen XZ, Quan DJ (1983) Two new oxoaporphine alkaloids isolated from *Aristolochia tuberosa*. I. Structures of tuberosinone and tuberosinone-*N*-β-D-glucoside. Acta Chim Sin 41:74–78
38. Ding LS, Zeng Q, Lou FC (1981) Active principles of Tong Cheng Hu Geng (roots of *Aristolochia tagala*). Chin Trad Herb Drugs 12:436–438
39. Zhou FX, Wen J, Liang PY, Ma Y (1982) Study on the chemical constituents of San Tong Guan (*Aristolochia championii*). Chin Trad Herb Drugs 13:3–5
40. Zhou FX, Wen J, Liang PY, Ma Y (1981) Studies on chemical constituents of *Aristolochia championii*. Chin Pharm Bull 17:243
41. Ho MS, Li XL, Lo JQ, Liu JC (1983) Studies on chemical constituents of *Aristolochia heterophylla*. Chin Trad Herb Drugs 14:158
42. Tian BZ, Zou WQ, Huang S, Tan TQ, Lu LQ (1982) Studies on the chemical constituents of Mu Ping Ma Dou Ling (*Aristolochia moupinensis*) and Yi Ye Ma Dou Ling (*Aristolochia heterophylla*) roots. Chin Trad Herb Drugs 13:10–12
43. Xu LZ, Sun NJ (1984) Chemical constituents of *Aristolochia moupinensis* Franch. Acta Pharm Sin 19:48–55
44. Xue HZ, Zhang J, He LX (1985) Isolation and identification of the constituents soluble in petrolether from *Aristolochia versicolar*. J Nanjing Pharm Coll 16:7–9

45. Zhang J, He LX (1986) Determination of the structures of versicolactone B and versicolactone C in the root of *Aristolochia versicolar* S.M. Hwang. Acta Pharm Sin 21:273–278
46. Zheng J, Li GP, Chen ZG, Tang YQ, Wei XC, He LX (1986) Crystal and molecular structure of acetyl versicolactone B. Acta Chim Sin 44:551–557
47. Zhang CZ, Lao JH, Wang ZW (1986) Studies of chemical components of domesticated and wild *Aristolochia cinabaria*. Chin J Pharm Anal 6:220–222
48. Xu FQ, Wu PE (1986) Studies on chemical constituents of *Aristolochia austrozechuanica*. Chin J Pharm Anal 6:55–56
49. Komatsu N, Nawata H, Kimino T, Tsunoda H, Nagumo N, Koike K, Shoji J, Tada A (1973) Biological activity of aristolochic acid. I. Antimicrobial spectrum and effects on bacterial toxins. Showa Igakkai Zasshi 33:769–775 (CA 82:3945m)
50. Komatsu N, Nawata H, Kimino T, Shoji J, Tada A (1973) Biological activities of aristolochic acid. II. Effect on experimental tumor, bacterial infection and RES (reticuloendothelial system) function. Showa Igakkai Zasshi 33:776–782 (CA 82:68199v)
51. Shunack W, Mutschler E, Rochelmeyer H (1967) Influence of aristolochic acids I and II on the survival rate of mice infected with pneumococci. Arzneimittelforschung 17:1215–1218
52. Möse JR (1975) The effect of aristolochic acid on the development of methylcholanthrene tumors in mice. Oesterr Z Onkol 2:151–153
53. Angeles LT (1975) Sex difference in response to a carcinostatic acid. J Philipp Med Assoc 48:81–95 (CA 78:79647h)
54. Viel C, Dore JC (1972) New synthetic cytotoxic and antitumor agents from aristolochic acid, a nitrophenanthrene acid with antitumor action, extracted from Aristolochiaceae. Farmaco [Sci] 27:257–312
55. Möse JR, Porta J (1974) Aristolochic acid. I. Arzneimittelforschung 24:52–54
56. Möse JR (1974) Aristolochic acid. II. Arzneimittelforschung 24:151–153
57. Wagner H, Porksch A, Vollmar A, Kreutzkamp B, Bauer J (1985) In vitro phagocytosis stimulation by isolated plant materials measured in the phagocytosis-chemoluminescence (CL) model. Planta Med 51:139–144
58. Siering H, Siering U (1984) Effect of prednisolone and aristolochic acid on the leukocyte adherence inhibition test. Folia Haematol (Leipz) 111:27–31
59. Stering H, Przesang R (1984) Effect of glucocorticoids and aristolochic acid on antibody formation in experimental animals. Z Gesamte Inn Med 39:81–84
60. Xing BH, Zhao LQ, Lou B, Jin L (1981) Pharmacology of aristolochic acid. Guanxi Yixue 2 4
61. Xing BH, Chen LL, Zhou LQ, Jin L, Liu F (1982) Experimental study on the carcinogenicity of aristolochic acid. Guangxi Yixue 4:118–121
62. Robisch G, Schimmer O, Göggelmann W (1982) Aristolochic acid is a direct mutagen in *Salmonella typhimurium*. Mutat Res 105:201–204
63. Schmeiser HH, Pool BL, Wiessler M (1984) Mutagenicity of the two components of commercially available carcinogenic aristolochic acid in *Salmonella typhimurium*. Cancer Lett 23:97–101
64. Schmeiser HH, Pool BL, Wiessler M (1986) Identification and mutagenicity of metabolites of aristolochic acid formed by rat liver. Carcinogenesis 7:59–63
65. Maier P, Schawalder HP, Weibel B, Zbinden G (1985) Aristolochic acid induces 6-thioguanine-resistant mutants in an extrahepatic tissue in rats after oral application. Mutat Res 108:143–148
66. Mengs U, Lang W, Poch JA (1982) The carcinogenic action of aristolochic acid in rats. Arch Toxicol 51:107–119
67. Mengs U (1983) On the histopathogenesis of rat forestomach carcinoma caused by aristolochic acid. Arch Toxicol 52:209–220
68. Wang WH, Zheng JH (1984) Pregnancy terminating effect and toxicity of an active component of *Aristolochia mollissima* Hance, aristolochic acid A. Acta Pharm Sin 19:405–409

Artemisia annua L. 22

22.1 Introduction

Qinghao, Herba Artemisae annuae, is the dry aboveground part of *Artemisia annua* L. (Asteraceae) collected in fall after the flowers are in full bloom. This officially listed herbal medicine is used mainly as an antimalarial agent.

22.2 Chemical Constituents

In 1977 a new type of sesquiterpene lactone was isolated as the active principle of *A. annua* and named Qinghaosu (*22-1*) [1]. This substance became known later by the names artemisinin and arteannuin [2]. The name artemisinin will be used here.

The chemical structure of artemisinin was ascertained by mass spectrometry, IR spectroscopy, ^{1}H NMR, chemical reactions [1, 2], and circular dichroism [3]. It is a new sesquiterpene lactone with a peroxy function in the ring. The absolute configuration of artemisinin was determined by X-ray crystallography and was established to be 3,6,9-trimethyl-octahydro-3,12-epoxy-pyrano[4,3-*j*]-1,2-benzodioxepin-10-one [4, 5]. The *trans* configuration of the lactone ring was confirmed by comparison of its optical rotation dispersion spectrum [2] with that of arteannuin B (*22-2*), a structurally related compound isolated from the same plant some years ago [6].

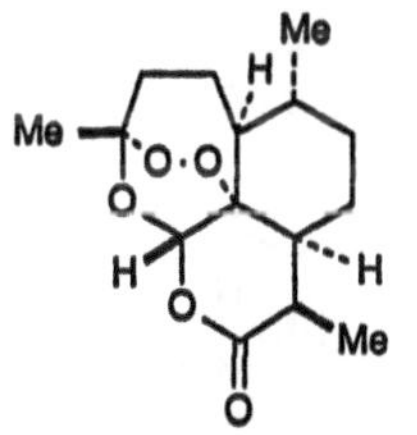

Artemisinin (Qinghaosu, Arteannuin) (*22-1*)

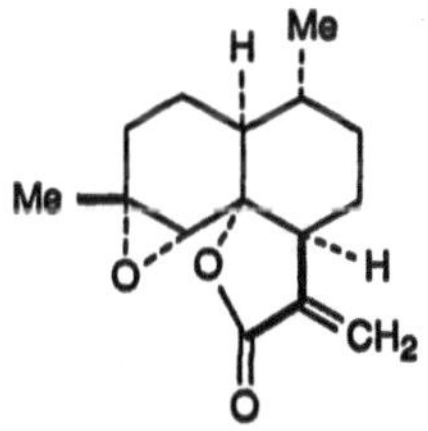

Arteannuin B (Qinghaosu II) (*22-2*)

The content of artemisinin in dry leaves of *A. annua* was 0.5%–0.6% determined after chromatographic separation on silica gel plates by spectrophotometry (580 nm) of the reaction product with *p*-dimethylaminobenzaldehyde [7]. Artemisinin was not detectable in plant callus tissue, but was detected in the regenerated shoot carrying callus and regenerated plantlets or plants. The artemisinin content in the regenerated plant and the original field plant was 0.9% and 0.6% (dry weight), respectively, indicating that plants regenerated from callus culture could provide a source with high artemisinin content [8].

In addition to artemisinin and arteannuin B (qinghaosu II), other sesquiterpenes and related compounds, artemisininic acid [9] (artemisic acid [10], arteannuic acid, qinghao acid [11]) (*22-3*), qinghaosu I (arteannuin A) (*22-4*) [10–12], qinghaosu III (deoxyartemisinin) (*22-5*) [13], qinghaosu IV (*22-6*) [11], qinghaosu V (arteannuin E) (*22-7*) [11], artemisilactone (arteannuin F) (*22-8*) [14], artemisinol (*22-9*) [15], artemisic acid methyl ester (*22-10*) [15], artemisitene (*22-11*) [16], arteannuin C (*22-12*) [17] and epoxyartemisic acid (*22-13*) [38] have been isolated from *A. annua*.

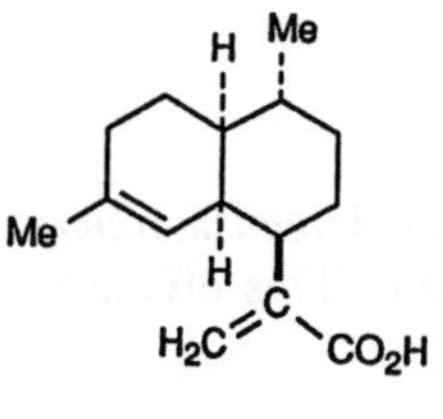

Artemisininic acid (Artemisic acid, Arteannuic acid, Qinghao acid) (*22-3*)

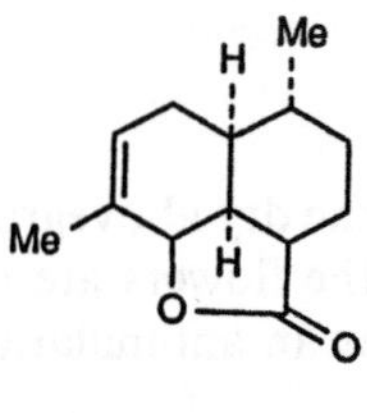

Qinghaosu I (Arteannuin A) (*22-4*)

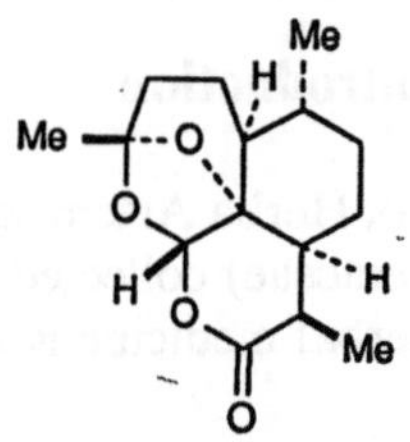

Qinghaosu III (Deoxyartemisinin) (*22-5*)

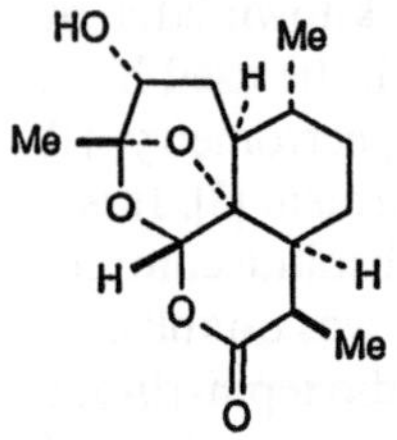

Qinghaosu IV (*22-6*)

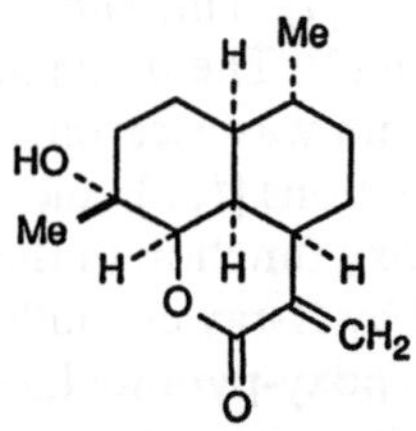

Qinghaosu V (Arteannuin E) (*22-7*)

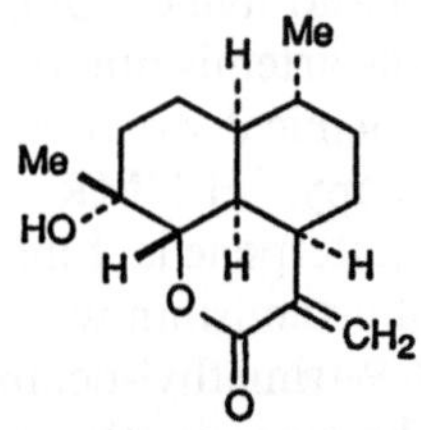

Artemisilactone (Arteannuin F) (*22-8*)

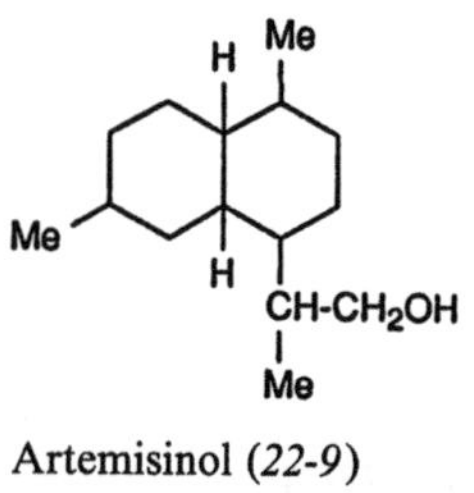

Artemisinol (*22-9*)

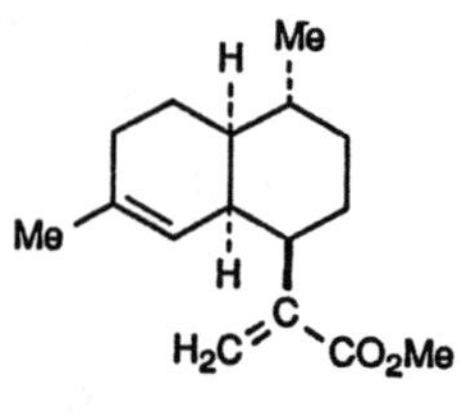

Artemisic acid methyl ester (*22-10*)

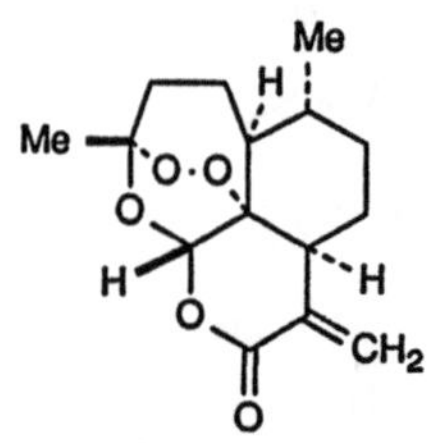

Artemisitene (*22-11*)

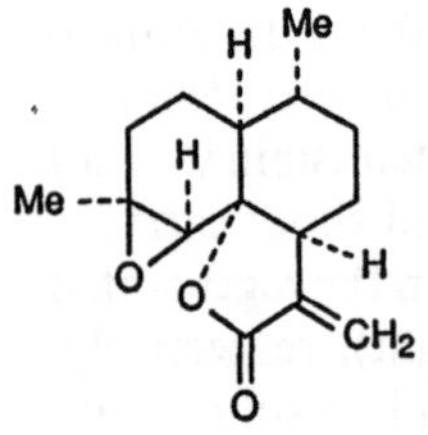

Arteannuin C (*22-12*)

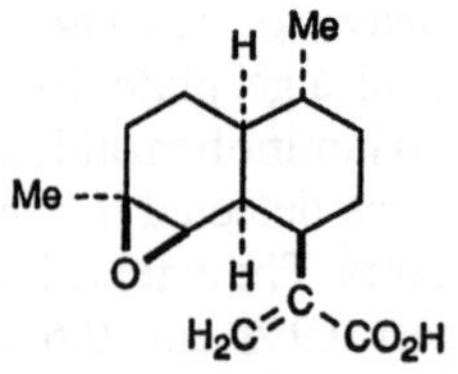

Epoxyartemisic acid (Epoxyarteannuic acid) (*22-13*)

Some flavone and coumarin derivatives were also isolated from *A. annua*: Coumarin [18, 19], scopoletin [18, 20], esculetin [21], 3,5-dihydroxy-6,7,3′,4′-tetramethoxyflavone [18, 22, 23], 5,4′-dihydroxy-3,6,7,3′-tetramethoxyflavone [24], and artemetin (*22-14*) [18–20] were identified.

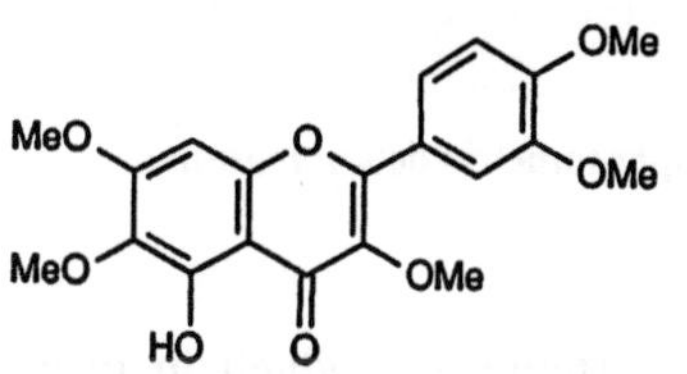

Artemetin (5-Hydroxy-3,6,7,3′,4′-pentamethoxyflavone) (*22-14*)

All parts of *A. annua* herbs contain essential oil, but the content is highest in the inflorescences. A maximal essential oil yield of 3.2% was obtained from plants in full bloom [25]. Camphor, 1,8-cineole, bornyl acetate, α-pinene, β-pinene, caryophyllene, *p*-cymol, β-myrcene, benzyl isovalerate, β-farnesene, artemisia ketone (*22-15*), and artemisia alcohol (*22-16*) were identified [9, 12, 25, 26]. The acid value of the essential oil of *A. annua* showed a seasonal variation with 2.5 for a June sample and 0.8 for an October sample [27].

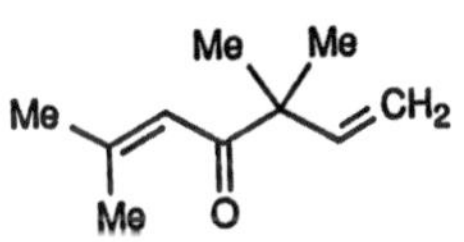

Artemisia ketone (*22-15*)

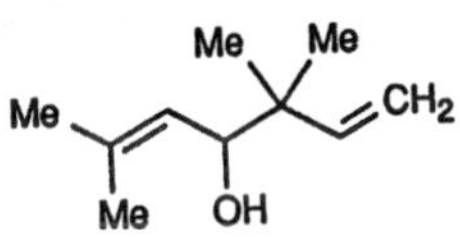

Artemisia alcohol (*22-16*)

The sesquiterpenes isolated from *A. annua* are all closely related compounds characterized by the presence of a *cis*-decalin skeleton with the isopropyl group *trans* to the hydrogen on the ring junction. There is no doubt that artemisinin is the most important component among the sesquiterpenes.

Analysis, chemical modification, and synthesis of artemisinin has been extensively studied. Artemisinin content can be determined quantitatively by volumetric titration using acidimetric [28] or iodometric [29] methods, by spectrophotometric methods [30, 31], or by HPLC using different columns and eluents [32–35]. A reductive electrochemical HPLC assay [36] and a polarographic determination [37] have been described. Since artemisinin only shows a UV absorption at $\lambda \leq 220$ nm, chemical modification has been proposed for UV spectrophotometric determination or HPLC analysis using UV detection. Alkaline treatment of artemisinin gives Q 292 (*22-17*) with a UV absorption maximum at 291–292 nm. Q 260 (*22-18*) (λ_{max} 258–260 nm) is formed after acidification [31, 34] (Fig. 22-1).

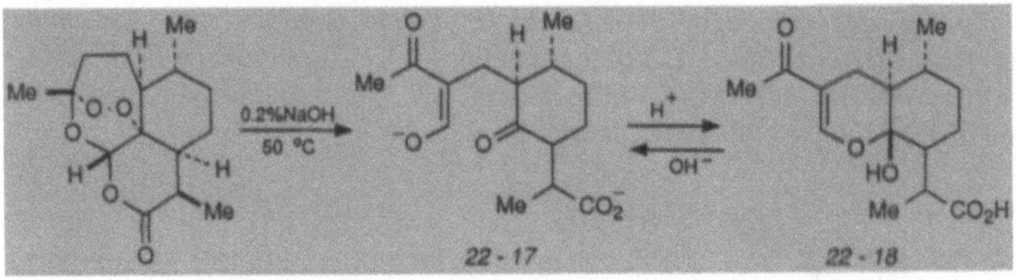

Fig. 22.1. Chemical modification of artemisinin for UV spectrophotometric determination

Catalytic hydrogenation of artemisinin gave deoxyartemisinin, which in turn yielded an α,β-unsaturated ketone (*22-19*) upon treatment with a 10% alcohol KOH solution [2]. This is explained by ring opening of deoxyartemisinin followed by intramolecular aldol condensation. A β-epoxide (*22-20*) was obtained when the α,β-unsaturated ketone was treated with 30% H_2O_2 and KOH (Fig. 22-2). In contrast, the α-epoxide (*22-21*) was obtained by treatment of artemisinin with K_2CO_3 in methanol [39] (Fig. 22-3). The lactonic carbonyl group in artemisinin can be reduced by sodium borohydride to give dihydroartemisinin (*22-22*) from which a hemiacetal and a number of ether and ester derivatives were synthesized and biologically investigated [40–46]. Reduction of artemisinin with lithium aluminum hydride under mild conditions gave a mixture of cyclohexanopyrane derivatives (*22-23*) and a substituted cyclohexane polyalcohol (*22-24*). The latter was the only product obtained after refluxing for more than 7 h [47] (Fig. 22-2).

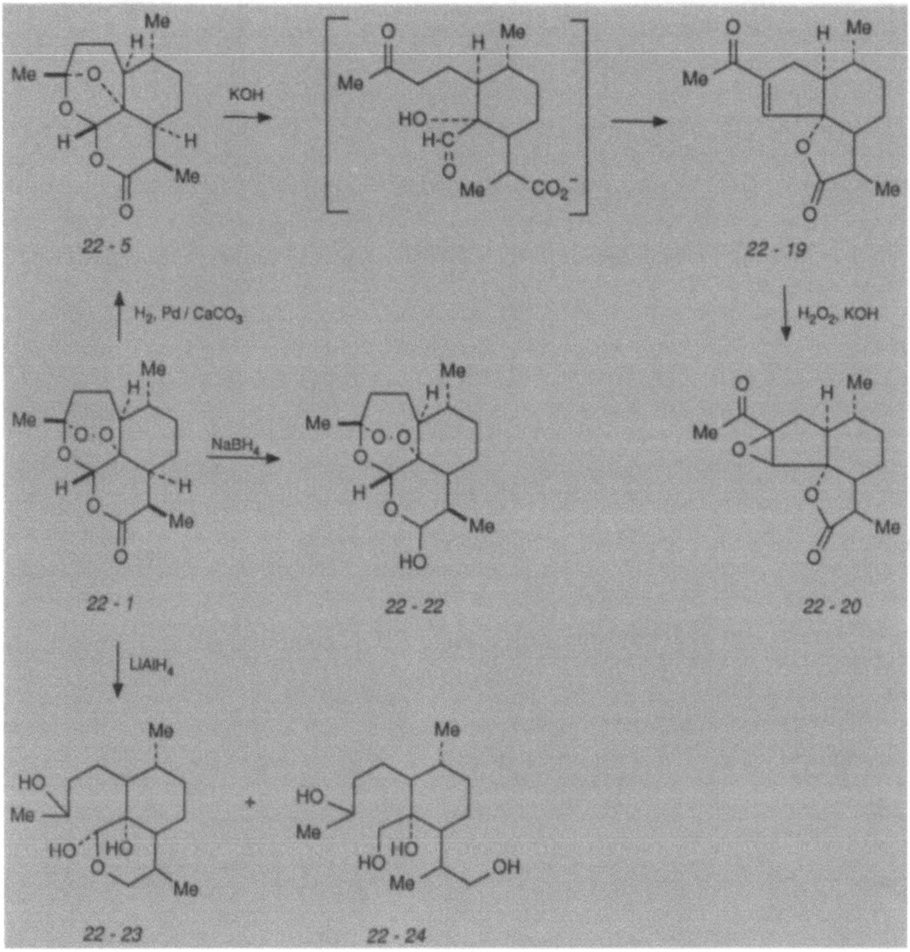

Fig. 22.2. Hydrogenation of artemisinin under different reaction conditions

By treatment with a strong acid such as H_2SO_4 in acetic acid, artemisinin gave a norsesquiterpene lactone (*22-25*). Hydrolysis of this product gave an acid, which converted into isodihydroartemisic acid methylester upon esterification, Wittig methylenation, and hydrogenation. Hydrolysis yielded the free acid (*22-26*) [48, 49] (Fig. 22-3).

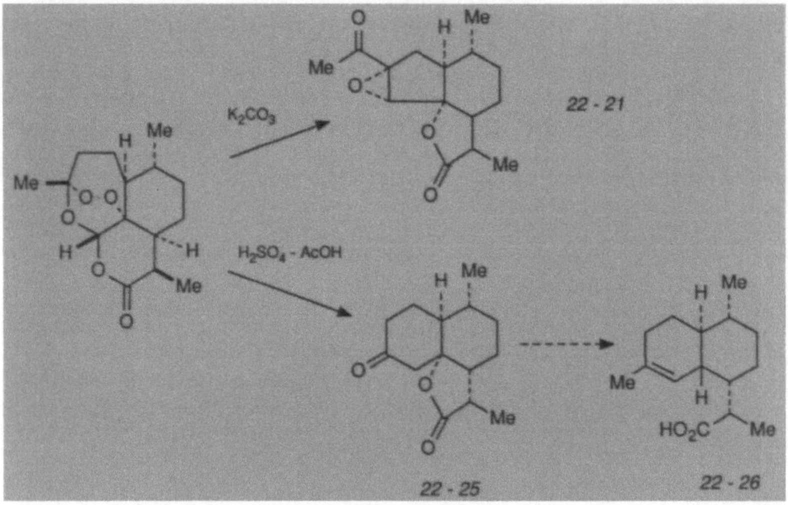

Fig. 22.3. Decomposition of artemisinin by treatment with K_2CO_3 or with a strong acid

After treatment of artemisinin with H_2SO_4/AcOH eight degradation compounds were detected, including epimeric oxolactones (*22-27*) and a hydrolactone (*22-28*) [50].

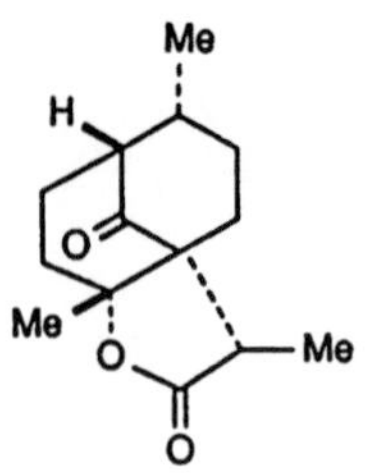

(*22-27*)

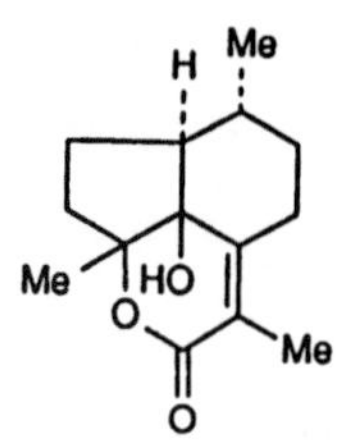

(*22-28*)

Products of heat-induced decomposition and rearrangement isolated after 10 min of heating artemisinin at 190 °C are shown in structures *22-29*–*22-31* [51, 52].

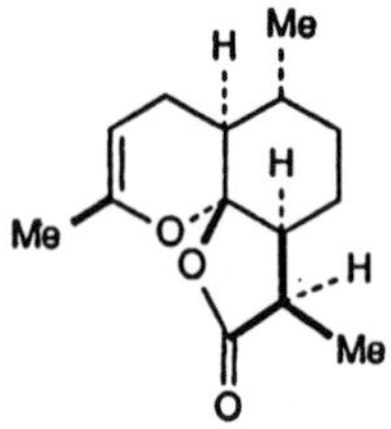

(*22-29*)

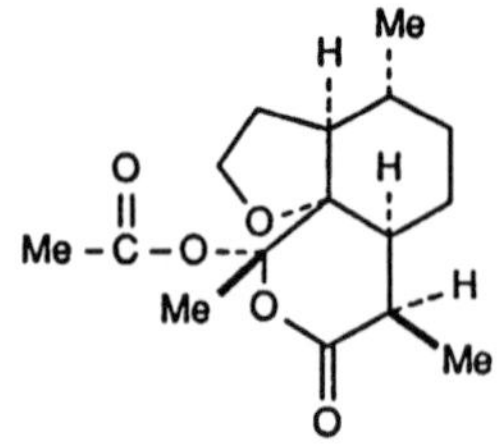

(*22-30*)

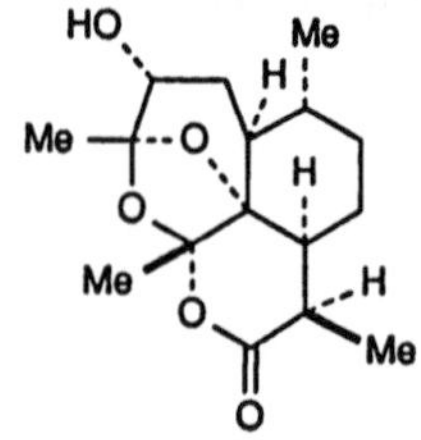

(*22-31*)

Thermolysis of dihydroartemisinin at 190 °C gave deoxyartemisinin and a diketonealdehyde (*22-32*) consisting of two epimers as the major decomposition product [53].

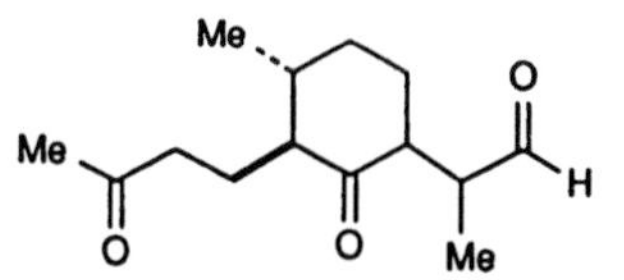

(*22-32*)

The total synthesis of artemisinin was reported in 1983 by Xu et al. from artemisic acid [54, 55] and by Schmid and Hofheinz from (−)isopulegol (*22-33*) [56]. The key step in the synthesis of artemisinin was the photooxygenation of a methyl enol ether (*22-34*) to obtain the assumed hydroperoxide intermediate (*22-35*) which could be converted into artemisinin (Fig. 22-4).

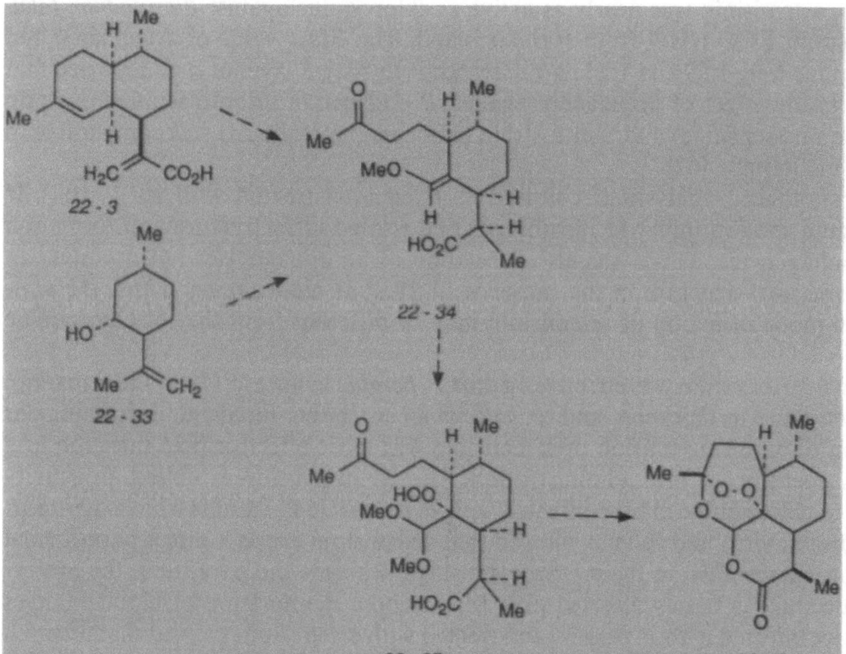

Fig. 22.4. Synthesis of artemisinin from artemisic acid and isopulegol

The direct synthesis of artemisinin and deoxyartemisinin from *R*-(+)-citronellal was also described [39, 57, 58].

Some highly effective antimalarial drugs have been developed from natural products and a number of plants such as *Cinchona* species, *Dichroa febrifuga, Brucea javanica* and *Artemisia annua* represent antimalarial principles that have been used in traditional Chinese medicine since ancient times. It is worth mentioning that a new peroxide compound with antimalarial activity was isolated from *Artabotrys uncinatus* (Annonaceae), a folk medicine used in China, and named Yingzhaosu A (*22-36*) [59, 60].

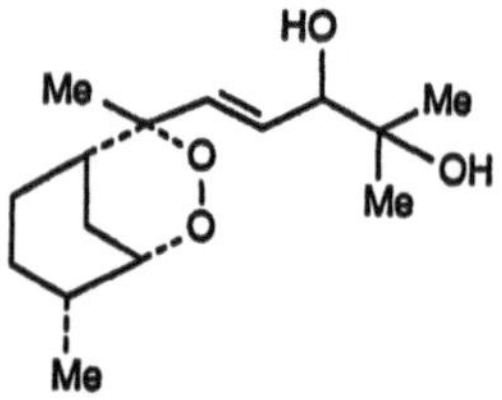
Yingzhaosu A (*22-36*)

22.3 Pharmacology

In vitro, artemisinin was nearly as active as chloroquine against *Plasmodium falciparum* strain FCC-1/HN from Hainan island. The ED_{50} value of artemisinin and chloroquine were 1.99 and 1.24 ng/ml, respectively [61]. Klayman et al. reported that the inhibiting effect of artemisinin against *P. falciparum* in both the Camp strain (chloroquine-sensitive) and Smith strain (chloroquine-resistant) was comparable to that of mefloquine [62].

Morphological analysis of cultured *P. falciparum* treated with 10^{-7}–10^{-6} *M* artemisinin showed injury to membranes and related ultrastructure and formation of autophagocytes. There was also formation of autophagocytes but no injury to membrane ultrastructure in the presence of 10^{-6} *M* chloroquine. Thus, the antimalarial mode of action of artemisinin may be different from that of chloroquine [63].

In vivo, artemisinin was effective against *P. berghei* in mice (ED_{50} = 138.8 mg/kg), *P. gallinaceum* in chickens, and *P. cynomolgi* in rhesus monkeys. Intramuscular injection of an oil suspension of artemisinin was more effective than injection of a water suspension or oral administration. The oil suspension exhibited an activity similar to chloroquine in mice infected with *P. berghei* [64]. Results from experimental studies in vitro and in vivo showed that artemisinin exerts a direct parasiticidal effect on *Plasmodium* in the erythrocytic stage, whereas the parasite in the preerythrocytic stage is barely affected [65]. Microscopic examination of blood samples from mice infected with *P. berghei* and treated with artemisinin showed that after 8 h the trophozoites began to change morphologically as indicated ley vacuolation, atrophy, and disappearance of the cytoplasm. After 20 h, the trophozoites showed extensive degeneration of their inner structures [65].

Rhesus monkeys infected with *P. cynomolgi* were treated with artemisinin or with chloroquine phosphate. The level of blood *P. cynomolgi* began to decrease markedly 6 h after artemisinin treatment and was undetectable after 13 h. In chloroquine treated animals the levels started to decrease after 8 h and was undetectable after 14 h. During a 10 day observation, *P. cynomolgi* reappeared in the blood of monkeys treated with artemisinin 9 days after treatment, whereas no reappearance of *P. cynomolgi* was observed in animals treated with chloroquine. The morphological changes in *P. cynomolgi* after artemisinin and chloroquine treatment were different which again points to different mechanisms of antimalarial action [66].

Disappearance of malaria symptoms and an increase in survival time was achieved in mice infected with *P. berghei* by combined use of artemisinin and primaquine as opposed to artemisinin or primaquine alone. The reoccurrence of malaria was also significantly delayed. Combination of artemisinin and primaquine resulted in a clear synergistic effect since the ED_{50} value for the combined use of artemisinin and primaquine was only $^1/_{10}$ that of artemisinin or primaquine alone; the acute toxicity of artemisinin or primaquine was not enhanced by combined use [67].

Artemisinin enhanced macrophage phagocytosis in mice. Both the in vivo phagocytic activity and the index of mouse abdominal macrophages were increased after intragastric administration of artemisinin. Phagocytic activity of macrophages for *Plasmodium* in an in vitro culture system was also increased in the presence of artemisinin. Acid phosphatase activity was elevated in abdominal macrophages from mice treated with artemisinin compared to those from control mice. The antimalarial activity of artemisinin appeared to be associated with its enhancing effect on macrophage phagocytic activity [68].

A pyronaridine-resistant line of *P. berghei* showed reduced sensitivity to six erythrocytic schizonticides including artemisinin, indicating the presence of cross resistance [69]. Resistance to artemisinin developed rapidly in a chloroquine-resistant line of *P. yoelii* passaged in mice but was not observed in chloroquine-sensitive *P. berghei* [70].

Treatment of mice infected with *P. falciparum* induced changes in membranes of the parasite together with alterations of ribosomal organization and endoplasmic reticulum. No changes were noted in digestive vacuoles or pigment, but nuclear membrane blebbing developed after 1 h and segregation of the nucleoplasm after 3 h. The morphological changes in ribosomes and endoplasmic reticulum in vivo correlated in time with the depression of protein synthesis observed in *P. falciparum* in vitro [71].

Artemisinin is a drug with low acute toxicity. In mice, an LD_{50} of about 4.2 g/kg after oral administration and 3.8 g/kg after i.m. injection of an oil suspension were obtained. In rats, the corresponding LD_{50} values were about 5.6 g/kg orally and 2.6 g/kg by i.m. injection. Restlessness, incoordination, tremors, diminished activity, lower respiration, and loss of the righting reflex appeared after administration of a toxic dose of artemisinin in dogs and in monkeys [64].

In monkeys, i.m. injections at a daily dose of 192 mg/kg artemisinin for 14 consecutive days caused severe ultrastructural changes in the myocardium which were visible on the third day after the last injection but apparently not after 35 days. This total dose caused the death of 75% of the animals within 3 days after 14 days of treatment. A dose of 24 mg/kg daily caused only mild myocardial lesions and diminution of circulating reticulocytes [72].

Intragastric administration of artemisinin at $^1/_{80}$ to $^1/_5$ the LD_{50} values (53–845 mg/kg) did not increase the number of micronuclei in mouse bone marrow polychromatic erythrocytes [73].

In clinical studies, artemisinin had good therapeutic effects on almost all patients treated and was without obvious side effects. All 2099 malaria patients infected with *P. vivax* or *P. falciparum* were clinically cured with artemisinin in different dosages.

Moreover, artemisinin was also effective in treatment of chloroquine-resistant *falciparum* malaria and cerebral malaria [74]. In general, the body temperature of patients became normal within 72 h after treatment and the asexual parasite forms were eliminated, as determined by blood films, within 120 h; however, the relapse rate within 1 month of patients treated with artemisinin was 21%. No relapse was observed in patients treated with chloroquine [74].

From a study on the effects of artemisinin and mefloquine on malaria patients infected with chloroquine-resistant *P. falciparum* it was concluded that clearance of parasitemia was more rapid with artemisinin than with mefloquine [75].

Treatment of cerebral malaria was one of the major successes of artemisinin as a new antimalarial. A clinical study of patients infected with cerebral malaria showed an average cure rate of about 90%, which was considerably higher than that of chloroquine and quinine. The time for recovery from the coma was about 21 h [74].

In 139 malaria patients treated with artemisinin, serum GPT activity, nonprotein nitrogen levels, and the electrocardiogram were observed before and after administration. No abnormalities were found except for a local pain at the site of injection of the aqueous suspension in some patients. Thus, artemisinin appeared to be a safe agent even in patients with complications involving heart, liver, and renal diseases or pregnancy [64, 65].

The liver appears to be the most important organ for metabolism of artemisinin. After a 1 h incubation of artemisinin with rat liver tissue 8.3% of the intact compound was recovered, kidney and lung were less active and blood was inactive. In mice, [^{3}H]artemisinin given orally was rapidly absorbed and the blood levels reached a maximum approximately 1 h after administration. The half-life of artemisinin was close to 4 h. About 80% of an oral dose was eliminated within 24 h [64].

In rats, artemisinin was completely and rapidly absorbed after oral administration; however, a very low plasma level was obtained even after a dose of 300 mg/kg. When given i.m., significantly higher and more persistent plasma levels were observed. Artemisinin was shown to pass the blood-brain and blood-placenta barriers after i.v. injection. Very little unchanged artemisinin was found in the urine and feces after 48 h regardless of the administration route [76].

In dogs, after i.m. injection of artemisinin at a dose of 10 mg/kg, rapid absorption was observed with a peak serum level of 0.2 μg/ml at 2 h and an elimination half-life of 1.6 h as detected by radioimmunoassay [77, 78].

After oral administration, four metabolites were isolated from the urines of malaria patients and healthy subjects. These metabolites were deoxyartemisinin, dihydrodeoxyartemisinin (*22-37*), an indenefuran derivative (*22-38*) [79], and 9,10-dihydroxyhydroartemisinin [80]. Artemisinin was also excreted unchanged in the feces [79].

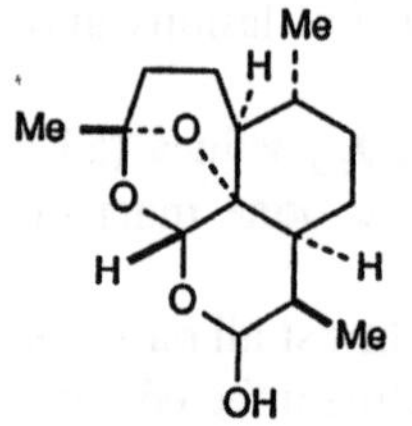

Dihydrodeoxyartemisinin (*22-37*)

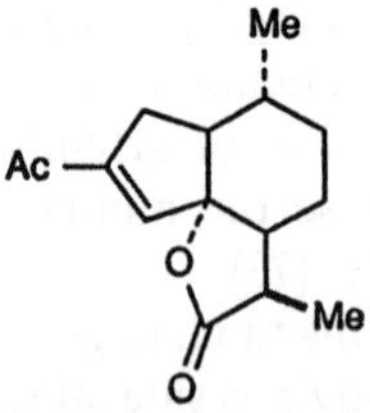

3,6-Dimethyl-2-oxo-8-acetyl-octahydroindene [3*a*, 4*b*] furan (*22-38*)

The isolated metabolites were inactive against *P. berghei* in mice, presumably due to loss of the peroxy function [79].

A series of artemisinin derivatives were synthesized and tested against experimental malaria, some of which were also used clinically. Most of these synthetic compounds were derivatives of dihydroartemisinin ethers or esters. Epimeric pairs of ethers of dihydroartemisinin were obtained by use of catalytic boron trifluoride etherate [42]. Most of the ether derivatives were more effective than the parent compound in experimental antimalarial activity [42, 81]. Epimeric esters of dihydroartemisinin could be prepared by reaction with acyl halides [43] or anhydrides in the presence of pyridine, 4-dimethylamino-pyridine, or dicyclohexylcarbodiimide [41, 42]. Esters, e.g., of formic acid, *o*-chlorobenzoic acid [81], valeric acid, cyclohexylcarboxylic acid, benzoic acid, and α-naphthylcarboxylic acid [40], were more than ten times as active as artemisinin in mice infected with *P. berghei*. Reaction with chloroformic esters in the presence of triethylamine and 4-dimethylamino-pyridine yielded carbonates of dihydroartemisinin, mainly in α-epimeric form. Most of these carbonates had antimalarial activity greater than that of artemisinin [40, 42].

Among the artemisinin derivatives, dihydroartemisinin, its methyl ether (artemether), and succinic acid hemiester (artesunic acid) have been thoroughly investigated. Artemether consists of a mixture of the α- and β-epimers and was more effective than artemisinin against *P. berghei* and *P. cynomolgi*. The toxicity of artemether was less than that of chloroquine, causing no irritation at the site of i.m. injection [82]. After i.v. administration of [^{3}H]artemether to mice and rats, radioactivity appeared rapidly in various tissues, reflecting systemic distribution after 30 min and decreased to very low levels after 24 h. In mice, urinary and fecal excretion of radioactivity were 41% and 27%, respectively, after 24 h; and 56% and 39%, respectively, after 72 h. In rats, 33% of the dose was excreted in the bile by 3 h. Following i.v. and i.m. injection of artemether into mice, 31% and 15% demethylation, respectively, were measured within 24 h. Demethylation could be stimulated by phenobarbital via induction of drug metabolizing enzymes in liver [83]. Appearance of ^{3}H radiolabel in various organs 30 min after intragastric administration of [^{3}H]artemether was in the following order: bile > kidneys > intestine > liver > bone > heart > spleen > brain [84]. The LD_{50} and LD_5 values of artemether in mice were 567 and 230 mg/kg, respectively; the therapeutic index calculated from LD_{50}/ED_{50} and the safety index calculated from LD_5/ED_{95} were 12.9 and 3.4, respectively [84].

A water soluble sodium salt of artesunic acid, known as sodium artesunate, can be administered i.v. It was found to be about five times more effective than artemisinin against both chloroquine-resistant and chloroquine-sensitive strains of *P. berghei* in mice. It was, however, also found to be more toxic than artemisinin but less toxic than chloroquine or artemether [74]. In a clinical study, treatment of cerebral malaria using sodium artesunate resulted in rapid recovery from coma within 12 h after administration, although patients treated with sodium artesunate had a high relapse rate [74].

Sodium artesunate was rapidly converted to dihydroartemisinin when administered to rats. The blood and organ concentrations of dihydroartemisinin were maximal at 10 min and dihydroartemisinin was no longer detectable 120 min after administration of sodium arsunate [85].

Artemisinin, artemether, dihydroartemisinin, and artesunic acid are able to bind to human plasma proteins [86]. In culture medium, dihydroartemisinin differentially accumulated into erythrocytes. Uninfected erythrocytes concentrated the drug less than two-fold, whereas erythrocytes infected with *P. falciparum* accumulated more than 300 times the medium concentration. The uptake process was reversible and saturable. Competition studies indicated that the receptor is the same as that for artemether. Chloroquine showed partial inhibition of uptake but was unable to release bound [^{3}H]dihydroartemisinin from infected cells [87].

An investigation of the incorporation of [^{3}H]hypoxanthine into nucleic acids of erythrocytes infected with *P. falciparum* showed a delayed inhibition by artemisinin and related compounds. It suggested that nucleic acid synthesis by *P. falciparum* was not the primary target of the drug [88]. In comparison, artemisinin, dihydroartemisinin, and artemether strongly inhibited [^{3}H]isoleucine incorporation by human erythrocytes infected with *P. falciparum*. Inhibition was observed at ≤ 1 h after incubation with these compounds at concentrations of 50 n*M*–5 μM. Inhibition of protein synthesis exerted by these compounds may be a direct effect [89].

Para-aminobenzoic acid did not antagonize the antimalarial action of artemisinin suggesting that folic acid metabolism was not disturbed by artemisinin [65].

The structure-activity relationships of artemisinin derivatives were also studied. A series of artemisinin derivatives with antimalarial activity were subjected to quantitative structure activity relationship (QSAR) according to Hansch analysis. The log P values (octanol/waters) of various compounds, as determined by HPLC, were consistent with the additive rule of Hansch. Variations in antimalarial activities were correlated to lipid solubility. In the series of ethereal compounds, electronic parameters of various substituents significantly affected antimalarial activity [90].

An analysis using the extended Hueckel molecular orbital method revealed a linear relationship between antimalarial activity and energy levels of the fourth molecular orbital [91].

Artemisinin and some related compounds were also tested in experimental schistosomiasis and clonorchiasis. Treatment of mice infected with *Schistosoma japonicum* with either an oral artemether suspension or a s.c. oily preparation and of infected dogs orally or i.m. with artemether resulted in significant worm reduction rates [92]. In schistosomiasis infected mice intragastric administration or s.c. injection of artemether shifted all or almost all of the worms into the liver. The majority of worms that survived after intragastric treatment with artemether migrated back between days 11 and 17, while after s.c. treatment the majority were still in the liver on day 14. Artemether appeared to be ineffective against the ova [93].

Artesunate was effective against experimental schistosomiasis. When infected mice were treated orally with artesunate suspension, the total worm reduction rates were 60%–70% and the female worm reduction rates were 80%–90%. When given i.p. to infected mice, the total worm reduction rate and the female worm reduction rate were 35%–45% and 55%–75%, respectively. Artesunate also appeared to be active in rabbits infected with *S. japonicum*. It had significant activity against 1-week-old immature worms without severe side effects to infected animals [94].

Artemether at an oral dose of 300 mg/kg daily for 2 days caused degeneration of the integument, intestine, and genital gland of *S. japonicum* in the liver of host mice, achieving a potent schistosomacidal effect. Activity was more rapid and marked for female worms than for males. The amounts of glycogen and RNA and the activities of alkaline phosphatase and phenolase in the worms were markedly decreased or even absent after treatment with artemether. The histochemical changes in female worms occurred earlier than those in male worms. Hydropic degeneration of the liver cells of infected mice was also observed [95]. The degeneration rate of schistosomulae in the livers of mice infected with *S. japonicum* following oral administration of artemether was 45%–80% [96].

Clonorchis sinensis was also found to be affected by artemisinin and some of its synthetic derivatives. In animal experiments, these agents showed significant clonorchicidal effects in rats with clonorchiasis. The compounds tested did not appear to cause organ damage or to affect normal activity and appetite of host animals, suggesting that they are effective clonorchicides with relatively low toxicity [97].

References

1. Coordinating Group for Research on the Structure of Qing Hao Su (1977) A new type of sesquiterpene lactone – Qing Hao Su. Kexue Tongbao 22:142
2. Liu JM, Ni MY, Fan JF, Tu YY, Wu ZH, Wu YL, Chou WS (1979) Structure and reaction of arteannuin. Acta Chim Sin 37:129–143
3. Liang XT (1982) Circular dichroism of the peroxidic linkage. Acta Chim Sin 40:287–288
4. Qinghaosu Research Group (1980) Crystal structure and absolute configuration of Qinghaosu. Sci Sin [Engl] 23:380–396
5. Zhang SD, Zhang JP, Wu BM, Yao JX, Lin XY (1981) Studies on the crystal structure of bromo Qinghaosu. Acta Phys Sin 30:976–982
6. Jeremic D, Jokic A, Behbud A, Stefanovic M (1973) New type of sesquiterpene lactone isolated from *Artemisia annua*. Tetrahedron Lett 3039–3042
7. Luo HM, Chao PP, Yu CC, Tai C, Liu CW (1980) TLC scanning determination of artemisinin in *Artemisia annua*. Chin Pharm Bull 15:8–10
8. He XC, Zeng MY, Li GF, Liang Z (1983) Callus induction and regeneration of plantlets from *Artemisia annua* and change of Qing Hao Su contents. Acta Bot Sin 25:87–90
9. Tu YY, Ni MY, Chung YY, Li LN (1981) Chemical constituents in *Artemisia annua* L. and the derivatives of artemisinine. Bull Chin Mat Met 6:31
10. Deng DA, Zhu DY, Jao YL, Dai JY, Xu RS (1981) Studies on the structure of artemisic acid. Kexue Tongbao 26:1209–1211
11. Tu YY, Ni MY, Zhong YR, Li LN, Cui SL, Zhang MQ, Wang XZ, Ji Z, Liang XT (1982) Studies on the constituents of *Artemisia annua*. Planta Med 44:143–145
12. Tu YY, Ni MY, Zhong YR, Li LN, Cui SL, Zhang MQ, Wang XZ, Liang XT (1981) Studies on the chemical constituents of *Artemisia annua* L. Acta Pharm Sin 16:366–370
13. Tian Y, Wei ZX, Wu ZH (1982) Studies on the chemical constituents of Qing Hao (*Artemisia annua*), a traditional Chinese herb. Chin Trad Herb Drugs 13:9–11
14. Zhu DY, Deng DA, Zhang SG, Xu RS (1984) Structure of artemisilactone. Acta Chim Sin 42:937–939
15. Zhu DY, Zhang SG, Liu BN, Fan GJ, Liu J, Xu RS (1982) Study on antibacterial constituents of Qing Hao (*Artemisia annua* L.) Chin Trad Herb Drugs 13:6
16. Acton N, Klayman DL (1985) Artemisitene, a new sesquiterpene lactone endoperoxide from *Artemisia annua*. Planta Med 51:441–442
17. Misra LN (1986) Arteannuin C, A sesquiterpene lactone from *Artemisia annua*. Phytochemistry 25:2892–2893
18. Liu HM, Li KL, Wo WC (1980) Studies on the constituents of *Artemisia annua*. Chin Pharm Bull 15:39
19. Tu YY, Yin JP, Gi L, Huang MM, Liang XT (1985) Studies on chemical constituents of sweet wormwood (*Artemisia annua*). III. Chin Trad Herb Drugs 16:200–201
20. Liu HM, Li GL, Wu HZ (1981) Studies on the chemical constituents of *Artemisia annua* L. Acta Pharm Sin 16:65–67

21. Aleskerova AN (1985) Detection of esculetin in *Artemisia annua* L. Izv Akad Nauk SSSR [Biol] 25–26 (CA 103:85125n)
22. Djermanovic M, Jokic A, Mladenovic S, Stefanovic M (1975) Quercetagetin 6,7,3′,4′-tetramethyl ether, a new flavonol from *Artemisia annua*. Phytochemistry 14:1873
23. Stefanovic M, Krstic L, Jokic A, Richter B, Mladenovic S (1979) Synthesis of 3,5-dihydroxy-3′,4′,6,7-tetramethoxy-flavone. Glas Khem Drush Beogr 44:497–502 (CA 92:198216r)
24. Bhardwaj DK, Jain RK, Jain SC, Manchanda CK (1985) Constituents of *Artemisia annua* flavone. Proc Indian Natl Sci Acad [A] 51:741–745 (CA 104:106190t)
25. Georgiev E, Genov N, Khristova N (1981) Changes in yield and quality of essential oil in annual wormwood (*Artemisia annua* L.) during growth. Rastenievd Nauki 18:95–102 (CA 97:52627g)
26. Georgiev E, Genov N, Lazarova R, Kuchkova D, Ganchev G (1979) On the distillation and extraction of annual wormwood (*Artemisia annua*). Congr Essent Oils (Pap) 7th 1977, 7:233–237 (CA 92:64527j)
27. Wu BK (1983) Acid values of Qing Hao volatile oils. Chin Trad Herb Drugs 14:298–299
28. Zhu HS (1980) Titrimetric analysis of artemisinin in *Artemisia annua*. Chin Pharm Bull 15:6–8
29. Zeng MY (1984) A modified iodometric method in determination of organic bridged peroxides-iodometric determination of Qing Hao Su. Chin J Pharm Anal 4:327–329
30. Shen XK, Yan KD, Luo ZY, Tian Y, Zeng MY (1983) UV spectrophotometric determination of Qing Hau Su. Chin J Pharm Anal 3:24–26
31. Zeng MY, Li LN, Liang XT (1985) Reaction mechanism for UV spectrophotometric quantitation of qinghaosu. Chin J Pharm Anal 5:268–271
32. Wang ZS, Zhu YH, Zhang SL, Lu XL (1981) Determination of artemisinin and its derivatives by high-pressure liquid chromatography. Acta Pharm Sin 16:466–470
33. Zhang RB, Chang JP, Xu SL, Li Y (1981) Separation of artemisinin and its derivatives by reversed-phase high-performance liquid chromatography. Acta Pharm Sin 16:460–465
34. Zhao SS, Zeng MY (1985) Studies on the analysis of Qinghaosu by high-pressure liquid chromatography and spectrometry (HPLC). Planta Med 51:233–237
35. Zhao SS, Zeng MY (1986) Determination of Qinghaosu in *Artemisia annua* L. by high-performance liquid chromatography. Chin J Pharm Anal 6:3–4
36. Acton N, Klayman DL, Rollman IJ (1985) Reductive electrochemical HPLC assay for artemisinin (Qinghaosu). Planta Med 51:445–446
37. Zhang XQ, Xu LX (1985) Determination of qinghaosu (arteannuin) in *Artemisia annua* L. by pulse polarography. Acta Pharm Sin 20:383–386
38. Wu ZH, Wang YY (1984) Stucture and synthesis of arteannuin and related compounds. XI. Identification of epoxyarteannuic acid. Acta Chim Sin 42:596–598
39. Zhou WS (1986) Total synthesis of arteannuin (Qinghaosu) and related compounds. Pure Appl Chem 58:817–824
40. Li Y, Yi PL, Chen YX, Ji RY (1982) Studies on analogs of arteannuin. II. Synthesis of some carboxylic esters and carbonates of dihydroarteannuin by using 4(N,N-diemthylamino)-pyridine as an active acylation catalyst. Acta Chim Sin 40:557–561
41. Li Y, Yu PL, Chen IH, Chi JY (1980) Synthesis of artemisinin derivatives by highly active acylation catalysts. Chin Pharm Bull 15:38
42. Li Y, Yu PL, Chen YX, Li LQ, Gai YZ, Wang DS, Zheng YP (1981) Studies on artemisinine analogs. I. Synthesis of ethers, carboxylates and carbonates of dihydroartemisinine. Acta Pharm Sin 16:429–439
43. Cao MZ, Hu SC, Li MH, Zhang S (1982) Synthesis of carboxylic ester of dihydroartemisinin. J Nanjing Coll Pharm 53–54
44. Zhang YL, Xu ML, Zhang LZ, Chen PC, Zhong JY, Yan CY, Ye DY, Lei XH (1985) Synthesis of derivatives of 12-dihydroartemisinine and their antischistosomal effects. Pharm Ind 16:254–257
45. Chen YX, Yu PL, Li Y, Ji RY (1985) Studies on analogs of arteannuin. III. Synthesis of diacid esters and mono esters of dihydroarteannuin. Acta Pharm Sin 20:105–111
46. Chen YX, Yu PL, Li Y, Ji RY (1985) Studies on analogs of arteannuin. VII. The synthesis of ethers of bis(dihydroarteannuin) and bis(dihydrodeoxyarteannuin). Acta Pharm Sin 20:470–473
47. Wu YL, Zhang JL (1986) Reduction of Qing Hao Su (Artemisinin) with lithium aluminum hydride. Org Chem 153–156

48. Gai YZ, Zheng YP, Li LQ (1984) Transformation of arteannuin to isodihydroartemisic acid and related reactions. Acta Chim Sin 42:259–263
49. Xu XX, Wu ZH, Shen JM, Cheng CH, Wu YL, Zhou WS (1984) Structure and synthesis of arteannuin and related compounds. III. Total synthesis of arteannuin degradation product, a norsesquiterpenoid lactone and determination of lactone ring configuration. Acta Chim Sin 42:333–339
50. Zhou WS, Wen YC (1984) Structure and synthesis of arteannuin and its related compounds. VI. Structures of arteannuin degradation products. Acta Chim Sin 42:455–459
51. Luo XD, Yeh HJC, Brossi A, Flippen-Anderson JL, Gilardi R (1985) The chemistry of drugs. VI. Thermal decomposition of Qing Hao Su. Heterocycles 23:881–887
52. Lin AJ, Klayman DL, Hoch JM, Silverton JV, George CF (1985) Thermal rearrangement and decomposition products of artemisinin (qinghaosu). J Org Chem 50:4504–4508
53. Lin AJ, Theoharides AD, Klayman DL (1986) Thermal decomposition products of dihydroartemisinin (dihydroquinghaosu). Tetrahedron 42:2181–2184
54. Xu XX, Zhu J, Huang DZ, Zhou WS (1983) Studies on the structure and synthesis of arteannuin and related compounds. X. Stereocontrolled synthesis of arteannuin and deoxyarteannuin from arteannuic acid. Acta Chim Sin 41:574–576
55. Xu XX, Zhu J, Chou WS (1982) Studies on the structure and synthesis of arteannuin and its related compounds. IX. Synthesis of some key compounds for the total synthesis of arteannuin. Org Chem 447–448
56. Schmid G, Hofheinz W (1983) Total synthesis of quinghaosu. J Am Chem Soc 105:624–625
57. Xu XX, Zhu J, Huang DZ, Zhou WS (1986) Total synthesis of arteannuin and deoxyarteannuin. Tetrahedron 42:819–828
58. Xu XX, Zhu J, Huang DZ, Zhou WS (1984) Structure and synthesis of arteannuin and related compounds. XVII. Stereocontrolled total synthesis of methyl dihydroarteannuate. Acta Chim Sin 42:940–942
59. Liang XT (1985) The chemistry of peroxidic anti-malarials. In: Chang HM, Yeung HW, Tso WW, Koo A (eds) Advances in Chinese Medicinal Material Research. World Scientific, Singapore, pp 427–432
60. Liang XT, Yu DQ, Wu WL, Deng HC (1979) The structure of Yingzhaosu A. Acta Chim Sin 37:215–230
61. Guan WB, Huang WJ, Zhou YC, Gong JZ (1982) Effect of artemisinin and its derivatives on *Plasmodium falciparum* in vitro. Acta Pharmacol Sin 3:139–141
62. Klayman DL, Lin AJ, Acton N, Scovill JP, Hoch JM, Milhous WK, Theoharides AD, Dobek AS (1984) Isolation of artemisinin (qinghaosu) from *Artemisia annua* growing in the United States. J Nat Prod 47:715–717
63. Ye ZG, Li ZL, Gao MX, Fu XQ, Lia HP (1982) Effect of arteannuin on the ultrastructure of intraerythrocytic forms of *Plasmodium falciparum* cultured in vitro. Chin Pharm Bull 17:247–248
64. Academy of Traditional Chinese Medicine (1979) Pharmacologic study of *Artemisia annua* L. J New Med Pharmacol 23–33
65. Qinghaosu Antimalaria Coordinating Research Group (1979) Antimalaria studies on qinghaosu. Chin Med J [Engl] 92:811–816
66. Chen HL, Cao ZH, Kang WM, Yang YH (1982) Experimental observation on the effect of qinghaosu on *P. cynomolgi*. New Chin Med 13:60–61
67. Wan YD, Zang QZ (1981) Study on the combined use of artemisinin and primaquine against shortterm malarial relapse. Chin Pharm Bull 16:9–12
68. Ye XS, Cheng DX, Wang YQ (1982) Effect of Qing-hao-su on macrophage phagocytosis in the mouse abdominal cavity. J Beijing Med Coll 14:141–142
69. Shao BR, Ye XY, Zheng H (1982) Development of pyronaridin-resistance in *Plasmodium berghei*. Acta Pharm Sin 17:566–571
70. Chawira AN, Warhurst DC, Peters W (1986) Qinghaosu resistance in rodent malaria. Trans R Soc Trop Med Hyg 80:477–480
71. Ellis DS, Li ZL, Gu HM, Peters W, Robinson BL, Tovey G, Warhurst DC (1985) The chemotherapy of rodent malaria. XXXIX. Ultrastructural changes following treatment with artemisinine of *Plasmodium berghei* infection in mice, with observations of the localization of (^{3}H)-dihydroartemisinine in *P. falciparum* in vitro. Ann Trop Med Parasitol 79:367–374

72. Wang DW, Liu XT (1983) Effect of Qing Hao Su (artemisinin) on the ultrastructure of the myocardium in rhesus monkeys. Acta Pharmacol Sin 4:191–194
73. Yang LX, Wu BF, Dai BQ (1984) Effect of Qing Hao Su on the frequency of micronucleus of mouse bone marrow polychromatic erythrocytes. Bull Chin Mat Med 9:132–134
74. China Cooperative Research Group on Qinghaosu and its derivatives as antimalarials (1982) J Trad Chin Med 2:9, 17, 25, 31
75. Jiang JB, Li GQ, Guo XB, Kong YC (1982) Antimalarial activity of mefloquine and qinghaosu. Lancet 2:285–288
76. Niu XY, Ho LY, Ren ZH, Song ZY (1985) Metabolic fate of qinghaosu in rats, a new TLC densitometric method for its determination in biological material. Eur J Drug Metab Pharmacokinet 10:55–59
77. Zhao KC, Chen QM, Song ZY (1986) Pharmacokinetics of qinghaosu and two of its active derivatives in dogs. Acta Pharm Sin 21:736–739
78. Song ZY, Zhao KC, Liang XT, Liu CX, Yi MG (1985) Radioimmunoassay of qinghaosu and artesunate. Acta Pharm Sin 20:610–614
79. Zhu DY, Huang BS, Chen ZL, Yin ML (1980) Isolation and identification of biotransformation metabolites of qinghaosu. I. Isolation and identification of biotransformation metabolites in humans. Acta Pharm Sin 15:509–512
80. Zhu DY, Huang BS, Chen ZL, Yin ML, Yang YM, Dai ML, Wang BD, Huang ZH (1983) Isolation and identification of the metabolite of artemisinin in human. Acta Pharmacol Sin 4:194–197
81. Yu PL, Chen IH, Lee Y, Chi JY (1980) Synthesis of artemisinin analogs containing halogen, nitrogen or sulfur atoms. Chin Pharm Bull 15:44
82. Gu HM, Liu MZ, Lu BF, Xu JY, Chen LJ, Wang MY, Sun WK, Yu B, Ji RY (1981) Antimalarial effect and toxicity of artemether in animals. Acta Pharmacol Sin 2:138–144
83. Jiang JR, Yan HY, Zhuang YH, Xu GY, Zeng YL, Ding SF (1983) Absorption, distribution and excretion of artemether in mice and rats. Acta Pharmacol Sin 4:197–201
84. Liu ZM, Yang YY, Hu MY, Li FW, Gan J, Li PS, Zhang SF, Wei ZQ, Shi WZ (1981) Arteannuin derivatives. Antimalarial activity and toxicity of arteannuin ether and the in vivo dynamics of ^{3}H-labeled arteannuin ether. Guangxi Med J 13–18
85. Li R, Liao TY, Huang KY, Chou LL (1981) Pharmacokinetic study of sodium qinghaosu hemisuccinate in vivo. Chin Trad Herb Drugs 12:20–22
86. Li WH, Shu HL, Xu GY, Zeng YL (1982) Binding of qinghaosu (artemisinin) and its derivatives to plasma protein. Acta Pharm Sin 17:783–786
87. Gu HM, Warhurst DC, Peters W (1984) Uptake of ^{3}H-dihydroartemisinin by erythrocytes infected with *Plasmodium falciparum* in vitro. Trans R Soc Trop Med Hyg 78:265–270
88. Li ZL, Gu HM, Warhurst DC, Peters W (1983) Effects of qinghaosu and related compounds on incorporation of (^{3}H)hypoxanthine by *Plasmodium falciparum*. Trans R Soc Trop Med Hyg 77:522–523
89. Gu HM, Warhurst DC, Peters W (1983) Rapid action of qinghaosu and related drugs on incorporation of (^{3}H)isoleucine by *Plasmodium falciparum* in vitro. Biochem Pharmacol 32:2463–2466
90. Wu JA, Ji RY (1982) A quantitative Hansch structure-activity study on artemisinin derivatives. Acta Pharmacol Sin 3:55–60
91. Wu JA, Chen KX, Ji RY (1981) Extended Hüeckel molecular orbital calculations of antimalarial drugs – artemisinin and its derivatives. Fenzi Kexue Xuebao 1:27–34
92. Le WJ, You JQ, Yang YQ, Mei JY, Guo HF, Yang HZ, Zhang CW (1982) Studies on the efficacy of artemether in experimental schistosomiasis. Acta Pharm Sin 17:187–193
93. Yue WJ, You JQ, Mei JY (1984) Effects of artemether on *Schistosoma japonicum* adult worms and ova. Acta Pharmacol Sin 5:60–63
94. Le WJ, You JQ, Mei JY (1983) Chemotherapeutic effect of artesunate in experimental schistosomiasis. Acta Pharm Sin 18:619–621
95. Wu LJ, Yang HZ, Yang YQ (1983) Histological and histochemical changes of *Schistosoma japonicum* and host liver caused by artemether. Acta pharm Sin 18:7–14
96. Yang YQ, Yang HZ, Zhang CW (1986) Histological observation on the effects of artemether, fuvinazole and niridazole on *Schistosoma japonicum* schistosomulae in mouse liver. Acta Pharmacol Sin 7:276–278
97. Chen RX, Qu ZQ, Zeng MA, Li JY (1983) Effect of qinghaosu and its derivatives on *Clonorchis sinensis* in rats. Chin Pharm Bull 18:410–411

Artemisia argyi Levl. et Vant.

23

23.1 Introduction

Aiye, Folium Artemisiae argyi, is the dry leaf of *Artemisia argyi* Levl. et Vant. (Asteraceae) collected in summer before the plant blooms. It is listed officially in the Chinese Pharmacopoeia and recommended for use as an analgesic and hemostatic.

23.2 Chemical Constituents

A. argyi contains a number of terpene compounds. From the leaves *trans*-carveol, α-terpineol, 4-terpineol, α-phellandrene, camphene, α-cedrene, bornyl acetate, elemol, isoborneol, and carvone were isolated and identified [1]. From the nonvolatile fraction ethyl palmitate, ethyl oleate, ethyl linoleate, lupenone, lupenyl acetate, α-amyrin acetate, β-amyrin acetate, glutinone, fernenone, 24-methylene-cycloartanone, simiarenol, and *trans*-phenylitaconic acid were found [2]. Flavones detected in the leaves of *A. argyi* were eupatilin and 5-hydroxy-6,7,3′,4′-tetramethoxyflavone [3]. The three bitter lactones isolated from the *A. argyi* herb were isoridentin, chrysartemin B, and its stereoisomer chrysartemin A (canin) [4].

Among the terpene compounds, *trans*-carveol (*23-1*), α-terpineol (*23-2*), 4-terpineol (*23-3*), and carvone (*23-4*) are oxygenated monocyclic monoterpenes. Elemol (*23-5*) and α-cedrene (*23-6*) are sesquiterpenes.

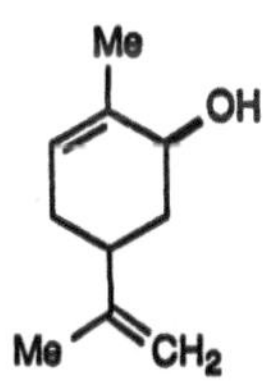

trans-Carveol (*23-1*)

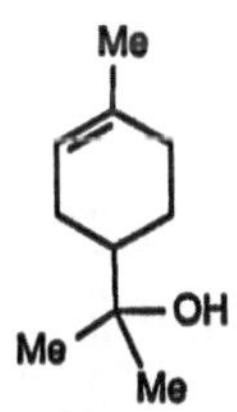

α-Terpineol (*23-2*)

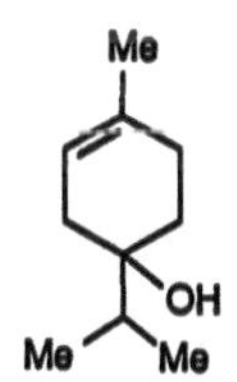

4-Terpineol (*23-3*)

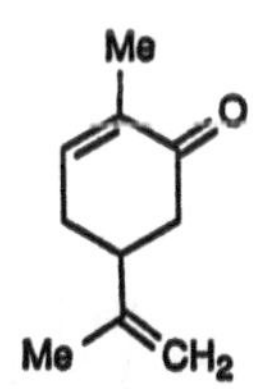

Carvone (*23-4*)

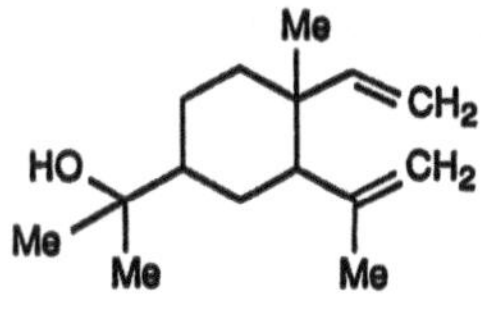

Elemol (*23-5*)

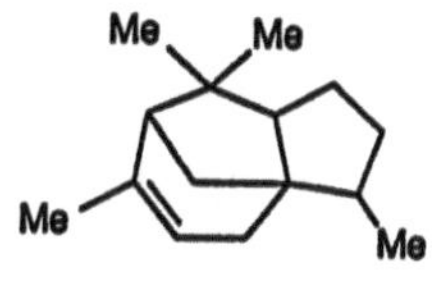

α-Cedrene (*23-6*)

Lupenone, lupenyl acetate, α-amyrin acetate, β-amyrin acetate, glutinone, fernenone, 24-methylene cycloartanone, and simiarenol are triterpene derivatives. They have different basic skeletons. Lupenone (*23-7*) and lupenyl acetate (*23-8*) are derived from lupane (*23-9*); 24-methylene cycloartanone (*23-10*) is derived from 9,19-cyclolanostane (*23-11*); fernenone (*23-12*) is derived from D:C-friedo-B′:A′-neogammacerane (*23-13*); simiarenol (*23-14*) is derived from D:B-friedo-B′:A′-neogammacerane (*23-15*); α-amyrin (*23-16*) possesses an ursane (*23-17*), β-amyrin (*23-18*) an oleanane (*2-4*), and glutinone (*23-19*) a D:B-friedooleanane (*23-20*) skeleton.

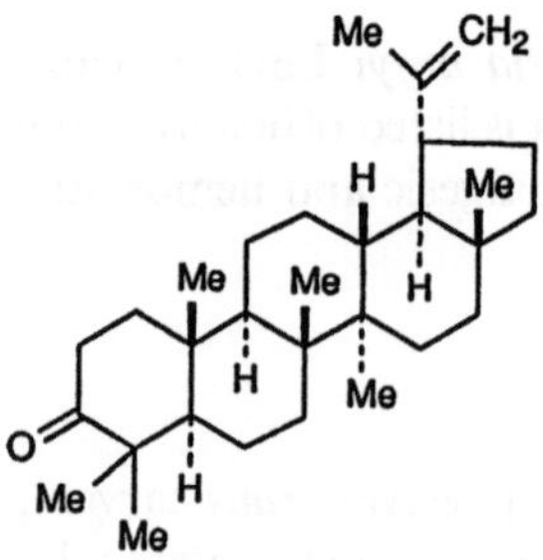

Lupenone (*23-7*)

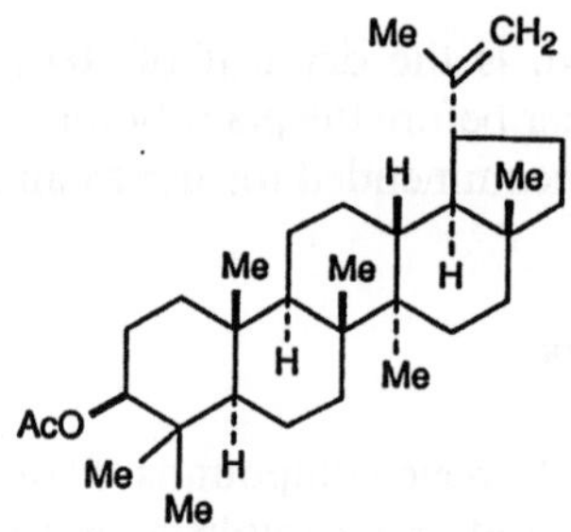

Lupenyl acetate (*23-8*)

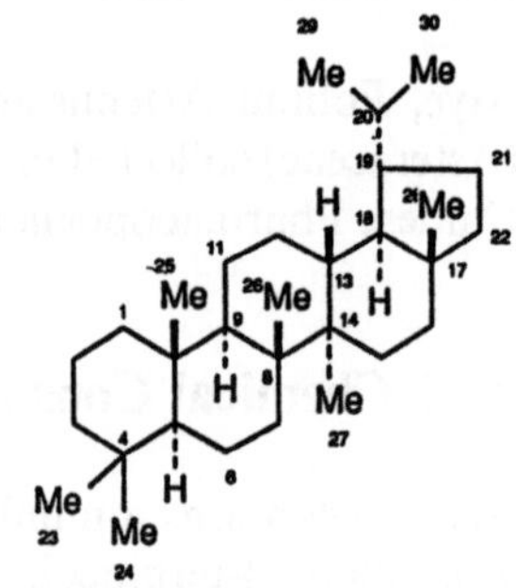

Lupane (*23-9*)

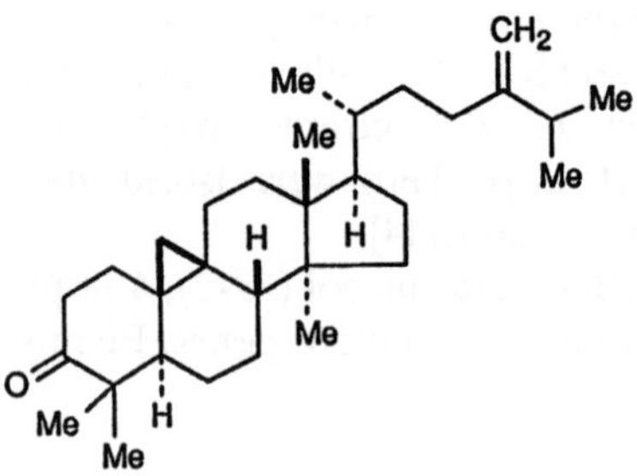

24-Methylene-cycloartanone (*23-10*)

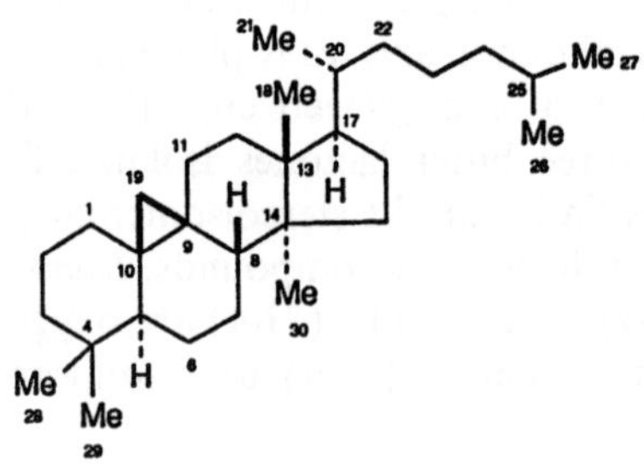

9,19-Cyclolanostane (*23-11*)

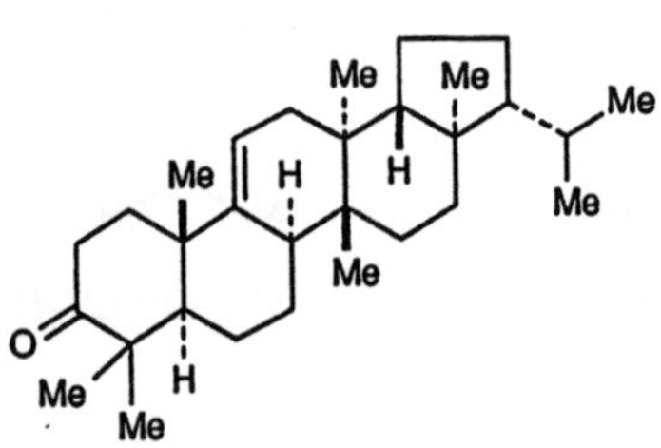

Fernenone (*23-12*)

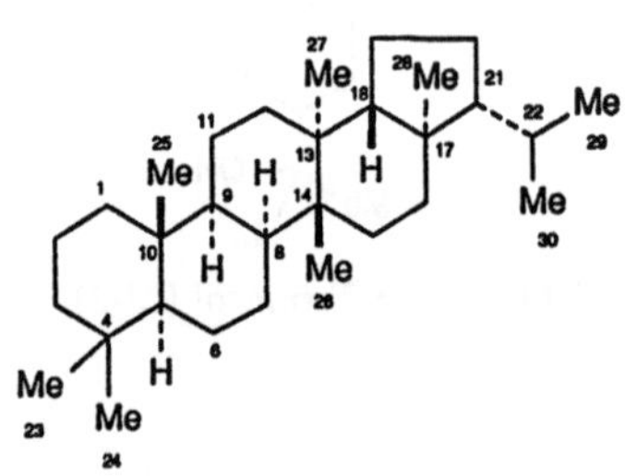

D:*C*-Friedo-*B*′:*A*′-neogammacerane (*23-13*)

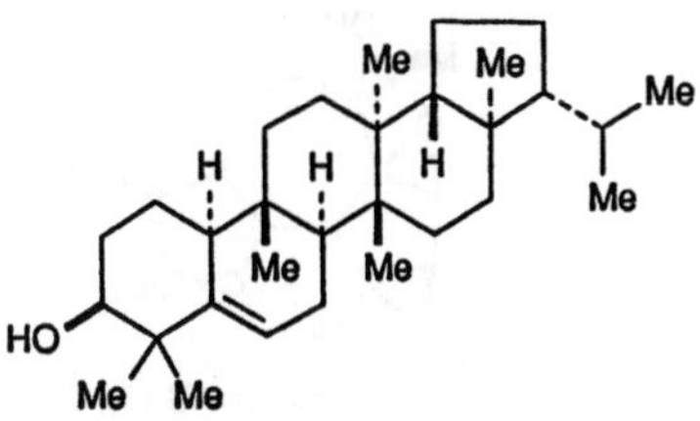

Simiarenol (*23-14*)

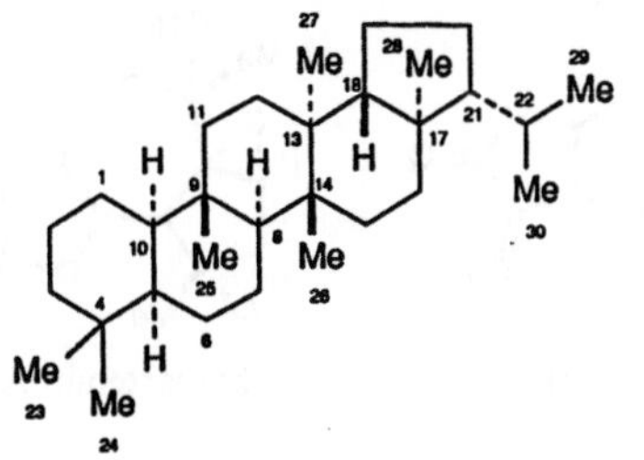

D:*B*-Friedo-*B′*:*A′*-neogammacerane (*23-15*)

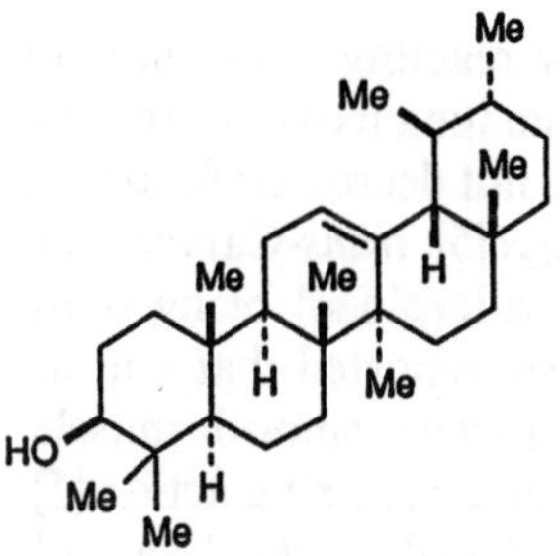

α-Amyrin (*23-16*)

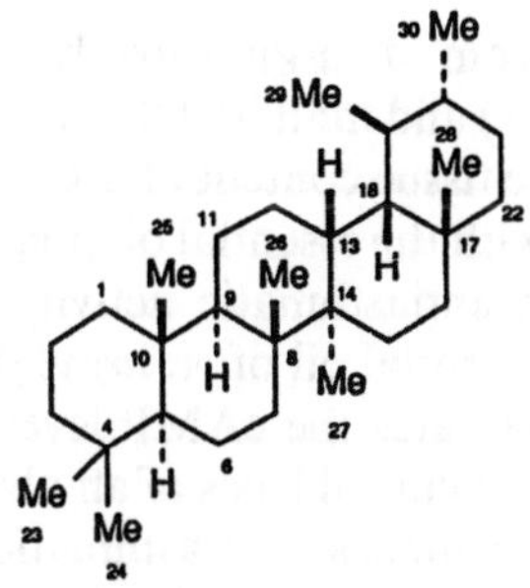

Ursane (*23-17*)

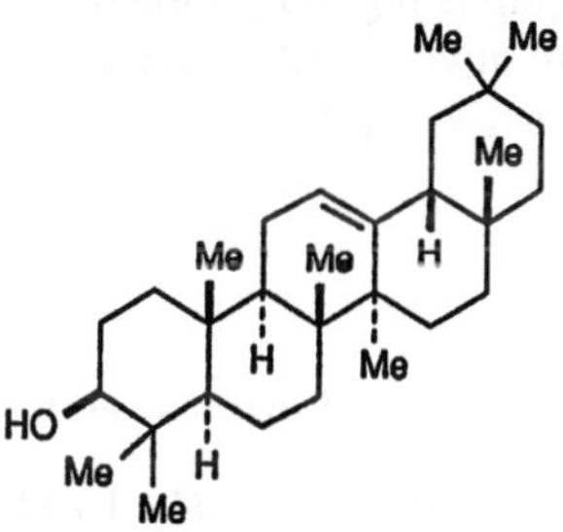

β-Amyrin (*23-18*)

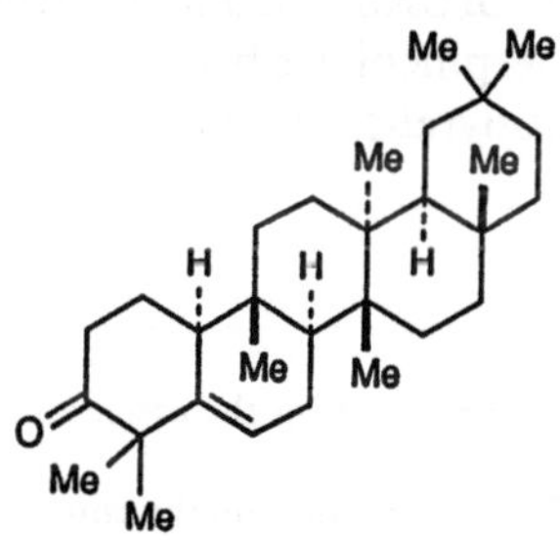

Glutinone (*23-19*)

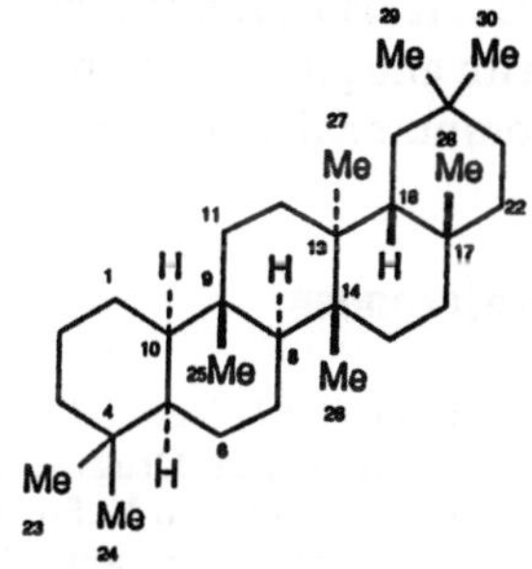

D:*B*-Friedoolaenane (*23-20*)

Phenylitaconic acid (*23-21*) is a butanedioic acid derivative, isolated for the first time from a natural source [2].

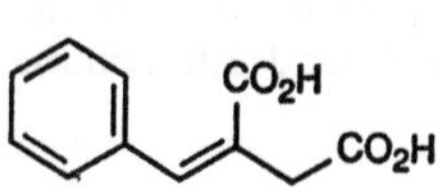

Phenylitaconic acid (*23-21*)

Eupatilin is 5,7-dihydroxy-6,3′,4′-trimethoxyflavone. The structures of the three lactones isoridentin (*23-22*), chrysartemin A (*23-23*), and chrysartemin B (*23-24*) isolated from *A. argyi* are shown below.

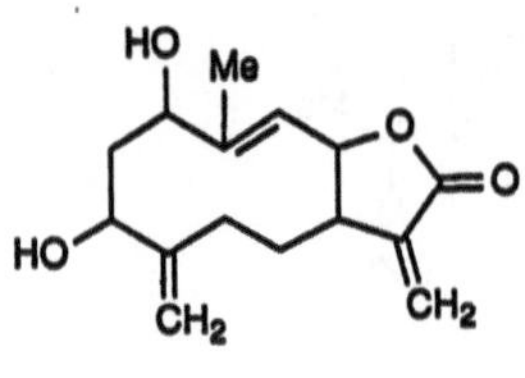

Isoridentin (*23-22*)

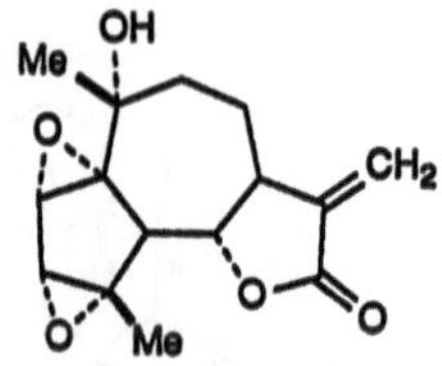

Chrysartemin A (*23-23*)

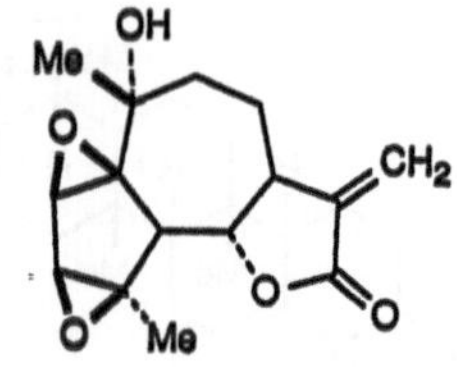

Chrysartemin B (*23-24*)

23.3 Pharmacology

The essential oil from *A. argyi* directly antagonized slow reacting substances of anaphylaxis (SRS-A) and inhibited the release of SRS-A from lung tissue or tracheal smooth muscle. The tissue content of SRS-A in the lung was not decreased following a single perfusion with the essential oil preparation of *A. argyi* [5]. *trans*-Carveol and α-terpineol had an antiasthmatic activity in guinea pigs, α-terpineol being more effective than the essential oil of *A. argyi* [1]. It has also been reported that α-terpineol significantly elevates the cAMP level in guinea pigs tracheal smooth muscle. This may be the biochemical basis of an observed smooth muscle relaxing action [5]. 4-Terpineol had the highest antiasthmatic efficacy and induced tracheal smooth muscle relaxation. At a concentration lower than that required for relaxation, it still blocked the allergic contraction of isolated guinea pig ileum sensitized with ovalbumin. It also inhibited the release of SRS-A from sensitized guinea pig lung challenged with antigen. Furthermore, 4-terpineol has been reported to have antitussive, expectorant, analgesic, sedative, antipyretic, and bacteriostatic activity [6].

References

1. Sun JY (1981) New antiasthmatic principles in the essential oil from leaves of *Artemisia argyi*. Chin Trad Herb Drugs 12:558
2. Lao AN, Fujimoto Y, Tatsuno T (1984) Studies on the constituents of *Artemisia argyi* Levl. and Vant. Chem Pharm Bull 32:723–727
3. Wu CM, Tu YY (1985) Studies on chemical constituents of *Artemisia* species. III. Isolation and identification of the lipophilic constituents from *Artemisia argyi*. Bull Chin Mat Med 10:31–32
4. Yusupov MI, Kasymov SZ, Sidyakin GP, Boiko EV (1985) *Artemisia argyi* lactones. Khim Prir Soedin 405–406 (CA 103:51216c)
5. Bian RL, Yang QH, Geng BQ, Yong DG, Chen LF (1982) Isolation and analysis of slow reacting substances of anaphylaxis (SRS-A) and the anti-SRS-A effect of essential oil of *Artemisia argyi*. Zhejiang Yike Daxue Xuebao 11:185–187
6. Bian RL, Yang QH, Geng BQ, Xie OM, Yang W (1981) Antiasthmatic constituents in the essential oil of *Artemisia argyi* – pharmacological study on terpineol-4. Chin J Tuberculosis Respir Dis 4:203–206

Artemisia scoparia Waldst. et Kit. and A. capillaris Thunb. 24

24.1 Introduction

Yinchen, Herba Artemisiae scopariae is the dry young sprout of *Artemisia scoparia* Waldst. et Kit. or *A. capillaris* Thunb. (Asteraceae) collected in spring when the sprouts have reached a height of 6–10 cm. It is listed officially in the Chinese Pharmacopoeia and used mainly as a choleretic, antiinflammatory, and diuretic agent in the treatment of epidemic hepatitis.

24.2 Chemical Constituents

24.2.1 Chemical Constituents of *Artemisia capillaris*

A number of polyacetylene derivatives were isolated from the essential oil of *A. capillaris*. Thus, capillene (*24-1*), capillone (*24-2*), capillin (*24-3*), capillarin (*24-4*), dehydrofalcarinone (*24-5*), dehydrofalcarinol (*24-6*) [1], norcapillene (5-phenyl-1,3-pentadiyne, *24-7*) [1, 2], capillanol (*24-8*) [3], methoxycapillene (*24-9*) [4], and neocapillene (*24-10*) [5] were detected. Neocapillene may be an artifact produced from capillene by UV irradiation [5]; capillarin is an isocoumarin derivative with an acetylenic side chain.

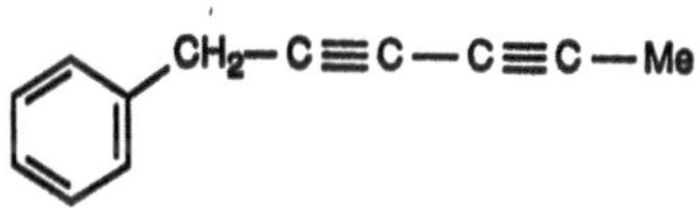

Capillene (*24-1*)

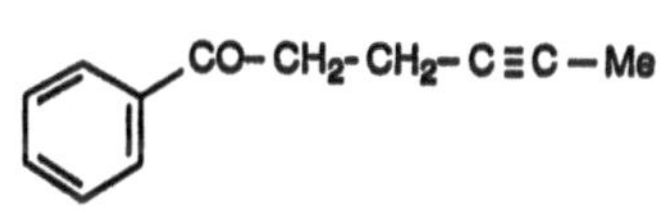

Capillone (*24-2*)

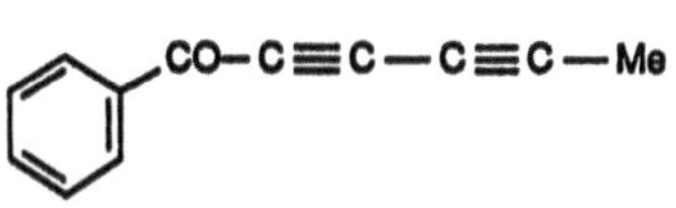

Capillin (*24-3*)

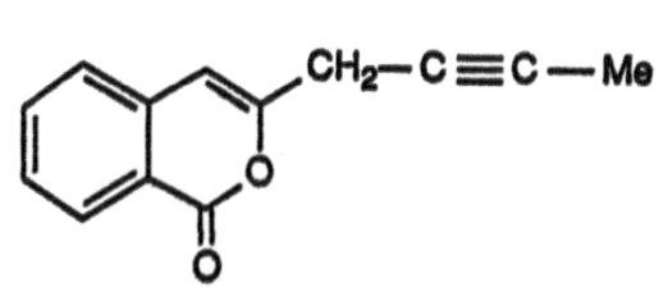

Capillarin (*24-4*)

$H_2C{=}CH{-}CO{-}C{\equiv}C{-}C{\equiv}C{-}CH_2{-}CH{=}CH{-}(CH_2)_5{-}CH{=}CH_2$

Dehydrofalcarinone (*24-5*)

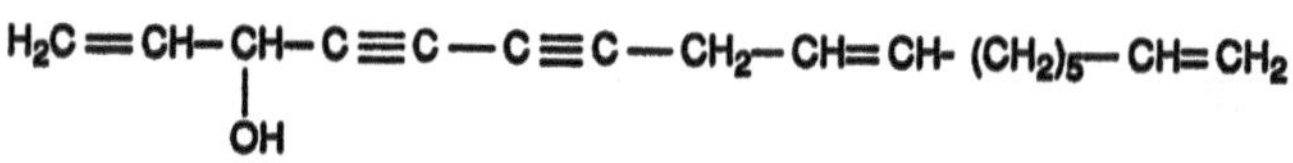

Dehydrofalcarinol (*24-6*)

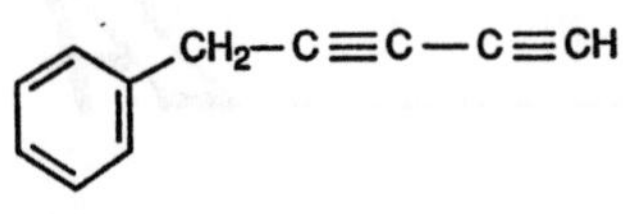

Norcapillene (*24-7*)

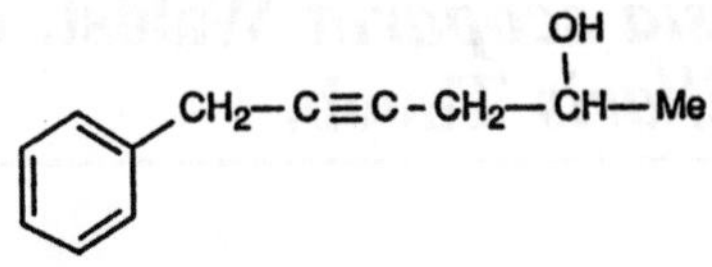

Capillanol (*24-8*)

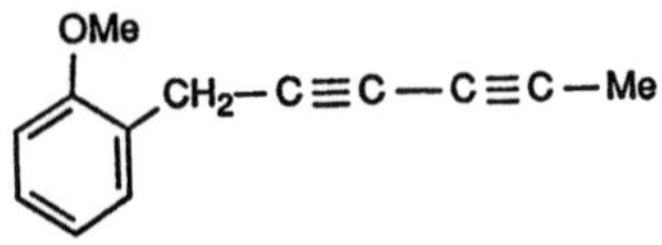

Methoxycapillene (*24-9*)

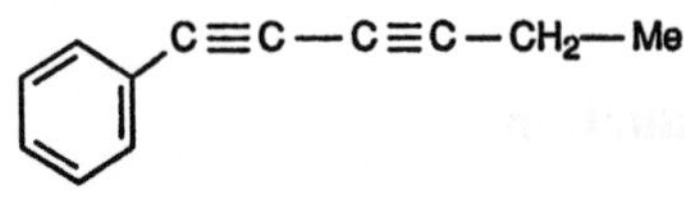

Neocapillen (*24-10*)

Besides the acetylenic derivatives, terpenes such as α- and β-pinene, *p*-cymene, Δ^3-carene, α-terpineol, bornyl acetate, methyleugenol, β-elemene, and β-caryophyllene were isolated and identified from *A. capillaris* [1]. The distribution of volatile components in different parts of the plant was also described. Thus, the main volatile components (% of total) in fine stem and leaves were capillene (26%), capillarin (14%), dehydrofalcarinol (9%), and β-caryophyllene (7%); the main volatile components in stem were dehydrofalcarinol (43%), dehydrofalcarinone (14%), and capillene (8%); the main volatile components in roots were dehydrofalcarinol (67%) and capillene (14%); the main volatile components in seeds were capillene (34%), capillarin (10%), Δ^3-carene (9%), and β-pinene (9%) [1].

Flavones isolated from *A. capillaris* were cirsilineol (5,4′-dihydroxy-6,7,3′-trimethoxyflavone), cirsimaritin (5,4′-dihydroxy-6,7-dimethoxyflavone), genkwanin (5,4′-dihydroxy-7-methoxyflavone), and rhamnocitrin (3,5,4′-trihydroxy-7-methoxyflavone) [6]. A new flavone, arcapillin (5,2′,4′-trihydroxy-6,7,5′-trimethoxyflavone (*24-11*), was isolated together with eupatolitin (3,5,3′,4′-tetrahydroxy-6,7-dimethoxyflavone) from *A. capillaris* and structurally elucidated by spectroscopic methods [7, 8].

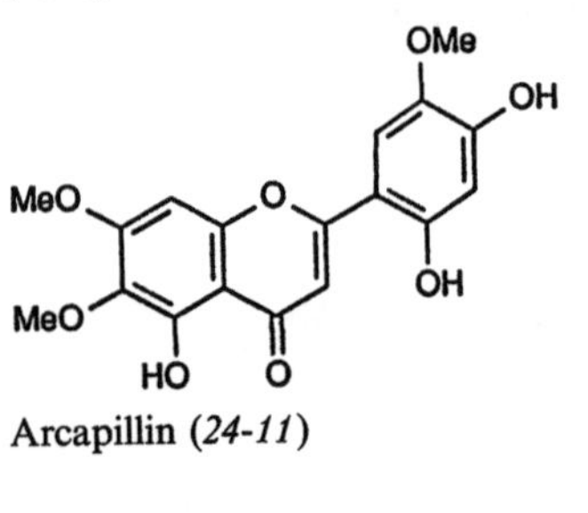

Arcapillin (*24-11*)

Capillarisin (*24-12*), a major constituent of *A. capillaris*, was shown to be a phenoxychromone derivative [9]. Some capillarisin related phenoxychromones with different substituents were also isolated [6].

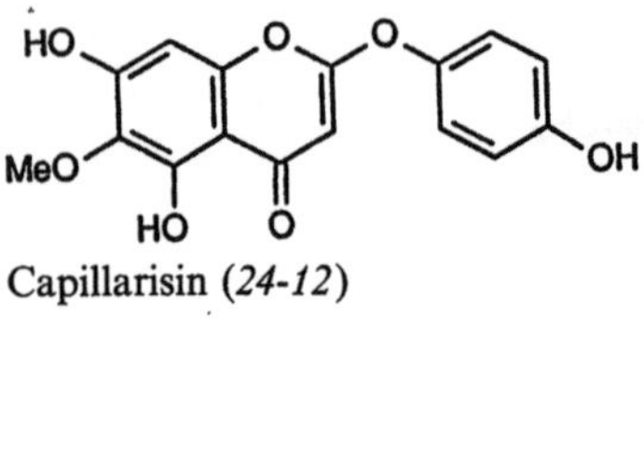

Capillarisin (*24-12*)

In addition to these constituents, the coumarin derivative scoparone (*24-13*) [1, 10, 11] and two new stereoisomeric constituents, capillartemisin A (*24-14*) and B (*24-15*) [12], were isolated from *A. capillaris*.

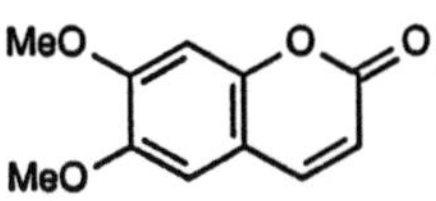

Scoparone (*24-13*)

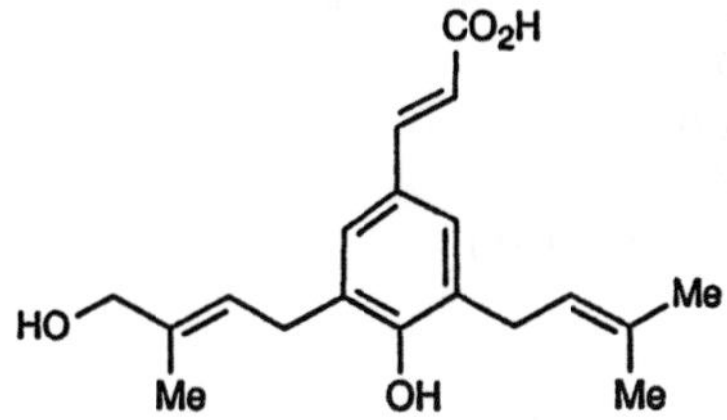

Capillartemisin A (*24-14*)

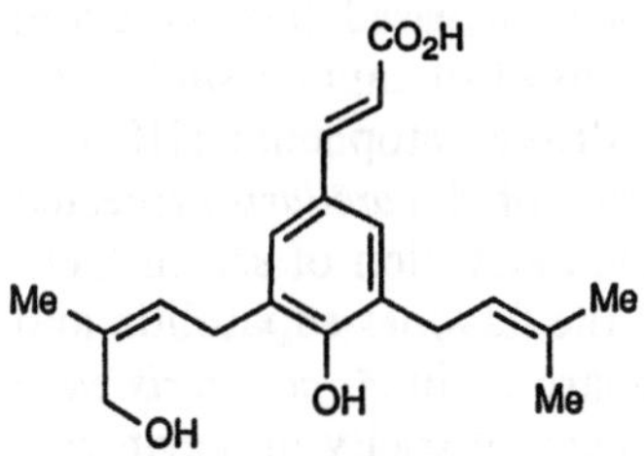

Capillartemisin B (*24-15*)

24.2.2 Chemical Constituents of *Artemisia scoparia*

From the essential oil of *A. scoparia*, the following compounds were isolated and identified: α- and β-pinene, myrcene, cineol, *p*-cymol, carvone, thujone, apiole (*24-16*), isoeugenol, cadinene [13], caryophyllene epoxide (*24-17*) [14], vanillin, capillin, and 1-phenyl-2,4-hexadiyne-1-ol [15].

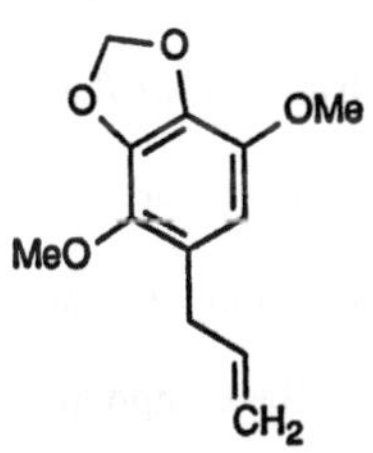

Apiole (*24-16*)

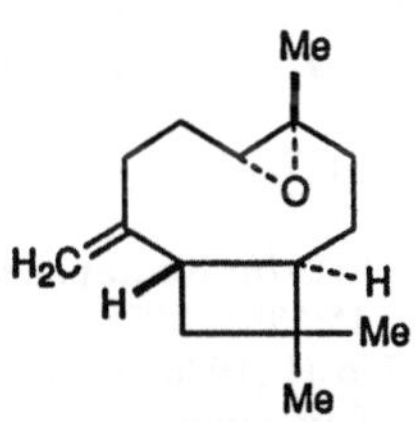

Caryophyllene epoxide (*24-17*)

The flavones 7-methylaromadendrin (3,5,4′-trihydroxy-7-methoxyflavanone), rhamnocitrin, eupalitin (3,5,4′-trihydroxy-6,7-dimethoxyflavone), cirsimaritin, and eupatolitin; the coumarin derivatives 7-methyl-esculetin, scoparone, and scopoletin [16]; and the isocoumarin derivative capillarin [17] were isolated from *A. scoparia*. In addition, *p*-hydroxyacetophenone and chlorogenic acid (*24-18*) were also found [18]. Large amounts of *p*-hydroxyacetophenone were found in the plant in June and July, of chlorogenic acid in July through October, and of scoparone in August through October [18].

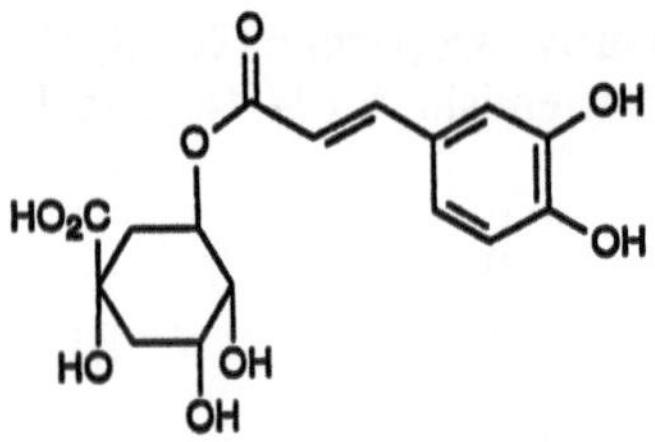

Chlorogenic acid (*24-18*)

24.3 Pharmacology

Essential oils from *A. capillaris* and its oil-free extract increased bile secretion; scoparone showed the same effect [19]. The choleretic effect of capillarisin [9, 20], capillartemisin A and B [12], chlorogenic acid, and *p*-hydroxyacetophenone [18] was also proven in experimental studies. The methanol extract of *A. capillaris* protected mice from CCl_4-induced hepatotoxicity [7] and inhibited elevation of serum GOT and GPT [22]. The active principles were shown to be the flavones eupatolitin and arcapillin [7]. Flavones and coumarins, especially scoparone, in *A. capillaris* were also able to prevent CCl_4- or galactosamine-induced hepatotoxicity in hepatocyte cell cultures [21, 23]. Moreover, scoparone had antiinflammatory and analgesic properties [10]. The decoction showed clinical effectiveness in treatment of biliary ascariasis [24].

References

1. Harada R, Iwasaki M (1982) Volatile components of *Artemisia capillaris*. Phytochemistry 21:2009–2011
2. Miyazawa M, Kameoka H (1976) Norcapillene, a new acetylenic hydrocarbon from the essential oil of *Artemisia capillaris*. Phytochemistry 15:223–224
3. Miyazawa M, Kameoka H (1975) Capillanol, a new acetylenic alcohol from *Artemisia capillaris*. Phytochemistry 14:1874
4. Miyazawa M, Kameoka H (1975) New polyacetylene from *Artemisia capillaris*. Phytochemistry 14:1126
5. Miyazawa M, Kameoka H (1976) Neocapillen, a new acetylenic hydrocarbon from *Artemisia capillaris*. Phytochemistry 15:1987–1988
6. Komiya T, Naruse Y, Oshio H (1976) Studies on "Inchinko". II. Studies on the compounds related to capillarisin and flavonoids. Yakugaku Zasshi 96:855–862
7. Kiso Y, Sasaki K, Oshima Y, Hikino H (1982) Liver-protective drugs. V. Validity of the oriental medicines. 42. Structure of arcapillin, an antihepatoxic principle of *Artemisia capillaris* herbs. Heterocycles 19:1615–1617
8. Namba T, Hattori M, Takehana Y, Tsunezuka M, Tomimori T, Kizu H, Miyaichi Y (1983) A flavone from *Artemisia capillaris*. Phytochemistry 22:1057–1058
9. Komiya T, Tsukui M, Oshio H (1975) Capillarisin, a constituent from *Artemisia capillaris* herba. Chem Pharm Bull 23:1387–1388
10. Yamahara J, Matsuda H, Sawada T, Mibu H, Fujimira H (1982) Biologically active principles of crude drugs. Pharmacological evaluation of *Artemisia capillaris* flos. Yakugaku Zasshi 102:285–291
11. Singh G, Nair GV, Aggarwal KP (1954) The structure of scoparone. Chem Ind 1294–1295
12. Kitagawa I, Fukuda Y, Yoshihara M, Yamahara J, Yoshikawa M (1983) Capillartemisin A and B, two new choleretic principles from *Artemisia capillaris* herba. Chem Pharm Bull 31:352–355

13. Mishurova SS, Abbasov RM, Mamedalieva FM (1983) Study of *Artemisia scoparia* essential oils. Izv Akad Nauk Az SSR [Biol] (2) 3–5 (CA 101:188066g)
14. Thappa RK, Vashisht VN, Singh J, Sharma BK (1970) Caryophyllene epoxide from the oil of *Artemisia scoparia, Elsholtzia polystachya, Piper hookeri*, and *Piper brachystachyum*. Curr Sci 39:182–183
15. Stefanovic M, Krstic L, Mladenovic S (1973) Extractives of *Artemisia scoparia*. Phytochemistry 12:2996–2997
16. Chandrasekharan I, Khan HA, Ghanim A (1981) Flavonoids from *Artemisia scoparia*. Planta Med 43:310–311
17. Nesmelova EF, Sidyakin GP (1971) Lactones of *Artemisia scoparia*. Khim Prir Soedin 7:376–377 (CA 75:115850t)
18. Xu QC, Yang LJ, Yang CS (1983) Studies on quality of *Artemisis scoparia* – determination and comparison of content variations of the three major choleretic constituents of *A. scoparia* collected in different seasons. Chin Trad Herb Drugs 14:35–40
19. Aburada M, Sasaki H, Harada M (1976) Pharmacological studies of Gardeniae Fructus. II. Contribution of the constituent crude drugs to choleretic activity of "Inchinko-to" in rats. Yakugaku Zasshi 96:147–153
20. Komiya T, Tsukui M, Oshio H (1976) Studies on "Inchinko". I. Capillarisin, a new choleretic substance. Yakugaku Zasshi 96:841–854
21. Kiso Y, Ogasawara S, Hirota K, Watanabe N, Oshima Y, Konno C, Hikino H (1984) Validity of Oriental medicines. LV. Liverprotective drugs. 10. Antihepatotoxic principles of *Artemisia capillaris* buds. Planta Med 50:81–85
22. Kimura Y, Okuda H, Okuda T, Hatano T, Agata I, Arichi S (1985) Studies on the activities of tannins and related compounds from medicinal plants and drugs. VII. Effects of extracts of leaves of *Artemisia* species, and caffeic acid and chlorogenic acid on lipid metabolic injury in rats fed peroxidized oil. Chem Pharm Bull 33:2028–2034
23. Hikino H (1985) Antihepatoxic constituents of Chinese drugs. Chin Pharm Bull 20:415–417
24. Wang SM, Tang SM (1985) Treatment of 15 cases of biliary ascariasis mainly with oriental wormwood. Zheijiang J Trad Chin Med 20:64–65

Asarum spp.

25

25.1 Introduction

Xixin, Herba Asari, is the dry whole plant of *Asarum heterotropoides* Fr. var. *mandshuricum* (Maxim.) Kitag., *A. sieboldii* Miq. var. *seoulense* Nakai, or *A. sieboldii* Miq. (Aristolochiaceae) collected in summer or fall when the fruits are ripe. It is listed officially in the Chinese Pharmacopoeia and used as an analgesic and antitussive agent for treatment of influenza, headache, rheumatic pain, and asthma.

25.2 Chemical Constituents

The chemical composition of the essential oil obtained from *A. sieboldii* was investigated by gas chromatography-mass spectroscopy (GC/MS). Thus, 1,8-cineole, asaricin, methyleugenol, croweacin, *β*-pinene, α-pinene [1], α-thujene, myrcene, terpinen-4-ol, α-terpineol, safrole, and myristicin [2] were identified.

From the essential oils of other *Asarum* species used in Chinese medicine such as *A. sieboldii* var. *soeulense, A. forbesii, A. inflatum, A. magnificum* var. *dinghugense*, and *A. caudigerum* var. *cardiophyllum* more than 100 peaks were found and some of them were identified. 3,5-Dimethoxytoluene, safrole, methyleugenol, and elemicin were found in all species [3, 4]. In addition, longifolene, α-phellandrene, *β*-pinene, limonene, bornyl acetate, α-bisabolene, α-himachalene, α-santalene, alloaromadendrene, *p*-methoxycarbonylbenzoic acid ethyl ester, 3-isopropyl-5,5-dimethylcyclohexanone, *o*-cymene, and the alcohol of acacipetalin were detected by GC/MS from the essential oil isolated from the whole plant of *A. himalaicum* [5].

Methyleugenol (*25-1*), elemicin (*25-2*), asaricin (*25-3*), croweacin (*25-4*), and safrole (*25-5*) are all allyl benzene derivatives, whereas α-santalene (*25-6*), longifolene (*25-7*), α-himachalene (*25-8*), and alloaromadendrene (*25-9*) are sesquiterpene compounds. Acacipetalin (*25-10*) is a hydroxybutenenitrile glucoside.

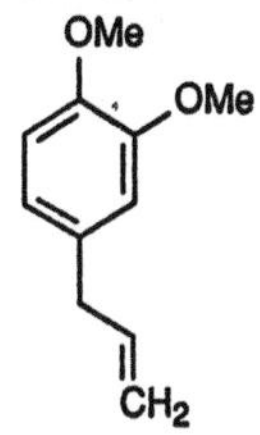

Methyleugenol (*25-1*)

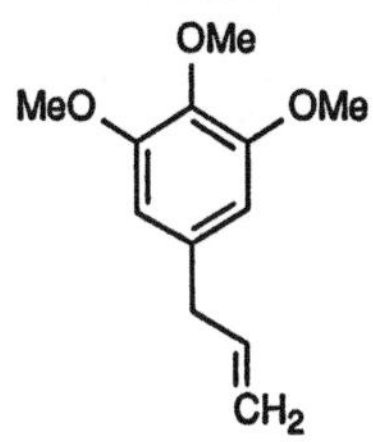

Elemicin (*25-2*)

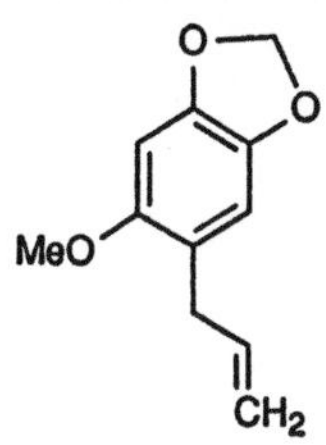

Asaricin (*25-3*)

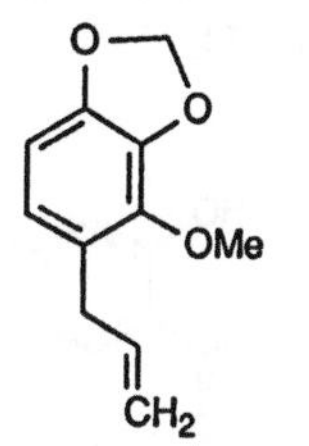

Croweacin (*25-4*)

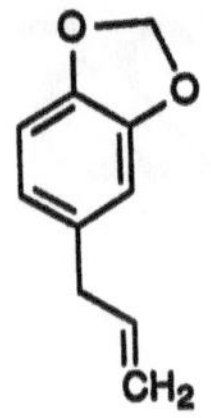

Safrole (*25-5*)

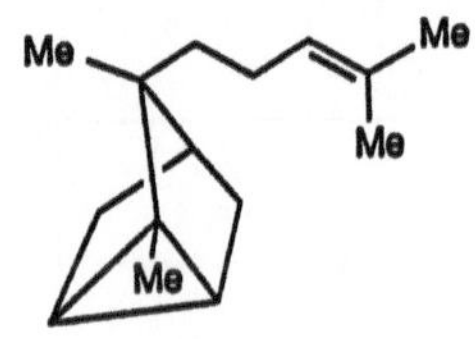

α-Santalene (*25-6*)

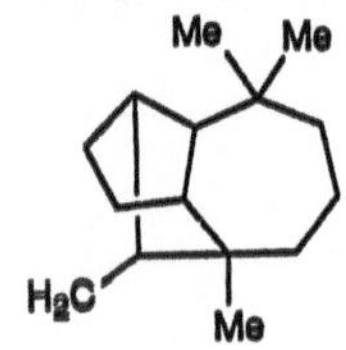

Longifolene (*25-7*)

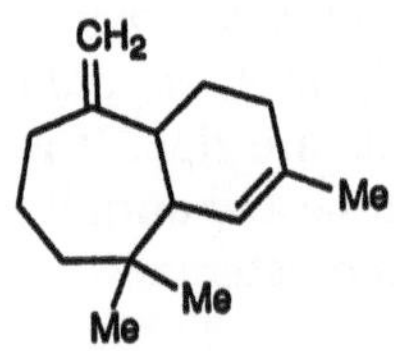

α-Himachalene (*25-8*)

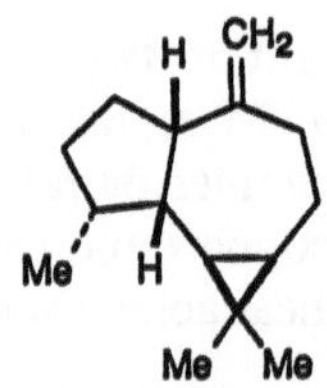

Alloaromadendrene (*25-9*)

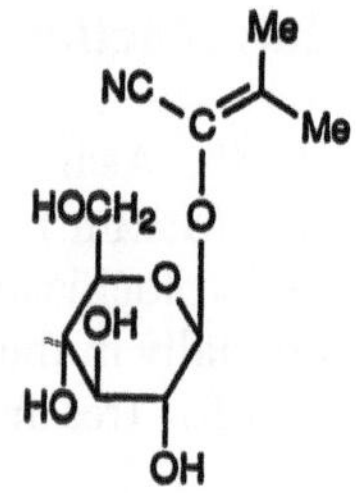

Acacipetalin (*25-10*)

The chemical compositions of the essential oils from other *Asarum* species have also been investigated including *A. caulescens, A. chinense, A. debile, A. ichangense, A. insigne, A. longerhizomatosum, A. magnificum, A. maximum, A. pulchellum*, and *A. wulingense*. A total of 53 components were determined by GC/MS. Safrole, methyleugenol, and elemicin were present in all ten species [6]. The essential oil content of *A. ichangense, A. wulingense*, and *A. longerhizomatosum* were analyzed and compared with that of the official species *A. heterotropoides* var. *manshuricum*. *A. longerhizomatosum* had the largest amount (3%) with methyleugenol as the major component; *A. ichangense* had the smallest amount (0.6%–0.9%) [7].

25.3 Pharmacology

The hemodynamic activity of *A. heterotropoides* in anesthesized dogs was compared with that of higenamine (*3-98*) and isoprenaline. *Asarum* extract had less heart rate increasing but more cardiac output stimulating effects than higenamine and isoprenaline [8, 9].

In mice, essential oil isolated from *A. forbesii* produced a dose dependent antihyperlipemic effect when given orally. The active principle was identified as kakuol (*25-11*).

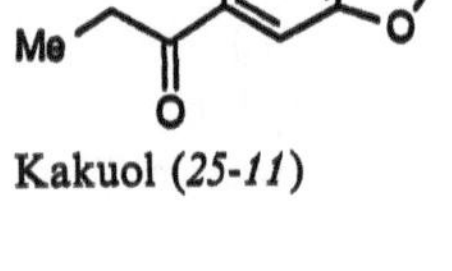

Kakuol (*25-11*)

The blood concentration time curve of kakuol in mice fitted a two compartment open model. It is rapidly absorbed and distributed in the body and slowly eliminated. The toxicity of kakuol appears to be very low [10].

Methyleugenol, isolated from *A. sieboldii,* completely inhibited toxin production by *Aspergillus versicolor* and three other *Aspergillus* strains at 100 and 200 µg/ml [11].

It is important to mention that safrole is known to be carcinogenic in mice and rats. It produces liver tumors after oral administration and liver and lung tumors in male infant mice following s.c. injection [12]. Innes et al. reported on the carcinogenicity of safrole to mice after oral administration. Male and female (C57BL/6 × C3H Anf)F_1 or (C57BL/6 × AKR)F_1 mice were given a daily dose of 464 mg/kg at 7 days of age by stomach tube until the animals were 28 days old. Subsequently, safrole was administered in the diet at a concentration of 1.1 g/kg of diet for up to 82 weeks. For each strain the difference in the occurrence of liver cell tumors between the experimental and control animals was significant [13]. Similar results were obtained in rats fed 0.1% or 0.5% safrole in the diet for 2 years [14]. Infant mice injected s.c. with a suspension of safrole in tricaprylin on days 1, 7, 14, and 21 after birth at a total dose of 0.66 mg or 6.6 mg developed hepatomas within 49–53 weeks. The animals in the higher dose group also developed pulmonary adenomas and pulmonary adenocarcinomas [15].

Studies on the metabolic activation of safrole in rats and mice revealed that the glucuronide of its 1′-hydroxy metabolite was excreted in the urine [16]. 1′-Hydroxysafrole (*25-12*) was considerably more hepatocarcinogenic than the parent substance, thus appearing to be a proximate carcinogen [17].

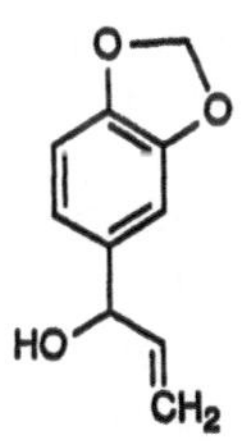

1′-Hydroxysafrole (*25-12*)

Evidence for metabolic activation of 1′-hydroxysafrole to reactive derivatives in vivo was provided by the formation of covalently bound adducts in liver DNA, RNA, and protein [18, 19]. Two adducts were characterized as N^2-(*trans*-isosafrol-3′-yl)deoxyguanosine (*25-13*) and N^6-(*trans*-isosafrol-3′-yl)-deoxyadenosine (*25-14*) [19].

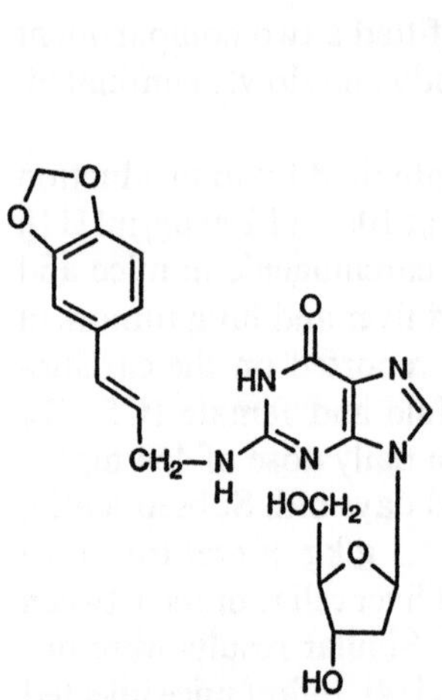

N^2-(*trans*-isosafrol-3′-yl)-
deoxyguanosine (*25-13*)

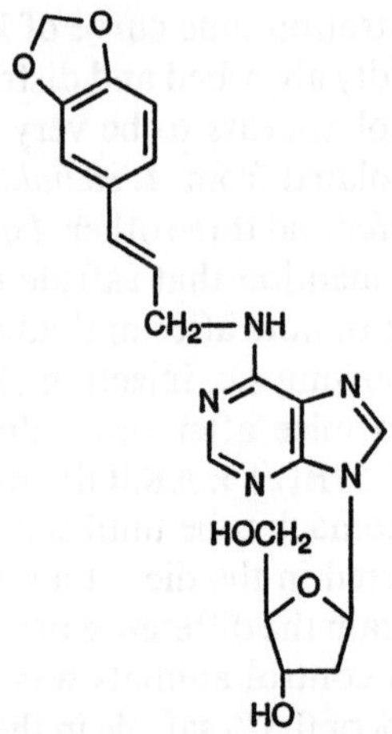

N^6-(*trans*-isosafrol-3′-yl)-
deoxyadenosine (*25-14*)

Rats, guinea pigs, and hamsters excreted 1%–3.5% of an i.p. dose of safrole as 1′-hydroxysafrole; male mice excreted 35% as 1′-hydroxysafrole and female mice 19% [17].

Following oral or i.p. administration of safrole to rats and guinea pigs, 3-*N,N*-dimethyl-amino-1-(3,4-methylenedioxyphenyl)-1-propanone (*25-15*) was the major urinary metabolite in guinea pigs and as a minor metabolite in rats. 3-Pyrrolidinyl-1-(3,4-methylenedioxyphenyl)-1-propanone, a further minor metabolite and 3-piperidyl-1-(3,4-methylenedioxyphenyl)-1-propanone were found in the urine of rats [20]. In male Sprague-Dawley rats and male guinea pigs given i.p. safrole, the main urinary metabolites were 1,2-dihydroxy-4-allyl-benzene, 1,2-methylenedioxy-4-(2,3-dihydroxypropyl)-benzene, and 1,2-dihydroxy-4-(2,3-dihydroxypropyl)-benzene (*25-16*) together with conjugated 1′-hydroxy-safrole. The diols were probably formed through their intermediate epoxides, since administration of 2′,3′-epoxy safrole to rats and guinea pigs yielded the same compounds [21].

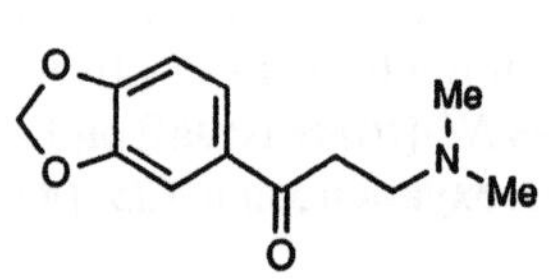

3-*N,N*-Dimethylamino-1-
(3,4-methylenedioxyphe-
nyl)-1-propanone (*25-15*)

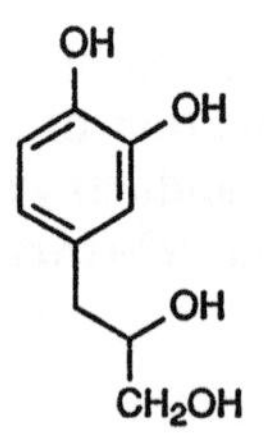

1,2-Dihydroxy-4-(2,3-di-
hydroxypropyl)-benzene
(*25-16*)

In mice given [^{14}C]safrole orally, 64% of the radioactivity was recovered in exhaled CO_2, 18% in urine, 6% in feces and intestine, 2% in liver, and 6% in the carcasses [22].

The oral LD_{50} values of safrole in mice and rats were 3.4 and 1.95 g/kg, respectively [23].

Methyleugenol also showed significant hepatocarcinogenicity in mice [24].

References

1. Tian Z, Dong SN, Wang BY, Lou ZC (1981) Identification of the constituents of volatile oil from Chinese *Asarum* species. II. Volatile oil from huaxixin, *Asarum sieboldii*. J Beijing Med Coll 13:282–284
2. Shen ZX (1982) GC/MS analysis of the essential oil of *Asarum sieboldii* Miq. Chin J Pharm Anal 2:335–338
3. Pan JG, Xu ZL, Wang GH, Yang CS, Zhang JJ (1984) GC/MS analysis of volatile oils from Chinese *Asarum* species. II. *Asarum sieboldii* forma *seoulense*, *A. forbesii*, *A. inflatum*, *A. magnificum* var. *dinghugense* and *A. caudigerum* var. *cardiophyllum*. Bull Chin Mat Med 9:175–177
4. Yang CS, Zhang JJ, Pan QG, Xu ZL, Zhu QC, Wang GF (1986) Gas chromatography-mass spectroscopic analysis of volatile oils from Chinese *Asarum* species. IV. Bull Chin Mat Med 11:423–427
5. Shen ZX (1981) Constituents of the essential oil of *Asarum himalaicum*. Chin Pharm Bull 16:695
6. Xu ZL, Pan JG, Zhu QC, Wang GH, Yang CS, Zhang JJ (1986) GC-MS of the volatile oils of *Asarum* species (III). Bull Chin Mat Med 11:46–49
7. Yang GM, Zhou TD (1986) Botanical resources and utilization of *Asarum* – the volatile oil contents and components of Xiang Xixin. Hunan Zhongyi Zazhi 40–42
8. Shen ZZ, Liu L, Zhou TJ, Li QN, Wang HX (1981) Comparative effects of *Asarum heterotropoides*, higenamine and isoprenaline on function of left ventricle in dogs. Acta Pharm Sin 16:721–727
9. Liu L, Li YE, Chen CC, Chou TC, Wang HH (1981) Comparative study of *Asarum heterotropoides*, higenamine and isoprenaline on the hemodynamics in anesthetized dogs. Chin Pharm Bull 16:50
10. Ling SS, Fang Q, Zhang J, Sun WL (1986) Experimental studies on the antihyperlipemic effect of *Asarum forbesii* Maxim. Chin Trad Herb Drugs 17:21–23
11. Ohmoto T, Sung YI (1982) Antimycotic substances in the crude drugs. II. Shoyakugaku Zasshi 36:307–314
12. International Agency for the Research on Cancer (1976) IARC Monogr Eval Carcinog Risk Chem Man 10:231–244
13. Innes JRM, Ulland BM, Valerio MG, Petrucelli L, Fishbein L, Hart ER, Pallota AJ, Bates RR, Falk HL, Garg JJ, Klein M, Mitchell I, Peters J (1969) Bioassay of pesticides and industrial chemicals for tumorigenicity in mice: a preliminary note. JNCI 42:1101–1114
14. Long EL, Nelson AA, Fitzhugh OG, Hansen WH (1963) Liver tumors produced in rats by feeding safrole. Arch Pathol 75:595–604
15. Epstein SS, Fujii K, Andrea J, Mantel N (1970) Carcinogenicity testing of selected food additives by parenteral administration to infant Swiss mice. Toxicol Appl Pharmacol 16:321
16. Borchert P, Wislocki PG, Miller JA, Miller EC (1973) The metabolism of the naturally occurring hepatocarcinogen safrole to 1′-hydroxysafrole and the electrophilic reactivity of 1′-acetoxysafrole. Cancer Res 33:575–589
17. Borchert P, Miller JA, Miller EC, Shires TK (1973) 1′-Hydroxysafrole, a proximate carcinogenic metabolite of safrole in the rat and mouse. Cancer Res 33:590–600
18. Wislocki PG, Borchert P, Miller JA, Miller EC (1976) The metabolic activation of the carcinogen 1′-hydroxysafrole in vivo and in vitro and the electrophilic reactivities of possible ultimate carcinogens. Cancer Res 36:1686–1695
19. Phillips DH, Miller JA, Miller EC, Adams B (1981) The N^2-atom of guanine and the N^6-atom of adenine residues as sites for covalent binding of metabolically activated 1′-hydroxysafrole to mouse-liver DNA in vivo. Cancer Res 41:2664–2671
20. Oswald EO, Fishbein L, Corbett BJ, Walker MP (1971) Identification of tertiary aminoethylenedioxypropion-phenones as urinary metabolites of safrole in the rat and guinea pig. Biochim Biophys Acta 230:237–247

21. Stillwell WG, Carman MJ, Bell L, Horning MG (1974) The metabolism of safrole and 2′,3′-epoxysafrole in the rat and guinea pig. Drug Metab Dispos 12:489–498
22. Kamienski FX, Casida JE (1970) Importance of demethylenation in the metabolism in vivo and in vitro of methylenedioxyphenyl synergists and related compounds in mammals. Biochem Pharmacol 19:91–112
23. Jenner PM, Hagen EC, Taylor JM, Cook EL, Fitzhugh OG (1964) Food flavourings and compounds of related structure. I. Acute oral toxicity. Food Cosmet Toxicol 2:327–343
24. Miller JA, Miller EC, Phillips DH (1982) The metabolic activation and carcinogenicity of alkenylbenzenes that occur naturally in many spices. In: Stich HF (ed) Carcinogens and Mutagens in the Environment, vol 1. CRC, Boca Raton, pp 83–95

Astragalus membranaceus (Fisch.) Bge. 26

26.1 Introduction

Huangqi, Radix Astragali, is the dried root of *Astragalus membranaceus* Bge. var. *mongholicus* (Bge.) Hsiao or *A. membranaceus* (Fisch.) Bge. (Fabaceae). *Astragalus* root is a very old and well known drug in traditional Chinese medicine. It is officially listed in the Chinese Pharmacopoeia and used mainly as a tonic and for treatment of nephritis and diabetes.

Another entry in the Chinese Pharmacopoeia concerning the *Astragalus* species is Shayuanzi, Semen Astragali complanati, the dry ripe seed of *A. complanatus* R. Br. collected in late fall to early winter. It is used as a tonic against polyuria and vertigo.

26.2 Chemical Constituents

The biologically active constituents of *Astragalus* roots represent two classes of chemical compounds, polysaccharides and saponins.

Fang et al. [1] isolated from the aqueous extract of the roots of *A. membranaceus* var. *mongholicus* three polysaccharides, astragalan I, II, and III. These three polysaccharides are homogeneous as judged by glass fiber paper electrophoresis and gel filtration on Sephadex G-150. Astragalan I is composed of D-glucose, D-galactose, and L-arabinose in the molar ratio 1.75:1.63:1. It also contains a trace of pentose. The average molecular weight of astragalan I is 36300. The sugar component of both astragalan II and III is D-glucose. Their average molecular weights are 12300 and 34600, respectively. Astragalan II and III, when treated by peroxidation and Smith degradation, give rise to glycerol in addition to a large amount of erythritol. These results suggest that both astragalan II and III consist mainly of $\alpha(1\rightarrow4)$ linked glucopyranosyl residues and also contain a small amount of $\alpha(1\rightarrow6)$ linked glucopyranosyl residues.

Two glucans (AG-1, AG-2) and two heterosaccharides (AH-1, AH-2) were further isolated and purified from a water extract of the roots of *A. membranaceus* var. *mongholicus* [2]. By electrophoresis and gel chromatography, these four polysaccharides were shown to be homogeneous. AG-1 was identified as an α-glucan, with a ratio of $\alpha(1\rightarrow4)$ and $\alpha(1\rightarrow6)$ linkages of about 5:2. AG-2 was identified as a $\alpha(1\rightarrow4)$ glucan. AH-1 is an acidic polysaccharide; the component sugars were identified as hexuronic acid (galacturonic acid and glucuronic acid), glucose, rhamnose, and arabinose in a ratio of approximately 1:0.04:0.02:0.01. AH-2 contains glucose and arabinose in a ratio of 1:0.15.

Kitagawa et al. [3] reported on triterpene oligoglycosides present in the roots of Korean *A. membranaceus*. By enzymatic and chemical degradation, two aglycones were separated and structurally elucidated. One of the two aglycones was the 9,19-cyclolanostane type triterpene cycloastragenol (*26-1*), which is the common genuine aglycone of 10 out of 11 glycosidic saponines called astragalosides. The second aglycone was the lanost-9(11)-ene type counterpart astragenol (*26-2*), which is formed as an artifact secondarily from cycloastragenol.

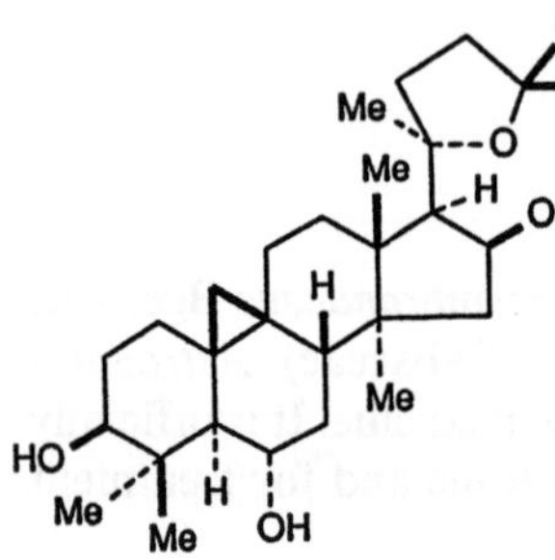

Cycloastragenol (*26-1*)

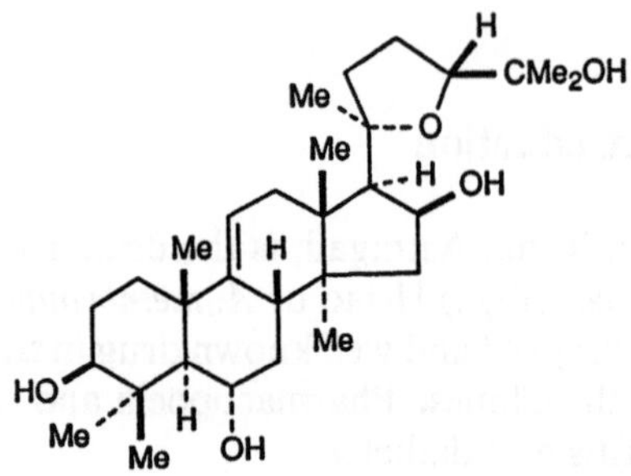

Astragenol (*26-2*)

The methanol extract of *Astragalus* roots was partitioned between n-butanol and water. The n-butanol soluble portion contained the total glycosidic constituents, which were further chromatographed on a reversed phase column. Eleven astragalosides and one soyasaponin were obtained. They are: astragaloside I–VIII (*26-3–26-10*), acetylastragaloside I (*26-11*), isoastragalosides I (*26-12*), II (*26-13*), and soyasaponin I (*26-14*) [4].

By chemical degradation and ^{13}C NMR examination, the structure of astragaloside IV was elucidated as 3-*O*-β-D-xylopyranosyl-6-*O*-β-D-glucopyranosylcycloastragenol. Astragaloside I, II, acetylastragaloside I, and isoastragaloside I, II are acetyl derivatives of astragaloside IV [4].

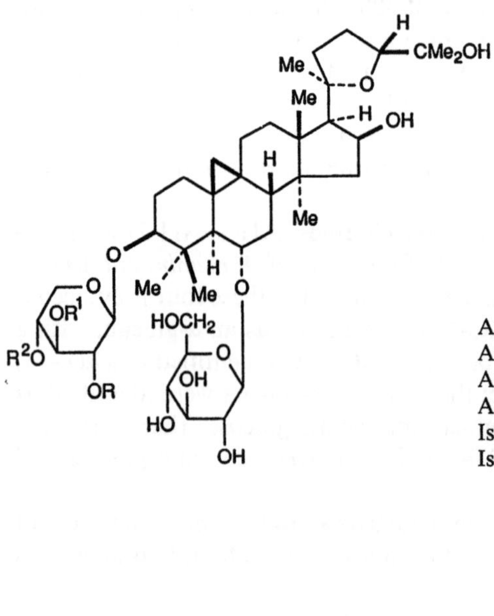

	R	R^1	R^2
Astragaloside I (*26-3*):	Ac	Ac	H
Astragaloside II (*26-4*):	Ac	H	H
Astragaloside IV (*26-6*):	H	H	H
Acetylastragaloside I (*26-11*):	Ac	Ac	Ac
Isoastragaloside I (*26-12*):	Ac	H	Ac
Isoastragaloside II (*26-13*):	H	Ac	H

The structures of astragalosides III, V, and VI were elucidated by ^{13}C NMR examination and by methylation [5]. Finally, the structures of astragalaside VII and VIII were also determined by enzymatic degradation, application of a selective cleavage method for the glucuronide linkage, and by ^{13}C NMR analysis [6]. The aglycone of astragaloside VIII and soyasaponine I is of oleanane type.

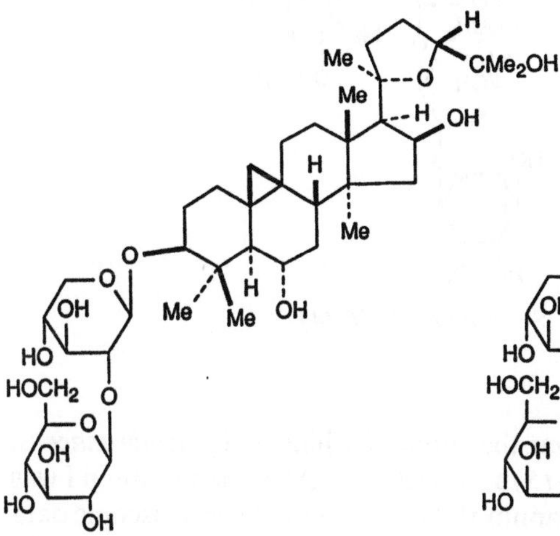

Astragaloside III (*26-5*)

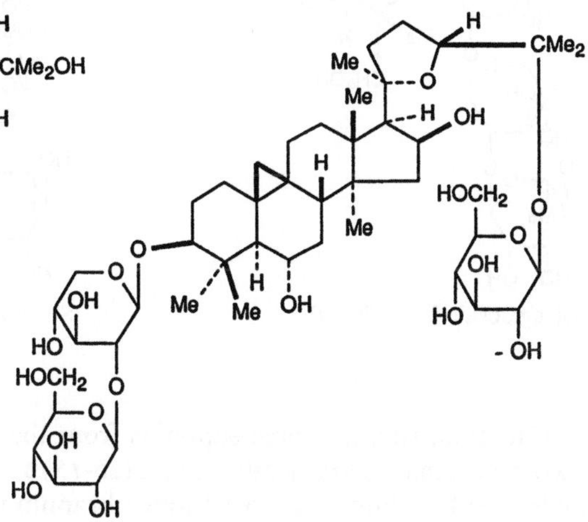

Astragaloside V (*26-7*)

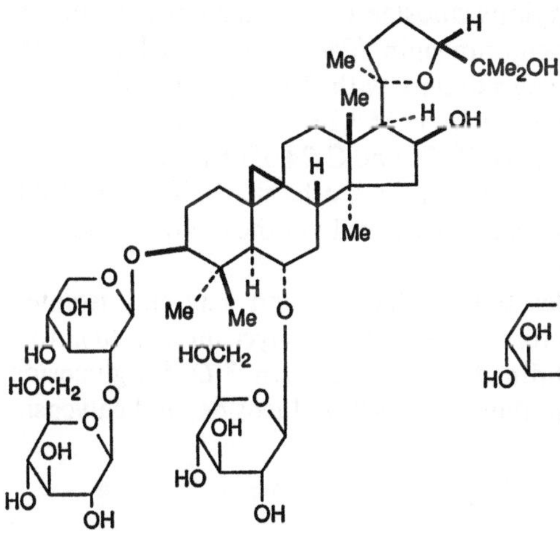

Astragaloside VI (*26-8*)

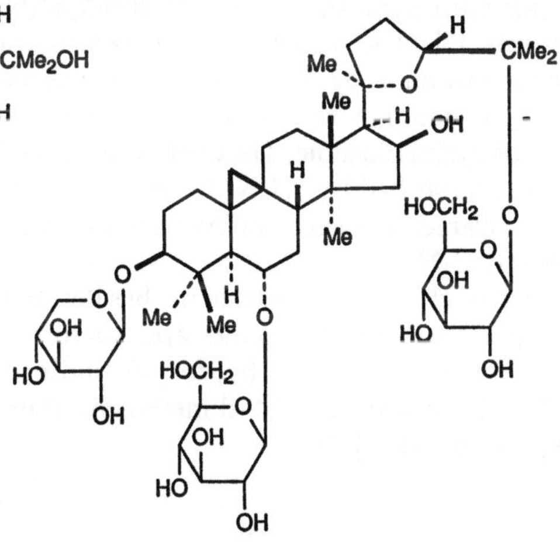

Astragaloside VII (*26-9*)

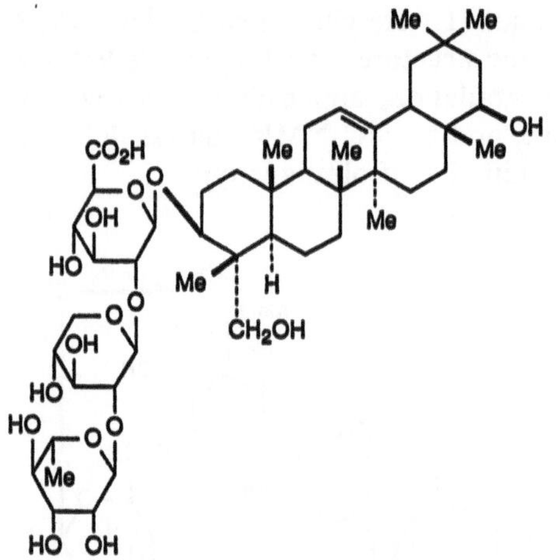

Astragaloside VIII (*26-10*)

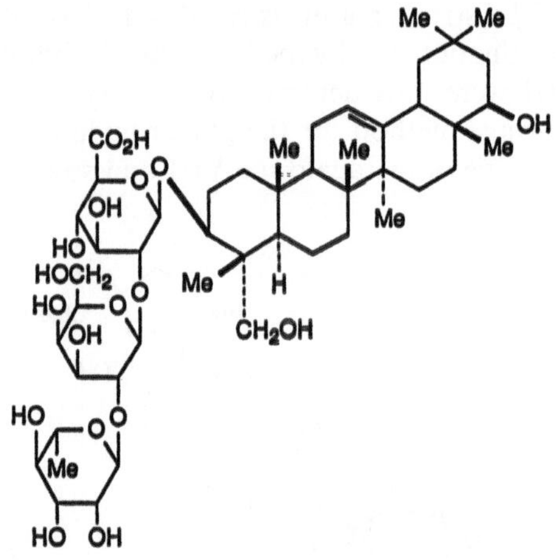

Soyasaponin I (*26-14*)

Cao et al. isolated three saponins from the roots of Chinese *A. membranaceus*. Two were named astramembrannin I (*26-15*) and II (*26-16*). Astramembrannin I was hydrolyzed by dilute acid to astramembrannin II. On the basis of spectroscopic data of astramembrannin I and II and of their peracetate and permethyl derivatives, the structures of astramembrannin I and II were established as 3-*O*-β-D-xylopyranosyl-6-*O*-β-D-glucopyranoside and 3-*O*-β-D-xylopyranoside of astramembrangenin, respectively [7]. The structure of astramembrangenin (*26-17*), obtained from astramembrannin I by Smith oxidative degradation with $NaIO_4$ and subsequent $NaBH_4$ reduction, was identified as (20*S*,24*R*)-3β,6α,16β,25-tetrahydroxy-20,24-epoxy-9,19-cyclolanostane. The configurations of C-20 and C-24 of astramembrangenin were established by ^1H NMR spectroscopy using a lanthanide shift reagent [8]. Thus, astramembragenin differs from cycloastragenol only in the configurations of C-20 and C-24.

Further chemical constituents besides polysaccharides and saponins isolated from the roots of *Astragalus* are: sucrose, β-sitosterol, calycosin (*26-18*), formononetin (*26-19*) [9], 3-hydroxy-9,10-dimethoxypterocarpan 3-*O*-β-D-glucoside (*26-20*); 2′,7-dihydroxy-3′,4′-dimethoxyisoflavone 7-*O*-β-D-glucoside; and calycosin 7-*O*-β-D-glucoside [10].

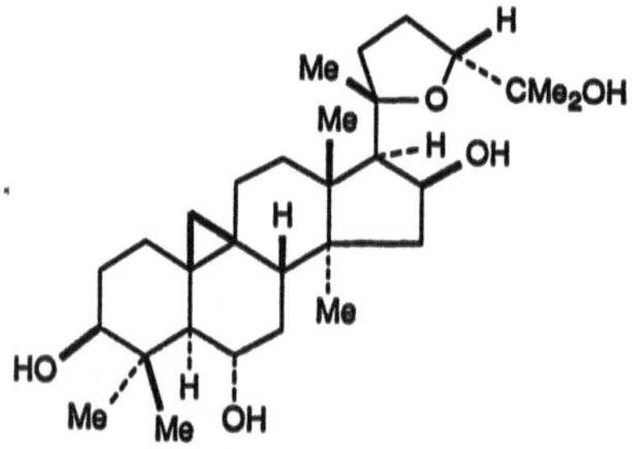

Astramembrangenin (*26-17*)

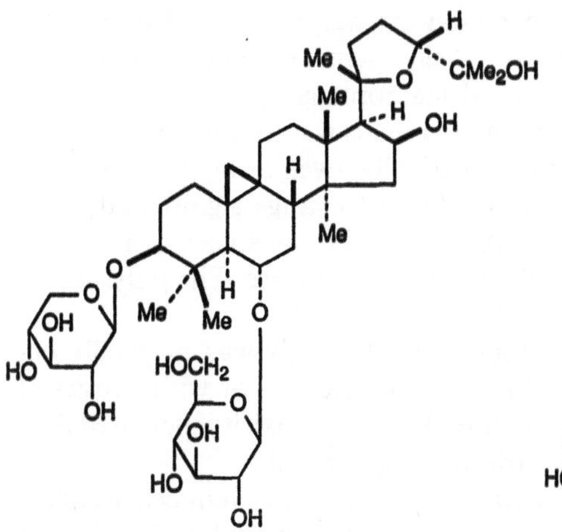

Astramembrainnin I (*26-15*)

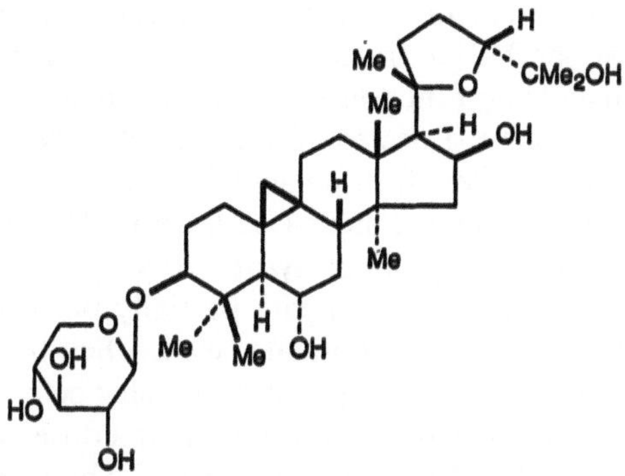

Astramembrainnin II (*26-16*)

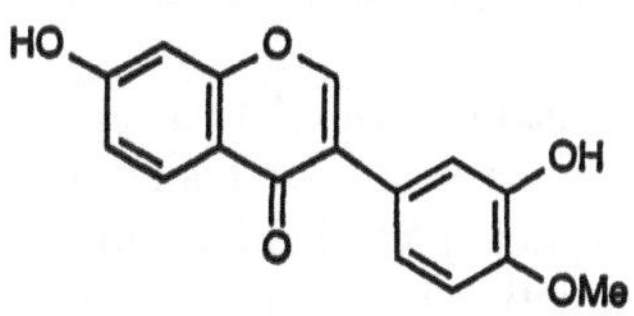

Calycosin (*26-18*)

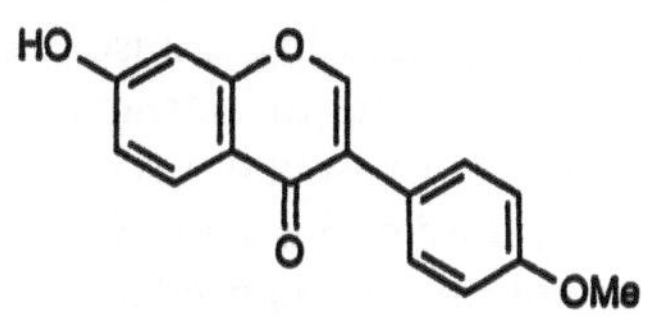

Formononetin (*26-19*)

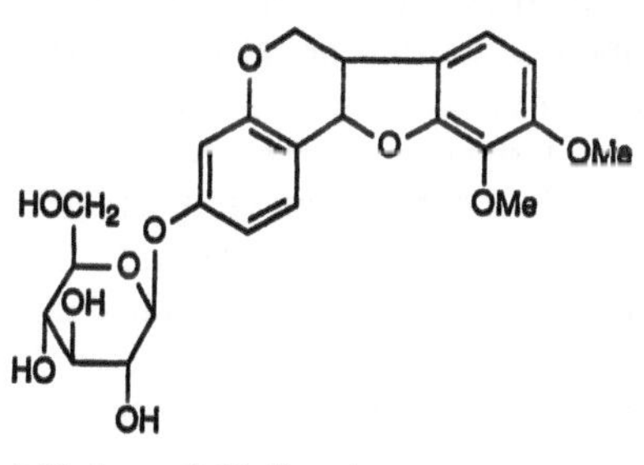

3-Hydroxy-9,10-dimethoxy-pterocarpan
3-*O*-β-D-glucopyranoside (*26-20*)

26.3 Pharmacology

The polysaccharides composed of glucose and arabinose extracted from *A. membránaceus* var. *mongholicus* were reported to increase the immune response when administered i.p. to mice. Moreover, they also caused an increase in the amount of RNA in the spleen and a decrease in the incorporation of [^{3}H]uridine into RNA [11]. Similar effects on other reticuloendothelial tissues but no effect on thymus, heart, or brain RNA or on DNA metabolism was noted [12].

The homogeneous fraction of polysaccharides obtained by water extraction, consisting mainly of astragalan I and II, exhibits a wide spectrum of immunological effects on mice. By i.p. administration it increased the weight and cell number of mouse spleen, elevated the response of mouse spleen against sheep red blood cells, and stimulated phagocytic activity of peritoneal macrophages [1]. The number of activated macrophages in the spleen of the treated animals was also increased. If the polysaccharide fraction was given i.v. or intragastrically, even at higher doses, the phagocytic function of peritoneal macrophages did not change significantly [13]. Astragalan II decreased the alkaline RNase activity in liver and spleen of mice and had a smaller effect on acid RNase but no effect on serum RNase. The polysaccharide fraction also increased hepatic RNase inhibitor activity [14].

The natural killer cytotoxicity of lymphocyte effector cells was markedly enhanced when treated with partially purified human interferon-α or with extract of *Astragalus*. They stimulated each other: the natural killer cytotoxicity increased five- to sixfold after treatment of effector cells with both agents [15].

Saponin astramembrannin I, at a dose of 10 mg/kg applied i.v., induced accumulation of cAMP in rabbit plasma. The increase in cAMP started after 30 min and reached a maximum in 0.5–4 h after a single injection. Saponin affected DNA biosynthesis in partially hepatectomized mice and increased incorporation of [^{3}H]thymidine into regenerating mouse liver [16].

Antiinflammatory effects of astramembrannin I were demonstrated in rats. It inhibited the increase in vascular permeability induced by serotonin or histamine when given i.v. at a dose of 5 mg/kg or orally at a dose of 50 mg/kg. Oral administration of astramembrannin I caused a dose dependent reduction in carrageenan-induced edema of the hind paw of rats. Hypotensive activity of astramembrannin I was observed after i.v. administration of 15 or 10 mg/kg to anesthetized cats or rats [17].

A clinical effect of *A. membranaceus* in the treatment of chronic hepatitis was also reported. Elevated levels of serum GPT returned to normal in 1–2 months, and symptoms were relieved. Patients had a good appetite and a sense of well-being after treatment, without showing significant side effects. Experiments on animals with toxic liver damage induced by CCl_4 indicated that the root of *A. membranaceus* might protect the liver, prevent decrease of hepatic glycogen contents, and raise the levels of total serum protein and albumin [18].

Phagocytosis of the reticuloendothelial cells of patients with chronic hepatitis was also stimulated, and cellular immunity was enhanced [18].

A decoction of the seed of *A. complanatus* given orally to mice increased the lymphocyte transformation rate and thus specific cellular immunity; however, the treatment had no effect on the spleen index [19].

References

1. Fang SD, Chen Y, Xu XY, Ye CQ, Zhai SK, Shen ML (1982) Studies of the active principles of *Astragalus mongholicus* Bunge. I. Isolation, characterization and biological effect of its polysaccharides. Org Chem 26–31
2. Huang QS, Lu GB, Li YC, Guo JH, Wang RX (1982) Studies on the polysaccharides of "Huang Qi" (*Astragalus mongholicus* Bunge). Acta Pharm Sin 17:200–206
3. Kitagawa I, Wang HK, Takagi A, Fuchida M, Miura I, Yoshikawa M (1983) Saponin and sapogenol. XXXIV. Chemical constituents of astragali radix, the root of *Astragalus membranaceus* Bunge. 1. Cycloastragenol, the 9,19-cyclolanostane type aglycone of astragalosides, and the artifact aglycone astragenol. Chem Pharm Bull 31:689–697

4. Kitagawa I, Wang HK, Saito M, Takagi A, Yoshikawa M (1983) Saponin and sapogenol. XXXV. Chemical constituents of astragali radix, the root of *Astragalus membranaceus* Bunge. 2. Astragalosides I, II and IV, acetylastragaloside I and isoastragalosides I and II. Chem Pharm Bull 31:698–708
5. Kitagawa I, Wang HK, Saito M, Yoshikawa M (1983) Saponin and sapogenol. XXXVI. Chemical constituents of astragali radix, the root of *Astragalus membranaceus* Bunge. 3. Astragalosides III, V and VI. Chem Pharm Bull 31:709–715
6. Kitagawa I, Wang HK, Yoshikawa M (1983) Saponin and sapogenol. XXXVII. Chemical constituents of astragali radix, the root of *Astragalus membranaceus* Bunge. 4. Astragalosides VII and VIII. Chem Pharm Bull 31:716–722
7. Cao ZZ, Yu JH, Gan LX, Chen YQ (1985) Structure of astramembrannins. Acta Chim Sin 43:581–585
8. Cao ZZ, Yu JH, Gan LX, Zhou WS (1983) The structure of astramembragenin. Acta Chim Sin 41:1137–1145
9. Wang ZX, Ma QF, Ho Q, Go JS (1983) Studies on chemical constituents of astragalus (*Astragalus membranaceus*). Chin Trad Herb Drugs 14:97–99
10. Lu GB, Lu SH, Zhang GQ, Xu SM, Li DY, Huang QS (1984) Isolation and identification of flavone-like constituents from Mongolian milkvetch (*Astragalus mongholicus*). Chin Trad Herb Drugs 15:452–454
11. Shanghai Institute of Materia Medica; Shanghai Second Medical College (1979) Immunopotentiating effects of *Astragalus* polysaccharide. Kexue Tongbao 24:764–768
12. Wang DY, Yang WY, Zhai SK, Shen ML (1980) Effect of *Astragalus* polysaccharide on ribonucleic acid metabolism. Acta Biochem Biophys Sin 12:343–348
13. Chen LJ, Shen ML, Wang MY, Zhai SK, Liu MZ (1981) Effect of *Astragalus* polysaccharides on phagocytic function in mice. Acta Pharmacol Sin 2:200–204
14. Wang DY, Li CY, Pong DW (1984) Effect of *Astragalus* polysaccharide on RNase and RNase inhibitor. Acta Biochem Biophys Sin 16:285–290
15. Jing JP, Lin WF (1983) Preliminary study on effects of mechanism of human umbilical cord blood derived interferon-α and of *Astragalus membranaceus* on neutral killer toxicity. Chin J Microbiol Immunol 3:293–296
16. Zhang YD, Shen JP, Song J, Wang YL, Shao YN, Li CF, Zhou SH, Li YF, Li DX (1984) Effects of *Astragalus* saponin 1 on cAMP and cGMP level in plasma and DNA synthesis in regenerating liver. Acta Pharm Sin 19:619–621
17. Zhang YD, Wang YL, Shen JP, Li DX (1984) Hypotensive and antiinflammatory effects of *Astragalus* saponin 1. Acta Pharm Sin 19:333–337
18. Zhou QJ (1985) Chinese medicinal herbs in the treatment of viral hepatitis. In: Chang HM, Yeung HW, Tso WW, Koo A (eds) Advances in Chinese Medicinal Materials Research. World Scientific, Singapore, pp 215–219
19. Wang JQ, Zhao XM, Yan HQ (1985) Effects of the seed of *Astragalus complanatus* on the immunological function of mouse splenocytes. Shaanxi Med J 14:47–49

Atractylodes macrocephala Koidz. 27

27.1 Introduction

Baizhu, Rhizoma Atractylodis macrocephalae, is the dry rootstock of *Atractylodes macrocephala* Koidz. (Asteraceae) collected in winter. It is listed officially in the Chinese Pharmacopoeia and is recommended in traditional Chinese medicine as a digestive, diuretic, and antihidrotic.

27.2 Chemical Constituents

From the essential oil of the rhizome of *A. macrocephala* atractylon (*27-1*) and two structurally related lactones, atractylenolide II (*27-2*) and atractylenolide III (*27-3*), were isolated and identified [1]. The content of essential oil was found to be 0.35% – 1.35% (w/w) with regional variations [1].

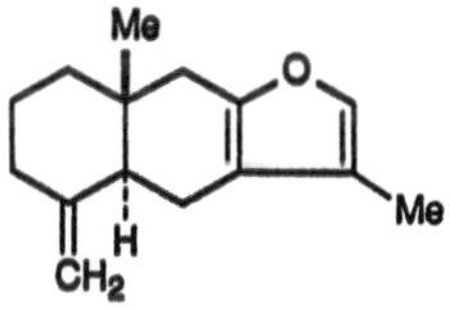

Atractylon (*27-1*)

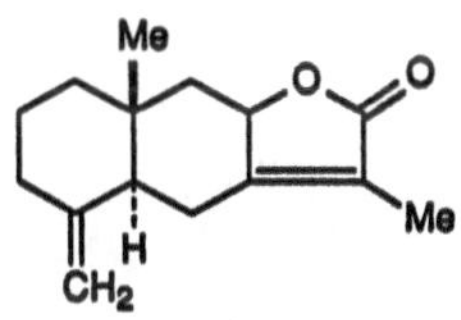

Atractylenolide II (*27-2*)

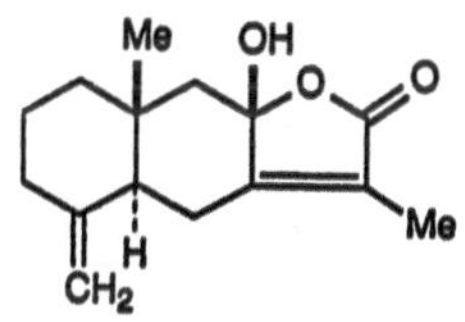

Atractylenolide III (*27-3*)

From the lipophilic fraction of *A. macrocephala* juniper camphor (*27-4*) was isolated together with atractylon and atractylenolides [2].

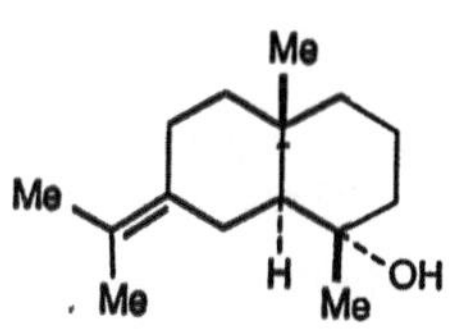

Juniper Camphor (*27-4*)

27.3 Pharmacology

Atractylon has antihepatotoxic activity; it inhibited CCl_4-induced cytotoxicity in primary cultured rat hepatocytes and CCl_4-induced lipid peroxidation in rat liver microsomes. Studies on its mechanism of action support the hypothesis that both CCl_4 and atractylon generate free radicals in rat liver microsomes. Free radicals from CCl_4 mediate lipid peroxidation and produce liver lesions, whereas atractylon forms free radicals which scavenge radicals induced by CCl_4 and thus inhibit lipid peroxidation by CCl_4 and suppresses CCl_4-induced liver damage [3].

27.4 *Atractylodes lancea* and *A. chinensis*

Cangzhou, Rhizoma Atractylodis, is another item listed in the Chinese Pharmacopoeia. It is the dry rootstock of *A. lancea* (Thunb.) DC. or *A. chinensis* (DC.) Koidz. collected in spring and fall. The rhizome of *A. lancea* or *A. chinensis* is recommended for treatment of digestive disorders, diarrhea, edema, beriberi, rheumatic diseases, influenza and nyctalopia.

From the rhizome of *A. lancea*, hydroxyatractylon (*27-5*) and acetoxyatractylon (*27-6*) were isolated and their structures determined [4]. In addition, hinesol (*27-7*) was isolated and its absolute configuration determined by spectral analyses [5]. From the rhizome of *A. lancea*, collected in Hokkaido, Japan, atractylodin (*27-8*) and β-eudesmol (*27-9*) were found together with hinesol; atractylon was not detected [6].

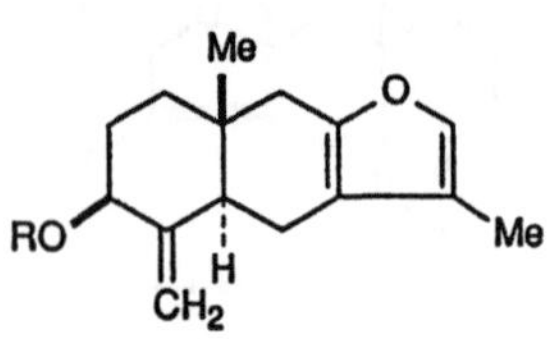

Hydroxyatractylon (*27-5*): R=H
Acetoxyatractylon (*27-6*): R=Ac

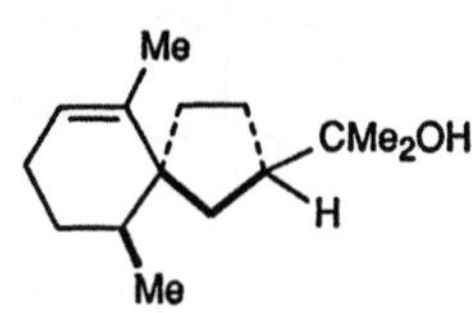

Hinesol (*27-7*)

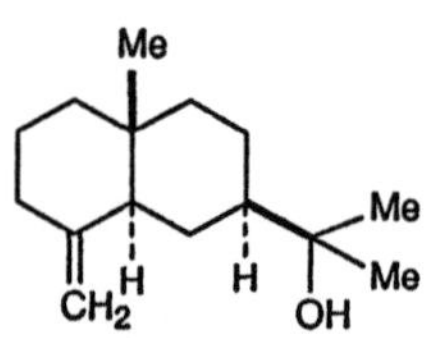

β-Eudesmol (*27-9*)

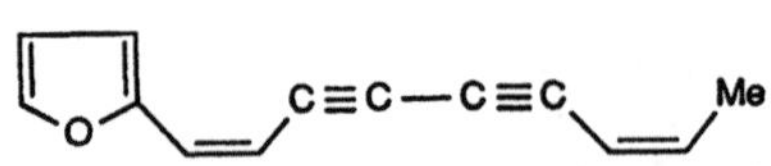

Atractylodin (*27-8*)

The essential oil of *A. lancea* contained *p*-cymene, β-selinene ar-curcumene, and elemol together with hinesol and β-eudesmol [7].

References

1. Liu GS (1980) Chemical constituents of the essential oil from the rhizome of *Atractylodes macrocephala* Koidz. Acta Bot Sin 22:395–396
2. Wang YS, Chang JC, Li KK, Wu CH, Tin K, Liu YH (1980) Studies on the constituents of *Atractylodes macrocephala* Koidz. Shanxi Med J 9:47–51
3. Kiso Y, Tohkin M, Hikino H (1985) Validity of the oriental medicines. LXXII. Liver-protective drugs. 17. Mechanism of antihepatotoxic activity of atractylon. I. Effect on free radical generation and lipid peroxidation. Planta Med 51:97–100
4. Nisikawa Y, Watanabe Y, Seto T, Yasuda I (1976) Studies on the components of Atractylodes. I. New sesquiterpenoids in the rhizome of *Atractylodes lancea* De Candolle. Yakugaku Zasshi 96:1089–1093
5. Yoshioka I, Kimura T (1969) Constituents of *Atractylodes*. XI. Structure and absolute configuration of hinesol. Chem Pharm Bull 17:856–857
6. Anetai M, Yamagishi T (1984) Chemical evaluation of *Atractylodes* rhizomes "Zhu" produced in Hokkaido. Hokkaidoritsu Eisei Kenkyushoho 19–23 (CA 102:209224v)
7. Hochmannova J, Novotny L, Herout V (1962) Terpenes 140. Composition of the oil from *Atractylodes lancea*. The structure of hinesol. Collect Czech Chem Commun 27:1914–1926

Bletilla striata (Thunb.) Reich. f. 28

28.1 Introduction

Baiji, Rhizoma Bletillae, is the dry tuber of *Bletilla striata* (Thunb.) Reich. f. (Orchidaceae) dug in summer and fall and dried after cooking or steaming. It is listed officially in the Chinese Pharmacopoeia and can be used in treatment of hemorrhagic and especially gastric and pulmonary diseases.

28.2 Chemical Constituents

The methanol extract from sliced tubers of *B. striata* exhibited antimicrobial activity in vitro. Subsequent fractionation of the extract revealed that the bioactivity was restricted to the phenolic part, from which five active principles were isolated. The structures of these five substances were determined by various spectroscopic methods. Two of them were identified as the dihydrophenanthrenes 4,7-dihydroxy-2-methoxy-9,10-dihydrophenanthrene (*28-1*) and its 1-*p*-hydroxybenzyl derivative (*28-2*). The other three compounds were the bibenzyl derivatives 3,3'-dihydroxy-5-methoxy- (*28-3*) and 3-hydroxy-3',5-dimethoxy-2,6-*bis*(*p*-hydroxybenzyl)bibenzyl (*28-4*) and 3,3'-dihydroxy-5-methoxy-2,5',6-*tris*-(*p*-hydroxybenzyl)bibenzyl (*28-5*) [1].

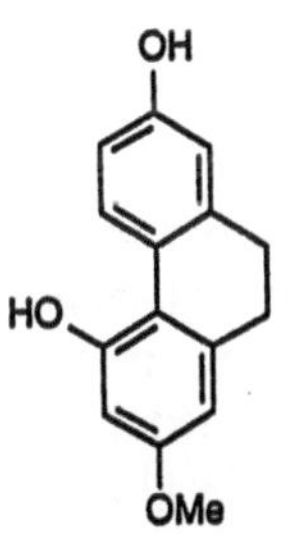

4,7-Dihydroxy-2-methoxy-9,10-dihydrophenanthrene (*28-1*)

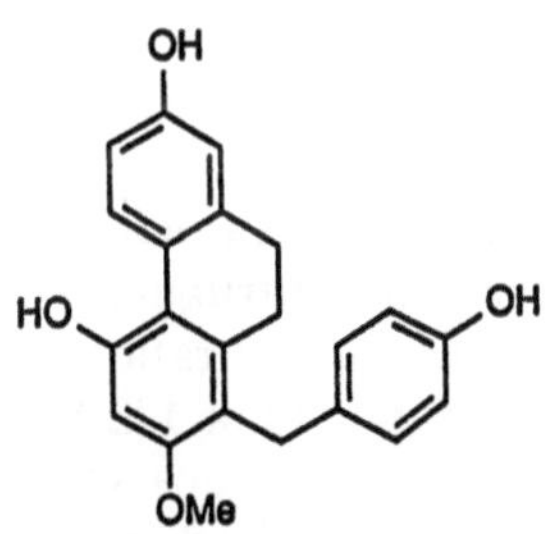

4,7-Dihydroxy-2-methoxy-1-*p*-hydroxy-benzyl-9,10-dihydrophenanthrene (*28-2*)

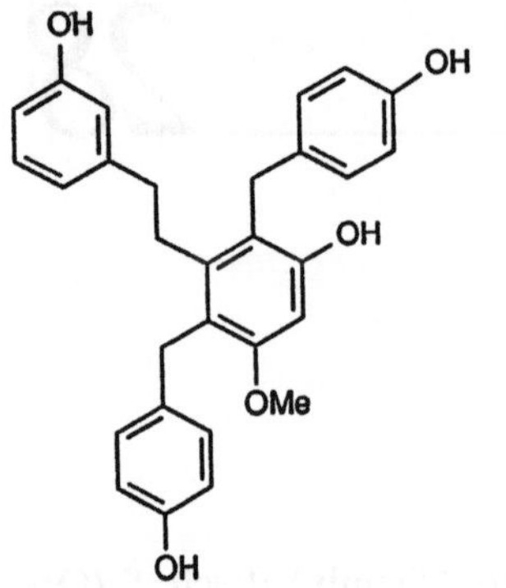

3,3′-Dihydroxy-5-methoxy-2,6-*bis*(*p*-hydroxybenzyl)-bibenzyl (*28-3*)

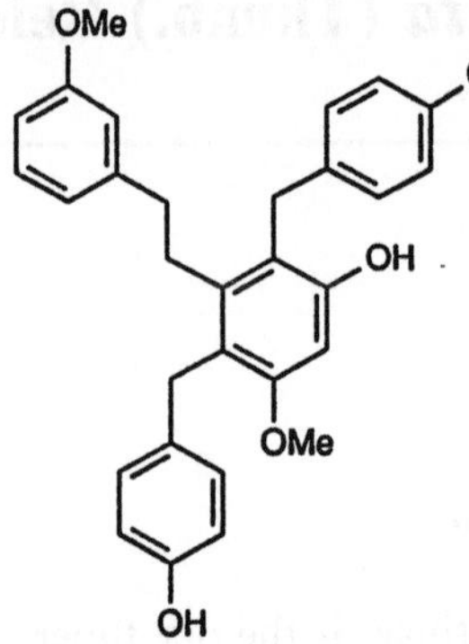

3-Hydroxy-5,3′-dimethoxy-2,6-*bis*(*p*-hydroxybenzyl)-bibenzyl (*28-4*)

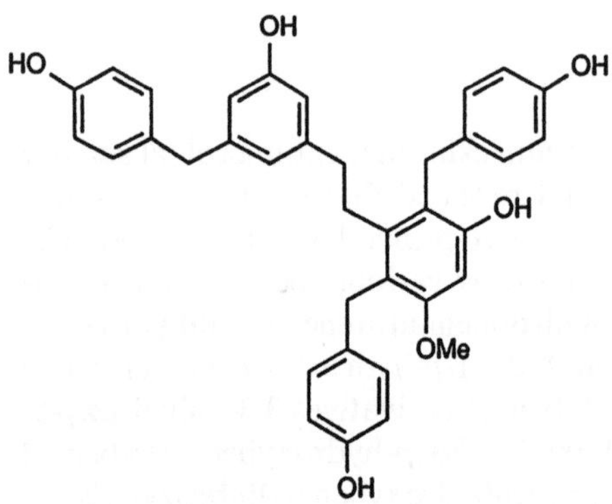

3,3′-Dihydroxy-5-methoxy-2,5′,6-*tris*(*p*-hydroxybenzyl)bibenzyl (*28-5*)

In addition, the tuber of *B. striata* is rich in plant mucilage. A mucous polysaccharide, bletilla-glucomannan, was isolated. It was homogeneous on glass fiber paper electrophoresis and by ultracentrifugal analysis. The component sugars were D-mannose and D-glucose with a molar ratio of 3:1. The molecular weight was estimated by osmometric determination to be 182000 [2]. Bletilla-glucomannan contained 4.2% O-acetyl groups. The O-acetyl groups were located in position 3 of most of the glucose units. Methylation studies provided evidence that the polysaccharide is mainly composed of $\beta(1\rightarrow4)$ linked aldohexopyranose units with $1\rightarrow2$ branch point at a part of mannose units [3]. Partial acetolysis of bletilla-glucomannan led to the isolation of 14 oligosaccharides. They were identified as: *O*-β-D-mannopyranosyl-(1→4)-D-mannopyranose; *O*-α-D-mannopyranosyl-(1→4)-D-mannopyranose; *O*-β-D-mannopyranosyl-(1→4)-D-glucopyranose; *O*-β-D-glucopyranosyl-(1→4)-D-mannopyranose; *O*-β-D-glucopyranosyl-(1→4)-D-glucopyranose; *O*-β-D-mannopyranosyl-(1→4)-*O*-β-D-mannopyranosyl-(1→4)-D-mannopyranose; *O*-β-D-glucopyranosyl-(1→4)-*O*-β-D-mannopyranosyl-(1→4)-D-mannopyranose; *O*-β-D-mannopyranosyl-(1→4)-*O*-β-D-mannopyranosyl-(1→4)-D-glucopyranose; *O*-β-D-mannopyranosyl-(1→4)-*O*-β-D-glucopyranosyl-(1→4)-D-glucopyranose; *O*-*β*-D-mannopyranosyl-(1→4)-*O*-*β*-D-glucopyranosyl-(1→4)-D-mannopyranose; *O*-*β*-D-glucopyranosyl-(1→4)-*O*-*β*-D-glucopyranosyl-(1→4)-D-mannopyranose; *O*-*β*-D-mannopyranosyl-(1→4)-*O*-*β*-D-mannopyranosyl-(1→4)-*O*-*β*-D-mannopyranosyl-(1→4)-D-mannopyranose; *O*-*β*-D-glucopyranosyl-(1→4)-*O*-*β*-D-mannopyranosyl-(1→4)-*O*-*β*-D-mannopyranosyl-(1→4)-D-mannopyranose; and *O*-*β*-D-mannopyranosyl-(1→4)-*O*-*β*-D-glucopyranosyl-(1→4)-*O*-*β*-D-mannopyranosyl-(1→4)-D-mannopyranose. Mannobiose, having an α-glyco- sidic linkage, might well be an artifact produced during the reaction [4].

28.3 Pharmacology

The dihydrophenanthrene and bibenzyl derivatives described above were tested for their antimicrobial properties against *Bacillus subtilis*, *B. cereus* var. *mycoides*, *Nocardia gardneri*, *Staphylococcus auerus*, *Candida albicans*, and *Trichophyton mentagrophytes*. The results showed that they are mainly active against gram-positive bacteria and very weakly active against certain fungi. Among them, 3,3′-dihydroxy-2,6-*bis*(*p*-hydroxybenzyl)-5-methoxy-bibenzyl and 3,3′-dihydroxy-2,5′,6-*tris*(*p*-hydroxybenzyl)-5-methoxybibenzyl were the most active against bacteria [1].

References

1. Takagi S, Yamaki M, Inoue K (1983) Antimicrobial agents from *Bletilla striata*. Phytochemistry 22:1011–1015
2. Tomoda M, Nakatsuka S, Tamai M, Nagata M (1973) Plant mucilages. VIII. Isolation and characterization of a mucous polysaccharide, Bletilla glucomannan from *Bletilla striata* tubers. Chem Pharm Bull 21:2667–2671
3. Tomoda M, Nakatsuka S, Satoh N (1974) Plant mucilages. IX. Location of the O-acetyl groups and the nature of the branches in Bletilla-glucomannan. Chem Pharm Bull 22:2710–2713
4. Tomoda M, Kimura S (1976) Plant mucilages. XII. Fourteen oligosaccharides obtained from Bletilla-glucomannan by partial acetolysis. Chem Pharm Bull 24:1807–1812

Brucea javanica (L.) Merr 29

29.1 Introduction

Yadanzi, Fructus Bruceae, is the dry mature fruit of *Brucea javanica* (L.) Merr (Simaroubaceae). It is used in traditional Chinese medicine for a long time as an antimalarial and antidysenteric agent. By external application, it was also used for treatment of warts and corns. The fruit of *B. javanica* is listed officially in the Chinese Pharmacopoeia.

In the last 20 years, a number of quassinoid constituents were isolated from the fruits of *B. javanica*. Interest in the quassinoids has increased vigorously because some display marked biological activities, particularly antileukemic activity.

29.2 Chemical Constituents

B. javanica is one of the most intensely investigated plants of the family of Simaroubaceae, which contain the bitter quassinoids as major constituents. The characteristic features of the quassinoids isolated from *B. javanica* are a diterpene ring system, picrasane (*6-1*), described for *Ailanthus altissima*. It is composed of four rings:

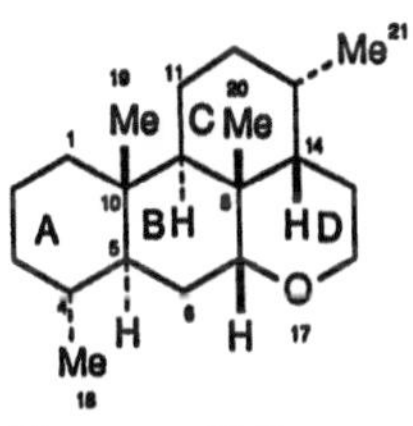

Picrasane (*6-1*)

The structure of ring A may be varied according to the degree of oxygenation, the positions of the functional groups containing oxygen, and the positions of the double bonds. Ring C possesses a hydroxymethyl group at position 8 that forms an ether bridge to C-13. Ring D may have a hydroxy function at C-15 esterified with different small fatty acids such as acetic acid, 2-methyl-butyric acid, isovaleric acid, senecioic acid, 2-hydroxy-2-methylbutyric acid, and 3,4-dimethyl-4-hydroxy-valeric acid. Furthermore, an oxo group is present at C-16 as part of a lactone function.

The following quassinoids and their glycosides were isolated from *B. javanica* and structurally elucidated: bruceines A–H and Q, dehydrobruceines A and B, 3,4-dihydrobruceine A, brusatol (yatansin), dehydrobrusatol, bruceantin, bruceantinol, dehydrobruceantinol, bruceantarin, bruceaketolic acid, bruceene, bruceosides A and B, yadanziolides A–C, yadanziosides A–K and M–P, and yadanzigan. These compounds may be classified according to their structural features concerning ring A:

- 2,3-dioxygenated picras-3-ene derivatives
- 2-monooxygenated picras-3-ene derivatives
- 1,2-dioxygenated picras-3-ene derivatives
- 2,3-dioxygenated picras-1-ene derivatives
- 2,3-dioxygenated picras-1,4-diene derivatives
- 1,2-dioxygenated picras-2-ene derivatives
- 1-monooxygenated picras-2,4(18)-diene derivatives
- 2,3-dioxygenated picrasane derivatives

The 2,3-dioxygenated picrasene derivatives occur in plants of the genus *Brucea* only. Under mild hydrolytic conditions the 2,3-dioxygenated picras-3-ene derivatives such as bruceine A, B, and C; brusatol; and bruceantin gave the same tetrol, bruceolide (*29-1*). The first structural elucidation of *Brucea* constituents was reported by Polonsky et al. in 1967 [1]. Since then, a great number of quassinoids have been isolated from *B. javanica* and structurally determined. Recently, a series of quassinoids and quassinoid glycosides named yadanziolides and yadanziosides, isolated from *B. javanica*, was described by Japanese scientists [2–5]. Two reviews describing extensively the chemistry of quassinoids were prepared by Polonsky [6, 7].

The quassinoids and quassinoid glycosides isolated from *B. javanica* are listed in Table 29-1. *B. sumatrana, B. ararissima,* and *B. gracilis* are all synonyms of *B. javanica.*

Table 29.1. Quassinoids isolated from *B. javanica*

Compound	Structure	R	Reference
Bruceolide (*29-1*)	2,3-Dioxygenated picras-3-ene derivatives	H	
Bruceine A (*29-2*)	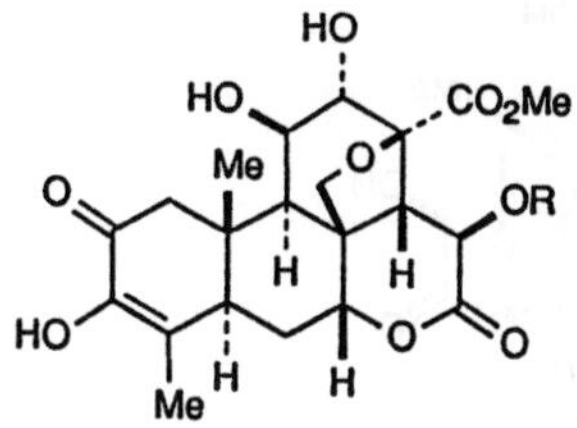	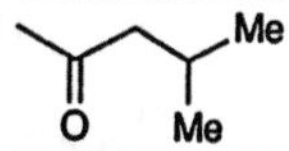	[1, 15, 51, 58]
Bruceine B (*29-3*)		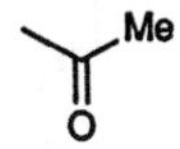	[1, 15, 51]
Bruceine C (*29-4*)		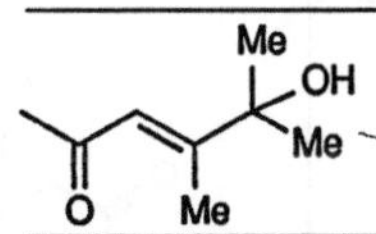	[1, 15, 49, 51]
Brusatol (Yatansin) (*29-5*)		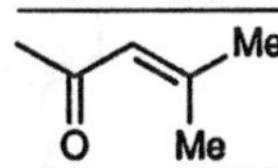	[46, 53, 57, 58]
Bruceantin (*29-6*)		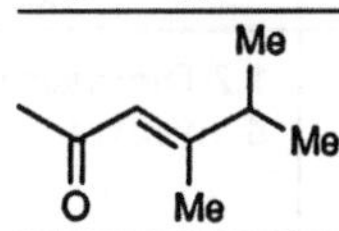	[13–15, 51]
Bruceantinol (*29-7*)		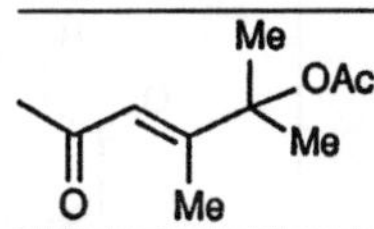	[14, 49, 51]
Bruceantarin (*29-8*)		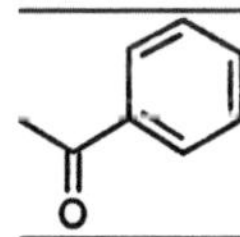	[13–15, 51]
Bruceoside B (*29-9*)	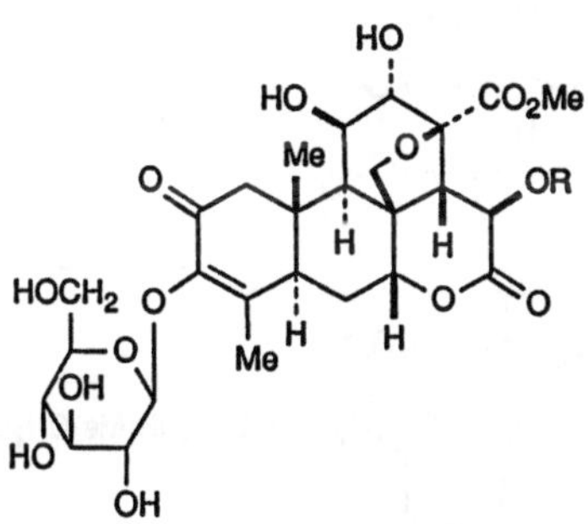	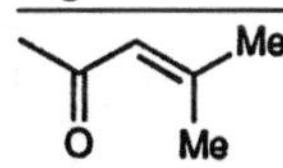	[48, 55]
Yadanzioside B (*29-10*)		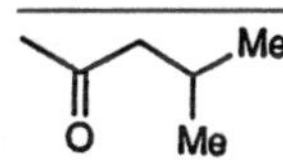	[4, 5, 58]
Yadanzioside I (*29-11*)		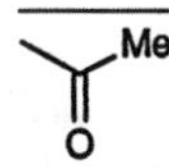	[3]
Yadanzioside L (*29-12*)		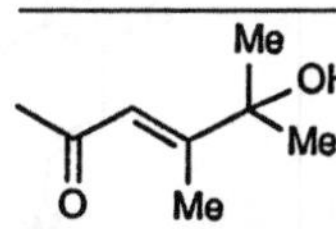	[3]
Yadanzioside K (*29-13*)		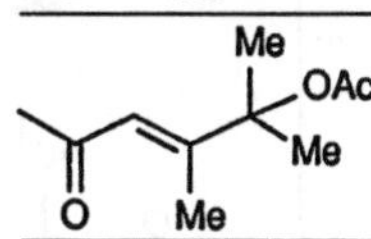	[62]
Yadanzioside P (*29-14*)		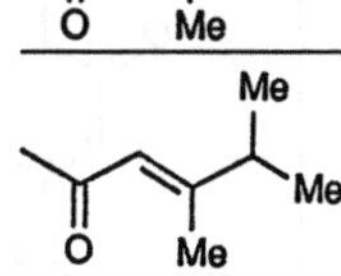	[63]

Table 29.1. (continued)

Compound	Structure	R	Reference
	2-Monooxygenated picras-3-ene derivatives		
Bruceine G (*29-15*)	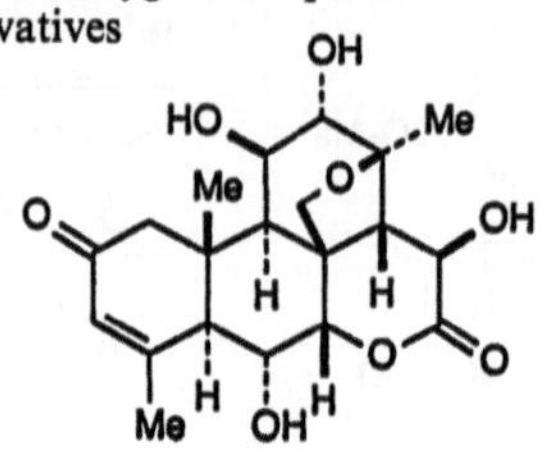		[45]
Bruceine H (*29-16*)	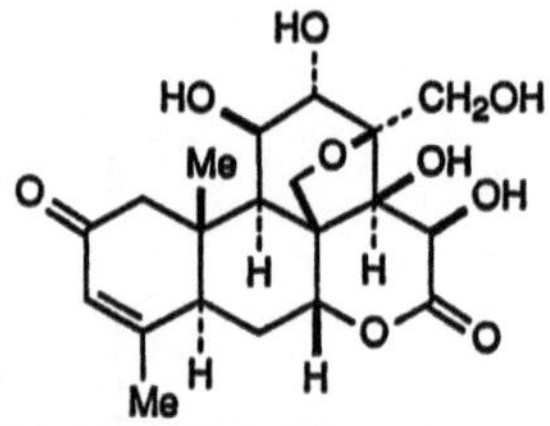		[61]
	1,2-Dioxygenated picras-3-ene derivatives		
Bruceine D (*29-17*)	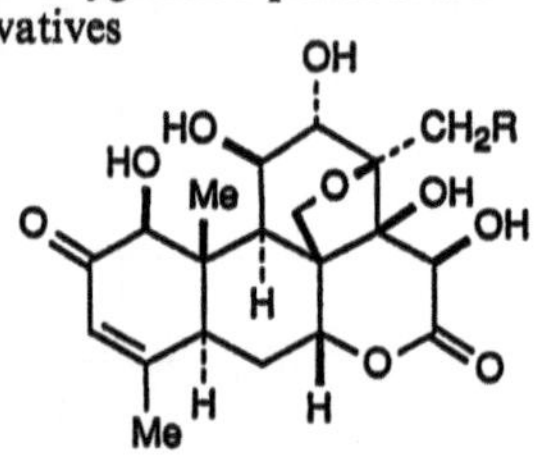	H	[43, 48, 56, 57]
Yadanziolide A (*29-18*)		OH	[2, 3]
Bruceine E (*29-19*)	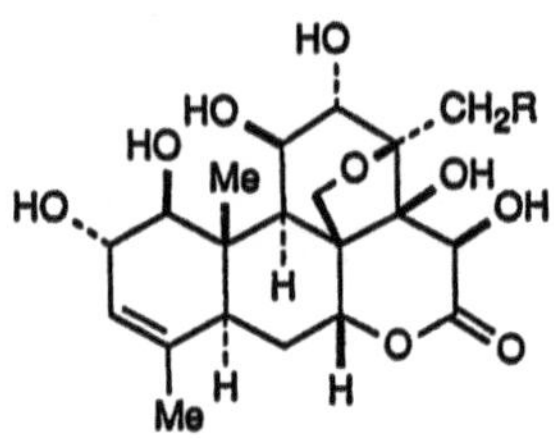	H	[43, 47, 48, 55–57]
Bruceine F (*29-20*)		OH	[44, 55]
Bruceine Q (*29-21*)	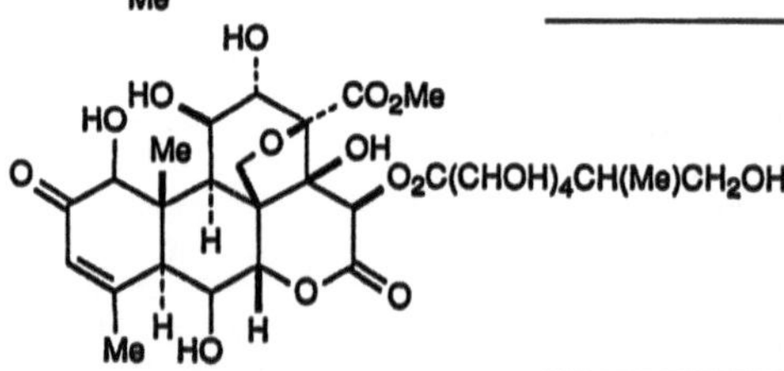		[52]
Yadanziolide B (*29-22*)	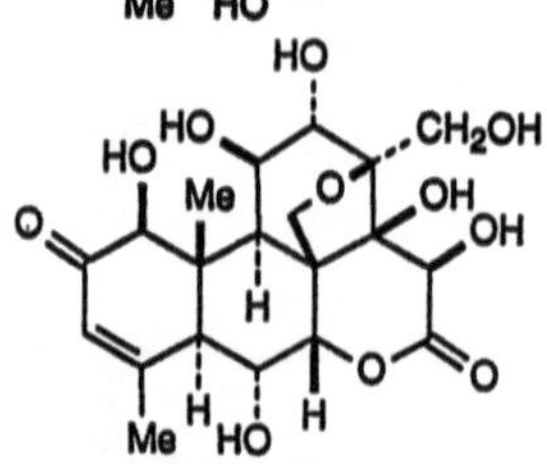		[2, 3]

Table 29.1. (continued)

Compound	Structure	R	Reference
Yadanziolide C (*29-23*)	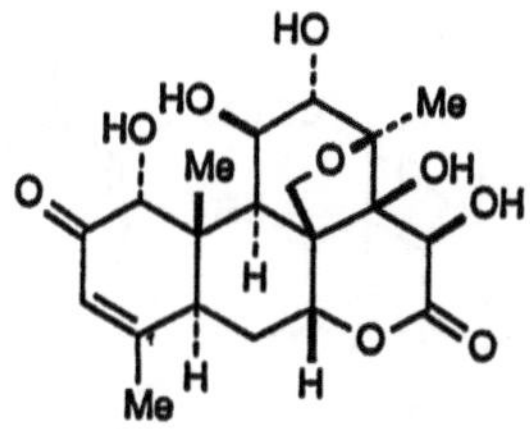		[2, 3]
Yadanzigan (*29-24*)	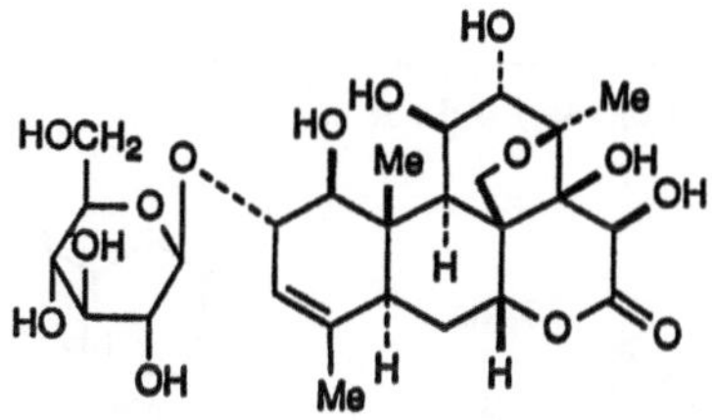		[59]
Yadanzioside D (*29-25*)	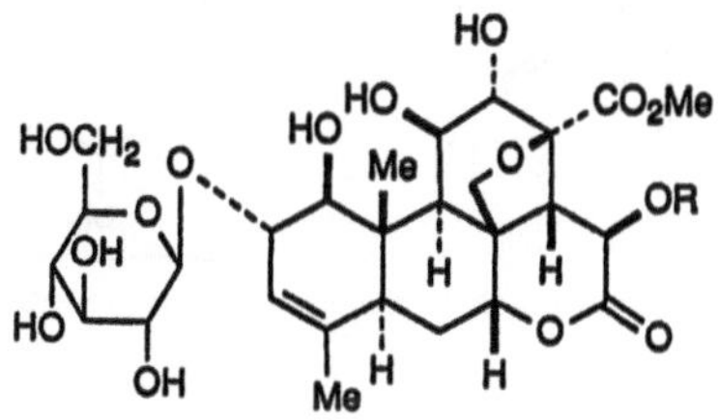	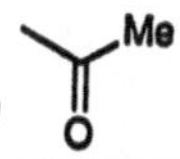	[4, 5]
Yadanzioside E (*29-26*)		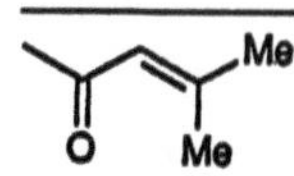	[4, 5]
Yadanzioside H (*29-27*)		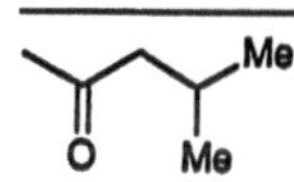	[4, 5]
	2,3-Dioxygenated picras-1-ene derivatives		
Bruceoside A (*29-28*)	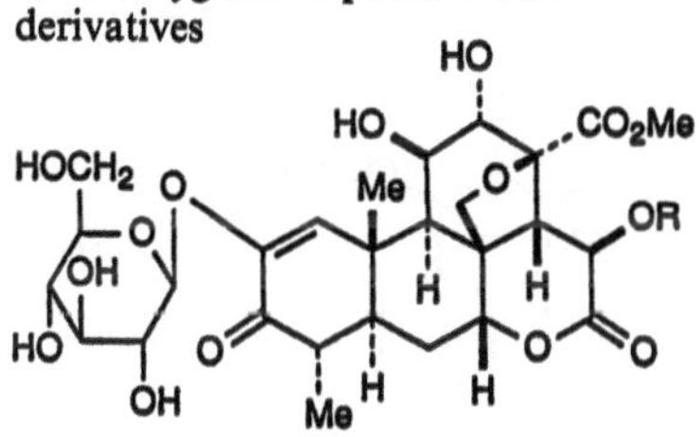	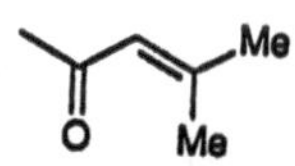	[16, 48, 55, 58]
Yadanzioside A (*29-29*)		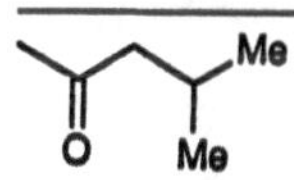	[4]
Yadanzioside C (*29-30*)		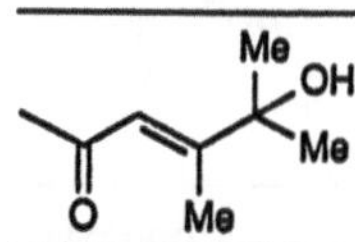	[5]
Yadanzioside F (*29-31*)		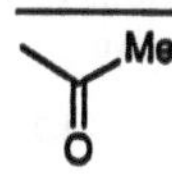	[3, 4]
Yadanzioside G (*29-32*)		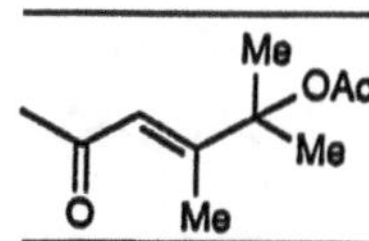	[4, 5]
Yadanzioside J (*29-33*)		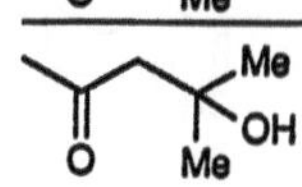	[3]

Table 29.1. (continued)

Compound	Structure	R	Reference
Yadanzioside M (*29-34*)	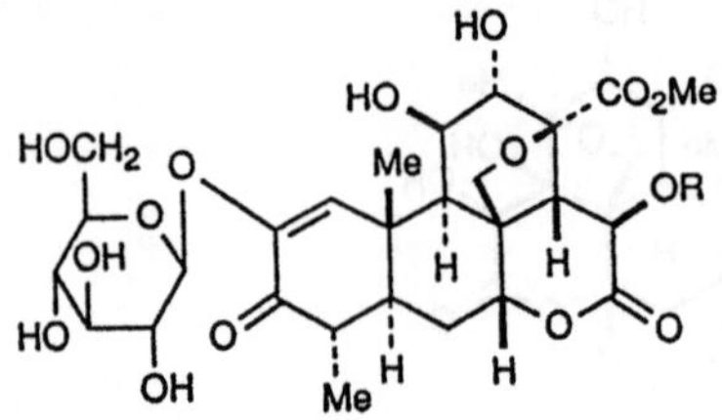	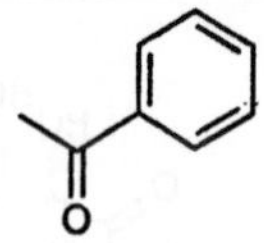	[62]
Yadanzioside O (*29-35*)		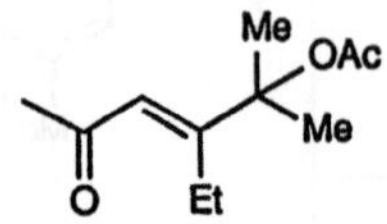	[62]
	2,3-Dioxygenated picras-1,4-diene derivatives		
Dehydrobruceine A (*29-36*)	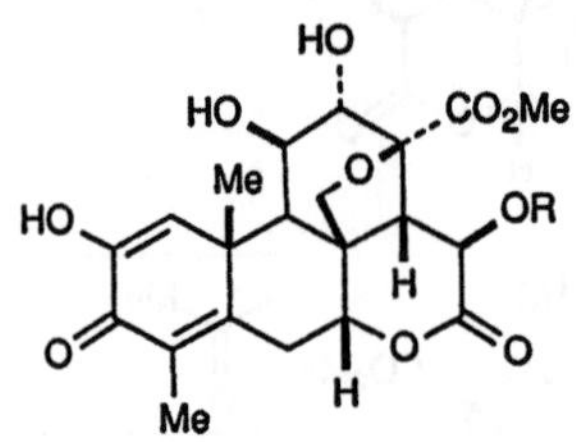	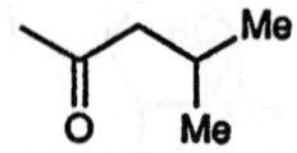	[15, 50, 51]
Dehydrobruceine B (*29-37*)		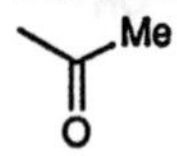	[14, 17, 50, 51, 57]
Dehydrobrusatol (*29-38*)		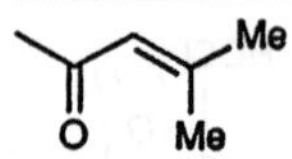	[5]
Dehydro-bruceantinol (*29-39*)		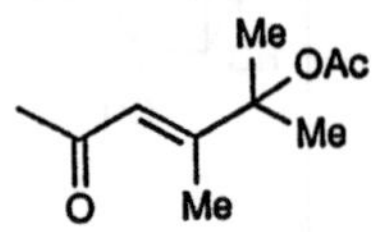	[5, 48]
	1,2-Dioxygenated picras-2-ene derivative		
Yadanzioside N (*29-40*)	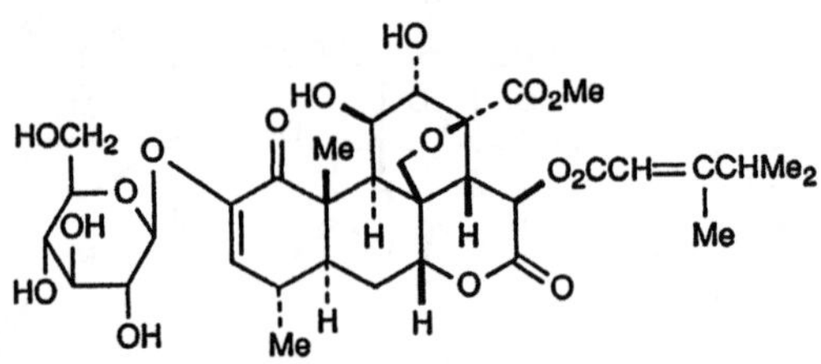		[62]
	1-Monooxygenated picras-2,4(18)-diene derivative		
Bruceene (*29-41*)	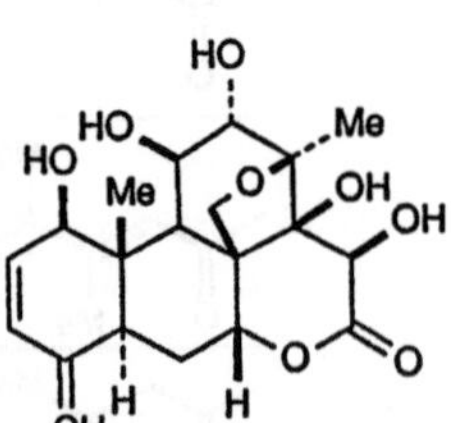		[54]

Table 29.1. (continued)

Compound	Structure R	Reference
	2,3-Dioxygenated picrasane derivative	
3,4-Dihydro-bruceine A (*29-42*)	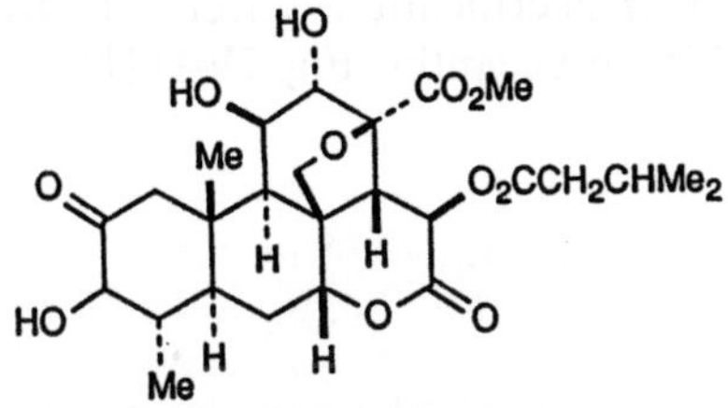	[50, 51]
Bruceaketolic acid (*29-43*)	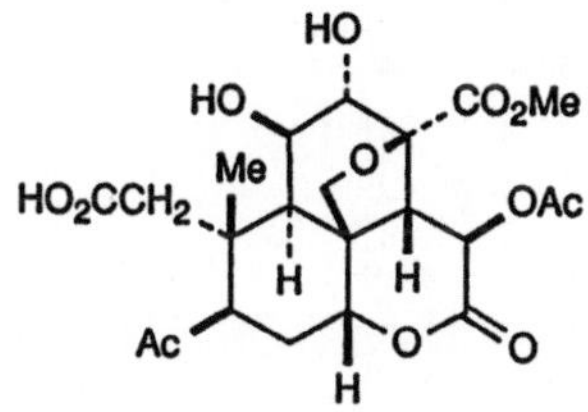	[57]

Bruceantin seems to be the most important and well-studied quassinoid compound of the *Brucea* species. Although efforts to achieve total synthesis of bruceantin have been made in the last years, only important intermediates have been obtained. For example, Heathcock et al. reported synthesis of the precursor (*29-44*) containing the ABCD ring system, starting from 2-methyl-cyclohexanone and 1-chloro-3-pentanone [8]; however, the group did not succeed in forming the tetrahydrofurane ring. The synthesis of another intermediate of bruceantin containing the ABC and tetrahydrofurane ring systems (*29-45*) was reported recently [9].

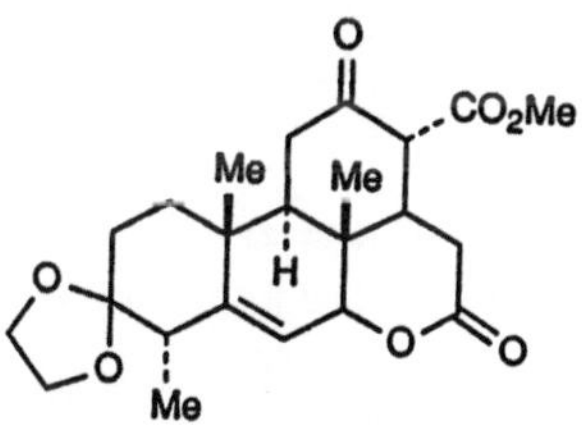

(*29-44*)

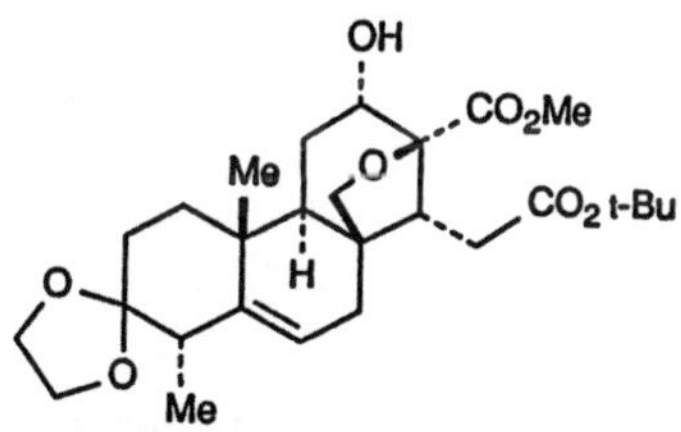

(*29-45*)

A pentacyclic system (*29-46*) as a model for bruceantin was also stereoselectively synthesized [10].

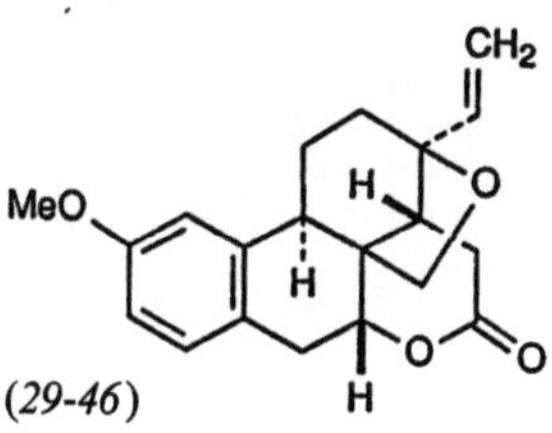

(*29-46*)

Bruceantin could be prepared by conversion of bruceoside A. Two methods have been described, one of them involving the hydrolysis of bruceoside A with potassium hydroxide followed by *p*-toluenesulfonic acid treatment in methanol to yield bruceolide (*29-1*). Esterification of bruceolide with 3,4-dimethyl-2-pentenoyl chloride (*29-47*) yielded the 3,15-diester (*29-48*) and the 3-monoester (*29-49*), which could be reconverted to the diester by further esterification. Selective hydrolysis of the diester with *p*-toluenesulfonic acid yielded bruceantin (Fig. 29-1) [11].

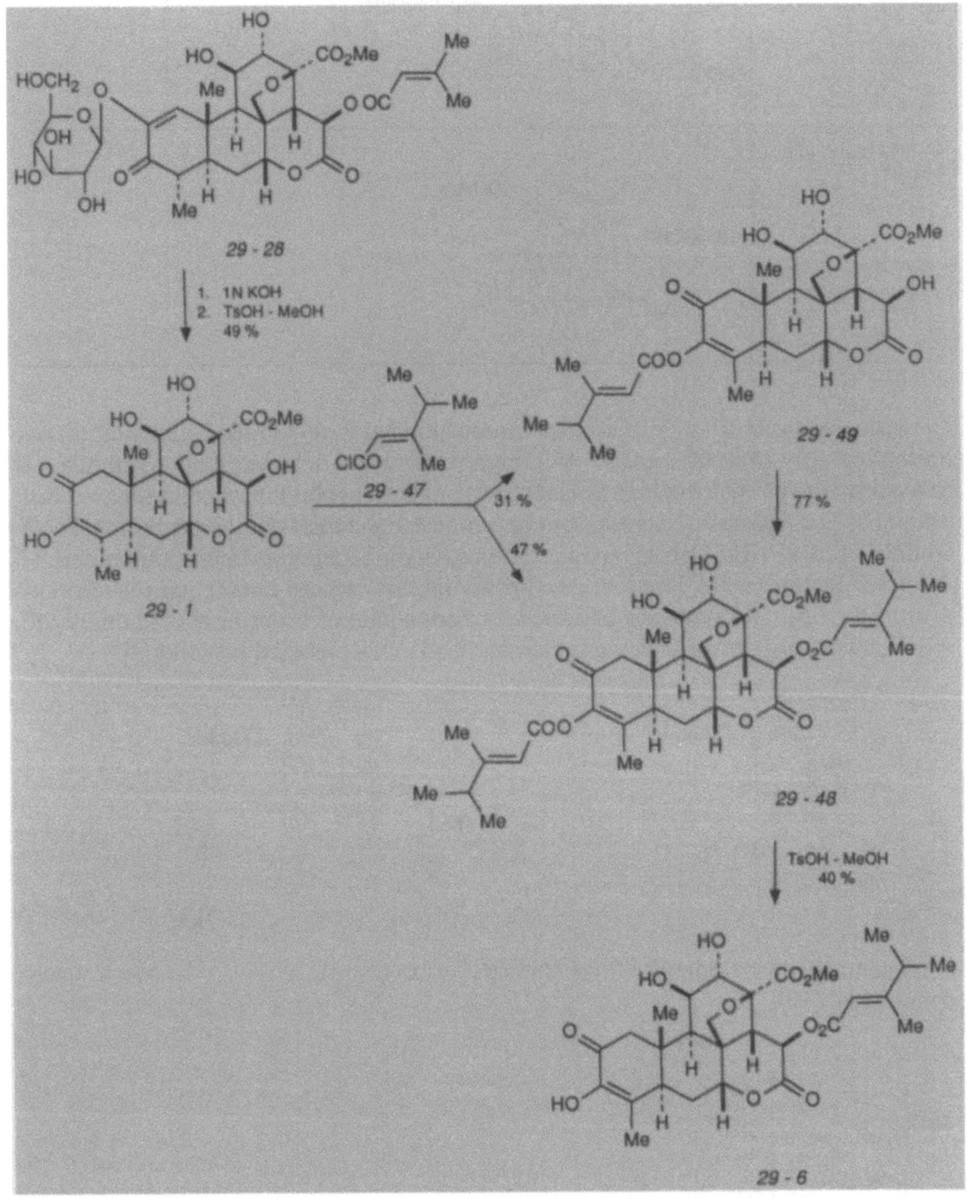

Fig. 29.1. Conversion of bruceoside A to bruceantin via bruceolide

The second method involves hydrolysis of bruceoside A with potassium hydroxide to yield 15-desenecioyl bruceoside A (*29-50*). Hydrolysis of the 3,4-dimethyl-2-pentenoyl ester, prepared by esterification of desenecioyl bruceoside A with the corresponding acid chloride, gave bruceantin (Fig. 29-2) [11, 12].

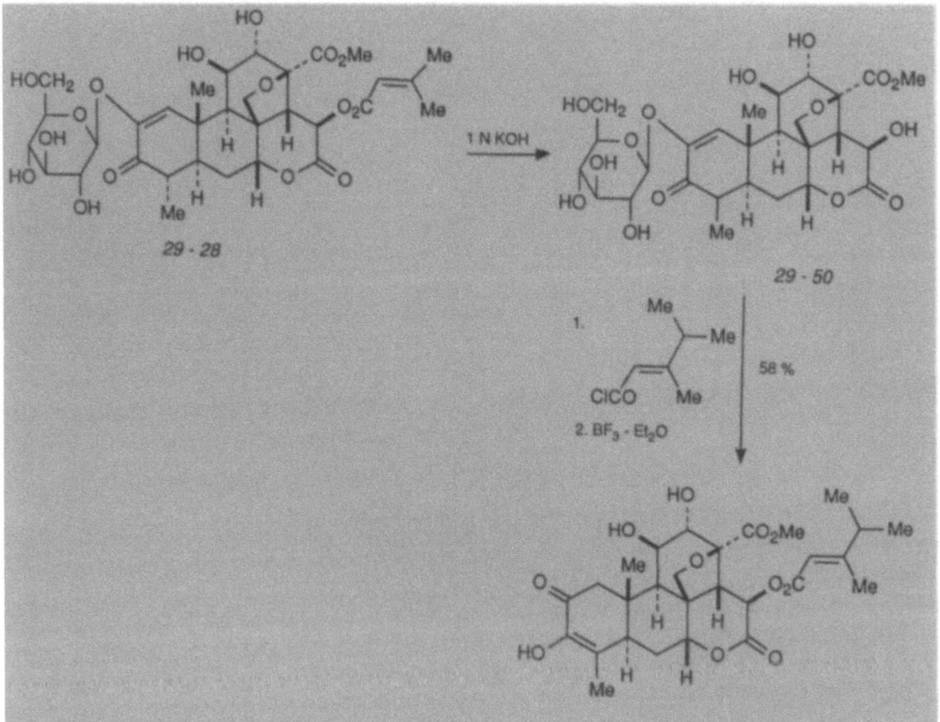

Fig. 29.2. Preparation of bruceantin from bruceoside A via desenecioyl bruceoside A

Furthermore, the conversion of bruceoside B into bruceantin was also reported [12]. Thus, bruceoside B or bruceoside A were hydrolyzed in $3N\,H_2SO_4$/MeOH (1 : 1) yielding brusatol (*29-5*), which was further hydrolyzed with potassium methylate in methanol after protecting the 3-hydroxyl group with *tert*-butyldimethylchlorosilane. After esterification of the hydroxyl group at position 15 with 3,4-dimethyl-2-pentenoyl chloride (*29-47*), deprotection with tetrabutylammonium fluoride in tetrahydrofuran (THF) yielded bruceantin (Fig. 29-3).

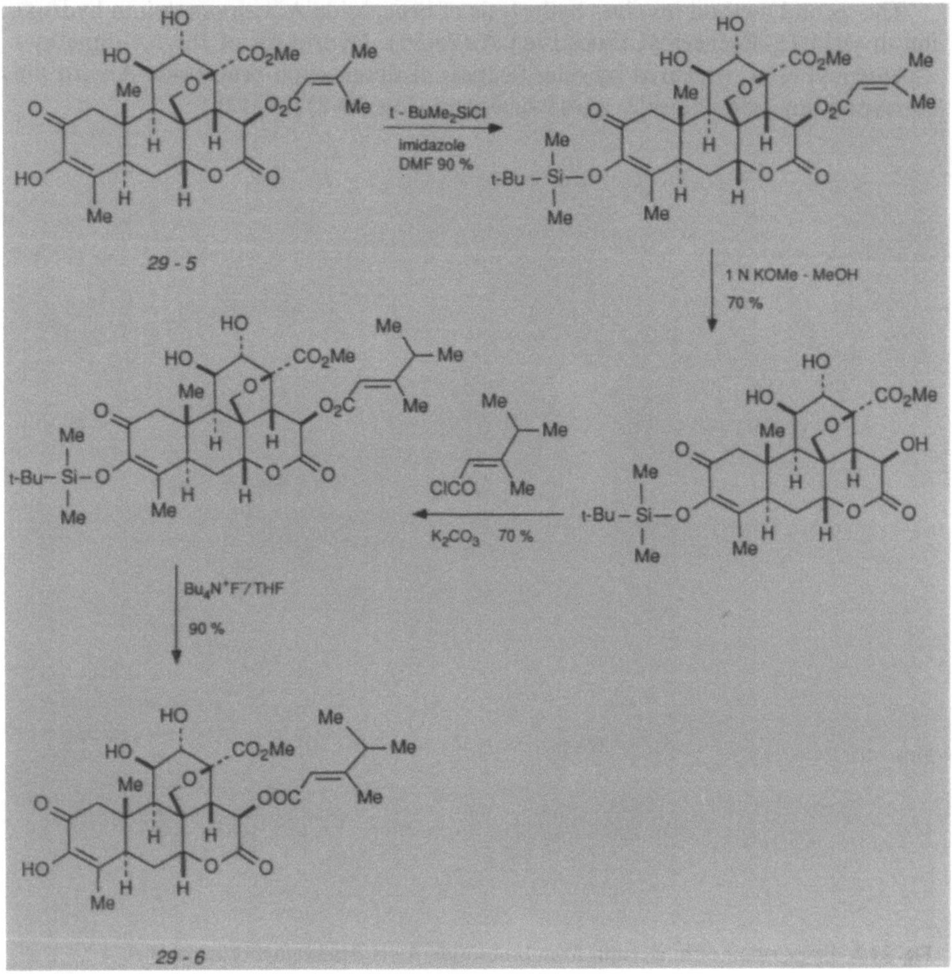

Fig. 29.3. Synthesis of bruceantin from brusatol

29.3 Pharmacology

The most important biological activity of some of the quassinoid bitter principles isolated from *Brucea* spp. is their antileukemic efficiency. Bruceantin was first isolated from *B. antidysenterica* and structurally identified by Kupchan et al. [13, 14]; it was first isolated from *B. javanica* by Darwish et al. [15]. Bruceantin is the bruceolide ester most intensely investigated for antileukemic activity.

In studies on experimental tumors in vitro and in vivo, bruceantin was active in a series of tumor systems including cells derived from human carcinoma of the nasopharynx, Walker 256 carcinoma in rats, and P388 lymphocytic leukemia in mice [13, 14, 16]. The T/C value (survival time) of bruceantin against P388 lymphocytic leukemia in mice was 200% [17]. A subline of P388 leukemia cells resistant to adriamycin did not show significant inhibition of nucleic acid synthesis after a dose of 10 mg/kg adriamycin; in contrast, it was sensitive to bruceantin [18]. Bruceantin therefore does not show cross-resistance to adriamycin.

The antileukemic activity of bruceantin and of brusatol in the P388 screen was different with regard to T/C values, depending on the P388 lymphocytic leukemia strain and the host strain of mice used. It was demonstrated that these substances caused different degrees of inhibition of nucleic acid and protein synthesis in the different P388 strains. The higher the inhibition of precursor incorporation into nucleic acids or protein, the higher was the observed T/C value [19].

The major mechanism responsible for antitumor activity of bruceantin on the molecular level might be the irreversible inhibition of protein synthesis as observed in HeLa cells, rabbit reticulocytes, and reticulocyte lysates. After addition of bruceantin to a reticulocyte lysate, a delay of 2–3 min before onset of inhibition of protein synthesis was observed. At a concentration of 0.1 m*M*, bruceantin induced sequential breakdown of polyribosomes to monosomes; concomitantly, nascent peptides were released from the ribosomes as completed chains of globin. This observation suggests that the principal inhibitory effect of bruceantin is on initiation of protein synthesis. The effects of bruceantin and its analogues on protein synthesis in HeLa cells, rabbit reticulocytes, and reticulocyte lysates correlated well with their antitumor activity in experimental animals. The side chain of bruceantin has been found to be important for transport of the drug into intact cells. The partial unsaturation of ring A is required for inhibition of protein synthesis [20].

Fresno et al. [21] used a model system from yeast and reticulocytes to study the effects of bruceantin on protein synthesis. Bruceantin was a potent inhibitor of polyphenylalanine synthesis as directed by poly U. Inhibition was less pronounced on protein synthesis directed by endogenous mRNA and the compound inhibited the poly U system only poorly when added after polyphenylalanine synthesis had been initiated.

The disposition and excretion of bruceantin were similar for normal and tumor-bearing mice. The concentration of bruceantin in blood decreased in a biphasic mode with $t_{1/2\alpha} = 5-6$ min and $t_{1/2\beta} > 0.5$ day after a single i.v. application. After 15 min, high concentrations of bruceantin were found in lung, pancreas, intestine, and spleen (1–6 µg/g) and low concentrations in liver, kidney, and tumor (0.3–0.5 µg/g). Concentrations of bruceantin in brain and peritoneal fat were below detectable limits (<0.1 µg/g). The excretion of bruceantin in urine and feces after 24 h was <2% of the dose applied. In vitro metabolism studies, using a postmitochondrial microsome fraction of liver, lung, and kidney, suggested that bruceantin was inactivated by an NADPH-dependent enzyme present in liver but not in lung or kidney [22].

Fong et al. [23] reported a radioimmunoassay for the quantitative detection of bruceantin using [^{3}H]acetyl-bruceantin prepared by reaction of bruceantin with [^{3}H]acetic anhydride. Antibody was induced by immunizing rabbits with succinyl-bruceantin-bovine serum albumin conjugates. With this simple and reproducible assay, drug levels were easily detected in tissues of experimental animals following administration of bruceantin. The lower limit of sensitivity of the assay was 1 µg/ml. During the phase I clinical trial of bruceantin, its pharmacology was studied in cancer patients by radioimmunoassay. The drug was infused i.v. for 3 h at doses ranging from 1–3.6 mg/m^2 per day for 5 days. Similar to the results in experimental studies, the drug disappeared from the plasma in a biphasic fashion with a fast initial

$t_{1/2\alpha}$ of less than 15 min. The $t_{1/2\beta}$ varied from 0.7–38 h among different patients and was not dose dependent. In patients with normal liver function, the main plasma concentration of bruceantin at the end of infusion was 22 ng/ml, whereas in patients with abnormal liver function the corresponding value was 115 ng/ml. Patients with abnormal liver function developed more severe toxicity, including fever, nausea, vomiting, and hypotension. Two patients with severe hepatic dysfunction received a reduced dose and developed no toxicity. The results showed the importance of liver function on drug metabolism.

The LD_{50} of i.v. administration in male and female B2D6/F1 mice was 1.95 and 2.85 mg/kg, respectively. Five daily injections of 0.025 mg or more of bruceantin produced mild to moderate irritation in rabbit muscle and in guinea pig subcutaneous tissue. Bruceantin produced clinical signs in dogs and monkeys that suggested an increase in vascular permeability. Except for absence of anemia, monkeys exhibited hematologic and blood chemical changes similar to those observed in dogs. The major target tissues in both dogs and monkeys were the liver, kidney, hematopoietic system, and lymphoid tissues [24]. The highest nontoxic dose was 0.0625 mg/kg bruceantin in dogs [25].

A phase I clinical trial on adult patients with various types of advanced solid tumors started with an initial daily dose of 0.2 mg/m^2 for 5 days repeated at 2 week intervals. The dose was increased progressively to a maximum daily dose of 4.5 mg/m^2. The dose limiting effect was hypotension, which was delayed, cumulative, and occurred more often in patients with abnormal liver function. Nausea, vomiting, and fever were common at high doses; diarrhea, stomatitis, alopecia, paresthesia, and rash were observed in some patients. The hematologic toxicity of bruceantin was moderate at high doses and was manifested mainly as thrombocytopenia. It was more severe in patients with abnormal hepatic and renal function. No objective tumor regressions were observed [26]. In another clinical phase I trial, adult patients with advanced solid tumors received bruceantin at doses of 1.6–6.0 mg/m^2 i.v. for 30 min 4 times a week, followed by a week rest. The dose limiting effect was nausea and vomiting, which was more severe in patients with hepatic metastases or liver function abnormalities. Other sporadic toxic effects included fever, chills, malaise, alopecia, and hypotension. Hematologic toxicity was insignificant [27]. The recommended dose of bruceantin for a phase II trial was 3.5 mg/m^2/day for 5 days [26] or 5 mg/m^2 four times a week every 6 weeks, with a reduction for patients with hepatic metastases [27].

Brusatol and its derivatives have also been investigated for their antitumor activity. Brusatol is chemically closely related to bruceantin; the difference is that the hydroxyl group at position 15 is esterified with 3-methyl-2-butenoic acid (senecioic acid) instead of 3,4-dimethyl-2-pentenoic acid in bruceantin. Bruceoside B is the 3-β-D-glucoside of brusatol.

A series of esters of brusatol, bruceantin, and bisbrusatolyl esters were synthesized. Bisbrusatolyl esters are formed from two molecules of brusatol *via* aliphatic dicarboxylic acids, connecting two brusatol molecules by esterification with the hydroxyl groups at position 3. In P388 lymphocytic leukemia cells the bisbrusatolyl esters of malonic, succinic, glutaric, adipic, and sebacic acid were as active or more active than the parent substance brusatol. The C-3 esters of brusatol or bruceantin were also found to be as active or more active than brusatol or bruceantin in general. The hydroxyl groups at C-11 and C-12 and the enone double bond in ring A of both bisbrusatolyl and brusatolyl esters are required for antileukemic activity. The presence of a double bond in the ester side chain contributes to the enhanced activity of the esters. Saturation of the side chain by catalytic hydrogenation decreased the antileukemic activity [28, 29]. A series of semisynthetic bruceolides with an ether function at position 3 and the hydroxyl group at position 15 esterified with C_{5-18} β-methyl-α,β-unsaturated fatty acids or with C_{1-18} alkylcarbamic acids were prepared and tested against experimental leukemia. The compounds showed an antileukemic effect at a dose of 0.1–5 mg/kg [30].

Brusatol and bisbrusatolyl malonate, like bruceantin, showed a strong correlation between their antileukemic activity and the ability to inhibit protein synthesis in P388 lymphocytic leukemia cells [31].

The antileukemic quassinoids from *Brucea* and their esters were, however, not universal protein synthesis inhibitors. Rather, they were selective for various types of cancer such as P388 lymphocytic leukemia, and L 1210 lymphoid leukemia, and for some normal tissues [32].

Brusatol, bisbrusatol esters, and bruceantin esters have also been shown to inhibit dihydrofolate reductase activity, basal respiration, and adenosine diphosphate stimulated respiration in P388 cells [33, 34]. RNA polymerase, thymidylate synthetase, phosphoribosyl pyrophosphate aminotransferase, and cathepsin protease activities were marginally inhibited in vitro by brusatol and bruceoside A. In vivo studies with brusatol and bruceoside A demonstrated similar inhibition and elevated cyclic AMP levels, correlating positively with their antileukemic activities. Purine synthesis was drastically inhibited by brusatol in vivo. One key inhibition site in purine synthesis was the regulatory enzyme phosphoribosyl pyrophosphate aminotransferase. Histone phosphorylation and ribonucleotide reductase activity also were inhibited marginally by brusatol [35].

B. javanica was used in traditional Chinese medicine as an antimalarial and antidysenteric agent. The chemical constituents isolated from the fruit were also tested against *Plasmodium* or *Entamoeba*. Guru et al. [36] reported that bruceantin was active against three chloroquine-resistant strains of *P. falciparum* in vitro. The concentration of bruceantin producing 50% inhibition was 0.0076 µg/ml. The 3-esters, 15-esters, and 3,15-diesters of bruceolide, including bruceolide, bruceantin, and brusatol, were reported to have generally inhibitory effects on chloroquine-resistant strains of *P. falciparum* [37].

In a study on the amoebicidal activity of certain quassinoids, bruceantin proved to be the most effective compound against *E. histolytica* with an LD_{50} of 0.018 µg/ml. This was 30-fold more active than metronidazole [38, 39].

The antiinflammatory activity of bruceolides should also be mentioned. In studies of the potential inhibitory activity of quassinoids against inflammation and arthritis induced in rodents, brusatol was the most potent followed by bruceine D. The 3-hydroxy-Δ^3-2-oxo moiety in brusatol or the 1-hydroxy-Δ^3-2-oxo moiety in bruceine D, the C-15 ester-bearing δ-lactone ring in brusatol, and the C-11 and C-12 free hydroxyl groups are required in both quassinoids for potent antiinflammatory activity. The studies also indicated that one of the modes of action is to stabilize lysosomal membranes, reducing the release of hydrolytic enzymes that cause damage to surrounding tissues [40].

After external application of the fruits of *B. javanica*, some cases of anaphylaxis were recently reported [41, 42]. For example, a patient using the fruit for local treatment of warts experienced numbness of the lips and generalized skin itch five minutes after chewing the fruit followed by tight chestedness, palpitation, severe abdominal cramps, and frequent vomiting [42].

References

1. Polonsky J, Baskevitch Z, Gandemer A, Das BC (1967) Constituants amers de *Brucea amarissima*, structures des bruceines A, B et C. Experientia 23:424–426
2. Yoshimura S, Sakaki T, Ishibashi M, Tsuyuki T, Takahashi T, Matsushita K, Honda T (1984) Structures of yadanziolides A, B and C, new bitter principles from *Brucea javanica*. Chem Pharm Bull 32:4698–4701
3. Yoshimura T, Sakaki T, Ishibashi N, Tsuyuki T, Takahashi T, Honda T (1985) Constituents of seeds of *Brucea javanica*. Structures of new bitter principles, yadanziolides A, B, C, Yadanziosides F, I, J and L. Bull Chem Soc Jpn 58:2673–2679
4. Sakaki T, Yoshimura S, Ishibashi M, Tsuyuki T, Takahashi T, Honda T, Nakanishi T (1984) New quassinoid glycosides, yadanziosides A-H, from *Brucea javanica*. Chem Pharm Bull 32:4702–4705
5. Sakaki T, Yoshimura S, Ishibashi M, Tsuyuki T, Takahashi T, Honda T, Nakanishi T (1985) Structures of new quassinoid glycosides, yadanziosides A, B, C, D, E, G, H and new quassinoids, dehydrobrusatol and dehydrobruceantinol from *Brucea javanica* (L.) Merr. Bull Chem Soc Jpn 58:2680–2686
6. Polonsky J (1973) Quassinoid bitter principles. In: Herz W, Grisebach H, Kirby GW (eds) Progress in the Chemistry of Organic Natural Products, Vol 30. Springer, Berlin Heidelberg New York, pp 101–150
7. Polonsky J (1985) Quassinoid bitter principles. II. In: Herz W, Grisebach H, Kirby GW, Tamm C (eds) Progress in the Chemistry of Organic Natural Products, Vol 47. Springer, Berlin Heidelberg New York, pp 221–264
8. Heathcock GH, Mahaim C, Schlecht MF, Utawanit T (1984) A synthetic approach to the quassinoids. J Org Chem 49:3264–3274
9. Kerwin SM, Paul AG, Heathcock CH (1987) Quassinoid synthesis. II. Preparation of a tetracyclic intermediate having the bruceantin tetrahydrofuran ring. J Org Chem 52:1686–1695
10. Shishido K, Saitoh T, Fukumoto K, Kametani T (1983) Stereoselective synthesis of the pentacyclic system as a model for bruceantin. J Chem Soc Chem Commun 852–853
11. Okano M, Lee KH (1981) Antitumor agents. XLIII. Conversion of bruceoside A into bruceantin. J Org Chem 46:1138–1141
12. Lee KH, Yamagishi T (1987) Antineoplastic terpenoids from Chinese medicinal plants. Abstr Chin Med 1:606–625
13. Kupchan SM, Britton RW, Ziegler MF, Sigel CW (1973) Bruceantin, a new potent antileukemic simaroubalide from *Brucea antidysenterica*. J Org Chem 38:178–179
14. Kupchan SM, Britton RW, Lacadie JA, Ziegler MF, Sigel CW (1975) The isolation and structural elucidation of bruceantin and bruceantinol, new potent antileukemic quassinoids from *Brucea antidysenterica*. J Org Chem 40:648–654
15. Darwish FA, Evans JF, Phillipson JD (1979) Cytotoxic bruceolides from *Brucea javanica*. J Pharm Pharmacol [Suppl] 31:10
16. Lee KH, Imakura Y, Huang HC (1977) Bruceoside A, a novel antileukemic quassinoid glycoside from *Brucea javanica*. J Chem Soc Chem Commun 69–70
17. Cassady JM, Suffness M (1980) Terpenoid antitumor agents. (Anticancer agents based on natural product models.) Med Chem (Academic) 16:201–269
18. Johnson RK, Chitnis MP, Embrey WM, Gregory EB (1978) In vivo characteristics of resistance and cross resistance of an adriamycin-resistant subline of P388 leukemia. Cancer Treat Rep 62:1535–1547
19. Hall IH, Lee KH, Okano M, Sims D, Ibuka T, Liou YF, Imakura Y (1981) Antitumor agents. XLII. Comparison of antileukemic activity of helenalin, brusatol and bruceantin and their esters on different strains of P388 lymphocytic leukemic cells. J Pharm Sci 70:1147–1150

20. Liao LL, Kupchan SM, Horwith SB (1976) Mode of action of the antitumor compound bruceantin, an inhibitor of protein synthesis. Mol Pharmacol 12:167–176
21. Fresno M, Gonzales A, Vazquez D, Jimenez A (1978) Bruceantin, a novel inhibitor of peptide bond formation. Biochim Biophys Acta 518:104–112
22. Suling WJ, Woolley CW, Shannon WM (1979) Disposition and metabolism of bruceantin in the mouse. Cancer Chemother Pharmacol 3:171–176
23. Fong KLL, Ho DHW, Carter CJK, Brown NS, Benjamin RS, Freireich EJ, Bodey GP (1980) Radioimmunoassay for the detection and quantitation of bruceantin. Anal Biochem 105:281–286
24. Castles TR, Bhandari JC, Lee CC, Guarino AM (1976) Preclinical toxicologic evaluation of bruceantin (NSC-165,563) in mice, dogs and monkeys. US National Technical Information Service, Springfield, p 348 (PB-257175)
25. Castles TR, Bhandari JE, Lee CC, Guarino AM, Cooney DA (1977) Preclinical toxicologic evaluation of bruceantin (NSC-165,563) in dogs. US National Technical Information Service, Springfield, p 56 (PB-269584)
26. Bedikian AY, Valdivieso M, Bodey GP, Murphy WK, Freireich E (1979) Initial clinical studies with bruceantin. Cancer Treat Rep 63:1843–1847
27. Liesmann J, Belt RJ, Haas CD, Hoogstraten B (1981) Phase I study on bruceantin administration on a weekly schedule. Cancer Treat Rep 65:883–885
28. Lee KH, Okano M, Hall IH, Brent DA, Soltmann B (1982) Antitumor agents. XLV. Bisbrusatolyl and brusatolyl esters and related compounds as novel potent antileukemic agents. J Pharm Sci 71:338–348
29. Kupchan SM, Lacadie JA, Howie GA, Sickles BR (1976) Structural requirements for biological activity among antileukemic glaucarubalone ester quassinoids. J Med Chem 19:1130–1133
30. Suntory Ltd (1985) Bruceolide derivatives. Jpn Kokai Tokkyo Koho JP 6032,793(8532793) (CA 102:221041 r)
31. Liou YF, Hall IH, Okano M, Lee KH, Chaney SG (1982) Antitumor agents. XLVIII. Structure-activity relationship of quassinoids as in vitro protein synthesis inhibitors of P388 lymphocytic leukemia tumor cell metabolism. J Pharm Sci 71:430–435
32. Hall IH, Liou YF, Lee KH, Chaney SG, Willingham W Jr (1983) Antitumor agents. LIX. Effects of quassinoids on protein synthesis of a number of murine tumors and normal cells. J Pharm Sci 72:626–630
33. Hall IH, Liou YF, Okano M, Lee KH (1982) Antitumor agents. XLVI. In vitro effects of esters of brusatol, bisbrusatol and related compounds on nucleic acid and protein synthesis of P388 lymphocytic leukemia cells. J Pharm Sci 71:345–348
34. Eigebaly SA, Hall IH, Lee KH, Sumida Y, Imakura Y, Wu RY (1979) Antitumor agents. XXXV. Effect of brusatol, bruceoside A and bruceantin on P388 lymphocytic leukemia cell respiration. J Pharm Sci 68:877–890
35. Hall IH, Lee KH, Eigebaly SA, Imakura Y, Sumida Y, Wu RY (1979) Antitumor agents. XXXIV. Mechanism of action of bruceoside A and brusatol on nucleic acid metabolism of P388 lymphocytic leukemia cells. J Pharm Sci 68:883–887
36. Guru PY, Warhurst DC, Harris A, Phillipson JD (1983) Antimalarial activity of bruceantin in vitro. Ann Trop Med Parasitol 77:433–435
37. Lee LK (1982) Bruceolides. US Patent 4,350,638
38. Gillin FD, Reiner DS (1982) In vitro activity of certain quassinoid antitumor agents against *Entamoeba histolytica*. Arch Invest Med [Suppl 3] 13:43–49
39. Gillin FD, Reiner DS, Suffness M (1982) Bruceantin, a potent amebicide from a plant, *Brucea antidysenterica*. Antimicrob Agents Chemother 22:342–345
40. Hall IH, Lee KH, Imakura Y, Okano M, Johnson A (1983) Antiinflammatory agents. III. Structure-activity relationships of brusatol and related quassinoids. J Pharm Sci 72:1282–1284
41. Zhang ZL, Wang GZ (1985) Report on a case of anaphylactic shock and atrial fibrillation after external application of Fructus Bruceae. Jilin Med J 6:674
42. Zheng GQ, Mao LZ, Xia WZ, Zhang GP (1986) A report on three cases of anaphylaxis caused by external application of the fruit of *Brucea javanica*. Bull Chin Mat Med 11:121
43. Polonsky J, Baskevitch Z, Das BC, Müller J (1968) Sur les constituants amers du *Brucea amarissima*: structures des bruceines D et E. C R Acad Sci [C] 267:1346–1349
44. Polonsky J, Baskevitch Z, Müller J (1969) Constituants amers du *Brucea amarissima*: structure de la bruceine F. C R Acad Sci [C] 268:1392

45. Duncan GR, Henderson GB (1968) Bruceines from *Brucea sumatrana*: the structure of bruceine G. Experientia 24:768–769
46. Sim KY, Sims J, Geissman TA (1968) Constituents of *Brucea sumatrana* Roxb. I. Brusatol. J Org Chem 33:429–431
47. Stöcklin W, Geissman TA (1968) A new bitter principle from *Brucea sumatrana* Roxb. Tetrahedron Lett 6007–6010
48. Lee KH, Imakura Y, Sumida Y, Wu RY, Hall JH, Huang HC (1979) Antitumor agents 33. Isolation and structural elucidation of bruceoside-A and -B, novel antileukemic quassinoid glycosides and bruceine-D and -E from *Brucea javanica*. J Org Chem 44:2180–2185
49. Polonsky J, Verenne J, Prange T, Pascard C (1980) Antileukemic quassinoids: structure (X-ray analysis) of bruceine C and revised structure of bruceantinol. Tetrahedron Lett 1853–1856
50. Phillipson JD, Darwish FA (1979) TLX-5-lymphoma cells in rapid screening for cytotoxicity in *Brucea* extract. Planta Med 35:308–315
51. Phillipson JD, Darwish FA (1981) Bruceolides from Fijian *Brucea javanica*. Planta Med 41:209–220
52. Chiang HC (1975) A new bitter principle from *Brucea sumatrana*. Proc Natl Sci Counc (Taiwan) [A] 8:139–143 (CA 86,34173z)
53. Sato Y, Hasegawa M, Suto N (1980) Identity of brusatol and yantansin, an antidysenteric agent. Agric Biol Chem 44:951–952
54. Zhang JS, Xu RS, Li YH, Chen ZL (1984) Chemical constituents of *Brucea javanica*. III. Bruceene – isolation and structure of a new quassinoid from *Brucea javanica*. Acta Chim Sin 42:684–687
55. Li X, Tso SH (1980) Studies on the anticancer principles in the seed of yadanzi (*Brucea javanica*). III. Isolation and structural elucidation of bruceoside A, bruceoside B, Bruceine E and bruceine F. Chin Trad Herb Drugs 11:530–532
56. Xie JX, Ji Z (1981) Chemical constituents of the Chinese medicinal herb *Brucea javanica*. I. Isolation and identification of daucosterol, bruceine D and bruceine E. Acta Pharm Sin 16:53–55
57. Lin LZ, Zhang JS, Chen ZL, Xu RS (1982) Studies on the chemical constituents of *Brucea javanica* (L.) Merr. I. Isolation and identification of bruceaketolic acid and other four quassinoids. Acta Chim Sin 40:73–78
58. Lin AQ, Liu SR, Ji Zheng, Xie JX (1982) Chemical constituents of the Chines drug "Yadanzi". II. Isolation and identification of brusatol and bruceoside A. Beijing Shifan Daxue Xuebao, Ziran Kexueban 71–73
59. Zhang JS, Lin LZ, Chen ZL, Xu RS, Sun XY (1983) Studies on the chemical constituents of *Brucea javanica*. II. Brucein-E-glucopyranoside. Acta Chim Sin 41:149–152
60. Fong KLL, Ho DHW, Benjamin RS, Brown NS, Bedikian A (1982) Clinical pharmacology of bruceantin by radioimmunoassay. Cancer Chemother Pharmacol 9:169–172
61. Li XA, Liu YN, Wang SX, Zhu YR (1986) The separation and identification of bruceine H in *Brucea javanica*. J Shenyang Coll Pharm 3:192–193
62. Sakaki T, Yoshimura S, Tsuyuki T, Takahashi T, Honda T, Nakanishi T (1986) Structures of yadanziosides K, M, N and O, new quassinoid glycosides from *Brucea javanica* (L.) Merr. Bull Chem Soc Jpn 59:3541–3546
63. Sakaki T, Yoshimura S, Tsuyuki T, Takahashi T, Honda T (1986) Yadanzioside P, a new antileukemic quassinoid glycoside from *Brucea javanica* (L.) Merr with the 3-O-(β-D-glucopyranosyl)-bruceantin structure. Chem Pharm Bull 34:4447–4450

Bupleurum spp.

30

30.1 Introduction

Chaihu, Radix Bupleuri, is the dry root of ***Bupleurum chinense*** DC. or ***B.*** *scorzonerifolium* Willd. (Apiaceae) collected in spring and fall. It is listed officially in the Chinese Pharmacopoeia and used in traditional Chinese medicine in treatment of influenza, fever, malaria, and menstrual disorders.

It is specifically stressed in the Chinese Pharmacopoeia that the root of ***B.*** *longiradiatum* Turcz. must not be used clinically because of its toxicity. In the appendix of the Chinese Pharmacopoeia ***B. marginatum*** Wall. ex DC. is accepted. Besides the official ***Bupleurum*** species, ***B. falcatum*** L. [1] has also been used.

30.2 Chemical Constitutents

30.2.1 Chemical Constituents of *Bupleurum falcatum*

A series of triterpene saponins and sapogenins, named saikosaponins and saikogenins, were isolated from the root of *B. falcatum* and structurally characterized. Thus, saikogenin A (*30-1*) [2–5], saikogenin B (*30-2*) [5–7], saikogenin C (*30-3*) [4, 5, 7], saikogenin D (*30-4*) [4, 5], saikogenin E (*30-5*) [8–10], saikogenin F (*30-6*) [9, 11], saikogenin G (*30-7*) [9, 10], saikosaponin A (*30-8*) [12, 13], saikosaponins B_1–B_4 (*30-9*–*30-12*) [14], saikosaponin C (*30-13*) [12], saikosaponin D (*30-14*) [12], saikosaponin E (*30-15*) [15], and saikosaponin F (*30-16*) [16] were isolated and their structure determined. All the saikogenins are derived from oleanane.

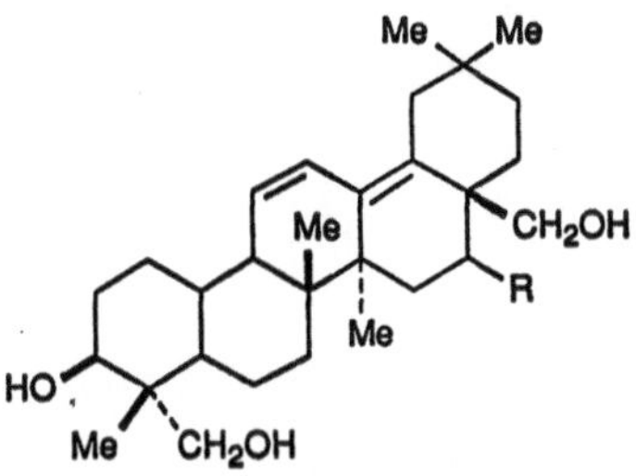

Saikogenin A (*30-1*): R = β-OH
Saikogenin D (*30-4*): R = α-OH

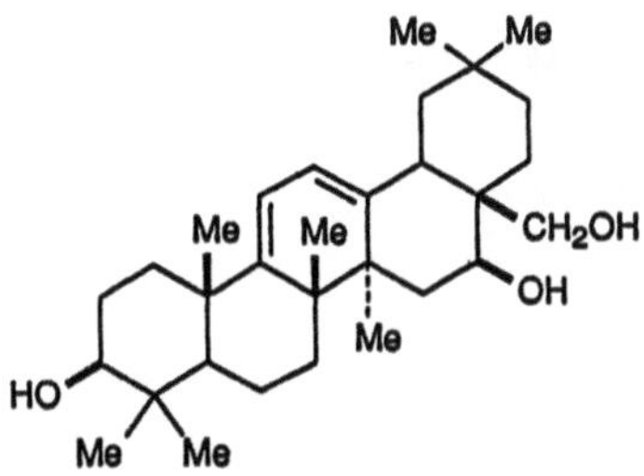

Saikogenin B (*30-2*)

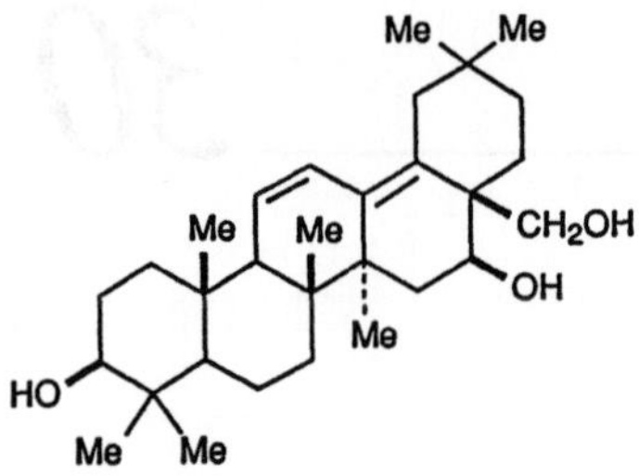

Saikogenin C (*30-3*)

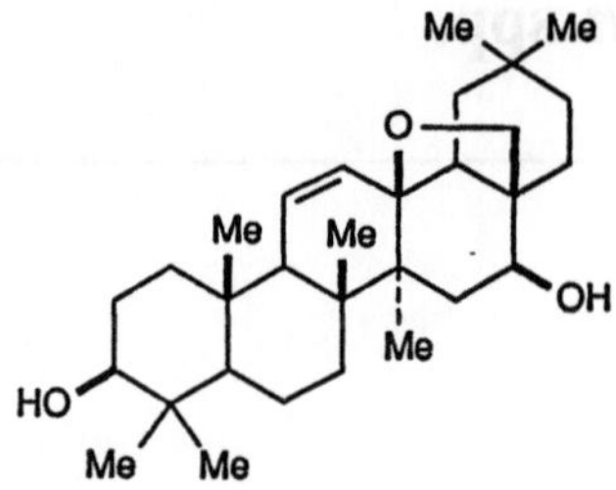

Saikogenin E (*30-5*)

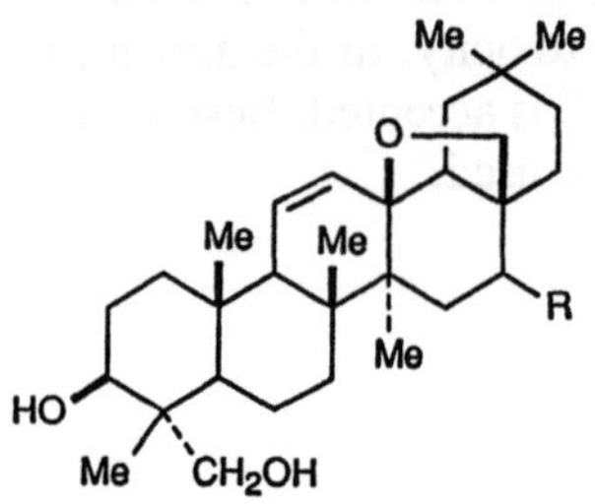

Saikogenin F (*30-6*): R = β-OH
Saikogenin G (*30-7*): R = α-OH

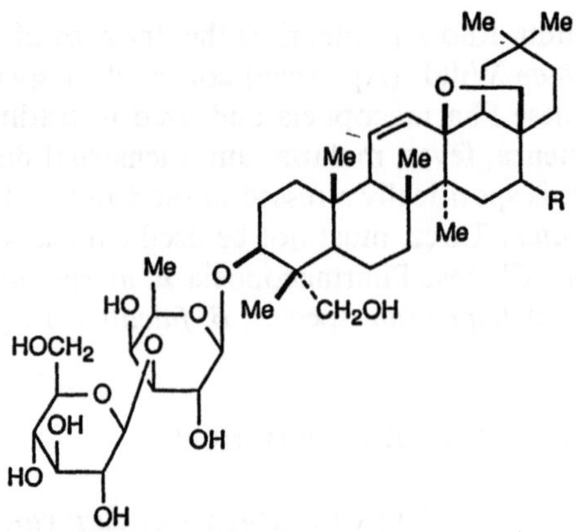

Saikosaponin A (*30-8*): R = β-OH
Saikosaponin D (*30-14*): R = α-OH

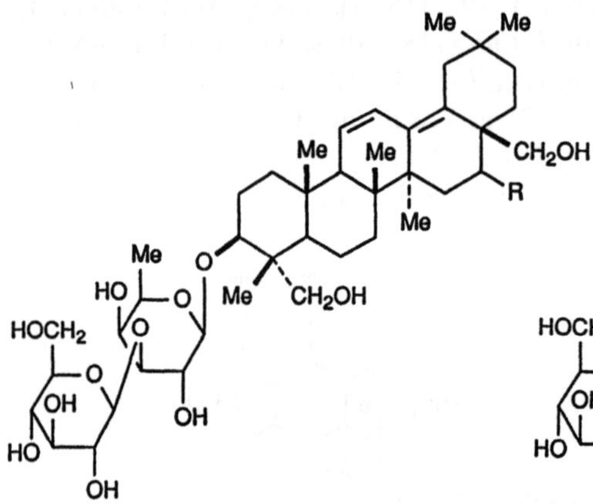

Saikosaponin B_1 (*30-9*): R = β-OH
Saikosaponin B_2 (*30-10*): R = α-OH

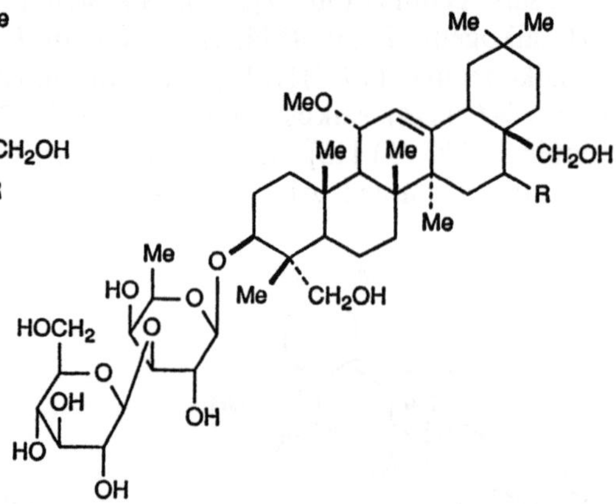

Saikosaponin B_3 (*30-11*): R = β-OH
Saikosaponin B_4 (*30-12*): R = α-OH

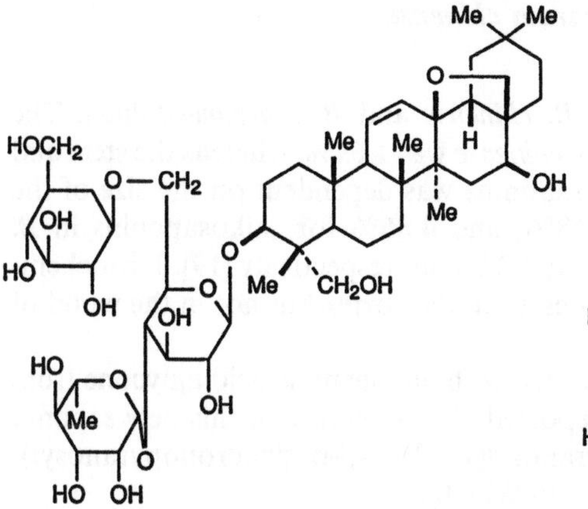

Saikosaponin C (*30-13*)

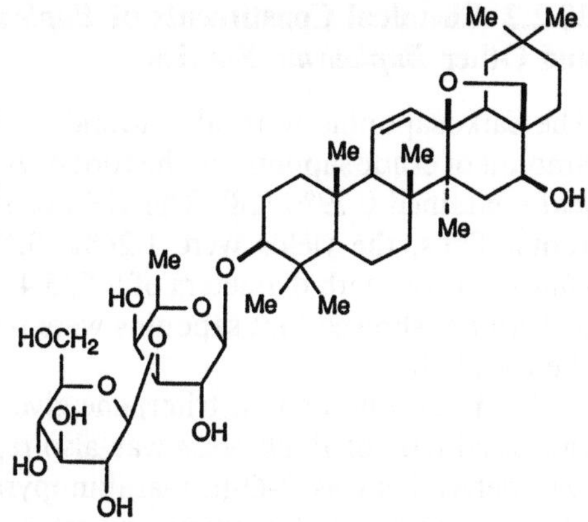

Saikosaponin E (*30-15*)

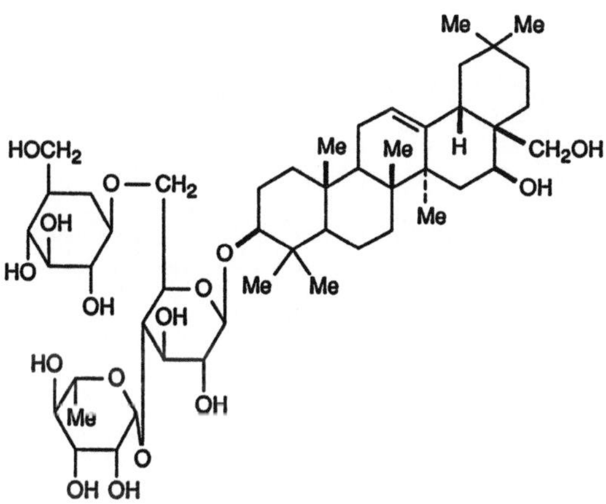

Saikosaponin F (*30-16*)

Besides the saikogenins, saikosaponins, and somc monoacctylsaikosaponins [15, 16], α-spinasterol (*30-17*) [17], a steroid derived from stigmasterol (*1-5*), and its β-D-glucopyranoside [16] were isolated and identified.

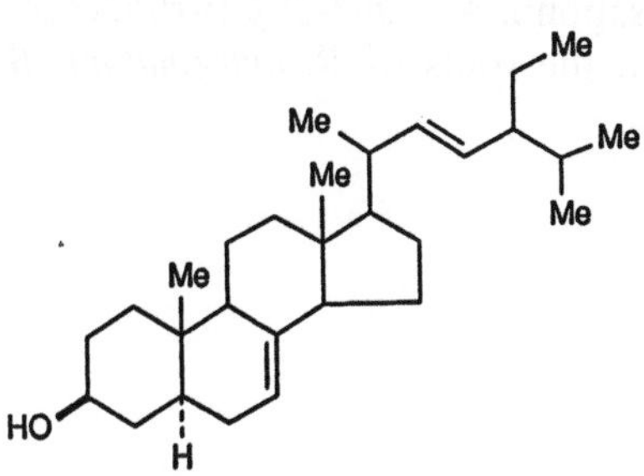

α-Spinasterol (*30-17*)

30.2.2 Chemical Constituents of *Bupleurum chinense* and Other *Bupleurum* Species

The saikosaponins were also found in *B. chinense* and *B. scorzonerifolium*. The amount of crude saponins in the root of *B. chinense* was 1.69%, whereas the stem and leaf contained 0.29% [18]. The yield of saponins was dependent on the size of the roots. Thus, the yields were 1.24%, 3.18%, and 4.86% for saikosaponins in *B. chinense* roots with diameters of 9.5, 5.4, and 2.7 mm, respectively [19]. Histochemical studies showed that saponins were present in the cortex but not in the wood of the root [19].

The isolation of a new triterpene glycoside with an oleanolic acid aglycone from the aerial part of *B. chinense* was also reported. The structure of this new saponin was determined as 3-*O*-(α-L-arabinopyranosyl(1→3)-*O*-β-D-glucuronopyranosyl)-oleanolic acid β-D-glucopyranosyl ester (*30-18*) [20].

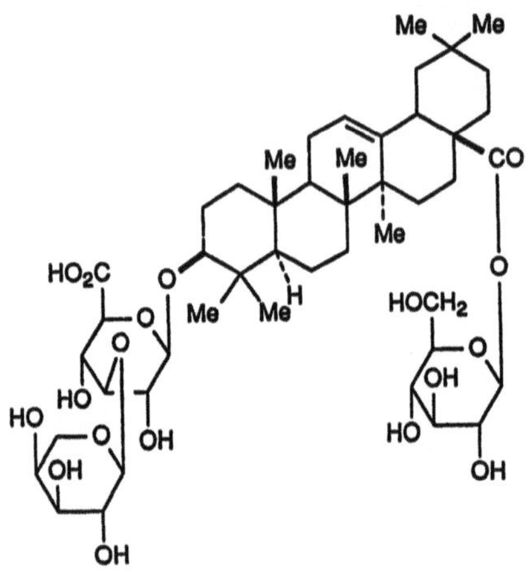

(*30-18*)

Five new triterpene glycosides were isolated from the root of *B. kunmingense*. Four of them were found to be acetylated saikosaponins, 3″-*O*-acetylsaikosaponin A, 4″-*O*-acetylsaikosaponin A, 2″-*O*-acetylsaikosaponin D, and 2″-*O*-acetylsaikosaponin A. In addition, 16-*epi*-chikusaikoside I (*30-19*) was identified [21].

2″-*O*-acetylsaikosaponin A, 3″-*O*-acetylsaikosaponin A, and 3-*O*-β-D-rhodeopyranosylsaikogenin F (*30-20*) were isolated from the roots of *B. marginatum, B. marginatum* var. *stenophyllum*, and *B. rockii* [22].

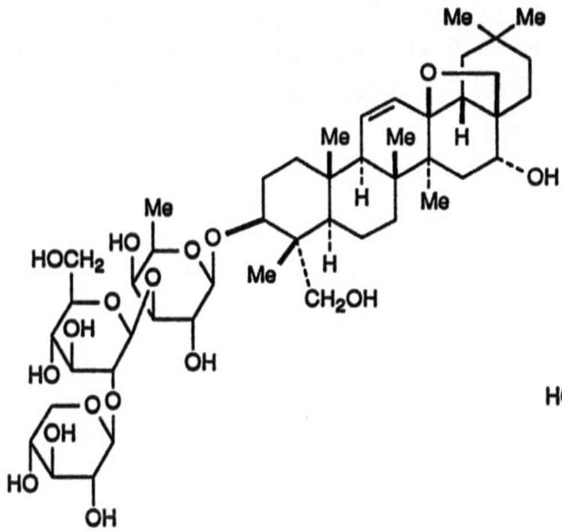

16-*epi*-Chikusaikoside (*30-19*)

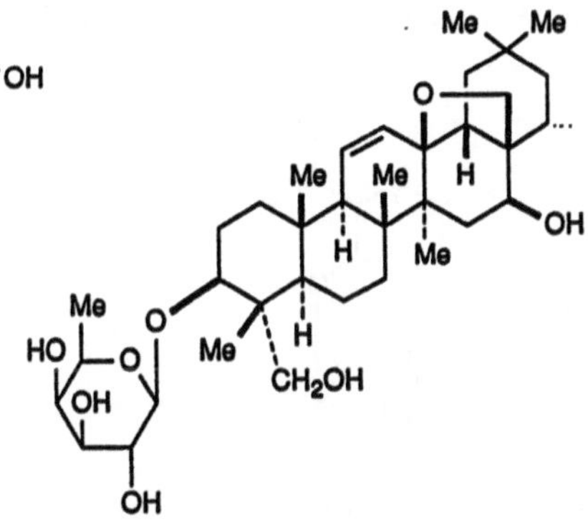

3-*O*-*β*-D-Rhodeopyranosylsaikogenin F (*30-20*)

A number of saponins were also isolated from the root of *B. polyclonum*: These included: 30-*β*-D-glucopyranosyloxy-saikosaponin B_2 (*30-21*), 2″-*O*-acetylsaikosaponin B_2, and 3″-*O*-acetylsaikosaponin B_2 [23].

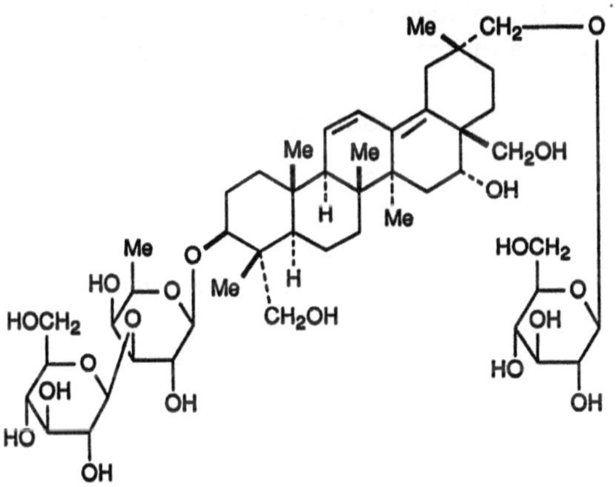

30-*β*-D-Glucopyranosyloxy-saikosaponin B_2 (*30-21*)

The saponin composition of the root of *B. tenue* was similar to that of the root of *B. chinense*, as determined by TLC comparison of the saponin fractions of both species indicating that *B. tenue* may be used as a substitute for *B. chinense* [24].

In addition to the saponins, a pharmacologically active essential oil was obtained from *B. chinense*. From the essential oil a number of components were isolated and identified as: pentanoic acid, hexanoic acid, heptanoic acid, 2-heptenoic acid, octanoic acid, 2-octenoic acid, nonanoic acid, 2-nonenoic acid, phenol, cresol, ethylphenol, thymol, eugenol, *o*-methoxyphenol, *γ*-heptalactone, *γ*-octalactone, *γ*-decalactone, *γ*-undecalactone, vanillin acetate [25], valeric acid, and *p*-methoxyacetophenone [26].

The amount of essential oil was 0.16% and 0.05% (v/w) in root and stems, respectively [18]. From the stems and leaves of *B. chinense* and *B. scorzonerifolium* the flavones kaempferitrin (*30-22*) (kaempferol 3,7-dirhamnoside) and kaempferol 7-rhamnoside were isolated and identified [27].

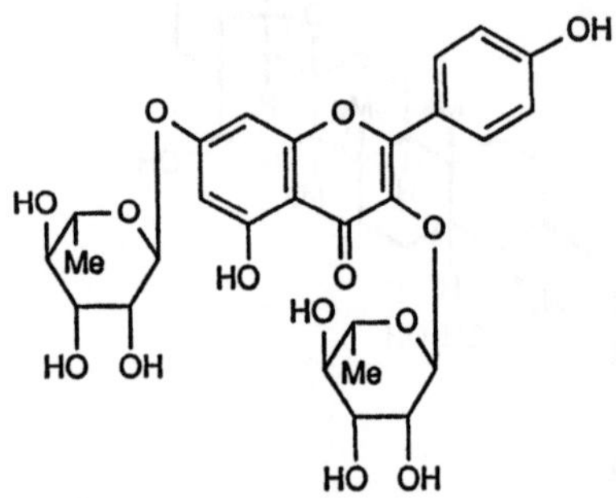

Kaempferitrin (*30-22*)

30.2.3 *Bupleurum longiradiatum*

The toxic principle of *B. longiradiatum* was isolated from the ether extract and identified as enanthotoxin, (*E,E,E*)-2,8,10-heptadecatriene-4,6-diyne-1,14-diol (*30-23*). The LD_{50} of enanthotoxin by i.p. administration in mice was 3 mg/kg [28].

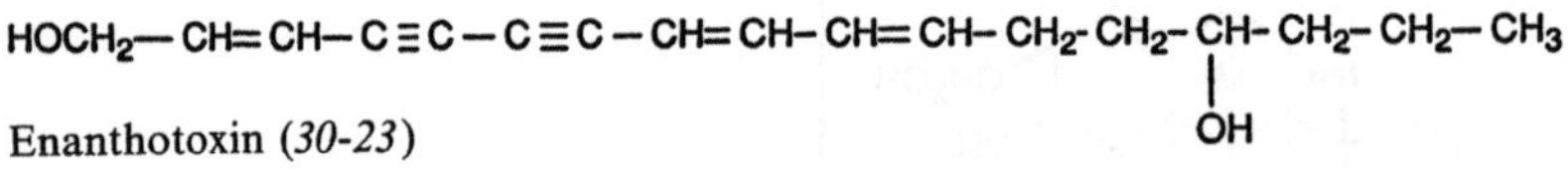

Enanthotoxin (*30-23*)

Furthermore, saponins were also isolated from the root of *B. longiradiatum*: saikosaponins A, C, and D and chikusaikoside I (*30-24*) were detected [29].

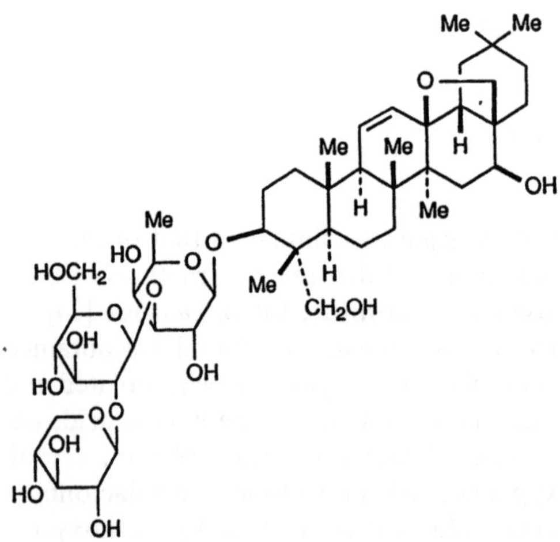

Chikusaikoside I (*30-24*)

30.3 Pharmacology

The crude saponin fraction extracted from the roots of *B. falcatum* had an LD_{50} of 4.7 g/kg in mice by oral administration and 58.3 mg/kg in guinea pigs by i.p. administration and was highly hemolytic in rats. Oral administration of 200–800 mg/kg of the crude saponin to mice exhibited sedative, analgesic, and antipyretic effects but no anticonvulsant activity or reduction of muscle tone. Moreover, the crude saponin was a potent antitussive with an ED_{50} of 9.1 mg/kg in guinea pigs by i.p. injection [30]. A local irritation by topical application was observed [31]. Given orally, this crude saponin fraction caused a significant decrease in dextran-induced edema of rat paw and an increased capillary permeability in the mouse peritoneal cavity and in the croton oil-induced granuloma pouch of the rat. The crude saponin fraction did not inhibit carrageenan-induced edema, acetic acid-induced edema, adjuvant arthritis, or experimental liver fibrosis of the rat. In addition, it had no protective effect on histamine shocked guinea pigs or anaphylactic shocked mice [31, 32]. Among saikosaponins, saikosaponins A and D, but not C, were active against inflammation [33].

The crude saponins did not affect gastric motility, but they were significantly effective in the prevention of stress ulcers in rats and showed a strong stimulating effect in the intestinal propulsion test in mice. The contraction of isolated guinea pig ileum induced by acetylcholine was potentiated by crude saponins but not the contraction induced by histamine. Furthermore, in dogs a transient hypotensive action and a decrease in heart rate caused by the crude saponins were also reported [31].

The crude saponins of *B. falcatum*, given orally to rats at a daily dose of 500 mg/kg for 3 days, normalized liver function, as indicated by serum alkaline phosphatase levels in CCl_4-treated rats [34]. Treatment of rats with saikosaponins 2 h after injection of D-galactosamine inhibited the increase in serum GOT and GPT levels and the histologic damage of liver tissues. Pretreatment with saikosaponins 2 h before administration of D-galactosamine was more effective against hepatic injury [35]. Unlike the activity against hepatic injury induced by D-galactosamine, saikosaponins did not affect the increase in serum GPT and experimental cirrhosis caused by CCl_4 intoxication [36].

Combined administration of adrenocortical hormones and root extract of *B. falcatum* enhanced the effectiveness of adrenocortical hormones. Thus, oral administration of corticosterone acetate together with the root extract to rats markedly elevated the hepatic activity of tyrosine transaminase as an indicator of corticosterone acetate activity. This enhancing effect was also clinically demonstrated [37]. When saikosaponins A, B_1, B_2, B_3, B_4, C, D, or F were injected alone i.p. into adrenalectomized rats, no induction of liver tyrosine aminotransferase activity was observed, whereas this enzyme was induced on administration of a low dose of cortisone [38].

Saikosaponin D was active against kidney damage in rats induced by aminonucleoside. It prevented the development of proteinuria and significantly lowered the electron microscopic abnormality in glomerular epithelial cells [39].

Saikosaponins and the corresponding sapogenins extracted from *B. falcatum* caused membrane stabilization and lytic effects on hypotonic and heat-induced hemolysis of rat erythrocytes. Low concentrations of saikosaponins protected or stabilized rat erythrocytes against both hypotonic and heat-induced hemolysis. Minor modifications of the aglycone of the saikosaponin altered the membrane stabilizing potency. Saikogenins also protected erythrocytes from hypotonic hemolysis but did not show any prevention of heat-induced hemolysis [40]. PGE_2 biosynthesis was stimulated by saikosaponins B_1, B_2, and D but inhibited by saikosaponins A and C [41].

The decoction of *B. chinense* antagonized caffeine-induced convulsions in mice and showed moderate analgesic action which was partly antagonized by atropine and naloxone. Crude saikosaponin isolated from *B. chinense* caused contracture of the isolated rectus abdominis muscle of toads and of the isolated ileum of guinea pigs. The former was antagonized by tubocurarine chloride, whereas the latter was partly antagonized by atropine. Crude saikosaponin from *B. chinense* also showed a significant inhibiting effect on the activity of acetylcholinesterase in mouse blood [42].

In rats, saponins isolated from *B. chinense* inhibited paw edema caused by dextran. The inhibition did not occur in bilaterally adrenalectomized rats [43]. The saponins antagonized serotonin- or histamine-induced increases in capillary permeability. The i.p. LD_{50} in mice was 1.53 g/kg [44].

In the essential oil of *B. chinense* the antipyretic constituents were identified as eugenol, hexanoic acid, γ-undecalactone, and *p*-methoxyacetophenone; the antiinflammatory constituents were identified as valeric acid, *p*-methoxyacetophenone, and 2-nonenoic acid [26].

α-Spinasterol, isolated from *B. chinense*, markedly suppressed the carrageenin-induced hind paw swelling of intact and adrenalectomized rats and mice. It also significantly suppressed edema of the hind paw caused by scalding and proliferation of granulation tissue in croton oil-induced air pouch granulomas in rats. Some inhibition of exudation of the granuloma pouch was observed. α-Spinasterol also significantly reduced the fever induced by s.c. injection of 15% yeast suspension, 20 ml/kg, in rats. The LD_{50} of α-spinasterol in mice was 479 mg/kg, i.p. The mechanism of antiinflammatory action of α-spinasterol is complicated. Marked inhibitory effects on the synthesis or release of PGE and bradykinin; the phlogistic effect of PGE, bradykinin, histamine, and serotonin; and inhibition of leukocyte migration were observed. These actions of α-spinasterol were not due to stimulation of the pituitary-adrenal axis; however, the weight of the adrenal gland was significantly increased by i.p. injection of α-spinasterol 48 mg/kg once daily for 7 days [29].

The polysaccharide isolated from *B. kunmingense* showed an immunomodulatory action on the proliferative response of mouse splenocytes to mitogens as measured by [^{3}H]thymidine incorporation [45].

References

1. Song SX (1984) Quantitative analysis of saikosaponin in *Bupleurum falcatum* L. and effects of environmental factors on its content. Zhiwu Shenglixue Tongxun 31–34
2. Shibata S, Kitagawa I, Fujimoto H (1965) Structure of saikogenin A, a sapogenin of the *Bupleurum* species. Tetrahedron Lett 3783–3788
3. Shibata S, Kitagawa I, Fujimoto H (1966) Chemical studies on oriental plant drugs. IV. Constituents of *Bupleurum*. 2. Structure of saikogenin A, a sapogenin of *Bupleurum falcatum*. Chem Pharm Bull 14:1023–1033

4. Kubota T, Tonami F, Hinoh H (1967) Triterpenoids from *Bupleurum falcatum*. I. The structure of saikogenins A, C and D. Tetrahedron 23:3333–3351
5. Kubota T, Tonami F, Hinoh H (1966) The structure of saikogenins A, B, C and D, triterpenoid alcohols of *Bupleurum falcatum* L. Tetrahedron Lett 701
6. Kubota T, Tonami F (1967) Triterpenoids from *Bupleurum falcatum*. II. Isolation of saikogenin B and longispinogenin. Tetrahedron 23:3353–3362
7. Aimi N, Shibata S (1966) Saikogenin E, a genuine sapogenin of *Bupleurum* roots. Tetrahedron Lett 4721–4724
8. Kubota T, Hinoh H (1966) Isolation of saikogenin E, a new triterpene from *Bupleurum falcatum*. Tetrahedron Lett 4725–4728
9. Kubota T, Hinoh H (1968) Triterpenoids from *Bupleurum falcatum* L. III. Isolation of genuine sapogenins, saikogenins E, F and G. Tetrahedron 24:675–686
10. Aimi N, Fujimoto H, Shibata S (1968) Chemical studies on oriental plant drugs. XVIII. Constituents of *Bupleurum* spp. 3. Saikogenins E and G. Chem Pharm Bull 16:641–646
11. Kubota T, Hinoh H (1966) Isolation of saikogenin F, another genuine sapogenin from *Bupleurum falcatum*. Tetrahedron Lett 5045–5048
12. Kubota T, Hinoh H (1968) The constitution of saponins isolated from *Bupleurum falcatum*. Tetrahedron Lett 303–306
13. Akahori A, Kagawa K, Shimaoka A (1975) Quantitative determination of saikosaponins in *Bupleurum* extract. III. Extraction of saikosaponins from *Bupleurum* root. Shoyakugaku Zasshi 29:99-105 (CA 85:10342 g)
14. Shimaoka A, Seo S, Hitoshi M (1975) Saponins isolated from *Bupleurum falcatum*. Components of saikosaponin B. J Chem Soc, Perkin Trans I:2043–2048
15. Ishii H, Seo S, Tori K, Tozyo T, Yoshimura Y (1977) The structures of saikosaponin-E and acetylsaikosaponins, minor components isolated from *Bupleurum falcatum* L., determined by carbon-13 NMR spectroscopy. Tetrahedron Lett 1227–1230
16. Ishii N, Nakamura M, Seo S, Tori T, Tozyo T, Yoshimura Y (1980) Isolation, characterization and nuclear magnetic resonance spectra of new saponins from the roots of *Bupleurum falcatum* L. Chem Pharm Bull 28:2367–2373
17. Nagoshi K, Tomimatsu T (1969) Constituents of *Bupleurum* plants. Shoyakugaku Zasshi 23:96–98 (CA 73:73845 w)
18. Gan HS, Chen SW (1982) Qualitative and quantitative comparison between root and stem and leaves of *Bupleurum chinense*. Bull Chin Mat Med 7:7–8
19. Zhang J (1985) Comparison on saikosaponin content in the root of *Bupleurum chinense* of different sizes. Bull Chin Mat Med 10:157–158
20. Seto H, Otake N, Luo SQ, Jin HF (1986) Studies on chemical constituents of *Bupleurum* genus. I. A new triterpenoid glycoside from *Bupleurum chinense* DC. Agric Biol Chem 50:939–942
21. Seto H, Otake N, Luo SQ, Qian FG, Xu GY, Pan SL (1986) Studies on chemical constituents of *Bupleurum* genus. II. Isolation of triterpenoid glycosides (saikosaponins) from *Bupleurum kunmingense* and their chemical structures. Agric Biol Chem 50:943–948
22. Ding JK, Fujino H, Kasai R, Fujimoto N, Tanaka O, Zhou J, Matsuura H, Fuwa T (1986) Chemical evaluation of *Bupleurum* species collected in Yunnan, China. Chem Pharm Bull 34:1158–1167
23. Seto H, Otake N, Kawai H, Luo SQ, Qian FG, Pan SL (1986) Studies on chemical constituents of *Bupleurum* genus. III. Isolation of triterpenoid glycosides (saikosaponins) from *Bupleurum polyclonum* Y. Li et S.L. Pan and their chemical structures. Agric Biol Chem 50:1607–1611
24. Feng SJ, Zhang PL (1981) Chemical constituents of *Bupleurum tenue*. Chin Trad Herb Drugs 12:30
25. Pu QL, Ji XD, Xu P, Huang L, Guo QY, Chen XX (1983) Studies on the constituents of the essential oil of *Bupleurum chinense* DC. Acta Chim Sin 41:559–561
26. Hubei Provincial Institute of Pharmaceutical Industry, Institute of Medical and Pharmaceutical Sciences of Guangxi Zhuang Autonomous Region (PR China) (1982) Studies of the antipyretic and antiinflammatory principles of the volatile oil of *Bupleurum chinense*. Chin Pharm Bull 17:202–205
27. Shi YN, Hsu L (1980) Isolation, separation and identification of kaempferitrin and kaempferol-7-rhamnoside from *Bupleurum scorzonerifolium* leaf and stem. Chin Trad Herb Drugs 11:241–243

28. Zhao JF, Guo YH, Meng XS (1985) Poisonous constituents of *Bupleurum longiradiatum* Turcz. J Shenyang Coll Pharm 2:301
29. Zhou CC, Sun XB, Liu JY, Luo SQ, Lu CY (1985) Anti-inflammatory effect of α-spinasterol. Acta Pharm Sin 20:257–261
30. Takagi K, Shibata M (1969) Pharmacological studies on *Bupleurum falcatum*. I. Acute toxicity and central depressant action of crude saikosides. Yakugaku Zasshi 89:712–720
31. Takagi K, Shibata M (1969) Pharmacological studies on *Bupleurum falcatum*. II. Antiinflammatory and other pharmacological actions of crude saikosides. Yakugaku Zasshi 89:1367–1378
32. Shibata M, Yoshida R, Motohashi S, Fukushima M (1973) Pharmacological studies on *Bupleurum falcatum*. IV. Pharmacological effects of crude saikosides, saikogenin A, and syrupy residue. Yakugaku Zasshi 93:1660–1667
33. Yamamoto M, Kumagai A, Yamamura Y (1975) Structure and actions of saikosaponins isolated from *Bupleurum falcatum*. I. Antiinflammatory action of saikosaponins. Arzneimittelforschung 25:1021–1023
34. Shibata M, Isomura A, Inoue T, Nagai M (1976) Some pharmacological studies on the crude drugs possessing antiinflammatory properties on the root of *Bupleurum* and the leaves of fig. Shoyakugaku Zasshi 30:62–66 (CA 87:62641 q)
35. Arichi S, Konishi H, Abe H (1978) Studies on the mechanism of action of saikosaponin. I. Effects of saikosaponin on hepatic injury induced by D-galactosamine. Kanzo 19:430–435 (CA 89:209044 z)
36. Zhao MQ, Han DW, Ma XH, Zhao YC, Yin L, Li CM (1983) Preventive and therapeutic effects of glycyrrhizin, glycyrrhetic acid and saikosides on experimental cirrhosis in rats. Acta Pharm Sin 18:325–331
37. Yuchi S, Uchida Y, Fujikawa A (1985) Enhancement of the effectiveness of corticosteroid by *Bupleurum falcatum* root extracts. Jpn Kokai Tokkyo Koho JP 60,166,623 (85,166,623) (CA 104:15803 d)
38. Hashimoto M, Inada K, Ohminami H, Kimura Y, Okuda H, Arichi S (1985) Effects of saikosaponins on liver tyrosine aminotransferase activity induced by cortisone in adrenalectomized rats. Planta Med 41:401–403
39. Abe H, Orita M, Konishi H, Arichi S, Odashima S (1986) Effects of saikosaponin D on aminonucleoside nephrosis in rats. Eur J Pharmacol 120:171–178
40. Abe H, Sakaguchi M, Anno M, Arichi S (1981) Erythrocyte membrane stabilization by plant saponins and sapogenins. Naunyn Schmiedebergs Arch Pharmacol 316:262–265
41. Ohuchi K, Watanabe M, Ozeki T, Tsurufuji S (1985) Pharmacological influence of saikosaponins on prostaglandin E_2 production by peritoneal macrophages. Planta Med 41:208–212
42. Wu CF, Yu QH (1984) Pharmacological studies on *Bupleurum chinense* and its active ingredient, crude saikosaponin. J Shenyang Coll Pharm 1:214–218
43. Yu QH, Wan LP (1986) Mechanism of the antiinflammatory effect of *Bupleurum chinense* crude saikosaponin. J Shenyang Coll Pharm 3:14–16
44. Wang BZ, Wang GZ, Liu AJ (1981) Antiinflammatory effect of Saikosaponins. Acta Pharmacol Sin 2:60–63
45. Zhang LX, Pan SL, Huang KX, Sun KX (1986) Immunomodulatory action of *Bupleurum kunmingense* polysaccharide on lymphocyte proliferation. Acta Acad Med Shanghai 13:20–23

Caesalpinia sappan L. 31

31.1 Introduction

Sumu, Lignum Sappan, is the dry heartwood of *Caesalpinia sappan* L. (Fabaceae). It is listed officially in the Chinese Pharmacopoeia and used in traditional Chinese medicine as an analgesic and antiinflammatory agent in treatment of traumatic diseases and menstrual disorders.

31.2 Chemical Constituents

The heartwood of *C. sappan* is known to contain a number of phenolic pigments. One of the major pigments, brazilin (*31-1*), was isolated in the last century. A sappanchalcone (*31-2*) was newly isolated from the heartwood of *C. sappan* and proposed as an intermediate in the biosynthesis of brazilin. The structure of sappanchalcone was elucidated on the basis of spectrometric data and synthesis from 2-*O*-methylresacetophenone and protocatechualdehyde [1].

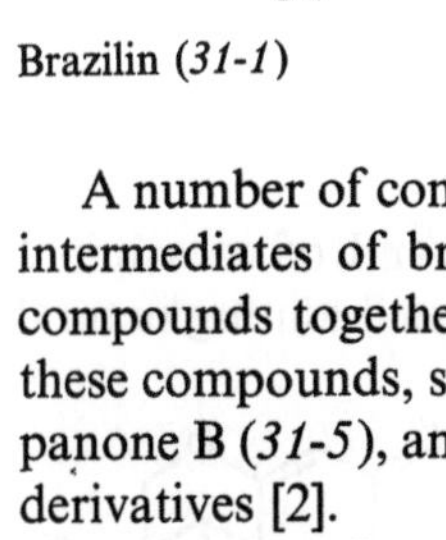

Brazilin (*31-1*)

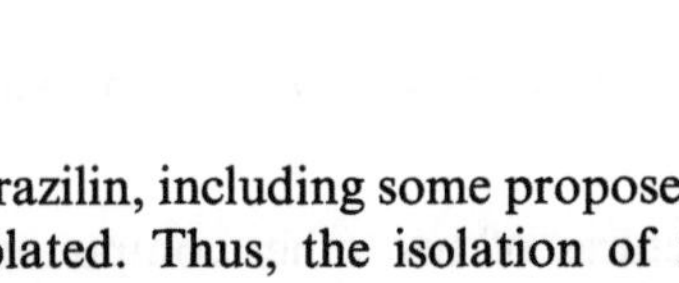

Sappanchalcone (*31-2*)

A number of compounds related to brazilin, including some proposed biogenetic intermediates of brazilin, were also isolated. Thus, the isolation of 17 aromatic compounds together with brazilin and sappanchalcone was reported [2]. Some of these compounds, such as sappanone B (*31-3*), sappanone A (*31-4*), 3-hydroxysappanone B (*31-5*), and sappanol (*31-6*), have been considered to be 3-benzylchroman derivatives [2].

Heller and Tamm [3] proposed a biogenetic pathway from chalcone via 3-benzylchroman analogues. Isolation of sappanol and its 3-methyl ether as key intermediates, which upon heating are converted partly into brazilin, together with sappanchalcone, sappanone A, sappanone B, and 3-hydroxysappanone B, which are considered to be precursors of brazilin, supported the above hypothesis [2]. The biogenetic pathway of brazilin from sappanchalcone is illustrated in Fig. 31-1.

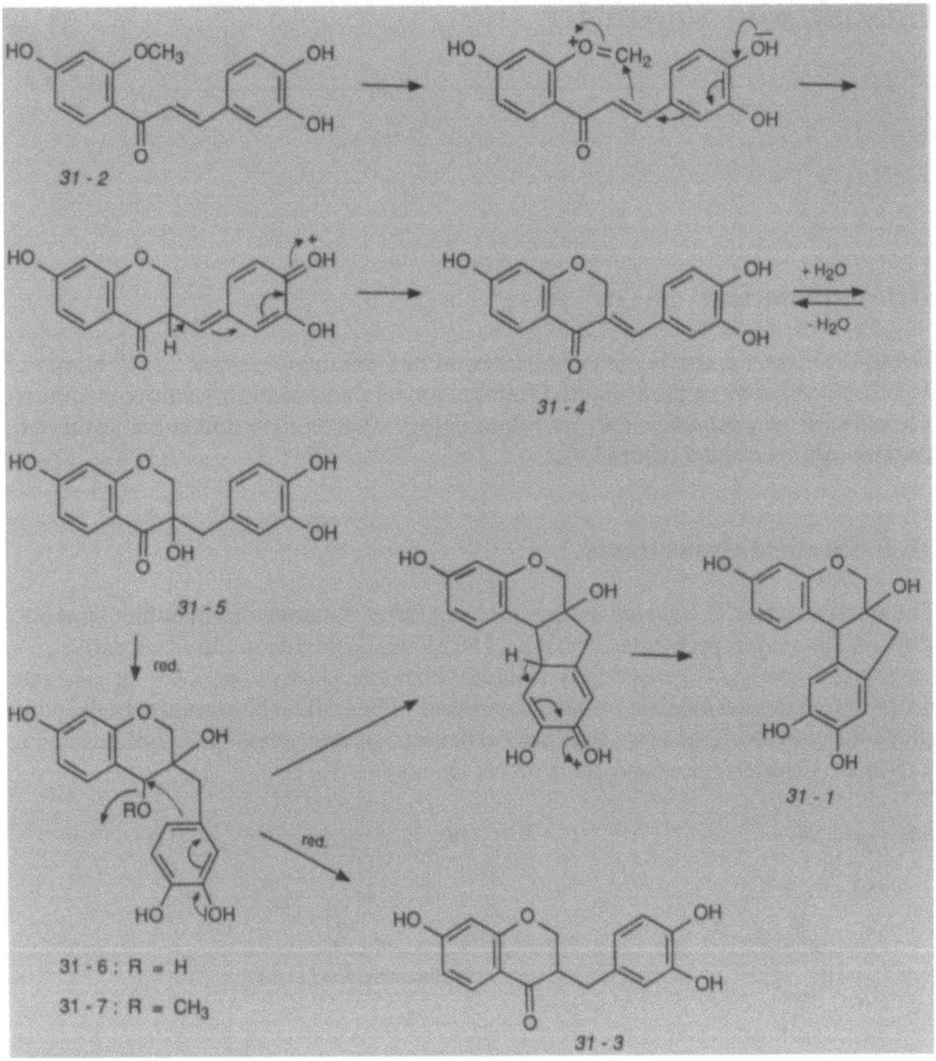

Fig. 31.1. Biogenetic pathway of brazilin from sappanchalcone

The structures of two other compounds (*31-8*, *31-9*) [2, 4] and of caesalpin J (*31-10*) and caesalpin P (*31-11*) [5] were found to be quite unusual among natural products.

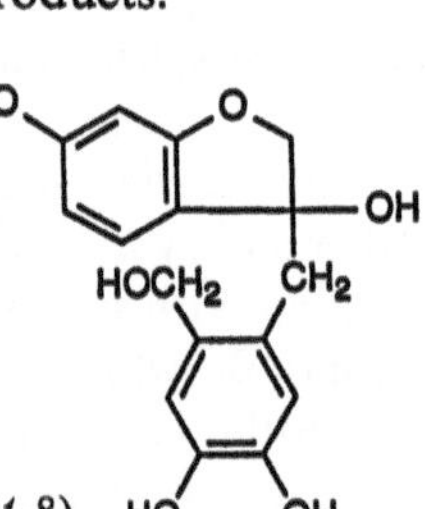

(*31-8*)

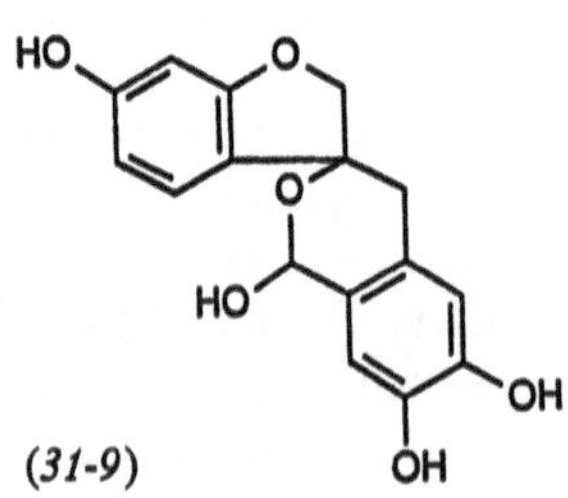

(*31-9*)

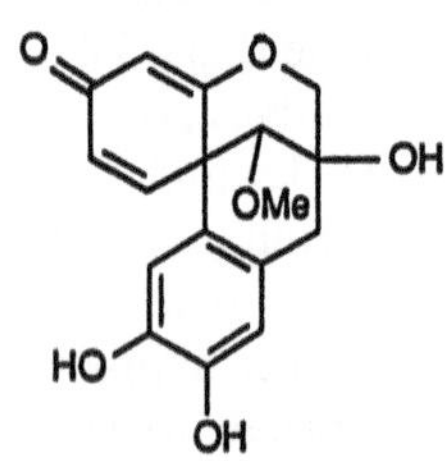

Caesalpin J (*31-10*)

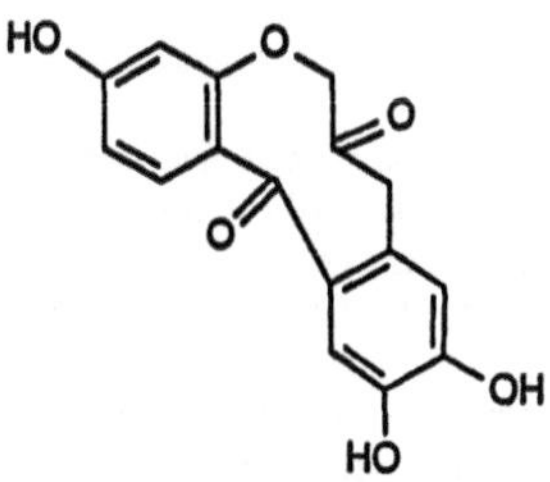

Caesalpin P (*31-11*)

In addition two dibenzoxocin derivatives, protosappanin A (*31-12*) [6] and B (*31-13*) [7], were isolated from sappan lignum and structurally elucidated by spectroscopic methods and by X-ray crystallography. They appear to be precursors of sappanin, 2,3′,4,4′-tetrahydroxybiphenyl (*31-14*), which has been known as an artificial compound derived from sappan lignum since 1872 and may be a metabolite of sappanchalcone. On alkali fusion, protosappanin A afforded sappanin along with small amounts of catechol and resorcinol [6].

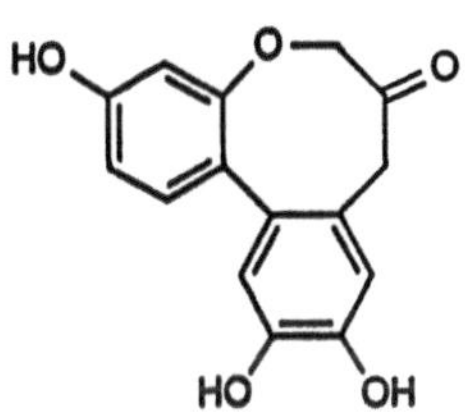

Protosappanin A (*31-12*)

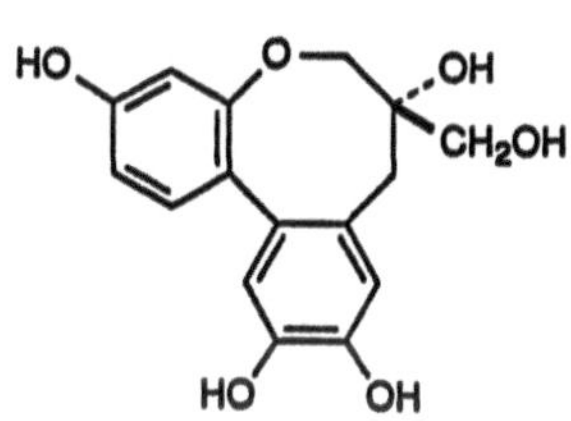

Protosappanin B (*31-13*)

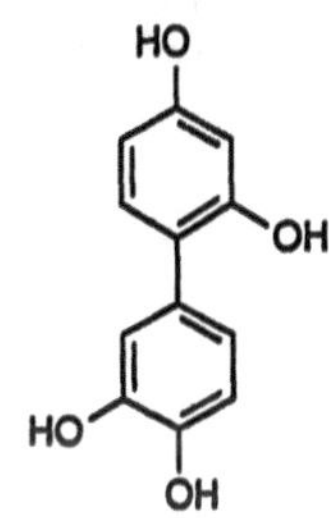

Sappanin (*31-14*)

31.3 Pharmacology

The methanol extract of the heartwood of *C. sappan* prolonged the duration of sleep in mice; protosapponin A had a weak sedative effect in mice [6]. The methanol extract also showed a significant antihypercholesteremic activity [2, 5].

Brazilin exhibited significant antiinflammatory activities in the carrageenin-induced rat paw edema test. Brazilin, at 10 mg/kg orally, was more effective than berberine chloride, at 100 mg/kg. It was approximately as active as berberine chloride in the granuloma pellet assay in rats [8].

In clinical studies some positive aspects of the administration of sappan wood, such as an improvement of visual disturbance, was observed in diabetic care suggesting that sappan wood might have an influence on the sorbitol pathway of the lens. Brazilin was examined for its effect on bovine lens aldose reductase activity. About 50% inhibition was observed, at a concentration of 10^{-4} *M* and about 95% inhibition was observed at a concentration of 10^{-3} *M*. Kinetic studies were also conducted with brazilin in which it was found to be a noncompetitive inhibitor [9].

References

1. Nagai M, Nagumo S, Eguchi I, Lee SM, Suzuki T (1984) Sappanchalcone from *Caesalpinia sappan* L., the proposed biosynthetic precursor of brazilin. Yakugaku Zasshi 104:935–938
2. Saitoh T, Sakashita S, Nakata H, Shimokawa T, Kinjo J, Yamahara J, Yamasaki M, Nohara T (1986) 3-Benzylchroman derivatives related to brazilin from Sappan lignum. Chem Pharm Bull 34:2506–2511
3. Heller W, Tamm C (1981) Homoisoflavanones and biogenetically related compounds. In: Herz W, Grisebach H, Kirby GW (eds) Progress in the Chemistry of Organic Natural Products, Vol 40. Springer, Berlin Heidelberg New York, pp 105–152
4. Fuke C, Yamahara J, Shimokawa T, Kinjo J, Tomimatsu T, Nahara T (1985) Two aromatic compounds related to brazilin from *Caesalpinia sappan*. Phytochemistry 24:2403–2405
5. Shimokawa T, Kinjo J, Yamahara J, Yamasaki M, Nohara T (1985) Two novel aromatic compounds from *Caesalpinia sappan*. Chem Pharm Bull 33:3545–3547
6. Nagai M, Nagumo S, Lee SM, Eguchi I, Kawai KI (1986) Protosappanin A, a novel biphenyl compound from sappan lignum. Chem Pharm Bull 34:1–6
7. Nagai M, Nagumo S (1986) Protosappanin B, a new dibenzoxocin derivative from sappan lignum. Heterocycles 24:601–605
8. Hikino H, Taguchi T, Fujimura H, Hiramatsu Y (1977) The validity of oriental medicine. III. Antiinflammatory principles of *Caesalpinia sappan* wood and of *Haematoxylon campechianum* wood. Planta Med 31:214–220
9. Moon CK, Yun YP, Lee JH, Wagner H, Shin YS (1985) Inhibition of lens-aldose reductase activity by brazilin and haematoxylin. Planta Med 51:66–67

Calvatia lilacina (Mont. et Berk.) Lloyd 32

32.1 Introduction

Mabo, Lasiosphaera seu Calvatia, is the dry fruiting body of *Calvatia gigantea* (Batsch ex Pers.) Lloyd, *C. lilacina* (Mont. et Berk.) Lloyd (Lycoperdaceae) or of *Lasiosphaera fenzlii* Reich., from the same family, collected in summer or fall. It is listed officially in the Chinese Pharmacopoeia and is used in treatment of laryngitis, cough and hoarseness and as a local hemostatic.

32.2 Chemical Constituents

From the culture broth of *C. lilacina* calvatic acid (*32-1*), an antibacterial and antifungal principle was isolated. Its structure was determined by spectroscopic methods as 4-(cyanoazoxy)-benzoic acid [1].

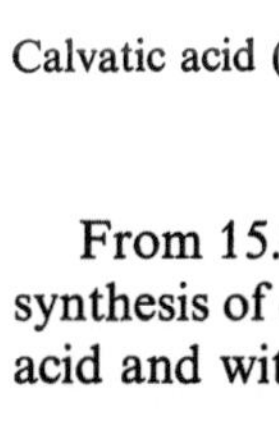

Calvatic acid (*32-1*)

From 15.9 liters of culture filtrate 30 mg calvatic acid was isolated [2]. The total synthesis of calvatic acid, carried out in four steps starting from *p*-hydrazinobenzoic acid and with a total yield of approximately 30%, is illustrated in Fig. 32-1 [2].

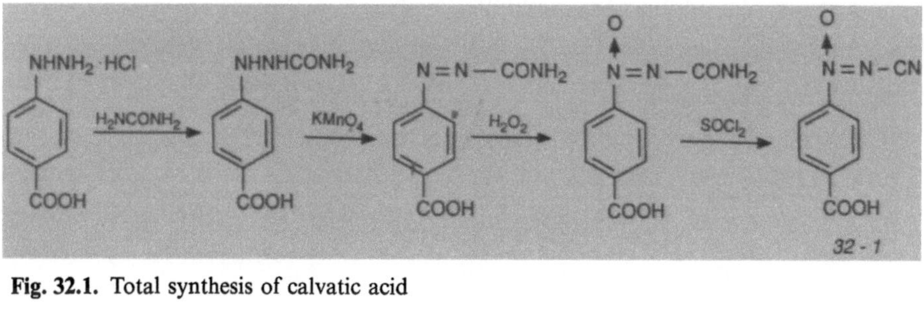

Fig. 32.1. Total synthesis of calvatic acid

32.3 Pharmacology

Calvatic acid inhibited growth of gram-positive and some gram-negative bacteria at a concentration of 3.1–6.2 μg/ml. It was inactive against most yeast and fungi at 100 μg/ml [2].

The LD_{50} of calvatic acid in mouse ranges between 125 and 250 mg/kg by i.p. administration [2]. Moreover, calvatic acid showed an cytotoxic activity in inhibiting cultured Yoshida sarcoma cell growth by 50% at 1.56 μg/ml. It also showed growth inhibition of mouse leukemia L 1210 cells. The i.p. dose of 400, 200, and 100 μg/animal of calvatic acid daily for 10 days to mice inoculated i.p. with 1×10^5 mouse leukemia L 1210 cells prolonged the survival time to 153%, 147%, and 138% of controls, respectively [2]. It was also found that *C. gigantea* contained antitumor active compounds effective against mouse sarcoma S 180 [3].

References

1. Gasco A, Serafino A, Mortarini V, Menziani E (1974) An antibacterial and antifungal compound from *Calvatia lilacina*. Tetrahedron Lett 3431–3432
2. Umezawa H, Takeuchi T, Iinuma H, Ito M, Ishizuka M, Kurakata Y, Umeda Y, Nakanishi Y, Nakamura T, Obayashi A, Tanabe O (1975) A new antibiotic, calvatic acid. J Antibiot (Tokyo) 28:87–90
3. Lucas EH, Byerrum RU (1959) Recovery of therapeutically effective extracts from fungi. German patent no. 1.048,675 (CA 55:8773c)

Camptotheca acuminata Decne

33

33.1 Introduction

Camptotheca acuminata Decne (Nyssaceae) is a tree native to and widely distrubuted in the southern part of China. The alkaloid component, camptothecin, with a new structural feature and an antileukemic activity, was isolated by Wall et al. in 1966 [1]. Since then, the chemical and biological properties of camptothecin and related alkaloids isolated from *C. acuminata* or synthesized have been investigated in great detail [2–5].

33.2 Chemical Constituents

As mentioned above, *C. acuminata* contains a number of alkaloids as main constituents. The alkaloid first isolated was camptothecin (*33-1*) [1]. The chemical structure of camptothecin was ascertained by its spectral and chemical properties and by X-ray crystallographic analysis of the 20-iodoacetate derivative [1, 6]. It has an indolizinoquinoline skeleton with a six-membered lactone ring.

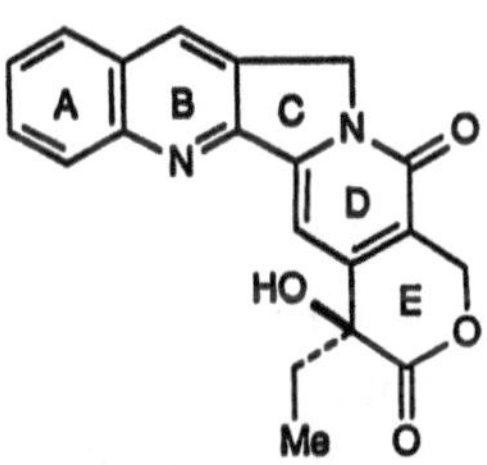

Camptothecin (*33-1*)

Camptothecin was first isolated from the stem wood and bark of *C. acuminata*. The yield of camptothecin from the bark was 0.005%–0.012%, depending on the extraction method [7]. From the wood and bark other camptothecin-related alkaloids were detected: 10-hydroxycamptothecin (*33-2*), 10-methoxycamptothecin (*33-3*) [8], 11-hydroxycamptothecin (*33-4*) [9], deoxycamptothecin (*33-5*), hexanoylcamptothecin (*33-6*), and hexanoyl-10-methoxycamptothecin (*33-7*) [7].

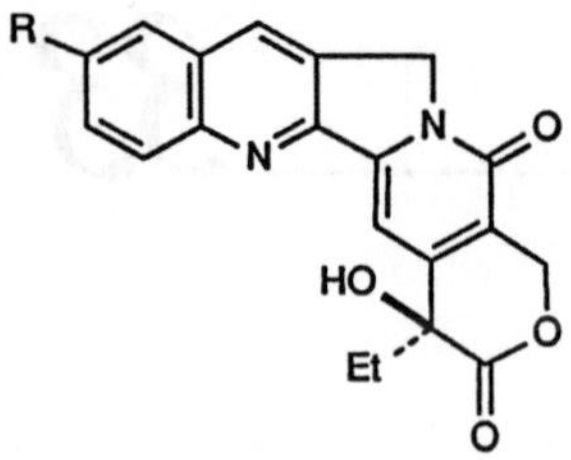

10-Hydroxycampothecin (*33-2*): R=OH
10-Methoxycamptothecin (*33-3*): R=OCH_3

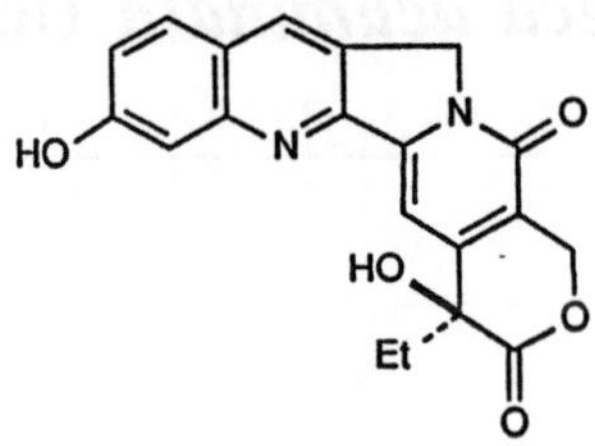

11-Hydroxycamptothecin (*33-4*)

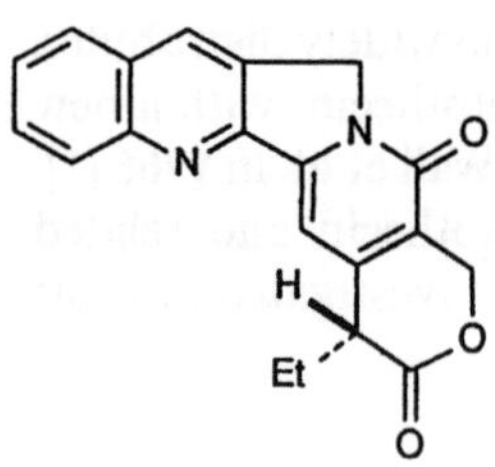

Deoxycamptothecin (*33-5*)

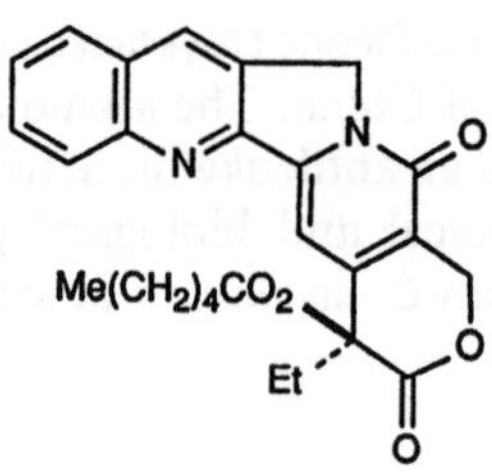

Hexanoylcamptothecin (*33-6*)

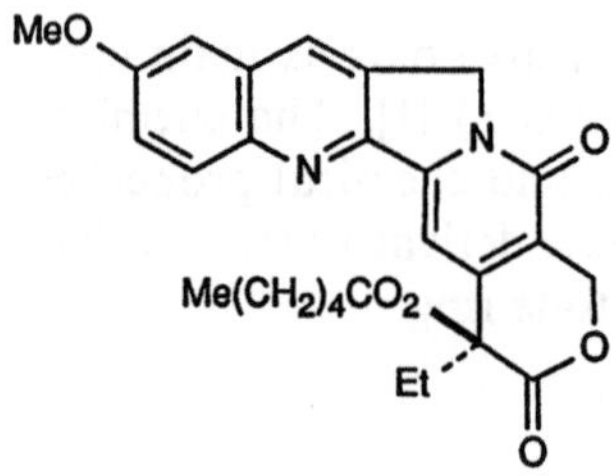

Hexanoyl-10-methoxycamptothecin (*33-7*)

Camptothecin was also isolated from the leaves [10] and roots [11] of *C. acuminata*. In the fruits of *C. acuminata*, camptothecin, 10-hydroxycamptothecin, deoxycamptothecin [12], 11-hydroxycamptothecin [12–15], and 10-methoxycamptothecin [13, 14] were isolated and identified.

Camptothecin has been subsequently found in some other plants such as *Nothapodytes foetida* (*Mappia foetida*, Icacinaceae) [16, 17], *Ophiorrhiza mungos* (Rubiaceae) [16, 18], and *Ervatamia heyneana* (Apocynaceae) [16, 19]. A new camptothecin analogue, 9-methoxycamptothecin (*33-8*) was also isolated. The amounts of camptothecin and 9-methoxycamptothecin in *N. foetida* were 0.3% and 0.045%, respectively [16]. These plants are generally found in the southern part of China.

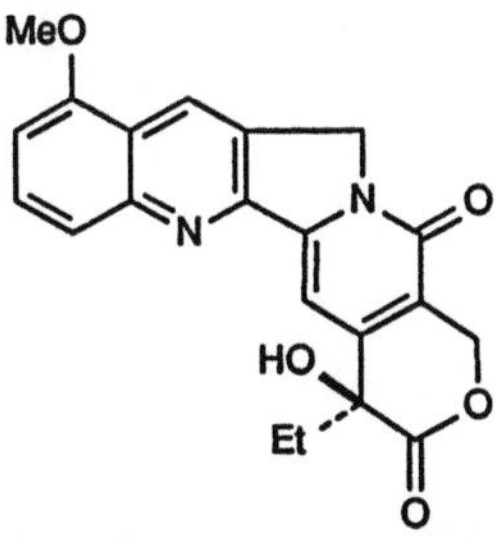

9-Methoxycampothecin (*33-8*)

Besides the camptothecin-related alkaloids, other chemical constituents were isolated from the fruits of *C. acuminata*: betulic acid [12], *β*-sitosterol [13], syringic acid (*33-9*) [20], abscisin II (*33-10*), syringaresinol [13], venoterpine (*33-11*), vincoside lactam (*33-12*) [12], and a number of ellagic acid derivatives: 2,3-*O*-dimethylellagic acid (*33-13*); 2,3,7-*O*-trimethylellagic acid (*33-14*); 2,3,7,8-*O*-tetramethyl-9-methoxyellagic acid (*33-15*): 2,3-*O,O*-methylene-ellagic acid (*33-16*): 2,3-*O,O*-methylene-7-*O*-methylellagic acid (*33-17*): 2,3-*O,O*-methylene-7,8-*O*-dimethylellagic acid (*33-18*); 2,3-*O,O*-methylene-7,8-*O*-dimethyl-9-hydroxyellagic acid (*33-19*); and 2,3-*O,O*-methylene-7,8-*O*-dimethyl-9-methoxyellagic acid (*33-20*) [14, 20].

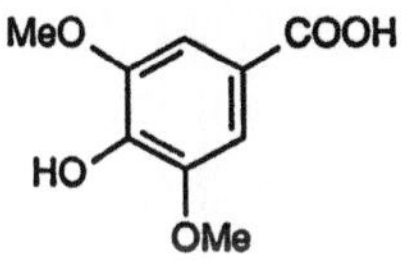

Syringic acid
(*33-9*)

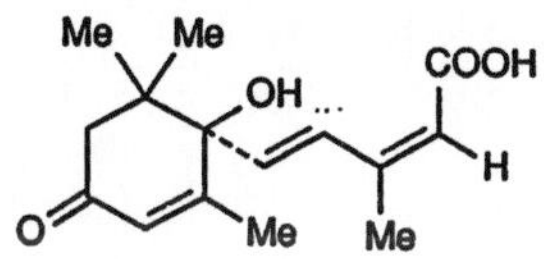

Abscisin II
(*33-10*)

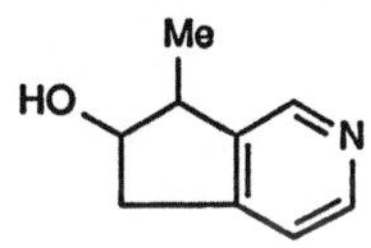

Venoterpine
(*33-11*)

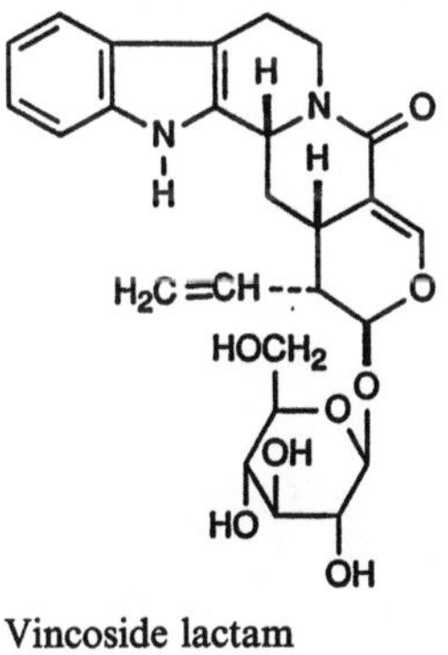

Vincoside lactam
(*33*-12)

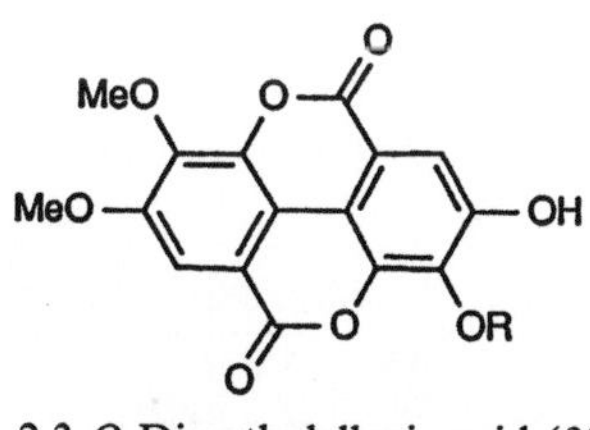

2,3-*O*-Dimethylellagic acid (*33-13*): R=H
2,3,7-*O*-Trimethylellagic acid (*33-14*): R=CH_3

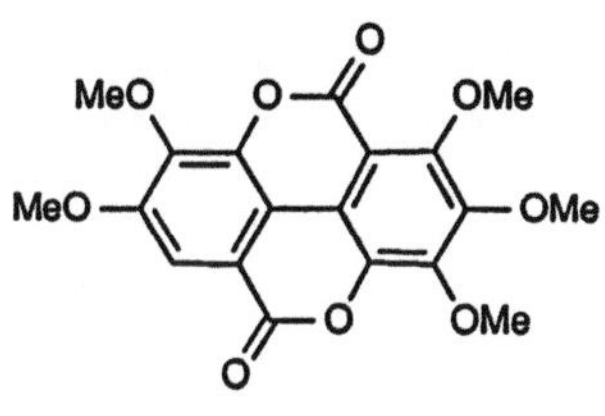

2,3,7,8-*O*-Tetramethyl-9-methoxyellagic acid (*33-15*)

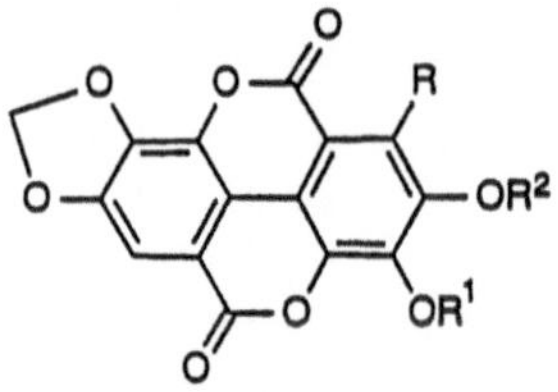

2,3-*O*,*O*-Methylene-ellagic acid (*33-16*): R, R^1, R^2=H
2,3-*O*,*O*-Methylene-7-*O*-methylellagic acid (*33-17*): R, R^2=H, R^1=CH_3
2,3-*O*,*O*-Methylene-7,8-*O*-dimethylellagic acid (*33-18*): R=H, R^1, R^2=CH_3
2,3-*O*,*O*-Methylene-7,8-*O*-dimethyl-9-hydroxyellagic acid (*33-19*): R=OH, R^1, R^2=CH_3
2,3-*O*,*O*-Methylene-7,8-*O*-dimethyl-9-methoxyellagic acid (*33-20*): R=OCH_3, R^1, R^2=CH_3

Since the yield of camptothecin from plant materials was very low, efforts were made to find a rational synthesis of camptothecin and its derivatives from different starting materials. The first total synthesis of racemic camptothecin was reported by Stork and Schultz [21]. They used *O*-aminobenzaldehyde and a pyrrolidone derivative as starting materials. The more important steps of the synthesis are summarized in Fig. 33-1. The overall yield of racemic camptothecin was 1%–2% [21].

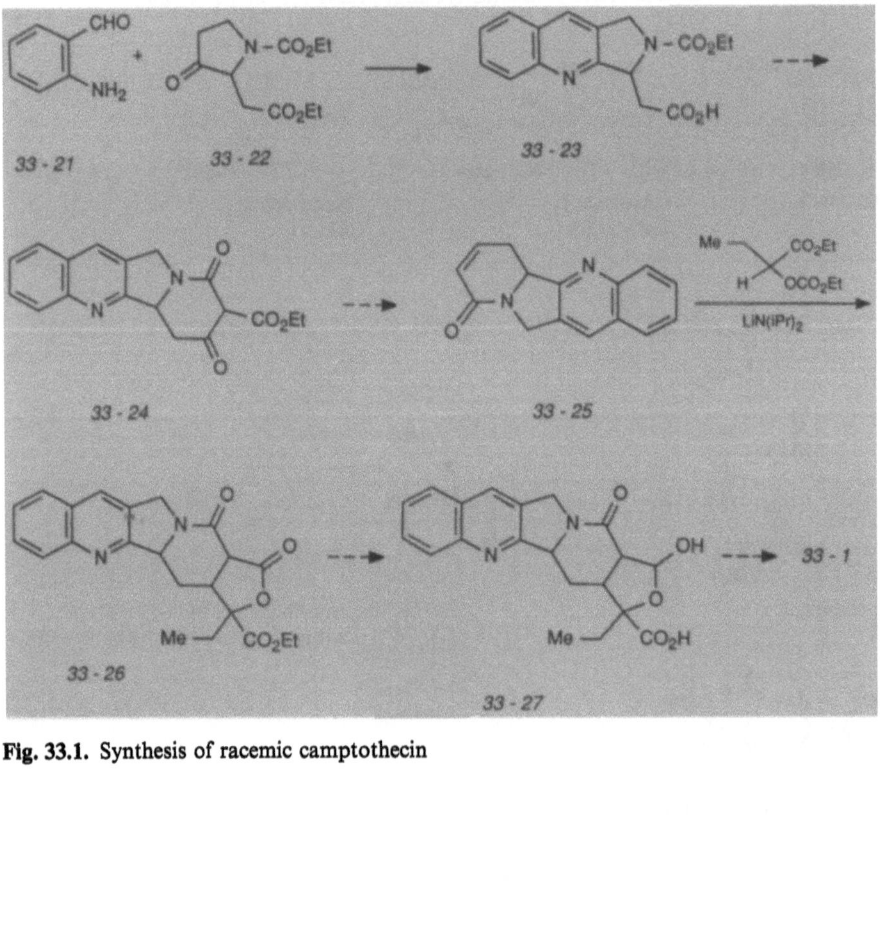

Fig. 33.1. Synthesis of racemic camptothecin

The tricyclic pyrrolo [3,4-*b*] quinoline acid (*33-23*) was obtained by base catalyzed condensation of *O*-aminobenzaldehyde (*33-21*) with pyrrolidone (*33-22*). Using a Dieckmann cyclization, the tricyclic compound could be converted into the tetracyclic *β*-ketoester (*33-24*) from which the unsaturated lactam (*33-25*) was obtained by hydrolysis, decarboxylation, reduction, and elimination of water from the resulting *β*-hydroxylactam. The key compound, a pentacyclic lactone (*33-26*) was obtained by reaction of the unsaturated lactam with the lithium salt of a protected α-hydroxybutyric acid ester at −70 °C. After hydrolysis of the pentacyclic lactone the free acid obtained could be converted into the hemiacetal (*33-27*) from which *dl*-camptothecin was finally obtained after dehydrogenation, acetate hydrolysis, hemiacetal reduction, and acidification [21].

A more effective synthesis of camptothecin was reported by Tang et al. [22] with an overall yield of 15%. This synthetic approach was based fundamentally on the bicyclic ketoacid 1-oxo-octahydroindolizine-6-carboxylic acid (*33-29*), obtainable in high yield from an inexpensive, commercially available starting diacid, isocinchomeronic acid (*33-28*). The important steps for this synthetic approach are shown in Fig. 33-2.

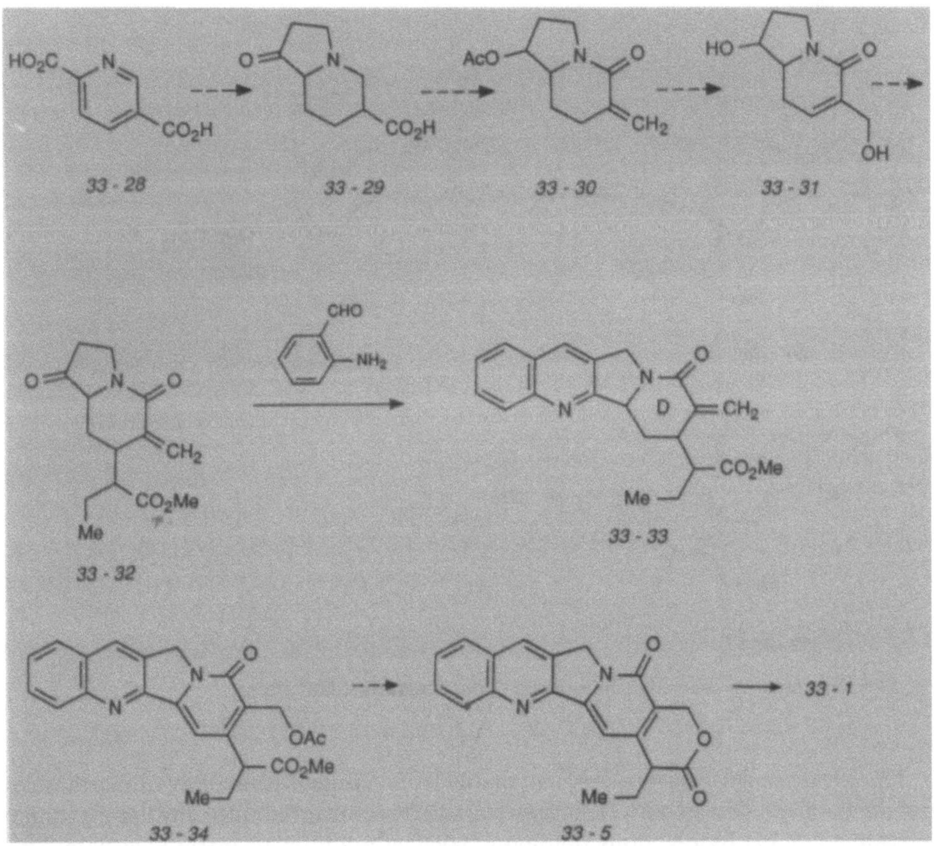

Fig. 33.2. Synthesis of camptothecin starting from isocinchomeronic acid

The bicyclic ketoacid (*33-29*) prepared from isocinchomeronic acid (*33-28*) was rearranged to the α-methylene lactam (*33-30*), which was then oxidized with SeO_2 and hydrolyzed to the lactamdiol (*33-31*). The keto-methylenelactam (*33-32*), obtained from the lactamdiol by Claisen orthoester rearrangement and subsequent oxidation, was condensed with *o*-aminobenzaldehyde to give the expected tetracyclic compound (*33-33*). Oxidation with SeO_2 in acetic acid simultaneously aromatized the D ring and introduced a C-17 acetoxy substituent (*33-34*). Finally, deoxycamptothecin, obtained by treatment of the acetoxy compound with acid, was oxidized to camptothecin with oxygen, $CuCl_2$, and aqueous dimethylamine in dimethylformamide [22].

Chinese scientists obtained racemic camptothecin with an impressive overall yield of 18% by using a pyridone derivative as the starting material [23, 24]. This synthesis is summarized in Fig. 33-3.

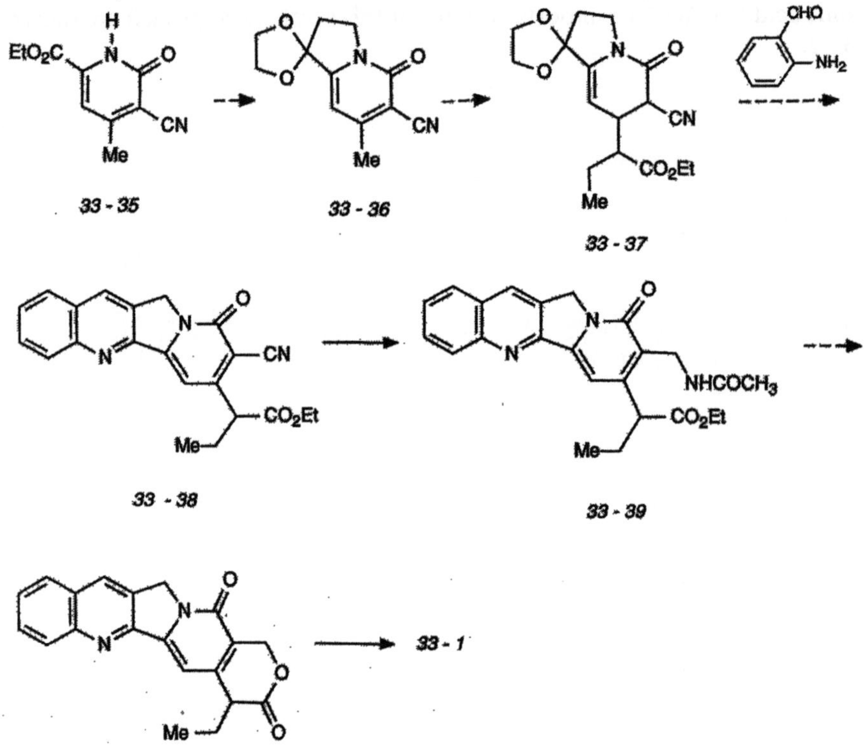

Fig. 33.3. Synthesis of racemic camptothecin from a pyridone derivative

The pyridone derivative (*33-35*) available from condensation of cyanoacetamide and the *O*-ethyl ether of ethyl acetopyruvate [25] is converted into another pyridone derivative (*33-36*) by reactions involving Michael addition to methyl acrylate, hydrolysis-decarboxylation, and ketalization. The C-4 methyl of the second derivative was sufficiently acidic to permit its carbethoxylation by reaction with NaH in diethyl carbonate. α-Ethylation of the resulting intermediate ester gives a new ester compound (*33-37*), which undergoes simultaneous deketalization and condensation with *o*-aminobenzaldehyde. The tetracyclic cyano compound (*33-38*) is then converted into an acetamide compound (*33-39*) which then is cyclized into deoxycamptothecin and camptothecin.

Wani et al. modified this synthetic approach and were able to prepare racemic camptothecin in an overall yield of 28%, and racemic 10-hydroxycamptothecin with a yield of 2% [26]. The key step used by this group was the construction of the complete CDE ring system from the cyanopyridone with the ester function (*33-37*) described above, from which camptothecin could be obtained by Friedländer condensation and subsequent standard transformation (Fig. 33-4) [26].

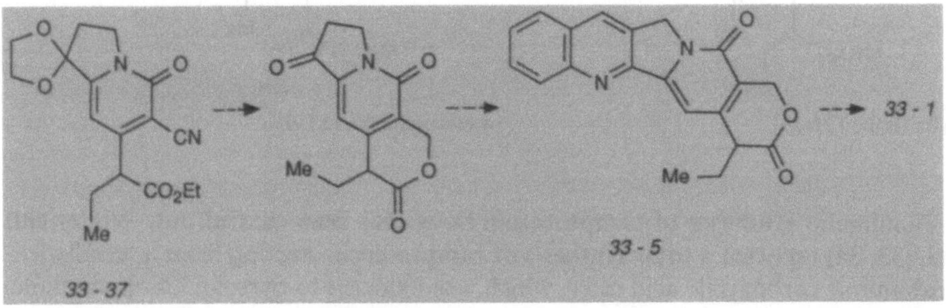

Fig. 33.4. Synthesis of racemic camptothecin via construction of the complete CDE ring system

This development allows synthesis of camptothecin and some of its derivatives on a large scale.

Thus, far the only synthesis of optically active 20(*S*)-camptothecin was reported by Corey et al. [27]. The strategy of this novel synthesis involves reaction of a chiral pseudoacid chloride (*33-40*), corresponding to ring E of camptothecin, with a tricyclic diamine (*33-41*), corresponding to the ABC ring system, to form the D ring by cyclization of an intermediate γ-aldehydo-*t*-amide (*33-42*) (Fig. 33-5) [27].

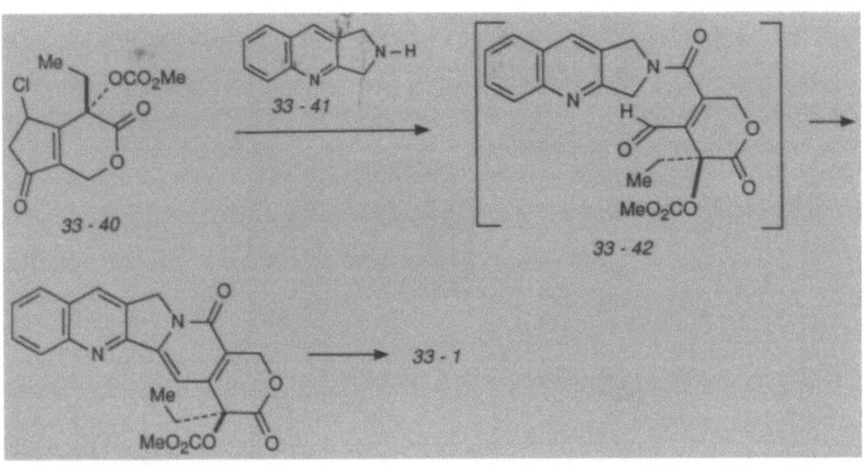

Fig. 33.5. Synthesis of 20(*S*)-camptothecin

For the biosynthesis of camptothecin, Wenkert proposed that camptothecin might result as a monoterpene indole alkaloid from isositsirikine (*33-43*) [28]. Winterfeldt proposed that geissoschizine (*33-44*), the didehydroderivative of isositsirikine, was a plausible biogenetic precursor of camptothecin [29, 30], whereas Hutchinson et al. [5, 31] and Cordell [32] independently postulated strictosamide as penult precursor in the biosynthesis of camptothecin. Strictosamide is the 13-epimeric mixture of vincoside lactam (*33-12*).

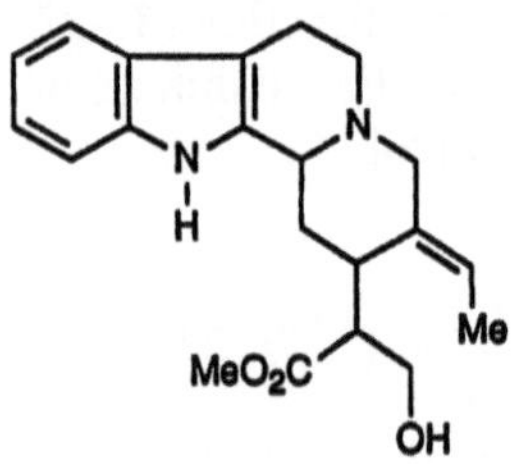

Isositsirikine (*33-43*)

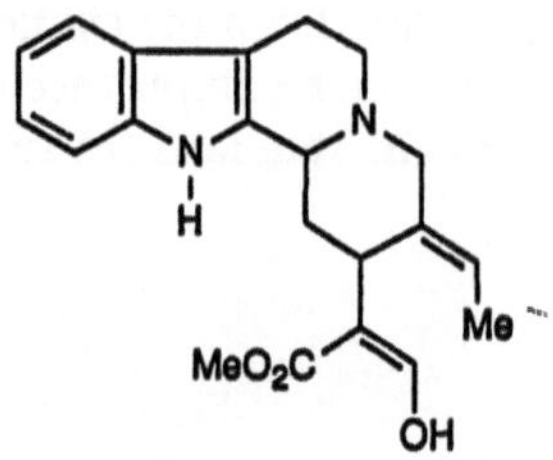

Geissoschizine (*33-44*)

Biomimetic syntheses of camptothecin have also been carried out. Winterfeldt et al. [33, 34] reported a total synthesis of camptothecin starting from a tetrahydropyridoindole carboxyclic acid ester, which was oxidized to pyrrolo[3,4-*b*]quinolone. This biomimetic synthesis is formulated in Fig. 33-6.

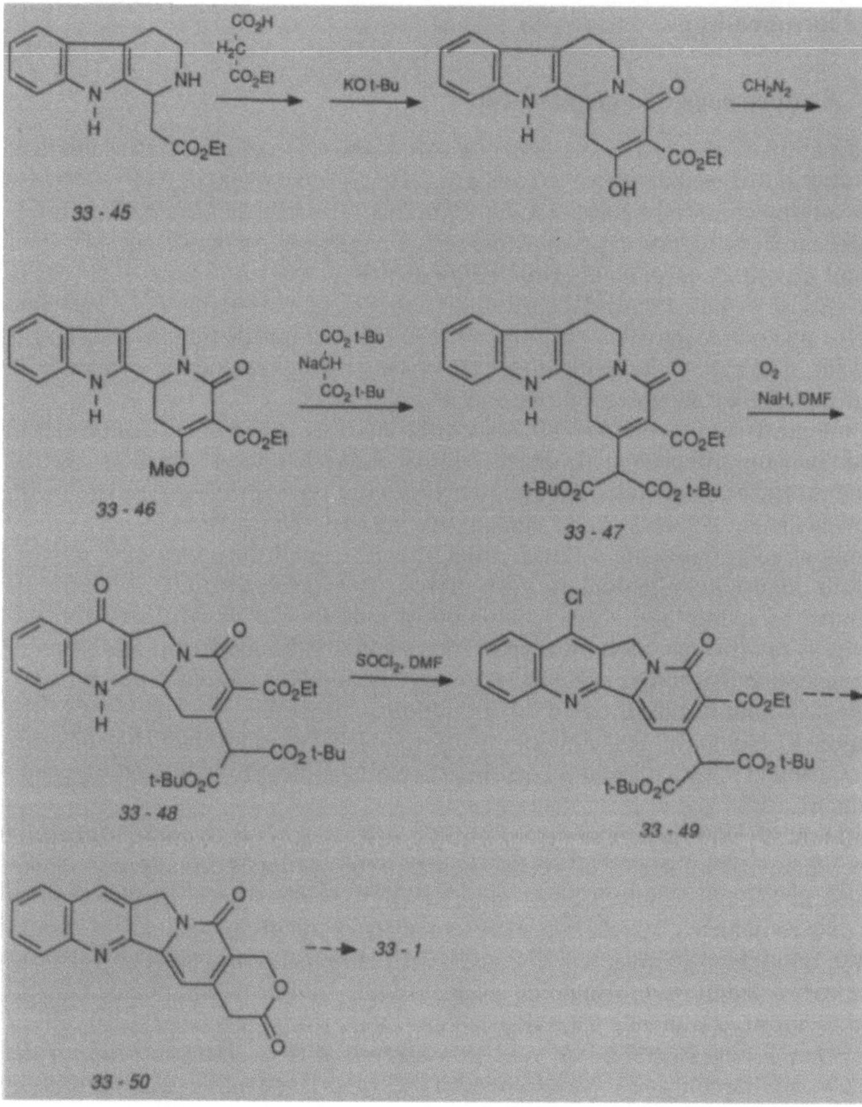

Fig. 33.6. Biomimetic synthesis of camptothecin

A tetracyclic lactam (*33-46*), obtainable from a tetrahydropyridoindole (*33-45*) was the keystone for formation of the pyrrolo[3,4-*b*]quinoline ring system. 1,4-Addition-elimination of sodium di-*t*-butylmalonate with the lactam followed by autoxidation of the resulting tetrahydro-β-carboline (*33-47*) and treatment of the quinolone (*33-48*) with thionyl chloride gave a chlorosubstituted compound (*33-49*), from which a pentacyclic lactone (*33-50*) was obtained via further steps. Regiospecific ethylation at C-20 of the lactone proceeded poorly to deoxycamptothecin, which was oxidized quantitatively to camptothecin [33, 34].

33.3 Pharmacology

33.3.1 Pharmacology of Camptothecin

The extract of *C. acuminata* was active in carcinoma 755 cells [1]. After isolation, camptothecin was tested against leukemia L1210 in mice and gave a 100% increase in survival time at a daily dose of 0.25–1.0 mg/kg. Treatment of Walker 256 intramuscular carcinosarcoma in rats at a dose of 1.25 mg/kg gave a significant inhibition of tumor growth. *C. acuminata* exhibited moderate cytotoxicity against KB cells in culture (ED_{50} = 0.07 μg/ml) [1]. Antitumor activity against murine P388 leukemia and B16 melanoma in vivo was also noticed [5]. Camptothecin, in a soft agar clonogenic assay, showed significant cytotoxic activity against human ovarian cancer and some other adenocarcinoma cells [80].

Pathological studies on the ultrastructural changes induced by camptothecin showed that camptothecin, at a concentration of 5 μg/ml, induces nucleolar segregation and produces microspherules and perichromatin type granules associated with fibrillar nucleolar elements in the human cell line ME-180 in vitro [35].

Camptothecin represents a novel structure with significant antitumor activity. Therefore, many investigations in vitro and in vivo have attempted to clarify its mechanism of action [36]. Potent inhibition of nucleic acid biosynthesis in mammalian cells in vitro has been observed. This is apparently not due to inhibition of either nucleotide biosynthesis or enzymatic activity of DNA and RNA polymerases. The effect of camptothecin on RNA formation is reversible after removal of the substance. It affects the biosynthesis of ribosomal RNA more than that of other types of cellular RNA. Protein biosynthesis is not significantly inhibited by camptothecin [37, 38].

Addition of camptothecin at a concentration of 2×10^{-5} *M* to cultures of HeLa cells or adenovirus-infected HeLa cells resulted in the partial degradation of cellular and viral DNA [39]. Camptothecin induced DNA single strand breaks in intact HeLa cells as detected by alkaline sucrose density gradient analysis. DNA breaks were not found after treatment of HeLa cells with camptothecin, when the DNA was sedimented in denaturing formamide gradients [40]. Camptothecin, at a concentration of 10 μg/ml, completely degraded cellular DNA from leukemic L1210 cells at 37 °C within 1 min. Degradation was also observed at 0 °C. Reaggregation of the DNA occurred rapidly at 37 °C and slowly at 0 °C [41]. The binding of camptothecin to isolated DNA was very weak [37, 42].

Lown and Chen [43] have observed that although camptothecin had no effect on PM2-ccc-DNA in vitro in the dark, it produced single strand breaks in the same DNA when irradiated with 360 nm light. Breakage was inversely dependent on oxygen concentration, and at least two mechanisms were indicated. Photosensitization of camptothecin may generate radicals to attack DNA or, in the presence of oxygen, may generate hydroperoxy radicals. The photolytic reaction of camptothecin itself was found to proceed via formation of racemic and photolabile camptothecin hemiacetal (*33-51*), isolated during the reaction. In the anaerobic pathway photodecarbonylation of the alkaloid might generate a diradical (*33-52*) which can be converted either into camptothecin hemiacetal or abstract H atoms from DNA, causing strand scission. In the presence of oxygen, the alternative aerobic pathway might become predominant, in which hydroperoxy radicals (*33-53*) are generated giving rise to hydrogen peroxide and then to reactive OH radicals, which attack DNA. The pathways postulated are summarized in Fig. 33-7 [43].

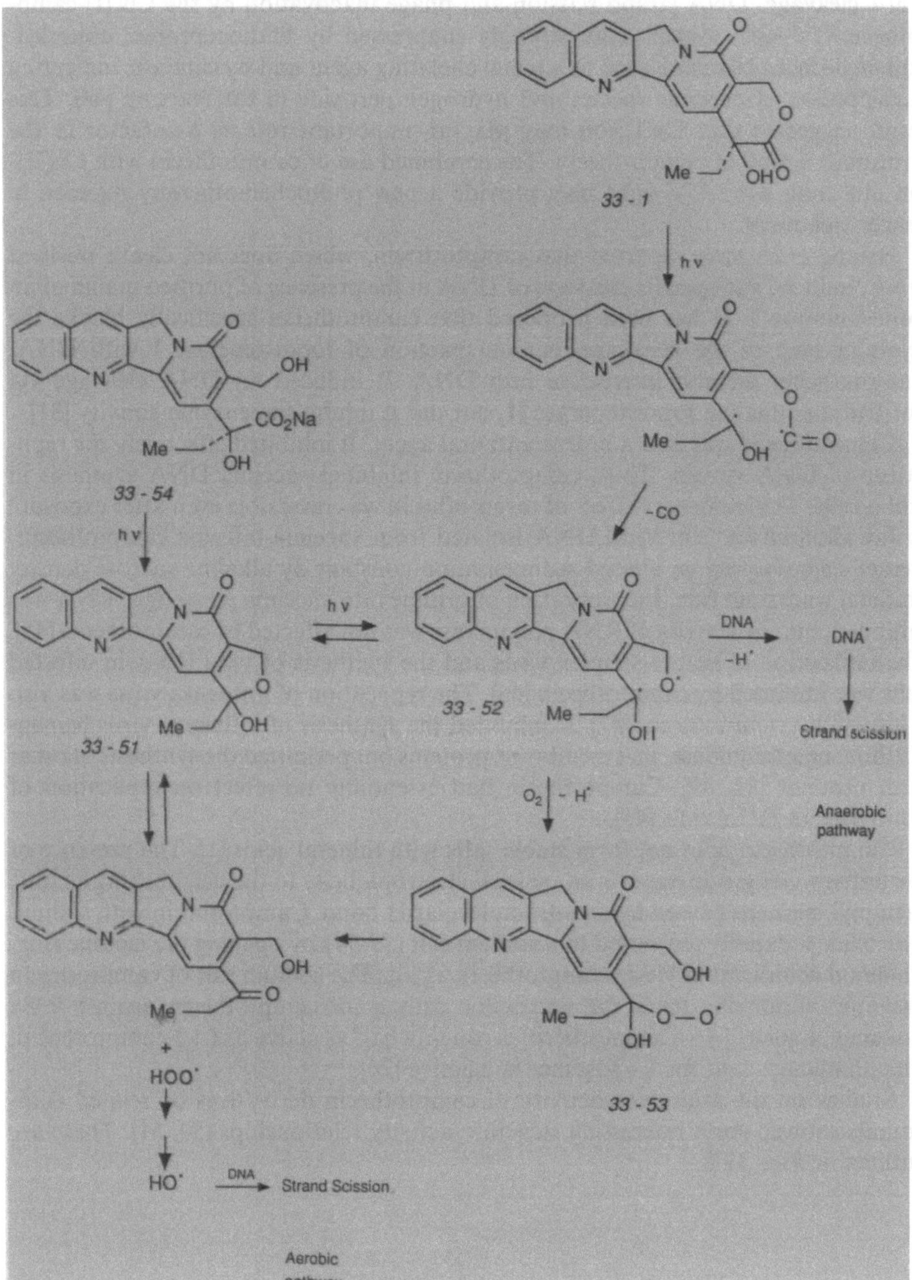

Fig. 33.7. Aerobic and anaerobic photosensitization of camptothecin

It was recently reported that camptothecin, in combination with Cu(II)ion and 365 nm light, produced single and double strand breaks in DNA and induced a marked inactivation of bacteriophage. Nucleotide sequence analysis showed random DNA cleavage. DNA strand scission and phage inactivation by the Cu(II)-camptothecin-UV light system were strongly suppressed by bathocuproine, dimethyldiphenyl-phenanthroline used as a metal chelating agent and by catalase, indicating participation of cuprous species and hydrogen peroxide in the reaction [44]. This result suggested that Cu(II)ion may play an important role as a cofactor in the antitumor action of camptothecin. The combined use of camptothecin with Cu(II)-ion and long wave UV light may provide a new photochemotherapy regimen in cancer treatment.

Hsiang et al. have reported that camptothecin, which does not cleave purified DNA, induced site-specific cleavage of DNA in the presence of purified mammalian topoisomerase I. It has been proposed that camptothecin specifically blocks the rejoining step of the breakage-reunion reaction of topoisomerase I with DNA. Camptothecin did not intercalate into DNA. It induced no DNA cleavage via purified mammalian topoisomerase II, nor did it inhibit the enzyme activity [81].

Camptothecin was also a potent antiviral agent. It inhibited effectively the replication of DNA viruses. Thus, camptothecin inhibited vaccinia DNA synthesis in HeLa cells. The inhibitory effect of camptothecin was reversible even after exposure to the alkaloid for 2 h. Viral DNA isolated from vaccinia-infected camptothecin-treated cells showed an altered sedimentation constant by alkaline sucrose density gradient centrifugation. Incorporation of uridine into vaccinia messenger RNA was inhibited, but the activity of RNA polymerase was not affected by camptothecin [45]. The replication of herpes simplex virus and the synthesis of viral DNA in infected cells was inhibited by camptothecin [46]. The replication of influenza virus was also inhibited by camptothecin [47]. It inhibited the synthesis of influenza virus hemagglutinin, neuraminidase, and membrane proteins but permitted the synthesis of other viral proteins [48, 49]. Camptothecin had essentially no effect on replication of poliovirus in HeLa cells [45].

Camptothecin does not form stable salts with mineral acids [1]. The presence of the tertiary OH group imparts an unusual electrophilicity to the neighboring lactone carbonyl, perhaps caused by an intramolecular H bond. Camptothecin with sodium hydroxide is rapidly converted to a sodium salt (*33-54*) by opening the lactone ring, which on acidification yields camptothecin again. The sodium salt of camptothecin has only about one tenth the antitumor activity of camptothecin against P388 leukemia in mice. (±)-Camptothecin is roughly half as active as (+)-camptothecin, which indicates that the (−)-isomer is inactive [26].

Studies on the antitumor activity of camptothecin derivatives or related compounds showed some interesting structure-activity relationships [50, 51]. These are outlined in Fig. 33-8.

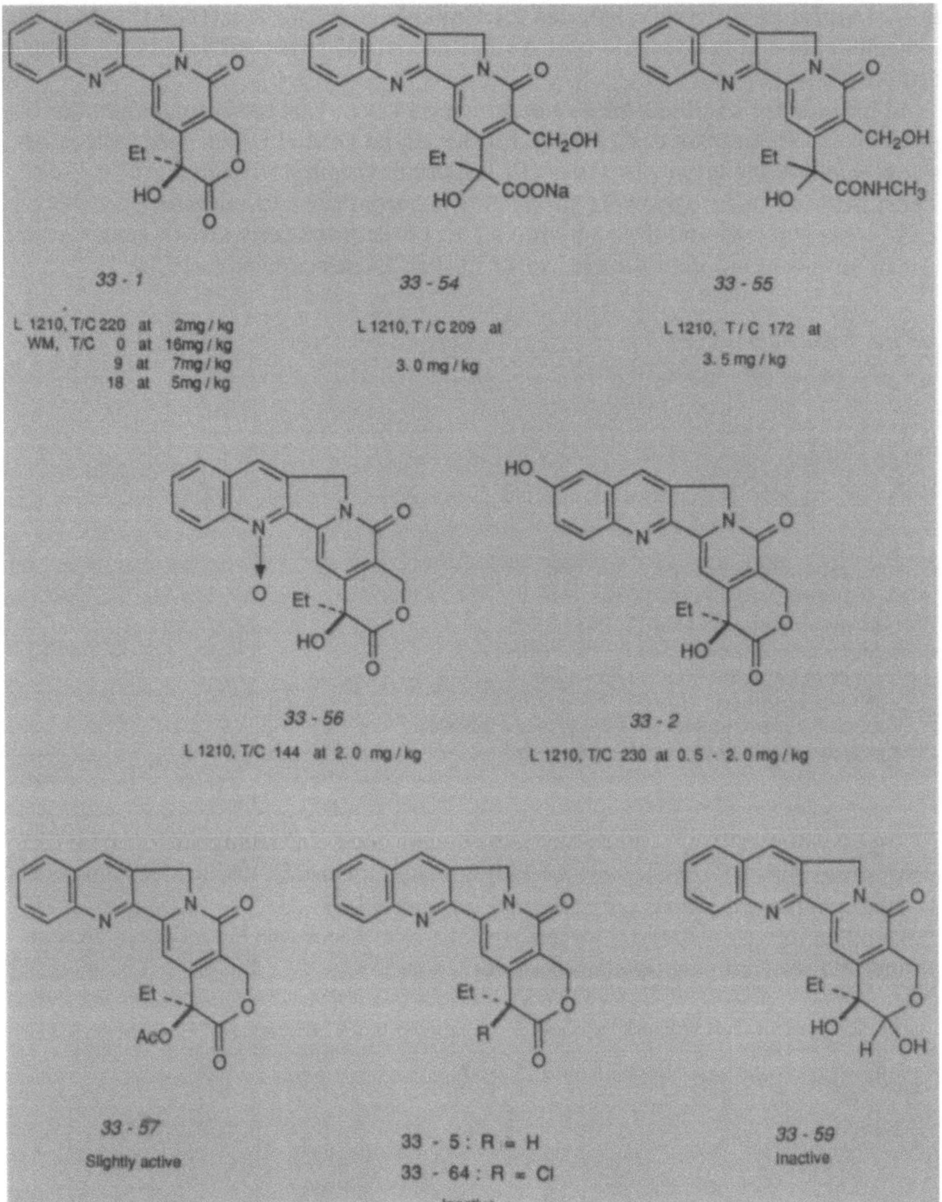

Fig. 33.8. Structure-activity relationship in the camptothecin series

The most important structural features may be summarized as follows: 10-hydroxycamptothecin and 11-hydroxycamptothecin [14] show the highest activity of all the various natural or synthetic analogues of camptothecin indicating a correlation between activity and the hydroxylation on ring A of camptothecin. Camptothecin-sodium is only marginally active after i.p. administration, probably due to limited recyclization. It is inactive when administered i.v. at pH 7.2. Marked reduction of antitumor activity was noted after modification of the lactone E ring. This indicates that the lactone E ring may be the critical structural feature of camptothecin.

In an attempt to obtain more potent congeners of camptothecin, a number of derivatives were prepared and biologically tested. Pan et al. [52] reported the synthesis and biological activity of a series of 12-substituted camptothecins (*33-60–33-67*). The 12-carboxy derivative was superior to camptothecin in inhibiting growth of L615 leukemia cells and the 12-hydroxy and 12-methoxy derivatives were superior to camptothecin in inhibiting growth of Ehrlich ascites carcinoma [52].

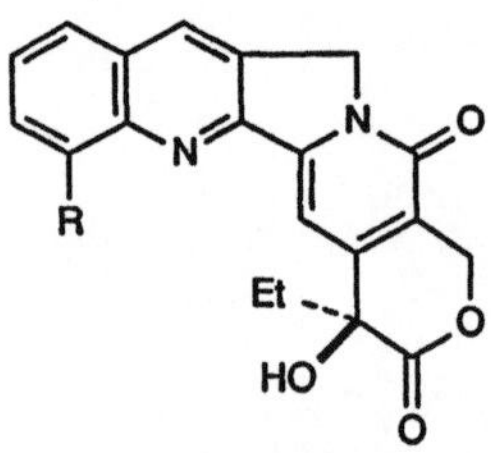

12-Nitrocamptothecin (*33-60*): R= $-NO_2$
12-Aminocamptothecin (*33-61*): R= $-NH_2$
12-Hydroxycamptothecin (*33-62*): R= $-OH$
12-Methoxycamptothecin (*33-63*): R= $-OCH_3$
12-Cyanocamptothecin (*33-64*): R= $-CN$
12-Carboxycamptothecin (*33-65*): R= $-COOH$
12-Carbomethoxycamptothecin (*33-66*): R= $-COOCH_3$
12-Mercaptocamptothecin (*33-67*): R= $-SH$

Sugusawa et al. [53] synthesized several camptothecin analogues in which the ethyl group at ring E of the parent compound was replaced by other side chains such as allyl, propargyl, benzyl, or phenacyl groups (*33-68–33-71*). Testing in L1210 leukemia in mice revealed that desethyl-allylcampothecin was the most active compound and was more active than camptothecin [53].

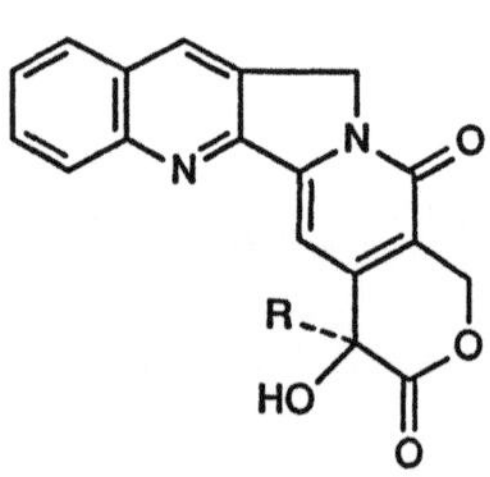

20-Desethyl-allylcamptothecin (*33-68*): R= $-CH_2-CH=CH_2$
20-Desethyl-propagylcamptothecin (*33-69*): R= $-CH_2-C\equiv CH$
20-Desethyl-benzylcamptothecin (*33-70*): R= $-CH_2-C_6H_5$
20-Desethyl-phenacylcamptothecin (*33-71*): R= $-CH_2COC_6H_5$

7-Substituted camptothecin derivatives (*33-72–33-75*) have been synthesized and tested by Winterfeldt et al. [33, 54].

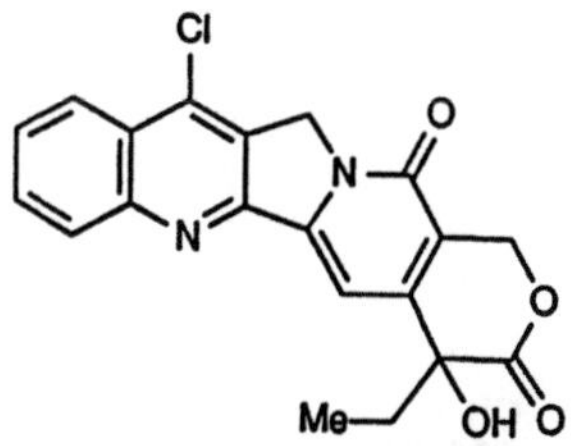

7-Chlorocamptothecin (*33-72*)

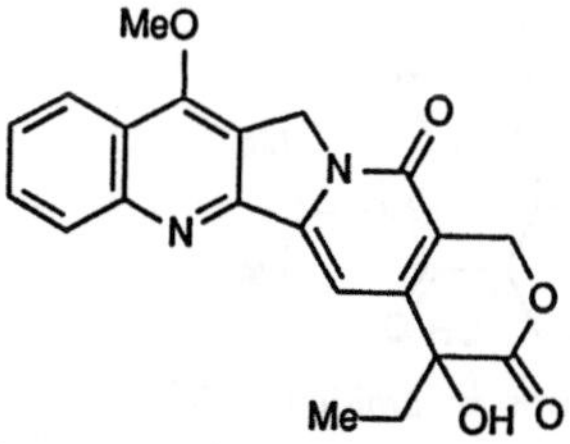

7-Methoxycamptothecin (*33-73*)

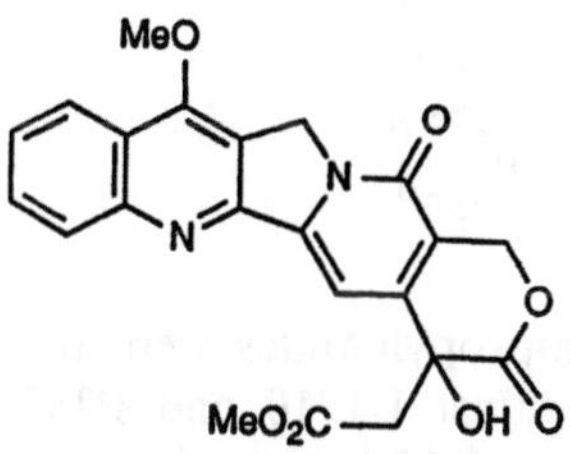

7-Methoxy-20-desethylcamptothecin-20-acetic acid methylester (*33-74*)

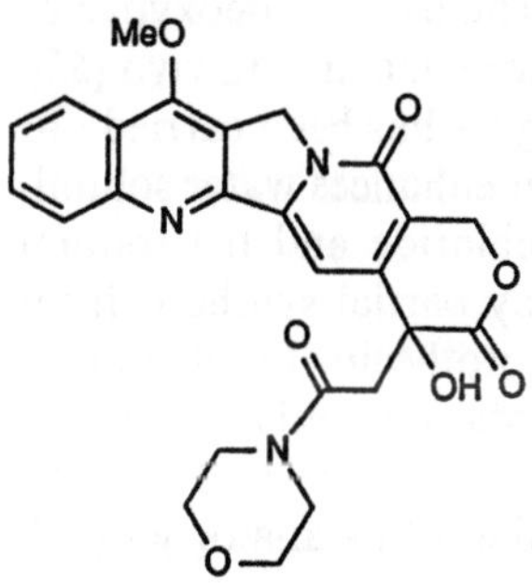

7-Methoxy-20-desethylcamptothecin-20-acetic acid morpholide (*33-75*)

In vivo, 7-chlorocamptothecin (*33-72*) showed good antitumor activity; the ester (*33-74*) and morpholide (*33-75*) moderate activity, and 7-methoxy-camptothecin (*33-73*) was inactive [5].

Adomovics and Hutchinson [55] described the preparation of compounds derived from isopropylamide of camptothecin (*33-76*). The rationale for preparation of these derivatives was based on the observation that the 21-methylamide of camptothecin had about 60% of the activity of camptothecin in the L 1210 assay [56]. The 17-O-esters, prepared by acylation of the isopropylamide of camptothecin, are lipophilic prodrug analogue that are potentially reconverted to camptothecin in vivo.

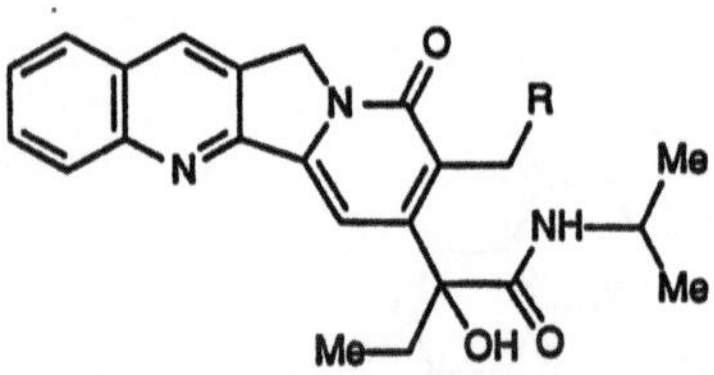

Camptothecin-21-isopropylamide (*33-76*): R = -OH
17-Acetylcamptothecin-21-isopropylamide (*33-77*): R = -$OCOCH_3$
17-Hexanoylcamptothecin-21-isopropylamide (*33-78*): R = -$OCOC_5H_9$.

17-Deoxy-pyrrolidinylcamptothecin-21-isopropylamide (*33-79*): R = 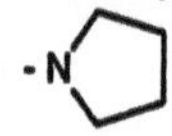

17-Deoxy-imidazolylcamptothecin-21-isopropylamide (*33-80*): R = 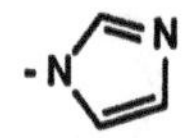

17-Acetyl- (*33-77*) and 17-acryloyl-camptothecin-21-isopropylamides were the most active compounds with 85% antitumor activity against L1210 and P388 leukemia in mice compared to camptothecin. 17-Acetyl- and 17-hexanoyl-camptothecin-21-isopropylamide (*33-78*) showed moderate activity against Lewis lung carcinoma cells but were inactive against melanoma B16. The basic 17 deoxypyrrolidinyl (*33-79*) and deoxy-imidazolyl (*33-80*) compounds were inactive in vivo [55].

The most comprehensive study on camptothecin analogues has been carried out by Wani et al. [26]. Analogues with a function in ring A that enhances water solubility, e.g., 10-(2-diethylamino-ethoxy)-camptothecin hydrochloride and the sodium salt of 10-carboxymethoxy-camptothecin, were prepared by partial synthesis from naturally occurring 10-hydroxycamptothecin. 12-Azacamptothecin, benz(*j*)camptothecin, and 18-methoxycamptothecin were prepared by total synthesis. Table 33-1 summarizes the antileukemic in vivo activity against P388 leukemia and in vitro cytotoxicity against 9KB cells of camptothecin and a number of its analogues [26].

Table 33-1. In vivo antileukemic activity against P388 leukemia in mice and in vitro cytotoxicity against 9KB cells of camptothecin and its analogues [26]

Compound	Dose range (mg/kg)	Optimal % T/C	Optimal dose (mg/kg)	Lowest toxic dose (mg/kg)	Therapeutic index	9KB ED_{50} (μg/ml)
Camptothecin	8.0–0.5	197	4.0	8.0	8	2×10^{-2}
Camptothecin · Na	80.0–2.5	212	40.0	80.0	4	2×10^{-1}
10-Hydroxycamptothecin	8.0–0.5	314	4.0	8.0	8	2×10^{-2}
10-Methoxycamptothecin	4.0–0.5	145	0.5	1.0	1	2×10^{-2}
rac. Camptothecin	16.0–1.0	222	8.0	16.0	8	1×10^{-1}
10-Carboxymethoxycampt · Na	23.0–4.0	Inactive	–	–	–	$>1\times10^{0}$
10-Diethylaminoethoxycampt · HCl	32.0–2.0	234	32.0	–	16	$>1\times10^{0}$
12-Azacamptothecin	32.0–2.0	175	32.0	–	2	$>1\times10^{0}$
18-Methoxycamptothecin	32.0–1.0	198	16.0	–	4	2×10^{-1}
Benz(*j*)camptothecin	8.0–0.5	160	4.0	8.0	2	2×10^{-1}

Toxicity of camptothecin in dogs and monkeys was characterized by emesis, hemorrhagic diarrhea, dehydration, coma, and death. Histopathological examination showed metaplasia of the intestinal epithelium and occasional necrotic cholecystitis in dogs. Intestinal toxicity may have resulted from biliary excretion of camptothecin after i.v. administration. In mice and rats, oral administration exhibited more severe intestinal toxicity and more severe morbidity than i.v. injection [57]. The monoammonium salt of glycyrrhetic acid, the sapogenin of glycyrrhizin isolated from *Glycyrrhiza* species, decreased the toxicity and increased the antitumor activity of camptothecin [58].

Distribution of camptothecin and of camptothecin sodium were compared in mice bearing ascitic hepatomas. Distribution of radioactivity in liver, kidney, and stomach after a single i.v. injection was higher after administration of [^{3}H] camptothecin in suspension than when given as the sodium salt. After 24 h, the radioactivity excreted in urine and feces was 21.3% and 22.6% in the suspension group and 22.8% and 17.2% in the sodium salt group, respectively. In normal mice, within 3 h of a single i.v. injection, radioactivity in the blood was higher after application of [^{3}H] camptothecin in suspension than in solution as a sodium salt [59].

Camptothecin-sodium has been subjected to preliminary clinical study [60]. Although initially camptothecin-sodium was active clinically against solid tumors of the gastrointestinal tract, subsequent clinical evaluation of this drug led to discouraging results [61, 62]: Camptothecin was extremely toxic and is therefore no longer of prime interest in clinical testing [4].

33.3.2 Pharmacology of 10-Hydroxycamptothecin

10-Hydroxycamptothecin and camptothecin are closely related chemical structures; however, there are some obvious differences in their toxicities and pharmacologic actions, which are summarized in Table 33-2 [63].

Table 33-2. Action properties of camptothecin and 10-hydroxycamptothecin [63]

	Camptothecin	10-Hydroxycamptothecin
Experimental antitumor activity	wide spectrum	wide spectrum
Effect on DNA synthesis	+	+
Effect on RNA synthesis	±	±
Effect on protein synthesis	−	±
Sister chromatid exchange	strong inducer	weak inducer
Effect on DNA polymerase	−	+
Effect on RNA polymerase	−	+
Cell cycle	G1 → S, S	S, G2/M
Immunosuppression	+	±
Plasma $t_{\frac{1}{2}}$	35 min	$t_{\alpha\frac{1}{2}}$: 4 min, $t_{\beta\frac{1}{2}}$: 29 min
Distribution tumor/liver	0.32	4.17
Excretion from urine	57%	23%
Excretion from feces	43%	77%
Hemorrhage cystitis	8%	1%

+: marked inhibition
±: slight effect

Antitumor activity of 10-hydroxycamptothecin in experimental investigations was first reported by Wani and Wall [8]. In the National Cancer Institute (NCI) screening program a variety of tumors including mouse leukemia P388, Lewis lung carcinoma, melanoma B16, CD8F1 breast cancer, colon 38, LX-1 lung cancer, and MX-1 breast cancer in mice and Walker carcinosarcoma 256 in rats were used to determine the antitumor spectrum of 10-hydroxycamptothecin [63]. The results have shown that 10-hydroxycamptothecin was active against leukemia P388 and melanoma B16 but inactive against Lewis lung carcinoma, MX-1 breast cancer, and LX-1 lung tumor. A moderate inhibition of L1210 was noticed [63]. Inhibition of ascites tumors such as Ehrlich carcinoma, hepatoma (H22), and reticulum cell sarcoma in mice and Yoshida sarcoma in rats was reported by Xu et al. [64]. Cell cycle studies using flow cytofluorometry indicated that 10-hydroxycamptothecin inhibits S phase of the tumor cells in vivo and S and G 2/M phases in vitro [65]. In most of the above experiments, 10-hydroxy-camptothecin-sodium was used.

In an in vitro study, it was found that murine erythroleukemia cells were very sensitive to 10-hydroxycamptothecin at a concentration of 100–200 ng/ml. It mainly delayed the progression of S phase of the subsequent cell cycle [66]. 10-Hydroxycamptothecin was also able to inhibit growth of a human stomach carcinoma cell line SGC-7901 at 0.2–100 μg/ml [67]. A similar inhibitory effect was observed in a human hepatoma cell line BEL-7404 [68]. 10-Hydroxycamptothecin (1–100 μg/ml) inhibited the growth of BEL-7404; the incorporation of [^{3}H] leucine into protein was inhibited by 31%–71%.

The transport of thymidine and uridine by mouse hepatoma cells in culture was competitively and reversibly inhibited by 10-hydroxycamptothecin at a concentration between 25 and 200 μ*M*. The concentration that gave 50% inhibition was 100 μ*M*. Since 10-hydroxycamptothecin did not inhibit hepatoma thymidine kinase or uridine kinase, its inhibition of nucleoside transport was probably not due to interference with phosphorylation but to inhibition of a membrane carrier system [69]. In a test with a human gastric carcinoma cell line, SGC-7901, the transport rate changes of nucleoside was dependent on the dose of and length of exposure to 10-hydroxycamptothecin. The transport of thymidine through the cell membrane was inhibited when the cells were exposed to 10-hydroxycamptothecin for a long time [67].

10-Hydroxycamptothecin was a potent inhibitor of nucleic acid synthesis in murine ascites hepatoma cells [69, 70], mouse hepatoma BW7755 cells [65], erythroleukemia cells [66], and human stomach cancer cells [67]. The incorporation of labeled uridine or cytidine into RNA and of acetate into lipids of murine P388 cells was also reduced significantly in the presence of 10-hydroxycamptothecin. The incorporation of thymidine into DNA was the parameter most sensitive to 10-hydroxycamptothecin in vitro with an ED_{50} of about 4 μg/ml [71], while the protein content was essentially unaffected as compared to the control group. 10-Hydroxycamptothecin inhibited synthesis of histone and nonhistone proteins in hepatoma cell nuclei of mice both in vitro and in vivo but had less or no effect on synthesis of cytoplasmic and nuclear sap protein suggesting that chromatin protein synthesis should be more sensitive to 10-hydroxycamptothecin than other kinds of proteins [72].

10-Hydroxycamptothecin caused a dose dependent growth inhibition in both the Chinese hamster cell line V79 and in Raji human lymphoid cells and induced a dose dependent increase in sister chromatid exchange frequencies. At the high concentration of 2 μg/ml about a twofold increase in sister chromatid exchange over the control was observed. Camptothecin caused a considerable increase in sister chromatid exchange up to elevenfold over the control under the same conditions [73]. In KB cell cultures, 10-hydroxycamptothecin also induced cellular chromatid breaks and increased the frequency of sister chromatid exchange, parallel to the inhibition of colony formation. By using the alkaline elution method, 10-hydroxycamptothecin was shown to induce protein associated DNA single strand breaks [74].

Mouse hepatoma RNA polymerase activity was also inhibited by 10-hydroxycamptothecin both in vitro and in vivo [75].

Morphological studies revealed that 10-hydroxycamptothecin caused degenerative changes of gastric carcinoma cells including disarrangement and pyknosis of the cells and coagulation and fragmentation of the nuclei [67]. Ultrastructural alterations such as swelling of mitochondria and variations in size of ascites hepatoma cells were seen when the tumor-bearing mice were treated with 10 mg/kg of 10-hydroxycamptothecin. The damage to mitochondria was followed by the appearance of many vacuoles, lipid droplets, and lysosomes in the cytoplasm [76].

Measurement of the cAMP and cGMP levels of tumor and normal tissue in mouse hepatomas showed that the cAMP content in the hepatoma was lower than that of normal liver. Treatment with 10-hydroxycamptothecin i.p. at a dose of 10 mg/kg inhibited tumor growth and induced an accumulation of cAMP in the hepatoma; however, 10-hydroxycamptothecin did not elevate cAMP in normal liver. The cGMP contents of the hepatoma and normal liver tissue were not significantly changed by 10-hydroxycamptothecin [70].

Biphasic pharmacokinetics of 10-hydroxycamptothecin were observed in mice given an i.v. dose of 10 mg/kg. The half-lives were 4.5 min and 29 min. High 10-hydroxycamptothecin levels were found in the bile and intestinal contents; moderate levels were found in carcinoma cells; and low levels were observed in bone marrow, thymus and spleen [77]. 10-Hydroxycamptothecin was excreted mainly in feces: 30% of the dose administered within 24 h and 48% within 48 h.

10-Hydroxycamptothecin was first used in China for clinical trials in 1974. In phase I and II clinical studies the results of 253 cancer patients have been reported [78]. 10-Hydroxycamptothecin is insoluble in water and therefore is given as the sodium salt in clinical applications. At the beginning, 10-hydroxycamptothecin (2 mg) was injected i.v. every other day. Later, doses of 4–8 mg per day or every other day were used. In the phase I study, the dose limiting factors of 10-hydroxycamptothecin were mild bone marrow depression, nausea, and vomiting. About 4% of patients treated with 10-hydroxycamptothecin developed toxic symptoms in the urinary tract including frequent micturition and mild albuminurea. Less than 1% of patients had a bleeding cystitis, indicating that the toxicity of 10-hydroxycamptothecin was much less than that of camptothecin.

In phase II clinical trials, the therapeutic dose of 4–8 mg 10-hydroxycamptothecin once daily or every other day was administered i.v. and 63 patients were evaluated. Tumor mass reduction greater than 50% and remission maintenance for more than 1 month occurred in 22 patients out of 52 patients (42.3%) with solid tumors. The responders included patients with primary liver carcinoma, cancers of head and neck, gastric carcinoma, or urinary bladder carcinoma. There were 8 patients with complete or partial remission in the group of 19 patients with cancer of head and neck, including 8 patients with malignant lymphoma [79]. 10-Hydroxy-camptothecin also induced partial remission of acute myelocytic leukemia and acute lymphocytic leukemia in adults or children.

References

1. Wall ME, Wani MC, Cook CE, Palmer KH, McPhail AT, Sim GA (1966) Plant antitumor agents. I. The isolation and structure of camptothecin, a novel alkaloidal leukemia and tumor inhibitor from *Camptotheca acuminata*. J Am Chem Soc 88:3888–3890
2. Wall ME, Wani MC (1980) Camptothecin. In: Cassady JM, Douros JD (eds) Anticancer Agents Based on Natural Product Models. Academic, New York, pp. 417–436
3. Cai JC, Hutchinson CR (1983) Camptothecin. Chem Heterocycl Comp 25:753–781
4. Schultz AG (1973) Camptothecin. Chem Rev 73:385–405
5. Hutchinson CR (1981) Camptothecin: chemistry, biogenesis and medicinal chemistry. Tetrahedron 37:1047–1065
6. McPhail AT, Sim GA (1969) The structure of camptothecin. X-ray analysis of camptothecin iodoacetate. J Chem Soc [B] 923–928
7. Adavovics JA, Cina JA, Hutchinson CR (1979) Minor alkaloids of *Camptotheca acuminata*. Phytochemistry 18:1085–1086
8. Wani MC, Wall ME (1969) Plant antitumor agents. II. The structure of two new alkaloids from *Camptotheca acuminata*. J Org Chem 34:1364–1367
9. Wall ME, Wani MC, Natschke SM, Nicholas AW (1986) Plant antitumor agents. XXII. Isolation of 11-hydroxycamptothecin from *Camptotheca acuminata* Decne: total synthesis and biological activity. J Med Chem 29:1553–1555
10. Tien HJ, Tien JM, Yeh MY, Wu TS, Huang CM (1977) Studies on the constituents of *Camptotheca acuminata* Decne. I. Constituents of leaves. Chemistry 51-4 (CA 92:37766e)
11. Lin LT, Chao TY, Hsu JS (1977) Chemical constituents of the anticancer plant *Camptotheca acuminata* Decne. I. Chemical constituents of the roots of *Camptotheca acuminata*. Acta Chim Sin 35:227–231
12. Hsu JS, Chao TY, Lin LT, Hsu CF (1977) Chemical constituents of the anticancer plant *Camptotheca acuminata* Decne. II. Chemical constituents of the fruits of *Camptotheca acuminata* Decne. Acta Chim Sin 35:193–200
13. Lin LZ, Song CQ, Xu RS (1982) Chemical constituents of the anticancer plant *Camptotheca acuminata* Decne. IV. Isolation and identification of 11-hydroxycamptothecin and four other compounds. Acta Chim Sin 40:85–89
14. Lin LT, Sung CC, Hsu JS (1978) Chemical constituents of the anticancer plant *Camptotheca acuminata* Decne. V. Other constituents of its fruits. Chemistry 327–328
15. Lin LT, Sung CC, Hsu JS (1979) A new anticancer alkaloid 11-hydroxycamptothecin. Kexue Tongbao 24:478–479
16. Ku KY, Tang TC (1980) Several botanical sources of camptothecin – an antitumor alkaloid. Chin Trad Herb Drugs 11:476–479
17. Govindachari TR, Viswanathan N (1972) 9-Methoxycamptothecin. New alkaloid from *Mappia foetida*. Indian J Chem 10:453–454
18. Tafur S, Nelson JD, DeLong DC, Svoboda GH (1976) Antiviral components of *Ophiorrhiza mungos*. Isolation of camptothecin and 10-hydroxycamptothecin. Lloydia 39:261–262
19. Gunasekera SP, Badawi MM, Cordell GA, Farnsworth NR, Chitnis M (1979) Plant anticancer agents. X. Isolation of camptothecin and 9-methoxycamptothecin from *Ervatamia heyneana*. J Nat Prod 42:475–477
20. Lin LZ, Song CQ, Xu RS (1979) Chemical constituents of the anticancer plant *Camptotheca acuminata* Decne. III. Ellagic acids from the fruits of *Camptotheca acuminata* Decne. Acta Chim Sin 37:207–214
21. Stork G, Schultz AG (1971) The total synthesis of *dl*-camptothecin. J Am Chem Soc 93:4074–4075
22. Tang CSF, Morrow CJ, Rapoport H (1975) A total synthesis of *dl*-camptothecin. J Am Chem Soc 97:159–167

23. Shanghai No. 5 Pharmaceutical Plant, Shanghai No. 12 Pharmaceutical Plant, Shanghai Institute of Pharmaceutical Industrial Research, Shanghai Institute of Materia Medica (1976) Total synthesis of *dl*-camptothecin. Kexue Tongbao 21:40–42
24. Shanghai No. 5 Pharmaceutical Plant, Shanghai No. 12 Pharmaceutical Plant, Shanghai Institute of Pharmaceutical Industrial Research, Shanghai Institute of Materia Medica (1978) The total synthesis of *dl*-camptothecin. Sci Sin 21:87–98
25. Henecke H (1949) Zur Kenntnis der β-Dicarbonyl-Verbindungen. VI. Die Kondensation von α-Aminomethylen-β-dicarbonyl-Verbindungen mit Ketonen und mit β-Dicarbonyl-Verbindungen. Chem Ber 82:41–46
26. Wani MC, Rouman PE, Lindley JT, Wall ME (1980) Plant antitumor agents. XVIII. Synthesis and biological activity of camptothecin analogs. J Med Chem 23:554–560
27. Corey EJ, Crouse DN, Anderson JE (1975) A total synthesis of natural 20(*S*)-camptothecin. J Org Chem 40:2140–2141
28. Wenkert E, Dave KG, Lewis RG, Spragne PW (1967) General methods of synthesis of indole alkaloids. VI. Syntheses of *dl*-corynantheidine and a camptothecin model. J Am Chem Soc 89:6741–6745
29. Winterfeldt E (1971) Reaktionen an Indolderivaten. XIII. Chinolon-Derivate durch Autoxidation. Liebigs Ann Chem 745:23–30
30. Warneke J, Winterfeldt E (1972) Reaktionen an Indolderivaten. XVI. Die autoxydative Indol-Chinolon-Umwandlung eines Camptothecin-Modells. Chem Ber 105:2120–2125
31. Hutchinson CR, Heckendorf AH, Straughn JL, Daddona PE, Cane DE (1979) Biosynthesis of camptothecin. III. Definition of strictosamide as the penultimate biosynthetic precursor assisted by carbon-13 and deuterium NMR-spectroscopy. J Am Chem Soc 101:3358–3369
32. Cordell GA (1974) Biosynthesis of indole alkaloids. Lloydia 37:219–298
33. Winterfeldt E, Korth T, Pike D, Boch M (1972) Biogenetically oriented total synthesis of camptothecin and 7-chlorocamptothecine. Angew Chem Int Ed Engl 11:289–290
34. Boch M, Korth T, Nelke JM, Pike D, Radunz H, Winterfeldt E (1972) Reaktionen an Indolderivaten. XVII. Die biogenetisch orientierte Totalsynthese von *dl*-Camptothecin and 7-Chlorcamptothecin. Chem Ber 105:2126–2142
35. Recher L, Chan H, Briggs L, Parry N (1972) Ultrastructural changes inducible with the plant alkaloid camptothecin. Cancer Res 32:2495–2501
36. Horwith SB (1975) Camptothecin. In: Corcoran JW, Hahn FE (eds) Mechanism of Action of Antimicrobial and Antitumor Agents. Springer, Berlin Heidelberg New York, pp 48–57 (Antibiotics III)
37. Horwith SB, Chang CK, Grollman AP (1971) Camptothecin. I. Effects on nucleic acid and protein synthesis. Mol Pharmacol 7:632–644
38. Bosmann HB (1970) Camptothecin inhibits macromolecular synthesis in mammalian cells but not in isolated mitochondria or Escherichia coli. Biochem Biophys Res Commun 41:1412–1420
39. Horwith MS, Horwith SB (1971) Intracellular degradation of HeLa and adenovirus type 2 DNA induced by camptothecin. Biochem Biophys Res Commun 45:723–727
40. Abelson HT, Penman S (1973) Induction of alkali labile links in cellular DNA by camptothecin. Biochem Biophys Res Commun 50:1048–1054
41. Spataro A, Kessel D (1972) Camptothecin-induced degradation and apparent reaggregation of DNA from L1210 cells. Biochem Biophys Res Commun 48:643–648
42. Li LH, Fraser TJ, Olin EJ, Buhyan BK (1972) Action of camptothecin on mammalian cells in culture. Cancer Res 32:2643–2650
43. Lown JW, Chen HH (1980) Studies related to antitumor antibiotics. XIX. Studies on the effects of the antitumor agent camptothecin and derivatives on deoxyribonucleic acid. Mechanism of the scission of deoxyribonucleic acid by photoactivated camptothecin. Biochem Pharmacol 29:905–915
44. Kuwahara J, Suzuki T, Funakoshi K, Sugiura Y (1986) Photosensitive DNA cleavage and phage inactivation by copper (II)-camptothecin. Biochemistry 25:1216–1221
45. Horwith SB, Chang CK, Grollman AP (1972) Antiviral action of camptothecin. Antimicrob Agents Chemother 2:395–401
46. Becker Y, Asher Y (1975) In vitro synthesis of DNA in nuclei isolated from herpes simplex virus-infected cells, untreated and treated with metabolic inhibitors. Virology 63:209–220
47. Kelly DC, Avery RJ, Dimmock NJ (1974) Camptothecin. Inhibitor of influenza virus replication. J Gen Virol 25:427–432

48. Minor PD, Dimmock NJ (1975) Inhibition of synthesis of influenza virus proteins. Evidence for two host-cell-dependent events during multiplication. Virology 67:114–123
49. Minor PD, Dimmock NJ (1977) Selective inhibition of influenza virus protein synthesis by inhibitors of DNA function. Virology 78:393–406
50. Wall ME (1969) Plant antitumor agents. V. Alkaloids with antitumor activity. In: Mothes K, Schreiber K, Schütte HR (eds) Symposiumberichte, 4. Internationales Symposium, Biochemie und Physiologie der Alkaloide. Akademie, Berlin, pp 77–87
51. Wall ME, Wani MC (1985) Antineoplastic structure activity relationships of camptothecin and related analogs. In: Chang HM, Yeung HW, Tso WW, Koo A (eds) Advances in Chinese Medicinal Material Research. World Scientific, Singapore, pp 391–405
52. Pan PC, Pan SY, Tu YH, Wang SY, Owen TY (1975) Studies on the derivatives of camptothecin. Acta Chim Sin 33:71–74
53. Sugasawa T, Toyoda T, Uchida N, Yamaguchi K (1976) Experiments on the synthesis of *d,l*-camptothecin. IV. Synthesis and antileukemic activity of *d,l*-camptothecin analogues. J Med Chem 19:675–679
54. Boxmann E, Winterfeldt E (1978) Totalsynthese des 7-Methoxycamptothecins. Chem Ber 111:3403–3411
55. Adamovics JA, Hutchinson CR (1979) Prodrug analogs of the antitumor alkaloid camptothecin. J Med Chem 22:310–314
56. Hartwell JL, Abbott BJ (1969) Antineoplastic principles in plants: recent development in the field. Adv Pharmacol Chemother 7:117–209
57. Schaeppi U, Fleischman RW, Cooney DA (1974) Toxicity of camptothecin (NCS-100880). Cancer Chemother Rep Part III 5:25–36
58. Yang JL, Han JX, Shen ZM, Xu B (1981) Study on the detoxification by monoammonium glycyrrhetate of anticancer drugs. Chin Pharm Bull 16:8–11
59. Chen RT, Hua Z, Lu Zy, Xu B (1980) Distribution and excretion of camptothecin suspension and sodium camptothecin in mice. Acta Pharmacol Sin 1:109–112
60. Gottlieb JA, Guarino AM, Call JB, Oliverio VT, Block JB (1970) Preliminary pharmacologic and clinical evaluation of camptothecin sodium (NSC-100880). Cancer Chemother Rep 54:461–470
61. Moertel CG, Schutt AJ, Reitmeier RJ, Hahn RG (1972) Phase II study of camptothecin (NSC-100880) in the treatment of advanced gastrointestinal cancer. Cancer Chemother Rep 56:95–101
62. Gottlieb JA, Luce JK (1972) Treatment of malignant melanoma with camptothecin (NSC-100880). Cancer Chemother Rep 56:103–105
63. Xu B, Yang JL (1985) Hydroxycamptothecin as an antitumor agent. In: Chang HM, Yeung HW, Tso WW, Koo A (eds) Advances in Chinese Medicinal Material Research. World Scientific, Singapore, pp 377–389
64. Shanghai Institute of Materia Medica, Academia Sinica (1979) Studies on the anticancer action of 10-hydroxycamptothecin. In: Fox BW (ed) Basis for Cancer Therapy. I. Pergamon, Oxford, pp. 105–109 (Advances in Medical Oncology, Vol 5)
65. Tsou KC, Xu B, Miller EE, Lo KW (1981) Antitumor effect of (*S*)-10-hydroxycamptothecin on mouse hepatoma BW 7756 and its possible mode of action. Anticancer Res 1:115–119
66. Xu B (1982) The influence of several anticancer agents on cell proliferation, differentiation and the cell cycle of murine erythroleukemia cells. Am J Chin Med 9:268–276
67. Yang JL, Han JX, Shen ZM, Zhou XJ (1982) Effect of 10-hydroxycamptothecin on the cell proliferation and ^{3}H-thymidine uptake into DNA of human gastric carcinoma cell (SGC-7901) in vitro. Tumor 2:21–23
68. Yang JL, Shen ZM, Sun YF, Han JX, Xu B (1985) Cultured human hepatoma cell (BEL-7404) for anticancer drug screening. Acta Pharmacol Sin 6:144–148
69. Ling YH, Yu WJ, Xu B (1983) Effects of 10-hydroxycamptothecin on the transport of pyrimidine nucleosides in murine hepatoma cells. Acta Biochim Biophys Sin 15:543–548
70. Yang JL, Han JX, Shen ZM, Xu B (1981) Effects of 10-hydroxycamptothecin on cAMP and cGMP levels in hepatoma and liver tissue of mice. Acta Pharmacol Sin 2:248–252
71. Creasey WA, Richards M, Gil D, Tsou KC (1983) Action of (*S*)-10-hydroxycamptothecin on P388 leukemia and distribution of the drug in mice. Cancer Treat Rep 67:179–182
72. Ling YH, Zhao CS, Xu B (1986) Effects of 10-hydroxycamptothecin on chromatin protein synthesis in murine hepatoma cells. Acta Pharmacol Sin 7:285–288

73. Huang CC, Han CS, Yue XF, Shen CM, Wang SW, Wu FG, Xu B (1983) Cytotoxicity and sister chromatid exchanges induced in vitro by six anticancer drugs developed in the People's Republic of China. JNCI 71:841–847
74. Wang XW, Shen ZM, Yang JL, Xu B (1986) Inhibitory effect of hydroxycamptothecin on colony formation of KB cells and DNA damage. Acta Pharm Sin 21:492–497
75. Ling YH, Yu WJ, Xu B (1984) Effect of 10-hydroxycamptothecin on nuclear RNA polymerase activity in mouse hepatoma cells. Acta Pharmacol Sin 5:211–213
76. Wang ZW, Shu RS, Xu B (1979) Effect of hydroxycamptothecin on the ultrastructure of ascitic hepatoma cells in mice. Chin J Oncol 1:183–185
77. Yang JL, Han JX, Xu B (1980) Distribution and excretion of 10-hydroxcycamptothecin and its effects on immune response. Acta Pharmacol Sin 1:44–48
78. Shanghai Institute of Materia Medica, Academia Sinica (1978) Studies on the anticancer action of 10-hydroxycamptothecin. Chin Med J 58:598–602
79. Department of Oral and Maxillo-Facial Surgery, 9th People's Hospital, Shanghai, 2nd Medical College (1978) China made camptothecin and hydroxycamptothecin in the treatment of oral and maxillofacial malignancies. Chin J Stomatol 13:75–77
80. Jiang TL, Salmon SE, Liu RM (1983) Activity of camptothecin, harringtonin, cantharidin and curcumae in the human tumor stem cell assay. Eur J Cancer Clin Oncol 19:263–270
81. Hsiang YH, Hertzberg R, Hecht S, Liu LF (1985) Camptothecin induces protein-linked DNA breaks via mammalian DNA topoisomerase I. J Biol Chem 260:14873–14878

Carpesium abrotanoides L. 34

34.1 Introduction

Heshi, Fuctus Carpesii, is the dry ripe fruit of *Carpesium abrotanoides* L. (Asteraceae) collected in fall when the fruits have ripened. It is listed officially in the Chinese Pharmacopoeia and is used as an anthelmintic.

34.2 Chemical Constituents

The essential oil, present at 0.25%–0.65% in the fruits of *C. abrotanoides*, has anthelminthic activity. A substance of lactone nature was isolated from the oil [1]. The first isolated sesquiterpene lactone was carpesialactone, a substance with powerful irritant properties and a bitter taste [2]. The structure of carpesialactone (*34-1*) was later elucidated by chemical reactions and IR spectral data [3, 4].

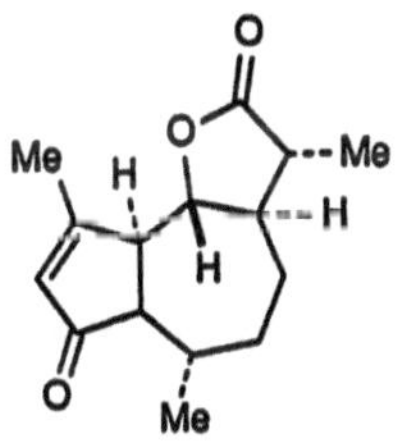

Carpesialactone (*34-1*)

Carabrone (*34-2*) was the second sesquiterpene lactone isolated from the ether extract of the *Carpesium* fruit. The structure of carabrone was determined by chemical reactions, IR spectral data [5, 6], and by total synthesis. The synthetic compound was identical with the natural product in IR spectral and gas chromatographic behavior [7].

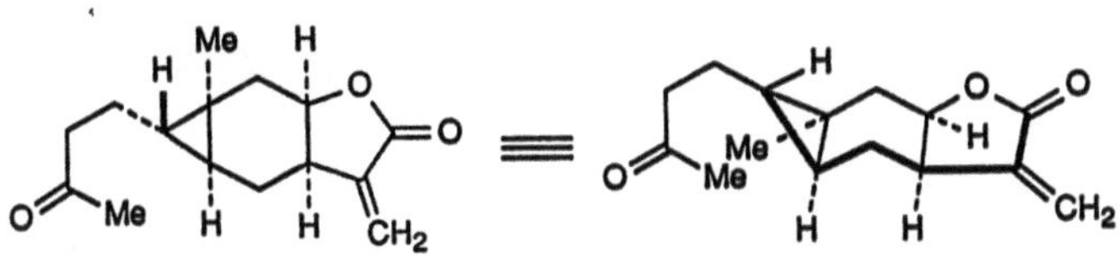

Carabrone (*34-2*)

Next, granilin (*34-3*) [8], carpesiolin (*34-4*) [9], carabrol (*34-5*), ivaxillin (*34-6*), dehydroivaxillin (*34-7*) [10], isoivaxillin (*34-8*), telekin, and dihydrotelekin (*34-9*) [11] were isolated and structurally characterized. Ivaxillin was found to be identical with eriolin [10]. Isoivaxillin and dihydrotelekin are two new sesquiterpenes.

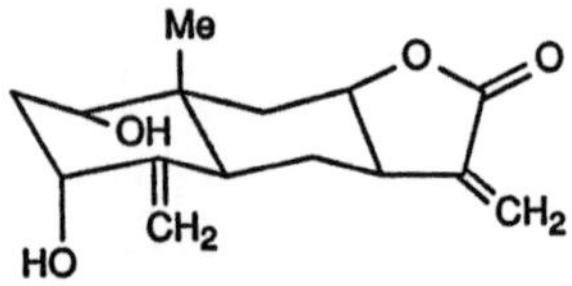
Granilin (*34-3*)

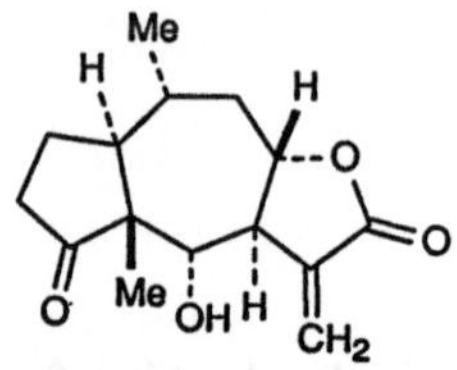
Carpesiolin (*34-4*)

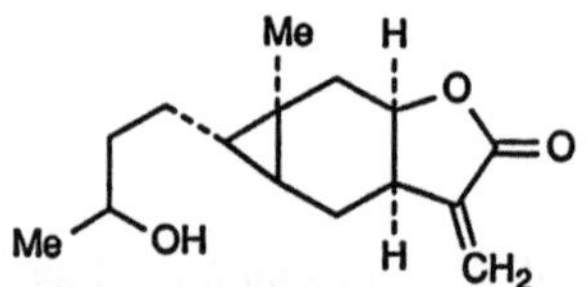
Carabrol (*34-5*)

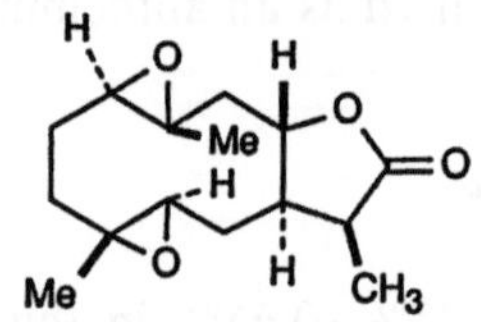
Ivaxillin (eriolin) (*34-6*)

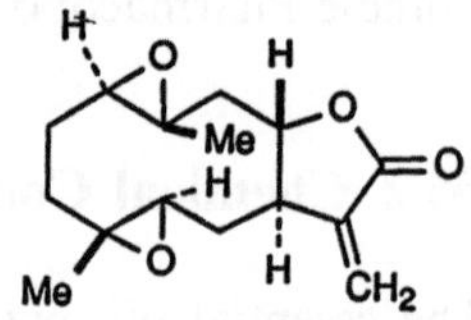
Dehydroivaxillin (*34-7*)

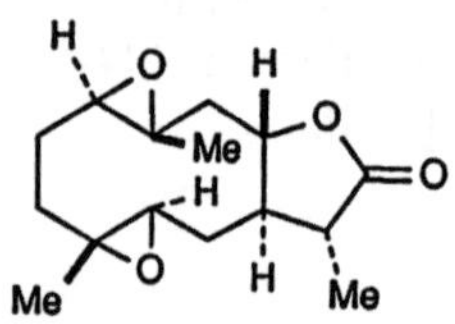
Isoivaxillin (*34-8*)

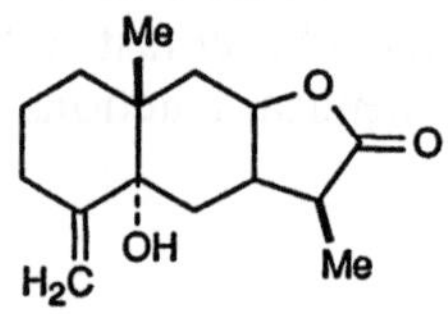
Dihydrotelekin (*34-9*)

Besides the sesquiterpene lactones, caproic, palmitic, stearic, oleic, and linoleic acids were also isolated from the seeds of *C. abrotanoides* [12].

34.3 Pharmacology

The alcohol extracts of *C. abrotanoides* had significant antifungal and antibacterial activity against *Cochliobolus miyabeanus* and *Xanthomonas oryzae*. Granilin [8], carpesiolin, and carabrone [9] were the isolated antifungal and antibacterial principles. Furthermore, caproic acid was demonstrated to be strongly toxic to water leeches [12].

References

1. Chu YL, Hsue CL, Liu PS, Tang TH (1957) A preliminary chemical study of the fruit of *Carpesium abrotanoides*. Acta Pharm Sin 5:155–156
2. Kariyone T, Kawano N (1949) Components of *Carpesium abrotanoides*. I. J Pharm Soc Jpn 69:317–318

3. Kariyone T, Naito S (1955) Components of *Carpesium abrotanoides*. III. Chemical structure of carpesia lactone. J Pharm Soc Jpn 75:39–43
4. Naito S (1955) Components of *Carpesium abrotanoides*. IV. Chemical structure of carpesia lactone. 3. J Pharm Soc Jpn 75:93–97
5. Minato H, Nosaka S, Horibe I (1964) Structure of carabrone from *Carpesium abrotanoides*. Proc Chem Soc 120–121
6. Minato H, Nosaka S, Horibe I (1964) Studies on sesquiterpenoids. VIII. The structure of carabrone, a new component of *Carpesium abrotanoides*. L. J Chem Soc 5503–5510
7. Minato H, Horibe I (1968) Studies on sesquiterpenoids. XVIII. Total synthesis of (±)-carabrone. J Chem Soc [C] 2131–2137
8. Maruyama M, Shibata F (1975) Stereochemistry of granilin isolated from *Carpesium abrotanoides*. Phytochemistry 14:2247–2248
9. Maruyama M, Omura S (1977) Carpesiolin from *Carpesium abrotanoides*. Phytochemistry 16:782–783
10. Maruyama M, Karube A, Sato K (1983) Sesquiterpene lactones from *Carpesium abrotanoides*. Phytochemistry 22:2773–2774
11. Dong YF, Ding YM (1988) Sesquiterpene lactones from *Carpesium abrotanoides*. Acta Bot Sin 30:71–75
12. Kaku T, Nakagawa K (1943) Component of "Kuan Shi". J Pharm Soc Jpn 63:252–255

Carthamus tinctorius L.

35

35.1 Introduction

Honghua, Flos Carthami, is the dry flower of *Carthamus tinctorius* L. (Asteraceae) collected in summer when the flowers have turned red. It is listed officially in the Chinese Pharmacopoeia and is used in traditional Chinese medicine as a hemokinetic and analgesic for treatment of menorrhalgia and traumatic diseases.

35.2 Chemical Constituents

The flower of *C. tinctorius* (safflower) is known to contain a number of coloring substances. The major red pigment is carthamin; however, structural determination of carthamin has required half a century. Since hydrolysis of carthamin yielded carthamidine (*35-1*) [1, 2] and isocarthamidine (*35-2*) [1–3], the structure was proposed to be the corresponding chalcone glycoside [1, 4] but this could not be proven by synthesis [5]. Reinvestigation [6–8] of carthamin (*35-3*) led to the structure shown below, consisting of two chalcone moieties with glucose bound directly at the C atom [7–9].

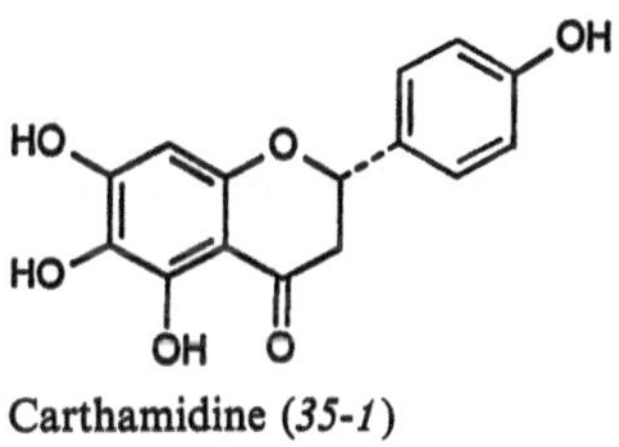

Carthamidine (*35-1*)

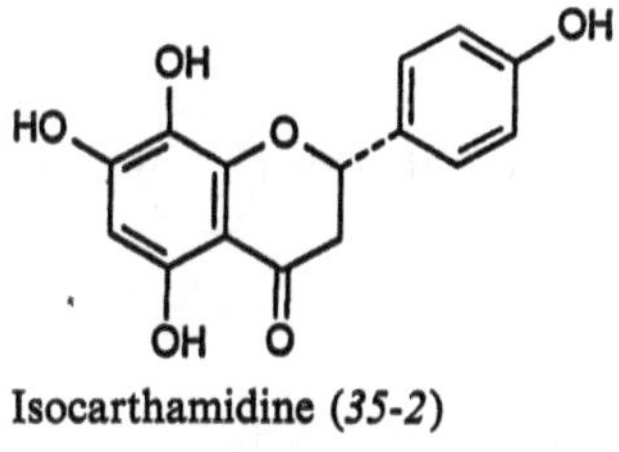

Isocarthamidine (*35-2*)

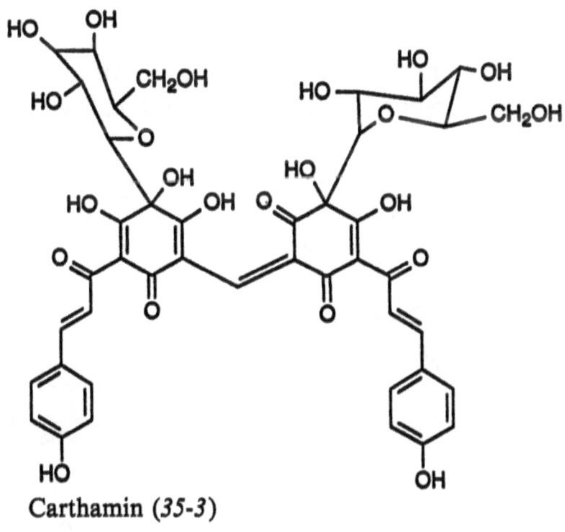

Carthamin (*35-3*)

Besides carthamin, a number of yellow pigments such as safflor yellow A (*35-4*) [8], safflor yellow B (*35-5*) [10], and safflomin A (*35-6*) [7], all related to carthamin were isolated.

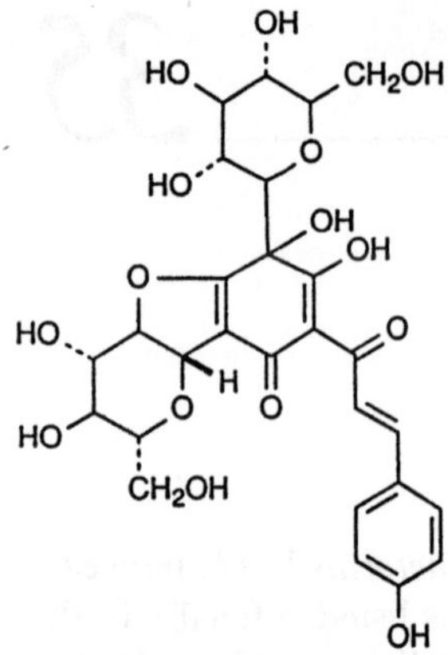

Safflor yellow A (*35-4*)

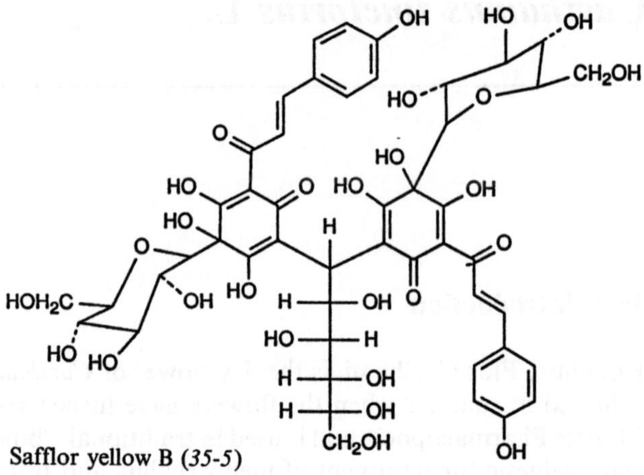

Safflor yellow B (*35-5*)

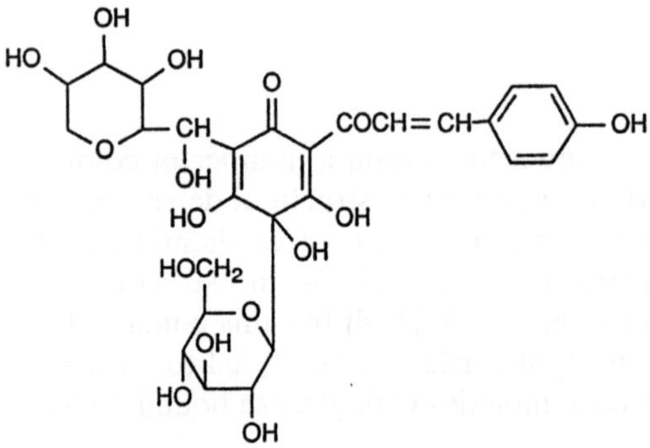

Safflomin A (*35-6*)

Furthermore, a novel phenolic amide, serotobenine (*35-7*), was isolated from safflower seeds together with *N*-feruloyltryptamine (*35-8*) and *N*-(*p*-coumaroyl)-tryptamine (*35-9*) [11].

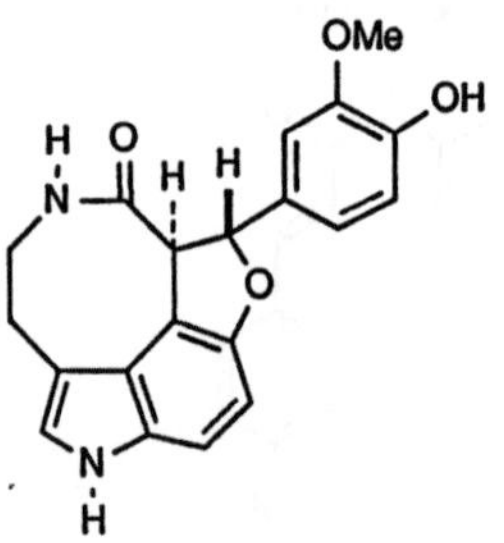

Serotobenine (*35-7*)

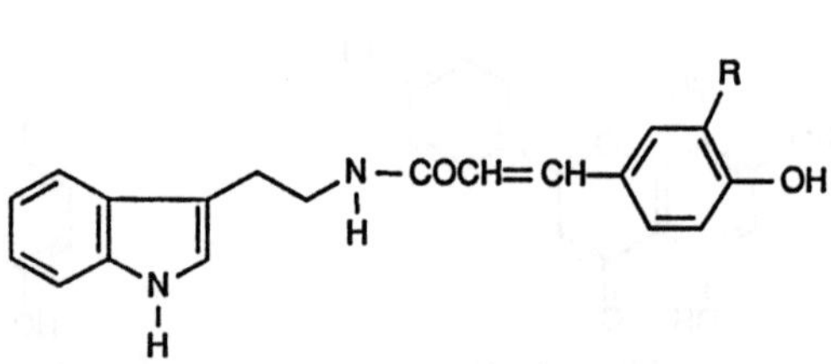

N-Feruloyltryptamine (*35-8*): R = OCH_3
N-(*p*-Coumaroyl)-tryptamine (*35-9*): R = H

In addition, the flavone luteolin and its 7-*O*-β-D-glucopyranoside [12, 13], β-sitosterol and its 3-*O*-β-D-glucopyranoside [12, 14] and a number of carboxylic acids such as lauric acid, myristic acid, palmitic acid, linoleic acid, arachidic acid and oleic acid [13–15] were also found in safflower. A sterol diglucoside with a pregnane skeleton (*35-10*), 15α,20β-dihydroxy-pregn-4-en-3-one diglucoside (*35-11*) was isolated from the seeds of *C. tinctorius* [16].

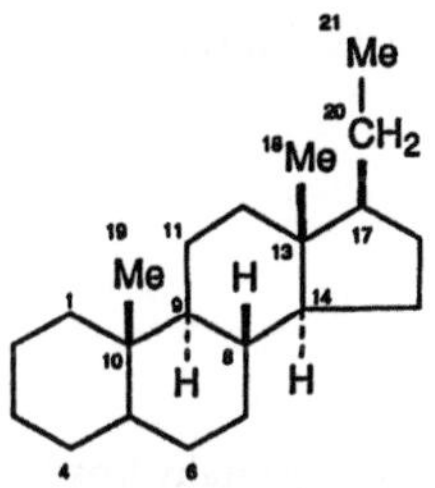

Pregnane (*35-10*)

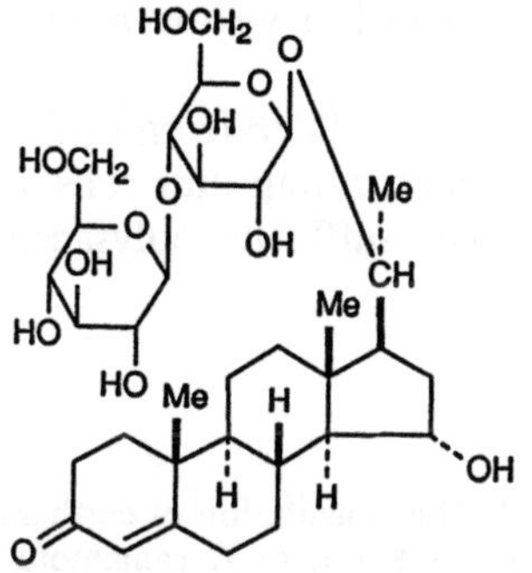

15α, 20β-Dihydroxy-pregn-4-en-3-one 20-*O*-β-D-glucopyranosyl-(1→4)-β-D-glucopyranoside (*35-11*)

A study on compounds in germinating safflower that stimulate fungal spore germination revealed the presence of some polyacetylene derivatives. These were identified as l-tridecene-3,5,7,9,11-pentayne; (11Z)-trideca-1,11-diene-3,5,7,9-tetrayne; (3Z,11Z)-trideca-1,3,11-triene-5,7,9-triyne: (3E,5Z,11E)-trideca-1,3,5,11-tetraene-7,9-diyne and (3Z,5E,11E)-trideca-1,3,5,11-tetraene-7,9-diyne [17]. Polyacetylenes rapidly increased during germination, but could not be detected in mature seeds [18]. From cultured cells of safflower, ubiquinone 9 (*35-12*) was isolated [19]. Ubiquinones are a group of compounds with a quinone ring and an isoprene side chain of different lengths. Ubiquinones act as coenzyme (coenzyme Q) in biological oxidation in the mitochondria.

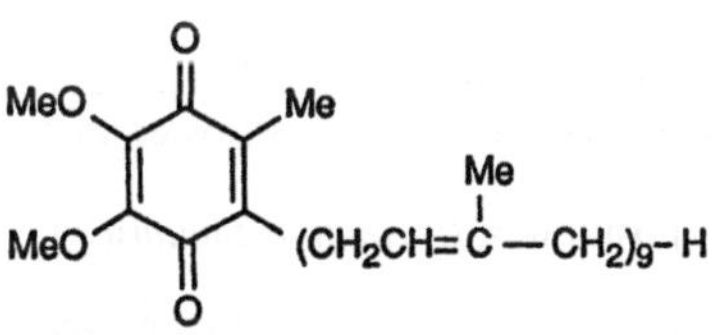

Ubiquinone 9 (*35-12*)

35.3 Pharmacology

Safflower is one of the most widely used herbal medicines in traditional Chinese medicine alone or in combination with other herbals for treatment of cardiovascular and hematological diseases, e.g., angina pectoris, cerebral hemorrhage, cerebral arteriosclerosis, rheumatalgia, amenorrhea, and menorrhalgia [20].

Intraperitoneal injection of safflow yellow A at doses of 550–1100 mg/kg in mice showed more sustained analgesic action than morphine. Moreover, safflor yellow A also markedly inhibited formaldehyde-induced foot swelling, histamine-stimulated capillary permeability, and formation of cotton ball granuloma in rats. The central inhibition induced by barbital or chloral in mice was significantly enhanced by safflor yellow A; coramine-induced convulsions and death were significantly reduced. The LD_{50} of safflor yellow A in mice by i.p. and oral administration was 5.5 g/kg [21].

The polysaccharide isolated from safflower was found to stimulate immune reactivity in vivo. Plaque-forming unit cells and spleen weight were both increased in mice injected i.p. with safflower polysaccharide [22].

References

1. Kuroda C (1930) The constitution of carthamin. I. J. Chem Soc 752–765
2. Obara H, Onodera J, Kurihara Y, Yamamota F (1978) Synthesis of 2′,3′,4,4′,6′-pentahydroxychalcone, an aglycone of carthamin and its isomerization into 4′,5,6,7- and 4′,5,7,8-tetrahydroxyflavanone, carthamidin and isocarthamidin. Bull Chem Soc Jpn 51:3627–3630
3. Zemplen G, Farkas L, Rakusa R (1958) Structure and synthesis of isocarthamidine. Acta Chim Acad Sci Hung 14:471–474
4. Seshadri TR, Thakur RS (1960) The coloring matter of the flowers of *Carthamus tinctorius*. Curr Sci 29:54–55
5. Saito N, Hatakeda K, Ito S, Asano T (1976) Synthesis of 4,4′-dibenzyloxy-2′-hydroxyquinochalcone. Tohoku Kogyo Gijutsu Shikensho Hokoku 7:55–57 (CA 88:38939t)
6. Obara H, Onodera J (1979) Structure of carthamin. Chem Lett 201–204
7. Onodera J, Obara H, Osone M, Maruyama Y, Sato S (1981) The structure of safflomin A, a component of safflower yellow. Chem Lett 433–436
8. Takahashi Y, Miyasaka N, Tasaka S, Miura I, Urano S, Ikura M, Hikichi K, Matsumoto T, Wada M (1982) Constitution of two coloring matters in the flower petals of *Carthamus tinctorius*. L. Tetrahedron Lett 23:5163–5166
9. Obara H, Onodera J, Abe S, Saito T (1980) Synthesis of 3-(*p*-hydroxycinnamoyl)-5-methylfilicinic acid and dehydro-3,3′-diacetyl-5,5′-methylenedifilicinic acid. Bull Chem Soc Jpn 53:289–290
10. Takahashi Y, Saito K, Yanagiya M, Ikura M, Hikichi K, Matsumoto T, Wada M (1984) Chemical constitution of safflor yellow B, a quinochalcone C-glycoside from the flower petals of *Cathamus tinctorius* L. Tetrahedron Lett 25:2471–2474
11. Sato H, Kawagishi H, Nishimura T, Yoneyama S, Yoshimoto Y, Sakamura S, Furusaki A, Katsuragi S, Matsumoto T (1985) Serotobenine, a novel phenolic amide from safflower seeds (*Carthamus tinctorius* L.). Agric Biol Chem 49:2969–2974
12. Ismailov NM, Kuliev AA (1985) Flavonoid and sterol composition of *Carthamus tinctorius* L. Dokl Akad Nauk Az SSR 41:61–63 (CA 103:34895f)
13. Takahashi M, Osawa K (1981) The components of *Carthamus tinctorius* L. Annu Rep Tohoku Coll Pharm 28:79–83 (CA 97:178747q)
14. Xu SX, Miao L (1986) Antiinflammatory active constituents of *Carthamus tinctorius* (II). Chin Trad Herbal Drugs 11:106–108
15. Xu SX, Qiu S, Zhang SJ (1984) Studies on the antiinflammatory principles in *Carthamus tinctorius*. I. Bull Chin Mat Med 9:31–32
16. Palter R, Lundin RE, Fuller G (1972) New steroid from safflower. Phytochemistry 11:819–822
17. Binder RG, Lundin RE, Kint S, Klisiewicz JM, Waiss AC, Jr (1978) Polyacetylenes from *Carthamus tinctorius*. Phytochemistry 17:315–317
18. Ichihara K, Noda M (1977) Distribution and metabolism of polyacetylenes in safflower. Biochim Biophys Acta 487:249–260
19. Hagimori M, Matsumoto T, Noguchi M (1978) Isolation and identification of ubiquinone 9 from culture cells of safflower (*Carthamus tinctorius*). Agric Biol Chem 42:499–500

20. Wang GM, Li YL (1985) Clinical application of safflower. Zhejiang J Trad Chin Med 20:42–43
21. Huang ZL, Gao QM, Cui ZM (1984) Pharmacological studies on safflor yellow from safflower (*Carthamus tinctorius*). Chin Trad Herb Drugs 15:348–350
22. Huang H, Yu ML, Qu SK, Shan ML, Song CQ (1984) Studies on the immunoactivity of the polysaccharide from safflower (*Carthamus tinctorius*). Chin Trad Herb Drugs 15:213–216

Centella asiatica (L.) Urb. 36

36.1 Introduction

Jixuecao, Herba Centellae, is the dry whole plant of *Centella asiatica* (L.) Urb. (Apiaceae) collected in summer and fall. It is listed officially in the Chinese Pharmacopoeia and used as an antipyretic, diuretic, and antidote in the treatment of icterus, heatstroke, diarrhea, ulcerations, eczema, and traumatic diseases.

36.2 Chemical Constituents

From the whole plant of *C. asiatica* several triterpene saponins and sapogenins were isolated. The major triterpene saponin is asiaticoside (*36-1*) [1, 2] with the sapogenin asiatic acid (*36-2*), a trihydroxyursenoic acid derivative. The structures of asiaticoside [1] and its aglycon [3] were elucidated by chemical and spectral methods. Thus, asiaticoside is a triglycoside of asiatic acid with one mannose and two glucose residues.

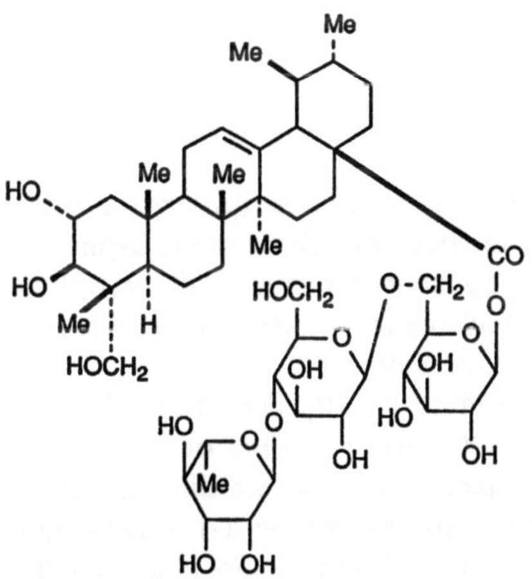

Asiaticoside (*36-1*)

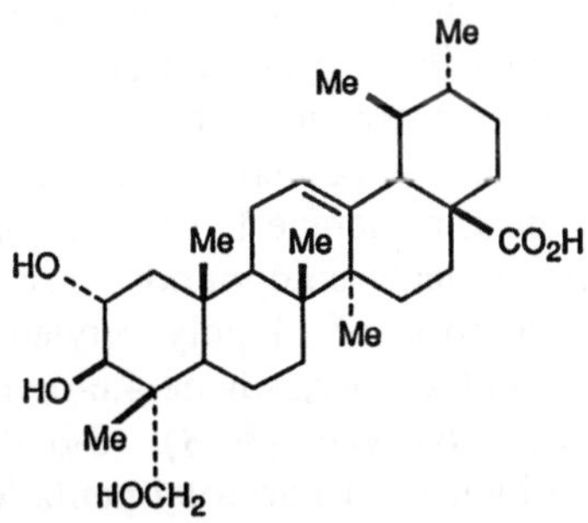

Asiatic acid (*36-2*)

From the whole plant of *C. asiatica*, collected in Madagascar, the triterpenes madasiatic acid (*36-3*) [4] and madecassic acid (*36-4*) [5] were isolated and structurally investigated. The latter was also isolated from *C. asiatica* collected in India and named brahmic acid [6]. A related saponin, madecassoside (*36-5*), with the aglycon madecassic acid was also isolated from *C. asiatica* collected in China [7]. All these triterpenes are derived from ursenoic acid (*36-6*) and have ursane (*23-17*) skeleton.

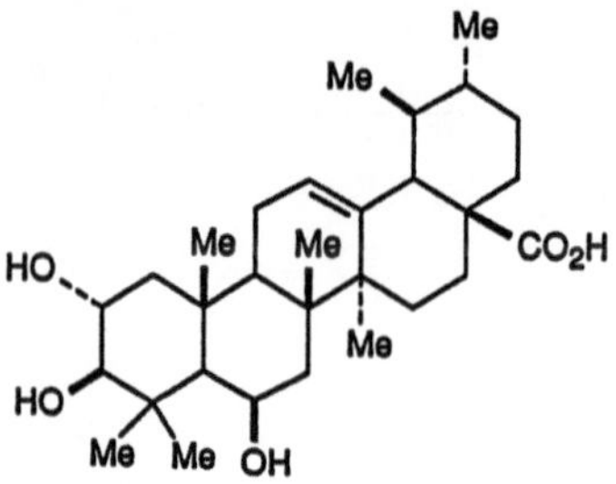

Madasiatic acid (*36-3*)

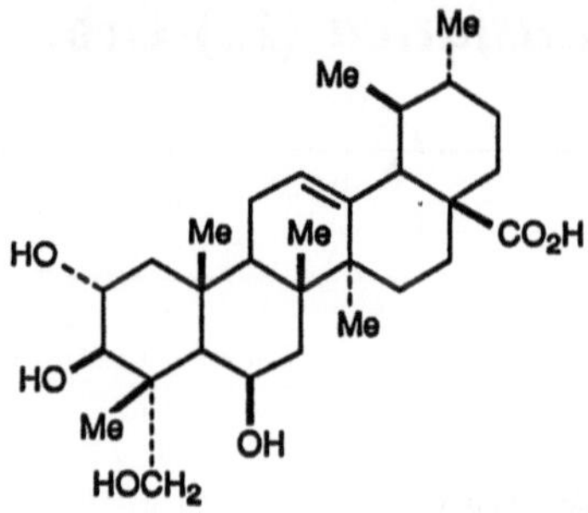

Madecassic acid (Brahmic acid) (*36-4*)

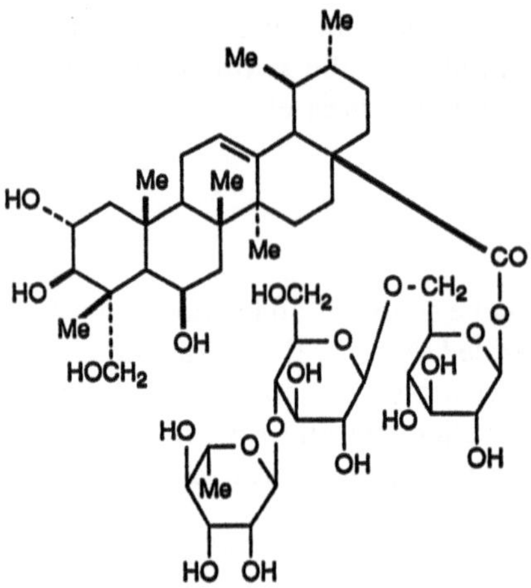

Madecassoside (*36-5*)

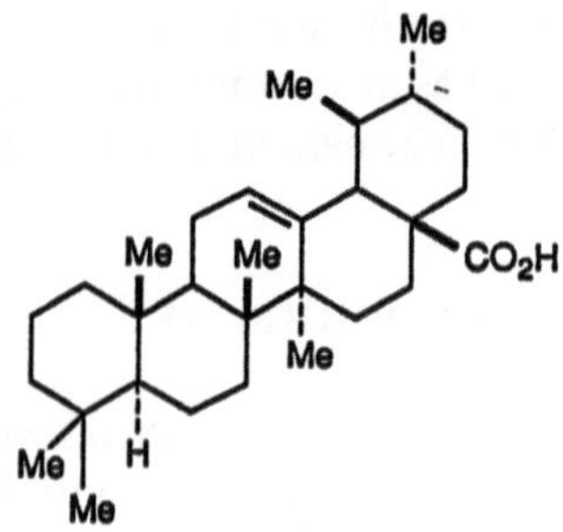

Ursenoic acid (*36-6*)

In addition to the triterpene constituents, some mono- and sesquiterpene compounds contained in *C. asiatica* were also investigated [8]. These were: α-pinene, β-pinene, myrcene, γ-terpinene, bornyl acetate, α-copaene, β-elemene, β-caryophyllene, *trans*-β-farnesene, germacrene-D, and bicycloelemene. β-Caryophyllene, *trans*-β-farnesene, and germacrene-D are the major components.

Some polyacetylenic compounds were also isolated [9, 10]. Schulte et al. have reported the isolation of 14 polyacetylenes from *C. asiatica*. Five of them were identified as: pentadeca-2,9-diene-4,6-diyn-1-ol acetate (*36-7*); 3,8-diacetoxypentadeca-1,9-diene-4,6-diyne (*36-8*); 3-hydroxy-8-acetoxy-pentadeca-1,9-diene-4,6-diyne (*36-9*); 3-hydroxy-10-acetoxy-pentadeca-1,8-diene-4,6-diyne (*36-10*); and pentadeca-1,8-diene-4,6-diyne-3,10-diol (*36-11*) [9]. The amounts of compounds *36-9* and *36-10* in dry plant material are 0.02% and 0.01%, respectively. Compound *36-9* is the major polyacetylene and common constituent of the *Centella* species [10].

$$CH_3(CH_2)_4CH{=}CH{-}CH_2{-}(C{\equiv}C)_2{-}CH{=}CH{-}CH_2{-}OCOCH_3$$

Pentadeca-2,9-diene-4,6-diyn-1-ol acetate (*36-7*)

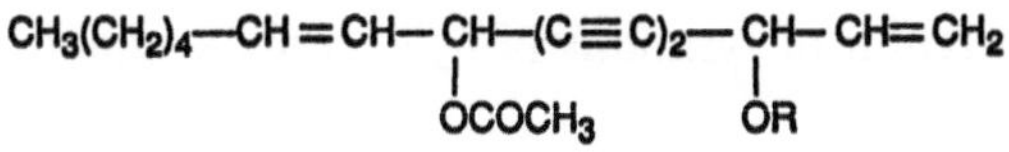

3,8-Diacetoxy-pentadeca-1,9-diene-4,6-diyne (*36-8*): R = Ac
3-Hydroxy-8-acetoxy-pentadeca-1,9-diene-4,6-diyne (*36-9*): R = H

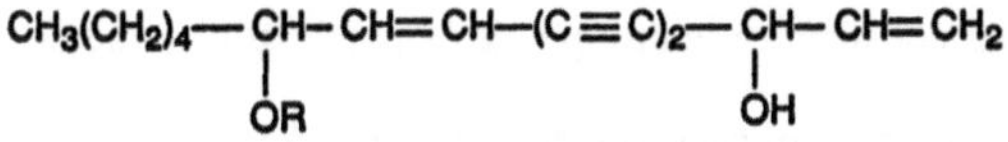

3-Hydroxy-10-acetoxy-pentadeca-1,8-diene-4,6-diyne (*36-10*): R = Ac
Pentadeca-1,8-diene-4,6-diyne-3,10-diol (*36-11*): R = H

Quercetin-3-glucoside and kaempferol-3-glucoside are flavone derivatives obtained from *C. asiatica* [11].

36.3 Pharmacology

Pharmacologic studies on *C. asiatica* in mice and rats indicated that the saponins exhibited a sedative action. The mode of action appeared to be mainly on the cholinergic mechanism in the central nervous system [12].

In rats, the extract of *C. asiatica* and three constituents, madecassic acid, asiatic acid, and asiaticoside, showed a healing effect when applied locally to wound. The area of skin necrosis induced by burn was decreased; the extract and the 3 constituents were equally effective [13]. Oral administration of *C. asiatica* extract, containing triterpenes, at a daily dose of 100 mg/kg shortened the lengthening phase of wound healing while having little effect on the contracting phase [14]. Intramuscular injection of asiaticoside into mice, rats, guinea pigs, and rabbits provoked a rapid thickening of the skin, local leukocytosis, increased vascularization of the connective tissue, mucous secretion, and hair and nail growth. Subcutaneous injection of asiaticoside, 0.04–0.05 g/kg, was toxic to mice and rabbits; 0.20–0.25 g/kg produced increased bleeding times and hemorrhage. A dose of 1 g/kg administered orally was tolerated [15].

In rats and mice good percutaneous absorption of tritiated madecassic acid and tritiated asiatic acid from reconstituted *C. asiatica* extract applied in a continuous oil phase or an oil-water emulsion was observed in rats, plasma levels of the triterpenes increased with time for the first 24 h. Levels of madecassic acid and asiatic acid in subcutaneous tissue and adjacent abdominal muscle were maximal by 1 and 3 h, respectively, decreasing slowly thereafter. In mice, 2.1% and 2.2% of madecassic acid and asiatic acid were absorbed by 3 h from oil phase and oil-water emulsion, respectively. The radioactivity after oral administration was greatest in the liver, kidney, and paw muscles [16].

Addition of the total triterpene fraction of *C. asiatica* to the culture medium of human embryo fibroblasts did not influence the growth rate at ≤ 25 µg/ml but inhibited growth at 50 µg/ml. The triterpene fraction stimulated the incorporation of proline without an extensive increase of total protein content, suggesting specific stimulation of collagen formation. A high incorporation of acetate into lipids and glycosaminoglycans was also observed. Analysis of glycosaminoglycans by ion exchange chromatography showed a stimulation of synthesis of hyaluronic acid and chondroitin sulfate [17].

References

1. Polensky J, Sach E, Lederer E (1959) Sur la constitution chimique de la partie glucidique de l'asiaticoside. Bull Soc Chim Fr 880–887
2. Luo SQ, Chin HF (1980) Isolation and identification of asiaticoside from *Centella asiatica* (L.) Urban. Chin Trad Herb Drugs 11:244–246
3. Polonsky J (1953) Sur la constitution chimique de l'acide asiatique, aglycone de l'asiaticoside. Rattachement de l'acide asiatique à la série de l'α-amyrine; formule développée de l'acide asiatique. Bull Soc Chim Fr 173–180
4. Pinhas H (1969) Structure de l'acide madasiatique, nouvel acide triterpénique isolé de *Centelle asiatica* L. Bull Soc Chim Fr 3592–3595
5. Pinhas H, Billet D, Heitz S, Chaigneau M (1967) Structure de l'acide madécassique, nouveau triterpène de *Centella asiatica* de Madagascar. Bull Soc Chim Fr 1890–1895
6. Singh B, Rastogi RP (1968) Chemical examination of *Centella asiatica.* III. Constitution of brahmic acid. Phytochemistry 8:1385–1393
7. Lou SQ, Jin HF (1981) Isolation and identification of madecassoside in *Centella asiatica.* Chin Trad Herb Drugs 12:5–6
8. Asakawa Y, Matsuda R, Takemoto T (1982) Mono- und sesquiterpenoids from *Hydrocotyle* and *Centella* species. Phytochemistry 21:2590–2592
9. Schulte KE, Ruecker G, Abdul Bary E (1973) Constituents of medical plants. XXVII. Polyacetylenes from *Hydrocotyle asiatica.* Arch Pharm (Weinheim) 306:197–209
10. Bohlmann F, Zdero C (1975) Polyacetylenic compounds. CXXX. A new polyene from *Centella* species. Chem Ber 108:511–514
11. Prum N, Illel B, Raynaud J (1983) Flavonoid glycosides from *Centella asiatica* L. (Umbelliferae). Pharmazie 38:423
12. Ramaswamy AS, Periyasamy SM, Basu NK (1970) Pharmacological studies on *Centella asiatica.* J Res Indian Med 4:160–175
13. Tsurumi K, Hiramatsu Y, Hayashi M, Fujimura H (1973) Effect of madecassol on wound healing. Oyo Yakuri 7:833–843 (CA 80:66674f)
14. Poizot A, Dumez D (1978) Modification of the healing kinetics after iterative exeresis in the rat. Action of titrated extract of *Centella asiatica* (TECA) on duration of healing. C R Acad Sci [D] 286:789–792
15. Boiteau P, Ratsimamanga AR (1956) Asiaticosid, extracted from *Centella asiatica,* its therapeutic uses in the healing of experimental or refractory wounds, leprosy, skin tuberculosis and lupus. Therapie 11:125–149
16. Viala A, Cano JP, Durand A, Paulin R, Roux F, Placidi M, Pinhas H, Lefournier C (1977) Animal study of the transcutaneous passage of tritium-labeled active principles from *Centella asiatica* L. applied in a continuous oil phase or an oil-water emulsion. Therapie 32:573–583
17. Del Vecchio A, Senni I, Cossu G, Molinaro M (1984) Effects of *Centella asiatica* on biosynthetic activity in cultured fibroblasts. Farmaco [Prat] 39:355–364

Centipeda minima (L.) A. Braun et Aschers. 37

37.1 Introduction

Ebushicao, Herba Centipedae, is the dry whole plant of *Centipeda minima* (L.) A. Braun et Aschers. (Asteraceae) collected in summer or fall when the plant blooms. This official medicinal herb is recommended in traditional Chinese medicine as an antiallergic, antitussive, and expectorant agent for treatment of cold, nasal allergy, and asthma.

37.2 Chemical Constituents

The n-hexane soluble fraction of the chloroform extract of *C. minima* contained a large amount of volatile compounds, from which heptan-2-ol, hepa-2,4-dien-1-ol, isobutyric acid, benzylalcohol, chrysanthenol (*37-1*), chrysanthenyl acetate (*37-2*), methyl linoleate, *β*-gurjunene (*37-3*), methyl palmitate, deca-2,4-dien-1-ol, ethyl palmitate, phytol, caryophyllane-2,6-oxide (*37-4*), and dihydroactinidiolide (*37-5*) were detected by GC/MS [1].

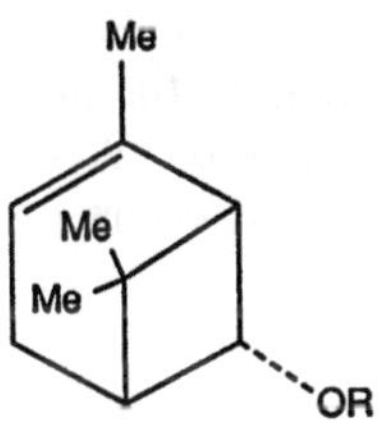

Chrysanthenol (*37-1*): R=H
Chrysanthenyl acetate (*37-2*): R=AC

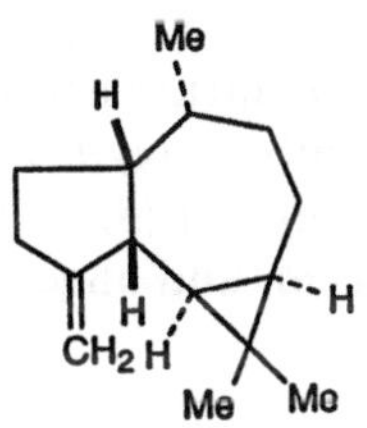

β-Gurjunene (*37-3*)

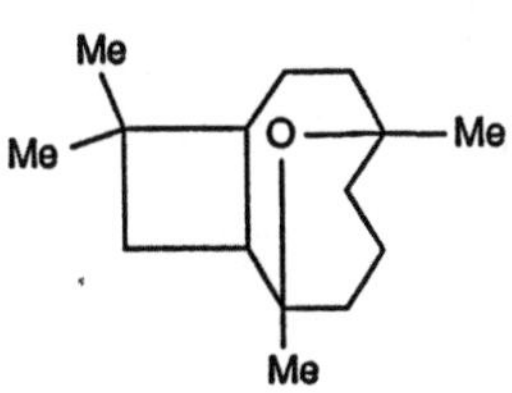

Caryophyllane-2,6-oxide (*37-4*)

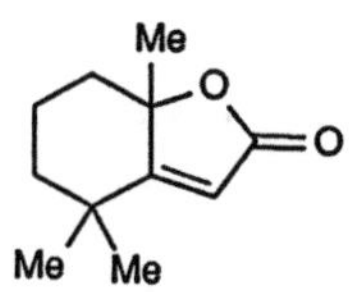

Dihydroactinidiolide (*37-5*)

From the methanol soluble fraction two sesquiterpene lactones, arnicolide C (*37-6*) and 6-O-senecioylplenolin (*37-7*) were isolated, together with aurantiamide acetate (*38-8*) [1]. The isolation of another structurally related sesquiterpene lactone helenalin (*37-9*) was also reported [2].

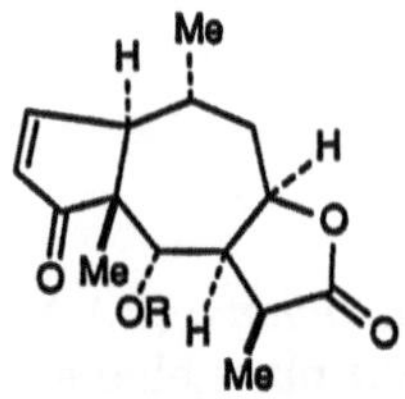

Arnicolide C (*37-6*): R= –CO–CH(CH$_3$)$_2$
6-O-Senecionylplenolin (*37-7*):
R= –CO–CH=C(CH$_3$)$_2$

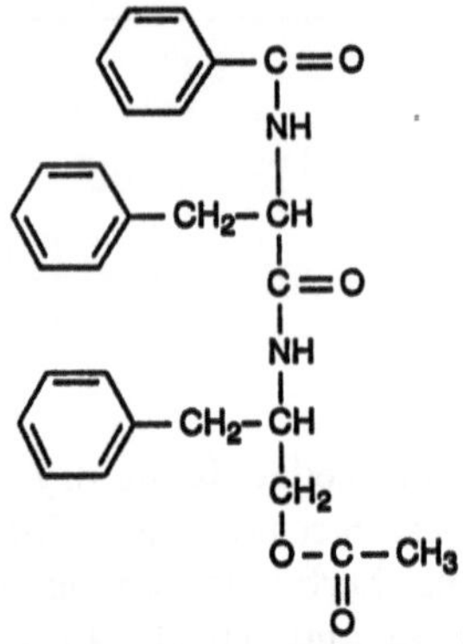

Aurantiamide acetate (*37-8*)

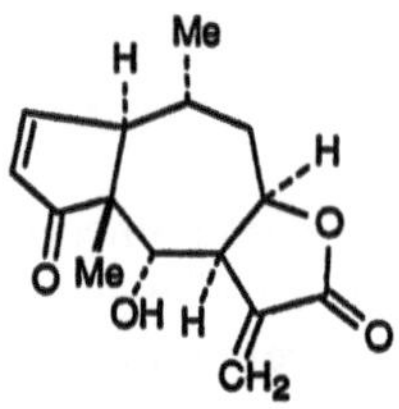

Helenalin (*37-9*)

In addition to the constituents described above, the flavones quercetin-3,3′-dimethylether, quercetin-3-methylether, and apigenin [1], the triterpene lupeyl acetate, *β*-sitosterol, taraxasterol, taraxasteryl palmitate, taraxasteryl acetate, stigmasterol, lupeol [2–4], 10-isobutyryloxy-8,9-epoxythymol isobutyrate (*37-10*), 9,10-di-isobutyryloxy-8-hydroxythymol (*37-11*), and florilenalin (*37-12*), and esters of isobutyric acid, isovaleric acid, and angelic acid [2] were isolated.

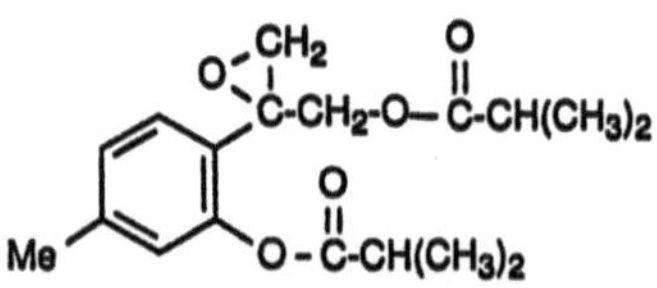

10-Isobutyryloxy-8,9-epoxy-thymol isobutyrate (*37-10*)

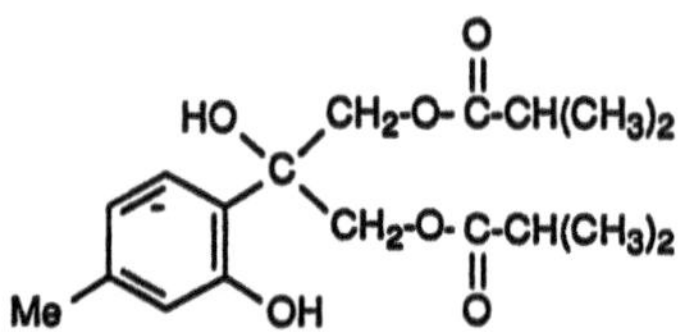

9,10-Diisobutyryloxy-8-hydroxythymol (*37-11*)

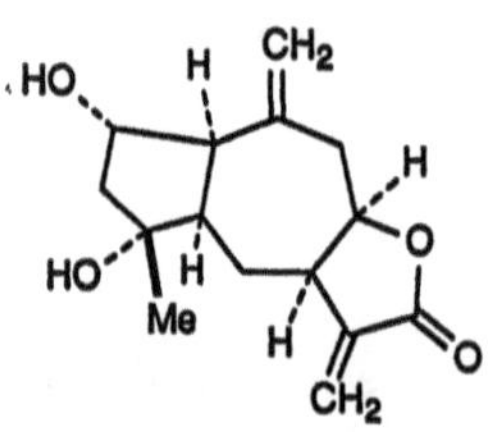

Florilenalin (*37-12*)

37.3 Pharmacology

The aqueous extract of *C. minima* exhibited a significant antiallergic activity in passive cutaneous anaphylaxis tests. The extract also showed potent inhibitory effects on histamine release from rat peritoneal mast cells induced by concanavalin A [1].

The chemical constituents arnicolide C, 6-*O*-senecioylplenolin, and three isolated flavones were all effective in inhibiting histamine release from rat peritoneal mast cells. The IC_{50} of arnicolide C, 6-*O*-senecioylplenolin and the flavones were 3.0×10^{-5} *M*, 1.8×10^{-5} *M*, and $0.5-1.0 \times 10^{-5}$ *M*, respectively. Aurantiamide acetate was less effective. Arnicolide C and 6-*O*-senecioylplenolin at doses of 50 mg/kg administered orally inhibited pigment leakage in passive cutaneous anaphylaxis tests by 44%–46% and 37%–41%, respectively. The isolated flavones also showed significant effects in passive cutaneous anaphylaxis tests by oral administration at the same dosage. Inhibition ratios of pigment leakage were in the range of 39%–67%. Quercetin 3,3′-dimethylether had the most potent antiallergic effect by oral administration [1].

References

1. Wu JB, Chun YT, Ebizuka Y, Sankawa U (1985) Biologically active constituents of *Centipeda minima*: isolation of a new plenolic ester and the antiallergy activity of sesquiterpene lactones. Chem Pharm Bull 33:4091–4094
2. Bohlmann F, Chen ZL (1984) New guaianolides from *Centipeda minima*. Kexue Tongbao [Foreign Lang] 29:900–903
3. Murakami T, Chen CM (1970) Constituents of *Centipeda minima*. I. Yakugaku Zasshi 90:846–849
4. Sen AB, Shukla YN (1970) Chemical constituents of *Centipeda minima*. J Indian Chem Soc 47:96

Cephalotaxus spp. 38

38.1 Introduction

Cephalotaxus, the only genus of the family Cephalotaxaceae with eight species and few varieties known so far, is mostly native to China. Seven of the eight known *Cephalotaxus* species growing in China are: *C. fortunei* Hook. f., *C. hainanensis* Li, *C. wilsoniana* Hay., *C. mannii* Hook. f., *C. oliveri* Mast., *C. lanceolata* K.M. Feng, and *C. sinensis* Li. *C. harringtonia* occurs in Japan and *C. mannii* is also found in India [1, 2].

After the antitumor activity of the ester alkaloids contained in *Cephalotaxus* against P388 and L1210 leukemias in mice was reported [3], great interest in the chemistry and pharmacology of *Cephalotaxus* was aroused and a number of studies were carried out worldwide, especially in China. In view of the interesting chemistry and pharmacologic properties of the alkaloid constituents from *Cephalotaxus*, they are described here, although the *Cephalotaxus* species are not listed officially in the Chinese Pharmacopoeia.

38.2 Chemical Constituents

38.2.1 Alkaloids

The presence of alkaloid components in *Cephalotaxus* species was already reported in the 1950s by different groups [4–6]. The first isolated alkaloid, cephalotaxine (*38-1*), together with some other alkaloids from *C. harringtonia* var. *drupacea* (*C. drupacea*), and *C. fortunei* were reported by Paudler et al. [7]. The powdered leaves and stems of *C. fortunei* and *C. harringtonia* var. *drupacea* yielded 0.39% and 0.35%, respectively, crude alkaloid material by conventional acid-base extraction of the concentrated alcohol extracts. Cephalotaxine was present in yields of 50% and 54%, respectively, in the crude alkaloid mixture of *C. fortunei* and *C. harringtonia* var. *drupacea*.

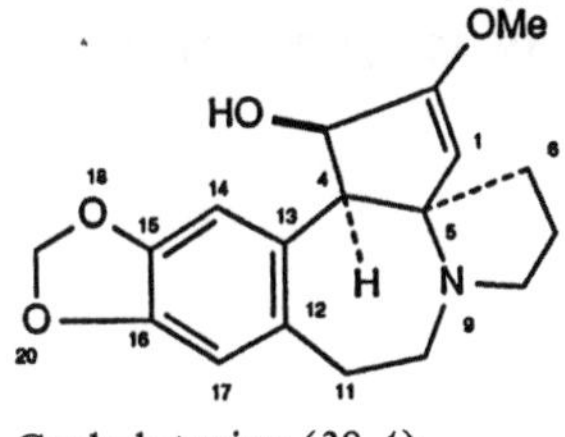

Cephalotaxine (*38-1*)

The structure of cephalotaxine, which is unique to the genus *Cephalotaxus*, was elucidated by spectral evidence [8] and by single crystal X-ray diffraction of its methiodide (*38-2*) [9]. In contrast to chromatographically pure cephalotaxine, the crystalline methiodide was optically inactive and racemic [8, 9].

The racemization of a molecule with four chiral centers is unusual and was suggested by breaking C-N bond to the azecine derivative (*38-3*) and subsequent reformation (Fig. 38-1) [10].

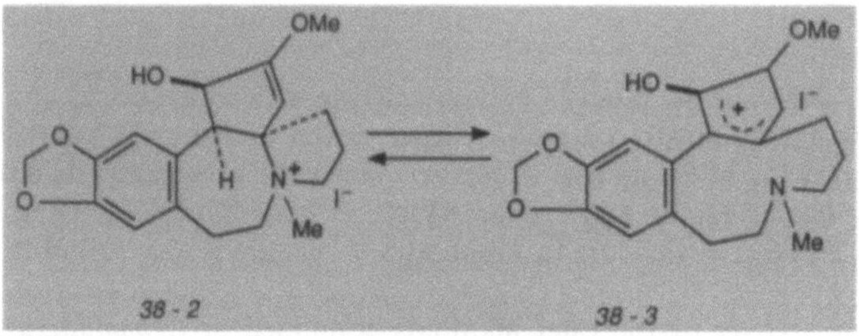

Fig. 38.1. Racemization of cephalotoxine methiodide

X-ray analysis of the optically active cephalotaxine *p*-bromobenzoate indicated the absolute configuration of cephalotaxine as 3*S*,4*S*,5*R* [10]. The X-ray study on cephalotaxine itself was then described [11].

The antitumor alkaloids harringtonine (*38-4*), homoharringtonine (*38-5*), isoharringtonine (*38-6*) [12] and deoxyharringtonine (*38-7*) [13], from *C. harringtonia* (Forbes) K. Koch var. *harringtonia* cv. Fastigiata, are esters of cephalotaxine with differently substituted succinic acid residues. All of these ester alkaloids yield cephalotaxine when transesterified by sodium methoxide in methanol along with the respective dimethyl esters of the substituted succinic acids. The stereochemistry of the acid moities in deoxyharringtonine, homoharringtonine [14], and isoharringtonine [15, 16] was also established.

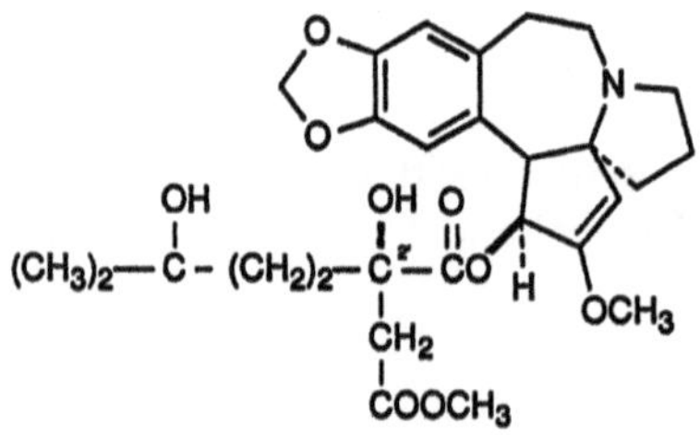

Harringtonine 3*S*, 4*S*, 5*R*, 2′*R* (*38-4*)

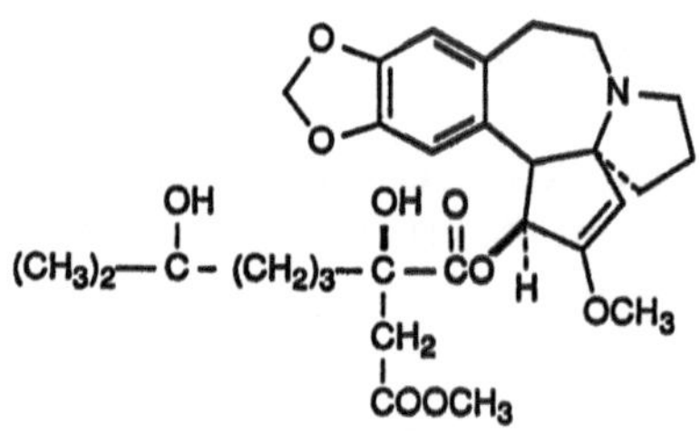

Homoharringtonine 3*S*, 4*S*, 5*R*, 2′*R* (*38-5*)

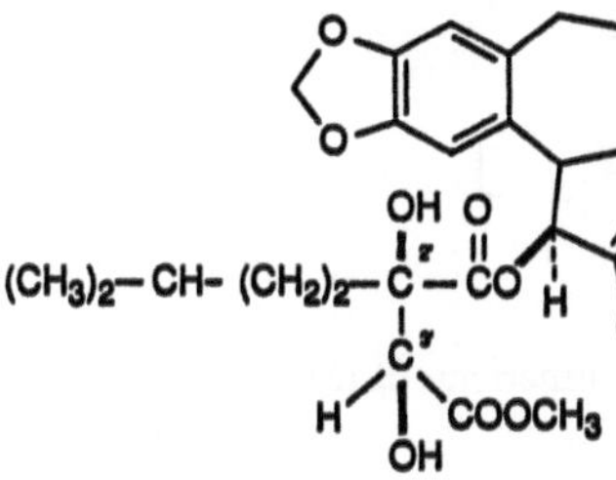

Isoharringtonine 3*S*, 4*S*, 5*R*, 2′*R*, 3′*S* (*38-6*)

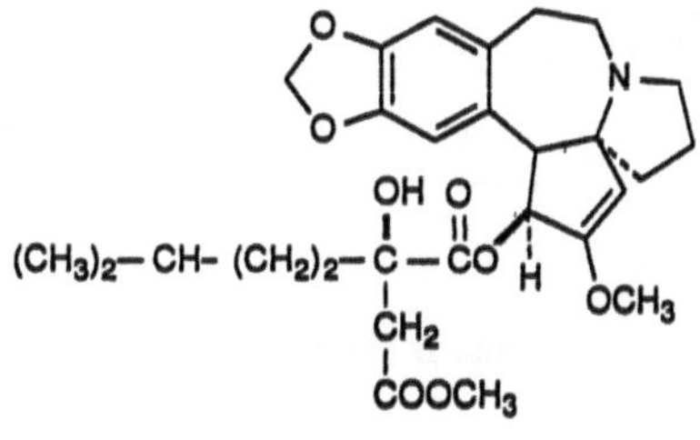

Deoxyharringtonine 3*S*, 4*S*, 5*R*, 2′*R* (*38-7*)

From *C. harringtonia* var. *harringtonia*, other alkaloids related to cephalotaxine, cephalotaxinone (*38-8*) [17], and demethylcephalotaxinone (*38-9*) [18] were isolated and structurally determined.

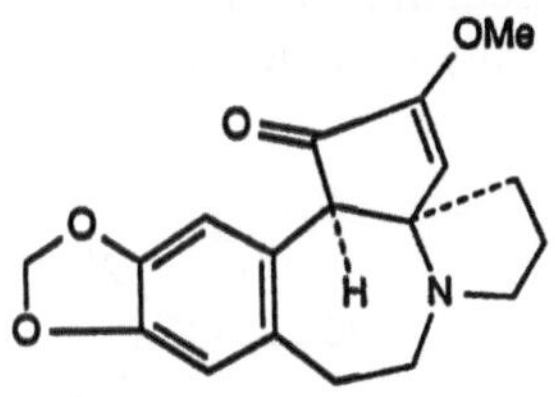

Cephalotaxinone (*38-8*)

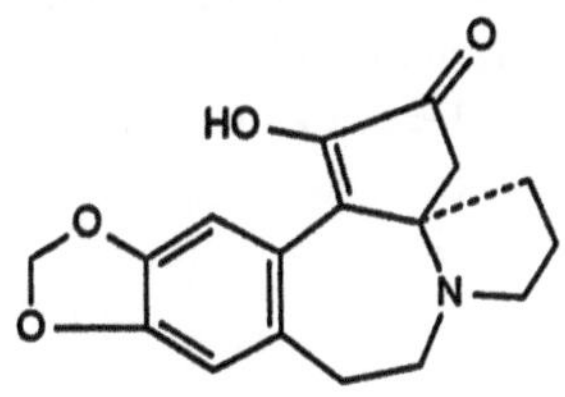

Demethylcephalotaxinone (*38-9*)

Besides the cephalotaxine related alkaloids, some minor alkaloids structurally derived from C-homoerythrinan (*38-10*), epischelhammericine (*38-11*), schelhammericine B (*38-12*), epischelhammericine B (*38-13*), methylschelhammericine B (*38-14*), and epimethylschelhammericine B (*38-15*) were isolated from the stem and root of *C. harringtonia* var. *harringtonia* with yields of 0.4%–1.1% of the total crude alkaloid [17]. Alkaloids possessing a C-homoerythrinan ring system are known in *Schelhammera pedunculata* (Liliaceae) [19–21]; however, epischelhammericine B and epimethylschelhammericine B were first isolated [21].

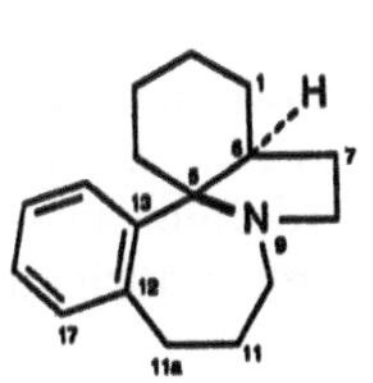

C-Homoerythrinan (*38-10*)

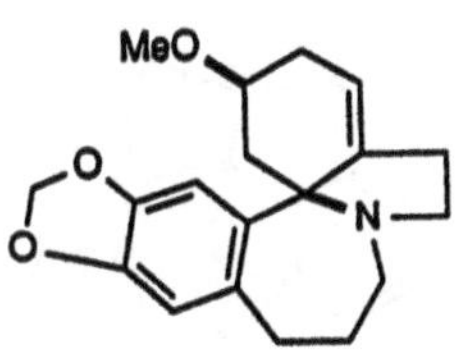

Epischelhammericine (*38-11*)

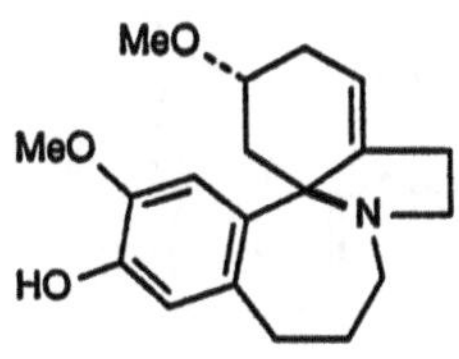

Schelhammericine B (*38-12*)

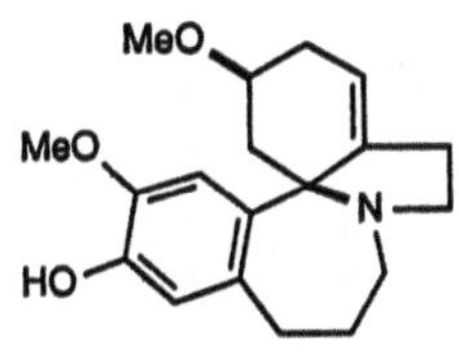

Epischelhammericine B (*38-13*)

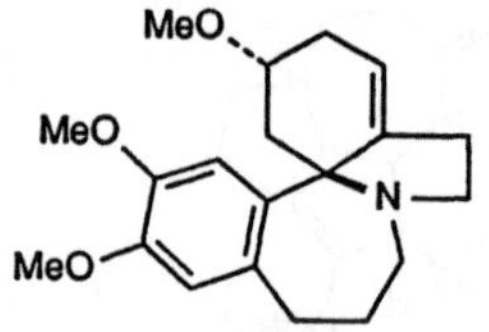

Methylschelhammericine B (*38-14*)

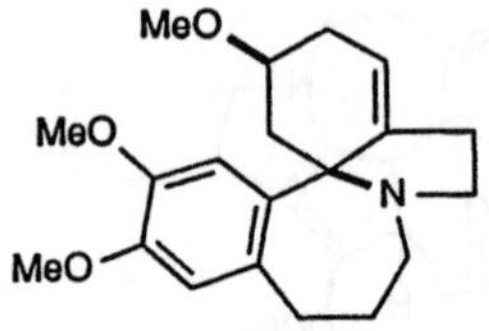

Epimethylschelhammericine B (*38-15*)

Thus, the alkaloids isolated from *Cephalotaxus* can be divided into two groups according to their structural features. One includes those compounds derived from cephalotaxine and the other, those from C-homoerythrinan.

From *C. harringtonia* var. *drupacea*, cephalotaxine [7, 22–24] and the related alkaloids 11-hydroxycephalotaxine (*38-16*), drupacine (*38-17*) [22, 23], demethylcephalotaxine (*38-18*), homoharringtonine and isoharringtonine [24], and the C-homoerythrinan-derived alkaloids epischelhammericine, epischelhammericine B, and epimethylschelhammericine B [24] were isolated and structurally investigated. Drupacine possesses a 2,11-epoxy linkage.

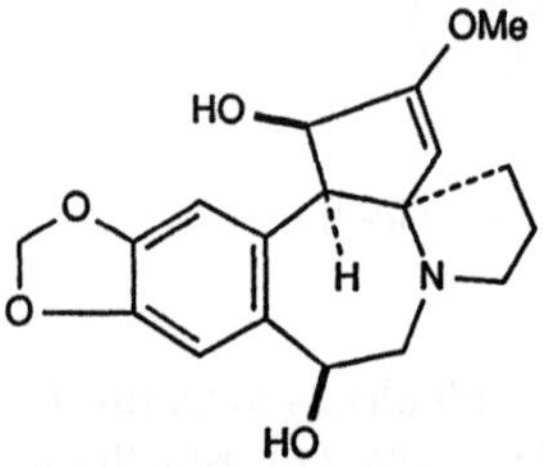

11-Hydroxycephalotaxine (*38-16*)

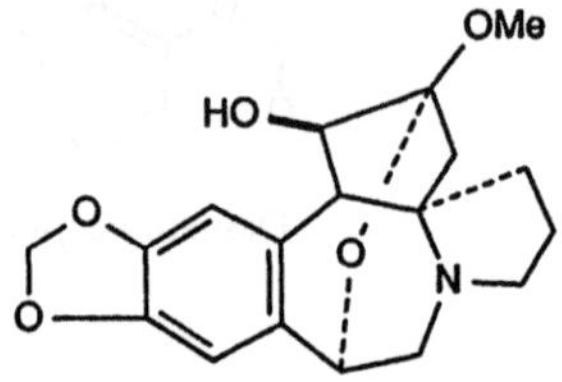

Drupacine (*38-17*)

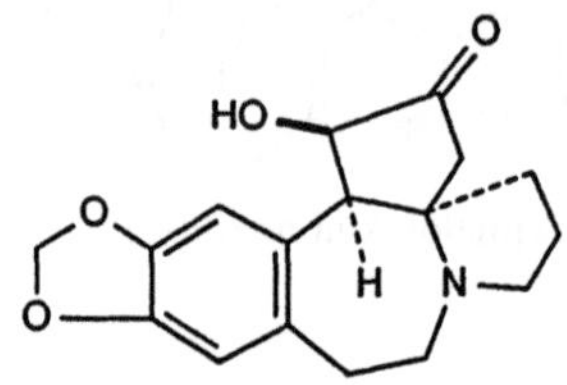

Demethylcephalotaxine (*38-18*)

Among the *Cephalotaxus* species native to China, *C. hainanensis* is the most thoroughly investigated. The first study on the alkaloid constituents of *C. hainanensis* led to isolation of 11 alkaloids, 9 of which were identified as cephalotaxine, epicephalotaxine (*38-19*), demethylcephalotaxinone, drupacine, harringtonine, isoharringtonine, homoharringtonine, deoxyharringtonine, and a homoerythrinan alkaloid, epischelhammericine [25]. Furthermore, cephalotaxinone, acetylcephalotaxine (*38-20*), cephalotaxinamide (*38-21*), demethylneodrupacine (*38-22*), deoxyharringtonic acid (*38-23*), isoharringtonic acid (*38-24*) [26], a new antitumor alkaloid of cephalotaxine type, hainanensine (*38–25*) [27, 28], and a C-homoerythrinan type alkaloid, cycloxyschelhammericine (*38-26*) [29], were isolated.

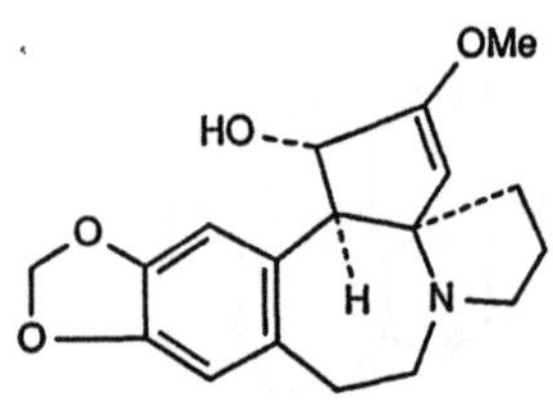

Epicephalotaxine (*38-19*)

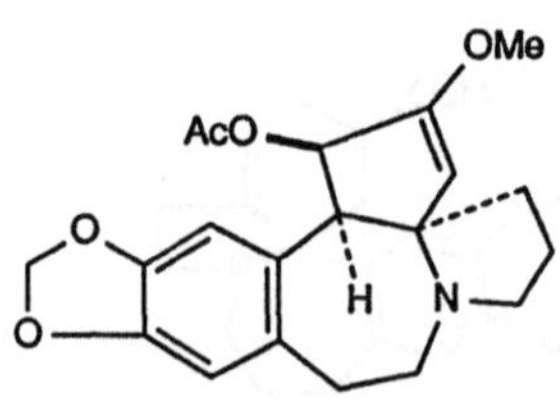

Acetylcephalotaxine (*38-20*)

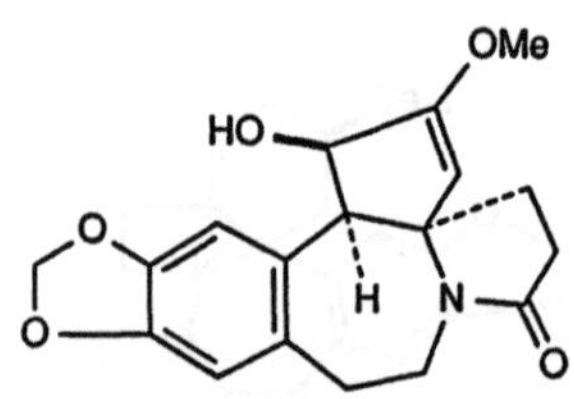

Cephalotaxinamide (*38-21*)

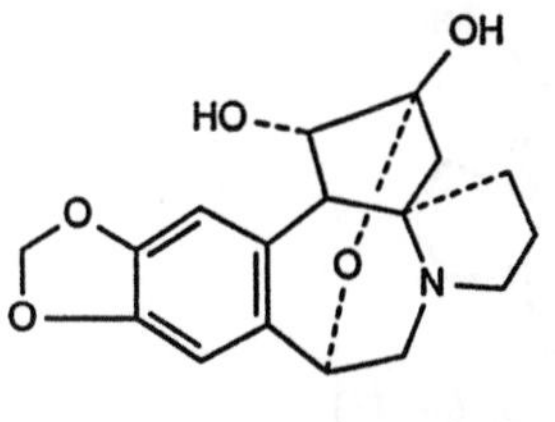

Demethylneodrupacine (*38-22*)

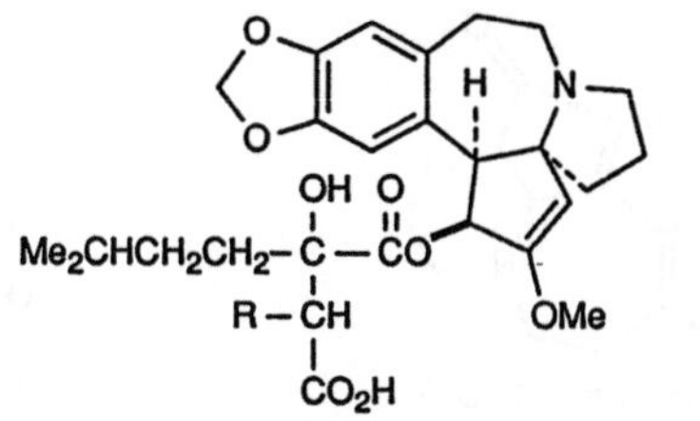

Deoxyharringtonic acid (*38-23*): R = H
Isoharringtonic acid (*38-24*): R = OH

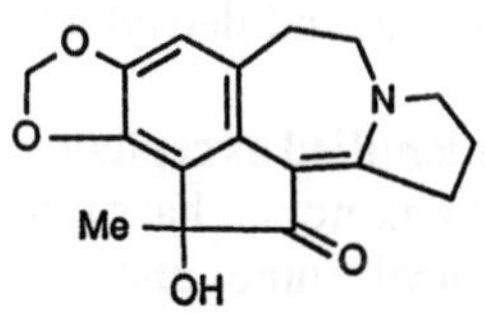

Hainanensine (*38-25*)

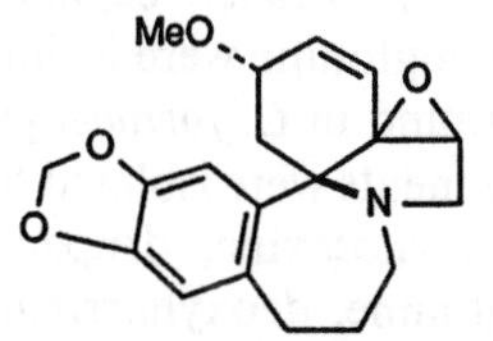

Cycloxyschelhammericine (*38-26*)

In addition to cephalotaxine [7], a number of other alkaloids were isolated from *C. fortunei* and chemically investigated. Paudler and McKay [30] isolated four minor alkaloids, epicephalotaxine, cephalotaxinone, demethylcephalotaxine and acetylcephalotaxine from *C. fortunei*. Ma et al. [31] reported the isolation of acetylcephalotaxine together with cephalotaxine, harringtonine, homoharringtonine, demethylcephalotaxinone, 11-hydroxycephalotaxine, and two homoerythrinan alkaloids, epiwilsonine (*38-28*) and cephalofortuneine (*38-30*) [32], from *C. fortunei*. Recently, isocephalotaxinone (*38-32*), epischelhammericine, epischelhammericine B, 2-epicephalofortuneine (*38-31*) [33], wilsonine (*38-27*), fortuneine (*38-29*) [34], and 4-hydroxycephalotaxine (*38-33*) [35] were isolated from *C. fortunei*. Fortuneine, cephalofortuneine, and 2-epicephalofortuneine are new natural products derived from C-homoerythrinan.

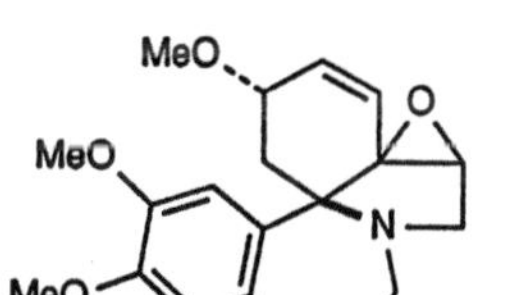

Wilsonine (*38-27*)

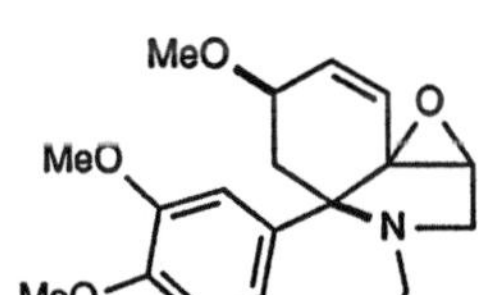

Epiwilsonine (*38-28*)

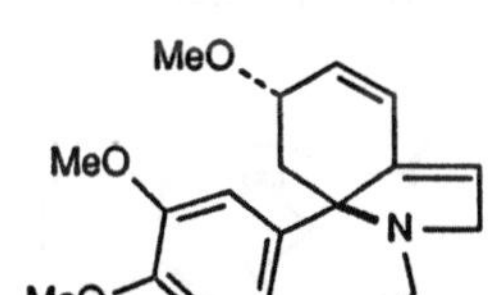

Fortuneine (*38-29*)

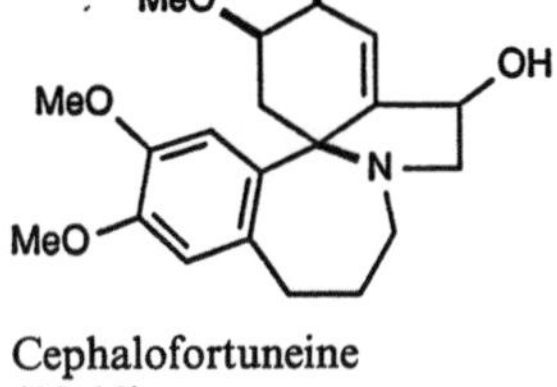

Cephalofortuneine (*38-30*)

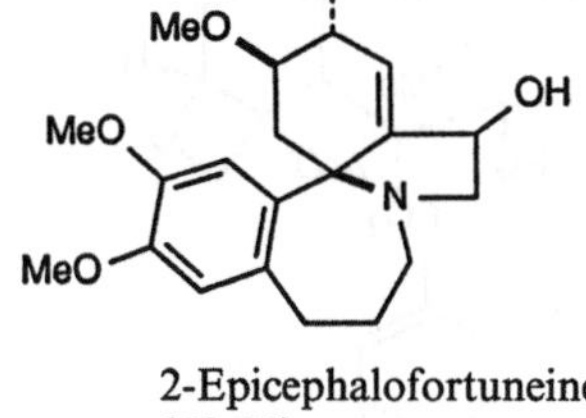

2-Epicephalofortuneine (*38-31*)

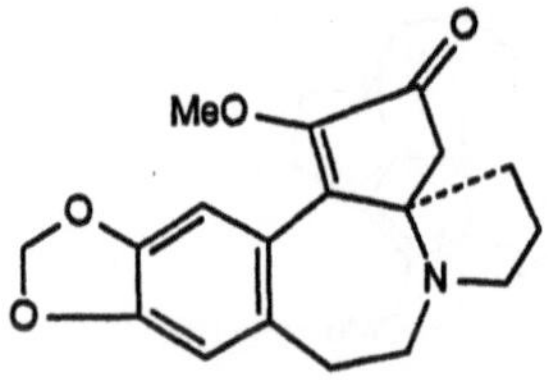

Isocephalotaxinone (*38-22*)

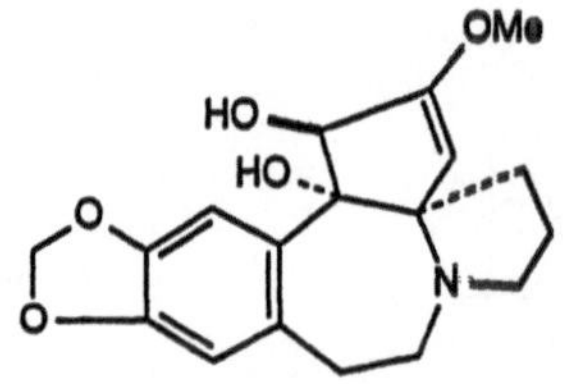

4-Hydroxycephalotaxine (*38-33*)

From the seed of *C. fortunei*, cephalotaxine, 11-hydroxycephalotaxine, drupacine, and homoharringtonine were isolated [35]. Isoharringtonine and deoxyharringtonine were not found in *C. fortunei* [36, 37].

Ten alkaloid components were isolated from *C. sinensis* and identified as cephalotaxine, 11-hydroxycephalotaxine, drupacine, demethylcephalotaxinone, harringtonine, homoharringtonine, deoxyharringtonine, wilsonine, epiwilsonine, and cycloxyschelhammericine [38, 39].

From *C. wilsoniana*, cephalotaxine, acetylcephalotaxine, wilsonine, epiwilsonine [40], drupacine, isoharringtonine, and epimethylschelhammericine B [41] were isolated and identified. Wilsonine is the major alkaloid component [40].

A study on the alkaloid component in *C. oliveri* showed that the plant contains cephalotaxine, harringtonine, and epischelhammericine B [42]. The content of harringtonine in the lower stem of *C. oliveri* was the highest in winter. Harringtonine content was highest in the root [43].

A new ester alkaloid with a diterpene skeleton of taxane (*38-34*) type esterified by an acid containing nitrogen atom was isolated from the stem and root of *C. mannii* and named cephalomannine (*38-35*) [44]. The structure of cephalomannine was determined by comparison of spectral data with taxol A (*38-36*) and baccatin III (*38-37*) and by hydrolysis of the side chain. Taxol A and baccatin III were also isolated from *C. mannii* together with cephalomannine. Taxane type diterpenes including taxol A and baccatin III have previously been reported only in the genus *Taxus* (Taxaceae).

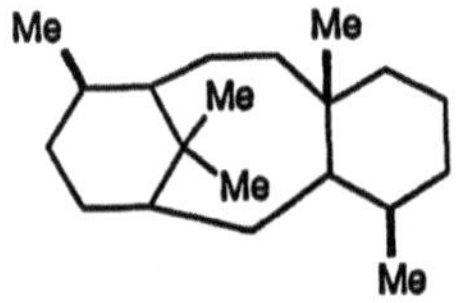

Taxane (*38-34*)

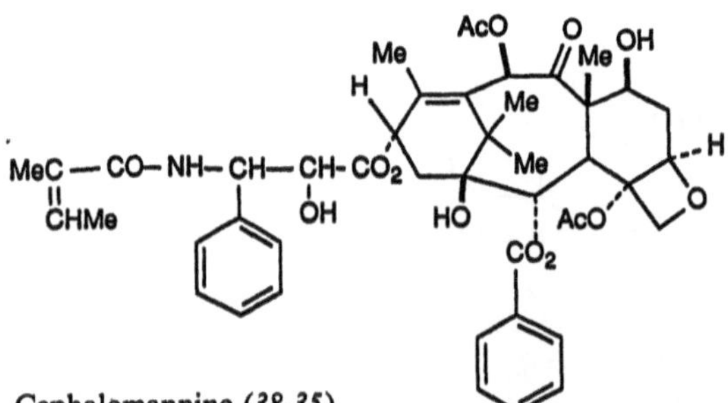

Cephalomannine (*38-35*)

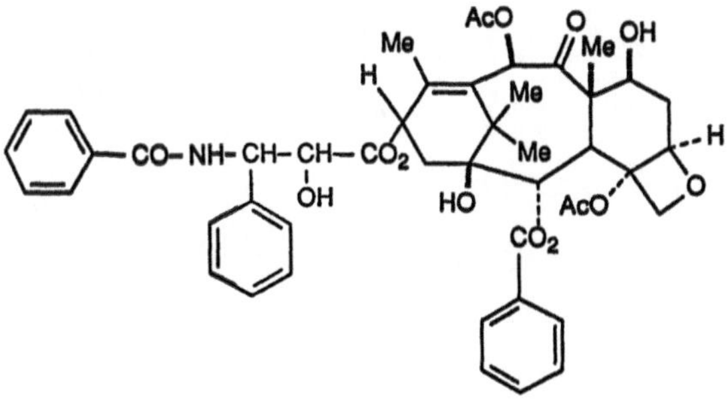

Taxol A (*38-36*)

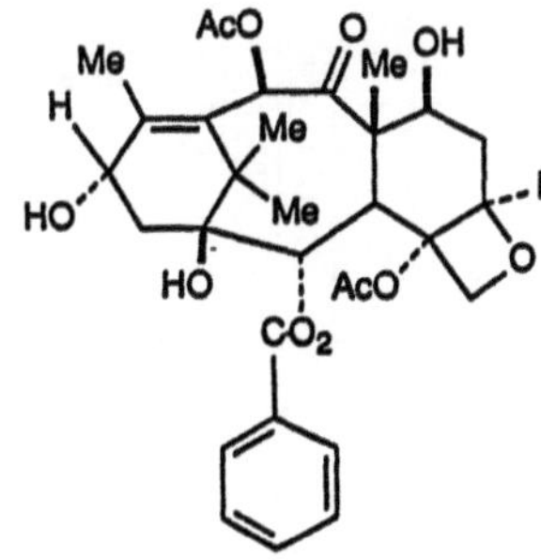

Baccatin III (*38-37*)

38.2.2 Synthesis of Cephalotaxine and Related Alkaloids

A number of studies on the total synthesis of cephalotaxine and its esterification to harringtonine and related alkaloids have been carried out. These studies were prompted by the low yield of cephalotaxine and its esters from plant materials, the interesting structural feature of cephalotaxine, and the chance to obtain new congeners of cephalotaxine and its esters. Huang and Xue [45] have discussed the synthetic approaches to *Cephalotaxus* alkaloids.

38.2.2.1 Total Synthesis of Cephalotaxine

The approach most often investigated for total synthesis of cephalotaxine is via a pyrrolo[2,1-*b*]-benzazepine three ring intermediate (*38-38*) converted into cephalotaxine. Dolby et al. [46], Weinstein and Craig [47], and Tse and Snieckus [48] established the bridge connecting the benzene ring with the pyrrolidine ring at position N-1 by using different starting materials. They completed the three ring system by formation of a C–C-bridge at position C-2 of the pyrrolidine ring. Auerbach and Weinreb [49, 50] established the C–C-bridge of benzene with pyrrolidine at position C-2 (Fig. 38-2).

Cyclization of the methyl dicarbonyl derivative (*38-39*) of the three ring intermediate resulted in formation of demethylcephalotaxinone (*38-9*). After conversion of demethylcephalotaxinone to cephalotaxinone (*38-8*) and subsequent reduction, racemic cephalotaxine was obtained [50] (Fig. 38-3).

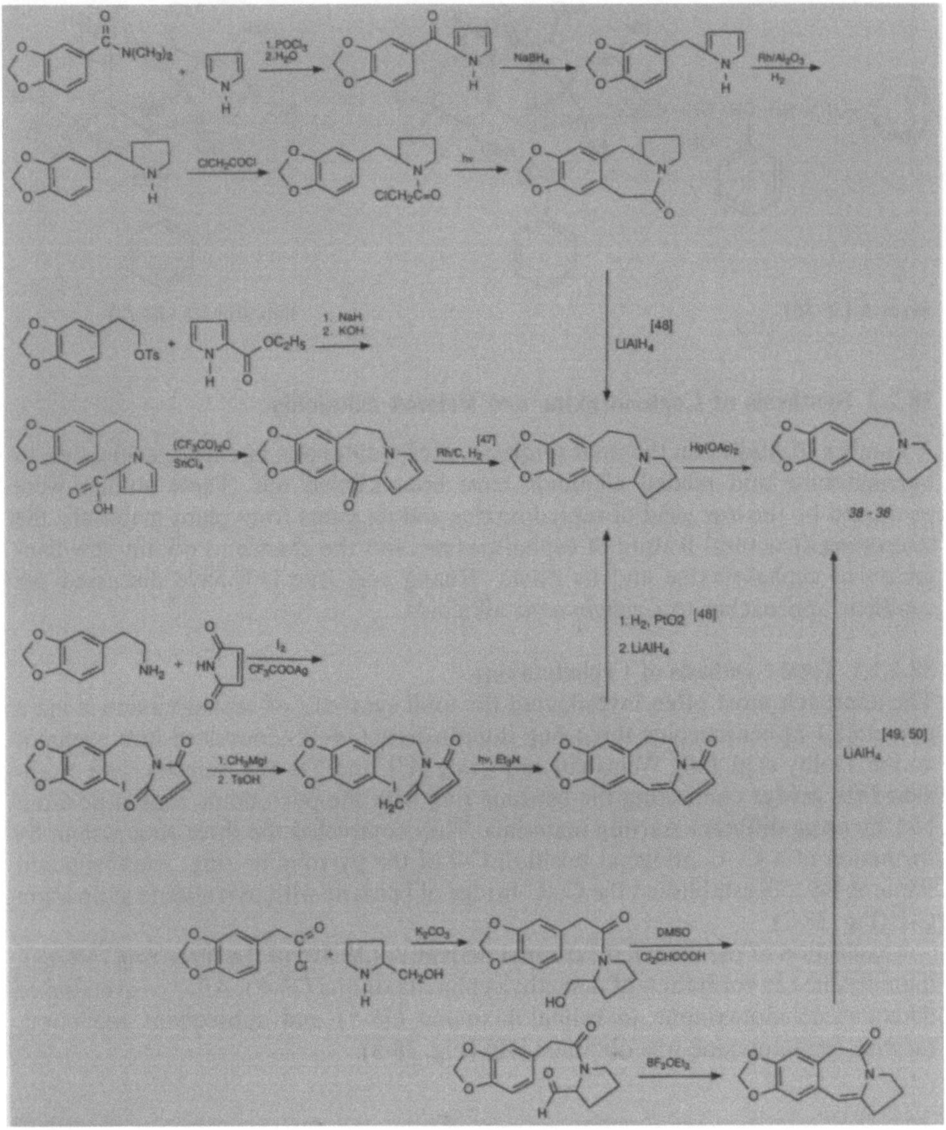

Fig. 38.2. Synthesis of pyrrolo[2,1-*b*]benzazepine three ring intermediate

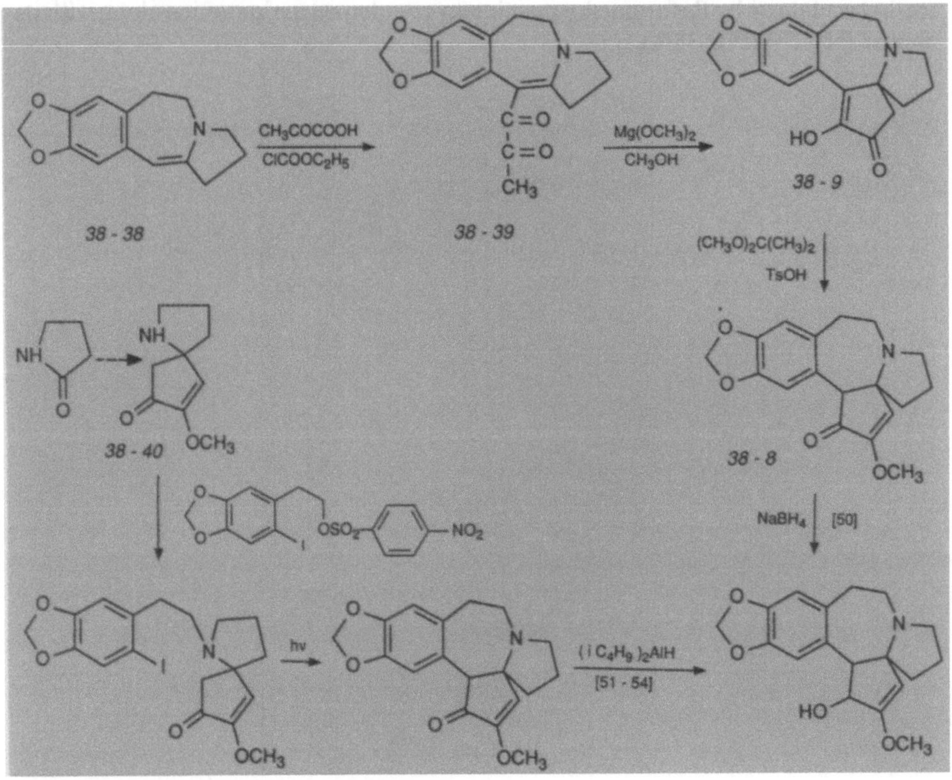

Fig. 38.3. Synthesis of racemic cephalotaxine

Another synthetic route to obtain cephalotaxine was reported by Semmelhack et al. [51–54]. They used the spiro CD ring moiety (*38-40*) of cephalotaxine as the key intermediate prepared from pyrrolidine as the starting material and then connected it with the aromatic moiety (Fig. 38-3).

38.2.2.2 Partial Synthesis of Ester Alkaloids Harringtonine, Isoharringtonine, Homoharringtonine, and Deoxyharringtonine

The amounts of the antitumor ester alkaloids harringtonine, isoharringtonine, homoharringtonine, and deoxyharringtonine in *Cephalotaxus* plants are much less than that of cephalotaxine. Much effort has been made, therefore, to achieve partial synthesis of these compounds. Attempts to acylate cephalotaxine directly with preformed acid gave no satisfactory results due to steric hindrance [13]. α-Keto acids were often used as the starting materials. Nonstereospecific formation of the chiral center C-2′ in the acid moiety led to a mixture of C-2′-epimers.

Deoxyharringtonine was the first ester alkaloid synthesized from cephalotaxine, due to its simple structure. Three groups reported independently on the synthesis of deoxyharringtonine, using 2-oxo-5-methyl-hexanoic acid chloride (*38-41*) as the starting material. The residual moiety was then introduced into the ester (*38-42*) formed from the acid and cephalotaxine either by treating the ester with lithium methyl acetate or by Reformatsky reaction to give deoxyharringtonine along with its epimer [55–58] (Fig. 38-4).

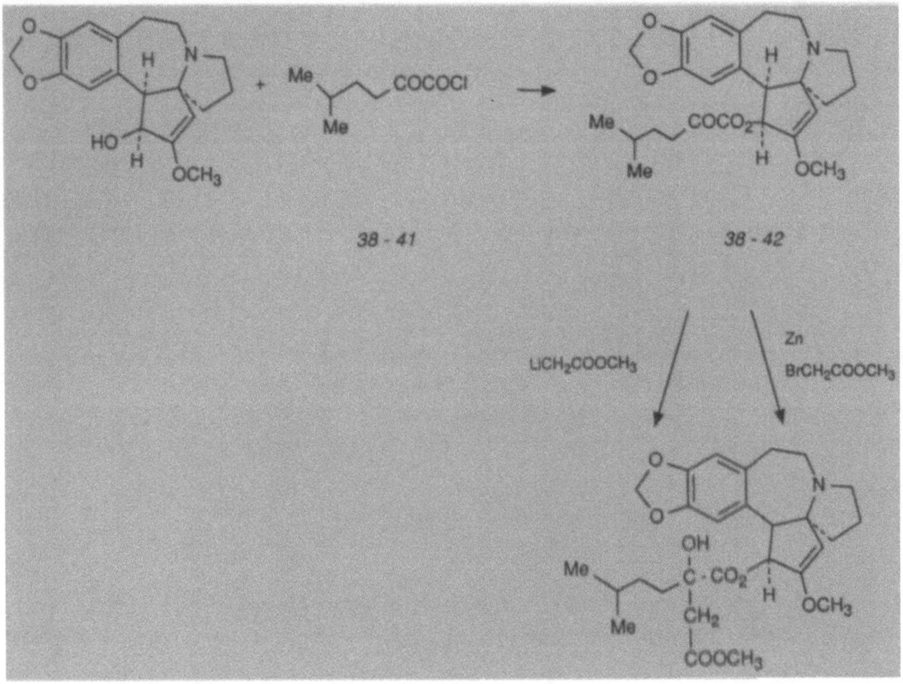

Fig. 38.4. Partial synthesis of deoxyharringtonine

The synthesis of harringtonine differs from that of deoxyharringtonine in the protection of the second hydroxyl group. The first successful synthesis of harringtonine was reported by Huang et al. [59, 60], using 4,4-dimethylbutyrolactone (*38-43*) as the starting material and esterifying cephalotaxine either with 5,5-dimethyl-dihydrofurane-2-carboxylic acid chloride (*38-44*) or with the corresponding ketal (*38-45*) in the presence of dicyclohexylcarbodiimide (DCC). The introduction of the side chain was carried out by condensation of the ester (*38-46*) with bromoacetic acid methyl ester to yield harringtonine, along with its epimer (Fig. 38-5).

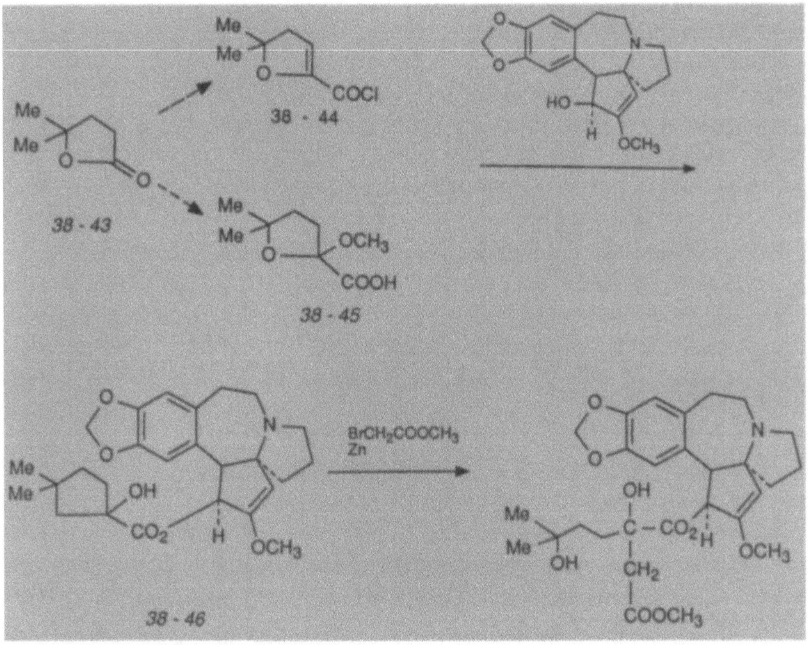

Fig. 38.5. Synthesis of harringtonine

Kelly et al. [61] reported a different method for the synthesis of harringtonine, using oxacycloheptane-carboxylic acid as a key intermediate. A special advantage of this synthesis was that natural harringtonine could be prepared when the optically active enantiomer of 4-benzyloxy-7,7-dimethyl-2-oxo-1-oxacycloheptane-4-carboxylic acid (*38-47*) with the *R* configuration was used (Fig. 38-6).

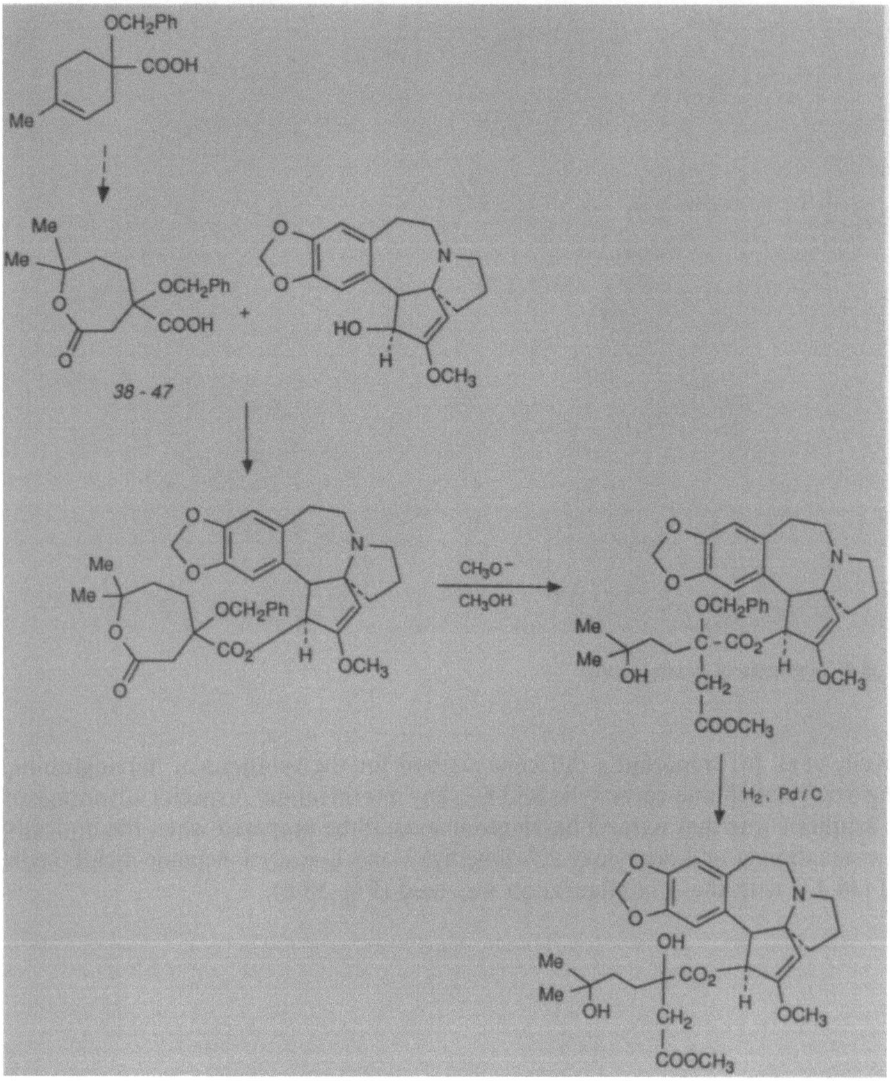

Fig. 38.6. Synthesis of harringtonine using oxacycloheptane-carboxylic acid as intermediate

Homoharringtonine could be synthesized either by esterification of cephalotaxine with 2-oxo-6-methyl-5-heptenoic acid, followed by condensation with bromoacetic acid methyl ester, hydroxylation, and reduction similar to the synthesis of harringtonine [62]. Alternatively, synthesis starting with oxoheptanoic acid and the geminal methyl group was introduced by treating with Grignard reagent [63, 64]. 6,6-Dimethyldihydropyrane-2-carboxylic acid, the homologue of 5,5-dimethyldihydrofurane-2-carboxylic acid, prepared from dimethylvalerolactone, was resistant to hydration due to steric hindrance in the six-membered ring and therefore was not suitable for the synthesis of homoharringtonine [62].

The synthesis of isoharringtonine was more complicated since two vicinal dihydroxyl groups and two chiral centers are present in the acid moiety. Isoharringtonine was prepared by using the same intermediate used for the synthesis of deoxyharringtonine. The residual part was introduced either by using methyl α-bromo-α-benzyloxyacetate in a Reformatsky reaction followed by hydrogenolysis [65] or by condensation with the lithium compound of the protected methyl α-hydroxyacetate and subsequent hydrolysis [66] to isoharringtonine and its three isomers (Fig. 38-7).

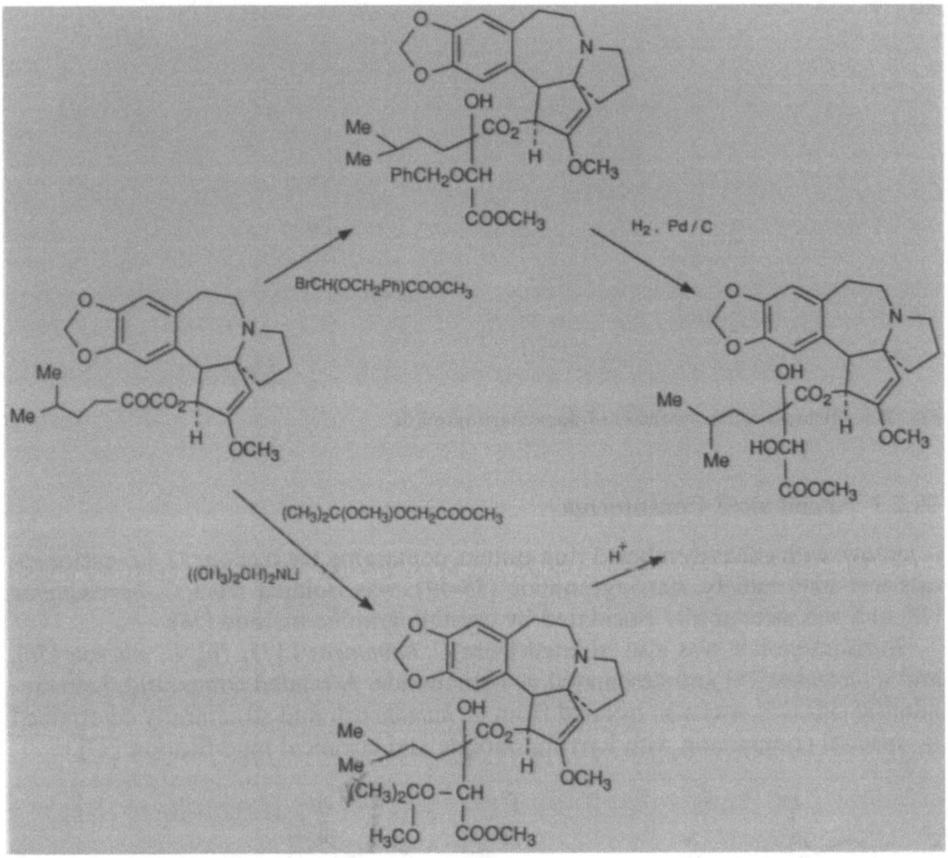

Fig. 38.7. Synthesis of isoharringtonine

Şeparation of the epimers of harringtonine [61, 67–69], deoxyharringtonine [55], and isoharringtonine [65, 70], obtained by partial synthesis from cephalotaxine, was carried out by chromatographic methods. Deoxyharringtonine and its epimer can also be separated by fractional recrystallization of their picrates [56, 57].

Cheng et al. [71] have recently reported the stereospecific synthesis of deoxyharringtonine and homoharringtonine using stereoselective condensation of cephalotaxyl α-ketocarboxylate with (+)-(*R*)-α-sulfinylester [72]. Deoxyharringtonine was prepared by condensation of the cephalotaxine ester of 2-oxo-5-methyl-hexanoic acid (*38-42*) with (+)-(*R*)-α-sulfinylester (*38-48*), followed by desulfurization, hydrolysis, and methylation with diazomethane (Fig. 38-8).

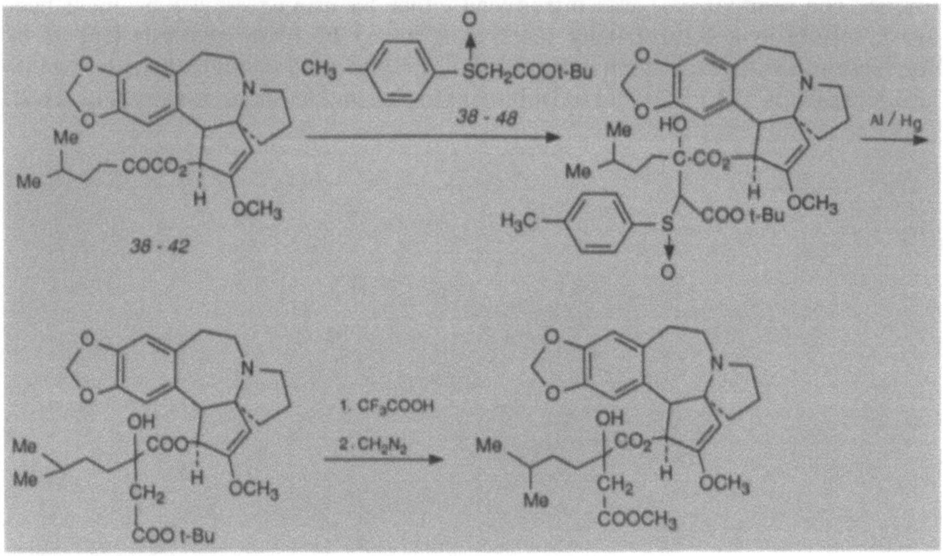

Fig. 38.8. Stereoselective synthesis of deoxyharringtonine

38.2.3 Nonalkaloid Constituents

A lactone with a five-membered ring system, containing the tropone (2,4,6-cycloheptatrien-1-one) moiety, harringtonolide (*38-49*), was isolated from *C. harringtonia* [73] and was structurally elucidated by crystallographic method [74].

Harringtonolide was also isolated from *C. hainanensis* [75, 76], *C. sinensis* [76], and *C. fortunei* [78] and designated as hainanolide. A related compound, hainanolidol (*38-50*) [75], was also isolated from *C. hainanensis* and structurally determined by spectral comparison with harringtonolide and chemical modification [77].

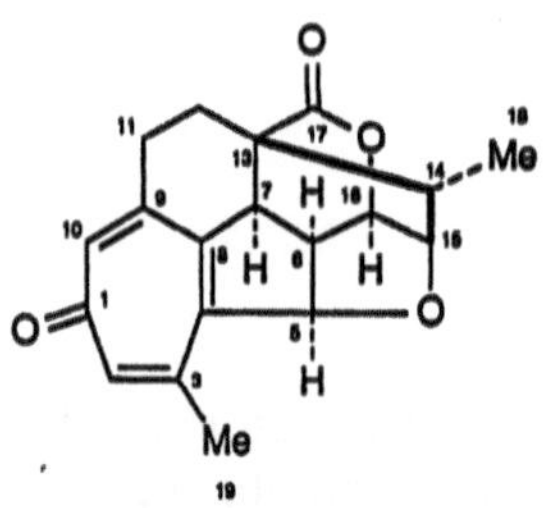

Harringtonolide (Hainanolide) (*38-49*)

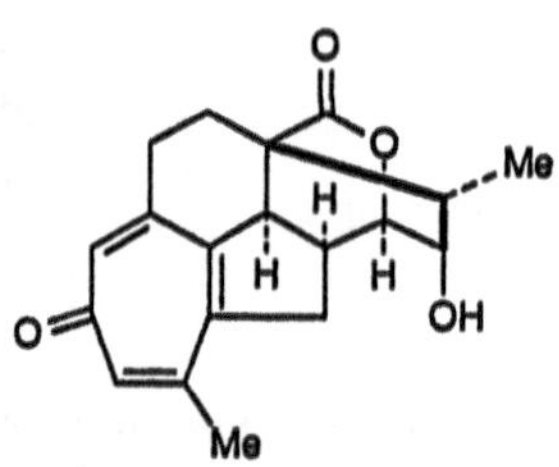

Hainanolidol (*38-50*)

From the leaves of *C. harringtonia* var. *drupacea*, a number of biflavones including amentoflavone (*38-51*), sciadopitysin (*38-52*), ginkgetin (*38-53*), and sequoiaflavone (*38-54*) were isolated and identified [79].

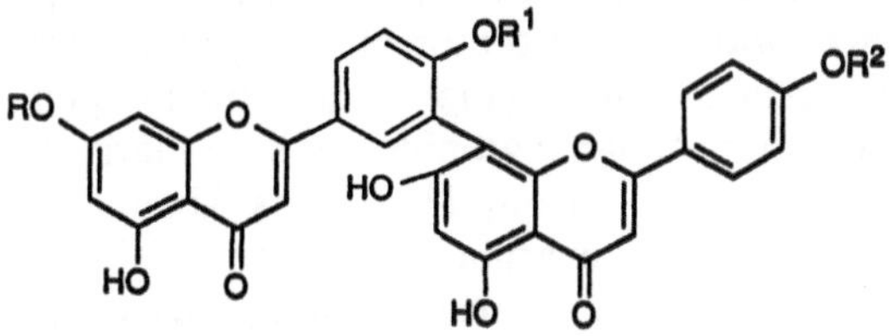

Amentoflavone (*38-51*): R, R^1, R^2=H
Sciadopitysin (*38-52*): R, R^1, R^2=CH_3
Ginkgetin (*38-53*): R, R^1=CH_3; R^2=H
Sequoiaflavone (*38-54*): R=CH_3; R^1, R^2=H

Besides sequoiaflavone, ginkgetin, and amentoflavone tetramethyl ether [80], a new C-methyl biflavone 6-C-methylsequoiaflavone, was isolated from *C. harringtonia* [81]. The structures were determined by spectral analysis and confirmed by synthesis [82]. From *C. fortunei*, the biflavones sciadopitysin and amentoflavone tetramethyl ether [83] and the flavone compound chrysoeriol (*38-55*) [84, 85] were isolated and identified.

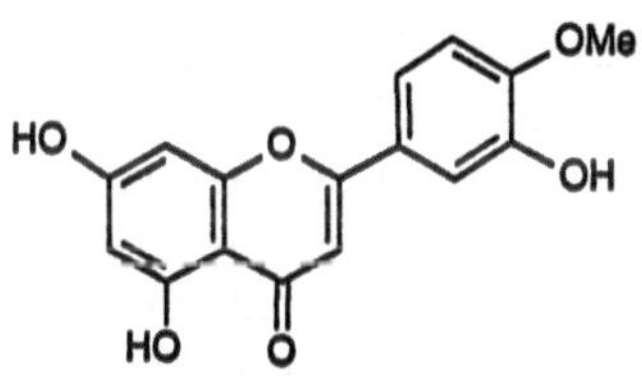

Chrysoeriol (*38-55*)

38.3 Pharmacology

The antitumor activity of the four ester alkaloids of cephalotaxine: harringtonine, homoharringtonine, isoharringtonine, and deoxyharringtonine was first tested in murine leukemias P388 and L1210. The life span of tumor-bearing animals [3] was increased significantly. Further tests were carried out on the following in vivo experimental tumor models: murine leukemia L615, rodent solid tumors, Lewis lung carcinoma, sarcoma 180, Walker 256 [45]; transplantable colon tumor 38 [86], and B16 melanoma [87]. Harringtonine and homoharringtonine showed significant activities against P388 leukemia, L1210 leukemia, and B16 melanoma by i.p. administration, comparable to vincristine and vinblastine. The therapeutic index (LD_{10}/ED_{50}) of harringtonine, homoharringtonine, vincristine, and vinblastine in B16 melanomas was 1.9, 2.3, 1.4 and <1.0, respectively. Harringtonine and homoharringtonine also demonstrated moderate activity against a P388 cell line resistant to vincristine [87].

In vitro, homoharringtonine was cytotoxic to HeLa, KB, and L cells growing in monolayer cell culture. The effect appeared to be cell cycle specific. In synchronized KB cells, protein synthesis was preferentially inhibited in the G_1 and G_2 phases, indicating that homoharringtonine might be an inhibitor of protein synthesis [88].

In a soft agar clonogenic assay system with human tumor biopsies, antitumor activity was observed for harringtonine against adenocarcinoma and sarcoma [89]. In a comparative study on the antitumor activity of harringtonine and homoharringtonine against solid tumors using fresh tumor cells from 23 patients, significant antitumor activity was noted for harringtonine in ovarian and endometrial carcinoma at 0.01 μg/ml. Homoharringtonine was active against sarcoma, breast cancer, and ovarian and endometrial carcinomas. In continuous exposure studies, homoharringtonine proved to be more potent than harringtonine [90]. Homoharringtonine was further tested against fresh surgical explants of human tumors in the 6 day in vivo subrenal capsule assay. Eighty tumors representing different histological types, including tumors from breast, lung, colon, ovarian, cervical, and neoplasms of undiagnosed origin, were screened simultaneously against homoharringtonine and five standard chemotherapeutics used clinically for treatment of the specific type of cancer. Homoharringtonine was active against 13 of 80 tumors, with a response rate of 16%, achieving an average of 45% reduction in tumor size. Of 80 tumors, 32 (40%) did not respond to any of the standard drugs, but two of them were sensitive to homoharringtonine [91].

Studies on the antitumor mechanisms of harringtonine and related compounds showed that harringtonine inhibited protein biosynthesis in HeLa cells, intact rabbit reticulocytes, and reticulocyte lysates. DNA synthesis was partially inhibited by a 200 n*M* concentration of harringtonine in HeLa cells; synthesis of RNA was unaffected. The inhibitory effects of harringtonine on protein and DNA synthesis were partially reversible. Harringtonine induced breakdown of polyribosomes to monosomes with concomitant release of completed globin chains. The principal effect of harringtonine appears to be inhibition of initiation of protein synthesis [92]. A similar activity in intact HeLa cells was also observed using homoharringtonine and isoharringtonine [93]. A study in a cell-free eukaryotic system showed that harringtonine, homoharringtonine, and isoharringtonine did not inhibit any of the actual steps of the initiation process but blocked poly(U)-directed polyphenylalanine synthesis and peptide bond formation in the fragment reaction assay. The enzymatic and nonenzymatic binding of Phe-tRNA to ribosomes were blocked, suggesting that the three alkaloids inhibit the elongation phase of translation in two ways: inhibiting both peptide bond formation and aminoacid-tRNA binding [94].

Evidence for hypothesis on the mechanism of action of harringtonine came from ultrastructural changes found in L1210 cells in mice treated with this agent [95]. Homoharringtonine stimulated enlargement of the hepatocyte nucleolus in mice and increased the number of nucleoli, indicating an inhibitory effect at the ribosomal level of protein synthesis [96].

The study of Wu et al. [97] showed that the synthesis of DNA and protein in L1210 cells was inhibited by harringtonine, while RNA synthesis was unaffected. The inhibitory effect of harringtonine on the incorporation of [^{14}C] leucine into protein was greater than that on the incorporation of [^{3}H] thymidine into DNA [98]. In an in vitro mouse Ehrlich ascites carcinoma cell system, harringtonine inhibited incorporation of [^{3}H]thymidine into DNA, but markedly increased the incorporation of [^{3}H]thymidine into the acid soluble fractions including thymidine diphosphate, thymidine monophosphate, and especially thymidine triphosphate. This suggested that harringtonine inhibits DNA formation by blocking polymerization of deoxynucleoside triphosphates into DNA, without influencing phosphorylation of thymidine. In addition, harringtonine did not significantly affect incorporation of [^{3}H]uridine and [^{3}H]hypoxanthine into nucleoside triphosphates and RNA [99]. The activity of DNA polymerase α, from Ehrlich ascites cells and L1210 leukemia cells, was inhibited by harringtonine in a concentration dependent manner. The results obtained from preincubation of harringtonine with the DNA template or the enzyme suggested that the inhibitory action of harringtonine appears to be on the enzyme rather than the DNA template. In addition, when the template or enzyme was present in excess, the inhibitory effect of harringtonine was decreased; thus harringtonine may be a competitive inhibitor of DNA polymerase α [100].

Harringtonine-related ester alkaloids were compared for their action on DNA polymerase α from Ehrlich ascites cells and DNA polymerase I from *Escherichia coli*. The inhibitory effect of homoharringtonine was stronger than that of harringtonine or isoharringtonine, but there was no significant difference between harringtonine and isoharringtonine at the same concentration. Epiharringtonine did not inhibit DNA polymerases from Ehrlich ascites cells or *E. coli* or enhance harringtonine-induced inhibition in cell-free systems [101]. In vivo, harringtonine at 20–30 μg/mouse had nonspecific effects on the L1210 leukemia cell cycle, but cells in S phase were sensitive to harringtonine [102].

A comparative study on harringtonine and related ester alkaloids in mice bearing leukemias P388 and L615 and normal mice showed that all four alkaloids tested, harringtonine, homoharringtonine, isoharringtonine, and deoxyharringtonine, caused initial inhibition of [^{3}H]thymidine incorporation into DNA in leukemic cells, spleen cells, intestinal mucosa, and bone marrow, with differences in the rates and extent of inhibition. The reduction in [^{3}H]thymidine incorporation usually started at 30 min after a single i.p. or i.v. dose; maximal inhibition was observed after 0.5–4.0 h and recovery after 24 h. The effects of deoxyharringtonine and isoharringtonine on DNA synthesis in leukemias L615 and P388 appeared to be more significant than those of harringtonine and homoharringtonine. The extent of inhibition of [^{3}H]thymidine incorporation in spleen cells and intestinal mucosa of normal mice caused by all of the alkaloids, except harringtonine, was similar to that observed in mice bearing L615 leukemias. The inhibitory effect of the alkaloids on the synthesis of DNA in bone marrow of normal mice, however, was greater than that observed in leukemic mice [103].

The inhibitory activity of harringtonine, homoharringtonine, isoharringtonine and deoxyharringtonine on protein synthesis was also studied in mouse leukemia L615 and P388 cells. When $^1/_5$ of the LD_{50} was injected into mice bearing leukemias L615 and P388, the incorporation rates of [^{14}C]DL-phenylalanine or [^{3}H]L-asparagine were reduced 30 min after single injection of the four alkaloids. Maximal inhibition was observed 30 min–4 h after administration with recovery after 4 h. Generally, the inhibitory effect of the four alkaloids was similar in terms of the time effect curve. Even if $^1/_{20}$ of the LD_{50} i.p. dose was used, the inhibitory effect on [^{3}H]DL-leucine incorporation by the four alkaloids in leukemia P388 cells was still significant; however, harringtonine did not markedly influence the incorporation of [^{14}C]DL-phenylalanine into protein in spleens of leukemic mice 24 h after the last of five consecutive doses. The inhibitory effects of the alkaloids on the incorporation of labeled amino acids into leukemic cell protein were significant, rapid, and reversible [104].

Harringtonine, isoharringtonine, homoharringtonine, and especially deoxyharringtonine increased the cAMP content in spleens of mice bearing leukemia L615: the content was very low in untreated controls. The increase closely paralleled the antitumor activity of the alkaloids. No changes in the splenic cAMP content of mice with P388 leukemia occurred after treatment with the ester alkaloids [105].

The acute toxicities of harringtonine, homoharringtonine, isoharringtonine, and deoxyharringtonine are listed in Table 38.1 [45]. Harringtonine and homoharringtonine are two to four times more toxic than isoharringtonine and deoxyharringtonine.

Table 38-1. Acute toxicity of harringtonine and related ester alkaloids in mice

Compound	LD_{50} (mg/kg)	
	i.p.	i.v.
Harringtonine	4.3±0.50	4.5±0.21
Homoharringtonine	3.3±0.44	2.4±0.25
Isoharringtonine	14.6±0.66	13.3±0.05
Deoxyharringtonine	16.0±2.40	8.8±0.52

The major target organs involved in toxicity in rabbits and dogs after 5 or 7 day treatments with harringtonine or homoharringtonine were gastrointestinal tract, heart, and hematopoietic organs. Most deaths were attributed to cardiac dysfunction. Hepatic and renal toxicities were seen only in individual cases at mild or moderate degree at the lethal dose. The cardiac and hematopoietic toxicities appeared to be moderately cumulative. All toxicities observed were dose dependent, predictable, and completely reversible upon discontinuation of treatment. No significant delayed toxicity was noted during an observation period of 6 weeks or more [106].

Intravenous administration of homoharringtonine in anesthetized dog at a dose of 4 mg/m^2 produced reduction in heart rate, cardiac output, and arterial blood pressure; however, myocardial contractility and total peripheral resistance were not altered. The bradycardia might account for decreases in both cardiac output and blood pressure. Homoharringtonine impaired tachycardia elicited during stimulation of cardiac sympathetic nerves. The positive chronotropic effects of either noradrenaline or isoproterenol were not altered by homoharringtonine. In animals in which both vagi and the right cardioaccelerator nerve were sectioned, the hypotensive and bradycardiac effects were smaller than in neurally intact animals. Thus, a dose of homoharringtonine comparable to that used in clinical trials produces hypotension and bradycardia which may result from inhibition of sympathetic nerve function [107]. An electron microscopic study indicated that homoharringtonine, at a dose of 1.5 mg/kg, caused a series of ultrastructural alterations in mouse cardiac muscle cells [96].

Semisynthetic harringtonine is a mixture of harringtonine and its epimer in a ratio of about 1:1.1. Epiharringtonine is biologically inactive compared to harringtonine. The LD_{50} of the mixture is about twice that of the natural isomer; the semisynthetic mixture demonstrated antitumor activity comparable to natural harringtonine at a dosage twice that of the latter. This was verified by the inhibitory activity shown by combined administration of pure harringtonine and its pure epimer in a ratio and doses corresponding to that of the synthetic mixture [45]. Nonetheless, the inhibitory action of harringtonine on DNA and protein synthesis was enhanced by the addition of epiharringtonine to tumor cell cultures, while synthesis of RNA was unaffected. The enhanced inhibition of protein synthesis was more pronounced than that of DNA synthesis, and it increased with time and reached a maximal value in 45 min. The reversibility of the inhibitory action of harringtonine was not changed by epiharringtonine [108].

A single i.p. injection of harringtonine into bone marrow donor mice at a dose of 1.5 mg/kg markedly decreased the number of colony-forming unit granulocyte-macrophage cells (CFU-GM) and colony-forming unit spleen cells (CFU-S) grown in recipient mice, while Li^+ treatment increased CFU-S and CFU-GM. Pretreatment of the donor or recipient mice with Li^+ inhibited or corrected, respectively, the myelosuppressive effect of harringtonine on hematopoietic progenitor cells [109].

Harringtonine showed genotoxic potential in inducing dose dependent chromosomal aberrations in vitro and nuclear fragmentation of bone marrow hemocytoblasts in mice [110].

Following i.v. injection of [^{3}H]harringtonine into normal rats, blood radioactivity levels decreased rapidly with $t_{1/2\alpha}$ about 3.5 min and $t_{1/2\beta}$ about 50 min [111]. The highest level of radioactivity was present in the kidneys [111, 112] 15 min after i.v. injection of harringtonine. Appreciable radioactivity was present in the liver, bone marrow, lung, heart, gastrointestinal tract, spleen and muscle; radioactivity in the testis, whole blood, and brain was low. The drug concentration in various tissues decreased rapidly 2 h after injection, but that of the bone marrow decreased more slowly, thus forming the highest concentration among the tissues tested. In all tissues only very low levels of activity were found 24 h after i.v. injection of harringtonine. The total urinary excretion of radioactivity was about 30%, whereas about 17% was present in feces. Approximately 15% of the adminstered harringtonine was excreted unchanged within a 24 h period. In tumor-bearing mice, a similar distribution pattern was observed. The absorption of harringtonine in normal mice following oral administration was rapid but incomplete [111].

In the case of homoharringtonine, the blood radioactivity level in normal rats after i.v. injection of [^{3}H]homoharringtonine decreased rapidly with two half-lives $t_{1/2\alpha} = 2.1$ min and $t_{1/2\beta} = 53.7$ min. The highest level of radioactivity 15 min after i.v. injection was present in the bone marrow, while appreciable radioactivity was present in the kidney, liver, lung, spleen, heart, and gastrointestinal tract. The radioactivity in muscle, whole blood, and brain was low. The radioactivity level in various tissues decreased rapidly 2 h after injection, except that of the gastrointestinal tract, which decreased more slowly. Of the administered radioactivity, 42% was excreted in the urine after 24 h and 6% was present in feces. It was also shown that only about 10% of the administered radioactivity was excreted as unchanged homoharringtonine within 24 h [113]. A similar distribution pattern of homoharringtonine was observed in tumor-bearing mice [113].

The first reports on clinical evaluations of harringtonine appeared in 1975 and 1977 [45, 114]. Results of 31 patients with different types of leukemia treated with a daily dose of 0.2–0.3 mg/kg of harringtonine for 5–7 days with a 1–2 week interval by slow i.v. infusion showed a complete remission in 7 patients and a partial remission in 18 patients. Among 26 patients with acute granulocytic leukemia, 7 complete remissions and 14 partial remissions occurred. The results of other clinical studies on harringtonine against nonlymphocytic leukemia, acute myelocytic leukemia, acute monocytic leukemia, and erythroleukemia are listed in Table 38-2 [45].

Table 38-2. Results of clinical trials of harringtonine against leukemias (from [45])

Leukemia	Number of patients	Remission	
		Complete	Partial
Nonlymphocytic leukemia	165	33	87
Acute myelocytic leukemia	102	21	57
Acute monocytic leukemia	33	5	15
Erythroleukemia	10	4	3

In a clinical trial of homoharringtonine on 94 patients with nonlymphocytic leukemia similar results were obtained with a complete remission of 22% and a partial remission of 42% [45]. Acute myelocytic leukemia and erythroleukemia were also sensitive to homoharringtonine.

Hypotension, disturbance, and myelosuppression were the main side effects. Of 102 patients using harringtonine, about 44% showed tachycardia during drug administration, 6 had ST segment changes, and 5 had arrhythmia [45].

Comparable results were obtained using semisynthetic harringtonine in the treatment of nonlymphocytic leukemia [45]. The dose of semisynthetic harringtonine has to be at least doubled since about half of the semisynthetic product is in an inactive form [115].

The complete remission rate in patients with acute nonlymphocytic leukemia was much higher, reaching 60%–80%, when harringtonine was used in combination with standard doses of vincristine, cytosine arabinoside, and prednisone [116].

A great number of unnatural cephalotaxine esters were synthesized [58, 117–120]. Among nearly 50 compounds tested, only 7 demonstrated some activity against P388 lymphocytic leukemia in mice, but none of them was more active than harringtonine (Table 38-3) [45].

Table 38.3. Antitumor activity of some unnatural cephalotaxine esters against P388 lymphocytic leukemia in mice

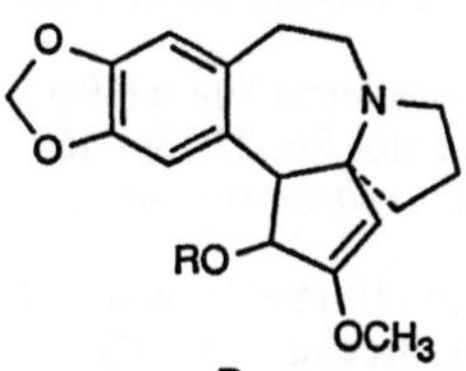

R	Dosage (mg/kg × 9)	T/C (%)
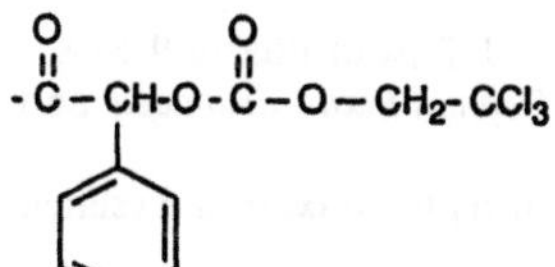	80	138
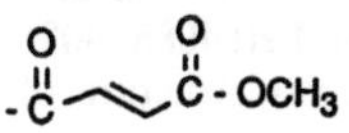	10 20	136 125
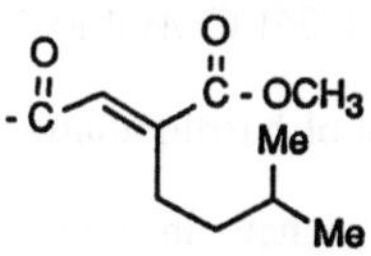	20	131
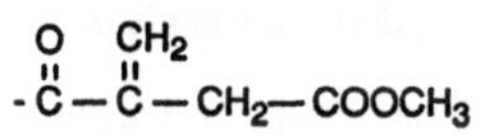	40 80 240	135 173 169
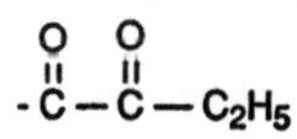	20 20	135 211
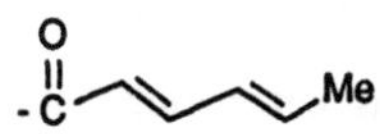	20 40	130 125
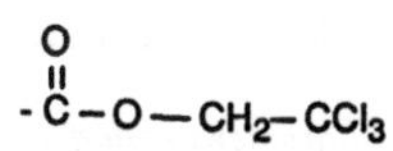	20 40 160	128–195 140–183 162
Harringtonine	0.5*	177

* mg/kg × 7

The lactone component harringtonine was also effective against sarcome 180 and L615 leukemia in an experimental study [75].

References

1. Zhu DY (1987) Recent advances on the active components in Chinese medicines. Abstr Chin Medicines 1:251–286
2. Alkaloids Group, Laboratory of Plant Chemistry, Institute of Botany, Academia Sinica (1980) Studies on alkaloids of *Cephalotaxus sinensis* (Rehd. et Will.) Li. Acta Bot Sin 22:156–160
3. Powell RG, Weisleder D, Smith CR Jr (1972) Antitumor alkaloids from *Cephalotaxus harringtonia*; structure and activity. J Pharm Sci 61:1227–1230
4. Well ME, Eddy CR, Williaman JJ, Correll DS, Schuber BG, Gentry HS (1954) Steroidal sapogenins. XII. Survey of plant for steroidal sapogenins and other constituents. J Am Pharm Assoc 43:503–505
5. Kariyone T, Takahashi M, Nitta A, Tsunehisa Y (1956) The alkaloid of coniferous plants. I. J Pharm Soc Jpn 76:611
6. Hsu HY (1957) Alkaloids in leaves of gymnosperms in Taiwan. I. J Taiwan Pharm 9:5–8
7. Paudler WW, Kerley GI, McKay J (1963) The alkaloids of *Cephalotaxus drupacea* and *Cephalotaxus fortunei*. J. Org Chem 28:2194–2197
8. Powell RG, Weisleder D, Smith CR Jr, Wolff IA (1969) Structure of cephalotaxine and related alkaloids. Tetrahedron Lett 4081–4084
9. Abraham DJ, Rosenstein RD, McGandy EL (1969) Single crystal X-ray structures of chemotherapeutic agents. II. The structure of cephalotaxine methiodide. Tetrahedron Lett 4085–4086
10. Arora SK, Bates RB, Grady RA, Powell RG (1974) Crystal and molecular structure of cephalotaxine *p*-bromobenzoate. J Org Chem 39:1269–1271
11. Arora SK, Bates RB, Grady RA, Germain G, Declercq P, Powell RG (1976) Crystal and molecular structure of cephalotaxine. J Org Chem 41:551–554
12. Powell RG, Weisleder D, Smith CR Jr, Rohwedder WK (1970) Structures of harringtonine, isoharringtonine and homoharringtonine. Tetrahedron Lett 815–818
13. Mikolajczak KL, Powell RG, Smith CR Jr (1972) Deoxyharringtonine, a new antitumor alkaloid from *Cephalotaxus*: structure and synthetic studies. Tetrahedron 28:1995–2001
14. Brandange S, Josephson S, Vallen S (1974) Absolute configurations of 2-alkylmalic acids. Acta Chem Scand [B]28:153–156
15. Brandange S, Josephson S, Vallen S, Powell RG (1974) Absolute configuration of 2,3-dihydroxy-2-isopentyl-butanedioic acid, a component of the alkaloid isoharringtonine. Acta Chem Scand [B]28:1237–1248
16. Ipaktchi T, Weinreb SM (1973) Relative configuration of the diacid sidechain of isoharringtonine. Tetrahedron Lett 3895–3898
17. Powell RG (1972) Structures of homoerythrina alkaloids from *Cephalotaxus harringtonia*. Phytochemistry 11:1467–1472
18. Powell RG, Mikolajczak KL (1973) Desmethylcephalotaxinone and its correlation with cephalotaxine. Phytochemistry 12:2987–2991
19. Fitzgerald JS, Johns SR, Lamberton JA, Sioumis AA (1969) Alkaloids of *Schelhammera pedunculata* (Liliaceae). I. Isolation of the alkaloids schelhammerine and schelhammeridine. Aust J Chem 22:2187–2201
20. Johns SR, Lamberton JA, Sioumis AA, Suares H (1969) Alkaloids of *Schelhammera pedunculata* (Liliaceae). II. Reactions of schelhammeridine. Aust J Chem 22:2203–2218
21. Johns SR, Lamberton JA, Sioumis AA (1969) Alkaloids of *Schelhammera pedunculata* (Liliaceae). III. The structures of schelhammericine and alkaloids A, B, E, G, H, J and K. Aust J Chem 22:2219–2231
22. Powell RG, Weisleder D, Smith CR Jr, Wolff IA (1969) Structure of cephalotaxine and related alkaloids. Tetrahedron Lett 4081–4084
23. Powell RG, Madrigal RV, Smith CR Jr, Mikolajczak KL (1974) Alkaloids of *Cephalotaxus harringtonia* var. *drupacea*. 11-Hydroxycephalotaxine and drupacine. J Org Chem 39:676–680
24. Asada S (1973) Alkaloids isolated from the *Cephalotaxus drupacea*. Yakugaku Zasshi 93:916–924
25. Chinese Academy of Medical Sciences, Chinese People's Liberation Army 187th Hospital (1976) Studies on the antitumor constituents of *Cephalotaxus hainanensis* Li. Acta Chim Sin 34:283–293
26. Xue Z, Xu LZ, Chen DH, Huang L (1981) Studies on minor alkaloids of *Cephalotaxus hainanensis* Li. Acta Pharm Sin 16:752–756

27. Sun NJ, Liang XT (1981) Studies on the structure of a new antitumor constituent, hainanensine. Acta Pharm Sin 16:24–26
28. Sun NJ, Liang HT (1980) Structure of a new antitumor constituent, hainanensine. Chin Pharm Bull 15:39–40
29. Xue Z, Sun NJ, Chen DH, Xu LZ, Chen WM, Ren LJ, Liang XT, Huang L (1982) The studies of chemical constituents of *Cephalotaxus hainanensis* Li. In: Wang Y (ed) Chem Nat Prod, Proc Sino-Am Symp 1980. Science Press, Beijing, 195–199
30. Paudler WW, McKay J (1973) The structures of some of the minor alkaloids of *Cephalotaxus fortunei*. J Org Chem 38:2110–2112
31. Ma GE, Lin LT, Chao TY, Fan HC (1977) Studies on the alkaloids of *Cephalotaxus*. I. Isolation and characterization of anticancer alkaloids and a new alkaloid (+)-acetylcephalotaxine from *Cephalotaxus fortunei*. Acta Chim Sin 35:201–208
32. Ma GE, Lin LT, Chao TY, Fan HC (1978) Studies on alkaloids of *Cephalotaxus*. II. Four minor alkaloids from *Cephalotaxus fortunei*. Hook. f. and the structure of cephalofortuneine. Acta Chim Sin 36:129–136
33. Lin W, Chen RT, Xue Z (1985) Studies on the minor alkaloids of *Cephalotaxus fortunei* Hook. f. Acta Pharm Sin 20:283–287
34. Ma GE, Lu CE, ElSohly HN, ElSohly MA, Turner CE (1983) Studies on the alkaloids of *Cephalotaxus*. IV. Fortuneine, a homoerythrina alkaloid from *C. fortunei*. Phytochemistry 22:251–253
35. Ma GE, Sun GQ, ElSohly MA, Turner CE (1982) Studies on the alkaloids of *Cephalotaxus*. III. 4-Hydroxycephalotaxine, a new alkaloid from *Cephalotaxus fortunei*. J Nat Prod 45:585–589
36. Ma GE, Lu CE, Fan GJ (1982) Isolation and identification of harringtonine and other alkaloids from seeds of *Cephalotaxus fortunei*. Chin Pharm Bull 17:205–206
37. Ma YQ, Guo MR, Zhu TP, Ma GE, Lu CE, Huang H, Yang YH (1984) Determination of harringtonine and homoharringtonine, antileukemic alkaloids in *Cephalotaxus fortunei* Hook. f. and *C. sinensis* (Rehd. et Wils) Li. Acta Bot Sin 26:405–410
38. Institute of Botany, Academia Sinica (1980) Studies on alkaloids of *Cephalotaxus sinensis* (Rehd. et Wils.) Li. Acta Bot Sin 22:156–160
39. Ren LJ, Xue Z (1981) Studies on the antitumor constituents of *Cephalotaxus sinensis*. Chin Trad Herb Drugs 12:241–244
40. Powell RG, Mikolajczak KL, Weisleder D, Smith CR Jr (1972) Alkaloids of *Cephalotaxus wilsoniana*. Phytochemistry 11:3317–3320
41. Furukawa H, Itoigawa M, Haruna M, Jinno Y, Ito K, Lu ST (1976) Alkaloids of *Cephalotaxus wilsoniana* Hay. Yakugaku Zasshi 96:1373–1377
42. Zhang FX, Wang ZH, Pan WD, Li YJ, Mai LT, Sun JQ, Ma GE (1978) A study on the antitumor plant *Cephalotaxus oliveri* Mast. Acta Bot Sin 20:129–134
43. Pan WD, Mai LT, Li YJ (1983) Studies on the variation of harringtonine content in *Cephalotaxus oliveri* with seasons, parts and ages of the tree. Bull Chin Mat Med 8:15–16
44. Powell RG, Miller RW, Smith CR Jr (1979) Cephalomannine: a new antitumor alkaloid from *Cephalotaxus mannii*. J Chem Soc Chem Commun 102–104
45. Huang L, Xue Z (1984) *Cephalotaxus* alkaloids. Alkaloids 23:157–226
46. Dolby LJ, Nelson SJ, Senkovich D (1972) Study of the synthesis of cephalotaxine. I. J Org Chem 37:3691–3695
47. Weinstein B, Craig AR (1976) A synthetic approach to the cephalotaxine skeleton. J Org Chem 41:875–878
48. Tse I, Snieckus V (1976) Photochemical preparation of dihydropyrrolo[2,1-*b*] [3] benzazepines. A *Cephalotaxus* alkaloid synthon. J Chem Soc Chem Commun 505–506
49. Auerbach J, Weinreb SM (1972) The total synthesis of cephalotaxine. J Am Chem Soc 94:7172–7173
50. Weinreb SM, Auerbach J (1975) Total synthesis of the *Cephalotaxus* alkaloids. Cephalotaxine, cephalotaxinone and demethylcephalotaxinone. J Am Chem Soc 97:2503–2506
51. Semmelhack MF, Chong BP, Jones LD (1972) Total synthesis of *Cephalotaxus* alkaloids. J Am Chem Soc 94:8629–8630
52. Semmelhack MF, Stauffer RD, Rogerson TD (1973) Nucleophilic aromatic substitution via a new nickel-catalyzed process and via the $S_{RN}1$ reaction. Improved synthesis of cephalotaxinone. Tetrahedron Lett 4519–4522

53. Semmelhack MF, Chong BP, Stauffer RD, Rogerson TD, Chong A, Jones LD (1975) Total synthesis of the *Cephalotaxus* alkaloids. A problem in nucleophilic aromatic substitution. J Am Chem Soc 97:2507–2516
54. Semmelhack MF, Margar TM (1977) Cyclizations of enolates onto aromatic rings via the photo-$S_{RN}1$ reaction. Preparative and mechanistic aspects. J Org Chem 42:1481–1482
55. Mikolajczak KL, Smith CR Jr, Weisleder D, Kelly TR, McKenna JC, Christenson PA (1974) Synthesis of deoxyharringtonine. Tetrahedron Lett 283–286
56. Huang WK, Li YL, Pan SF (1976) Crystallization of (oxoacyl)cephalotaxine, a key intermediate for deoxyharringtonine synthesis. Kexue Tongbao 21:178
57. Huang WK, Li YL, Pan SF (1980) Synthesis of deoxyharringtonine and separation of its stereomers. Sci Sin [Engl] 23:835–846
58. Li SW, Dai JY (1975) A study on the deoxyharringtonine and its analogs. Acta Chim Sin 33:75–78
59. Institute of Pharmacology (1976) Partial synthesis of harringtonine. Kexue Tongbao 21:512
60. Huang L, Xi YG, Guo JY, Liu DK, Xu SP, Wu KM, Chen JC, Jiang YZ, Cao YS, Guo Z, Li L, Zhang M, Chu F (1979) Partial synthesis of harringtonine. Sci Sin [Engl] 22:1333–1345
61. Kelly TR, McNutt RW Jr, Montury M, Tosches NP, Mikolajczak KL, Smith CR Jr, Weisleder D (1979) Preparation of harringtonine from cephalotaxine. J Org Chem 44:63–67
62. Zhao ZZ, Xi YG, Zhao HF, Hou JY, Zhang JY, Wang ZH (1980) Partial synthesis of homoharringtonine from cephalotaxine. Acta Pharm Sin 15:46–49
63. Wang YK, Li YL, Pan HF, Li CP, Huang WK (1980) Synthesis of antitumor alkaloids. Kexue Tongbao 25:576
64. Hiranuma S, Hudlicky T (1982) Synthesis of homoharringtonine and its derivative by partial esterification of cephalotaxine. Tetrahedron Lett 23:3431–3434
65. Li YN, Wu KM, Huang L (1982) Synthesis of isoharringtonine and separation of the isomers. Acta Pharm Sin 17:866–867
66. Pan XF, Li YL, Li SB, Tian J, Zhang DN, Cui YX, Cheng PG, Wang YK, Huang WK (1982) Synthesis of isoharringtonine. Kexue Tongbao 27:1048
67. Chinese Academy of Medical Sciences, Institute of Materia Medica (1978) Separation and identification of harringtonine and epiharringtonine. Kexue Tongbao 23:696–699
68. Zhang GD, Zhou ZH, Guo YS, Guo JY, Liu DK, Chu FM (1981) Isolation and determination of the two epimers of partially synthesized harringtonine. Anal Chem 291–295
69. Huang L, Fang QN, Cheng JC, Jiang ZY (1983) Semisynthetic diastereoisomers – separation and identification of harringtonine and epiharringtonine by preparative high-performance liquid chromatography. Anal Chem 158–159
70. Li YN, Wu KM, Huang L (1984) Synthesis of isoharringtonine and separation of the isomers. Acta Pharm Sin 19:582–589
71. Cheng JC, Zhang JH, Zhang QB, Yang J, Huang L (1984) Stereospecific synthesis of deoxyharringtonine and homoharringtonine. Acta Pharm Sin 19:178–183
72. Mioskowski C, Solladie G (1980) Synthésis asymétriques de β-hydroxy acids par condensation d'anions énolates d'esters α-sulfinyle chiraux sur des composés carbonyles. Tetrahedron 36:227–236
73. Steffens GL, Buta JG, Gregory LE, Momdava NB, Meudt WJ, Worley JF (1979) New plant growth regulators isolated from higher plants. In: Geissbuehler H (ed) Advances in Pestic Sciences, Vol. 2. Pergamon, Oxford, pp 343–346
74. Flippen-Anderson JL, Duesler E (1979) Harringtonolide: a complex tropone. Acta Crystallogr [B]35:1253–1254
75. Sun NJ, Xue Z, Liang XT, Huang L (1979) Studies on the structure of a new antitumor agent – hainanolide. Acta Pharm Sin 14:39–44
76. Bao GH, He CH, Xu CF (1983) Crystal structure of hainanolide. Acta Acad Med Sin 5:89–93
77. Xue Z, Sun NJ, Liang XT (1982) Studies on the structure of hainanolidol. Acta Pharm Sin 17:236–237
78. Sun NJ, Zhao ZF, Chen RT, Lin W, Zhou YZ (1981) Isolation and identification of the antitumor component – hainanolide from *Cephalotaxus fortunei*. Acta Pharm Sin 16:233–234
79. Khan NU, Ilyas M, Rahman W, Okigawa M, Kawano N (1971) Occurrence of biflavones in *Cephalotaxus*. Phytochemistry 10:2541–2542
80. Ishratullah K, Rahman W, Okigawa M, Kawano N (1981) Biflavones from the leaves of *Cephalotaxus harringtonia* C. Koch (Cephalotaxacaeae). Indian J Chem [B] 20:935–936

81. Aqil M, Rahman W, Okigawa M, Kawano N (1976) Isolation of C-methylbiflavone from *Cephalotaxus* and use of lanthanide shift reagent in deciding 6- or 8-C-methylflavone. Chem Ind 567–568
82. Aqil M, Rahman W, Hasaka N, Okigawa M, Kawano N (1981) A C-methylbiflavone from *Cephalotaxus harringtonia* K. Koch. J Chem Soc Perkin Trans I 1389–1392
83. Ma ZW, He GF, Yin WF (1984) Biflavonoids in the leaves of *Cephalotaxus fortunei* Hook. f. var. *alpina* native to China. Acta Bot Sin 26:416–418
84. Yang WW, Xing YQ, Xu JY, Shi XF (1983) Isolation of chrysoeriol from *Cephalotaxus fortunei* Hook. f. J Lanzhou Univ [Natl Sci] 19:114
85. Yang WW, Xing YQ, Xu JY, Shi XF (1983) Isolation of chrysoeriol from *Cephalotaxus fortunei*. Chin Trad Herb Drugs 14:164
86. Corbett TH, Griswold DP Jr, Roberts BJ, Peckham JC, Schabel FM Jr (1977) Evaluation of single agents and combinations of chemotherapeutic agents in mouse colon carcinoma. Cancer 40:2660–2680
87. Takeda S, Yajima N, Kitazato K, Unemi N (1982) Antitumor activities of harringtonine and homoharringtonine, *Cephalotaxus* alkaloids which are active principles from plant by intraperitoneal and oral administration. J Pharmacobiodyn 5:841–847
88. Baaske DM, Heinstein P (1977) Cytotoxicity and cell cycle specificity of homoharringtonine. Antimicrob Agents Chemother 12:298–300
89. Jiang TL, Salmon SE, Liu RM (1983) Activity of camptothecin, harringtonine, cantharidin and cucumae in the human tumor stem cell assay. Eur J Cancer Clin Oncol 19:263–270
90. Jiang TL, Liu RH, Salmon SE (1983) Comparative in vitro antitumor activity of homoharringtonine and harringtonine against clonogenic human tumor cells. Invest New Drugs 1:21–25
91. Cobb WR, Bogden AE, Reich SD, Griffin TW, Kelton DE, LePage DJ (1983) Activity of two phase I drugs, homoharringtonine and tricyclic nucleotide, against surgical explants of human tumor in the 6-day subrenal capsule assay. Cancer Treat Rep 67:173–178
92. Huang MT (1975) Harringtonine, an inhibitor of initiation of protein biosynthesis. Mol Pharmacol 11:511–519
93. Tscherne JS, Pestka S (1975) Inhibition of protein synthesis in intact HeLa cells. Antimicrob Agents Chemother 8:479–487
94. Fresno M, Jimenez A, Vazquez D (1977) Inhibition of translation in eukaryotic systems by harringtonine. Eur J Biochem 72:323–330
95. Si JY, Son KX, Hu WW, Men SS, Han R, Pan ZK, Li ZR (1979) Ultrastructural study on the effect of harringtonine on mouse leukemia L-1210 cells. Acta Biochem Biophys Sin 11:333–338
96. Wang ZW, Hsu B (1980) Electron microscopic studies of the effect of homoharringtonine on hepatocyte nucleolus and myocardiac cells. Acta Biochim Biophys Sin 12:231–235
97. Wu GY, Fang FD, Han R, Sun ZR (1981) Studies on the mechanism of the inhibition of harringtonine on DNA synthesis. I. Effects of harringtonine on DNA template and metabolism. Acta Biochim Biophys Sin 13:509–515
98. Wu GY, Fang FD, Zuo J, Han R, Sun ZR (1982) Studies on mechanism of inhibition of harringtonine on DNA and on protein synthesis. Acta Acad Med Sin 4:78–81
99. Yang SR, Fang FD, Wu GY (1982) Effects of harringtonine on nucleotide metabolism in tumor cells. Acta Pharm Sin 17:721–727
100. Wu GY, Qiou CC (1982) Studies on the mechanism of inhibition of harringtonine on DNA biosynthesis. II. Effect of harringtonine on DNA polymerase. Acta Biochim Biophys Sin 14:553–558
101. Qiu CC, Wu GY (1984) Studies on the mechanism of inhibition of harringtonine on DNA synthesis. III. Kinetics of inhibition by harringtonine and comparison with its analogs. Acta Biochim Biophys Sin 16:645–649
102. Pan ZK, Han R, Wang YC (1980) The cytokinetic effects of harringtonine on leukemia L1210 cells. I. Autoradiographic studies. Acta Biochim Biophys Sin 12:13–20
103. Xu YT, Du CZ (1981) Effect of harringtonine and its related alkaloids on DNA synthesis in mice bearing leukemias P388, L615 and normal mice. Acta Pharmacol Sin 2:252–256
104. Xu YT, Du CZ, Zhang FR, Ji XJ (1981) Effect of harringtonine alkaloids on the protein biosynthesis in mouse leukemias L615 and P388 cells. Acta Pharm Sin 16:661–666
105. Li ZR, Sun ZR, Han R (1983) Effects of harringtonine and related alkaloids on the cAMP levels of L615 and P388 leukemic cells. Acta Pharm Sin 18:303–306

106. Zhang YJ, Yu H, Luo XY, Zheng YX, Li WJ, Liu XL, Yuan YY (1979) Toxicological studies of harringtonine and homoharringtonine. Acta Pharm Sin 14:135–140
107. Lokhandwala MF, Sabouni MH, Jandhyala BS (1985) Cardiovascular actions of an experimental antitumor agent, harringtonine, in anesthetized dogs. Drug Dev Res 5:157–163
108. Wu GY, Fang FD, Zuo J, Han R, Sun ZR (1983) Epiharringtonine-induced enhancement of harringtonine inhibition on protein and DNA formation in tumor cells. Acta Acad Med Sin 5:157–160
109. Jiang DZ, Xu YH, Li MF, Yuan WD, Wang QR (1985) Lithium protects mouse bone marrow cells against the myelosuppressive effect of harringtonine. Bull Hunan Med Coll 10:307–311
110. Huang NJ, Chen SM (1986) Induction of induced chromosome aberrations and micronuclei by harringtonine. Acta Pharmacol Sin 7:173–175
111. Ji XJ, Liu JS, Liu ZM (1979) Metabolism of harringtonine. Acta Pharm Sin 14:234–240
112. Chinese Academy of Medical Sciences (1979) Chemical, pharmacological and clinical studies on the antitumor active principle of *Cephalotaxus hainanensis* Li. Chin J Oncol 1:176–182
113. Ji XJ, Liu Y, Lin H, Liu ZM (1982) Metabolism of homoharringtonine in rats and mice. Acta Pharm Sin 17:881–888
114. Harringtonine Cooperative Group (1975) Preliminary clinical evaluation of cephalotaxine ester in the treatment of acute leukemia. Natl Med J China 55:712
115. Zhang ZY (1981) Clinical analysis of the therapeutic effect of semisynthetic harringtonine in treating 55 cases of nonlymphocytic leukemia. Chin J Intern Med 20:667
116. Chang CN (1985) Antileukemia Chinese herbs and the effective ingredients. In: Chang HM, Yeung HW, Tso WW, Koo A (eds) Advances in Chinese Medical Material Research. World Scientific, Singapore, pp 369–375
117. Mikolajczak KL, Smith CR Jr, Powell RG (1974) Partial synthesis of harringtonine analogs. J Pharm Sci 63:1280–1283
118. Mikolajczak KL, Powell RG, Smith CR Jr (1975) Preparation and antitumor activity of a rearranged ester of cephalotaxine. J Med Chem 18:63–66
119. Mikolajczak KL, Smith CR Jr, Weisleder D (1977) Synthesis of cephalotaxine esters and correlation of their structures with antitumor activity. J Med Chem 20:328–332
120. Li SW, Sun HL, Lu SJ, Zhang SY, Lu FL, Dai JY, Xu YB (1981) Synthesis of cephalotaxine esters and their antitumor activity. Acta Pharm Sin 16:821–827

Choerospondias axillaris (Roxb.) Burtt et Hill 39

39.1 Introduction

Guangzao, Fructus Choerospondiatis, is the dry ripe fruit of *Choerospondias axillaris* (Roxb.) Burtt et Hill (Anacardiaceae), which is collected in the fall when the fruits have become ripe. It is a herbal medicine especially used by the Mongolians as a sedative and cardiotonic. The Chinese Pharmacopoeia requires a qualitative determination of the flavone compound of this official crude drug detected with boric acid and citric acid and detected with $AlCl_3$ after extraction.

39.2 Chemical Constituents

Two flavanone compounds were isolated from the bark of *C. axillaris*, one identified as naringenin and the second, named choerospondin (*39-1*) was determined as the 4′-β-D-glucopyranoside of naringenin [1,2].

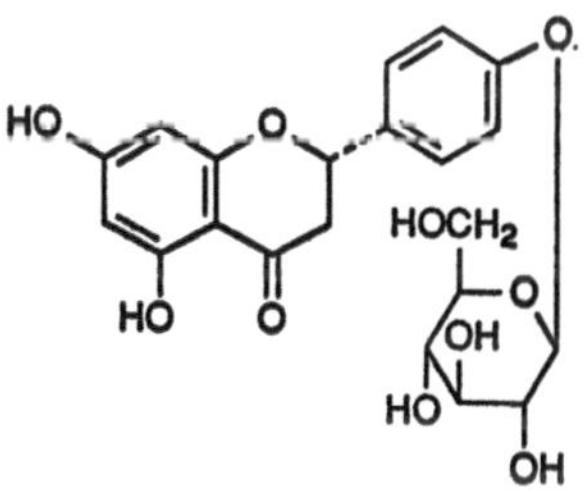

Choerospondin (*39-1*)

39.3 Pharmacology

The total flavone extract obtained from the fruit of *C. axillaris* at i.p. doses of 5.1–11.2 mg/kg markedly reduced the rate and amount of oxygen consumption in rats and markedly enhanced the tolerance of mice to hypoxia. Changes in the electrocardiogram induced by pituitrin treatment of rats were markedly reduced by administration of the total flavone extract. Additionally, the total flavone extract antagonized arrhythmia induced by acute myocardial ischemia [3].

In mice, the antiarrhythmic intravenous dose of a total flavone extract from *Choerospondias* fruits was 11.2 mg/kg and the LD_{50} 112 mg/kg [4].

References

1. Lu YZ (1982) Isolation and structure identification of naringenin and choerospondin from the bark of *Choerospondia axillaris*. Chin Pharm Bull 17:120
2. Lu YZ, Wang YL, Lou ZX, Zu JY, Liang HQ, Zhou ZL (1983) Isolation and structural determination of naringenin and choerospondin from the bark of *Choerospondias axillaris*. Acta Pharm Sin 18:199–202
3. Li ZX, Tian FJ, Wu XY, Zhang YP, Tian L, Shi S (1984) Effects of total flavones of axillary choerospondias (*Choerospondias axillaris*) on the tolerance of animals to hypoxia and the protection of animals from acute myocardial ischemia. Chin Trad Herb Drugs 15:265–267
4. Li ZX, Tian FJ, Wu XY, Zhang YP, Tian L, Shi S (1984) Antiarrhythmic effect of total flavones from fruits of *Choerospondias axillaris*. Acta Pharmacol Sin 5:251–254

Chrysanthemum indicum L. and *C. morifolium* Ramat. 40

40.1 Introduction

Two *Chrysanthemum* species are officially listed in the Chinese Pharmacopoeia:

- Yejuhua, Flos Chrysanthemi indici, is the dry composite inflorescence of *Chrysanthemum indicum* L. (Asteraceae), which is collected in the fall and winter. It is used in traditional Chinese medicine as an antiphlogistic for the treatment of carbuncle, furuncle, conjunctivitis, headache, and vertigo.
- Juhua, Flos Chrysanthemi, is the dry composite inflorescence of *C. morifolium* Ramat, which is collected from September to November. It is used as an antiphlogistic and against cold, headache, vertigo, and conjunctivitis.

40.2 Chemical Constituents

40.2.1 Chemical Constituents of *Chrysanthemum indicum*

Three sesquiterpene guaianolides, lactones with a guaiane (*40-1*) skeleton were isolated from the aerial parts of *C. indicum* and structurally elucidated as angeloylcumambrin B (*40-2*), arteglasin A (*40-3*), and angeloylajadin (*40-4*) [1].

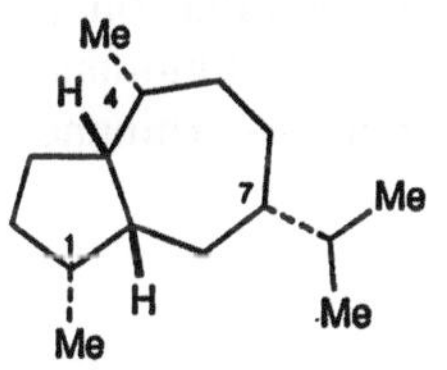

Guaiane (*40-1*)

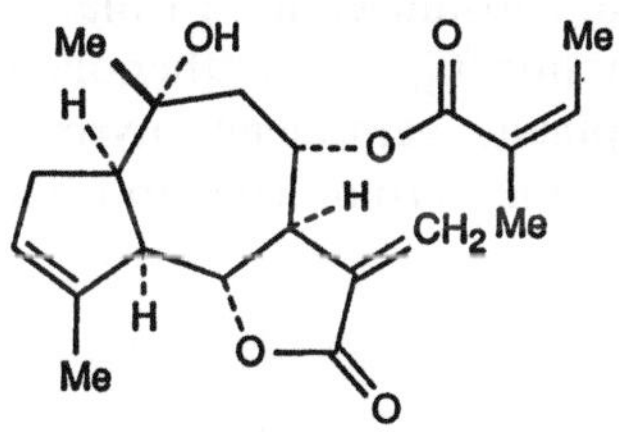

Angeloylcumambrin B (*40-2*)

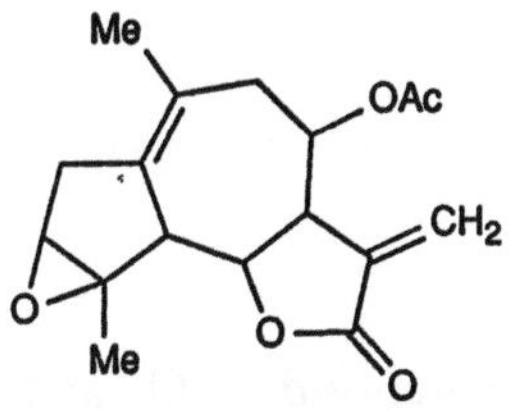

Arteglasin A (*40-3*)

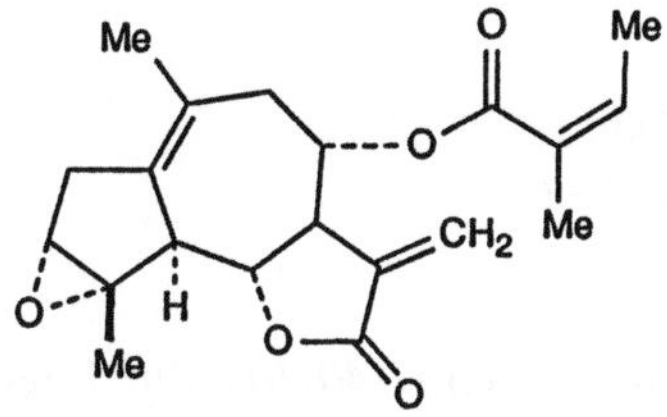

Angeloylajadin (*40-4*)

Another sesquiterpene lactone named yejuhua lactone isolated from *C. indicum* was reported in 1963 [2], the structure of which was recently determined as being identical to that of handelin (*40-5*), a diguaianolide [3].

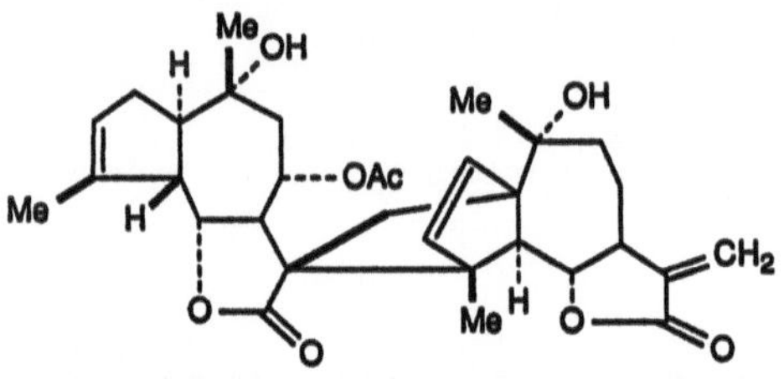

Handelin (*40-5*)
(Yejuhua lactone)

The isolation of further new sesquiterpenes chrysetunone (*40-6*), tunefulin (*40-7*) [4], and cumambrin A (*40-8*) [5] from *C. indicum* var. *Tuneful* and *C. indicum* was recently reported.

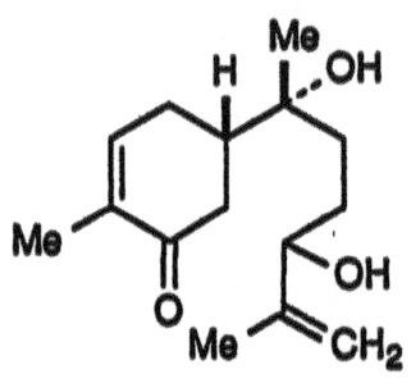

Chrysetunone (*40-6*)

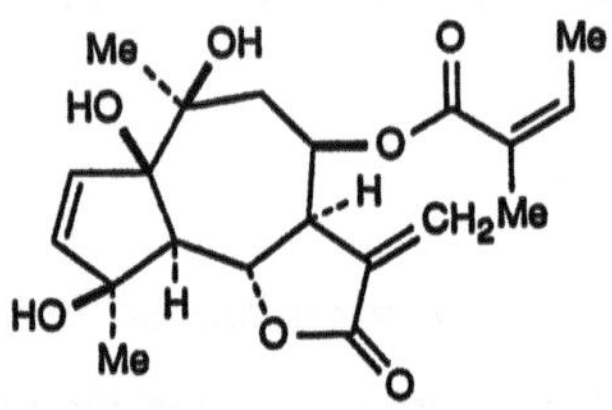

Tunefulin (*40-7*)

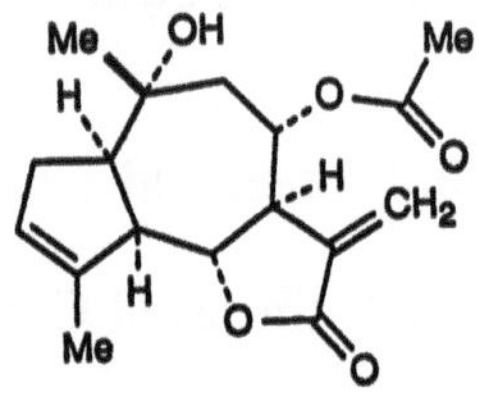

Cumambrin A (*40-8*)

The following substances were isolated and identified from the essential oil of *C. indicum*, camphor [6], chrysanthenone (*40-9*), *β*-caryophyllene oxide, limonene, *α*-pinene [7], *β*-pinene, camphene, myrcene, 3-carene, *α*-phellandrene, *α*-terpinene, *γ*-cymene, *β*-ocimene, terpinolene, and borneol [8].

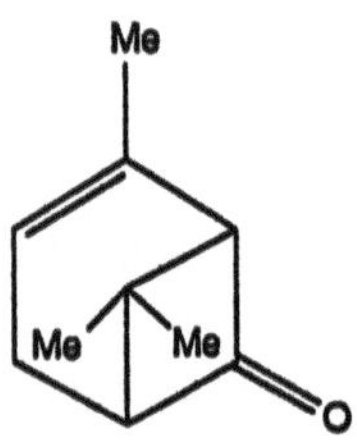

Chrysanthenone (*40-9*)

The flavones acaciin (*40-10*) [9], luteolin 7-*O*-*β*-D-glucopyranoside [10], and acacetin 7-*O*-*β*-D-galactopyranoside (*40-11*) [11] were isolated from *C. indicum*.

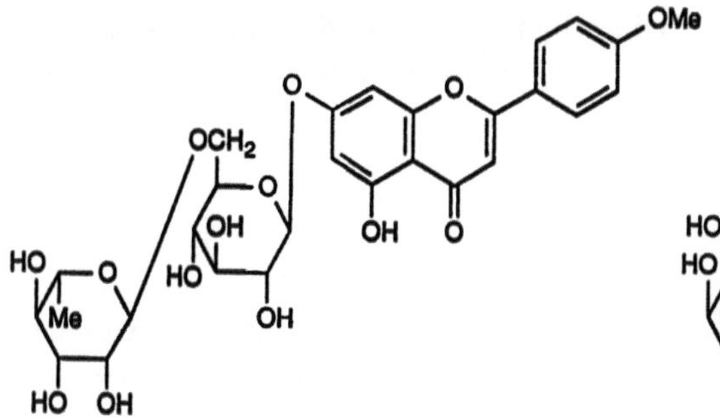

Acaciin (*40-10*)

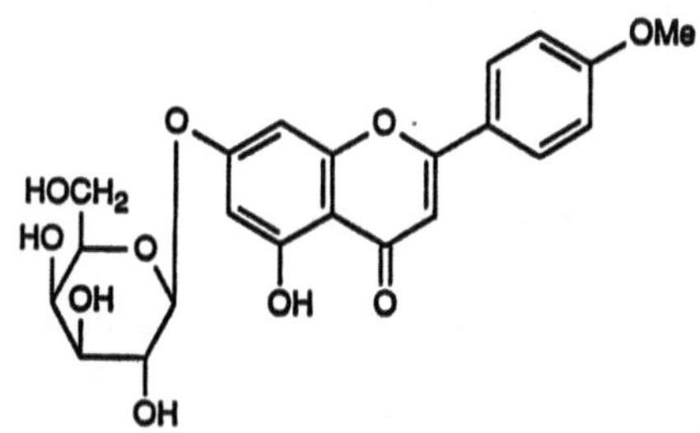

Acacetin 7-*O*-β-D-galactopyranoside (*40-11*)

Recently, a new bisabolane ketodiol named indicumenone (*40-12*) was also isolated from *C. indicum* [12].

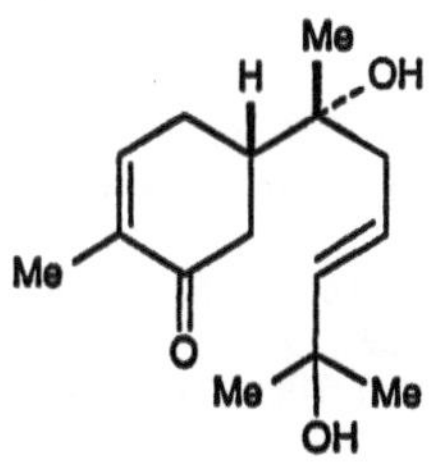

Indicumenone (*40-12*)

In addition, β-sitosterol, α- and β-amyrin, and their acetates, friedelin and sesamin, were identified from the nonvolatile fraction of *C. indicum* [7].

40.2.2 Chemical Constituents of *Chrysanthemum morifolium*

Two novel sesquiterpenes chlorochrymorin [13] and chrysandiol [14, 15] were isolated from *C. morifolium*. Chlorochrymorin (*40-13*) is the first natural sesquiterpene lactone with a 1-methyl-6-isopropyl-4-ethylperhydroindene skeleton containing a chlorine atom, whereas chrysandiol (*40-14*) is a sesquiterpene diol with a germacrane skeleton containing a cyclodecane ring. Furthermore, two known sesquiterpene lactones, chrysartemin A (*23-23*) and chrysartemin B (*23-24*), first isolated from *C. parthenium* [16], were found in *C. morifolium* [17].

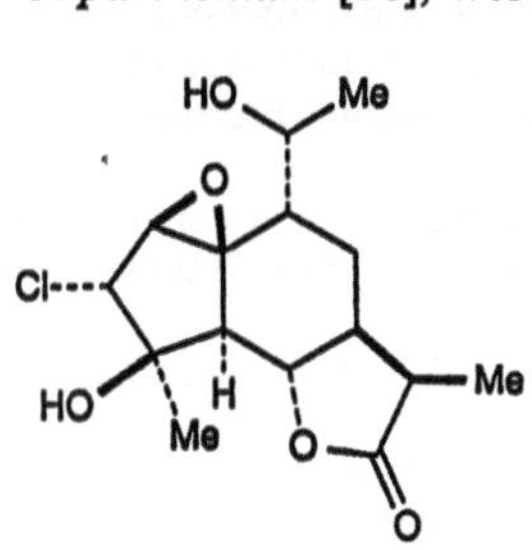

Chlorochrymorin (*40-13*)

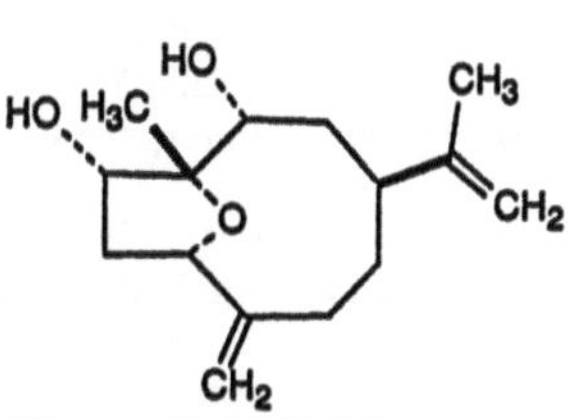

Chrysandiol (*40-14*)

In addition, the flavone glycosides acaciin [18], luteolin 7-*O*-*β*-D-glucopyranoside, and apigenin 7-*O*-*β*-D-glucopyranoside [19] were isolated from the flowers of *C. morifolium*.

40.3 Pharmacology

Arteglasin A was detected as a contact allergen of *C. indicum* [20]. It exhibited anaphylactic activity by skin tests in guinea pigs [21, 22] and caused allergic contact dermatitis in humans [23].

The flower of *C. indicum* also showed inhibitory activity in vitro against platelet aggregation of blood platelets from experimental animal (rabbits and rats) induced by bacteria such as *Staphylococcus aureus*, *Escherichia coli*, *Bacillus subtilis*, and *Pseudomonas aeruginosa* [24].

References

1. Mladenova K, Tsankova E, Stoianova-Ivanova B (1985) Sesquiterpene lactones from *Chrysanthemum indicum*. Planta Med 51:284–285
2. Chien MK, Chen CH, Tseng KF (1963) The constituents of Yejuhua, the flower of *Chrysanthemum indicum*. II. The structure of yejuhua lactone. Acta Pharm Sin 10:129–134
3. Chen ZN, Xu PJ (1987) Structural determination of yejuhua lactone, isolated from *Chrysanthemum indicum* L. Acta Pharm Sin 22:67–69
4. Mladenova K, Tsankova E, Dinh VH (1988) New sesquiterpenoides from *Chrysanthemum* var. *Tuneful*. Planta Med 54:553–555
5. Yu DQ, Xie FZ (1987) Chemical constituents of *Chrysanthemum indicum* L. Acta Pharm Sin 22:837–840
6. Tsumaki T, Hattori S, Kuramoto M (1947) Aromatic components of *Chrysanthemum* flowers. I. Repts Res Sci Dept Kyushu Univ 1:10–16 (CA 46:4507a)
7. De Pascual TJ, Bellido IS, Salado VJR, Moliner F, Alberdi MR (1980) Components of *Chrysanthemum indicum*, Linnaeus (var. cult.). Riv Ital EPPOS 62:236–238 (CA 94:52674h)
8. Stoianova-Ivanova B, Budzikiewicz H, Koumanova B, Tsoutsoulova A, Mladenova K, Brauner A (1983) Essential oil of *Chrysanthemum indicum*. Planta Med 49:236–239
9. Xu LX, Liu AR (1987) Coulometric titration of acaciin in *Chrysanthemum indicum*. Acta Pharm Sin 22:318–320
10. He YQ, Li RZ, Shen L (1982) Separation and identification of flavonoids from the flower of *Chrysanthemum indicum* L. J Beijing Med Coll 14:259–261
11. Chatterjee A, Sarkar S, Saha SK (1981) Acacetin 7-*O*-*β*-D-galactopyranoside from *Chrysanthemum indicum*. Phytochemistry 20:1760–1761
12. Mladenova K, Tsankova E, Kostova I, Stoianova-Ivanova B (1987) Indicumenone, a new bisabolene ketodiol from *Chrysanthemum indicum*. Planta Med 53:118–119
13. Osawa T, Suzuki A, Tamura S, Ohashi Y, Sasada Y (1973) Structure of chlorochrymorin, a novel sesquiterpene lactone from *Chrysanthemum morifolium*. Tetrahedron Lett 5135–5138
14. Osawa T, Suzuki A, Tamura S (1974) Structure of chrysandiol, a novel sesquiterpene diol from *Chrysanthemum morifolium*. Agric Biol Chem 38:685–686
15. Osawa T, Suzuki A, Tamura S, Ohashi Y, Sasada Y (1974) Molecular structure and stereochemistry of chrysandiol. Novel sesquiterpene diol from *Chrysanthemum morifolium*. Tetrahedron Lett 1569–1572
16. Osawa T, Suzuki A, Tamura S (1971) Isolation of chrysartemins A and B as rooting cofactor in *Chrysanthemum morifolium*. Agric Biol Chem 35:1966–1972
17. Romo J, Romo de Vivar A, Trevino R, Joseph-Nathan P, Diaz E (1970) Constituents of *Artemisia* and *Chrysanthemum* species. Structures of chrysartemins A and B. Phytochemistry 9:1615–1621

18. Hsu KK, Hong WH (1964) The constituents of the flowers of *Chrysanthemum morifolium*. Taiwan Ke Hsueh 18:102–104 (CA 62:12159h)
19. Arisawa M, Ishiwari Y, Nakaoki T, Sekino S, Takakuwa T (1969) Unutilized resources. III. Components of *Juncus* genus plants, the leaves of *Aesculus turbinata* and the petals of *Chrysanthemum morifolium*. Shoyakugaku Zasshi 23:49–52 (CA 73:73843u)
20. Hausen BM, Schulz KH (1976) *Chrysanthemum* allergy. III. Identification of the allergens. Arch Dermatol Res 255:111–121
21. Hausen BM, Schulz KH (1977) Isolation and identification of *Chrysanthemum* allergens. Z Immunitaetsforsch [Suppl]2:133–134
22. Hausen BM, Schulz KH, Jarchow O, Klaska KH, Schmalle H (1975) First allergenic sesquiterpene lactone from *Chrysanthemum indicum*. Arteglasin A. Naturwissenschaften 62:585–586
23. Schulz KH, Hausen BM, Wallhöfer L, Schmidt-Löffler P (1975) *Chrysanthemum* allergy. II. Experimental studies on the causative agents. Arch Dermatol Forsch 251:235–244
24. Yang JL, Liu WF (1982) Bacteria induced platelet aggregation and the effects of some drugs. Acta Acad Med Sin 4:306–309

Cimicifuga dahurica (Turcz.) Maxim. 41

41.1 Introduction

Shengma, Rhizoma Cimicifugae, is the dry rootstock of *Cimicifuga heracleifòlia* Kom., *C. dahurica* (Turcz.) Maxim. or *C. foetida* L. (Ranunculaceae) which is collected in the fall. This officially listed traditional Chinese medicine is used as an antipyretic and analgesic.

41.2 Chemical Constituents

The main constituents of the rhizome of *C. dahurica* are the triterpenes with a total content of 4.3% [1]. Triterpenes first isolated from the rhizome of *C. dahurica* were cimigenol (*41-1*), cimigenol 3-*O*-*β*-D-xylopyranoside (*41-2*), and dahurinol (*41-3*) [2], all derived from 9,19-cyclolanostane (*23-11*).

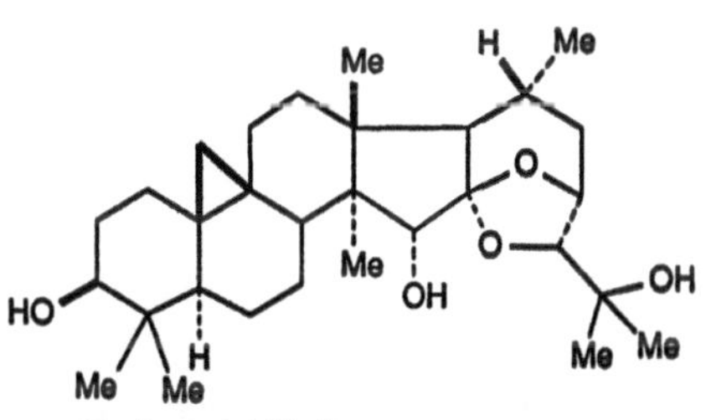

Cimigenol (*41-1*)

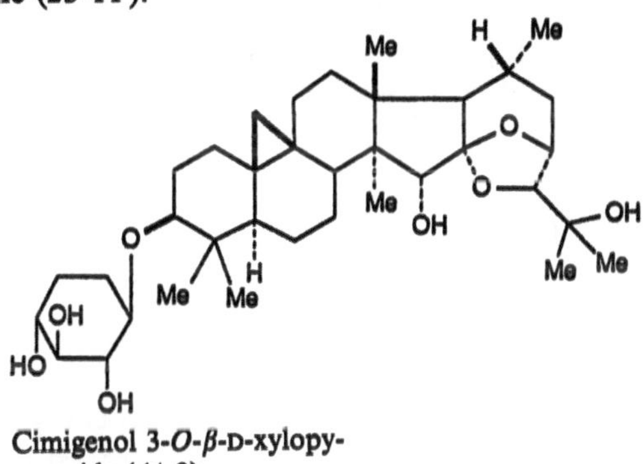

Cimigenol 3-*O*-*β*-D-xylopyranoside (*41-2*)

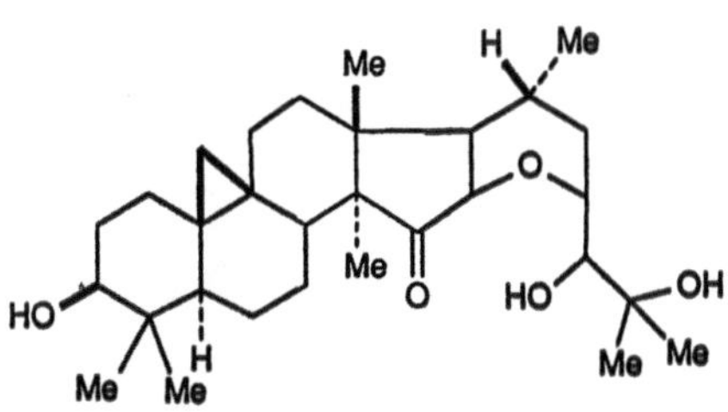

Dahurinol (*41-3*)

Acetylshengmanol 3-*O*-*β*-D-xylopyranoside (*41-4*) and 24-*O*-acetylhydroshengmanol 3-*O*-*β*-D-xylopyranoside (*41-5*) were isolated from the rhizome of *C. dahurica*.

Shengmanol 3-*O*-*β*-D-xylopyranoside (*41-6*), which was first isolated from the rhizome of *C. japonica* [3], was also detected in the rhizome of *C. dahurica* by high-performance liquid chromatography [4].

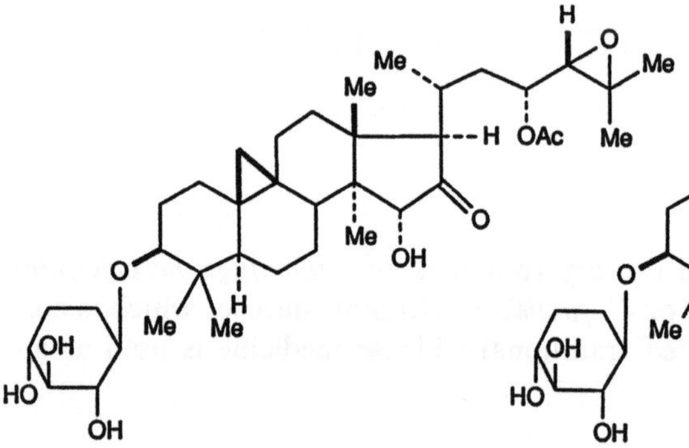

Acetylshengmanol
3-*O*-*β*-D-xylopyranoside (*41-4*)

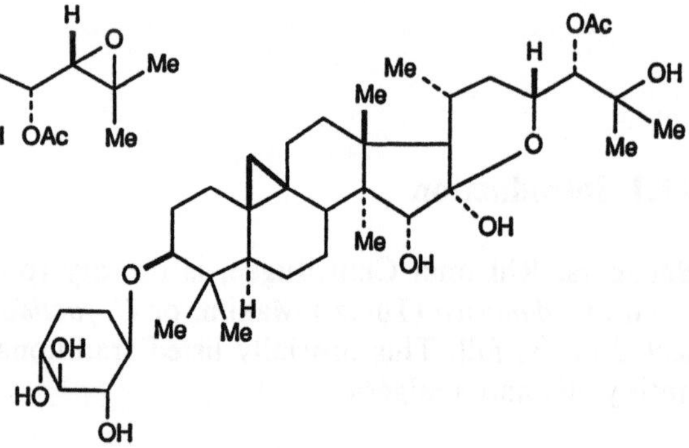

24-*O*-acetylhydroshengmanol
3-*O*-*β*-D-xylopyranoside (*41-5*)

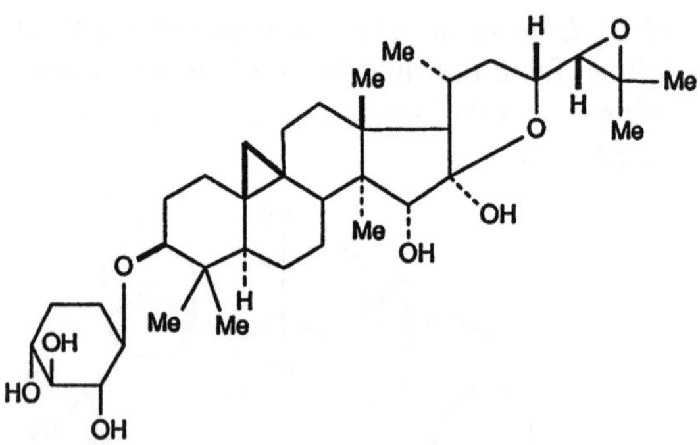

Shengmanol 3-*O*-*β*-D-xylopyranoside (*41-6*)

Besides the triterpenes and triterpene glycosides described, some phenolic carboxylic acids were isolated and identified from the rhizome of *C. dahurica.* Thus, caffeic acid (*41-7*), ferulic acid (*41-8*) and isoferulic acid (*41-9*) were detected [5].

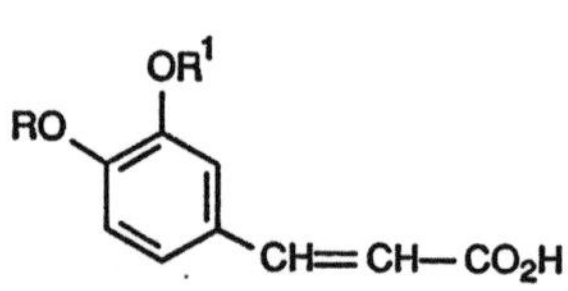

Caffeic acid (*41-7*): $R=R^1=H$
Ferulic acid (*41-8*): $R=H$, $R^1=CH_3$
Isoferulic acid (*41-9*): $R=CH_3$, $R^1=H$

Furthermore, the furochromones visamminol (*41-10*), visnagin (*41-11*), norvisnagin (*41-12*) [6], and two yellow pigments, 3-(3′-methyl-2′-butenylidene)-2-indolinone as geometrical isomers were isolated and identified [7, 8].

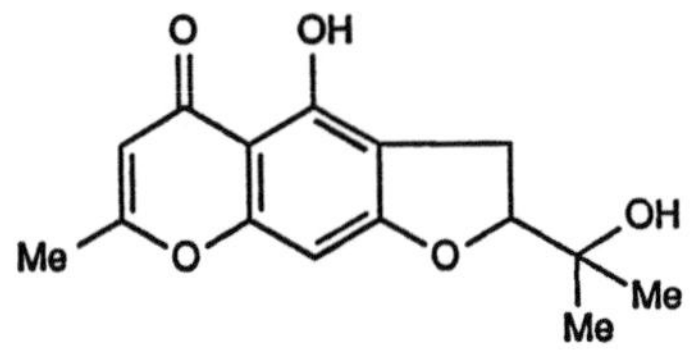

Visamminol (*41-10*)

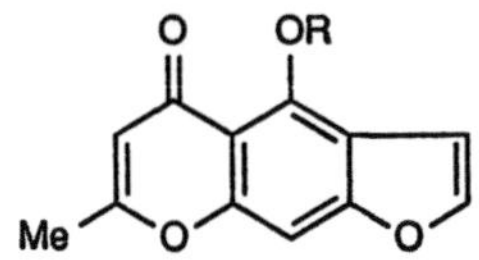

Visnagin (*41-11*): R=CH_3
Norvisnagin (*41-12*): R=H

41.3 Pharmacology

Visamminol and visnagin showed spasmolytic effectiveness in the isolated guinea pig jejunum, reaching about 10%–30% that of papaverine hydrochloride. Norvisnagin was inactive [6]. In rats with hyperlipidemia induced by vitamin D_2 and cholesterol, oral treatment with cyclolanostan triterpenes from *C. dahurica* decreased blood cholesterol and triglyceride levels. It was supposed that the triterpenes are metabolized to a compound of sterol-like structure that might competitively inhibit cholesterol formation [1]. The methanolic extract of *Cimicifuga* rhizome is effective in preventing liver disorders of mice induced by carbon tetrachloride. Based on biochemical and histological examinations, cimigenol 3-*O*-*β*-D-xylopyranoside was also effective in preventing liver damage [9].

References

1. Muravev IA, Vasilenko YK, Basharov AY, Parfenteva EP, Skulte IV (1985) Hypolipidemic properties of cimicilen, isolated from *Cimicifuga dahurica* (Turcz.) Maxim. Farmatsiya (Moscow) 34:38–42 (CA 103:206007m)
2. Sakurai N, Inoue T, Nagai M (1972) Chinese crude drug shoma. II. Triterpenes of *Cimicifuga dahurica*. Yakugaku Zasshi 92:724–728
3. Kumura O, Sakurai N, Nagai M, Inoue T (1982) Studies on the Chinese crude drug "Shoma". VI. Shengmanol xyloside, a new genuine natural product of some *Cimicifuga* glycosides. Yakugaku Zasshi 102:538–545
4. Kimura O, Sakurai N, Inoue T (1983) Studies on the Chinese crude drug shoma. VII. Isolation and determination of genuine natural products, acetylshengmanol xyloside, 24-*O*-acetylhydroshengmanol xyloside and shengmanol xyloside in *Cimicifuga dahurica* and the other *Cimicifuga* plants. Yakugaku Zasshi 103:293–299
5. Inoue T, Nakata C, Izawa K (1970) Chinese crude drug shoma. I. Isolation of phenolic carboxylic acids from the rhizomes of *Cimicifuga dahurica* and *Cimicifuga simplex*. Shoyakugaku Zasshi 24:76–80
6. Ito M, Kondo Y, Takemoto T (1976) Studies on the constituents of *Cimicifuga* spp. X. Spasmolytic substances from *Cimicifuga dahurica* Maxim. Chem Pharm Bull (Tokyo) 24:580–583
7. Hata K, Baba K, Kozawa M (1978) The structure of the yellow pigment from the rhizomes of *Cimicifuga dahurica* Maxim. Chem Pharm Bull (Tokyo) 26:2279–2280
8. Baba K, Kozawa M, Hata K, Ishida T, Inoue M (1981) The structure of yellow pigments from the rhizomes of *Cimicifuga dahurica*. Maxim. Chem Pharm Bull (Tokyo) 29:2182–2187
9. Yamahara J, Kobayashi M, Kimura H, Miki K, Kozuka M, Sawada K, Fujimura H (1985) Biologically active principles of crude drugs. The effect of Cimicifugae Rhizoma and its constituents in preventive action of the carbon tetrachloride-induced liver disorder in mice. Shoyakugaku Zasshi 39:80–84

Cinnamomum cassia Presl

42

42.1 Introduction

There are three entries for the medicinal herb *Cinnamomum cassia* Presl (Lauraceae) in the Chinese Pharmacopoeia:

- Rougui, Cortex Cinnamomi, is the dry stem bark of *C. cassia* peeled in the fall. It is used in traditional Chinese medicine as an analgesic, stomachic, and anti-inflammatory agent.
- Rouguiyou, Oleum Cinnamomi, is the essential oil of *C. cassia* obtained by steam distillation of the dry stems and leaves. It is used as an analgesic and stomachic. The cinnamaldehyde content in this essential oil should not be less than 85% (ml/ml).
- Guizhi, Ramulus Cinnamomi, is the dry young branches of *C. cassia* collected in spring and summer and is used as an analgesic and antipyretic against influenza and rheumatic pain.

42.2 Chemical Constituents

Besides cinnamaldehyde (*42-1*) as the main component in the essential oil [1, 2], coumarin, cinnamic acid (*42-2*), β-sitosterol, choline, protocatechuic acid, vanillic acid, and small amounts of syringic acid were isolated and identified from the aqueous extract of *C. cassia* [3–5].

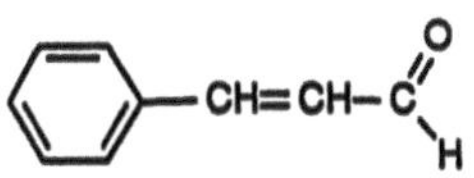

Cinnamaldehyde (*42-1*)

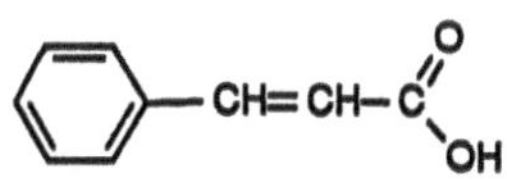

Cinnamic acid (*42-2*)

Because the stem bark of *C. cassia* is one of the most widely used crude drugs, the chemical composition has been investigated in detail. Thus, in the past several years a number of new closely related diterpenes were isolated from the fraction exhibiting anticomplement activity. These diterpenes were named cinncassiols. The following substances were isolated and structurally elucidated by various spectral methods as well as by X-ray crystallographic studies: cinncassiol A (*42-3*) and its glucoside (*42-4*) [6], cinncassiol B (*42-5*) and its glucoside (*42-6*) [7], cinncassiol C_1 (*42-7*) [8] and its glucoside (*42-8*) [9], cinncassiols C_2 (*42-9*) and C_3 (*42-10*) [9], cinncassiol D_1 (*42-11*) and its glucoside (*42-12*) [10, 11], cinncassiol D_2 (*42-13*) and its glucoside

(*42-14*) [11], cinncassiol D_3 (*42-15*) [11], cinncassiol D_4 (*42-16*) and its glucoside (*42-17*) [12], cinncassiol E (*42-18*) [13], cinnzeylanol (*42-19*) [6, 14], cinnzeylanin (*42-20*), anhydrocinnzeylanol (*42-21*) [6], anhydrocinnzeylanin (*42-22*). Cinncassiol B and cinnzeylanin as well as cinnzeylanol are derived from ryanodol (*42-23*), whereas cinncassiols C_1, C_2, and C_3 possess a securyanodol (*42-24*) skeleton. These diterpene compounds can also be divided into lactone type, ketal-type, and diketone type (Fig. 42.1) [11].

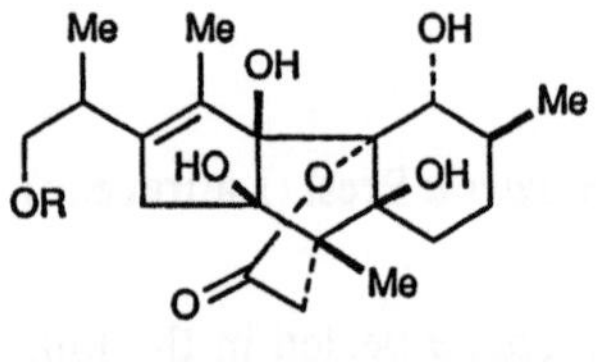

Cinncassiol A (*42-3*): R=H

Cinncassiol A
O-*β*-D-glucopyranoside (*42-4*): R =

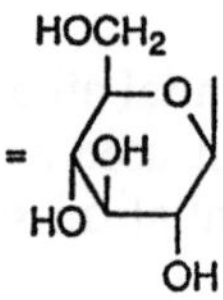

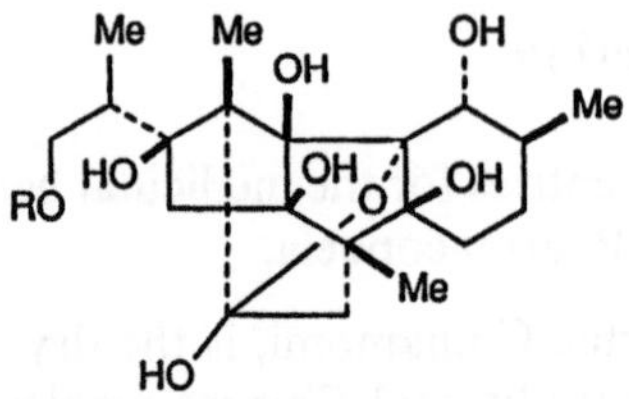

Cinncassiol B (*42-5*): R=H

Cinncassiol B
O-*β*-D-glucopyranoside (*42-6*): R =

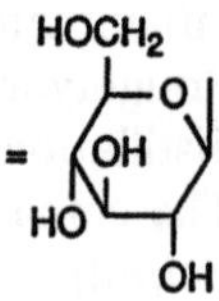

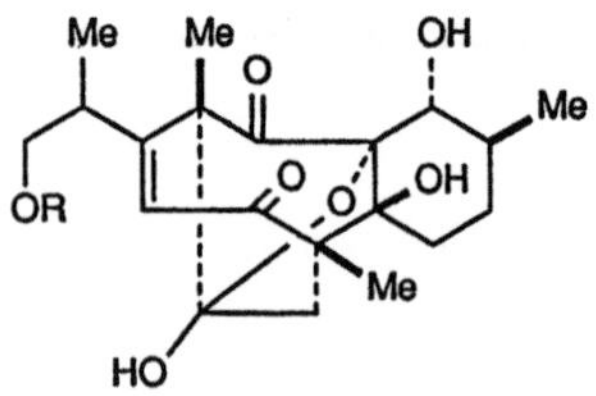

Cinncassiol C_1 (*42-7*): R=H

Cinncassiol C_1
O-*β*-D-glucopyranoside (*42-8*): R =

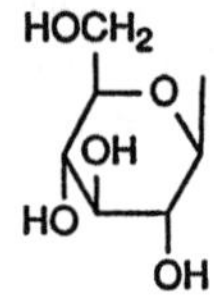

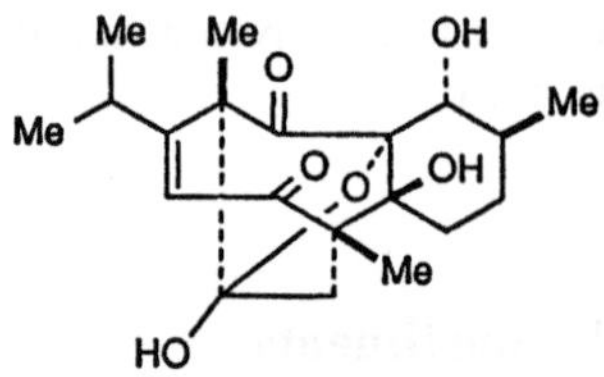

Cinncassiol C_2 (*42-9*)

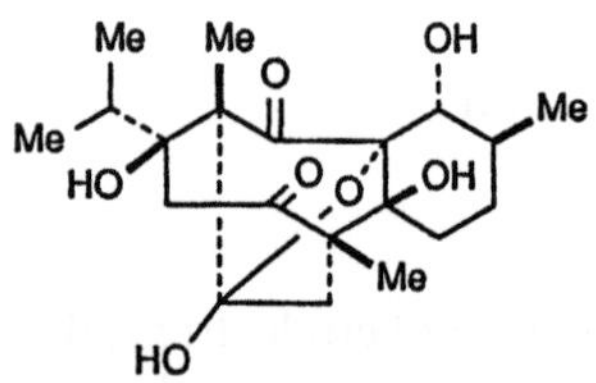

Cinncassiol C_3 (*42-10*)

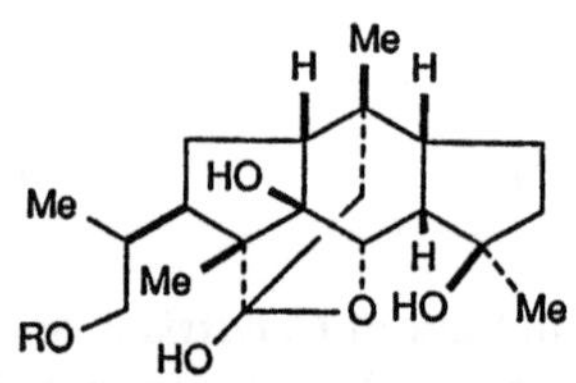

Cinncassiol D_1 (*42-11*): R=H

Cinncassiol D_1
O-*β*-D-glucopyranoside (*42-12*): R =

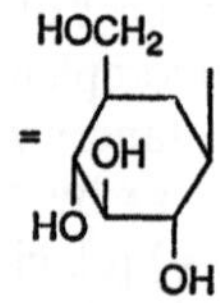

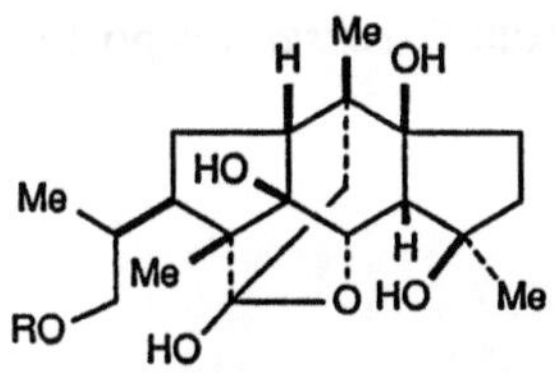

Cinncassiol D_2 (*42-13*): R=H

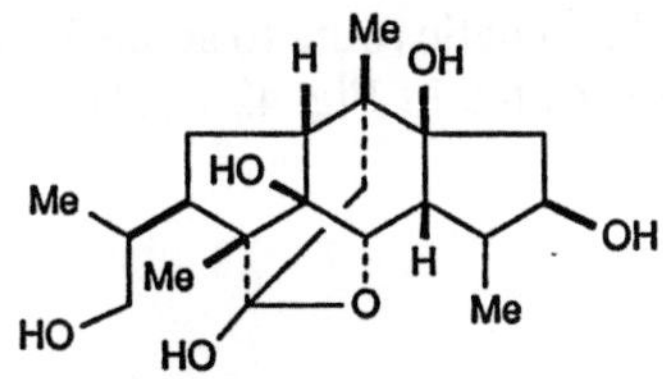

Cinncassiol D_3 (*42-15*)

Cinncassiol D_2
O-β-D-glucopyranoside (*42-14*): R =

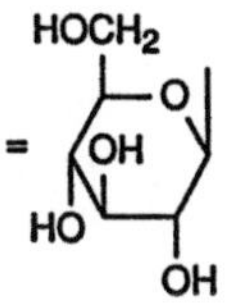

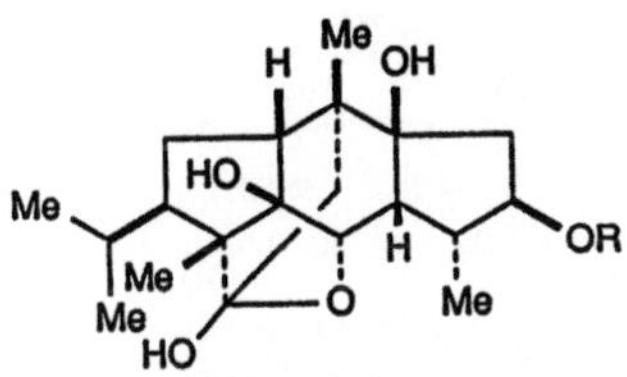

Cinncassiol D_4 (*42-16*): R=H

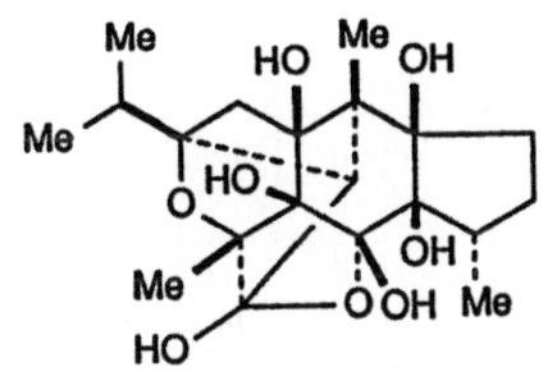

Cinncassiol E (*42-18*)

Cinncassiol D_4
O-β-D-glucopyranoside (*42-17*): R =

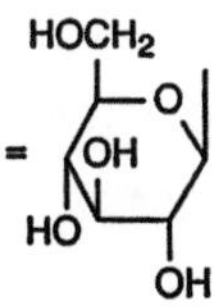

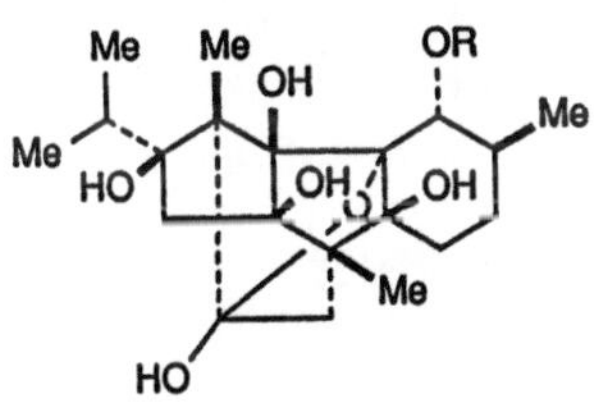

Cinnzeylanol (*42-19*): R=H
Cinnzeylanin (*42-20*): R=Ac

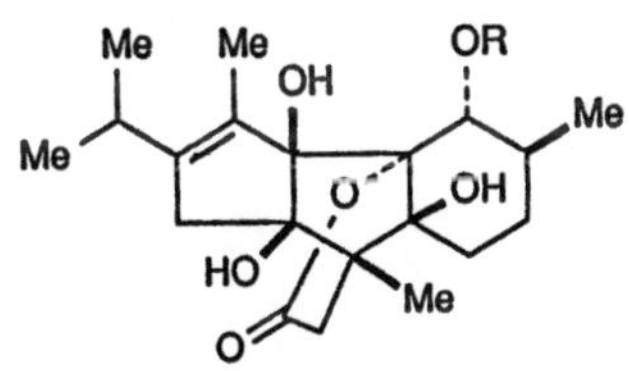

Anhydrocinnzeylanol (*42-21*): R=H
Anhydrocinnzeylanin (*42-22*): R=Ac

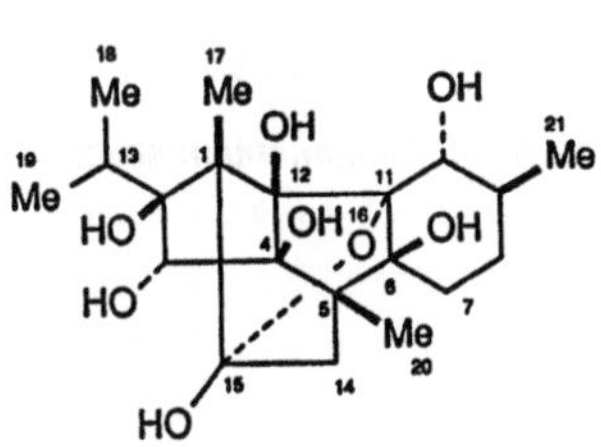

Ryanodol (*42-23*)

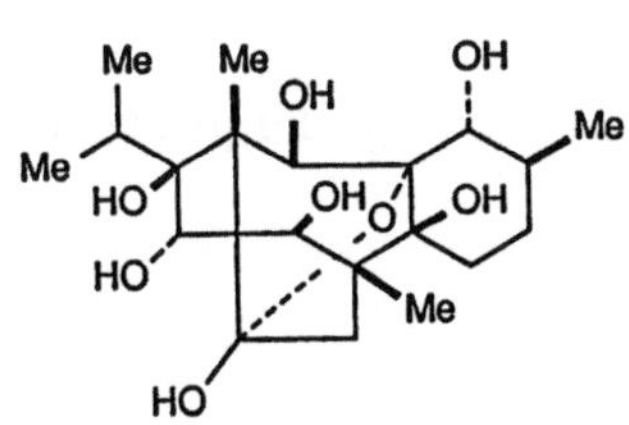

Securyanodol (*42-24*)

A possible biogenetic route to some diterpenes isolated from *C. cassia* was postulated as demonstrated in Fig. 42.1 [11].

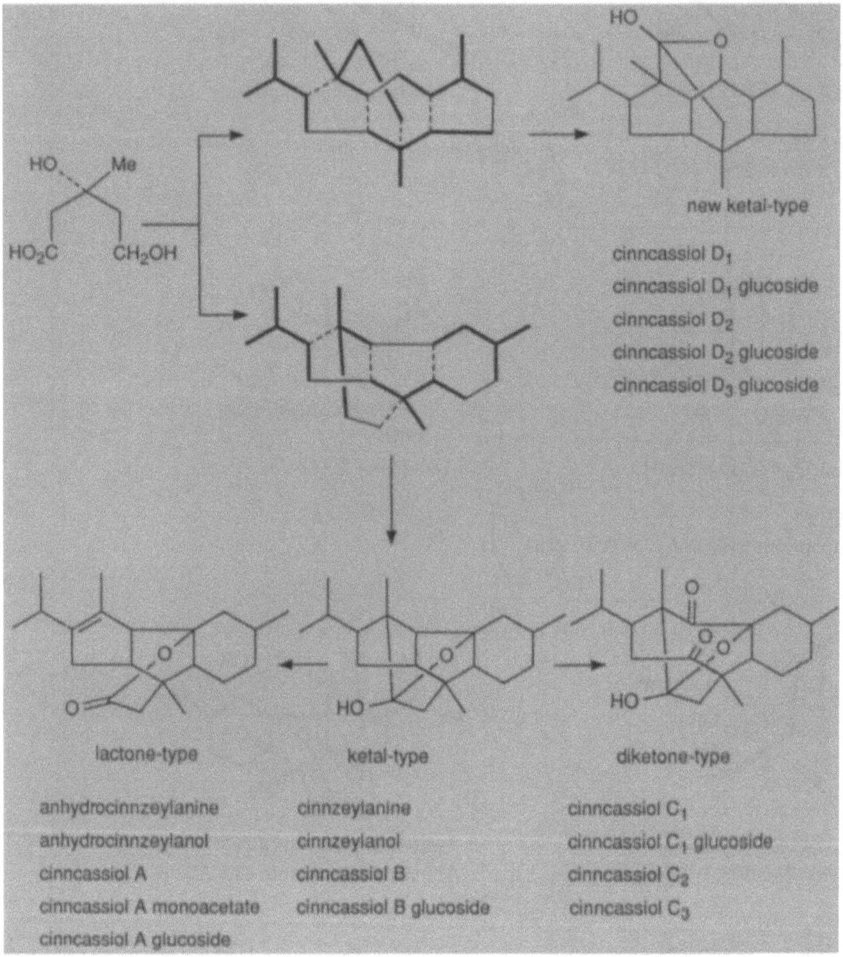

Fig. 42.1. Possible biogenesis of diterpenes isolated from the stem bark of *Cinnamomum* species

In connection with the isolation of the diterpenes described above, the aromatic compounds lyoniresinol 2α-*O*-β-D-glucopyranoside (*42-25*), 3,4,5-trimethoxyphenol *O*-β-D-apiofuranosyl-(1→6)-β-D-glucopyranoside (*42-26*), syringaresinol, two epicatechins [5,7,3′(or 4′)-trimethyl-epicatechin, 5,7-*O*-dimethyl-3′,4′-di-*O*-methylene-epicatechin (*42-27*), and two cinnamic aldehyde cyclic glycerol 1,3-acetals (*42-28*, *42-29*) were further isolated from the stem bark of *C. cassia* [15].

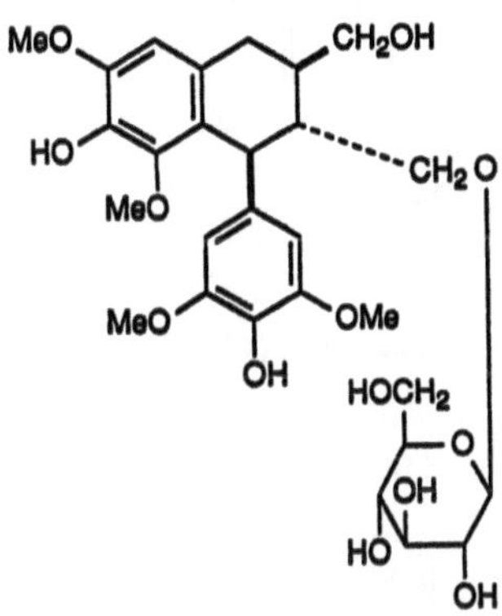

Lyoniresinol 2α-*O*-β-D-glucopyranoside (*42-25*)

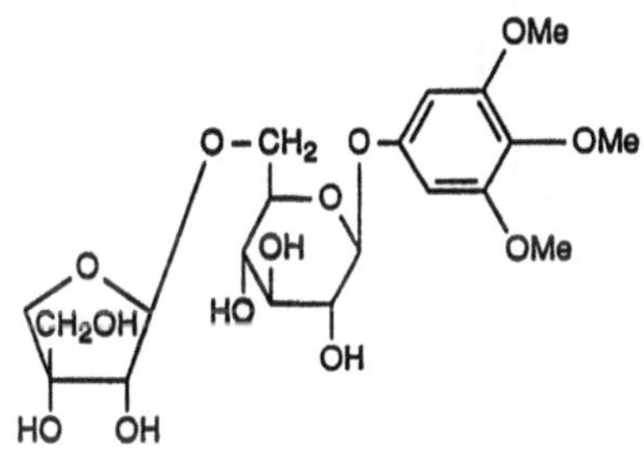

3,4,5-Trimethoxyphenol *O*-β-D-apiofuranosyl-(1→6)-*O*-β-D-glucopyranoside (*42-26*)

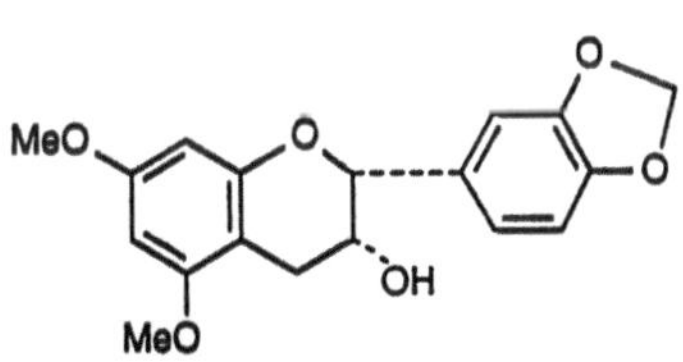

5,7-*O*-Dimethyl-3′,4′-di-*O*-methylene-epicatechin (*42-27*)

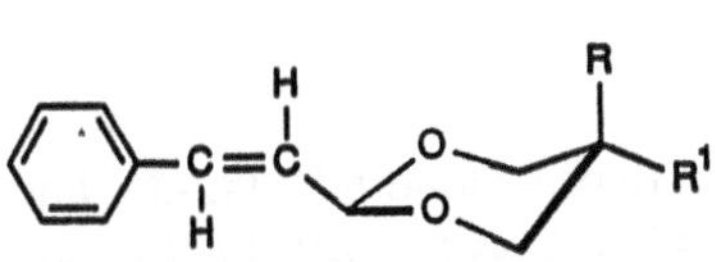

(9,2′-*trans*)-Cinnamic aldehyde cyclic glycerol 1,3-acetal (*42-28*): R=H, R^1=OH
(9,2′-*cis*)-Cinnamic aldehyde cyclic glycerol 1,3-acetal (*42-29*): R=OH, R^1=H

Additionally, a number of flavon-3-ol derivatives, including procyanidins, were isolated from the bark of *C. cassia*. The momomeric flavan-3-ols isolated were 3′-*O*-methyl-, 5,3′-di-*O*-methyl-, 5,7,3′-tri-*O*-methyl-epicatechin (*42-30*), and 4′-*O*-methyl-, 7,4′-di-*O*-methyl-, 5,7,4′-tri-*O*-methyl-catechin (*42-31*) [16], as well as the epicatechin 3-*O*-, 8-*C*- and 6-*C*-*β*-D-glucopyranosides (*42-32*–*42-34*) [17].

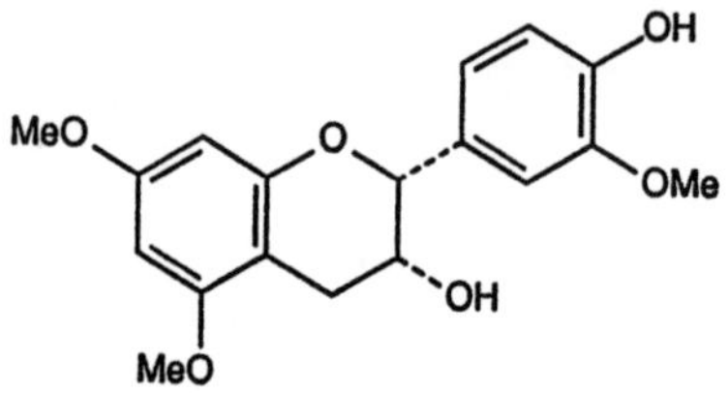

5,7,3′-*O*-Trimethyl-(−)-epicatechin (*42-30*)

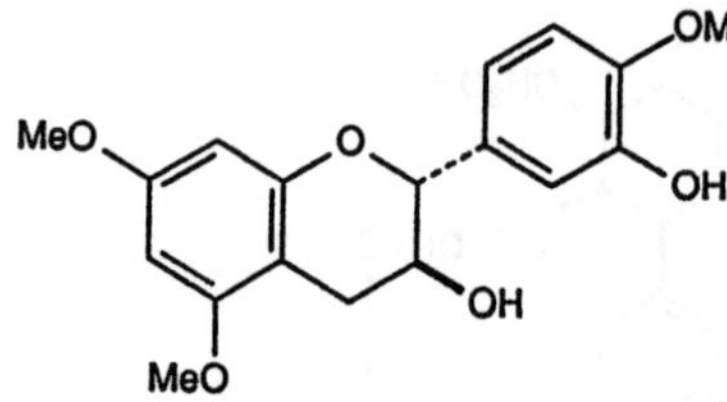

5,7,4′-*O*-Trimethyl-(+)-catechin (*42-31*)

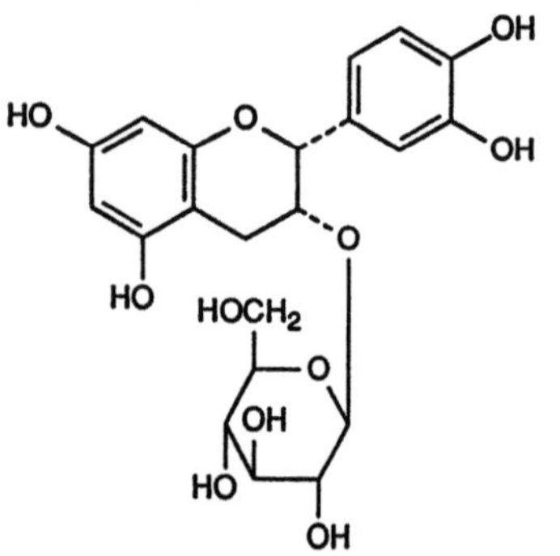

Epicatechin-3-*O*-*β*-D-glucopyranoside (*42-32*)

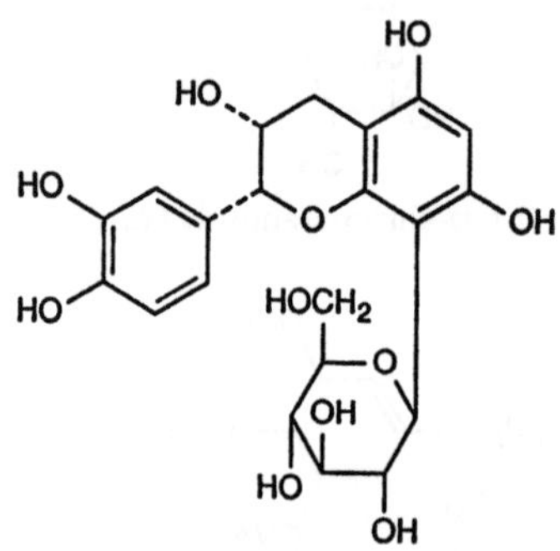

Epicatechin-8-*C*-*β*-D-glucopyranoside (*42-33*)

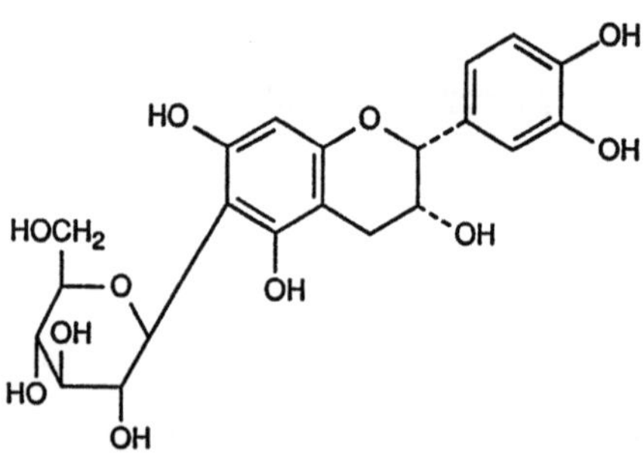

Epicatechin-6-*C*-*β*-D-glucopyranoside (*42-34*)

Procyanidins isolated from the bark of *C. cassia* were procyanidin B_1 (*42-35*) and its epimer B_2 (*42-36*), B_5 (*42-37*) and its epimer B_7 (*42-38*), A_2 (*42-39*) [17, 18], and two procyanidin glucosides procyanidin B_2 8-*C*- and 6-*C*-*β*-D-glucopyranoside (*42-40, 42-41*) [18]. Oligomeric procyanidin C_1 and cinnamtannins A_2 (*42-42*), A_3, and A_4 were also isolated and structurally elucidated as tri-, tetra-, penta-, and hexameric procyanidins, respectively. The latter consist exclusively of epicatechin units linked linearly through C(4*β*)–C(8) bonds [17, 18].

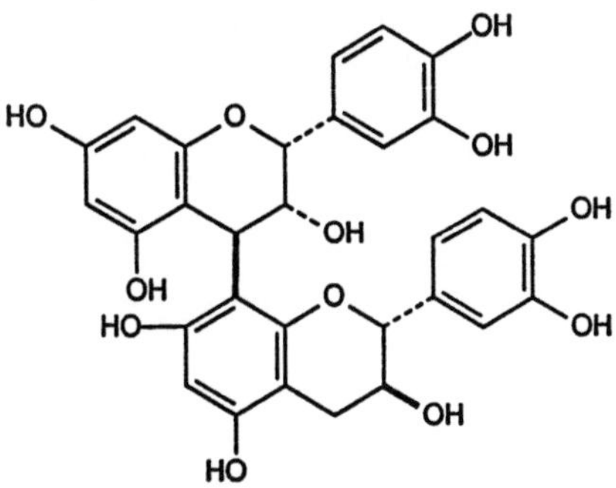

Procyanidin B_1 (*42-35*)

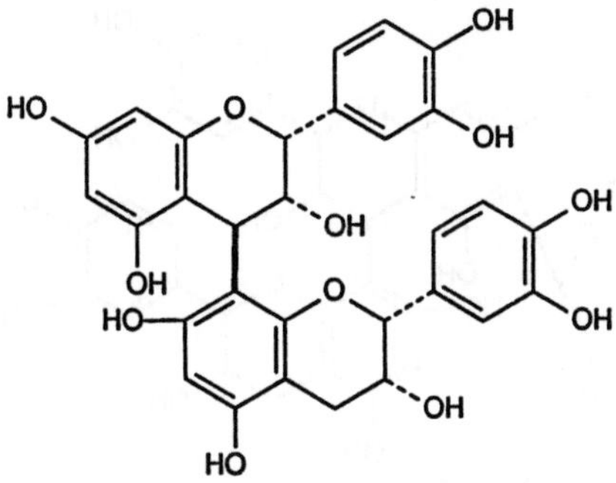

Procyanidin B_2 (*42-36*)

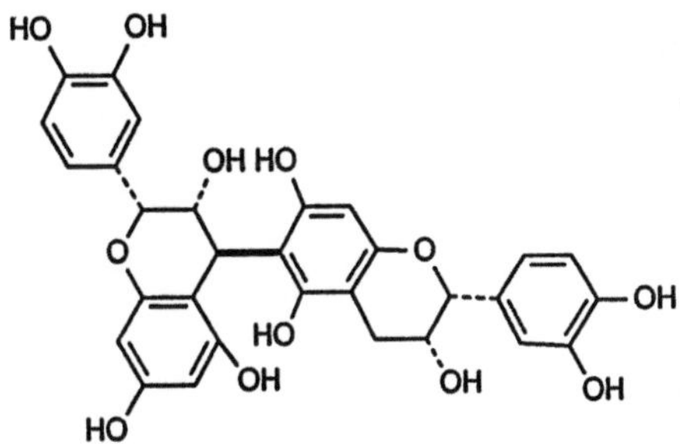

Procyanidin B_5 (*42-37*)

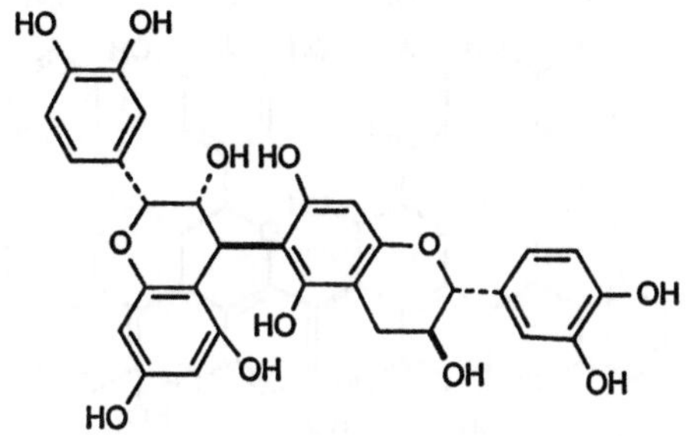

Procyanidin B_7 (*42-38*)

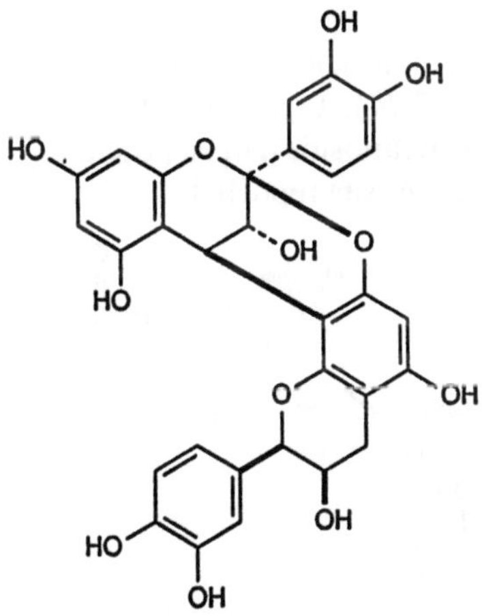

Procyanidin A_2 (*42-39*)

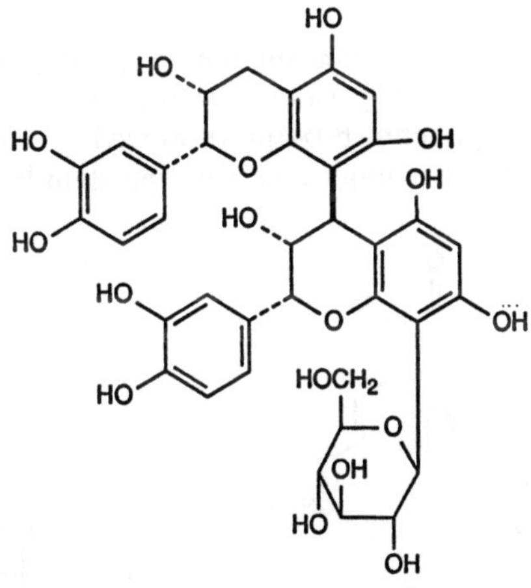

Procyanidin B_2-8-*C*-*β*-D-glucopyranoside (*42-40*)

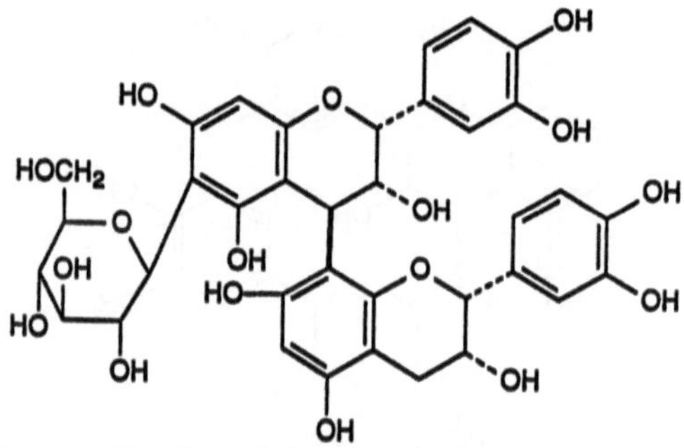

Procyanidin B_2-6-*C*-β-D-glucopyranoside (*42-41*)

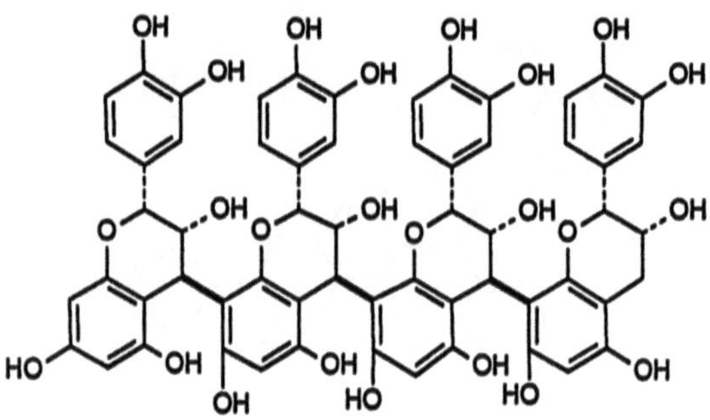

Cinnamtannin A_2 (*42-42*)

Recently, the isolation of 3-(2-hydroxyphenyl)-propanoic acid and its *O*-β-D-glucopyranoside [19], cassioside (*42-43*), cinnamoside (*42-44*), together with 3,4,5-trimethoxyphenol-β-D-apiofuranosyl-(1→6)-β-D-glucopyranoside [20] as potent antiulcerogenic compounds from the stem bark of *C. cassia* was reported.

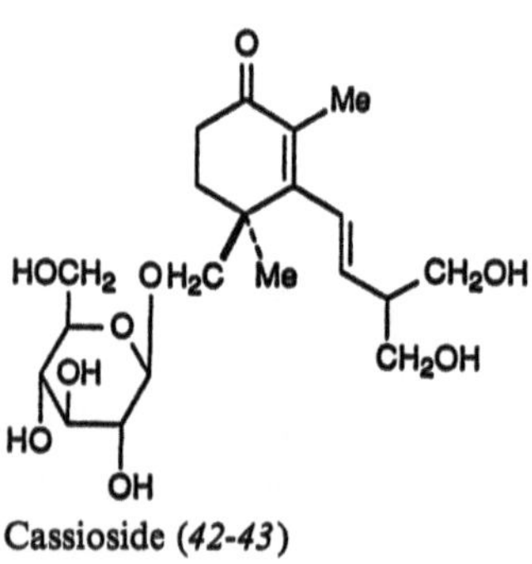

Cassioside (*42-43*)

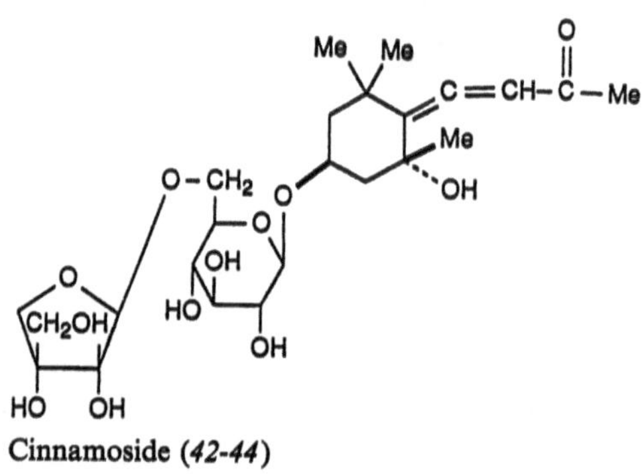

Cinnamoside (*42-44*)

42.3 Pharmacology

Most of the pharmacological studies on the chemical constituents of *C. cassia* have been limited to the effects of the essential oil and its major component cinnamaldehyde. Cinnamaldehyde exhibited bactericidal activity against some bacteria species in high dilution using the agar-diffusion method [21]. The cinnamon oil was found to have fungicidal activity against *Trichophyton mentagrophytes*. The active component of the oil was cinnamaldehyde [22]. Cinnamaldehyde was also active against *Microsporum audouini*. The minimal fungicidal concentration of cinnamaldehyde against these two dermatophytes was less than 0.09 mg/ml [23]. A study on the antifungal activity of cinnamaldehyde against a number of fungi showed that cinnamaldehyde, at a concentration of 0.33 m*M*, has inhibitory effects on *Aspergillus nidurans*, *Cryptococcus neoformans*, *Penicillium rugulosum*, *Sporothrix schenckii*, *Trichophyton rubrum*, *T. mentagrophytes*, *T. violaceum*, *Microsporum gypseum*, *Histoplasma capsulatum*, and *Blastomyces dermatitidis* [24]. The growth-inhibiting activity of cinnamaldehyde against fungi was reduced or abolished in the presence of cysteine or glutathione in some cases, indicating that the inhibition was mainly due to reactions of the aldehyde with thiol groups involved in the fungal growth [24]. A synergistic effect of cinnamaldehyde with sodium chloride was observed [25].

Cinnamaldehyde (>30 mg/kg, i.p.) has sedative activity, including a decrease in spontaneous motor activity in mice, and has an antagonistic effect to locomotor activity induced by apomorphine or methamphetamine in mice at doses of 125 and 250 mg/kg (i.p.). Prolongation of hexobarbital-induced hypnosis was also observed [26, 27]. Cinnamaldehyde also showed hypothermic and antipyretic activities in mice [26]. The central inhibitory and excitatory effects of cinnamaldehyde may be ascribed to interaction with monoaminergic neurons in the CNS and may play an important role in the therapeutic effect of the stem bark or branches of *C. cassia* [26, 27].

In anesthetized dogs and guinea pigs, cinnamaldehyde produced a hypotensive effect, which appeared to be due mainly to peripheral vasodilatation. The vasodilatation induced by cinnamaldehyde in dogs persisted through the period of blood pressure recovery.

In the isolated ileum from the guinea pig and the mouse, cinnamaldehyde exhibited a papaverine-like activity apparently contributing to vasodilatation. An increase in cardiac contractile force and beating rate was exerted by cinnamaldehyde in isolated guinea pig heart preparations. Repeated administrations of cinnamaldehyde, however, led to a progressive decrease in such effects and to cardiac inhibition [28]. Cinnamaldehyde induced a release of catecholamines from the adrenal gland of dogs. The positive inotropic and chronotropic effects of cinnamaldehyde on isolated perfused guinea pig heart were presumably due to release of endogenous catecholamine [29].

The aqueous extract of *C. cassia* administered orally to rats with nephritis prevented the increase in protein level in the urine [30], but had no effect on the recovery of anemia induced by an acute blood loss and on the total leukocyte and lymphocyte counts [31]. Immunosuppressive properties have been ascribed to this extract [3–5].

The i.p. administration of an aqueous extract of *C. cassia* to rats at a dose of 100 mg/kg prevented the occurrence of stress ulcers under exposure to a cold atmosphere (3°–5 °C) or by water immersion. This extract also strongly inhibited gastric ulcers induced by a subcutaneous injection of serotonin in rats. Pharmacological studies have shown that this extract not only inhibits gastric secretion, but also promotes gastric mucosal blood flow, indicating that the antiulcerogenic effect of *C. cassia* is due to both the inhibition of aggressive factors and the potentiation of defensive factors [32].

Cinnamaldehyde at a concentration of 4.8 μg/ml inhibited the growth of L1210 leukemia cells in culture by 50%. The aldehyde group of cinnamaldehyde was responsible for the inhibition. Incorporation of [^{3}H]uridine, [^{3}H]thymidine, and [^{3}H]leucine by the cells was suppressed by the presence of cinnamaldehyde. The inhibitory effect of cinnamaldehyde on glycolysis was insignificant. Direct reaction between aldehyde groups of cinnamaldehyde and thiol groups of cell components was demonstrated, indicating that the inhibitory effect of cinnamaldehyde on the growth of L1210 cells might be ascribed to such reactions [33]. Cinnamaldehyde inhibited the growth of SV40-induced tumor W2K-11 in mice [34].

Cinnamaldehyde was also tested for antimutagenic activity toward chemical mutagens or UV irradiation. Antimutagenic effects of cinnamaldehyde on chemical mutagenesis were tested in *Escherichia coli*. Cinnamaldehyde greatly suppressed the SOS repair-dependent mutagenesis induced by 4-nitroquinoline-*1*-oxide, furylfuramide, or captan. It was less effective against the SOS repair-independent mutagenesis by alkylating agents such as *N*-methyl-*N′*-nitro-*N*-nitrosoguanidine and ethyl methanesulfonate [35]. Despite the decrease in the number of revertants, a remarkable increase was observed in the survival of *E. coli* cells treated with 4-nitroquinoline-*1*-oxide after exposure to cinnamaldehyde. The reactivation of survival suggested the promotion of some DNA repair system by cinnamaldehyde [36].

In UV-irradiated *E. coli*, cinnamaldehyde did not affect the surviving colonies up to a concentration of 80 μg/ml, but the number of revertant colonies was reciprocal to the concentration of cinnamaldehyde. The results clearly showed that cinnamaldehyde had an antimutagenic effect on UV-induced mutation in the microbial assay system with *E. coli* [37].

Cinnamaldehyde is generally recognized as safe for food additive applications and is an important fragrance raw material used in perfumery. Cinnamaldehyde at the concentrations contained in consumer products and fragrances had a very low potential for the induction of hypersensitivity or the elicitation of sensitization reactions in the general population [38].

After i.p. administration of cinnamaldehyde in rats, 6.5% of the dose was excreted as thioether in urine. Two sulfur-containing metabolites 3-*S*-(*N*-acetylcysteinyl)-3-phenylpropanol (*42-45*) and 3-*S*-(*N*-acetylcysteinyl)-3-phenylpropionic acid (*42-46*) were identified at a ratio of 4:1 [39, 40].

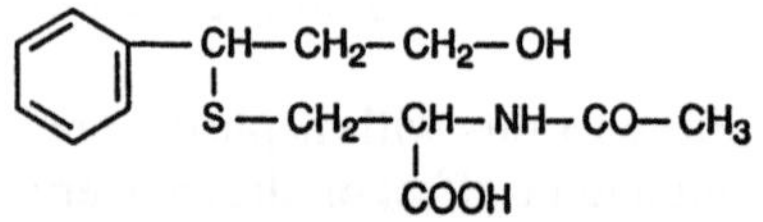

3-*S*-(*N*-Acetylcysteinyl)-3-phenylpropanol (*42-45*)

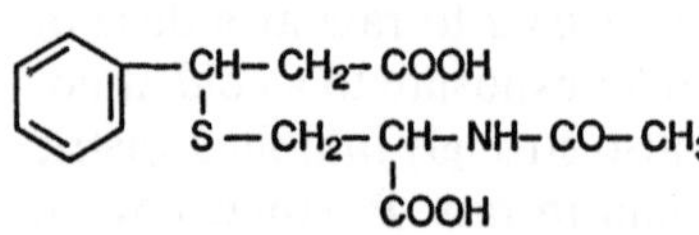

3-*S*-(*N*-Acetylcysteinyl)-3-phenylpropionic acid (*42-46*)

References

1. Wu TM, Chen YP, Hsu HY (1972) Comparison on the extraction methods of Chinese cinnamon. Tai-Wan Yao Hsueh Tsa Chih 24:86–90
2. Shen SJ, Yu CH, Chen YP, Hsu HY (1981) Studies on processing of Chinese herbs and analysis of their active principles. I. Quantitative analysis of cinnamaldehyde and cinnamic acid in the cinnamon crude drugs and their preparations. Chung-ku Nung Yeh Hua Hsueh Hui Chih 19:46–51 (CA 95:209738e)
3. Yuan AX, Tan L, Jiang DG (1982) Studies on chemical constituents of Rou Gui (*Cinnamomom cassia* Presl). I. Bull Chin Mat Med 7:26–28
4. Yuan AX, Qin L, Jiang DG (1981) Studies on the chemical constituents of *Cinnamomum cassia.* Chin Pharm Bull 16:631
5. Yuan AX, Than L, Wei SX, Kang SH, Jiang DG (1984) Chemical constituents of Guizhi (*Cinnamomum cassia*), a traditional Chinese medicine. Bull Chin Mat Med 9:127–128
6. Yagi A, Tokubuchi N, Nohara T, Nonaka G, Nishioka I, Koda A (1980) The constituents of Cinnamomi Cortex. I. Structures of cinncassiol A and its glucoside. Chem Pharm Bull (Tokyo) 28:1432–1436
7. Nohara T, Tokubuchi N, Kuroiwa M, Nishioka I (1980) The constituents of Cinnamomi Cortex. III. Structures of cinncassinol B and its glucoside. Chem Pharm Bull (Tokyo) 28:2682–2686
8. Nohara T, Nishioka I, Tokubuchi N, Miyahara K, Kawasaki T (1980) The constituents of Cinnamomi Cortex. II. Cinncassiol C, a novel type of diterpene from Cinnamomi Cortex. Chem Pharm Bull (Tokyo) 28:1969–1970
9. Kashiwada Y, Nohava T, Tomimatsu T, Nishioka I (1981) Constituents of Cinnamomi Cortex. IV. Structures of cinncassiols C_1 glucoside, C_2 and C_3. Chem Pharm Bull (Toyko) 29:2686–2688
10. Nohara T, Kashiwada Y, Tomimatsu T, Kido M, Tokubuchi N, Nishioka I (1980) Cinncassiol D_1 and its glucoside, novel pentacyclic diterpenes from Cinnamomi Cortex. Tetrahedron Lett 21:2647–2648
11. Nohara T, Kashiwada Y, Murakami K, Tomimatsu T, Kido M, Yagi A, Nishioka I (1981) Constituents of Cinnamomi Cortex. V. Structures of five novel diterpenes, cinncassiols D_1, D_1 glucoside, D_2, D_2 glucoside and D_3. Chem Pharm Bull (Tokyo) 29:2451–2459
12. Nohara T, Kashiwada Y, Tomimatsu T, Nishioka I (1982) Studies on the constituents of Cinnamomi Cortex. VII. Two novel diterpenes from bark of *Cinnamomum cassia.* Phytochemistry 21:2130–2132
13. Nohara T, Kashiwada Y, Nishioka I (1985) Cinncassiol E, a diterpene from the bark of *Cinnamomum cassia.* Phytochemistry 24:1849–1850
14. Nohara T, Tomimatsu T, Tokubuchi N, Yagi A, Nishioka I, Miyahara K (1979) Chemical studies on the diterpenoids from Cinnamomi Cortex. Koen Yoshishu-Tennen Yuki Kagobutsu Toroukai 22:87–93 (CA 92:215578e)
15. Miyamura M, Nohara T, Tomimatsu T, Nishioka I (1983) Studies on the constituents of Cinnamomi Cortex. VIII. Seven aromatic compounds from bark of *Cinnamomum cassia.* Phytochemistry 22:215–218
16. Morimoto S, Nonaka G, Nishioka I, Ezaki N, Takizawa N (1985) Tannins and related compounds. 29. Seven new methyl derivatives of flavan-3-ols and a 1,3-diarylpropan-2-ol from *Cinnamomum cassia, C. obtusifolium* and *Lindera umbellata* var. *membranacea.* Chem Pharm Bull (Tokyo) 33:2281–2286
17. Morimoto S, Nonaka G, Nishioka I (1986) Tannins and related compounds. 38. Isolation and characterization of flavan-3-ol glucosides and procyanidin oligomers from cassia bark (*Cinnamomum cassia* Blume). Chem Pharm Bull (Tokyo) 34:633–642
18. Morimoto S, Nonaka G, Nishioka I (1986) Tannins and related compounds 39. Procyanidin *C*-glucosides and an acylated flavan-3-ol glucoside from the barks of *Cinnamomum cassia* Blume and *C. obtusifolium* Ness. Chem Pharm Bull (Tokyo) 34:643–649
19. Tanaka S, Yoon YH, Fukui H, Tabata M, Akira T, Okan K, Iwai M, Iga Y, Yokoyama K (1989) Antiulcerogenic compounds isolated from Chinese cinnamon. Planta Med 55:245–248
20. Shiraga Y, Okano K, Akira T, Fukaya C, Yokoyama K, Tanaka S, Fukui H, Tabata M (1988) Structures of potent antiulcerogenic compounds from *Cinnamomum cassia.* Tetrahedron 44:4703–4711

21. Yousef RT, Tawil GG (1979) Antimicrobial activity of volatile oil components. Manuf Chem Aerosol News 50:59 (CA 91:151896b)
22. Korbely I, Florian E (1971) Effect of essential oils on *Trichophyton mentagrophytes* and *Candida albicans*. Gyogyszereszet 15:462–471 (CA 77:1182n)
23. Yousef RT, Aggag ME, Tawil GG (1978) Evaluation of the antifungal activity of some components of volatile oils against dermatophytes. Mykosen 21:190–193 (CA 89:141139b)
24. Kurita N, Miyaji M, Kurane R, Takahara Y, Ichimura K (1979) Antifungal activity and molecular orbital energies of aldehyde compounds from oils of higher plants. Agric Biol Chem 43:2365–2371
25. Kurita N, Koike S (1982) Synergistic antimicrobial effect of sodium chloride and essential oil components. Agric Biol Chem 46:159–165
26. Harada M, Ozaki Y (1972) Pharmacological studies on Chinese cinnamon. I. Central effects of cinnamaldehyde. Yakugaku Zasshi 92:135–140
27. Watanabe H, Hagiwara M, Tohda M, Hiyama Y, Terasawa K, Watanabe K (1984) Central effect of cinnamaldehyde. Yakugaku Zasshi 104:1095–1100
28. Harada M, Yano S (1975) Pharmacological studies on Chinese cinnamon. II. Effects of cinnamaldehyde on the cardiovascular and digestive systems. Chem Pharm Bull (Tokyo) 23:941–947
29. Harada M, Saito A (1978) Pharmacological studies on Chinese cinnamon. IV. Effect of cinnamaldehyde on the isolated heart of guinea pigs and its catecholamine releasing effect from the adrenal gland of dogs. J Pharmocobiodyn 1:89–97
30. Yagi A, Nohara T, Nishioka I, Koda A, Nagai H, Noda K, Tokubuchi N (1980) Chemical studies on aqueous extract of cinnamoni cortex and its therapeutic trial from experimental nephritis. Wakanyaku Simpojumu (Kiroku) 13:72–78 (CA 95:35420a)
31. Chung SH, Paik SY (1981) An experimental study on the effects of several phytohaemagglutinins, prednisolone acetate and testosterone propionate on the peripheral blood picture of rabbits in acute blood loss. Karyo Taehakkyo Uikwa Taehak Chapchi 18:257–269 (CA 95:214989q)
32. Akira T, Tanaka S, Tabata M (1986) Pharmacological studies on the antiulcerogenic activity of Chinese cinnamon. Planta Med 52:440–443
33. Moon KH, Pack MY (1983) Cytotoxicity of cinnamic aldehyde on leukemia L1210 cells. Drug Chem Toxicol 6:521–535
34. Institute of Physical and Chemical Research Higashikaze, Mutsuyuki; Kaken Chemical (1981) Aldehydes and neoplasm inhibitors. Jpn Kokai Tokkyo Koho 8112, 313 (CA 94:168054k)
35. Ohta T, Watanabe K, Moriya M, Shirasu Y, Kada T (1983) Antimutagenic effects of cinnamaldehyde on chemical mutagenesis in *Escherichia coli*. Mutat Res 107:219–227
36. Ohta T, Watanabe K, Moriya M, Shirasu Y, Kada T (1983) Analysis of the antimutagenic effect of cinnamaldehyde on chemically induced mutagenesis in *Escherichia coli*. Mol Gen Genet 192:309–315
37. Kakimuma K, Koike J, Kotani K, Ikekawa W, Kado T, Nomoto M (1984) Cinnamaldehyde: identification of an antimutagen from a crude drug, cinnamomi cortex. Agric Biol Chem 48:1905–1906
38. Danneman PJ, Broman KA, Dorsky J, Kohrman KA, Rothenstein AS, Sedlak RI, Steltenkamp RJ (1983) Cinnamic aldehyde: a survey of consumer patch-test sensitization. Fd Chem Toxicol 21:721–725
39. Delbressine LPC, Klippert PJM, Renvers JTA, Seutter-Berlage F (1980) Identification of two sulfur containing urinary metabolites of cinnamic aldehyde in the rat. Br J Pharmacol 68:165 P
40. Delbressine LPC, Klippert PJM, Renvers JTA, Seutter-Berlage F (1981) Isolation and identification of mercapturic acid of cinnamic aldehyde and cinnamyl alcohol from urine of female rats. Arch Toxicol 49:57–64

Cissampelos pareira L. var. *hirsuta* (Buch. ex DC.) Formen

43

43.1 Introduction

Yahunu, Herba Cissampelotis, is the dry whole plant of *Cissampelos pareira* L. var. *hirsuta* (Buch. ex DC.) Formen (Menispermaceae) collected in spring and summer. It is officially listed in the Chinese Pharmacopoeia and is used as an analgesic and hemostatic for reduction of swelling and for external treatment of traumatic pain and bleeding.

43.2 Chemical Constituents

Cissampelos pareira contains a number of alkaloids, especially bisbenzylisoquinoline alkaloids. The major alkaloid is hayatine (*43-1*) with a content of 0.1% [1]. Hayatine has a tubocurarane (*43-2*) skeleton [2–4] and is racemic. The optically active compound, named curine [5, 6] and its stereoisomer bebeerine [7, 8], were also isolated from *C. pareira*. Further alkaloids with a tubocurarane skeleton isolated from *C. pareira* are 12′-*O*-methylcurine (*43-3*) [9–11] and its stereoisomer hayatinine [12].

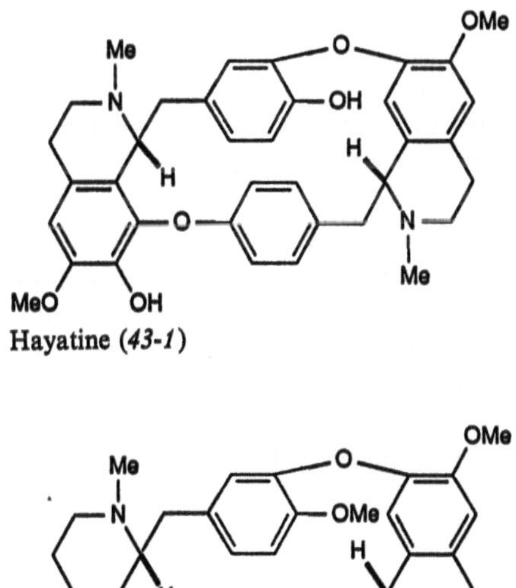
Hayatine (*43-1*)

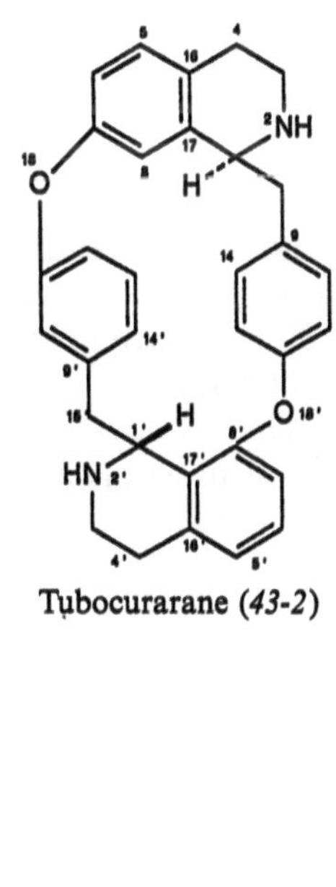
Tubocurarane (*43-2*)

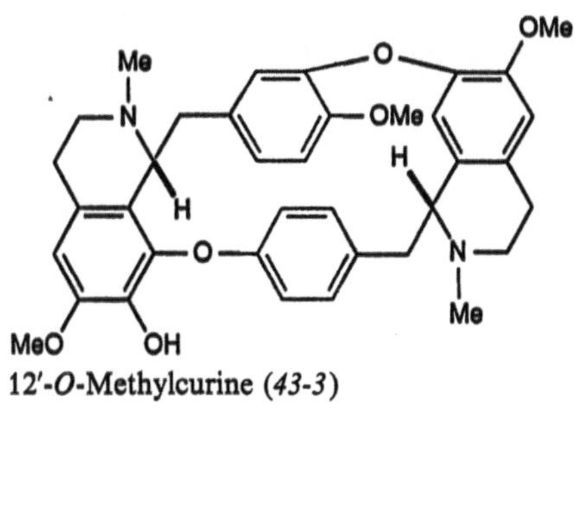
12′-*O*-Methylcurine (*43-3*)

In addition, isochondrodendrine (*43-4*) [5,7], cycleanine (*43-5*) [8, 13], cissampareine (*43-6*) [14, 15], and cissamine (cyclanoline, *43-7*) [16, 17], an alkaloid derived from dibenzo[*a*,*g*]-quinolizine, were also isolated from *C. pareira*.

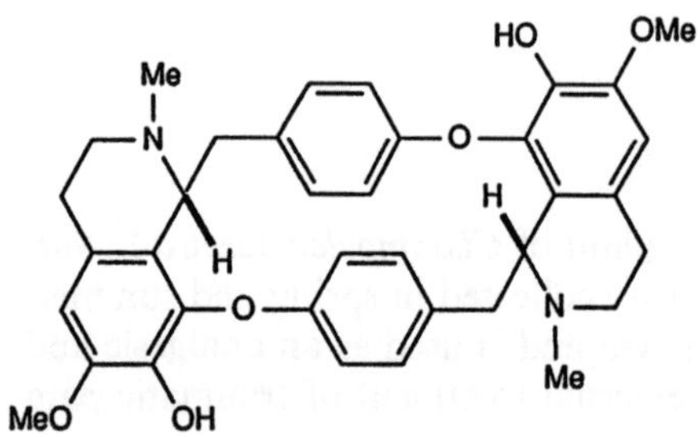

Isochondrodendrine (*43-4*)

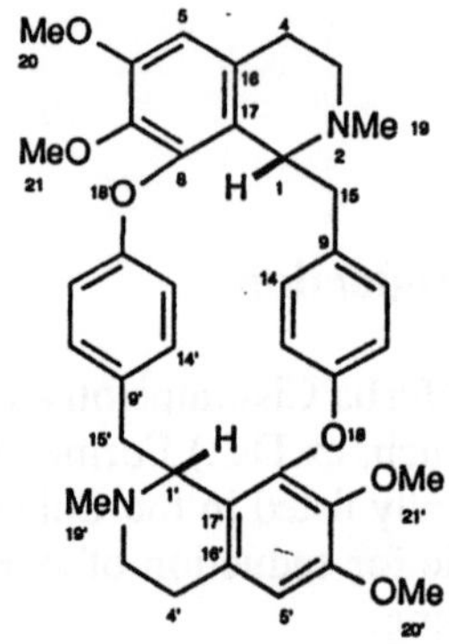

Cycleanine (*43-5*)

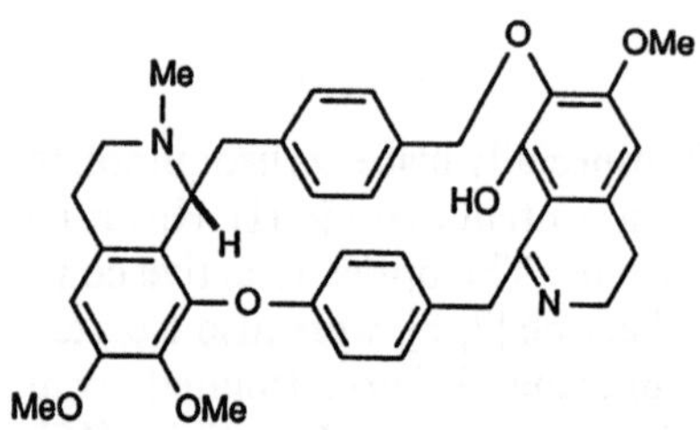

Cissampareine (*43-6*)

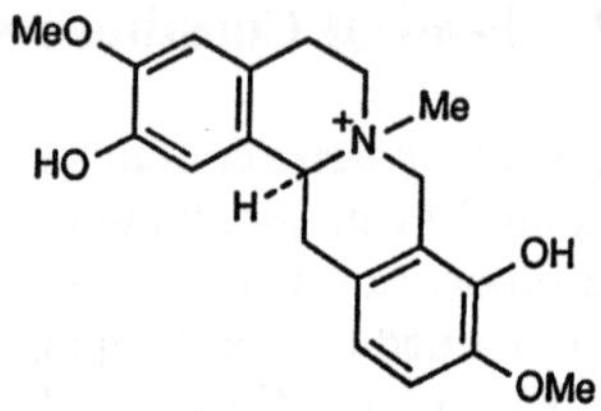

Cissamine (*43-7*)

Cyclea species, belonging also to the menispermaceous plant family, contain a number of isoquinoline, especially bisbenzylisoquinoline alkaloids. They are used in folk medicine as analgesic and muscle relaxant; however, they are not officially listed in the Chinese Pharmacopoeia. The alkaloid constituents of *Cyclea densiflora*, *C. barbata*, *C. hypoglauca*, *C. migoana*, *C. racemosa*, *C. polypetala*, and *C. hainanensis* were investigated. Homoaromoline (*43-8*), tetrandrine, isotetrandrine, berbamine, fangchinoline, isochondrodendrine, curine, cycleanine, cycleanorine (*43-9*), and magnoflorine (*21-3*) were detected. Curine and isochondrodendrine were the major constituents [18]. Homoaromoline possesses an oxyacanthan (*43-10*) skeleton, and cycleanorine, a berbaman skeleton (*43-11*). Tetrandrine, isotetrandrine, fangchinoline, and berbamine all having the berbaman skeleton, are the main alkaloid constituents contained in *Stephania* species and will be described in that chapter in detail.

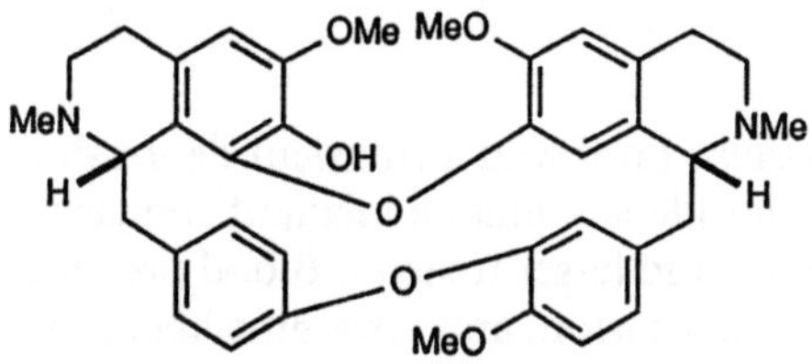

Homoaromoline (*43-8*)

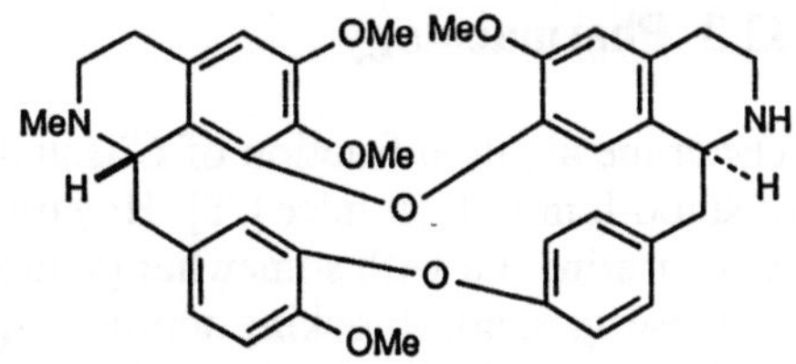

Cycleanorine (*43-9*)

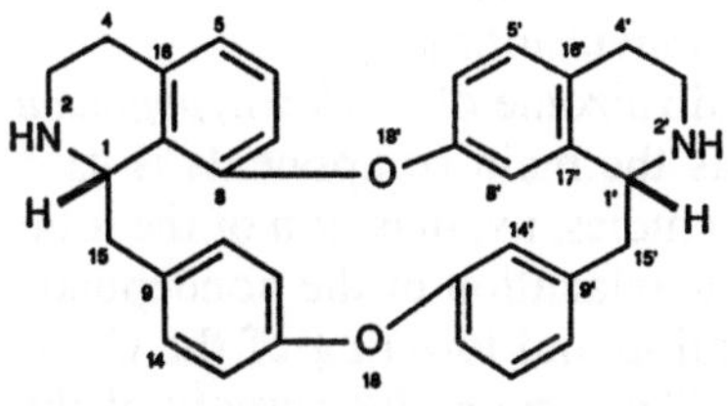

Oxyacanthan (*43-10*)

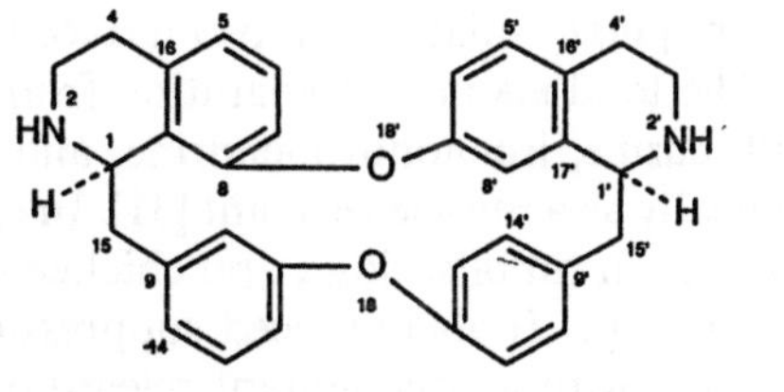

Berbaman (*43-11*)

Among the *Cyclea* species *C. barbata* [19–21] *C. hainanensis* [21–23], *C. tonkinensis* [24, 25], and *C. hypoglauca* [26] are the most investigated. From *C. hainanensis*, 12′-*O*-methylcurine, hayatine, cyclanoline, cissamine, and a new alkaloid named α-hainanine [22] were isolated. α-Hainanine (*43-12*) is like cissamine, a quaternary alkaloid possessing the dibenzo[*a,g*]quinolizinium skeleton; the latter was also found in *C. barbata* [19] and *C. tonkinensis* [24, 25]. Insularine (*43-13*) and its 7-*O*-demethyl analog insulanoline (*43-14*) are further alkaloids isolated from *C. hypoglauca* [26]. A new bisbenzylisoquinoline alkaloid named cycleaneonine (*43-15*) was recently isolated from the roots of *C. racemosa* [27].

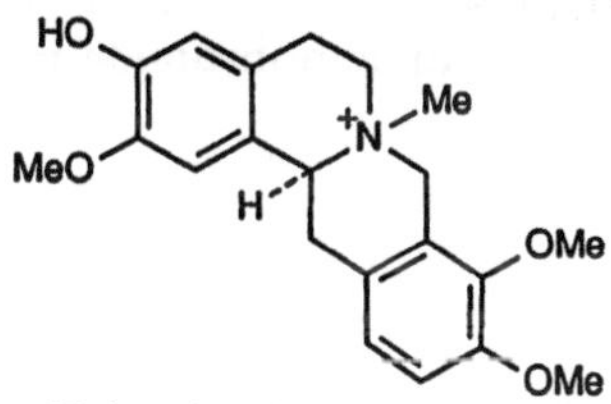

α-Hainanine (*43-12*)

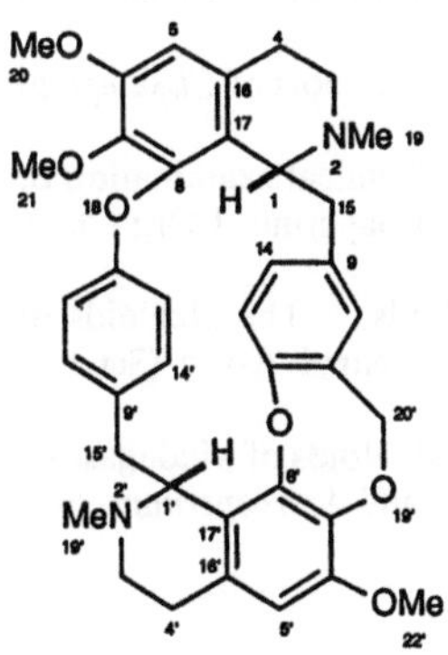

Insularine (*43-13*)

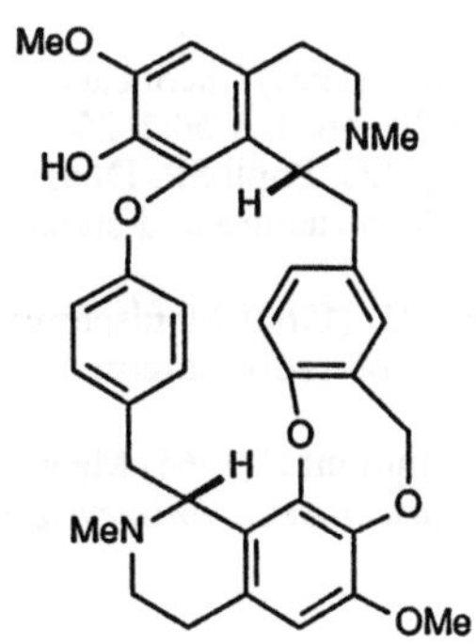

Insulanoline (*43-14*)

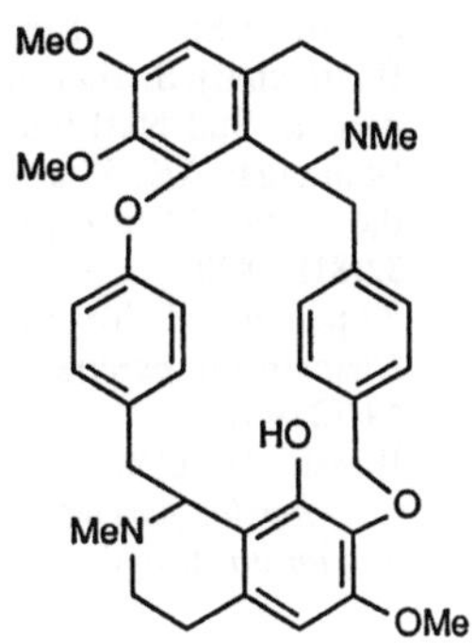

Cycleaneonine (*43-15*)

43.3 Pharmacology

The crude alkaloid fraction of *Cissampelos pareira* produced a transient depression of smooth muscle in mice [28]. Hayatine methiodide is a muscle relaxant similar to tubocurarine but with somewhat greater potency and lesser toxicity. Blood pressure decrease, ganglion-blocking action, respiratory depression, and histamine liberation are less marked compared to tubocurarine in equipotent dosage [29]. Hayatin methiodide injected intraventricularly, intrathecally, or intracisternally in cats or dogs increased the rate and depth of respiration, somatic reflexes, salivation, lacrimation, and pupillary dilation. It also induced generalized convulsions [30].

The total alkaloid preparation from the root and rhizome of *Cyclea hypoglauca* with curine, isochondrodendrine, and cycleanine as the main components is used clinically as a muscle relaxant [31]. In experimental studies, i.v. injection of the total alkaloid in rabbits, dogs, and chicken caused muscle relaxation of the nondepolarization type. It also induced suppression of respiration and lowering of the blood pressure with a concomitant release of histamine. The hypotensive activity of the total alkaloid was antagonized by antihistaminergics such as diphenhydramine. In high doses the total alkaloid caused respiratory paralysis. The LD_{50} value of i.v. injected total alkaloid in mice was 0.53 mg/kg [32]. The muscle relaxant activity of curine methochloride was found to be within the range of that of tubocurarine chloride. Both have a nondepolarizing mode of action, and the blocking site is at the cholinergic receptor in the postsynaptic membrane. The muscle relaxant action and the inhibition of breathing could be antagonized by neostigmine. It is considered a safe muscle relaxant for surgery [33].

Cycleanine dimethobromide given i.v. at a dose of 0.5–1.0 mg/kg had a hypotensive effect in dogs. Blood pressure returned to normal 30 min later [34]. Heart rate, cardiac output, left ventricular performance and total peripheral resistance were decreased [35]. Ganglionic blockade plays a role in the hypotensive mechanism of cycleanine dimethobromide [36]. The i.p. LD_{50} of cycleanine dimethobromide in mice was 2.9 mg/kg [34].

References

1. Chen ZL (1980) Structure of Daijisong. Acta Chim Sin 38:567–572
2. Bhattacharji S, Sharma VN, Dhar ML (1952) Chemical examination of the roots of *Cissampelos pareira*. J Sci Ind Res 11B:81–82
3. Bhattacharji S, Sharma VN, Dhar ML (1955) Chemical constituents of the roots of *Cissampelos pareira*. Bull Natl Inst Sci India 4:39–46 (CA 50:2626c)
4. Bhatnagar AK, Bhattacharji S, Roy AC, Popli SP, Dhar ML (1967) Chemical examination of the roots of *Cissampelos pareira*. IV. Structure and stereochemistry of hayatine. J Org Chem 32:819–820
5. Kupchan SM, Yokoyama N, Beal JL (1960) Menispermaceae alkaloids. I. The alkaloids of *Cissampelos pareira* and the origin of radix pareira bravae. J Am Pharm Assoc (Sci Ed) 49:727–731
6. Boissier JR, Combes G, Pernet R, Dumont C (1965) Menispermaceae alkaloids of Madagascar: *Cissampelos pareira*, *Cycela madagascariensis*, *Anisocycla grandidieri*, and *Spirospermum penduliflorum*. Lloydia 28:191–198

7. Srivastava RM, Khare MP (1963) Water-soluble alkaloids of the root bark of *Cissampelos pareira*. Curr Sci 32:114
8. Chowdhury AR (1972) Chemical investigation on *Cissampelos pareira*. Sci Cult 38:358–359 (CA 78:82058x)
9. Haynes LJ, Herbert EF, Plimmer JR (1986) (+ +)-4″-*O*-methylcurine from *Cissampelos pareira*. J Chem Soc (C Org) 615–617
10. Bhattacharji S, Sharma VN, Dhar ML (1956) Chemical examination of the roots of *Cissampelos pareira*. J Sci Ind Res 15B:363–368
11. Bhattacharji S, Roy AC, Dhar ML (1962) Chemical examination of the roots of *Cissampelos pareira*. II. Isolation and structure of hayatinine. J Sci Ind Res 21B:428–433
12. Bhatnagar K, Popli SP (1967) Chemical examination of the roots of *Cissampelos pareira*. III. Structure and stereochemistry of hayatinine. Indian J Chem 5:102–104
13. Don SQ, Guan ZX, Shan GJ, Zhou Y, Xie R (1987) The crystal structure of alkaloid B from *Cissampelos pareira* L. J Struct Chem 6:84–88
14. Kupchan SM, Patel AC, Fujita E (1965) Tumor inhibitors. VI. Cissampareine, new cytotoxic alkaloid from *Cissampelos pareira*. Cytotoxicity of bisbenzylisoquinoline alkaloids. J Pharm Sci 54:580–583
15. Kupchan SM, Kubota S, Fujita E, Kobayashi S, Block JH, Telang SA (1966) Tumor inhibitors. XV. The structure and configuration of cissampareine, a novel bisbenzylisoquinoline alkaloid. J Am Chem Soc 88:4212–4218
16. Srivastava RM, Khare MP (1964) Water-soluble alkaloids from the root bark of *Cissampelos pareira*. Chem Ber 97:2732–2741
17. Anwer F, Popli SP, Srivastava RM, Khare MP (1968) Medicinal plants. III. Protoberberine alkaloids from the roots of *Cissampelos pareira*. Experientia 24:999
18. Zhu ZY, Feng YX, He LY, Wang YC (1983) Study on the utilization of medicinal plant resources of the genus *Cyclea* of Menispermaceae in China. Acta Pharm Sin 18:535–540
19. Klughardt G, Zymalkowski F (1982) Magnoflorine and protoquercitol as contents of *Cyclea barbata* Miers. Arch Pharm (Weinheim) 315:7–11
20. Dahmen K, Pachaly P, Zymalkowski F (1977) Alkaloids from the Thai drug Krung Kha Mao (*Cyclea barbara*, Menispermaceae). V. Isolation and structural elucidation of further bisbenzylisoquinoline alkaloids. Arch Pharm (Weinheim) 310:95–102
21. Shanghai Inst Mat Med, Acad Sinica (1979) Active principles of neuromuscular blocking actions from *Cyclea barbata*, *Cyclea hainanensis* and *Stephania epigaea* (Menispermaceae). Koxue Tongbao 24:574–476
22. Zhang XX, Tang ZJ, Gao YL, Chen R, Lao AN, Wang CG (1981) Studies on the alkaloids of *Cyclea hainanensis* Merr. Acta Bot Sin 23:216–221
23. Tang XC, Feng J, Wang YE, Liu MZ (1980) Neuromuscular blocking activity of alkaloids of *Cyclea hainanensis*. Acta Pharmacol Sin 1:17–22
24. Chou FH, Chen CC, Liang PY, Wen C (1981) Chemical constituents of *Cyclea tonkinensis* Gagnep. Chin Pharm Bull 16:50–51
25. Zhou XF, Chen ZJ, Liang PY, Wen J (1981) Chemical constituents of *Cyclea tonkinensis*. Chin Trad Herb Drugs 12:439–440
26. Fang XD, Qian LJ, Shen PZ, Shi ZY (1985) Alkaloids of *Cyclea hypoglauca*. Chin Trad Herb Drugs 16:536
27. Lai S, Zhao TF, Wang XK (1988) Cycleaneonine, a new bisbenzylisoquinoline alkaloid from *Cyclea racemosa* Oliv. Acta Pharm Sin 23:356–360
28. Roy PK, Dutta AT, Ray GK, Mukerji B (1952) Preliminary note on the pharmacological action of the total alkaloids isolated from *Cissampelos pareira*. Indian J Med Res 40:95–99
29. Mukerji B, Bhandari PR (1959) *Cissampelos pareira*, source of a new curariform drug. Planta Med 7:250–259
30. Sur RN, Pradhan SN (1964) *Cissampelos* alkaloids. I. Action of hayatine derivatives on the central nervous system of cats and dogs. Arch Int Pharmacodyn 152:106–114
31. Zhou FX, Liang F, Fang JS, Zhang KS, Liang C, Tian AP, Fang XD, Shi ZY, Gu ZH (1982) Preparation of the neuromuscular blocking agent, Guang Ji Song. Chin Pharm Bull 17:135–137
32. Huang SL, Zhao CS, Wei HY, Bong XX, Li YD, Li M, Yang WL, Chen DZ, Wu YQ (1983) Neuromuscular blocking effect of Guang Ji Song. Chin Pharm Bull 18:669–672
33. Cooperative group on clinical trials of dimethyl-1-curine dimethochloride (1981) Chin Trad Herb Drugs 12:26–29

34. Sun Z, Yang XY, Dai ZL, Jiu GZ, Zhang ZD (1980) Hypotensive effect of dimethyl cycleanine bromide. Acta Pharmacol Sin 1:23–27
35. Chenz WZ, Dong YL, Ding GS (1980) Effects of dimethyl cycleanine bromide on cardiac hemodynamics of anesthetized dogs. Acta Pharmacol Sin 1:27–30
36. Cao ZE, Shen JF, Song JZ, Hu ZM, Zhu XZ (1981) Mechanism of hypotensive effect of cycleanine dimethobromide. Acta Pharmacol Sin 2:160–163

Citrus spp. 44

44.1 Introduction

A number of *Citrus* species have been used in traditional Chinese medicine. The following official entries involving with *Citrus* species are in the Chinese Pharmacopoeia:

- Huajuhong, Exocarpium Citri grandis, is the dry external pericarp of the fruits of *Citrus grandis* "Tomentosa" or *C. grandis* (L.) Osbeck (Rutaceae). It should be collected in summer when the fruits are still unripe or almost half ripe. It is used as an expectorant and stomachic.
- Foshou, Fructus Citri sarcodactylis, is the dry fruit of *C. medica* L. var. *sarcodactylis* Swingle, collected in the fall and dried after cutting into thin slices. It is used mainly as a stomachic.
- Chenpi, Pericarpium Citri reticulatae, is the dry pericarp of the ripe fruits of *C. reticulata* Blanco and its cultivated variants. It is used as an expectorant and stomachic. The main cultivated variants mentioned in the Chinese Pharmacopoeia are: *C. reticulata* "Chachi," *C. reticulata* "Dahongpao," *C. reticulata* "Unshiu," and *C. reticulata* "Tangerina."
- Qingpi, Pericarpium Citri reticulatae viride, is the pericarp of the unripe fruits of *C. reticulata* Blanco, which is gathered in July and August. It is used as a stomachic and also in the treatment of hernia.
- Zhiqiao, Fructus Aurantii, is the dry unripe fruits of *C. aurantium* L. and its cultivated variants collected in July when the pericarp of the fruits is still green. It is used as a digestant, expectorant, and in the treatment of anal prolapse. The cultivated variants listed in the Chinese Pharmacopoeia are *C. aurantium* "Huangpi," *C. aurantium* "Daidai," *C. aurantium* "Chuluan," and *C. aurantium* "Tangcheng."
- Zhishi, Fructus Aurantii immaturus, is the immature fruit of *C. aurantium* L. and its cultivated variants and of *C. sinensis* Osbeck collected from May to June. It is used as an expectorant and digestant, and in the treatment of diarrhea, anal prolapse, and frank prolapse.
- Xiangyuan, Fructus Citri, is the dry ripe fruit of *C. medica* L. or *C. wilsonii* Tanaka, collected in the fall. It is used for stimulation of digestive function and as an expectorant.
- Juhong, Exocarpium Citri rubrum, is the dry external pericarp of the ripe fruits of *C. reticulata* Blanco and its cultivated variants, collected in the late fall and early winter. It is used as an expectorant.
- Juhe, Semen Citri reticulatae, is the seeds of *C. reticulata* Blanco and is used mainly in the treatment of hernia.

44.2 Chemical Constituents

The chemical composition of the *Citrus* fruits or fruit peel shows a wide variability because the *Citrus* species officially listed in the Chinese Pharmacopoeia are mainly cultivated fruits such as oranges, mandarins, and shaddocks from different areas with a great number of variants. Flavones, limonins, alkaloids, essential oil with terpene and nonterpene constituents, and some other compounds were isolated and identified from fruits and fruit peel of different *Citrus* species.

Naringin (*44-1*), a flavanone glycoside with naringenin (*44-2*) as aglycone, is the main flavone constituent of the fruit peel of *C. grandis* and its variant *C. grandis* "Tomentosa" [1]. In a study with 26 samples, the naringin contents were 1.8%–6.0% in 24 samples and as high as 32%–33% in two samples [1]. Naringin was isolated from *Citrus* species more than 100 years ago, and the structure of the aglycon naringenin and the composition and position of the sugar moiety were determined in 1928 and 1929, respectively [2, 3].

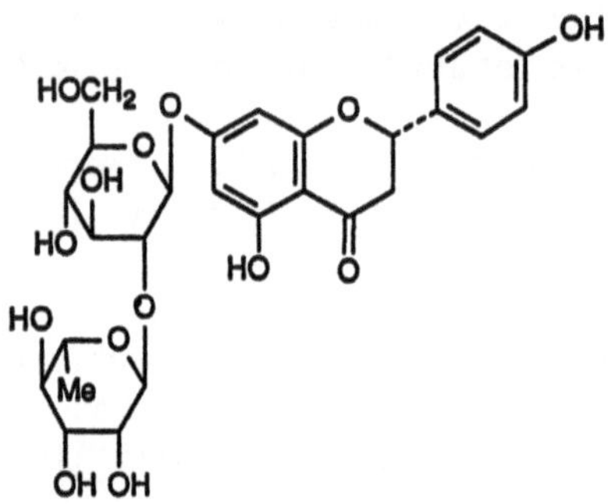

Naringin (*44-1*)

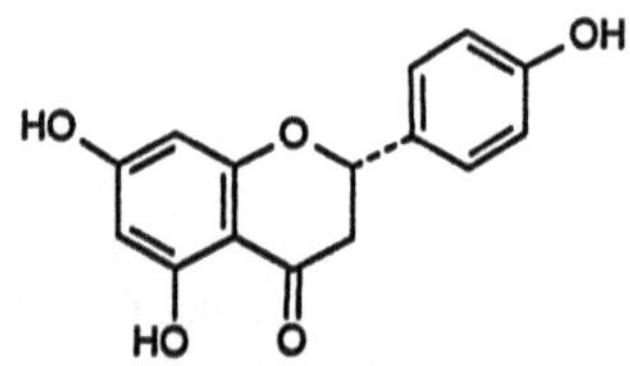

Naringenin (*44-2*)

Hesperidin (*44-3*) is the major flavone derivative in the peel of *C. reticulata* and its variants. It is also a flavanone derivative with hesperetin as the aglycon and rutinose as the sugar moiety. Hesperidin was isolated from *Citrus* species more than 150 years ago. The structure of its aglycon [4] and the position [5] and structure of the sugar moiety [6] were determined in the early 1930s.

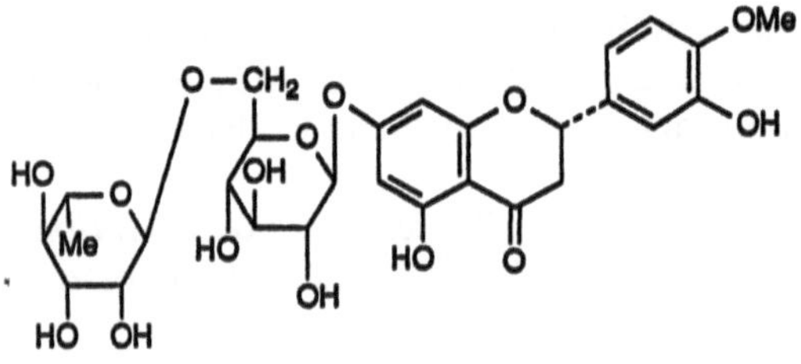

Hesperidin (*44-3*)

The hesperidin content in the peels of *C. reticulata* is dependent on species. The peel of *C. reticulata* "Unshiu," *C. reticulata* "Tangerina," and *C. reticulata* "Chachi" contained 7%–10% hesperidin, and that of *C. sinensis* contained 6% hesperidin [7, 8]. The hesperidin content in the peel decreased with the ripening of the fruit [9].

Besides hesperidin, flavanone citromitin (*44-4*) [10] and a number of flavones [11] such as nobiletin (*44-5*), 5-demethylnobiletin, 5,7,8,4′-tetramethoxyflavone [11, 12], and 5,4′-dihydroxy-6,7,8,3′-tetramethoxyflavone [13] were isolated and identified.

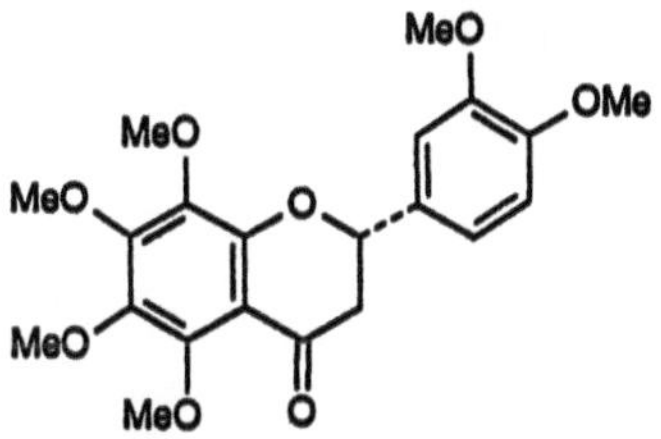

Citromitin (*44-4*)

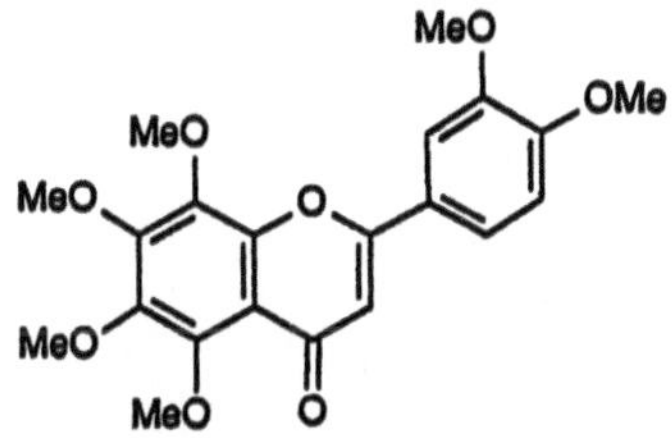

Nobiletin (*44-5*)

Hesperidin was also detected in the peel of *C. aurantium* [14] together with neohesperidin (*44-6*), another hesperetin glycoside carrying the sugar neohesperidose instead of rutinose in position 7 [15, 16].

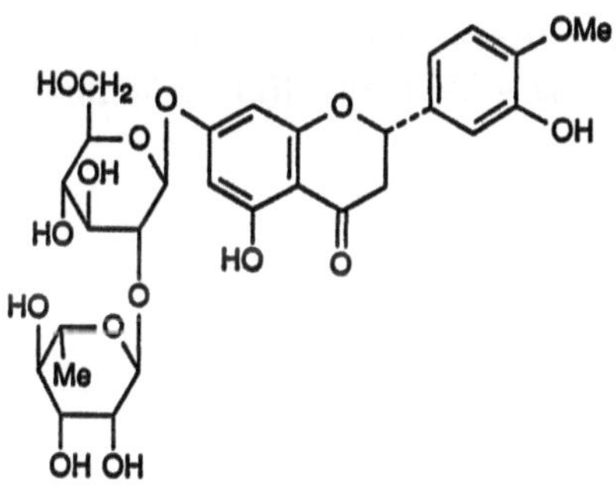

Neohesperidin (*44-6*)

The flavones tangeretin (*44-7*), tetra-*O*-methylscutellarin, 3,5,6,7,8,3′,4′-heptametoxyflavone, nobiletin, sinensetin (*44-8*) [17], auranetin (*44-9*), and 5-hydroxyauranetin (*44-10*) [18] were found in the peel of *C. aurantium*. 8-*C*-*β*-D-Glucopyranosyldiosmetin (*44-11*) was isolated from the peel of *C. sinensis* [19].

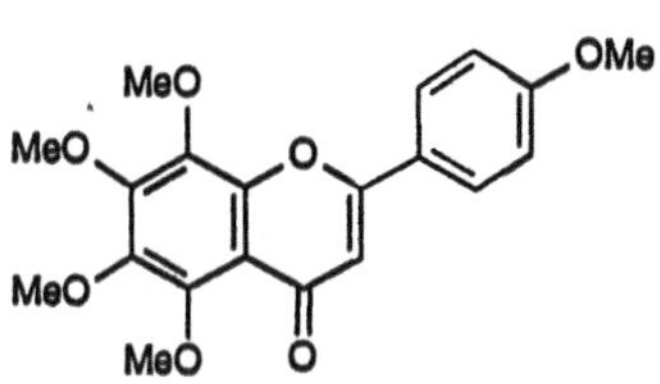

Tangeretin (*44-7*)

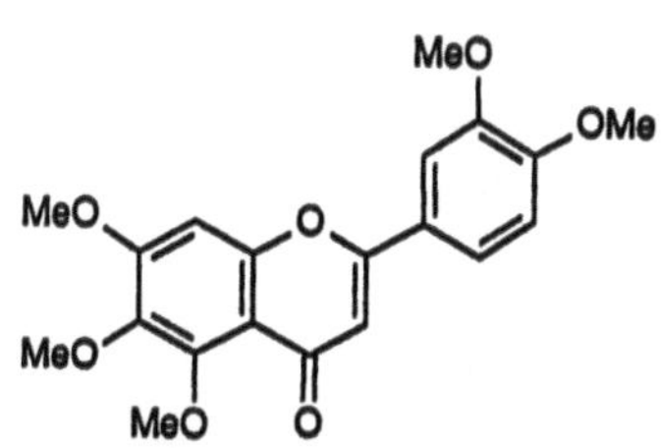

Sinensetin (*44-8*)

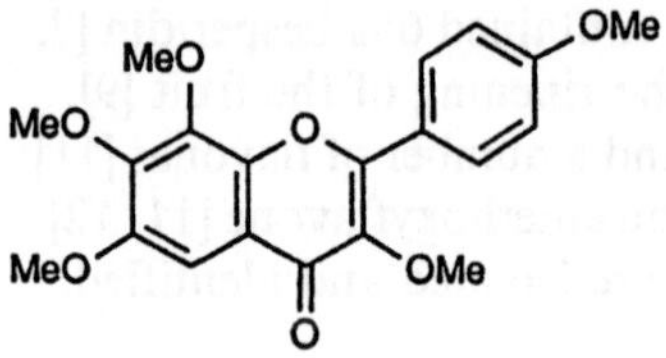

Auranetin (*44-9*)

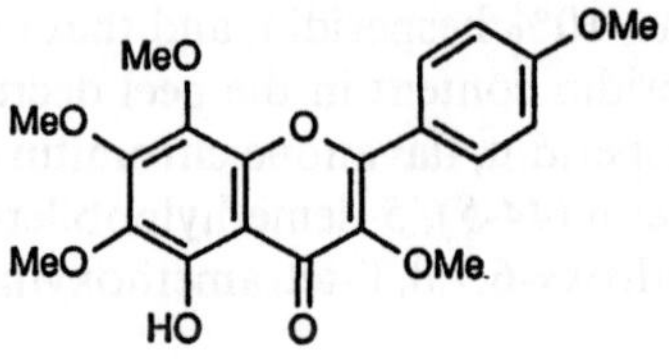

5-Hydroxyauranetin (*44-10*)

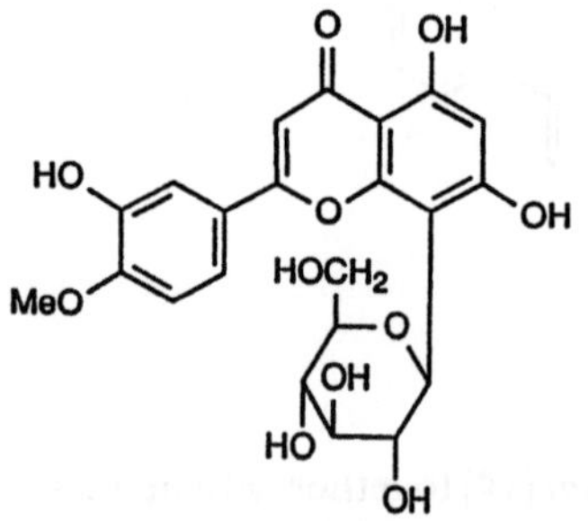

8-*C*-*β*-D-Glucopyranosyl-diosmetin (*44-11*)

Another active principle in the peel of *C. aurantium* is the alkaloid synephrine (*44-12*), the structure of which was determined as 4-hydroxy-α-[(methylamino)-methyl]-benzenemethanol [20].

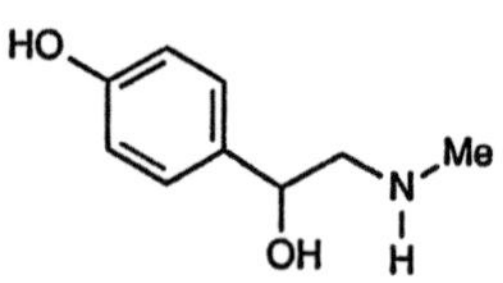

Synephrine (*44-12*)

Synephrine was also found in the peel of *C. reticulata* [21, 22], *C. sinensis* [21], and *C. wilsonii* [23] and in the peel and fruit of *C. aurantium*, *C. reticulata*, and *C. sinensis* from various locations in China at contents of 0.1%–2% [21, 24]. In addition to synephrine, *N*-methyltyramine (*44-13*) was also isolated [22]. Synephrine is formed biosynthetically in *Citrus* by a pathway involving tyramine and *N*-methyltyramine [25].

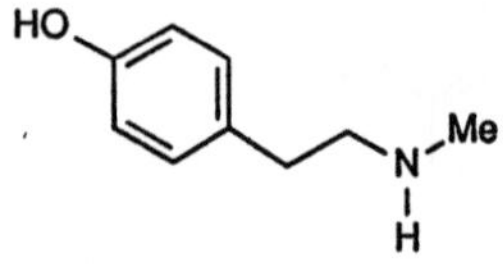

N-Methyltyramine (*44-13*)

Carotene pigments in the peel of *Citrus* species were also studied. Most of carotene pigments of *C. sinensis* isolated from the peel are derived from *β*,*β*-carotene

(*44-14*) such as cryptoxanthin (*44-15*), 5,5′,6,6′-tetrahydro-*β*,*β*-carotene, luteoxanthin (*44-16*), mutatochrome (*44-17*), auroxanthin (*44-18*), and zeaxanthin (*44-19*); some are derived from *ψ*,*ψ*-carotene (*44-20*) such as phytoene (*44-21*) and phytofluene (*44-22*), as well as the polyene compounds sintaxanthin (*44-23*) and *β*-apo-10′-carotenal (*44-24*) [26]. Cryptoxanthin is the main carotene pigment.

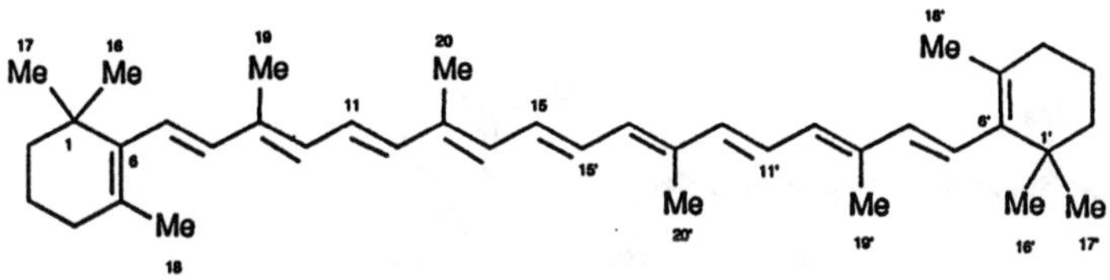

β,*β*-Carotene (*44-14*)

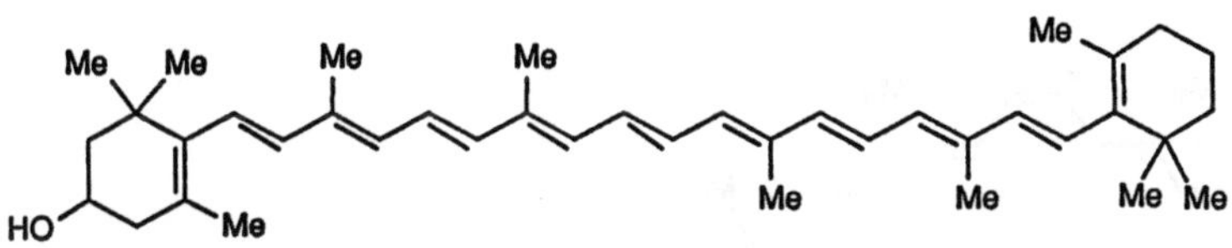

Cryptoxanthin (*44-15*)

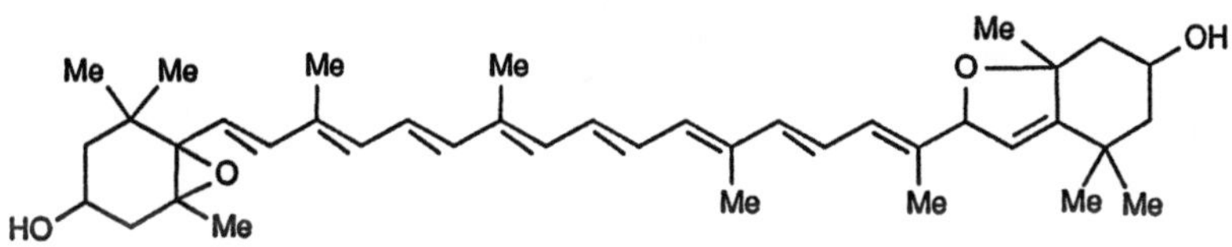

Luteoxanthin (*44-16*)

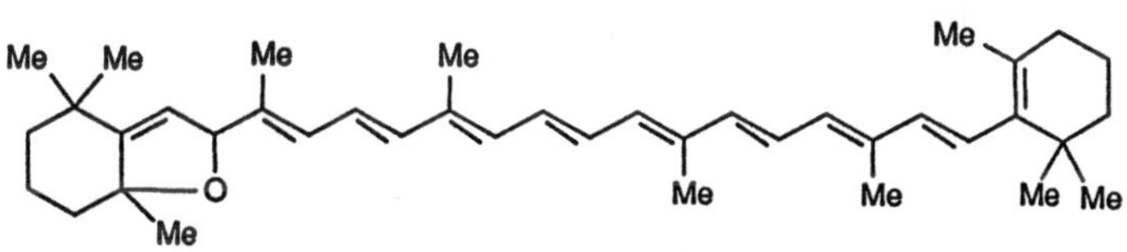

Mutatochrome (*44-17*)

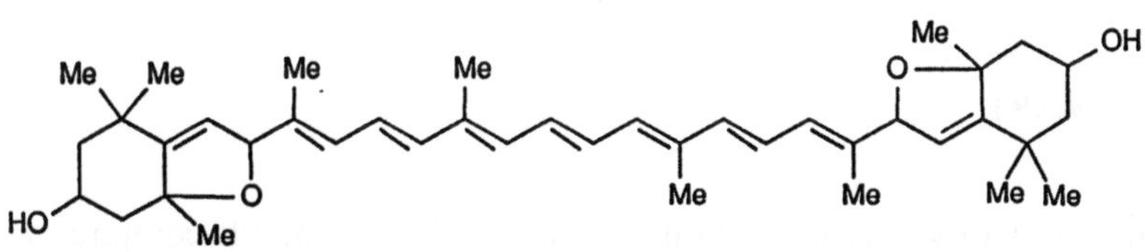

Auroxanthin (*44-18*)

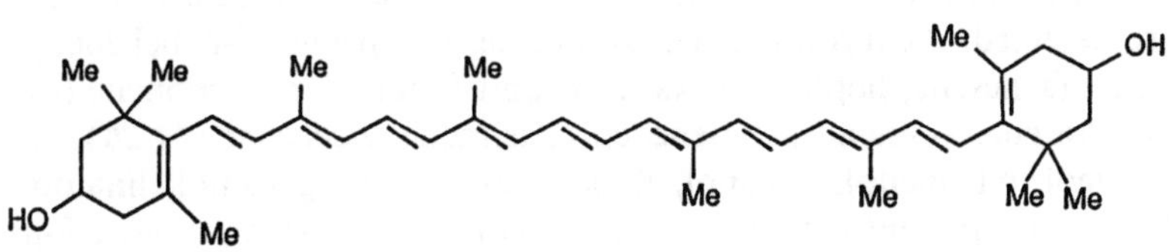

Zeaxanthin (*44-19*)

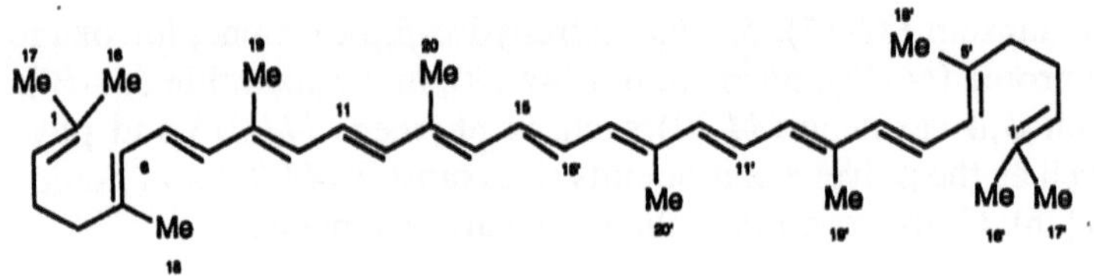

ψ,ψ-Carotene (*44-20*)

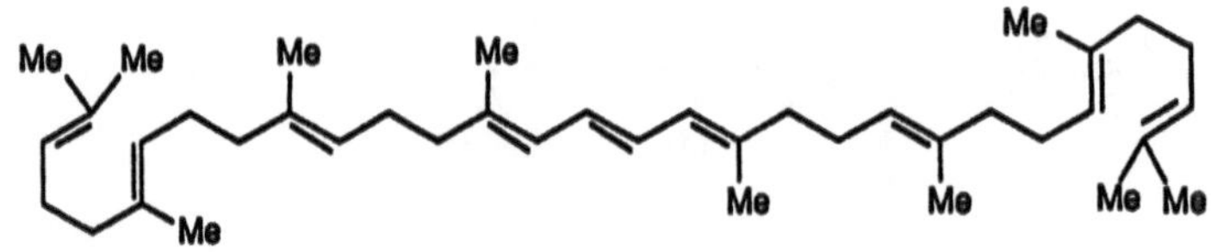

Phytoene (*44-21*)

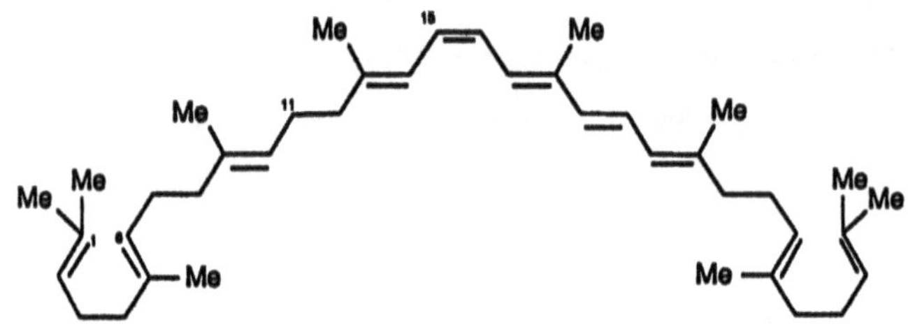

Phytofluene (*44-22*)

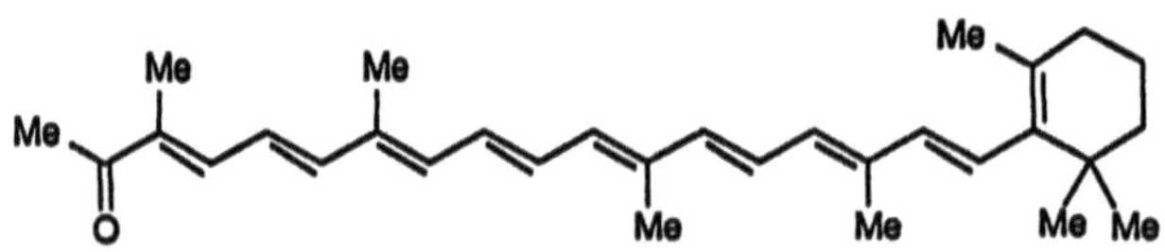

Sintaxanthin (*44-23*)

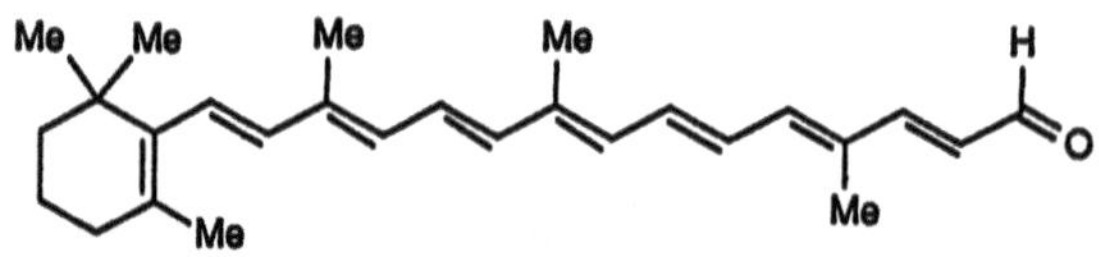

β-Apo-10′-carotenal (*44-24*)

The essential oil of *Citrus* species contains a great number of volatile compounds including hydrocarbons, alcohols, aldehydes, ketones, and esters of a terpene or nonterpene nature. From the essential oil of *C. aurantium* "Valencia", the following compounds were detected: β-caryophyllene, α-copaene, β-copaene, β-cubebene, *p*-cymene, β-elemene, farnesene, heptane, hexane, α- and β-humulene, limonene (*44-25*), myrcene, nootkatene, α-pinene, sabinene, epi-α-selinene, valencene (*44-26*), *cis*-carveol, *trans*-carveol, citronellol, decanol, dodecanol, elemol, geraniol, linalool, intermedeol (*44-27*), 1,8-*p*-menthadien-9-ol, *cis*-2,8-*p*-methadien-1-ol, *trans*-2,8-*p*-menthadien-1-ol, nerol, nonanol, octanol, α-terpineol, undecanol, acetaldehyde, decanal, dodecanal, geranial, octanal, perillaldehyde (*44-28*), α-sinensal (*44-29*), β-sinensal (*44-30*), acetone, carvone, ethyl vinyl ketone, nootkatone, piperitenone (*44-31*), ethyl acetate, ethyl butyrate, ethyl propionate, methyl butyrate, 1,8-*p*-menthadien-9-yl acetate, octyl acetate, 1,1-diethoxyethane, and *trans*-linalool oxide [27]. Limonene is the major compound in the essential oil, with a content of ca. 93% [27].

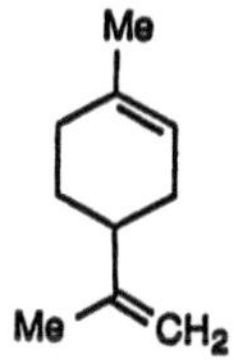
Limonene (*44-25*)

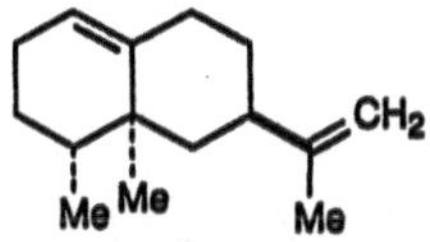
Valencene (*44-26*)

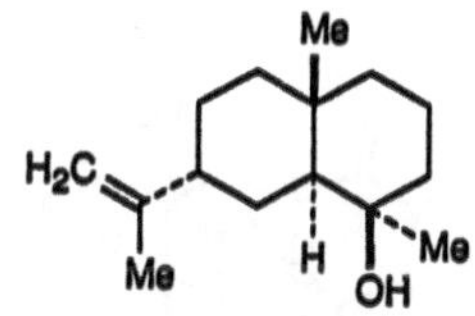
Intermedeol (*44-27*)

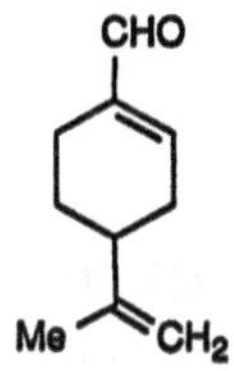
Perillaldehyde (*44-28*)

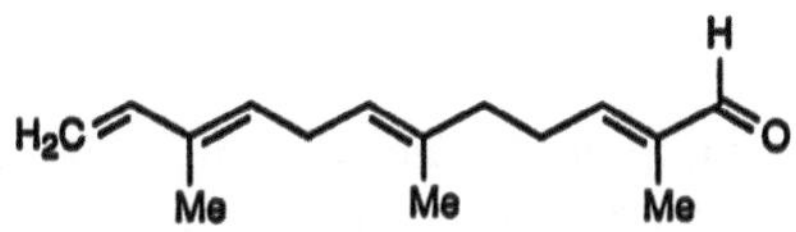
α-Sinensal (*44-29*)

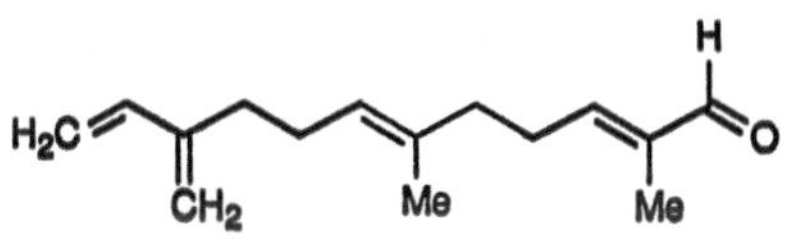
β-Sinensal (*44-30*)

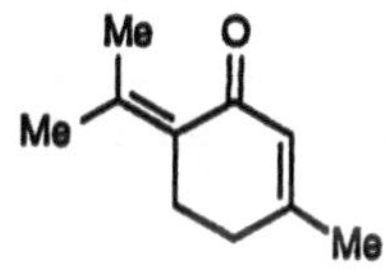
Piperitenone (*44-31*)

The monoterpene composition of essential oil from *C. reticulata* "Mandarin" and *C. reticulata* "Tangarine" is different from that of *C. aurantium* "Valencia"; however, limonene is still the major constituent. The monoterpenes of *C. reticulata* oil have been shown to include α-pinene, β-pinene, camphene, myrcene, 3-carene, α-phellandrene, β-phellandrene, α-terpinene, β-terpinene, limonene, and the aromatic hydrocarbon *p*-cymene [28].

Recently, the isolation of two new ketones *cis*- and *trans*-4,7-dimethyl-bicyclo-[3.2.1]oct-3-en-6-one (*44-32*) from the essential oil of orange was reported [29].

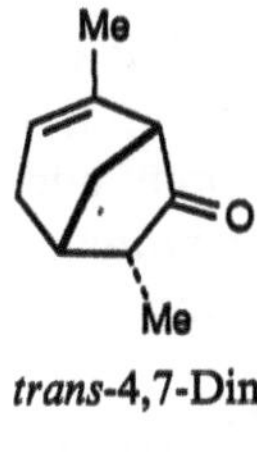
trans-4,7-Dimethyl-bicyclo[3,2,1]oct-3-en-6-one (*44-32*)

The seeds of *Citrus* species are rich in limonin bitter substances, which are a class of C_{26}-degraded triterpenes believed to arise as oxidation products from tetracyclic triterpenes. Limonin (*44-33*), the representative of this substance class and the major bitter principle in *Citrus* seeds, has been known for over 100 years [30]. As a reasonably accessible material, it is the most extensively studied member of this class. The structure was determined in 1960 [31] and was confirmed by X-ray crystallographic studies with the iodoacetate of epilimonol [32, 33].

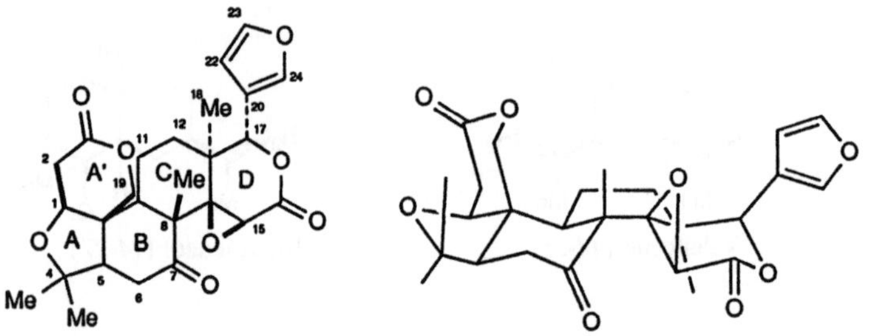

Limonin (*44-33*)

Limonin-related compounds deoxylimonin (*44-34*) [34], nomilin (*44-36*) [35], obacunone (*44-35*) [36], and deacetylnomilin (*44-37*) [34] were isolated [37]. From a chemotaxonomic study on the distribution of the limonin bitter substances in different *Citrus* species, the presence of obacunone, limonin, and deacetylnomilin in *C. reticulata* and *C. sinensis*, obacunone and limonin in *C. medica* and limonin and deacetylnomilin in *C. aurantium* was described [38].

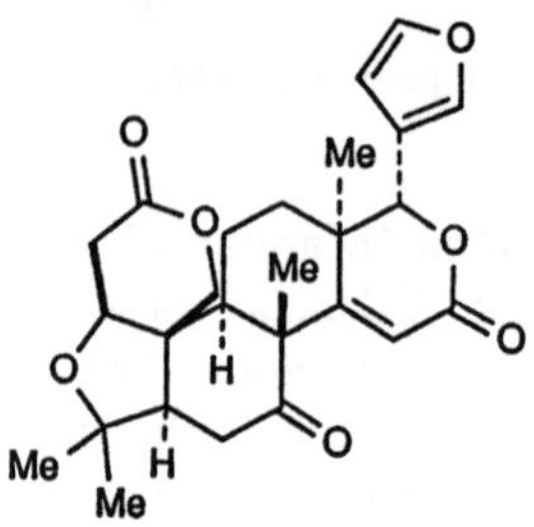

Deoxylimonin (*44-34*)

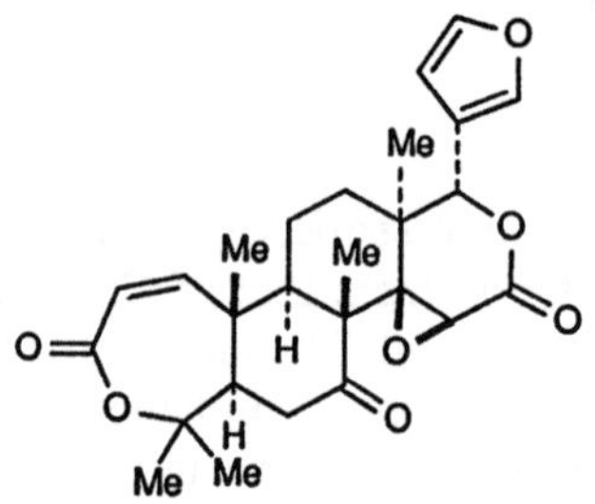

Obacunone (*44-35*)

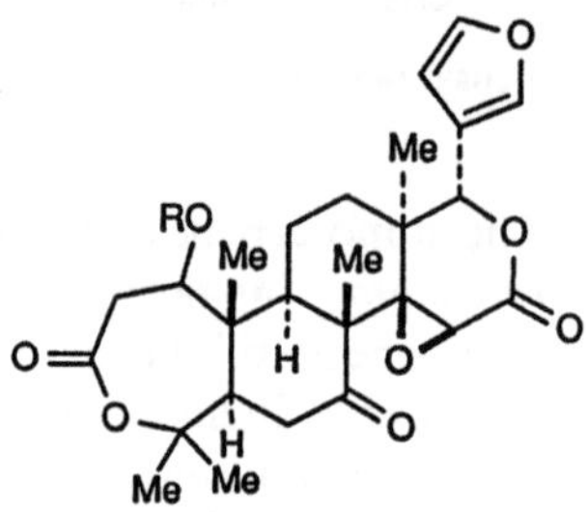

Nomilin (*44-36*): R=Ac
Deacetylnomilin (*44-37*): R=H

The fruit pulp of *Citrus* species also contains limonin bitter principles, but the content is less than that in the seed. Thus, nomilin, limonin, deoxylimonin, and obacunone were reported to be present in seeds of *C. grandis* at 70–600 ppm, and in fruit pulp at 0.02–3 ppm, respectively [37]. Biochemical experiments with radioactive tracer have shown that *Citrus* seeds accumulate limonins by translocation from fruit tissue during growth of the fruit [39].

Studies on the chemical constituents in *C. medica* var. *sarcodactylis* resulted in the isolation of 3,5,6-trihydroxy-4′,7-dimethoxyflavone, 3,5,6-trihydroxy-3′,4′,7-trimethoxyflavone [40], the flavone glycoside diosmin (*44-38*), and the coumarin derivative limettin (*44-39*) together with hesperidin [41].

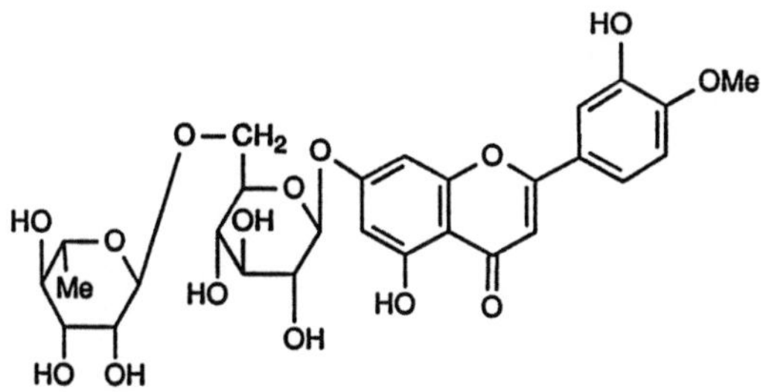

Diosmin (*44-38*)

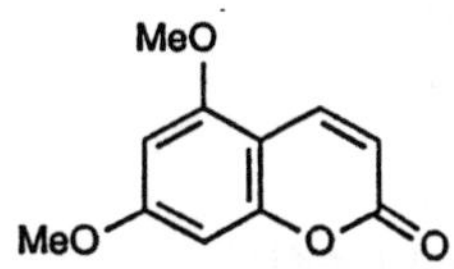

Limettin (*44-39*)

Two dimeric limettins, *cis*-head-to-tail-limettin dimer (*44-40*) and *cis*-head-to-head-limettin dimer (*44-41*), were also isolated from the fruit of *C. medica* var. *sarcodactylis*. Compounds *44-40* and the *trans*-isomer of *40-41* were synthesized from limettin upon UV irradiation [42].

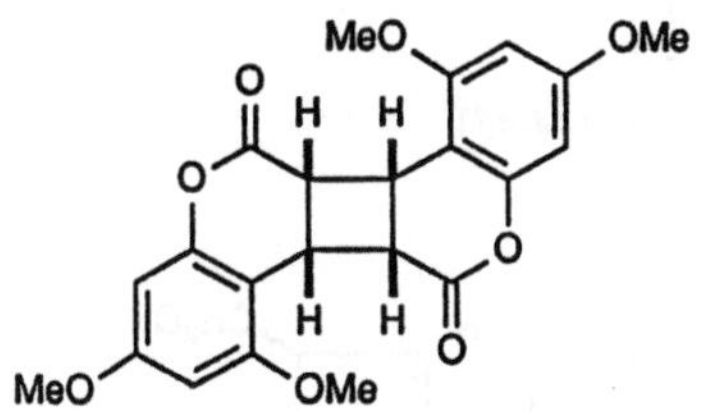

(*44-40*)

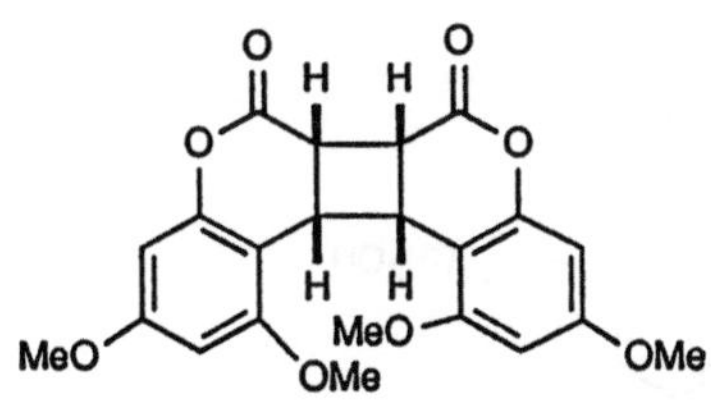

(*44-41*)

Nomilin, limonin, and a new related compound named citrusin (*44-42*) were isolated from the fruit of *C. medica* [43, 44].

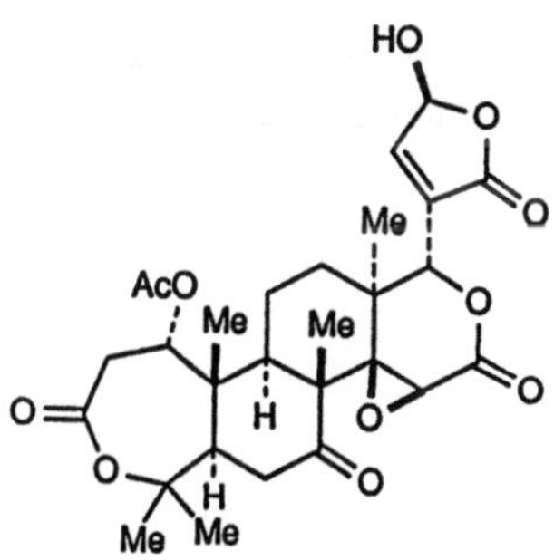

Citrusin (*44-42*)

The fruit peel of *C. sinensis* was reported to contain some phenyl propanoids and their glycosides, including three new compounds named citrusin A (*44-43*), citrusin B (*44-44*), and citrusin C (*44-45*) together with known compounds identified as syringin, coniferin (*44-46*), and dehydrodiconiferyl alcohol (*44-47*).

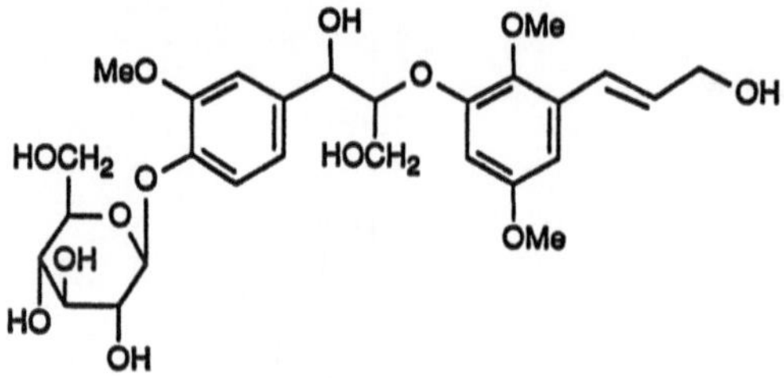

Citrusin A (*44-43*)

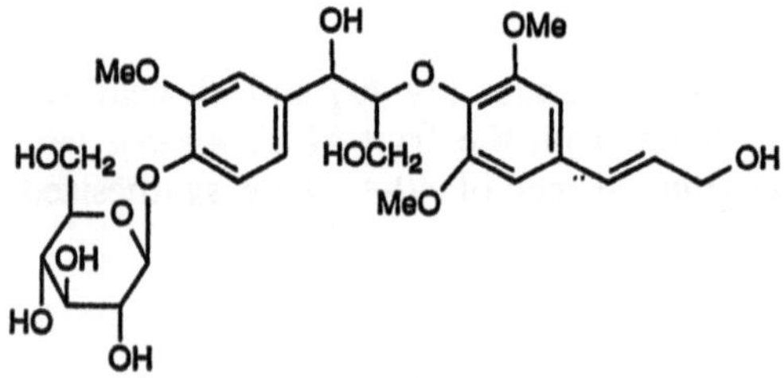

Citrusin B (*44-44*)

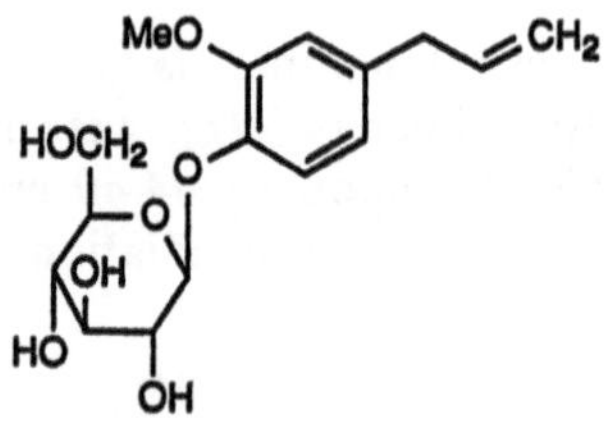

Citrusin C (*44-45*)

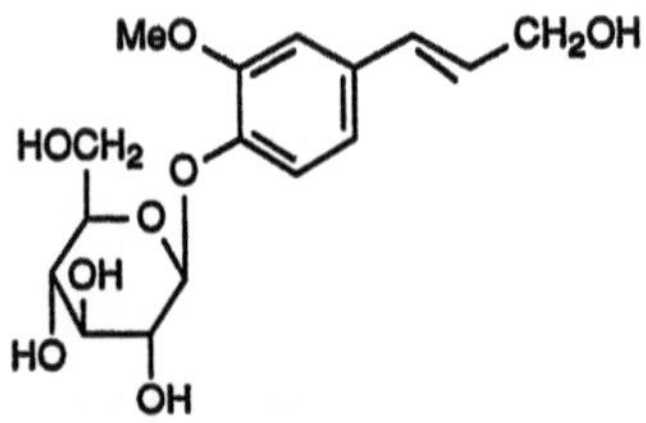

Coniferin (*44-46*)

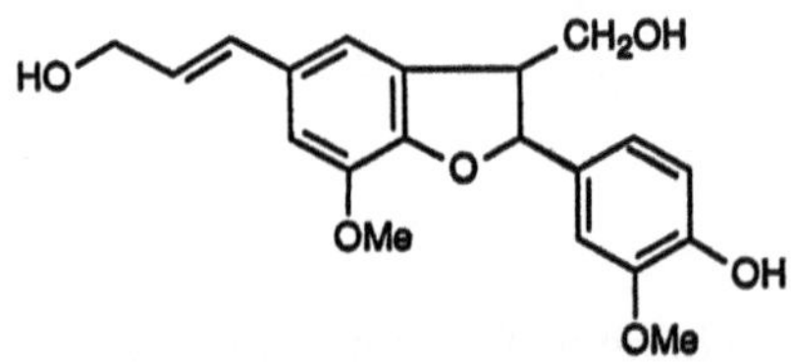

Dehydrodiconiferyl alcohol (*44-47*)

Recently, a series of acridone alkaloids was isolated from the root bark of *C. grandis* [45–47] and of other domestic citrus plants [48]. The structure of buntanine (*44-48*) as a representative is demonstrated [45].

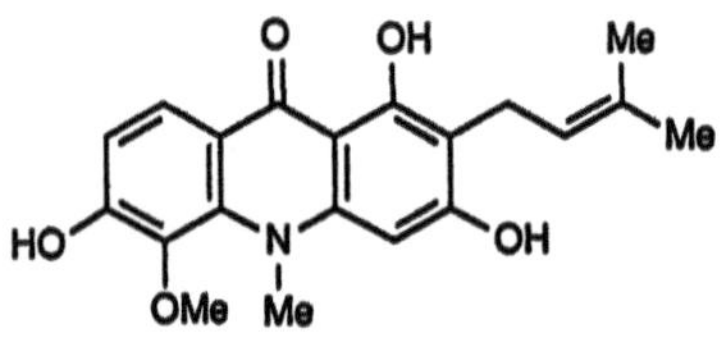

Buntanine (*44-48*)

44.3 Pharmacology

Intraperitoneal administration of 5 mg naringin per rat caused a depression of cardiac action immediately after injection. Continuous administration for 9 days led to maximal depressive action on the 3rd day. Naringin also showed some tendency to reduce blood pressure [49]. A similar effect was observed with methylhesperidin [50]. The flavanone glycosides hesperidin and naringin and their aglycones hesperetin and naringenin are excreted via the bile. After oral administration of 50 mg/kg of these flavanones, 12% of naringenin, 8% of hesperetin, 2% of naringin, and 3% of hesperidin, respectively, were found to be excreted within 48 h in rat bile and 2%, 2%, 3%, and 1%, respectively, in urine as metabolites containing the flavanone ring system. In comparison, after i.p. administration of 10 mg/kg naringenin, hesperetin, naringin, or hesperidin, biliary excretion was rapid, constituting 99%, 83%, 94%, and 97% of the dose, respectively. In urine flavanone conjugates, mainly glucuronides, were detected only after administration of hesperetin [51, 52].

Synephrine was first prepared as a synthetic sympathomimetic drug, being later isolated from the leaves and fruits of *Citrus* species. Continuous i.v. infusion of 4 mg/min synephrine into healthy volunteers significantly increased the systolic and mean arterial blood pressure, but did not affect the diastolic pressure and heart rate. The cardiac index increased and the peripheral vascular resistance decreased significantly [53].

After the i.v. administration of [^{3}H]synephrine to humans, about 80% of the radioactivity was recovered in the urine, two-thirds of the activity being associated with *p*-hydroxymandelic acid and 10% with unchanged synephrine. After the oral administration of [^{3}H]synephrine, total urinary radioactivity was similar to that in an i.v. injection; however, the quantity of unchanged synephrine was only 2.5%. The biological half-life of synephrine was about 2 h [54].

In mice and in rats, synephrine increased the plasma and myocardial cGMP levels, whereas *N*-methyltyramine increased plasma and myocardial cyclic GMP as well as cAMP levels [55]. The increase in cGMP in plasma and myocardium after administration of *N*-methyltyramine was also observed in rabbits [56]. Injection of *N*-methyltyramine increased renal and cerebral vascular resistance in dog, resulting from α-receptor stimulation [57].

Coniferin and syringin showed hypertensive activity in rats on i.v. administration at a dose of 10 mg/kg, whereas citrusin A, citrusin C, and dehydrodiconiferyl alcohol showed antihypertensive action in rats at the same dosage [58].

References

1. Huang MS, Zhou ZC (1984) Determination of naringin contents in pummelo (*Citrus grandis* or *C. grandis* var *tomentosa*) peel. Chin Trad Herbal Drugs 15:203–208
2. Asahina Y, Inbuse M (1928) Flavanone glucoside. II. Constitution of naringenin. Chem Ber 61:1514–1516
3. Asahina Y, Inbuse M (1929) Flavanone glucoside. IV. Naringin and hesperidin. J Pharm Soc Japan 49:128–134
4. Hoffmann E (1876) Über Hesperidin. Chem Ber 9:685–690
5. King FE, Robertson A (1931) Natural glucosides, III. The position of the biose residue in hesperidin. J Chem Soc 1704–1709
6. Zemplen G, Tettamanti AK (1938) Über die Biose des Hesperidins and des Neohesperidins. Chem Ber 71:2511–2520
7. Huang HY, He ZY (1982) Assay of hesperidin in Chen Pi (peels of some Chinese citrus fruit). Chin Trad Herbal Drugs 13:21–24
8. Xu LX, Liu AR (1983) Analysis of active principles in Chinese herbal drugs. VII. Determination of hesperidin in *Citrus reticulata* Blanco. Chin J Pharm Anal 3:203–206

9. Iwasaki Y (1936) Changes of the contents of hesperidin in the peel of mandarin orange during ripening. The content of hesperidin in the juice, the imbedded fiber and the endocarp of the ripened fruits. J Agr Chem Soc Japan 12:279–280
10. Chaliha BP, Sastry GP, Rao PR (1967) Chemical investigation of *Citrus reticulata*. Indian J Chem 5:239–241
11. Iinuma M, Matsuura S, Kurogochi K, Tanaka T (1980) Studies on the constituents of useful plants. V. Multisubstituted flavones in the fruit peel of *Citrus reticulata* and their examination by gas-liquid chromatography. Chem Pharm Bull 28:717–722
12. Lu ST, Wu TS, Chang SI (1977) Studies on the constituents of the peels of *Citrus reticulata* Blanco. J Taiwan Pharm Assoc 29:1–6
13. Wagner H, Maurer G, Hörhammer L, Farkas L (1971) Zur Struktur eines Flavones aus *Citrus reticulata* Blanco. Synthese von 5,4′-dihydroxy-6,7,8,3′-tetramethoxyflavon. Chem Ber 104:3357–3360
14. Tseng KF, Yu RD (1936) The studies of flavone derivatives in Chinese drugs. I. The isolation method of hesperidin from Chen-pi. J Chin Pharm Assoc 1:1–23
15. Karrer W (1949) Presence of hesperidin and neohesperidin in unripe oranges and in the seed vessels and stigmas of orange flowers. Helv Chim Acta 32:714–717
16. Horowitz RM, Gentile B (1963) Flavonoids of citrus. VI. The structure of neohesperidose. Tetrahedron 19:773–782
17. Swift LJ (1967) TLC-Spectrophotometric analysis for neutral fraction. Flavones in orange peel juice. J Agric Food Chem 15:99–101
18. Sarin PS, Seshadri TR (1960) New components of *Citrus aurantium*. Tetrahedron 8:64–66
19. Gentile B, Horowitz RM (1968) Flavonoids of citrus. IX. Some new *C*-glycosylflavones and a nuclear magnetic resonance method for differentiating 6- and 8-*C*-glycosyl isomers. J Org Chem 33:1571–1577
20. Stewart I, Newhall WF, Edwards GJ (1964) The isolation and identification of synephrine in the leaves and fruits of citrus. J Biol Chem 239:930–932
21. Chen FQ, Hou L (1984) Determination of synephrine in citrus plants. Chin J Pharm Anal 4:169–171
22. Fang XD, Qian LJ, She BZ, Shi ZY (1985) Chemical constituents in the fruit of *Citrus reticulata* Blanco. Bull Chin Mat Med 10:77–78
23. He CQ (1981) Alkaloid contents in Zhike (immature fruit of some species of *Poncirus* and *Citrus*). Chin Trad Herbal Drugs 12:345–348
24. Zheng HJ, Liu YF, Zhou YY (1983) Determination of synephrine and *N*-methyltyramine in citrus crude drugs and Zhi Shi (immature citrus fruit) injection by reversed-phase ion-pairing HPLC. Chin Trad Herbal Drugs 14:200–204
25. Wheaton TA, Stewart I (1969) Biosynthesis of synephrine in citrus. Phytochemistry 8:85–92
26. Kudritskaya SE, Fishman GM, Zagorodskaya LM (1977) Identification of orange peel (*Citrus sinensis* Osb.) carotenoids. Subtrop Kult 154–158 (CA 89:211937m)
27. Moshonas MG, Shaw PE (1979) Composition of essence oil from overripe oranges. J Agric Food Chem 27:1337–1339
28. Ashoor SHM, Bernhard RA (1967) Isolation and characterization of terpenes from *Citrus reticulata* Blanco and their comparative distribution among other *Citrus* species. J Agr Food Chem 15:1044–1047
29. Koepsel M, Surburg H (1988) Two new naturally occurring ketones from the essential oil of orange peel. Flavour Flagrance J 3:135–136
30. Bernay F (1841) Limonin. Liebigs Ann Chem 40:317
31. Arigoni D, Barton DHR, Corey EJ, Jeger O, Caglioti L, Dev S, Ferrini PG, Glazier ER, Melera A, Pradhan SK, Schaffner K, Sternhall S, Templeton JF, Tobinaya S (1960) The constitution of limonin. Experientia 16:41
32. Arnott S, Davie AW, Robertson JM, Sim GA, Watson DG (1960) The structure of limonin. Experientia 16:49
33. Arnott S, Davie AW, Robertson JM, Sim GA, Watson DG (1961) The structure of limonin: X-ray analysis of epilimonol iodoacetate. J Chem Soc 4183–4200
34. Dreyer DL (1965) Citrus bitter principles. III. Isolation of deacetylnomilin and deoxylimonin. J Org Chem 30:749–751
35. Emerson OH (1948) The bitter principles of citrus fruit. I. Isolation of nomilin, a new bitter principle from the seeds of oranges and lemons. J Am Chem Soc 70:545–549

36. Emerson OH (1951) Bitter principles of citrus. II. Relation of nomilin and obacunone. J Am Chem Soc 73:2621
37. Morishita T, Jinnai H, Nakachi H (1985) The isolation of limonoids of banpeiyn (*Citrus grandis* Osbeck). Nippon Shokuhin Kogyo Gakkaishi 32:590–591 (CA 104:145558p)
38. Dreyer DL (1966) *Citrus* bitter principles. V. Botanical distribution and chemotaxonomy in the Rutaceae. Phytochemistry 5:367–378
39. Hasegawa S, Bennett RD, Verdon CP (1980) Limonoids in citrus seeds: origin and relative concentration. J Agric Food Chem 28:922–925
40. He HY, Ling LQ (1985) Chemical studies on a Chinese traditional drug, fingered citron (*Citrus medica* L. var *sarcodactylis* (Noot.) Swingle). Acta Pharm Sin 20:433–435
41. Matsuno T (1959) Components of *Citrus* species. V. Components of *Citrus medica* var *sarcodactylus*. Yakugaku Zasshi 79:540–541
42. He HY, Ling LQ, Zhou MH (1987) Isolation and structure elucidation of two dimeric limettins from fingered citron. Org Chem 193–196
43. He HY, Ling LQ, Shi GP, Zhang N, Mao QM (1988) Studies on the chemical constituents of Chinese medicinal plant, fingered citron. Bull Chin Mat Med 13:352–354
44. He HY, Ling LQ, Qin HZ (1986) A new triterpenoid from traditional Chinese medicine Xiangyuan (*Citrus medica*). Chin Trad Herbal Drugs 17:530–532
45. Wu TS (1988) Alkaloids and coumarins of *Citrus grandis*. Phytochemistry 27:3717–3718
46. Wu TS (1987) Baiyumine A and B, two acridone alkaloids from *Citrus grandis*. Phytochemistry 26:871–872
47. Wu TS, Huang SC, Jong TT, Lai JS, Kuoh CS (1988) Coumarins, acridone, alkaloids and a flavone from *Citrus grandis*. Phytochemistry 27:585–587
48. Juichi M, Inone M, Ito C, Matsuoka M, Furnkawa H, Kajiura I (1987) Constituents of domestic *Citrus* plants. Part V. Acridone alkaloids. Part XVII. Two new acridone alkaloids from genus *Citrus* plants. Heterocycles 26:1873–1876
49. Oshiba J, Kato M (1981) Nutritional regulation. III. Depression of cardiac action by isolated naringin. Mukogawa Joshi Daigaku Kiyo, Shokumotsu-hen 29:1–8 (CA 97:37927m)
50. Feng MG, Feng GH, Zhou QG (1988) Effects of methylhesperidin on coronary, renal and cerebral circulation in dogs. Acta Pharmacol Sin 9:548–550
51. Hackett AM, Marsh I, Barrow A, Griffths LA (1979) The biliary excretion of flavanones in the rat. Xenobiotica 9:491–502
52. Hackett AM, Griffths LA (1979) Selective excretion of flavanones in bile. Biochem Soc Trans 7:645–646
53. Hofstetter R, Kreuder J, Bernuth G von (1985) Effect of oxedrine on the left ventricle and peripheral vascular resistance. Arzneimittelforschung 35:1844–1846
54. Hengstmann JH, Aulepp H (1978) Pharmacokinetics and metabolism of ^{3}H-synephrine. Arzneimittelforschung 28:2326–2331
55. Yen YF, Chung WP, Liu ZS, Ye YW (1983) Mechanism of the antishock effect of the active principles of the Chinese medicine Zhi Shi. Effect of *N*-methyltyramine and synephrine on plasma and myocardial cyclic nucleotide levels. Bull Hunan Med Coll 8:145–147
56. Yen YF, Chung HP (1981) Effect of *N*-methyltyramine, the active principle of *Citrus aurantium* on cardiac α-receptor and cGMP content. Chin Pharm Bull 16:51
57. Chen X, Lin LY, Deng HW, Fang YX, Ye YW (1981) Effects of *Citrus aurantium* and its active principle *N*-methyltyramine on the cardiovascular receptors. Acta Pharm Sin 16:253–259
58. Sawabe A, Matsubara Y, Kumamoto H, Iizuka Y, Okamoto K (1986) Structure and physiological activity of phenyl propanoid glycosides of hassaku (*Citrus hassaku* Hort.) and orange (*Citrus sinensis* Osbeck.) peels. Nippon Nogei Kaguku Kaishi 60:593–599 (CA: 105:222728u)

Clematis spp. 45

45.1 Introduction

Weilingxian, Radix Clematidis, is the dry rootstock and root of *Clematis chinensis* Osbeck, *C. manshurica* Rupr, or *C. hexapetala* Pall. (Ranunculaceae) collected in the fall. This official crude drug was used in Chinese traditional medicine as an antirheumatic and analgesic.

Another item from the *Clematis* plant listed in the Chinese Pharmacopoeia is Chuanmutong, Caulis Clematidis armandii. It is the dry stem of *C. armandii* Franch. or *C. montana* Buch.-Ham. collected in spring and fall and used as an antipyretic and diuretic in the treatment of edema and rheumatism.

45.2 Chemical Constituents

45.2.1 Chemical Constituents of *Clematis chinensis*

It has been reported that saponin appeared to be the major constituent of the roots of *C. chinensis*. The crude saponin obtained from the butanol soluble fraction of a methanol extract and precipitated on treatment with methanol and acetone contained more than ten components by thin-layer chromatography. A number of prosapogenins tentatively namend CP_1 [1], CP_2 [2], CP_3, CP_4, CP_5 [1], CP_6, CP_7, CP_8 [2], CP_9, CP_{10} [3], CP_0 [4], CP_{2a} [4], CP_{2b} [3], CP_{3a} [4], CP_{3b} [3], CP_{7a}, CP_{8a}, CP_{9a} and CP_{10a} [5] obtained by alkaline hydrolysis of the crude saponin, without further purification.

These prosapogenins have either hederagenin or oleanolic acid as the aglycon with the exception of CP_{3a}, which has 4-epihederagenin (*45-1*) as the aglycon. The sugar moieties of the prosapogenins are composed from glucose, arabinose, rhamnose, ribose, and xylose. The aglycon and the sugar sequence of the prosapogenins reported are listed in Table 45.1. Only the structures CP_0 (*45-2*), CP_8 (*45-3*), CP_9 (*45-4*) and CP_{3a} (*45-5*) are given as examples.

Table 45.1. Chemical composition of the prosapogenins isolated from *C. chinensis*

Prosapogenin	Aglycon	Sugar moiety	Ref.
CP_0	Hederagenin	23-*O*-α-L-arabinopyranoside	[4]
CP_1	Hederagenin	3-*O*-α-L-arabinopyranoside	[1]
CP_2	Oleanolic acid	3-*O*-α-L-rhamnopyranosyl-(1→2)-α-L-arabinopyranoside	[2]
CP_{2a}	Hederagenin	23-*O*-β-D-glucopyranoside	[4]
CP_{2b}	Oleanolic acid	3-*O*-β-D-xylopyranosyl-(1→2)-α-L-arabinopyranoside	[3]
CP_3	Oleanolic acid	3-*O*-β-D-xylopyranosyl-(1→3)-α-L-rhamnopyranosyl-(1→2)-α-L-arabinopyranoside	[1]
CP_{3a}	Olean-12-ene-28-oic acid 3β, 24-diol (4-epihederagenin)	3-*O*-α-L-rhamnopyranosyl-(1→2)-α-L-arabinopyranoside	[4]
CP_{3b}	Hederagenin	3-*O*-α-L-rhamnopyranosyl-(1→2)-α-L-arabinopyranoside	[3]
CP_4	Oleanolic acid	3-*O*-β-D-ribopyranosyl-(1→3)-α-L-rhamnopyranosyl-(1→2)-α-L-arabinopyranoside	[1]
CP_5	Hederagenin	3-*O*-β-D-xylopyranosyl-(1→3)-α-L-rhamnopyranosyl-(1→2)-α-L-arabinopyranoside	[1]
CP_6	Hederagenin	3-*O*-β-D-ribopyranosyl-(1→3)-α-L-rhamnopyranosyl-(1→2)-α-L-arabinopyranoside	[2]
CP_7	Oleanolic acid	3-*O*-β-D-glucopyranosyl-(1→4)-β-D-ribopyranosyl-(1→3)-α-L-rhamnopyranosyl-(1→2)-α-L-arabinopyranoside	[2]
CP_{7a}	Oleanolic acid	3-*O*-β-D-glucopyranosyl-(1→4)-β-D-xylopyranosyl-(1→3)-α-L-rhamnopyranosyl-(1→2)-α-L-arabinopyranoside	[5]
CP_8	Hederagenin	3-*O*-β-D-glucopyranosyl-(1→4)-β-D-ribopyranosyl-(1→3)-α-L-rhamnopyranosyl-(1→2)-α-L-arabinopyranoside	[2]
CP_{8a}	Hederagenin	3-*O*-β-D-glucopyranosyl-(1→4)-β-D-xylopyranosyl-(1→3)-α-L-rhamnopyranosyl-(1→2)-α-L-arabinopyranoside	[5]
CP_9	Oleanolic acid	3-*O*-β-D-glucopyranosyl-(1→4)-β-D-glucopyranosyl-(1→4)-β-D-ribopyranosyl-(1→3)-α-L-rhamnopyranosyl-(1→2)-α-L-arabinopyranoside	[3]
CP_{9a}	Oleanolic acid	3-*O*-β-D-glucopyranosyl-(1→4)-β-D-glucopyranosyl-(1→4)-β-D-xylopyranosyl-(1→3)-α-L-rhamnopyranosyl-(1→2)-α-L-arabinopyranoside	[5]
CP_{10}	Hederagenin	3-*O*-β-D-glucopyranosyl-(1→4)-β-D-glucopyranosyl-(1→4)-β-D-ribopyranosyl-(1→3)-α-L-rhamnopyranosyl-(1→2)-α-L-arabinopyranoside	[3]
CP_{10a}	Hederagenin	3-*O*-β-D-glucopyranosyl-(1→4)-β-D-glucopyranosyl-(1→4)-β-D-xylopyranosyl-(1→3)-α-L-rhamnopyranosyl-(1→2)-α-L-arabinopyranoside	[5]

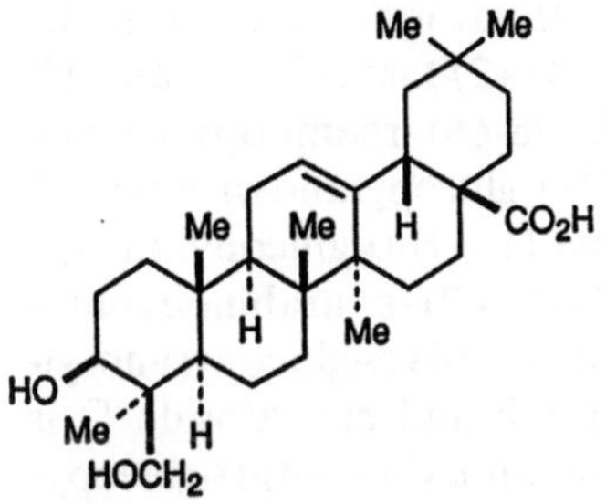

4-Epihederagenin (*45-1*)

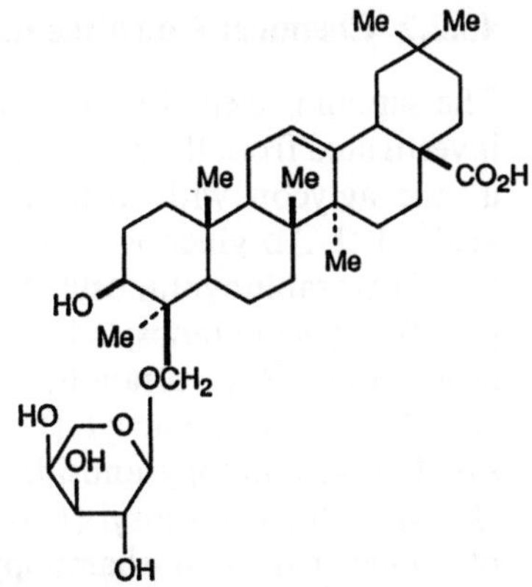

Prosapogenin CP$_0$ (*45-2*)

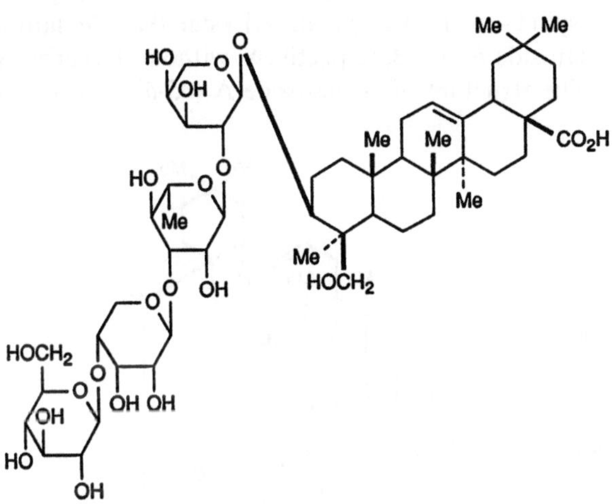

Prosapogenin CP$_8$ (*45-3*)

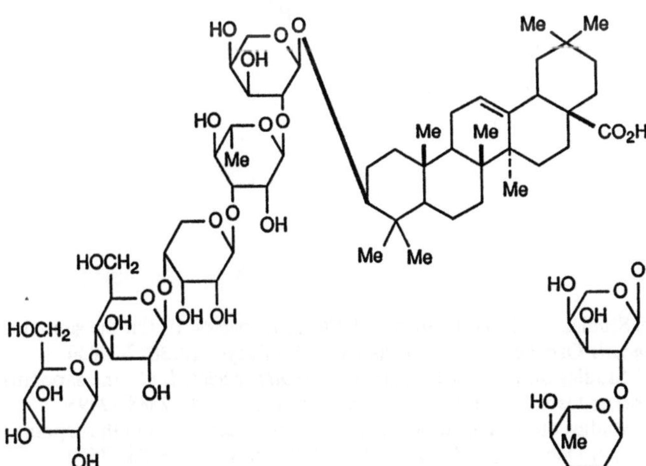

Prosapogenin CP$_9$ (*45-4*)

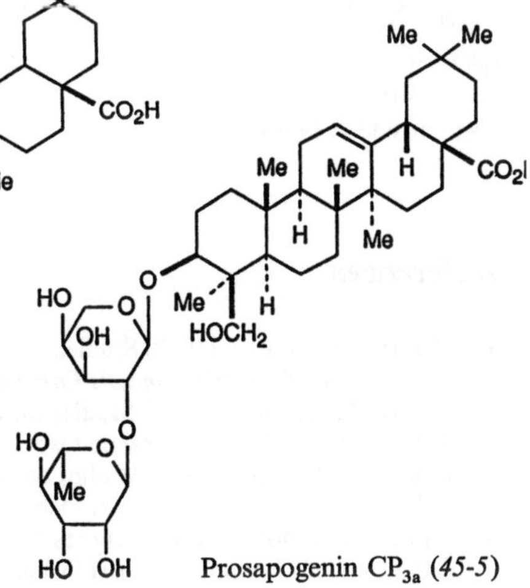

Prosapogenin CP$_{3a}$ (*45-5*)

45.2.2 Chemical Constituents of *Clematis manshurica*

The saponins clematoside A, A′, B, B′, and C have been isolated and structurally investigated from the root of *C. manshurica*. The clematosides all have oleanolic acid as the aglycon with different sugar moieties. Thus, clematoside A is oleanolic acid 3-*O*-D-glucopyranosyl-(1→4)-D-xylopyranosyl-(1→2)-L-arabinopyranosyl-(1→2)-L-arabinoypranosyl-(1→4)-L-rhamnopyranoside 28-*O*-L-rhamnopyranosyl-(1→4)-D-glucopyranosyl-(1→4)-D-glucopyranosyl-(1→6)-D-glucopyranosyl ester [6]; clematoside B is oleanolic acid 3-*O*-D-glucopyranosyl-(1→4)-D-glucopyranosyl-(1→4)-D-xylopyranosyl-(1→2)-L-arabinopyranosyl-(1→2)-L-arabinopyranosyl-(1→4)-rhamnopyranoside 28-*O*-L-rhamnopyranosyl-(1→4)-D-glucopyranosyl-(1→4)-D-glucopyranosyl-(1→6)-D-glucopyranosyl ester [7]; and clematoside C is oleanolic acid 3-*O*-L-rhamnopyranosyl-(1→6)-D-glucopyranosyl-(1→4)-D-glucopyranosyl-(1→4)-D-xylopyranosyl-(1→2)-L-arabinopyranosyl-(1→2)-L-arabinopyranosyl-(1→4)-L-rhamnopyranoside 28-*O*-L-rhamnopyranosyl-(1→4)-D-glucopyranosyl-(1→4)-D-glucopyranosyl-(1→6)-D-glucopyranosyl ester [8]. Clematoside A′ and B′ are derived from clematoside A and B, respectively, without the carbohydrate chain at position C-28 [9]. The structure of clematoside A (*45-6*) is demonstrated below.

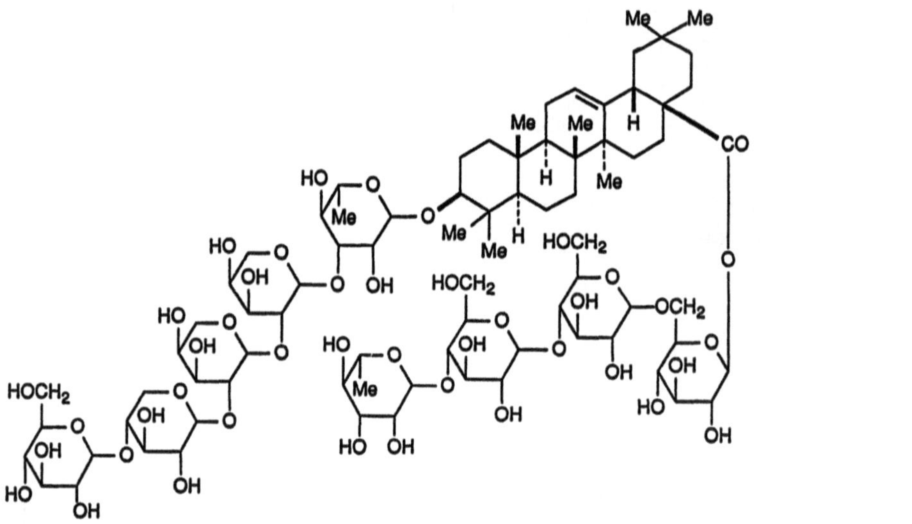

Clematoside A (*45-6*)

References

1. Kizu H, Tomimori T (1980) Studies on the constituents of *Clematis* species. II. On the saponins of the root of *Clematis chinensis* Osbeck. (2). Chem Pharm Bull (Tokyo) 28:2827–2830
2. Kizu H, Tomimori T (1979) Studies on the constituents of *Clematis* species. I. On the saponins of the root of *Clematis chinensis* Osbeck. (1). Chem Pharm Bull (Tokyo) 27:2388–2393
3. Kizu H, Tomimori T (1980) Studies on the constituents of *Clematis* species. III. On the saponins of the root of *Clematis chinensis* Osbeck. (3). Chem Pharm Bull (Tokyo) 28:3555–3560
4. Kizu H, Tomimori T (1982) Studies on the constituents of *Clematis* species. V. On the saponins of the root of *Clematis chinensis* Osbeck. (5). Chem Pharm Bull (Tokyo) 30:3340–3346

5. Kizu H, Tomimori T (1982) Studies on the constituents of *Clematis* species. IV. On the saponins of the root of *Clematis chinensis* Osbeck. (4). Chem Pharm Bull (Tokyo) 30:859–865
6. Chirva VY, Konyukhov VP (1968) Structure of Clematoside A. Khim Prir Soedin 4:140–141
7. Chirva VY, Konyukhov VP (1968) Structure of Clematoside B. Khim Prir Soedin 4:141–142
8. Kochetkov NK, Khorlin AJ, Chirva VJ (1965) Clematoside *C*-triterpene oligoside from *Clematis manshurica*. Tetrahedron Lett 2201–2205
9. Chirva VY, Usov AI, Konykhov VP (1968) Carbohydrate chain structures of clematosides A′, A and B. Izv Akad Nank SSSR, Ser Khim 2541–2545

Codonopsis pilosula (Franch.) Nannf. 46

46.1 Introduction

Dangshen, Radix Codonopsis pilosulae, is the dry root of *Codonopsis pilosula* (Franch.) Nannf. (Campanulaceae), which is collected in the fall. It is officially listed in the Chinese Pharmacopoeia and is used mainly as a tonic.

46.2 Chemical Constituents

The root of *C. pilosula* is one of the best-known traditional Chinese medicines and has sometimes been used as a substitute for ginseng in traditional Chinese medicine. Comparative thin-layer chromatographic studies on the butanol extracts of the roots of *C. pilosula* and their hydrolysates were carried out. The thin-layer chromatogram of the root extracts of *C. pilosula* from different areas were found be similar to each other but quite different from those of the roots of *P. ginseng* and *P. pseudoginseng* [1].

The water-soluble components of the root of *C. pilosula* contain sugars, amino acids, and inorganic elements [2]. Essential mineral elements for humans detected in the root include K, Na, Ca, Mg, Fe, Cu, Co, Zn, Mn, Cr, and Mo [2, 3]. Inulin and fructose were the two major sugars. The monosaccharide content in the root was 4.3 g/10 g [4]. At least 17 amino acids were detected in the root [2] including threonine, aspartic acid, isoleucine, alanine, asparagine, glutamic acid, glycine, serine, valine, proline, and glutamine [5, 6].

The sterols and triterpenes in the root of *C. pilosula* include α-spinasterol [7] and its β-D-glucopyranoside, Δ^7-stigmasterol and its β-D-glucopyranoside [8], $\Delta^{5,22}$-stigmasterol [9], taraxerol, taraxeryl acetate [10], and friedelin [10].

The following substances were also isolated and identified: butyloxycarbonylurea [10], 5-(hydroxymethyl)-2-furaldehyde and 5-(methoxymethyl)-2-furaldehyde (*46-1*) [11], 2-furancarboxylic acid, atractylenolide II (*27-2*), atractylenolide III (*27-3*), syringin, 4-hydroxy-3,5-dimethoxybenzaldehyde [12], alkanol- or alkenolglycosides ethyl-α-D-fructofuranoside, *n*-hexyl-β-D-glucopyranoside [12], (Z)-3-hexenyl-β-D-glucopyranoside, and (E)-2-hexenyl-β-D-glucopyranoside [13]. The major components of the essential oil of *C. pilosula* were identified as methyl palmitate, octadecane, nonadecane, heptadecane, and long-chain carboxylic acid [14].

Antiulcerous polysaccharides named CP-1, CP-2, CP-3, and CP-4 were also isolated from the root of *C. pilosula*. They are primarily composed of glucose, fructose, galactose, arabinose, mannose, rhamnose, xylose, and ribose in different ratios [15, 16].

Isolation of the *β*-carboline alkaloid perlolyrine (*46-2*) from the roots of *C. pilosula* was recently reported [17].

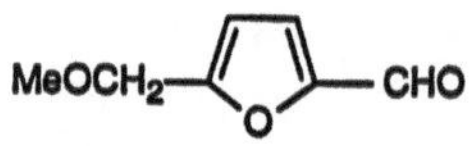

5-(Methoxymethyl)-2-furaldehyde (*46-1*)

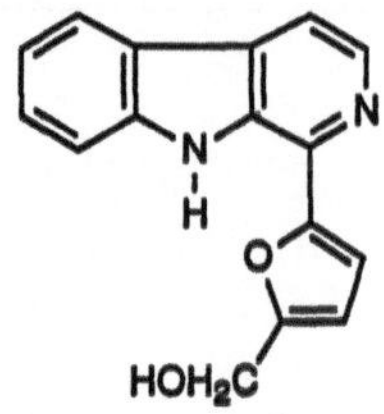

Perlolyrine (*46-2*)

Besides *C. pilosula*, a series of other *Codonopsis* species such as *C. pilosula* var. *volubilis* [6, 18], *C. tangshen*, *C. clematidea*, *C. moestra*, *C. subglobosa*, *C. convolvulacea*, and *C. mollis* are also used in Chinese medicine [6]. From the root of *C. tangshen*, two new phenylpropanoid glucosides named tangshenoside I (*46-3*) and tangshenoside II (*46-4*) were isolated and structurally determined [19].

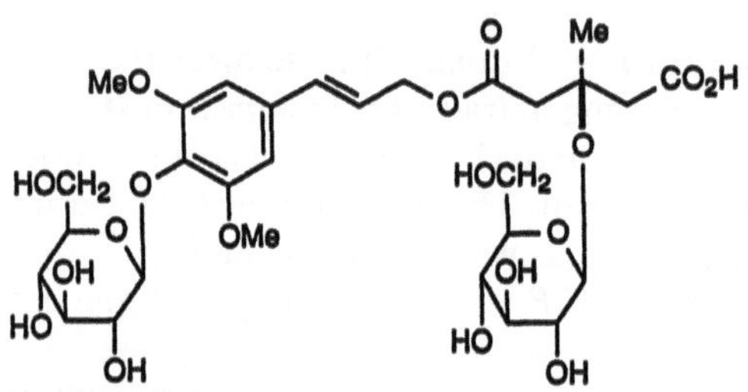

Tangshenoside I (*46-3*)

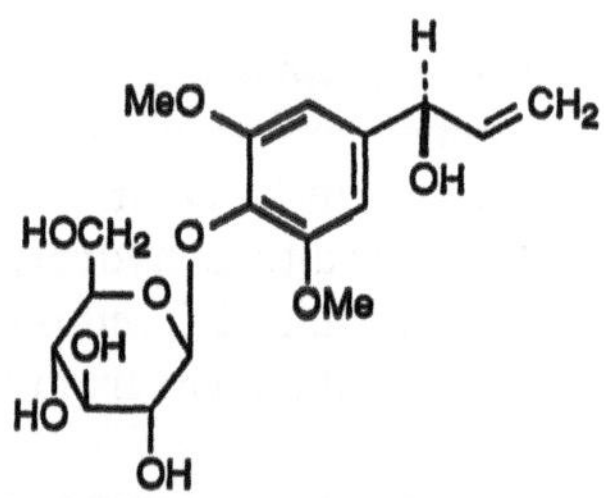

Tangshenoside II (*46-4*)

From the root of *C. clematidea*, two alkaloids named codonopsine (*46-5*) [20] and codonopsinine (*46-6*) [21] were isolated. The absolute configuration of these two alkaloids was recently revised [22].

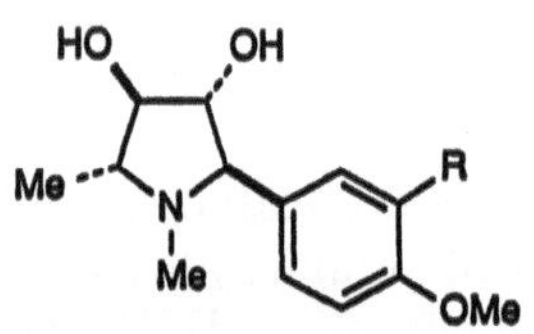

Codonopsine (*46-5*): R=H
Codonopsinine (*46-6*): R=OCH_3

References

1. Wong MP, Chiang TC, Chang HM (1983) Chemical studies on dang-shen, the root of *Codonopsis pilosula*. Planta Med 49:60
2. Cai DG, Wang YZ, Han C, Zhao WW (1982) Studies on chemical constituents of Dang Shen (*Codonopsis pilosula*). II. Chin Trad Herbal Drugs 13:442–444

3. Wang SM, Yang Y, Li CZ, Lin WZ (1987) Comparison of trace element content in 11 species of Dangshen (*Codonopsis pilosula, C. tangshen* and *C. tubulosa*). Chin Trad Herbal Drugs 18:16–17
4. Liu ZY, Wang YH (1983) Comparison of monosaccharides content in several *Codonopsis* species. Chin Trad Herbal Drugs 8:16–17
5. Jiang Y, Huang MY, Zhu GB, Zhou LA, Wang MQ, Zheng YD (1986) Analysis of free amino acids in silkworm excrement, *Dioscorea opposita, Astragalus membranacenus, Angelica sinensis* and *Codonopsis pilosula*. Chin Trad Herbal Drugs 17:105–107
6. Sha DZ, Lu YR, Shen LS (1988) Determination of amino acids and inorganic elements in eight species of medicinal Dangshen. Chin J Pharm Anal 8:143–147
7. Lee IR, Jung MH (1979) A phytochemical study on the component of *Codonopsis pilosula* roots. Yakhak Hoe Chi 23:57–61 (CA 92:169112y)
8. Lee IR, Kim YH, Park SB (1982) Sterols and steryl glycosides from the root of *Codonopsis pilosula*. Saengyak Hakhoe Chi (Hanguk Saengyak Hakhoe) 13:129–131 (CA 98:221673x)
9. Chen HS, Wang YZ, Han C, Cai DG (1985) Studies on the chemical constituents of pilose asiabell (*Codonopsis pilosula*). Chin Trad Herbal Drugs 16:295–296
10. Wang YZ, Cai DG, Zhao WW (1982) Chemical constituents of *Codonopsis pilosula*. Part I. Chin Trad Herbal Drugs 13:1–3
11. Lee IR (1978) A phytochemical study on components of *Codonopsis pilosula* radix. Yakhak Hoe Chi 22:1–7 (CA 90:83635t)
12. Wang ZT, Xu GJ, Hattori M, Namba T (1988) Constituents of the roots of *Codonopsis pilosula*. Shoyakugaku Zasshi 42:339–342
13. Mizutani K, Yuda M, Tanaka O, Saruwatari Y, Fuwa T, Jia MR, Ling YK, Pu XF (1988) Chemical studies on the Chinese traditional medicine dangshen. I. Isolation of (Z)-3- and (E)-2-hexenyl β-D-glucopyranosides. Chem Pharm Bull (Tokyo) 36:2689–2690
14. Liao J, Lu YQ (1987) Chemical constituents of pilosa asiabell (*Codonopsis pilosula*) V. Studies of the essential oils. Chin Trad Herbal Drugs 18:386–388
15. Zhang SJ, Zhang SY (1987) Polysaccharides of Dangshen (*Codonopsis pilosula*). Chin Trad Herbal Drugs 18:98–100
16. Cui JC, Zhang W, Zhang YQ, Xu YJ (1988) Anti-ulcer action of the polysaccharides from pilose asiabell (*Codonopsis pilosula*). Chin Trad Herbal Drugs 19:357–359
17. Liu T, Liang WZ, Tu GS (1988) Perlolyrine: a β-carboline alkaloid from *Codonopsis pilosula*. Planta Med 54:472–473
18. Sha DZ, Lu YR, Shen LS (1989) Chemical studies on Chanraodangshen (the root of *Codonopsis pilosula* var. *volubilis*). Chin J Pharm Anal 9:13–17
19. Mizutani K, Yuda M, Tanaka O, Saruwatari Y, Jia MR, Ling YK, Pu XF (1988) Tangshenosides I and II from Chuandangshen, the root of *Codonopsis tangshen* Oliv. Chem Pharm Bull (Tokyo) 36:2726–2729
20. Matkhalikova SF, Malikov VM, Yunosov SY (1969) Alkaloids of *Codonopsis clematidea*. Khim Priv Soedin 5:30 (CA71:13245z)
21. Matkhalikova SF, Malikov VM, Yunosov SY (1969) Structure of codonopsinine. Khim Priv Soedin 5:607 (CA 73:25712d)
22. Iida H, Yamazaki N, Kibayashi C, Nagase H (1986) Stereochemical revision and absolute configuration of codonopsinine. Tetrahedron Lett 27:5393–5396

Coptis spp.

47

47.1 Introduction

Huanglian, Rhizoma Coptidis, is the dry rhizome of *Coptis chinensis* Franch., *C. deltoidea* C.Y. Cheng et Hsiao, or *C. teetoides* C.Y. Cheng (Ranunculaceae), which is collected in the fall. *Coptis* rhizome is one of the most well known and widely used herbs in traditional Chinese medicine. It is officially listed in the Chinese Pharmacopoeia and is used as a bacteriostatic, antipyretic, and antiphlogistic for the treatment of gastroenteritis, diarrhea, vomiting, icterus, fever, insomnia, hematemesis, nose bleeding, conjunctivitis, toothache, carbuncle, and abscess and as a bitter digestive for the treatment of indigestion. It is used for treatment of eczema by external application.

47.2 Chemical Constituents

A number of alkaloids, especially quaternary protoberberine type alkaloids, have been isolated from the rootstocks of *Coptis* species [1]. Protoberberine is 5,6-dihydro-dibenzo[*a*,*g*]-quinolizinium (*47-1*).

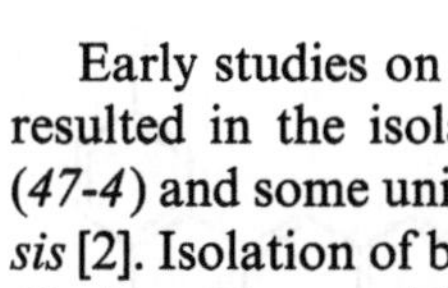

Protoberberine (*47-1*)

Early studies on the chemical constituents in the *Coptis* species native to China resulted in the isolation of berberine (*47-2*), palmatine (*47-3*), and jatrorrhizine (*47-4*) and some unidentified alkaloids from the rhizome of *C. chinensis* var. *omeiensis* [2]. Isolation of berberine, jatrorrhizine, and worenine (*47-5*) from the rhizome of *C. chinensis* var. *shihchuensis* [3], as well as the isolation of berberine, palmatine, and jatrorrhizine from the rhizome of *C. teeta* were also described [4]. From the rhizome of *C. quinquefolia*, berberine, jatrorrhizine, coptisine (*47-6*), columbamine (*47-7*), and magnoflorine were isolated and identified [5]. Berberine, jatrorrhizine, palmatine, worenine, coptisine, and columbamine are all protoberberine-type alkaloids.

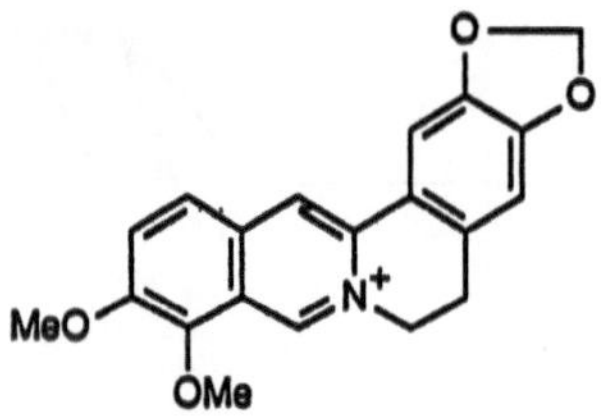

Berberine (*47-2*)

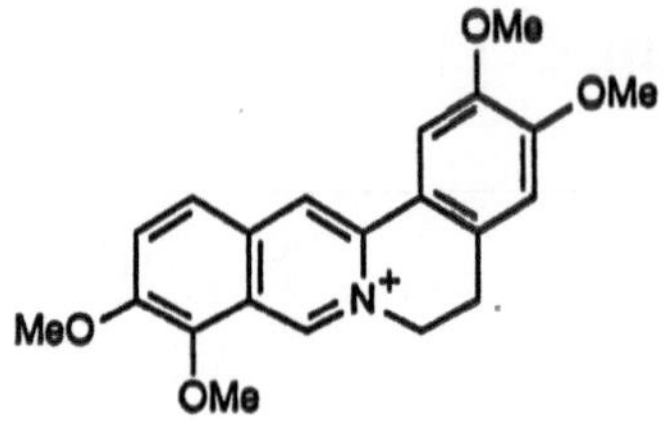

Palmatine (*47-3*)

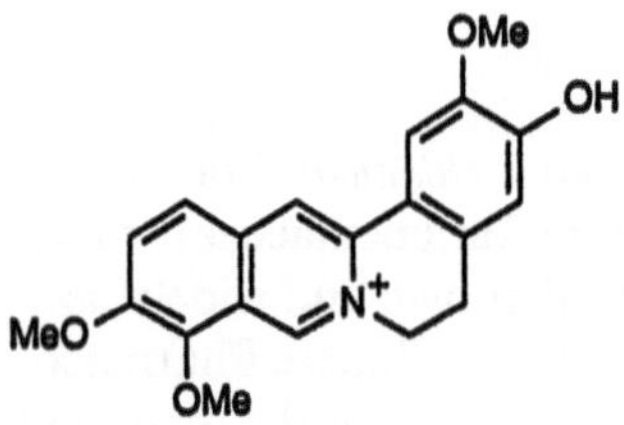

Jatrorrhizine (*47-4*)

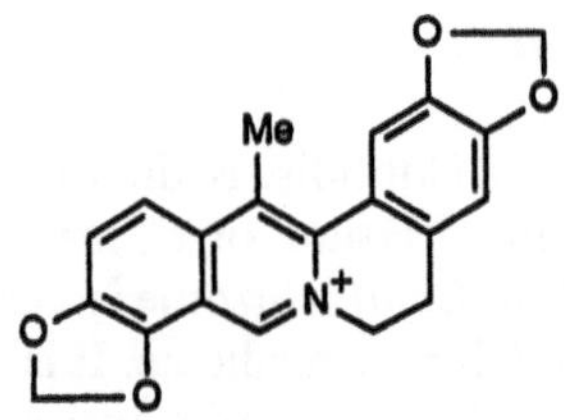

Worenine (*47-5*)

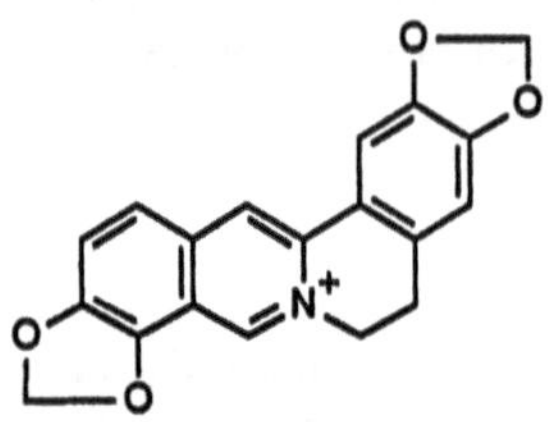

Coptisine (*47-6*)

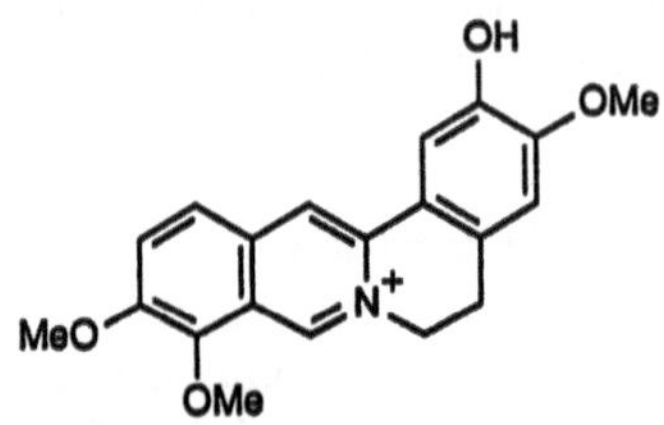

Columbamine (*47-7*)

A study on the alkaloids in the rhizome of *C. deltoidea* revealed the presence of four alkaloids identified as berberine, coptisine, jatrorrhizine, and palmatine [6]. The *Coptis* species officially listed in the Chinese Pharmacopoeia *C. chinensis*, *C. deltoidea*, and *C. teetoides* are from different growing areas, but have the same basic chemical constituents [7–10]. In addition, epiberberine (*47-8*) [11], groenlandicine (*47-9*) [12], and berberastine (*47-10*) [13] were isolated from *C. chinensis* and *C. deltoidea*.

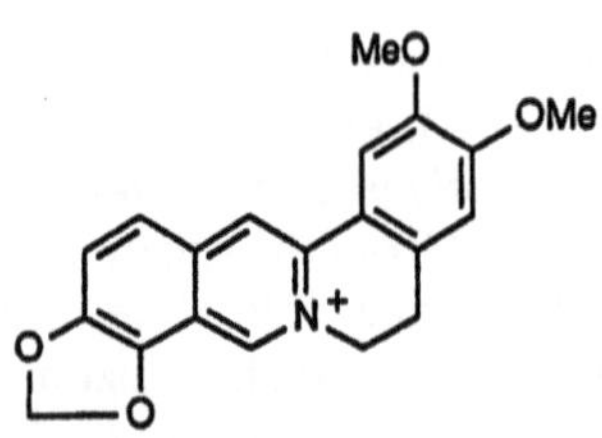

Epiberberine (*47-8*)

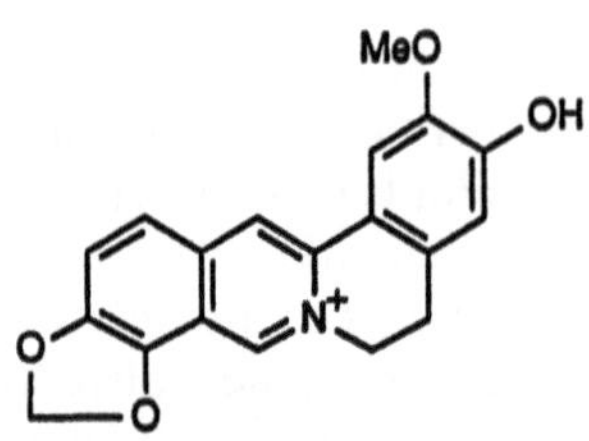

Groenlandicine (*47-9*)

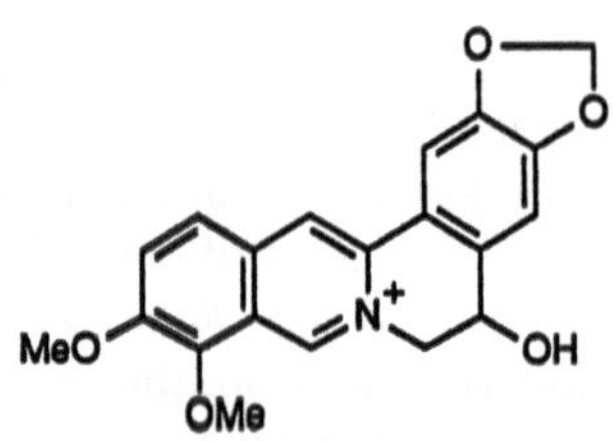

Berberastine (*47-10*)

47.3 Pharmacology

47.3.1 Antimicrobial Activities

The extract of *C. chinensis* rootstock showed a high antibacterial activity against numerous gram-positive and gram-negative bacteria including *Shigella*, *Brucella*, *Staphylococcus*, and *Streptococcus* species [14]. Berberine chloride was found to be active against a number of gram-positive as well as gram-negative bacteria, such as *Staphylococcus aureus*, *S. hemolyticus*, *Salmonella typhosa*, *Shigella dysenteriae*, *S. paradysenteriae*, *Escherichia coli*, *Neisseria gonorrhoeae*, and *Diplococcus pneumoniae* in different media. It had about the same antibacterial activity as some sulfonamides. However, berberine chloride also had an effect in broth and serum, where the sulfonamides were antagonized [15]. The antibacterial effect of berberine chloride at pH 5 was the same as at pH 9. The microorganisms acquired resistance when left in contact with berberine chloride for a long time [16].

Berberine sulfate was shown to be bactericidal to *Vibrio cholerae* and bacteriostatic to *Staphylococcus aureus* at concentrations of 35 and 50 μg/ml, respectively. In both these organisms berberine at the concentrations mentioned above inhibited RNA and protein synthesis almost immediately after the addition of the drug [17]. Berberine sulfate at concentrations of 50–600 μg/ml in culture medium showed bactericidal activity against *S. aureus*. The growth of 70% from 196 strains of *S. aureus* was inhibited. Cross-resistance between berberine sulfate and antibiotics used in therapy was not observed except for streptomycin [18].

Cell-free preparations made from vibrios pretreated with berberine did not produce choleraic symptoms in infant rabbits, suggesting that the toxin was either inactivated or neutralized [19, 20]. Oral administration of berberine to infant rabbits 18–24 h before a single fatal intraintestinal dose of choleragenic toxin prevented toxin-induced diarrhea and consequently a prolonged survival time when compared with untreated choleragenic animals [21]. Berberine also markedly inhibited the secretory response of *E. coli* heat-stable enterotoxin in rabbits and mice [22] and was dose dependently effective in reducing water and electrolyte secretions induced by *E. coli* heat-stable enterotoxin [23]. Mild changes in mucosal histology due to cholera toxin were also reversed by berberine. Berberine did not significantly alter normal ileal water and electrolyte transport [24]. The antisecretory effects of berberine may be explained by stimulation of NaCl-coupled absorptive transport process [25].

Berberine chloride appeared to be a weak tuberculostatic agent. It inhibited the multiplication of 7 of 16 strains of *Mycobacterium tuberculosis* at a concentration of 0.5 μg/ml in nutrient [26]. Berberine sulfate was also studied for antitrachoma activity [27].

The fecal flora of rats after oral treatment of berberine chloride, coptisine chloride, and a methanol extract of *C. japonica* for 10 days showed no significant change in the counts of bacteria such as *Escherichia coli*, *Staphylococcus*, and *Streptococcus* as compared with controls [28].

The quaternary ammonium group in berberine is necessary for its antibacterial activity. Derivatives without the quaternary ammonium group, such as tetrahydroberberine, showed little antibacterial effect [29].

Berberine sulfate in concentrations of 10–25 mg/ml inhibited the growth of 11 of 13 fungi: *Alternaria* sp., *Aspergillus flavus*, *A. fumigatus*, *Candida albicans*, *Curvula-*

ria sp., *Drechslera* sp., *Fusarium* sp., *Mucor* sp., *Penicillium* sp., *Rhizopus oryzae*, and *Scopulariopsis* sp. At 50 mg/ml the growth of *Syncephalastrum* sp. was inhibited as well, but *Aspergillus niger* remained unaffected [30]. Oral administration of berberine sulfate at doses of 350–700 mg/kg was effective in treating *Candida albicans* infections of the intestine in mice [31]. Coptisine showed a greater antimicrobial effect against *Saccharomyces carlsbergensis* than berberine, palmatine, and jatrorrhizine [32].

Berberine sulfate administered to rats at doses of 100 mg/kg 10 days after experimentally induced intestinal amebiasis was effective in 80% of the animals [33]. It completely inhibited the growth to trophozoites of *Entamoeba histolytica* at concentrations of 0.5–1 mg/ml in vitro, and was active in vivo against infections with *E. histolytica* in hamsters and rats [34].

47.3.2 Antiinflammatory Effect

The antiinflammatory activity of berberine was confirmed by some test methods [35, 36]. Subcutaneous injection of berberine sulfate together with cholera toxin dose dependently inhibited toxin-induced inflammation in the neck of rats; separate injection of berberine sulfate also had antiinflammatory effects [37].

47.3.3 Antitumor Activity

Berberine sulfate (50 μg/ml) and berberine chloride (25 μg/ml) showed growth inhibition of Ehrlich and lymphoma ascites tumor cells [38]. The presence of berberine in granules inside the cells was detected by its fluorescence [39]. Berberine and tetrahydroberberine inhibited oxidation of DL-alanine in rat kidney homogenate [40]. Berberine was the most cytotoxic alkaloid isolated from *Thalictrum minus elatum* with 70% inhibition of protein synthesis in KB cells at a concentration of 1 μg/ml [41]. The cytotoxic ED_{50} values of berberine in HeLa cell cultures were 3.5–30.0 μg/ml [42].

Berberine and tetrahydroberberine showed no antitumor activity against sarcoma 180 ascites tumors in mice, but berberrubine chloride, the 9-*O*-demethyl analog of berberine, and its esters had strong antitumor activity [43]. Berberine chloride inhibited the formation of DNA, RNA, proteins, and lipids as well as the oxidation of [^{14}C]glucose to $^{14}CO_2$ when incubated with S180 cells in vitro. Protein and RNA syntheses were most sensitive to berberine [44].

Berberine chloride and berberine sulfate activated macrophage-mediated inhibition of [^{3}H]thymidine incorporation into EL4 leukemia cells. The activation of macrophages by berberine was not augmented by the addition of a suboptimal dose of macrophage-activating factor. It was concluded that berberine is a potent activator for macrophages to induce inhibition tumor cells in vitro [45].

47.3.4 Cardiovascular Actions

Intravenous infusion of berberine sulfate to rats lowered the blood pressure in a dose-dependent manner. A significant hypotensive effect was followed by bradycardia. These effects were also observed in bilaterally vagotomized rats [46]. Berberine chloride at doses of 0.5–5.0 mg/kg administered to rabbits anesthetized with urethane produced a long-lasting, dose-related decrease in blood pressure. Intraventricular administration of berberine did not elicit the hypotensive response. Berberine-induced hypotension is attributable to α-adrenoceptor blockade and not to a direct relaxant effect of vascular smooth muscle [47]. A comparative study of the effect of berberine and its substituted derivatives on blood pressure of rats showed that substitution of phenoxy or *n*-pentoxy groups for a methoxy group prolonged the transient hypotensive activity of berberine, whereas substitution by a hydroxy group caused a slight hypertension [48].

Arrhythmia induced by chloroform could be antagonized by i.p. administered berberine chloride and jatrorrhizine chloride. Palmatine chloride was ineffective. Berberine chloride antagonized aconitine-induced ventricular fibrillation but was not effective against ouabain-induced arrhythmias [49]. Standard cardiovascular function tests in dogs suggested that small i.v. injected doses of berberine may excite the myocardium, accelerate heart rate, and slightly increase coronary blood flow and cardiac contractility. With increasing doses it may directly dilate peripheral blood vessel, inhibit myocardium, and weaken cardiac contractility. The heart cells became seriously damaged and stopped in diastole [50].

Berberine exerted no direct vasodilatory effects on isolated rabbit pulmonary and cat coronary arteries. However, berberine reversed vasoconstriction mediated by α-adrenergic agents in both preparations [51]. The results obtained from a study on the cardiac effects of berberine in isolated right and left arterial preparations from guinea pigs showed that berberine has a unique profile of action in isolated guinea pig arterial tissue, showing both positive inotropic and negative chronotropic activity [52].

47.3.5 Choleretic Action

Increased bile bilirubin output and concentration were observed after oral and subcutaneous administration of berberine to rats. Intravenous injection of berberine, even at a smaller dose, produced strong responses [53]. In rats, berberine chloride increased the secretion of bilirubin in experimental hyperbilirubinemia without affecting the liver bilirubin uridine diphosphoglucuronyltransferase activity [54]. Berberine sulfate at a dose of 10 mg/kg administered i.v. to dogs increased the bile flow and the output of bile acid and bilirubin. Berberine acted as a mild choleretic agent [55].

47.3.6 Influence on Enzymes

The oxidation of ethanol by liver alcohol dehydrogenase was competitively inhibited by berberine. Fluorometric and spectrometric equilibrium measurements showed the formation of a 1:2 complex between the enzyme and berberine [56]. The highest enzyme-inhibiting effect was shown by 13-ethylberberine, which was bound more firmly to the enzyme at pH 10 than NAD and NADH [57, 58]. Substantially higher concentrations of berberine and related compounds are needed for the inhibition of the oxidation of succinate [59]. Berberine sulfate and tetrahydropalmatine inhibited the NADH oxidase system of electron transport particles from beef heart by 90%–

100%. The concentrations for 50% inhibition were 50 μ*M* for berberine sulfate and 0.55 m*M* for tetrahydropalmatine [60].

Berberine and palmatine inhibited specific cholinesterase in rabbit spleen and pseudocholinesterase in normal horse serum. Both compounds were less effective inhibitory agents than neostigmine, but palmatine exhibited lower toxicity than berberine. Tetrahydropalmatine and tetrahydroberberine had no anticholinesterase effect, suggesting that the quaternary ammonium group is crucial for the effect of isoquinoline alkaloids on this enzyme [61]. The interaction of berberine and 13-ethylberberine with acetylcholinesterase from *Electrophorus electricus* was compared with that of known inhibitors of acetylcholinesterase. Electrostatic interaction of the quaternary alkaloids with the γ-anionic site of the enzyme could be confirmed, with two molecules of the substances being bound to the enzyme [62]. Berberine and coptisine were weak inhibitors of butyrylcholinesterase [63].

The cholinesterase-inhibiting action of berberine was also observed in homogenates of frog rectus abdominis and rabbit brain: addition of berberine kept acetylcholine undecomposed. The contractions caused by curare drugs were counteracted by a large dose of berberine, but this effect was only transient. Reoccurring contractions were only slowly removed [64].

Inhibitory effects on Na^+/K^+-ATPase, and Mg^{2+}-ATPase in rat brain microsomal preparations by isoquinoline alkaloids including berberine were also studied. In most cases Na^+/K^+-ATPase was more sensitive to the alkaloids than Mg^{2+}-ATPase [65].

Palmatine chloride, berberine chloride, and some related derivatives showed inhibition of reverse transcriptase activity of RNA tumor viruses. It was suggested that the alkaloids caused inhibition of the enzyme activity by interacting with the template primer, particularly of the adenine-thymine base pair. Furthermore, the alkaloids competed with the template primer-binding site of the enzyme. The time course of inhibition indicated that the alkaloids stopped the DNA synthesis instantly when added after the initiation of the polymerization processes. Inhibition of reverse transcriptase activity was correlated with the structure and antileukemic activity of the protoberberine alkaloids [66].

47.3.7 Interaction with DNA

The absorption maximum of berberine at 3475 Å shifted to 3500 Å when mixed with DNA. The height of the extinction maximum was markedly diminished depending on the DNA concentration used and on the degree of polymerization of DNA. Similar but less pronounced changes also occurred in the region of the extinction maximum at 4250 Å, which shifted to 4300 Å when mixed with DNA [67]. Calf thymus DNA produced systematic changes in the absorption spectrum of berberine which suggest that berberine forms a complex with DNA and binds up to the extent of one alkaloid molecule per two base pairs [68]. Flow dichroism of purines and pyrimidines and of berberine in the complex with DNA had the same signs and magnitudes. Berberine shifted the thermal strand separation profile of DNA to higher temperatures. This indicates that the alkaloid apparently forms a complex with DNA, probably by intercalation [69]. The viscometric titrations and flow-polarized fluorescence results also indicated that berberine binds to DNA by intercalation [70]. The DNA-alkaloid complexes formed were very stable due to the high affinity and specificity of berberine chloride for domains rich in adenine and thymine [71]. Unsubstituted protoberberine had the highest association constant [72]. The binding properties of protoberberine alkaloids with respect to DNA were influenced not only by the position and type of substituent but also by the presence of charge on the protoberberine skeleton [73].

47.3.8 Other Pharmacological Actions

Berberine produced sedation when given i.p. to cats and mice, or intraventricularly to cats, and potentiated the pentobarbital sleeping time. Berberine had no tranquilizing, anticonvulsant, or analgesic action [74], and it decreased rectal temperature and spontaneous motor activity and had local anesthetic action on guinea pig corneal reflex [75, 76].

Oral and intramuscular administration of berberine sulfate decreased serum cholesterol levels [77, 78]. A decrease in the anticoagulant action of heparin by berberine sulfate in dog and human blood in vitro was also noted [79]. Berberine formed a strongly fluorescent complex with the heparin of fixed mast cells and can be used for cytofluorometric measurement of heparin in both mast cells and individual mast cell granules [80].

The water extract of *C. chinensis* decreased serum glucose level in mice. Berberine, the major component of the extract, also showed a hypoglycemic effect in normal, spontaneously diabetic and alloxan-induced diabetic mice. Berberine also antagonized the hyperglycemic effect induced by i.p. injection of glucose or adrenaline in normal mice and inhibited the aggregation of rabbit platelets in vitro [78].

No significant changes in monoamines in rat brain were observed with berberine chloride and palmatine chloride at doses of 60 mg/kg 30 min after i.p. injection [81]. Berberine was phototoxic to mosquito (*Aedes atropalpus*) larvae. In the presence of near UV the median lethal concentration for acute 24 h toxicity was 8.8 ppm, compared to 250 ppm for dark controls. A slight increase in chromosome aberrations in Chinese hamster cells was also observed with berberine plus near UV treatment [82].

47.3.9 Metabolism of Berberine In Vivo

Berberine when administered orally to rabbits was absorbed through the gastrointestinal tract. The maximal blood level was reached after 8 h and berberine was still detectable 72 h later. Berberine levels were highest in the heart, followed by pancreas and liver. It was excreted through feces and urine [83]. The blood level of orally administered [^{3}H]berberine chloride in rats plateaued at 4–24 h, and maximal levels in the liver and muscles were achieved at 12 h. Urinary berberine excretion reached a maximum at 12–24 h. Excretion in the urine and feces at 48 h amounted to 2.7% and 86% of the administered dose, respectively [84]. Fecal elimination as the main excretion route indicates that berberine is not readily absorbed by the gastrointestinal tract [85]. The biological half-life of berberine chloride in rats was 5.2 and 5.4 h, after i.p. and oral administration, respectively [86]. Following subcutaneous injection of berberine chloride in rabbits and dogs, a metabolite was isolated from the urine by paper chromatography, which was identical with a compound obtained by oxidation of berberine chloride with Ag_2O_2. Perfusion experiments indicated oxidation of berberine chloride in liver [87].

47.3.10 Toxicity

Berberine sulfate administered orally at a dose of 25 mg/kg to cats caused a general depression 50–60 min after ingestion, lasting 6–8 h. A dose of 50 mg/kg caused salivation and occasional vomiting with recovery the next day. A dose of 100 mg/kg induced vomiting for 6–8 h and caused the death of all animals 8–10 days later. Intravenous injections of 10–15 mg/kg produced a short-lasting increase in respiratory frequency and agitation with a subsequent depression. A dose of 25 mg/kg injected i.v. caused an immediate increase in breathing, dyspnea, salivation, defecation, and micturation [87]. In experiments with rats, the LD_{50} of berberine sulfate was >1 g/kg after oral and about 90 mg/kg after i.p. administration. Histopathological examination revealed no changes in tissues and organs even in cases when berberine sulfate was given for 6 weeks at daily oral doses of 500 mg/kg [88].

References

1. Schramm G, Tang W (1959) Pharmacognosy of *Coptis*, Chinese Pharmacopoeia 1953. Pharmazie 14:405–408
2. Wang HK, Yang PC, Chen HM (1964) Alkaloids from roots of *Coptis chinensis* var. *omeiensis*. IV. Acta Pharm Sin 11:382–388
3. Wang HK, Yang PC, Chen HM (1964) Alkaloids from roots of *Coptis chinensis* var. *shihchuensis*. V. Acta Pharm Sin 11:389–392
4. Chatterjee A (1950) Constituents of *Coptis teeta*. Sci Cult 15:330
5. Yang TH, Chen KT (1969) Alkaloids of Formosan *Coptis quinquefolia*. Taiwan Ko Hsueh 23:30–36 (CA 73:106340h)
6. Yang MK, Liao RH, Huang ZM (1983) Study on the alkaloids of Ya Lian (*Coptis deltoidea*). Chin Trad Herb Drugs 14:151–153
7. Zhang XC, Yuan XL, Sun YQ (1985) Identification of species of *Coptis* and *Phellodendron* by TLC-DM. J Shenyang Coll Pharm 2:280–283
8. Qin XX, Wu CY, Chen BL (1986) Analysis of berberine-type alkaloids in Rhizoma coptidis and Chinese patent medicines by micellar chromatography. Acta Pharm Sin 21:458–465
9. Wang MZ, Zhu M (1984) Determination of alkaloids in *Coptis* by high performance thin-layer chromatography. Chin J Pharm Anal 4:12–15
10. Fang XP, Wang TZ, Zhang H, Shuai H, Li D, Xie CK (1989) Quantitative determination of five alkaloids in plants of *Coptis* from China. J Chin Mat Med 14:97–99
11. Li XJ, Tian XH, Chen YX, Li F, Duan MS, Zhou YY (1986) Separation and identification of epiberberine in coptis (*Coptis chinensis*). Chin Trad Herb Drugs 17:2–3
12. Ikuta A, Itokawa H (1983) Estimation of *Coptis* rhizomes on TLC. Shoyakugaku Zasshi 37:195–197
13. Ikuta A, Kobayashi A, Itokawa H (1984) Studies on the quantitative analysis of protoberberine alkaloids in Japanese, Chinese and other countries *Coptis* rhizomes by thin layer chromatograph-densitometry. Shoyakugaku Zasshi 38:279–282
14. Chang NC (1948) In vitro antibacterial action of extracts from *Coptis* roots. Proc Soc Exp Biol Med 69:141–143
15. Ukita T, Mizuno D, Tamura T (1949) The antibacterial properties of berberinium chloride. Jpn J Exp Med 20:103–108 (CA 44:3087e)
14. Chang NC (1948) In vitro antibacterial action of extracts from *Coptis* roots. Proc Soc Exp Biol Med 69:141–143
15. Ukita T, Mizuno D, Tamura T (1949) The antibacterial properties of berberinium chloride. Jpn J Exp Med 20:103–108 (CA 44:3087e)
16. Sado S (1947) Pharmacological study of berberine and its derivatives. II. J Pharm Soc Jpn 67:166–170
17. Amin AH, Subbaiah TV, Abbasi KM (1969) Berberine sulfate: antimicrobial activity, bioassay and mode of action. Can J Microbiol 15:1067–1076

18. Kowalewski Z, Kedzia W, Mirska I (1972) Effect of berberine sulfate on *Staphylococci*. Arch Immunol Ther Exp 20:353–360
19. Modak S, Modak MJ, Venkataraman A (1970) Mechanism of action of berberine on *Vibrio cholerae* and *V. cholerae* biotype eltor. Indian J Med Res 58:1510–1522
20. Modak MJ, Modak S, Venkataraman A (1970) Effect on berberine on the fatty acid composition of *Vibrio cholerae* and *V. cholerae* biotype eltor. Indian J Med Res 58:1523–1527
21. Dutta NK, Marker PH, Rao NR (1972) Berberine in toxin-induced experimental cholera. Appendix on berberine in toxin-induced experimental cholera. Br J Pharmacol 44:153–159
22. Sack RB, Froehlich JL (1982) Berberine inhibits intestinal secretory response of *Vibrio cholerae* and *Escherichia coli* enterotoxins. Infect Immun 35:471–475
23. Zhu BL, Ahrens FA (1982) Effect of berberine on intestinal secretion mediated by *Escherichia coli* heat-stable enterotoxin in jejunum of pigs. Am J Vet Res 43:1594–1598
24. Swabb EA, Tai YH, Jordan L (1981) Reversal of cholera toxin-induced secretion in rat ileum by luminal berberine. Am J Physiol 241:G248–G252
25. Tai YH, Feser JF, Marnane WG, Desjeux JF (1981) Antisecretory effects of berberine in rat ileum. Am J Physiol 241:G253–G258
26. Jezowa L, Kaczmarek F, Rafinski T (1966) The synergic action of berberine hydrochloride and tuberculostatic preparations upon the growth of *Mycobacterium tuberculosis* in vitro. Herba Pol 12:266–272 (CA 67:88503h)
27. Sabir M, Mahajan VM, Mohapatra LN, Bhide NK (1976) Experimental study of the antitrachoma action of berberine. Indian J Med Res 64:1160–1167
28. Yamahara J, Kawakita J, Kuraba K, Sawada T (1972) Biological studies of *Coptis japonica* and berberine type alkaloids. 3. Influence of berberine, coptisine and the extract of *Coptis* rhizoma on the fecal flora. Shoyakugaku Zasshi 26:84–89
29. Pitea M, Margineanu C (1972) Correlations between chemical structure and antibacterial activity of berberine. Clujul Med 45:465–469 (CA 80:78706f)
30. Mahajan VM, Sharma A, Rattan A (1982) Antimycotic activity of berberine sulfate: an alkaloid from an Indian medicinal herb. Sabouraudia 20:79–81 (CA 98:2666u)
31. Mirska I, Kedzia H, Kowalewski Z, Kedzia W (1972) Effect of berberine sulfate on healthy mice infected with *Candida albicans*. Arch Immunol Ther Exp 20:921–929
32. Sawada T, Yamahara J, Goto K, Yamamura M (1971) Evaluation of crude drugs by bioassay. IV. Antibacterial activities of the components of Coptidis rhizoma. Shoyakugaku Zasshi 25: 74–78
33. Kulkarni SK, Dandiya PC, Varandani NL (1972) Pharmacological investigations of berberine sulfate. Jpn J Pharmacol 22:11–16
34. Sabbaiah TV, Amin AH (1967) Effect of berberine sulfate on *Entamoeba histolytica*. Nature 215:527–528
35. Otsuka H, Tsukui M, Matsuoka T, Goto M, Fujimura H, Hiramatsu Y, Sawada T (1974) Antiinflammatory agents. Antiinflammatory screening by a fertile egg method. Yakugaku Zasshi 94:796–801
36. Otsuka H, Fujimura H, Sawada T, Goto M (1981) Studies on antiinflammatory agents. II. Antiinflammatory constituents from rhizome of *Coptis japonica* Makino. Yakugaku Zasshi 101:883–890
37. Akhter MH, Sabir M, Bhide NK (1977) Antiinflammatory effect of berberine in rats injected locally with cholera toxin. Indian J Med Res 65:133–141
38. Shvarev LF, Tsetlin AL (1972) Antiblastic properties of berberine and its derivatives. Farmakol Toksikol 35:73–75 (CA 76:148880b)
39. Schmitz H (1951) Storage of berberine in the granules of tumor cells. Naturwissenschaften 38:405
40. Honda F, Oka M, Akashi A, Nishino Y (1960) Pharmacological studies on the metabolism of cancer tissues. XXX. Influence of berberine and related compounds on the oxidation of DL-alanine in rat ascites hepatoma. Yakugaku Kenkyu 32:830–835 (CA 55:10695e)
41. Mollov NM, Duchevska KB, Silyanovska K, Stoichev S (1968) Cytotoxic effect of alkaloids from *Thalictrum minus elatum* and their derivatives. CR Acad Bulg Sci 21:605–608 (CA 69:58087z)
42. Hladon B, Kowalewski Z, Bobkiewicz T, Gronostaj K (1978) Cytotoxic activity of some *Chelidonium maius* alkaloids on human and animal tumor cell cultures in vitro. Ann Pharm (Poznan) 13:61–68 (CA 92:15307d)

43. Hoshi A, Ikekawa T, Ikeda Y, Shirakawa S, Iigo M, Kuretani K (1976) Antitumor activity of berberrubine derivatives. Gann 67:321–325
44. Creasey WA (1979) Biochemical effects of berberine. Biochem Pharmacol 28:1081–1084
45. Kumazawa Y, Itagaki A, Fukumoto M, Fujisawa H, Nishimura C, Nomoto K (1984) Activation of peritoneal macrophages by berberine-type alkaloids in terms of induction of cytostatic activity. Int J Immunopharmacol 6:587–592
46. Chun YT, Yip TT, Lau KL, Kong YC, Sankawa U (1979) A biochemical study on the hypotensive effect of berberine in rats. Gen Pharmacol 10:177–182
47. Ko ST, Lim DY (1980) Influence of berberine on the blood pressure of rabbits. Arch Pharmacol Res 3:23–30 (CA 95:35556z)
48. Fukuda H, Watanabe K, Kudo Y (1970) Cardiovascular effect of 9-substituted berberines. Chem Pharm Bull (Tokyo) 18:1299–1304
49. Hou DH, Li L, Wang QJ, Chen GF (1983) Studies on the antihypoxic and antiarrhythmic effect of five protoberberine compounds. J Nanjing Pharm Coll 30–33
50. Tang QY, Li GH, Li JL, Xu Y (1985) Effects of intravenous infusion of berberine hydrochloride on cardiovascular function. Pharm Ind 16:130–133
51. Greco RJ, Lefer AM, Maroko PR (1983) Actions of a new inotropic agent, berberine, on vascular smooth muscle. IRCS Med Sci Libr Compend 11:570–571
52. Shaffer JE (1985) Inotropic and chronotropic activity of berberine on isolated guinea pig atria. J Cardiovasc Pharmacol 7:307–315
53. Hobara N, Watanabe A (1984) Berberine-induced bile bilirubin secretion in the rat. Curr Ther Res 35:663–667
54. Chan MY (1977) The effect of berberine on bilirubin excretion in the rat. Comp Med East West 5:161–168
55. Oshiba S, Ueno H, Mihara H, Okamoto S (1974) Effect of berberine on bile secretion. Nihon Univ J Med 16:69–79 (CA 81:114662v)
56. Skursky L, Kovar J (1970) Inhibition of LADH (liver alcohol dehydrogenase) by berberine and some related alkaloids. Wenner-Gren Cent Int Symp Ser 18:653–662 (CA 80:142267t)
57. Pavelka S, Kovar J (1975) Interaction of liver alcohol dehydrogenase with protoberberine alkaloids. Collect Czech Chem Commun 40:753–768
58. Kovar J, Duerrova E, Skursky L (1979) Differences in the interactions of liver alcohol dehydrogenase with probes binding into the substrate pocket. Eur J Biochem 101:507–514
59. Mikes V, Yaguzhinskij LS (1985) Interaction of fluorescent berberine alkyl derivatives with respiratory chain of rat liver mitochondria. J Bioenerg Biomembr 17:23–32
60. Schewe T, Müller W (1976) Inhibition of the respiratory chain by the alkaloids berberine sulfate, alpinigenine and tetrahydropalmatine. Acta Biol Med Ger 35:1010–1021
61. Berezhinskaya VV, Aleshinskaya EE, Trutneva (1968) Anticholinesterase activity of some isoquinoline alkaloids. Farmakol Toksikol 31:296–299
62. Ulrichova J, Kovar J, Simanek V (1985) Interaction of quaternary aromatic isoquinoline alkaloids with acetylcholinesterase from *Electrophorus electricus*. Collect Czech Chem Commun 50:978–983
63. Ulrichova J, Walterova D, Preininger V, Simanek V (1983) Inhibition of butyrylcholinesterase activity by some isoquinoline alkaloids. Planta Med 48:174–177
64. Shimamoto K, Uchizumi S, Inoue K (1957) The anticurare action of berberine. Nippon Yakurigaku Zasshi 53:75–80 (CA 52:7520a)
65. Meyerson LR, McMurtrey KD, Davis VE (1978) Isoquinoline alkaloids. Inhibitory actions on cation-dependent ATP-phosphohydrolases. Neurochem Res 3:239–357
66. Sethi ML (1983) Enzyme inhibition. VI. Inhibition of reverse transcriptase activity by protoberberine alkaloids and structure-activity relationships. J Pharm Sci 72:538–541
67. Klimek M, Hnilica L (1959) Influence of deoxyribonucleic acid on ultraviolet and visible light absorption of berberine. Arch Biochem Biophys 81:105–110
68. Hahn FE, Krey AK (1971) Interactions of alkaloids with DNA. Prog Mol Subcell Biol 2:134–151
69. Krey AK, Hahn FE (1969) Berberine: complex with DNA. Science 166:755–757
70. Davidson MW, Lopp I, Alexander S, Wilson WD (1977) The interaction of plant alkaloids with DNA. Nucleic Acids Res 4:2697–2712
71. Maiti M, Chaudhuri K (1981) Interaction of berberine chloride with naturally occurring deoxyribonucleic acids. Indian J Biochem Biophys 18:245–250

72. Smekal E, Kubova N (1984) Alkaloids of protoberberine and benzophenanthridine groups as intercalating drugs with DNA. Stud Biophys 101:121–123 (CA 102:263n)
73. Smekal E, Koudelka J, Hung MA (1980) Fluorescence investigation of berberine-nucleic acid complexes. Stud Biophys 81:89–90 (CA 94:204201h)
74. Shanbhag SM, Kulkarni HJ, Gaitonde BB (1970) Pharmacological actions of berberine on the central nervous system. Jpn Pharmacol 20:482–487
75. Sabir M, Bhide NK (1971) Pharmacological actions of berberine. Indian J Physiol Pharmacol 15:111–132
76. Yamahara J (1976) Behavioral pharmacology of berberine-type alkaloids. I. Central depressive action of *Coptis rhizoma* and its constituents. Nippon Yakurigaku Zasshi 72:899–908 (CA 87:193774h)
77. Vad BG, Raman PH, Desmukh VK (1971) Effect of berberine on serum cholesterol and pentobarbitone sleeping time in rats. Indian J Pharm 33:23–24
78. Chen QM, Xie MZ (1986) Hypoglycemic effect of *Coptis chinensis* extract and berberine. Acta Pharm Sin 21:401–406
79. Sabir M, Bhide NK (1971) In vitro antiheparin action of berberine on the dog and human blood. Indian J Physiol Pharmacol 15:97–100
80. Berlin G, Enerbaeck L (1983) Fluorescent berberine binding as a marker of secretory activity in mast cells. Int Arch Allergy Appl Immunol 71:332–339
81. Liu GQ, Jiang Y, Pang G, Ibrahim, Zhou LQ (1985) Effect of some isoquinoline alkaloids on monoamines in rat brain. Acta Pharm Sin 20:566–570
82. Philogene BJR, Arnason JT, Towers CHN, Abramowski Z, Champagne D, McLachlan D (1984) Berberine: a naturally occurring phototoxic alkaloid. J Chem Ecol 10:115–123
83. Bhide MB, Chavan SR, Dutta NK (1969) Absorption, distribution and excretion of berberine. Indian J Med Res 57:2128–2131
84. Sakurai S, Tezuka M, Tamemasa O (1976) Studies on the absorption, distribution and excretion of ^{3}H-berberine chloride. Oyo Yakuri 11:351–355 (CA 88:145908d)
85. Yamahara J, Goto K, Sawada T (1972) Biological studies of *Coptis japonica* and berberine-type alkaloids. 2. Metabolism and chemotherapeutic effect. Shoyakugaku Zasshi 26:53–57 (CA 78:11673h)
86. Mrozikiewicz A, Kowalewski Z, Drost-Karwowska K, Bobkiewicz T, Klimaszewska O (1980) Pharmacokinetics of berberine chloride. Herba Pol 26:123–127 (CA 94:185343w)
87. Turova AD, Leskov AI, Bichevina VI (1962) Berberine. Lekarstv Sredstva iz Rast 303–307 (CA 58:2763b)
88. Hladon B (1975) Toxicity of berberine sulfate. Acta Pol Pharm 32:113–120 (CA 83:91108u)

Cordyceps sinensis (Berk.) Sacc. 48

48.1 Introduction

Dongchongxiacao, *Cordyceps*, is the dry complex composed of the sclerotium of the fungus *Cordyceps sinensis* (Berk.) Sacc. (Clavicipitaceae) and the larva corpses of insects of the family Hepialidae, on which the fungus is parasitic. It is officially listed in the Chinese Pharmacopoeia and is used as hemostatic, mycolytic, antiasthmatic, and expectorant in the treatment of respiratory diseases and as a tonic.

48.2 Chemical Constituents

The chemical constituents of *C. sinensis* were first studied by Chatterjee et al. in 1957 [1]. A crystalline substance was isolated and named "cordycepic acid", and was then identified by Sprecher and Sprinson as D-mannitol [2]. Further studies on the chemical constituents of *Cordyceps* revealed the presence of a series of known substances, but new structures or compounds with significant pharmacological efficacy were not found. The chemical constituents isolated from *C. sinensis* were amino acids, stearic acid, D-mannitol, mycose, ergosterol, uracil, adenine, adenosine [3, 4], palmitic acid, cholesterol palmitate and 5α-8α-epidioxy-5α-ergosta-6,22-dien-3β-ol (*48-1*) [5].

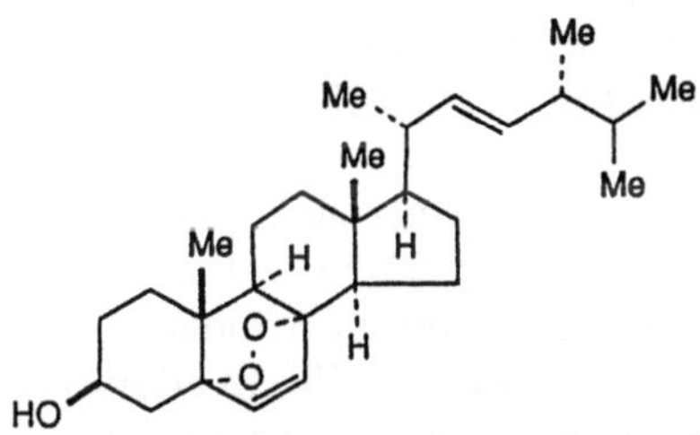

5α,8α-Epidioxy-5α-ergosta-6,22-dien-3β-ol (*48-1*)

Since the production of *C. sinensis* has not satisfied demand, *C. hawkesii* was found as a substitute for *C. sinensis*. Comparison of the chemical constituents between *C. hawkesii* and *C. sinensis* based on thin-layer chromatography has shown that amino acid, alkaloid, sterol, and organic acid contents are similar [6, 7]. Moreover, *Cordyceps* can be cultured submerged.

Ergosterol, stearic acid, D-mannitol, mycose, uracil, uridine, adenine, adenosine, and 13 amino acids were separated in both the cultured broth and mycelium of *C. sinensis* [8].

In addition, *C. barnesii* [9, 10], *C. liangshanensis* [11], *C. militaris* [12], and *C. shanxiensis* [13] were investigated botanically and chemically. They were all used as substitutes for *C. sinensis*. The D-mannitol contents in *C. barnesii* and *C. sinensis* were 8.7% and 7.8%, respectively, and the ergosterol contents in *C. barnesii* and *C. sinensis* were 1.2% and 1.1%, respectively. The contents of proteins, total amino acids, and alkaloids in both species were also similar [9]. Cultivation of *C. barnesii* was also reported [14].

From ascocarps of *C. sinensis*, a water-soluble polysaccharide was isolated and purified by ethanol fractionation and gel filtration. D-Galactose and D-mannose in a molar ratio of 1:1 were obtained by acid hydrolysis of purified polysaccharide. Chemical degradation and carbon 13 NMR spectrometric analysis showed that this polysaccharide is a highly branched galactomannan with a mannan core and galactosyl oligomer branches [15]. A water-soluble, minor galactomannan containing a small proportion of protein was further isolated from a 5% Na_2CO_3 extract of *C. sinensis*. It showed a homogeneous pattern in gel filtration and the molecular weight was estimated to be about 23 000. This minor polysaccharide was mainly composed of D-mannose and D-galactose in a molar ratio of 3:5 and had a highly branched structure [16].

48.3 Pharmacology

Both natural *Cordyceps* and cultured mycelia of *C. sinensis* have significant effects on the immune system of mice. They can increase the size of the spleen, decrease the size of the thymus, and prevent atrophy of spleen and liver and hypertrophy of the thymus in mice induced by cyclophosphamide [17]. The DNA, RNA, and protein contents in the enlarged spleen were significantly increased. The effect on the thymus was abolished by adrenalectomy. The extract increased the incorporation of [^{3}H]thymidine into spleen DNA in vivo and the proliferation of splenocytes in vitro. The active principle was found to be present in the stroma rather than in the larva [18].

The aqueous extracts of natural *Cordyceps* and the cultured mycelia of *C. sinensis* can enhance the production of macrophages and activate the functions of the phagocytic system. They not only enhance the phagocytic activity of the macrophages, but also increase the alkaline phosphatase activity of the macrophages [19]. In addition, serum hemolysin and splenocytic immunohemolytic activities were elevated in mice immune suppressed by hydrocortisone. In normal mice, however, no such regulatory effect on humoral immunity was observed [20].

The polysaccharide of *C. sinensis* also showed immunostimulating activity in mice. It activated the phagocytic function of the reticuloendothelial system and of macrophages in the abdominal cavity and increased blood serum IgG and plasma corticosterone levels and spleen weight. The polysaccharide also antagonized spleen atrophy and leukocyte decrease induced by cortisone and cyclophosphamide, and the reduction of phagocytic function of macrophages in the abdominal cavity, but did not inhibit the antiinflammatory function of cortisone [21].

Some results of clinical studies with *C. sinensis* for treatment of tinnitus [22], chronic nephritis [23, 24], arrhythmia [25], and sexual hypofunction [26] were recently reported. For instance, treatment with *C. sinensis* significantly decreased proteinuria of the patients with chronic nephritis in 24 h. Two patients with hematuria as the major clinical manifestation responded with strongly decreased erythrocyte counts in urine [23].

References

1. Chatterjee R, Srinivasan KS, Maiti PC (1957) *Cordyceps sinensis*: structure of cordycepic acid. J Am Pharm Assoc 46:114–118
2. Sprecher M, Sprinson DB (1963) A reinvestigation of the structure of "cordycepic acid". J Org Chem 28:2490–2491
3. Lu RM, Yang YC, Yang YP, Wang SF (1981) Study on chemical constituents in *Cordyceps sinensis* Sacc. Chin Pharm Bull 16:55
4. Xu WH, Xue Z, Ma JM (1988) Water-soluble constituents of *Cordyceps sinensis* (Berk.) Sacc. – the nucleosides. Bull Chin Mat Med 13:226–228
5. Xiao YQ, Liu JM, Tu YY (1983) Studies on chemical constituents in *Cordyceps sinensis*. I. Bull Chin Mat Med 8:32–33
6. Huang HY, Chou SH, Ho HL (1980) Comparison of the chemical constituents between yaxianbangchoncao (*Cordyceps hawkesii*) and dongchonxiacao (*C. sinensis*). Chin Trad Herbal Drugs 11:435–439
7. Huang HT, Chou SH, Ho HL (1981) Comparison of chemical constituents between *Cordyceps hawkesii* and *C. sinensis*. Chin Pharm Bull 16:53
8. Lu RM, Yang YC, Yue DC, Wang SF, Fan TJ, Huo ZM, Wang CF, Yang YP (1982) Chemical composition of the submerged culture of *Cordyceps* sp. No 1. Weishengwuxue Tongbao 9:166–168
9. Yang HD, Ma ZL, Sun TQ, Zhang XC, Cai JG (1985) Comparative study on the chemical constituents between xiangbangchongcao (*Cordyceps barnesii*) and cordyceps (*C. sinensis*). Chin Trad Herbal Drugs 16:194–195
10. Guo YW, Wang SM, Gao JD, Zhou YZ, Ma XF, Zhuang X, Cheng XB, Liu JH (1985) Preliminary study on *Cordyceps barnesii* – comparison of chemical constituents between *Cordyceps barnesii* and *Cordyceps sinensis*. Bull Chin Mat Med 10:129–131
11. Tan ZY, Gao R, Gao Y, Xie CZ (1985) Comparative study on the chemical constituents between liangshanchongcao (*Cordyceps liangshanensis*) and cordyceps (*C. sinensis*) Chin Trad Herbal Drugs 16:196–198
12. Gao HA, Chenz SZ, Wang LR, Zhang LS, Li JR (1987) Comparison of some chemical constituents of *Cordyceps militaris* and *Cordyceps sinensis*. Bull Chin Mat Med 12:108–109
13. Liu B, Rong FX, Jin HS (1985) A new species of the genus *Cordyceps*. J Wuhan Bot Res 3:23–24
14. Ma XF, Chai QY, Hou X (1986) Survey of the ecology of *Cordyceps barnesii*. Bull Chin Mat Med 11:13–14
15. Miyazaki T, Oikawa N, Yamada H (1977) Studies on fungal polysaccharides. XX. Galactomannan of *Cordyceps sinensis*. Chem Pharm Bull 25:3324–3328
16. Kiho T, Tabata H, Ukai S, Hara C (1986) Polysaccharides in fungi. XVIII. A minor protein-containing galactomannan from a sodium carbonate extract of *Cordyceps sinensis*. Carbohydr Res 156:189–197
17. Chen DM, Zhang SL, Li ZN, Cheng ZQ, Liu XP (1985) Effects of natural cordyceps and the cultured mycelia of *Cordyceps sinensis* on murine immune organs and functions of mononuclear phagocyte system. Chin J Integr Trad West Med 5:42–44
18. Liu GT, Xu RL (1985) Immunopharmacology of *Cordyceps sinensis*. Chin J Integr Trad Western Med 5:622–624
19. Zhang SL, Chen DM, Cheng ZQ, Liu XP (1985) Activation of murine peritoneal macrophage by the natural cordyceps and the cultured mycelia of *Cordyceps sinensis*. Chin J Integr Trad West Med 5:45–47
20. Tang RJ, Wang ZP, Min ZH, Zhang J (1986) Pharmacology of natural cordyceps and cultured mycelia of *Cordyceps sinensis*. II. Effects on immunologic function. Chin Trad Herbal Drugs 17:214–216

21. Zang QZ, He GX, Zheng ZY, Xu JH, Liu JZ, Wang SY, Huang JZ, Du DJ, Zeng QT (1985) Pharmacological action of the polysaccharide from cordyceps (*Cordyceps sinensis*). Chin Trad Herbal Drugs 16:306–311
22. Zhuang JM, Chen HL (1985) Treatment of tinnitus with *Cordyceps* infusion: a report of 23 cases. Fujian Med J 7:42
23. Shen LM, Chen YP (1985) Treatment of 18 cases of chronic nephritis mainly with cultivated cordyceps. Liaoning J Trad Chin Med 9:32–33
24. Chen YP, Liu WZ, Shen LM, Xu SN (1986) Clinical effects of natural cordyceps and cultured mycelia of *Cordyceps sinensis* in kidney failure. Chin Trad Herbal Drugs 17:256–258
25. Yu HS (1985) Treatment of arrhythmia with *Cordyceps sinensis*. J Zhejiang Trad Chin Med Coll 9:28
26. Yang WZ, Deng XA, Hu W (1985) Treatment of sexual hypofunction with *Cordyceps sinensis*. Jiangxi Zhongyiyao 5:46–47

Corydalis turtschaninovii Bess. f. *yanhusuo* Y.H. Chou et C.C. Hsü

49

49.1 Introduction

Yanhusuo, Rhizoma Corydalis, is the dry tuber of *Corydalis turtschaninovii* Bess. f. *yanhusuo* Y.H. Chou et C.C. Hsü (Papaveraceae), which is collected in early summer after the stems and leaves have withered. It is officially listed in the Chinese Pharmacopoeia and is used as an analgesic for the treatment of abdominalgia, menorrhalgia, menostasia, and traumatic pain.

Tetrahydropalmatine is listed in the Chinese Pharmacopoeia (1985), Vol II.

In addition to the officially listed *C. turtschaninovii* f. *yanhusuo* (*C. yanhusuo*), there are many species of the subgenus Capnites known to be used in traditional Chinese medicine or folk medicine, mainly as an analgesic, antirheumatic, and emmenagogue. These medicinal species are *Corydalis ambigua*, *C. decumbens*, *C. glaucescens*, *C. humosa*, *C. ledebouriana*, *C. remota*, *C. repens*, *C. schanginii*, and *C. ternata* [1]. Thirty species of *Corydalis* subgenus Capnoides in China were noted to be of medicinal value for the treatment of fever, bleeding, pain, infection, jaundice, irregular menstruation, hypertension, diarrhea, and tumors. These plants are *Corydalis adunca*, *C. balansae*, *C. bungeana*, *C. conspersa*, *C. curviflora*, *C. davidii*, *C. delavayi*, *C. denticulato-bracteata*, *C. edulis*, *C. hendersonii*, *C. impatiens*, *C. incisa*, *C. linearioides*, *C. melanochlora*, *C. mucronifera*, *C. ochotensis*, *C. ophiocarpa*, *C. pachypoda*, *C. pallida*, *C. racemosa*, *C. saxicola*, *C. scaberula*, *C. sheareri*, *C. speciosa*, *C. stricta*, *C. taliensis*, *C. temulifolia*, *C. thyrsiflora*, *C. tomentella*, and *C. trachycarpa* [2]. *C. stricta* Steph. and *C. bungeana* Turcz. are included in the appendix of the Chinese Pharmacopoeia.

49.2 Chemical Constituents

49.2.1 Chemical Constituents of *Corydalis ambigua*

The chemical constituents of the tuber of *C. ambigua* have been intensively investigated for the past 60 years [3]. Chou isolated a number of alkaloids from the tuber and named them Corydalis A, B, C, D, E [4], F, G, H [5], I [6], J, K [7], L, and M [8]. Corydalis A, B, C, D, E, G, I, K, and M were then identified as corydaline, tetrahydropalmatine, protopine, *l*-tetrahydrocoptisine [9], *dl*-tetrahydrocoptosine, corybulbine, glaucine, tetrahydrocolumbamine, and allocryptopine, respectively. Corydaline (*49-1*), corybulbine (*49-2*), tetrahydropalmatine (*49-3*), tetrahydrocolumbamine (*49-4*) , and tetrahydrocoptisine (*49-5*) are structurally all derived from 6*H*-dibenzo[*a*,*g*]-quinolizine (*49-6*), whereas glaucine (*49-7*) is derived from 4*H*-dibenzo[*de*,*g*]quinoline (*49-8*). The protopine-type alkaloids protopine (*49-9*) and allocryptopine (*49-10*) possess an azecine ring structure. Tetrahydropalmatine, tetrahydrocolumbamine, and tetrahydrocoptisine differ from the corresponding dehydro analogs palmatine, columbamine, and coptisine by the presence of a tertiary nitrogen atom instead of a quaternary one. Corydalis J was shown to be identical with corydalis I and F identical with K [10].

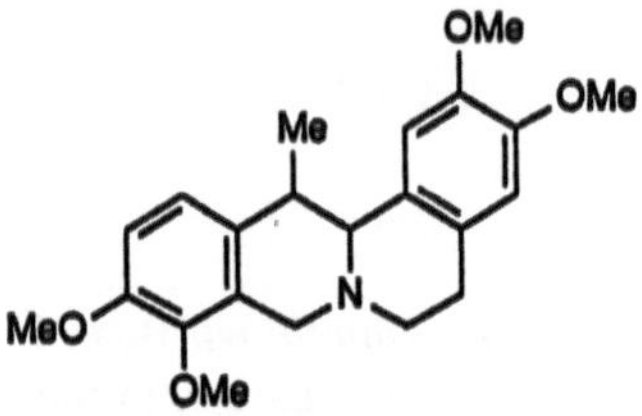

Corydaline (*49-1*)

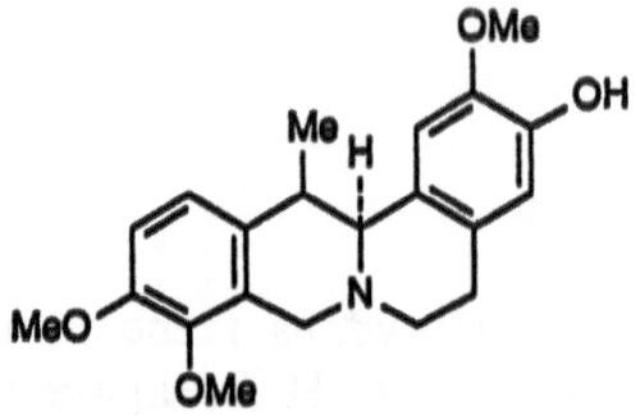

Corybulbine (*49-2*)

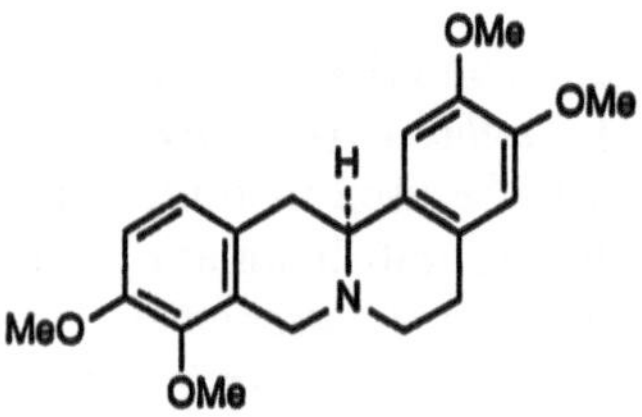

Tetrahydropalmatine (*49-3*)

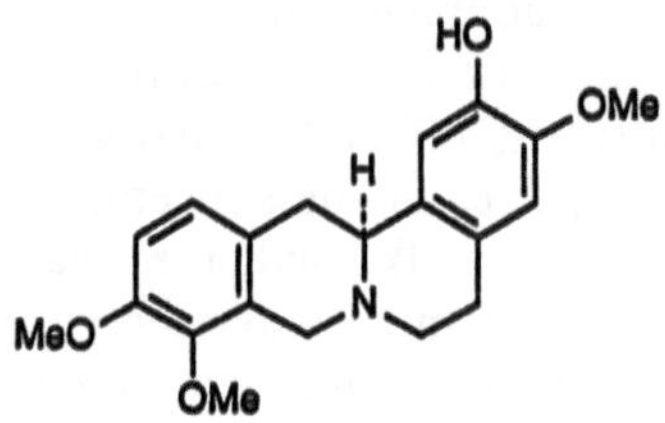

Tetrahydrocolumbamine (*49-4*)

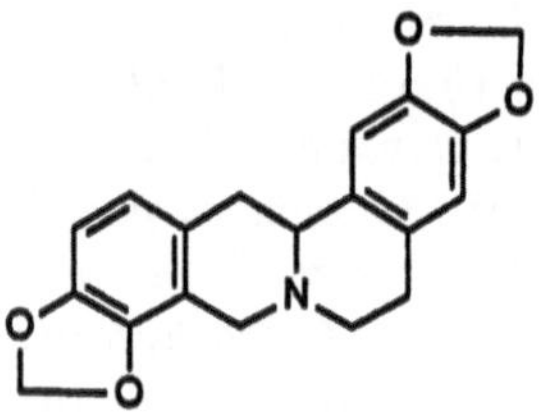

Tetrahydrocoptisine (*49-5*)

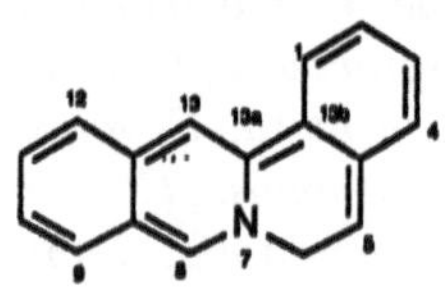

6*H*-Dibenzo[*a*, *g*]-quinolizine (*49-6*)

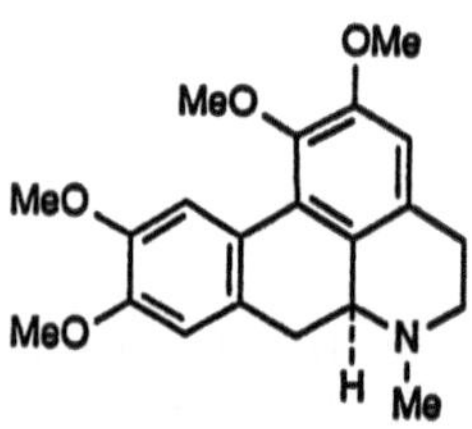

Glaucine (*49-7*)

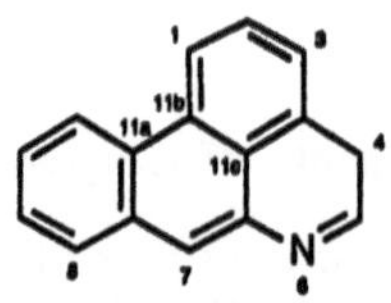

4*H*-Dibenzo[*de*, *g*]-quinoline (*49-8*)

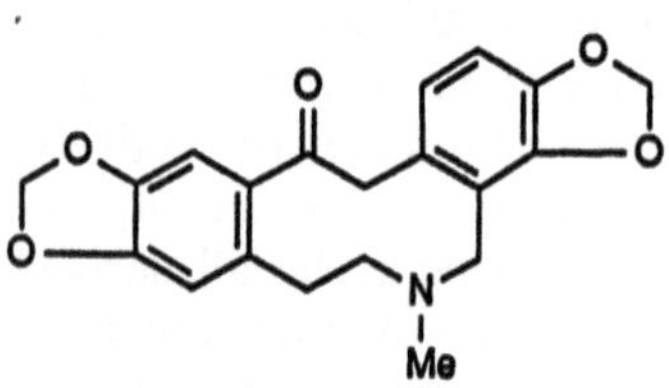

Protopine (*49-9*)

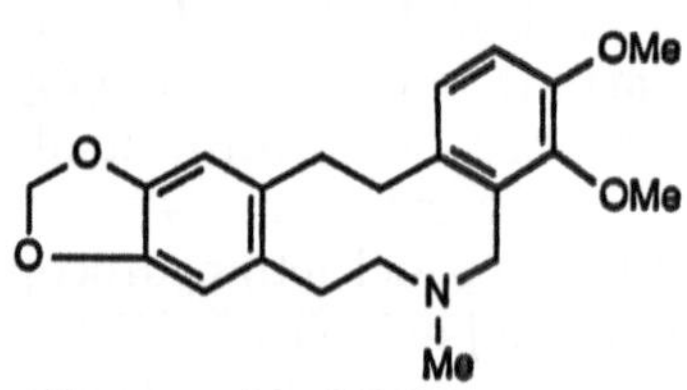

Allocryptopine (*49-10*)

Further studies on the chemical constituents of the tuber of *C. ambigua* resulted in the isolation of the alkaloids scoulerine (*49-11*), tetrahydrojatrorrhizine (*49-12*), noroxyhydrastinine (*49-13*) [11], methylcorypalline (*49-14*), cavidine (*49-15*), dehydrothalictrifoline (*49-16*) [12], ambinine (*49-17*) [13], corydalmine (*49-18*), and dehydrocorydalmine (*49-19*) [11, 14]. Among these, noroxyhydrastinine and methylcorypalline are two isoquinoline alkaloids, whereas ambinine possesses a benzo[*c*]phenanthridine skeleton. The corydaline and corybulbine contents reached their maximum in plants approximately 4 and 5 years old, whereas the cavidine content did not change with age of the cultivated plants [15].

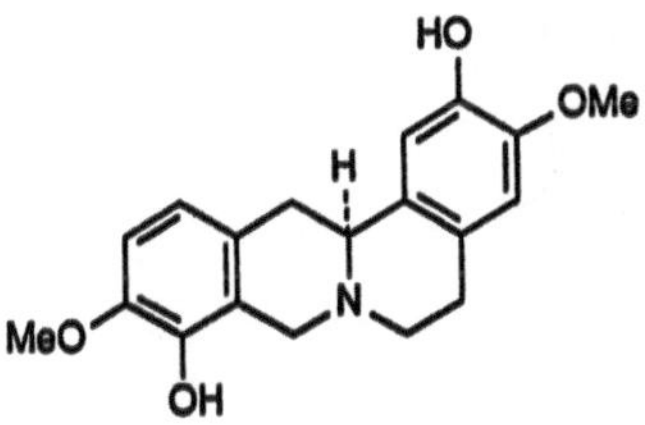

Scoulerine (*49-11*)

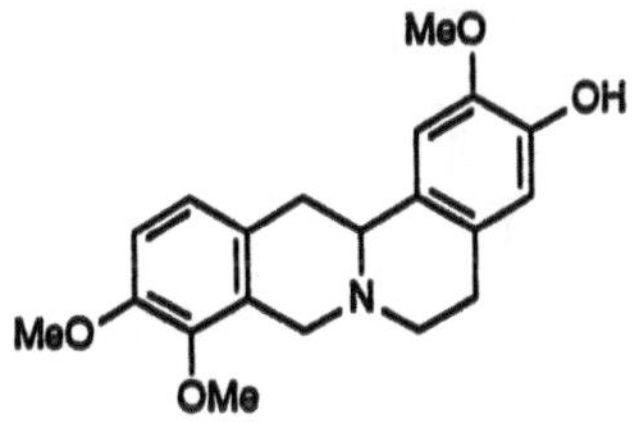

Tetrahydrojatrorrhizine (*49-12*)

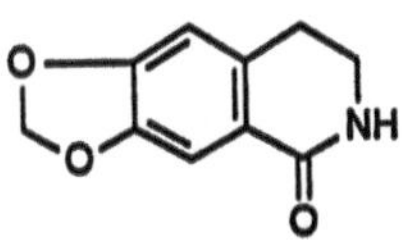

Noroxyhydrastinine (*49-13*)

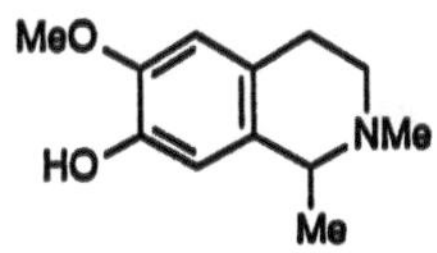

Methylcorypalline (*49-14*)

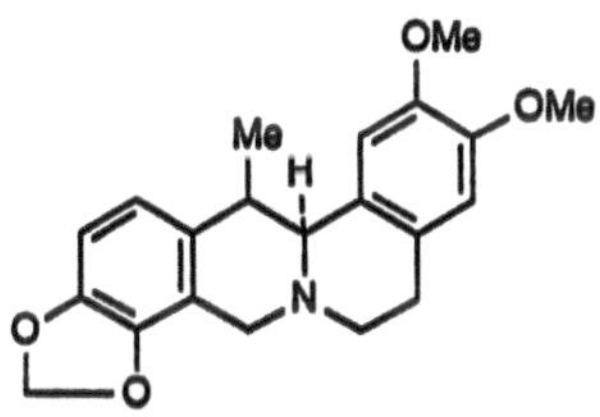

Cavidine (*49-15*)

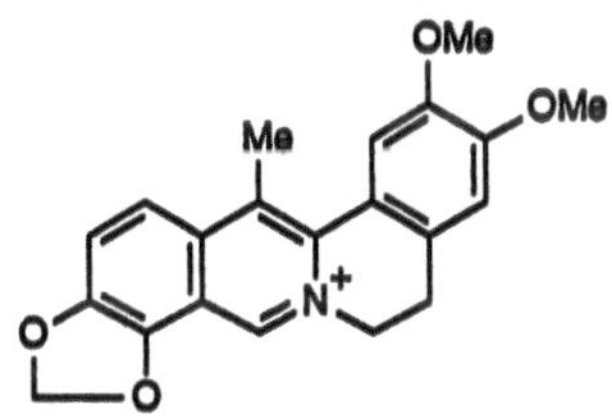

Dehydrothalictrifoline (*49-16*)

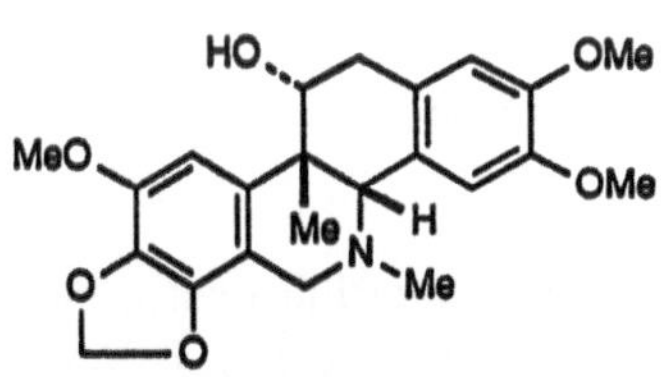

Ambinine (*49-17*)

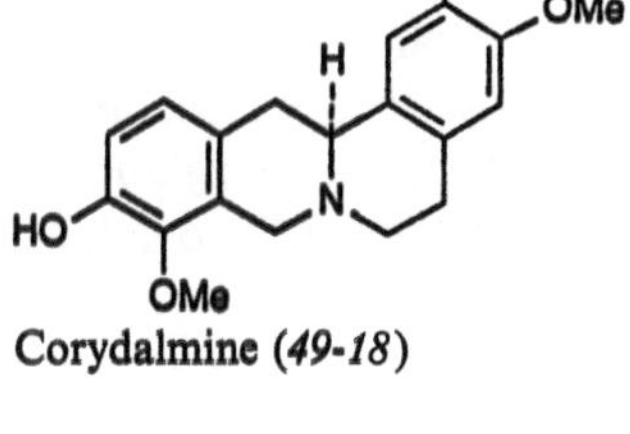

Corydalmine (*49-18*)

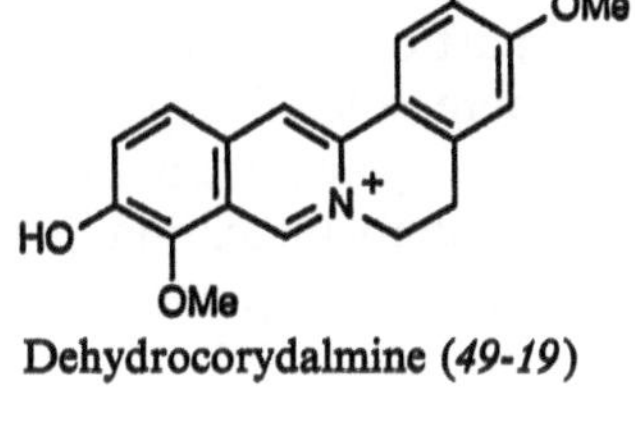

Dehydrocorydalmine (*49-19*)

49.2.2 Chemical Constituents of *Corydalis turtschaninovii* f. *yanhusuo*

A number of alkaloids have been isolated and identified from the tuber of *C. turtschaninovii* f. *yanhusuo*. They are corydaline, tetrahydropalmatine, tetrahydrocoptisine, glaucine, protopine, allocryptopine, tetrahydrocolumbamine, corybulbine, dehydrocorydaline [16–19], tetrahydroberberine (*49-20*), palmatine, columbamine, lauroscholtzine (*49-21*), dehydroglaucine, yuanhunine (*49-22*), a new protoberberine-type alkaloid [17, 20], leonticine (*49-23*), and dihydrosanguinarine (*49-24*) [21].

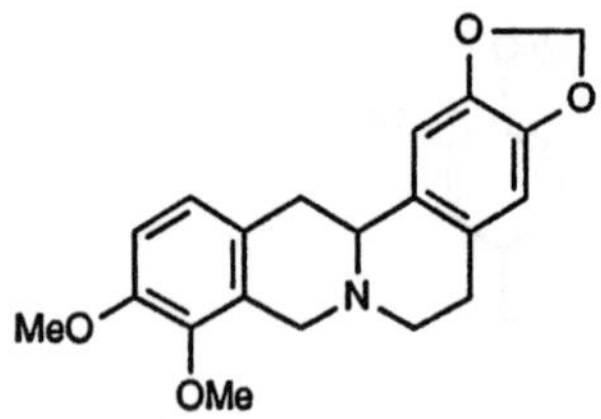

Tetrahydroberberine (*49-20*)

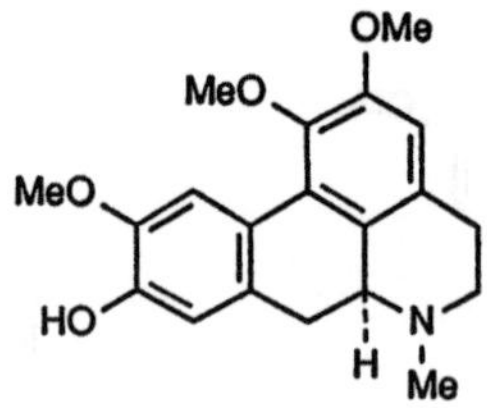

Lauroscholtzine (*49-21*)

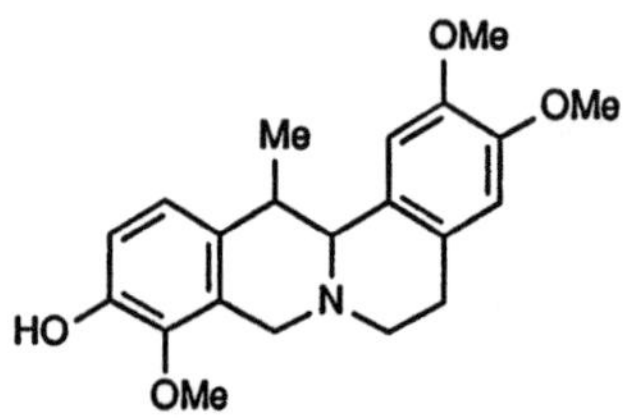

Yuanhunine (*49-22*)

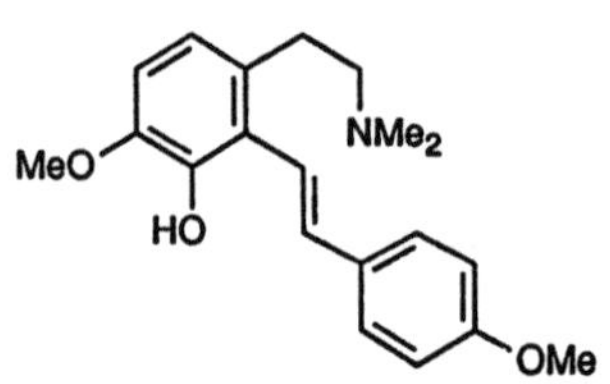

Leonticine (*49-23*)

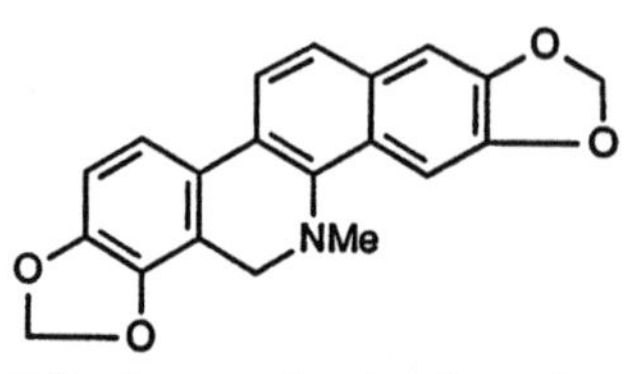

Dihydrosanguinarine (*49-24*)

The contents of tetrahydrocoptisine, corydaline, tetrahydropalmatine, protopine, coptisine, palmatine, and glaucine were found to be 0.02%–0.04%, 0.1%, 0.06%, 0.05%, 0.03%–0.04%, 0.03%, and 0.06%–0.08%, respectively, in the tuber of *C. turtschaninovii* f. *yanhusuo* [22]. Contents also depend on the localities where samples are collected [21].

From the aerial part of *C. turtschaninovii* f. *yanhusuo*, 12 alkaloids and a long-chain fatty alcohol were isolated. They were identified as tetrahydrocoptisine, dehydroglaucine, protopine, glaucine, nantenine (*49-25*), allocryptopine, norglaucine (*49-26*), tetrahydrocolumbamine, thaliporphine (*49-27*), lirioferine (*49-28*), lauroscholtzine, isoboldine (*49-29*), and the fatty alcohol nonacosan-10-ol.

A new alkaloid named coryphenanthrene (*49-30*) was isolated from the aerial part of *C. turtschaninovii* f. *yanhusuo* and structurally determined as 1-methylaminoethyl-3,4,6,7-tetramethoxyphenanthrene [24]. Glaucine and *d*-nantenine are the main constituents in the aerial part [25]. The glaucine content was more than 0.25% [23]. Nantenine (domestine), thaliporphine, lirioferine, and isoboldine are all alkaloids of the dibenzo[*de,g*]-quinoline type.

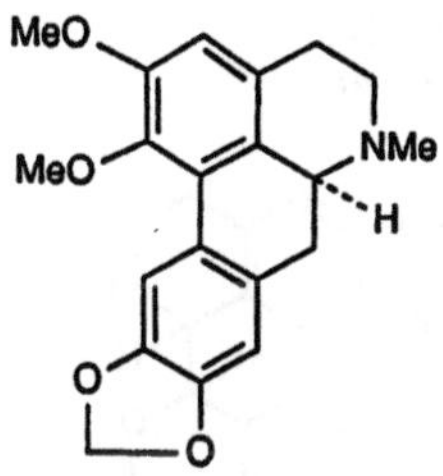

Nantenine (*49-25*)

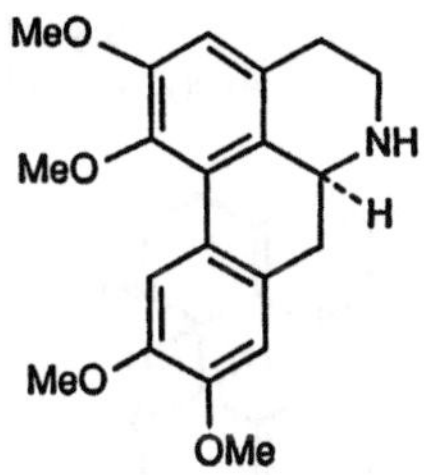

Norglaucine (*49-26*)

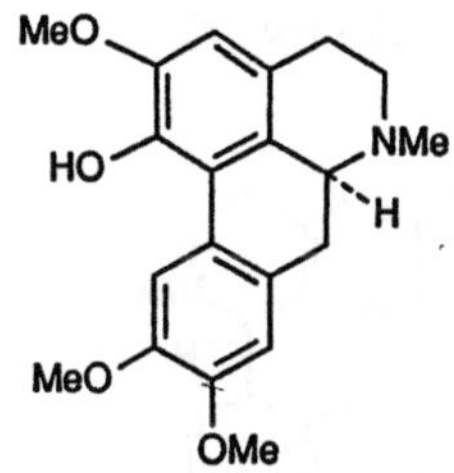

Thaliporphine (*49-27*)

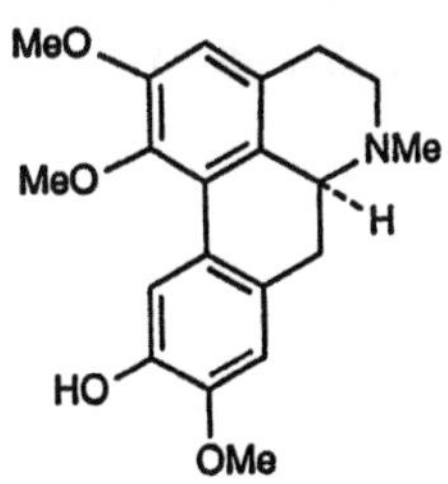

Lirioferine (*49-28*)

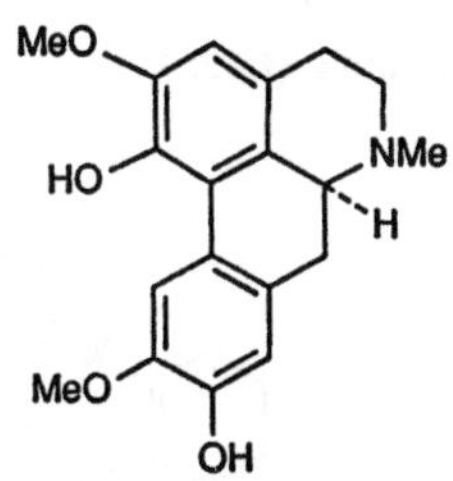

Isoboldine (*49-29*)

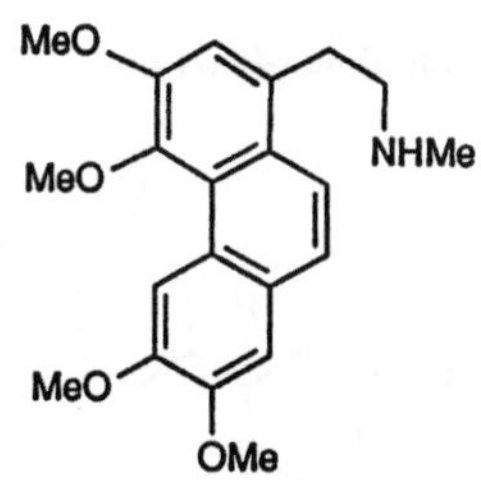

Coryphenanthrene (*49-30*)

49.2.3 Chemical Constituents of Other *Corydalis* Species

Corydalis stricta, *C. melanochlora*, *C. linearioides*, *C. mucronifera*, and *C. conspersa* are five *Corydalis* species native to the Qinghai plateau, China, and are used in folk medicine. Studies on the alkaloidal constituents have given the following results: hydrastine (*49-31*) was isolated from *C. stricta*; corynoline (*49-32*) and acetylcorynoline (*49-33*) were isolated from *C. melanochlora*; corynoline, corlumidine (*49-34*), protopine, and adlumine (*49-35*) were isolated from *C. linearioides*; protopine, adlumine, bicuculline (*49-36*) and cheilanthifoline (*49-37*) were isolated from *C. mucronifera*; and acetylcorynoline, corynoline, and corynoxine (*49-38*) were isolated from *C. conspersa* [26, 27].

These alkaloids are different in structure from the alkaloids of *C. ambigua* and *C. turtschaninovii* f. *yanhusuo*. Corynoline is a chelidonine (*49-39*) type alkaloid derived from benzo[*c*]phenanthridine; hydrastine, corlumidine, adlumine, and bicuculline (adlumidine) have a benzylisoquinoline structure with a lactone ring and corynoxine has a corynoxane (*49-40*) skeleton.

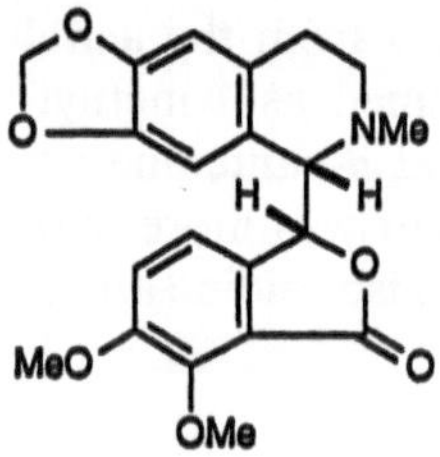

Hydrastine (*49-31*)

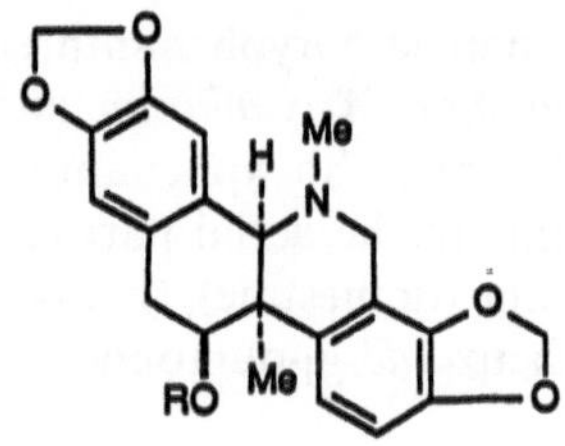

Corynoline (*49-32*): R=H
Acetylcorynoline (*49-33*): R=Ac

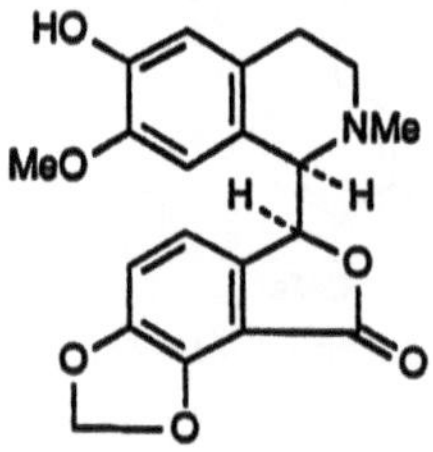

Corlumidine (*49-34*)

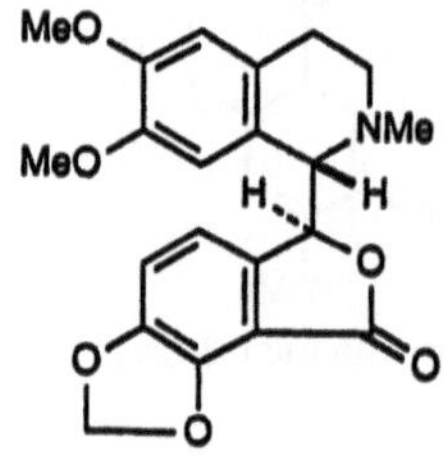

Adlumine (*49-35*)

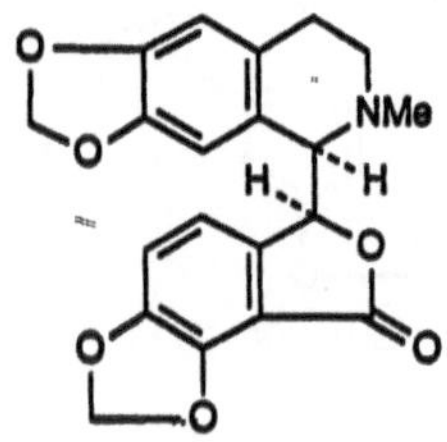

Bicuculline (*49-36*)

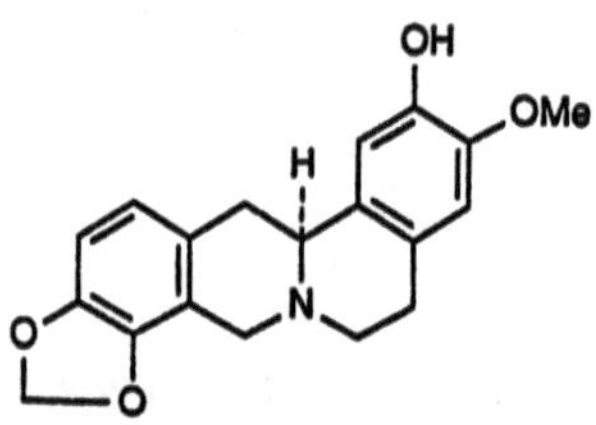

Cheilanthifoline (*49-37*)

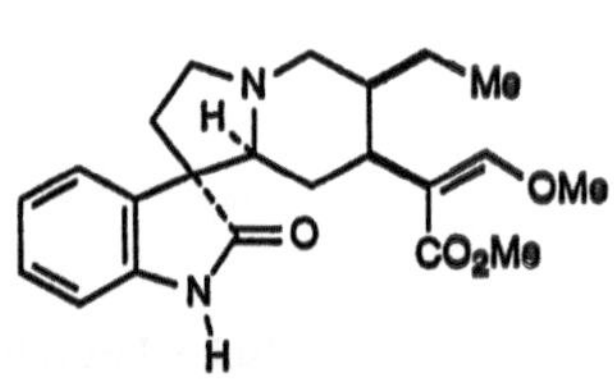

Corynoxine (*49-38*)

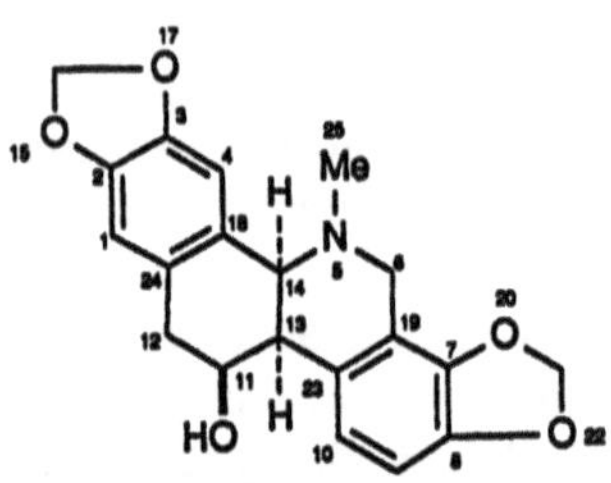

Chelidonine (*49-39*)

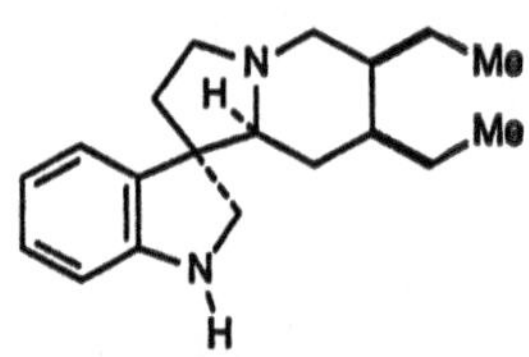

Corynoxane (*49-40*)

Corydalis bungeana is a folk medicine used as an antipyrogenic and an antiinflammatory agent. From the above-ground part of *C. bungeana*, the alkaloids protopine, isocorynoline (*49-41*), acetylisocorynoline, corynoline, tetrahydrocoptisine, acetylcorynoline, tetrahydrocorysamine (*49-42*), and 11-epicorynoline were isolated and identified [28–31].

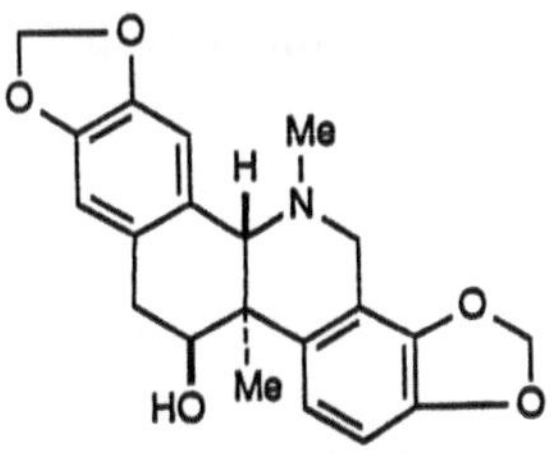

Isocorynoline (*49-41*)

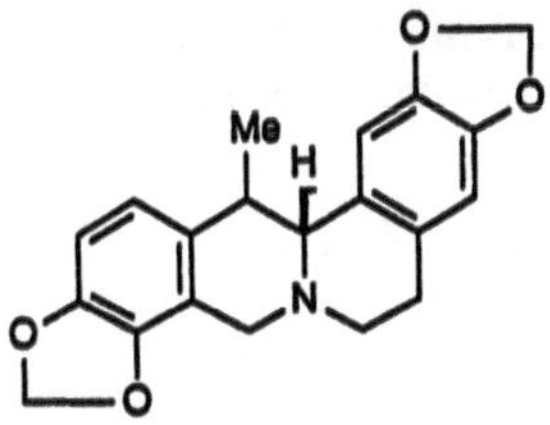

Tetrahydrocorysamine (*49-42*)

Besides corynoline and acetylcorynoline [26, 27], alkaloids of the benzo[*c*]phenanthridine type, corynoloxine (*49-43*), and consperine (*49-44*) were isolated from the whole plant of *C. conspersa* [32]. Consperine was structurally elucidated as acetonylacetylcorynoline on the basis of spectral analysis [32].

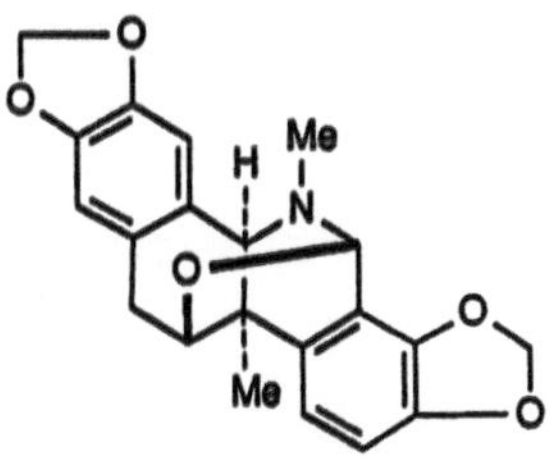

Corynoloxine (*49-43*)

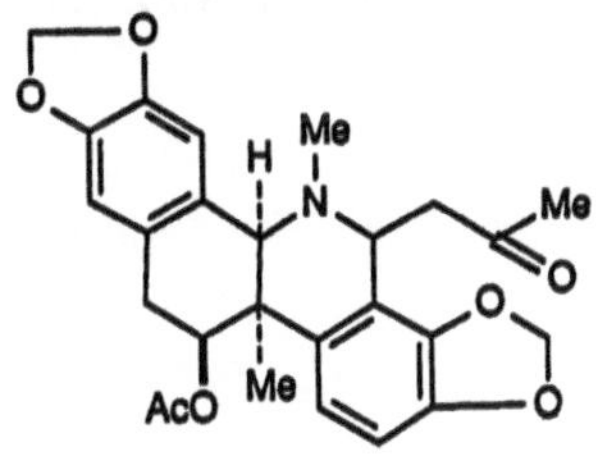

Consperine (*49-44*)

The alkaloid components isolated from the whole plant of *C. decumbens* were identified as dihydropalmatine, hydrastinine (*49-45*), 3,4-dehydrohydrastine [33], corlumidine, *d*-tetrahydropalmatine, bicuculline, bulbocapnine (*49-46*), protopine, palmatine, corydaline, hydroxyhydrastine, berberine, jatrorrhizine, and the alkaloids decumbenine (*49-47*) [34, 35], decumbenine C (*49-48*) [36], decumbesine (*49-49*), and epidecumbesine [37].

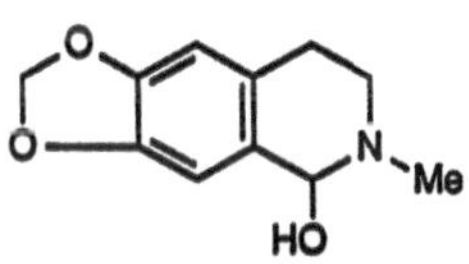

Hydrastinine (*49-45*)

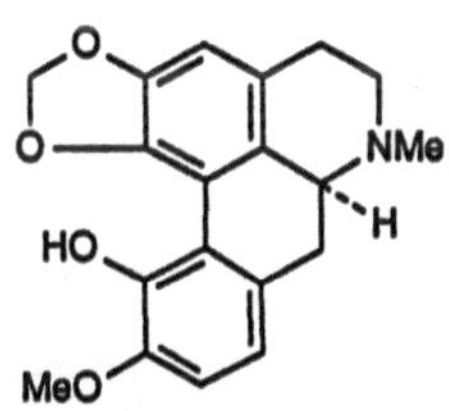

Bulbocapnine (*49-46*)

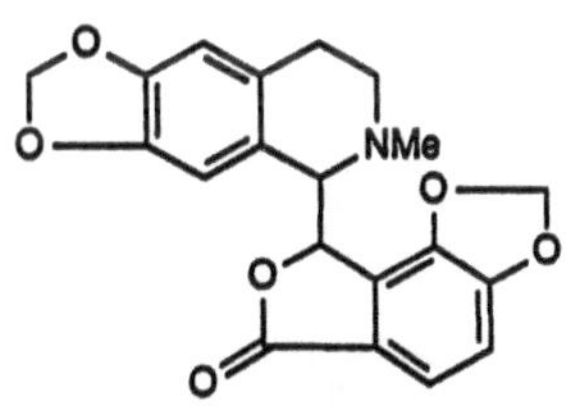

Decumbenine (*49-47*)

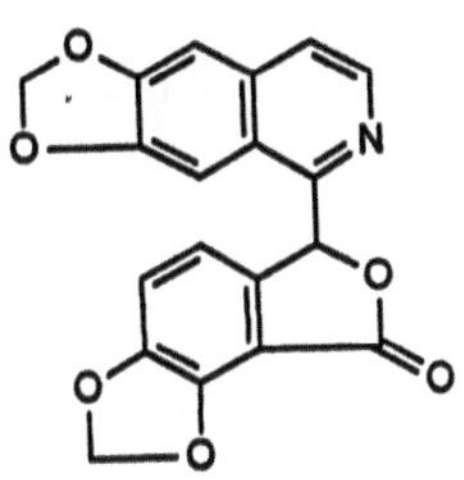

Decumbenine C (*49-48*)

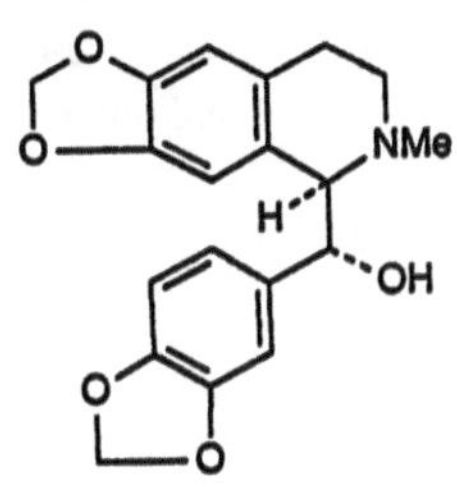

Decumbesine (*49-49*)

Protopine, corycavine (*49-50*), corynoline, acetylcorynoline [38], and acetonylcorynoline [39] were isolated from *C. delavayi.*

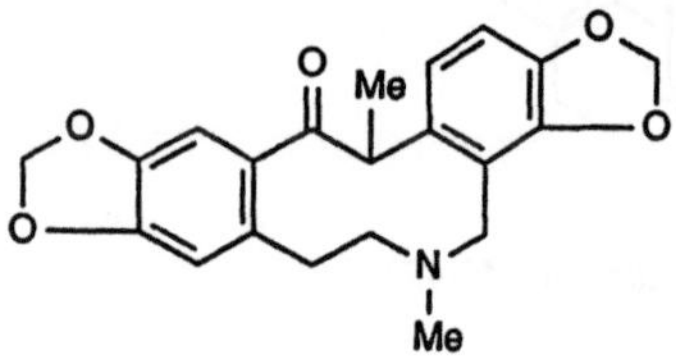

Corycavine (*49-50*)

From the whole plant of *C. hendersonii*, a new alkaloid named henderine (*49-51*) was isolated together with allocryptopine, protopine, stylopine, and cheilanthifoline [40]. Stylopine is the (*S*)-configuration of tetrahydrocoptisine.

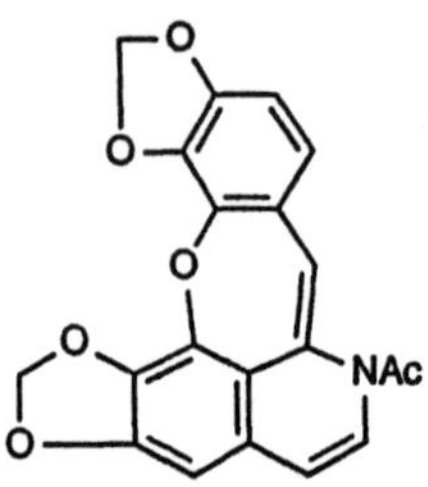

Henderine (*49-51*)

New alkaloids corytensine (*49-52*) [41] and isoochotensine (*49-53*) [42] were isolated from *C. ochotensis* together with the known alkaloids ochotensine (*49-54*), ochotensimine (*49-55*), lienkonine (*49-56*), and cheilanthifoline [42]. Ochotensine, ochotensimine, and isoochotensine are alkaloids possessing a spiroindenoisoquinoline structure.

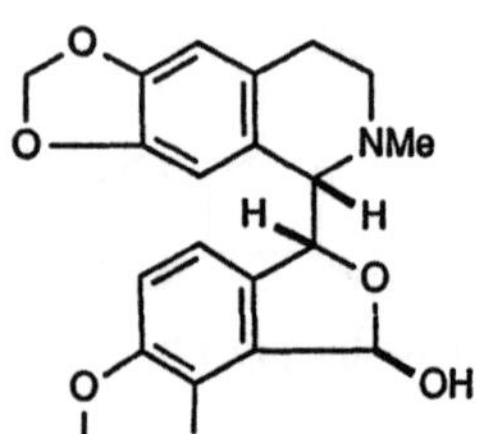

Corytensine (*49-52*)

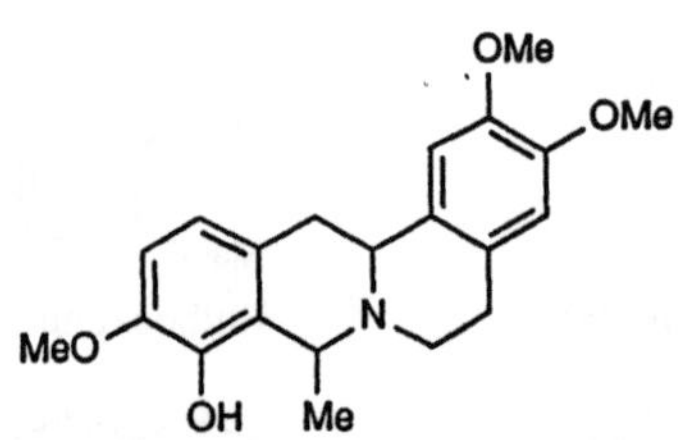

Lienkonine (*49-56*)

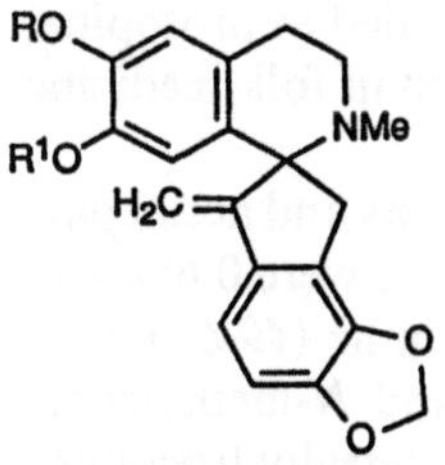

Isoochotensine (*49-53*): $R = CH_3$, $R^1 = H$
Ochotensine (*49-54*): $R = H$, $R^1 = CH_3$
Ochotensimine (*49-55*): $R = R^1 = CH_3$

From *C. pallida* var. *speaose* six constituents were isolated and identified as *trans*-3-ethylidene-2-pyrrolidone, protopine, dihydrosanguinarine, tetrahydrocoptisine, and nonacosan-10-ol [43].

The new alkaloid corymotine (*49-57*) was isolated from *C. remota*, in addition to the known alkaloids tetrahydrocorysamine, allocryptopine, corybulbine, acetylcorynoline, cavidine, tetrahydrocoptisine, corydaline, tetrahydropalmatine, corycavine, protopine, sinactine (*49-58*), and corynoline [44].

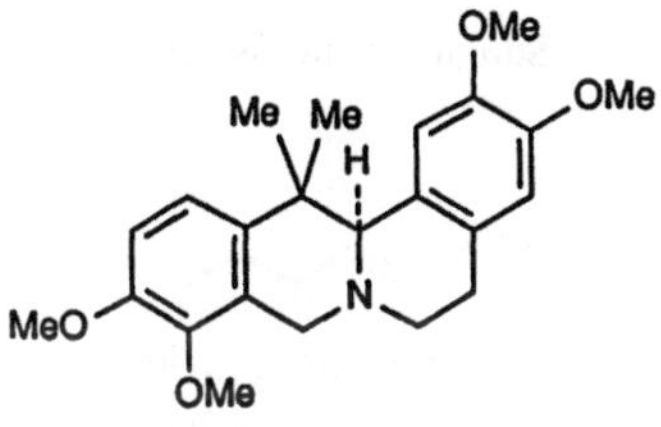

Corymotine (*49-57*)

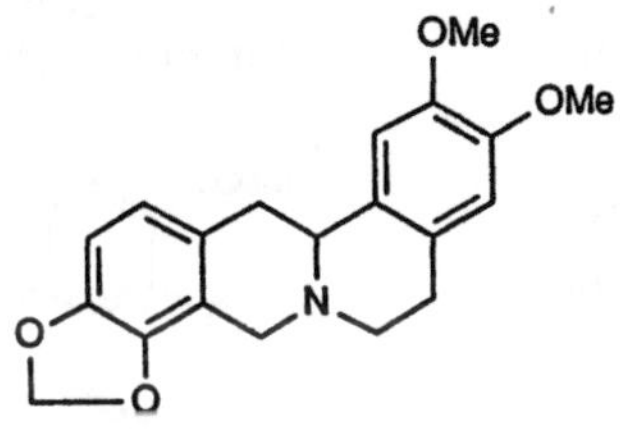

Sinactine (*49-58*)

From the bulbs of *Corydalis repens* 0.2% total alkaloid was isolated, from which protopine, scoulerine, bicuculline, cheilanthifoline, and stylopine were identified by comparison of their mass spectra with those of the authentic compounds [45].

Ten alkaloids were isolated from the medicinal plant *C. saxicola*. They were identified as berberine, cavidine, thalictrifoline, dehydrocavidine, 13β-hydroxystylopine, tetrahydropalmatine, scoulerine, chelerythrine (*49-59*), tetrahydrocolumbamine, and protopine [46].

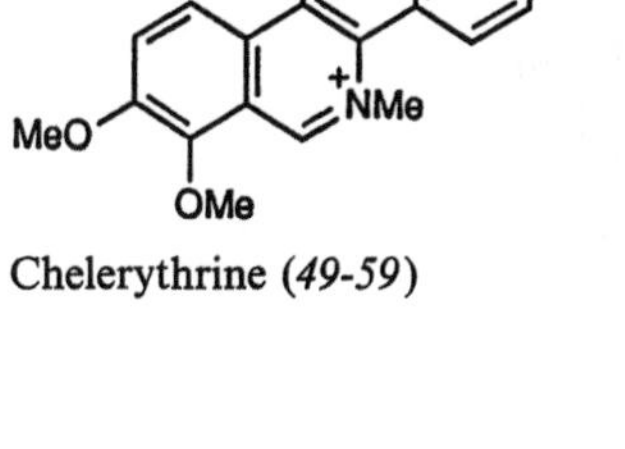

Chelerythrine (*49-59*)

Three alkaloids were isolated from *C. sheareri*. They were identified as protopine, corynoline, and isocorynoline. The tubers of *C. sheareri* were used in folk medicine for the treatment of arthritis and for induction of abortion [47].

Irgashev et al. reported the isolation of 23 alkaloids from the roots and aerial part of *C. stricta*. The total alkaloid contents in roots and the aerial part were 0.6% and 1.7%, respectively. From the roots, pancoridine (*49-60*), pancorinine (*49-61*), sanguinarine (*49-62*), bicuculline, protopine, wilsonirine (*49-63*), and *N*-methylcorypalline were identified. From the aerial part, cheilanthifoline, tetrahydrocolumbamine, scoulerine, reticuline (*49-64*), *N*-methylcoclaurine (*49-65*), bicuculline, adlumine, protopine, stylopine, sanguinarine, dihydrosanguinarine, coreximine (*49-66*), isoboldine, juziphine (*49-67*), corypalline (*49-68*), hydroxymethylstylopine, pycnarrhine (*49-69*), and hydrastine were identified [48].

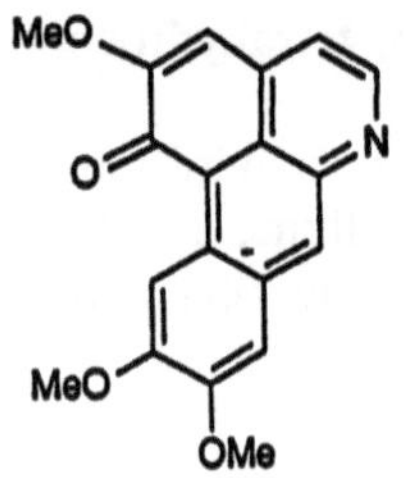

Pancoridine (*49-60*)

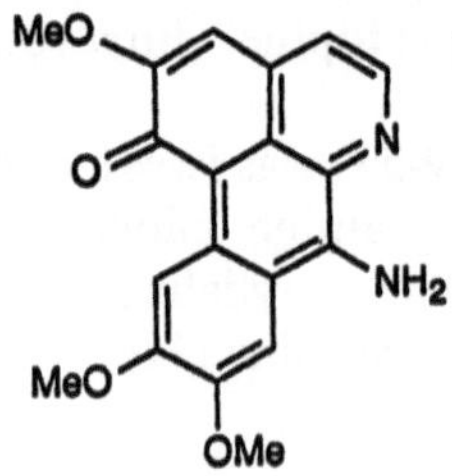

Pancorinine (*49-61*)

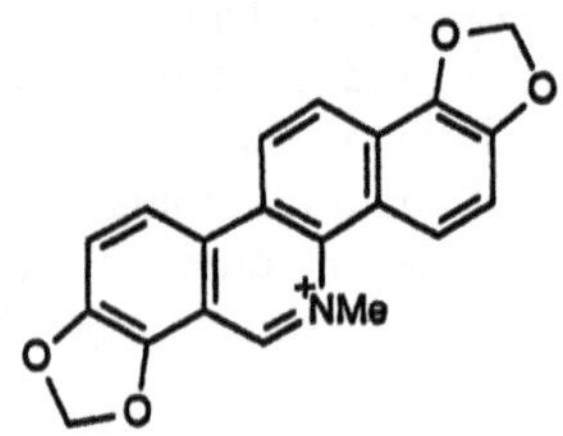

Sanguinarine (*49-62*)

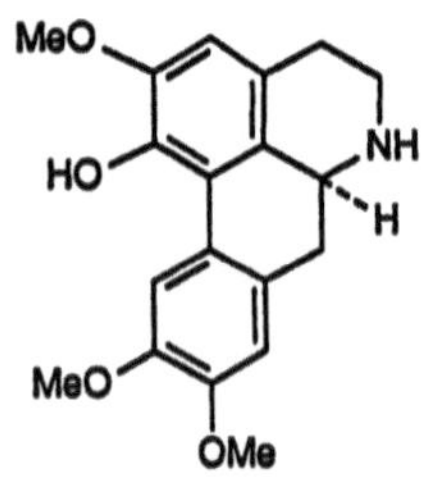

Wilsonirine (*49-63*)

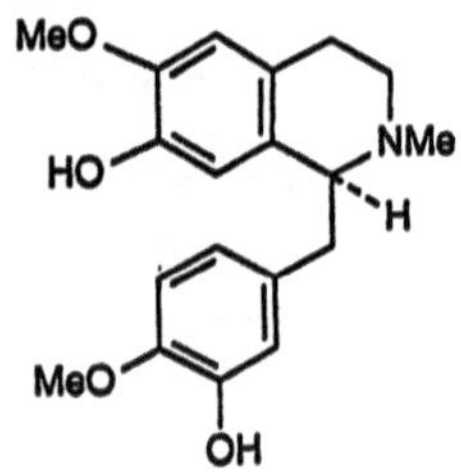

Reticuline (*49-64*)

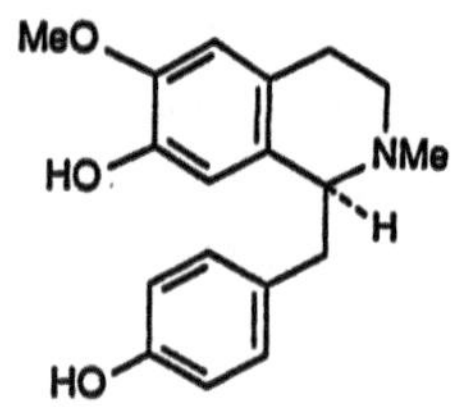

N-Methylcoclaurine (*49-65*)

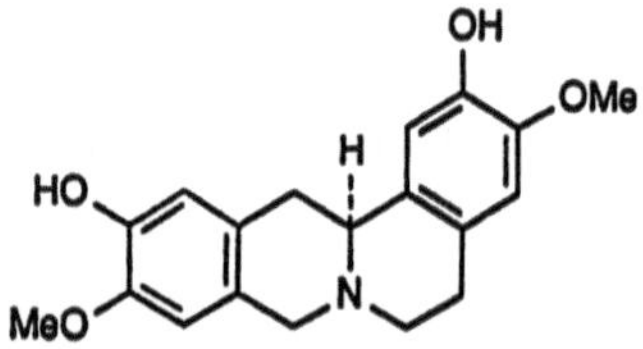

Coreximine (*49-66*)

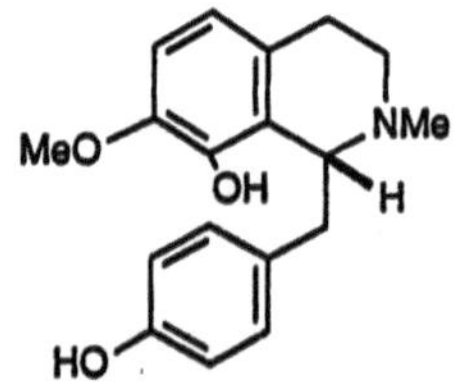

Juziphine (*49-67*)

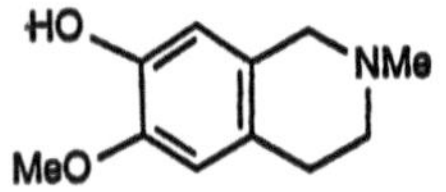

Corypalline (*49-68*)

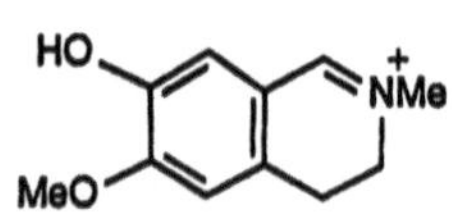

Pycnarrhine (*49-69*)

The seeds of *C. stricta* also contain a number of alkaloids. The total alkaloid content of the seeds was found to be 1%. The alkaloids identified were hydrastine, adlumidine (*49-70*), protopine, scoulerine, stylopine, *N*-methylstylopinium hydroxide, and pycnarrhine [49].

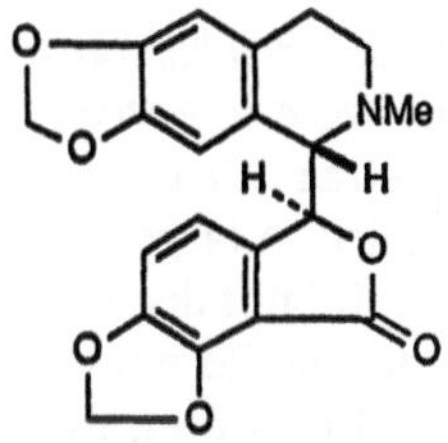

Adlumidine (*49-70*)

From the rootstock of *C. suaveolens*, another folk medicine used in China, the following alkaloids were isolated: domesticine (*49-71*), protopine, bicuculline, allocryptopine, and corytuberine (*49-72*) [50].

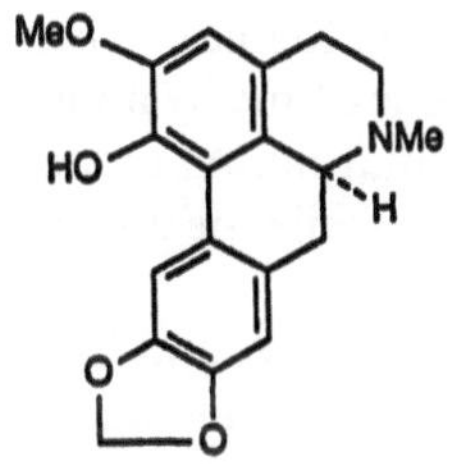

Domesticine (*49-71*)

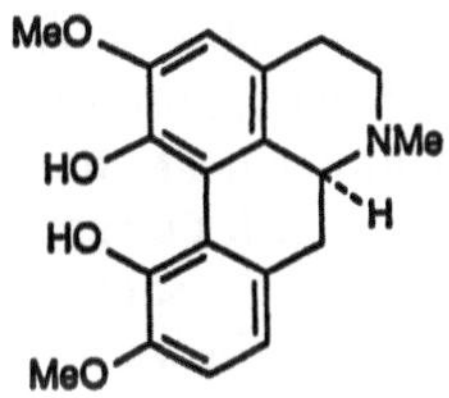

Corytuberine (*49-72*)

From *C. taliensis*, a folk medicine used in treatment of pain and to decrease swelling, nine constituents were isolated. Seven of them were identified as acetylcorynoline, bicuculline, corynoline, corycavine, protopine, and corydamine hydrochloride (*49-73*). The other two compounds were identified as nonacosan-10-ol and KNO_3 [51].

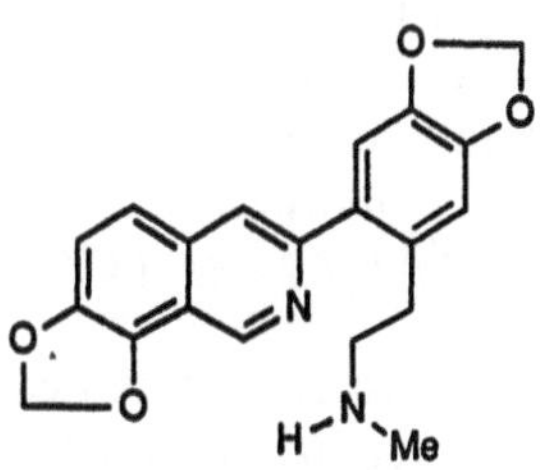

Corydamine (*49-73*)

49.3 Pharmacology

49.3.1 Pharmacology of *Corydalis turtschanivovii* f. *yanhusuo* and Its Main Active Constituent Tetrahydropalmatine

Tetrahydropalmatine is the main active component of the tuber of *C. turtschaninovii* f. *yanhusuo*. The pharmacological actions of tetrahydropalmatine were intensively investigated [52, 53]. Intraperitoneal injection of *l*-tetrahydropalmatine at a dose of 20–40 mg/kg caused a marked sedative and tranquilizing effect in mice. A large dose of *l*-tetrahydropalmatine also caused ataxia. *l*-Tetrahydropalmatine did not decrease the brain serotonine level [54]. Unlike reserpine, *l*-tetrahydropalmatine did not deplete stored monoamines. Pharmacological studies in rats indicated that *l*-tetrahydropalmatine blocked postsynaptic dopaminergic receptors but not receptors of γ-aminobutyric acid (GABA). Thus, an important mechanism in the tranquilizing action of *l*-tetrahydropalmatine seems to be the blocking of dopaminergic receptors without depletion of monoamine stores or an increase in GABA function [55]. These effects of *l*-tetrahydropalmatine are similar to those observed with tranquilizers or dopamine antagonists [56].

No marked influence on the motor defense conditioned reflex was observed when tetrahydropalmatine in doses of 0.1–1.0 mg/kg was administered i.v. to mice. When the dosage was raised to 5 mg/kg, it prolonged the conditioned reflex time by about 2 s and increased the reinforcement but did not alter the differentiation. At a dose of 10 mg/kg the conditioned reflex disappeared almost completely, while the unconditioned reflex time remained the same [57].

l-Tetrahydropalmatine was shown to have marked analgesic, sedative, and hypnotic effects. These effects are not observed with *d*-tetrahydropalmatine. Side effects of tetrahydropalmatine were rare and thus its therapeutic index appeared to be high [58].

l-Tetrahydropalmatine at a dose of 5–30 mg/kg given i.v. to rabbits altered the electroencephalogram, causing the appearance of slow high-amplitude waves and sleep spindles [59].

dl-Tetrahydropalmatine was found to decrease the firing rate of the neurones in the nucleus accumbens. Intracerebroventricular administration of noradrenaline reversed the depression of the firing in the nucleus accumbens induced by tetrahydropalmatine 30 min after injection. Serotonine given intracerebroventricularly only produced a 50% reversal of the depression of firing 15 min after injection, whereas dopamine showed no effect on the tetrahydropalmatine-induced inhibition. Noradrenaline and serotonin, rather than dopamine, may be responsible for the tetrahydropalmatine-induced inhibition [60]. The effect of tetrahydropalmatine could also be antagonized by pretreatment with the monoamine oxidase inhibitor pargyline [61, 62].

Injection of *dl*-tetrahydropalmatine at a dose of 60 mg/kg into rats decreased brain dopamine, noradrenaline, and serotonine contents by 70%, 50% and 30%, respectively, and increased the contents of their metabolites homovanillic acid, dihydroxyphenylacetic acid, 5-hydroxyindole-acetic acid, and 3-methoxy-5-hydroxyphenyleneglycol sulfate in the striatum and limbic area. However, the methylated metabolite of dopamine, 3-methoxytyramine, was decreased [63–65].

The catalepsy induced in rats by tetrahydropalmatine at i.p. doses of 70–100 mg/kg was associated with increased brain 5-hydroxyindoleacetic acid levels, indicating that the serotoninergic nervous system seems to be involved in tetrahydropalmatine-induced catalepsy [66, 67]. The central inhibitory effects of *l*-tetrahydropalmatine do not appear to be directly related to the prostagladins [68], because the endogenous biosynthesis of prostaglandins in cortex homogenates of mice and rats was not influenced by *l*-tetrahydropalmatine in vivo or in vitro.

Tetrahydropalmatine was rapidly and almost completely absorbed in mice when a single dose of 60 mg/kg was given orally. Fifteen minutes after subcutaneous injection of 150 mg/kg tetrahydropalmatine into rats, appreciable amounts of tetrahydropalmatine were detected in various tissues. One hour after injection, tetrahydropalmatine tended to concentrate in the fat, lung, kidneys, liver, and spleen at the level of 40–50 μg/g, while lower levels occurred in blood, brain, and heart. Later still, the high concentration of 73–103 μg/g was attained in fat tissue and this level persisted for more than 8 h. In contrast, the concentration of tetrahydropalmatine in viscera began to decrease in the 2nd h, reaching a negligible level 8 h later.

After i.v. injection of 40 mg/kg tetrahydropalmatine into rabbits the drug passes readily from the blood into the visceral organs and was found to penetrate the blood-brain barrier easily, to reach a maximal level in brain tissue within 2–5 min.

Equilibrium between blood and brain or viscera was rapidly reached (<15 min). The amount in the cerebral cortex and diencephalomesencephalon was slightly greater than that in the cerebral cortex and the disappearance of tetrahydropalmatine occurred first in the latter. Palmatine was found only in trace amounts in all parts of the brain. The results suggest, therefore, that palmatine is hardly able to penetrate the blood-brain barrier. Two hours after treatment, no biotransformation from tetrahydropalmatine to palmatine was found in the brain and liver of the rabbit in vivo. Tetrahydropalmatine, however, could be oxidized to palmatine in the air. Eighty percent of tetrahydropalmatine was bound to blood protein in a reversible manner. The partition coefficient was greater than that of palmatine. In rabbits receiving i.p. injections of 60 mg/kg tetrahydropalmatine, only a very small fraction of the drug was recovered from the urine and much less from the feces [69].

The total alkaloid isolated from *C. turtschaninovii* had no significant therapeutic effect against cardiac arrhythmia induced by aconitine in rats. The water-insoluble alkaloid fraction showed a stronger antiarrhythmic activity than the total alkaloid or the water-soluble alkaloid fraction [70].

49.3.2 Pharmacology of Other *Corydalis* Species and Some Alkaloids Isolated Therefrom

The alkaloid dehydrocorydaline isolated from several *Corydalis* species caused dilatation of the coronary artery [13]. In addition, dehydrocorydaline decreased ulceration, gastric secretion, and pepsin output in rats and protected against ulceration induced by restraint, fasting, and chemical agents such as aspirin, cortisone, ACTH, and reserpine. It antagonized histamine-induced gastric secretion in dogs. The gastrin-induced secrection in rats could be inhibited by dehydrocorydaline, but the gastric gastrin content was not affected. The antisecretory action of dehydrocorydaline was observed in vagotomized rats. Rats pretreated with reserpine showed a significant decrease in the antisecretory action [71].

Dehydrocorydaline was more effective than tetrahydropalmatine in increasing the tolerance of mice to normobaric and hypobaric hypoxia. It enhanced the coronary venous flow, but slightly lowered the heart rate. Dehydrocorydaline showed protective action on the change of electrocardiogram induced by pituitrin and on experimental myocardial necrosis induced by isoproterenol [72].

The disposition and metabolism of dehydrocorydaline was also studied in mice and rats, using 13-[^{14}C]dehydrocorydaline. After oral administration of 50 mg/kg dehydrocorydaline, maximal blood levels were as low as 0.5 and 0.06 μg/ml for mice and rats, respectively. This might be due to poor absorption from the digestive tract. Radiometric and autoradiographic results after oral administration revealed appreciable radioactivity in the gastrointestinal tract, liver, and kidneys. Comparison of the distribution with that after i.v. administration suggested that the compound was subject to a first-pass effect by the liver. The distribution of the radioactivity in liver was hetereogeneous, regardless of the route of administration, and did not differ between normal rats and rats with experimentally induced gastric ulcer. Most of the activity was excreted in the feces. Excretion via urine and bile was quite minor. Metabolite analysis suggested that dehydrocorydaline was metabolized by *O*-demethylation and by subsequent conjugation with glucuronic acid. However, metabolite excretion was low compared with the unchanged compound [73].

The total alkaloids extracted from the tubers of *C. decumbens* produced no significant effect on the blood flow and vascular resistance of the lower limbs but significantly decreased the blood pressure. Vascular tension of the lower limbs induced by noradrenaline was also observed [74].

Protopine, an active constituent of the tuber of *C. decumbens*, has direct smooth muscle-relaxing action. It inhibited the spastic contraction of isolated guinea pig ileum induced by acetylcholine and barium chloride; the IC_{50} were 3.4 and 5.1 μM, respectively. Protopine antagonized the contracting effect of physostigmine-acetylcholine on isolated cat ciliary muscle with an IC_{50} of 12.7 mg/ml. The IC_{50} of papaverine was comparable to that of protopine; however, atropine was more potent than protopine [75].

The i.p. injection of protopine in mice caused an analgesic effect, with an ED_{50} of 27 mg/kg [76].

References

1. Su ZY, Wu CY (1985) A study on the classification, distribution, evolutionary trends and uses of Chinese *Corydalis* subgenus Capnites DC. Acta Bot Yunnan 7:253–276
2. Lian WY (1985) Medicinal uses of the plants in *Corydalis* subgenus Capnoides. Zhongyaocai 5:14–19
3. Chou TQ (1954) Investigation of some Chinese drugs. Pharmazie 9:688–691
4. Chou TQ (1928) Alkaloids of Chinese *Corydalis ambigua* Cham. et Sch. (yen-hu-so). I. Chin J Physiol 2:203–217
5. Chou FG, Chou TQ (1929) Alkaloids of *Corydalis ambigua* (yen-hu-so). II. Corydalis. Chin J Physiol 3:69–74
6. Chou TQ (1929) Alkaloids of Chinese *Corydalis ambigua*. Cham. et Sch. (yen-hu-so). III. Corydalis I and monomethylether of corydalis F and G. Chin J Physiol 3:301–305
7. Chou TQ (1933) The alkaloids of Chinese *Corydalis ambigua* Cham et Sch. (yen-hu-so). IV. Corydalis J and K. Chin J Physiol 7:35–40

8. Chou TQ (1934) The alkaloids of the Chinese *Corydalis ambigua* Cham. et Sch. (yen-hu-so). V. Corydalis L and corydalis M. Chin J Physiol 8:155–160
9. Chou TQ (1936) The alkaloids of the Chinese *Corydalis ambigua* Cham. et Sch. (yen-hu-so). VI. Identification of corydalis D and corydalis M. Chin J Physiol 10:507–511
10. Zhu DY, Song CQ, Gao YL, Xu RS (1981) Identification of the structures of corydalis F, I, J and K. Acta Chim Sin 39:280–283
11. Kaneko H, Naruto S (1969) Studies on the constituents of *Corydalis* sp. VI. Alkaloids from Chinese *Corydalis* and the identity of *d*-corydamine with *d*-corybulbine. J Org Chem 34:2803–2805
12. Naruto S, Kaneko H (1973) Constituents of *Corydalis* species. X. Alkaloids from *Corydalis ambigua*. Phytochemistry 12:3008–3009
13. Cui Z, Qi ML, Lin L, Yu DQ (1984) Chemical structure of ambinine isolated from *Corydalis ambigua* Cham et Schltd. Acta Pharm Sin 19:904–908
14. Imaseki I, Taguchi H (1962) Components of *Corydalis* species. I. Alkaloids of the Chinese *Corydalis*. New bases corydalmine and dehydrocorydalmine. Yakugaku Zasshi 82:1214–1219
15. Yasuda M, Kano Y, Konoshima M (1988) Pharmacognostical and botanical studies on the crude drugs in Hokkaido. (IX). Cultivation of *Corydalis ambigua* Cham et Schlecht. (V). Annual variation of alkaloid contents. Shoyakugaku Zasshi 42:214–215
16. Fu XY, Liang WZ, Tu GS (1985) Chemical studies on the alkaloids from yuanhu (*Corydalis turtschaninovii* Bess. f. *yanhusuo* Y.H. Chou et C.C.Hsü). III. Quantitative determination of the tertiary alkaloids in yuanhu by TLC-densitometry. Chin J Pharm Anal 5:194–198
17. Fu XY, Liang WZ, Tu GS (1986) Chemical studies on the alkaloids isolated from the tuber of yuanhu (*Corydalis turtschaninovii* Bess. f. *yanhusuo* Y.H. Chou et C.C. Hsü). Acta Pharm Sin 21:447–453
18. Feng WS, Ji CR, Yu ZL, Zhang JX (1987) Comparison of the constituents of Rhizoma corydalis (*Corydalis yanhusuo*) cultured in Hunan Province Deng Country with those cultured in Zhejiang Province. Bull Chin Mat Med 12:295–297
19. Yu Q, Xu GJ, Jin RL, Xu LS, Cong XD (1988) Pharmacognostical studies on the Chinese drug Yanhusuo. I. Comparison of alkaloid contents in the Yanhusuo tubers of four species. J China Pharm Univ 19:4–7
20. Fu XY, Liang WZ, Fang DM, Tu GS (1986) Chemical studies on the alkaloids from Yuanhu (*Corydalis turtschaninovii* Bess. f. yanhusuo Y.H. Chou et C.C. Hsü). V. HPLC assay for determination of quaternary alkaloids in Yuanhu. Chin J Pharm Anal 6:195–198
21. Fu XY, Liang WZ, Tu GS (1986) Chemical studies on the alkaloids from Yuanhu (*Corydalis yanhusuo*). VIII. Alkaloids from the tuber of Yuanhu. Chin Trad Herbal Drugs 17:533–534
22. Chu M, Chen BZ, Lian WY, Feng RZ, Fu GX, Xiao PG (1986) Studies on traditional Chinese drugs "yuanhu" II. Evaluation of the constituents and quality. Chin Trad Herbal Drugs 17:150–152
23. Fu XY, Liang WZ, Tu GS (1986) Chemical studies on the alkaloids from yuanhu (*Corydalis turtschaninovii* Bess. f. *yanhusuo* Y.H. Chow et C.C. Hsü). II. Alkaloids from the aerial part of yuanhu. Chin J Pharm Anal 6:6–9
24. Hu TM, Zhao SX (1985) Alkaloids from the aerial parts of yanhusuo. J Nanjing Coll Pharm 16:7–11
25. Fu XY, Liang WZ, Tu GS (1986) Chemical studies on the alkaloids from Yuanhu (*Corydalis turtschaninovii* Bess. f. *yanhusuo* Y.H. Chow et C.C. Hsü) IV. TLC-densitometric determination of the main aporphine alkaloids in the aerial part of Yuanhu. Chin J Pharm Anal 6:66–69
26. Fang CC, Lin M, Weng CM, Chu CT, Liu H (1981) Chemical study on alkaloids of five *Corydalis* species in Qinghai Plateau, China. Chin Pharm Bull 16:49
27. Fang QC, Lin M, Weng QM, Zhu CD, Liu X (1981) Chemistry of alkaloids of *Corydalis* – chemical studies of four *Corydalis* species from Qinghai plateau (China). Acta Pharm Sin 16:798–800
28. Pan HC, Kuan MC, Fan CC (1981) Preliminary study on the alkaloids in *Corydalis bungeana* Turcz. Chin Pharm Bull 16:57
29. He LY, Zhang YB (1985) TLC separation and densitometric determination of six isoquinoline alkaloids in *Corydalis bungeana*. Acta Pharm Sin 20:377–382
30. Zeng WG, Liang WZ, He CH, Zheng QT, Tu CS (1988) An alkaloid from *Corydalis bungeana*. Phytochemistry 27:599–602

31. Zeng WG, Liang WZ, Tu GS (1987) Assay of five corynoline-type alkaloids in *Corydalis bungeana* by reversed phase high-performance liquid chromatography. J Chromatogr 408:426–429
32. Fang QC, Lin M, Weng QM (1984) Chemical study of alkaloids from *Corydalis conspersa*. Planta Med 50:25–27
33. Zhang JS, Xu RS, Chen ZL, Zhu DY, Deng SS (1987) Chemical constituents of *Corydalis decumbens*. Chin Trad Herb Drugs 18:8
34. Chu TY, Sung SC, Kao YL, Hsu JS, Tai PH, Chen L, Tang SS (1980) Isolation and structural elucidation of decumbenine and eleven other chemical constituents. Chin Trad Herb Drugs 11:341–342
35. Wang MH, Deng HX (1983) Colorimetric determination of the total alkaloids in Xia Tian Wu (*Corydalis decumbens*) injection. Chin Trad Herb Drugs 14:536–537
36. Zhang JS, Yu HG, Liu LZ, Chen ZL, Xu RS, Deng SS (1988) Decumbenine, a new phthalide isoquinoline alkaloid. Acta Chim Sin 46:595–597
37. Zhang JS, Xu RS (1988) Decumbensine and epi-α-decumbensine: two new alkaloids isolated from *Corydalis decumbens*. J Nat Prod 51:1245–1246
38. Luo SD (1981) Alkaloid constituents of *Corydalis delavayi* Franch. Acta Bot Yunnan 3:185–188
39. Luo SD (1982) A new alkaloid from *Corydalis delavayi* Franch. Acta Bot Yunnan 4:69–70
40. Lin M, Liu X, Fang QC (1986) A chemical study on *Corydalis hendersonii* (=*C. nepalensis*). Acta Bot Sin 28:91–94
41. Wu TS, Huang SC, Lu ST, Wu YC, McPhail DR, McPhail AT, Lee KH (1988) Structure and stereochemistry of corytensine, a new phthalideisoquinoline alkaloid from *Corydalis ochotensis*. Heterocycle 27:1565–1568
42. Wu TS, Huang SC, Lu ST, Wu YC (1987) Studies on the alkaloids of Formosan *Corydalis* species. Part VII. Isoochotensine, a new spirobenzylisoquinoline alkaloid from *Corydalis ochotensis* Turcz. J Chin Chem Soc (Taipei) 34:157–160
43. Luo SD, Gong XH, Gao ZY, Tan JQ (1983) Studies on the chemical constituents of *Corydalis pallida* var. *speaose* Kom. Acta Bot Yunnan 5:315–316
44. Fu XY, Liang WZ, Tu GS (1988) Alkaloids from *Corydalis remota*. J Nat Prod 51:262–264
45. Fang QC, Lin M, Zhou JY, Liu X (1982) Chemical studies on alkaloids of *Corydalis repens*. Chin Pharm Bull 17:3–4
46. Ke MM, Zhang XD, Wu LZ, Zhao Y, Zhu DY, Skong CQ, Xu RS (1982) Studies on the active principles of *Corydalis saxicola* Bunting. Acta Bot Sin 24:289–291
47. Zhou JM, Yu CC, Tsao Y, Chou PC, Liu CY (1981) Studies on active principles of *Corydalis sheareri* S. Moore. Chin Trad Herb Drugs 12:3–4
48. Irgashev T, Israilov IA, Batsuren D, Yunusov MS (1983) *Corydalis stricta* alkaloids. Khim Prir Soedin 490–493 (CA 100:82695a)
49. Veznik F, Israilov IA (1985) Alkaloids from the seeds of *Corydalis stricta*. Planta Med 51:469
50. Xin WF, Lin M (1981) Chemical studies on alkaloids in genus *Corydalis* (Papaveraceae) plants. Part II. Chemical study on alkaloids of *Corydalis suaveolens* Hance. Chin Trad Herb Drugs 12:1–4
51. Luo SD, Wu SB (1982) Studies on chemical constituents of *Corydalis taliensis* Fr. Acta Pharm Sin 17:699–702
52. Ding GS (1980) Trials of some Chinese medicinal herbs. In: Burns JJ, Tsuchitani PJ (eds) Proceedings of the US-China Pharmacological Symposium 1979. NAS, Washington DC, pp 103–121
53. Chin KC, Tang HT, Chou K, Hsü P (1959) Pharmacologic actions of *Corydalis* B. Kexue Tongbao 737–738
54. Chin KC, Wang YN, Shü P (1964) Some neuropharmacological actions of rotundine. Acta Pharm Sin 11:754–761
55. Jin GZ, Xu JA, Zhang FT, Yu LP, Li JH, Wang XL (1983) Sedative-tranquilizing effect of *l*-tetrahydropalmatine in relation to brain monoaminergic neurotransmitters. Acta Pharmacol Sin 4:4–10
56. Buda C, Sallanon M, Jin GZ (1984) Effect of *l*-tetrahydropalmatine on sleep-waking cycle of cats. Acta Pharmacol Sin 5:5–8
57. Hsu B, Hsu KL (1958) Pharmacological actions of *Corydalis*. III. The influence of Corydalis B (*dl*-tetrahydropalmatine) on the higher nervous activity. Acta Physiol Sin 22:40–46

58. Hsu B, Kin KC (1962) Pharmacological study of tetrahydropalmatine and its analogs. A new type of central (nervous system) depressant. Arch Int Pharmacodyn 139:318–327
59. Bantsekina MM, Berezhinskaya VV (1971) Effect of gindarine on the bioelectrical activity of the brain. Tr Vses Nauch Issled Inst Lek Rast 297–301 (CA 79:374f)
60. Liu GQ, Wu HQ, Xie L, Mei XY (1985) Effect of monoamines on the *dl*-tetrahydropalmatine-induced inhibition of single unit activity in the nucleus accumbens. Acta Pharm Sin 20:500–504
61. Liu GQ (1983) Effect of pargyline on depletion of catecholamines in rat brain by *dl*-tetrahydropalmatine. Acta Pharm Sin 18:472–474
62. Wu HQ, Liu GQ, Xie L, Jiang Y (1984) Effect of tetrahydropalmatine on single unit activity of the nucleus accumbens in the rat. Acta Pharm Sin 19:495–498
63. Liu GQ, Algeri S, Garattini S (1983) Depletion of monoamines in the rat by *dl*-tetrahydropalmatine. Acta Pharm Sin 18:641–647
64. Liu GQ, Algeri S, Garattini S (1982) *dl*-Tetrahydropalmatine as monoamine depletor. Arch Int Pharmacodyn Ther 258:39–50
65. Liu GQ, Jiang Y, Pang G, Ibraham, Zhou LQ (1985) Effect of some isoquinoline alkaloids on monoamines in rat brain. Acta Pharm Sin 20:566–570
66. Jin GZ, Liu XJ, Yu LP, Xu J (1980) Serotoninergic implication in *l*-tetrahydropalmatine-induced catalepsy in rats. Acta Pharmacol Sin 1:12–17
67. Xu J, Jin GZ, Yu LP, Liu XJ (1981) Differentiation between catalepsies induced by *l*-tetrahydropalmatine and by haloperidol and morphine. Acta Pharmacol Sin 2:152–155
68. Xu JA, Zheng LZ, Jin GZ (1982) Relation of brain prostaglandines to central effects of *l*-tetrahydropalmatine. Acta Pharmacol Sin 3:217–220
69. Chin KC, Chang CT, Hsu P (1963) Pharmacological actions of *corydalis* XI. Physiologic disposition and fate of Corydalis B. Acta Physiol Sin 26:347–359
70. Ma SX, Chen KJ, Ma YL, Bao XF (1985) Preliminary study on the antiarrhythmic action of the alkaloids of *Corydalis yanhusuo*. Bull Chin Mat Med 10:41–42
71. Soji Y, Kadokawa T, Masuda Y, Kawashima K, Nakamura K (1969) Influence of *Corydalis* alkaloid on gastric ulcers in experimental animals. Nippon Yakurigaku Zasshi 65:196–209 (CA 73:33759q)
72. Jiang XR, Wu Q, Shi H, Chen W, Chang S, Zhao S, Tian X, Zhou L, Guo S, Lu Y (1982) Pharmacological action of dehydrocorydaline on the cardiovascular system. Acta Pharm Sin 17:61–65
73. Fujii T, Miyazaki H, Nambu K, Kagemoto A, Hashimoto M (1984) Disposition and metabolism of ^{14}C-dehydrocorydaline in mice and rats. Radioisotopes 33:519–525
74. Wang DY, Cheng MZ, Wang CG, Zhang SE, Chen TF, Dai PX (1986) Effects of the alkaloids of *Corydalis decumbens* on cerebral and peripheral circulation in dogs. Chin J Integr Trad Wes Med 6:477–479
75. Zhong RX, Shi RR, Huang LX, Liu H, Tao JY, Zhu XY, Jiang MH, Yang XY, Zhao HM, Deng SS (1986) Spasmolytic effect of protopine and *Corydalis decumbens* on isolated cat ciliary muscle and guinea pig ileum. Chin Trad Herb Drugs 17:303–306
76. Yue KL (1981) Analgesic and antispastic effects of protopine. Acta Pharmacol Sin 2:16–18

Crocus sativus L. 50

50.1 Introduction

Xihonghua, Stigma Croci, is the dry stigmata of *Crocus sativus* L. (Iridaceae). It is officially listed in the Chinese Pharmacopoeia and is used in traditional Chinese medicine for treatment of hematoma, menostasis, melancholia, and convulsion, and as a sedative.

50.2 Chemical Constituents

Saffron, the stigma of *C. sativa*, has been known to contain a pigment named crocin since the beginning of the nineteenth century. Crocin gave two components by hydrolysis, one of them being a pigment named crocetin and the other glucose. The structure of crocetin (*50-1*) was determined as 8,8′-diapo-ψ,ψ-carotenedioic acid [1–5]. Two minor pigments, β-crocetin and γ-crocetin, were also isolated by the hydrolysis of crocin and were structurally elucidated as the methylester and dimethylester of crocetin, respectively [1, 2].

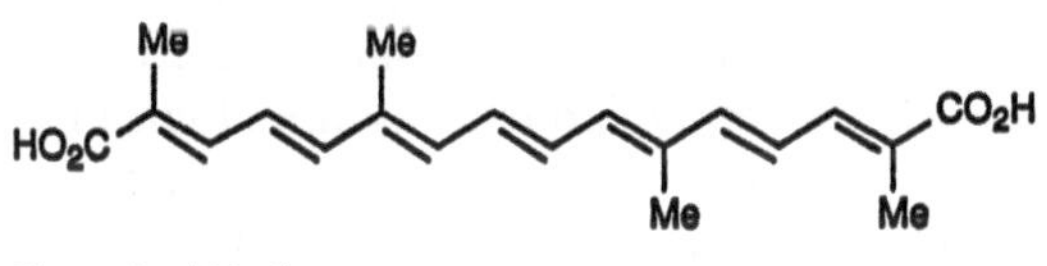

Crocetin (*50-1*)

The sugar in crocin was determined as gentiobiose and the structure of crocin (*50-2*) as crocetin bis(β-D-gentiobiosyl)-ester [6, 7]. Five other crocetin glycosyl esters, crocetin β-D-gentiobiosyl β-D-glucopyranosyl ester (crocin 2), crocetin gentiobiosyl ester (crocin 3) [8–11], crocetin bis(β-D-glucopyranosyl)-ester, crocetin β-D-glucopyranosyl ester [9–11], and crocetin β-D-glucopyranosyl methyl ester (crocin 4, *50-3*) [8], were isolated from dry or fresh [12] saffron. Recently, the isolation of 13-*cis*-crocin, a stereoisomer of crocin, was also reported [13].

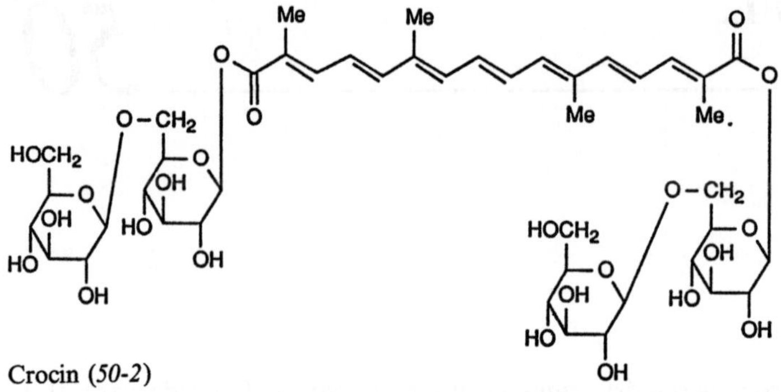

Crocin (*50-2*)

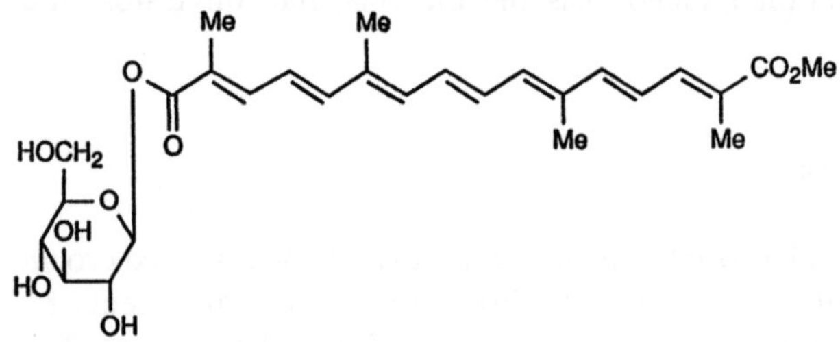

Crocin 4 (*50-3*)

The naturally occurring crocetin bis(β-D-glucopyranosyl)-ester can be synthesized from crocetin bis-imidazolide or crocetin bis(1,2,4-triazolide) and β-D-glucose in pyridine in the presence of a base [12, 14]. In general, the content of crocetin glycosyl esters in saffron was more than 23% and that of crocin was more than 14% [15]. Another type of compound isolated from saffron is the bitter principle picrocrocin (*50-4*), a glycoside which produced by hydrolysis glucose and safranal (*50-5*) or 3-hydroxy-β-cyclocitral [16, 17]. The absolute configuration of picrocrocin was also determined [18].

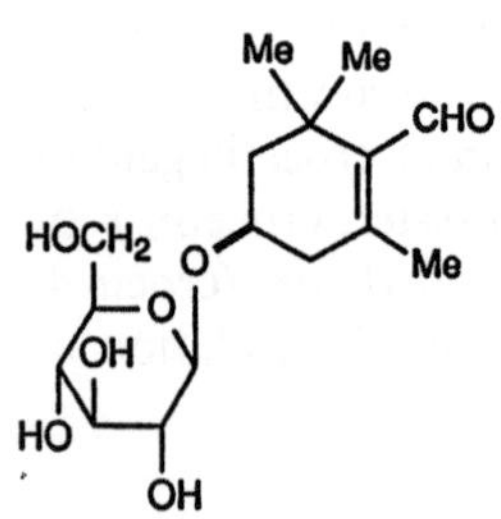

Picrocrocin (*50-4*)

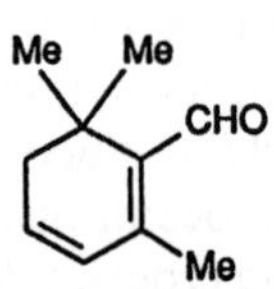

Safranal (*50-5*)

In addition, ψ,ψ-carotene, β-carotene, γ-carotene, zeaxanthine, and some other C_{20} carotenoids were isolated from saffron [19, 20]. It was discussed whether crocin and picrocrocin might be decomposition products of C_{40} carotenoids such as zeaxanthin [19, 21]. Picrocrocin can be extracted with ether, whereas crocin and related pigments are extracted with methanol. Safranal has the specific odor of saffron.

Compounds identified from the volatile fraction of saffron were safranal [22], phenylethanol, naphthalene, 2-butenoic acid lactone, palmitic, stearic, oleic, and linoleic acid [23] and a number of monoterpene aldehydes and isophorone (*50-6*) analogs 2,6,6-trimethyl-4-hydroxy-l-cyclohexen-l-carboxaldehyde, 2,4,4-trimethyl-3-formyl-6-hydroxy-2,5-cyclohexandien-l-one, isophorone, 3,5,5-trimethyl-4-hydroxy-2-cyclohexen-l-one, 3,5,5-trimethyl-1,4-cyclohexadione, 3,5,5-trimethyl-2-cyclohexene-1,4-dione, and 3,5,5-trimethyl-2-hydroxy-2-cyclohexene-1,4-dione [24]. Safranal is the major component of the volatile fraction of saffron [22].

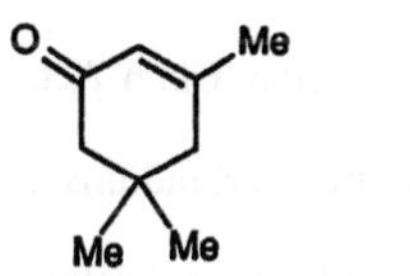

Isophorone (*50-6*)

Recently, a novel xanthone-carotenoid glycosidic conjugate, named mangicrocin (*50-7*) has been isolated from saffron and structurally determined [25].

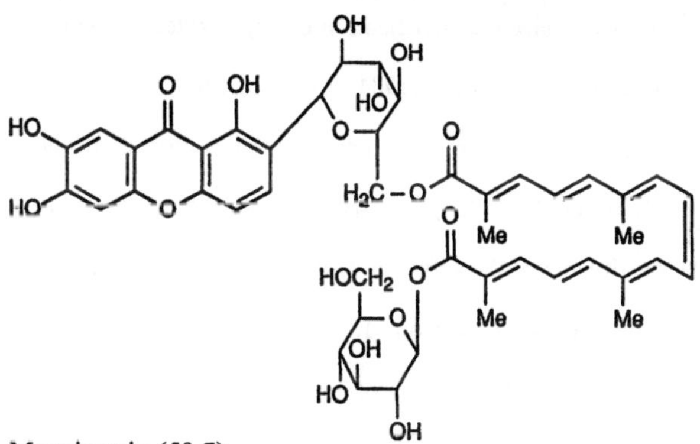

Mangicrocin (*50-7*)

50.3 Pharmacology

Various extracts of saffron showed stimulating action on uteri of guinea pigs, rabbits, and dogs, both gravid and nongravid. The effect appeared to be both myogenic and neurogenic. When mice were fed a diet containing 0.22%–2% saffron powder for 3 weeks, the duration of complete cornification of vaginal epithelium was prolonged from the normal 1–2 days to 3–4 days. The oral LD_{50} of saffron in mice was 20.7 g/kg administered as decoction [26]. Intramuscular injection of crocetin into rabbits fed a cholesterol-containing diet resulted in greatly reduced severity of atherosclerosis. Serum cholesterol levels were reduced by 50% [27, 28].

References

1. Karrer P, Salomon H (1927) Über Pflanzenfarbstoffe III. Zur Kenntnis der Safranfarbstoffe I. Helv Chim Acta 10:397–405
2. Karrer P, Salomon H (1928) Über Pflanzenfarbstoffe IV. Über die Safranfarbstoffe II. Helv Chim Acta 11:513–525
3. Kuhn R, L'Orsa F (1931) Über konjugierte Doppelbindungen. XVIII. Mitteilung: zur Konstitution des Safran-Farbstoffes. Chem Ber 64:1732–1736
4. Karrer P, Benz P, Morf R, Raudnitz H, Stoll M, Takahashi T (1932) Konstitution des Safranfarbstoffes Crocetin, Synthese des Perhydrobixinäthylesters and Perhydronorbixins, vorläufige Mitteilung. Helv Chim Acta 15:1218–1219
5. Karrer P, Benz P, Morf R, Raudnitz A, Stoll M, Takahashi T (1932) Pflanzenfarbstoffe XLVI. Konstitution des Crocetins und Bixins. Synthese des Perhydronorbixins. Helv Chim Acta 15:1399–1419
6. Karrer P, Miki K (1929) Pflanzenfarbstoffe XV. Der Zucker des α-Crocins. Helv Chim Acta 12:985–986
7. Kuhn R, Wang Y (1939) Synthese des Tetradecacetyl-crocins und verwandter Verbindungen. Chem Ber 72:871–878
8. Dhingra VK, Seshadri TR, Mukerjee SK (1975) Minor carotenoid glycosides from Saffron (*Crocus sativus*). Indian J Chem 13:339–341
9. Pfander, H, Wittwer F (1975) Carotinoid-Glycoside. 2. Mitteilung: Untersuchungen zur Carotinoid-Zusammensetzung im Safran. Helv Chim Acta 58:1608–1620
10. Pfander H, Wittwer F (1975) Carotinoid-Glycoside. 3. Mitteilung: Untersuchung zur Carotinoidzusammensetzung im Safran. Helv Chim Acta 58:2233–2236
11. Pfander H (1976) Carotinoid glycoside. Pure Appl Chem 47:121–128
12. Pfander H, Wittwer F (1979) Carotenoid glycosyl ester, part 3. The synthesis of crocetin di(β-D-glycosyl)ester. A new method for the selective esterification of unprotected β-D-glucose. Helv Chim Acta 62:1944–1951
13. Speranza G, Dada G, Manitto P, Monti D, Gramatica P (1984) 13-*cis*-Crocin: a new crocinoid of saffron. Gazz Chim Ital 114:189–192
14. Pfander H (1979) Synthesis of carotenoid glycosylesters and other carotenoids. Pure Appl Chem 51:565–580
15. Jin RL, Qiao ZP, Zhou SD, Ye YL, Zhou JX (1986) Quality analysis of *Crocus sativus*. J Nanjing Coll Pharm 17:247–251
16. Winterstein E, Teleczky J (1922) Constituents of the saffrons. 1. Picrocrocin. Helv Chim Acta 5:376–381
17. Winterstein E, Teleczky J (1922) Constituents of saffron. 1. Picrocrocin. Z Physiol Chem 120:141–166
18. Buchecker R, Eugster CH (1973) Absolute configuration of picrocrocin. Helv Chim Acta 56:1121–1124
19. Kuhn R, Winterstein A (1934) Über die Konstitution des Pikrocrocins und seine Beziehung zu den Carotin-Farbstoffen des Safrans. Chem Ber 67:344–357
20. Pfander H, Schurtenberger H (1982) Biosynthesis of C_{20}-carotenoids in *Crocus sativus*. Phytochemistry 21:1039–1042
21. Schurtenberger H, Vögeli U, Pfander H (1983) Synthese von Diterpenen als mögliche biogenetische Vorläufer des C_{20}-Carotinoids Crocetin. Helv Chim Acta 66:2346–2356
22. Curro P, Lanuzza F, Micali G (1986) Evaluation of the volatile fraction of saffron by headspace gas chromatography. Rass Chim 38:331–334 (CA 107:132735f)
23. Zarghami NS, Heinz DE (1971) Volatile constituents of saffron. Lebensm Wiss Technol 4:43–45
24. Zarghami NS, Heinz DE (1971) Monoterpene aldehydes and isophorone-related compounds of saffron. Phytochemistry 10:2755–2761
25. Ghosal S, Singh SK, Battacharya SK (1989) Mangicrocin, an adaptogenic xanthone carotenoid glycosidic conjugate from saffron. J Chem Res Synop 70–71
26. Chang PY, Wang CK, Liang CT, Kuo W (1964) The pharmacological action of Zang Hong Hua (*Crocus sativus* L.). I. Effect on the uterus and estrous cycle. Acta Pharm Sin 11:94–100
27. Gainer JL, Jones JR (1975) Use of crocetin in experimental atherosclerosis. Experientia 31:548–549
28. Gainer JL, Chisolm GM, III (1974) Oxygen diffusion and atherosclerosis. Atherosclerosis 19:135–138

Cucurbita moschata Duch. 51

51.1 Introduction

The seeds of *Cucurbita moschata* Duch. (Cucurbitaceae) were used originally in traditional Chinese medicine as an ascaricidal agent for treatment of human ascariasis. It was then found to be active in inhibiting the growth of immature *Schistosoma japonicum* and was used clinically for treatment of schistosomiasis. It is not officially listed in the Chinese Pharmacopoeia.

51.2 Chemical Constituents

In 1961, an active principle of *C. moschata* was first isolated and structurally investigated as a new amino acid named cucurbitin (*51-1*), which was also the active substance against ascaris [1]. The structure and configuration of this amino acid was determined as (*R*)-3-amino-3-pyrrolidinecarboxylic acid by X-ray diffraction of the perchlorate [2].

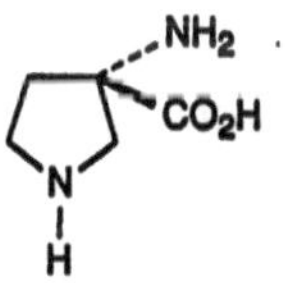

Cucurbitin (*51-1*)

51.3 Pharmacology

As an anthelmintic, cucurbitin induced paralysis in earthworms. The inhibitory effect was in the same order of magnitude as that observed with piperazine [3]. In animal experiments, cucurbitin hydrochloride or perchlorate showed a similar prophylactic activity against schistosomes given orally to mice on the day of infection. The rates of worm number reduction were 47% and 50% respectively. Cucurbitin also exhibited some therapeutic effect on mature schistosomes, but no killing effect was found. A combination of cucurbitin with tartar emetic increased the schistosomacidal activity compared to cucurbitin or tartar emitic alone [4].

Cucurbitin had an inhibitory effect on glutamic-pyruvic transaminase [5] and arginase [6] of schistosomes. However, the glutamic-pyruvic transaminase and glutamic-oxalacetic transaminase levels were not affected in vivo, when the worms were removed from the infected mice treated with cucurbitin [5].

The acute oral LD_{50} values of cucurbitin hydrochloride and perchlorate in mice were 1.25 and 1.1 g/kg. No apparent histological damage to the tissues was observed at therapeutic doses on mice, except that hepatic cells exhibited a slight atrophy, and showed a slight infiltration with fatty globules after 3 weeks. In vitro, cucurbitin showed a prolonged depressive action on the contraction of isolated strips of rabbit and guinea pig ileum. In vivo it caused a transient rise in blood pressure in urethane-anesthetized rabbits after intravenous injection, accompanied by a slight increase in respiration [4].

Autoradiographic study on tissue localization of [^{14}C] cucurbitin in mice showed that after intravenous injection the radioactivity was predominantly localized in the liver, kidneys, dorsal root ganglion, tracheal cartilage, and pancreas. Intense activity persisted for 24 h in these tissues, except for the pancreas, in which the concentration decreased with time. A moderate level of radioactivity was found in nasal epithelium, gastrointestinal mucosa, esophagus, salivary glands, thymus, bone marrow, and Harder's gland. After 3 days, the highest concentration was in cartilage. Because of the structural similarity of cucurbitin with proline, the retention in the cartilage may involve incorporation into the protein [7].

The cucurbitin content in seeds of some *Cucurbita* species varied widely even within the same species, e.g., *C. maxima* 0.53%–0.94%, *C. moschata* 0.4%–0.84%, and *C. pepo* 0.18%–0.66%. No cucurbitin could be detected in the fruit pulp [8]. A number of N- or N′-substituted cucurbitin derivatives as well as proline derivatives were synthesized and their schistosomicidal activities were tested. All compounds showed less activity against schistosomes than the parent compound cucurbitin [9].

References

1. Fang ST, Li LC, Niu CI, Tseng KF (1961) Chemical studies on *Cucurbita moschata.* I. The isolation and structural studies of cucurbitin, a new amino acid. Sci Sin 10:845–851
2. Fan HF, Lin CC (1965) The crystal structure of cucurbitin perchlorate and the absolute configuration of the cucurbitin molecule. Acta Phys Sin 21:253–262
3. Gonzalez AE, Bravo OR, Garcia MH, Santos RM de la, Tomas ML del (1974) Pharmacological (anthelmintic) study of *Cucurbita maxima* seeds and their active principle, cucurbitin. An R Acad Farm 40:47–86
4. Shiao SH, Shao BJ, Ho YH, Yang YC, Mao CP (1962) Prophylactic therapeutic effect of cucurbitin in experimental schistosomiasis in mice. Sci Sin 11:1527–1534
5. Hsiao SH, Hsu YC (1965) The effect of some schistosomacides on glutamic-pyruvic transaminase and glutamic-oxalacetic transaminase of *Schistosoma japonicum.* Acta Pharm Sin 12:242–248
6. Tao IH, Huang TY (1964) Arginase of *Schistosoma japonicum.* Acta Biochim Biophys Sin 4:161–166
7. Liang Y, Marlowe C, Waddell WJ (1982) Autoradiographic study on tissue localization of (C-14)cucurbitin in mice. Acta Pharmacol Sin 3:267–269
8. Misranian VH, Abou-Chaar CI (1968) Extraction, detection and estimation of cucurbitin in *Cucurbita* seeds. Lloydia 31:23–29
9. Sun CJ, Ji RY (1985) Synthesis of some cucurbitin derivatives and related compounds. Acta Pharm Sin 20:214–218

Curcuma spp.

52

52.1 Introduction

Three items from *Curcuma* plants are officially listed in the Chinese Pharmacopoeia:

- Yujin, Radix Curcumae, is the dry tubers of *Curcuma aromatica* Salisb., *C. kwangsiensis* S. Lee et C. F. Liang, *C. longa* L., or *C. zedoaria* Rosc. (Zingiberaceae) collected in winter when the aerial part of the plant has withered. It is used as a choleretic, analgesic, and sedative and in the treatment of hepatitis, menstrual disorders and epilepsy.
- Jianghuang, Rhizoma Curcumae longae, is the dry rhizome of *C. longa* collected in winter. It is used as an analgesic in the treatment of menstrual disorders, rheumatism, and traumatic diseases.
- Ezhu, Rhizoma Zedoariae, is the dry rhizome of *C. zedoaria, C. aromatica,* or *C. kwangsiensis* collected in winter. It is used as an analgesic and digestive and also for treatment of cervical cancer.

52.2 Chemical Constituents

52.2.1 Chemical Constituents of *Curcuma longa*

Turmeric from the rhizome of *C. longa* has been known for its coloring, flavoring, and digestive properties since ancient times. It is a constituent of curry powder and contributes to its characteristic color and odor. From the rhizome or from turmeric, three major phenolic pigments were isolated and structurally determined as diarylheptane derivatives: the bisferuloyl-methane commonly named curcumin (*52-1*), bis(4-hydroxy-cinnamoyl)-methane (*52-2*), and 4-hydroxycinnamoyl feruloyl methane (*52-3*) [1–4]. An assymmetrical derivative, dihydrocurcumin (*52-4*), was also isolated [5].

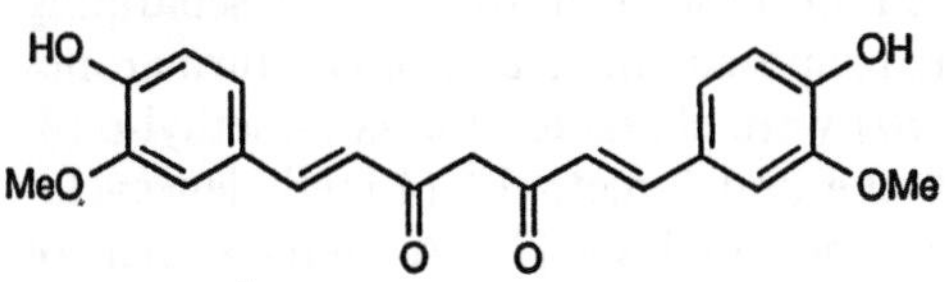

Curcumin (*52-1*)

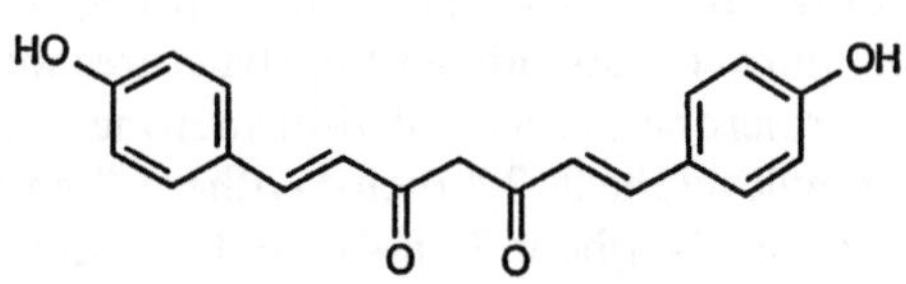

Bis(4-hydroxy-cinnamoyl)methane (*52-2*)

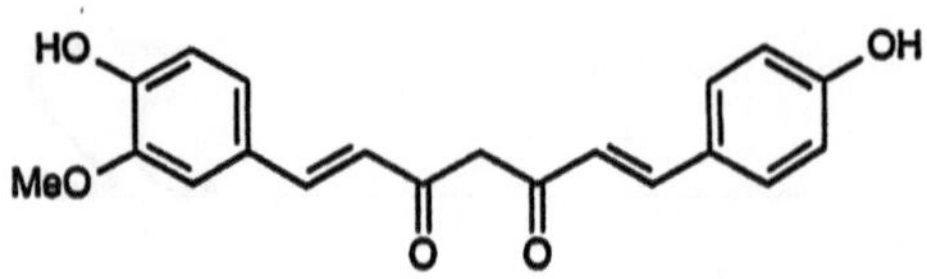

4-Hydroxycinnamoyl feruloyl methane (*52-3*)

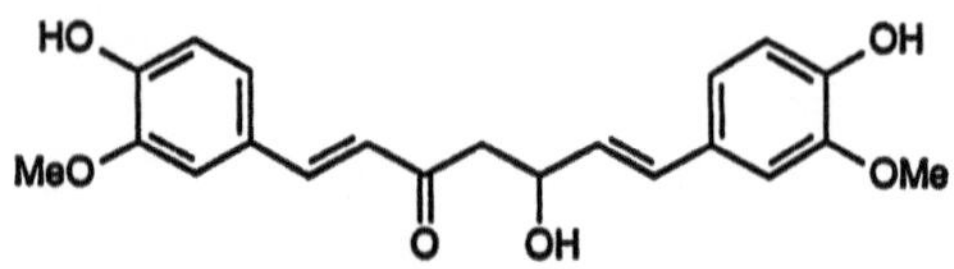

Dihydrocurcumin (*52-4*)

Three C9-chain homologs of curcumin derivatives were also isolated and assigned as curcumins I (*52-5*), II (*52-6*), and III (*52-7*) [6].

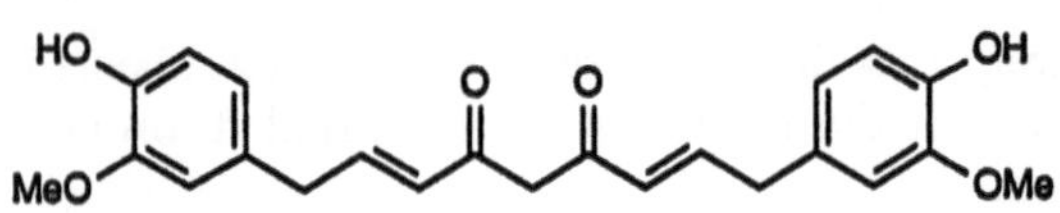

Curcumin I (*52-5*)

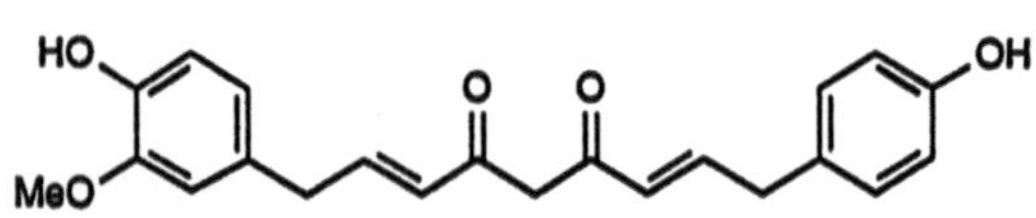

Curcumin II (*52-6*)

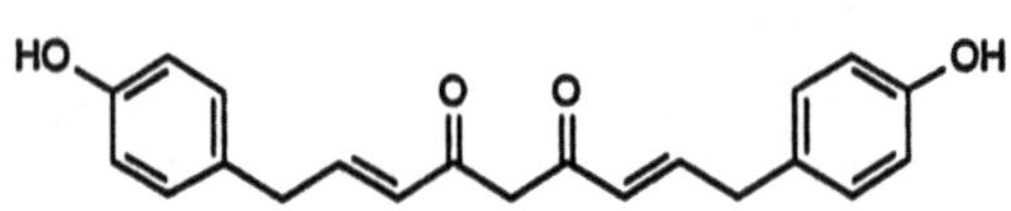

Curcumin III (*52-7*)

The major constituents of the essential oil of the rhizome of *C. longa* were established to be sesquiterpene ketones, ar-turmerone (*52-8*) with an aromatic ring, and turmerone, considered to have the enone and carbon skeleton of ar-turmerone but the aromatic ring partially hydrogenated. Because of the difficulty in separating the pure compounds several structures were proposed. Structures of two turmerones α-turmerone (*52-9*) and β-turmerone (*52-10*) were characterized as 2-methyl-6-(4-methylcyclohexa-2,4-dien-1-yl)hept-2-en-4-one and 2-methyl-6-(4-methylenecyclohex-2-en-1-yl)hept-2-en-4-one by spectroscopic analysis [7]. A stereoisomer of β-turmerone named curlone (*52-11*) was also isolated and structurally defined [8].

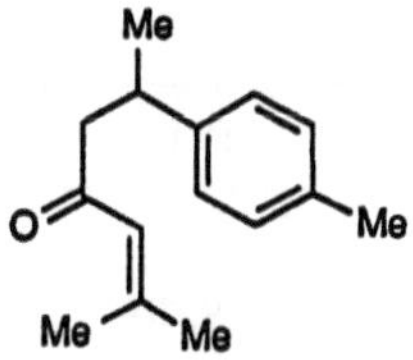
ar-Turmerone (*52-8*)

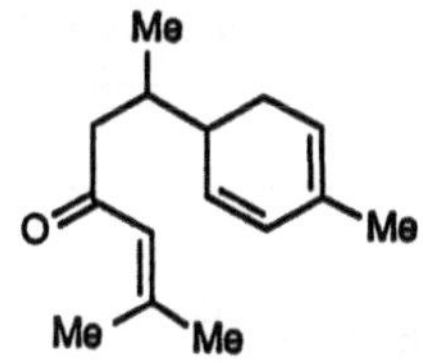
α-Turmerone (*52-9*)

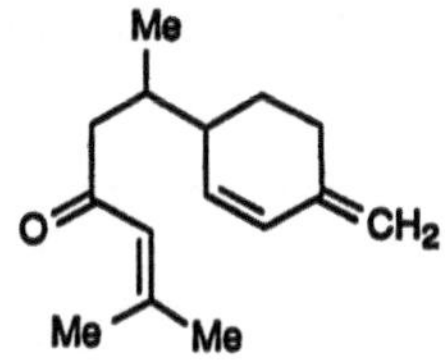
β-Turmerone (*52-10*)

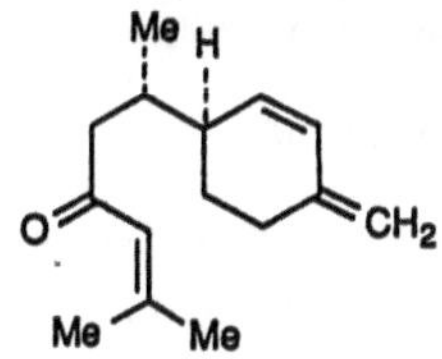
Curlone (*52-11*)

Additionally, a number of terpene compounds were isolated from the essential oil of *C. longa*. They were identified as α- and β-pinene, camphene, limonene, terpinene, caryophyllene, linalool, borneol, isoborneol, camphor, eugenol, cineole, curdione, curzerenone, and curcumenes (*52-12–52-14*) [4].

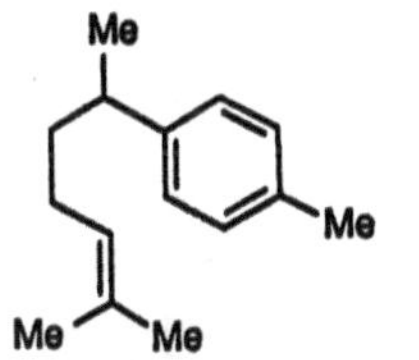
ar-Curcumene (*52-12*)

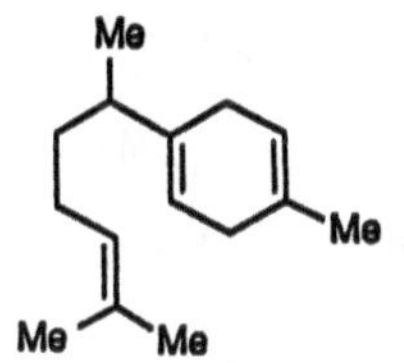
β-Curcumene (*52-13*)

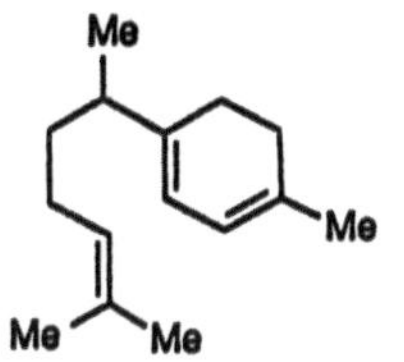
γ-Curcumene (*52-14*)

52.2.2 Chemical Constituents of *Curcuma zedoaria*

From the rhizome of *C. zedoaria*, especially from its essential oil, a great number of sesquiterpenes were isolated and structurally investigated. The compounds determined are listed in Table 52.1. They are derived from germacrane (*52-15*), elemane (*52-16*), cadinane (*52-17*), eudesmane (*52-18*), guaiane (*52-19*), and other type skeletons.

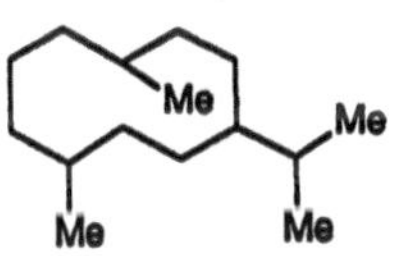
Germacrane (*52-15*)

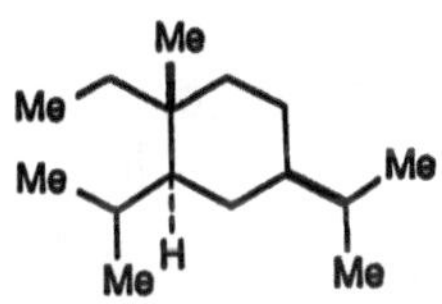
Elemane (*52-16*)

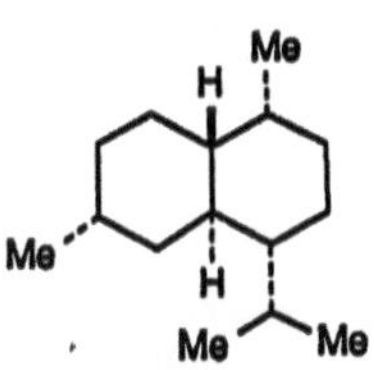
Cadinane (*52-17*)

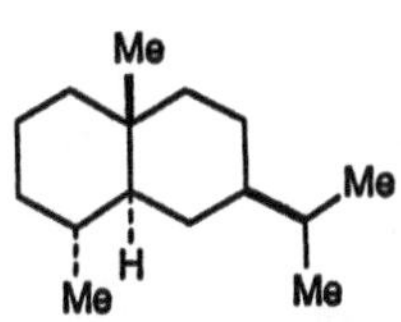
Eudesmane (*52-18*)

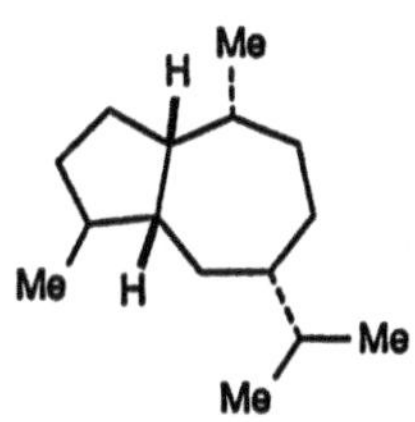
Guaiane (*52-19*)

In addition, curcumin and its related compounds bis(4-hydroxycinnamoyl)-methane and 4-hydroxycinnamoyl feruloyl methane were also detected in the rhizome of *C. zedoaria* [31]. As an antifungal principle, ethyl *p*-methoxycinnamate was detected in zedoary rootstock [32].

Table 52.1. Sesquiterpenes isolated from the essential oil of *Curcuma zedoaria*

Compound	Structure	Ref.
Germacrane-type compounds		
Curdione (*52-20*)	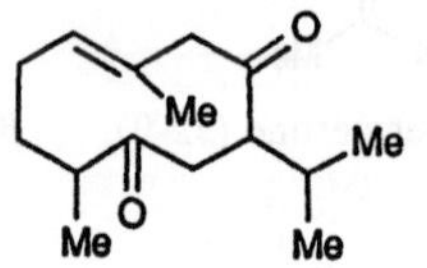	[9, 10]
Dehydrocurdione (*52-21*)	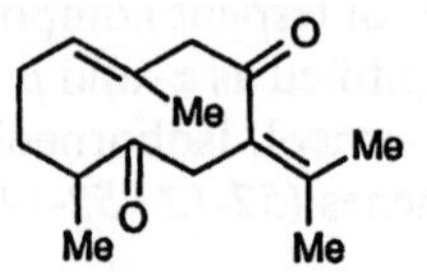	[11]
Germacrone-7,8-epoxide (*52-22*)	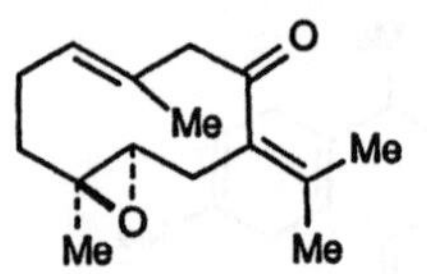	[12]
13-Hydroxygermacrone (*52-23*)	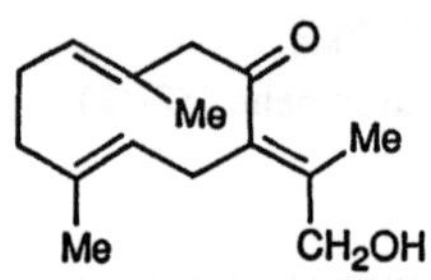	[13]
Zederone (*52-24*)	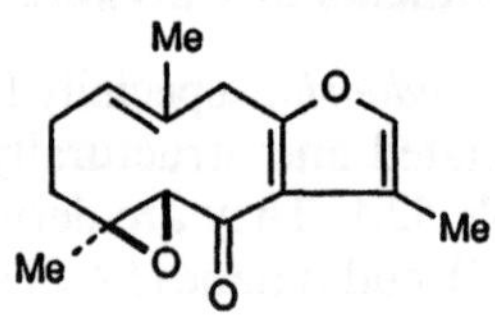	[14–17]
Furanodienone (*52-25*)	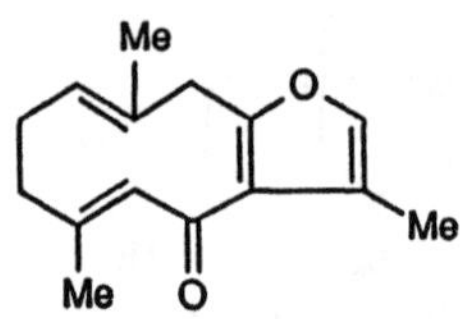	[18]
Furanogermenone (*52-26*)	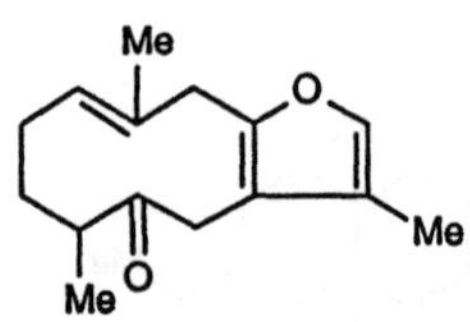	[19]
Furanodiene (*52-27*)	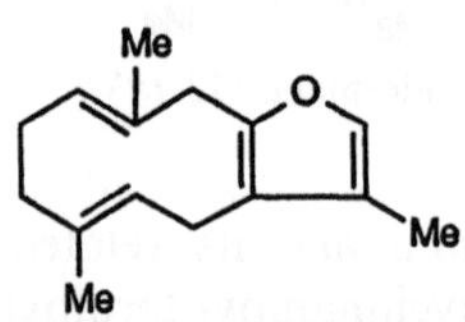	[20]
Isofuranodienone	(E, Z)-Isomer of furanodienone	[19]

Table 52.1. (continued)

Compound	Structure	Ref.
Elemane-type compounds		
Zedoarone (*52-28*)	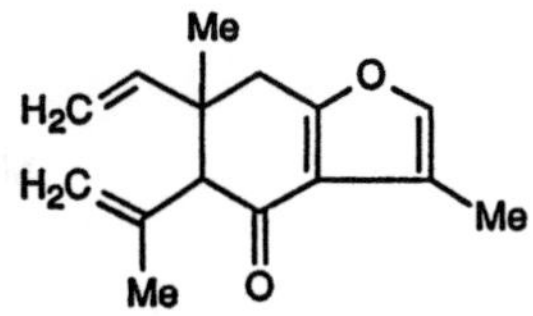	[21]
Curzerenone	*trans*-Isomer of zedearone	[18, 22]
Epicurzerenone	*cis*-Isomer of zedearone	[18]
Curzerene (isofuranogermacrene) (*52-29*)	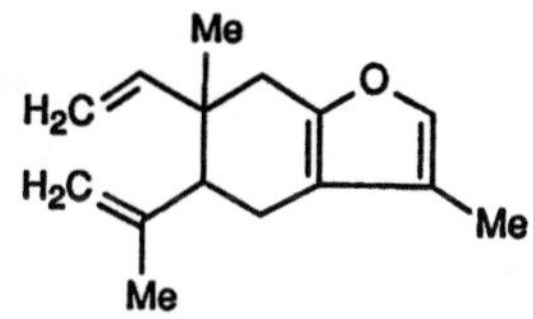	[20]
Cadinane-type compounds		
Pyrocurzerenone (*52-30*)	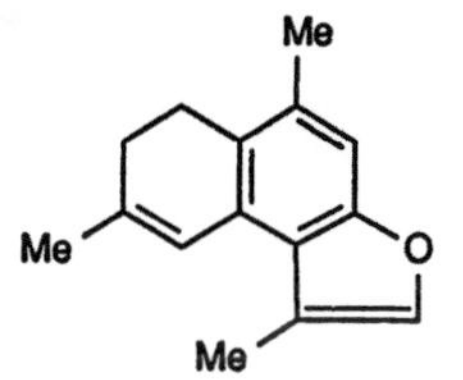	[18, 23]
Curzeone (*52-31*)	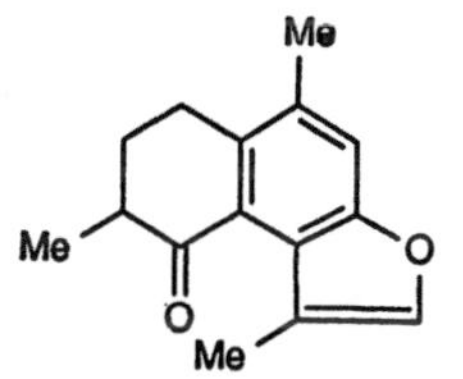	[13]
Eudesmane-type compound		
Curcolone (*52-32*)	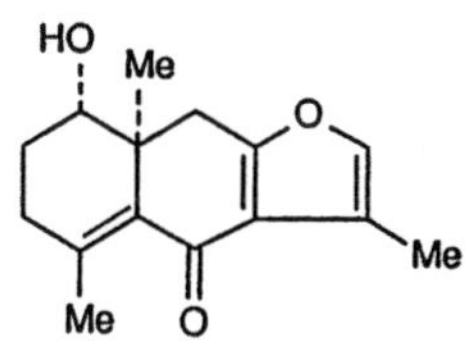	[24]
Guaiane-type compounds		
Procurcumenol (*52-33*)	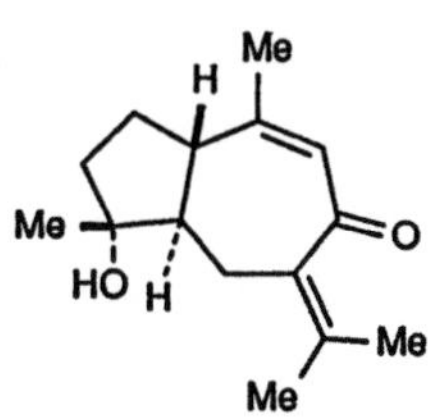	[25]

Table 52.1. (continued)

Compound	Structure	Ref.
Guaiane-type compounds		
Curcumadiol (*52-34*)	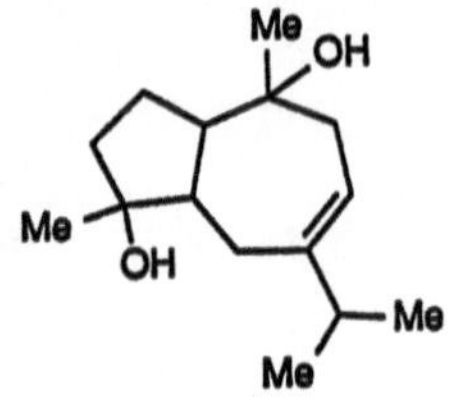	[26]
Isocurcumenol (*52-35*)	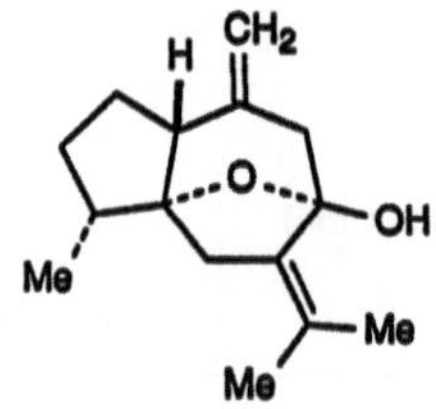	[27]
Zedoarondiol (*52-36*)	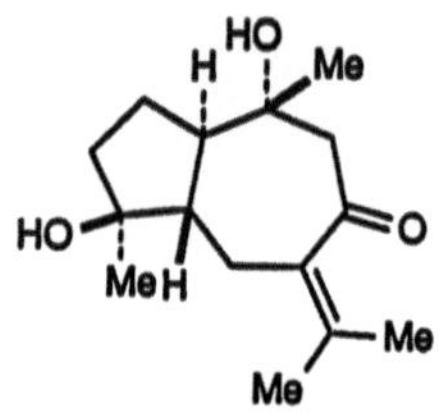	[28, 29]
Zedoarol (*52-37*)	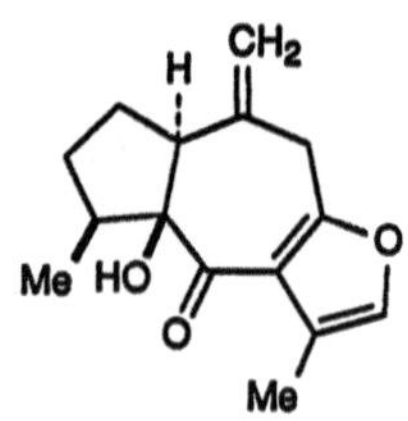	[13]
Other type compounds		
Curcumenone (*52-38*)	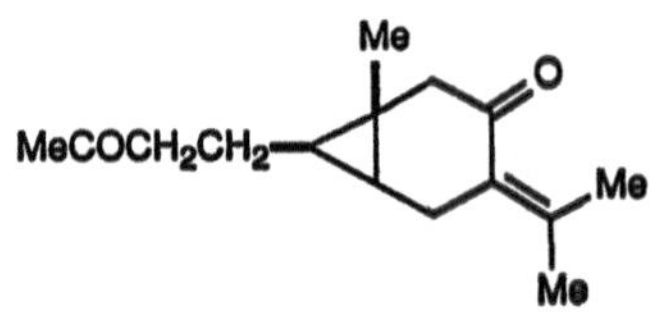	[30]
Curcumanolide A (*52-39*)	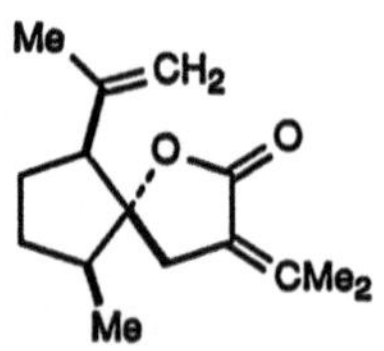	[30]
Curcumanolide B (*52-40*)	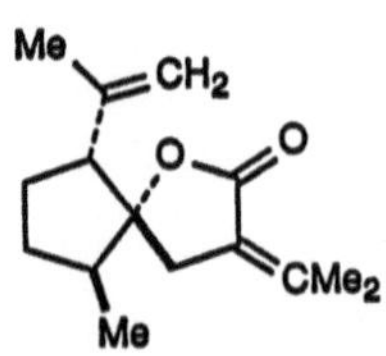	[30]

52.2.3 Chemical Constituents of *Curcuma aromatica*

From the essential oil of the rhizome of *C. aromatica* α-pinene, β-pinene, camphene, cineole, curzerene, borneol, isoborneol, camphor, germacrone, tetramethylpyrazine [33], and curcumol (*52-41*) [34] were isolated. The contents of curcumol in two samples of *C. aromatica* essential oil were 7.4% and 6.5% [35]. Recently, new sesquiterpenes isozedoarondiol (*52-42*), methylzedoarondiol (*52-43*), and neocurdione (*52-44*) along with curdione, dehydrocurdione, procurcumenol, zedoarondiol, and curcumenone were isolated from *C. aromatica* [10, 29, 36].

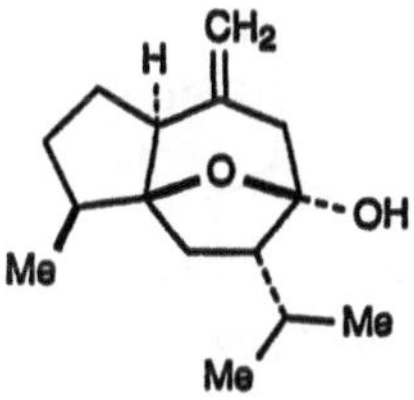

Curcumol (*52-41*)

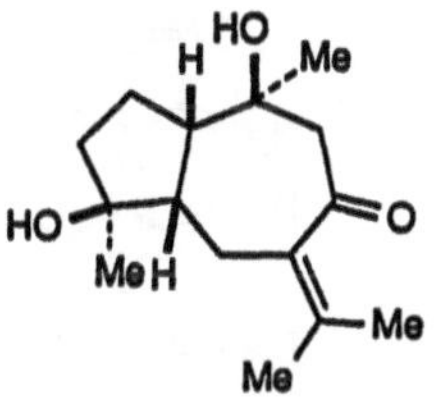

Isozedoarondiol (*52-42*)

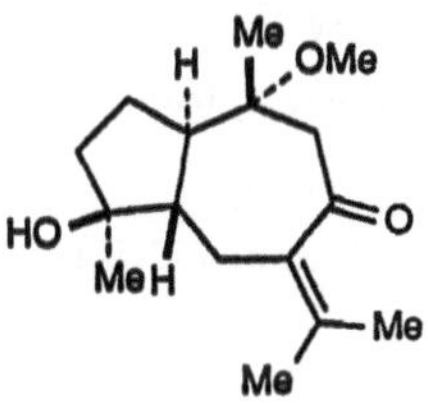

Methylzedoarondiol (*52-43*)

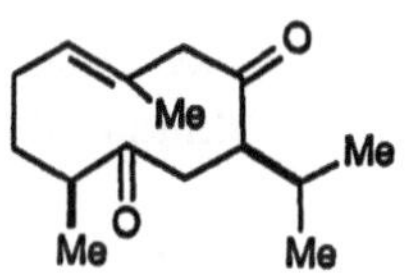

Neocurdione (*52-44*)

52.2.4 Chemical Constituents of *Curcuma kwangsiensis*

From the essential oil of *C. kwangsiensis* the following components were determined: β-pinene, 1,8-cineole, linalool, camphor, isoborneol, borneol, 4-terpineol, α-terpineol, δ-elemene, β-elemene, caryophyllene, humulene, germacrene (*52-45*) [37], linderazulene (*52-46*), germacrone, and isocurcumenol [38].

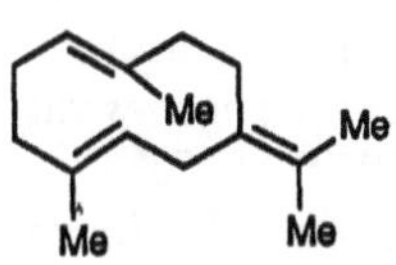

Germacrene (*52-45*)

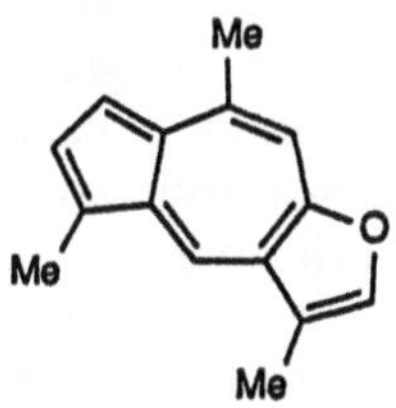

Linderazulene (*52-46*)

Curcumin and related compound bis(4-hydroxycinnamoyl)methane and 4-hydroxycinnamoyl feruloyl methane were also isolated from the rhizome of *C. kwangsiensis*, but their content was not so high as that in rhizome of *C. longa* [39].

52.2.5 Chemical Constituents of *Curcuma wenyujin*

A series of studies were carried out on the constituents of *C. wenyujin*. Besides the known sesquiterpenes curdione, curcumol [40], neocurdione, germacrene D (*52-47*), elemol (*52-48*) [41–43], *β*-elemene (*52-49*), *γ*-elemene (*52-50*), *δ*-elemene (*52-51*) [44], new compounds curcumalactone (*52-52*) [45, 46], wenjine (*52-53*) [47], germacrone-1(10),4-diepoxide (*52-54*) [47] and 1,10-dihydrocurdione (*52-55*) [48] were isolated from the rhizome of *C. wenyujin* and structurally investigated.

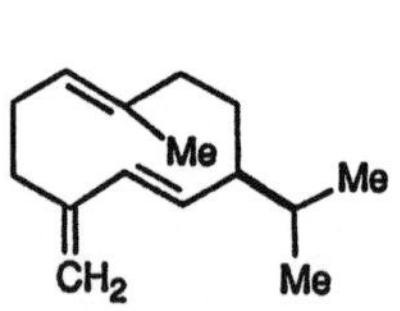

Germacrene D (*52-47*)

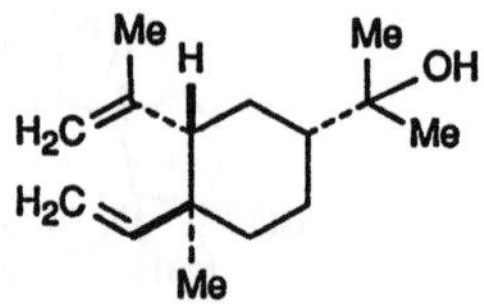

Elemol (*52-48*)

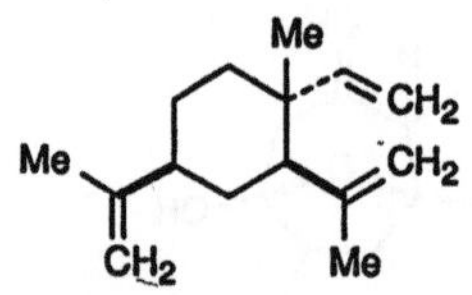

β-Elemene (*52-49*)

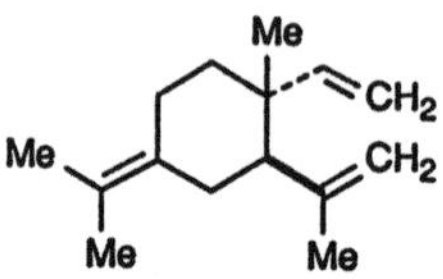

γ-Elemene (*52-50*)

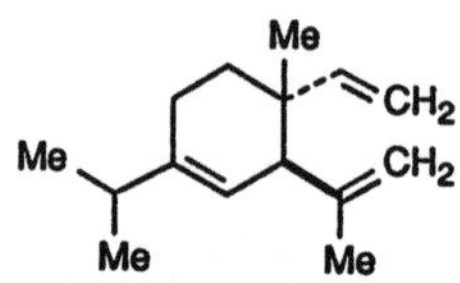

δ-Elemene (*52-51*)

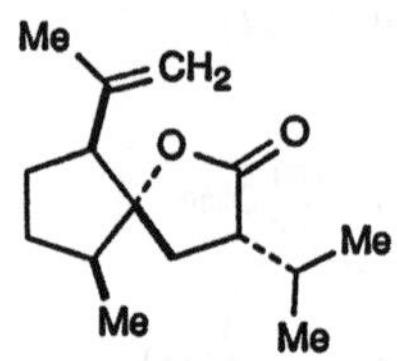

Curcumalactone (*52-52*)

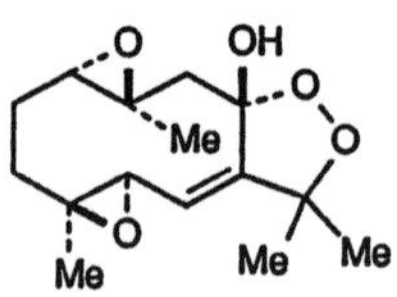

Wenjine (*52-53*)

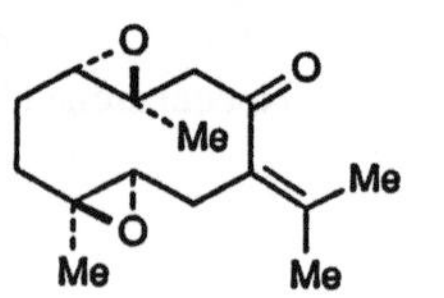

Germacrone-diepoxide (*52-54*)

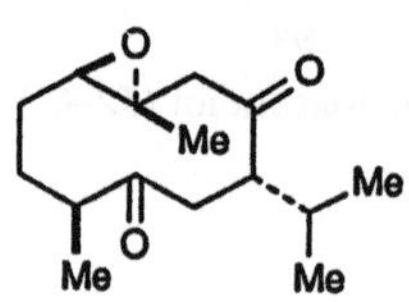

Epoxy-dihydrocurdione (*52-5*

52.3 Pharmacology

The ethanolic extract of *C. longa* as well as curcumin and essential oil of *C. longa* rhizome were found to inhibit the growth of most microorganisms occurring in cholecystitis. Curcumin showed a bacteriostatic effect against *Staphylococcus*, whereas the alcoholic extract and essential oil were bactericidal [49]. Curcumin sodium and essential oil of rhizome of *C. longa* inhibited *Micrococcus pyrogenes* var. *aureus*. The effective concentration of curcumin sodium was 1×10^{-6} *M* [50].

The rhizome of *C. longa* inhibited in vitro growth and acid production by *Lactobacillus acidophilus* and *L. plantarum*, but it stimulated these parameters in *Streptococcus faecalis*, *S. lactis*, and *Escherichia coli*. It has been suggested that *C. longa* may alter the intestinal microflora by inhibiting or stimulating the growth of different bacteria. A concentration of 0.5% turmeric was the minimal effective level for control of *Lactobacillus* growth [51]. The gas formation by *Clostridium perfringens* of intestinal origin decreased gradually as the curcumin concentration increased and there was no gas formation at 0.05% curcumin, the level at which bacterial growth was inhibited completely [52].

The alcoholic extract as well as the essential oil of rhizome of *C. longa* induced morphological changes in *Streptococcus*, *Lactobacillus*, and *Staphylococcus* [53].

A distillation fraction at 80°–110° of the essential oil of *C. longa* administered orally to rats, was found to have antiinflammatory activity [54].

Naturally occurring curcumin analogs, curcumin, bis-(4-hydroxycinnamoyl)-methane, and 4-hydroxycinnamoyl feruloyl methane, from *C. longa* exhibited an inflammatory action against carrageenin-induced paw edema of rats. 4-Hydroxycinnamoyl feruloyl methane was the most potent compound among the three analogs. Curcumin analogs revealed a dose-dependent effect up to the dose of 30 mg/kg. A further increase in the dose of curcumin analogs resulted in decreased antiinflammatory activity [55]. Water-soluble curcumin sodium and curcumin potassium showed better antiinflammatory activity than curcumin as well as hydrocortisone acetate in inflammation induced by carrageenin and formalin in rats [56]. Curcumin sodium reversibly inhibited contractions induced by nicotine, acetylcholine, serotonin, histamine, and $BaCl_2$ on isolated guinea pig ileum in order of decreasing potency.

Oral dose of 3 g/kg curcumin sodium did not cause mortality in rats within 24 h and subacute toxicity experiments for 6 weeks at an oral dose level of 550 mg/kg curcumin sodium showed no undesirable side effects [56].

The antiinflammatory activity of diacetylcurcumin, triethylcurcumin, and tetrahydrocurcumin was also tested in comparison to that of curcumin, curcumin sodium, and phenylbutazone. Maximal activity was observed with triethylcurcumin, whereas curcumin, curcumin sodium, and phenylbutazone were almost half as effective as triethylcurcumin [57].

Inflammation induced by carrageenin in mice was accompanied by an increase in the in vitro formation of lipid peroxides in the liver. Pretreatment of mice with curcumin at a dose of 500 mg/kg administered orally or phenylbutazone at a dose of 80 mg/kg injected intraperitoneally prevented both edema development and lipid peroxide formation [58]. The effect of curcumin on hyaluronidase of the kidney, liver, spleen, serum, and testes of male rats was also studied. Except for liver, the in vitro effect of curcumin on the lysosomal enzyme was very marked [59].

Moreover, curcumin was a more potent inflammation inhibitor than ibuprofen in comparative studies on inflammation-induced changes in rats. In vitro, curcumin was found to be more potent than ibuprofen as a stabilizer of lysosomal membranes and as an uncoupler of oxidative phosphorylation, but was much less effective than ibuprofen in inhibiting prostaglandin synthesis in the inflammatory exudate [60]. However, curcumin was reported to exhibit a strong inhibitory effect on cyclooxygenase system from sheep seminal vesicles [61].

Curcumin sodium given intravenously to anesthetized dogs at a dose of 24 mg/kg caused an increase in bile flow by nearly 100% but without any appreciable disturbance in blood pressure and respiration [62]. Absolute values for the entire period of choleresis indicated increased total excretion of bile salts, bilirubin, and cholesterol, whereas fatty acid contents remained almost constant [63].

The plasma secretin level was markedly stimulated by duodenal infusion of curcumin in dogs and human subjects in a concentration-dependent fashion. These increases in secretin levels were comparable to those induced by duodenal infusion with hydrochloric acid [64].

Curcumin administered orally at a dose of 100 mg/kg for six consecutive days produced gastric ulceration in rats due to the marked reduction in the mucin content of gastric juice. Pretreatment with adrenergic, cholinergic, tryptaminergic, and histaminergic receptor antagonists provided partial protection against curcumin-induced gastric ulcers [65].

Oral administration of paracetamol at doses of 100, 200, and 400 mg/kg in rats caused increases in serum glutamic pyruvic transaminase, serum alkaline phosphatase activity, and serum cholesterol level in the liver; serum bilirubin levels were not increased. Pretreatment of the rats 1 h before the administration of paracetamol with curcumin protected the rats from the induced changes [66]. Curcumin and related compounds present in *C. longa* also possessed significant antihepatotoxic action against CCl_4-induced liver damage in mice and against CCl_4 or galactosamine toxicity in primary cultured rat hepatocytes [67].

In rats fed cholesterol and curcumin, levels of serum and liver cholesterol decreased to one-half or one-third of those in rats fed cholesterol alone [68, 69]. Curcumin given intraperitoneally to mice showed an antithrombotic effect in a dose-dependent manner. Furthermore, curcumin inhibited the cyclooxygenase activity of platelets in vitro [70]. The curcumin-related derivatives bis-(4-hydroxycinnamoyl)methane and 4-hydroxycinnamoyl feruloyl methane were found to have an anticoagulant activity similar to that of curcumin [71]. In addition, curcumin also inhibited the production of 5-hydroxyeicosatetraenoic acid in intact human neutrophils [72].

Curcumin also inhibited ADP-, epinephrine-, and collagen-induced platelet aggregation in monkey plasma [73]. Rapid intravenous injection of curcumin produced transient hypotension and bradycardia in anesthetized dogs and cats. This hypotensive effect of curcumin may be due to its myocardial depressant action. Curcumin exhibited a negative inotropic and chronotropic effect on isolated perfused rabbit heart, an antispasmodic effect on smooth muscle of dog intestine in vivo and of vas deferens of guinea pig in vitro, but no effect on the rectus abdominis muscle of frog or its response to cholinergic stimulation [74].

Toxicological studies showed that the rhizome of *C. longa* or curcumin fed to rats at doses up to 125-fold those corresponding to normal human intake caused no adverse effects on growth, feed efficiency ratio, blood counts, or clinical blood chemistry.

At a concentration of 10% in food, feed efficiency was much lower because of reduced food intake [75]. Acute toxicity studies on *C. longa* have also been conducted in different species of animals, including nonrodents. Neither the rhizome nor its alcoholic extract was toxic even at doses of 2.5 g/kg and 300 mg/kg respectively [76]. Turmeric oleoresin containing curcumin was fed for 102–109 days to pigs at a daily dose of 60, 296, and 1551 mg/kg. The highest dose group showed a reduction in weight gain and in feed efficiency. Statistically significant dose-related increases in the weight of the liver and thyroid were recorded at all dose levels. Pericholangitis, hyperplasia of the thyroid, and epithelial changes in the kidney and urinary bladder were observed in the two higher dose groups [77].

Turmeric oleoresin and curcumin showed no mutagenic activity in tests using *Salmonella typhimurium* strains TA 1535, TA 100, and TA 98 [78]. *C. longa* also did not induce the mitotic gene conversion in *Saccharomyces cerevisiae* in diploid yeast [79]. Addition of either the rhizome of *C. longa* (0.5%) or curcumin (0.015%) into diets of mice showed no significant effects on the incidence of micronucleated polychromatic erythrocytes, structural and numerical aberrations in bone marrow chromosomes, pregnancy rate, number of live and dead embryos, total implants, and mutagenic index. Rats fed cooked diets containing 0.5% or 0.05% turmeric showed no significant differences in the incidence of chromosomal aberrations in bone marrow [80].

Cytotoxic activity of the rhizomes of *C. longa* was tested in vitro. *C. longa* extract inhibited cell growth in Chinese hamster ovary cell culture at a concentration of 0.4 mg/ml and was cytotoxic to lymphocytes and Dalton's lymphoma cells at the same concentration. The constituent found to be responsible for cytotoxicity was curcumin, being toxic to lymphocytes and Dalton's lymphoma cells at a concentration of 4 μg/ml [81].

The metabolism of curcumin was thoroughly investigated [82–87]. Curcumin administered orally to rats was found to be excreted in the feces to about 75%, while negligible amounts appeared in the urine. Measurements of blood plasma levels and biliary excretion showed that curcumin was poorly absorbed from the gut. Curcumin injected intravenously or added to the perfusate of the isolated liver was actively transported into the bile. The major biliary metabolites were glucuronides of tetrahydrocurcumin, hexahydrocurcumin, and dihydroferulic acid; traces of ferulic acid were also detected. In suspensions of isolated hepatocytes or liver microsomes, 90% of the added curcumin was metabolized within 30 min.

Ethyl p-methoxycinnamate isolated from the rhizome of *C. zedoariae* possesses antifungal activity. It inhibited the growth of *Trichophyton rubrum*, *Aspergillus niger*, *Saccharomyces cerevisiae*, and *Epidermorphyton floccosum* at <10 μg/ml and *A. fumigatus*, *Penicillium purpurogenum*, *Microsporum gypseum*, and *Fusarium oxysporum* at <25 μg/ml [32].

A new sesquiterpene, furanogermenone, isolated from *C. zedoaria* suppressed the increase of serum glutamic oxalacetic and glutamic pyruvic transaminases resulting from CCl_4-induced liver damage in rats. It showed no significant effect on liver damage induced by other hepatotoxins, including *d*-galactosamine, thioacetamide, phalloidin, and α-amanitin [88]. It also prevented the prolongation of sleeping time in mice caused by the administration of CCl_4 [89].

A protein-bound polysaccharide with antitumor effectiveness was also isolated from the rhizome of *C. zedoaria* containing glucose, arabinose, xylose and rhamnose in a ratio of 45–55:2–4:4–6:4–6. The major amino acids of the protein were aspartic acid, glutamic acid, leucine, valine, and glycine. Antitumor activity was demonstrated against Ehrlich ascites tumor in mice [90]. Moreover, the sesquiterpene derivatives curcumol and curdione also showed an inhibitory effect on sarcoma-180 in mice [91].

References

1. Srinivasan KR (1953) Chromatographic study of the curcuminoids in *Curcuma longa*. J Pharm Pharmacol 5:448–457
2. Yang MK, Dong XP, Tang YS (1984) Studies on the chemical constituents of common turmeric (*Curcuma longa*). Chin Trad Herb Drugs 15:197–198

3. Toda S, Miyase T, Arichi H, Tanizawa H, Takino Y (1985) Natural antioxidants. III. Antioxidative components isolated from rhizome of *Curcuma longa* L. Chem Pharm Bull (Tokyo) 33:1725–1728
4. Chen YH, Yu JG, Fang HJ (1983) Studies on Chinese *Curcuma*. III. Comparison of the volatile oil and phenolic constituents from the rhizome and the tuber of *Curcuma longa*. Bull Chin Mat Med 8:27–29
5. Ravindranath V, Satyanarayana MN (1980) An unsymmetrical diarylhepatanoid from *Curcuma longa*. Phytochemistry 19:2031–2032
6. Gorchakova NK, Grinkevich NI, Fogal AN (1984) *Curcuma longa* L. as a source of bile expelling drugs. Farmatsiya (Moscow) 33:12–13
7. Golding BT, Pombo E, Samuel CJ (1982) Tumerones: isolation from turmeric and their structure determination. J Chem Soc, Chem Commun 363–364
8. Kiso Y, Suzuki Y, Oshima Y, Hikino H (1983) Stereostructure of curlone, a sesquiterpenoid of *Curcuma longa* rhizomes. Phytochemistry 22:596–597
9. Hikino H, Sakurai Y, Takahashi H, Takemoto T (1967) Structure of curdione. Chem Pharm Bull (Tokyo) 15:1390–1394
10. Ohkura T, Gao JF, Harimaya K, Hikichi M, Iitaka Y, Kawamata T, Kuroyañagi M, Fukushima S, Inayama S (1986) The absolute configuration and conformation of neocurdione. Chem Pharm Bull (Tokyo) 34:4435–4438
11. Hikino H, Konno C, Takemoto T (1972) Sesquiterpenoids. 43. Structure of dehydrocurdione, a sesquiterpenoid of *Curcuma zedoaria*. Chem Pharm Bull (Tokyo) 20:987–989
12. Yoshihara M, Shibuya H, Kitano E, Yanagi K, Kitagawa I (1984) The absolute stereostructure of (4S,5S)-(+)-germacrone 4,5-epoxide from Zedoariae rhizoma cultivated in Yakushima Island. Chem Pharm Bull (Tokyo) 32:2059–2062
13. Shiobara Y, Asakawa Y, Kodama M, Takemoto T (1986) Zedoarol, 13-hydroxygermacrone and curzeone, three sesquiterpenoids from *Curcuma zedoaraia*. Phytochemistry 25:1351–1353
14. Hikino H, Takahashi H, Sakurai Y, Takemoto T, Bhacca NS (1966) Structure of Zederone. Chem Pharm Bull (Tokyo) 14:550–551
15. Hikino H, Takahashi, Shuji S, Sakurai Y, Takemoto T, Bhacca NS (1968) Sesquiterpenoids, 23. Structure of zederone. Chem Pharm Bull (Tokyo) 16:1081–1087
16. Hikino H, Tori K, Horibe I, Kuriyama K (1971) Sesquiterpenoids 37. Absolute configuration and conformation of zederone, a sesquiterpenoid of *Curcuma zedoaria*. J Chem Soc 688–691
17. Shibuya H, Hamamoto Y, Cai Y, Kitagawa I (1987) A reinvestigation of the structure of zederone, a furanogermacrane-type sesquiterpene from zedoary. Chem Pharm Bull (Tokyo) 35:924–927
18. Hikino H, Konno C, Agatsuma K, Takemoto T, Horibe I, Tori K, Ueyama M, Takeda K (1975) Sesquiterpenoids. 47. Structure, configuration, conformation and thermal rearrangements of furanodienone, isofuranodienone, curzerenone, epicurzerenone and pyrocurzerenone, sequiterpenoids of *Curcuma zedoaria*. J Chem Soc, Perkin Trans 1:478–484
19. Shibuya H, Yamamoto Y, Miura I, Kitagawa I (1982) Absolute stereostructure of furanogermenone, a biologically active sesquiterpene from Zedoariae rhizoma. Heterocycles 215–218
20. Hikino H, Agatsuma K, Konno C, Takemoto T (1970) Sesquiterpenoids. 35. Structure of furanodiene and isofuranogermacrene (curzerene), Chem Pharm Bull (Tokyo) 18:752–755
21. Fukushima S, Kuroyanagi M, Akahori Y, Saiki Y, Ueno A (1968) Structure of Zedoarone, a new sesquiterpene from *Curcuma zedoaria*. Yakugaku Zasshi 88:792–794
22. Fukenshima S, Kuroyanagi M, Ueno A, Akahori Y, Saiki Y (1970) Structure of curzerenone, a new sesquiterpene from *Curcuma zedoaria*. Yakugaku Zasshi 90:863–869
23. Viswanatha V, Rao GSK (1974) Terpernoids. 31. Synthesis of pyrocurzerenone, a furosesquiterpenoid from *Curcuma zedoaria*. J Chem Soc, Perkin Trans 1:450–453
24. Hikino H, Sakurai Y, Takemoto T (1968) Sesquiterpenoids. 20. Structure and absolute configuration of curcolone. Chem Pharm Bull (Tokyo) 16:827–831
25. Hikino H, Sakurai Y, Takemoto T (1968) Sesquiterpenoids. XXVI. Structure of procurcumenol. Chem Pharm Bull (Tokyo) 16:1605–1607
26. Hikino H, Konno C, Takemoto T (1971) Sesquiterpenoids. 39. Structure of curcumadiol, a sesquiterpenoid of *Curcuma zedoaria*. Chem Pharm Bull (Tokyo) 19:93–96
27. Hikino H, Agatsuma K, Takemoto T (1969) Sesquiterpenoids. 32. Structure of isocurcumenol. Chem Pharm Bull (Tokyo) 17:959–960

28. Konno I, Kawano N (1985) Structure of a guaiane from *Curcuma zedoaria*. Phytochemistry 24:1845–1847
29. Kuroyanagi M, Ueno A, Ujiie K, Sato S (1987) Structures of sesquiterpenes from *Curcuma aromatica* Salisb. Chem Pharm Bull (Tokyo) 35:53–59
30. Shiobara Y, Asakawa Y, Kodama M, Yasuda K, Takemoto T (1985) Curcumenone, curcumenolide A and curcumenolide B, three sesquiterpenoids from *Curcuma zedoaria*. Phytochemistry 24:2629–2633
31. Kuroyanagi M, Natori S (1970) Curcuminoids from Zingiberaceae plants. Yakugaku Zasshi 90:1467–1470
32. Gupta SK, Banejee AB, Achari B (1976) Isolation of ethyl *p*-methoxycinnamate, the major antifungal principle of *Curcuma zedoaria*. Lloydia 39:218–222
33. Guo YT, Chu XK, Chen YR, Wu XY, Chen G, Lu XR (1980) Studies on the constituents of Wenezhu (*Curcuma aromatica* Salisb). Acta Pharm Sin 15:251–252
34. Yang SD, Ghen JM, Ghen YH (1979) Determination of curcumol in the volatile oil of *Curcuma aromatica*. Acta Pharm Sin 14:356–361
35. Gu MX, Yang Y (1982) Gas chromatographic determination of curcumol in oil of *Curcuma aromatica* Salisb and menthol in oil of *Mentha hyplocalyx* Brig. Chin J Pharm Anal 2:75–78
36. Kuroyanagi M, Ujiie K, Ueno A, Sato S (1987) Sesquiterpenoids of *Curcuma aromatica* and transannular reaction of germacrone 4,5-epoxide. Tennen Yuki Kagobutsu Toronkai Koen Yoshishu 29:528–535 (CA 109:190589p)
37. Tian SJ, Liang WF (1985) Essential oil of *Zedoaria*. III. Separation and identification of the essential oil from *Curcuma kwangsiensis* and comparison with the essential oil from *C. wenyujin* by gas chromatography and TLC. Chin J Pharm Anal 5:325–328
38. Chen YH, Yu JG, Fang HJ (1983) Studies on Chinese curcumas. VI. Isolation and identification of three chemical constituents from Kwangsi turmeric (*Curcuma kwangsiensis*). Chin Trad Herb Drugs 14:534–535
39. Chen JM, Chen YH, Yu JG (1983) Studies on Chinese curcuma plants. IV. Assay of curcumonoids in the root and tuber of *Curcuma* spp. Chin Trad Herbal Drugs 14:59–63
40. Inayama S, Gao JF, Hariyama K, Kawamata T, Iitaka Y, Guo YT (1984) The absolute stereostructure of curcumol isolated from *Curcuma wenyujin*. Chem Pharm Bull 32:3783–3786
41. Ohkura T, Gao JF, Nishishita T, Harimaya K, Kawamata T, Inayama S (1987) Identification of sesquiterpenoid constituents in the essential oil of *Curcuma wenyujin* by capillary gas chromatographic mass spectrometry. Shoyakugaku Zasshi 41:102–107
42. Harimaya K, Ohkura T, Gao JF, Iitaka Y, Osawa E, Inayama S (1987) Conformational analysis of curdione and neocurdione. Chem Pharm Bull (Tokyo) 35:3866–3869
43. Inayama S, Gao JF, Harimaya K, Iitaka Y, Gua YT, Kawamata T (1985) Stereostructure of curdione. Chem Pharm Bull (Tokyo) 33:1323–1326
44. Guo YT, Wu XY, Chen YR (1983) Separation and identification of elemenes from the volatile oil of *Curcuma wenyujin*. Bull Chin Mat Med 8:31
45. Inayama S, Gao JF, Harimaya K, Hikichi M, Iitaka Y, Guo YT, Kawamata T (1985) The absolute configuration of curdione and the stereostructure of curcumalactone from *Curcuma wenyujin*. Chem Pharm Bull (Tokyo) 33:2179–2182
46. Gao JF, Ohkura T, Harimaya K, Hikichi M, Kawamata T, Ying WX, Iitaka Y, Inayama S (1986) The absolute configuration of curcumalactone. Chem Pharm Bull (Tokyo) 34:5122–5132
47. Gao JF, Xie JH, Iitaka Y, Inayama S (1989) The stereostructure of wenjine and related (1S,10S),(4S,5S)-germacrone-1(10),4-diepoxide isolated from *Curcuma wenyujin*. Chem Pharm Bull (Tokyo) 37:233–236
48. Gao JF, Xie JH, Iitaka Y, Inayama S (1988) The stereostructure of (1R,10R)-epoxy-(−)1,10-dihydrocurdione from *Curcuma wenyujin*. Shoyakugaku Zasshi 42:347–349
49. Lutomski J, Kedzia B, Debska W (1974) Effect of an alcohol extract and active ingredients from *Curcuma longa* on bacteria and fungi. Planta Med 26:9–19
50. Ramprasad C, Sirsi M (1956) Indian medicinal plants: *Curcuma longa*; in vitro antibacterial activity of curcumin and the essential oil. J Sci Ind Research 15C:239–241
51. Shankar TNB, Murthy VS (1978) Effect of turmeric (*Curcuma longa*) on the growth of some intestine bacteria in vitro. J Food Sci Technol 15:152–153
52. Bhavanishankar TN, Murthy VS (1985) Inhibitory effect of curcumin on intestinal gas formation by *Clostridium perfringens*. Nutr Rep Int 32:1285–1292

53. Shankar TNB, Murthy VS (1979) Effect of turmeric (*Curcuma longa*) fractions on the growth of some intestinal and pathogenic bacteria in vitro. Indian J Exp Biol 17:1363–1366
54. Chandra D, Gupta SS (1972) Antiinflammatory and antiarthritic activity of volatile oil of *Curcuma longa*. Indian J Med Res 60:138–142
55. Rao TS, Basu N, Siddiqui HH (1982) Antiinflammatory activity of curcumin analogs. Indian J Med Res 75:574–578
56. Ghatak N, Basu N (1972) Sodium curcuminate as an effective antiinflammatory agent. Indian J Exp Biol 10:235–236
57. Mukhopadhyay A, Basu N, Ghatak N, Gujral PK (1982) Antiinflammatory and irritant activities of curcumin analogs in rats. Agents Actions 12:508–515 (CA 97:174653h)
58. Sharma SC, Mukhtar H, Sharma SK, Krishna MCR (1972) Lipid peroxide formation in experimental inflammation. Biochem Pharmacol 21:1210–1214
59. Kushwa A, Amma MKP, Sareen KN (1978) Effect of some antiinflammatory agents on lysosomal and testicular hyaluronidases. Indian J Exp Biol 16:222–224
60. Srivastava R, Srimal RC (1985) Modification of certain inflammation-induced biochemical changes by curcumin. Indian J Med Res 81:215–223
61. Wagner H, Wierer M, Bauer R (1986) Antiinflammatory drugs. Part 3. In vitro inhibition of prostaglandin biosynthesis by essential oils and phenolic compounds. Planta Med 52:184–187
62. Ramprasad C, Sirsi M (1956) Indian medicinal plants: *Curcuma longa* – effect of curcumin and the essential oils of *C. longa* on bile secretion. J Sci Ind Res 15C:262–265
63. Ramprasad C, Sirsi M (1957) *Curcuma longa* and bile secretion: quantitative changes in the bile constituents induced by sodium curcuminate. J Sci Ind Res 16C:108–110
64. Matsuo H (1979) Mechanism of secretin secretion. Shokakan Horumon No Shinpo 72–76 (CA 95:18922y)
65. Gupta B, Kulshrestha VK, Srivastava RK, Prasad DN (1980) Mechanisms of curcumin induced gastric ulcer in rats. Indian J Med Res 71:806–814
66. Banerjee M, Sengupta M, Das J, Sikdar S, Mitra NK (1976) Effects of paracetamol on livers in rats. Calcutta Med J 73:119–123 (CA 88:130867r)
67. Kiso Y, Suzuki Y, Watanabe N, Oshima Y, Hikino H (1983) Antihepatotoxic principles of *Curcuma longa* rhizomes. Planta Med 49:185–187
68. Rao DS, Sekhara NC, Satyanarayana MN, Srinivasan M (1970) Effect of curcumin on serum and liver cholesterol levels in the rat. J Nutr 100:1307–1316
69. Patil TN, Srinivasan M (1971) Hypocholesteremic effect of curcumin in induced hypercholesteremic rats. Indian J Exp Biol 9:167–169
70. Srivastava R, Dikshit M, Srimal RC, Dhawan BN (1985) Antithrombotic effect of curcumin. Thromb Res 40:413–417
71. Kosuge T, Ishida H, Yamazaki H (1985) Studies on active substances in the herbs used for Oketsu ("stagnant blood") in Chinese medicine. III. On the anticoagulative principles in curcumae rhizoma. Chem Pharm Bull (Tokyo) 33:1499–1502
72. Flynn DL, Rafferty MF, Boctor AM (1986) Inhibition of 5-hydroxyeicosatetraenoic acid (5-HETE) formation in intact human neutrophils by naturally occurring diarylheptanoids: inhibitory activities of curcuminoids and yakuchinones. Prostaglandins Leukotrienes Med 22:357–360
73. Srivastava R, Puri V, Srimal RC, Dhawan BN (1986) Effect of curcumin on platelet aggregation and vascular prostacyclin synthesis. Arzneimittelforschung 36:715–717
74. Rao TS, Basu N, Seth SD, Siddioui HH (1984) Some aspects of pharmacological profile of sodium curcuminate. Indian J Physiol Pharmacol 28:211–215
75. Sambaiah K, Katankumar S, Kamanna VS, Satyanarayana MN, Rao MVL (1982) Influence of turmeric and curcumin on growth, blood constituents and serum enzymes in rats. J Food Sci Technol 19:187–190
76. Shankar TNB, Shantha NV, Ramesh HP, Murthy IAS, Murthy VS (1980) Toxicity studies on turmeric (*Curcuma longa*): acute toxicity studies in rats, guinea pigs and monkeys. Indian J Exp Biol 18:73–75
77. Bille N, Larsen JC, Hansen EV, Würtzen G (1985) Subchronic oral toxicity of turmeric oleoresin in pigs. Food Chem Toxicol 23:967–973
78. Juul Jensen N (1982) Lack of mutagenic effect of turmeric oleoresin and curcumin in the *Salmonella*/mammalian microsome test. Mutat Res 105:393–396

79. Sankaranarayanan N, Murthy MSS (1979) Testing of some permitted food colors for the induction of gene conversion in diploid yeast. Mutat Res 67:309–314
80. Vijayalaxmi (1980) Genetic effects of turmeric and curcumin in mice and rats. Mutat Res 79:125–132
81. Kuttan R, Bhanumathy P, Nirmala K, George MC (1985) Potential anticancer activity of turmeric (*Curcuma longa*). Cancer Lett 29:197–202
82. Ravindranath V, Chandrasekhara N (1981) In vitro studies on the intestinal absorption of curcumin in rats. Toxicology 20:251–257
83. Wahlstrom B, Blennow G (1978) A study on the fate of curcumin in the rat. Acta Pharmacol Toxicol 43:86–92
84. Holder GM, Plummer JL, Ryan AJ (1978) The metabolism and excretion of curcumin [1,7-bis(4-hydroxy-3-methoxyphenyl)-1,6-heptadiene-3,5-dione] in the rat. Xenobiotica 8:761–768
85. Ravindranath V, Chandrasekhara N (1982) Metabolism of curcumin – studies with (^{3}H)-curcumin. Toxicology 22:337–344
86. Khanna M, Singh S, Sarin JPS (1981) The metabolic disposition of curcumin in rats. Indian Drugs 19:31 (CA 96:135289d)
87. Ravindranath V, Chandrasekhara N (1980) Absorption and tissue distribution of curcumin in rats. Toxicology 16:259–265
88. Yamahara J, Matsuda H, Sawada T, Sakurai H, Matsumura A, Toyama S, Suzuki I (1982) Effect of crude drugs on experimental liver damages. II. Effect of new sesquiterpenoid "furanogermenone". Yakugaku Zasshi 102:272–277
89. Yamahara J, Matsuda H, Sawada T, Kushida H, Shibuya H, Kitagawa I (1982) Effect of crude drugs on experimental liver damages I. The active principle of Zedoariae rhizoma. Yakugaku Zasshi 102:306–309
90. Kokan, Takuo Tsumura Juntendo, Inc (1985) Antitumor protein-bound polysaccharides from *Curcuma* plants. Jpn Kokai Tokkyo Koho JP 60 67,428(85 67,428) (CA 103:92834w)
91. Lien EJ, Li WY (1985) Anticancer Chinese drugs: structure-activity relationships. In: Chang HM, Yeung HW, Tso WW, Koo A (eds) Advances in Chinese Medicinal Materials Research. World Scientific Publ, Singapore, pp 433–452

Cynanchum glaucescens (Decne.) Hand.-Mazz. 53

53.1 Introduction

Baiqian, Rhizoma Cynanchi stauntonii, is the dry root and rootstock of *Cynanchum stauntonii* (Decne.) Schltr. ex Levl. or *C. glaucescens* (Decne.) Hand.-Mazz. (Asclepiadaceae), which is collected in the fall. It is officially listed in the Chinese Pharmacopoeia.

This crude drug was used in traditional Chinese medicine as an antitussive, expectorant, and antiasthmatic agent.

53.2 Chemical Constituents

53.2.1 Chemical Constituents of *Cynanchum glaucescens*

It is mainly the chemical constituents of the rootstock of *C. glaucescens* that have been investigated. The chloroform extract of *C. glaucescens* rhizome showed positive Liebermann-Burchard and Keller-Killiani reactions, indicating the presence of steroidal glycosides, containing 2-deoxysugars. From the extract, some glycosides named glaucosides were isolated, which by mild hydrolysis gave the aglycones glaucogenin A, glaucogenin B, and glaucogenin C D-thevetoside. The structures of glaucogenins A and B were characterized chemically and spectroscopically. Glaucogenin C D-thevetoside was determined structurally by X-ray crystallography [1–3]. In these structures, a novel 13,14:14,15-disecopregnane-type was found. For comparison, the structures of pregnane (*53-1*), and the 13,14:14,15-disecopregnane-type skeleton (*53-2*) characteristic of the glycosides of *C. glaucescens* are demonstrated. The structures of glaucogenins and glucosides are listed in Table 53.1.

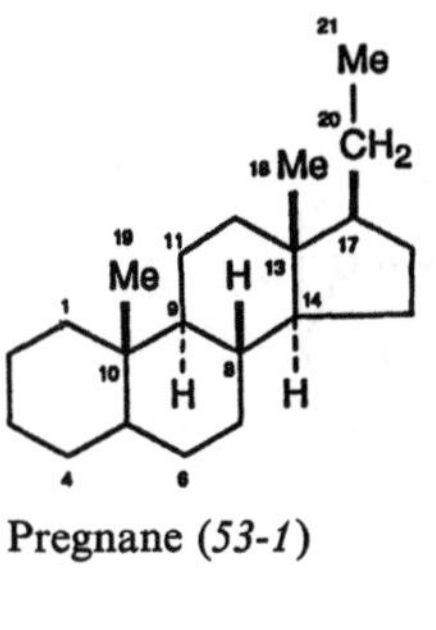

Pregnane (*53-1*)

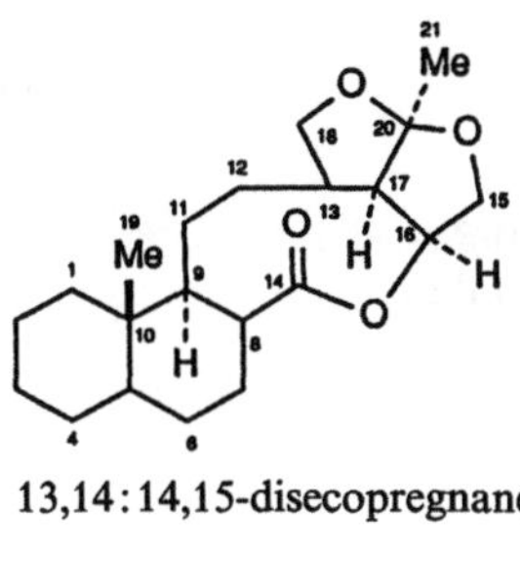

13,14:14,15-disecopregnane-type skeleton (*53-2*)

Table 53.1. Glaucogenins and glaucosides from *Cynanchum glaucescens*

Compound	R	Ref.
	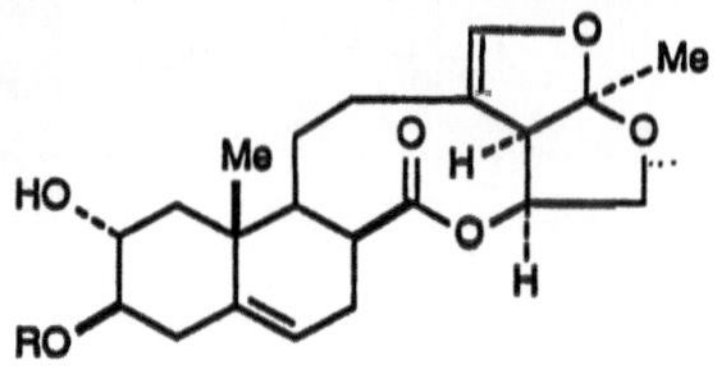	
Glaucogenin A (*53-3*)	H	[1–3]
Glaucoside A (*53-4*) (Glaucogenin A 3-*O*-β-D-oleandropyranoside)	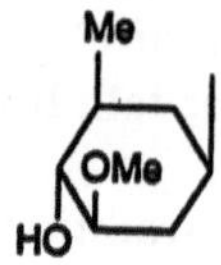	[4]
Glaucoside B (*53-5*) (Glaucogenin A 3-*O*-α-L-cymaropyranosyl-(1→4)-β-L-cymaropyranosyl-(1→4)-β-L-cymaropyranoside)	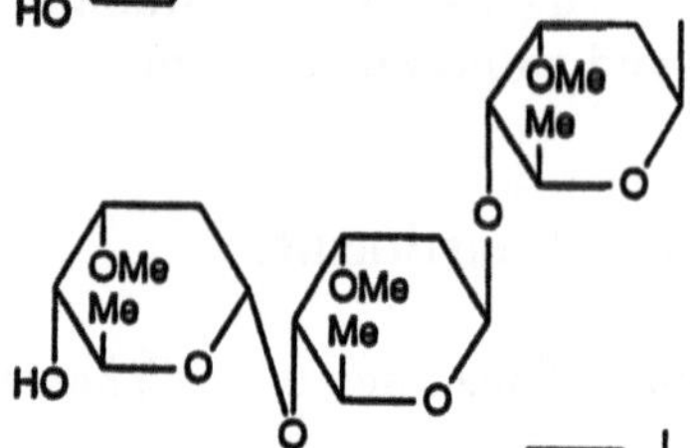	[4]
Glaucoside C (*53-6*) (Glaucogenin A 3-*O*-α-L-cymaropyranosyl-(1→4)-β-D-digitoxopyranosyl-(1→4)-β-L-cymaropyranoside)	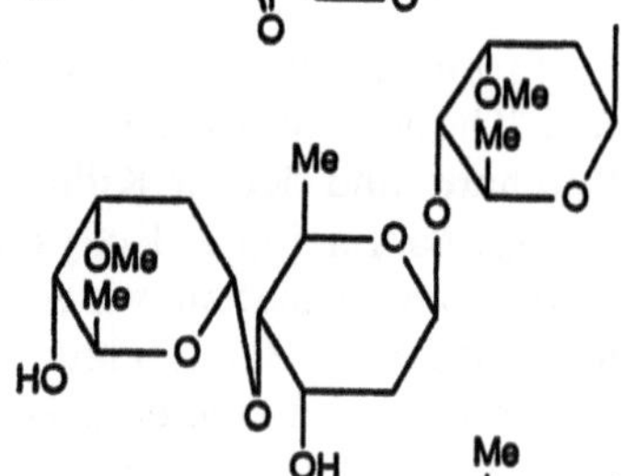	[4]
Glaucoside D (*53-7*) (Glaucogenin A 3-*O*-α-L-cymaropyranosyl-(1→4)-β-D-digitoxopyranosyl-(1→4)-β-D-oleandropyranoside)	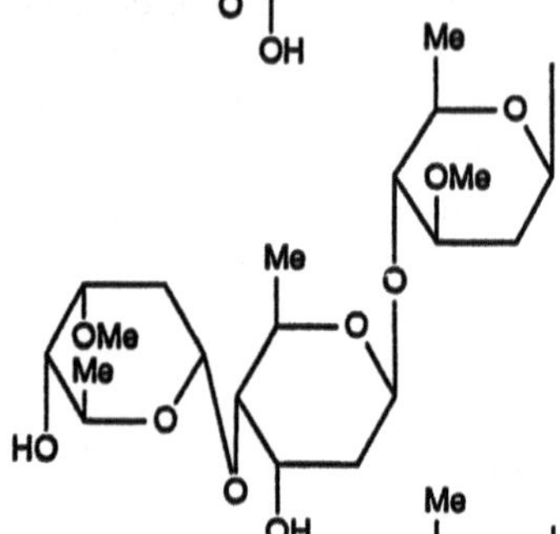	[4]
Glaucoside F (*53-8*) (Glaucogenin A 3-*O*-α-L-cymaropyranosyl-(1→4)-β-L-cymaropyranosyl-(1→4)-β-D-oleandropyranoside)	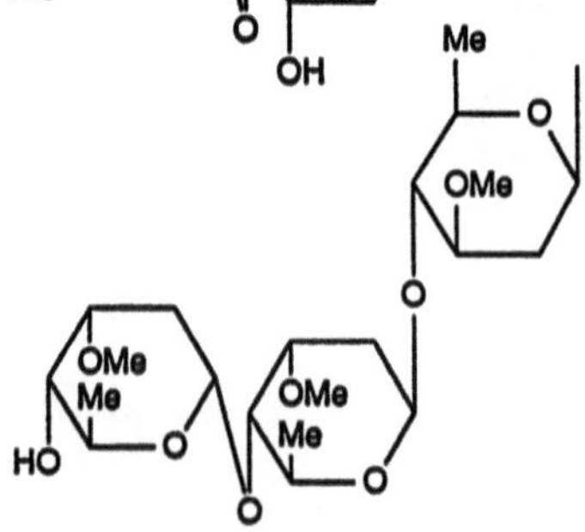	[5]

Table 53.1. (continued)

Compound	R	Ref.
	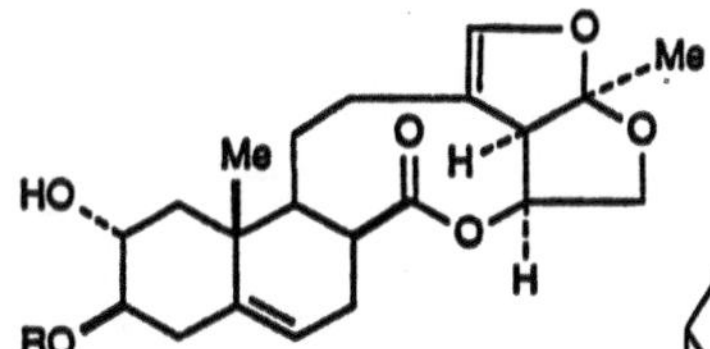	
Glaucoside H (*53-9*) (Glaucogenin A 3-*O*-β-D-glucopyranosyl-(1→4)-α-L-cymaropyranosyl-(1→4)-β-D-digitoxopyranosyl-(1→4)-β-L-cymaropyranoside)	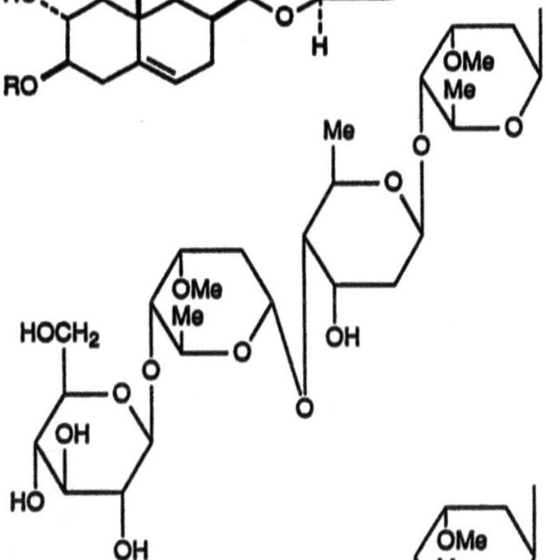	[6]
Glaucoside I (*53-10*) (Glaucogenin A 3-*O*-β-D-glucopyranosyl-(1→4)-α-L-cymaropyranosyl-(1→4)-β-L-cymaropyranosyl-(1→4)-β-L-cymaropyranoside)	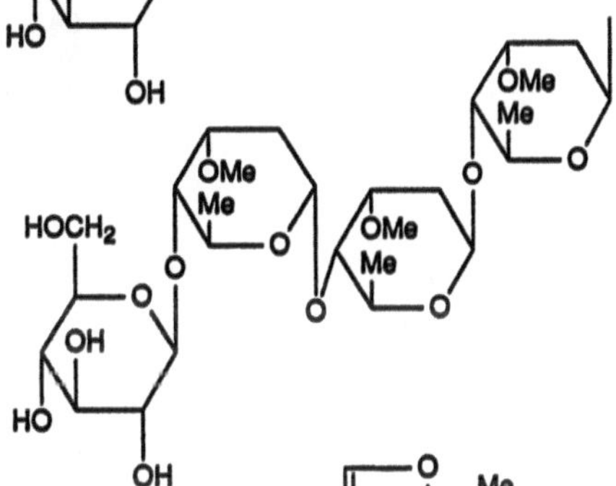	[6]
	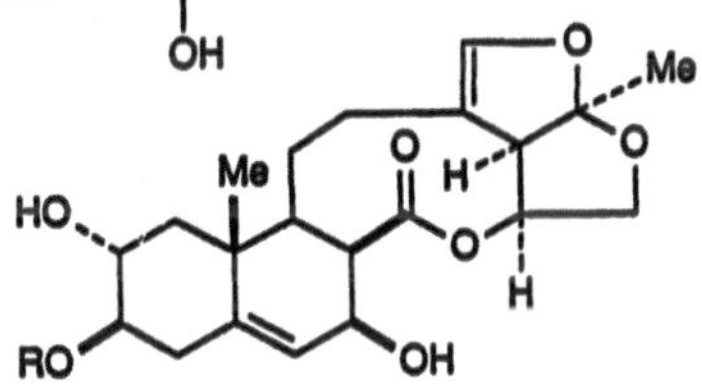	
Glaucogenin B (*53-11*)	H	[1–3]
Glaucoside J (*53-12*) (Glaucogenin B 3-*O*-β-D-glucopyranosyl-(1→4)-α-L-cymaropyranosyl-(1→4)-β-D-digitoxopyranosyl-(1→4)-β-D-oleandropyranoside)	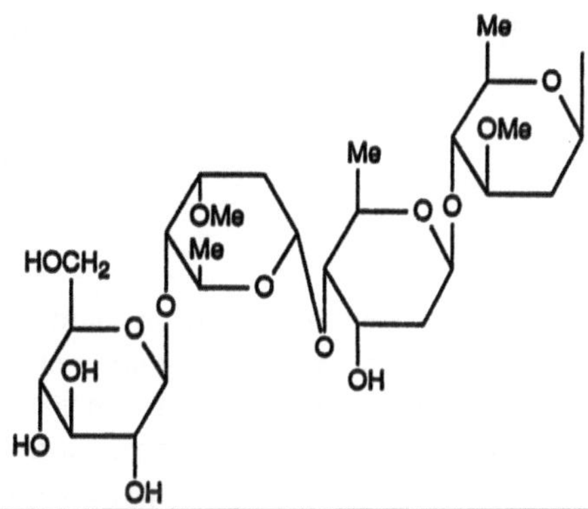	[6]

Table 53.1. (continued)

Compound	R	Ref.
	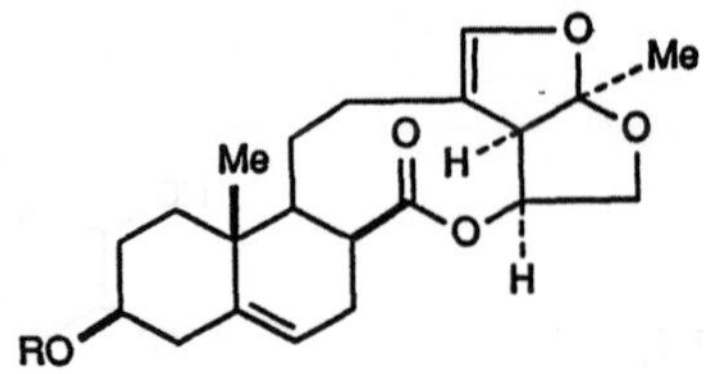	
Glaucogenin C (*53-13*)	H	[1–3]
Glaucoside E (*53-14*) (Glaucogenin C 3-*O*-α-L-cymaropyranosyl-(1→4)-β-L-cymaropyranosyl-(1→4)-β-D-thevetopyranoside)	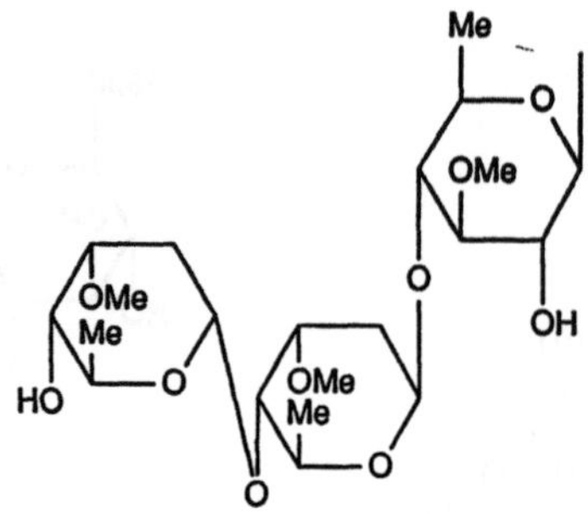	[4]
Glaucoside G (*53-15*) (Glaucogenin C 3-*O*-α-L-cymaropyranosyl-(1→4)-β-D-digitoxopyranosyl-(1→4)-β-D-thevetopyranoside)	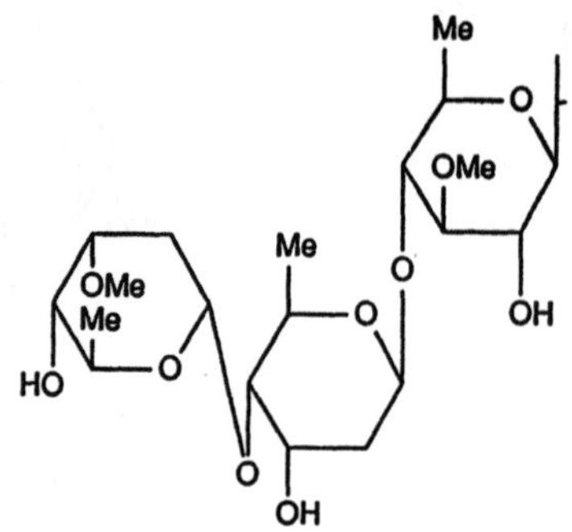	[5]
Glaucogenin C 3-*O*-β-D-thevetopyranoside (*53-16*)	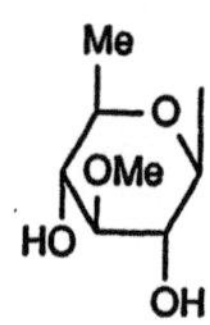	[1–3]

In addition, a new disaccharide named glaucobiose (*53-17*) was found in *C. glaucescens*. On hydrolysis using snail enzyme (β-glucosidase) glaucobiose produced cymarose and glucose. The structure of glaucobiose was determined by chemical and spectral methods and by comparison with its stereoisomer strophanthobiose from *C. candatum* as 4-*O*-β-D-glucopyranosyl-L-cymaropyranose [7].

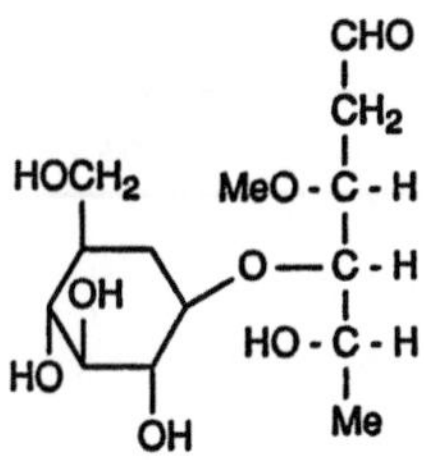
Glaucobiose (*53-17*)

53.2.2 Chemical Constituents of *Cynanchum atratum*

Another item of the *Cynanchum* genus listed in the Chinese Pharmacopoeia is Baiwei, Radix Cynanchi atrati. It is the dry root and rootstock of *Cynanchum atratum* Bge. and *C. versicolor* Bge. collected in the spring and fall. It is used as an antipyretic and diuretic.

From the rootstock of *C. atratum*, glycosides named cynatratosides A, B, C, D, E [8], and F [9] and atratosides A, B, C, and D [10] were isolated and structurally determined. Cynatratosides have been found to possess glaucogenin A and glaucogenin C as aglycones. New sapogenins atratogenin A, atratogenin B, and cynajaponin were found to be the aglycones of atratosides. Cynajaponin was first isolated from *C. japonicum* [11]. The structures of cynatratosides and atratosides are listed in Table 53.2. In addition, glaucogenin A, glaucoside C, and glaucoside H have also been found in *C. atratum* [9].

Table 53.2. Cynatratosides and atratosides of *Cynanchum atratum*

Compound	R	Ref.
	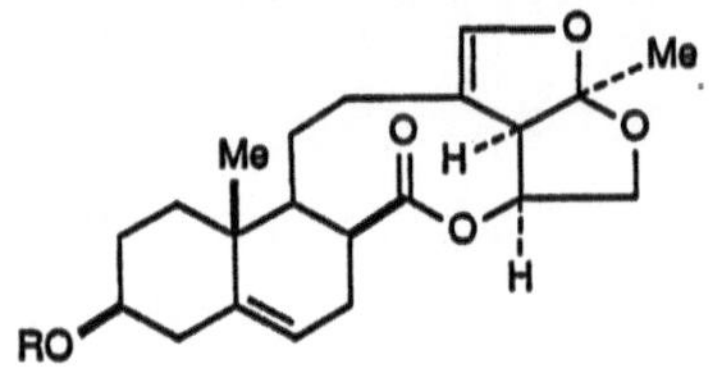	
Cynatratoside A (*53-18*) (Glaucogenin C 3-*O*-*β*-D-oleandropyranoside)	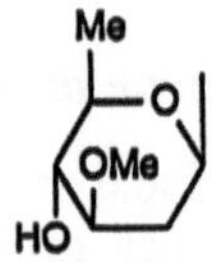	[8]
Cynatratoside B (*53-19*) (Glaucogenin C 3-*O*-α-L-cymaropyranosyl-(1→4)-*β*-D-digitoxopyranosyl-(1→4)-*β*-D-oleandropyranoside)	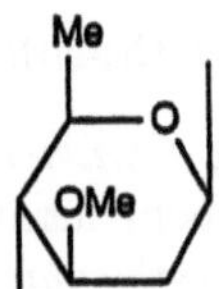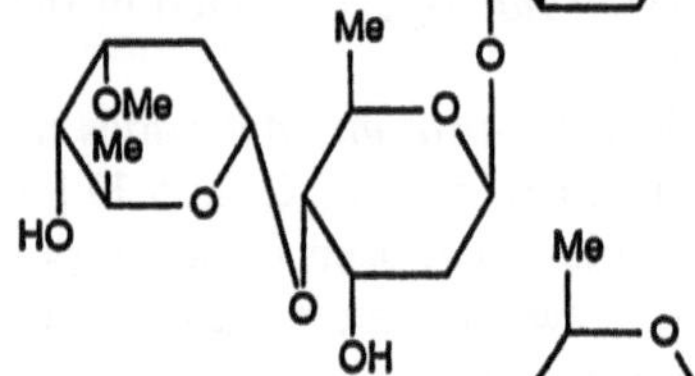	[8]
Cynatratoside C (*53-20*) (Glaucogenin C 3-*O*-α-D-oleandropyranosyl-(1→4)-*β*-D-digitoxopyranosyl-(1→4)-*β*-D-oleandropyranoside)	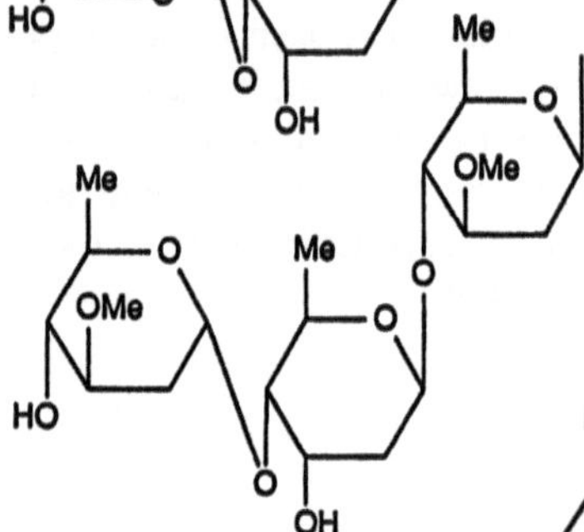	[8]
Cynatratoside D (*53-21*) (Glaucogenin C 3-*O*-*β*-D-glucopyranosyl-(1→4)-α-L-cymaropyranosyl-(1→4)-*β*-D-digitoxopyranosyl-(1→4)-*β*-D-oleandropyranoside)	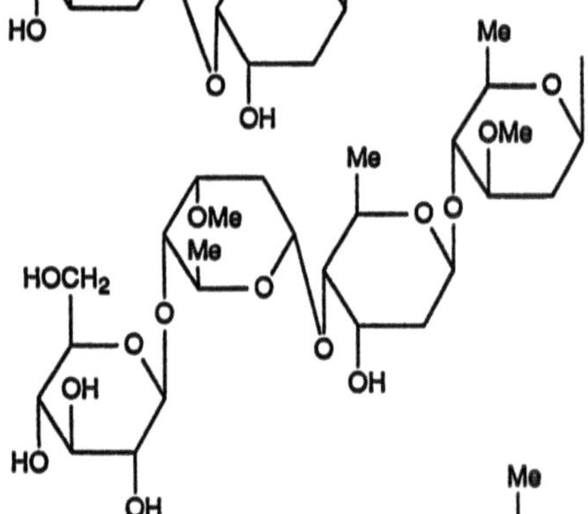	[8]
Cynatratoside E (*53-22*) (Glaucogenin C 3-*O*-*β*-D-glucopyranosyl-(1→4)-α-D-oleandropyranosyl-(1→4)-*β*-D-digitoxopyranosyl-(1→4)-*β*-D-oleandropyranoside)	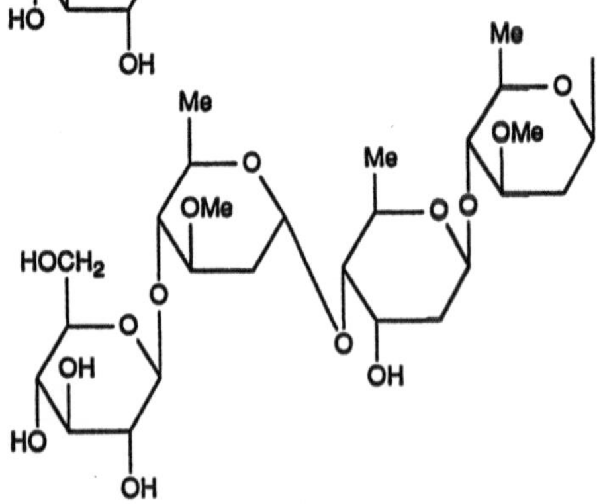	[8]

Table 53.2. (continued)

Compound	R	Ref.
Cynatratoside F (*53-23*) (Glaucogenin A 3-*O*-β-D-cymaropyranosyl-(1→4)-α-L-diginopyranosyl-(1→4)-β-D-cymaropyranoside)	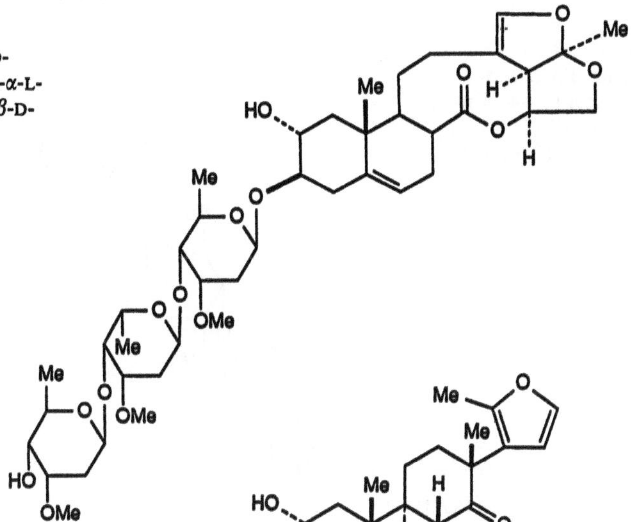	[9]
Atratoside A (*53-24*) (Atratogenin A 3-*O*-β-D-cymaropyranosyl-(1→4)-α-L-diginopyranosyl-(1→4)-β-D-cymaropyranoside)	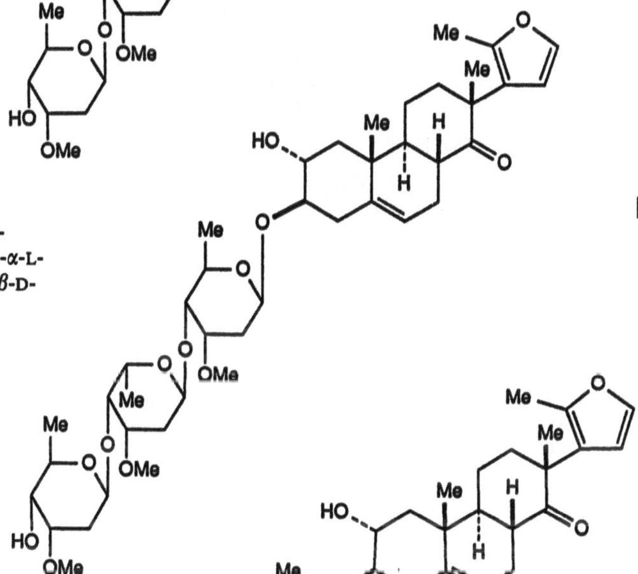	[10]
Atratoside B (*53-25*) (Atratogenin A 3-*O*-β-D-glucopyranosyl-(1→4)-β-D-cymaropyranosyl-(1→4)-α-L-diginopyranosyl-(1→4)-β-D-cymaropyranoside)	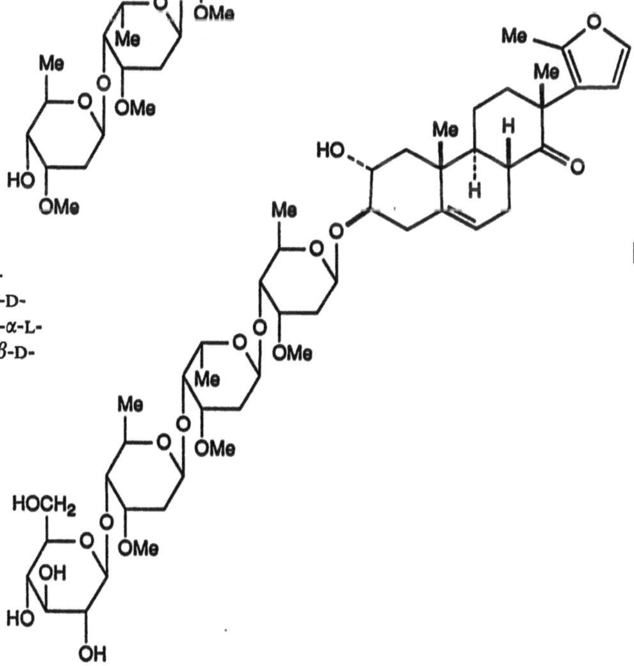	[10]

Table 53.2. (continued)

Compound	R	Ref.
Atratoside C (*53-26*) (Atratogenin B 3-*O*-*β*-D-glucopyranosyl-(1→4)-*β*-D-cymaropyranosyl-(1→4)-α-L-diginopyranosyl-(1→4)-*β*-D-cymaropyranoside)	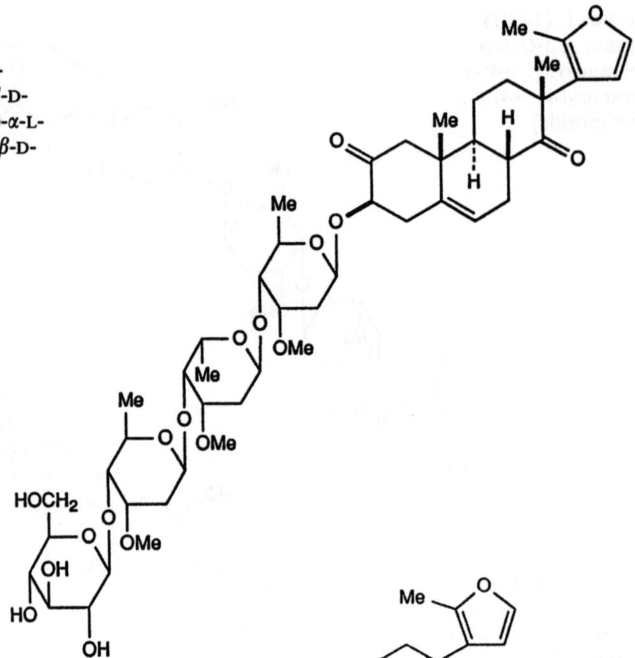	[10]
Atratoside D (*53-27*) (Cynajapogenin A 3-*O*-α-D-oleandropyranosyl-(1→4)-*β*-D-digitoxopyranosyl-(1→4)-*β*-D-cymaropyranoside)	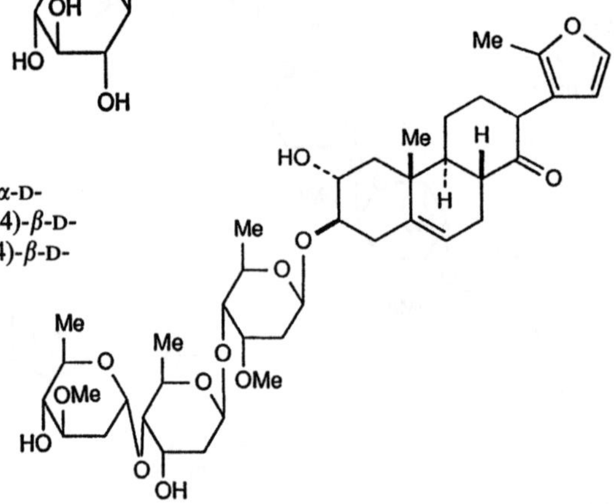	[10]

53.2.3 Chemical Constituents of Other *Cynanchum* Species with Medicinal Use

53.2.3.1 Chemical Constituents of *Cynanchum otophyllum*

Two new C_{21} steroidal glycosides named otophylloside A (*53-28*) and otophylloside B (*53-29*) were isolated from the roots of *C. otophyllum* and their structures determined by X-ray crystallographic, spectrometric, and chemical methods [12, 13].

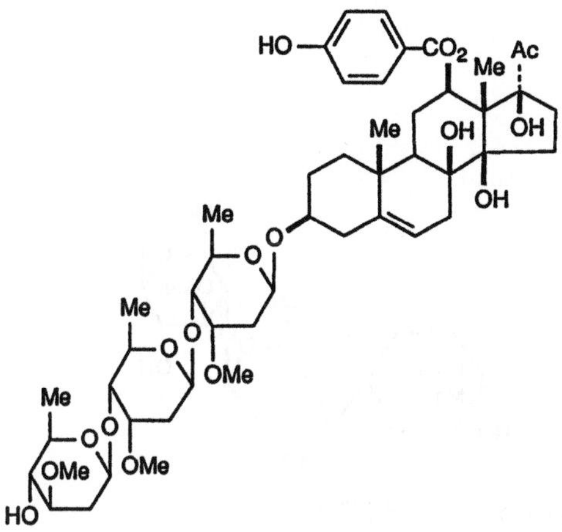

Otophylloside A (*53-28*)

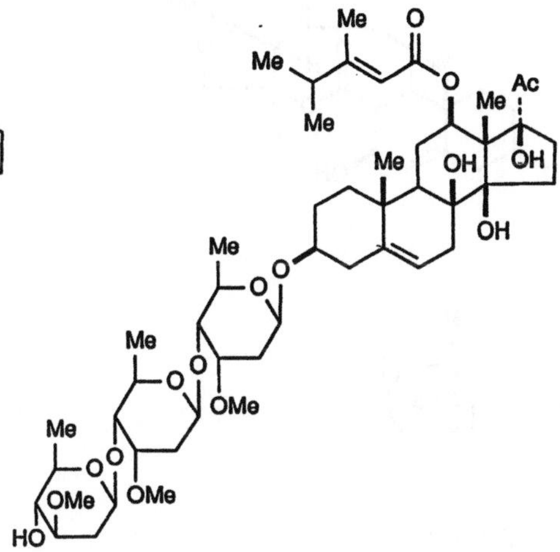

Otophylloside B (*53-29*)

Besides the otophyllosides, steroidal esters rostratamine (*53-30*), qingyangshengenin (*53-31*), and caudatin (*53-32*) were also isolated from the roots of *C. otophyllum* together with methyl palmitate, *β*-sitosterol, vanillic acid, and digitoxose [14].

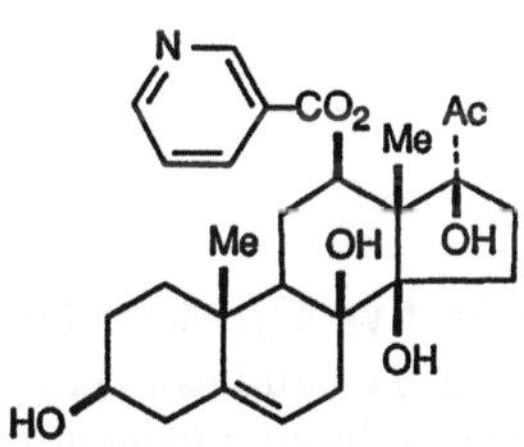

Rostratamine (*53-30*)

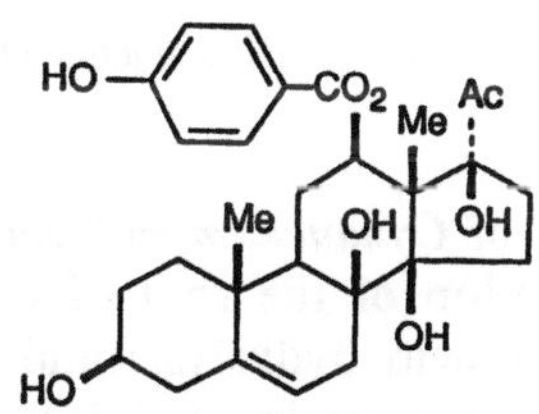

Qingyangshengenin (*53-31*)

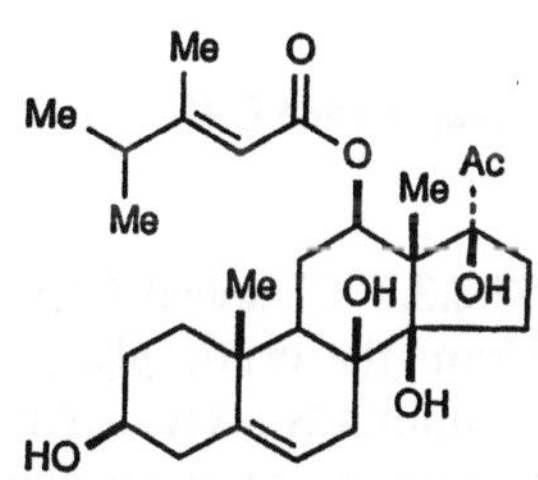

Caudatin (*53-32*)

53.2.3.2 Chemical Constituents of *Cynanchum paniculatum*

Three new glycosides named cynapanosides A (*53-33*), B (*53-34*), and C (*53-35*) which contain a new aglycone, glaucogenin D (*53-36*), were isolated from *C. paniculatum* in addition to the known compounds cynatratoside B and 3β,14-dihydroxy-14β-pregn-5-en-20-one [15].

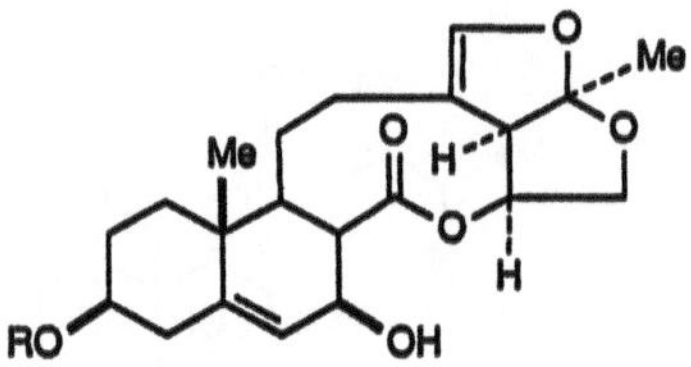

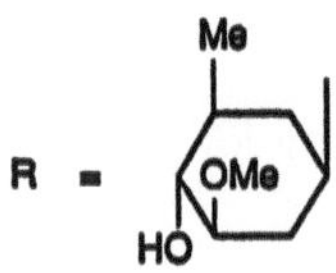

Cynapanoside A (*53-33*)

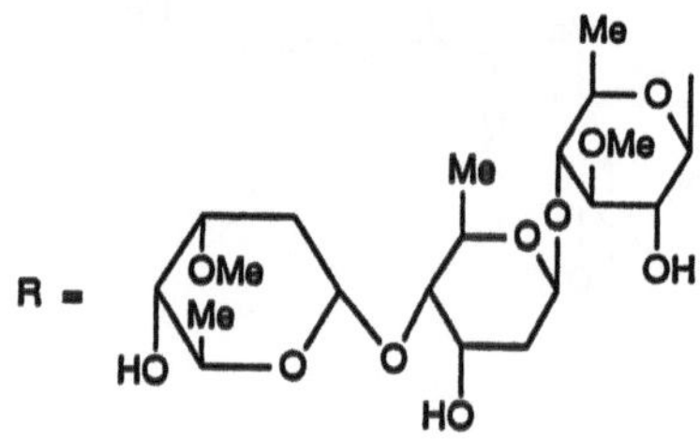

Cynapanoside B (*53-34*)

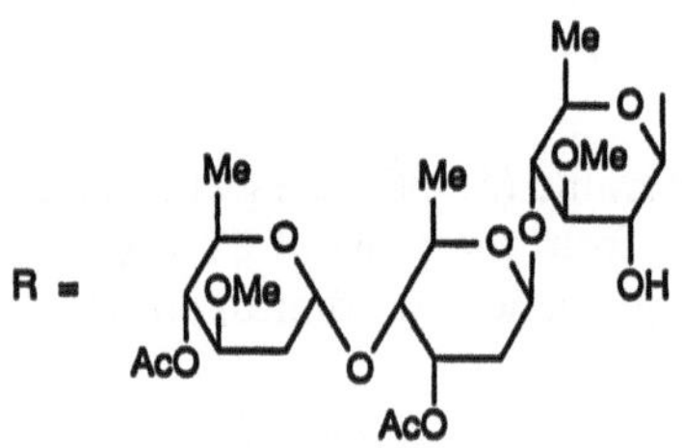

Cynapanoside C (*53-35*)

R=H

Glaucogenin D (*53-36*)

53.2.3.3 Chemical Constituents of *Cynanchum wallichii*

From the crude glycoside fraction of the root of *C. wallichii* five C_{21}-steroidal compounds were isolated after acid hydrolysis and identified as caudatin, rostratamine, qingyangshengenin, gagaminine (*53-37*), and deacetylmetaplexigenin (*53-38*) [16]. A glycoside named wallicoside (*53-39*) was also isolated from the crude glycoside mixture of *C. wallichii* and structurally elucidated as caudatin 3-*O*-β-D-glucopyranosyl-(1→4)-β-D-glucopyranosyl-(1→4)-β-D-oleandropyranosyl-(1→4)-β-D-cymaropyranosyl-(1→4)-β-D-cymaropyranoside [17].

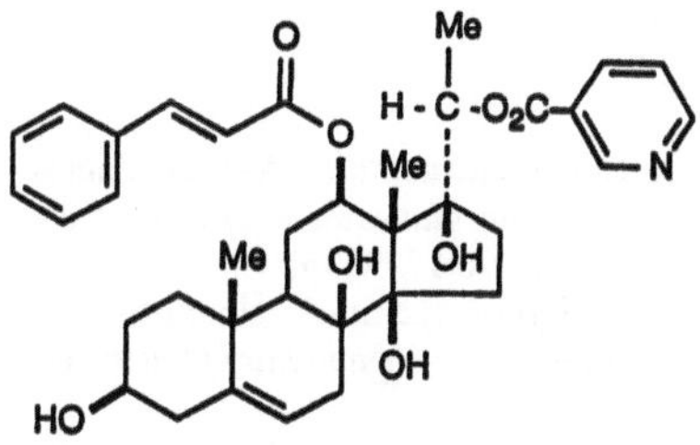

Gagaminine (*53-37*)

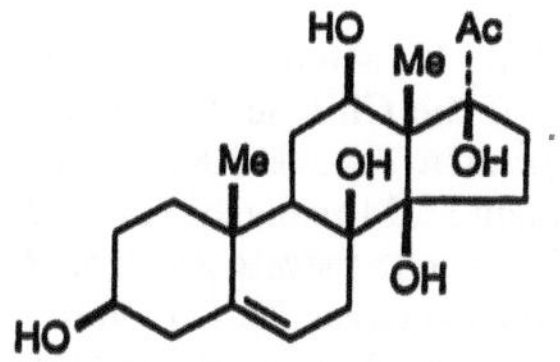

Deacetylmetaplexigenin (*53-38*)

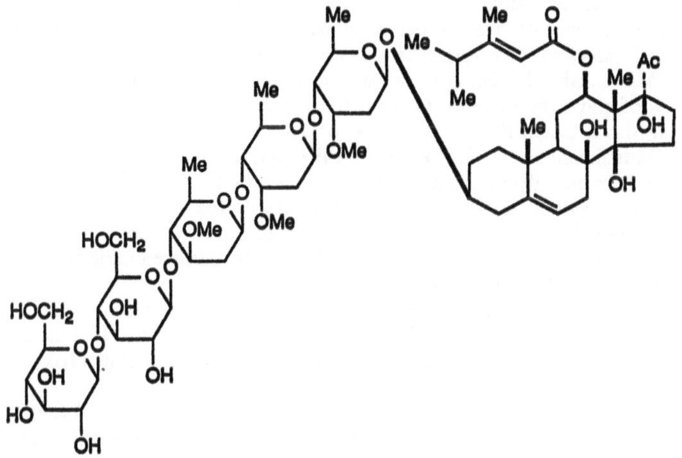

Wallicoside (*53-39*)

53.3 Pharmacology

The roots of *C. otophyllum* is a traditional Chinese medicine used for treatment of epilepsy and chronic hepatitis [14]. The C_{21} steroidal glycosides otophyllosides A and B were found to be active against epilepsy, protecting rats from audiogenic seizures with an ED_{50} of 10.2 mg/kg [12]. Intraperitoneal injection of the total glycosides of *C. otophyllum* roots to mice did not affect glutamate decarboxylase activity in the brain of normal mice. However, a decrease in brain glutamate decarboxylase activity was antagonized by the glucosides in mice with seizure induced by thiosemicarbazide [18]. It was reported that the root or the whole plant of *C. paniculatum* was very effective in the treatment of urticaria and atophic rhinitis. Pharmacological investigation indicated the inhibitory effects on the anaphylaxis of guinea pigs sensitized with ovalbumin. The majority of the animal were protected from death due to anaphylactic shock [19].

References

1. Nakagawa T, Hayashi K, Mitsuhashi H (1981) Studies on the constituents of Asclepiadaceae plants. Glycosides of the Chinese drug pai-ch'ien from *Cynanchum glaucescens* Hand.-Mazz. Tennen Yuki Kagobutsu Toronkai Koen Yoshishu 24:79–86 (CA 96:177925p)
2. Nakagawa T, Hayashi K, Mitsuhashi H (1982) The structures of glaucogenin-A, glaucogenin-B and glaucogenin-C mono D-thevetoside from Chinese drug "pai-ch'ien" *Cynanchum glaucescens* Hand-Mazz. Tetrahedron Lett 23:757–760
3. Nakagawa T, Hayashi K, Mitsuhashi H (1983) Studies on the constituents of Asclepiadaceae plants. LIII. The structures of glaucogenin-A, -B and -C mono-D-thevetoside from the Chinese drug "pai-ch'ien", *Cynanchum glaucescens* Hand-Mazz. Chem Pharm Bull (Tokyo) 31:870–878
4. Nakagawa T, Hayashi K, Wada K, Mitsuhashi H (1983) Studies on the constituents of Asclepiadaceae plants. LII. The structures of five glycosides glaucoside A, B, C, D and E from Chinese drug "pai-ch'ien", *Cynanchum glaucescens* Hand-Mazz. Tetrahedron 39:607–612
5. Nakagawa T, Hayashi K, Mitsuhashi H (1983) Studies on the constituents of Asclepiadaceae plants. LIV. The structures of glaucoside-F and -G from the Chinese drug "pai-ch'ien", *Cynanchum glaucescens* Hand-Mazz. Chem Pharm Bull (Tokyo) 31:879–882
6. Nakagawa T, Hayashi K, Mitsuhashi H (1983) Studies on the constituents of Asclepiadaceae plants. LV. The structures of three new glycosides, glaucoside-H, -I and -J from the Chinese drug "pai-ch'ien", *Cynanchum glaucescens* Hand-Mazz. Chem Pharm Bull (Tokyo) 31:2244–2253
7. Nakagawa T, Hayashi K, Wada K, Mitsuhashi H (1982) A new disaccharide, glaucobiose from Chinese drug "pai-ch'ien": a comparison of carbon-13 NMR with its diastereomeric isomer, strophanthobiose. Tetrahedron Lett 23:5431–5434
8. Zhang ZX, Zhou J, Hayashi K, Mitsuhashi H (1985) Studies on the constituents of Asclepiadaceae plants. LVIII. The structures of five glycosides, cynatratoside-A, -B, -C, -D and -E, from the Chinese drug "pai-wei", *Cynanchum atratum* Bunge. Chem Pharm Bull (Tokyo) 33:1507–1514
9. Zhang Z, Zhou J, Hayashi K, Mitsuhashi H (1985) Studies on the constituents of Asclepiadaceae plants LXI. The structure of cynatratoside-F from the Chinese drug "pai-wei" dried root of *Cynanchum atratum*. Chem Pharm Bull (Tokyo) 33:4188–4192
10. Zhang ZX, Zhuo J, Hayashi K, Kameko K (1988) Studies on the constituents of Asclepiadaceae plants. Part 68. Atratosides A, B, C, and D steroid glycosides from the root of *Cynanchum atratum*. Phytochemistry 27:2935–2941
11. Hayashi K, Sugama K, Zhang ZX, Tsukamoto S, Nakaya H, Sasaki K, Nakagawa T, Mitsuhashi H, Kaneko K (1986) On the pregnane glycosides from the plants belonging to the genus *Cynanchum* (Asclepiadaceae). Tennen Yuki Kagobutsu Toronkai Koen Yoshishu 28:216–223 (CA 106:135258t)
12. Mu QZ, Lu JR, Zhou QL (1986) Two new antiepilepsy compounds – otophyllosides A and B. Sci Sin, Ser B (Engl Ed) 29:295–301
13. Mu QZ, Zhou QL (1983) Study on chemical constituents of Qing Yang Shen (*Cynanchum otophyllum* Schneid.) Acta Pharm Sin 18:356–362
14. Mu QZ, Zhou QL (1983) Studies on constituents of *Cynanchum otophyllum* Schneid. roots. Acta Bot Yunnan 5:99–103
15. Sugama K, Hayashi K, Mitsuhashi H, Kaneko K (1986) Studies on the constituents of Asclepiadaceae plants. LXVI. The structures of three new glycosides, cynapanosides A, B and C from the Chinese drug "Xu-Chang-Qing", *Cynanchum paniculatum* Kitagawa. Chem Pharm Bull (Tokyo) 34:4500–4507
16. Zhang ZX, Zhou J (1982) Chemical components of *Cynanchum wallichii*. Acta Bot Yunnan 4:413–418
17. Zhang ZX, Zhou J (1983) Structure of wallicoside. Acta Chim Sin 41:1058–1064
18. Li JY, Cai XL, Zhao TR, Nan GH, Zhou X (1987) Effect of total glucosides of Qingyangshen (*Cynanchum otophyllum*) on glutamate decarboxylase and GABA transaminase in mouse brain. Chin Trad Herbal Drugs 18:264–266
19. Qiu G, Wu AR (1986) Chinese material medica with anti-atophy effect. Abst Chin Med 1:113–129

Daphne genkwa Sieb. et Zucc. 54

54.1 Introduction

Yuanhua, Flos Genkwa, is the dry flower buds of *Daphne genkwa* Sieb. et Zucc. (Thymelaeaceae) collected in spring before blossom. It is officially listed in the Chinese Pharmacopoeia and is a traditional Chinese medicine used as a diuretic in treatment of ascites, edema, and asthma. Externally it is used against scabies and ulcer. In addition, the root bark of *D. genkwa* has also been used as a diuretic, especially in treatment of ascites in late-stage schistosomiasis.

54.2 Chemical Constituents

54.2.1 Flavones

From the flowers and buds of *D. genkwa* a series of flavones was isolated and identified as apigenin, luteolin, luteolin-7-methyl ether, genkwanin (*54-1*) [1], 3′-hydroxygenkwanin, and yuankanin (*54-2*) [2]. Yuankanin is a genkwanin-5-bioside, the sugar moiety being composed of xylose and glucose. Yuankanin was first isolated from the root bark of *D. genkwa* with a yield of about 0.1% [3]. Galuteolin (*54-3*) and its 7-methyl ether named yuanhuanin (*54-4*) [4], yuankanin, luteolin, genkwanin, isoquercetin, and 3′-hydroxygenkwanin were also isolated from the leaves of *D. genkwa* [5]. The 3′-hydroxygenkwanin content in dried leaves was 0.33% [6].

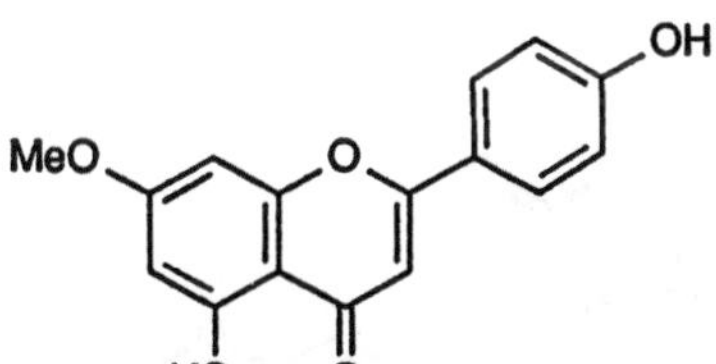

Genkwanin (*54-1*)

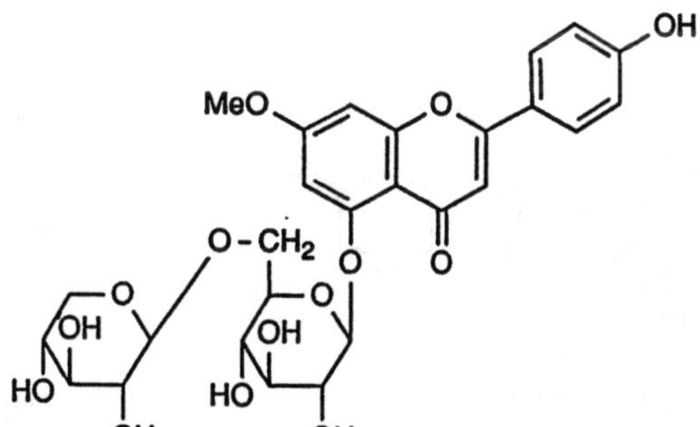

Yuankanin (*54-2*)

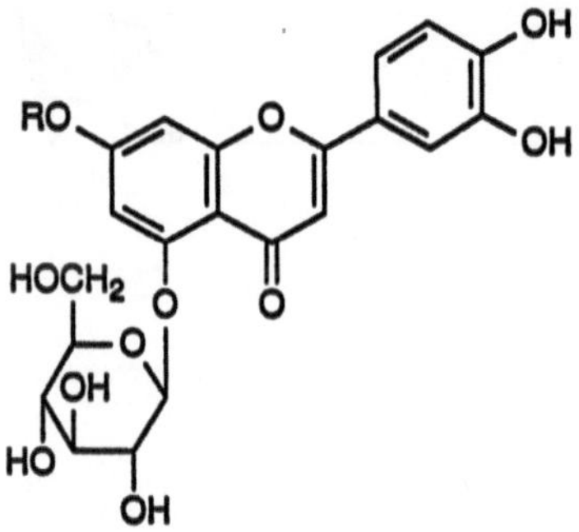

Galuteolin (*54-3*): R=H
Yuanhuanin (*54-4*): R=CH_3

A new spiro compound named genkwanol A (*54-5*) was isolated from the root of *D. genkwa* together with the known compounds daphnodorin B (*54-6*), umbelliferone, daphnin (*54-7*), daphnoretin (*54-8*), syringin, and yuankanin [7]. Daphnodorin B is a flavan derivative first isolated from *D. odora* [8, 9], and daphnoretin is a bicoumarin compound.

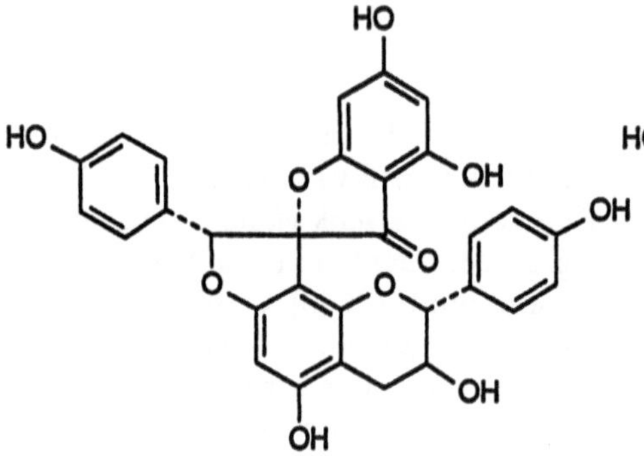

Genkwanol A (*54-5*)

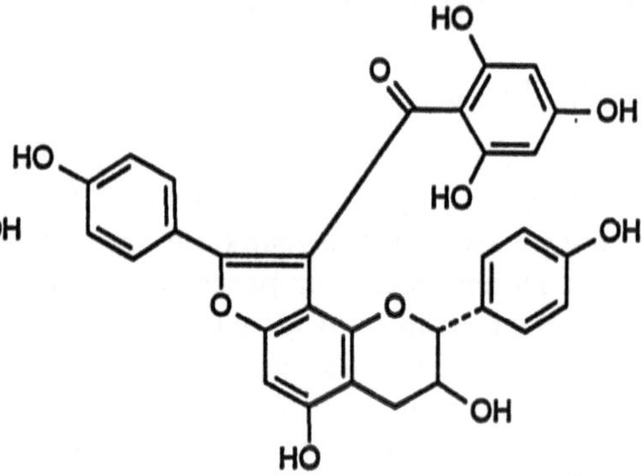

Daphnodorin B (*54-6*)

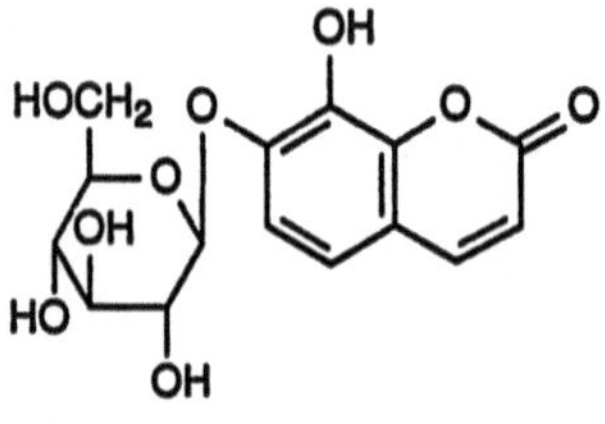

Daphnin (*54-7*)

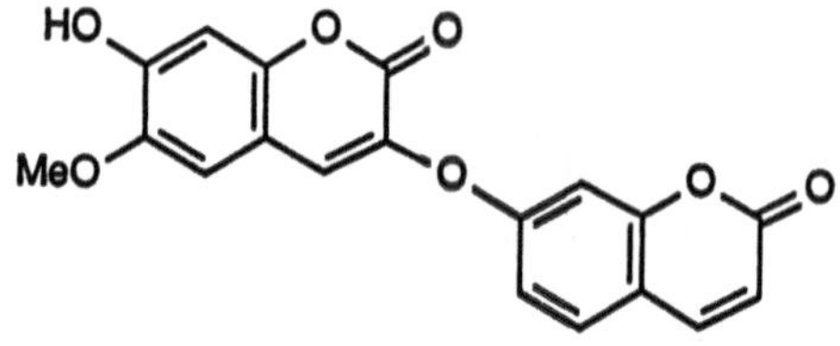

Daphnoretin (*54-8*)

54.2.2 Diterpene Orthoester

A series of diterpene orthoesters were isolated from the flowers of *D. genkwa*. Thus, the isolation and structure elucidation of genkwadaphnin (*54-9*) [10–12], yuanhuafin (*54-10*) [11, 12], yuanhuacin (*54-11*) [13], yuanhuapin (*54-12*) [14] and yuanhuatin (*54-13*) [15] were reported, whereas yuanhuacin [16] and yuanhuadin (*54-14*) [17, 18] were isolated from the roots. All these constituents are orthoesters structurally derived from daphnetoxin (*54-15*) or simplexin (*54-16*).

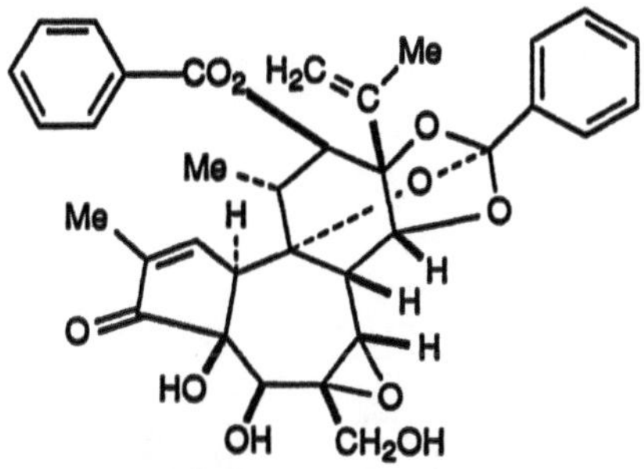

Genkwadaphnin (*54-9*)

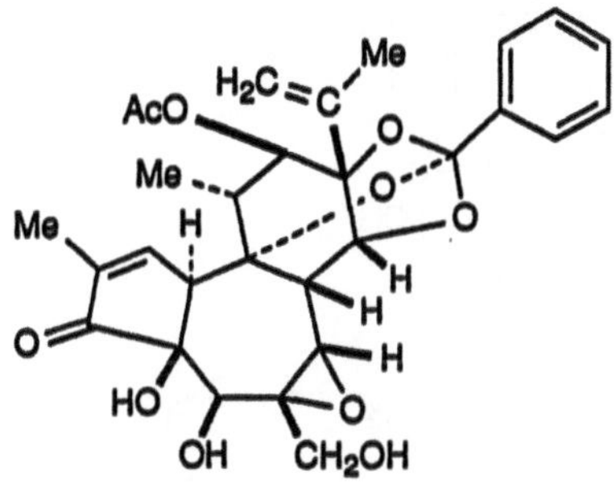

Yuanhuafin (*54-10*)

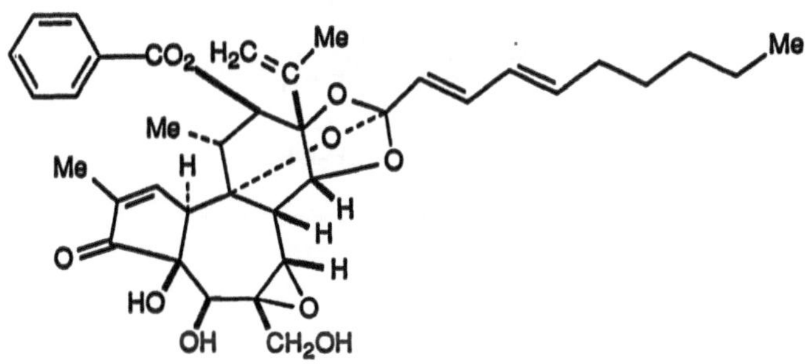

Yuanhuacin (*54-11*)

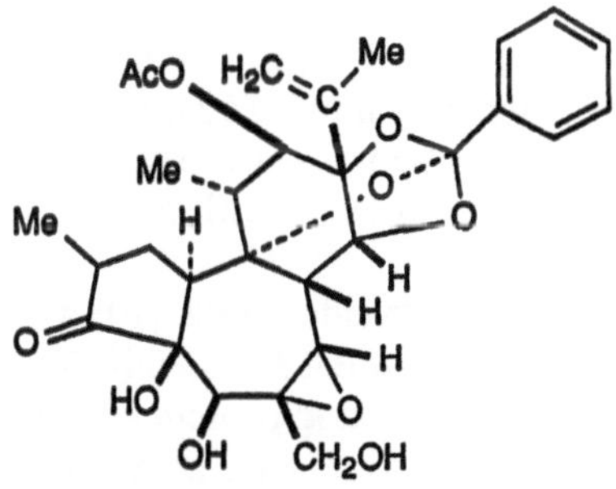

Yuanhuapin (*54-12*)

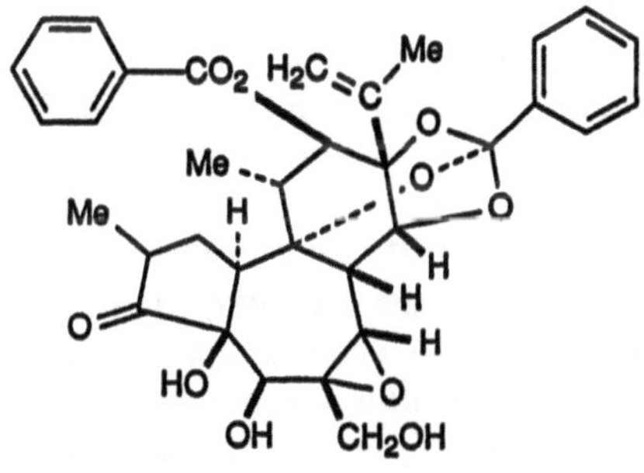

Yuanhuatin (*54-13*)

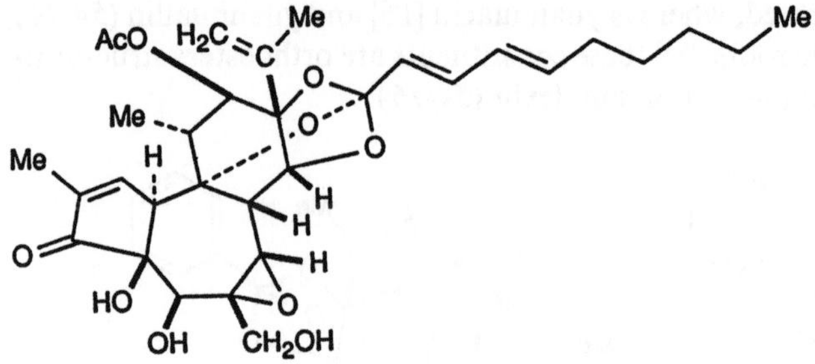

Yuanhuadin (*54-14*)

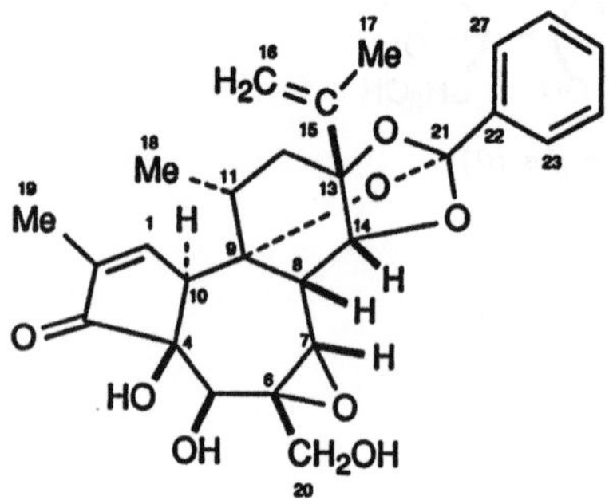

Daphnetoxin (*54-15*)

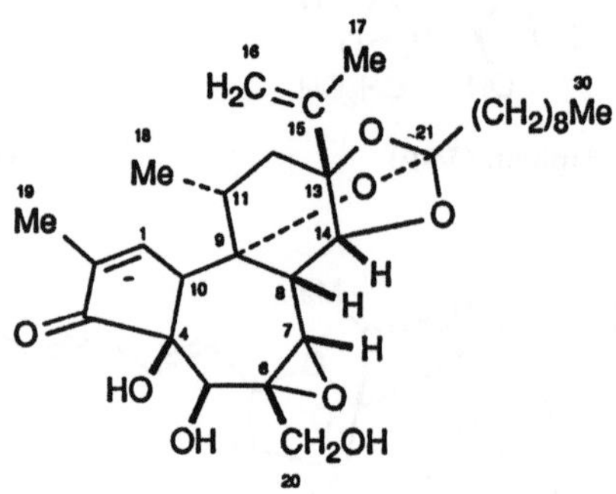

Simplexin (*54-16*)

54.3 Pharmacology

The flavones genkwanin, apigenin, yuankanin, and 3′-hydroxygenkwanin isolated from the flowers of *D. genkwa* were tested for antitussive activity. Genkwanin showed a greater antitussive activity than the other three flavones [2]. The flavones genkwanin, apigenin, luteolin, and luteolin-7-methyl ether also showed an inhibitory effect on xanthine oxidase [1]. Apigenin and luteolin showed particularly strong inhibitory activity. The modes of inhibition by apigenin and luteolin with respect to xanthine as substrate were of mixed type [1].

In contrast to the flavones, the diterpene orthoester genkwadaphnine [10] and yuanhuacin [19] showed significant antileukemic activity against P388 leukemia. The major effects of genkwadaphnin and yuanhuacin were on DNA and protein synthesis [20, 21]. Inhibitory effects on DNA synthesis in vitro were seen at a lower concentration than that required for protein synthesis inhibition. Targets affected in DNA synthesis were DNA polymerase on the one hand and de novo purine synthesis on the other. In the latter pathway, enzyme activities inhibited were phosphoribosyl aminotransferase, inosinic acid dehydrogenase and dihydrofolate reductase. In vivo administration of the diterpene orthoesters in mice bearing P388 leukemia at 0.8 mg/kg afforded identical types of effects on purine and DNA synthesis and in addition suppressed histone phosphorylation; this treatment reduced the number of surviving tumor cells. The in vivo effects on purine and DNA synthesis were evident as early as 6 and 24 h after administration of a single dose of the diterpene orthoesters [20]. The effects of genkwadaphnin and yuanhuacin on protein synthesis consisted in blocking of the elongation process and interference with the peptidyl transferase reaction. The latter reaction was suppressed at concentrations of the diterpene orthoesters which were commensurate with concentrations that inhibited whole cell in vitro protein synthesis in P388 cells [21]. In vitro DNA synthesis of mouse embryos was also decreased by yuanhuacin [22].

It is worth mentioning that a related compound, mezerein, was isolated from *D. mezereum*. Mezerein (*54-17*) is structurally related to genkwadaphnin, the only difference being in the substituent at C-12 of daphnetoxin; thus, mezerein possesses a phenylpentadienoyloxy group instead of the benzoyloxy group in genkwadaphnin. It was first isolated as an antileukemic active principle [23, 24]. On the other hand, mezerein is also structurally related to the phorbol esters known as tumor promotors such as TPA (tetradecanoylphorbolacetate, *54-18*) isolated from some *Croton* species (Euphorbiaceae) [25]. Mezerein showed weak tumor promotion activity and was classified as a "deactivated" promotor [26, 27].

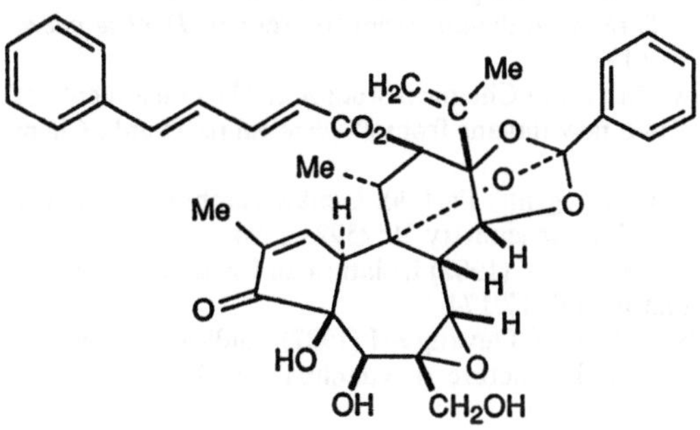

Mezerein (*54-17*)

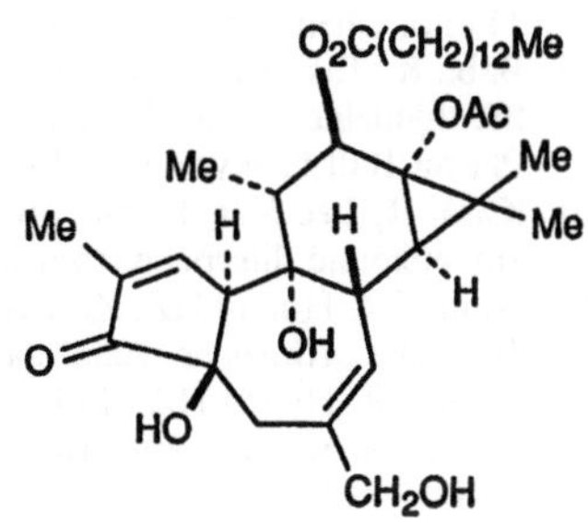

TPA (*54-18*)

Yuanhuafin, yuanhuatin, yuanhuadin, and yuanhuacin [18] all showed abortive activity. The dose of yuanhuafin inducing abortion in monkeys was 200–300 μg/animal [11, 12], whereas the minimal effective dose of yuanhuatin [14] and yuanhuadin [17, 18] inducing abortion in monkeys was 50 μg/animal. The LD_{50} of yuanhuatin was 3.0 mg/kg. Yuanhuafin caused severe irritation of the skin [12].

The contractions of rat and guinea pig uteri were increased by yuanhuacin both in vivo and in vitro. The effect was greater in pregnant than in nonpregnant uteri and greater in the uterus body than in the cervix. The rectal temperature of guinea pigs increased by 0.9 °C after intrauterine injections of yuanhuacin. The temperature of rabbits increased by 0.8 °C after intravenous administration of yuanhuacin and by 1.05 °C after yuanhuacin was administered subcutaneously (granulome pouch). Intragastric administration of acetylsalicylic acid to rabbits prevented the pyrogenic effect of yuanhuacin. Serum of rabbits with yuanhuacin-induced fever contained endogenous pyrogens which were destroyed in part by pepsin treatment [28, 29].

Intraamniotic, extraamniotic, or intravenous administration of yuanhuacin to pregnant women during mid-pregnancy induced labor. Inflammation and necrosis of the decidual membrane was found and, as a result, the synthesis and release of prostaglandins were markedly enhanced. These effects may cause abortion [30].

References

1. Noro T, Oda Y, Miyase T, Ueno A, Fukushima S (1983) Studies of enzyme inhibitors. II. Inhibitors of xanthine oxidase from the flowers and buds of *Daphne genkwa*. Chem Pharm Bull (Tokyo) 31:3984–3987
2. Li SF, Wang ZX (1983) Isolation and identification of Yuan Hua (*Daphne genkwa*) flavonoids. Chin Trad Herbal Drugs 14:392–394
3. Chen CL, Tseng KF (1965) The flavonoids in Chinese drugs. XI. The chemical composition of the root bark of *Daphne genkwa*. Acta Pharm Sin 12:119–122
4. Wang MT, Zhao TZ, Zhang ZW, Li CR (1985) Flavonoid glycosides from lilac daphne (*Daphne genkwa*) leaf. Chin Trad Herbal Drugs 16:98–100
5. Ji CR, Liu YZ, Feng WS, Wang MT, Zhao TZ (1986) Flavonoids in the Yuanhua leaf (*Daphne genkwa*). Chin Trad Herbal Drugs 17:487–489
6. Ji CR, Feng WS, Liu YZ, Zheng XK, Yuan WM (1986) Determination of hydroxygenkwanin in leaves of *Daphne genkwa*. Bull Chin Mat Med 11:427–428
7. Baba K, Takeuchi K, Tabata Y, Taniguchi M, Kozawa M (1987) Chemical studies on the constituents of the thymelaeaceous plants. IV. Structure of a new spiro biflavonoid, genkwanol A, from the root of *Daphne genkwa*. Sieb. et Zucc. Yakugaku Zasshi 107:525–529
8. Baba K, Takeuchi K, Hamasaki F (1985) Three new flavans from the root of *Daphne odora* Thunb. Chem Pharm Bull (Tokyo) 33:416–419
9. Baba K, Takeuchi K, Hamasaki F, Kozawa M (1986) Chemical studies on the constituents of the thymelaeaceous plants I. Structures of two new flavans from *Daphne odora* Thunb. Chem Pharm Bull (Tokyo) 34:595–602
10. Kasai R, Lee KH, Huang HC (1981) Antitumor agents. Part 40. Genkwadaphnine, a potent antileukemic diterpene from *Daphne genkwa*. Phytochemistry 20:2592–2594
11. Wang CR, Huang HZ, Xu RS, Dou YY, Wu XC, Li Y (1982) Isolation and structure of a new diterpene orthoester, yuanhuafine. Chin Pharm Bull 17:174
12. Wang CR, Huang HZ, Xu RS, Dou YY, Wu XC, Li Y, Ouyang SH (1982) Studies on the active principles of Yuan-Hua roots. III. Isolation and structure of yuanhuafine. Acta Chim Sin 40:835–839
13. Lin LW, Mi GT, Hou ZY, Shao RQ (1983) Studies on active principles in the flower of *Daphne genkwa*. Isolation and structure of yuanhuacin ester A (a brief report). Acta Acad Med Shandong 46–47
14. Hu BH, Sha H, Wang CR, Yu DF, Wu XC, Yu XG (1985) Antifertility constituent of the flower Yuan-Hua – isolation and structure of yuanhuatine. Acta Chim Sin 43:460–462
15. Sha H, He ZW, Wu XC (1986) Constituents of the Yuanhua's flower buds – isolation and structure of yuanhuapine. Acta Chim Sin 44:843–845
16. Ying BP, Wang CR, Chou PN, Pan PC, Liu JS (1977) Studies on the active principles of the root of Yuan-Hua (*Daphne genkwa*). I. Isolation and structure of yuanhuacine. Acta Chim Sin 35:103–108
17. Wang CT, Chen CH, Yin PP, Pan PC (1980) Studies on the active principle of the root of *Daphne genkwa*. II. Isolation and structure of a new antifertile diterpene orthoester, yuanhuadine. Chin Pharm Bull 15:39
18. Wang CR, Chen ZX, Ying BP, Zhou BN, Liu JS, Pan BC (1981) Studies on active principles in the root of Yuan-Hua (*Daphne genkwa*). II. Isolation and structure of a new antifertile diterpene yuanhuadine. Acta Chim Sin 39:421–426
19. Borris RP, Cordell GA (1984) Studies of the Thymelaeaceae. II. Antineoplastic principles of *Gnidia kraussiana*. J Nat Prod 47:270–278
20. Hall IH, Kasai R, Wu RY, Tagahara K, Lee KH (1982) Antitumor agents. LV. Effects of genkwadaphnine and yuanhuacine on nucleic acid synthesis of P388 lymphocytic leukemia cells. J Pharm Sci 71:1263–1267
21. Liou YF, Hall IH, Lee KH (1982) Antitumor agents. LVI. The protein synthesis inhibition by genkwadaphnin and yuanhuacine of P388 lymphocytic leukemia cells. J Pharm Sci 71:1340–1344
22. Yang HY, Li L (1984) Effects of three polypeptide hormones and yuanhuacine on DNA synthesis in preimplantation mouse embryos cultured in vitro. J Bethune Univ Med Sci 10:126–130

23. Lotter H, Jones A, Sturm M (1977) X-ray structure analysis of mezerein from *Daphne mezereum* L. Z Naturforsch [C] 32C:678–682
24. Kupchan SM, Baxter RL (1975) Mezerein. Antileukemic principle isolated from *Daphne mezereum*. Science 187:652–653
25. Hecker E (1981) Cocarcinogenesis and tumor promotors of the diterpene ester type as possible carcinogenic risk factors. J Cancer Res Clin Oncol 99:103–124
26. Marks F (1983) Stufen der Krebsentstehung – am Beispiel experimenteller Hauttumoren. In: Krebsforschung heute – Berichte aus dem Deutschen Krebsforschungszentrum. Steinkopff, Darmstadt, pp 52–57
27. Fürstenberger G, Berry DL, Sorg B, Marks F (1981) Skin tumor promotion by phorbol esters is a two-stage process. Proc Natl Acad Sci USA 78:7722–7726
28. Lu XR, Chen SM (1981) Uterine contraction and pyrogenic effect of yuanhuacine. Acta Pharmacol Sin 2:186–188
29. Wang WC, Shen SR (1988) Effects of yuanhuacin and yuanhuadin on in vitro contraction of rat uterus. Reprod Contracept 8:60–61
30. Yang BY, Lin ZM, Wang SX, Yang SZ (1981) Mechanism of action of yuanhuacine to induce labor during mid pregnancy. Natl Med J China 61:613–616

Datura metel L.

55

55.1 Introduction

Yangjinhua, Flos Daturae, is the dry flowers of *Datura metel* L. (Solanaceae) collected from April to November when the plants are blooming. It is officially listed in the Chinese Pharmacopoeia and is used as an antiasthmatic, spasmolytic, and anesthetic in surgical treatment. It should not be used for patients with glaucoma or hypertension.

55.2 Chemical Constituents

The major alkaloid constituents in the flower of *D. metel* are scopolamine (*55-1*) and hyoscyamine (*55-2*) [1]. The scopolamine content in the flowers was found to be 0.26% by thin-layer chromatography and densitometry [2]. The highest scopolamine contents in *D. metel* were 1.1% in branches and 0.6% in the leaves during the flowering period [3]. A new tropane alkaloid named datumetine (*55-3*) was isolated from the leaves of *D. metel* [4]. Atropine is the racemic mixture of L- and D-hyoscyamine.

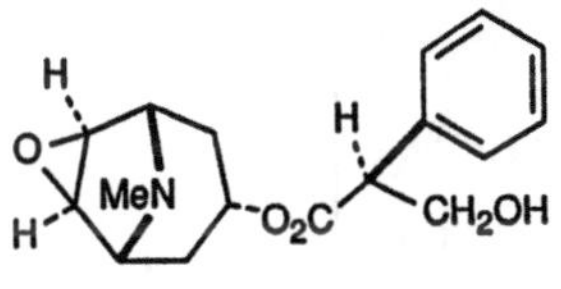

Scopolamine (*55-1*)

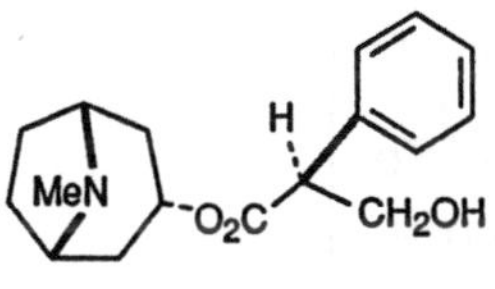

Hyoscyamine (*55-2*)

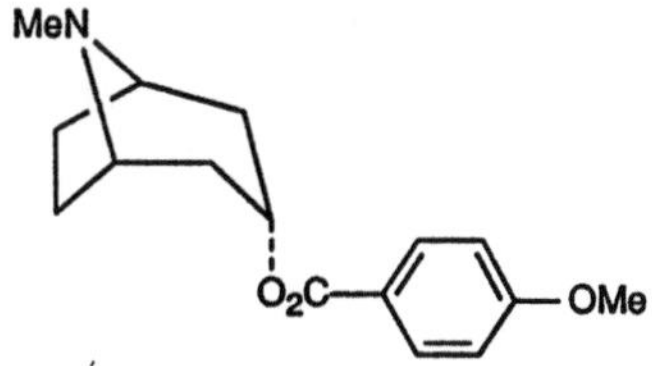

Datumetine (*55-3*)

The leaves of *Datura* species are rich with withanolides [5]. Withanolides are steroidal compounds of plant origin with a 22-hydroxy-ergostan-26-oic acid δ-lactone (withanolide, *55-4*) skeleton. Recently, a series of withanolides were isolated from the leaves of *D. metel* and structurally determined. They are datumetelin (*55-5*)

[6, 7], datumelin (*55-6*), daturilin (*55-7*) [9], daturilinol (*55-8*) [10], withametelin (*55-9*) [11], and the withanolide glycosides daturametelin A (*55-10*) and daturametelin B (*55-11*) [12].

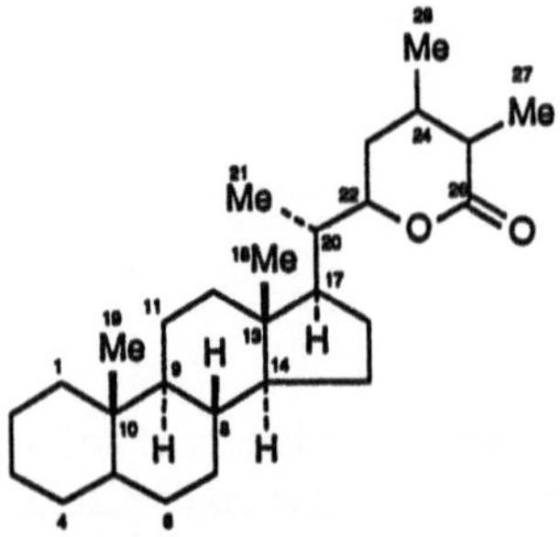

Withanolide (*55-4*)

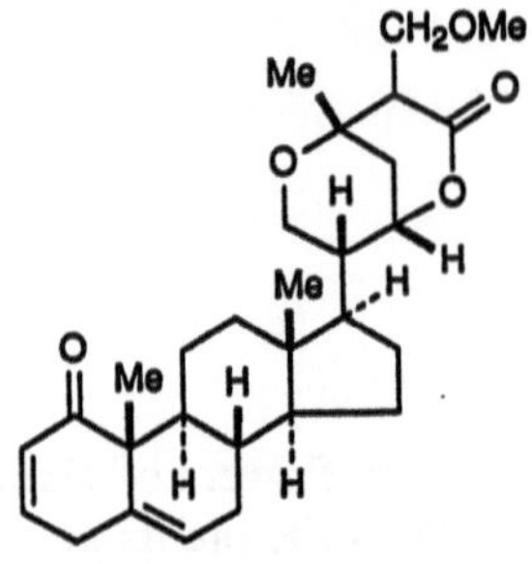

Datumetelin (*55-5*)

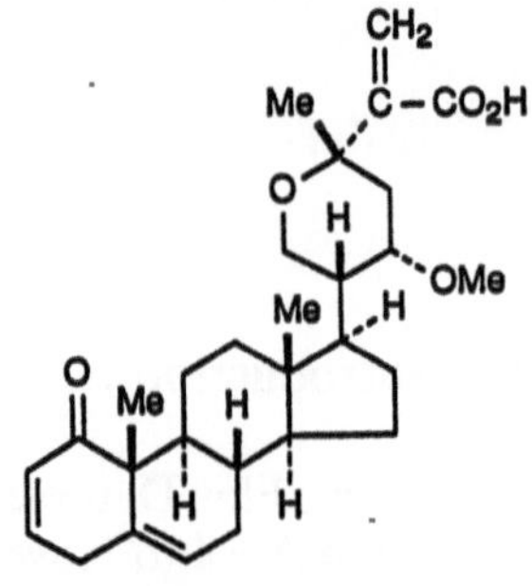

Datumelin (*55-6*)

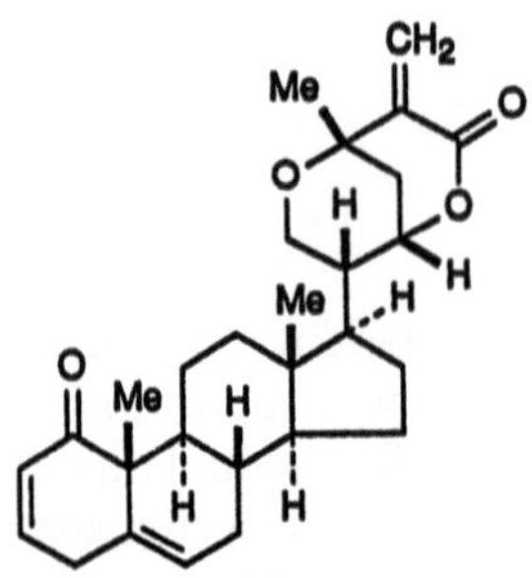

Daturilin (*55-7*)

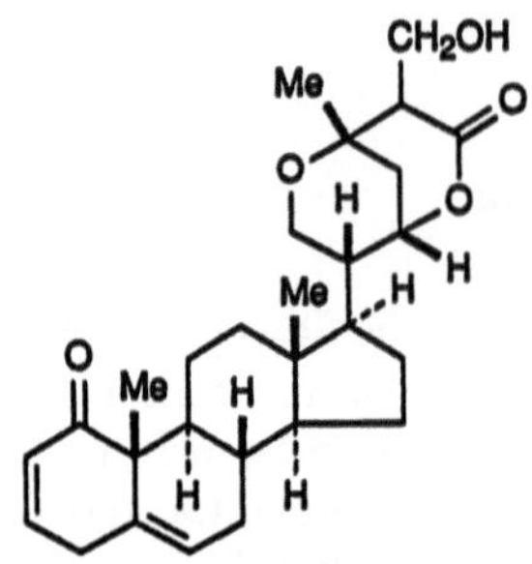

Daturilinol (*55-8*)

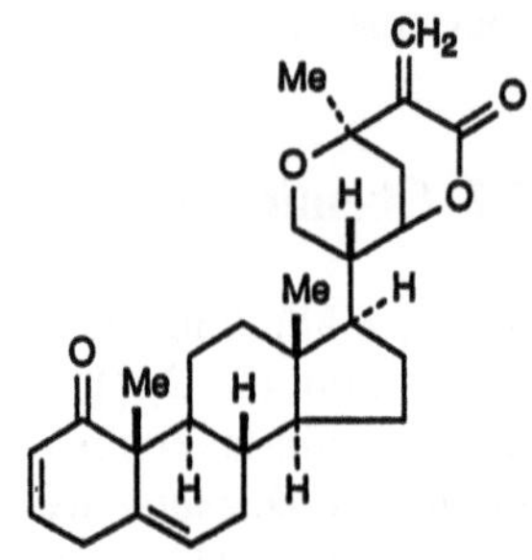

Withametelin (*55-9*)

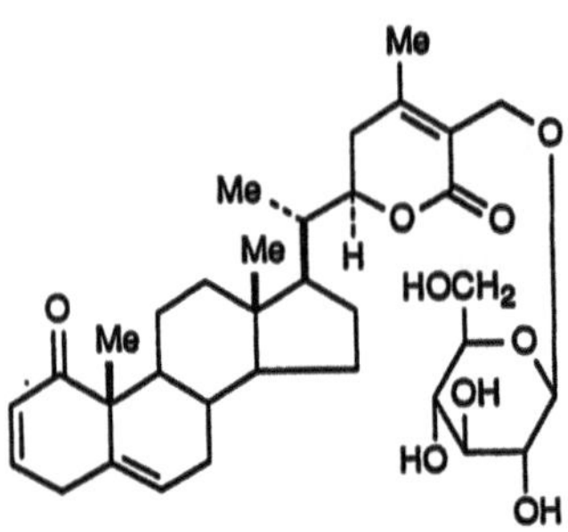

Daturametelin A (*55-10*)

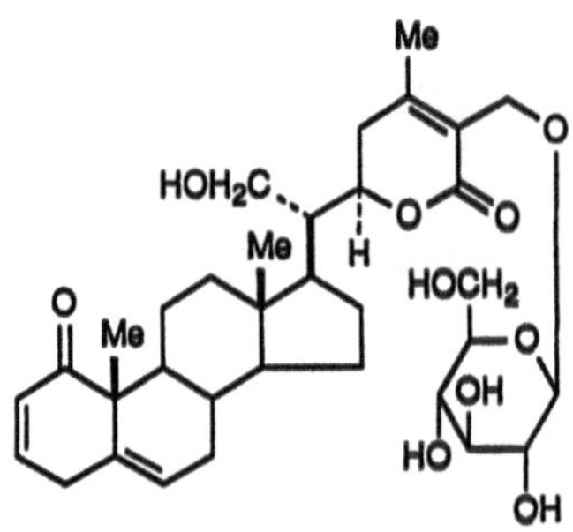

Daturametelin B (*55-11*)

55.3 Pharmacology

Atropine and scopolamine are well known as parasympatholytic agents which have long been used clinically for treatment of gastrointestinal spasm as a result of spastic gastritis or enteritis, ulcus ventriculi, hyperacidity, and for treatment of bronchial asthma and bradycardic arrhythmia. They can also be used for premedication in anesthesiology and as an antidote in the treatment of different intoxications. Atropine is used additionally in ophthalmology for diagnostic purposes because of its dilatational activity on the pupilla. In general, atropine is half as active as L-hyoscyamine; D-hyoscyamine exhibits only 1/10–1/20 the biological activity of the L-form [13, 14].

55.3.1 Pharmacological Actions on the Nervous System

Atropine inhibits the muscarinic activity of acetylcholine released at the parasympathic terminal. The following pharmacological actions appear with increasing doses: inhibition of salivary, bronchial, and sweat secretions; dilatation of the pupilla; inhibition of parasympathic control on the urinary bladder and the intestines with dysfunction in urinary excretion and decrease of intestinal tonus and motility; and inhibition of gastric motility and secretion. At high doses the nicotinic activity of acetylcholine can also be inhibited by atropine. All the atropine effects can be antagonized by cholinesterase inhibitors. Scopolamine differs from atropine mainly in the intensity of its action [15]. Recently, the muscarinic cholinergic antagonists, including atropine and scopolamine, were investigated in rat heart membrane preparations [16], isolated guinea pig atria and ileum [16, 17], and the structure-activity relations analyzed. All results confirmed the competitive antagonism of atropine and its derivatives at muscarinic receptors. Scopolamine given intracerebroventricularly to rabbits produced a loss of the righting reflex. The CNS-inhibiting effects of scopolamine were antagonized by adrenergic agonists [18]. Intravenous injection of scopolamine and atropine caused high electroencephalographic synchronization and blocked the electroencephalographic arousal response [19].

A series of atropine analogs were also compared for their receptor-binding affinities and antitremor and anticatalepsy potencies. The affinity constants correlated well with their antitremor activities [20] and the compounds with the highest partition coefficients possessed the greatest central selectivity [21].

55.3.2 Analgesic Effect

Scopolamine showed analgesic activity in the rat tail flick test when administered intraperitoneally or intraventricularly. Scopolamine produced a similar degree of analgesia in rats by intracerebroventricular injection of 1 mg/kg as by intraperitoneal injection of 25 or 50 mg/kg. The analgesia persisted for 32–36 h and showed a circadian rhythm, being more pronounced in the afternoon and early evening than at night and in the early morning [22]. Scopolamine administered intraperitoneally, intravenously, or intraventricularly to rabbits increased the pain threshold [23]. The analgesic activity of scopolamine was not affected by naloxane, was temporarily potentiated by L-dopa, and was antagonized by physostigmine, suggesting the involvement of the muscarinic cholinergic system [22].

Atropine and scopolamine are also used as anesthetics or as premedications in anesthesia [24]. The levels of serotonin and 5-hydroxyindolacetic acid in rat brain were increased during anesthesia induced by intracerebroventricularly administered scopolamine. The duration of anesthesia from scopolamine in mice was prolonged by L-tryptophan. Dopamine levels in midbrain were increased but no significant change in norepinephrine levels was observed [25]. Treatment of mice with L-dopa or of rabbits with hydroxydopamine prolonged subsequent anesthesia from scopolamine [26].

55.3.3 Cardiovascular Action

The most noted effect of atropine or scopolamine on the cardiovascular system is the change in heart rate. While tachycardia is the expected response this can be preceded by bradycardia [15, 27]. This biphasic response has been found to occur in spite of a steadily rising concentration of atropine in the blood [14, 28]. Following intravenous injection of scopolamine at a dose of 30 mg/kg to dogs, the rate of ventricular pressure increase, the mean arterial pressure, and the left ventricular systolic pressure all started to decrease within 1 min and reached a maximal decrease in 3–5 min, started to recover after 10 min, and were restored to pretreatment levels after 60 min [29]. Atropine and scopolamine at minimal intraperitoneal doses markedly increased the foodpad skin temperature of rats due to vasodilatory action. Both adrenaline and acetylcholine antagonized the vasodilatory action of atropine and scopolamine [30].

Combined use of atropine and scopolamine by intravenous administration had a greater protective effect than propranolol against myocardial infarction in rabbits induced by ligation of the left anterior descending coronary artery. Infarct size and height and number of elevated ST segment and pathologic Q waves in electrocardiogram were decreased and the changes in plasma cAMP and cGMP were inhibited [31]. Scopolamine prolonged the duration of the action potential and the postrepolarization refractory period in isolated guinea pig myocardial cells. It also decreased the spontaneous contraction rate of the cells [32].

In hemorrhagic shocked rabbits treated with scopolamine, the glomerular and renal tubule capillaries were dilated and the proximal tubule epithelia appeared nearly normal, whereas the shocked rabbits without treatment showed constriction and occlusion of glomerular and renal tubule capillaries, swelling of epithelial mitochondria, and disappearance of membrane organization in renal tissue [33].

55.3.4 Action on Respiratory Tract

Scopolamine given intravenously in anesthetized rabbits led to an abrupt increase in the pulmonary vascular resistance, but with a mild drop in the femoral arterial pressure, which maintained a more constant value. Scopolamine, given during a 20-min period of hypoxic gas inhalation, prevented the posthypoxia pulmonary hypotension [34].

Inhibition of bronchoconstrictor responses to inhaled acetylcholine and to acetylcholine released by electric stimulation of the vagus nerves in anesthetized dogs [35] was similar after intravenous scopolamine or inhaled atropine. Atropine inhalation also resulted in a marked decrease in sensitivity against histamine inhalation in monkeys [36]. Bronchodilation was also produced in normal subjects by inhalation of atropine. The specific airway conductance and forced expiratory flows were both significantly increased by atropine inhalation [37]. In anesthetized guinea pig with bronchoconstriction induced by histamine, bradykinin, and leukotriene C_4, atropine prevented both the increased airway resistance and release of thromboxane A_2-like substances in blood [38].

55.3.5 Antisecretory Activity

Atropine blocked equally effectively the basal and the stimulated acid secretion in conscious gastric fistula rats [39]. Intestinal secretion and hypermotility of the rat jejunum in vivo induced by bethanechol were inhibited by atropine [40]. Acetylcholine-stimulated pepsin output by the rat isolated whole stomach preparation was inhibited by atropine. This dose-related inhibition of the pepsin output was completely reversed by increasing the concentration of acetylcholine, indicating that the inhibition was mediated by muscarinic receptors [41].

Intraluminal perfusion of pig jejunum with *Escherichia coli* heat-stable enterotoxin reversed net absorption of water and electrolytes to net secretion. Addition of atropine to the perfusate reduced the secretory response to enterotoxin and enhanced Na^+ and Cl^- absorption in control segments [42]. Apparently, blockade of cholinergically mediated secretion in the small intestine attenuates the enterosorptive effects of heat-stable enterotoxin and atropine may be useful therapeutically in the treatment of secretory diarrhea. Atropine given orally to rats and mice reduced the incidence, number, size and severity of duodenal ulcers in both species induced by cysteamine [43].

55.3.6 Antidotal Action

Atropine and scopolamine are used clinically as antidotes, especially against organophosphorus anticholinesterase compounds. The lethality of organophosphorus compounds methylphosphonofluoridic acid l-methylethyl ester (sarin) and methyl-*S*-[2-[bis(1-methylethyl)amino]ethyl]phosphonothioic acid *O*-ethyl ester (VX) was diminished when rats were pretreated with atropine. Pretreated animals receiving sarin showed significant recovery of morphological and functional properties of the neuromuscular junction compared with the damage of structures in animals without pretreatment. Blood cholinesterase inhibition was slightly decreased whereas brain and muscle acetylcholinesterase levels recovered significantly with pretreatment [44]. Poisoning with cyolane, 1,3-dithiolan-2-ylidene-phosphoramidic acid diethylester, in experimental animals, can also be treated with atropine when given immediately after the intoxication [45].

55.3.7 Toxicity

The toxicity of atropine and scopolamine in experimental animals is species dependent. Cats, dogs, and birds are very sensitive to these alkaloids compared with goats, sheep, and rabbits [46, 47]. The LD_{50} values of atropine in rats and mice with oral administration are 622 and 400 mg/kg, respectively, and the minimal lethal dose in rabbits is 1450 mg/kg. The LD_{50} values of scopolamine in mice with intravenous or subcutaneous administration are 163 and 1700 mg/kg, respectively [48]. Twenty-five to 50% of all rabbits possess an atropinesterase, which hydrolyzes L-hyoscyamine to tropic acid and tropine. Scopolamine is also a substrate for this enzyme. Atropinesterase occurs in serum and in almost all organs, and the liver shows the highest content [49–51].

The toxicity of atropine and scopolamine in humans shows great individual differences. A quantity of 1–10 mg atropine can be fatal for children and the minimal lethal dose for adults is 100 mg [15]. However, survival of children after intoxication with 400–600 mg atropine and survival of adults after intoxication with 1000 mg atropine was reported [15, 52].

55.3.8 Pharmacokinetics

Atropine and scopolamine administered orally are rapidly absorbed from the intestinal tract but not from the stomach [53]. The poor absorption through the mucous membranes of the stomach is because the drugs are almost completely ionized in the acidic gastric contents [54]. In the circulation, up to 50% of the atropine administered is bound to the plasma proteins and the plasma half-life of atropine is 2.5 h [52]. The main route of excretion is the urine [28]. The kinetics of elimination of atropine in human subjects were found to be first order and there was evidence that the kinetics of distribution of the drug were dose dependent [55]. After administration of [^{3}H]atropine sulfate to a normal volunteer, noratropine (24%), atropine *N* oxide (15%), topine (2%), and tropic acid (3%) appear to be the major urinary metabolites, while 50% of the administered dose is excreted as apparently unchanged atropine. No conjugates were detected and the presence of D-hyoscyamine suggested the occurrence of stereoselective metabolism [56].

The mean serum concentrations of inhaled atropine in healthy subjects were comparable to those with intramuscular administration. The concentrations increased as the inhaled dose increased. The observed bronchodilating, anticholinergic, and other pharmacological effects were seen after all dose concentrations and were typical of atropine [57].

Scopolamine was rapidly and completely absorbed from the rat intestine. After intravenous injection scopolamine showed biphasic half-lives of 11 and 95 min. The highest levels of scopolamine were found in lung. In the brain, the highest levels of scopolamine were found in the striatum, cerebral cortex, and hippocampus. After intravenous administration of [^{3}H]scopolamine, 62% of the dose was excreted in urine and 25% in feces within 48 h [58].

In addition, atropine and scopolamine can cross the placenta and affect the fetus. A number of studies have shown that placental transfer is sufficiently rapid to cause changes of the fetal pulse within 5 min after an intravenous injection into the mother [59]. The rapid transfer of atropine across the placenta has also been confirmed by studies with [^{3}H]atropine in both early [60] and late pregnancy [61].

References

1. Ye CY, Zhang SX (1981) Gas chromatographic assay of scopolamine and hyoscyamine in *Datura metel.* Chin Trad Herbal Drugs 12:493–498
2. He LY (1982) TLC separation and densitometric determination of topine alkaloids. Chin Trad Herbal Drugs 13:13–16
3. Yasuda K, Nishijima M, Saito K, Kamimura H, Ibe A, Nagayama T, Ushiyama H, Naoi Y, Tanaka K, Yoshizawa M (1981) Seasonal variation of alkaloid contents in *Datura* and the effect of cooking. Shokuhin Eiseigaku Zasshi 22:397–403 (CA 96:33562v)
4. Siddiqui S, Sultana N, Ahmed SS, Haider SI (1986) Isolation and structure of a new alkaloid datumetine from the leaves of *Datura metel.* J Nat Prod 49:511–513
5. Christen P (1989) Withanolide – Naturstoffe mit vielversprechendem Wirkungsspektrum. Pharm Unserer Zeit 18:129–139

6. Mahmood T, Ahmad SS, Fazal A (1988) A new withanolide datumetelin from the leaves of *Datura metel* L. J Indian Chem Soc 65:526–527
7. Mahmood T, Ahmad SS, Fazal A (1988) A new withanolide, datumetelin, from the leaves of *Datura metel*. Planta Med 54:468–469
8. Siddiqui S, Ahmad SS, Mahmood T (1987) Datumelin – a new withanolide from *Datura metel* L. Pak J Sci Ind Res 30:567–568
9. Siddiqui S, Sultana N, Ahmad SS, Haider SI (1987) A novel withanolide from *Datura metel*. Phytochemistry 26:2641–2643
10. Mahmood T, Ahmad SS, Siddiqui S (1988) Daturilinol. A new withanolide from the leaves of *Datura metel*. Heterocycles 27:101–103
11. Oshima Y, Bagchi A, Hikino H, Sinha SC, Sahai M, Ray AB (1987) Steroids. Part 35: C_{28}-steroidal lactone. Part 16: Withametelin, a hexacyclic withanolide of *Datura metel*. Tetrahedron Lett 28:2025–2028
12. Shingu K, Kajimoto T, Furusawa Y, Nahara T (1987) Studies on the constituents of solanaceous plants. Part 9. The structures of daturametelin A and B. Chem Pharm Bull (Tokyo) 35:4359–4361
13. Greeff K, Wirth KE (1987) Pharmakotherapeutische Anwendungen von Parasympathomimetika, Parasympatholytika and peripheren Muskelrelaxantien. In: Forth W, Henschler D, Rummel W (eds) Pharmakologie und Toxikologie, 5th edn. Wissenschaftsverlag, Mannheim, pp 103–121
14. Seeger R (1986) Hyoscyamine – atropine. Hyoscine (scopolamine). Dtsch Apoth Ztg 126:1930–1934
15. Goodman LS, Gilman A (1985) In: Goodman LS, Rall TW, Murad F (eds) The Pharmacological Basis of Therapeutics.7th edn. MacMillan, New York, pp 130–144
16. Shen SY, Xu J, Jin GZ (1985) Effect of 13 cholinergic antagonists on M-cholinergic receptors in rat hearts. Acta Pharmacol Sin 6:158–161
17. Dai DZ, Lin J, Wang LY, Zhang DL, Huang J, Xing WF (1986) The M-receptor antagonism by a new cholinolytic compound TBBB in comparison with hyoscine butyl bromide (HBB) and atropine. Acta Pharm Sin 21:853–856
18. Bian CF, Duan SM (1981) Relation between central inhibitory effect of scopolamine and its adrenergic antagonistic activity. Acta Pharmacol Sin 2:78–81
19. Peng JZ, Jin LR, Chen XY, Chen ZX (1983) Central effects of anisodamine, atropine, anisodine and scopolamine after intraventricular injection. Acta Pharmacol Sin 4:81–87
20. Xu JA, Jin GZ, Yu LP, Li JH, Yu AY (1983) Central antimuscarinic cholinergic effects of scopolamine and other cholinergic antagonists. Acta Pharmacol Sin 4:156–162
21. Niu XY, Ren ZH (1984) Structure-activity relations of interaction between atropine derivatives and muscarinic receptors. Acta Pharm Sin 19:326–332
22. Hong GX, Li JH, Jin GZ (1984) Analgesic effect of scopolamine. Acta Physiol Sin 36:149–156
23. Bian CF, Xing SH, Jin SJ, Zhang JF (1979) Effect of scopolamine on pain and analgesic activity. Acta Pharm Sin 14:397–401
24. Shutt LE, Bowes JB (1979) Atropine and hyoscine. Anaesthesia 34:476–490
25. Dai DZ, Ma JR, Wang YZ, Zhang H (1983) Scopolamine induced anesthesia and monamine transmitters in the brain. J Nanjing Coll Pharm 14:1–6
26. Dai DZ, Li JY, Yang PB (1984) The dopaminergic system in relation to scopolamine-induced anesthesia. J Nanjung Coll Pharm 15:53–56
27. Grainger SL, Smith SE (1983) Dose-response relationships of intravenous hyoscine butylbromide and atropine sulfate on heart rate in healthy volunteers. Br J Clin Pharmacol 16:623–626
28. Kalser SC (1971) The fate of atropine in man. Ann NY Acad Sci 179:667
29. Wang ZM, Peng BL, Wan QS, Shen YT (1983) Effects of scopolamine on the cardiac contractility in dogs. Chin J Anesth 3:138–140
30. Wang WJ (1985) Vasodilatory effects of four atropine-like drugs on rat footpad. Acta Pharmacol Sin 6:26–29
31. Yang GD, Miao JL, Liu MX, Zheng Y, Wang ZL, Ye MZ, Shen JH (1987) Effects of scopolamine plus atropine against experimental myocardial infarction in rabbits. Acta Pharm Sin 8:128–131
32. Li ZY (1984) Effect of scopolamine on the action potential of guinea pig myocardial cells. Acta Pharmacol Sin 5:170–173
33. Yang ML, Shi Y (1983) Effects of anesthesia with scopolamine on the kidney of rabbits in hemorrhagic shock. An optical and electron microscopic study. Chin J Anesth 3:136–137

34. Fan YL, Yang L, Wang XQ, Sun HL, Sun YW, Hu F, Zhou WQ, Xiao YZ (1986) Effects of anisodamine and scopolamine on pulmonary circulation in rabbits. Acta Pharmacol Sin 7:117–121
35. Holtzman MJ, McNamara MP, Sheppard D, Fabbri LM, Hahn HL, Graf PD, Nadel JA (1983) Intravenous versus inhaled atropine for inhibiting bronchoconstrictor responses in dogs. J Appl Physiol 54:134–139
36. Pare PD, Nicholls I (1982) Bronchial response to histamine after inhaled propranolol and atropine in monkeys. J Allergy Clin Immunol 69:213–220
37. Gal TJ, Suratt PM, Lu JY (1984) Glycopyrrolate and atropine inhalation: comparative effects on normal airway function. Am Rev Respir Dis 129:871–873
38. Folco G, Omini C, Rossoni G, Vigano T, Berti F (1982) Anticholinergic agents prevent guinea pig airway constriction induced by histamine, bradykinin and leukotriene C_4: relationship to circulating TXA_2. Eur J Pharmacol 78:159–165
39. Ekelund M, Haakanson R, Vallgren S (1987) Effects of cimetidine, atropine and pirenzepine on basal and stimulated gastric acid secretion in the rat. Eur J Pharmacol 138:225–232
40. Greenwood B, Read NW, Hardcastle PT, Hardcastle J (1984) Effect of secoverine and atropine on intestinal secretion and motor activity in the rat small intestine in vivo. J Pharm Pharmacol 36:100–106
41. Parsons ME, Price C, Wintermeyer D (1983) Inhibition of acetylcholine-stimulated secretion in the isolated whole stomach of the rat. J Pharm Pharmacol 35:104–109
42. Ahrens FA, Zhu BL (1982) Effects of indomethacin, acetazolamide, ethacrynate sodium and atropine on intestinal secretion mediated by *Escherichia coli* heat-stable enterotoxin pig jejunum. Can J Physiol Pharmacol 60:1281–1286
43. Smith JA (1983) The effect of atropine, cimetidine and FPL 52694 on duodenal ulcers in mice. Eur J Pharmacol 88:215–221
44. Albuquerque EX, Deshpande SS, Kawabuchi M, Aracava Y, Idriss M, Rickett DL, Boyne AF (1985) Multiple actions of anticholinesterase agents on chemosensitive synapses: molecular basis for prophylaxis and treatment of organophosphate poisoning. Fundam Appl Toxicol 5 [6,Pt2]:182–203
45. Shen SY, Zheng HX, Deng H, Peng SL, Liu ZY, Huang H, Zhang LZ (1983) Toxicity of cyolane. Chin J Prev Med 17:216–218
46. Korb H (1960) Über Atropinvergiftungen. Münch Med Wochenschr 39:1858–1860
47. Albanus L, Sundwall A, Vangbo B, Winbladh B (1968) The fate of atropine in the dog. Acta Pharmacol Toxicol 26:571–582
48. Virtanen K, Kanto J, Iisalo E (1980) Radioimmunoassay for atropine and L-hyoscyamine. Acta Pharmacol Toxicol 47:208–212
49. Stormont C, Suzuki Y (1970) Atropinesterase and cocainesterase of rabbit serum: localisation of the enzyme activity in isoenzymes. Science 167:200–202
50. Herz A (1963) Über die Abhängigkeit der zentralen und peripheren Wirkung der Belladonnaalkaloide; vom Vorkommen einer Hyoscyaminesterase bei einem Teil der Kaninchen. Arch Int Pharmacodyn 141:595–604
51. Cauthen SE, Ellis RD, Larrison SB, Kidd MR (1976) Resolution, purification and characterization of rabbit serum atropinesterase and cocainesterase. Biochem Pharmacol 25:181–185
52. Arthurs GJ, Davies R (1980) Atropine – a safe drug. Anaesthesia 35:1077–1079
53. Beerman B, Hellström K, Rosen A (1971) The gastric-intestinal absorption of atropine in man. Clin Sci 40:95–106
54. Travell J (1940) The influence of the hydrogen ion concentration on the absorption of alkaloids from the stomach. J Pharmacol Exp Ther 69:21
55. Hinderling PH, Gundert-Remy U, Schmidlin O (1985) Integrated pharmacokinetics and pharmacodynamics of atropine in healthy humans. I. Pharmacokinetics. J Pharm Sci 74:703–710
56. Meer MJ, van der Hundt HKL, Muller FO (1986) The metabolism of atropine in man. J Pharm Pharmacol 38:781–784
57. Harrison LI, Smallridge RC, Lasseter KC, Goldlust MB, Shamblen EC, Gam VW, Chang SF, Kvam DC (1986) Comparative absorption of inhaled and intramuscularly administered atropine. Am Rev Respir Dis 134:254–257
58. Yue TL, Wang GF, Song ZY (1979) Metabolism of ^{3}H-scopolamine and ^{3}H-anisodamine. Acta Pharm Sin 14:208–217

59. John AH (1965) Placental transfer of atropine and the effect on foetal heart rate. Br J Anaesth 37:57
60. Kivalo I, Saarikosi S (1970) Quantitative measurements of placental transfer and distribution of radioactive atropine in fetus. Ann Chir Gynaecol Fenn 59:80
61. Kivalo I, Saarikosi S (1977) Placental transmission of atropine at full term pregnancy. Br J Anaesth 49:1017

Daucus carota L. 56

56.1 Introduction

Nanheshi, Fructus Carotae, is the dry ripe fruits of *Daucus carota* L. (Apiaceae) collected in the fall when the fruits have become ripe. It is officially listed in the Chinese Pharmacopoeia and used in traditional Chinese medicine mainly as an anthelmintic.

56.2 Chemical Constituents

From the essential oil of carrot seed some terpenes were isolated and identified. They are α-thujene, α-pinene, camphene, β-pinene, β-phellandrene, limonene, β-caryophyllene, β-bisabolene (*56-1*), geranyl acetate, terpinyl acetate, bornyl acetate, carotol (*56-2*), daucol (*56-3*), and a mixture of coumarins with tarry materials. The major components were carotol, daucol, and terpinyl acetate with contents of 36%, 13%, and 18.6%, respectively [1]. The presence of esters, particularly terpinyl acetate, is interesting. The isolation of a new sesquiterpene, carota-1,4-β-oxide (*56-4*), from the seeds of *D. carota* was reported [2].

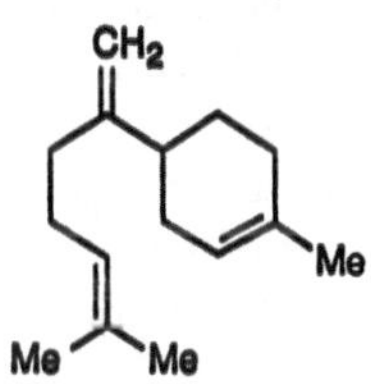

β-Bisabolene (*56-1*)

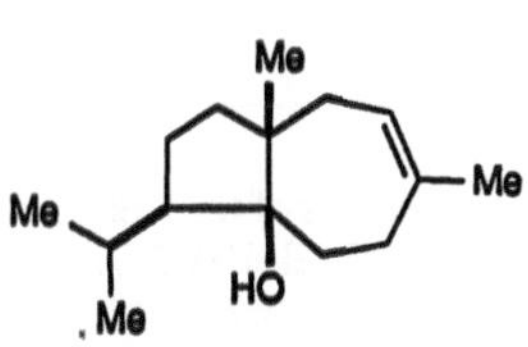

Carotol (*56-2*)

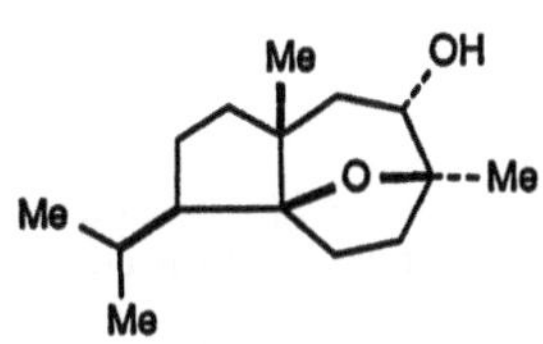

Daucol (*56-3*)

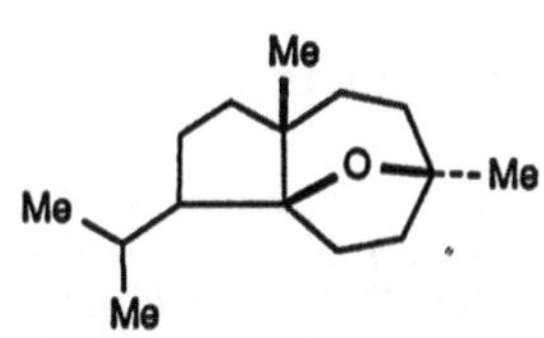

Carota-1,4-β-oxide (*56-4*)

The flavone glycosides apigenin-4′-*O*-β-D-glucoside, kaempferol-3-*O*-β-D-glucoside, and apigenin-7-*O*-β-D-galactopyranosyl-(1→4)-β-D-mannopyranoside were also isolated from carrot seeds [3].

56.3 Pharmacology

Administration of an alcoholic extract of carrot seeds at doses of 50, 150, and 500 mg/kg either alone or in combination with estradiol dipropionate on alternate days for 42 days to ovariectomized rats showed a definite but relatively mild estrogenic activity on the basis of alterations in the reproductive organ weight and histology and chemistry of the uterus [4]. Ovarian compensatory hypertrophy in the remaining ovary of rats following unilateral ovariectomy was prevented by administration of carrot seed extract, which was comparable to daily administration of 10 μg estradiol [5].

Administration of carrot seed oil to mice inhibited fertility and caused abortions. One of the active components was separated by column chromatography and identified as β-bisabolene [6]. The antifertility effect of the carrot seed oil may be related to the reduction of the plasma level of progesterone and to the inhibition of decidual reaction. The vaginal cornification test of ovariectomized mice demonstrated that the effective fraction had neither estrogenic nor antiestrogenic activity. In isolated early pregnant rat uteri contractions were inhibited by a high concentration of this effective fraction [7].

The essential oil of *D. carota* seeds showed an antibacterial effect against *Bacillus subtilis* and *Salmonella typhimurium* at concentrations of 0.2%, but was totally inactive against *Escherichia coli*, *Klebsiella pneumoniae*, and *Salmonella stanley* [8].

Three milliliters of a 2% aqueous emulsion of the seed oil of *D. carota* produced a transient fall in arterial blood pressure in the anesthetized dog without influencing respiration, but higher doses (3–5 ml of a 4% suspension) produced a persistent hypotension also causing depression of respiration. The oil was found to exert a direct cardiac depressant action on frog heart and dog heart in the open chest preparation. The oil showed no analgesic effect in rats, but exerted a marked CNS depressant action in rats. The oil offered moderate protection to frogs from the convulsant effect of strychinine. In isolated intenstine of rat and rabbit, as well as on the isolated uterus of rat, the aqueous emulsion of carrot seed oil produced a relaxant effect. In these preparations, as well as in the isolated skeletal muscle of frog, the contractions induced by acetylcholine were also decreased by this emulsion [9].

References

1. Ashraf M, Zaidi SA, Mahmood S, Bhatty MK (1979) Studies on the essential oils of the Pakistani species of the family Umbelliferae. Part 31. Wild *Dauces carota* (carrot) seed oil. Pak J Sci Ind Res 22:258–259 (CA 92:90947r)
2. Dhillon RS, Gantam VK, Kalst PS, Chhabra BR (1989) Carota-1,4-β-oxide, a sesquiterpene from *Daucus carota*. Phytochemistry 28:639–640
3. Gupta KR, Niranjan GS (1982) A new flavone glycoside from seeds of *Daucus carota*. Planta Med 46:240–241
4. Kant A, Jacob D, Lohiya NK (1986) The estrogenic efficacy of carrot (*Daucus carota*) seeds. J Adv Zool 7:36–41
5. Kaliwal BB, Rao MA (1981) Inhibition of ovarian compensatory hypertrophy by carrot seed (*Daucus carota*) extract or 17 β-estradiol in hemicastrated albino rats. Indian J Exp Biol 19:1058–1060

6. Tung CY, Hsu LC, Chu HW, Chou Y (1981) Studies on the antifertility constituents in carrot seeds (*Daucus carota* L.). Chin Trad Herbal Drugs 12:13
7. Chu YH, Zhou MH, Li QA, Bao YM (1985) Antifertility effect of volatile oil from *Daucus carota* seeds. Reprod Contracept 5:37–40
8. Grover GS, Rao JT (1978) In vitro antimicrobial studies of the essential oil of *Daucus carota*. Indian Drugs Pharm Ind 13:39–40 (CA 89:100872s)
9. Bhargava AK, Ali SM, Chauhan CS (1967) Pharmacological investigation of the essential oil of *Daucus carota* var. *sativa*. Indian J Pharm 29:127–129

Dendrobium nobile Lindl. 57

57.1 Introduction

Shihu, Herba Dendrobii, is the dry stems of *Dendrobium loddigesii* Rolfe., *D. fimbriatum* Hook. var. *oculatum* Hook., *D. chrysanthum* Wall., *D. candidum* Wall. ex Lindl. or *D. nobile* Lindl. (Orchidaceae) and is officially listed in the Chinese Pharmacopoeia. It can be collected throughout the year. *Dendrobium* stems are used in traditional Chinese medicine as a tonic to improve the digestive function after recovery from ailments.

57.2 Chemical Constituents

The main constituents obtained from *Dendrobium* species are sesquiterpene alkaloids [1]. A number of sesquiterpene alkaloids were isolated from *D. nobile*. Dendrobine (*57-2*), one of the major alkaloid components, was the first to be isolated [2] and structurally determined [3, 4]. The sesquiterpene skeleton for dendrobine and most of the *Dendrobium* alkaloids was designated as dendrobane (*57-1*). Then other alkaloids, nobiline (*57-3*) [5, 6], dendroxine (*57-4*) [7], dendramine (6-hydroxydendrobine, *57-5*) [8, 9], dendrine (*57-6*) [10], 8-hydroxydendroxine (*57-7*) [11], 3-hydroxy-2-oxodendrobine (*57-8*) [12], and 6-hydroxydendroxine (*57-9*) [9] were successively isolated and structurally elucidated.

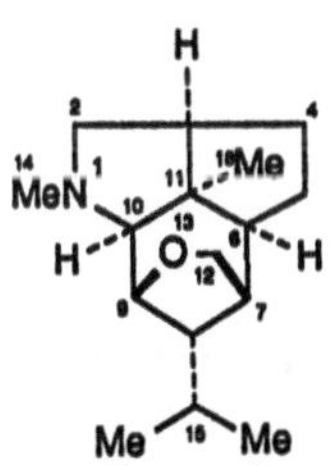

Dendrobane (*57-1*)

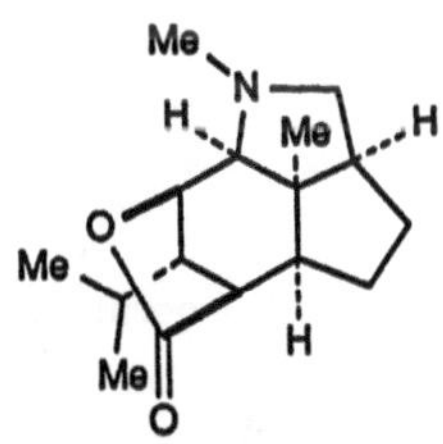

Dendrobine (*57-2*)

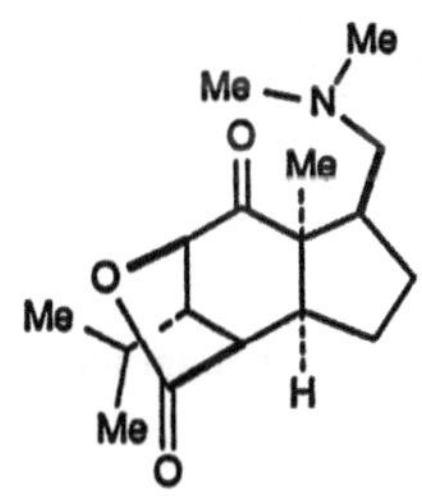

Nobiline (*57-3*)

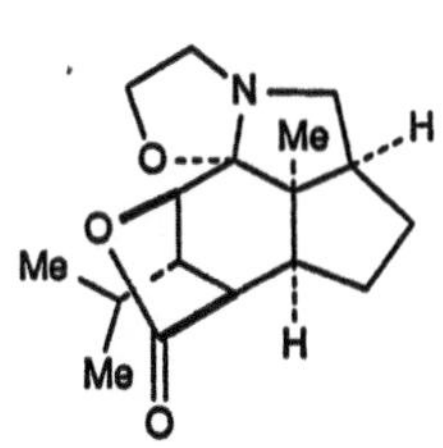

Dendroxine (*57-4*)

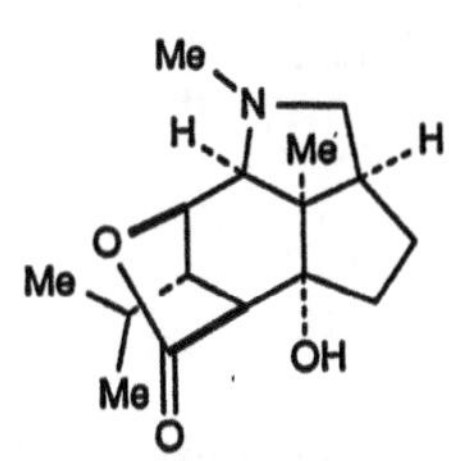

Dendramine (*57-5*)

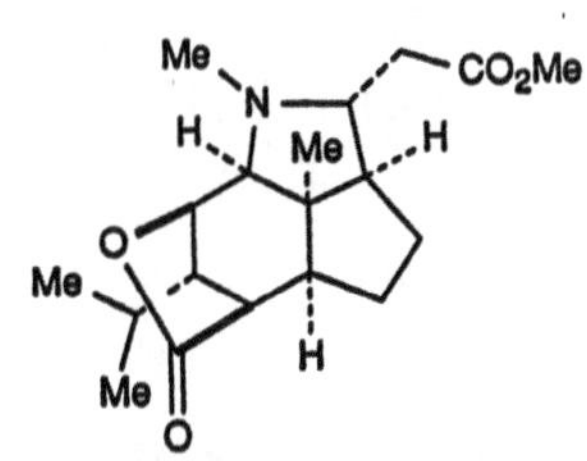

Dendrine (*57-6*)

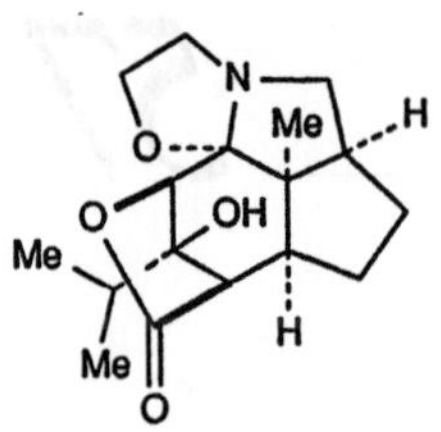

8-Hydroxydendroxine (*57-7*)

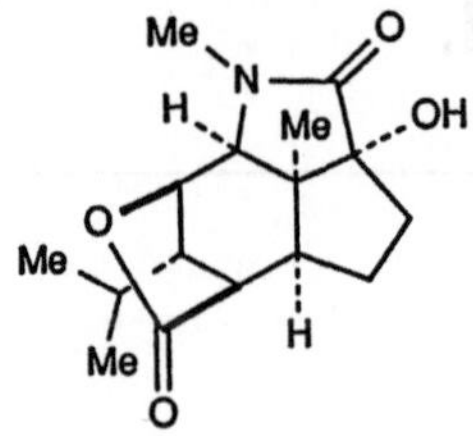

3-Hydroxy-2-oxodendrobine (*57-8*)

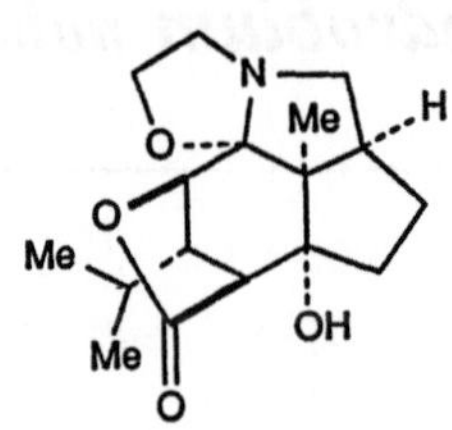

6-Hydroxydendroxine (*57-9*)

Dendrobium nobile can be cultivated on trees. *Sapium sebiferum*, *Prunus salicina*, *Pyrus* sp., *Phoebe nanmu*, *Camptotheca acuminata*, *Aleurites fardii*, *Toona sinensis*, *Pterocarya stenoptera*, *Diaspyros kaki*, *Zizyphus jujuba*, and *Prunus armeniaca* were used as host trees. *D. nobile* cultivated on *P. armeniaca* and on *C. acuminata* contained one to three unknown alkaloids not found in wild *D. nobile*, indicating that the host trees have some effect on alkaloids of *D. nobile* grown on them and could be useful in the investigation of methods of cultivation of *D. nobile* for medicinal use. The dendrobine contents were lower in *D. nobile* cultivated on trees than in wild plants [13]. The average contents of dendrobine in stems and leaves of *D. nobile* cultivated on trees were 0.58% and 0.6%, respectively, and the contents in wild *D. nobile* were 3.2% in stems, 0.8% in leaves and 0.08% in roots [14].

Besides the sesquiterpene alkaloids mentioned above, some related quaternary salts of the dendrobine type *N*-methyldendrobinium iodide (*57-10*), *N*-isopentenyldendrobinium bromide (*57-11*), dendrobine *N*-oxide (*57-12*), *N*-isopentenyldendroxinium chloride (*57-13*), and *N*-isopentenyl-6-hydroxydendroxinium chloride (*57-14*) were also isolated from *D. nobile* [15].

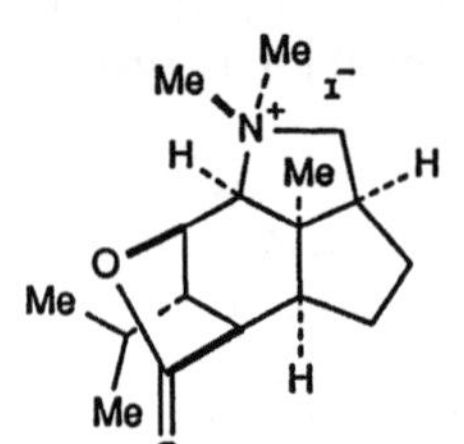

N-Methyldendrobinium iodide (*57-10*)

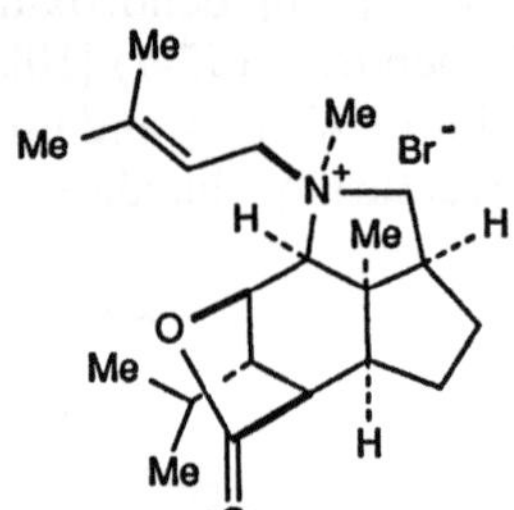

N-Isopentenyldendrobinium bromide (*57-11*)

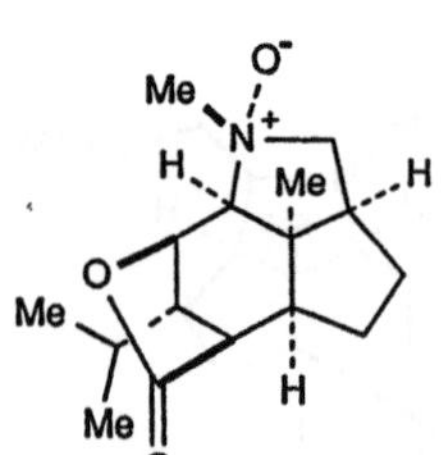

Dendrobine *N*-oxide (*57-12*)

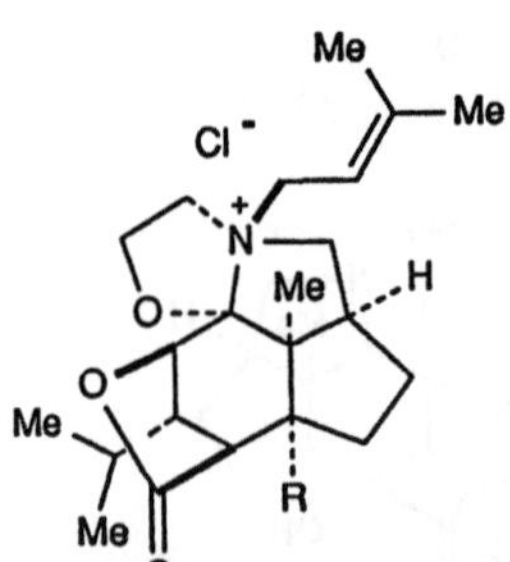

N-Isopentenyldendroxinium chloride (*57-13*): R=H
N-Isopentenyl-6-hydroxy-dendroxinium chloride (*57-14*): R=OH

In addition to the alkaloids, a related sesquiterpene nobilomethylene (*57-15*) [11] and a phenanthraquinone denbinobin (*57-16*) [16] were also found in *D. nobile*.

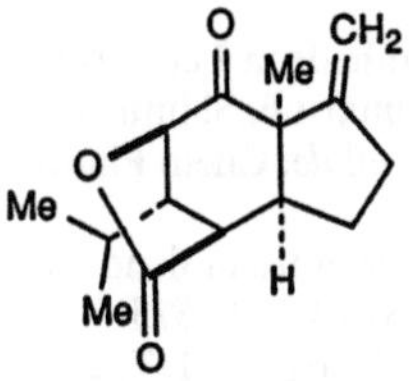

Nobilomethylene (*57-15*)

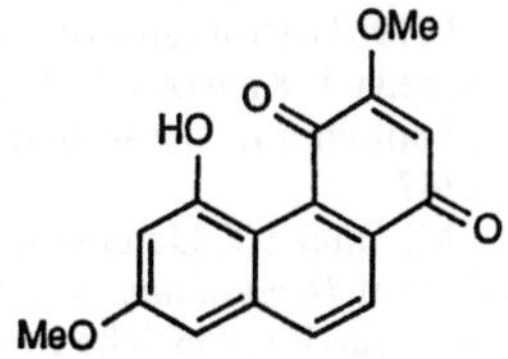

Denbinobin (*57-16*)

57.3 Pharmacology

Dendrobine hydrochloride showed a slight but demonstrable analgesic and antipyretic action, produced moderate hyperglycemia, diminished cardiac activity in large doses, lowered blood pressure, depressed respiration, inhibited isolated rabbit intestines, and contracted isolated guinea pig uteri. The minimal lethal dose of dendrobine by intravenous injection was 20 mg/kg in mice and rats, 22 mg/kg in guinea pigs, and 17 mg/kg in rabbits. Death was preceded by convulsions, which appear to be central in origin [17].

The effects of dendrobine and nobiline on the electric activity and on amino acid induced depolarizations of primary afferent terminals were studied in the frog spinal cord and compared with those of picrotoxinin and strychnine [18]. Picrotoxinin is a sesquiterpene isolated from some menispermaceous plants such as *Anamirta cocculus* and *Tinomiscium philippinensis* and strychnine is an alkaloid obtained from *Strychnos* species. Both picrotoxinin and strychnine are potent convulsant agents. Moreover, the sesquiterpene lactone picrotoxinin (*57-17*) shows a structural similarity with the sesquiterpene alkaloid dendrobine. The effects of dendrobine were qualitatively similar to those of strychnine but were somewhat different from those of picrotoxinin [18].

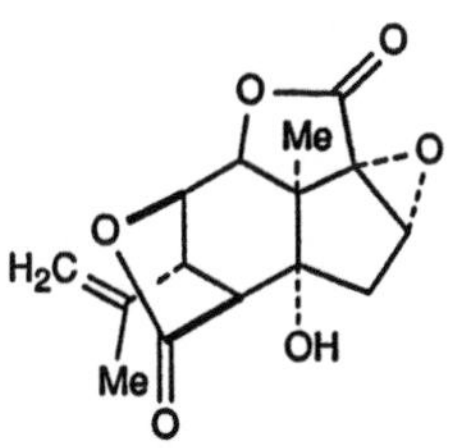

Picrotoxinin (*57-17*)

References

1. Lars G (1976) Studies on some *Dendrobium* (Orchidaceae) constituents. Chem Commun (Univ Stockholm) 2:14
2. Chen KK, Chen Ling A (1935) The alkaloid of chin-shih-hu. J Biol Chem 111:653–658

3. Inubushi Y, Ishii H, Yasui B, Konita T, Harayama T (1964) Isolation and characterization of the Chinese drug "Chin-Shih-Hu". Chem Pharm Bull (Tokyo) 12:1175–1180
4. Inubushi Y, Sasaki Y, Tsuda Y, Yasui B, Konita T, Matsumoto J, Katarao E, Nakano J (1964) Structure of dendrobine. Tetrahedron 20:2007–2023
5. Yamamura S, Hirata Y (1964) Structures of nobiline and dendrobine. Tetrahedron Lett 79–87
6. Onaka T, Kamata S, Maeda T, Kawazoe Y, Natsume M, Okamoto T, Uchimaru F, Shimuzu M (1965) The structure of nobilonine, the second alkaloid from *Dendrobium nobile*. Chem Pharm Bull (Tokyo) 13:745–747
7. Okamoto T, Natsume M, Onaka T, Uchimaru R, Shimizu M (1966) The structure of dendroxine, the third alkaloid from *Dendrobium nobile*. Chem Pharm Bull (Tokyo) 14:672–675
8. Inubushi Y, Tsuda Y, Katarao E (1966) The structure of dendramine. Chem Pharm Bull (Tokyo) 14:668–671
9. Okamoto T, Netsume M, Onaka T, Uchimaru F, Shimizu M (1966) The structure of dendramine (6-hydroxydendrobine) and 6-hydroxydendroxine, the fourth and fifth alkaloid from *Dendrobium nobile*. Chem Pharm Bull (Tokyo) 14:676–680
10. Inubushi Y, Nakano J (1965) Structure of dendrine. Tetrahedron Lett 2723–2728
11. Ikamoto T, Natsume M, Onaka T, Uchimaru F, Shimizu M (1972) Alkaloidal constituents of *Dendrobium nobile* (Orchidaceae). Structure determination of 4-hydroxy-dendroxine and nobilomethylene. Chem Pharm Bull (Tokyo) 20:418–421
12. Wang XK, Zhao TF, Che CT (1985) Dendrobine and 3-hydroxy-2-oxodendrobine from *Dendrobium nobile*. J Nat Prod 48:796–801
13. Wang XK, Zhao TF, Wang M (1985) Gas chromatographic-mass spectroscopic investigation of the alkaloids of *Dendrobium nobile* cultivated on eleven trees. Bull Chin Mat Med 10:367–369
14. Wang XK, Zhao TF (1985) TLC determination of dendrobine in *Dendrobium nobile* Lindl. cultivated on trees. Acta Acad Med Sichuan 16:249–253
15. Hedman K, Leander K (1972) Studies on orchidaceae alkaloids. 27. Quaternary salts of the dendrobine type from *Dendrobium nobile* Lindl. Acta Chem Scand 26:3177–3180
16. Talapatra B, Mukhopadhyay P, Chaudhury P, Talapatra SK (1982) Denbinobin, a new phenanthraquinone from *Dendrobium nobile* Lindl (Orchidaceae). Indian J Chem [B] 21B:386–387
17. Chen KK, Chen Ling A (1935) The pharmacological action of dendrobine, the alkaloid of Chin-shih-hu. J Pharmacol Exp Ther 55:319–325
18. Kudo Y, Tanaka A, Yamada K (1983) Dendrobine, an antagonist of β-alanine, taurine and of presynaptic inhibition in the frog spinal cord. Br J Pharmacol 78:709–715

Dichroa febrifuga Lour.

58

58.1 Introduction

Changshan, Radix Dichroae, is the dry root of *Dichroa febrifuga* Lour. (Saxifragaceae) collected in the fall. It is officially listed in the Chinese Pharmacopoeia and used as an antimalarial agent [1, 2].

58.2 Chemical Constituents

The active principles of the roots of *D. febrifuga* are alkaloids, isolated for the first time by Fu, Jang et al. [3–5], and Koepfli et al. [6]. These alkaloids were named by Jang as dichroine A, B, C [7], which were substituted later by α-dichroine, β-dichroine, and γ-dichroine, whereas Koepfli called these alkaloids febrifugine and isofebrifugine. Febrifugine is identical with β-dichroine and isofebrifugine with α-dichroine.

Febrifugine (*58-1*) and isofebrifugine (*58-2*) are isomeric quinazolin derivatives [8]. They are readily interconverted under the influence of heat, acid, alkali, and solvents [3], e.g. isofebrifugine isomerizes to febrifugine by heat [5]. The stereochemistry of febrifugine [9, 10] and its absolute configuration [11] has also been determined.

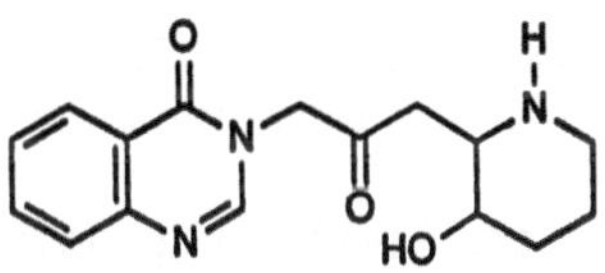

Febrifugine (β-dichroine) (*58-1*)

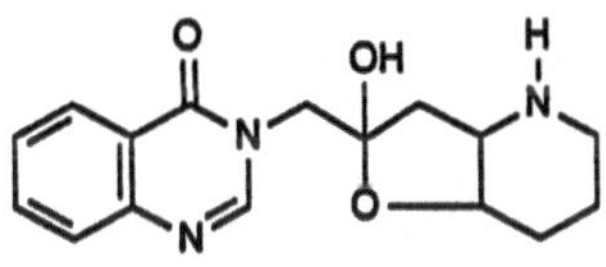

Isofebrifugine (α-dichroine) (*58-2*)

The total febrifugine and isofebrifugine content in roots was approximately 0.1%, whereas the total content in leaves was found to be 0.7% [12]. Moreover, umbelliferone and 4-quinazolone were also detected in roots of *D. febrifuga* [5]. The synthesis of febrifugine was also reported [13].

58.3 Pharmacology

In experimental studies, febrifugine showed an antimalarial activity 50–100 times higher than that of quinine, whereas isofebrifugine was only slightly active. Thus, febrifugine was approximately 100 times as active as quinine against *Plasmodium lophurae* in ducks [6, 14], 50 times as active as quinine against *P. gallinaceum* in chicks or against *P. cynomolgi* infection in rhesus monkeys [15]. In anesthetized dogs, the *Dichroa* alkaloids produced hypotension in all animals treated and depressed cardiac contractions. Splenic and kidney volumes were frequently increased, but respiration was not affected [16]. Febrifugine showed an LD_{50} of 2.5–3.0 mg/kg in mice after oral administration and had delayed toxic manifestations. In rhesus monkeys, febrifugine is more than 300 times as toxic than quinine in subacute experiments, causing loss of body weight at doses of 0.6 mg/kg daily [15]. A major toxic side effect of febrifugine is its emetic activity. The emetic effect of febrifugine has been suggested mainly as a reflex action induced by stimulating the afferent vagi and sympathetic nerves of the gastrointestinal tract [17].

A pharmacokinetic study showed that after oral administration to rats, febrifugine was effectively absorbed from the gastrointestinal tract 1 h after administration, and 58% of the administered dose could be recovered from the gastrointestinal tract. However, 30% was still detectable 4 h after administration of the drug. One hour after febrifugine was given intravenously to a rat, the highest concentration was found in the kidneys. Moderate levels were found in the heart, spleen, liver, fat, and muscle, whereas very low levels were present in blood. Only 16% of an administered dose was excreted unchanged in the urine in a 24-h period. Very little was excreted in the feces and no febrifugine was found in the bile [18]. Febrifugine was also tested for cytostatic [19] and anticoccidial [20] activities. In vitro, febrifugine was active in killing Ehrlich ascites cells [19].

Because the high antimalarial activity was accompanied by a high gastrointestinal toxicity in comparison with quinine, the structure of febrifugine was used as a lead to synthesize some active congeners with lower toxicity, such as methylenedioxy analogs [21] and amino ketones of 4-quinazolone as analogs of febrifugine [22, 23]. Some of these synthetic analogs of febrifugine showed a significant antimalarial activity in animal experiments.

It is worthwhile mentioning here that in the studies on the febrifugine analogs a quinazoline derivative 4-[3,5-bis(*N*-pyrrolidinylmethyl)-4-hydroxyanilino]quinazoline named changrolin (*58-3*) was obtained, which showed a novel type of antiarrhythmic activity in pharmacological and clinical investigations [24].

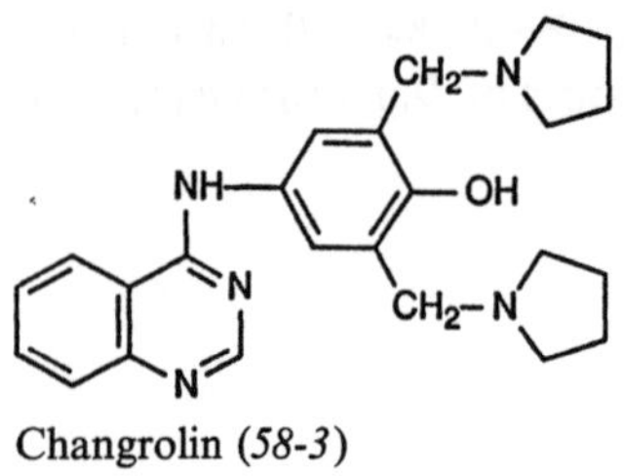

Changrolin (*58-3*)

References

1. Tang W, Beyrich T (1961) *Dichroa febrifuga*, ein Antimalariamittel aus China. Pharmazie 16:482–485
2. Ding GS (1980) Trials of some Chinese medicinal herbs. In: Burns JJ, Tsuchitani PJ (eds) Proceedings of the US-China Pharmacological Symposium. NAS, Washington DC, 1979, pp 103–121
3. Fu FY, Jang CS (1948) Chemotherapeutic studies on chang shan *Dichroa febrifuga*. III. Potent antimalarial alkaloids from chang shan. Sci Technol China 1:56–61
4. Jang CS, Fu FY, Wang CY, Huang KC, Lu G, Chou TC (1946) Chang Shan, a Chinese antimalarial herb. Science 103:59
5. Chou TQ, Fu FY, Kao YS (1948) Antimalarial constituents of Chinese drug, Chang Shan, *Dichroa febrifuga* Lour. J Am Chem Soc 70:1765–1767
6. Koepfli JB, Mead JF, Brockman JA (1947) Alkaloid with high antimalarial activity from *Dichroa febrifuga*. J Am Chem Soc 69:1837
7. Jang CS, Fu FY, Huang KC, Wang CY (1948) Pharmacology of chang shan (*Dichroa febrifuga*), a Chinese antimalarial herb. Nature 161:400–401
8. Koepfli JB, Brockman JA Jr, Moffat J (1950) Structure of febrifugine and isofebrifugine. J Am Chem Soc 72:3323
9. Barringer DF Jr, Berkelhammer G, Carter SD, Goldman L, Lanzilotti AE (1973) Stereochemistry of febrifugine. I. Equilibrium between *cis*- and *trans*-(3-substituted 2-piperidyl)-2-propanones. J Org Chem 38:1933–1937
10. Barringer DF Jr, Berkelhammer G, Wayne RS (1973) Stereochemistry of febrifugine. II. Evidence for the *trans* configuration in the piperidine ring. J Org Chem 38:1937–1940
11. Hill RK, Edwards AG (1962) Absolute configuration of febrifugine. Chem Ind (Lond) 858
12. Li LH, Qu CW, Hsieh SM (1965) Colorimetric method for determination of febrifugine and isofebrifugine. Acta Chim Sin 31:482–485
13. Baker BR, Schaub RE, McEvoy FJ, Williams JH (1952) An antimalarial alkaloid from hydrangea. XII. Synthesis of 3-(β-keto-γ-(3-hydroxy-2-piperidyl)propyl)-4-quinazolone. J Org Chem 17:132–140
14. Hewitt RI, Wallace WS, Gill ER, Williams JH (1952) An antimalarial alkaloid from hydrangea. XIII. The effects of various synthetic quinazolones against *Plasmodium lophurae* in ducks. Am J Trop Med Hyg 1:768–772
15. Koepfli JB, Mead JF, Brockman JA Jr (1950) Obtaining febrifugine alkaloids. US patent no. 2,504,847 (CA 44:6086d)
16. Jang CS, Huang KC (1956) Pharmacological studies on dichroines-3 isomeric alkaloids from Chang Shan. Acta Physiol Sin 20:30–36
17. Chiang W (1961) The mechanism of the emetic action of β-dichroine in dog. Acta Physiol Sin 24:180–186
18. Sung CY, Ho CF (1964) The physiological disposition of β-dichroine, an alkaloid of *Dichroa febrifuga*. Acta Pharm Sin 11:437–443
19. Vermel EM, Kruglyak-Syrkina SA (1960) Anticancer activity of the alkaloid febrifugine in animal experiments. Vopr Onkol 6:56–61 (CA 54:23040d)
20. Reid WM (1969) Efficacy studies on some new anticoccidial drugs. Acta Vet (Brno) 38:137–147 (CA 72:97858h)
21. Chien PL, Cheng CC (1970) Structural modification of febrifugine. Some methylenedioxy analogs. J Med Chem 13:867–870
22. Magidson OY, Lu YK (1959) Amino ketones of the 4-quinazolone series as analogs of febrifugine. I. Derivatives of acetone and methyl ethyl ketone. Zhur Obshchei Khim 29:2843–2851
23. Lu YK, Magidson OY (1959) Amino ketones of the 4-quinazolone series as analogs of febrifugine. II. Methyl alkyl ketone derivatives. Zhur Obshchei Khim 29:3299–3305
24. Ding GS (1987) Anti-arrhythmia agents in traditional Chinese medicine. Abstr Chin Med 1:287–308

Dioscorea spp. 59

59.1 Introduction

The following items from *Dioscorea* species are listed in the Chinese Pharmacopoeia:

- Shanyao, Rhizoma Dioscoreae, is the dry rhizome of *Dioscorea opposita* Thunb. (Dioscoreaceae) collected in winter and used mainly for treatment of diarrhea, asthma, polyuria, and diabetes.
- Fenbixie, Rhizoma Dioscoreae hypoglaucae, is the dry rhizome of *D. hypoglauca* Palibin collected in the fall and winter and used against rheumatic diseases.
- Mianbixie, Rhizoma Dioscoreae septemlobae, is the dry rhizome of *D. septemloba* Thunb. or *D. futschauensis* Uline collected in the fall and winter and has the same medicinal use as *D. hypoglauca.*

In addition, *D. nipponica* Makino is listed in the appendix of the Chinese Pharmacopoeia, and a number of *Dioscorea* species are used in folk medicine for different purposes. The main sapogenin diosgenin is an important starting material in the synthesis of steroid hormones.

The contents of sapogenin were determined in 16 *Dioscorea* species, *D. althaeoides, D. biformifolia, D. chingii, D. collettii, D. collettii* var. *hypoglauca, D. deltoidea, D. futschauensis, D. gracillima, D. nipponica, D. nipponica* subsp. *rothornii, D. panthaica, D. poilanei, D. septemloba, D. tenuipes, D. tokoro,* and *D. zingiberensis* native in China. The *D. zingiberensis* rhizome had the highest sapogenin content at 5.9%, whereas the tuber of *D. poilanei* had the lowest content at only 0.07% [1].

59.2 Chemical Constituents

59.2.1 Chemical Constituents of *Dioscorea hypoglauca*

Nine steroid sapogenins were isolated from the rhizome of *D. hypoglauca* and identified as $\Delta^{3,5}$-deoxytigogenin, $\Delta^{3,5}$-deoxyneotigogenin, β-sitosterol, diosgenin, yamogenin, diosgenin acetate, yamogenin acetate, diosgenin palmitate, and yamogenin palmitate [2]. Diosgenin (*59-1*), tigogenin (*59-2*), and yamogenin (*59-3*) are all derived from spirostan. Diosgenin and yamogenin as well as $\Delta^{3,5}$-deoxytigogenin (*59-4*) and $\Delta^{3,5}$-deoxyneotigogenin (*59-5*) are epimers. Whereas diosgenin and tigogenin possess a 25R configuration, yamogenin and neotigogenin show a 25S configuration.

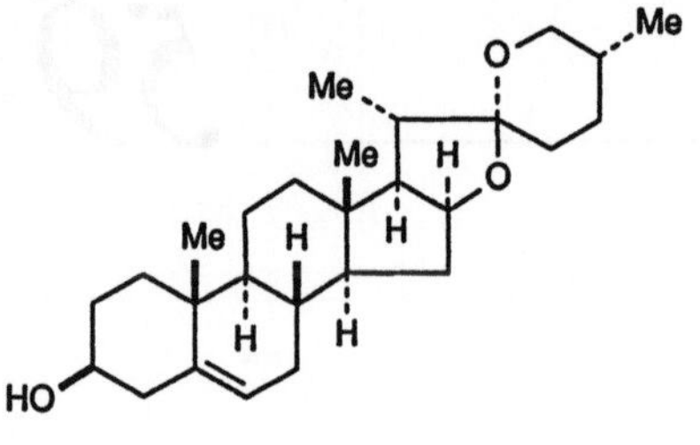

Diosgenin (*59-1*)

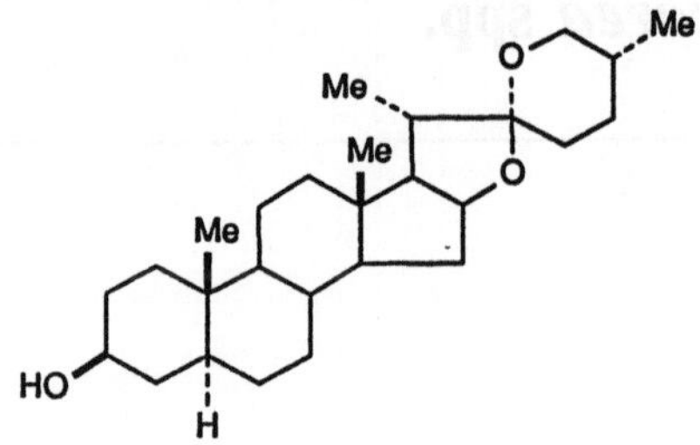

Tigogenin (*59-2*)

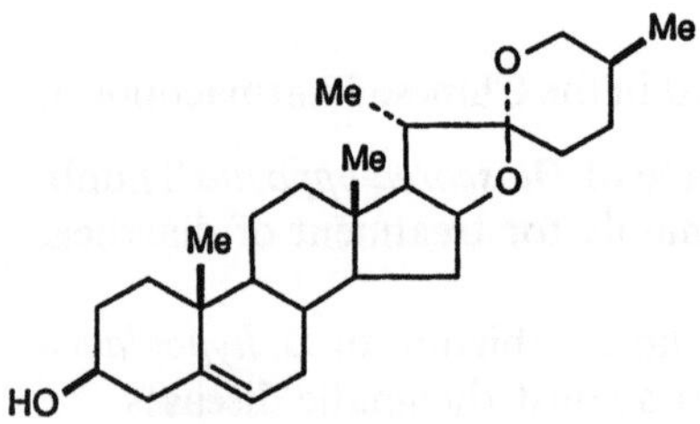

Yamogenin (*59-3*)

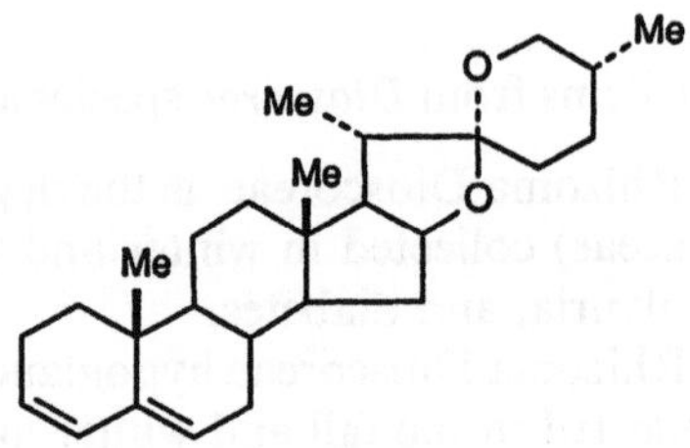

$\Delta^{3,5}$-Deoxytigogenin (*59-4*)

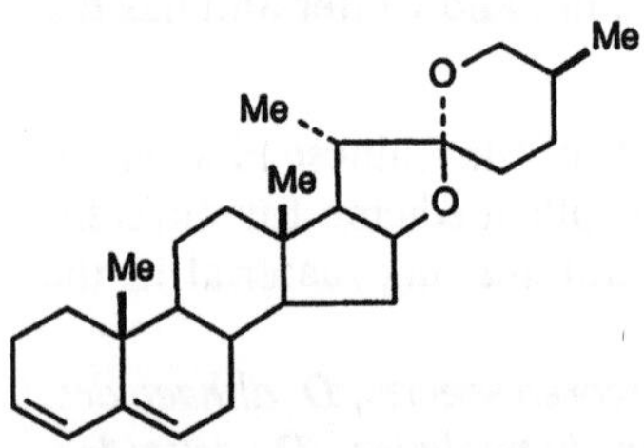

$\Delta^{3,5}$-Deoxyneotigogenin (*59-5*)

Two new saponins named hypoglaucine A (*59-6*) and protohypoglaucine A (*59-7*) were isolated from the rhizome of *D. hypoglauca* [3]. Hypoglaucin A is a yamogenin glycoside and protohypoglaucin A is the corresponding dihydrofurostadiene glycoside with the same sugar moiety.

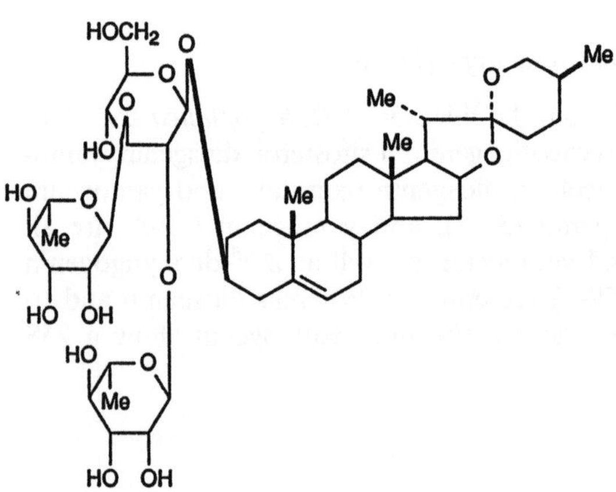

Hypoglaucin A (*59-6*)

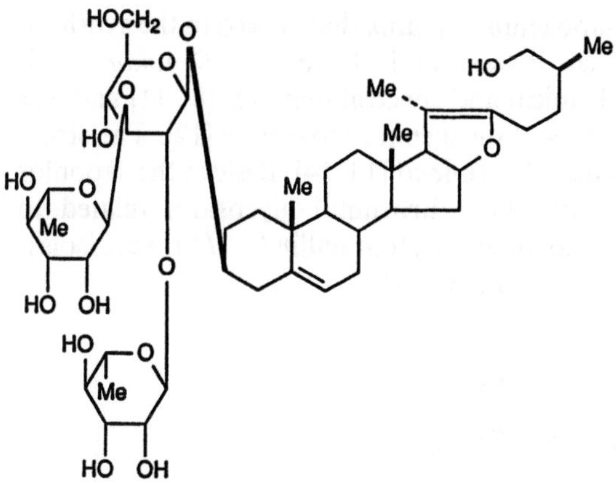

Protohypoglaucin A (*59-7*)

59.2.2 Chemical Constituents of *Dioscorea septemloba*

Two sapogenins, diosgenin and $\Delta^{3,5}$-deoxytigogenin, and two saponins dioscin (*59-8*) and gracillin (*59-9*), were initially isolated from the rhizome of *D. septemloba* collected in Japan [4]. Diosgenin, β-sitosterol, $\Delta^{3,5}$-deoxytigogenin, diosgenin palmitate, and yamogenin were detected in the hydrolysate of the extract from the rhizome of *D. septemloba* collected in China [5]. The yield of diosgenin from the rhizome of *D. septemloba* was 0.5% [5].

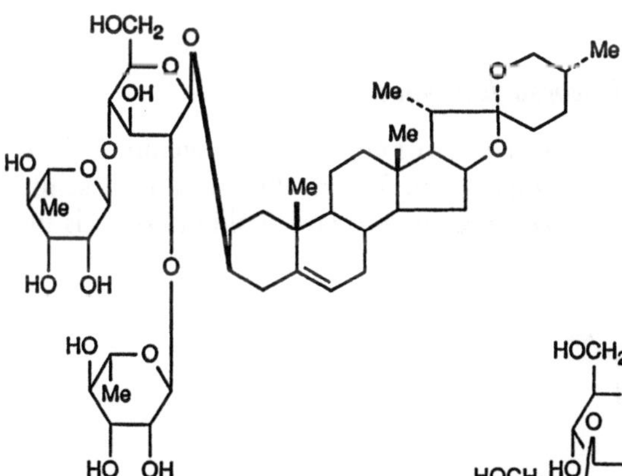

Dioscin (*59-8*)

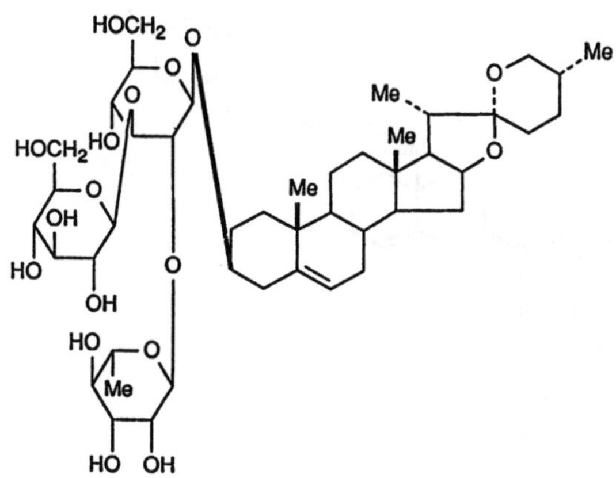

Gracillin (*59-9*)

Diosgenin is an important steroid sapogenin of plants that is used in the synthesis of steroid hormones as starting material. It was first isolated from *D. tokoro* [6, 7] and its structure was determined by chemical and spectral methods [8–11] and was proven by X-ray crystallographic analysis of diosgenin iodoacetate [12]. The structure of the major saponin dioscin was also determined [13, 14]. Besides the saponins and sapogenins derived from spirostane, two furostanol glycosides related to gracillin, named protogracillin (*59-10*) and methylprotogracillin (*59-11*), were isolated from *D. septemloba* and strucurally investigated [15].

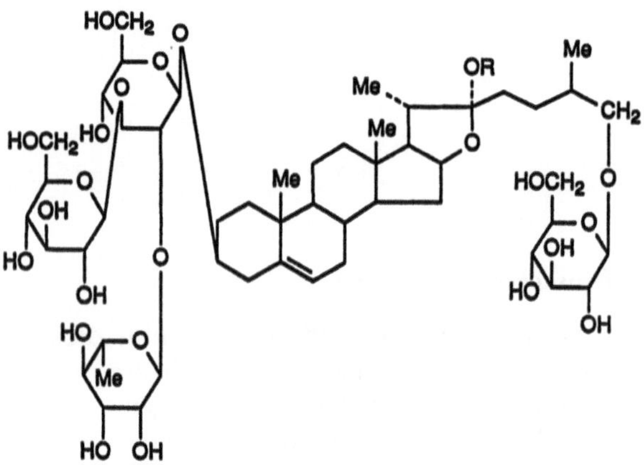

Protogracillin (*59-10*): R=H
Methylprotogracillin (*59-11*): R=CH_3

59.2.3 Chemical Constituents of *Dioscorea futschauensis*

The acid hydrolysate of the extract of *D. futschauensis* contained diosgenin, β-sitosterol, diosgenin palmitate, and $\Delta^{3,5}$-deoxytigogenin. Trillin, dioscin, and gracillin were saponins detected in the rhizome of *D. futschauensis* [16]. Trillin (*59-12*) is the 3-*O*-β-D-glucopyranoside of diosgenin.

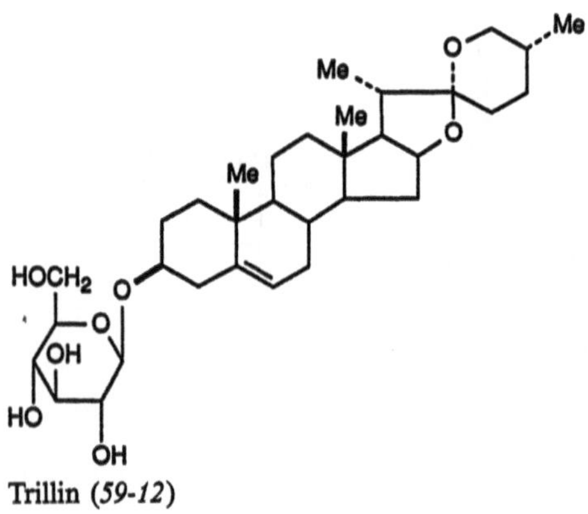

Trillin (*59-12*)

59.2.4 Chemical Constituents of *Dioscorea opposita*

Two percent diosgenin was obtained from the bulbs of *D. opposita* by an improved technique [17]. Besides diosgenin, a number of phenolic compounds designated as batatasins I (*59-13*) [18, 19], II (*59-14*) [20], III (*59-15*), IV (*59-16*), and V (*59-17*) [21, 22] were detected from *D. opposita*.

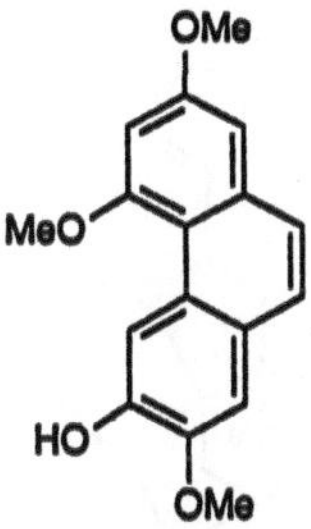

Batatasin I (*59-13*)

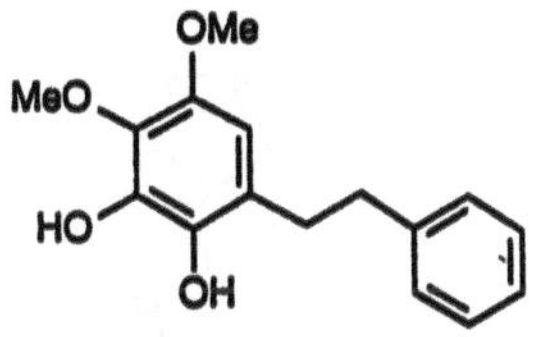

Batatasin II (*59-14*)

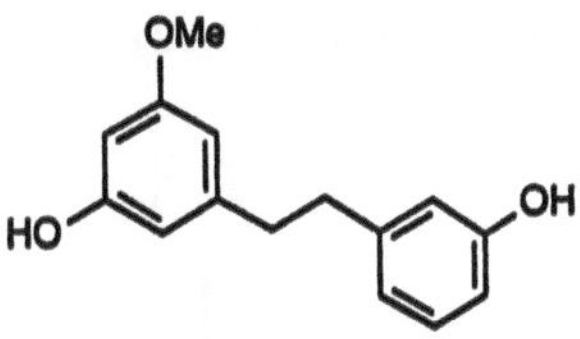

Batatasin III (*59-15*)

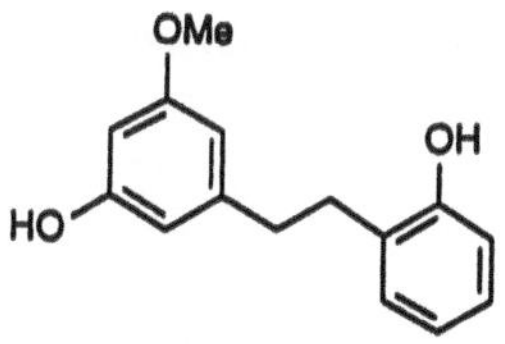

Batatasin IV (*59-16*)

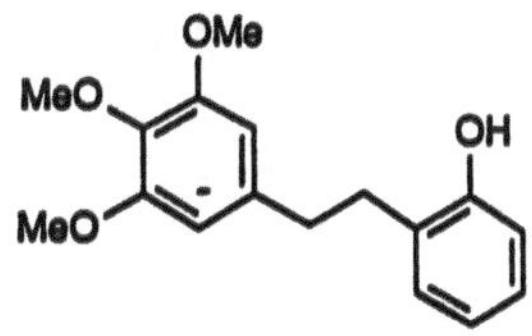

Batatasin V (*59-17*)

59.2.5 Chemical Constituents of Other *Dioscorea* Species

59.2.5.1 Chemical Constituents of *Dioscorea althaeoides*

In addition to dioscin and gracillin, four sapogenins diosgenin, diosgenin palmitate, $\Delta^{3,5}$-deoxytigogenin, and β-sitosterol were isolated from the rhizome of *D. althaeoides* [23].

59.2.5.2 Chemical Constituents of *Dioscorea bulbifera*

The rhizome of *D. bulbifera* is a folk medicine in China used as an antipyretic and antiphlogistic. The saponin fraction of *D. bulbifera* gave three sapogenins, diosgenin, yamogenin, and cryptogenin (*59-18*), a sapogenin derived from cholestane. Diosgenin is the major sapogenin with a content of 0.45% [24].

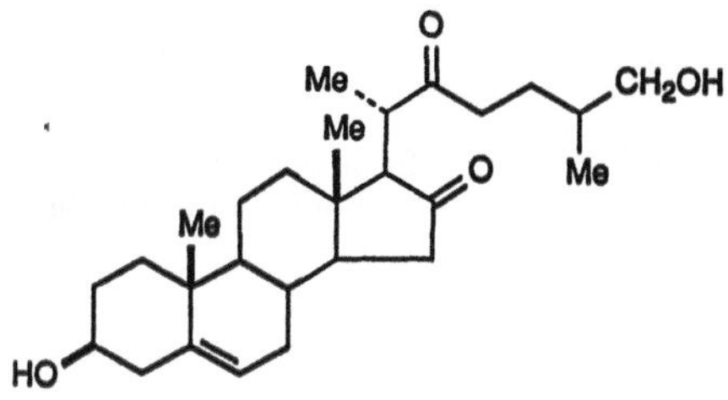

Cryptogenin (*59-18*)

Besides the steroid components, a series of furanoid norditerpenes named diosbulbins A (*59-19*) [25, 26], B (*59-20*), C (*59-21*) [25], D (*59-22*), E (*59-23*), F (*59-24*), G (*59-25*), and H (*59-26*) [27] were isolated. The structures of diosbulbins were determined by chemical correlations and spectral analysis and the absolute configuration of diosbulbins A and B was established by crystal analysis of 8-bromo-9-dehydroxy-9-oxo-tetrahydrodiosbulbin A (*59-27*) [28, 29].

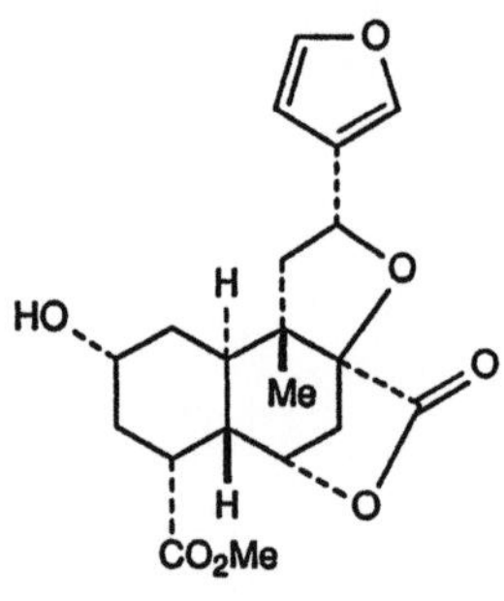

Diosbulbin A (*59-19*)

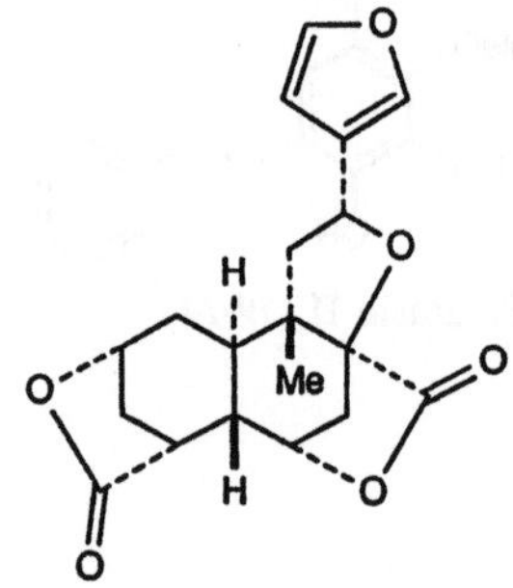

Diosbulbin B (*59-20*)

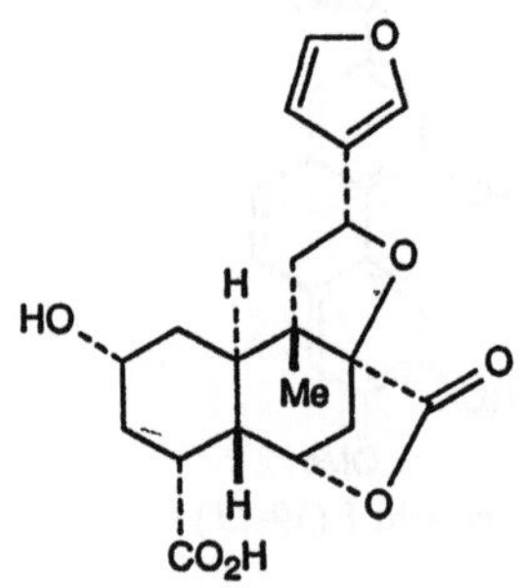

Diosbulbin C (*59-21*)

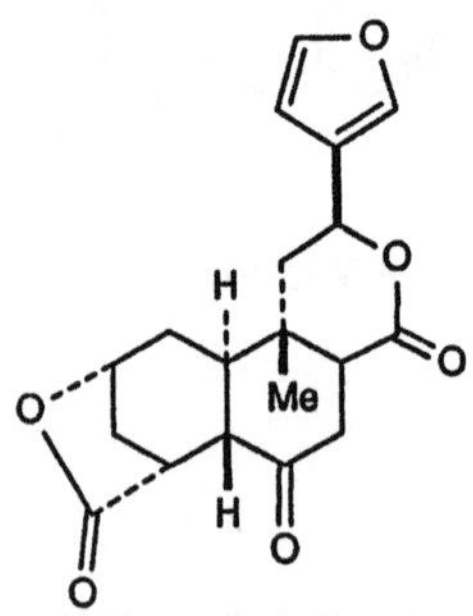

Diosbulbin D (*59-22*)

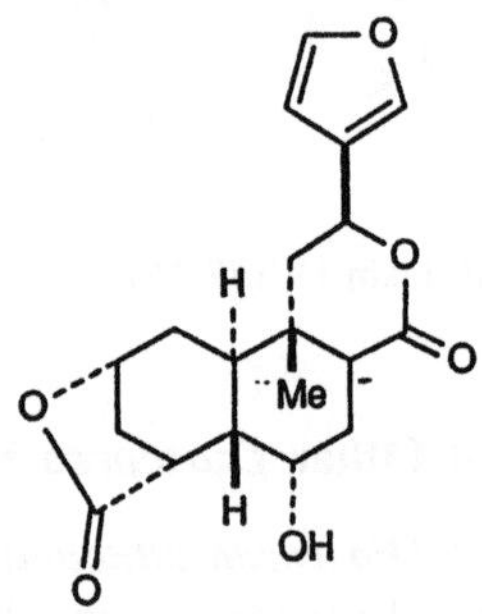

Diosbulbin E (*59-23*)

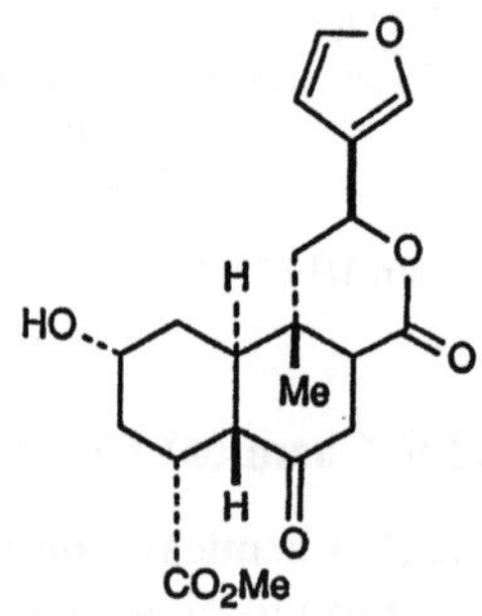

Diosbulbin F (*59-24*)

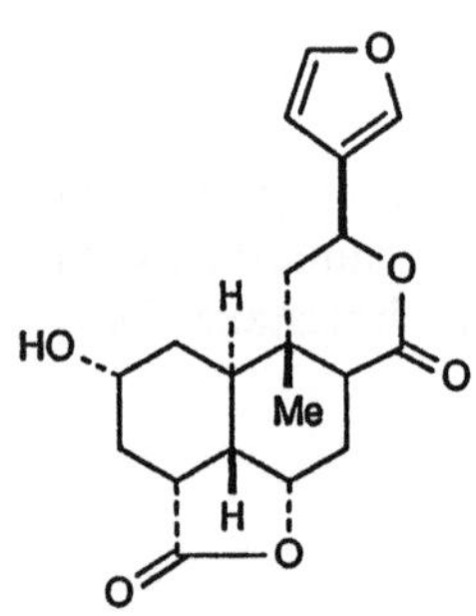

Diosbulbin G (*59-25*)

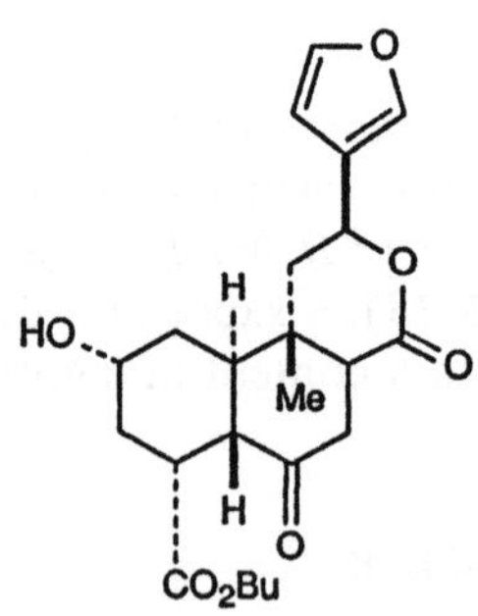

Diosbulbin H (*59-26*)

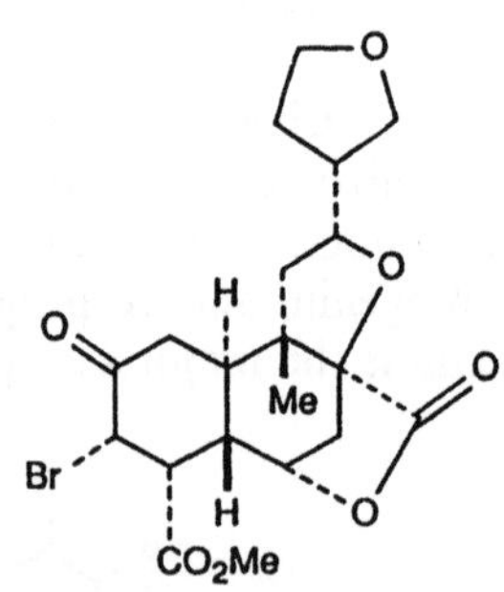

8-Bromo-9-dehydroxy-9-oxo-tetrahydrodiosbulbin A (*59-27*)

Furthermore, two glycosides, diosbulbinoside D (*59-28*) and diosbulbinoside F (*59-29*), were isolated and their structures determined [30]. The isolation of two new acetophenone derivatives, 4-hydroxy-2-(*trans*-3,7-dimethylocta-2,6-dienyl)-6-methoxyacetophenone and 4,6-dihydroxy-2-(4-hydroxybutyloxy)-acetophenone, was reported recently [31].

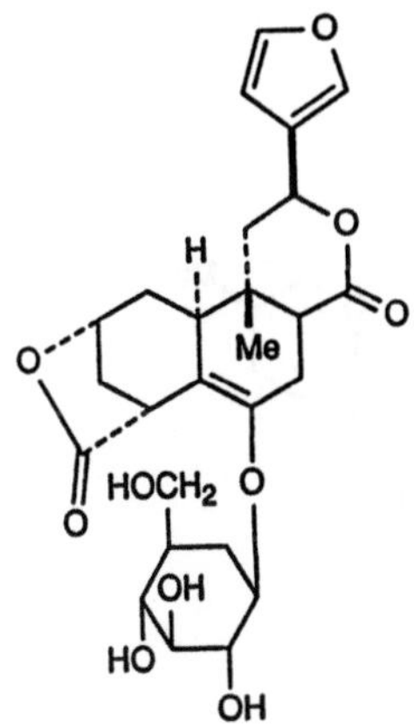

Diosbulbinoside D (*59-28*)

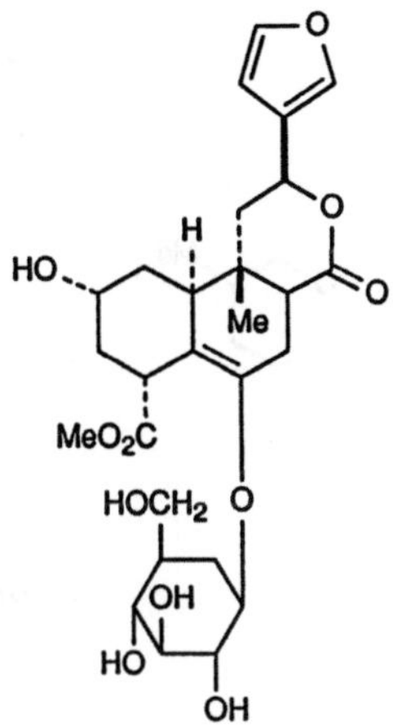

Diosbulbinoside F (*59-29*)

59.2.5.3 Chemical Constituents of *Dioscorea cirrhosa*

The rhizome of *D. cirrhosa* is a folk medicine used as a hemostatic. The isolation of dimeric, trimeric, and tetrameric procyanidins together with catechin and epicatechin from the rhizome of *D. cirrhosa* was reported [32, 33].

59.2.5.4 Chemical Constituents of *Dioscorea collettii*

Nine steroid sapogenins and their esters or glycosides diosgenin, yamogenin, $\Delta^{3,5}$-deoxytigogenin, $\Delta^{3,5}$-deoxyneotigogenin, β-sitosterol, isonarthogenin (*59-30*), diosgenin palmitate, yamogenin palmitate, and yamogenin-3-*O*-β-D-glucopyranoside were isolated from the rhizome of *D. collettii* [34–36]. Isonarthogenin is a sapogenin derived from furostane.

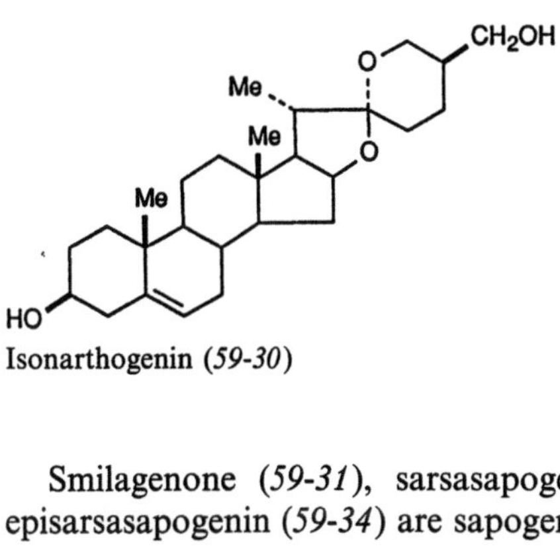

Isonarthogenin (*59-30*)

Smilagenone (*59-31*), sarsasapogenone (*59-32*), epismilagenin (*59-33*), and episarsasapogenin (*59-34*) are sapogenins further isolated from *D. collettii* [37].

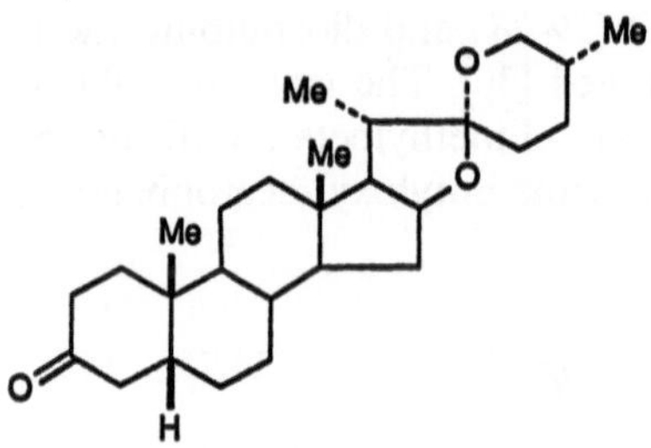

Smilagenone (*59-31*)

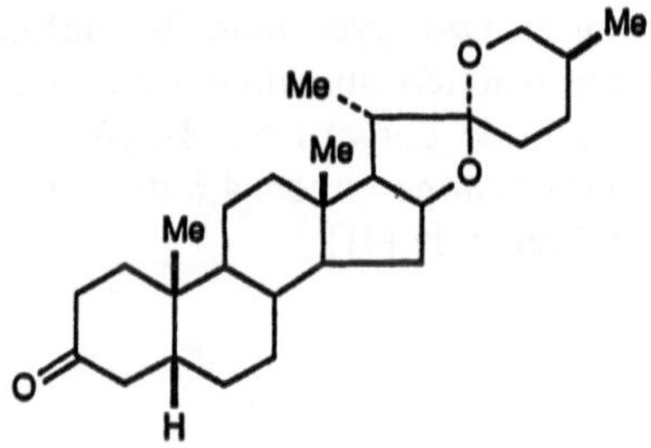

Sarsasapogenone (*59-32*)

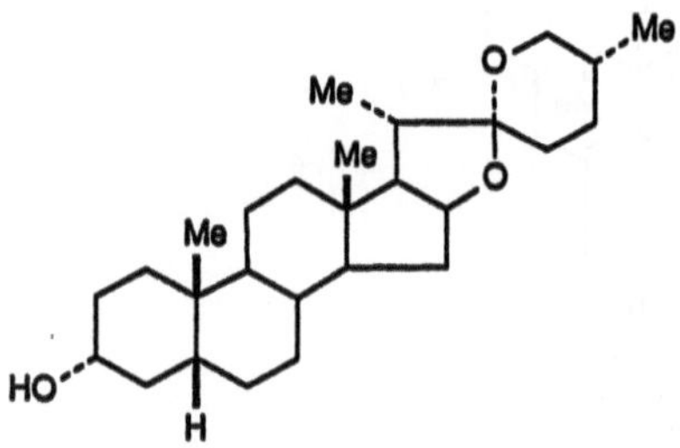

Epismilagenin (*59-33*)

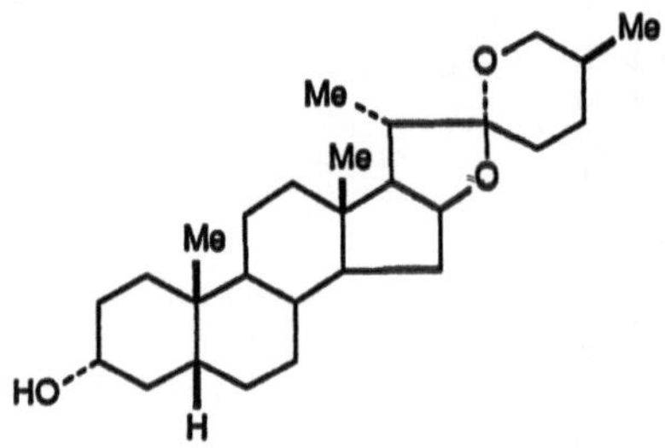

Episarsasapogenin (*59-34*)

Some glycosides named collettinsides I, II, III, and IV with yamogenin as aglycon were also isolated. Collettinside I is identical with yamogenin-3-*O*-*β*-D-glucopyranoside, and collettinsides II (*59-35*), III (*59-36*), and IV (*59-37*) are three new saponins [38].

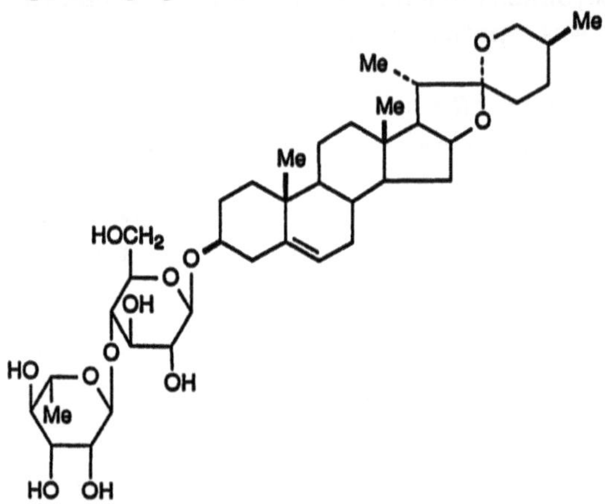

Collettinside II (*59-35*)

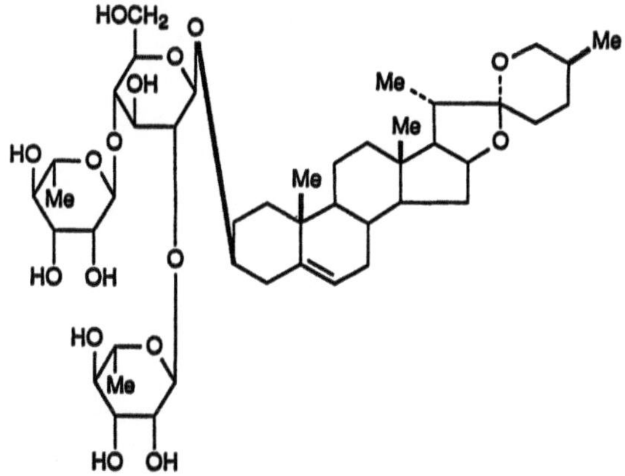

Collettinside III (*59-36*)

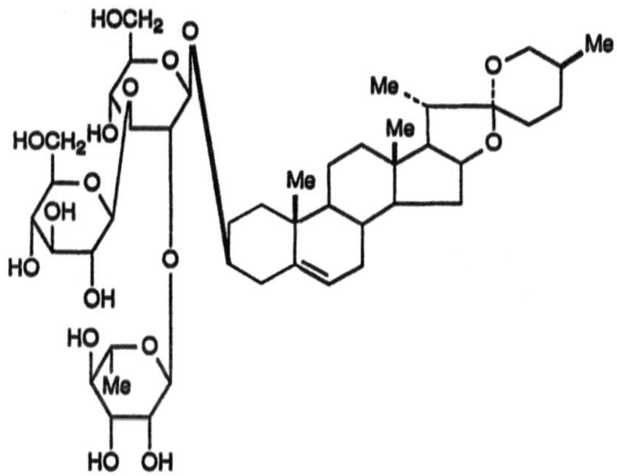

Collettinside IV (*59-37*)

The diosgenin/yamogenin ratio for *D. collettii* was 62/38 [39].

59.2.5.5 Chemical Constituents of *Dioscorea deltoidea*

The *D. deltoidea* rhizome contained 1.87% of the total sapogenins [1]. Besides diosgenin and yamogenin, the new saponins deltonin (*59-38*) [40] deltoside (*59-39*) [40, 41] and diosgenin-3-*O*-β-D-glucopyranosyl-(1→3)-[β-D-glucopyranosyl-(1→4)]-β-D-glucopyranoside (*59-40*) [42] were isolated from the rhizome and deltofolin (*59-41*) [43] from the leaves of *D. deltoidea*.

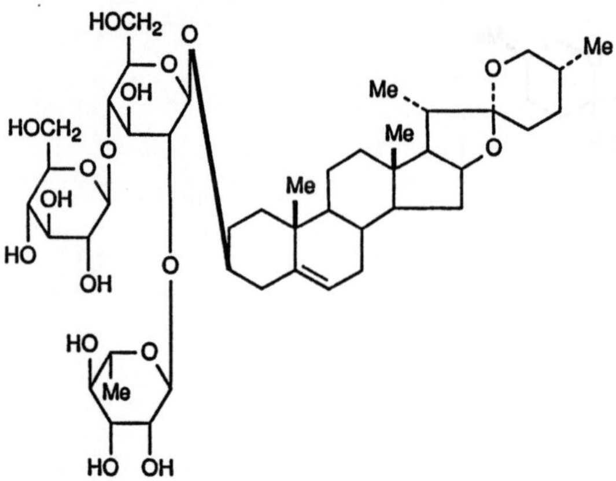

Deltonin (*59-38*)

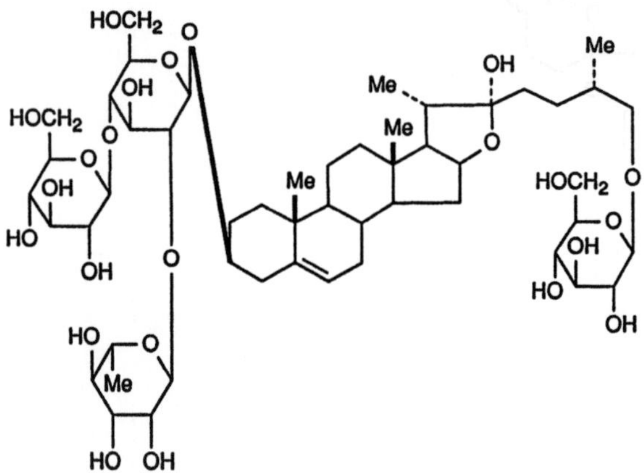

Deltoside (*59-39*)

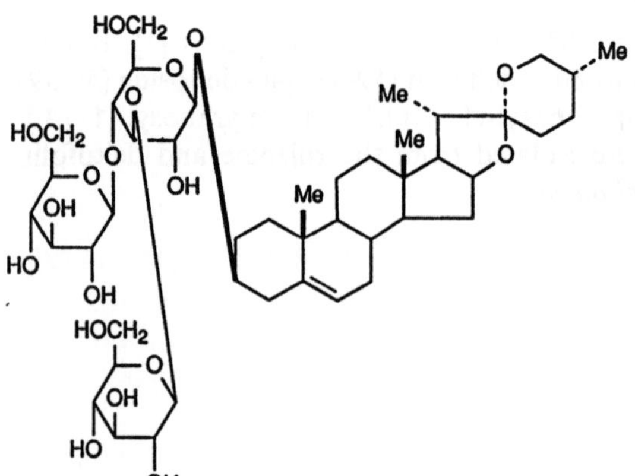

Diosgenin-3-*O*-*β*-D-glucopyranosyl-(1→3)-[*β*-D-glucopyranosyl-(1→4)]-*β*-D-glucopyranoside (*59-40*)

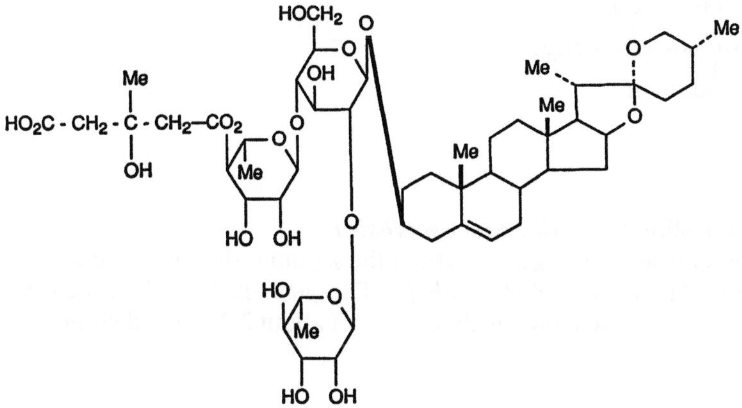

Deltofolin (*59-41*)

59.2.5.6 Chemical Constituents of *Dioscorea gracillima*

Dioscin and gracillin were isolated and identified from the rhizome of *D. gracillima* in a ratio of 3:1 [44]. The total sapogenin content in the rhizome of *D. gracillima* was 2.4% [1]. In addition, furostanol bisglycoside corresponding to dioscin, named protodioscin (*59-42*) and its artifact methylprotodioscin (*59-43*), were also isolated [15].

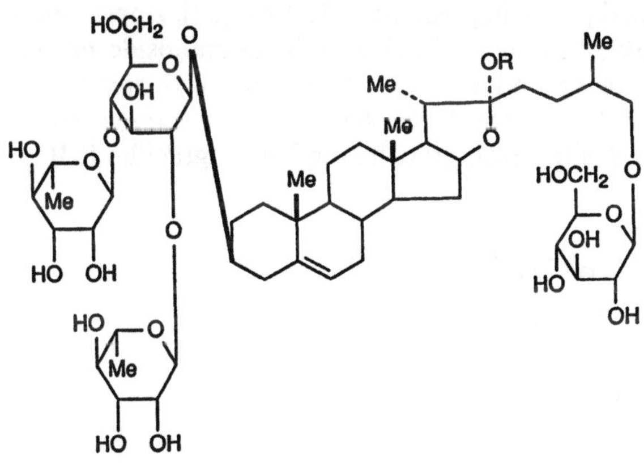

Protodioscin (*59-42*): R=H
Methylprotodioscin (*59-43*): R=CH_3

59.2.5.7 Chemical Constituents of *Dioscorea nipponica*

The rhizome of *D. nipponica* is one of the most extensively used resources for extraction of diosgenin and dioscin [45]. The sapogenin content in the rhizome was found to be about 1.6% [1]. Dioscin and gracillin were detected as saponins [46]. Besides the steroid saponins and their aglycones, piscidic acid (*59-44*) was isolated and identified [47, 48].

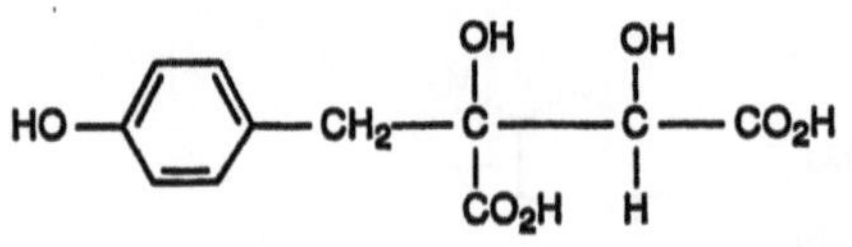

Piscidic acid (*59-44*)

59.2.5.8 Chemical Constituents of *Dioscorea panthaica*

The sapogenins diosgenin and yamogenin [39] and the saponins dioscin and gracillin [49] were isolated from the rhizome of *D. panthaica*. Diosgenin is the major sapogenin component in the rhizome and a sapogenin content of about 2.2% was determined [1] in the rhizome.

59.2.5.9 Chemical Constituents of *Dioscorea parviflora*

The rhizome of *D. parviflora* is used as a plant resource for diosgenin with an average content of 3.0%–3.5% [50]. Further investigation led to the isolation of four saponins, which were identified as gracillin, deltonine, protogracillin, and deltoside [41].

59.2.5.10 Chemical Constituents of *Dioscorea zingiberensis*

Epismilagenin (*59-33*), trillin (*59-12*), gracillin, diosgenin 3-*O*-β-D-glucopyranosyl-(1→4)-β-D-glucopyranoside [51], diosgenin 3-*O*-β-D-glucopyranosyl-(1→2)-[β-D-rhamnopyranosyl-(1→3)]-β-D-glucopyranoside named zingiberenin A (*59-45*) and its 26-*O*-β-D-glucopyranoside named protozingiberenin A (*59-46*) [52], zingiberenin B, identical to collettinside IV (*59-37*), and its 26-*O*-β-D-glucopyranoside named protozingiberenin B (*59-47*) [52] were isolated from the dry rhizome of *D. zingiberensis*. Protozingiberenin A was also isolated from the fresh rhizome of *D. zingiberensis* together with diosgenin palmitate, β-sitosterol, gracillin, and protogracillin [53].

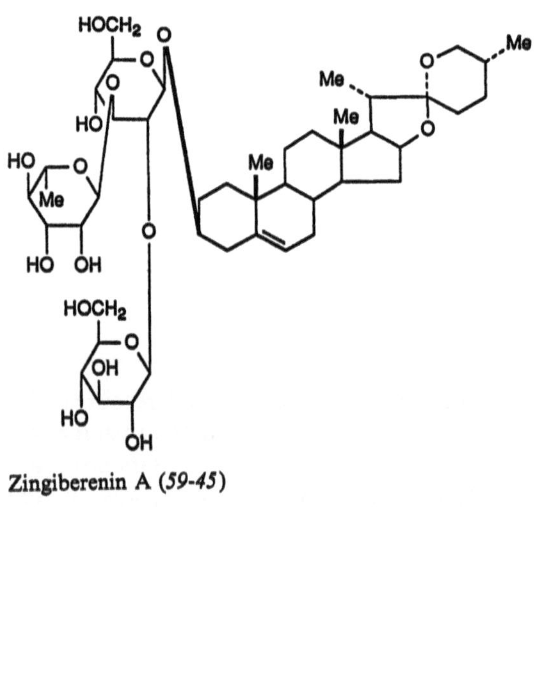

Zingiberenin A (*59-45*)

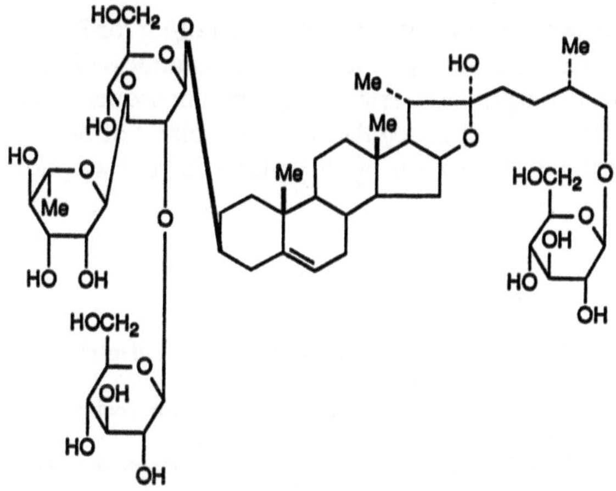

Protozingiberenin A (*59-46*)

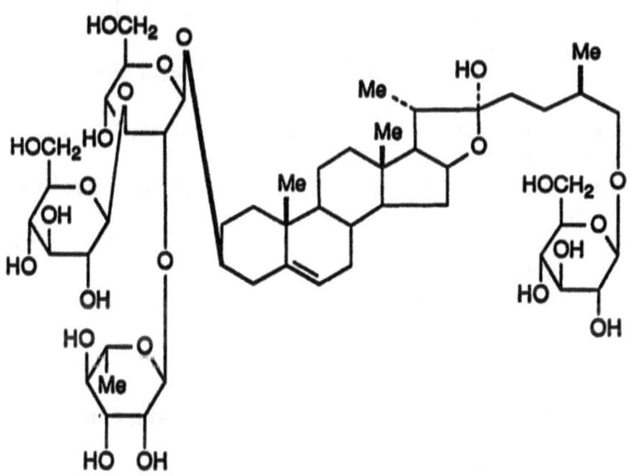

Protozingiberenin B (*59-47*)

Dioscorea zingiberensis showed the highest diosgenin content (4%–8%). The diosgenin contents were higher in older than in younger rhizomes and were maximal during flowering [54].

59.3 Pharmacology

In male Wistar rats fed diets containing different plant steroids including diosgenin, the secretion of biliary cholesterol, bile salt, and phospholipid was investigated. The theoretical maximal biliary cholesterol output significantly increased by 500% in diosgenin-fed animals. No changes were found in the kinetic characteristics of biliary phospholipid output. Adding 2% cholesterol to the diosgenin diet abolished the increment of biliary cholesterol output induced by diosgenin [55].

The intraperitoneal injection of diosgenin at a daily dose of 45 μmol/kg for 3 days significantly increased biliary cholesterol output by 70%. The results indicated that diosgenin-induced biliary cholesterol output was apparently the consequence of direct effects of diosgenin on the intrahepatocytic regulatory mechanisms of biliary cholesterol secretion [55]. The intestinal absorption of cholesterol in rats was not affected by diosgenin [56].

References

1. Tang SR, Zhang HC, Dong YF, Li HI, Ding ZZ (1979) Quantitative analysis and identification of steroidal sapogenins of the *Dioscorea* plants. Acta Bot Sin 21:171–176
2. Lou W, Yang YQ, Chen YY (1984) Isolation and identification of steroidal sapogenins from rhizome of *Dioscorea collettii* var. *hypoglauca*. Acta Bot Yunnan 6:461–462
3. Tang SR, Pang ZJ (1984) Isolation and identification of steroid saponins of *Dioscorea collettii* Hook. f. var. *hypoglauca* (Palibin) Pei et Ting. Acta Bot Sin 26:419–424
4. Tsukamoto T, Kawasaki T, Shimauchi Y (1958) Saponins of Japanese Dioscoreaceae. VII. Sapogenins and saponins from *Dioscorea septemloba*. Yakugaku Zasshi 77:1221–1224
5. Lou W, Chen YY (1983) Isolation and identification of steroidal sapogenins from rhizome of *Dioscorea septemloba* Thunb. Acta Bot Sin 25:352–355
6. Fujii K, Matuskawa T (1936) Saponin of *Dioscorea tokoro* Makino. J Pharm Soc Japan 56:408–414
7. Tsukamoto T, Ueno Y (1936) Glucosides of *Dioscorea tokoro* Makino. I. Dioscin, dioscorea-sapotoxin and diosgenin. J Pharm Soc Jpn 56:802
8. Fujii K, Matsukawa T (1937) Saponin and sterols XII. Saponin of the root of *Dioscorea tokoro* Makino. 2. J Pharm Soc Jpn 57:114–120
9. Tsukomoto T, Ueno Y, Ohta Z (1936) Constitution of diosgenin I. Glucoside of *Dioscorea tokoro* Makino. J Pharm Soc Jpn 56:931–940
10. Tsukamoto T, Ueno Y, Ohta Z (1937) Diosgenin II. Glucoside of *Dioscorea tokoro* Makino. 3. Constitution of diosgenin. J Pharm Soc Jpn 57:985–991
11. Marker RE, Tsukamoto T, Turner DL (1940) Sterols. C. Diosgenin. J Am Chem Soc 62:2525–2536
12. O'Donnell EA, Ladd MFC (1963) The crystal structure of diosgenin iodoacetate. Chem Ind 1984–1985
13. Tsukamoto T, Kawasaki T (1953) Sugar part of dioscin. J Pharm Soc Jpn 73:415
14. Kawasaki T, Yamauchi T, Yamauchi R (1962) Saponins of Japanese Dioscoreaceae. XI. Structures of dioscin, gracillin and kikuba saponin. 2. Chem Pharm Bull (Tokyo) 10:703–708
15. Kawasaki T, Komori T, Miyahara K, Nohara T, Hosokawa I, Mihashi K (1974) Furostanol bisglycosides corresponding to dioscin and gracillin. Chem Pharm Bull (Tokyo) 22:2164–2175
16. Liu CL, Chen YY (1984) Chemical constituents of *Dioscorea* plants VIII. Isolation and identification of steroidal saponins and sapogenins from Foochow yam (*Dioscorea futschauensis*). Chin Trad Herbal Drugs 15:10–12
17. Zhou ZQ, Feng YX (1983) Improved production technique of diosgenin. Pharm Ind (12):1–2
18. El-Olemy MM, Reisch J (1979) Natural product chemistry, part 67. Isolation of batatasin I from nondormant bulbils of *Dioscorea opposita*. An improved method for isolation of batatasin I. Planta Med 37:70-72
19. Letcher RM (1973) Structure and synthesis of the growth inhibitor batatasin I from *Dioscorea batatas*. Phytochemistry 12:2789–2790
20. Hashimoto T, El-Olemy MM (1980) The structure of batatasin II from *Dioscorea opposita* Thunb. Egypt J Pharm Sci 21:249–253 (CA 98:68856n)
21. Ireland CR, Schwabe WW, Coursey DG (1981) The occurrence of batatasins in the Dioscoreaceae. Phytochemistry 20:1569–1571
22. Hashimoto T, Tajima M (1978) Structures and synthesis of the growth inhibitors batatasins IV and V and their physiological activities. Phytochemistry 17:1179–1184
23. Liu CL, Chen YY (1984) Studies on chemical constituents of *Dioscorea* plants. X. Isolation and identification of steroidal saponins and sapogenins from *Dioscorea althaeoides* Kunth. Acta Pharm Sin 19:799–801

24. Marker RE, Wagner RB, Ulshofer PR, Wittbecker EL, Goldschmith PJ, Ruof CH (1943) Sterols 157. Sapogenins. 69. Isolation and structures of thirteen new steroidal sapogenins. New sources for known sapogenins. J Am Chem Soc 65:1199–1209
25. Komori T, Setoguchi S, Kawasaki T (1968) Furanoid-Norditerpene aus Pflanzen der Familie Dioscoreaecae, II. Konstitution and Konfiguration von Diosbulbin-A, -B and -C. Chem Ber 101:3096–3112
26. Institute of Military Medicine (1980) Isolation and identification of diosbulbin. Chin Trad Herbal Drugs 11:522
27. Ida Y, Kubo S, Fujita M, Komori T, Kawasaki T (1978) Furanoid-Norditerpene aus Pflanzen der Familie Dioscoreacea, V. Struktur der Diosbulbine-D, -E, -F, -G and -H. Liebigs Ann Chem 818–833
28. Kamiya K, Wada Y (1972) Furanoid-Norditerpene aus Pflanzen der Familie Dioscoreaceae, III. Die Kristall-Analyse von 2-Keto-3-bromo-tetrahydrodiosbulbin-A. Tetrahedron Lett 1869–1872
29. Komori T, Kawasaki T, Kamiya K, Wada Y (1977) Structure and absolute configuration of diosbulbin-A, -B and -C. Crystal analysis of 3-bromo-2-oxo-tetrahydrodiosbulbin-A. Chem Pharm Bull (Tokyo) 25:1701–1707
30. Ida Y, Noda K, Kubo S, Komori T, Kawasaki T (1978) Furanoid norditerpenes from *Dioscorea* plants. VII. Structures of diosbulbinosides D and F. Chem Pharm Bull (Tokyo) 26:435–439
31. Gupta D, Singh J (1989) *p*-Hydroxyacetophenone derivatives from *Dioscorea bulbifera*. Phytochemistry 28:947–949
32. Hsu FL, Nonaka G, Nishioka I (1985) Tannins and related compounds. 33. Isolation and characterization of procyanidins in *Dioscorea cirrhosa*. Chem Pharm Bull (Tokyo) 33:3293–3298
33. Jiang KN, Luo QY, Sun DW, Foo LY (1988) Studies on condensed tannins. IV. Tannins from *Dioscorea cirrhosa* and *Rosa* spp. bark. Chem Ind Forest Prod 8:32–38
34. Yang MH, Chen YY (1982) Steroidal sapogenins in *Dioscorea collettii*. Chin Pharm Bull 17:175
35. Yang MH, Chen YY (1983) Steroidal sapogenins in *Dioscorea collettii*. Planta Med 49:38–42
36. Liu Ch, Chen YY (1984) Studies on the chemical constituents of *Dioscorea* plants. V. Isolation and identification of esters of steroidal sapogenin from collett yam (*Dioscorea collettii*) rhizome. Chin Trad Herbal Drugs 15:193–194
37. Liu CL, Chen YY (1985) Steroid sapogenins from mildewed *Dioscorea collettii* Hook. f. Acta Pharm Sin 20:143–145
38. Liu CL, Chen YY, Ge SB, Li BG (1983) Studies on the chemical constituents of *Dioscorea* plants. II. Isolation and identification of steroidal saponins from *Dioscorea collettii* Hook. f. Acta Pharm Sin 18:597–606
39. Li DG, Ruan YC (1980) Diosgenin and yamogenin from four species of *Dioscorea* L. in Yunnan (China). Acta Bot Yunnan 2:476–479
40. Paseshnichenko VA, Guseva AR (1975) Isolation and properties of saponins from *Dioscorea deltoidea* rhizome. Prikl Biokhim Mikrobiol 11:94–101 (CA 82:125560g)
41. Liu CL, Chen YY, Ge SB (1985) Isolation and identification of steroidal saponins from *Dioscorea parviflora* Ting. Acta Bot Sin 27:635–639
42. Rajaraman K, Seshadri V, Rangaswami S (1976) Prazerigenin-*A*-3-*O*-*β*-D-glucopyranoside and prazerigenin-*A*-3-O-α-L-rhamnopyranosyl-(1→6)-*β*-D-glucopyranoside, from *Dioscorea prazeri* and diosgenin-3-*O*-*β*-D-glucopyranosyl-(1→3)-*O*-[*β*-D-glucopyranosyl-(1→4)]-*β*-D-glucopyranoside from *D. deltoidea*. Indian J Chem [B] 14B:735–738
43. Paseshnichenko VA, Guseva AR, Strigina LI, Isakov VV (1981) Structure of deltofolin – a steroidal saponin from leaves of *Dioscorea deltoidea* Wall. Dokl Akad Nauk SSSR 256:742–745
44. Tang SR, Wu YF (1984) Isolation and identification of steroidal saponin from *Dioscorea gracillima* Miq. Acta Bot Sin 26:630–633
45. Zhou ZQ, Feng YX (1985) Technological studies on the dioscin extraction from *Dioscorea*. Chin Trad Herbal Drugs 16:303–305
46. Fang YW, Zhao JJ, Ho YZ, Li BG, Xu CJ (1982) Elucidation of the chemical structures of two steroid saponins of *Dioscorea nipponica* Makino. Acta Pharm Sin 17:388–391
47. Ho PC, Liu CY, Chin KS, Li YM (1980) Isolation and identification of *p*-hydroxybenzyltartaric acid (piscidic acid): a water-soluble active constituent of *Dioscorea nipponica*. Chin Pharm Bull 15:39

48. He BJ, Liu ZY, Jiu GS, Li YM (1980) Studies on an aqueous soluble active constituent of *Dioscorea nipponica* Makino. I. Isolation and identification of *p*-hydroxybenzyltartaric acid (piscidic acid). Acta Pharm Sin 15:764–765
49. Li BG, Tang YF, Shi Y (1986) Isolation and identification of steroidal saponins from *Dioscorea panthaica* Prain et Burkill. Acta Bot Sin 28:409–414
50. Xu CF, Zhou J, Dou YZ, He MF (1983) Studies on the cultivation of *Dioscorea* – an important resource for steroidal hormones. Bull Chin Mat Med 8:3–5
51. Liu CL, Chen YY, Tang YF, Li BG (1984) Isolation and identification of steroidal saponins from *Dioscorea zingiberensis*. Acta Bot Sin 26:283–289
52. Tang SR, Wu YF, Pang ZJ (1983) Identification and isolation of steroidal saponins of *Dioscorea zingiberensis* Wright. Acta Bot Sin 25:556–562
53. Liu CL, Chen YY (1985) Isolation and identification of protosaponins from fresh rhizomes of *Dioscorea zingiberensis* Wright. Acta Bot Sin 27:68–74
54. Dring ZZ, Zhou XL, Wang YC, Tang SR (1981) Factors influencing diosgenin content of *Dioscorea zingiberensis* Wright. Chin Trad Herbal Drugs 12:34–35
55. Ulloa N, Nervi F (1985) Mechanism and kinetic characteristics of the uncoupling by plant steroids of biliary cholesterol from bile salt output. Biochim Biophys Acta 837:181–189
56. Malinow MR (1985) Effects of synthetic glycosides on cholesterol absorption. Ann NY Acad Sci 454 (Atherosclerosis):23–27

Ecklonia kurome Okam. 60

60.1 Introduction

Kunbu, Thallus Eckloniae or Thallus Laminariae, is the dry thallus of the brown alga *Ecklonia kurome* Okam. (Alariaceae) or *Laminaria japonica* Aresch. (Laminariaceae) collected in summer and fall. It is officially listed in the Chinese Pharmacopoeia and used in treatment of scrofula, goiter, hypertyroidism, and edema.

60.2 Chemical Constituents

Some novel phlorotannines with a dibenzo-*p*-dioxin skeleton have been isolated from *E. kurome*. They are generally constructed of phloroglucinol moieties linked by aryl-aryl, aryl-ether bonds or of a mixed type. The structures of the phlorotannins eckol (*60-1*), phloroeckol (*60-2*), dieckol (*60-3*) [1, 2], 6,6′-bieckol (*60-4*), 8,8′-bieckol (*60-5*) [2], and a phlorotannin with a furane ring (*60-6*) [3] were determined on the basis of spectral data. These compounds are the first examples of naturally occurring phlorotannins having a dibenzo-*p*-dioxin skeleton. The structure of eckol was recently established by X-ray analysis [4].

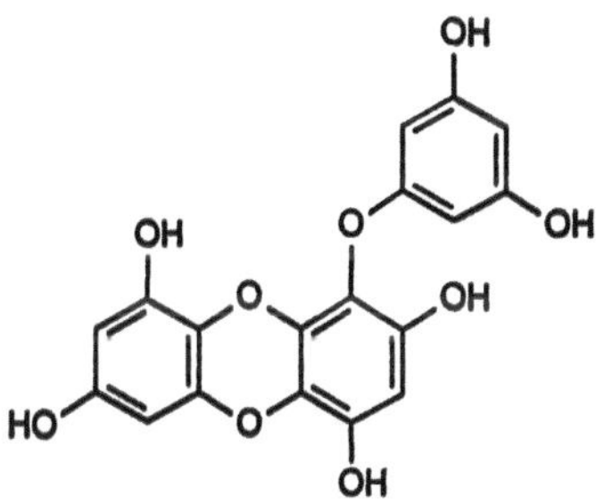

Eckol (*60-1*)

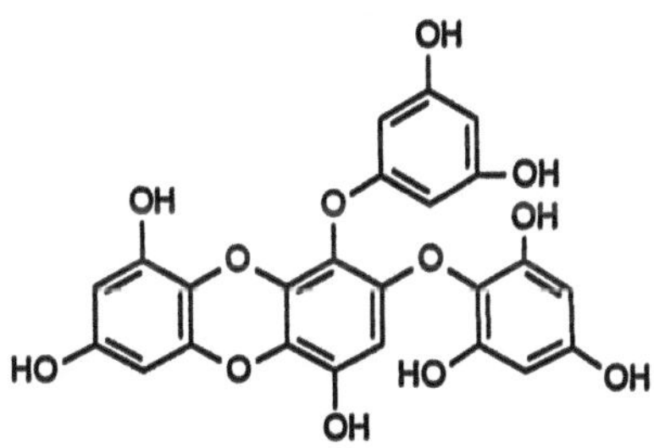

Phloroeckol (*60-2*)

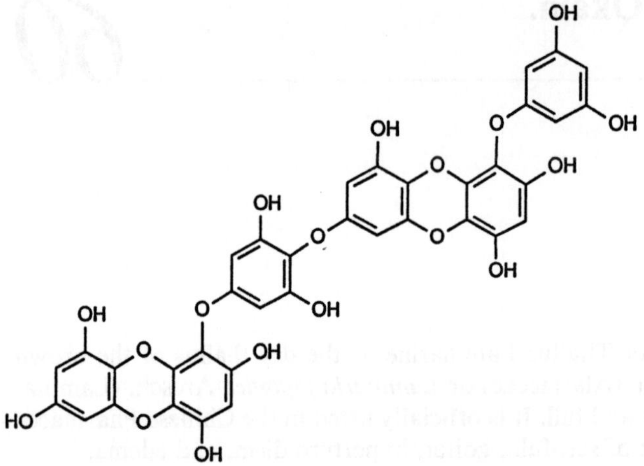

Dieckol (*60-3*)

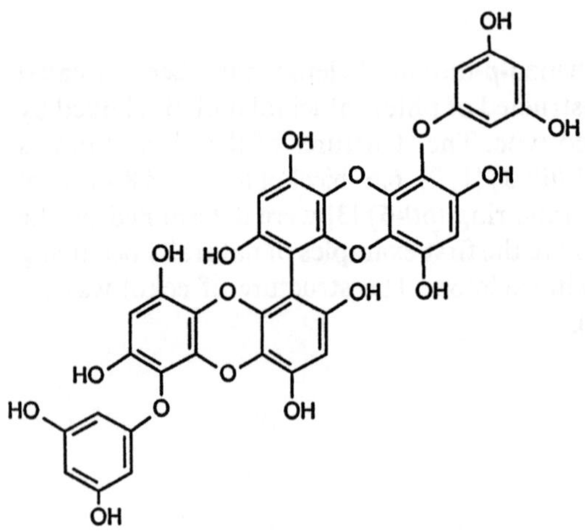

6,6′-Bieckol (*60-4*)

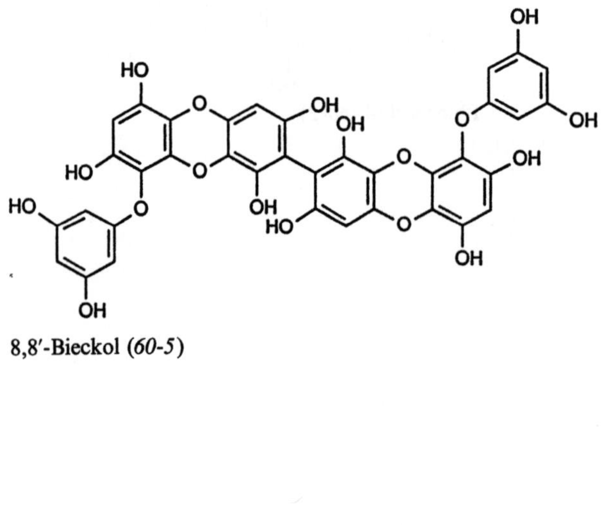

8,8′-Bieckol (*60-5*)

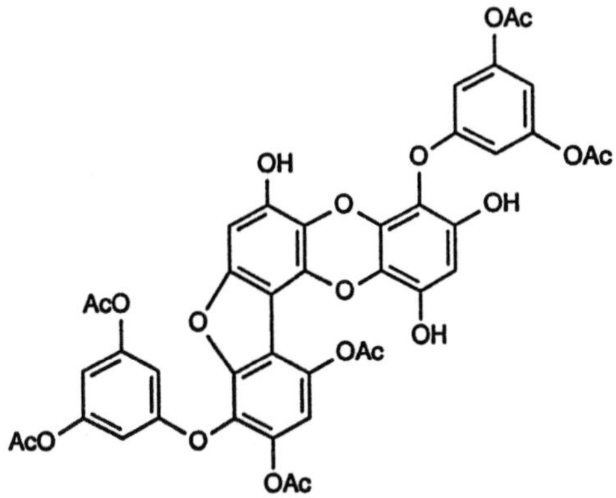

1-(3,5-Diacetoxyphenoxy)-6,7-[2,4-diacetoxy-5-(3,5-diacetoxyphenoxy)-benzo[*b*]furan]-dibenzo-*p*-dioxin-2,4,9-triol (*60-6*)

60.3 Pharmacology

The methanolic extract of *E. kurome* has been found to inhibit the action of α_2-macroglobulin and α_2-plasmin inhibitor which contribute to suppressing the fibrinolytic enzyme system. The phlorotannins were isolated as the active principles responsible for inhibition of the α_2-macroglobulin and α_2-plasmin inhibitors [1, 2, 4]. The ID_{50} of the phloroeckol sodium [5], 8,8′-bieckol sodium [6] and of the 1-(3,5-diacetoxyphenoxy)-6,7-[2,4-diacetoxy-5-(3,5-diacetoxyphenoxy)benzo[*b*]furan]dibenzo-*p*-dioxin-2,4,9-triol sodium [3] against plasmin α_2-inhibitor (3 μg/ml) were 1.9, 2, and 0.3 μg/ml, respectively.

References

1. Fukuyama Y, Miura I, Kinzyo Z, Mori H, Kido M, Nakayama Y, Takahashi M, Ochi M (1985) Eckols, novel phlorotannins with a dibenzo-*p*-dioxin skeleton possessing inhibitory effects on α_2-macroglobulin from the brown alga *Ecklonia kurome* Okamura. Chem Lett 739–742
2. Fukuyama Y, Miura I, Kinzyo Z, Nakayama Y, Takahashi M, Kido M (1983) Anti-plasmin inhibitors, polyhydroxydibenzo-*p*-dioxins isolated from *Ecklonia kurome* Okamura. Tennen Yuki Kagobutsu Toronkai Koen Yoshishu 26:126–133 (CA 100:144874u)
3. Otsuka Pharmaceutical Co (1983) Dibenzo-*p*-dioxin derivatives as antiplasmin inhibitors. Jpn Kokai Tokkyo Koho Jp 58, 118, 591 (83, 118, 591) (CA 100:12637j)
4. Fukuyama Y, Kodama M, Miura I, Kinzyo Z, Kido M, Mori M, Nakayama Y, Takahashi M (1989) Anti-plasmin inhibitor, part III. Structure of an anti-plasmin inhibitor, eckol, isolated from the brown alga *Ecklonia kurome* Okamura and inhibitory activities of its derivatives on plasma plasmin inhibitors. Chem Pharm Bull (Tokyo) 37:349–353
5. Otsuka Pharmaceutical Co (1983) Dibenzo-*p*-dioxin derivatives as inhibitors of antiplasmin agents. Jpn Kokai Tokkyo Koho JP 58, 118, 580 (83, 118, 580) (CA 100:12635g)
6. Otsuka Pharmaceutical Co (1983) Dibenzo-*p*-dioxin derivatives as antiplasmin inhibitors. Jpn Kokai Tokkyo Koho Jp 58, 118, 581 (83, 118, 581) (CA 100:12636h)

Eleutherine americana Merr. et Heyne 61

61.1 Introduction

Eleutherine americana Merr. et Heyne (Iridaceae), is a herbal plant cultivated on the island of Hainan, the rhizome of which was used as a folk medicine for the treatment of coronary diseases.

61.2 Chemical Constituents

From the ethanol extract of the rhizome of *E. americana* three naphthalene derivatives eleutherol (*61-1*), eleutherin (*61-2*), and isoeleutherin (*61-3*) were isolated [1]. Stearic acid, aceritannin, and gallic acid were also isolated [1].

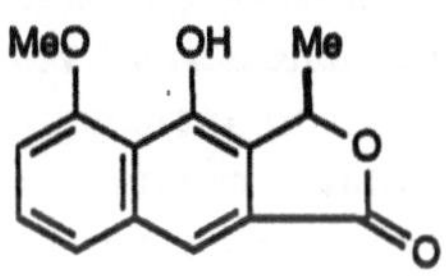

Eleutherol (*61-1*)

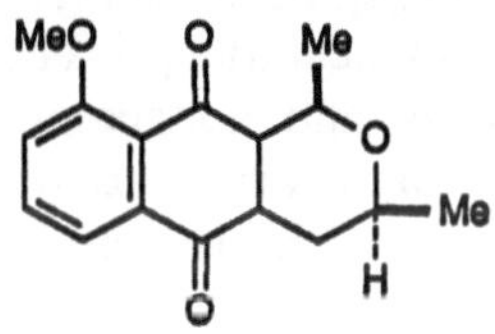

Eleutherin (*61-2*)

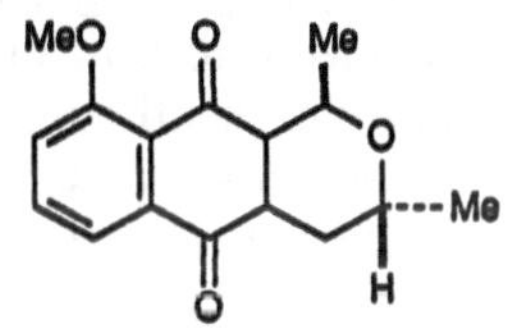

Isoeleutherin (*61-3*)

A new naphthalene derivative named hongconin (*61-4*) was further isolated. Its structure was elucidated on the basis of chemical and spectral studies, as well as by X-ray analysis [2, 3].

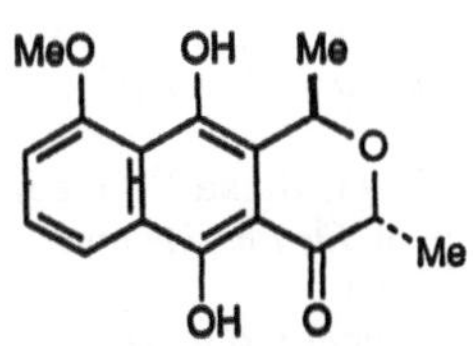

Hongconin (*61-4*)

The racemic eleutherin and isoeleutherin could be synthesized by intramolecular cyclization to the naphthopyrans of 2-acetyl-3-allyl-8-methoxy-1,4-hydronaphthoquinone dimethylether (*61-6*) which is itself obtained by the Lewis acid-mediated allylation of 2-acetyl-8-methoxy-1,4-naphthoquinone (*61-5*) with allyltrimethylstannane. Oxidative demethylation of the cyclization product led to eleutherin and isoeleutherin (Fig. 61.1) [4, 5].

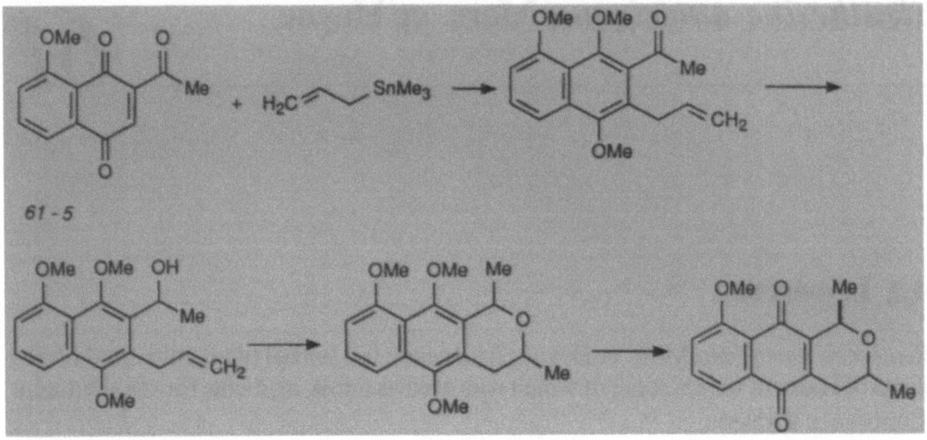

Fig. 61.1. Synthesis of racemic eleutherine and isoeleutherine

61.3 Pharmacology

The biological activity of the rhizome of *E. americana* was confirmed by pharmacological testing using isolated guinea pig heart and it was also shown to be effective in angina pectoris in a preliminary clinical trial [2]. The four naphthalene derivatives eleutherol, eleutherin, isoeleutherin, and hongconin were effective in increasing coronary flow on isolated guinea pig heart [2]. It was reported that a mixture of naphthalene derivatives isolated from the rhizome of *E. americana* prepared as tablets for clinical use against angina pectoris was as effective as dipyridamol in clinical trials [6].

References

1. Chen ZX, Huang HZ, Wang CR, Li YH, Ding JM (1981) Studies on the active constituents of Hong Cong (rhizome of *Eleutherine americana*). Chin Trad Herb Drugs 12:484
2. Chen ZX, Huang HZ, Wang CR, Li YH, Ding JM, Sankawa U, Noguchi H, Iitaka Y (1984) Hongconin, a new naphthalene derivative from the rhizome of *Eleutherine americana* (Hong-Cong). Heterocycles 22:691–694
3. Chen ZX, Huang HZ, Wang CR, Li YH, Ding JM, Sankawa U, Noguchi H, Iitaka Y (1986) Hongconin, a new naphthalene derivative from Hong-Cong, the rhizome of *Eleutherine americana* Merr. and Heyne (Iridaceae). Chem Pharm Bull (Tokyo) 34:2743–2746
4. Naruta Y, Uno H, Maruyama K (1981) Synthesis of (±)-eleutherin, (±)-isoeleutherin, and their demethoxy analogues. A novel synthetic approach. J Chem Soc, Chem Commun 1277–1278
5. Uno H (1986) Allylation of 2-alkanoyl 1,4-quinones with allylsilanes and allylstannanes. Efficient synthesis of pyranonaphthoquinone antibiotics. J Org Chem 51:350–358
6. Ding JM, Huang HZ (1982) Preparation of Hong Cong Su tablet. Chin Trad Herb Drugs 13:499–501

Ephedra spp.

62

62.1 Introduction

Mahuang, Herba Ephedrae, is the dry haulms of *Ephedra sinica* Stapf, *E. intermedia* Schrenk et C.A. Mey. or *E. equisetina* Bge. (Ephedraceae). The green haulms are harvested in the fall and dried. The *Ephedra* herb is one of the oldest and most widely used traditional Chinese medicines and is used as a diaphoretic, antiasthmatic, and diuretic. It is officially listed in the Chinese Pharmacopoeia.

Mahuanggen, Radix Ephedrae, is the dry root of *E. sinica* Stapf or *E. intermedia* collected in the fall. It is also officially listed in the Chinese Pharmacopoeia and used as an antisudorific.

62.2 Chemical Constituents

The major constituent in *Ephedra* species is the alkaloid ephedrine, which was isolated about 100 years ago [1] and was structurally determined as α-[1-(methylamino)-ethyl]-benzenemethanol [2–6]. Because this compound possesses two asymmetrical carbon atoms, two diastereomeric series are possible. One of them was designated as ephedrine, the other as pseudoephedrine. The configuration of ephedrine and pseudoephedrine were formerly the subject of a number of studies [7–9] which demonstrated the absolute configuration of (−)-ephedrine (*62-1*) and (+)-ephedrine (*62-2*) as well as of (+)-pseudoephedrine (*62-3*) and (−)-pseudoephedrine (*62-4*) [10–13]. Ephedrine shows a 1(R),2(S)(erythro) configuration, whereas pseudoephedrine has a 1(S),2(S)(threo) configuration.

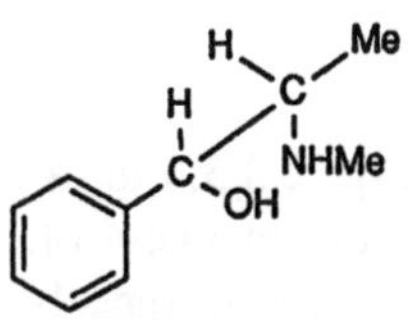

(−)-Ephedrine (*62-1*)

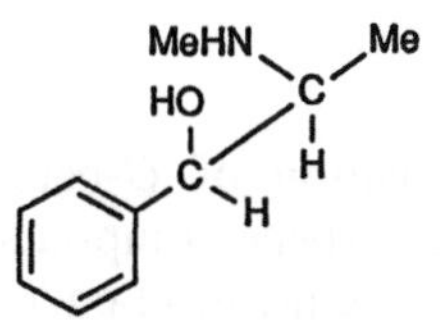

(+)-Ephedrine (*62-2*)

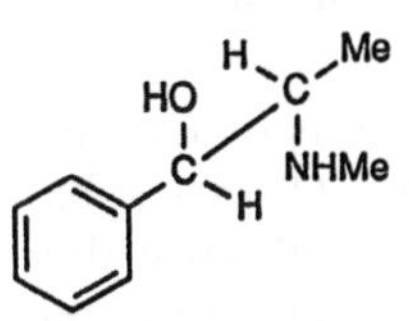

(+)-Pseudoephedrine (*62-3*)

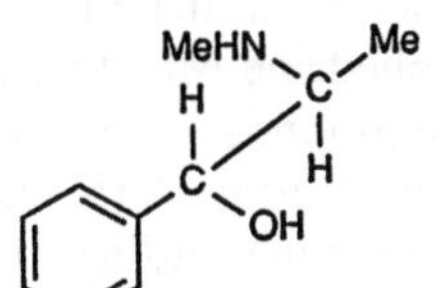

(−)-Pseudoephedrine (*62-4*)

(−)-Ephedrine and (+)-pseudoephedrine occur in plant material; (+)-ephedrine and (−)-pseudoephedrine are synthetic products. From the *Ephedra* species native to China such as *E. equisetina* and *E. sinica* [14] about 1% total alkaloid containing 63%−80% ephedrine and 20%−34% pseudoephedrine was obtained [15−17]. The contents showed great variation, depending on the origin of the plant materials [18].

Besides (−)-ephedrine and (+)-pseudoephedrine, the minor alkaloids (−)-*N*-methylephedrine (*62-5*) [19−22], (−)-norephedrine [21, 23] (*62-6*), (+)-*N*-methylpseudoephedrine (*62-7*) [20], and (+)-norpseudoephedrine (*62-8*) [24, 25] were isolated from *Ephedra* species [26, 27].

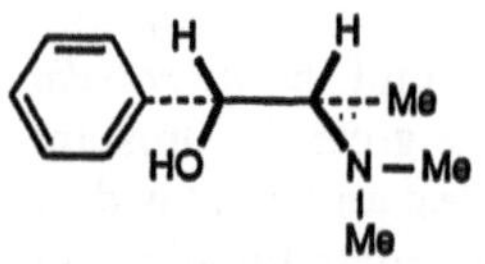

(−)-*N*-Methylephedrine (*62-5*)

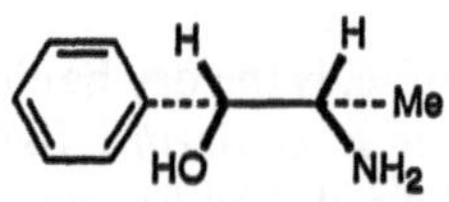

(−)-Norephedrine (*62-6*)

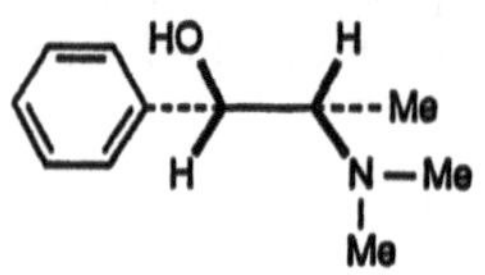

(+)-*N*-Methylpseudoephedrine (*62-7*)

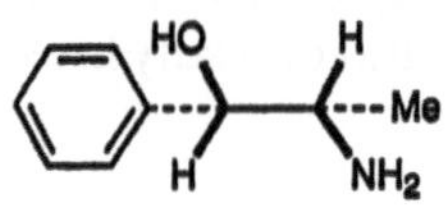

(+)-Norpseudoephedrine (*62-8*)

Recently, a new alkaloid with an oxazolidin ring named ephedroxane (*62-9*) was isolated from *E. intermedia, E. equisetina, E. sinica* and some other *Ephedra* species. Its structure was elucidated as 4(S),5(R)-3,4-dimethyl-5-phenyloxazolidone [28].

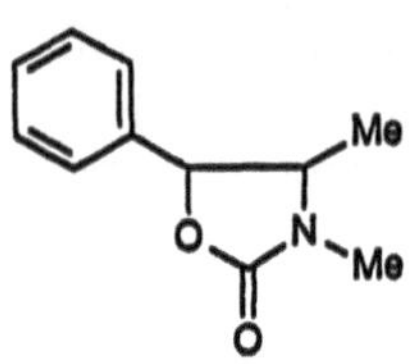

Ephedroxane (*62-9*)

Furthermore, the isolation of *O*-benzoylpseudoephedrine [29], 2,3,5,6-tetramethylpyrazine [30] and terpene compounds α-terpineol, β-terpineol, terpine-4-ol, myrcene, dihydrocarveol, *p*-menth-2-en-7-ol and 1,3,4-trimethyl-3-cyclohexene-1-carboxaldehyde [30] from *Ephedra* stems was also reported. Usually ephedrine and other alkaloid components have been obtained from *Ephedra* herb by extraction with suitable solvents such as toluene or xylene [31, 32]. A membrane process for extracting the natural ephedrine was described [33].

A number of new compounds were recently isolated from the roots of *Ephedra* species. Among them, ephedradine A (*62-10*) [34], ephedradine B (*62-11*) [35], ephedradine C (*62-12*) [36] and ephedradine D (*62-13*) [37] are related alkaloids with a macrocyclic spermine (*N,N'*-bis(3-aminopropyl)-1,4-butanediamine) moiety in the structure.

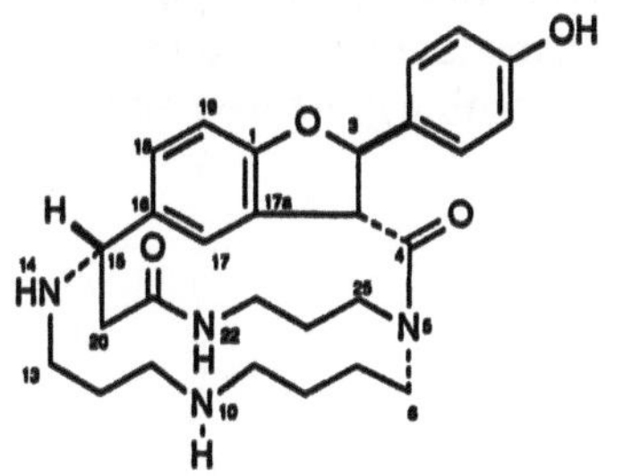

Ephedradine A (*62-10*)

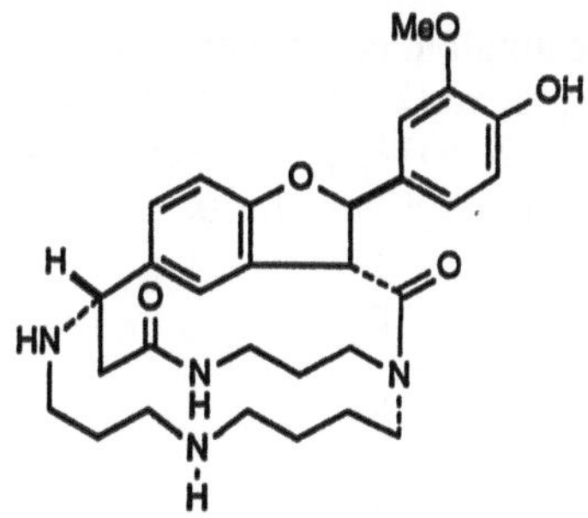

Ephedradine B (*62-11*)

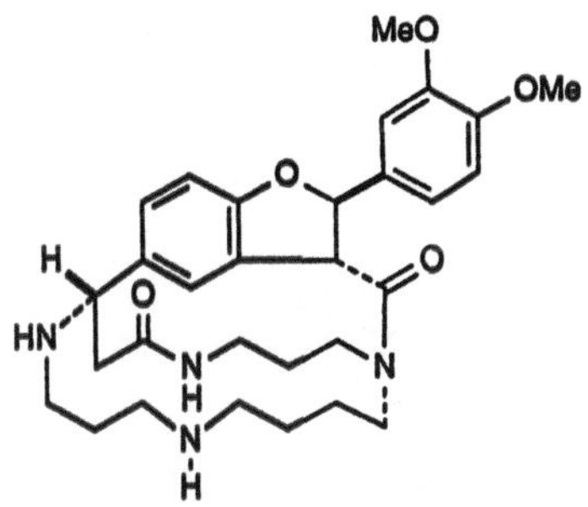

Ephedradine C (*62-12*)

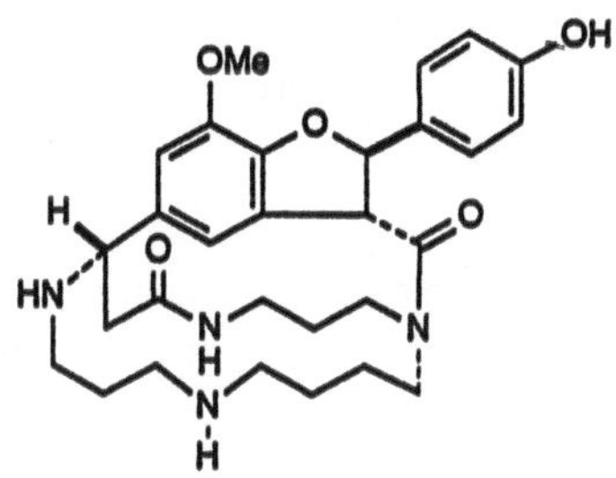

Ephedradine D (*62-13*)

Another alkaloid from the *Ephedra* roots was determined as feruloylhistamine (*62-14*) [38]. Feruloylhistamine and ephedradines are all hypotensive principles of the root.

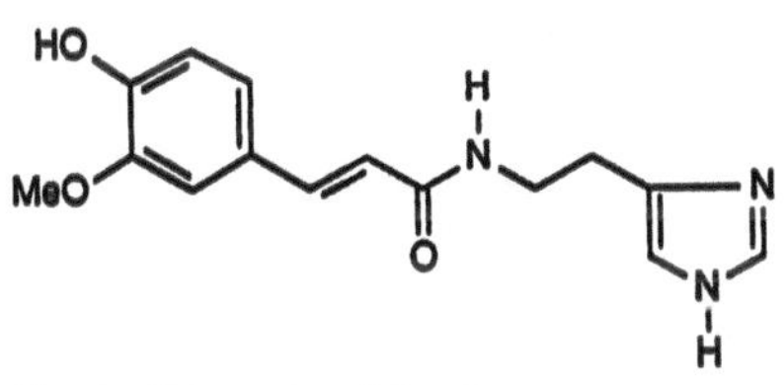

Feruloylhistamine (*62-14*)

An alkaloid with hypertensive activity named maokonine (*62-15*) was shown by conventional chemical and spectral methods and by comparison with a synthetic sample to be an inner salt of α-carboxy-4-hydroxy-*N,N,N*-trimethylbenzeneethan-ammonium hydroxide [39].

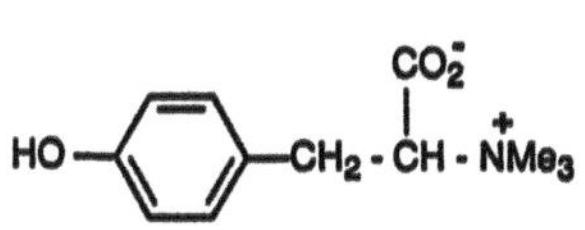

Maokonine (*62-15*)

Besides the alkaloid constituents, a flavonoflavanol compound named ephedrannin A (*62-16*) [40] and bisflavanols mahuannin A (*62-17*), mahuannin B (*62-18*) [41], mahuannin C (*62-19*) [42], and mahuannin D (*62-20*) [43] with hypotensive activity were isolated from *Ephedra* roots.

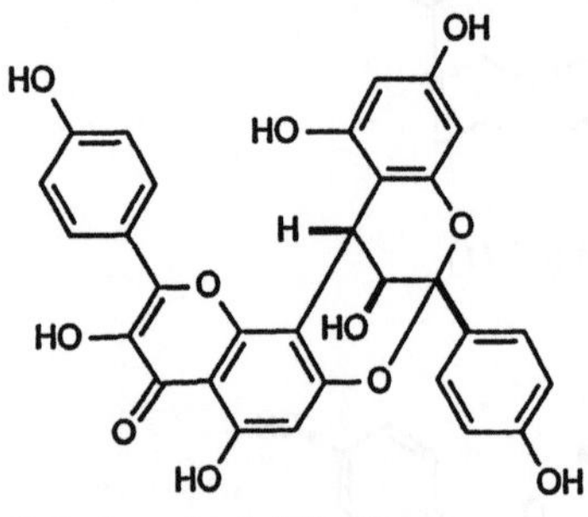

Ephedrannin A (*62-16*)

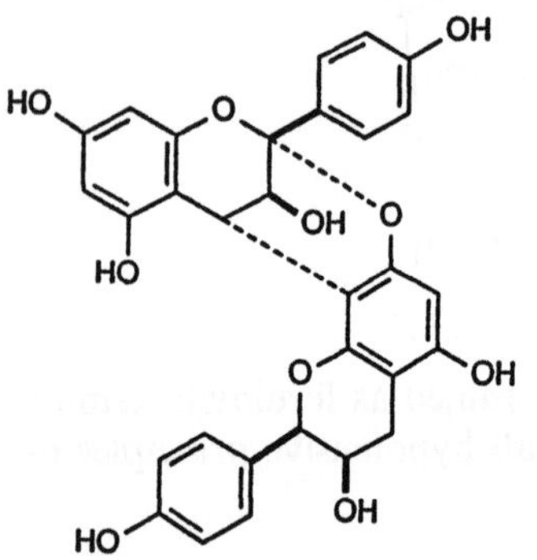

Mahuannin A (*62-17*)

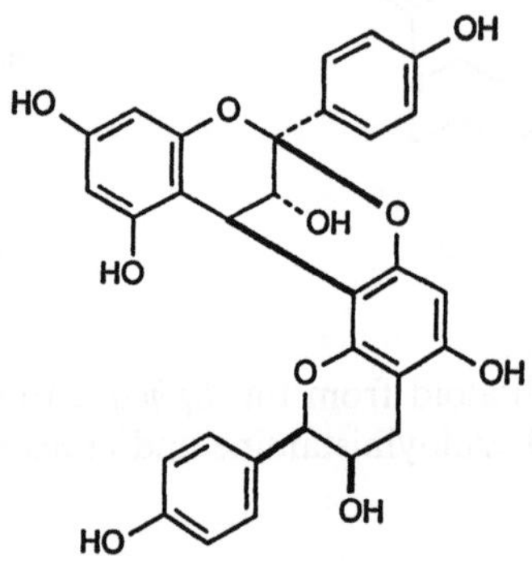

Mahuannin B (*62-18*)

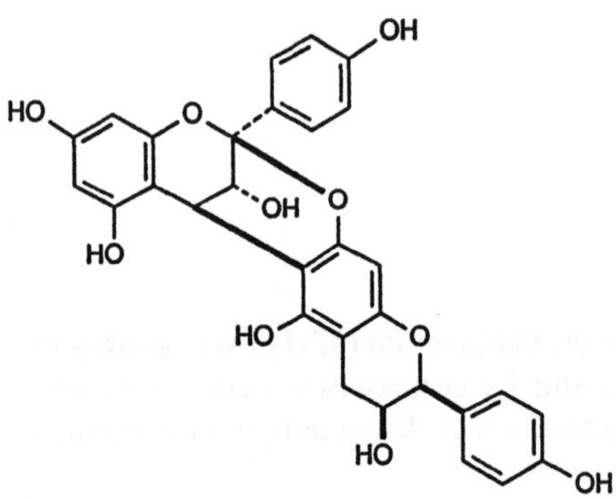

Mahuannin C (*62-19*)

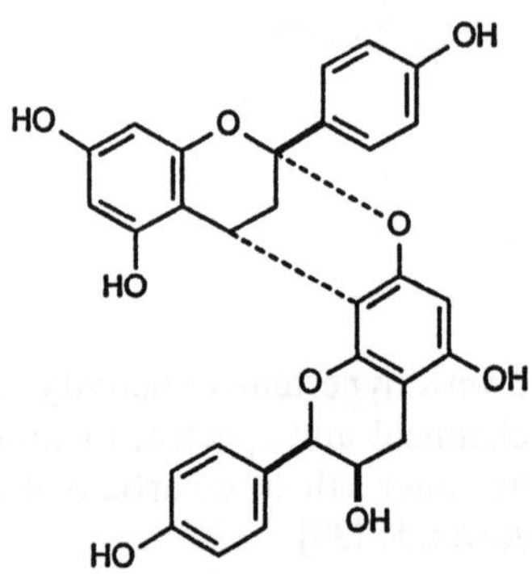

Mahuannin D (*62-20*)

62.3 Pharmacology

Ephedrine, the main active principle of the *Ephedra* herb, is a well-known sympathomimetic used clinically for the treatment of bronchial asthma, anaphylactic reaction, and hypotension.

(−)-Ephedrine, (+)-pseudoephedrine, and (±)-ephedrine showed a relaxant activity on isolated guinea pig trachea previously contracted by acetylcholine in vitro and partially inhibited the bronchoconstriction of guinea pig caused by acetylcholine in vivo due to stimulation of β-adrenoceptors [44]. The affinities of (−)-ephedrine, (+)-pseudoephedrine, (−)-methylephedrine, and (+)-norpseudoephedrine to β-adrenoceptors were investigated on isolated rat lung cell membranes by ligand-binding assay. All the compounds tested competed with the binding of [^{3}H]dihydroalprenolol to the rat lung cell membrane preparations in the following order: (−)-ephedrine > (+)-norpseudoephedrine > (+)-pseudoephedrine ≫ (−)-methylephedrine. The ability of (−)-ephedrine to activate adenylate cyclase was greater than that of (+)-pseudoephedrine and (−)-methylephedrine, which appeared to be consistent with the orders of affinity for the β-adrenoceptors. However, (+)-norpseudoephedrine had no effect on adenylate cyclase activity and antagonized the stimulating action of isoproterenol on adenylate cyclase activity, suggesting (+)-norpseudoephedrine to be a β-adrenoceptor blocker [45].

Intravenous injection of ephedrine dose dependently increased pulmonary artery blood pressure and blood vessel resistance in cats. An increase in intratracheal pressure induced by acetylcholine was inhibited by ephedrine by concentrations higher than those to increase pulmonary artery pressure. Results obtained in isolated pulmonary artery strips of rabbits indicated that ephedrine induces the contraction of the artery by indirect and direct actions [46].

Ephedrine inhibited the electrically stimulated twitch response in preparations of isolated guinea pig ileum muscle [47, 48], its spontaneous contraction and contractions induced by acetylcholine or serotonin [47].

Both ephedrine and norepinephrine produced concentration-dependent contractile effects on aortic strips of rabbits in vitro. The effect of norepinephrine was significantly increased, that of ephedrine markedly decreased by cocaine. In strips isolated from reserpine-pretreated rabbits, the actions of both norepinephrine and ephedrine were potentiated [49].

Ephedrine exhibited an anticholinesterase activity studied in vitro with acetylcholinesterase of human erythrocytes and butyrylcholinesterase of equine blood serum. The erythro derivatives of ephedrine are more active than the threo derivatives. Evidently, an erythro isomer is better adapted for binding to an anionic site of acetylcholinesterase than a threo isomer. The anionic site of the active center of butyrylcholinesterase appears to be less specific with regard to stereoisomerism [50].

Ephedrine, pseudoephedrine, and ephedroxane showed antiinflammatory activity measured by inhibition of hind-paw edema induced by histamine, serotonin, and bradykinin in mice [51, 52]. These alkaloids inhibited arachidonic acid release and prostaglandin production of rat peritoneal macrophages stimulated with zymosan, suggesting that a part of the antiinflammatory activities of these alkaloids is attributable to their inhibitory effects on arachidonate metabolism [53].

In contrast to ephedrine and pseudoephedrine showing stimulatory activity in the CNS, ephedroxane exerted inhibitory actions. Additionally, ephedroxane possessed only a weak antihistaminic activity compared with that of ephedrine [54].

By intravenous injection, the minimum lethal doses of ephedrine in rats, rabbits, cats, and dogs were 135–140, 66–70, 75, and 70–75 mg/kg, respectively. In rabbits, ephedrine is least toxic when given orally. The minimum lethal doses of ephedrine in rabbits by oral, intramuscular, subcutaneous, and intraperitoneal administrations were 590, 340, 320–400, and 310–400 mg/kg, respectively. Evidence from direct observation, perfusion experiments, and electrocardiographic studies indicates death after intravenous injection in rabbits, cats, and dogs to be due primarily to cardiac collapse. Respiratory failure follows immediately [55]. In chronic toxicologic studies, ephedrine administered in doses of 25 mg/kg intravenously, intramuscularly, or orally in rabbits daily for 4 weeks produced no observable toxic effects. All the animals, with few exceptions, showed a large gain in body weight after a period of about 140 days. The visceral organs from treated animals showed no definite or uniform lesion that can be attributed to the effect of ephedrine. In rats, ephedrine given in doses of 100 mg/kg every other day for 7 days produced no damage to the visceral organs [56].

Ephedrine is a secondary amine and can be N-nitrosated. Formation of *N*-nitrosephedrine was studied by using the WHO Nitrosation Assay Procedure [57] and under conditions simulating those within the normal fasting stomach with a nitrite concentration of 25 μM [58]. The mutagenicity of *N*-nitrosoephedrine was shown using *Salmonella typhimurium* TA 100 and TA 98 and *Escherichia coli* WP_2 uvrA/pkM in the presence of hamster S9 cells [59]. Ephedrine itself showed no mutagenic activity [60]. In experimental studies in mice and rats, ephedrine was not found to be carcinogenic. Furthermore, the survival of ephedrin-exposed rats during the 2-year study was greater than that of controls [60, 61].

A study of the carcinogenicity of *N*-nitrosoephedrine by oral administration on male Sprague-Dawley rats at doses of 120 mg/kg, twice weekly, showed that 50% of treated animals died with preneoplastic and malignant lesions mainly in the liver, lung, and forestomach. The observation of hyperkeratosis, papillomas, and one squamous cell carcinoma of the forestomach suggests that the *N*-nitrosoephedrine not only exhibits systemic effects but is probably also a weak local carcinogen [62].

Following an oral dose of [^{14}C](−)-ephedrine to human subjects, 97% of the dose was excreted in the urine within 48 h, and 88% in the first 24 h. Unchanged drug was the major urinary excretory product (53%–74%), with *N*-demethylation to norephedrine occurring to a variable extent (8%–20%). Oxidative deamination was also variable (4%–13%); the main products identified were benzoic acid and 1-phenylpropane-1,2-diol both free and conjugated [63]. After oral administration of phenyl-$^2H_5$-labeled (−)-ephedrine, about 75% of the intact ephedrine was recovered in the urine. Major metabolites isolated from the urine were norephedrine and hippuric acid. Benzoic acid was only detected in trace amounts [64–66]. (−)-Ephedrine administered to human subjects was excreted in urine at the same rate whether given orally or percutaneously, but (±)-norephedrine and (−)-methylephedrine were excreted more slowly after percutaneous than after oral administration [67].

The metabolism of ephedrine showed great species differences. After subcutaneous administration of [^{14}C](−)-ephedrine to mice, rabbits, guinea pigs, and rats, about 70%–90% of the administered radioactivity was found in the urine 24 h after injection. The major urinary metabolites were norephedrine and benzoic acid in guinea pigs, *p*-hydroxyephedrine in mice and rats, and 1-phenyl-propane-1,2-diol, and hippuric acid in rabbits. Approximately 86% of the urinary radioactivity in mice was unchanged ephedrine. Ephedrine and *p*-hydroxyephedrine were excreted as their glucuronides [68, 69]. Urinary excretion of ^{14}C activity in rats or rabbits given α-[^{14}C]phenyl-$^{2}H_5$-labeled ephedrine subcutaneously was not significantly different from that in animals given the non-deuterated compund [70].

In rat livers, norephedrine, *p*-hydroxyephedrine, ephedrine glucuronide, and *p*-hydroxyephedrine glucuronide were identified as the metabolites of [^{14}C](−)-ephedrine and [^{14}C](+)-ephedrine after subcutaneous administration. The amount of radioactivity in various organs at 2.5 h after injection decreased in the order: liver > kidney > brain > spleen [71]. A stereoselective reaction in the formation of the glucuronides of ephedrine, norephedrine, and *p*-hydroxyephedrine was observed. The (−)-isomers were more easily subjected to glucuronide formation than the (+)-isomers [72].

Feruloylhistamine, the imidazole alkaloid from *Ephedra* roots and its synthetic analogs cinnamoylhistamine, *p*-coumaroylhistamine, caffeoylhistamine, and sinapoylhistamine exerted hypotensive, histidine decarboxylase inhibiting, and antihepatotoxic action in mice or rats [73].

Ephedradines A, B, C, and D elicited hypotensive effects in rats. Administration of ephedradine B at a dose of 0.1–3 mg/kg intravenously to normal rats and to spontaneously hypertensive rats reduced the blood pressure in a dose-dependent manner. The hypotensive mechanism of ephedradine B was postulated to occur mainly by ganglion block [74].

References

1. Ladenburg A, Oelschägel C (1889) Über das „Pseudo-Ephedrin". Chem Ber 22:1823–1827
2. Emde H (1907) Ephedrine and pseudoephedrine: a case of unlike-halved asymmetry. Arch Pharm (Weinheim) 245:662–679
3. Fourneau E (1907) Synthetic ephedrines. J Pharm Chim 25:593–602
4. Schmidt E (1908) Ephedrine and pseudoephedrine. Arch Pharm (Weinheim) 246:210–214
5. Schmidt E (1914) Ephedrine and pseudoephedrine. Arch Pharm (Weinheim) 252:98–138
6. Rabe P (1911) Über das Ephedrin und das Pseudoephedrin. Chem Ber 44:824–827
7. Späth E, Göhring R (1920) Die Synthese des Ephedrins, des Pseudoephedrins, ihrer optischen Antipoden und Razemkörper. Monath Chem 41:319–338
8. Nagai WN, Kanao S (1929) Über die Synthese der isomeren Ephedrine und ihre Homologe. Liebigs Ann Chem 470:157–182
9. Emde H (1929) Über Diastereomerie. I. Konfiguration des Ephedrins. Helv Chim Acta 12:365–376
10. Leithe W (1932) Die Konfiguration der Ephedrin-Basen. Chem Ber 65:660–666
11. Skita A, Keil F, Meiner H (1933) Kernhydrierte optisch aktive Ephedrine. Chem Ber 66:974–984
12. Freudenberg K, Nikolai F (1934) Die Konfiguration des Ephedrins. Liebigs Ann Chem 510:223–230
13. Freudenberg K, Schoeffel E, Braun E (1932) Study on the configuration of ephedrine. J Am Chem Soc 54:234–236
14. Read BE, Lire JC (1928) Chinese botanical sources of ephedrine and psdeudoephedrine. J Am Pharm Assoc 17:339–344

15. Read BE (1927) Note on Chinese *Ephedra*. Pharm J 118:681
16. Read BE, Feng CT (1927) The ephedrine content of Chinese *Ephedra*. Pharm J 119:356–357
17. Fan TC (1958) Quantitative determination of *l*-ephedrine, pseudoephedrine and their combined weight in *Ephedra equisetina*. Aptechn Delo 7 (5):9–17 (CA 54:7979f)
18. Yamazaki K (1985) Chemical components of Ma-Huang. Wakan Iyaku Gakkaishi 2:93–94 (CA 104:56500w)
19. Smith S (1927) *l*-Methylephedrine, an alkaloid from *Ephedra* species. J Chem Soc 2056–2059
20. Nagai WN, Kanao S (1928) Constituents of Chinese drug "Ma Huang" VI. J Pharm Soc Jpn 48:845–851
21. Sagara K, Oshima T, Misaki T (1983) A simultaneous determination of norephedrine, pseudoephedrine, ephedrine and methylephedrine in *Ephedra* herb and oriental pharmaceutical preparations by ion-pair high-performance liquid chromatography. Chem Pharm Bull (Tokyo) 31:2359–2365
22. Iwanami N, Ohtsuka Y, Kubo H (1985) Determination of ephedrine alkaloids in *Ephedra* herb and oriental pharmaceutical preparations by HPLC. Chin Pharm Bull 20:149–153
23. Kanao S (1930) *l*-Norephedrin, über die Bestandteile der chinesischen Droge "Ma Huang". VII. Mitteilung. Chem Ber 63:95–98
24. Smith S (1928) Nor-*d*-pseudoephedrine, an alkaloid from *Ephedra* species. J Chem Soc 51–53
25. Yue N (1983) Extraction and transformation of (+)-norpseudoephedrine. Pharm Ind 2:45–46
26. Endo M, Kanazawa R, Hashimoto Y, Kato A, Mizumo M (1984) High performance liquid chromatographic determination of organic substances by metal chelate derivatization. III. Analysis of *Ephedra* bases. Chem Pharm Bull (Tokyo) 32:7447
27. Zhang JA, Tian Z, Lou ZC (1988) Simultaneous determination of six alkaloids in Ephedrae Herba by high-performance liquid chromatography. Planta Med 54:69–70
28. Konno C, Taguchi T, Tamada T, Hikino H (1979) Ephedroxane, anti-inflammatory principle of *Ephedra* herbs. Phytochemistry 18:697–698
29. Cheng DL, Wang DM, Li S, Chu TT (1985) A minor alkaloid from the extract of *Ephedra* herb. Chem J Chin Univ 6:609–612
30. Sun JY (1983) Novel active constituents of *Ephedra sinica*. Chin Trad Herb Drugs 14:345–346
31. Su YF, Wang G, Fang YX, Zhi YZ, Zhon KC (1958) Extraction of ephedrine from *Ephedra sinica* leach liquor. I. Selection of solvents and determination of distribution data. Hua Kung Hsüeh Pao 24–33
32. Su YF, Wang G, Fang YZ, Zhi YZ, Zhon KC (1958) Extraction of ephedrine from *Ephedra sinica* leach liquor. II. Performance of extraction column. Hua Kung Hsüeh Pao 34–50
33. Wang Y, Yang HL, Quyang CG, Xu XZ, Gao YQ, Jing W, Jia XR (1986) Extraction of natural ephedrine by a membrane process. Membr Sep Sci Technol 6(3):47–56
34. Tamada M, Endo K, Hikino H, Kabuto C (1979) Structure of ephedradine A, a hypotensive principle of *Ephedra* roots. Tetrahedron Lett 873–876
35. Tamada M, Endo K, Hikino H (1979) Structure of ephedradine B, a hypotensive principle of *Ephedra* roots. Heterocycles 12:783–786
36. Konno C, Tamada M, Endo K, Hikino H (1980) Structure of ephedradine C, a hypotensive principle of *Ephedra* roots. Heterocycles 14:295–298
37. Hikino H, Ogata M, Konno C (1982) Structure of ephedradine D, a hypotensive principle of *Ephedra* roots. Heterocycles 17:155–158
38. Hikino H, Ogata M, Konno C (1983) Structure of feruloylhistamine, a hypotensive principle of *Ephedra* roots. Planta Med 48:108–110
39. Tamada M, Endo K, Hikino H (1978) Studies on the constituents of *Ephedra*. II. Maokonine, hypertensive principle of *Ephedra* roots. Planta Med 34:291–293
40. Hikino H, Takahashi M, Konno C (1982) Studies on the constituents of *Ephedra*. 10. The validity of oriental medicines. 33. Structure of ephedrannin A, a hypotensive principle of *Ephedra* roots. Tetrahedron Lett 23:673–676
41. Hikino H, Shimoyama N, Kasahara Y, Takahashi M, Konno C (1982) Studies on the constituents of *Ephedra*. 11. Validity of the oriental medicines. 35. Structures of mahuannin A and B, hypotensive principle of *Ephedra* roots. Heterocycles 19:1381–1384
42. Kasahara Y, Shimoyama N, Konno C, Hikino H (1983) Studies on the constituents of *Ephedra*. 14. Validity of the oriental medicines. 59. Structure of mahuannin C, a hypotensive principle of *Ephedra* roots. Heterocycles 20:1741–1744

43. Kasahara Y, Hikino H (1983) Studies on the constituents of *Ephedra*. 15. Structure of mahuannin D, a hypotensive principle of *Ephedra* roots. Heterocycles 20:1953–1956
44. Floch A, Lagente V, Feslon JC, Blanc M, Advenier C (1985) Study of the bronchomotor effects of (−)ephedrine, (±)-ephedrine and (+)-pseudoephedrine on the guinea pig. Ann Pharm Fr 43:31–38
45. Jiang MH, Liu L, Wang Q, Zhan WX, Shu HD (1987) Effects of ephedrine and its analogs on the β-adrenoceptors of the rat lung cell membrane. Acta Pharmacol Sin 8:318–320
46. Zhu MY, Zhao NC, Zhang KY, Hao MS (1985) Effects of ephedrine on pulmonary vessels. J Chin Med Univ 14:437–439
47. Shu HD, Huang GP, Yang ZC (1987) Effect of ephedrine on the myenteric plexus-longitudinal muscle of the guinea pig ileum in vitro. Acta Pharmacol Sin 8:213–216
48. Shu HD, Zhang LH, Zhong ZD, Shen DL, Yang ZC (1987) Effects of ephedrine on neuromuscular transmission in vitro. Acta Pharm Sin 8:313–317
49. Gong QY, Yang ZC (1984) Studies on the mechanism of ephedrine's action on rabbit aorta and atrium. Acta Physiol Sin 36:367–373
50. Dalimov DN, Vaizburg GM, Abdullaeva LK, Abduvakhabov AA, Sadykov AS (1986) Specificity of interaction of ephedrine- and pseudoephedrine-based choline analogs with acetylcholinesterase and butyrylcholinesterase. Dokl Akad Nauk SSSR 289:227–230
51. Hikino H, Konno C, Takata H, Tamada M (1980) Studies on the constituents of *Ephedra*. VI. Antiinflammatory principle of *Ephedra* herbs. Chem Pharm Bull (Tokyo) 28:2900–2904
52. Kasahara Y, Hikino H, Tsurufuji S, Watanabe M, Ohuchi K (1985) Oriental medicines. Part 75. Studies on the constituents of *Ephedra*. 18. Antiinflammatory actions of ephedrines in acute inflammations. Planta Med 51:325–331
53. Sugawara T, Ohuchi K, Watanabe M, Hirasawa N, Tsurufuji S, Kasahara Y, Hikino H (1986) Arachidonate metabolism in macrophages and the effect of "mao" alkaloids. Ensho 6:245–249 (CA 106:95783 b)
54. Hikino H, Ogata K, Kasahara Y, Konno C (1985) Pharmacology of ephedroxanes. J Ethnopharmacol 13:175–191
55. Chen KK (1926) Acute toxicity of ephedrine. J Pharmacol 27:61–76
56. Chen KK (1926) Effect of repeated administration of ephedrine. J Pharmacol 27:77–86
57. Gillatt PN, Hart RJ, Walters CL, Reed PI (1984) Susceptibilities of drugs to nitrosation under standardized chemical conditions. Food Chem Toxicol 22:269–274
58. Gillatt PN, Palmer RC, Smith PLR, Walters CL, Reed PI (1985) Susceptibilities of drugs to nitrosation under simulated gastric conditions. Food Chem Toxicol 23:849–855
59. Shimizu H, Takemura N, Ando H, Morita M, Machida K (1984) The in vivo formation of mutagenic *N*-nitroso compounds from methamphetamine and ephedrine. Nippon Eiseigaku Zasshi 39:573–580 (CA 102:39787 m)
60. National Toxicology Program (1986) Toxicology and carcinogenesis studies of ephedrine sulfate (CAS No. 134-72-5) in F 344/N rats and B6C3F_1 mice (Feed studies). Natl Toxicol Program Tech Rep Ser 307
61. National Toxicology Program (1986) Toxicology and carcinogenesis studies of ephedrine sulfate (CAS 134-72-5) in F344/N rats and B6C3F_1 mice (feed studies). Report NTP-TR-307, NIH/PUB-86-2563; order no PB 86-247285/GAR
62. Eisenbrand C, Peussmann R, Schmaehl D (1978) Carcinogenicity of *N*-nitrosoephedrine in rats. Cancer Lett 5:103–106
63. Sever PS, Dring LG, Williams RT (1975) The metabolism of (−)-ephedrine in man. Eur J Clin Pharmacol 9:193–198
64. Baba S, Kawai K, Horie M, Shida Y (1972) Metabolic studies in man using deuterium-labeled drugs. Quantitative determination of the metabolites of *l*-ephedrine labeled in the aromatic ring. Yakugaku Zasshi 92:1569–1571
65. Kawai K, Baba S (1976) Studies on drug metabolism by use of isotopes. XVII. Mass spectrometric quantification of urinary metabolites of deuterated *l*-ephedrine in man. Chem Pharm Bull (Tokyo) 24:2728–2732
66. Baba S (1973) Analysis of drug metabolites in human bodies using deuterium and carbon-13 labeled drugs. Nippon Aisotopu Kaigi Hobunshu 11:171–176 (CA 84:83892 a)
67. Beckett AH, Gorrod JW, Taylor DC (1972) Comparison of oral and percutaneous routes in man for the systemic administration of ephedrines. J Pharm Pharmacol 24:[Suppl] 65P–70P
68. Baba S, Enogaki K, Matsuda A, Nagase Y (1972) Analysis of drugs using radioisotopes. VIII. Species differences in *l*-ephedrine metabolism. Yakugaku Zasshi 92:1270–1274

69. Kawai T, Baba S (1975) Studies of drug metabolism using isotopes. XIV. Mass spectrometric quantification of urinary metabolites of deuterated *l*-ephedrine in rabbits. Chem Pharm Bull (Tokyo) 23:289–293
70. Kawai K, Baba S (1974) Studies on the drug metabolism by use of isotopes. XIII. Isotope effect on metabolism of deuterated *l*-ephedrine. Chem Pharm Bull (Tokyo) 22:2372–2376
71. Baba S, Kawai K (1972) Analysis of drugs using radioisotopes. IX. Distribution and identification of metabolites of *d*- and *l*-ephedrine in the rat. Yakugaku Zasshi 92:1534–1539
72. Baba S, Kuroda Y, Horie M (1986) Studies on drug metabolism by use of isotopes. XXIX. Studies of the differences in biological fates of ephedrine isomers by use of a pseudoracemic mixture. Biomed Environ Mass Spectrom 13:141–143
73. Hikino H, Kiso Y, Ogata M, Konno C, Aisaka K, Kubota H, Hirosa N, Ishihara T (1984) Pharmacological actions of analogues of feruloylhistamine, an imidazole alkaloid of *Ephedra* roots. Planta Med 50:478–480
74. Hikino H, Ogata K, Konno C, Sato S (1983) Hypotensive actions of ephedradines, macrocyclic spermine alkaloids of *Ephedra* roots. Planta Med 48:290–293

Epimedium spp. 63

63.1 Introduction

Yinyanghuo, Herba Epimedii, is the dry above ground part of *Epimedium brevicornum* Maxim., *E. sagittatum* (Sieb. et Zucc.) Maxim., *E. pubescens* Maxim, or *E. koreanum* Nakai (Berberidaceae) collected in summer and fall when the plant is mature. The Chinese Pharmacopoeia requires a qualitative determination of the main flavone constituent icariin for this official herbal medicine by thin-layer chromatographic comparison with an authentic sample after extraction of the powdered herb with ethanol. It is used in traditional Chinese medicine as a tonic and in the treatment of rheumatic and paralytic diseases and involutional hypertension.

63.2 Chemical Constituents

The main constituents in *Epimedium* plants are the prenylflavone glycosides. The isolation of icariin (*63-1*) from *E. brevicornum* [1–3] and *E. koreanum* [2, 3], and its demethyl analog epimedoside A (*63-2*) from *E. koreanum* [4] was reported.

The glycoside icariin was first isolated from *E. macranthum* [5]. By acidic hydrolysis of icariin, icariside I (*63-3*) was obtained by loss of a rhamnose [6]. Icariside I was also isolated from *E. brevicornum* [7] and *E. sagittatum* [8].

Icariin and other prenylflavone derivatives isolated from some *Epimedium* species are listed in Table 63.1.

Table 63.1. Prenylflavone glycosides isolated from some *Epimedium* species used in Chinese medicine

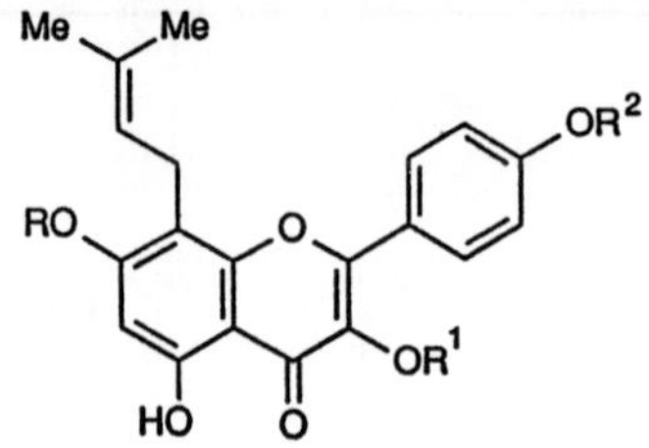

Compound	R	R^1	R^2	Plant origin	Ref.
Icariin (*63-1*)	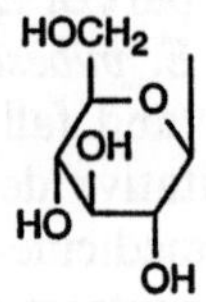	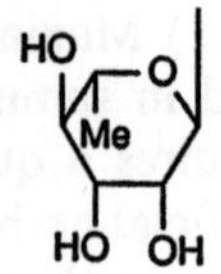	CH_3	*E. brevicornum* Other *Epimedium* species	[1-3] [2, 3, 9]
Epimedoside A (*63-2*)	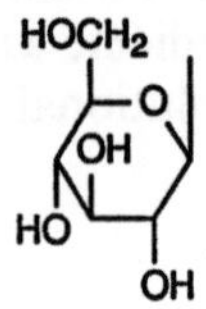	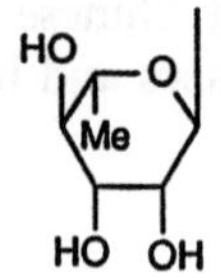	H	*E. koreanum* *E. davidii*	[4] [16]
Icariside I (*63-3*)	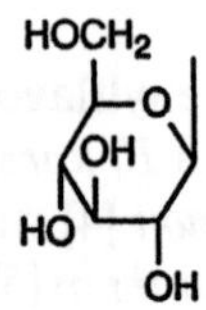	H	CH_3	*E. brevicornum* *E. sagittatum*	[7] [8]
Anhydroicaritin 3-*O*-α-L-rhamnopyranoside (*63-4*)	H	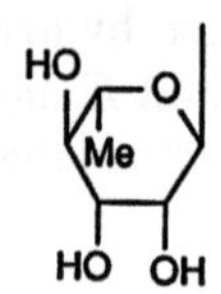	CH_3	*E. sagittatum*	[8]
Sagittatoside A (*63-6*)	H	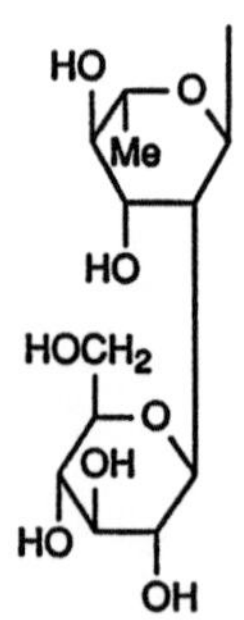	CH_3	*E. sagittatum*	[10]

Table 63.1. (continued)

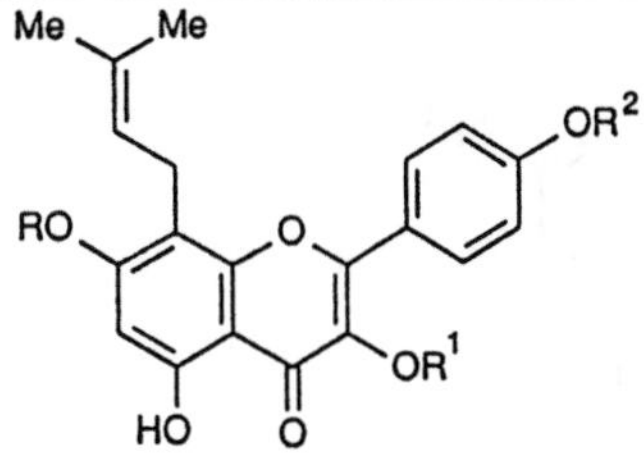

Compound	R	R¹	R²	Plant origin	Ref.
Sagittatoside B (*63-7*)	H	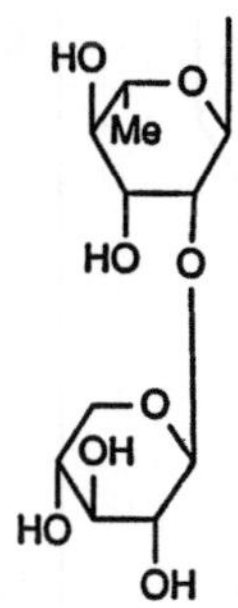	CH_3	*E. sagittatum*	[10]
Sagittatoside C (*63-8*)	H	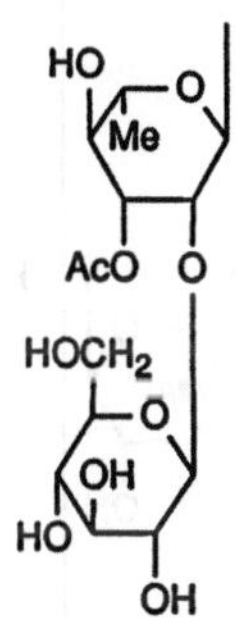	CH_3	*E. sagittatum*	[10]
Epimedin A (*63-11*)	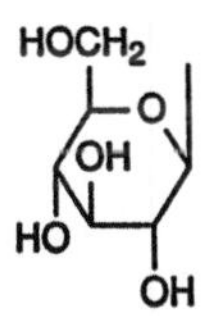	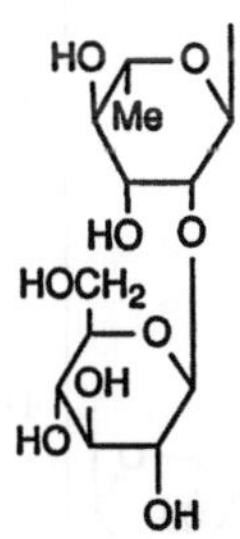	CH_3	*E. koreanum*	[12]
Epimedin B (*63-12*)	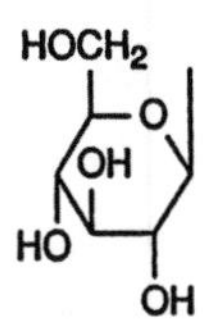	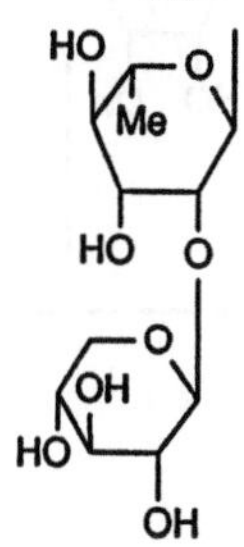	CH_3	*E. koreanum*	[12]

Table 63.1. (continued)

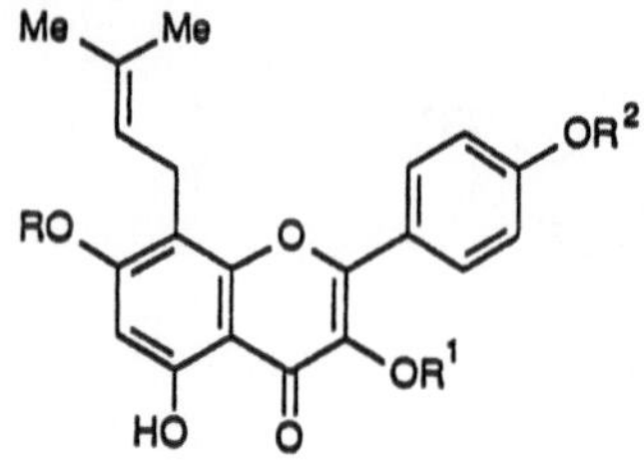

Compound	R	R¹	R²	Plant origin	Ref.
Epimedin C (*63-13*)	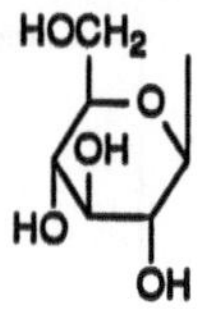	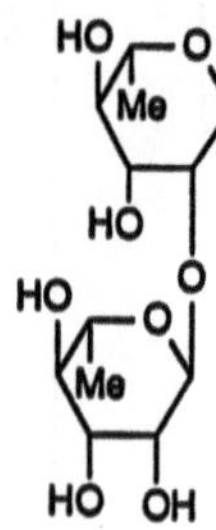	CH_3	*E. koreanum*	[12]
Baohuoside II (*63-14*)	H	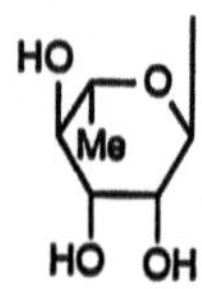	H	*E. davidii*	[16]
Baohuoside III (*63-15*)	H	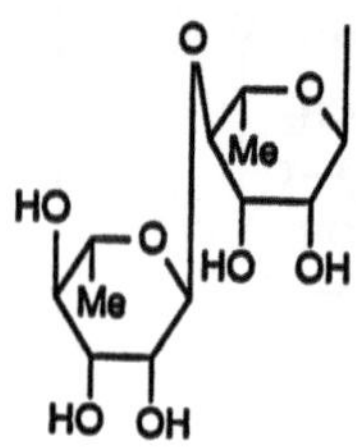	H	*E. davidii*	[16]
Baohuoside IV (*63-16*)		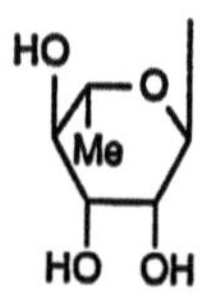	H	*E. davidii*	[16]
Baohuoside V (*63-17*)	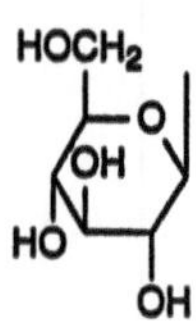	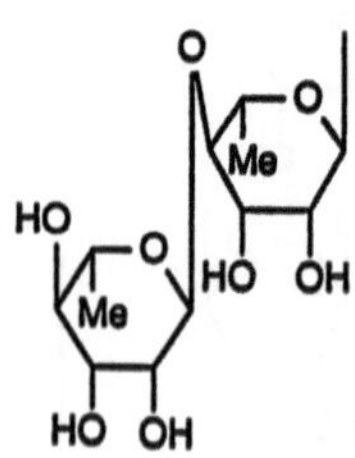	H	*E. davidii*	[16]

Table 63.1. (continued)

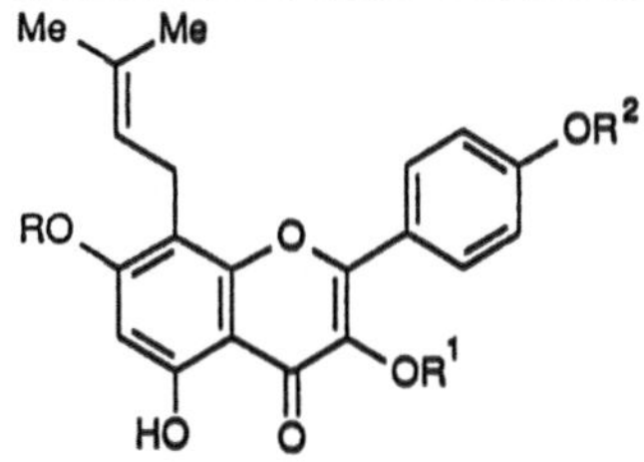

Compound	R	R¹	R²	Plant origin	Ref.
Baohuoside VI (*63-18*)	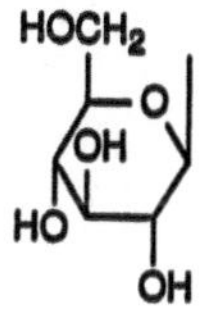	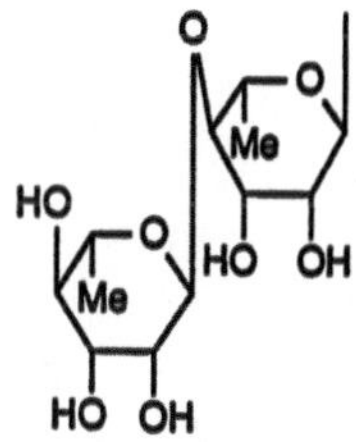	CH_3	*E. davidii*	[15]
Baohuoside VII (*63-19*)	H	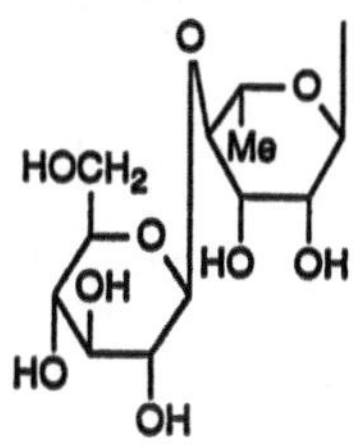	CH_3	*E. davidii*	[15]
Epimedoside C (*63-20*)	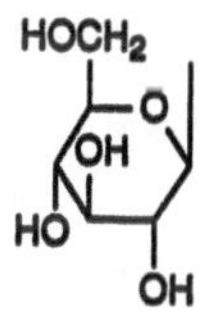	H	H	*E. davidii*	[16]

Eight species of *Epimedium* are the botanical sources for use in Chinese medicine, *E. acuminatum, E. brevicornum, E. davidii, E. hunanense, E. koreanum, E. pubescens, E. sagittatum*, and *E. wushanense*. They all contain icariin as detected thin-layer chromatographically, but other flavones might vary in different species [9]. Furthermore, new glycosides, anhydroicaritin-3-*O*-rhamnoside (*63-4*), icaritin-3-*O*-rhamnoside (*63-5*) [8], sagittatosides A (*63-6*), B (*63-7*), and C (*63-8*) [10], sagittatin A (*63-9*), and sagittatin B (*63-10*) [11] were isolated from *E. sagittatum* and three new glycosides named epimedins A (*63-11*), B (*63-12*), and C (*63-13*) from *E. koreanum* [12]. Icaritin is a flavone with an 8-hydroxyisopentyl group and sagittatin A and sagittatin B are flavones without a substituent at position 8.

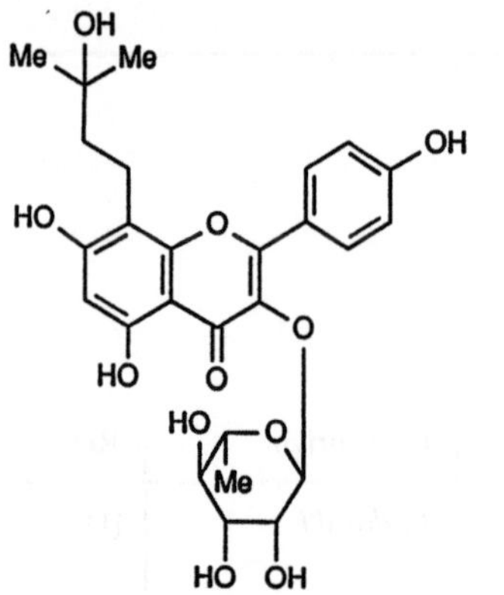

Icaritin 3-*O*-α-L-rhamnopyranoside (*63-5*)

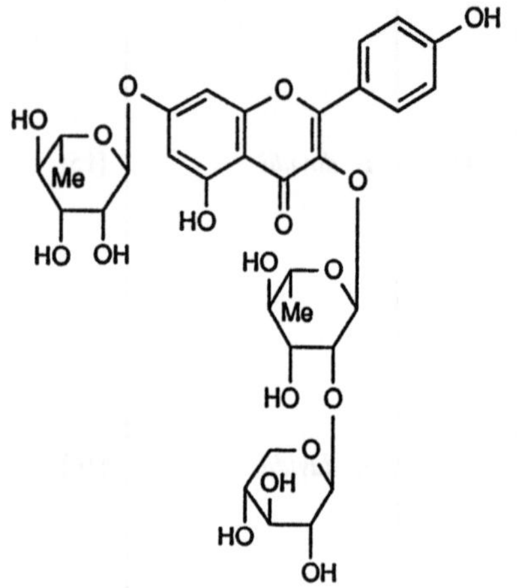

Sagittatin A (*63-9*)

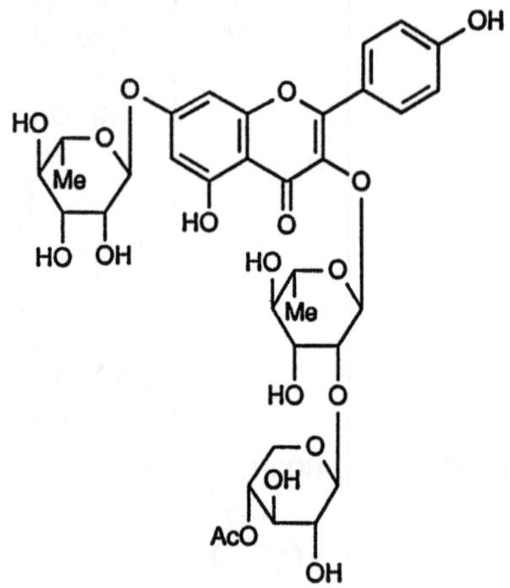

Sagittatin B (*63-10*)

Sagittatosides A and B [13] and anhydroicaritin 3-*O*-rhamnoside [14] were also isolated from *E. koreanum*, whereas epimedins A, B, and C were also found in *E. sagittatum* [10].

A series of prenylflavone glycosides was also isolated from *E. davidii* [15, 16] and named baohuosides I–VII. Baohuoside I was determined as identical to *63-4*, and baohuosides II–VII (*63-14–63-19*) are new prenylflavone glycosides. In addition, epimedosides A and C (*63-20*) were isolated from *E. davidii* [16]. Baohuosu (*63-21*) is another flavone from *E. davidii* with a highly oxygenated 2-phenyl ring [15].

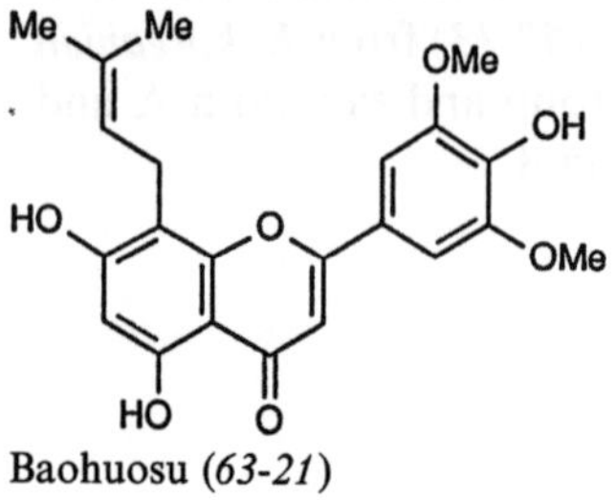

Baohuosu (*63-21*)

The isolation of prenylflavones from *E. wushanense* was also reported [17]. One was determined as icariin. Another prenylflavone glycoside named wushanicariin (*63-22*) has a prenyl group at position 6.

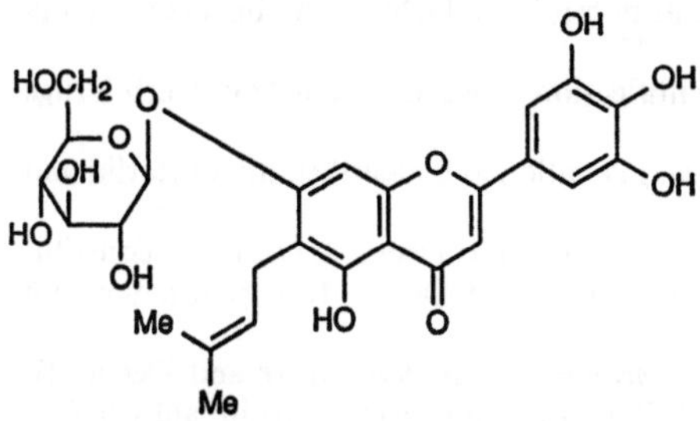

Wushanicariin (*63-22*)

63.3 Pharmacology

Epimedium herb has been shown to regulate the immunological functions. It enhanced the functions of antibody-forming cells as well as the excitatory state of the lymphatic cells; it also increased the phagocytic activity of the monocytes and the number of T cells [18]. In an in vitro study, the methanolic extract of the leaf of *E. pubescens* markedly inhibited the proliferation of mouse lymphocytes induced by mitogens and the mixed lymphocyte reaction [19].

The polysaccharide isolated from *E. sagittatum* was found to accelerate the production of T-suppressor cells of immunized mice and to inhibit the antibody production in recipient mice; icariin, on the other hand, attenuated the production of T-suppressor cells and the antibody titer was therefore markedly elevated [20]. A polysaccharide with a molecular weight of about 75000 was isolated from *E. koreanum* and was used as an immune adjuvant [21].

In a clinical study, 22 cases of leukopenia unresponsive to conventional treatment were treated with an infusion made of the leaf of *E. sagittatum*. Of the 14 patients who complied with the treatment, 3 were cured and 9 were significantly improved. The leukocyte count was increased from $2440 \pm 992/mm^3$ to $4060 \pm 966/mm^3$. The immune complex titers and serum Zn^{2+} and Mg^{2+} were decreased. The lymphocyte transformation rate was also significantly increased [22].

The effect of icariin on myocardial activities was studied in rabbits and compared with an aqueous extract of *E. grandiflorum* containing icariin. Intravenously injected icariin (1 mg/kg) and the extract at a dose equal to 1 g plant/kg did not significantly change the heart rate and electrocardiogram. Other myocardial and circulatory indices indicated that both icariin and the extract decreased the peripheral resistance, suggesting their usefulness in treatment of hypertension-complicated coronary disease [23].

The clinical applications of the aerial part of *Epimedium* species in Chinese medicine have been summarized [24]. They include treatment of paraplegia and hemiplegia, neurasthenia, coronary disease, hypertension, hyperlipidemia, aplastic anemia, leukopenia, hepatitis B, nephritis, urolithiasis, hematuria, enuresis, chronic prostatitis, infertility, chronic bronchitis and asthma, hypothyroidism, menopausal syndrome, adolescent dysfunctional uterine bleeding, and osteohyperplasia [24].

References

1. Yang CH, Liu HK, Wu CL (1980) Chemical constituents of Xinyeyinyanghuo (*Epimedium brevicornum*). Chin Trad Herb Drugs 11:444
2. Liu AR, Xu LX (1984) Analysis of active ingredients in traditional Chinese herbal drugs. Assay of icariin in *Epimedium*. Chin J Pharm Anal 4:81–84
3. Liu BQ, Ma HS, Mou P (1980) Isolation and identification of icariin. Chin Trad Herb Drugs 11:201
4. Xu SC, Wang ZX, Wu LJ, Wang NB, Chen YJ (1982) Isolation and identification of icariin and epimedoside A. Chin Trad Herb Drugs 13:9–11
5. Akai S (1935) Constituents of *Epimedium macranthum* Morr and Decne. I. Chemical constitution of a new glucoside of *Epimedium macranthum* Morr and Decne. 1. J Pharm Soc Jpn 55:537–599
6. Akai S, Matsukawa T (1935) Constituents of *Epimedium macranthum* Morr and Decne. II. Constitution of a new flavone glucoside, icariin. 2. Relationship between icaritin, anhydroicaritin and β-anhydroicaritin and oxidation of anhydroicaritin. J Pharm Soc Jpn 55:705–719
7. Xu GW, Xu BJ, Wang MT (1987) Isolation and identification of icariin and icariside I. Chin Pharm Bull 22:129–130
8. Mizuno M, Hanioka S, Suzuki N, Iinuma M, Tanaka T, Liu XS, Min ZD (1987) Flavonol glycosides from *Epimedium sagittatum*. Phytochemistry 26:861–863
9. Shi DW, Huang DJ, Bi ZQ, Wang ZW, Yang XY (1986) Comparison of the chemical constituents and identification of the Chinese kidney-tonic yinyanghuo. Acta Acad Med Shanghai 13:13–19
10. Mizuno M, Sakakihara N, Hanioka S, Iinuma M, Tanaka T, Liu XS, Shi DW (1988) Flavonol glycosides from *Epimedium sagittatum*. Phytochemistry 27:3641–3643
11. Oshima Y, Okamoto M, Hikino H (1989) Sagittatins A and B, flavonoid glycosides of *Epimedium sagittatum* herbs. Planta Med 55:309–311
12. Oshima Y, Okamoto M, Hikino H (1987) Validity of oriental medicines. Part 122. Epimedins A, B and C, flavonoid glycosides of *Epimedium koreanum* herbs. Heterocycles 26:935–938
13. Ito Y, Hirayama F, Suto K, Sagara K, Toshida T (1988) Three flavonol glycosides from *Epimedium koreanum*. Phytochemistry 27:911–913
14. Kang SS, Shin KH, Chung SG, Cho EH (1988) Flavonoids from *Epimedium koreanum*. Saengyak Hakhoechi 19:93–96
15. Li F, Liu YL (1988) Isolation and structures of baohuosides I, VI, VII, and baohuosu. Acta Pharm Sin 23:739–748
16. Li F, Liu YL (1988) Isolation and structures of baohuoside II, III, IV, and V. Acta Pharm Sin 23:672–681
17. Liang HR, Yan WM, Li JS, Yang CS (1988) Chemical studies of *Epimedium wushanense*. T.S. Ying. Acta Pharm Sin 23:34–37
18. Chen KJ, Zhang WP (1987) Advances on antiageing herbal medicines in China. Abstr Chin Med 1:309–330
19. Wang YC, He QZ (1986) In vivo and in vitro studies on the effect of *Epimedium* extract on mouse immune response. Shanghai J Immunol 6:6–9
20. Wang TR, Xing ST, Zhou JH (1986) Action of *Epimedium sagittatum* polysaccharide and icariin on T suppressor cells. Chin J Immunol 2:74–76
21. Zenyaku Kogyo Co (1982) Polysaccharide PS-A as immune adjuvant from *Epimedium*. Jpn Kokai Tokkyo Koho JP 57.181.017 (82.181.017) (CA 98:40576x)
22. Liu FC, Li JX, Ding GX, Zhang JY, Zhou SH, Guo F, Wu YC, Hu TH (1985) Correlation between trace elements and immunological function in patients with vital energy deficiency. J Trad Chin Med 26:856–857
23. Liu CM, Yu QH, Zhang LM (1982) Effect of icariin on heart. Chin Trad Herb Drugs 13:414–416
24. Yang ZZ (1985) Clinical applications of Yinyanghuo. Zhejiang J Trad Chin Med 20:478–480

Erycibe obtusifolia Benth. 64

64.1 Introduction

Dinggongteng, *Erycibe obtusifolia* Benth. (Convolvulaceae), is a medicinal plant listed in the appendix of the Chinese Pharmacopoeia. Its vines could be used in traditional Chinese medicine in the treatment of rheumatism and is the major component of the official preparation Dinggongteng Fengshi Yaojiu, a medicinal wine made of 25 herbal medicines and used in the treatment of rheumatic diseases.

64.2 Chemical Constituents

A new alkaloid baogongteng A with miotic activity was isolated from *E. obtusifolia* [1]. Its structure was determined by ^{1}H and ^{13}C NMR and high-resolution mass spectral data as 2β-hydroxy-6β-acetoxynortropane (*64-1*) [2, 3]. The corresponding 2β, 6β-dihydroxynortropane named baogongteng C (*64-2*) without miotic activity was also isolated [4].

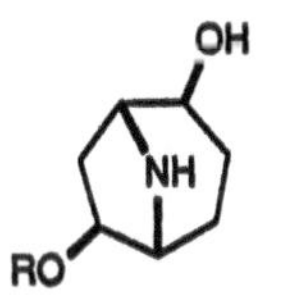

Baogongteng A (*64-1*): R=AC
Baogongteng C (*64-2*): R=H

In addition to the alkaloids, scopoletin and its glucoside were isolated and identified [5]. Baogongteng A and C were also isolated from *E. elliptilimba*, which contains a further new alkaloid named erycibelline (*64-3*), which was identified as 2β,7β-dihydroxynortropan [6].

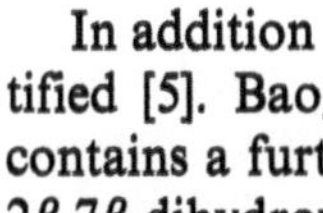

Erycibelline (*64-3*)

64.3 Pharmacology

Baogongteng A showed miotic activity 100-fold greater than that of pilocarpine [7]. Baogongteng A benzoate solution can lower ocular pressure and improve the aqueous outflow and has been used in the treatment of glaucoma [8]. A hypotensive effect of baogongteng A was also reported [9]. Scopoletin and scopolin have been suggested to be responsible for the antirheumatic and antiinflammatory activity of *E. obtusifolia* [5].

References

1. Yao TR, Chen ZN (1979) Chemical studies on *Erycibe obtusifolia*, Bao Gong Teng. I. Isolation and preliminary study on a new miotic constituent, Bao Gong Teng A. Acta Pharm Sin 14:731–735
2. Yao TR, Chen ZN, Yi DN, Xu GY (1981) Chemical study on Bao Gong Teng (*Erycibe obtusifolia* Benth.). II. Structure of baogongteng A – a new miotic agent. Acta Pharm Sin 16:582–588
3. Fang YW, Zhao JJ, Bian ZL (1981) Determination of the structure of *Erycibe obtusifolia* Benth's Base II – a new medicine for glaucoma. Chemistry 209–210
4. Chen ZI, Xu PJ, Yao TR (1986) Chemical investigation of Baogongteng (Dinggongteng; *Erycibe obtusifolia*). III. The identification of boagongteng B and studies on baogongteng C. Chin Trad Herb Drugs 17:386–387
5. Ye HZ, Fan YH, Liu CW, Chin IS (1981) Study of the antirheumatic active principle of Ding Gong-teng (*Erycibe obtusifolia* Benth.). Chin Trad Herb Drugs 12:5–7
6. Lu Y, Yao TR, Chen ZI (1986) Constituents of *Erycibe elliptilimba*. Acta Pharm Sin 21:829–835
7. Shanghai Second Medical College (1981) A new miotic agent to treat glaucoma – baogongteng A. Chin Pharm Bull 16:55
8. Li TA, Hsu H (1980) Comparison of bioavailability between two miotic solutions of baogongteng A. Chin Pharm Bull 15:44–45
9. Shanghai Department of Pharmacology (1981) Preliminary study on the cardiovascular effect of Baogongteng A, an alkaloid from *Erycibe obtusifolia*. Chin Pharm Bull 16:51–52

Eucommia ulmoides Oliv.

65

65.1 Introduction

Duzhong, Cortex Eucommiae, is the dry stem bark of *Eucommia ulmoides* Oliv. (Eucommiaceae) peeled in April to June. It is officially listed in the Chinese Pharmacopoeia and is one of the oldest tonic herbs in traditional Chinese medicine and used in the treatment of hypertension.

65.2 Chemical Constituents

The main constituents isolated from the stem bark of *E. ulmoides* so far have been irridoid glycosides and lignan glucosides. Iridoids are monoterpene cyclic ethers with a basic skeleton of hexahydrodimethylcyclopenta[*c*]pyran (*65-1*). At first, six iridoids were detected in the ethanolic extract of the stem bark of *E. ulmoides* and five of them were identified as the known compounds aucubin (*65-2*), harpagide acetate (*65-3*), ajugoside (*65-4*), reptoside (*65-5*), and a new compound named eucommiol (*65-6*) [1]. Aucubin was the major constituent, with a content of 0.1%–4.0% in stem bark and 1.6%–1.7% in the leaves [2].

An aucubigenin diglycoside named ulmoside (aucubigenin-1-β-isomaltoside, *65-7*) [3], eucommioside I (*65-8*), geniposidic acid (*65-9*), and geniposide (*65-10*) [4] were further isolated and structurally determined. Recently, the isolation of 4-deoxyeucommiol from the stems as well as from the leaves [6] of *E. ulmoides* was also reported.

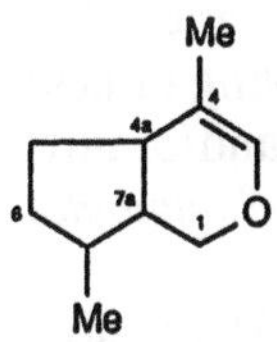

Hexahydrodimethyl-cyclopenta[*c*]pyran (*65-1*)

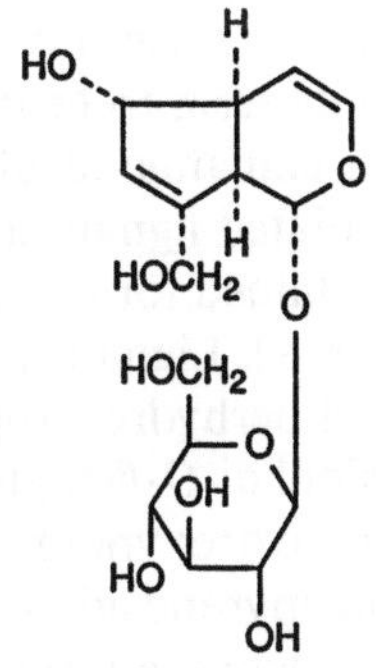

Aucubin (*65-2*)

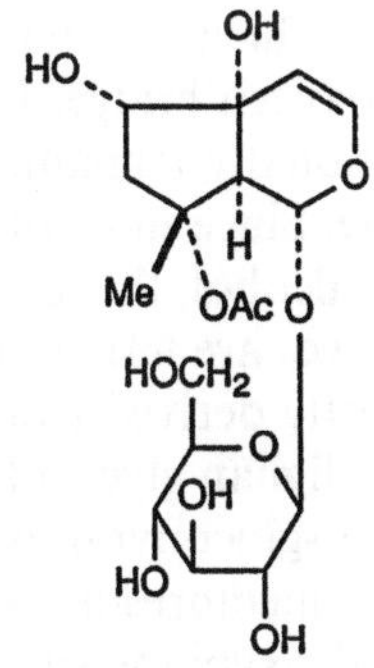

Harpagide acetate (*65-3*)

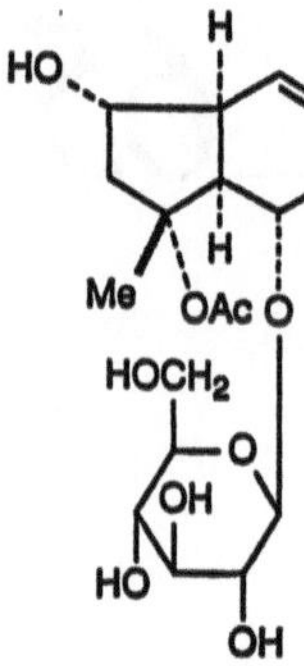
Ajugoside (*65-4*)

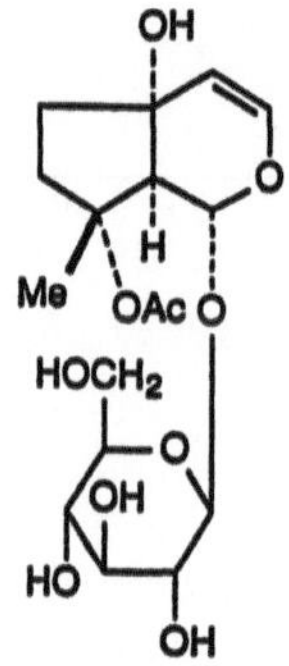
Reptoside (*65-5*)

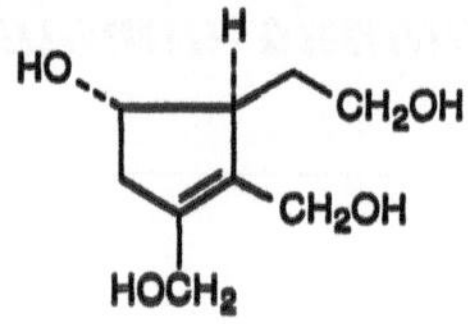
Eucommiol (*65-6*)

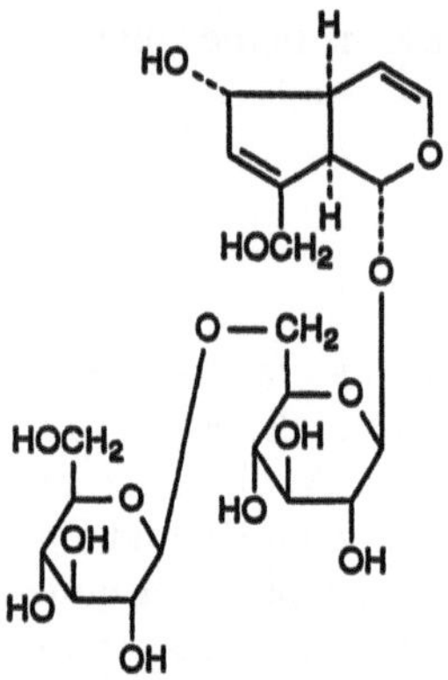
Ulmoside (*65-7*)

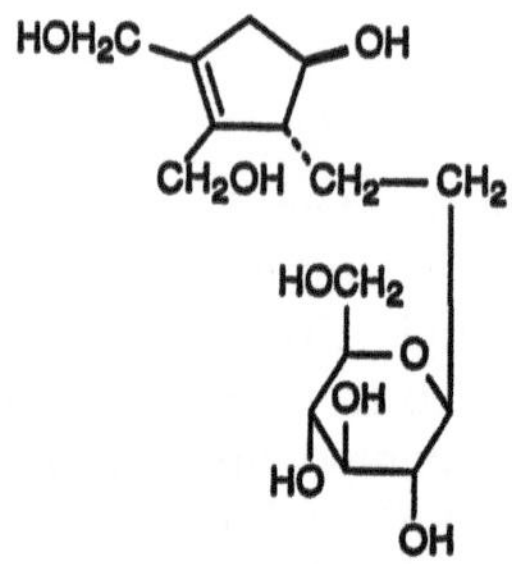
Eucommioside I (*65-8*)

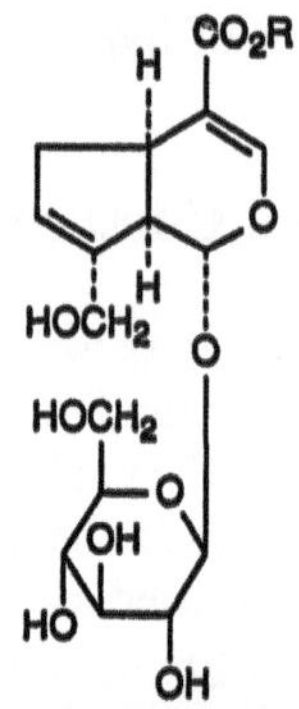
Geniposidic acid (*65-9*): R=H
Geniposide (*65-10*): $R=CH_3$

Lignans were first defined as plant products with a carbon skeleton having two *n*-propylbenzene residues linked by the *β*-carbon atoms of the side chain [7]. The term was later extended to cover all natural products of low molecular weight that arise primarily from the oxidative coupling of *p*-hydroxyphenylpropene units, a concept which also refers to variants of skeletons in which the two units are linked by an oxygen bridge. Four such units seem to be involved. They are: cinnamic acid (exceptionally cinnamic aldehyde), cinnamic alcohol, propenylbenzene, and allylbenzene. Since most of the early isolated lignans are derived from coupling of acid and/or alcohol, the term lignan is retained for this group. The propenyl and/or allyl derivatives are termed neolignans [8, 9]. Lignans present in the bark of *E. ulmoides* are mostly derived from a bisbenzyl-perhydrofuro[3,4-*c*]furan skeleton.

New lignan glycosides medioresinol di-*O*-*β*-D-glucopyranoside (*65-11*) [10], olivil di-*O*-*β*-D-glucopyranoside (*65-12*), hydroxypinoresinol di-*O*-*β*-D-glucopyranoside (*65-13*), medioresinol 4′-*O*-*β*-D-glucopyranoside named eucommin A (*65-14*) [11], and hydroxypinoresinol 4″-*O*-*β*-D-glucopyranoside (*65-15*) [12] were isolated from the bark of *E. ulmoides*, in addition to the known lignan compounds pinoresinol di-*O*-*β*-D-glucopyranoside (*65-16*) [10, 13], liriodendrin (*65-17*), pinoresinol *O*-*β*-D-glucopyranoside [4], syringaresinol *O*-*β*-D-glucopyranoside (*65-18*) [11], hydroxypinoresinol 4′-*O*-*β*-D-glucopyranoside (*65-19*), cycloolivil (*65-20*), and olivil

[12]. The structures of the new lignan derivatives were elucidated chemically and spectroscopically.

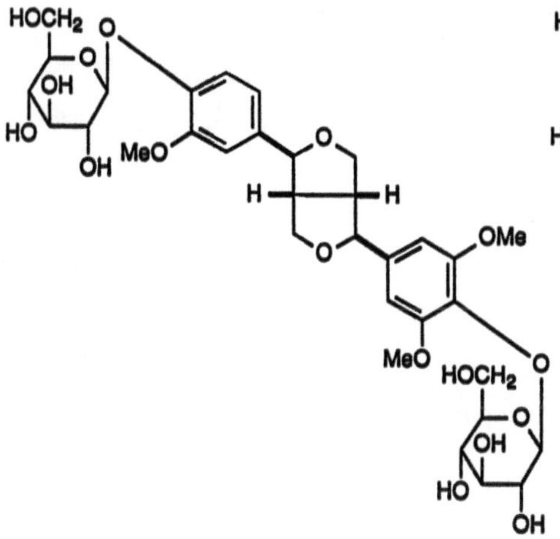

Medioresinol di-*O*-β-D-glucopyranoside (*65-11*)

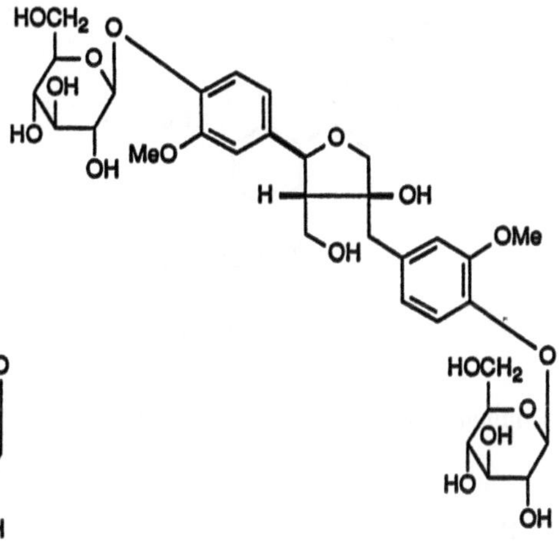

Olivil di-*O*-β-D-glucopyranoside (*65-12*)

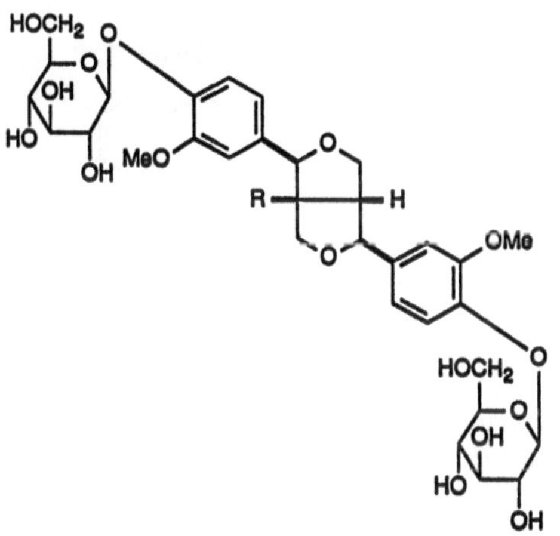

Pinoresinol di-*O*-β-D-glucopyranoside (*65-16*): R=H
Hydroxypinoresinol di-*O*-β-D-glucopyranoside (*65-13*): R=OH

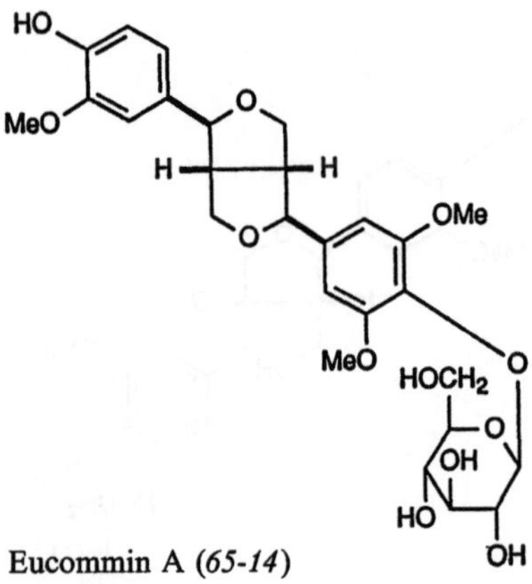

Eucommin A (*65-14*)

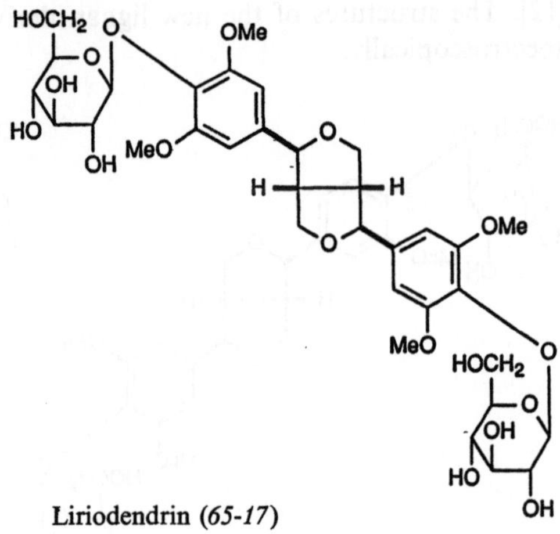

Liriodendrin (*65-17*)

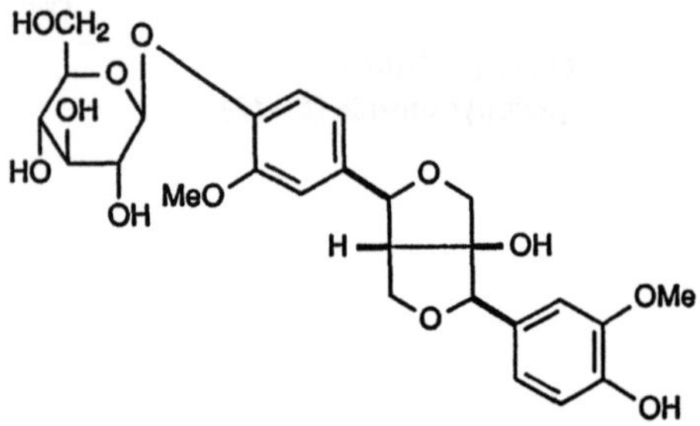

Hydroxypinoresinol-4″-*O*-*β*-D-glucopyranoside (*65-15*)

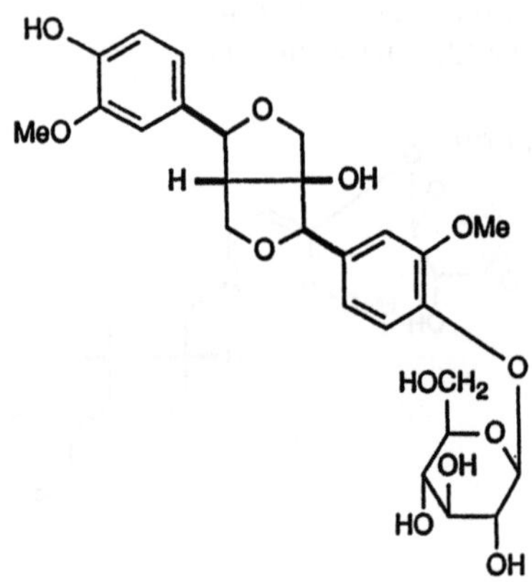

Hydroxypinoresinol-4′-*O*-*β*-D-glucopyranoside (*65-19*)

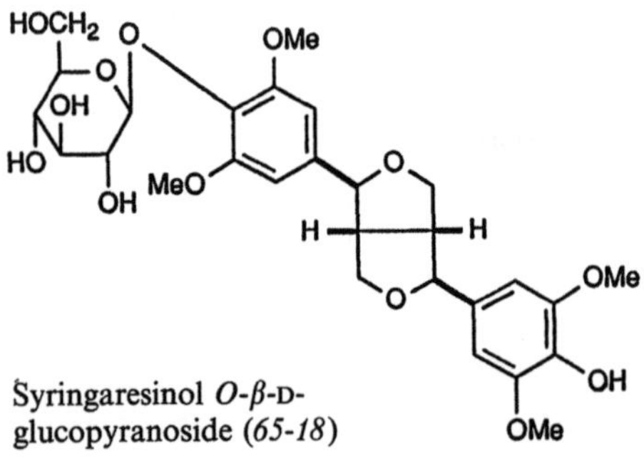

Syringaresinol *O*-*β*-D-glucopyranoside (*65-18*)

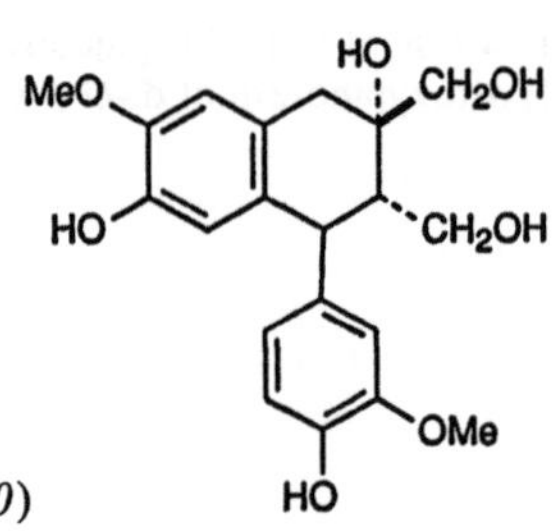

Cycloolivil (*65-20*)

Sih et al. described pinoresinol di-*O*-β-D-glucopyranoside as the major antihypertensive principle of *E. ulmoides* and reported a synthesis of this glucoside [9]. Thus, coniferyl alcohol (*65-21*), prepared by oxidation of eugenol acetate (*65-22*) with mercuric acetate, was dimerized to pinoresinol by enzymatic catalysis using the chloroperoxidase-containing microorganism, *Caldariomyces fumago* (Fig. 65.1). Reaction of pinoresinol with α-bromoacetoglucose, in the presence of Ag_2O, followed by alkaline hydrolysis, produced pinoresinol di-*O*-β-D-glucopyranoside.

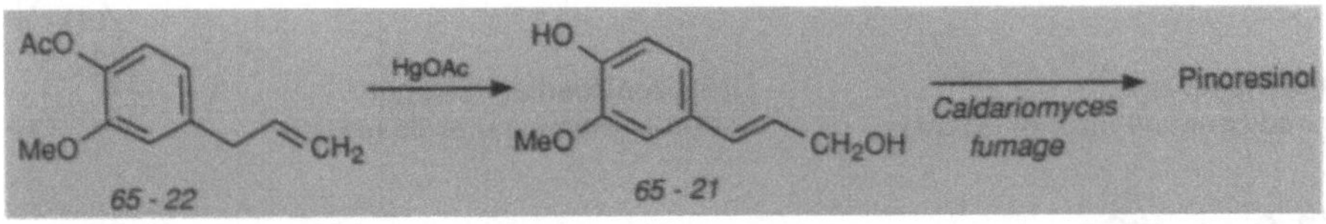

Fig. 65.1.

Pinoresinol diglucoside is mainly present in the phloem of the bark. The contents of pinoresinol diglucoside found in the phloem of various samples were up to 0.55%, whereas other parts of the bark did not contain the diglucoside [14].

Recently, a number of new lignans and lignan glycosides have been found and characterized such as guaiacylglycerol-β-medioresinol ether di-*O*-β-D-glucopyranoside (*65-23*) [4], *erythro*-dihydroxydehydrodiconiferyl alcohol (*65-24*) and its *threo* isomer [15], syringylglycerol-β-syringaresinol ether di-*O*-β-D-glucopyranoside (*65-25*) and dehydrodiconiferyl alcohol di-*O*-β-D-glucopyranoside (*65-26*) [16] together with the known compounds olivil 4′-*O*-β-D-glucopyranoside, olivil 4″-*O*-β-D-glucopyranoside [4], and citrusin B [16].

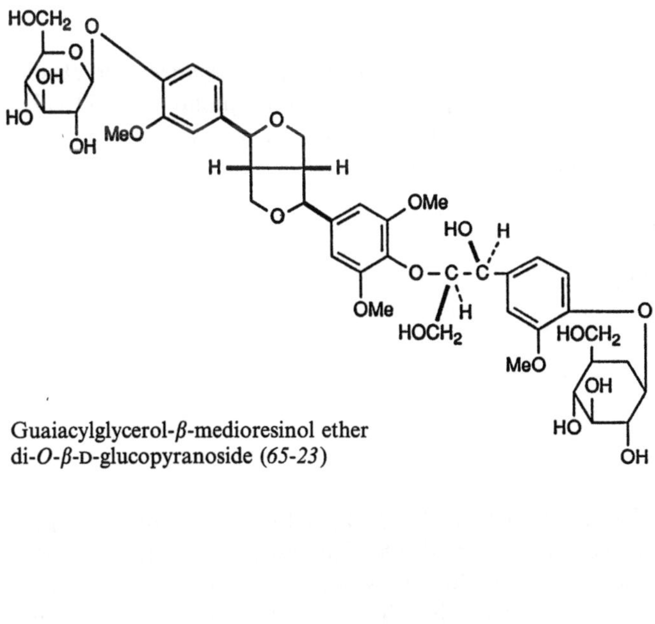

Guaiacylglycerol-β-medioresinol ether di-*O*-β-D-glucopyranoside (*65-23*)

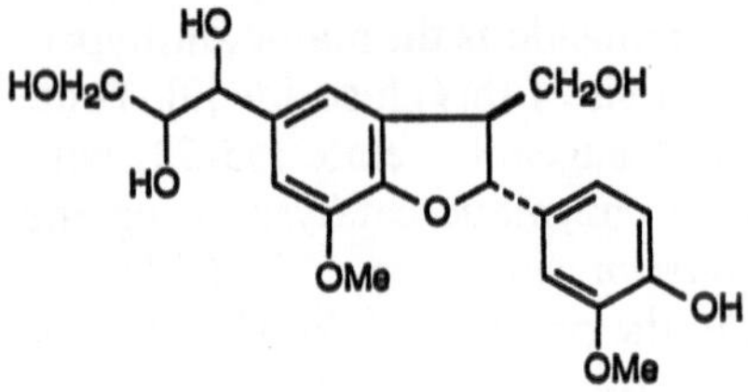

erythro-Dihydroxy-
dehydrodiconiferyl alcohol (*65-24*)

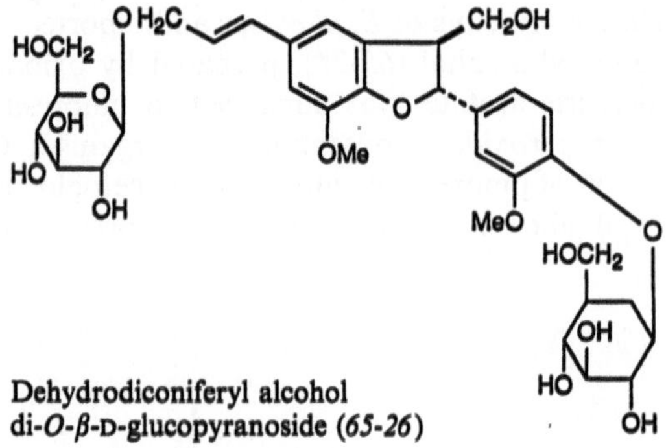

Dehydrodiconiferyl alcohol
di-*O*-β-D-glucopyranoside (*65-26*)

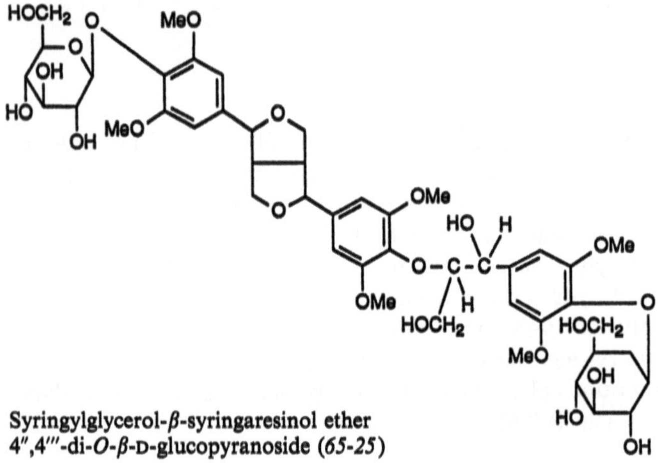

Syringylglycerol-β-syringaresinol ether
4″,4‴-di-*O*-β-D-glucopyranoside (*65-25*)

In addition to the iridoid glycosides and lignan derivatives, *erythro*- and *threo*-guaiacylglycerol [12], a new type C_{30}-polyprene named ulmoprenol (*65-27*) [17], nonacosane, *n*-triacontanol, β-sitosterol, betulin, betulic acid, ursolic acid, and vanillic acid [18, 19] were also found in *E. ulmoides*.

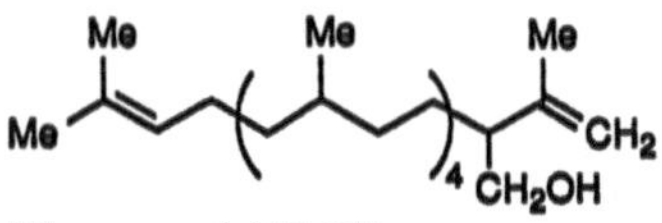

Ulmoprenol (*65-27*)

65.3 Pharmacology

The aqueous extract of dried stem bark of *E. ulmoides* lowered blood pressure, dilated ear veins, and raised tension of excited intestine and uterus in rabbits [20]. Three hypertensive dogs showed a drop in blood pressure, accompanied by a slowing down of the heart rate, when treated orally with aqueous extract of leaves of

E. ulmoides for 4 weeks [21]. Intraperitoneal injection of the aqueous extract of the bark of *E. ulmoides* at a dose of 800 mg/kg decreased the blood pressure and heart rate of anesthetized rats. At 200-, 400-, or 800-mg/kg doses, the extract increased cardiac perfusion by 8.4%, 11.9%, and 15%, respectively. Saline-loaded rats were given 400 mg/kg of the extract intravenously. Urine output increased within 30 min but electrolyte concentration and urine pH were unaltered [22].

Oral administration of the bark of *E. ulmoides* as tea or wine to 62 hypertensive patients resulted in improvement after 2–4 months in 94% of cases. The systemic arterial hypotension caused by the extract of *E. ulmoides* is apparently the result of peripheral vasodilation by direct action on the vascular smooth muscle. The synthetic pinoresinol di-*O*-*β*-D-glucopyranoside was found to be indistinguishable from the natural product in antihypertensive activity [13]. Oral administration of extract of the bark of *E. ulmoides* increased the plasma levels of cAMP and cGMP from 70.9 and 54.5 pmol/ml in the control group to 100.4 and 137.8 pmol/ml, respectively [23], determined by radioimmunoassay.

References

1. Bianco A, Iavarone C, Trogolo C (1974) Structure of eucommiol, a new cyclopentanoid-tetrol from *Eucommia ulmoides*. Tetrahedron 30:4117–4121
2. Li JS, Yan YN (1986) Chemical analysis of duzhong (*Eucommia ulmoides*) bark and leaves. Bull Chin Mat Med 11:489–490
3. Bianco A, Bonini C, Guiso M, Iavarone C, Trogolo C (1978) Iridoids 26. Ulmoside (aucubigenin-1-*β*-isomaltoside), a new iridoid from *Eucommia ulmoides*. Gazz Chim Ital 108:17–20
4. Deyama T, Ikawa T, Kitagawa S, Nishibe S (1986) The constitutents of *Eucommia ulmoides* Oliv. IV. Isolation of a new sesquilignan glycoside and iridoids. Chem Pharm Bull (Tokyo) 34:4933–4938
5. Gewali MB, Hattori M, Namba T (1988) Constituents of the stems of *Eucommia ulmoides* Oliv. Shoyakugako Zasshi 42:247–248
6. Hattori M, Che QM, Gewali MB, Nomura Y, Tezuka Y, Kikuchi T, Namba T (1988) Studies on Du-Zhong leaves (III). Constituents of the leaves of *Eucommia ulmoides* (1). Shoyakugaku Zasshi 42:76–80
7. Haworth RD (1942) The chemistry of the lignan group of natural products. J Chem Soc 448
8. Gottlieb OR (1974) Lignans and neolignans. Rev Latinoam Quim 5:1
9. Gottlieb OR (1978) Neolignans. Prog Chem Org Nat Prod 35:1–72
10. Deyama T (1983) The constituents of *Eucommia ulmoides* Oliv. I. Isolation of (+)-medioresinol di-*O*-*β*-D-glucopyranoside. Chem Pharm Bull (Tokyo) 31:2993–2997
11. Deyama T, Ikawa T, Nishibe S (1985) The constituents of *Eucommia ulmoides* Oliv. II. Isolation and structures of three new lignan glycosides. Chem Pharm Bull (Tokyo) 33:3651–3657
12. Deyama T, Ikawa T, Kitagawa S, Nishibe S (1986) The constituents of *Eucommia ulmoides* Oliv. III. Isolation and structure of new lignan glycoside. Chem Pharm Bull (Tokyo) 34:523–527
13. Sih CJ, Ravikumar PR, Huang FC, Buckner C, Whitlock H jr (1976) Isolation and synthesis of pinoresinol diglucoside, a major antihypertensive principle of Tu-chung (*Eucommia ulmoides* Oliver). J Am Chem Soc 98:5412–5413
14. Sha ZF, Sun WJ (1986) High performance liquid chromatography of pinoresinol diglucoside in *Eucommia ulmoides* Oliv. bark. Acta Pharm Sin 21:708–711
15. Deyama T, Ikawa T, Kitagawa S, Nishibe S (1987) The constituents of *Eucommia ulmoides* Oliv. V. Isolation of dihydroxydehydrodiconiferyl alcohol isomers and phenolic compounds. Chem Pharm Bull (Tokyo) 35:1785–1789
16. Deyama T, Ikawa T, Kitagawa S, Nishibe S (1987) The constituents of *Eucommia ulmoides* Oliv. VI. Isolation of new sesquilignan and neolignan glycosides. Chem Pharm Bull (Tokyo) 35:1803–1807

17. Horii ZI, Ozaki Y, Nagao K, Kim SW (1978) Ulmoprenol, a new type C_{30}-polyprenoid from *Eucommia ulmoides* Oliver. Tetrahedron Lett 5015–5016
18. Li D, Wang HL, Chen JM, Xu JW (1986) Chemical constituents of Duzhong (the bark of *Eucommia ulmoides*). Acta Bot Sin 28:528–532
19. Wang HL, Li D, Chen JM, Xu JW (1986) Chemical constituents of the bark of *Eucommia ulmoides* (II). Chin Trad Herbal Drugs 17:232
20. Chien TH (1957) Pharmacological action of *Eucommia ulmoides* Oliv. Jpn J Pharmacol 6:122–137
21. Kin KC, Ting KS (1956) Drugs for the treatment of hypertension. II. Toxicity and experimental therapy of *Eucommia ulmoides*. Acta Physiol Sin 20:247–254
22. Yu C, Yang CP, Huang CJ, Liu HJ (1986) The diuretic and cardiovascular effects of cortex eucommiae. Bull Taipei Med Coll (Abstr Chin Med 870966)
23. Xu SL, Zeng QZ, Huang WG, Yin Q (1986) Effects of Duzhong on plasma cAMP and cGMP levels in mice. Chin Trad Herbal Drugs 17:204

Evodia rutaecarpa (Juss.) Benth.

66

66.1 Introduction

Wuzhuyu, Fructus Evodiae, is the dry, nearly ripe fruits of *Evodia rutaecarpa* (Juss.) Benth., *E. rutaecarpa* (Juss.) Benth. var. *officinalis* (Dode) Huang, or *E. rutaecarpa* (Juss.) Benth. var. *bodinieri* (Dode) Huang (Rutaceae) collected in August to November before the fruits burst. It is officially listed in the Chinese Pharmacopoeia and is used as an analgesic, antiemetic, and astringent, and in the treatment of hypertension.

66.2 Chemical Constituents

The fruits of *E. rutaecarpa* contain a number of alkaloids with an indolo[2′,3′:3,4]pyrido[2,1-*b*]quinazolin (*66-1*) skeleton as the main constituent. The first alkaloid isolated was evodiamine (*66-2*) [1]. It is the main alkaloid component and the active principle and was isolated by Asahina et al. more than 70 years ago. Asahina also isolated another alkaloid, rutaecarpine (*66-3*), from the fruits of *E. rutaecarpa* [2, 3].

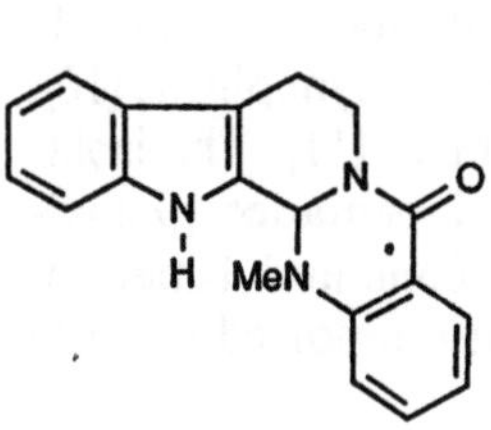

Indolo[2′,3′:3,4]pyrido[2,1-*b*]quinazolin (*66-1*)

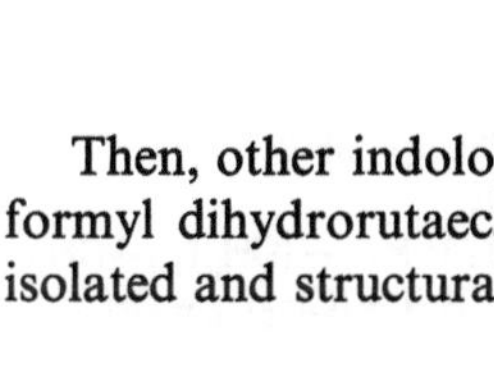

Evodiamine (*66-2*)

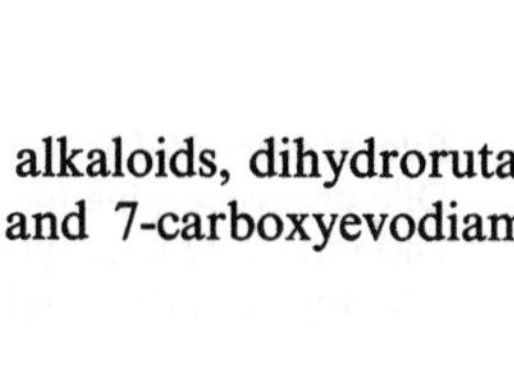

Rutaecarpine (*66-3*)

Then, other indolopyridoquinazoline alkaloids, dihydrorutaecarpine (*66-4*), 14-formyl dihydrorutaecarpine (*66-5*) [4], and 7-carboxyevodiamine (*66-6*) [5] were isolated and structurally investigated.

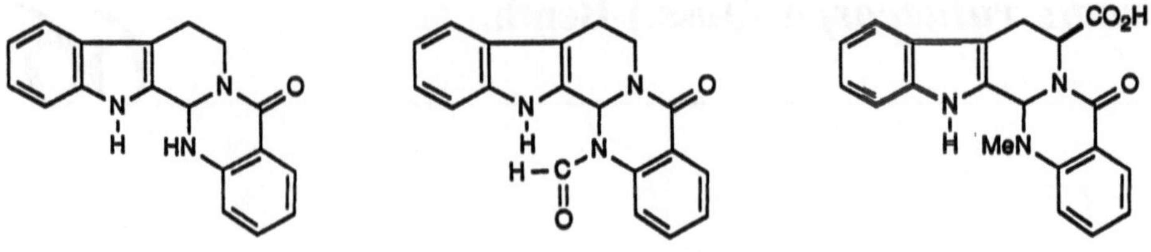

Dihydrorutaecarpine (*66-4*) 14-Formyl-dihydrorutaecarpine (*66-5*) 7-Carboxyevodiamine (*66-6*)

Dehydroevodiamine (*66-7*) was the major alkaloid isolated from the leaves of *E. rutaecarpa* together with rhetisinine (*66-8*), an alkaloid without a quinazoline moiety [6]. Rhetisinine was later also found in the fruits of *E. rutaecarpa* [7, 8].

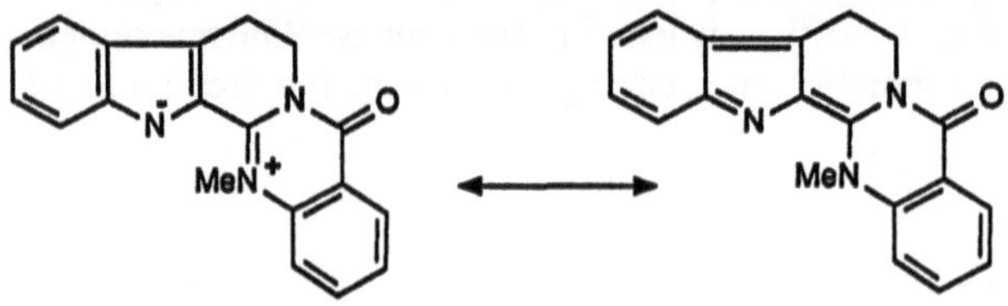

Dehydroevodiamine (*66-7*)

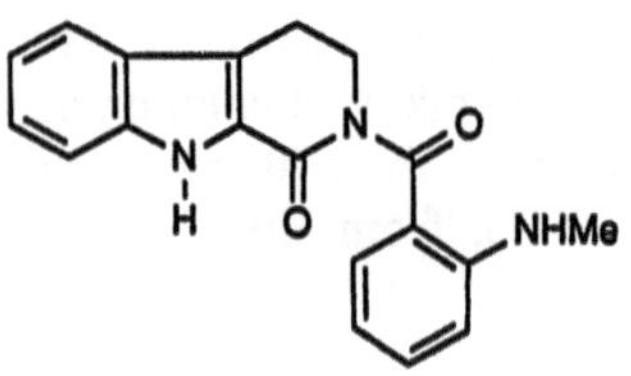

Rhetisinine (*66-8*)

In addition to the indolopyridoquinazoline alkaloids, some quinolone alkaloids and other nitrogen-containing compounds were isolated and structurally investigated. These include evocarpine (*66-9*) [7], the first quinolone alkaloid of *E. rutaecarpa*, isolated from the fruits, its dihydro analog dihydroevocarpine (*66-10*), the homologs of dihydroevocarpine 1-methyl-2-pentadecyl-4(1H)-quinolone (*66-11*) and 1-methyl-2-undecyl-4(1H)-quinolone (*66-12*) from the fresh leaves [9], synephrine [10], *N,N*-dimethyl-5-methoxytryptamine, and *N*-methylanthranilamide [11]. The isolation of 1-methyl-2-pentadec-6-enyl-4(1H)-quinolone, 1-methyl-2-pentadec-10-enyl-4(1H)-quinolone, 1-methyl-2-pentadeca-6,9-dienyl-4(1H)-quinolone and 1-methyl-2-trideca-4,7-dienyl-4(1H)-quinolone from *E. rutaecarpa* was also reported recently [12].

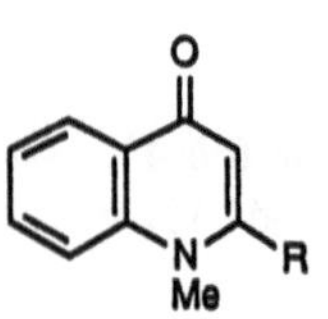

Evocarpine (*66-9*): $R=(CH_2)_7-CH=CH-(CH_2)_3-CH_3$
Dihydroevocarpine (*66-10*): $R=(CH_2)_{12}-CH_3$
1-Methyl-2-pentadecyl-4(1H)-quinolone (*66-11*): $R=-(CH_2)_{14}-CH_3$
1-Methyl-2-undecyl-4(1H)-quinolone (*66-12*): $R=-(CH_2)_{10}-CH_3$

Furthermore, a new indole alkaloid named evodiamide (*66-13*) was isolated from *E. rutaecarpa* [13].

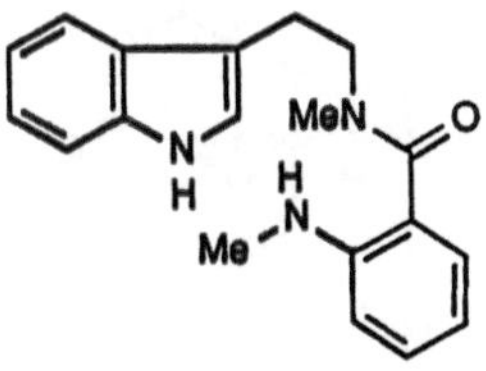

Evodiamide (*66-13*)

Bitter principles isolated from the fruits of *E. rutaecarpa* were limonin derivatives evodin [14], evodol (*66-14*), and evodinon (*66-15*) [3]. Evodin has been found to be identical with limonin [15], and evodol and evodinon were found to be identical with limonin diosphenol and rutaevin, respectively. Limonin (*44-33*) [16], limonin-diosphenol [17], and rutaevin [18] were also isolated from *Citrus* species.

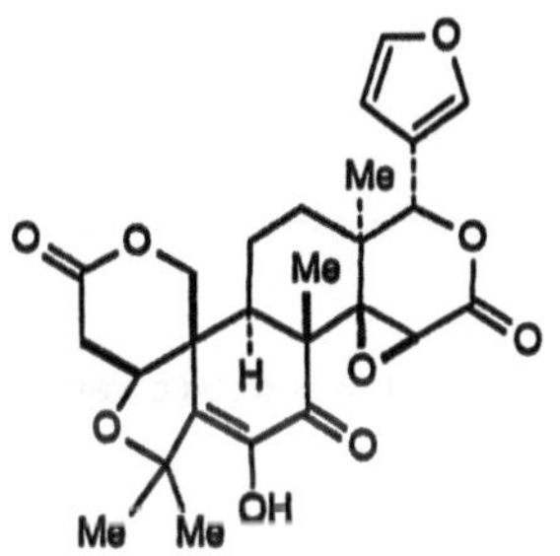

Evodol (Limonin diosphenol) (*66-14*)

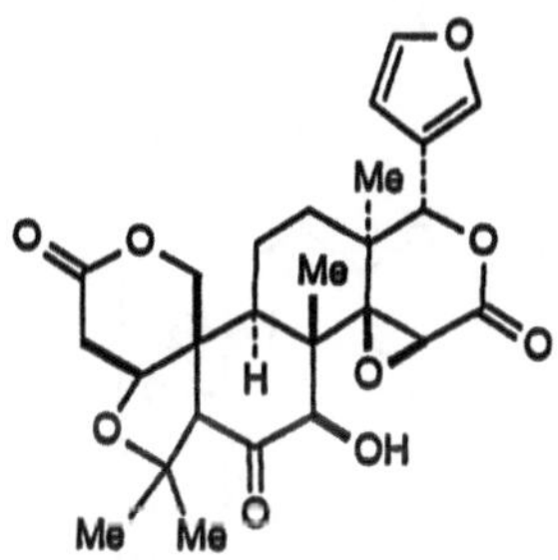

Evodinon (Rutaevin) (*66-15*)

Further limonoids isolated from *E. rutaecarpa* are obacunone (*44-35*), jangomolide (*66-16*), rutaevin acetate, graucin A (*66-17*), 12α-hydroxylimonin (*66-18*), 12α-hydroxyevodol [12], 6α-acetoxy-5-epilimonin (*66-19*), and 6β-acetoxy-5-epilimonin (*66-20*) [12, 19]. This is the first time the latter four substances have been isolated from plant tissue.

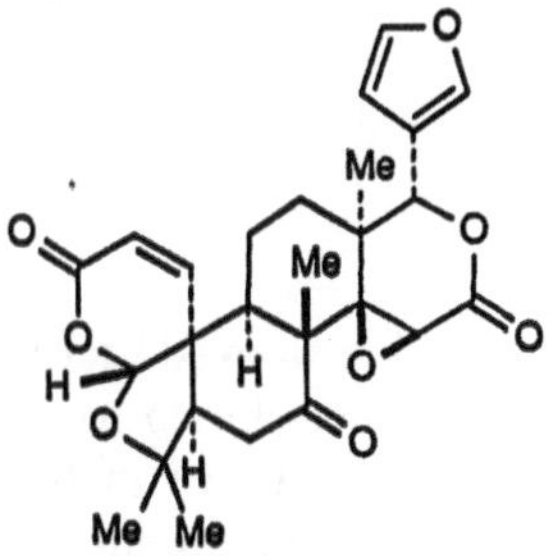

Jangomolide (*66-16*)

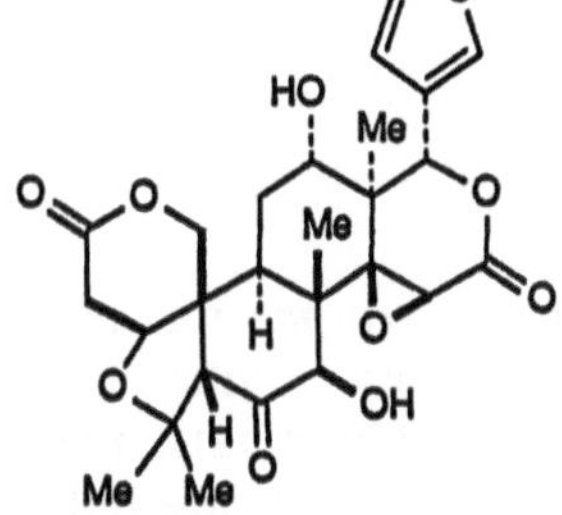

Graucin A (*66-17*)

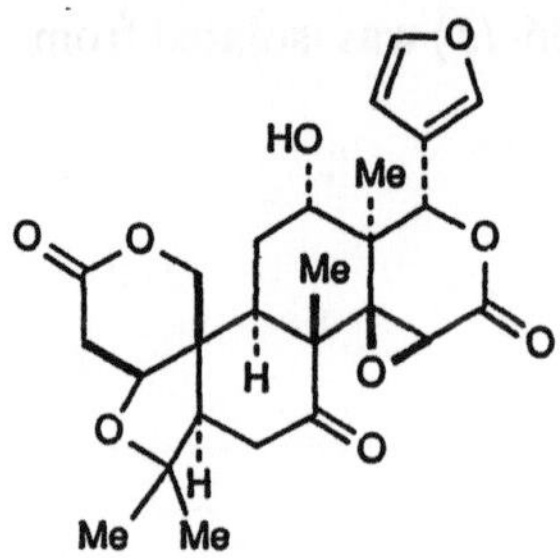

12α-Hydroxylimonin (*66-18*)

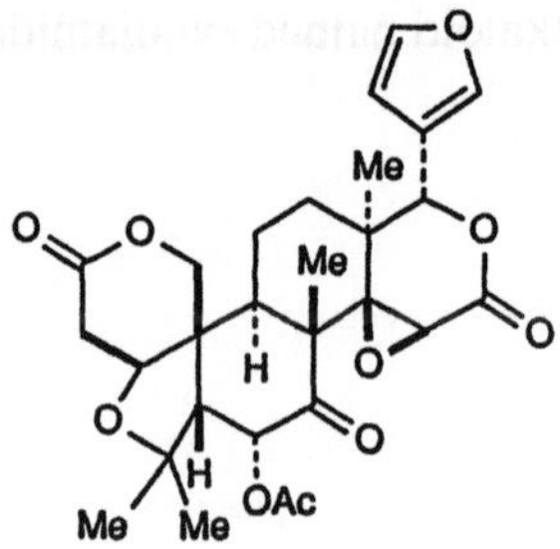

6α-Acetoxy-5-epilimonin (*66-19*)

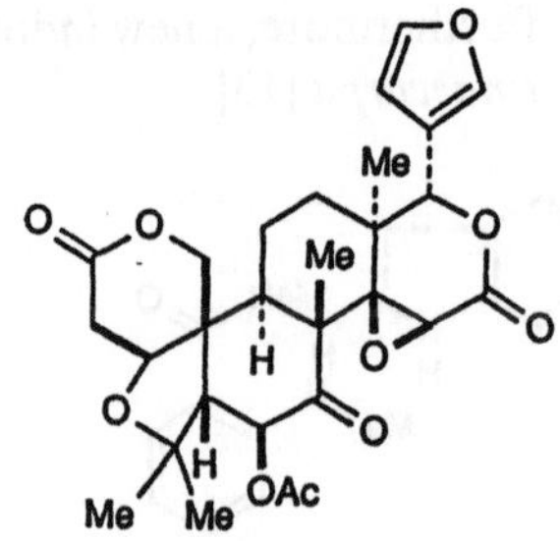

6β-Acetoxy-5-epilimonin (*66-20*)

The bitter principle in fruits of *E. rutaecarpa* obtained during July to August was identified as evodin and the component in fruits obtained during September to October was identified as rutaevin. Fruits collected from the latter part of August to September contained rutaevin and evodol [20].

An isopentenylflavonone glycoside was also isolated from *E. rutaecarpa* and identified as 4′,5,7-trihydroxy-6(or 8)-(3-methylbut-2-enyl)flavanone 7,4′-di*O*-β-D-glucopyranoside (*66-21*) [21].

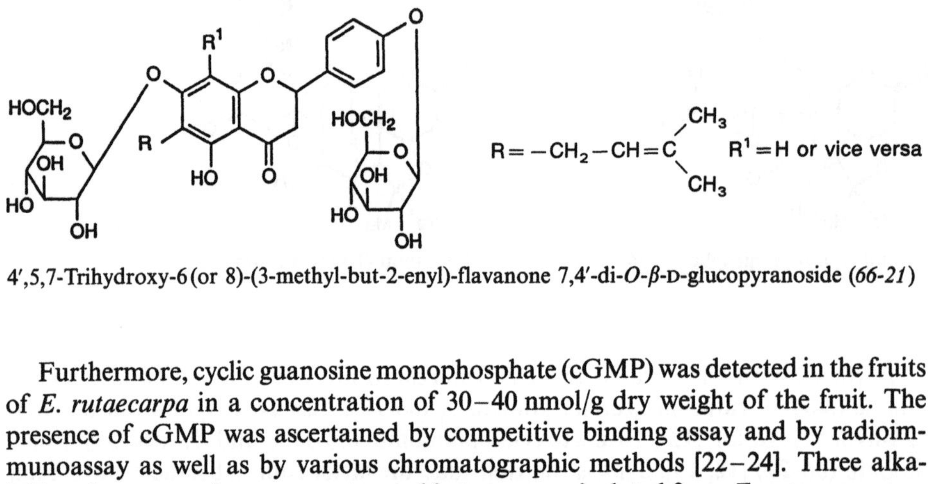

4′,5,7-Trihydroxy-6(or 8)-(3-methyl-but-2-enyl)-flavanone 7,4′-di-*O*-β-D-glucopyranoside (*66-21*)

Furthermore, cyclic guanosine monophosphate (cGMP) was detected in the fruits of *E. rutaecarpa* in a concentration of 30–40 nmol/g dry weight of the fruit. The presence of cGMP was ascertained by competitive binding assay and by radioimmunoassay as well as by various chromatographic methods [22–24]. Three alkaloids, a lactone and a new unsaturated ketone, were isolated from *E. rutaecarpa* var. *officinalis*. According to physical, chemical, and spectral analyses the alkaloids were determined as evodiamine, rutaecarpine, and hydroxyevodiamine, and the lactone was identified as evodin [25].

66.3 Pharmacology

The alkaloids of *E. rutaecarpa* rutaecarpine and dehydroevodiamine showed uterotonic activity on rat uterus in vitro. Dehydroevodiamine was further shown to be active in an in vivo study in rats. Evodiamine was found to be inactive in the uterotonic activity assay [26]. A threshold dose of dehydroevodiamine can potentiate the uterine-contracting action of acetylcholine, serotonin, oxytocin, prostaglandin

F_2, and $BaCl_2$. The uterotonic action and the potentiation effect of dehydroevodiamine can be blocked by indomethacin and mepacrine, suggesting that the dehydroevodiamine action may be mediated through prostaglandin synthesis [27].

Dehydroevodiamine was also shown to lower blood pressure and to produce bradycardia in anesthetized rats. At a cumulative dose of 22.5 mg/kg within 30 min, there was a very significant decrease in blood pressure and heart rate. A stronger decrease in diastolic pressure than in systolic pressure was observed, implying vasodilation. The hypotensive activity of dehydroevodiamine may be mediated by prostaglandin synthesis [28] and may involve its antihistaminic or calcium channel-blocking properties [29].

Injection of evodiamine into the anesthetized dog produced by prompt hypotension with bradycardia and apnea followed by a marked increase in blood pressure with increased depth and rate of respiration before returning to previous levels [30].

References

1. Asahina Y, Ohta T (1928) Eine Synthese des Evodiamins. Chem Ber 61:319–321
2. Boit HG (1960) Ergebnisse der Alkaloid-Chemie bis 1960. Akademie, Berlin, p 741
3. Chu JH (1951) Constituents of the Chinese drug Wu-Chu-Yu, *Evodia rutaecarpa*. Sci Record (China) 4:279–284
4. Kamikado T, Murakoshi S, Tamura S (1978) Structure elucidation and synthesis of alkaloids isolated from fruits of *Evodia rutaecarpa*. Agric Biol Chem 42:1515–1519
5. Danieli B, Lesma G, Palmisano G (1979) A new tryptophan derived alkaloid from *Evodia rutaecarpa* (Juss.) Benth. et Hook. Experientia 35:156
6. Nakasato T, Asada S, Marui K (1962) Dehydroevodiamine, main alkaloid from the leaves of *Evodia rutaecarpa*. Yakugaku Zasshi 82:619–626
7. Tschesche R, Werner W (1967) Evocarpin, ein neues Alkaloid aus *Evodia rutaecarpa*. Tetrahedron 23:1873–1881
8. Yamazaki M, Kawana T (1967) Isolation of hydroxyevodiamine (rhetisinine) from the fruits of *Evodia rutaecarpa*. Yakugaku Zasshi 87:608–610
9. Kamikado T, Chang CF, Murakoshi S, Sakurai A, Tamura S (1976) Isolation and structure elucidation of three quinolone alkaloids from *Evodia rutaecarpa*. Agric Biol Chem 40:605–609
10. Takagi S, Kinoshita T, Sameshima M, Akiyama T, Kobayashi S, Sankawa U (1979) Isolation of synephrine from *Evodia* fruits. Shoyakugaku Zasshi 33:35–37
11. Takagi S, Akiyama T, Konoshita T, Sankawa U, Shibata S (1979) Minor basic constituents of *Evodia* fruits. Shoyakugaku Zasshi 33:30–34
12. Sugimoto T, Miyase T, Kuroyanagi M, Ueno A (1988) Limonoids and quinolone alkaloids from *Evodia rutaecarpa* Bentham. Chem Pharm Bull (Tokyo) 36:4453–4461
13. Shoji N, Umeyama A, Iuchi A, Saito N, Takemoto T, Nomoto K, Ohizumi Y (1988) Isolation of a new alkaloid from *Evodia rutaecarpa*. J Nat Prod 51:791–792
14. Maeda S (1935) Constituents of *Evodia danielli* Hemsl. J Pharm Soc Jpn 55:531–537 (CA 29:5831)
15. Fujita A, Akatsuka M (1949) Obakulactone. V. Evodin. J Pharm Soc Jpn 69:322–325
16. Arnott S, Davie AW, Robertson JM, Sim GA, Watson DG (1961) The structure of limonin: X-ray analysis of epilimonin iodoacetate. J Chem Soc 4183–4200
17. Barton DHR, Pradhan SK, Sternhell S, Templeton JF (1961) Triterpenoids. Part XXV. The constitutions of limonin and related bitter principles. J Chem Soc 255–275
18. Dreyer DL (1967) Citrus bitter principles VII. Rutaevin. J Org Chem 32:3442–3445
19. Sugimoto T, Ueno A, Kadota S, Cut CB, Kikuchi T (1988) New 5β-H limonoids from *Evodia rutaecarpa* Bentham. Chem Pharm Bull 36:1237–1240
20. Hirose Y, Kondo K, Arita H, Fujita A (1967) Components of *Evodia* fruit. II. Shoyakugaku Zasshi 21:126–127
21. Grimshaw J, Lamer-Zarawaka E (1975) Isopentenylflavanone from *Evodia rutaecarpa*. Phytochemistry 14:838–839

22. Cyong JC, Takahashi M (1982) Purification and identification of guanosine 3′,5′-monophosphate from higher plants (*Evodiae fructus*). Chem Pharm Bull (Tokyo) 30:2463–2466
23. Cyong JC, Takahashi M, Hanabusa K, Otsuka Y (1982) Guanosine 3′,5′-monophosphate in fruits of *Evodia rutaecarpa* and *E. officinalis*. Phytochemistry 21:777–778
24. Takahashi M, Cyong JC, Hanabusa K (1980) Cyclic GMP in *Evodiae fructus*. Wakanyaku Shinojumu (Kiroku) 13:96–100 (CA 94:80245n)
25. Li MT, Huang HI (1966) Studies on the chemical constituents of the Chinese drug Shih-Hu (*Evodia rutaecarpa* var. *officinalis*). Acta Pharm Sin 13:265–272
26. King CL, Kong YC, Wong NS, Yeung HW, Fong HHS, Sankawa U (1980) Uterotonic effect of *Evodia rutaecarpa* alkaloids. J Nat Prod 43:577–582
27. Kong YC (1982) *Evodia rutaecarpa*, from Pents'ao to action mechanism. Adv Pharmacol Ther 6:239–243
28. Xu SB, Huang YM, Lau CNB, Wat CKH, Kong YC (1982) Hypotensive effect of dehydroevodiamine from *Evodiae fructus*. Am J Chin Med 10:75–85
29. Yang HY, Li SY, Chen CF (1988) Hypotensive effects of dehydroevodiamine, a quinazolinocarboline alkaloid isolated from *Evodia rutaecarpa*. Asia Pac J Pharmacol 3:191–196
30. Hamet R (1962) Pressure and respiratory effects of evodiamine of Yasuhiko Asahima. Compt Rend 255:1152–1154

Forsythia suspensa (Thunb.) Vahl. 67

67.1 Introduction

Lianqiao, Fructus Forsythiae, is the dry fruits of *Forsythia suspensa* (Thunb.) Vahl. (Oleaceae) collected in the fall when the fruits have become ripe. It is a well-known traditional Chinese medicine and officially listed in Chinese Pharmacopoeia. The fruits are widely used as an antipyretic and antiinflammatory in the treatment of bacterial infections.

67.2 Chemical Constituents

Phenolic glycosides, lignans, triterpenes, and some C_6-C_2 natural alcohols were isolated and identified from the fruits of *F. suspensa*. Among the lignans, phillygenin (*67-1*) and pinoresinol and their corresponding glucosides phillyrin (*67-2*) and pinoresinol glucoside are compounds of the diphenyl-perhydrofurotetrahydrofurane type [1–5]. Matairesinol (*67-3*) and arctigenin (*67-5*) and their glycosides arctiin (*67-6*) and matairesinoside (*67-4*) [2, 4] as well as *O*-methylarctigenen (*67-7*) [6] are compounds of the 2,3-dibenzylbutyrolactone type.

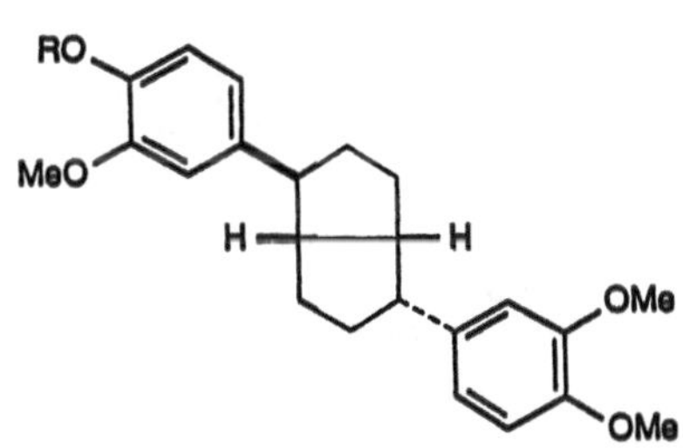

Phillygenin (*67-1*): R=H

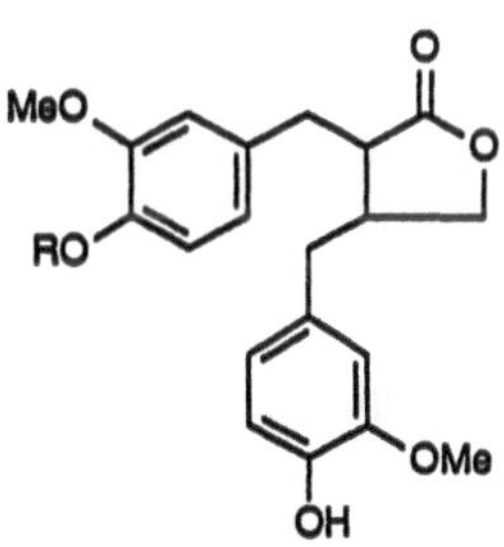

Matairesinol (*67-3*): R=H

Phillyrin (*67-2*): R=

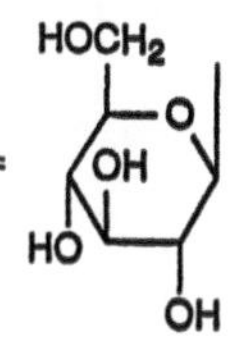

Matairesinoside (*67-4*): R=

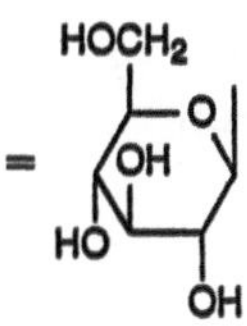

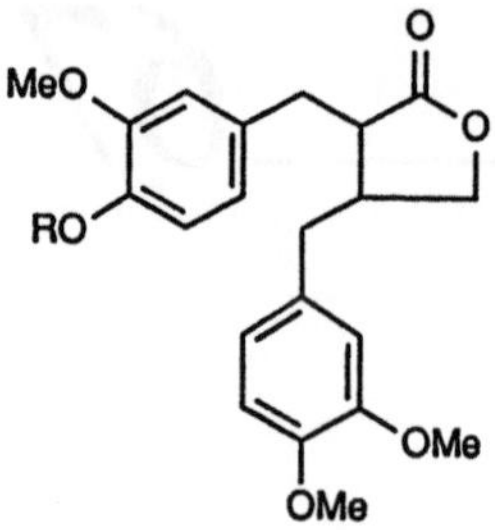

Arctigenin (*67-5*): R=H

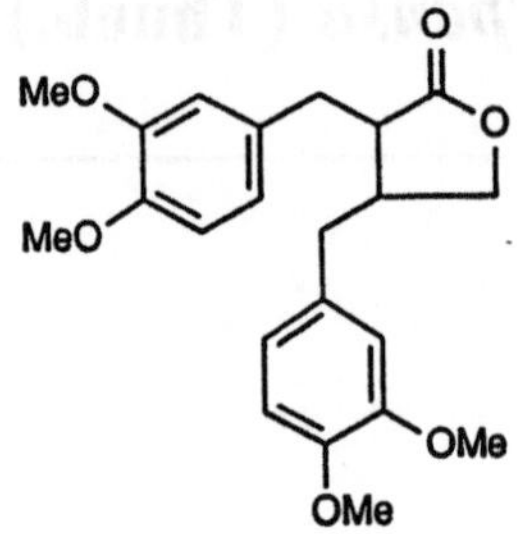

O-Methylarctigenin (*67-7*)

Arctiin (*67-6*): R=

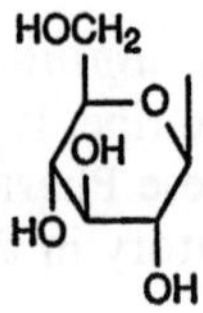

Epipinoresinol-4″-glucoside was obtained from the bark of *F. suspensa* collected in the Federal Republic of Germany [7]. A number of phenolic glycosides from *F. suspensa* were named forsythosides. Forsythoside A (*67-8*) was isolated from the leaves of *F. suspensa*. The structure of forsythoside A was determined on the basis of chemical degradation and spectral analyses [8]. It was also isolated from the fruits of *F. suspensa* and was named forsythiaside [9]. Then, forsythosides C (*67-9*) (suspensaside [10]), D (*67-10*) [11], and E (*67-11*) [12] were also isolated from the fruits of *F. suspensa*. Forsythoside B (*67-12*) did not occur in *F. suspensa*. It was isolated from *F. koreana* [13].

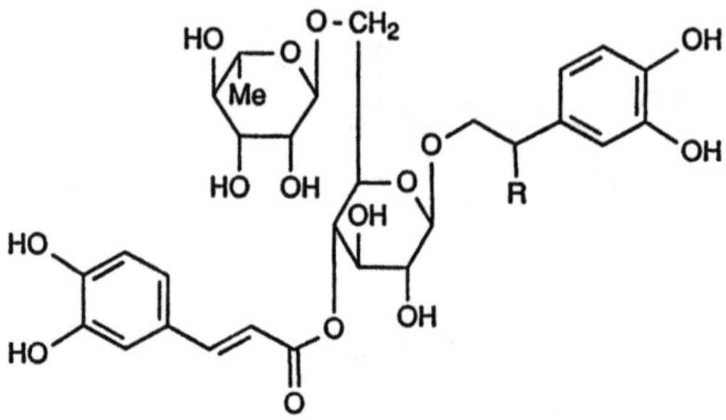

Forsythoside A (Forsythiaside) (*67-8*): R=H
Forsythoside C (Suspensaside) (*67-9*): R=OH

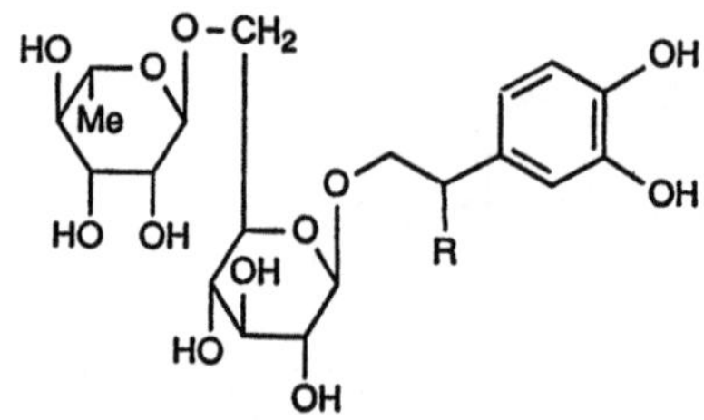

Forsythoside D (*67-10*): R=OH
Forsythoside E (*67-11*): R=H

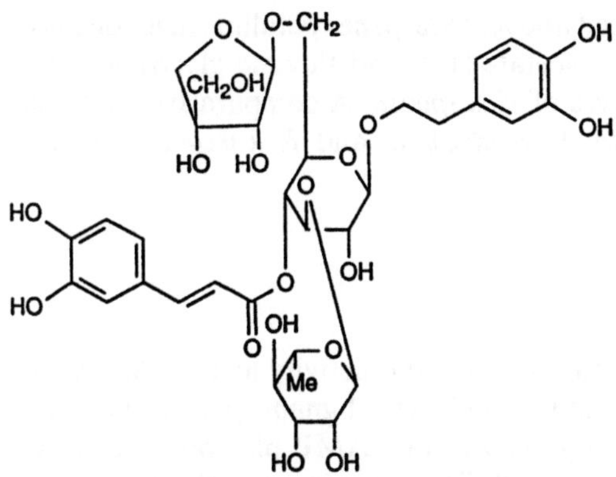

Forsythoside B (*67-12*)

In addition to the forsythosides, two caffeoyl glycosides of 3,4-dihydroxyphenethyl alcohol were also detected from the leaves of *F. suspensa* [14].

The fruits of *F. suspensa* produced, besides the lignans and glycosides, a number of new natural alcohols and their glucosides containing the cyclohexylethane skeleton. The compounds isolated were rengyol (*67-13*) and its glucoside rengyoside A (*67-14*), rengyoxide (*67-15*), rengyolone (*67-16*), and the known glycosides cornoside (*67-17*) and salidroside (*67-18*) [12, 15]. The isolation of a related polyol named suspenol (*67-19*) [16] was recently reported.

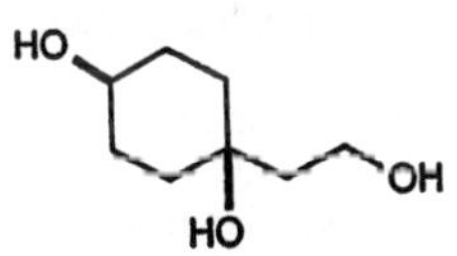

Rengyol (*67-13*)

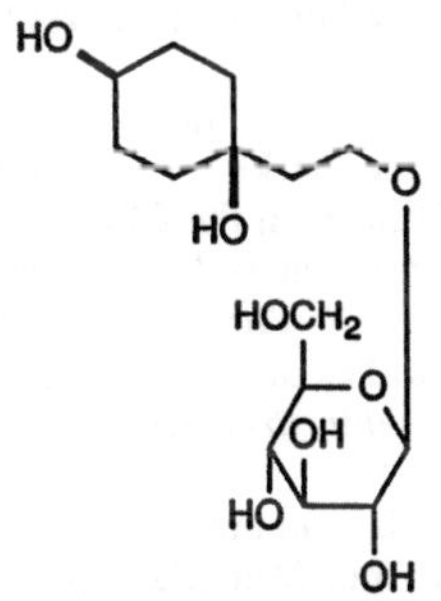

Rengyoside A (*67-14*)

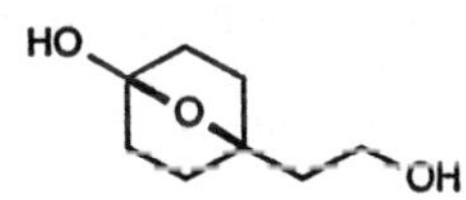

Rengyoxide (*67-15*)

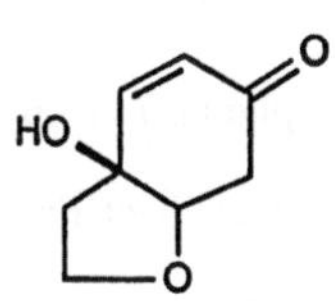

Rengyolone (*67-16*)

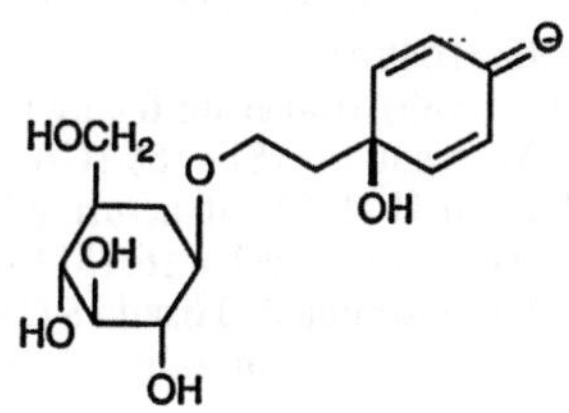

Cornoside (*67-17*)

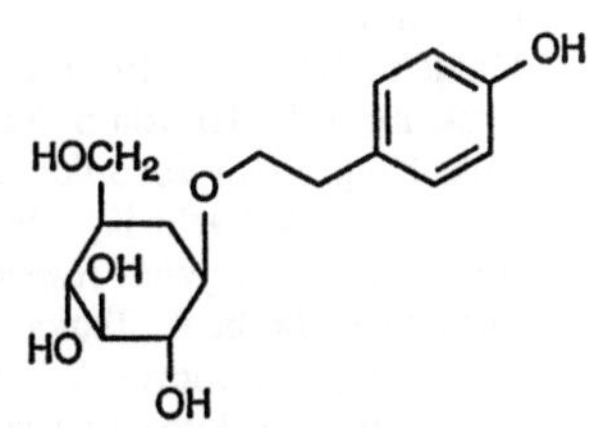

Salidroside (*67-18*)

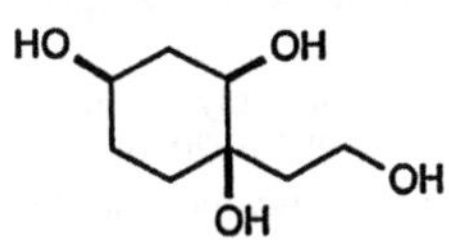

Suspenol (*67-19*)

In addition to the constituents described above, triterpenes betulinic acid, oleanolic acid [2], ursolic acid [4, 17], β-amyrin acetate [18], and flavone glycoside rutin [2, 4, 19] were found in fruits and/or leaves of *F. suspensa*. A comparative study on the chemical constituents of *F. suspensa, F. viridissima*, and *F. koreana* was also reported [14].

67.3 Pharmacology

Forsythoside A [8], C, and D [11] exhibited antibacterial activity against *Staphylococcus aureus* at a concentration less than 2 m*M*. The lignans pinoresinol and pinoresinol glucoside showed an inhibitory activity on cAMP phosphodiesterase. A structure–activity relationship was also noted. Thus, in pinoresinol analogs, the configuration of the two benzene rings is very important with respect to inhibitory activity. Substitution of *p*-hydroxyl groups in matairesinol analogs by a methyl or glucosyl group leads to decrease of activities in comparison with those of the unsubstituted compounds [20]. Phosphodiesterase was also found to be inhibited by the chloroform extract of *Forsythia* fruits [21]. The caffeoylglycosides of *Forsythia* fruits also possessed an inhibitory activity on the formation of 5-hydroxy-6,8,11,14-eicosatetraenoic acid from arachidonic acid in rat peritoneal cells [22].

References

1. Yee CL (1960) Bacteriostatic principle isolated from *Forsythia suspensa* (Lien Chiao). Acta Pharm Sin 8:241–244
2. Nishibe S, Chiba M, Hisada S (1977) Studies on the Chinese crude drug "Forsythiae fructus". I. Constituents of Forsythiae fruits on the market. Yakugaku Zasshi 97:1134–1137
3. Nishibe S, Chiba M, Hisada S (1977) Studies on the Chinese crude drug "Forsythiae fructus". II. Comparative examination on the lignin glucosides of *Forsythia* fruits of the original plants listed in the Japanese Pharmacopoeia Ed IX. Shoyakugaku Zasshi 31:131–135
4. Chiba M, Hisada S, Nishibe S (1978) Studies on the Chinese crude drug "Forsythiae fructus". III. On the constituents of fruits of *Forsythia viridissima* and *F. suspensa*. Shoyakugaku Zasshi 32:194–197
5. Kuang HX, Zhang N, Lu ZB (1988) Studies on antibacterial constituents of the unripe fruit of *Forsythia suspensa* (Thunb.) Vahl. Bull Chin Mat Med 13:416–418
6. Takizawa Y, Suzuki E, Mitsuhashi T (1981) Naturally occurring antioxidant. (I). Isolation and determination of natural phenolic antioxidants from *Forsythia suspensa* Vahl. Tokyo Gakugei Daigaku Kiyo 33:119–123 (CA 96:31648d)
7. Tsukamoto H, Hisada S, Nishibe S (1983) Studies on the lignans from Oleaceae plants. Tennen Yuki Kagobutsu Toronkai Koen Yoshishu 26:181–188 (CA 100:117805r)
8. Endo K, Takahashi K, Abe T, Hikino H (1981) Structure of forsythoside A, an antibacterial principle of *Forsythia suspensa* leaves. Heterocycles 16:1311–1314
9. Nishibe S, Okabe K, Tsukamoto H, Sakushima A, Hisada S (1982) Studies on the Chinese crude drug *Forsythia* fructus. V. The structure of forsythiaside isolated from *Forsythia suspensa*. Chem Pharm Bull (Tokyo) 30:1048–1050
10. Nishibe S, Okabe K, Tsukamoto H, Sakushima A, Hisada S, Baba H, Akisada T (1982) Studies on the Chinese drug "forsythiae fructus" VI. The structure and antibacterial activity of suspensaside from *Forsythia suspensa*. Chem Pharm Bull (Tokyo) 30:4538–4553
11. Endo K, Hikino H (1982) Validity of oriental medicine, part 44. Structures of forsythoside C and D, antibacterial principles of *Forsythia suspensa* fruits. Heterocycles 19:2033–2036
12. Endo K, Hikino H (1984) Structures of rengyol, rengyoxide and rengyolone, new cyclohexylethane derivatives from *Forsythia suspensa*. Can J Chem 62:2011–2014

13. Endo K, Takahashi K, Abe T, Hikino H (1982) Structure of forsythoside B, an antibacterial principle of *Forsythia koreana* stems. Heterocycles 19:261–264
14. Kitagawa S, Hisada S, Nishibe S (1984) Phenolic compounds from *Forsythia* leaves. Phytochemistry 23:1635–1636
15. Endo K, Seya K, Hikino H, Akiyama M, Ogasawara K, Takano S (1985) Structures and reactions of rengyol and the related natural products. Tennen Yuki Kagobutsu Toronkai Koen Yoshishu 27:656–663 (CA 104:164877h)
16. Endo K, Seya K, Hikino H (1987) Structure and enantioselective synthesis of suspenol, a new polyol of Forsythia suspensa. Tennen Yuki Kagobutsu Toronkai Koen Yoshishu 29:660–667 (CA 109:110788q)
17. Liang WZ, Zhou JG, Tu GS (1985) Analysis of constituents of *Forsythia suspensa.* III. Isolation, identification and determination of ursolic acid. Chin J Pharm Anal 5:67–69
18. He M, Wang JC (1983) Studies on chemical constituents in the seeds of *Forsythia suspensa.* I. Isolation and identification of lipid soluble constituents. Bull Chin Mat Med 8:34
19. Liang WZ, Zhou JG, Tu GS (1985) Analysis of constituents of *Forsythia suspensa.* II. Isolation, identification and determination of rutin. Chin J Pharm Anal 5:79–80
20. Nikaido T, Ohmoto T, Kinoshita T, Samkawa U, Nishibe S, Hisada S (1981) Inhibitors of cyclic AMP phosphodiesterase in medicinal plants. II. Inhibition of cyclic AMP phosphodiesterase by lignans. Chem Pharm Bull (Tokyo) 29:3586–3592
21. Sankawa U (1980) Screening of bioactive compounds in oriental medicinal drugs. Saengyak Hakhoe Chi (Hanguk Saengyak Hakhoe) 11:125–132 (CA 95:125747s)
22. Kimura Y, Okuda H, Nishibe S, Arichi S (1987) Effects of caffeoylglycosides on arachidonate metabolism in leukocytes. Planta Med 53:148–153

Fraxinus spp.

68

68.1 Introduction

Qinpi, Cortex Fraxini, is the dry barks from stem or branches of the following *Fraxinus* species: *F. rhynchophylla* Hance, *F. chinensis* Roxb., *F. chinensis* Roxb. var. *acuminata* Lingelsh., or *F. stylosa* Lingelsh. (Oleaceae) collected in spring and the fall. It is officially listed in the Chinese Pharmacopoeia and is used in traditional Chinese medicine as an astringent for the treatment of diarrhea and as an antiphlogistic in ophthalmology for the treatment of conjunctivitis.

68.2 Chemical Constituents

The first report about coumarin components in *Fraxinus* bark as a traditional Chinese medicine was given in 1962. Esculin (*68-1*) and esculetin (*68-2*) were isolated as active principles against dysentery from *F. rhychophylla* [1].

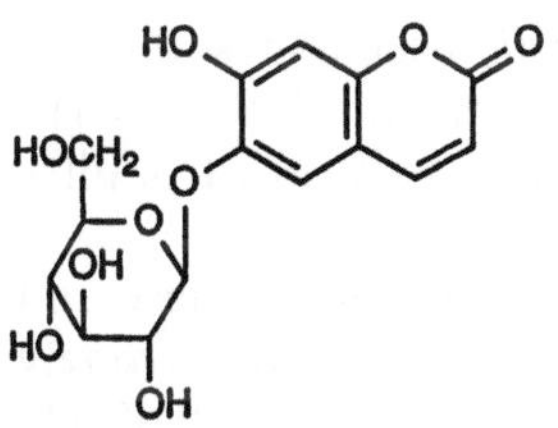

Esculin (*68-1*)

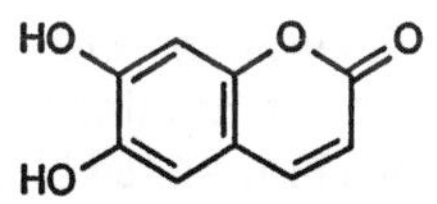

Esculetin (*68-2*)

Esculin is a coumarin glucoside isolated initially from the bark of *Aesculus hippocastanum* (Hippocastanaceae) and was structurally determined in 1930 by Head and Robertson [2]. The aglycone esculetin was first obtained by hydrolysis of esculin and structurally investigated more than 100 years ago [3, 4]. Recently, the X-ray crystallographic analysis of esculetin demonstrated that the molecules of esculetin are nearly planar [5]. The esculin and esculetin content in the bark of *F. rhynchophylla* was 3.4% [6] and was highest at the flowering stage [7]. From the bark of *F. chinensis*, two further compounds, syringin (*1-8*) and fraxin (*68-3*) were detected together with esculin and esculetin [8]. Fraxin is another coumarin glycoside isolated first from *F. excelsior* [9]. The bark of *F. stylosa* contains a new coumarin glycoside named stylosin, in addition to the known components esculin, esculetrin, fraxin, and syringin. The structure of stylosin was determined by spectral analysis as 6-methoxy-7-hydroxy-8-rhamnosyl-rhamnosyl-glucosyloxycoumarin [10].

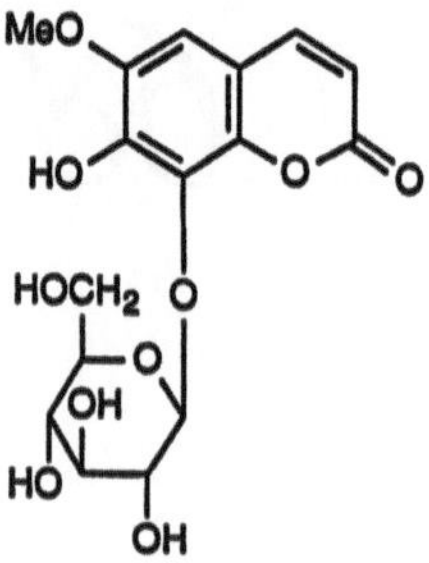

Fraxin (*68-3*)

Stylosin was also isolated from the bark of *F. chinensis*. The contents of esculetin, fraxin, esculin, and stylosin in the bark of *F. stylosa* were 0.1%, 0.5%, 2.6%, and 0.3%, and in the bark of *F. chinensis* 0.1%, 0.6%, 3.1%, and 0.1%, respectively [11]. Seasonal variations in the coumarin contents were observed. The values were lowest in January and highest in July [11]. A study of the coumarin contents in various *Fraxinus* species showed that *F. stylosa, F. fallax, F. paxiana, F. ornus, F. bungeana, F. rhychophylla, F. rhychophylla* var. *huashanensis, F. caudata, F. chinensis*, and *F. sargentiana* had high esculin contents (more than 1.75%) [12].

68.3 Pharmacology

The major constituents esculin and esculetin from the *Fraxinus* bark both inhibited the growth of *Shigella flexneri, S. sonnei*, and *S. schmitzii*. Esculetin showed a significant curative effect when administered orally to patients with dysentery [1]. Esculetin effectively inhibits human red blood cell glyoxalase I activity in vitro. Fifty percent inhibition was obtained at 0.03 m*M* [13].

Studies of the effects of coumarin derivatives on rat platelet lipoxygenase and cyclooxygenase activities showed that esculetin inhibited lipoxygenase more strongly than cyclooxygenase. The concentrations needed for 50% inhibition were 0.65 μ*M* for platelet lipoxygenase and 0.45 m*M* for platelet cyclooxygenase. Esculin also inhibited lipoxygenase, though less strongly (50% inhibition at 290 μ*M*). The inhibition of lipoxygenase by esculetin was noncompetitive [14]. The 50% inhibition concentrations of esculetin on 5- and 12-lipoxygenase of cloned mastocytoma cells were 4 μ*M* and 2.5 μ*M*, respectively. No inhibition at all, but rather a stimulating effect on prostaglandin synthesis, was observed, especially at higher doses of esculetin. Leukotriene synthesis by mouse mast tumor cells was also reduced in the presence of esculetin, in parallel with 5-lipoxygenase inhibition [15]. An inhibitory activity of esculetin on 5-lipoxygenase of human polymorphonuclear leukocytes was also observed [16].

References

1. Mei PF, Hsu CC, Wang Y (1962) Active principles of the Chinese drug Chin Pie (*Fraxinus rhychophylla*). Acta Chim Sin 28:25–30
2. Head FSH, Robertson A (1930) Natural glucosides, part II. The constitution of aesculin. J Chem Soc 2434–2444
3. Tiemann F, Will W (1882) Zur Konstitution des Aesculetins. Chem Ber 15:2072–2083
4. Will W, Albrecht K (1884) Über einige Pyrogallussäure- und Phloroglucinderivate und die Beziehungen derselben zu Daphnetin und Aesculetin. Chem Ber 17:2098–2109
5. Ueno K, Saito N (1977) Esculetin; 6,7-dihydroxycoumarin. Acta Crystallogr [B] B33:283–285
6. Zhang XQ, Xu LX (1982) Polarographic determination of aesculetin in Qin Pi (*Fraxinus rhychophylla* Hance or *F. chinensis* Roxb.). Acta Pharm Sin 17:305–308
7. Zeng MY, Fu GL, Wu JL (1982) Assay of aesculin and aesculetin in Qin Pi. Bull Chin Mat Med 7:26–29
8. Guo XS, Zhang YZ (1983) HPLC separation and determination of active principles in Chin Pie (*Fraxinus chinensis*). Acta Pharm Sin 18:525–528
9. Wessely F, Demmer E (1929) Konstitution und Eigenschaften des Fraxins. Chem Ber 62:120
10. Guo XS, Zhang YZ (1983) Chemical studies on the Chinese drug Qin Pi, the bark of *Fraxinus stylosa*. Acta Pharm Sin 18:434–439
11. Guo XS, Zahng YZ (1983) Quantitative TLC-densitometry of coumarins in Qin Pi (*Fraxinus stylosa*). Acta Pharm Sin 18:446–452
12. Wu JL, Fu GL, Zeng MY (1983) Studies on the quality and resource of the Chinese crude drug "Qin Pi" (*Cortex fraxini*). Chin J Pharm Anal 3:12–18
13. Brandt RB, Brandt ME, April ME (1983) Inhibitors of glyoxalase I in vitro. Biochem Med 29:385–391
14. Sekiya K, Okuda H, Arichi S (1982) Selective inhibition of platelet lipoxygenase by esculetin. Biochim Biophys Acta 713:68–72
15. Neichi T, Koshihara Y, Murota T (1983) Inhibitory effect of esculetin on 5-lipoxygenase and leukotriene biosynthesis. Biochim Biophys Acta 753:130–132
16. Panossian AG (1984) Inhibition of arachidonic acid 5-lipoxygenase of human polymorphonuclear leukocytes by esculetin. Biomed Biochim Acta 43:1351–1355

Fritillaria spp. 69

69.1 Introduction

Beimu, Bulbus Fritillariae, is one of the most widely used Chinese medicines. The following items from *Fritillaria* species (Liliaceae) are officially listed in the Chinese Pharmacopoeia:

- Chuanbeimu, Bulbus Fritillariae cirrhosae, is the dry bulbs of the following *Fritillaria* species: *F. cirrhosa* D. Don, *F. unibracteata* Hsiao et K. C. Hsia, *F. przewalskii* Maxim. and *F. delavayi* Franch. The bulbs should be dug and collected in summer or the fall. They are used in the treatment of diseases of the respiratory tract, especially as an expectorant.
- Yibeimu, Bulbus Fritillariae pallidiflorae, is the dry bulbs of *F. walujewii* Regel and *F. pallidiflora* Schrenk. dug and collected from May to July. It is also used in the treatment of respiratory diseases.
- Zhebeimu, Bulbus Fritillariae thunbergii, is the dry bulbs of *F. thunbergii* Miq. collected in early summer when the aboveground part has withered. The bulbs of *F. thunbergii* have the same medical use as those of the other *Fritillaria* species mentioned above.

69.2 Chemical Constituents

69.2.1 Chemical Constituents of *Fritillaria thunbergii*

Fritillaria bulbs were known to contain alkaloids. The *Fritillaria* alkaloids are characterized by having a cevane (*69-1*) skeleton, which is composed of an isosteroid structure with a quinolizidine system.

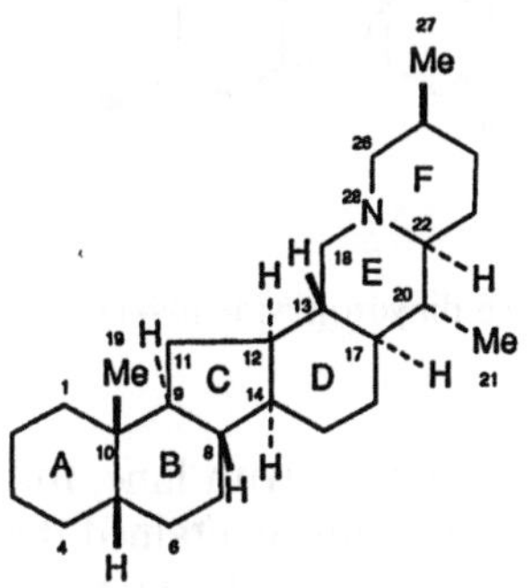

Cevane (*69-1*)

The first two alkaloids named peimine (*69-2*) and peiminine (*69-3*) were isolated more than 50 years ago [1]. A number of studies to clarify their structures were performed [2–12]. Peimine was also designated as verticine and peiminine as verticinone. The structures of peimine and peiminine were elucidated only 30 years after their isolation [13]. Their stereochemistry was determined by X-ray analysis using peiminine methobromide [14].

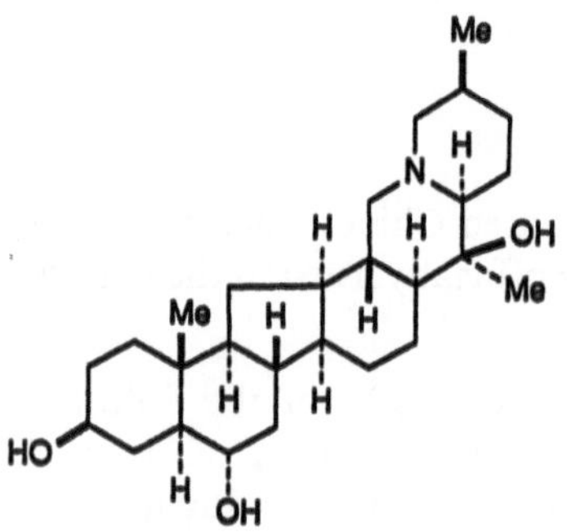

Peimine (*69-2*)

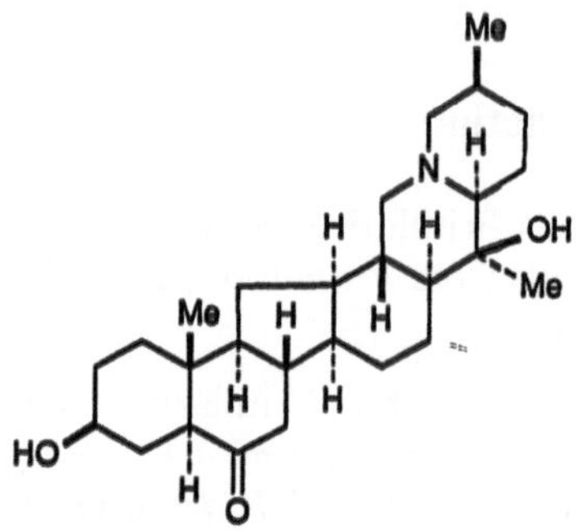

Peiminine (*69-3*)

Peimine and peiminine were both found in fresh and processed bulbs of *F. thunbergii* [15]. From the fresh bulb of *F. thunbergii* another alkaloid related to jervine (*69-5*), a constituent of *Veratrum nigrum* derived from veratraman (*69-4*), 11-deoxo-6-oxo-5α,6-dihydrojervine (*69-6*) was isolated [15].

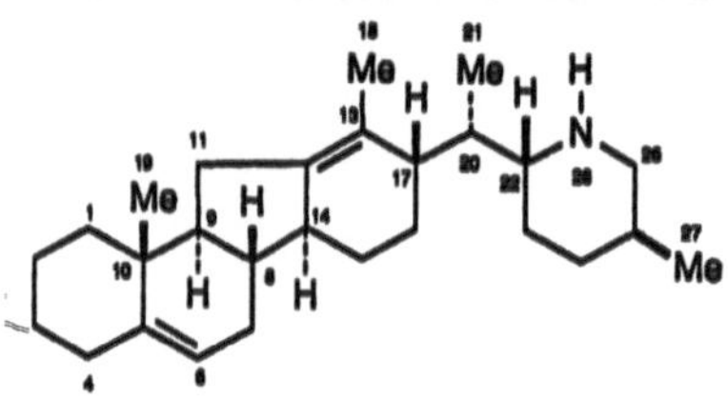

Veratraman (*69-4*)

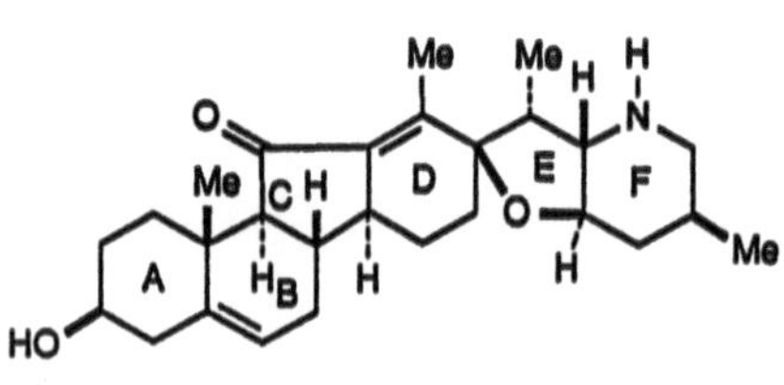

Jervine (*69-5*)

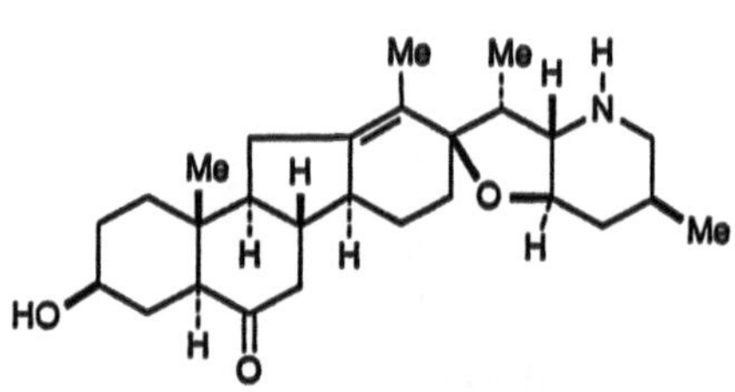

11-Deoxo-6-oxo-5α,6-dihydrojervine (*69-6*)

In contrast to the fresh bulbs, the bulbs processed by treatment with lime, followed by bleaching in the sun in addition to peimine and peiminine, contained the peimine *N*-oxide and peiminine *N*-oxide together with 12,13-epoxy-11-deoxo-6-oxo-5,6-dihydrojervine *N,O*-diacetate (*69-7*) and another compound with an open

E ring, 12,13-epoxy-5α,6-dihydro-veratraman-3β,17,23α-triol-6-one *N*,*O*(3)-diacetate (*69-8*) [15].

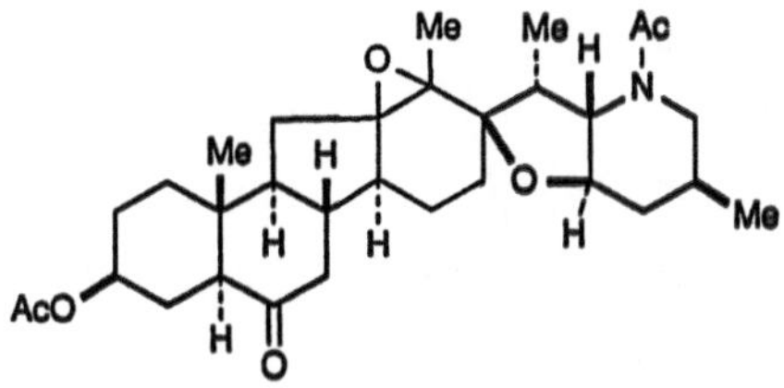

12,13-Epoxy-11-deoxo-6-oxo-5α,6-dihydrojervine *N*,*O*-diacetate (*69-7*)

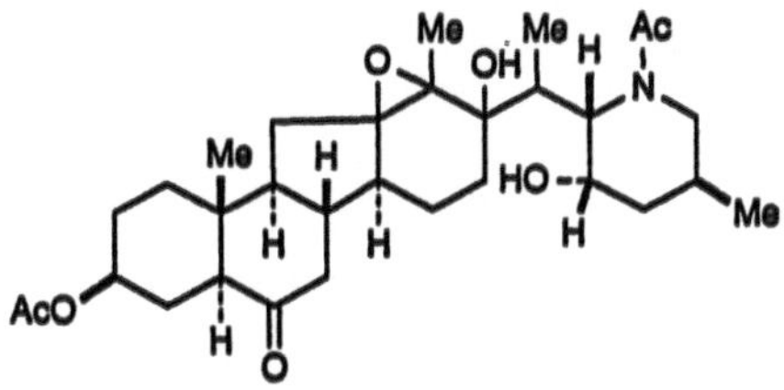

12,13-Epoxy-5α,6-dihydro-veratraman-3β,17,23α-triol-6-one *N*,*O* (3)-diacetate (*69-8*)

Furthermore, the isolation of another new cevane alkaloid, isoverticine (*69-9*), from the bulb of *F. thunbergii* (*F. verticillate* Wild var. *thunbergii* (Miq.) Baker [15]) was also reported [16]. It is the 6-epimer of peimine.

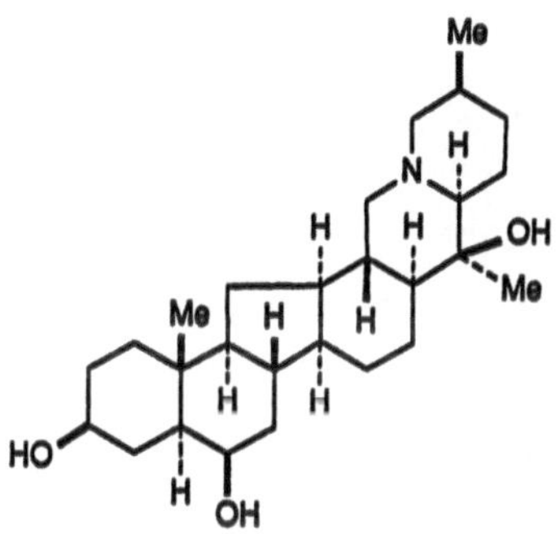

Isoverticine (*69-9*)

In addition to the alkaloid components, a number of diterpene constituents were isolated from the processed bulbs of *F. thunbergii.* They were kaurane derivatives ent-15,16-epoxykauran-17-ol (*69-10*), ent-16-hydroxykauran-17-yl ent-kaur-15-en-17-oate (*69-11*), ent-kauran-16,17-diol (*69-12*), ent-17-norkauran-16-one (*69-13*), phenanthrene derivatives isopimaran-19-ol (*69-14*), and isopimaric acid (*69-15*) and *trans*-communol (*69-16*), *trans*-communic acid (*69-17*) as well as ent-(16S)-atisan-13,17-oxide (*69-18*) [17]. Ent-15β,16-epoxykauran-17-ol, ent 17β-hydroxykauran-17-yl ent-kaur-15-en-17-oate, ent-(16S)-atisan-13,17-oxide, and isopimaric acid were not found in fresh bulbs. Lime treatment and bleaching in the sun oxidized most of the diterpenes in the fresh bulbs [17].

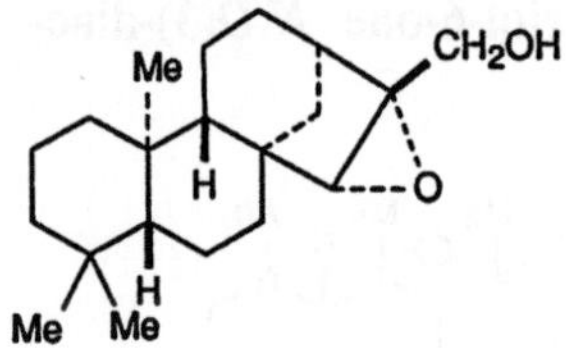

15,16-Epoxykauran-17-ol (*69-10*)

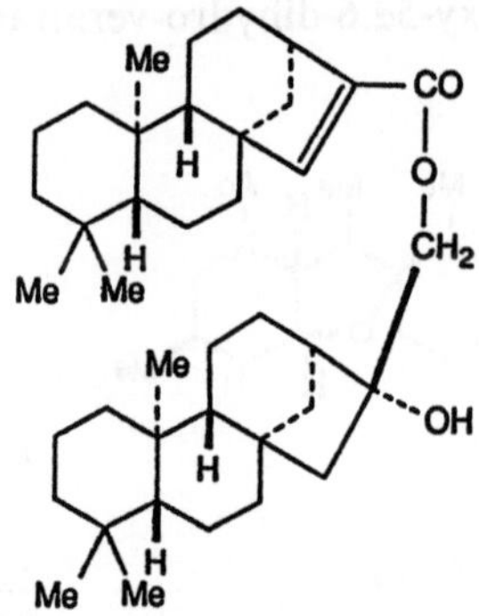

16-Hydroxykauran-17-yl kaur-15-en-17-oate (*69-11*)

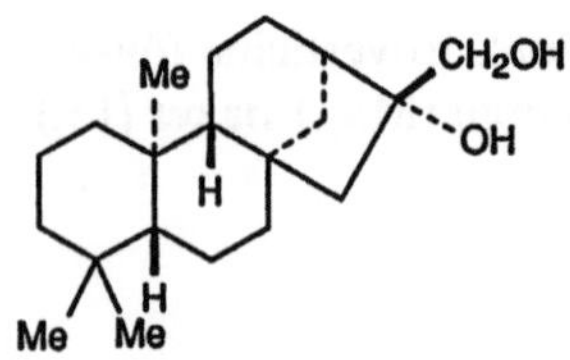

Kauran-16,17-diol (*69-12*)

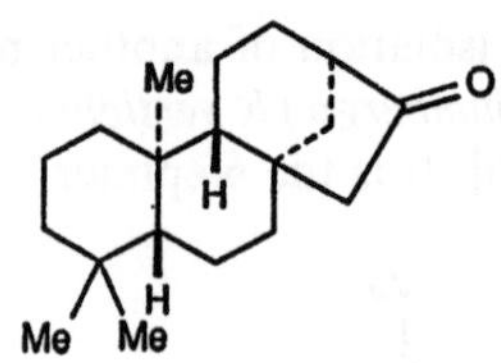

17-Norkauran-16-one (*69-13*)

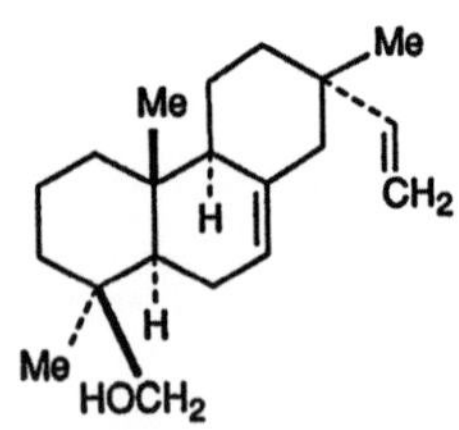

Isopimaran-19-ol (*69-14*)

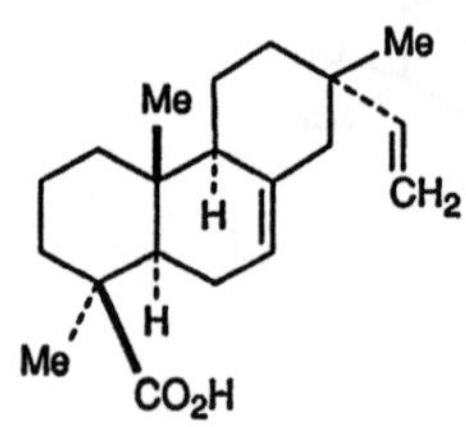

Isopimaric acid (*69-15*)

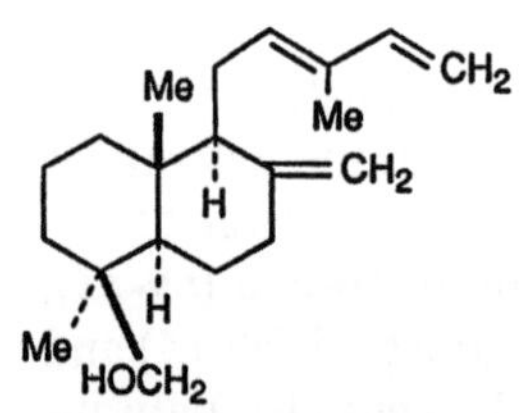

trans-Communol (*69-16*)

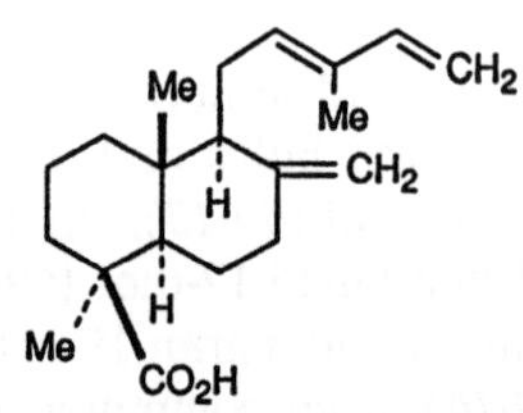

trans-Communic acid (*69-17*)

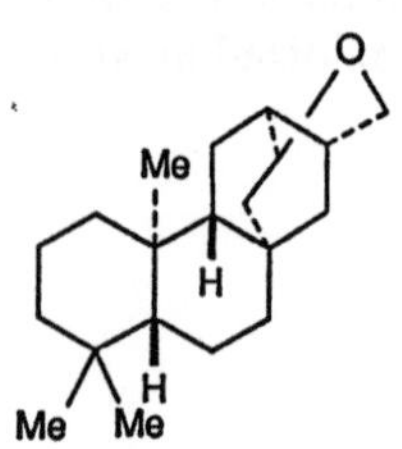

(16S)-Atisan-13,17-oxide (*69-18*)

Steroid components stigmasterol, campesterol, β-sitosterol, β-sitosterol glucoside as well as succinic acid and thymidine diacetate were further isolated from the fresh bulbs of *F. thunbergii*. 5α,6β-Dihydroxycholestan derivatives were found in the processed bulbs [18]. In addition to peimine and peiminine, two new epimers of peimine and isoverticine, named baimonidine (*69-19*) [16] and isobaimonidine (*69-20*) [19], were isolated from the aerial part of *F. thunbergii*. The structures of these four epimers are given below.

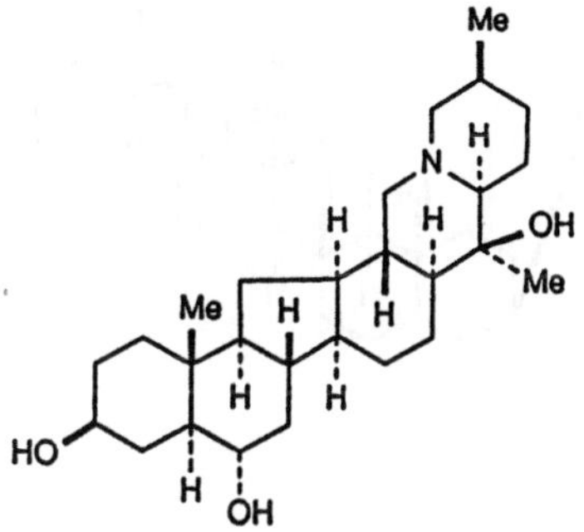

Peimine (*69-2*)

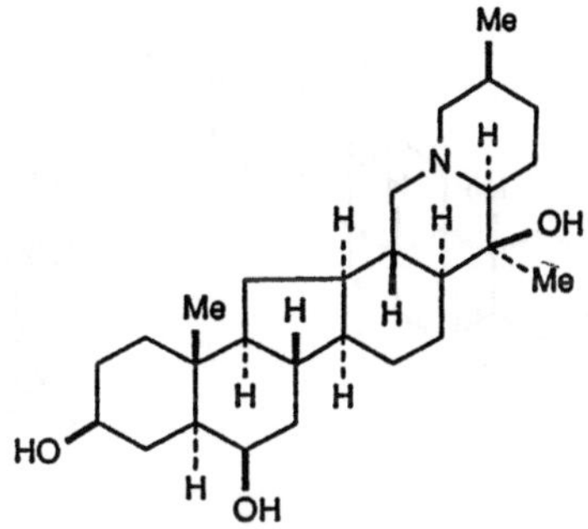

Isoverticine (*69-9*)

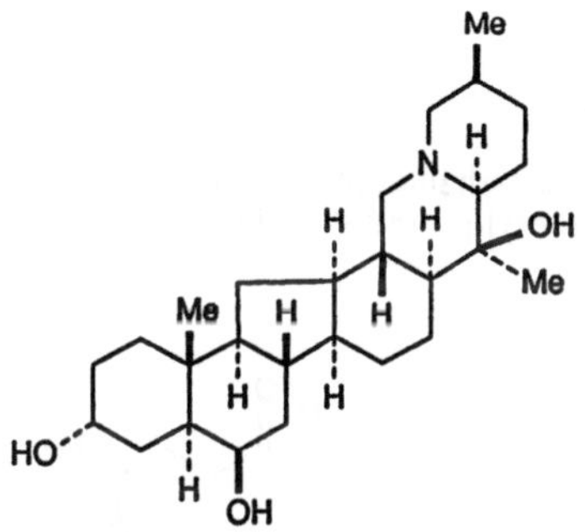

Baimonidine (*69-19*)

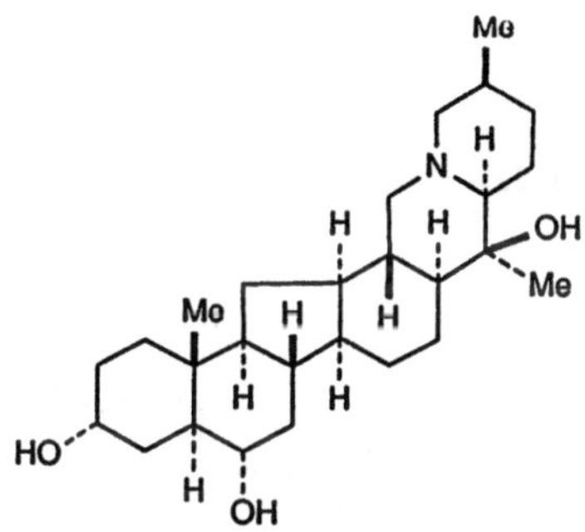

Isobaimonidine (*69-20*)

The 3β-hydroxy epimers peimine and isoverticine mainly accumulate in the bulb, whereas the α-hydroxy epimers baimonidine and isobaimonidine are contained mainly in the aerial parts of the plant. Another alkaloid named fritillarizine (*69-21*) without the substituent at C-6 was also found [20].

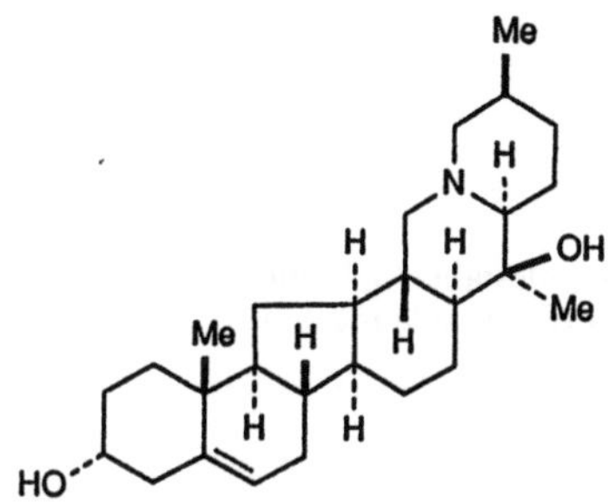

Fritillarizine (*69-21*)

Besides the alkaloids mentioned above, some alkaloid glycosides and saponins were also isolated from the aerial part of *F. thunbergii*. They are β-chaconine (*69-22*), solanidine 3-*O*-α-L-rhamnopyranosyl-(1→2)-[β-D-glucopyranosyl-(1→4)-β-D-glucopyranoside (*69-23*), hapepunine-3-*O*-α-L-rhamnopyranosyl-(1→2)-β-D-glucopyranoside (*69-24*) [21], 6,22-dioxo-5α-cholestane-3β,26-diol bis-*O*-β-D-glucopyranoside (*69-25*), and 3β-[α-L-rhamnopyranosyl-(1→2)-β-D-glucopyranosyloxy]-6,22-dioxo-5α-cholestan-26-ol 26-*O*-β-D-glucopyranoside (*69-26*) [22].

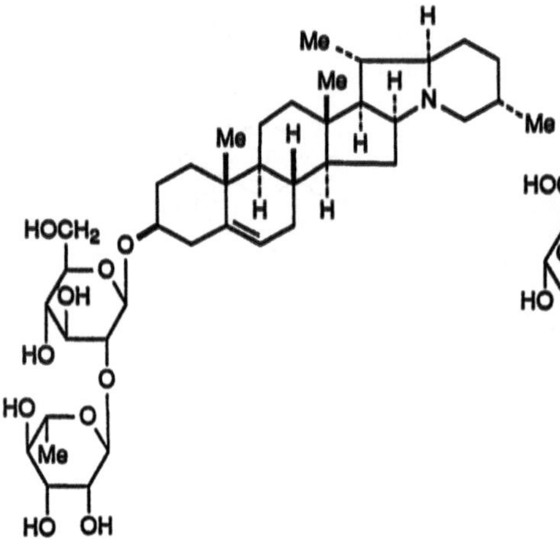

β-Chaconine (*69-22*)

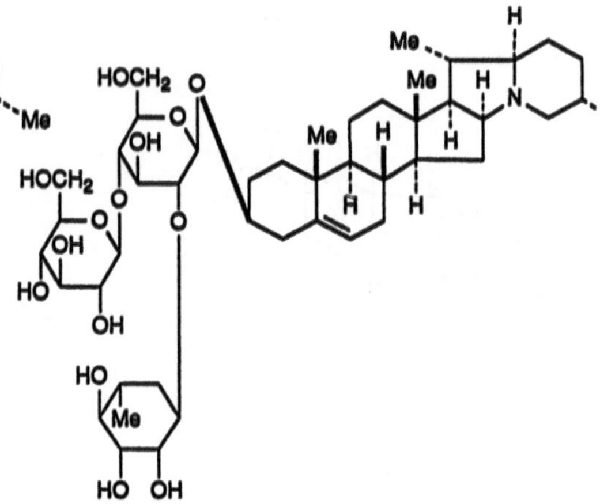

Solanidine 3-*O*-α-L-rhamnopyranosyl-(1→2)-[β-D-glucopyranosyl-(1→4)]-β-D-glucopyranoside (*69-23*)

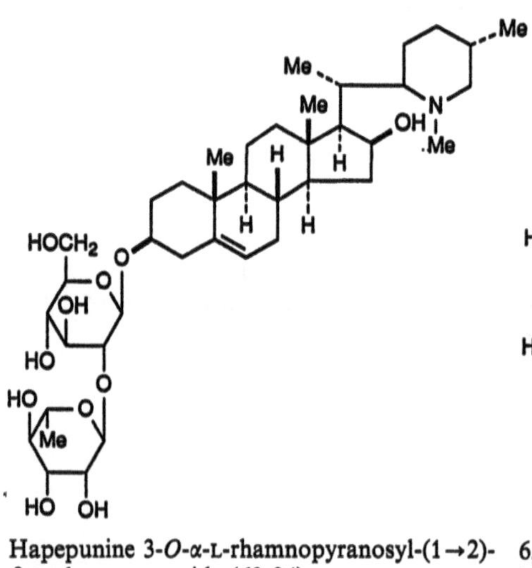

Hapepunine 3-*O*-α-L-rhamnopyranosyl-(1→2)-β-D-glucopyranoside (*69-24*)

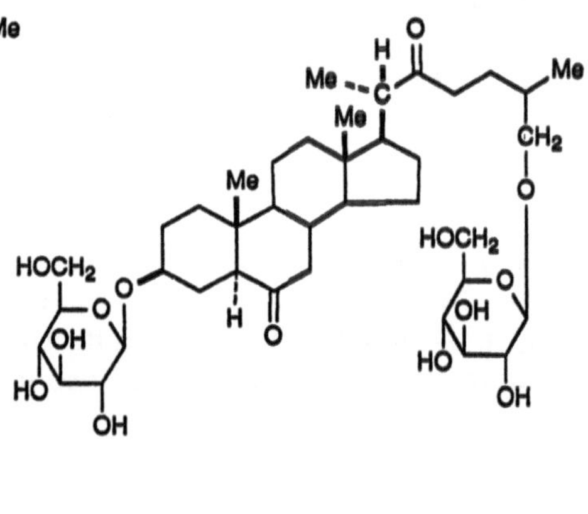

6,22-dioxo-5α-cholestane-3β,26-diol bis-*O*-β-D-glucopyranoside (*69-25*)

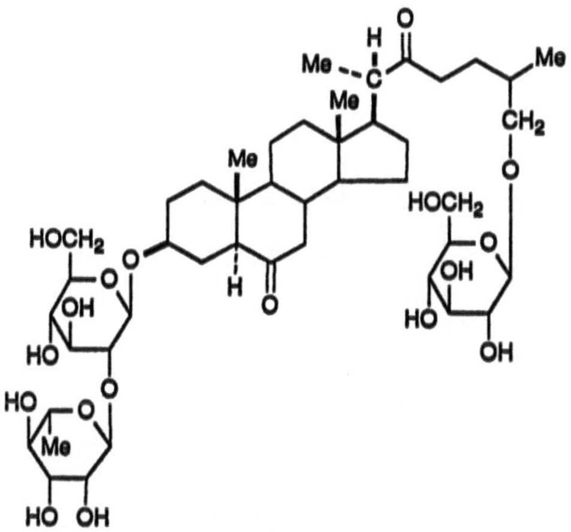

3β-[α-L-rhamnopyranosyl-(1→2)-O-β-D-glucopyranosyloxy-6,22-dioxo-5α-cholestan-26-ol 26-O-β-D-glucopyranoside (*69-26*)

69.2.2 Chemical Constituents of *Fritillaria delavayi*

The bulbs of *F. delavayi* contain some new alkaloids of the cevane type, delavine (*69-27*), delavinone (*69-28*) [23], chuanbeinone (*69-29*) [24], delafrinone (*69-30*), delafrine (*69-31*), and solanidane-3β,5α,6β-triol (*69-32*) [25] together with the known alkaloid imperialine (*69-33*) [26, 27]. The chuanbeinone alkaloids are all D/E *cis*-(22*R*,25*S*)-5α-cevane alkaloids. The absolute configuration of chuanbeinone was confirmed by X-ray crystallographic analysis [24].

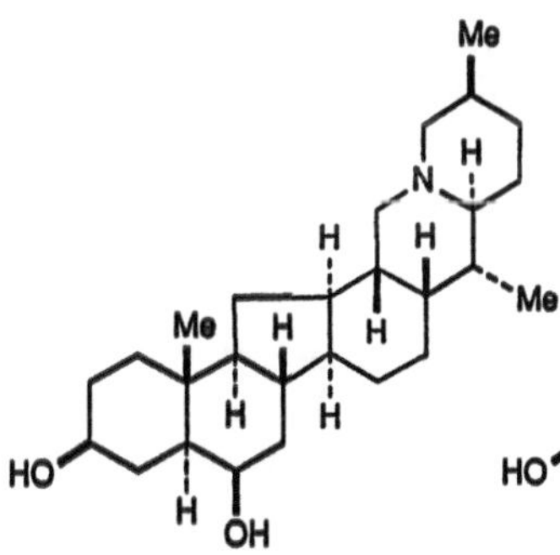

Delavine (*69-27*)

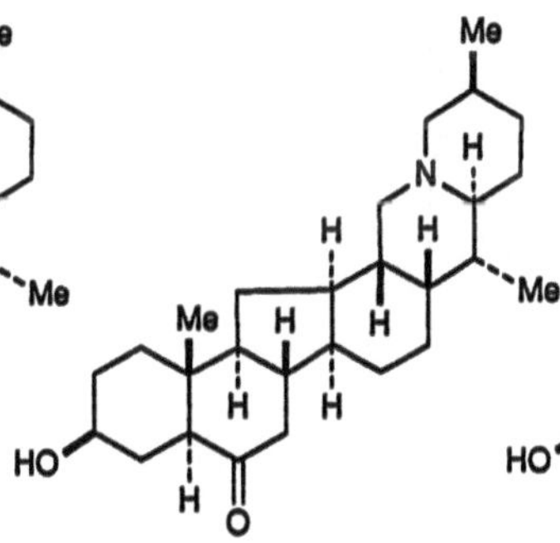

Delavinone (*69-28*)

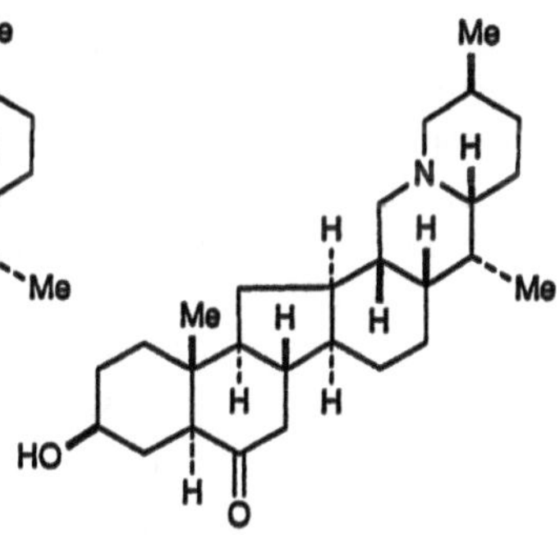

Chuanbeinone (*69-29*)

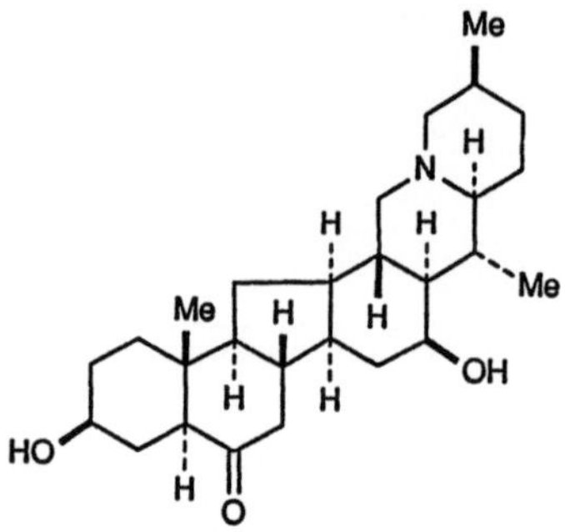

Delafrinone (*69-30*)

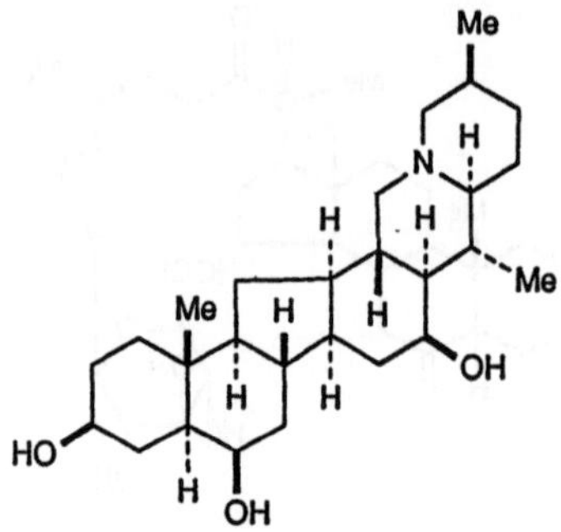

Delafrine (*69-31*)

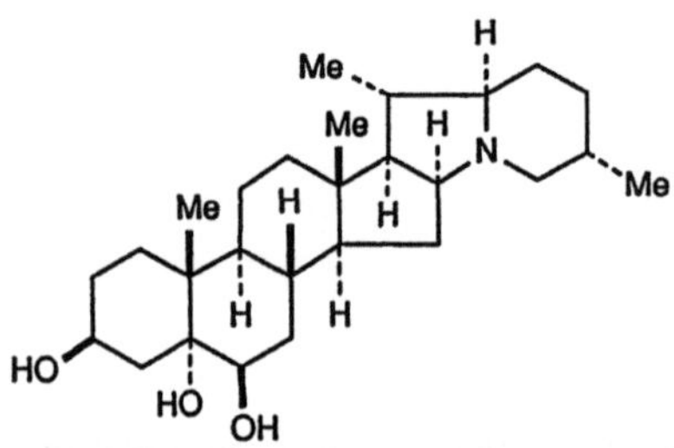

Solanidane-3β,5α,6β-triol (*69-32*)

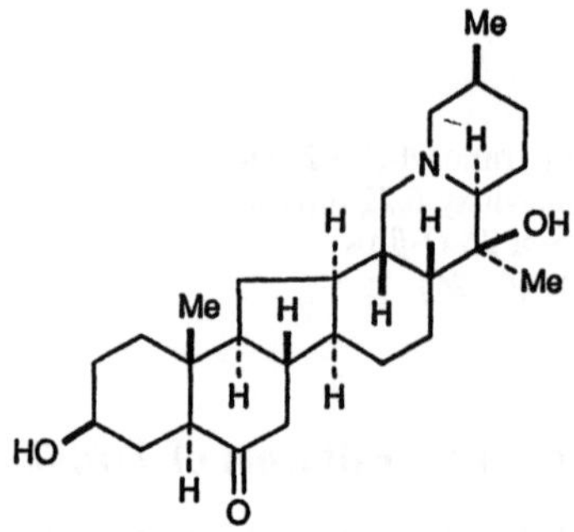

Imperialine (*69-33*)

69.2.3 Chemical Constituents of Unofficial *Fritillaria* Species with Medicinal Use

From the bulbs of *F. hupehensis* some new alkaloids were isolated together with peimine and peiminine. They were designated as hupehenine (*69-34*) [28], hupehenirine (*69-35*), and hupehenizine (*69-36*) [29]. Hupehenine is a 3,6-diol derivative of cevane. Its oxidation yielded hupehenirine and hupehenizine, by conversion of one of the two hydroxy groups into an oxo function. The corresponding dione derivative is hupehenidione (*69-37*) [29].

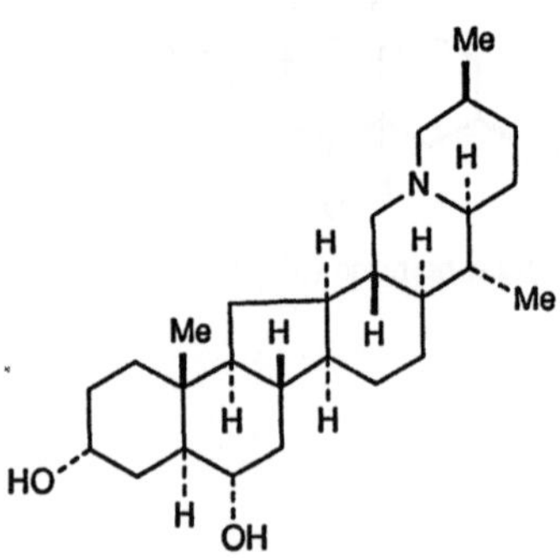

Hupehenine (*69-34*)

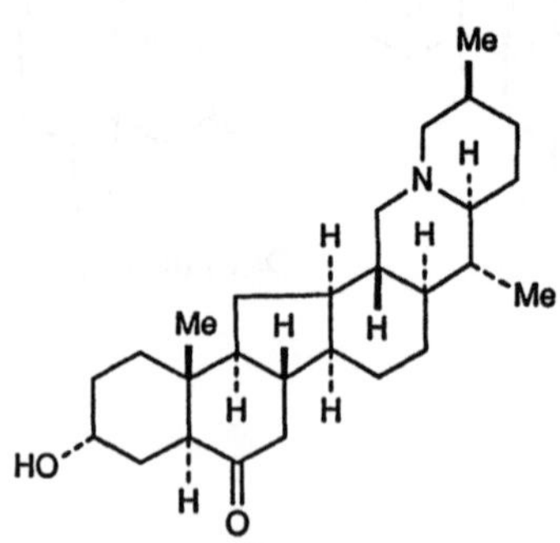

Hupehenirine (*69-35*)

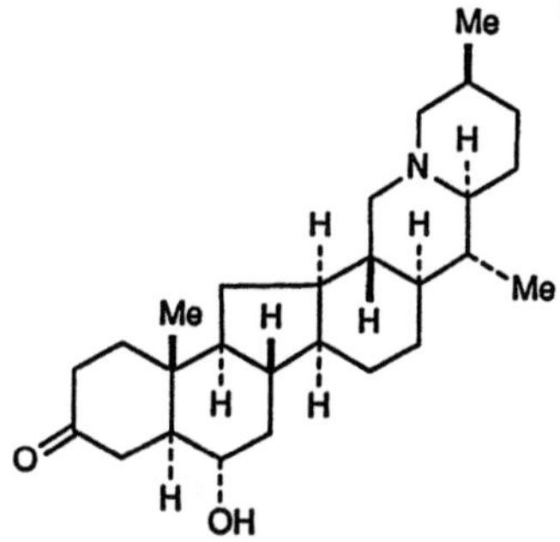

Hupehenizine (*69-36*)

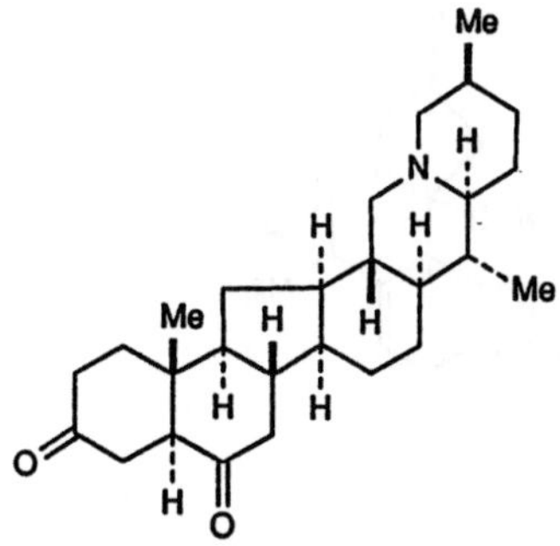

Hupehenidione (*69-37*)

Additionally, a jervine-type alkaloid named hupehenisine (*69-38*) [30] and an alkaloid glycoside named hupeheninoside (*69-39*) [31] were isolated from *F. hupehensis*.

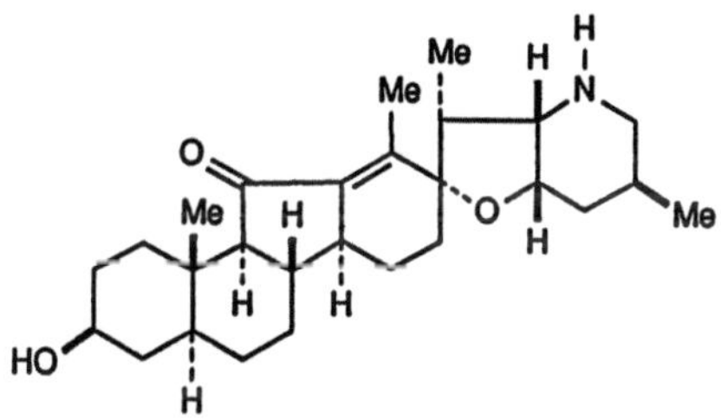

Hupehenisine (*69-38*)

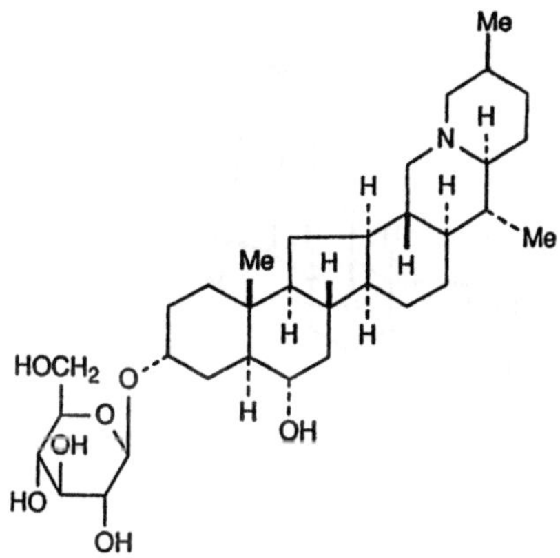

Hupeheninoside (*69-39*)

From the bulb of *F. ussuriensis* Maxim. peimisine, sipeimine [32], sipeimine-3-glucoside [33, 34] and new alkaloids pingpeimine A (*69-40*) [32], pingpeimine B (*69-41*) [35] and ussurienine (*69-42*) [36] with an aromatic D ring and an 18,25-methylene bridge were isolated. Sipeimine was proven to be identical to imperialine [37].

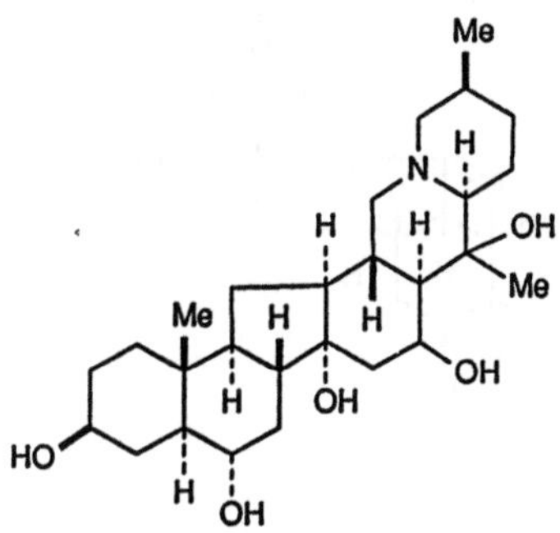

Pingpeimine A (*69-40*)

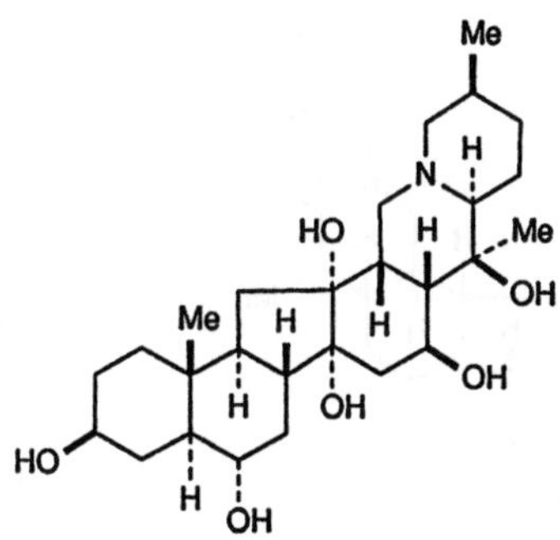

Pingpeimine B (*69-41*)

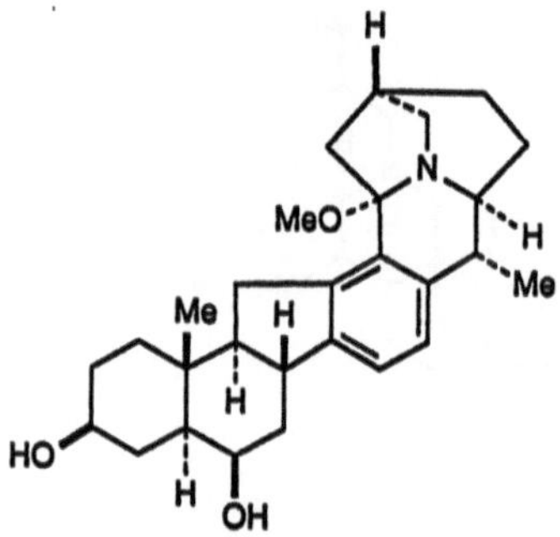

Ussurienine (*69-42*)

Four alkaloids were isolated from the bulb of *F. anhuiensis*: peiminine, peimisine, isoverticine, and a new alkaloid of the cevane type, named wanpeinine (*69-43*). Wanpenine was structurally elucidated as 5α,22β,25α-cevane-3β,6α,20β-triol [38].

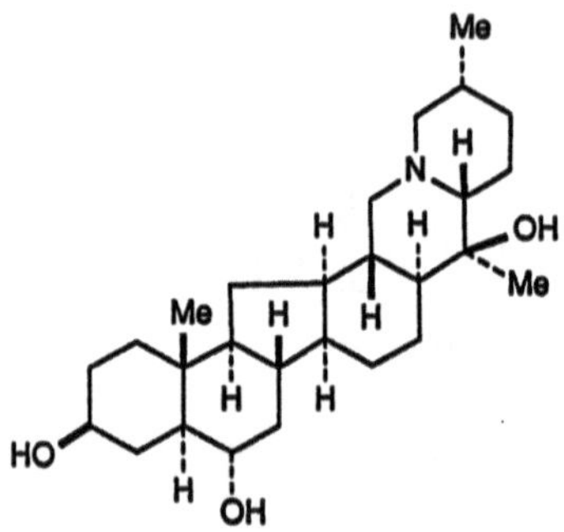

Wanpeinine (*69-43*)

A new cevane alkaloid named harepermine (*69-44*) and its 3-*O*-β-D-glucopyranoside (*69-45*) named harepermiside were isolated from the bulbs of *F. harelinii* together with the known alkaloid peiminine. The structure of harepermine was determined to be 3β,6β-dihydroxy-5α,17β,25α-cevane [39].

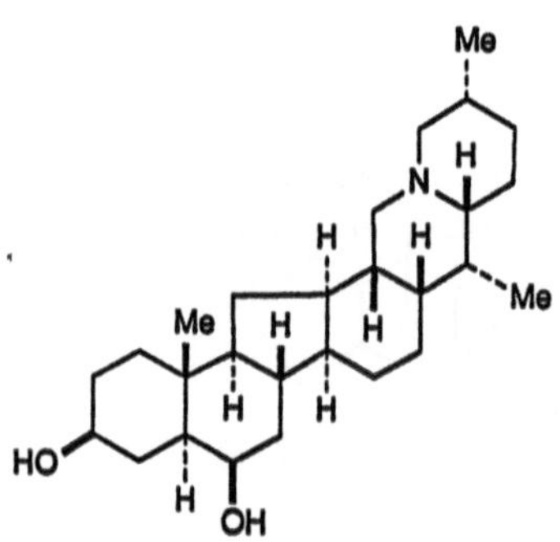

Harepermine (*69-44*)

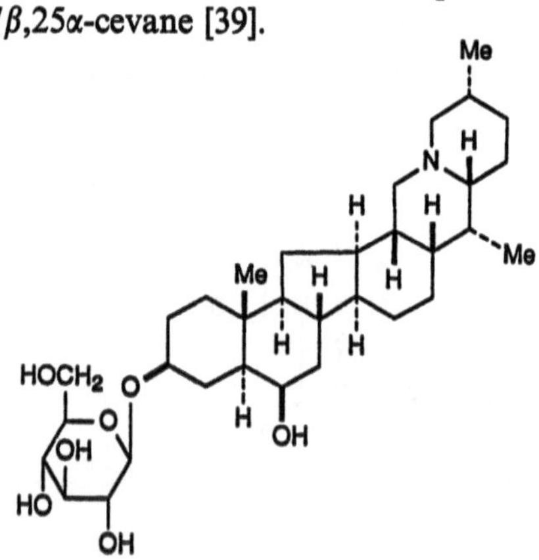

Hareperminside (*69-45*)

Five alkaloids were isolated from the bulbs of *F. walujewii*. One of them was identified as imperialine. A new steroid alkaloid named sinpeinine A (*69-46*) was structurally elucidated on the basis of spectral analysis and chemical correlations [40].

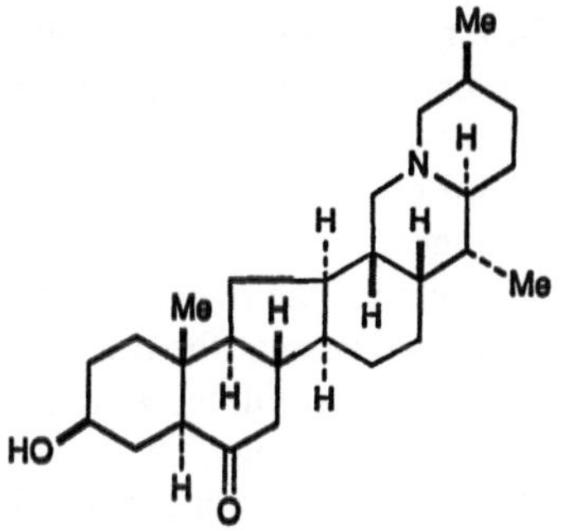

Sinpeinine A (*69-46*)

A new veratraman alkaloid named ningpeisine (*69-47*) was isolated from the bulb of *F. ninggnoensis* together with peimine, peiminine, isoverticine, and peimisine [41].

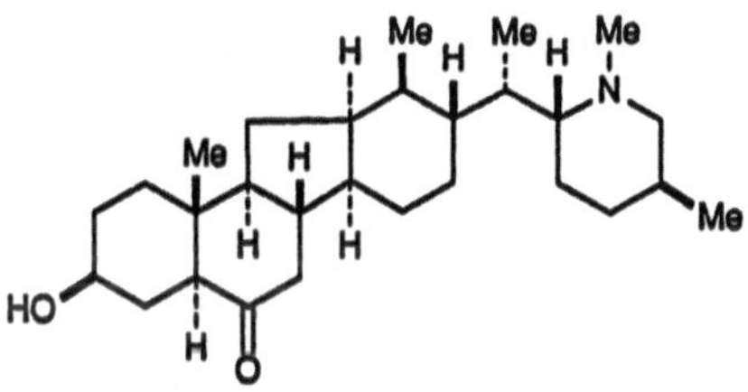

Ningpeisine (*69-47*)

Tortifoline (*69-48*) is a new cevane-type alkaloid isolated from the dried bulb of *F. tortifolia* used as an antitussive, expectorant, or sedative in Chinese folk medicine. Known alkaloids from *F. tortifolia* are solanidine and imperialine [42].

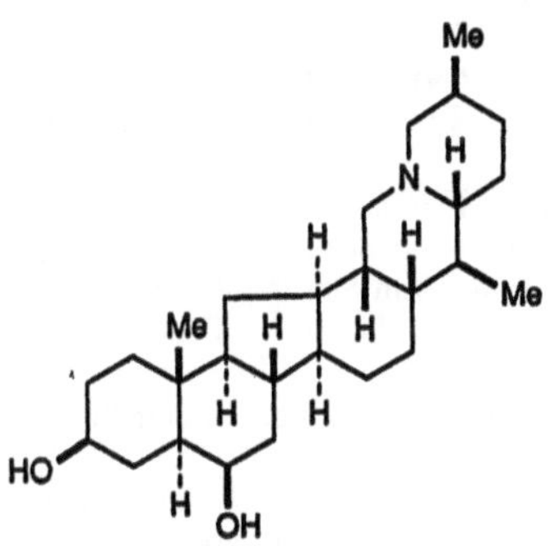

Tortifoline (*69-48*)

69.3 Pharmacology

Peimine and peiminine showed hypotensive activity. Both alkaloids have a similar physiological action and a minimal lethal dose of 0.9 mg/kg by intravenous injection into mice [1]. Peimine *N*-oxide and peiminine *N*-oxide, isolated from processed bulb, were much less toxic and more potent hypotensives than peimine and peiminine in mice, indicating that the processing is of pharmacological significance [15].

The antitussive and sedative effects of peimine and peiminine were also studied. At oral doses of 4 mg/kg, peimine and peiminine prolonged the time needed to induce 50% mice to cough by ammonia to 130% of the control. The cough amplitude of anesthetized guinea pig induced by stimulation of the mucosa at the tracheal bifurcation with a bristle was reduced by 45% by peimine or peiminine, 4 mg/kg subcutaneously. Peak action appeared 30–60 min after the injection. They also inhibited the cough of cats induced by electrical stimulation of the superior laryngeal nerve. The spontaneous activities of mice were significantly decreased by peimine or peiminine, given subcutaneously at 2 mg/kg. They antagonized the central stimulating action of caffeine and potentiated the sedative action of chlorpromazine. Their antitussive action was suspected to be central in nature. Peimine and peiminine exhibited the same pharmacological actions with equal potency [43].

The bulb of *F. anhuiensis* was compared clinically with the bulb of *F. delavayi* in the treatment of chronic bronchitis and showed no difference in the effect [44].

References

1. Chou TQ, Chen KK (1932) The alkaloids of the Chinese drug pei-mu, *Fritillaria roylei*. I. Peimine and peiminine. Chin J Physiol 6:265–270
2. Chou TQ, Chu TT (1941) Preparation and properties of peimine and peiminine. J Am Chem Soc 63:2936–2938
3. Chi YF, Kao YS, Chang KJ (1936) Alkaloids of *Fritillaria roylei*. I. Isolation of peimine. J Am Chem Soc 58:1306–1307
4. Chi YF, Kao YS, Chang KJ (1940) Alkaloids of *Fritillaria roylei*. II. Isolation of peiminine. J Am Chem Soc 62:2896–2897
5. Fukuda M (1948) Alkaloids of *Fritillaria verticillata*. II. Constitution of verticine. J Chem Soc Jpn (Pure Chem Sect) 69:165–167
6. Wu YH (1944) Constituents of *Fritillaria roylei*. J Am Chem Soc 66:1778–1780
7. Chu TC, Lo JY, Huang WK, Ho FC (1957) *Fritillaria* alkaloids XII. The dehydrogenation of peimine and oxidation and reduction of peiminine. Kexue Tongbao 371–372
8. Ho FC, Lo JY, Liu CC, Chu TC (1957) *Fritillaria* alkaloids. XIII. Function of the third O atom of peimine, peiminine and sipeimine. Kexue Tongbao 372
9. Chu TC, Lu JY, Huang WK, Ho FC, Liu CC (1958) A study of *Fritillaria* alkaloids. XII. The function of the third oxygen atom of peimine, peiminine and sipeimine. Selenium dehydrogenation of peimine and oxidation and reduction of peiminine. Acta Chim Sin 24:377–382
10. Morimoto H, Kimata S (1960) Components of *Fritillaria thunbergii*. I. Isolation of peimine and its new glycoside. Chem Pharm Bull (Tokyo) 8:302–307
11. Morimoto H, Kimata S (1960) Components of *Fritillaria thunbergii* II. Peimine. 1. Chem Pharm Bull (Tokyo) 8:871–874
12. Ito S, Kato M, Shibata K, Nozoe T (1961) The alkaloid of *Fritillaria verticillata*. I. The position of the secondary hydroxyls and the skeleton of verticine. Chem Pharm Bull (Tokyo) 9:253–255
13. Ito S, Kato M, Shibata K, Nozoe T (1963) The alkaloid of *Fritillaria verticillata*. II. The structure of verticine. Chem Pharm Bull (Tokyo) 11:1337–1340
14. Ito S, Fukazawa Y, Okuda T, Iitaka Y (1968) Structure of verticinone methobromide. Tetrahedron Lett 5373–5375

15. Kitajima J, Noda N, Ida Y, Miyahara K, Kawasaki T (1981) Steroid alkaloids of fresh bulbs of *Fritillaria thunbergii* Miq. and of crude drug "Bai-Mo" prepared therefrom. Heterocycles 15:791–796
16. Kaneko K, Tanaka M, Hiruki K, Naruse N, Mitsuhashi H (1979) ^{13}C-NMR studies on the cevanine alkaloids: the application of ^{13}C-NMR spectrum for structure elucidation of new alkaloids, baimonidine and isoverticine. Tetrahedron Lett 3737–3740
17. Kitajima J, Nada N, Ida Y, Komori T, Kawasaki T (1982) Studies on the constituents of the crude drug "Fritillariae Bulbus". IV. On the diterpenoid constituents of the crude drug "Fritillariae Bulbus". Chem Pharm Bull (Tokyo) 30:3922–3931
18. Kitajima J, Ida Y, Nada N, Komori T, Kawasaki T (1982) Studies on the constituents of crude drug "Fritillariae Bulbus". VI. On the sterol, organic acid and nucleoside of the fresh bulbs of *Fritillaria thunbergii* Miq. and crude drug "Bai-mo". Yakugaku Zasshi 102:1016–1022
19. Kaneko K, Naruse N, Haruki K, Mitsuhashi H (1980) Isobaimonidine, a new *Fritillaria* alkaloid from the aerial part of *Fritillaria verticillata.* Chem Pharm Bull (Tokyo) 28:1345–1346
20. Kaneko K, Naruse N, Tanaka M, Yoshida N, Mitsuhashi H (1980) Fritillarizine, a new *Fritillaria* alkaloid isolated from the aerial part of mature *Fritillaria verticillata.* Chem Pharm Bull (Tokyo) 28:3711–3713
21. Kitajima J, Komori T, Kawasaki T, Schulten HR (1982) Field desorption mass spectrometry of natural products, part 9. Basic steroid saponins from aerial parts of *Fritillaria thunbergii.* Phytochemistry 21:187–192
22. Kitajima J, Komori T, Kawasaki T (1982) Studies on the constituents of crude drug "Fritillariae Bulbus". New sterol glycosides from aerial parts of *Fritillaria thunbergii* Miq. Yakugaku Zasshi 102:1009–1015
23. Kaneko K, Katushara T, Mitsuhashi H, Chen YP, Hsu HY, Shiro M (1985) Isolation and structure elucidation of new alkaloids from *Fritillaria delavayi* Franch. Chem Pharm Bull (Tokyo) 33:2614
24. Kaneko K, Takao K, Mitsuhashi H, Chen YP, Hsu HY, Shiro M (1986) Chuanbeinone, a novel D/E *cis*-(22R,25S)-5-cevanine alkaloid from Chinese herbal drug, Chuan-Bei-Mu. Tetrahedron Lett 27:2387–2390
25. Kaneko K, Katsuhara T, Kitamura Y, Nishizawa M, Chen YP, Hsu HY (1988) New steroidal alkaloids from the Chinese herb drug "Bei-mu". Chem Pharm Bull (Tokyo) 36:4700–4705
26. Boit HG (1954) Imperialine. Chem Ber 87:472–475
27. Ito S, Fukazawa Y, Miyashita M (1976) Structure of imperialine. Tetrahedron Lett 3161–3164
28. Wu JZ (1982) Studies on chemical constituents of Hubei Bei Mu (*Fritillaria hupehensis*). I. Studies on its alkaloids. Chin Trad Herbal Drugs 13:339–342
29. Wu JZ, Pu QL (1986) Studies on the Hubei Beimu (*Fritillaria hupehensis*) IV. Isolation and identification of hupehenirine, hupehenizine and conversion into hupehenine. Chin Trad Herbal Drugs 17:101–104
30. Wu JZ, Wang YY, Ling DK (1986) Chemical constituents of hubei-beimu (*Fritillaria hupehensis* Hsiao et K. C. Hsia). V. Isolation and identification of hupehenisine. Acta Pharm Sin 21:546–550
31. Wu JZ, Pu QL (1985) Studies on the chemical constituents of *Fritillaria hupehensis*. III. Isolation and identification of hupeheninoside. Acta Pharm Sin 20:372–376
32. Xu DM, Zhang B, Li HR, Xu ML (1982) Isolation and identification of alkaloids from *Fritillaria ussuriensis* Maxim. Acta Pharm Sin 17:355–359
33. Xu DM, Zhang B, Huang WZ, Qi Y, Ma JL (1982) Isolation and identification of sipeimine-3-D-glucoside. Chin Trad Herbal Drugs 13:337–338
34. Xu DM, Zhang B, Qu Y, Ma JL (1983) Studies on alkaloids of Ping Bei Mu (*Fritillaria ussuriensis* Maxim.). II. Isolation and identification of sipeimine glycoside. Chin Trad Herbal Drugs 14:55–56
35. Xu DM, Wang SQ, Huang EX, Xu ML, Zhang YX, Wen XG (1988) Isolation and identification of pingpeimine B. Acta Pharm Sin 23:902–905
36. Kitamura Y, Nishizawa M, Kaneko K, Ikura M, Hikichi K, Shiro M, Chen YP, Hsu HY (1988) Ussurienine, a novel 5α-cevanine alkaloid from *Fritillaria ussuriensis* Maxim. Tetrahedron Lett 29:1959–1962
37. Chu TT, Huang WK, Loh JY (1957) *Fritillaria* alkaloids. XI. Proof of the identity of sipeimine with imperialine. Kexue Tongbao 207–208

38. Li QH, Wu ZH (1986) Isolation and identification of alkaloids from *Fritillaria anhuiensis* S. C. Chen et S. F. Yin. Acta Pharm Sin 21:767–771
39. Min ZD, Qian JF, Iinuma M, Tanaka T, Mizuno M (1986) Two steroidal alkaloids from *Fritillaria harelinii*. Phytochemistry 25:2008–2009
40. Lin QH, Jia XG, Ren YF, Muhatal, Liang XT (1984) Studies on the chemical constituents of *Fritillaria walujewii*. Acta Pharm Sin 19:894–898
41. Li QH, Wu ZH, Zhang LL, Shao L (1988) Isolation and identification of alkaloids from *Fritillaria ninggnoensis* S. C. Chen et S. F. Yin. Acta Pharm Sin 23:415–421
42. Kitamura Y, Kaneko K, Shiro M, Chen YP, Hsu HY, Lee P, Xu CJ (1989) Tortifoline, a novel (20S,22R)-5α-cevanine alkaloid from *Fritillaria tortifolia*. Chem Pharm Bull (Tokyo) 37:1514–1516
43. Qian BC, Xu HJ (1985) Antitussive and sedative effects of peimine and peiminine. Acta Pharm Sin 20:306–308
44. Zhang SH, Xia MX, Tang DQ, Cui XL (1986) Clinical effects of wanbei and chuanbei. Chin J Integr Trad West Med 6:416–418

Gardenia jasminoides Ellis 70

70.1 Introduction

Zhizi, Fructus Gardeniae, is the dry ripe fruits of *Gardenia jasminoides* Ellis (Rubiaceae) collected in September to November when the fruits have become ripe. It is officially listed in the Chinese Pharmacopoeia and is used in traditional Chinese medicine as an antiphlogistic, diuretic, laxative, choleretic, and hemostatic in the treatment of traumatosis by external application.

70.2 Chemical Constituents

The fruits of *G. jasminoides* contain a number of iridoid glycosides. The first isolated two iridoid glycosides were gardenoside (*70-1*) and geniposide (*70-2*). Whereas the major iridoid geniposide was elucidated as genipin-1-glucoside, gardenoside was found to be a related compound with a further hydroxyl group [1].

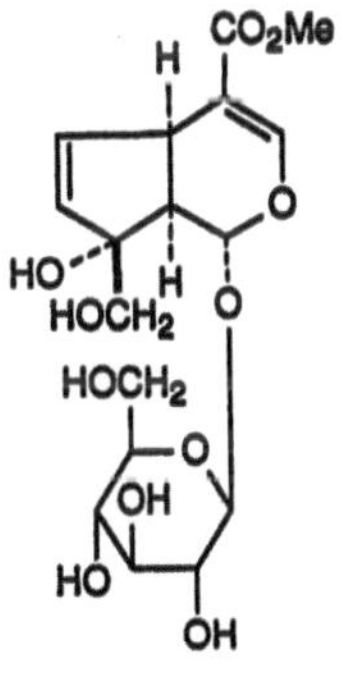

Gardenoside (*70-1*)

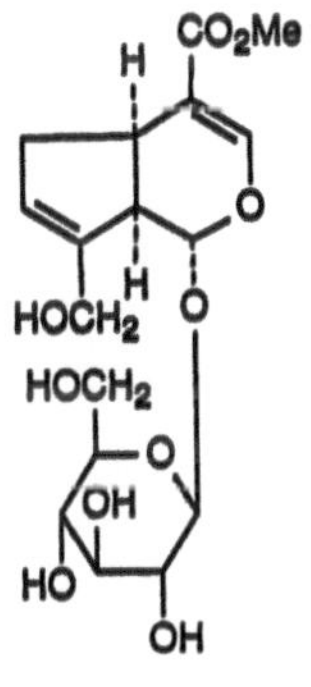

Geniposide (*70-2*)

A number of other iridoid glycosides were isolated from the fruits of *G. jasminoides* successively and structurally investigated. Thus, the isolation of shanzhiside (*70-3*) [2], genipin gentiobioside (*70-4*) [3, 4], gardoside (*70-5*) [5], scandoside (*70-6*) methyl ester [5], and geniposidic acid (*70-7*) [6] were reported. 5β-Hydroxygeniposide (*70-8*) [7] and 10-acetylgeniposide (*70-9*) [8] were isolated from *G. jasminoides* forma *grandiflora*.

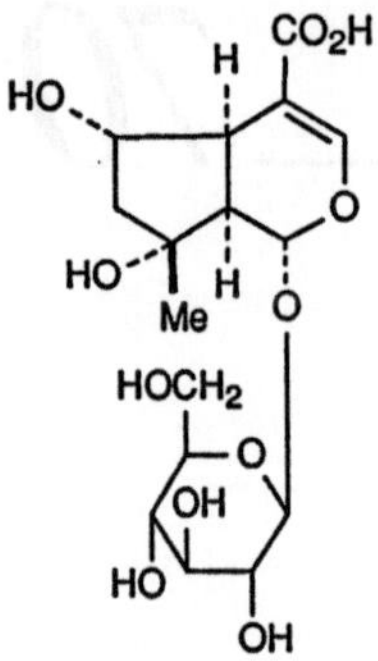

Shanzhiside (*70-3*)

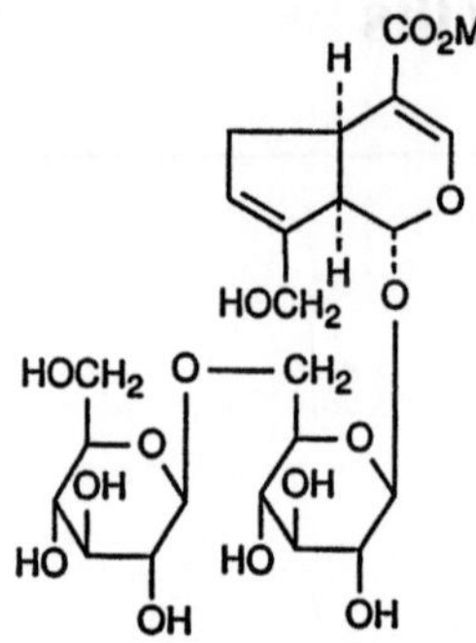

Genipin gentiobioside (*70-4*)

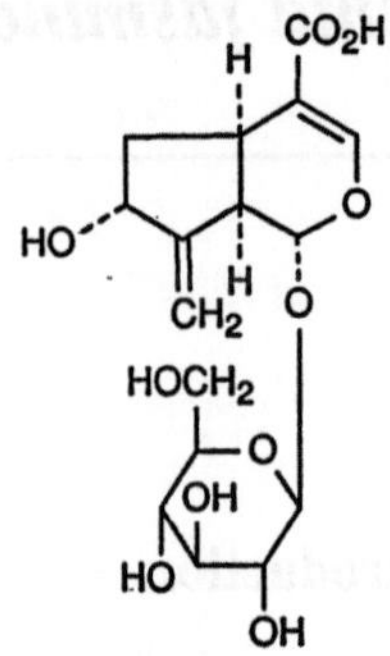

Gardoside (*70-5*)

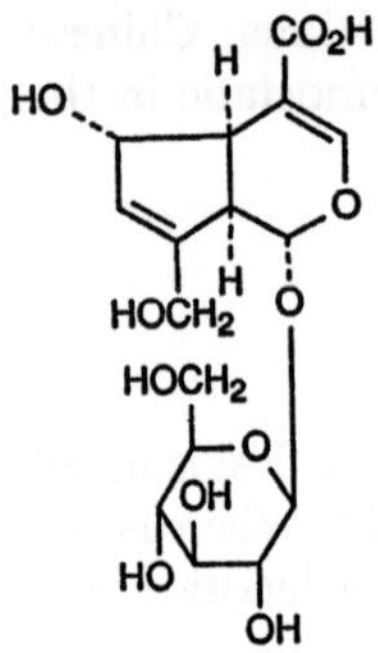

Scandoside (*70-6*)

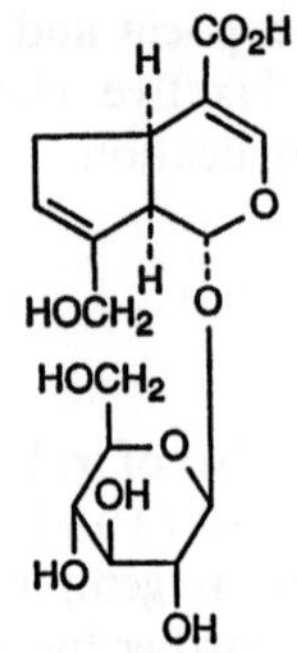

Geniposidic acid (*70-7*)

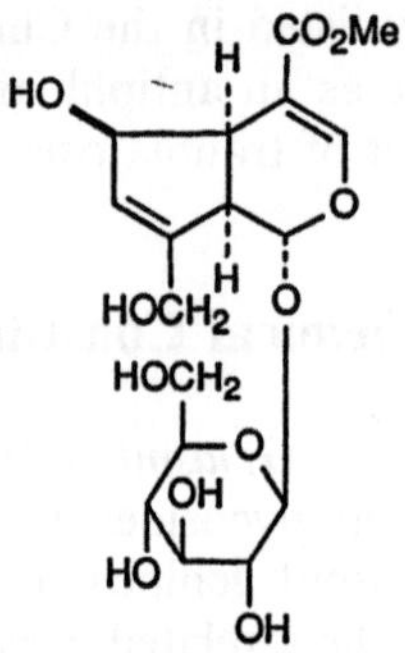

5β-Hydroxygeniposide (*70-8*)

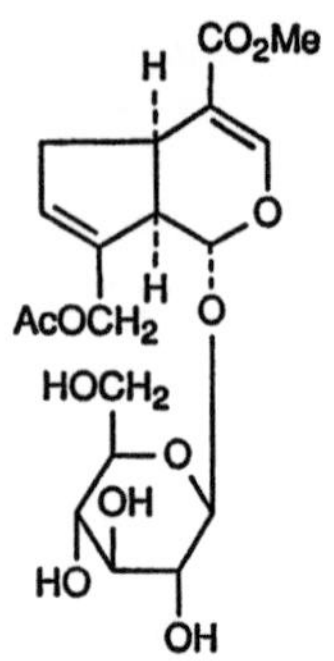

10-Acetylgeniposide (*70-9*)

In addition, the pigment crocetin (*50-1*), its glycoside crocin (*50-2*) [9], and a glycosidic bitter substance, picrocrocinic acid (*70-10*) [8], were isolated from the fruits of *G. jasminoides*.

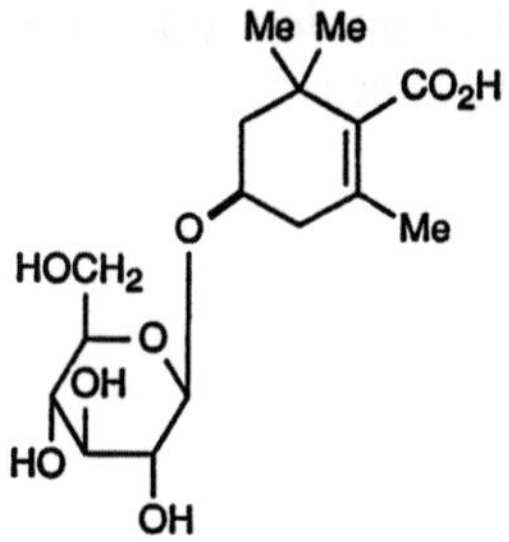
Picrocrocinic acid (*70-10*)

Two new lipoxygenase inhibitors were also isolated from the fruits of *G. jasminoides* and structurally determined as 3,4-dicaffeoyl-5-(3-hydroxy-3-methylglutaroyl)quinic acid (*70-11*) [10] and 3-caffeoyl-4-sinapoylquinic acid (*70-12*) [11].

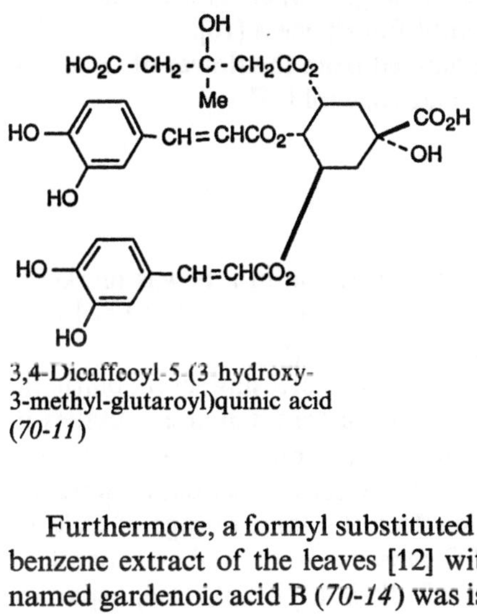
3,4-Dicaffeoyl-5-(3 hydroxy-3-methyl-glutaroyl)quinic acid (*70-11*)

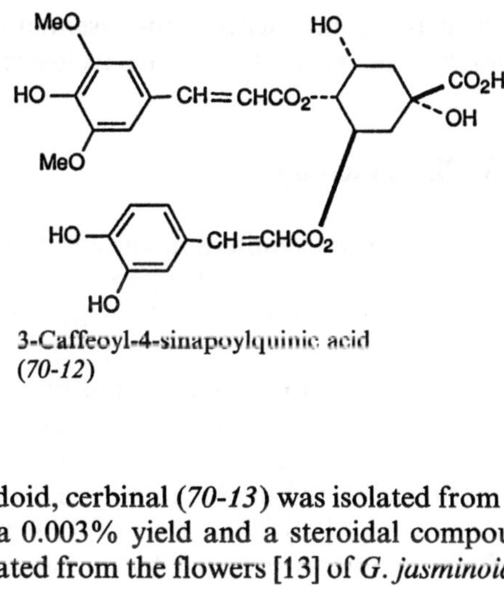
3-Caffeoyl-4-sinapoylquinic acid (*70-12*)

Furthermore, a formyl substituted iridoid, cerbinal (*70-13*) was isolated from the benzene extract of the leaves [12] with a 0.003% yield and a steroidal compound named gardenoic acid B (*70-14*) was isolated from the flowers [13] of *G. jasminoides*.

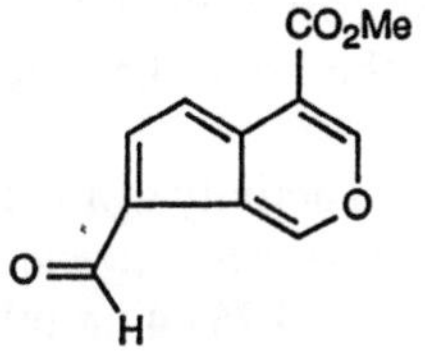
Cerbinal (*70-13*)

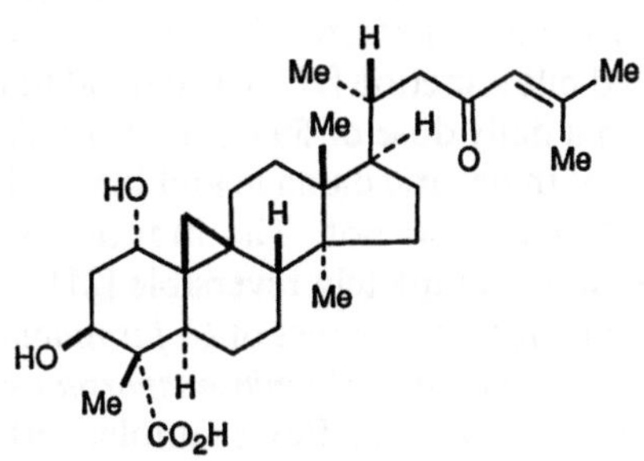
Gardenoic acid B (*70-14*)

Oleanolic acid acetate, D-mannitol, and stigmasterol were isolated and identified from the stem and root of *G. jasminoides* [14]. Enzymatic hydrolysis of gardenoside yielded its aglycone gardenogenin A (*70-15*). Acid treatment of gardenoside gave scandoside methyl ester and deacetylasperulosidic acid methyl ester [15].

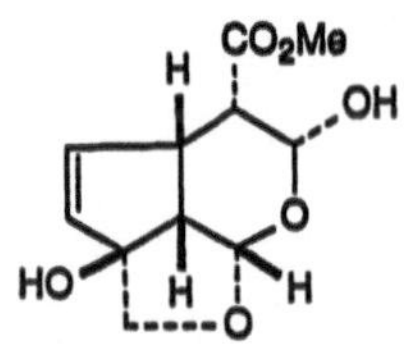

Gardenogenin A (*70-15*)

The fruit growth of *G. jasminoides* forma *grandiflora* and the formation of crocin and geniposide can be divided into two stages. In the first stage, 1–6 weeks after flowering, fruit weight and the geniposide content were found to increase rapidly and no crocin was detected. In the second stage, 8–23 weeks postflowering, the geniposide content per fresh weight of fruit barely changed, whereas crocin accumulation began and increased linearly with time until full ripeness [16].

The major constituents of the essential oil obtained from the flower of *G. jasminoides* were benzyl acetate, hydroxycitronellal, and eugenol [17].

70.3 Pharmacology

Pharmacological examination showed that the bile secretion in rats was markedly increased by administration of genipin, the aglycone of geniposide, but hardly increased by administration of the extract of *Gardenia* fruits [18].

The hepatotoxic activity of α-naphthylisothiocyanate, increasing serum bilirubin, glutamic pyruvic transaminase, and glutamic oxalacetic transaminase activities in rats, was significantly reduced by geniposide administered orally. Histopathological observations of the liver gave good agreement with the serological data. However, geniposide appeared unable to reduce the toxic effect of a large dosage of CCl_4 or D-galactosamine [18]. The extract of the fruit of *G. jasminoides* showed no hepatotoxic effects in rats as detected by measurement of alkaline phosphatase, aspartate aminotransferase, and lactate dehydrogenase activities in serum and liver [19].

Intravenous injection of crocin or crocetin at a dose of 0.1 g/kg into rabbits increased bile secretion [20]. Crocin did not affect hepatic function when given orally to rats in a daily dose of 50 mg/kg for 8 days. A high dose of 100 mg/kg for 2 weeks induced both hepatic damage and black pigmentation, but a lower dosage of 10 mg/kg for 40 days did not. The induced black pigmentation and the acute hepatic damage were completely reversible [21].

Cerbinal from the leaves of *G. jasminoides* exhibited an antifungal activity against *Bipolaris sorokiniana, Helminthosporium* sp., *Pyricularia* sp., *Colletotrichum lagenarium*, and *Puccinia* sp. For example, cerbinal at a concentration of 0.75–4 μg/ml caused complete inhibition of germination of spores of *Puccinia* sp. on oats, wheat, and white clover [12].

Gardenoic acid B from the flowers of *G. jasminoides* showed significant effects on terminating early pregnancy in rats. The flowers are used in Chinese folk medicine for birth control [13].

References

1. Inouye H, Saito S, Taguchi H, Endo T (1969) Zwei neue Iridoidglucoside aus *Gardenia jasminoides*: Gardenosid and Geniposid. Tetrahedron Lett 2347–2350
2. Inouye H, Saito S, Shingu T (1970) Ein weiteres neues Iridoidglucosid aus *Gardenia jasminoides*. Shanzhisid. Tetrahedron Lett 3581–3584
3. Endo T, Taguchi H (1970) New iridoid glycoside from *Gardenia jasminoides*: genipin 1-β-gentiobioside. Chem Pharm Bull (Tokyo) 18:1066–1067
4. Endo T, Taguchi A (1973) Constituents of *Gardenia jasminoides* geniposide and genipin gentiobioside. Chem Pharm Bull (Tokyo) 21:2684–2688
5. Inouye H, Takeda Y, Nishimura H (1974) Monoterpene glucosides and related natural products. XXVI. Two new iridoid glucosides from *Gardenia jasminoides* fruits. Phytochemistry 13:2219–2224
6. Tsumura Juntendo Co. Ltd (1981) Choleretic geniposidic acid aglycone. Jpn Kokai Tokkyo Koho 81, 92, 211 (CA 95:209643t)
7. Inouye H, Takeda Y, Saito S, Nishimura H, Sakuragi R (1974) Monoterpene glucosides and related natural products. XXV. Iridoid glucosides of *Gardenia jasminoides grandiflora*. I. Yakugaku Zasshi 94:577–586
8. Takeda Y, Nishimura H, Kadota O, Inouye H (1976) Studies on monoterpene glucosides and related natural products. XXXIV. Two further new glucosides from the fruits of *Gardenia jasminoides* Ellis forma *grandiflora* (Lour.) Makino. Chem Pharm Bull (Tokyo) 24:2644–2646
9. Kamikura M, Nakazato K (1985) Studies on the quality of natural coloring matters. II. Natural yellow colors extracted from gardenia fruit (*Gardenia jasminoides* Ellis) and colors found in commercial gardenia fruit extract color. Analysis of natural yellow colors by high performance liquid chromatography. Shokuhin Eiseigaku Zasshi 26:150–159 (CA 103:177142u)
10. Nishizawa M, Fujimoto Y (1986) Isolation and structural elucidation of a new lipoxygenase inhibitor from gardeniae fructus. Chem Pharm Bull (Tokyo) 34:1419–1421
11. Nishizawa M, Izuhara R, Kaneko K, Fujimoto Y (1987) 3-Caffeoyl-4-sinapoylquinic acid, a novel lipoxygenase inhibitor from *Gardenia* fructus. Chem Pharm Bull (Tokyo) 35:2133–2135
12. Ohashi H, Tsurushima T, Ueno T, Fukami H (1986) Cerbinal, a pseudoazulene iridoid, as a potent anti-fungal compound isolated from *Gardenia jasminoides* Ellis. Agric Biol Chem 50:2655–2657
13. Xu RS, Qin GW, Zhu DY, Fan ZY, Jiang FX, Zhang BX, Wang JC, Wang YL (1987) Chemical constituents of the antifertility plant *Gardenia jasminoides* Ellis. I. Structure of gardenoic acid B, an early pregnancy terminating component. Acta Chim Sin 45:301–304
14. Wang XF, Chen JY, Zhang GL (1986) Studies on the chemical constituents from the stems and roots of *Gardenia jasminoides*. Bull Chin Mat Med 11:620–621
15. Ishiguro K, Yamaki M, Takagi S (1983) Studies on iridoid-related compounds. II. The structure and antimicrobial activity of aglucones of galioside and gardenoside. J Nat Prod 46:532–536
16. Umetani YX, Fukui H, Tabata M (1980) Changes in the crocin and geniposide contents in the developing fruits of *Gardenia jasminoides* forma *grandiflora*. Yakugaku Zasshi 100:920–924
17. Wang DJ (1979) Studies on the constituents of the essential oils of four aromatic flowers. Ko Hsueh Fa Chan Yueh Kan 7:1036–1048 (CA 92:124929d)
18. Chang HM, Cheng YY, Chan YS, Choang KF (1985) Active component from a Chinese composite prescription for the treatment of liver diseases. In: Chang HM, Yeung HW, Tso WW, Koo A (eds) Advances in Chinese Medicinal Materials Research. World Scientific, Singapore, pp 221–237
19. Kong YC, Che CT, Yip TT, Chang HM (1977) Effect of fructus Gardeniae extract on hepatic function. Comp Med East West 5:241–255 (CA 92:157854d)
20. Miwa T (1954) *Gardenia florida* as a remedy for icterus. IV. Action of active principle and extract of fructus gardeniae on bile secretion of rabbits, blood bilirubin and peripheral lymph bilirubin of common bile-duct ligated rabbits. Jpn J Pharmacol 4:69–81
21. Lin JK, Wang CJ (1986) Reversible hepatic toxic effect of crocin dyes in rats. Saengyak Hakhoechi 16:227–232 (CA 105:41394m)

Gastrodia elata Bl.

71

71.1 Introduction

Tianma, Rhizoma Gastrodiae, is the dry tuber of *Gastrodia elata* Bl. (Orchidaceae). It has to be collected from late fall to early spring and dried at room temperature after heating in a steam bath. It is officially listed in the Chinese Pharmacopoeia and used as an anticonvulsant, analgesic, and sedative against vertigo, general paralysis, epilepsy, and tetanus.

71.2 Chemical Constituents

Gastrodin (*71-1*), a new phenolic glucoside, was isolated as the first active principle from *G. elata*. The structure was determined spectroscopically and by synthesis from acetobromoglucose and *p*-hydroxybenzaldehyde via Koenigs-Knorr glycoside synthesis followed by reduction and hydrolysis [1, 2].

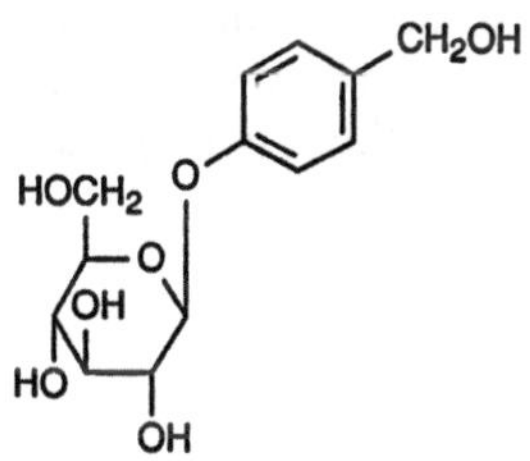

Gastrodin (*71-1*)

Gastrodin was the major constituent, accompanied by its aglycone 4-hydroxybenzyl alcohol, 4-hydroxybenzaldehyde, succinic acid, citric acid and its mono methyl ester, palmitic acid, sucrose, *β*-sitosterol, daucosterol [3, 4]. Gastrodin content ranged from 0.16% to 1.18%, as determined in *G. elata* samples from various areas [5]. Average contents of gastrodin and *p*-hydroxybenzyl alcohol in *G. elata* were 0.41% and 0.14%, respectively [6].

The gastrodin contents in *G. elata* samples varied, depending on the collection seasons. Thus, the average content of gastrodin was 0.31% in September samples, 0.23% in December samples, and 0.93% in July samples [7]. The cultivated plant contained less gastrodin than the wild-growing plant [8].

Another new glucoside, named gastrodioside (*71-2*), was also isolated from the rhizome of *G. elata* and its structure was elucidated as bis(4-hydroxy-benzyl)ether-mono-*β*-D-glucopyranoside [9].

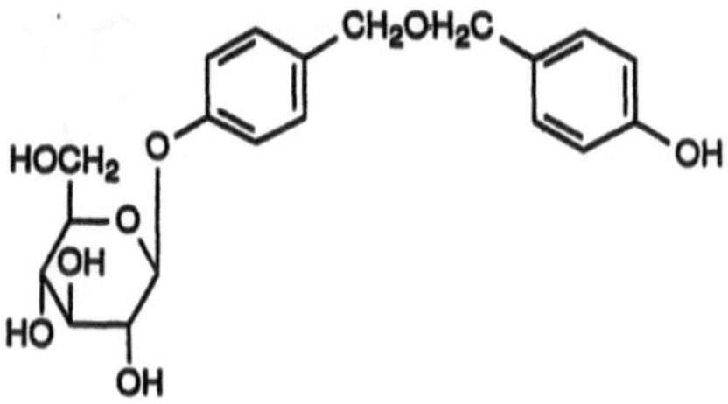

Gastrodioside (*71-2*)

p-Hydroxybenzyl methyl ether, 4-(4-hydroxybenzyloxy)benzyl methyl ether, bis(4-hydroxybenzyl)-ether, 4-(β-D-glucopyranosyloxy)-benzyl alcohol, and tris[4-(β-D-glucopyranosyloxy)benzyl] citrate were also isolated [9]. From the fresh rootstock of *G. elata* five phenolic compounds were isolated and identified as bis(4-hydroxyphenyl) methane, bis(4-hydroxybenzyl) ether, 4-ethoxymethylphenyl 4-hydroxybenzyl ether, 4-ethoxymethylphenol, and 3,4-dihydroxybenzaldehyde [10]. A number of other *Gastrodia* species were studied for their chemical constituents. Thus, gastrodin, 4-hydroxybenzyl alcohol, 4-hydroxybenzaldehyde, 3,4-dihydroxybenzaldehyde, bis(4-hydroxyphenyl)-methane, bis(4-hydroxybenzyl) ether, 4-hydroxybenzyl ethyl ether, and 4-ethoxymethylphenyl 4-hydroxybenzyl ether were detected in the rhizome of *G. elata glauca*. Bis(4-hydroxyphenyl)methane and 4-ethoxymethylphenyl 4-hydroxybenzyl ether were not detected, however, in the rhizome of *G. elata elata*; gastrodin and 3,4-dihydrobenzaldehyde were not detected in the rhizome of *G. elata alba* [11]. In rhizomes of *G. tuberculata* and *G. angista*, gastrodin, 4-hydroxybenzyl alcohol, and 4-hydroxybenzaldehyde were found in considerable amounts. With the exception of a small amount of bis(4-hydroxybenzyl) ether in *G. tuberculata*, no other phenolic compounds were detectable in these two species [11].

71.3 Pharmacology

Gastrodin and its genin, 4-hydroxybenzyl alcohol, were not toxic to mice when given orally or intravenously at doses below 5 g/kg. Both compounds showed sedative activity in mice, monkeys, rabbits, and human subjects. In addition, intravenously administered gastrodin and its genin had anticonvulsant activity in mice [12]. The rhizome of *G. elata* also showed activity in treatment of experimental epilepsy of the guinea pig [13]. Synthetic gastrodin showed antiepileptic activity against experimental seizures in rabbits. It was less active than diazepam but had no side effect [14]. A number of gastrodin analogs [15] and the aglycone analogs and homologs [16] were synthesized and tested for anticonvulsant activity.

In mice, the aqueous extract of *G. elata* dose dependently increased the uptake of ^{86}Rb into the myocardium, the survival time during hypoxia, and the duration of pentobarbital-induced sleep time and decreased spontaneous activity [17]. It was further reported that the DNA, RNA, and glycogen contents of the heart cells of neonatal rats as well as their succinate dehydrogenase, lactate dehydrogenase, and

ATPase activities were increased, indicating that gastrodin can promote the energy metabolism of the heart especially under hypoxia conditions [18].

The physiological disposition of ^{3}H-labeled gastrodin was investigated in rats. The decline in radioactivity from the gastrointestinal tract was rapid following oral administration of gastrodin and only 1.1% of the dose was recovered from the gastrointestinal tract after 8 h. In rats given gastrodin intragastrically, the radioactivity level in blood was moderate at 5 min and reached its peak at 50 min after administration. Radioactivity was highest in kidneys, moderate in liver, lung, and uterus, and relatively lower in the brain, reaching a maximum at 2 h in the brain. Elimination of radioactivity via urine, feces, and bile within 24 h was 66%, 0.6% and 3.1%, respectively, of the oral dose. Drug plasma protein binding of [^{3}H]gastrodin was 4.3%, whereas that of its genin, 4-hydroxybenzyl alcohol, was 69%. The main metabolite of gastrodin detected by thin-layer chromatography was the genin [19]. In rats following oral administration of [^{3}H]gastrodin at 8 a.m. and 8 p.m., the maximal time to reach peak blood radioactivities was 1.1 and 0.7 h, respectively. The area under the plasma radioactivity-time curve (AUC) was the lowest when [^{3}H]gastrodin was given at 2 a.m., as compared with AUCs obtained after administration at 8 a.m., 2 p.m., and 8 p.m., respectively. Thus, the pharmacokinetics of gastrodin in rats obviously reflect a circadian rhythm [20].

References

1. Chow J, Yang YB, Yang TR (1979) A new phenolic glucoside of *Gastrodia elata* Blume – gastrodin. Kexue Tongbao 24:335–336
2. Pang QJ, Zong YG (1983) Improved synthesis of gastrodin. Pharm Ind 3–4
3. Feng XZ, Chen YW, Yang JS (1979) Studies on constituents of Tian-Ma (*Gastrodia elata* Bl.). Acta Chim Sin 37:175–182
4. Zhou J, Yang YB, Yang TR (1979) Chemistry of *Gastrodia elata* Bl. I. Isolation and identification of chemical constituents of *Gastrodia elata* Bl. Acta Chim Sin 37:183–189
5. Zhang GD, Liu HY (1983) Assay of gastrodin in *Gastrodia elata*. Chin Trad Herb Drugs 14:353–355
6. Sha ZF, Sun WJ (1985) HPLC determination of gastrodin and *p*-hydroxybenzyl alcohol in *Gastrodia elata*. Chin J Pharm Anal 5:218–221
7. Meng ZM, Shen LJ, Shen LX (1985) Determination of gastrodin in *Gastrodia elata* at different harvest times. J Nanjing Coll Pharm 16:15–20
8. Ma GJ, Wang LF, Gao YZ (1982) Preliminary comparison of the constituents of *Gastrodia elata*, cultivated and wild, from Zhaotong, Yunnan, China. Chin J Pharm Anal 2:280–283
9. Taguchi H, Yosioka I, Yamasaki K, Kim IH (1981) Studies on the constituents of *Gastrodia elata* Blume. Chem Pharm Bull 29:55–62
10. Zhou J, Yang YB, Pu XY (1980) Phenolic constituents of fresh *Gastrodia elata* Blume. Acta Bot Yunnan 2:370
11. Zhou J, Pu XY, Yang YB, Yang TR (1983) Chemical studies on *Gastrodia elata* Bl. IV. Chemical constituents of some Chinese species of *Gastrodia*. Acta Bot Yunnan 5:443–444
12. Deng SX, Mo YT (1979) Pharmacological studies on *Gastrodia elata* Blume. I. Sedative and anticonvulsant effects of gastrodin and its genin. Acta Bot Yunnan 1:66–73
13. Koang NK, Wu YJ, Chen CL, Chou J (1958) Experimental epilepsy of the guinea pig: therapeutic action of procain, sodium diphenylhydantoin, *Gastrodia elata, Uncaria sinensis* and vanillin. Natl Med J China 44:582–585
14. Chai HX, Zeng HD, Xie YG, Xu JG, Chen QX (1983) Preliminary observations on the effect of synthetic gastrodin against epilepsy in rabbits induced by *Coriaria* lactone. Acta Acad Med Sichuan 14:288–292

15. Zhou J, Yang YB, Yang CR (1980) Chemical study on *Gastrodia elata* Bl. II. Synthesis of gastrodin and related compounds. Acta Chim Sin 38:162–166
16. Zhong YG, Pang QJ, Zhang HY, Tao AQ, Zhang ZQ, Gao MY, Song YL (1984) Synthesis and anticonvulsive effect of gastrodigenin homologs and analogs. Acta Acad Med Sichuan 15:17–22
17. Huang JH, Wang GL (1985) Some pharmacological effects of gastrodia injection and gastrodin. Acta Acad Med Sin 7:399–402
18. Huang XF, Xiao Y, Lei PH (1986) Effect of synthetic gastrodin on the beating of cultured heart cell of neonatal rat and histochemical changes. Bull Chin Mat Med 11:307–309
19. Lu GW, Zou YJ, Mo QZ (1985) Absorption, distribution, metabolism and excretion of ^{3}H-gastrodin in rats. Acta Pharm Sin 20:167–172
20. Lu GW, Zhou YJ, Mo QZ, Huang JA, Chu DQ, Ye DY (1986) Circadian rhythm of (^{3}H)gastrodin pharmacokinetics in rats. Acta Pharmacol Sin 7:190–191

Gentiana spp.

72

72.1 Introduction

Longdan, Radix Gentianae, is the dry roots and rootstocks of the following *Gentiana* species: *Gentiana manshurica* Kitag., *G. scabra* Bge., *G. triflora* Pall. and *G. regescens* Franch. (Gentianaceae), that are collected in the spring and fall. It is officially listed in the Chinese Pharmacopoeia and used in the treatment of hepatic and cholesteric diseases.

Qinjiao, Radix Gentianae macrophyllae, is the dry roots of the following *Gentiana* species: *G. macrophylla* Pall., *G. straminea* Maxim., *G. crassicaulis* Duthie ex Burk., and *G. dahurica* Fisch. collected in the spring and fall. It is also officially listed in the Chinese Pharmacopoeia and used mainly against rheumatism.

72.2 Chemical Constituents

72.2.1 Chemical Constituents of *Gentiana scabra*

The plants of the genus *Gentiana* usually contain bitter principles as the main constituents. The major bitter principle of these plants is gentiopicroside (gentiopicrin *72-1*), a secoiridoidglucoside, first isolated from *G. lutea* more than 100 years ago. But the structure determination of gentiopicroside ran into some difficulties, so that the structure was only elucidated in 1968 [1]. In contrast to the iridoids, which have a hexahydrodimethylcyclopenta[c]pyran skeleton, the secoiridoids are derived from 3,4-diethyl-pyran.

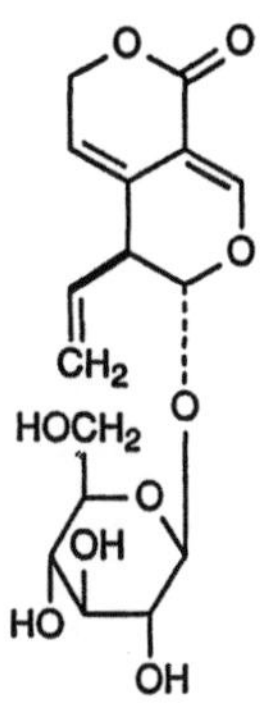

Gentiopicroside (*72-1*)

Gentiopicroside was also detected in *G. scabra* in large amounts. The gentiopicroside content in the root of *G. scabra* was dependent on the season of collection and on the years of cultivation. It was highest in the fall during the 3rd year of cultivation (7.8%). The gentiopicroside content in the above ground part of the plant was about 1% [2]. In addition to gentiopicroside, the secoiridoid glycosides sweroside (*72-2*) and swertiamarin (*72-3*) were also isolated from the root of *G. scabra* [3].

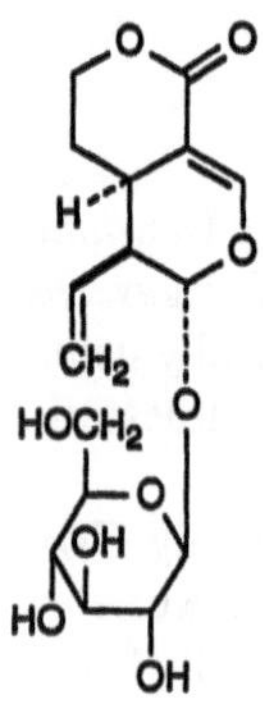

Sweroside (*72-2*)

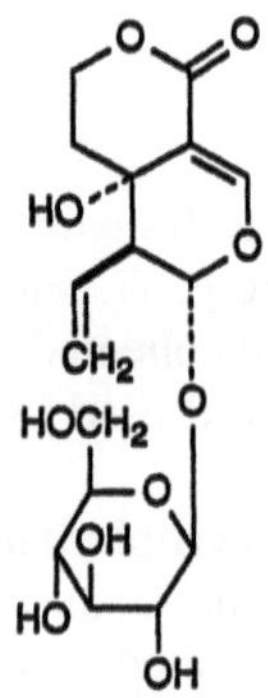

Swertiamarin (*72-3*)

Results of quantitative determination by thin-layer chromatography-densitometry of the three bitter glycosides in different *Gentiana* species showed that gentiopicroside and total bitter glycoside content is higher (4.1%–6.7%) in the roots of *G. rigescens, G. triflora, G. atuntsiensis, G. manshurica,* and *G. scabra* compared with the roots of *G. cephalantha* and *G. suffrutescens* (1.8%–3.7%) [4, 5].

The roots of *G. atuntsiensis* and *G. rigescens* also contain amarogentin (*72-4*), another secoiridoid glucoside, having a tri-hydroxy-biphenyl-carboxylic acid moiety esterified with glucose at C-2 [5].

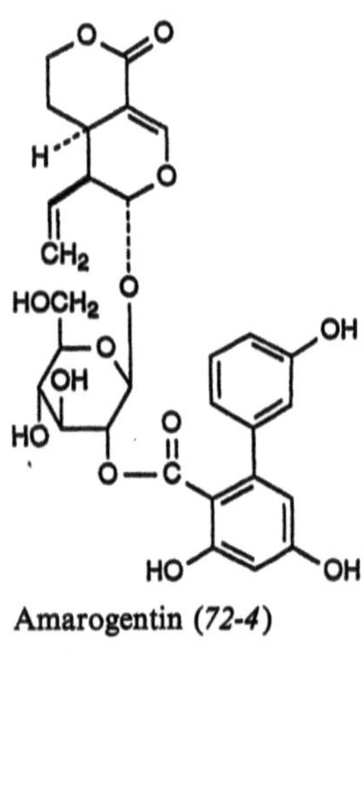

Amarogentin (*72-4*)

Furthermore, two new secoiridosid glycosides trifloroside (*72-5*) [6] and scabraside (*72-6*) [7] were isolated from the roots of *G. triflora* var. *japonica* and *G. scabra* var. *Buergeri*, respectively. As a common structural feature, they contain 2,3-dihydroxybenzoic acid-3-glycoside esterified to another glucose molecule that is acetylated and linked to the secoiridoid aglycone.

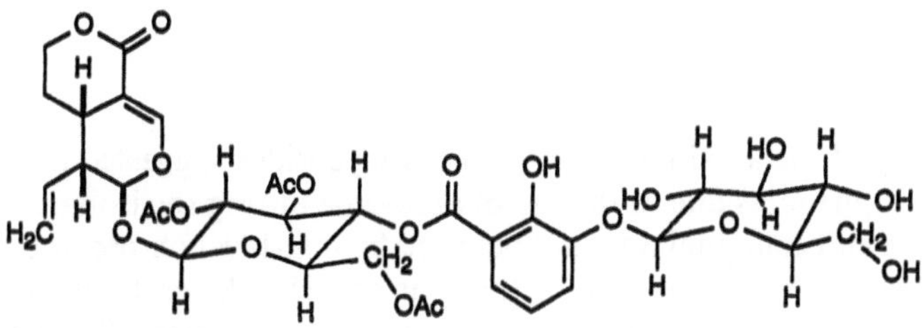

Trifloroside (*72-5*)

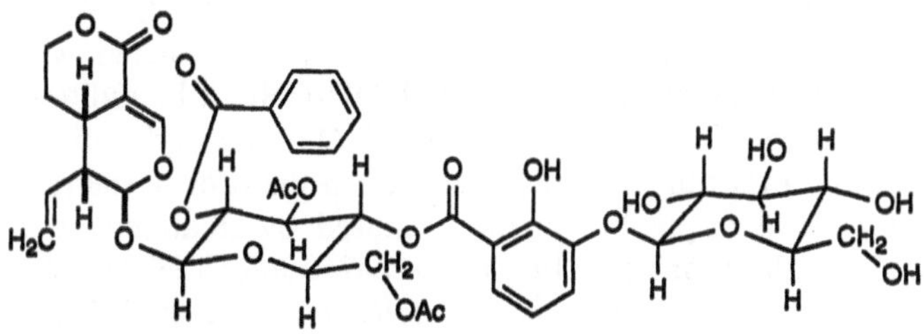

Scabraside (*72-6*)

72.2.2 Chemical Constituents of *Gentiana macrophylla*

In contrast to the bitter secoiridoid glycosides found in *G. scabra*, compounds first isolated from *G. macrophylla* are of alkaloid nature. The first alkaloid isolated from the root of *G. macrophylla* and structurally elucidated was gentianine (*72-7*) [8]. Gentianine is a structurally relative simple lactonic tertiary base derived from nicotinic acid. The second alkaloid isolated from *G. macrophylla* was gentianidine (*72-8*) [9], a gentianine-related alkaloid also derived from nicotinic acid.

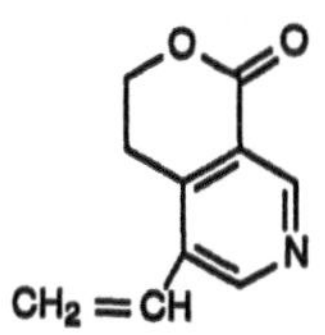

Gentianine (*72-7*)

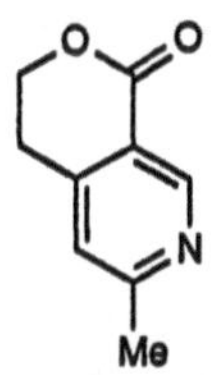

Gentianidine (*72-8*)

The third compound containing nitrogen from *G. macrophylla* was gentianal (*72-9*) [7, 8], which has a methylpyrone instead of a lactone ring.

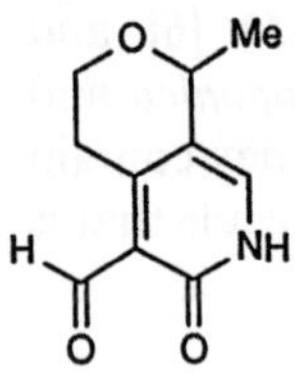

Gentianal (*72-9*)

The structural similarity of the alkaloids with the bitter glycosides suggested that the alkaloids might arise as artifacts due to the use of ammonia during the isolation procedure [10]. Gentianine and gentianal were not detected by thin-layer chromatography in the ethanol extract of the dried root of the four officially listed *Gentiana* species used as qinjiao, including *G. macrophylla, G. straminea, G. crassicaulis*, and *G. dahurica*. However, gentianine and gentianal were detected when the ethanol extract was mixed with an NH_4OH solution and allowed to stand for 24 h. Similarly, no gentianine and gentianal were detected when the root was extracted with Na_2CO_3-$CHCl_3$, but were detected by extraction with NH_4OH-$CHCl_3$. In addition, gentianine and gentianal were produced on treatment of *O*-tetraacetylgentiopicroside with NH_4OH, indicating that the alkaloids detected in the root of the four *Gentiana* spp. are not of natural origin but artifacts caused by treatment with NH_4OH [11]. A technical procedure for extraction and separation of the alkaloids gentianine and gentianal from the root of *G. macrophylla* was described. Thus, powdered root was extracted with ethanol (95%)-NH_4OH solution (10%) (50:7). The filtrate was concentrated, treated with H_2O, titrated to pH 2–3, and filtered again. The filtrate was then passed through cation-exchange resin. Gentianal was obtained by eluting the resin with 50% ethanol, whereas gentianine was obtained by using a solution of 95% ethanol, water, and concentrated NH_4OH (20:3:2). The pure alkaloids were obtained after $CHCl_3$ extraction and recrystallization from ether and absolute ethanol. Yields of gentianal and gentianine by this method were 0.7 and 0.9 g/kg, respectively [12].

Gentianine could be synthesized by treatment of 4-methyl-5-vinyl-nicotinic acid with formaldehyde [13]. Similarly, gentianidine could be obtained by reaction of 4,6-dimethylnicotinic acid with formaldehyde [9].

The gentiopicroside content in the roots of *G. macrophylla* and *G. dahurica* was 0.2%–1.4% [10].

72.3 Pharmacology

72.3.1 Pharmacology of the Bitter Principle

Gentiopicroside isolated from various *Gentiana* species including *G. scabra* and *G. macrophylla* showed antiinflammatory activity in carrageenin-induced foot edema in rats [14].

72.3.2 Pharmacology of Gentianine and Related Compounds

Antiinflammatory activity of gentianine was also demonstrated. Intraperitoneal injection of gentianine at a dose of 90 mg/kg reduced the swelling of the ankle joint of the rat hind leg induced by injection of formalin [15] or caused by egg white [16]. The effect of gentianine was not observed in bilateral adrenalectomized or hypophysectomized rats and in normal rats anesthetized with pentobarbital [15, 16], suggesting that it is mediated via the nervous and hypophyseal system [16]. Hypotensive activity in cats was also reported for gentianine [17]. The minimal lethal dose of gentianine in mice by intraperitoneal injection was 400 mg/kg [18].

References

1. Inouye H, Yoshida T, Nakamura Y, Tobita S (1968) Die Stereochemie einiger Secoiridoidglucoside und die Revision der Struktur des Gentiopicrosids. Tetrahedron Lett 4429–4432
2. Lu YR, Yang XH, Shao AX (1986) Comparison of gentiopicroside contents in the root and its preparation of *Gentiana scabra* collected in different seasons at different years of cultivation. Bull Chin Mat Med 11:298–300
3. Luo JP, Lou ZC (1986) Separation and identification of gentiopicroside, swertiamarin and sweroside in the traditional drug longdan, Radix Gentianae. Chin Trad Herb Drugs 17:145–149
4. Luo JP, Lou ZC (1986) TLC-densitometry determination of bitter glycosides in the Chinese drug Longdan, Radix Gentianae and its quality evaluation. Acta Pharm Sin 21:40–46
5. Luo JP, Lou ZC (1985) Silica gel thin layer and polyamide sheet chromatographic identification of the secoiridoid glucosides in certain *Gentiana* species used in the Chinese traditional medicine Long Dan. Chin J Pharm Anal 5:7–10
6. Inouye H, Ueda S, Nakamura Y, Inoue K, Hayano T, Matsumura H (1974) Über die Monoterpenglycoside und verwandte Naturstoffe XXIV. Triflorosid, ein neues Secoiridoidglucosid aus *Gentiana triflora* var. *japonica*. Tetrahedron 30:571–577
7. Ikeshiro Y, Tomita Y (1983) A new bitter secoiridoid glucoside from *Gentiana scabra* var. *Buergeri*. Planta Med 48:169–173
8. Fu FY, Sun NC (1958) The chemical constituents of *Gentiana macrophylla*. Acta Pharm Sin 6:198–203
9. Liang HT, Yu TC, Fu FY (1964) Investigation of the chemical constituents of *Gentiana macrophylla*. II. The structure and synthesis of gentianidine. Acta Pharm Sin 11:412–416
10. Hayashi T, Higashino M (1976) Studies on crude drugs originated from gentianaceous plants. III. The bitter principle of the Chinese crude drug Qinjiao and its contents. Yakugaku Zasshi 96:362–365
11. Guo YJ, Lu YR (1983) Studies on the transformation of gentiopicroside to gentianal. Chin J Pharm Anal 3:268–271
12. Cai BY, Mu FF, Li WL, Jin JH, Zhong JF (1986) New technology for extraction of Qinjiao alkaloids. Chin Trad Herb Drugs 17:111–112
13. Govindachari TR, Nagarajan K, Rajappa S (1957) Synthesis of gentianine. J Chem Soc 2725–2726
14. Hayashi T, Kubo M (1979) Antiinflammatory secoididoids. Jpn Kokai Tokkyo Koho 79,26,323 (CA 91:9485y)
15. Sung CY, Chi HC, Liu KT (1958) Pharmacology of gentianine. I. Anti-inflammatory effect and action of pituitary-adrenal function of the rat. Acta Physiol Sin 22:201–205
16. Chi HC, Liu KT, Sung CY (1959) The pharmacology of gentianine. II. The antiphlogistic effect of gentianine and its comparison with some clinically effective drugs. Acta Physiol Sin 23:151–157
17. Sadritdinov FS, Tulyaganov N (1967) Pharmacology of the alkaloids gentianine and gentianidine. Farmakol Alkaloidov Glikozidov 128–137 (CA 70:2217v)
18. Natarajan PN, Wan ASC, Zaman V (1974) Antimalarial, antiamebic and toxicity tests on gentianine. Planta Med 25:258–260

Ginkgo biloba L.

73

73.1 Introduction

Baiguo, Semen Ginkgo, is the dry ripe seeds of *Ginkgo biloba* L. (Ginkgoaceae) collected in the fall when the seeds are ripe. This crude drug is officially listed in the Chinese Pharmacopoeia and used in traditional Chinese medicine as an antiasthmatic and against polyuria. The Chinese Pharmacopoeia also contains a note on the toxicity of the raw seeds.

Ginkgo biloba is the sole representative of its family and is not linked to any other living plant. The order Ginkgoales was once widely distributed throughout the world but in the past few million years all species except *G. biloba* have become extinct, the other species being found only as fossils, as that the ginkgo tree is called the "fossil tree." *G. biloba* was unknown outside the Orient before the eighteenth century but is now rather commonly distributed in Europe, America, and other continents.

73.2 Chemical Constituents

The purified lipid fraction, obtained from the nuts of *G. biloba* with a yield of 1.7% on a wet weight basis, is composed of ca. 90% neutral lipids, ca. 7% polar lipids [1], and ca. 3% glycolipids [2]. Main fatty acids in the triglyceride fraction are oleic and linoleic acid, and also palmitic acid in the phospholipid fraction. Enzymatic hydrolysis of the triglyceride and phosphatidylcholine fractions demonstrated relatively large amounts of unsaturated acids in the β-position. Analysis by gas chromatography-mass spectrometry of the steroid ester fraction indicated the presence of tetracosanoic, hexacosanoic, octacosanoic, and triacontanoic acids, of a lactone and of compounds suspected to be phenolic acids connected to long-chain diols [1]. The largest fraction of glycolipids was found to be digalactosyldiglyceride (64.1%), followed by monogalactosyldiglyceride (31.2%), and cerebroside (4.7%). The main component, amounting to 85% of total fatty acids, in the glycolipid fraction was α-hydroxypalmitic acid. The sugar component in the cerebroside was glucose [2].

In addition, a number of phenolic acids and phenols were isolated and identified as toxic principles of the raw fruit pulp of *G. biloba* [3, 4]. They are ginkgolic acid (*73-1*), hydroginkgolic acid (*73-2*), hydroginkgolinic acid (*73-3*), ginkgol (*73-4*), and bilobol (*73-5*) [5]. A further toxic principle from the seed of *G. biloba* was determined as 4-*O*-methylpyridoxine. It is proposed that this toxic principle causes food poisoning through not only antagonizing vitamin B_6 in the body, but also inhibiting the formation of 4-aminobutyric acid from glutamate in the brain [6].

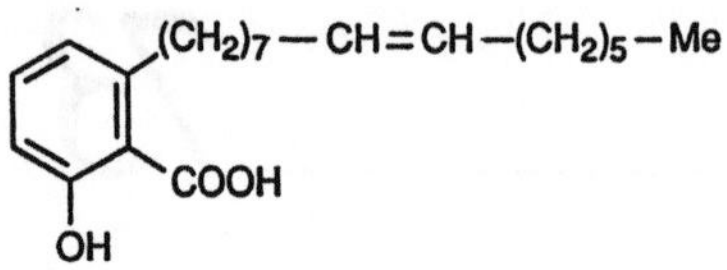

Ginkgolic acid (*73-1*)

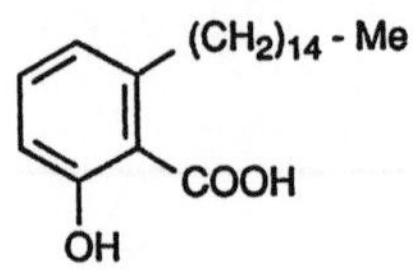

Hydroginkgolic acid (*73-2*)

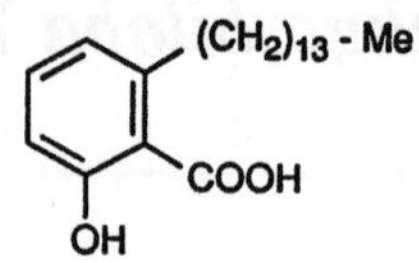

Hydroginkgolinic acid (*73-3*)

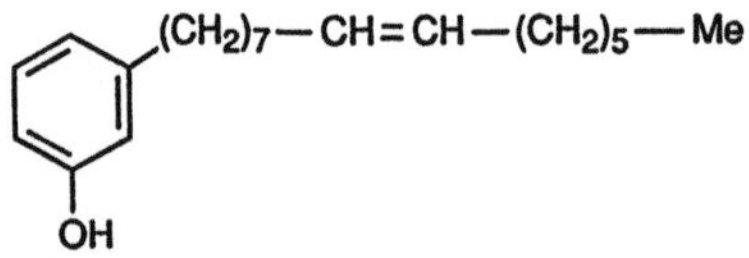

Ginkgol (*73-4*)

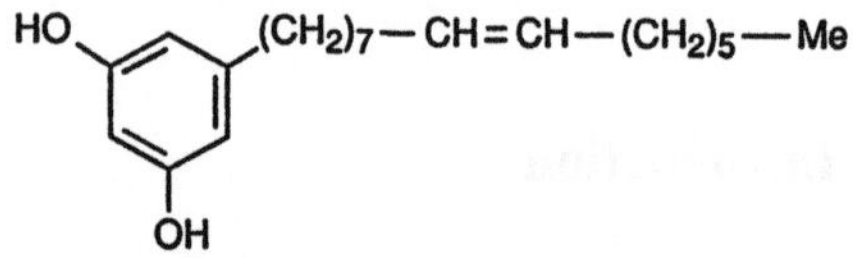

Bilobol (*73-5*)

More important was the discovery of ginkgolides isolated from the root bark [7] and leaves [8] of *G. biloba*. Structural investigations on ginkgolides revealed that the ginkgolides are unique cage molecules, representing diterpene lactones incorporating a tertiary butyl group and six five-membered rings [7–13]. Thus, ginkgolides A (*73-6*), B (*73-7*), C (*73-8*), and M (*73-9*) were isolated from the root bark of *G. biloba* and ginkgolides A, B, and C from the leaves. The leaves of *G. biloba* are used in Chinese folk medicine for the treatment of cardiovascular diseases.

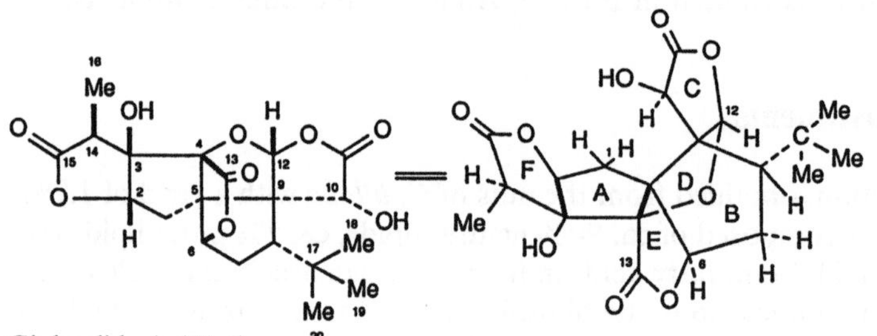

Ginkgolide A (*73-6*)

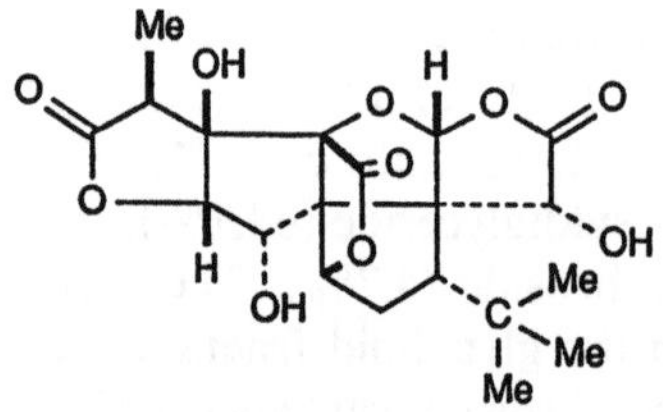

Ginkgolide B (*73-7*)

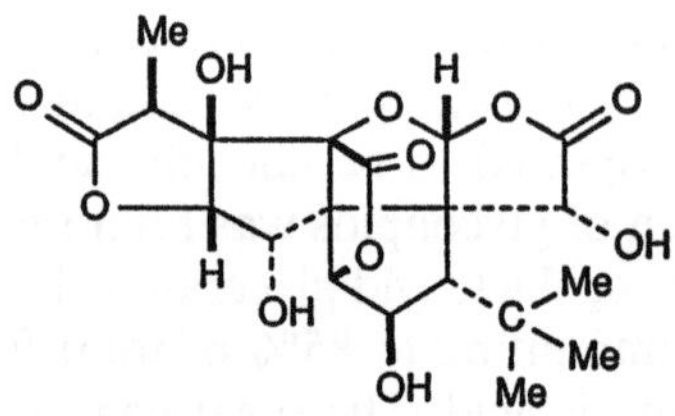

Ginkgolide C (*73-8*)

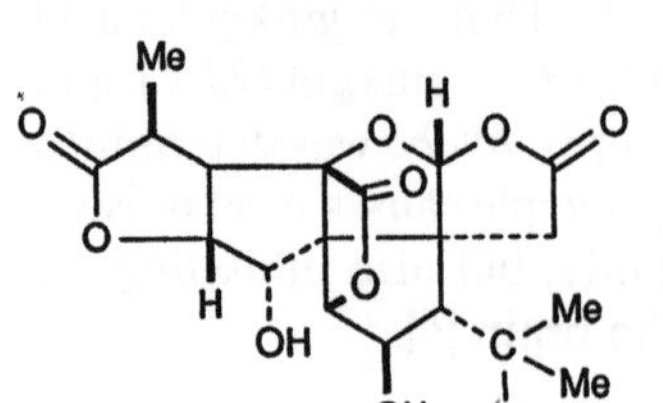

Ginkgolide M (*73-9*)

The ginkgolides were extracted from the root bark with methanol. The syrupy material obtained after concentration of the methanolic extract was then distributed in water and benzene. Solid residue from the aqueous layer was recrystallized from ethanol and then gave a mixture of ginkgolides, which were separated on a silica gel column to give ginkgolides C, M, and a mixture of ginkgolides A and B. Ginkgolides A and B were separated by fractional recrystallization. From 100 kg root bark of *G. biloba* (five trees 30 cm in diameter were used) 10 g (0.01%) ginkgolide A, 10 g (0.01%) ginkgolide B, 20 g (0.02%) ginkgolide C, and 200 mg (0.0002%) ginkgolide M were obtained [7]. The yields of ginkgolides A, B, and C from 200-kg ginkgo leaves were 800 mg, 10 g, and 150 mg, respectively [8].

The structure of ginkgolide A was confirmed by X-ray analysis of its mono-*p*-bromobenzate [8]. The terpene nature of ginkgolides was derived from biosynthetic evidence [14].

Recently, a new member of the ginkgolides, ginkgolide J (*73-10*), was isolated from the leaves of *G. biloba* and structurally elucidated [15].

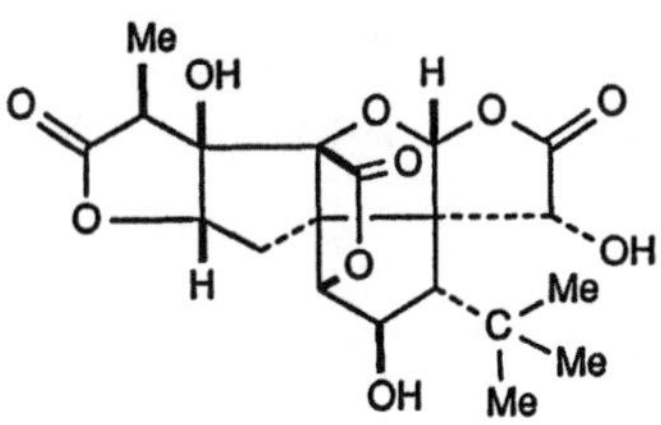

Ginkgolide J (*73-10*)

In addition to the diterpene ginkgolides, a new sesquiterpene compound named bilobalide (*73-11*) closely related to the diterpene ginkgolides from leaves of *G. biloba* was isolated [16, 17] and structurally determined [17–19]. Bilobalide also has a very unique cage molecule containing three lactone rings and a tertiary butyl group.

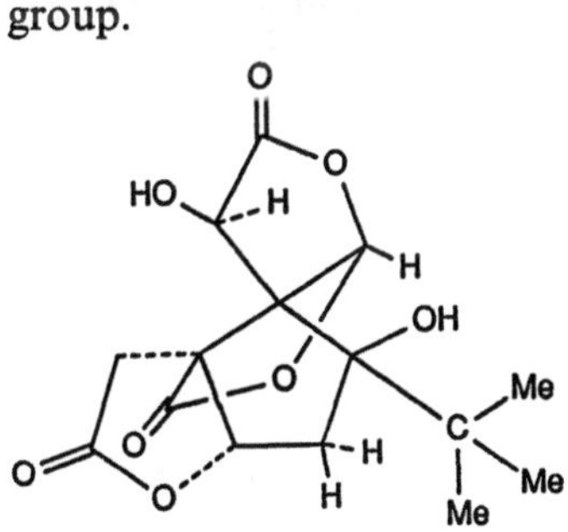

Bilobalide (*73-11*)

The terpene nature of bilobalide was also derived from biosynthetic evidence [14]. The relationship between the ginkgolides and bilobalide is apparent from scheme 73.1.

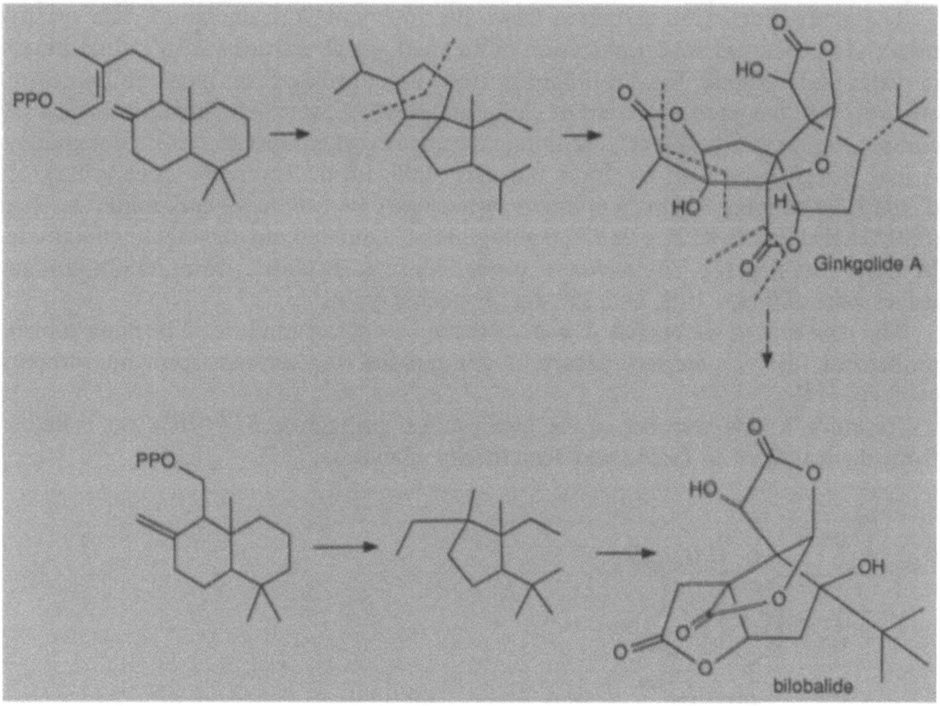

Fig. 73.1. Relationship between ginkgolide A and bilobalide

Besides the terpene constituents (ginkgolides and bilobalide) characteristic of ginkgo, the leaves of ginkgo contain a further number of compounds, such as the flavone glycosides kaempferol-3-rutinoside, quercetin-3-rutinoside, isorhamnetin-3-rutinoside [20], and octaacetylquercitrin [21], the flavones quercetin and its pentaacetyl derivative [21], the biflavones sciadopitysin (*73-16*), ginkgetin (*73-14*), isoginkgetin (*73-15*), bilobetin (*73-13*), and amentoflavone (*73-12*) [22–25].

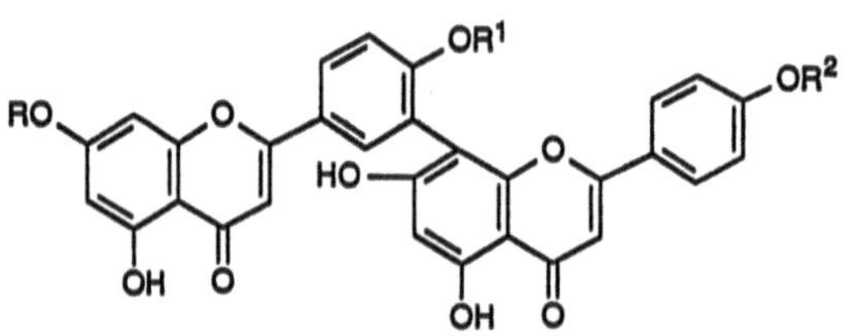

	R	R¹	R²
Amentoflavone (*73-12*):	H	H	H
Bilobetin (*73-13*):	H	CH_3	H
Ginkgetin (*73-14*):	CH_3	CH_3	H
Isoginkgetin (*73-15*):	H	CH_3	CH_3
Sciadopitysin (*73-16*):	CH_3	CH_3	CH_3

The phenolic compounds catechin pentaacetate, epicatechin pentaacetate, gallocatechin hexaacetate, and epigallocatechin hexaacetate have also been isolated [21].

The flavone glycoside, 3′-*O*-methylmyricitrin (*73-17*), was isolated from the autumnal leaf of *G. biloba* [20]. A new flavone glycoside was determined as quercetin 3-*O*-β-D-(6-*p*-coumaroylglucopyranosyl)-(1→4)-α-L-rhamnopyranoside (*73-18*) [26, 27]. The isolation of β-sitosterol, containing a trace amount of stigmasterol, from the ginkgo leaf was also reported [28].

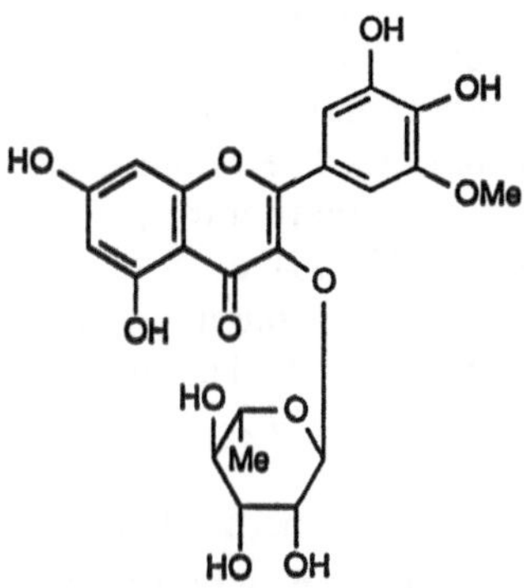

3′-*O*-Methylmyricitrin (*73-17*)

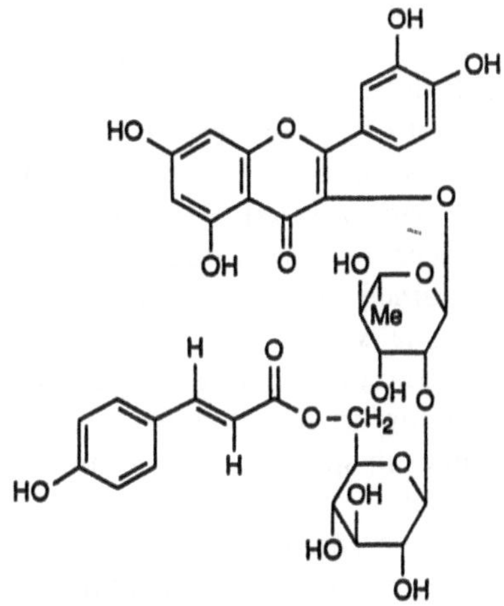

Quercetin 3-*O*-β-D-(6-coumaroylglucopyranosyl)-(1→4)-α-L-rhamnopyranoside (*73-18*)

Recently, an unusual natural product (Z,Z)-4,4′-(1,4-pentadien-1,5-diyl)-di-phenol (*73-19*) was isolated from the leaf of *G. biloba*. The structure was determined by synthesis (Fig. 73.2) starting from the bis(phosphorane) (*73-20*) and *p*-hydroxybenzaldehyde and by X-ray analysis of its *O,O*′-bis(4-nitrobenzoyl) derivative [29]. 6-Hydroxykynurenic acid (*73-21*) was the first nitrogen-containing compound isolated from the leaves of *G. biloba* [27, 30].

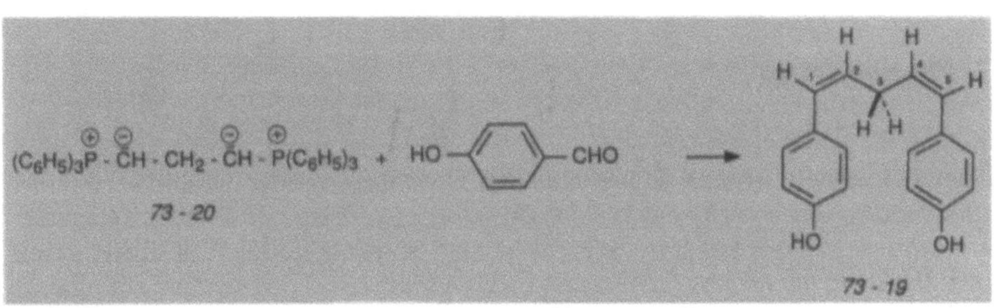

Fig. 73.2.

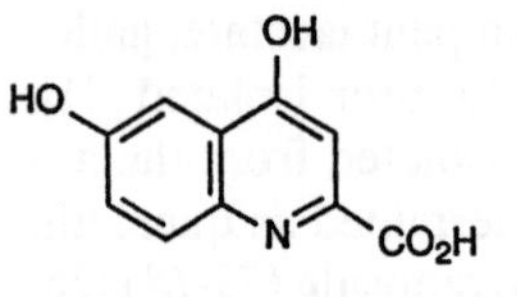

6-Hydroxykynurenic acid (*73-21*)

Water- and alkali-soluble polysaccharides, which could be separated into low- and high-molecular weight fractions, were also isolated from the leaves of *G. biloba*. After hydrolysis, it was shown that the polysaccharides with high molecular weight were composed of D-galactose, D-arabinose, L-rhamnose, and minor amounts of D-glucose, D-mannose, and D-xylose. Furthermore, D-galacturonic acid and D-glucuronic acid were detected [31]. Fractionation of the polysaccharides on DEAE cellulose resulted in a separation of seven distinct polysaccharides. Their homogeneity was ascertained by high-voltage electrophoresis and gel chromatography.

The biflavone content in the ginkgo leaf was determined. The leaves in the autumn season contained about 10 mg/g (dry wt.) sciadopitysin, 3.6–4.6 mg/g ginkgetin, 2.4–2.9 mg/kg isoginkgetin, and 0.9–1.4 mg/g bilobetin. The total contents of these biflavones in spring, summer, and autumn were 5.2, 4.4, and 17.2–19.0 mg/g respectively [32].

A sesquiterpene named bilobanone (*73-22*) was isolated and structurally determined from the heartwood of the ginkgo tree. It is a sesquiterpene with an isobutyl moiety [33].

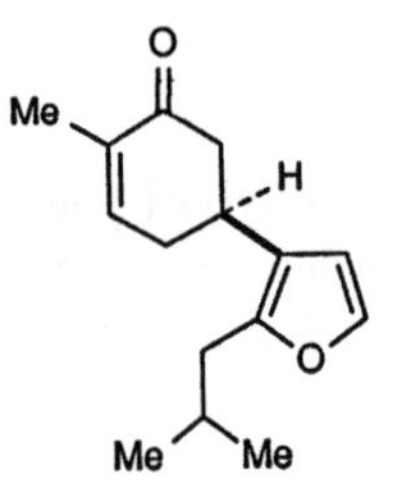

Bilobanone (*73-22*)

73.3 Pharmacology

Phenolic acids as the major components in the extracts of ginkgo fruits showed good antibacterial activity against *Mycobacterium smegmatis*, moderate activity against *Bacillus subtilis*, low activity against *Staphylococcus aureus*, and no activity against gram-negative microorganisms [34]. They also showed good antifungal action against *Trichophyton mentagrophytes* and *Saccharomyces cerevisiae* but not against *Candida albicans* or *Aspergillus niger* [34].

Bilobol showed a paralytic effect upon isolated rabbit intestine and contractive activity upon isolated uterus, but produced no pharmacological effect on frog heart even at a dosage of 500 mg/kg. The LD_{50} of bilobol in mice was 761 mg/kg. Bilobol produced a transitional hypotensive effect in rabbits. Repeated administration increased respiration in rabbits. Bilobol-induced capillary vessel permeability appeared to be strongest in guinea pigs, followed by rats and rabbits [35].

Quercetin, kaempferol, and isorhamnetin extracted from the green leaves of *G. biloba* showed a vasodilating and spasmolytic activity in the isolated intestine of quinea pigs against histamine and $BaCl_2$-induced spasms. The alcoholic extract showed the same effects as the pure flavonols. Neither hypotensive activity nor influence on heart and respiratory frequency was observed in vivo in cats and rabbits. In guinea pigs, doses 100- to 1000-fold the vasodilatatory dose led to moderate hypotension, a rise in the respiratory rate and a drop in the heart rate [36]. The biflavones from ginkgo leaves showed 75% of the activity of papaverine in peripheral vasodilation in rat hind limb preparations [37].

Recently, a number of studies were performed on the pharmacology and clinical use of a well-defined extract of the green leaves of *G. biloba*, designated EGb 761. Its main uses are treatment of psychic and behavioral disorders of the elderly, of peripheral vascular deficiency, and of functional disorders of ischemic origin in the cerebrovascular, rhinolaryngeal, and ophthalmic areas.

This extract contains flavone compounds, such as the ginkgo flavone glycosides and terpenes, e.g., ginkgolides and bilobalide, characteristic for ginkgo [38].

Numerous experiments over several years in different pathological models of cerebral ischemia to evaluate the effects of *G. biloba* extract and experiments at both cellular and molecular levels to explain its mechanisms of action have been performed. In experimental models of ischemia, edema, and hypoxia, ginkgo extract reduced vascular and metabolic disturbances as well as the resulting neurological and behavioral consequences. The pharmacological effects of ginkgo extract apply to vascular, rheological, and metabolic processes. Membrane effects also appear to be involved, including protection of the membrane ultrastructure against free radicals and modulation of membrane-located enzymes and ionic pumps [39–42].

The antiradical properties of *G. biloba* extract have been determined on several in vitro models. It was found that the extract is as active as uric acid. In addition, the extract prevented the formation of the adriamycinyl radical by phenobarbital-induced rat liver microsomes, a process not influenced by uric acid. *G. biloba* extract inhibited lipid peroxidation not influenced by rat liver microsomes and also stimulated cyclooxygenase activity in vitro, evidently due to trapping of radicals that damage the enzyme [43, 44].

Unilateral ligation of the right carotid artery in gerbils caused a well-definable brain ischemia. It was less severe when the animals were pretreated with *G. biloba* extract [45].

Edema is one of the major complications of cerebral ischemia. *G. biloba* extract limited the formation of cerebral edema and suppressed its neurological consequences irrespective of whether the edema was of cytotoxic origin, such as with triethyltin [46, 47] or of vasogenic origin such as from unilateral traumatic edema [47]. Several membrane mechanisms have been implicated in the protective action manifested by *G. biloba* extract against cerebral edema [48].

Ginkgo biloba extract exerted a specific effect on the noradrenergic system and on β-receptors. No variation was found in the α_2-receptors and serotonin uptake [49].

Perfusion with the *G. biloba* extract had beneficial effects on ischemic-reperfused rat and guinea pig hearts in vitro and on ischemic rat hearts in vivo. Arrhythmic and electrocardiographic dysfunctions were decreased, with no effect on basal cardiovascular parameters [50].

After oral administration of a ^{14}C-labeled *G. biloba* extract to rats, the amount of radioactivity in the expired air and urine indicated an absorption of ≥60%, probably from the upper digestive tract. The pharmacokinetics fitted a two-compartment model with a first-order absorption phase having a half-life of 4.5 h. Radioactivity was mainly present in the plasma for the first 3 h, but after 48 h the specific activity of the red cells equaled that of the plasma. Glandular and neuronal tissues and the eyes showed especially high uptake of the labeled material [51]. The bioavailability and kinetics of flavones from standardized *G. biloba* extract were also studied in rats and humans. After oral administration of 400 mg extract to volunteers, the calculated half-life of the flavone constituents was about 3 h [52].

Pharmacological studies on pure ginkgolides A (BN 52020), B (BN 52021) and C (BN 52022) specifically inhibited the binding of platelet-activating factor to its receptor in isolated rabbit platelet membranes and inhibited platelet-activating factor-dependent in vitro platelet aggregation. Ginkgolide B was the most active ginkgolide with an IC_{50} of ca. 10^{-7} *M*. After oral administration, ginkgolide B inhibited platelet-activating factor-dependent platelet aggregation in rabbits and inhibited thrombus formation induced by electric stimulation of the carotid artery in rats. Ginkgolide B also inhibited platelet-activating factor-dependent human polymorphonuclear leukocyte aggregation in vitro. Ginkgolides did not significantly affect arachidonic acid metabolism and did not bind to neurotransmitter receptors [53]. Ginkgolides may have a clinical use and might also be of value as pharmacological tools for studying biological activities of platelet-activating factor [54].

Experiments on the interaction between ginkgolide B and the effects of platelet-activating factor on the bronchopulmonary system of the guinea pig showed that ginkgolide B, at an intravenous dose of 1 mg/kg or an oral dose of 10 mg/kg, inhibited bronchoconstriction and the thrombopenia and leukopenia induced by platelet-activating factor [55]. Ginkgolide B also inhibited bronchoconstriction in the guinea pig induced by antigen in passively sensitized animals [56] and protected the animals from lethal immunological reaction, suggesting that platelet-activating factor must play a role in the expression of anaphylactic bronchoconstriction. The antianaphylactic activity of ginkgolide A was weaker than that of ginkgolide B [57, 58].

Isolated perfused hearts of guinea pigs sensitized passively with antiovalbumin rabbit serum responded to the specific antigen with a marked decrease in contractile force, an increase in perfusion pressure, and rhythm disturbances. All these impairments, except tachycardia, were decreased by ginkgolide B. Apparently, platelet-activating factor plays an important role as mediator in cardiac anaphylaxis and the antagonist ginkgolide B may be a valuable therapeutic agent in allergic conditions [59].

Rats with experimental cirrhosis of the liver induced by phenobarbital combined with CCl_4 showed a hyperdynamic status with enhanced cardiac output, decreased main arterial pressure and peripheral vascular resistance, and increased vascular permeability, similar to some of the known effects of the systemic infusion of low doses of synthetic platelet-activating factor into the systemic circulation of normal rats. The levels of platelet-activating factor in samples of blood were found to be higher in cirrhotic than in control rats. Ginkgolide B decreased the cardiac output, increased the peripheral vascular resistance, and slightly increased the main arterial pressure in cirrhotic animals, but induced no significant hemodynamic changes in normal animals [60].

Ginkgo biloba extract (EGb 761) was also studied clinically in the treatment of old age disorder [61], arteritis of the pelvic limbs [62], senile macula degeneration [63], acute cochlear deafness [64], and tinnitus [65]. In the treatment of arteritis, *G. biloba* extract gave a significant improvement in pain relief and walking tolerance which continued throughout the treatment period [62]. The condition of tinnitus patients was improved by treatment with *G. biloba* extract [65]. In a rare but severe case of hypovolemic shock, intravenous infusion of *G. biloba* extract resulted in dramatic recovery [66].

A study of the quantitative structure-activity relationship for antitumor activity of long-chain phenol from *G. biloba* or synthesized was also carried out. 4-Undecylcatechol, selected on the basis of the results, exhibited strong antitumor activity against sarcoma 180 ascites and P-388 lymphocytic leukemia in mice [67].

References

1. Urakami C, Oka S, Han JS (1976) Composition of the neutral and phospholipid fractions from ginkgo fruits (*Ginkgo biloba*) and fatty acid composition of individual lipid classes. J Am Oil Chem Soc 53:526–529
2. Kemeyama H, Urakami C (1979) Glycolipids isolated from ginkgo nuts (*Ginkgo biloba*) and their fatty acid compositions. J Am Oil Chem Soc 56:549–551
3. Lin QS (1977) Chemistry of the Constituents in Chinese Medicinal Herbs. Science Press, Beijing, pp 148–149
4. Morimoto H, Kawamatsu Y, Sugihara H (1968) Steric structure of the toxic substances from the fruit pulp of *Ginkgo biloba*. Chem Pharm Bull (Tokyo) 16:2282–2286
5. Itokawa H, Totsuka N, Nakahara K, Takeya K, Lepoittevin JP, Asakawa Y (1987) Antitumor principles from *Ginkgo biloba* J. Chem Pharm Bull (Tokyo) 35:3016–3020
6. Wada K, Ishigaki S, Ueda K, Take Y, Sasaki K, Sakata M, Haga M (1988) Studies on the constitution of edible and medicinal plants. I. Isolation and identification of 4-*O*-methylpyridoxine, toxic principle from seed of *Ginkgo biloba* L. Chem Pharm Bull (Tokyo) 36:1779–1782
7. Nakanishi K (1967) The ginkgolides. Pure Appl Chem 14:89–113
8. Okabe K, Yamada K, Yamamura S, Takada S (1967) Ginkgolides. J Chem Soc [C] 2201–2206
9. Maruyama M, Terahara A, Itagaki Y, Nakanishi K (1967) The ginkgolides. I. Isolation and characterization of the various groups. Tetrahedron Lett 299–302
10. Maruyama M, Terahara A, Itagaki Y, Nakanishi K (1967) The ginkgolides. II. Derivation of partial structures. Tetrahedron Lett 303–308
11. Maruyama M, Terahara A, Nakadaira Y, Woods MC, Nakanishi K (1967) The ginkgolides. III. The structure of the ginkgolides. Tetrahedron Lett 309–313
12. Maruyama M, Terahara A, Nakadaira Y, Woods MC, Takagi Y, Nakanishi K (1967) The ginkgolides. IV. Stereochemistry of the ginkgolides. Tetrahedron Lett 315–319
13. Woods MC, Miura I, Nakadaira Y, Terahara A, Maruyama M, Nakanishi K (1967) The ginkgolides. V. Some aspects of their NMR spectra. Tetrahedron Lett 321–326
14. Nakanichi K, Habaguchi K (1971) Biosynthesis of ginkgolide B, its diterpenoid nature and origin of the tert.-butyl group. J Am Chem Soc 93:3546–3547
15. Weinges K, Hepp M, Jaggy H (1987) Chemistry of ginkgolides. II. Isolation and structural elucidation of a new ginkgolide. Liebigs Ann Chem 521–526
16. Major RT (1967) The ginkgo, the most ancient living tree. Science 157:1270–1273
17. Weinges K, Bähr W (1969) Kondensierte Ringsysteme II. Bilobalid A, ein neues Sesquiterpen mit tert.-Butyl Gruppe aus den Blättern von *Ginkgo bioloba* L. Liebigs Ann Chem 724:214–216
18. Nakanishi K, Habaguchi K, Nakadaira Y, Woods MC, Maruyama M, Major RT, Alauddin M, Patel AR, Weinges K, Bähr W (1971) Structure of bilobalide, a rare tert.-butyl containing sesquiterpenoid related to the C_{20}-ginkgolides. J Am Chem Soc 93:3544–3546

19. Weinges K, Bähr W (1972) Natural products from medicinal plants. XVI. Comparison of the NMR and mass spectra of bilobalide $C_{15}H_{18}O_8$ and of the ginkgolides $C_{20}H_{24}O_{9-11}$. Liebigs Ann Chem 759:158–172
20. Geiger H (1979) 3′-*O*-Methylmyricetin-3-rhamnoglucoside, a new flavonoid from the autumnal leaf of *Ginkgo biloba* L. Z Naturforsch [C] 34C:878–879
21. Weinges K, Bähr W, Kloss P (1968) Natural phenolic compounds. X. Phenolic components of *Ginkgo biloba* leaves. Arzneimittelforschung 18:539–543
22. Briancon-Scheid F, Guth A, Anton R (1982) High-performance liquid chromatography of biflavones from *Gingko biloba* L. J Chromatogr 245:261–267
23. Briancon-Scheid F, Lobstein-Guth A, Anton R (1983) HPLC separation and quantitative determination of biflavones in leaves from *Ginkgo biloba*. Planta Med 49:204–207
24. Piette P, Mauri P, Rava A (1988) Reversed-phase high-performance liquid chromatographic method for the analysis of biflavones in *Ginkgo biloba* L. extract. J Chromatogr 437:453–456
25. Lobstein-Guth A, Briancon-Scheid F, Victoire C, Haag-Berrurier M, Anton R (1988) Isolation of amentoflavone from *Ginkgo biloba*. Planta Med 54:555–556
26. Nasr C, Lobstein-Guth A, Haag-Berrurier M, Anton R (1987) Quercetin coumaroyl glucorhamnoside from *Ginkgo biloba*. Phytochemistry 26:2869–2870
27. Victoire C, Haag-Berrurier M, Lobstein-Guth A, Balz JP, Anton R (1988) Isolation of flavonol glycosides from *Ginkgo biloba* leaves. Planta Med 54:245–247
28. Kircher HW (1970) *β*-Sitosterol in *Ginkgo biloba* leaves. Phytochemistry 9:1879
29. Plieninger H, Schwarz B, Jaggy H, Huber-Patz U, Rodewald H, Irngartinger H, Weinges K (1986) Naturstoffe aus Arzneipflanzen, XXIV. Isolierung, Strukturauflärung und Synthese von (Z,Z)-4,4′-(1,4-Pentadien-1,5-diyl)diphenol, einen ungewöhnlichen Naturstoff aus den Blättern des Ginkgo-Baumes (*Ginkgo biloba* L.). Liebigs Ann Chem 1772–1778
30. Schennen A, Hoelzl J (1986) 6-Hydroxykynurenic acid, the first nitrogen-containing compound from the *Ginkgo biloba* leaf. Planta Med 52:235–236
31. Hoellriegl H, Koehler H, Franz G (1986) High-molecular-weight polysaccharides of *Ginkgo biloba* leaves. Sci Pharm 54:321–330
32. Song YF (1986) Chemical composition and utilization of *Ginkgo biloba* L. Linchan Huaxue Yu Gongye 6:42–45
33. Kimura H, Irie H, Ueda K, Ueda S (1968) Constituents of the heartwood of *Ginkgo biloba*. V. Structure and absolute configuration of bilobanone. Yakugaku Zasshi 88:562–572
34. Adawadkar PD, El Sohly MA (1981) Isolation, purification and antimicrobial activity of anacardic acids from *Ginkgo biloba* fruits. Fitoterapia 52:129–135
35. Han DS (1964) Chemical and pharmacological study on bilobol, a component of *Ginkgo biloba* fruits. Soul Uidae Chapchi 5:281–290 (CA 65:7861 c)
36. Peter H (1968) Vasoactivity of *Ginkgo biloba* preparations. In: Dumbovich B (ed) Conf Hung Ther Invest Pharmacol, Soc Pharmacol Hung, 4th 1966, Akad Kiado, Budapest, Hung, pp 177–181
37. Natarajan S, Murti WS, Seshadri T, Ramaswamy AS (1970) Pharmacological properties of flavonoids and biflavonoids. Curr Sci 39:533–534
38. Drieu K (1986) Preparation and definition of *Ginkgo biloba* extract. Presse Med 15:1455–1457
39. Clostre F (1986) From the body to cell membranes: the different levels of action of *Ginkgo biloba* extract. Presse Med 15:1529–1538
40. Chatterjee SS (1985) Effects of *Ginkgo biloba* extract on cerebral metabolic processes. In: Agnoli A, Rapin JR, Scapagnini V, Weitbrecht WV (eds) Effects of *Ginkgo biloba* Extract on Organic Cerebral Impairment. Libbey, London, pp 5–15
41. Karcher L, Zagermann P, Krieglstein J (1984) Effect of an extract of *Ginkgo biloba* on rat brain energy metabolism in hypoxia. Naunyn-Schmiedebergs Arch Pharmacol 327:31–35
42. Attella MJ, Hoffmann SW, Stasio MJ, Stein DG (1989) *Ginkgo biloba* extract facilitates recovery from penetrating brain injury in adult male rats. Exp Neurol 105:62–71
43. Deby PC (1986) Antiradical properties of *Ginkgo biloba* extract. Presse Med 15:1475–1479
44. Deby C, Deby-Dupont G, Noel FX, Lavergne L (1981) In vitro and in vivo arachidonic acid conversions into biologically active derivatives are enhanced by uric acid. Biochem Pharmacol 30:2243–2249
45. Spinnewyn B, Blavet N, Clostre F (1986) Effects of *Ginkgo biloba* extract on a cerebral ischemia model in gerbils. Press Med 15:1511–1515

46. Borzeix MG, Weber S, Akimjak JP (1978) Cerebral edema induced by triethyltin (TET) in the rabbit and in the rat: In: Proceedings of the 7th International Congress of Pharmacology, Satellite Symposium of Reims
47. Borzeix MG (1985) Effect of *Ginkgo biloba* extract on 2 types of cerebral edema. In: Agnoli A, Rapin JR, Scapagnini V, Weitbrecht WV (eds) Effects of *Ginkgo biloba* Extract on Organic Cerebral Impairment. Libbey, London, pp 51–56
48. Etienne A, Hecquet F, Clostre F (1986) Mechanism of action of *Ginkgo biloba* extract on experimental cerebral edema. Press Med 15:1506–1510
49. Brunello N, Racagni G, Clostre F, Drieu K, Braquet P (1985) Effects of an extract of *Ginkgo biloba* on noradrenergic systems of rat cerebral cortex. Pharmacol Res Commun 17:1063–1072
50. Guillon JM, Rochette L, Baranes J (1986) Effects of *Ginkgo biloba* extract on two models of experimental myocardial ischemia. Presse Med 15:1516–1519
51. Moreau JP, Eck CR, McCabe J, Skinner S (1986) Absorption, distribution and elimination of radiolabeled *Ginkgo biloba* leaf extract in the rat. Presse Med 15:1458–1461
52. Drieu K, Moreau JP, Eck CR, McCabe J, Skinner S (1986) Animal distribution and preliminary human kinetic studies of the flavonoid fraction of a standardized *Ginkgo biloba* extract (GBE 761). Stud Org Chem (Amsterdam) 23:351–359
53. Braquet PG, Spinnewyn B, Braquet M, Bourgain RH, Taylor JE, Etienne A, Drieu K (1985) BN 52021 and related compounds: a new series of highly specific PAF-acether receptor antagonists isolated from *Ginkgo biloba* L. Ketsueki to Myakkan 16:558–572 (CA 105:58959x)
54. Braquet P (1986) Proofs of involvement of PAF-acether in various immune disorders using BN 52021 (ginkgolide B): a powerful PAF-acether antagonist isolated from *Ginkgo biloba* L. Adv Prostaglandin Thromboxane Leukotriene Res 16:179–198
55. Desquand S, Touvay C, Randon J, Lagente V, Vilain B, Maridouneau-Parini I, Etienne A, Lefort J, Braquet P, Vargaftig BB (1986) Interference of BN 52021 (ginkgolide B) with the bronchopulmonary effects of PAF-acether in the guinea pig. Eur J Pharmacol 127:83–95
56. Vilain B, Lagente V, Touvay C, Desquand S, Randon J, Lefort J, Braquet P, Vargaftig BB (1986) Pharmacological control of the in vivo passive anaphylactic shock by the PAF-acether antagonist compound BN 52021. Pharmacol Res Commun 18 [Suppl]:119–126
57. Berti F, Omini C, Rossoni G, Braquet P (1986) Protection by two ginkgolides, BN-52020 and BN-52021, against guinea pig lung anaphylaxis. Pharmacol Res Commun 18:775–793
58. Braquet P, Bougain RH (1987) Anti-anaphylactic properties of BN 52021: a potent platelet activating factor antagonist. Adv Exp Med Biol 215:215–235
59. Koltai M, Lepran I, Szekeres L, Viossat I, Chabrier E, Braquet P (1986) Effect of BN 52021, a specific PAF-acether antagonist, on cardiac anaphylaxis in Langendorff hearts isolated from passively sensitized guinea pigs. Eur J Pharmacol 130:133–136
60. Villamediana LM, Sanz E, Fernandez-Gallardo S, Caramelo C, Sanchez Crespo M, Braquet P, Lopez-Novoa JM (1986) Effects of the platelet-activating factor antagonist BN 52021 on the hemodynamics of rats with experimental cirrhosis of the liver. Life Sci 39:201–205
61. Allard M (1986) Treatment of old age disorders with *Ginkgo biloba* extract. Presse Med 15: 1540–1545
62. Bauer U (1986) *Ginkgo biloba* extract in the treatment of lower limb arteritis. A sixty-five week trial. Presse Med 15:1546–1549
63. Lebnisson DA, Leroy L, Rigal G (1986) Treatment of senile macula degeneration with *Ginkgo biloba* extract. A preliminary double-blind, drug versus placebo study. Presse Med 15:1556–1558
64. Dubreuil C (1986) Comparative therapeutic trial of *Ginkgo biloba* extract and nicergoline in acute cochlear deafness. Presse Med 15:1559–1561
65. Meyer B (1986) A multicenter, randomized, double-blind drug versus placebo study of *Ginkgo biloba* extract in the treatment of tinnitus. Presse Med 15:1562–1564
66. Lagrue G, Rahbar K, Behar A, Sobel A, Laurent J (1986) Recurrent shock associated with monoclonal gammapathy. Acute and chronic treatment with parenteral and oral *Ginkgo biloba* extract. Presse Med 15:1554–1555
67. Itokawa H, Tobsuka N, Nakahara K, Maezuru M, Takeya K, Kondo M, Inamatsu M, Morta H (1989) A quantitative structure-activity relationship for antitumor activity of long-chain phenols from *Ginkgo biloba* L. Chem Pharm Bull (Tokyo) 37:1619–1621

Glycyrrhiza spp.

74

74.1 Introduction

Gancao, Radix Glycyrrhizae, is the dry root and rhizome of *Glycyrrhiza uralensis* Fisch., *G. inflata* Bat. or *G. glabra* L. (Fabaceae), collected in the spring and fall. It is officially listed in the Chinese Pharmacopoeia. *Glycyrrhiza* root is one of the oldest traditional Chinese medicines and is used as a tonic, antiphlogistic, mucolytic, expectorant, analgesic for treatment of gastrointestinal and respiratory disorders, and also to alleviate the toxicity of some other drugs. However, it is reported not to be suitable in combined use with *Euphorbia kansui, E. pekinensis*, and *Daphne genkwa.*

Two galenic preparations of *Glycyrrhiza* root are listed in the Chinese Pharmacopeia:

- Gancao Jingao, Extractum Glycyrrhizae, is the dry extract of *Glycyrrhiza* root, prepared by extracting the root with boiling water.
- Gancao Liujingao, Extractum Glycyrrhizae liquidum, is the fluid extract of *Glycyrrhiza* root, prepared from the dry extract by adding concentrated ammonia solution, ethanol, and water.

The extract, as well as the fluid extract of *Glycyrrhiza* root, is used as an alleviative often in combination with mucolytic and expectorant preparations for the treatment of asthma, laryngitis, and bronchitis. Furthermore, they show an inhibitory effect on smooth muscle contraction in the gastrointestinal tract. Because of some desoxycorticosterone-like activity it could be used for treatment of chronic dysfunction of adrenal cortex. Successive administration of the extract or fluid extract of *Glycyrrhiza* root for a long time might cause side effects such as edema and hypertension, which disappear after withdrawal.

74.2 Chemical Constituents

The major constituent of the roots of *G. glabra, G. uralensis* [1], and *G. inflata* [2] is the triterpene saponin glycyrrhizin (glycyrrhizic acid, glycyrrhizinic acid) with the sapogenin glycyrrhetic acid (glycyrrhetin, glycyrrhetinic acid). Glycyrrhetic acid is a triterpene with an oleanan skeleton. Glycyrrhizin was first isolated from *Glycyrrhiza* root by Robiquet in 1809 [3]. After hydrolysis an aglycone and two molecules of sugar were obtained [4, 5]. Structure [6–8] and configuration [9] of the aglycone glycyrrhetic acid were shown to correspond to 3β,20β-3-hydroxy-11-oxo-olean-12-en-29-oic acid (*74-1*).

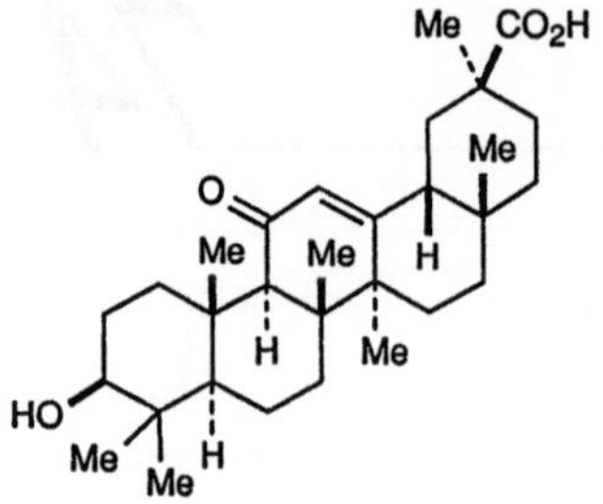

Glycyrrhetic acid (*74-1*)

The sugar moiety of glycyrrhizin was found to be β-D-glucopyranuronosyl-(1→2)-α-D-glucopyranosiduronic acid. Thus, glycyrrhizin is structurally 3-*O*-β-D-glucopyranuronosyl(1→2)-α-D-glucopyranuronosyl-(3β,20β)-20-carboxy-11-oxo-30-norolean-12-ene (*74-2*) [10].

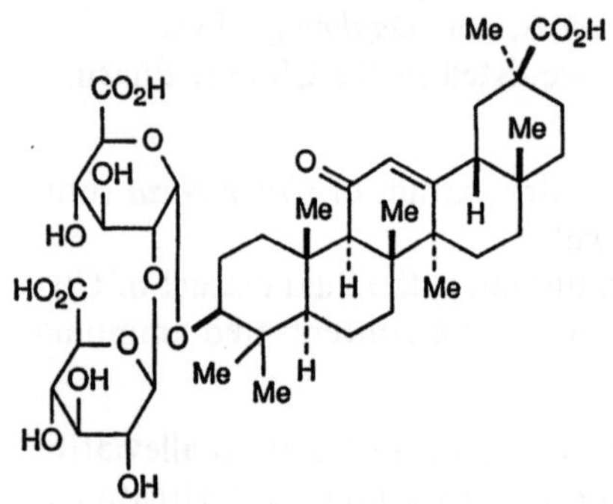

Glycyrrhizin (*74-2*)

Further triterpene components isolated from the roots of *G. glabra, G. uralensis,* and *G. inflata* are summarized in Table 74.1. Besides the triterpenes, a number of flavone, isoflavone, chalcone, and related compounds as well as their glycosides were isolated from *Glycyrrhiza* species. The flavone, isoflavone, chalcone and related compounds detected in *G. glabra, G. inflata,* and *G. uralenis* are summarized in Table 74.2.

Table 74.1. Triterpene constituents from the roots of *Glycyrrhiza glabra*, *G. inflata*, and *G. uralensis*

Compound	Structure	Plant origin	Ref.
Liquiritic acid (*74-3*)	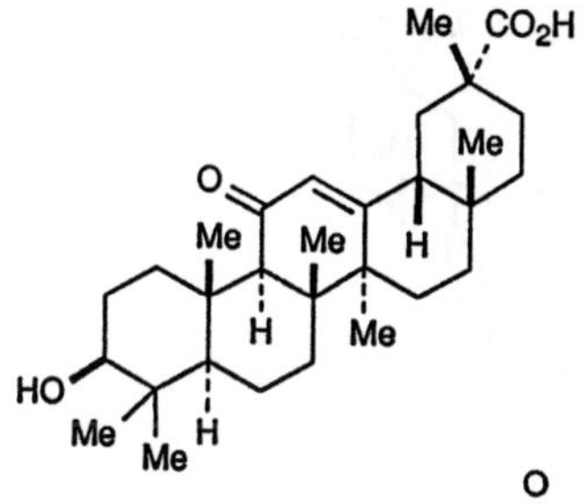	*Glycyrrhiza glabra*	[44]
Glabrolide (*74-4*)	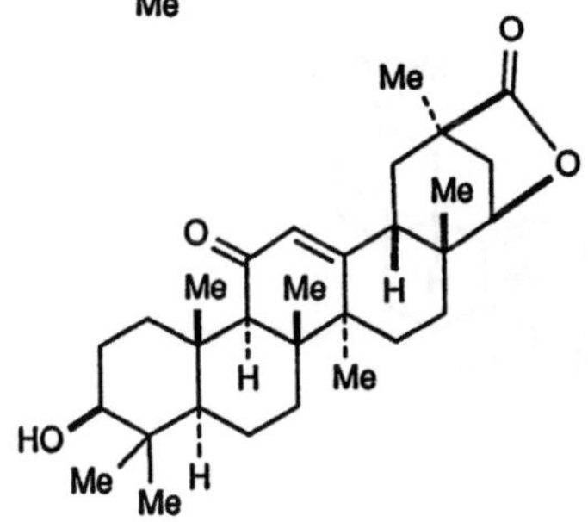	*G. glabra* *G. uralensis*	[45] [46]
Isoglabrolide (*74-5*)	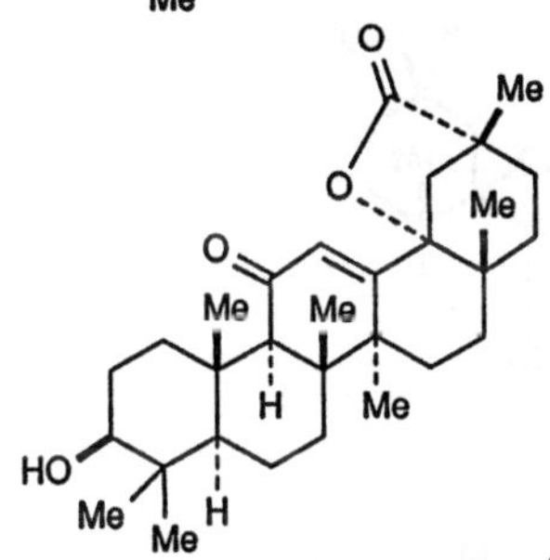	*G. glabra*	[45, 47]
Deoxoglabrolide (*74-6*)	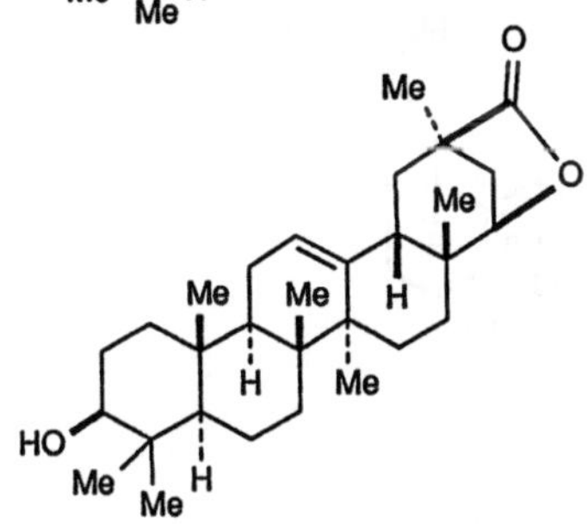	*G. glabra*	[45]
Glabric acid (*74-7*)	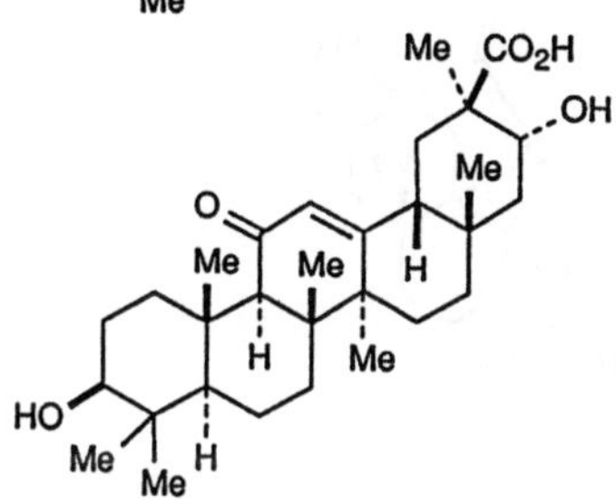	*G. glabra*	[48, 49]

Table 74.1. (continued)

Compound	Structure	Plant origin	Ref.
Deoxoglycyrrhetic acid (*74-8*)	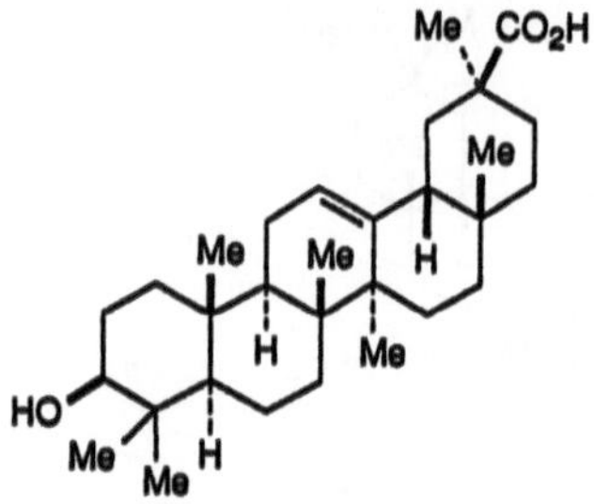	*G. glabra*	[45]
18α-Glycyrrhetic acid (*74-9*)	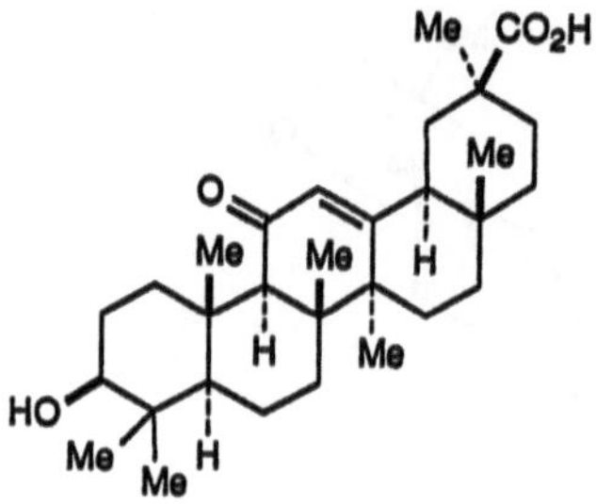	*G. glabra*	[9]
18α-Hydroxy-glycyrrhetic acid (*74-10*)	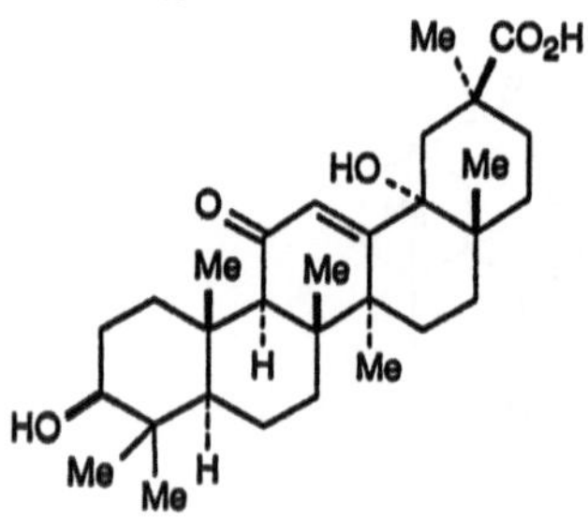	*G. glabra* *G. uralensis*	[50] [51]
Glycyrrhetol (*74-11*)	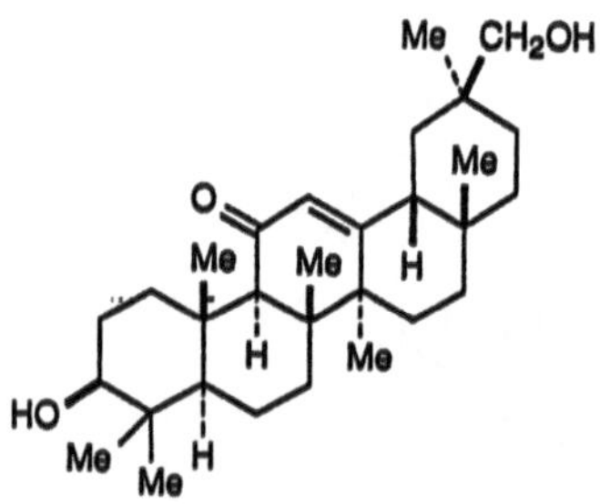	*G. glabra*	[52]
21α-Hydroxy-isoglabrolide (*74-12*)	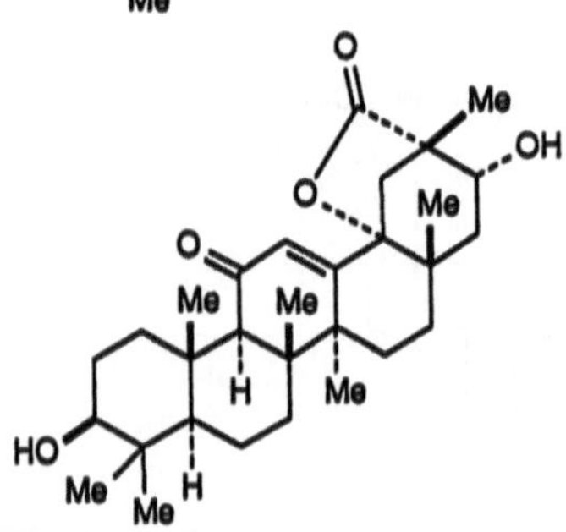	*G. glabra*	[52]

Table 74.1. (continued)

Compound	Structure	Plant origin	Ref.
23-Hydroxy-glycyrrhetic acid		*G. glabra* *G. uralensis*	[53] [46]
24-Hydroxy-11-deoxy-gly-cyrrhetic acid		*G. glabra*	[53]
24-Hydroxy-liquiritic acid		*G. glabra*	[54]
Liquiridiolic acid (*74-13*)	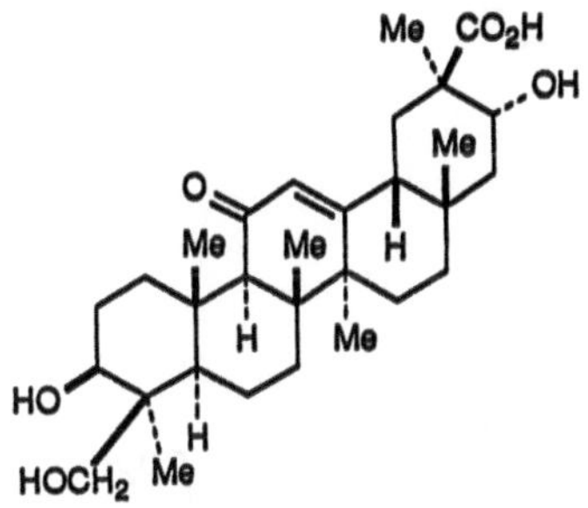	*G. glabra*	[54]
28-Hydroxygly-cyrrhetic acid (*74-14*)	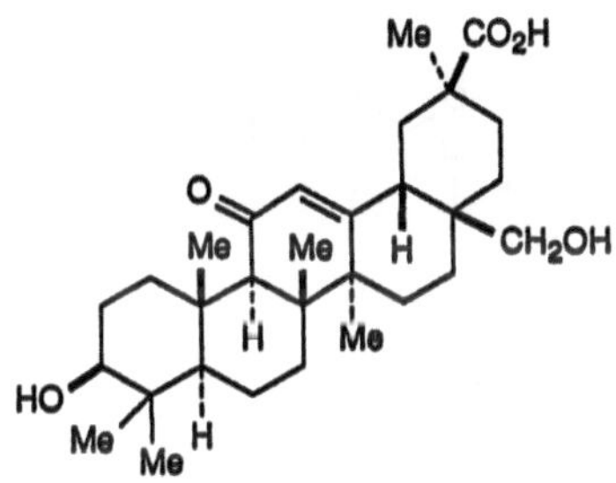	*G. glabra*	[46, 49]
Methyl 3*β*, 24-dihydroxyolean-11, 13(18)-diene-30-oate (*74-15*)	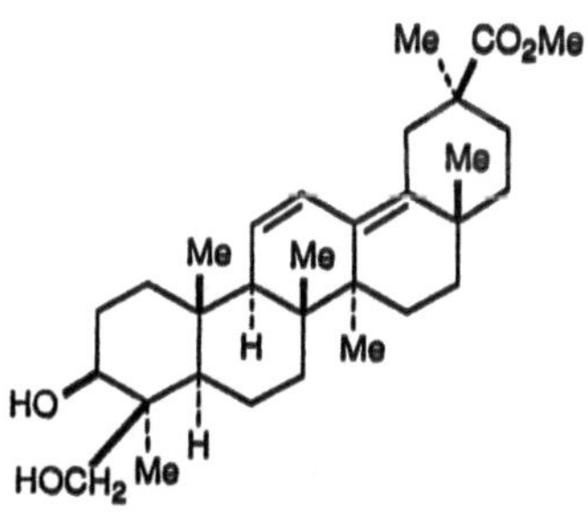	*G. uralensis*	[46]
24-Hydroxy-glabrolide (*74-16*)	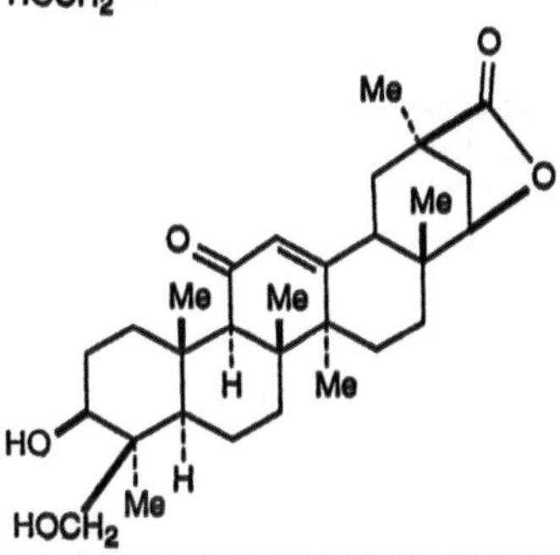	*G. uralensis*	[46]

Table 74.1. (continued)

Compound	Structure	Plant origin	Ref.
Uralsaponin A (*74-17*)	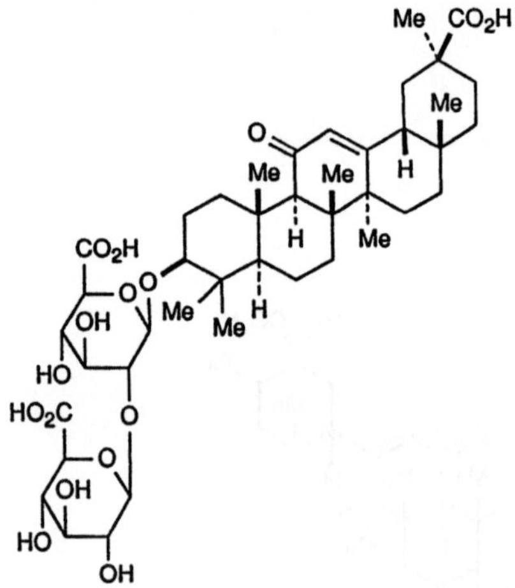	*G. uralensis*	[55]
Uralsaponin B (*74-18*)	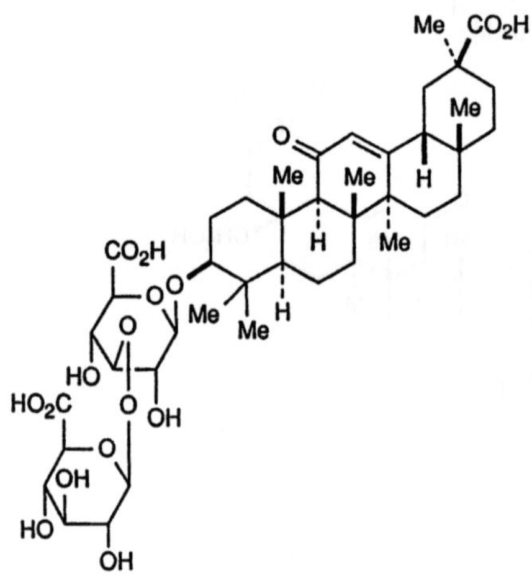	*G. uralensis*	[55]
Uralenolide (*74-19*)	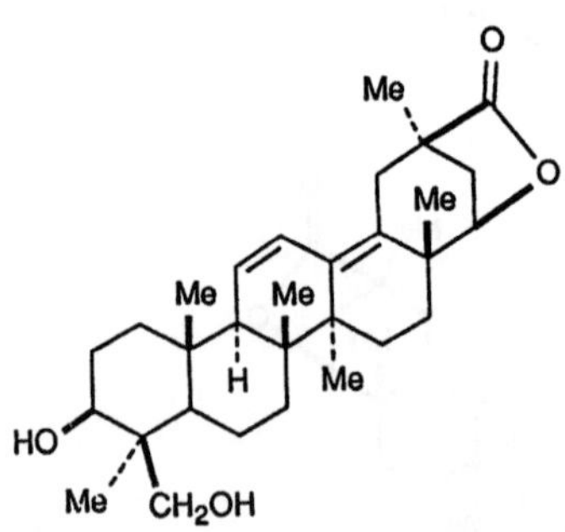	*G. uralensis*	[51]

Table 74.1. (continued)

Compound	Structure	Plant origin	Ref.
Licorice-saponin A3 (*74-20*)	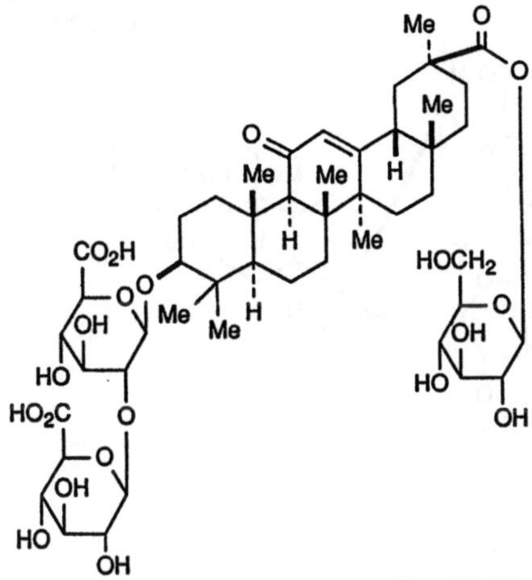	*G. uralensis*	[56]
Licorice-saponin B2 (*74-21*)	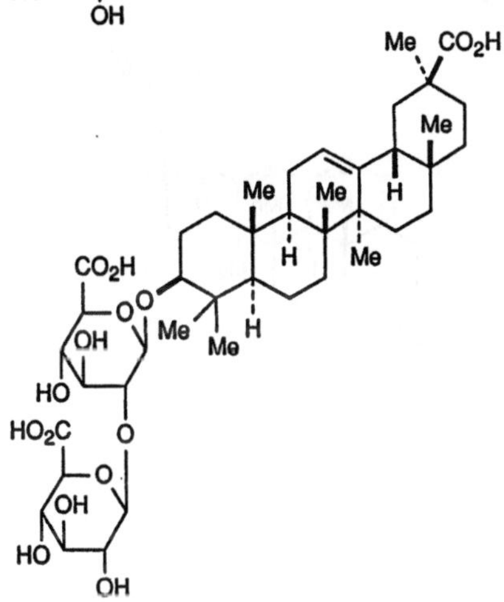	*G. uralensis*	[56]
Licorice-saponin C2 (*74-22*)	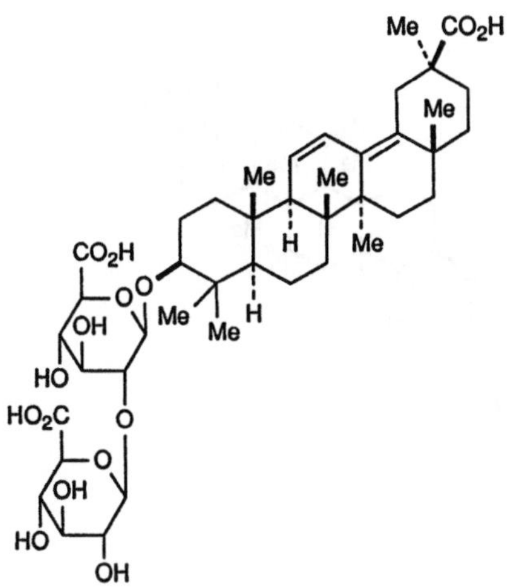	*G. uralensis*	[56]

Table 74.1. (continued)

Compound	Structure	Plant origin	Ref.
Licorice-saponin D3 (*74-23*)	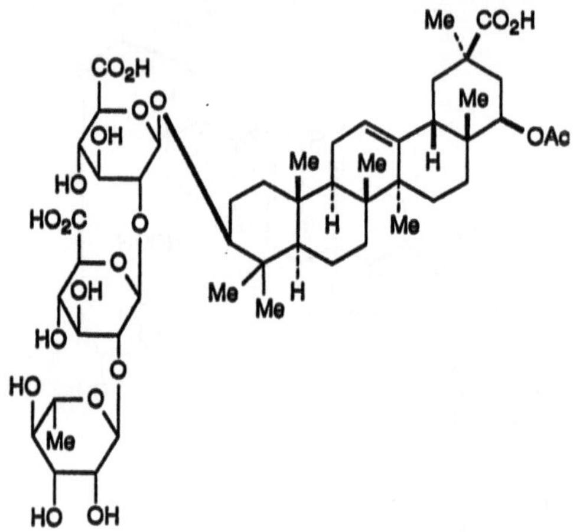	*G. uralensis*	[56]
Licorice-saponin E2 (*74-24*)	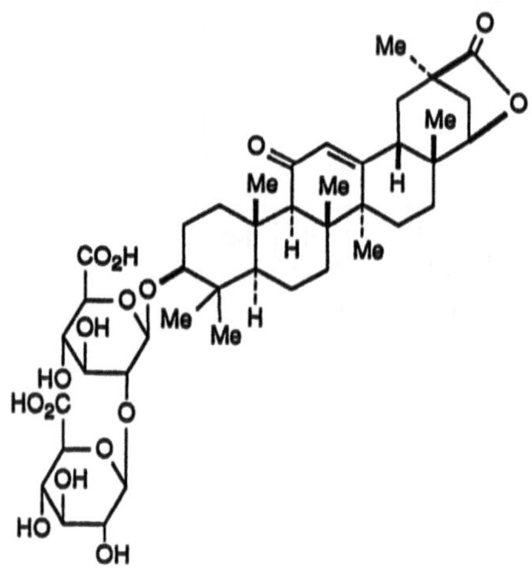	*G. uralensis*	[56]
Araboglycyrrhizin (*74-25*)	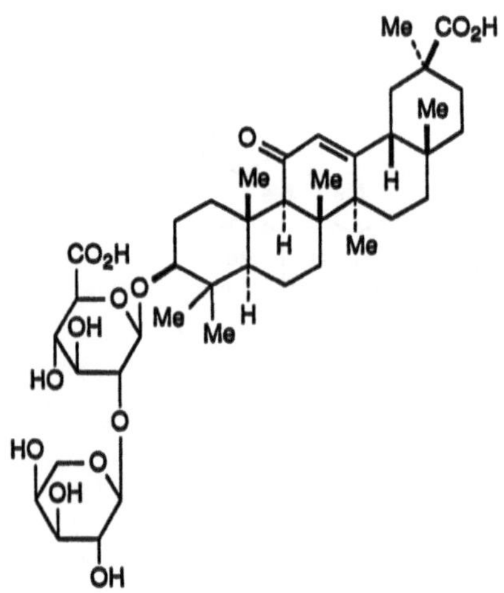	*G. inflata*	[57]

Table 74.1. (continued)

Compound	Structure	Plant origin	Ref.
Apioglycyrrhizin (*74-26*)	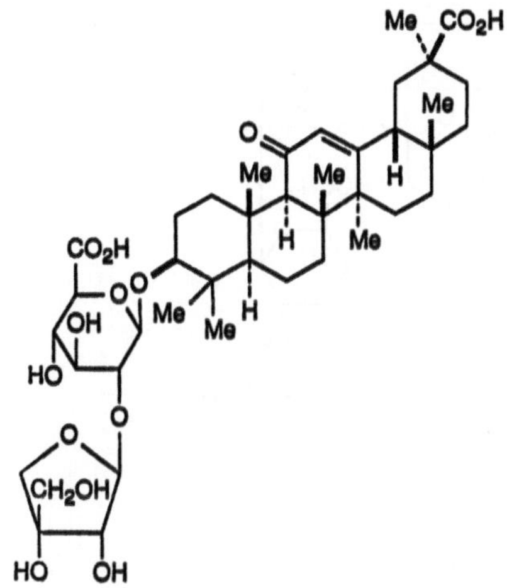	*G. inflata*	[57]

Table 74.2. Flavone, isoflavone, chalcone, and related compounds from *Glycyrrhiza glabra, G. inflata,* and *G. uralensis*

Compound	Structure	Plant origin	Ref.
Liquiritin (*74-27*)	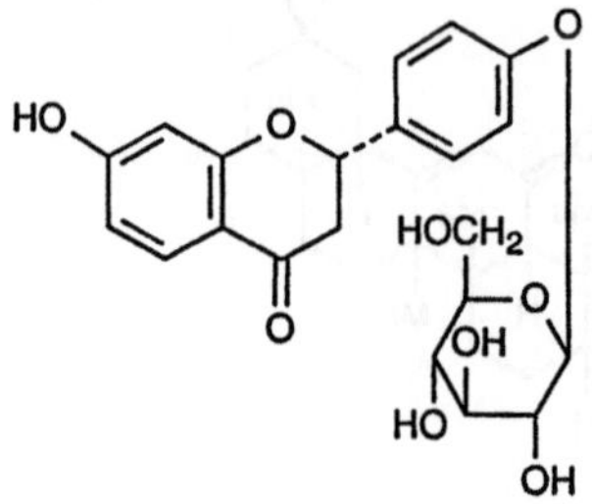	*Glycyrrhiza glabra* *G. uralensis*	[58, 59] [60]
Liquiritigenin (*74-28*)	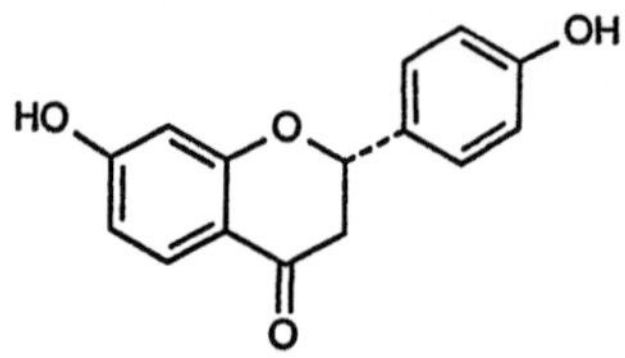	*G. glabra* *G. uralensis*	[58, 61] [60]
Isoliquiritin (*74-29*)	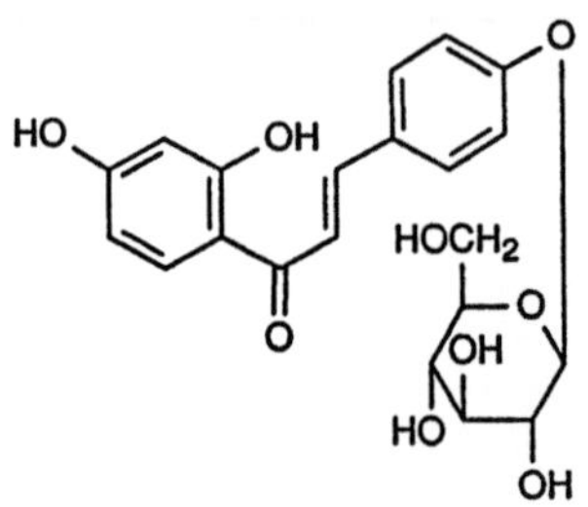	*G. glabra*	[62]
Isoliquiritigenin (*74-30*)	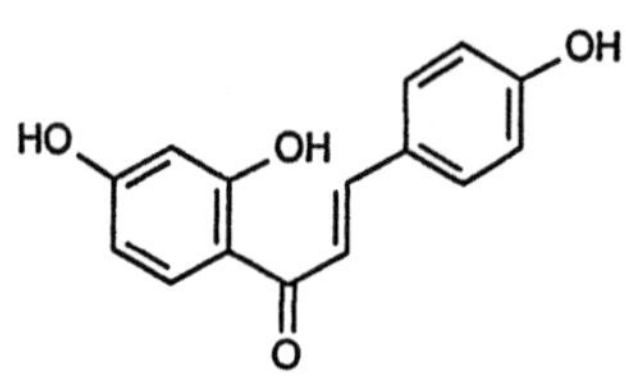	*G. glabra* *G. uralensis*	[61, 62] [60]
Neoisoliquiritin (*74-31*)	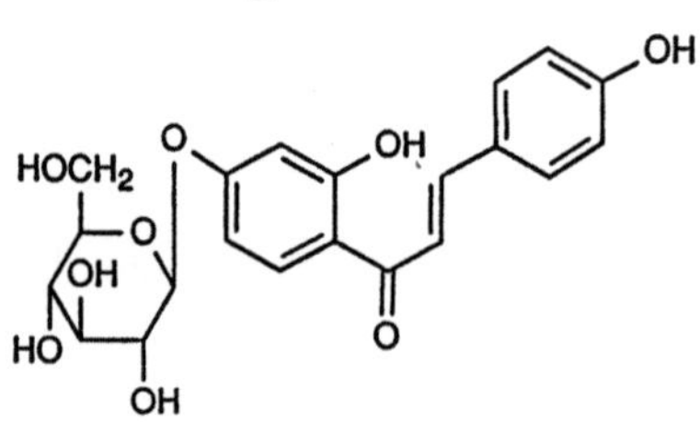	*G. glabra* *G. uralensis*	[61] [60]
Neoliquiritin (*74-32*)	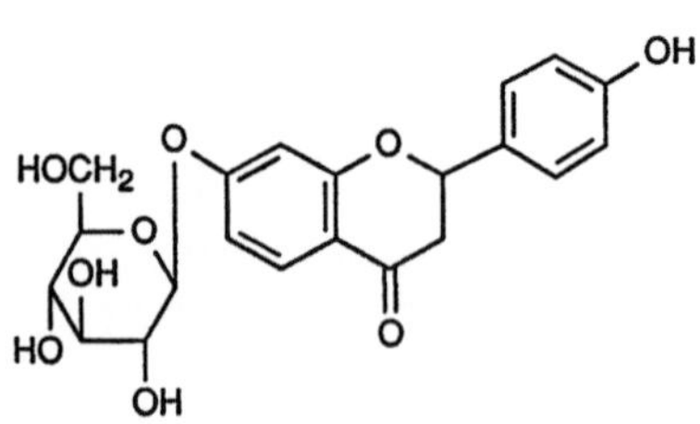	*G. glabra* *G. uralensis*	[63] [60]

Table 74.2. (continued)

Compound	Structure	Plant origin	Ref.
Licuroside (*74-33*)	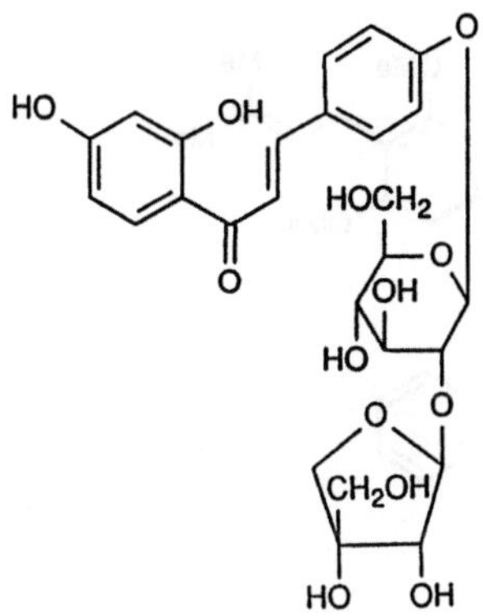	*G. glabra* *G. uralensis*	[64, 65] [60, 64]
Saponaretin (*74-34*)	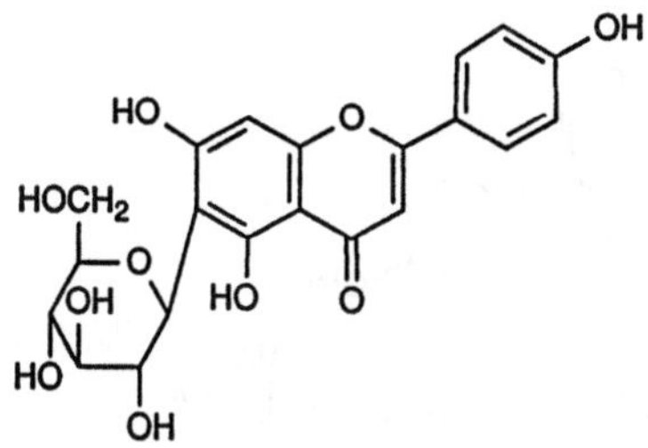	*G. glabra*	[66]
Vitexin (*74-35*)	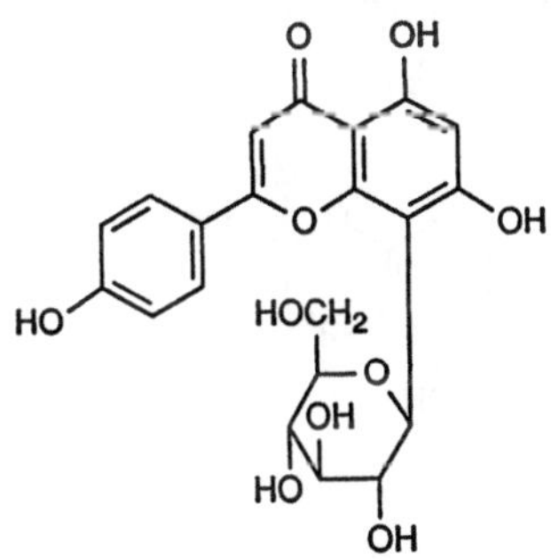	*G. glabra*	[66]
Pinocembrin (*74-36*)	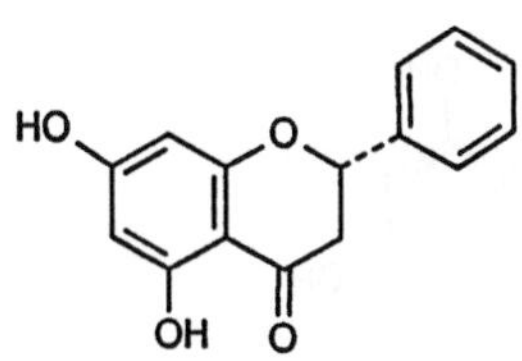	*G. glabra*	[67]
Prunetin (*74-37*)	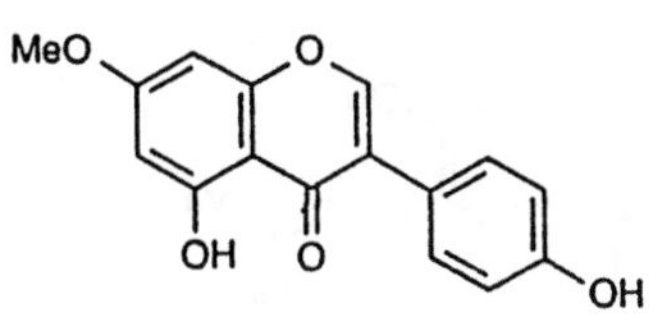	*G. glabra*	[67]

Table 74.2. (continued)

Compound	Structure	Plant origin	Ref.
Licoricone (*74-38*)	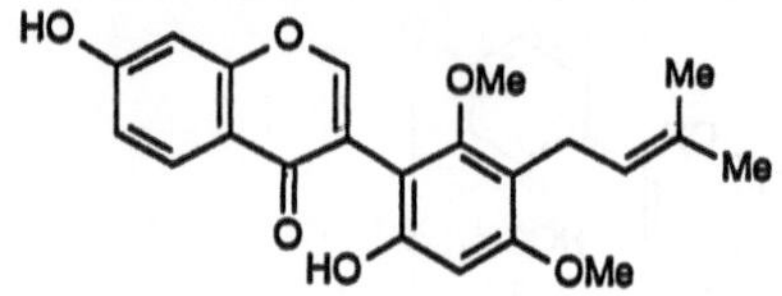	*G. uralensis*	[68, 69]
Glabranin (*74-39*)	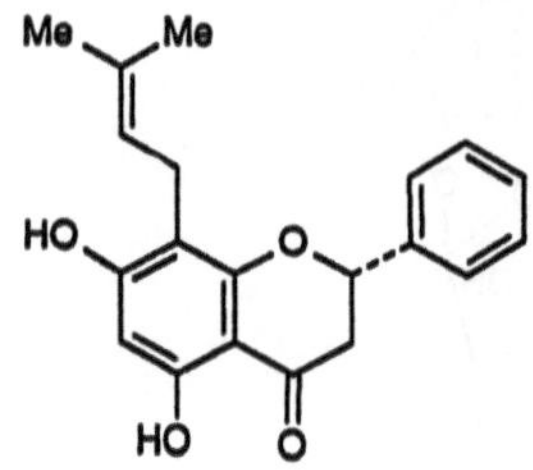	*G. glabra*	[70]
Formononetin (*74-40*)	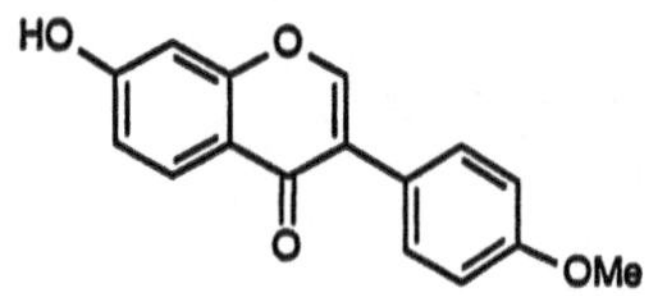	*G. glabra*	[71]
Glabrone (*74-41*)	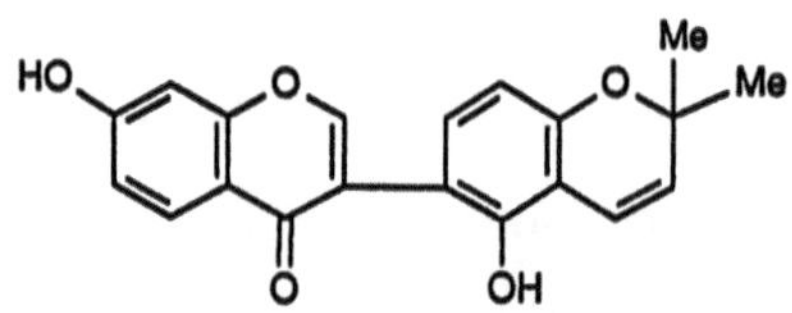	*G. glabra*	[72]
Glabrene (*74-42*)	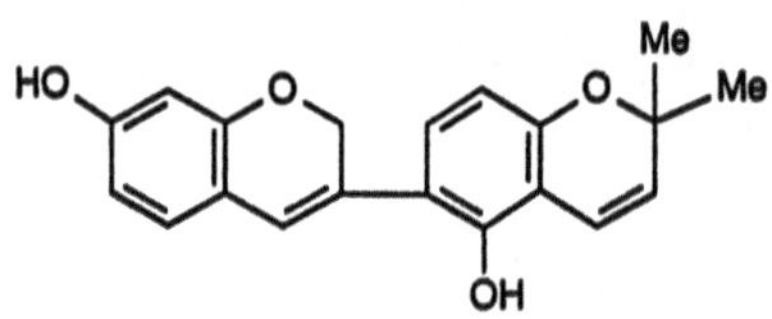	*G. glabra*	[72]
Glabridin (*74-43*)	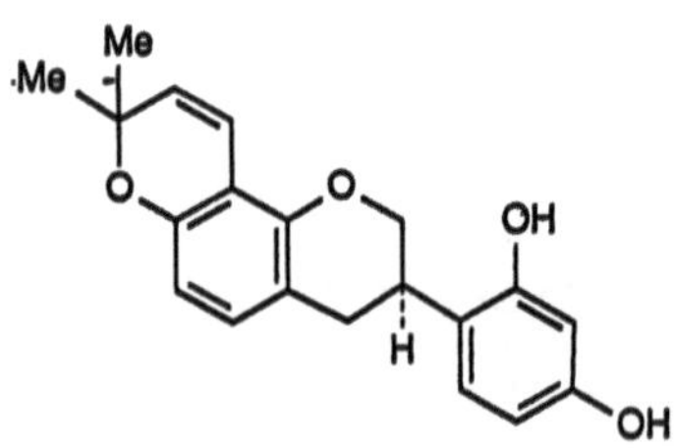	*G. glabra*	[73]
Glabrol (*74-44*)	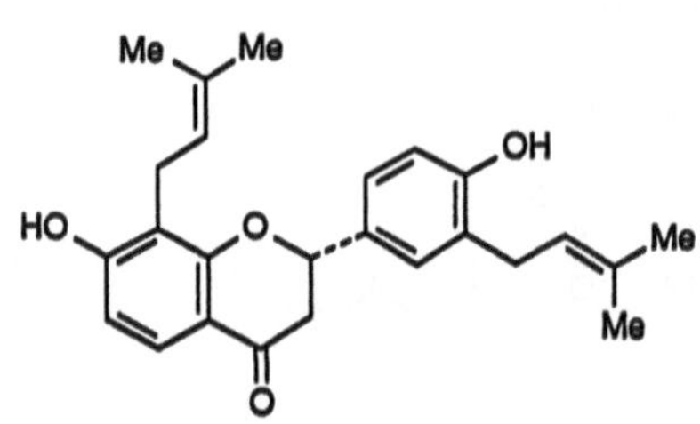	*G. glabra*	[73]

Table 74.2. (continued)

Compound	Structure	Plant origin	Ref.
7-Acetoxy-2-methyl-isoflavone (*74-45*)	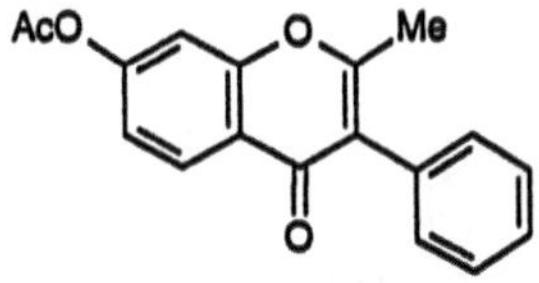	*G. glabra*	[74]
7-Methoxy-2-methyl-isoflavone (*74-46*)	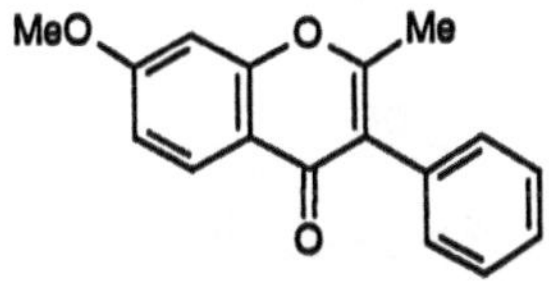	*G. glabra*	[74]
7-Hydroxy-2-methyl-isoflavone (*74-47*)	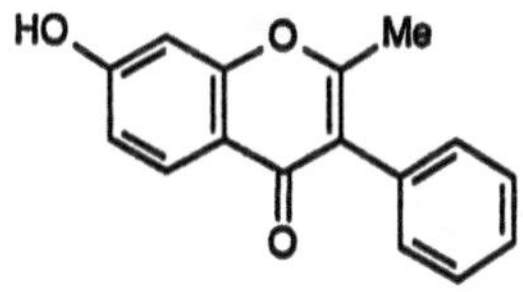	*G. glabra*	[74]
Licochalcone A (*74-48*)	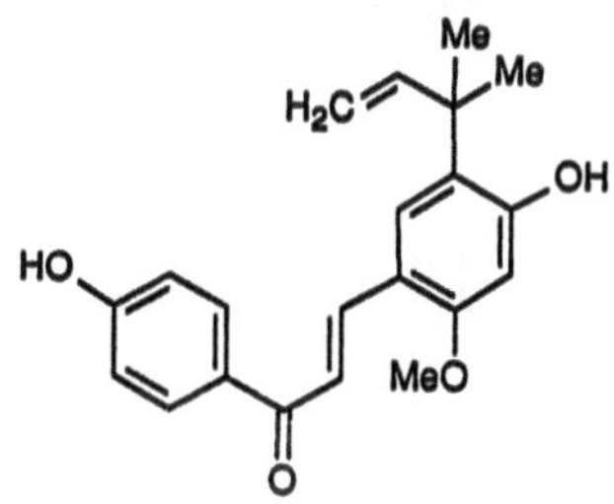	*G. glabra* *G. inflata*	[75] [76]
Licochalcone B (*74-49*)	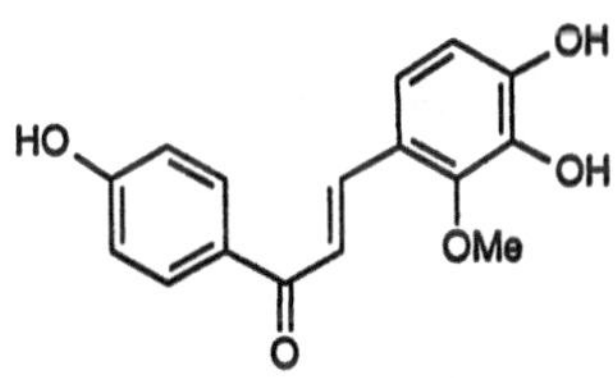	*G. glabra*	[75]
4-Hydroxychalcone (*74-50*)	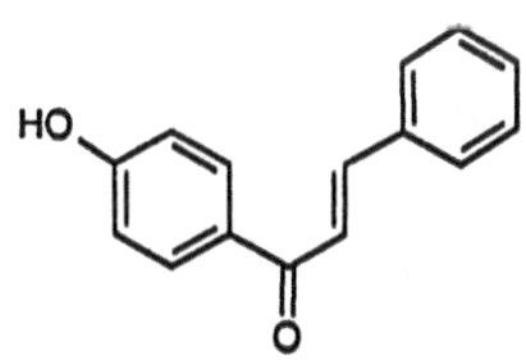	*G. glabra*	[77]
Liqcoumarin (*74-51*)	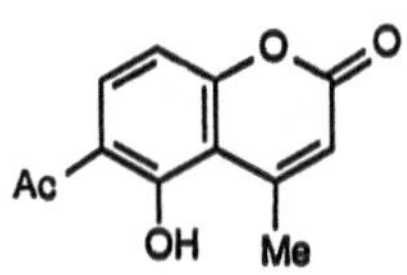	*G. glabra*	[78]
Glycyrol (*74-52*)	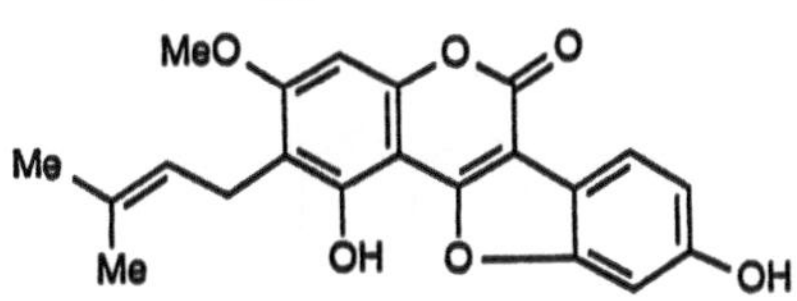	*G. uralensis*	[79, 80]

Table 74.2. (continued)

Compound	Structure	Plant origin	Ref.
Isoglycyrol (*74-53*)	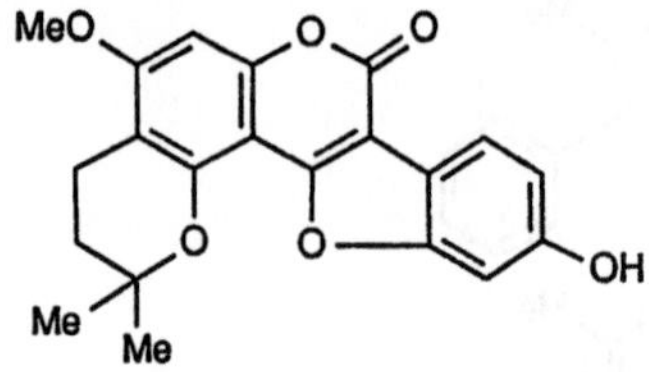	*G. uralensis*	[79, 80]
Isolicoflavonol (*74-54*)	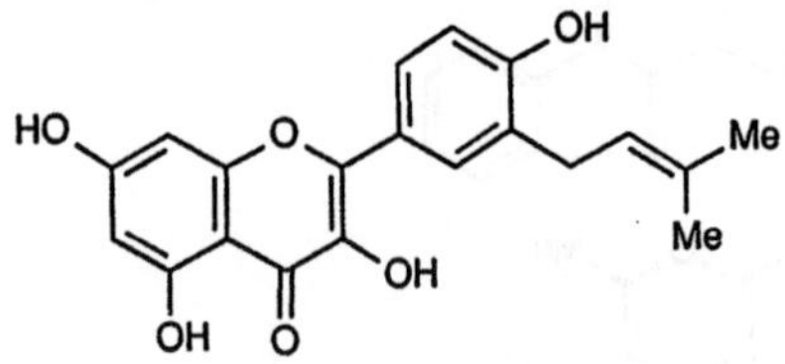	*G. uralensis*	[79]
Glycycoumarin (*74-55*)	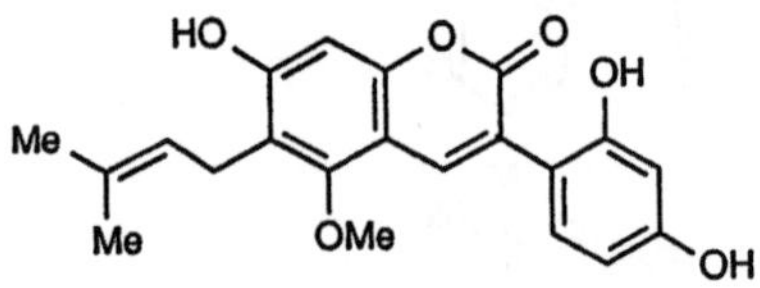	*G. uralensis*	[79]
Licoricidin (*74-56*)	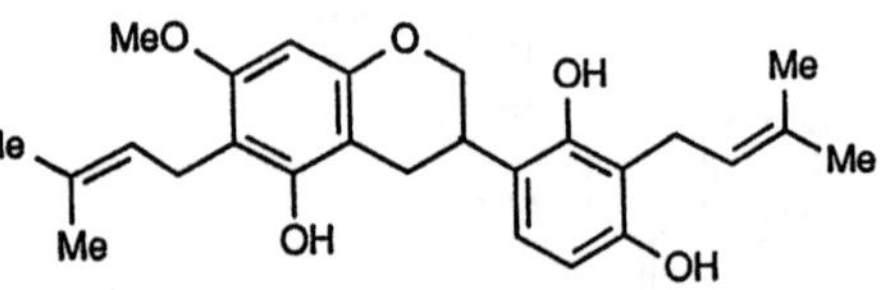	*G. uralensis*	[81, 82]
Liconeolignan (*74-57*)	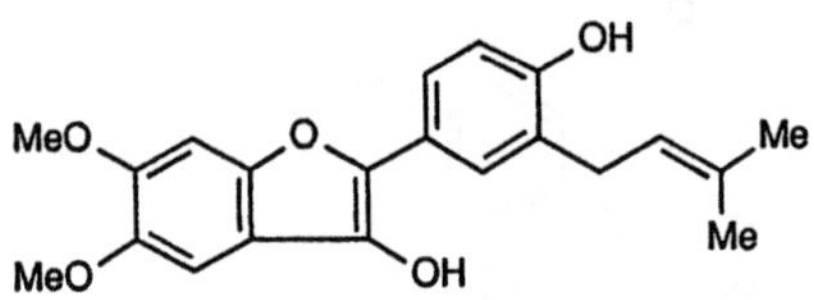	*G. uralensis*	[81, 82]
4′-*O*-*β*-D-apio-D-furanosyl-(1→2)-*β*-D-glucopyranosyl liquiritigenin (*74-58*)	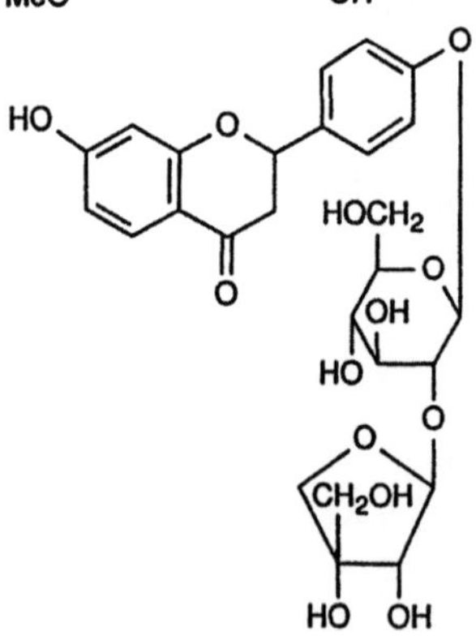	*G. uralensis*	[84]
Ononin (*74-59*)	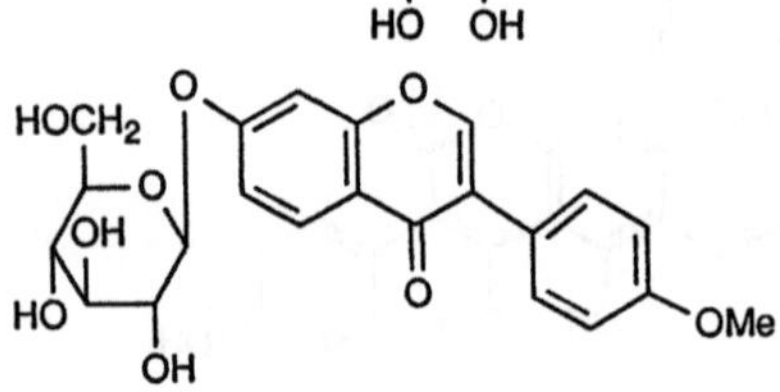	*G. uralensis*	[84]

Table 74.2. (continued)

Compound	Structure	Plant origin	Ref.
Neolicuroside (*74-60*)	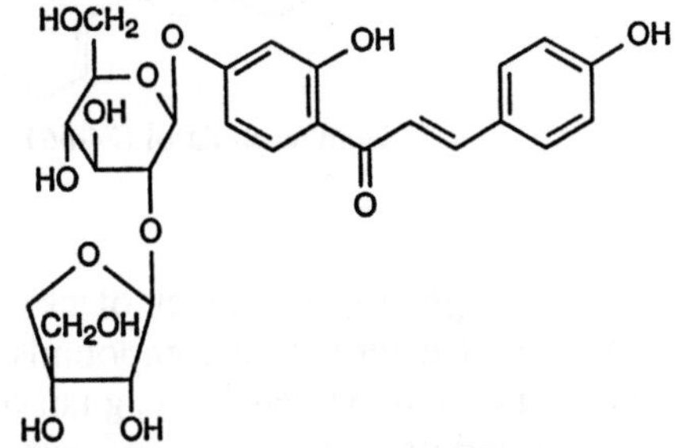	*G. glabra*	[85]
Licoflavanone (*74-61*)	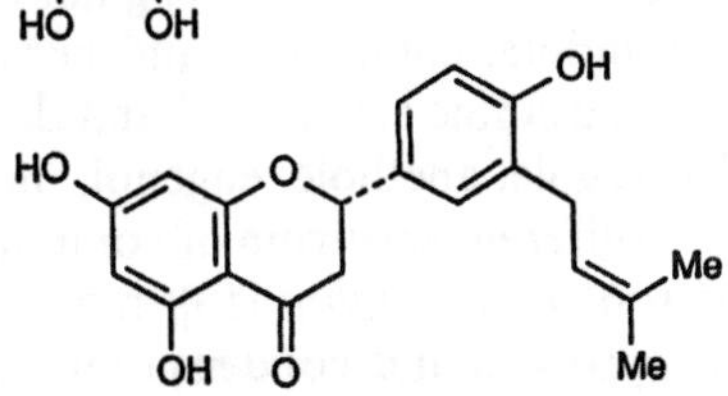	*G. glabra*	[86]
Glycyrrhiso-flavone (*74-62*)	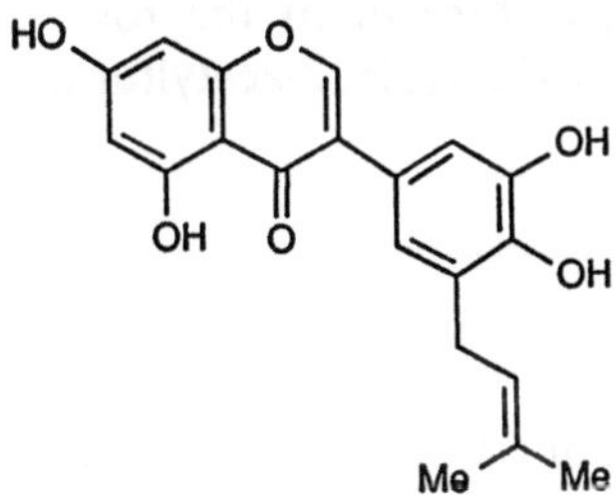	*Glycyr-rhiza* sp.	[87]
Glycyrrhiso-flavanone (*74-63*)	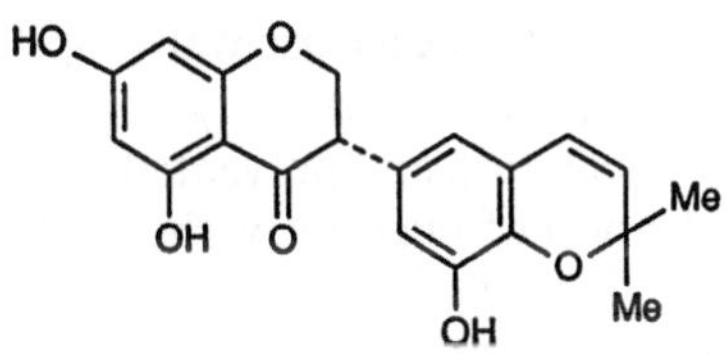	*Glycyr-rhiza* sp.	[87]
Licocoumarone (*74-64*)	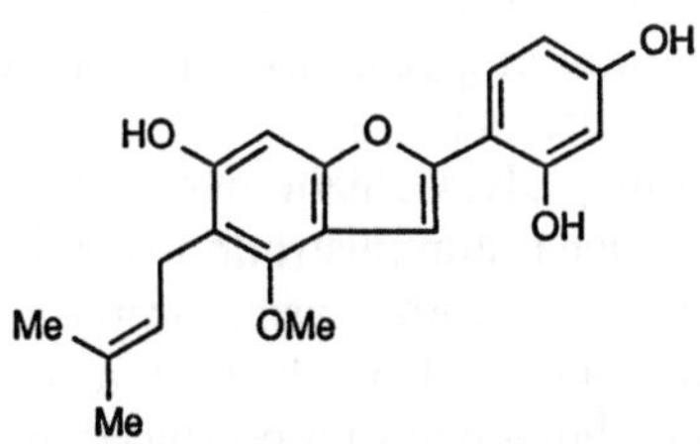	*Glycir-rhiza* sp.	[88]

Flavones and related compounds such as quercetin, isoquercitrin, kaempferol, astragalin, astragalin monoacetate, isorhamnetin, saponaretin, liquiritigenin, isoliquiritigenin, genkwanin, folerogenin (*74-65*) [11], and isomucronulatol (*74-66*) [12] were isolated and identified from the leaves and aerial parts of *G. glabra*. Folerogenin is the first flavanonol of a *cis* configuration found in nature [11].

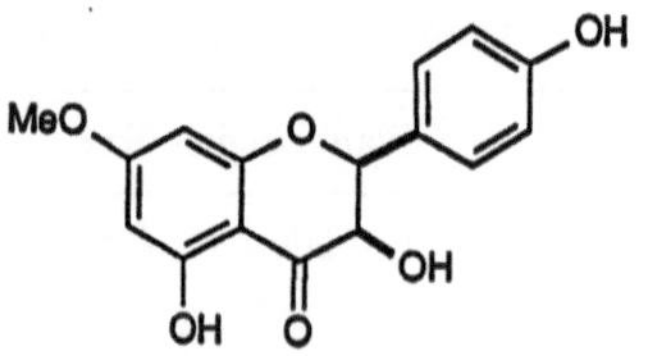

Folerogenin (*74-65*)

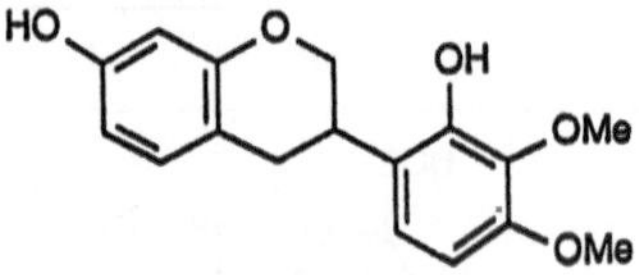

Isomucronulatol (*74-66*)

From the volatile fraction of *G. glabra*, a number of esters, alcohols, ethers, hydrocarbons, ketones, phenols, and heterocyclic compounds such as γ-nonalactone, linalool, α-terpineol, *p*-cymene, thujone, fenchone, guaiacol, thymol, geraniol [14], eugenol, estragole, anethole, indole, cumic acid and hexanoic acid [15] were detected. The main component was hexanoic acid (32%) and the main characteristic aroma was due to a mixture of estragole, anethole, eugenol, indole, γ-nonalactone, and cumic acid [15]. In addition, a different spectrum of compounds were identified in the essential oil from heated root of *G. glabra* [13], many of which are furan derivatives. This may be due to pyrolysis and condensation reactions which may occur during heating between sugars present in the root. The most abundant components are propionic acid, 2-acetylpyrrole, 2-acetylfuran, and furfuryl alcohol [13].

74.3 Pharmacology

The glycyrrhizin-containing fraction of *Glycyrrhiza* root showed a significant therapeutic effect at oral doses of 200 and 400 mg/kg on chronic gastric ulcers in rats induced by the injection of acetic acid into the gastric wall [16]. The glycyrrhizin-containing fraction and the deglycyrrhizinized fraction, given intraduodenally to pylorus-ligated rats, inhibited gastric secretion and prevented the formation of ulcer by aspirin administration. Administration of pure glycyrrhizin showd no effect on the ulcer. The antiulcer activity of the deglycyrrhizinized fraction was less than that of the glycyrrhizin-containing fraction, indicating that, although glycyrrhizin itself had no antiulcer activity, the combination of glycyrrhizin and some factors in the fraction had a synergistic effect on the ulcers [17].

The methanolic extract containing glycyrrhizin given intraduodenally inhibited gastric secretion in rats. Intraduodenal administration of this extract in different doses to dogs resulted in increases of both plasma secretin concentrations and pancreatic bicarbonate secretion in a dose-dependent manner. The plasma secretin concentrations and pancreatic bicarbonate output produced by the extract correlated with each other [18].

Glycyrrhetic acid showed mineralocorticoid-like effects [19]. The direct mineralocorticoid effect of *Glycyrrhiza* root is probably due to the affinity of glycyrrhetic acid for kidney aldosterone receptors. The relative low affinity of glycyrrhetic acid for mineralocorticoid receptors is in good agreement with the high doses of *Glycyrrhiza* root required for the development of hypertension [20, 21].

In addition, glycyrrhetic acid inhibited 5β-reduction of cortisol, aldosterone, and testosterone by rat liver preparations in vitro. Since 5β-reductase is involved in cortisol and aldosterone metabolisms, glycyrrhetic acid may delay clearance of corticosteroids and prolong their biological effects in the body [22]. Urinary cortisol excretion by normal volunteers given *Glycyrrhiza* extract was elevated [23].

Glycyrrhizin at a concentration of 8 m*M* completely inhibited growth and the cytopathic effects of vaccinia, herpes simplex type 1, Newcastle disease, and vesicular stomatitis viruses in cultures of human aneuploid HEP2 cells [24]. At the same concentration, glycyrrhizin inactivated herpes simplex virus irreversibly [24, 25]. The action mechanism of glycyrrhizin is probably based on the interaction with sensitive virus proteins at the virionic stage and during a later phase when these proteins are synthesized in host cells [24]. Antiviral activity of glycyrrhizin against varicella zoster virus in human embryonic fibroblast cells in vitro was also described and ascribed to inhibition of penetration, uncoating, or release of virus particle [26].

Glycyrrhizin was found completely to inhibit human immunodeficiency virus-induced plaque formation in MT-4 cells at the concentration of 0.5 mg/ml. The ID_{50} was 0.125 mg/ml. It also completely inhibited the human immunodeficiency virus-induced cytoplasmic effect and virus-specific antigen expression in MT-4 cells [27]. In comparison to glycyrrhizin, glycyrrhizin sulfate showed a stronger activity against human immunodeficiency virus [28].

Furthermore, glycyrrhizin showed an antiallergic activity [29]. It inhibited the passive cutaneous anaphylaxis response in rats and concentration dependently inhibited the contraction of rabbit ileum and guinea pig trachea induced by histamine, acetylcholine, or slow-reacting substances of anaphylaxis [30]. Ammonium glycyrrhizinate inhibited PGE_2 and $PGF_{2\alpha}$ formation by mouse lung and kidney in vivo and in vitro [31].

Glycyrrhizin and glycyrrhetic acid were able to prevent the development of experimental cirrhosis. In rats intoxicated with CCl_4, the elevation of serum glutamic oxalacetic transaminase and the accumulation of triglyceride in the liver were decreased. Histopathological studies revealed that liver lesions in rats induced by CCl_4 were less severe in glycyrrhizin- and glycyrrhetic acid-treated animals than in controls. The liver glycogen level in treated rats was markedly decreased [32]. Glycyrrhizin significantly increased the weight of the spleen and thymus of mice and also increased the leukocyte count and the clearance rate of charcoal particles [33]. The therapeutic significance of glycyrrhizin alone or in combination with methionine against acute or chronic liver injury induced by CCL_4 or D-galactosamine was also described [34–36]. The combined use of glycyrrhizin and methionine had greater therapeutic effects than glycyrrhizin alone.

The LD_{50} values of a crude extract from *Glycyrrhiza* root containing about 50% glycyrrhizin were 1.4–1.7 g/kg by intraperitoneal and 14–18 g/kg by oral administration in rats and mice. Rats given orally a daily dose of 2.5 g/kg for 3 months showed decreased body weight gain, blood cell count, and thymus weight. Atrophic cortex and sporadic lymphofollicle formation were noted in the medulla of the thymus. All changes disappeared after discontinuation of *Glycyrrhiza* extract administration. Oral administration of 0.3–0.6 g/kg for 90 days in rats had no toxic effect [37]. Rats exposed to a dietary level of 4% ammonium salt of glycyrrhizin, corresponding to a daily dose of 2.6 g/kg for 4–6 months, produced hypertension, increase in relative weights of kidney, and a slight decrease in body weight and growth [38].

Glycyrrhizin was absorbed from rat small intestine according to an apparent first-order process. After oral administration of glycyrrhizin, glycyrrhetic acid was detected in blood. Since glycyrrhetic acid was not detectable in blood when glycyrrhizin was injected into the portal vein, glycyrrhizin was probably absorbed in the small intestine in the form of glycyrrhetic acid. With the decline of glycyrrhetic acid in the blood, there was a rise in the blood level of a substance which appeared to be a glucuronic acid conjugate. Glycyrrhizin injected into the portal vein was eliminated from the blood slowly [39]. After intravenous injection of glycyrrhizin into normal subjects, its serum concentration decreased rapidly for 6 h, its metabolite, glycyrrhetic acid, appeared in the serum at about 6 h after administration and increased gradually, reaching a maximum at 24 h. The patterns of the glycyrrhetic acid increase after oral administration of glycyrrhizin were similar, except that one additional peak was observed between 1 and 4 h [40].

A glycyrrhetic acid derivative, the hemisuccinate of glycyrrhetic acid named carbenoxolone (*74-67*) was used clinically for ulcus treatment [41].

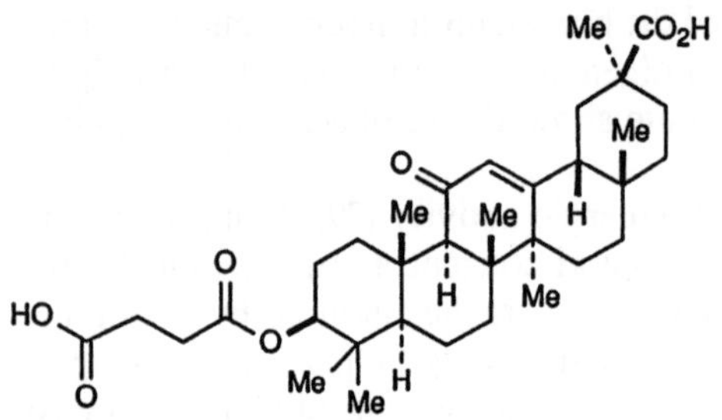

Carbenoxolone (*74-67*)

The flavones liquiritigenin, liquiritin, licuroside, and the total flavones from *G. glabra* and *G. uralensis* administered to rats inhibited the passage of $BaSO_4$ suspension from the stomach into the intestine by reducing stomach motility. They showed a spasmolytic action on isolated rat and guinea pig ileum preparations constricted by $BaCl_2$, acetylcholine, and histamine. In addition, an inhibitory effect on the development of exudative processes in inflammation caused by formalin, albumin, and dextran in rats was also described [42].

A polysaccharide from the root of *G. uralensis* was observed to have mitogenic activity on murine spleen cell proliferation in vitro. This mitogenic activity was directed to B lymphocytes. B lymphocytes stimulated by this polysaccharide were macrophage dependent and T cell independent. On the contrary, production of antibodies and interleukin-2 was inhibited by this polysaccharide in vivo [43].

References

1. Liu LJ, Pan JY, Yin XH, Lu YS (1985) Quantitative determination of glycyrrhizin in *Glycyrrhiza* and its preparations by macro-reticular resin. Chin Trad Herb Drugs 16:202–206
2. Shu YH, Xu GX, Tang WJ (1986) Determination of glycyrrhizic acid in *Glycyrrhiza uralensis* Fisch. and *G. inflata* Batal. by TLC-UV. Chin Trad Herb Drugs 17:202–203
3. Karrer W (1976) Konstitution und Vorkommen der organischen Pflanzenstoffe, 2nd ed. Birkhäuser, Basel

4. Gorup-Besanez E v (1861) Zur Kenntnis des Glycyrrhizins. Liebigs Ann Chem 118:236–247
5. Tschirch A, Cederberg H (1907) On glycyrrhizin. Arch Pharm (Weinheim) 245:97–111
6. Ruzicka L, Cohen SL (1937) Polyterpene und Polyterpenoide CXIII, Oxydation in der Reihe der Oleanolsäure ohne Sprengung des Ringsystems. Über die Natur des vierten Sauerstoffatoms der Glycyrrhetinsäure. Helv Chim Acta 20:804–808
7. Ruzicka L, Jeger O (1942) Zur Kenntnis der Triterpene. 67. Mitteilung: zur Lage der Carboxylgruppe bei der Glycyrrhetinsäure. Helv Chim Acta 25:775–785
8. Ruzicka L, Jeger O, Ingold W (1943) Zur Kenntnis der Triterpene. 84. Mitteilung: neuer Beweis für die verschiedene Lage der Carboxylgruppe bei der Oleanolsäure und der Glycyrrhetinsäure. Helv Chim Acta 26:2278–2282
9. Beaton JM, Spring FS (1955) Triterpenoids Part XLII. The configuration of the carboxyl group in glycyrrhetic acid. J Chem Soc 3126–3129
10. Lythgoe B, Trippett S (1950) The constitution of the disaccharide of glycyrrhizic acid. J Chem Soc 1983–1991
11. Litvinenko VI, Nadezhina TP (1972) Flavonoids from the above-ground part of *Glycyrrhiza glabra*. Rast Resur 8:35–42 (CA 76:124123f)
12. Ingham JL (1977) An isoflavan phytoalexin from leaves of *Glycyrrhiza glabra*. Phytochemistry 16:1457–1458
13. Frattini C, Bicchi C, Barettini C, Nano GM (1977) Volatile flavor components of licorice. J Agric Food Chem 25:1238–1241
14. Toulemonde B, Mazza M, Bricout J (1977) Composition of the aroma of *Glycyrrhiza glabra* L. rhizoma. Ind Aliment Agric 94:1179–1182
15. Kameoka H, Nakai K (1987) Components of essential oil from the root of *Glycyrrhiza glabra*. Nippon Nogei Kagaku Kaishi 61:1119–1121 (CA 108:62302g)
16. Takagi K, Okabe S, Kawashima K, Hirai T (1971) Therapeutic effect of FM 100, a fraction of licorice root, on acetic ulcer in rats. Jpn J Pharmacol 21:832–833
17. Okabe S, Kunimi H, Nosaka A, Ishii Y, Fujii Y, Nakamura K (1979) Effects of FM100, FM100-deG and glycyrrhizin on gastric secretion and experimental gastric ulcers in rats. Oyo Yakuri 18:469–474 (CA 92:157697e)
18. Shin I, Chey WY, Lee KY, Chang TM (1986) Release of secretin by licorice extract in dogs. Pancreas 1:449–454
19. Kumagai A (1973) Hormone-like action of licorice and glycyrrhizin. Taisha 10:632–645 (CA 81:130969f)
20. Corval P, Ulmann A, Michaud A, Menard J (1975) Mineralcorticoid-like effect of glycyrrhetinic acid. Res Steroids 6:65–75
21. Tamaya T, Sato S, Okada HH (1986) Possible mechanism of steroid action of the plant herb extracts glycyrrhizin, glycyrrhetinic acid and paeoniflorin: inhibition by plant herb extract of steroid protein binding in the rabbit. Am J Obstet Gynecol 155:1134–1139
22. Tamura Y, Nishikawa T, Yamada K, Yamamoto M, Kumagai A (1979) Effects of glycyrrhetinic acid and its derivatives on Δ^4-5α- and 5β-reductase in rat liver. Arzneimittelforschung 29:647–649
23. Epstein MT, Espiner EA, Donald RA, Hughes H, Cowles RJ, Lun S (1978) Licorice raises urinary cortisol in man. J Clin Endocrinol Metab 47:397–400
24. Pompei R, Flore O, Marccialis MA, Pani A, Loddo B (1979) Glycyrrhizic acid inhibits virus growth and inactivates virus particles. Nature 281(5733):689–690
25. Pompei R (1979) Activity of *Glycyrrhiza glabra* extracts and glycyrrhizic acid on virus growth and infectivity. Riv Farmacol Ter 10:281–284 (CA 92:104967m)
26. Baba M, Shigeta S (1987) Antiviral activity of glycyrrhizin against varicella-zoster virus in vitro. Antiviral Res 7:99–107
27. Ito M, Nakashima H, Baba M, Pauwels R, DeClercq E, Shigeta S, Yamamoto N (1987) Inhibitory effect of glycyrrhizin on the in vitro infectivity and cytopathic activity of the human immunodeficiency virus [HIV (HTLV-III/LAV)]. Antiviral Res 7:127–137
28. Nakashima H, Matsui T, Yoshida O, Isowa Y, Kido Y, Motoki Y, Ito M, Shigeta S, Mori T, Yamamoto N (1987) A new anti-human immunodeficiency virus substance, glycyrrhizin sulfate: endowment of glycyrrhizin with reverse transcriptase-inhibitory activity by chemical modification. Jpn J Cancer Res (GANN) 78:767–771
29. Inoue H, Mori T, Shibata S, Saito H (1987) Pharmacological activities of glycyrrhetinic acid derivatives: analgesic and anti-type IV allergic effects. Chem Pharm Bull (Tokyo) 35:3888–3893

30. Zhang HQ, Liu F, Sun B, Li GH (1986) Antiallergic action of glycyrrhizin. Acta Pharmacol Sin 7:175–177
31. Ren J, Cheng JX, Wang ZG (1988) The effect of ammonium glycyrrhizinate on the biosynthesis of prostaglandins E_2 and $F_{2\alpha}$. Kexue Tongbao (Foreign language edn) 33:1561–1564
32. Zhao MQ, Han DW, Ma XH, Zhao YC, Yin L, Li CM (1983) Preventive and therapeutic effects of glycyrrhizin, glycyrrhetic acid and sikosides on experimental cirrhosis in rats. Acta Pharm Sin 18:325–331
33. Zhang HQ, Liu F, Zheng HQ, Li GH (1984) Some pharmacologic effect of glycyrrhizin in mice. Acta Pharm Sin 19:926–927
34. Fujita H, Sakurai T, Toyoshima S (1978) Studies on the regulation by drugs against experimental hepatitis. 1. The therapeutic effects of glycyrrhezinic acid, DL-methionine, their mixture and tablets against acute injury induced by carbon tetrachloride or D-galactosamine. Oyo Yakuri 16:637–645 (CA 91:13665z)
35. Fujita H, Sakurai T, Toyoshima S (1978) Study on the regulation by drugs against experimental hepatitis. 2. The therapeutic effects of glycyrrhizinic acid, DL-methionine and their complex against chronic liver injury induced by D-galactosamine. Oyo Yakuri 16:647–657 (CA 91:13666a)
36. Fujita H, Sakurai T, Toyoshima S (1978) Study on the regulation by drugs against experimental hepatitis. 3. The therapeutic effects of glycyrrhizinic acid, DL-methionine and their mixture against chronic liver injury induced by carbon tetrachloride in rats. Oyo Yakuri 16:659–669 (CA 91:13667b)
37. Komiyama K, Kawakubo Y, Fukushima T, Sugimoto K, Takeshima H, Ko Y, Sato T, Okamoto M, Umezawa I, Nishiyama Y (1977) Acute and subacute toxicity test on the extract from *Glycyrrhiza*. Oyo Yakuri 14:535–548 (CA 88:69205e)
38. Sobotka TJ, Spaid SL, Brodie RE, Reed GF (1981) Neurobehavioral toxicity of amminiated glycyrrhizin, a licorice component, in rats. Neurobehav Toxicol Teratol 3:37–44
39. Sakiya Y, Akada Y, Kawano S, Miyauchi Y (1979) Rapid estimation of glycyrrhizin and glycyrrhetic acid in plasma by high-speed liquid chromatography. Chem Pharm Bull (Tokyo) 27:1125–1129
40. Nakano N, Kato H, Suzuki H, Nakao K, Yano S, Kaneoka M (1980) Enzyme immunoassay for glycyrrhetic acid and glycyrrhizin. 2. Measurement of glycyrrhetic acid and glycyrrhizin in blood. Yakuri To Chiryo 8:4171–4173
41. Forth W, Henschler D, Rummel W (1983) Allgemeine und spezielle Pharmakologie und Toxikologie. BI-Wissenschaftsverlag, Mannheim, p 310
42. Obolentseva GV, Khadzhai YI (1966) Pharmacological data on certain aspects of the action of *Glycyrrhiza* flavonoids. Vop Izuch Ispol's Solodki Nauka 163–166 (CA 67:107314r)
43. Shi Y, Yang GZ (1986) Effect of GPS on stimulating murine lymphocyte proliferation. Chin J Immunol 2:295–300
44. Canonica L, Russo G, Bombardelli E (1966) Triterpenes of *Glycyrrhiza glabra*. II. Liquiritic acid. Gazz Chim Ital 96:832–842
45. Canonica L, Russo G, Bonati A (1966) Triterpenes of *Glycyrrhiza glabra*. I. Two new lactones with an oleanane structure. Gazz Chim Ital 96:772–785
46. Shu YH, Zhao YY, Zhang RY (1985) Isolation and structural identification of triterpene sapogenins from *Glycyrrhiza uralensis* Fisch. Acta Pharm Sin 20:193–197
47. Canonica L, Danielli B, Manitto P, Russo G (1966) Triterpenes of *Glycyrrhiza glabra*. III. Structure of isoglabrolide. Gazz Chim Ital 96:843–851
48. Beaton JM, Spring FS (1956) Triterpenoids Part LI. The isolation and characterisation of glabric acid, a new triterpenoid acid from liquorice root. J Chem Soc 2417–2419
49. Elgamal MHA, Fayer MBE (1968) Structure of glabric acid, a further triterpenoid constituent of *Glycyrrhiza glabra*. Acta Chim (Budapest) 58:75–84
50. Canonica L, Danielli B, Russo G, Bonati A (1967) Triterpenes from *Glycyrrhiza glabra* IV. 18α-Hydroxyglycyrrhetic acid. Gazz Chim Ital 97:769–786
51. Shu YH, Zhang RY, Zhao YY, Zhang JW, Tong WD (1987) Isolation and structure determination of new triterpene sapogenins from *Glycyrrhiza uralensis* Fisch. Acta Pharm Sin 22:512–514
52. Canonica L, Danielli B, Manitto P, Russo G, Bombardelli E (1967) *Glycyrrhiza glabra* triterpenes V. Glycyrrhetol and 21α-hydroxyisoglabrolide. Gazz Chim Ital 97:1347–1358
53. Canonica L, Danielli B, Manitto P, Russo G, Bonati A, Bombardelli E (1967) *Glycyrrhiza glabra* triterpenes VI. 23-Hydroxyglycyrrhetic acid and 24-hydroxy-11-deoxyglycyrrhetic acid. Gazz Chim Ital 97:1359–1369

54. Canonica L, Danielli B, Manitto P, Russo G, Bombardelli E, Bonati A (1968) *Glycyrrhiza glabra* triterpenes VII. 24-Hydroxyliquiritic acid (3β,24-dihydroxy-11-oxo-olean-12-en-29-oic acid) and liquiridolic acid (3β,21α,24-trihydroxy-olean-12-en-29-oic acid). Gazz Chim Ital 98:712–728
55. Zhang RY, Zhang JH, Wang MT (1986) Saponins from the root of *Glycyrrhiza uralensis* Fisch. Acta Pharm Sin 21:510–515
56. Kitagawa I, Zhou JL, Sakagami M, Taniyama T, Yoshikawa M (1988) Licorice saponins A3, B2, C2, D3 and E2, five new oleanene-type triterpene oligoglycoside from Chinese glycyrrhizae radix. Chem Pharm Bull (Tokyo) 36:3710–3713
57. Kitagawa I, Sakagami M, Hashiuchi F, Zhou JL, Yoshikawa M, Ren TL (1989) Apioglycyrrhizin and araboglycyrrhizin, two new sweet oleanene-type triterpene oligoglycosides from the root of *Glycyrrhiza inflata*. Chem Pharm Bull (Tokyo) 37:551–553
58. Shinoda J, Ueeda S (1934) Über das Flavanonglucosid in *Glycyrrhiza glabra* L. var. *glandulifera* Regel et Herder. Chem Ber 67:434–440
59. Paris RA, Guillot M (1955) Liquiritoside, flavonoid of the root of *Glycyrrhiza glabra*. Ann Pharm Fr 13:592–595
60. Litvinenko VI (1963) Flavonoid compounds of *Glycyrrhiza uralensis*. Farmatsevt Zh (Kiev) 18(5):20–25 (CA 60:6700g)
61. Litvinenko VI, Maksyutina NP, Kolesnikov IG (1963) Flavonoid compounds of *Glycyrrhiza glabra*. Zh Obshch Khim 33:296–299
62. Puri B, Seshadri TR (1954) Survey of anthoxanthins. V. Coloring matter of licorice roots. J Sci Ind Res 13B:475–480
63. Litvinenko VI, Maksyutina NP, Kolesnikov IG (1963) Flavonoidal compounds of *Glycyrrhiza glabra*. I. Flavonoid L-1. Zh Obshch Khim 33:4014–4018
64. Litvinenko VI, Obolentseva GV (1964) Chemical and pharmacological investigations of flavonoids from licorice (*Glycyrrhiza glabra* and *G. uralensis*). Med Prom SSSR 18(10):20–23 (CA 62:8286b)
65. Litvinenko VI, Kovalev IP (1966) Licuroside, a new flavonoid glycoside of *Glycyrrhiza glabra*. Dokl Akad Nauk SSSR 169:347–350
66. Litvinenko VI, Kovalev IP (1967) Glycoflavonoids of *Glycyrrhiza glabra*. I. Saponaretin and vitexin. Khim Prir Soedin 10:56–57
67. Kattaev NS, Nikonov GK (1974) Flavonoids of *Glycyrrhiza glabra*. Khim Prir Soedin 10:93
68. Kaneda M, Saitoh T, Iitaka Y, Shibata S (1973) Chemical studies on the oriental plant drugs. XXXVI. Structure of licoricone, an isoflavone from licorice root. Chem Pharm Bull (Tokyo) 21:1338–1341
69. Kaneda M, Iitaka Y, Shibata S (1973) Crystal and molecular structure of licoricone monobromoacetate. Acta Crystallogr [B] 29:2827–2832
70. Kattaev NS, Nikonov GK (1972) Glabranin, a new flavanone from *Glycyrrhiza glabra*. Khim Prir Soedin 805–806
71. Elgamal MHA, Fayez MBE (1972) Isolation of formononetin from the roots of *Glycyrrhiza glabra* collected locally. Indian J Chem 10:128
72. Kinoshita T, Saitoh T, Shibata S (1976) Chemical studies on the oriental plant drugs. XL. The occurrence of an isoflavane and the corresponding isoflavone in licorice root. Chem Pharm Bull (Tokyo) 24:991–994
73. Saitoh T, Kinoshita T, Shibata S (1976) Chemical studies on oriental plant drugs. XXXIX. New isoflavane and flavanone from licorice root. Chem Pharm Bull (Tokyo) 24:752–755
74. Bharadwaj DK, Murari R, Seshadri TR, Singh R (1976) Occurrence of 2-methylisoflavones in *Glycyrrhiza glabra*. Phytochemistry 15:352–353
75. Saitoh T, Shibata S (1975) New type chalcone from licorice root. Tetrahedron Lett 4461–4462
76. Xu RS, Wen KL, Jiang SF, Wang CG, Jiang FX, Xie YY, Gao YS (1978) Isolation, structure and total synthesis of licochalcone. Acta Chim Sin 37:289–297
77. Hoton-Dorge M (1974) Identification of some flavonoid aglycone extracts of *Glycyrrhiza glabra* roots. J Pharm Belg 29:560–572
78. Bhardwaj DK, Murari R, Seshadri T, Singh R (1976) Liqcoumarin, a novel coumarin from *Glycyrrhiza glabra*. Phytochemistry 15:1182–1183
79. Zhu DY, Song GQ, Jian FX, Chang XR, Guo WB (1984) Chemical constituents of *Glycyrrhiza uralensis* Fisch. – Structures of isolicoflavonal and glycycoumarin. Acta Chim Sin 42:1080–1084

80. Saitoh T, Shibata S (1969) Chemical studies on the oriental plant drugs. XXII. Some new constituents of licorice root. 2. Glycyrol, 5-*O*-methylglycyrol and isoglycyrol. Chem Pharm Bull (Tokyo) 17:729–734
81. Chang XR, Xu QH, Zhu DY, Song GQ, Xu RS (1983) Isolation and structural elucidation of liconeolignan from *Glycyrrhiza uralensis*. Acta Pharm Sin 18:45–50
82. Shibata S, Saitoh T (1968) Chemical studies on the oriental plant drugs. XIX. New constituents of licorice root. 1. Structure of licoricidin. Chem Pharm Bull (Tokyo) 16:1932–1936
83. Chang XR, Xu QH, Zhu DY, Song GJ, Xu RS (1981) Studies on the chemical constituents of *Glycyrrhiza*. I. Isolation and structural elucidation of the antibacterial constituent, licobenzofuran. Chin Trad Herb Drugs 12:530
84. Nakanishi T, Inada A, Kambayashi K, Yoneda K (1985) Flavonoid glycosides of the roots of *Glycyrrhiza uralensis*. Phytochemistry 24:339–341
85. Miething H, Speicher-Brinker A (1989) Neolicuroside, a new chalcone glycoside from the roots of *Glycyrrhiza glabra*. Arch Pharm (Weinheim) 322:141–143
86. Fukui H, Goto K, Tabata M (1988) Two antimicrobial flavanones from the leaves of *Glycyrrhiza glabra*. Chem Pharm Bull (Tokyo) 36:4174–4176
87. Hatano T, Kagawa H, Yasuhara T, Okuda T (1988) Two new flavonoids and other constituents in licorice root: their relative astringency and radical scavenging effects. Chem Pharm Bull (Tokyo) 36:2090–2097
88. Demizu S, Kajiyama K, Takahashi K, Hiraga Y, Yamamoto S, Tamura Y, Okada K, Kinoshita T (1988) Antioxidant and antimicrobial constituents of licorice: isolation and structure elucidation of a new benzofuran derivative. Chem Pharm Bull (Tokyo) 36:3474–3479

Houttuynia cordata Thunb. 75

75.1 Introduction

The whole plant of *Houttuynia cordata* Thunb. (Saururaceae) is a folk medicine in China used as an antibacterial agent especially against infections of the respiratory tract.

75.2 Chemical Constituents

Early studies on the chemical composition of the essential oil of *H. cordata* revealed that it contains aldehyde and ketone derivatives such as methyl *n*-nonyl ketone, 1-decanal, and 1-dodecanal. These compounds showed no antimicrobial activity [1]. An antimicrobial component was isolated from the essential oil of the rhizome of *H. cordata* and structurally determined as 3-oxododecanal (*75-1*) [2].

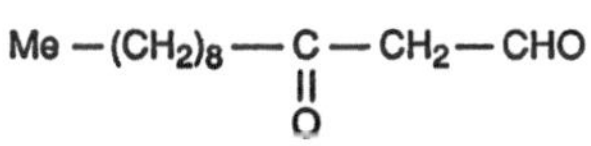

3-Oxododecanal (*75-1*)

Recently, Liu and Deng [3] reported the determination of a number of terpene and nonterpene derivatives from the essential oil of *H. cordata* from Guangdong province by GC-MS. Ten compounds were identified: the monoterpenes α-pinene, camphene, myrcene, and limonene, two oxygenated monoterpenes linalool and bornyl acetate, the sesquiterpene caryophyllene and methyl *n*-nonyl ketone, 3-oxododecanal and 1-dodecanal. This result is different from that obtained by Tutupalli and Chaubal [4], who studied the chemical composition of the essential oil isolated from *H. cordata* collected in Japan and found that the essential oil is very rich in lipids and is devoid of 3-oxododecanal.

From the benzene fraction of *H. cordata*, β-sitosterol, palmitic acid, linoleic acid, oleic acid, and stearic acid were detected [5]. In addition, the flavone glycosides afzelin (*75-2*), hyperin (*75-3*), rutin [5], isoquercitrin (*75-4*), and quercitrin [6] were isolated from *H. cordata*.

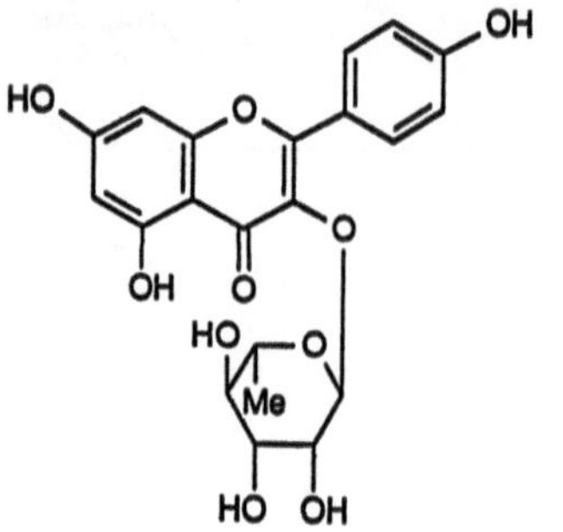

Afzelin (*75-2*)

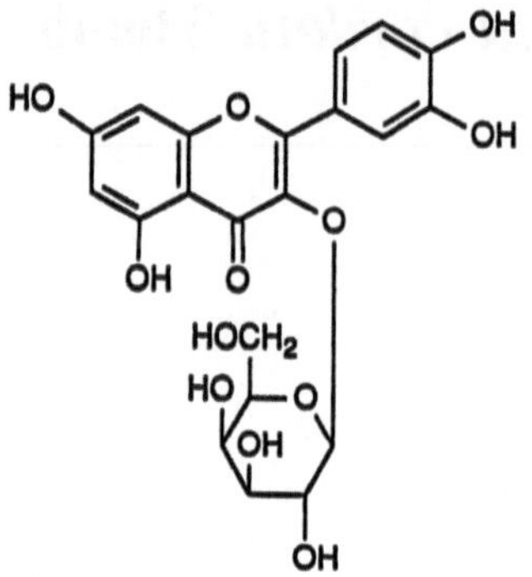

Hyperin (*75-3*)

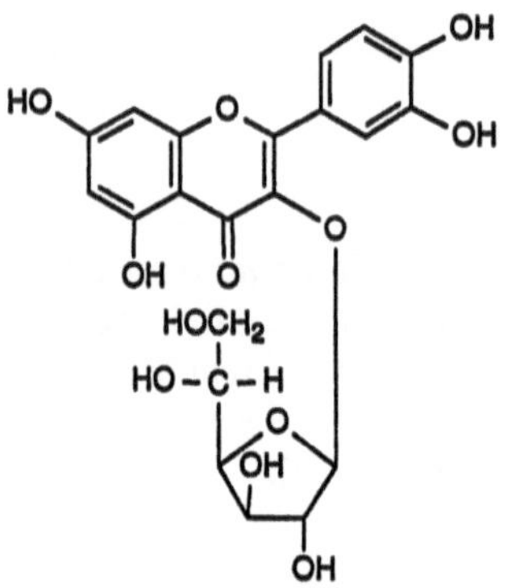

Isoquercitrin (*75-4*)

3-Oxododecanal can be synthesized by formylation of methyl *n*-nonyl ketone with ethyl formate in benzene in the presence of metallic sodium [7]. The sodium bisulfite compound of 3-oxododecanal, a crystalline substance, was also synthesized for medicinal use against some infectious diseases, because the free compound polymerizes very easily [2]. A number of hydrazine derivatives produced from 3-oxododecanal and substituted hydrazines were also synthesized and tested for antimicrobial activity. As an example, the bis(isonicotinic hydrazide) of 3-oxododecanal (*75-5*) was synthesized by treating 3-oxododecanal with isonicotinic hydrazide. It showed tuberculostatic activity [7].

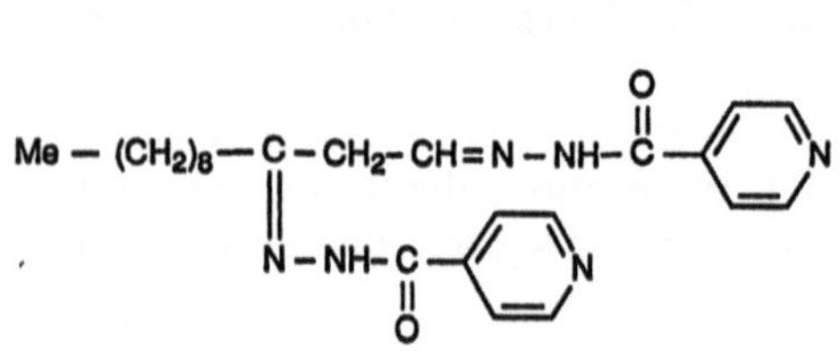

3-Oxododecanal bis(isonicotinic hydrazide) (*75-5*)

75.3 Pharmacology

A study on antimicrobial activity of some constituents of *H. cordata* showed that linalool exhibited inhibitory activity against *Staphylococcus aureus, Escherichia coli, Bacillus dysenteriae, Streptococcus pneumoniae*, and *S. pyrogenes*; α-pinene was active against *Staphylococcus aureus, Streptococcus pneumoniae*, and *S. pyrogenes*; 1-dodecanal showed no antimicrobial activity [3].

Rectal administration of 3-oxododecanal at a dose of 50 mg/suppository to rabbits for 11 days did not cause local irritation. A slight local irritation was observed when the dose was increased to 100–200 mg. 3-Oxododecanal was rapidly absorbed by rectal administration [8].

After intravenous administration of 3-oxododecanal sodium bisulfite to mice, 3-oxododecanal was found to be bound completely to plasma protein; free 3-oxododecanal was not detectable in the serum (M. Z. Xie 1978, personal communication).

References

1. Shinozaki Y (1921) Essential oil of "dokudame". J Chem Ind (Jpn) 24:557–562 (CA 16:990)
2. Kosuge T (1952) Structure of an antimicrobial substance isolated from *Houttuynia cordata* Thunb. J Pharm Soc Jpn 72:1227–1231
3. Liu YL, Deng ZF (1979) Investigation of the chemical constituents of the essential oil of *Houttuynia cordata* Thunb. Acta Bot Sin 21:244–249
4. Tutupalli LV, Chaubal MG (1975) Saururaceae. V. Composition of essential oil from foliage of *Houttuynia cordata* and chemosystematics of Saururaceae. Lloydia 38:92–96
5. Takagi S, Yamaki M, Masuda K, Kubota M (1978) On the constituents of the terrestrial part of *Houttuynia cordata* Thunb. Shoyakugaku Zasshi 32:123–125
6. Kimura Y, Nishikawa Y (1953) Standardization of crude drugs. III. 2. Component of *Houttuynia cordata*. J Pharm Soc Jpn 73:194–195
7. Zhao ZZ, Jiang XJ, Wang L, Wei Z (1979) Studies on antimicrobial and antiviral compounds – synthesis of derivatives of decanoyl acetaldehyde. Acta Pharm Sin 14:428–433
8. Li YM (1981) LD_{50} and local irritability studies of decanoyl acetaldehyde suppositories. Chin Trad Herb Drugs 12:516

Ilex pubescens Hook et Arn. 76

76.1 Introduction

Ilex pubescens Hook et Arn. (Aquifoliaceae), is a folk medicine in China used in the treatment of cardiovascular diseases and as an antiinflammatory and analgesic agent. It is not officially listed in the Chinese Pharmacopoeia.

76.2 Chemical Constitutents

Homovanillic acid (*76-1*), vomifoliol (*76-2*) [1–3], 3,4-dihydroxyacetophenone, scopoletin, esculetin [2, 3], and a new compound named glaberide I [2–4] were isolated from the leaves of *I. pubescens* var. *glaber*.

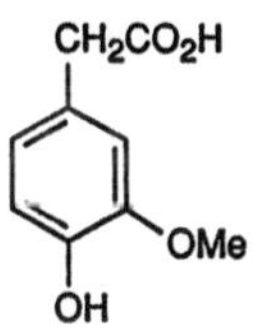

Homovanillic acid (*76-1*)

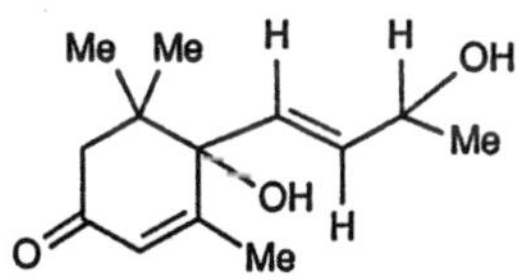

Vomifoliol (*76-2*)

The structure of glaberide I (*76-3*) was determined on the basis of spectral and X-ray analyses. It represents a condensed furofuranone derivative [2, 5].

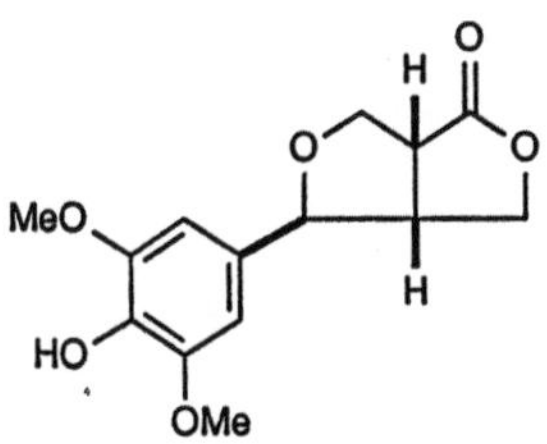

Glaberide I (*76-3*)

From the root of *I. pubescens*, a triterpene glycoside named ileoxolide A (*76-4*), containing the aglycon ilexalic acid connected to xylose, was isolated and structurally determined [6]. Recently, new saponins named ilexsaponin A (*76-5*) [7], ilexsaponin B_1 (*76-6*), ilexsaponine B_2 (*76-7*), and ilexsaponin B_3 (*76-8*) were isolated from the root of *I. pubescens.*

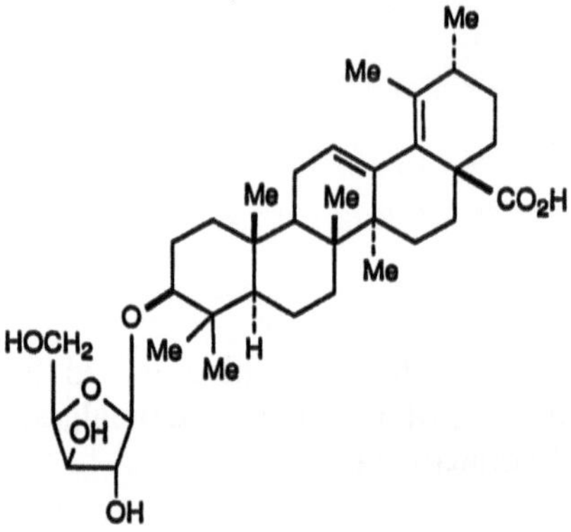

Ilexolide A (*76-4*)

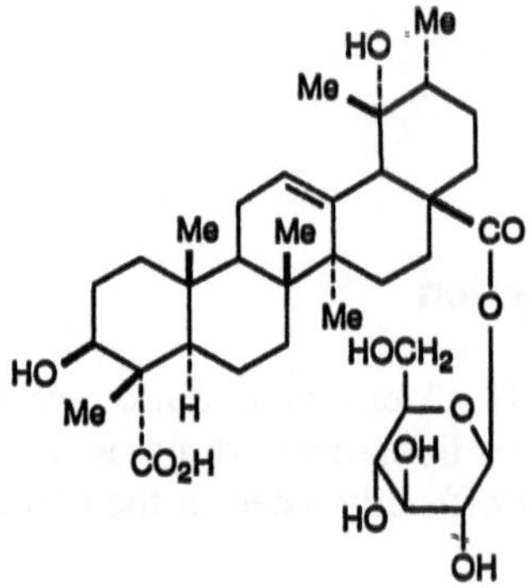

Ilexsaponin A (*76-5*)

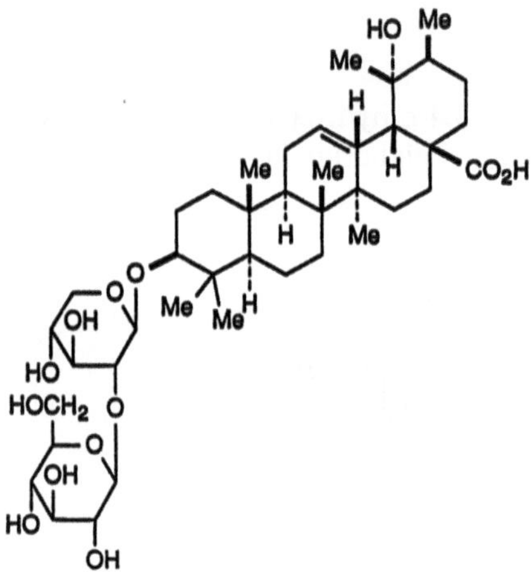

Ilexsaponin B_1 (*76-6*)

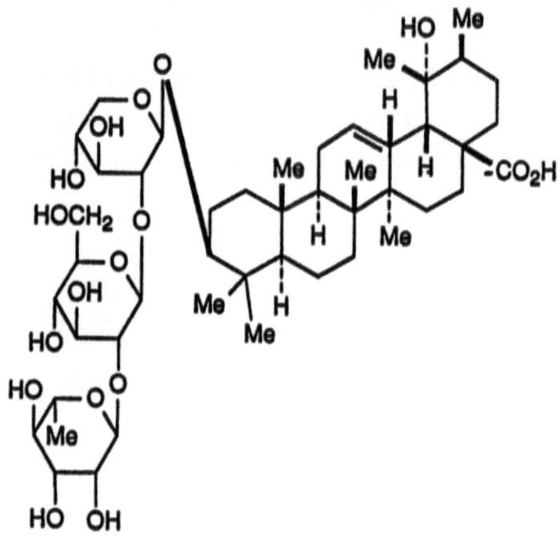

Ilexsaponin B_2 (*76-7*)

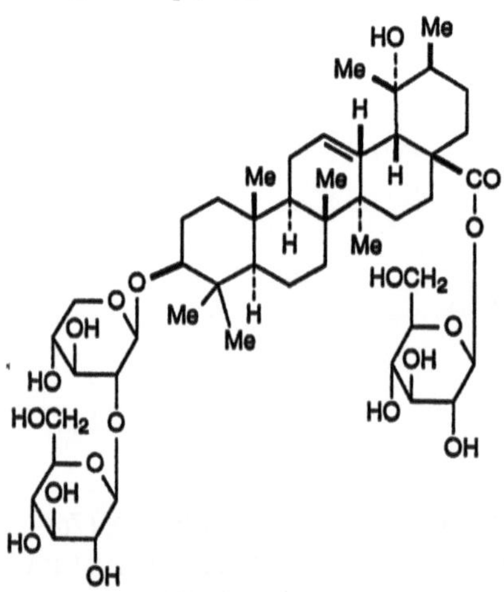

Ilexsaponin B_3 (*76-8*)

Structurally related saponins ilexoside A (*76-9*) and ilexoside B (*76-10*) methyl ester were isolated from the leaves of *I. chinensis*, another species used in Chinese medicine [9].

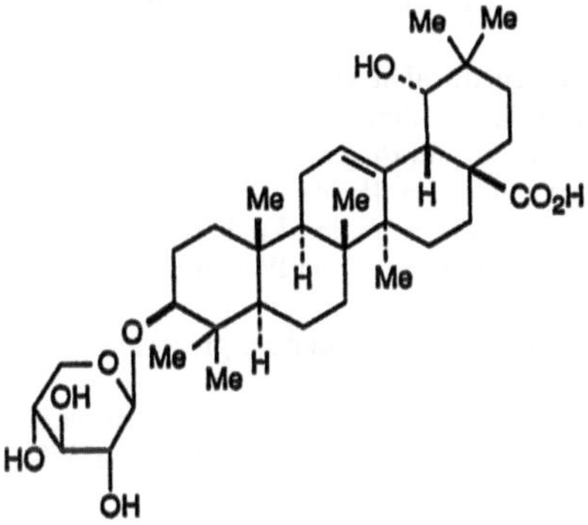

Ilexoside A (*76-9*)

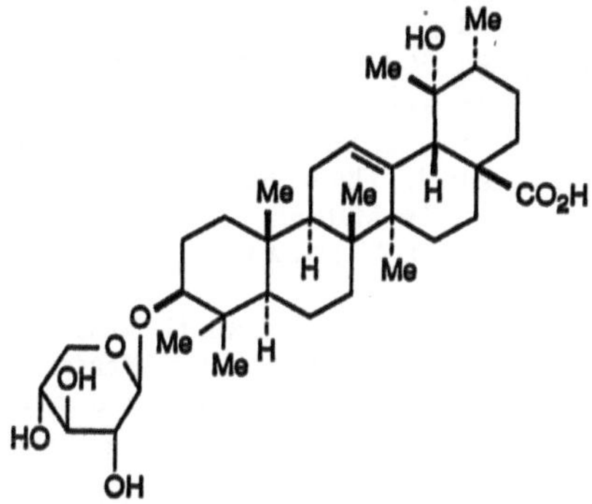

Ilexoside B (*76-10*)

76.3 Pharmacology

The extract of *I. pubescens* var. *glaber* and its active ingredients, especially 3,4-dihydroxyacetophenone, showed cardiovascular and circulatory activities in experimental animals, including rabbits and dogs. Coronary circulation was stimulated and coronary resistance was decreased, as was myocardial oxygen consumption. Cerebral circulation was increased and cerebral circulatory resistance decreased. The vasoconstriction induced by potassium chloride and the platelet aggregation caused by ADP were inhibited, while the myocardial uptake of ^{86}Rb was stimulated. Thus, the extract of *I. pubescens* var. *glaber* and 3,4-dihydroxyacetophenone are useful for treatment of the coronary diseases. 3,4-Dihydroxyacetophenone appears to be the most active pharmacological component. The LD_{50} of 3,4-dihydroxyacetophenone in mice was 235 mg/kg.

76.4 *Ilex rotunda* and *I. chinensis*

Ilex rotunda and *I. chinensis* are two other *Ilex* species used in the folk medicine for treatment of cardiovascular diseases. From the bark of *I. rotunda* a hemostatic constituent was isolated and identified as syringin [11, 12]. Protocatechualdehyde was detected in *I. chinensis* at a content of 0.03%–0.05% [13].

References

1. Qin WJ, Chiao CY, Fan TT, Chang KS, Pei LC (1980) Studies on the active constituents in the leaves of *Ilex pubescens* Hook. et Arn. var. *glaber* Chang, part II. Chin Trad Herb Drugs 11:97–98
2. Qin WJ, Jiao ZY, Fan ZT, Chen BQ, Lin XY, Yao JX (1980) Studies on the chemical constituents of the leaves of *Ilex pubescens* Hook. et Arn. var. *glaber* Chang. III. Structural elucidation of glaberide I. Acta Pharm Sin 15:669–673

3. Peking Institute of Pharmaceutical Industry (1980) Chemical constituents of the leaves of *Ilex pubescens* var. *glaber*. Chin Pharm Bull 15:42
4. Qin WJ, Chiao CX, Fan CT, Chen PC, Lin HY, Yao CH (1980) Studies on the active principles of *Ilex pubescens* Hook. et Arn. var. *glaber* Chang leaf, part III. Chin Trad Herb Drugs 11:199
5. Lin XY, Yao JX, Zhang SD (1982) Crystal and molecular structure of glaberide I. Acta Chim Sin 40:469–475
6. Zeng LM, Su JY, Liu DW, Ma FE (1980) Studies on the chemical structure of ilexolide A. In: Wang Y (ed) Chem Nat Prod Proc Sino-Am Symp (Pub 1982). Beijing, pp 280–284
7. Qin GW, Chen ZX, Xu RS, Jiang ZF, Liang JG (1987) Studies on the chemical constituents of *Ilex pubescens*. II. The structure of ilexsaponin A. Acta Chim Sin 45:249–255
8. Hidaka K, Ito M, Matsuda Y, Kohda H, Yamasaki K, Yamahara J, Chisaka T, Kawakami Y, Sato T, Kagei K (1987) New triterpene saponins from *Ilex pubescens*. Chem Pharm Bull (Tokyo) 35:524–529
9. Inada A, Kobayashi M, Murata H, Nakanishi T (1987) Two new triterpenoid glycosides from leaves of *Ilex chinensis* Sims. Chem Pharm Bull (Tokyo) 35:841–845
10. Peking Institute of Pharmaceutical Industry (1980) Pharmacological studies of constituents from *Ilex pubescens* var. *glaber* leaf in coronary heart disease. Chin Trad Herb Drugs 11:358–361
11. Chu JH, Hung SH, Wang YH (1956) Glucosides of the Chinese drug, Chiu-Pi-Ying (Ilex). Acta Chim Sin 22:128–132
12. Xie PS, Yang ZX (1980) Isolation and identification of a hemostatic constituent from Chinese drug-Jiu Bi Ying, the bark of *Ilex rotunda* Thunb. Acta Pharm Sin 15:303–305
13. Yang SD, Zhou TH (1981) Studies on the analysis of protocatechualdehyde. II. Determination of protocatechualdehyde in *Ilex chinensis* Sims and *Salvia miltiorrhiza* Bge. Acta Pharm Sin 16:530–534

Inula spp. 77

77.1 Introduction

The following entries representing *Inula* species are listed in the Chinese Pharmacopoeia:

- Tumuxiang, Radix Inulae, is the dry roots of *Inula helenium* L. or *I. racemosa* Hook f. (Asteraceae) collected in the fall. It is used as a stomachic and analgesic and is recommended for the treatment of emesis, diarrhea, and prevention of miscarriage.
- Jinfocao, Herba Inulae, is the dry aboveground part of *I. linariifolia* Turcz. and *I. japonica* Thunb., collected in the summer and fall. It is used as a diuretic and mucolytic for the treatment of cough and asthma.
- Xuanfuhua, Flos Inulae, is the dry inflorescence of *I. japonica* and *I. britannica* L., collected in the summer and fall when the flowers come into bloom. It is used as a diuretic, antiemetic, and mucolytic.

In addition, the dry root of *I. cappa* DC is listed in the appendix of the Chinese Pharmacopoeia.

77.2 Chemical Constituents

77.2.1 Chemical Constituents of *Inula helenium*

Among the *Inula* species, *I. helenium* is the most investigated. A number of sesquiterpene lactones were isolated from the root of *I. helenium*. The first, designated as helenin and later as alantolactone, was isolated about 150 years ago [1]. The structure elucidation of alantolactone (*77-1*) was a topic of a number of studies [2–5]. Stereoselective synthesis [6] and X-ray crystallographic determination [7] of this eudesmane derivative have also been reported.

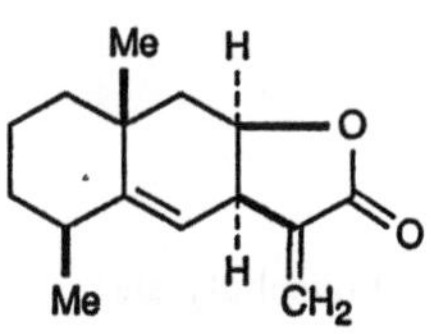

Alantolactone (Helenin, *77-1*)

Other sesquiterpenes isolated from *I. helenium* were isoalantolactone (*77-2*) [8, 9], dihydroalantolactone (*77-3*), dihydroisoalantolactone (*77-4*) [10, 11], and tetrahydroalantolactone [12]. Total sesquiterpene lactone contents, including alantolactone, dihydroalantolactone, tetrahydroalantolactone, isoalantolactone, and dihydroisoalantolactone in five samples of roots of *I. helenium* were 4.3%–5.2%: Alantolactone and isoalantolactone contents were 1.6%–2.0% and 2.1%–2.7%, respectively [13].

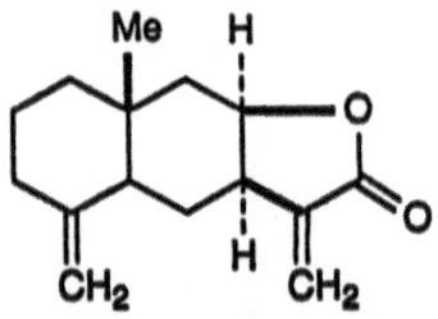
Isoalantolactone (*77-2*)

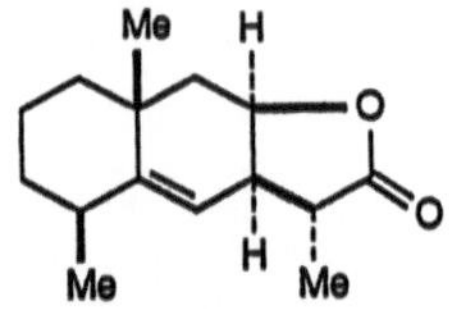
Dihydroalantolactone (*77-3*)

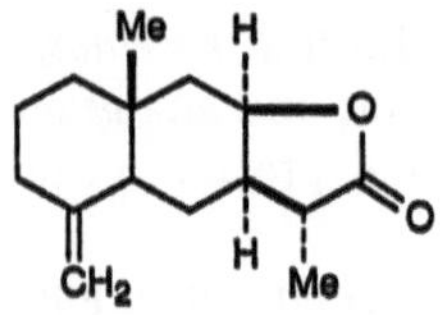
Dihydroisoalantolactone (*77-4*)

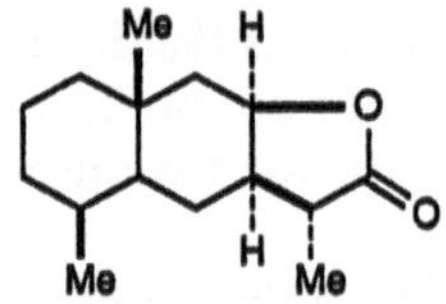
Tetrahydroalantolactone (*77-5*)

Besides alantolactone and related compounds, deoxyepiivangustin (*77-6*), germacrene-D-lactone (*77-7*), pentaynene (*77-8*), isohumulene (*77-9*), 8,9-epoxy-10-hydroxy-thymol diisobutyrate (*77-10*) [14], dammaradienol (*77-11*) [15] and its acetate (*77-12*) [14, 16], and α- and β-sitosterol [15] were isolated from the root of *L. helenium*.

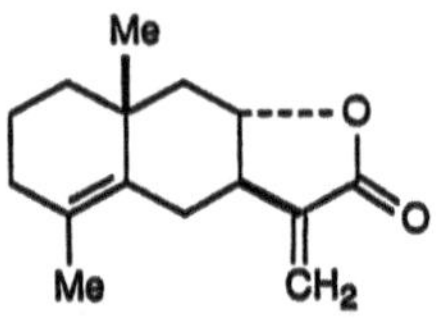
Deoxyepiivangustin (*77-6*)

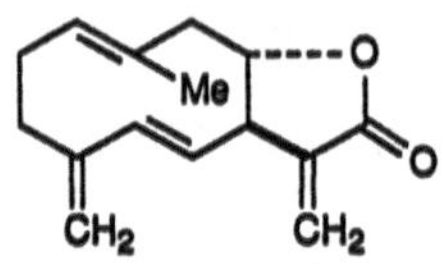
Germacrene-D-lactone (*77-7*)

$H_3C(C{\equiv}C)_5CH{=}CH_2$

Pentaynene (*77-8*)

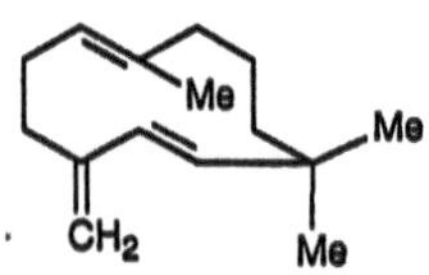
Isohumulene (*77-9*)

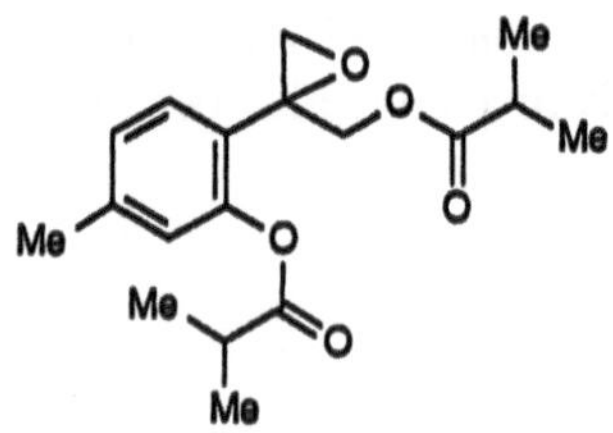
8,9-Epoxy-10-hydroxythymol diisobutyrate (*77-10*)

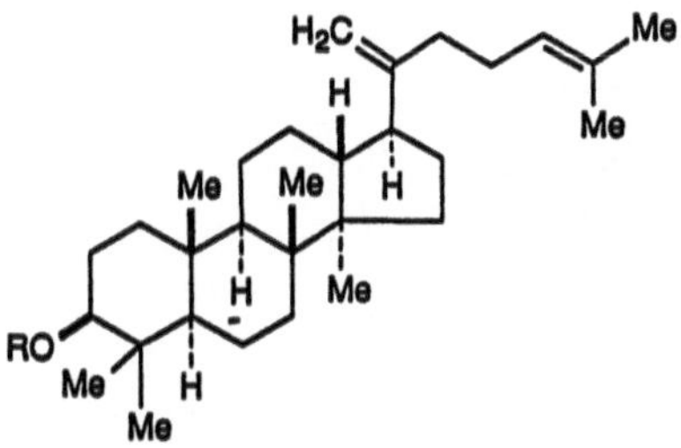

Dammaradienol (*77-11*): R=H
Dammaradienol acetatet (*77-12*): R=Ac

The following substances were isolated and identified from the aerial part of *I. helenium* [14]: alantolactone, isoalantolactone, hydroxyalantolactone (*77-13*), hydroxyisoalantolactone (*77-14*), oxoalantolactone (*77-15*), epiisovangustin (*77-16*), costunolide (*77-17*), 5β-hydroxycostunolide (*77-18*), 5β-(2-methyl-butyryloxy)-costunolide (*77-19*), 5β-isovaleryloxycostunolide (*77-20*), 5β-isobutyryloxycostunolide (*77-21*), 5α,6β-epoxyinunolide (*77-22*), 9β,10α-epoxyinunolide (*77-23*), epiisoinuviscolide (*77-24*), 4aα,5α-epoxyinuviscolide (*77-25*), 5-deoxo-5α-hydroxyconfertin (*77-26*), tomentosin (*77-27*), carabrone (*77-28*), and epitomentosin (*77-29*).

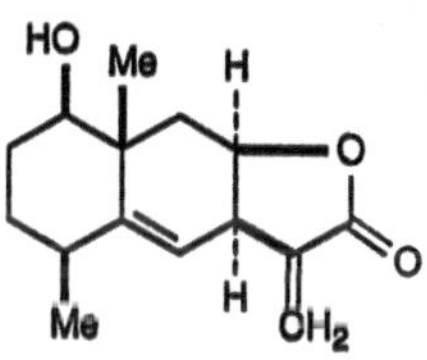

8β-Hydroxyalantolactone (*77-13*)

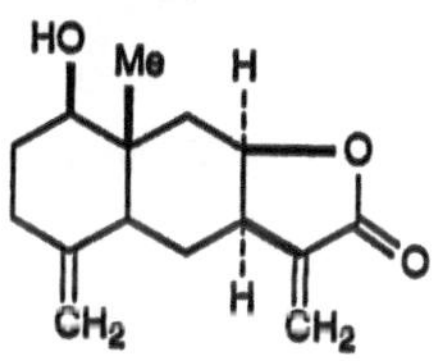

8β-Hydroxyisoalantolactone (*77-14*)

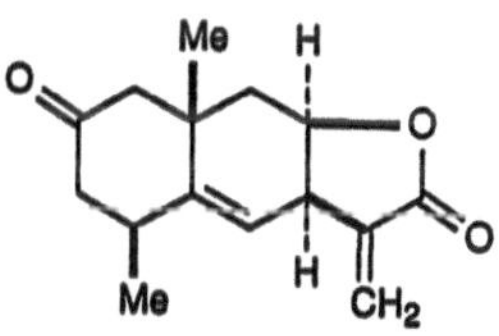

7-Oxoalantolactone (*77-15*)

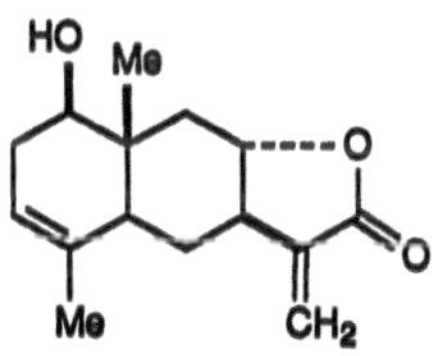

Epiisovangustin (*77-16*)

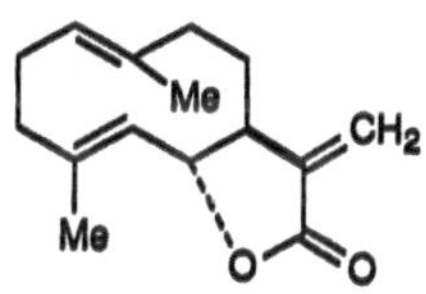

Costunolide (*77-17*)

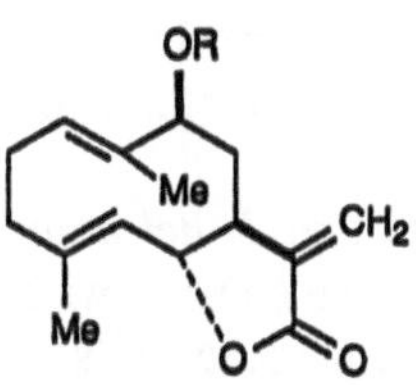

5β-Hydroxycostunolide (*77-18*): R=H
5β-(2-Methylbutyryloxy)-costunolide (*77-19*): $R=CH_3-CH_2-CH(CH_3)-CO-$
5β-Isovaleryloxy-costunolide (*77-20*): $R=(CH_3)_2-CH_2-CH_2-CO-$
5β-Isobutyryloxy-costunolide (*77-21*): $R=(CH_3)_2-CH_2-CO-$

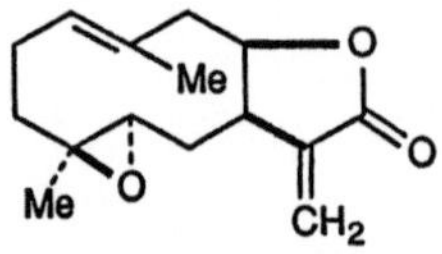

5α,6β-Epoxyinunolide
(*77-22*)

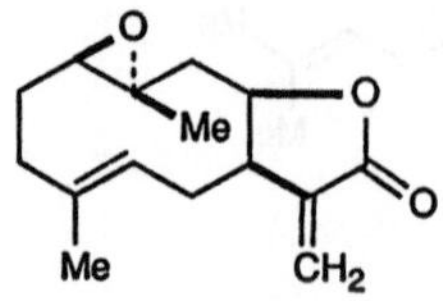

9β,10α-Epoxyinunolide
(*77-23*)

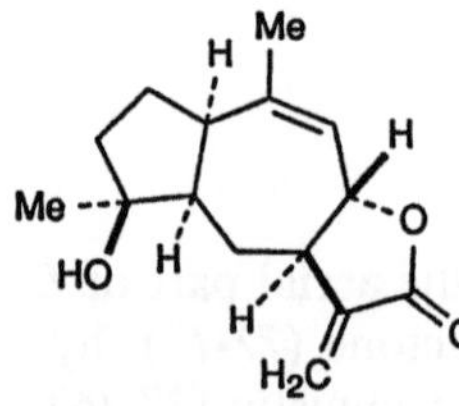

Epiisoinuviscolide (*77-24*)

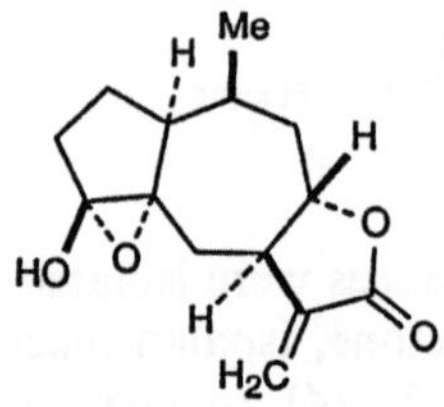

4aα,5α-Epoxyinuviscolide (*77-25*)

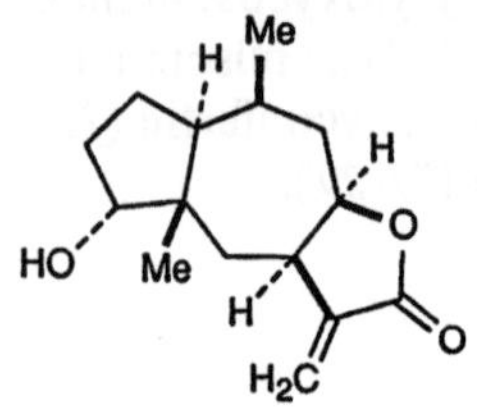

5-Deoxo-5α-hydroxyconfertin (*77-26*)

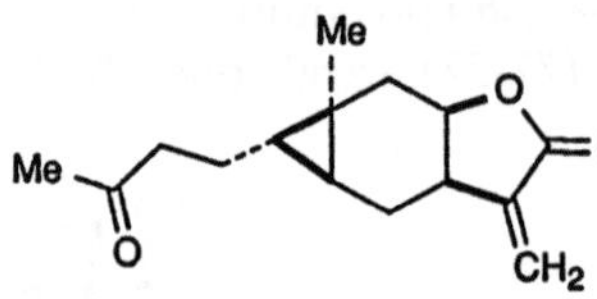

Carabrone (*77-28*)

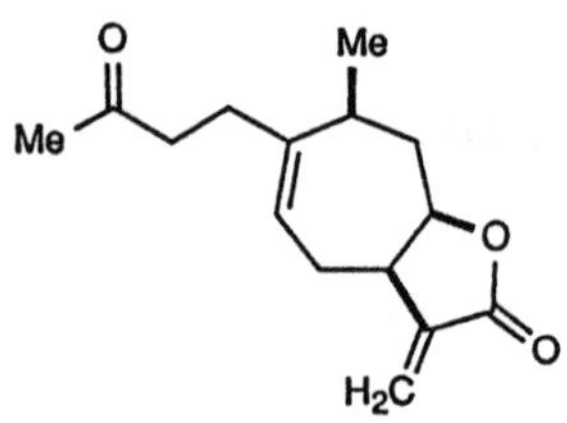

Tomentosin (*77-27*)

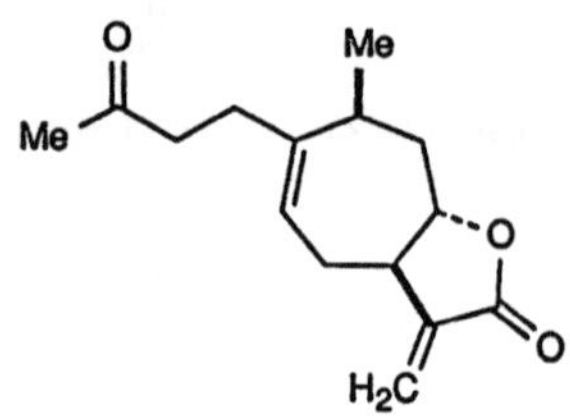

Epitomentosin (*77-29*)

The inflorescences of *I. helenium* contained quercetin, methylquercetin, and a quercetin triglucoside [16]. From the aerial part of *I. helenium*, alantolactone, isoalantolactone, and their dihydro analogs as well as scopoletin and umbelliferone were isolated [17].

A well-known plant polysaccharide first isolated from the root of *I. helenium* about 2 centuries ago is inulin [18]. Inulin is a high-molecular polysaccharide with a molecular weight of about 5200, containing about 30 β-D-fructofuranose units in (2→1)-glycosidic linkage, with a β-D-glucopyranose terminus [19]. It can be extracted with water and precipitated with alcohol [20].

77.2.2 Chemical Constituents of *Inula britannica*

From the flowers of *I. britannica*, the sesquiterpene gaillardin (*77-30*) was isolated as the main constituent [21].

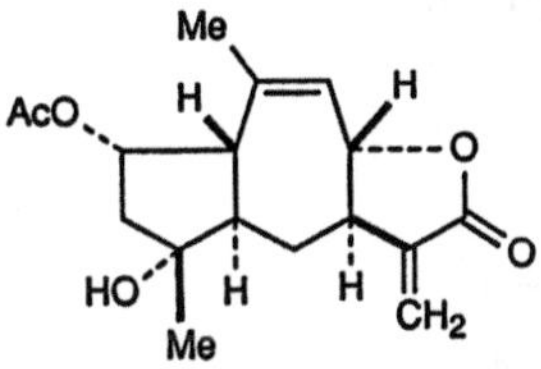

Gaillardin (*77-30*)

From the aerial part of *I. britannica*, a sesquiterpene lactone named britanin (*77-31*) was isolated [22] and structurally investigated [23].

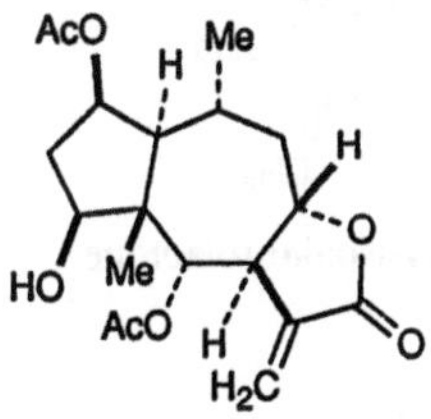

Britanin (*77-31*)

Other components in the aerial part of *I. britannica* reported are esculetin, scopoletin, chlorogenic acid, isochlorogenic acid, salicylic acid, *p*-hydroxybenzoic acid, protocatechuic acid, vanillic acid, syringic acid, *p*-hydroxyphenylacetic acid, *p*-coumaric acid, caffeic acid, ferulic acid [24], luteolin, 6-hydroxy-luteolin 7-glucoside, isoquercitrin, quercimeritrin (*77-32*), quercetagitrin (*77-33*), nepitrin (*77-34*) and aglycone, patulitrin (*77-35*) and aglycone, quercetin 7-glucoside, 6-hydroxyluteolin 7-diglucoside [24, 25], 4-hydroxycostunolide (*77-36*), 3-epiisotelekin (*77-37*), deoxyartemisiifolin (*77-38*), and 7β-hydroxy-6α-senecioyloxy-isoalantolactone (*77-39*) [26].

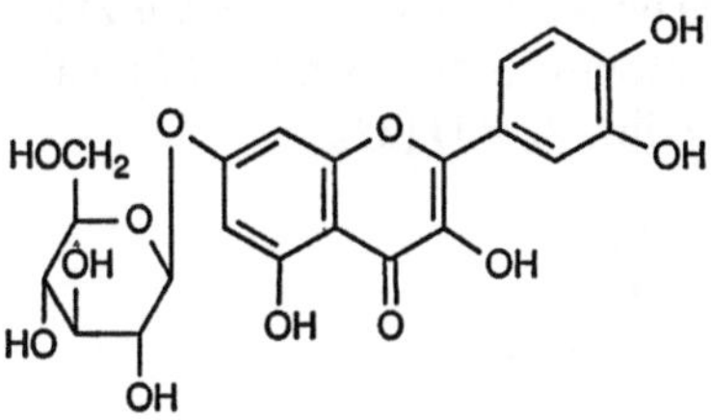

Quercimeritrin (*77-32*)

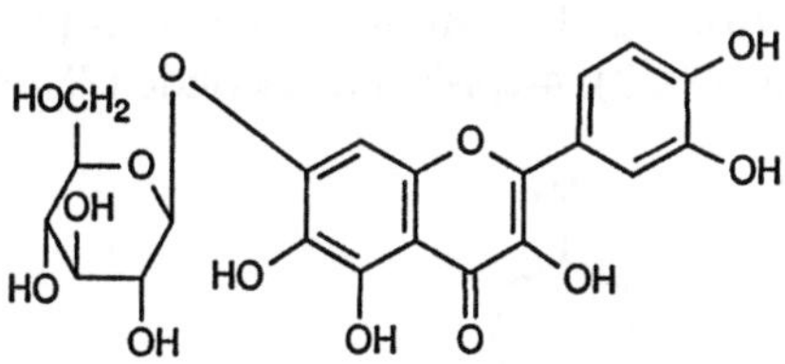

Quercetagitrin (*77-33*)

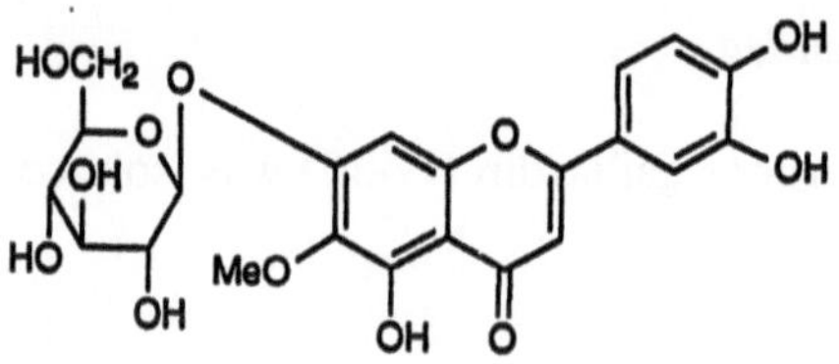

Nepitrin (*77-34*)

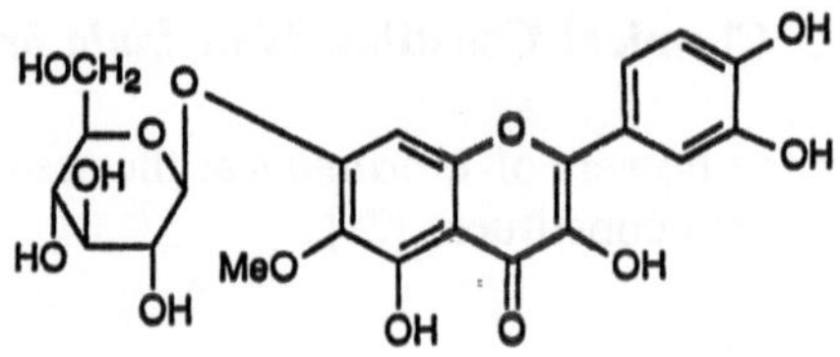

Patulitrin (*77-35*)

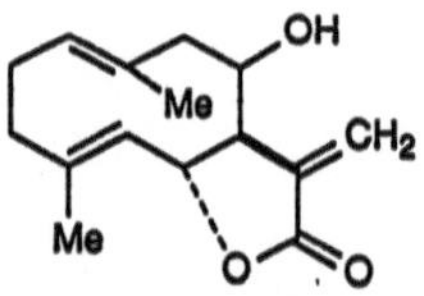

4-Hydroxycostunolide (*77-36*)

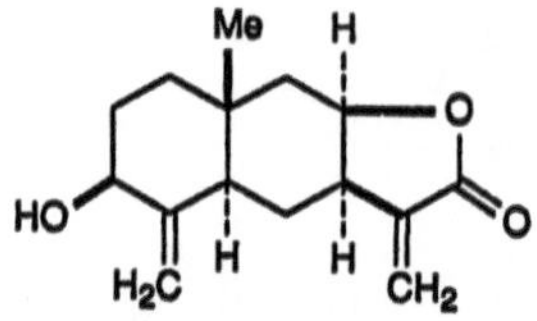

3-Epiisotelekin (*77-37*)

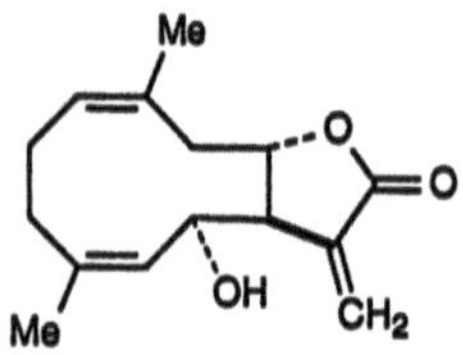

Deoxyartemisiifolin (*77-38*)

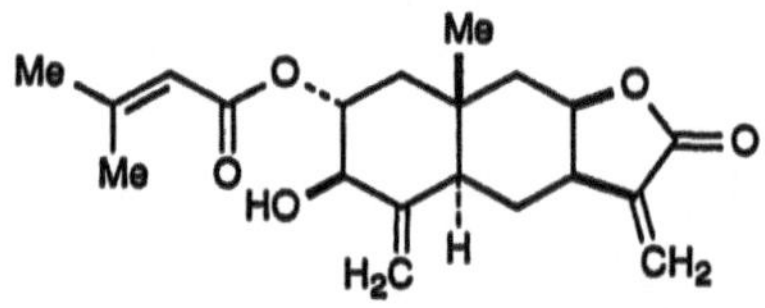

7β-Hydroxy-6α-senecioyloxy-isoalantolactone (*77-39*)

The root of *I. britannica* contained pentaynene, nerolisobutyrate (*77-40*), nerolisovalerate (*77-41*), and thymol derivatives 8,9-epoxy-10-hydroxythymolisobutyrate, 8,9-epoxy-10-hydroxythymoldiisobutyrate (*77-10*), and 8,9-epoxy-10-isovaleryloxythymolisobutyrate [26].

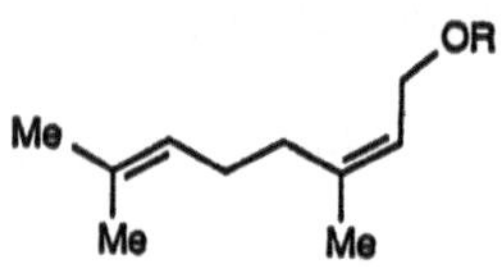

Nerolisobutyrate (*77-40*): R = $-CO-CH-(CH_3)_2$
Nerolisovalerate (*77-41*): R = $-CO-CH_2-CH-(CH_3)_2$

Furthermore, three new sesquiterpene lactones, named inuchinenolides A (*77-42*), B (*77-43*), and C (*77-44*), were isolated from the whole plant of *I. britannica* var. *chinensis*, together with the known sesquiterpene lactones tomentosin (*77-27*), gaillardin (*77-30*), 4-epi-isoinuviscolide (*77-24*), and ivalin (*77-45*) [27].

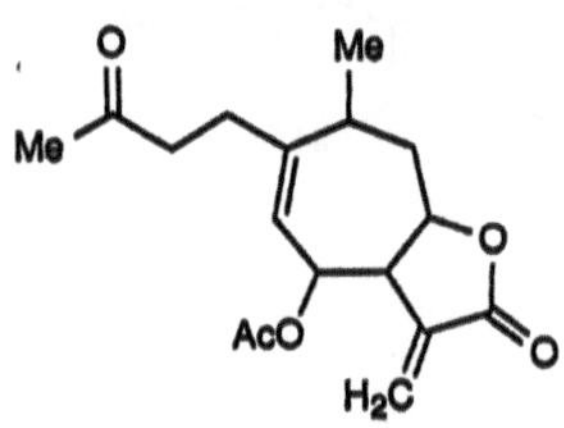

Inuchinenolide A (*77-42*)

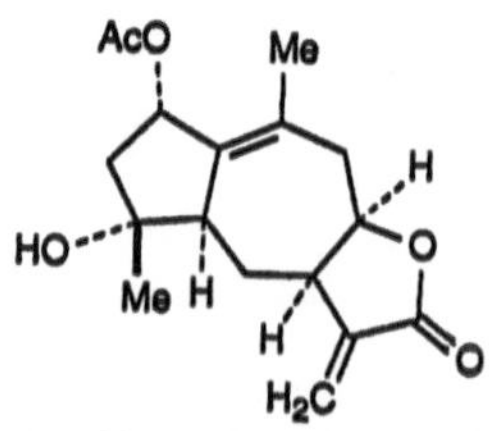

Inuchinenolide B (*77-43*)

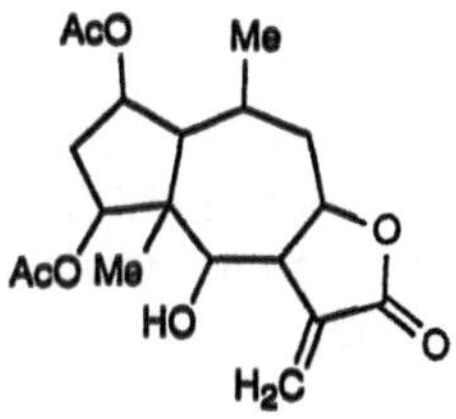

Inuchinenolide C (*77-44*)

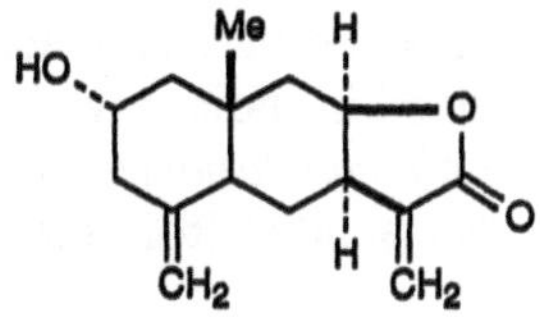

Ivalin (*77-45*)

77.2.3 Chemical Constituents of *Inula linariaefolia*

From the flowers of *I. linariaefolia*, a sesquiterpene lactone was isolated and identified as the known compound britanin. Another compound isolated was taraxasteryl palmitate [28].

77.2.4 Chemical Constituents of *Inula racemosa*

The essential oil of the root of *I. racemosa* contained about 60% sesquiterpenes [29]. Among them, inunolide (*77-46*) [30], 1-deoxy-8-epi-ivangustin (alloalantolactone) [31], inunal (*77-47*), and isoalloalantolactone (*77-48*) [32] were isolated.

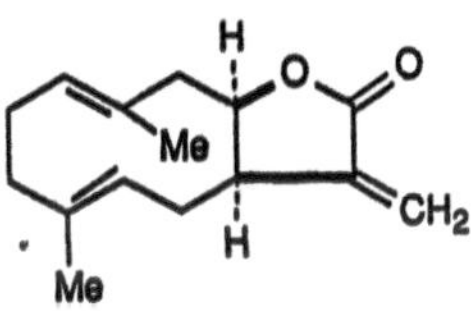

Inunolide (*77-46*)

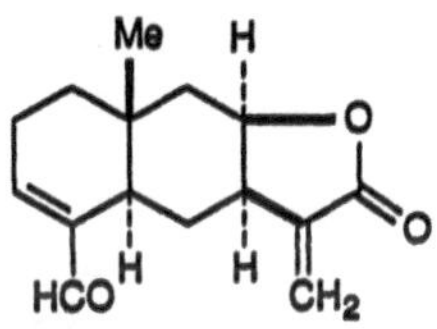

Inunal (*77-47*)

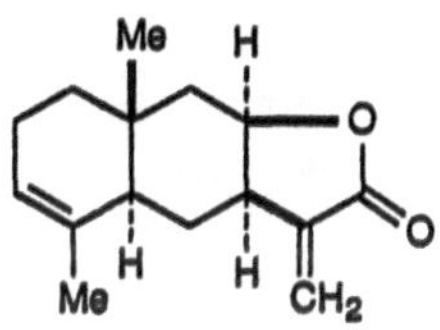

Isoalloalantolactone (*77-48*)

The most abundant constituent in the essential oil of *I. racemosa* root was heptadeca-1,8,11,14-tetraene (aplotaxene) at 22%. Other components isolated from the essential oil were *p*-cymene, 2-furfural, norbornyl acetate, benzaldehyde, *β*-elemene, *α*-pinene oxide, *α*-humulene, *α*-farnesene, ar-curcumene, heptadeca-1,8,11-triene, *α*-ionone (*77-49*), *β*-ionone (*77-50*), and phenylacetonitrile [29].

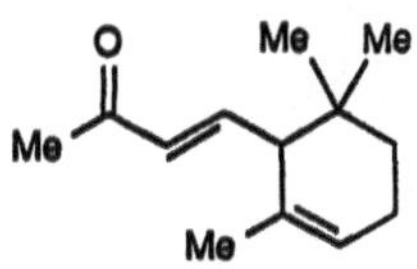

α-Ionone (*77-49*)

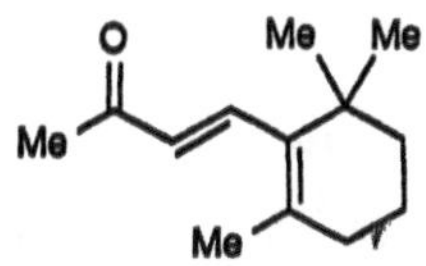

β-Ionone (*77-50*)

Dammara-20,24-dien-3*β*-yl acetate was also isolated from the root of *I. racemosa* [33].

77.2.5 Chemical Constituents of *Inula cappa*

The isolation of a number of new germacranolides (*77-51*–*77-57*) from the aerial parts of *I. cappa*, collected in India, together with a known germacranolide (*77-58*) previously isolated from *I. eupatorioides*, was reported [34]. In addition, some inositol tetraangelates were also isolated from the aerial parts of *I. cappa* [35].

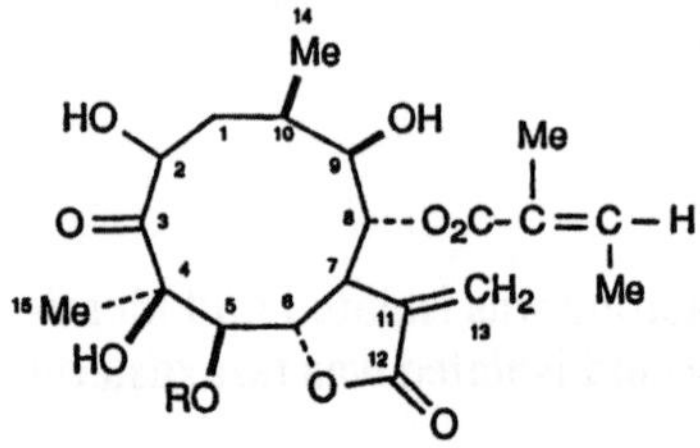

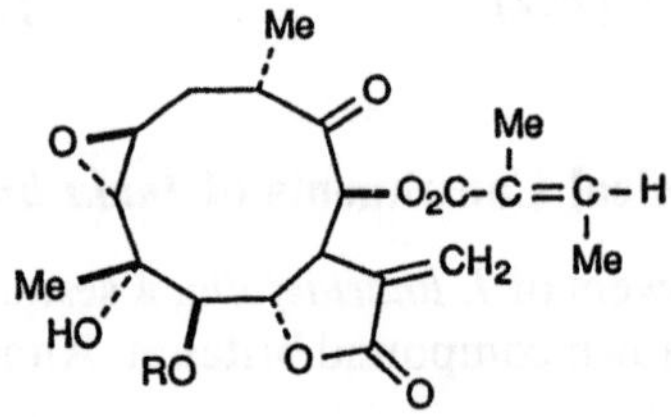

(*77-51*): R =

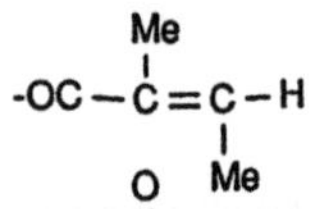

(*77-52*): R =

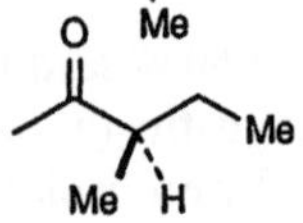

(*77-53*): R =

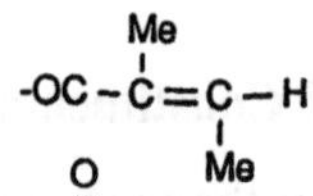

(*77-54*): R =

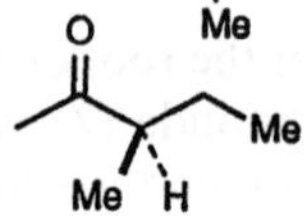

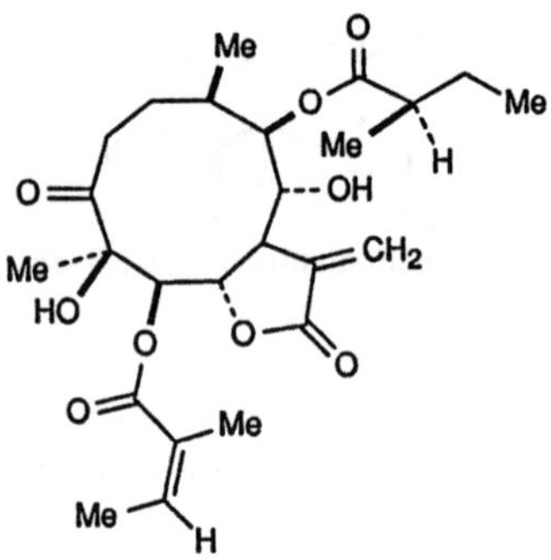

(*77-55*)

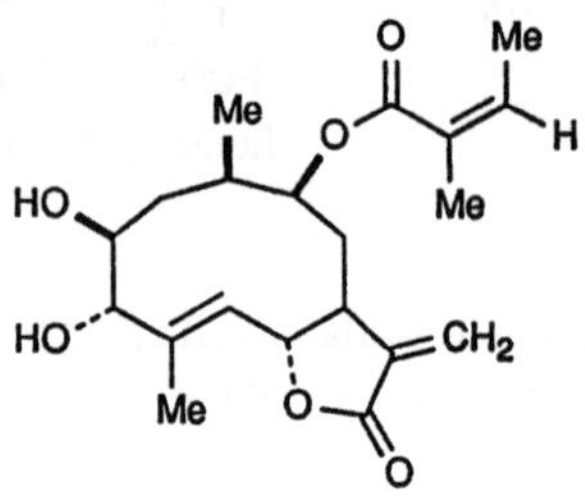

(*77-56*)

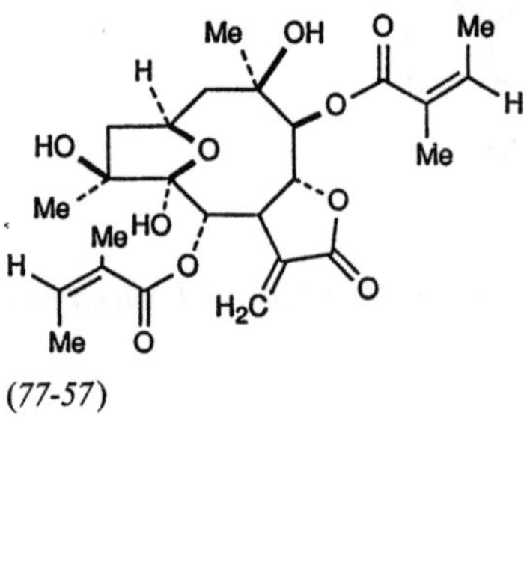

(*77-57*)

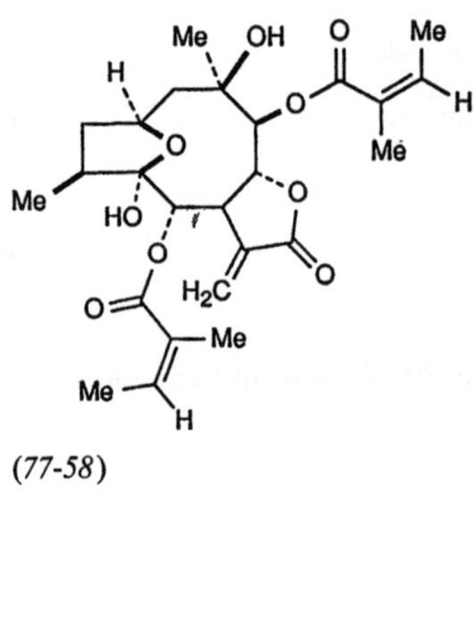

(*77-58*)

77.3 Pharmacology

Alantolactone and isoalantolactone were tested against a number of fungi. Their antifungal activity varied greatly with the fungi species. At a concentration of 10 µg/ml, the lactones strongly inhibited the growth of *Microsporum cookei, Trichophyton mentagrophytes,* and *Trichothecium roseum* [36]. Alantolactone showed an anthelmintic activity against *Hymenolepis nana* var. *fraterna* and *Syphacia obvelata* in vitro. However, their activity against these parasites in mice was poor, due to considerable absorption in the digestive tract before reaching the intestinal site of parasitization [37]. Alantolactone administered to dogs or rabbits at a dose of 3 mg/kg by subcutaneous injection increased the thrombocyte adhesiveness and aggregation capacity and shortened the recalcification time of oxalated plasma. Thus, alantolactone has hemostatic properties [38].

References

1. Gerhardt C (1840) Chemische Untersuchungen über das Helenin. Liebigs Ann Chem 34:192–204
2. Ruzicka L, Pieth P, Reichstein T, Ehmann T (1933) Polyterpene und Polyterpenoide LXXX, Zur Kenntnis der Alantolactone. Synthese des 1,4-Dimethyl-6-isopropyl und des 1,5-Dimethyl-7-isopropylnaphthalins. Helv Chim Acta 16:268–275
3. Tsuda K, Tanabe K, Iwai I, Funakoshi K (1957) The structure of alantolactone. J Am Chem Soc 79:5721–5724
4. Asselineau C, Bory S (1958) The separation and structure of alantolactone and isoalantolactone. Compt Rend 246:1874–1877
5. Marshall JA, Cohen N (1964) The structure of alantolactone. J Org Chem 29:3727–3729
6. Marshall JA, Cohen N (1965) The stereoselective total synthesis of alantolactone. J Am Chem Soc 87:2773–2774
7. Schmalle HW (1986) Structure of 4β,10β-dimethyleudesma 5,11(13)-diene-7β-lactone (alantolactone). Acta Crystallogr [C] C42:705–708
8. Wunderlich W (1959) Isolierung von reinem Isoalantolacton aus *Inula helenium.* J Pr Chem 9:107
9. Ruzicka L, Melsen JA van (1931) Höhere Terpenverbindungen XLV. Zur Kenntnis des Alantolactons und des Isoalantolactons. Helv Chim Acta 14:397–410
10. Hansen KFW (1931) Über Bitterstoffe aus der Alantwurzel (vorläufige Mitteilung). Chem Ber 64:67–71
11. Hansen KFW (1931) Über Bitterstoffe aus der Alantwurzel (II. Mitteilung über Bitterstoffe). Chem Ber 64:943–947
12. Ukita C, Nakazawa S (1960) Santonin analogs. IV. Structure of "iso"-tetrahydroalantolactone. J Am Chem Soc 82:2224–2228
13. Rosik GG, Zinchenko AA, Reznichenko AA, Kovalev IP (1987) Quantitative determination of sesquiterpene lactones of *Inula helenium* L. by gas-liquid chromatography. Khim-Farm Zh 21:632–634
14. Bohlmann F, Mahanta PK, Jakupovic J, Rastogi C, Natu AA (1978) New sesquiterpene lactones from *Inula* species. Phytochemistry 17:1165–1172
15. Yoshioka I, Yamada Y (1963) Isolation of dammaradienyl acetate from *Inula helenium.* Yakugaku Zasshi 83:801–802
16. Kowalewska K, Lutomski J (1978) Flavonoids in the inflorescences of *Inula helenium* L. Herba Pol 24:107–113 (CA 90:183169n)
17. Khvorost PP, Komissarenko NF (1976) Lactones of *Inula helenium.* Khim Prir Soedin 820–821
18. Karrer W (1976) Konstitution und Vorkommen der organischen Pflanzenstoffe, 2nd edn. Birkhäuser, Basel
19. Beyer H, Walter W (1984) Lehrbuch der organischen Chemie, 20th edn. Hirzel, Stuttgart, p 429

20. Bell DJ, Palmer A (1952) Structural studies on inulin from *Inula helenium* and on levans from *Dactylis glomerata* and *Lolium italicum*. J Chem Soc 3763–3770
21. St. Pyrek J (1977) Terpenes of Compositae plants. IV. Isolation of gaillardin from flowers of *Inula britannica* L. Rocz Chem 51:1277–1279 (CA 87:197230a)
22. Rybalko KS, Sheichenko VI, Maslova GA, Kiseleva EY, Gubanov IA (1968) Britanin, a lactone from *Inula britannica*. Khim Prir Soedin 4:251–252
23. Chugunov PV, Sheichenko VI, Bankovskii AI, Rybalko KS (1971) Structure of britanin, a sesquiterpenoid lactone from *Inula britannica*. Khim Prir Soedin 7:276–280
24. Krolikowska M, Wolbis M (1981) Polyphenolic compounds in *Inula britannica*. Acta Pol Pharm 38:107–114 (CA 95:183890u)
25. Krolikowska M, Wolbis M (1979) Polyphenolic compounds in *Inula britannica*. Acta Pol Pharm 36:395 (CA 92:160563p)
26. Bohlmann F, Zdero C (1977) Naturally occurring terpene derivatives, part 104. New sesquiterpene lactones and thymol derivatives from *Inula* species. Phytochemistry 16:1243–1245
27. Ito K, Iida T (1981) Seven sesquiterpene lactones from *Inula britannica* var. *chinensis*. Phytochemistry 20:271–273
28. Chien MK, Chen ZN, Qin GW, Han J, Wang BD, Xu RS (1983) Studies on the chemical constituents of *Inula linariaefolia* Turcz. Acta Chim Sin 41:254–261
29. Bokadia MM, MacLeod AJ, Mehta SC, Mehta BK, Patel H (1986) The essential oil of *Inula racemosa*. Phytochemistry 25:2887–2888
30. Raghavan R, Ravindranath KR, Trivedi GK, Paknikar SK, Bhattacharyya SC (1969) Inunolide. A sesquiterpene lactone from *Inula racemosa* root. Indian J Chem 7:310
31. Bhandari P, Rastogi RP (1983) Alloalantolactone, a sesquiterpene lactone from *Inula racemosa*. Indian J Chem [B] 22B:286–287
32. Kaur B, Kalsi PS (1985) Stereostructures of inunal and isoalloalantolactone, two biologically active sesquiterpene lactones from *Inula racemosa*. Phytochemistry 24:2007–2010
33. Paknikar SK, Naik US, Raghavan R (1982) Occurrence of (−)-dammara-20,24-dien-3β-yl acetate in *Inula racemosa*. Indian J Chem [B] 21B:894
34. Goswami AC, Baruah RN, Sharma RP, Baruah JN, Kulanthaivel P, Herz W (1984) Germacranolides from *Inula cappa*. Phytochemistry 23:367–372
35. Bohlmann F, Ahmed M, Jakupovic J (1982) Inositol angelates from *Inula cappa*. Phytochemistry 21:780–782
36. Pieman AK (1983) Antifungal activity of helenin and isohelenin. Biochem Syst Ecol 11:183–186
37. Azoulay P, Reynier JP, Balansard G, Gasquet M, Timon-David P (1986) Biogalenic and pharmacological study on three sesquiterpenes with antiparasitic actions: helenin, santonin and 12-carboxy-3,11(13)-endesma-diene. Pharm Acta Helv 61:345–352
38. Mansurov MM (1973) Helenin action on blood coagulation. Farmakol Toksikol (Moscow) 36:335–338

Leonurus heterophyllus Sweet 78

78.1 Introduction

Yimucao, Herba Leonuri, is the dry aboveground part of *Leonurus heterophyllus* Sweet (Lamiaceae) collected in the summer when the plant is growing rapidly but has not bloomed. It is officially listed in the Chinese Pharmacopoeia. Herba Leonuri is a traditional Chinese medicine most used in the treatment of gynecologic disorders such as menorrhagia, menostasia, and other irregular menstruation. It is also used as a diuretic in the treatment of edema and acute nephritis.

Two galenic preparations of Herba Leonuri are entered in the Chinese Pharmacopoeia: Yimucao Liujingao, Extractum Leonuri liquidum, and Yimucao Gao, Extracum Leonuri inspissatum. Chemical examination of alkaloid content as described for Herba Leonuri should be carried out for both preparations, which have the same medicinal uses as Herba Leonuri.

Chongweizi, Fructus Leonuri, another official entry in the Chinese Pharmacopoeia, is the dry ripe fruits of *L. heterophyllus* collected in the fall when the fruits have ripened. It is used mainly in the treatment of menstrual disorders.

78.2 Chemical Constituents

From the aboveground part of *L. heterophyllus*, two alkaloids with a known structure, leonurine and stachydrine, were isolated and identified [1].

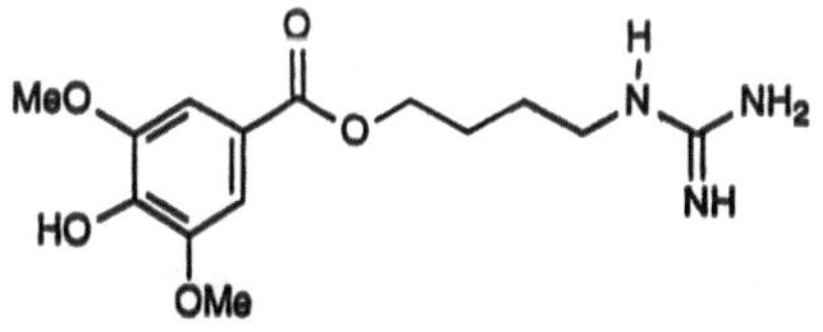

Leonurine (*78-1*)

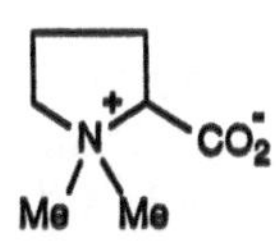

Stachydrine (*78-2*)

The leonurine content was the highest during the early fruiting stage and was the lowest before the flowering stage. In contrast, the stachydrine content was the highest before the flowering stage and was the lowest during the early fruiting stage [1]. The total alkaloid content in Herba Leonuri ranged from 0.1% to 2.1%. Younger and more succulent plants showed higher alkaloid contents [2]. The highest alkaloid content was determined in overwintering young seedlings and decreased with advancing maturation [3].

Leonurine is chemically a guanidinobutanolester of syringic acid. A number of studies were carried out on the determination of this rare structure [4–9]. Crystallographic structure elucidation was performed with nitroleonurine monohydrate [10]. In contrast to the planar, symmetric structure of nitroguanidine, the nitroguanidyl moiety in the nitroleonurine molecule assumes the hitherto unobserved unsymmetric form, containing a formal imino group, deviating significantly from exact planarity. Extensive H bonding, involving nitroleonurine and water molecules, results in a layer structure with weak van der Waals interactions between neighboring layers [10].

Methods for the synthesis of leonurine have been described [10, 11].

78.3 Pharmacology

Leonurine possesses stimulative activity on uterus contraction [12] through a process not involving known neural transmitters. It also showed an initial inhibitory effect on rabbit small intestine, followed by a prolonged stimulatory action. Its effect on the isolated rabbit heart was inhibitory [13].

References

1. Luo SR, Mai L (1986) Analysis of alkaloids in *Leonurus heterophyllus*. Chin J Pharm Anal 6:47–48
2. Xu LQ, Dong YW (1985) Quantitative determination of total alkaloids in Herba leonuri by colorimetric method with ammonium chromium thiocyanate. Chin Trad Herb Drugs 16:487–488
3. Hui YM, Zhang CY, Sun DQ, Wu YP (1982) Qualitative and quantitative analysis of Yi Mu Cao (*Leonurus heterophyllus*) in different growth stages. Bull Chin Mat Med 7:8–9
4. Ho CS, Yu CF, Wang H (1962) Chemical studies on the Chinese drug, I-mu-tsao. I. Structure of alkaloid A. Sci Sin 11:1341–1352
5. Goto T, Kato N, Hirata Y (1962) The structure of leonurine. Tetrahedron Lett 545–548
6. Hayashi Y (1962) Ingredients of *Leonurus sibiricus*. I. Yakugaku Zasshi 82:1020–1024
7. Hayashi Y (1962) Ingredients of *Leonurus sibiricus*. II. Structure of leonurine. 2. Yakugaku Zasshi 82:1025–1027
8. Hayashi Y (1962) Ingredients of *Leonurus sibiricus*. III. Structure of leonurine. 3. Yakugaku Zasshi 83:271–274
9. Sugiura S, Inoue S, Hayashi Y, Kishi Y, Goto T (1969) Structure and synthesis of leonurine. Tetrahedron 25:5155–5161
10. Camerman N, Chan LYY, Yeung HW, Mak TCW (1979) The crystal and molecular structure of nitroleonurine monohydrate. Acta Crystallogr [B] B35:3004–3007
11. Cheng KF, Yip CS, Yeung HW, Kong YC (1979) Leonurine, an improved synthesis. Experientia 35:571–572
12. Yu CF (1981) Studies on the chemical constituents of the Chinese traditional medicine I-Mu-Tsao. II. Structure of "alkaloid A". Acta Pharm Sin 39:94–96
13. Kong YC, Hu SY, Hwang JC, Chang SHH, Yeung HW, Ng CKH, Yip TT, Keung WM (1974) I-mu tsao. J Chin Univ Hong Kong 2:345–364

Ligusticum chuanxiong Hort. 79

79.1 Introduction

Some plants of the genus *Ligusticum* are used as herbal medicines in China. Two items including three *Ligusticum* species are listed in the Chinese Pharmacopoeia:

Chuanxiong, Rhizoma Chuanxiong, is the dry rootstock of *Ligusticum chuanxiong* Hort. (Apiaceae), which is collected in summer. It is used as an analgesic agent in the treatment of rheumatic and traumatic diseases and against menstrual disorder.

Gaoben, Rhizoma Ligustici, is the dry root and rootstock of *L. sinensis* Oliv. or *L. jeholense* Nakai et Kitag., which is collected in the fall, when the aboveground part of the plant has withered, or in spring, when the seedlings have grown. It may be used as an analgesic in the treatment of cold, headache, and rheumatic pains.

79.2 Chemical Constituents

The rhizome of *L. chuanxiong* is known to contain phthalides, alkaloids, acids, esters, and neutral substances [1]. A number of phthalides were isolated and identified from the rhizome of *L. chuanxiong*: ligustilide (*79-1*) [2], chuanxingol (*79-2*) [3], butylphthalide (*79-3*), butylidene phthalide (*79-4*), senkyunolide (*79-5*), neocnidilide (*79-6*) [4], 3-butyl-3-hydroxy-4,5,6,7-tetrahydro-6,7-dihydroxyphthalide (*79-7*) [5], and ligustilidiol (*79-8*) [6].

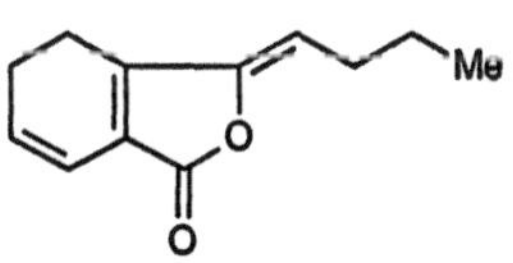

Ligustilide (*79-1*)

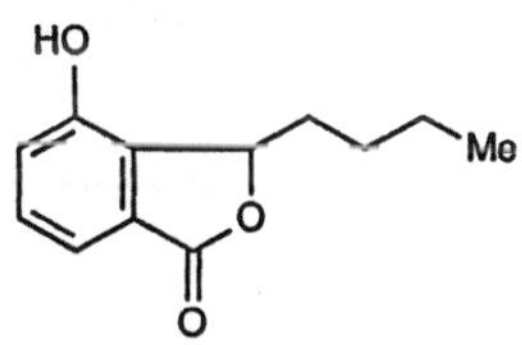

Chuanxingol (*79-2*)

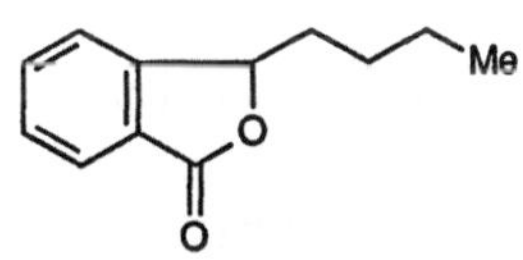

Butylphthalide (*79-3*)

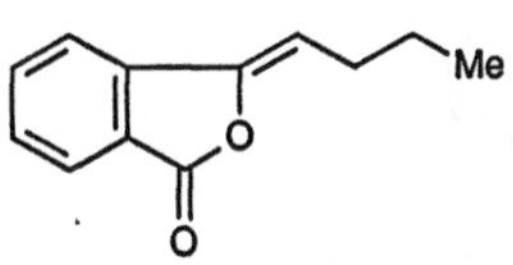

Butylidene phthalide (*79-4*)

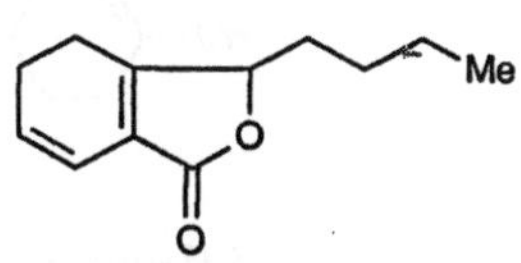

Senkyunolide (*79-5*)

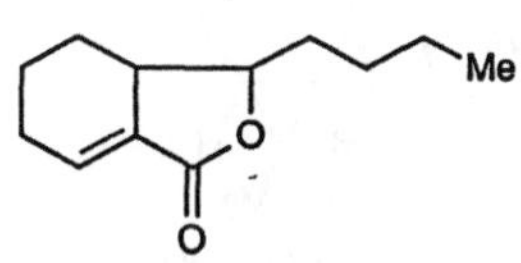

Neocnidilide (*79-6*)

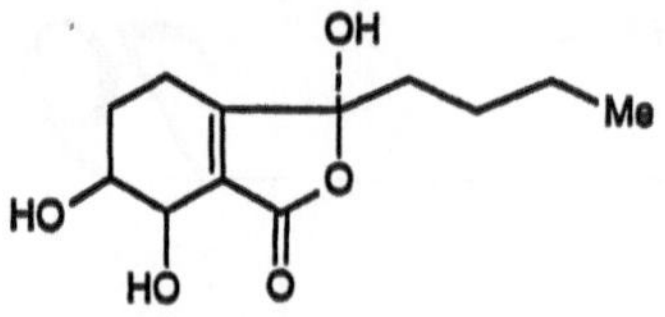

3-Butyl-3-hydroxy-4,5,6,7-tetra-hydro-6,7-dihydroxyphthalide (*79-7*)

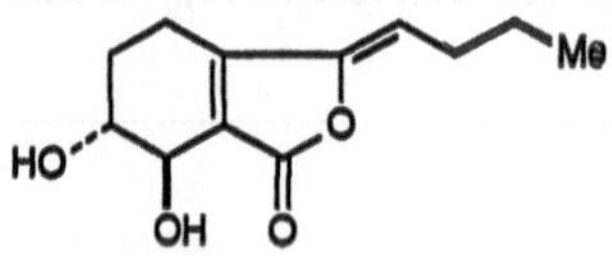

Ligustilidiol (*79-8*)

In addition to the phthalides, some nitrogen-containing substances were isolated and identified: tetramethylpyrazine, named ligustrazine or chuanxiongzine (*79-9*), L-isoleucyl-L-valine anhydride, perlolyrine (*79-10*) [3, 7], L-valyl-L-valine anhydride, uridine, trimethylamine hydrochloride, choline chloride, and 1-acetyl-β-carboline [3]. Compounds without nitrogen were also isolated and identified: ferulic acid, chrysophanic acid (*79-11*), sedanonic acid (*79-12*), palmitic acid, vanillin, bis-5,5-formyl-furfuryl-ether [3], spathulenol (*79-13*), linoleic acid, β-sitosterol, dilinoleoyl palmitoyl glyceride [8], and 5-hydroxymethyl-3'-methoxy-4'-hydroxyphenyl-8-oxabicyclo[3,2,1]-oct-3-en-2-one (*79-14*) [5].

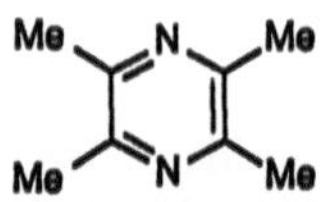

Ligustrazine (*79-9*)

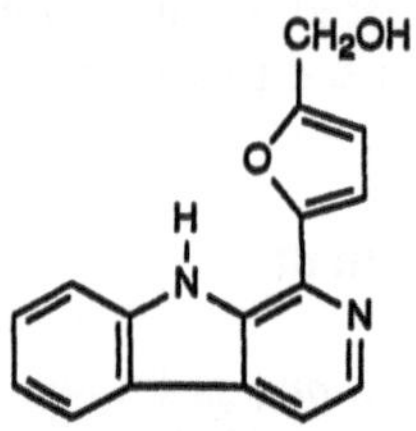

Perlolyrine (*79-10*)

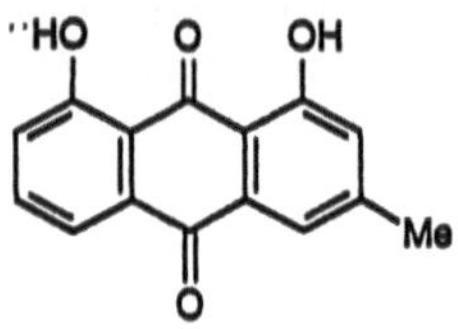

Chrysophanic acid (*79-11*)

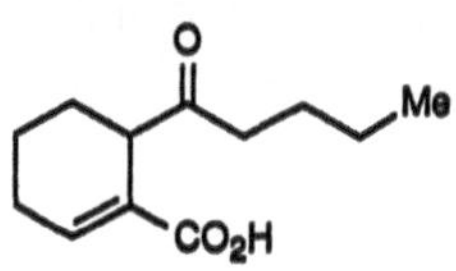

Sedanonic acid (*79-12*)

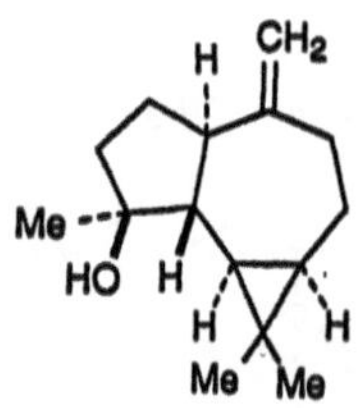

Spathulenol (*79-13*)

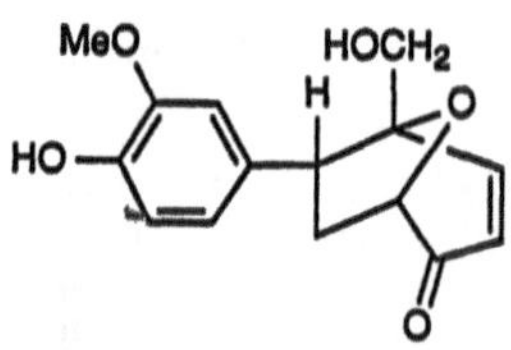

5-Hydroxymethyl-3'-methoxy-4'-hdroxyphenyl-8-oxabicyclo[3,2,1]-oct-3-en-2-one (*79-14*)

Recently, a new dimeric phthalide named diligustilide (*79-15*) was also isolated from the rhizome and structurally investigated. It was converted into ligustilide by thermolysis [9]. A related dimeric phthalide named riligustilide (*79-16*) was isolated from the root of *L. acutilobum* [10].

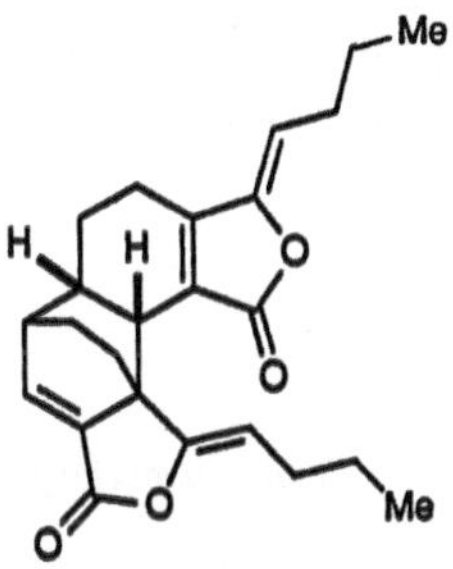

Diligustilide (*79-15*)

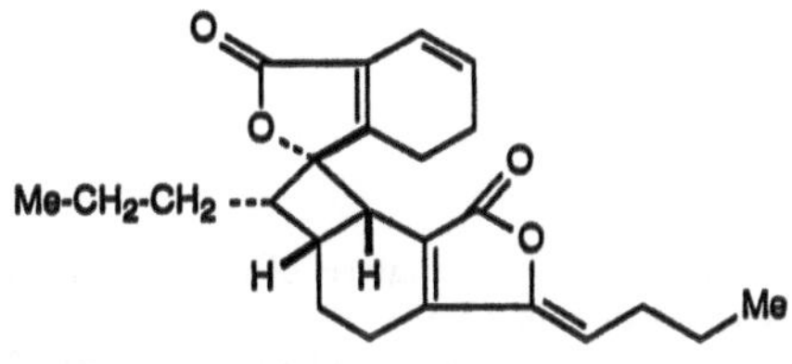

Riligustilide (*79-16*)

The major components in the essential oil from *L. jeholense* are β-phellandrene, 4-terpineol acetate, and ligustilide, whereas in that from *L. sinensis* the major components are limonene, neocnidilide, and cnidilide (*79-17*) [11]. Isolation of ligustilide, butylphthalide, sabinene [12], and butylidene phthalide [13] was also reported.

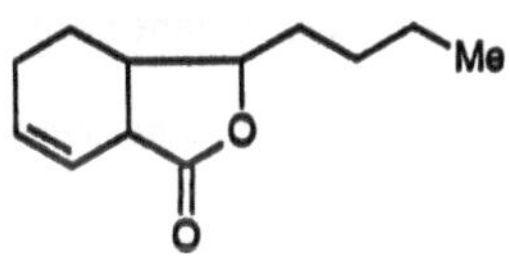

Cnidilide (*79-17*)

79.3 Pharmacology

The phthalides butylphthalide, butylidene phthalide, and ligustilide isolated from *Ligustrum* plant all showed an inhibitory effect on the contraction of nonpregnant rat uterus in vitro induced by prostaglandin $F_{2\alpha}$. Alkyl substitution at C-3 of the phthalide structure increased the inhibitory activity, probably due to an increase in lipid solubility. The Δ^3-double bond was also an important structural feature for the inhibitory effect. Butylidene phthalide was more active than ligustilide, possibly due to the aromatic character of butylidene phthalide [14]. The phthalides have also been shown to exhibit antiarrhythmic effects and dilatating activity on coronary arteries [15].

Ligustrazine inhibited platelet aggregation and might have the ability to displace Ca^{2+} from platelet membranes [16]. It prevented arterial thrombus formation, probably by inhibiting platelet aggregation [17]. Ligustrazine administered intravenously to guinea pigs at a dose of 10 mg/kg increased the coronary flow and decreased the myocardial contractile force in isolated heart. In anesthetized dogs, it lowered the arterial blood pressure and increased the left ventricular pressure, heart rate, coronary blood flow, and myocardial oxygen consumption [18].

Clinical investigations with the extract of *L. chuanxiong* for treatment of coronary disease patients was also reported [19]. Before treatment, platelet aggregation, β-thromboglobulin, platelet factor IV, and thromboxane B_2 were much higher whereas 6-ketoprostaglandin $F_{1\alpha}$ was lower in these patients than in subjects of comparable age and sex. After treatment, platelet aggregation, β-thromboglobulin, platelet factor IV, and thromboxane B_2 were decreased and 6-keto-prostaglandin $F_{1\alpha}$ was increased, indicating that this herbal medicine is a good inhibitor of platelet aggregation and thromboxane synthesis [19].

References

1. Xu ZM (1980) TLC analysis of the ethereal extract of the rhizome of *Ligusticum chuanxiong*. Chin Pharm Bull 15:5–6
2. Lu RM, He LY, Fang HJ, Zhang XQ (1980) Thin layer chromatography and densitometry of ligustilide in Umbelliferae plants. Acta Pharm Sin 15:371–374
3. Cao FY, Liu WX, Wen YS, He ZR, Qin WJ (1983) Studies on chemical constituents of *Ligusticum chuanxiong*. Chin Trad Herb Drugs 14:241–242
4. Wang PS, Gao XL, Fukuyama Y, Kanbara M (1985) Chemical constituents of rhizomes of *Ligusticum chuanxiong* Hort. – five lactones. Chin Trad Herb Drugs 16:137–138
5. Xue KF, Cao FY (1986) Chemical components of *Ligusticum chuanxiong*. Chin Trad Herb Drugs 17:122
6. Kaouadji M, Puech-Baronnat M, Mariotte AM (1983) (Z)-Ligustilidiol, a new hydroxy phthalide isolated from *Ligusticum wallichii* Franch. Tetrahedron Lett 24:4675–4676
7. Peking Institute of Pharmaceutical Industry (1980) Chemical studies on the components of *Ligusticum chuanxiong*. Chin Pharm Bull 15:39
8. Wang PS, Gao XL, Fukuyama Y, Kanbara M (1985) Chemical constituents of *Ligusticum chuanxiong* Hort. – a terpenoid compound. Chin Trad Herb Drugs 16:174
9. Kaouadji M, Reutenauer H, Chulia AJ, Marsura A (1983) (Z,Z′)-Diligustilide, a new dimeric phthalide isolated from *Ligusticum wallichii* Franch. Tetrahedron Lett 24:4677–4678
10. Meng YM, Wang QG, Zhang HD, Chen YZ, Fan YG, Wang FS (1983) Crystal and molecular structure of riligustilide. J Lanzhou Univ Nat Sci 19:76–83
11. Dai B (1988) Comparison of chemical constituents of essential oil from four species of gaoben (*Ligusticum*) by GC-MS analysis. Acta Pharm Sin 23:361–369
12. Huang YZ, Pu FD (1988) Studies on the chemical components of the essential oil from the rhizome of *Ligusticum sinense* Oliv. cv. *chuanxiong* Hort. Acta Pharm Sin 23:426–429
13. Xi YG, Sun MI, Li WM (1987) Chemical constituents of gaoben (*Ligusticum sinense*). Chin Trad Herb Drugs 18:54–55
14. Ko WC, Lin SC, Yeh CY, Wang YT (1977) Alkylphthalides isolated from *Ligusticum wallichii* Franch and their in vitro inhibitory effect on rat uterine contraction induced by prostaglandin F_2. Taiwan I Hsueh Hui Tsa Chih 76:669–677 (CA 88:130721p)
15. Fukuyama Y, Hiroyoshi O, Nobuaki K, Yoshio O (1985) *Ligusticum wallichii* phthalides as medicines. Jpn Kokai Tokkyo Koho JP 60,155,175 (85,155,175) (CA 105:85184x)
16. Nie SQ, Xie ZC, Lin KC (1985) Effects of tetramethylpyrazine on membrane fluidity and electrophoretic mobility of platelets and their relation to antiaggregation effect. Acta Pharm Sin 20:689–692
17. Zhou XB, Salganicoff L, Sevy R (1985) Pharmacological effect of ligustrazine on human platelets. Acta Pharm Sin 20:334–339
18. Chen LF, Wang GX, Li QA, Liu GH (1987) Comparison of the effect of narcissin and ligustrazin on cardiac hemodynamics. Acta Pharmacol Sin 8:123–127
19. Yu Z, Chen KJ, Qian ZH, Weng WL, Yu YQ, Tu XH, Ma HM, Wu YS, Zhang H (1987) Effect of chuanxiong granule on platelet function and prostaglandin metabolism in coronary disease patients. Chin J Intergrated Trad West 7:8–11

Lithospermum erythrorhizon Sieb. et Zucc. 80

80.1 Introduction

Zicao, Radix Lithospermi or Radix Arnebiae, is the dry root of *Lithospermum erythrorhizon* Sieb. et Zucc. or *Arnebia euchroma* (Royle) Johnst. (Boraginaceae), respectively, collected in the spring or fall. It is officially listed in the Chinese Pharmacopoeia and is used as an antiinflammatory and antipyretic agent in the treatment of measles, eczema, and thermal burn.

80.2 Chemical Constituents

80.2.1 Chemical Constituents of *Lithospermum erythrorhizon*

The root of *L. erythrorhizon* contains a number of naphthoquinone pigments as the main constituents. The average content of total naphthaquinone pigments, calculated as alkannin, was 0.47% [1]. Of the naphthaquinone pigments, shikonin is the main representative. It is isolated from the root of *L. erythrorhizon* and is structurally a naphthaquinone with an unsaturated side chain and an asymmetric centrum bearing a hydroxy group. The enantiomer alkannin was isolated first from another boraginaceous plant *Alkanna tinctoria* [2]. The absolute configurations of shikonin (*80-1*) and alkannin (*80-2*) were elucidated [3]. Shikonin possesses the (R)- and alkannin the (S)-configuration.

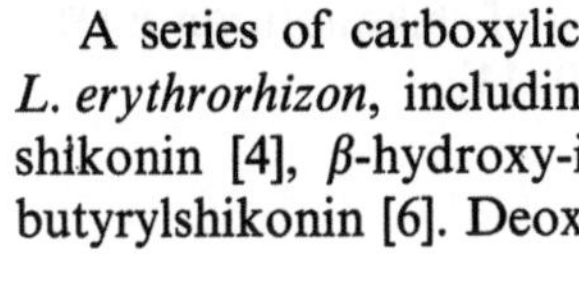

Shikonin (*80-1*)

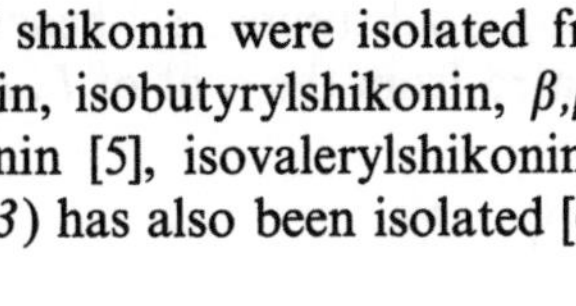

Alkannin (*80-2*)

A series of carboxylic acid esters of shikonin were isolated from the root of *L. erythrorhizon*, including acetylshikonin, isobutyrylshikonin, β,β-dimethylacrylshikonin [4], β-hydroxy-isovalerylshikonin [5], isovalerylshikonin and α-methylbutyrylshikonin [6]. Deoxyshikonin (*80-3*) has also been isolated [6].

Lithospermidins A (*80-4*) and B (*80-5*) are two new minor pigments isolated from the root of *L. erythrorhizon* [7]. Their structures were determined by comparing their spectral data with those of other pigments. They are isomeric esters of a shikonin-like structure, which differs from shikonin by side-chain modification.

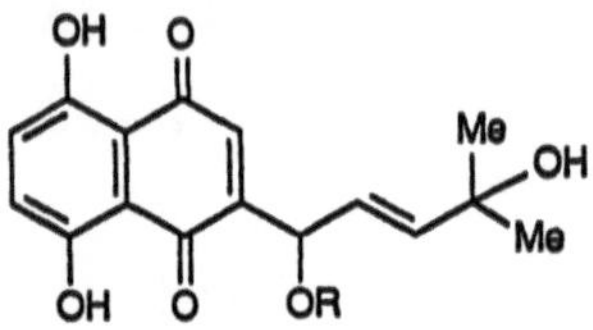

Lithospermidin A (*80-4*): R = $CH_3\text{-}CH_2\text{-}CH(CH_3)\text{-}CO\text{-}$

Lithospermidin B (*80-5*): R = $CH_3\text{-}CH(CH_3)\text{-}CH_2\text{-}CO\text{-}$

Alkannin was also isolated from *L. erythrorhizon* by extraction of the root powder with hot petroleum ether [8]. Further substances isolated from the root of *L. erythrorhizon* are furylhydroquinone derivatives. Thus, shikonofurans A (*80-6*), B (*80-7*), C (*80-8*), D (*80-9*), and E (*80-10*) were found and their structures determined [9].

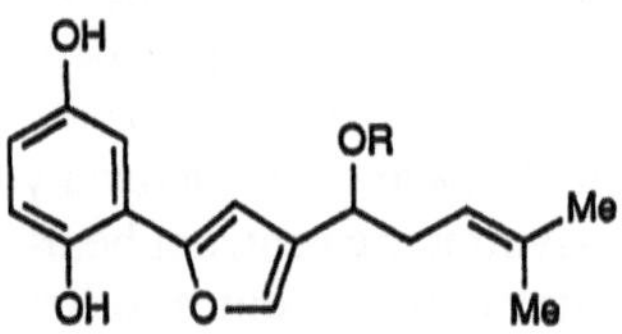

Shikonofuran A (*80-6*): R = $CH_3\text{-}CO\text{-}$

Shikonofuran B (*80-7*): R = $CH_3\text{-}CH_2\text{-}CH(CH_3)\text{-}CO\text{-}$

Shikonofuran C (*80-8*): R = $CH_3\text{-}CH(CH_3)\text{-}CH_2\text{-}CO\text{-}$

Shikonofuran D (*80-9*): R = $CH_3\text{-}CH(CH_3)\text{-}CO\text{-}$

Shikonofuran E (*80-10*): R = $CH_3\text{—}C(CH_3)\text{=}CH\text{—}CO\text{-}$

A corresponding 2-substituted 1,4-benzoquinone analog of shikonofuran E (*80-11*) was also isolated from the root of *L. erythrorhizon* [10, 11]:

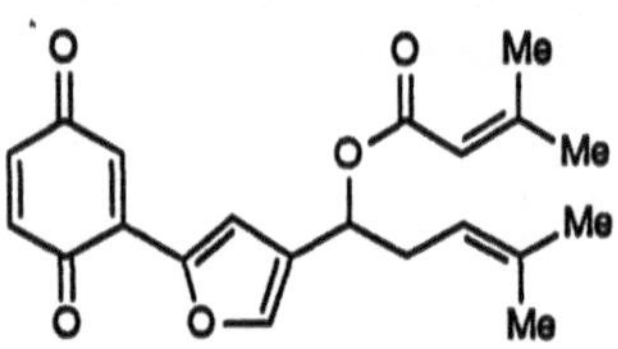

1,4-Benzoquinone analog of shikonofuran E (*80-11*)

Shikonin and its esters were also produced from callus cultures [12, 13] or cell suspension cultures of *L. erythrorhizon* [14–16]. Different strains of callus cultures showed wide variations in the production of shikonin derivatives. Mizukami et al. [13] reported that two high pigment-producing strains with stable shikonin, β,β-dimethylacrylshikonin, and β-hydroxyisovalerylshikonin contents, similar to that of intact plant root, were established by repeated selection. Deoxyshikonin, acetylshikonin, isovalerylshikonin, and β-hydroxyisovalerylshikonin were also formed by undifferentiated callus tissues of *L. erythrorhizon* [14]. A study on the localization of shikonin formation in cultured cells of *L. erythrorhizon*, demonstrated that shikonin biosynthesis is localized in the endoplasmic reticulum [17]. The isolation of dihydroshikonofuran (*80-12*) from the cell culture was also reported [18].

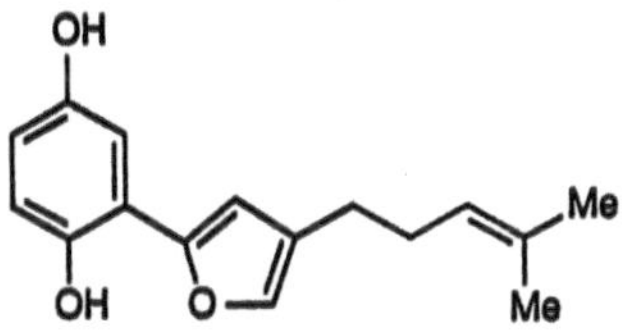

Dihydroshikonofuran (*80-12*)

Two phenolic acids were isolated from colorless cell cultures of *L. erythrorhizon* and identified as rosmarinic acid (*80-13*) and lithospermic acid (*80-14*) [19]. Lithospermic acid was first isolated from *L. ruderale* [20], a plant with gonadotropin-inhibitory activity [21]. The structure of lithospermic acid was determined 10 years later [22]. The aliphatic alcohols in the colorless mixture of caffeic acid esters isolated from *L. erythrorhizon* were identified after hydrolysis as stearyl alcohol, eicosanol, docosanol, and tetracosanol [23].

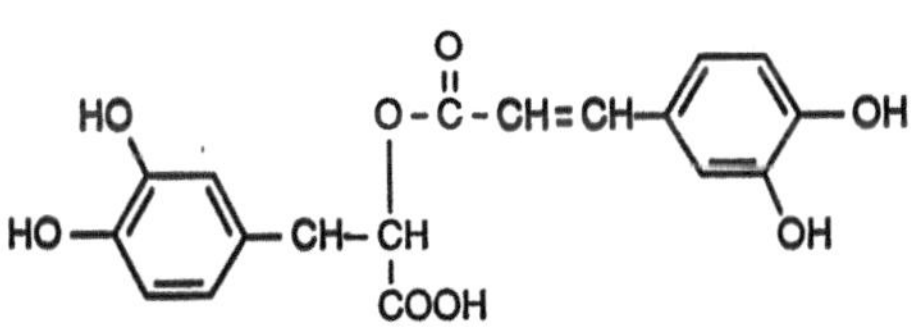

Rosmarinic acid (*80-13*)

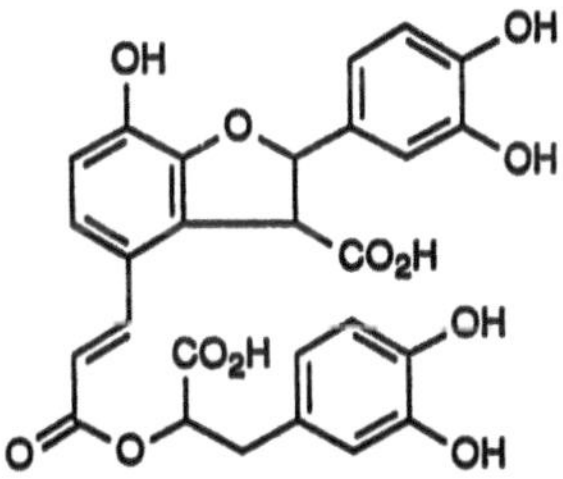

Lithospermic acid (*80-14*)

Three glycans, named lithospermans A, B, and C, were isolated from the water extract of *L. erythrorhizon* roots [24]. The analytical results showed that lithosperman A contains rhamnose, fucose, arabinose, and galactose as neutral sugar components in a molar ratio of 2.3 : 0.7 : 1.8 : 1.0; lithosperman B contains rhamnose, fucose, arabinose, xylose, mannose, galactose, and glucose in a ratio of 0.9 : 0.7 : 0.1 : 0.1 : 0.5 : 1.0 : 1.2; and lithosperman C contains rhamnose, fucose, arabinose, xylose, mannose, galactose, and glucose in a ratio of 0.8 : 0.6 : 0.1 : 0.1 : 0.4 : 1.0 : 1.1. Furthermore, lithospermans A, B, and C were found to contain acetoxy groups and a peptide moiety. In addition, galacturonic acid and glucuronic acid were present in lithosperman A.

80.2.2 Chemical Constituents of *Arnebia euchroma*

The root of *A. euchroma*, like the root of *L. erythrorhizon*, contains a number of naphthaquinone pigments. Shikonin, deoxyshikonin, acetylshikonin [25], *β,β*-dimethylacrylalkannin [25, 26], *β*-acetoxyisovalerylalkannin, *β*-hydroxyisovaleryl-alkannin [25], teracrylshikonin [27], and alkannin [8] were found. Alkannin contents varied from 5% in the top-quality material from *A. euchoma* to 0.8% in the low-quality material from *A. guttata* [8]. A mixture of shikonofurans B and C was also found in *A. euchroma* as a slightly yellow oil [28].

Arnebifuranone (*80-15*), a new monoterpenylbenzoquinone, was isolated from *A. euchroma*. It was found to be a 2,3-dimethoxy-1,4-benzoquinone with a monoterpene side chain containing a furan ring [29].

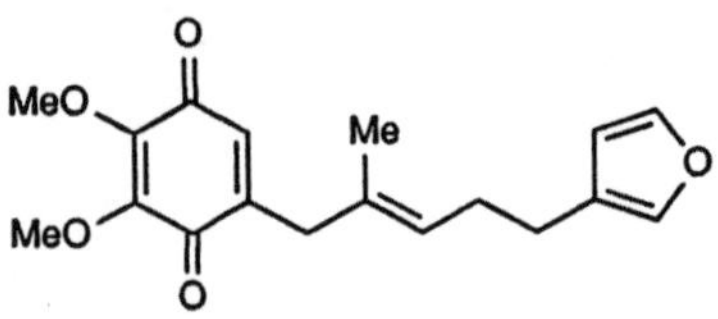

Arnebifuranone (*80-15*)

Arnebinone (*80-16*), isolated from the root of *A. euchroma*, is another novel monoterpenylbenzoquinone, in which the monoterpenyl moiety forms a six-membered ring condensed to benzoquinone. The structure of arnebinone was elucidated by spectroscopic studies [28].

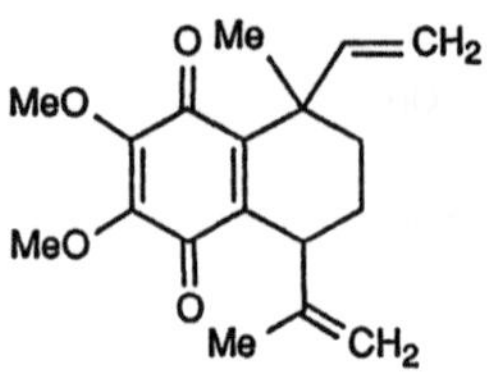

Arnebinone (*80-16*)

Furthermore, *O*-demethyllasiodiplodin (*80-17*) and a new monoterpenylbenzoid of the ansa type named arnebinol (*80-18*) were isolated from the roots of *A. euchroma* [30]. Their structures were determined by spectral and X-ray analysis.

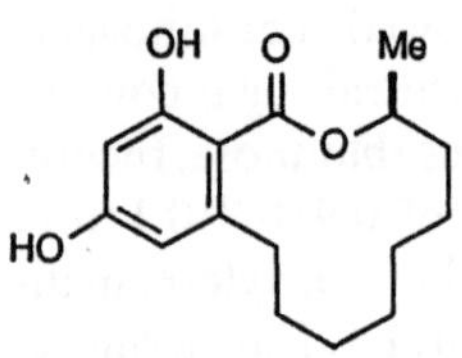

O-Demethyllasiodiplodin (*80-17*)

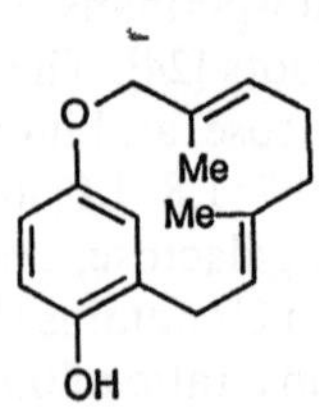

Arnebinol (*80-18*)

Several other plants, e.g. *Arnebia guttata, Onosma paniculatum*, and *O. hookeri*, are used under the name of the traditional Chinese medicine Zicao. All these plants contain naphthaquinone pigments with β,β-dimethylacrylalkannin and acetylshikonin [25] as the main constituents.

80.3 Pharmacology

The ether extract of *L. erythrorhizon*, the fraction soluble in petroleum ether and the subfraction soluble in acetone of the petroleum ether fraction were inhibitory to the growth of *Staphylococcus aureus, S. epidermidis, Sarcina lutea*, and *Bacillus subtilis*. The growth of *Saccharomyces cerevisiae* was slightly inhibited but not the growth of *Escherichia coli* and *Pseudomonas aeruginosa* [31]. β,β-Dimethylacrylshikonin and hydroxyisovalerylshikonin inhibited the growth of *Bacillus subtilis, Staphylococcus aureues*, and *Sarcina lutea*, but not that of *E. coli* [32]. In concentrations of 20–30 μg/ml, shikonin had a bactericidal effect on *Lactobacillus* [33].

Shikonin showed a significant antiamebic action on *Entamoeba histolytica* at 0.5–10 μg/ml in the culture medium. However, when administered orally at 0.25–0.50 mg/animal per day for 6 days to rats with experimental intestinal amebiasis, shikonin only showed a weak therapeutic effect [34].

LD_{50} values of shikonin and acetylshikonin were 20 and 41 mg/kg respectively by intraperitoneal administration and >1000 mg/kg oral administration in mice. Weak analgesic effects and moderate antipyretic actions were also observed [35].

Shikonin and acetylshikonin administered orally to rats inhibited histamine-induced vascular permeability [35]. Acetylshikonin also showed antiinflammatory effects in adrenalectomized rats [36]. β,β-Dimethylacrylshikonin showed antiinflammatory activity in the following tests: inhibition of histamine-induced capillary permeability in rats, inhibition of rat paw edema in intact and adrenalectomized rats, and inhibition of cotton pellet granuloma. The intraperitoneal LD_{50} of β,β-dimethylacrylshikonin in mice was 48 mg/kg [37].

The aqueous extract of the herb was found to exhibit gonadotropin antagonistic activity and was effective in terminating early pregnancy in mice and rabbits [38, 39].

Alkannin was fed to mice for 15 weeks at 1% of the diet with no evidence of toxicity. The average total intake per animal was 3.4 g. Alkannin was excreted via urine and was not deposited in abdominal fat. The LD_{50} of alkannin was 3 g/kg for mice and >1.0 g/kg for rats by oral administration [40].

The pharmacokinetics of shikonin in mice were studied using [^{3}H]shikonin. The plasma concentration time curve obtained after intravenous injection was shown to fit a three-compartment open model. The drug concentration was found to be high in bile and liver; moderate in lung, spleen, kidney, heart, and skin; and low in testis, muscle, and brain. The absorption of shikonin was rapid after oral or intramuscular administration. In both cases radioactivity could be detected from the plasma about 1 min after administration. The bioavailability was 34% for the oral route and 65% for the intramuscular injection. Within 96 h following intravenous injection, the total radioactivity excreted in both urine and feces was 40% of the dose, of which only 3.6% and 7.7% were unchanged drug [41]. Shikonin showed high antitumor activity against ascites cells of sarcoma 180 in mice. It completely inhibited tumor growth at a dose of 5–10 mg/kg per day [42]. The naphthoquinones with a prenyl side chain were not mutagenic in the *Salmonella*/microsome test [43].

Shikonofurans B and C and *O*-demethyllasiodiplodin isolated from the roots of *A. euchroma* possess significant inhibitory effect on prostaglandin biosynthesis. Arnebinol and arnebinone had less inhibitory activity [44].

References

1. Li ZL, Zhang M, Guo LJ (1983) Assay of total pigments in *Lithospermum erythrorhizon*. Chin Trad Herb Drugs 14:476
2. Brockmann H (1936) The constitutions of alkannin, shikonin and alkannan. Liebigs Ann Chem 521:1–47
3. Arakawa H, Makazaki M (1961) Absolute configuration of shikonin and alkannin. Chem Ind 947
4. Morimoto I, Kishi T, Ikegami S, Hirata Y (1965) Naphthoquinone derivatives from *Lithospermum erythrorhizon*. Tetrahedron Lett 4737–4739
5. Morimoto I, Hirata Y (1966) Naphthoquinone derivatives from *Lithospermum erythrorhizon*. Tetrahedron Lett 3677–3680
6. Kyogoku K, Terayama H, Tachi Y, Suzuki T, Komatsu M (1973) Constituents of shikon. I. Structure of three new shikonin derivatives and isolation of anhydroalkannin. Shoyakugaku Zasshi 27:24–30
7. Hisamichi S, Yoshizaki F (1982) Studies on the shikon. I. Structures of new minor pigments and isolation of two isomers of shikonin derivatives from *Lithospermum erythrorhizon* Sieb. et Zucc. Shoyakugaku Zasshi 36:154–159
8. Li ZL, Zhang M, Guo LJ (1986) Determination of alkannin in Zicao (radix arnebiae or lithospermi). Chin J Pharm Anal 6:41–42
9. Yoshizaki F, Hisamichi S, Kondo Y, Sato Y, Nozoe S (1982) Studies on shikon. III. New furylhydroquinone derivatives, shikonofurans A, B, C, D and E, from *Lithospermum erythrorhizon* Sieb. et Zucc. Chem Pharm Bull (Tokyo) 30:4407–4411
10. Krivoshchekova OE, Fedorev SA, Denisenko VA, Maksimov OB (1977) New quinoid pigment from *Lithospermum erythrorhizon*. Khim Prir Soedin 702–703
11. Denisenko VA, Fedoreev SA, Mishchenko NP (1979) Determination of the structure of a new quinoid pigment from *Lithospermum erythrorhizon* by proton and carbon-13 NMR spectroscopic methods. Deposited Doc VINITI 2544, 8 pp (CA 94:80234h)
12. Inouye H, Uedo S, Inoue K, Matsumura H (1979) Quinones and related compounds in higher plants. VIII. Biosynthesis of shikonin in callus cultures of *Lithospermum erythrorhizon*. Phytochemistry 18:1301–1308
13. Mizukami H, Konoshima M, Tabata M (1978) Variation in pigment production in *Lithospermum erythrorhizon* callus cultures. Phytochemistry 17:95–97
14. Tabata M, Mizukami H, Hirakoa N, Konoshima M (1974) Pigment formation in callus cultures of *Lithospermum erythrorhizon*. Phytochemistry 13:927–932
15. Fujita Y, Hara Y, Ogino T, Suga C (1981) Production of shikonin derivatives by cell suspension cultures of *Lithospermum erythrorhizon*. I. Effects of nitrogen sources on the production of shikonin. Plant Cell Rep 1:59–60 (CA 96:119129p)
16. Fujita Y, Hara Y, Suga C, Morimoto T (1981) Production of shikonin derivatives by cell suspension cultures of *Lithospermum erythrorhizon*. II. A new medium for the production of shikonin derivatives. Plant Cell Rep 1:61–63 (CA 96:119130q)
17. Tabata M, Yoshikawa N, Tsukada M, Fukui H (1982) Localization and regulation of shikonin formation in the cultured cells of *Lithospermum erythrorhizon*. In: Fujiwara A (ed) Plant Tissue Cultures. Proc Int Congr Plant Tissue Cell Cult, 5th, Maruzen, Tokyo, pp 335–336
18. Yazaki K, Fukui H, Tabata M (1987) Dihydroshikonofuran, an unusual metabolite of quinone biosynthesis in *Lithospermum* cell cultures. Chem Pharm Bull (Tokyo) 35:898–901
19. Fukui H, Yazaki K, Tabata M (1984) Two phenolic acids from *Lithospermum erythrorhizon* cell suspension cultures. Phytochemistry 23:2398–2399
20. Johnson G, Sunderwirth SG, Gibian H, Coulter AW, Grassner FX (1963) *Lithospermum ruderale*, partial characterization of the principal polyphenyl isolated from the roots. Phytochemistry 2:145–150

21. Breneman WR, Zeller FJ, Carmack M, Kelley CJ (1976) In vivo inhibition of gonadotropins and thyrotropin in the chick by extracts of *Lithospermum ruderale*. Gen Comp Endocrinol 28:24–32
22. Kelley CJ, Mahajan JR, Brooks LC, Neubert LA, Breneman WR, Carmack M (1975) Polyphenolic acids of *Lithospermum ruderale* (Boraginaceae). I. Isolation and structure determination of lithospermic acid. J Org Chem 40:1804–1815
23. Hisamichi S, Yoshizaki F, Kondo Y (1982) Studies on the shikon. II. The colorless substances isolated from *Lithospermum erythrorhizon* Sieb. et Zucc. Shoyakugaku Zasshi 36:170–172
24. Konno C, Mizuno T, Hikino H (1985) Validity of oriental medicines. LXXIX. Antidiabetes drugs. 3. Isolation and hypoglycemic activity of lithospermans A, B and C, glycans of *Lithospermum erythrorhizon* roots. Planta Med 51:157–158
25. Fu SL, Shang TM, Xiao PG (1984) Analysis of naphthaquinone pigments in some Chinese medicinal "Zi Cao". Acta Pharm Sin 19:921–925
26. Liu GS (1981) Isolation and identification of alkannin β,β′-dimethylacrylate, a new naphthaquinone component in *Arnebia euchroma*. Johnst. Acta Pharm Sin 16:14–15
27. Zhu FC, Lu FS, Xiang GQ (1984) Isolation of shikonin and its derivatives by HPLC. Sepu 1:131–133
28. Yao XS, Ebizuka Y, Noguchi H, Kiuchi F, Sankawa U, Seta H (1983) Structure of anebinone, a novel monoterpenylbenzoquinone with inhibitory effect to prostaglandin biosynthesis. Tetrahedron Lett 24:3247–3250
29. Yao XS, Ebizuka Y, Noguchi H, Kiuchi F, Seto H, Sankawa U (1984) Structure of arnebifuranone, new monoterpenylbenzoquinone from *Arnebia euchroma*. Tetrahedron Lett 25:5541–5542
30. Yao XS, Ebizuka Y, Noguchi H, Kiuchi F, Iitaka Y, Sankawa U, Seto H (1983) Structure of arnebinol, a new ansa-type monoterpenylbenzoid with inhibitory effect on prostaglandin biosynthesis. Tetrahedron Lett 24:2407–2410
31. Kyogoku K, Tarayama H, Tachi Y, Suzuki T, Komatsu M (1973) Constituents of shikon. II. Comparison of contents, constituents and antibacterial effect of fat soluble fraction between nanshikon and koshikon. Shoyakugaku Zasshi 27:31–36
32. Tanaka Y, Odani T (1972) Pharmacodynamic study on Shiunko. I. Antibacterial effect of Shiunko. Yakugaku Zasshi 92:525–530
33. Sherbanivskii LP (1971) Presence of the shikonin in some species of the family Boraginaceae and its effect on lactic acid bacteria. Ukr Bot Zh 28:504–508 (CA 76:68687d)
34. Rubinchik MA (1972) Antiamoebic properties of shikonin. In: Turova AD (ed) Mater Vses Konf Issled Lek Rast Perspekt Ikh Ispolz Proizvod Lek Prep 1970 Bittsa USSR 236–237 (CA 83:671h)
35. Hayashi M (1977) Pharmacological studies on crude plant drugs, Shikon and Tooki. II. Shikonin and acetylshikonin. Nippon Yakurigaku Zasshi 73:193–203
36. Liu ZB, Chai BL, Wang P, Guo QX, Lu FS, Xiang GQ (1980) Studies on the antiinflammatory effect of chemical principle of Zi-Cao (*Arnebia euchroma* (Royle) Johnst.). J Beijing Med Coll 12:101–106
37. Liu ZB, Wang P, Ruan Y, Guo QX (1980) Antiinflammatory effect of β,β-dimethylacrylshikonin. Acta Pharmacol Sin 1:60–63
38. Gao JH (1986) Resources, chemistry, pharmacology and clinical application of zicao. Chin Trad Herb Drugs 17:268–271
39. Winterhoff H, Gumbinger HG, Sourgens H (1988) On the antigonadotropic activity of *Lithospermum* and *Lycopus* species and some of their phenolic constituents. Planta Med 54:101–106
40. Majlathova L (1971) Feeding trial with alkannin in mice. Nahrung 15:505–508
41. Wang WJ, Yi MG, Zhu XY (1988) A study on absorption, distribution and excretion of [^{3}H]shikonin in mice. Acta Pharm Sin 23:246–251
42. Sankawa U, Ebizuka Y, Miyazaki T, Isomura Y, Otsuka H, Shibata S, Inomata M, Fukuoka F (1977) Antitumor activity of shikonin and its derivatives. Chem Pharm Bull (Tokyo) 25:2392–2395
43. Tikkanen L, Matsushima T, Natori S, Yoshihira K (1983) Mutagenicity of natural naphthoquinones and benzoquinones in the *Salmonella*/microsome test. Mutat Res 124:25–34
44. Yao XS, Ebizuka Y, Noguchi H, Kiuchi F, Iitaka Y, Sankawa U, Seto H (1983) Prostaglandin biosynthesis inhibitors of nan-shikon, the root of *Arnebia euchroma*. Tennen Yuki Kagobutsu Toronkai Koen Yashishu 26:134–141 (CA 100:17999r)

Lonicera japonica Thunb. 81

81.1 Introduction

Two items from *Lonicera* species are officially listed in the Chinese Pharmacopoeia:

- Rendongteng, Caulis Lonicerae, is the dry cane of *Lonicera japonica* Thunb. (Caprifoliaceae) collected in the fall and winter. It is used in traditional Chinese medicine as an antiphlogistic and bacteriostatic in the treatment of fever, dysentery, abscess, and rheumatic swelling.
- Jinyinhua, Flos Lonicerae, is the dry flower buds of *L. japonica, L. hypoglauca* Miq., *L. confusa* DC. or *L. dasystyla* Rehd. collected in early summer before the buds bloom. It is used as an antibacterial and antiphlogistic agent in the treatment of abscess, laryngeal catarrh, erysipelas, dysentery, cold, and fever.

81.2 Chemical Constituents

Using chromatography on silica gel, two compounds were isolated from the essential oil of the flowers of *L. japonica*. They were determined as linalool and 2,6,6-trimethyl-2-vinyl-5-hydroxytetrahydropyran (*81-1*).

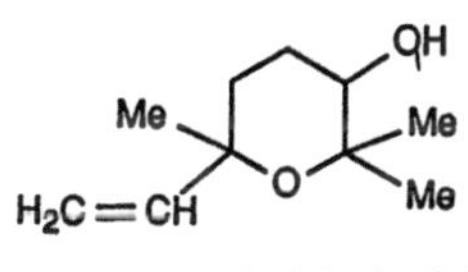

2,6,6-Trimethyl-2-vinyl-5-hydroxytetrahydropyran (*81-1*)

In addition, 12 compounds of the essential oil were identified by gas-liquid chromatography-mass spectroscopy as pinene, hex-1-ene, hex-3-en-1-ol, *cis-* and *trans*-2-methyl-2-vinyl-5-(α-hydroxyisopropyl)-tetrahydrofuran, geraniol, α-terpineol, benzyl alcohol, *β*-phenylethyl alcohol, carvacrol, eugenol, and an isomer of the above-mentioned pyran derivative [1].

The flower of *L. japonica* is also rich in chlorogenic acid (*81-2*). The chlorogenic acid content in the flower was about 6% [2, 3–6]. The contents in root, cane, and leaves were 1.4%, 0.9%, and 2.6%, respectively [2]. Isochlorogenic acid was also found in flower and cane of *L. japonica* and some other *Lonicera* species [7, 8]. The so-called isochlorogenic acid is a mixture [9] composed of three isomers, isochlorogenic acids a, b, and c. Isochlorogenic acid a (*81-3*) represents 3,5-dicaffeoyl quinic acid, whereas isochlorogenic acids b (*81-4*) and c (*81-5*) are two isomers of 3,4-dicaffeoyl quinic acid. The leaves of some varieties of *L. japonicca* were found to contain more chlorogenic and isochlorogenic acids than the corresponding flowers. The acid levels in the leaves peaked in August and September [10].

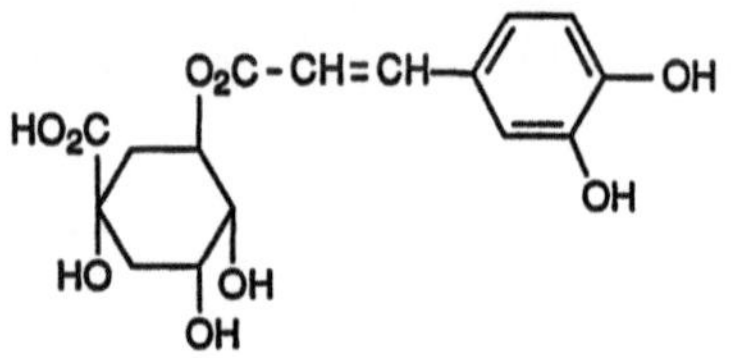

Chlorogenic acid (*81-2*)

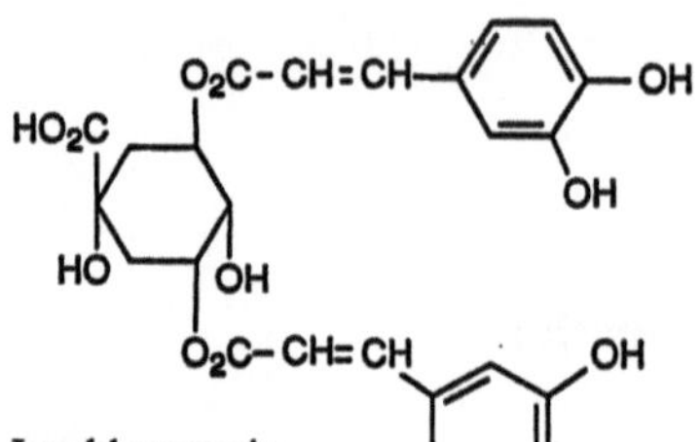

Isochlorogenic acid a (*81-3*)

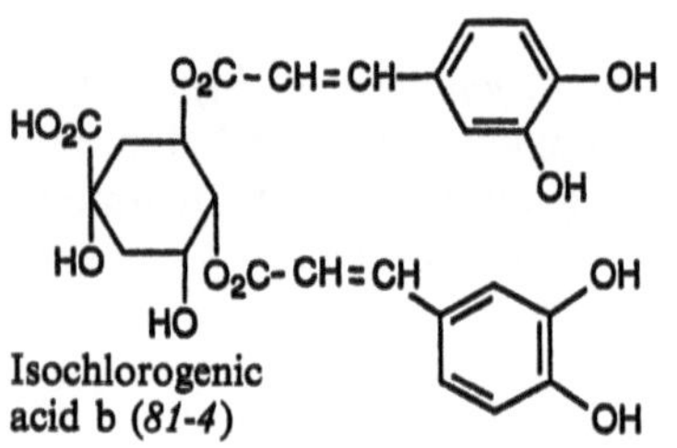

Isochlorogenic acid b (*81-4*)

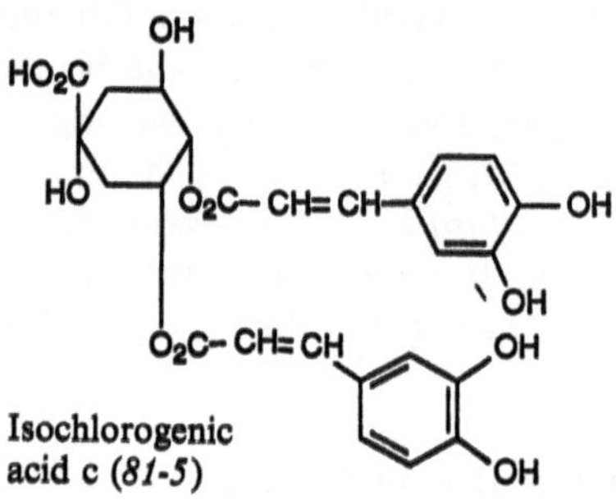

Isochlorogenic acid c (*81-5*)

Moreover, two flavones lonicerin (*81-6*) [11] and loniceraflavone [12] were found in *L. japonica*. Lonicerin is a neohesperidoside of luteolin, whereas loniceraflavone corresponds to 5,6,4′-trihydroxyflavone.

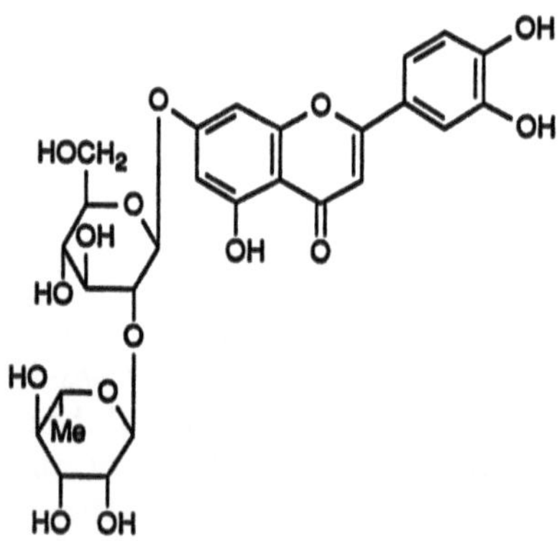

Lonicerin (*81-6*)

A number of iridoid glycosides were isolated from *L. japonica* and identified as loganin (*81-7*) [13], secoxyloganin (*81-8*) [14], secologanin dimethylacetal (*81-9*), and vogeloside (*81-10*) [15]. A new iridoid glycoside, epivogeloside (*81-11*) was also isolated and structurally determined [15].

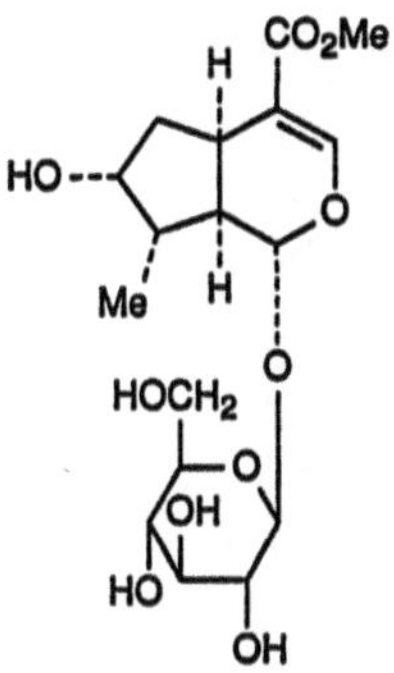

Loganin (*81-7*)

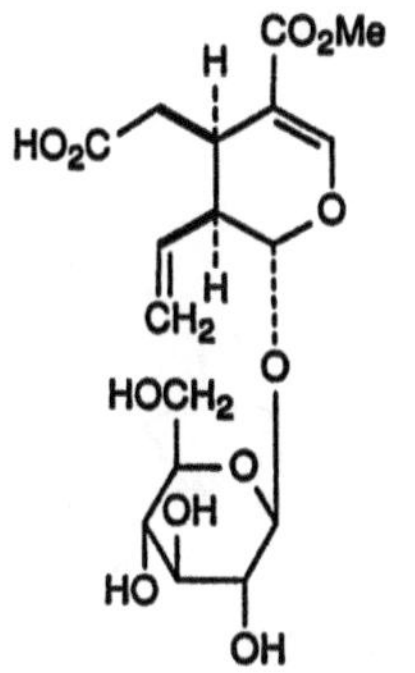

Secoxyloganin (*81-8*)

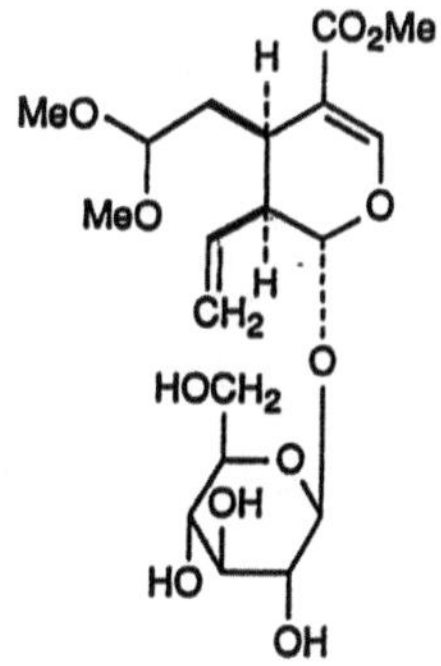

Secologanin dimethylacetal (*81-9*)

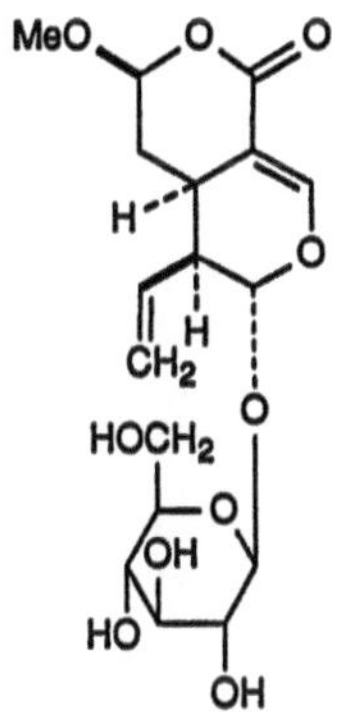

Vogeloside (*81-10*)

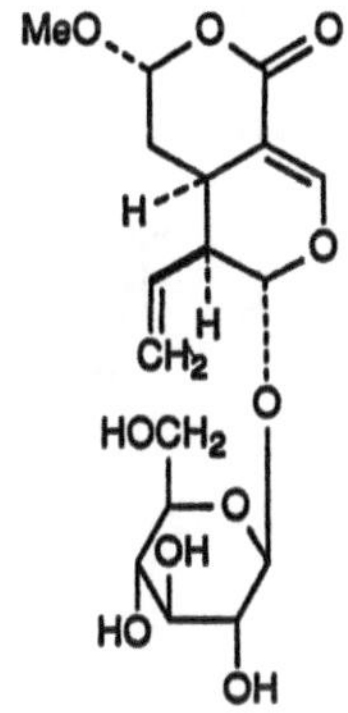

Epivogeloside (*81-11*)

Twelve saponins having oleanolic acid or hederagenin as aglycones were isolated from the aerial part of *L. japonica* and identified as: 3-*O*-α-L-arabinopyranosyl hederagenin, 3-*O*-β-D-glucopyranosyl-(1→2)-α-L-arabinopyranosyl hederagenin, 3-*O*-α-L-rhamnopyranosyl-(1→2)-α-L-arabinopyranosyl hederagenin, 3-*O*-α-L-arabinopyranosyl hederagenin 28-β-D-glucopyranosyl-(1→6)-β-D-glucopyranosyl ester, 3-*O*-β-D-glucopyranosyl-(1→2)-α-L-arabinopyranosyl hederagenin 28-β-D-glucopyranosyl-(1→6)-β-D-glucopyranosyl ester, 3-*O*-α-L-rhamnopyranosyl-(1→2)-α-L-arabinopyranosyl hederagenin 28-β-D-glucopyranosyl ester (*81-12*), 3-*O*-α-L-rhamnopyranosyl-(1→2)-α-L-arabinopyranosyl hederagenin 28-β-D-glucopyranosyl-(1→6)-β-D-glucopyranosyl ester, 3-*O*-α-L-rhamnopyranosyl-(1→2)-α-L-arabinopyranosyl hederagenin 28-*O*-β-D-6-*O*-acetyl-glucopyranosyl-(1→6)-β-D-glucopyranosyl ester (*81-13*), 3-*O*-β-D-glucopyranosyl-(1→2)-α-L-arabinopyranosyl oleanolic acid, 3-*O*-α-L-arabinopyranosyl oleanolic acid 28-*O*-β-D-glucopyranosyl-(1→6)-β-D-glucopyranosyl ester (*81-14*), 3-*O*-β-D-glucopyranosyl-(1→2)-α-L-arabinopyranosyl oleanolic acid 28-β-D-glucopyranosyl-(1→6)-β-D-glucopyranosyl ester, and 3-*O*-α-L-rhamnopyranosyl-(1→2)-α-L-arabinopyranosyl oleanolic acid 28-β-D-glucopyranosyl-(1→6)-β-D-glucopyranosyl ester (*81-15*). The structures of four new saponins (*81-12–81-15*) were determined [16].

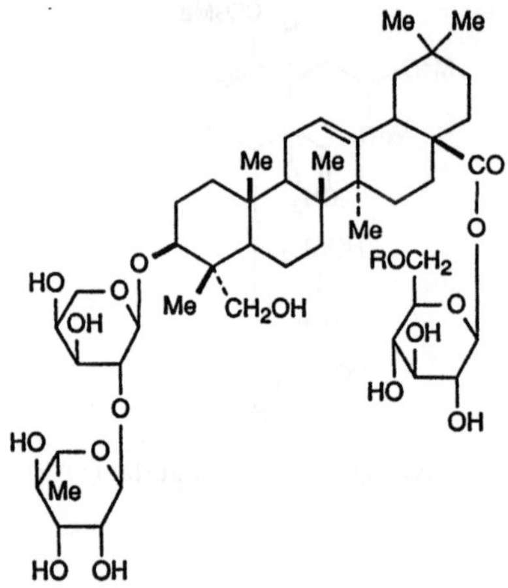

(*81-12*): R=H

(*81-13*): R= 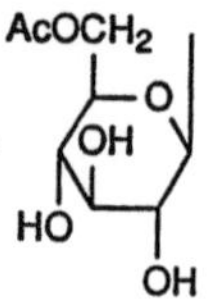

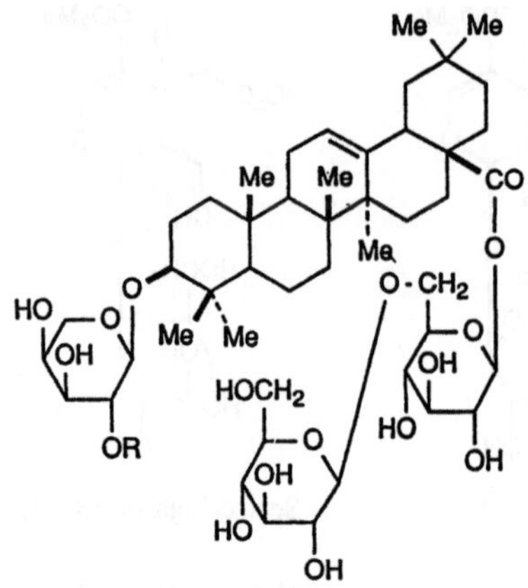

(*81-14*): R=H

(*81-15*): R= 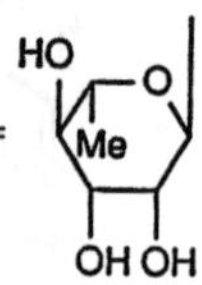

81.3 Pharmacology

Water extracts of *L. japonica* showed a bacteriostatic effect on *Staphylococcus albus* ranked in the order leaf ≥ flower bud > root > stem extract. Leaf extract also had a bacteriostatic effect on *S. aureus* [2]. Control of influenza virus infection in mice by chlorogenic acid has been demonstrated [13].

Intraperitoneal injection of an aqueous-ethanolic extract of *L. japonica* to mice on day 8 after successful mating reduced pregnancy in the test animals dose dependently. Intrauterine and intraamniotic administration of the extract killed the fetuses in dogs and caused abortion in monkeys, respectively [17].

Lonicera chrysantha and *L. maackii* had chlorogenic acid contents comparable to those of *L. japonica* and thus could be employed medicinally [18].

References

1. Wu YL, Fang HJ (1980) Constituents of the essential oil from the flowers of *Lonicera japonica* Thunb. Acta Chim Sin 38:573–580
2. Song GY, Cai MQ, Liu QQ (1985) Chlorogenic acid contents in different organs of Japanese honeysuckle (*Lonicera japonica*) and its bacteriostatic effect. Chin Trad Herb Drugs 16:229–230
3. Zhang XQ, Xu LX (1987) A rapid pulsed polarographic method for the determination of total chlorogenic acid in *Lonicera japonica* Thunb. Chin J Pharm Anal 7:162–164
4. Huang XF (1988) Separation of components in honeysuckle by reversed-phase HPLC and determination of chlorogenic acid. Chin Trad Herb Drugs 19:206–208

5. Peng GF, Lin BB, Zhong FX (1988) The comparison of chlorogenic acid and volatile oil during the various development stages in Japanese honeysuckle (*Lonicera japonica*). Chin Trad Herb Drugs 19:558–559
6. Gao WL, Liu QL (1989) Quantitative determination of chlorogenic acid and isochlorogenic acid by TLC-densitometry. J Shenyang Coll Pharm 6:99–103
7. Li BT (1986) Comparative analysis of the chlorogenic acid and isochlorogenic acid in flower and cane of Japanese honeysuckle (*Lonicera japonica* Thunb.) Chin Trad Herb Drugs 17:10–11
8. Ding J, Li CH, Kung HC (1981) Comparative determination of isochlorogenic acid and chlorogenic acid in 14 species of honeysuckle. Chin Trad Herb Drugs 12:10–14
9. Mialoundama F, Paulet P (1975) Change in chlorogenic acid and isochlorogenic acid contents during cold treatment of *Cichorium intybus* roots. Physiol Plant 35:39–44
10. Dong XG, Ding L, Cui YZ (1985) Levels of chlorogenic acid and isochlorogenic acid in the leaves of *Lonicera japonica*. Bull Chin Mater Med 10:223–225
11. Inagaki I, Sakushima A, Hisada S, Nishibe S, Morita N (1974) Comparison of lonicerin and veronicastroside. Yakugaku Zasshi 94:524–525
12. Lin QS (1977) Chemistry of the Constituents in Chinese Medicinal Herbs. Science, Beijing, pp 325–326
13. Taguchi H, Iketani Y, Niitsu K, Mori K (1985) Pharmaceuticals for treatment of influenza virus infection. Jpn Kokai Tokkyo Koho JP 60,243,016 [85,243,016]. (CA 104:116118q)
14. Mehrotra R, Singh C, Popli SP (1988) Isolation of secoxyloganin from *Lonicera japonica* and its conversion into secologanin. J Nat Prod 51:319–321
15. Kawai H, Kuroyanagi M, Ueno A (1988) Iridoid glucosides from *Lonicera japonica* Thunb. Chem Pharm Bull (Tokyo) 36:3664–3666
16. Kawai H, Kuroyanagi M, Umehara K, Ueno A, Satake M (1988) Studies on the saponins of *Lonicera japonica* Thunb. Chem Pharm Bull (Tokyo) 36:4769–4775
17. Cao CP, Huang ZN, Qian PL, Pan XX (1986) Fertility regulating effect of the flower bud of *Lonicera japonica*. Pharm Ind 17:115–117
18. Zhang DZ, Lu QW, Po MZ, Wang YQ, Dong MH, Duan WH (1984) New sources of jinyinhua. Jilin Med J 5:56–57

Luffa cylindrica (L.) Roem. 82

82.1 Introduction

Sigualuo, Retinervus Luffae fructus, is the fibrovascular bundle of the dry ripe fruits of *Luffa cylindrica* (L.) Roem. (Cucurbitaceae) collected in the summer and fall when the fruits are ripe and the pericarp has turned yellow. It is officially listed in the Chinese Pharmacopoeia and used for treatment of paralytic diseases. In folk medicine it is used as an antitussive and in the treatment of chronic bronchitis.

82.2 Chemical Constituents

In recent years, a number of saponins named lucyosides A–M [1–3] together with oleanolic acid [4], oleanolic acid 3-*O*-β-D-glucopyranoside, hederagenin 3-*O*-β-D-glucopyranoside [3], and ginsenosides Re and Rg_1 [1] were isolated from the fruits or leaves and stem of *L. cylindrica*. Oleanolic acid, hederagenin, 21β-hydroxyhederagenin, arjunolic acid, machaerinic acid, maslinic acid, gypsogenin, 2α-hydroxygypsogenin, and 21β-hydroxygypsogenin were identified as sapogenins for the lucyosides. The structures of lucyosides A–M (*81-1–81-13*) are summarized in Table 82.1.

Table 82.1. Structures of the triterpene saponins lucyosides from *Luffa cylindrica*

Saponin	Structure
Lucyoside A (*82-1*) 3-*O*-*β*-D-Glucopyranosyl-21*β*-hydroxyhederagenin *β*-D-glucopyranosyl ester	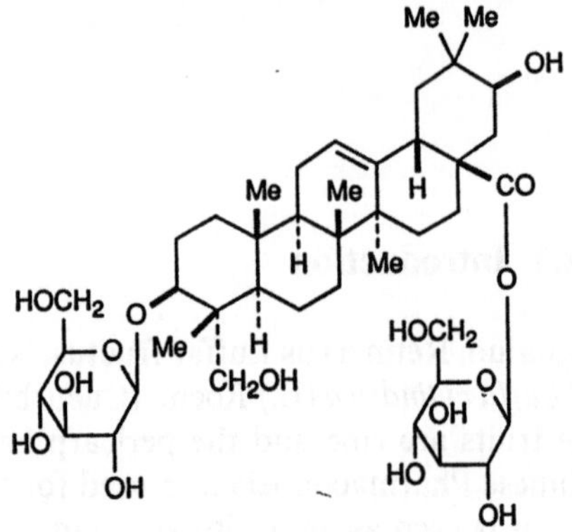
Lucyoside B (*82-2*) 3-*O*-*β*-D-Glucopyranosyl-arjunolic acid *β*-D-glucopyranosyl ester	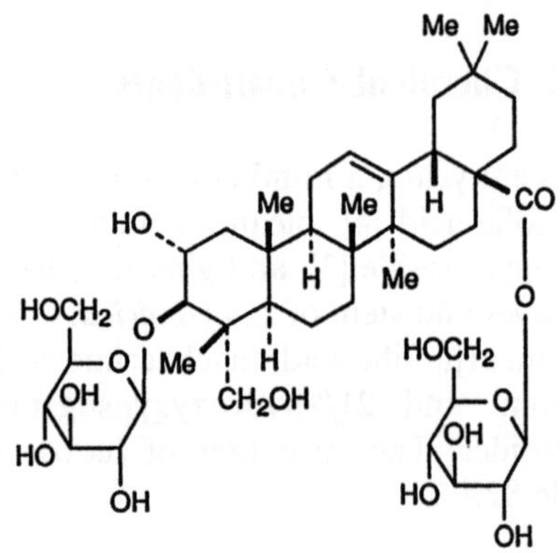
Lucyoside C (*82-3*) 3-*O*-*β*-D-Glucopyranosyl-machaerinic acid *β*-D-glucopyranosyl ester	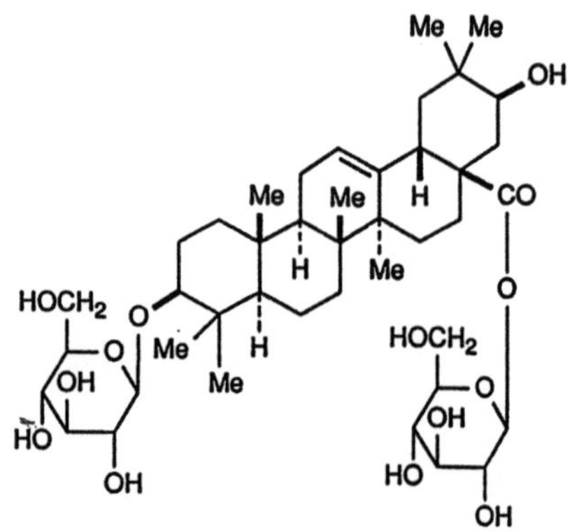

Table 82.1. (continued)

Saponin	Structure
Lucyoside D (*82-4*) 3-*O*-*β*-D-Glucopyranosyl- 2α-hydroxygypsogenin *β*-D-glucopyranosyl ester	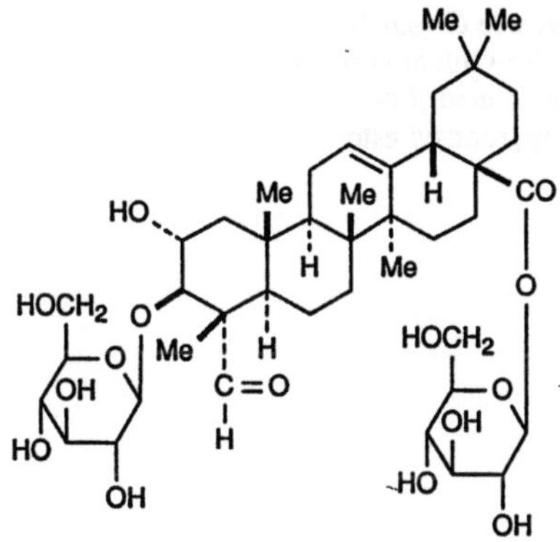
Lucyoside E (*82-5*) 3-*O*-*β*-D-Glucopyranosyl- hederagenin *β*-D- glucopyranosyl ester	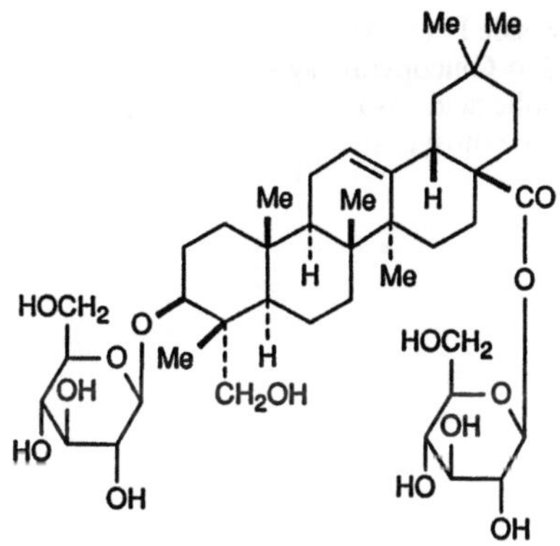
Lucyoside F (*82-6*) 3-*O*-*β*-D-Glucopyranosyl- gypsogenin *β*-D- glucopyranosyl ester	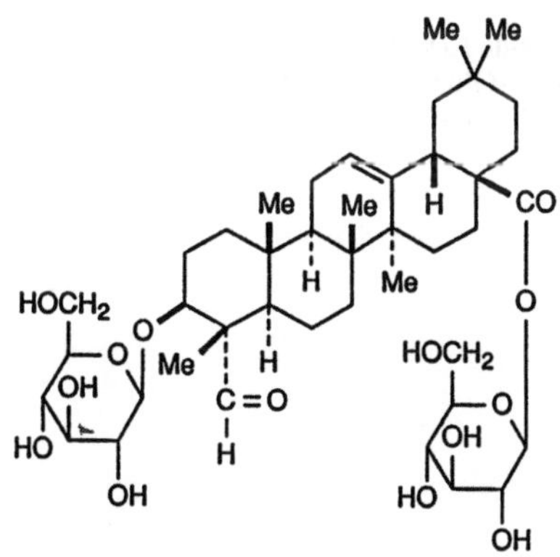

Table 82.1. (continued)

Saponin	Structure
Lucyoside G (*82-7*) 3-*O*-β-D-Glucopyranosyl-maslinic acid β-D-glucopyranosyl ester	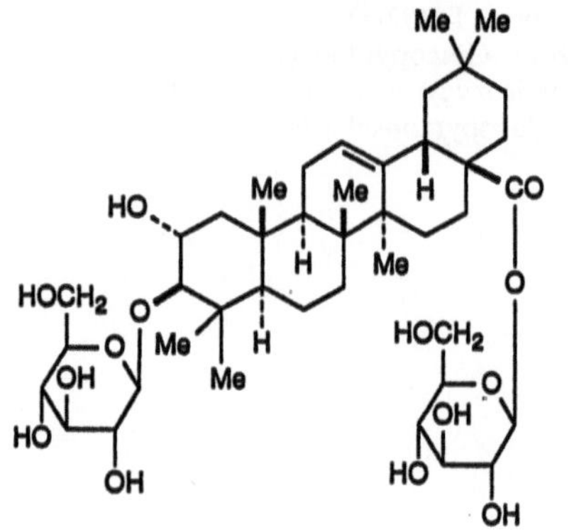
Lucyoside H (*82-8*) 3-*O*-β-D-Glucopyranosyl-oleanolic acid β-D-glucopyranosyl ester	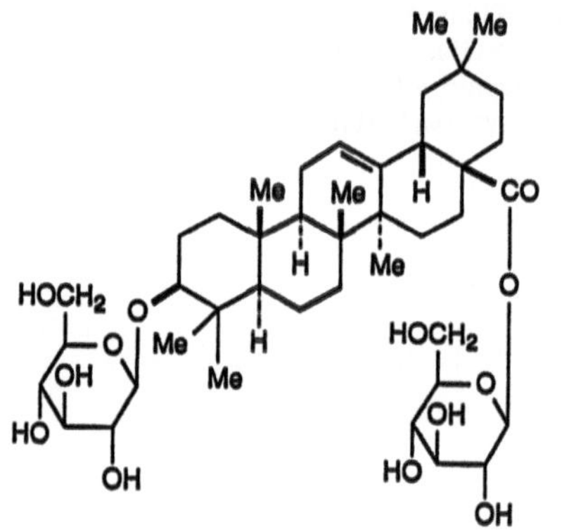
Lucyoside I (*82-9*) 3-*O*-β-D-Glucopyranosyl-arjunolic acid	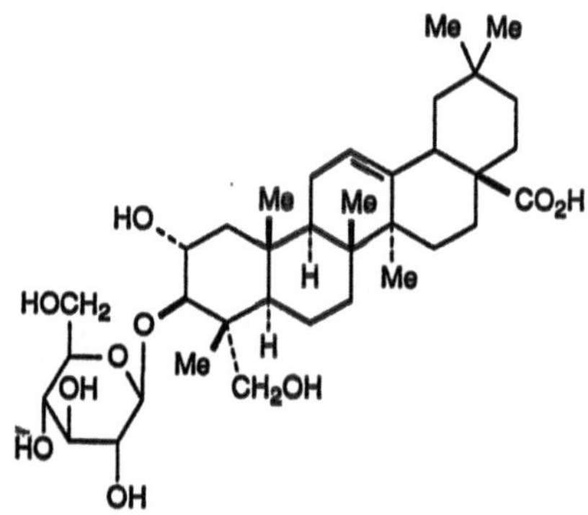

Table 82.1. (continued)

Saponin	Structure
Lucyoside J (*82-10*) 3-*O*-β-D-Glucopyranosyl-21β-hydroxy-gypsogenin β-D-glucopyranosyl ester	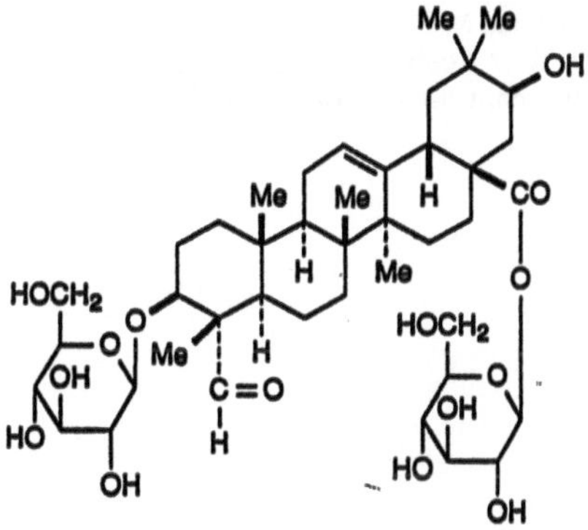
Lucyoside K (*82-11*) 3-*O*-β-D-Glucopyranosylgypsogenin	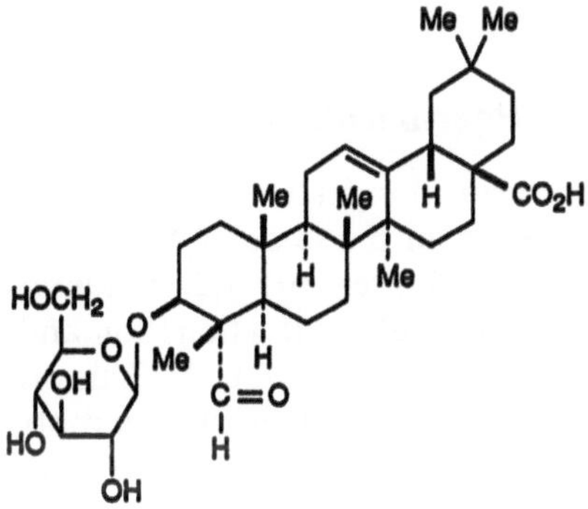
Lucyoside L (*82-12*) 3-*O*-β-Sophorosyl-gypsogenin β-D-glucopyranosyl ester	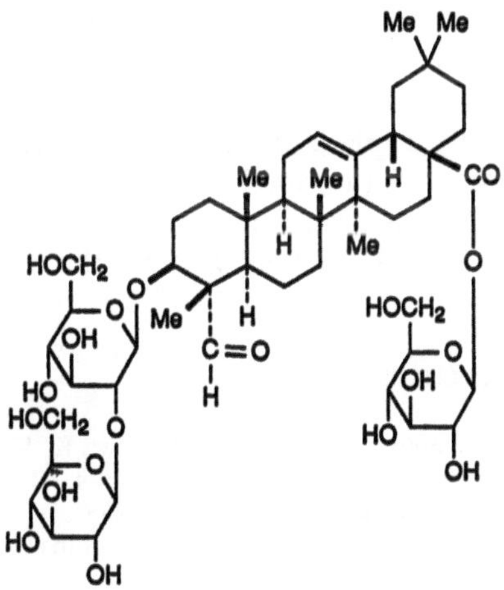

Table 82.1. (continued)

Saponin	Structure
Lucyoside M (*82-13*) 3-*O*-(6′-*O*-Acetyl)-β-D-glucopyranosyl-gypsogenin β-D-glucopyranosyl ester	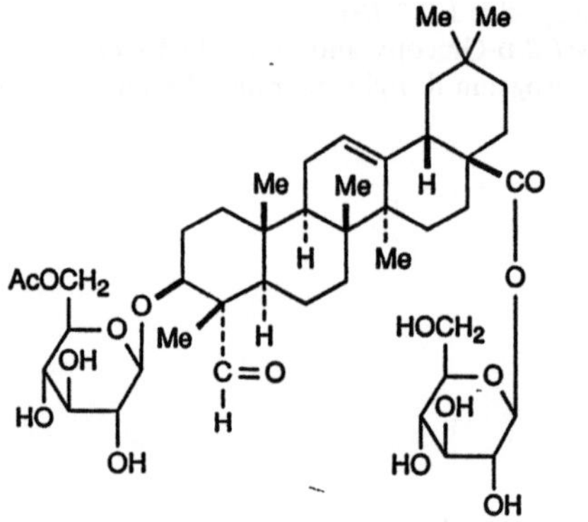

82.3 Pharmacology

A protein, designated luffin, with an apparent molecular weight of 26000, was isolated from the seeds of *L. cylindrica*. Seeds were extracted with 20 m*M* sodium phosphate buffer, pH 7.2, containing 0.2 *M* NaCl. The extract was purified by ammonium sulfate fractionation and chromatographic separation on Sephacryl S-200 and CM-Sephadex C-25. This protein exhibited ten times the inhibitory activity against protein synthesis in the rabbit reticulocyte lysate test (IC_{50}, 0.42 ng/ml) as that of ricin A chain. In contrast, cytotoxicity against murine leukemia L 1210 cells was very weak, corresponding to $1 \times 10^{-6} - 1 \times 10^{-5}$ that of ricin [5].

Interferon synthesis was induced in rabbit by intravenous injection of an extract from the sprout of *L. cylindrica*. The serum interferon level peaked 2 or 4 h after administration of 6 mg/kg or 1.2 mg/kg of the extract, respectively. Based on its stability at pH 2 and thermostability, this interferon was identified as type I interferon. Under optimal conditions, all mice challenged with a lethal dose of Japanese encephalitis virus were protected. The active substance of the extract was shown to be the RNA fraction, whereas the polysaccharide fraction was not active [6].

References

1. Takemoto T, Arihara S, Yoshikawa K, Kusumoto K, Yano I, Hayashi T (1984) Studies on the constituents of Cucurbitaceae plants. VI. On the saponin constituents of *Luffa cylindrica* Roem. (1). Yakugaku Zasshi 104:246–255
2. Takemoto T (1985) Lucyoside saponins from a gourd. Jpn Kokai Tokkyo Koho JP 60 48,995 [85 48,944] (CA 103:166146x)
3. Takemoto T, Arihara S, Yoshikawa K, Tanaka R, Hayashi T (1985) Studies on the constituents of Cucurbitaceae plants. XIII. On the saponin constituents of *Luffa cylindrica* Roem. (2). Yakugaku Zasshi 105:834–839
4. Nanking Institute of Materia Medica (1980) A note on the chemical study on the vines (leaves) of *Luffa cylindrica* (L.) Roem. Chin Trad Herb Drugs 11:55
5. Kishida K, Masuho Y, Hara T (1983) Protein synthesis inhibitory protein from seeds of *Luffa cylindrica* Roem. FEBS Lett 153:209–212
6. Xu ZX, Zhou ZQ, Qu FZ, Tong LL, Li LQ (1985) Extract of *Luffa cylindrica* sprout (L042) as an interferon inducer. Chin J Microbiol Immunol 5:130–132

Lycium barbarum L. and *L. chinensis* Mill. 83

83.1 Introduction

Gouqizi, Fructus Lycii, is the dry ripe fruits of *Lycium barbarum* L. (Solanaceae). The orange red fruits are harvested in summer and fall. It is a tonic used in traditional Chinese medicine and officially listed in the Chinese Pharmacopoeia.

Digupi, Cortex Lycii, is the dry root bark of *L. barbarum* and *L. chinensis* Mill. The roots are dug in early spring or in late fall and the bark collected. The root bark of *Lycium* is also officially listed in the Chinese Pharmacopoeia and is used as a tonic and hemostyptic.

83.2 Chemical Constituents

83.2.1 Chemical Constituents of *Lycium chinensis*

The half-dried fruits of *L. chinensis* have a characteristic odor. Study of the volatile components led to the isolation of two sesquiterpenes [1]. One of them is a new naturally occurring sesquiterpene dehydro-α-cyperone (*83-1*) and the other is a known compound, solavetivone (*83-2*), first isolated from potato tuber [2]. The absolute configuration of 1,2-dehydro-α-cyperone was also determined [1].

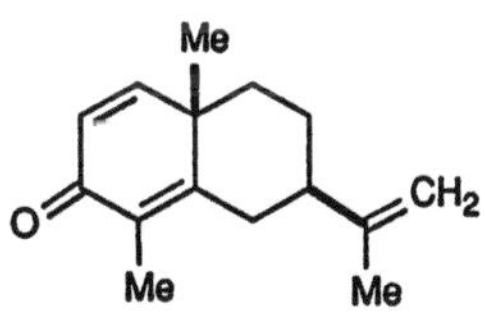

Dehydro-α-cyperone (*83-1*)

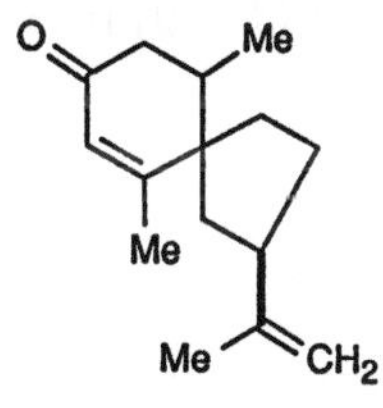

Solavetivone (*83-2*)

Furthermore, 36 neutral volatile compounds were detected by GC-MS. Some components were isolated by preparative GC and analyzed by spectroscopic methods. This neutral fraction of volatile compounds represents most of the odor aspects of the dried berries [3]. The 36 neutral volatiles include 2 hydrocarbons, 5 alcohols, 1 aldehyde, 13 ketones, 2 ketals, 9 esters, 2 lactones, 1 pyrrole, and 1 pyridine. Methyl linoleate was the predominant component of the neutral volatiles (18%). Esters of C_{14}, C_{16}, and C_{18} fatty acids were also found in large quantities [3]. From the nonvolatile fraction, betaine was isolated from the fruits and leaves of *L. chinensis*, but was not detected in its roots [4]. A polyene alcohol, zeaxanthine, and its dipalmitate, named physalien, were also isolated from the fruits of *L. chinensis* [5]. It was reported that the seeds of some solanaceous plants, including *L. chinensis* contain a number of steroid compounds. Among the 4,4-dimethylsterols (triterpene 3β-monols), cycloartanol (*83-3*), cycloartenol (*83-4*) and 24-methylene-cycloartanol (*83-5*) were isolated as major compounds, whereas other triterpene 3β-monoalcohols such as lanosterol derivatives were only found in trace amounts [6].

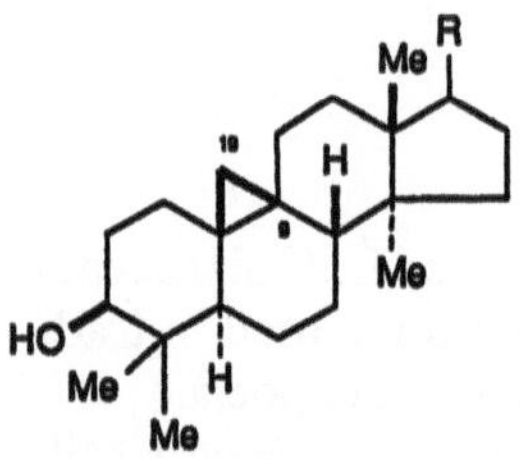

Cycloartanol (*83-3*): R =

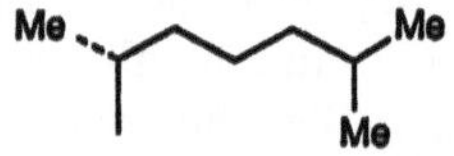

Cycloartenol (*83-4*): R =

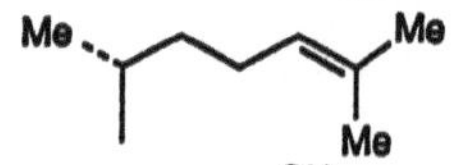

24-Methylenecycloartanol (*83-5*): R =

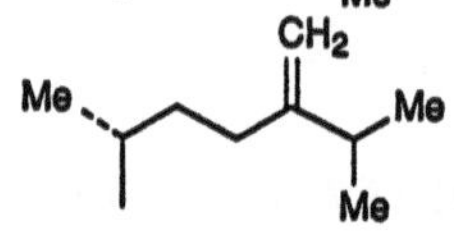

Besides the 4,4-dimethylsterols, the presence of 4-desmethylsterol derivatives in the seeds of solanaceous plants including *L. chinensis* was also investigated [7]. Among them, gramisterol (*83-6*) is the major compound (44%) in the seeds of *L. chinensis*. Four other α-methylsterols in large quantities from the seeds of *L. chinensis* are citrostadienol (*83-7*) (18%), lophenol (*83-8*) (9%), cycloeucalenol (*83-9*) (8%), norcycloartenol (*83-10*) (6%), and obtusifoliol (*83-11*) (6%) [8].

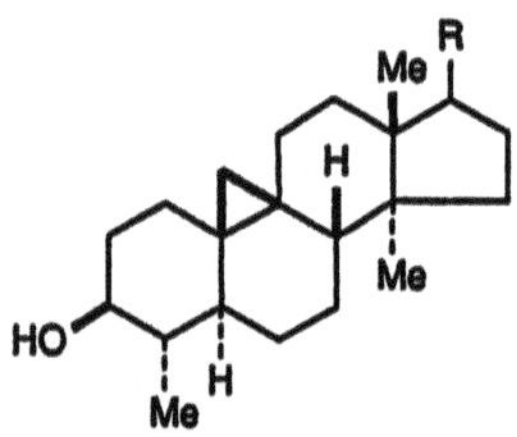

Cycloeucalenol (*83-9*): R =

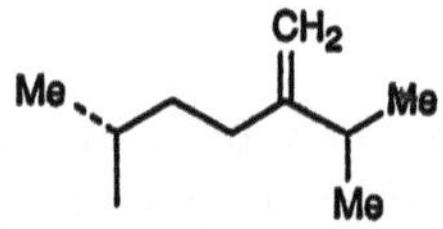

Norcycloartenol (*83-10*): R =

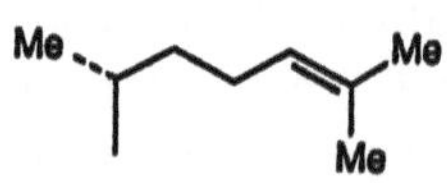

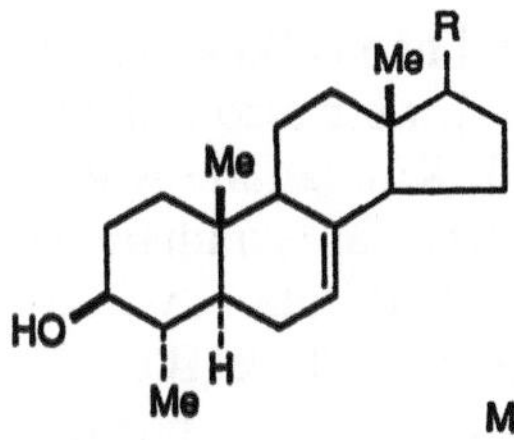

Gramisterol (*83-6*): R =

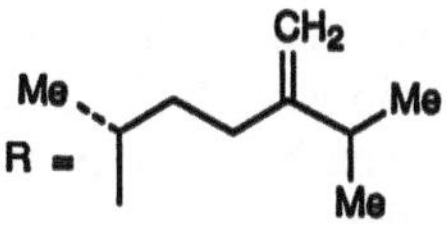

Citrostandienol (*83-7*): R =

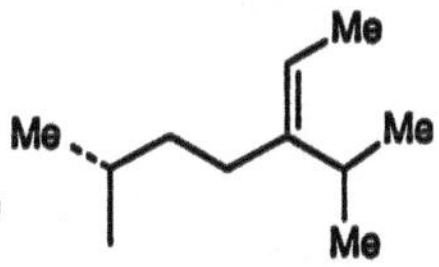

Lophenol (*83-8*): R =

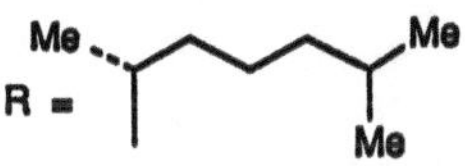

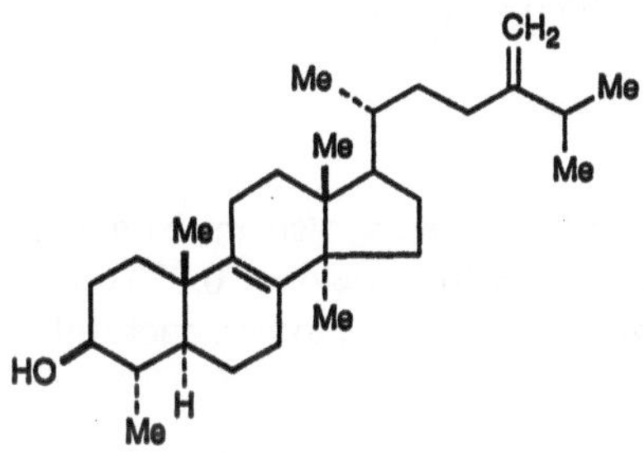

Obtusifoliol (*83-11*)

β-Sitosterol and melissic acid ($CH_3-(CH_2)_{28}-COOH$) were also isolated and identified from the fruits and bark of *L. chinensis* [9].

The bark of *L. chinensis* also contained linoleic acid [10] and 5α-stigmastan-3,6-dione (*83-12*) [11] in addition to a phenolic compound sugiol (*83-13*) [11].

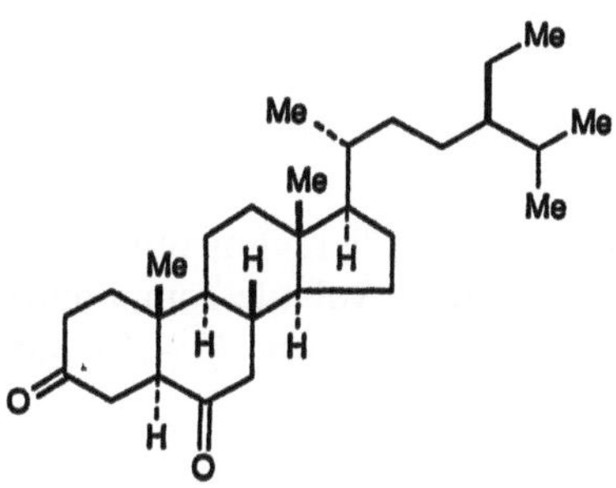

5-α-Stigmastan-3,6-dione (*83-12*)

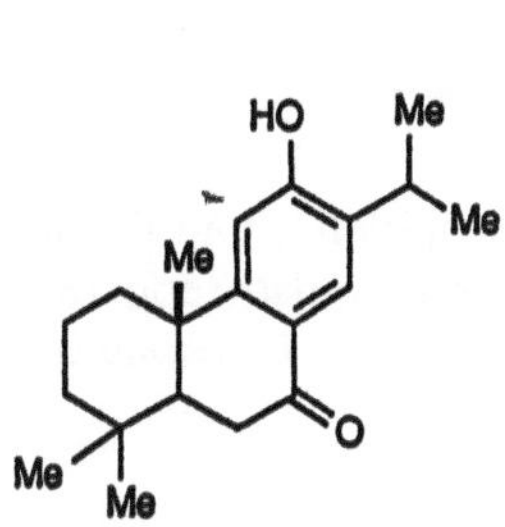

Sugiol (*83-13*)

Moreover, two nitrogen-containing components, an alkaloid named kukoamine A (*83-14*) [12] and a dipeptide named lyciumamide (*83-15*) [13], were also isolated from the root bark of *L. chinensis* and structurally elucidated. Kukoamine A was considered to be the spermine bisamide of dihydrocaffeic acid, whereas lyciumamide was determined as *N*-benzoyl-L-phenylalanyl-L-phenylalaninol acetate. Lyciumamide is the first natural product of the phenylalanine-phenylalaninol type. It could be synthesized by a reaction of phenylalaninol with *N*-carbobenzoxy-L-phenylalanine *p*-nitrophenylester. Elimination of the carbobenzoxy group by catalytic hydrogenation, benzoylation, and acetylation yielded *83-15*.

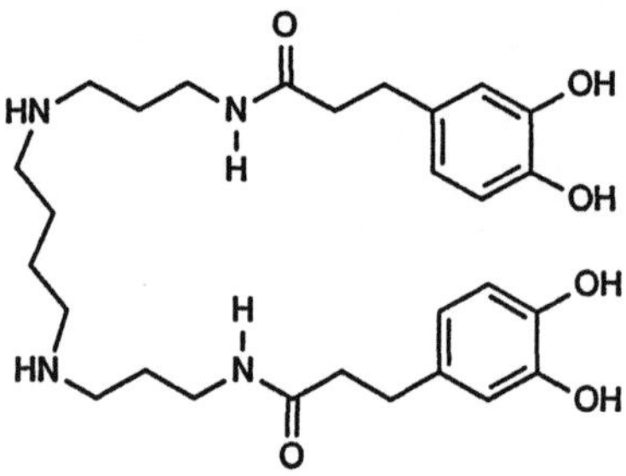

Kukoamine A (*83-14*)

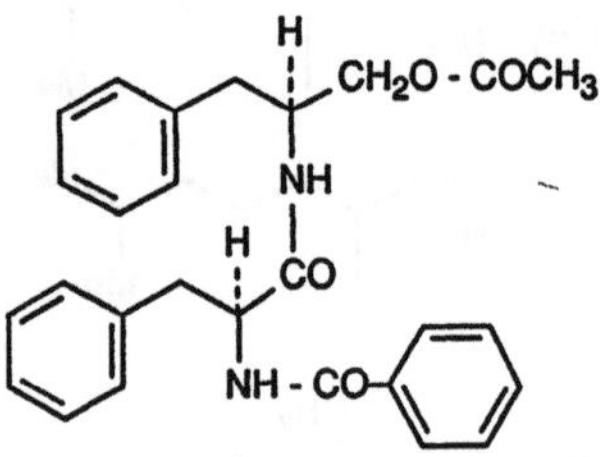

Lyciumamide (*83-15*)

The chemical constituents of the leaves of *L. chinensis* were also investigated. Thus, two lyciumwithanolides A and B were isolated with a yield of 0.02% and 0.03%, respectively [14]. Lyciumwithanolides A and B were proven to be structurally identical with withanolide A (*83-16*) and B (*83-17*) first isolated from *Withania coagulans* [15].

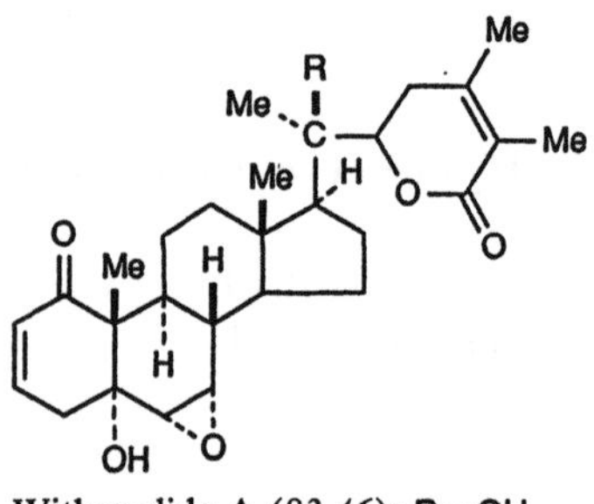

Withanolide A (*83-16*): R=OH
Withanolide B (*83-17*): R=H

The isolation of scopoletin, vanillic acid [16], betaine [10, 17], and nicotinamine [18] from *L. chinensis* was also reported. Among the neutral volatile compounds isolated from the leaves of *L. chinensis*, hydroxy-dehydro-*β*-ionone (*83-18*) was identified [19].

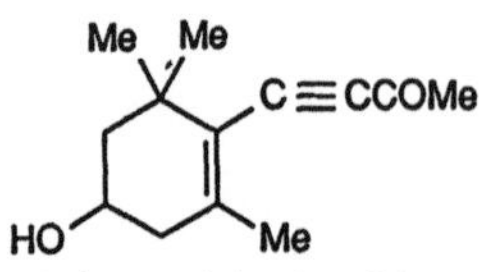

Hydroxy-dehydro-*β*-ionone (*83-18*)

83.2.2 Chemical Constituents of *Lycium barbarum*

The chemical components of the fruit and leaves of *L. barbarum* were examined. Results revealed that 100 g fruit contained 3.1 g protein, 1.9 g lipid, 9.1 g carbohydrate, 1.6 g fiber, 22.5 mg calcium, 56 mg phosphorus, 1.3 mg iron, 19.6 mg carotene, 0.08 mg thiamine, 0.14 mg riboflavin, 0.67 mg nicotinic acid, and 42.6 mg ascorbic acid. The contents of 100 g leaves was 1087–4579 mg calcium, 111–340 mg phosphorus, 64.6–89.2 mg iron, and 10.6 mg nicotinic acid [20, 21]. The leaves of *L. barbarum* contained 0.2% total sterol components. Free sterols such as sitosterol, isofucosterol, campesterol, and stigmasterol comprised 73%, whereas the other 27% consisted of esters, sterylglycosides, or acylated sterylglycosides, with cholesterol as aglycone [22].

83.3 Pharmacology

The water-soluble component of the fruit of *L. chinensis* decreased blood pressure and stimulated respiration when given intravenously to rabbits at a dose of 20 mg/kg [23]. The alkaloid component kukoamine A isolated from the root bark of *L. chinensis* induced hypotension in rats when given intravenously at a dose of 5 mg/kg [12]. The extract of *L. barbarum* caused a sharp long-lasting reduction in the blood sugar content, with an increase in carbohydrate tolerance [24]. Additionally, a water-soluble, nondialyzable substance extracted from the leaves of *L. chinensis* induced ovulation in the adult female rabbit. The substance was eluted relatively early on gel filtration with Sephadex G-25 and G-50, and the molecular weight appeared to be relatively large and the fraction containing the active component was ninhydrin negative [25].

References

1. Sannai A, Fujimori T, Kato K (1982) Isolation of (–)-1,2-Deydro-α-cyperone and solavetivone from *Lycium chinense*. Phytochemistry 21:2986–2987
2. Coxon DT, Keith RP, Howard B (1974) Two new vetispirine derivatives: stress metabolites from potato (*Solanum tuberosum*) tubers. Tetrahedron Lett 2921–2924
3. Sannai A, Fujimori T, Kato K (1983) Neutral volatile components of "Kukoshi" (*Lycium chinense* M.). Agric Biol Chem 47:2397–2399
4. Nishiyama R (1963) Betaine of *Lycium chinense*. Nippon Shokuhin Kogyo Gakkaishi 10:517–519 (CA 63:4660 h)
5. Lin QS (1977) Chemistry of the Constituents in Chinese Medicinal Herbs. Science, Beijing, p 568
6. Itoh T, Tamura T, Matsumoto T (1977) Triterpene alcohols in the seeds of Solanaceae. Phytochemistry 16:1723–1726
7. Itoh T, Tamura T, Matsumoto T (1977) 4-Desmethylsterols in the seeds of Solanaceae. Steroids 30:425–433
8. Itoh T, Ishii T, Tamura T, Matsumoto T (1978) Four new and other 4α-methylsterols in the seeds of Solanaceae. Phytochemistry 17:971–977
9. Mizobuchi K, Kobayashi H, Nagai M, Shimaoka T, Higashi J (1965) Constituents of box thorn. IV. Components of Ti-ku-pi and Kou-chi-tzu. Tokushima Daigaku Yakugakubu Kenkyu Nempo 14:10–13 (CA 67:105978 m)
10. Mizobuchi K, Inoue Y, Kiuchi T, Higashi J (1963) Constituents of box thorn. II. Chemical components of root bark of box thorn. Shoyakugaku Zasshi 17:16–18

11. Noguchi M, Mochida S, Shingu T, Fujitani K, Kozuka M (1985) Sugiol and 5α-stigmastane-3,6-dione from the Chinese drug Ti-ku-pi (Lycii radicis cortex). J Nat Prod 48:342–343
12. Funayama S, Yoshida K, Konno C, Hikino H (1980) Structure of kukoamine A, a hypotensive principle of *Lycium chinense* root barks. Tetrahedron Lett 21:1355–1356
13. Noguchi M, Mochida K, Shingu T, Kozuka M, Fujitani K (1984) Constituents of a Chinese drug, Ti Ku Pi. I. Isolation and structure of lyciumamide, a new dipeptide. Chem Pharm Bull (Tokyo) 32:3584–3587
14. Haensel R, Huang JT (1977) *Lycium chinense*. II. Semiquantitative determination of withanolides. Arch Pharm 310:35–38
15. Haensel R, Huang JT (1975) Two withanolids from *Lycium chinense*. Arch Pharm (Weinheim) 308:653–654
16. Haensel R, Huang JT (1977) *Lycium chinense*. III. Isolation of scopoletin and vanillic acid. Arch Pharm (Weinheim) 310:38–40
17. Mizobuchi K, Inoue Y, Kiuchi T, Higashi J (1963) Constituents of box thorn. I. Chemical components of the leaves of Japanese *Lycium chinense*. Shoyakugaku Zasshi 17:14–15
18. Noma M, Noguchi M (1976) Occurrence of nicotinamine in higher plants. Phytochemistry 15:1701–1702
19. Sannai A, Fujimori T, Uegaki R, Akaki T (1984) Isolation of 3-hydroxy-7,8-dehydro-β-ionone from *Lycium chinense* M. Agric Biol Chem 48:1629–1630
20. Qi ZS, Li SF (1981) Determination of the chemical composition of Fructus lycii. Ningxia Daxue Xuebao Ziran Kexueban 67–72
21. Qi ZS, Li SF, Wu JP, Qu R, Yang YF (1986) Chemical constituents of Fructus lycii and Folium lycii. I. Nutrients in Fructus lycii and Folium lycii. Bull Chin Mater Med 11:169–171
22. Duperon R, Thiersault M, Duperon P (1984) High level of glycosylated sterols in species of *Solanum* and sterol changes during the development of the tomato. Phytochemistry 23:743–746
23. Kurokawa S (1962) Pharmacological properties and lipotropic action of various components derived from the fruit of *Lycium chinense* and betaine hydrochloride. I. General pharmacological studies of the water and various organic solvent-soluble components derived from *Lycium chinense* and betaine hydrochloride. Shikoku Igaku Zasshi 18:127–136 (CA 57:11822e)
24. Lapinina LO, Sisoeva TF (1964) Investigation of some plants to determine their sugar-lowering action. Farmatsevt Zh 52–58 (CA 66:1451e)
25. Suzuki M, Osawa S, Hirano M (1972) *Lycium chinense* Miller component inducing ovulation in adult female rabbits. Tohoku J Exp Med 106:219–231 (CA 77:28992p)

Magnolia spp. 84

84.1 Introduction

There are three official entries containing *Magnolia* species in the Chinese Pharmacopoeia:

- Xinyi, Flos Magnoliae, is the dry flower buds of *Magnolia biondii* Pamp., *M. denudata* Desr., or *M. sprengeri* Pamp. (Magnoliaceae) collected from late winter to early spring before the buds have flowered. The flower buds are used in traditional Chinese medicine for the treatment of cold and nasal catarrh.
- Houpohua, Flos Magnoliae officinalis, is the dry flower buds of *M. officinalis* Rehd. et Wils. or *M. officinalis* Rehd. et Wils. var. *biloba* Rehd. et Wils., collected in the spring before the buds flower. It is mainly used as a stomachic.
- Houpo, Cortex Magnoliae officinalis, is the dry bark of stem, branch, and root of *M. officinalis* and *M. officinalis* var. *biloba* collected in April to June. It is used as a stomachic and antiasthmatic agent.

84.2 Chemical Constituents

84.2.1 Chemical Constituents of *Magnolia officinalis*

The bark of *M. officinalis* contains as major components magnolol (*84-1*) and honokiol (*84-2*) as mentioned above, two substituted biphenols with a propenyl side chain.

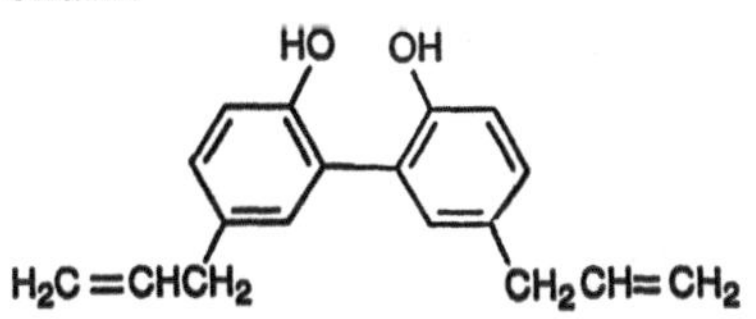

Magnolol (*84-1*)

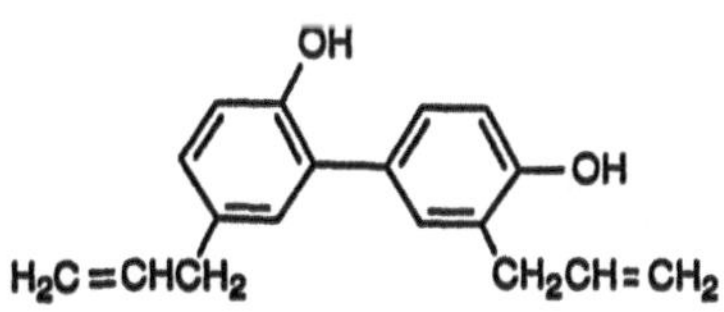

Honokiol (*84-2*)

Magnolol was isolated from *M. officinalis* and *M. obovata* in 1930 and structurally determined [1], whereas honokiol was first isolated more than 40 years later from *M. obovata* [2] and then from *M. officinalis* [3]. According to crystal and molecular structure data, the benzene rings have a conformation with a 45° dihedral angle, maximizing strong intra- and intermolecular H-binding. The solid structure corresponds to a helix stabilized through H-bonds parallel to the *b*-axis [4].

Quantitative determination of magnolol and honokiol by HPLC [5, 6], TLC spectrophotometry [5, 6], and GC [7] has been described. Samples from various areas showed magnolol contents of 2%–11% and honokiol contents of 0.3%–4.6% [5]. The root bark contains more magnolol and honokiol than the stem bark [8]. Magnolol and honokiol can be extracted from the bark with NaOH solution, precipitated with hydrochloric acid, and purified by recrystallization [9].

Besides magnolol and honokiol, a number of hydroxybiphenyls such as 8,9-dihydroxydihydrohonokiol (*84-3*), 8,9-dihydroxy-7-methoxy-dihydrohonokiol (*84-4*), 8,9-dihydroxymagnolol (*84-5*) [10], and bornylmagnolol (*84-6*) [11] were isolated from the bark of *M. officinalis*.

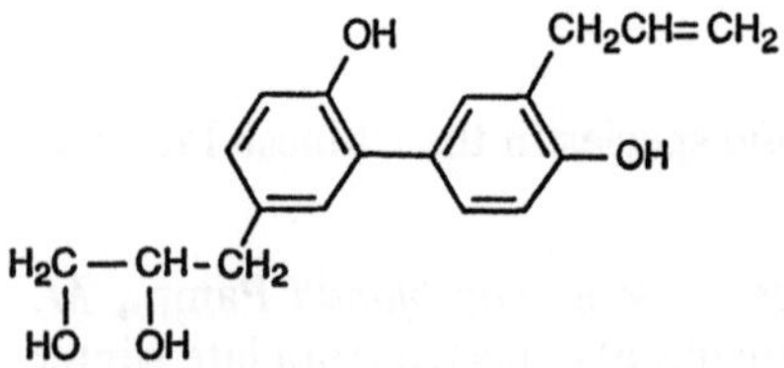

8,9-Dihydroxydihydro-
honokiol (*84-3*)

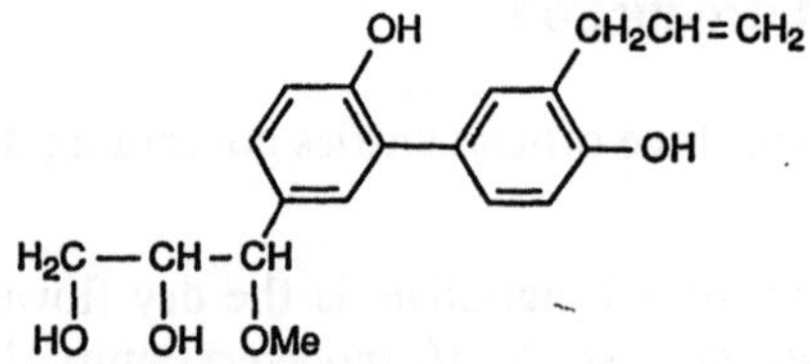

8,9-Dihydroxy-7-methoxy-
dihydrohonokiol (*84-4*)

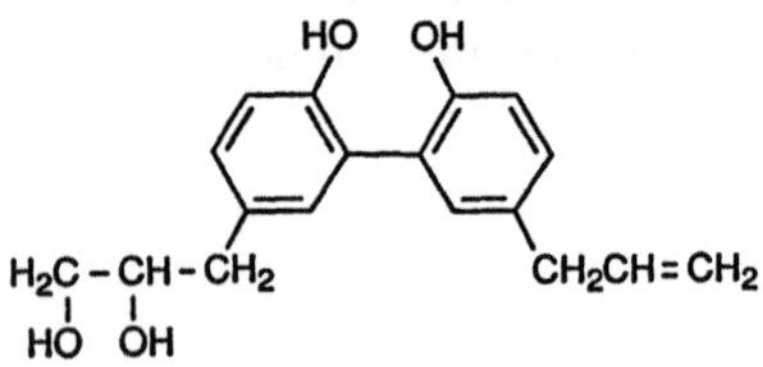

8,9-Dihydroxydihydro-
magnolol (*84-5*)

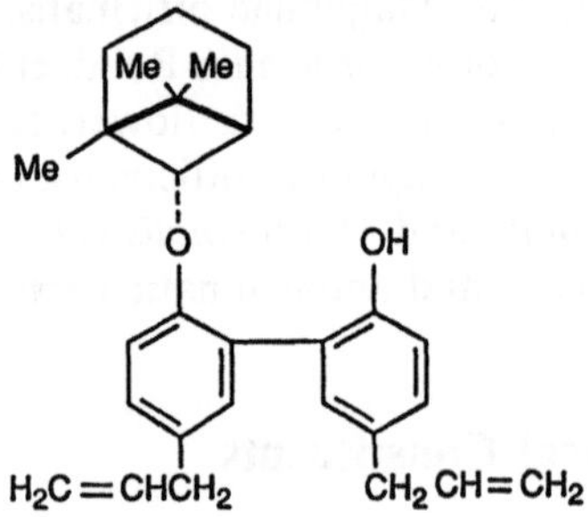

Bornylmagnolol (*84-6*)

From the bark of *M. officinalis* var. *biloba*, α-eudesmol (*84-7*) and β-eudesmol (*84-8*) were isolated together with magnolol and honokiol [12].

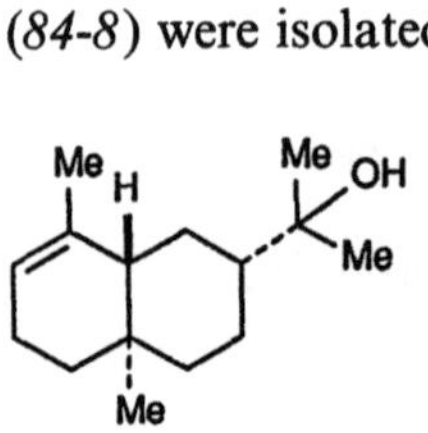

α-Eudesmol (*84-7*)

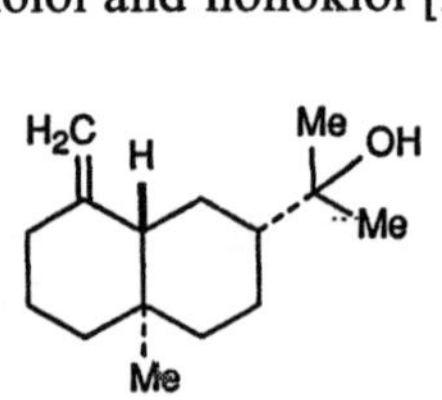

β-Eudesmol (*84-8*)

84.2.2 Chemical Constituents of Other *Magnolia* Species

From the essential oil of the buds of *M. biondii* collected in Hubei and Henan, a series of compounds were identified by gas chromatographic retention indices and mass spectral data. Both essential oils contain 1,8-cineol, sabinene, limonene, and terpinen-4-ol as the major components [13]. Gas chromatographic and thin-layer chromatographic determination indicated that the bark of *M. biondii, M. denudata, M. sprengeri,* and the unofficial *M. liliflora* did not contain magnolol and honokiol and that the total eudesmol content in these barks was lower than that in the bark of *M. officinalis* [14]. Moreover, some alkaloids were isolated from the bark of *M. sprengeri* and identified as salicifoline (*84-9*) and magnocurarine (*84-10*) [15]. A new alkaloid named magnosprengerine (*84-11*) was also isolated and structurally elucidated [16]. The major components of the volatile oil from the flower buds and twigs of *M. sprengeri* have been identified as sabinene, *β*-pinene, *p*-cymene, bornyl acetate, *trans*-caryophyllene, caryophyllene oxide, and *β*-eudesmol [17]. Similar composition was found in the volatile oil of *M. sargentiana* [18].

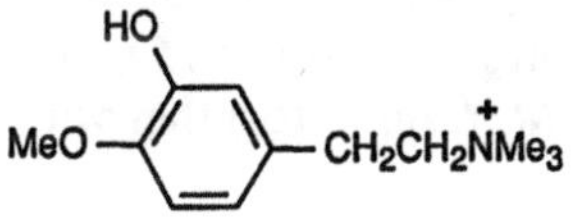

Salicifoline (*84-9*)

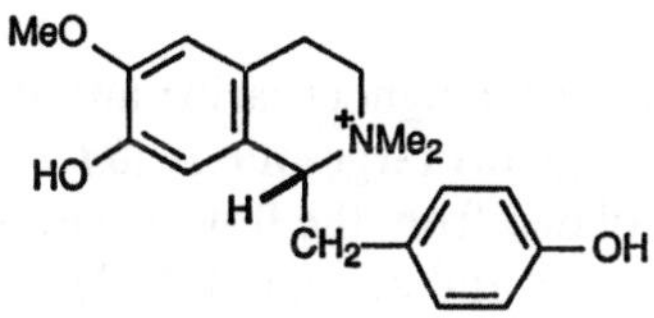

Magnocurarine (*84-10*)

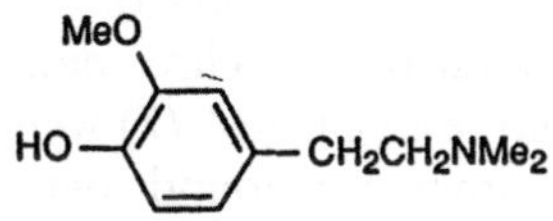

Magnosprengerine (*84-11*)

Whereas magnolol and honokiol are active ingredients of *M. officinalis*, the bark of *M. liliflora* contains the alkaloid D-coclaurine as the active principle [19]. Three percent essential oil was isolated from the flowers of *M. liliflora*, containing mainly cineole together with α-pinene, methylchavicol, and citral. In addition to the essential oil, the lignans pinoresinol dimethyl ether, lirioresinol B dimethyl ether, magnolin (*84-12*), and fargesin (*84-13*) were also isolated from the flowers of *M. liliflora* [20].

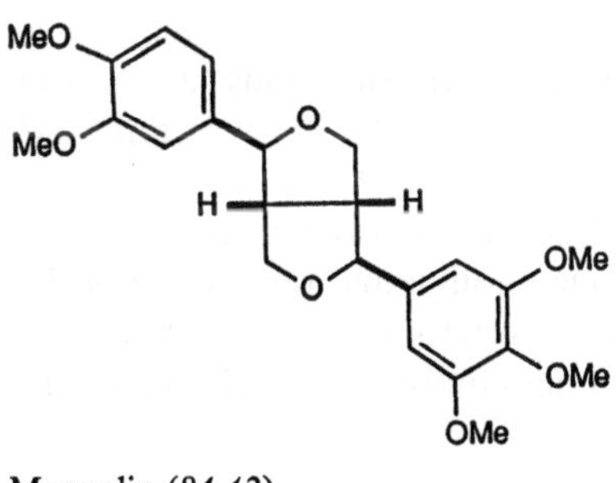

Magnolin (*84-12*)

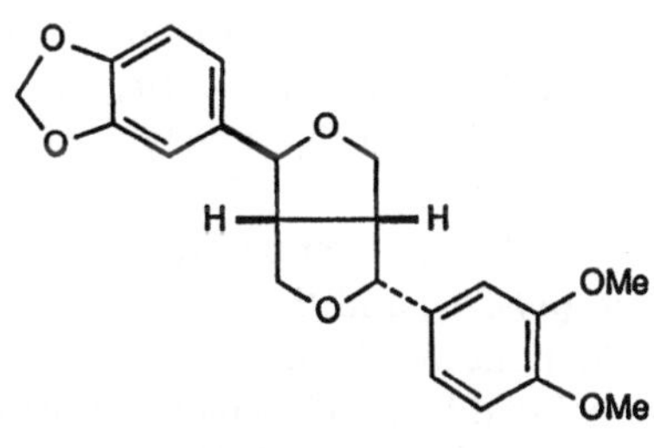

Fargesin (*84-13*)

The bark of *M. rostrata*, a substitute for the bark of *M. officinalis*, contains magnolol, honokiol [21], and magnocurarine [22]. A comparative study revealed that the bark of *M. officinalis* contained 3.6% magnolol and 0.8% honokiol and that of *M. rostrata* contained 2.6% magnolol and 2.3% honokiol [23]. Recently, the lignans lirioresinol B dimethylether, pinoresinol dimethylether, magnolin, fargesin, aschantin (*84-14*), and demethoxyaschantin were isolated from the flowers of *M. biondii* [24].

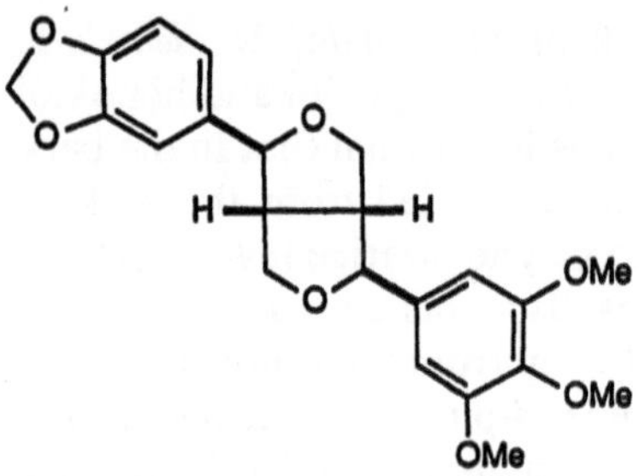

Aschantin (*84-14*)

From the bark of *M. wilsonii*, magnocurarine and 0.3% (v/w) essential oil were isolated [25]. Three new neolignans fargesone A (*84-15*), fargesone B (*84-16*), and fargesone C (*84-17*) were isolated from the flower buds of *M. fargesii* together with known compounds denuatine B and syringin [26, 27].

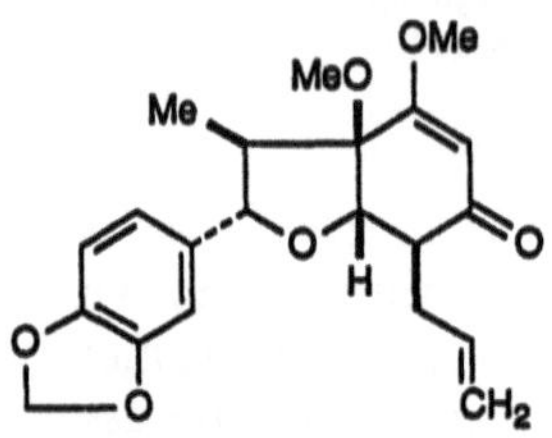

Fargesone A (*84-15*)

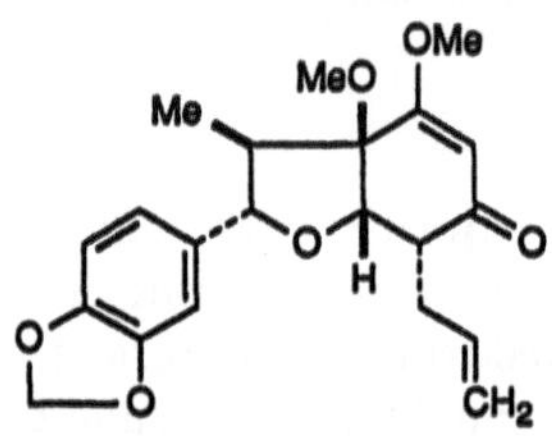

Fargesone B (*84-16*)

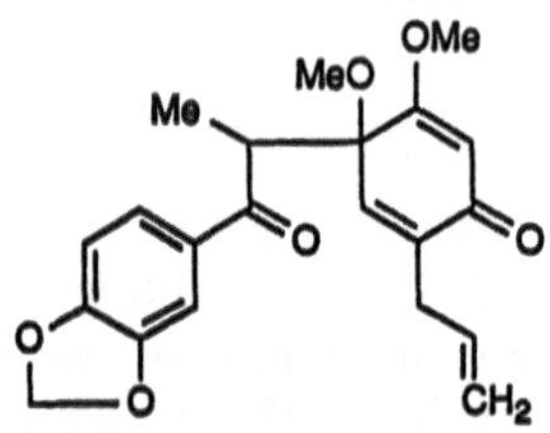

Fargesone C (*84-17*)

The leaves of some *Magnolia* species contain a number of neolignan compounds. For example, the isolation of four new neolignans, the dihydrobenzofurane neolignans liliflols A (*84-18*) and B (*84-19*), the hexahydrobenzofurane neolignan liliflone (*84-20*), and the bicyclo(3,2,1)octane neolignan liliflodione (*84-21*) was reported, together with that of seven other neolignans, denudatins A (*84-22*) and B (*84-23*), burchellin (*84-24*), piperenone (*84-25*), the 2,5-diaryltetrahydrofurane neolignan veraguensin (*84-26*), and the spirocyclohexadienone neolignans futoenone (*84-27*) and maglifloenone (*84-28*) [28]. Maglifloenone was proven to be identical to denudatone, first isolated from *M. denudata* together with denudatins A and B, burchellin, veraguensin, and futoenone [29].

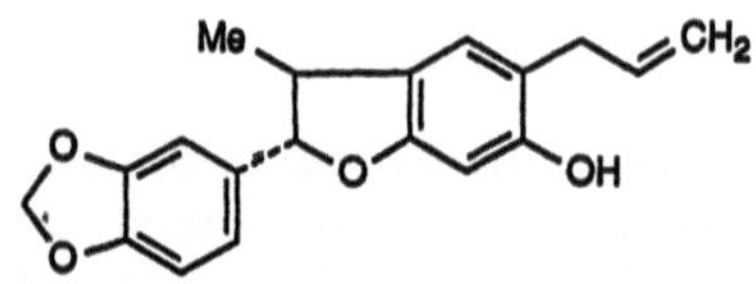

Liliflol A (*84-18*)

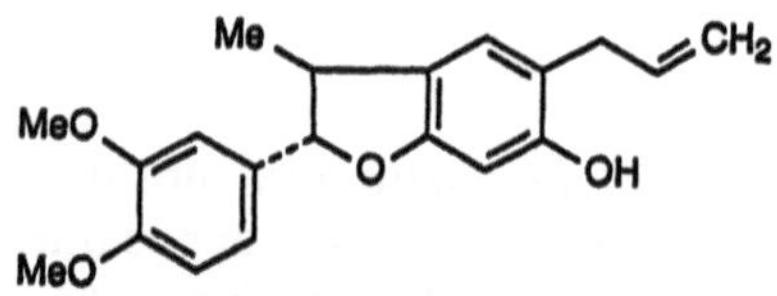

Liliflol B (*84-19*)

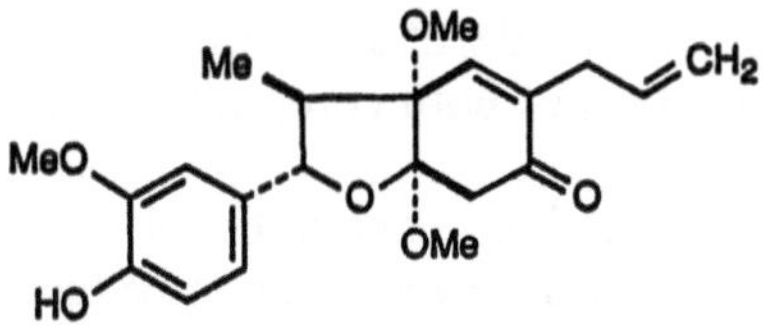

Liliflone (*84-20*)

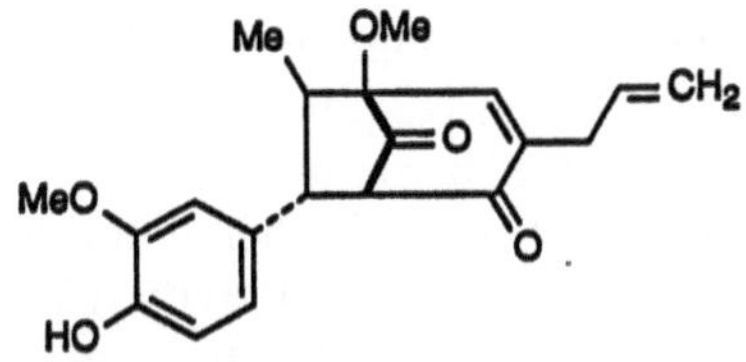

Liliflodione (*84-21*)

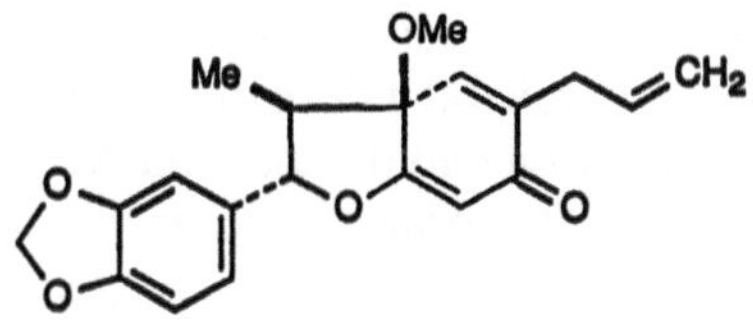

Denudatin A (*84-22*)

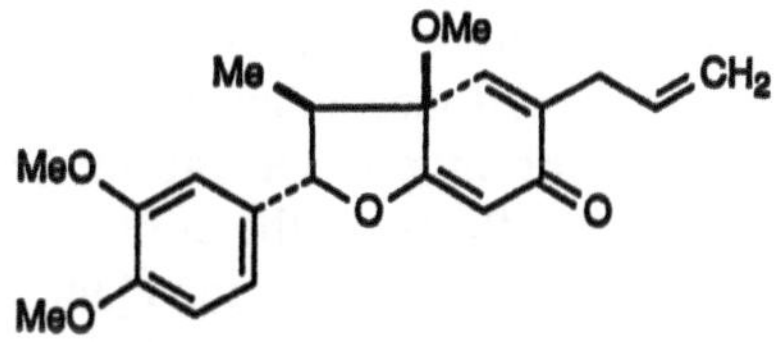

Denudatin B (*84-23*)

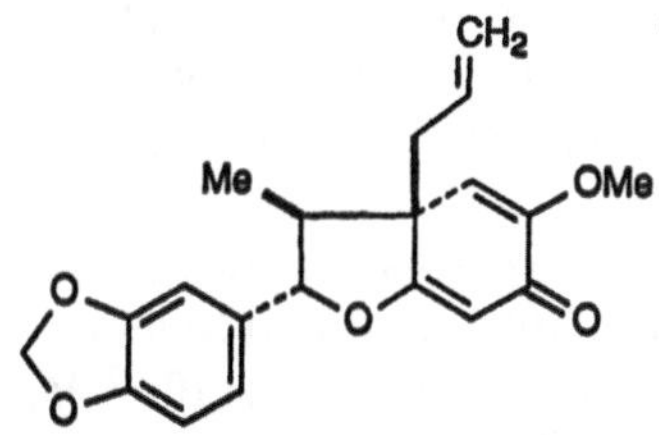

Burchellin (*84-24*)

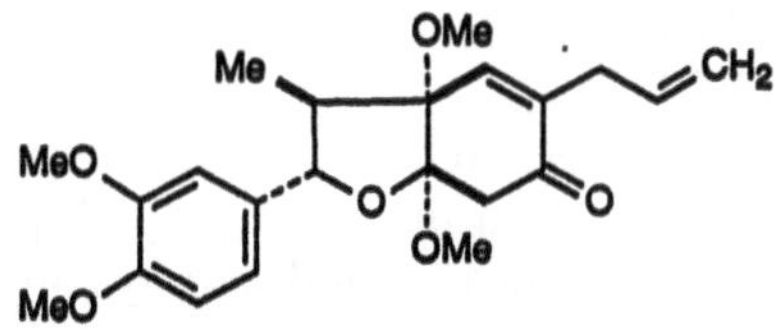

Piperenone (*84-25*)

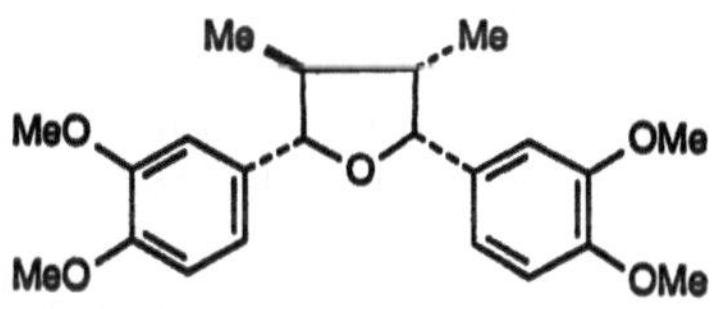

Veraguensin (*84-26*)

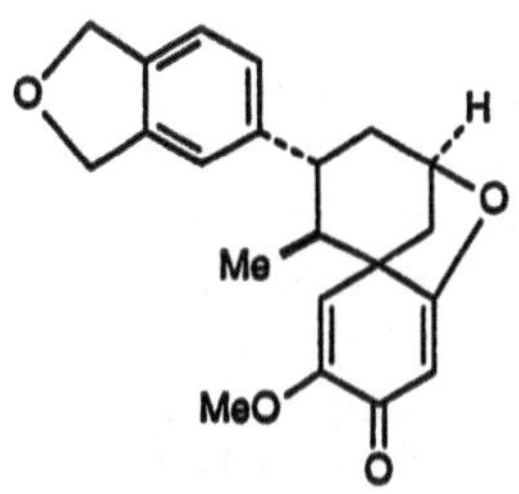

Futoenone (*84-27*)

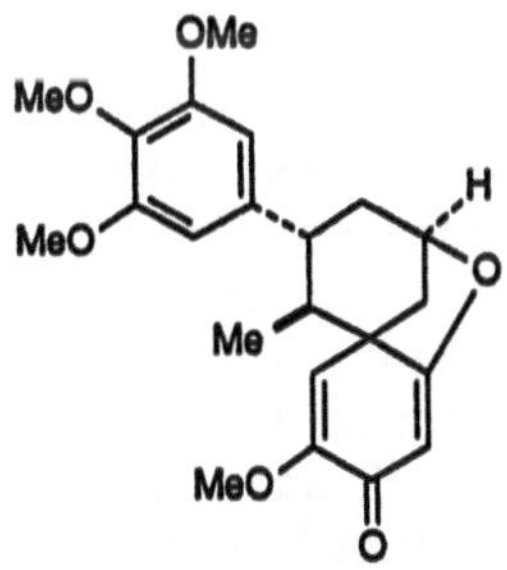

Denudatone (Maglifloenone) (*84-28*)

84.3 Pharmacology

Magnolol and honokiol exhibited significant activity against gram-positive bacteria and fungi [30]. The ether and methanolic extracts with the active principles magnolol and honokiol showed potent antibacterial action against a cariogenic bacterium

Streptococcus mutans. Magnolol and honokiol were bactericidal at a minimal inhibitory concentration of 6.3 μg/ml. The antibacterial action of both compounds was stronger than that of berberine [31].

Some synthetic hydroxybiphenyl derivatives produced from phenylphenols and biphenols showed a structure-activity relationship against *S. mutans*. The introduction of allyl groups to the aromatic rings increased their antibacterial activities, but their position did not affect activities. Furthermore, the introduction of hydroxyl groups to aromatic rings also increased the activity [32].

Magnolol and honokiol both showed a significant and long-lasting muscle relaxant activity. Intraperitoneal administration of magnolol to mice at a dose of 100 mg/kg produced strong muscle relaxation for 2 h [33]. Both magnolol and honokiol showed inhibitory effects on the contractile response of rat thoracic aorta to high concentration of K^+ and Ca^{2+} ions [34].

Studies on the structure-activity relationship of hydroxybiphenyls including magnolol and honokiol on grip strength in mice and spinal reflexes in young chicken showed that a hydroxyl substituent accelerates the onset and shortens the muscle relaxant activity of biphenyl and an allyl substituent influences the activity in the opposite direction [35]. Central depressant effects were also reported for magnolol and honokiol [36].

The lignans isolated from various *Magnolia* species showed antagonistic activity against platelet-activating factor [24]. Magnolol and honokiol were found to inhibit rabbit platelet aggregation induced by collagen with an IC_{50} value of 1.9×10^{-6} *M*, which is about three times more potent than aspirin [37, 38].

Furthermore, magnolol showed significant preventive effects against the water-immersion stress ulceration and stress-induced gastric hemorrhage. Because the antiulcer and antisecretory activity of magnolol was dissimilar to those of atropine, cimetidine and dimethylprostaglandin E_2, it was suggested that the prophylactic action of magnolol may be partly attributed to its central depressant effects [39].

After single oral administration of ring-^{14}C-labeled magnolol into rats, the blood levels of radioactivity showed 2 peaks at 15 min and 8 h, respectively, suggesting an enterohepatic circulation of magnolol and its metabolites. Radiolabel was mainly found in the gastrointestinal tract and liver, with some radioactivity also in the kidney, pancreas, and lung. A major metabolite excreted in the bile was ring-^{14}C-labeled magnolol-2-*O*-glucuronide. Most of the radioactivity was eliminated into the feces and urine within the first 12 h after oral or intraperitoneal administration. The oral dose was recovered to a greater extent from the feces (72% of the administered radioactivity) than from the urine (7.4%) in 144 h and the intraperitoneal dose was similarly recovered from the feces (67%) and from the urine (12%). On repeated oral administration of cold and ring-^{14}C-labeled magnolol, the composition of the fecal metabolites significantly changed. Tetrahydromagnolol, 5-(prop-1-enyl)-5′-propyl-2,2′-dihydroxybiphenyl, 5-allyl-5′-propyl-2,2′-dihydroxybiphenyl, isomagnolol, and 5-allyl-5′-(prop-1-enyl)-2,2′-dihydroxybiphenyl were detected [40]. On the other hand, magnolol was transformed to isomagnolol and 5-allyl-5′-(prop-1-enyl)-2,2′-dihydroxybiphenyl together with small amounts of 5-(prop-1-enyl)-5′-propyl-2,2′-dihydroxyphenyl and 5-allyl-5′-propyl-2,2′-dihydroxybiphenyl by anaerobic incubation with rat feces [41].

References

1. Sugii Y (1930) Constituents of the bark of *Magnolia officinalis* Rehd. et Wils. and *M. obovata* Thunb. J Pharm Soc Jpn 50:183–217
2. Fujita M, Itokawa H, Sashida Y (1972) Honokiol, a new phenolic compound isolated from the bark of *Magnolia obovata*. Chem Pharm Bull (Tokyo) 20:212–213
3. Fujita M, Itokawa H, Sashida Y (1973) Components of *Magnolia obovata*. III. Occurrence of magnolol and honokiol in *M. obovata* and other allied plants. Yakugaku Zasshi 93:429–434
4. Wang Y, Cheng MC, Lee JS, Chen FC (1982) Molecular and crystal structure of magnolol-$C_{18}H_{18}O_2$. J Chin Chem Soc (Taipei) 30:215–221
5. Li AJ, Guo XF, Wang XM, Chen CB, Shi YH, Sui NH, Du JQ (1983) Determination of phenolic substances in Hou Po by HPLC. Chin J Pharm Anal 3:1–3
6. Cui JF, Zhang GD, Song WZ (1988) Analysis of magnolol and honokiol in the traditional Chinese drug Houpo. Chin J Pharm Anal 8:274–277
7. Ou XC, Zhang L, Xi YG (1984) Gas chromatographic assay of magnolol and honokiol in *Magnolia officinalis* and its extracts. Chin Pharm Bull 19:421–423
8. Li AJ, Guo XF, Feng HL, Chen CB, Fang ZX (1985) Contents of magnolol and honokiol in different parts of *Magnolia officinalis* and the effect of processing on their contents. Bull Chin Mater Med 10:154–157
9. He QY (1985) Extraction of medicinal magnolol and its isomer from *Magnolia* bark. Faming Zhuanli Shenqing Gongkai Shuomingshu CN 85,103,128 (CA 109:216000s)
10. Tsumura Juntendo, Inc (1984) Hydroxybiphenyls from *Magnolia officinalis* for dermatitis treatment. Jpn Kokao Tokkyo Koho JP 59,139,335(84,139,335) (CA 102:12348g)
11. Tsumura Juntendo Inc (1984) Antiallergy agents from magnolia bark. Jpn Kokai Tokkyo Koho JP 59,95,229(84,95,229) (CA 101:198183g)
12. Song WZ, Liu YL, Ji QY (1984) Studies on medicinal plants of Magnoliaceae. IV. Study on active constituents of rootbark of "Ao Ye" magnolia (*Magnolia biloba*). Chin Trad Herb Drugs 15:450–451
13. Zhang J, Mao SZ, Wu CY, Zeng SR (1986) FFAP (Carbowax 20m-2-nitro-terephthalic acid) glass capillary column for the separation and identification of the components in the essential oils of Chinese herb "Xinyi". Anal Chem 14:325–329
14. Zhou GZ, Zhu ZF (1985) Comparison tests of Shanxi's Jiang Po (*Magnolia*) and Hou Po (*Magnolia officinalis* and *M. officinalis* var. *biloba*) by TLC and GC. Chin Trad Herb Drugs 16:104–106
15. Chen FQ, Wang HZ (1981) Studies on the water-soluble alkaloids of *Magnolia sprengeri*. Chin Trad Herb Drugs 12:389–391
16. Cao ZQ, Li HG, Tian YJ, Mu FF, Yang JP, Wang ML, Zhao RN (1985) Chemical constituents of Sprenger magnolia (*Magnolia sprengeri*) Isolation and structural determination of magnosprengerine. Chin Trad Herb Drugs 16:386–388
17. Fang HJ, Song WZ, Yan YP (1987) Analysis and comparison of the constituents of the volatile oil from the flower buds and twigs of *Magnolia sprengeri* Pamp. Acta Pharm Sin 22:908–912
18. Fang HJ, Song WZ, Yuan ZM, Ni JH (1988) Medicinal plants of Chinese Magnoliaceae. III. Analysis of the constituents of the volatile oil from flower buds of *Magnolia sargentiana*. Chin J Pharm Anal 8:266–269
19. Watanabe K (1985) Pharmacology of the constituents of the cortex of *Magnolia officinalis* and *M. liliflora*. Chin Pharm Bull 20:522–524
20. Hsu HY (1987) Advances of research on medicinal plants in Taiwan. Abstr Chin Med 1:455–474
21. Yan WM (1979) Chemical components in the bark of *Magnolia rostrata* W.W. Smith. Acta Bot Sin 21:54–56
22. Chen DC, Yen WM, Shih TH (1981) Studies on *Magnolia rostrata*, a substitute of *Magnolia officinalis*. II. Isolation and identification of magnocurarine. Chin Trad Herb Drugs 12:10–11
23. Chen DC, Liu JW (1982) Quantitative analysis of magnolol and honokiol in the bark of *Magnolia officinalis* Rehd. et Wils. and *Magnolia rostrata*. Acta Pharm Sin 17:360–364
24. Pan JX, Hensens OD, Zink DL, Chang MN, Hwang SB (1987) Lignans with platelet activating factor antagonist activity from *Magnolia biondii*. Phytochemistry 26:1377–1379
25. Wang HZ (1982) Study on the constituents of *Magnolia wilsonii* Rehd. Chin J Pharm Anal 2:95–97

26. Chen CC, Huang YL, Chen YP, Hsu HY, Kuo YH (1988) Three new neolignans, fargesones A, B and C from the flower buds of *Magnolia fargesii*. Chem Pharm Bull (Tokyo) 36:1791–1795
27. Chen CC, Huang YL, Chen HT (1988) On the calcium-antagonistic principles of the flower buds of *Magnolia fargesii*. Planta Med 54:438–440
28. Iida T, Ito K (1983) Four neolignans from *Magnolia liliflora*. Phytochemistry 22:763–766
29. Iida T, Ichino K, Ito K (1982) Neolignans from *Magnolia denudata*. Phytochemistry 21:2939–2941
30. Clark AM, El-Feraly FS, Li WS (1981) Antimicrobial activity of phenolic constituents of *Magnolia grandiflora* L. J Pharm Sci 70:951–952
31. Namba T, Tsunezuka M, Hattori M (1982) Dental caries prevention by traditional Chinese medicines. II. Potent antibacterial action of Magnoliae Cortex extracts against *Streptococcus mutans*. Planta Med 44:100–106
32. Bae K, Koo S, Seo W (1986) Antimicrobial activities of hydroxybiphenyl derivatives. Korean J Pharmacogn 17:85–90 (CA 105:168755k)
33. Watanabe K, Watanabe HY, Goto Y, Yamamoto N, Yoshizaki M (1975) Studies on the active principles of magnolia bark. Centrally acting muscle relaxant activity of magnolol and honokiol. Jpn J Pharmacol 25:605–607
34. Yamahara J, Miki S, Matsuda H, Fujimura H (1986) Screening test for calcium antagonists in natural products. The active principles of *Magnolia obovata*. Yakugaku Zasshi 106:888–893
35. Watanabe H, Watanabe K, Hagino K (1983) Chemostructural requirement for centrally acting muscle relaxant effect of magnolol and honokiol, neolignan derivatives. J Pharmacobiodyn 6:184–190
36. Watanabe K, Watanabe H, Goto Y, Yamaguchi M, Yamamoto N, Hagino K (1983) Pharmacological properties of magnolol and honokiol extracted from *Magnolia officinalis*: central depressant effects. Planta Med 49:103–108
37. Teng CM, Chen CC, Ko FN, Lee LG, Huang TF, Chen YP, Hsu HY (1988) Two antiplatelet agents from *Magnolia officinalis*. Thromb Res 50:757–765
38. Chen CC, Huang YL, Chen HT, Chen YP, Hsu HY, Sheu SJ (1988) The platelet aggregation inhibiting activity of honokiol, magnolol, and related compounds. Chung Hua Yao Hsueh Tsa Chih 40:15–19
39. Watanabe K, Goto Y, Hara N, Kanaoka S (1981) Neuronal factors involved in the pathogenesis of experimental ulceration and antiulcer effect of magnolol, an active component of *Magnolia* bark. Wakanyaku Shinpojumu, (Kiroku) 14:1–6 (CA 96:110706t)
40. Hattori M, Endo Y, Takebe S, Kobashi K, Fukasaku N, Namba T (1986) Metabolism of magnolol from Magnoliae Cortex. II. Absorption, metabolism and excretion of (ring-^{14}C)-magnolol in rats. Chem Pharm Bull (Tokyo) 34:158–167
41. Hattori M, Sakamoto T, Endo Y, Kakiuchi N, Kobashi K, Mizuno T, Namba T (1984) Metabolism of magnolol from magnoliae cortex. I. Application of liquid chromatography-mass spectrometry to the analysis of metabolites of magnolol in rats. Chem Pharm Bull (Tokyo) 32:5010–5017

Melia azedarach L. and *M. toosendan* Sieb. et Zucc.

85

85.1 Introduction

Kulianpi, Cortex Meliae, is the dry stem bark or root bark of *Melia toosendan* Sieb. et Zucc. or *M. azedarach* L. (Meliaceae). The bark is peeled in the spring or fall. It is officially listed in the Chinese Pharmacopoeia and can be used as an anthelmintic and externally to treat scabies.

Chuanlianzi, Fructus Toosendan, the dry ripe fruits of *M. toosendan* is also officially listed in the Chinese Pharmacopoeia. It is harvested in winter when the fruits have ripened. It is also used as an anthelmintic.

85.2 Chemical Constituents

85.2.1 Chemical Constituents of *Melia azedarach*

The first compound of *M. azedarach* isolated from the fruits was melianone [1]. Structural investigation showed that melianone (*85-2*) is a triterpene compound [2] derived from euphane (13α,14β,17α-lanostane, *85-1*) showing 23R, 24S configuration [3]. Related compounds melianol (*85-3*) [2], melianodiol (*85-4*) [4], and meliantriol (*85-5*) were then isolated and structurally investigated.

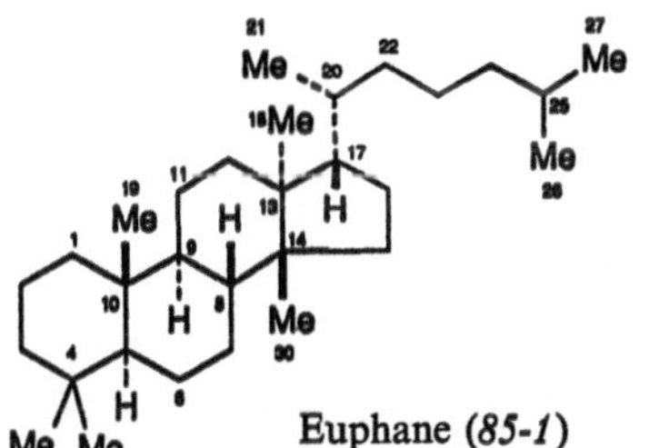

Euphane (*85-1*)
(13α,14β,17α-Lanostane)

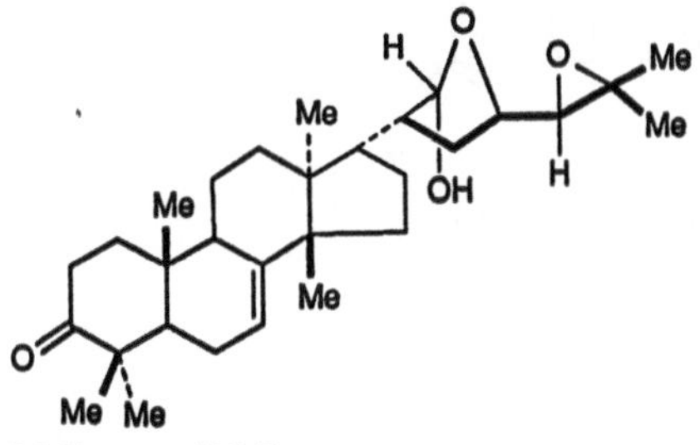

Melianone (*85-2*)

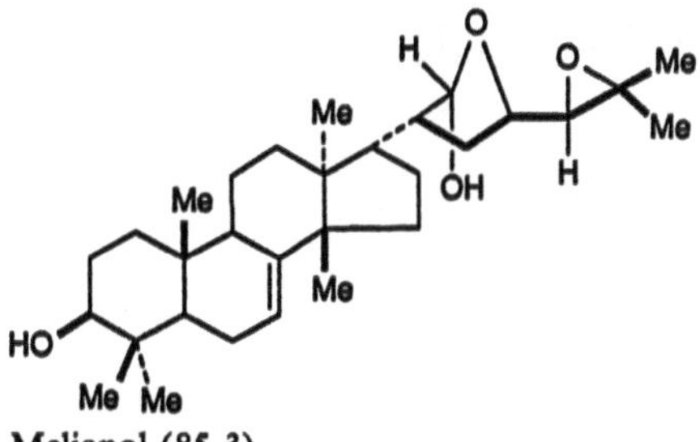

Melianol (*85-3*)

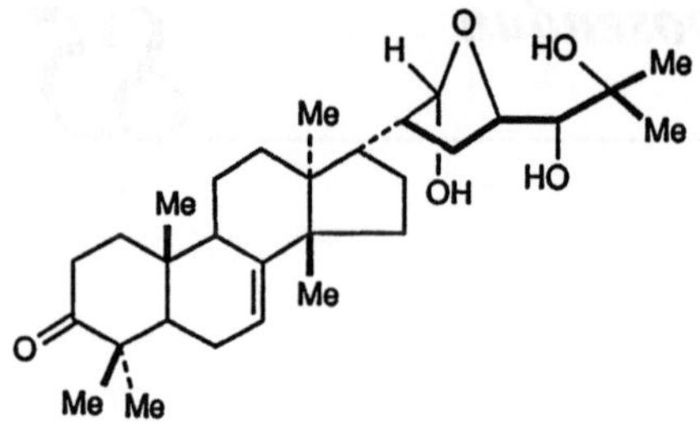

Melianodiol (*85-4*)

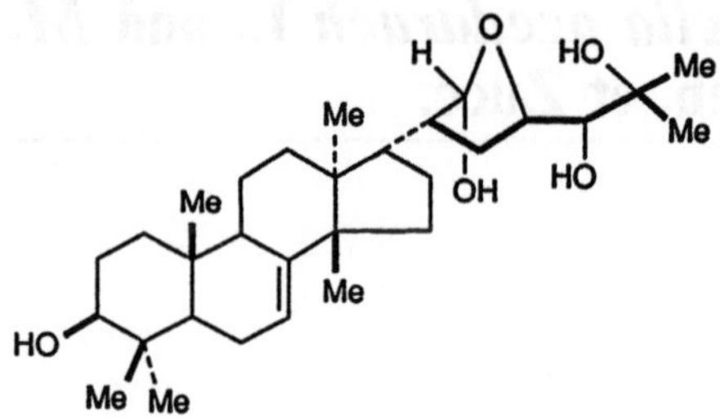

Meliantriol (*85-5*)

A number of new triterpene compounds were further isolated from the fruits of *M. azedarach*, including 1-cinnamoylmelianolone (*85-6*) [5] and a triterpene compound (*85-7*) [6], tetranortriterpene derivatives ohchinolides A (*85-8*), B (*85-9*) isolated first from the fruits of *M. azedarach* var. *japonica* [7], nimbolidines A (*85-10*), B (*85-11*), nimbolinine B (*85-12*), 1-deacetylnimbolinine B (*85-13*) [8] and a triterpene compound with the proposed structure of 21,23:24,25-diepoxyeuph-7-ene-21-ol [9]. The isolation of a new tetranortriterpene (*85-14*) from the fruits of *M. azedarach* cultivated in Egypt was reported recently [10].

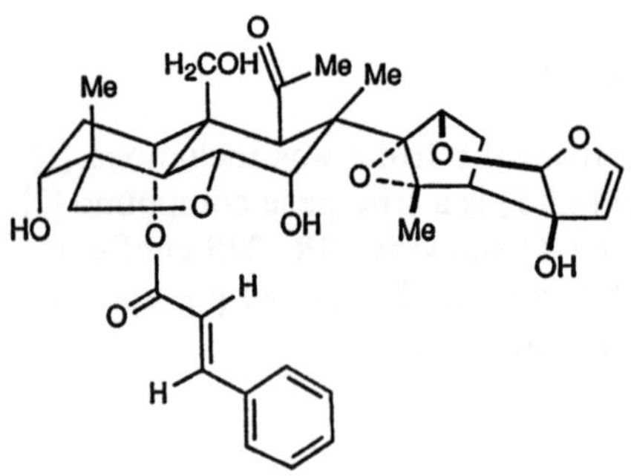

1-Cinnamoylmelianolone (*85-6*)

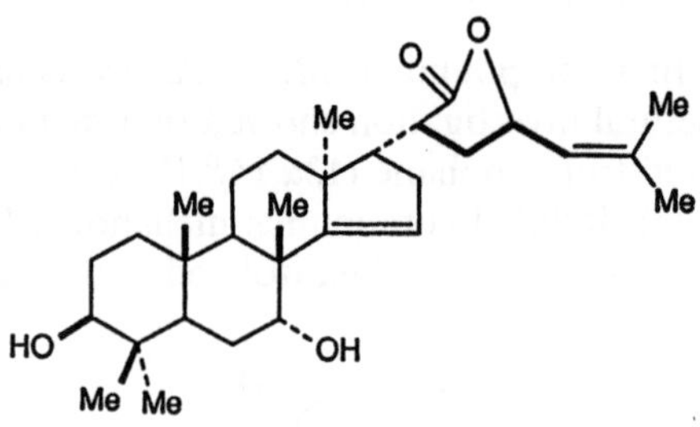

Triterpene compound (*85-7*)

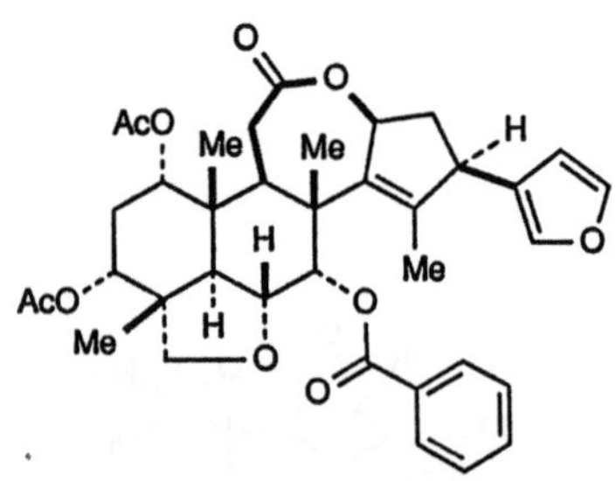

Ohchinolide A (*85-8*)

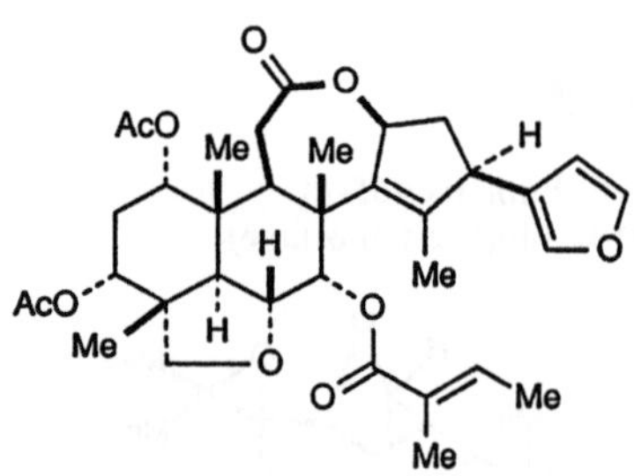

Ohchinolide B (*85-9*)

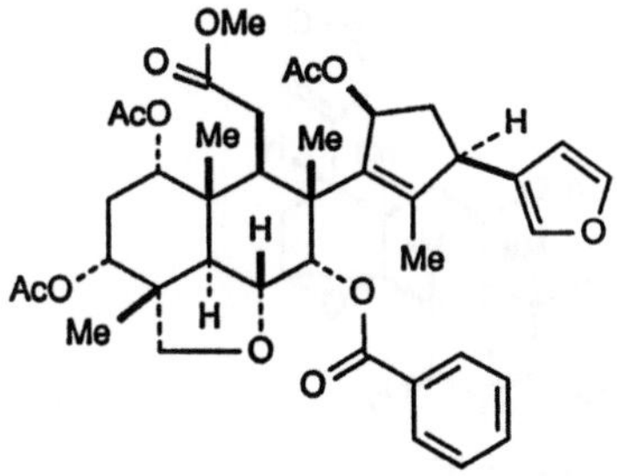

Nimbolidine A (*85-10*)

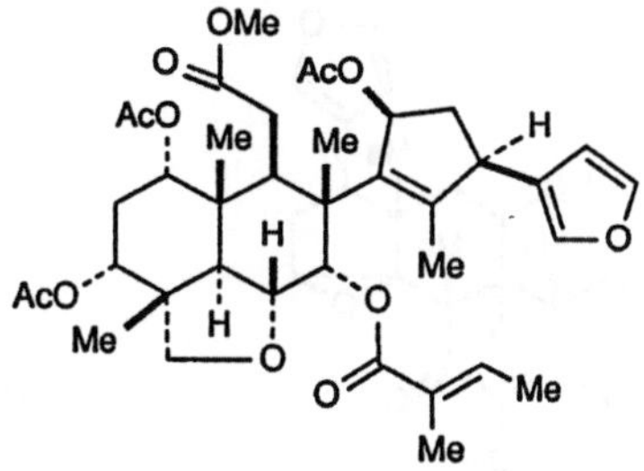

Nimbolidine B (*85-11*)

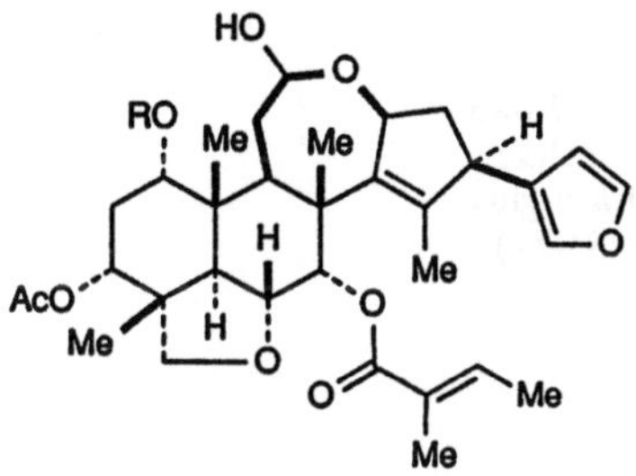

Nimbolinine B (*85-12*): R=Ac
1-Deacetylnimbolinine B (*85-13*): R=H

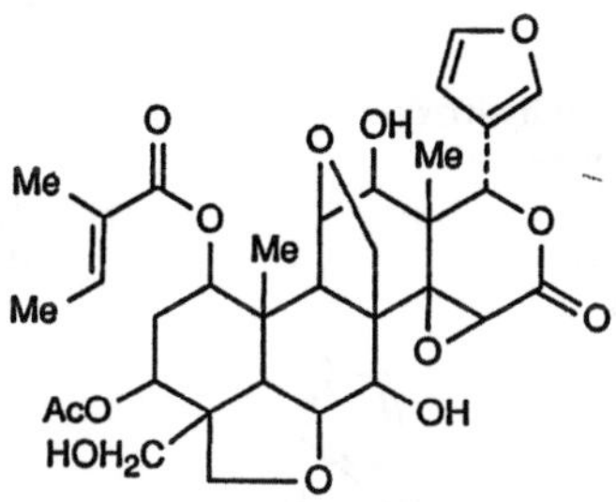

Tetranortriterpene (*85-14*)

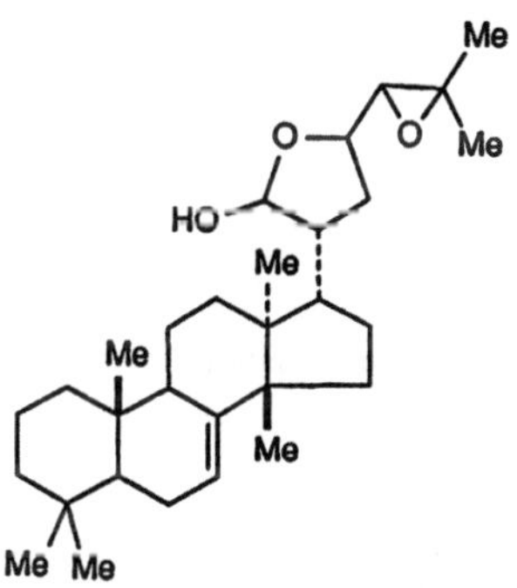

21,23:24,25-Diepoxy-euph-7-ene-21-ol (*85-15*)

From the seeds of *M. azedarach* two glycosides 6-acetoxy-3β,11α-dihydroxy-7-oxo-14β,15β-epoxymeliac-1,5-diene 3-*O*-α-L-rhamnopyranoside (*85-16*) [11], 6,11-diacetoxy-7-oxo-14β,15β-epoxymeliac-1,5-diene 3-*O*-β-D-glucopyranoside (*85-17*) [12] and 6-acetoxy-3β-hydroxy-7-oxo-14β,15β-epoxymeliac-1,5-diene 3-*O*-β-D-xylopyranoside (*85-18*) [13] were isolated together with the known compounds salannin (*85-19*) [14] and meldenin (*85-20*) [15], which were first found in the oil of *M. azadirachta*.

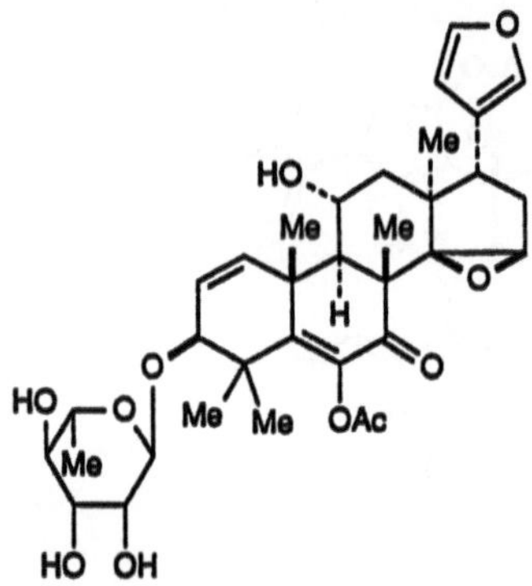

6-Acetoxy-3β,11α-dihydroxy-7-oxo-14β,15β-epoxy-meliac-1,5-diene 3-*O*-α-L-rhamnopyranoside (*85-16*)

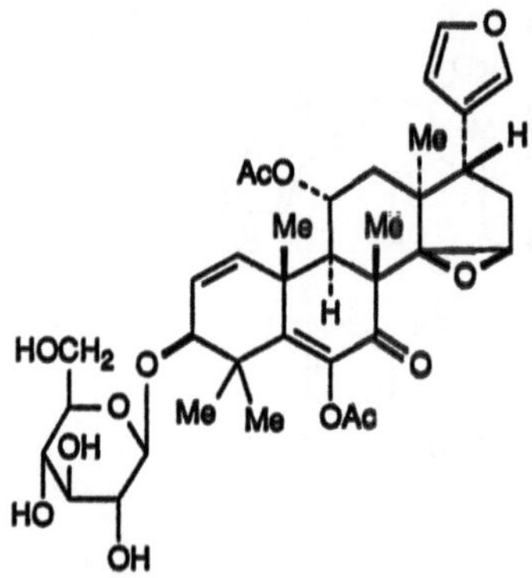

6,11α-Diacetoxy-7-oxo-14β,15β-epoxymeliac 1,5-diene 3-*O*-β-D-glucopyranoside (*85-17*)

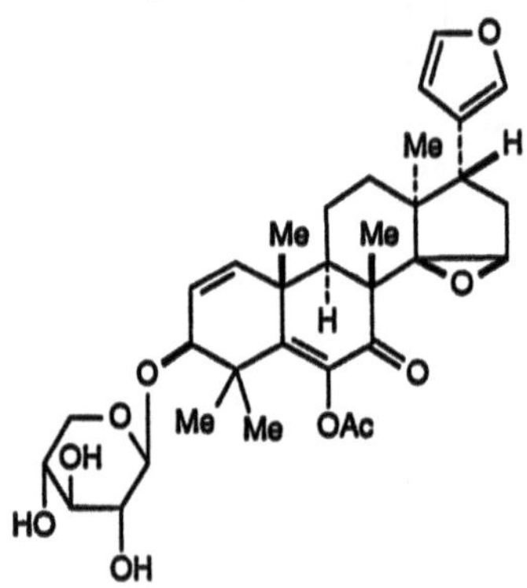

6-Acetoxy-3β-hydroxy-7-oxo-14β,15β-epoxymeliac-1,5-diene 3-*O*-β-D-xylopyranoside (*85-18*)

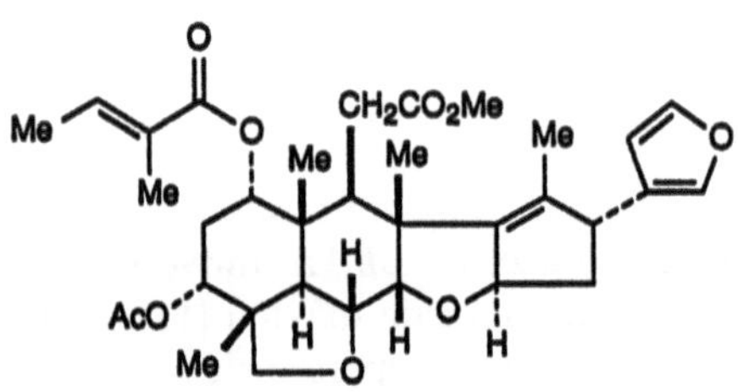

Salannin (*85-19*)

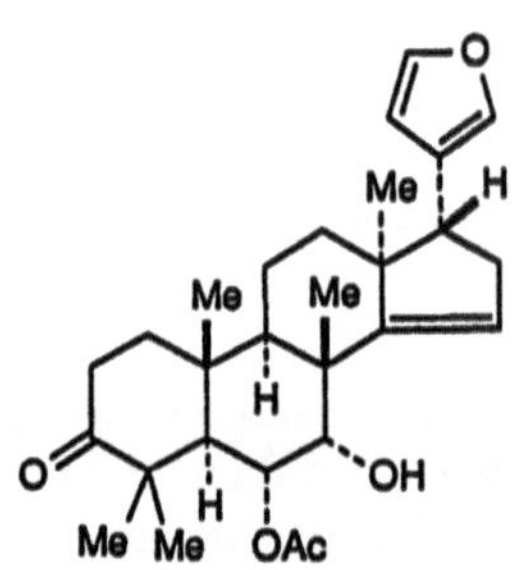

Meldenin (*85-20*)

A number of studies undertaken on the chemical constituents of the bark of *M. azedarach* led to the isolation of a series of triterpene compounds related to those isolated from the fruits. The first compound isolated from the bark was kulinone (*85-21*), characterized by a hydroxy function at C-16 [16].

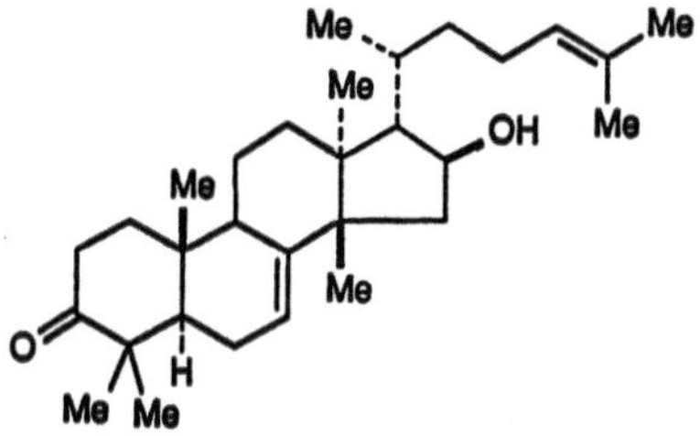

Kulinone (*85-21*)

The isolation of two related compounds named kulactone (*85-22*) and kulolactone (*85-23*) from *M. azedarach*, was further reported. They are characterized by a lactone group [17].

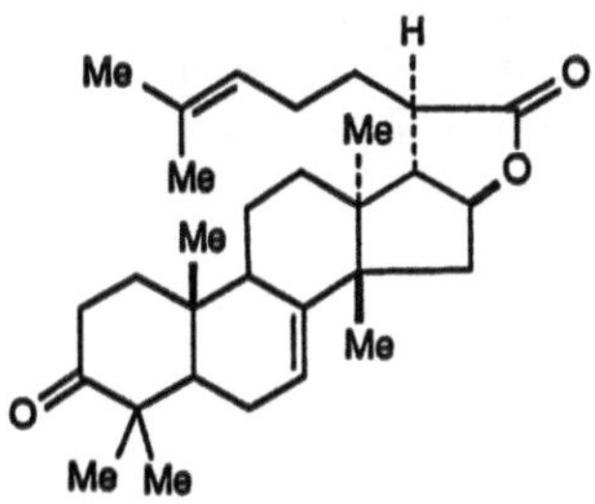

Kulactone (*85-22*)

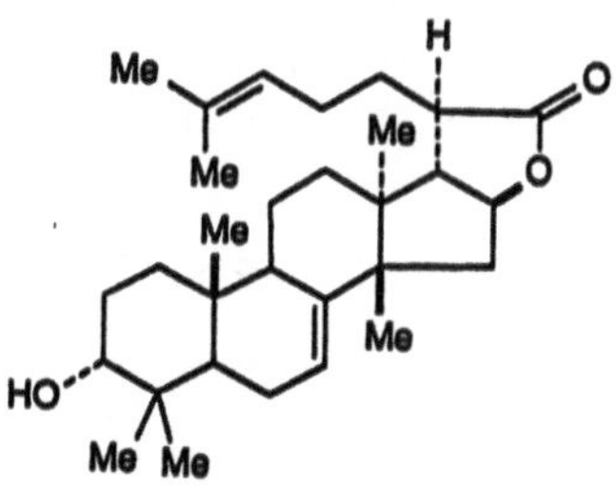

Kulolactone (*85-23*)

Another compound related to kulinone from the bark of *M. azedarach* is methyl kulonate (*85-24*) [18].

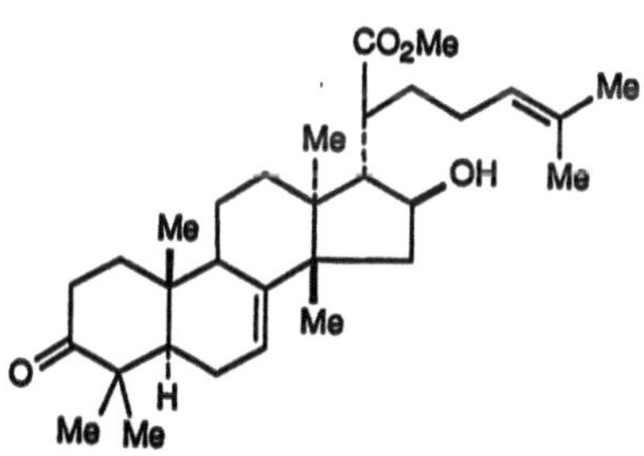

Methyl kulonate (*85-24*)

Two 6-hydroxylated steroids 6β-hydroxy-4-stigmasten-3-one (*85-25*) and its close analog, 6β-hydroxy-4-ergosten-3-one (*85-26*), were also isolated from the bark of *M. azedarach*. They are the first naturally occurring 6-hydroxylated Δ^4-3-oxo-steroids with an intact side chain [19].

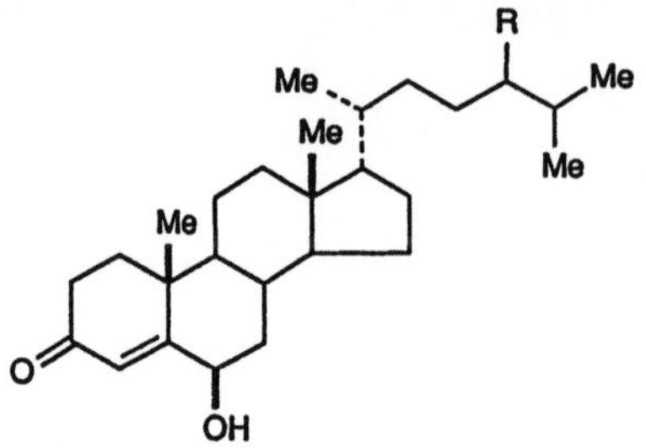

6β-Hydroxy-4-stigmasten-3-one (*85-25*): $R=CH_2-CH_3$
6β-Hydroxy-4-ergosten-3-one (*85-26*): $R=CH_3$

In addition to the tetracyclic triterpene compounds, the flavone glycoside apigenin 7-*O*-α-L-rhamnoypranosyl-(1→4)-β-D-glucopyranoside (*85-27*) [20] and the anthraquinone pigments 1,3,8-trihydroxy-2-methyl-anthraquinone 3-*O*-β-D-galactopyranoside (*85-28*) and 1,3,5-trihydroxy-8-methoxy-2-methylanthraquinone 3-*O*-α-L-rhamnopyranoside (*85-29*) [21] were isolated and identified from the bark of *M. azedarach*.

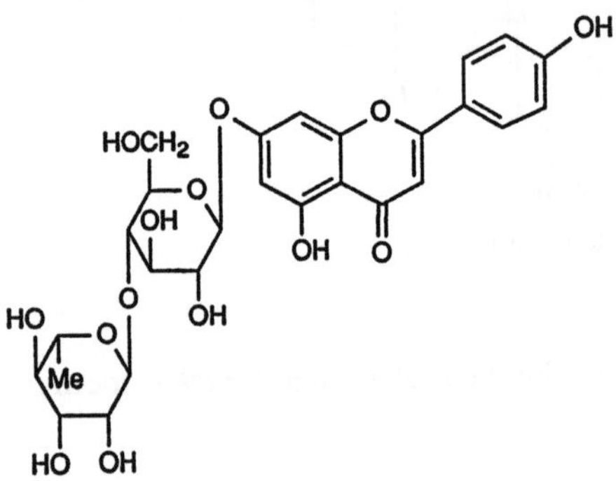

Apigenin 7-*O*-α-L-rhamnopyranosyl-(1→4)-
β-D-glucopyranoside (*85-27*)

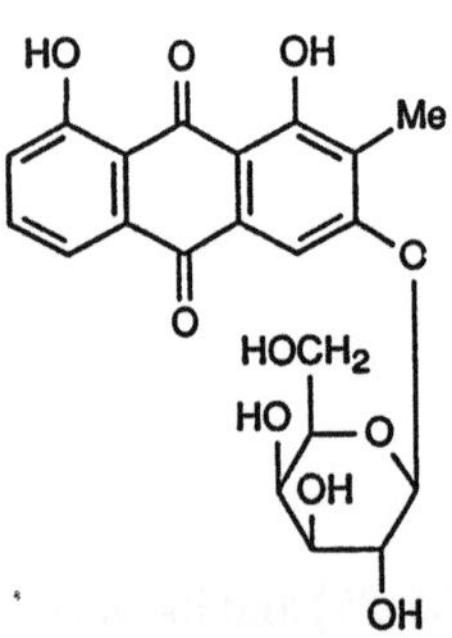

1,3,8-Trihydroxy-2-methyl-
anthraquinone 3-*O*-β-D-
galactopyranoside (*85-28*)

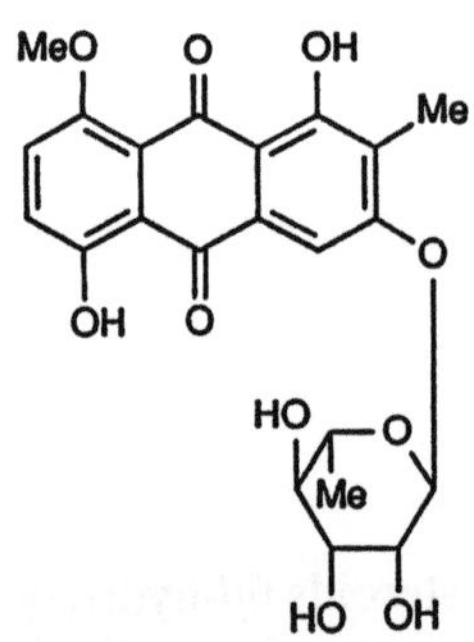

1,3,5-Trihydroxy-8-methoxy-
2-methylanthraquinone
3-*O*-α-L-rhamnopyra-
noside (*85-29*)

A new limonoide glycoside 3-deoxo-3β-hydroxy-gedunin 3-*O*-β-D-glucopyranoside (*85-30*) from the stem bark of *M. azedarach* was also found [22].

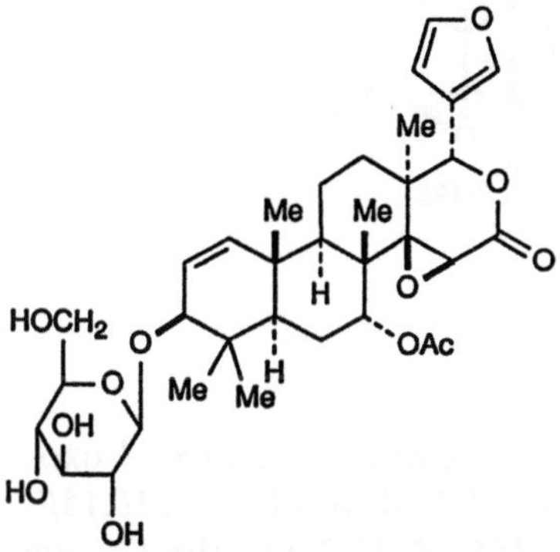

3-Deoxo-3β-hydroxy-gedunin 3-*O*-β-D-glucopyranoside (*85-30*)

The wood of *M. azedarach* was found to contain nimbolins A (*85-31*), B (*85-32*), fraxinellone, gedunin, 24-methylenecycloartanone [23], and melianins A (*85-33*) and B (*85-34*) [24]. Fraxinellon (*85-35*) and gedunin (*85-36*) are both known natural products. Fraxinellon was first isolated from *Dictamnus alba* (Rutaceae) [25], whereas gedunin was first isolated from another meliaceous plant *Entandrophragma angolense* [26].

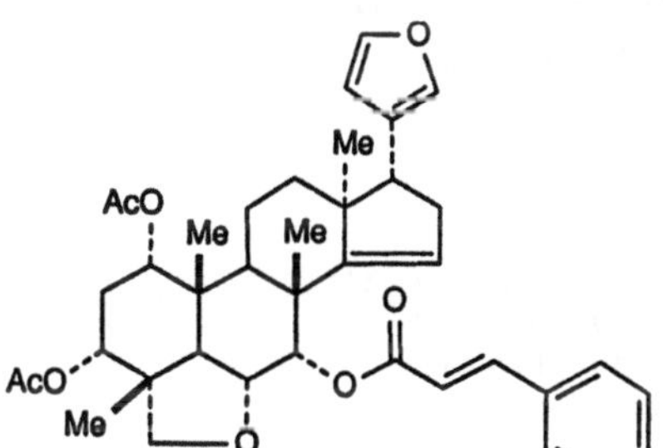

Nimbolin A (*85-31*)

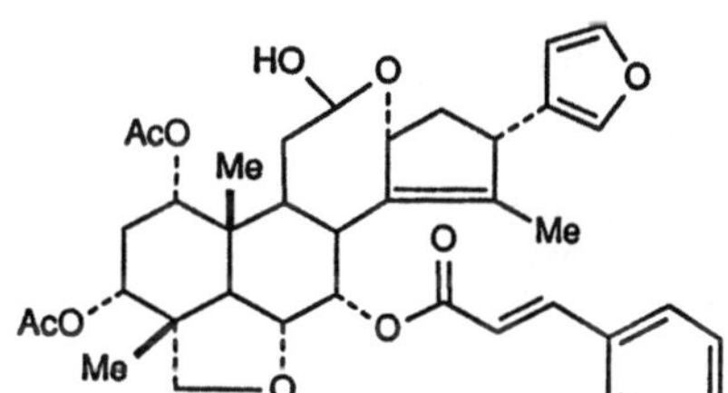

Nimbolin B (*85-32*)

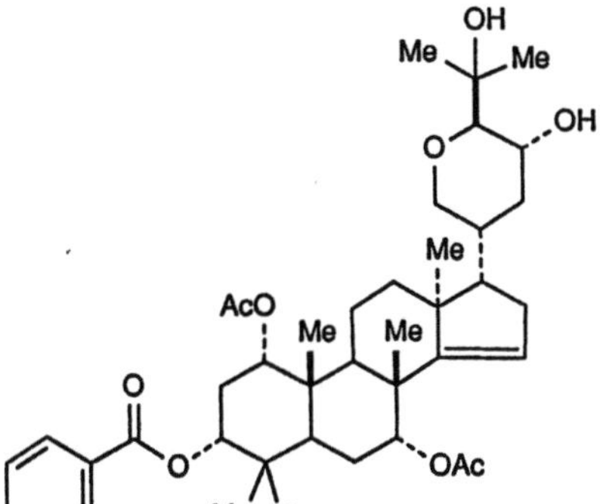

Melianin A (*85-33*)

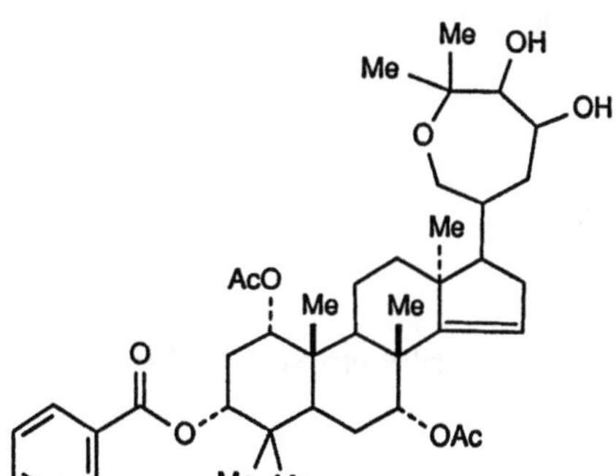

Melianin B (*85-34*)

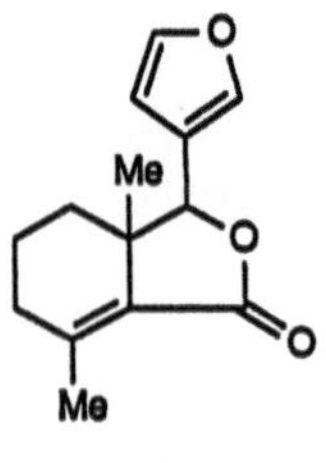

Fraxinellone (*85-35*)

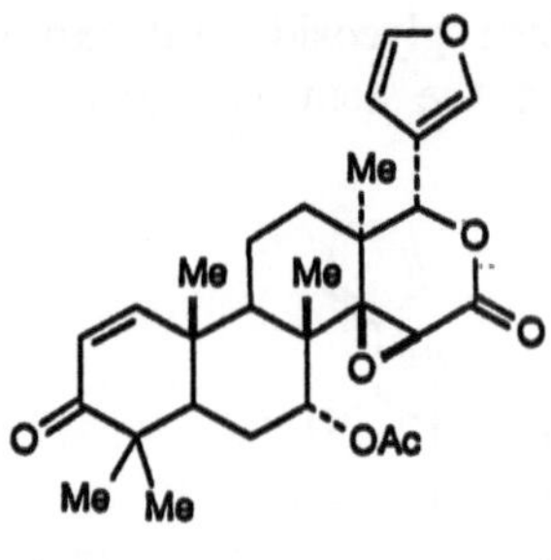

Gedunin (*85-36*)

Salannin and two new meliacan derivatives 6-acetoxy-7α-hydroxy-3-oxo-14β,15β-epoxymeliac-1,5-diene (*85-37*) and 6-acetoxy-3β-hydroxy-7-oxo-14β,15β-epoxymeliac-1,5-diene 3-*O*-β-D-glucuronopyranoside (*85-38*) [27] and the flavone apigenin 5-*O*-β-D-galactopyranoside (*85-39*) were further isolated [28] from the root of *M. azedarach*.

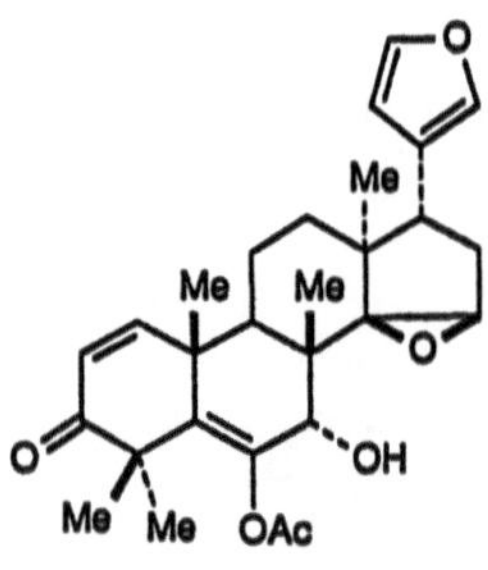

6-Acetoxy-7α-hydroxy-3-oxo-14β,15β-epoxymeliac-1,5-diene (*85-37*)

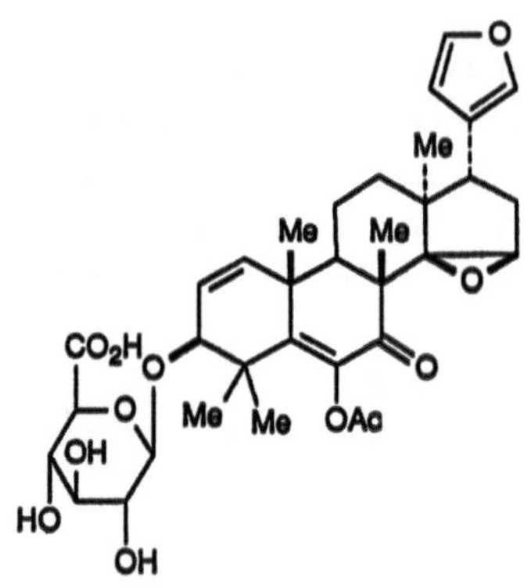

6-Acetoxy-3β-hydroxy-7-oxo-14β,15β-epoxymeliac-1,5-diene 3-*O*-β-D-glucuronopyranoside (*85-38*)

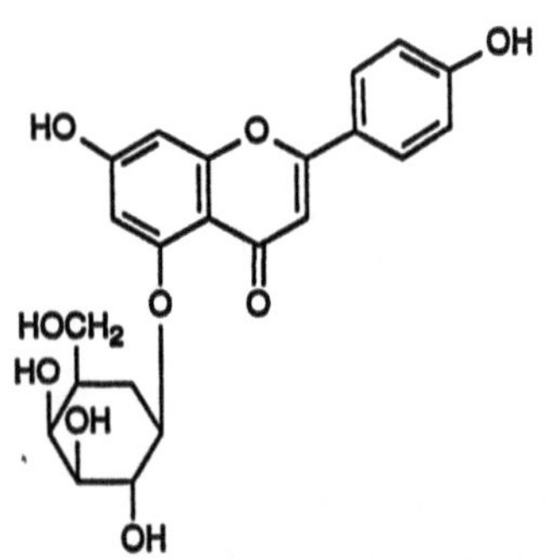

Apigenin 5-*O*-β-D-galactopyranoside (*85-39*)

Flavone glycosides rutin, quercitrin [29], and kaempferol 3-*O*-rutinoside [30] were isolated and identified from the leaves of *M. azedarach*.

85.2.2 Chemical Constituents of *M. toosendan*

An ascaricidal compound named chuanliansu (toosendanin, *85-40*) was isolated from the bark of *M. toosendan* [31]. The structure of chuanliansu was ascertained to correspond to deacetylsendanin [32]. The crystal structure of chuanliansu $C_{30}H_{38}O_{11} \cdot 3\ H_2O$ was also determined [33]. The content of chuanliansu in the bark of *M. toosendan* was dependent upon the season and age of the trees [34]. Highest contents were found in December to March, lowest in July to September. The contents of chuanliansu decreased from about 0.2% to <0.1% from 6- to 19-year-old plants. Chuanliansu was also isolated from the bark of *M. azedarach* [34]. The levels of chuanliansu in *M. toosendan* ranged from 0.15% to 0.24% and those in *M. azedarach* ranged from 0.06% to 0.1%.

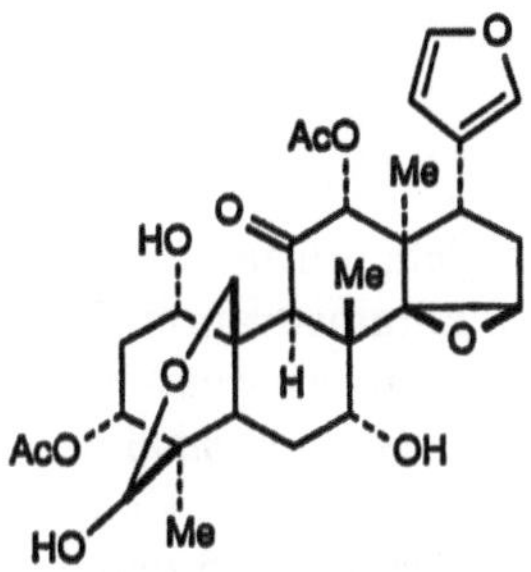

Chuanliansu (*85-40*)

A structurally related compound designated isochuanliansu (isotoosendanin, *85-41*) was isolated from the bark of *M. toosendan* and *M. azedarach* [35]. Chuanliansu and isochuanliansu are both compounds with a hemiacetal structural feature. Treatment of chuanliansu with 0.1% HCl yielded the isomer. Isochuanliansu differs from chuanliansu only in the replacement of the 14β,15β-epoxide group by a carbonyl group situated at position 15.

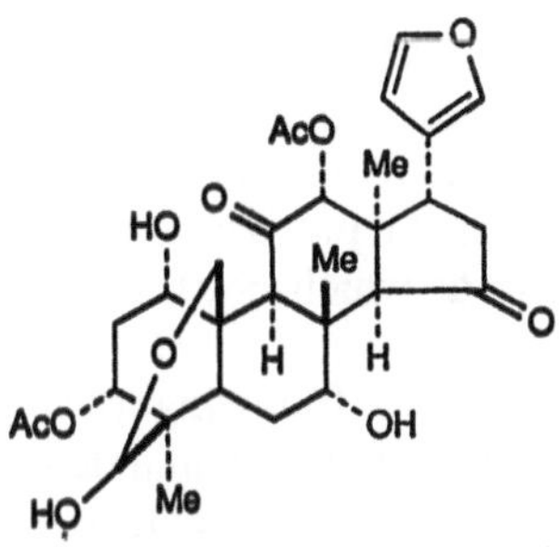

Isochuanliansu (*85-41*)

A euphane-type triterpene compound, 21-*O*-acetyl toosendantriol (*85-42*), was isolated from the fruits of *M. toosendan* and its structure detected by X-ray analysis [36].

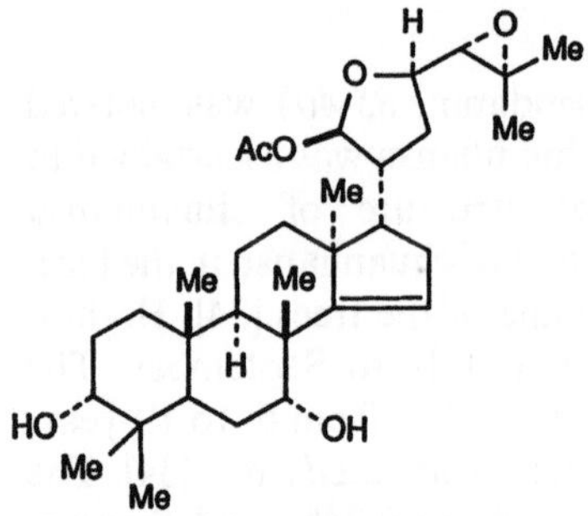

21-*O*-Acetyl-toosendantriol (*85-42*)

85.3 Pharmacology

Extracts of the fruits of *M. azedarach* caused mortality in piglets, buffalo calves, rabbits, and fowls after subcutaneous, intraperitoneal, or oral administration. Degenerative changes and hemorrhages in the liver, kidney, and gastrointestinal tract were found in all four species [37]. Chuanliansu from the bark of *M. toosendan* was reported to show anthelmintic and possibly antibotulism activity. It did not significantly affect the activities of other nervous systems and the cardiovascular system in experimental animals. The LD_{50} value of chuanliansu in mice was 13.8 mg/kg by intraperitoneal, 14.6 mg/kg by intravenous, 14.3 mg/kg by subcutaneous, and 244 mg/kg by oral administration; the LD_{50} in rats was 98 mg/kg by intraperitoneal and in rabbits 4.2 mg/kg by intravenous injection. An increase in the serum glutamic-pyruvic transaminase activity was the most predominant indication for subacute toxicity. The toxic effects of chuanliansu were reversible [38].

The survival rate of mice poisoned with lethal dosages of botulinum toxin was increased by more than 80% after intravenous, subcutaneous, or oral administration of chuanliansu within 6 h after intoxication. More than 50% of monkeys poisoned with A, B, or C type botulinum toxin survived when chuanliansu was administered 24 h after intoxication. The detoxifying effect of chuanliansu was comparable regardless of the type of toxin. Combined use of chuanliansu and antiserum of botulinum toxin A markedly decreased the amount of antiserum required for detoxication [39]. Experiments with mouse phrenic nerve-diaphragm preparations revealed that the neuromuscular junction is the primary site at which chuanliansu exerts its antagonistic effect on botulinum toxin [40].

The pharmacokinetics of chuanliansu were studied in monkeys following intravenous, intramuscular, and intragastric administration of [^{3}H]chuanliansu at a dose of 0.25 mg/kg [41]. Chuanliansu showed two-compartment kinetics. The half-lives $t_{1/2\alpha}$ and $t_{1/2\beta}$ following intravenous, intraperitoneal, and intragastric administration were 0.2, 0.6, 1.3, and 6.6, 18.0, 25.4 h, respectively [41].

The results indicated that the half-life of chuanliansu was much longer after intragastric than after intravenous administration. Other kinetic parameters showed that chuanliansu was rapidly absorbed and distributed in various organs, but its elimination was very slow. The combined urinary and fecal excretion of 24 h after administration was about 51% by intravenous, 27% by intramuscular, and 47% by intragastric application, and the corresponding excretion values 11 days after administration were 75%–80%. The distribution of chuanliansu resulted in the highest concentrations in the gallbladder and liver, followed by the spleen, stomach, kidney, and intestine, with lowest values in the bone [41].

References

1. Lavie D, Jain MK, Kirson I (1966) Terpenoids. V. Melianone from *M. azedarach* L. Tetrahedron Lett 2049–2052
2. Lavie D, Jain MK, Kirson I (1967) Terpenoids. VI. The complete structure of melianone. J Chem Soc [C] 1347–1351
3. Lyons CW, Taylor DR (1975) Sterochemistry of melianone and sapelin F. Correlation with bourjotinolone A. J Chem Soc Chem Commun 517–518
4. Lavie D, Jain MK, Shpan-Gabrielith SR (1967) A locust phagorepellent from two *Melia* species. Chem Commun 910–911
5. Lee SM, Klocke JA, Balandrin MF (1987) The structure of 1-cinnamoylmelianolone, a new insecticidal tetranortriterpenoid, from *Melia azedarach* L. (Meliaceae). Tetrahedron Lett 28:3543–3546
6. Lavie D, Levy EC (1969) A compound linking melianes with meliacins. Tetrahedron Lett 3525–3528
7. Ochi M, Kotsuki H, Ido M, Nakai H, Shiro M, Tokoroyama T (1979) The structures of ohchinolide A and B, two new limonoids from *Melia azedarach* Linn. var. *japonica* Makino. Chem Lett 1137–1140
8. Kraus W, Bokel M (1981) Neue Tetranortriterpenoide aus *Melia azedarach* Linn. (Meliaceae). Chem Ber 114:267–275
9. Schulte KE, Ruecker G, Matern HU (1979) Some constituents of the fruits and root of *Melia azedarach* L. Planta Med 35:76–83
10. Sabri NN, Abou-Donia AH, Ghazy NA (1987) On the constituents of the fruits of *Melia azedarach* L. cultivated in Egypt. Alexandria J Pharm Sci 1:28–31 (CA 110:132168k)
11. Srivastava SD (1986) Limonoids from the seeds of *Melia azedarach*. J Nat Prod 49:56–61
12. Srivastava SD (1987) Further constituent from the seeds of *Melia azedarach*. Planta Med 53:100–101
13. Rusia K, Srivastava SK (1988) Structure of a new limonoid glycoside from the seeds for *Melia azedarach* Linn. Proc Natl Acad Sci India [A] 58:33–36
14. Henderson R, McCrindle R, Overton KH, Meleva A (1964) Salannin. Tetrahedron Lett 3969–3974
15. Connolly JD, Handa KL, McCrindle R (1968) Further constituents of nim oil: the constitution of meldenin. Tetrahedron Lett 437–440
16. Chang FC, Chiang CK (1968) Kulinone, a euphane-type triterpenoid from *Melia azedarach*. Chem Commun 1156–1158
17. Chang FC, Chiang CK (1969) Tetracyclic triterpenoids from *Melia azedarach* L. II. 2-oxa-*trans*-bicyclo 3,3,*O*-octanones. Tetrahedron Lett 891–894
18. Chiang CK, Chang FC (1973) Tetracyclic triterpenoids from *Melia azedarach* L. III. Tetrahedron 29:1911–1929
19. Nair MG, Chang FC (1973) 6β-Hydroxy-4-stigmasten-3-one and 6β-hydroxy-4-campesten-3-one. Phytochemistry 12:903–906
20. Mishra M, Srivastava SK (1984) A new flavone glycoside from *Melia azedarach* Linn. Curr Sci 53: 694–695
21. Srivastava SK, Mishra M (1985) New anthraquinone pigments from the stem bark of *Melia azedarach* Linn. Indian J Chem [B] 24:793–794
22. Saxena M, Srivastava SK (1986) A new limonoid glycoside from the stem bark of *Melia azedarach* Linn. Indian J Chem [B] 25:1087–1088
23. Ekong DEU, Fakunle CO, Fasina AK, Okogun JI (1969) The meliacins (limonoids). Nimbolin A and B, two new meliacin cinnamates from *Azadirachta indica* L. and *Melia azedarach* L. Chem Commun 1166–1167

24. Okogun JI, Fakunle CO, Ekong DEU, Connolly JD (1975) Chemistry of the meliacins (limonoids). The structure of melianin A, a new protomeliacin from *Melia azedarach*. J Chem Soc Perkin Trans I 1352–1356
25. Pailer M, Schaden G, Spiteller G, Fenzl W (1965) Die Konstitution des Fraxinellons (aus *Dictamnus albus* L.). Monatsh 96:1324–1346
26. Akisanya A, Bevan CWL, Hirst J, Halsall TG, Taylor DAH (1960) West African timbers. III. Petroleum extracts from the genus *Entandrophragma*. J Chem Soc 3827–3829
27. Srivastava SK, Gupta HO (1985) New limonoids from the roots of *Melia azedarach* Linn. Indian J Chem [B] 24:166–170
28. Gupta HO, Srivastava SK (1985) Apigenin-5-*O*-galactoside from the roots of *Melia azedarach* Linn. Curr Sci 54:570–571
29. Nair AGR, Subramanian SS (1975) Quercetin glycosides from the leaves of *Soymida febrifuga* and *Melia azedarach*. Indian J Chem 13:527
30. Marco JA, Barbera O, Sanz JF, Sanchez-Parareda J (1986) Flavonol diglycosides from *Melia azedarach*. J Nat Prod 49:170
31. Chung CC, Hsie TH, Chen SF, Liang HT (1975) The structure of chuanliansu. Acta Chim Sin 33:35–47
32. Shu GX, Liang XT (1980) Correction of the structure of chuanliansu. Acta Chim Sin 38:196–198
33. Guo F, Hou YG, Zhu NJ, Fu H (1984) Crystal structure of chuanliansu $C_{30}H_{38}O_{11} \cdot 3\ H_2O$. Jiegou Huaxue 3:91–94
34. Zhuo JM, Gu YC, Ran CX, Zhao PP, Yu CJ (1981) Study on toosendanin dynamics in the bark of *Melia toosendan* Set Z. Chin Pharm Bull 16:10–12
35. Xie JX, Yuan AX (1985) Molecular structure of isochuanliansu isolated from traditional Chinese medicine – the bark of *Melia toosendan* and *Melia azedarach*. Acta Pharm Sin 20:188–192
36. Nakanishi T, Inada A, Nishi M, Miki T, Hino R, Fujiwara T (1986) The structure of a new natural apotirucallane-type triterpene and the stereochemistry of the related terpene. X-ray and carbon-13 NMR spectral analyses. Chem Lett 69–72
37. Hothi DS, Singh B, Kwatra MS, Chawla RS (1976) A note on the comparative toxicity of *Melia azedarach* (Dhrek) berries to piglets, buffalo calves, rabbits and fowls. J Res Punjab Agric Univ 13:233–234 (CA 86:184305h)
38. Li PZ, Shi XY, Xu ZH, Li JF, Sun GZ, Qin BY (1982) Pharmacological and toxicological studies on toosendanin. Chin Trad Herb Drugs 13:29–32
39. Li PZ, Zou J, Miao WY, Ding FH, Meng JY, Jai GR, Li JF, Ye HJ, He XY (1982) Efficacy of the treatment of botulinum toxin poisoning by toosendanin. Chin Trad Herb Drugs 13:28–30
40. Li PZ, Sun GZ (1983) Antagonistic effect of toosendanin to botulin toxin on neuromuscular preparations of mice. Acta Physiol Sin 35:480–483
41. Zou J, Jia GR, He XY (1982) Pharmacokinetic study of toosendanin. Chin Trad Herb Drugs 13:408–410

Menispermum dauricum DC.

86

86.1 Introduction

Beiduogen, Rhizoma Menispermi, is the dry rootstock of *Menispermum dauricum* DC. (Menispermaceae), which occurs widely in China. It is a traditional Chinese medicine and is now officially listed in the Chinese Pharmacopoeia. Beiduogen Pian, Tabellae Menispermi, is another official item produced from the total alkaloidal extract of *M. dauricum*. The rhizome as well as the tablets are used as an analgesic and antipyretic.

86.2 Chemical Constituents

Menispermum dauricum contains mainly alkaloids as active ingredients. A number of alkaloids of different structural types were isolated and identified. Dauricine (*86-1*), an alkaloid of the bisbenzylisoquinoline type, is the major alkaloid. Two benzylisoquinolines are connected via an ether bridge between the benzyl moieties [1, 2].

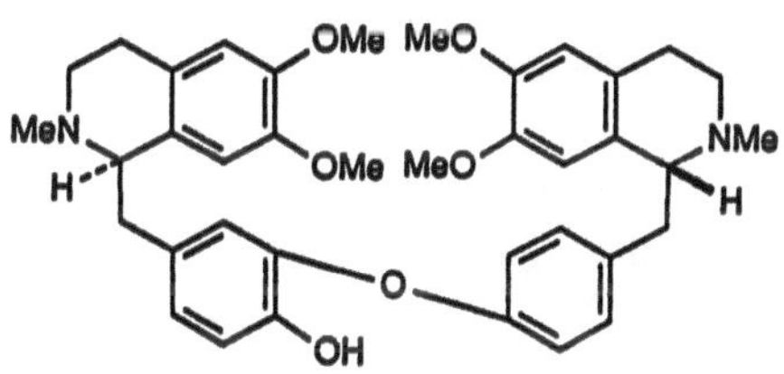

Dauricine (*86-1*)

Some related alkaloids, derived by partial *O*-demethylation of dauricine, were also isolated and identified: dauricinoline (*86-2*) [3], dauricoline (*86-3*) [4], daurinoline (*86-4*) [5], and daurisoline (*86-5*) [6].

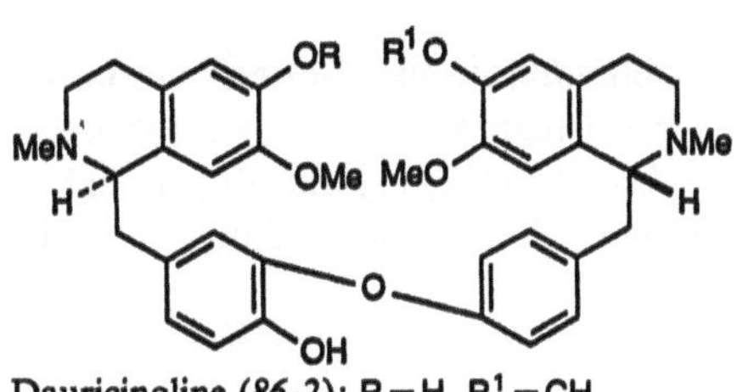

Dauricinoline (*86-2*): R=H, $R^1=CH_3$
Dauricoline (*86-3*): $R=R^1=H$
Daurinoline (*86-4*): $R=CH_3$, $R^1=H$

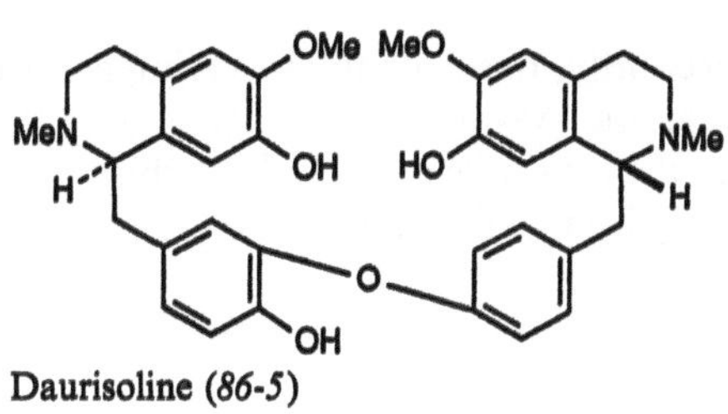

Daurisoline (*86-5*)

O,O-Dimethyl-corytuberine, isocorydine, and the quaternary alkaloids menisperine (*86-8*), *O*-methylmenisperine [7], and magnoflorine [3] are alkaloids of the aporphine type, isolated from *M. dauricum*.

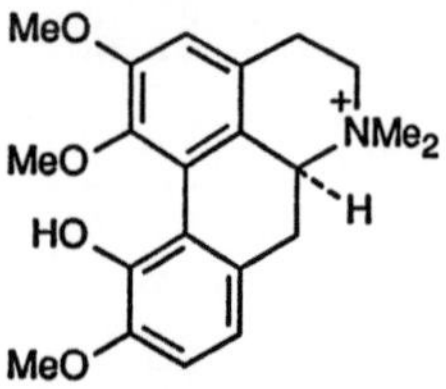

Menisperine (*86-6*)

In addition, some alkaloids of different chemical types, sinomenine (*86-7*) [8], cheilantifoline (*86-8*), stepharine (*86-9*), and stepholidine (*86-10*) [9] were also isolated. Sinomenine possesses a morphinan skeleton, and cheilanthifoline and stepholidine are two alkaloids derived from dibenzoquinolizin.

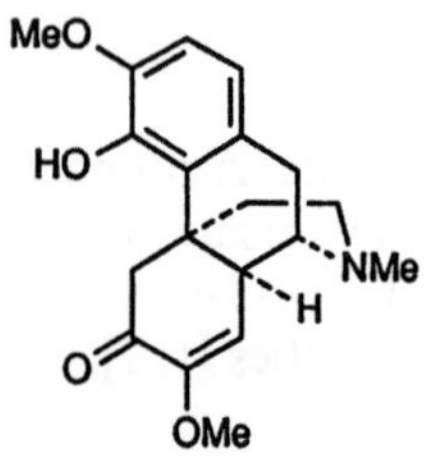

Sinomenine (*86-7*)

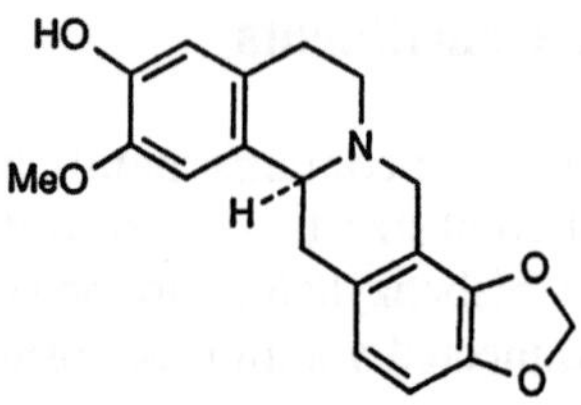

Cheilanthifoline (*86-8*)

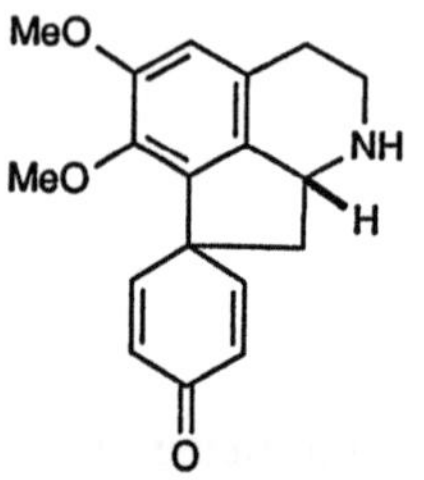

Stepharine (*86-9*)

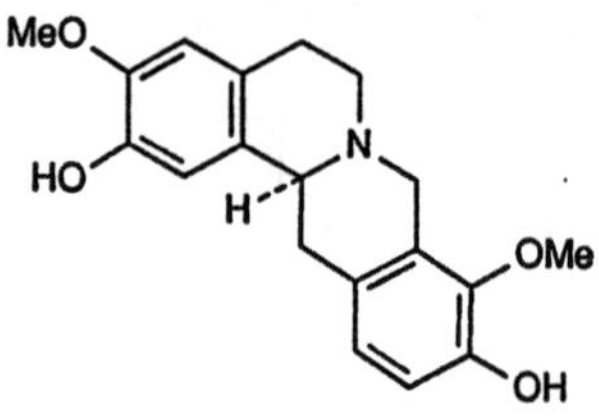

Stepholidine (*86-10*)

Recently, three alkaloids having a 7H-dibenzo[*de,h*]quinolin-7-one skeleton, bianfugecine (*86-11*), bianfugedine (*86-12*), and bianfugenine (*86-13*), were isolated from the ethanolic extract of the rhizome of *M. dauricum* [10, 11]. Bianfugenine had already been isolated from the vines of *M. dauricum* [12] and named dauriporphine.

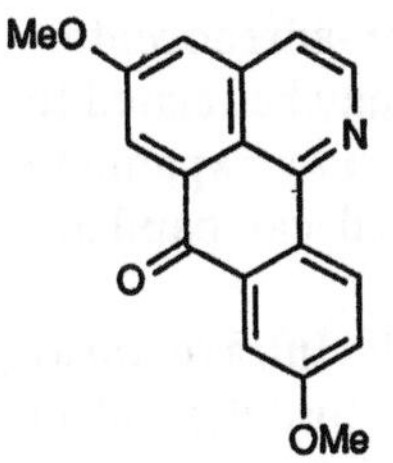

Bianfugecine (*86-11*)

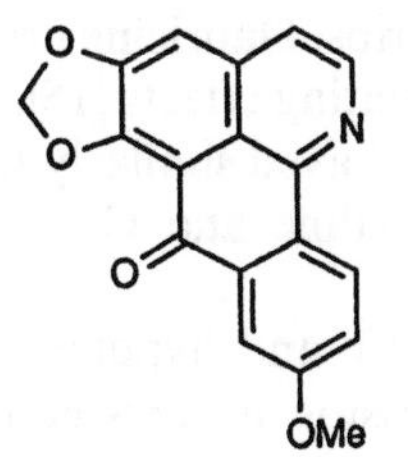

Bianfugedine (*86-12*)

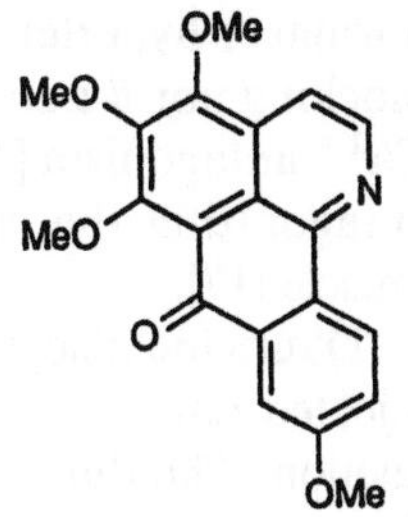

Bianfugenine (*86-13*)

A structurally related alkaloid isolated from *M. dauricum* was described by Japanese scientists and named menisporphine (*86-14*) [13].

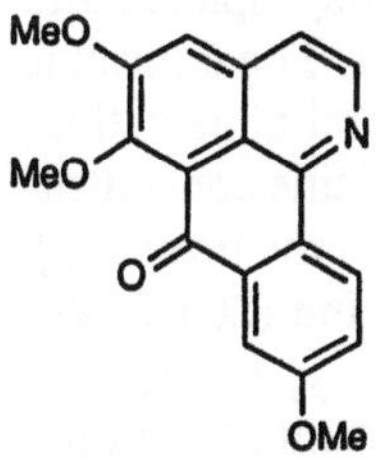

Menisporphine (*86-14*)

Two alkaloids containing chlorine with a novel skeleton were isolated from *M. dauricum*, acutumidine (*86-15*) and its *N*-methyl analog acutumine (*86-16*). Structures were elucidated by degradative and spectroscopic methods [3, 14]. The absolute configuration of the alkaloids was ascertained by X-ray examination [14].

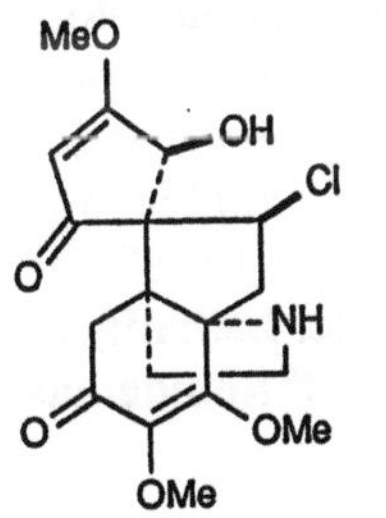

Acutumidine (*86-15*)

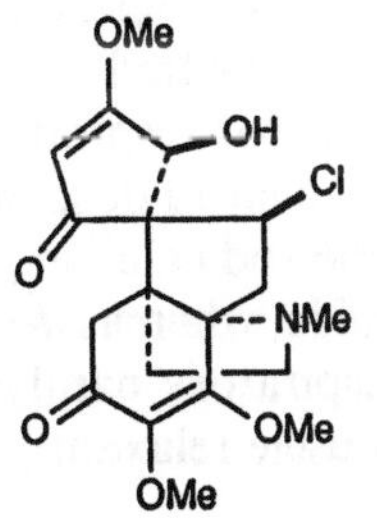

Acutumine (*86-16*)

86.3 Pharmacology

86.3.1 Pharmacology of Dauricine

Dauricine inhibits contractions of rabbit thoracic aortic strips induced by epinephrine or high K^+ concentration and those induced by norepinephrine in normal or hypertensive rat aortic strips. Dauricine produces neither α-adrenoceptor-blocking nor β-adrenoceptor-stimulating effects [15]. The activity may be related to Ca^{2+} antagonism [16]. Dauricine behaved similarly to tetrandrine and verapamil in antagonizing the effects of isoprenaline and Ca^{2+} on the isolated cat papillary muscle [17].

Dauricine also has antiarrhythmic and hypotensive actions [18]. Intravenously injected dauricine induced hypotension in anesthetized cats in a dose-dependent fashion. The duration of dauricine-induced hypotension was also dose dependent. Dauricine did not decrease epinephrine- or carotid ganglion-induced hypertension, indicating that dauricine is neither a receptor blocker nor a ganglionic blocker [19, 20]. The lowering effect of dauricine on arterial blood pressure in cats after intravenous administration of 5 mg/kg might be caused by arterial dilation [21].

One hour after a single oral administration of dauricine at a dose of 150 mg/kg to rats, about 50% of the administered dose had been absorbed from the gastrointestinal tract and rapidly appeared in organs including liver, lung, kidney, spleen, and brain, indicating that dauricine can also cross the blood-brain barrier. The dauricine content in the liver, lung, and brain reached peak values 1 h after the administration, whereas that in the spleen reached a peak value only after 12 h. Dauricine concentration was relatively high in the liver, spleen, and kidney and was low in the brain [22]. The kinetics of disappearance of intravenously injected [^{3}H]dauricine (20 μCi/kg) from the rat circulation fit an open two-compartment model with $t_{1/2\alpha}$ and $t_{1/2\beta}$ of 0.27 and 2.9 h, respectively. The urinary and fecal ^{3}H activity accounted for only 9.2% and 9.8%, respectively, of the total [^{3}H]dauricine administered after 72 h. The fate of the remaining ^{3}H in the body is not known [23]. The pharmacokinetics of dauricine were also studied in rabbits [24].

86.3.2 Pharmacology of Daurisoline

The didemethyl derivative of dauricine, daurisoline, was also studied pharmacologically. The DL_{50} of daurisoline methylbromide in mice was 1.25 mg/kg intravenously. In rabbits, the mean dose for respiratory paralysis was 3.7 mg/kg and that for cardiac arrest 153 mg/kg. The mean dose for induction of head drop for rabbits was 1.7 mg/kg. Both neostigmine and calcium gluconate were antagonistic to the effects of daurisoline methyl bromide, whereas *d*-tubocurarine was synergistic. Lethality in chickens was caused by respiratory paralysis [25]. Daurisoline methylbromide has been used clinically as a muscle relaxant [25].

References

1. Tomita M, Okamoto Y (1964) Alkaloids of menispermaceous plants. CCVIII. Alkaloids of *Menispermum dauricum*. Yakugaku Zasshi [Suppl 3] 84:1030–1031
2. Sun YQ, Gao CY, Zhang LX, Zhang AJ, Li FM, Wang LQ, Cai H (1984) Study of chromatography and spectrometry for Chinese medicines. II. TLC analysis and HPLC separation for the total alkaloid and injections of the root of *Menispermum dauricum* DC. J Shenyang Coll Pharm 1:223–227

3. Tomita M, Okamoto Y, Nagai Y, Tanaka S, Hayata T (1970) Alkaloids of menispermaceous plants. CCLVIII. Alkaloids of *Menispermum dauricum*. Basic components of Siberian *Menispermum dauricum*. Yakugaku Zasshi 90:1182–1186
4. Tomita M, Okamoto Y, Nagai Y, Kitayama K, Yanagawa H (1970) Alkaloids of menispermaceous plants. CCLVII. Alkaloids of *Menispermum dauricum*. Structure of a new tertiary phenolic biscoclaurine type alkaloid dauricoline. Yakugaku Zasshi 90:1178–1181
5. Tomita M, Okamoto Y (1965) Alkaloids of menispermaceous plants. CCIX. Alkaloids of *Menispermum dauricum*. 4. The structure of a new tertiary biscoclaurine type alkaloid "daurinoline". Yakugaku Zasshi 85:456–459
6. Zheng XW, Min ZD, Zhao SX (1979) A new alkaloid of *Menispermum dauricum*, daurisoline. Kexue Tongbao 24:285–288
7. Tomità M, Kikuchi T (1955) Alkaloids of menispermaceous plants. CXXIV. Alkaloids of *Menispermum dauricum*. Pharm Bull Jpn 3:100–104
8. Ilinskaya TN (1956) Alkaloids of *Menispermum dauricum*. Dokl Akad Nauk SSSR 108:1081–1082
9. Okamoto Y, Tanaka S, Kitayama K, Isomoto M, Masaishi M, Yanagawa H, Kunimoto J (1971) Alkaloids of menispermaceous plants. CCLXI. Alkaloids of *Menispermum dauricum*. 8. Yakugaku Zasshi 91:684–687
10. Hou CY, Xue H (1985) Studies on chemical constituents of *Menispermum dauricum* DC. Acta Pharm Sin 20:112–117
11. Hou CY, Xue H (1984) Studies on the chemical constituents of *Menispermum dauricum* DC. Acta Pharm Sin 19:471–472
12. Takani M, Takasu Y, Takahashi K (1983) Studies on constituents of medicinal plants. XXIII. Constituents of the vines of *Menispermum dauricum* DC. 2. Chem Pharm Bull (Tokyo) 31:3091–3093
13. Kumitomo J, Satoh M, Shingu T (1983) Studies on the alkaloids of menispermaceous plants. CCLXXIX. Alkaloids of *Menispermum dauricum* DC. 9. Structure and synthesis of menisporphine, a new type of isoquinoline alkaloid. Tetrahedron 39:3261–3265
14. Tomita M, Okamoto Y, Kikuchi T, Osaki K, Nishikawa M, Kamiya K, Sasaki Y, Matoba K, Goto K (1971) Alkaloids of menispermaceous plants. CCLIX. Alkaloids of *Menispermum dauricum*. Structure of acutumine and acutumidine, chlorine-containing alkaloids with a novel skeleton. Chem Pharm Bull (Tokyo) 19:770–791
15. Chen SH, Hu CJ (1982) Effect of dauricine on aortic stripe. Acta Pharmacol Sin 3:178–182
16. Li GR, Fang DC, Hu CJ, Lu FH (1984) Effects of dauricine on physiologic properties of cat myocardium papillary muscle. Acta Pharmacol Sin 5:20–22
17. Li GR, Fang DC, Hu CJ, Lu FH (1983) Effects of dauricine on the concentration-response curve of isoprenaline and calcium and the electromechanical activity of the cat papillary muscle. Acta Pharm Sin 18:660–664
18. Li GR, Hu CJ, Lu FH (1983) Effects of dauricine on ECG and its local anesthetic activity. Acta Acad Med Wuhan 12:280–282
19. Chen SH, Wu CJ (1981) Antihypertensive action of dauricine. Chin Trad Herb Drugs 12:450
20. Chen SH, Hu CJ (1982) Hypotensive effect of dauricine and preliminary elucidation of its mechanism. Acta Acad Med Wuhan 11:75–78
21. Zeng FD, Zeng WC, Leng DM, Hu CJ (1984) Effects of dauricine and verapamil on systolic time intervals and blood pressure in anesthetized cats. Acta Acad Med Wuhan 13:205–208
22. Dai ZS, Hou SX, Guo LY, Hu CJ (1983) Absorption, distribution and excretion of dauricine in rats. Chin Pharm Bull 18:278–279
23. Dai ZS, Yi MG, Li J, Hu CJ (1983) Study on the pharmacokinetics of ^{3}H-dauricine. Acta Acad Med Wuhan 12:290–291
24. Hou SX, Dai ZS, Wang FJ, Guo LJ, Hu CJ (1982) Pharmacokinetic studies on dauricine. Acta Pharm Sin 17:863–865
25. Gong T, Wu ZY (1979) Pharmacological effects of daurisoline methyl bromide, a preliminary report. Acta Pharm Sin 14:439–442

Momordica cochinchinensis (Lour.) Spreng. and *M. grosvenori* Swingle

87

87.1 Introduction

Mubiezi, Semen Momordicae, is the dry ripe seeds of *Momordica cochinchinensis* (Lour.) Spreng. (Cucurbitaceae) collected in winter when the fruits have ripened. It is officially listed in the Chinese Pharmacopoeia and used for treatment of ulcer, carbuncle, anal fistula, tetter, and favus by external administration.

Luohanguo, Fructus Momordicae, is the dry fruits of *M. grosvenori* Swingle collected in the fall when the fruits have turned dark green. It is also officially listed in the Chinese Pharmacopoeia and used for treatment of cough and pharyngitis and as a laxative.

87.2 Chemical Constituents

87.2.1 Chemical Constituents of *Momordica cochinchinensis*

Two saponins named momordica saponons I (*87-1*) and II (*87-2*) were isolated from the seeds of *M. cochinchinensis*. The aglycone of saponon I is gypsogenin, whereas that of saponin II is quillaic acid. Both saponins possess the same sugar moiety [1]. Gypsogenin and quillaic acid are both triterpenes derived from oleanane characterized by a carbonyl group at C-23.

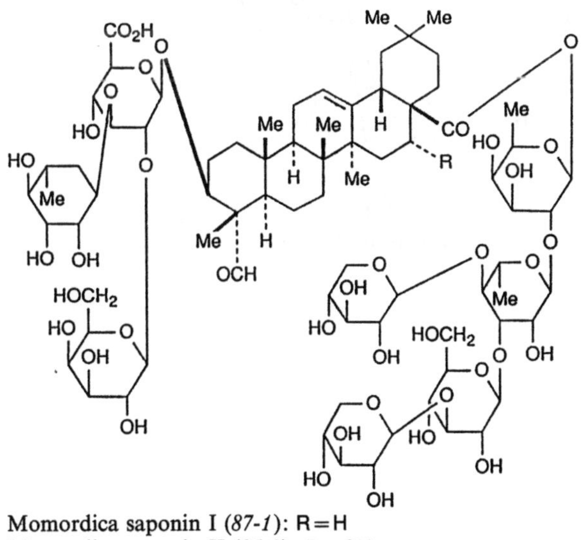

Momordica saponin I (*87-1*): R=H
Momordica saponin II (*87-2*): R=OH

Seeds of *M. cochinchinensis* contain up to 41% of fat oil. Besides palmitic, stearic, oleic, and linolic acids, a rare fatty acid, α-eleostearic acid, was also identified in the seed oil [2]. α-Eleostearic acid is chemically (E,Z,E)-9,11,13-octadecatrienoic acid. Furthermore, oleanolic acid [3] and its glycosides named momordins I (*87-3*), II (*87-4*), and III (*87-5*) were isolated from the root of *M. cochinchinensis* in addition to momordic acid (*87-6*) [4], the first example of a triterpenic acid with an 1-oxo group in the oleanane series.

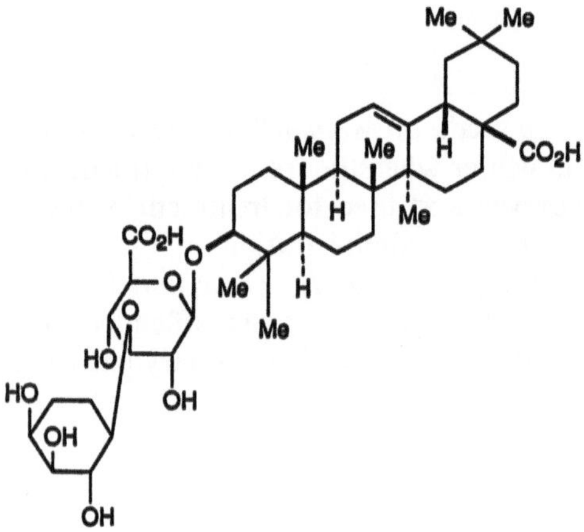

Momordin I (*87-3*)

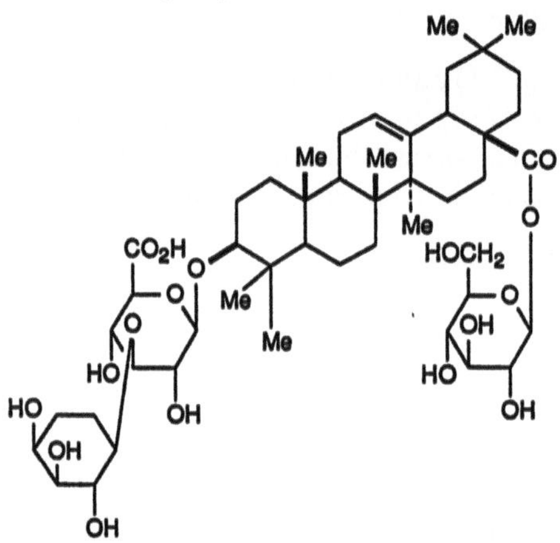

Momordin II (*87-4*)

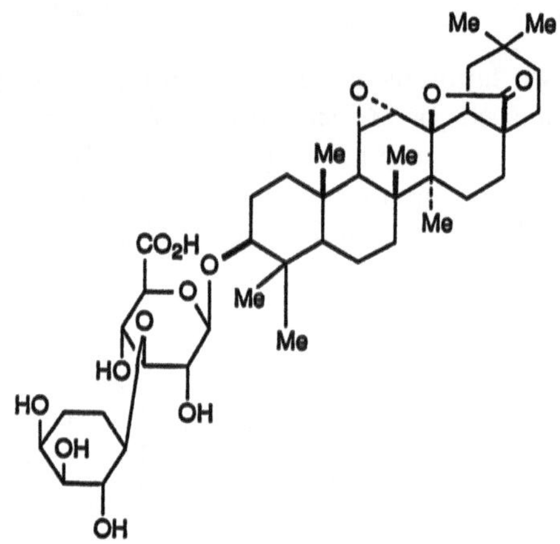

Momordin III (*87-5*)

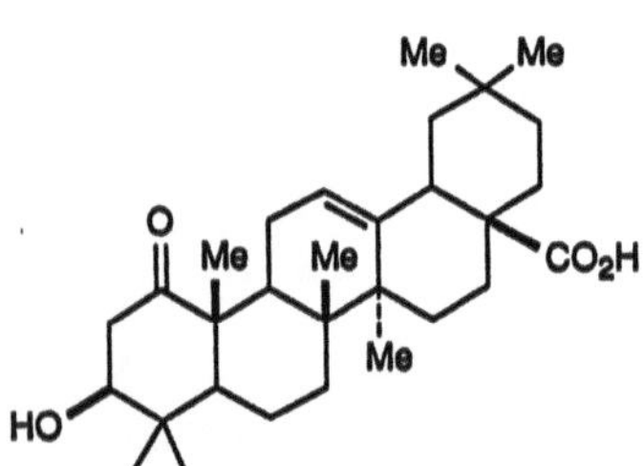

Momordic acid (*87-6*)

A bitter diterpene, columbin (*87-7*) [5], and the sterol spinasterol [6–8] were also isolated from the root of *M. cochinchinensis.*

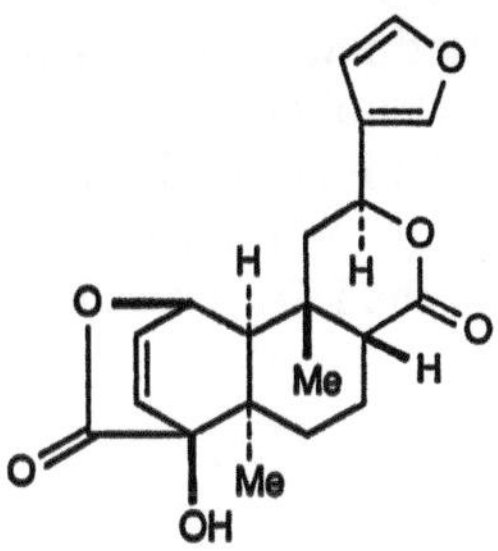

Columbin (*87-7*)

The isolation of a glycoprotein named momorcochin from the fresh root of *M. cochinchinensis* has been reported. It is characterized by an abundance of aspartic acid and glutamic residues and an absence of cysteine residues in its amino acid sequence. The molecular weight of momorcochin was 32 000 [9]. Momorcochin has been reported to have abortifacient activity.

87.2.2 Chemical Constituents of *Momordica grosvenori*

The fruit of *M. grosvenori* is well known to have an intense sweet taste. The sweet principle can be extracted by water or by 50% ethanol and can be separated from the bitter fractions by chromatography or Sephadex G-25 or Amberlite XAD-2. The sweet fraction obtained had an Rg value (relative to glucose) of 0.67. Stevioside, the sweetener from *Stevia rebaudiana*, under the same conditions, had an Rg value of 1.1, while glycyrrhizin as a free acid had an Rg value of 0.7 [10].

The major component of the sweet fraction, named mogroside V (*87-8*) is a glycoside containing the aglycone mogrol [11, 12]. Hydrolysis of mogroside V by acid or using maltase yielded a number of degradation products [13].

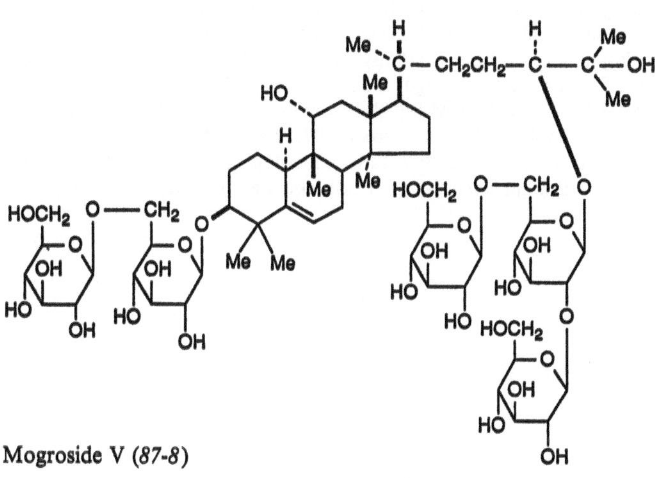

Mogroside V (*87-8*)

87.3 Pharmacology

The saponin isolated from the seeds of *M. cochinchinensis* inhibited both the carageenin-induced rat paw edema and intestine motility of rabbit duodenum. It also potentiated the contractile response of guinea pig ileum to acetylcholine and abolished the antiacetylcholine action of papaverine in vitro [14]. Preliminary toxicological studies on a lyophilized extract of the fruits of *M. grosvenori* with sweetness intensity equal to that of ammonium glycyrrhizinate showed that the LD_{50} exceeded 10 g/kg in mice [10].

References

1. Iwamoto M, Okabe H, Yamauchi T, Tanaka M, Rokutani Y, Hara S, Mihashi K, Higuchi R (1985) Studies on the constituents of *Momordica cochinchinensis* Spreng. I. Isolation and characterization of the seed saponins, momordica saponins I and II. Chem Pharm Bull (Tokyo) 33:464–478
2. Huang MQ (1986) Fatty acids in the kernel oil of *Momordica cochinchinensis*. Guihaia 6:297–299
3. Kuwada S, Fuwa Y (1935) Chemical investigation of saponins. XI. Sapogenin of the roots of *Momordica cochinchinensis* (Lour.) Spreng. J Pharm Soc Jpn 55:467–473
4. Murakami T, Nagasawa M, Itokawa H, Tachi Y, Tanaka K (1966) The structure of a new triterpene, momordic acid, obtained from *Momordica cochinchinensis* Sprenger. Tetrahedron Lett 5137–5140
5. Waterman PG, Reza-ul-Zalil, Hasan CM, Jabbar A (1985) Columbin from the root of *Momordica cochinchinensis*: high field NMR studies. Planta Med 51:181–182
6. Kuwada S, Yoshiki S (1937) Sterols. VII. Sterol of *Momordica cochinchinensis* Spreng. J Pharm Soc Jpn 57:695–708
7. Kuwada S, Yoshiki S (1940) Sterols. XXIII. A sterol of the root of *Momordica cochinchinensis* Spreng. J Pharm Soc Jpn 60:581–585
8. Kuwada S, Yoshiki S (1940) Sterols. XXII. Identity of bessisterol and spinasterol. J Pharm Soc Jpn 60:407–419
9. Yeung HW, Ng TB, Wong NS, Li WW (1987) Isolation and characterization of an abortifacient protein, momorcochin, from root tubers of *Momordica cochinchinensis* (family Cucurbitaceae). Int J Pept Protein Res 30:135–140
10. Lee CH (1975) Intense sweetener from Lo Han Kuo (*Momordica grosvenori*). Experientia 31:533–534
11. Takemoto T, Arihara S, Nakajima T, Okuhira M (1983) Studies on the constituents of Fructus Momordicae. II. Structure of sapogenin. Yakugaku Zasshi 103:1155–1166
12. Takemoto T, Arihara S, Nakajima T, Okuhira M (1983) Studies on the constituents of Fructus Momordicae. III. Structure of mogrosides. Yakugaku Zasshi 103:1167–1173
13. Takemoto T, Arihara S, Nakajima T, Okuhira M (1983) Studies on the constituents of Fructus Momordicae. I. On the sweet principle. Yakugaku Zasshi 103:1151–1154
14. Kubota K, Sato M, Murakami T, Yamagishi T (1971) Pharmacological studies on the saponin isolated from the seed of *Momordica cochinchinensis*. Yakugaku Zasshi 91:174–179

Morus alba L.

88

88.1 Introduction

The following four items on *Morus alba* L. (Moraceae) appear in the Chinese Pharmacopoeia:

- Sangye, Folium Mori, is the dry leaves of *M. alba* collected in the fall and used as an antiphlogistic.
- Sangbaipi, Cortex Mori, is the dry root bark of *M. alba*. It could be used as an antiinflammatory and a diuretic agent.
- Sangzhi, Ramulus Mori, is the dry young branches of *M. alba* collected in the late spring and early summer. It is used for treatment of arthritis and rheumatism.
- Sangren, Fructus Mori, is the ripe aggregate fruit of *M. alba*, used as a tonic and sedative.

88.2 Chemical Constituents

88.2.1 Chemical Constituents of the Leaves of *Morus alba*

88.2.1.1 Volatile Components

The essential oil from mulberry leaves could be separated into neutral (32%), acidic (26%), phenolic (28%), carbonyl (11%), and basic (4.4%) fractions [1]. Some neutral components were identified as isobutanol, isoamyl alcohol, isoamyl acetate, and acetophenone [2]. In the acidic fraction, acetic, propionic, butyric, isobutyric, isovaleric, caproic, isocaproic, and lactic acids were found [2, 3]. Phenol, *o*-, *m*-, and *p*-cresol, guaiacol, eugenol, and methyl salicylate were isolated and identified from the phenolic fraction [3] and benzaldehyde and phenylacetaldehyde were detected in the carbonyl fraction [4].

88.2.1.2 Nonvolatile Components

Besides amino acids, saccharides, and vitamins, a number of compounds of different types were isolated from mulberry leaves and structurally determined. Thus, oxalic, succinic, malic, tartaric, citric, fumaric [5], and palmitic acids, and ethyl palmitate [6] were isolated from mulberry leaves. Rutin, quercetin [7], and quercetin-3-triglucoside were isolated as flavones from the mulberry leaves [8]. The sterols β-sitosterol, campesterol [9], β-sitosterolglycoside [10], β-ecdysone, and inokosterone [11] were also isolated from the mulberry leaves. Moreover, a polyprenoid alcohol moraprenol-11 [12] and a polypyranoid ketone bombiprenone [13] were also reported to be isolated and identified. Moraprenol-11 was structurally elucidated as

3,7,11,15,19,23,27,31,35,39,43-undecamethyl-2,6,10,14,18,22,26,30,34,38,42-tetratetracontaundecaen-1-ol with a Z,Z,Z,Z,Z,Z,Z,E,E,E configuration. Bombiprenone was identified as 6,10,14,18,22,26,30,34-octamethyl-5,9,13,17,21,25,29,33-pentatriacontaoctaen-2-one, with an all-E configuration.

88.2.2 Chemical Constituents of the Root Bark of *M. alba*

The chemical constituents of mulberry root bark have been thoroughly investigated and a number of compounds of different structural features were isolated and structurally determined. Prenylflavones represent an interesting substance group isolated from mulberry root bark [14]. The initial research on prenylflavones of *M. alba* was carried out by Desphande and coworkers. This group isolated four flavone derivatives from the stem and root bark of *M. alba*, mulberrin (*88-1*), cyclomulberrin (*88-2*), mulberrochromene (*88-3*), and cyclomulberrochromene (*88-4*) [15]. Structures were elucidated by spectral analysis and chemical evidence. Since then, a number of further prenylflavones were isolated and structurally investigated. Among them, prenylflavones, prenylflavanones, and Diels-Alder adducts of two chalcones, or of a chalcone with a flavone or flavanone were found. The flavone constituents isolated from the root bark of *M. alba* and other *Morus* species are listed in Table 88.1.

Table 88.1. Flavone and related compounds isolated from the root bark of *Morbus alba* and other *Morus* species

Compound	Structure	Plant origin	Reference
Mulberrin (Kuwanone C, *88-1*)	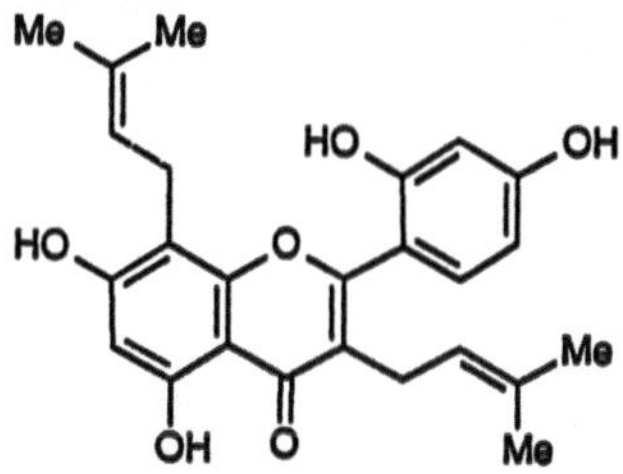	*M. alba*	[15–19]
Cyclomulberrin (*88-2*)	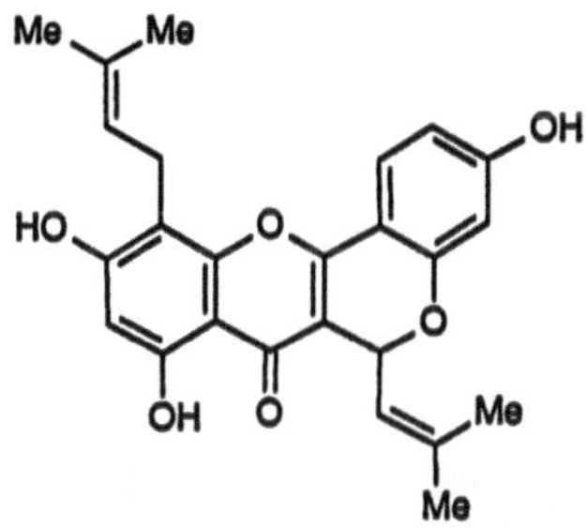	*M. alba*	[15, 18, 19]
Mulberro-chromene (Morusin, *88-3*)	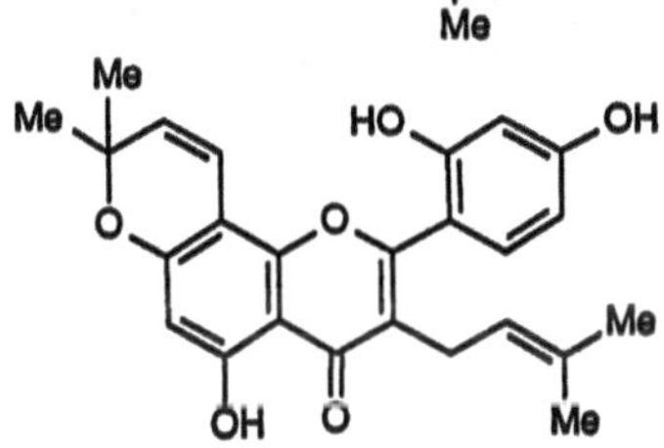	*M. alba*	[15, 18–21]
Cyclomulberro-chromene (Cyclomorusin, *88-4*)	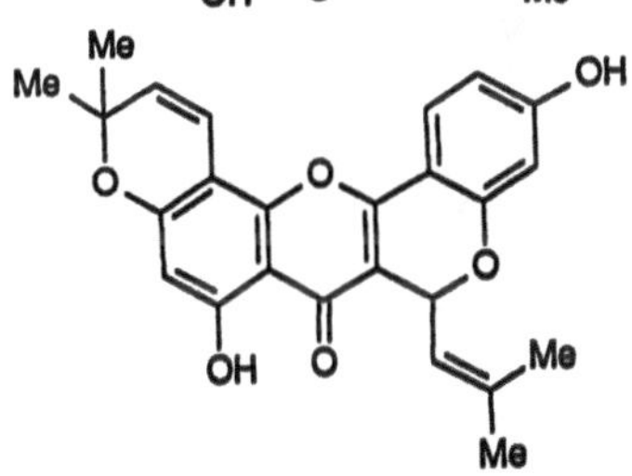	*M. alba*	[15, 18–21]
Mulberranol (*88-5*)	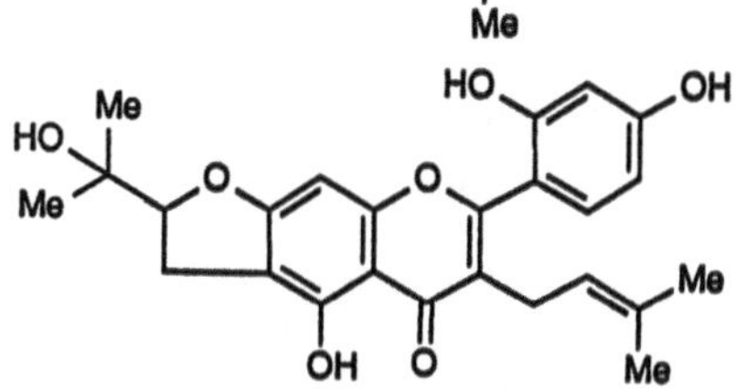	*M. alba*	[22]
Oxydihydro-morusin (Morusinol, *88-6*)	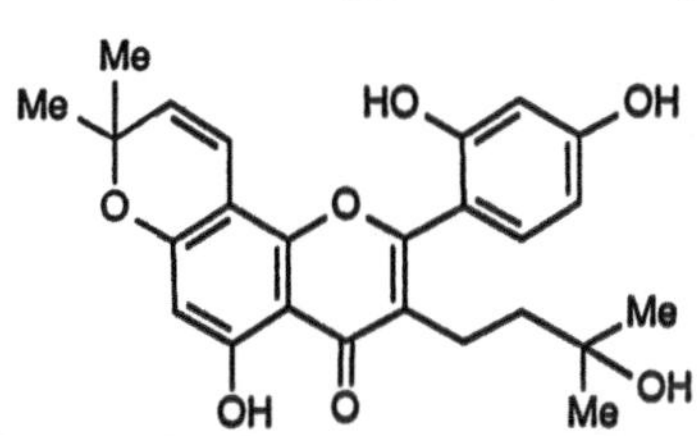	*M. alba*	[16, 17, 23]

Table 88.1. (continued)

Compound	Structure	Plant origin	Reference
Compound A (*88-7*)	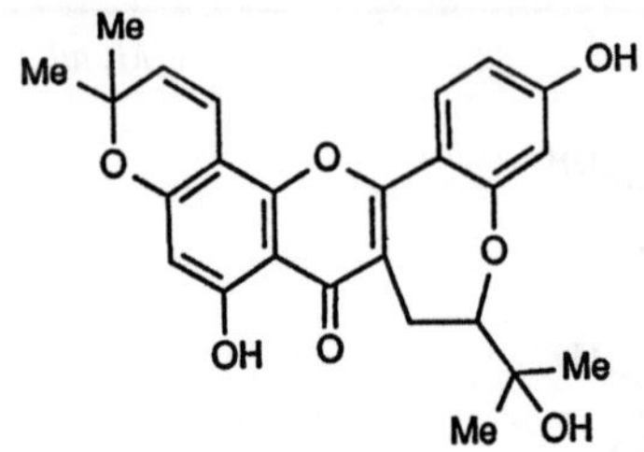	*M. alba*	[20, 21]
Kuwanon A (*88-8*)	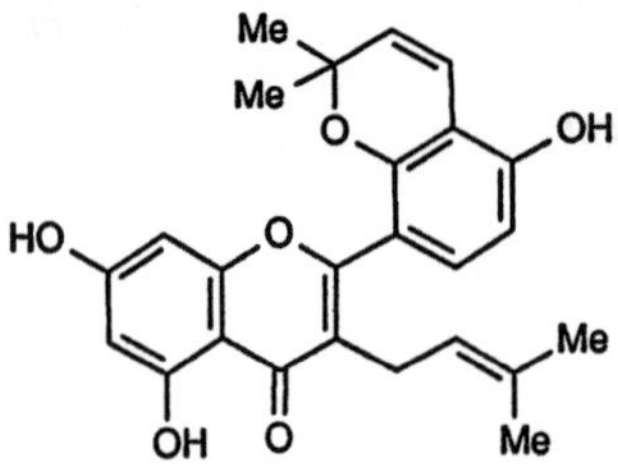	*M. alba*	[16, 17]
Kuwanon B (*88-9*)	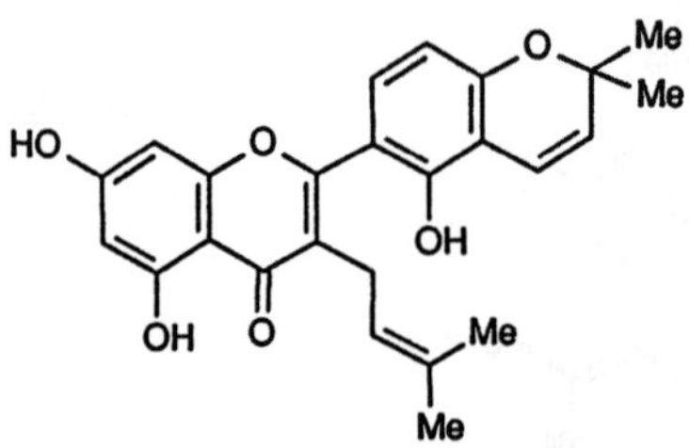	*M. alba*	[16, 17]
Kuwanon D (*88-10*)	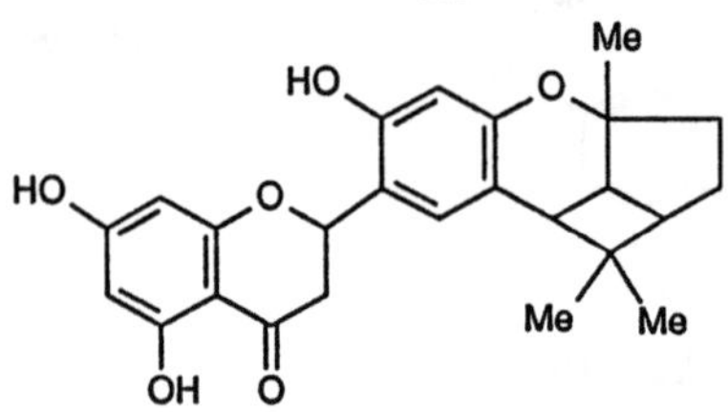	*M. alba*	[24, 25]
Kuwanon E (*88-11*)	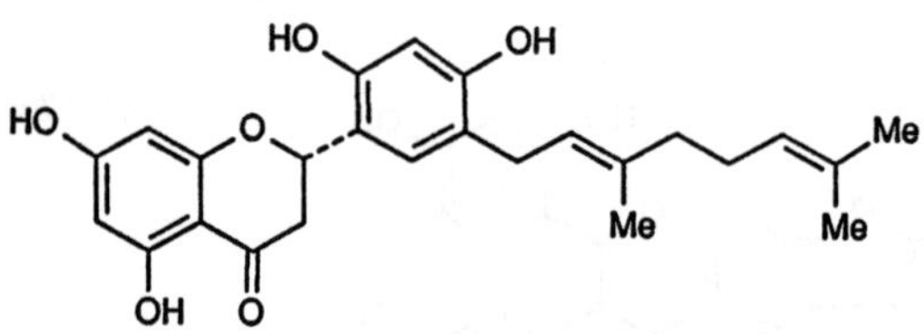	*M. alba*	[24, 26]
Kuwanon F (*88-12*)	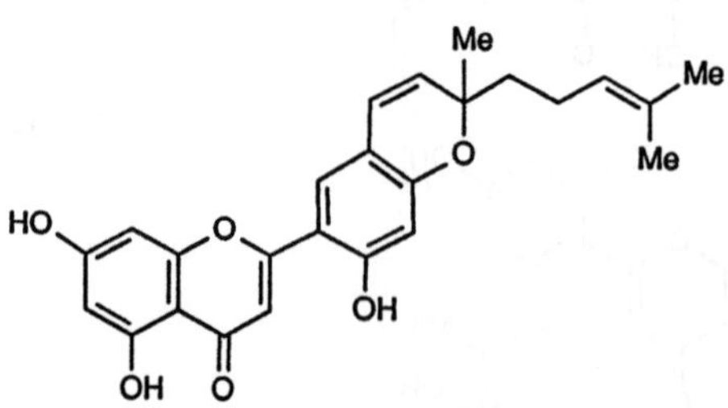	*M. alba*	[24, 27]

Table 88.1. (continued)

Compound	Structure	Plant origin	Reference
Kuwanon G (Albanin F, Moracenin B, *88-13*)	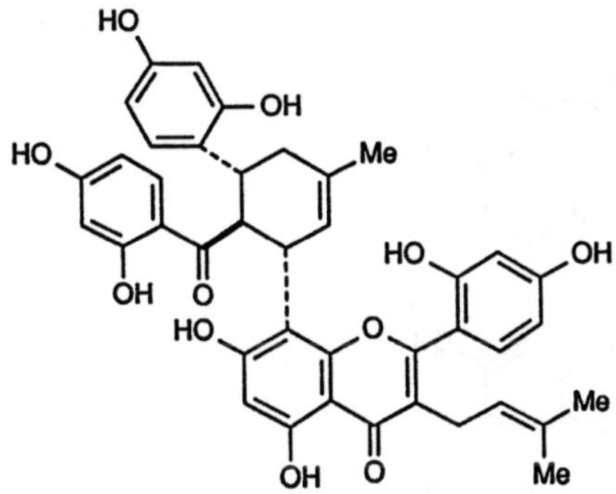	*M. alba*	[28–31]
Kuwanon H (Albanin G Moracenin A, *88-14*)	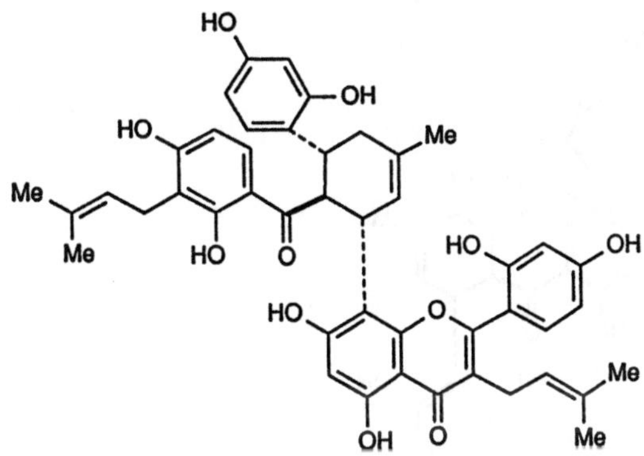	*M. alba*	[29, 30, 32, 33]
Kuwanon I (*88-15*)	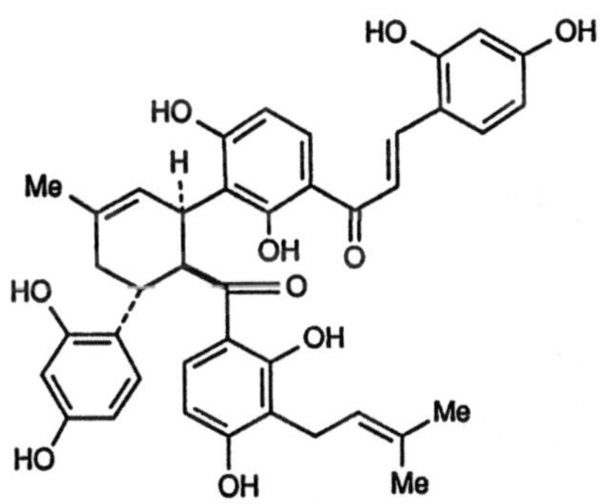	*M. alba*	[34, 35]
Kuwanon J (*88-16*)	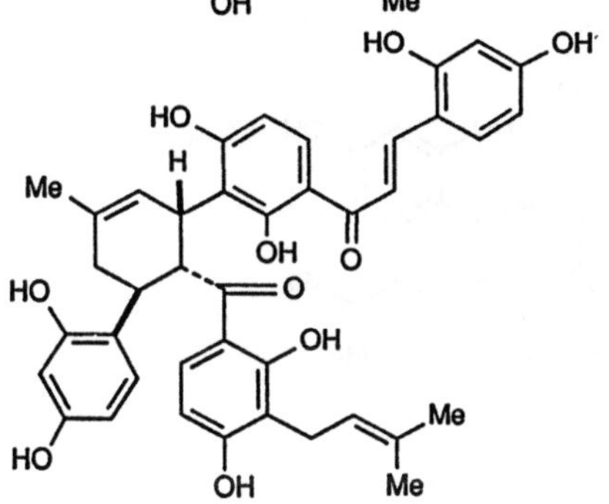	Callus culture of *M. alba*	[36, 37]

Table 88.1. (continued)

Compound	Structure	Plant origin	Reference
Kuwanon K (*88-17*)	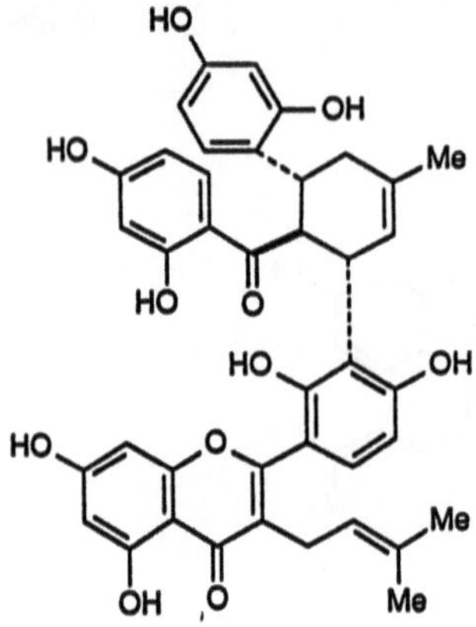	*M. alba*	[38]
Kuwanon L (*88-18*)	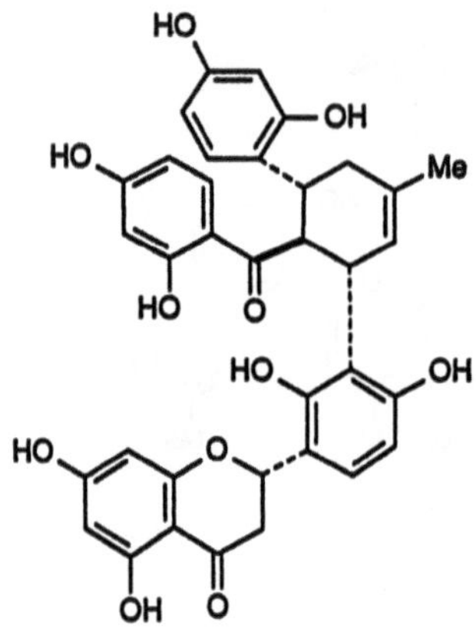	*M. alba*	[38–40]
Kuwanon M (*88-19*)	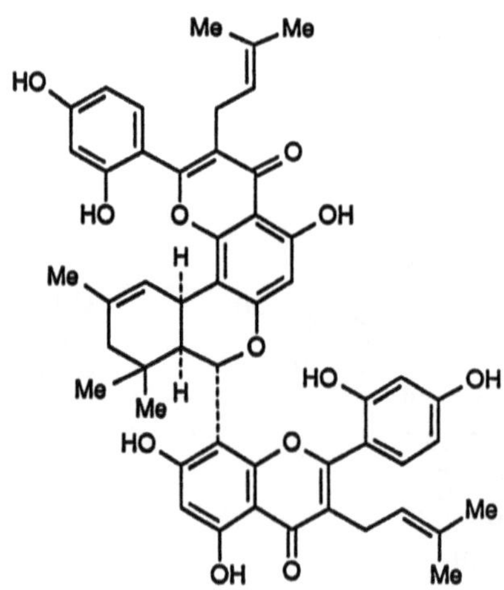	*M. lhou*	[40, 41]

Table 88.1. (continued)

Compound	Structure	Plant origin	Reference
Kuwanon N (*88-20*)	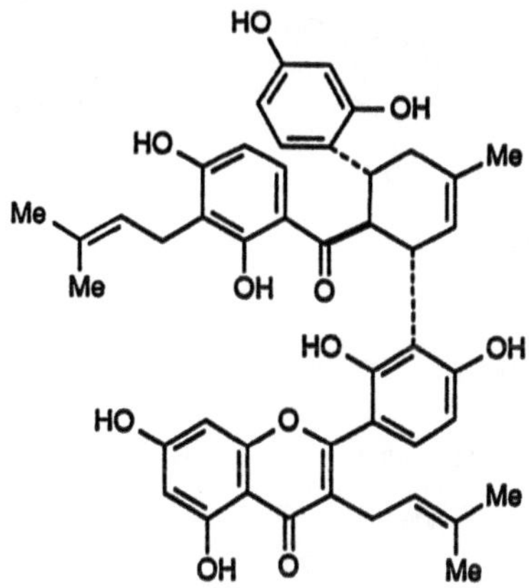	*M. lhou*	[40, 42]
Kuwanon O (*88-21*)	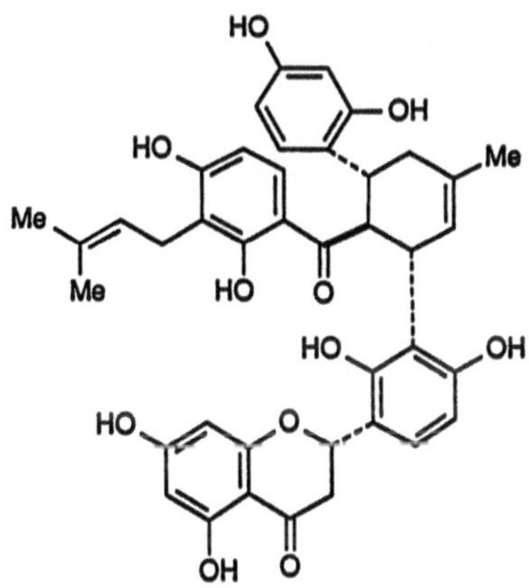	*M. lhou*	[40, 42]
Kuwanon P (*88-22*)	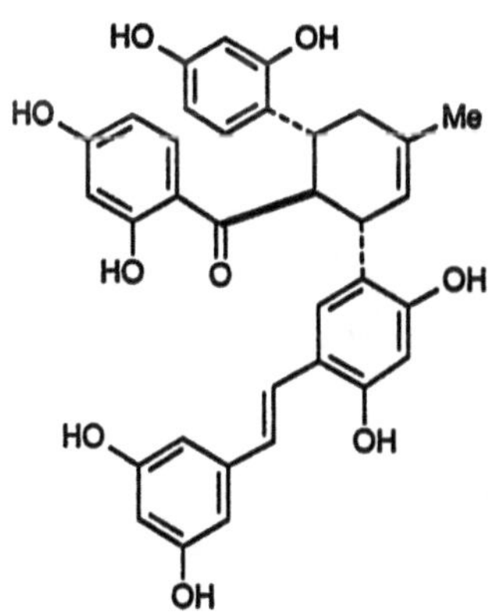	*M. lhou*	[40, 43]

Table 88.1. (continued)

Compound	Structure	Plant origin	Reference
Kuwanon Q (*88-23*)	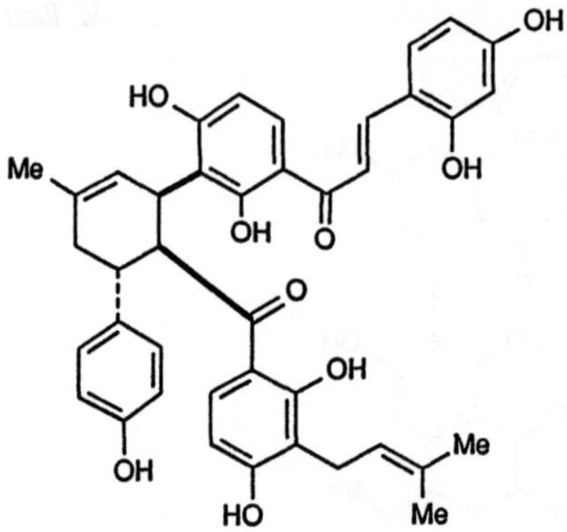	Callus culture of *M. alba*	[37, 44]
Kuwanon R (*88-24*)	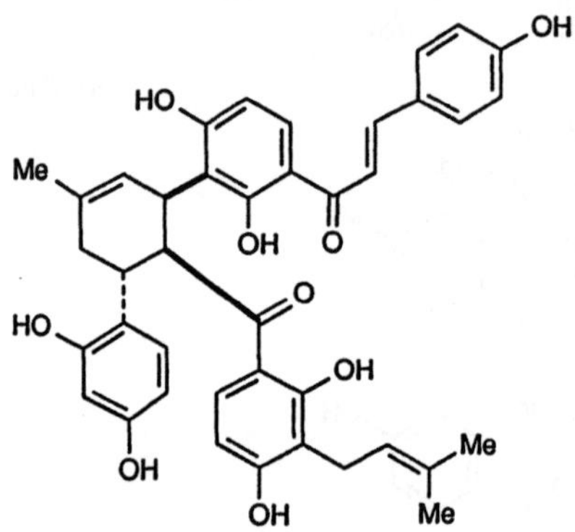	Callus culture of *M. alba*	[37, 44]
Kuwanon S (*88-25*)	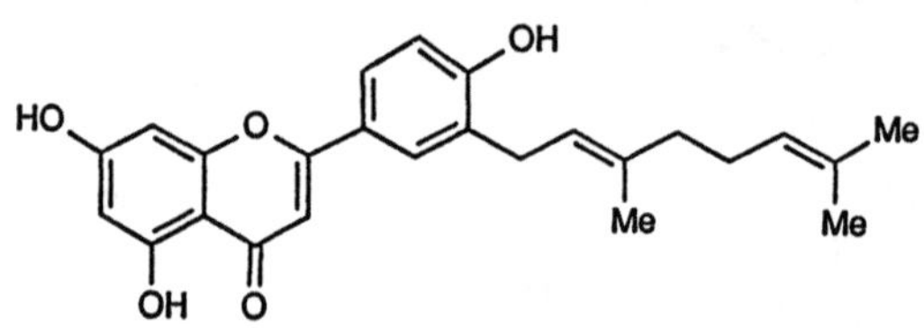	*M. lhou*	[37, 44]
Kuwanon T (*88-26*)	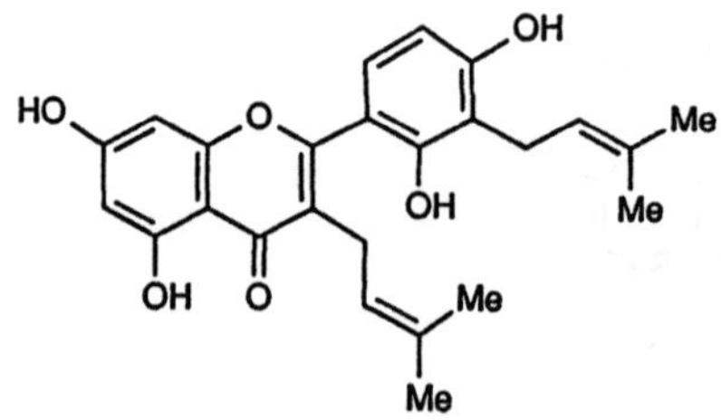	*M. lhou*	[45]

Table 88.1. (continued)

Compound	Structure	Plant origin	Reference
Kuwanon V (*88-27*)	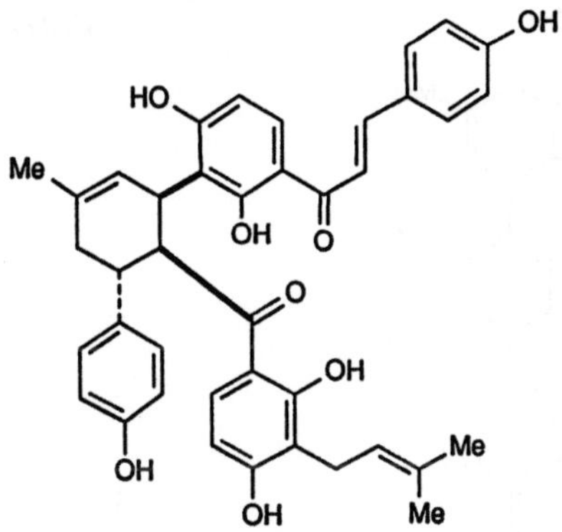	Callus culture of *M. alba*	[37, 44]
Kuwanon W (*88-28*)	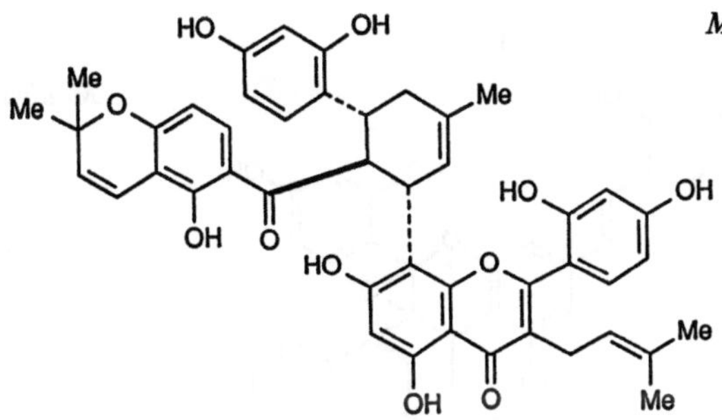	*M. lhou*	[46]
Kuwanon X (*88-29*)	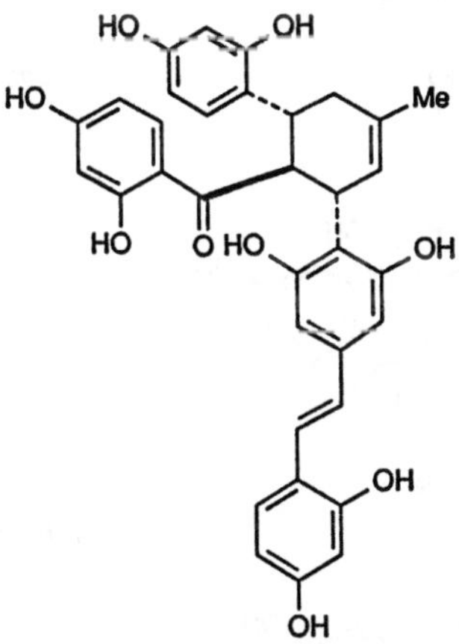	*M. lhou*	[43]

Table 88.1. (continued)

Compound	Structure	Plant origin	Reference
Kuwanon Y (*88-30*)	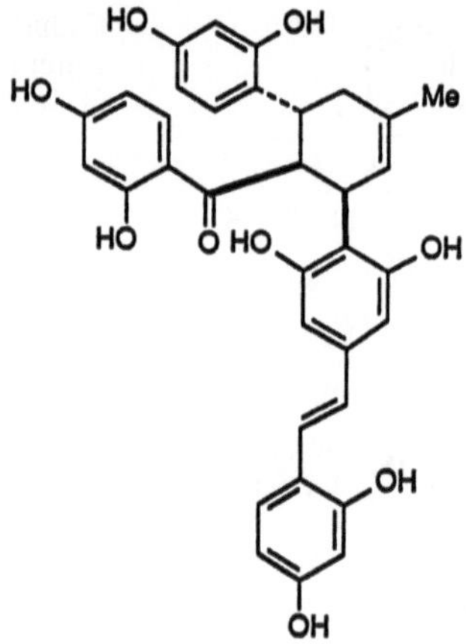	*M. alba*	[47]
Kuwanon Z (*88-31*)	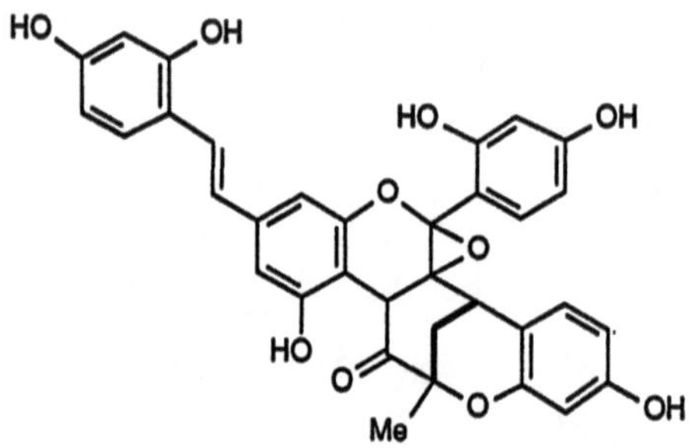	*M. alba*	[46]
Kuwanol A (*88-32*)	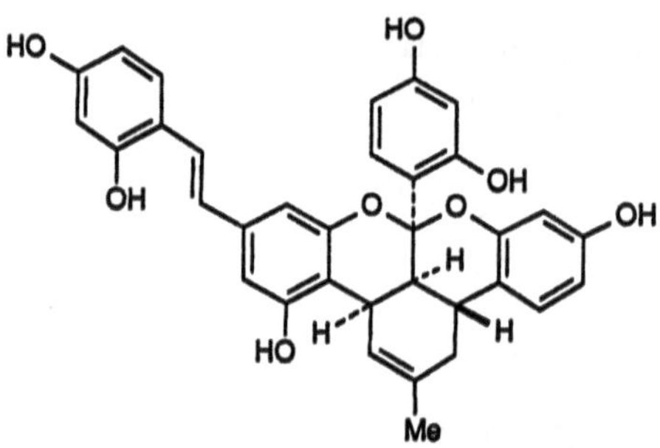	*M. bombycis*	[48]
Kuwanol B (*88-33*)	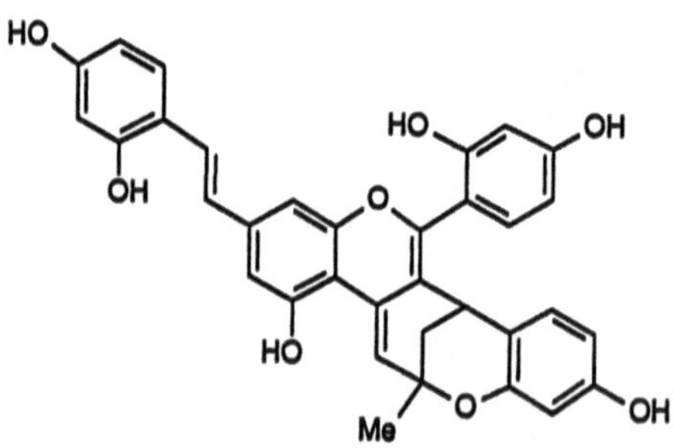	*M. bombycis*	[48]

Table 88.1. (continued)

Compound	Structure	Plant origin	Reference
Sanggenone A (*88-34*)	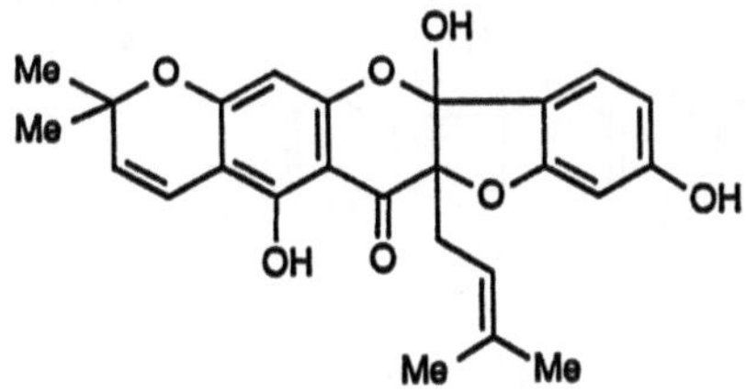	*Morus* sp.	[49–51]
Sanggenone B (*88-35*)	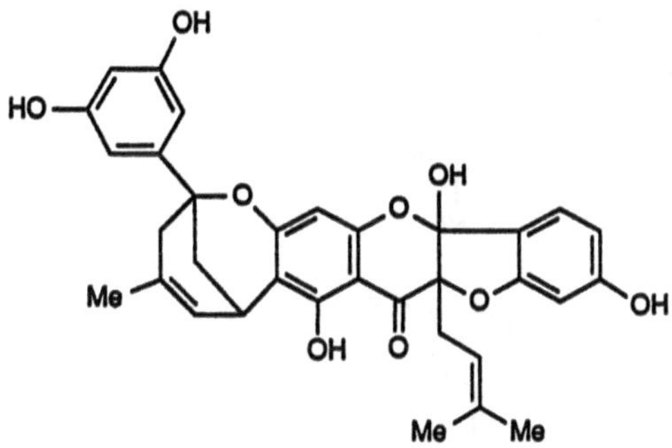	*Morus* sp.	[49, 52]
Sanggenone C (*88-36*)	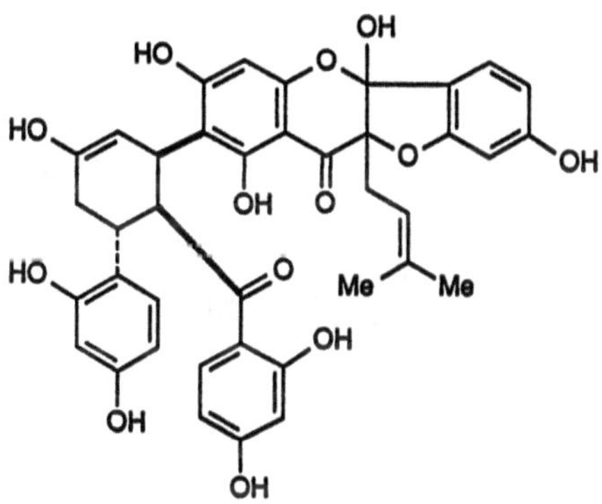	*Morus* sp.	[49, 53]
Sanggenone D (*88-37*)	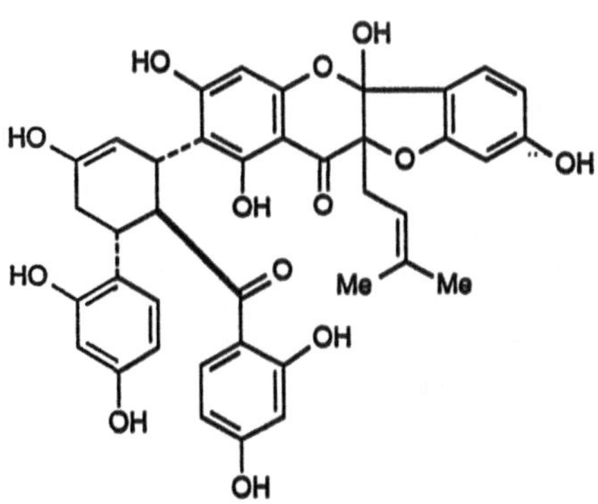	*Morus* sp.	[49, 54]

Table 88.1. (continued)

Compound	Structure	Plant origin	Reference
Sanggenone E (*88-38*)	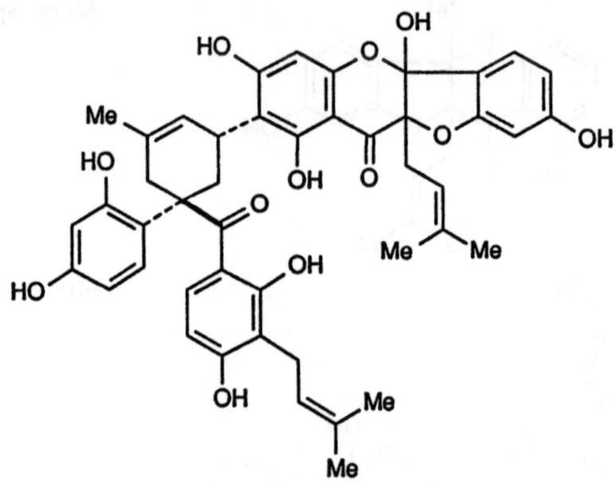	*Morus* sp.	[49, 55]
Sanggenone F (*88-39*)	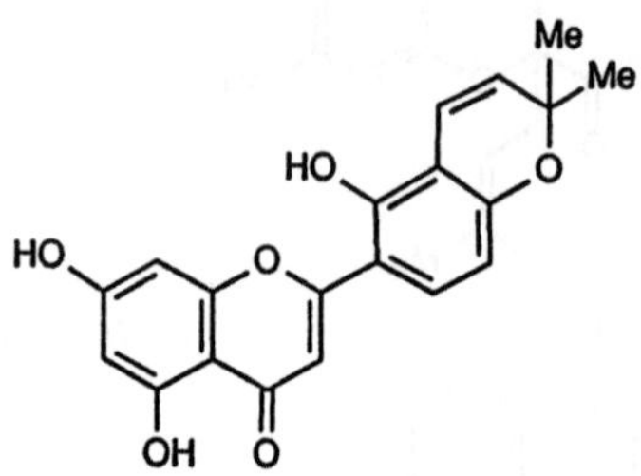	*Morus* sp.	[56]
Sanggenone G (*88-40*)	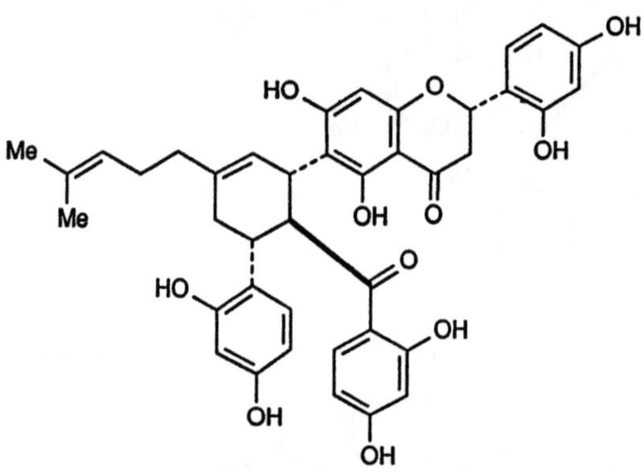	*Morus* sp.	[57]
Sanggenone H (*88-41*)	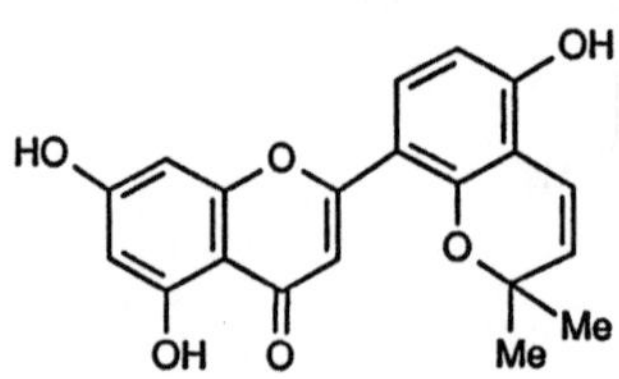	*Morus* sp.	[58]

Table 88.1. (continued)

Compound	Structure	Plant origin	Reference
Sanggenone I (*88-42*)	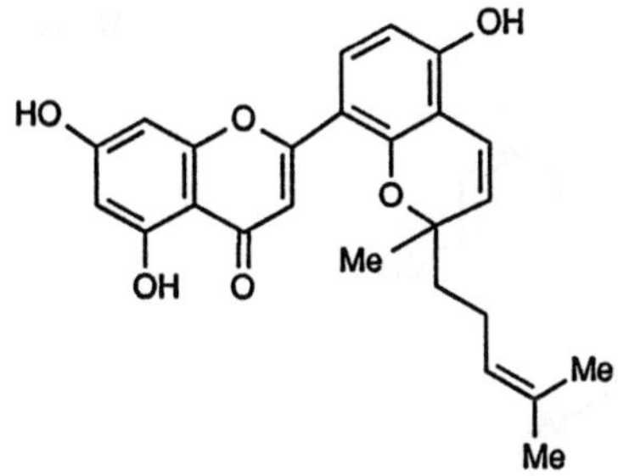	*Morus* sp.	[58]
Sanggenone J (*88-43*)	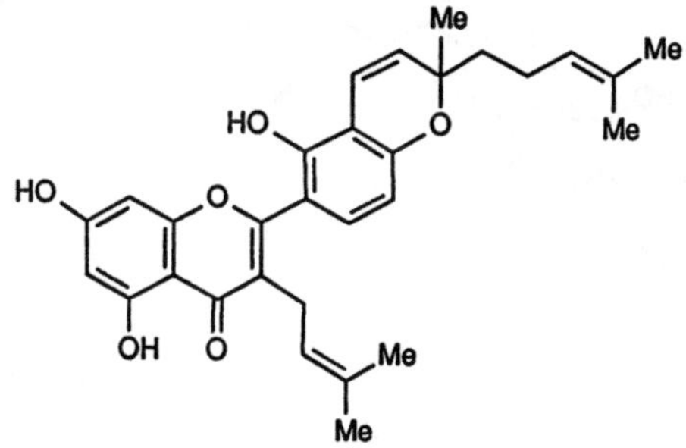	*Morus* sp.	[58]
Sanggenone K (*88-44*)	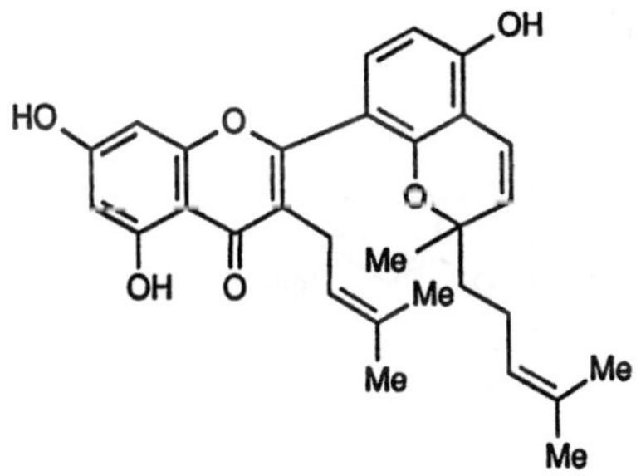	*Morus* sp.	[58]
Sanggenone L (*88-45*)	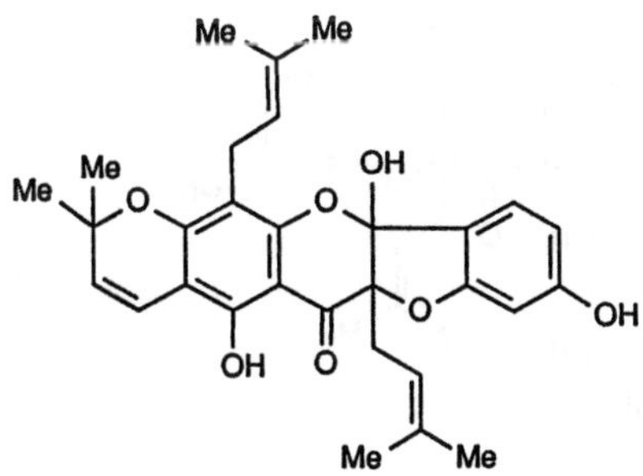	*Morus* sp.	[59]

Table 88.1. (continued)

Compound	Structure	Plant origin	Reference
Sanggenone M (*88-46*)	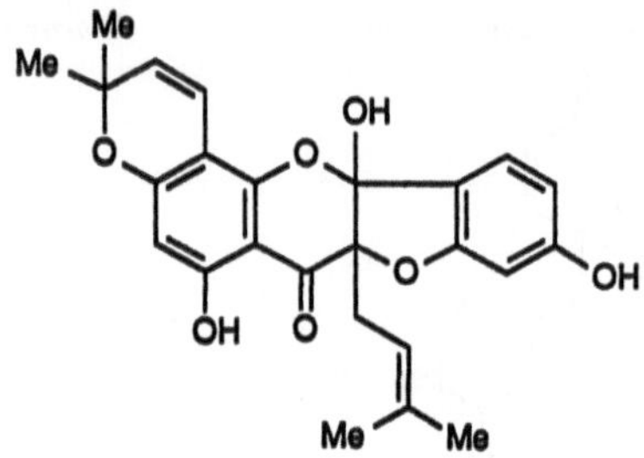	*Morus* sp.	[59]
Sanggenone N (*88-47*)	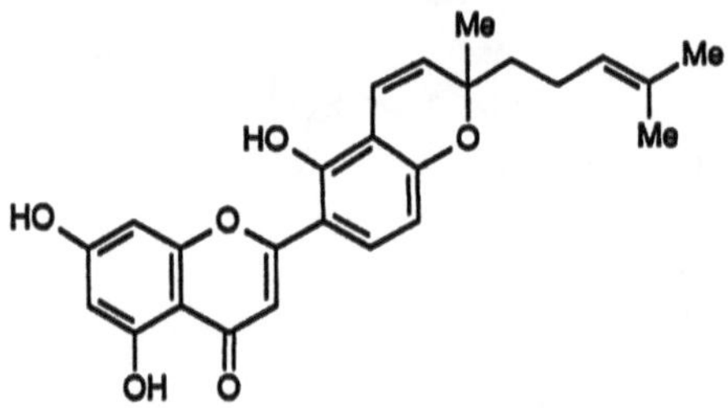	*Morus* sp.	[59]
Sanggenone O (*88-48*)	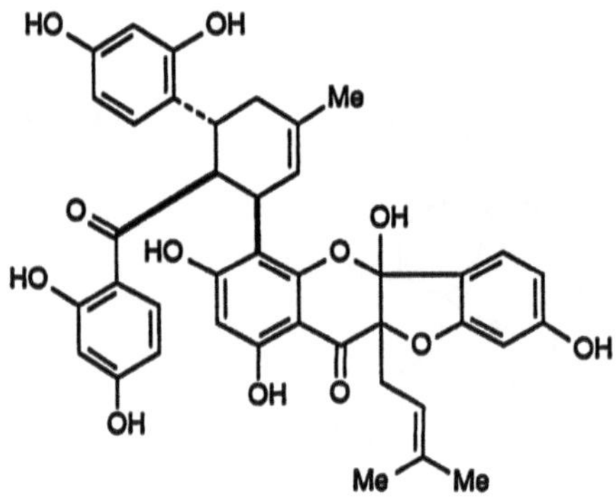	*Morus* sp.	[60]
Sanggenone P (*88-49*)	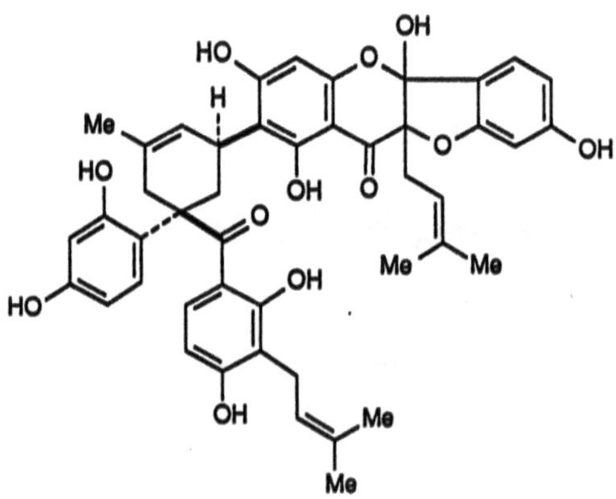	*Morus* sp.	[55]

Table 88.1. (continued)

Compound	Structure	Plant origin	Reference
Sanggenone Q (*88-50*)	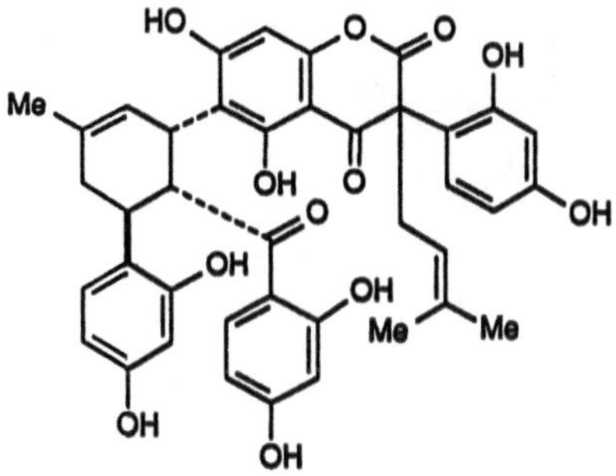	*M. mongolica*	[61]
Moracenin C (*88-51*)	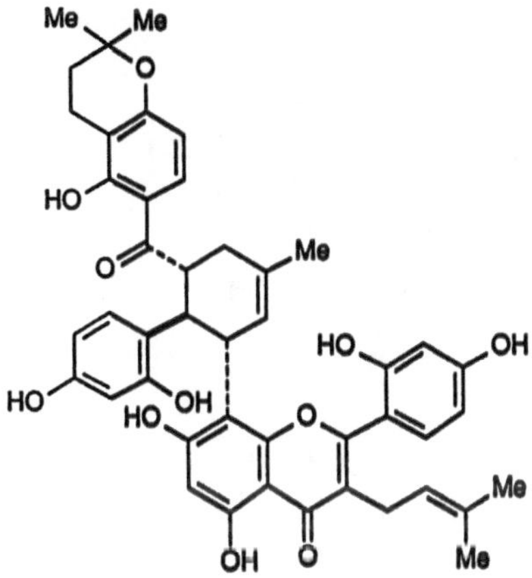	*Morus* sp.	[62]
Moracenin D (*88-52*)	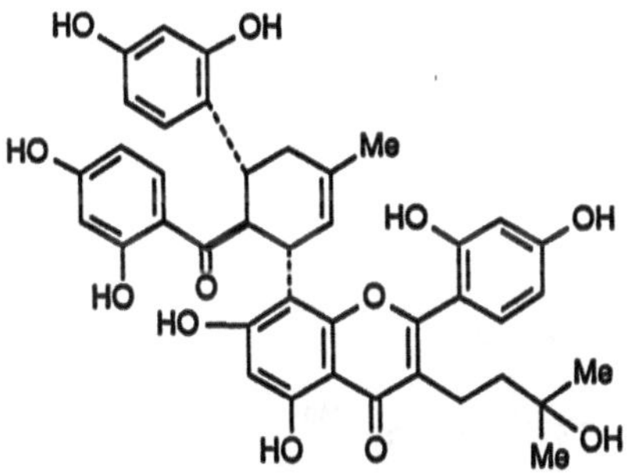	*Morus* sp.	[63, 64]
Mulberro-furan A (*88-53*)	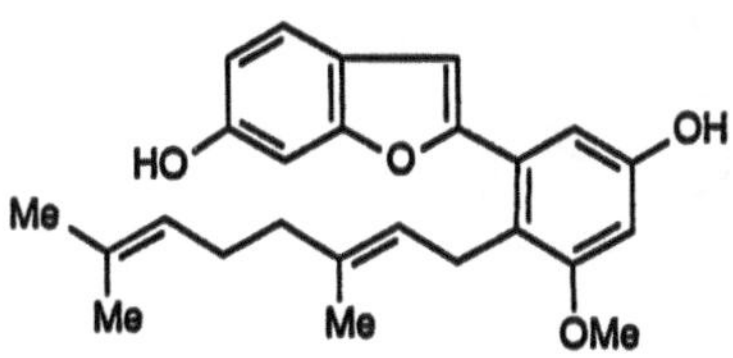	*M. alba*	[65]

Table 88.1. (continued)

Compound	Structure	Plant origin	Reference
Mulberro-furan B (*88-54*)	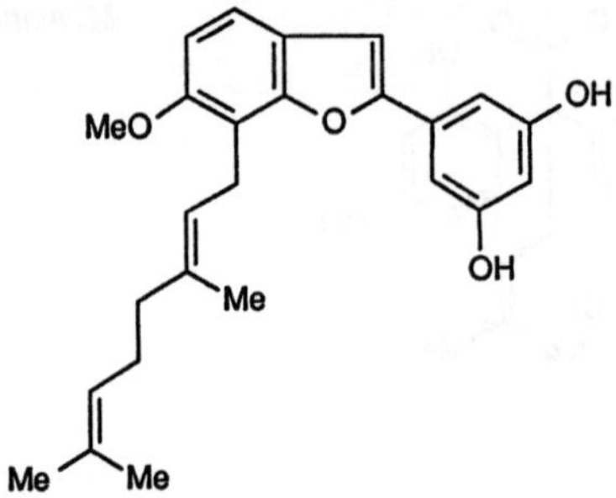	*Morus* sp.	[66]
Mulberro-furan C (*88-55*)	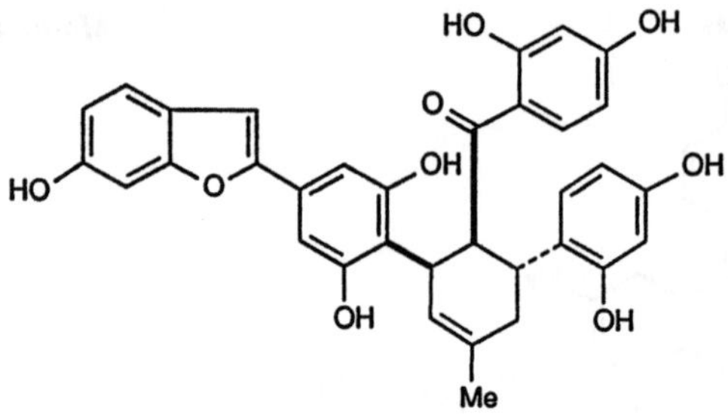	*M. bombycis*	[67]
Mulberro-furan D (*88-56*)	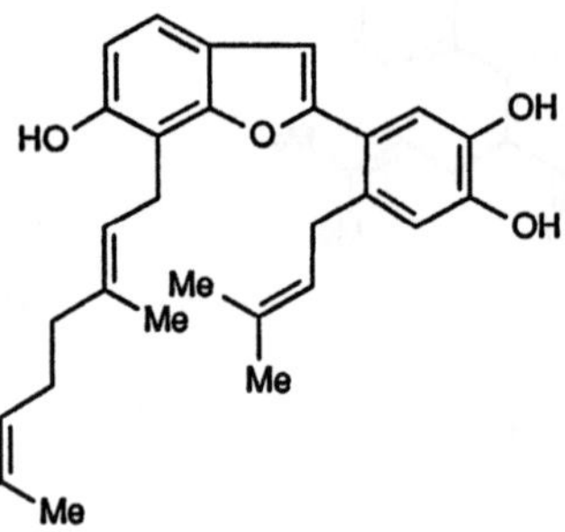	*M. australis*	[68]
Mulberro-furan E (*88-57*)	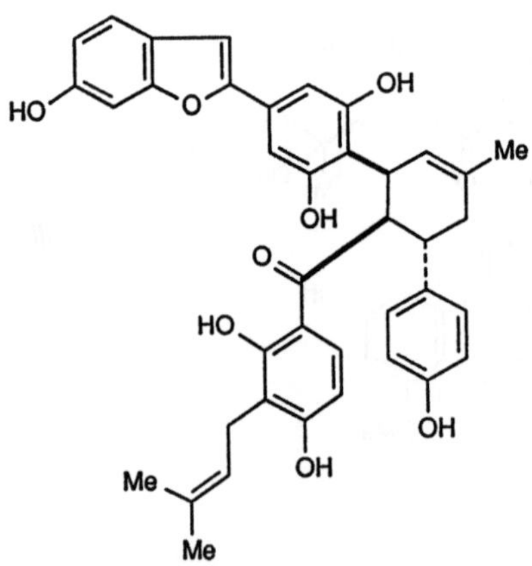	Callus tissue of *M. alba*	[44]

Table 88.1. (continued)

Compound	Structure	Plant origin	Reference
Mulberro-furan F (*88-58*)	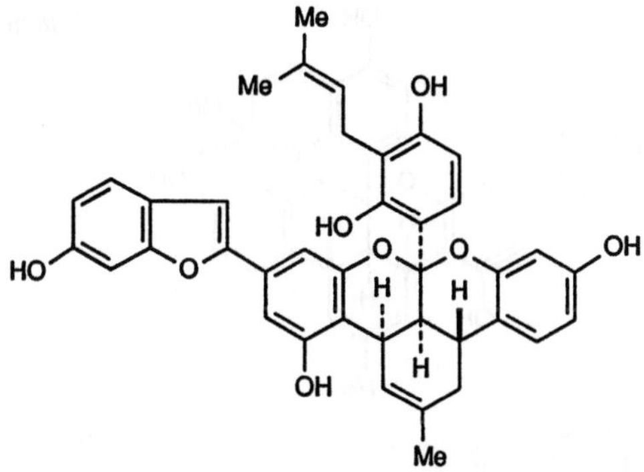	*M. lhou*	[69, 70]
Mulberro-furan G (Albanol A, *88-59*)	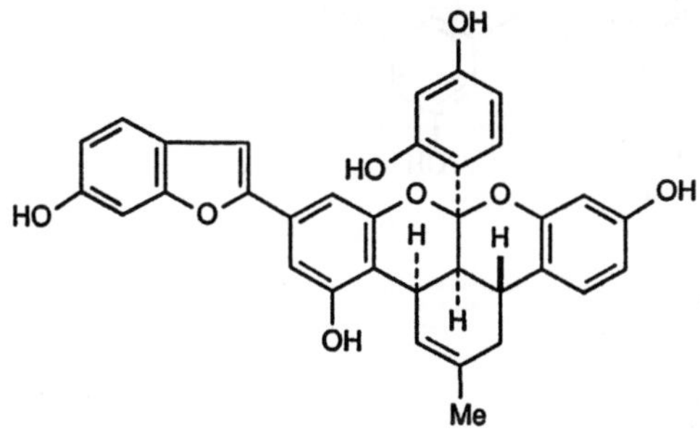	*M. lhou*	[69–71]
Mulberro-furan H (*88-60*)	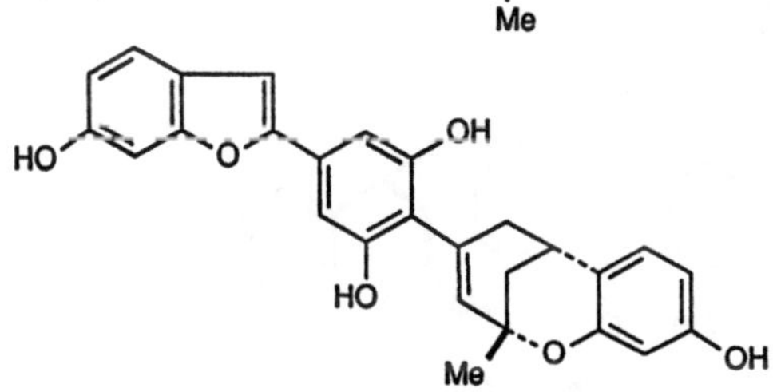	*M. lhou*	[72]
Mulberro-furan I (*88-61*)	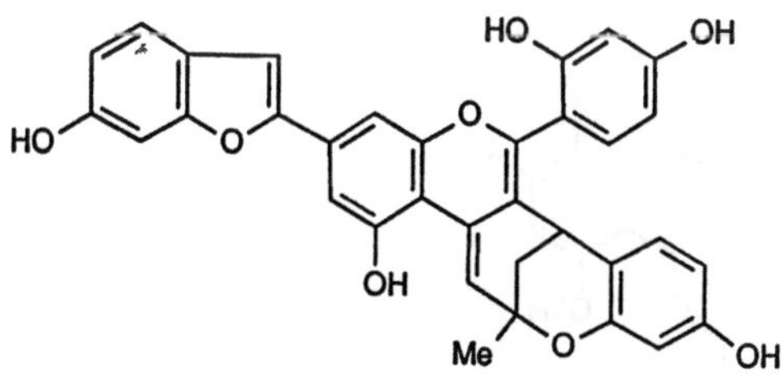	*M. lhou*	[40, 73]
Mulberro-furan J (*88-62*)	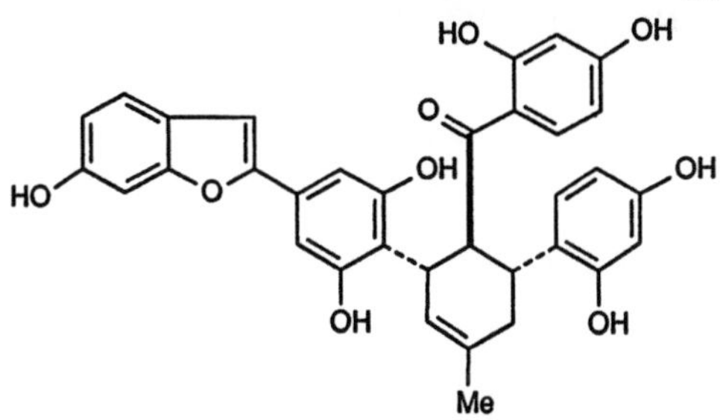	*M. lhou*	[43, 74]

Table 88.1. (continued)

Compound	Structure	Plant origin	Reference
Mulberro-furan K (*88-63*)	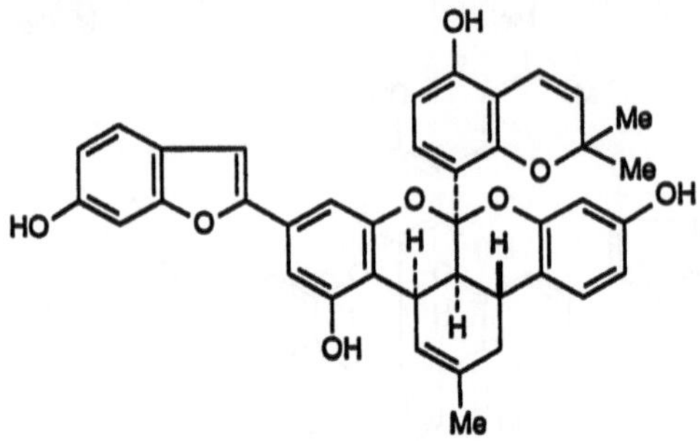	*M. lhou*	[75, 76]
Mulberro-furan L (*88-64*)	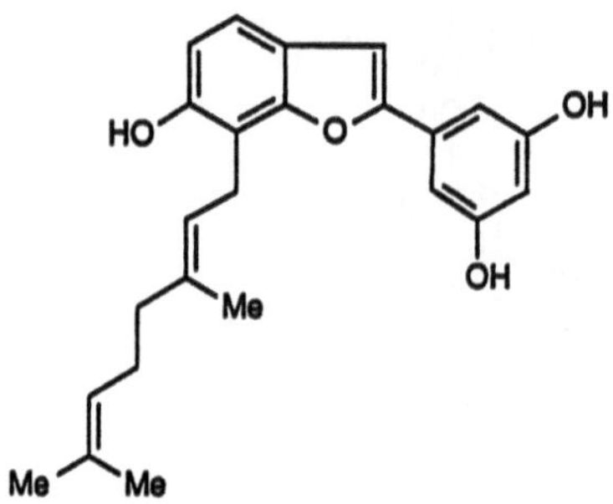	*M. lhou*	[77]
Mulberro-furan M (*88-65*)	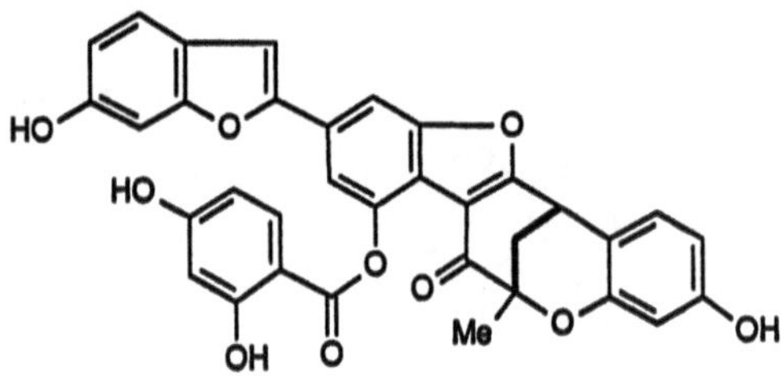	*M. alba*	[78]
Mulberro-furan N (*88-66*)	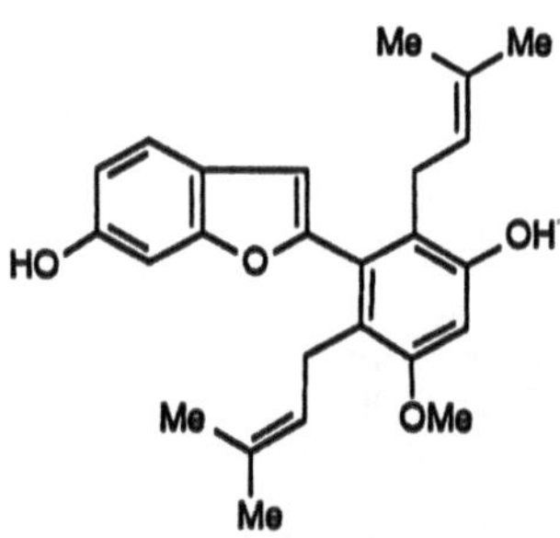	*Morus* sp.	[76]

Table 88.1. (continued)

Compound	Structure	Plant origin	Reference
Mulberro-furan O (*88-67*)	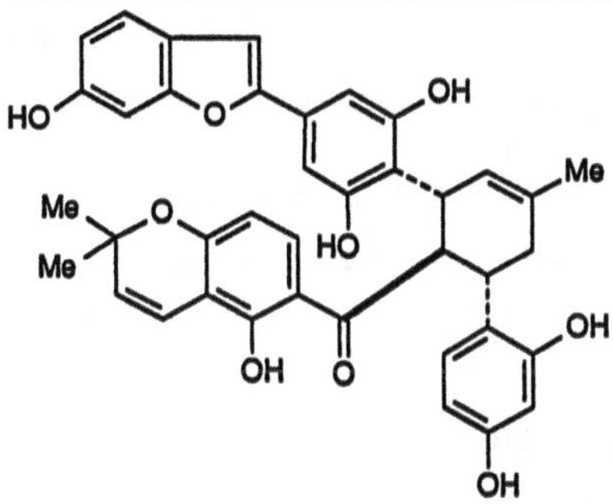	*Morus* sp.	[76]
Mulberro-furan P (*88-68*)	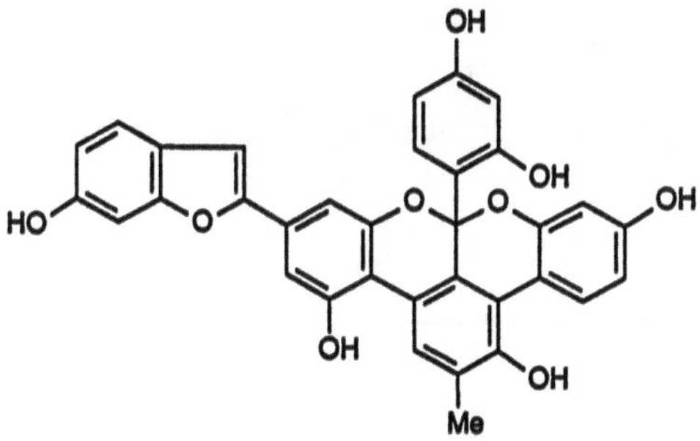	*M. alba*	[79]
Mulberro-furan Q (*88-69*)	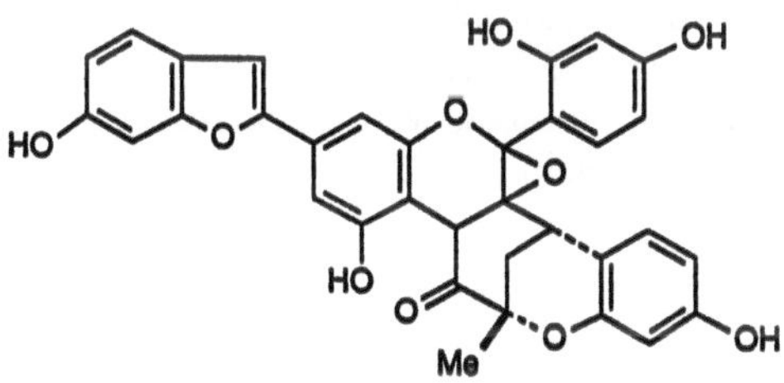	*M. alba*	[80]
Albanol B (*88-70*)	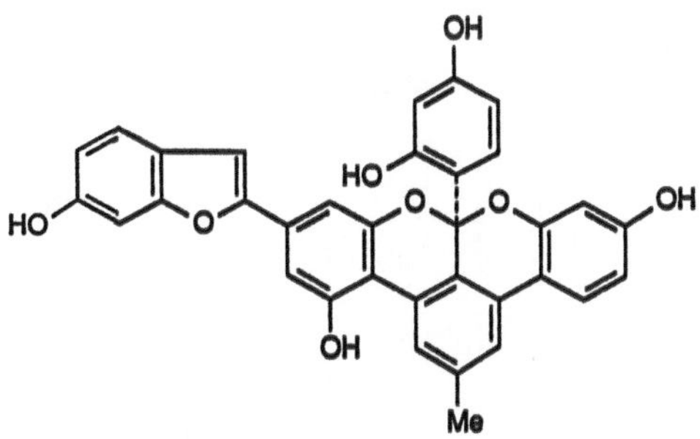	*M. alba*	[71]
Albafuran A (*88-71*)	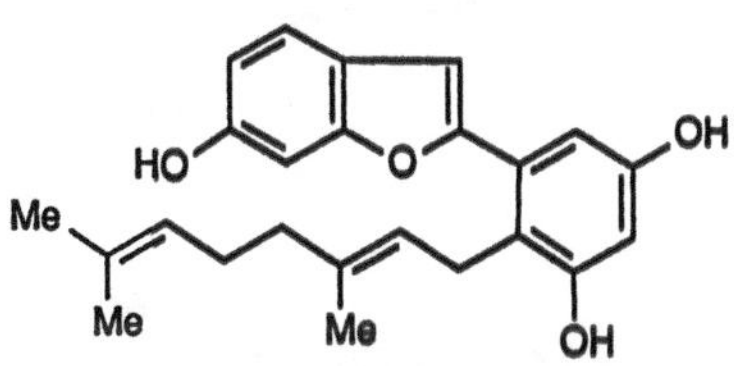	*M. alba*	[81]

Table 88.1. (continued)

Compound	Structure	Plant origin	Reference
Albafuran B (*88-72*)	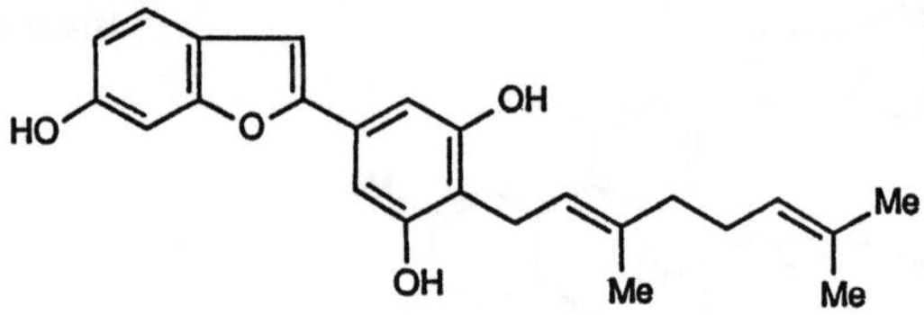	*M. alba*	[81]
Albafuran C (*88-73*)	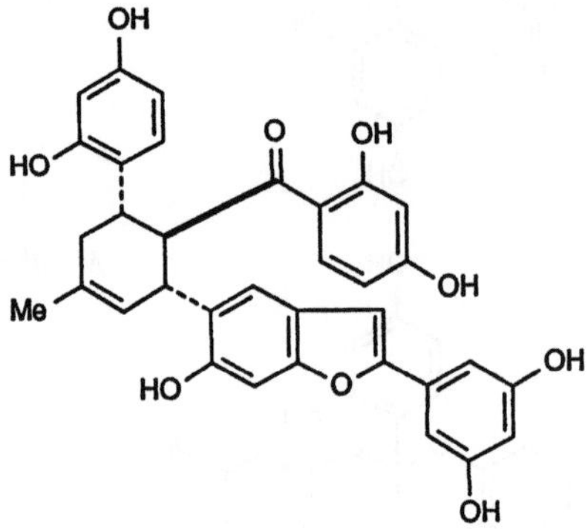	*M. alba*	[82]

A number of antifungal phenylbenzofurane derivatives, named moracins (*88-74*–*88-86*) were isolated from diseased *M. alba* xylem or bark infected with *Fusarium solani* f. sp. *mori*. The structures of moracins were elucidated [83–87]. A Diels-Alder adduct of moracin C named dimoracin (*88-87*) was additionally isolated from the diseased shoot of *M. alba* [88].

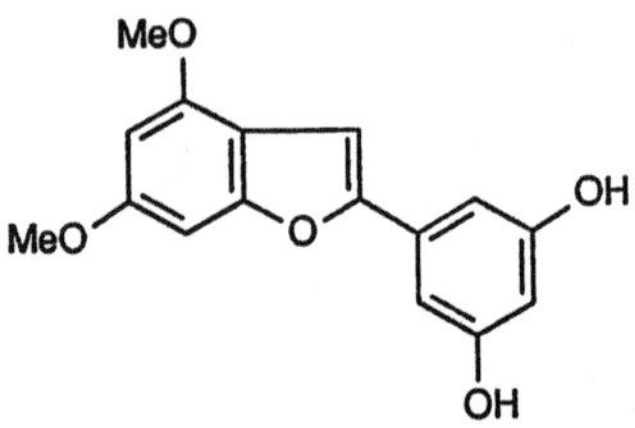

Moracin A (*88-74*)

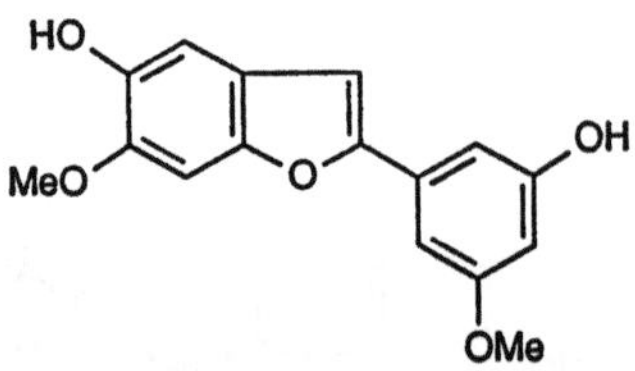

Moracin B (*88-75*)

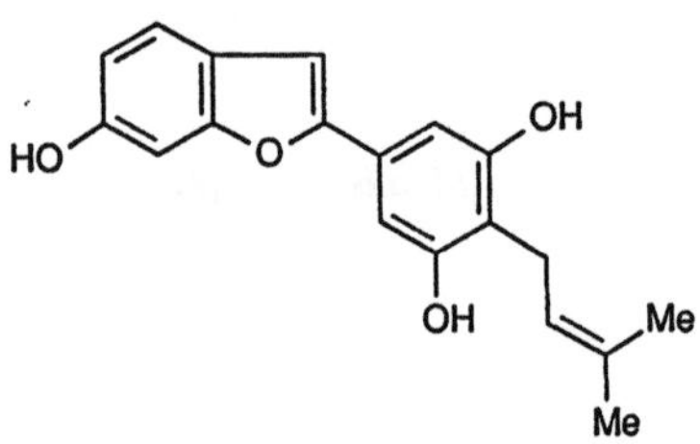

Moracin C (*88-76*)

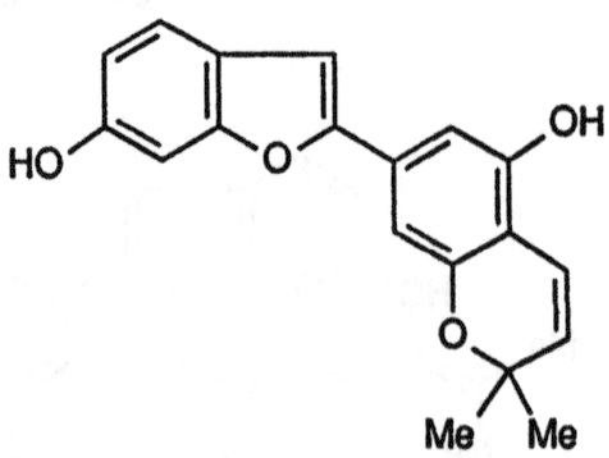

Moracin D (*88-77*)

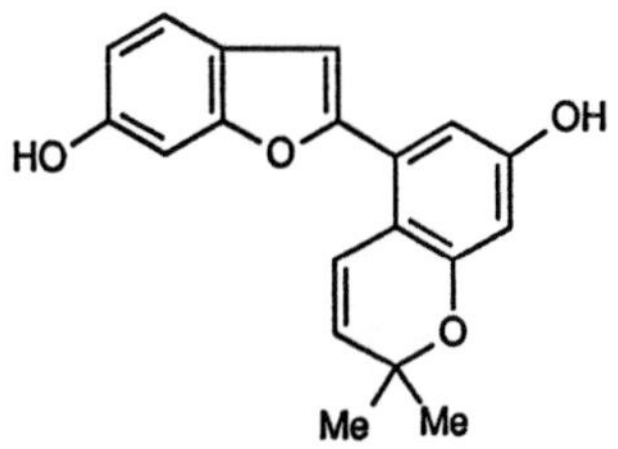

Moracin E (*88-78*)

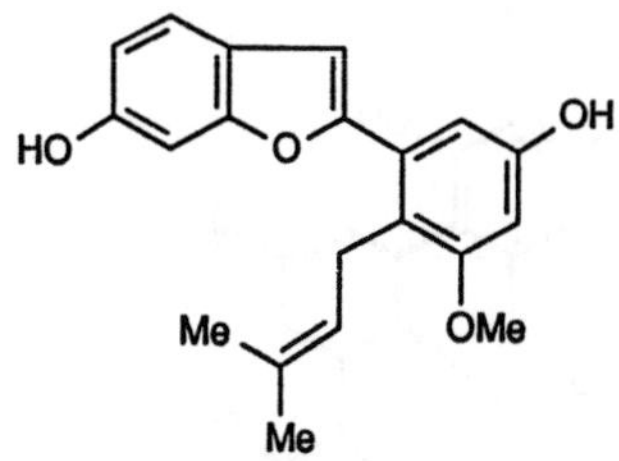

Moracin I (*88-82*)

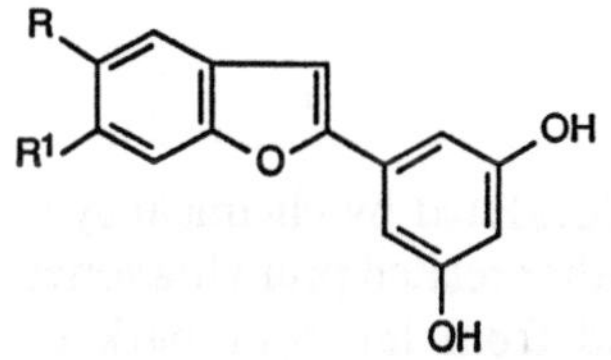

Moracin F (*88-79*): $R=R^1=OCH_3$
Moracin J (*88-83*): $R=OH$, $R^1=OCH_3$
Moracin M (*88-86*): $R=H$, $R^1=OH$

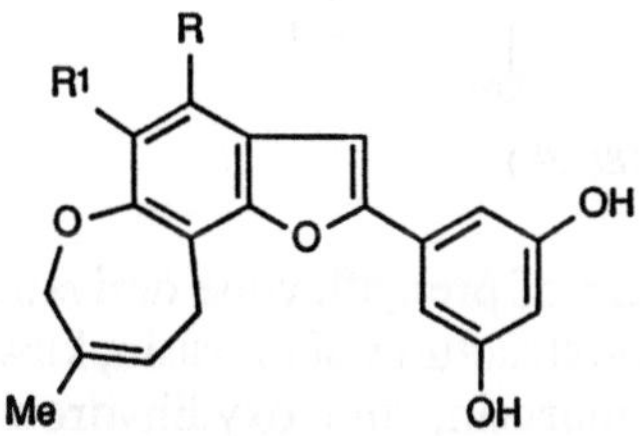

Moracin G (*88-80*): $R=R^1=H$
Moracin H (*88-81*): $R=OHC_3$, $R^1=H$
Moracin L (*88-85*): $R=H$, $R^1=OH$

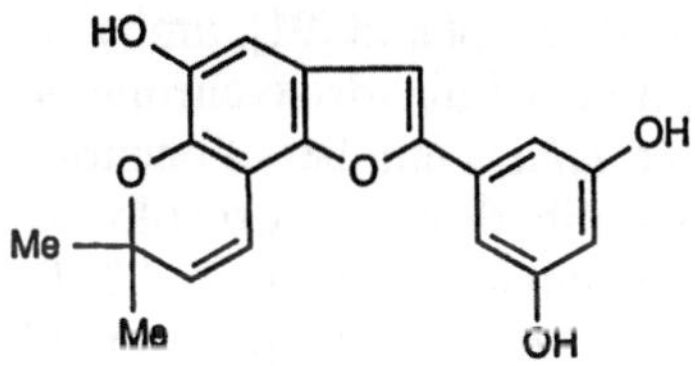

Moracin K (*88-84*)

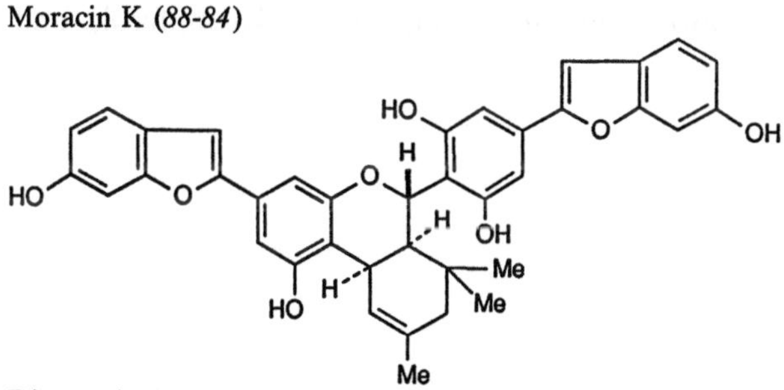

Dimoracin (*88-87*)

From the leaves of diseased *M. alba*, an antifungal Diels-Alder adduct named chalcomoracin (*88-88*) was isolated and structurally detected [89].

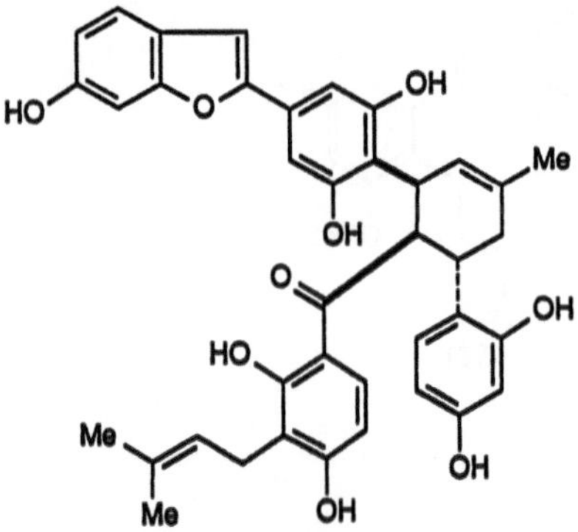

Chalcomoracin (*88-88*)

The structure of prenylflavone derivatives was also elucidated by chemical synthesis. Thus, the structures of morusin, kuwanon C, and other related prenylflavones such as cyclomorusin, and oxydihydromorusin isolated from the root bark of *M. alba* were confirmed by synthesis of tetrahydrokuwanon C tetramethyl ether: catalytic hydrogenation of morusin or kuwanon C and subsequent methylation yielded tetrahydrokuwanon C tetramethyl ether [90].

In addition to the prenylflavones a number of other compounds were isolated from the root bark of *M. alba*. They were identified as β-tocopherol [91], umbelliferone, scopoletin [92], ethyl 2,4-dihydroxybenzoate, and 5,7-dihydroxychromone [93]. From the wood of some *Morus* species including *M. alba*, a number of phenolic compounds were isolated. They were identified as 3,4′-dihydroxydihydrostilbene, 3,2′,4′-trihydroxydihydrostilbene, 6,3′,5′-trihydroxy-2-phenylbenzofuran, 5,7,2′,4′-tetrahydroxyflavanone, resorcinol, kaempferol, quercetin, resveratrol, piceatannol (*88-89*), dihydroxyresveratrol, morin, dihydromorin, dihydrokaempferol, and oxyresveratrol [94].

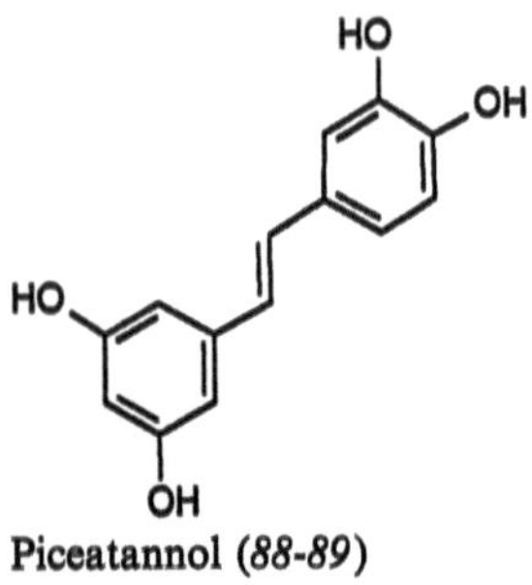

Piceatannol (*88-89*)

A piperidine alkaloid named moranoline (*88-90*) was isolated from *Morus* species and structurally elucidated by spectral analysis [95].

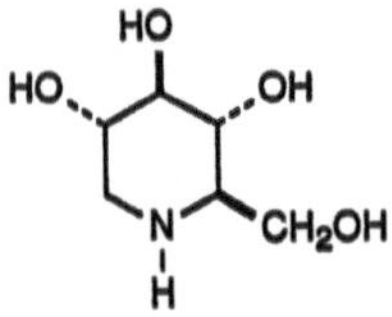

Moranoline (*88-90*)

88.2.3 Chemical Constituents of the Fruits of *M. alba*

A carbon dioxide extract of the air-dried fruits of *M. alba* contained lipids (63%), organic acids (27%), alcohols (1.6%) and essential oil. The essential oil was composed of cineol, geraniol, linalool and its acetate, camphor, α-pinene, limonene, and some nonidentified compounds, probably of a terpene nature [96]. Cineole was the major constituent of the essential oil of mulberry fruits with a content of 69%, followed by geraniol (17%). The fatty acids in the lipid fraction were identified as heptanoic, caprylic, pelargonic, capric, myristic, palmitic, palmitoleic, methyl-heptadecanoic, stearic, oleic, linoleic, nonadecanoic, and linolenic acids.

The major constituent (68%) was linoleic acid, followed by oleic acid (13%) and palmitic acid (12%) [97]. Cyanidin and its 3-*O*-β-D-glucopyranoside chrysanthemin were also isolated from the fruits of *M. alba* [98].

88.3 Pharmacology

Both the water-soluble and butanol-soluble fractions of the root bark of *M. alba* showed analgesic, diuretic, antitussive, antiedemic, sedative, anticonvulsive, and hypotensive activities in rodents and dogs. These effects were similar to the clinical observations reported in the literature concerning traditional Chinese medicine [99].

Kuwanons G [28, 29] and H [29, 32], sanggenone C [53], and mulberrofuran C [67] all showed significant hypotensive activity when given intravenously to rabbits at a dose of 1 mg/kg. Mulberrofuranes F and G also showed marked hypotensive activity in rabbits [69, 70]. Sanggenone C [49] (0.5 mg/kg) and sanggenone D [54, 99] (0.5–2.0 mg/kg) caused hypotension in rats. Moracenins A [33], B [31], C [62], and D [63] also possessed a marked hypotensive effect in rats.

Mulberrofuran A was effective against gram-positive bacteria but inactive against gram-negative bacteria. The minimum inhibiting concentrations against *Staphylococcus aureus*, *Streptococcus faecalis*, *Bacillus subtilis*, and *Mycobacterium* sp. were 6.3, 3.1, 3.1, and 1.6 μg/ml, respectively [65]. The minimum inhibiting concentrations of kuwanons I [35] and L [39] against *Staphylococcus aureus* were 3.1 and 6.1 μg/ml, respectively.

The moracins A–H [83–85] isolated from diseased cortex and xylem tissues of *M. alba* and albanins F and G [30] as well as the stilbene derivatives oxyresveratrol (trans-2,3′,4,5′-tetrahydroxystilbene) and its 4′-(3-methyl-2-butenyl)-analog [100] were found to possess antifungal activity.

The root bark of *M. alba* reduced the plasma sugar level in mice greatly. The active component was shown to be a glycoprotein named moran A, which elicited marked hypoglycemic effects in normal and alloxan-induced hyperglycemic mice when given intraperitoneally [101]. The neutral sugar fraction in moran A was composed of rhamnose, arabinose, mannose, galactose, and glucose in a molar ratio of 0.1:1.1:0.2:1.0:0.5. Amino acid components were glycine, glutamic acid, aspartic acid, alanine, proline, threonine, serine, hydroxyproline, cystine, and other minor amino acids. An interferon-inducing polysaccharide named morusan was further isolated from the root of *M. alba* and *M. bombycis*. The molecular weight of morusan was 60 000 [102].

References

1. Yamazaki M (1967) Studies on the volatile components of mulberry leaves by gas chromatography. IV. Relative quantities of the major components of the volatiles. Nippon Sanshigaku Zasshi 36:67–70 (CA 69:103806w)
2. Yamazaki M (1964) The volatile components of mulberry leaves by gas chromatography. I. On the neutral and acidic components. Nippon Nogei Kagaku Kaishi 38:95–99 (CA 62:9461f)
3. Yamazaki M (1966) Volatile components of mulberry leaves by gas chromatography. II. Acidic and phenolic components. Nippon Nogei Kagaku Kaishi 40:456–460 (CA 67:776r)
4. Yamazaki M (1967) Studies on the volatile components of mulberry leaves by gas chromatography. III. Carbonyl components. Nippon Sanshigaku Zasshi 36:63–65 (CA 69:103805v)
5. Yoshida T, Matsuzaki K (1961) The nonvolatile organic acids in mulberry leaves, their decomposition by the enzymes of the alimentary tract of the silk moth larvae, and their physiological effects on the silk worm. Sanshi Skiensho Hokoku 17:225–268 (CA 61:7423a)
6. Naito K, Nishina J, Hamamura Y (1963) Micro constituents in mulberry leaves. V. Isolation of palmitic acid, ethyl palmitate, hentriacontane, fumaric acid and hydroxycoumarine from mulberry leaves. Nippon Nogei Kagaku Kaishi 37:449–452 (CA 63:4572d)
7. Naito K (1968) Trace components in mulberry leaves. VII. Isolation of rutin and quercetin from mulberry leaves. Nippon Nogei Kagaku Kaishi 42:423–425 (CA 69:93633w)
8. Naito K (1968) Trace constituents of mulberry leaves. VIII. Isolation of quercetin glycoside from mulberry leaves. Nippon Nogei Kagaku Kaishi 42:450–453 (CA 70:26376x)
9. Moriya S (1977) Sterol composition in mulberry leaves. Sanshi Kenkyu 103:32–36 (CA 91:52726q)
10. Goto M, Kamada M, Imai S, Murata T, Fujioka S, Fujita E, Hamamura Y (1965) Mulberry and related materials. I. Annu Rep Takeda Res Lab 24:55–65 (CA 64:7045c)
11. Takemoto T, Ogawa S, Nishimoto N, Hirayama H, Tangiguchi S (1967) Isolation of the insect moulting hormones from mulberry leaves. Yakugaku Zasshi 87:748
12. Verunova GI, Glukhoded IS, Danilov LL, Eliseeva GI, Kotchetkov NK, Troitsskii MF, Usov AI, Shashkov AS, Shibaev VN (1977) Structure of moraprenol and the synthesis of moraprenyl phosphate. Bioorg Khim 3:1484–1492 (CA 88:89882b)
13. Nisshin Flour Milling Co Nisshin Kagaku KK (1981) Isolation of polyprenoid ketones from plants. Jpn Kokai Tokkyo Koho 81, 25,129 (CA 95:150964t)
14. Nomura T, Fukai T (1981) Prenylflavonoids from the root bark of the cultivated mulberry tree. Heterocycles 15:1531–1567
15. Deshpande VH, Parthasarathy PC, Venkataraman K (1968) Four analogs of artocarpin and cycloartocarpin from *Morus alba*. Tetrahedron Lett 1715–1719
16. Nomura T, Fukai T, Katayanagi M (1977) Kuwanon A, B, C and oxydihydromorusin, four new flavones from the root bark of the cultivated mulberry tree (*Morbus alba* L.). Chem Pharm Bull (Tokyo) 25:529–532
17. Nomura T, Fukai T, Katayanagi M (1978) Studies on the constituents of the cultivated mulberry tree. II. Isolation of four new flavones, kuwanon A, B, C and oxydihydromorusin from the root bark of *Morus alba* L. Chem Pharm Bull (Tokyo) 26:1453–1458
18. Chari VM, Ahmad S, Oesterdahl BG (1978) Carbon-13 NMR spectra of chromeno- and prenylated flavones. Structure revision of mulberrin, mulberrochromene, cyclomulberrin and cyclomulberrochromene. Z Naturforsch 33B:1547–1549
19. Nomura T, Fukai T (1979) On the structure of mulberrin, mulberrochromene, cyclomulberrin, and cyclomulberrochromene. Heterocycles 12:1289–1295
20. Nomura T, Fukai T, Yamanda S, Katayanagi M (1976) Phenolic constituents of the cultivated mulberry tree (*Morus alba* L.). Chem Pharm Bull (Tokyo) 24:2898–2900
21. Nomura T, Fukai T, Yamanda S, Katayanagi M (1978) Studies on the constituents of the cultivated mulberry tree. I. Three new prenylflavones from the root bark of *Morus alba* L. Chem Pharm Bull (Tokyo) 26:1394–1402
22. Deshpande VH, Wakharkar PV, Rao AV (1976) Wood phenolics of *Morus* species: V. Isolation of new flavone, mulberranol and a novel phenol, albotalol from *Morus alba*. Indian J Chem Sect B 14B:647–650
23. Hikino H (1977) Morusinol, isoprenoid flavone from *Morus* root barks. Planta Med 32:118–124

24. Nomura T, Fukai T (1981) Constituents of the cultivated mulberry tree. VII. Isolation of three new isoprenoid flavanones, kuwanon D, E and F from the root bark of *Morus alba* L. Planta Med 42:79–88
25. Nomura T, Fukai T, Katayanagi M (1978) Kuwanon D, new isoprenoid flavanone from the root bark of the cultivated mulberry tree (*Morus alba* L.). Heterocycles 9:745–752
26. Nomura T, Fukai T (1978) Kuwanon E, a new flavanone derivative from the root bark of the cultivated mulberry tree (*Morbus alba* L.). Heterocycles 9:1295–1300
27. Nomura T, Fukai T (1979) Kuwanon F, a new flavanone derivative from the root bark of the cultivated mulberry tree (*Morus alba* L.). Heterocycles 12:943–946
28. Nomura T, Fukai T (1980) Kuwanon G, a new flavone derivative from the root bark of the cultivated mulberry tree (*Morus alba* L.). Chem Pharm Bull (Tokyo) 28:2548–2552
29. Zenyaku Kogyo Co (1981) 8-Cyclohexenylflavones from mulberry species. Jpn Kokai Tokkyo Koho JP 81, 123, 979 (CA 96:19970v)
30. Takasugi M, Ishikawa S, Nagao S, Masamune T, Shirata A, Takahashi K (1980) Studies on phytoalexins of the Moraceae. 8. Albanins F and G, natural Diels-Alder adducts from mulberry. Chem Lett 1577–1580
31. Oshima Y, Konno C, Hikino H, Matsushita K (1980) Validity of oriental medicines. XXIII. Structure of moracenin B, a hypotensive principle of *Morus* root barks. Tetrahedron Lett 21:3381–3384
32. Nomura T, Fukai T, Narita T (1980) Hypotensive constituent, kuwanon H, a new flavone derivative from the root bark of the cultivated mulberry tree (*Morus alba* L.). Heterocycles 14:1943–1951
33. Oshima Y, Konno C, Hikino H, Matsushita K (1980) The validity of oriental medicines. XXIV. Structure of moracenin A, a hypotensive principle of *Morus* root barks. Heterocycles 14:1287–1290
34. Zenyaku Kogyo Co (1983) Kuwanon I. Jpn Kokai Tokkyo Koho JP 58 43,990 [83 43,990] (CA 99:164007c)
35. Zenyaku Kogyo Co (1983) Kuwanon I. Jpn Kokai Tokkyo Koho JP 58, 150, 538 [83, 150, 538] (CA 100:39588f)
36. Ueda S, Nomura T, Fukai T, Matsumoto J (1982) Kuwanon J, a new Diels-Alder adduct, and chalcomoracin from callus culture of *Morus alba* L. Chem Pharm Bull (Tokyo) 30:3042–3045
37. Ikuta J, Fukai T, Nomura T, Ueda S (1986) Constituents of the cultivated mulberry tree. XXXV. Constituents of *Morbus alba* L. cell cultures. I. Structures of four new natural Diels-Alder type adducts, Kuwanons J, Q, R, and V. Chem Pharm Bull (Tokyo) 34:2471–2478
38. Nomura T, Fukai T, Hano Y, Nemoto K, Terada S, Kuramochi T (1983) Constituents of cultivated mulberry tree. XII. Isolation of two new natural Diels-Alder adducts from root bark of *Morus alba*. Planta Med 47:151–156
39. Zenyaku Kogyo Co (1982) Isolation of mikrobicidal kuwanon L. Jpn Kokai Tokkyo Koho JP 58, 150, 584 [83, 150, 584] (CA 100:73968q)
40. Fukai T, Hano Y, Ikuta H, Fujimoto T, Hirakura K, Nomura T, Uzawa J, Komatsu K, Fukushima K (1983) Structure of new phenolic constituents of *Morus* root barks. Tennen Yuki Kagobutsu Toronkai Koen Yoshishu, 26th, 150–157 (CA 100:135777n)
41. Nomura T, Fukai T, Hano Y, Ikuta H (1983) Kuwanon M, a new Diels-Alder adduct from the root bark of the cultivated mulberry tree (*Morus lhou* (ser.) Koidz). Heterocycles 20:585–591
42. Hano Y, Hirakura K, Nomura T, Terada S, Fukushima K (1984) Constituents of the cultivated mulberry tree. XVI. Components of root bark of *Morus lhou*. I. Structures of two new natural Diels-Alder adducts, kuwanons N and O. Planta Med 50:127–130
43. Hirakura K, Hano Y, Fukai T, Nomura T, Uzawa J, Fukushima K (1985) Constituents of the cultivated mulberry tree. XXI. Constituents of the root bark of *Morus lhou* Koidz. III. Structures of three natural Diels-Alder type adducts, kuwanon P and X, and mulberrofuran J, from the cultivated mulberry tree (*Morbus lhou* Koidz.). Chem Pharm Bull (Tokyo) 33:1088–1096
44. Ueda S, Matsumoto J, Nomura T (1984) Four new natural Diels-Alder adducts, mulberrofuran E, kuwanon Q, R, and V from callus culture of *Morus alba* L. Chem Pharm Bull (Tokyo) 32:350–353
45. Fukai T, Hano Y, Hirakura K, Nomura T, Uzawa J (1985) Structures of a novel 2-arylbenzofuran derivative and two flavone derivatives from the cultivated mulberry tree (*Morus lhou* Koidz.). Chem Pharm Bull (Tokyo) 33:4288–4295

46. Hirakura K, Fukai T, Hano Y, Nomura T (1985) Constituents of the cultivated mulberry tree. XX. Constituents of the root bark of *Morus lhou*. II. Kuwanon W, a natural Diels-Alder type adduct from the root bark of *Morus lhou*. Phytochemistry 24:159–161
47. Hano Y, Tsubura H, Nomura T (1986) Constituents of the cultivated mulberry tree. XXXVIII. Structures of kuwanons Y and Z, two new stilbene derivatives from the cultured mulberry tree (*Morus alba* L.). Heterocycles 24:2603–2610
48. Hano Y, Itoh M, Nomura T (1985) Structures of kuwanol A and B, two novel stilbene derivatives from the cultivated mulberry tree (*Morbus bombycis* Koidz.). Heterocycles 23:819–824
49. Nomura T, Fukai T, Hano Y, Matsumoto J, Imashimizu A, Uzawa J, Urano S, Fukushima K (1981) Structures of natural Diels-Alder adducts of *Morus* root barks. Tennen Yuki Kagobutsu toronkai Koen Yoshishu 24:151–158 (CA 96:214241k)
50. Nomura T, Fukai T, Hano Y, Sugaya Y, Hosoya T (1980) Sanggenone A, a new flavanone derivative from Chinese crude drug "Sang-Bai-Pi" (*Morus* root bark). I. Heterocycles 14:1785–1790
51. Nomura T, Fukai T, Hano Y (1983) Constituents of the cultivated mulberry tree. IX. Constituents of the Chinese crude drug "Sang-Bai-Pi" (*Morus* root barks). I. Structure of a new flavanone derivative, sanggenone A. Planta Med 47:30–34
52. Nomura T, Fukai T, Hano Y, Urano S (1983) Constituents of the cultivated mulberry tree. XI. Constituents of the Chinese crude drug "Sang-Bai-Pi" (*Morus* root barks). II. Structure of a new flavanone derivative, sanggenone B. Planta Med 47:95–99
53. Nomura T, Fukai T, Hano Y, Uzawa J (1981) Structure of sanggenone C, a natural hypotensive Diels-Alder adduct from Chinese crude drug "Sang-Bai-Pi" (*Morus* root barks). Heterocycles 16:2141–2148
54. Nomura T, Fukai T, Hano Y, Uzawa J (1982) Structure of sanggenone D, a natural hypotensive Diels-Alder adduct from Chinese crude drug "Sang-Bai-Pi" (*Morus* root barks). Heterocycles 17:381–389
55. Hano Y, Kohno H, Suziki S, Nomura T (1986) Constituents of the cultivated mulberry tree. XXXVII. Constituents of the Chinese crude drug "Sang-Bai-Pi" (*Morus* root bark). VIII. Structures of sanggenons E and P, two new Diels-Alder-type adducts from the Chinese crude drug "Sang-Bai-Pi" (*Morus* root bark). Heterocycles 24:2285–2291
56. Nomura T, Fukai T, Hano Y, Tsukamoto K (1983) Constituents of the Chinese crude drug "Sang-Bai-Pi" (*Morus* root barks). III. Structure of a new flavanone derivative, sanggenone F. Heterocycles 20:661–666
57. Fukai T, Hano Y, Fujimoto T, Nomura T (1983) Structure of sanggenone G, a new Diels-Alder adduct from the Chinese crude drug "Sang-Bai-Pi" (*Morus* root barks). Heterocycles 20:611–615
58. Hano Y, Nomura T (1983) Constituents of the Chinese crude drug "Sang-Bai-Pi" (*Morus* root barks). IV. Structures of four new flavonoids, sanggenone H, I, J, and K. Heterocycles 20:1071–1076
59. Hano Y, Itoh M, Koyama N, Nomura T (1984) Constituents of the Chinese crude drug "Sang-Bai-Pi" (*Morus* root barks). V. Structures of three new flavanones, sanggenone L, M, and N. Heterocycles 22:1791–1800
60. Hano Y, Nomura T (1985) Structure of sanggenone O, a natural Diels-Alder type adduct from the Chinese crude drug "Sang-Bai-Pi" (*Morua* root bark). Heterocycles 23:2499–2503
61. Sun JY, Hano Y, Nomura T (1989) Constituents of cultivated mulberry tree. XL. The structure of sanggenone Q, a new Diels-Alder type adduct from *Morus mongolica* Schneider. Heterocycles 29:195–202
62. Oshima Y, Konno C, Hikino H, Matsushita K (1980) Validity of oriental medicines. XXVI. Structure of moracenin C, a hypotensive principle of *Morus* root barks. Heterocycles 14:1461–1464
63. Oshima Y, Konno C, Hikino H (1981) Validity of oriental medicines. XXVIII. Structure of moracenin D, a hypotensive principle of *Morus* root barks. Heterocycles 16:979–982
64. Nomura T, Fukai T, Sato E, Fukushima K (1981) The formation of moracenin D from kuwanon G. Heterocycles 16:983–986
65. Nomura T, Fukai T, Uno J, Arai T (1978) Mulberrofuran A, a new isoprenoid 2-arylbenzofuran from the root bark of the cultivated mulberry tree (*Morus alba* L.). Heterocycles 9:1593–1601

66. Nomura T, Fukai T (1981) Mulberrofuran B, a new isoprenoid 2-arylbenzofuran from the root bark of the cultivated mulberry tree. Planta Med 42:197–199
67. Nomura T, Fukai T, Matsumoto J, Fukushima K, Momose Y (1981) Structure of mulberrofuran C, a natural hypotensive Diels-Alder adduct from root barks of the cultivated mulberry tree (*Morus bombycis* Koidzumi). Heterocycles 16:759–765
68. Nomura T, Fukai T, Shimada T, Chen IS (1983) Constituents of the cultivated mulberry tree. XIII. Components of the root bark of *Morus australis*. I. Structure of a new 2-acrylbenzofuran derivative, mulberrofuran D. Planta Med 49:90–94
69. Fukai T, Hano Y, Hirakura K, Nomura T, Uzawa J, Fukushima K (1985) Constituents of the cultivated mulberry tree. XXV. Constituents of the root bark of *Morus lhou* Koidz. V. Structures of two natural hypotensive Diels-Alder type adducts, mulberrofuran F and G, from the cultivated mulberry tree (*Morus lhou* Koidz.). Chem Pharm Bull (Tokyo) 33:3195–3204
70. Fukai T, Hano Y, Hirakura K, Nomura T, Uzawa J, Fukushima K (1984) Structures of mulberrofurans F and G, two natural hypotensive Diels-Alder type adducts from the cultivated mulberry tree (*Morus lhou* (ser.) Koidz.). Heterocycles 22:473–477
71. Rao AVR, Deshpande VH, Shastri RK, Tavale SS, Dhaneshwar NN (1983) Structures of albanols A and B, two novel phenols from *Morus alba* bark. Tetrahedron Lett 24:3013–3016
72. Fukai T, Hano Y, Hirakura K, Nomura T, Uzawa J (1984) Structure of mulberrofuran H, a novel 2-arylbenzofuran derivative from the cultivated mulberry tree (*Morus lhou* (ser.) Koidz.). Chem Pharm Bull (Tokyo) 32:808–811
73. Nikaido T, Ohmoto T, Nomura T, Fukai T, Sankawa U (1984) Inhibitors of cyclic AMP phosphodiesterase in medicinal plants. IX. Inhibition of adenosine 3′,5′-cyclic monophosphate phosphodiesterase by phenolic constituents of mulberry tree. Chem Pharm Bull (Tokyo) 32:4929–4934
74. Fukai T, Hano Y, Hirakura K, Nomura T, Uzawa J, Fukushima K (1984) Structure of mulberrofuran J, a 2-arylbenzofuran derivative from the cultivated mulberry tree (*Morus lhou* (ser.) Koidz.). Heterocycles 22:1007–1011
75. Hano Y, Fukai T, Kohno H, Hirakura K, Nomura T, Uzawa J (1984) Structure of mulberrofuran K, an optically active natural Diels-Alder type adduct from the Chinese crude drug "Sang-Bai-Pi" (*Morus* root barks). Heterocycles 22:2729–2733
76. Hano Y, Kohno H, Itoh M, Nomura T (1985) Constituents of the cultivated mulberry tree. XXIX. Structures of three new 2-arylbenzofuran derivatives from the Chinese crude drug "Sang-Bai-Pi" (*Morus* root barks). Chem Pharm Bull (Tokyo) 33:5294–5300
77. Fukai T, Fujimoto T, Hano Y, Nomura T, Uzawa J (1984) Structures of mulberrofurans B and L, 2-arylbenzofuran derivatives from the root bark of the cultivated mulberry tree (*Morus lhou* (ser.) Koidz.) Heterocycles 22:2805–2814
78. Hano Y, Hirakura K, Someya T, Nomura T (1986) Structure of mulberrofuran M, a novel 2-arylbenzofuran derivative from the cultivated mulberry tree (*Morus alba* L.). Heterocycles 24:1251–1255
79. Hano Y, Nomura T (1986) Constituents of the cultivated mulberry tree. XXXVI. Structure of mulberrofuran P, a novel 2-arylbenzofuran derivative from the cultivated mulberry tree (*Morus alba* L.). Heterocycles 24:1381–1386
80. Hano Y, Tsubura H, Nomura T (1986) Structure of mulberrofuran Q, a novel 2-arylbenzofuran derivative from the cultivated mulberry tree (*Morus alba* L.). Heterocycles 24:1807–1813
81. Takasugi M, Ishikawa S, Masamune T (1982) Stuides on phytoalexins of the Moraceae. XI. Albafurans A and B, geranyl 2-phenylbenzofurans from mulberry. Chem Lett 1221–1222
82. Takasugi M, Ishikawa S, Nagao S, Masamune T (1982) Studies on the phytoalexins of the Moraceae. 12. Albafuran C, a natural Diels-Alder adduct of a dehydroprenyl-2-phenylbenzofuran with a chalcone from mulberry. Chem Lett 1223–1224
83. Takasugi M, Nagao S, Masamune T, Shirata A, Takahashi K (1978) Structure of moracin A and B, new phytoalexins from diseased mulberry. Tetrahedron Lett 797–798
84. Takasugi M, Nagao S, Masamune T, Shirata K (1978) Moracin C and D, new phytoalexins from diseased mulberry. Chem Lett 1239–1240
85. Takasugi M, Nagao S, Masamune T, Shirata A, Takahashi K (1979) Studies on phytoalexins of the Moraceae. IV. Structures of moracins E, F, G, and H, new phytoalexins from diseased mulberry. Tetrahedron Lett 4675–4678

86. Takasugi M, Nagao S, Munoz L, Ishikawa S, Masamune T, Shirata A, Takahashi K (1979) The structure of phytoalexins produced in diseased mulberry. Koen Yoshishu Tennen Yuki Kagobutsu Toronkai, 22:275–282 (CA 92:160540d)
87. Takasugi M, Munoz L, Masamune T, Shirata A, Takahashi K (1978) Studies on phytoalexins of the Moraceae. III. Stilbene phytoalexins from diseased mulberry. Chem Lett 1241–1242
88. Takasugi M, Nagao S, Masamune T (1982) Studies on phytoalexins of the Moraceae. X. Structure of dimoracin, a new natural Diels-Alder adduct from diseased mulberry. Chem Lett 1217–1220
89. Takasugi M, Nagao S, Masamune T, Shirata A, Takahashi K (1980) Chalcomoracin, a natural Diels-Alder adduct from diseased mulberry. Chem Lett 1573–1576
90. Nomura T, Sawaura Y, Fukai T, Yamada S, Tamura S (1978) Studies on the constituents of the cultivated mulberry tree. V. The synthesis of tetrahydrokuwanon C tetramethyl ether. Heterocycles 9:1355–1366
91. Shibata H, Mikoshiba I, Shimizu S (1974) Constituents of the mulberry tree. I. Isolation of β-tocopherol from the root bark of the mulberry tree. Agric Biol Chem 38:1745
92. Uno T (1970) Isolation of umbelliferone and scopoletin from the root bark of the mulberry tree. Sanshi Shikenjo Hokoku 24:437–442 (CA 76:56567b)
93. Uno T, Isogai A, Suzuki A, Shirata A (1981) Isolation and identification of ethyl 2,4-dihydroxybenzoate and 5,7-dihydroxychromone from the root bark of mulberry tree (*Morus alba*) and their biological activity. Nippon Sanshigaku Zasshi 50:422–427 (CA 96:82739s)
94. Deshpande VH, Srinivasan R, Rao AVR (1975) Wood phenolics of *Morus* species. IV. Phenolics of the heartwood of five *Morus* species. Indian J Chem 13:453–457
95. Yagi M, Kouno T, Aoyagi Y, Murai H (1976) The structure of moranoline, a piperdine alkaloid from *Morus* species. Nippon Nogei Kagaku Kaishi 50:571–572 (CA 86:167851r)
96. Kasyanov GI, Pekhov AV, Shaftan EA, Klimova ES, Shishkov GZ (1976) Carbon dioxide extract from *Morus alba* L. fruit. Rastit Resur 12:597–599 (CA 86:40143f)
97. Shishkov GZ, Golberg ND, Marchenko PS (1978) Fatty acid composition of a carbon dioxide extract from *Morus alba* fruits. Rastit Resur 14:585–586 (CA 90:19061a)
98. Ikuta H, Fukai T, Nomura T, Uzawa J (1985) Constituents of the cultivated mulberry tree. XXIII. Carbon-13 NMR spectra of cyanidin and chrysanthemin. Heteracycles 23:121–126
99. Yamatake Y, Shibata M, Nagai M (1976) Pharmacological studies on root bark of mulberry tree (*Morus alba* L.). Jpn J Pharmacol 26:461–469
100. Takasugi M, Munoz L, Masamune T, Shirata A, Takahashi K (1978) Stilbene phytoalexins from diseased mulberry. Chem Lett 1241–1242
101. Hikino H, Mizuno T, Oshima Y, Konno C (1985) Validity of the oriental medicines. LXXX. Antidiabetes drugs. IV. Isolation and hypoglycemic activity of moran A, a glycoprotein of *Morus alba* root barks. Planta Med 51:159–160
102. Kitasato Institute (1984) Interferon-inducing polysaccharide from mulberry roots. Jpn Tokkyo Koho JP 59 40,135 [84 40,135] (CA 102:100795f)

Nelumbo nucifera Gaertn.

89

89.1 Introduction

Nelumbo nucifera Gaertn. (Nymphaeaceae), is a medical plant used in traditional Chinese medicine. Officially listed in the Chinese Pharmacopoeia are:

- Lianzi, Semen Nelumbinis, the dried mature seeds of *N. nucifera*, harvested in the fall, used as a sedative and tonic.
- Lianzixin, Plumula Nelumbinis, the dried embryos from the mature seeds, used as a sedative and hemostatic.
- Lianfang, Receptaculum Nelumbinis, the dried receptacles collected in the fall when the fruits have ripened, used as a hemostatic.
- Lianxu, Stamen Nelumbinis, the dry stamen collected in summer when the lotus flowers bloom, used as an adstringent.
- Heye, Folium Nelumbinis, the dry leaves harvested in summer or fall, used as a hemostatic.
- Oujie, Nodus Nelumbinis rhizomatis, the dry nodes of lotus rootstock collected in the fall or winter used as a hemostatic. After charring the receptacles, leaves, and rootstock nodes are also used against bleeding.

89.2 Chemical Constituents

The chemical constituents of the leaves and embryos of *N. nucifera* were investigated. Both leaves and embryos contain a number of isoquinoline alkaloids of different types as the main constituents.

89.2.1 Chemical Constituents of the Leaves of *Nelumbo nucifera*

The first alkaloid isolated from the leaves of *N. nucifera* was nuciferine (*89-1*) [1]. From 33 kg leaves about 0.2 g nuciferine, 8 g remerine (*89-2*), and 11 g nornuciferine (*89-3*) [2, 3] were isolated.

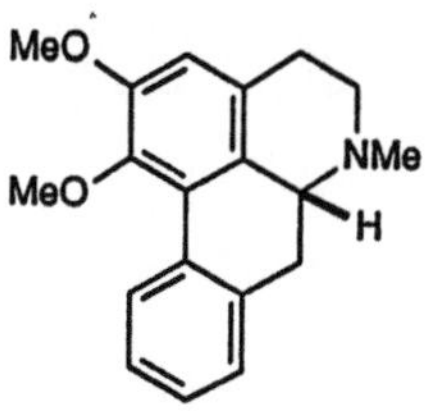

Nuciferine (*89-1*)

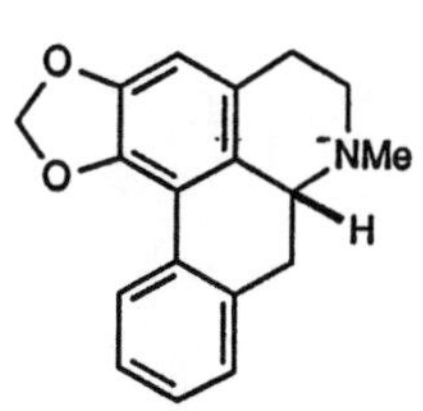

Remerine (*89-2*)

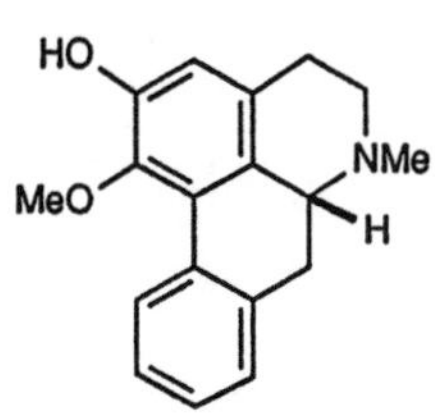

Nornuciferine (*89-3*)

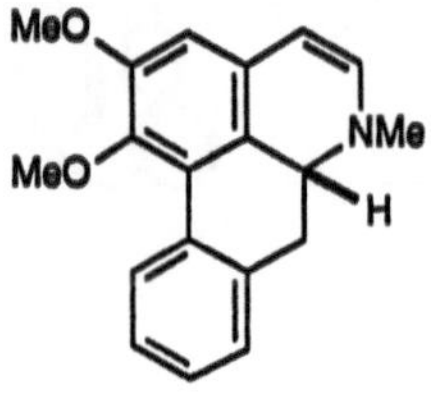

Dehydronuciferine (*89-4*)

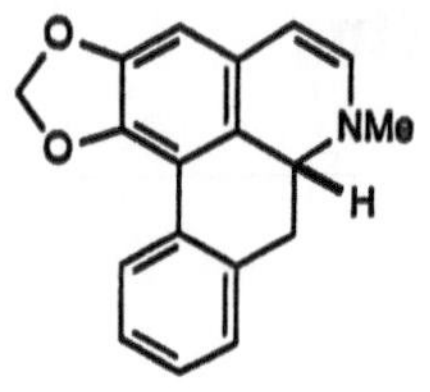

Dehydroremerine (*89-5*)

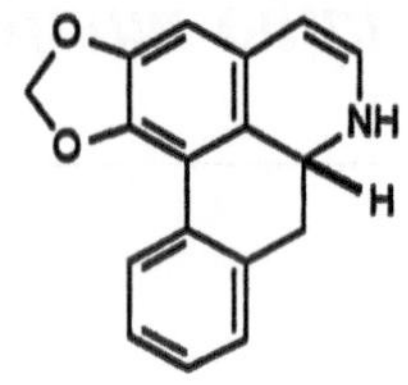

Dehydroanonaine (*89-6*)

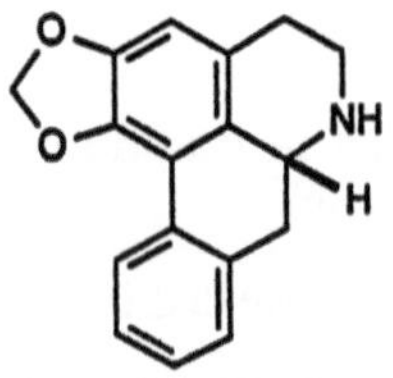

Anonaine (*89-7*)

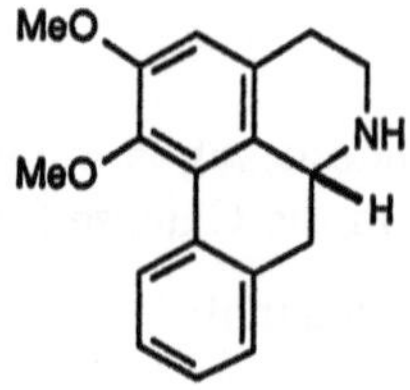

N-Nornuciferine (*89-8*)

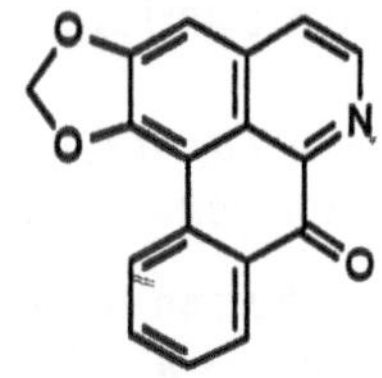

Liriodenine (*89-9*)

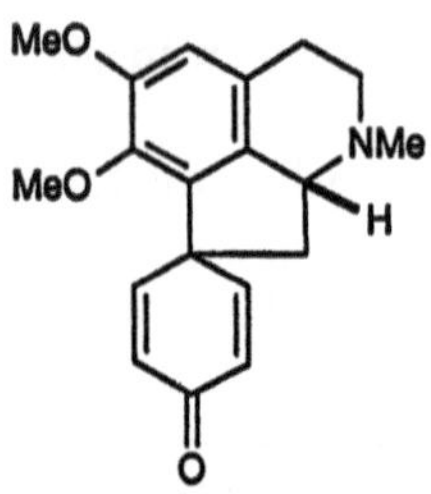

Pronuciferine (*89-10*)

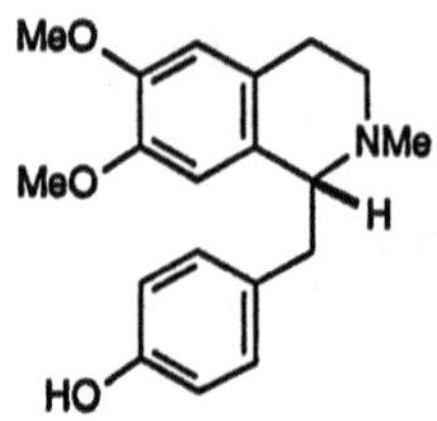

Armepavine (*89-11*)

Furthermore, the alkaloids armepavine (*89-11*) [4, 5], anonaine (*89-7*), pronuciferine (*89-10*), *N*-nornuciferine (*89-8*), liriodenine (*89-9*), methylcoclaurine [6], dehydroremerine (*89-5*), dehydronuciferine (*89-4*), dehydroanonaine (*89-6*), and *N*-methyl- coclaurine [7] were isolated. Armepavine is an alkaloid of the 1-benzylisoquinoline type; remerine, nuciferine, nornuciferine, anonaine, liriodenine are all aporphine-type alkaloids, whereas pronuciferine is a proaporphine-type alkaloid, which can be converted into nuciferine by treatment with acid after reduction (Fig. 89.1) [8].

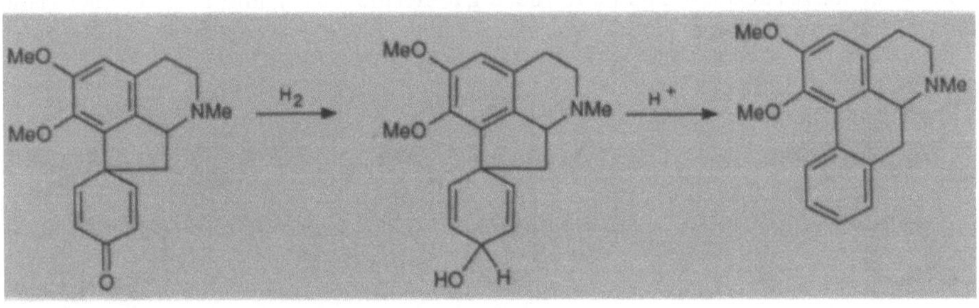

Fig. 89.1. Conversion of pronuciferine into nuciferine

Recently, asimilobine (*89-12*) and lirinidine (*89-13*) were also isolated from the leaves of *N. nucifera* [9].

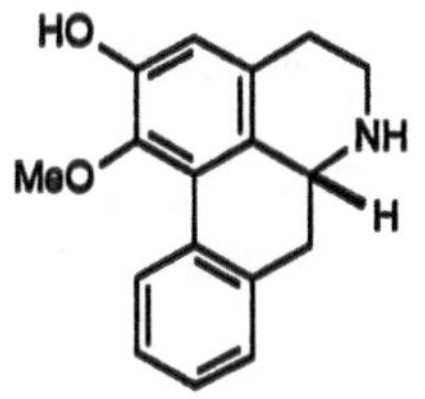

Asimilobine (*89-12*)

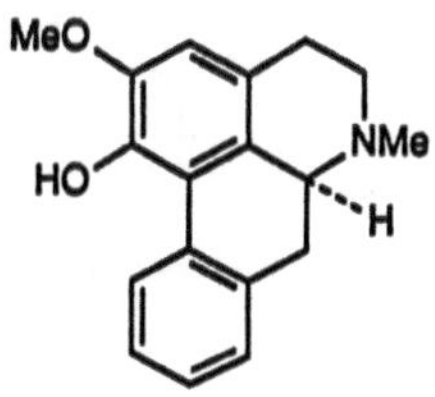

Lirinidine (*89-13*)

89.2.2 Chemical Constituents of the Embryos of *Nelumbo nucifera*

Besides the alkaloids isolated from the leaves, such as nuciferine, pronuciferine, and anonaine [10, 11], a number of bisbenzylisoquinoline alkaloids were isolated from the lotus embryos and structurally determined [12]. The first alkaloid of this type isolated from the embryos of *N. nucifera* was liensinine (*89-14*) [13]. Its structure and absolute configuration were determined by chemical reactions [14, 15]. Isoliensinine (*89-15*) [16, 17] and neferine (*89-16*) [18] are other alkaloids related to liensinine isolated from lotus embryos.

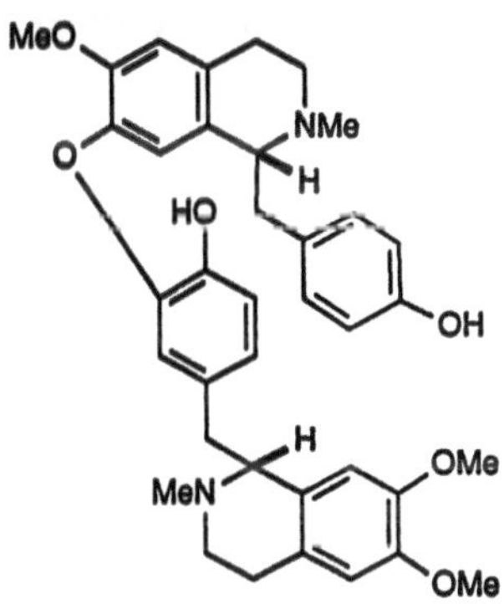

Liensinine (*89-14*)

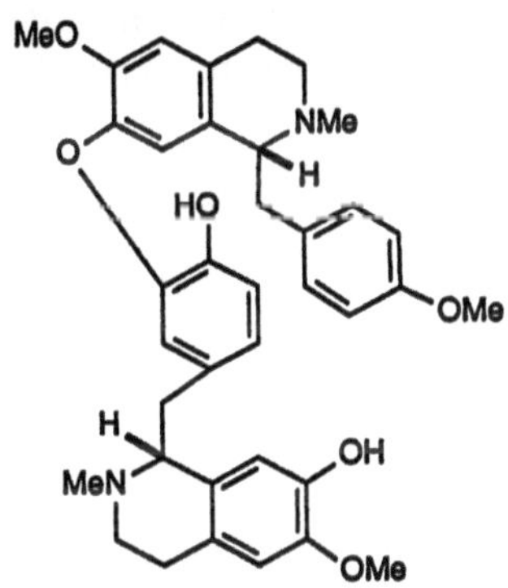

Isoliensinine (*89-15*)

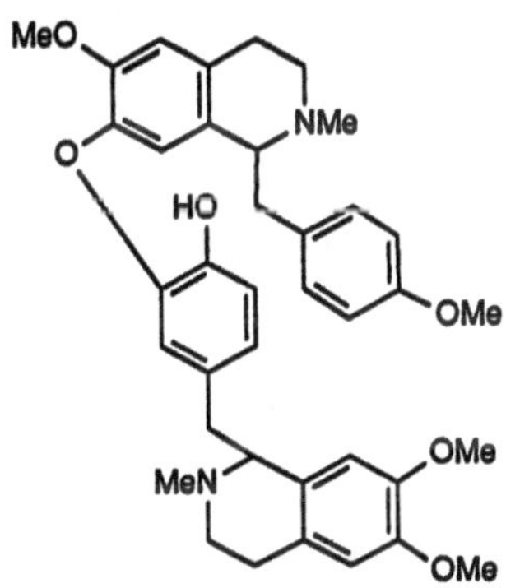

Neferine (*89-16*)

Liensinine was also totally synthesized by Ullmann condensation of 1-(3-bromo-4-benzyloxy-benzyl)-2-methyl-6,7-dimethoxy-1,2,3,4-tetrahydroisoquinoline and 1-(4-benzyloxy-benzyl)-2-methyl-6-methoxy-7-hydroxy-1,2,3,4-tetrahydroisoquinoline [19]. In addition, lotusine (*89-17*), a water-soluble quarternary base [20], demethylcoclaurine (*89-18*) [21], and methylcorypalline (*89-19*) [22, 23] were also isolated from the lotus embryos and structurally elucidated. Recently, the isolation of 4′-*O*-methyl-*N*-methylcoclaurine from the lotus embryo was also reported [24].

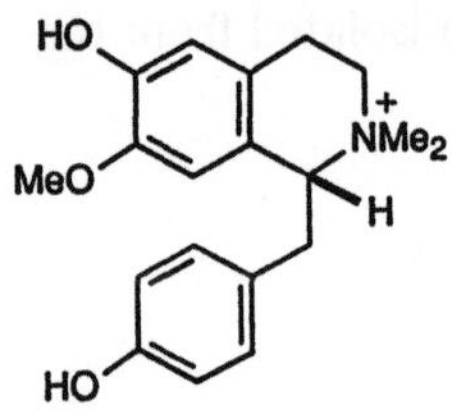

Lotusine (*89-17*)

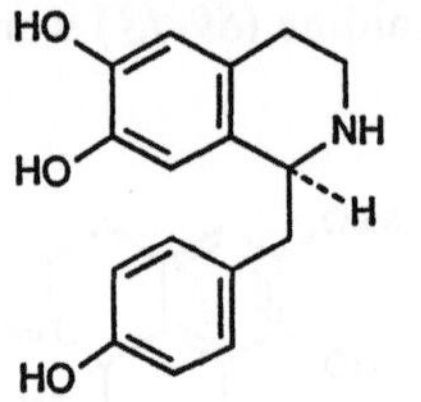

Demethylcoclaurine (*89-18*)

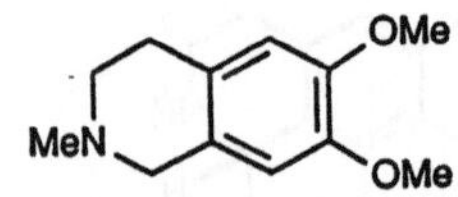

Methylcorypalline (*89-19*)

From the lotus receptacle, four alkaloids, nuciferine, *N*-nornuciferine, liriodenine, and *N*-norarmepavine were isolated and identified [25].

89.3 Pharmacology

Nuciferine had depressant properties on the CNS in rodents, as well as weak antiinflammatory, analgetic, antitussive, antiserotonin, and adrenergic blocking activities [26]. Nornuciferine derivatives administered intraperitoneally to mice as hydrobromide or hydrochloride provoked intense clonic convulsions. Unlike apomorphine, they did not induce emesis in dogs. *N*-Propylnornuciferine was the most effective convulsant agent [27]. Nornuciferine was found to be metabolized by rat and rabbit liver microsomes to lysicamine (*89-20*). Rat, rabbit, and guinea pig liver microsomes also dealkylated *N*-alkylated analog of nornuciferine [28].

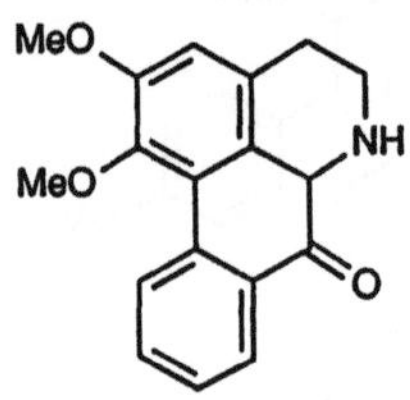

Lysicamine (*89-20*)

Quercetin was identified as the antihemorrhagic principle in the receptacle of *N. nucifera* [29].

References

1. Arthur HR, Cheung HT (1959) An aporphine alkaloid, nuciferine, from asiatic lotus cultivated in Hong Kong. J Chem Soc 2306
2. Tomita M, Watanabe Y, Tomita M, Furukawa H (1961) Alkaloids of *Nelumbo nucifera*. I. Yakugaku Zasshi 81:469–473
3. Tomita M, Watanabe Y, Furukawa H (1961) Alkaloids of *Nelumbo nucifera*. II. Structure of nornuciferine. Yakugaku Zasshi 81:942–947
4. Tomita M, Watanabe Y, Furukawa H (1961) Alkaloids of *Nelombo nucifera*. IV. Isolation of *dl*-armepavine. Yakugaku Zasshi 81:1644–1647

5. Tomita M, Furukawa H (1962) Alkaloids of *Nelumbo nucifera*. V. Alkaloids of Ohga-Hasu. Yakugaku Zasshi 82:1458–1460
6. Kunitomo J, Nagai Y, Okamoto Y, Furukawa H (1970) Alkaloids of *Nelumbo nucifera*. XIV. Tertiary base. Yakugaku Zasshi 90:1165–1169
7. Kunimoto J, Yoshikawa Y, Tanaka S, Imori Y, Isoi K, Masada Y, Hashimoto K, Inoue T (1973) Alkaloids of *Nelumbo nucifera*. XVI. Phytochemistry 12:699–701
8. Bernauer K (1963) Pronuciferin, ein Benzylisochinolin-Alkaloid mit *para*-Cyclohexadienon-Gruppierung. Helv Chim Acta 46:1783–1785
9. Shoji N, Umeyama A, Saito N, Iuchi A, Takemoto T, Kajiwara A, Ohizumi Y (1987) Asimilobine and lirinidine, serotoninergic receptor antagonists, from *Nelumbo nucifera*. J Nat Prod 50:773–774
10. Bernauer K (1964) Über die Isolierung von (+)-Pronuciferin und (−)-Anonain aus den Keimlingen von *Nelumbo nucifera* Gaertn. Helv Chim Acta 47:2119–2122
11. Furukawa H (1966) Alkaloids of *Nelumbo nucifera*. XII. Alkaloids of loti embryo. V. Yakugaku Zasshi 86:75–77
12. Guo MD, Chen LG (1984) Studies on the alkaloids constituents of the embryo nelumbinis (*Nelumbo nucifera*) produced in China. Chin Trad Herb Drugs 15:291–293
13. Chao YC, Chou YL, Yang PC, Chao CK (1962) Alkaloids of embryo of loti *Nelumbo nucifera*. I. Isolation and characterization of liensinine. Sci Sin 11:215–219
14. Pan PC, Chou YL, Sun TC, Kao IS (1962) Alkaloids of embryo of loti *Nelumbo nucifera*. II. Structure of liensinine. Sci Sin 11:321–336
15. Hsieh YY, Chen WC, Kao YS (1964) Alkaloids of *Nelumbo nucifera*. III. Absolute configuration of liensinine. Sci Sin 12:2018–2019
16. Tomita M, Furukawa H, Yang TH, Lin TJ (1965) Alkaloids of *Nelumbo nucifera* Gaertn. VIII. Alkaloids of loti embryo. 2. Structure of isoliensinine, a new biscoclaurine type alkaloid. Chem Pharm Bull (Tokyo) 13:39–43
17. Tomita M, Furukawa H, Yang TH, Lin TJ (1964) Studies on the alkaloids of loti embryo. I. Structure of isoliensinine. Tetrahedron Lett 2637–2642
18. Furukawa H (1965) Alkaloids of *Nelumbo nucifera*. IX. Alkaloids of loti embryo. 2. Structure of neferine, a new biscoclaurine type alkaloid. Yakugaku Zasshi 85:335–338
19. Hsieh YY, Pan PC, Chen WC, Kao YS (1964) Alkaloids of *Nelumbo nucifera*. IV. Total synthesis of liensinine. Sci Sin 12:2020–2025
20. Furukawa H, Yang TH, Lin TJ (1965) Alkaloids of *Nelumbo nucifera*. XI. Alkaloids of *Nelumbo nucifera* embryo. 4. Structure of lotusine, a new water soluble quarternary base. Yakugaku Zasshi 85:472–475
21. Koshiyama H, Ohkuma H, Kawaguchi H, Hsu HY, Chen YP (1970) Isolation of 1-(*p*-hydroxybenzyl)-6,7-dihydroxy-1,2,3,4-tetrahydroisoquinoline (demethylcoclaurine), an active alkaloid from *Nelumbo nucifera*. Chem Pharm Bull (Tokyo) 18:2564–2568
22. Yang TH, Chen CM (1970) Isolation of methylcorypalline from embryo loti. J Chin Chem Soc (Taipei) 17:54–56 (CA 73:99072s)
23. Yang TH, Chen CM (1970) Alkaloids of *Nelumbo nucifera*. Alkaloids of the lotus embryo. J Chin Chem Soc (Taipei) 17:235–242 (CA 74:100254g)
24. Nishibe S, Tsukamoto H, Kinoshita H, Kitagawa S, Sakushima A (1986) Alkaloids from embryo of the seed of *Nelumbo nucifera*. J Nat Prod 49:548
25. Yang TH, Chen CM, Lu CS, Liao CL (1972) Alkaloids of lotus receptacle. J Chin Chem Soc (Taipei) 19:143–147 (CA 77:161937r)
26. Macko E, Douglas B, Weisbach JA, Waltz DT (1972) Pharmacology of nuciferine and related aporphines. Arch Int Pharmacodyn Ther 197:261–273
27. Burkman AM, Cannon JG (1972) Screening nornuciferine derivatives for apomorphine-like activity. J Pharm Sci 61:813–814
28. Smith RV, Sood SP (1971) In vitro metabolism of certain nornuciferine derivatives. J Pharm Sci 60:1654–1658
29. Ishida H, Umino T, Tsuji K, Kosuge T (1988) Studies on the antihemorrhagic substances in herbs classified as hemostatics in Chinese medicine. VIII. On the hemorrhagic principle in Nelumbinis receptaculum. Chem Pharm Bull (Tokyo) 36:4585–4587

Paeonia spp. 90

90.1 Introduction

Chishao, Radix Paeoniae rubra, is the dry roots of *Paeonia lactiflora* Pall. and *P. veitchii* Lynch (Ranunculaceae). It is officially listed in the Chinese Pharmacopoeia and collected in the spring and fall.

Baishao, Radix Paeoniae alba, is the dry roots of *P. lactiflora*, which are collected in the summer or fall, peeled after cooking in water, and dried.

Mudanpi, Cortex Moutan, is the dry root bark of *P. suffruticosa* Andr., which is collected in the fall.

Paeonia roots as well as moutan root bark are used in traditional Chinese medicine as analgesic, hemostyptic, and bacteriostatic agents.

90.2 Chemical Constituents

90.2.1 Chemical Constituents of *Paeonia lactiflora*

The most important constituent of the root of *P. lactiflora* is paeoniflorin (*90-1*), a glucoside of a monoterpene of the pinane series. The structure of paeoniflorin was elucidated by chemical and physicochemical methods [1]. Phylogenetically, plant species with higher paeoniflorin contents were mainly in the genus *Paeonia*, *P. lactiflora* and *P. veitchii* being the two with the highest contents [2]. The root of *P. lactiflora* contained 0.05%–5.8% paeoniflorin [3]. *Paenoia* roots from plants cultivated for 1 year contained more paeoniflorin than those cultivated for 3 years [4]. *Paeonia tenuifolia*, another species, contained the highest amount of paeoniflorin [5].

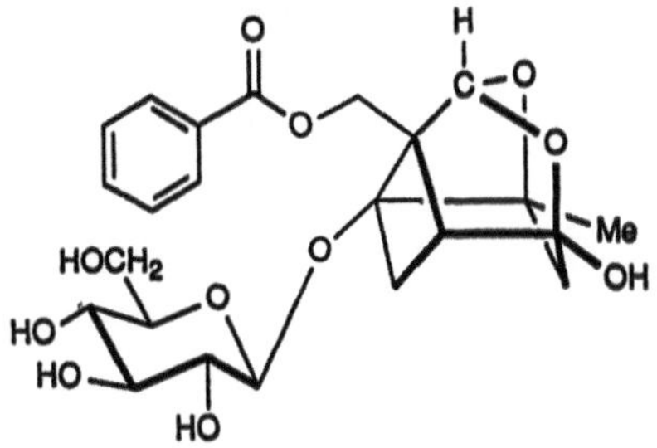

Paeoniflorin (*90-1*)

The structure of paeoniflorin, which involves a cage-like pinane skeleton, is unique among natural products. An X-ray crystallographic investigation, using a single crystal of the bromo derivative, was undertaken to confirm the structure and establish the absolute configuration [6]. Chromatographic separation of the methanolic extracts of the paeony roots, in addition to the major constituent paeoniflorin, demonstrated three minor principles named albiflorin (*90-2*), oxypaeoniflorin (*90-3*), and benzoylpaeoniflorin (*90-4*), which were isolated and structurally determined on the basis of paeoniflorin structure [6].

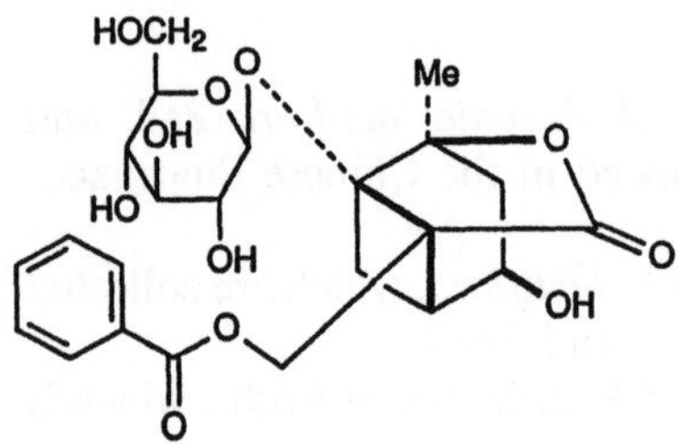

Albiflorin
(*90-2*)

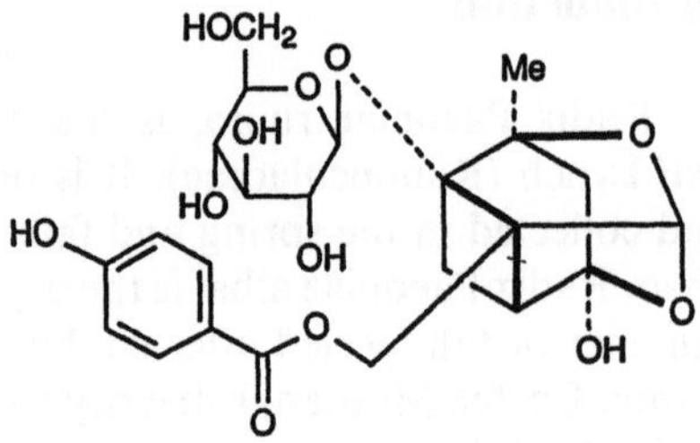

Oxypaeoniflorin
(*90-3*)

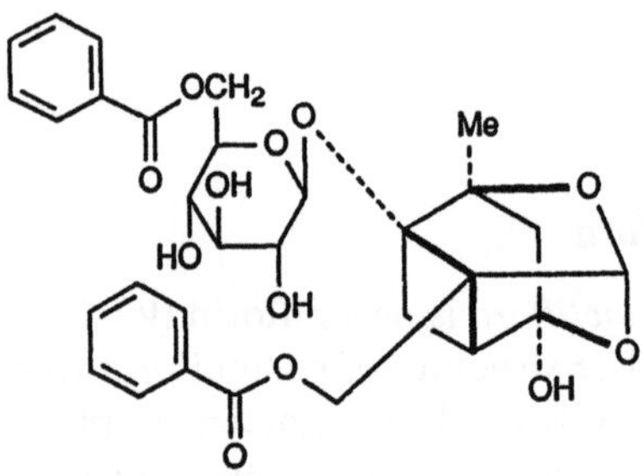

Benzoylpaeoniflorin
(*90-4*)

Furthermore, the monoterpenes paeoniflorigenone (*90-5*) [7, 8], and the paeonilactones A (*90-6*), B (*90-7*), and C (*90-8*) [9] were isolated from paeony roots.

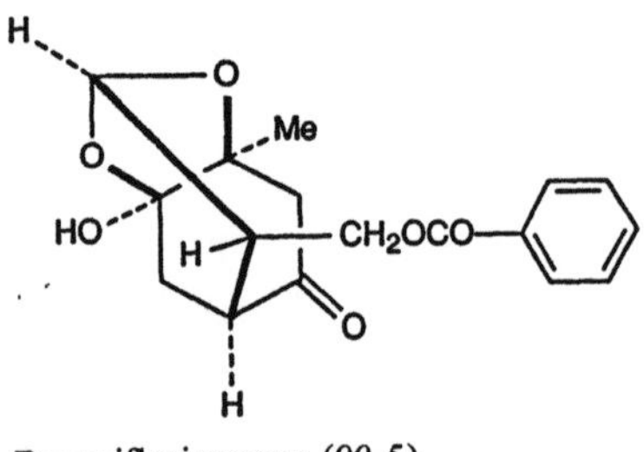

Paeoniflorigenone (*90-5*)

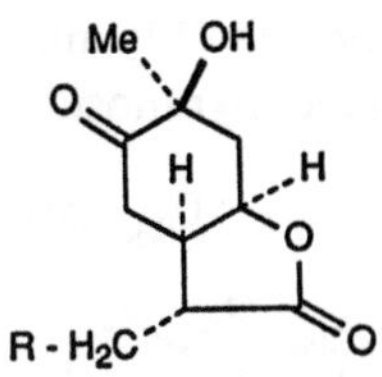

Paeonilactone A (*90-6*): R=H

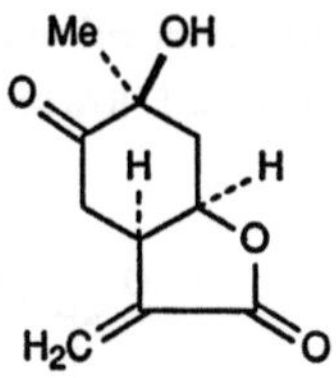

Paeonilactone B (*90-7*)

Paeonilactone C (*90-8*): R = 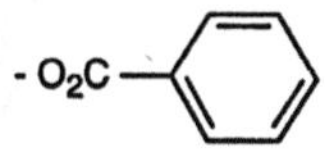

Paeoniflorigenone was not detected in a hot extract of paeony roots. Obviously, paeoniflorigenone decomposes during thermal treatment [10].

Two further monoterpenes isolated from the roots of *P. lactiflora* were *β*-pinen-10-yl vicianoside (*90-9*) and lactiflorin (*90-10*) [11, 12].

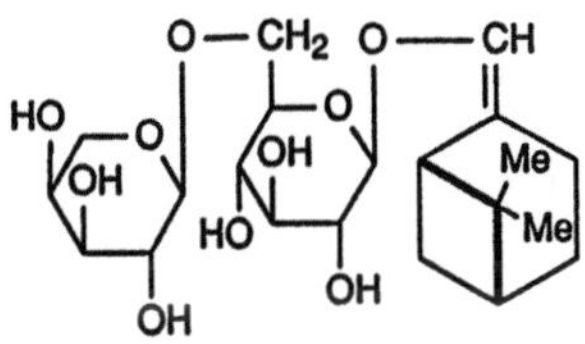

β-pinen-10-yl vicianoside (*90-9*)

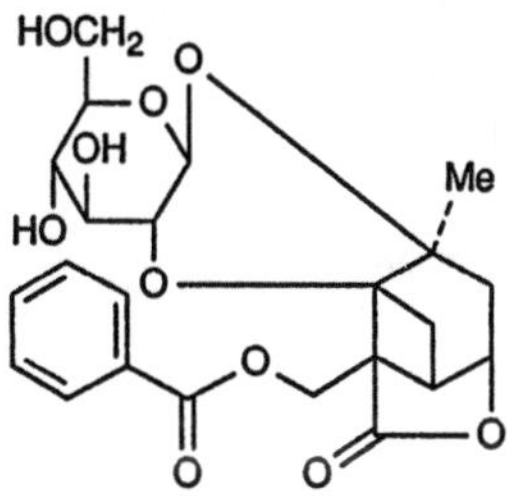

Lactiflorin (*90-10*)

Besides monoterpenes and monoterpene glycosides, benzoic acid and its esters [13] and a number of gallotannins [14] were isolated from paeony roots. From the flowers of *P. lactiflora*, phyrethrin and flavones astragalin (kaempferol-3-glucoside) and kaempferol-3,7-diglucoside [15] were isolated. Pyrethrin gives a 1:1 mixture of chrysanthemic acid (*90-11*) and chrysanthemum-dicarboxylic acid (*90-12*) [16] by alkaline hydrolysis.

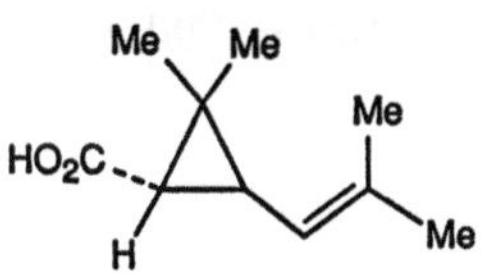

Chrysanthemic acid (*90-11*)

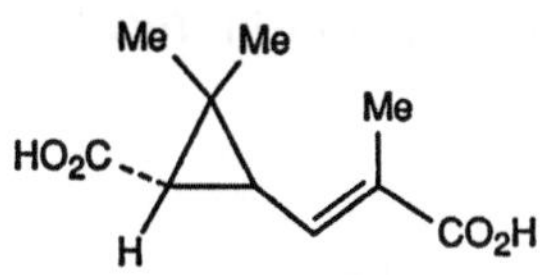

Chrysanthemumdicarboxylic acid (*90-12*)

90.2.2 Chemical Constituents of *Paeonia suffruticosa*

The most important constituent of the root bark of *P. suffruticosa* is paeonol (*90-13*) [17, 19]. Glycosides of paeonol were also isolated from moutan root bark.

Paeonoside (*90-14*) was characterized as paeonol-β-D-glucopyranoside [17] and paeonolide (*90-15*) as paeonol α-L-arabinopyranosyl-(1→6)-β-D-glucopyranoside [20, 21].

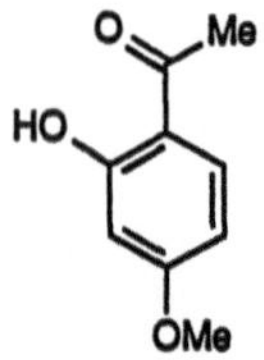
Paeonol (*90-13*)

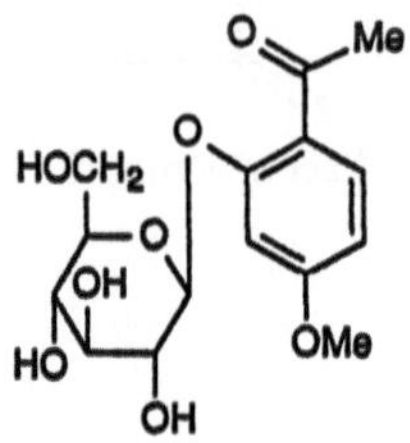
Paeonoside (*90-14*)

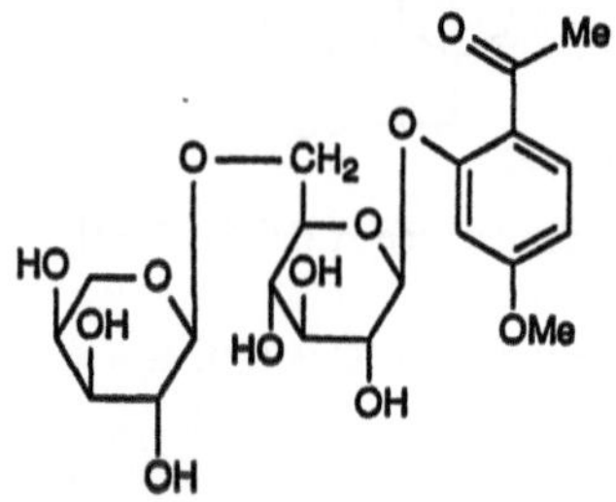
Paeonolide (*90-15*)

Paeonol D-apio-β-D-furanosyl(1→6)-β-D-glucopyranoside, named apiopaeonoside (*90-16*), was also isolated from the root of *P. suffruticosa* [22].

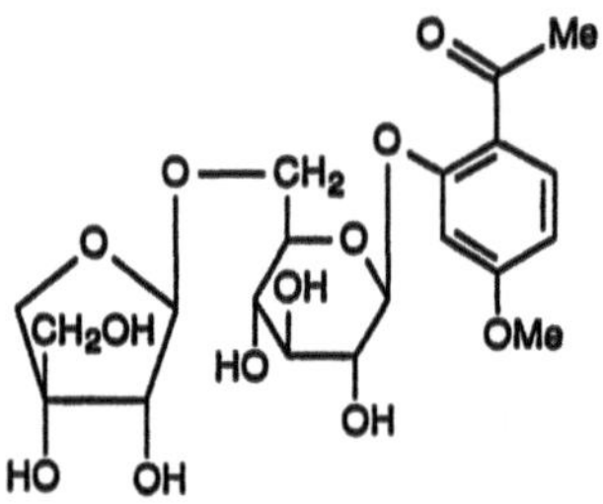
Apiopaeonoside (*90-16*)

Paeonol was found mostly in periderm, cambium, and neighboring tissues of the root and was also found in xylem [23]. Waste from processing of the herbal medicine was found to contain more than 1.9% paeonol and can thus be utilized for isolation of paeonol [24]. Paeonol in moutan root bark was proven to be formed via phenylalanine and cinnamic acid. Feeding of [^{14}C]phenylalanine and [^{14}C]cinnamic acid to the plant resulted in efficient incorporation of ^{14}C into the corresponding position of paeonol. On the other hand, neither acetate nor malonate were incorporated significantly into paeonol [25]. Moutan root bark also contained paeoniflorin, oxypaeoniflorin, and benzoylpaeoniflorin. These monoterpene glycosides existed mainly in the periderm and neighboring tissues of the root [26]. New monoterpene glucosides benzoyloxypaeoniflorin (*90-17*) [27] and galloylpaeoniflorin [28] were further isolated.

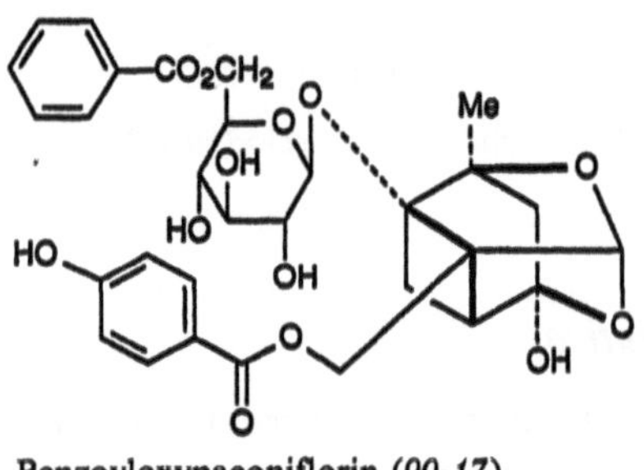
Benzoyloxypaeoniflorin (*90-17*)

Paeoniflorin occurs ubiquitously as a characteristic constituent in all *Paeonia* species, whereas paeonol and its glycoside were found only in the root of *P. suffruticosa* and its substitutes. This difference may be employed as a chemical criterion to distinguish *P. suffruticosa* from *P. lactiflora* [5].

An antiviral substance was isolated from the root bark of *P. suffruticosa* and was identified as 1,2,3,4,6-pentagalloylglucose [29]. The flowers of *P. suffruticosa* also contained astragalin [30].

90.3 Pharmacology

90.3.1 Pharmacology of Paeoniflorin

Paeoniflorin had depressant, antispasmodic, and antiinflammatory effects [30]. The acute toxicity of paeoniflorin was found to be very low. It exhibited a sedative activity in rats. Loss of righting reflex was obtained in rats by intravenous administration. Sleeping duration induced by hexobarbital was prolonged. It was further found that paeoniflorin inhibited the writhing symptoms in mice induced by intraperitoneal administration of acetic acid. It also showed weak hypothermic and weak anticonvulsive activities [32].

An antiinflammatory effect of paeoniflorin was demonstrated by significant inhibition of rat paw edema caused by carrageenin and also by a tendency to inhibit dextran- or chymotrypsin-induced edema and exudation of dye into the abdominal cavity of the mouse. Paeoniflorin also had a preventive effect on stress ulcer in the rat [33]. Paeoniflorin, desbenzoylpaeoniflorin, and paeonone (*90-18*), a monoterpene glucoside obtained by treatment of paeoniflorin with potassium carbonate in methanol [34], inhibited experimental contact hypersensitivities and passive cutaneous anaphylaxis reactions [35].

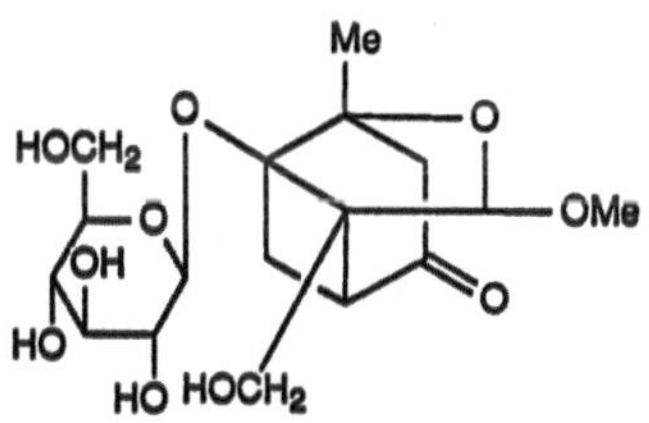

Paenone (*90-18*)

Further pharmacological studies showed that paeoniflorin exhibited a hypotensive effect in guinea pigs, possibly based on peripheral vasodilation. Paeoniflorin produced vasodilation of the coronary vessels and hind limb vessels of the dog. Relaxation and inhibition of movement and tonus of smooth muscle organs such as rat stomach or rat uterus were also observed [36]. Furthermore, paeoniflorin and benzoylpaeoniflorin inhibited blood platelet coagulation [37], and also showed inhibitory effects on plasminogen and plasmin [26].

90.3.2 Pharmacology of Paeonol

Paeonol, the main constituent of moutan root bark, was found to inhibit the growth of *Escherichia coli* and *Bacillus subtilis* at a 1:1500 dilution and that of *Staphylococcus aureus* and *Streptococcus faecalis* at a 1:2000 dilution [38].

Moutan root bark and its glycosidic and aqueous fractions showed antiinflammatory effects after oral administration in rats. The methanolic extract prevented disseminated intravascular coagulation induced by enterotoxin in the rat. The methanolic extract, the glycosidic fraction, and paeonol inhibited blood platelet aggregation [37, 39]; ADP- or collagen-induced human blood platelet aggregation was suppressed by the aqueous extract and by paeonol in a concentration-dependent fashion. The formation of thromboxane B_2 was also inhibited but the formation of 12-hydroxy-5,8,10,14-eicosatetraenoic acid from [^{14}C]arachidonate in human blood platelets was stimulated. In addition, paeonol also inhibited the formation of prostanoids such as prostaglandins and thromboxanes from [^{14}C]arachidonate in rat peritoneal macrophages. Thus, the antithrombotic and antiinflammatory actions of moutan root bark appeared to be due to the inhibitory effects of paeonol on prostanoid synthesis [40, 41]. However, although oral administration of paeonol to rats inhibited carrageenin-induced edema [42], other fractions without paeonol also showed strong inhibitory activity [43].

Paeonol administered orally or intraperitoneally to mice showed sedative effects on the CNS. The acute toxicity was low. A decrease in spontaneous motor activity and caffeine-induced hyperactivity and a loss of righting reflex were recognized. Analgesic activity against writhing symptoms induced by intraperitoneal administration of acetic acid and against tail pain induced by pressing were found in mice [44].

The extract of the root bark of *P. suffruticosa* and its main constituent paeonol showed antimutagenic activity, decreasing the frequency of mutations induced by 4-nitroquinoline-1-oxide in *Escherichia coli* WP2s [45].

References

1. Aimi N, Inaba M, Watanabe M, Shibata S (1969) Chemical studies on the oriental plant drugs. XXIII. Paeoniflorin, a glucoside of Chinese paeony root. Tetrahedron 25:1825–1838
2. Ho LY, Feng RZ, Xiao PG (1980) Studies of the medicinal plants of the family Ranunculaceae in China. IV. The occurrence of paeoniflorine in the genus *Paeonia*. Acta Pharm Sin 15:429–433
3. He LY (1983) Assay of paeoniflorin. Chin Pharm Bull 18:230–231
4. Shimizu M, Hashimoto T, Ishikawa S, Kurosaki F, Morita N (1979) Analysis of constituents in crude drugs by high-speed liquid chromatography. I. Quantitative analysis of paeoniflorin in peony roots. Yakugaku Zasshi 99:432–435
5. Yu J, Lang HY, Xiao PG (1985) The occurrence of paeniflorins and paeonols in Paeoniaceae. Acta Pharm Sin 20:229–234
6. Kaneda M, Iitaka Y, Shibata S (1972) Chemical studies on the oriental plant drugs. XXXIII. Absolute structure of paeoniflorin, albiflorin, oxypaeoniflorin and benzoylpaeoniflorin isolated from Chinese peony root. Tetrahedron 28:4309–4317
7. Shimizu M, Hayashi T, Morita N, Kiuchi F, Noguchi H, Iitaka Y, Sankawa U (1983) The structure of paeoniflorigenone, a new monoterpene isolated from *Paeonia* radix. Chem Pharm Bull (Tokyo) 31:577–583
8. Shimizu M, Hayashi T, Morita N, Kimura I, Kimura M, Kiuchi F, Noguchi H, Iitaka Y, Sankawa U (1981) Paeoniflorigenone, a new monoterpene from paeony roots. Tetrahedron Lett 22:3069–3070

9. Hayashi T, Shinbo T, Shimizu M, Arisawa M, Morita N, Kimura M, Matsuda S, Kikuchi T (1985) Paeonilactone-A, -B and -C, new monoterpenoids from paeony root. Tetrahedron Lett 26:3699–3702
10. Hayashi T, Kurosawa S, Shimizu M, Morita N (1985) Studies on muscle relaxants in Paeoniae radix. Effect of heat on stability of paeoniflorigenone. Shoyakugaku Zasshi 39:214–217
11. Lang HY, Li SZ, Liang XT (1983) Studies on the chemical constituents of *Paeonia lactiflora* Pall. roots, a traditional Chinese medicine. Acta Pharm Sin 18:551–552
12. Lang HY, Li SZ, McCabe T, Clardy J (1984) A new monoterpene glycoside of *Paeonia lactiflora.* Planta Med 50:501–504
13. Ling CS, Lee C, Wang MC, Li C (1956) Chemical constituents of two commercial varieties of paeony root, Pai-Shao and Cheh-shao. Acta Pharm Sin 4:17–24
14. Nishizawa M, Yamagishi T, Nonaka G, Nishioka I (1980) Structure of gallotannins in Paeoniae radix. Chem Pharm Bull (Tokyo) 28:2850–2852
15. Egger K (1961) Astragalin and kaempferol 3,7-diglucoside, the flavonol glycosides of white paeonies. Z Naturforsch [B] 16:430–431
16. Chmielewska I, Kasprzyk Z (1962) Pyrethrin and gallotannin in the flowers of paeony. Nature 196:776
17. Kariyone T, Takahashi M, Takaishi K (1956) Glycosides. VII. Components of *Paeonia suffruticosa.* 1. J Pharm Soc Jpn 76:917–919
18. Chen WD (1980) Studies on the coupled colorimetric determination of paeonol. Chin Pharm Bull 15:4–7
19. Xu LX, Liu AR (1980) Determination of active constituents in Chinese herb drugs. I. Colormetric titration of paeonol in *Paeonia suffruticosa* Andr. and *Pycnostelma paniculatum* (Bunge) K. schum. Acta Pharm Sin 15:184–188
20. Kariyone T, Takahashi M, Takaishi K (1956) Glycosides. VIII. Components of *Paeonia suffruticosa.* 2. J Pharm Soc Jpn 76:920–921
21. Takahashi M, Osawa K, Sato T, Abe K, Nakaguchi O (1970) Components of *Paeonia suffruticosa.* III. Structure of paeonolide. Tohoku Yakka Daigaku Kenyu Nempo 25-35 (CA 75:148478f)
22. Yu J, Lang HY, Xiao PG (1986) A new compound, apiopaenoside, isolated from the root of *Paeonia suffruticosa.* Acta Pharm Sin 21:191–197
23. Tani T, Katsuki T, Kubo M, Arichi S, Kitagawa I (1980) Studies of moutan cortex. IV. Distribution of paeonol in the root of *Paeonia moutan* from Nara Prefecture. Shoyakugaku Zasshi 34:292–298
24. Shang MY, Kang YD, Ke GT, Liu JG, Hu YQ, Cui WJ (1982) Paeonol in waste from *Paeonia suffruticosa* root bark processing. Bull Chin Mater Med 7:27–28
25. Yamasaki K, Tamaki T, Sankawa U, Shibata S (1976) Biosynthesis of paeonol. Yakugaku Zasshi 96:1103–1106
26. Tani T, Katsuki T, Matsuda H, Kubo M, Arichi S, Yoshikawa M, Kitagawa I (1980) Studies of moutan cortex. V. Distribution of monoterpene glucosides in the root of *Paeonia moutan* from Nara Prefecture. Shoyakugaku Zasshi 34:299–305
27. Kitagawan I, Yoshikawa M, Tsunaga K, Tani T (1979) Studies on Moutan cortex. II. On the chemical constituents. Shoyakugaku Zasshi 33:171–177
28. Kang SS, Shim KH, Chi HJ (1989) Galloylpaenoflorin, a new monoterpene glycoside from peony roots. Korean J Pharmacogn 20:48–49 (CA III:160062K)
29. Takechi M, Tanaka Y (1982) Antiviral substances from the root of *Paeonia* species. Planta Med 45:252–253
30. Egger K, Keil M (1969) Flavone glycosides in the flowers of *Paeonia arborea* and *P. suffruticosa.* Planta 88:154–156
31. Harada M (1969) Pharmacological studies on herb peony root. IV. Analysis of therapeutic effects of peony- and licorice-containing common prescriptions in Chinese medicine and comparison with effects of experimental pharmacological tests. Yakugaku Zasshi 89:899–908
32. Takagi K, Harada M (1969) Pharmacological studies on herb peony root. I. Central effects of peoniflorin and combined effects with licorice component F_M 100. Yakugaku Zasshi 89: 879–886
33. Takagi K, Harada M (1969) Pharmacological studies on herb peony roots. II. Antiinflammatory effect, inhibitory effect on gastric juice secretion, preventive effect on stress ulcer, antidiuretic effect of paeniflorin and combined effects with licorice component F_M 100. Yakugaku Zasshi 89:887–892

34. Institute for Production Development Science (1983) Paeonone. Jpn Kokai Tokkyo Koho JP 5818,398 (8318,398) (CA 99:16561p)
35. Yamahara J, Yamada T, Kimura H, Sawada T, Fujimura H (1982) Biologically active principles of crude drugs. II. Antiallergic principles in "Shoseiryu-To", antiinflammatory properties of paeoniflorin and its derivatives. J Pharmacobiodyn 5:921–929
36. Takagi K, Harada M (1969) Pharmacological studies on herb peony root. III. Effects of peoniflorin on circulatory and respiratory systems and isolated organs. Yakugaku Zasshi 89:893–898
37. Tani T, Katsuki T, Kosoto H, Arichi S, Kubo M, Matsuda H, Kimura Y, Kitagawa I, Yoshikawa M (1981) Studies on the medicinal history, chemical constituents and pharmacological effects of Moutan Cortex. Wakayaku Shinpojumu 14:86–92 (CA 97:426w)
38. Ohta T, Mihashi S, Saheki Y, Wakabayashi K (1961) Bacteriostatic effect of paeonol, the main component of paeony root, on some bacteria causing appendicitis. Tokyo Yakka Daigaku Kenkyu Nempo 79–81 (CA 58:1824f)
39. Shi L, Fan PS, Fang JX, Han ZX (1988) Inhibitory effects of paeonol on experimental atherosclerosis and platelet aggregation in rabbit. Acta Pharmacol Sin 9:555–558
40. Hirai A, Tamura Y, Yoshida H (1985) Mechanism of antithrombotic and anti-inflammatory actions of moutan cortex. Studies on arachidonic acid cascade. Wakan Iyaku Gakkaishi 2:63–65 (CA 104:45497z)
41. Hirai A, Terano T, Hamazaki T, Sajiki J, Saito H, Tahara K, Tamura Y, Kumagai A (1983) Studies on the mechanism of antiaggregatory effect of Moutan cortex. Thromb Res 31:29–40
42. Haraka M, Yamashita A, Aburada M (1972) Pharmacological studies on the root bark of *Paeonia moutan*. II. Antiinflammatory effect, preventive effect on stress-induced gastric erosion, inhibitory effect on gastric juice secretion and other effects on paeonol. Yakugaku Zasshi 92:750–756
43. Arichi S, Kubo M, Matsuda H, Tani T, Tsunaga K, Yoshikawa M, Kitagawa I (1979) Studies on Moutan cortex. III. On antiinflammatory activities. I. Shoyakugaku Zasshi 33:178–184
44. Harada M, Yamashita A (1969) Pharmacological studies on the root bark of *Paeonia moutan*. I. Central effects of paeonol. Yakugaku Zasshi 89:1205–1211
45. Fukuhara Y, Yoshida D (1987) Paeonol: a bio-anti-mutagen isolated from a crude drug, Moutan cortex. Agric Biol Chem 51:1441–1442

Panax ginseng C.A. Mey. 91

91.1 Introduction

Renshen, Radix Ginseng, is the dry root of *Panax ginseng* C.A. Mey. (Araliaceae), a worldwide well-known traditional Chinese medicine with the popular name "ginseng". The wild-growing or cultivated ginseng root, which is collected in the fall, is officially listed in the Chinese Pharmacopoeia and used as a tonic.

Ginseng was formerly a wild plant growing in the northeastern region of China. Wild Ginseng, the mountain ginseng, is called "Shanshen" in Chinese and should be dried in the sun. Nowadays, wild ginseng is rarely available, and almost all of the commercially available ginseng root is cultivated in the northeastern district and other regions of China, where the growing conditions for ginseng plant are favorable. Cultivated ginseng, the garden ginseng, is called "Yuanshen" in Chinese and should be dried either in the sun or after steaming. The steamed root has a caramel color and is also called "red ginseng."

91.2 Chemical Constituents

Although ginseng has been used for a long time and may be the best-known traditional Chinese medicine, isolation and characterization of the chemical constituents only became successful during the 1960s. The major constituents of ginseng are the saponins. The first sapogenin of ginseng saponins isolated and structurally elucidated was oleanolic acid. It was obtained by hydrolysis of a 50% ethanolic extract of ginseng with sulfuric acid [1]. Then, another compound named panaxadiol (*91-1*) was isolated from the crude ginseng saponin mixture by hydrolysis with dilute mineral acid in boiling aqueous ethanol. It was supposed to be a sapogenin [2], but was later shown to be formed as an artifact during acid hydrolysis. Panaxadiol was recovered unchanged after treatment with mineral acid, whereas a chlorine-containing compound (*91-2*), described first by Katake [3], could be obtained directly by hydrolysis of the crude ginseng saponins with concentrated hydrochloric acid [4].

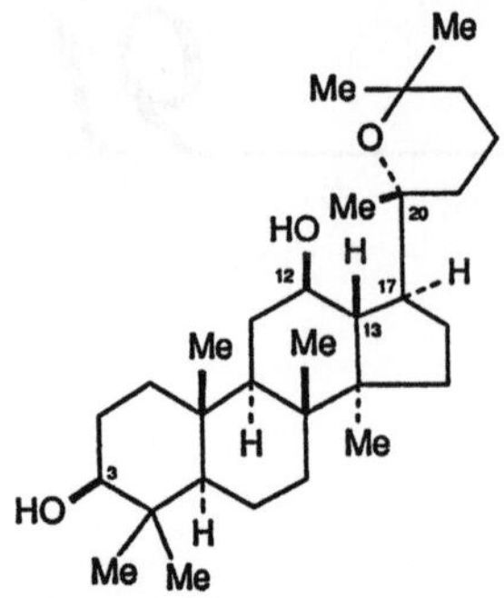

Panaxadiol (*91-1*)

Dehydrochlorination of the chlorine-containing compound yielded an unsaturated derivative, which gave panaxadiol by treatment with dilute mineral acid in boiling aqueous ethanol [4]. The structure of the new compound was determined as 12*β*-hydroxydammarenediol I, named protopanaxadiol [5–7]. However, further investigation of the acid-catalyzed isomerization of dammarane-type triterpenes and subsequent study of the mild hydrolysis of the purified ginseng saponins led to the conclusion that the sapogenin was not 12*β*-hydroxydammarenediol I but its C-20 epimer 12*β*-hydroxydammarenediol II or 20(*S*)-protopanaxadiol (*91-3*). 12*β*-Hydroxydammarenediol I was then designated as 20(R)-protopanaxadiol (*91-4*) [8, 9].

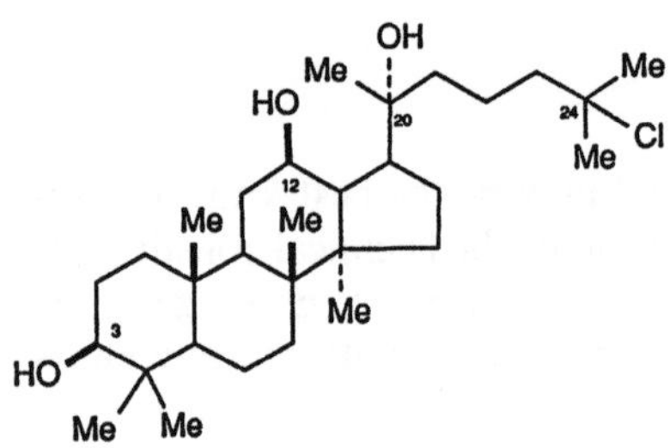

Chlorine-containing compound (*91-2*)

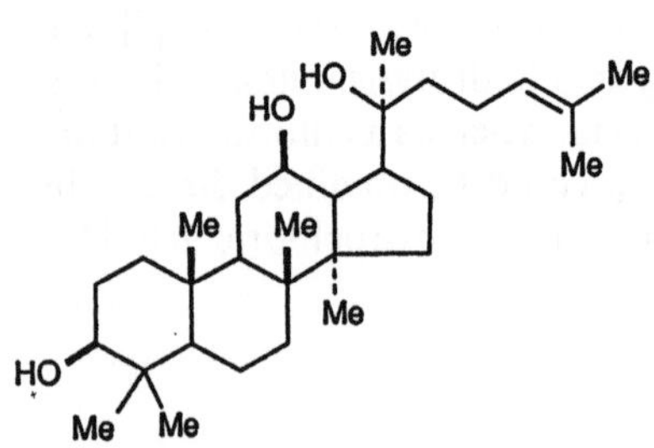

20(S)-Protopanaxadiol
(12*β*-Hydroxydammarene-
diol II, *91-3*)

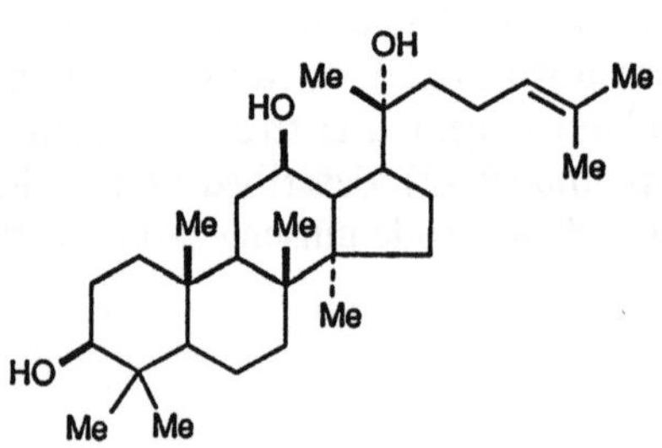

20(R)-Protopanaxadiol
(12*β*-Hydroxydammarene-
diol I, *91-4*)

On the other hand, the pure ginsenoside Rg_1 on hydrolysis with dilute mineral acid yielded glucose and a crystalline compound named panaxatriol (*91-5*) [10]. The structure of panaxatriol was established as 6α-hydroxypanaxadiol by comparison of its spectral data with that of panaxadiol. The crystal structures of panaxadiol and panaxatriol monohydrate were also determined [11].

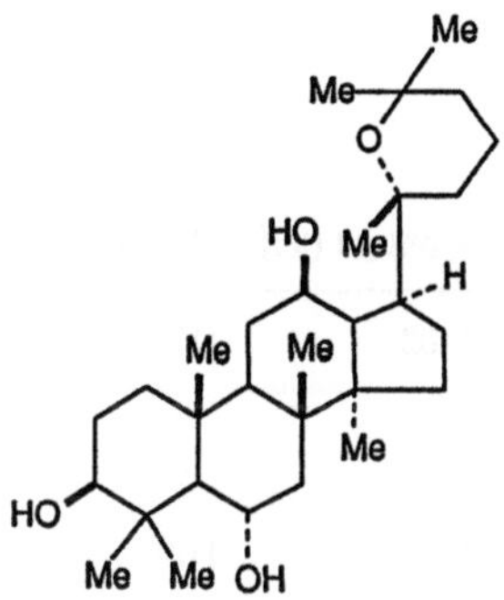

Panaxatriol (*91-5*)

Smith degradation of ginsenoside Rg_1 gave a sapogenin, which afforded panaxatriol on acid treatment with dilute mineral acid. Further studies on the hydrochlorination and dehydrochlorination and on the isomerization of the sapogenin revealed that the sapogenin of ginsenoside Rg_1 can be designated as 20(*S*)-protopanaxatriol (*91-6*).

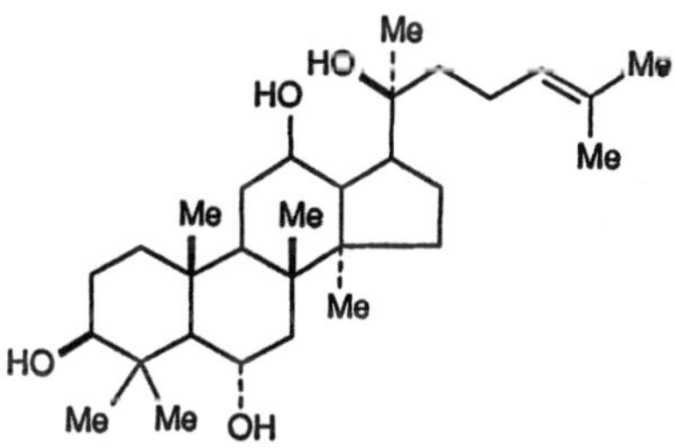

20(*S*)-protopanaxatriol (*91-6*)

Thus, the three sapogenins 20(*S*)-protopanaxadiol, 20(*S*)-protopanaxatriol, and oleanolic acid were identified from the ginseng saponins. Protopanaxadiol and protopanaxatriol are both dammarane (*91-7*) derivatives.

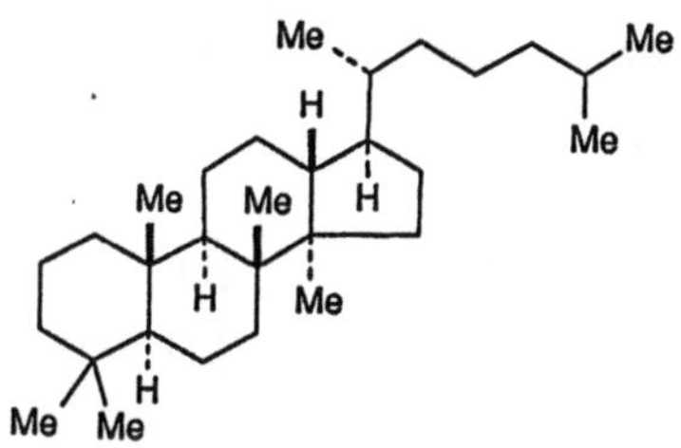

Dammarane (*91-7*)

Table 91.1. Protopanaxadiol-type ginsenosides from ginseng root

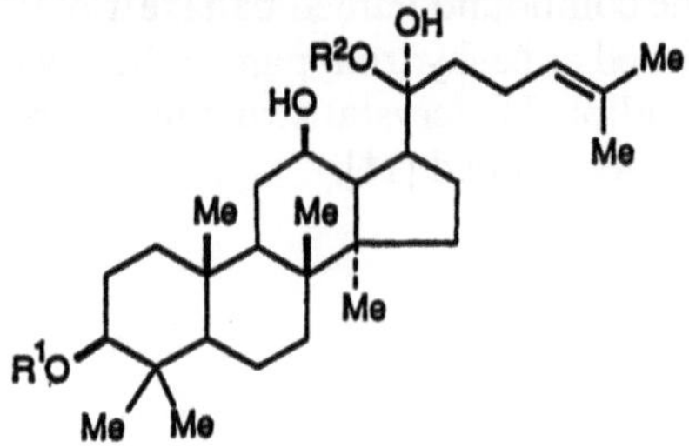

Ginsenoside	R^1	R^2	Content (%)	Reference
Ra_1	*O*-β-D-Glucopyranosyl-(1→2)-β-D-glucopyranosyl	*O*-β-D-Xylopyranosyl-(1→4)-α-L-arabinopyranosyl-(1→6)-*O*-β-D-glucopyranosyl	0.02	[12; 13]
Ra_2	*O*-β-D-Glucopyranosyl-(1→2)-β-D-glucopyranosyl	*O*-β-D-Xylopyranosyl-(1→2)-α-L-arabinofuranosyl-(1→6)-*O*-β-D-glucopyranosyl	0.03	[12]
Ra_3	*O*-β-D-Glucopyranosyl-(1→2)-β-D-glucopyranosyl	*O*-β-D-Xylopyranosyl-(1→3)-β-D-glucopyranosyl-(1→6)-*O*-β-D-glucopyranosyl	0.005	[14]
Rb_1	*O*-β-D-Glucopyranosyl-(1→2)-β-D-glucopyranosyl	*O*-β-D-Glucopyranosyl-(1→6)-β-D-glucopyranosyl	0.37–0.4	[15, 16]
Rb_2	*O*-β-D-Glucopyranosyl-(1→2)-β-D-glucopyranosyl	*O*-α-L-Arabinopyranosyl-(1→6)-β-D-glucopyranosyl	0.18–0.21	[15, 16]
Rb_3	*O*-β-D-Glucopyranosyl-(1→2)-β-D-glucopyranosyl	*O*-β-D-Xylopyranosyl-(1→6)-β-D-glucopyranosyl	0.005–0.01	[16, 17]
Rc	*O*-β-D-Glucopyranosyl-(1→2)-β-D-glucopyranosyl	*O*-α-L-Arabinofuranosyl-(1→6)-β-D-glucopyranosyl	0.13–0.15	[15, 16, 18]
Rd	*O*-β-D-Glucopyranosyl-(1→2)-β-D-glucopyranosyl	*O*-β-D-Glucopyranosyl	0.13–0.15	[15, 16]
Rg_3	*O*-β-D-Glucopyranosyl-(1→2)-β-D-glucopyranosyl	H	0.003–0.014	[15, 16]
Notoginsenoside R_4	*O*-β-D-Glucopyranosyl-(1→2)-β-D-glucopyranosyl	*O*-β-D-Xylopyranosyl-(1→6)-β-D-glucopyranosyl-(1→6)-β-D-glucopyranosyl	0.002	[14]
Rh_2	*O*-β-D-Glucopyranosyl-	H	0.001	[19]
Rs_1	6-*O*-Acetyl-β-D-glucopyranosyl-(1→2)-β-D-glucopyranosyl-	*O*-α-L-Arabinopyranosyl-(1→6)-β-D-glucopyranosyl	0.008	[16]

Table 91.1. (continued)

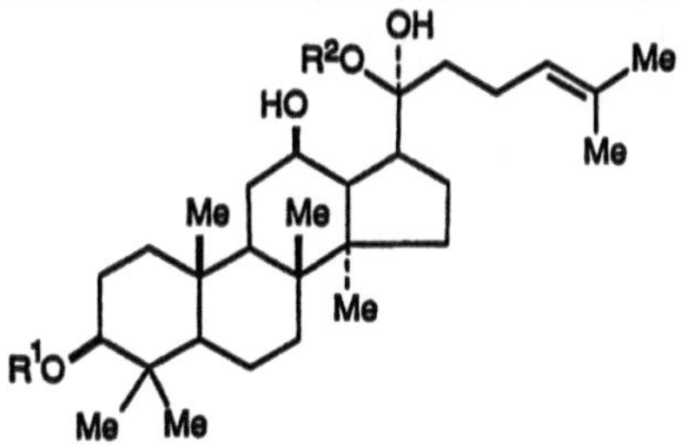

Ginsenoside	R^1	R^2	Content (%)	Reference
Rs_2	6-*O*-Acetyl-β-D-glucopyranosyl-(1→2)-β-D-glucopyranosyl-	*O*-α-L-Arabinofuranosyl-(1→6)-β-D-glucopyranosyl	0.01	[16]
Quinquenoside R_1	6-*O*-Acetyl-β-D-glucopyranosyl-(1→2)-β-D-glucopyranosyl-	*O*-β-D-Glucopyranosyl-(1→6)-β-D-glucopyranosyl	0.002–0.015	[16, 20]
Malonylginsenoside Rb_1	6-*O*-Malonyl-β-D-glucopyranosyl-(1→2)-β-D-glucopyranosyl	*O*-β-D-Glucopyranosyl-(1→6)-β-D-glucopyranosyl	Trace–0.82	[21]
Malonylginsenoside Rb_2	6-*O*-Malonyl-β-D-glucopyranosyl-(1→2)-β-D-glucopyranosyl	*O*-α-L-Arabinopyranosyl-(1→6)-β-D-glucopyranosyl	Trace–0.41	[21]
Malonylginsenoside Rc	6-*O*-Malonyl-β-D-glucopyranosyl-(1→2)-β-D-glucopyranosyl	*O*-α-L-Arabinofuranosyl-(1→6)-β-D-glucopyranosyl	Trace–0.30	[21]
Malonylginsenoside Rd	6-*O*-Malonyl-β-D-glucopyranosyl-(1→2)-β-D-glucopyranosyl	*O*-β-D-Glucopyranosyl	Trace–0.12	[21]

Table 91.2. Protopanaxatriol-type ginsenosides from ginseng root

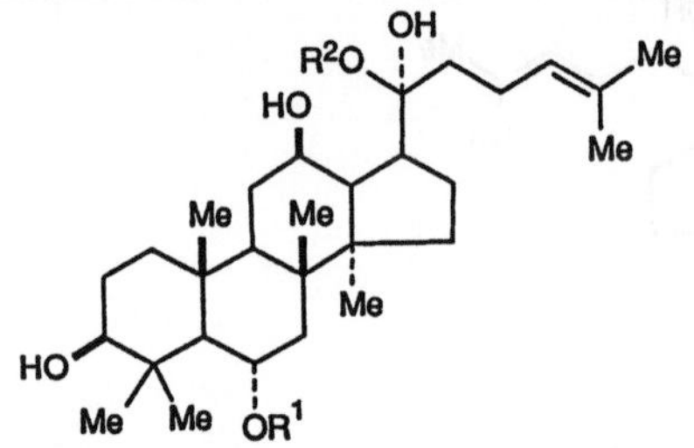

Ginsenoside	R^1	R^2	Content (%)	Reference
Re	*O*-α-L-Rhamnopyranosyl-(1→2)-β-D-gluco-pyranosyl-	*O*-β-D-Glucopyranosyl	0.15–0.20	[16, 22]
Rf	*O*-β-D-Glucopyranosyl-(1→2)-β-D-gluco-pyranosyl-	H	0.05	[16, 22]
20-Gluco-ginsenoside Rf	*O*-β-D-Glucopyranosyl-(1→2)-β-D-gluco-pyranosyl-	*O*-β-D-Glucopyranosyl	0.005	[16, 17]
Rg_1	*O*-β-D-Glucopyranosyl-	*O*-β-D-Glucopyranosyl	0.21	[16, 23–25]
Rg_2	*O*-α-L-Rhamnopyranosyl-(1→2)-β-D-gluco-pyranosyl-	H	0.01–0.02	[16, 22]
20(R) Rg_2			0.003	[19]
Rh_1	*O*-β-D-Glucopyranosyl-	H	0.0015–0.0023	
20(R) Rh_1			0.007	[19]
Notogin-senoside R_1	*O*-β-D-Xylopyranosyl-(1→2)-β-D-gluco-pyranosyl-	*O*-β-D-Glucopyranosyl	0.002–0.007	[16]

Since then, a number of ginseng saponins have been isolated and structurally determined. They may be divided into three groups, depending on their aglycones. Ginseng saponins named ginsenosides isolated and identified to date are listed in Tables 91.1 and 92.2. Thus, ginsenosides Rc, Rg_1, and Ro are representatives of the ginseng root saponins of the protopanaxadiol, protopanaxatriol, and oleanolic acid type, respectively. They are used as reference substances for qualitative determination of ginseng root in the Chinese Pharmacopoeia. Usually, ginseng saponins may be extracted with methanol or aqueous methanol at room temperature or under reflux. The extract is then suspended in water, washed with ether, and extracted with *n*-butanol saturated with water. The representative structures of ginsenosides Ra_1, Rc, Rs_1, Rg_1, and malonylginsenoside Rb_1 are demonstrated below:

Ginsenoside Ra_1 ((3β,12β)-12-hydroxy-20-[(*O*-β-D-xylopyranosyl-(1 → 4)-α-L-arabinopyranosyl-(1 → 6)-β-D-glucopyranosyl)oxy]dammar-24-en-3-yl 2-*O*-β-D-glucopyranosyl-β-D-glucopyranoside) (*91-8*)

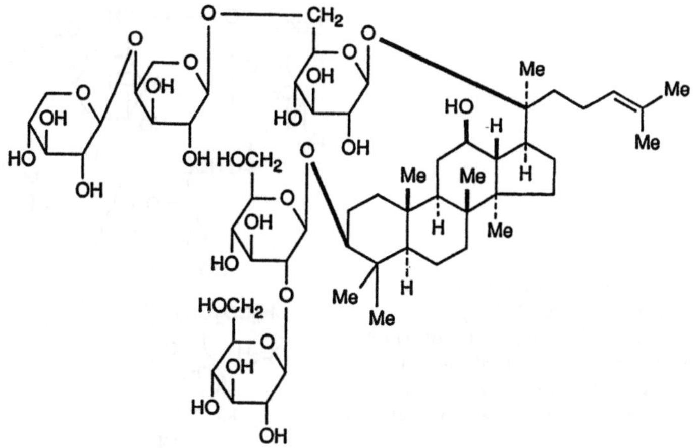

Ginsenoside Rc ((3β,12β)-20-[(6-*O*-α-L-arabinofuranosyl-β-D-glucopyranosyl)oxy]-12-hydroxydammar-24-en-3-yl 2-*O*-β-D-glucopyranosyl-β-D-glucopyranoside) (*91-9*)

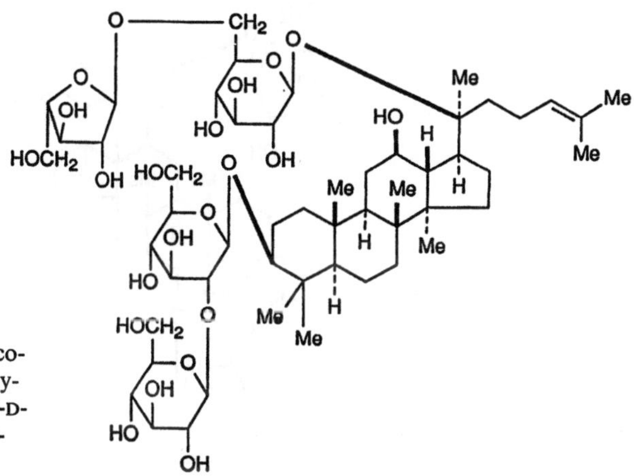

Ginsenoside Rs_1 ((3β,12β)-20-[(6-*O*-α-L-arabinopyranosyl-β-D-glucopyranosyl)oxy]-12-hydroxydammar-24-en-3-yl 2-*O*-(6-*O*-acetyl-β-D-glucopyranosyl)-β-D-glucopyranoside (*91-10*)

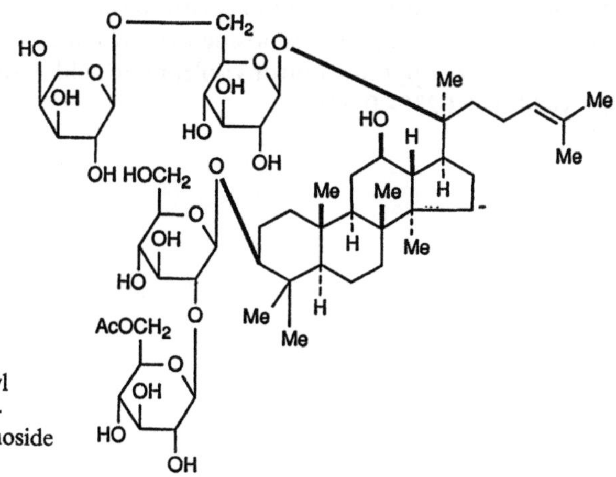

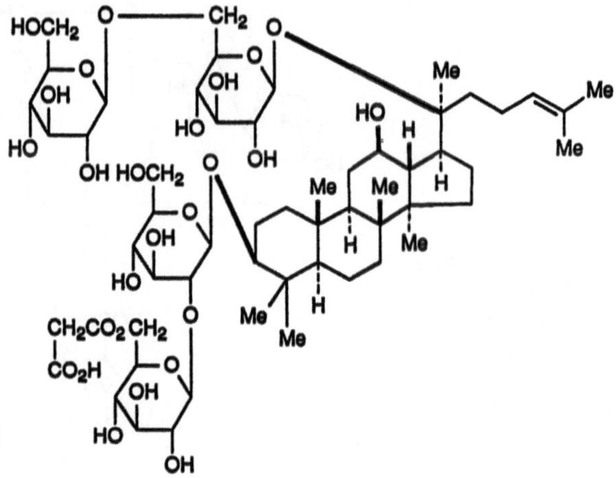

Malonylginsenoside Rb_1
((3β,12β)-20-[(6-*O*-β-D-gluco-pyranosyl-β-D-glucopyranosyl)oxy]-12-hydroxydammar-24-en-3-yl 2-*O*-[6-*O*-(carboxy-acetyl)-β-D-glucopyranosyl]-β-D-glucopyranoside (*91-11*)

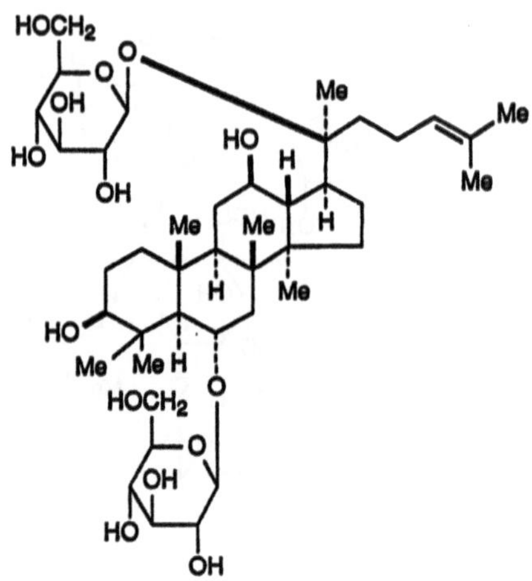

Ginsenoside Rg_1
((3β,6α,12β)-3,12-dihydroxy-dammar-24-ene-6,20-diyl bis-*O*-β-D-glucopyranoside) (*91-12*)

The only oleanolic acid-type saponin identified in the root of *P. ginseng* is the ginsenoside Ro (*91-13*) [15, 25]. Ginsenoside Ro is chemically 3β-*O*-β-D-Glucopyranosyl-(1→2)-*O*-β-D-glucopyranosiduronyl-oleanolic acid 28-β-D-glucopyranosyl ester. It has a yield of 0.02%–0.04%.

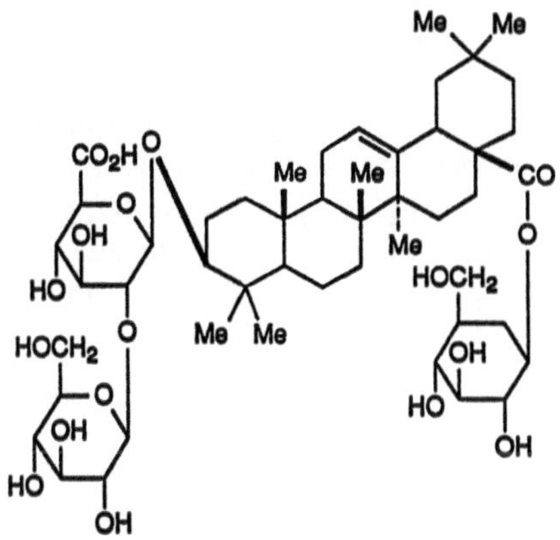

Ginsenoside Ro (*91-13*)

The butanol-soluble fraction is referred to as total saponin or crude saponin fraction. Two typical thin-layer chromatograms of saponin fraction of ginseng root compared with pure ginsenosides are shown Fig. 91.1 [26].

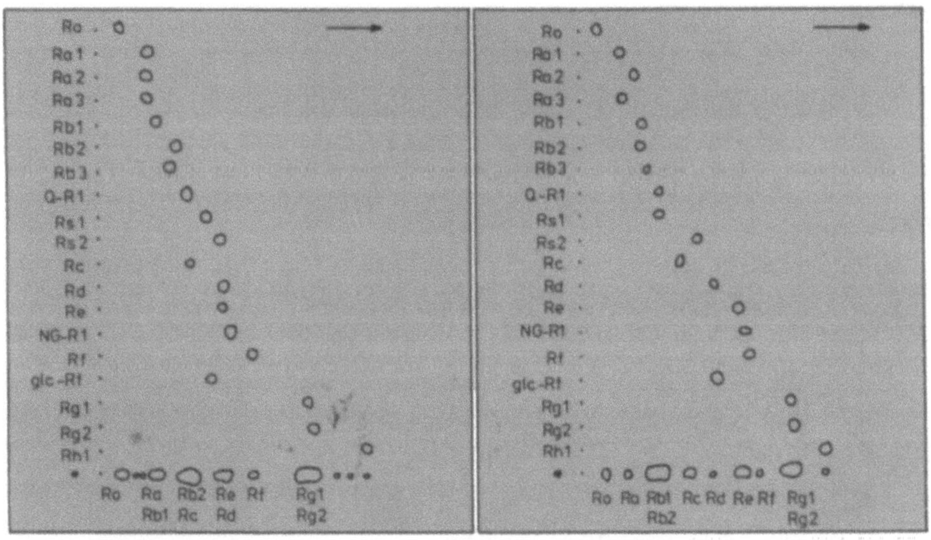

Fig. 91.1. Thin-layer chromatograms of the saponins of ginseng root [25]

Almost all the ginsenosides isolated from the ginseng root are also found in red ginseng [16, 19, 27, 28]. But some ginsenosides such as 20(R)-ginsenoside Rg_2, ginsenoside Rg_3, 20(R)-ginsenoside Rh_1, and ginsenoside Rh_2 are characteristic saponins found in red ginseng. They are considered to be degradation products by heating and hydrolysis during the processing of red ginseng [19]. The ginsenoside composition of the ginseng rhizome is similar to that of the ginseng root [18, 29–31].

However, the ginsenoside content was found to be higher in the root than in the rhizome.

The saponin constituents of the aboveground parts of *P. ginseng*, including stems and leaves, flowers and buds, and fruits have also been investigated. Besides ginsenosides Rb_1, Rb_2, Rc, Rd, Re, Rg_1, Rg_2, Rg_3, and 20-glucoginsenoside Rf, ginsenosides F_1, F_2 and F_3 were isolated from the leaves [32–37]. Ginsenoside F_2 (*91-14*) is a protopanaxadiol-type saponin, whereas ginsenosides F_1 (*91-15*) and F_3 (*91-16*) are saponins of the protopanaxatriol-type [32, 33].

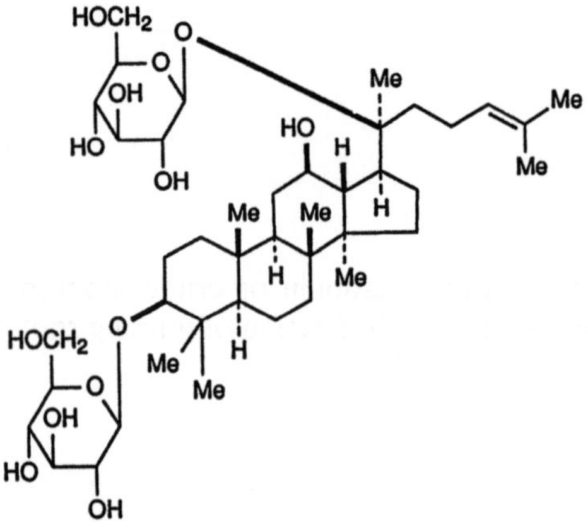

Ginsenoside F_2
(protopanaxadiol-3,20-bis-
O-*β*-D-glucopyranoside)
(*91-14*)

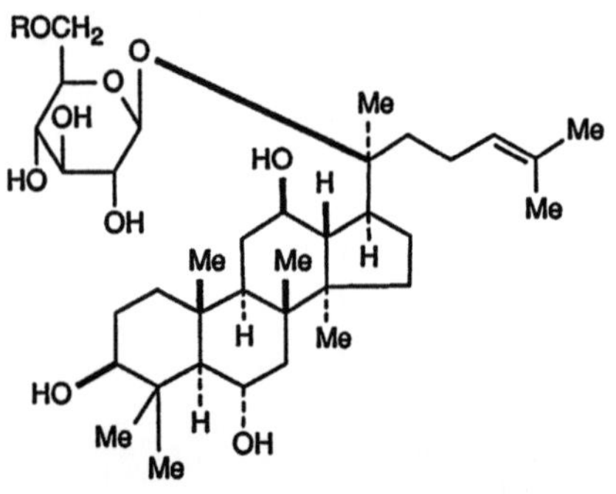

Ginsenoside F_1 (protopanaxatriol-20-*O*-*β*-D-glucopyranoside) (*91-15*): R=H
Ginsenoside F_3 (protopanaxatriol-20-*O*-α-L-arabinopyranosyl-(1 → 6)-*O*-*β*-D-glucopyranoside) (*91-16*): R=

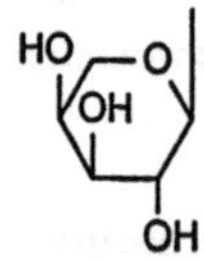

From flowers and flower buds of *P. ginseng*, the isolation of ginsenosides Rb_1, Rb_2, Rb_3, Rc, Rd, Re, Rg_1, Rg_2, and 20-glucoginsenoside Rf was reported [32, 37–41], and in fruits ginsenosides Rb_2, Rc, Rd, Re, and Rg_1 were found [37].

Quantitative determination of the saponin content in different parts of *P. ginseng* showed that the contents were in the following order: flower bud > fruit > stem > leaf > root > seed. Leaves appeared to be a good source for the extraction of ginseng saponins. After defatting with petroleum, leaves are extracted with 90% ethanol. The supernatant of the ethanolic extract is dried under reduced pressure to give the saponin fraction [42].

The processing methods of ginseng root also influence the saponin content. Quantitative determination of the total ginsenoside content in processed ginseng root showed that the total ginsenoside contents were the highest in freeze-dried ginseng roots, followed by roots dried in the sun and steamed roots, and were the highest in the fibrous root, followed by the lateral root and the main root with a total ginsenoside content of 9.7%, 6.4%, and 3.3%, respectively [43]. The saponin and ginsenoside contents were higher in wild than in cultivated ginseng [44, 45].

The quantitative determination of ginsenoside could be carried out by HPLC [46–48] TLC [49, 50], or colorimetric method [51–53]. Colorimetric determinations after hydrolysis and reaction with vanillin-sulphuric acid or with vanillin-perchloric acid in acetic acid enable a rapid comparison of different ginseng products to be made but they are not specific because only the total amount of the formed panaxadiol and panaxatriol can be determined. TLC or HPLC methods are more specific and can be used to determine the individual ginsenoside [54, 55].

The saponin content of some commercial ginseng preparations was also reported, e.g., Renshenlu, steam distillate of ginseng, contained 58 mg ginsenosides/l [50]; Renshenjin, aqueous solution of ginseng, contained 0.2%–0.3% total saponin [57]; Renshengao, ginseng extract, contained 3.89% total saponin [58], and ginseng tea showed a total ginsenoside content of 1.1% [59].

Furthermore, ginseng root contained a number of chemical constituents other than saponins. The volatile component of ginseng root contained a series of sesquiterpenes such as eremophilene (*91-17*), β-gurjunene (*91-18*), *trans-* and *cis*-caryophyllene, ε-muurolene (*91-19*), γ-patchoulene (*91-20*), β-eudesmol, β-farnesene, β-bisabolene, aromadendrene (*91-21*), alloaromadendrene (*91-22*), β-guaiene (*91-23*), γ-elemene, mayurone (*91-24*), pentadecane, 2,5-dimethyltridecane, and palmitic acid [60].

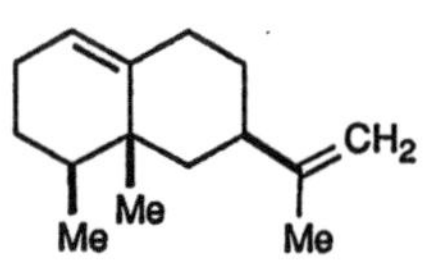

Eremophilene (*91-17*)

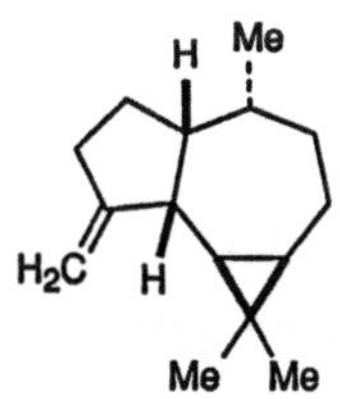

β-Gurjunene (*91-18*)

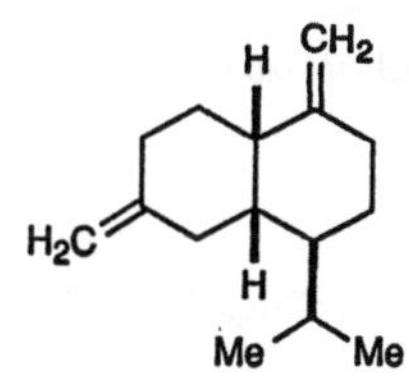

ε-Muurolene (*91-19*)

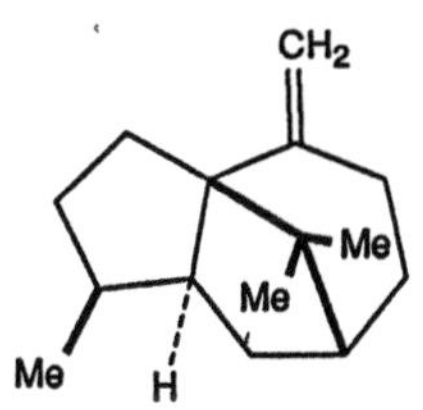

γ-Patchoulene (*91-20*)

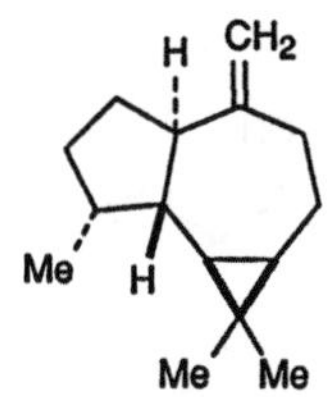

Aromadendrene (*91-21*)

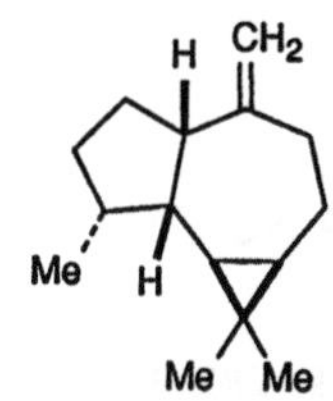

Alloaromadendrene (*91-22*)

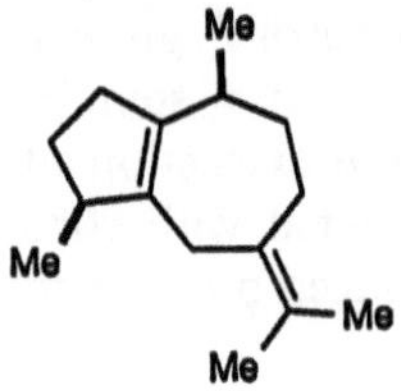

β-Guaiene (*91-23*)

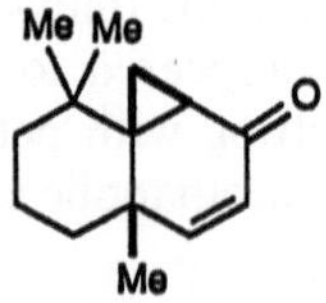

Mayurone (*91-24*)

From the basic fraction of the volatile extract of ginseng root, some pyrazine derivatives such as 3-*sec*-butyl-2-methoxy-5-methylpyrazine and 3-isopropyl-2-methoxy-5-methylpyrazine were identified by gas chromatography-mass spectrometry. 3-*sec*-Butyl-2-methoxy-5-methylpyrazine (*91-25*) exhibitis a characteristic floral, moldy aroma with an odor threshold concentration of 2 ppb (2×10^{-9}) in water [61].

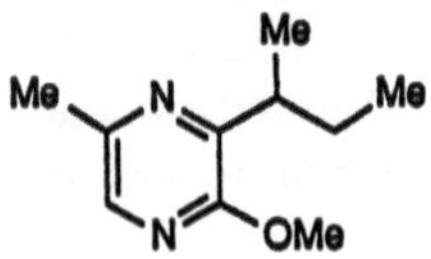

3-*sec*-Butyl-2-methoxy-5-methylpyrazine (*91-25*)

Some acetylenic compounds were also isolated from ginseng root. Panaxynol (*91-26*) was obtained from the higher boiling fraction of volatile oil isolated from ginseng root [62, 63]. Panaxydol (*91-27*), a polyacetylenic epoxide [64], panaxytriol (*91-28*) [65], and heptadeca-1-ene-4,6-diyn-3,9-diol (*91-29*) were also isolated and identified [66].

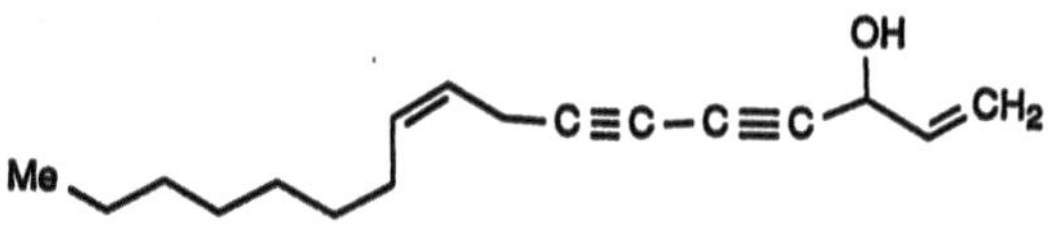

Panaxynol (*91-26*)

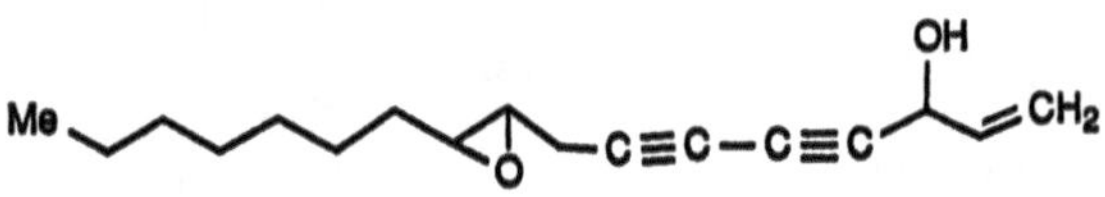

Panaxydol (*91-27*)

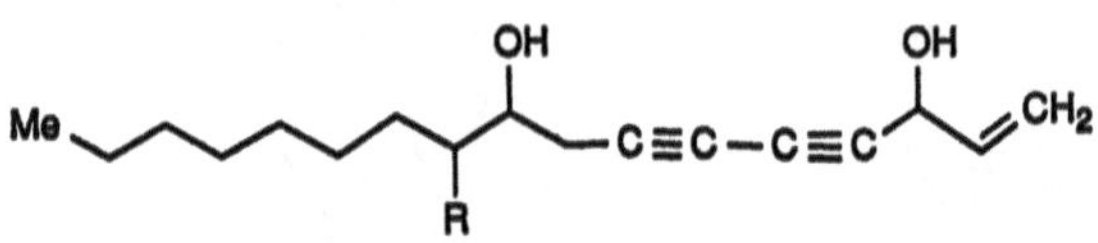

Panaxytriol (*91-28*): R=OH
Heptadeca-1-ene-4,6-diyn-3,9-diol (*91-29*): R=H

Ginseng contains about 5% of a sugar fraction, which comprises two monosaccharides, D-glucose and D-fructose, two disaccharides, sucrose and maltose, and three trisaccharides, maltosyl-β-D-fructofuranose, O-α-D-glucopyranosyl(1→2)-O-β-D-fructofuranosyl(1→2)-β-D-fructofuranose, and O-α-D-glucopyranosyl(1→6)-O-α-D-glucopyranosyl-(1→4)-α-D-glucopyranose [67].

A number of hypoglycemic peptide glycans named panaxans were also isolated from ginseng root. Isolation of panaxans A, B, C, D, E [68], F, G, H [69], I, J, K, L [70], Q, R, S, T, and U [71] has been reported. Panaxan A was shown to have a molecular weight of ca. 14 000. Spectroscopic and degradation studies revealed that panaxan A is mainly composed of α-1→6-linked D-glucopyranose residues with a branching at the C-3 position, showing a ratio of terminals, branching positions, and intermediate units of about 1:1:2 [72]. Panaxan B was shown to be a peptidoglycan with a molecular weight of ca. 1 800 000, mainly composed of α-1→6-linked D-glucopyranose residues with branching at the C-3 position and a ratio of terminals, branching positions, and intermediate units of approximately 1:1:1.8 [73].

A crude polysaccharide preparation was extracted with boiling water at a yield of 8%–10% from ginseng roots, containing 89% sugars and 5% proteins. The polysaccharide fraction was composed of 80% starch and 20% pectin [74]. By chromatography on Sephadex 4B, ginseng pectin gave two acidic polysaccharides. One contained galactose, arabinose, and rhamnose in a molar ratio of 4.7:2.6:1 and β-(1→3)D-galactan was obtained after partial hydrolysis. The other contained galactose, arabinose, and rhamnose in a molar ratio of 3.3:1.8:1 and α-(1→4)-D-galacturonan was obtained after partial hydrolysis with acid [75]. The protein fraction contained 15 amino acids including aspartic acid, threonine, serine, glutamic acid, glycine, alanine, valine, methionine, isoleucine, leucine, tyrosine, phenylalanine, lysine, histidine, and arginine [74].

A pyran-4-one derivative 3-hydroxy-2-methyl-pyran-4-one [76] and its glucoside (*91-30*) [77], a glucoside of hydroxyacetone, and an α-D-glucopyranoside of dihydroxypyran-3-one (*91-31*) [78] were also found in red ginseng. However, they may be artifacts formed during the steaming process.

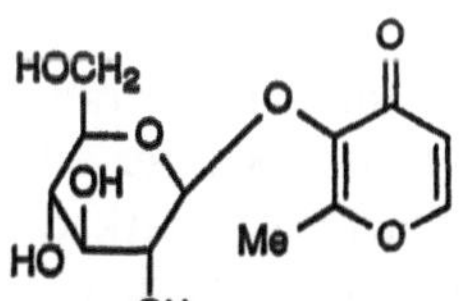

3-Hydroxy-2-methyl-pyran-4-one 3-O-β-D-glucopyranoside (*91-30*)

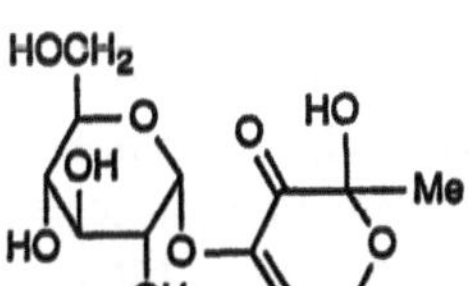

α-D-Glucopyranoside of dihydroxy-pyran-3-one derivative (*91-31*)

Three flavone constituents, kaempferol, trifolin (*91-32*), and panasenoside (*91-33*), were isolated from the leaves and stems of *P. ginseng* [79].

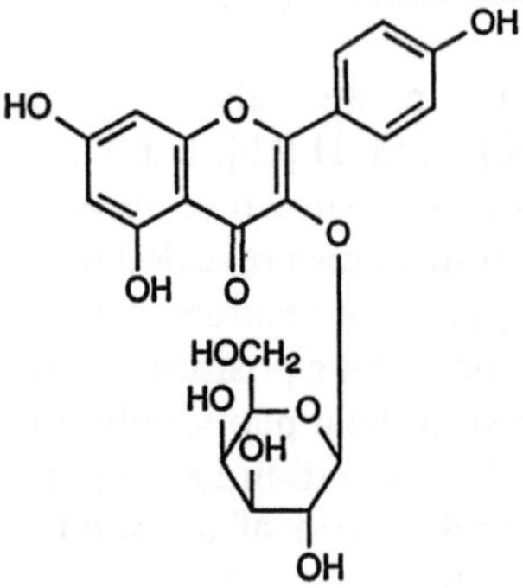

Trifolin (*91-32*)

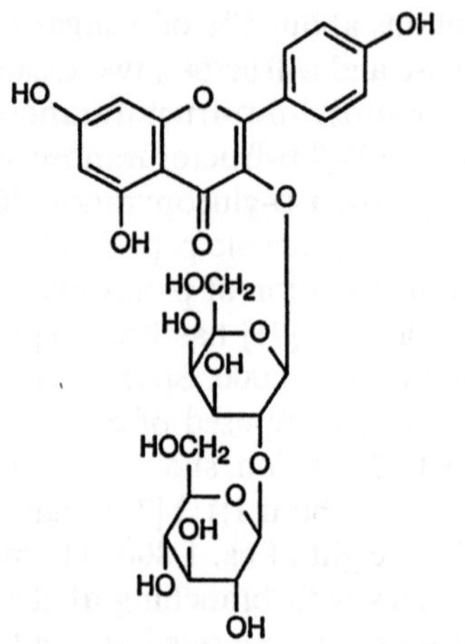

Panasenoside (*91-33*)

It is noteworthy that a cucurbitaceous plant *Gynostemma pentaphyllum* also contained a number of ginsenosides such as ginsenosides Rb_1, Rb_2, Rd, and F_2 [80–82].

91.3 Pharmacology

Ginseng was used in traditional Chinese medicine for a long time as a general tonic and cardiotonic. Systematic pharmacological investigations revealed a multifaceted biological activity of ginseng, including effects on the cardiovascular, immune, and nervous systems, and activity as an antidote, antitumor agent, or antitumor adjuvant and as an antidiabetic [83].

91.3.1 Activity as a General Tonic

When ginsenosides of ginseng stems and leaves were given intraperitoneally or orally to young mice or rats, a marked increase in body weight of the animals was observed. The protein and RNA contents of muscles and liver in treated rats increased markedly, but no increase of the weight of the prostata and seminal vesicles was noted. The anabolic action of ginsenosides is different from that of androgenic steroids. The promotion of animal growth by ginsenosides was suggested to be mediated by influence on the synthesis of RNA and protein [84].

Administration of ginseng root extract to rats increased the incorporation of labeled precusor into rat renal nuclear RNA. This stimulating effect was dose dependent, and maximal activity was observed at 8 h after treatment. Sequential cytoplasmic RNA synthesis was stimulated maximally at 10 h after administration. Stimulation of RNA synthesis by ginseng extract in turn increased the incorporation of leucine into rat renal protein, maximal activity being observed at 12 h after intraperitoneal administration [85].

The stimulating effect of ginsenosides isolated from the leaves and stems on DNA, RNA, and protein in liver and kidney was also studied in normal and reserpinized mice [86, 87]. The incorporation of [^{3}H]uridine into the liver and kidney RNA was increased in mice receiving 25 mg/kg ginsenosides daily orally for 7 days. In contrast, oral administration did not alter incorporation of [^{3}H]thymidine into kidney and liver DNA. Incorporation of [^{3}H]leucine into protein was also increased in mice. However, RNA and protein formation in liver and kidney reserpinized animals were not affected by ginsenoside [86]. Although synthesis of DNA in liver and kidney remained unchanged, it was increased in bone marrow cells [87]. Bone marrow mitosis was enhanced, as well as numbers of total nucleated cells in bone marrow and of reticulocytes in peripheral blood [88]. Pure ginsenosides Rb_2, Rc, Re, or Rg_1 given intraperitoneally at a dose of 5–10 mg/kg to rats increased DNA, RNA, protein, and lipid formation in bone marrow cells. Ginsenosides Rb_2, Rc, and Rg_1 decreased cAMP, but increased cGMP in bone marrow 20 min after injection [89].

The aqueous extract of *P. ginseng* inhibited intracellular protein degradation and stimulated protein synthesis in confluent cultures of IMR-90 human diploid fibroblasts, indicating that ginseng extract is capable of acting directly on human cells to promote protein accumulation [90].

91.3.2 Effects on Immunological Functions

Mice treated with a total aqueous ginsenoside extract from ginseng root at a dosage of 10, 50, and 250 mg/kg orally for 5–6 days responded in a dose-dependent manner by enhanced formation of antibody to either a primary or a secondary challenge with sheep red cells. At the highest dosage, the primary IgM response was increased by 50% and the secondary IgG and IgM responses by 50% and 199%, respectively. An even more pronounced effect was obtained for natural killer cell activity [91].

In vitro, an enhancement of interferon production particularly in nonstimulated spleen cells was observed. The immunostimulating effects of ginsenoside observed in vivo were in agreement with the stimulation of interferon production observed in vitro [91]. The phagocytic activity of guinea pig reticuloendothelial system was increased after intragastric infusion of ginsenosides from the leaves and steams of *P. ginseng* at a daily dose of 400 mg/kg for 3 days. Serum levels of specific antibodies and of IgG, IgA, and IgM were also increased in mice after intraperitoneal injection of ginsenosides [92].

Gastric infusion of ginseng polysaccharide into mice caused stimulation of the phagocytic function of the reticuloendothelial system, an increase in serum-specific antibodies and IgG contents, and also an increase in the relative percentages of B cells. In addition, gastric infusion of ginseng polysaccharide also increased serum complement in guinea pigs [93]. Ginseng polysaccharide given orally, intraperitoneally, or subcutaneously was effective against immunodeficiency induced by cyclophosphamide in mice. Suppressed macrophage phagocytosis and hemolysin formation, as well as delayed hypersensitivity reaction, could be normalized [94, 95].

91.3.3 Effect on Cardiovascular System

Extracts of *P. ginseng* by ether, ethanol, or water showed different effects on cardiovascular function in dogs after intravenous administration (40 mg/kg). Following the administration of the ether extract, the heart rate and the central venous pressure decreased significantly. The administration of ethanol extract caused a significant decrease in the heart rate and the mean arterial pressure. After the administration of the aqueous extract, the cardiac output, stroke volume, and central venous pressure were significantly decreased whereas the total peripheral resistance was significantly increased [96].

Intravenous injection of total ginsenosides into dogs rapidly decreased the peak value of left ventricular pressure and arterial systolic pressure. The heart rate and renal arterial blood flow were also decreased, whereas renal vasoresistance was markedly increased by treatment with total ginsenosides. The vasoconstrictory effect of ginsenoside was not blocked by α-adrenoceptor blocking or serotonin receptor blocking [97, 98].

The cardiovascular responses to the saponins from leaves and stems of ginseng in dog with significant decrease in blood pressure, left ventricular pressure, and heart rate were similar to those observed for saponins from the root of ginseng [99]. Intraperitoneal injection of total ginsenosides in rabbits with experimental myocardial infarction caused a marked decrease in elevated total amplitude of S-T potential and the appearance of an abnormal Q wave. Ginsenosides also increased the myocardial tolerance of animals to hypoxia, including dogs, guinea pigs, rats, and mice. This increased tolerance to hypoxia appeared to be associated with a decrease in myocardial oxygen consumption during the hypoxia [100]. Treatment with ginseng extract normalized the anoxia-induced alterations in lactic dehydrogenase and succinic dehydrogenase activities in the ischemic mouse myocardium [101].

Ginsenosides from various parts of ginseng, including stem, leaf, fruit, and root, were effective against mouse myocardial necrosis induced by isoproterenol. Changes in electrocardiogram and serum creatine phosphate kinase, lactic dehydrogenase, and γ-glutamyl-transferase were normalized. These effects were comparable to those of propranolol [102]. An ultrastructure study in the myocardium of rats aged 18 months showed that oral administration of ginseng at 100 mg/kg for 3 months decreased degradative changes associated with aging [103]. Ginsenosides caused different responses in isolated, contracted ring preparations of various blood vessels from rabbits, dogs, and humans, a finding that may explain biphasic actions on blood pressure [104].

91.3.4 Effect on Lipid Metabolism

In normal rats, serum cholesterol was decreased in proportion to the administrated dose of purified total saponin of ginseng [105]. In rats fed on a diet high in cholesterol, administration of a high dose of ginseng saponins also decreased the serum cholesterol level [105, 106]. Ginsenoside Rb_2 given intraperitoneally at a single dose to rats fed on a high cholesterol diet produced a significant decrease in serum levels of total cholesterol, free cholesterol, low-density lipoprotein cholesterol, triglyceride, 3-hydroxybutyrate, and acetoacetate [107]. In contrast, a significant increase in high-density lipoprotein cholesterol level in the serum was observed after the treatment [105, 107].

A time course study on the effect of ginsenoside Rb_2 on lipid metabolism in rats showed that the first phenomenon to be observed was a stimulation of glucose-6-phosphate dehydrogenase activity beginning 4 h after the administration of saponins. At the same time, a slight increase in acetyl-CoA carboxylase activity became apparent and significantly enhanced activity was noted 16 h after treatment. An increase in the lipolytic activity of lipoprotein lipase was observed 4 h after intraperitoneal administration of ginsenoside Rb_2, reaching a maximum 16 h after treatment, whereas an inhibitory effect was observed on the hormone-sensitive lipase activity throughout the experimental period [108].

Similar effects of ginseng in lowering the serum total cholesterol level and serum low-density lipoprotein cholesterol level were also observed in white leghorn females. In addition, the activities of β-hydroxy-β-methylglutaryl-CoA reductase and cholesterol 7α-hydroxylase in liver were inhibited [109].

91.3.5 Effect of Ginseng on Alcohol Metabolism

Pretreatment with ginseng delayed anesthesia induction by alcohol in rabbits and dogs; recovery time and duration of the anesthesia were shortened [110]. Pretreatment with ginseng extract in rats significantly stimulated the elimination of ^{14}C-labeled ethanol [111]. Alcohol dehydrogenase of rat liver mitochondria was stimulated by ginseng saponins. Aldehyde dehydrogenase was stimulated at a moderate concentration of ginseng saponin, but was inhibited at a higher concentration [112]. In mice, the ginseng saponin elevated the incorporation of ^{14}C-labeled ethanol into liver lipids, indicating that the saponins appear to stimulate the oxidation of ethanol in liver [112]. The decrease in aldehyde dehydrogenase activity caused by disulfiram was abrogated by injection of the ginseng saponins in mice [113].

Rats treated orally with ethanol showed liver damage as reflected by various degrees of cellular swelling, congestion, and bile pigmentation. Serum glutamic-oxalacetic transaminase, glutamic-pyruvic transaminase, and alkaline phosphatase levels as well as bilirubin levels were elevated. This histopathological liver damage and the elevation of enzyme and bilirubin levels caused by ethanol were reduced by oral administration of ginseng saponins [114].

91.3.6 Hypoglycemic Activity

Total saponins of ginseng root showed hypoglycemic activity in alloxan-induced diabetic mice, when given before and after alloxan injection. When treatment with ginseng saponin started 24 h after injection of alloxan, no hypoglycemic effect was observed [115]. The pure ginsenoside Rb_2 decreased the blood glucose levels in streptozotocin-induced diabetic rats. In addition, the animals treated with ginsenoside Rb_2 showed a significant rise in glucokinase activity in the liver and a significant decrease in the glucose-6-phosphatase activity [116].

Panaxans from ginseng root exhibited significant hypoglycemic actions in normal and alloxan-induced hyperglycemic mice [68–70]. Furthermore, ginseng root contained an unidentified component, which lowered the blood glucose level in normal and diabetic rats and mice and stimulated insulin release in diabetic animals. It did not increase the incorporation of radioactive leucine in pancreatic preparations of normal animals. However, in preparations from animals with hyperglycemia, an 1.5- to 1.8-fold increase in incorporation of radioactive leucine into insulin was observed [117].

91.3.7 Stimulating Effect on Pituitary-Adrenocortical System

Intraperitoneal administration of ginseng saponin to rats caused an increase in plasma ACTH and corticosterone 30, 60, and 90 min after the treatment as detected by radioimmunoassay and competitive protein-binding methods. The kinetic pattern of the increase in plasma ACTH was almost parallel to that in plasma corticosterone. The separated ginsenosides of the protopanaxadiol or protopanaxatriol type also increased plasma corticosterone. 20(S)-Panaxadiol also showed significant inducing activity on corticosterone secretion whereas the effect of the epimer 20(R)-panaxadiol was not significant [118].

Intraperitoneal injection of total ginsenosides of stems and leaves, of panaxadiol or ginsenoside Rd into rats increased adrenal intracellular cAMP concentrations [119, 120]. The increase was potentiated by ACTH and was prevented by hypophysectomy, suggesting that the ACTH-like effect in ginseng saponin is due to the direct effect on the hypophysis [119].

In contrast to the results observed in the study on hypoglycemic activity of ginseng saponins, it was also reported that ginseng saponins injected intraperitoneally into rats increased plasma corticosterone and plasma glucose, and decreased plasma immunoreactive insulin [121]. It was further described that the extract of red ginseng given subcutaneously to mice had a protective effect against insulin shock. This protective effect may be related to serum glucose elevation during the shock. In contrast, saponins isolated from ginseng root, stem, or leaves potentiated insulin shock [122].

Saponins isolated from ginseng fruit given intraperitoneally into rats inhibited the depletion of the ascorbic acid content of the adrenals caused by stress conditions or ACTH. However, the adrenal ascorbic acid regulating effect of the saponins disappeared in hypophysectomized rats, suggesting that the saponins possess a stimulatory action on the hypophyseal-adrenal system. In addition, saponin from ginseng fruit exhibited a significant antihypoxia effect and antifatigue action, and raised the ability to tolerate temperature stress in mice [123]. In rats subjected to hyperthermia stress, ginseng root saponins decreased the rectal temperature, reversed the decrease in acetylcholine levels in the brain, and markedly increased plasma corticosterone levels [124].

91.3.8 Antitumor Activity

The petroleum ether fraction and ethyl acetate fraction extracted from ginseng root inhibited the growth of murine leukemia L5178Y cells and murine sarcoma 180 cells in vitro in exponential dependence on concentration. The cytotoxicity correlated with an inhibitory effect on biosynthesis of macromolecules. This applied especially to the inhibitory action on protein synthesis of the petroleum-ether fraction [125], whereas biosynthesis of a particular RNA species was inhibited by the ethyl acetate fraction [126].

Ginsenoside Rh_2 at a concentration of 5–15 μM inhibited the growth of B16 melanoma cells in culture. In contrast, 20(S)-ginsenoside Rh_1 and 20(R)-ginsenoside Rh_1 had no growth inhibitory effect in B16 melanoma, but rather stimulated melanin synthesis in a concentration-dependent manner [127].

In contrast to untreated control cells, the subcellular organelles of Morris hepatoma cells cultured in medium containing ginsenosides were well developed with a well-organized distribution, suggesting that ginsenosides could reverse transformation of Morris hepatoma cells [128, 129]. Treatment of mice bearing sarcoma 180 with 50 mg/kg ginsenoside Rg_1 for 7 days resulted in a 52% tumor inhibition [130].

The use of ginseng as adjuvant in tumor chemotherapy is much more relevant. Thus, total saponin from *P. ginseng* prevented the stem cell killing effects of harringtonine in treatment of L1210 leukemia in mice [131, 132]. The water extract of ginseng inhibited the side effects of anticancer agents 5-fluorouracil and mitomycin C in decreasing the number of leukocytes, urine flow, renal plasma flow, glomerular filtration rate, and urinary excretion of sodium [133]. However, cytosin arabinoside-induced damages to the hematopoietic precursor cells in mice were enhanced by intraperitoneal or oral pretreatment of total saponins of ginseng [132].

91.3.9 Toxicity of Ginseng Saponins

Acute toxicities of purified ginseng saponin in mice and subacute toxicities of ginseng saponin in rats were reported. The LD_{50} of purified ginseng saponin in male mice was 270 mg/kg by intravenous, 342 mg/kg by intraperitoneal, 505 mg/kg by intramuscular, 950 mg/kg by subcutaneous, and 5000 mg/kg by oral administration. In subacute toxicity studies in rats, body weight was markedly increased by subcutaneous administration of 7.7 mg/kg saponin. Side effects were seen after administration of doses exceeding 77 mg/kg. Ginseng saponin administered at 240 mg/kg caused a marked decrease in the albumin : globulin ratio, and a 28% increase in urea nitrogen in serum and a 33% increase in liver weight/body weight ratio [134]. The intraperitoneal LD_{50} of total ginsenosides from stems and leaves in mice was 670 mg/kg [135].

The total ginseng saponin did not show hemolytic activity. Generally, ginsenosides of the protopanaxadiol type are not hemolytic but exhibit an antihemolytic effect, whereas the ginsenosides of protopanaxatriol- and oleanolic acid types are strongly hemolytic [136, 137].

91.3.10 Pharmacokinetic Studies

The pharmacokinetic study was carried out on ginsenoside Rg_1 as a representative of protopanaxatriol-type ginsenosides. Ginsenoside Rg_1 was absorbed rapidly from the upper parts of the digestive tract in rats. The serum level reached its peak at 30 min, and maximal levels in tissues were reached within 1.5 h. Ginsenoside Rg_1 was not found in brain. Excretion into urine and bile occurred at a ratio of 2:5. It was not metabolized to a significant extent in the liver, but decomposition and metabolism in rat stomach and large intestine were confirmed [138]. In a simulated gastric acid medium, ginsenoside Rg_1 was degraded to some prosapogenins [139].

The pharmacokinetics of ginsenoside Rb_1, a representative of protopanaxadiol-type ginsenosides, were quite different from those of ginsenoside Rg_1. Little ginsenoside Rb_1 was absorbed from the digestive tract after oral administration of 100 mg/kg to rats. After intravenous injection of 5 mg/kg in rats, the serum level of ginsenoside Rb_1 decreased biexponentially, and the half-life of β-phase was 14.5 h. The high persistence of ginsenoside Rb_1 in serum and tissues in rats was assumed to result from a high degree of plasma protein binding. Ginsenoside Rb_1 was gradually excreted into urine, but not bile. In the digestive tract, unabsorbed ginsenoside Rb_1 was rapidly decomposed and/or metabolized in the large intestine [140]. Ginsenoside Rb_1 was also degraded to some prosapogenins in simulated gastric acid [139].

Following oral administration of labeled ginseng saponins to rats, saponins distributed preferentially into the liver, kidney, blood serum, and gastrointestinal tract. Total recovery of radioactivity was only about 30%, suggesting the binding of the saponins to macromolecules and membrane structures [141].

A synthesis of ^3H-labeled ginsenoside Rg_1 was described. Ginsenoside Rg_1 after sequential partial acetylation, oxidation, and saponification gave 12-ketoginsenoside Rg_1. The latter was tritiated with T_2O by keto-enol tautomerization and the product was reduced to give [^3H]ginsenoside Rg_1 (*91-34*) [142].

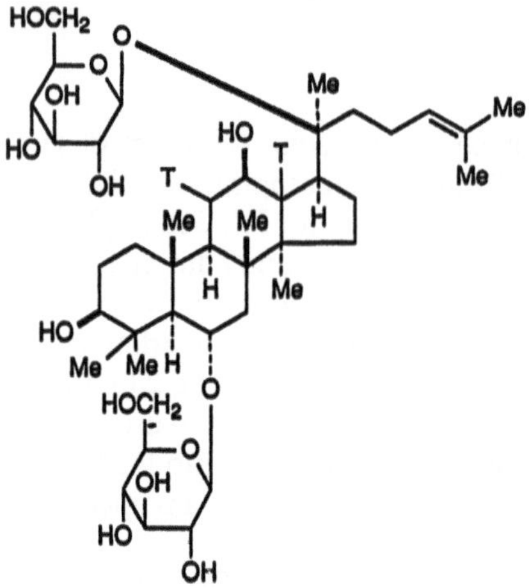

[^3H]Ginsenoside Rg_1 (*91-34*)

The pharmacokinetics of ginsenosides Rb_2, Re, and Rg_1 were also studied in rabbits. Ginsenoside Rb_2 of the protopanaxadiol type showed a significantly longer half-life, higher plasma protein binding, and lower metabolic and renal clearance than the protopanaxatriol ginsenosides Re and Rg_1. All the three ginsenosides were slowly absorbed after intraperitoneal administration. Ginsenosides were not found in rabbit plasma or urine samples after oral administration. The observed differences in the pharmacokinetics of the ginsenosides may be caused by different extents of protein binding. Ginsenoside Rb_2 was more toxic than ginsenoside Rg_1 after intraperitoneal administration to mice. After oral administration, no toxicity of any of the ginsenosides was observed [143]. Panaxadiol and panaxatriol were more toxic and had larger volumes of distribution than the ginsenosides [143]. Ginsenoside Rg_1 of the protopanaxatriol type showed an extremely short half-life of 27 min after intravenous administration into minipigs. The volume of distribution was 360 ml/kg, and pharmacokinetics were best described by a one-compartment model. In contrast to ginsenoside Rg_1, the protopanaxadiol type ginsenoside Rb_1 showed a biexponential decline of the blood level, with a half-life of the β-phase of 16 h, and a volume of distribution of 160 ml/kg [144]. These results correlated with the pharmacokinetic results in rats and in rabbits.

Table 91.3 summarizes the half-lives of the two different types of ginsenosides in different animals. It can clearly be seen that protopanaxatriol saponins have completely different intravenous pharmacokinetics than protopanaxadiol saponins.

Table 91.3. Different half-lives of protopanaxadiol-type and protopanaxatriol-type ginsenosides

Protopanaxadiol type

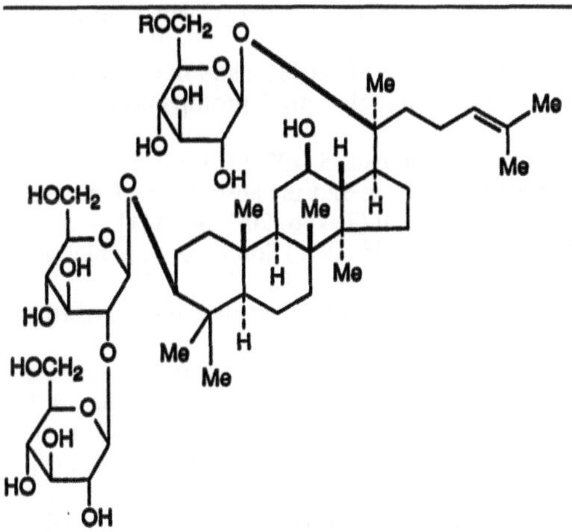

Protopanaxatriol type

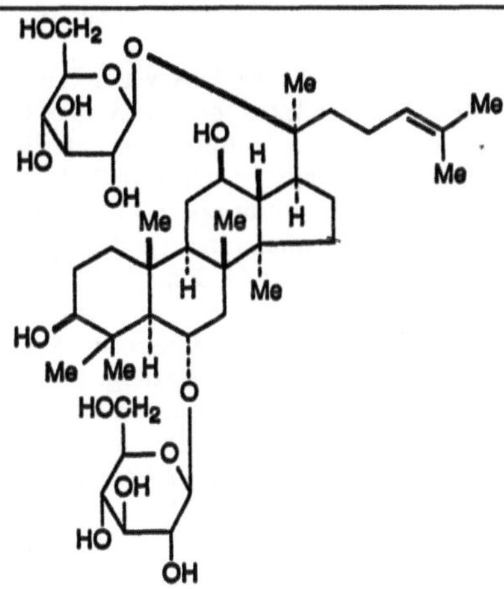

Ginsenoside Rb_1: R=

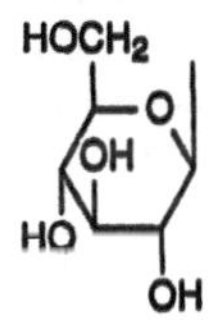

Ginsenoside Rg_1

Ginsenoside Rb_2: R=

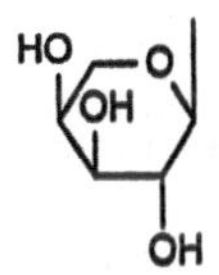

Ginsenoside	Animal		
	Rat [138, 140]	Rabbit [143]	Minipig [144]
Rb_1	14.5 h		16 h
Rb_2		8 h	
Rg_1	6.3 min	69.5 min	27 min

References

1. Höhrhammer L, Wagner H, Loy B (1961) Contents of the *Panax ginseng* root. Preliminary report. Pharm Z 106:1307–1311
2. Shibata S, Fujita M, Itokawa H, Tanaka O, Ishii T (1963) Studies on the constituents of Japanese and Chinese crude drugs. XI. Panaxadiol, a sapogenin of Ginseng roots (1). Chem Pharm Bull (Tokyo) 11:759–761

3. Kotake M (1930) Glucosides. I. Glucoside of *Panax ginseng*. J Chem Soc Jpn 51:357–360
4. Shibata S, Tanaka O, Sado M, Tsushima S (1963) On the genuine sapogenin of Ginseng. Tetrahedron Lett 795–800
5. Shibata S, Tanaka O, Ando T, Sado M, Tsushima S, Ohsawa T (1966) Chemical studies on the oriental plant drugs. XIV. Protopanaxadiol, a genuine sapogenin of Ginseng saponins. Chem Pharm Bull (Tokyo) 14:595–600
6. Tanaka O, Nagai M, Shibata S (1964) The stereochemistry of protopanaxadiol, a genuine sapogenin of Ginseng. Tetrahedron Lett 2291–2297
7. Tanaka O, Nagai M, Shibata S (1966) Chemical studies on the oriental plant drugs. XVI. The stereochemistry of protopanaxadiol, a genuine sapogenin of Ginseng. Chem Pharm Bull (Tokyo) 14:1150–1156
8. Tanaka O, Nagai M, Ohsawa T, Tanaka N, Shibata S (1967) Stereochemistry of protopanaxadiol. Acid catalyzed epimerization of C-20 hydroxy of betulafolienetriol, protopanaxadiol, and their derivatives. Tetrahedron Lett 391–396
9. Tanaka O, Nagai M, Ohsawa T, Tanaka N, Kawai K, Shibata S (1972) Chemical studies on the oriental plant drugs. XXVII. The acid catalyzed reaction and the absolute configuration at C(20) of dammarane type triterpenes. Chem Pharm Bull (Tokyo) 20:1204–1211
10. Shibata S, Tanaka O, Soma K, Iida Y, Ando T, Nakamura H (1965) Studies on saponins and sapogenins of Ginseng. The structure of panaxatriol. Tetrahedron Lett 207–213
11. Puff H, Friedrichs E, Habscheid M, Quante G, Chen WS (1986) Panaxadiol and panaxatriol monohydrate. Acta Crystallogr [C] 42:576–579
12. Besso H, Kaisai R, Saruwatari Y, Fuwa T, Tanaka O (1982) Ginsenoside-Ra_1 and ginsenoside-Ra_2, new dammarane-saponins of Ginseng root. Chem Pharm Bull (Tokyo) 30:2380–2385
13. Koizumi H, Sanada S, Ida Y, Shoji J (1982) Studies on the saponins of Ginseng. IV. On the structure and enzymatic hydrolysis of ginsenoside Ra_1. Chem Pharm Bull (Tokyo) 30:2393–2398
14. Matsuura H, Kasai R, Tanaka O, Saruwatari Y, Kunihiro K, Fuwa T (1984) Further studies on dammarane-saponins of Ginseng root. Chem Pharm Bull (Tokyo) 32:1188–1192
15. Sanada S, Kondo N, Shoji J, Tanaka O, Shibata S (1974) Studies on the saponins of Ginseng. I. Structures of ginsenoside-Ro, -Rb_1, -Rb_2, -Rc and -Rd. Chem Pharm Bull (Tokyo) 22:421–428
16. Kasai R, Besso M, Tanaka O, Saruwatari Y, Fuwa T (1983) Saponins of red Ginseng. Chem Pharm Bull (Tokyo) 31:2120–2125
17. Sanada S, Shoji J (1978) Studies on the saponins of Ginseng. III. Structures of ginsenoside-Rb_3 and 20-glucoginsenoside-Rf. Chem Pharm Bull (Tokyo) 26:1694–1697
18. Luo SD, Liu AY, Liu YY (1983) Comparison of active constituents and pharmacological effects of ginseng root and ginseng rhizome. Chin Pharm Bull 18:468–470
19. Kitagawa I, Yoshikawa M, Yashihara M, Hayashi T, Taniyama T (1983) Chemical studies on crude drug procession. I. On the constituents of Ginseng Radix Rubra. Yakugaku Zasshi 103:612–622
20. Besso H, Kasai R, Wei J, Wang JF, Saruwatari Y, Fuwa T, Tanaka O (1982) Further studies on dammarane-saponins of American Ginseng, root of Panax quinquefolium L. Chem Pharm Bull (Tokyo) 30:4534–4538
21. Kitagawa I, Taniyama T, Hayashi T, Yoshikawa M (1983) Malonyl-ginsenosided-Rb_1, -Rb_2, -Rc and Rd, four new malonylated dammarane-type triterpene oligosaccharides from Ginseng radix. Chem Pharm Bull (Tokyo) 31:3353–3356
22. Sanada S, Kondo N, Shoji J, Tanaka O, Shibata S (1974) Studies on the saponins of Ginseng. II. Structures of ginsenoside-Re, -Rf, and -Rg_2. Chem Pharm Bull (Tokyo) 22:2407–2412
23. Iida Y, Tanaka O, Shibata S (1968) Studies on saponins of Ginseng: the structure of ginsenoside-Rg_1. Tetrahedron Lett 5449–5453.
24. Nagai Y, Tanaka O, Shibata S (1971) Chemical studies on the oriental plant drugs. XXIV. Structure of ginsenoside-Rg_1, a neutral saponin of Ginseng root. Chem Pharm Bull (Tokyo) 27:881–892
25. Li XG, Teng FT (1979) Studies on the triterpenoid in *Panax ginseng*. Acta Bot Sin 21:181–185
26. Tanaka O, Kasai R (1984) Saponins of Ginseng and related plant. In: Herz W, Grisebach H, Kirby GW, Tamm C (eds) Progress in the Chemistry of Organic Natural Products Vol. 46. Springer, Vienna New York, pp 1–76
27. Xu SX, Wang NL, Li YH (1986) Chemical constituents of red ginseng. II. Acta Pharm Sin 21:356–360

28. Xu SX, Zhang FX, Yao NX (1986) Chemical constituents of ginseng. IX. Comparative analysis of the contents of ginseng saponins in Chinese and Korean red ginseng. J Shenyang Coll Pharm 3:95–99
29. Xu JD (1982) Separation and identification of ginsenosides in the rootstock of Jilin Ginseng. Chin Pharm Bull 17:501
30. Kuang HX, Xu JD, Jiang XK (1986) Studies on the rhizomes of ginseng. II. Isolation and identification of saponins D_1 and D_2 from the rhizome of ginseng. Bull Chin Mater Med 11:104–106
31. Kuang HX, Xu JD, Jiang XK, Cheng GR (1986) Studies on the rhizoma ginseng. III. Isolation and identification of saponins E_1 and E_2. Bull Chin Mater Med 11:166–168
32. Yahara S, Kaji K, Tanaka O (1979) Further study on dammarane-type saponins of roots, leaves, flower-buds and fruits of *Panax ginseng*. C.A. Meyer. Chem Pharm Bull (Tokyo) 27:88–92
33. Yahara S, Tanaka O, Komori T (1976) Saponins of the leaves of *Panax ginseng*. C.A. Meyer. Chem Pharm Bull (Tokyo) 24:2204–2208
34. Cai PL, Xu JD, Gu YQ (1982) Isolation and identification of ginsenosides in the leaves of Jilin Ginseng. Chin Pharm Bull 17:500
35. Cai PL, Xu JD, Gu YQ (1986) Chemical composition of the home-grown ginseng leaves. I. Isolation and identification of ginsenosides. J Bethune Univ Med Sci 12:16–20
36. Xu SX, Wang NL, Shen M, Lu XK (1986) Chemical constituents of saponins of the stems and leaves of *Panax ginseng*. C.A. Meyer. Acta Bot Sin 28:95–101
37. Yahara S, Matsuura S, Kasai R, Tanaka O (1976) Saponins of buds and flowers of *Panax ginseng* C.A. Meyer. I. Isolation of ginsenoside-Rd, -Re and -Rg_1. Chem Pharm Bull (Tokyo) 24:3212–3213
38. Shao CJ, Xu JD (1982) Isolation and identification of ginsenosides in the flowers and flower buds of Jilin Ginseng. Chin Pharm Bull 17:500
39. Shao CJ, Xu JD, Jiang XK, Cheng GR (1984) Studies on chemical constituents of flower buds of *Panax ginseng* – isolation and identification of ginsenoside-Rb_3 and -Rc. Bull Chin Mater Med 9:172–173
40. Shao CJ, Xu JD, Jiang XK, Cheng GR (1984) Chemical studies on the tetracyclic triterpenic saponins in flower buds of *Panax ginseng* C.A. Meyer. Chem J Chin Univ 5:674–676
41. Shao CJ, Xu JD, Jiang XK, Cheng GR (1985) Studies on the chemical constituents of flower buds of *Panax ginseng* C.A. Meyer of Jilin. I. Isolation and identification of ginsenosides Rg_2, Re and Rd. J Bethune Univ Med Sci 11:142–145
42. Tu ES, Wang YS, Liu YC (1980) Studies on the extraction technology and the saponin content in aerial parts of Ginseng. Chin Trad Herb Drugs 11:515–516
43. Zhang YH (1983) Different processings of ginseng and saponin content in different parts of plant for medicinal use. Chin Trad Herb Drugs 14:211–212
44. Wu GX, Wei YD, Song CC, Wang CR, Ma XY, Xu JD, Zhang DX, Jiang XK (1988) Comparative analysis of the contents of ginseng saponins in wild ginseng and cultivated ginseng. Chin Pharm Bull 23:397–398
45. Yamaguchi H, Matsuura H, Kasai R, Tanaka O, Satake M, Koda H, Izumi H, Nuno M, Katsuki S, et al. (1988) Analysis of saponins of wild *Panax ginseng*. Chem Pharm Bull (Tokyo) 36:4177–4181
46. Sollorz G (1985) Quality evaluation of ginseng roots. Quantitative HPLC determination of ginsenosides. Dtsch Apoth Z 125:2052–2055
47. Meier B, Meier-Bratschi A, Dallenbach-Toelke K, Sticher O (1985) Quantitative analysis of ginseng by HPLC. In: Chang HM, Yeung HW, Tso WW, Koo A (eds) Advances in Chinese Medicinal Material Research. World Science Singapore, pp 471–484
48. Yamaguchi H, Kasai R, Matsuura H, Tanaka O, Fuwa T (1988) High-performance liquid chromatographic analysis of acidic saponins of ginseng and related plants. Chem Pharm Bull (Tokyo) 36:3468–3473
49. Shao CJ, Xu JD (1982) Thin-layer scanning and colorimetric determination of ginseng saponins. Chin Trad Herb Drugs 13:355–359
50. Zhang GD, Zhou ZH, Liu HY (1983) Analysis of ginseng. III. Isolation and TLC-densitometry determination of ginseng saponins. Acta Pharm Sin 18:607–611
51. Wang MZ, Gao FY, Zhang GD, Zhang SR (1979) Analysis of ginseng. I. Extraction and colorimetric determination of ginseng saponins. Acta Pharm Sin 14:309–315

52. Tai BS, Geng CG, Li J (1983) Colorimetric determination of ginsenosides in ginseng preparations. Chin J Pharm Anal 3:234–236
53. Xu SX, Wang NL, Li B, Yao NX (1986) Studies on the chemical composition of ginseng. X. Studies on the comparative quantitative analysis of ginseng saponins. J Shenyang Coll Pharm 3:173–177
54. Xie PS, Yan YZ (1987) HPTLC fingerprint identification of commercial ginseng drugs – reinvestigation of HPTLC of ginsenosides. J High Resolut Chromatogr Commun 10:607–613
55. Xie PS, Yan YZ (1988) Differentiation and evaluation of ginsengs and their preparations by means of HPTLC fingerprint analysis. J Planar Chromatogr Mod TLC 1:29–34
56. Song CC, Ma XY (1983) Analysis of the chemical constituents of Ren Shen Lu. J Bethune Univ Med Sci 9:56–59
57. Shao CJ, Kuang HX, Xu JD (1983) Quantitative determination of saponins of ginseng extract. Chin Pharm Bull 18:734–736
58. Wang CR, Wei YD, Xu JD (1983) Analysis of the chemical constituents of Ren Shen Gao. J Bethune Univ Med Sci 9:60–62
59. Shou CJ, Xu JD (1983) Quantitative and qualitative determination of ginseng tea – a ginseng tonic. J Bethune Univ Med Sci 9:93–95
60. Zhang HX, Sun YX, Wang SQ, Jiang WP, Yang LR (1985) Analysis of the volatile constituents of Jilin ginseng. Kexue Tongbao [Foreign Lang] 30:195–199
61. Iwabuchi H, Yoshikura M, Obata S, Kamisako W (1984) Studies on the aroma constituents of crude drugs. I. On the aroma constituents of Ginseng radix. Yakugaku Zasshi 104:951–958
62. Takahashi M, Yoshikura M (1966) Studies on the components of *Panax ginseng* C. A. Meyer. IV. On the structure of a new acetylene derivative "panaxynol" (2). Synthesis of 1,7,9-heptadecatrien-4-yn-3-ol. Yakugaku Zasshi 86:1051–1053
63. Takahashi M, Yoshikura M (1966) Studies on the components of *Panax ginseng*. C.A. Meyer. V. On the structure of a new acetylene derivative, "panaxynol" (3). Synthesis of 1,9-(*cis*)-heptadecadiene-4,6-diyn-3-ol. Yakugaku Zasshi 86:1053–1056
64. Poplawski J, Wrobel JT, Glinka T (1980) Panaxydol, a new polyacetylenic expoxide from *Panax ginseng* roots. Phytochemistry 19:1539–1541
65. Shim SC, Koh HY, Han BH (1983) Polyacetylenes from *Panax ginseng* roots. Phytochemistry 22:1817–1818
66. Dabrowski Z, Wrobel JT, Wojtasiewicz K (1980) Structure of an acetylenic compound from *Panax ginseng*. Phytochemistry 19:2464–2465
67. Shoji J (1985) Recent advances in the chemical studies on ginseng. In: Chang HM, Yeung HW, Tso WW, Koo A (eds) Advances in Chinese Medicinal Material Research. World Science, Singapore, pp 455–469
68. Konno C, Sugiyama K, Kano M, Takahashi M, Hikino H (1984) Validity of the oriental medicines. LXX. Antidiabetes drugs. 1. Isolation and hypoglycemic activity of panaxans A, B, C, D and E, glycans of *Panax ginseng* roots. Planta Med 50:434–436
69. Hikino H, Oshima Y, Suzuki Y, Konno C (1985) Isolation and hypoglycemic activity of panaxans F, G and H, glycans of *Panax ginseng* roots. Shoyakugaku Zasshi 39:331–333
70. Oshima Y, Konno C, Hikino H (1985) Antidiabetes drugs. XIV. Isolation and hypoglycemic activity of panaxans I, J, K and L, glycans of *Panax ginseng* roots. J Ethnopharmacol 14:255–259
71. Konno C, Murakami M, Oshima Y, Hikino H (1985) Validity of the Oriental medicines. CVI. Antidiabetes drugs. 19. Isolation and hypoglycemic activity of panaxans Q, R, S, T and U, glycans of *Panax ginseng* roots. J Ethnopharmacol 14:69–74
72. Tomoda M, Shimada K, Konno C, Sugiyama K, Hikino H (1984) Validity of Oriental medicines. LXXI. Antidiabetes drugs. 2. Partial structure of panaxan A, a hypoglycaemic glycan of *Panax ginseng* roots. Planta Med 50:436–438
73. Tomoda M, Shimada K, Konno C, Hikino H (1985) Validity of Oriental medicines. XCIV. Antidiabetes drugs. 12. Structure of panaxan B, a hypoglycemic glycan of *Panax ginseng* roots. Phytochemistry 24:2431–2433
74. Li RQ, Zhang YS (1985) Analysis of the chemical constituents of ginseng polysaccharide. Chin Trad Herb Drugs 16:389–391
75. Li RQ, Zhang YS (1984) Purification and characterization of *Panax ginseng* C.A. Meyer pectin. Acta Pharm Sin 19:764–768
76. Wei JX (1982) Studies on the constituents of Korean red ginseng – isolation and identification of 3-hydroxy-2-methyl-4-pyrone. Acta Pharm Sin 17:549–550

77. Xu SX, Wang NL, Li YH (1986) The isolation and identification of 2-methyl-pyrone-3-*O*-β-D-glucoside from Chinese red ginseng. Acta Pharm Sin 21:71–72
78. Matsuura H, Hirao Y, Yoshida S, Kunihiro K, Fuwa T, Kasai R, Tanaka O (1984) Study of red ginseng: new glucosides and a note on the occurrence of maltol. Chem Pharm Bull (Tokyo) 32:4674–4677
79. Wang ZX, Zang YY, Liu MS, Shang FC, Li YQ, Chi JP, Lin L, Jiao Q, Zhao HB (1985) Flavonoid constituents of the stems and leaves of *Panax ginseng*. C.A. Meyer. J Sheuyang Coll Pharm 2:284–287
80. Cheng GR (1986) A review of the studies on *Gynostemma pentaphyllum*. Guihaia 6:305–308
81. Yu ZY, Yang H (1988) Recent developments in the study of *Gynostemma pentaphyllum*. Chin Pharm Bull 23:12–14
82. Zhou HP (1988) The saponin constituents and pharmacology of *Gynostemma pentaphyllum*. Chin Pharm Bull 23:720–724
83. Wang BX (1980) Recent progress of pharmacological studies of *Panax ginseng*. Acta Pharm Sin 15:312–320
84. Wang BX, Cui JC, Liu AJ (1982) Effect of ginsenosides extracted from the stems and leaves of *Panax ginseng* in promoting animal growth. Acta Pharm Sin 17:899–904
85. Nagasawa T, Oura H, Hiai S, Nishinaga K (1977) Effect of ginseng extract on ribonucleic acid and protein synthesis in rat kidney. Chem Pharm Bull (Tokyo) 25:1665–1670
86. Zhang BF, Pan WJ, Wu CF, Chen S, Wang LY, Tan YZ (1985) Effects of saponin (GSL) on DNA, RNA, and protein synthesis of liver and kidney in normal and reserpinized mice. J Shenyang Coll Pharm 2:121–124
87. Wen SR, Li MY, Wang SX, Liu DX (1982) Effect of total saponins from aerial parts of ginseng on the biosynthesis of nucleic acid and protein. Chin Pharm Bull 17:362–363
88. Yamamoto M, Takeuchi N, Kumagai A, Yamamura Y (1977) Stimulatory effect of *Panax ginseng* principles on DNA, RNA, protein and lipid synthesis in rat bone marrow. Arzneimittelforschung 27:1169–1173
89. Yamamoto M, Masaka M, Yamada K, Hayashi Y, Hirai A, Kumagai A (1978) Stimulatory effect of ginsenosides on DNA, protein and lipid synthesis in rat bone marrow and participation of cyclic nucleotides. Arzneimittelforschung 28:2238–2241
90. Lu ZQ, Dice JF (1985) Ginseng extract inhibits protein degradation and stimulates protein synthesis in human fibroblasts. Biochem Biophys Res Commun 126:636–640
91. Jie YH, Cammisuli S, Baggiolini M (1984) Immunomodulatory effects of *Panax ginseng* C.A. Meyer in the mouse. Agents Actions 15:386–391
92. Cui JC, Liu AJ, Wang BX (1982) Effect of ginsenosides from stems and leaves of ginseng on immunological functions. Chin Trad Herb Drugs 13:29–31
93. Wang BX, Cui JC, Liu AJ (1981) Effect of polysaccharide of ginseng root on immunological function. Chin Pharm Bull 16:53
94. Yuan WX, Shang XH, Jin ML, Shang K (1986) Effects of ginseng root polysaccharides on cyclophosphamide-suppressed immunity. J Shenyang Coll Pharm 3:162–165
95. Wang BX, Cui JC, Liu AJ (1985) The effect of ginseng on immune responses. In: Chang HM, Yeung HW, Tso WW, Koo A (eds) Advances in Chinese Medicinal Material Research. World Science, Singapore, pp 519–527
96. Lee DC, Lee MO, Kim CY, Clifford DH (1981) Effect of ether, ethanol and aqueous extracts of ginseng on cardiovascular function in dogs. Can J Comp Med 45:182–187
97. Chen X, Zhu QY, Liu LY, Tang XL (1981) Effects of ginsenosides on cardiac performance and hemodynamics in dogs. Chin Pharm Bull 16:757
98. Chen X, Zhu QY, Liu LY, Tang XL (1982) Effect of ginsenosides on cardiac performance and hemodynamics of dogs. Acta Pharmacol Sin 3:236–239
99. Pan WJ, Zhang BF, Wu CF, Wang LY, Chen S, Tan YZ (1985) Effect of ginseng leaf and stem saponins on cardiac performance and hemodynamics of dogs. J Shenyang Coll Pharm 2:51–55
100. Chen H, Fang YH, Teng HW, Chen N, Liu LY, Hsiao H, Tang HL (1981) Effects of active constituents of ginseng and *Citrus aurantium* and clonidine on experimental myocardial infarction. Chin Pharm Bull 16:50–51
101. Shi LR, Zhang Y (1984) Histochemistry and electron microscopic cytochemistry on SDH and LDH in the myocardium of experimental anoxia mice. Acta Anat Sin 15:210–215
102. Chen YZ, Huo YS, Yu ZS, Zhang W (1986) Mouse myocardial necrosis preventive effects of ginsenosides from various parts of ginseng. Bull Chin Mater Med 11:368–370

103. Wang HP, Zhang SF, Zhao ZH, Fan GY, Xie RH (1986) The effect of ginseng leaf saponins on the myocardium of rats of various ages. Acta Zool Sin 32:101–105
104. Chen X, Gillis CN, Moalli R (1984) Vascular effects of ginsenosides in vitro. Br J Pharmacol 82:485–491
105. Lim CJ, Park EH, Rhee DK (1981) Studies on the effect of ginseng total saponin on serum cholesterol and triglyceride levels in rats. Hanguk Saenghwa Hakhoe Chi 14:188–198 (CA 96:28471k)
106. Moon CK, Kang NY, Yun YP, Lee SH, Lee HA, Kang TL (1984) Effects of red ginseng-crude saponin on plasma lipid levels in rats fed on a diet high in cholesterol and triglyceride. Arch Pharmacol Res 7:41–45 (CA 101:204131g)
107. Yokozawa T, Kobayashi T, Kawai A, Oura H, Kawashima Y (1985) Hyperlipemia-improving effects of ginsenoside-Rb_2 in cholesterol-fed rats. Chem Pharm Bull (Tokyo) 33:722–729
108. Yokozawa T, Kobayashi T, Kawai A, Oura H, Kawashima Y (1984) Stimulation of the lipogenic pathway in ginsenoside-Rb_2 treated rats. Chem Pharm Bull (Tokyo) 32:4490–4496
109. Qureshi AA, Abuirmeileh N, Din ZZ, Ahmad Y, Burger WC, Elson CE (1983) Suppression of cholesterogenesis and reduction of LDL cholesterol by dietary ginseng and its fraction in chicken liver. Arterosclerosis 48:81–94
110. Shin MR (1976) Studies on the antidotal effect of *Panax ginseng*. Therapeutic effect of ginseng on acute alcoholism. Koryo Taehakkyo Uikwa Taehak Chapchi 13:231–246 (CA 88:70260a)
111. Petkov V, Koushev V, Panova Y (1977) Accelerated elimination under the effect of ginseng (experiments in rats). Acta Physiol Pharmacol Bulg 3:46–50 (CA 88:70259g)
112. Joo CN, Koo JD, Kim DS, Lee SJ (1977) Biochemical studies of ginseng saponins. XI. The effect of ginseng saponins on aldehyde dehydrogenase. Hanguk Saenghwa Hakhoe Chi 10:109–120 (CA 88:130977b)
113. Huh K, Choi CW (1978) Effect of *Panax ginseng* on the hepatic alcohol and aldehyde dehydrogenase activity in mouse and rat. Yongnam Taehakkyo Chonyonmul Hwahak Yonguso Yongu Pogo 5:1–13 (CA 91:83306g)
114. Yu HY, Kim CM, Koo KH (1982) An experimental study on the effect of ginseng saponin upon alcoholic liver injury. Hanyang Uidae Haksulchi 2:287–304 (CA 98:174615t)
115. Bao TT (1981) Effect of total saponin of *Panax ginseng* on alloxan diabetes in mice. Acta Pharm Sin 16:618–620
116. Yokozawa T, Kobayashi T, Oura H, Kawashima Y (1985) Studies on the mechanism of the hypoglycemic activity of ginsenoside-Rb_2 in streptozotocin-induced rats. Chem Pharm Bull (Tokyo) 33:869–872
117. Waki I, Kyo H, Yasuda M, Kimura M (1982) Effects of a hypoglycemic component of ginseng radix on insulin biosynthesis in normal and diabetic animals. J Pharmacobiodyn 5:547–554
118. Hiai S, Yokoyama H, Oura H, Kawashima Y (1983) Evaluation of corticosterone secretion-inducing activities of ginsenosides and their prosapogenins and sapogenins. Chem Pharm Bull (Tokyo) 31:168–174
119. Zong RY, Zheng SQ, Liu J (1985) Adrenocorticotropic hormone-like effects of ginseng saponin. J Bethune Univ Med Sci 11:254–258
120. Hiai S, Sasaki S, Oura H (1979) Effect of ginseng saponin on rat adrenal cyclic AMP. Planta Med 37:15–19
121. Hiai S, Yokoyama H, Oura H (1979) Features of ginseng saponin-induced corticosterone secretion. Endocrinol Jpn 26:737–740
122. Yuan WX, Wu XJ, Han YJ, Shang XH, Zhang LL (1985) Effects of extracts of ginseng and saponins of ginseng on insulin shock and serum glucose. J Shenyang Coll Pharm 2:207–211
123. Zhang SC, Jiang XL (1981) Antistress effect and the effect on functions of hypophyseal-adrenal system of saponins from *Panax ginseng* fruit. Acta Pharm Sin 16:860–863
124. Cheng XJ, Liu YL, Lin GF, Luo XT (1986) Effects of ginseng root saponins on central transmitters and plasma corticosterone in hypothermia-stressed rats. Acta Pharmacol Sin 7:6–8
125. Yun YS, Lee SY, Kim BS, Yun TK (1980) Studies on the mechanism of action of the cytotoxic fraction from Korean ginseng roots. I. Effects of petroleum ether fraction from Korean ginseng roots on the biosynthesis of macromolecules in mammalian neoplastic cells. Hanguk Saenghwa Hakhoe Chi 13:203–217 (CA 95:126682d)
126. Yun YS, Lee SY, Kim BS, Yun TK (1980) Studies on the mechanism of action of the cytotoxic fraction from Korean ginseng roots. II. Effects of ethyl acetate fraction from Korean ginseng

roots on the biosynthesis of macromolecules in mammalian neoplastic cells. Hanguk Saenghwa Hakhoe Chi 13:219–234 (CA 95:126683e)
127. Tahara M, Kono H, Mune S, Odashima S (1985) Action of ginsenosides on tumor cells. 1. Growth inhibition and redifferentiation of neoplasia. Wakan Iyaku Gakkaishi 2:170–171 (CA 104:28514k)
128. Abe H, Arichi S, Hayashi T, Odashima S (1979) Ultrastructural studies of Morris hepatoma cells reversely transformed by ginsenosides. Experientia 35:1647–1649
129. Odashima S, Nakayabu Y, Honjo N, Abe H, Arichi S (1979) Induction of phenotypic reverse transformation by ginsenosides in cultured Morris hepatoma cells. Eur J Cancer 15:885–892
130. Arichi S (1980) Antitumor agent containing ginsenosides. Fr Demande 2,430,234 (CA 93:53959w)
131. Fan IJ, Han R (1979) Effect of harringtonine on L-1210 leukemia cells and normal bone marrow cells in mice. Acta Pharm Sin 14:467–473
132. Zhou QX, Han R (1983) Effects of the total saponins in *Panax ginseng* on bone morrow hemopoietic stem cells in mice. Chin Trad Herb Drugs 14:27–29
133. Kim KJ, Kim HS (1982) Effects of *Panax ginseng* on the side effects of anticancer agent. Chungnam Uidae Chapchi 9:157–164
134. Rhee DK, Lim CJ, Kim DH, Hong SK, Park EH, Han YN (1982) Studies on the acute and subacute toxicity of ginseng saponin. Yakhak Hoe Chi 26:209–214 (CA 98:100786c)
135. Wang BX, Cui JC, Liu AJ (1980) Antidiuretic effect of ginsenosides of the stems and leaves of *Panax ginseng*. Acta Pharmacol Sin 1:126–130
136. Namba T, Yoshizaki M, Tomimori T, Kobashi K, Mitsui K, Hase J (1973) Hemolytic and its protective activity of ginseng saponin. Chem Pharm Bull (Tokyo) 21:459
137. Namba T, Yoshizaki M, Tomimori T, Kobashi K, Mitsui K, Hase J (1974) Fundamental studies on the evaluation of the crude drugs. I. Hemolytic and its protective activity of ginseng saponins. Planta Med 25:28
138. Odani T, Zanizawa H, Takino Y (1983) Studies on the absorption, distribution, excretion and metabolism of ginseng saponins. II. The absorption, distribution and excretion of ginsenoside Rg_1 in the rat. Chem Pharm Bull (Tokyo) 31:292–298
139. Han BH, Park MH, Han YN, Woo LK, Sankawa U, Yahara S, Tanaka O (1981) The transformation of ginsenosides by acid catalysis in gastric pH. Soul Taehakkyo Saengyak Yonguso Opjukjip 20:7–13 (CA 98:65039n)
140. Odani T, Tanizawa H, Takino Y (1983) Studies on the absorption, distribution, excretion and metabolism of ginseng saponin. III. The absorption, distribution and excretion of ginsenoside-Rb_1 in the rat. Chem Pharm Bull (Tokyo) 31:1059–1066
141. Joo CN, Koo JH, Lee HB, Yoon JB, Byun YS (1982) Biochemical studies on the absorption of ginseng saponin and its effect on metabolism in animal body. Hanguk Saenghwa Hakhoe Chi 15:189–199 (CA 97:207793j)
142. Han BH, Woo LK (1979) Synthesis of ^{3}H-labeled dammarane triterpene glycosides of Korean ginseng. Soul Taehakkyo Saengyak Yonguso Opjukjip 18:25–29 (CA 94:65992k)
143. Chen SE, Sawchuk RJ, Staba EJ (1980) American ginseng. III. Pharmacokinetics of ginsenosides in the rabbit. Eur J Drug Metab Pharmacokinet 5:161–168
144. Jenny E, Soldati F (1985) Pharmacokinetics of ginsenosides in the mini pig. In: Chang HM, Yeung HW, Tso WW, Koo A (eds) Advances in Chinese Medicinal Material Research. World Science, Singapore, pp 499–507

Panax japonicus C. A. Mey. 92

92.1 Introduction

Zhujieshen, Rhizoma Panacis japonici, the dry rootstocks of *Panax japonicus* C.A. Mey. (Araliaceae), and Zhuzishen, Rhizoma Panacis majoris, the dry rootstocks of *P. japonicus* C.A. Mey. var. *major* (Burk.) C.Y. Wu et K.M. Feng or *P. japonicus* C.A. Mey. var. *bipinnatifidus* (Seem.) C.Y. Wu et K.M. Feng, are further *Panax* species officially listed in the Chinese Pharmacopoeia. They can be collected in the fall and are used as tonic and hemostatic drugs.

92.2 Chemical Constituents

The major constituents contained in the rootstock of *P. japonicus* and its variant are also saponins. The sapogenins in the saponins of *P. japonicus* are protopanaxadiol, protopanaxatriol, and oleanolic acid, like those in the saponins of *P. ginseng*. The main saponins of *P. japonicus* are oleanolic acid-type saponins. Hydrolysis yields oleanolic acid, panaxadiol, and panaxatriol. The saponins of *P. japonicus* isolated and identified to date are listed in Table 92.1.

Table 92.1. Saponins isolated from *Panax japonicus* and variants

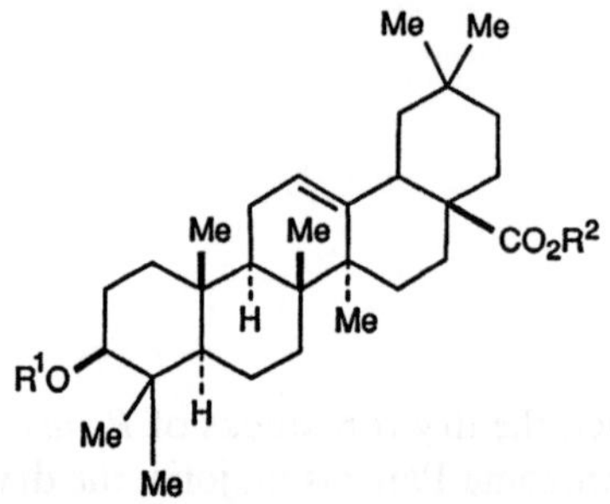

Oleanolic acid type

Saponin	R^1	R^2	Origin	Reference
Ginsenoside Ro (Chikusetsu-saponin V)	*O*-*β*-D-Glucopyranosyl-(1→2)-*β*-D-glucopyranosiduronic acid	*β*-D-Glucopyranosyl	*Panax japonicus* (rhizome) *P. japonicus* var. *major* (rhizome) *P. japonicus* var. *bipinnatifidus* (rhizome)	[1–3] [4, 5] [6]
Chikusetsu-saponin IV	*O*-*α*-L-Arabinofuranosyl-(1→4)-*β*-D-glucopyranosiduronic acid	*β*-D-Glucopyranosyl	*P. japonicus* (rhizome) *P. japonicus* var. *bipinnatifidus* (rhizome)	[2, 3, 7] [6]
Chikusetsu-saponin IVa	*O*-*β*-D-Glucopyranosiduronic acid	*β*-D-Glucopyranosyl	*P. japonicus* (rhizome) *P. japonicus* var. *major* (rhizome) *P. japonicus* var. *bipinnatifidus*	[3, 8] [4, 9] [6]
Chikusetsu-saponin IVa methyl ester	*O*-*β*-D-Glucopyranosiduronic acid methyl ester	*β*-D-Glucopyranosyl	*P. japonicus* var. *major* (rhizome)	[9]
Chikusetsu-saponin Ib	*O*-*α*-L-Arabinofuranosyl-(1→4)-*β*-D-glucopyranosiduronic acid 6-*β*-D-glucopyranosylester	H	*P. japonicus* (rhizome)	[8]
Oleanolic acid 28-*β*-D-glucopyranosyl ester	H	*β*-D-Glucopyranosyl	*P. japonicus* (rhizome) *P. japonicus* var. *major* (rhizome)	[10] [5]
Pjs-2	*O*-*β*-D-Xylopyranosyl-(1→2)-*β*-D-glucopyranosiduronic acid	*β*-D-Glucopyranosyl	*P. japonicus* (rhizome)	[11]
3-*O*-*β*-D-Glucopyranosiduronic acid-(methylester)-oleanolic acid	*O*-*β*-D-Glucopyranosiduronic acid methyl ester	H	*P. japonicus* (root, stem)	[12]
Zingiberoside R_1	*O*-*β*-D-Glucopyranosyl-(1→2)-*β*-D-glucopyranosiduronic	H	*P. japonicus* var. *bipinnatifidus* (rhizome)	[6]

Table 92.1. (continued)

Saponin	R^1	R^2	Origin	Reference
	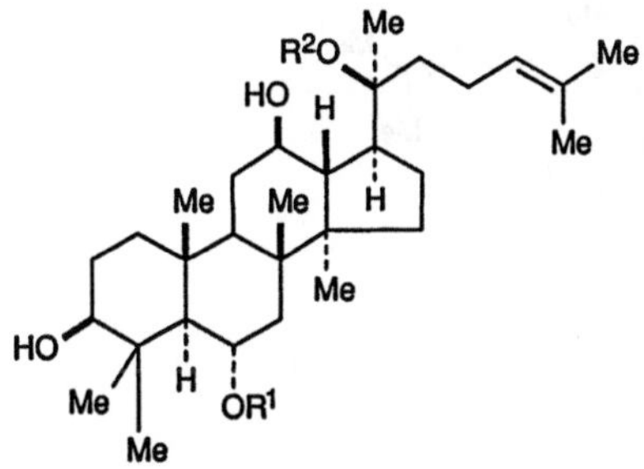 Protopanaxatriol type			
Ginsenoside Re	*O*-α-L-Rhamnopyranosyl-(1→2)-β-D-glucopyranosyl	*O*-β-D-Glucopyranosyl	*P. japonicus* (rhizome, leaf) *P. japonicus* var. *major* (rhizome, leaf) *P. japonicus* var. *bipinnatifidus* (rhizome)	[3, 10, 13] [5, 9] [6]
20-Glucoginsenoside Rf	*O*-β-D-Glucopyranosyl-(1→2)-β-D-glucopyranosyl	*O*-β-D-Glucopyranosyl	*P. japonicus* var. *major* (rhizome)	[4]
Ginsenoside Rg_1	*O*-β-D-Glucopyranosyl	*O*-β-D-Glucopyranosyl	*P. japonicus* (rhizome) *P. japonicus* var. *major* (leaf) *P. japonicus* var. *bipinnatifidus* (rhizome)	[3] [9] [6]
Ginsenoside Rg_2	*O*-α-L-Rhamnopyranosyl-(1→2)-β-D-glucopyranosyl	H	*P. japonicus* (rhizome) *P. japonicus* var. *major* (rhizome, leaf) *P. japonicus* var. *bipinnatifidus* (rhizome)	[3, 8] [5, 9] [6]
Ginsenoside F_1	H	*O*-β-D-Glucopyranosyl	*P. japonicus* (leaf)	[13]
Notoginsenoside R_2	*O*-β-D-Xylopyranosyl-(1→2)-β-D-glucopyranosyl	H	*P. japonicus* (rhizome) *P. japonicus* var. *major* (rhizome)	[3] [4]
Chikusetsusaponin L_5	H	*O*-β-D-Xylopyranosyl-(1→4)-α-L-arabinopyranosyl-(1→6)-β-D-glucopyranosyl	*P. japonicus* (leaf)	[13]
Chikusetsusaponin L_{10}	12β-*O*-β-D-glucopyranosyl-protopanaxatriol		*P. japonicus* (leaf)	[13]

Table 92.1. (continued)

Saponin	R^1	R^2	Origin	Reference
	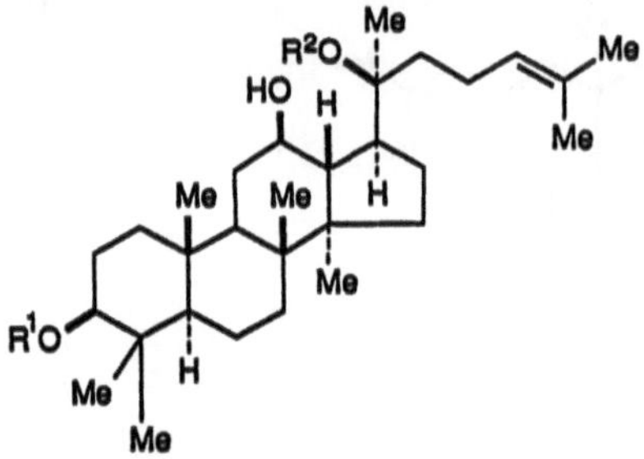			
	Protopanaxadiol type			
Ginsenoside Rb_1	O-β-D-Glucopyranosyl-(1→2)-β-D-glucopyranosyl	O-β-D-Glucopyranosyl-(1→6)-β-D-glucopyranosyl	*P. japonicus* var. *major* (leaf) *P. japonicus* var. *bipinnatifidus* (rhizome)	[14] [6]
Ginsenoside Rb_3	O-β-D-Glucopyranosyl-(1→2)-β-D-glucopyranosyl	O-β-D-Xylopyranosyl-(1→6)-β-D-glucopyranosyl	*P. japonicus* var. *major* (leaf)	[14]
Ginsenoside Rc	O-β-D-Glucopyranosyl-(1→2)-β-D-glucopyranosyl	O-α-L-Arabinofuranosyl-(1→6)-β-D-glucopyranosyl	*P. japonicus* var. *major* (leaf)	[14]
Ginsenoside Rd	O-β-D-Glucopyranosyl-(1→2)-β-D-glucopyranosyl	O-β-D-Glucopyranosyl	*P. japonicus* (rhizome) *P. japonicus* var. *major* (leaf) *P. japonicus* var. *bipinnatifidus* (rhizome)	[3] [9, 14] [6]
Chikusetsusaponin Ia	O-β-D-Xylopyranosyl-(1→6)-β-D-glucopyranosyl	H	*P. japonicus* (rhizome)	[8]
Chikusetsusaponin III	O-β-D-Glucopyranosyl-(1→2)-[β-D-xylopyranosyl-(1→6)]-O-β-D-glucopyranosyl	H	*P. japonicus* (rhizome)	[2, 15]
	Saponins with other sapogenins			
Majoroside F_1	20(s)-β-D-Glucopyranosyloxy-dammar-25-ene-3β, 12β, 24β-triol 3-O-β-D-glucopyranosyl-(1→2)-β-D-glucopyranoside		*P. japonicus* (leaf)	[9]
Majoroside F_2	20(s)-β-D-Glucopyranosyloxy-dammar-25-ene-3β, 12β, 24α-triol 3-O-β-D-glucopyranoside		*P. japonicus* (leaf)	[9]
Majoroside F_3	20(s)-β-D-Glucopyranosyloxy-dammar-23-ene-3β, 12β, 22-triol 3-O-β-D-glucopyranoside		*P. japonicus* (leaf)	[9]
Majoroside F_4	20(s)-β-D-Glucopyranosyloxy-dammar-23-ene-3β, 12β, 25-triol 3-O-β-D-glucopyranoside		*P. japonicus* (leaf)	[9]
24(S)-Pseudoginsenoside	3β, 12β, 25-20, 24-Epoxytrihydroxydammaran-6α-yl-O-α-L-rhamnopyra-		*P. japonicus* var. *bipinnatifidus* (rhizome)	[6]

References

1. Kondo N, Marumoto Y, Shoji J (1971) Studies on the constituents of Panacis japonici rhizoma. IV. The structure of chikusetsusaponin V. Chem Pharm Bull (Tokyo) 19:1103–1107
2. Kondo N, Shoji J (1968) Studies on the constituents of panacis japonici Rhizoma. I. Isolation and purification of the saponins. 1. Yakugaku Zasshi 88:325
3. Morita T, Kasai R, Kohda H, Tanaka O, Zhou J, Yang T (1983) Chemical and morphological study on Chinese *Panax japonicus* C.A. Meyer (Zhujie-shen). Chem Pharm Bull (Tokyo) 31:3205
4. Morita T, Kasai R, Tanaka O, Zhou J, Yang TR, Shoji J (1982) Saponins of Zutiseng, rhizomes of *Panax japonicus* C.A. Meyer var. *major* (Burk.) C. Y. Wu et K. M. Feng, collected in Yunnan, China. Chem Pharm Bull (Tokyo) 30:4341–4345
5. Wang DQ, Fan J, Wang XB, Feng BS, Yang CR, Zhou J, Ning YC, Tao JX (1988) Saponins of the rhizome of *Panax japonicus* C.A. Meyer var. *major* (Burk.) Wu et Feng collected in Qinling Mountain (Shaanxi). Acta Bot Sin 30:403–408
6. Wang DQ, Fan J, Li SR, Wang XB, Feng BS, Yang CR, Zhou J, Ning YC, Feng YP, Tao JX (1988) Saponins of the rhizome of *Panax japonicus* var. *bipinnatifidus* and their significance for chemotaxonomy. Acta Bot Yunnan 10:101–104
7. Kondo N, Shoji J, Nagumo N, Komatsu N (1969) Studies on the constituents of Panacis japonici rhizoma. II. The structure of chikusetsusaponin IV and some observations on the structural relationship with araloside A. Yakugaku Zasshi 89:846–850
8. Lin TD, Kondo N, Shoji J (1976) Studies on the constituents of Panacis japonici rhizoma, V. The structures of chikusetsusaponin I, Ia, Ib, IVa and glycoside P_1. Chem Pharm Bull (Tokyo) 24:253–261
9. Feng BS, Wang XB, Wang DQ, Yang CR, Zhou J (1987) Dammarane saponins of leaves of *Panax japonicus* var. *major* collected in Qinling Mountain, China. 1. Acta Bot Yunnan 9:477–484
10. Cai P, Xiao ZY, Wei JX (1982) Studies on the chemical constituents of Zhu Jie Shen (*Panax japonicus*). I. Chin Trad Herb Drugs 13:1–2
11. Cai P, Xia ZY (1984) Chemical constituents of Japanese ginseng (*Panax japonicus*). II. Chin Trad Herb Drugs 15:241–246
12. Peng SL, Xiao R, Xiao ZY (1987) Chemical constituents of Daye Zhuzishen (*Panax japonicus*). I. Chin Trad Herb Drugs 18:346–347
13. Yahara S, Kasai R, Tanaka O (1977) New dammarane type saponins of leaves of *Panax japonicus* C.A. Meyer. I. Chikusetsusaponin-L5, -L9a and -L10. Chem Pharm Bull (Tokyo) 25:2041–2047
14. Yang TR, Wu MZ, Zhou J (1984) Triterpenoid saponins of leaves of *Panax japonicus* C.A. Meyer var. *major* (Burk.) Wu et Feng. Acta Bot Yunnan 6:118–120
15. Kondo N, Aoki K, Ogawa H, Kasai R, Shoji J (1970) Studies on the constituents of Panacis japonici rhizoma. III. The structure of chikusetsusaponin III. Chem Pharm Bull (Tokyo) 18:1558–1562

Panax notoginseng (Burk.) F. H. Chen 93

93.1 Introduction

Sanqi, Radix Notoginseng, is the dry roots of *Panax notoginseng* (Burk.) F. H. Chen (Araliaceae), another plant of the genus *Panax* used in traditional Chinese medicine and officially listed in the Chinese Pharmacopoeia. It should be collected in the fall before the plants bloom. The medicinal use of notoginseng roots is different from that of ginseng roots. Notoginseng roots are used mainly as a hemostatic drug in the treatment of different types of bleeding.

93.2 Chemical Constituents

The main constituents in notoginseng are also saponins, especially of the protopanaxadiol and protopanaxatriol types. Thus, a number of ginsenosides such as Rb_1, Rb_2, Rb_3, Rc, Rd, Re, Rg_1, Rg_2-Rh_1, F_2, and glucoginsenoside R_f were isolated from the underground part or above ground part of *P. notoginseng* [1–5]. In addition to the ginsenosides, a number of new saponins named notoginsenosides have been isolated and structurally investigated, which are listed in Table 93.1.

Table 93.1. Notoginsenosides isolated from *Panax notoginseng*

Notoginsenoside	R[1]	R[2]	Origin	Reference
	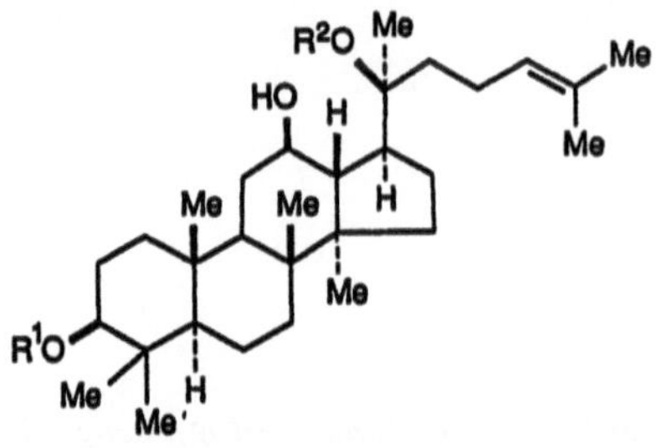 Protopanaxadiol type			
R4	*O*-β-D-Glucopyranosyl-(1→2)-β-D-glucopyranosyl	*O*-β-D-Xylopyranosyl-(1→6)-β-D-glucopyranosyl-(1→6)-β-D-glucopyranosyl	Root	[3]
Fa	*O*-β-D-Xylopyranosyl-(1→2)-β-D-glucopyranosyl-(1→2)-β-D-glucopyranosyl	*O*-β-D-Glucopyranosyl-(1→6)-β-D-glucopyranosyl	Leaf, seed	[5]
Fc	*O*-β-D-Xylopyranosyl-(1→2)-β-D-glucopyranosyl-(1→2)-β-D-glucopyranosyl	*O*-β-D-Xylopyranosyl-(1→6)-β-D-glucopyranosyl	Leaf, seed	[5]
Fe	*O*-β-D-Glucopyranosyl	*O*-β-D-Arabinofuranosyl-(1→6)-β-D-glucopyranosyl	Leaf	[5]
Gypenoside XVII	*O*-β-D-Glucopyranosyl	*O*-β-D-Glucopyranosyl-(1→6)-β-D-glucopyranosyl	Root	[3]
Gypenoside IX	*O*-β-D-Glucopyranosyl	*O*-β-D-Xylopyranosyl-(1→6)-β-D-glucopyranosyl	Leaf, seed	[5]

Table 93.1. (continued)

Notoginsenoside	R^1	R^2	Origin	Reference
	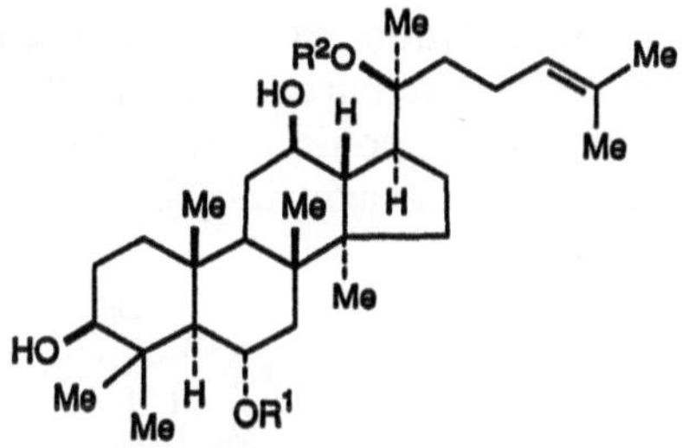			
	Protopanaxatriol type			
R1	*O*-β-D-Xylopyranosyl-(1→2)-β-D-gluco-pyranosyl	*O*-β-D-Glucopyranosyl	Root Corm	[6] [3]
R2	*O*-β-D-Xylopyranosyl-(1→2)-β-D-gluco-pyranosyl	H	Root	[6]
R3	*O*-β-D-Glucopyranosyl	*O*-β-D-Glucopyranosyl-(1→6)-β-D-gluco-pyranosyl	Root	[3]
R6	*O*-β-D-Glucopyranosyl	*O*-α-D-Glucopyranosyl-(1→6)-β-D-gluco-pyranosyl	Root	[3]

The structure of the major notoginsenoside, notoginsenoside R_1 (*93-1*), is given below.

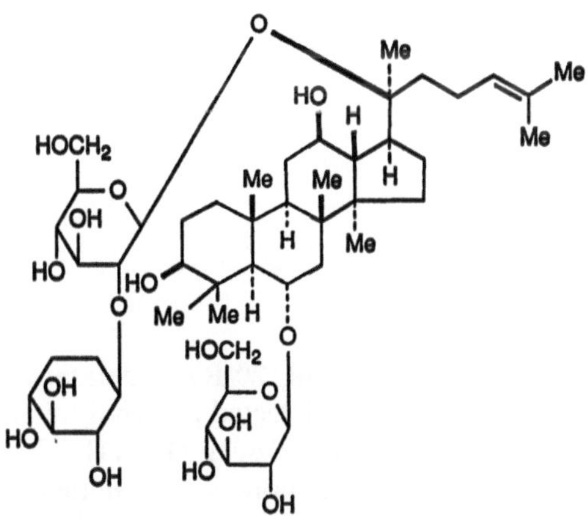

Notoginsenoside R_1 (*93-1*)
(Protopanaxatriol 20-*O*-β-D-xylopyranosyl-(1 → 2)-β-D-glucopyranosyl 6-*O*-β-D-glucopyranoside)

Ginsenosides Rb_1, Rg_1, and notoginsenoside R_1 are representatives of protopanaxadiol and protopanaxatriol types which are used in the qualitative determination of notoginseng root and in the differentiation of notoginseng root from ginseng root in the Chinese Pharmacopoeia. In addition, two new sapogenins, dammar-20(22)ene-3β,12β,26-triol (*93-2*) and 20(R)-dammarane-3β,12β,20,25-tetrol (*93-3*), were isolated from the hydrolysate of crude saponin extracted from the leaves of *P. notoginseng* together with panaxadiol and panaxatriol [7].

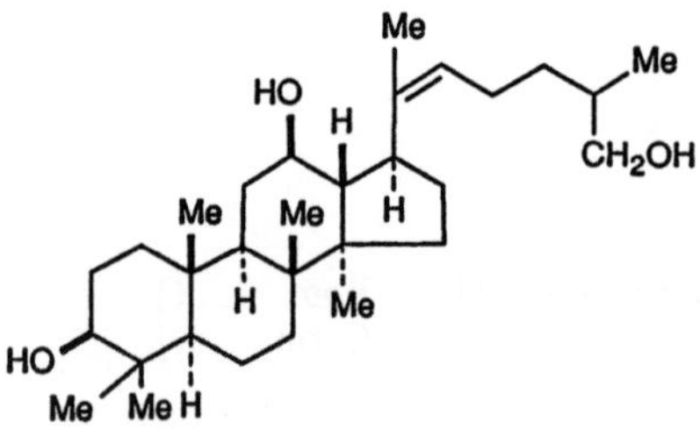

Dammar-20(22)-ene-
3β,12β,26-triol (*93-2*)

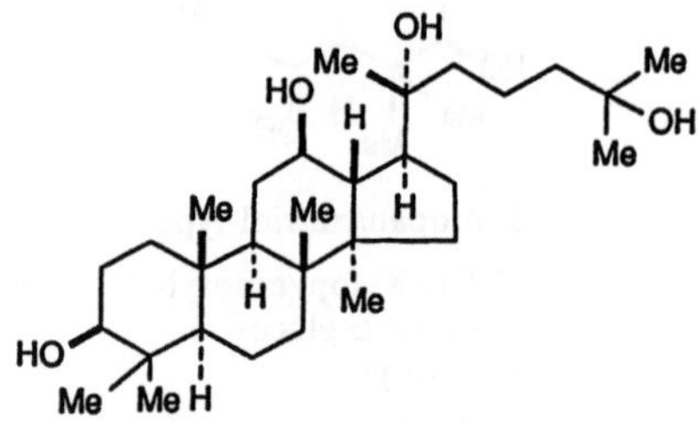

20(R)-Dammarane-
3β,12β,20,25-tetrol (*93-3*)

Sapogenins from the hydrolysate of the saponin fraction isolated from notoginseng flowers were identified as panaxadiol, dammar-20(22)-ene-3β,12β,26-triol, 20(R)-dammarane-3β,12β,20,25-tetrol, and a new sapogenin with a cyclic ether structural feature [8]. This oxepane derivative (*93-4*) was also found in the leaves of notoginseng [9].

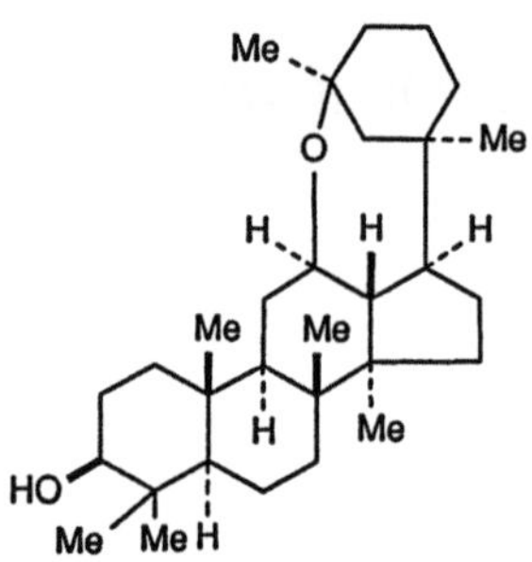

Oxepane sapogenin (*93-4*)

Panaxadiol, panaxatriol, dammar-20(22)-ene-3β,12β,26-triol, 20(R)-dammarane-3β,12β,20,25-tetrol, and 20(R)-protopanaxatriol have been detected as sapogenins of the saponin fraction isolated from the rootlets of *P. notoginseng* [10, 11]. The isolation of a new saponin, sanchinoside B_1 (*93-5*), from the rootlets of *P. notoginseng* was also reported [12].

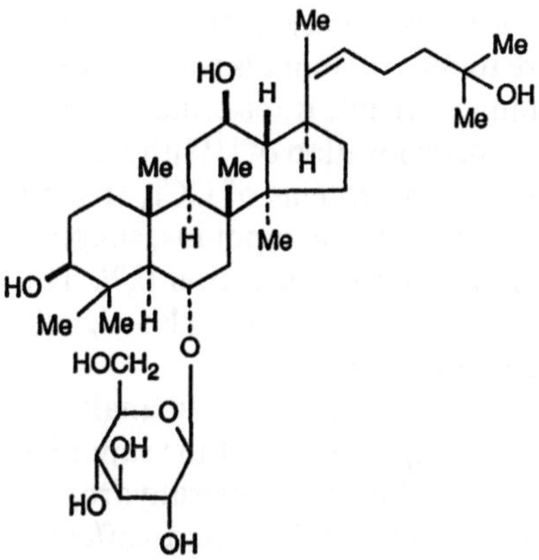

Sanchinoside B_1 (*93-5*)

The major components in the essential oil of the root of *P. notoginseng* have been identified as α-guaiene, β-guaiene, and octadecane [13] and in that of the flower γ-elemene, heptacosane, and pentacosane [14].

93.3 Pharmacology

Panax notoginseng showed a wide spectrum of pharmacological activities, such as hemostatic activity, platelet aggregation inhibitory activity, antiinflammatory activity, and therapeutic effects in cardiac infarction, cardiac ischemia, and angina pectoris [15]. The saponin fraction of notoginseng given intravenously to dogs decreased blood pressure and peripheral vascular resistance. The hypotensive effect of the saponins appeared to be due primarily to direct dilation of the blood vessels [16] caused by inhibition of Ca^{2+}-influx [1, 17–20]. Heart rate was not affected [16]. Antiarrhythmic effects of the total saponions [21] and of panaxatriol saponins [22] of notoginseng were recorded on chloroform-induced ventricular fibrillation in mice, $BaCl_2$- or aconitine-induced arrhythmia in rats, and epinephrine-induced arrhythmia in rabbits.

The extract of *P. notoginseng* consisting mostly of saponins and flavones decreased myocardial cAMP and cGMP, when given orally to mice for 1 week. In rabbits with experimental heart ischemia, the ST segment of the T wave was normalized after treatment with the extract in comparison to the control animals [23]. Intravenous injection of the saponins from notoginseng was also found to be effective in protecting rabbits against hemorrhagic shock due to the improvement of heart function [24].

The total saponins of *P. notoginseng* were effective against several experimental inflammations in mice and rats. The antiinflammatory effect was stronger in normal mice than in adrenalectomized rats [25]. The saponins decreased the ascorbic acid content in adrenals of intact but not of hypophysectomized rats [25, 26]. Thus, in addition to the direct antiinflammatory effect, the saponins may act indirectly via the pituitary-adrenocortical system.

The increase in the serum level of glutamic-pyruvic transaminase induced by CCl_4 in mice was markedly inhibited by subcutaneous treatment of animals with the total saponins from *P. notoginseng*. In contrast, the glutamic-pyruvic transaminase activity in serum of normal animals was not affected. The incorporation of [^{3}H]thymidine into liver DNA and [^{3}H]leucine into both liver and serum proteins of CCl_4-treated mice was significantly increased by subcutaneous injection of the saponins, suggesting that the total saponins of *P. notoginseng* stimulate liver regeneration [27]. The synthesis of protein and DNA in liver, kidney, and testis of mice was also significantly increased by oral treatment of notoginseng extract [23].

Intraperitoneal injection of the total saponins of *P. notoginseng* showed analgesic activity in mice comparable to that of aminopyrine. The appearance of the saponin-induced analgesia was faster, but shorter than that of morphine and 1-tetrahydropalmatine. The total saponins also induced a sedative effect and inhibited caffeine-induced locomotive excitation [28]. The subcutaneous LD_{50} of the total saponins of *P. notoginseng* in mice was 1667 mg/kg [27].

References

1. Sanada S, Shoji J (1978) Comparative studies on the saponins of Ginseng and related crude drugs. I. Shoyakugaku Zasshi 32:96
2. Wu MZ (1979) Studies on the saponin components of plants in Yunnan. IV. Two saponins of *Panax notoginseng*. Acta Bot Yunnan 1:119–124
3. Matsuura H, Kasai R, Tanaka O, Saruwatari Y, Fuwa T, Zhou J (1983) Further studies on dammarane-saponins of Sanchi-Ginseng. Chem Pharm Bull (Tokyo) 31:2281–2287
4. Taniyasu S, Tanaka O, Yang T, Zhou J (1982) Dammarane saponins of flower buds of *Panax notoginseng* (Sanchi Ginseng). Planta Med 44:124–125
5. Yang TR, Kasai R, Zhou J, Tanaka O (1983) Dammarane saponins of leaves and seeds of *Panax notoginseng*. Phytochemistry 22:1473–1478
6. Zhou J, Wu MZ, Taniyama S, Besso H, Tanaka O, Saruwatari Y, Fuwa T (1981) Dammarane-saponins of Sanchi-Ginseng, roots of *Panax notoginseng* (Burk.) F. H. Chen (Araliaceae): structures of new saponins, notoginsenosides-R_1 and -R_2 and identification of ginsenosides-Rg_2 and Rh_1. Chem Pharm Bull (Tokyo) 29:2844–2850
7. Wei JX, Chang LY, Wang JF, Friedrichs E, Jores M, Puff H (1982) Two new dammaran sapogenins from leaves of *Panax notoginseng*. Planta Med 45:167–171
8. Wei JX, Liu K, Xu RX (1984) Studies on the sapogenins in flower buds of *Panax notoginseng*. Bull Chin Mater Med 9:223–225
9. Wei JX, Chang LY, Wang JF, Chen WS, Friedrichs E, Puff H, Breitmaier E (1984) An oxepane derivative of panaxadiol from the leaves *Panax notoginseng*. Planta Med 50:47–52
10. Wei JX, Wang JF, Chang LY, Du YC (1980) Chemical studies on San-chi, *Panax notoginseng* (Burk.) F.H. Chen. I. Studies on the constituents of San-chi root hairs. Acta Pharm Sin 15:359–364
11. Wei JX, Li B, Ma XB (1984) Studies on the sapogenins from the rootlets of *Panax notoginseng*. Bull Chin Mater Med 9:267–269
12. Wei JX, Wang LA, Du H, Li R (1985) Isolation and identification of sanchinoside B_1 and B_2 from rootlets of *Panax notoginseng* (Burk.) F.H. Chen. Acta Pharm Sin 20:288–293
13. Lu Q, Li XG (1988) Studies on the chemical constituents of the essential oil in the Renshen Sanqi (*Panax notoginseng*). Chin Trad Herb Drugs 19:5–7
14. Shuai F, Li XG (1986) Studies and comparison of chemical constituents of essential oil of *Panax notoginseng* flowers. Chin Pharm Bull 21:513–514
15. Zhang BH (1984) Recent advances in the pharmacological studies on San chi (*Panax notoginseng*). Chin Trad Herb Drugs 15:514–518
16. Wang JD, Chen JX (1984) Cardiac and hemodynamic effects of the total saponins of *Panax notoginseng*. Acta Pharmacol Sin 5:181–185

17. Guan YY, He H, Chen JX (1985) Effect of the total saponins of *Panax notoginseng* on contraction of rabbit aortic strips. Acta Pharmacol Sin 6:267–269
18. Wu JX, Chen JX (1988) Negative chronotropic and inotropic effects of *Panax notoginseng* saponins. Acta Pharmacol Sin 9:409–412
19. Xiong ZG, Chen JX, Sun JJ (1989) Effects of *Panax notoginseng* saponins on cardiac action potentials and slow inward current. Acta Pharmacol Sin 10:122–125
20. Wu JX, Chen JX (1988) Depressant actions of *Panax notoginseng* saponins on vascular smooth muscles. Acta Pharmacol Sin 9:147–152
21. Liu S, Chen JX (1984) Antiarrhythmic effects of total saponins of *Panax notoginseng*. Acta Pharmacol Sin 5:100–103
22. Li XJ, Zhang BH (1988) Studies on the antiarrhythmic effects of panaxatriol saponins (PTS) isolated from *Panax notoginseng*. Acta Pharm Sin 23:168–173
23. Zhang BH, Wang T, Zhang SX, Su HD, Cui JR, Shang JH, Wei JX (1984) Effects of *Panax notoginseng* (76017) on myocardial ischemia, cyclic nucleotides, nucleic acid and protein. Hejishu 69
24. Li LX, Wang ZC, Li SQ, Zhao WD, Zhao WJ, Zhang XF, Wei JX (1988) Effects on *Panax notoginseng* saponins on hemorrhagic shock in rabbits. Acta Pharmacol Sin 9:52–55
25. Hao CQ, Yang F (1986) Antiinflammatory effects of total saponins of *Panax notoginseng*. Acta Pharmacol Sin 7:252–255
26. Wang JD, Chen JX (1984) Effect of total saponins of *Panax notoginseng* on adrenocortical function of rats and guinea pigs. Acta Pharmacol Sin 5:50–52
27. Song LC, Lui J, Wu MZ, Zang QZ, Zhou J, Chang Y (1982) Effects of total saponins of *Panax notoginseng* on DNA and protein metabolism in mice intoxicated with carbon tetrachloride. Chin Pharm Bull 17:67–69
28. Lei WY, Shi QT, Yu SC (1984) Analgesic and CNS inhibitory effects of total saponins from the leaves of *Panax notoginseng*. Bull Chin Mater Med 9:134–137

Peucedanum praeruptorum Dunn

94

94.1 Introduction

Qianhu, Radix Peucedani, is the dry root of *Peucedanum praeruptorum* Dunn or *P. decursivum* Maxim. (Apiaceae) collected in the winter or spring. It is officially listed in the Chinese Pharmacopoeia and could be used as an expectorant and mucolytic.

94.2 Chemical Constituents

94.2.1 Chemical Constituents of *Peucedanum praeruptorum*

A number of pyranocoumarins were isolated from the root of *P. praeruptorum*. The first two were designated as praeruptorin A (*94-1*) and praeruptorin B (*94-2*) [1]. The isolation of praeruptorin E (*94-3*) was also reported [2, 3]. A series of pyranocoumarin glycosides, praeroside II (*94-4*), praeroside III (*94-5*), praeroside IV (*94-6*), and praeroside V (*94-7*) [4], furanocoumarin glycosides praeroside I (*94-8*), isorutarin (*94-9*), rutarin (*94-10*), and marmesinin (*94-11*) and coumarin glycosides scopoline and skimmin [5] were also isolated from the root of *P. praeruptorum*.

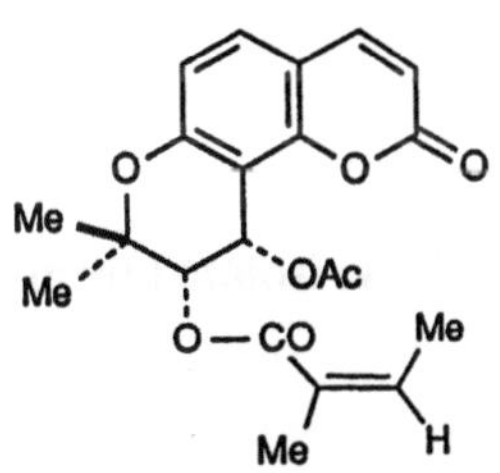
Praeruptorin A (*94-1*)

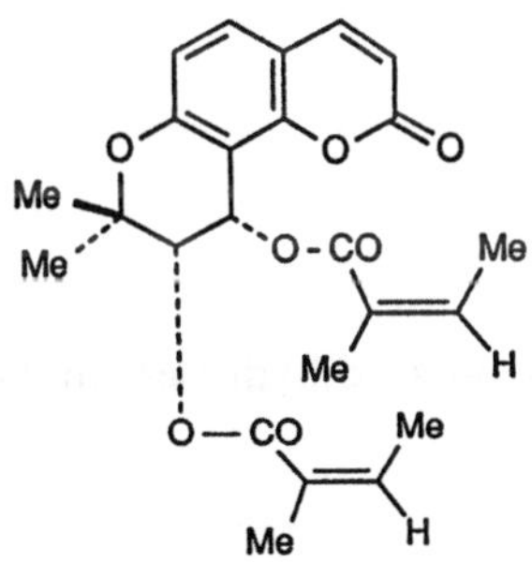
Praeruptorin B (*94-2*)

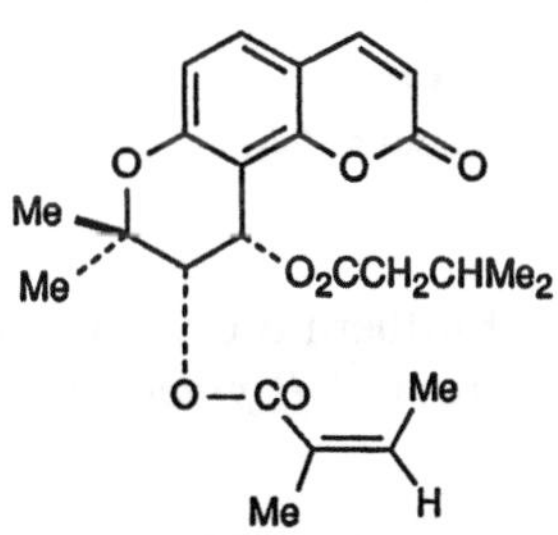
Praeruptorin E (*94-3*)

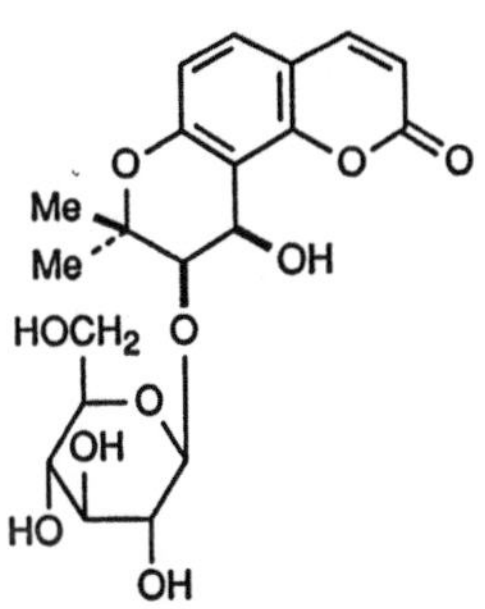
Praeroside II (*94-4*)

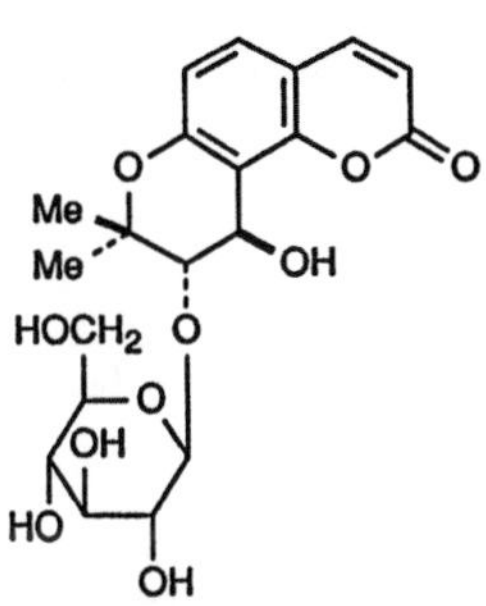
Praeroside III (*94-5*)

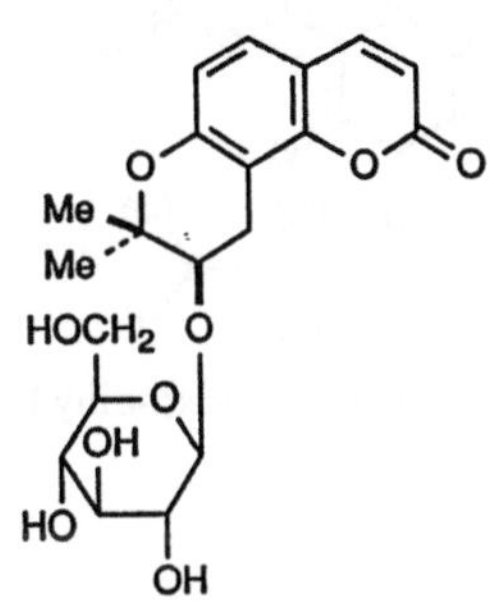
Praeroside IV (*94-6*)

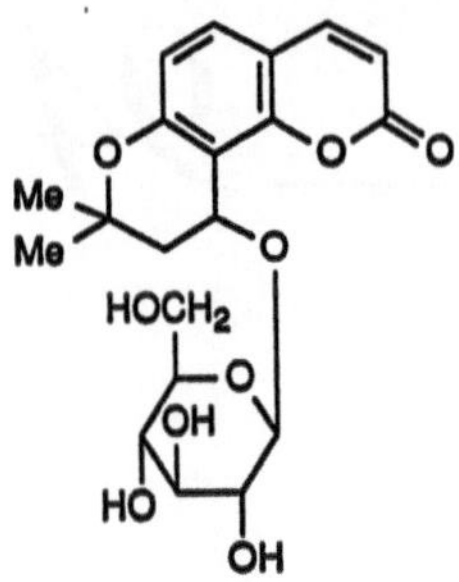

Praeroside V (*94-7*)

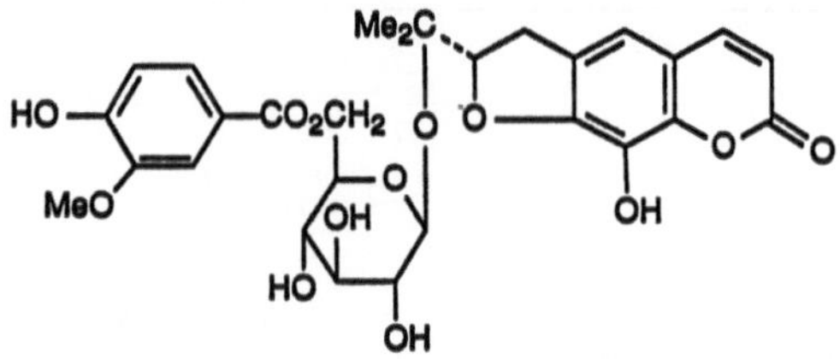

Praeroside I (*94-8*)

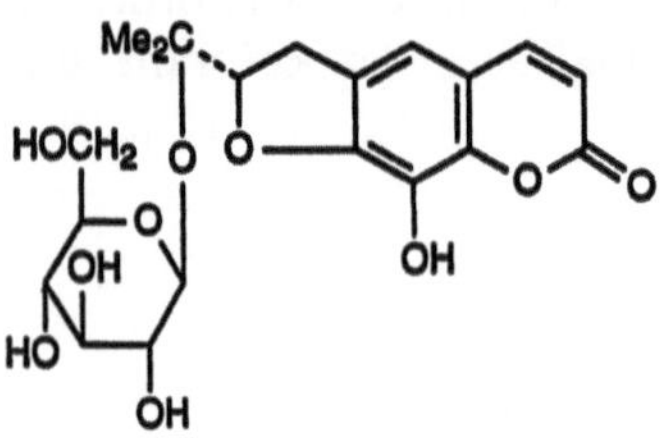

Isorutarin (*94-9*)

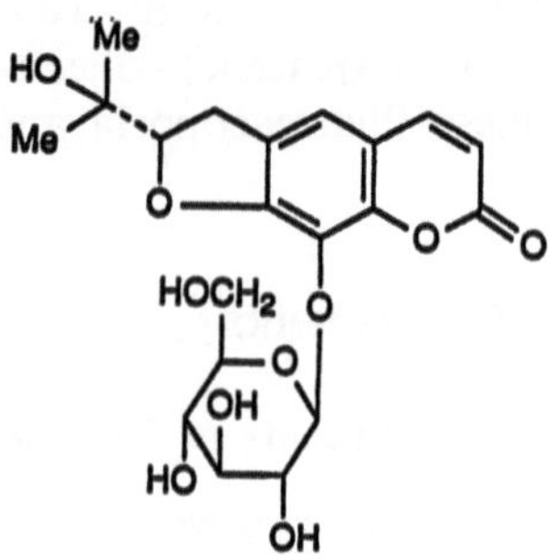

Rutarin (*94-10*)

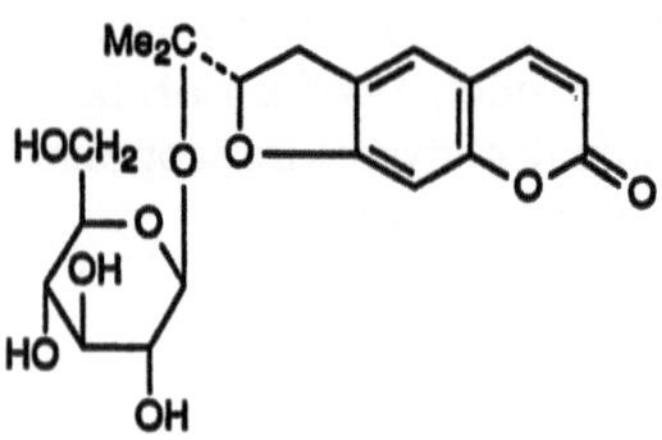

Marmesinin (*94-11*)

Furthermore, 9-angeloyloxy-10-oxo-dihydroseselin (*94-12*) [3] was isolated from the root of *P. praeruptorum*.

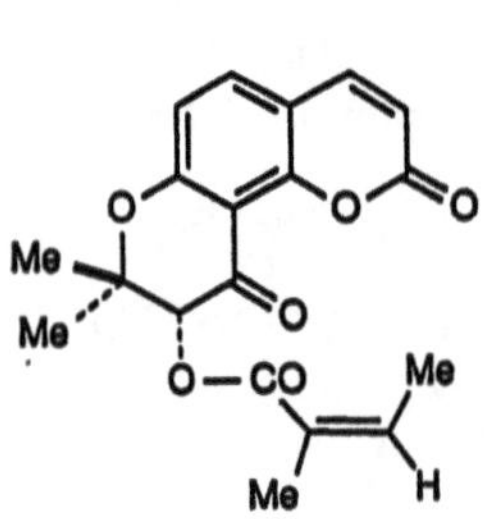

9-Angeloyloxy-10-oxo-dihydroseselin (*94-12*)

94.2.2 Chemical Constituents of *Peucedanum decursivum*

A number of known or new furocoumarins or pyranocoumarins were isolated from the root of *P. decursivum* and structurally determined. Among the furocoumarins, bergapten (*94-13*), nodakenetin (*94-14*) and its glycoside nodakenin (*94-15*) [6], furocoumarin glycosides decurosides I (*94-16*), II (*94-17*), III (*94-18*) [7], IV (*94-19*), and V (*94-20*) [8] were isolated.

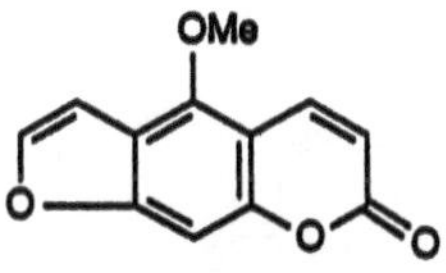

Bergapten (*94-13*)

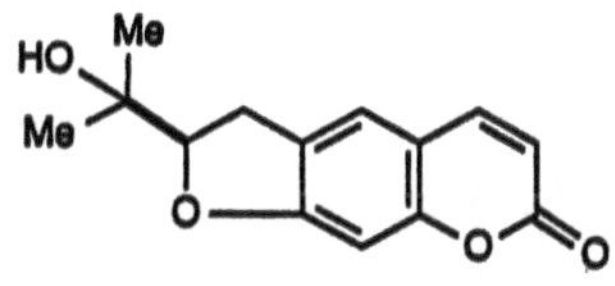

Nodakenetin (*94-14*)

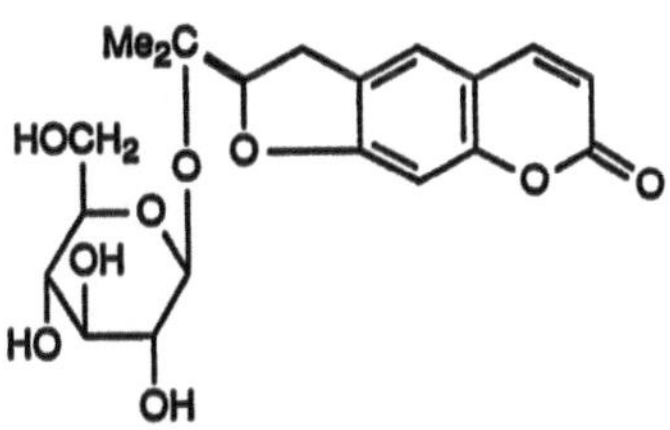

Nodakenin (*94-15*)

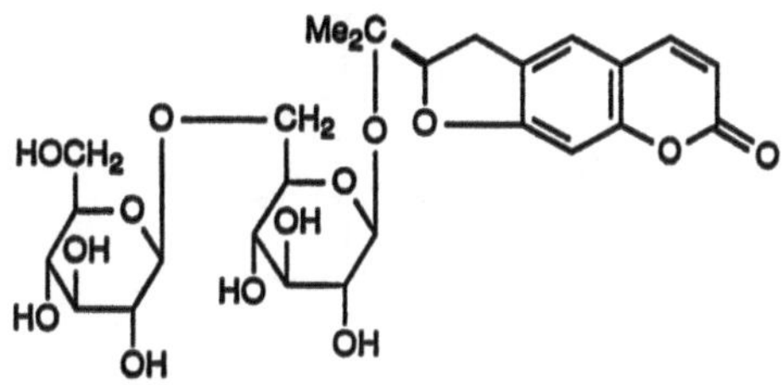

Decuroside I (*94-16*)

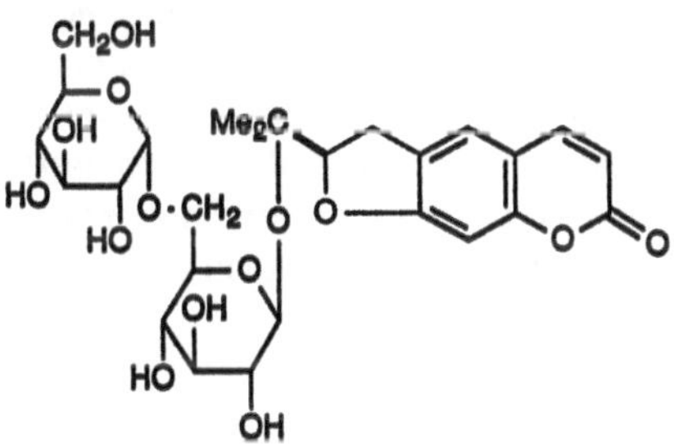

Decuroside II (*94-17*)

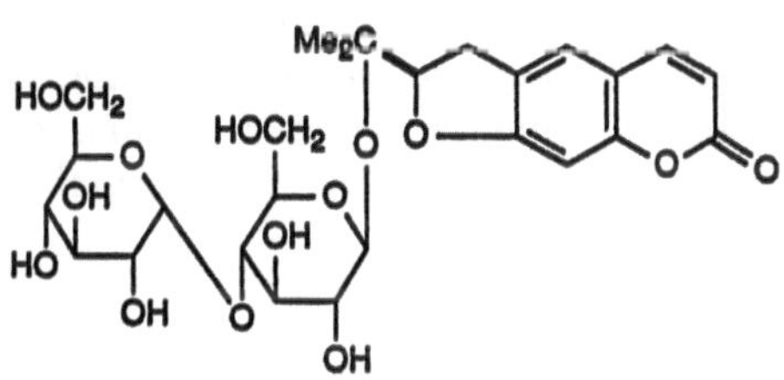

Decuroside III (*94-18*)

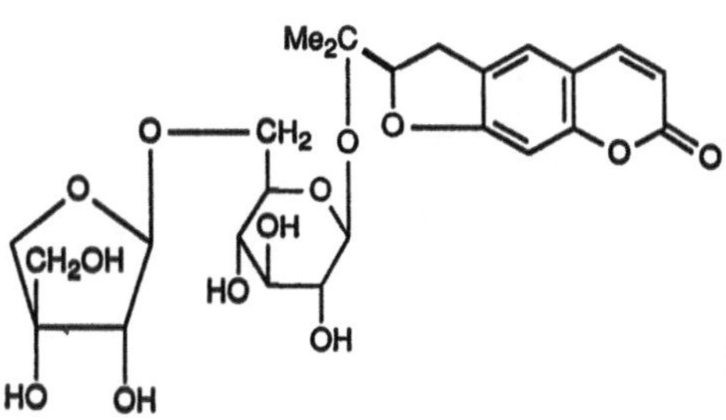

Decuroside IV (*94-19*)

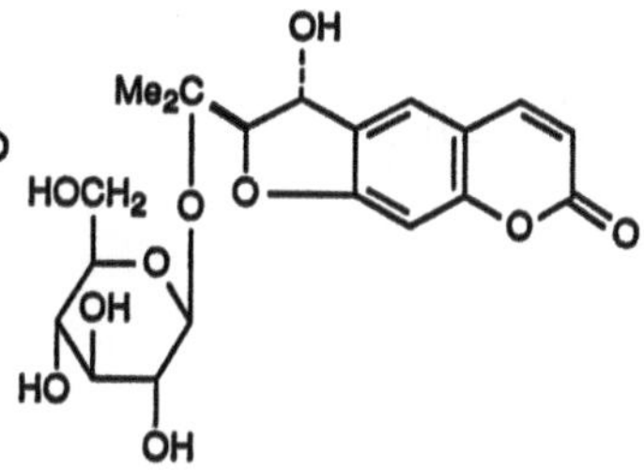

Decuroside V (*94-20*)

Pyranocoumarins isolated from the root of *P. decursivum* (*Angelica decursiva*) [8] have been identified as xanthyletin (*94-21*) derivatives decursin (*94-22*) [6, 9–11], decursidin (*94-23*) [9–11], 7-angeloyloxy-6-isovaleroyloxy-6,7-dihydroxanthyletin, 7-angeloyloxy-6-senescioyloxy-6,7-dihydroxanthyletin [11], 7-senecioyloxy-6-hydroxy-6,7-dihydroxanthyletin, and 7-hydroxy-6-senecioyloxy-6,7-dihydroxanthyletin [6]. Decursin is 7-senecioyloxy-6,7-dihydroxanthyletin, whereas decursidin is 6,7-disenecioyloxy-6,7-dihydroxanthyletin [9, 10].

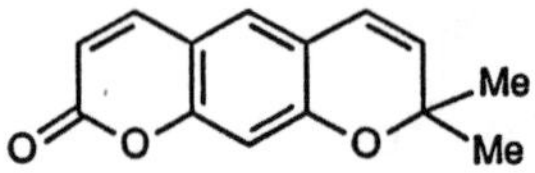

Xanthyletin (*94-21*)

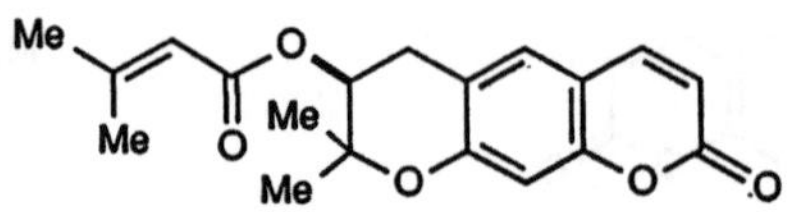

Decursin (*94-22*)

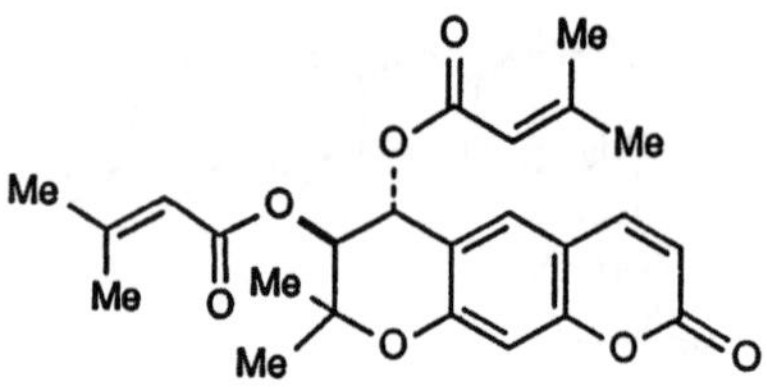

Decursidin (*94-23*)

The isolation of 7-acetoxy-6-isovaleroyloxy-6,7-dihydroxanthyletin and 7-acetoxy-6-angeloyloxy-6,7-dihydroxanthyletin was also reported [12]. Coumarin contents in the root of *P. praeruptorum* and *P. decursivum* were 0.6% and 1.1%, respectively [13]. *P. terebinthaceum* [14], *P. rubricaule* [15], and *P. medium* [16] are also used medicinally. From the root of *P. terebinthaceum*, the coumarin compound suberosin (*94-24*) was detected [14], whereas from the root of *P. medium* praeruptorin F (*94-25*) was isolated [16].

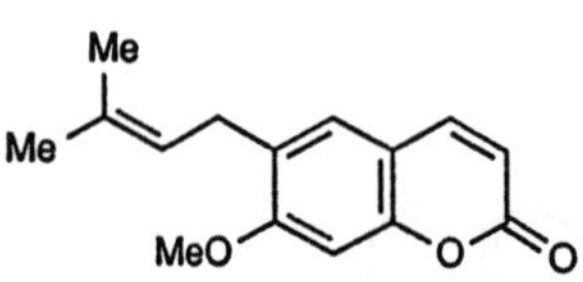

Suberosin (*94-24*)

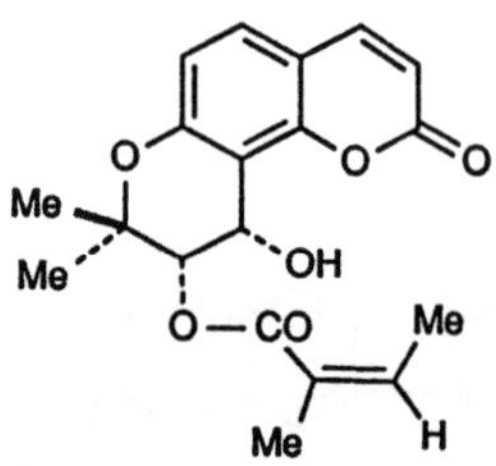

Praeruptorin F (*94-25*)

94.3 Pharmacology

The ether-soluble fraction of the root of *P. praeruptorum*, from which praeruptorins A, B, E, and 9-angeloyloxy-10-oxo-dihydroseselin are isolated, showed an antispasmodic activity on smooth muscle of guinea pig ileum and taenia coli contracted by acetylcholine and histamine [3]. Praeruptorin A was the most effective in antagonizing the action of acetylcholine noncompetitively in small intestine [17]. A study on the effect of coumarine against human platelet aggregation induced by ADP in vitro showed that nodakenin and nodakenetin were most active in inhibiting both the primary and secondary wave of aggregation [18]. Among the decurosides, decuroside III and decuroside IV showed the strongest inhibiting activity against aggregation of human platelets [7].

References

1. Chen ZX, Huang BS, She QL, Zeng GF (1979) Study on the chemical constituents of the Chinese medicinal plant, *Peucedanum praeruptorum* Dunn. Structures of four new coumarins. Acta Pharm Sin 14:486–496
2. Ye JS, Zhang HQ, Yuan CQ (1982) Isolation and identification of coumarin praeruptorin E from the root of the Chinese drug *Peucedanum praeruptorum* Dunn (Umbelliferae). Acta Pharm Sin 17:431–434
3. Okuyama T, Shibata S (1981) Studies on coumarins of a Chinese drug "Qian-Hu". Planta Med 42:89–96
4. Takata M, Okuyama T, Shibata S (1988) Studies on coumarins of a Chinese drug, Qian-Hu: VIII. Structures of new coumarin glycosides of Bai-Hua-Qian-Hu. Planta Med 54:323–327
5. Okuyama T, Takata M, Shibata S (1989) Studies on coumaringlycosides of a Chinese drug Qian-Hu. IX. Structures of linear furano- and simple-coumarin glycosides of Bai-Hua-Qian-Hu. Planta Med 55:64–67
6. Sakakibara I, Okuyama T, Shibata S (1982) Studies on coumarins of a Chinese drug "Qian-Hu". III. Coumarins from "Zi-Hua Qian-Hu". Planta Med 44:199–203
7. Matano Y, Okuyama T, Shibata S, Hoson M, Kawada T, Osada H, Noguchi T (1986) Studies on coumarins of a Chinese drug "Qian-Hu". VII. Structures of new coumarin-glycosides of Zi-Hua Qian-Hu and effect of coumarin-glycosides on human platelet aggregation. Planta Med 52:135–138
8. Asahara T, Sakakibara I, Okuyama T, Shibata S (1984) Studies on coumarins of a Chinese drug "Qian-Hu" V. Coumarin-glycosides from "Zi-Hua Qian-Hu". Planta Med 50:488–492
9. Hata K, Sano K (1967) Coumarins from the root of *Angelica decursiva*. I. Structure of decursin and decursidin. Yakugaku Zasshi 89:549–557
10. Sano K, Yosioka I, Kitagawa I (1973) Stereostructures of decursin, decursidin, and a new coumarin isolated from *Angelica decursiva*. Chem Pharm Bull (Tokyo) 21:2095–2097
11. Sano K, Yosioka I, Kitagawa I (1975) Studies on coumarins from the root of *Angelica decursiva*. II. Stereostructures of decursin, decursidin and other new pyranocoumarin derivatives. Chem Pharm Bull (Tokyo) 23:20–28
12. Sakakibara I, Okuyama T, Shibata S (1984) Studies on coumarins of a Chinese drug "Qian Hu". IV. Coumarins from "Zi Hua Qian Hu". Planta Med [Suppl] 50:117–120
13. Zhang XQ, Xu LX (1984) Determination of coumarins in Qian Hu. Bull Chin Mater Med 9:79–81
14. Ye JS, Zhang HQ, Yuan CQ (1983) Chemical components of the Chinese drug *Peucedanum terebinthaceum*. Bull Nanjing Bot Garden Mem Sun Yat Sen 89–91
15. Hu RY (1986) The original plant and anatomy of Yunqianhu produced in Sichuan. Bull Chin Mater Med 11:81–82
16. Zhang HQ, Yuan CQ, Chen GY (1984–5) A preliminary study of the chemical constituents of *Peucedanum medium* (Umbelliferae). Bull Nanjing Bot Garden Mem Sun Yat Sen 136–137
17. Kozawa T, Sakai K, Uchida M, Okuyama T, Shibata S (1981) Calcium antagonistic action of a coumarin isolated from "Qian-Hu", a Chinese traditional medicine. J Pharm Pharmacol 33:317–320
18. Okuyama T, Kawasaki C, Shibata S, Hoson M, Kawada T, Osada H, Noguchi T (1986) Effect of oriental plant drugs on platelet aggregation. II. Effect of Qian-Hu coumarins on human platelet aggregation. Planta Med 52:132–134

Phellodendron amurense Rupr. 95

95.1 Introduction

Huangbai, Cortex Phellodendri, is the dry stem bark of *Phellodendron amurense* Rupr. or *P. chinense* Schneid. (Rutaceae). It is officially listed in the Chinese Pharmacopoeia and used as an antiphlogistic, antibacterial, antiinflammatory agent for the treatment of diarrhea, icterus, ulcer, carbuncle, and eczema.

95.2 Chemical Constituents

The major chemical constituents in the bark of *P. amurense* are alkaloids of the isoquinoline type. Alkaloids isolated from the bark of *P. amurense* are berberine (*47-2*), palmatine (*47-5*) [1], magnoflorine, phellodendrine (*95-1*) [2], candicine (*95-2*) [3], and jatrorrhizine (*47-3*) [4]. Phellodendrine is an alkaloid of the protoberberine type [5–8], whereas candicine is an aliphatic quaternary ammonium compound.

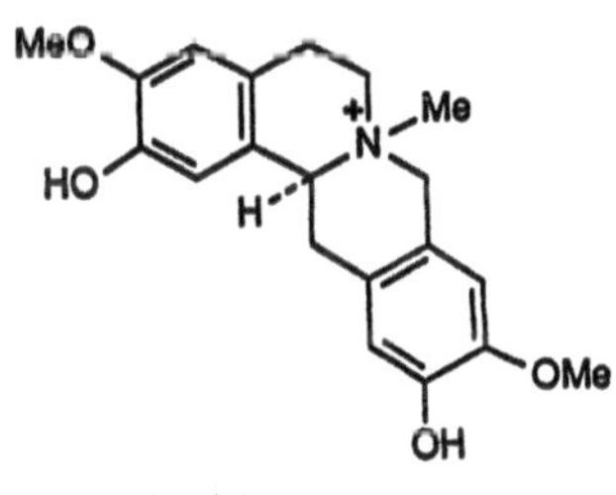

Phellodendrine (*95-1*)

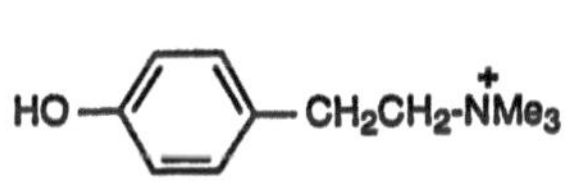

Candicine (*95-2*)

Berberine, jatrorrhizine, phellodendrine, and candicine were also isolated from the root bark, whereas the fruits including seeds contained berberine and jatrorrhizine and only berberine was isolated from the wood [9]. An aqueous acid-lime method for the isolation of berberine from the bark of *P. amurense* with a yield of 1.7% was described [10]. Comparative study on the alkaloid content in the bark of *P. chinense* and *P. amurense* showed that the bark of *P. chinense* contained much more berberine than that of *P. amurense*, but the latter had higher palmatine and jatrorrhizine contents [4]. In addition to the alkaloids, limonin and related compounds (*95-3*, *95-4*) with a γ-hydroxybutenolide moiety [11] were isolated and structurally determined.

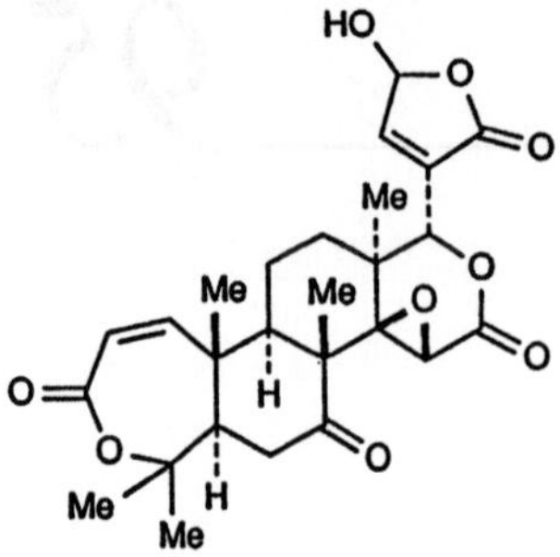

(*95-3*)

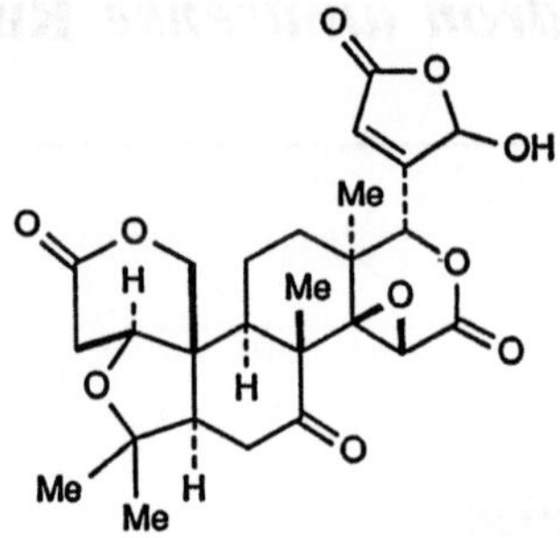

(*95-4*)

The leaves of *P. amurense* contained a number of flavone derivatives. The isolation of phellodendroside (*95-5*) with the aglycone phellamuretin [12], hyperin, phelloside (*95-6*) and 2,3-dihydrophelloside [13], noricariside (*95-7*) [14], phellatin (*95-8*), and phellavin (*95-9*) [15] was reported.

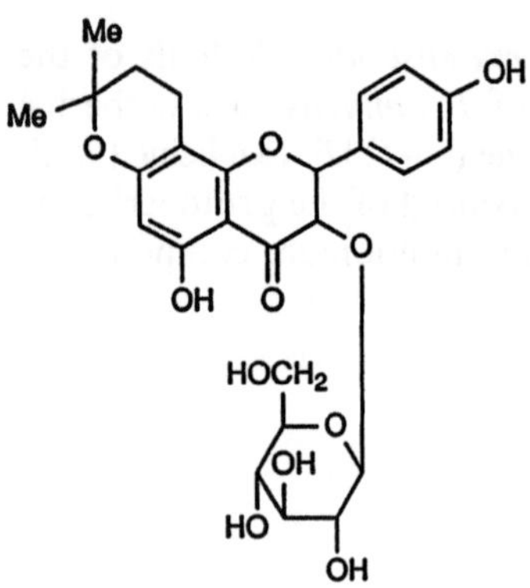

Phellodendroside (*95-5*)

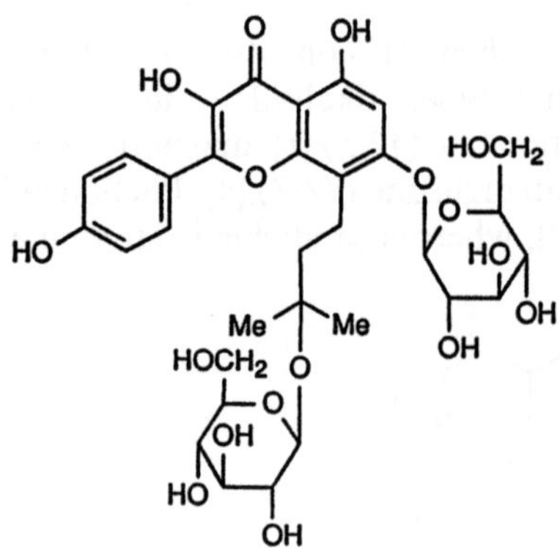

Phelloside (*95-6*)

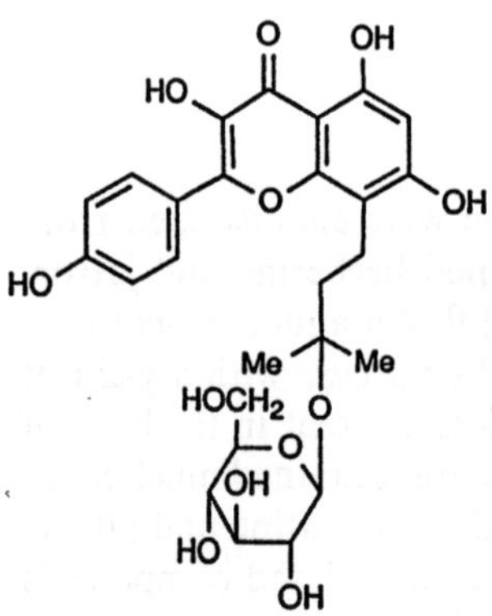

Noricariside (*95-7*)

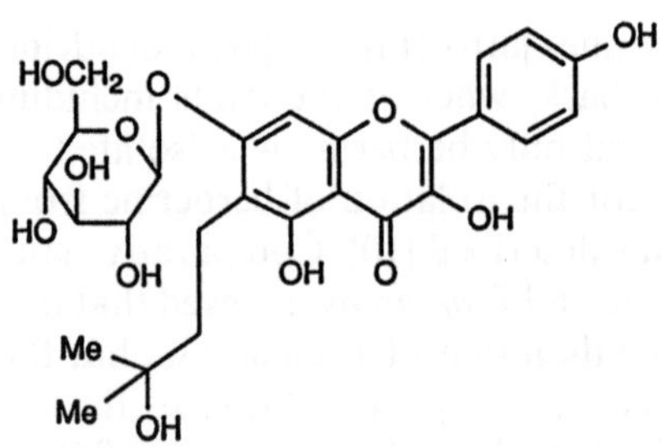

Phellatin (*95-8*)

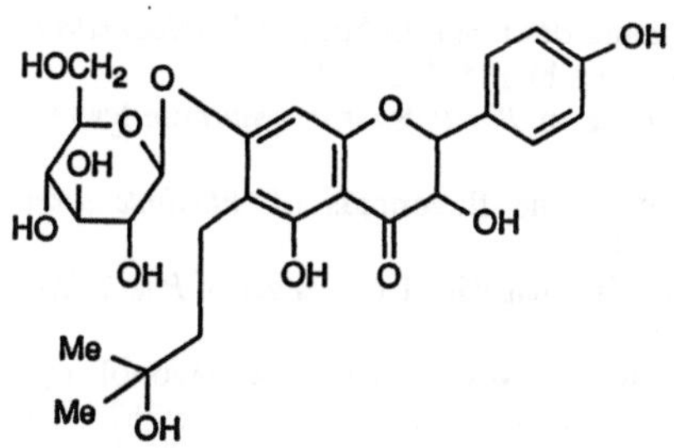

Phellavin (*95-9*)

The fruit of *P. amurense* contained essential oil, from which myrcene was isolated [16].

95.3 Pharmacology

The biological activity of berberine and related alkaloids is described under *Coptis*. Gastric infusion of the essential oil from the fruit of *P. amurense* and myrcene or direct infusion of the essential oil and myrcene into the respiratory tract of normal and vagotomized mice showed significant expectorant effects, but not in mice injected with atropine. Gastric infusion of essential oil markedly decreased total protein levels in respiratory tract lavages; this decrease in protein levels, however, was not observed in animals treated with atropine. In mice exposed to ammonia, gastric infusion of the essential oil from the fruit of *P. amurense* and myrcene prolonged the cough-inducing time [16].

References

1. Murayama Y, Shinozaki K (1926) Alkaloids of *Coptis* root and bark of *Phellodendron amurense* Rupr. J Pharm Soc Jpn 530:299–302 (CA 21:2049)
2. Tomita M, Nakano T (1957) Alkaloids of rutaceous plants. I. Alkaloids of *Phellodendron amurense*. 1. Pharm Bull (Tokyo) 5:10–12
3. Tomita M, Kunitomo J (1960) Alkaloids of rutaceous plants. XI. Alkaloids of *Phellodendron amurense*. 6. Isolation of candicine. Yakugaku Zasshi 80:1300–1301
4. Zhu M, Xiao PG (1985) Quality evaluation of Chinese traditional drug "Huang Bo". Chin Trad Herb Drugs 16:34–35
5. Tomita M, Kunitomo J (1960) Alkaloid of rutaceous plants. VII. Alkaloids of *Phellodendron amurense*. 2. Structure of phellodendrine. a. Yakugaku Zasshi 80:880–884
6. Tomita M, Kunitomo J (1960) Alkaloid of rutaceous plants. VIII. Alkaloids of *Phellodendron amurense*. 3. Structure of phellodendrine. b. Yakugaki Zasshi 80:885–887
7. Tomita M, Kunitomo J (1960) Alkaloids of rutaceous plants. IX. Alkaloids of *Phellodendron amurense*. 4. Synthesis of *dl*-phellodendrine and its isomers. Yakugaku Zasshi 80:1238–1244
8. Tomita M, Kunitomo J (1960) Alkaloids of rutaceous plants. X. Alkaloids of *Phellodendron amurense*. 5. Synthesis of phellodendrine isomers. Yakugaku Zasshi 80:1245–1248
9. Kunitomo J (1962) Alkaloids of Rutaceae. XVII. Alkaloids of *Phellodendron amurense*. 7. Yakugaku Zasshi 82:611–613
10. He R, Chen Y (1982) Isolation of berberine from Huang Bo (*Phellodendron amurense* Rupr.) by the aqueous acid-lime method. Chin Trad Herb Drugs 13:26

11. Kondo Y, Suzuki H, Nozoe S (1985) Two γ-hydroxybutenolides from the bark of *Phellodendron amurense* and photooxidation of limonoids. Yakugaku Zasshi 105:742–746
12. Bodalski T, Lamer E (1965) Phellodendroside occurrence in *Phellodendron amurense* leaves. Acta Pol Pharm 22:281–284 (CA 63:16767b)
13. Shevchuk OI, Maksyutina NP, Litvinenko VI (1968) The flavonoids of *Phellodendron sachalinense* and *P. amurense*. Khim Prir Soedin 4:77–82
14. Bodalski T, Lamer E (1969) Isolation and identification of noricariside from leaves of *Phellodendron*. Diss Pharm Pharmacol 21:181–184 (CA 71:70452p)
15. Glyzin VI, Bankovskii AI, Sheichenko VI, Molodozhnikov MM (1970) New flavonol glycosides from *Phellodendron lavallei* and *Phellodendron amurense*. Khim Prir Soedin 6:762–763
16. Li LY, Ye JM (1982) Expectorant activity of volatile oil and myrcene in the fruit of *Phellodendron amurense*. Chin Pharm Bull 17:304–305

Physochlaina infundibularis Kuang 96

96.1 Introduction

Huashanshen, Radix Physochlainae, is the dry root of *Physochlaina infundibularis* Kuang (Solanaceae) collected in the spring. It is officially listed in the Chinese Pharmacopoeia and used as an antiasthmatic, expectorant, and sedative, but is not to be used for glaucoma patients. Therapy of pregnant women must be carried out with great care.

A galenic preparation of the roots of *P. infundibularis*, Huashanshen Pian, Tabellae Physochlainae, the tablets of the root extract is also officially listed in the Chinese Pharmacopoeia. The latter is produced by percolation of the root powder with ethanol containing 0.1% hydrochloric acid and concentration in vacuum.

96.2 Chemical Constituents

From the root of *P. infundibularis* the coumarin fabiatrin (*96-1*) and the alkaloids scopoline, atropine, anisodamine, scopolamine, and aposcopolamine (*96-2*) were isolated and identified [1]. The major alkaloids in *Physochlaina* tablets are anisodamine, hyoscyamine, and scopolamine [2].

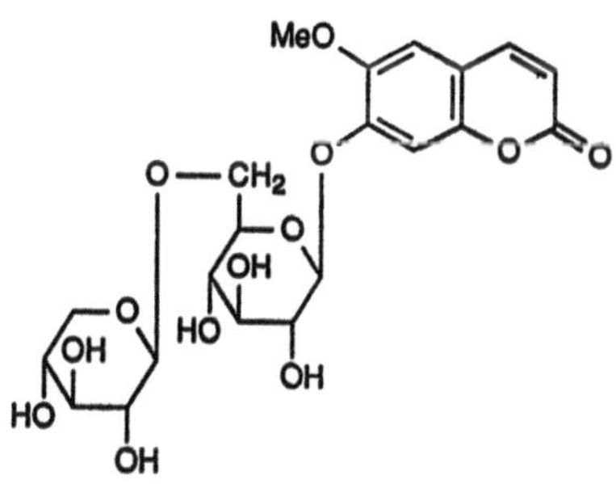

Fabiatrin (*96-1*)

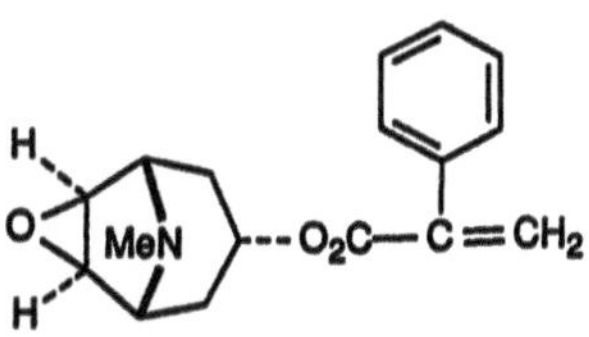

Aposcopolamine (*96-2*)

Physochlaina physaloides is another *Physochlaina* species with medicinal use occurring in China. From the root of *P. physaloides* hyposcyamine and anisodamine were isolated and identified [3]. It is a plant growing wild can also be cultivated. The anisodamine content in the root of cultivated *P. physaloides* was 0.03%–0.2% [4].

96.3 Pharmacology

The pharmacological activities of *P. infundibularis* are based on the anisodamine and scopolamine content described under *Anisodus tanguticus* and *Datura metel*, respectively.

References

1. Chen ZA, Chang XR, Qin GW, Wang BD, Xu RS (1981) Studies on the chemical constituents of the root of *Physochlaina infundibularis* Kuang. Chin Trad Herb Drugs 12:1–6
2. Hua ZL, Lin NQ, Yang XA (1986) Determination of total alkaloids in physochlaine tablets by a modified acid dye colorimetry. Chin J Pharm Anal 6:95–96
3. Lin YL, Xie FZ (1979) Study on the alkaloids of *Physochlaina physaloides*. G. Don. Acta Pharm Sin 14:497–501
4. Chen LS, Si DZ (1984) Introduction of *Physochlaina physaloides* (L.) G. Don in Beijing and observations of its biological characteristics. Acta Pharm Sin 19:869–875

Phytolacca americana L. and *P. acinosa* Roxb. 97

97.1 Introduction

Shanglu, Radix Phytolaccae, is the dry root of *Phytolacca acinosa* Roxb. or *P. americana* L. (Phytolaccaceae) collected in the spring and fall. This officially in the Chinese Pharmacopoeia listed herbal medicine is toxic and to be used for treatment of edema or treatment of abscess by external administration.

97.2 Chemical Constituents

97.2.2 Chemical Constituents of *Phytolacca americana*

The major constituents of *P. americana* are saponins. The first representative was isolated in 1949 from the roots and was designated phytolaccatoxin [1]. The structure of the aglycone Phytolaccagenin (*97-1*) obtained by hydrolysis of the saponin with either methanol-hydrochloric acid or dioxane-hydrochloric acid was determined by X-ray analysis [2]. Unusual structural features of phytolaccagenin are a high oxidation status and a carbomethoxy group. Since phytolaccagenin can be prepared with no exposure to methanol, its ester function is apparently of natural origin [2].

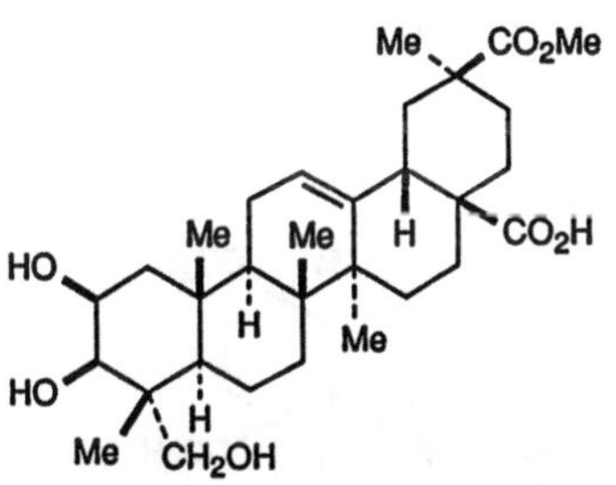

Phytolaccagenin (*97-1*)

Structures of saponins from the root of *P. americana* were determined. They were named phytolaccoside A (*97-2*) [3], phytolaccoside B (*97-3*) [4] (phytolaccasaponin G [5]), phytolaccoside D (*97-4*) [3], phytolaccoside D_2 (*97-5*) [6], phytolaccoside E (*97-6*) [3] (phytolaccasaponin E [5]), phytolaccoside G (*97-7*) [7], and phytolaccasaponin B (*97-8*) [5].

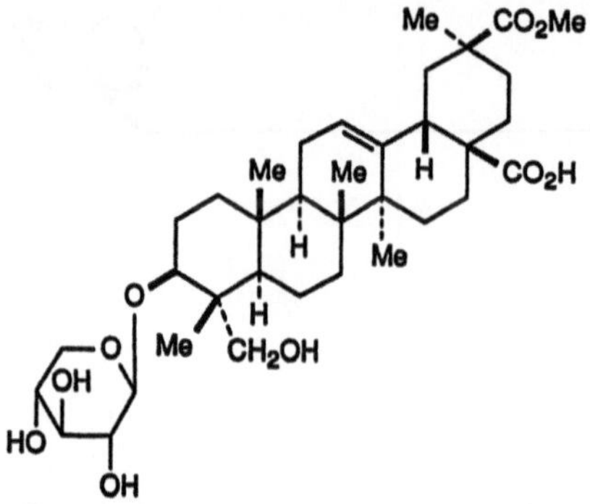

Phytolaccoside A (*97-2*)

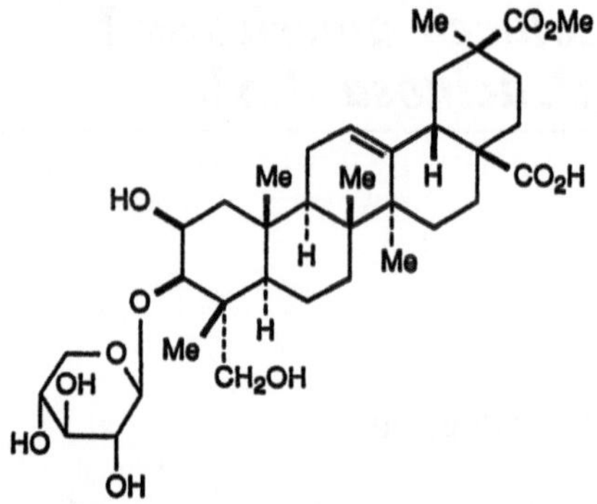

Phytolaccoside B (*97-3*)
(Phytolaccasaponin G)

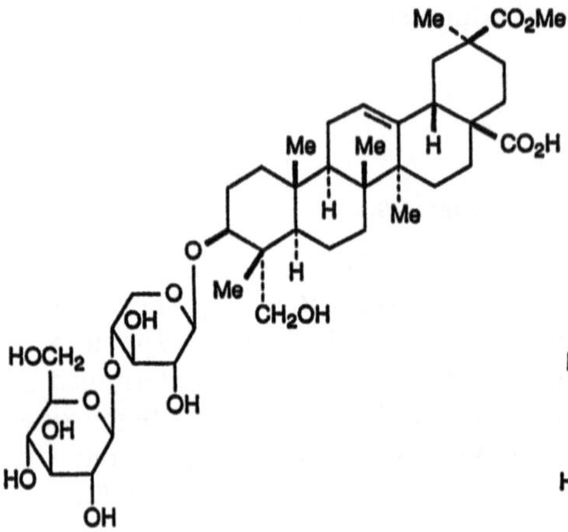

Phytolaccoside D (*97-4*)

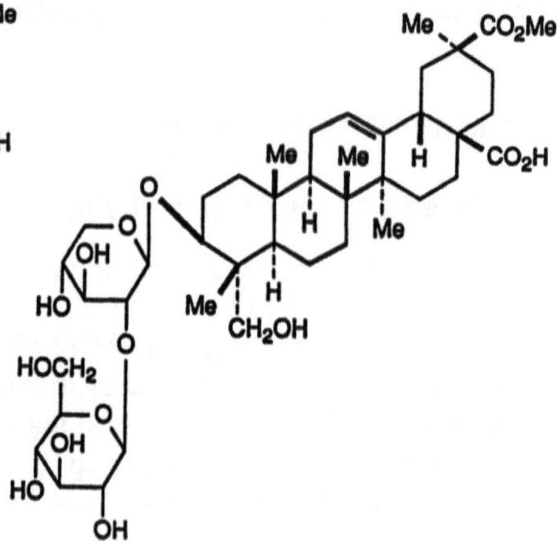

Phytolaccoside D_2 (*97-5*)

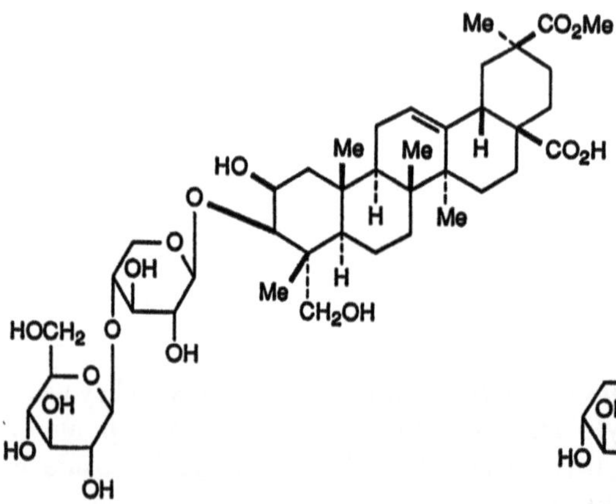

Phytolaccoside E (*97-6*)
(Phytolaccasaponin E)

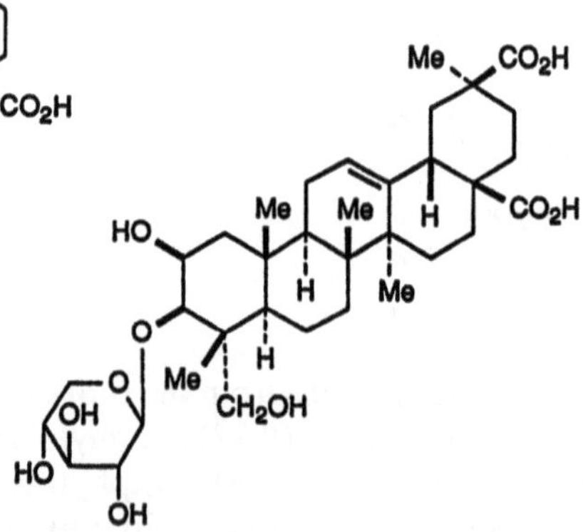

Phytolaccoside G (*97-7*)

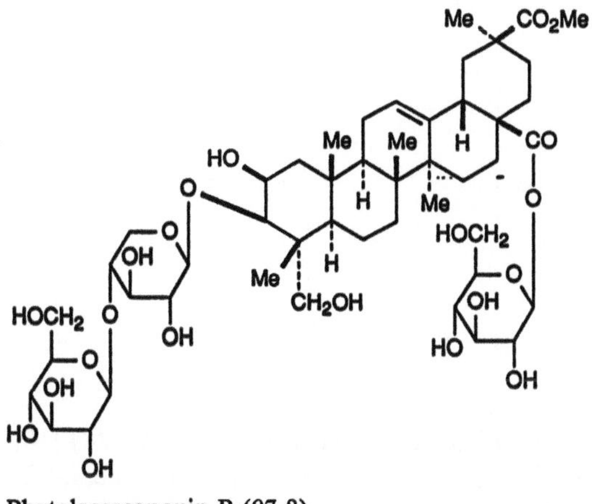

Phytolaccasaponin B (*97-8*)

A number of aglycones structurally related to phytolaccagenin were isolated from the berries of *P. americana* and identified as pokeberrygenin (*97-9*), jaligonic acid (*97-10*), phytolaccagenic acid (*97-11*), esculentic acid (*97-12*), and a trace amount of acinosolic acid (*97-13*) [8].

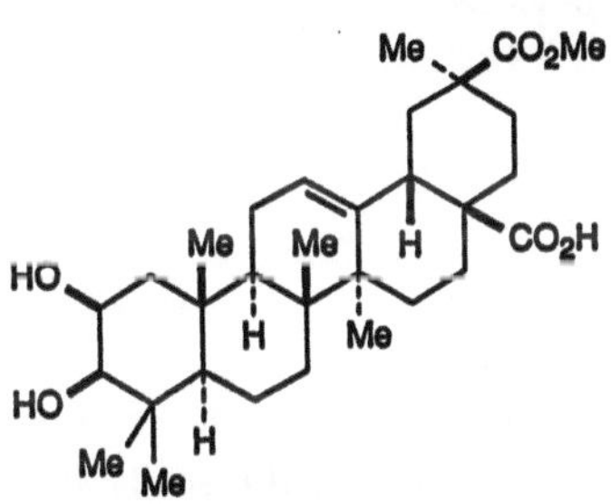

Pokeberrygenin (*97-9*)

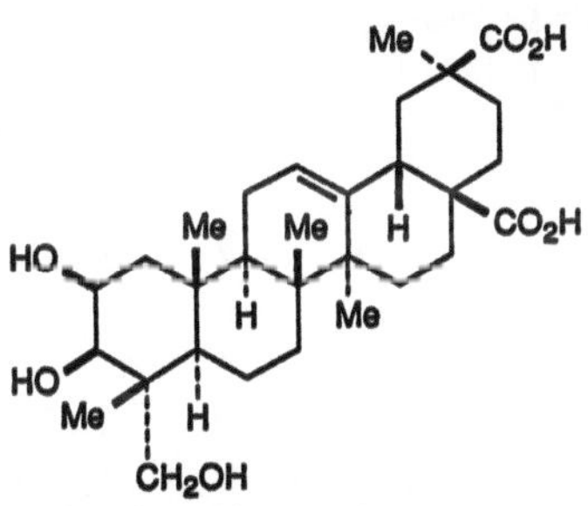

Jaligonic acid (*97-10*)

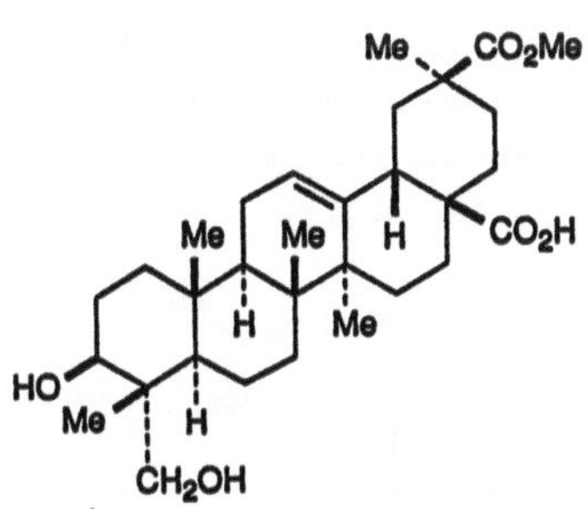

Phytolaccagenic acid (*97-11*)

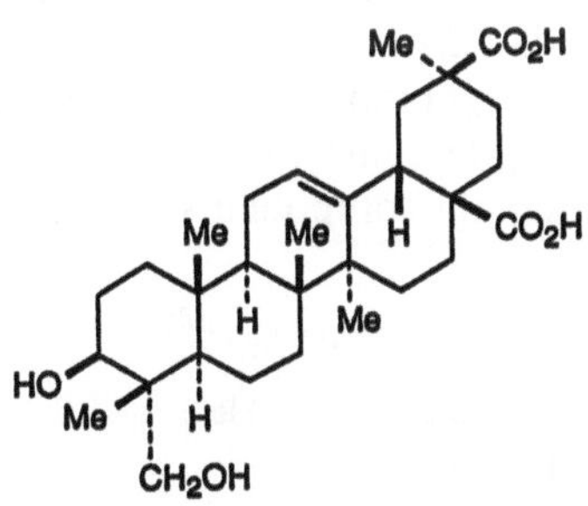

Esculentic acid (*97-12*)

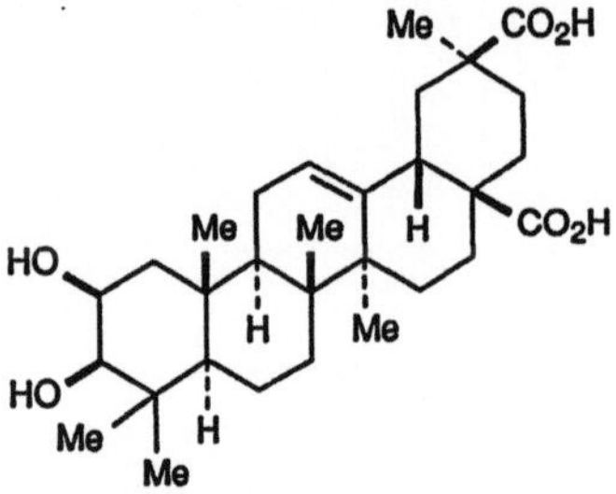

Acinosolic acid (*97-13*)

Furthermore, 3-acetyloleanolic acid [9] and 3-acetylaleuritolic acid (*97-14*) [10] were found in the seeds of *P. americana*. 3-Oxo-30-carbomethoxy-23-norolean-12-en-28-oic acid (*97-15*), obtained by the hydrolysis of the saponins from the root of *P. americana*, was determined as an artifact [11].

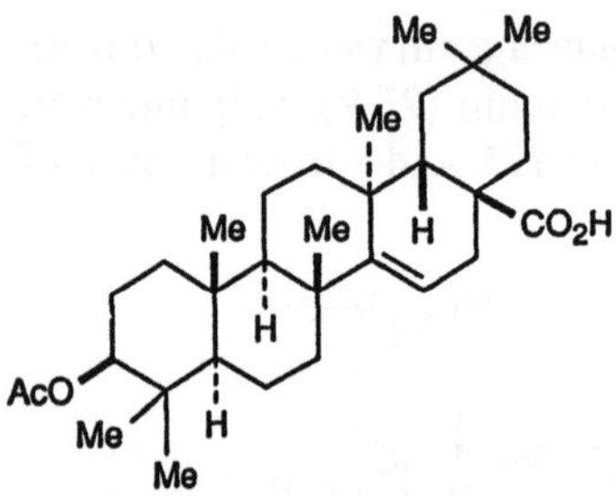

Acetylaleuritolic acid (*97-14*)

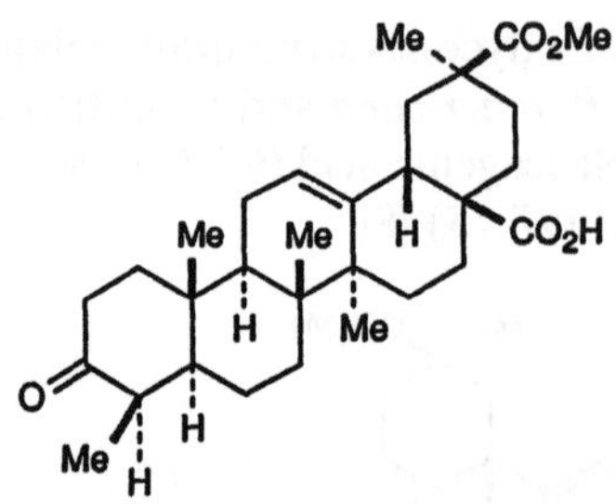

3-Oxo-30-carbomethoxy-23-norolean-12-en-28-oic acid (*97-15*)

Steroid compounds α-spinasterol and Δ^7-stigmasterol were found in the root of *P. americana* [10]. A number of triterpene alcohols were detected in the seeds and identified as cycloartenol, 24-methylenecycloartenol, parkeol (*97-16*), 24-methylene-24-dihydroparkeol, lanosterol, euphol (*97-17*), butyrospermol (*97-18*), tirucallol (*97-19*), tirucalla-7,24-dienol, dammaradienol, 24-methylenedammarenol, α-amyrin, β-amyrin, lupeol, germanicol (*97-20*), taraxasterol (*97-21*), ψ-taraxasterol (*97-22*), taraxerol (*97-23*), and myricadiol (*97-24*). Tirucalla-7,24-dienol and butyrospermol are the predominant components of the seed oil of *P. americana* [12].

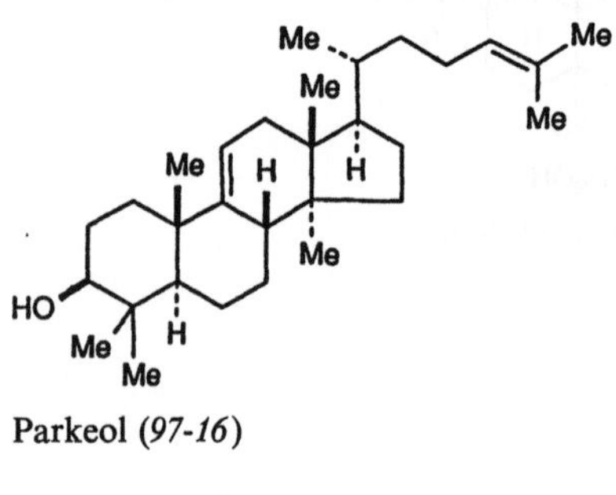

Parkeol (*97-16*)

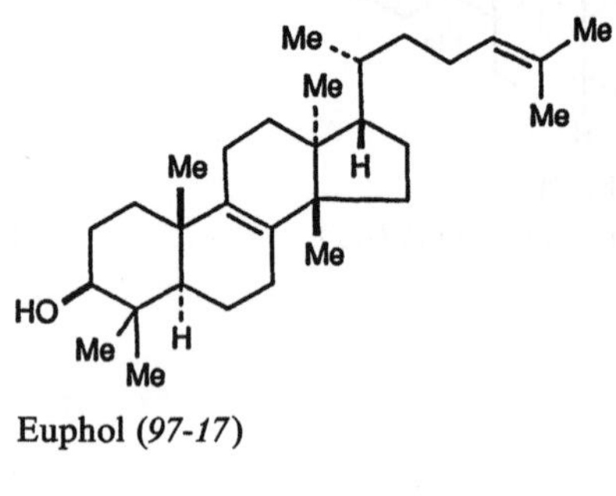

Euphol (*97-17*)

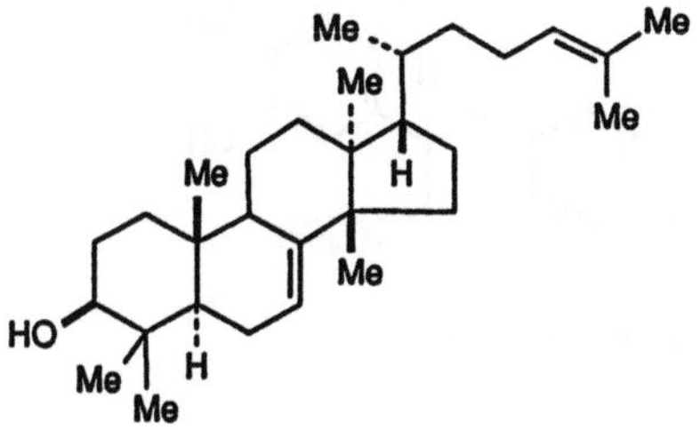

Butyrospermol (*97-18*)

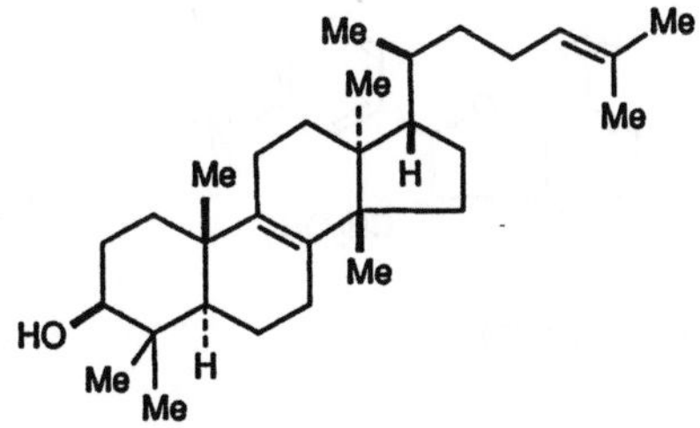

Tirucallol (*97-19*)

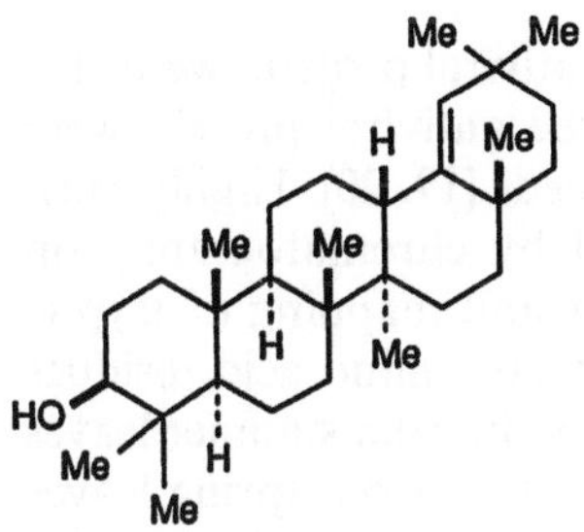

Germanicol (*97-20*)

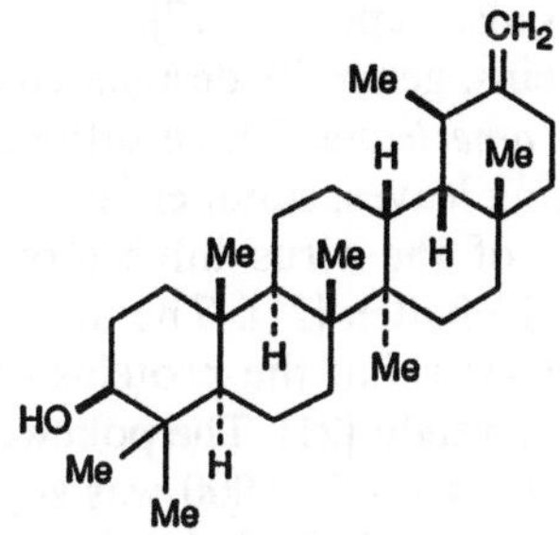

Taraxasterol (*97-21*)

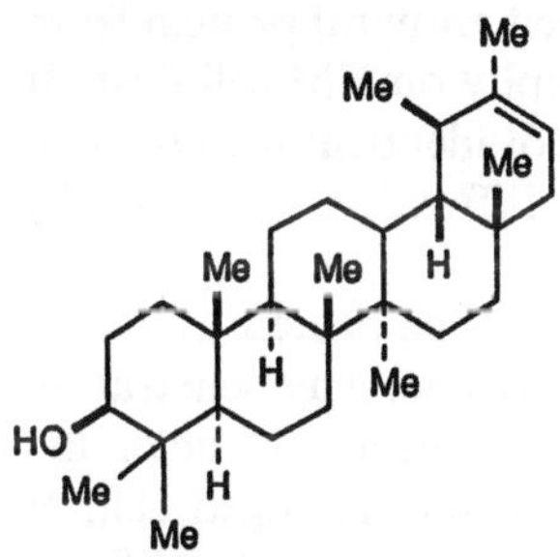

ψ-Taraxasterol (*97-22*)

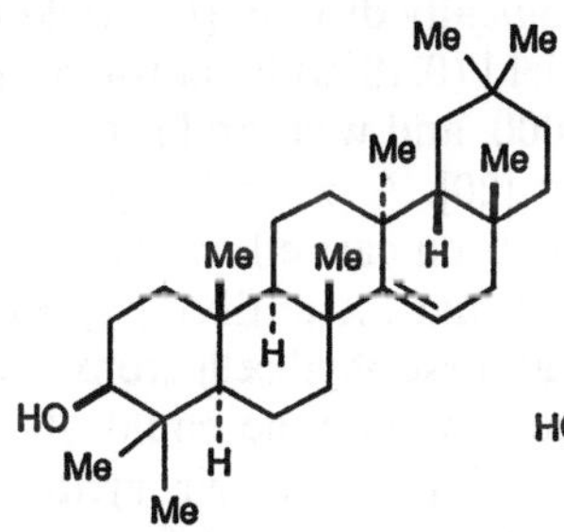

Taraxerol (*97-23*)

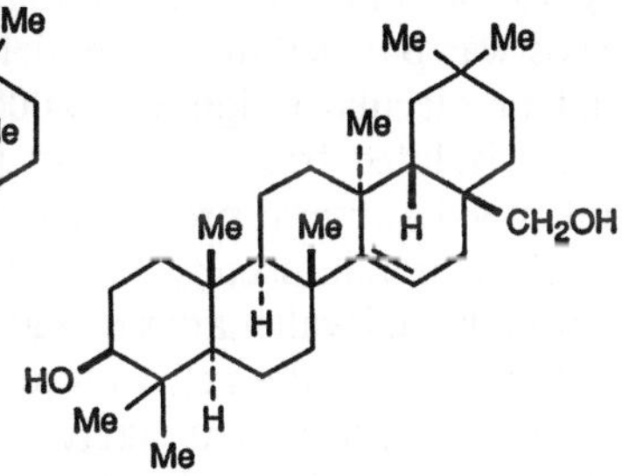

Myricadiol (*97-24*)

Besides the saponins and sapogenins, some new neolignans named americanins A (*97-25*) [13, 14], B (*97-26*), D (*97-27*) [15], and isoamericanin A (*97-28*) [16] were isolated from the seeds and structurally identified.

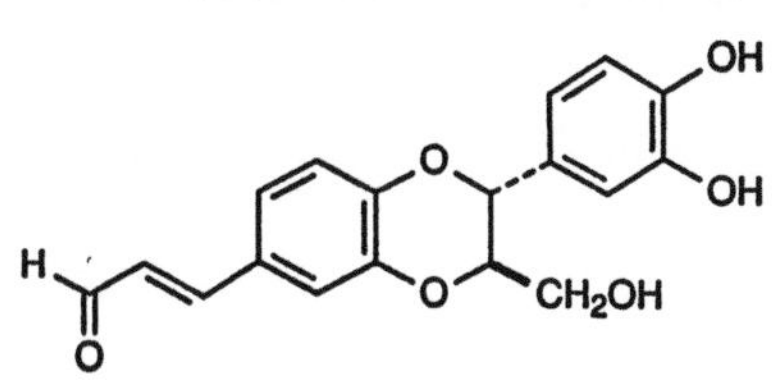

Americanin A (*97-25*)

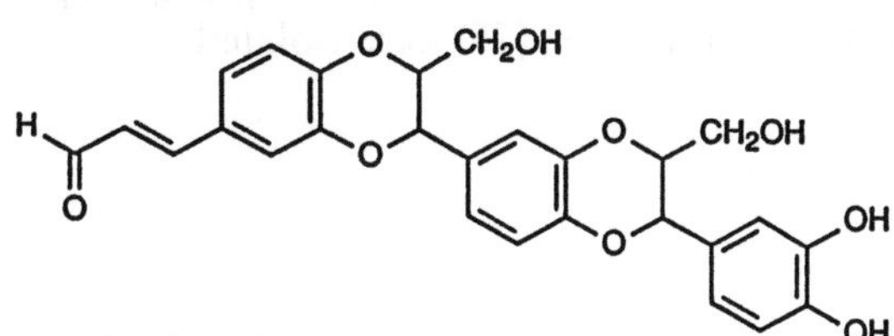

Americanin B (*97-26*)

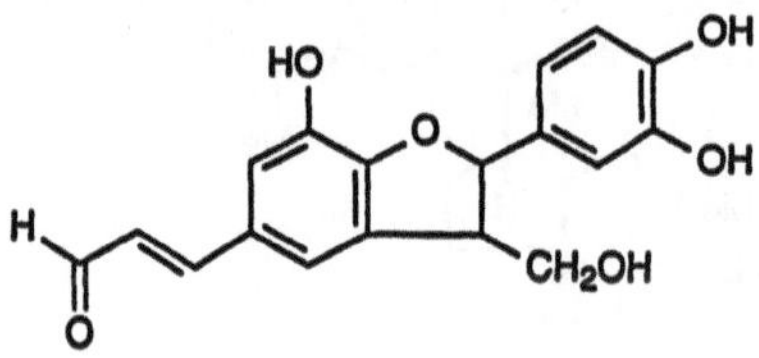

Americanin D (*97-27*)

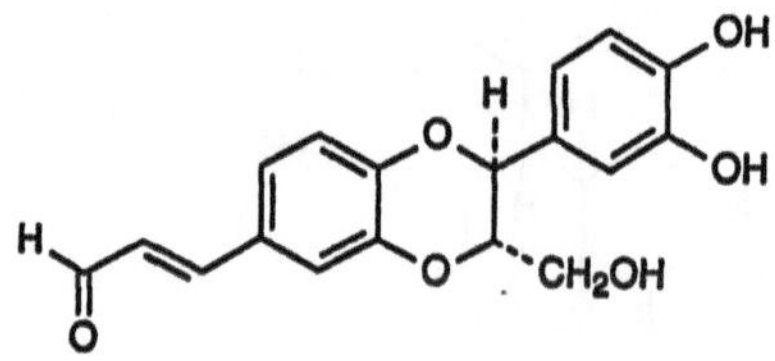

Isoamericanin A (*97-28*)

In addition, the isolation of γ-aminobutyric acid and histamine from the roots of *P. americana* was also reported [17].

Antiviral proteins, generally designated as pokeweed antiviral protein, were also isolated from *P. americana*. Three different species of the antiviral protein were isolated from spring leaves, summer leaves [18, 19], and seeds [19, 20]. Highly purified preparations of the virus inhibitors were obtained by chromatography on columns of CM Sephadex [21]. The amino acid analyses and mapping of tryptic peptides demonstrated that the proteins consist of about 116 amino acid residues with 12% lysine in weight [21]. The pokeweed antiviral protein from summer leaves with a molecular weight of 30 000 was slightly larger than that from spring leaves with a molecular weight of 29 000 by electrophoresis. These two proteins had similar amino acid compositions but yielded different peptide distributions on tryptic hydrolysis and were immunologically distinct [18]. Pokeweed antiviral protein from seeds was purified at a high yield (0.18%) by chromatography on CM cellulose. It had a molecular weight of 30 000, and was similar to, but not identical with the two forms isolated from the leaves [20]. A preliminary X-ray diffraction study on the pokeweed antiviral protein has been carried out [22].

Electron microscopy showed that the antibody specific for the protein is bound within the cell wall matrix of leaf mesophyll cells from *P. americana*. Any penetration or breakage of the cell wall and membrane could allow the enzyme to enter the cytoplasm, where it is likely to inhibit protein synthesis in the damaged cell. It appears that pokeweed antiviral protein is a defensive agent whose principal function is probably antiviral [23].

97.2.2 Chemical Constituents of *Phytolacca acinosa*

From the root of *P. acinosa* (*P. esculenta*) saponins and phytolaccosides B, D, E [24], and G (esculentoside E) [25], 3-*O*-β-D-glucopyranosyl-jaligonic acid [24], and esculentoside F (*97-29*) [25] were isolated.

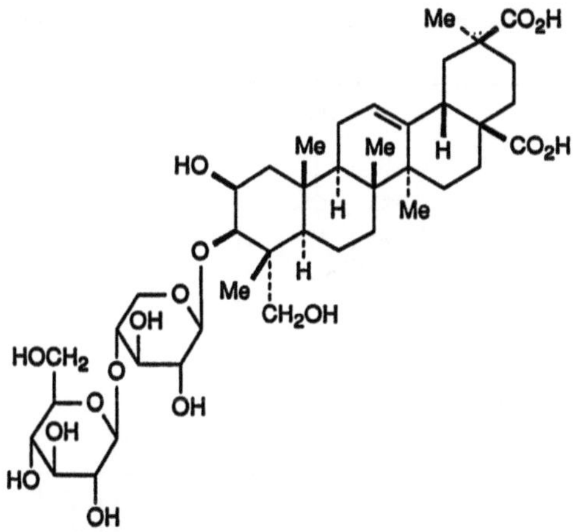

Esculentoside F (*97-29*)

A number of sapogenins related to phytolaccagenin were isolated from the seeds of *P. acinosa*. They are spergulagenic acid (*97-30*), jaligonic acid, acinosolic acid [26], phytolaccanol (*97-37*), epiacetylaleuritolic acid [27], spergulagenic acid A (*97-31*), isophytolaccagenic acid A (*97-33*), isophytolaccagenin A (*97-32*) [28], phytolaccagenin, phytolaccagenic acid, esculentic acid, phytolaccagenin A (*97-34*), acinosolic acid A (*97-35*), and acinosolic acid B (*97-36*) [29]. Besides phytolaccanol and epiacetylaleuritolic acid, all other triterpenes from the seeds of *P. acinosa* are derived from the olean-12-ene skeleton. Epiacetylaleuritolic acid is the 3-epimer of acetylaleuritolic acid (*94-14*) and phytolaccanol is also derived from D-friedooleane.

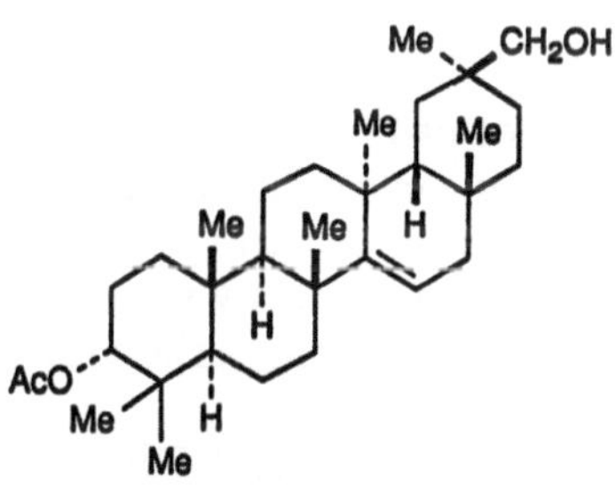

Phytolaccanol (*97-37*)

The structural characteristics of the triterpenes with the olean-12-ene skeleton from *P. americana* and *P. acinosa* are listed in Table 97.1.

Table 97.1. Structural characteristics of triterpenes with olean-12-ene skeleton isolated from *P. americana* and *P. acinosa*

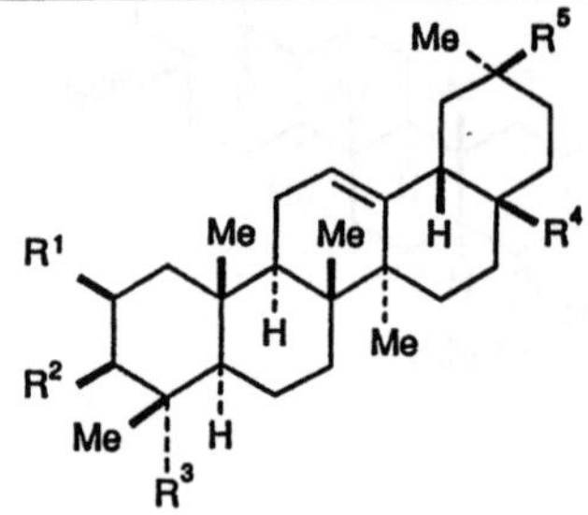

Sapogenin	R^1	R^2	R^3	R^4	R^5
Phytolaccagenin (*97-1*)	OH	OH	CH_2OH	COOH	$COOCH_3$
Pokeberrygenin (*97-9*)	OH	OH	H	COOH	$COOCH_3$
Jaligonic acid (*97-10*)	OH	OH	CH_2OH	COOH	COOH
Phytolaccagenic acid (*97-11*)	H	OH	CH_2OH	COOH	$COOCH_3$
Esculentic acid (*97-12*)	H	OH	CH_2OH	COOH	COOH
Acinosolic acid (*97-13*)	OH	OH	H	COOH	COOH
Spergulagenic acid (*97-30*)	H	OH	H	COOH	COOH
Spergulagenic acid A (*97-31*)	H	OAc	H	COOH	$COOCH_3$
Isophytolaccagenin A (*97-32*)	OAc	OAc	CH_2OAc	$COOCH_3$	COOH
Isophytolaccagenic acid A (*97-33*)	H	OAc	CH_2OAc	$COOCH_3$	COOH
Phytolaccagenin A (*97-34*)	OH	OAc	CH_2OH	COOH	$COOCH_3$
Acinosolic acid A (*97-35*)	OH	OAc	H	$COOCH_3$	COOH
Acinosolic acid B (*97-36*)	OAc	OH	H	$COOCH_3$	COOH

97.3 Pharmacology

The crude saponin and phytolaccagenin isolated from root of *P. americana*, given intraperitoneally, exhibited potent inhibitory effects on acute edema in rats and in mice. At 15–30 mg/kg, crude saponin showed 50% inhibition of carrageenin-induced paw edema in rats [30, 31]. The LD_{50} value of crude saponins by intraperitoneal administration was 181 mg/kg in mice and 208 mg/kg in rats. The crude saponin produced a severe hemolysis at high doses [30].

The liver and spleen tissue incorporation of intraperitoneal-injected [^{3}H]thymidine into DNA and the survival of mice fed a diet containing hydroxyurea were both increased by oral administration of a total saponin preparation from *P. acinosa* at a dose of 1.5 mg/animal daily. It has been suggested that the saponin preparation may act as an activator for nucleotide reductase. In addition, the saponin preparation also enhanced the freezing tolerance of animals [32].

The two pokeweed antiviral proteins isolated from the leaves of *P. americana* were both efficient inhibitors of eukaryotic protein synthesis, tobacco mosaic virus transmission, and ribosome elongation factor interaction. The primary mechanism of the inhibition of protein synthesis by pokeweed antiviral protein is the reduction of ribosomal affinity for elongation factor 2, which results in the inhibition of the translocation step in protein synthesis [18]. The pokeweed antiviral protein from the seed inhibited protein formation in rabbit reticulocyte lysate with an ID_{50} value of

1.1 ng/ml, but had much less effect on protein formation in whole cells, as reflected by an ID_{50} value of 1 mg/ml. Replication of herpes simplex virus type 1 was also inhibited [20].

Treatment of poliovirus with pokeweed antiviral protein resulted in more than 96% inhibition of infectious virus replication in HeLa cells. The inhibitor could be quantitatively removed from the virus by centrifugation in sucrose solution. The antiviral protein may bind reversibly to poliovirus and thus be transported into the cell, where it enzymatically inactivates the HeLa 60S ribosomal subunits [33].

Pokeweed antiviral protein has functional similarities to ricin A chain, even the sequences of the pokeweed proteins showing little similarity with ricin A chain. Ricin B chain is responsible for helping ricin A chain across the plasma membrane; since pokeweed antiviral protein has no counterpart to ricin B chain, it is not nearly as cytotoxic as ricin. However, when pokeweed antiviral protein was covalently coupled to ricin B chain, a strongly cytotoxic species was formed. Pokeweed antiviral protein, however, fails to interact noncovalently with the ricin B chain to produce a cytotoxic species equivalent in function to ricin [19].

The pokeweed antiviral protein has been used as a cytotoxic group in conjugation with antibody [34–39]. For example, it was conjugated with the Fab′ fragment of IgG from a rabbit antiserum against murine leukemia L1210 cells via a disulfide bond employing *N*-succinimidyl-3-(2-pyridyldithio)propionate as the coupling agent. The conjugate showed a potent in vitro cytotoxicity against L1210 cells which was competitively blocked by $F(ab')_2$ directed against L1210 cells. Pokeweed antiviral protein itself did not exhibit the cytotoxicity at the concentration corresponding to its content in the conjugate concurrently tested [34].

Monoclonal antibodies against human T-cell antigen 3Al, human transferrin receptor, and mouse Thy 1.1 antigen were linked to pokeweed antiviral protein by a disulfide bond. Blood samples collected from rabbits at different times after the injection of immunotoxin showed that the immunotoxin did not dissociate in circulation immediately, about 90% of the initial concentration of the conjugate being present for more than 4 h. The half-life of the conjugate in the circulation ranged between 17 and 24 h. Immunotoxins remained intact immunologically, retaining their biological activity as measured by the ability of blood-borne immunotoxins efficiently to block protein synthesis of target cells in vitro. Apparently, the disulfide linkage of toxin to antibody is reasonable stable [38]. Because of its stability, ease of purification, and absence of a ricin B chain analog, pokeweed antiviral protein may be more useful than ricin A for immunotoxin synthesis [36].

References

1. Ahmed ZF, Zufall CJ, Jenkins GL (1949) A contribution to the chemistry and toxicology of the root of *Phytolacca americana*. J Am Pharm Assoc 38:443–448
2. Stout GH, Malofsky BM, Stout VF (1964) Phytolaccagenin: a light atom X-ray structure proof using chemical information. J Am Chem Soc 86:957–958
3. Woo WS, Kang SS, Wagner H, Seligmann O, Chari VM (1978) Triterpenoid saponins from the roots of *Phytolacca americana*. Planta Med 34:87–92
4. Woo WS, Kang SS (1976) Phytolaccoside B: triterpene glycoside from *Phytolacca americana*. Phytochemistry 15:1315–1317

5. Suga Y, Maruyama Y, Kawanishi S, Shoji J (1978) Studies on the constituents of phytolaccaceous plants. I. On the structures of phytolaccasaponin B, E and G from the roots of *Phytolacca americana* L. Chem Pharm Bull (Tokyo) 26:520–525
6. Kang SS, Woo WS (1987) Two new saponins from *Phytolacca americana*. Planta Med 53:338–340
7. Woo WS, Kang SS (1977) The structure of phytolaccoside G. Yakhak Hoe Chi 21:159–162 (CA 88:191324h)
8. Kang SS, Woo WS (1980) Triterpenes from the berries of *Phytolacca americana*. J Nat Prod 43:510–513
9. Burke DE, Le Quesne PW (1971) 3-Acetyloleanolic acid from *Phytolacca americana* seeds. Phytochemistry 10:3319–3320
10. Woo WS, Wagner H (1977) 3-Acetylaleuritolic acid from the seeds of *Phytolacca americana*. Phytochemistry 16:1845–1846
11. Woo WS (1974) Steroids and pentacyclic triterpenoids from *Phytolacca americana*. Phytochemistry 13:2887–2889
12. Itoh T, Uetsuki T, Tamura T, Matsumoto T (1980) Characterization of triterpene alcohols of seed oils from some species of Theaceae, Phytolaccaceae and Sapotaceae. Lipids 15:407–411
13. Woo WS, Kang SS, Wagner H, Chari VM (1978) The structure of americanin, a new neolignan from *Phytolacca americana*. Tetrahedron Lett 3239–3242
14. Lin LJ, Meksuriyen D, Cordell GA, Woo WS, Lee CK (1987) The structure of americanin A. Saengyak Hakhoechi 18:94–98 (CA 107:242496m)
15. Woo WS, Kang SS, Seligmann O, Chari VM, Wagner H (1980) The structure of new lignans from the seeds of *Phytolacca americana*. Tetrahedron Lett 21:4255–4258
16. Hasegawa T, Fukuyama Y, Koshino K, Nakagawa K, Tori M, Asakawa Y (1987) Studies of isoamericanin A, a prostaglandin I_2 inducer, isolated from the seeds of *Phytolacca americana* L. Chem Lett 329–332
17. Funayama S, Hikino H (1979) Hypotensive principles of *Phytolacca* roots. J Nat Prod 42:672–674
18. Irvin JD, Kelly T, Robertus JD (1980) Purification and properties of a second antiviral protein from *Phytolacca americana* which inactivates eukaryotic ribosomes. Arch Biochem Biophys 200:418–425
19. Houston LL, Ramakrishnan S, Hermodson MA (1983) Seasonal variations in different forms of pokeweed antiviral protein, a potent inactivator of ribosomes. J Biol Chem 258:9601–9604
20. Barbieri L, Aron GM, Irvin JD, Stirpe F (1982) Purification and partial characterization of another form of the antiviral protein from the seeds of *Phytolacca americana* L. (pokeweed). Biochem J 203:55–59
21. Wyatt SD, Shepherd RJ (1969) Isolation and characterization of a virus inhibitor from *Phytolacca americana*. Phytopathology 59:1787–1794
22. Robertus JD, Monzingo AF, Irvin JD (1977) Preliminary X-ray diffraction studies on a antiviral protein. Biochem Biophys Res Commun 74:775–779
23. Ready MP, Brown DT, Robertus JD (1986) Extracellular localization of pokeweed antiviral protein. Proc Natl Acad Sci USA 83:5053–5056
24. Yi YH, Wang ZL (1984) Active principles of *Phytolacca esculenta*. I. Isolation and identification of the triterpene saponins. Chin Trad Herb Drugs 15:55–59
25. Wang ZL, Yi YH (1984) Studies on the active principles of the traditional Chinese drug "Shang Lu" (*Phytolacca esculenta* Van Houtte). II. Isolation and identification of esculentoside E and F. Acta Pharm Sin 19:825–829
26. Glombitza KW, Gielsdorf W, Eckhardt G, Koch ML (1975) Triterpenoid sapogenins from the fruits of *Phytolacca acinosa*. Planta Med 27:367–371
27. Razdan TK, Harkar S, Kachroo V, Koul GL (1982) Phytolaccanol and epiacetylaleuritolic acid, two triterpenoids from *Phytolacca acinosa*. Phytochemistry 21:2339–2342
28. Razdan TK, Harkar S, Kachroo V, Koul GL, Waight ES (1983) Triterpenoids from *Phytolacca*, three oleanane derivatives. Phytochemistry 22:1797–1800
29. Harkar S, Razdan TK, Waight ES (1984) Further triterpenoids and carbon-13 NMR spectra of oleanane derivatives from *Phytolacca acinosa*. Phytochemistry 23:2893–2898
30. Woo WS, Shin KH, Kang SS (1976) Constituents of *Phytolacca* species. I. Antiinflammatory saponins. Soul Taehakkyo Saengyak Yonguso Opjukjip 15:103–106 (CA 88:146072p)

31. Woo WS, Shin KH (1976) Antiinflammatory action of *Phytolacca* saponin. Soul Taehakkyo Saengyak Yonguso Opjukjip 15:90–96 (CA 88:69156g)
32. Liu FC, Ding GX, Li JX (1984) Effects of the total saponin from *Phytolacca* root on the incorporation ^{3}H-thymidine into liver and spleen tissue. Chin Trad Herb Drugs 15:506–508
33. Ussery MA, Irvin JD, Hardesty B (1977) Inhibition of poliovirus replication by a plant antiviral peptide. Am NY Acad Sci 284:431–440
34. Masuho Y, Kishida K, Hara T (1982) Targeting of the antiviral protein from *Phytolacca americana* with an antibody. Biochem Biophys Res Commun 105:462–469
35. Ramakrishnan S, Houston LL (1984) Inhibition of human acute lymphoblastic leukemia cells by immunotoxins: potentiation by chloroquine. Science 223:58–61
36. Ramakrishnan S, Houston LL (1984) Comparison of the selective cytotoxic effects of immunotoxins containing ricin A chain or pokeweed antiviral protein and anti-Thy 1.1 monoclonal antibodies. Cancer Res 44:201–208
37. Bjorn MJ, Larrick J, Piatak M, Wilson KJ (1984) Characterization of translational inhibitors from *Phytolacca americana*. Amino-terminal sequence determination and antibody-inhibitor conjugates. Biochim Biophys Acta 790:154–163
38. Ramakrishnan S, Houston LL (1985) Immunological and biological stability of immunotoxins in vivo as studied by the clearance of disulfide-linked pokeweed antiviral protein-antibody conjugates from blood. Cancer Res 45:2031–2036
39. Lambert JM, Senter PD, Yau-Young A, Blattler WA, Goldmacher VS (1985) Purified immunotoxins that are reactive with human lymphoid cells. Monoclonal antibodies conjugated to the ribosome-inactivating proteins gelonin and the pokeweed antiviral proteins. J Biol Chem 260:12035–12041

Pinellia ternata (Thunb.) Breit.

98

98.1 Introduction

Banxia, Rhizoma Pinelliae, is the dry tubers of *Pinellia ternata* (Thunb.) Breit. (Araceae) collected in the summer and fall. It is officially listed in the Chinese Pharmacopoeia and used in traditional Chinese medicine as an antiemetic, mucolytic, and antiasthmatic.

98.2 Chemical Constituents

The tuber of *P. ternata* has a pungent taste. An irritating glycoside was isolated and found to be composed of 3,4-dihydroxybenzaldehyde and D-glucose. The aglycone had a strong acrid taste [1]. The alkaloid ephedrine hydrochloride was isolated at a yield of 0.002% as an active principle of the *Pinellia* tuber [2]. Pinellin, a crystalline plant protein isolated from the *Pinellia* tubers, was found to have a low cystein content and a relatively low molecular weight of 10500 and did not contain hexose [3]. A trypsin inhibitor was also isolated from the tubers. It inhibited trypsin, but not chymotrypsin, kallikrein, or papain, and was estimated to have a molecular weight of 40800 [4]. In addition to the constituents described above, palmitic acid, stearic acid, 6,7-dihydroxystearic acid [5], β-sitosterol [6], β-sitosteryl-D-glucoside [7], and choline [6] were detected in the root. The amino acids aspartic acid, glutamic acid, arginine, and β-aminobutyric acid were also isolated and identified from *Pinellia* tuber [8].

Pinellia pedatisecta is a herbal drug used in folk medicine in China. The tubers are used as an antiarrhythmic. A number of cyclodipeptide alkaloids with a piperazinedione structure were isolated from the tuber of *P. pedatisecta*. They were identified as L-alanyl-L-valine anhydride (*98-1*), L-analyl-L-leucine anhydride (*98-2*), L-alanyl-L-isoleucine anhydride (*98-3*), L-alanyl-L-phenylalanine anhydride (*98-4*), L-valyl-L-valine anhydride (*98-5*), L-valyl-L-α-amino-β,β-dimethylbutyric acid anhydride (*98-6*), L-valyl-L-leucine anhydride (*98-7*), L-valyl-L-tyrosine anhydride (*98-8*), L-leucyl-L-tyrosine anhydride (*98-9*), L-prolyl-glycine anhydride (*98-10*), and L-prolyl-L-proline anhydride (*98-11*) [9–11].

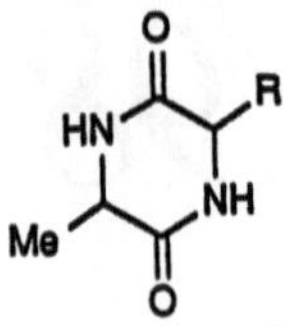

L-Alanyl-L-valine anhydride (*98-1*): R = $—CH(CH_3)_2$

L-Alanyl-L-leucine anhydride (*98-2*): R = $—CH_2—CH(CH_3)_2$

L-Alanyl-L-isoleucine anhydride (*98-3*): R = $—CH(CH_3)—CH_2—CH_3$

L-Alanyl-L-phenylalanine anhydride (*98-4*): R = 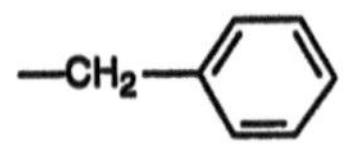

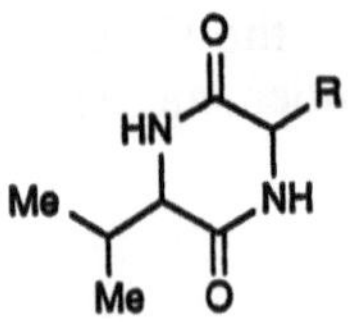

L-Valyl-L-valine anhydride (*98-5*): R = $—CH(CH_3)_2$

L-Valyl-L-α-amino-β,β-dimethyl-butyric acid anhydride (*98-6*): R = $—C(CH_3)_3$

L-Valyl-L-leucine anhydride (*98-7*): R = $—CH_2—CH(CH_3)_2$

L-Valyl-L-tyrosine anhydride (*98-8*): R = 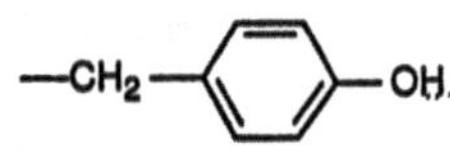

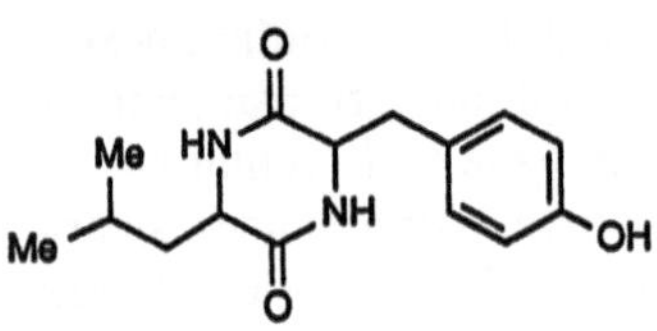

L-Leucyl-tyrosine anhydride (*98-9*)

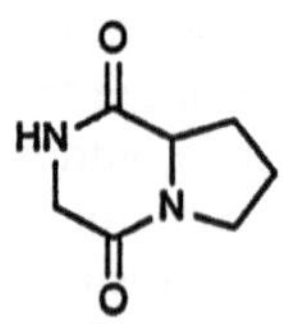

L-Prolyl-glycine anhydride (*98-10*)

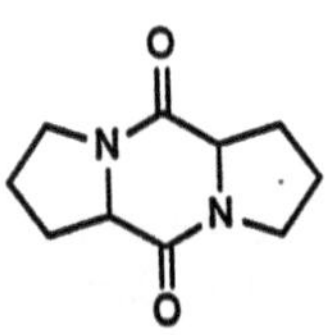

L-Prolyl-L-proline anhydride (*98-11*)

In addition to the piperazine-2,5-dione alkaloids, the following compounds were isolated and identified from the tubers of *P. pedatisecta*: β-carboline, 1-acetyl-β-carboline, pedatisectine B (*98-12*) [12], pedatisectine A (*98-13*) [13], uracil, 5-methyluracil, nicotinamide, 2-methyl-3-hydroxypyridine, β-sitosterol, palmitic acid [12], choline, and mannitol [10].

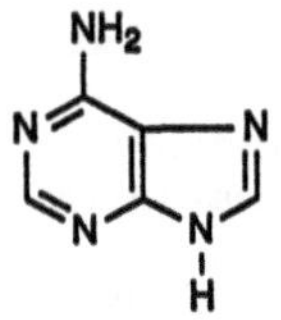

Pedatisectine B (*98-12*)

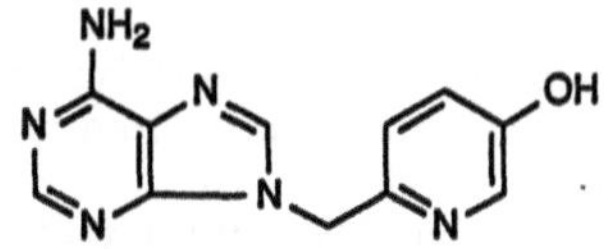

Pedatisectine A (*98-13*)

98.3 Pharmacology

A decoction of *P. ternata* not only prevented early vomiting caused by deslanoside but also prevented emesis caused by orally administered copper sulfate [14]. The basic fraction of methanol and water extracts of *Pinellia* tuber showed a relaxant action and antihistamine-like action upon an isolated rectum of Japanese quails. The active principle of this fraction might have been ephedrine [2].

The crystalline protein pinellin exhibited cell-agglutinating and mitogenic activity. It agglutinated erythrocytes from sheep, dog, cat, rabbit, guinea pig, rat, mouse, and pigeon but did not agglutinate erythrocytes of man, monkey, pig, chicken, duck, goose, tortoise, toad, and eel. Hemagglutination by pinellin is thus species specific. In addition to erythrocytes, pinellin also agglutinated splenocytes, Ehrlich ascites cells, Hep A hepatoma ascitic cells from mouse, and cultured human hepatoma cells. Pinellin did not agglutinate epididymal adipose cells of the rat or cells of omentum majus of the pig. Therefore, the cell agglutination of pinellin exhibited not only species specificity, but also cell type specificity. Pinellin was a mitogen for rabbit peripheral lymphocytes and mouse spleen lymphocytes, but not for human peripheral lymphocytes, demonstrating species specificity in mitogenic action [15]. Pinellin given subcutaneously at a dose of 30 mg/kg effectively inhibited pregnancy at an early stage in mice [16].

Denaturation of pinellin in 6 *M* guanidine hydrochloride solution was reversible when the denaturing agent was removed by dialysis [17]. Pepsin digestion of pinellin destroyed its activity [4]. Fluorescence results and CD measurements indicated the presence of intermediates along the path of denaturation [18]. The chloroform extract of *P. pedatisecta* has been found to possess quinidine-like antiarrhythymic activity on isolated porcine heart [19].

References

1. Suzuki M (1969) Irritating substance of *Pinellia ternata.* Arzneimittelforschung 19:1307–1309
2. Oshio H, Tsukui M, Matsuoka T (1978) Isolation of l-ephedrine from "Pinellia Tuber". Chem Pharm Bull (Tokyo) 26:2096–2097
3. Tao ZJ, Xu QY, Wu KZ, Lian SH, Sun D (1980) Isolation, crystallization, biological activities and some chemical characteristics of pinellin. In: Shen ZW (ed) Nucleic Acids, Proteins. Science, Peking, pp 123–126
4. Wu KZ, Tao ZJ (1981) Isolation and characterization of a trypsin inhibitor from the rhizome of *Pinellia ternata.* Acta Biochim Biophys Sin 13:267–274
5. Yajima M, Ozeki S (1954) Components of rhizome of *Pinellia ternata.* Bull Nagoya City Univ Pharm School 37–41 (CA 50:11611i)
6. Oziki S (1961) Constituents of *Pinellia ternata.* II. Sterol and bases of *Pinellia ternata.* Yakugaku Zasshi 81:1706–1708

7. Oziki S (1962) Constituents of *Pinellia ternata*. III. Steryl glucoside of *Pinellia ternata*. Yakugaku Zasshi 82:766–768
8. Murakami T, Nagasawa M, Itokawa H, Inatomi H (1965) Water-soluble constituents of crude drugs. I. Free amino acids isolated from tubers of *Pinellia ternate* and *Arisaema ringens*. Yakugaku Zasshi 85:832–835
9. Chin WC, Kung CF, Su HY, Fan CT (1981) Studies on chemical constituents of *Pinellia pedatisecta* Schott. Chin Pharm Bull 16:51
10. Qin WJ, Kung CF, Fan CT (1981) Studies on chemical constituents of *Pinellia pedatisecta* Schott. I. Chin Trad Herb Drugs 12:5–9
11. Qin WJ, Kong QF, Fan ZT, Su XY, Li LP (1984) Studies on the chemical constituents of pedate pinellia (*Pinellia pedatisecta*). III. Chin Trad Herb Drugs 15:490–492
12. Qin WJ, Wang SX, Fan ZT, Zhang L, Li LP (1983) Studies on the chemical constituents of Zhang Yie Ban Xia (*Pinellia pedatisecta*). II. Chin Trad Herb Drugs 14:443–445
13. Qin WJ, Kong QF, Fan ZT, Su XY, Lin XY, Li LP (1986) Elucidation of the structure of pedatisectine A. Chin Trad Herb Drugs 17:197–199
14. Ho ST, Liu TS (1975) Studies on *Pinellia ternata* Breitenbach. Report I. Antiemetic action of *Pinellia ternata* Breitenbach and its components. Taiwan Yao Hsueh Tsa Chih 27:41–46 (CA 86:11845f)
15. Sun C, Xu JH, Zhai SK, Tao ZJ, Yau TY, Zhu Z, Shen ZW (1983) Some biological properties of pinellin. Acta Biochim Biophys Sin 15:333–338
16. Xia LN, Li CJ (1985) Effects of *Pinellia ternata* Breit. protein on the termination of early pregnancy in mice and its action mechanism. Acta Acad Med Primae 12:193–198
17. Xu QY, Sun D, Tao ZJ (1981) Reversible denaturation of crystalline pinellin in 6M guanidine hydrochloride. Acta Biochim Biophys Sin 13:153–158
18. Lu ZX, Shi QL, Xu QY (1986) Conformational changes of pinellin in solution using intrinsic fluorescence and CD as probes. Biopolymers 25:393–405
19. Zhou RP, Liu TF (1983) Effect of Ning Xin (alkaloids from *Pinellia pedatisecta*) on the action potential of porcine ventricular fibers. Chin Trad Herb Drugs 14:117–118

Polygala tenuifolia Willd. 99

99.1 Introduction

Yuanzhi, Radix Polygalae, is the dry root of *Polygala tenuifolia* Willd. or *P. sibirica* L. (Polygalaceae) collected in the spring and fall. It is officially listed in the Chinese Pharmacopoeia and is used in traditional Chinese medicine as a mucolytic for the treatment of cough. It is further used as a sedative and for antiswelling purposes.

Two galenic preparations are also listed in the Chinese Pharmacopoeia: Yuanzhi Liujingao, Extractum Polygalae liquidum; and Yuanzhi Ding, Tinctura Polygalae. They are used only as mucolytics.

99.2 Chemical Constitution

The root of *Polygala* species is known to contain triterpene saponins. The first prosapogenin identified from *P. tenuifolia* was tenuifolin (*99-1*). It was obtained by alkaline hydrolysis of the butanol extract containing the saponin. It has been shown to be 2β,27-dihydroxy-23-carboxyoleanolic acid 3-*O*-β-D-glucopyranoside [1].

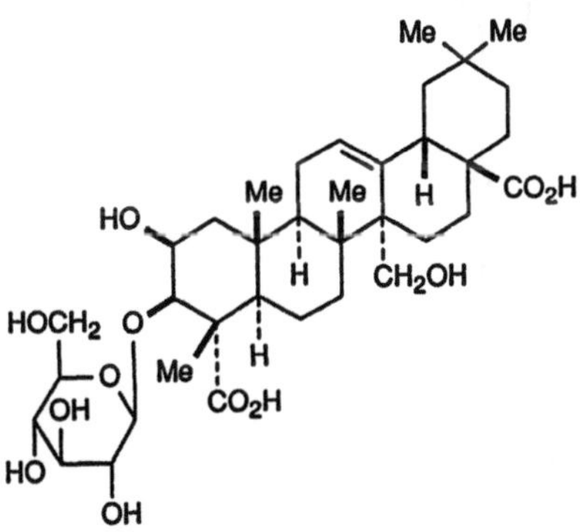

Tenuifolin (*99-1*)

Acidic hydrolysis of the whole saponin from *P. tenuifolia* with diluted hydrochloric acid gave mostly senegenin (*99-2*). With phosphoric acid, hydroxysenegenin (*99-3*) was obtained [1].

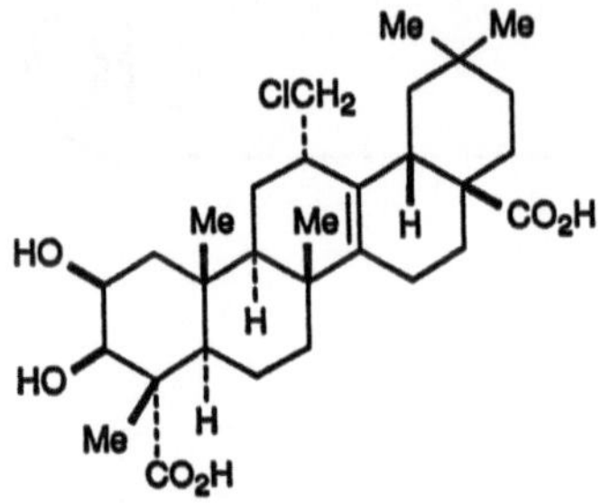

Senegenin (*99-2*)

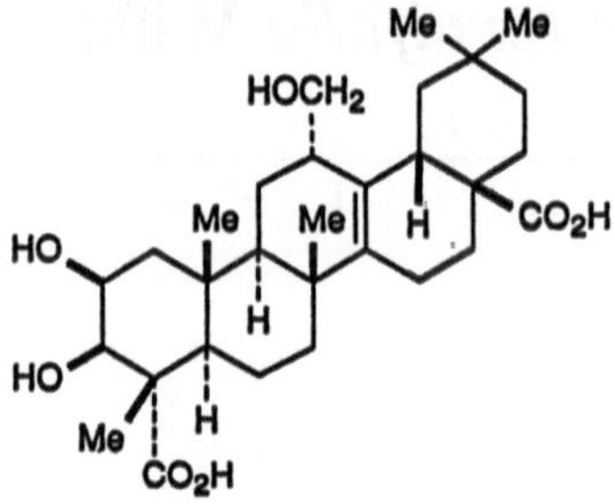

Hydroxysenegenin (*99-3*)

Seven intact saponins were isolated and designated as onjisaponins A–G [2, 3]. Five of them were structurally determined. They are all tenuifolin-28-esters of different oligosaccharides. The oligosaccharide moieties of onjisaponins A (*99-4*), B, E (*99-5*) [3], F, and G [2] are listed in Table 99.1. Onjisaponin A and onjisaponin E structures are given as examples.

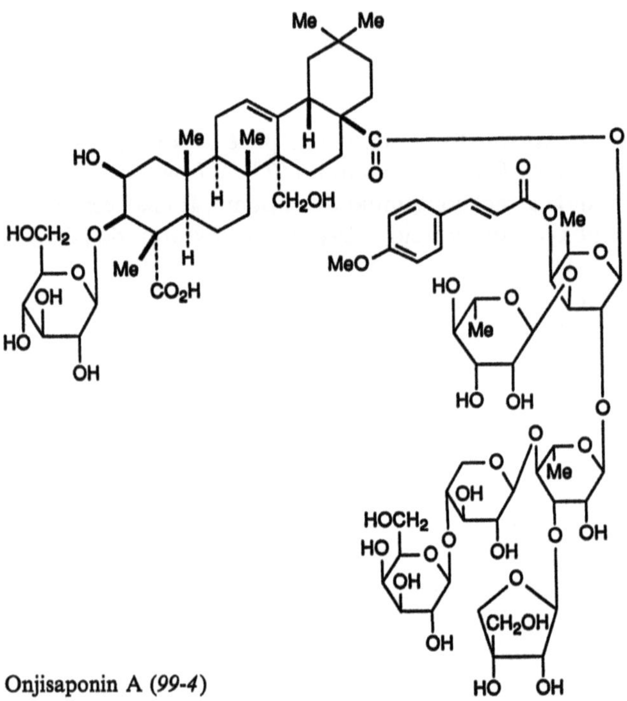

Onjisaponin A (*99-4*)

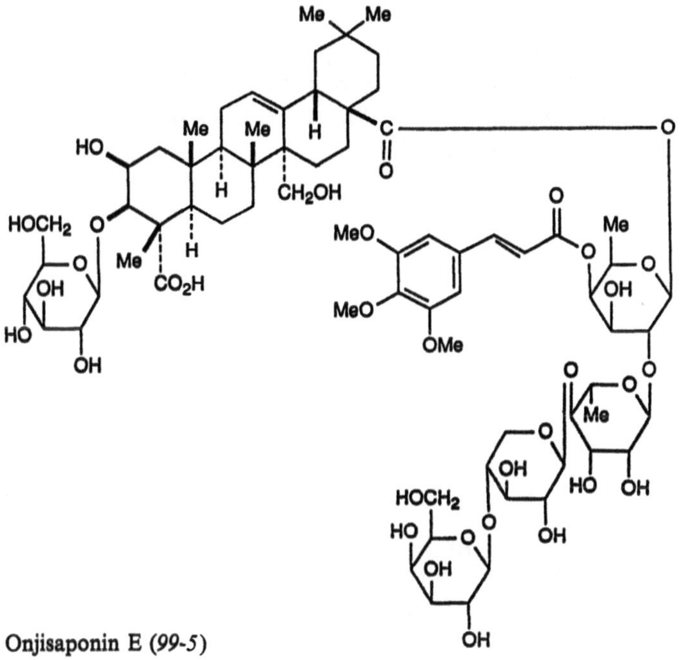

Onjisaponin E (*99-5*)

Table 99.1. Sugar moieties in onjisaponins esterified at position 28 of tenuifolin

Saponin	Oligosaccharide moiety
Onjisaponin A (*99-4*)	*O*-D-Apio-*β*-D-furanosyl-(1→3)-*O*-[*O*-*β*-D-galactopyranosyl-(1→4)-*β*-D-xylopyranosyl-(1→4)]-*O*-6-deoxy-*α*-L-mannopyranosyl-(1→2)-*O*-[6-deoxy-*α*-L-mannopyranosyl-(1→3)]-6-deoxy-4-*O*-[3-(4-methoxyphenyl)-1-oxo-2-propenyl]-*β*-D-galactopyranosyl
Onjisaponin B	*O*-6-Deoxy-*α*-L-mannopyranosyl-(1→3)-*O*-[*O*-*β*-D-galactopyranosyl-(1→4)-*O*-*β*-D-xylopyranosyl-(1→4)-6-deoxy-*α*-L-mannopyranosyl-(1→2)]-6-deoxy-4-*O*-[3-(4-methoxyphenyl)-1-oxo-2-propenyl]-*β*-D-galactopyranosyl
Onjisaponin E (*99-5*)	*O*-*β*-D-Galactopyranosyl-(1→4)-*O*-*β*-D-xylopyranosyl-(1→4)-*O*-6-deoxy-*α*-L-mannopyranosyl-(1→2)-6-deoxy-4-*O*-[1-oxo-3-(3,4,5-trimethoxyphenyl)-2-propenyl]-*β*-D-galactopyranosyl
Onjisaponin F	*O*-D-Apio-*β*-D-furanosyl-(1→3)-*O*-[*O*-*α*-L-arabinopyranosyl-(1→3)-*β*-D-xylopyranosyl-(1→4)]-*O*-6-deoxy-*α*-L-mannopyranosyl-(1→2)-6-deoxy-4-*O*-[1-oxo-3-(3,4,5-trimethoxy-phenyl)-2-propenyl]-*β*-D-galactopyranosyl
Onjisaponin G	*O*-D-Apio-*β*-D-furanosyl-(1→3)-*O*-[*β*-D-xylopyranosyl-(1→4)]-*O*-6-deoxy-*α*-L-mannopyranosyl-(1→2)-6-deoxy-4-*O*-[1-oxo-3-(3,4,5-trimethoxyphenyl)-2-propenyl]-*β*-D-galactopyranosyl

As seen in their structures, the onjisaponins contain a substituted cinnamic acid esterified to the oligosaccharide moiety. Besides the saponins, 3,4,5-trimethoxycinnamic acid (*99-6*) and three xanthone derivatives 6-hydroxy-1,2,3,7-tetramethoxyxanthone (*99-7*), 1,2,3,7-tetramethoxyxanthone (*99-8*), and 1,2,3,6,7-pentamethoxyxanthone (*99-9*) were isolated from *P. tenuifolia* [4].

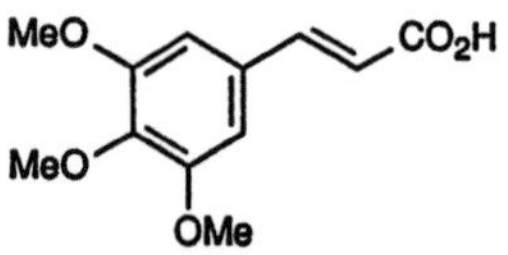

3,4,5-Trimethoxy-
cinnamic acid (*99-6*)

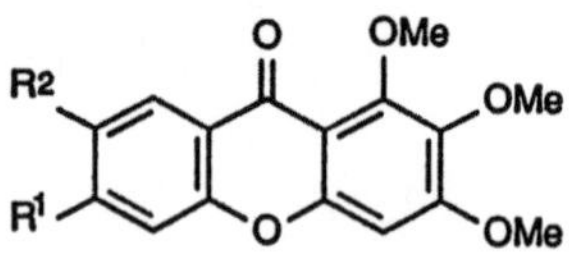

6-Hydroxy-1,2,3,7-tetramethoxyxanthone (*99-7*): $R^1 = OH$, $R^2 = OCH_3$
1,2,3,7-Tetramethoxyxanthone (*99-8*): $R^1 = H$, $R^2 = OCH_3$
1,2,3,6,7-Pentamethoxyxanthone (*99-9*): $R^1 = R^2 = OCH_3$

Xanthone derivatives were also isolated from *P. caudata* and designated as wubangzisides A, B, and C. Wubangzisides A (*99-10*) and B (*99-11*) are euxanthone 7-*O*-glycosides [5], whereas wubangziside C (*99-12*) is 1,3,7-trihydroxy-xanthone 2,4-*C*-di-*β*-D-glucopyranoside [6].

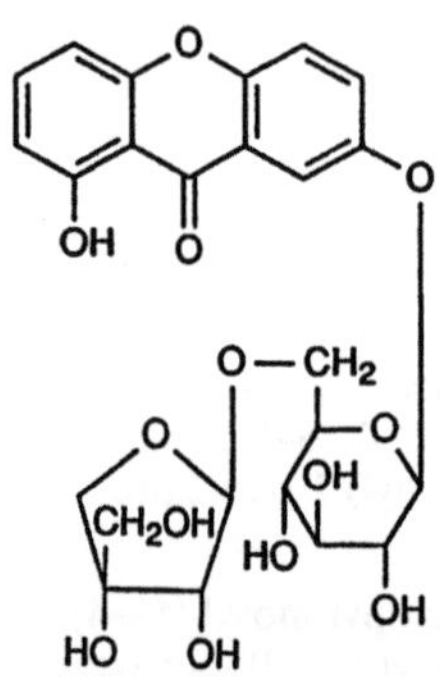
Wubangziside A (*99-10*)

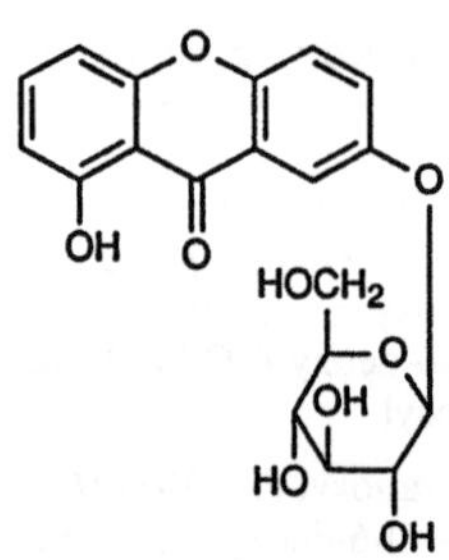
Wubangziside B (*99-11*)

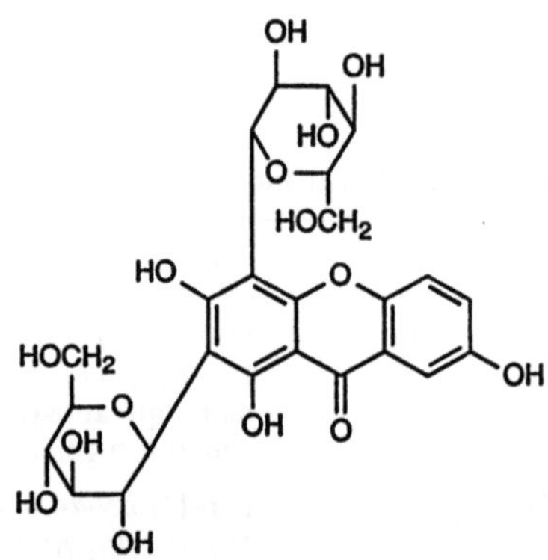
Wubangziside C (*99-12*)

Polygala japonica is another *Polygala* species used in folk medicine as a mucolytic agent. From the aerial part a new triterpene saponin B (*99-13*) was isolated. Acidic hydrolysis of this saponin gave a sapogenin identified as 2α,3α,24-trihydroxyolean-12-en-28-oic acid [7]. The structure determination of the saponin was reported [8].

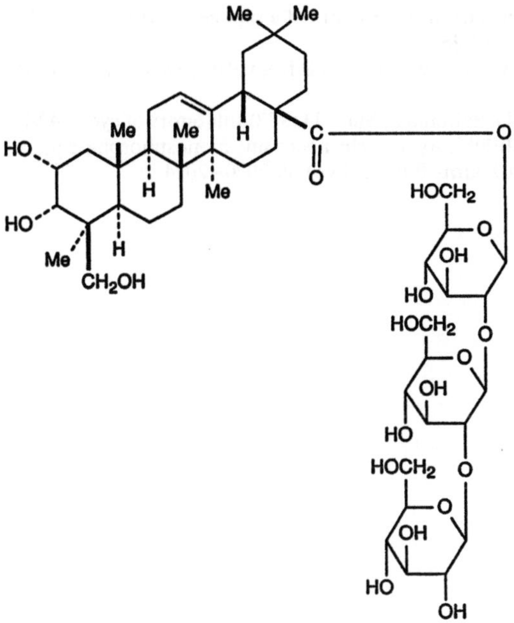

Saponin B from *P. japonica* (*99-13*)

99.3 Pharmacology

Saponins isolated from *P. tenuifolia* showed inhibitory activity against cAMP phosphodiesterase. The concentrations of onjisaponins E, F, and G required to give 50% inhibition were of the same order as that of papaverine. A kinetic study revealed that onjisaponin F acts noncompetitively against cAMP phosphodiesterase, like papaverine. Onjisaponin F exhibited a prolongation effect on hexobarbital sleeping time in mice [9].

References

1. Pelletier SW, Nakamura S, Soman R (1971) Constituents of *Polygala* species. The structure of tenuifolin, a prosapogenin from *P. senega* and *P. tenuifolia*. Tetrahedron 27:4417–4427
2. Sakuma S, Shoji J (1981) Studies on the constituents of the root of *Polygala tenuifolia* Willdenow. I. Isolation of saponins and the structures of onjisaponins G and F. Chem Pharm Bull (Tokyo) 29:2431–2441
3. Sakuma S, Shoji J (1982) Studies on the constituents of the root of *Polygala tenuifolia* Willdenow. II. On the structures of onjisaponins A, B, and E. Chem Pharm Bull (Tokyo) 30:810–821
4. Ito H, Taniguchi H, Kita T, Matsuki Y, Tachikawa E, Fujita T (1977) Xanthones and a cinnamic acid derivative from *Polygala tenuifolia*. Phytochemistry 16:1614–1616
5. Pan MD, Mao Q (1984) Isolation and identification of wubangziside A and B from *Polygala caudata* Rehd et Wils. Acta Pharm Sin 19:899–903
6. Pan MD, Mao Q (1985) Isolation and identification of wubangziside C from *Polygala caudata* Rehd et Wils. Acta Pharm Sin 20:662–665

7. Fang ZP, Yu JZ, Yu DQ (1983) Isolation and chemical structure of a sapogenin from Gua Zi Jin (*Polygala japonica* Houtt). Acta Pharm Sin 18:266–271
8. Fang ZP, Ying GJ (1986) Structure of a new triterpene saponin B from *Polygala japonica* Houtt. Acta Bot Sin 28:196–200
9. Nikaido T, Ohmoto T, Saitoh H, Sankawa U, Sakuma S, Shoji J (1982) Inhibitors of cyclic AMP phosphodiesterase in medicinal plants. IV. Inhibitors of cyclic adenosine monosphosphate phosphodiesterase in *Polygala tenuifolia*. Chem Pharm Bull (Tokyo) 30:2020–2024

Polygonum spp. 100

100.1 Introduction

A number of *Polygonum* species are used in traditional Chinese medicine. The following eight entries with six species are officially listed in the Chinese Pharmacopoeia:

- Bianxu, Herba Polygoni avicularis, is the dry aboveground part of *Polygonum aviculare* L. (Polygonaceae) collected in the summer. It is used mainly as a diuretic and against eczema by external application.
- Quanshen, Rhizoma Bistortae, is the dry rhizome of *P. bistorta* L. collected in early spring and fall. It is used as an antiphlogistic, antidote, and hemostatic for treatment of dysentery and hemorrhagic diseases.
- Huzhang, Rhizoma Polygoni cuspidati, is the dry root and rhizome of *P. cuspidatum* Sieb. et Zucc. collected in spring and fall. It is used for treatment of rheumatism, icterus, menostasis, cough, burn, trauma, and ulcer.
- Heshouwu, Radix Polygoni multiflori, is the dry tuber of *P. multiflorum* Thunb. collected in the fall and winter. It is used as an antidote and laxative for treatment of carbuncle, abscess, and for treatment of hyperlipidemia.
- Zhiheshouwu, Radix Polygoni multiflori preparata, is the processed root of *P. multiflorum*. It is used as a tonic and against hyperlipidemia.
- Shouwuteng, Caulis Polygoni multiflori, is the dry canes of *P. multiflorum* collected in the fall and winter. It is used as a sedative and antirheumatic.
- Shuihonghuazi, Fructus Polygoni orientalis, is the dry ripe fruits of *P. orientale* L. collected in the fall. It is used as an analgesic and stomachic as well as for the treatment of abscess by external application.
- Liaodaqingye, Folium Polygoni tinctorii, is the dry leaves of *P. tinctorium* Ait. collected in the summer and fall. It is mainly used as an antipyretic for treatment of infectious diseases such as measles, laryngitis, mumps, and carbuncle.

100.2 Chemical Constituents

100.2.1 Chemical Constituents of *Polygonum aviculare*

A number of flavone derivatives and their glycosides were isolated and identified from *P. aviculare*, together with the coumarins umbelliferone and scopoletin [1]. The flavones isolated were quercetin, avicularin (*100-1*), quercitrin [1], vitexin, isovitexin, luteolin, rhamnetin-3-*O*-β-D-galactopyranoside, and quercetin-3-*O*-β-D-galactopyranoside [2]. Avicularin is the major flavone glycoside in *P. aviculare*, with an average content of about 0.2% [3].

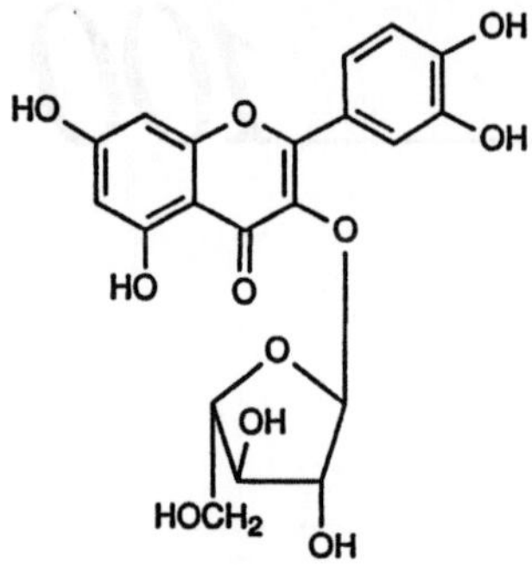

Avicularin (*100-1*)

100.2.2 Chemical Constituents of *Polygonum cuspidatum*

From the rhizome of *P. cuspidatum*, a number of anthraquinone derivatives have been isolated. They are emodin (*100-2*), physcion (*100-3*), chrysophanol (*100-4*) [4], fallacinol (*100-6*), citreorosein (*100-5*), questin (*100-7*), and questinol (*100-8*) [5] as well as their glycosides emodin 8-*O*-β-D-glucopyranoside (*100-9*) and physcion 8-*O*-β-D-glucopyranoside (*100-10*) [6].

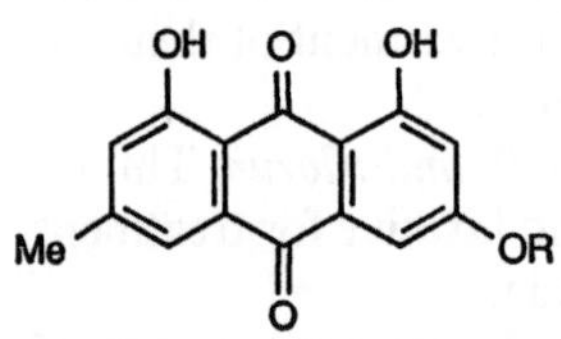

Emodin (*100-2*): R=H
Physcion (*100-3*): R=CH_3

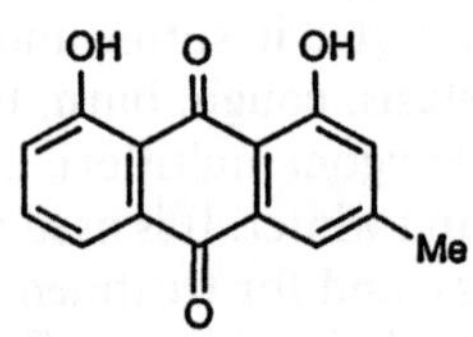

Chrysophanol (*100-4*)

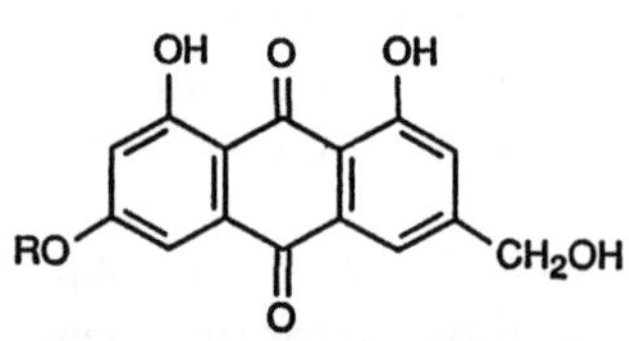

Citreorosein (*100-5*): R=H
Fallacinol (*100-6*): R=CH_3

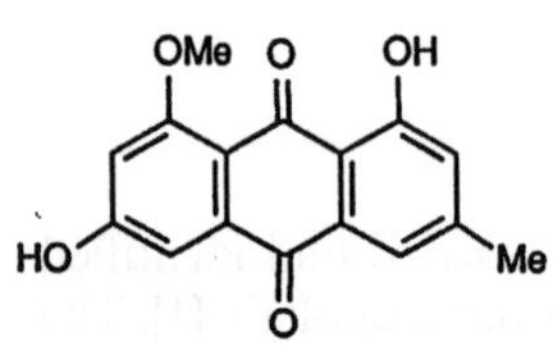

Questin (*100-7*)

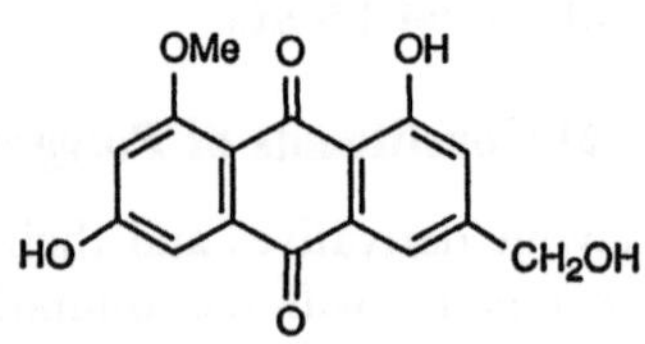

Questinol (*100-8*)

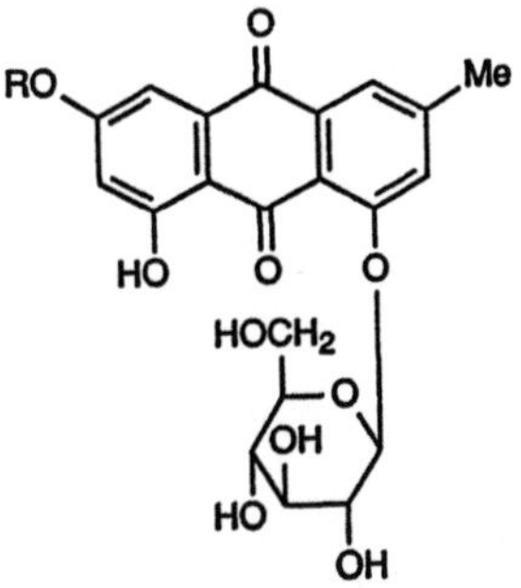

Emodin 8-*O*-β-D-glucopyranoside (*100-9*): R=H
Physcion 8-*O*-β-D-glucopyranoside (*100-10*): R=CH_3

In addition to the anthraquinones, the stilbene compound resveratrol (*100-11*) and its glucoside piceid (*100-12*) [7], protocatechuic acid, catechin, 2,5-dimethyl-7-hydroxy-chromone, 7-hydroxy-4-methoxy-5-methylcoumarin, and the naphthoquinone compound 2-methoxy-6-acetyl-7-methyljuglone (*100-13*) [5] were isolated from the rhizome of *P. cuspidatum*.

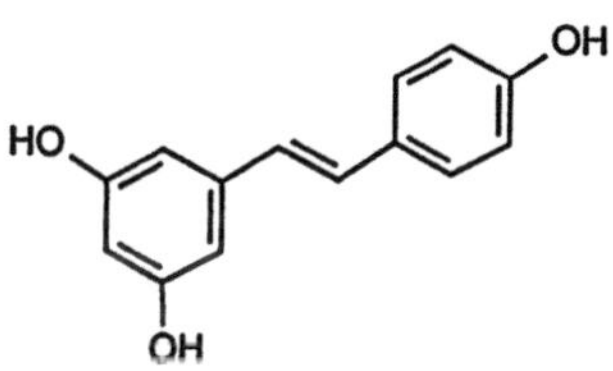

Resveratrol (*100-11*)

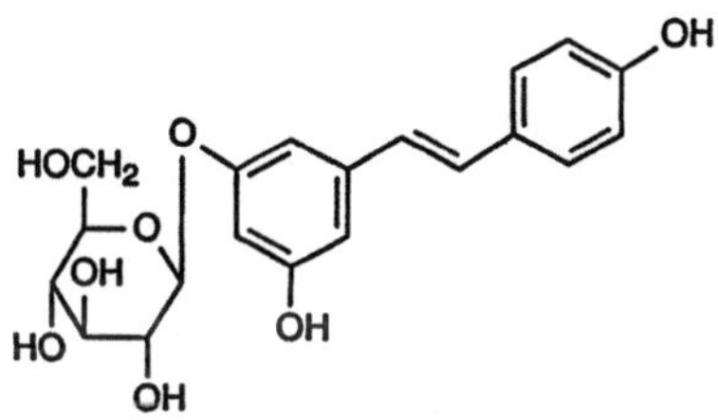

Piceid (*100-12*)

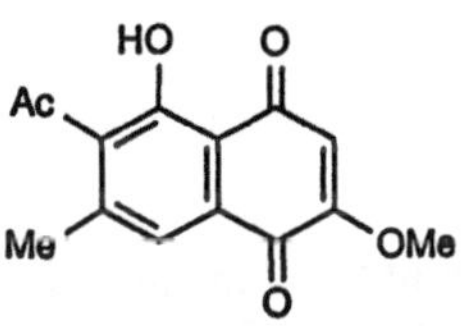

2-Methoxy-6-acetyl-7-methyljuglone (*100-13*)

100.2.3 Chemical Constituents of *Polygonum multiflorum*

Studies on the chemical constituents of *P. multiflorum* showed that the root contains as major constituents the anthraquinones emodin and physcion [8, 9] together with hydroxylated stilbene glycosides. The stilbene derivatives isolated from the root of *P. multiflorum* were identified as 2,3,5,4′-tetrahydroxystilbene 2-*O*-β-D-glucopyranoside (*100-14*) [10] and its 2″- and 3″-*O*-monogalloyl esters [11].

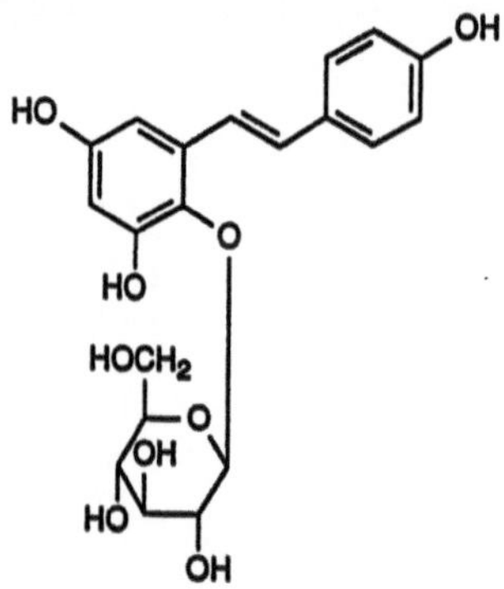

2,3,5,4′-Tetrahydroxystilbene 2-*O*-β-D-glucopyranoside (*100-14*)

The average content of 2,3,5,4′-tetrahydroxystilbene 2-*O*-β-D-glucopyranoside in samples from seven locations ranged from 1% to 3.8% [12]. A new hydroxylated acetophanone glycoside, 2,3,4,6-tetrahydroxyacetophenone 3-*O*-β-D-glucopyranoside named polygoacetophenoside (*100-15*), was also isolated from *P. multiflorum* [18].

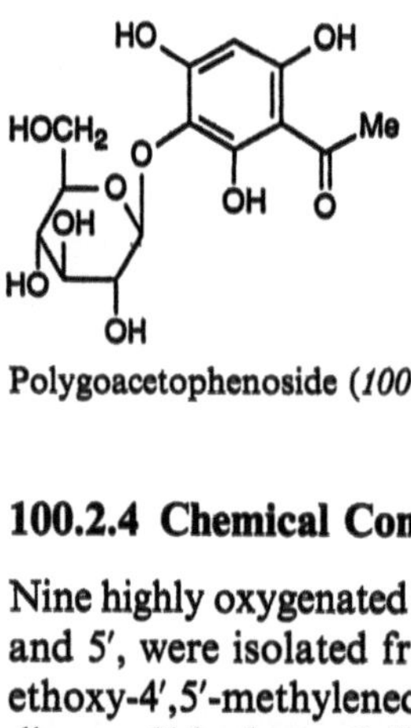

Polygoacetophenoside (*100-15*)

100.2.4 Chemical Constituents of *Polygonum orientale*

Nine highly oxygenated flavones, having oxygen functional groups at C-3,5,6,7,8,3′,4′, and 5′, were isolated from *P. orientale*. They were identified as 3,3′,5,6,7,8-hexamethoxy-4′,5′-methylenedioxy-, 5-hydroxy-3,3′,6,7,8-pentamethoxy-4′,5′-methylenedioxy-, 3′-hydroxy-3,4′,5,5′,6,7,8-heptamethoxy-, 3,3′,5,8-tetramethoxy-4′,5′,6,7-bis (methylenedioxy)-, 3′-hydroxy-3,4′,5,5′,8-pentamethoxy-6,7-methylenedioxy-, and 3′,5-dihydroxy-3,4′,5′,8-tetramethoxy-6,7-methylenedioxyflavone [13].

100.3 Pharmacology

Treatment of rats with *P. multiflorum* extract increased the plasma and liver lipoprotein lipase activities and the ratio of lipoprotein lipase to total lipase activity, but decreased the plasma total cholesterol and high-density lipoprotein cholesterol levels [14, 15]. Resveratrol and piceid inhibited the deposition of triglyceride and cholesterol in the liver of rats fed corn oil-cholesterol-cholic acid mixture. Piceid reduced the serum triglyceride and low-density lipoprotein-cholesterol levels of the treated rats. Intraperitoneal or oral administration of resveratrol or piceid reduced triglyceride synthesis from [^{14}C]palmitate in the liver of mice, but did not affect hormone-induced lipolysis in fat cells from rat epididymal adipose tissue [16].

Piceid and 2,3,5,4′-tetrahydroxystilbene 2-*O*-*β*-D-glucopyranoside partly inhibited the deposition of lipid peroxides in the liver of rats fed with peroxidized oil. The stilbene glycosides reduced the elevation of glutamic-oxalacetic transaminase and glutamic-pyruvic transaminase levels in the serum of the rats. Furthermore, resveratrol, piceid, and 2,3,5,4′-tetrahydroxystilbene 2-*O*-*β*-D-glucopyranoside inhibited lipid peroxidation induced by ADP and NADPH in rat liver microsomes [17].

References

1. Khvorost PP (1980) Flavonoids of *Polygonum aviculare*. Khim Prir Soedin 840
2. Panosyan AG, Barikyan ML, Grigoryan RT, Amroyan EA, Gabrielyan ES (1986) The effect of *Polygonum aviculara* L. flavonoids on platelet aggregation. Khim Farm Zh 20:190–194
3. Xu LX, Liu AR (1983) Assay of avicularin in *Polygonum aviculara* L. Acta Pharm Sin 18:700–704
4. Tsukida K, Yonemichi M (1954) Constituents of polygonaceous plants. III. Constituents of Ko-jo-kon (*Polygonum cuspidatum*). J Pharm Soc Jpn 74:379–382
5. Kimura Y, Kozawa M, Baba K, Hata K (1983) New constituents of roots of *Polygonum cuspidatum*. Planta Med 48:164–168
6. Murakami T, Ikeda K, Takido M (1968) Structure of anthraglycosides from the rhizomes of *Polygonum cuspidatum*. Chem Pharm Bull (Tokyo) 16:2299–2300
7. Nonomura S, Kanagawa H, Makimoto A (1963) Chemical constituents of polygonaceous plants. I. Components of *Polygonum cuspidatum*. Yakugaku Zasshi 83:988–990
8. Yan XZ (1981) Isolation and identification of stilbene glycoside from *Polygonum multiflorum* Thunb. Acta Acad Med Primae Shanghai 8:123–126
9. Zhang XQ, Xa LX (1984) Pulse polarographic determination of anthraquinones in *Polygonum multiflorum* Thunb. Chin J Pharm Anal 4:347–350
10. Hata K, Kozawa M, Baba K (1975) New stilbene glucoside from Chinese crude drug Heshouwn, roots of *Polygonum multiflorum*. Yakugaku Zasshi 95:211–213
11. Nonaka G, Miwa N, Nishioka I (1982) Tannins and related compounds. Stilbene glycoside gallates and proanthocyanidins from *Polygonum multiflorum*. Phytochemistry 21:429–432
12. Yao GG, Sun XP, Zhou GH, Qui ZL (1984) Studies on the quality control of *Polygonum multiflorum* Thunb. VI. Assay of stilbene glycoside in *Polygonum multiflorum* Thunb. Chin J Pharm Anal 4:28–31
13. Kuroyanagi M, Fukushima S (1982) Highly oxygenated flavonoids from *Polygonum orientale*. Chem Pharm Bull (Tokyo) 30:1163–1168
14. Zhang Z, Zhuang QQ, Mei MZ (1983) Effects of some drugs on plasma and liver lipoprotein lipase activities and plasma cholesterol levels in rats. Acta Pharm Sin 18:468–471
15. Mei MZ, Zhuang QQ, Liu GZ, Xie WJ (1979) Rapid screening method for hypocholesterolemic agents. Acta Pharm Sin 14:8–11
16. Arichi H, Kimura Y, Okuda H, Baba K, Kozawa M, Arichi S (1982) Effects of stilbene components of the roots of *Polygonum* cuspidatum Sieb. et Zucc. on lipid metabolism. Chem Pharm Bull (Tokyo) 30:1766–1770
17. Kimura Y, Ohminami H, Okuda H, Baba K, Kozawa M, Archi S (1983) Effects of stilbene components of roots of *Polygonum* spp. on liver injury in peroxidized oil-fed rats. Planta Med 49:51–54
18. Yoshida M, Fujino H, Arise A, Ohmura K, Arisawa M, Morita N (1987) Polygoacetophenoside, a new acetophenone glucoside from *Polygonum multiflorum*. Planta Med 53:273–275

Pseudolarix kaempferi Gord. 101

101.1 Introduction

Tujingpi, Cortex Pseudolaricis, is the dry root bark or stem bark from the stem near to the root of *Pseudolarix kaempferi* Gord. (Pinaceae) shelled and collected in May. It is officially listed in the Chinese Pharmacopoeia and due to its toxicity is used for extradermal treatment of scabies and as an antifungal agent.

101.2 Chemical Constituents

From the root bark of *P. kaempferi*, a number of antifungal acidic compounds named pseudolaric acids A (*101-1*) [1–3], B (*101-2*) [1–6], C (*101-3*), C_2 (*101-4*) [1–3], D (*101-5*) [7, 8], and E (*101-6*) [8] as well as pseudolaric acid A β-D-glucopyranosylester (*101-7*) and pseudolaric acid B β-D-glucopyranosylester (*101-8*) [9] were isolated and chemically investigated. These pseudolaric acids are compounds of a diterpene nature. Pseudolaric acids A, B, C, and C_2 possess the same carbon skeleton and are closely correlated to each other. Pseudolaric acid C is identical with deacetylpseudolaric acid B and pseudolaric acid C_2 with the free acid of pseudolaric acid B. Oxidation of pseudolaric acid A yielded pseudolaric acid B [10, 11]. The structure of pseudolaric acid A was also elucidated by X-ray crystallography [12]. Pseudolaric acids D and E are structurally derived from kaurane.

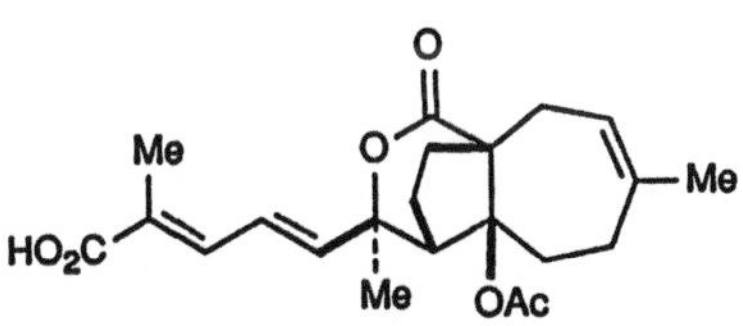

Pseudolaric acid A (*101-1*)

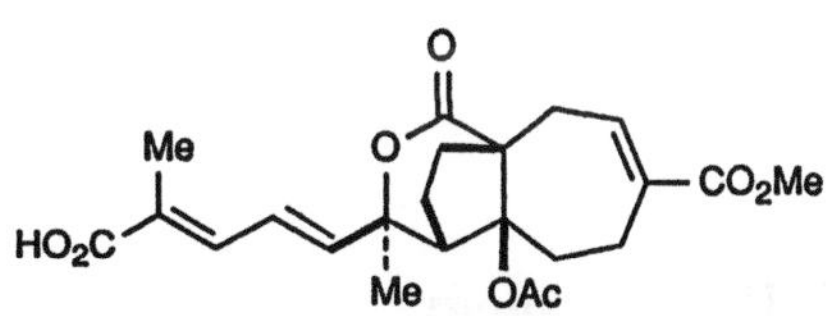

Pseudolaric acid B (*101-2*)

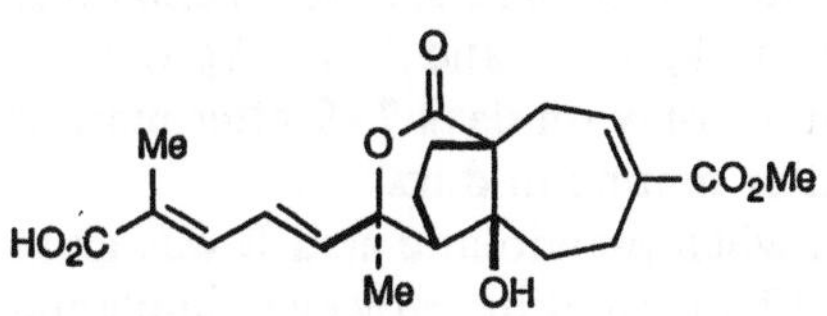

Pseudolaric acid C (*101-3*)

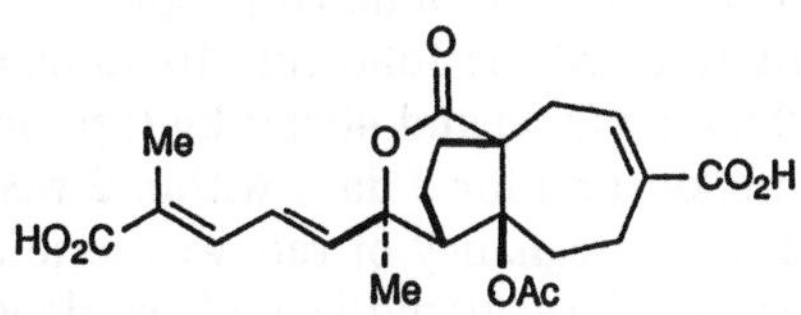

Pseudolaric acid C_2 (*101-4*)

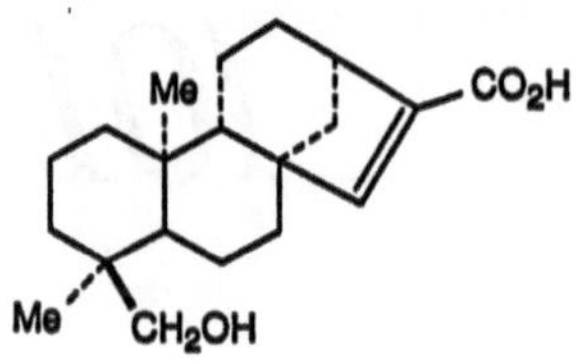

Pseudolaric acid D (*101-5*)

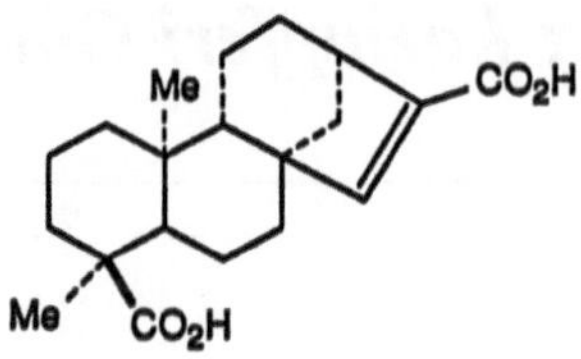

Pseudolaric acid E (*101-6*)

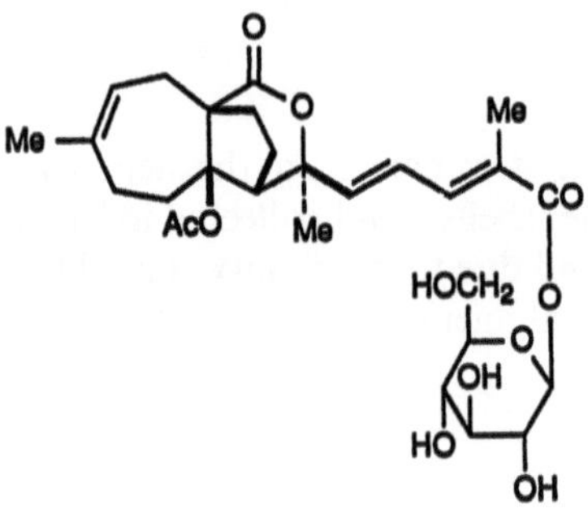

Pseudolaric acid A β-D-glucopyranosylester (*101-7*)

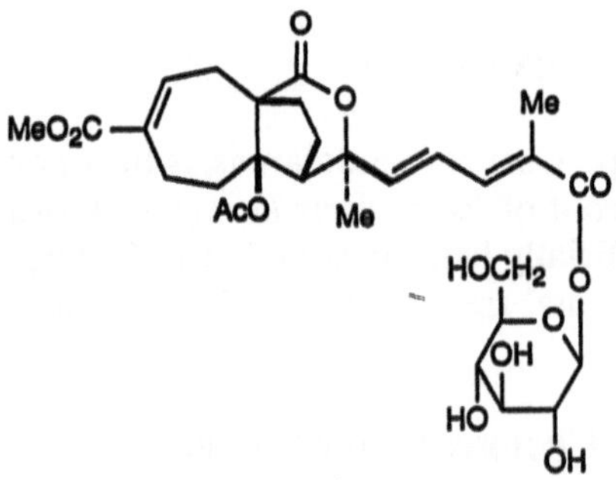

Pseudolaric acid B β-D-glucopyranosylester (*101-8*)

The first attempt at total synthesis of pseudolaric acid A was made by Pan et al. [13] with (*101-9*) as the key intermediate.

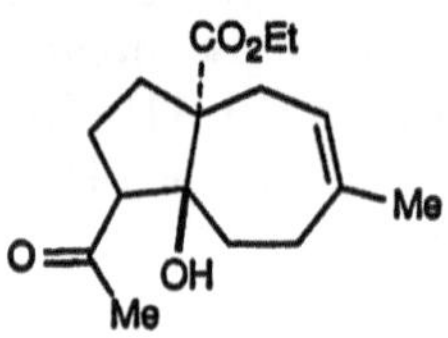

101-9

101.3 Pharmacology

Pseudolaric acid B terminated pregnancy at an early stage in rats, rabbits, and dogs. Animals received the following treatments: 25 mg/kg (s.c.) and 15 mg/kg (i.p.) in rats; 30 mg/kg (s.c.) and 40 mg/kg (i.p.) in rabbits, only on days 7–9 after mating; and 5 mg/kg (i.p.) for 3 days within 2 weeks after mating in dogs.

Mid-term pregnancy of rats was terminated when pseudolaric acid B was given intraperitoneally at 10 mg/kg daily on days 10–12 after mating. However, implantation was not prevented in rats when 40 mg/kg pseudolaric acid B was injected subcutaneously or intraperitoneally daily on days 1–3 after mating. Pseudolaric acid B showed no estrogenic activity. It lowered the plasma progesterone levels, but progesterone did not antagonize the effects of pseudolaric acid B on early pregnancy in rats [14].

The ED_{50} of a single dose of pseudolaric acid A and B in terminating the early pregnancy in rats 7 days after mating was determined to be 14.5 and 9.3 mg/kg, respectively. The LD_{50} values of pseudolaric acid A and B given by gavage on day 7 of pregnancy in rats were 220 and 130 mg/kg, respectively. The therapeutic index (LD_{50}/ED_{50}) for pseudolaric acid A is 15 and that for pseudolaric acid B 14. Therefore, pseudolaric acid A is less active but also less toxic than pseudolaric acid B [15].

References

1. Li ZL, Pan TC, Wu CL, Hsu KI (1980) Studies on the antifungal constituents of *Pseudolarix kaempferi* Gord. Acta Acad Med Primae Shanghai 7:386
2. Zhou BN, Ying BP, Song GQ, Chen ZX, Han J, Yan YF (1983) Pseudolaric acids from *Pseudolarix kaempferi*. Planta Med 47:35–38
3. Li ZL, Pan DJ, Hu CQ, Wu QL, Xu GY, Zhou BN, Yin BP, Sun GJ, Chen ZX (1982) Studies on the antifungal constituents of tu jin pi. The structures of novel diterpenoids, pseudolaric acid A, B, C, and C_2. In: Wang Y (ed) Chemistry in Natural Products. Science, Beijing, pp 150–155
4. Ying BP, Yu HG, Han J (1988) Study on the NMR spectra of pseudolaric acid B. Org Chem 8:273–275
5. Ying BP, Xu RS, Mi JF, Han J (1988) Configuration of pseudolaric acid B. Acta Chim Sin 46:85–86
6. Yu HG, Song GQ (1988) Determination of relative signs of scalar coupling constants in pseudolaric acid derivatives by proton COSY-45. Acta Chim Sin 46:76–77
7. Wu JH, Wang BY, Zheng PJ, Li ZL, Chen K, Pan DJ, Xu GY (1986) The crystal structure of pseudolaric acid D. J Struct Chem 5:190–192
8. Li ZL, Chen K, Pan DJ, Wu GY (1989) New diterpenic constituents of Tu-Jin-Pi. IV. Isolation and identification of pseudolaric acid D and pseudolaric acid E. Acta Chim Sin 47:258–261
9. Li ZL, Chen K, Pan DJ, Xu GY (1985) New diterpenic constituents of Tu-Jin-Pi. III. Isolation and identification of pseudolaric acid A-β-D-glucoside and pseudolaric acid B-β-D-glucoside. Acta Chim Sin 43:786–788
10. Li ZL, Pan DJ, Wu QL, Xu GY (1982) Studies on the novel diterpenic constituents of Tu-Jin-Pi. II. Identification of pseudolaric acid C_2 and structural correlation of pseudolaric acid A, B, C, and C_2. Acta Chim Sin 40:757–761
11. Li ZL, Pan DJ, Hu CQ, Wu QL, Yang SS, Xu GY (1982) Studies on the novel diterpenic constituents of Tu-Jin-Pi. I. Determination of chemical structures of pseudolaric acid A and pseudolaric acid B. Acta Chim Sin 40:447–457
12. Yao JX, Lin XY (1982) Crystal and molecule structure of pseudolaric acid A. Acta Chim Sin 40:385–393
13. Pan BC, Chang HY, Cai GL, Guo YS (1989) Synthetic studies on pseudolaric acid A. Pure Appl Chem 61:389–392
14. Wang WC, Lu RF, Zao SX, Zhu YZ (1982) Antifertility effect of pseudolaric acid B. Acta Pharmacol Sin 3:188–192
15. Wang WC, Lu RF, Zhao SX, Gu ZP (1988) Comparison of early pregnancy-terminating effect and toxicity between pseudolaric acids A and B. Acta Pharmacol Sin 9:445–448

Pueraria lobata (Willd.) Ohwi 102

102.1 Introduction

Gegen, Radix Puerariae, is the dry root of *Pueraria lobata* (Willd.) Ohwi or *P. thomsonii* Benth. (Fabaceae) collected in the fall and winter. It is officially listed in the Chinese Pharmacopoeia and used as a muscle relaxant, antipyretic, antidysenteric, and for treatment of hypertension.

102.2 Chemical Constituents

The major active principles in the root of *P. lobata* are isoflavone derivatives, puerarin (*102-1*), daidzein (*102-2*), daidzin (*102-3*), and daidzein-7,4′-diglucoside (*102-4*) [1, 2]. Daizein and daidzin are well-known compounds isolated from soybean (*Phaseolus vulgaris*) and some other plants. Daidzin and daidzein-7,4′-diglucoside are *O*-glucosides of daidzein and puerarin is a *C*-glucoside of daidzein.

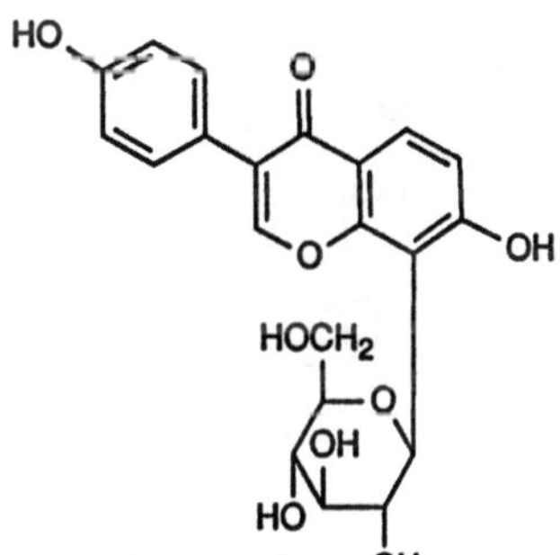

Puerarin (*102-1*)

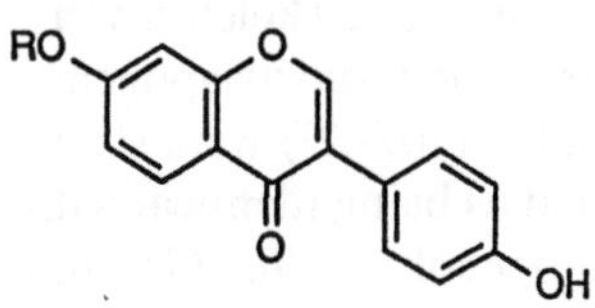

Daidzein (*102-2*): R=H

Daidzin (*102-3*): R=

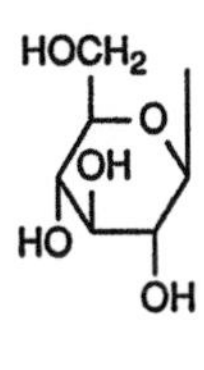

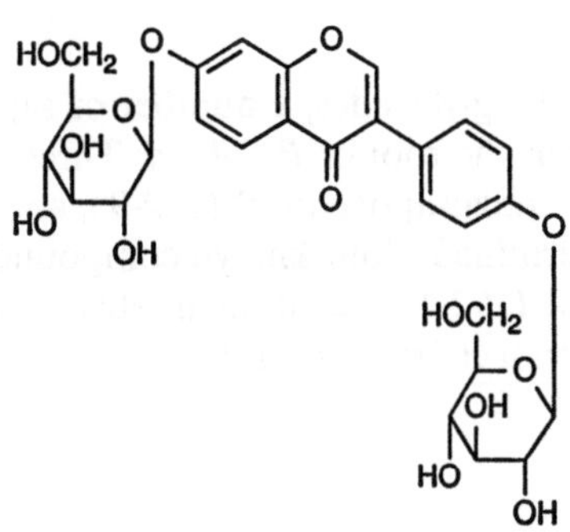

Daidzein-7,4′-di-*O*-*β*-D-glucopyranoside (*102-4*)

The contents of daidzein, daidzin, daidzein-7,4′-diglucoside, puerarin, and 4′-methylpuerarin in roots of *P. lobata* from various locations in China ranged from 0.02% to about 2%, on the basis of HPLC determination [3]. A series of studies on the determination of isoflavones in *Pueraria* roots with HPLC, TLC [4–7], or polarography [8] were reported. Formononetin [9], 3′-hydroxypuerarin, 6″-*O*-D-xylosylpuerarin, 3′-methoxypuerarin, puerarin 4′-*O*-D-glucoside [9], and 8-*C*-apiosyl-(1→6)-glucosides of daidzein and genistein [10] were further isoflavones and isoflavone glycosides isolated from the root of *P. lobata*. A coumestan derivative named puerarol (*102-5*) was also isolated from the root of *P. lobata* [9].

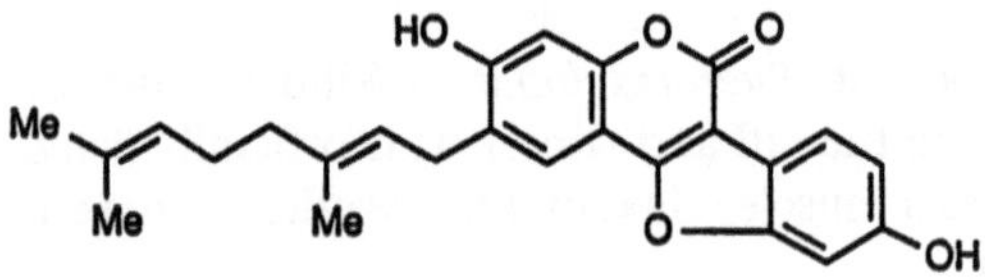

Puerarol (*102-5*)

The isolation of two new aromatic glycosides named pueroside A (*102-6*) and pueroside B (*102-7*) from the root of *P. lobata* was also reported [11].

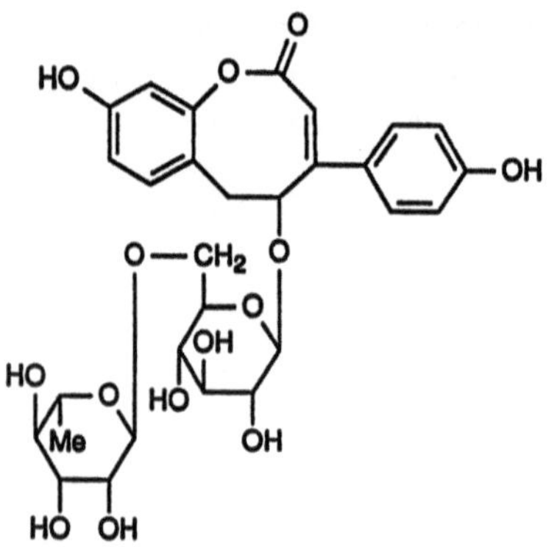

Pueroside A (*102-6*)

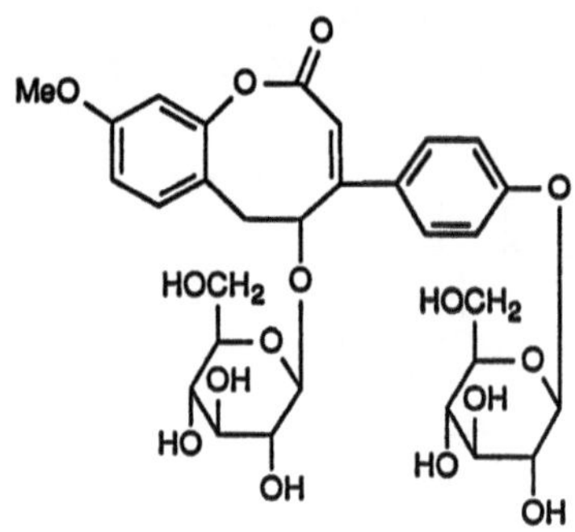

Pueroside B (*102-7*)

Besides the glycosides, a number of sapogenins with an oleanane skeleton were isolated from the root of *P. lobata*. Three new compounds named kudzusapogenol A (*102-8*), kudzusapogenol B (*102-9*), and kudzusapogenol C (*102-10*) were structurally determined. Four known compounds were elucidated as being identical with sophoradiol (*102-11*), cantoniensistriol (*102-12*), soyasapogenol A (*102-13*), and soyasapogenol B (*102-14*) [12].

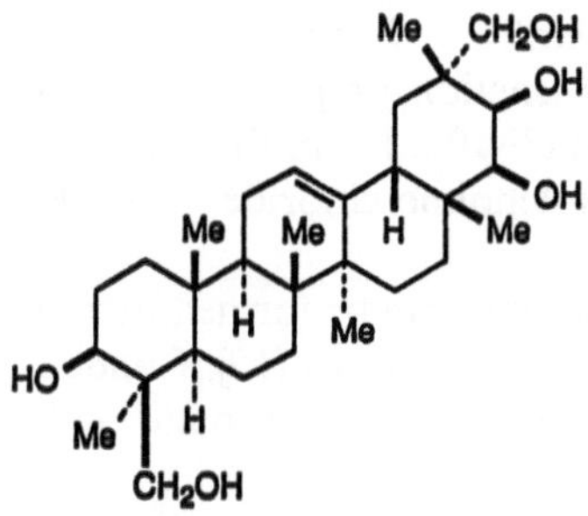

Kudzusapogenol A (*102-8*)

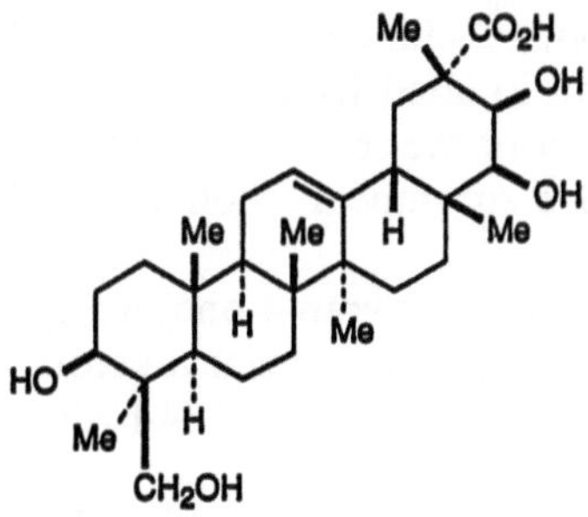

Kudzusapogenol B (*102-9*)

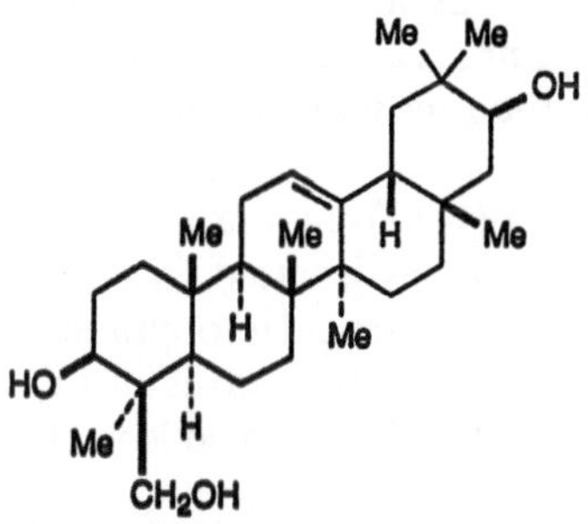

Kudzusapogenol C (*102-10*)

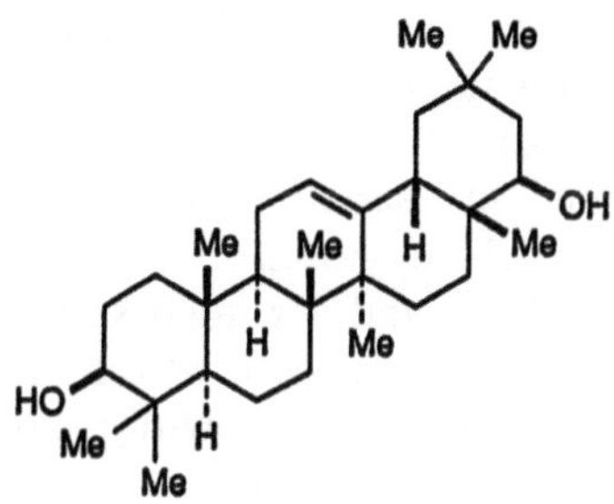

Sophoradiol (*102-11*)

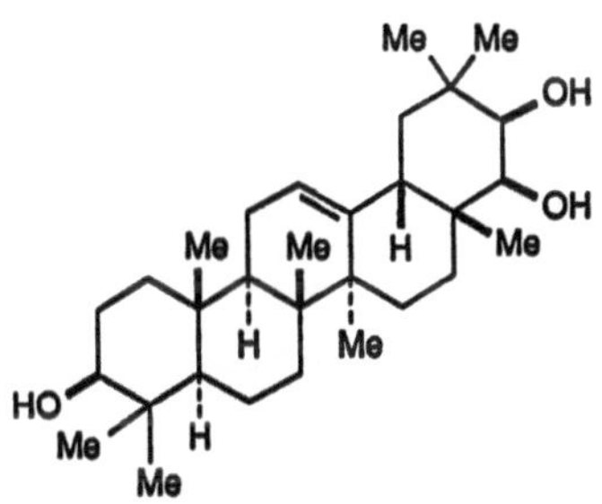

Cantoniensistriol (*102-12*)

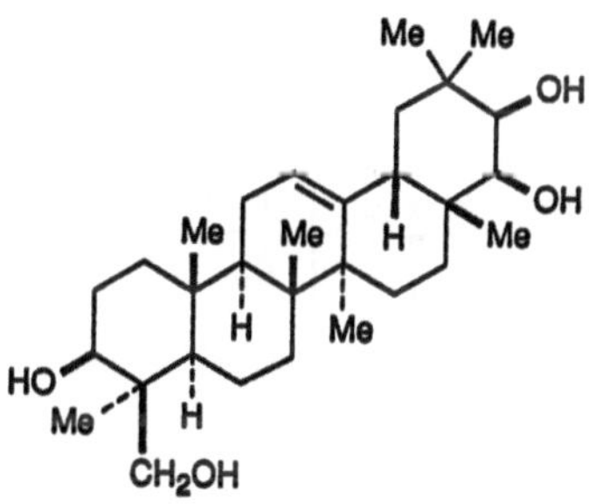

Soyasapogenol A (*102-13*)

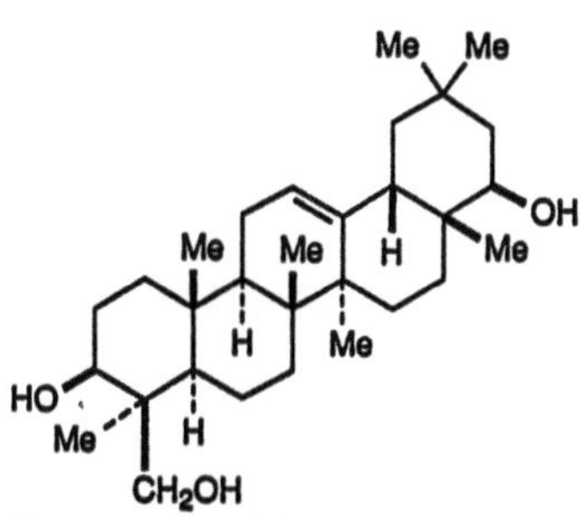

Soyasapogenol B (*102-14*)

In addition, 6,7-dimethoxycoumarin, 5-methylhydantoin, β-sitosterol, and its glucoside were isolated from *P. lobata* [13]. A polar fraction prepared from the aqueous extract of the root of *P. lobata* was termed MTF-101. From this fraction, two cholinergic substances, choline chloride and acetylcholine chloride, were detected [14].

The 3-phenylbenzopyran-4-one ring system of isoflavones can be formed by ring closure of *O*-hydroxyphenylbenzyl ketones under various conditions [15]. Condensation of 2,4-dihydroxyphenyl-(4-hydroxybenzyl)-ketone in ethylformate with sodium yielded daidzein [16].

Studies with cell cultures revealed that the biosynthetic pathway of isoflavones from malonyl CoA and *p*-coumaroyl CoA in *P. lobata* is catalyzed by chalcone synthase, chalcone-flavanone isomerase, and isoflavone synthase. Kinetic experiments indicated that liquiritigenin, the precursor of daidzein, is a much more favorable substrate for the isoflavone synthase in *P. lobata* than naringenin, the corresponding precursor of genistein [17].

A new triterpene saponin 3-*O*-α-L-rhamnopyranosyl-(1→2)-α-L-arabinopyranosyl-(1→2)-β-D-glucopyranuronosyl sophoradiol (*102-15*) was isolated from the flowers of *P. lobata,* and a related saponin (*102-16*) with a β-D-galactopyranose instead of the arabinopyranose was isolated from both flowers and leaves of this plant [18]. Furthermore, a glucosyltryptophan derivative (*102-17*) was also isolated from the flower of *P. lobata* [19].

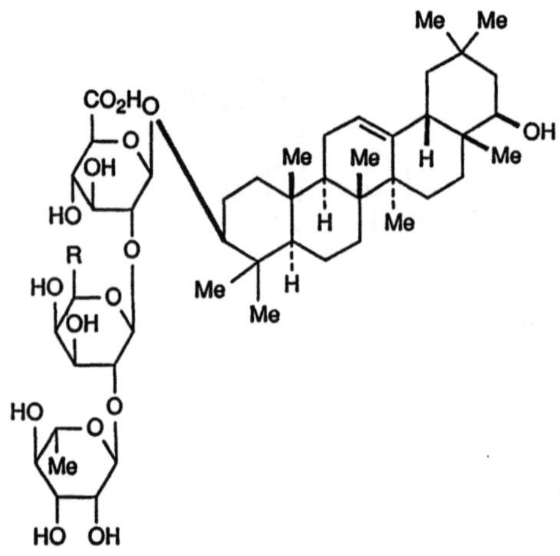

Sophoradiol 3-*O*-α-L-rhamnopyranosyl-(1→2)-α-L-arabinopyranosyl-(1→2)-β-D-glucopyranuronoside (*102-15*): R=H

Sophoradiol 3-*O*-α-L-rhamnopyranosyl-(1→2)-β-D-galactopyranosyl-(1→2)-β-D-glucopyranuronoside (*102-16*): $R=CH_2OH$

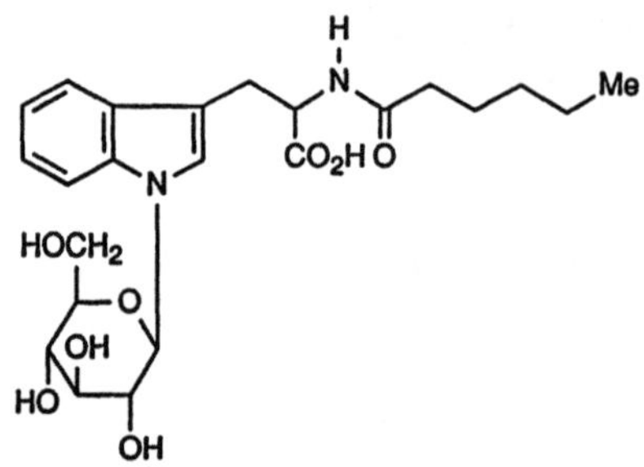

(*102-17*)

102.3 Pharmacology

Studies on the pharmacological activities of different fractions obtained by extracting the root of *P. lobata* with different solvents revealed a coexistence of substances having a mutually reverse pharmacological effect in the root. Some fractions showed a decrease in body temperature in mice, while others showed an increase; some fractions exerted a relaxant effect on the isolated quinea pig ileum like that of papaverine, whereas some fractions possessed a contractive effect [20].

The total isoflavones from the ethanolic extract of roots of *P. lobata* had a hypotensive effect on anesthetized dogs and unanesthetized hypertensive dogs [21]. In dogs, the intraarterial or intravenous injection of total isoflavones of *P. lobata* roots or puerarin dilated coronary arteries to a greater extent than other arteries examined. Heart rate, blood pressure, and total peripheral resistance decreased while cardiac output remained unchanged.

After intravenous injection of total isoflavones, oxygen, lactate, and pyruvate content of the blood of the coronary sinus increased and the myocardial consumption of these substances decreased. These effects may be beneficial in the treatment of human coronary artery disease. The effects of the isoflavones may be due to a direct relaxation of coronary vessels [22]. Daidzein, daidzin, and puerarin have been reported to relieve headaches and other symptoms associated with hypertension [2].

In dogs, intravenous administration of puerarin decreased the size of acute myocardial infarction. Puerarin was administered 5 min after acute coronary ligation. Beneficial effects of puerarin were demonstrated by epicardial electrocardiogram, plasma creatine kinase, and radiocardiogram [23–26]. A clinical study in patients with hypertension or angina pectoris showed that puerarin applied intravenously at a dose of 100–200 mg decreased blood catecholamine levels, blood pressure, and heart rate [27].

In anesthetized rats, arrhythmia induced by aconitine and $BaCl_2$ was antagonized by puerarin administered intravenously or orally or by daidzein administered orally. Furthermore, ventricular fibrillation induced by $CaCl_2$ in rats and by chloroform in mice was prevented by daidzein or the alcoholic extract of *P. lobata* given orally [28]. Puerarin also antagonized the cardiac arrhythmia induced by chloroform in rabbits [29]. The intravenous LD_{50} value of puerarin in mice was 738 mg/kg [29]. Blocking of isoprenaline-induced tachycardia in cats and a decrease in normal heart rate and blood pressure in anesthetized cats by intravenous administration of puerarin was mediated by its β-adrenergic receptor-blocking activity [30]. Puerarin showed a hypoglycemic effect in alloxan-induced diabetic mice, and also decreased serum cholesterol levels [31]. Daidzein exerted a very high antihemolytic activity on peroxidative hemolysis in sheep erythrocytes, whereas only little activity was found in rat erythrocytes. No antihemolytic effect at all was noted in rabbit erythrocytes [32].

Absorption, distribution, and elimination of ^{14}C-labeled daidzein were studied in rats. Radioactivity appeared in the blood 30 min after oral administration of daidzein and reached its peak within 6–8 h, therefore decreasing steadily. About 65% of the radioactivity was absorbed from the gastrointestinal tract within 24 h. After intravenous injection of daidzein, the distribution phase and the elimination phase in the blood had half-lives of 13 and 42 min, respectively. Radioactivity was highest in the kidney and liver, moderate in plasma, lung, and heart, and low in skeletal muscle, spleen, testis, and brain. After intravenous injection of daidzein, about 70% of the radioactivity was excreted in urine within 24 h, while only 17% was recovered from the feces. After oral administration, the amounts excreted in the urine and feces were about equal. Radioactivity recovered from the gastrointestinal tract, urine, and bile was mainly attributable to metabolites of daidzein, indicating that daidzein was metabolized rapidly [33].

The metabolism of puerarin was also studied in rats, using polyamide thin-layer chromatography and UV spectrophotometry. The blood level of puerarin following intravenous administration in rats decreased in two phases with half-lives of 3 and 18 min, respectively. Puerarin was widely distributed in the body and eliminated rapidly. The highest level of puerarin was found in the kidney, moderate levels in plasma, liver, and spleen, and lowest in the brain. Absorption of puerarin from the gastrointestinal tract was rapid but incomplete. About 40% of the dose could still be recovered from the gastrointestinal content and feces 24 h after administration. About 1.8% and 36% was excreted in urine and feces within 24 h, respectively, after oral administration. After intravenous administration, about 37% was found in urine and 7% in feces. Puerarin was stable in the gastrointestinal tract, but appears to be metabolized in the blood, liver, lung, and kidney [34].

References

1. Shibata S, Murakami T, Nishikawa Y, Harada M (1959) The constituents of *Pueraria* root. Chem Pharm Bull (Tokyo) 7:134–136
2. Fang CC, Lin M, Sun CM, Liu HM, Lang HY (1974) Flavones of Radix puerariae. Natl Med J China 54:271–274
3. Zhang YH, Yang F (1984) HPLC determination of isoflavones in Ge Gen (Radix puerariae) and its tablets. Chin J Pharm Anal 4:67–70
4. Fang QN, Wu P, Yang L (1983) Separation and determination of puerarin by high-performance liquid chromatography. Acta Pharm Sin 18:695–699
5. Hayakawa J, Noda N, Yamada S, Uno K (1984) Studies on physical and chemical quality evaluation of crude drug preparations. I. Analysis of pueraria radix and pueraria species. Yakugaku Zasshi 105:50–56
6. Tseng TH, Sheu SJ (1983) Analysis and processing of Chinese herbs. IV. Quantitative study of pueraria root. J Taiwan Pharm Assoc 35:177–183
7. Zhao SP, Zhang YZ (1985) TLC-densitometry of isoflavones in *Pueraria lobata* (Willd.) Ohwi. Acta Pharm Sin 20:203–208
8. Xu LX, Liu AR, Zhang XQ (1987) Differential pulse polarographic determination of flavonoids in *Pueraria lobata.* Acta Pharm Sin 22:208–211
9. Ohshima Y, Okuyama T, Takahashi K, Takizawa T, Shibata S (1988) Isolation and high performance liquid chromatography (HPLC) of isoflavonoids from the *Pueraria* root. Planta Med 54:250–254
10. Kinjo J, Kurusawa J, Baba J, Takeshita T, Yamasaki M, Nohara T (1987) Studies on the constituents of *Pueraria lobata.* III. Isoflavonoids and related compounds in the roots and the voluble stems. Chem Pharm Bull (Tokyo) 35:4846–4850
11. Kinjo J, Furusawa J, Nohara T (1985) Two novel aromatic glycosides, Pueroside-A and -B, from puerariae radix. Tetrahedron Lett 26:6101–6102
12. Kinjo J, Miyamoto I, Murakami K, Kida K, Tomimatsu T, Yamasaki M, Nohara T (1985) Oleanene-sapogenols from puerariae radix. Chem Pharm Bull (Tokyo) 33:1293–1296
13. Chen MH, Zhang SJ (1985) Studies on the chemical constituents of *Pueraria lobata.* Chin Pharm Bull 10:274–276
14. Nakamoto H, Miyamura S, Inada K, Nakamura N (1975) Aqueous extract of pueraria radix. I. Preparation and the components of the active extract. Yakugaku Zasshi 95:1123–1127

15. Wagner H, Farkas L (1975) Synthesis of flavonoids. In: Harborne JB, Mabry TJ, Mabry H (eds) The Flavonoids. Chapman and Hall, London, pp 182–183
16. Wessely F, Kornfeld L, Lechner F (1933) Über die Synthese von Daidzein und von 7-oxy-4′-methoxy-isoflavon. Chem Ber 66:685–687
17. Takashi H, Hiroshi N, Yutaka E, Ushio S (1989) Isoflavone synthase fron cell suspension cultures of *Pueraria lobata*. Chem Pharm Bull (Tokyo) 37:249–252
18. Kinjo J, Takeshita T, Abe Y, Terada N, Yamashita H, Yamasaki M, Takeuchi K, Murakami K, Tomimatsu T, Nohara T (1988) Studies on the constituents of *Pueraria lobata*. IV. Chemical constituents in the flowers and the leaves. Chem Pharm Bull (Tokyo) 36:1174–1179
19. Kinjo J, Takeshita T, Nohara T (1988) Constituents of *Pueraria lobata*. V. A tryptophan derivative from Puerariae flos. Chem Pharm Bull (Tokyo) 36:4171–4173
20. Ueno K (1975) Pharmacological studies on *Pueraria* root. I. Fractional extraction of *Pueraria* root and identification of its pharmacological effects. Chem Pharm Bull (Tokyo) 23:1798–1805
21. Tseng KY, Chou YP, Chang LY, Fan LL (1974) Pharmacologic studies on Radix puerariae. I. Effects on dog arterial pressure, vascular reactivity, cerebral and peripheral circulation. Natl Med J China 54:265–270
22. Fan L, Zeng GY, Zhou YP, Zhang LY, Cheng YS (1982) Pharmacologic studies on Radix puerariae. Effects of pueraria flavones on coronary circulation, cardiac hemodynamics, and myocardial metabolism in dogs. Chin Med J [Engl] (Beijing) 95:145–150
23. Li XY, Weng PR, Shao JH, Zhu XL, Zhu JF (1984) Effect on intravenous puerarin injection on the size of experimental acute myocardial infarction in dogs. Acta Acad Med Shandong 22:9–17
24. Fan LL, O'Keefe DD, Powell WJ Jr (1984) Effect of puerarin on regional myocardial blood flow and cardiac hemodynamics in dogs with acute myocardial ischemia. Acta Pharm Sin 19:801–807
25. Fan LL, O'Keefe DD, Powell WJ Jr (1985) Pharmacologic studies on radix puerariae. Effect of puerarin on regional myocardial blood flow and cardiac hemodynamics in dogs with acute myocardial ischemia. Chin Med J [Engl] (Beijing) 98:821–832
26. Zhu XL, Wang KP, Liu SJ, Zhang XL, Liu J (1985) Effect of puerarin on heart hemodynamics and effective refractory periods in dogs. Acta Acad Med Shandong 23:48–50
27. Zeng GY, Zhang LY, Zhou YP, Fan LL (1979) Pharmacological studies on Radix puerariae. IV. Effects of puerariae flavones on plasma catecholamine levels in patients with hypertension and angina pectoris. Natl Med J China 59:238–241
28. Fan LL, Zhao DH, Zhao MQ, Zeng GY (1985) The antidysrhythmic effect of *Pueraria* isoflavones. Acta Pharm Sin 20:647–651
29. Chai XS, Wang ZX, Chen PP, Wang LY, Lu XR, Kang B (1985) Antiarrhythmic action of puerarin. Acta Pharmacol Sin 6:166–168
30. Lu XR, Gao E, Xu LZ, Li HZ, Kang B, Chen WN, Chen SM, Chai XS (1987) Puerarin β-adrenergic receptor blocking effect. Chin Med J [Engl] (Beijing) 100:25–28
31. Shen ZF, Xie MZ (1985) Hypoglycemic effect of the combined use of puerarin and aspirin in mice. Acta Pharm Sin 20:863–865
32. Naim M, Gestetner B, Bondi A, Birk Y (1976) Antioxidative and antihemolytic activities of soybean isoflavones. J Agric Food Chem 24:1174–1177
33. Su CY, Zhu XY (1979) Metabolic fate of the effective components of Radix puerariae. II. Absorption, distribution and elimination of ^{14}C-daidzein. Acta Pharm Sin 14:129–134
34. Zhu XY, Su GY, Li ZH, Yue TL, Yan XZ, Wei HL (1979) Metabolic fate of the effective components of Radix puerariae. III. Metabolism of puerarin. Acta Pharm Sin 14:349–355

Qingdai 103

103.1 Introduction

Qingdai, Indigo naturalis, is the natural dye obtained from plant material of *Baphicacanthus cusia* (Nees) Bremek. (Acanthaceae), *Indigofera suffruticosa* Mill. (Fabaceae), *Polygonum tinctorium* Ait. (Polygonaceae), or *Isatis indigotica* Fort. (Brassicaceae). The natural indigo is officially listed in the Chinese Pharmacopoeia and is to be used in traditional Chinese medicine as a hemostatic, antipyretic, anti-inflammatory, and sedative in the treatment of bacterial and viral infections.

Isatis indigotica and *Polygonum tinctorium* are also separately entered in the Chinese Pharmacopoeia:

- Daqingye, Folium Isatidis, is the dry leaves of *I. indigotica* collected in the summer and fall.
- Banlangen, Radix Isatidis, is the dry roots of *I. indigotica* collected in the fall.
- Liaodaqingye, Folium Polygoni tinctorii, is the dry leaves of *P. tinctorium* collected in the summer and fall.

These three herbal medicines are used as antipyretic, antibacterial, and antiviral agents against infectious diseases.

103.2 Chemical Constituents

103.2.1 Chemical Constituents of *Baphicacanthus cusia*

From the natural indigo isolated from *B. cusia*, indirubin (*103-1*) [1–3], indigo (*103-2*) [3], and isoindigo (*103-3*) were isolated and identified. The discovery of the antileukemic activity of natural indigo and the isolation of indirubin as the effective ingredient have greatly stimulated interest in the study of antitumor agents from natural sources [4–6].

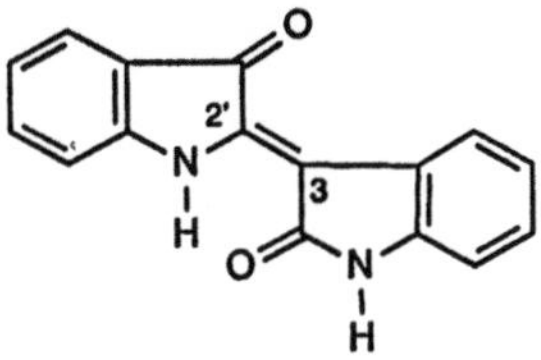

Indirubin (*103-1*)

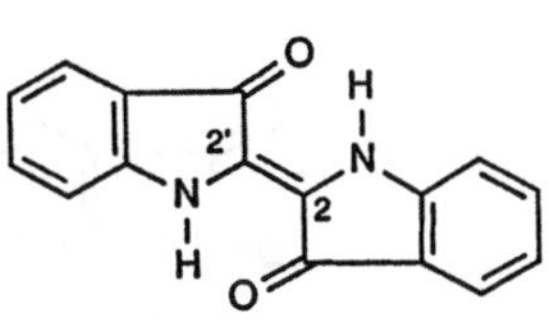

Indigo (*103-2*)

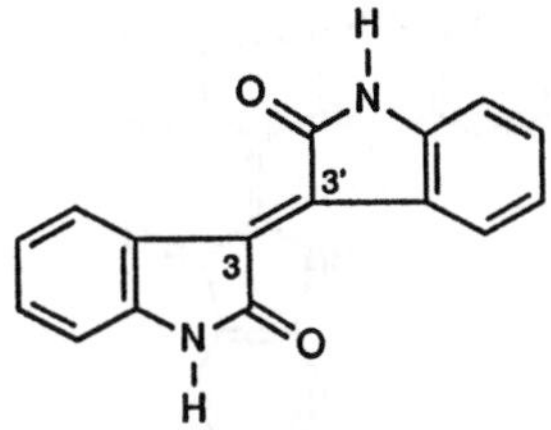

Isoindigo (*103-3*)

103.2.2 Chemical Constituents of *Indigofera suffruticosa*

Besides indigo and indirubin [3], a new compound louisfieserone (*103-4*) which may be formed from an 8-methyl-5-hydroxymethyl flavanone and a mevalonic acid derivative, was reported from Mexican *I. suffruticosa* [7].

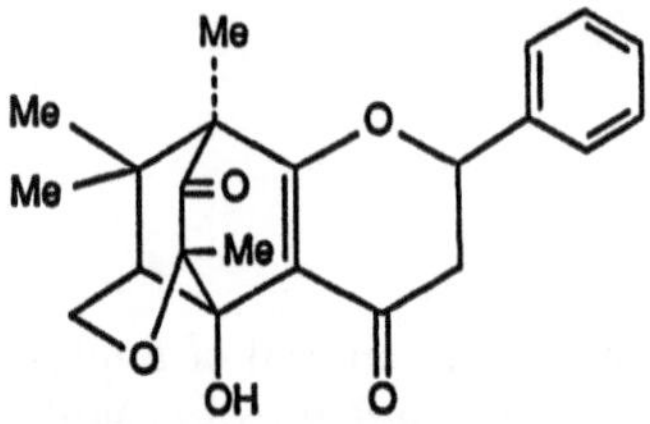

Louisfieserone (*103-4*)

103.2.3 Chemical Constituents of *Polygonum tinctorium*

The natural indigo isolated from *P. tinctorium* also contained indigo and indirubin [3]. In addition, an antidermatophytic component tryptanthrine (*103-5*) was isolated and identified [8].

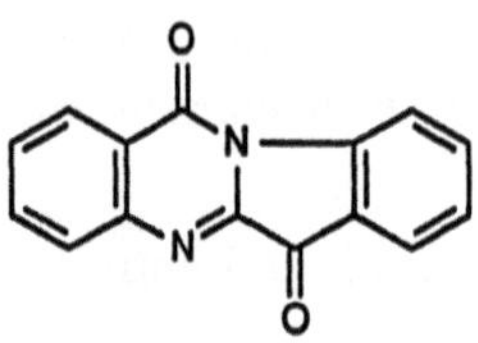

Tryptanthrine (*103-5*)

103.2.4 Chemical Constituents of *Isatis indigotica*

From the leaves and aerial part of *I. indigotica*, indirubin [9, 10] and indigo [10] were isolated as two major constituents. In addition, pigments named qingdainone (*103-6*), tryptanthrin, isatin (*103-7*), and *n*-nonacosane were isolated [10]. The structure of qingdainone was confirmed by condensation of tryptanthrin with indoxyl [10].

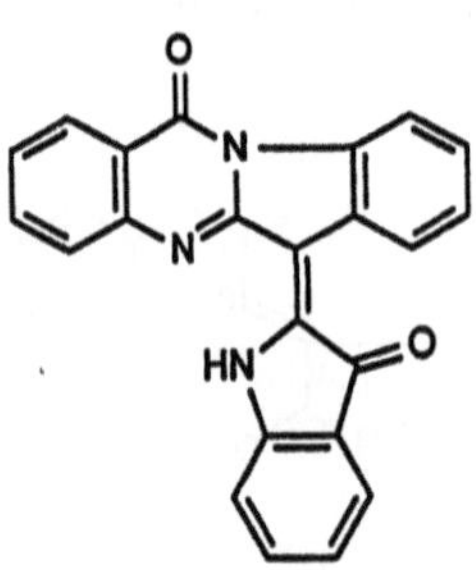

Qingdainone (*103-6*)

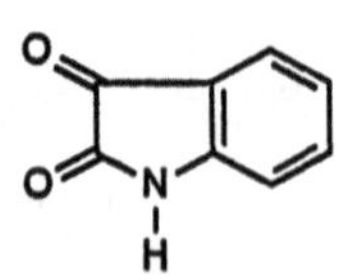

Isatin (*103-7*)

The indirubin contents in the leaves of *I. indigotica* varied with harvest time and were the highest in September. They were also correlated with the color of leaf, being highest in purple or gray-purple leaves and lowest in green to gray-green leaves [11]. The isatin contents in the dry leaves were about 1 μg/2.5 g leaves [12]. In addition, tryptanthrine [13] was isolated from the leaves of *I. indigotica*. Two sulfur-containing compounds epigoitrin (*103-8*) and 2-hydroxy-3-butenyl thiocyanate were isolated from its root [14, 15]. Indigo, indirubin, β-sitosterol, γ-sitosterol, and certain amino acids were further constituents detected in the root [16].

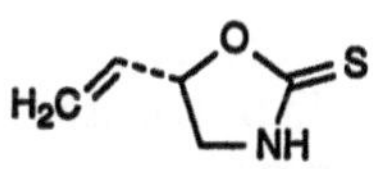

Epigoitrin (*103-8*)

Because indirubin is the antileukemic effective principle of the natural indigo of different origins, a number of studies on the quantitative analysis of indirubin in natural indigo, including TLC [17], spectrophotometric [18–22], and HPLC [20] methods, have been described. Regional variations of indirubin content (0.2%–0.4%) in natural indigo were observed [21]. Contents were also dependent on the plant origin of the natural indigo. For example, indirubin in natural indigo obtained from *B. cusia* contained 0.1%–0.3% indirubin, *Indigofera suffruticosa* 0.05%–0.1%, *P. tinctorium* 0.3%–0.5%, and *Isatis indigotica* 0.1%–0.3% [23].

A number of indirubin derivatives were synthesized as potentially antineoplastic agents. N-Substituted derivatives of indirubin were prepared by alkylating indirubin with the appropriate alkyl bromide [24–26]. Dihydroindirubin has been prepared by hydrogenation of indirubin [27]. The synthesis of acyl derivatives [27] and oximes of indirubin was also reported [28].

103.3 Pharmacology

Indirubin inhibited Lewis lung carcinoma in mice and Walker carcinosarcoma 256 in rats, but not leukemia L 7212 or P 388 in mice [29]. In a clinical study, treatment with indirubin of patients with chronic myelocytic leukemia at doses of 300–450 mg daily showed 26% complete remissions and 33.4% partial remissions [30]. In patients with chronic myelocytic leukemia, treatment with indirubin for 1.5–6 months markedly increased 5′-nucleotidase activity of white cells in cases with a palliative effect [31].

Cell-free DNA synthesis catalyzed by partially purified DNA-dependent DNA polymerase from Ehrlich ascites tumor cells was inhibited by indirubin in a concentration-dependent manner. This inhibition could not be reversed by increasing the DNA concentration, but could be abolished by increasing the enzyme concentration. The strongest inhibition resulted after preincubation with the enzyme and DNA, which might indicate that indirubin, DNA, and enzyme form a tertiary complex [32].

Incorporation of [^{3}H]thymidine into DNA of sarcoma tissue of rats bearing Walker 256 sarcoma was partially inhibited by intraperitoneal or subcutaneous injection of indirubin at a daily dose of 200 mg/kg. Incorporation of [^{3}H]uridine into RNA and of [^{14}C]phenylalanine into protein was not affected by indirubin [33].

[^{3}H]Thymidine incorporation into DNA of liver and spleen of healthy control mice was not affected by intraperitoneal injection of nitrogen mustard, cyclophosphamide, and indirubin. In contrast, hepatic and splenic [^{3}H]thymidine incorporation was inhibited in mice bearing L 7212 leukemia by nitrogen mustard, cyclophosphamide, and indirubin [34].

After incubation of indirubin with calf thymus DNA, the λ_{max} 207 nm of indirubin shifted toward a longer wavelength with decreasing absorbance. The binding between DNA and indirubin was rather weak, as indirubin molecules could be released easily during precipitation with alcohol or gel filtration. The binding was not affected by NaCl even at high concentration, but was greatly decreased in the presence of 8 *M* urea. The amount of bound [^{3}H]indirubin was directly proportional to the concentration of indirubin. Calf thymus DNA bound 46 indirubin molecules/1000 nucleotides [35].

Observations on the action of indirubin on the surface of white blood cells of patients with chronic myelocytic leukemia showed that the interaction of indirubin with chronic myelocytic leukemia cell surface was related to neuraminic acid but unrelated to other groups which carry negative charges such as phosphate. It appears that indirubin binds to certain receptors on the surface of the cell membrane. Structural changes in the membrane surface of the cell might influence the permeability of the membrane and alter cell metabolism and the transduction properties of the membrane accounting for the therapeutic action of indirubin in chronic myelocytic leukemia [36].

Electron microscopic examination of peripheral blood samples and leukocytes of the bone marrow from patients with chronic granulocytic leukemia treated with indirubin showed a swelling of the cell nucleus membrane, swelling and degeneration of the rough endoplasmic reticulum, and condensation of chromosomes [37].

Subacute toxicity tests with indirubin administered orally at 100–400 mg/kg per day to rats for 30 days showed no effects on leukocytes or on liver or renal function. Oral administration of 80 mg/kg per day of indirubin for 1–3 months to rats caused anorexia and diarrhea, but no effects on bone marrow were observed [29]. Long-term administration of high doses of indirubin did not affect hematopoietic stem cell production or DNA formation. Neither the volume of bone marrow nor the blood picture was affected [38]. A subacute toxicity study of indirubin in dogs receiving three different doses was also reported. No adverse effect was shown at the low dosage (20 mg/kg per day). At the middle dosage (100 mg/kg per day), mild diarrhea occurred within 10–30 days. Serum glutamic-pyruvic transaminase (SGPT) levels were slightly elevated after 5 months in one of three dogs treated. At a high dosage (200 mg/kg per day), serious diarrhea and hematofecia occurred during 40–60 days and SGPT levels were elevated after 3 months of treatment. Pathological changes were also observed in all tissue sections obtained after 6 months. The bone marrow and blood indexes, renal function, and electrocardigram were not affected by any of the three dosages. Indirubin was reported to be an apparently safe antitumor drug without serious side effects [39]. In human peripheral lymphocytes, indirubin did not increase sister chromatid exchanges, whether activated with microsomal enzyme S-9 or not [40].

The absorption half-life of [^{3}H]indirubin in mice was 5.9 h by gastric infusion of 45.7 mg/kg and the time to reach maximal absorption was 15 h, indicating slow absorption. The elimination half-life of intravenously injected indirubin was 17.5 h and 21 h after gastric infusion. Intravenously administered [^{3}H]indirubin was found mostly in liver and bile. Within 96 h, 15% and 60% of the administered substance were excreted via urine and feces, respectively. The liver seems to be the major organ for indirubin metabolism [41].

Antitumor activity of various indirubin derivatives against rat carcinosarcoma W256 and mouse leukemia L7212 was examined and compared with that of indirubin. Of all compounds tested, *N,N'*-dimethylindirubin (*103-9*), *N*-methylindirubin monooxime (*103-10*), *N*-acetylindirubin (*103-11*), and *N*-methylindirubin monomethyloxime (*103-12*) had equal or slightly greater activity against W256 as compared with that of indirubin. They also showed activity against L7212, but were not effective against mouse reticulocyte leukemia L615 or lymphocyte leukemia L7712 [28]. In an in vitro study, *N*-carboxymethylindirubin (*103-13*) inhibited CH–Cl and U_{27} malignant cell growth and [^{3}H]thymidine incorporation into DNA, and induced cell degeneration similar to indirubin [42].

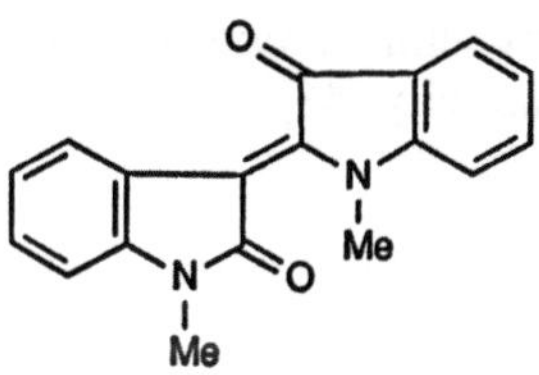

N,N'-Dimethylindirubin (*103-9*)

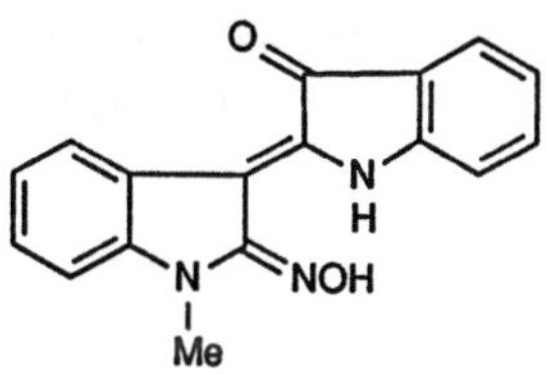

N-Methylindirubin monooxime (*103-10*)

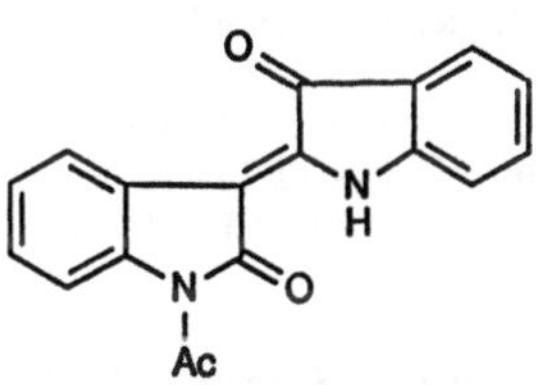

N-Acetylindirubin (*103-11*)

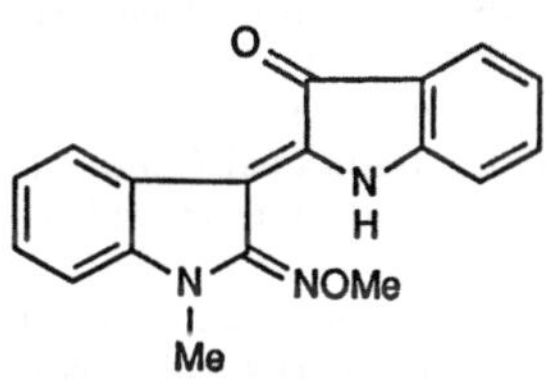

N-Methylindirubin monomethyloxime (*103-12*)

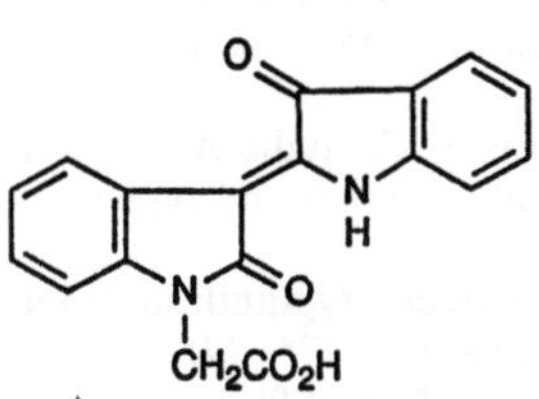

N-carboxymethylindirubin (*103-13*)

Derivatives of bisindole with 2,2'; 3,3', or 3,2' linkages were compared for their effect on nucleic acid and protein formation. They formed complexes with calf thymus DNA as measured by UV and visible spectrometry and inhibited nucleic acid and protein formation on cultured W256 carcinoma cells and in vivo. Many bisindoles also inhibited cell-free DNA formation. 3,3′-bisindoles showed the highest potency in inhibiting the DNA formation in cancer cells and in cell-free systems, and the 2,2′-bisindoles had the lowest activity [43]. *N*-Methylisoindigo (*103-14*) was more potent than indirubin against W256 carcinoma in rats [26].

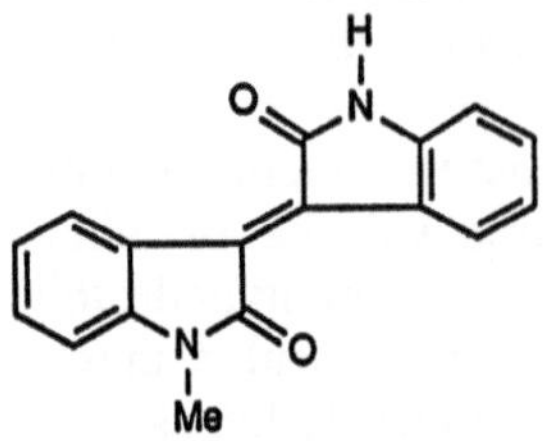

N-Methylisoindigo (*103-14*)

In both rats and mice transplanted with W256 carcinoma or Lewis lung tumor cells, *N*′-ethylindirubin showed the highest antitumor activity, followed by indirubin, and *N*′-octadecylindirubin had the lowest effect [44]. Tritiated *N*′-ethylindirubin given orally accumulated in neoplastic lung tissue about twice as high as indirubin- or *N*-′octadecylindirubin-associated radioactivity. Obviously chemical modification of indirubin can change its oral absorption and antitumor activity [45].

References

1. Tang Y (1987) Determination of indirubin in Qingdai (*Baphicacanthus* cusia Bremek.) and Chinese medicines containing it. Chin J Pharm Anal 7:40–42
2. Chen DH, Xie JX (1984) Chemical constituents of traditional Chinese medicine Qing Dai. Chin Trad Herb Drugs 15:534–536
3. Ben BL (1981) Column chromatographic-spectrophotometric determination of indigo and indirubin in Qingdai, a traditional Chinese medicine. Chin Trad Herb Drugs 12:11–15
4. Fang FD, Wu GY (1982) Advances in studies on the anticancer effect of indirubin. Chin J Intern Med 21:312–314
5. Chang CN (1985) Anti-leukemia Chinese herbs and the effective ingredients. In: Chang HM, Yeung HW, Tso WW, Koo A (eds) Advances in Chinese Medicinal Materials Research. World Science, Singapore, p 373
6. Lien EJ, Li WY (1985) Anticancer Chinese drugs: structure-activity relationships. In: Chang HM, Yeung HW, Tso WW, Koo A (eds) Advances in Chinese Medicinal Materials Research. World Science; Singapore, p 447
7. Dominguez XA, Martinez C, Calero A, Dominguez XA Jr, Hinojosa M, Zamudio A, Watson WH, Zabel V (1978) Mexican medicinal plants. XXXI. Chemical components from "Jiquelite" *Indigofera suffruticosa* Mill. Planta Med 34:172–175
8. Honda G, Tosirisuk V, Tabata M (1980) Isolation of an antidermatophytic, tryptanthrin, from the indigo plants, *Polygonum tinctorium* and *Isatis tinctoria*. Planta Med 38:275–276
9. Chen FK, Chen KM, Guo YZ (1986) The extraction and isolation of indirubin from leaves of *Isatis indigotica* Fort. J Shenyang Coll Pharm 3:194
10. Li QH (1987) The chemical constituents of Qing-Dai. Acta Bot Sin 29:67–72
11. Wang XP, Zhou TY, Yuan CQ, Ding ZZ (1985) Comparison of the content of indirubin in *Isatis indigotica* at different harvest time. Bull Chin Mater Med 10:178–179
12. Guo YZ, Chen FK (1986) TLC-UV spectrophotometric and TLC-scanning determination of isatin in leaf of *Isatis*. Chin Trad Herb Drugs 17:8–11
13. Li QH, Jin JS, Chong MC, Song ZY (1983) Studies on the antifungal constituents of Qing Dai (*Isatis indigotica*). Chin Trad Herb Drugs 14:440–441

14. Huang CS, Yoshihira K, Natori N (1981) Study on chemical constituents of *Isatis indigotica* Fort. Chin Pharm Bull 16:54–55
15. Huang QS, Yoshihira K, Natori S (1981) Isolation of 2-hydroxy-3-butenyl thiocyanate, epigoitrin, and adenosine from 'Banlangen', *Isatis indigotica* root. Planta Med 42:308–310
16. Zhang SH (1983) Studies on the chemical constituents of *Isatis indigotica* root. Chin Trad Herb Drugs 14:247–248
17. Zhao PP, Luo HM, Yu CQ, Liu CW (1981) Determination of indirubin by dual wavelength TLC scanner. Bull Chin Mater Med 6:28–30
18. Gu YC, Yang QL, Fu JP, Yang MK (1980) Determination of indirubin. Acta Pharm Sin 15:308–311
19. Zhang SH, Wang BL (1985) Determination of indigo and indirubin in Qing Dai (*Isatis indigotica* Fort.) by dual wavelength spectrophotometry. Acta Pharm Sin 20:301–305
20. Dai FB, Qiao CZ, Li L (1986) Determination of indigo and indirubin in Qingdai by HPLC. Acta Pharm Sin 21:868–871
21. Zhao LQ (1984) Paper chromatographic separation and spectrophotometric determination of indirubin in Qing Dai (crude natural indigo). Bull Chin Mater Med 9:78–79
22. Lu RG (1986) Determination of indirubin and indigo in natural indigo (Qingdai) with dual wavelength spectrometry. Chin Pharm Bull 21:72–74
23. Deng BL (1986) Direct colometric method for determination of indigo and indirubin in Qingdai. Chin Trad Herb Drugs 17:163–164
24. Wu KM, Zhang MY, Fang Z, Huang L (1984) Synthesis of N_1-substituted derivatives of indirubin, an antileukemic compound. Acta Pharm Sin 19:513–518
25. Isukura Sangyo KK (1982) Indirubin derivatives. Jpn Kokai Tokkyo Koho JP 57,209,271 (82,209,271) (CA 98:143702q)
26. Ji XJ, Zhang FR (1985) Antineoplastic effect of indirubin derivatives and their structure-activity relationship. Acta Pharm Sin 20:137–139
27. Isukura Sangyo KK (1982) Indirubin derivatives. Jpn Kokai Tokkyo Koho JP 57,209,272 (82,209,272) (CA 98:143701p)
28. Zeng QT, Du DJ, Xie DC, Wang XP, Rau CQ (1982) Antitumor activities of indirubin derivatives. Chin Trad Herb Drugs 13:24–30
29. Ji XJ, Zhang FR, Lei JL, Xu YT (1981) Studies on the antineoplastic effect and toxicity of synthetic indirubin. Acta Pharm Sin 16:146–148
30. Indirubin Cooperative Group (1980) Clinical study of indirubin in the treatment of 314 patients with chronic granulocytic leukemia. Chin J Hematol 1:132
31. Gan WJ, Yang TY, Wen SD, Liu YY, Tan Z, Deng CA, Wu JX, Liu MP (1985) Studies on the mechanism of indirubin action in treatment of chronic myelocytic leukemia (CML). II. 5'-Nucleotidase in the peripheral white blood cells of CML. Chin J Hematol 6:611–613
32. Zhang L, Wu GY, Qiu CC (1985) Effect of indirubin on DNA synthesis in vitro. Acta Acad Med Sin 7:112–116
33. Du DJ, Ceng QT (1981) Effect of indirubin on the incorporation of isotope labeled precursors into nucleic acid and protein of tumor tissues. Chin Trad Herb Drugs 12:406–409
34. Du DG, Ceng QT, Weng ZJ, Wan XP (1982) Study on the incorporation of ^{3}H-TdR into DNA of liver and spleen of L7212, a lymphatic leukemia in mice. Chin J Hematol 3:277–280
35. Wu GY, Liu JZ, Fang FD, Zuo J (1982) Studies on the mechanism of indirubin action in the treatment of chronic granulocytic leukemia. V. Binding between indirubin and DNA and identification of the type of binding. Sci Sin [B] [Engl] 25:1071–1079
36. Wang XQ, Gan WJ, Yang TY, Wang ZC, Qiau LS, Qi RB (1984) Effect of indirubin on the cell surface of chronic myelocytic leukemia. Tianjin Med J 12:707–710
37. Lee K, Shih CY, Yang TY, Chen LS, Chao WM, Sun CS, Wang TC, Pien SK, Sung KH (1979) Ultrastructural study on the mechanism of the therapeutic effect of indirubin for human chronic granulocytic leukemia. Natl Med J China 59:129–132
38. Wang JH, You YC, Mi JX, Ying HG (1981) Effect of indirubin on hematopoietic cell production. Acta Pharmacol Sin 2:241–244
39. Sichuan Institute of Traditional Chinese Medicine (1981) Subacute toxicity of indirubin in dogs. Chin Trad Herb Drugs 12:27–29
40. Cai YY, Xu CL, Li SH, Liu Y (1983) Studies on sister chromatid exchanges induced by harringtonine, indirubin and pyquiton before or after activation with microsome enzyme. Acta Acad Med Sin 5:161–164

41. Qi SB, He GX, Wang YD (1981) Pharmacological studies on indirubin. III. Pharmacokinetic studies on indirubin in mice. Chin Trad Herb Drugs 12:23–27
42. Fang FD, Cai YY, Zuo J, Wu GY (1984) Effects of indirubin derivatives no. 2 on two malignant cell lines. Chin Pharm Bull 19:664–665
43. Wu GY, Liu JZ, Zhang L (1985) Effect of bisindole compounds with three different kinds of linkage on the synthesis of nucleic acid and protein. Shengwu Huaxue Yu Shengwu Wuli Jinzhan 61:48–51
44. Li CL, Ji XJ (1984) The relationship between antitumor activity and intestine absorption in vivo of indirubin and its derivatives. J Beijing Med Coll 16:326–328
45. Li CL, Ji XJ (1983) A comparative study on the physiological disposition of indirubin and its ethyl and octadecyl derivatives in animals. Acta Pharm Sin 18:332–338

Quisqualis indica L. 104

104.1 Introduction

Shijunzi, Fructus Quisqualis, is the dry fruits of *Quisqualis indica* L. (Combretaceae) collected in the fall. This herbal medicine is officially listed in the Chinese Pharmacopoeia and is used mainly as an anthelmintic.

104.2 Chemical Constituents

The anthelmintic principle in fruits of *Q. indica* was isolated and determined as an amino acid quisqualic acid (*104-1*). The structure of quisqualic acid was elucidated by spectral analysis, chemical synthesis [1, 2], and crystallographic investigation [3]. It is an L-alanin derivative substituted with a dioxooxadiazolidin ring. Isolation from the fruits yields potassium quisqualate [4].

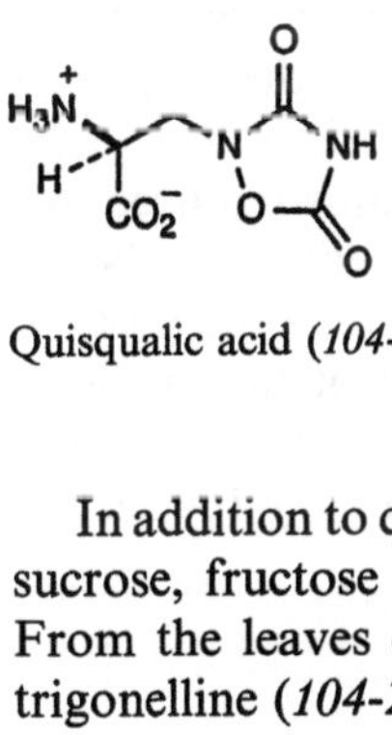

Quisqualic acid (*104-1*)

In addition to quisqualic acid, linoleic, oleic, palmitic, stearic, and arachidic acids, sucrose, fructose [5], and D-mannitol [4] were detected in the fruits of *Q. indica*. From the leaves of *Q. indica* potassium quisqualate, L-proline, L-asparigine, and trigonelline (*104-2*) were isolated and identified [6].

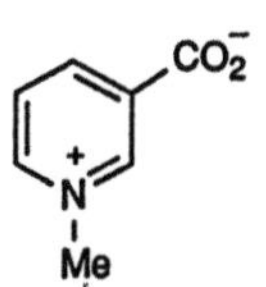

Trigonelline (*104-2*)

Because of the relative inaccessibility of quisqualic acid from natural sources or by a convenient method of synthesis, investigations on the synthesis of this enantiomeric compound have been carried out. Synthesis of (RS)-quisqualic acid at an overall yield of 20% from an enol α-benzoylamino-β-hydroxy-acrylic acid ethyl ester

(*104-3*) was reported (Fig. 104.1). Enzymatic resolution of the *N*-acyl derivatives of quisqualic acid using either hog renal or *Escherichia coli* acylases led to the L-(S)-isomer, which is identical to the natural product [7].

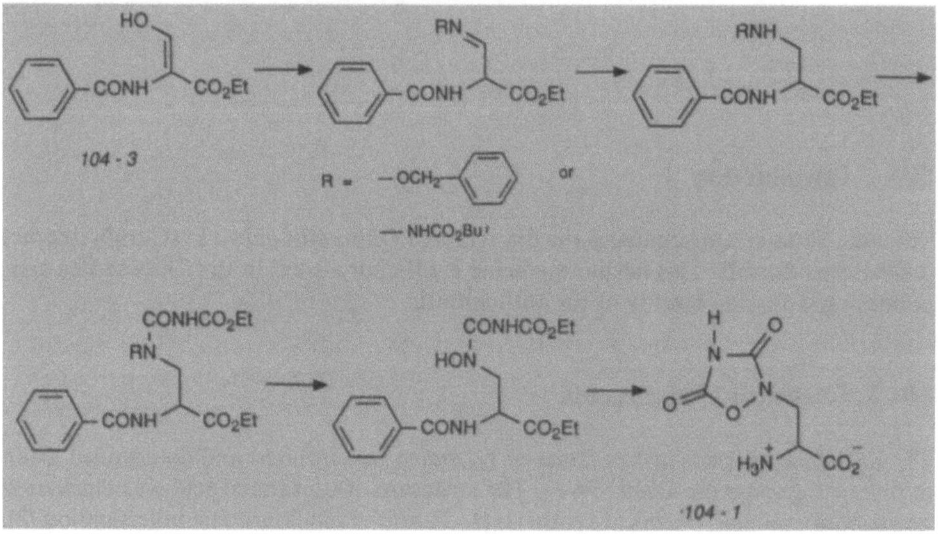

Fig. 104.1. Synthesis of (RS)-quisqualic acid

A general method for enantio-efficient synthesis of β-amino-alanine derivatives, which involves intramolecular transfer of the amino substituent from the α-carboxy to the β-carbon atom *via* an azetidinone, was applied to the synthesis of quisqualic acid in an optically pure state [8]. Thus, L-serine was converted by known procedures into 3*S*-azetidinone (*104-4*), which was isomerized to the isooxazolidinone (*104-5*). Treatment of this substance with ethoxycarbonylisocyanate gave the urea (*104-6*), which was immediately ring opened to the salt. Upon treatment of the salt with trifluoracetic acid and subsequent ion-exchange chromatography, L-quisqualic acid was obtained with an overall yield of 89% from isooxazolidinone (Fig. 104.2).

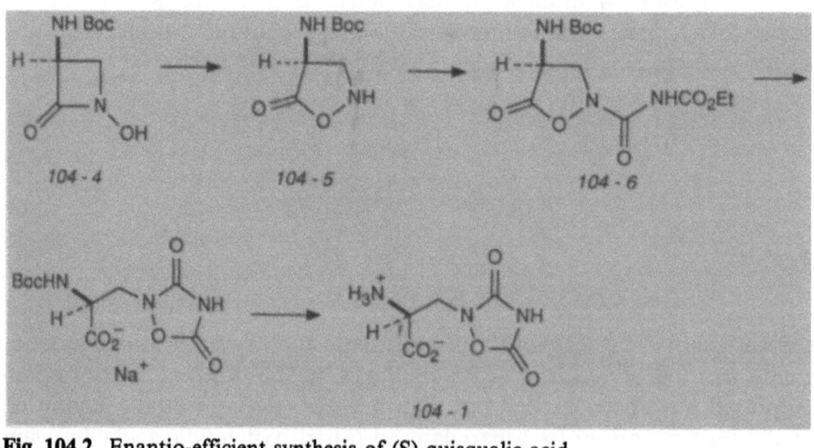

Fig. 104.2. Enantio-efficient synthesis of (S)-quisqualic acid

104.3 Pharmacology

L-Quisqualic acid is one of the most potent agonists of the neurotransmitter amino acid L-glutamate in both the central and peripheral nervous systems of vertebrates and invertebrates [9]. Quisqualic acid added to incubation media containing *Ascaris suum* isolated from the pig inhibited the forward movement of the worms but did not have a parasiticidal effect [10]. In a clinical study on potassium quisqualate of natural origin it was found that 125 mg of the drug used without a laxative was effective as an anthelminthic with activity similar to that of santonin [11]. The synthesized DL- and L-quisqualic acids and the natural L-quisqualic acid were capable of paralysing earthworms, whereas D-quisqualic acid was inactive [12].

References

1. Pan PC, Fang SD, Tsai CC (1976) The chemical constituents of Shihchuntze, *Quisqualis indica* L. II. Structure of quisqualic acid. Sci Sin 19:691–701
2. Takemoto T, Nakajima T, Arihara S, Koike K (1975) Constituents of Quisqualis fructus. II. Structure of quisqualic acid. Yakugaku Zasshi 95:326–332
3. Flippen JL, Gilardi RD (1976) Quisqualic acid. Acta Crystallogr [B] 32:951–953
4. Zhang RW, Guan BQ (1981) Chemical constituents of *Quisqualis indica* L. Chin Trad Herb Drugs 12:40
5. Hsu CF, King PH (1940) Chemical study of the seed of *Quisqualis indica* (shih-chun-tze). I. Composition of the crude oil. J Chin Pharm Assoc 2:132–156
6. Fang ST, Chu JH (1964) The chemical constituents of leaves of *Quisqualis indica*. Acta Chim Sin 30:226–229
7. Bycroft BW, Chhabra SR, Grout RJ, Crowley PJ (1984) A convenient synthesis of the neuroexcitatory amino acid quisqualic acid and its analogues. J Chem Soc Chem Commun 1156–1157
8. Baldwin JE, Adlington RM, Birch DJ (1985) Synthesis of the L-quisqualic acid: a general method for enantio-efficient synthesis of β-aminoalanine derivatives. J. Chem Soc Chem Commun 256–257
9. Gration KAF, Lambert JJ, Ramsey RL, Rand RP, Usherwood PNR (1981) Agonist potency determination by patch clamp analysis of single glutamate receptors. Brain Res 230:400–405
10. Ishizaki T, Kato K, Kumata M, Takemoto T, Nakajima T, Takagi N, Koike K (1973) Effect of quisqualic acid upon *Ascaris suum* in vitro in comparison with those of kainic acid, α-allokainic acid, and pyrantel pamoate. Kiseichugaku Zasshi 22:181–186 (CA 81:33173j)
11. Tuan YC, Li CH, Chen TC (1957) Preliminary study of anthelminthic action of potassium quisqualate. Acta Pharm Sin 5:87–91
12. Gu XQ, Pan BC, Gao YS (1985) The synthesis of quisqualic acid. Acta Chim Sin 43:675–679

Rabdosia spp.

105

105.1 Introduction

Several plants of the genus *Rabdosia* (Lamiaceae) that occur in China have been used in folk medicine as antitumor or antiinflammatory agents. None of these has been officially listed in the Chinese Pharmacopoeia. Thorough studies on the chemical constituents of the leaves and stems of various species and on their biological activities have been carried out by Chinese and Japanese scientists. An extensive review of the chemistry and biological activities of the diterpene constituents of *Rabdosia* species was published by Fujita and Node [1]. This showed that, up to the beginning of 1983, 108 different diterpenes had been isolated from 22 species of *Rabdosia* and had been structurally elucidated.

In recent years, further *Rabdosia* species have been investigated in China and several new diterpenes have been isolated. To date, more than 30 *Rabdosia* species have been studied with regard to their biological activity, with special emphasis on antitumor activity. More than 100 diterpene components have been isolated from these species, demonstrating great interest in this plant genus. The species are: *Rabdosia adenantha* [2, 3], *R. amethystoides* [4–8], *R. angustifolia* [9], *R. bulleyana* [10], *R. coetsa* [11], *R. coetsoides* [12], *R. eriocalyx* [13–19], *R. eriocalyx* var. *laxiflora* [20], *R. excisa* [21], *R. flexicaulis* [22], *R. forrestii* [23], *R. glutinosa* [24], *R. henryi* [25–29], *R. japonica* [30–34], *R. japonica* var. *glaucocalyx* [35–37], *R. kunmingensis* [38], *R. lasiocarpa* [39], *R. latifolia* var. *reniformis* [40], *R. liangshanica* [41], *R. lophantoides* [42, 43], *R. lungshengensis* [44], *R. macrocalyx* [45–51], *R. macrocalyx* var. *jiuhua* [52], *R. macrophylla* [8, 53–61], *R. nervosa* [62–66], *R. phyllostachys* [67], *R. rosthornii* [68, 69], *R. rubescens* [67, 70–76], *R. rubescens* f. *lushanensis* [77–80], *R. sculponeata* [67, 81], *R. serra* [82–85], *R. ternifolia* [86–90], and *R. weisiensis* [91, 92].

105.2 Chemical Constituents

The main chemical constituents of the *Rabdosia* species are the diterpene compounds. The first compound of this series was enmein (*105-1*), isolated from *R. japonica* and *R. trichocarpa*. It was used in Japan as a folk medicine for the treatment of gastrointestinal disorders [93, 94]. Its structure was determined by chemical reactions and spectral analyses as a diterpene with six oxygen atoms [95]. The absolute configuration was elucidated by X-ray analysis of dihydroenmein-3-acetate-6-bromoacetate [96].

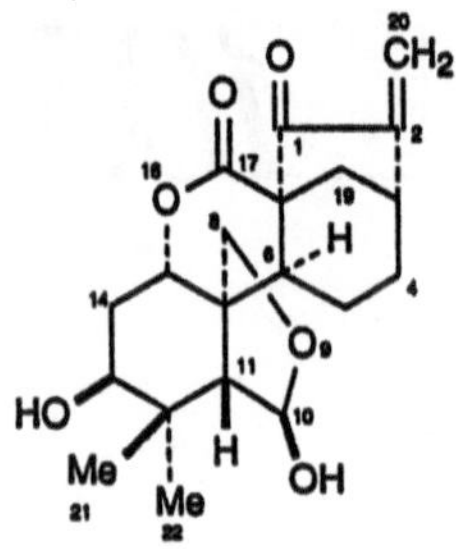

Enmein (*105-1*)

An initial test on the antitumor activity of enmein as a pure *Rabdosia* constituent was carried out by Arai et al. against Ehrlich ascites carcinoma [97]. The antitumor activity reported for enmein initiated investigations into other *Rabdosia* species and constituents. Studies on chemical constituents of Chinese *Rabdosia* species and their potential antitumor activity have also been made in China. The diterpene constituents isolated from Chinese *Rabdosia* species are listed in Tabel 105.1, most of which are derived from kaurane (*105-2*) or enmein. The basic skeleton of enmein can be regarded as 6,7-secokaurane (*105-3*) [1]. Some diterpenes isolated from *R. lophanthoides* are abietane quinones [42, 43].

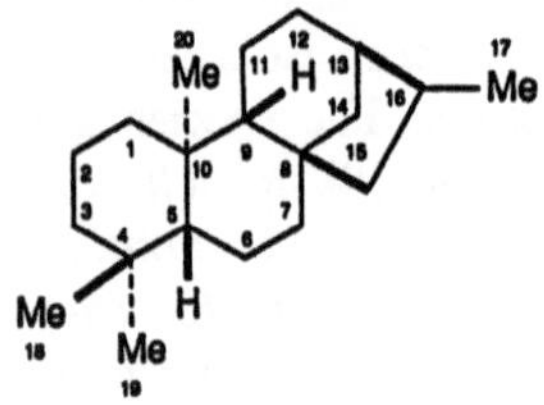

Kaurane (*105-2*)

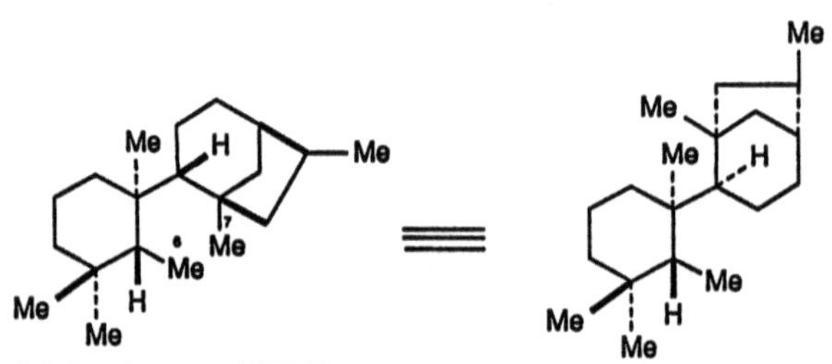

6,7-Secokauran (*105-3*)

Table 105.1. Diterpenes isolated from *Rabdosia* species occurring in China

Rabdosia sp.	Compound	Structure	Reference
Rabdosia adenantha	Adenanthin (*105-4*)	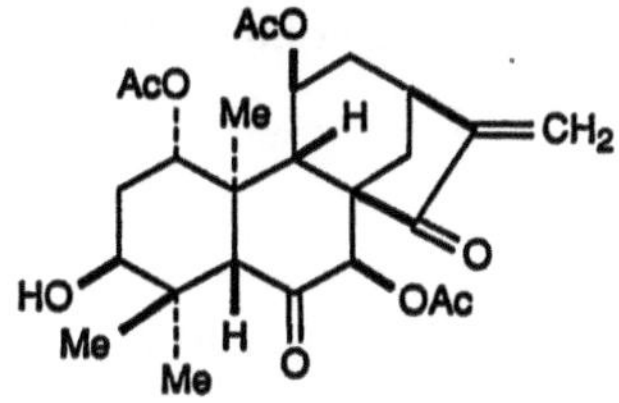	[2, 3]
Rabdosia amethystoides	Umbrosin A (*105-5*)	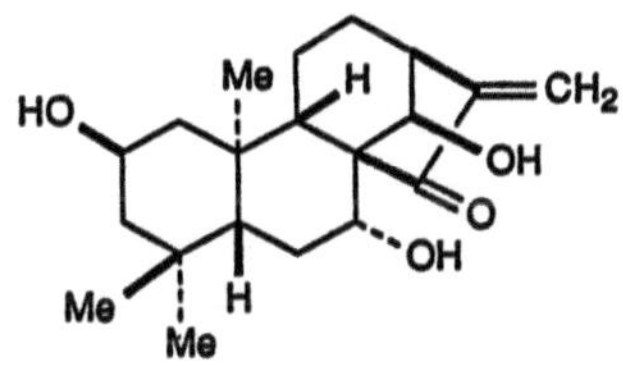	[4]
	Umbrosin B (*105-6*) R=H 14-Acetylumbrosin B (*105-7*) R=Ac	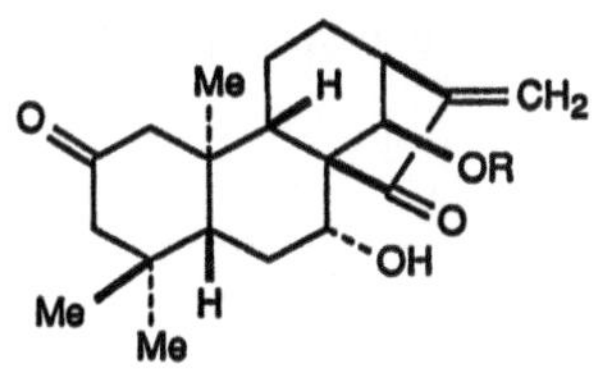	[4]
	Glaucocalyxin A (*105-8*) (Wangzaozin B): R=H Glaucocalyxin B (*105-9*) (Wangzaozin C): R=Ac	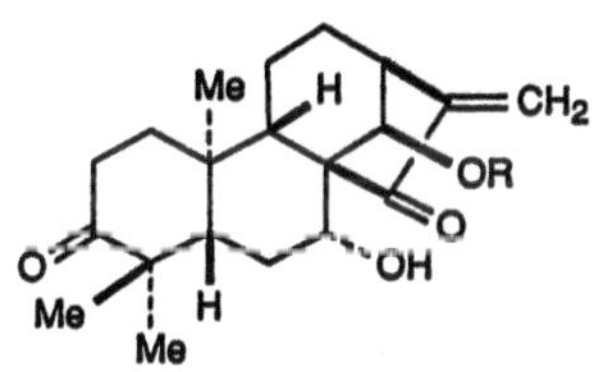	[5]
	Wangzaozin A (*105-10*)	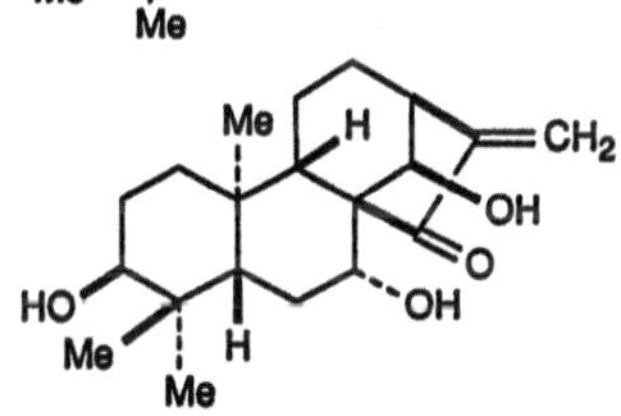	[5]
	Amethystoidin A (*105-11*)	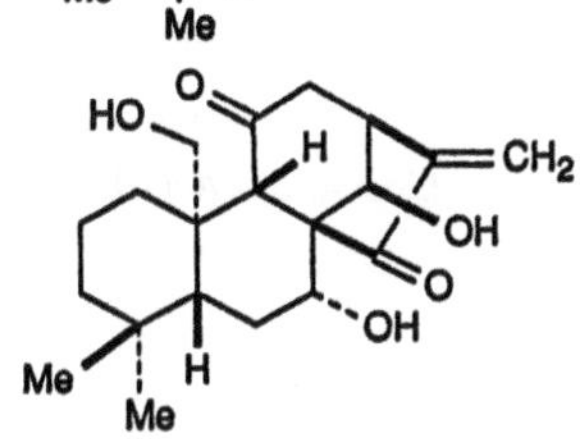	[6, 8]
	Amethystonal (*105-12*): R=CHO Amethystonoic acid (*105-13*): R=COOH	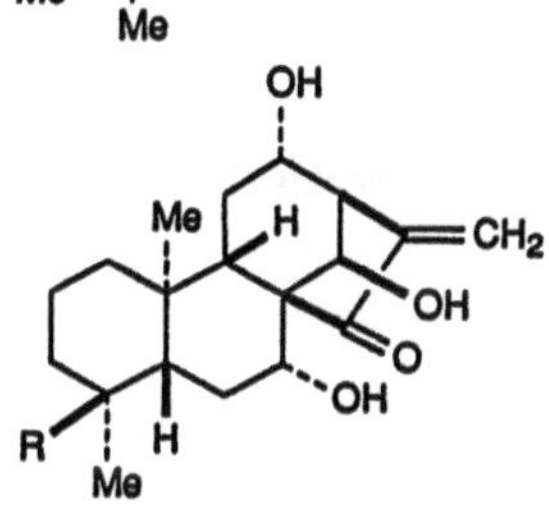	[7]

Table 105.1. (continued)

Rabdosia sp.	Compound	Structure	Reference
Rabdosia angustifolia	Angustifolin (*105-14*)	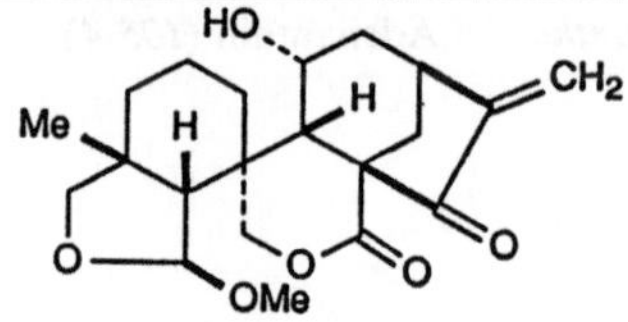	[9]
	Isodonal (*105-15*)	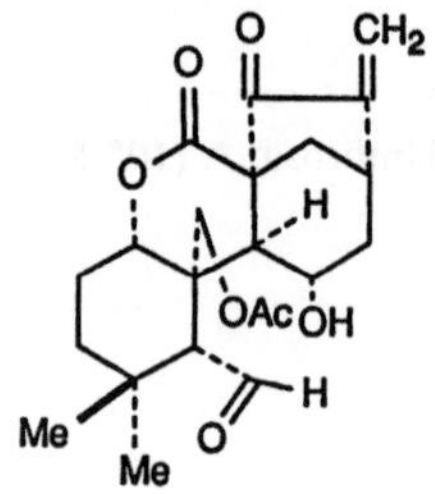	[9, 34]
Rabdosia bulleyana	Bulleyanin (*105-16*)	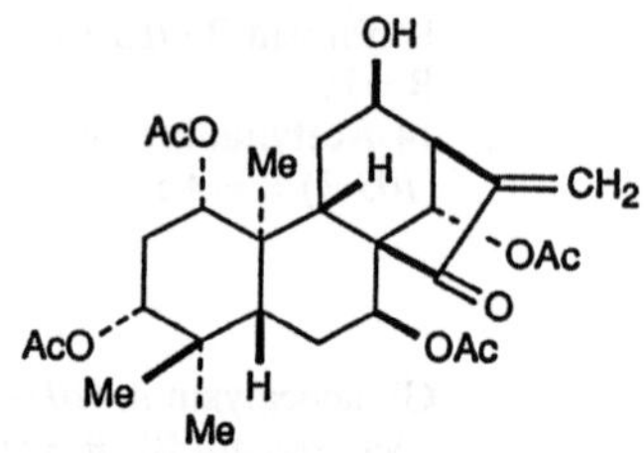	[10]
Rabdosia coetsa	Coetsin A (*105-17*)	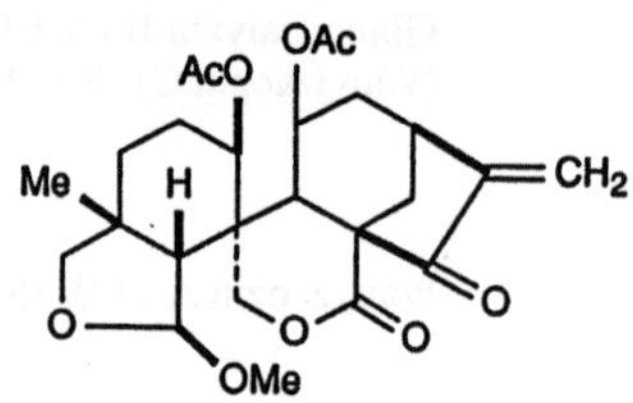	[11]
	Coetsin B (*105-18*)	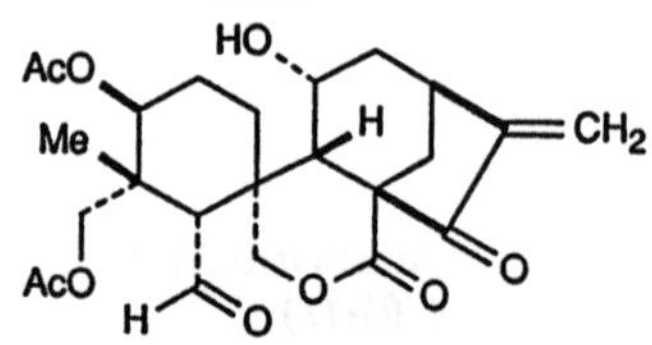	[11]
Rabdosia coetsoides	Coetsoidin A (*105-19*)	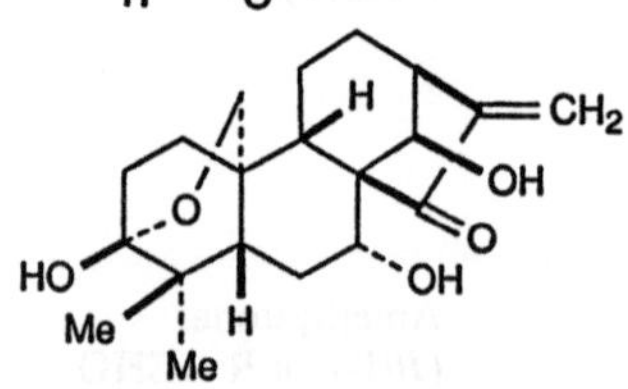	[12]
	Coetsoidin B (*105-20*)	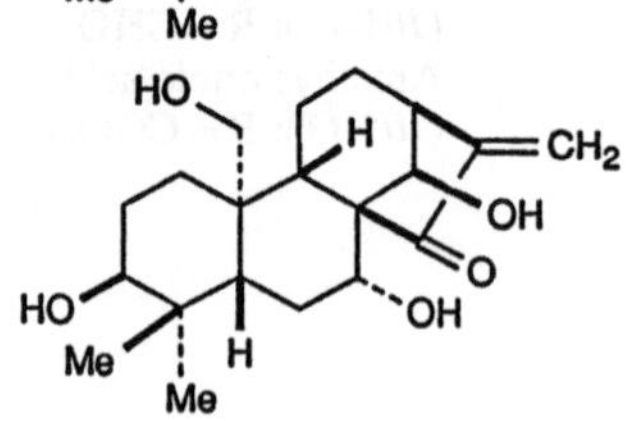	[12]

Table 105.1. (continued)

Rabdosia sp.	Compound	Structure	Reference
Rabdosia coetsoides	Coetsoidin C (*105-21*)	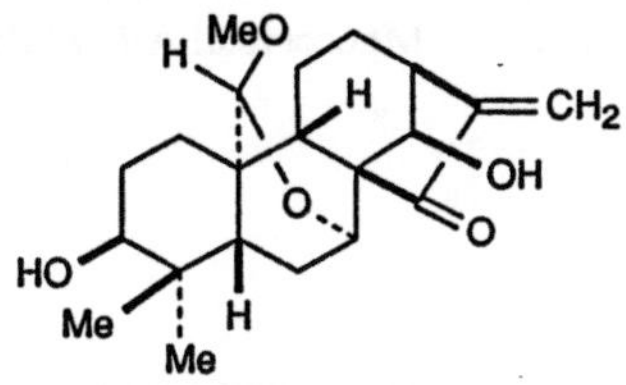	[12]
	Coetsoidin D (*105-22*)	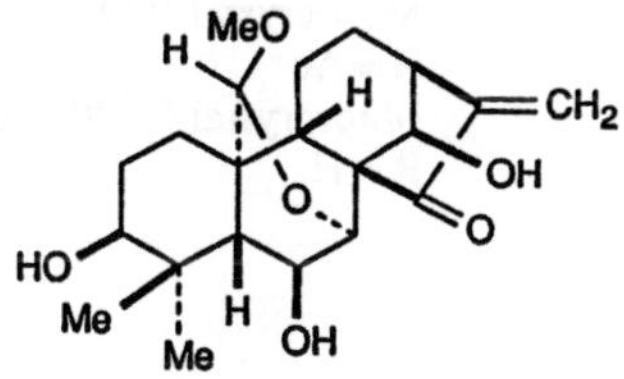	[12]
	Coetsoidin E (*105-23*)	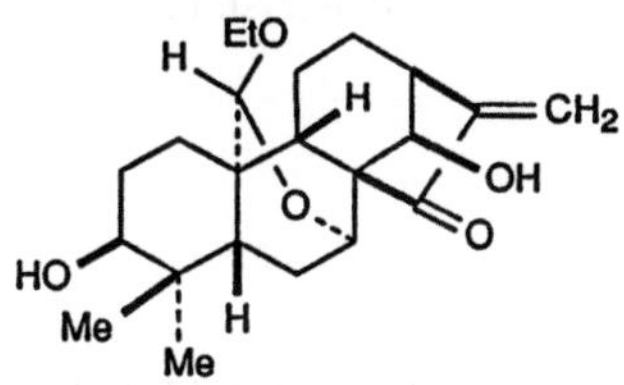	[12]
	Coetsoidin F (*105-24*)	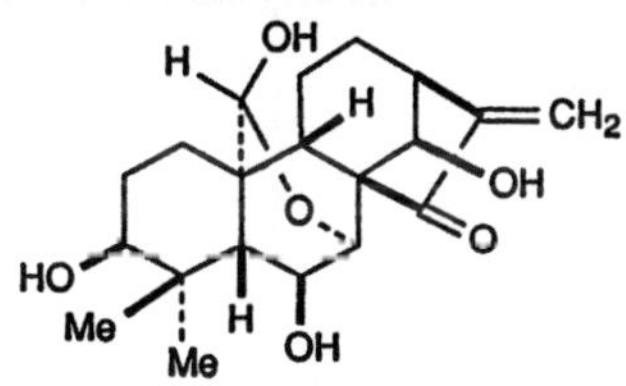	[12]
	Coetsoidin G (*105-25*)	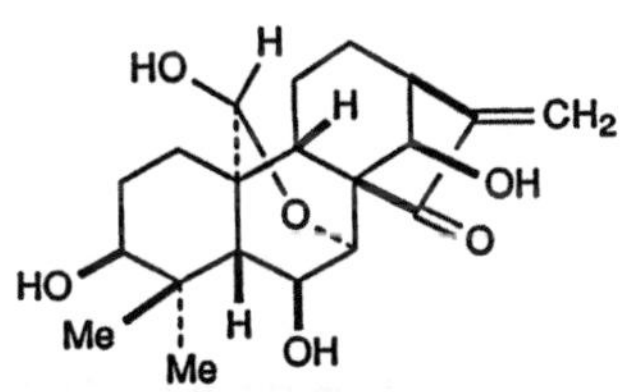	[12]
Rabdosia eriocalyx	Eriocalyxin A (*105-26*)	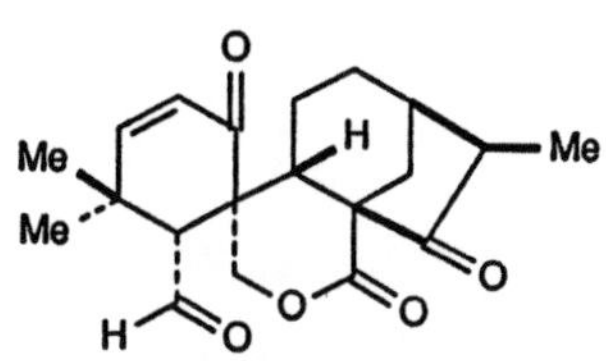	[13–15]
	Eriocalyxin B (*105-27*)	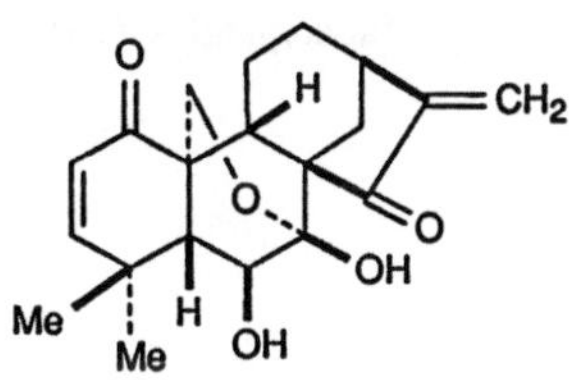	[13–15]

Table 105.1. (continued)

Rabdosia sp.	Compound	Structure	Reference
Rabdosia eriocalyx	Maoecrystal A (*105-28*)	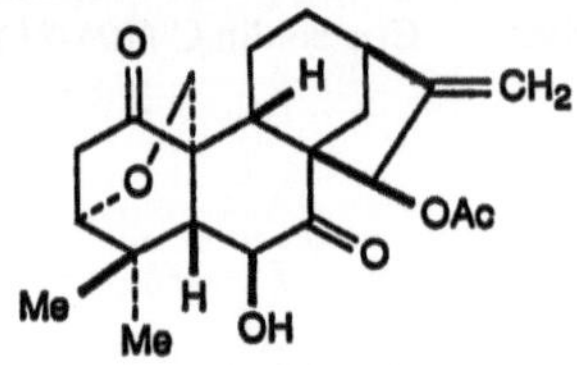	[14, 16]
	Maoecrystal B (*105-29*): R=Ac Maoecrystal C (*105-30*): R=H	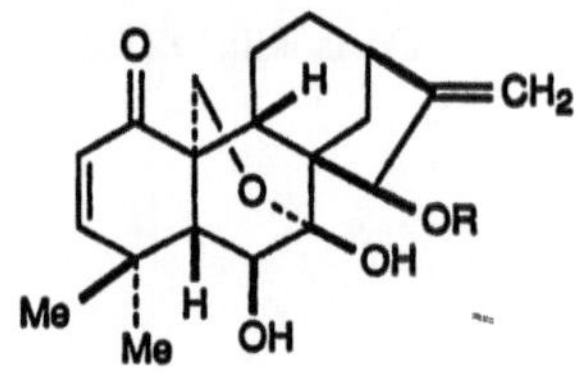	[14–16]
	Maoecrystal D (*105-31*)	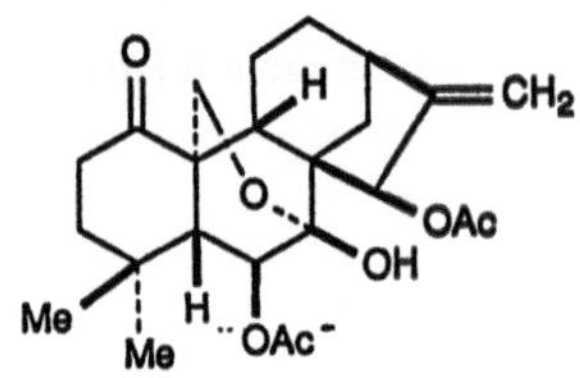	[14]
	Maoecrystal E (*105-32*)	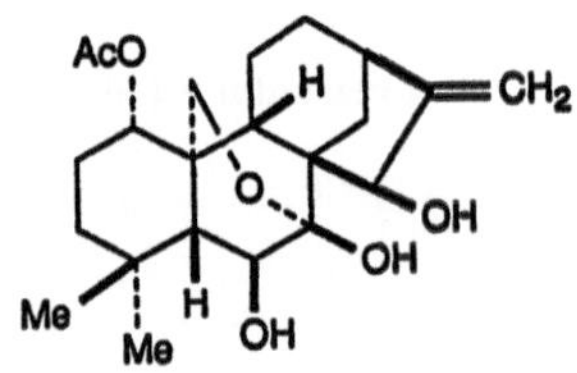	[14]
	Maoecrystal I (*105-33*)	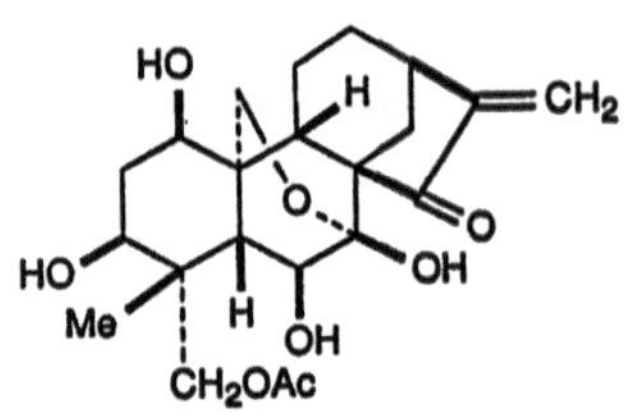	[17]
	Maoecrystal J (*105-34*)	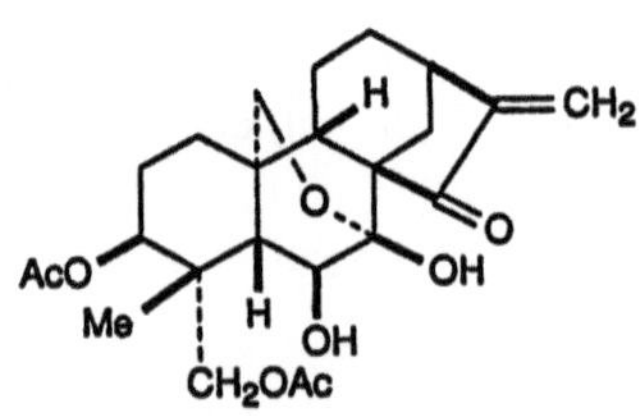	[17]
	Maoecrystal K (*105-35*)	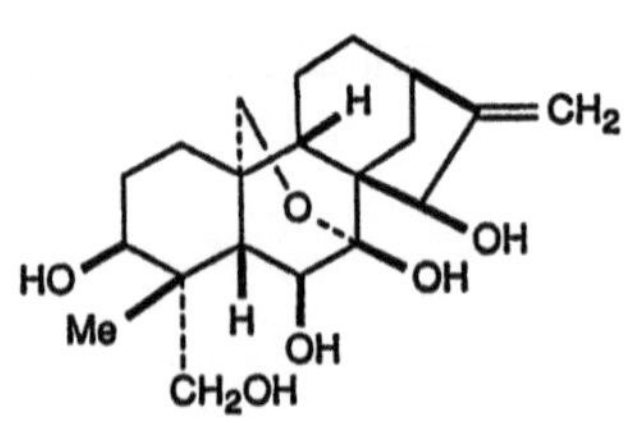	[19]

Table 105.1. (continued)

Rabdosia sp.	Compound	Structure	Reference
Rabdosia eriocalyx	Rabdoside 1 (*105-36*)	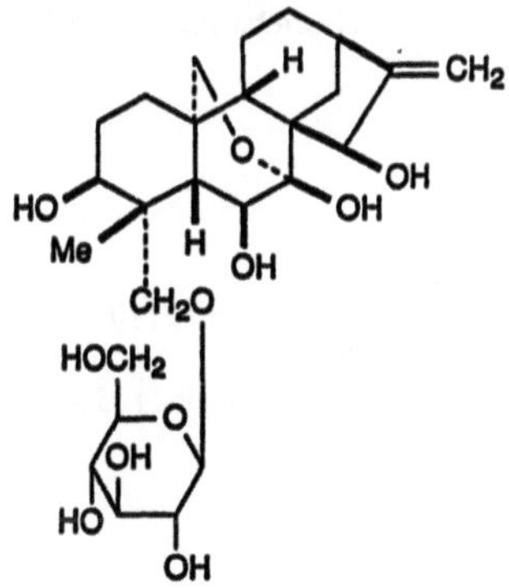	[19]
	Rabdoside 2 (*105-37*)	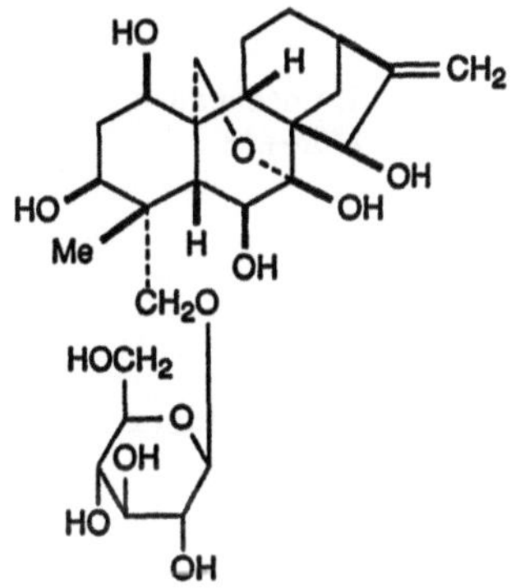	[19]
	Odonicin (*105-38*)	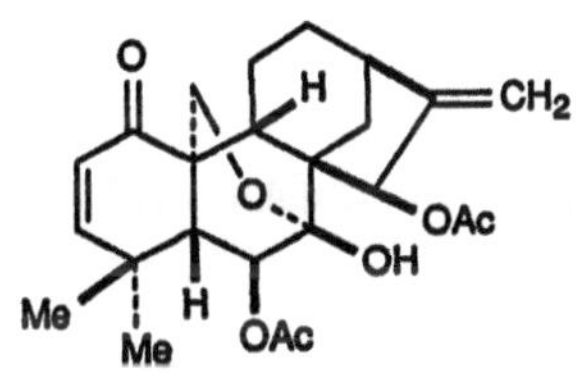	[15]
Rabdosia eriocalyx var. *laxiflora*	Oridonin (*105-39*)	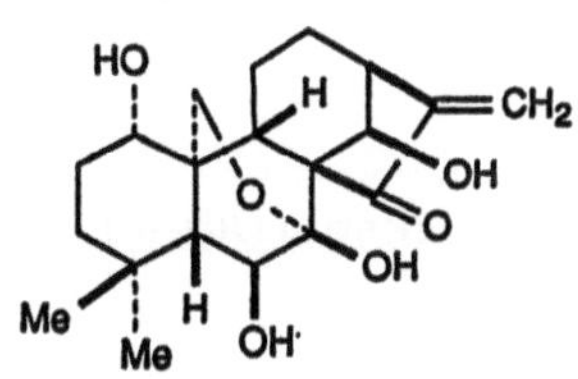	[20]
	Eriocalyxin B (*105-27*)		[20]
	Maoecrystal A (*105-28*)		[20]
	Maoecrystal B (*105-29*)		[20]

Table 105.1. (continued)

Rabdosia sp.	Compound	Structure	Reference
Rabdosia excisa	Kamebanin (*105-40*)	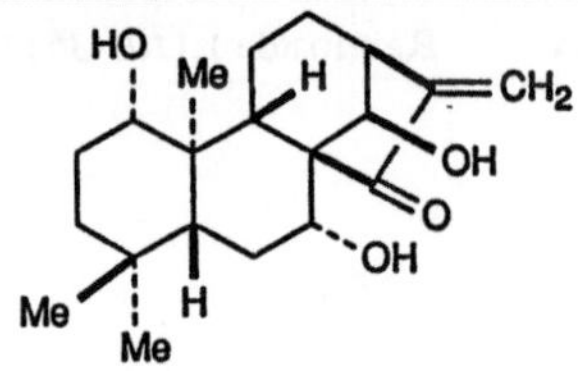	[21]
	Excisanin A (*105-41*): R=H Excisanin B (*105-42*): R=Ac	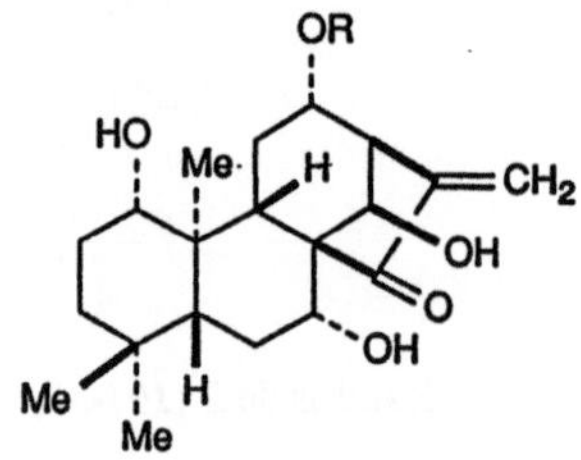	[21] [21]
	Kamebakaurin (*105-43*)	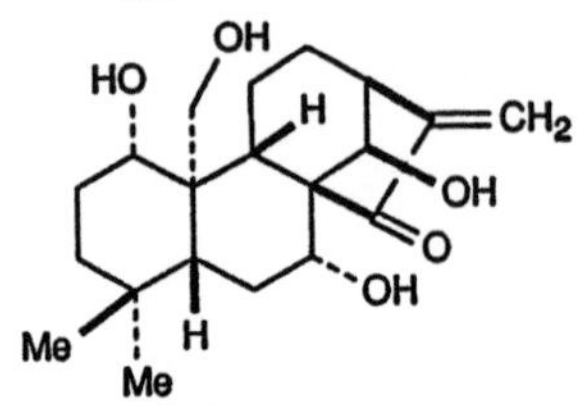	[21]
	Kamebacetal B (*105-44*)	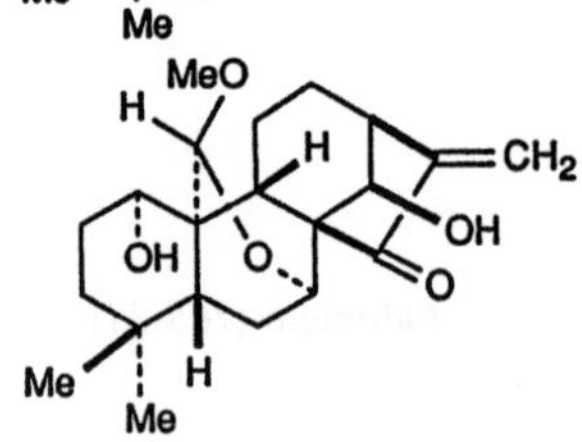	[21]
Rabdosia flexicaulis	Flexicaulin A (*105-45*)	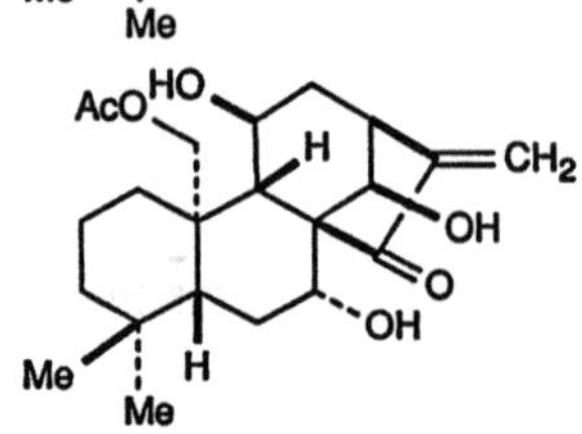	[22]
	Henryin (*105-46*)	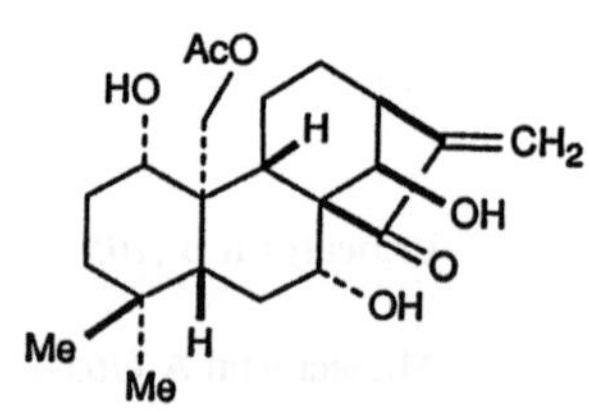	[22]

Table 105.1. (continued)

Rabdosia sp.	Compound	Structure	Reference
Rabdosia flexicaulis	Rabdoloxin B (*105-47*)	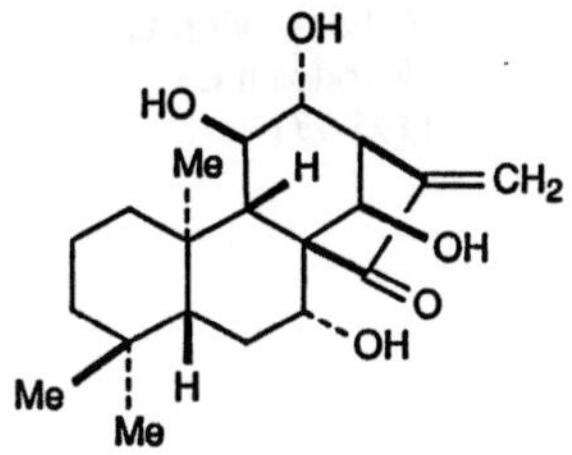	[22]
Rabdosia forrestii	Rabdoferrestin A (*105-48*)	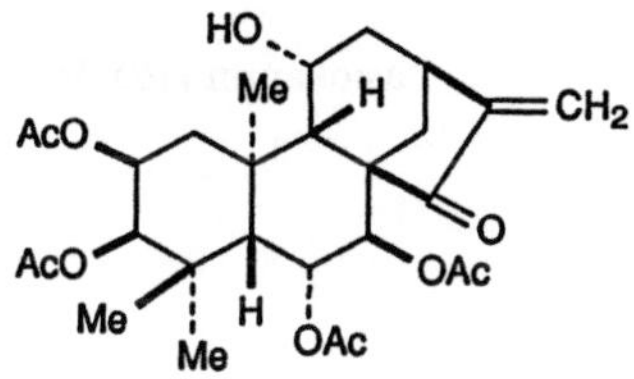	[23]
Rabdosia glutinosa	Glutinosin (*105-49*)	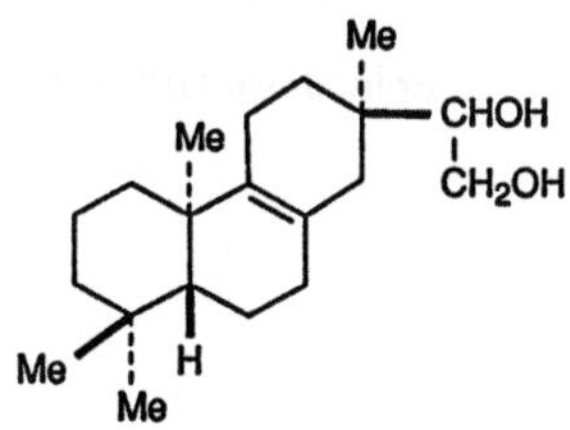	[24]
Rabdosia henryi	Kamebacetal A (*105-50*)	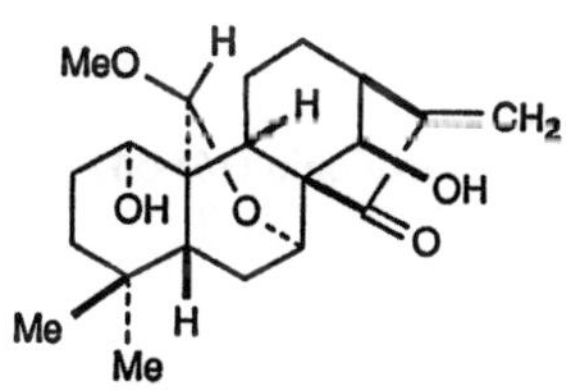	[26]
	Henryin A (*105-51*)	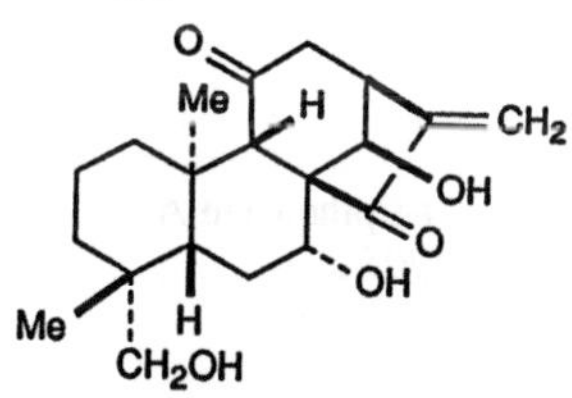	[25]
	Macrocalyxoformin A (*105-52*)	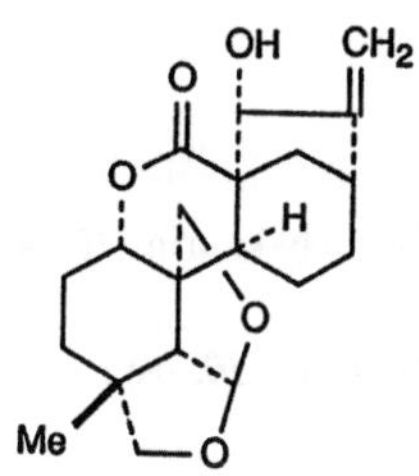	[28]

Table 105.1. (continued)

Rabdosia sp.	Compound	Structure	Reference
Rabdosia henryi	Rabdophyllin G (Rabdosin C) (*105-53*)	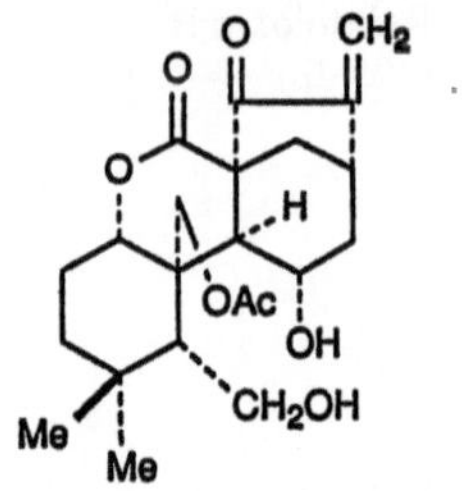	[27]
	Lasiokaurin (*105-54*)	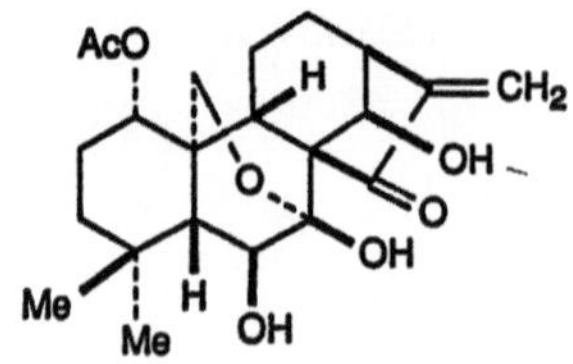	[27]
	Epinodosin (*105-55*)	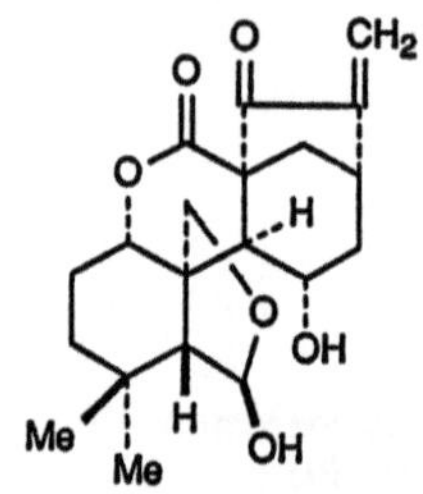	[27]
	Exidonin (*105-56*)	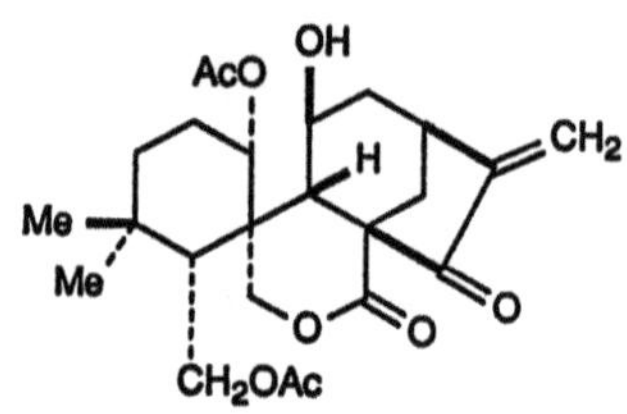	[27]
	4-Epihenryin A (*105-57*)	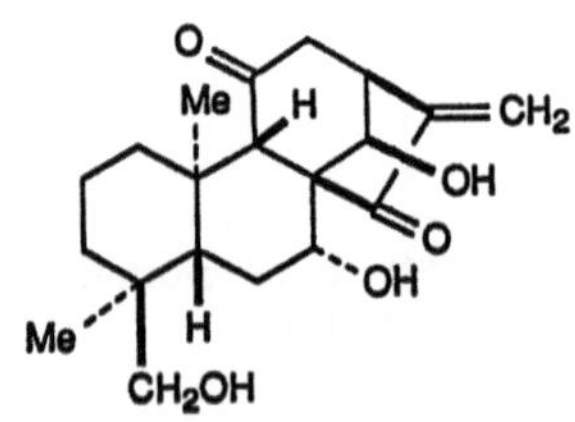	[29]
	Kamebakaurin (*105-43*)		[26]
	Henryin (*105-46*)		[26]

Table 105.1. (continued)

Rabdosia sp.	Compound	Structure	Reference
Rabdosia japonica	Enmenol (*105-58*)	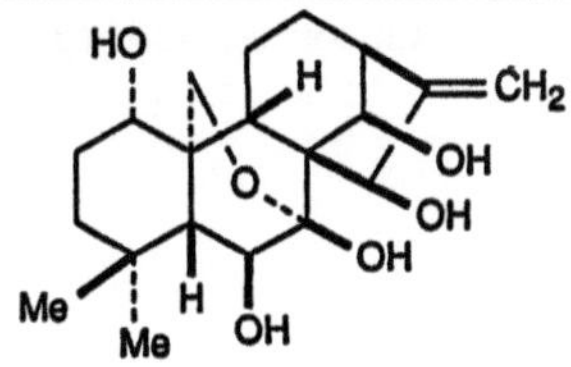	[30]
	Epinodosinol (*105-59*)	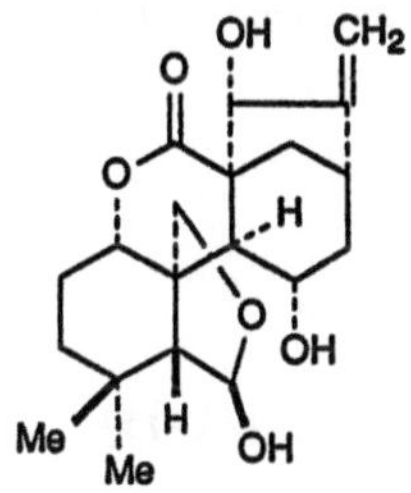	[30]
	Rabdosin A (*105-60*)	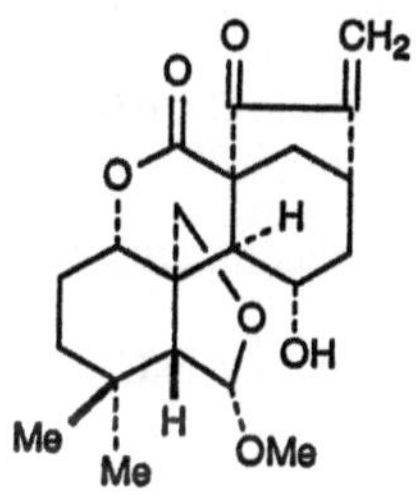	[30]
	Rabdosin B (*105-61*)	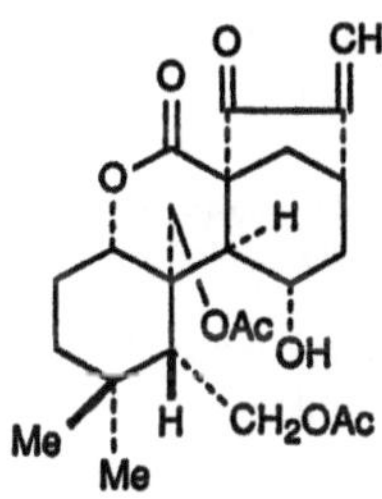	[30, 34]
	Epinodosin (*105-55*)		[31]
	Isodonoiol (Rabdophyllin G) (*105-53*)		[32]
	Maoyerabdosin (*105-62*)	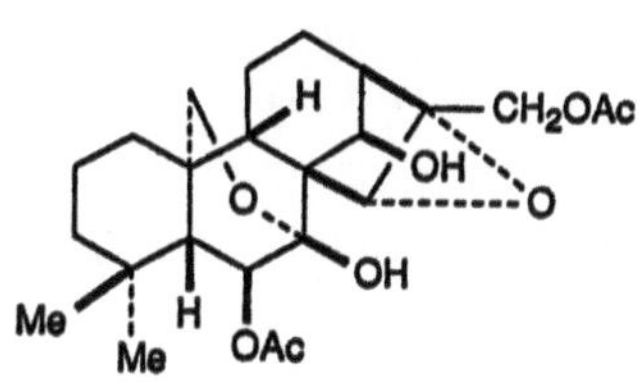	[32]

Table 105.1. (continued)

Rabdosia sp.	Compound	Structure	Reference
Rabdosia japonica	Rabdosinate (*105-63*)	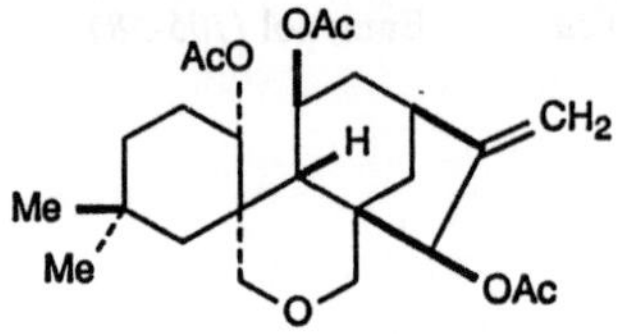	[33]
	Rabdosinatol (*105-64*)	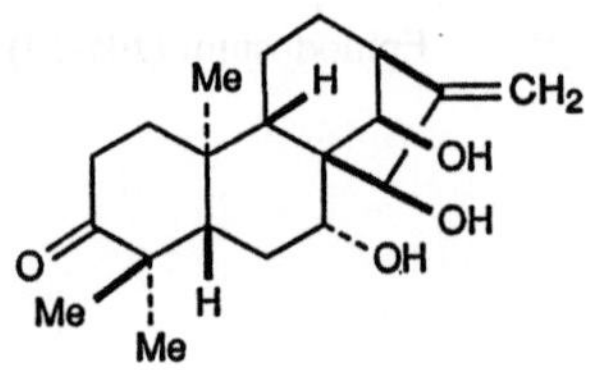	[33]
	Isodonal (*105-15*)		[32]
	Oridonin (*105-39*)		[31]
	Rabdosin C (*105-53*) (Rabdophyllin G)		[31]
	Lasiokaurin (*105-54*)		[31]
Rabdosia japonica var. *glaucocalyx*	Glaucocalyxin C (Rabdosinatol) (*105-64*)		[37]
	Glaucocalyxin A (*105-8*)		[35, 36]
	Glaucocalyxin B (*105-9*)		[35, 36]
Rabdosia kunmingensis	Rabdokunmin A (*105-65*)	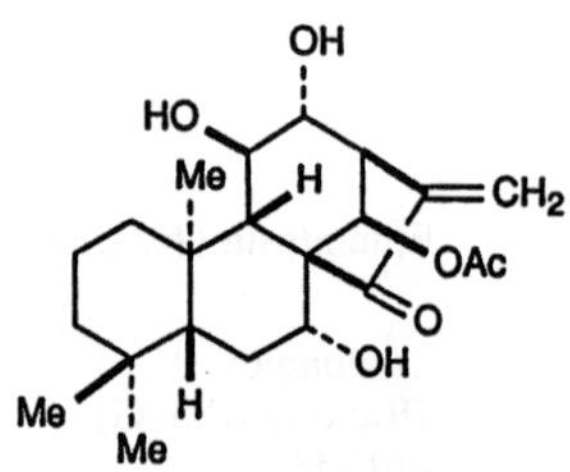	[38]
	Rabdokunmin B (*105-66*)	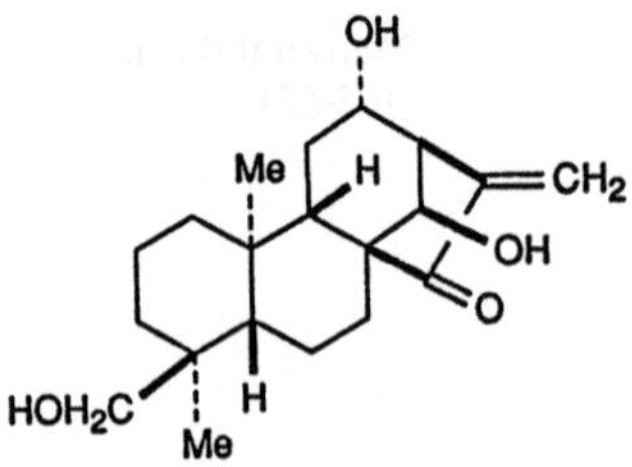	[38]

Table 105.1. (continued)

Rabdosia sp.	Compound	Structure	Reference
Rabdosia kunmingensis	Rabdokunmin C (*105-67*)	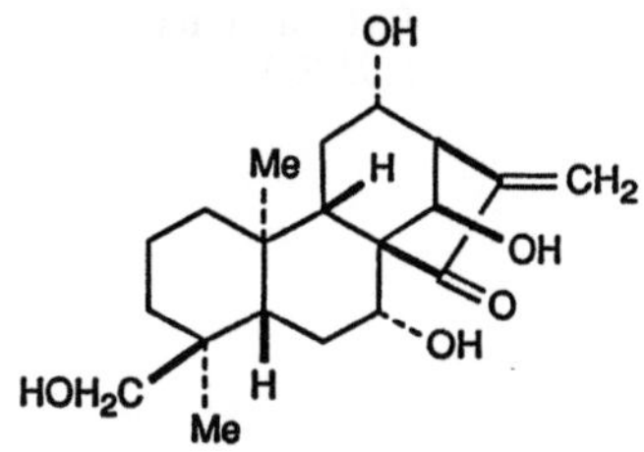	[38]
	Rabdokunmin D (*105-68*)	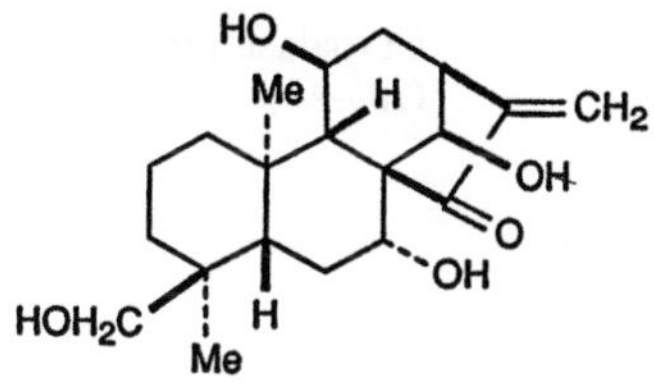	[38]
	Rabdokunmin E (*105-69*)	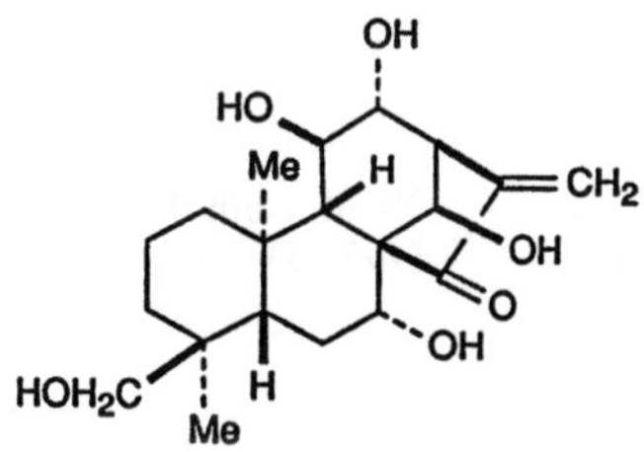	[38]
	Rabdoloxin B (*105-47*)		[38]
	4-*epi*-Isopimaric acid (*105-70*)	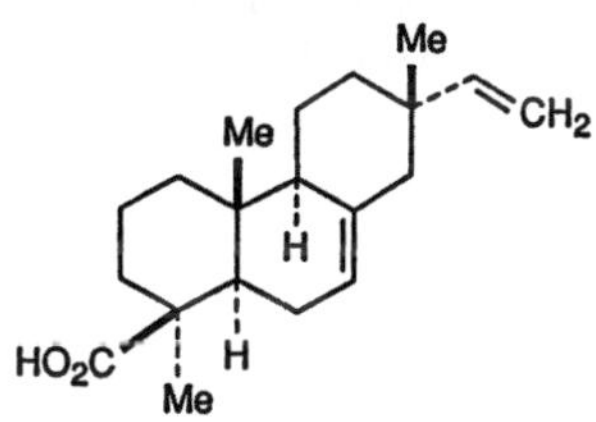	[38]
	Callitrisic acid (*105-71*)	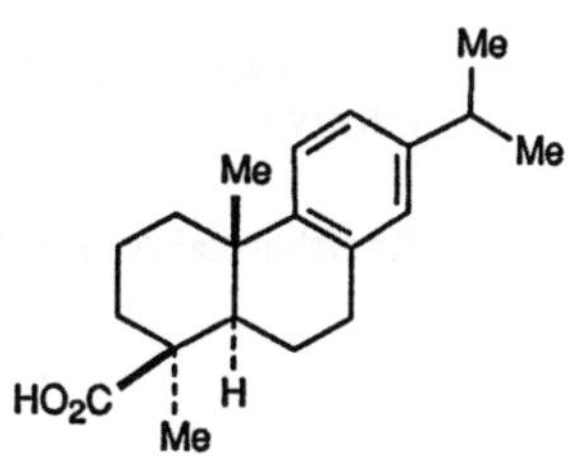	[38]
Rabdosia lasiocarpa	Lasiocarpanin (*105-72*)	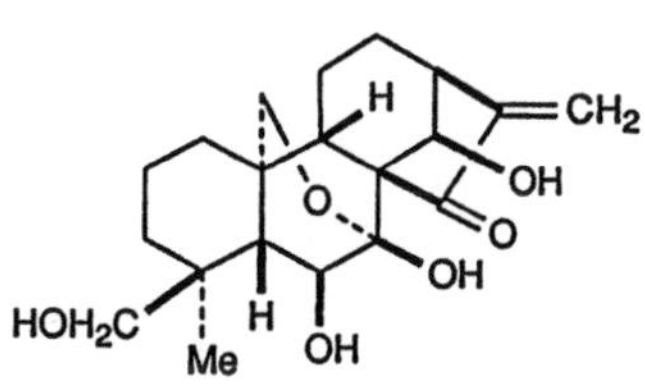	[39]

Table 105.1. (continued)

Rabdosia sp.	Compound	Structure	Reference
Rabdosia lasiocarpa	Rabdolasional (*105-73*)	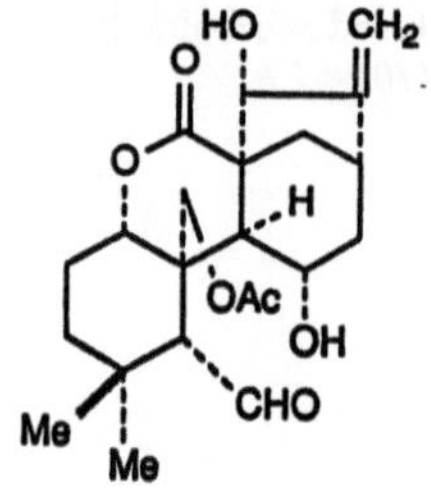	[39]
	Carpalasionin (*105-74*)	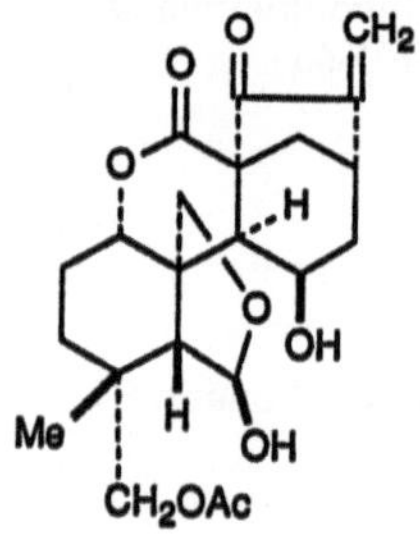	[39]
Rabdosia latifolia var. *reniformis*	Reniformin B (*105-75*)	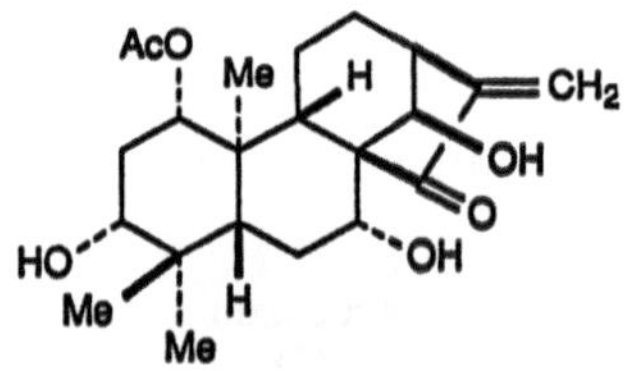	[40]
	Reniformin C (*105-76*)	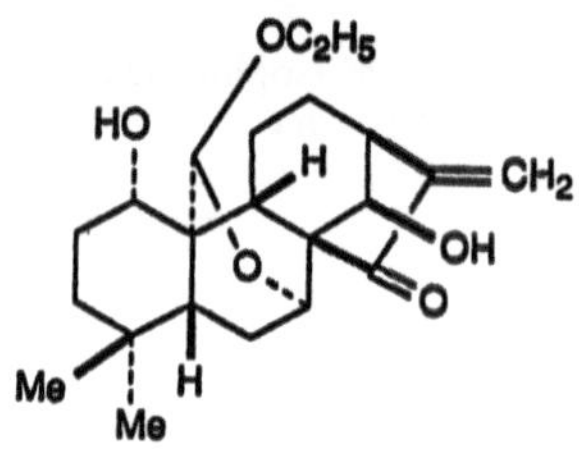	[40]
	Reniformin A (*105-46*) (Henryin)		[40]
	Kamebacetal A (*105-50*)		[40]

Table 105.1. (continued)

Rabdosia sp.	Compound	Structure	Reference
Rabdosia liangshanica	Liangshanin A (*105-77*)	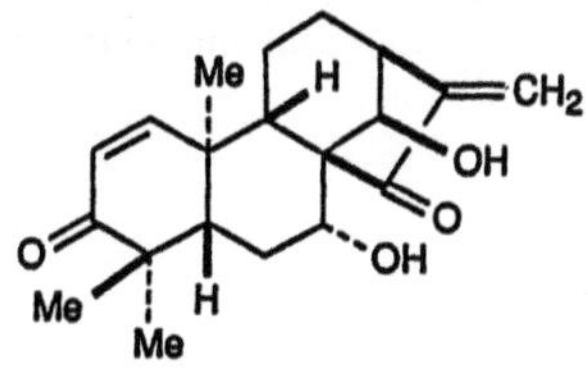	[41]
	Liangshanin B (*105-78*)	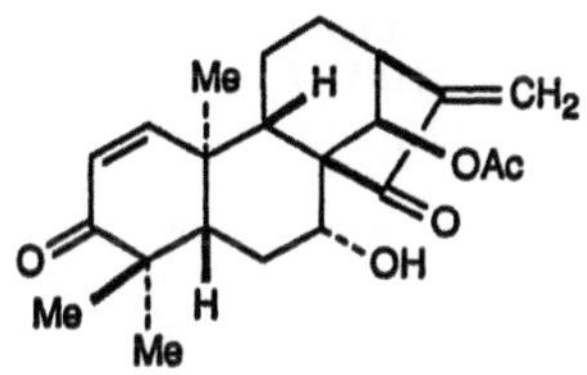	[41]
	Liangshanin C (*105-79*)	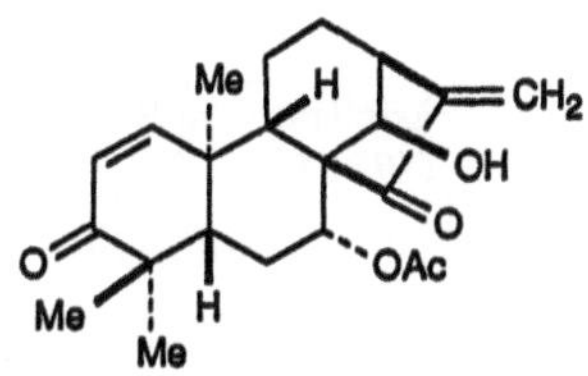	[41]
	Liangshanin D (*105-80*)	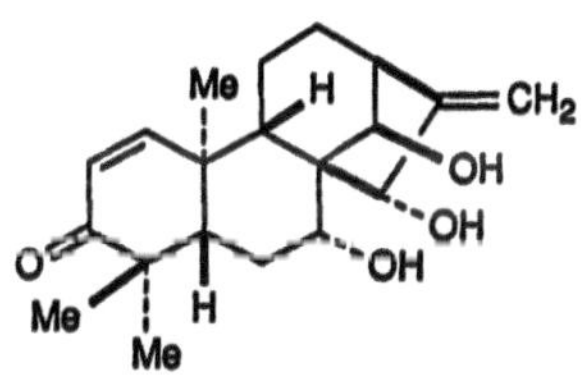	[41]
	Liangshanin E (*105-81*)	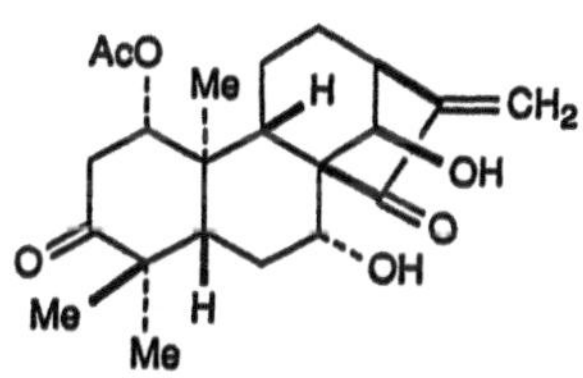	[41]
	Liangshanin F (*105-82*)	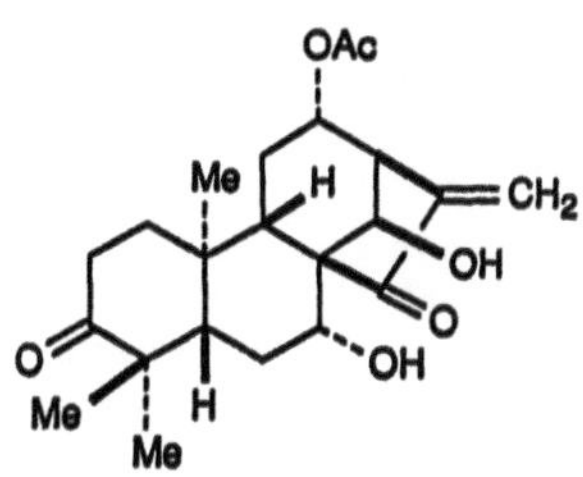	[41]

Table 105.1. (continued)

Rabdosia sp.	Compound	Structure	Reference
Rabdosia liangshanica	Liangshanin G (*105-83*)	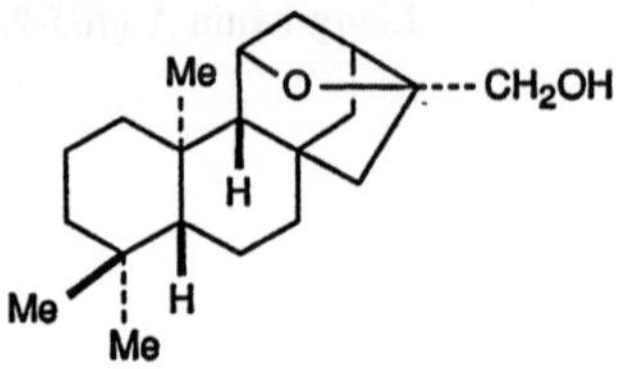	[41]
Rabdosia lophantoides	Lophanthoidin A (*105-84*)	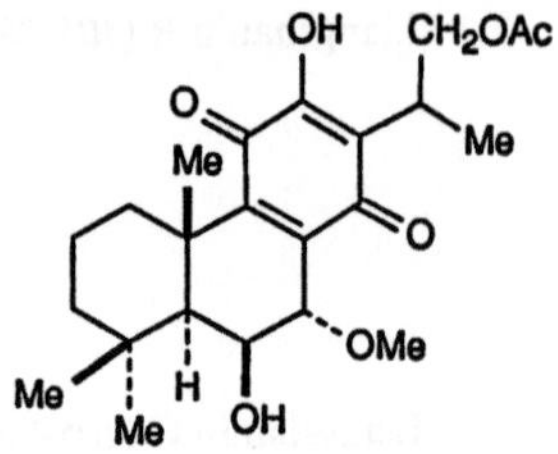	[42]
	Lophantoidin B (*105-85*)	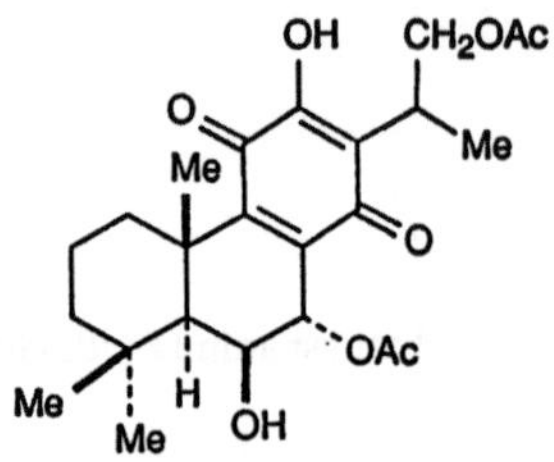	[42]
	Lophantoidin C (*105-86*)	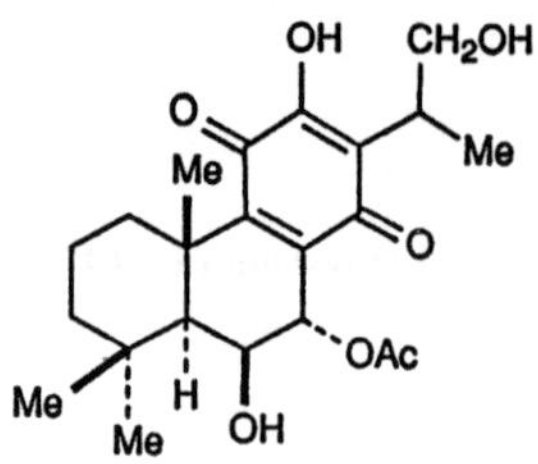	[42]
	Lophantoidin D (*105-87*)	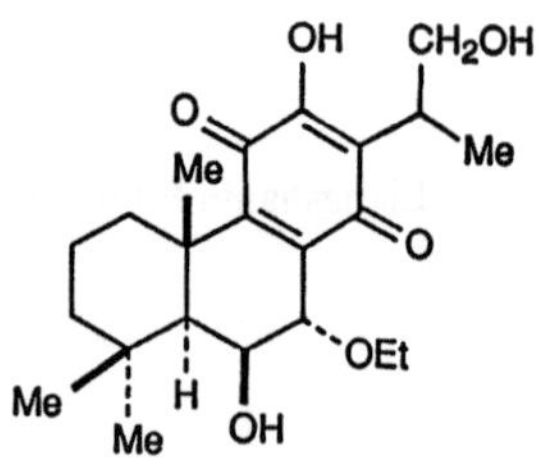	[42]

Table 105.1. (continued)

Rabdosia sp.	Compound	Structure	Reference
Rabdosia lophantoides	Lophanthoidin E (*105-88*)	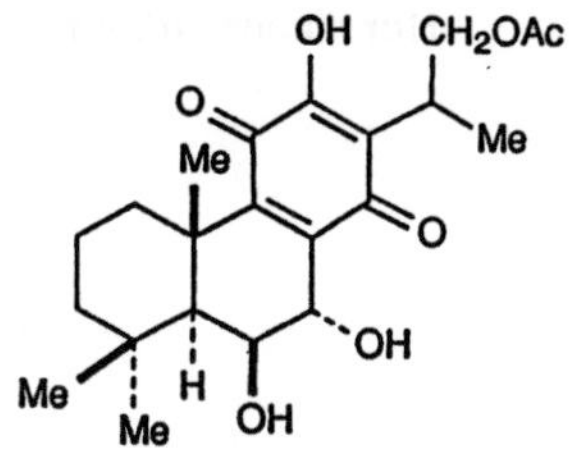	[42]
	Lophanthoidin F (*105-89*)	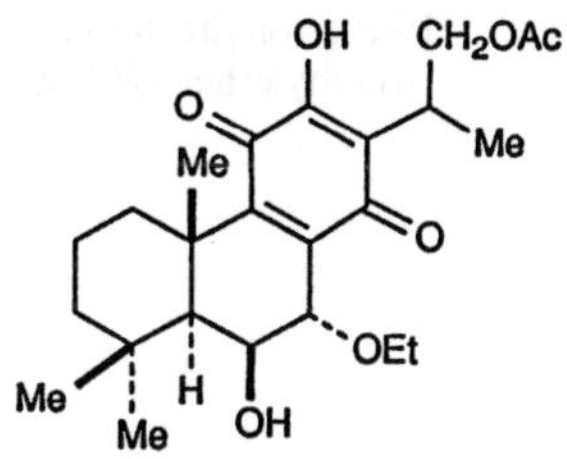	[42]
Rabdosia lophantoides var. *gerardiana*	16-Acetoxy-horminone-7-acetate (*105-90*)	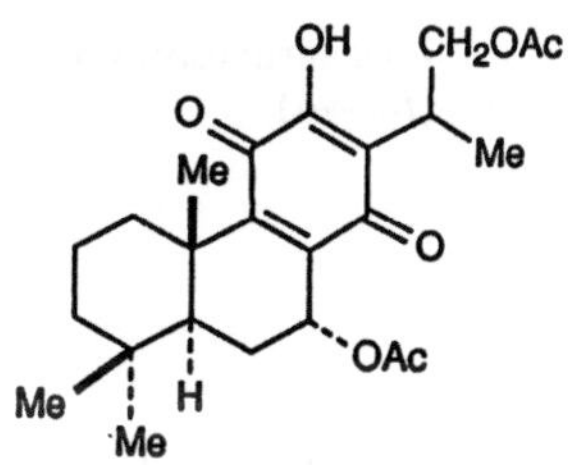	[43]
	16-Acetoxy-horminone-12-acetate (*105-91*)	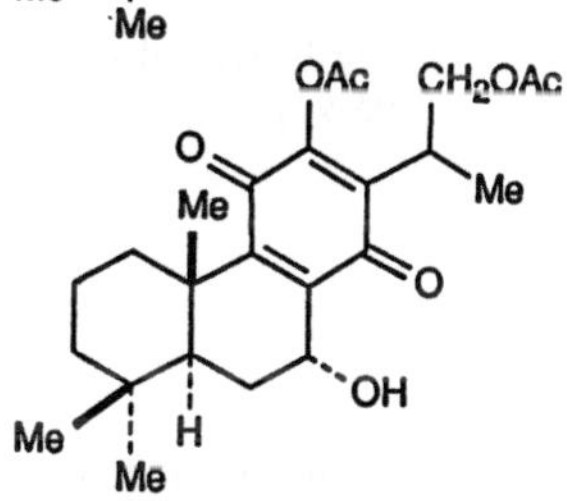	[43]
	Royleanone (*105-92*)	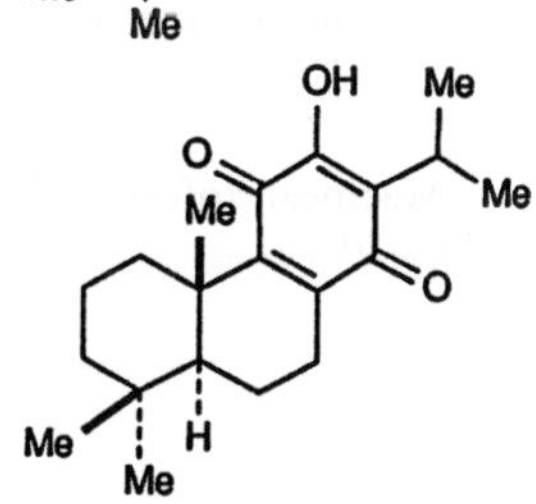	[43]
	Dehydroroyleanone (*105-93*)	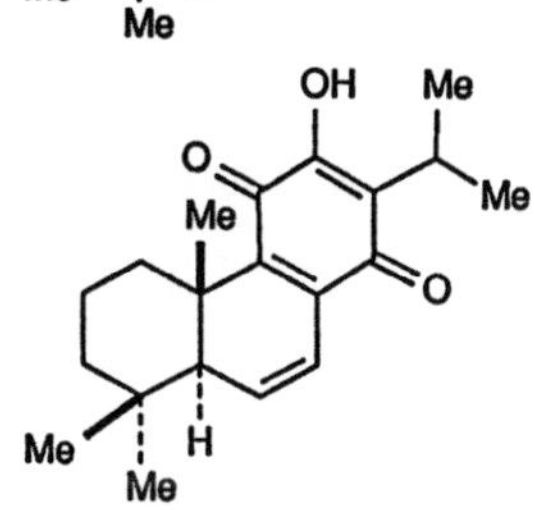	[43]

Table 105.1. (continued)

Rabdosia sp.	Compound	Structure	Reference
Rabdosia lophantoides var. *gerardiana*	Horminone (*105-94*)	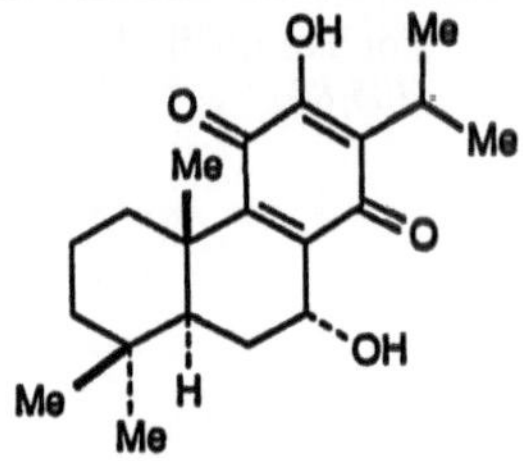	[43]
	16-Acetoxyhorminone-7-methylether (*105-95*)	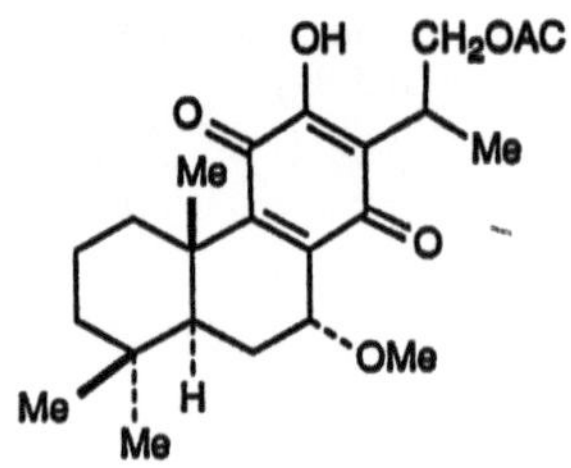	[43]
Rabdosia lungshengensis	Lushanrubescensin C (*105-96*)	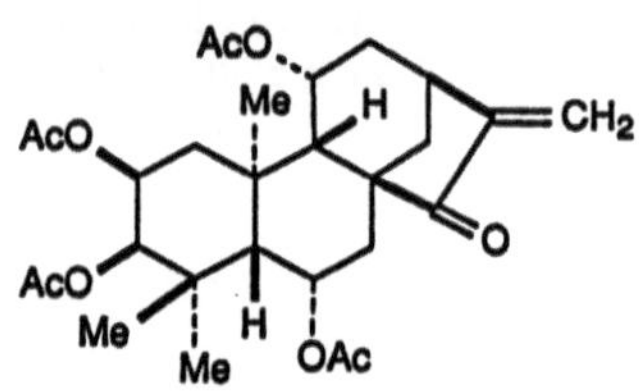	[44]
	Lungshengrabdosin (*105-97*)	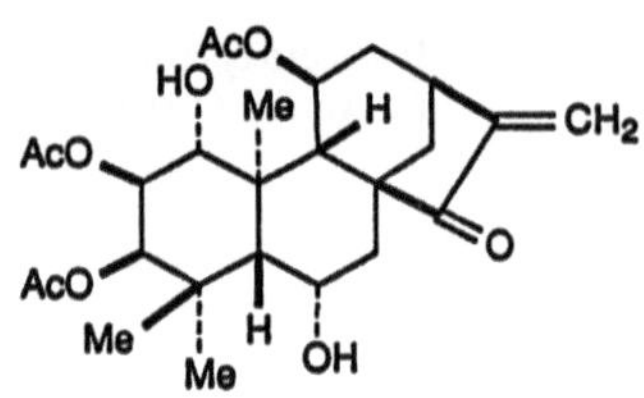	[44]
Rabdosia macrocalyx	Macrocalyxoformin A (*105-52*)		[45]
	Macrocalyxoformin B (*105-98*)	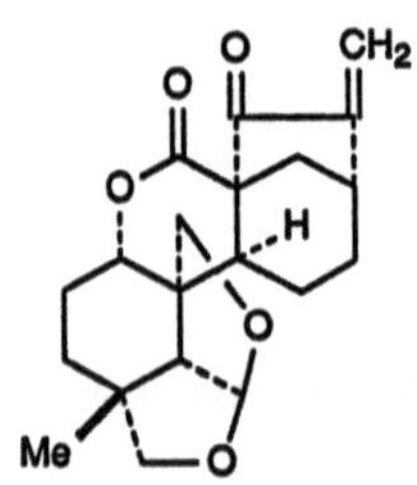	[50]

Table 105.1. (continued)

Rabdosia sp.	Compound	Structure	Reference
Rabdosia macrocalyx	Macrocalyxoformin C (*105-99*)	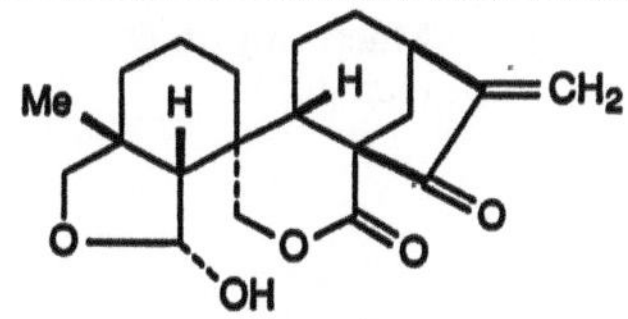	[50]
	Macrocalyxoformin D (*105-100*)	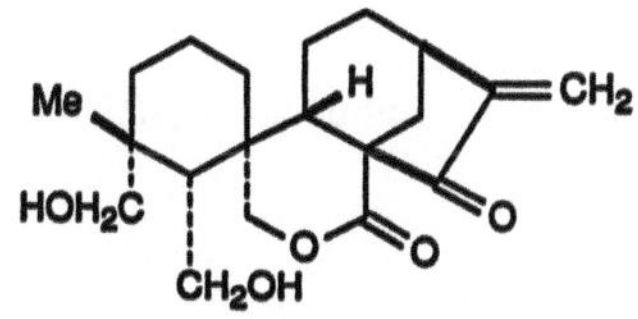	[49]
	Macrocalyxoformin E (*105-101*)	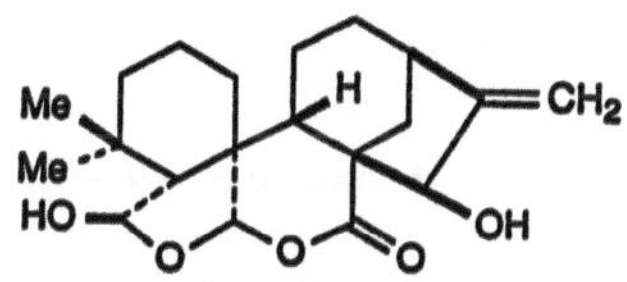	[50]
	Macrocalin A (*105-102*)	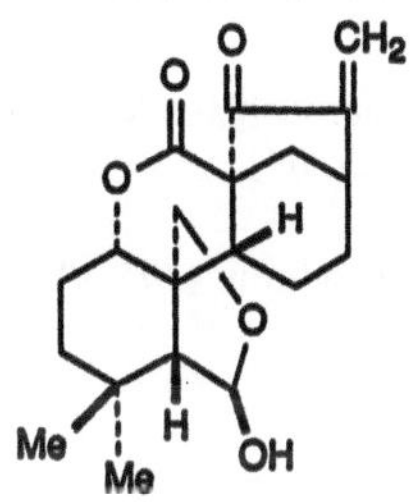	[46]
	Macrocalin B (*105-103*)	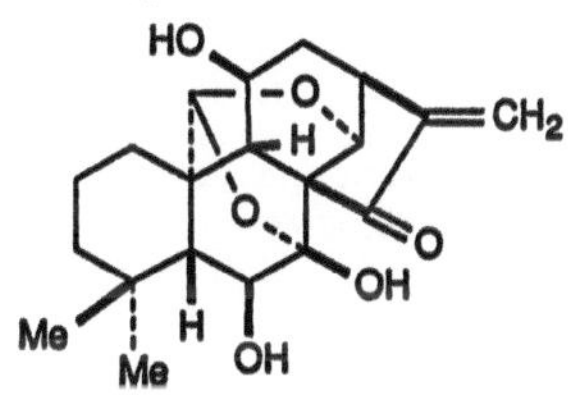	[46]
	Macrocalyxin A (*105-104*)	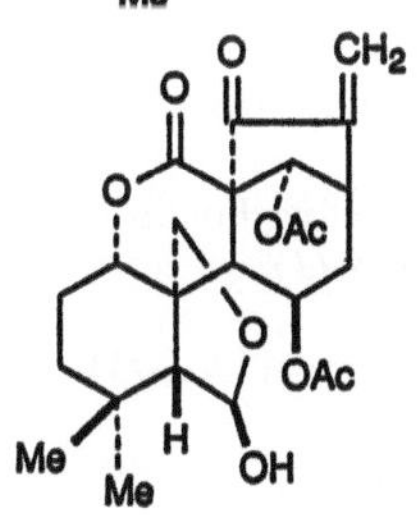	[47]
	Macrocalyxin C (*105-12*) (Amethystonal)		[51]

Table 105.1. (continued)

Rabdosia sp.	Compound	Structure	Reference
Rabdosia macrocalyx	Macrocalyxin D (*105-105*)	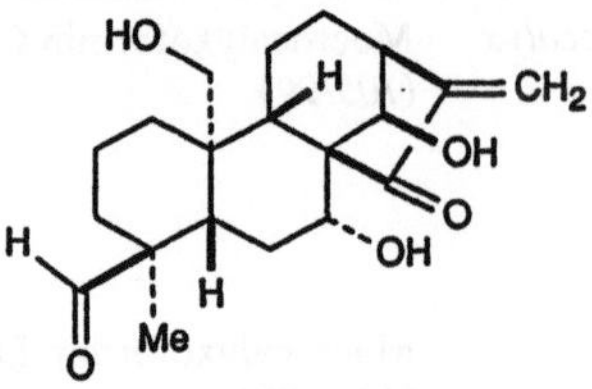	[48]
Rabdosia macrocalyx var. *jiuhua*	Jiuhuanin A (*105-106*)	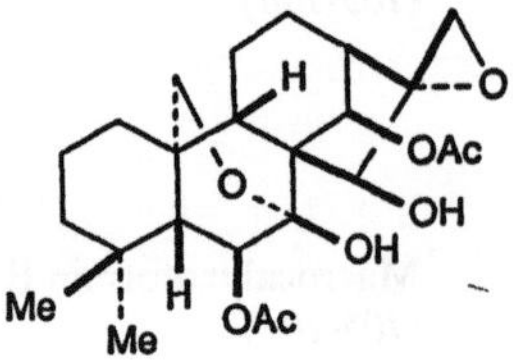	[52]
	Excisanin A (*105-41*)		[52]
	Excisanin B (*105-42*)		[52]
	Macrocalyxin A (*105-104*)		[52]
Rabdosia macrophylla	Lasiodonin (*105-107*)	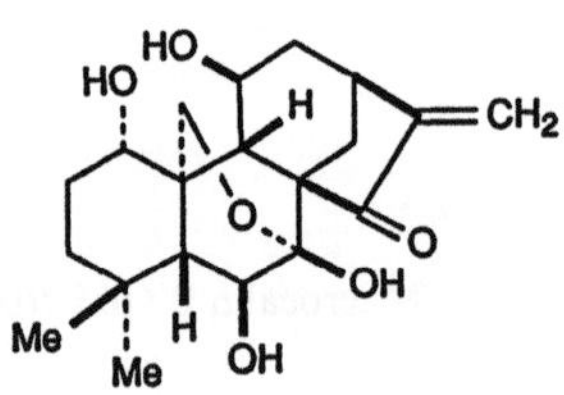	[57]
	Rabdophyllin H (*105-108*)	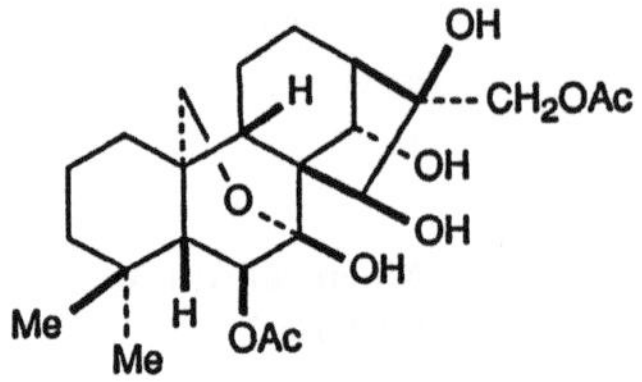	[60, 61]
	Amethysthoidin A (*105-11*)		[55, 56, 59]
	Isodonal (*105-15*)		[53, 54, 59]
	Oridonin (*105-39*)		[53, 54, 59]
	Rabdophyllin G (*105-53*)		[55, 56, 58, 59]
	Lasiokaurin (*105-54*)		[57]
	Enmenol (*105-58*)		[55, 56]

Table 105.1. (continued)

Rabdosia sp.	Compound	Structure	Reference
Rabdiosa nervosa	Nervosin (*105-109*)	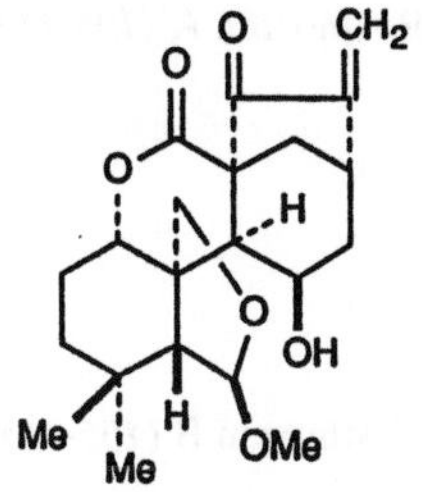	[62, 63]
	Novelrabdosin (*105-110*)	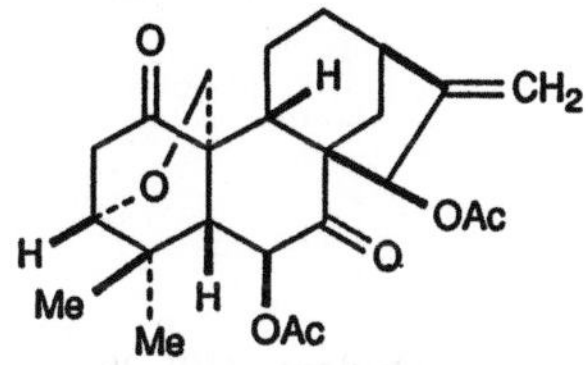	[63, 65]
	Ganervosin A (*105-111*)	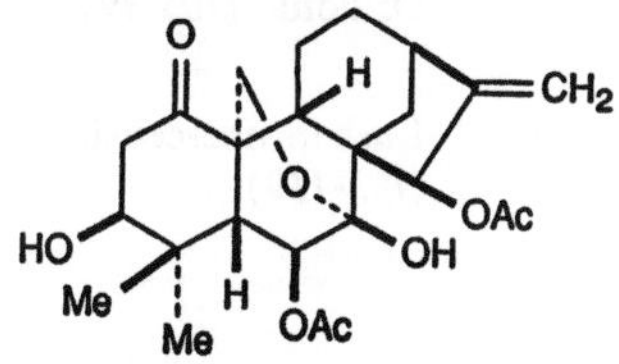	[66]
	Effusanin A (*105-112*): R=H Effusanin E (*105-113*): R=OH	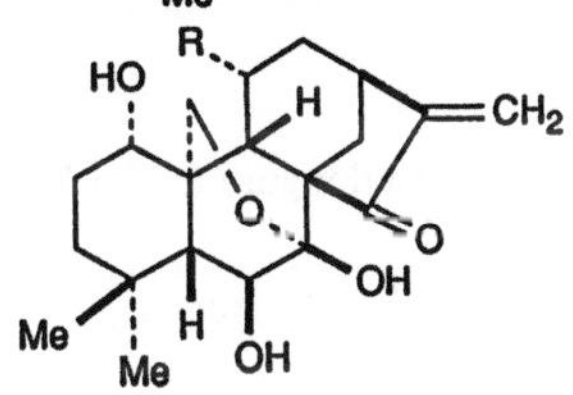	[64]
Rabdosia phyllostachys	Phyllostachysin A (*105-114*)	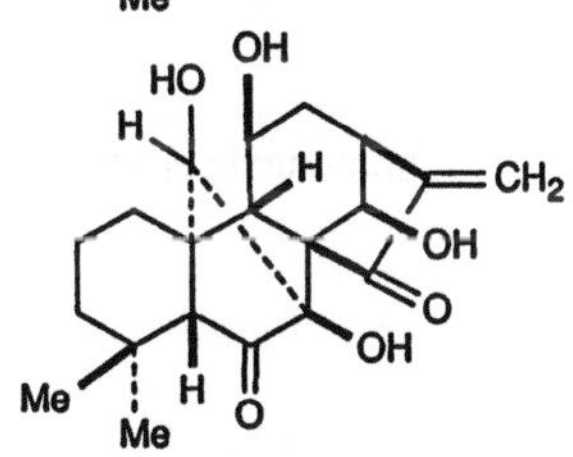	[67]
Rabdosia rosthornii	Ponicidin (*105-115*)	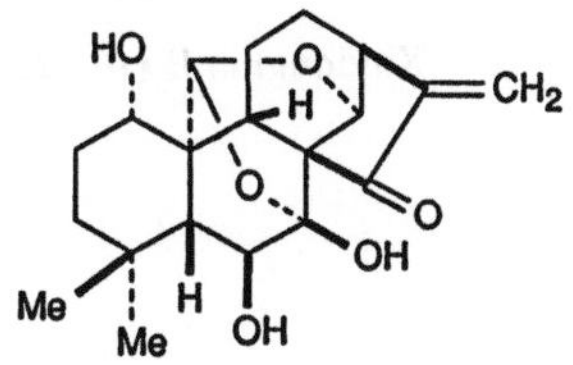	[68]
	Rosthorin A (*105-116*)	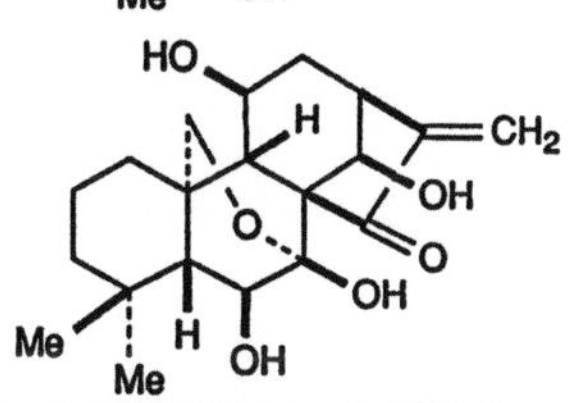	[68]

Table 105.1. (continued)

Rabdosia sp.	Compound	Structure	Reference
Rabdosia rosthornii	Rosthornin A (*105-117*)	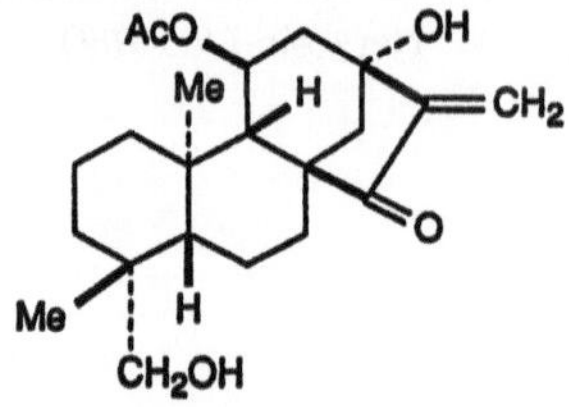	[69]
	Rosthornin B (*105-118*)	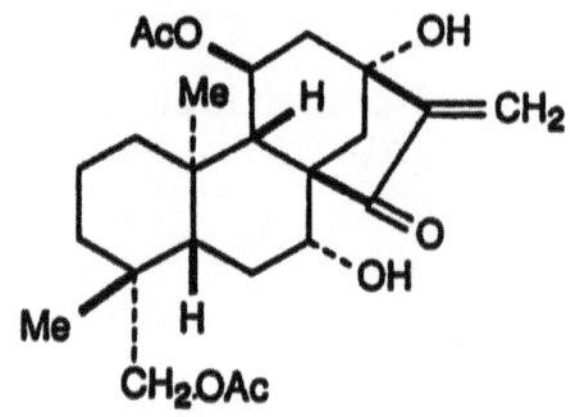	[69]
	Oridonin (*105-39*)		[68]
Rabdosia rubescens	Lushanrubescensin A (*105-119*)	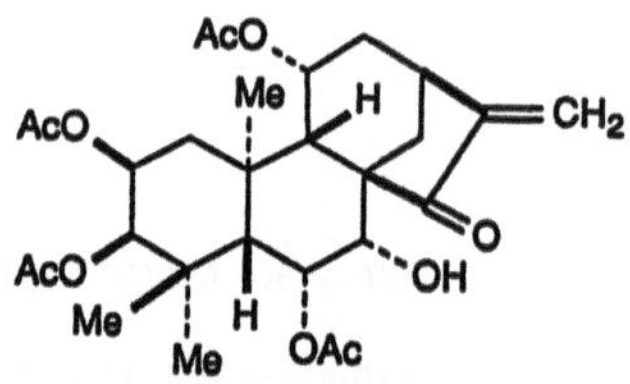	[73]
	Ludongnin (*105-120*)	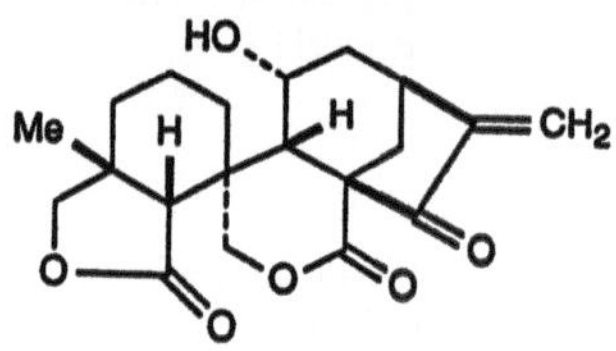	[74]
	Xindongnin A (*105-121*)	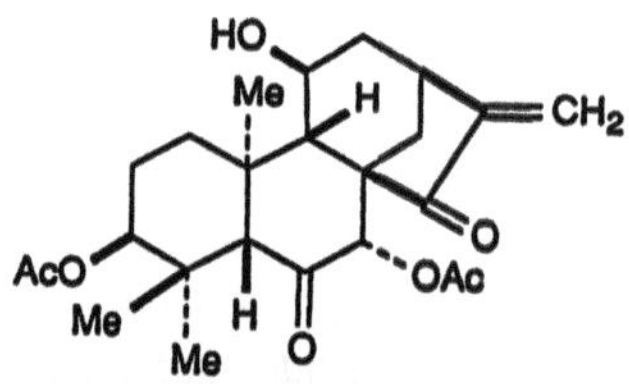	[75]
	Xindongnin B (*105-122*)	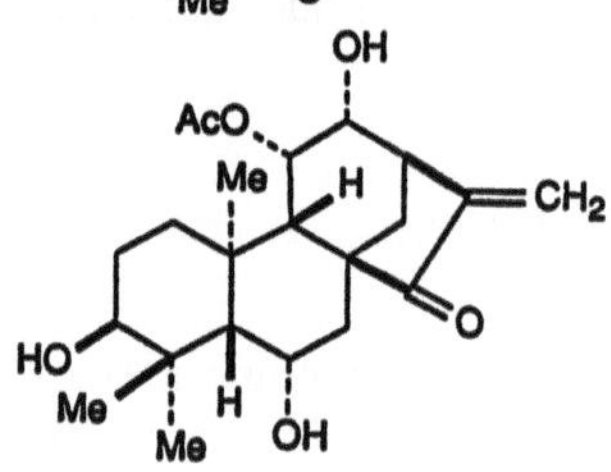	[75]

Table 105.1. (continued)

Rabdosia sp.	Compound	Structure	Reference
Rabdosia rubescens	Guidongnin (*105-123*)	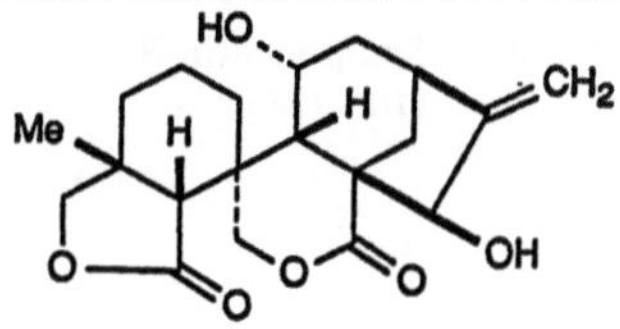	[76]
Rabdosia rubescens f. *lushanensis*	Ludongnin B (*105-124*)	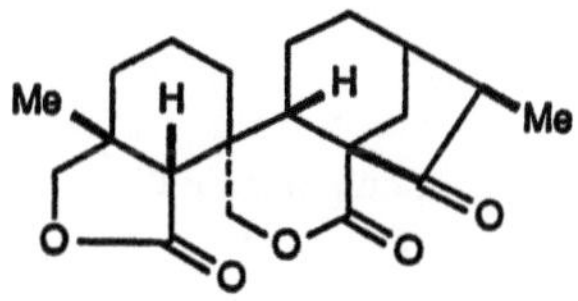	[77]
	Lushanrubescensin A (*105-119*)		[78]
	Lushanrubescensin B (*105-125*)	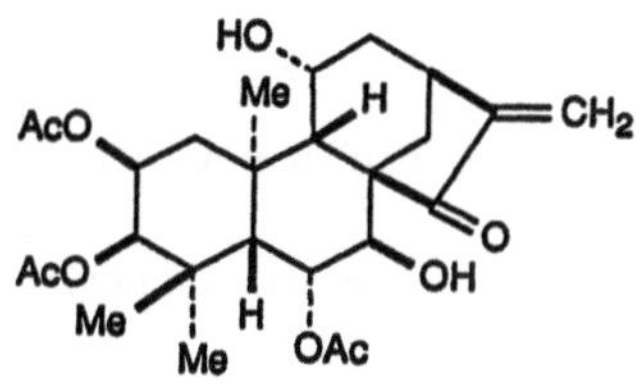	[78]
	Lushanrubescensin C (*105-96*)		[78]
	Lushanrubescensin D (*105-126*)	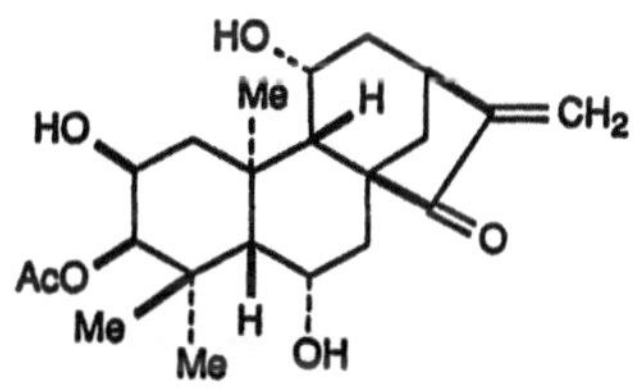	[79]
	Lushanrubescensin E (*105-127*)	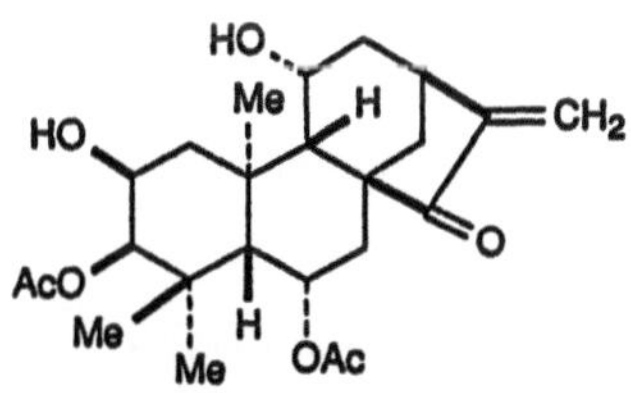	[80]
Rabdosia sculponeata	Enmein (*105-1*)		[81]
	Sculponeatin A (*105-52*) (Macrocalyxoformin A)		[81]

Table 105.1. (continued)

Rabdosia sp.	Compound	Structure	Reference
Rabdosia sculponeata	Sculponeatin B (*105-128*)	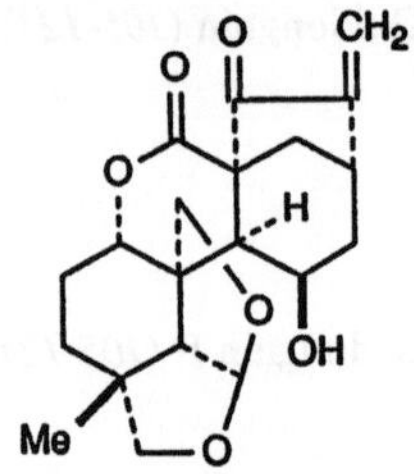	[81]
	Sculponeatin C (*105-129*)	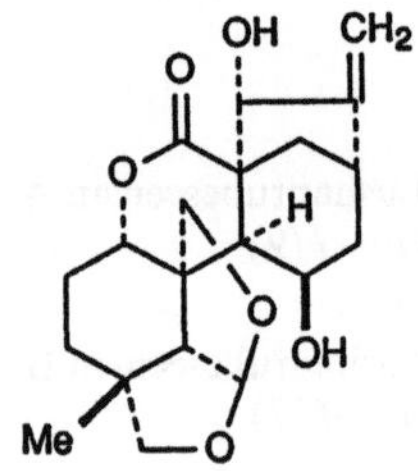	[81]
Rabdosia serra	Excisanin A (*105-41*)		[82]
	Kamebakaurin (*105-43*)		[82]
	Rabdoserrin B (*105-130*)	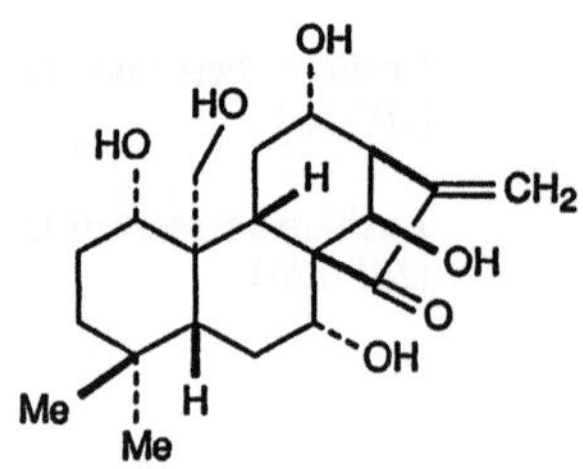	[83, 85]
	Rabdoserrin A (*105-131*)	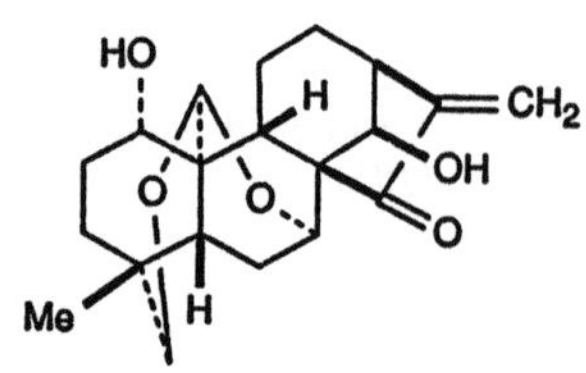	[84]
Rabdosia ternifolia	Isodonal (*105-15*)		[86]
	Isodonoic acid (*105-132*)	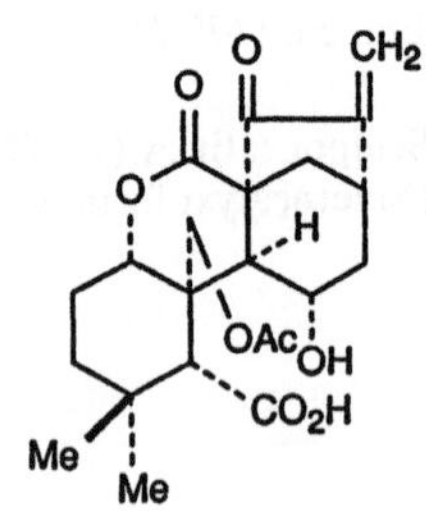	[86]

Table 105.1. (continued)

Rabdosia sp.	Compound	Structure	Reference
Rabdosia ternifolia	Longikaurin A (*105-133*)	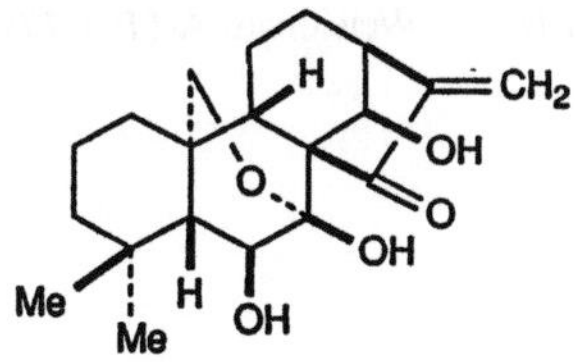	[86, 88]
	Longikaurin E (*105-134*)	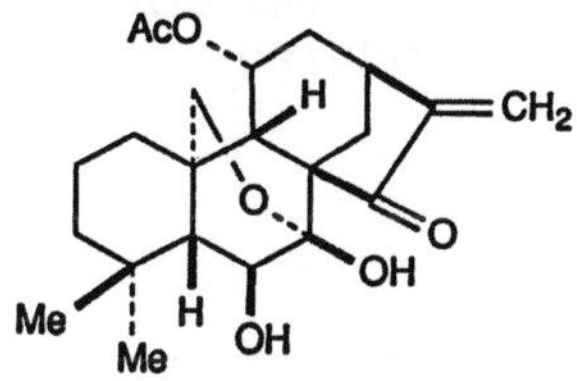	[86]
	Oridonin (*105-39*)		[87]
	Ponicidin (*105-115*)		[87]
	Rabdoternin A (*105-135*)	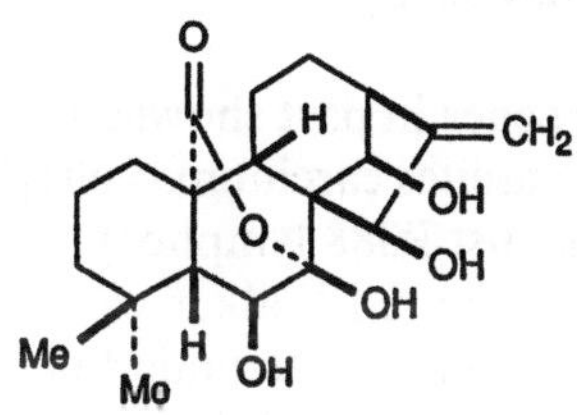	[87]
	Effusanin B (*105-136*)	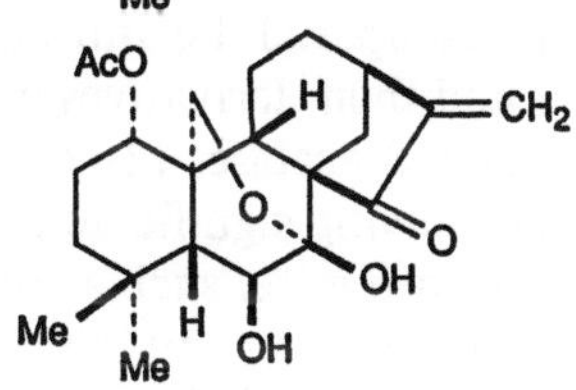	[88]
	Effusanin E (*105-113*)		[88]
	Sodoponin (*105-137*)	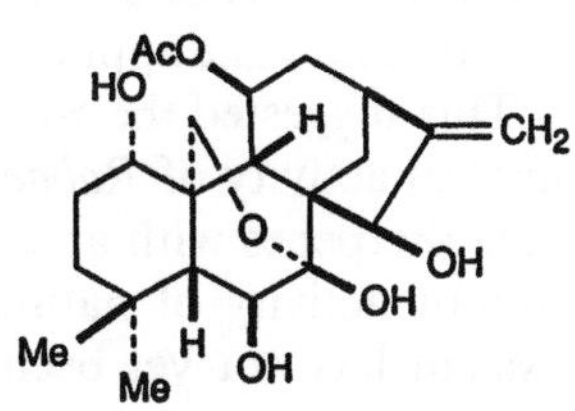	[89, 90]
	Ternifolin (*105-138*)	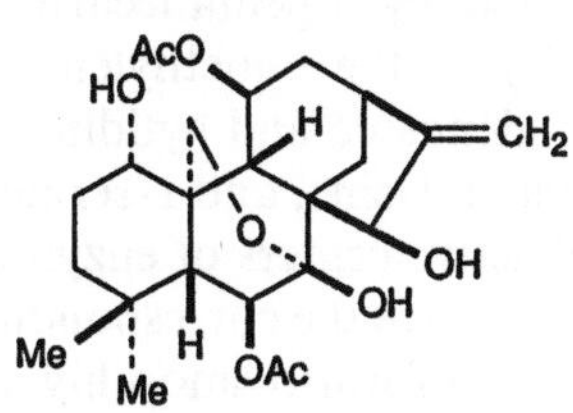	[89, 90]

Table 105.1. (continued)

Rabdosia sp.	Compound	Structure	Reference
Rabdosia weisiensis	Weisiensin A (*105-139*)	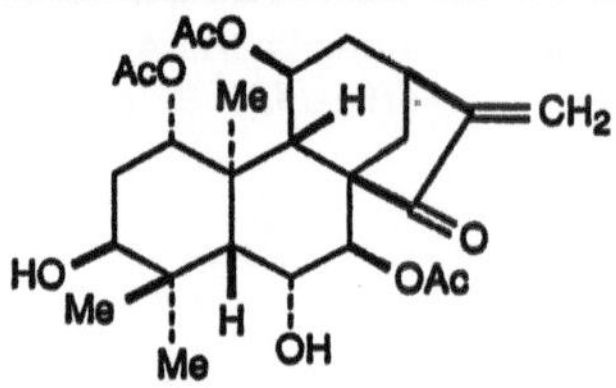	[91, 92]
	Trichorabdal A (*105-140*)	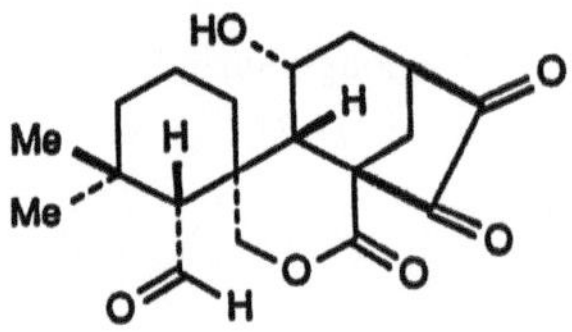	[91, 92]

105.3 Pharmacology

Rabdosia diterpenes in part showed marked cytotoxic activity against HeLa cells [4, 25, 49]. Ehrlich ascites carcinoma cells [21, 26, 59], human OGT 7703 hepatoma cells [59], in vivo against P388 lymphocytic leukemia [30], Ehrlich ascites carcinoma [56, 59], and i.m. implanted Walker carcinosarcoma [98] in mice. A clinical trial with oridonin and ponicidin was undertaken [70].

Fujita et al. [1] compared the antitumor and antibacterial activities of some *Rabdosia* diterpenes and their derivatives including oridonin (*105-39*), lasiokaurin (*105-54*), enmein, enmein-3-acetate, 14-deoxyoridonin, 16,17-dihydrooridonin.

All compounds having significant antitumor activity against Ehrlich ascites carcinoma inoculated into mice, such as oridonin, lasiokaurin, and enmein, possess an α-methylene-cyclopentanone unit in the molecule. Specific activity of such structures against gram-positive bacteria has also been indicated. Analogs without the α-methylene function such as dihydrooridonin and dihydroenmein showed no antitumor or antibacterial activity. The same applied to compounds lacking the ketone group such as trichokaurin. This suggested the α-methylene-cyclopentanone system to be essential for the biological activity of *Rabdosia* diterpenes.

Antitumor sesquiterpenes with an α-methylene-γ-lactone system are well known [99, 100]. Antitumor activity of natural products containing the α-methylene-cyclopentanone system has not yet been taken into account except for sarcomysin (2-methylene-3-oxo-cyclopentanecarboxylic acid) [101].

The reactivity of the α-methylene group was examined using the reaction of oridonin with adenosine and cytidine as nucleic acid model compounds and with thiols, L-cysteine, L-lysine, and L-serine as model compounds representing relevant constituents of active centers of enzymes. Oridonin reacted with thiols easily under mild conditions to give the corresponding thioether adducts. The adduct with L-cysteine (*105-141*) was formed smoothly and practically quantitatively. However, oridonin did not react with adenosine and cytidine under the same conditions [102].

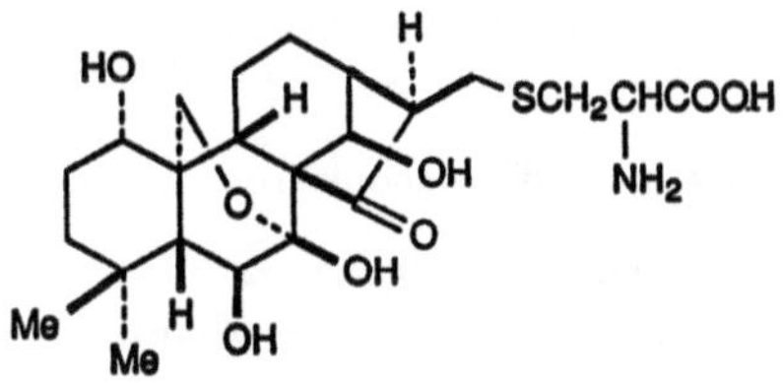

Adduct of oridonin with L-cysteine (*105-141*)

A hypothetical transition state between oridonin and a specific site of an enzyme in tumor cells has been proposed [102, 103]. According to this hypothesis, hydroxy groups in the molecule might play an important role in biological activities. For example, antitumor activity of 14-deoxyoridonin was found to be lower than that of oridonin. The 14β-hydroxy group has been speculated to influence biological activity by increasing the binding affinity to specific sites of target enzymes. A similar function could be ascribed to the 7-hydroxy group. Likewise, hydrogen bonding between the C-6 hydroxy and C-15 carbonyl groups might possibly contribute to antitumor activity [102].

Oridonin given i.p. to mice at a dose of 15 mg/kg on days 5 or 8 after implantation of L1210 leukemia cells exhibited 73% and 39% cell killing, respectively. The G_2 and S phases of L1210 cells were prolonged, while the G_1 phase was unchanged [104].

Decreased incorporation of [^{3}H]TdR, [^{3}H]UR, and [^{3}H]leucine into Ehrlich ascites carcinoma cells showed that oridonin inhibits DNA, RNA, and protein synthesis in Ehrlich ascites carcinoma cells in a dose-dependent manner. Inhibition of DNA and RNA synthesis was fast but reversible upon removal of oridonin. Inhibition of protein synthesis, however, was strong and long lasting [105]. Intraperitoneal administration of oridonin at a dose of 15 mg/kg to mice inhibited incorporation of [^{3}H]TdR, [^{3}H]UR, and [^{3}H]leucine into DNA, RNA, and protein, respectively, also in vivo. Inhibition of DNA synthesis by oridonin preceded that of RNA and protein synthesis [106].

References

1. Fujita E, Node M (1984) Diterpenoids of *Rabdosia* species. In: Herz W, Grisebach H, Kirby GW, Tamm C (ed) Progress in the Chemistry of Organic Natural Products, Vol 46. Springer, Berlin Heidelberg New York, pp 77–157
2. Xu YL, Sun HD, Wang DZ (1982) Chemical structure of adenanthin. In: Wang Y (ed) Chemistry of Natural Products. Science, Beijing, pp 275–279
3. Xu YL, Sun HD, Wang DZ, Iwashita T, Komura H, Kozuka M, Naya K, Kubo I (1987) Structure of adenanthin. Tetrahedron Lett 28:499–502
4. Li GY, Wang YL, Xu ZP, Zhang PL, Zhao W (1981) Studies on chemical constituents of *Isodon amethystoides* (Benth) C Y Wu et Husan. Acta Pharm Sin 16:667–671
5. Wang XR, Wang ZQ, Shi PC, Zhou BN (1982) Wangzaozine A, a novel diterpenoid from *Isodon amethystoides*. Chin Trad Herb Drugs 13:12–13
6. Chang PY, Xu MJ, Lin YL, Shi JC (1981) Study on antitumor constituents of *Rabdosia amethystoides* Hara. Chin Pharm Bull 16:57–58
7. Zhao QZ, Chao JH, Wang HQ, Sun HD, Minami Y (1983) Two new diterpenoids from *Rabdosia amethystoides*. Acta Bot Yunnan 5:305–309

8. Hu JH, Chen ZW, Xue XH, Lin LX (1984) Determination of two diterpenes in *Rabdosia macrophylla* and *Rabdosia amethystoides* in different growing stages. J Nanjing Coll Pharm 36–39
9. Sun HD, Lin ZW, Minami Y, Marunaka T (1984) Studies on the diterpenoids from *Rabdosia angustifolia.* Acta Pharm Sin 19:633–635
10. Xu YL (1985) The chemical structure of bulleyanin. Acta Bot Yunnan 7:437–442
11. Wang XR, Wang ZQ, Wang HP, Hu HP, Wang DQ (1987) Chemical structures of coetsin A and B. Acta Bot Sin 29:412–415
12. Huang H, Xu YL, Sun HD (1989) Diterpenoids from *Rabdosia coetsoides.* Phytochemistry 28:2753–2757
13. Wang ZY, Xu YL (1982) New diterpenoid constituents of *Rabdosia eriocalyx* I. Acta Bot Yunnan 4:407–411
14. Li CB, Zhou J, Sun HD (1985) Structures of new diterpenoids, maoecrystal A–E from *Rabdosia eriocalyx.* Acta Bot Yunnan 7:115–116
15. Ruan DC, Wang DZ, Li CB (1986) Quantitative analysis of diterpenoids from *Rabdosia eriocalyx* by HPLC. Acta Bot Yunnan 8:175–179
16. Xu MJ, Cheng PY, Liu YL, Hu JH (1987) Chemical constituents of *Rabdosia eriocalyx.* J China Pharm Univ 18:113–116
17. Shen XY, Isogai A, Furihata K, Kaniwa H, Sun HD, Suzuki A (1989) ent-Kaurene diterpenoids from *Rabdosia eriocalyx.* Phytochemistry 28:855–858
18. Wang ZY, Ruan DC, Chen ZL, Wang CF (1988) The variation and ecochemistry of *Rabdosia eriocalyx.* Acta Bot Sin 30:528–533
19. Isogai A, Shen XY, Furihata K, Kaniwa H, Sun HD, Suzuki A (1989) Diterpene and diterpene glycosides from *Rabdosia eriocalyx.* Phytochemistry 28:2427–2432
20. Shen PQ, Sun HD, Lin ZQ (1986) Diterpenoids of *Rabdosia eriocalyx* var. *laxiflora.* Acta Bot Yunnan 8:163–166
21. Sun HD, Sun XC, Lin ZW, Xu YL, Minami Y, Muranaka T, Fujita T (1981) Excisanin A and B, new diterpenoids from *Rabdosia excisa.* Chem Lett 753–756
22. Zhang HJ, Sun HD (1989) Diterpenoids from *Rabdosia flexicaulis.* Phytochemistry 28:3534–3536
23. Sun HD, Lin ZW (1988) Structure of rabdoforrestin A. Acta Bot Yunnan 10:215–218
24. Sun HD, Lin ZW, Shen PQ (1987) The chemical structure of glutinosin. Acta Bot Yunnan 9:247–252
25. Wang ZQ, Wang XR, Dong JG, Xue ZW, Wang XW (1984) Chemical structure of henryine A. Acta Bot Sin 26:634–638
26. Li JC, Lin CJ, An XZ, Sun HD, Lin ZW (1984) Structure of henryin. Acta Bot Yunnan 6:453–456
27. Meng XJ, Chen YZ, Cui YX, Chen JL (1988) Diterpenoids from *Rabdosia henryi.* J Nat Prod 51:812–816
28. Wang ZQ, Wang YR, Xu FM, Hu HP, Wang HP (1987) Active constituents of *Rabdosia henryi* (Hemsl) Hara. Bull Chin Mater Med 12:609–610
29. Zhou BN, Chen JB, Wang CY, Blasko G, Cordell GA (1989) 4-*epi*-henryine A, a diterpene from *Rabdosia henryi.* Phytochemistry 28:3536–3538
30. Li JC, Liu CJ, An XZ, Wang MT, Zhao TZ, Yu SZ, Zhao GS, Chen RF (1982) Studies on the antitumor constituent of *Rabdosia japonica* (Burm. f) Hara. I. Structures of rabdosin A and B. Acta Pharm Sin 17:682–687
31. Liu CJ, Li JC, An XZ, Cheng RM, Shen FZ, Xu YL, Wang DZ (1982) Studies on the antitumor constituents of *Rabdosia japonica* (Burm. f) Hara II. Structure of rabdosin C. Acta Pharm Sin 17:750–754
32. Zhao QZ, Chao JH, Wang HQ, Sun HD (1984) Chemical constituents in *Rabdosia japonica.* Chin Trad Herb Drugs 15:49–52
33. Wang MT, Zhao TZ, Li JC, Liu CJ, An XZ (1987) Structure of rabdosinate and rabdosinatol. Acta Chim Sin 49:871–874
34. Wang MT, Liu CJ, Li JC (1990) Structural revision of rabdosin B from *Rabdosia japonica.* Phytochemistry 29:664–665
35. Xu YL, Sun XC, Sun HD, Lin ZW, Wang DZ (1981) Chemical structures of glaucocalyxin A and B. Acta Bot Yunnan 3:283–286
36. Zhao QC, Li CS (1987) Chemical constituents of *Rabdosia japonica* var. *glaucocalyx.* Bull Chin Mater Med 12:294–295

37. Liu CJ, Zhao ZH, Wang QD, Sun HD, Lin ZW (1988) Structure of glaucocalyxin C. Acta Bot Yunnan 10:471–474
38. Zhang HJ, Sun HD (1989) Diterpenoids from *Rabdosia kunmingensis.* Phytochemistry 28:3405–3409
39. Takeda Y, Fujita T, Chen CC (1982) Structures of lasiocarpanin, rabdolasional and carpalasionin, new diterpenoids from *Rabdosia lasiocarpa.* Chem Lett 833–836
40. Wang XR, Wang ZQ, Dong JG, Wang XW (1986) Chemical structure of reniformins A, B and C. Acta Bot Sin 28:292–298
41. Zhang FL, Xu YL, Sun HD (1989) Diterpenoid constituents of *Rabdosia liangshanica.* Phytochemistry 28:1671–1674
42. Xu YL, Wang D, Li XJ, Fu J (1989) Abietane quinones from *Rabdosia lophanthoides.* Phytochemistry 28:189–191
43. Xu YL, Ma YB, Zhou L, Sun HD (1988) Abietane quinones from *Rabdosia lophanthoides.* Phytochemistry 27:3681–3682
44. Luo Y, Luo JT, Sun HD, Lin ZW (1988) Constituents of *Rabdosia lungshengensis.* Structure of a new diterpenoid, lungshengrabdosin. Acta Chim Sin 46:763–767
45. Wang ZQ, Wang XR, Dong JG (1983) A new antibacterial diterpene from *Rabdosia macrocalyx* – macrocalyxoformin A. Chin Trad Herb Drugs 14:481–485
46. Cheng PY, Lin YL, Xu GY (1984) New diterpenoids of *Rabdosia macrocalyx*: structure of macrocalin A and macrocalin B. Acta Pharm Sin 19:593–598
47. Wang XR, Wang ZQ, Dong JG, Xue ZW (1984) Chemical structures of macrocalyxin A and C. Acta Bot Sin 26:425–431
48. Wang XR, Wang ZQ, Dong JG (1985) Chemical structure of macrocalyxin D. Acta Bot Sin 27:285–289
49. Wang ZQ, Wang XR, Dong JG, Xu GY (1985) Chemical structure of macrocalyxoformin D. Acta Bot Sin 27:171–176
50. Wang ZQ, Wang XR, Dong JG (1986) Chemical structures of macrocalyxoformin B, C and E. Acta Bot Sin 28:79–85
51. Wang XR, Wang ZQ, Dong JG (1986) Chemical structure of macrocalyxin E. Acta Bot Sin 28:415–418
52. Wang ZQ, Wang XR, Dong JG, Wang XW (1986) Diterpenoids of *Rabdosia macrocalyx* var. *jiuhua* Z.W. Xue et X.W. Wang. Acta Bot Sin 28:185–191
53. Cheng PY, Hsu MC, Lin YT, Shih CC (1980) Studies on the antitumor components of *Rabdosia macrophylla.* I. Chin Pharm Bull 15:43–44
54. Cheng PY, Xu MJ, Lin YL, Shi JC (1981) Antitumor constituents of *Rabdosia macrophylla* (Migo). Acta Pharm Sin 16:796–797
55. Cheng PY, Xu MJ, Lin YL, Shi JC (1982) Structure of rabdophyllin G, an antitumor constituent of *Rabdosia macrophylla.* Chin Pharm Bull 17:174
56. Cheng PY, Xu MJ, Lin YL, Shi JC (1982) Studies on rabdophyllin G, an antitumor constituent of *Rabdosia macrophylla.* II. Acta Pharm Sin 17:917–921
57. Cheng PY, Xu MJ, Lin YL (1984) Antitumor principle of large leaf rabdosia (*Rabdosia macrophylla*). III. Chin Trad Herb Drugs 15:53–54
58. Chen YZ, Wu ZW, Cheng PY (1984) Crystal and molecular structure of rabdophyllin G. Acta Chim Sin 42:645–649
59. Yao JS, Sun FZ, Su TD, Zhou GL, Qian HM (1984) Preliminary evaluation of antitumor effect of chemical constituents of *Rabdosia macrophylla.* Jiangsu Med J 10:575–576
60. Chen YZ, Xu BL, Yao XW, Cheng PY, Xu MJ (1985) Crystal and molecular structure of rabdophyllin H. Acta Chim Sin 43:1190–1194
61. Cheng PY, Xu MJ, Lin YL, Shi JC, Xu GY (1986) Structure of rabdophyllin H. Acta Pharm Sin 21:109–113
62. Cha JH, Zhao QZ, Wang HQ, Sun HD (1983) Studies on the diterpenoids from *Rabdosia nervosa* (Hemsl.) C.Y. Wu et H.W. Li. I. Structure of nervosin. Acta Bot Yunnan 5:311–314
63. Sun HD, Zhao QZ, Cha JH, Wang HP, Lin ZW, Wang DZ, Gong YH (1984) Studies on the diterpenoids from *Rabdosia nervosa.* II. Structure of novelrabdosin, an unique 3,20-epoxy-entkaurene diterpenoid. Acta Bot Yunnan 6:235–236
64. Xu MJ, Tang M, Cheng PY (1985) Studies on the antitumor constituents of *Rabdosia nervosa.* Chin Trad Herb Drugs 16:51–52
65. Sun HD, Lin ZW, Wang DZ, Gong YH, Zhao QZ, Cha JH (1985) Structure of neorabdosin. Acta Chim Sin 43:481–483

66. Wang QG, Hua SM, Bai G, Chen YZ (1988) X-ray crystal structure of a new diterpene, ganervosin A. J Nat Prod 51:775–778
67. Sun H, Lin Z, Minami Y, Marunaka T, Fujita T, Takeda T (1982) Structure of new diterpenoids isolated from *Rabdosia* plants, *R. phyllostachys, R. rubescens* and *R. sculponeata* in China (Abstr). 25th Symposium on the Chemistry of Natural Products, (Japan) p 266
68. Li GY, Wang YL (1984) Studies on the diterpenoids of *Rabdosia rosthornii* (Diels) Hara. Acta Pharm Sin 19:590–592
69. Xu YL, Ma YB (1989) Diterpenoid constituents from *Rabdosia rosthornii.* Phytochemistry 28:3235–3237
70. Sun HD, Lin ZW, Qin CQ, Chao JH, Zhao QZ (1981) Studies on the chemical constituents of antitumor plant *Rabdosia rubescens* (Hemsl.) Hara. Acta Bot Yunnan 3:95–100
71. Chang TL, Chen CY, Miau CH, Chao CT, Sun HT, Liu CW (1980) Rubescensine B – another effective antitumor agent in *Rabdosia rubescens* Hemsl. Kexue Tongbao 25:1051–1054
72. Sun HD, Chao JH, Lin ZW, Marunaka T, Minami Y, Fujita T (1982) The structure of rubescensin C: a new minor diterpenoid isolated from *Rabdosia rubescens.* Chem Pharm Bull (Tokyo) 30:341–343
73. Qin CQ, Liu CJ, Li JC, An XZ, Sun HD, Lin ZW (1984) Structure of lushanrubescensin. Acta Bot Yunnan 6:333–337
74. Zheng XR, Gao ZY, Sun HD, Lin ZW (1984) Structure of ludongnin. Acta Bot Yunnan 6:316–320
75. Sun HD, Lin ZW, Fu JA, Zheng XR, Gao ZY (1985) Xindongnin A and B, new diterpenoids from *Rabdosia rubescens.* Acta Chim Sin 43:353–359
76. Sun HD, Pan LT, Lin ZW, Niu FD (1988) Structure of guidongnin. Acta Bot Yunnan 10:325–330
77. Zhang XR, Gao ZI, Tang JQ, Sun HD, Lin ZW (1986) Structure of ludongnin B. Acta Bot Yunnan 8:161–162
78. Li JC, Liu CJ, Sun HD, Lin ZW (1986) Structures of lushanrubescensin B and C: new diterpenoids from *Rabdosia rubescens* f. *lushanensis.* Acta Bot Yunnan 8:93–97
79. Qin CQ, Li FQ, Li HL, Sun HD, Lin ZW (1986) Structure of lushanrubescensin D. Acta Bot Yunnan 8:99–102
80. Li JC, Sun HD, Lin ZW (1987) The structure of lushanrubescensin E. Acta Bot Yunnan 9:485–488
81. Wang XR, Wang ZQ, Dong JG (1982) New diterpenes from Huang Hua Xiang Cha Cai (*Rabdosia sculponeata*). Chin Trad Herb Drugs 13:491–492
82. Jin RL, Cheng PY, Xu GY (1985) The structure of rabdoserrin A, isolated from *Rabdosia serra* (Maxim.) Hara. Acta Pharm Sin 20:366–371
83. Jin RL, Cheng PY, Xu GY (1987) Structure of rabdoserrin B, isolated from *Rabdosia serra.* J China Pharm Univ 18:172–174
84. Wu ZW, Chen YZ, Jin RL, Chen PY (1986) The crystal and molecular structure of rabdoserrin A. Acta Chim Sin 44:1139–1143
85. Huang DY, Lu GY, Gu XC, Han YZ, Li GP, Jin RL, Cheng PY (1988) Crystal and molecular structure of rabdoserrin B. Struct Chem 7:17–21
86. Sun HD, Lin ZW, Minami Y, Takeda Y, Fujita T (1982) On the constituents of *Rabdosia ternifolia* (D. Don) Hara: the structure of a new diterpenoid, isodonoic acid. Yakugaku Zasshi 102:887–890
87. Fujita T, Takeda Y, Sun H, Minami Y (1983) The chemical structure of a new diterpene of *Rabdosia ternifolia* (Abst). 103rd Annual Meeting of Japan Pharmaceutical Society, Tokyo, p 242
88. Li GY, Wang YL, Song WZ, Ji ZY (1983) Studies on the diterpenoids of *Rabdosia ternifolia.* Bull Chin Mater Med 8:28–29
89. Wang XF, Wei RF, Lu SC, Xu RS (1984) Chemical constituents of Xi Ye Xiang Cha Cai (*Rabdosia ternifolia* Hara). Chin Trad Herb Drugs 15:102
90. Wang XF, Wei RF, Lu WJ, Wu LZ, Zhu DY, Song GQ, Feng SC, Xu RS (1985) Diterpenoids of *Rabdosia ternifolia.* Acta Bot Sin 27:520–524
91. Xu YL, Wu M (1989) Diterpenoid constituents from *Rabdosia weisiensis.* Phytochemistry 28:1978–1979
92. Xu YL, Wu M (1988) Diterpenoids of *Rabdosia weisiensis.* Acta Bot Yunnan 10:357–358
93. Ikeda T, Kanamoto S (1958) Study of bitter principles of *Isodon trichocarpus.* I. Yakugaku Zasshi 78:1128–1132

94. Fujita E, Fujita T, Shibuya M (1966) Isolation of enmein and its 3-acetate from *Isodon japonicus*. Chem Commun 297
95. Kubota T, Matsuura T, Tsutsui T, Uyeo S, Irie H, Numata A, Fujita T, Suzuki T (1966) Constitution and stereochemistry of enmein, a diterpene from *Isodon trichocarpus* Kudo. Tetrahedron 22:1659–1699
96. Iitaka Y, Natsume M (1966) The crystal and molecular structure of acetylbromoacetyldihydroenmein. Acta Crystallogr 20:197–210
97. Arai T, Koyama Y, Suenaga T, Morita T (1963) Antitumor activity of the components of *Isodon trichocarpus* and related plant. J Antibiot [A] (Tokyo) 16:132–138
98. Kubo I, Miura I, Kamikawa T, Isobe T, Kubota T (1977) The structure of kamebnin, a new antitumor ent-kauranoid from *Isodon kameba* Okuyama. Chem Lett 1289–1292
99. Kupchan SM, Davies VH, Fujita T, Cox R, Restivo RJ, Bryan RF (1973) The isolation and structural elucidation of liatrin, a novel antileukemic sesquiterpene lactone from *Liatris chapmanii*. J Org Chem 38:1853–1858
100. Kupchan SM (1970) Recent advances in the chemistry of terpenoid tumor inhibitors. Pure Appl Chem 21:227–246
101. Hill RK, Foley PJ Jr, Gardella LA (1967) The absolute configuration of sarcomycin. J Org Chem 32:2330–2335
102. Fujita E, Nagao Y, Kaneko K, Nakazawa S, Kuroda H (1976) The antitumor and antibacterial activity of the *Isodon* diterpenoids. Chem Pharm Bull (Tokyo) 24:2118–2127
103. Fujita E, Nagao Y, Node M, Kaneko K, Nakazawa S, Kuroda H (1976) Antitumor activity of the *Isodon* diterpenoids; structural requirements for the activity. Experientia 32:203–206
104. Wang MY, Lin C, Zhang TM (1985) Cytokinetic effects of oridonin on leukemia L1210 cells. Acta Pharmacol Sin 6:195–198
105. Wang MY, Shou MG, Wang QD, Wang Q, Zhang TM (1985) Effect of oridonin on DNA, RNA and protein syntheses in vitro. Acta Acad Med Henan 20:15–18
106. Wang MY, Lin C, Zhang TM (1987) Autoradiographic study on the effects of oridonin on DNA, RNA and protein syntheses of leukemia L1210 cells. Acta Pharmacol Sin 8:164–165

Rehmannia glutinosa Libosch.

106

106.1 Introduction

Dihuang, Radix Rehmanniae, is the fresh or dry root of *Rehmannia glutinosa* Libosch. (Scrophulariaceae). This officially in the Chinese Pharmacopoeia listed herbal drug is to be used in fresh or dried form or after processing. Processing is carried out by boiling in the rice wine or steaming. The fresh and dry roots of *R. glutinosa* are used in traditional Chinese medicine as an antipyretic and hemostatic. The processed roots are used mainly as a tonic and sedative.

106.2 Chemical Constituents

The main constituents in the root of *R. glutinosa* are iridoid glycosides. The first iridoid glycoside isolated from fresh tubers was catalpol (*106-1*) [1].

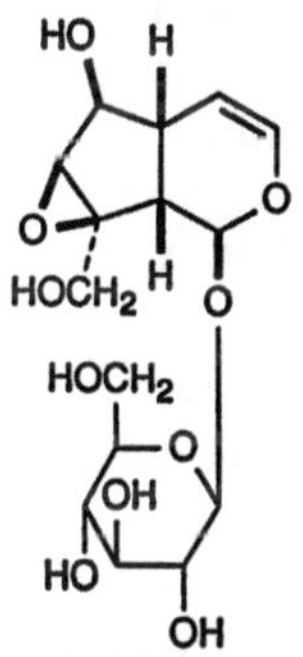

Catalpol (*106-1*)

A systematic study of the iridoid glycosidic constituents in the root of *R. glutinosa* led to the isolation of four new iridoid glycosides named rehmannioside A (*106-2*), rehmannioside B (*106-3*), rehmannioside C (*106-4*), and rehmannioside D (*106-5*) together with known compounds of the same class, catalpol, ajugol (*106-6*), aucubin, and melittoside (*106-7*). The structures of rehmanniosides A, B, C, and D were determined as 6′-*O*-α-D-galactopyranosylcatalpol, 6-*O*-α-D-galactopyranosylcatalpol, 7-*O*-α-D-galactopyranosyl-ajugol, and sophorosylmonomelittoside, respectively [2].

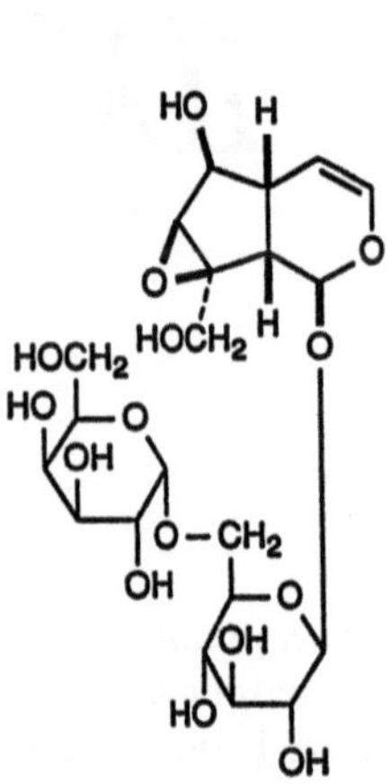

Rehmannioside A (*106-2*)

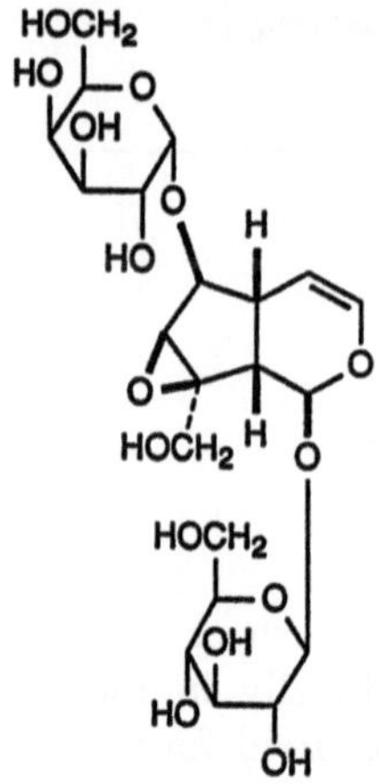

Rehmannioside B (*106-3*)

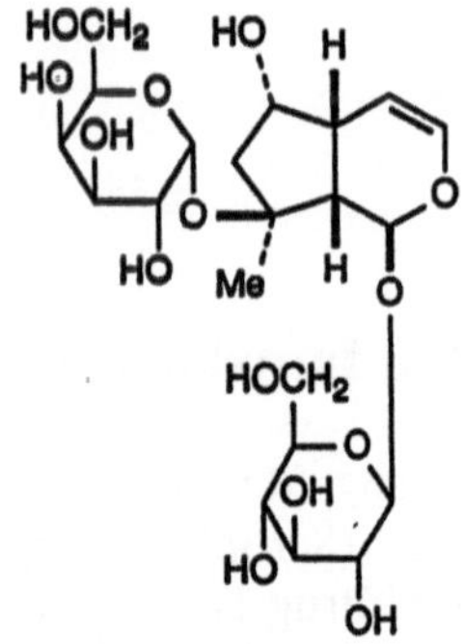

Rehmannioside C (*106-4*)

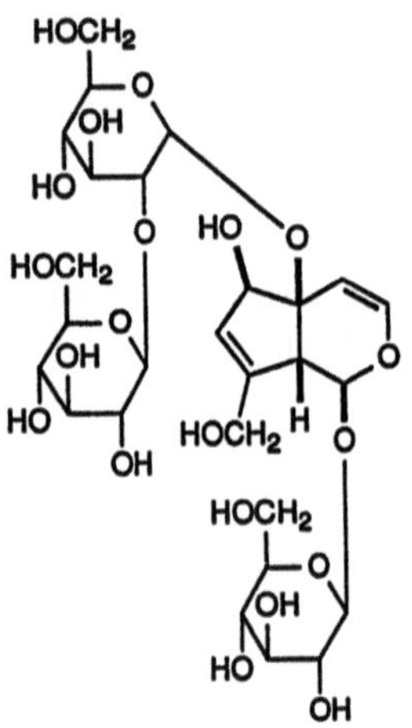

Rehmannioside D (*106-5*)

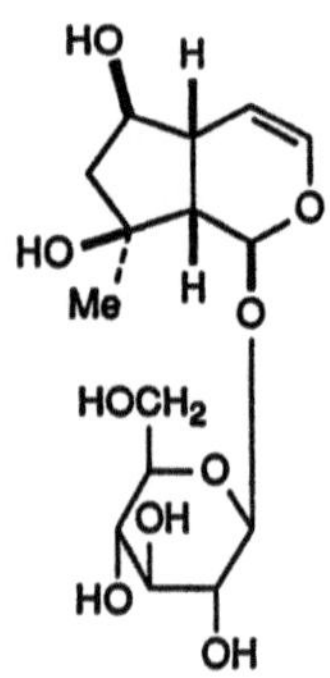

Ajugol (*106-6*)

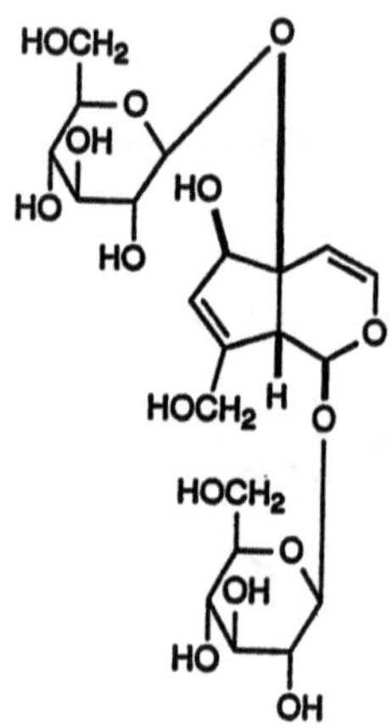

Melittoside (*106-7*)

Recently, the isolation of six ajugol esters (*106-8*–*106-13*) from the root of *R. glutinosa* var. *purpurea* [3] and two new iridoid glycosides named jioglutoside A (*106-14*) and jioglutoside B (*106-15*) from the root of *R. glutinosa* var. *hueichingensis* [4] was reported.

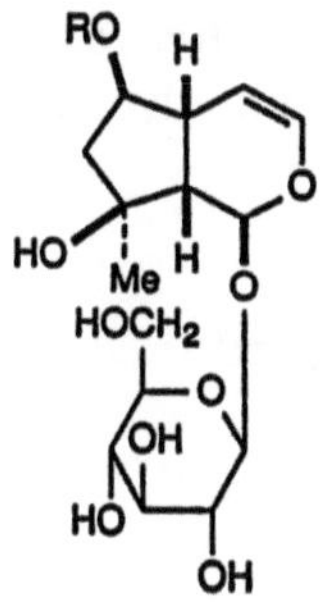

E-Feruloylajugol (*106-8*): R=

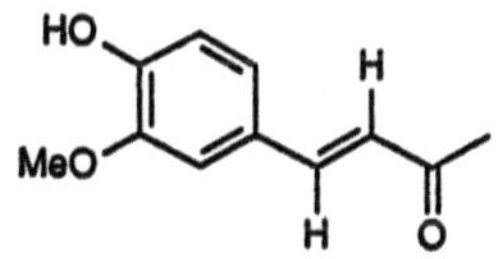

Z-Feruloylajugol (*106-9*): R=

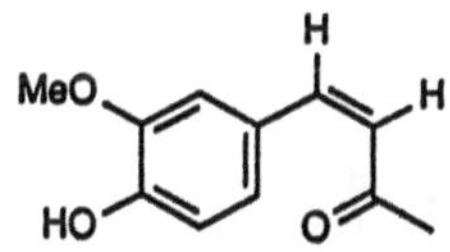

p-Coumaroylajugol (*106-10*): R=

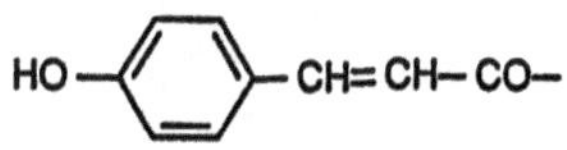

p-Hydroxybenzoylajugol (*106-11*): R=

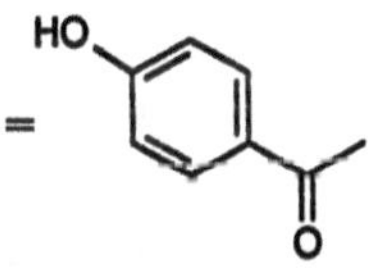

Vanilloylajugol (*106-12*): R=

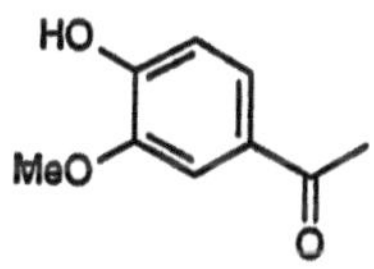

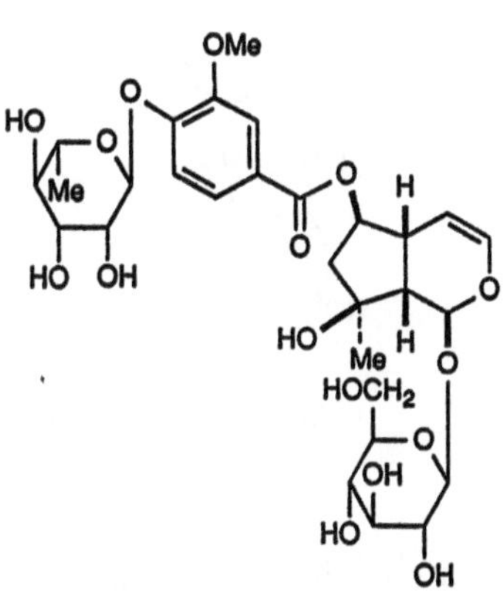

4-(α-L-Rhamnopyranosyloxy)-3-methoxybenzoylajugol (*106-13*)

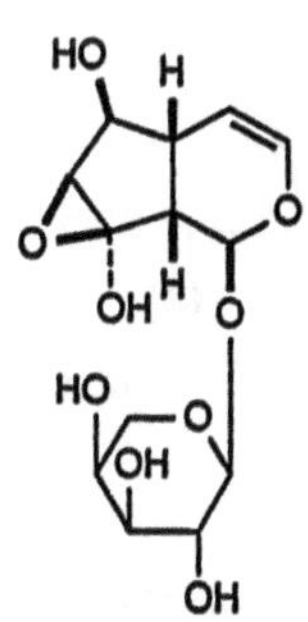

Jioglutoside A (*106-14*)

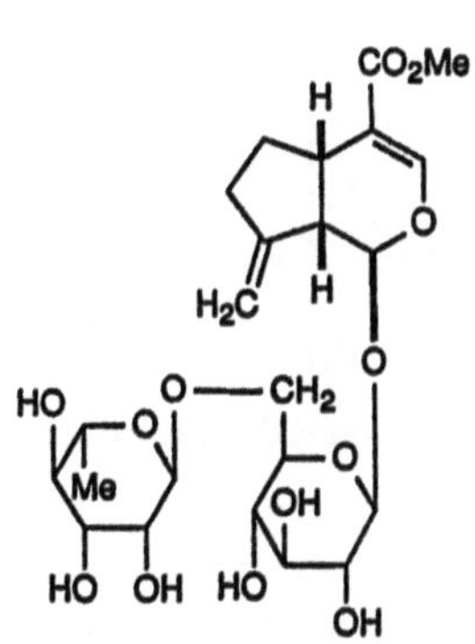

Jioglutoside B (*106-15*)

Furthermore, iridoids named rehmaglutins A (*106-16*), B (*106-17*), C (*106-18*), and D (*106-19*), and a chlorine-containing iridoid glycoside, glutinoside (*106-20*), were isolated and structurally determined [5, 6].

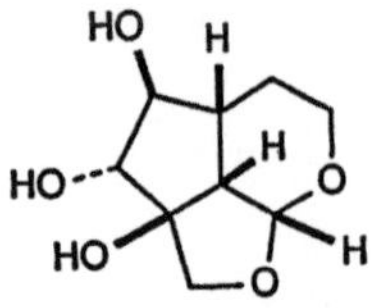

Rehmaglutin A (*106-16*)

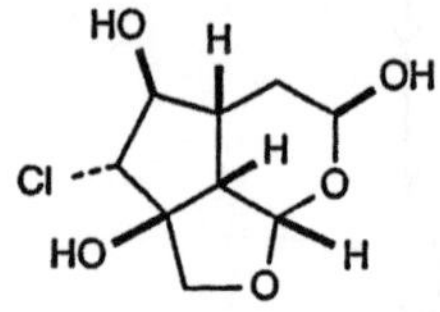

Rehmaglutin B (*106-17*)

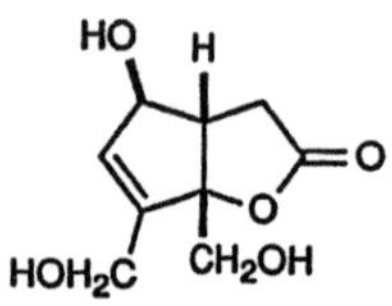

Rehmaglutin C (*106-18*)

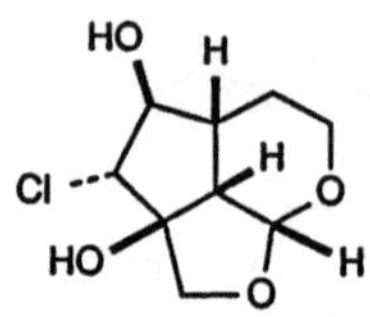

Rehmaglutin D (*106-19*)

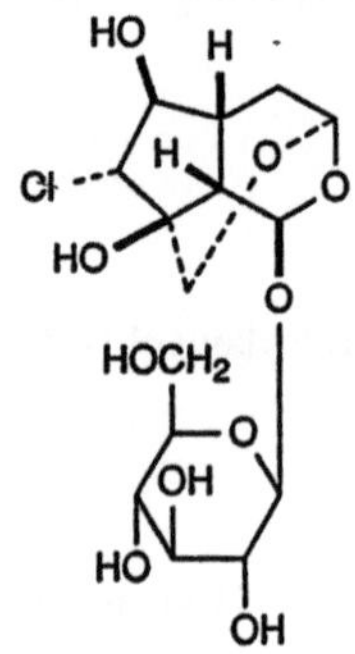

Glutinoside (*106-20*)

Three ionone glucosides, rehmaionosides A (*106-21*), B, and C (*106-22*), and a monoterpene glucoside rehmapicroside (*106-23*) were further constituents isolated from the roots of *R. glutinosa* [7]. Rehmaionosides A and B are stereoisomers. Rehmaionoside A shows 3′S configuration, whereas rehmaionoside B possesses 3′R configuration.

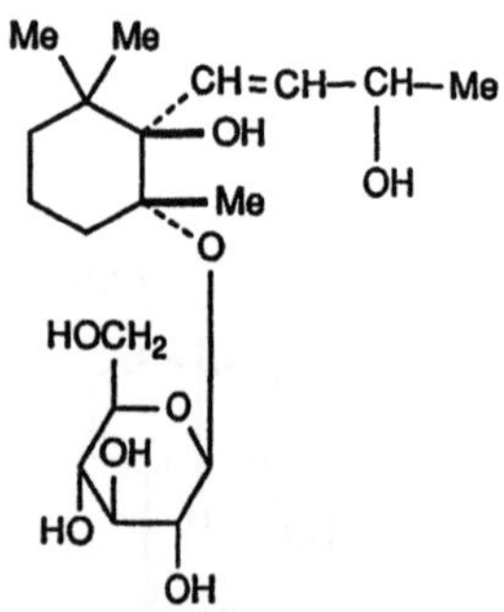

Rehmaionoside A (*106-21*)

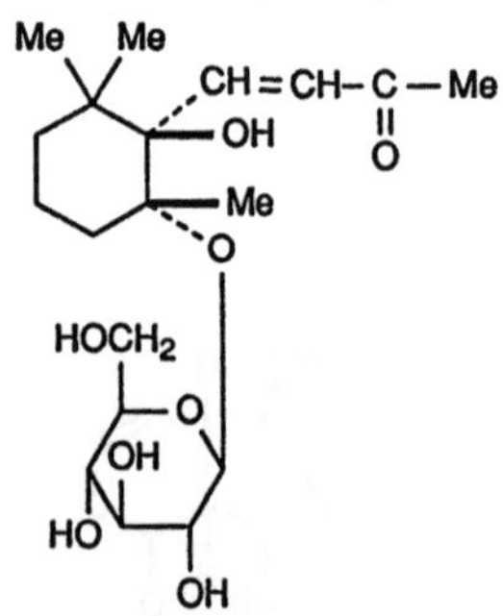

Rehmaionoside C (*106-22*)

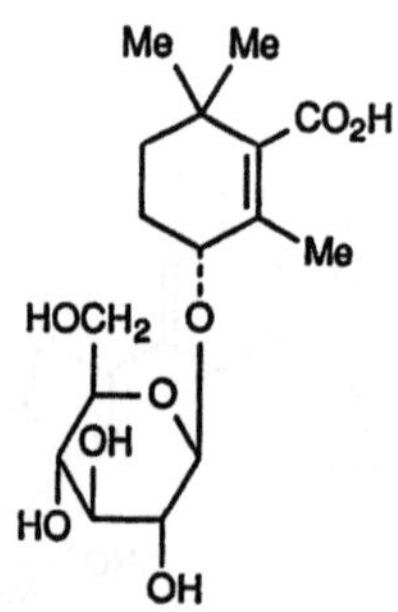

Rehmapicroside (*106-23*)

The aboveground part of *R. glutinosa* was found to contain catalpol, ajugol, aucubin, dihydrocatalpol, and monomelittoside [5]. The flavones chrysoeriol (*106-24*) and luteolin were isolated from the leaves [8].

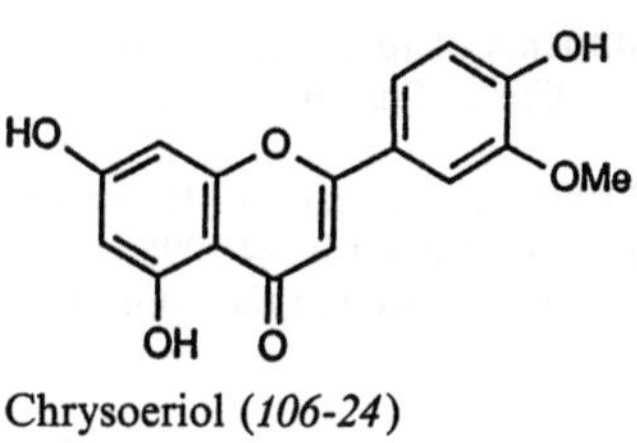

Chrysoeriol (*106-24*)

In addition to the iridoid glycosides and related compounds, the aqueous extract of the root of *R. glutinosa* was found to contain a number of amino acids, D-glucosamine, phosphoric acid, and carbohydrates, including D-glucose, D-fructose, sucrose, manninotriose, raffinose, stachyose, verbascose, and D-mannitol. Stachyose is the main component of the aqueous extract [9, 10]. The isolation of 1-ethyl-β-D-galactopyranoside from *R. glutinosa* was also reported [11].

Monosaccharide contents in raw root and in processed root were found to be quite different. In raw and processed root decoctions of *R. glutinosa* contents of 16% and 52% were found, respectively [12].

106.3 Pharmacology

Of the iridoid glycosides isolated from *R. glutinosa*, rehmannioside D showed weak hypoglycemic activity in spontaneously diabetic mice [2].

References

1. Kitagawa I, Nishimura T, Furubayashi A, Yosioka I (1971) Constituents of rhizome of *Rehmannia glutinosa* f. *hueichingensis*. Yakugaku Zasshi 91:593–596
2. Oshio H, Inouye H (1981) Iridoid glycosides of *Rehmannia glutinosa*. Phytochemistry 21:133–138
3. Nishimura H, Sasaki H, Morota T, Chin M, Mitsuhashi H (1989) Six iridoid glycosides from *Rehmannia glutinosa*. Phytochemistry 28:2705–2709
4. Morota T, Sasaki H, Nishimura H, Sugama K, Chin M, Mitsuhashi H (1989) Two iridoid glycosides from *Rehmannia glutinosa*. Phytochemistry 28:2149–2153
5. Kitagawa I, Fukuda Y, Taniyama T, Yoshikawa M (1986) Absolute stereostructures of rehmaglutins A, B, and D, three new iridoids isolated from Chinese rehmanniae radix. Chem Pharm Bull (Tokyo) 34:1399–1402
6. Yoshikawa M, Fukuda Y, Taniyama T, Kitagawa I (1986) Absolute stereostructures of rehmaglutin C and glutinoside, a new iridoid glucoside from Chinese rehmanniae radix. Chem Pharm Bull (Tokyo) 34:1403–1406
7. Yoshikawa M, Fukuda Y, Taniyama T, Cha BC, Kitagawa I (1986) Absolute configurations of rehmaionosides A, B, and C and rehmapicroside, three new ionone glucosides and a new monoterpene glucoside from Rehmanniae radix. Chem Pharm Bull (Tokyo) 34:2294–2297
8. Hasegawa T, Koike K, Takahashi S, Ariyoshi U (1982) Constituents of leaves and roots of Kaikei Jio (*Rehmannia glutinosa* Libosch. forma *hueichingensis* Hsiao). Shoyakugaku Zasshi 36:1–5
9. Tomoda M, Kato S, Onuma M (1971) Water-soluble constituents of rehmanniae radix. I. Carbohydrates and acids of *Rehmannia glutinosa* f. *hueichingensis*. Chem Pharm Bull (Tokyo) 19:1455–1460

10. Tomoda M, Tanaka M, Kondo N (1971) Water-soluble constituents of rehmannia radix. II. Constituents of roots of *Rehmannia glutinosa* var. *purpurea*. Chem Pharm Bull (Tokyo) 19:2411–2413
11. Wu SJ, Xu SM, Li YC, Li DY, Zhao GZ (1984) Studies on the chemical constituents of Huaiqing rehmannia (*Rehmannia glutinosa* f. *huaichingensis*). Chin Trad Herb Drugs 15:294–296
12. Liu ZY (1984) Monosaccharide content in raw and processed roots of *Rehmannia glutinosa*. Bull Chin Mater Med 9:17–18

Rheum spp. 107

107.1 Introduction

Dahuang, Radix et Rhizoma Rhei, is the dry root and rootstock of *Rheum palmatum* L., *R. tanguticum* Maxim. ex Balf. or *R. officinale* Baill. (Polygonaceae), collected in the late fall, when the aboveground part has withered or in early spring, before sprouting. It is officially listed in the Chinese Pharmacopoeia. The rhubarb root is one of the oldest and best-known Chinese herbal medicines and is used or recommended as a laxative, antiphlogistic, and hemostatic in the treatment of obstipation, gastrointestinal indigestion, diarrhea, and jaundice. Further indications include bleeding of the gastrointestinal tract, menstrual disorders, conjunctivitis, traumatic diseases, carbuncle, and ulcer. It can also be used for treatment of thermal burn by external application. It is utilized as raw material or after processing.

Dahuang Liujingao, Extractum Rhei liquidum, the fluid extract prepared by percolating the root powder with 60% ethanol, is also officially listed in the Chinese Pharmacopoeia and is mainly used as a laxative and stomachic.

107.2 Chemical Constituents

The most important constituents from rhubarb root are the anthraquinone derivatives. The first anthraquinone, chrysophanol (*107-1*) [1], was reported about one and half centuries ago. The anthraquinones isolated from *R. palmatum, R. officinale, R. tanguticum*, and other medicinal *Rheum* species are listed in Table 107.1. The anthraquinone derivatives occur in the rhubarb root mainly as glycosides. Names and structures are given in Table 107.2 together with their plants of origin.

Table 107.1. Anthraquinone derivatives isolated from *Rheum palmatum*, *R. officinale*, *R. tanguticum*, and other *Rheum* species

Compound	Structure	Plant origin	Reference
Chrysophanol (*107-1*)	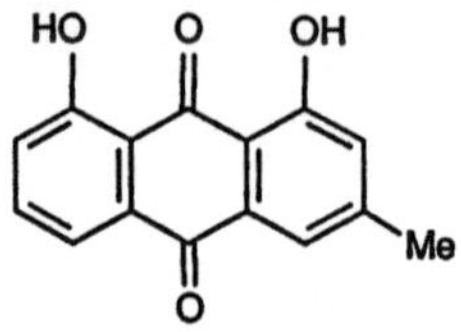	*Rheum* sp. *R. palmatum* *R. officinale* *R. tanguticum*	[1–7] [8, 9] [8, 10] [11]
Aloe emodin (*107-2*)	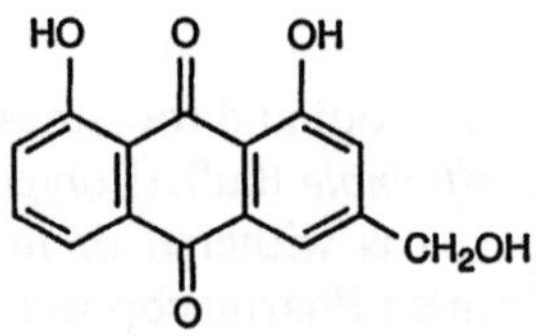	*Rheum* sp. *R. palmatum* *R. officinale* *R. tanguticum*	[5, 7] [8, 9] [8, 10] [11]
Emodin (*107-3*)	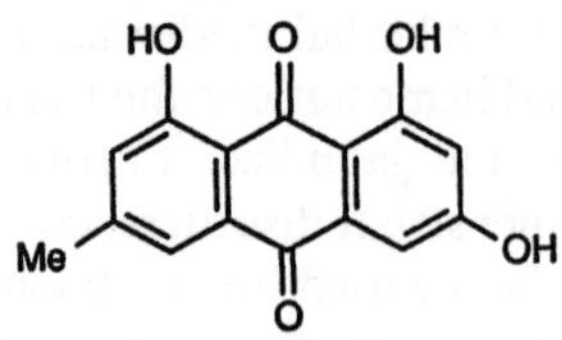	*Rheum* sp. *R. palmatum* *R. officinale* *R. tanguticum*	[5, 7, 12] [8, 9] [10] [11]
Physcion (*107-4*)	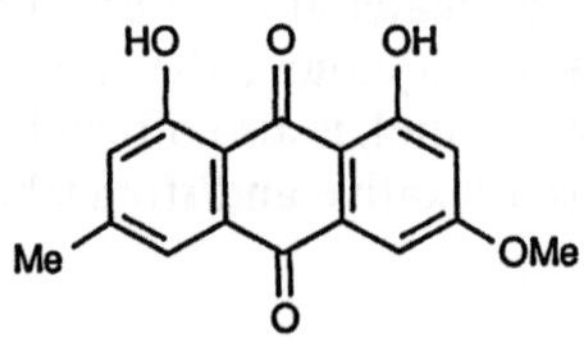	*Rheum* sp. *R. palmatum* *R. officinale* *R. tanguticum*	[5, 7, 13] [9] [10] [11]
Rhein (*107-5*)	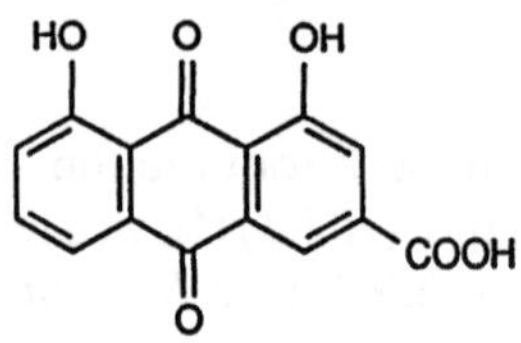	*Rheum* sp. *R. palmatum* *R. officinale* *R. tanguticum*	[5, 7] [8, 9, 14–16] [8, 10] [11]
Citreorosein (*107-6*)	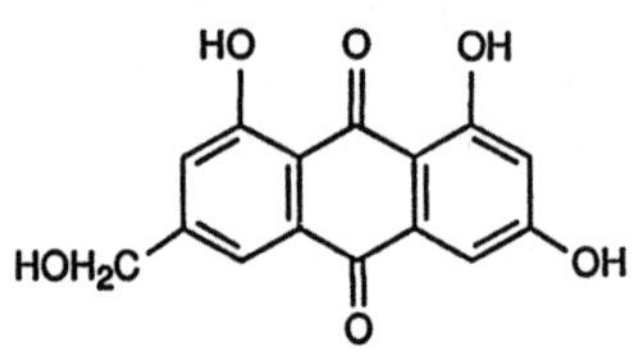	*Rheum* sp.	[17]
Alizarin (*107-7*)	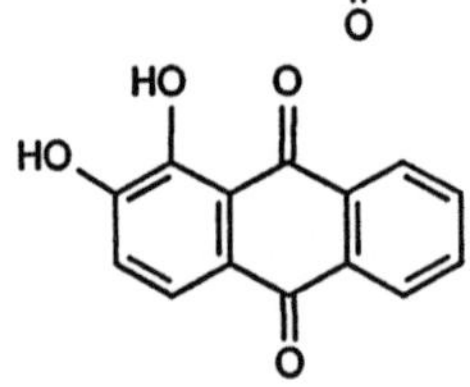	*Rheum* sp.	[18]
Laccaic acid D (*107-8*)	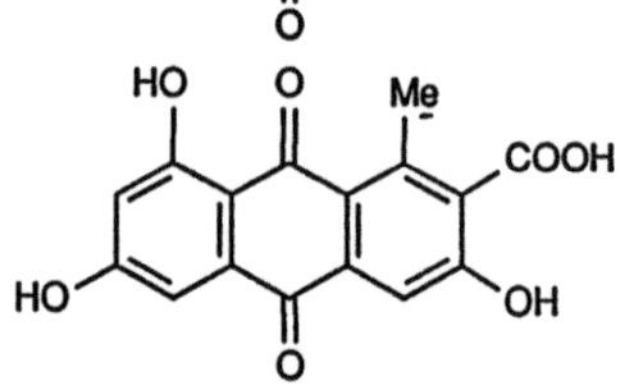	*Rheum* sp.	[17]

Table 107.2. Anthraquinone glycosides from *Rheum palmatum* and other *Rheum* species

Compound	Structure	Plant origin	Reference
Physcion 1-*O*-β-D-glucopyranoside (*107-9*)	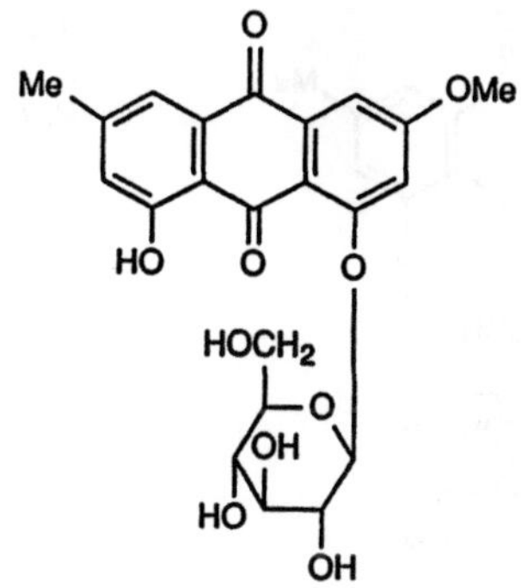	*Rheum* sp. *R. palmatum* *R. tanguticum*	[19, 20] [21–23] [24, 25]
Aloe emodin 1-*O*-β-D-glucopyranoside (*107-10*)	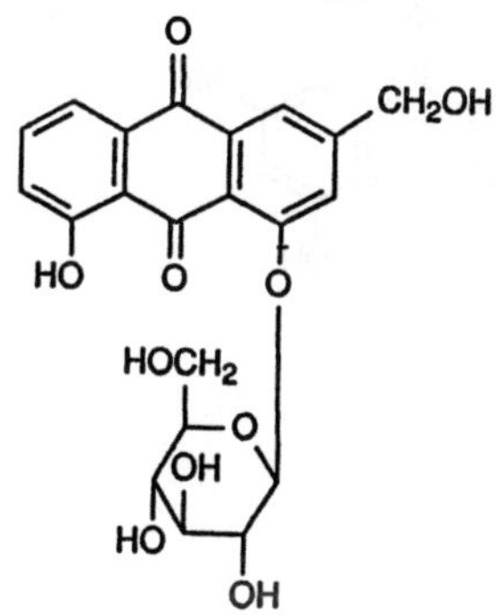	*Rheum* sp. *R. palmatum* *R. tanguticum*	[20] [21–23] [24–27]
Emodin 8-*O*-β-D-glucopyranoside (*107-11*)	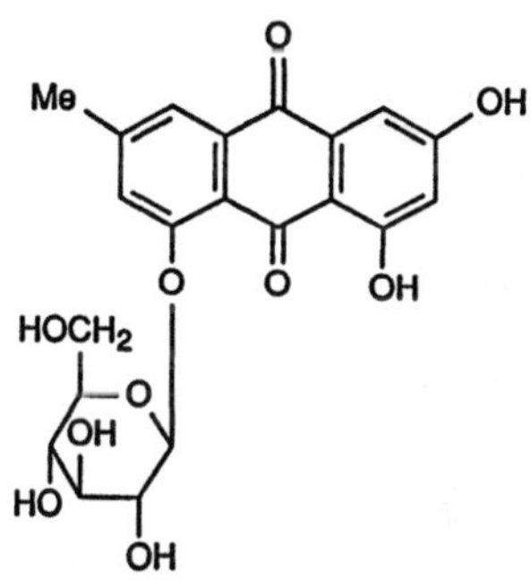	*Rheum* sp. *R. tanguticum*	[19, 20] [26]
Emodin 1-*O*-β-D-glucopyranoside (*107-12*)	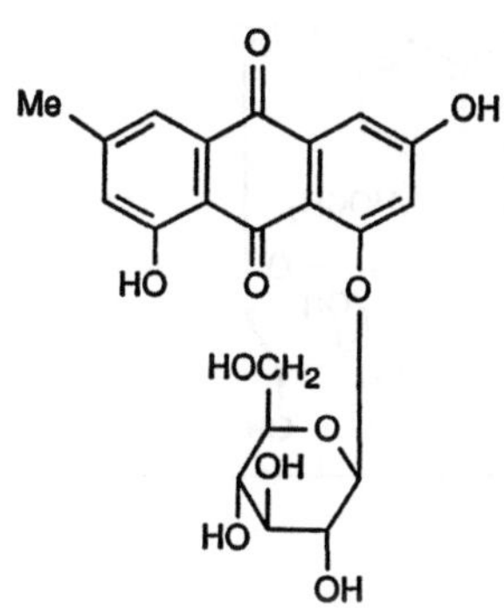	*Rheum* sp.	[20]

Table 107.2. (continued)

Compound	Structure	Plant origin	Reference
Chrysophanol 1-*O*-β-D-glucopyranoside (Chrysophanein, *107-13*)	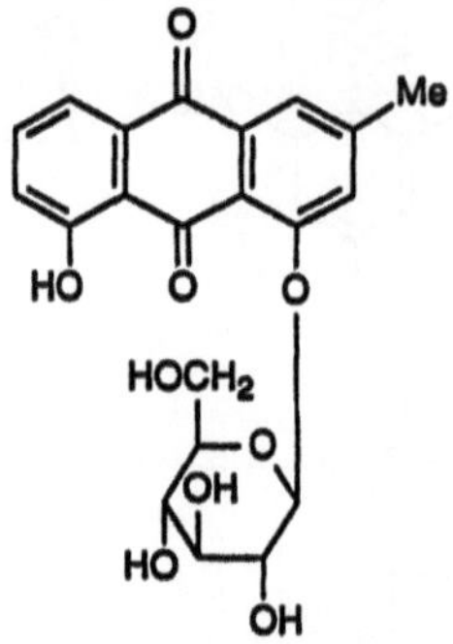	*Rheum* sp. *R. palmatum* *R. tanguticum*	[19, 20] [21–23, 28] [24, 25, 29]
Chrysophanol 8-*O*-β-D-glucopyranoside (*107-14*)	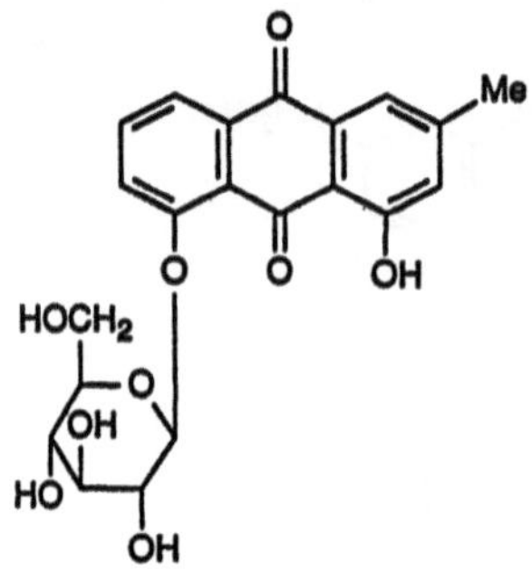	*Rheum* sp.	[20]
Rhein 1-*O*-β-D-gluco-pyranoside (*107-15*)	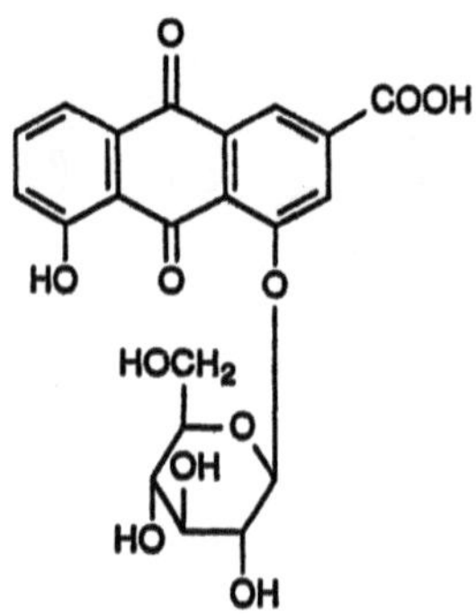	*Rheum* sp. *R. tanguticum*	[19, 20] [25, 29]
Aloe emodin 1′-*O*-β-D-glucopyranoside (*107-16*)	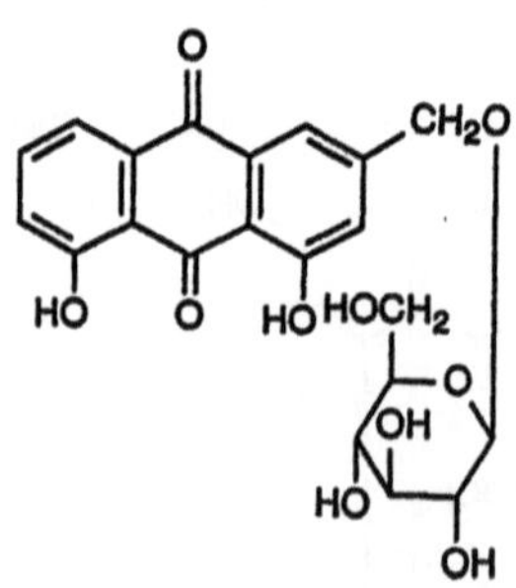	*Rheum* sp.	[20]

Table 107.2. (continued)

Compound	Structure	Plant origin	Reference
Emodin 3-*O*-β-D-glucopyranoside (Glucoemodin, *107-17*)	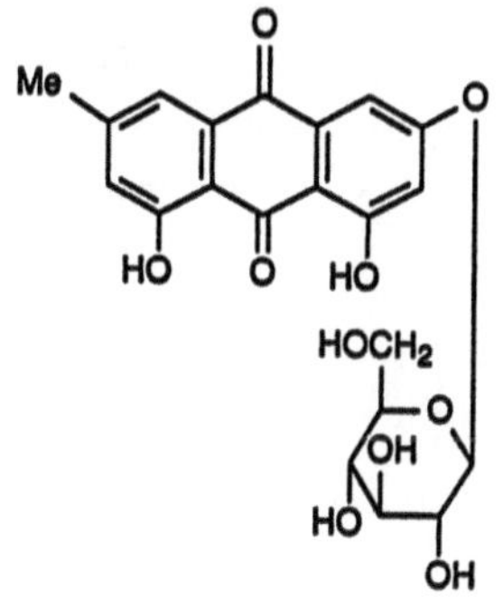	*R. palmatum* *R. tanguticum*	[30] [25]
Physcion 8-*O*-β-D-gentiobioside (*107-18*)	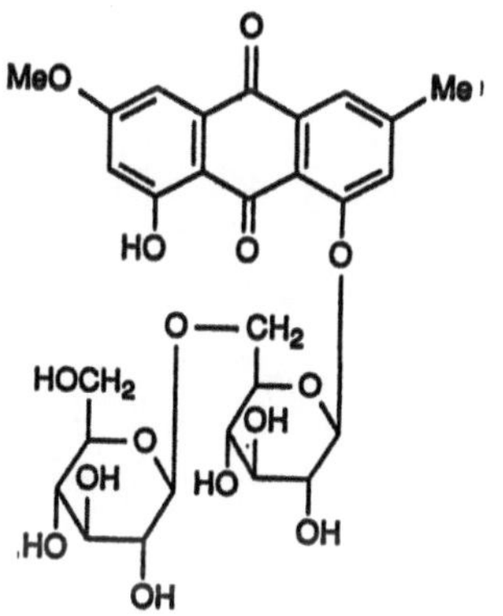	*Rheum* sp. *R. palmatum*	[31] [32]

Bianthrones (dianthrones) are another group of compounds isolated from rhubarb root, which are derived from the five major anthraquinone derivatives physicon, rhein, emodin, aloe emodin, and chrysophanol together with isobianthrones and heterobianthrones. The isolation of a number of bianthrones from the root of *R. palmatum* was reported [33]. These bianthrone compounds also occur mainly as glycosides and can be converted to the corresponding aglycones by acid hydrolysis. Catalytic hydrogenation of bianthrone glycosides yields two molecules of corresponding anthrone glycosides. Oxidative decomposition of bianthrone aglycones yields two molecules of the corresponding anthraquinones (Fig. 107.1) [34]. Bianthrones and their glycosides isolated from the root of *R. palmatum* are listed in Table 107.3.

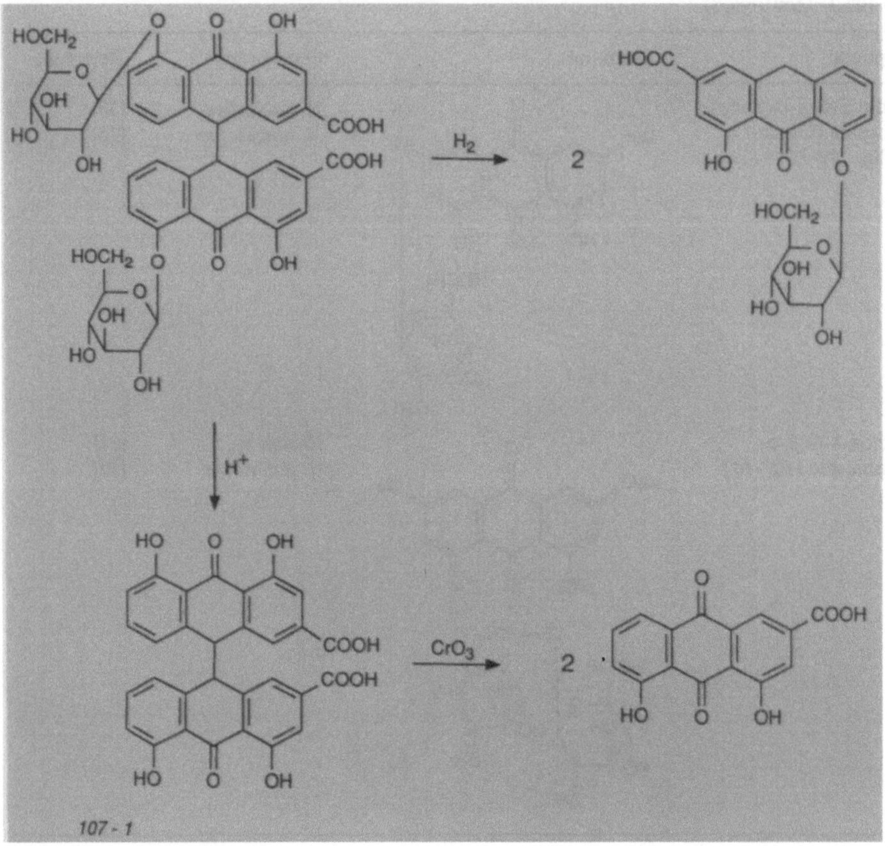

Fig. 107.1. Catalytic hydrogenation of sennoside A and oxidative decomposition of sennidin A

Table 107.3. Bianthrones and their glycosides from *Rheum palmatum*

Compound	Structure	Reference
Chrysophanol bianthrone (*107-19*)	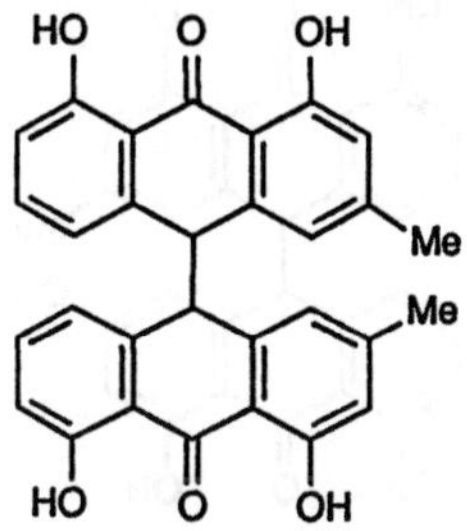	[35]
Aloe emodin bianthrone (*107-20*)	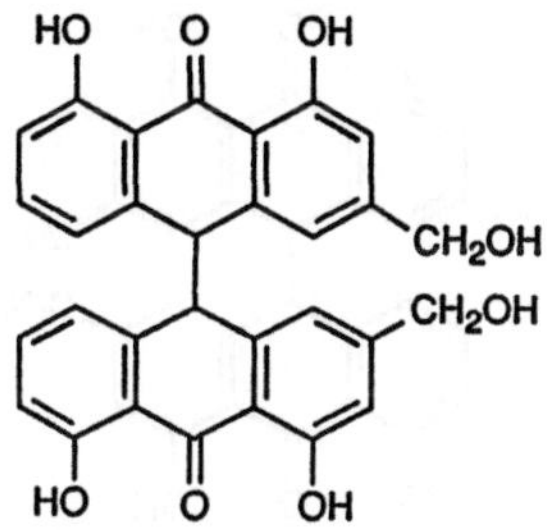	[32]
Sennidin A (R*, R*) (*107-21*)	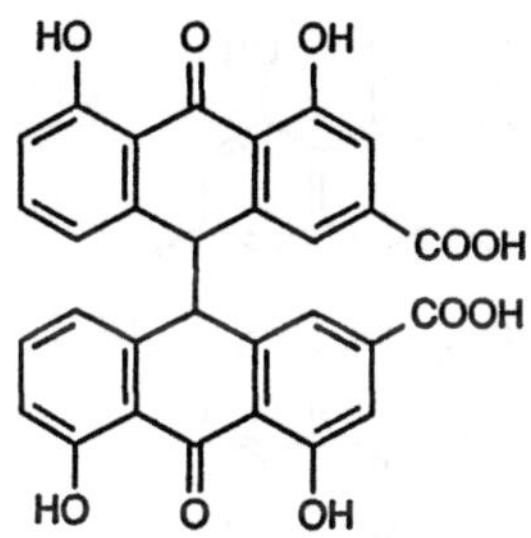	[41]
Sennidin B (R*, S*)		
Palmidin A (*107-22*)	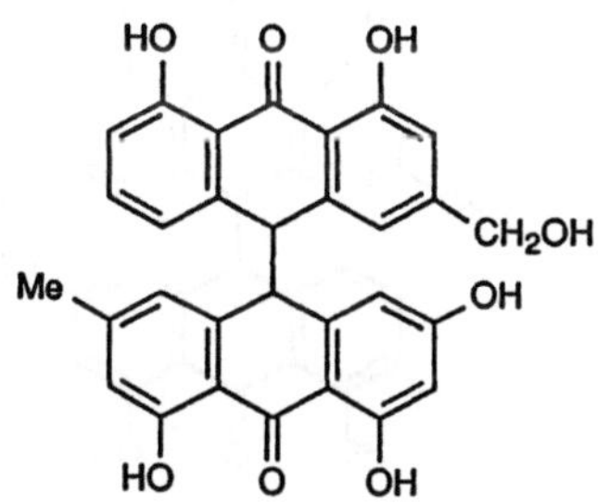	[36]
Palmidin B (*107-23*)	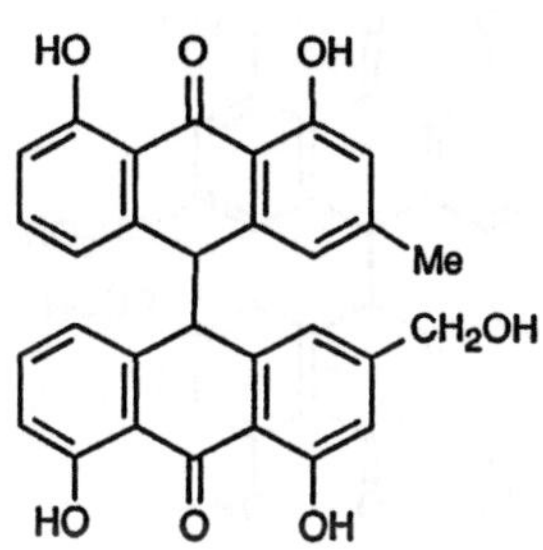	[36]

Table 107.3. (continued)

Compound	Structure	Reference
Palmidin C (*107-24*)	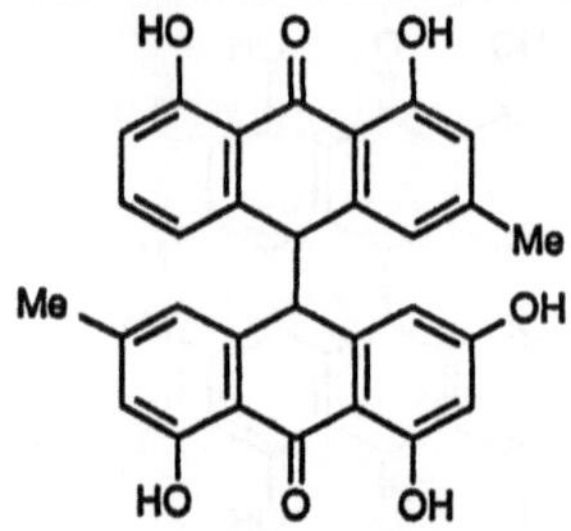	[36]
Rheidin A (*107-25*)	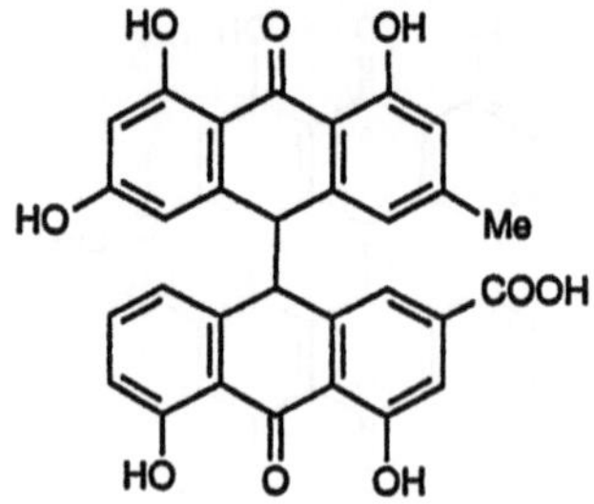	[14, 37]
Rheidin B (*107-26*)	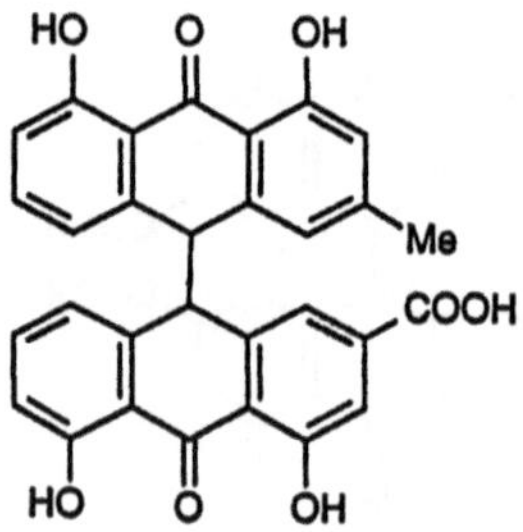	[14, 38]
Rheidin C (*107-27*)	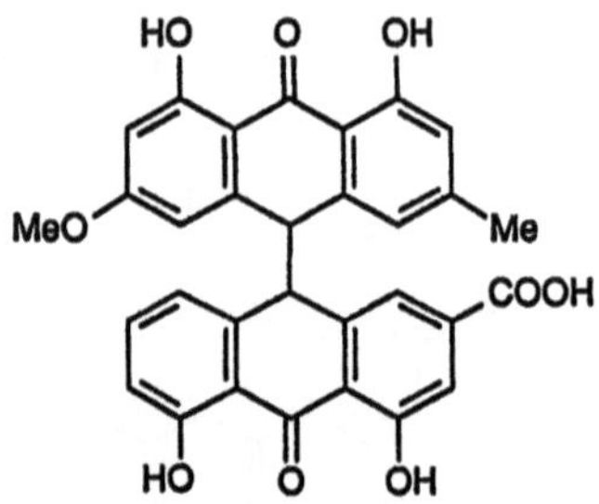	[14, 38]
Sennidin C (*107-28*)	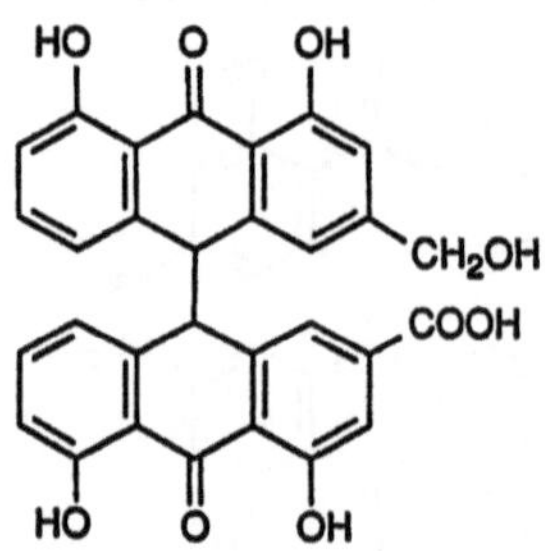	[14, 38]

Table 107.3. (continued)

Compound	Structure	Reference
Sennoside A (R*, R*) (*107-29*)	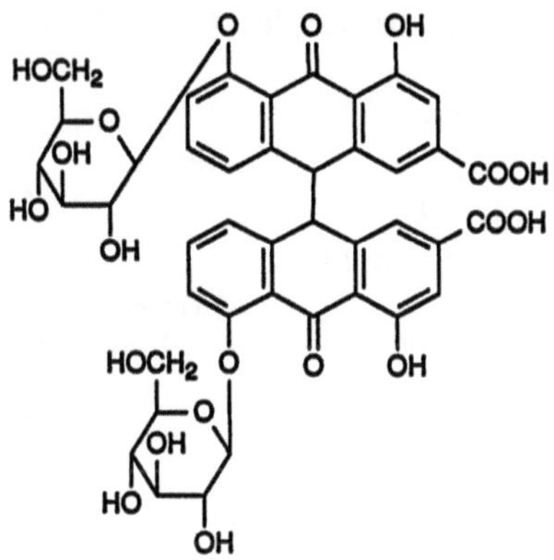	[39, 40] [32, 41]
Sennoside B (R*, S*)		[39, 41] [32]
Sennoside C (R*, R*) (*107-30*)	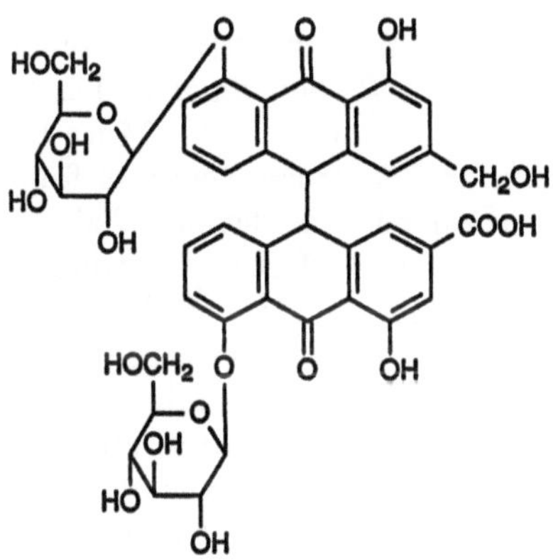	[32, 41]
Sennoside D (R*, S*)		[32, 41] [17]
Sennoside E (R*, R*) (*107-31*)	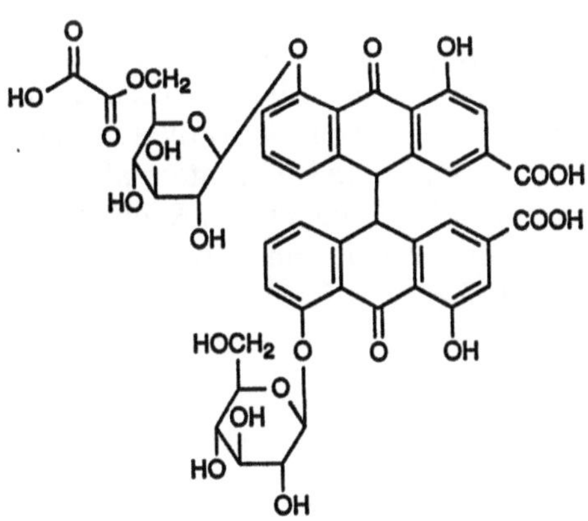	[42, 43]
Sennoside F (R*, S*)		[43]

Four anthrone *C,O*-diglycosides, named rheinosides A, B, C, and D, were isolated from rhubarb root. Rheinosides A (*107-32*) and B are steroisomeric 5-*O*-*β*-D-glucopyranosyl-9-*C*-*β*-D-glucopyranosyl 9-hydroxyrheinanthrones. Rheinosides C (*107-33*) and D differ from rheinosides A and B by the absence of the 9-hydroxy group [44].

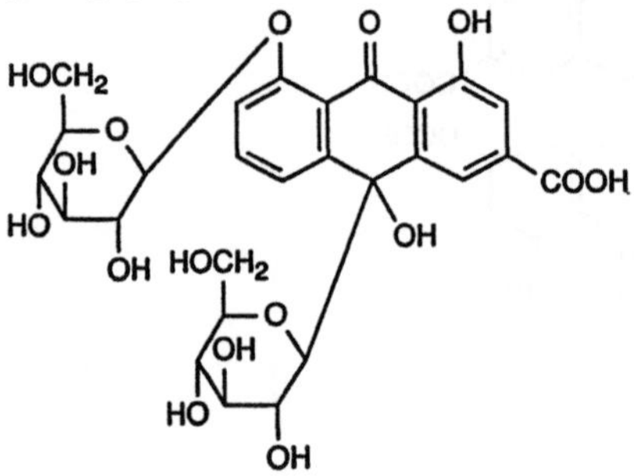

Rheinoside A (*107-32*)

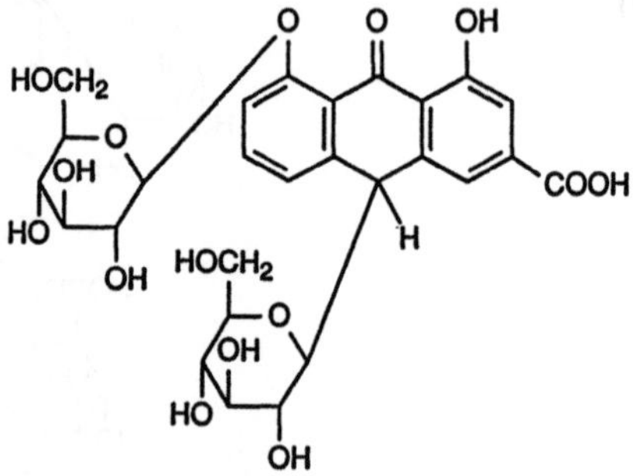

Rheinoside C (*107-33*)

Great variability in anthraquinone derivative contents was found within different species of Chinese rhubarb. *R. palmatum* had a total content of 3.4%, while *R. tanguticum* was found to contain 1.2%. More emodin than aloe emodin was found in *R. palmatum*, and *R. tanguticum* was found to contain a considerable amount of glycosidic rhein but was devoid of aloe emodin. Duration of storage appeared not to have any major effect on the contents [45].

The contents of rhein, aloe emodin, emodin, physcion, chrysophanol, and sennoside in processed root of *R. palmatum* were lower than those in raw material, the type of processing also being of some influence [46]. For the extraction of anthraquinones from the root of *R. palmatum*, ethanol was a more effective solvent than water [47]. The rhein content in *R. palmatum* and *R. tanguticum* was found to be 0.9%–1.7%, *R. officinale* contained 0.6%, whereas *R. hotaoense* contained only 0.1%. No rhein was detected in *R. emodi, R. franzenbachii, R. mobile, R. acuminatum, R. wittrochii*, and *R. lhasaense* [48].

Tannins are also present in rhubarb. The chemically rather heterogeneous class of tannins includes hydrolysable tannins, containing ester or glycosidic bonds composed of gallic acid, glucose, and other monosaccharides [49] and condensed tannins, derived primarily from the flavane derivatives catechin and leucocyanidin [50]. Samples of *R. palmatum* were found to contain about 11% tannin, whereas those of *R. emodi, R. franzenbachii, R. hotaoense, R. officinale*, and *R. tanguticum* contained 4%–7% [51]. Tannins isolated from *R. palmatum* or other medicinal *Rheum* species are listed in Table 107.4.

Table 107.4. Tannins isolated from rhubarb roots

Compound	Structure	Reference
3,3′-Di-*O*-galloyl-procyanidin B-2 (*107-34*)	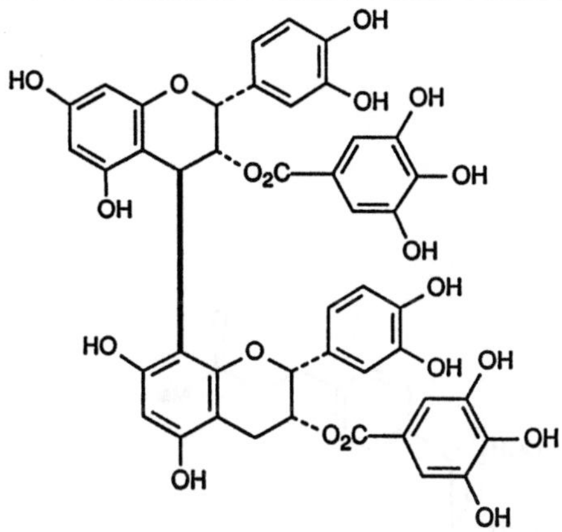	[52]
3-*O*-Galloylprocyanidin B-1 (*107-35*)	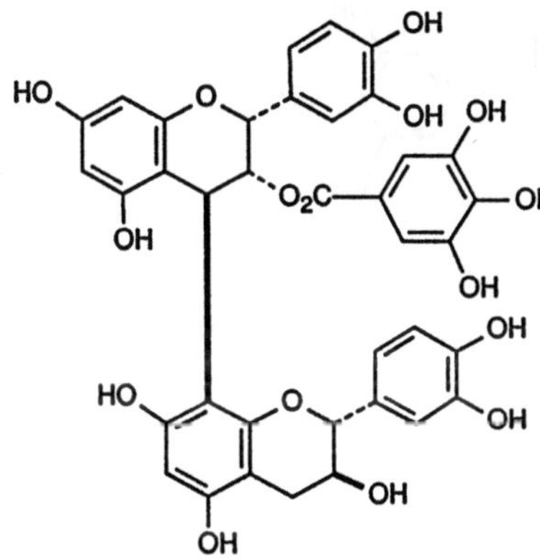	[52, 55]
1,2,6-Tri-*O*-galloyl-glucose		[52, 53]
Lindleyin (*107-36*)	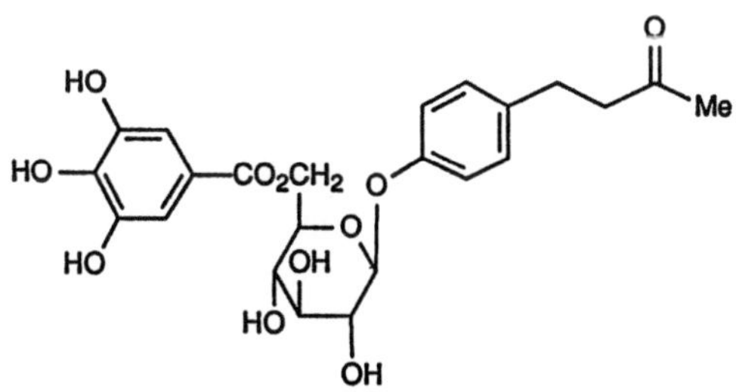	[52, 54]
3-*O*-Galloyl-epi-catechin		[52, 53]
Gallic acid		[52]
1-*O*-Galloylglycerol		[55]
Gallic acid 3-*O*-*β*-D-(6′-*O*-galloyl)-glucopyranoside		[55]

Table 107.4. (continued)

Compound	Structure	Reference
Gallic acid 4-*O*-*β*-D-(6′-*O*-galloyl)-glucopyranoside		[55]
1,6-Di-*O*-galloyl-D-glucose		[53, 55]
6-*O*-Galloylglucose		[53, 55]
Isolindleyin (*107-37*)	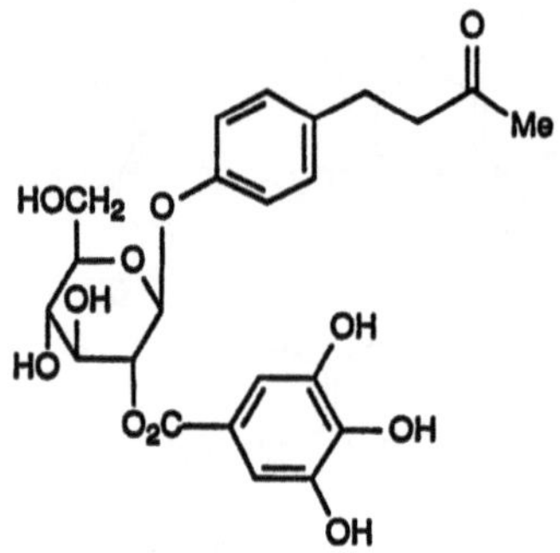	[54, 55]
Catechin-5-*O*-*β*-D-glucopyranoside		[56, 57]
Catechin-7-*O*-*β*-D-glucopyranoside		[56, 57]
2-*O*-Cinnamoyl-D-glucose		[53]
2-*O*-Cinnamoyl-1,6-di-*O*-galloyl-D-glucose		[53]
2-*O*-(*p*-Coumaroyl)-1-*O*-galloyl-D-glucose		[53]
1-*O*-Galloylfructose		[53]
2-*O*-Cinnamoyl-1-*O*-galloyl-D-glucose		[53]
1-*O*-Galloyl-D-glucose		[53]
2,6-Di-*O*-galloyl-D-glucose		[53]
3,5-Dihydroxyphenol 1-*O*-*β*-D-(6′-*O*-galloyl)-glucopyranoside		[53]

Table 107.4. (continued)

Compound	Structure	Reference
Procyanidin B2 6-*C*-*β*-D-glucopyranoside (*107-38*)	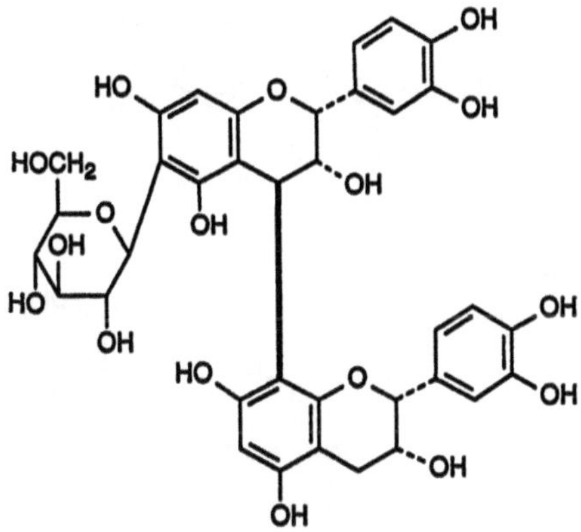	[57]
Procyanidin B2 8-*C*-*β*-D-glucopyranoside (*107-39*)	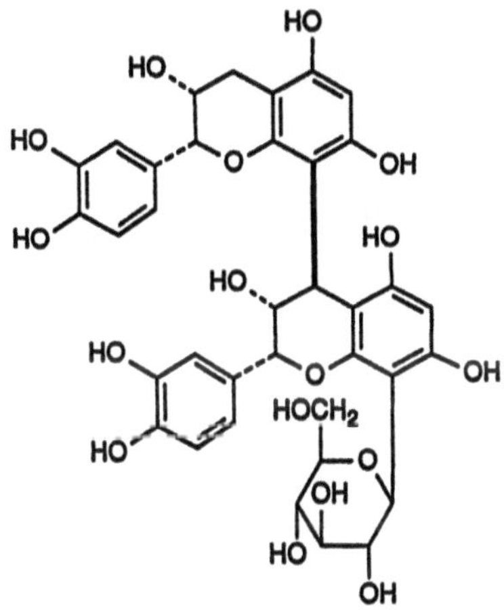	[57]
Catechin 3′-*O*-*β*-D-glucopyranoside		[57]
Catechin 4′-*O*-*β*-D-glucopyranoside		[57]
Catechin 7,3′-di-*O*-*β*-D-glucopyranoside		[57]
Catechin 5,3′-di-*O*-*β*-D-glucopyranoside		[57]
Catechin 3′,4′-di-*O*-*β*-D-glucopyranoside		[57]
Catechin 5,4′-di-*O*-*β*-D-glucopyranoside		[57]
Catechin 8-*C*-*β*-D-glucopyranoside		[57]
Catechin 6-*C*-*β*-D-glucopyranoside		[57]

Table 107.4. (continued)

Compound	Structure	Reference
Procyanidin B3 7-*O*-*β*-D-glucopyranoside (*107-40*)	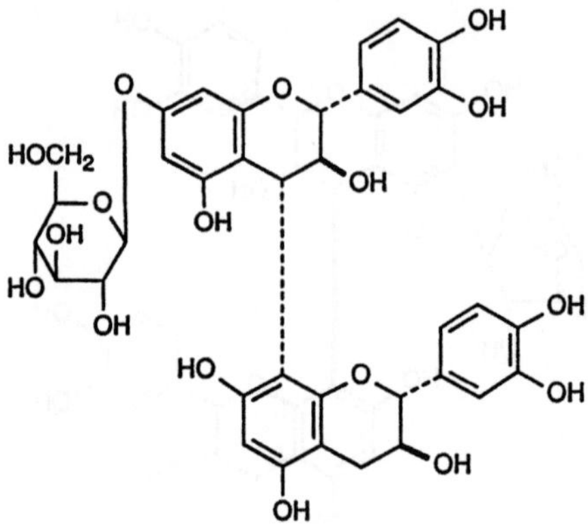	[57]
Procyanidin B1 8-*C*-*β*-D-glucopyranoside		[57]
Procyanidin B1 6-*C*-*β*-D-glucopyranoside		[57]
4-(4-Hydroxyphenyl)-2-butanone 4′-*O*-*β*-D-glucopyranoside		[54]
Gallic acid 3-*O*-*β*-D-glucopyranoside		[54]
Gallic acid 4-*O*-*β*-D-glucopyranoside		[54]
6-Galloylisolindleyin (*107-41*)	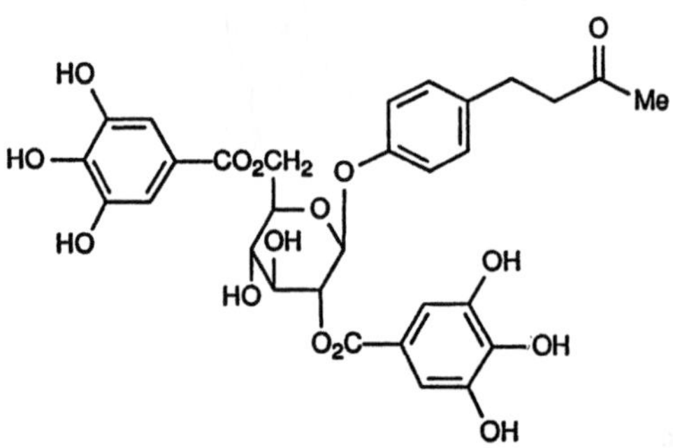	[54]
6-Cinnamoyliso-lindleyin (*107-42*)	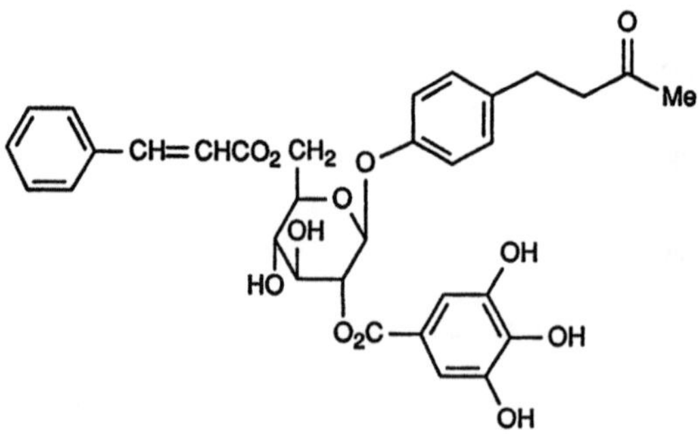	[54]

Table 107.4. (continued)

Compound	Structure	Reference
6-*p*-Hydroxycinnamoylisolindleyin (*107-43*)	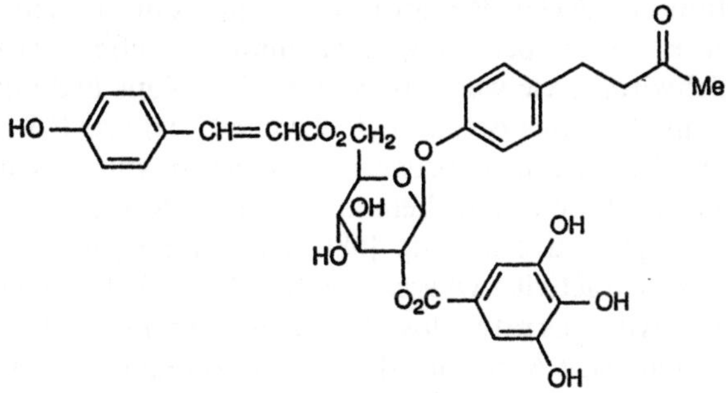	[54]

Chromone and chromanone derivatives were also isolated from rhubarb, including 2,5-dimethyl-7-hydroxychromone, 2-methyl-5-carboxymethyl-7-hydroxychromone, 2-(2-hydroxypropyl)-5-methyl-7-hydroxychromone, 2-(2-hydroxypropyl)-5-methyl-7-hydroxychromone 7-*O*-*β*-D-glucopyranoside, and 2-methyl-5-carboxymethyl-7-hydroxychromanone, together with torachrysone 8-*O*-*β*-D-glucopyranoside (*107-44*) [58].

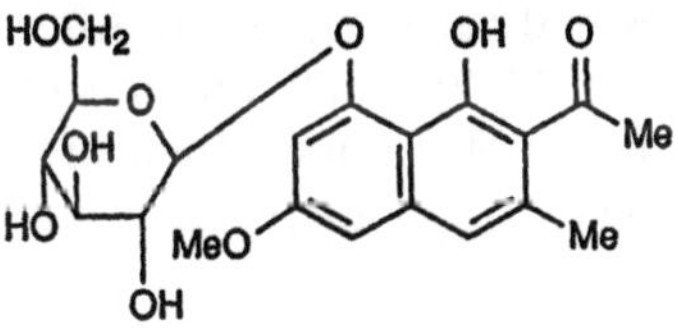

Torachrysone 8-*O*-*β*-D-glucopyranoside (*107-44*)

Besides anthraquinones and tannins, a number of stilbene compounds were isolated from rhubarb and were identified as piceid (*100-12*) [59], 3,5,4′-trihydroxystilbene 4′-*O*-*β*-D-glucopyranoside [60, 61], 3,5,4′-trihydroxystilbene 4′-*O*-*β*-D-(6″-*O*-galloyl)-glucopyranoside [52, 60], and 3,5,4′-trihydroxystilbene 4′-*O*-*β*-D-(2″-*O*-galloyl)-glucopyranoside [54].

107.3 Pharmacology

The anthraquinones aloe emodin, emodin, and rhein were found to inhibit in vitro the growth of *Bacillus subtilis* and *Staphylococcus aureus*, with rhein being the most effective compound [62]. The antimicrobial activity of rhein was also observed against *Escherichia coli*, *Micrococcus luteus*, *Candida albicans*, *Clostridium perfringens*, *Fusobacterium varium* and *Bacteriodes fragilis*, a major anaerobic microorganism in human intestinal flora [63, 64]. The antibacterial activity of the anthraquinone derivatives appears to result from inhibition of the mitochondrial respiratory chain of microorganisms. Respiration of *S. aureus* in ordinary broth was found to be strongly inhibited by emodin, aloe emodin, and rhein at a minimal growth inhibitory concentration [65]. Rhein, emodin, and aloe emodin specifically interfered with the redox function of NADH dehydrogenase. Inhibition of membrane-bound NADH dehydrogenase was competitive whereas inhibitory effects of the three compounds on malate dehydrogenase were noncompetitive. None had any effect on coenzyme Q – cytochrome C reductase and cytochrome C oxidase [66, 67].

A study on the correlation between chemical constituents and therapeutic effects of *Rheum* species showed that rhein and sennosides in *R. palmatum* appeared to be responsible for purgative activity [68]. Sennoside content in rhubarb correlated highly with laxative activity, whereas the correlation between anthraquinone content and laxative activity was rather low [69]. In mice, sennoside C had a stronger laxative activity than sennoside A. Sennoside C was also found to exert a potentiating effect on the activity of sennoside A [70]. In mice, the laxative ED_{50} of rhubarb was not influenced by sex [71]. Studies on the oxidized products of sennosides suggested that sennosides act predominantly on large intestine motility after their degradation by colonic microorganisms [72].

Studies on the metabolism of sennoside A and sennoside B in the intestinal contents and feces of rats and humans showed that sennosides are hydrolyzed by microbial β-glycosidase in a stepwise fashion to the corresponding sennidins via 8-monoglycosides. The resulting metabolites, sennidins A and B, were found to be interconvertible under the experimental conditions used, and were futher reduced to give rheinanthrone as the laxative principle. The latter reduction is possibly performed by a reductase bound to cell membranes of intestinal bacteria [73, 74]. After oral administration of sennoside A [75, 76] or sennoside B [77], small amounts of different anthracene derivatives were found in urine and feces. After oral administration of rhein to rats, rhein was detected together with the corresponding sulfate and glucuronide conjugates [75, 78].

After administration of a single oral dose of 91 mg/kg emodin to mice, approximately 30% of the dose was excreted as unchanged emodin or its derivatives in urine, and 21% in feces within 24 h. Only traces were excreted within 24–48 h (2%). The major free anthraquinone metabolites of emodin in 24 h urine were found to be 6-hydroxyaloe emodin, 6-hydroxyrhein, chrysophanol, and physicon [79]. Rhein, aloe emodin, and possibly 4-hydroxychrysophanol, 4-hydroxyemodin, 4-hydroxyrhein, 4,6-dihydroxyaloe emodin, and 4,6-dihydroxyemodin were also detected [80].

Urinary excretion of [^{14}C]emodin administered orally to rats at a dose of 50 mg/kg amounted to 18% and 22% after 24 and 72 h, respectively. The free anthraquinones emodin and 6-hydroxyrhein predominated in urine compared with only 3% of conjugates. In feces, 48% (24 h) and 68% (120 h) of the dose was excreted. Biliary excretion reached a maximum at about 6 h. Radioactivity levels in most organs decreased between days 3 and 5. In kidneys, however, they remained constantly high for 5 days after administration [81].

Physcion was metabolized by mouse and rat into emodin, chrysophanol, aloe emodin, rhein, 6-hydroxyaloe emodin, 6-hydroxyrhein, and, possibly, 6-methoxyaloe emodin and 6-methoxyrhein [80].

In addition to laxative and antimicrobial activities, several other pharmacological effects of rhubarb anthraquinones were observed in experimental studies. Oral administration of emodin and rhein provoked marked diuretic, natriuretic, and kaliuretic effects in rabbits [82]. The plasma corticosterone level was found to be increased in rats suffering from diarrhea induced by feeding rhubarb anthraquinones, and the relative adrenal weight increased in these animals [83]. Rabbits with fever induced by s.c. inoculation of pneumococci responded with decreased body temperature after oral administration of a decoction of rhubarb [84].

The rhizome of *R. palmatum* has also been used as a hemostatic in traditional Chinese medicine. This hemostatic activity has also been demonstrated experimentally and clinically [85, 86]. Rhubarb rhizome was also found to be effective in treatment and prevention of experimental gastric bleeding and ulcer formation in rats [87]. Significant therapeutic effects of the powdered rhizome of *R. palmatum* were also reported in the treatment of clinical gastrointestinal bleeding [88, 89].

No mutagenicity was found in an Ames test using *Salmonella typhimurium* strains TA97, TA98, TA100, and TA102 at different concentrations of a decoction of *R. palmatum* rhizome [90]. In contrast, emodin, chrysophanol, and 6-hydroxyrhein showed strong mutagenic activity in strain TA2637 with metabolic activation. Sennoside A was nonmutagenic in this test system with or without metabolic activation [91]. Mutagenicity was also reported for chrysophanol in strain TA1537 after metabolic activation [92] and for emodin in strains TA97, TA100, and TA1537 in the presence of liver homogenate [93]. No genotoxicity of emodin was found with or without metabolic activation in Chinese V79 hamster cells as a mammalian test system for sister chromatid exchange (SCE) and using the hypoxanthine guanine phosphoribosyltransferase forward mutation assay (HGPRT test) [93].

With an ED_{50} value of 20 µg/ml, emodin was a strong inhibitor of respiration in Ehrlich ascites carcinoma cells [94]. This was confirmed in leukemia L1210 cells [95]. Rhein showed moderate inhibition of respiration in Ehrlich ascites carcinoma cells, whereas aloe emodin and chrysophanol were only rather weak inhibitors [94]. It was further reported that the anthraquinone derivatives rhein, emodin, and aloe emodin had an in vivo inhibitory effect toward P388 leukemia in mice. The survival time of the animals tested was markedly increased and the ascites volume and tumor cell number were decreased [96].

References

1. Schlossberger J, Doepping O (1844) Chemische Untersuchung der Rhabarberwurzel. Liebigs Ann Chem 50:196–223
2. Liebermann C, Fischer O (1875) Über Chrysophansäure. Chem Ber 8:1102–1107
3. Fischer O, Falco F, Gross H (1911) Beitrag zur Kenntnis der Chrysophansäure. J Prakt Chem 83:208–214
4. Leger E (1912) Constitution of chrysophanic acid. Compt Rend 154:281–283
5. Tutin F, Clewer HWB (1908) Constituents of rhubarb. J Chem Soc 99:946–967
6. Oesterle OA, Johann U (1910) Chysophanic acid. Arch Pharm 248:492
7. He LY, Luo SR (1980) Studies on the analysis of anthraquinone derivatives of Chinese medicinal herbs. I. Separation and determination of constituents of Chinese rhubarb. Acta Pharm Sin 15:555–562
8. Takido M, Kojima Y (1957) Quantitative evaluation of anthraquinone pigments in *Rheum palmatum* and *Rheum officinale* by paper chromatography. Nippon Daigaku Yakugaku Kenkyu Hokoku 1:20–22 (CA 52:5554g)
9. He LY (1985) Determination of anthraquinone derivatives in rhubarb for export. Bull Chin Mater Med 10:33–35

10. Wang NL, Chen YJ (1983) Improvement in the separation method of anthraquinones from rhubarb. Bull Chin Mater Med 8:28–30
11. Romanova AS, Bankovskii AI (1963) Anthracene derivatives of *Rheum palmatum* var. *tanguticum*. I. Aglycones. Med Prom SSSR 17(3):13–17 (CA 59:7318c)
12. Hesse O (1908) Rhubarb and related substances. J Prakt Chem 77:383–390
13. Oesterle OA, Johann U (1910) Methylchrysophanic acid. Arch Pharm 248:476
14. Dequeker R, Lemli J, Cuveele J (1964) Anthraquinone drugs. VII. Densitometric determination of the sennidins, rheidins and rhein in the root of rhubarb. Planta Med 12:51–56
15. Hörhammer L, Wagner H, Köhler I (1958) The analysis of rhein from *Rheum palmatum*. Naturwissenschaften 45:389–390
16. Hörhammer L, Wagner H, Köhler I (1959) The analysis of the rheins. Arch Pharm (Weinheim) 292:591–601
17. Oshio H (1978) Investigation of rhubarbs. IV. Isolation of Sennoside D, citreorosein and laccaic acid D. Shoyakugaku Zasshi 32:19–23
18. Müller H (1908) Occurrence of alizarin in rhubarb. J Chem Soc 99:967–968
19. Bellart AC (1958) Studies on rhubarb roots grown in Holland. Pharm Weekbl 93:1077–1083
20. Okabe H, Matsuo K, Nishioka I (1973) Rhubarb (Rhei rhizoma). II. Anthraquinone glycosides. Chem Pharm Bull (Tokyo) 21:1254–1260
21. Bellart AC (1954) Rhubarb grown in the Netherlands. III. The structure of rhein glucosides. Pharm Weekbl 89:579–581
22. Hörhammer L, Wagner H, Leue HJ (1959) *Rheum palmatum*. II. Separation of anthraquinone glycosides by polyamide chromatography. Dtsch Apoth Ztg 99:1043–1044
23. Wagner H, Hörhammer L, Farkas L (1963) Genuine anthraquinone glycosides from the rhizome of *Rheum palmatum*. Z Naturforsch 18b:89
24. Romanova AS, Bankovski AI (1963) Anthracene derivatives of *Rheum palamatum* var. *tanguticum*. II. Glycosides. Med Prom SSSR 17:21–23 (CA 60:8116c)
25. Romanova AS, Bankovskii AI (1969) Chemical study of anthraglycosides of *Rheum palmatum* var. *tanguticum*. Tr Vses Nauchno Issled Inst Lek Aromat Rast 15:504–510 (CA 75:31304e)
26. Romanova AS, Bankovski AI (1963) Anthracene derivatives of *Rheum palmatum* var. *tanguticum*. III. Glycosides. Med Prom SSSR 17:29–30 (CA 60:11849c)
27. Hörhammer L, Farkas L, Wagner H, Müller E (1964) Synthesis and proof of structure of 8-(β-D-glucopyrnaosyloxy)-aloe emodin from *Rheum palmatum* var. *tanguticum*. Chem Ber 97:1662–1665
28. Hörhammer L, Wagner H, Müller E (1965) Definitive structure and proof for the chrysophanol 1-mono-β-D-glucoside (chrysophanein) from *Rheum palmatum*. Chem Ber 98:2859–2861
29. Romanova AS, Bankovski AI (1964) Anthracene derivatives from *Rheum palmatum* var. *tanguticum*. IV. Glycosides. Med Prom SSSR 18:18–20 (CA 62:5497e)
30. Romanova AS, Bankovski AI, Perelson ME, Kiryanov AA (1966) Glucoemodin from *Rheum palmatum*. Khim Prir Soedin Akad Nauka Uz SSR 2(2):83–85 (CA 65:9241h)
31. Holzschuh L, Kopp B, Kubelka W (1982) Physicon-8-*O*-β-D-gentiobioside, a new anthraquinone glycoside from rhubarb roots. Planta Med 46:159–161
32. Zwaving JH (1974) Qualitative and quantitative content of anthracene derivatives in *Rheum palmatum*. II. Composition of the mixture of anthracene derivatives in the root of *R. palmatum*. Pharm Weekbl 109:1169–1177
33. Engelshowe R (1985) Rhabarber. Pharm Unserer Zeit 14:40–49
34. Lin QS (1977) Chemical Constituents of Chinese Herbal Medicine. Science, Beijing, p 203
35. Lemli J, Dequecker R, Cuveele J (1963) Anthraquinone drugs. I. The presence of dianthrones in rhubarb root. Pharm Weekbl 98:500–502
36. Lemli J, Dequeker R, Cuveele J (1964) The palmidins, a new group of dianthrones from the fresh root of *Rheum palmatum*. Planta Med 12:107–111
37. Lemli J, Dequecker R, Cuveele J (1963) Anthraquinone drugs. III. Rheidin A, a new dianthrone from the roots of *Rheum palmatum*. Pharm Weekbl 98:655–659
38. Lemli J, Dequeker R, Cuveele J (1964) Anthraquinone drugs. VI. The heterodianthrones sennidin C, rheidin B and rheidin C from fresh rhubarb roots. Pharm Weekbl 99:613–616
39. Zwaving JH (1965) Separation and isolation of anthraquinone glycosides of *Rheum palmatum*. Planta Med 13:474–484
40. Miyamoto H, Imai S, Shinohara M, Fujioka S, Goto M, Matsuoka T, Fujimura H (1967) Purgatives. I. Isolation of sennoside A, one of the most active principles from rhubarb. Yakugaku Zasshi 87:1040–1043

41. Zwaving JH (1972) Sennoside content of *Rheum palmatum*. Planta Med 21:254–262
42. Oshio H, Imai S, Fujioka S, Sugawara T, Miyamoto M, Tsukui M (1972) Rhubarbs. II. Isolation of sennoside E, a new purgative compound. Chem Pharm Bull (Tokyo) 20:621–622
43. Oshio H, Imai S, Fujioka S, Sugawara T, Miyamoto M, Tsukui M (1974) Rhubarb III. New purgative constituents, sennosides E and F. Chem Pharm Bull (Tokyo) 22:823–831
44. Yamagishi T, Nishizawa M, Ikura M, Hikichi K, Nonaka G, Nishioka I (1987) Studies on rhubarb (Rhei rhizoma). XI. New laxative constituents of rhubarb, isolation and characterization of rheinosides A, B, C and D. Chem Pharm Bull (Tokyo) 35:3132–3138
45. Su HL, Chen CH (1962) Chinese rhubarb. II. Paper chromatography of anthraquinone derivatives. Acta Pharm Sin 10:725–730
46. Zhao YS, Liu FQ, Li QH (1984) Effect of wine on the chemical constituents during processing of rhubarb. Chem Pharm Bull (Tokyo) 19:417–420
47. Luo SD, Cai HS, Su W (1987) Effect of extraction methods on the anthraquinone content of rhubarb. Bull Chin Mater Med 12:286–288
48. Tan YX, Yang Y, Yan YH (1983) Quality control of ten species of rhubarb – qualitative and quantitative analysis of aglycone of anthraquinone glucosides. Chin J Pharm Anal 3:74–78
49. Haslam E (1982) The metabolism of gallic acid and hexahydroxydiphenic acid in higher plants. In: Herz W, Grisebach H, Kirby GW (eds) Progress in the Chemistry of Organic Natural Products, Vol 41. Springer, Berlin Heidelberg New York, pp 1–46
50. Roux DG, Ferreira D (1982) The direct biomimetic synthesis, structure and absolute configuration of angular and linear condensed tannins. In: Herz W, Grisebach H, Kirby GW (eds) Progress in the Chemistry of Organic Natural Products, Vol 41. Springer, Berlin Heidelberg New York, pp 47–76
51. Luo WY, Zhang YZ (1986) Studies on the determination of tannin in Dahuang. I. Casein method. Chin J Pharm Anal 6:15–18
52. Nonaka G, Nishioka I, Nagasawa T, Oura H (1981) Tannins and related compounds. I. Rhubarb. I. Chem Pharm Bull (Tokyo) 29:2862–2870
53. Kashiwada Y, Nonaka G, Nishioka I (1984) Tannins and related compounds. XXIII. Rhubarb. 4. Isolation and structures of new classes of gallotannins. Chem Pharm Bull (Tokyo) 32:3461–3470
54. Kashiwada Y, Nonaka G, Nishioka I (1986) Tannins and related compounds. XLVII. Rhubarb. 6. Isolation and characterization of new *p*-hydroxyphenylbutanones, stilbenes and gallic acid glycosides. Chem Pharm Bull (Tokyo) 34:3237–3243
55. Nonaka G, Nishioka I (1983) Tannins and related compounds. X. Rhubarb. 2. Isolation and structures of a glycerol gallate, gallic acid glucoside gallates, galloylglucose and isolindleyin. Chem Pharm Bull (Tokyo) 31:1652–1658
56. Nonaka G, Ezaki E, Hayashi K, Nishioka I (1983) Tannins and related compounds. XI. Flavanol glucosides from rhubarb and *Rhaphiolepis umbellata*. Phytochemistry 22:1659–1661
57. Kashiwada Y, Nonaka G, Nishioka I (1986) Tannins and related compounds. XLV. Rhubarb. 5. Isolation and characterization of flavan-3-ol and procyanidin glucosides. Chem Pharm Bull (Tokyo) 34:3208–3222
58. Kashiwada Y, Nonaka G, Nishioka I (1984) Studies on rhubarb (rhei rhizoma). V. Isolation and characterization of chromone and chromanone derivatives. Chem Pharm Bull (Tokyo) 32:3493–3500
59. Yagi A, Koizumi Y, Nishioka I (1971) Rhubarb (rhei rhizoma). I. Stilbene derivatives from "dudaioo" (Chinese inferior rhubarb). Shoyakugaku Zasshi 25:52–54
60. Nonaka G, Minami M, Nishioka I (1977) Studies on rhubarb (Rhei rhizoma). III. Stilbene glycosides. Chem Pharm Bull (Tokyo) 25:2300–2305
61. Murakami T, Tanaka K (1972) New phenolic glycosides in Rhei palmati rhizomes. Tetrahedron Lett 2965–2968
62. Fuzellier MC, Mortier F, Girard T, Payen J (1981) Study of antibiotic properties of some anthraquinones using chromatographic microplates. Ann Pharm Fr 39:313–318
63. Matsumoto O, Kiyohara H, Yamada A, Arakawa K, Cho M, Ohtsuka Y (1985) Anti-*Bacteroides fragilis* component in rhubarb and senna. Wakan Iyaku Gakkaishi 2:248–249 (CA 104:39569h)
64. Cyong JC, Matsumoto T, Arakawa K, Kiyohara H, Yamada H, Otsuka Y (1987) Anti-*Bacteroides fragilis* substance from rhubarb. J Ethnopharmacol 19:279–283

65. Chen CH, Li TT, Su HL, Wang CI (1963) Chinese rhubarb. VII. Mechanism of antibiotic action of anthraquinone derivatives. Effects on the respiration of *Staphylococcus aureus*. Acta Biochim Biophys Sin 3:426–433
66. Zhang Y, Chen QH (1986) Biochemical study of Chinese rhubarb. XIV. Inhibitory effects of anthraquinone derivatives on some NAD linked dehydrogenase. Acta Biochim Biophys Sin 18:239–245
67. Chen CL, Chen QH (1987) Biochemical study of Dahuang. XIX. Localization of inhibition of anthraquinone derivatives on the mitochondrial respiratory chain. Acta Pharm Sin 22:12–18
68. Xiao PG, Chen BZ, Wang LW, Ho LY, Luo SR, Guo HZ (1980) Preliminary study on the correlation between phylogeny, chemical constituents and therapeutic effects of *Rheum* species. Acta Pharm Sin 15:33–39
69. Oshio H, Kawamura N (1985) Determination of the laxative components in rhubarb by high performance liquid chromatography. Shoyakugaku Zasshi 39:131–138
70. Kisa K, Sasaki K, Yamauchi K, Kuwano S (1981) Potentiating effect of sennoside C on purgative activity of sennoside A in mice. Planta Med 42:302–303
71. Tsukui M, Yamazaki M, Toyosato T, Matsuoka T, Fujimura H (1974) Biological assay of rhubarb. Yakugaku Zasshi 94:1095–1100
72. Garcia-Villar R, Leng-Peschlow E, Ruckebusch Y (1980) Effect of anthraquinone derivatives on canine and rat intestinal motility. J Pharm Pharmacol 32:323–329
73. Hottori M, Kim G, Motoike S, Kobashi K, Namba T (1982) Metabolism of sennosides by intestinal flora. Chem Pharm Bull (Tokyo) 30:1338–1346
74. Kobashi K, Nishimura T, Kusaka M, Hattori M, Namba T (1980) Metabolism of sennosides by human intestinal bacteria. Planta Med 40:225–236
75. Lemli J, Lemmens L (1980) Metabolism of sennosides and rhein in the rat. Pharmacology [Suppl 1] 20:50–57
76. Lemmens L (1979) The laxative action of anthracene derivatives. II. Absorption, metabolism and excretion of sennoside A and B in the rat. Pharm Weekbl [Sci] 1:178–185
77. Hietala P, Lainonen H, Marvola M (1988) New aspects on the metabolism of the sennosides. Pharmacology 36:138–143
78. Lemmens L, Dreessen M (1979) The laxative action of anthracene derivatives. III. Absorption, metabolism and excretion of rhein in the rat. Pharm Weekbl [Sci] 1:950–956
79. Sun Y, Li Q, Chen QH (1986) Excretion of emodin and its metabolites in mice. J Nanjing Coll Pharm 17:132–135
80. Sun Y, Chen QH (1985) Isolation and identification of the metabolites of emodin and physcion in rat and mouse urine. J Nanjing Coll Pharm 16:72
81. Bachmann M, Schlatter C (1981) Metabolism of ^{14}C-emodin in the rats. Xenobiotica 11:217–225
82. Zhou XM, Chen QH (1988) Biochemical study of Chinese rhubarb. XXII. Inhibitory effect of anthraquinone derivatives on sodium-potassium-ATPase of rabbit renal medulla and their diuretic action. Acta Pharm Sin 23:17–20
83. Zhou A, Wei KY (1985) Changes of plasma corticosterone and plasma sodium and potassium during diarrhea induced by genuine rhubarb in rats. J Beijing Norm Univ [Nat Sci] 62–63
84. Guo CY, Wang BE, Zhao SY, Li ZJ (1986) Effects of *Rheum palmatum* on body temperature and cerebral PGE level. Chin J Integrat Trad West Med 6:106–107
85. Liang ZJ, Chen YJ, Shao YD, Jiao DH, Liu XC (1986) Effects of *Rheum palmatum* on the hemorheology of rabbits. Chin J Integrat Trad West Med 6:294–296
86. Weng WL, Cui J, Chen YC, Wang JC (1986) Effect of Dahuang on platelet aggregation and thrombus formation in vitro. Chin Trad Herb Drugs 17:507–508
87. Jiang WJ, Mao SJ, Wu LY, Yu MM, Ma YL, Pei DK (1985) Effects of raw and processed Dahuang on experimental gastric ulcer in rats. Bull Chin Mater Med 10:65–67
88. Jin YC, Hu GH, Zhu ZZ, Dai PX, Kang FC, Zhang XB, Feng SC, Fang ZQ, Chen ZC, Qi GB (1986) Experimental and clinical studies on the effects of *Cassia angustifolia* in acute gastric and duodenal hemorrhage. Chin J Integrat Trad West Med 6:455–457
89. Sun DA, Zhuang HQ, Deng A, Wang LZ (1986) Comparison on the clinical effects of *Rheum palmatum* and cimetidine in upper gastrointestinal bleeding. Chin J Integrat Trad West Med 6:458–459
90. Dong GB, Fang YM, Jin TY (1985) Ames test of Dahuang decoction. Guizhouyiyao 9:54–55

91. Tikkanen L, Matsushima T, Natori S (1983) Mutagenicity of anthraquinones in the *Salmonella* preincubation test. Mutat Res 116:297–304
92. Liberman DF, Schaefer FL, Fink RC, Ramgopal M, Ghosh AC, Mulcahy R (1980) Mutagenicity of islandicin and chrysophanol in the *Salmonella/microsome* system. Appl Environ Microbiol 40:476–479
93. Bruggeman IM, van der Hoeven JCM (1984) Lack of activity of the bacterial mutagen emodin in HGPRT and SCE assay with V79 Chinese hamster cells. Mutat Res 138:219–224
94. Chen QH, Liu CY, Qiu CH (1980) Studies on Chinese rhubarb. XII. Effect of anthraquinone derivatives on the respiration and glycolysis of Ehrlich ascites carcinoma cells. Acta Pharm Sin 15:65–70
95. Kawai K, Kato T, Mori H, Kitamura J, Nozawa Y (1984) A comparative study on cytotoxicities and biochemical properties of anthraquinone mycotoxins emodin and skyrin from *Penicillium islandicum* Sopp. Toxicol Lett 20:155–160
96. Lu M, Chen QH (1989) Biochemical study of Chinese rhubarb. XXIX. Inhibitory effects of anthraquinone derivatives on P388 leukemia in mice. J China Pharm Univ 20:155–157

Rhododendron dauricum L. 108

108.1 Introduction

Rhododendron dauricum L. (Ericaceae) is a folk medicine used in China as an expectorant and mucolytic in the treatment of acute and chronic bronchitis. The essential oil Manshanhongyou, Oleum Rhododendri daurici, obtained by steam distillation of the dry leaves of *R. dauricum* and its galenic preparation Manshanhongyou Jiaowan, Capsulae Olei Rhododendri daurici, are officially listed in the Chinese Pharmacopoeia.

108.2 Chemical Constituents

108.2.1 Nonvolatile Components

In studies to determine the active principles of *R. dauricum*, a number of flavone derivatives such as farrerol (*108-1*), 8-methyl-farrerol, hyperoside (*108-2*), kaempferol, quercetin, myricetin, and isohyperoside (quercetin-3-*O*-α-D-galactopyranoside [1] were isolated and identified.

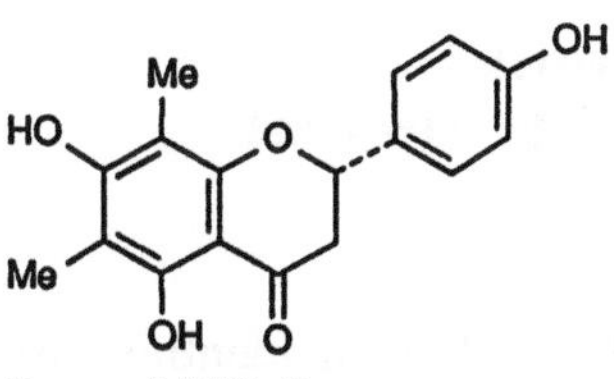

Farrerol (*108-1*)

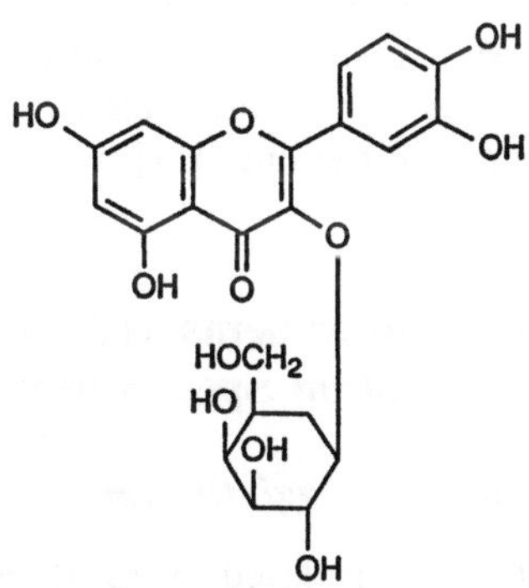

Hyperoside (*108-2*)

With a content of 0.08%–0.15% in the dry leaves of *R. dauricum*, farrerol is the most prevalent constituent [2, 3]. It has been snythesized by cyclocondensation of dimethylphloroglucinol with *p*-propionyloxycinnamyl chloride or of acetyldimethylphloroglucinol with *p*-hydroxybenzaldehyde [4, 5].

Besides the flavone derivatives, the coumarins scopoletin and umbelliferone [6], the phenolcarboxylic acids vanillic acid and *p*-hydroxybenzoic acid [7], and andromedotoxin (*108-3*) [6], a toxic diterpene derivative, were isolated and identified. A new chromene derivative, daurichromenic acid (*108-4*), was also isolated from the leaves of *R. dauricum* and structurally characterized on the basis of spectral data [8].

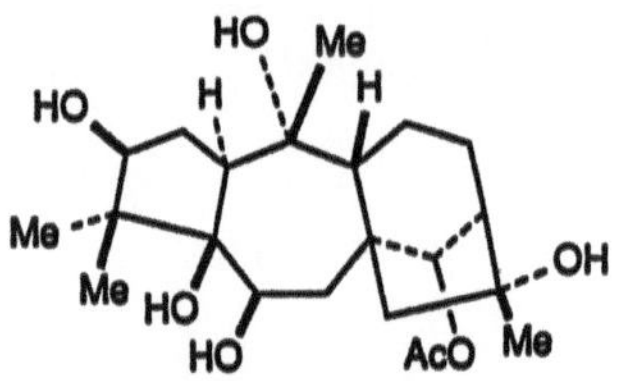

Andromedotoxin (*108-3*)

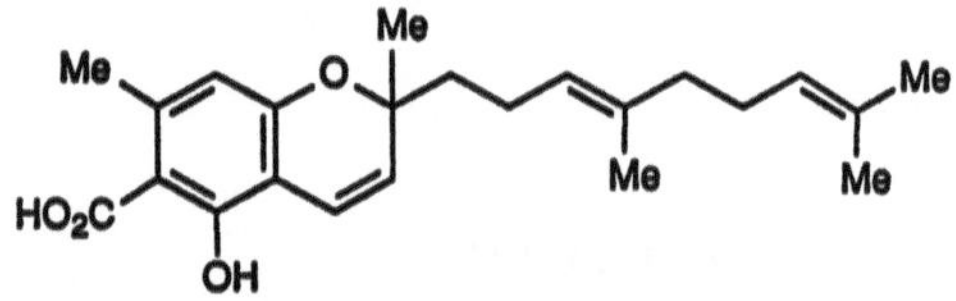

Daurichromenic acid (*108-4*)

108.2.2 Volatile Components

Eight constituents were isolated from the essential oil of *R. dauricum* by chromatography. Six were identified as germacrone, menthol, juniper champor, and α-, β-, and γ-eudesmol [9].

108.3 Pharmacology

Farrerol given orally or i.p. at doses of 0.2–1 g/kg to rats increased the output of phenol red from the respiratory tract. This effect is dose dependent and probably due to direct action on the respiratory tract. By repeated oral administration, farrerol decreased protein content of respiratory tract washings in control mice and in mice or rats exposed to sulfur dioxide [10].

Within 6–12 h about 70%–80% farrerol given orally (200 mg/kg) to rats was found to be cleared from the gastrointestinal tract. About 30% of the dose was excreted in the feces. One hour after i.v. administration, the farrerol content was found to be highest in the lung [11].

108.4 Chemical Constituents and Biological Activities of Other *Rhododendron* Species used in Chinese Folk Medicine

108.4.1 *Rhododendron agglutinaum*

Six components were isolated from *R. agglutinaum*, four of them identified as quercetin, hyperoside, rhododendrol (betuligenol, *108-5*), and rhododendrine (betuloside, *108-6*). *R. agglutinaum* is used in folk medicine for the treatment of arthritis [12].

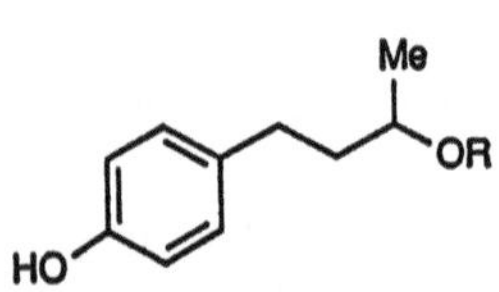

Rhododendrol (*108-5*): R=H

Rhododendrine (*108-6*): R=

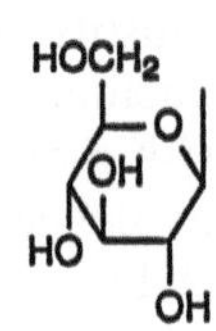

108.4.2 *Rhododendron anthopogonoides*

Volatile oil [13], gossypetin-3-*O*-*β*-D-galactopyranoside, 8-methoxyquercetin, and hyperoside [14] were isolated from the leaves of *R. anthopogonoides*. Intravenous doses of 50–80 mg/kg of the volatile oil decreased the arterial pressure in anesthetized rabbits [13].

108.4.3 *Rhododendron chrysanthum*

The following 11 constituents were isolated and identified from the leaves of *R. chrysanthum*: campanulin (*108-7*), simiarenol (*108-8*), *β*-sitosterol, uvaol (*108-9*), oleanolic acid, hyperoside, polystachoside (*108-10*), avicularin (quercetin-3-L-arabinoside), quercetin, rhododendrol, and rhododendrine [15].

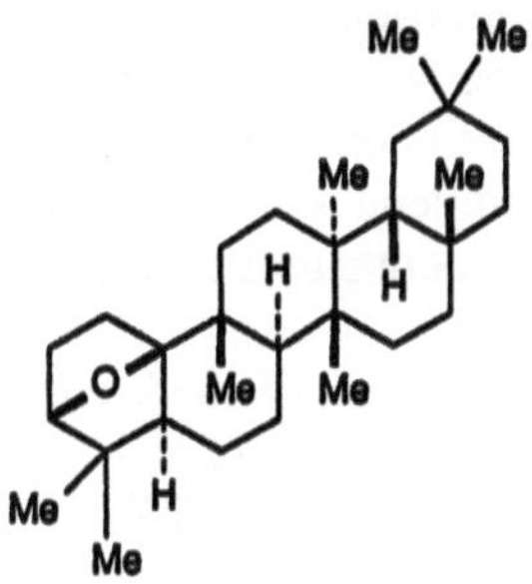

Campanulin (*108-7*)

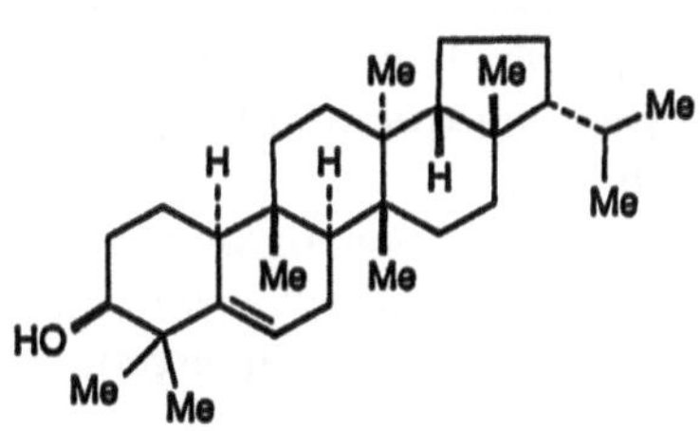

Simiarenol (*108-8*)

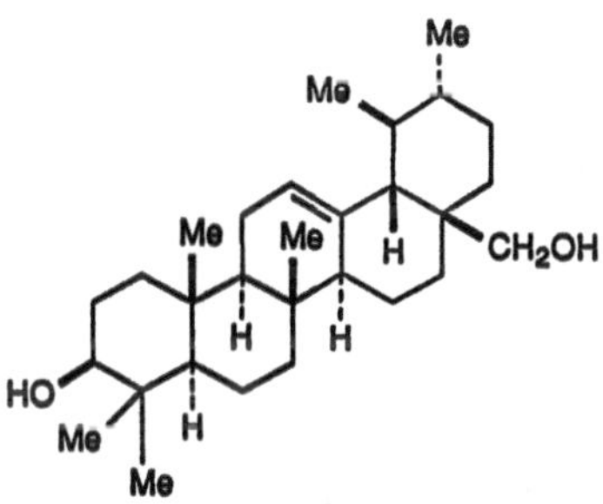

Uvaol (*108-9*)

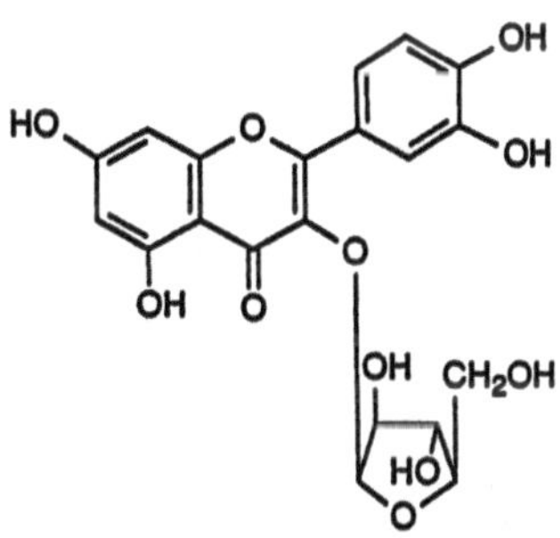

Polystachoside (*108-10*)

108.4.4 *Rhododendron dabanshanense*

The following constituents were isolated and identified from the aqueous extract of *R. dabanshanense*: rhododendrol, rhododendrine, quercetin, avicularin, catechin, hyperoside, and grayanotoxins I, II (*108-11*), and IV (*108-12*) [16, 17]. Grayanotoxin I is identical to andromedotoxin [18].

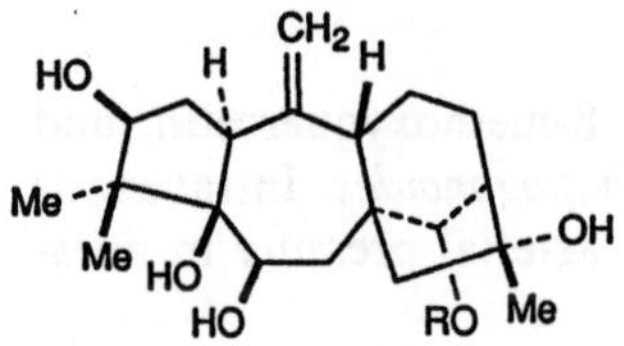

Grayanotoxin II (*108-11*): R=H
Grayanotoxin IV (*108-12*): R=Ac

108.4.5 *Rhododendron capitatum*

A number of terpene derivatives were isolated from the essential oil of *R. capitatum* and identified as: α-pinene, β-pinene, β-myrcene, δ-cadinene, selinenes, *trans*-β-farnesene, preisocalamendiol (*108-13*), juniper champhor, *cis*-α-ocimene, β-gurjunene, α-humulene, kauran-16α-ol (*108-14*), borneol acetate, α̃-terpineol, *trans*-pinocarveol, linalool, and humulene epoxide [19–21]. The essential oil showed some antimicrobial activity [20]. α-Pinene, β-pinene, β-myrcene, δ-cadinene, selinenes, and α-humulene showed expectorant activity and α-pinene was also active as an antitussive agent [19].

Ten nonvolatile compounds were isolated from the aqueous extract of the leaves of *R. capitatum* and identified as scopoletin, fraxetin (*108-15*), grayanotoxins I, II, and IV, raffinose, hyperoside, quercetin, myricetin, and capitatin (*108-16*) [22].

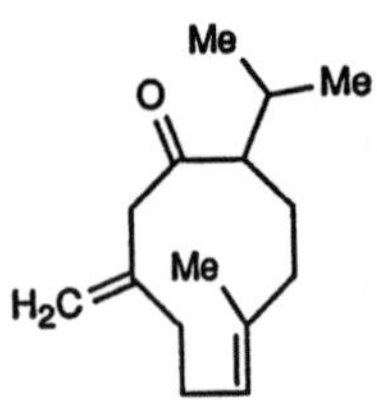

Preisocalamendiol (*108-13*)

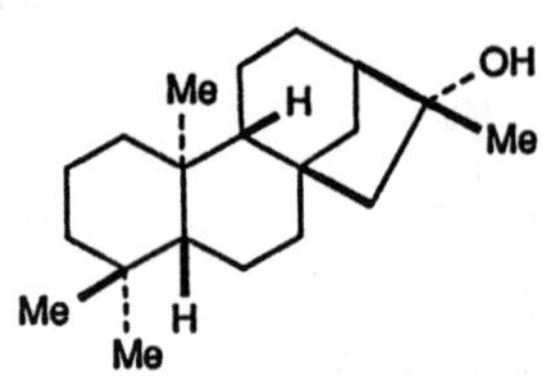

Kauran-16α-ol (*108-14*)

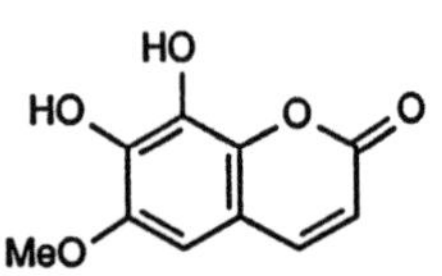

Fraxetin (*108-15*)

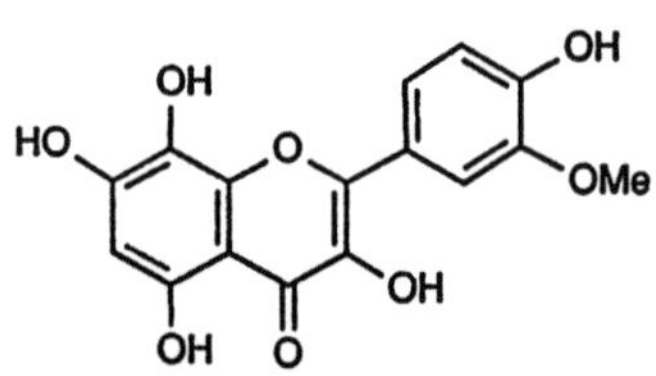

Capitatin (*108-16*)

108.4.6 *Rhododendron micranthum*

The flavone derivatives hyperoside, astragalin, quercetin, and kaempferol were isolated and identifed from the leaves of *R. micranthum*, a folk medicine used for the treatment of bronchitis [23].

108.4.7 *Rhododendron molle*

Rhododendron molle is listed in the appendix of the Chinese Pharmacopoeia and the fruits are used as an analgesic. From the ethanolic extract of the fruits of *R. molle*, rhododendrotoxin (rhomotoxin *108-17*) was isolated and was proven to be identical to rhodojaponin III [24]. Rhododendrotoxin has been reported to have hypotensive and antitachycardial activity [25].

In addition, rhododendrotoxin has been found selectively to act on the myocardial Na^+ channel, causing an increase in Na^+ influx. It has been recommended as a physiological tool in the study of the cardiac Na^+ channel [26].

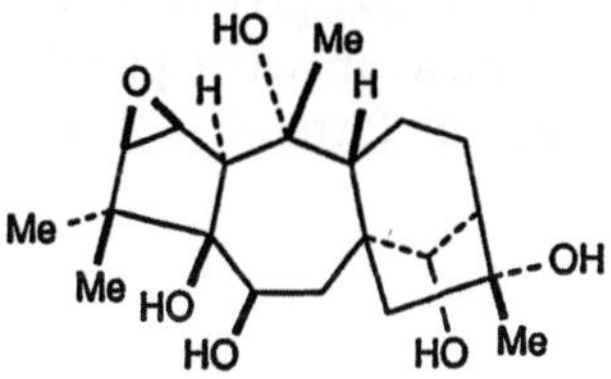

Rhododendrotoxin (*108-17*)

108.4.8 *Rhododendron racemosum*

The constituents of the essential oil obtained from the leaves of *R. racemosum* were studied by gas chromatography-mass spectroscopy. The following components were detected: α-pinene, β-pinene, *p*-cymene, 1,8-cineole, bornyl acetate, *trans*-caryophyllene, and caryophyllene. The essential oil of *R. racemosum* is used for the treatment of bronchitis [27].

108.4.9 *Rhododendron seniavinii*

Rhododendron seniavinii is a Chinese medicinal herb widely distributed in the province of Fujian. In clinical trials it showed efficacy in the treatment of chronic bronchitis. The isolation of seven compounds was reported from the alcoholic extract of the leaves of *R. seniavinii*: α-amyrin, β-sitosterol, hyperoside, quercitrin, quercetin, friedelin, and epifriedelanol (*108-18*) [28].

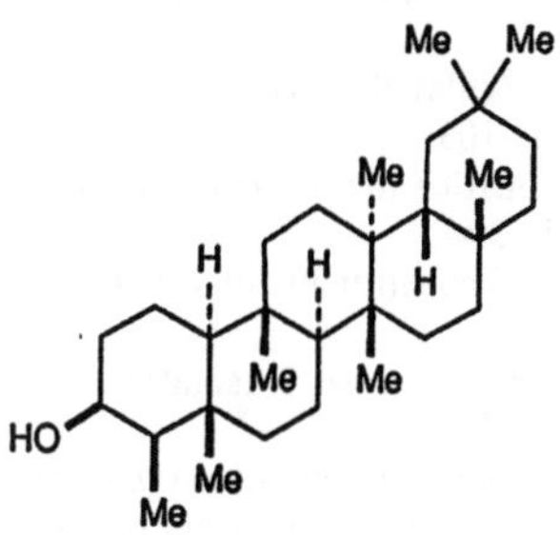

Epifriedelanol (*108-18*)

108.4.10 *Rhododendron thymifolium*

Rhododendron thymifolium is another Chinese medicinal herb used to treat chronic bronchitis. Raffinose, scopoletin, fraxetin, hyperoside, isohyperoside, and quercetin have been isolated from the alcoholic extract [29]. Ten constituents were isolated from the essential oil of the leaves: germacron, juniper camphor, nonaldehyde, myrcene, limonene, humulene, camphene, farnesene, and α- and β-pinenes [30].

108.4.11 *Rhododendron simsii*

The total flavone fraction of *R. simsii* was found to be effective in the treatment of chronic bronchitis by i.m. administration to rats at a dose of 300 mg/kg [31]. Matteucinol (4′-methylfarrerol), isolated from the total flavone fraction as an active principle, was found to have expectorant activity similar to that of farrerol [32]. Its synthesis was also reported [32].

References

1. Liu YL, Fu FY, Hsie TH, Chang SL (1976) Studies on the chemical constituents of Man-Shan-Hong (*Rhododendron dauricum* L.). I. Acta Chim Sin 34:211–221
2. Chang KT, Wang MC, Chang SJ (1980) Studies on the quantitative determination of farrerol in the leaves of *Rhododendron dauricum*. Chin Pharm Bull 15:45
3. Zhang GD, Wang MZ, Zhang SR (1980) Studies on the determination of farrerol content in *Rhododendron dauricum* leaves. Acta Pharm Sin 15:736–740
4. Isukura Sangyo KK (1983) Farrerol. Jpn Kokai Tokkyo Koho JP 5821,678 [8321,678] (CA 98:197878 v)
5. Liang XT, Chen SF, Lu YH, Ge DL (1985) Studies on synthesis of 6,8-dimethyl-5,7,4′-trihydroxyflavanone. Acta Pharm Sin 20:33–38
6. Fu FY, Liu YL, Liang HT, King PY, Chen YY, Yu CK (1976) Studies on the constituents of Man-Shang-Hong (*Rhododendron dauricum* L.). II. Acta Pharm Sin 34:223–227
7. Liu YL, Fu FY, Chin PY (1980) Studies on the chemical constituents of *Rhododendron dauricum* L. IV. Identification of phenolcarboxylic acids. Chin Trad Herb Drugs 11:152–153
8. Suntory Ltd (1982) Daurichromenic acid. Jpn Kokai Tokkyo Koho JP 82 28, 080 (CA 97:3777 x)
9. Hsu CC, Yu TC (1976) Studies on the chemical constituents of Man-Shang Hong (*Rhododendron dauricum* L.). III. Acta Chim Sin 34:275–281
10. Chinese Academy of Medical Sciences (1977) Expectorant action of farrerol. Chin Med J [Engl] 3:259–265
11. Feng YS, Zhu DZ (1979) Metabolism of farrerol. Acta Pharm Sin 14:149–155
12. Chang CL, Chang CC (1981) Study on chemical constituents in *Rhododendron agglutinaum*. Chin Pharm Bull 16:53–54
13. Du JZ, Li QF, Liu LQ, Gong SX, Li MX, Xing DJ (1980) Cardiovascular effects of the essential oil of *Rhododendron anthopogonoides*. Acta Pharmacol Sin 1:105–109
14. Zhang ZL, Chuan FY, Liu YL (1980) Studies on the chemical constituents of liexiangdujuan (*Rhododendron anthopogonoides*). Chin Trad Herb Drugs 11:393–394
15. Jin ZX, Chen RY, Shang TM, Liang XT (1981) Studies on chemical constituents of *Rhododendron chrysanthum* Pallas. Acta Pharm Sin 16:869–871
16. Yang HR, Wang SX (1978) Chemical studies of *Rhododendron dabanshanense*. I. Isolation and identification of four phenolic components. Acta Bot Sin 20:355–359
17. Wang SX, Yang HR (1981) Studies on chemical constituents of *Rhododendron dabanshanense*. II. Isolation and identification of (+)-catechin, hyperin and toxicant components. Acta Bot Sin 23:47–51
18. Humphreys DJ, Stodulski JBJ (1986) Detection of andromedotoxins for the diagnosis of rhododendron poisoning in animals. J Appl Toxicol 6:121–122

19. Lu IC, Hsing CP, Wei SL (1981) Studies on chemical constituents of *Rhododendron capitatum* Maxim. I. Chin Pharm Bull 16:54
20. Lu IC, Hsing CP (1981) Studies on chemical constituents of *Rhododendron capitatum* Maxim. II. Chin Pharm Bull 16:54–55
21. Lu YC, Xing JP (1982) Studies on the constituents of the essential oil of *Rhododendron capitatum* Maxim. Acta Chim Sin 40:531–538
22. Wang SX, Ji LJ, Yi FS, Jiang SH, Zhou BN (1984) Chemical constituents of *Rhododendron capitatum* Maxim. Acta Bot Sin 26:71–75
23. Zhang YZ, Cui JF, Zhao SP (1984) TLC densitometric determination of flavonoid contents in the leaves and tablets of *Rhododendron micranthum* Turcz. Chin J Pharm Anal 4:1–4
24. Pu QL (1983) Structural elucidation of rhodojapanin III from *Rhododendron molle* G. Don. Chin Trad Herb Drugs 14:293–294
25. Wuhan Medical College (1980) Extraction and isolation of rhododendrotoxin from *Rhododendron molle*. Chin Pharm Bull 15:39–40
26. Jin MW, Zong XG, Fang DC, Jiang MX (1985) Effect of rhomotoxin on the membrane potential and contractile force of guinea pig papillary muscles. Acta Pharm Sin 20:481–484
27. Fang HJ, Chen LS, Zhou TH (1980) Studies on the component of essential oils. III. Studies of chemical constituents of the essential oil from *Rhododendron racemosum* Franch. Comparison of the constituents of *Vitex negundo* L. var. *cannabifolia* (Sieb. et Zucc) Hand-Mazz. and *V. negundo* L. var. *heterophylla* (Franch.) Rehd. Acta Pharm Sin 15:284–287
28. Deng FX, Lin GT, Gui XM, Zhou BN (1982) Studies on the chemical constituents of Man Shan Bai (*Rhododendron seniavinii* Maxim). Acta Chim Sin 40:178–180
29. Chinghai Sheng Institute of Biology, Academia Sinica, Institute of Botany (1978) Chemical constituents of the essential oil of *Rhododendron thymifolium* Maxim. Acta Bot Sin 20:135–139
30. Yang HR, Wang SX (1982) Studies on the chemical constituents of *Rhododendron thymifolium* Maxim. Acta Bot Sin 24:65–67
31. Zhang YF, Tan JX, Zhou YF (1981) Change of action potentials in vagus nerve (efferent) of chronic bronchitic rats and the effect of treatment with flavonoids of *Rhododendron simsii*. Chin Pharm Bull 16:4–6
32. Xie JX, Wang L, Liu CX, Ge DL (1986) Synthesis of matteucinol, an expectorant principle. Acta Acad Med Sin 8:84–87

Rubia cordifolia L. 109

109.1 Introduction

Qiancao, Radix Rubiae, is the dry root and rhizome of *Rubia cordifolia* L. (Rubiaceae) collected in spring and fall. It is an officially listed herbal medicine in the Chinese Pharmacopoeia and is recommended as a hemostatic agent for the treatment of hematorrhea, hematemesis, nose bleeding, traumatic bleeding, dysmenorrhea, and arthritis.

109.2 Chemical Constituents

The root of *R. cordifolia* contains a number of anthraquinone pigments. The first pigment named purpurin (*109-1*) was reported more than 100 years ago [1–3]. Later the related pigments munjistin (*109-2*) [3, 4], pseudopurpurin (*109-3*) [5], and alizarin [6] were isolated. Pseudopurpurin is purpurin-3-carboxylic acid. On sublimation it undergoes decarboxylation to yield purpurin [5].

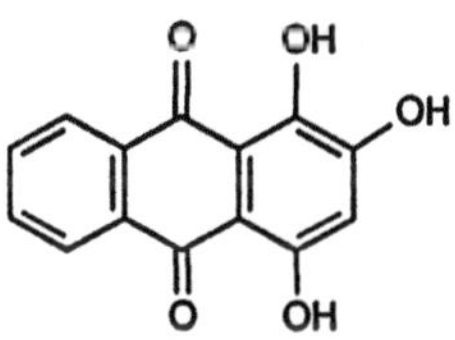

Purpurin (*109-1*)

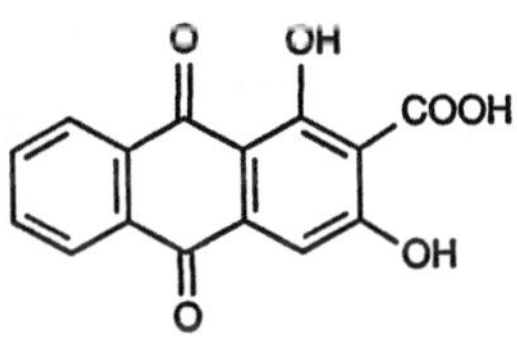

Munjistin (*109-2*)

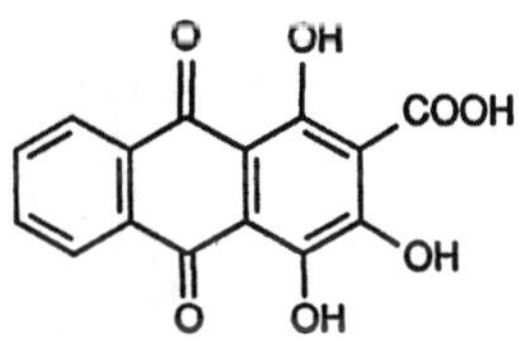

Pseudopurpurin (*109-3*)

A number of further known and new anthraquinone pigments were isolated from the root of *R. cordifolia* and identified. They include: 1-hydroxy-2-methylanthraquinone, nordamnacanthal (*109-4*), physcion, 1,4-dihydroxy-6-methylanthraquinone [7], 1,4-dihydroxy-2-methylanthraquinone, 1,5-dihydroxy-2-methylanthraquinone [8], 1-hydroxy-2-methoxyanthraquinone, 1,3-dimethoxy-2-carboxyanthraquinone, rubiadin (*109-5*) [9], 1,3-dihydroxy-2-ethoxymethylanthraquinone, lucidin primeveroside (*109-6*), ruberythric acid (*109-7*), 1,3,6-trihydroxy-2-methylanthraquinone, 1,3,6-trihydroxy-2-methylanthraquinone 3-*O*-α-L-rham-nopyranosyl-(1→2)-β-D-(6-*O*-acetyl)-glucopyranoside (*109-8*) and its deacetyl derivative [10] 1-acetoxy-3,6-dihydroxy-2-methylanthraquinone 3-*O*-α-L-rhamnopyranosyl-(1→4)-α-D-glucopyranoside (*109-9*) [11], 1,4-dihydroxy-2-carboethoxyanthraquinone, and 1-hydroxy-2-carboxy-3-methoxyanthraquinone [12].

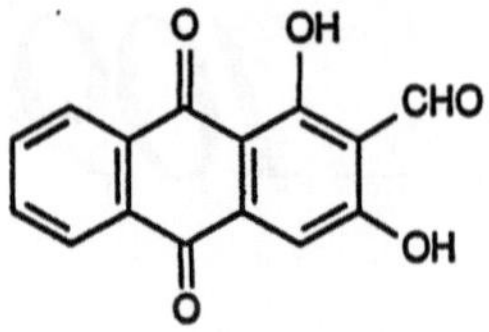

Nordamnacanthal (*109-4*)

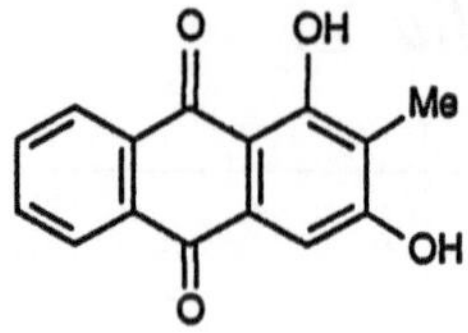

Rubiadin (*109-5*)

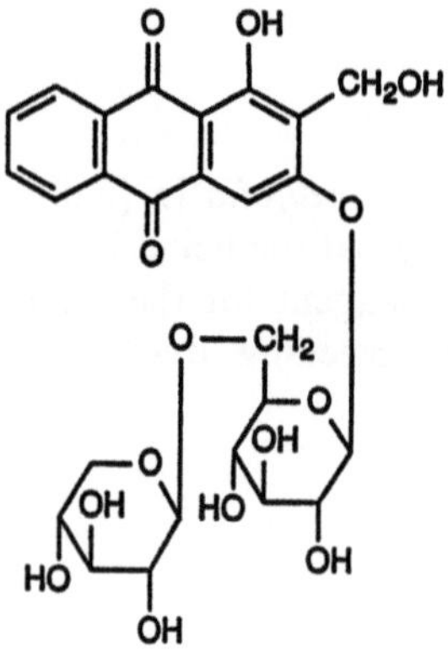

Lucidin primeveroside (*109-6*)

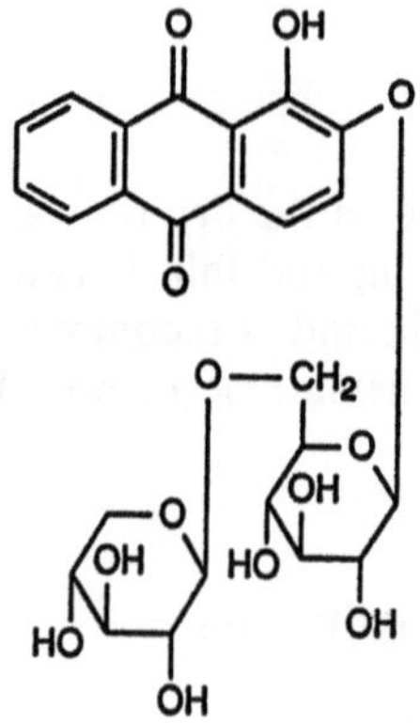

Ruberythric acid (*109-7*)

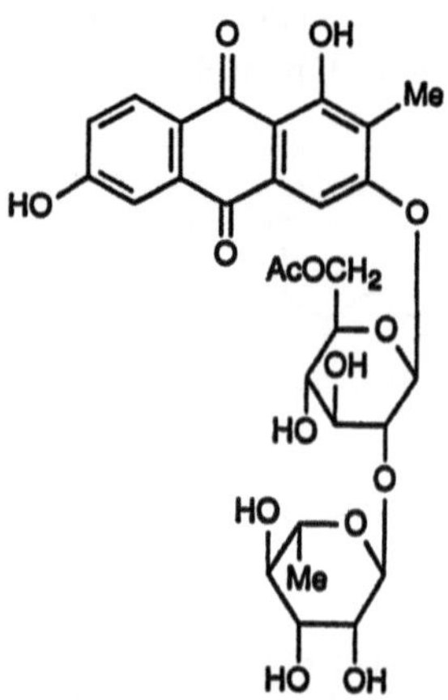

1,3,6-Trihydroxy-2-methylanthraquinone 3-*O*-α-L-rhamnopyranosyl-(1→2)-β-D-(6-*O*-acetyl)-glucopyranoside (*109-8*)

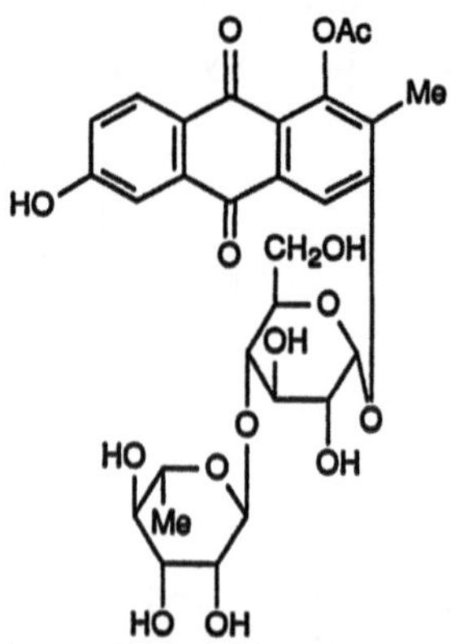

1-Acetoxy-3,6-dihydroxy-2-methylanthraquinone-3-*O*-α-L-rhamnopyranosyl-(1→4)-*O*-α-D-glucopyranoside (*109-9*)

In addition to anthraquinone derivatives the naphthalene compounds mollugin (*109-10*) [10] and rubidate (*109-11*) [13] were isolated.

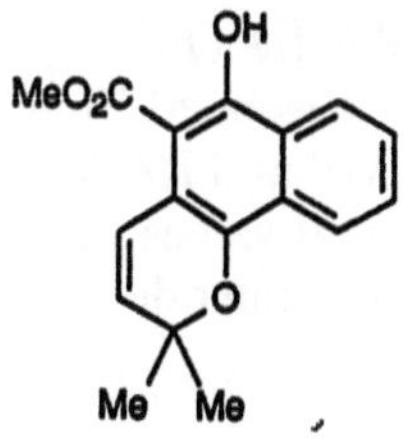

Mollugin (*109-10*)

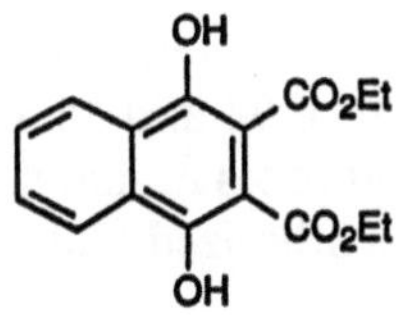

Rubidate (*109-11*)

New triterpenes isolated from the root of *R. cordifolia* are rubifolic acid (*109-12*), rubicoumaric acid (*109-13*) [14], and rubiatriol (*109-14*) [15] together with oleanolic acid acetate and sitosterol [12].

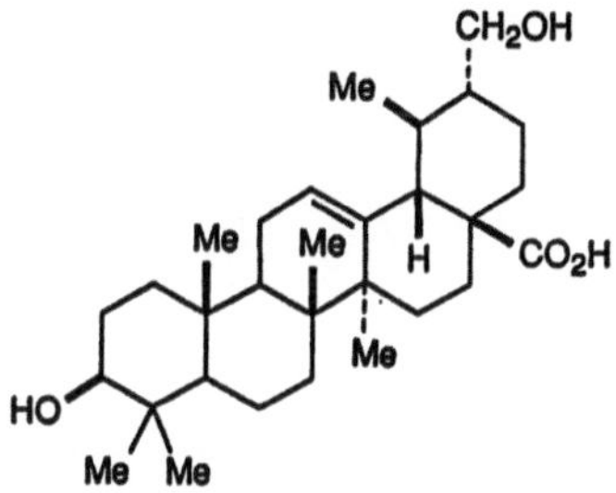

Rubifolic acid (*109-12*)

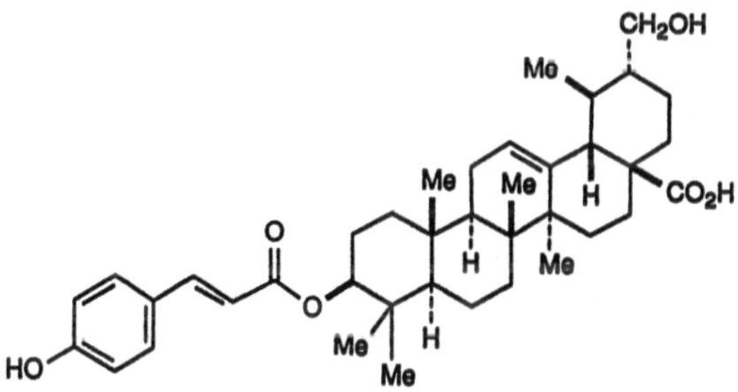

Rubicoumaric acid (*109-13*)

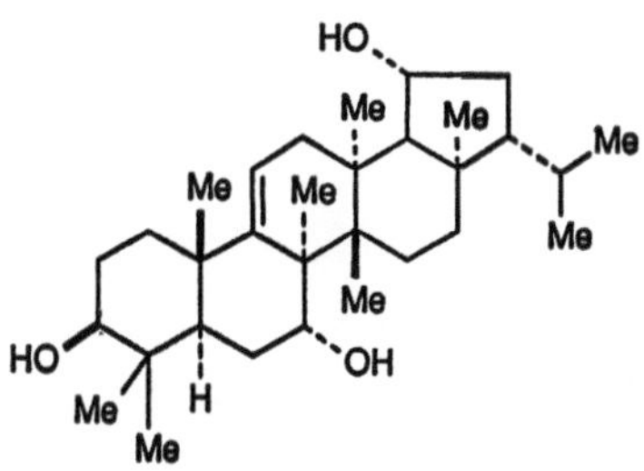

Rubiatriol (*109-14*)

Cyclic hexapeptides named RA-I (*109-15*), RA-II (*109-16*), RA-III (*109-17*), RA-IV (*109-18*), RA-V (*109-19*), and RA-VII (*109-20*) [16–18] were isolated from the methanolic extract of the roots of *R. cordifolia* and *R. akane*. RA-V and RA-VII showed an average content of 94 and 36 mg/kg, respectively. Maximal values were found in the aerial part and roots in mid-summer [19]. Alkyl ether and ester derivatives of RA-V were also synthesized [20].

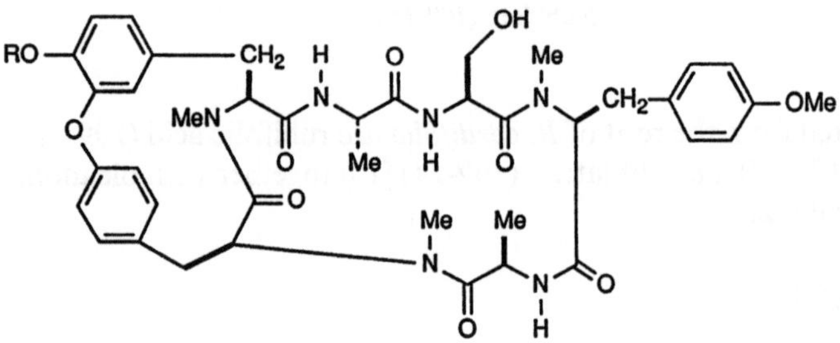

RA-I (*109-15*): R=H
RA-III (*109-17*): R=CH_3

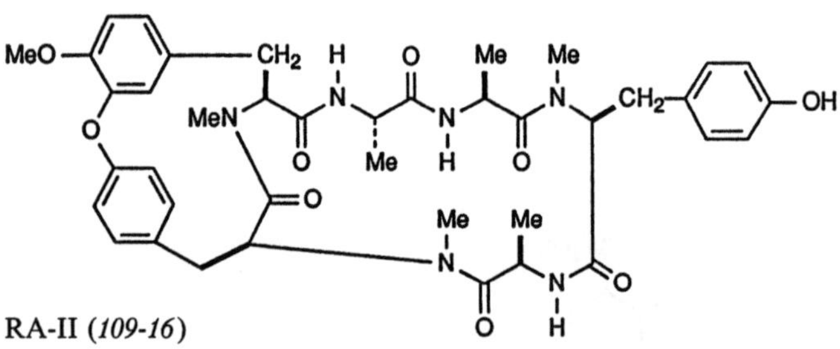

RA-II (*109-16*)

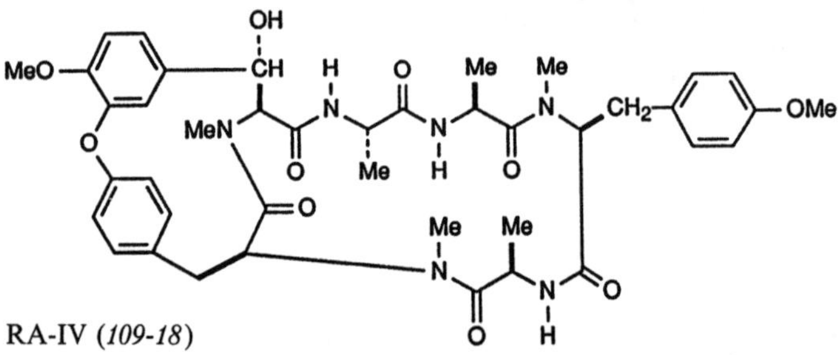

RA-IV (*109-18*)

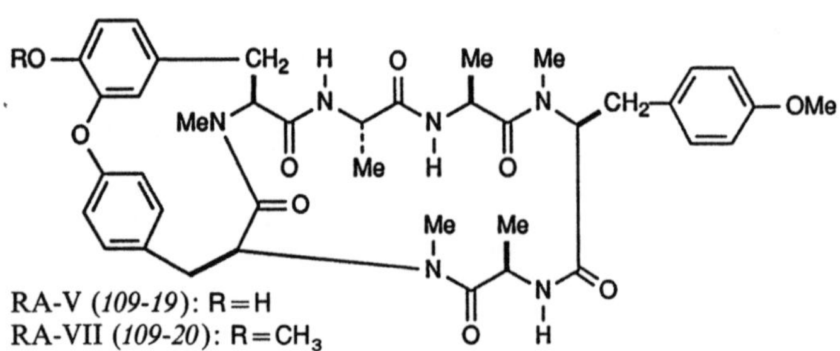

RA-V (*109-19*): R=H
RA-VII (*109-20*): R=CH_3

109.3 Pharmacology

Rubidate, 1,4-naphthalenediol-2,3-dicarboxylic acid diethylester, was found to stimulate hematopoiesis in mice and dogs. It increased peripheral leukocyte counts and stimulated proliferation and differentiation of hematopoietic stem cells. It has been reported to prevent leukopenia induced by cyclophosphamide in mice. The LD_{50} of rubidate in mice was about 3 g/kg by i.p. administration [14].

The cyclic hexapeptides exhibited significant antineoplastic activity in mice transplanted with leukemias P388 and L1210, B16 melanoma and solid tumor colon 38, Lewis lung carcinoma, and Ehrlich carcinoma. RA-V was found to be highly effective on MM2 mammary carcinoma in mice [21]. Acute LD_{50} values of RA-VII were 10 mg/kg by i.p., 16.5 mg/kg by i.v., and 63 mg/kg by oral administration. The optimal dose of RA-VII against P388 leukemia was 4 mg/kg [22].

References

1. Stenhouse J (1864) Untersuchung der *Rubia munjista*, des ostindischen Krapps oder des Munjeet des Handels. Liebigs Ann Chem 130:325
2. Gräbe C, Liebermann C (1968) Über Farbstoffe aus der Anthracengruppe. Chem Ber 1:104–106
3. Takagi Y (1961) The pigments in the root of *Rubia cordifolia* variety *munjista*. Nippon Kaguku Zasshi 82:1561–1563 (CA 57:6296e)
4. Mitter PC, Biswas H (1932) Über die Synthese des Munjistins. Chem Ber 65:622–625
5. Wada M (1941) Coloring matter of *Rubia cordifolia*. Science (Japan) 11:415 (CA 44:10817d)
6. Murti VVS, Seshadri TR, Sivakumaran S (1972) Anthraquinones of *Rubia cordifolia*. Phytochemistry 11:1524
7. Tessier AM, Delaveau P, Champion B (1981) New anthraquinones in *Rubia cordifolia* root. Planta Med 41:337–343
8. Dosseh C, Tessier AM, Delaveau P (1981) *Rubia cordifolia* roots. II. New quinones. Planta Med 43:141–147
9. Dosseh C, Tessier AM, Delaveau P (1981) New quinones in *Rubia cordifolia* L. roots. III. Planta Med 43:360–366
10. Itokawa H, Mihara K, Takeya K (1983) Studies on a novel anthraquinone and its glycosides isolated from *Rubia cordifolia* and *R. akane*. Chem Pharm Bull (Tokyo) 31:2353–2358
11. Varma N, Painuly P, Sharma SC, Tandon JS (1985) A new anthraquinone glycoside from *Rubia cordifolia*. Indian J Chem [B] 24:791–792
12. Vidal-Tessier AM, Delaveau P, Champion B (1986) New quinones of *Rubia cordifolia* L. roots. Ann Pharm Fr 44:117–122
13. Song SY, Ding LM, Chen Y, Bai YZ (1985) Studies on the hematopoietic effect and toxicity of rubidate. Chin J Integrat Trad West Med 5:625–626
14. Talapatra SK, Sarkar AC, Talapatra B (1981) Terpenoid and related compounds. Part XVIII. Two pentacyclic triterpenes from *Rubia cordifolia*. Phytochemistry 2:1923–1927
15. Arisawa M, Ueno H, Nimura M, Hayashi T, Morita N (1986) Rubiatriol, a new triterpenoid from the Chinese drug Qian Cao Gen, *Rubia cordifolia*. J Nat Prod 49:1114–1116
16. Itokawa H, Takeya K, Mori N, Sonobe T, Mihashi S, Hamanaka T (1986) Studies on antitumor cyclic hexapeptides RA obtained from *Rubia* radix, Rubiaceae. VI. Minor antitumor constituents. Chem Pharm Bull (Tokyo) 34:3762–3768
17. Itokawa H, Takeya K, Mihara K, Mori N, Hamanaka T, Sonobe T, Iitaka Y (1983) Studies on the antitumor cyclic hexapeptides obtained from Rubiae radix. Chem Pharm Bull (Tokyo) 31:1424–1427
18. Itokawa H, Takeya K, Mori N, Sonobe T, Hamanaka T, Mihara K, Iitaka Y (1983) Studies on the antineoplastic cyclic hexapeptides obtained from Rubiae radix. I. Chemical structures of RA-VII, RA-V, RA-IV and RA-III. In: Spitzy KG, Karrer K (eds) Proceedings of the 13th International Congress on Chemotherapy. Egermann, Vienna, 284/100–284/113

19. Itokawa H, Takeya K, Mori N, Kidokoro S, Yamamoto H (1984) Studies on antitumor cyclic hexapeptides RA obtained from Rubiae Radix, Rubiaceae. IV. Quantitative determination of RA-VII and RA-V in commercial Rubiae Radix and collected plants. Planta Med 50:313–316
20. Itokawa H, Takeya K, Mori N, Takanashi M, Yamamoto H, Sonobe T, Kidokoro S (1984) Cell growth-inhibitory effects of derivatives of antitumor cyclic hexapeptide RA-V obtained from Rubiae radix. V. Gann 75:929–936
21. Itokawa H, Takeya K, Mori N, Hamanaka T, Sonobe T, Mihara K (1984) Isolation and antitumor activity of cyclic hexapeptides isolated from rubiae radix. Chem Pharm Bull (Tokyo) 32:284–290
22. Itokawa H, Takeya K, Mori N, Hamanaka T, Sonobe T, Mihara K, Tsukagoshi S (1983) Studies on the antineoplastic cyclic hexapeptides obtained from Rubiae Radix. II. Biological activities of RA-VII, RA-V, RA-IV and RA-III. In: Spitzy KG, Karrer K (eds) Proceedings of the 13th International Congress on Chemotherapy. Egermann, Vienna, 284/114–284/116

Salvia miltiorrhiza Bge. 110

110.1 Introduction

Danshen, Radix Salviae miltiorrhizae, is the dry root and rhizome of *Salvia miltiorrhiza* Bge. (Lamiaceae), collected in spring and fall. It is officially listed in the Chinese Pharmacopoeia and used for treatment of menstrual disorder, menostasis, menorrhalgia, insomnia, blood circulation diseases, and angina pectoris as well as against inflammation.

110.2 Chemical Constituents

The major chemical constituents of the root of *S. miltiorrhiza* are diterpene pigments with a phenanthrenequinone structure, especially phenanthrofurane quinone derivatives. They have been classified into two series, a phenanthro[1,2-*b*]furan-10,11-dione (*110-1*) series and a phenanthro[3,2-*b*]furan-7,11-dione (*110-2*) series.

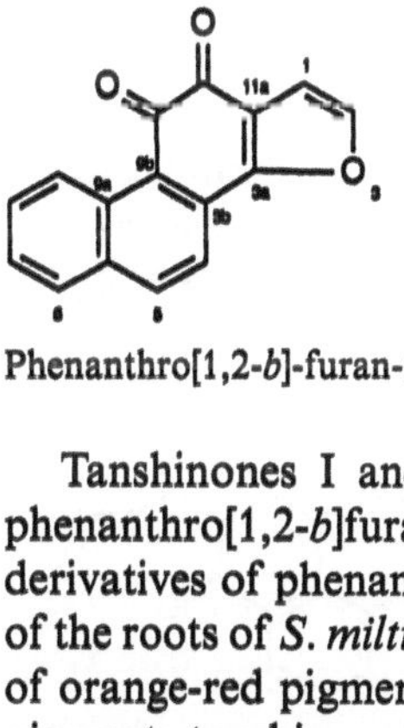

Phenanthro[1,2-*b*]-furan-10,11-dione (*110-1*)

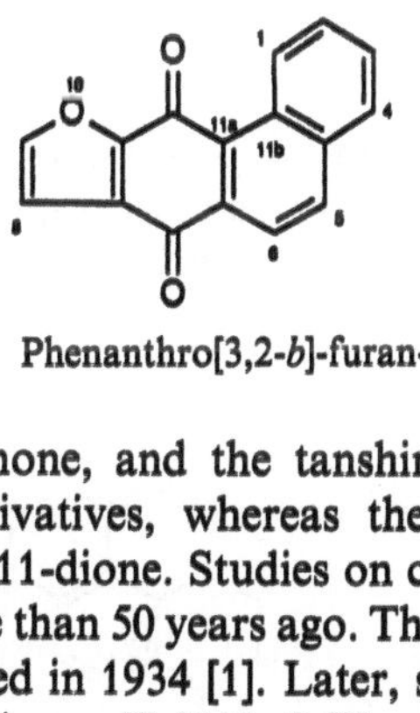

Phenanthro[3,2-*b*]-furan-7,11-dione (*110-2*)

Tanshinones I and II, cryptotanshinone, and the tanshindiols belong to the phenanthro[1,2-*b*]furan-10,11-dione derivatives, whereas the isotanshinones are derivatives of phenanthro[3,2-*b*]furan-7,11-dione. Studies on chemical components of the roots of *S. miltiorrhiza* began more than 50 years ago. The presence of a group of orange-red pigments was first reported in 1934 [1]. Later, structures of isolated pigments tanshinone I (*110-3*) [2], tanshinone II (*110-4*) [3], and cryptotanshinone (*110-5*) [4] were determined.

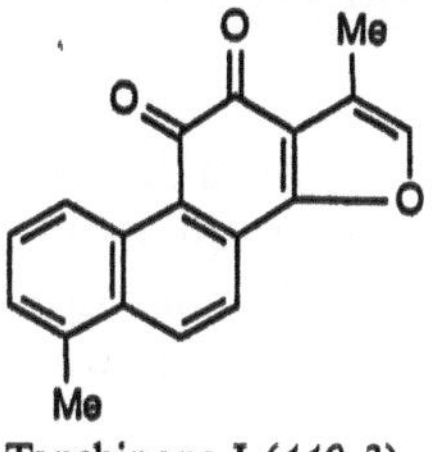

Tanshinone I (*110-3*)

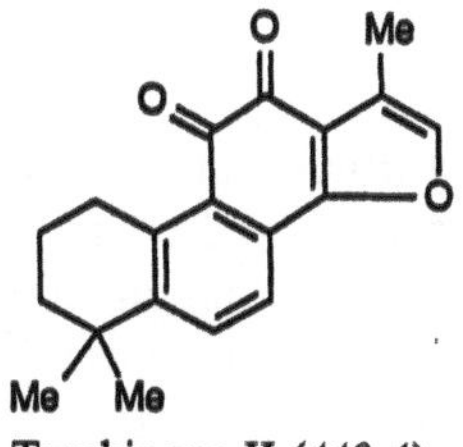

Tanshinone II (*110-4*)

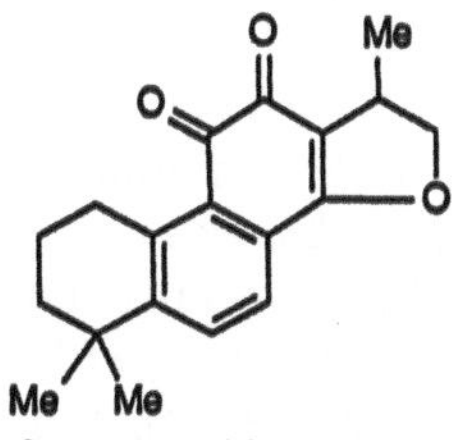

Cryptotanshinone (*110-5*)

Some years later, tanshinone IIB (*110-6*) [5], 9-hydroxytanshinone II (*110-7*) [6], and methyl tanshinonate (*110-8*) [7] were successively isolated and structurally identified.

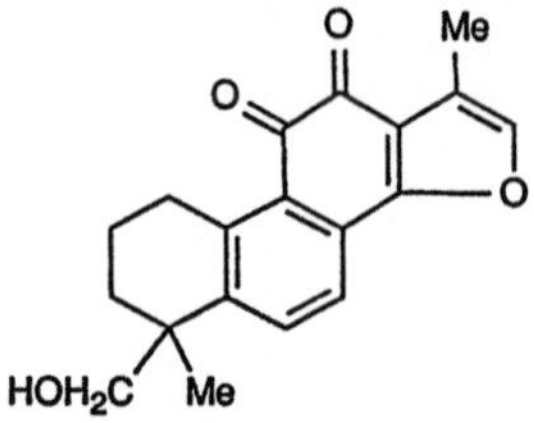

Tanshinone IIB (*110-6*)

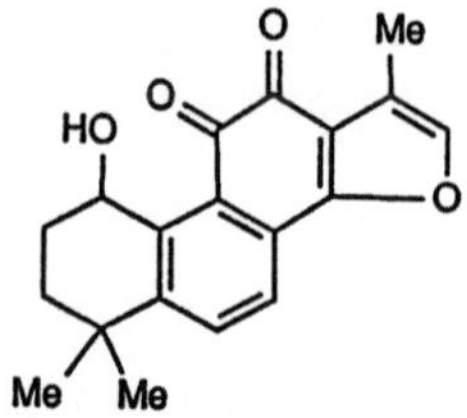

9-Hydroxytanshinone II (*110-7*)

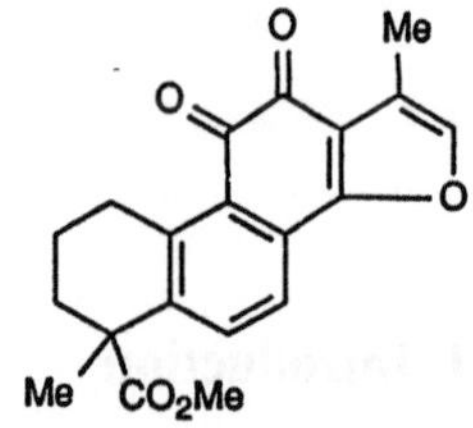

Methyl tanshinonate (*110-8*)

A related compound with a methylene group at position 6, named methylene tanshinquinone (*110-9*) [7], was further isolated and its structure determined by spectral analysis.

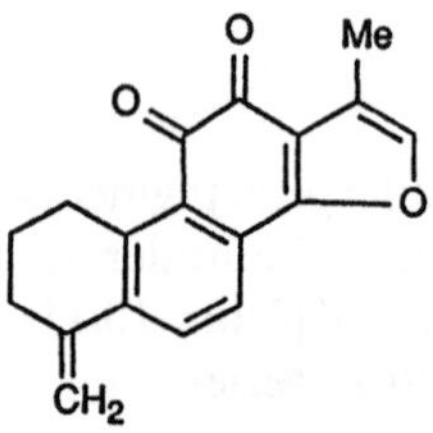

Methylene tanshinquinone (*110-9*)

Pigments of the phenanthro[3,2-*b*]furan-7,11-dione type are isotanshinone I (*110-10*), isotanshinone II (*110-11*), and isocryptotanshinone (*110-12*) [8]. Brieskorn et al. [9] discussed the biogenesis of tanshinones and classified these compounds as diterpenes.

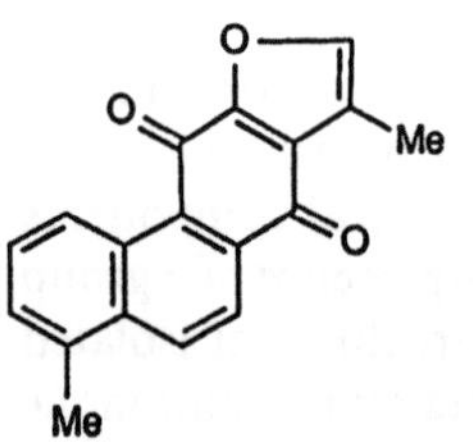

Isotanshinone I (*110-10*)

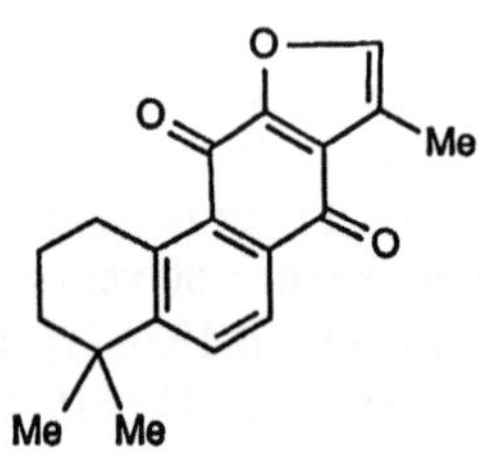

Isotanshinone II (*110-11*)

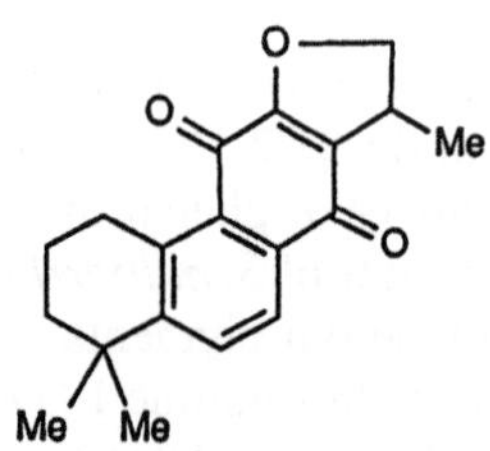

Isocryptotanshinone (*110-12*)

Minor compounds with a phenanthrofurandione structure isolated from the root of *S. miltiorrhiza* are tanshindiol A (*110-13*), tanshindiol B (*110-14*), tanshindiol C (*110-15*), nortanshinone (*110-16*), 7α-hydroxytanshinone II (*110-17*) [10], dihydrotanshinquinone (*110-18*) [11], didehydrotanshinone II [12], dihydroisotanshinone I (*110-19*) [13, 14], and isotanshinone IIB (*110-20*) [15].

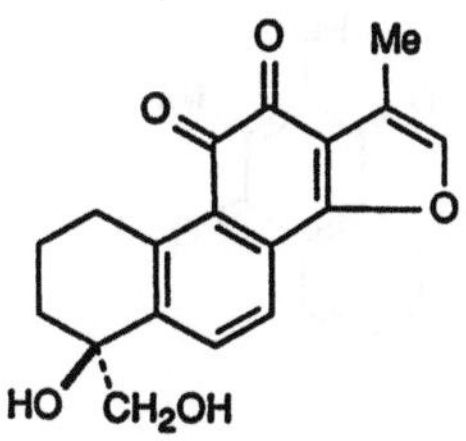

Tanshindiol A (*110-13*)

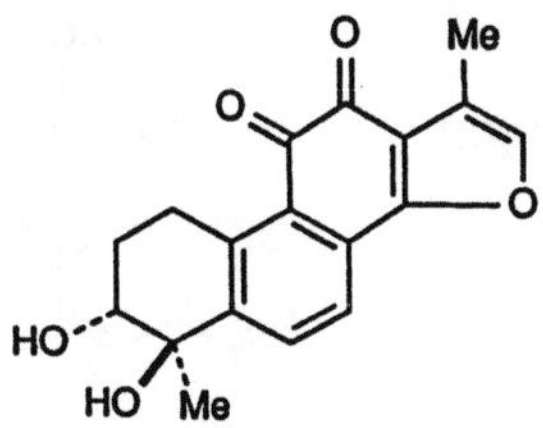

Tanshindiol B (*110-14*)

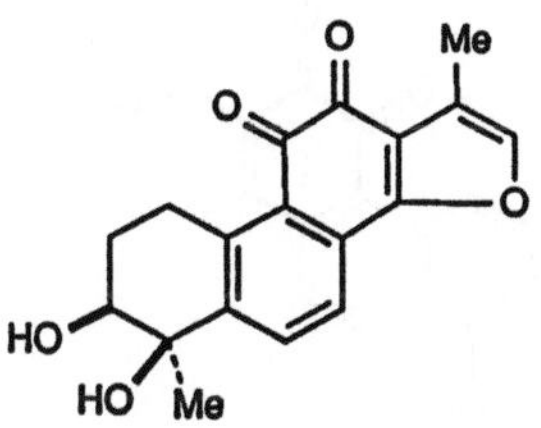

Tanshindiol C (*110-15*)

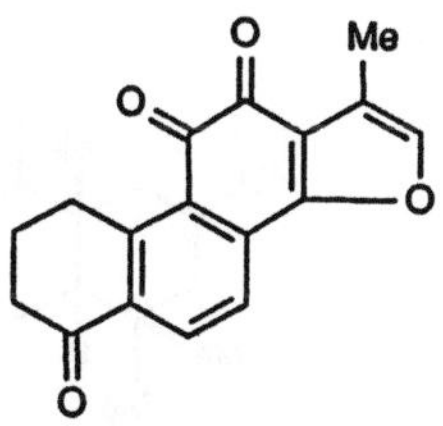

Nortanshinone (*110-16*)

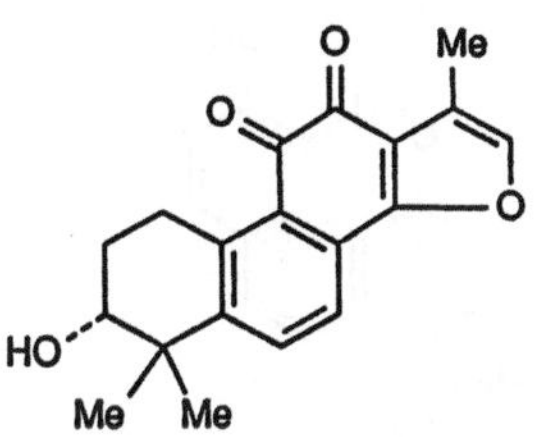

7α-Hydroxytanshinone II (*110-17*)

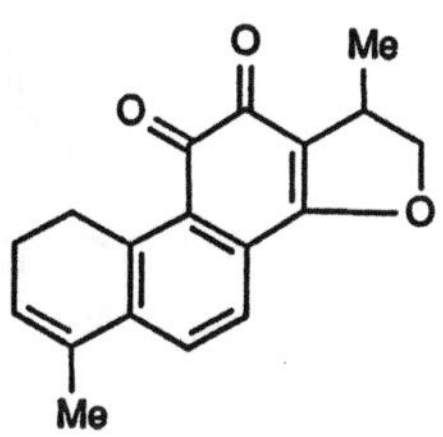

Dihydrotanshinquinone (*110-18*)

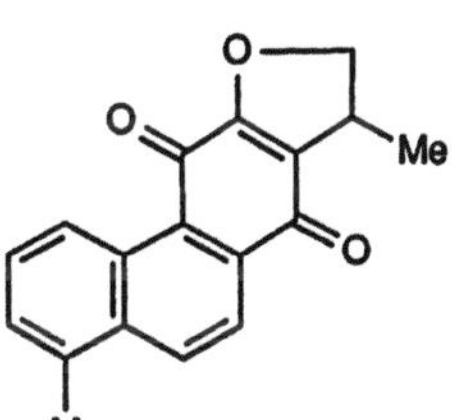

Dihydroisotanshinone I (*110-19*)

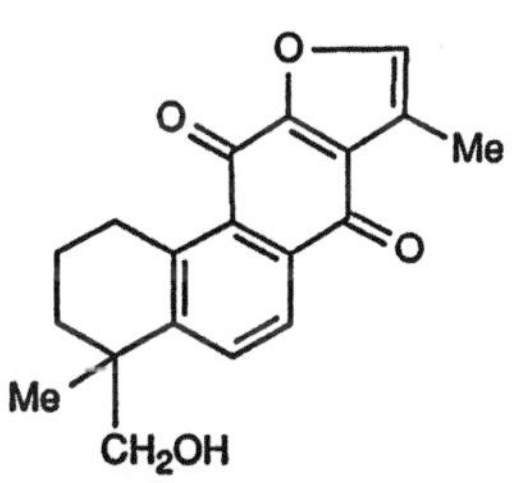

Isotanshinone IIB (*110-20*)

Further minor components derived from a phenanthrene skeleton without a furan ring were isolated from *S. miltiorrhiza*. These are danshenxinkun A (*110-21*), danshenxinkun B (*110-22*), danshenxinkun C (*110-23*) [16], miltirone (*110-24*) [17], salviol (*110-25*) [18], Ro 09-0680 (*110-26*) [19], ferruginol (*110-27*) [12, 20], dehydromiltirone (*110-28*) [12, 21], miltiodiol (*110-29*) [22], and miltionone I (*110-30*) [23].

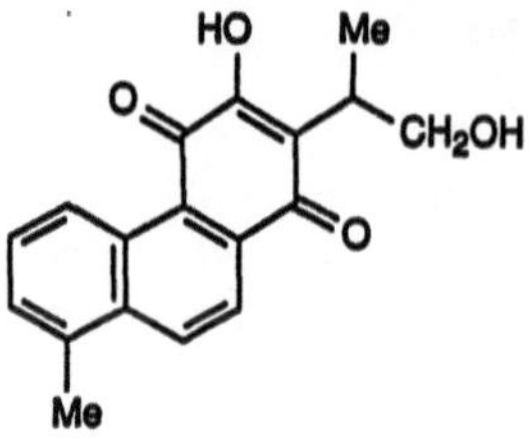

Danshenxinkun A (*110-21*)

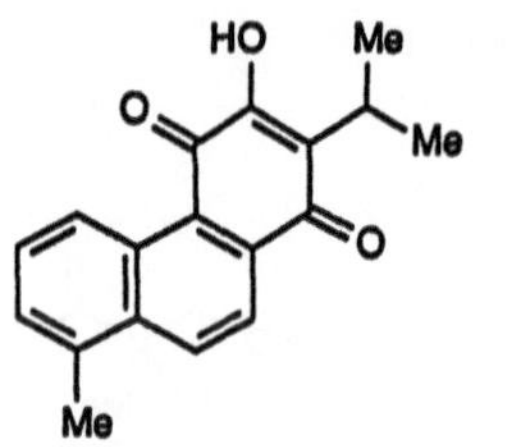

Danshenxinkun B (*110-22*)

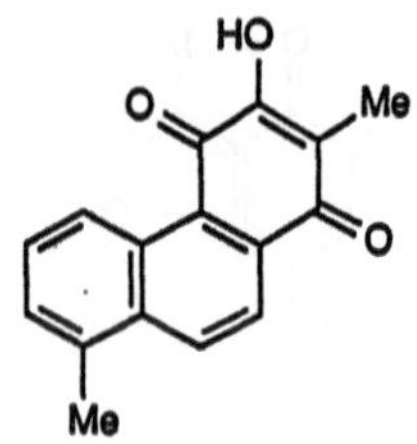

Danshenxinkun C (*110-23*)

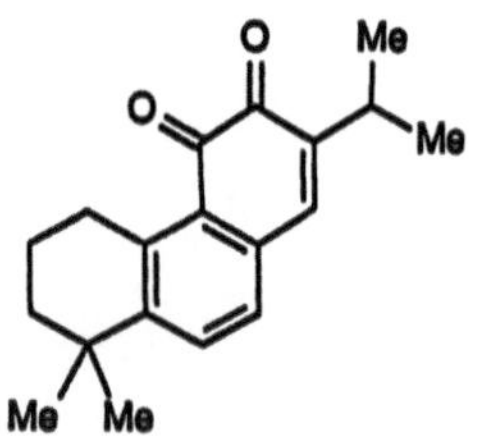

Miltirone (*110-24*)

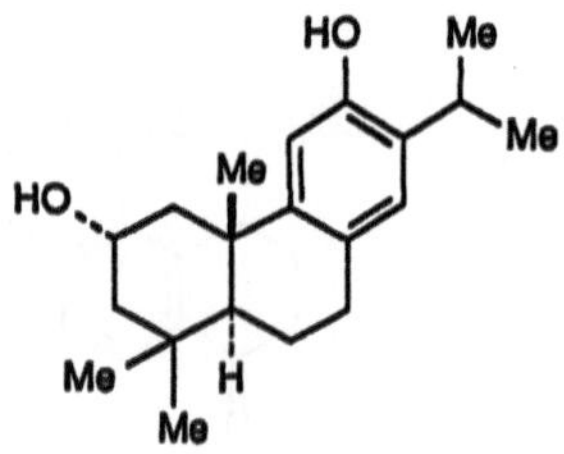

Salviol (*110-25*)

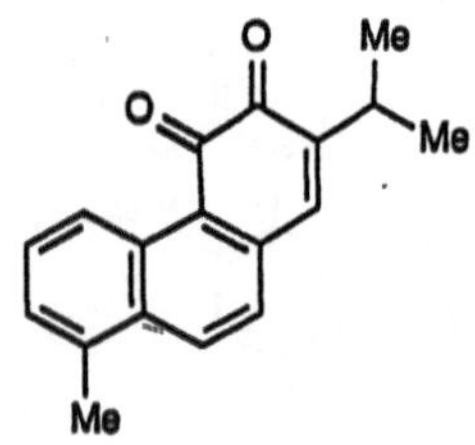

Ro 09-0680 (*110-26*)

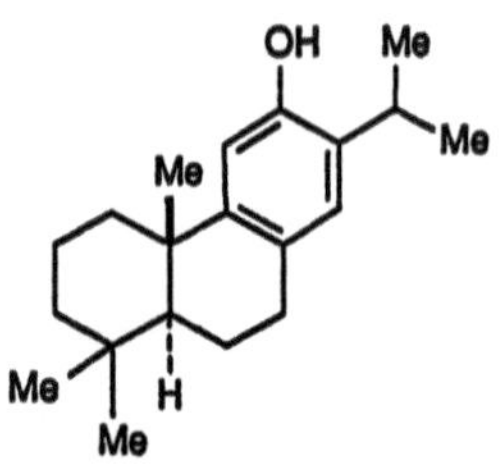

Ferruginol (*110-27*)

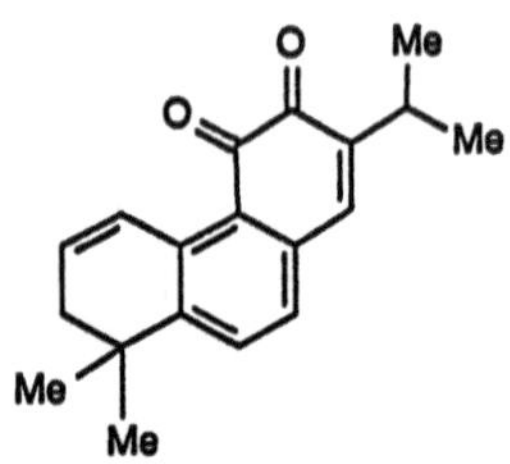

Dehydromiltirone (*110-28*)

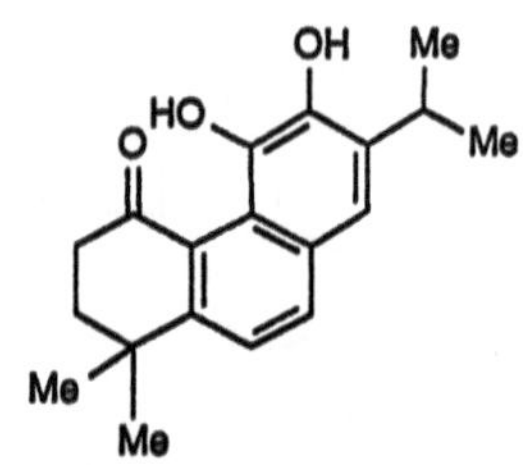

Miltiodiol (*110-29*)

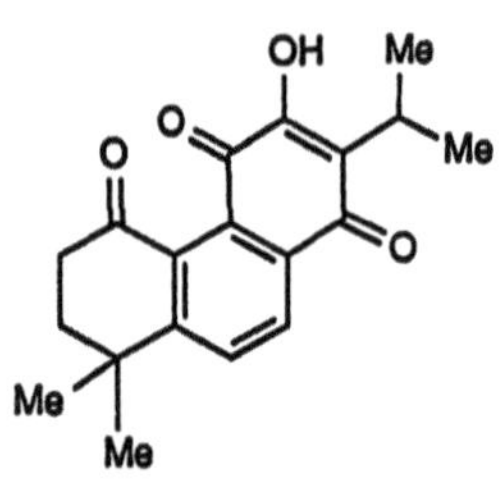

Miltionone I (*110-30*)

Furthermore, a diterpene derivative with a phenalenofuran skeleton named salvilenone (*110-31*) [24], spiro ketal lactones named danshenspiroketal lactone (*110-32*) [14, 25, 26], and epi-danshenspiroketal lactone (*110-33*) [27], danshenxinkun D (*110-34*), a phenanthropyrandione derivative [28], tanshinlactone (*110-35*), a furonaphthopyrane derivative [29], and miltionone II (*110-36*), a phenanthro[1,2-*b*]-furan-5,11-dione [23] were isolated.

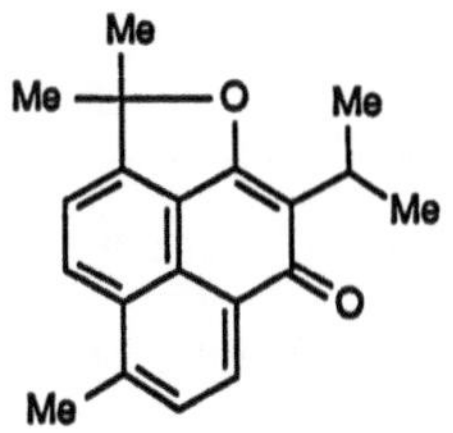

Salvilenone (*110-31*)

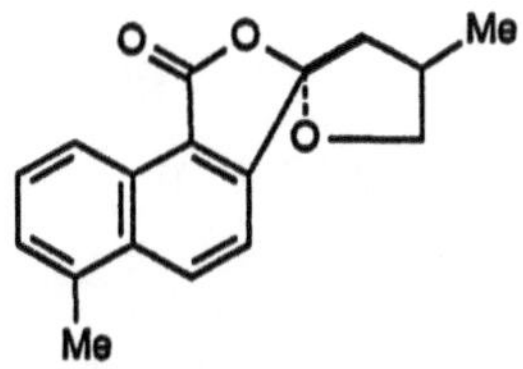

Danshenspiroketal lactone (*110-32*)

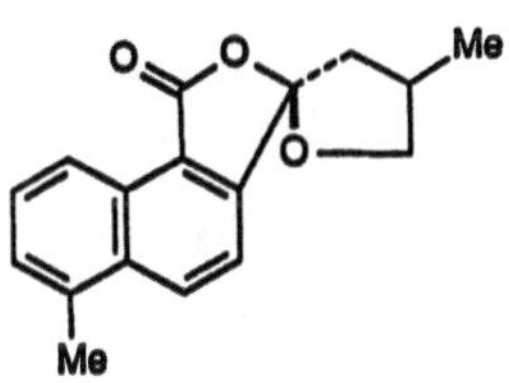

epi-Danshenspiroketal lactone (*110-33*)

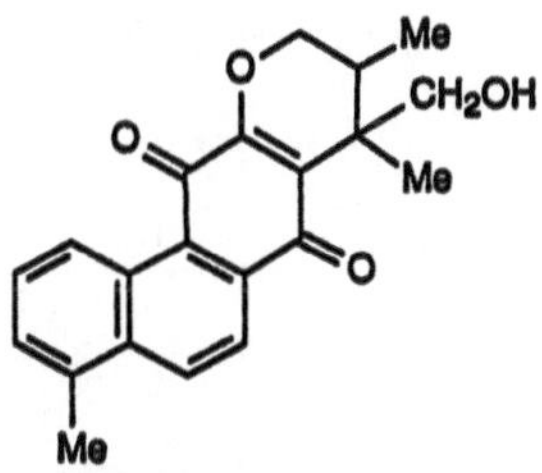

Danshenxinkun D (*110-34*)

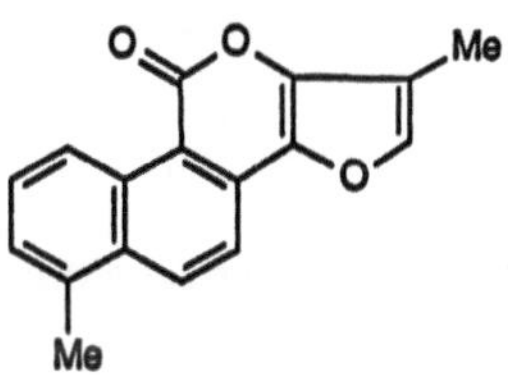

Tanshinlactone (*110-35*)

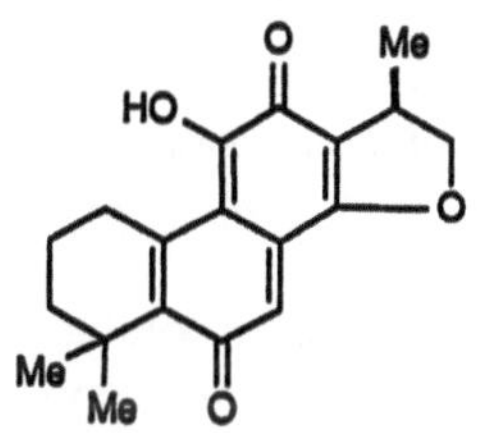

Miltionone II (*110-36*)

Several studies described the total syntheses of tanshinones [5, 30, 31]. The product of the Diels-Alder reaction (*110-39*) of 3-methylbenzofuran-4,7-quinone (*110-37*) with 6,6-dimethyl-1-vinyl-cyclohexene (*110-38*) was converted into isotanshinone II (*110-11*) by air oxidation in alcoholic KOH solution. The dihydro derivative of isotanshinone II was formed by catalytic hydrogenation over palladium charcoal and rearranged to cryptotanshinone (*110-5*) by alkaline hydrolysis followed by acidic cyclization. Oxidation of cryptotanshinone gave tanshinone II (*110-4*) (Fig. 110.1) [18]. The same procedures were applied to the synthesis of tanshinone I using *O*-methylstyrene instead of 6,6-dimethyl-1-vinylcyclohexene as the starting material [18].

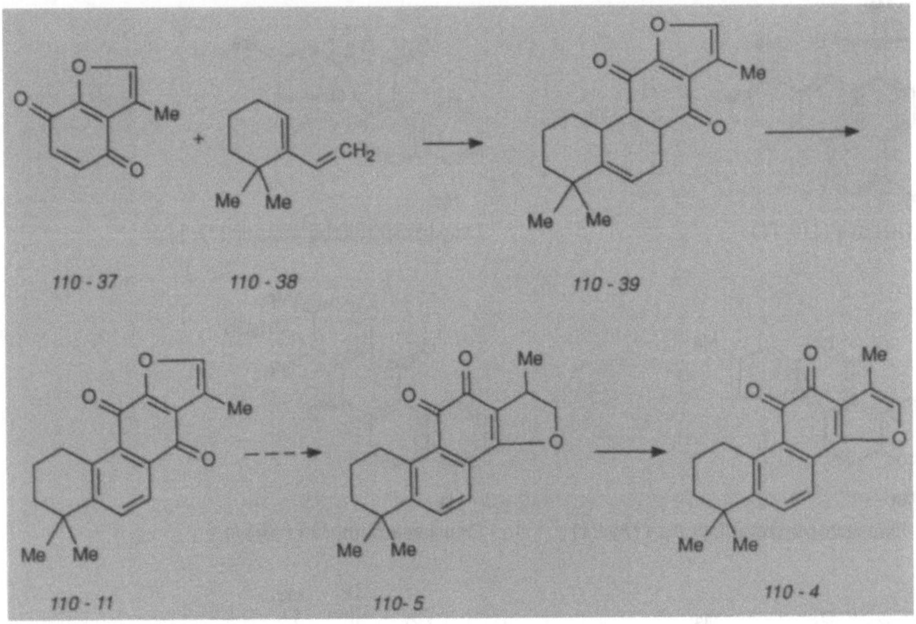

Fig. 110.1. Total synthesis of tanshinones

Furthermore, a series of water-soluble compounds was isolated from the aqueous extract of *S. miltiorrhiza*. The more important compounds were identified as protocatechualdehyde [32], danshensu (*110-40*) [33, 34], salvianolic acid A (*110-41*) [35], salvianolic acid B (*110-42*) [36, 37], and salvianolic acid C (*110-43*) [36].

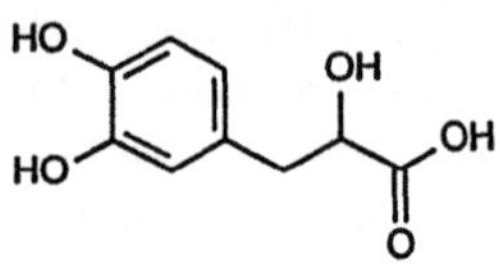

Danshensu (*110-40*)

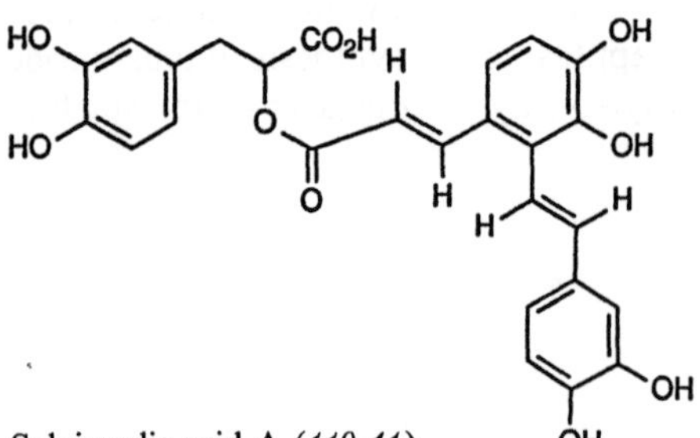

Salvianolic acid A (*110-41*)

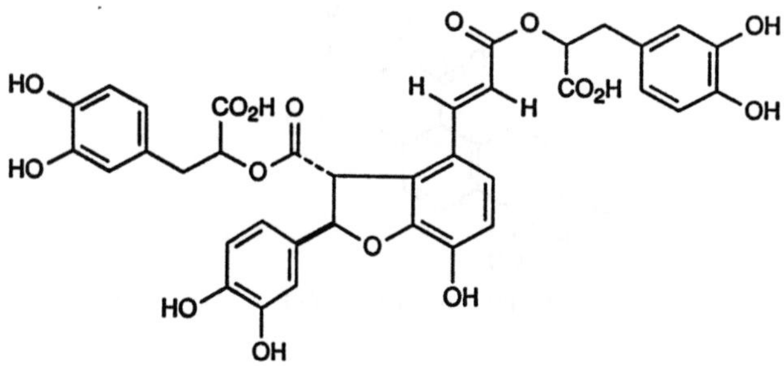

Salvianolic acid B (*110-42*)

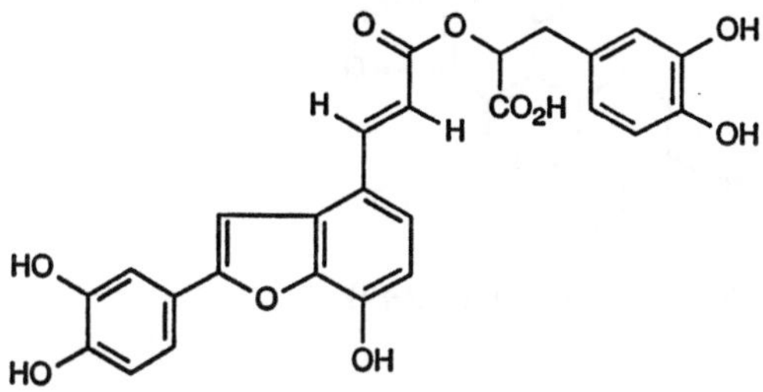

Salvianolic acid C (*110-43*)

A study of the occurrence of phenanthroquinones in the genus *Salvia* found those of the tanshinone type in 39 of 89 species. The occurrence of quinones can be used as a taxonomic characteristic for the genus *Salvia* [38].

In a search for plant material containing tanshinones, the contents of tanshinone II, methylene tanshinquinone, and tanshinone I of 13 *Salvia* species or varieties collected from 10 provinces in China were investigated [39]. The contents of these three tanshinones ranged from 0.1% to 1.0%. *S. kiaometiensis* had the highest content of tanshinone II, *S. przewalskii* had the highest content of methylene tanshinquinone, and *S. castanea* had the highest content of tanshinone I [40]. Tanshinone contents in *Salvia* species were highest in November [41]. The root of *S. miltiorrhiza* from different areas showed varying contents of cryptotanshinone and phenolic components [42].

Salvia przewalskii is a traditional Tibetan drug. It was demonstrated to be an appropriate source for tanshinones. Besides the known compounds, tanshinone I, tanshinone II, cryptotanshinone, methyl tanshinonate [43], methylene tanshinquinone, and 1,2-dihydrotanshinquinone [44], new phenanthrofurandione pigments named przewaquinones A (*110-44*), B (*110-45*) [44], C (*110-46*) [45], D (*110-47*), E (*110-48*) [45, 46], and F (*110-49*) [45] were isolated and structurally determined.

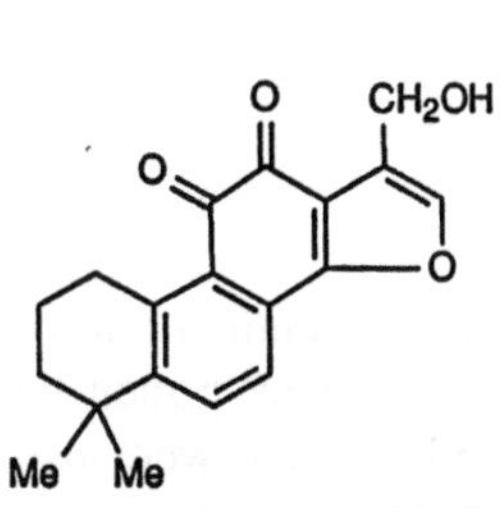

Przewaquinone A (*110-44*)

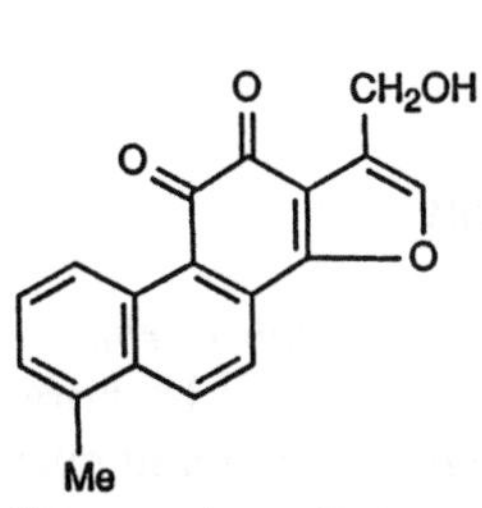

Przewaquinone B (*110-45*)

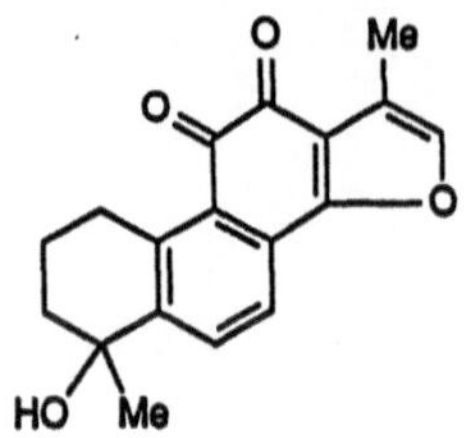

Przewaquinone C (*110-46*)

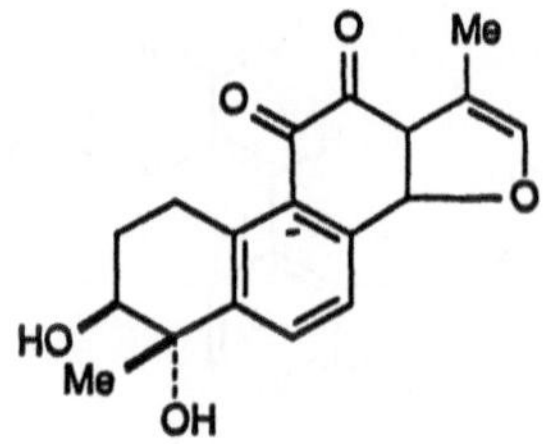

Przewaquinone D (*110-47*)

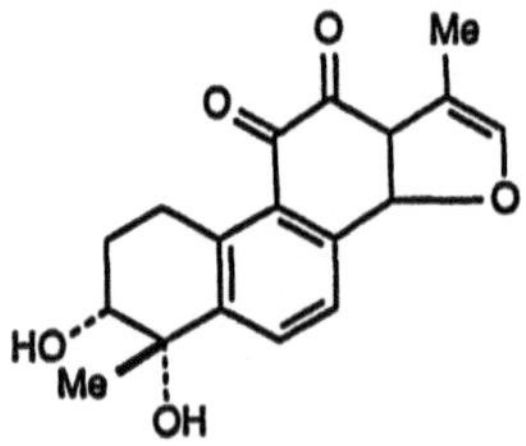

Przewaquinone E (*110-48*)

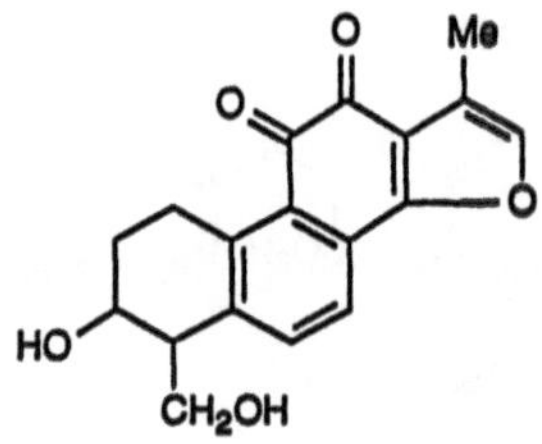

Przewaquinone F (*110-49*)

Przewaquinone A and przewaquinone B are characterized by the hydroxymethyl group at position 1. These two compounds were also detected in the root of *S. miltiorrhiza* [47].

Nine diterpene quinones were isolated from the roots of *S. trijuga*, which were found to be tanshinone I, tanshinone II, tanshinone IIB, cryptotanshinone, methylene tanshinquinone, dihydrotanshinone I, hydroxytanshinone II, and methyl tanshinonate. A new compound was identified as 7-hydroxymethylene tanshinquinone (*110-50*) [48]. In addition, *S. paramiltiorrhiza* f. *purpureorubra* was tested as a substitute for *S. miltiorrhiza*. No significant differences in contents of tanshinone II and protocatechualdehyde in the roots of either species were found [49].

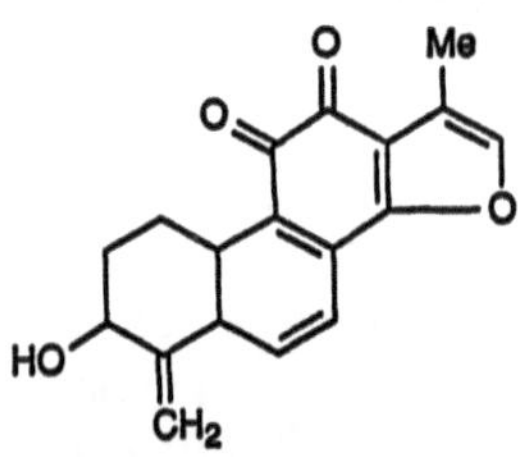

7-Hydroxymethylene tanshinquinone (*110-50*)

110.3 Pharmacology

The extract of the root of *S. miltiorrhiza* was reported to be effective against microcirculation disturbances induced by noradrenaline in the hamster cheek pouch by increasing the arteriolar diameter and microcirculation rate. Similar effects were also seen in the venous and capillary microcirculation [50].

Tanshinone is one of the active principles of *S. miltiorrhiza*. The pure substance has limited clinical use because it is poorly soluble in water. A water-soluble derivative, tanshinone II sodium sulfonate, was prepared [7] and tested biologically. In dogs, injection of tanshinone II sodium sulfonate at a dose of 1 mg/kg into the distal end of the descending coronary artery beyond the occlusion significantly reduced the size of acute myocardial infarct 24 h after administration. The vascular deficit areas markedly diminished or disappeared. The results suggest that the beneficial effects of tanshinone II sodium sulfonate on ischemic hearts may be related to an acceleration of the opening of coronary collaterals [51, 52].

A clinical investigation of tanshinone sodium sulfonate in 180 patients with coronary heart disease having abnormal electrocardiograms showed a significant effect as evidenced by electrocardiograms as well as improvement of angina pectoris and chest oppression [7].

In vitro, tanshinone II sodium sulfonate showed a membrane-stabilizing effect on red blood cells as evidenced by increased resistance against hemolysis caused by hypotonic solution, heat, low pH, or saponin [53].

No toxic effects were observed in mice and rats after oral or s.c. administration of the extract of *S. miltiorrhiza*. Antipyretic activity of the extract in rabbits and antiinflammatory activity in rats with infective arthritis and in mice with croton oil induced inflammation of the ear were also reported [54]. Furthermore, tanshinone-related pigments showed bacteriostatic activity against *Staphylococcus aureus* [54] and *Mycobacterium* sp. [55]. However, these pigments were bound by plasma protein to varying degrees and protein binding significantly decreased their bacteriostatic activity [55].

Results of pharmacokinetic studies with [^{3}H]tanshinone II sodium sulfonate in rats were reported. Radioactivity was highest in liver, followed by spleen, kidney, and lung after i.v. administration. Peak levels of radioactivity in organs were reached 2 h after injection. The half-lives of the substance for the fast and slow phase were 27 and 199 min, respectively. Within 72 h, 75% of the administered activity was excreted in the feces and 18% in the urine. Three undefined metabolites were found by thin-layer chromatography and autoradiography from bile, urine, and feces [56]. Biotransformation of cryptotanshinone into tanshinone II by hydrogenation in the liver was observed after administration of cryptotanshinone into the duodenum of rats [57].

Danshensu, 3,4-dihydroxyphenyllactic acid, is another active compound isolated from the root of *S. miltiorrhiza*. It was found to dilate isolated swine coronary artery and to antagonize the constricting response elicited by morphine and propranolol. These interactions have been considered of practical importance in cases when propranolol or morphine might be administered concomitantly with danshensu for treatment of severe angina attacks. On the other hand, danshensu showed no antagonistic effect on the contracture response of coronary artery elicited by high potassium medium [58].

Moreover, decoctions made from the root of *S. miltiorrhiza* were found to be effective in reducing enhanced serum glutamic pyruvic transaminase and pathological changes in rabbits with acute liver damage induced by CCl_4 [59]. They were also reported to be effective in restoring liver function and in preventing liver fibrosis in clinical studies [60].

References

1. Nakao H, Fukushima T (1934) Chemical composition of *S. miltiorrhiza* I. Yakugaku Zasshi 54:844–858
2. Von Wessely F, Wang S (1940) Über einen neuartigen natürlichen Chinonfarbstoff aus der Klasse eines Phenanthrofurans. Chem Ber 73:19–24
3. Okumura Y, Kakisawa H, Kato M, Hirata Y (1961) Structure of tanshinone II. Bull Chem Soc Jpn 34:895–897
4. Takiura K, Koizumi K (1962) Components of the Chinese drug tanshin. IV. The structure of tanshinone IIA. Chem Pharm Bull (Tokyo) 10:112–116
5. Baillie AC, Thomson RH (1968) Naturally occurring quinones. Part XI. The tanshinones. J Chem Soc [C] 48–52
6. Kakisawa H, Hayashi T, Okazaki I, Ohashi M (1968) Isolation and structures of new tanshinones. Tetrahedron Lett 3231–3234
7. Chen MK, Young PT, Ku WH, Chen ZX, Chen HT, Yeh HC (1978) Studies on the active principles of Dan-Shen. I. The structure of sodium tanshinone II A sulfonate and methylene tanshinquinone. Acta Chim Sin 36:199–206
8. Kakisawa H, Hayashi T, Yamazaki T (1969) Structures of isotanshinones. Tetrahedron Lett 301–304
9. Brieskorn CH, Fuchs A, Bredenberg JB, McChesney JD, Wenkert E (1964) The structure of carnosol. J Org Chem 29:2293–2298
10. Luo H, Wu BJ, Wu MY, Yong ZG, Niwa M, Hirata Y (1985) Pigments from *Salvia miltiorrhiza*. Phytochemistry 24:815–817
11. Feng BS, Li SR (1980) Studies of chemical components of Danshen – discovery of two new isomers. Acta Pharm Sin 15:489–494
12. Luo HW, Hu XJ, Wang N, Ji J (1988) Platelet aggregation inhibitors from *Salvia miltiorrhiza* Bunge. Acta Pharm Sin 23:830–834
13. Kong DY, Liu XJ (1984) Structure of dihydroisotanshinone I from Dan Shen. Acta Pharm Sin 19:755–759
14. Kong DY, Liu XJ (1983) Studies on the chemical components of "Dan Shen". Acta Acad Med Primae Shanghai 10:313–315
15. Lee AR, Wu WL, Chang WL, Lin HC, King ML (1987) Isolation and bioactivity of new tanshinones. J Nat Prod 50:157–160
16. Fang CN, Chang PL, Hsu TP (1976) The antibacterial components of Dan-shen. Acta Chim Sin 34:197–209
17. Hayashi T, Kakisawa H, Hsu HY, Chen YP (1970) Structure of miltirone, a new diterpenoid quinone. J Chem Soc [D] 299
18. Kakisawa H, Hayashi T, Handa T, Ohashi M, Hsu HY, Chen YP (1971) Structure of salviol, a new phenolic diterpene. J Chem Soc [D] 541–542
19. Onitsuka M, Fujiu M, Shinma N, Maruyama HB (1983) New platelet aggregation inhibitors from Tan-Shen, *Salvia miltiorrhiza* Bunge root. Chem Pharm Bull (Tokyo) 31:1670–1675
20. Tomita Y, Annaka M, Ikeshiro Y (1989) Biosynthesis of the abietane type diterpene ferruginol in cell cultures of *Salvia miltiorrhiza*: synthesis of (15S)- and (15R)-(16-$^{2}H_{1}$)-12-*O*-methylferruginol by enzymic resolution of 12-*O*-methyl-16-hydroxyferruginol and stereochemistry of 1,2-methyl migration in the formation of the isopropyl group. J Chem Soc Chem Commun 108–110
21. Lin LZ, Wang XM, Huang XL, Huang Y, Yang BJ (1988) A new diterpenoid quinone dehydromiltirone. Acta Pharm Sin 23:273–275
22. Ginda H, Kusumi T, Ishitsuka MO, Kakisawa H, Zhao WJ, Chen J, Guo YT (1988) Salviolone, a cytotoxic bisnorditerpene with a benzotropolone chromophore from a Chinese drug Dan-Shen (*Salvia miltiorrhiza*). Tetrahedron Lett 29:4603–4606
23. Ikeshiro Y, Mase I, Tomita Y (1989) Abietane type diterpenoids from *Salvia miltiorrhiza*. Phytochemistry 28:3139–3141
24. Kusumi T, Ooi T, Hayashi T, Kakisawa H (1985) A diterpenoid phenalenone from *Salvia multiorrhiza*. Phytochemistry 24:2118–2120
25. Kong DY, Liu XJ, Teng MK, Rao ZH (1985) The structure of Dan Shen spiroketallactone of Dan Shen (*Salvia miltiorrhiza* Bunge). Acta Pharm Sin 20:747–751
26. Teng MK, Rao ZH, Kong DY, Liu XJ (1985) Crystal and molecular structure of Danshen spiroketal lactone (crystal IX). Zhong-guo Kexue Jishu Daxue Xuebao 15:92–99

27. Luo HW, Chen SX, Lee JN, Snyder JK (1988) Epi-danshenspiroketal lactone from *Salvia miltiorrhiza*. Phytochemistry 27:290–292
28. Luo HW, Wu BJ, Wu MY, Yong ZG, Jin Y (1985) Isolation and structures of danshenxinkun D. Acta Pharm Sin 20:542–544
29. Luo HW, Ji J, Wu MY, Yong ZG, Niwa M, Hirata Y (1986) Tanshinlactone, a novel secoabietanoid from *Salvia miltiorrhiza*. Chem Pharm Bull (Tokyo) 34:3166–3168
30. Kakisawa H, Tateishi M, Kusumi T (1986) Synthesis of tanshinone II and cryptotanshinone. Tetrahedron Lett 3783–3786
31. Kakisawa H, Inouye Y (1968) Total synthesis of tanshinone I, tanshinone II and cryptotanshinone. Chem Commun 1327
32. Chen W, Liu HC (1980) Isolation and identification of catecholaldehyde from danshen (*Salvia miltiorrhiza*). Chin Trad Herb Drugs 11:442
33. Zhang DC, Wu CL, Liu HC, Pan TC, Yang CH, Chang PT (1980) Studies on the water-soluble active principles in *Salvia miltiorrhiza* Bunge. II. Structure of D-(+)-(3,4-dihydroxyphenyl)lactic acid. Acta Acad Med Primae Shanghai 7:384–385
34. Chen ZX, Gu WH, Huang HZ, Yang XM, Sun CJ, Chen WZ, Dong YL, Ma HL (1981) Studies on the water-soluble phenolic acid constituents of *Salvia miltiorrhiza*. Chin Pharm Bull 16:24–25
35. Li LN, Tan R, Chen WM (1984) Salvianolic acid A, a new depside from roots of *Salvia miltiorrhiza*. Planta Med 50:227–228
36. Ai CB, Li LN (1988) Stereostructure of salvianolic acid B and isolation of salvianolic acid C from *Salvia miltiorrhiza*. J Nat Prod 51:145–149
37. Yokozawa T, Chung HY, Oura H, Nonaka G, Nishioka I (1988) Isolation of the active component having the uremic-preventive effect from *Salvia miltiorrhiza* radix extract. Chem Pharm Bull (Tokyo) 36:316–320
38. Patudin AV, Romanova AS, Sokolov VS, Pribylova GF (1974) Occurrence of phenanthrenquinones in the genus *Salvia*. Planta Med 26:201–207
39. Huang XL, Yang BJ, Huang HZ, Xu Y, Hu ZB (1980) Studies of the active principles of Dan Shen. III. Search for plant resources containing tanshinone IIA. Acta Bot Sin 22:98–100
40. Huang XL, Yang BJ, Hu ZB (1981) Studies on the active principles of Dan-Shen. II. Searching for plant resources containing tanshinone IIA. Acta Bot Sin 23:70–71
41. Huang XL, Yang PC, Hu CP (1980) A brief note on the active principles of *Salvia miltiorrhiza* Bunge in relation to the sources and seasons. Chin Trad Herb Drugs 11:276
42. Lu HB (1986) Comparison of cryptotanshinone and phenolic components in Danshen from different areas. Bull Chin Mater Med 11:750
43. Xu XL, Xiao ZY (1982) Studies on chemical constituents of Hong Qin Jiu (*Salvia przewalskii* Maxim.) a Tibetan drug. I. Chin Trad Herb Drugs 13:485–488
44. Xu XM, Xiao ZY (1984) Studies on the chemical constituents of the Tibetan traditional drug Hong Qin Jiao (*Salvia przewalskii*). II. Chin Trad Herb Drugs 15:1–4
45. Yang BJ, Huang XL, Zhou QR (1984) Structures of four minor diterpene quinones from the root of *Salvia przewalskii* Maxim. var. *mandarinorum* (Diels) Stib. Acta Pharm Sin 19:274–281
46. Yang BJ, Hong SH (1985) Stereochemistry of przewaquinone D and przewaquinone E. Acta Chim Sin 43:898–900
47. Yang BJ, Qian MK, Qin GW, Chen ZX (1981) Studies on active principles of Dan-shen (*Salvia miltiorrhiza* Bunge). V. Isolation and chemical structure of przewaquinone A and przewaquinone B. Acta Pharm Sin 16:837–841
48. Yang BJ, Huang TL, Hu ZB, Chen ZX (1982) Study on chemical constituents of *Salvia trijuga* Diels. Chin Pharm Bull 17:242
49. Wang XM (1981) Preliminary comparison of the quality between *Salvia paramiltiorrhiza* f. *purpureo-rubra* and *Salvia miltiorrhiza* Bunge. Chin Pharm Bull 16:8–9
50. Xue QF, Dai SL, Wu YQ, Yuan SY, Zhu LX, Liu CG, Wang ZY (1986) Effect of ligustrazine and *Salvia miltiorrhiza* on microcirculation in the hamster cheek pouch. Natl Med J China 66:334–337
51. Wang YP, Chen YH, Xu DZ, Jiang WD (1980) Effect of "tanshinone IIA" sodium sulfonate on the hemodynamics and extent of myocardial infarct in coronary occluded dogs. Acta Acad Med Primae Shanghai 7:347–353
52. Jiang WD, Yu YZ, Liu WW, Chen YH, Wang YP, Huang TC (1981) Effects of sodium tanshinone IIA sulfonate and propranolol on coronary collaterals in infarcted dogs. Acta Pharmacol Sin 2:29–33

53. Chen WZ, Dong DL, Wang CG, Ting GS (1979) Pharmacological studies of sodium tanshinone IIA sulfonate. Acta Pharm Sin 14:277–283
54. Gao YG, Song YM, Yang YY, Liu WF, Tang JX (1979) Pharmacology of tanshinone. Acta Pharm Sin 14:75–82
55. Luo HW, Zheng JR, Jiang BL, Xu LF (1982) Relation between the RM value and biological activity of tanshinones. J Nanjing Coll Pharm 42–48
56. Shao HS, Jing XA, Ying LQ, Zhao MM, Gu JE (1981) Distribution and metabolism of ^{3}H-tanshinone IIA sulfonate in rats. Nucl Tech 55–57
57. Luo HW, Sheng LS, Zhang SQ, Xu LF, Wei P (1983) Tanshinones: antimicrobacterial agents – bile excretion and biotransformation in rat liver. Acta Pharm Sin 18:1–6
58. Dong ZT, Jiang WD (1982) Effect of Dan Shen Su on isolated swine coronary artery perfusion preparation. Acta Pharm Sin 17:226–228
59. Cheng YX, Gan SX, Shu SF, Wang QQ, Guo X, Jing FY (1985) Treatment for acute toxipathic hepatitis in rabbits with variously processed danshen. Bull Chin Mater Med 10:161–162
60. Wang ZL (1985) Progress in clinical and experimental studies of *Salvia miltiorrhiza* in the treatment of chronic hepatitis. Chin J Integrat Trad West Med 5:509–510

Schisandra chinensis (Turcz.) Baill. 111

111.1 Introduction

Wuweizi, Fructus Schisandrae, is the dry fruits of *Schisandra chinensis* (Turcz.) Baill. or *S. sphenanthera* Rehd. et Wils. (Magnoliaceae) collected in the fall when the fruits have become ripe. It is officially listed in the Chinese Pharmacopoeia and is used as a tonic and sedative.

111.2 Chemical Constituents

The fruits of *Schisandra* have been used in traditional Chinese medicine for over 2000 years as a tonic and sedative. Some time ago they were found to lower elevated serum glutamic pyruvic transaminase (SGPT) levels [1] and were introduced in the treatment of human hepatitis [2]. A series of dibenzo[*a*,*c*]cyclooctene (*111-1*) derivatives were elucidated as the active principles. The lignan content of *Schisandra* seed varies between 7.2% and 19.2% according to species and geographical origin. Contents are highest between May and July, and stems have a lignan content of between 1.3% and 10.3% [3].

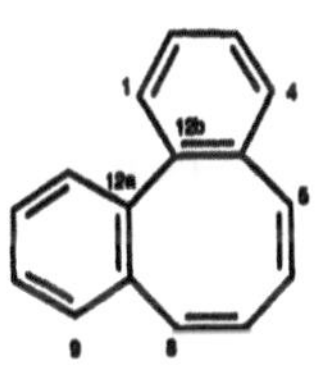

Dibenzo[*a*, *c*]cyclooctene (*111-1*)

In the past 10 years, a series of dibenzo[*a*,*c*]cyclooctene derivatives have been isolated from various *Schisandra* species, some of which showed significant biological and biochemical activity in experimental and clinical studies. The compounds isolated are listed in Table 111.1.

Table 111.1. Dibenzo[*a,c*]cyclooctene derivatives isolated from *Schisandra* species

Compound	Structure	Plant origin	Reference
Schisandrin A (Wuweizisu A, *111-2*)	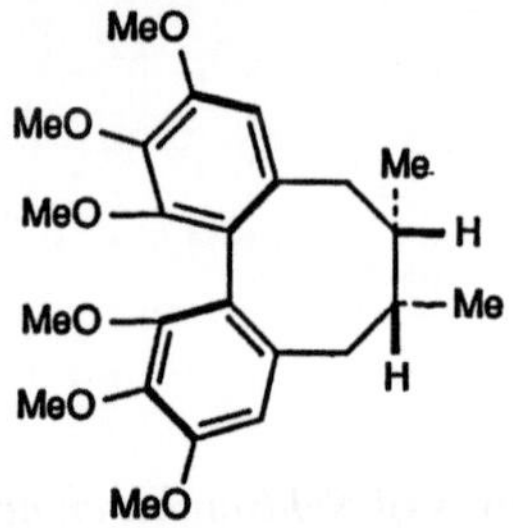	*S. chinensis* *S. sphenanthera* *S. rubriflora*	[3, 4]
Schisandrin B (Wuweizisu B, γ-Schisandrin, *111-3*)	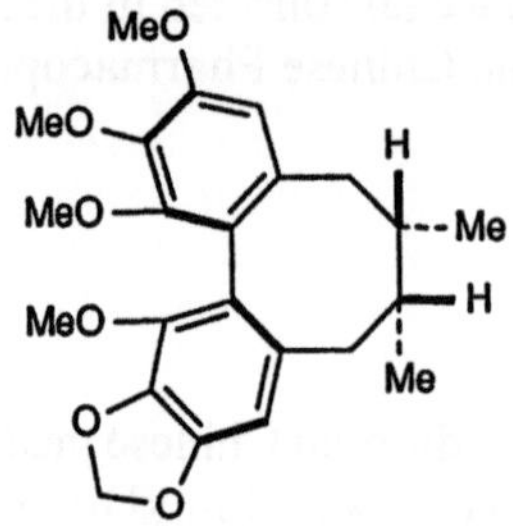	*S. chinensis* *S. rubriflora*	[3–6]
Schisandrin C (Wuweizisu C, *111-4*)	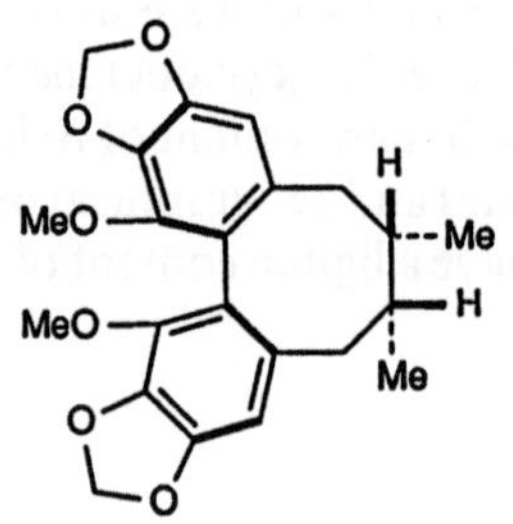	*S. chinensis* *S. rubriflora*	[3, 4, 7]
Schisandrol A (Wuweizi alcohol A, Schisandrin, *111-5*)	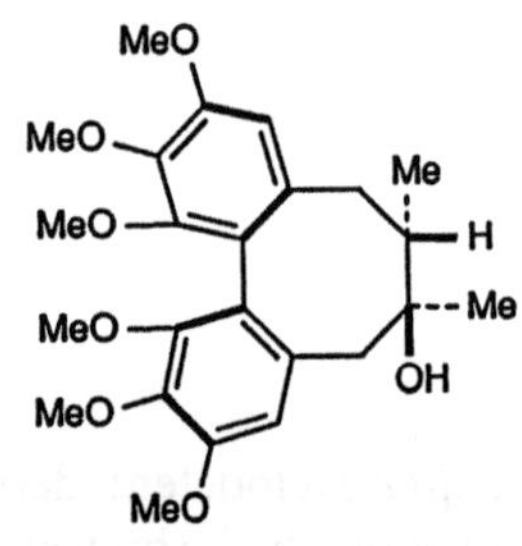	*S. chinensis*	[8, 9]
Schisandrol B (Wuweizi alcohol B, Gomisin A, *111-6*)	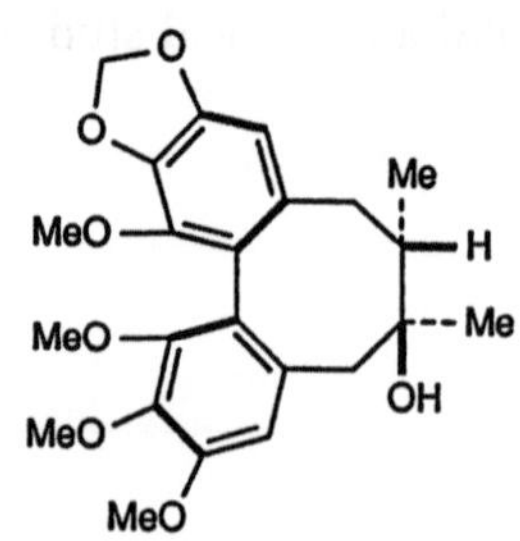	*S. chinensis* *S. rubriflora*	[9–12]

Table 111.1. (continued)

Compound	Structure		Plant origin	Reference
Schisantherin A (Wuweizi ester A, Gomisin C, *111-7*)	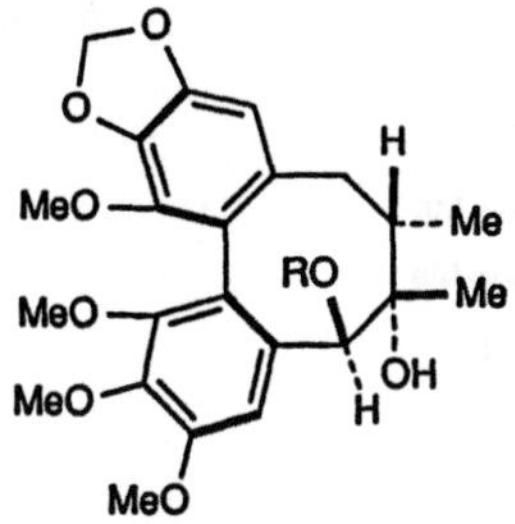	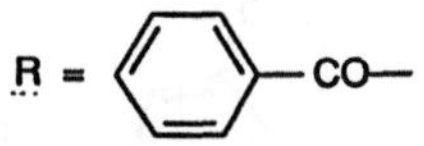	*S. chinensis* *S. sphenanthera*	[9–11, 13–16]
Schisantherin B (Wuweizi ester B, Gomisin B, *111-8*)		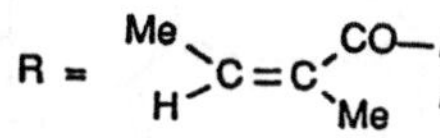	*S. chinensis* *S. sphenanthera*	[9–11, 14–16]
Schisantherin C (*111-9*)		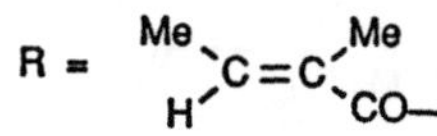	*S. sphenanthera*	[16]
Schisantherin D (*111-10*)	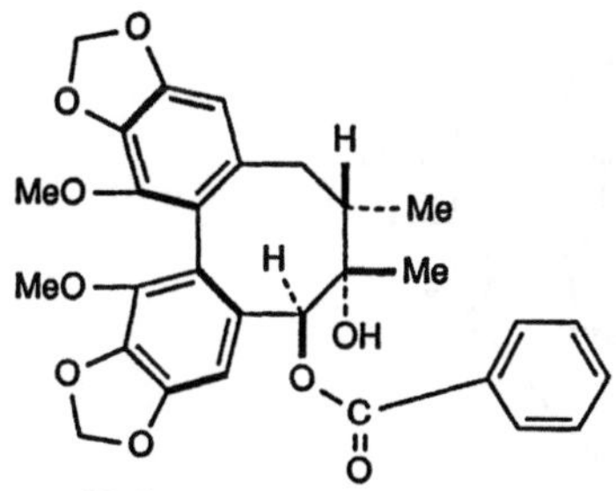		*S. sphenanthera*	[7, 15, 16]
Schisantherin E (*111-11*)	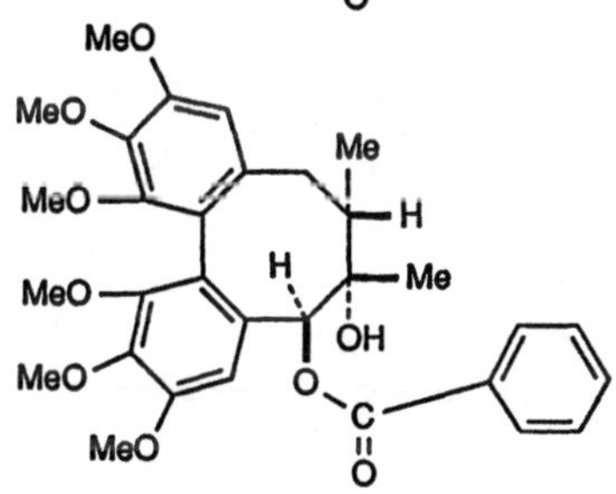		*S. sphenanthera*	[15, 16]
Isoschisandrin (*111-12*)	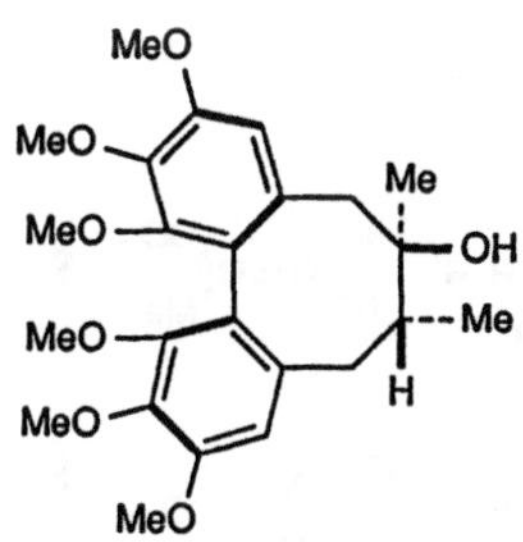		*S. chinensis*	[17]

Table 111.1. (continued)

Compound	Structure	Plant origin	Reference
Gomisin D (*111-13*)	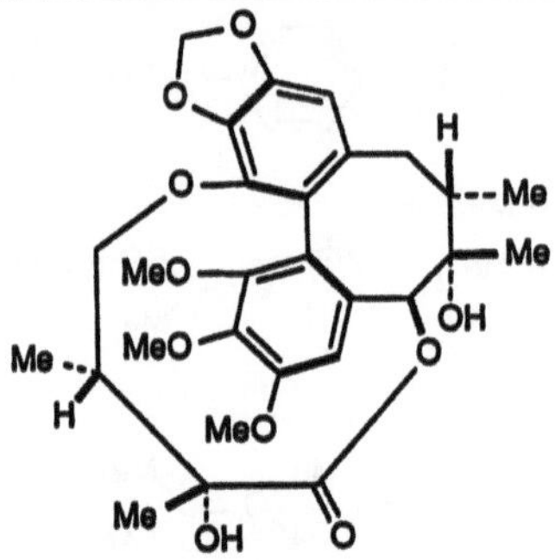	*S. chinensis*	[18, 19]
Gomisin E (*111-14*)	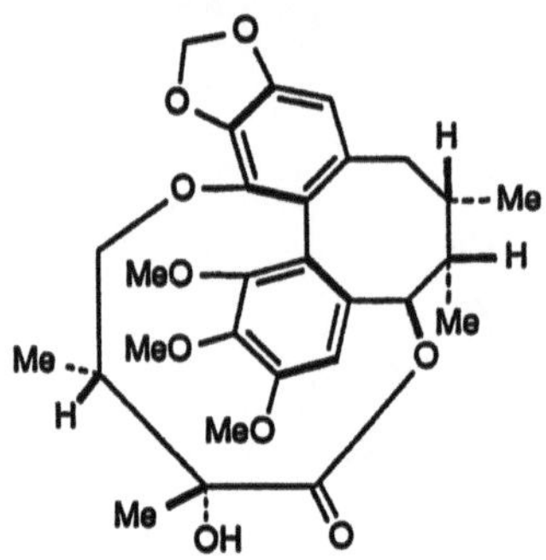	*S. chinensis*	[20, 21]
Gomisin F (*111-15*)	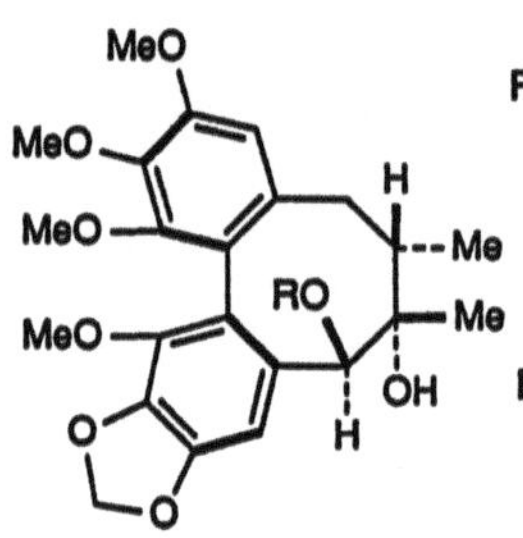 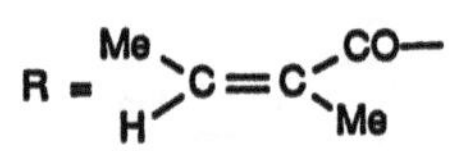	*S. chinensis*	[9]
Gomisin G (*111-16*)	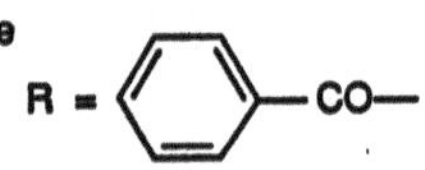	*S. chinensis*	[9]
Gomisin H (*111-17*)	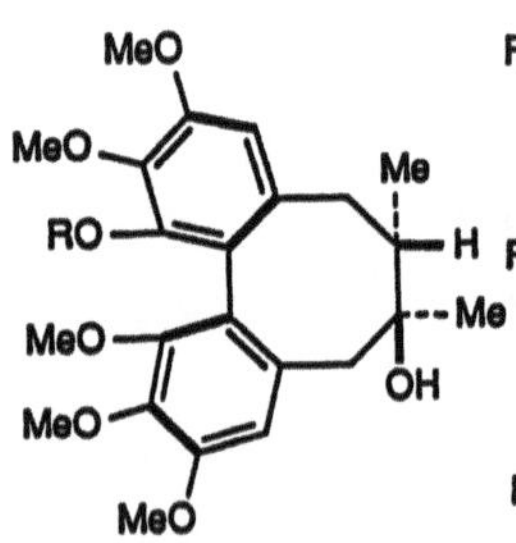 R = H	*S. chinensis*	[22]
Angeloylgomisin H (*111-18*)	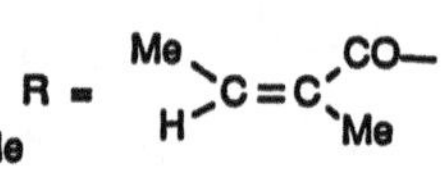	*S. chinensis*	[22, 23]
Benzoylgomisin H (*111-19*)	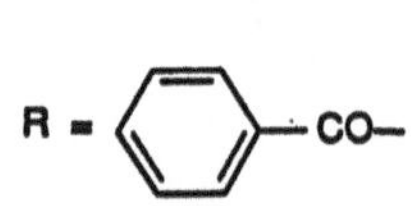	*S. chinensis*	[22, 23]
Tigloylgomisin H (*111-20*)	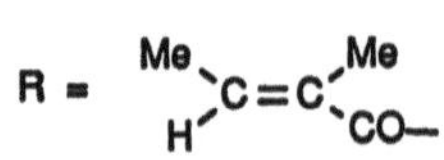	*S. chinensis*	[22, 23]

Table 111.1. (continued)

Compound	Structure	Plant origin	Reference
Gomisin J (*111-21*)	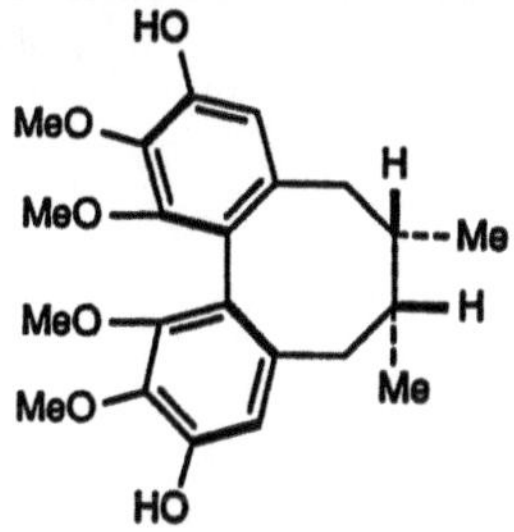	*S. chinensis*	[24, 25]
Gomisin K_1 (*111-22*)	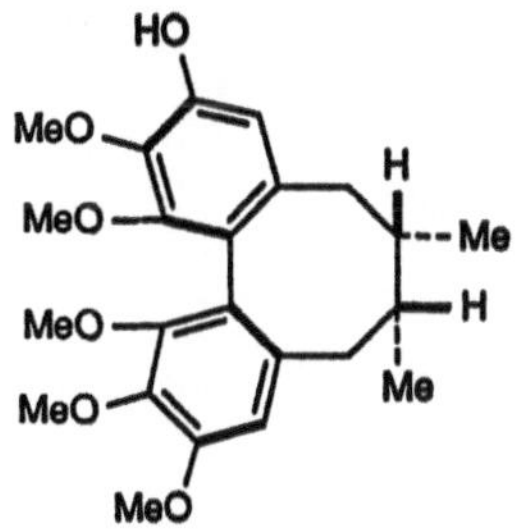	*S. chinensis*	[26]
Gomisin K_2 (*111-23*)	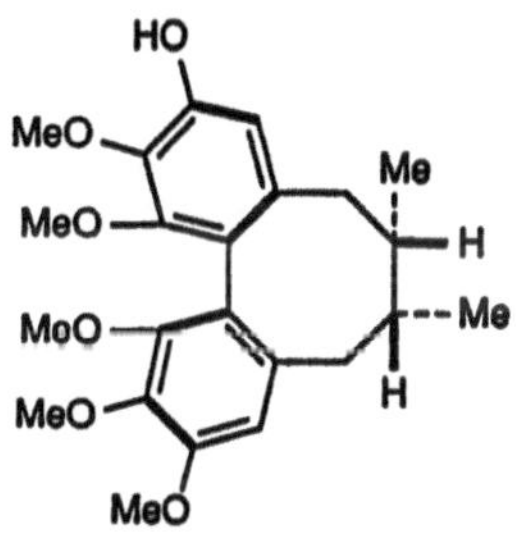	*S. chinensis*	[26]
Gomisin K_3 (schisanhenol, *111-24*)	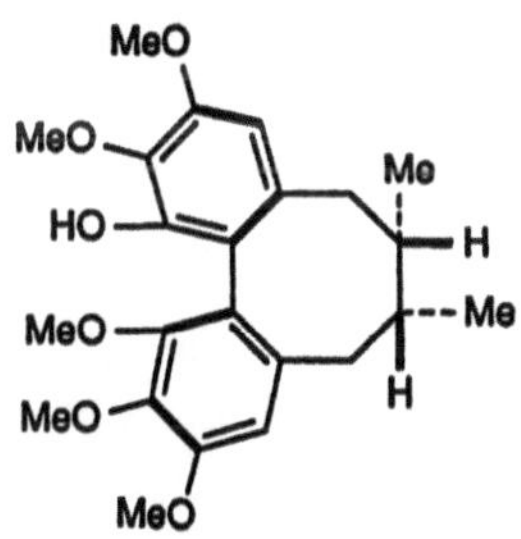	*S. chinensis* *S. rubriflora*	[26, 27]
Gomisin L_1 (*111-25*)	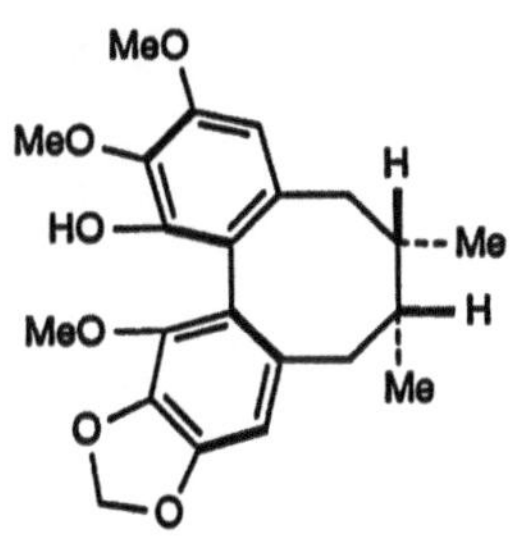	*S. chinensis*	[5]

Table 111.1. (continued)

Compound	Structure	Plant origin	Reference
Gomisin L_2 (*111-26*)	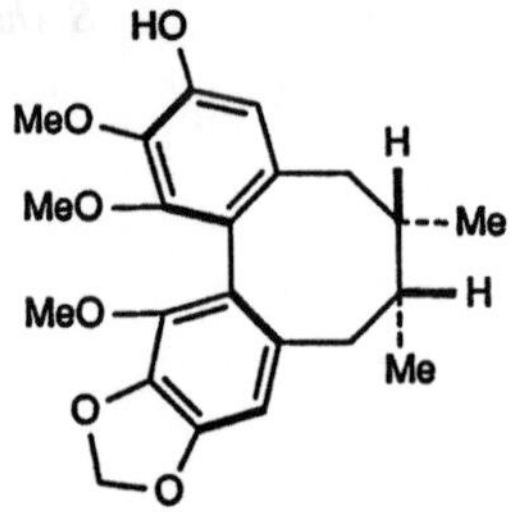	*S. chinensis*	[5]
Gomisin M_1 (*111-27*)	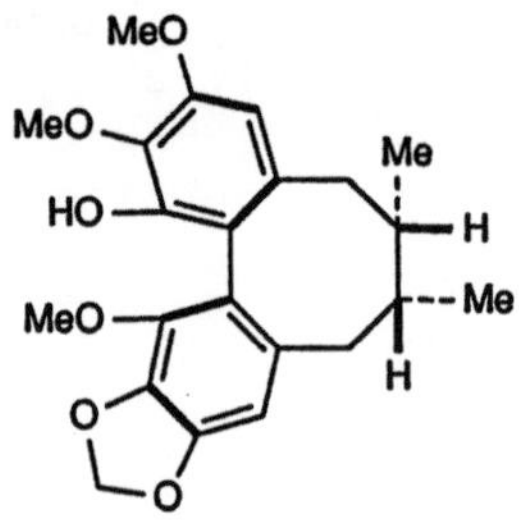	*S. chinensis*	[5]
Gomisin M_2 (*111-28*)	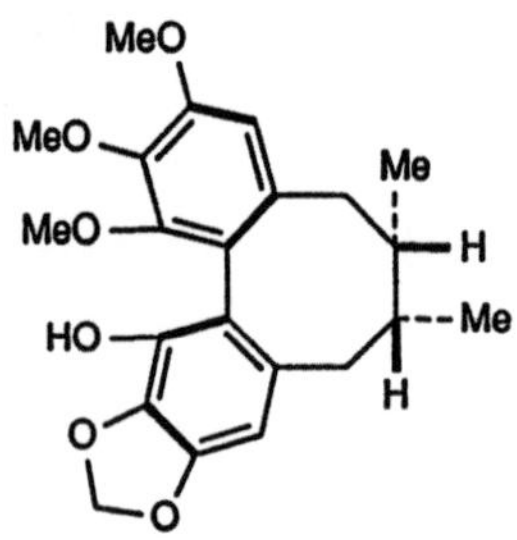	*S. chinensis*	[5]
Gomisin N (*111-29*)	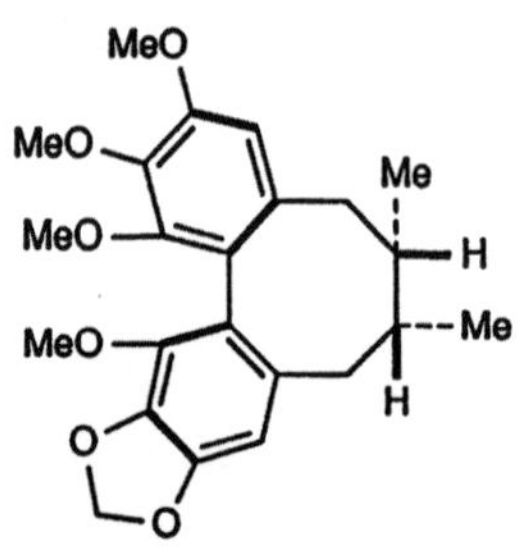	*S. chinensis*	[28, 29]
Gomisin O (*111-30*)	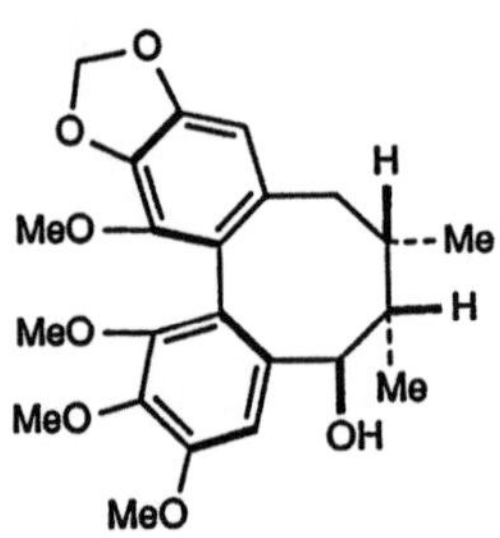	*S. chinensis*	[20]

Table 111.1. (continued)

Compound	Structure	Plant origin	Reference
Angeloylgomisin O (*111-31*)	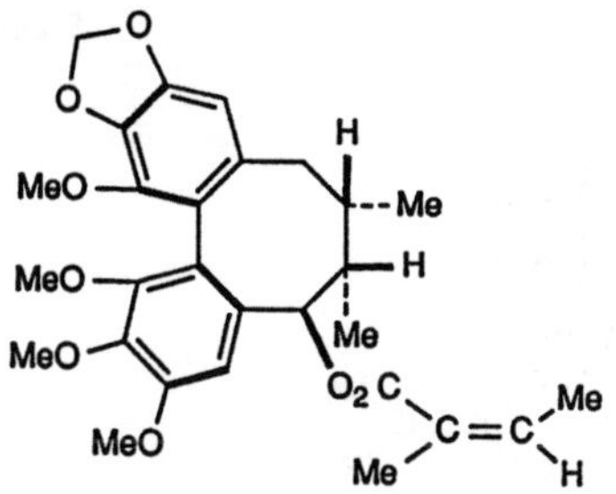	*S. chinensis*	[30]
Benzoyliso-gomisin O (*111-32*)	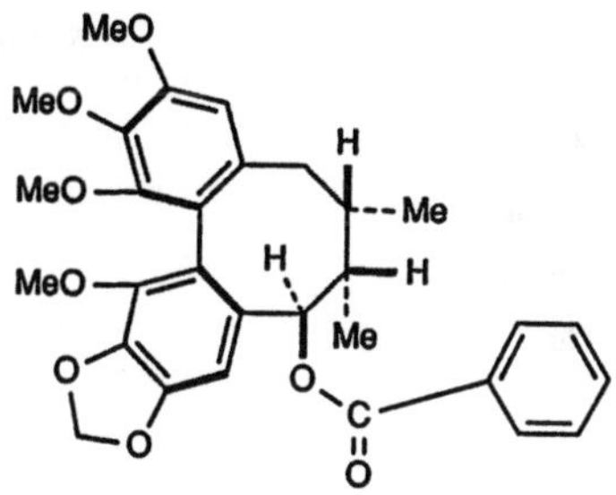	*S. chinensis*	[30]
Angeloyliso-gomisin O (*111-33*)	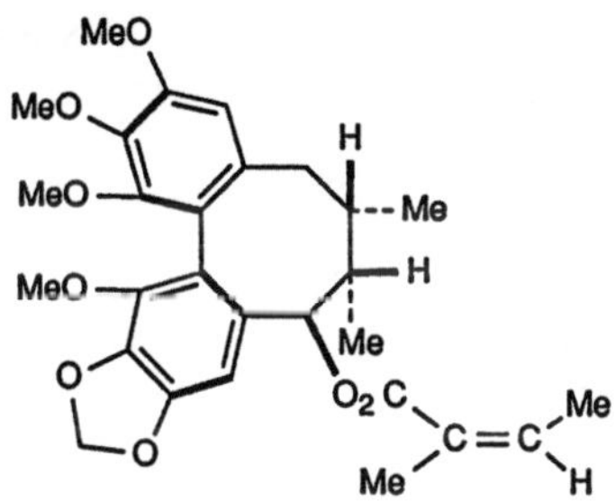	*S. chinensis*	[30]
Epigomisin O (*111-34*)	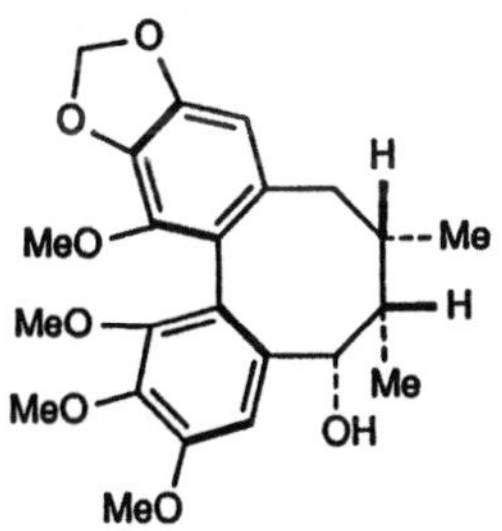	*S. chinensis*	[20]
Angeloylgomisin P (*111-35*)	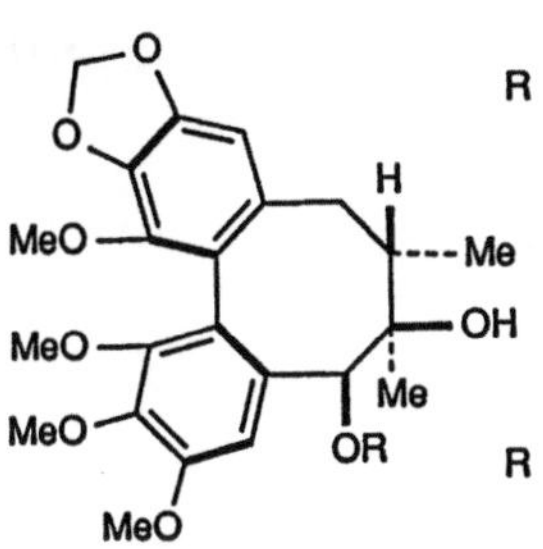 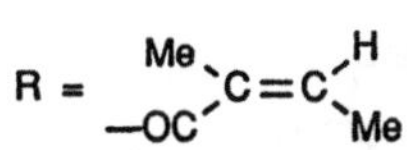	*S. chinensis*	[31]
Tigloylgomisin P (*111-36*)	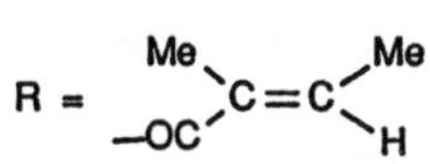	*S. chinensis*	[28, 31]

Table 111.1. (continued)

Compound	Structure	Plant origin	Reference
Angeloylgomisin Q (*111-37*)	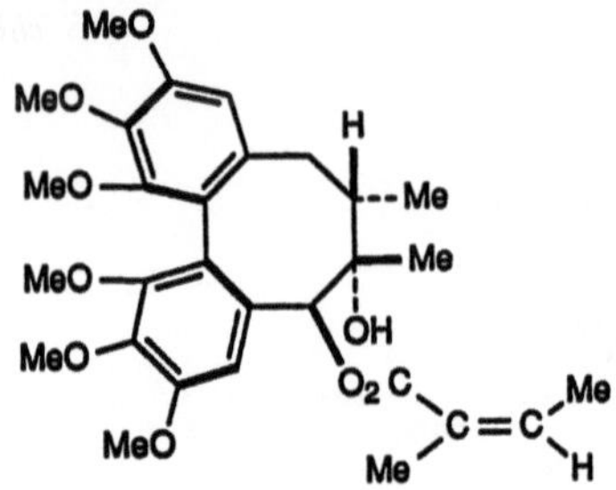	*S. chinensis*	[21, 32]
Gomisin R (*111-38*)	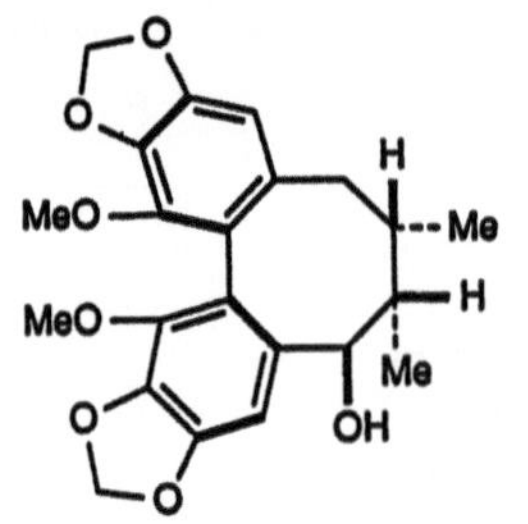	*S. chinensis*	[7]
Gomisin S (*111-39*)	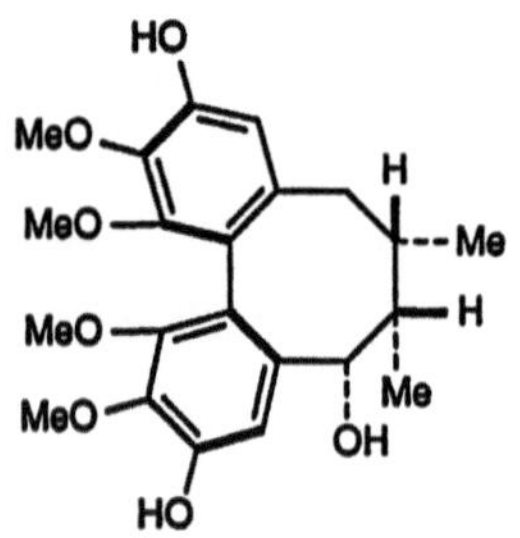	*S. chinensis*	[33]
Gomisin T (*111-40*)	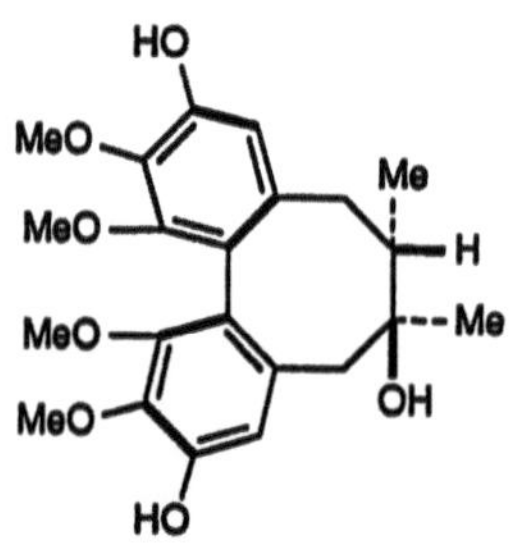	*S. chinensis*	[33]
Deoxygomisin A (*111-41*)	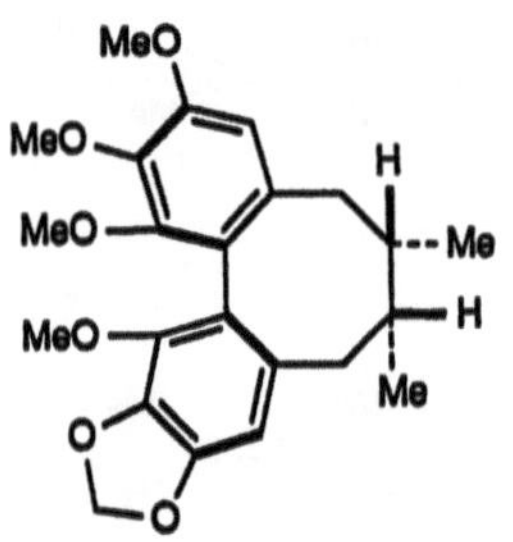	*S. chinensis*	[29]

Table 111.1. (continued)

Compound	Structure	Plant origin	Reference
Rubschisantherin (*111-42*)	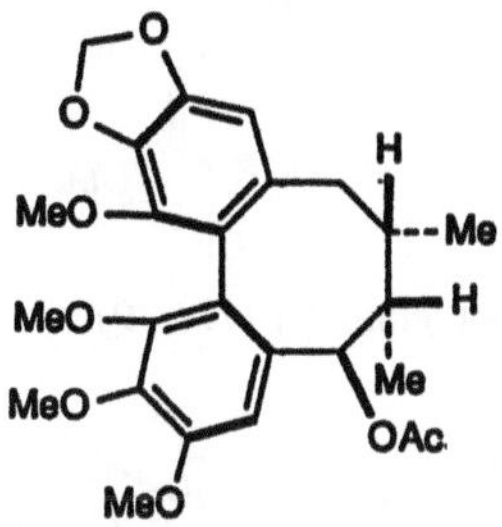	*S. rubriflora*	[34]
Schisanhenol B (*111-43*)	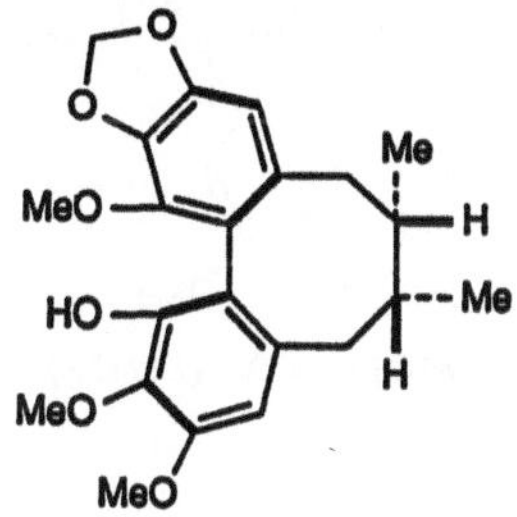	*S. rubriflora*	[34]

In addition several lignan derivatives without a dibenzo[*a,c*]cyclooctene structure were isolated from *Schisandra* species, such as pregomisin (*111-44*) from *S. chinensis* [24], epigalbacin (*111-45*) [35], anwulignan (*111-46*) [36] and ganschisandrine (*111-47*) [37] from *S. sphenanthera* and henricine (*111-48*) [38], wulignan A_1 (*111-49*), wulignan A_2 (*111-50*), epiwulignan A_1 (*111-51*), and epischisandrone (*111-52*) [39] from *S. henryi*.

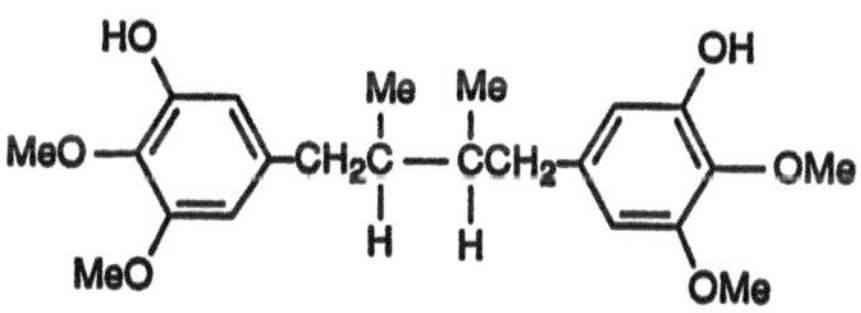

Pregomisin (*111-44*)

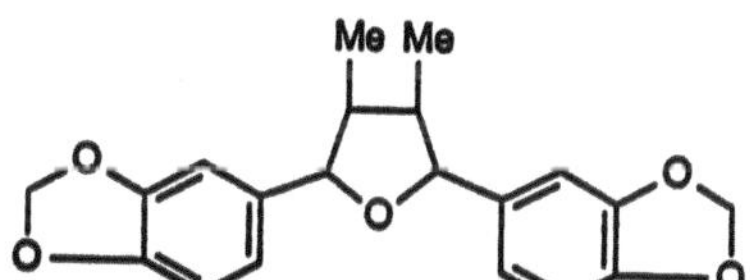

Epigalbacin (*111-45*)

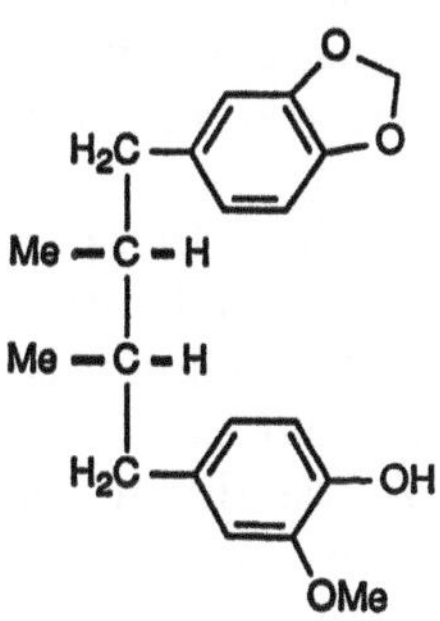

Anwulignan (*111-46*)

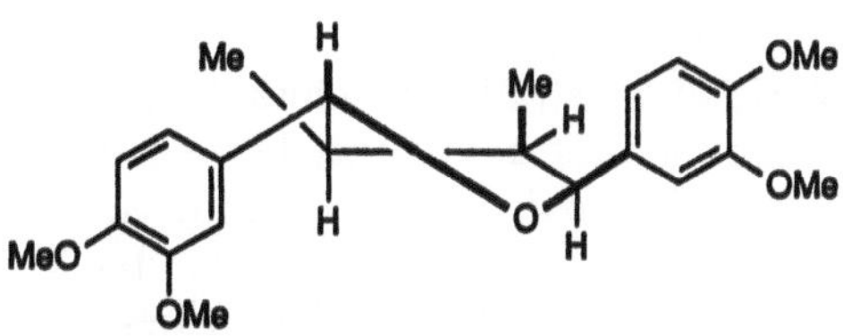

Ganschisandrine (*111-47*)

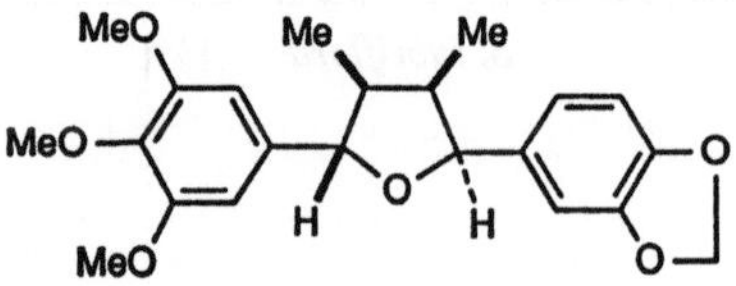

Henricine (*111-48*)

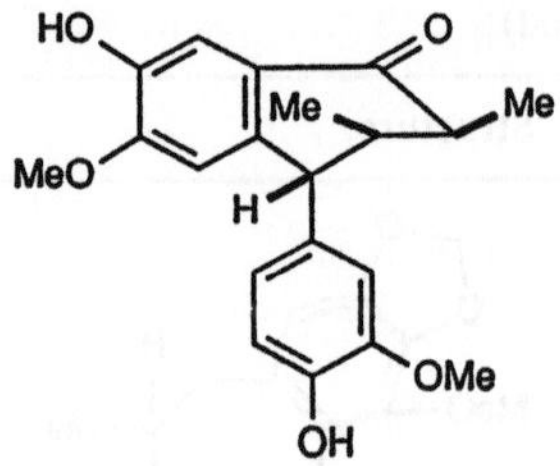

Wulignan A_1 (*111-49*)

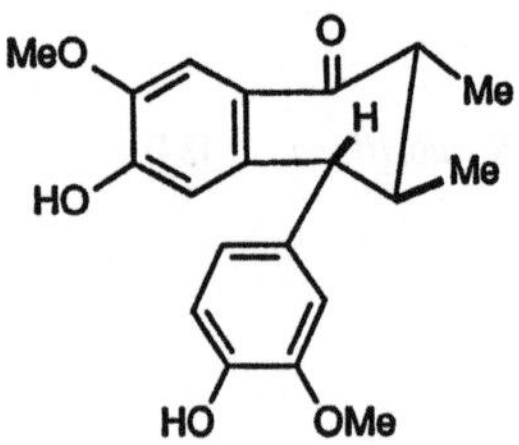

Wulignan A_2 (*111-50*)

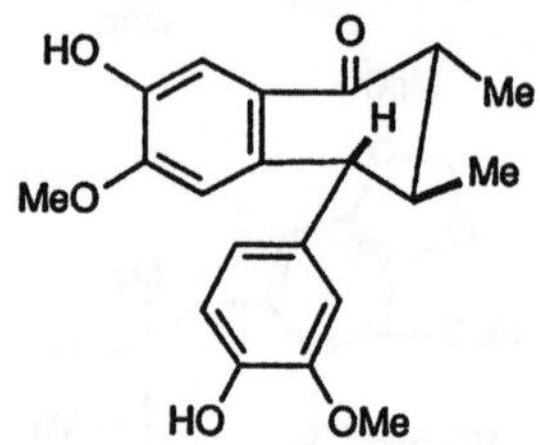

Epiwulignan A_1 (*111-51*)

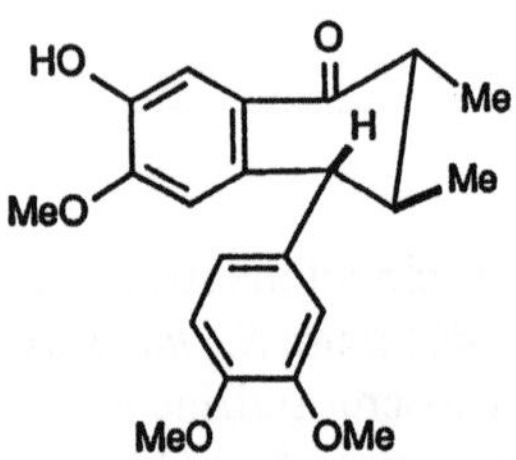

Epischisandrone (*111-52*)

A series of dibenzo[*a,c*]cyclooctene lignans were also isolated from *Kadsura* species [40, 41].

111.3 Pharmacology

In experimental pharmacological studies, increased SGPT activities induced by CCl_4 or acetaminophen in mice, thiacetamide in rats, and ethinylestradiol-3-cyclopentylether in rabbits were reported to be reduced by oral administration of the ethanol extract of seeds of *S. chinensis* prior to and after the administration of the hepatotoxic agents [1, 42].

Schisantherins A, B, C, and D were also found to decrease SGPT levels in experimental animals [43] and of patients with chronic viral hepatitis [15, 43]. Intragastric administration of schisantherin A gave protection against CCl_4-induced liver damage in mice, as evidenced by electron microscopy. Biochemical studies revealed that the lowered blood transaminase activity by schisantherin A was due to an inhibition of the enzyme activity in the liver [43]. Schisandrin A showed not only a decreasing effect on SGPT activities, but also on those of liver glutamic pyruvic transaminase (LGPT) [44].

After administration of schisandrin B to mice, an enlargement of the liver was observed. Relative contents of water, protein, glycogen, total lipid, and RNA per gram liver remained similar to those in untreated controls, but total amounts increased markedly. In contrast, the total amount of DNA in the liver remained constant and this resulted in a decrease of relative DNA content per gram liver tissue. Moreover, administration of schisandrin B increased the incorporation of [^{14}C]phenylalanine into liver protein. Concomitantly with hepatic protein content, hepatic microsomal cytochrome P-450 content was significantly increased. Schisandrin B therefore is an inducing agent for the hepatic metabolizing enzyme system. The enlargement of the liver following administration of the agent has been ascribed mainly to hypertrophy associated with some hyperplasia [45, 46].

Cytochrome P-450 from rat liver microsomes after treatment with schisandrin B or schisandrol B was compared with cytochrome P-450 obtained from untreated or phenobarbital-treated animals. A cytochrome P-450 species cochromatographed on DEAE cellulose with a species having a common antigenic site with phenobarbital-induced cytochrome P-450 exhibited a high benzphetamine demethylase activity. This isoenzyme is induced by schisandrin B and schisandrol B, whereas the other cytochrome P-450 isoenzymes are not affected [47].

Pretreatment of rats with gomisin A decreased the activities of glutamic oxalacetic transaminase (GOT), GPT, lactate dehydrogenase, total bilirubin, and total cholesterol in the serum elevated by CCl_4. Alkaline phosphatase activity was also found to be slightly decreased. Histologically, hepatocellular necrosis by CCl_4 was inhibited by pretreatment with gomisin A, the action of which was reported to be even stronger than that of schisandrol A [48, 49].

Schisanhenol was also found to protect mice against CCl_4-induced liver damage, and to increase hepatic microsomal cytochrome P-450, NADPH cytochrome C reductase, benzo[*a*]pyrene hydroxylase activities, and microsomal protein content in rats. It also stimulated the proliferation of endoplasmic reticulum of liver cells [27, 50].

Schisandrins B and C, and schisandrol B were also found to inhibit CCl_4-induced lipid peroxidation and [^{14}C]CCl_4 covalent binding to lipids of liver microsomes from phenobarbital-treated mice. These compounds also decreased CO production and cofactor (NADPH, O_2) utilization during CCl_4 metabolism by liver microsomes. The hepatoprotective effect of these compounds was concluded to be due to their inhibitory activity on CCl_4-induced lipid peroxidation and on binding of CCl_4 metabolites to lipids of liver microsomes [51]. Radioprotective effects of schisandrin B have also been related to its antioxidizing activity [52].

In vitro, schisandrin B and schisandrol B have been found to decrease the mutagenicity of benzo[*a*]pyrene in the Ames test [53].

Incubation of ascites hepatoma cells with schisandrin B was reported to inhibit the formation of DNA, ATP, and nucleoprotein. In mouse sarcoma 180 cells and human embryo lung fibroblasts, DNA formation was also inhibited by schisandrin B [54]. An inhibitory effect of *S. chinensis* on xenobiotic metabolizing enzymes of liver and small intestine has also been demonstrated after feeding to mice and rats [55, 56].

On the basis of these studies, simple compounds with a structural relationship to *Schisandra* lignans such as dimethyl biphenyldicarboxylate were discovered as being active against liver damage induced by CCl_4 in mice [2, 57].

Gomisin A has been further found effectively to prolong hexobarbital-induced sleeping time in mice and showed a hypothermic, and at high doses an analgesic, effect. Gomisin A was also found to inhibit gastric contraction after i.v. administration in mice. After oral administration, stress-induced gastric ulceration was prevented. Schisandrol A showed similar activity, also inhibiting gastric secretion. It was also found to have choleretic activity in experimental animals [48, 58, 59].

After oral administration to rats, schisandrol A was readily absorbed from the gastrointestinal tract with a half-life of 58 min. Following i.v. injection the blood level showed a biphasic decline. The half-life of the distribution phase was 1.4 and of the elimination phase 42 min. Schisandrol A was detectable in urine 1 h after oral administration. Five minutes after i.v. injection, high levels of schisandrol A were found in the lungs, moderate amounts in the liver, heart, brain, and kidneys, and low amounts in the ileum and spleen. Thus, schisandrol A is distributed widely in the body and is eliminated rather rapidly. In the brain, the highest amounts were found in the hypothalamus, striatum, and hippocampus and moderate amounts in the cerebral cortex and cerebellum. These differences may be relevant to the neuroleptic and anticonvulsant properties of schisandrol A [60].

References

1. Pao TT, Liu KT, Hsu KF, Sung CY (1974) Fructus schizandrae. I. Effect on increased SGPT (serum glutamic pyruvic transaminase) levels in animals caused by hepatotoxic chemical agents. Natl Med J China 54:275–278
2. Liu G (1983) From the study of Fructus schizandrae to the discovery of biphenyl dicarboxylate. Acta Pharm Sin 18:714–720
3. Song W, Tong Y (1983) Occurrence and assay of some important lignans in Wu Wei Zi (*Schisandra chinensis*) and its allied species. Acta Pharm Sin 18:138–143
4. Chen YY, Shu ZB, Li LN (1976) Studies on Fructus schisandrae. IV. Isolation and determination of the active compounds (in lowering high SGPT levels) of *Schisandra chinensis* Baill. Sci Sin 19:276–290
5. Ikeya Y, Taguchi H, Yosioka I (1982) The constituents of *Schisandra chinensis* Baill. X. The structure of γ-schizandrin and four new lignans, (−)-gomisins L_1 and L_2, (±)-gomisin M_1 and (+)-gomisin M_2. Chem Pharm Bull (Tokyo) 30:132–139
6. Tan R, Li LN, Fang QC (1986) The stereostructure of wuweizisu B. Planta Med 52:49–51
7. Ikeya Y, Taguchi H, Yosioka I (1982) The constituents of *Schisandra* chinensis Baill. XII. Isolation and structure of a new lignan, gomisin R, the absolute structure of wuweizisu C and isolation of schisantherin D. Chem Pharm Bull (Tokyo) 30:3207–3211
8. Kochetkov NK, Khorlin AY, Chizhov OS, Sheichenko VI (1961) Schizandrin, a lignan of unusual structure. Tetrahedron Lett 730–734
9. Ikeya Y, Taguchi H, Yosioka I, Kobayashi H (1979) The constituents of *Schisandra chinensis* Baill. I. Isolation and structure determination of five new lignans, gomisin A, B, C, F and G, and the absolute structure of schisandrin. Chem Pharm Bull (Tokyo) 27:1383–1394
10. Taguchi H, Ikeya Y (1975) The constituents of *Schisandra chinensis* Baill. I. The structures of gomisin A, B and C. Chem Pharm Bull (Tokyo) 23:3296–3298
11. Taguchi H, Ikeya Y (1977) The constituents of *Schisandra chinensis* Baill. The structures of two new lignans, gomisin F and G, and the absolute structures of gomisin A, B and C. Chem Pharm Bull (Tokyo) 25:364–366
12. Chinese Academy of Medical Science (1977) X-ray Diffraction Group: the crystal structure of *p*-bromobenzoyloxy ester of schisandrin. Ko Hsueh Tung Pao 22:504–506
13. Fang SD, Huang MF, Liu JS, Gao YL, Gsu JS (1975) Studies on the constituents of Hua-Zhong-Wu-Wei-Zi (*Schisandra sphenanthera* Rehd. et Wils.). I. The isolation and structure of schisantherin A. Acta Chim Sin 33:57–60

14. Chen YY, Li LN (1976) Studies on fruit of *Schisandra*. Structure determination of wuweizi ester A and wuweizi ester B. Acta Chim Sin 34:45–52
15. Liu JS, Fang SD, Huang MF, Gao YL, Hsu JS (1976) Studies on the constituents of Hua-Zhong-Wu-Wei-Zi (*Schisandra sphenanthera* Rehd. et Wils.). II. The structures of schisantherin A, B, C, D, E and related compounds. Acta Chim Sin 34:229–240
16. Liu CS, Fang SD, Huang MF, Kao YL, Hsu JS (1978) Studies on the active principles of *Schisandra sphenanthera* Rehd. et Wils. The structures of schisantherin A, B, C, D, E and related compounds. Sci Sin [Engl] 21:483–502
17. Ikeya Y, Taguchi H, Mitsuhashi H, Takeda S (1988) The constituents of *Schizandra chinensis* Baill. XIV. A lignan from *Schizandra chinensis*. Phytochemistry 27:569–573
18. Ikeya Y, Taguchi H, Itaka Y (1976) The constituents of *Schisandra chinensis* Baill. The structure of a new lignan, gomisin D. Tetrahedron Lett 1359–1362
19. Ikeya Y, Taguchi H, Yosioka I, Iitaka Y, Kobayashi H (1979) The constituents of *Schisandra chinensis* Baill. II. The structure of a new lignan, gomisin D. Chem Pharm Bull (Tokyo) 27:1395–1401
20. Ikeya Y, Taguchi H, Yosioka I, Kobayashi H (1979) The constituents of *Schisandra chinensis* Baill. V. The structures of four new lignans, gomisin N, gomisin O, epigomisin O and gomisin E and transformation of gomisin N to deangeloylgomisin B. Chem Pharm Bull (Tokyo) 27:2695–2709
21. Taguchi H, Ikeya Y, Yosioka I (1979) Structures and reactions of new dibenzocyclooctadiene lignans from the fruits of *Schisandra chinensis* Baill. Koen Yoshishu – Tennen Yuki Kagobutsu Toronkai, 22nd, 299–306 (CA 93:46238 g)
22. Ikeya Y, Taguchi H, Yosioka I, Kobayashi H (1979) The constituents of *Schisandra chinensis* Baill. III. The structures of four new lignans, gomisin H and its derivatives, angeloyl-, tigloyl- and benzoyl-gomisin H. Chem Pharm Bull (Tokyo) 27:1576–1582
23. Ikeya Y, Taguchi H, Yosioka I (1978) The constituents of *Schisandra chinensis* Baill. The structures of three new lignans, angeloylgomisin H, tigloylgomisin H and benzoylgomisin H and the absolute structure of schisandrin. Chem Pharm Bull (Tokyo) 26:328–332
24. Ikeya Y, Taguchi H, Yosioka I (1978) The constituents of *Schisandra chinensis* Baill. The structures of two new lignans, pre-gomisin and gomisin J. Chem Pharm Bull (Tokyo) 26:682–684
25. Ikeya Y, Taguchi H, Yosioka I, Kobayashi H (1979) The constituents of *Schisandra chinensis* Baill. IV. The structures of two new lignans, pre-gomisin and gomisin J. Chem Pharm Bull (Tokyo) 27:1583–1588
26. Ikeya Y, Taguchi H, Yosioka I (1980) The constituents of *Schisandra chinensis* Baill. VII. The structures of three new lignans, (−)-gomisin K_1 and (+)-gomisin K_2 and K_3. Chem Pharm Bull (Tokyo) 28:2422–2427
27. Liu GT, Wei HL (1985) Induction of hepatic microsomal monooxygenases by schisandrol in rats. Acta Pharmacol Sin 6:41–44
28. Ikeya Y, Taguchi H, Yosioka I, Kobayashi H (1978) The constituents of *Schisandra chinensis* Baill. The structure of two new lignans, gomisin N and tigloylgomisin P. Chem Pharm Bull (Tokyo) 26:3257–3260
29. Hikino H, Kiso Y, Taguchi H, Ikeya Y (1984) Validity of the oriental medicines. LX. Liver protective drugs. 11. Antihepatotoxic actions of lignoids from *Schisandra chinensis* fruits. Planta Med 50:213–218
30. Ikeya Y, Ookawa N, Taguchi H, Yosioka I (1982) The constituents of *Schisandra chinensis* Baill. XI. The structures of three new lignans, angeloylgomisin O and angeloyl- and benzoyl-isogomisin O. Chem Pharm Bull (Tokyo) 30:3202–3206
31. Ikeya Y, Taguchi H, Yosioka I, Kobayashi H (1980) The constituents of *Schisandra chinensis* Baill. VIII. The structures of two new lignans, tigloylgomisin P and angeloylgomisin P. Chem Pharm Bull (Tokyo) 28:3357–3361
32. Ikeya Y, Taguchi H, Yosioka I (1979) The constituents of *Schisandra chinensis* Baill. The cleavage of the methylenedioxy moiety with lead tetraacetate in benzene and the structure of angeloylgomisin Q. Chem Pharm Bull (Tokyo) 27:2536–2538
33. Ikeya Y, Kanatani H, Hakozaki M, Taguchi H, Mitsuhashi H (1988) The constituents of *Schizandra chinensis* Baill. XV. Isolation and structure determination of two new lignans gomisin S and gomisin T. Chem Pharm Bull (Tokyo) 36:3974–3979
34. Wang HJ, Chen YY (1985) Lignans from *Schisandra rubriflora* Rehd et Wils. Acta Pharm Sin 20:832–841

35. Huang MF, Fang SD, Kao YL (1982) Studies on the constituents of *Schisandra sphenanthera* Rehd et Wils. III. Constituents of the lignan in different districts and the structure of *d*-epigalbacin. Acta Bot Sin 24:451–455
36. Liu J, Huang M (1984) Constituents of *Schisandra sphenanthera* Rehd. et Wils. IV. Structures of anweizic acid and *dl*-anwulignan and the absolute configurations of *d*-epigalbacin. Acta Chim Sin 42:264–270
37. Yue JM, Chen YZ, Hua SM, Cheng JL, Cui YX (1989) Ganschisandrine, a lignan from *Schisandra spheranthera*. Phytochemistry 28:1774–1776
38. Li LN, Xue H (1986) Henricine, a new tetrahydrofuran lignan from *Schisandra henryi*. Plant Med 52:493–494
39. Liu JS, Tao Y, Huang MF (1988) Studies on the constituents of *Schisandra henryi*. V. Structures of wulignan A_1, A_2, epiwulignan A_1 and epischisandrone. Acta Chim Sin 46:483–488
40. Liu JS, Ma YT (1988) Constituents of Schisandraceae plants in Shennongjia District. II. Isolation and structure of schisantherin F. Acta Chim Sin 46:460–464
41. Liu JS, Ma YT (1988) Studies on the constituents of Schisandraceae plants in Shennongjia District. III. Isolation and structures of schisantherin G, H and I. Acta Chim Sin 46:465–471
42. Liu GT, Wei HL (1987) Protection by Fructus schisandrae against acetaminophen hepatotoxicity in mice. Acta Pharm Sin 22:650–654
43. Shanghai Institute of Materia Medica (1976) Studies on the transaminase-lowering principles of *Schisandra spheranthera* Rehd. et Wils. Acta Biochim Biophys Sin 8:333–340
44. Bao TT, Xu GF, Liu GT, Sun RH, Song ZY (1979) Comparison of the pharmacological effects of seven constituents isolated from the fruits of Schisandrae. Acta Pharm Sin 14:1–7
45. Liu GT, Bao TT, Wei HL, Song ZY (1980) Induction of hepatocyte microsomal cytochrome P-450 by schisandrin B in mice. Acta Pharm Sin 15:206–211
46. Liu GT, Bao TT, Wei HL, Song ZY (1980) Induction of hepatic microsomal cytochrome P-450 by Schizandrin B in mice. In: Burns JJ, Tsuchitani PJ (eds) Proceedings of the US-China Pharmacology Symposium 1979. NAS, Washington, pp 301–313
47. Liu KT, Cresteil T, Le Provost E, Lesca P (1981) Specific evidence that schisandrins induce a phenobarbital-like cytochrome P-450 form separated from rat liver. Biochem Biophys Res Commun 103:1131–1137
48. Maeda S, Sudo K, Aburada M, Ikeya Y, Taguchi H, Yosioka I, Harada M (1981) Pharmacological studies on *Schizandra* fruit. I. General pharmacological effects of gomisin A and schisandrin. Yakugaku Zasshi 101:1030–1041
49. Maeda S, Sudo K, Miyamoto Y, Takeda S, Shinbo M, Aburada M, Ikeya Y, Taguchi H, Harada M (1982) Pharmacological studies on *Schizandra* fruits. II. Effects of constituents of *Schizandra* fruits on chemical-induced hepatic damage in rats. Yakugaku Zasshi 102:579–588
50. Liu KT, Cresteil T, Columelli S, Lesca P (1982) Pharmacological properties of dibenzo[*a,c*]cyclooctene derivatives isolated from Fructus Schisandrae chinensis. II. Induction of phenobarbital-like hepatic monooxygenases. Chem Biol Interact 39:315–330
51. Liu KT, Lesca P (1982) Pharmacological properties of dibenzo[*a,c*]cyclooctene derivatives isolated from fructus Schisandrae chinensis. III. Inhibitory effects on carbon tetrachloride-induced lipid peroxidation, metabolism and covalent binding of carbon tetrachloride to lipids. Chem Biol Interact 41:39–47
52. Liu L, Zheng R, Zhang H, Zhang X, Liu Y, Liu M (1983) Radiation protection mechanism of γ-schizandrin. VI. Effect of γ-schizandrin on the formation of lipid peroxide in homogenates of liver and brain. J Lanzhou Univ [Nat Sci] 19:99–104
53. Liu KT, Lesca P (1982) Pharmacological properties of dibenzo[*a,c*]cycloocytene derivatives isolated from Fructus Schisandrae chinensis. I. Interaction with rat liver cytochrome P-450 and inhibition of xenobiotic metabolism and mutagenicity. Chem Biol Interact 39:301–314
54. Liu L, Xiao X, Wang X, Zheng R, Zhou X (1984) Effect of γ-schisandrin on the metabolism of DNA, ATP and nucleoprotein of cancer cells. Acta Pharmacol Sin 5:130–132
55. Hendrich S, Bjeldanes LF (1983) Effects of dietary cabbage, brussels sprouts, *Illicium verum*, *Schisandra chinensis* and alfalfa on the benzo[*a*]pyrene metabolic system in mouse liver. Food Chem Toxicol 21:479–486
56. Salbe AD, Bjeldanes LF (1985) The effects of dietary brussels sprouts and *Schisandra chinensis* on the xenobiotic-metabolizing enzymes of the rat small intestine. Food Chem Toxicol 23:57–65
57. Liu GT, Wei HL, Song ZY (1982) Further studies on the protective action of dimethyl biphenyldicarboxylate against experimental liver injury in mice. Acta Pharm Sin 17:101–106

58. Taguchi H, Iketani Y, Maeda S, Aburada M, Yosioka I (1979) Gomisin derivatives. Jpn Kokai Tokkyo Koho 79, 52, 066 (CA 91:78900s)
59. Niu XY, Wang WJ, Bian ZJ, Ren ZH (1983) Effects of schizandrol A, an active principle from the dried fruits of *Schizandra chinensis* Baill. on the central nervous system. Acta Pharm Sin 18:416–421
60. Niu XY, Bian ZJ, Ren ZH (1983) Metabolism of schisandrol A in rats and its distribution in brain determined by TLC-UV. Acta Pharm Sin 18:491–495

Scutellaria baicalensis Georgi 112

112.1 Introduction

Huangqin, Radix Scutellariae, is the dry roots of *Scutellaria baicalensis* Georgi (Lamiaceae), which is collected in spring and fall. It is officially listed in the Chinese Pharmacopoeia and is one of the most widely used Chinese herbal medicines against bacterial infections of the respiratory and the gastrointestinal tract.

Banzhilian, Herba Scutellariae berbatae, is another item for *Scutellaria* listed in the Chinese Pharmacopoeia. It is the dry whole plant of *Scutellaria barbata* D. Don, collected in summer and fall, and is used as an antipyretic and diuretic agent.

112.2 Chemical Constituents

The root of *S. baicalensis* is known to contain a number of flavone derivatives. The first flavone isolated from its root was wogonin (*112-1*), although its structure was determined only many years later [1]. Wogonin is present only in small amounts in the root; the flavone glycoside baicalin (*112-2*) predominates by far [2]. Baicalin was extracted from the roots of *S. baicalensis* with 50% alcohol. Acid hydrolysis yielded glucuronic acid and a flavone aglycone named baicalein (*112-3*) [3]. Glucuronic acid was established to be linked at position 7 in baicalin [4].

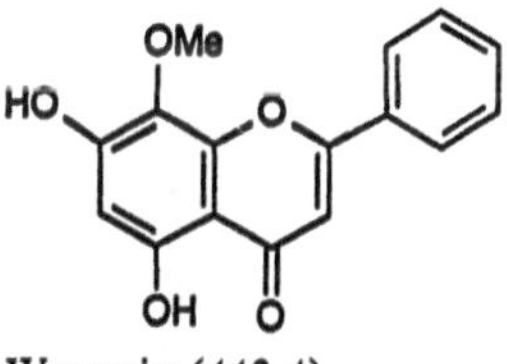

Wogonin (*112-1*)

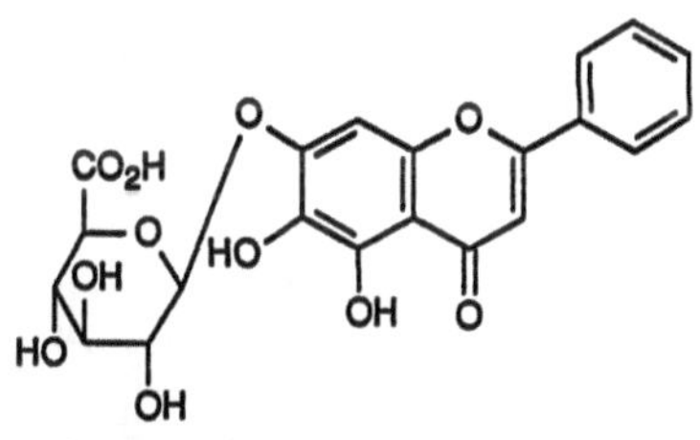

Baicalin (*112-2*)

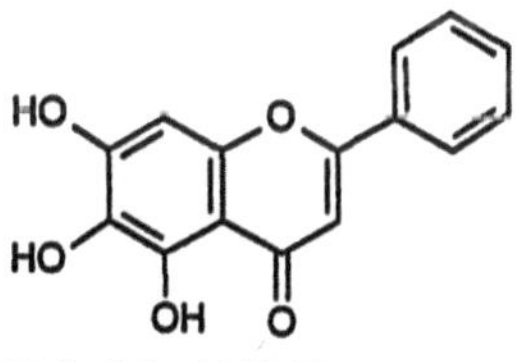

Baicalein (*112-3*)

Further flavones and related compounds isolated from *S. baicalensis* are listed in Table 112.1. Baucalin and wogonoside contents in samples of *S. baicalensis* roots from China were found to vary between 12%–17% and about 3%–4%, respectively [23–26].

Table 112.1. Flavone and related compounds isolated from the root of *Scutellaria baicalensis*

Compound	Structure	Reference
Flavones		
Chrysin (*112-4*)	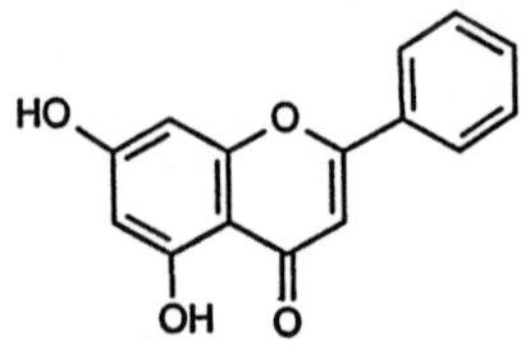	[5]
Baicalein (*112-3*)		[2, 6]
Norwogonin (*112-5*)	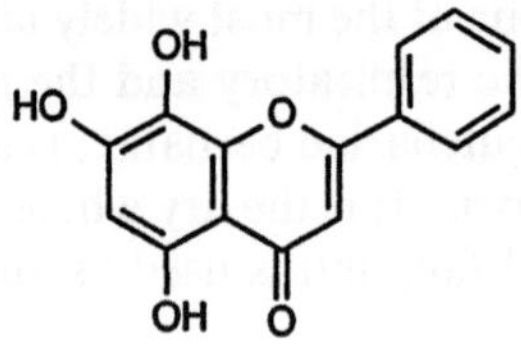	[7]
Wogonin (*112-1*)		[1, 6]
Oroxylin A (*112-6*)	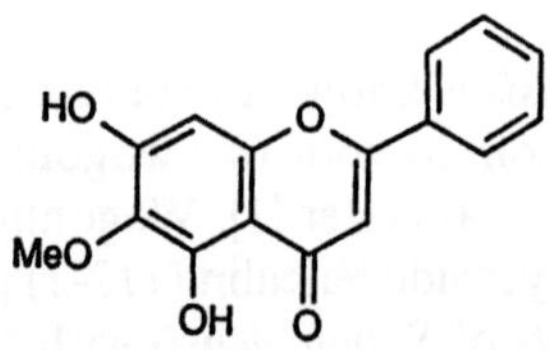	[8]
Baicalein-7-methylether (*112-7*)	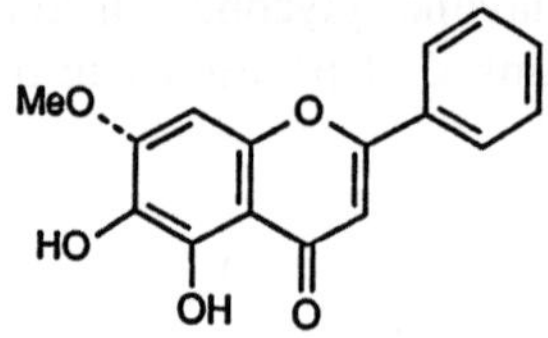	[8]
5-Hydroxy-7,8-dimethoxyflavone (*112-8*)	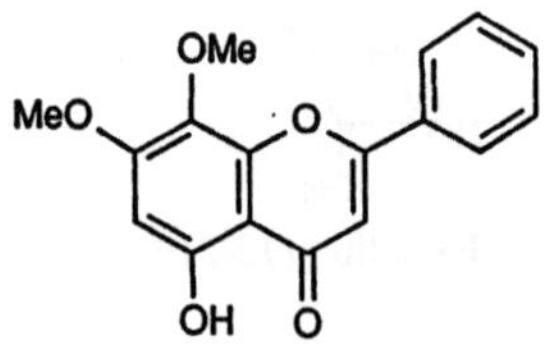	[7]

Table 112.1. (continued)

Compound	Structure	Reference
5,8-Dihydroxy-6,7-dimethoxyflavone (*112-9*)	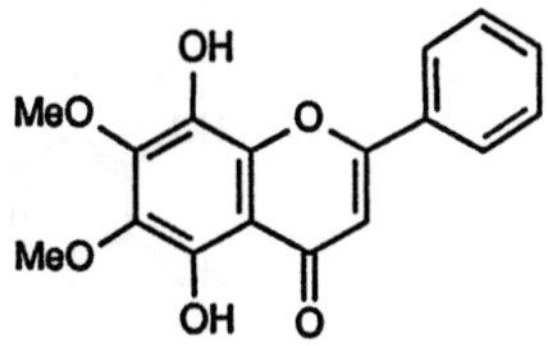	[9]
5,7,2′-Trihydroxy-8-methoxyflavone (*112-10*)	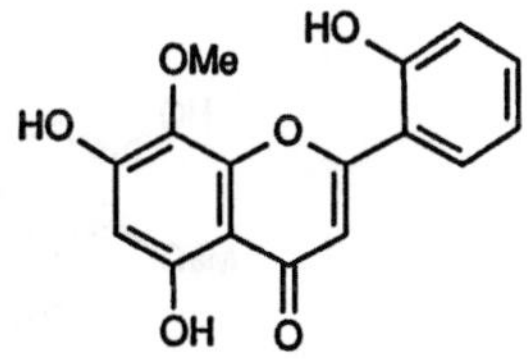	[10]
5,8,2′-Trihydroxy-7-methoxyflavone (*112-11*)	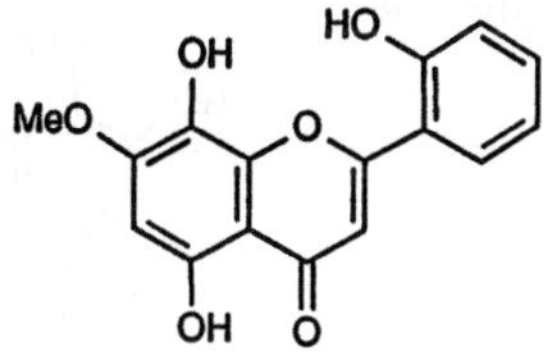	[11]
5,7,2′-Trihydroxy-6-methoxyflavone (*112-12*)	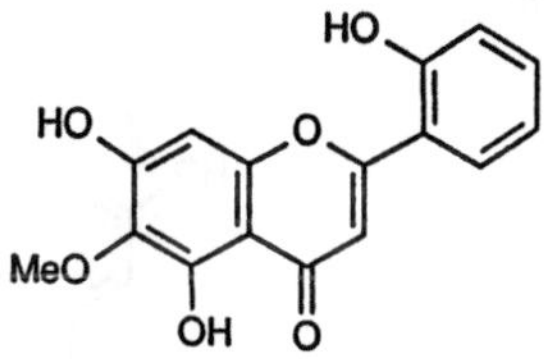	[7]
Skullcapflavone I (*112-13*)	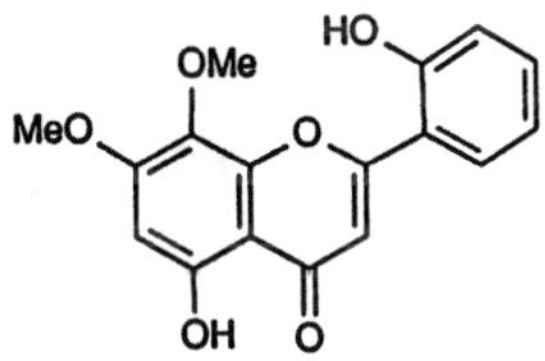	[12–15]
5,8,2′-Trihydroxy-6,7-dimethoxyflavone (*112-14*)	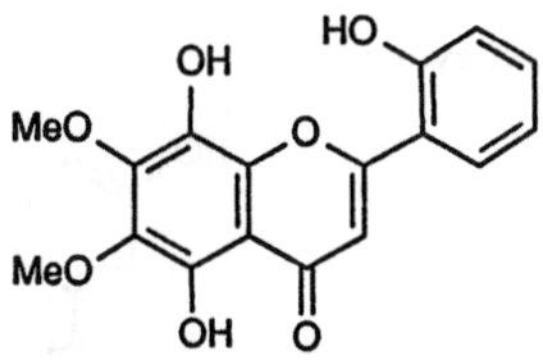	[11]
Tenaxin I (*112-15*)	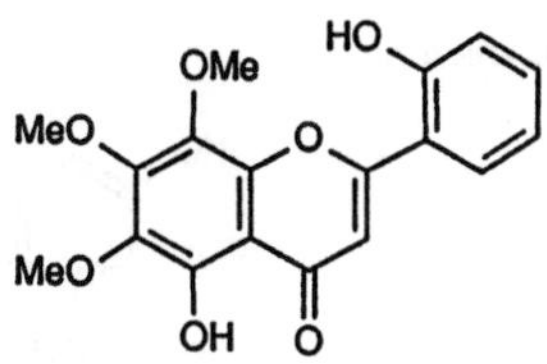	[7]

Table 112.1. (continued)

Compound	Structure	Reference
Scutellarein (*112-16*)	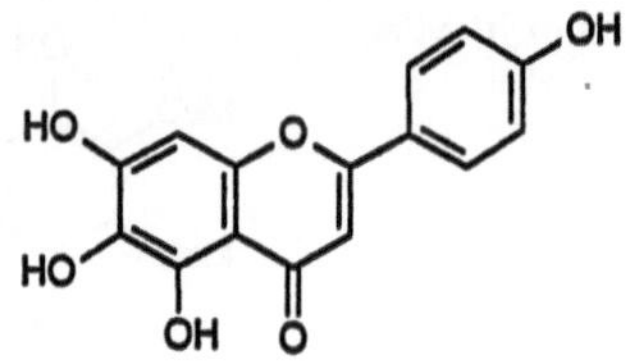	[16]
Hispidulin (*112-17*)	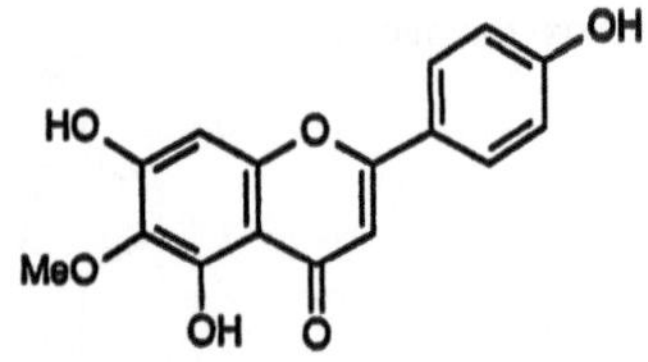	[11]
5,7,4′-Trihydroxy-8-methoxyflavone (*112-18*)	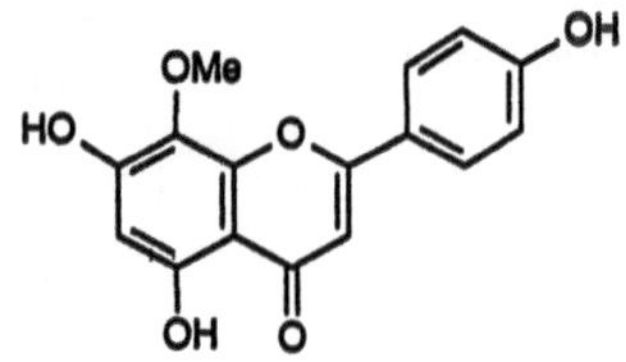	[9]
5,7,2′,3′-Tetrahydroxyflavone (*112-19*)	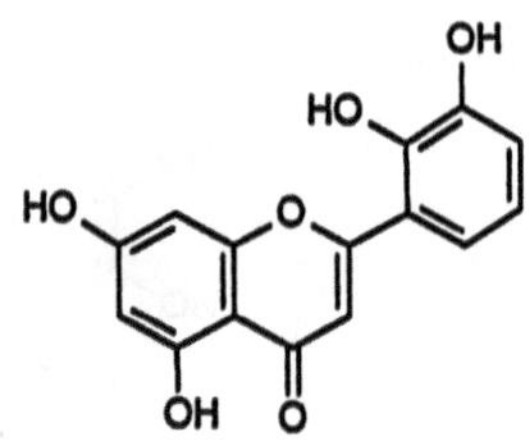	[10]
5,7,2′,6′-Tetrahydroxyflavone (*112-20*)	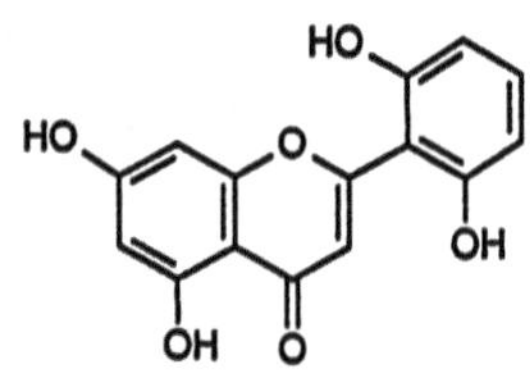	[9]
7,2′,6′-Trihydroxy-5-methoxyflavone (*112-21*)	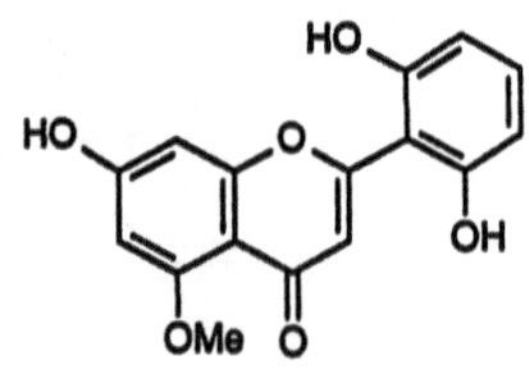	[10]
5,7,2′-Trihydroxy-6′-methoxyflavone (*112-22*)	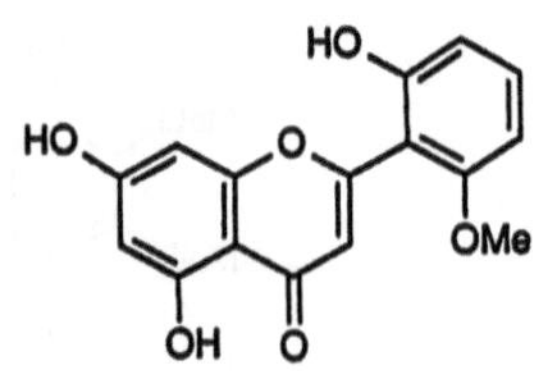	[10]

Table 112.1. (continued)

Compound	Structure	Reference
Viscidulin II (*112-23*)	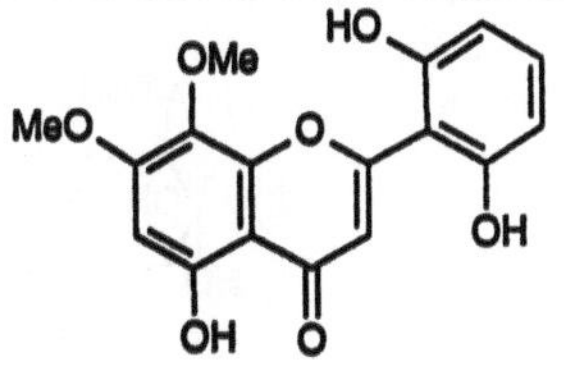	[10]
Skullcapflavone II (Neobaicalein) (*112-24*)	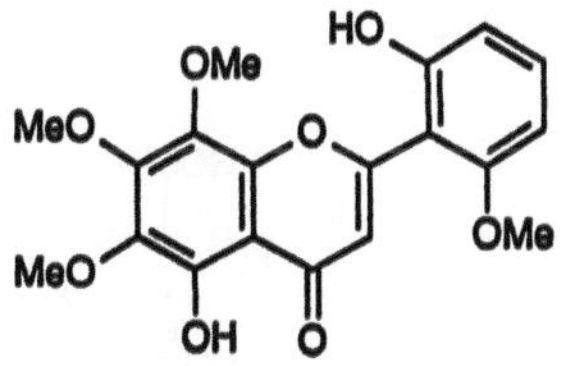	[12, 15, 17]
Ganhuangenin (*112-25*)	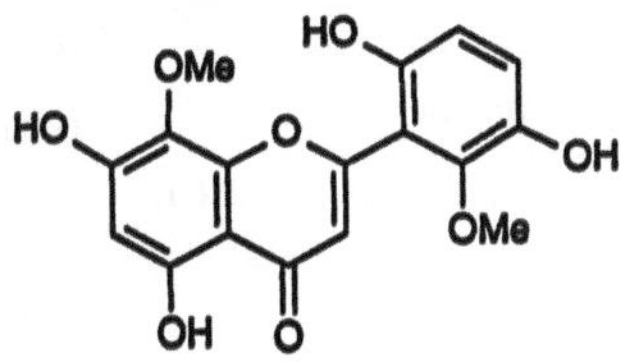	[18]
Viscidulin I (*112-26*)	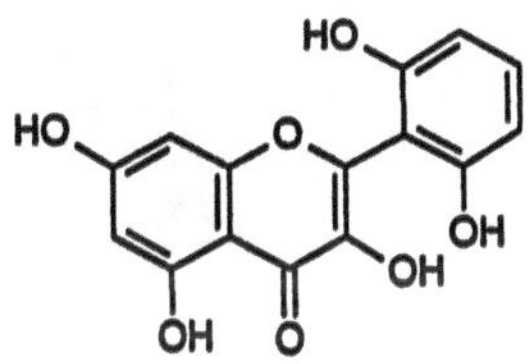	[10]
Flavone *O*-glycosides		
Baicalin (*112-2*)		[2, 4, 6]
Wogonoside (*112-27*)	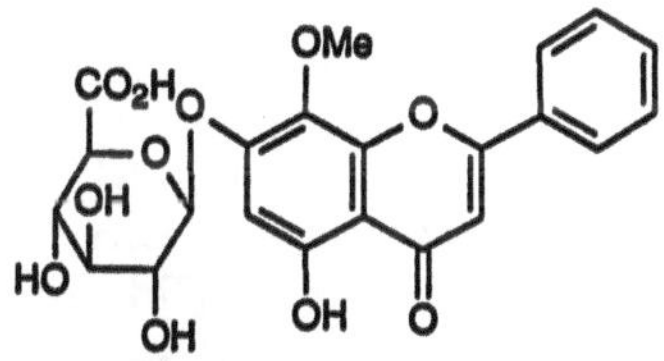	[6, 17]
Oroxylin A 7-*O*-β-D-glucopyranosiduronide (*112-28*)	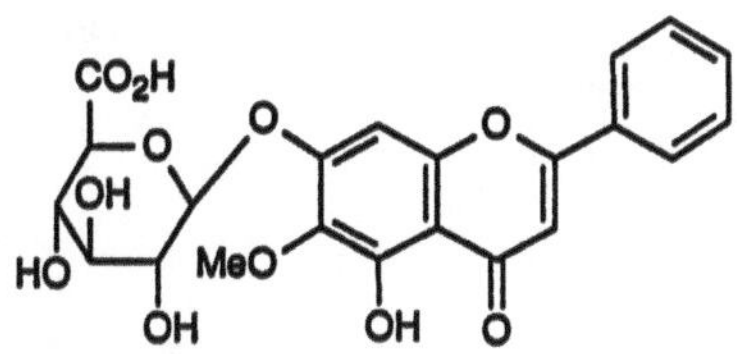	[5]

Table 112.1. (continued)

Compound	Structure	Reference
Scutellarin (*112-29*)	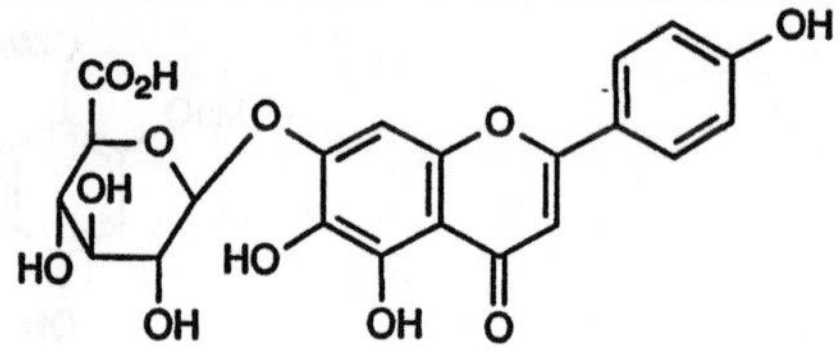	[19]
Baicalein 7-*O*-*β*-D-glucopyranoside (*112-30*)	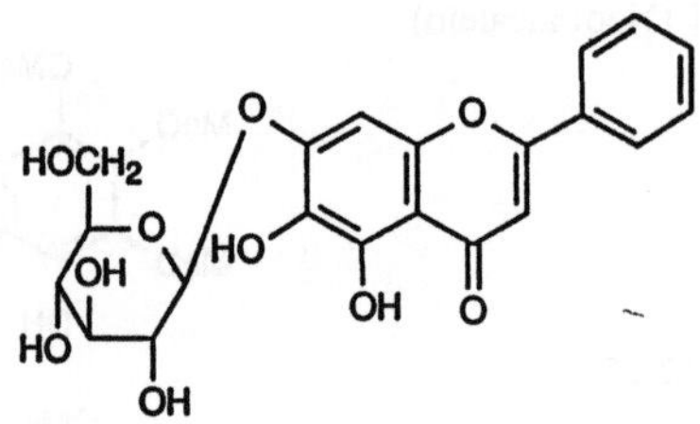	[5]
Wogonin 5-*O*-*β*-D-glucopyranoside (*112-31*)	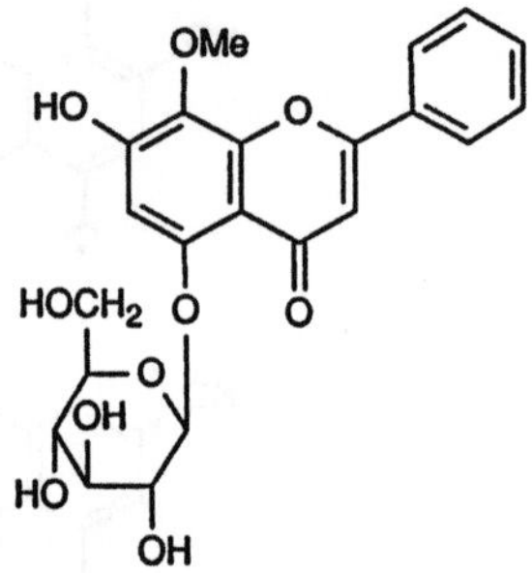	[20]
Flavone C glycosides		
6-*C*-*β*-D-Glucopyranosyl-8-*C*-α-L-arabinopyranosylchrysin (*112-32*)	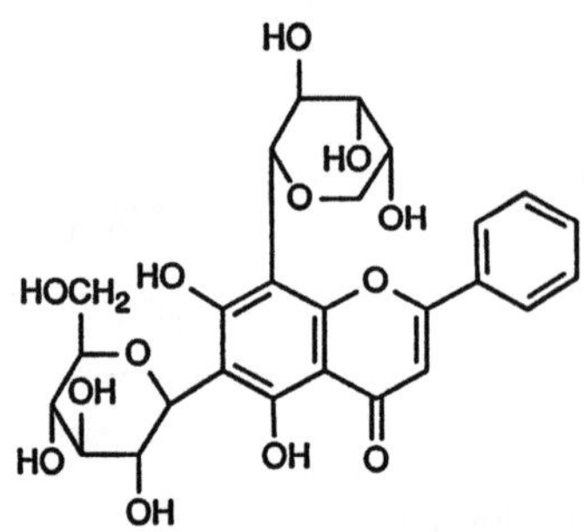	[21]
6-*C*-α-L-Arabinopyranosyl-8-*C*-*β*-D-glucopyranosylchrysin (*112-33*)	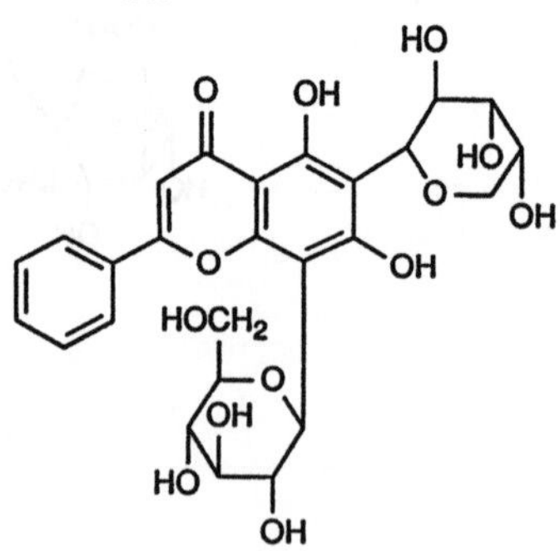	[21]

Table 112.1. (continued)

Compound	Structure	Reference
Flavanones		
Dihydrobaicalein (*112-34*)	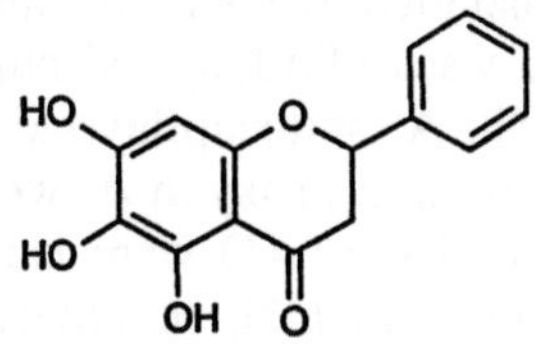	[7]
Dihydrooroxylin A (*112-35*)	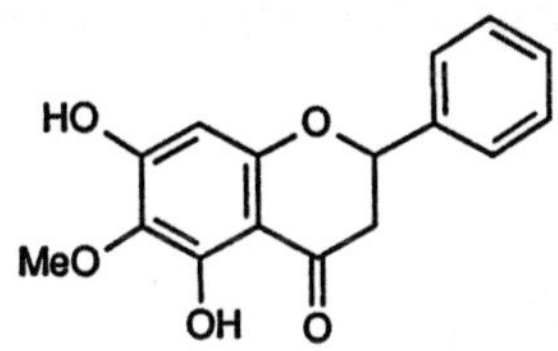	[5]
Carthamidin (*112-36*) (from leaves)	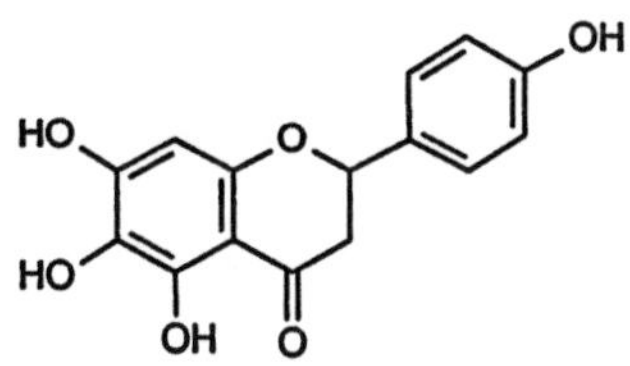	[22]
Isocarthamidin (*112-37*) (from leaves)	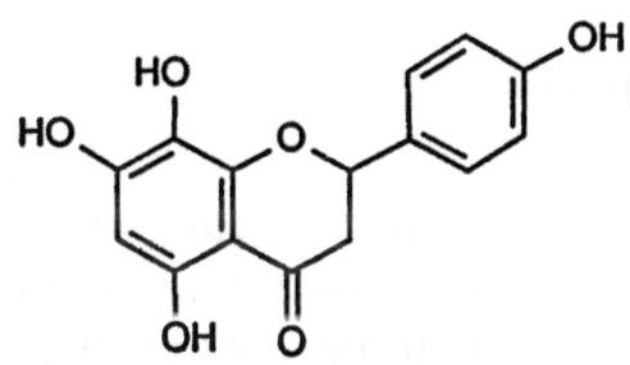	[22]
5,7,2′,6′-Tetrahydroxyflavanone (*112-38*)	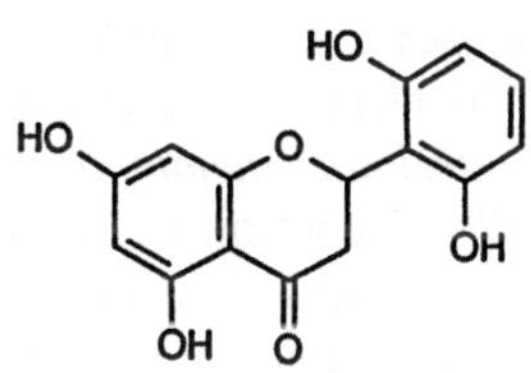	[5, 14]
3,5,7,2′,6′-Pentahydroxyflavanone (*112-39*)	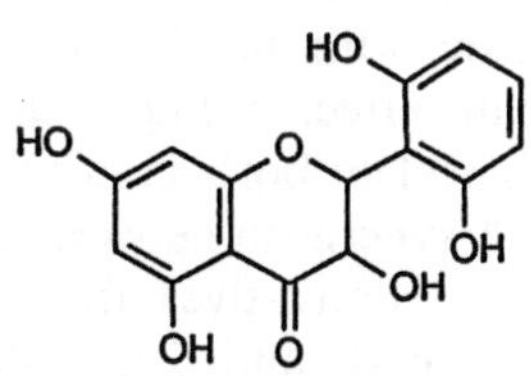	[20]
Chalcone		
2,6,2′,4′-Tetrahydroxy-6′-methoxy-chalcone (*112-40*)	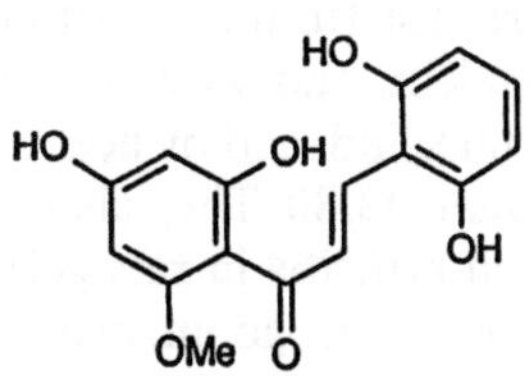	[10]

A number of *Scutellaria* species have been described as substitutes for *S. baicalensis*. They are identified as *S. amoena*, *S. hypericifolia*, *S. likianensis*, *S. rehderiana*, *S. tenax*, and *S. vircidula* [27]. The main components in all species are baicalin, baicalein, and wogonin. In addition to the three major flavone derivatives, hispidulin and wogonoside were isolated from *S. amoena* [28]; skullcapflavones I and II, wogonoside, oroxylin A, viscidulin I, and visculin II were isolated from *S. viscidula* [29]; wogonoside, oroxylin A, skullcapflavone II, viscidulin I, and tenaxin I were isolated from *S. tenax* [30]; and oroxylin A, wogonoside, ganhuangenin (viscidulin III) [31], and rehderianin I (*112-41*) were isolated from *S. rehderiana* [32]. The flavone compounds scutellarin [33], scutellarein, and the flavonone derivatives carthamidin and isocarthamidin [34] were isolated from the whole plant of *S. barbata*.

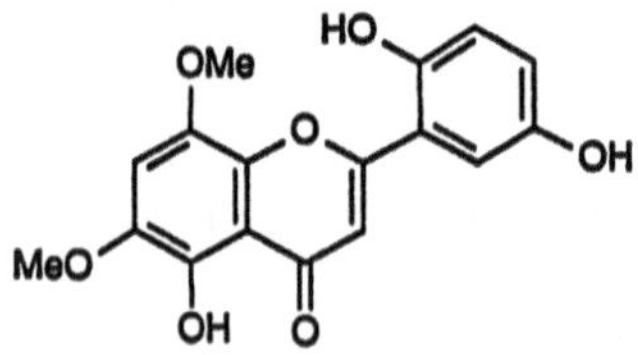

Rehderianin I (*112-41*)

112.3 Pharmacology

The ether extract of *S. baicalensis* root was found to possess marked antibacterial activity against gram-positive bacteria. One of the active principles was elucidated as 5,7,2′,6′-tetrahydroxyflavanone [14]. Of the flavones, wogonin showed inhibitory activity against *Vibrio comma* and *Staphylococcus aureus* at dilutions of 1:800 and 1:400, respectively [35]. In tests with selected oral bacteria, including suspected periodontopathogens, *Bacterioides melamnogenicus intermedius* was found to be the most sensitive against a 2% decoction of *S. baicalensis* [36].

The *Scutellaria* flavones baicalin, baicalein, wogonin, and oroxylin A showed no significant cytotoxic activity against L1210 leukemia cells in vitro. However, skullcapflavone II was cytotoxic, with an ED_{50} of 1.5 μg/ml [37, 38].

In tests for antiasthmatic effects using isolated tracheal muscle from guinea pigs as model, baicalin and baicalein both showed antihistaminic, anticholinergic, and papaverine-like activity. A synergism between baicalin and ephedrine was observed [39]. Many of the flavone derivatives from *S. baicalensis* including wogonin, wogonoside, skullcapflavone II, and 5,7,2′5′-tetrahydroxy-8,6′-dimethoxyflavone, 5,7,2′,6′-tetrahydroxyflavanone, 3,5,7,2′,6′-pentahydroxyflavanone were active in inhibiting the histamine release from rat peritoneal mast cells induced by compound 48/80 [40]. Baicalin, baicalein, and wogonin were found to inhibit the increase in vascular permeability in mice induced by acetic acid and to reduce acute paw edema in rats induced by compound 48/80. They also suppressed development of secondary lesion in adjuvant-induced arthritis in rats [41].

It was reported that baicalin, wogonin, oroxylin A, skullcapflavone II, and chrysin showed an inhibitory effect in vitro on collagen-induced blood platelet aggregation at a concentration of 0.1 m*M*. Chrysin was also found to inhibit ADP-induced blood platelet aggregation, whereas baicalein and wogonin had an inhibitory action on arachidonic acid induced blood platelet aggregation. Baicalein and baicalin showed inhibitory activity on the conversion of fibrinogen to fibrin by thrombin. They also were found to prevent the decrease in blood platelets and fibrinogen in rats with experimental disseminated intravascular coagulation induced by endotoxin [42]. Several flavone compounds isolated from *S. baicalensis* were also studied for their effects on arachidonate metabolism on leukocytes [43, 44].

Effects of flavone components of the root of *S. baicalensis* on mammalian lipid metabolism were also studied. Wogonin inhibited deposition of liver triglycerides and increased serum high-density lipoprotein-cholesterol levels in rats fed on a corn oil-cholesterol-sodium cholate mixture. Skullcapflavone II reduced total cholesterol levels and liver triglyceride content in serum and increased serum high-density lipoprotein-cholesterol. Baicalein and baicalin reduced serum-free fatty acid and triglyceride levels and liver glyceride content [45]. Various flavone derivatives of *S. baicalensis* inhibited lipid peroxidation in rat liver homogenate stimulated by a mixture of $FeCl_2$ and ascorbic acid or by a mixture of NADPH and ADP [46, 47]. When given 3 and 20 h before galactosamine treatment, an extract of *S. baicalensis* also showed protection against galactosamine-induced lethal toxicity [48].

Mutagenicity of wogonin in *Salmonella typhimurium* TA98 and TA100 [49] as well as mutagenicity of norwogonin in *S. typhimurium* TA100 with metabolic activation [50] was reported. On the other hand, baicalein showed a marked inhibition of 12-*O*-tetradecanoylphorbol-13-acetate(TPA)-induced epidermal ornithine decarboxylase activity and also inhibited promotion of dimethylbenzanthracene carcinogenesis by TPA [51].

Absorption of baicalin by the rat small intestine in situ was 20%–40% within half-times of 11–16 h. Removal of the biliary tract did not affect absorption [52]. After i.m. injection to humans, baicalin appeared in the urine at 0.4 h. Bioavailability was assessed to be 89%, the mean half-life was 0.6 h, indicating a rapid elimination. Bioavailability from tablets or solutions after oral administration was 22%–36% [53].

References

1. Hattori S (1930) Spectrography of the flavone series. III. The constitution of wogonin. Acta Phytochim 5:99–116 (CA 25:1525)
2. Shibata K, Iwata S, Nakamura M (1923) Baicalin, a new flavone-glucuronic acid compound from the roots of *Scutellaria baicalensis*. Acta Phytochim 1:105–39 (CA 17:3506)
3. Bargellini G (1919) 1,2,3-Trihydroxyflavone. Contribution to the knowledge of the constitution of scutellarein. Gazz Chim Ital 49:47–63
4. Tseng KF, Chang TY (1958) Studies on the flavonoids in Chinese drugs. VII. Chemical composition of huang-chin (*Scutellaria baicalensis*). 2. Position of glucuronic acid linking in baicalin. Acta Pharm Sin 6:21–25
5. Tamimori T, Jin H, Miyaichi Y, Toyofuku S, Namba T (1985) Studies on the constituents of *Scutellaria* species. VI. On the flavonoid constituents of the root of *Scutellaria baicalensis* Georgi. 5. Quantitative analysis of flavonoids in *Scutellaria* roots by high-performance liquid chromatography. Yakugaku Zasshi 105:148–155
6. Liu ML, Gao W (1982) Analysis of flavonoids in *Scutellaria baicalensis* G. Chin J Pharm Anal 2:134–140

7. Tomimori T, Miyaichi Y, Imoto Y, Kizu H, Tanabe Y (1983) Studies on the constituents of *Scutellaria* species. II. On the flavonoid constituents of the root of *Scutellaria baicalensis* Georgi 2. Yakugaku Zasshi 103:607–611
8. Popova TP, Litvinenko VI, Kovalev IP (1973) Flavones of *Scutellaria baicalensis* roots. Khim Prir Soedin 729–733
9. Tomimori T, Miyaichi Y, Kizu H (1982) On the flavonoid constituents from the roots of *Scutellaria baicalensis* Georgi. I. Yakugaku Zasshi 102:388–391
10. Tomimuri T, Miyaichi Y, Imoto Y, Kizu H, Suzuki C (1984) Studies on the constituents of *Scutellaria* species. IV. On the flavanoid constituents of the root of *Scutellaria baicalensis* Georgi. 4. Yakugaku Zasshi 104:529–534
11. Takagi S, Yamaki M, Inoue K (1980) Studies on the water-soluble constituents of the roots of *Scutellaria baicalensis* Georgi (Wogon). Yakugaku Zasshi 100:1220–1224
12. Horie T, Tsukayama M, Masumura M, Nakayama M, Hayashi S (1979) The synthesis of 5,2′-dihydroxy-6,8-dimethoxyflavone and its isomers: a revised structure of skullcapflavone. I. Bull Chem Soc Jpn 52:2950–2952
13. Takido M, Yasukawa K, Matsuura S, Iinuma M (1979) On the revised structure of skullcapflavone I, a flavone compound in the roots of *Scutellaria baicalensis* Georgi (Wogon). Yakugaku Zasshi 99:443–444
14. Kubo M, Kimura Y, Odani T, Tani T, Namba K (1981) Studies on *Scutellaria* radix. II. The antibacterial substance. Planta Med 43:194–201
15. Takido M, Aimi M, Takahashi S, Yamanouchi S, Torii H, Dohi S (1975) Constituents in the water extracts of crude drugs. I. Roots of *Scutellaria baicalensis*. Yakugaku Zasshi 95:108–113
16. Zhang XQ, Xu LY (1985) Pulse polarographic determination of flavone glycosides and their aglycones in *Scutellaria baicalensis* Georgi. Chin Trad Herb Drugs 16:216–217
17. Institute of Chinese Materia Medica, Academy of Traditional Chinese Medicine (1973) Chemical constituents of traditional drug *Scutellaria baicalensis*. Natl Med J China 417–420
18. Kimura Y, Okuda H, Taira Z, Shoji N, Takemoto T, Arichi S (1984) Studies on Scutellariae radix. IX. New component inhibiting lipid peroxidation in rat liver. Planta Med 50:290–295
19. Xiao HA (1981) Study on the stability of scutellarin in aqueous solution. Chin Trad Herb Drugs 12:21–22
20. Takagi S, Yamaki M, Inoue K (1981) On the minor constituents of the roots of *Scutellaria baicalensis* Georgi (Wogon). Yakugaku Zasshi 101:899–903
21. Takagi S, Yamaki M, Inoue K (1981) Flavone di-*C*-glycosides from *Scutellaria baicalensis*. Phytochemistry 20:2443–2444
22. Takido M, Aimi M, Yamanouchi S, Yasukawa K, Torii H, Takahashi S (1976) Studies on the constituents in the water extracts of crude drugs. II. On the leaves of *Scutellaria baicalensis* Georgi. Yakugaku Zasshi 96:381–383
23. Yu LR, Liu ML, Zhang YZ (1983) TLC densitometry of baicalin and wogonoside in *Scutellaria*. Chin J Pharm Anal 3:18–21
24. Katsura E, Yamagishi T (1982) Quantitative determination of flavonoids in scutell radix. Hokkaidoritsu Eisei Kenkyushoho 17–20 (CA 101:12273h)
25. Liang GF, Li LP, Li ZL (1985) A preliminary investigation on the technology of baicalin extraction. Bull Chin Mater Med 10:364–366
26. Jin RJ (1981) Experiments on improving the yield of baicalin from *Scutellaria baicalensis*. Chin Trad Herb Drugs 12:401
27. Song WZ (1981) Study on the resource of the Chinese herb *Scutellaria baicalensis* Georgi. Acta Pharm Sin 16:139–145
28. Liu YL, Li NW, Sung WC, Wu C (1980) Study on the flavonoid constituents in *Scutellaria amoena* C.H. Wright. Chin Trad Herb Drugs 11:337–340
29. Yu LR, Liu ML, Wang XT (1984) Studies on the flavonoids of *Scutellaria viscidula* Bunge. Acta Pharm Sin 19:397–398
30. Liu ML, Li ML, Wang FH (1984) Studies on flavonoids in *Scutellaria tenax*. Acta Pharm Sin 19:545–546
31. Liu YL, Song WZ, Ji QY, Bai YL (1984) Structure of ganhuangenin, a new flavanoid from the root of Gau Su Huang Qin (*Scutellaria rehderiana*). Acta Pharm Sin 19:830–835
32. Liu ML, Li ML, Wang FH, Yang CS, Liang XT (1984) Studies on the flavonoids of medicinal *Scutellaria*. I. *Scutellaria rehderiana*. Bull Chin Mater Med 9:76–77

33. Wang CC (1981) Brief report on the study of chemical constituents of *Scutellaria barbata* D. Don. Chin Trad Herb Drugs 12:19
34. Xiang RD, Zheng JF, Yao ZC (1982) Study on chemical constituents of Ban Zhi Lian (*Scutellaria barbata* D. Don). Chin Trad Herb Drugs 13:345–348
35. Hsu HY (1954) Chinese drugs Huang-Lien (*Coptis teeta*) Huang-Chin (*Scutellaria baicalensis*), and Ko-Kun (*Pueraria thunberiana*). IV. Antibacterial activity of wogonin from *Scutellaria baicalensis*. J Taiwan Pharm Assoc 6:7–10
36. Tsao TF, Newman MG, Kwok YY, Horikoshi AK (1982) Effect of Chinese and Western antimicrobial agents on selected oral bacteria. J Dent Res 61:1103–1106
37. Ryu SH, Ahn BZ, Pack MY (1985) The cytotoxic principle of Scutellariae radix against L1210 cell. Planta Med 51:355
38. Ryu SH, Ahn BZ, Pack MY (1985) The cytotoxic principle of Scutellariae radix against L1210 cell. Planta Med 51:462
39. Koda A, Nagai H, Yoshida Y, Lauw KH (1970) Pharmacologic actions of baicalin and baicalein. III. Experimental asthma. Nippon Yakubutsugaku Zasshi 66:471–86 (CA 76:21262w)
40. Kubo M, Matsuda H, Kimura Y, Okuda H, Arichi S (1984) Scutellariae radix. X. Inhibitory effects of various flavonoids on histamine release from rat peritoneal mast cells in vitro. Chem Pharm Bull (Tokyo) 32:5051–5054
41. Kubo M, Matsuda H, Tanaka M, Kimura Y, Okuda H, Higashino M, Tani T, Namba K, Arichi S (1984) Studies on Scutellariae radix. VII. Anti-arthritic and anti-inflammatory actions of methanolic extract and flavonoid components from Scutellaria radix. Chem Pharm Bull (Tokyo) 32:2724–2729
42. Kubo M, Matsuda H, Tani T, Arichi S, Kimura Y, Okuda H (1985) Studies on Scutellariae radix. XII. Anti-thrombic actions of various flavonoids from Scuttelariae Radix. Chem Pharm Bull (Tokyo) 33:2411–2415
43. Kimura Y, Okuda H, Arichi S (1985) Studies on Scutellariae Radix. XIII. Effects of various flavonoids on arachidonate metabolism on leukocytes. Planta Med 51:132–136
44. Sekiya K, Okuda H (1982) Selective inhibition of platelet lipoxygenase by baicalein. Biochem Biophys Res Commun 105:1090–1095
45. Kimura Y, Kubo M, Tani T, Arichi S, Ohminami H, Okuda H (1981) Studies on Scutellariae Radix. III. Effects on lipid metabolism in serum, liver and fat cells of rats. Chem Pharm Bull (Tokyo) 29:2308–2312
46. Kimura Y, Kubo M, Tani T, Arichi S, Okuda H (1981) Studies on Scutellariae radix. IV. Effects on lipid peroxidation in rat liver. Chem Pharm Bull (Tokyo) 29:2610–2617
47. Kimura Y, Okuda H, Tani T, Arichi S (1982) Studies on Scutellariae radix. VI. Effects of flavanone compounds on lipid peroxidation in rat liver. Chem Pharm Bull (Tokyo) 30:1792–1795
48. Ihara N, Arichi S (1981) Protective effect of oriental crude drugs and kampo medicament on a fulminant type of galactosamine-induced hepatitis. Wakanyaku Shinpojumn (Kiroku) 14:45–55 (CA 97:16912n)
49. Nagao M, Morita N, Yahagi T, Shimizu M, Kuroyanagi M, Fukuoka M, Yoshihara K, Natori S, Fujino T, Sugimura T (1981) Mutagenicities of 61 flavonoids and 11 related compounds. Environ Mutagen 3:401–419
50. Elliger CA, Henika PR, MacGregor JT (1984) Mutagenicity of flavones, chromones and acetophenones in *Salmonella typhimurium*. New structure-activity relationship. Mutat Res 135:77–86
51. Kato R, Nakadate A, Yamamoto S (1985) Lipoxygenase inhibition and tumor promotor inhibition by medicinal plant compounds. Wakan Iyaku Gakkaishi 2:162–163 (CA 104:61607r)
52. Mao FF, In XD, Zhu JB, Zhao LH, Hu ZZ (1984) Absorption of baicalin in the small intestine of rats. J Nanjing Pharm Coll 15:61–66
53. Mao FF, Tu XD, Zhu JB, Zhao LH, Shen GM, Philippe CC, Fan BT (1983) Absorption of baicalin in humans. J Nanjing Pharm Coll 14:61–66

Sophora flavescens Ait. 113

113.1 Introduction

Kushen, Radix Sophorae flavescentis, is the dry roots of *Sophora flavescens* Ait. (Fabaceae) collected in spring and fall. It is officially listed in the Chinese Pharmacopoeia and is a well-known Chinese herbal medicine used as a diuretic and for the treatment of diarrhea, gastrointestinal hemorrhage, and eczema.

113.2 Chemical Constituents

113.2.1 Chemical Constituents of *Sophora flavescens*

The root of *S. flavescens* is known to contain a number of quinolizidine alkaloids as major constituents. A total alkaloid content of more than 2%, calculated as matrine, is required by the Chinese Pharmacopoeia as a qualitative characteristic for the dry root. Matrine (*113-2*) [1–4] and oxymatrine (*113-3*) [1–5] are the main alkaloids in the root of *S. flavescens*. Matrine represents a saturated four-ring system containing two condensed quinolizidine moieties with matridine (*113-1*) as the basic structure [6].

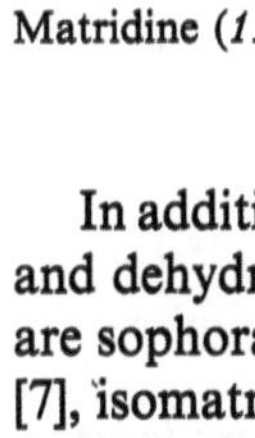

Matridine (*113-1*)

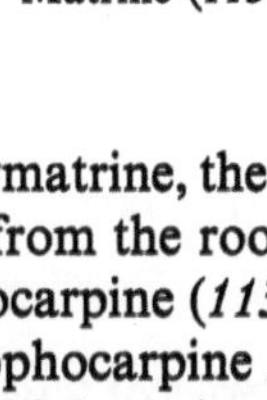

Matrine (*113-2*)

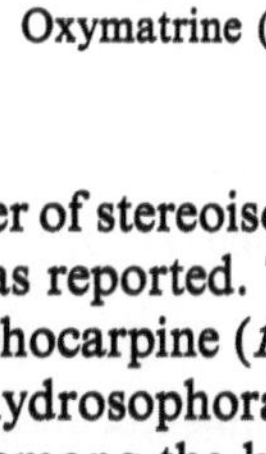

Oxymatrine (*113-3*)

In addition to matrine and oxymatrine, the isolation of a number of stereoisomers and dehydroanalogs of matrine from the root of *S. flavescens* was reported. These are sophoranol (*113-4*) [1], sophocarpine (*113-5*) [2] and 5-episophocarpine (*113-6*) [7], isomatrine (*113-7*) [8], and sophocarpine *N*-oxide [9]. 7,8-Dehydrosophoramine (*113-8*), the compound with the highest degree of unsaturation among the known matridine-type alkaloids, was isolated from the plant [10] and 7,11-dehydromatrine (*113-9*) was isolated from the fresh flowers [11] of *S. flavescens*.

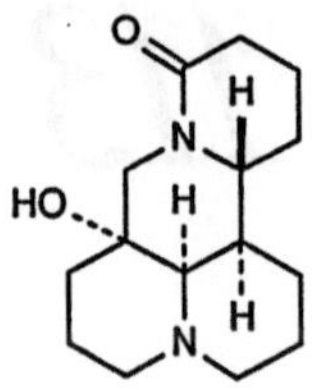

Sophoranol (*113-4*)

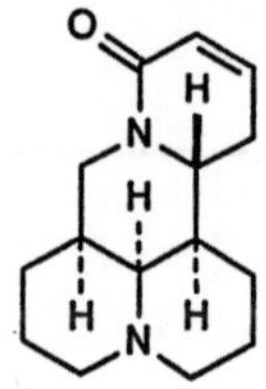

Sophocarpine (*113-5*)

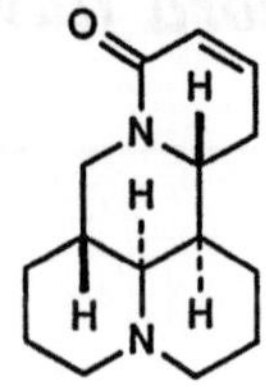

5-Episophocarpine (*113-6*)

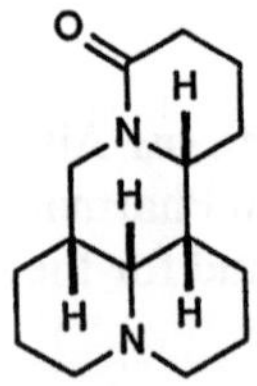

Isomatrine (*113-7*)

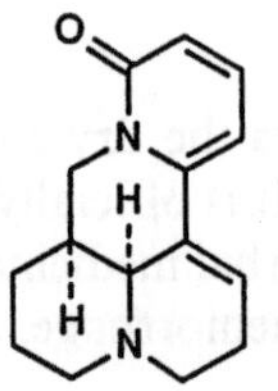

7,8-Dehydrosophramine (*113-8*)

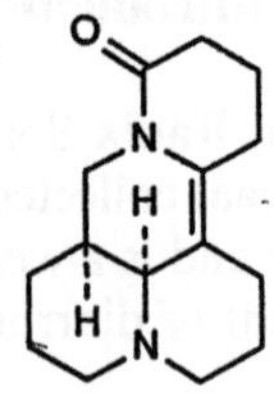

7,11-Dehydromatrine (*113-9*)

Other types of quinolizidine alkaloids without the matridine skeleton isolated from *S. flavescens* are methylcytisine (*113-10*), anagyrine (*113-11*), and baptifoline (*113-12*) [1, 2]. Methylcytisine, anagyrine, mamanine (*113-13*), kuraramine (*113-14*) [12], and isokuraramine (*113-15*) [11] were isolated from the flower of *S. flavescens*. Mamanine and kuraramine are possible metabolites of methylcytisine and anagyrine [12].

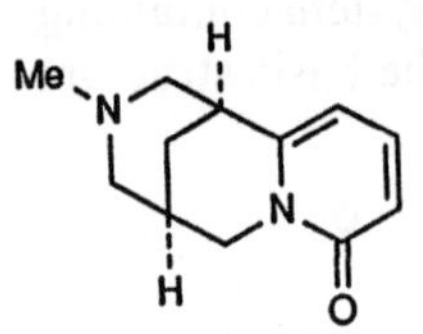

Methylcytisine (*113-10*)

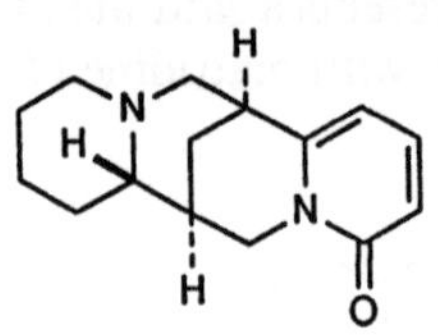

Anagyrine (*113-11*)

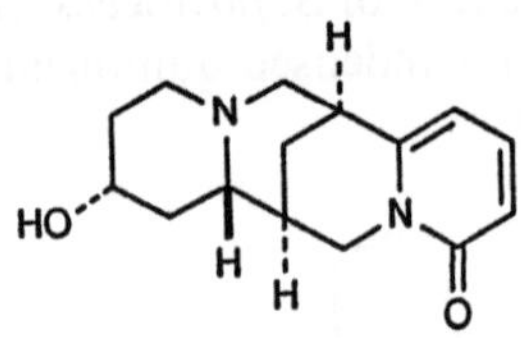

Baptifoline (*113-12*)

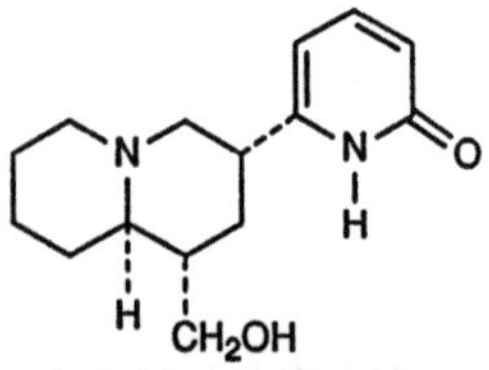

Mamanine (*113-13*)

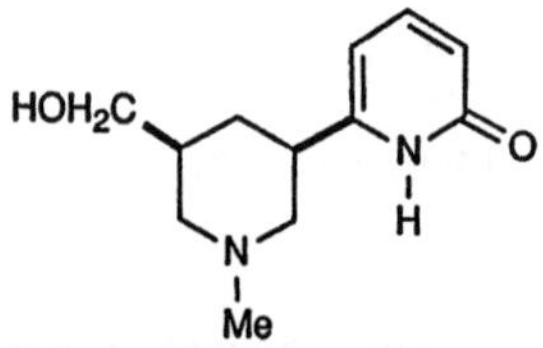

Kuraramine (*113-14*)

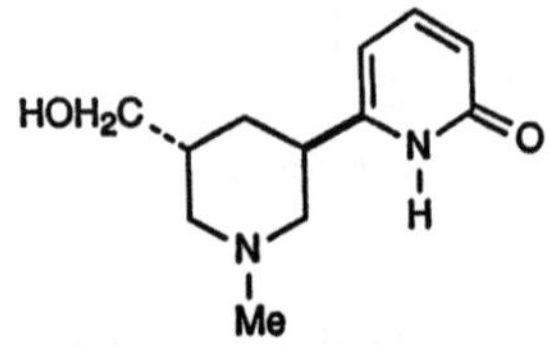

Isokuraramine (*113-15*)

Besides the alkaloid constituents, a series of flavone and related compounds were isolated from the root of *S. flavescens*, which are summarized in Table 113.1.

Table 113.1. Flavone and related compounds isolated from *Sophora flavescens*

Compound	Structure	Origin	Reference
Kuraridinol (*113-16*)	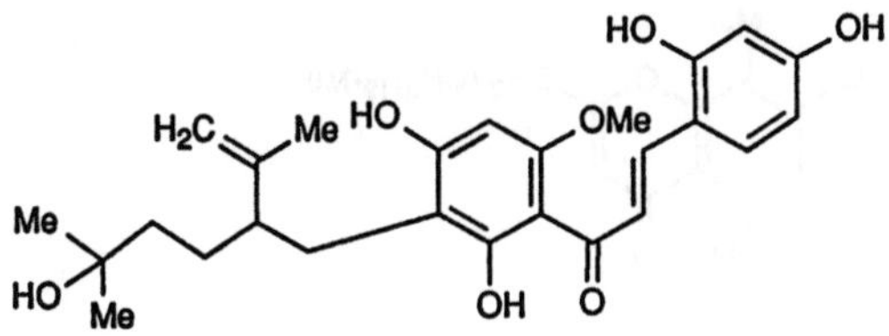	Root	[13]
Kurarinol (*113-17*)	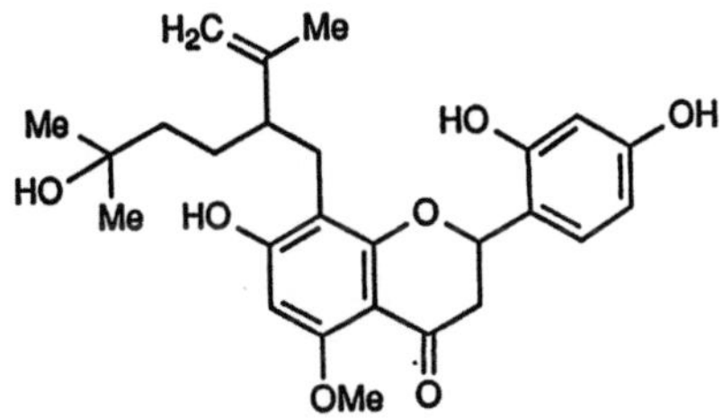	Root	[13]
Neokurarinol (*113-18*)	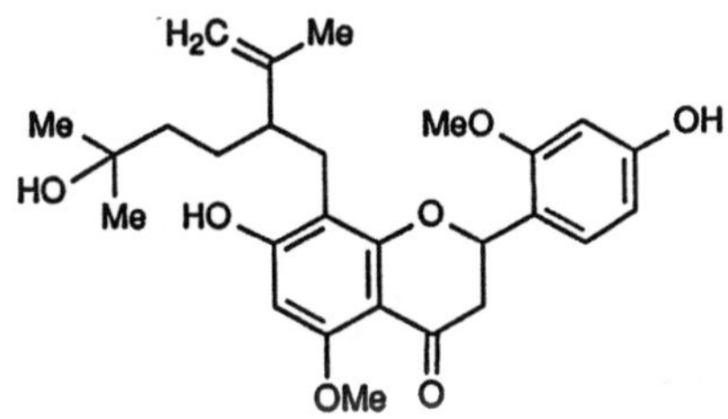	Root	[13]
Norkurarinol (*113-19*)	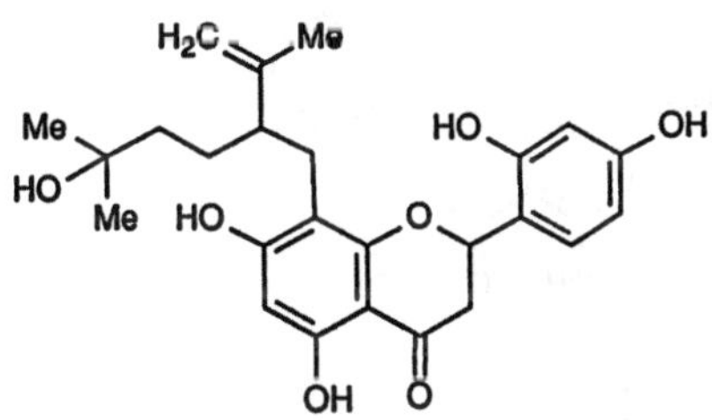	Root	[13]
Isokurarinone (*113-20*)	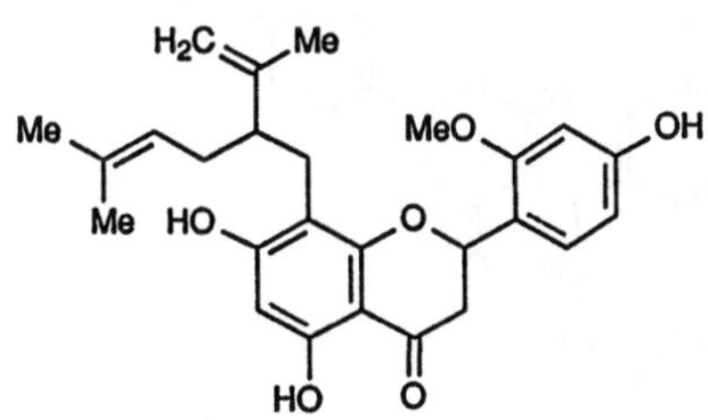	Root	[13]
Formononetin (*113-21*)	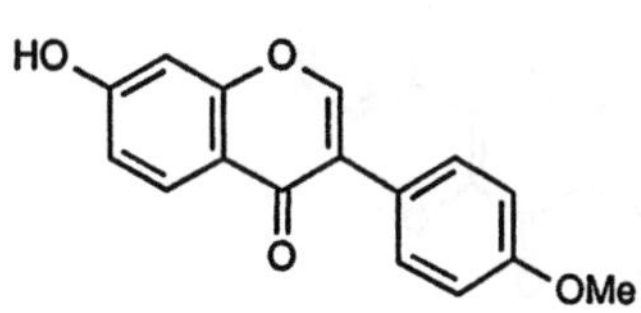	Root	[13]

Table 113.1. (continued)

Compound	Structure	Origin	Reference
2-Heneicosyl-5,7-dihydroxy-6,8-dimethylchromone (*113-22*)	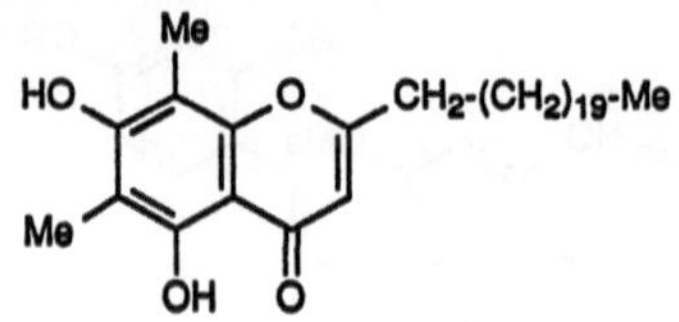	Plant	[14]
2-Undecyl-5,7-dihydroxy-6,8-dimethylchromone		Plant	[14]
2-Tridecyl-5,7-dihydroxy-6,8-dimethylchromone		Plant	[14]
2-Pentadecyl-5,7-dihydroxy-6,8-dimethylchromone		Plant	[14]
2-Heptadecyl-5,7-dihydroxy-6,8-dimethylchromone		Plant	[14]
2-Tricosyl-5,7-dihydroxy-6,8-dimethylchromone		Plant	[14]
Kushenol A (*113-23*)	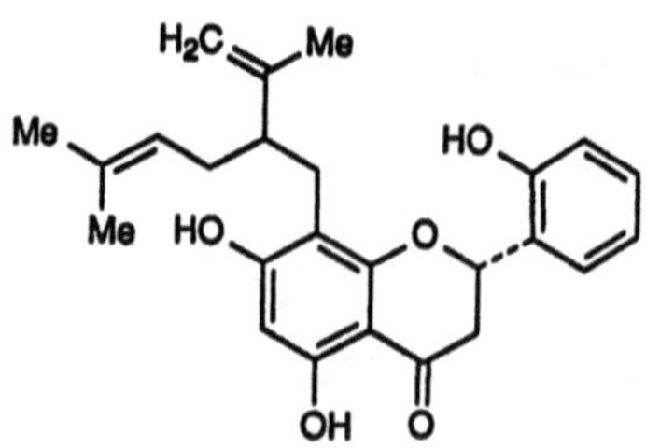	Root	[15]
Kushenol B (*113-24*)	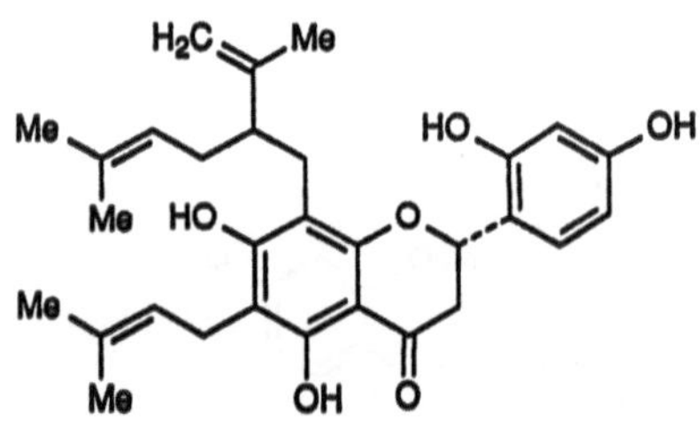	Root	[15]
Kushenol C (*113-25*)	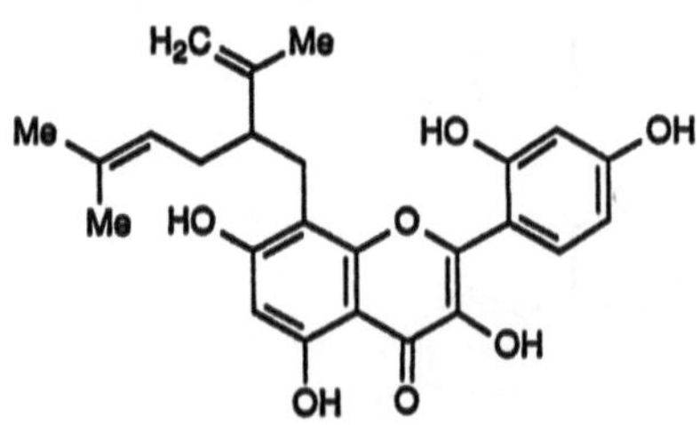	Root	[15]

Table 113.1. (continued)

Compound	Structure	Origin	Reference
Kushenol D (*113-26*)	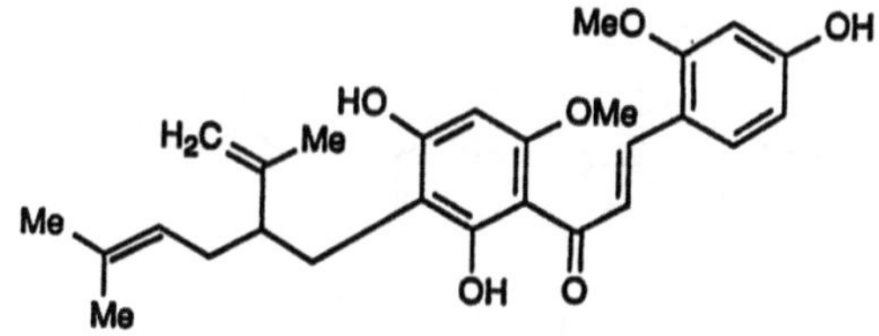	Root	[15]
Kushenol E (*113-27*)	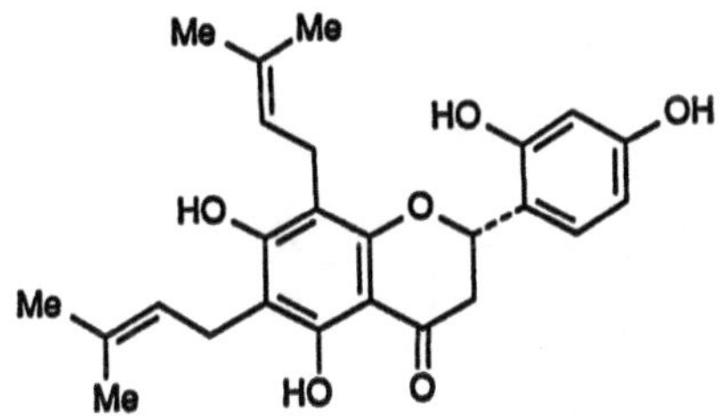	Root	[16]
Kushenol F (*113-28*)	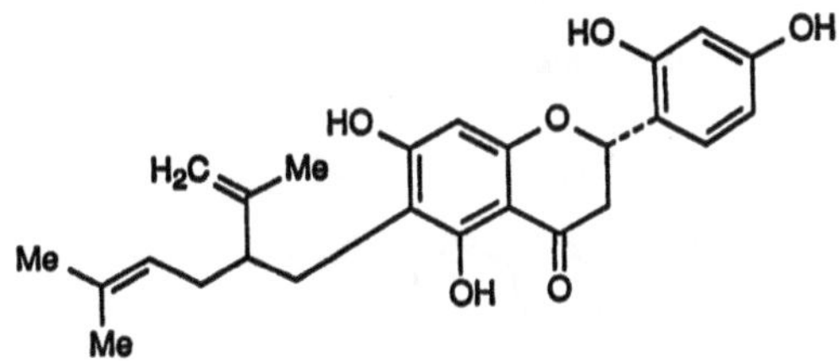	Root	[16]
Kushenol G (*113-29*)	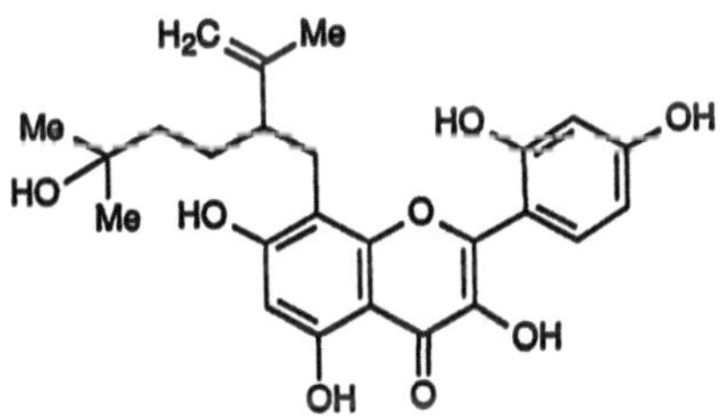	Root	[16]
Kushenol H (*113-30*)	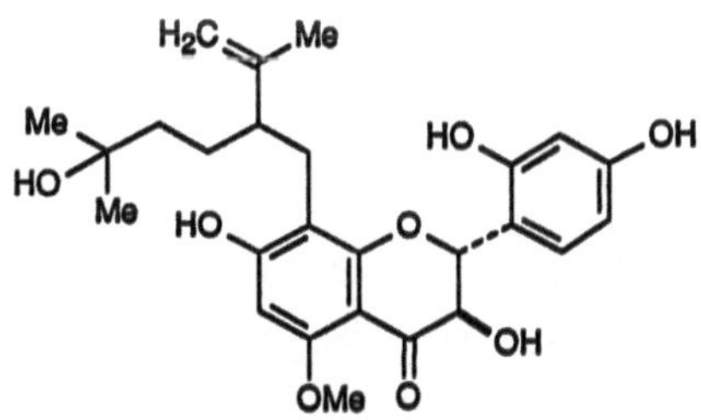	Root	[16]
Kushenol I (*113-31*)	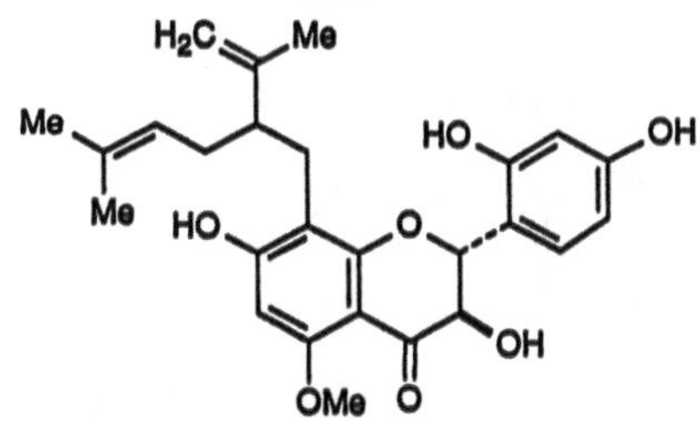	Root	[16]

Table 113.1. (continued)

Compound	Structure	Origin	Reference
Kushenol J (*113-32*)	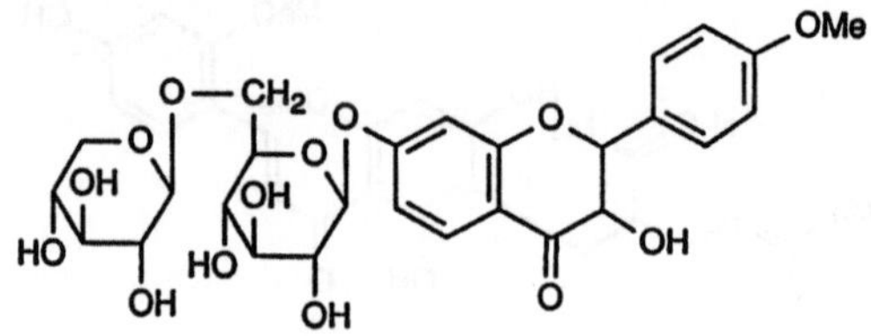	Root	[17]
Kushenol K (*113-33*)	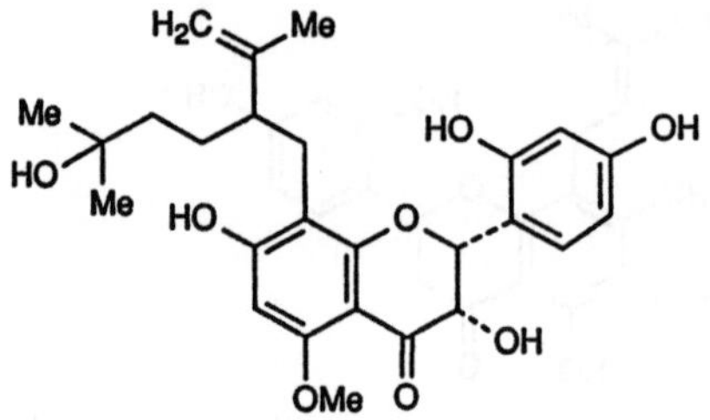	Root	[17]
Kushenol L (*113-34*)	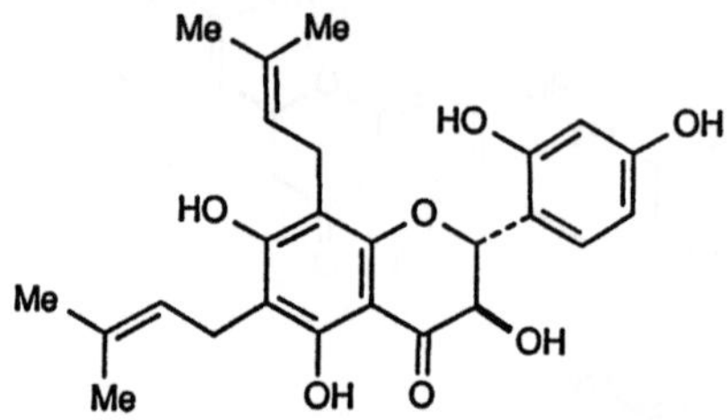	Root	[17]
Kushenol M (*113-35*)	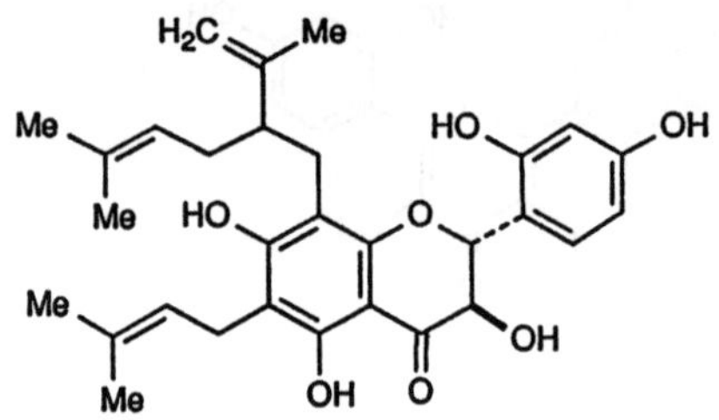	Root	[17]
Kushenol N (*113-36*)	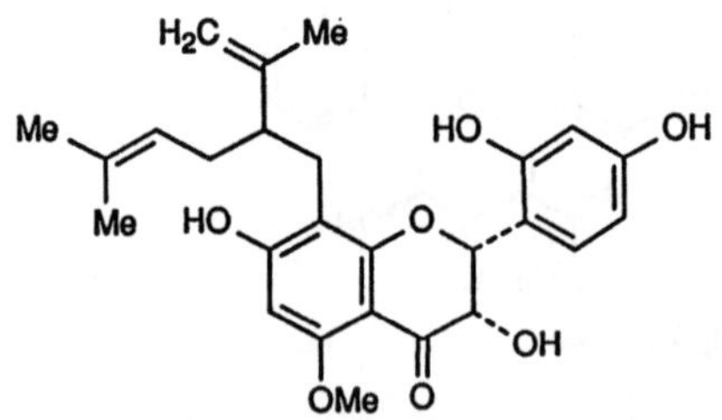	Root	[18]
Kushenol O (*113-37*)	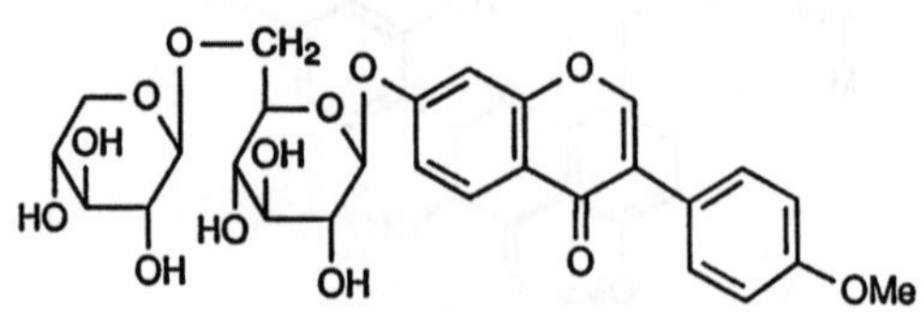	Root	[18]

Table 113.1. (continued)

Compound	Structure	Origin	Reference
Kushequinone A (*113-38*)	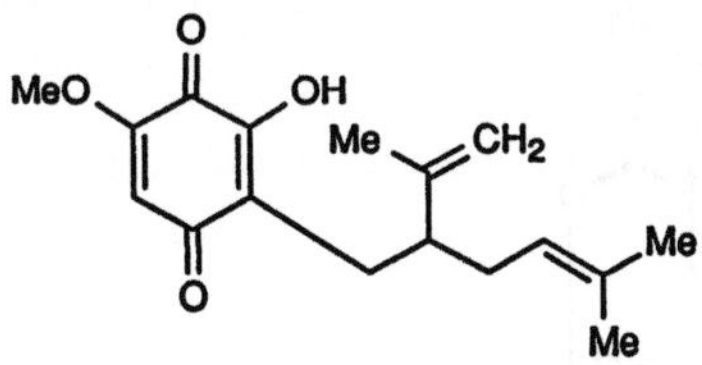	Root	[18]
Kushenin (*113-39*)	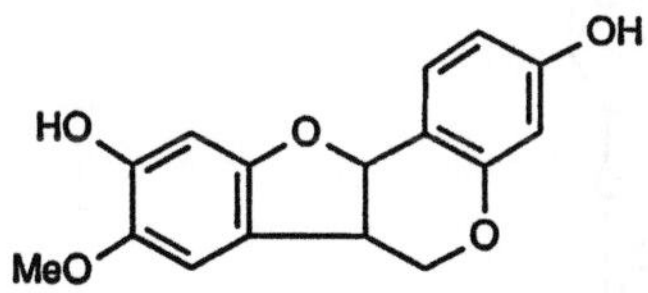	Root	[15]
Kuraridin (*113-40*)	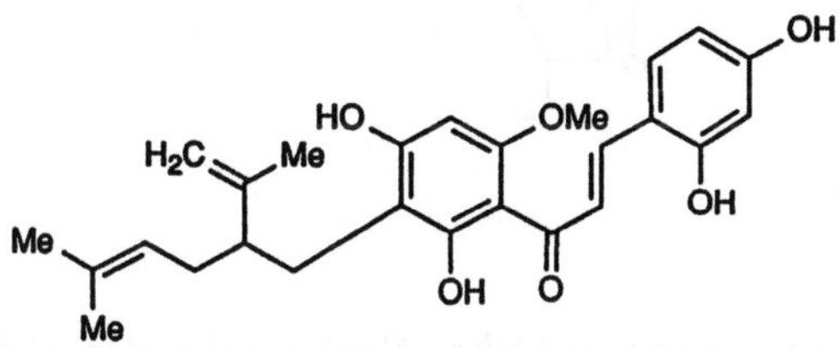	Root	[19]
Kurarinone (*113-41*)	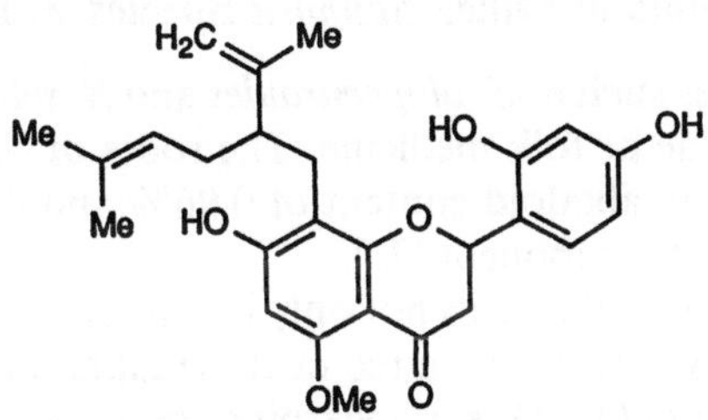	Root	[19]
Norkurarinone (*113-42*)	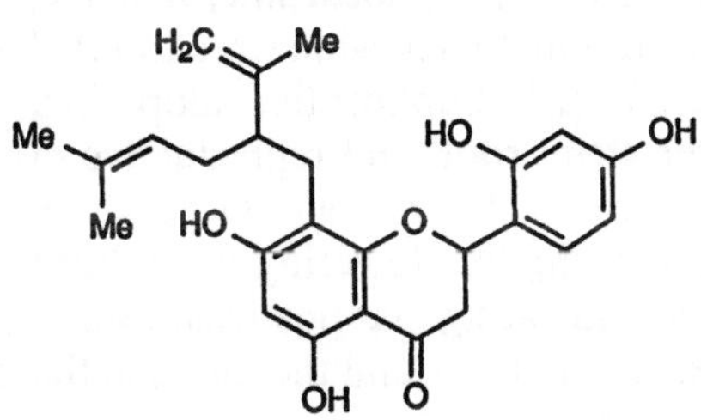	Root	[19]

A soponin named sophoraflavoside I (*113-43*) with soyasapogenol B as aglycone was isolated together with soyasaponin I (*26-14*) from the root of *S. flavescens* [20].

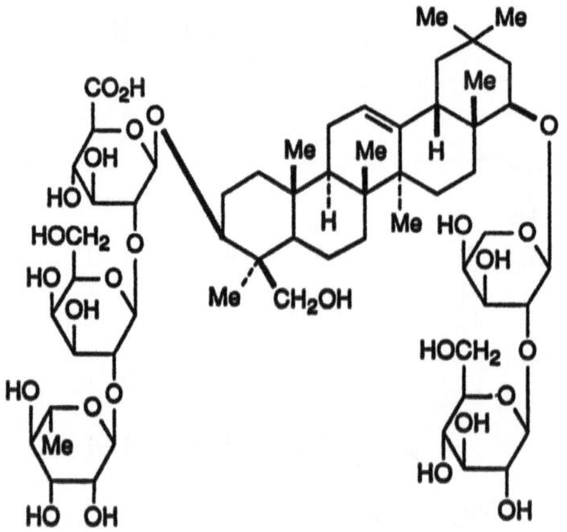

Sophoraflavoside I (*113-43*)

113.2.2 Chemical Constituents of Other *Sophora* Species with Medicinal Use

Several other *Sophora* species such as *S. alopecuroides* and *S. subprostrata* have also been used in Chinese medicine or folk medicine. The roots of *S. alopecuroides* and *S. subprostrata* have shown an alkaloid content of 0.96% and 0.66%, respectively, with oxymatrine as the major component [21].

Sophra alopeculoides contains alkaloids not only in the root, but also in the whole plant and may be used as an industrial source of these alkaloids [22]. Sophoridine (5-epimatrine), sophoramine (*113-44*), sophocarpine, matrine, aloperine (*113-45*), cytisine (*113-46*), oxymatrine, and methylcytisine were isolated and identified from the aerial part of *S. alopeculoides* [23]. Sophoridine, aloperine, sorphocarpine, and matrine were all present in the stem, root, and reproductive organs, and especially in the seeds, whereas in the leaves exclusively aloperine was found. The aloperine content in leaves was maximal during the flowering stage, but the alkaloid contents in seeds were highest after frost, indicating that postfrost seeds may be a good source of sophoridine, sophocarpine, and matrine, and the leaves in the flowering stage may be a good source of aloperine [24].

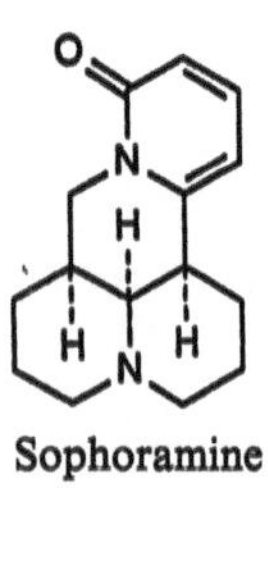

Sophoramine (*113-44*)

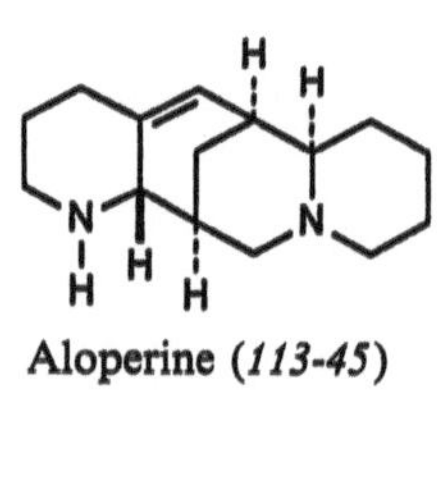

Aloperine (*113-45*)

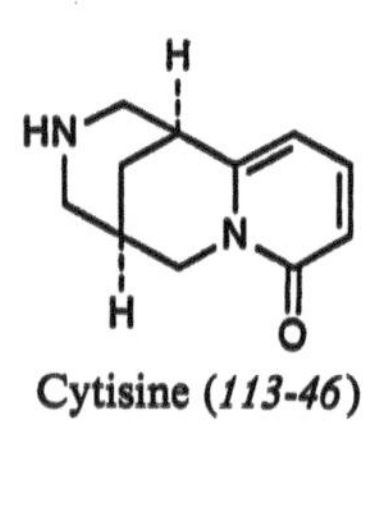

Cytisine (*113-46*)

In addition to the alkaloids mentioned above, the isolation of *N*-(2-hydroxyethyl)-cytisine, baptifoline, 3α-hydroxysophoridine [25], and *N*-allylaloperine [26] from *S. alopeculoides* was also reported. Another species, namely *S. viciifolia*, may also be a good source of industrial production of sophocarpine, since the leaves have been found to contain 0.4%–0.5% of the alkaloid [27, 28].

113.3 Pharmacology

113.3.1 Antiarrhythmic Activity

Total alkaloids from *S. flavescens* at a dose of 200 mg/kg exhibited a therapeutic and preventive effect against arrhythmia in rats and guinea pigs induced by $BaCl_2$, aconitine, or ouabain [29, 30]. They were also found to be active against chloroform- or epinephrine-induced arrhythmia in cats [31].

Of the pure alkaloids isolated from *S. flavescens*, sophocarpine was studied with regard to its antiarrhythmic activity. It showed effects against $CaCl_2$-induced ventricular arrhythmia in mice, aconitine-induced arrhythmia in rats, ouabain-induced arrhythmia in rabbits, and arrhythmia in dogs by coronary occlusion and reperfusion. In contrast, sophocarpine was not effective against $CaCl_2$-acetylcholine-induced atrial fibrillation in mice or against arrhythmias in rabbits and isolated rabbit hearts induced by chloroform, epinephrine, or ouabain. The effect of sophocarpine on myocardial cAMP content in mice was also negligible. Thus, sophocarpine is a substance for treatment of ventricular arrhythmia. Its effect was not found to be mediated by myocardial β-receptors, but was rather on the myocardium and nervous system that regulates heart rhythm [32]. The antiarrhythmic activity of sophoridine [33] and sophoramine [34, 35] was also reported.

The total flavone from *S. flavescens* has also shown antiarrhythmic activity. The incidence of spontaneous or ouabain-induced arrhythmias of cultured rat heart cells was reduced by the addition of flavones of *S. flavescens* root at a concentration of 125–250 μm/ml [36, 37].

In vivo, the total flavone of *S. flavescens* root inhibited chloroform-induced ventricular fibrillation in mice after i.v. administration. Intravenous administration of total flavone also decreased aconitine-induced arrhythmia in rats [38].

113.3.2 Antiulcer Activity

It was reported that the root of *S. flavescens* exhibited a potent antiulcer effect after oral administration similar to that observed after matrine and oxymatrine administration [39]. Oxymatrine inhibited the formation of ulcers induced by pylorus ligation or by indomethacin administration. This effect was considered to relate to inhibition of acid secretion [40]. When administered intraduodenally, it decreased acid secretion in rats and inhibited gastric motility induced by experimental stress. Thus, the protective effect of oxymatrine on stress ulcer appears to be due to a decrease in acid secretion and inhibition of gastric motility. On the other hand, matrine showed only a weak inhibition of gastric acid secretion, but was fairly effective at preventing stress ulcer after intravenous administration [41].

113.3.3 Antiasthmatic and Antitussive Activity

Oxymatrine has been used as an antiasthmatic agent when administered orally. A pharmacokinetic study showed that, after intramuscular injection of oxymatrine into rats, oxymatrine appeared at high concentrations in the tissues, bile, and urine. In contrast, when oxymatrine was given orally, concentrations of matrine exceeded those of oxymatrine in all samples, indicating that oxymatrine was converted to matrine. Intravenous injection of oxymatrine was ineffective against experimental asthma in guinea pigs, whereas oral administration of oxymatrine relieved asthma symptoms. Matrine is probably the pharmacologically active metabolite for asthma treatment arising from oxymatrine. In healthy volunteers given 100 mg oxymatrine orally, about 40% of the dose was excreted in the urine, 13%–33% representing oxymatrine [42, 43].

The metabolism of sophocarpine has also been studied. Sophocarpine is well absorbed from the gastrointestinal tract. Distribution in different organs after oral administration was similar to that after i.v. or i.m. administration with some delay in appearance of maxima. Sophocarpine was eliminated mainly via the renal route [44].

Sophocarpine, sophocarpine hydrobromide, and the aqueous extract of *S. alopeculoides* all showed antitussive activity. This action has been reported to be mediated via the central β-receptors [45]. Flavones and related compounds of *S. flavescens*, including kushenol A, kurarinone, and kuraridin, showed inhibitory activity on cAMP phosphodiesterase. The prenyl group in the structure has been found to be important for high inhibitory activity. Kinetic study revealed that norkurarinone, kurarinol, and kuraridin noncompetitively inhibited cAMP phosphodiesterase [46]. The antiinflammatory and antiallergic action of aloperine was also reported [47].

113.3.4 Antineoplastic Activity

Matrine and oxymatrine exhibited significant antitumor activity against sarcoma 180, and matrine also showed antitumor activity against Ehrlich ascites tumor in mice [19].

Sophocarpine moderately inhibited the transplanted tumors S180, U14, Lio-1, Walker 256, and L615. When sophocarpine was given i.p. to mice inoculated with S180 ascites sarcoma, Ehrlich ascites carcinoma, or Walker 256, mitotic indices of cancer cells were moderately or slightly reduced. When normal or tumor-bearing mice were given sophocarpine intragastrically at 24 mg/kg daily for 10 days, RNA and DNA contents in the tumor and spleen were decreased slightly. In dogs given sophocarpine orally at a dose of 45 mg/kg, mild thrombopenia was observed. No significant changes in immune activity in mice were caused by sophocarpine treatment [48, 49]. RNA and DNA contents in Ehrlich ascites tumor cells of mice were decreased by 7%–9% and 20–30%, respectively, following administration of 60–120 mg/kg sophocarpine. Sophocarpine treatment also inhibited the incorporation of [^{3}H]thymidine into tumor cell DNA by 21%–34% [50].

In contrast, matrine showed immunosuppressive activity in vivo. Murine spleen cell proliferation and interleukin 2 formation were reduced by 50% in culture at fairly high concentrations of matrine (0.6 and 0.1 mg matrine/ml, respectively) [51].

113.3.5 Other Pharmacological Actions

Oxymatrine exhibited a protective effect on experimental liver damage in animals. Hepatic necrosis, glycogen depletion, and elevation of serum glutamic pyruvic transaminase activity in rabbits and mice induced by CCl_4 or D-glucosamine were all significantly decreased by oxymatrine at a dose of 3.6 mg/kg by i.m. administration.

Repeated i.p. administration of oxymatrine at a dosage of 100–150 mg/kg per day for 2–4 weeks caused no significant damage to organs including heart, spleen, and kidneys in mice. The i.p. LD_{50} of oxymatrine in mice was 521 mg/kg [52].

Matrine, given i.p. or orally at doses of 20 or 30 mg/kg, inhibited yeast-provoked elevation of body temperature in rats in a dose-dependent manner. The antipyretic activity of matrine has been considered to be mediated by dopamine release or through dopaminergic receptor blockade [53].

Intramuscular administration of matrine at a dose of 25 mg/kg in rats markedly decreased carrageenan-induced hind paw inflammation. Matrine at a daily dose of 15 and 25 mg/kg for 8 days in rats significantly reduced ear inflammation induced by croton oil. Adrenolectomy did not influence the antiinflammatory effect of matrine in mice. Thus, matrine exhibited characteristics of nonsteroidal antiinflammatory agents and its effect is probably not related to the hypothalamic-adrenal axis [54].

References

1. Bohlmann F, Rahtz D, Arndt C (1958) Lupinen-Alkaloide. XI. Die Alkaloide aus *Sophora flavescens*. Chem Ber 91:2189–2193
2. Okuda S, Murakoshi I, Kamata H, Kashida Y, Haginiwa J, Tsuda K (1965) Lupine alkaloids. I. Minor alkaloids of Japanese *Sophora flavescens*. Chem Pharm Bull (Tokyo) 13:482–487
3. Zhang YZ, Zhang SR, Cui JF, Wu XM, Yan AC (1981) Thin layer chromatographic separation and determination of alkaloids in *Sophora flavescens* Ait. and its preparation. Acta Pharm Sin 16:283–288
4. Bai SZ, He JH, Yang ZQ, Rao EC (1982) Studies on alkaloidal components of Ku Shen (*Sophora flavescens* Ait.) II. Isolation of oxymatrine and other alkaloids. Chin Trad Herb Drugs 13:8–9
5. Ibragimov BT, Talipov SA, Tishchenko GN, Kushmuradov YK, Aripov TF (1979) Molecular and crystal structure of matrine *N*-oxide. Kristallografiya 24:45–51 (CA 90:204319d)
6. Bohlmann F, Weise W, Rahtz D, Arndt C (1958) Lupinen-Alkaloide. X. Die Konfiguration des Matrins. Chem Ber 91:2176–2189
7. Morinaga K, Ueno A, Fukushima S, Namikoshi M, Iitaka Y, Okuda S (1978) Studies on lupin alkaloids. VIII. A new stereoisomer of sophocarpine. Chem Pharm Bull (Tokyo) 26:2483–2488
8. Ueno A, Morinaga K, Fukushima S, Iitaka Y, Koiso Y, Okuda S (1975) Lupine alkaloids. VI. Isolation and structure of (+)-isomatrine. Chem Pharm Bull (Tokyo) 23:2560–2566
9. Ota N, Mino Y (1979) Pharmacognostical studies on the Chinese crude drug "Sophorae radix". I. High-speed liquid chromatographic studies on the alkaloid constituents of Sophorae radix. Shoyakugaku Zasshi 33:140–145
10. Ueno A, Morinaga K, Fukushima S, Okuda S (1978) Studies on the lupine alkaloids. VII. Isolation and structure of (−)-Δ7-dehydrosophoramine. Chem Pharm Bull (Tokyo) 26:1832–1836
11. Murakoshi I, Kidoguchi E, Haginiwa J, Ohmiya S, Higashiyama K, Otomasu H (1982) Isokuraramine and (−)-7,11-dehydromatrine, lupine alkaloids from flowers of *Sophora flavescens*. Phytochemistry 21:2379–2384
12. Murakoshi I, Kidoguchi E, Haginiwa J, Ohmiya S, Higashiyama K, Otomasu H (1981) (+)-Kuraramine, a possible metabolite of (−)-*N*-methylcytisine in flowers of *Sophora flavescens*. Phytochemistry 20:1407–1409

13. Kyogoku K, Hatayama K, Komatsu M (1973) Constituents of a Chinese crude drug Kushen (the root of *Sophora flavescens*). Isolation of five new flavonoids and formononetin. Chem Pharm Bull (Tokyo) 21:2733–2738
14. Ueno A, Hirakawa K, Fukushima S, Noro T, Moninaga K (1978) Studies on the constituents of *Sophora flavescens*. Aiton. Chem Pharm Bull (Tokyo) 26:2407–2410
15. Wu LJ, Miyase T, Ueno A, Kuroyanagi M, Noro T, Fukushima S (1985) Studies on the constituents of *Sophora flavescens* Aiton. II. Chem Pharm Bull (Tokyo) 33:3231–3236
16. Wu LJ, Miyase T, Ueno A, Kuroyanagi M, Noro T, Fukushima S (1985) Studies on the constituents of *Sophora flavescens* Ait. III. Yakugaku Zasshi 105:736–741
17. Wu LJ, Miyase T, Ueno A, Kuroyanagi M, Noro T, Fukushima S, Sasaki S (1985) Studies on the constituents of *Sophora flavescens* Ait. IV. Yakugaku Zasshi 105:1034–1039
18. Wu LJ, Miyase T, Ueno A, Kuroyanagi M, Noro T, Fukushima S, Sasaki S (1986) Studies on the constituents of *Sophora flavescens* Ait. V. Yakugaku Zasshi 106:22–26
19. Kojima R, Fukushima S, Ueno A, Saiki Y (1970) Antitumor activity of Leguminosae plants constituents. I. Antitumor activity of constituents of *Sophora subprostrata*. Chem Pharm Bull (Tokyo) 18:2555–2563
20. Yoshikawa M, Wang HK, Kayakiri H, Taniyama T, Kitagawa I (1985) Soponin and sapogenol. XI. Structure of sophoraflavoside I, a bisdesmoside of soyasapogenol B, from Sophorae Radix, the root of *Sophora flavescens* Aiton. Chem Pharm Bull (Tokyo) 33:4267–4274
21. Cui JF, Zhang GD (1986) Analysis of alkaloid constituents in four *Sophora* species. Bull Chin Mater Med 11:102–103
22. Zhao BG, Yen CC, Yang H, Shih CH, Chuang M, Yang C, Chang YF, Wang CT, Wang W (1980) Study on the extraction and isolation of alkaloids from *Sophora alopeculoides* L. for industrial production. Chin Trad Herb Drugs 11:350–352
23. Zhao BG (1980) Studies on the alkaloids of *Sophora alopeculoides* L. Acta Pharm Sin 15:182–183
24. Zhao BG, Liu XH, Yan ZJ (1984) TLC determination of the main alkaloids in foxtail-like sophora (*Sophora alopeculoides*) and its accumulation trends. Chin Trad Herb Drugs 15:559–560
25. Monakhova TE, Proskurnina NF, Tolkachev ON, Kavanov VS, Perel'son ME (1973) Alkaloids of *Sophora alepeculoides*. 3α-hydroxysophoridine. Khim Prir Soedin 9:59–64 (CA 78:15960v)
26. Tolkachev ON, Monakhova TE, Sheichenko VI, Kabanov VS, Fesenko OH, Proskurinina NF (1975) New types of alkaloids from *Sophora alopeculoides*. Khim Prir Soedin 11:30–37 (CA 83:97667x)
27. Li S, Chang HJ (1981) Study on alkaloids of *Sophora viciifolia* Hance. Chin Pharm Bull 16:52
28. Li S, Chang XR, Yan JY, Chen YL (1984) Studies on alkaloids of *Sophora viciifolia* Hance. I. A resource for the isolation of sophocarpine. Shaanxi Med J 13:61–62
29. Zhang BF, Song LY, Zhu LG, Liu YL, Li XQ (1985) Effect of *Sophora* alkaloids on experimental arrhythmia. Chin Trad Herb Drugs 10:229–230
30. Cha L, Qian JQ, Lu FH (1981) Antiarrhythmic effect of tetrandrine and the total alkaloids of *Sophora flavescens* root. Acta Pharmacol Sin 2:26–28
31. Cha L, Chien CC, Lu FH (1981) Antiarrhythmic effect of *Angelica sinensis* root, tetrandrine and *Sophora flavescens* root. Chin Pharm Bull 16:53–54
32. Zhao ZY, Li YQ, Lin ZY (1983) Antiarrhythmic effect of sophocarpine hydrobromide. Acta Pharmacol Sin 4:173–176
33. Cui LH, Zhang BH (1986) The antiarrhythmic effects and mechanism of sophoridine. Chin J Pharmacol Toxicol 1:3–10
34. Wu QH, Long YD, Yao LY, Luo W, Xiong HF, Zhang JM, Zeng JY (1988) Effects of sophoramine on ventricular fibrillation threshold in dogs. Acta Pharmacol Sin 9:137–139
35. Li H, Zhang BH (1989) The antiarrhythmic effect of sophoramine. Acta Pharm Sin 24:147–150
36. Wang JZ, Sun FL, Han HW, Zhang BH, Yuan SH, Ma L (1983) Effect of total flavones of *Sophora flavescens* root on arrhythmia of cultured rat heart cells. Acta Pharmacol Sin 4:32–35
37. Wang CC, Sun FL, Han HW, Chang PH, Yuan SH, Ma L (1981) Experimental study on the effect of total flavone of *Sophora flavescens* root on the arrhythmia heart cell. Chin Pharm Bull 16:53
38. Zhang BH, Su Y, Ma L, Li QH (1979) Experimental study on antiarrhythmic activity of total flavones from "Ku Shen" (*Sophora flavescens* Ait.). Acta Pharm Sin 14:449–454

39. Yamazaki M, Shirota H, Suzuki S (1981) Protective effect of Kujin and Sanzukon on experimental stress ulcer in mice. Wakanyaku Shinpojumu (Kiroku) 14:7–10 (CA 79:528f)
40. Yamazaki M (1983) Inhibition of gastric acid secretion by matrine and oxymatrine. Jpn Kokai Tokkyo Koho JP 58 57,316 (83 57,316) (CA 99:16563r)
41. Yamazuki M, Arai A, Suzuki S, Takeuchi T (1984) Protective effects of matrine and oxymatrine on stress ulcer in relation to their effects on the central nervous system. Yakugaku Zasshi 104:293–301
42. Xie MZ, Zhou WZ, Zhang Y (1981) Metabolism of oxymatrine. Acta Pharm Sin 16:481–487
43. Xie MZ, Zhou WZ, Zhang Y (1983) Oxymatrine metabolic fate. Chin Med J [Engl] 96:145–150
44. Li YQ, Yao DF, Yu SL, Wang JX, Liu CR, Cheng Y (1982) Neural mechanism of bronchospasmolytic effect of sophocarpine hydrobromide. Acta Pharmacol Sin 3:169–171
45. Li YQ, Yao TF, Yu SL, Chao ML, Cheng Y, Wang CH (1980) Pharmacological effects of kugancao (*Sophora alopeculoides*) and sophocarpine. Chin Trad Herb Drugs 11:555–557
46. Ohmoto T, Aikawa R, Nikaido T, Sankawa U, Wu LJ (1986) Inhibition of cyclic AMP phosphordiesterase in medicinal plants. Part XI. Inhibition of adenosine 3′,5′-cyclic monophosphate phosphodiesterase by components of *Sophora flavescens* Aiton. Chem Pharm Bull (Tokyo) 34:2094–2099
47. Zhou CC, Gao HB, Sun XB, Shi HB, Lin W, Yuan HN, Wang ZX (1989) Anti-inflammatory and antiallergic action of aloperine. Acta Pharmacol Sin 10:360–365
48. Li XM, Li LZ, Chen SL, Quin XH, Yu YH, Pan DX, Zou LL, Yang ZL, Han YS (1984) Inhibitory effects of sophocarpine on animal tumors. Acta Pharmacol Sin 5:125–130
49. Li XM, Wu YG, Chen SL, Pan DX, Wu JN, Yu YH (1987) Antitumor action of sophoridine. Acta Pharmacol Sin 8:153–158
50. Cai XL (1981) The effects of sophocarpine on nucleic acid contents and DNA synthesis from (^{3}H)thymidine in cancer cells. Nucl Tech 62
51. Xie DW, Wu Q, Zhang HH, Chen WF, Li RZ, He YQ (1985) Effect of the polysaccharide component BN_3C from *Ganoderma* and *d*-matrine on murine T lymphocytes. Chin J Microbiol Immunol 5:8–13
52. Chen JY, Xiang WY, Zhou GL, Yu G, Wang ZL (1983) Studies on the protective effect of oxymatrine in experimental liver damage. Chin Pharm Bull 18:407–409
53. Cho CH, Chuang CY, Chen CF (1986) Study of the antipyretic activity of matrine. A lupine alkaloid isolated from *Sophora subprostrata*. Planta Med 52:343–345
54. Tan HR, Zhang BH (1985) Antiinflammatory effects of matrine. Chin J Integrat Trad West Med 5:108–110

Sophora japonica L.

114

114.1 Introduction

Sophora japonica L. (Fabaceae) is a well-known herbal medicine which has been used in traditional Chinese medicine for a long time. Three entries are listed in the Chinese Pharmacopoeia:

Huaimi, Flos Sophora immaturus, is the dry flower buds of *S. japonica* collected in summer. The rutin content in the flower buds should not be less than 20%. It is used mainly as a hemostatic agent for the treatment of different hemorrhagic diseases.

Huaihua, Flos Sophorae, is the dry flowers of *S. japonica* collected in summer when the plant has flowered. The rutin content in flowers should not be less than 8%. The medicinal indications of the flower are similar to those of flower buds for treatment of different hemorrhagic diseases.

Huaijiao, Fructus Sophorae, is the dry ripe fruits of *S. japonica* collected in winter. It is used for treatment of intestinal hemorrhage.

114.2 Chemical Constituents

Sophora japonica is one of the richest sources of rutin (*114-1*) with a dry weight content of more than 20% in flower buds and more than 8% in flowers. Chemically, rutin is a flavone glycoside with quercetin as the aglycone and rutinose as the sugar moiety. The aglycone quercetin was also isolated from the flower of *S. japonica* [1].

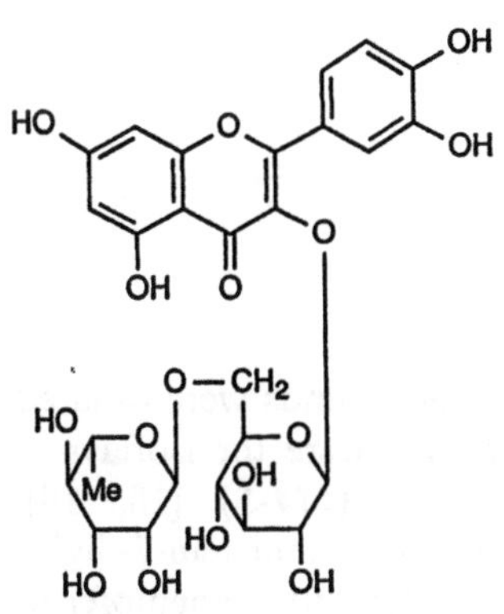

Rutin (*114-1*)

Rutin is a well-known and widely distributed flavone glycoside first isolated from *Ruta graveolens* (Rutaceae) in 1842 [2, 3]. Its isolation from *S. japonica* was also reported in 1853 [3]. The structure of quercetin [4–6] and rutin [7] was elucidated by classical chemical methods. In *S. japonica* flower buds rutin contents were highest at the initial stage, and decreased with development. However, the total yield of rutin per flower was highest at the fully blooming stage [8]. On the other hand, rutin content was highest in carpels, followed by petals, sepals, filaments, and anthers. Petals contained 70% of the rutin in flowers [8]. Other organs of *S. japonica* also contain rutin. Rutin contents were found to be 24% in buds, 4%–11% in pericarps, 0.5%–2% in seeds, 5%–6% in leaflets, and 0.5%–2% in small branches [9].

Besides rutin and quercetin, triterpene glycosides were isolated from the mother liquor by the extraction of rutin from the flowers or flower buds. After hydrolysis, the aglycones betulin and sophoradiol were isolated [10]. The structure of sophoradiol (*114-4*) was identified as olean-12-ene-3,22-diol [11, 12].

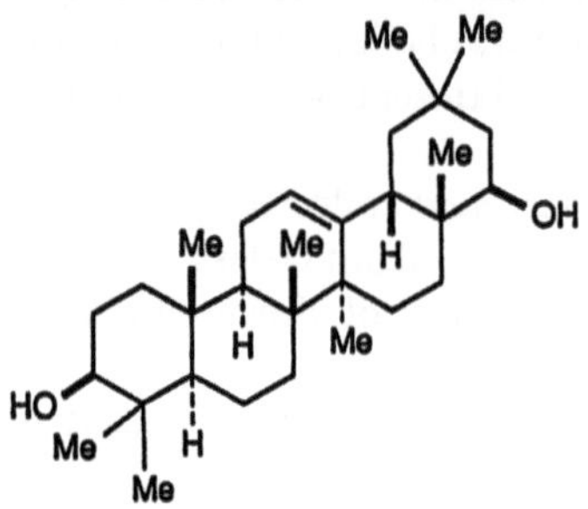

Sophoradiol (*114-2*)

A disaccharide named sophorose (*114-3*) was isolated from the buds of *S. japonica* and identified as 2-*O*-*β*-D-glycopyranosyl-D-glucose [13].

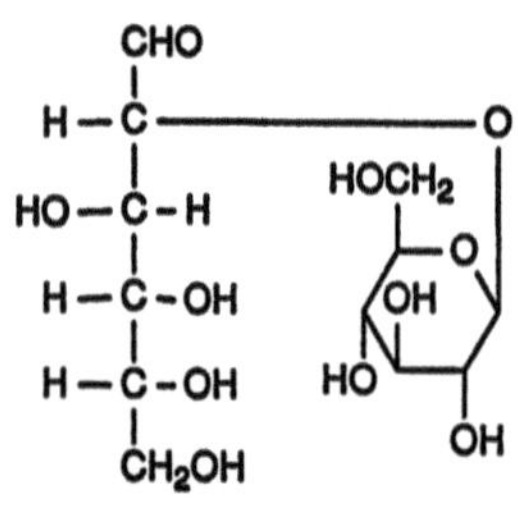

Sophorose (*114-3*)

In addition to rutin, ten other flavone and isoflavone compounds were isolated from the fruits of *S. japonica* [14–16]. Compounds identified were the isoflavone genistein (*114-4*) [17] and its glycosides sophoricoside (*114-5*) [18, 19], sophorabioside (*114-6*) [20, 21], genistein-7-diglucoside, and genistein-7-diglucorhamnoside [16] as well as the flavone kaempferol and its glycosides, kaempferol-3-sophoroside (*114-7*) [15], and kaemperol-3-rhamnodiglucoside [16]. The isolation of quercetin from the fruit of *S. japonica* was also reported [22].

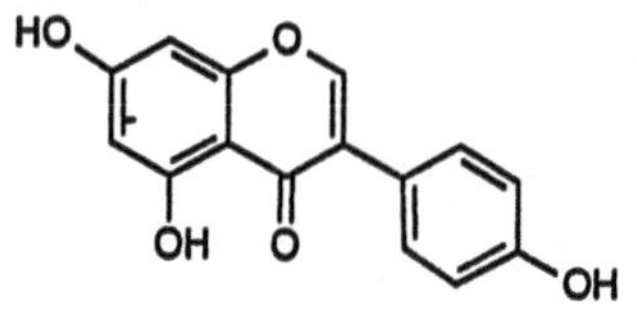

Genistein (*114-4*)

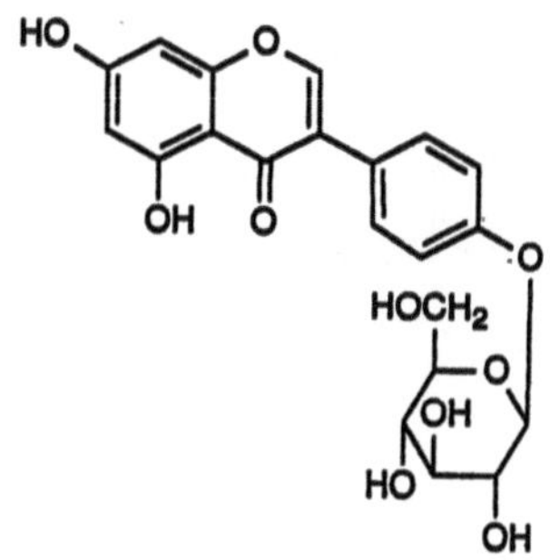

Sophoricoside (*114-5*)

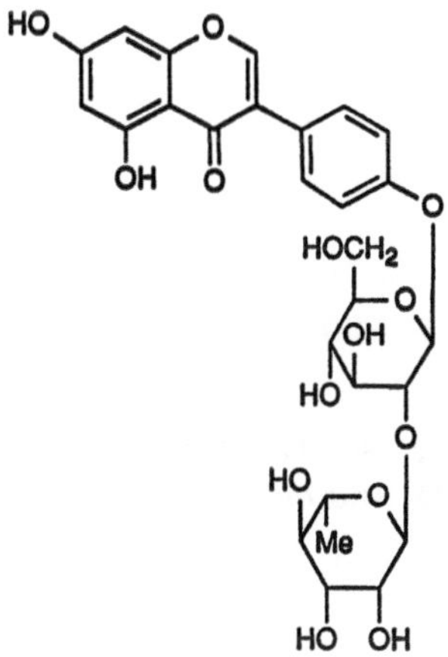

Sophorabioside (*114-6*)

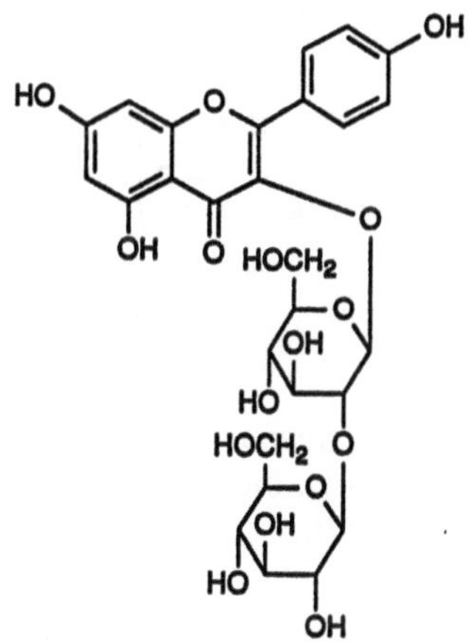

Kaempferol-3-sophoroside (*114-7*)

Sophorose was also isolated from the fruits of *S. japonica* [23]. Moreover, the alkaloids cytisine, methylcytisine, sophocarpine, and matrine were isolated from the seeds of *S. japonica*. The total alkaloid content of dried seed was determined to about 0.05% [24].

From the root of *S. japonica*, a number of isoflavone and related compounds were isolated and identified as irisolidone (*114-8*), 5,7-dihydroxy-3′,4′-methylenedioxyisoflavone (*114-9*), biochanin A (*114-10*), flemmichapparin B (*114-11*), maackiain (*114-12*) [25], and its glycoside sophojaponicin (*114-13*) [26].

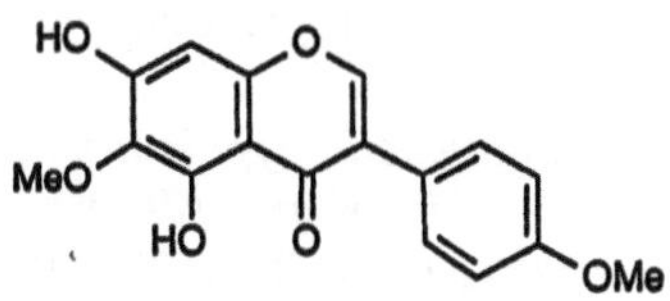

Irisolidone (*114-8*)

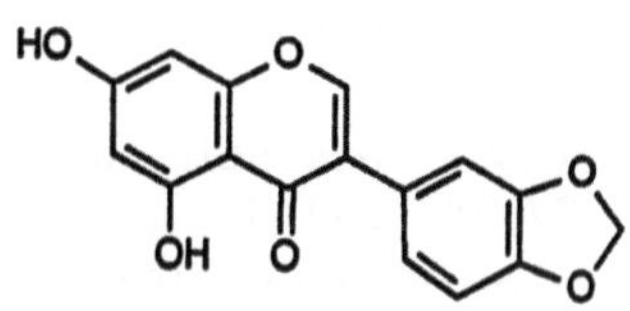

5,7-Dihydroxy-3′,4′-methylenedioxyisoflavone (*114-9*)

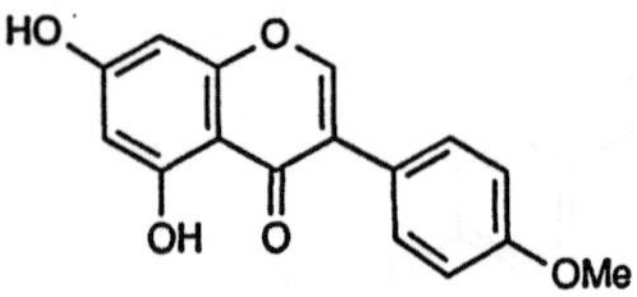

Biochanin A (*114-10*)

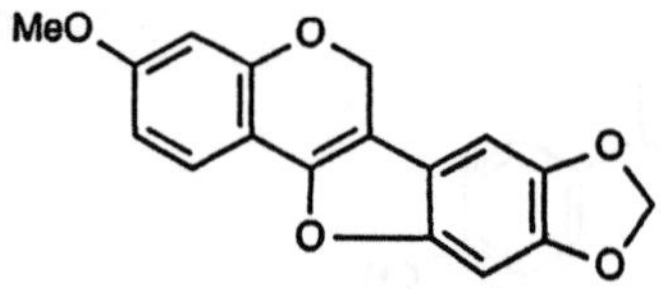

Flemichapparin B (*114-11*)

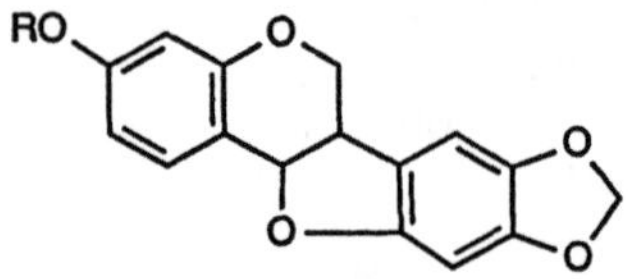

Maackiain (*114-12*): R = H

Sophojaponicin (*114-13*): R =

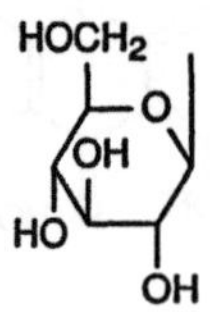

The wood of *S. japonica* besides rutin contained the isoflavone glycosides irisolidone 7-D-glucoside, biochanin A 7-D-xylosylglucoside, and biochanin A 7-D-glucoside (sissotrin) [27].

114.3 Pharmacology

In view of the ubiquitous occurrence of rutin and quercetin in herbal medicines and even more in vegetables, the discovery of the mutagenic activity of quercetin attracted great attention [28]. At first, the mutagenicity of rutin and quercetin was detected in the histidine reversion system with the *Salmonella* frameshift strain TA98. The flavonol quercetin was mutagenic without metabolic activation and mutagenicity increased when a rat liver microsomal preparation (S-9) was included. In contrast, rutin required metabolic activation for its mutagenic activity [28]. In addition to pure compounds, mutagenicity of crude extracts from commonly available nutritional items containing rutin has been detected and was found to correlate with the flavone content [29]. Among the flavonols, quercetin showed the highest mutagenic activity [30, 31].

The mutagenicity of flavonols was also tested with *Salmonella typhimurium* strains TA1535, TA100, TA1537, and TA1538 [30, 32]. *S. typhimurium* TA97 was detected to be far more susceptible to quercetin mutagenesis than strain TA1537 [33].

A study of structure-activity relationships revealed that without a hydroxyl group in the 2-phenyl ring or with only one hydroxyl group there was an absolute requirement for microsomal activation. Flavonols with two adjacent hydroxyl groups at the 2-phenyl ring, such as in the case of quercetin, did not require microsomal activation [32]. For strong mutagenicity, a double bond between positions 2 and 3 and a hydroxyl group at position 3 are required. An exception is wogonin, which does not possess a hydroxyl group at position 3 but is strongly mutagenic in the presence of S9 mixture. In contrast, its 3-*O*-methylether was not found to be mutagenic [34, 35].

Combinations of oxygen and alkaline pH irreversibly inactivated the mutagenicity of quercetin in *S. typhimurium* TA98 [36].

The mutagenic activity of quercetin was also inhibited by adding metal salts [36, 37]. $MnCl_2$ was a potent inhibitor, followed by $CuCl_2$, $FeSO_4$, and $FeCl_3$, the probable mechanism being facilitated catalytic oxidation of quercetin. When an oxygen-saturated solution of quercetin was exposed to polyphenol oxidase at various pH values, the UV absorption maximum of quercetin near 370 nm decreased to an extent that correlated with the decrease in mutagenicity of quercetin under these conditions [36, 38].

DNA single-strand breakage in L5178 mouse lymphoma cells induced by quercetin was observed using the alkaline elution technique [39]. In addition, quercetin was also mutagenic in *Drosophila melanogaster*, as measured by the recessive sex-linked lethality test [40].

In contrast, quercetin and rutin administered i.p. to normal mice at daily doses of 50, 100, or 150 mg/kg for 7 days, showed no effect on the frequency of micronuclei in mouse bone marrow polychromatic erythrocytes [36]. Quercetin also showed no mutagenic activity when given orally to mice at concentrations that were about 10^3 times greater than the estimated average human intake of total flavonols, using the micronucleus test and the host-mediated Ames test [41].

Mutagenic activity of pyrolyzates of albumin was reduced by quercetin or several other polyphenols [42]. Quercetin administered i.p. at daily doses of 100 and 150 mg/kg for 7–10 days to mice treated with cyclophosphamide decreased the frequency of micronuclei induced by cyclophosphamide alone but rutin had no such effect [43].

The mutagenicity of benz[*a*]pyrene [44, 45] and its ultimate mutagenic and carcinogenic metabolite $(\pm)$-7β,8α-dihydroxy-9α,10α-epoxy-7,8,9,10-tetrahydrobenz[*a*]pyrene [44] were inhibited by quercetin, probably by affecting the metabolic pathway of benz[*a*]pyrene [45] or by directly interacting with the activated diol-epoxide, since the rate of disappearance of the diol-epoxide from cell-free solutions was markedly stimulated by quercetin [44].

In contrast to quercetin's inhibitory activity on the mutagenicity of benz[*a*]pyrene, it enhanced the mutagenicity of 2-acetylaminofluorene [44, 46]. In the presence of quercetin, the total metabolic rate of 2-acetylaminofluorene decreased, whereas the formation of *N*-hydroxyacetylaminofluorene and 2-aminofluorene increased. It is suggested that the comutagenic effect of quercetin on 2-acetylaminofluorene is due to the inhibition of aryl-hydroxylation and the promotion of *N*-hydroxylation and deacetylation in S9 mixture [46].

Because quercetin was mutagenic in a variety of experimental models, a number of studies on the potential carcinogenic activity of rutin and quercetin as well as some further flavone compounds was carried out. ACI rats that obtained a diet containing 10% rutin and 10% quercetin for 850 days showed no significant difference in incidence of tumors between the experimental and the control groups [47]. In experiments with mice, quercetin administration started at the age of 6 weeks by feeding a diet containing 2% quercetin and this was continued throughout the life span of the animals. No difference in the tumor incidence of the test and the control groups was observed [48]. Likewise, no carcinogenic activity was observed in an experiment with golden hamsters fed with 10% rutin or 10% quercetin in the diet [49]. No evidence was obtained that quercetin was carcinogenic in Fischer rats [50, 51]. The sodium salt of the rutin sulfate ester was devoid of carcinogenic activity in Sprague-Dawley rats [52].

On the other hand, quercetin was found to inhibit tumor promotion by 12-*O*-tetradecanoylphorbol 13-acetate (TPA) (*114-14*). For example, quercetin suppressed the effect of TPA on skin tumor formation in CD-1 mice initiated with 7,12-dimethyl-benz[*a*]anthracene (DMBA) [53]. TPA-induced epidermal ornithine decarboxylase activity was also inhibited by quercetin. It did, however, not inhibit the stimulation of epidermal DNA synthesis by TPA [53]. Activity of lipoxygenase, as measured by the formation of 12-hydroxyeicosatetraenoate in mouse epidermis, was clearly inhibited by quercetin. Lipoxygenase appeared to be involved in the mechanism of TPA-caused epidermal ornithine decarboxylase induction and tumor promotion [54]. Inhibition of lipoxygenase by quercetin has been proposed as one of the major actions of quercetin responsible for its inhibitory effects on TPA-induced tumor promotion and ornithine decarboxylase activity [53]. In contrast to lipoxygenase, epidermal cyclooxygenase activity was not inhibited by quercetin [55].

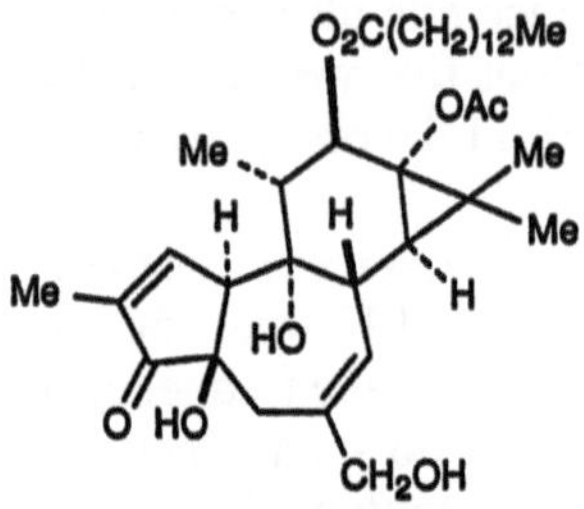

12-*O*-Tetradecanoylphorbol-13-acetate (TPA, *114-14*)

In addition, quercetin inhibited TPA-induced vascular permeability in mice [55], and also inhibited the increase in incorporation of inorganic phosphate into phospholipids of human embryo fibroblasts induced by TPA. Moreover, TPA-induced increases of sugar transport and RNA synthesis in chick embryo fibroblasts as well as TPA-induced aggregation of human platelets were also inhibited [56]. Quercetin very effectively suppressed alkaline phosphatase activity stimulated by tumor promotor [57].

The pharmacological effect of quercetin that has been most intensively investigated was its activity against anaphylactic reactions. A study on the effect of quercetin on anaphylactic smooth muscle contraction of ileum from guinea pigs sensitized to egg albumin showed that quercetin inhibited the contraction in a concentration-dependent manner with an IC_{50} of about 10 μM [58].

Quercetin was the most active of various naturally occurring flavones in inhibiting antigen-induced histamine release from human basophils. The presence of both the keto group at position 4 and hydroxysubstituent(s) of the benzopyrane moiety was found to be necessary for this effect. Of the transitional metals, copper salts effectively blocked the inhibitory activity of quercetin, probably through a chelation mechanism [59]. In contrast, rutin lacked the inhibitory activity in antigen-induced histamine release of human basophils [60].

In vitro, quercetin and several other flavone derivatives affected various pathways involved in the metabolism of arachidonic acid. A comparative study on the effect of quercetin on arachidonic acid metabolism by cyclooxygenase and lipoxygenase in intact human platelets revealed that quercetin is effective in vitro as an inhibitor of cyclooxygenase and lipooxygenase [61, 62]. Inhibition of neutrophil phospholipase A_2 by quercetin was also reported [63].

Furthermore, quercetin and several other natural flavone derivatives have been found to inhibit washed human platelet aggregation, and serotonin release paralleled the inhibition of thromboxane synthesis induced by arachidonic acid [64]. The inhibitory effect of flavone derivatives on human platelet functions was diminished by saturation of the double bond between positions 2 and 3, by lack of the carbonyl function, by glycosylation of the 3-hydroxyl group, and by a high number of hydroxyl substituents [65].

High concentrations of quercetin increased platelet cAMP levels. At not so high concentrations the increase in cAMP caused by PGI_2 was potentiated [66]. Modification of platelet cAMP metabolism through inhibition of phosphodiesterase activity was discussed as a probable mechanism for the antiaggregating effect [67]. The inhibitory effect of quercetin on the aggregation of human platelets has also been found in a manner similar to calmodulin inhibitors. The possibility that calmodulin antagonistic activity of quercetin is involved in the inhibition of human platelet aggregation was also discussed [68].

The inhibitory effect of flavone derivatives on cyclic nucleotide phosphodiesterases was tested in the rat cerebrum, rabbit myocardium, and Ehrlich ascites tumor cells. It was found that both cAMP phosphodiesterase and cGMP phosphodiesterase activity were inhibited, depending on the organ source of the phosphodiesterase and on the structure of flavone derivatives. Quercetin was one of the compounds with a maximal inhibitory effect on both cAMP and cGMP phosphodiesterases. In general, inhibitory activities of flavone aglycones on phoshodiesterases were greater than those of the corresponding flavone glycosides [69]. A kinetic study on the inhibition of cAMP phosphodiesterase of rabbit heart by several flavone derivatives including quercetin indicated that the inhibition was competitive [70].

Quercetin showed cardiotonic action in the isolated toad heart perfusion system. This effect was, however, different from that of isoprenaline sulfate and was not associated with β-receptors. Perfusion of isolated toad heart with 50–100 μM quercetin increased the myocardial content of cAMP by 40%, whereas that of cGMP remained unchanged, indicating that quercetin selectively inhibited cAMP phosphodiesterase in toad myocardial cells. The cardiotonic action of quercetin is probably the result of its inhibitory effect on cardiac cAMP phosphodiesterase [71].

Oral administration of quercetin has been found to exhibit hypolipidemic activity in rats fed with 1% cholesterol in the diet for 6 and 12 weeks, but had no effect in rats fed a stock diet [72]. The serum triglyceride levels in both mice and rats were depressed by quercetin [73]. Anesthetized dogs given rutin i.v. at 3 mg/kg within 45 min showed decreased total cholesterol levels in their arterial and venous blood, increased thoracic lymph cholesterol, and decreased cholesterol in the liver [74].

In a study of the metabolism of 4-[^{14}C]quercetin in ACI rats, 20% absorption from the digestive tract was found after oral administration. More than 30% of the dose was exhaled as $^{14}CO_2$ and about 30% was excreted unchanged in the feces. Absorbed quercetin was rapidly excreted into the bile and urine as glucuronide and sulfate conjugates of quercetin, 3′-*O*-methyl quercetin, and 4′-*O*-methyl quercetin [75].

A pharmacokinetic study on quercetin in men showed a bisphasic elimination life, with $t_{1/2\alpha} = 8.8$ min at $t_{1/2\beta} = 2.4$ h after a single i.v. dose. Protein binding was more than 98%. About 7.4% of the i.v. dose was excreted in the urine in conjugated form, about 0.6% being excreted unchanged. After oral administration, no measurable plasma concentrations could be detected [76].

Urinary metabolites in rat urine of orally administered [^{2}H]rutin were identified as 3-hydroxyphenylacetic acid, 3-methoxy-4-hydroxyphenylacetic acid, 3,4-dihydroxyphenylacetic acid, 3,4-dihydroxytoluene, and 3-(3-hydroxyphenyl)propionic acid. Unchanged rutin and quercetin were not present in urine. Results from intraperitoneal injection in rats, oral administration to neomycin-treated rats, and incubation in vitro with the intestinal contents of rats suggested the involvement of intestinal microflora in the metabolism of orally administered rutin [77].

After intraperitoneal administration of rutin to bile duct-cannulated rats, three conjugates were detected as biliary metabolites. One was found to have 3′-*O*-methyl quercetin as the aglycone. The other two conjugates were not fully characterized but each was degraded by acid hydrolysis to give a mixture of quercetin and 3′-*O*-methyl quercetin [78].

References

1. Shimizu M, Ohta G (1952) Solubilization of flavonoids. VIII. Glucosides in flos sophorae japonicae. J Pharm Soc Jpn 72:331–333
2. Schindler H (1954) *Sophora japonica*, an East Asian drug rich in rutin. The history of rutin. Dtsch Apoth Ztg 94:995–997
3. Karrer W (1976) Konstitution und Vorkommen der organischen Pflanzenstoffe, 2nd edn. Birkhäuser, Basel, pp 614–615
4. Foerster P (1882) Zur Identitätsfrage der Farbstoffe der chinesischen Gelbbeeren, der Kapern und der Raute mit dem Quercitrin und Quercetin. Chem Ber 15:214–217
5. Hlasiwetz H (1859) Ueber das Quercitrin. Liebigs Ann Chem 112:96–117
6. Kostanecki S, Tambor J (1895) Über die Constitution des Fisetins. Chem Ber 28:2302–2309
7. Attree GF, Perkin AG (1927) The position of the sugar nucleus in the quercetin glucosides. J Chem Soc 234–240
8. Tabata M (1975) Quantitative distribution of rutin in the panicles and flower organs of *Sophora japonica*. Shoyakugaku Zasshi 29:93–95
9. Balbaa SI, Zaki AY, El Shamy AM (1974) Total flavonoid and rutin content of the different organs of *Sophora japonica*. J Assoc Off Anal Chem 57:752–755
10. Kariyone T, Ishimasa S, Shiomi T (1956) Triterpenoids. VIII. Triterpenoids contained in *Sophora japonica*. J Pharm Soc Jpn 76:1210–1211
11. Kimura K, Takahashi M, Ishimasa S, Kodama Y (1958) Components of *Sophora japonica*. II. Structure of sophoradiol. Yakugaku Zasshi 78:1090–1094
12. Ishimasa S (1960) Components of *Sophora japonica*. III. Stereochemistry of sophoradiol. Yakugaku Zasshi 80:304–310
13. Clancy MJ (1960) Sophorose and its derivatives. J Chem Soc 4213–4217
14. Ho LX, Xu YX, Xue HZ, Wang WH, Zhou ZR, Chen PY, Fan MF (1982) Studies on the antifertility constituents of Huai Jiao. I. Isolation of constituents I–XI and identification of I–IV and IX. Reprod Contracept 2:23–27
15. Ho LX, Xu YX, Xue HZ (1984) Studies on antifertility constituents of Huai Jiao. II. Separation and identification of four compounds V–VIII. Reprod Contracept 4:51–53
16. Liu JS, Tiu CM, Huang MF (1980) Study on the active constituents of *Sophora japonica* L. Chin Trad Herb Drugs 11:145–146

17. Szabo V, Bognar R, Farkas E, Litkei G (1967) The glycosides of the fruit of *Sophora japonica*. VII. Acta Univ Debrecen Ludovico Kossuth Nominatae, Ser Phys Chim 13:129–144 (CA 69:16776t)
18. Charaux C, Rabate J (1938) Biochemical study of the fruits of *Sophora japonica* L. I. Presence of sophoricoside. Bull Soc Chim Biol 20:454–458
19. Zemplen G, Bognar R, Farkas L (1943) Determination of the structure of sophoricoside, an isoflavone glycoside of *Sophora japonica* L. Chem Ber 76B:267–272
20. Zemplen G, Bognar R (1942) Sophorabioside, an new glucoside from *Sophora japonica*. L. Chem Ber 75B:482–489
21. Farkas L, Nogradi M, Wagner H, Hoerhammer L (1968) Isoflavone glycosides. X. Final structure determination and total synthesis of sophorabioside, a glycoside from *Sophora japonica*. Chem Ber 101:2758–2761
22. Akhmedkhodzhaeva NM, Svechnikova AN, Bandyukova VA, Kambarova DM (1986) Determination of flavones in *Sophora japonica* fruits. Farmatsiya (Mosk) 35:60–61
23. Freudenberg K, Knauber H, Cramer F (1951) Die Übereinstimmung der Sophorose mit 2-(β-Glucosido)-glucose. Chem Ber 84:144–146
24. Abdusalamov BA, Aslanov KA, Sadykov AS, Khoroshkova OA (1972) Level of alkaloids in *Sophora japonica*. Khim Prir Soedin 658
25. Komatsu M, Yokoe I, Shirataki Y (1976) Studies on the constituents of *Sophora* species. X. Constituents of the root of *Sophora japonica* L. Yakugaku Zasshi 96:254–257
26. Shibata S, Nishikawa Y (1963) Constituents of Japanese and Chinese crude drugs. VII. Constituents of the roots of *Sophora subprostrata* and *Sophora japonica*. Chem Pharm Bull (Tokyo) 11:167–177
27. Takeda T, Ishiguro I, Masegi M, Ogihara Y (1977) New isoflavone glycosides from the woods of *Sophora japonica*. Phytochemistry 16:619–620
28. Stan HJ, Hunt W (1984) Flavonols – mutagens in our daily nutrition. Dtsch Lebensm Rundsch 80:85–87
29. Hardigree AA, Epler JL (1978) Comparative mutagenesis of plant flavonoids in microbial system. Mutat Res 58:231–239
30. Brown JP, Dietrich PS, Brown RJ (1977) Frameshift mutagenicity of certain naturally occurring phenolic compounds in the *Salmonella*/microsome test. Activation of anthraquinone and flavanol glycosides by gut bacterial enzymes. Biochem Soc Trans 5:1489–1492
31. Sugimura T, Minako M, Matsushima T, Yahagi T, Seino Y, Shirai A, Sawamura M, Natori S, Yoshihira K (1977) Mutagenicity of flavone derivatives. Proc Jpn Acad [B] 53:194–197
32. Brown JP, Dietrich PS (1979) Mutagenicity of plant flavonols in the *Salmonella*/mammalian microsome test. Activation of flavonol glycosides by mixed glycosidase from rat cecal bacteria and other sources. Mutat Res 66:223–240
33. Busch DB, Hatcher JF, Bryan GT (1986) Urine recovery experiments with quercetin and other mutagens using the Ames test. Environ Mutagen 8:393–399
34. Nagao M, Marita N, Yahagi T, Shimizu M, Kuroyanagi M, Fukuoka M, Yoshihara K, Natori S, Fujino T, Sugimura T (1981) Mutagenicities of 61 flavonoids and 11 related compounds. Environ Mutagen 3:401–419
35. MacGregor JT, Jurd L (1978) Mutagenicity of plant flavonoids: structural requirements for mutagenic activity in *Salmonella typhimurium*. Mutat Res 54:297–309
36. Hatcher JF, Bryon GT (1985) Factors affecting the mutagenic activity of quercetin for *Salmonella typhimurium* TA 98: metal ions, antioxidants and pH. Mutat Res 148:13–23
37. Friedman M, Smith GA (1984) Inactivation of quercetin mutagenicity. Food Chem Toxicol 22:535–539
38. Friedman M, Smith GA (1984) Factors which facilitate inactivation of quercetin mutagenicity. Adv Exp Med Biol 177:527–544
39. Meltz ML, MacGregor JT (1981) Activity of the plant flavonol quercetin in the mouse lymphoma L 5178y $TK^{+/-}$mutation, DNA single-strand break and Balb/c3T3 chemical transformation assays. Mutat Res 88:317–324
40. Watson WAF (1982) The mutagenic activity of quercetin and kaempferol in *Drosophila melanogaster*. Mutat Res 103:145–147
41. Aeschbacher HU, Meier H, Ruch E (1982) Nonmutagenicity in vivo of the food flavonol quercetin. Nutr Cancer 4:90–98
42. Fukuhara Y, Yoshida D, Goto F (1981) Reduction of mutagenic products in the presence of polyphenols during pyrolysis of protein. Agric Biol Chem 45:1061–1066

43. Hang BQ, Wu Y, Hang S, Yang Y, Wang MS (1985) Effect of quercetin and rutin on the occurrence of micronuclei in mouse bone-marrow polychromatic erythrocytes. J Nanjing Coll Pharm 16:52–55
44. Huang MT, Wood AW, Newmark HL, Sayer JM, Yagi H, Jerina DM, Conney AH (1983) Inhibition of the mutagenicity of bay-region diol-epoxides of polycyclic aromatic hydrocarbons by phenolic plant flavonoids. Carcinogenesis 4:1631–1637
45. Ogawa S, Hirayama T, Nohara M, Tokuda M, Hirai K, Fukui S (1985) The effect of quercetin on the mutagenicity of 2-acetylaminofluorene and benz[*a*]-pyrene in *Salmonella typhimurium* strains. Mutat Res 142:103–107
46. Ogawa S, Hirayama T, Tokuda M, Hirai K, Fukui S (1986) The effect of quercetin, a mutagenicity-enhancing agent, on the metabolism of 2-acetylaminofluorene with mammalian metabolic activation systems. Mutat Res 162:179–186
47. Hirano I, Ueno I, Hosaka S, Takanashi H, Matsushima T, Sugimura T, Natori S (1981) Carcinogenicity examination of quercetin and rutin in ACI rats. Cancer Lett 13:15–21
48. Saito D, Shirai A, Matsushima T, Sugimura T, Hirono I (1980) Test of carcinogenicity of quercetin, a widely distributed mutagen in food. Teratogenesis Carcinog Mutagen 1:213–221
49. Morino K, Matsukura N, Kawachi T, Ohgaki H, Sugimura T, Hirono I (1982) Carcinogenicity test of quercetin and rutin in golden hamsters by oral administration. Carcinogenesis 3:93–97
50. Stoewsand GS, Anderson JL, Boyd JN, Hrazdina G, Babish JG, Walsh KM, Losco P (1984) Quercetin: a mutagen, not a carcinogen, in Fisher rats. J Toxicol Environ Health 14:105–114
51. Takanashi H, Aiso S, Hirono I, Matsushima T, Sugimura T (1983) Carcinogenicity test of quercetin and kaempferol in rats by oral administration. J Food Safety 5:55–60
52. Habs M, Habs H, Berger MR, Schmaehl D (1984) Negative dose-response study for carcinogenicity of orally administered rutin sulfate in Sprague-Dawley rats. Cancer Lett 23:103–108
53. Kato R, Nakadate T, Yamamoto S, Sugimura T (1983) Inhibition of 12-*O*-tetradecanoylphorbol 13-acetate-induced tumor promotion and ornithine decarboxylase activity by quercetin: possible involvement of lipoxygenase inhibition. Carcinogenesis 4:1301–1305
54. Kato R, Nakadate T, Yamamoto S (1984) Involvement of lipoxygenase products of arachidonic acid in tumor-promoting activity of TPA. In: Thaler-Dao H, Crastes de Paulet A, Paoletti R (eds) Icosanoids Cancer. Raven, New York, pp 101–103
55. Nakadate T, Yamamoto S, Aizu E, Kato R (1984) 12-*O*-Tetradecanoylphorbol 13-acetate-caused increase in vascular permeability in mouse skin. II. Effects of caffeic acid derivatives, chalcone derivatives and calcium blockers on it. Ensho 4:554–556 (CA 102:99302c)
56. Nishino H, Nishino A, Iwashima A, Tanaka K, Matsuura T (1984) Quercetin inhibits the action of 12-*O*-tetradecanoylphorbol 13-acetate, a tumor promoter. Oncology 41:120–123
57. Gschwendt M, Kittstein W, Marks F (1984) Stimulation of alkaline phosphatase activity in mouse epidermis by tumor promoters. Cancer Lett 22:219–225
58. Fanning MJ, Macander P, Drzewiecki G, Middleton E Jr (1983) Quercetin inhibits anaphylactic contraction of guinea pig ileum smooth muscle. Int Arch Allergy Appl Immunol 71:371–373
59. Middleton E Jr, Drzewiecki G (1982) Effects of flavonoids and transitional metal cations on antigen-induced histamine release from human basophils. Biochem Pharmacol 31:1449–1453
60. Middleton E Jr, Drzewiecki G, Krishnarao D (1981) Quercetin: an inhibitor of antigen-induced human basophil histamine release. J Immunol 127:546–550
61. Michel F, Mercklein L, Rey R, Crastes de Paulet A (1986) Comparative effects of some flavonoids on cyclooxygenase and lipoxygenase activities in different cell systems or subfractions. Stud Org Chem (Amsterdam) 23:389–401
62. Van Wauwe J, Goosens J (1983) Effects of antioxidants on cyclooxygenase and lipoxygenase activities in intact human platelets: comparison with indomethacin and ETYA. Prostaglandins 26:725–730
63. Lanni C, Becker EL (1985) Inhibition of neutrophil phospholipase A_2 by *p*-bromophenylacyl bromide, nordihydroguaiaretic acid, 5,8,11,14-eicosatetraenoic acid and quercetin. Int Arch Allergy Appl Immunol 76:214–217
64. Corvazier E, Maclouf J (1985) Interference of some flavonoids and nonsteroidal anti-inflammatory drug with oxidative metabolism and arachidonic acid by human platelets and neutrophils. Biochim Biophys Acta 835:315–321
65. Beretz A, Cazenave JP, Anton R (1982) Inhibition of aggregation and secretion and human platelets by quercetin and other flavonoids: structure-activity relationship. Agents Actions 12:382–387

66. Beretz A, Stierle A, Anton R, Cazenave JP (1982) Role of cyclic AMP in the inhibition of human platelet aggregation by quercetin, a flavonoid that potentiated the effect of prostacyclin. Biochem Pharmacol 31:3597–3600
67. Landolfi R, Mower RL, Steiner M (1984) Modification of platelet function and arachidonic acid metabolism by bioflavonoids. Structure-activity relationship. Biochem Pharmacol 33:1525–1530
68. Nishino H, Tanaka K, Matusura T (1983) Quercetin inhibits the aggregation of human platelets. Kyoto-furitsu Ika Daigaku Zasshi 92:165–168 (CA 98:191481p)
69. He SP, Wei MC, Li RZ, He YQ, Zhang RY, Chen YY, Yue BZ, Lu YX (1982) Study on the inhibitory effect of flavonoids in some traditional Chinese medicines on phosphodiesterase. J Beijing Med Coll 14:253–257
70. He SP, Qiao J (1985) The inhibition kinetic study of flavonoids on cAMP phosphodiesterase. Shenwu Huaxue Zazhi 1:55–58
71. He SP, Wei MC, Pei YQ, Wang XY (1984) Mechanism of cardiotonic effect of flavonoids from some traditional Chinese drugs. Hejishu 41–43
72. Basarker PW, Nath N (1983) Hypocholesterolemic and hypolipidemic activity of quercetin – a vitamin P-like compound in rats. Indian J Med Res 77:122–126
73. Kato N, Tosa N, Doudou T, Imamura T (1983) Effects of dietary quercetin on serum lipids. Agric Biol Chem 47:2119–2120
74. Shaikemeleva US (1983) Effect of rutin on cholesterol content in dog lymph, blood, and tissues. Byull Eksp Biol Med 95:53–57 (CA 98:196824n)
75. Ueno I, Nakano N, Hirono I (1983) Metabolic fate of ^{14}C-quercetin in the ACI rat. Jpn J Exp Med 53:41–50
76. Gugler R, Leschik M, Dengler HJ (1975) Disposition of quercetin in man after single oral and intravenous doses. Eur J Clin Pharmacol 9:229–234
77. Baba S, Futura T, Fujioka M, Goromaru T (1983) Studies on drug metabolism by use of isotopes. XXVII. Urinary metabolites of rutin in rats and the role of intestinal microflora in the metabolism of rutin. J Pharm Sci 72:1155–1158
78. Griffiths LA, Brown S (1983) New metabolites of naturally occurring flavonols; identification by E.I. mass spectrometry and chromatographic techniques. Anal Chem Symp Ser 14:121–124

Stemona spp. 115

115.1 Introduction

Baibu, Radix Stemonae, is the dry bulbs of *Stemona sessilifolia* (Miq.) Miq., *S. japonica* (Bl.) Miq. or *S. tuberose* Lour. (Stemonaceae) collected in spring and fall. It is officially listed in the Chinese Pharmacopoeia and used for treatment of cough and whooping cough.

115.2 Chemical Constituents

The isolation of alkaloids from the root of *S. sessilifolia* was reported first in 1913 [1]. Suzuki [2, 3] reported later on the isolation of two alkaloids from the root of *S. japonica* which were named stemonidine and stemonine, but the plant material was subsequently shown to correspond to the root of *S. ovata* [4]. An alkaloid named tuberostemonine was isolated from the root of *S. tuberosa* [5–7]. A series of alkaloids were then successively isolated from the root of *Stemona* species and structurally investigated. They have rather complex structures, and most of them were determined by crystallographic analysis. More than 15 alkaloids have been isolated so far from *Stemona* species. Their structures are given in Table 115.1.

Table 115.1. Alkaloids isolated from *Stemona* species

Compound	Structure	Plant origin	Reference
Tuberostemonine (*115-1*)	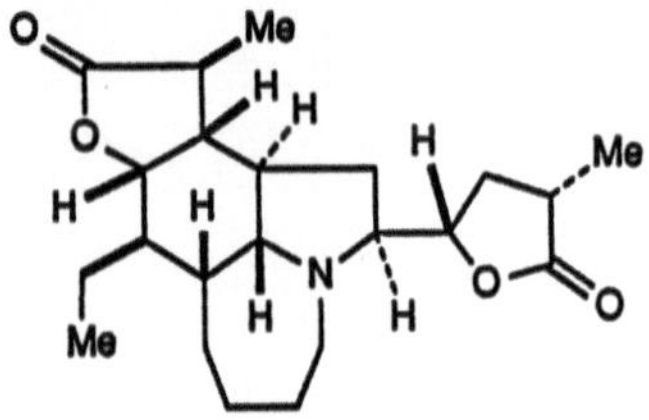	*S. tuberosa* *S. sessilifolia*	[8–20] [20–22]
Stenine (*115-2*)	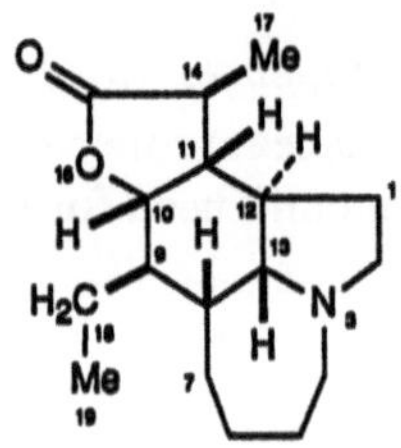	*S. tuberosa*	[20, 23]
Oxotuberostemonine (*115-3*)	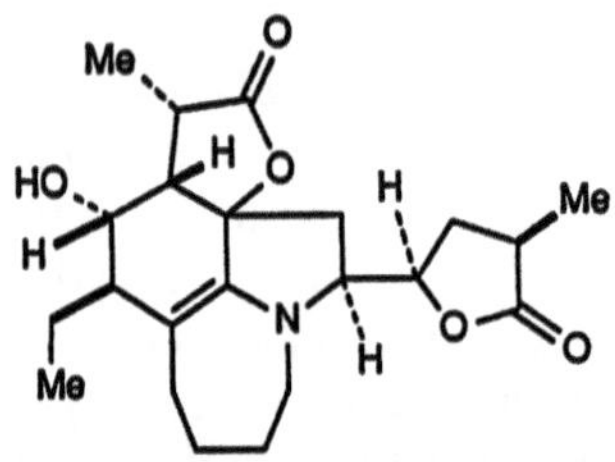	*S. tuberosa* *S. sessilifolia*	[11, 24–26] [21]
Protostemonine (*115-4*)	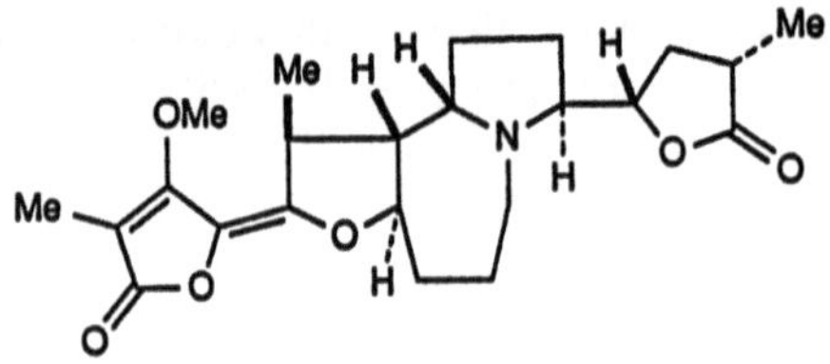	*S. japonica*	[27–32]
Stemonine (C_{17} alkaloid) (*115-5*)	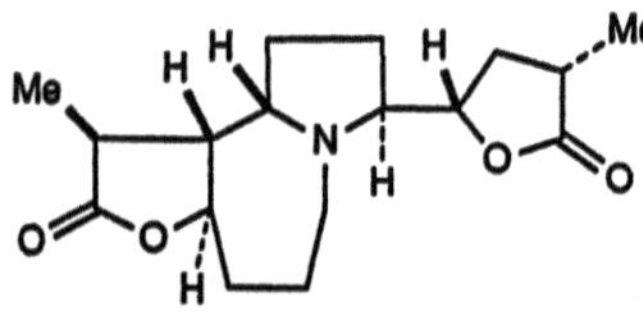	*S. ovata*	[2–4, 31–33]
Stemonine (C_{22} alkaloid)	Stereoisomer of tuberostemonine	*S. tuberosa*	[5]
Stemonamine (*115-6*)	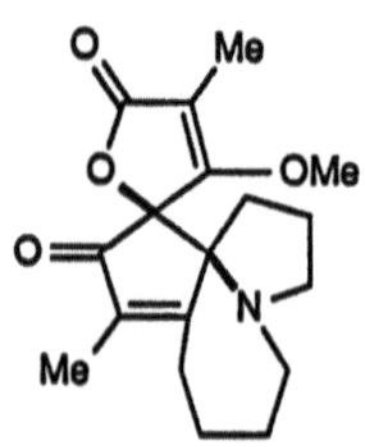	*S. japonica*	[34]

Table 115.1. (continued)

Compound	Structure	Plant origin	Reference
Isostemonamine (*115-7*)	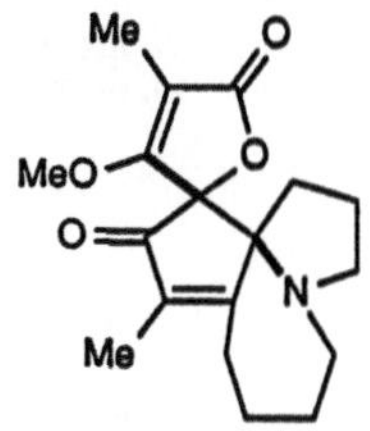	*S. japonica*	[34]
Stemonidine (*115-8*)	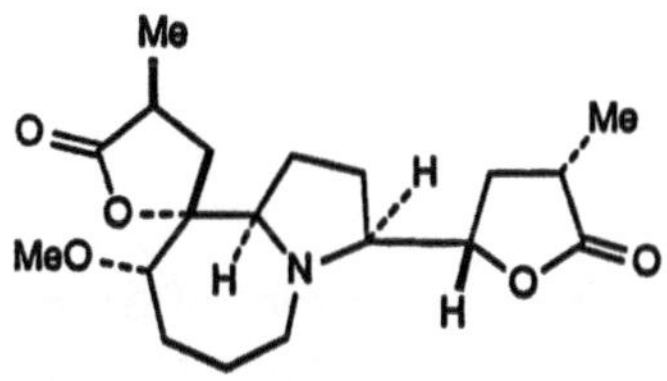	*S. ovata* *Stemona* sp.	[2] [35, 36]
Stemotinine (*115-9*)	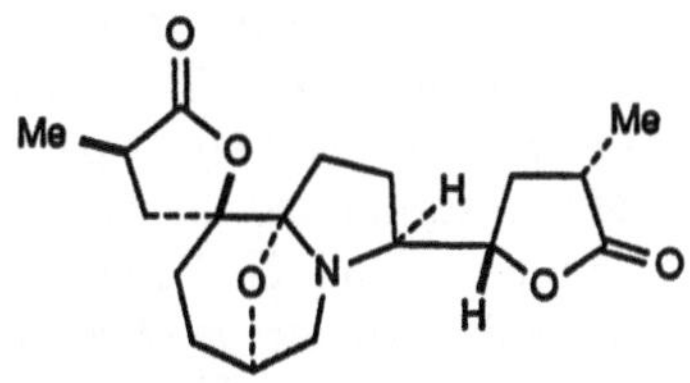	*S. tuberosa*	[36]
Isostemotinine (*115-10*)	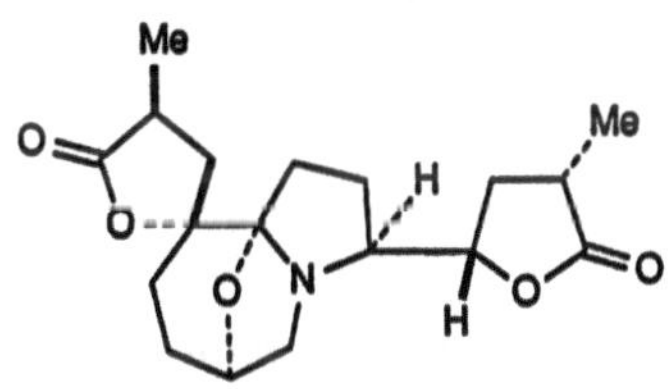	*S. tuberosa*	[36]
Croomine (*115-11*)	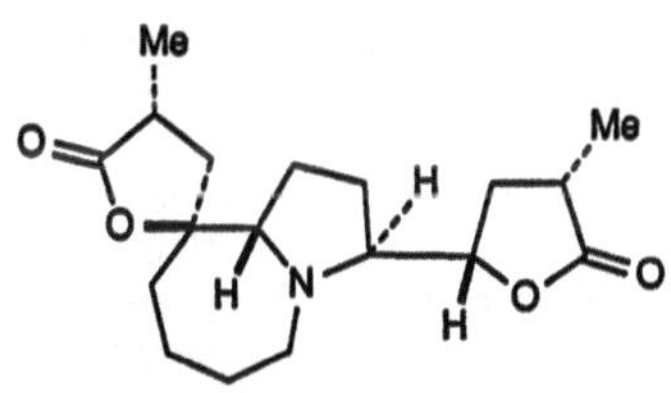	*Stemona* sp.	[36, 37]
Stemofoline (*115-12*)	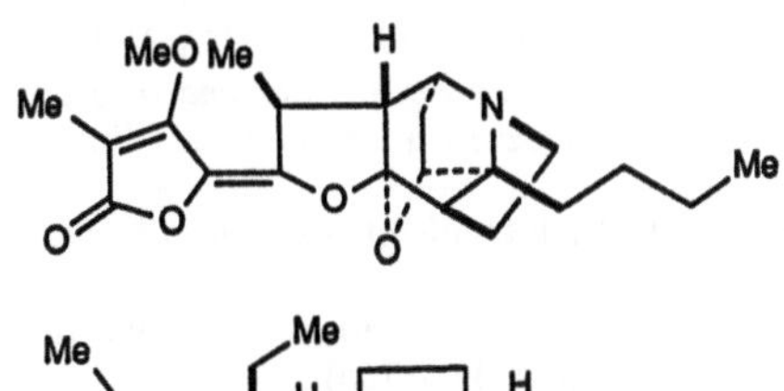	*S. japonica* leaves, stem	[38, 39]
Stemoninine (*115-13*)	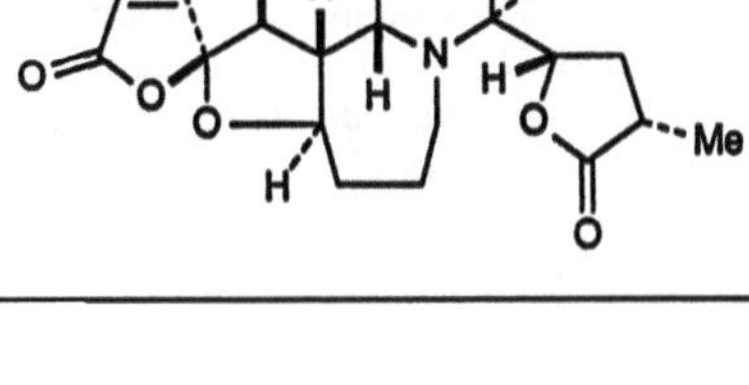	*Stemona* sp.	[40, 41]

Table 115.1. (continued)

Compound	Structure	Plant origin	Reference
Stemospironine (*115-14*)	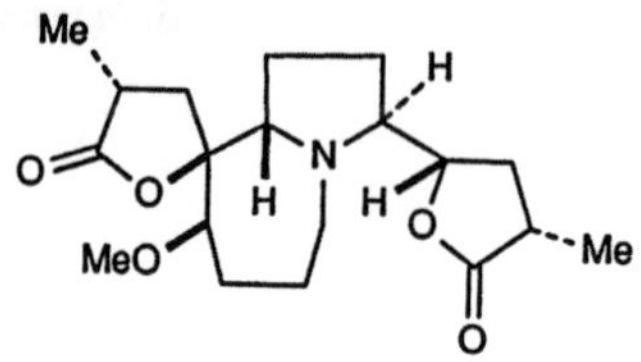	*S. japonica* stem, leaves	[39]

115.3 Pharmacology

Only few studies on the biological activity of the root of *Stemona* species and its alkaloid constituents have been reported. Thus, tuberostemonine showed an inhibitory effect on the motility of *Angiostrongylus cantonensis*, *Dipylidium caninum*, and *Fasciola hepatica* in vitro at a concentration of $6.7 \times 10^{-6} - 6.7 \times 10^{-5} M$, but had little effect on the motility of *Schistosoma japonicum* [42]. It paralyzed the motility of isolated mouse ileum at $6.7 \times 10^{-5} M$ and stimulated the twitch response induced by guanidine in the isolated frog rectus preparation [42]. Stemospironine and stemofoline showed insecticidal activity against silkworm larvae (*Bombyx mori*). Stemofoline had a much stronger activity than stemospironine [39].

References

1. Furuya T (1913) The constituents contained in the roots of *Stemona sessilifolia* Miquel, Stemonaceae. Arb Pharm Inst Univ Berlin 9:112–116
2. Suzuki K (1929) Alkaloid of *Stemona japonica.* I. J Pharm Soc Jpn 49:457–464 (CA 23:4221)
3. Suzuki K (1931) Alkaloid of *Stemona japonica.* II, III. J Pharm Soc Jpn 51:419–429 (CA 25:4551)
4. Suzuki K (1934) Alkaloids of *Stemona.* V. A new alkaloid, isostemonidine, from the roots of *Stemona ovata* Nakai. J Pharm Soc Jpn 54:567–572 (CA 31:106)
5. Lobstein E, Grumbach J (1932) An alkaloid extracted from the root of *Stemona tuberosa* Loureiro (sinoannamite drug). C R Hebd Seances Acad Sci 194:386–389
6. Suzuki K (1934) Alkaloids of *Stemona.* VI. An alkaloid, tuberostemonine, from the roots of *Stemona tuberosa* Loureiro. J Pharm Soc Jpn 54:573–579 (CA 31:106)
7. Kondo H, Suzuki K, Satomi M (1939) The bases of *Stemona tuberosa* Loureiro. VII, VIII. Tuberostemonine of the root of *Stemona tuberosa* Loureiro. 2. J Pharm Soc Jpn 59:443–450 (CA 34:767)
8. Kondo H, Suzuki K, Satomi M (1941) Alkaloids of *Stemona tuberosa.* X. Tuberostemonine. J Pharm Soc Jpn 61:369–374 (CA 44:9457e)
9. Kondo H, Suzuki K (1943) *Stemona* bases. XI. Tuberostemonine, the alkaloid of *Stemona tuberosa.* J Pharm Soc Jpn 63:334–338 (CA 45:5170d)
10. Kondo H, Satomi M, Odera T (1954) *Stemona* alkaloids. XV. Structure of tuberostemonine. 6. Dehydrogenation reaction. Annu Rep ITSUU Lab 5:95–98 (CA 49:15932a)
11. Kondo H, Satomi M, Odera T (1954) *Stemona* alkaloids. XVI. Alkaloid from root of *Stemona tuberosa.* Annu Rep ITSUU Lab 5:99–102 (CA 49:15932e)
12. Kondo H, Satomi M, Kaneko T (1955) *Stemona* alkaloids. XVII. Structure of tuberostemonine. 7. Annu Rep ITSUU Lab 6:63–66 (CA 50:10111a)

13. Kondo H, Suzuki K, Satomi M (1940) Alkaloids of *Stemona tuberosa* Loureiro; tuberostemonine. III. J Pharm Soc Jpn 60:389–398 (CA 35:459)
14. Kondo H, Satomi M, Kaneko T (1956) *Stemona* alkaloids. XVIII. Structure of tuberostemonine. 8. Annu Rep ITSUU Lab 7:19–23 (CA 51:2826i)
15. Kondo H, Satomi M, Kaneko T (1957) *Stemona* alkaloids. XX. Tuberostemonine. 9. Annu Rep ITSUU Lab 8:15–17 (CA 51:16499e)
16. Kondo H, Satomi M, Kaneko T (1957) *Stemona* alkaloids. XXI. Examination of the catalytic reduction of tuberostemonine. Annu Rep ITSUU Lab 8:17–22 (CA 51:16499h)
17. Kondo H, Satomi M, Kaneko T (1958) *Stemona* alkaloid. XXIII. Tuberostemonine. 10. ITSUU Kenkyushu Nempo 9:48–54 (CA 54:599c)
18. Kaneko T (1960) Studies on *Stemona* alkaloids. XXIV. Tuberostemonine. 13. ITSUU Kenkyusho Nempo 11:39–51 (CA 55:27394g)
19. Götz M, Bögri T, Gray AH (1969) The structure of tuberostemonine. Tetrahedron Lett 707–715
20. Harada H, Irie H, Masaki K, Osaki K, Uyeo S (1967) The stereochemistry and absolute configuration of stenin and tuberostemonine. Chem Commun 460–462
21. Edwards OE, Feniak G, Handa KL (1962) The alkaloids of *Stemona sessilifolia*. Can J Chem 40:455–462
22. Edwards OE, Feniak G (1962) The structure of tuberostemonine. Can J Chem 40:2416–2418
23. Uyeo S, Irie H, Harada H (1967) The structure of stenine, a new alkaloid occurring in *Stemona tuberosa*. Chem Pharm Bull (Tokyo) 15:768–770
24. Kondo H, Satomi M, Kaneko T (1956) New alkaloids from the root of *Stemona tuberosa*. Annu Rep ITSUU Lab 7:24–29 (CA 51:1540f)
25. Huber CP, Hall SR, Maslen EN (1968) The crystal structure of oxotuberostemonine. Tetrahedron Lett 4081–4084
26. Pfeiffer S, Nastewa W (1968) Alkaloids of Vietnamese *Stemona tuberosa*. Pharmazie 23:342–343
27. Kondo H, Satomi M (1947) *Stemona* alkaloids. XII. Alkaloids from the roots of *Stemona japonica* and *S. sessilifolia*. 1. J Pharm Soc Jpn 67:182–184
28. Kondo H, Satomi M (1947) *Stemona* alkaloids. XIII. Alkaloids from the roots of *Stemona japonica* and *S. sessilifolia*. 2. J Pharm Soc Jpn 67:185–187
29. Kondo H, Satomi M (1947) *Stemona* alkaloids. XIV. Alkaloids from the roots of *Stemona japonica* and *S. sessilifolia*. 3. J Pharm Soc Jpn 67:188–190
30. Irie H, Harada H, Ohno K, Mizutani T, Uyeo S (1970) The structure of the alkaloid protostemonine. Chem Commun 268–269
31. Irie H, Ohno K, Osaki K, Taga T, Uyeo S (1973) X-ray crystallographic determination of the structure of the alkaloid prostostemonine. Chem Pharm Bull (Tokyo) 21:451–452
32. Irie H, Harada H, Ohno K, Mizutani T, Uyeo S (1970) Structure of the alkaloid protostemonine. J Chem Soc [D] 268–269
33. Koyama H, Oda K (1970) Crystal and molecular structure of stemonine hydrobromide hemihydrate. J Chem Soc [B] 1330–1333
34. Iizuka H, Irie H, Masaki N, Osaki K, Uyno S (1973) X-ray crystallographic determination of the structure of stemonamine, a new alkaloid from *Stemona japonica* Miq.: isolation of isostemonamine. J Chem Soc Chem Commun 125–126
35. Chu JH (1949) The alkaloids of the Chinese drug, Pai-pu. Sci Rec 2:310–314
36. Xu RS, Lu YJ, Chu JH, Iwashita T, Naoki H, Naya Y, Nakanishi K (1982) Studies on some new *Stemona* alkaloids. Tetrahedron 38:2667–2670
37. Noro T, Fukushima S, Ueno A, Iitaka Y, Saiki Y (1979) A new alkaloid, croomine, from *Croomia heterosepala* Okuyama. Chem Pharm Bull (Tokyo) 27:1495–1497
38. Irie H, Masaki N, Ohno K, Osaki K, Taga T, Uyeo S (1970) The crystal structure of a new alkaloid, stemofoline, from *Stemona japonica*. Chem Commun 1066
39. Sakata K, Aoki K, Chang CF, Sakurai A, Tumura S, Murakoshi S (1978) Stemospironine, a new insecticidal alkaloid of *Stemona japonica* Miq., isolation, structural determination and activity. Agric Biol Chem 42:457–463
40. Kuo C, Chu TT (1978) A study of *Stemona* alkaloids I. Acta Chim Sin 36:291–296
41. Guo J (1981) Study of *Stemona* alkaloids. II. Acta Chim Sin 39:865–868
42. Terada M, Sano M, Ishii AI (1982) Studies on chemotherapy of infestation with parasitic helminths. III. Effects of tuberostemonine from *Stemona japonica* on the motility of parasitic helminths and isolated host tissues. Nippon Yakurigaku Zasshi 79:93–103 (CA 96:135368d)

Stephania tetrandra S. Moore 116

116.1 Introduction

Fangji, Radix Stephaniae tetrandrae, is the dry roots of *Stephania tetrandra* S. Moore (Menispermaceae) which are collected in the fall. It is officially listed in the Chinese Pharmacopoeia and is used in traditional Chinese medicine as an analgesic and diuretic agent and for the treatment of hypertension.

116.2 Chemical Constituents

The main chemical constituents in the roots of *S. tetrandra* are the alkaloids tetrandrine (*116-1*) and fangchinoline (*116-2*) [1–3]. Tetrandrine and fangchinoline are alkaloids of the bisbenzylisoquinoline type, in which the two benzylisoquinoline moieties are connected via two ether linkages. Berbamine (*116-3*), another alkaloid of the bisbenzylisoquinoline type, was also found in the roots of *S. tetrandra* [4]. The tetrandrine content in the roots of *S. tetrandra* ranged from 0.7% to 1.3% [5]. The Chinese Pharmacopoeia requires a tetrandrine content of not less than 0.7% for this officially listed herbal medicine.

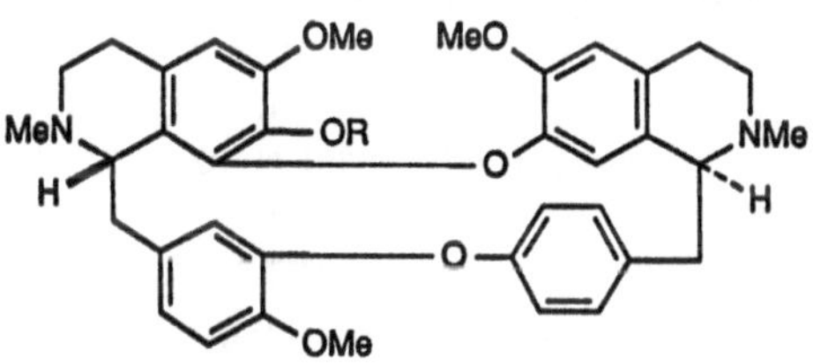

Tetrandrine (*116-1*): R = CH_3
Fangchinoline (*116-2*): R = H

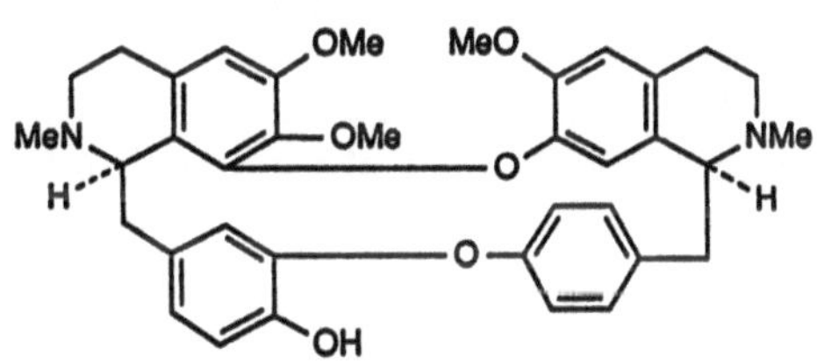

Berbamine (*116-3*)

In addition to the bisbenzylisoquinoline alkaloids, a quaternary alkaloid of the protoberberine type, cyclanoline (*116-4*), was isolated from the roots of *S. tetrandra* [6]. Recently, two new alkaloids, oxofangchirine (*116-5*) and stephenanthrine (*116-6*), were isolated. Oxofangchirine is an alkaloid of the bisbenzylisoquinoline type, whereas stephananthrine possesses a phenanthrene skeleton with a tertiary amine side chain [7].

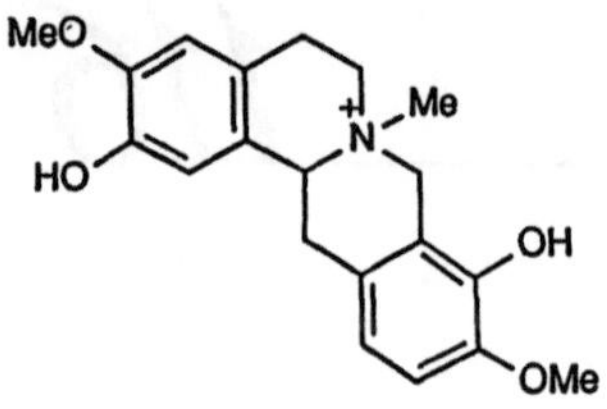

Cyclanoline (*116-4*)

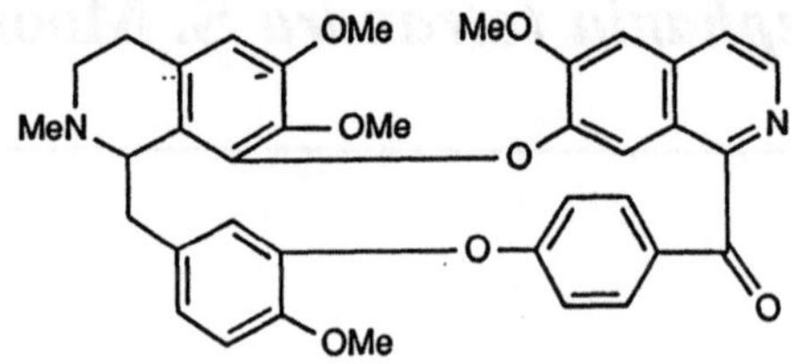

Oxofanchirine (*116-5*)

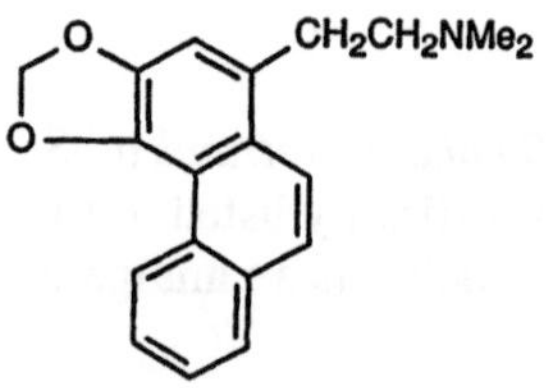

Stephenanthrine (*116-6*)

116.3 Pharmacology

116.3.1 Pharmacology of Tetrandrine

The biological activities of *S. tetrandra* depend on the presence of the chemical constituents, especially of tetrandine [4]. Tetrandrine in a concentration of 10 μ*M*–1.2 m*M* blocked the contractions of pig coronary artery strips induced by ouabain. The synergistic effect of ouabain and Ca^{2+} on contraction was also antagonized by tetrandrine. All the actions of tetrandrine were reversed by excess Ca^{2+}, indicating that tetrandrine appears to be a new Ca^{2+} antagonist. The effects of tetrandrine were similar to those of the clinically used drug verapamil, but less potent [8, 9].

The effects of tetrandrine on contractility and oxygen consumption were also compared with those of verapamil in isolated guinea pig left atrial strips and cat papillary muscles. Both compounds diminished Ca^{2+}-dependent contractility and oxygen consumption. These effects were antagonized by increasing the extracellular Ca^{2+} concentration [10, 11]. The contraction of pig coronary artery strips could be relaxed by tetrandrine and verapamil and their effect was reversible by increasing Ca^{2+} concentration. Neither drug inhibited contractions induced by norepinephrine [12]. Furthermore, tetrandrine did not antagonize the isoprenaline-induced increased in cAMP and contractility of isolated rabbit atrial strips, and therefore appeared not to be a β-receptor blocker [13]. Tetrandrine also inhibited Ca^{2+}-induced contractions of the isolated rat uterus [14].

Tetrandrine showed a significant effect on acute experimental myocardial infarction in the dog. The increase in the ST segment of the electrocardiogram and the release of myocardial infarction induced by ligation of the left anterior descending coronary artery were both significantly decreased by pretreatment of the animals with tetrandrine at an i.v. dose of 5 mg/kg, administered 5 min before ligation. Tetrandrine caused only a slight decrease in the blood pressure and heart rate [15]. The acute and subacute effect of isoprenaline was found to be inhibited by tetrandrine [16].

It was further reported that duration and ventricular action potential of the effective refractory period of isolated pig heart were markedly increased by perfusion with a Tyrode's buffer containing tetrandrine. A high concentration of Ca^{2+} in Tyrode's buffer partially antagonized tetrandrine-induced effects [17].

In isolated guinea pig papillary muscle tetrandrine produced concentration-dependent negative inotropic effects without affecting resting potential or the amplitude of the action potential [18]. Tetrandrine was also described to have antiarrhythmic [19, 20] action and to inhibit the platelet aggregation [21, 22].

Tetrandrine may inhibit the development of experimental silicosis in rats when given to the animal immediately after dusting or after the formation of silicotic nodules. These effects are characterized by lowering of total weight and total collagen content of the lungs compared with that of the silicotic control animals. Some degradation of collagen fibers was also indicated from histopathological examinations [23].

Rabbits were given powdered quartz in intratracheally administered saline to induce silicosis, which was detected by X-ray 4 months later. The silicotic animals were then treated i.m. with tetrandrine at a dose of 40 mg/animal, three doses a week for 3 months. After treatment the symptoms had almost disappeared [24].

The lungs of normal rats, silicotic rats, and silicotic rats treated with tetrandrine were different in total glycosaminoglycan content and in relative individual glycosaminoglycan content. The glycosaminoglycan content in silicotic rat lung was raised much more than that of normal rat. Tetrandrine inhibited the increase in glycosaminoglycan content in silicotic rat lungs [25].

It has also been reported that the extract of *S. tetrandra* containing tetrandrine was highly effective against *Mycobacterium tuberculosis* in mice even when the strain used was resistant to streptomycine, isoniazid, or *p*-aminosalicyclic acid [26]. Studies on the antiallergic activities of tetrandrine indicated that tetrandrine not only antagonized allergenic activities, but also blocked the release of allergens. Thus, tetrandrine suppressed both the passive cutaneous allergic response in rats and the allergic contraction of isolated ileum from sensitized guinea pigs. The asthmatic response and the contraction of ileum in guinea pigs induced by histamine or acetylcholine, as well as the increase in cutaneous vessel permeability induced by serotonin, were antagonized by tetrandrine. Release of the slow-reacting substance of anaphylaxis from the lung of sensitized guinea pigs and the release of histamine from mast cells of dextrane-treated rats were inhibited by tetrandrine [27].

Tetrandrine administered orally to patients and intragastrically into rats was recovered predominantly unchanged in human urine and from rat liver, lung, and urine. In addition to tetrandrine, tetrandrine 2′-oxide and 2′-nortetrandrine [28] were detected in human and rat urine and rat organs. *O*-Demethyltetrandrine was isolated as a metabolite in the liver and excreta after oral administration of tetrandrine [29].

116.4 Chemical Constituents of Other *Stephania* Species

A number of other *Stephania* species are used in traditional Chinese medicine or in folk medicine. Total alkaloid contents in the roots of a series of *Stephania* species with medicinal use have been determined. The species studied (total alkaloid contents) were: *S. delavayi* (3.1%), *S. japonica* (0.2%), *S. longa* (0.9%), *S. tetrandra* (2.9%), *S. excentrica* (19.6%), *S. cepharantha* (1.5%), *S. epigeae* (2.9%), *S. brachyandra* (6.6%), *S. sinica* (2.0%), *S. dielsiana* (2.6%), *S. yunnanensis* (5.1%), *S. succifera* (2.3%), *S. hainanensis* (2.4%), *S. mashanica* (2.0%), *S. micrantha* (1.0%), *S. dicentrinifera* (4.2%), *S. kwangsiensis* (2.6%), *S. viridiflavens* (6.0%) [30].

116.4.1 *Stephania japonica*

Stephania japonica is a plant with medicinal use. A number of alkaloids of different chemical types were isolated from the roots.

Epistephanine (*116-7*) [31, 32], hypoepistephanine (*116-8*) [33], stepholine (*116-9*) [34, 35], stebisimine (*116-10*) [36], and insularine (*116-11*) [37] were alkaloids of the bisbenzylisoquinoline type isolated from the roots of *S. japonica*, and stepinonine (*116-12*) is an alkaloid with a benzazepine structural element [38, 39]. The alkaloid insularine showed curare-like activity [37].

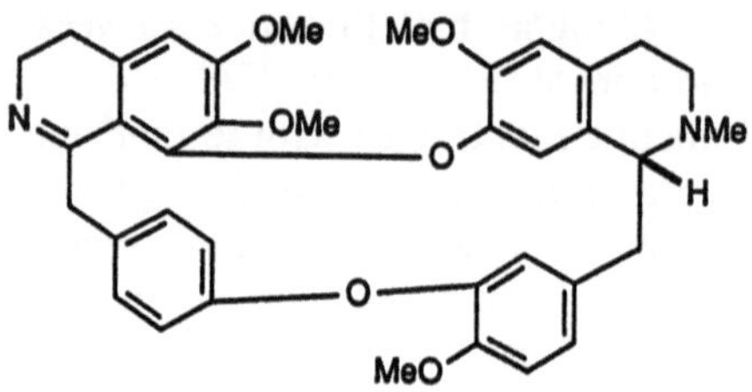

Epistephanine (*116-7*)

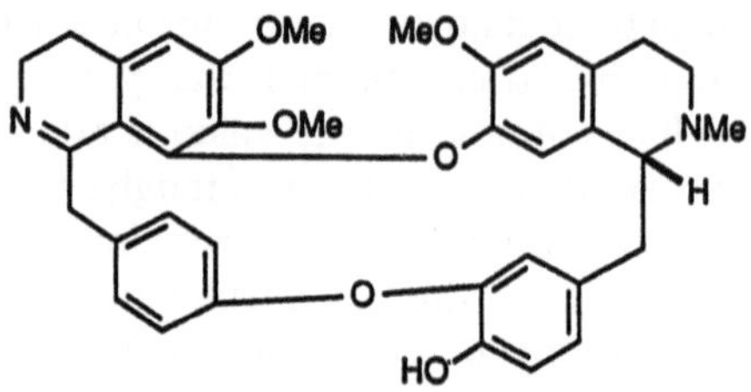

Hypoepistephanine (*116-8*)

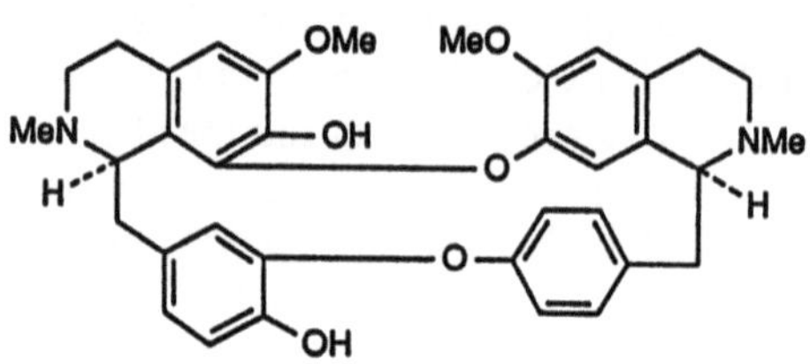

Stepholine (*116-9*)

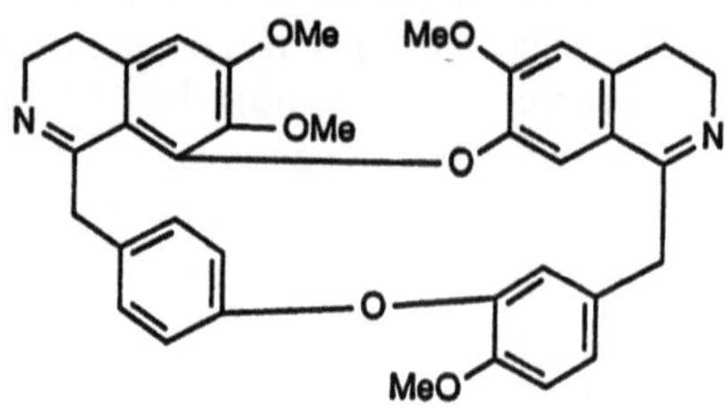

Stebisimine (*116-10*)

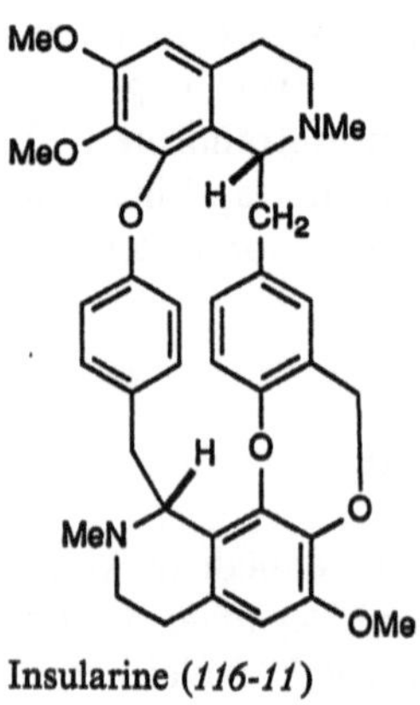

Insularine (*116-11*)

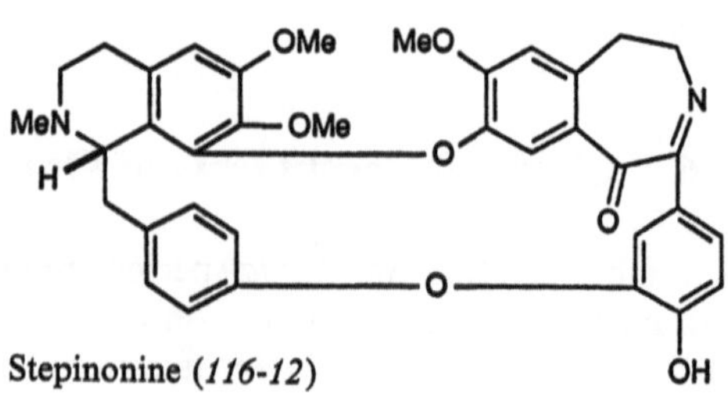

Stepinonine (*116-12*)

A series of alkaloids derived from hasubanan were also isolated from the roots of *S. japonica*. Hasubanan (*116-13*) is a structural isomer of morphinan. Further alkaloids isolated from the roots of *S. japonica* were: hasubanonine (*116-14*) [40–43], metaphanine (*116-15*) [44, 45], homostephanoline (*116-16*) [43, 46], prometaphanine (*116-17*) [47], stephamiersine (*116-18*), epistephamiersine (*116-19*), oxostephamiersine (*116-20*), stephasunoline (*116-21*) [48], oxoprometaphanine (*116-22*), oxohasubanonine (*116-23*) [49], oxoepistephmiersine (*116-24*) [50], and oxostephasunoline (*116–25*) [51].

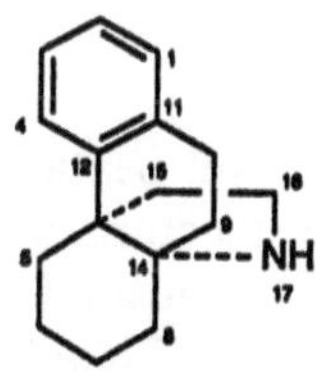

Hasubanan (*116-13*)

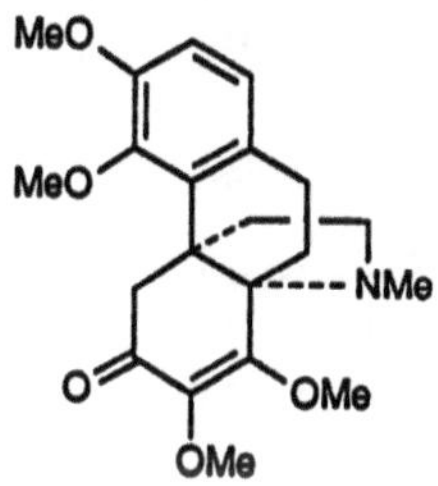

Hasubanonine (*116-14*)

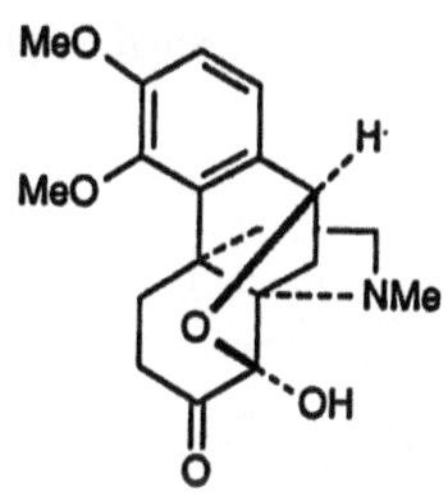

Metaphanine (*116-15*)

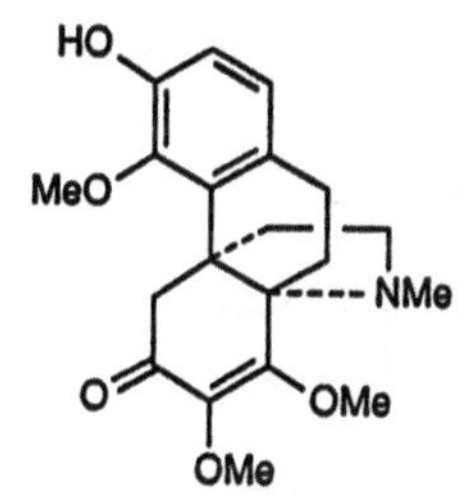

Homostephanoline (*116-16*)

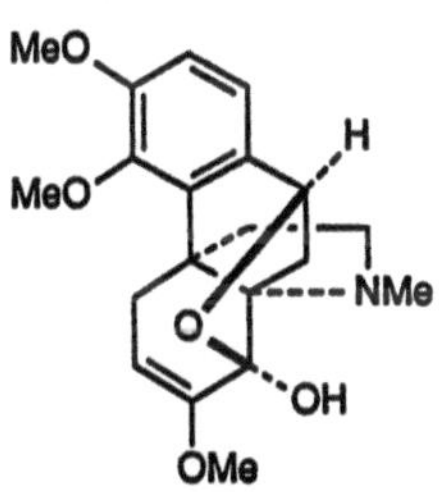

Prometaphanine (*116-17*)

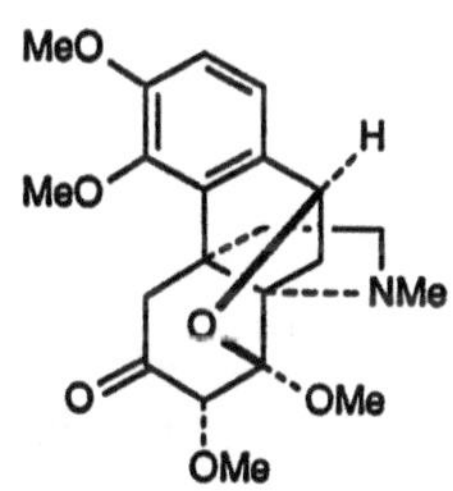

Stephamiersine (*116-18*)

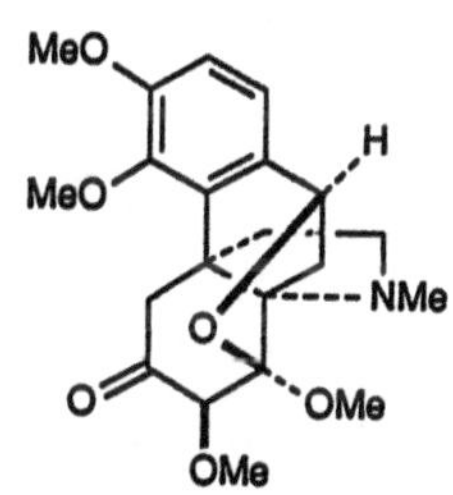

Epistephamiersine (*116-19*)

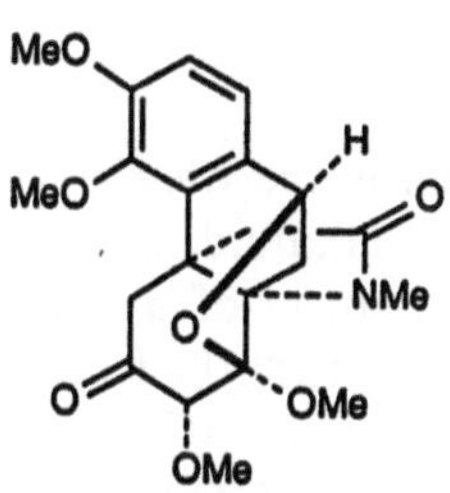

Oxostephamiersine (*116-20*)

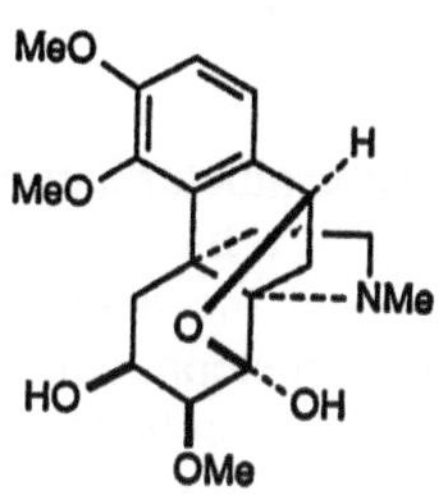

Stephasunoline (*116-21*)

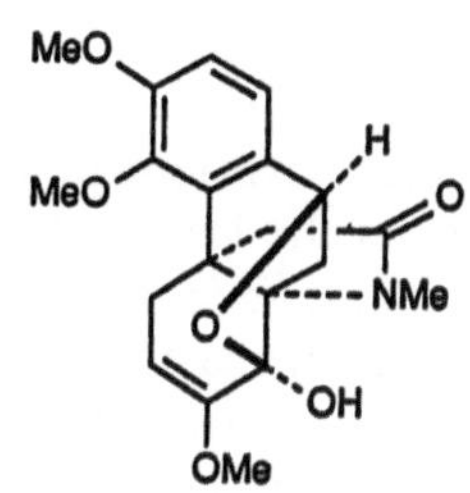

Oxoprometaphanine (*116-22*)

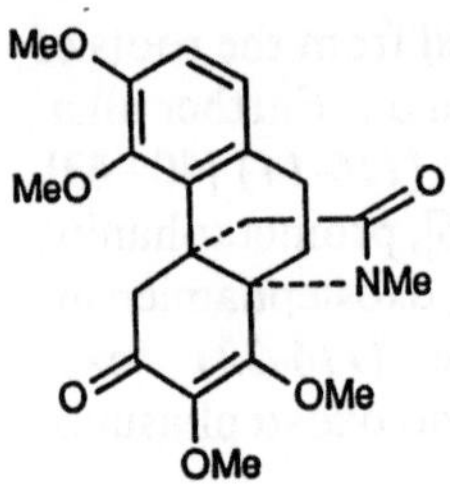
Oxohasubanonine (*116-23*)

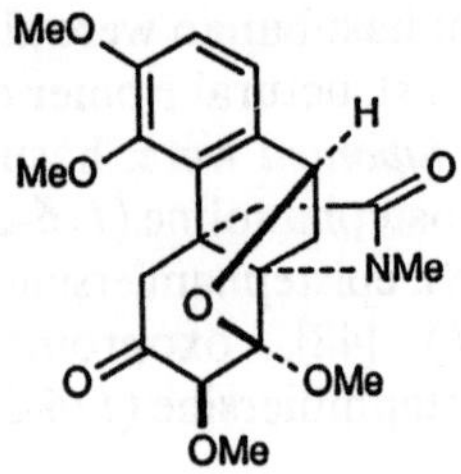
Oxoepistephamiersine (*116-24*)

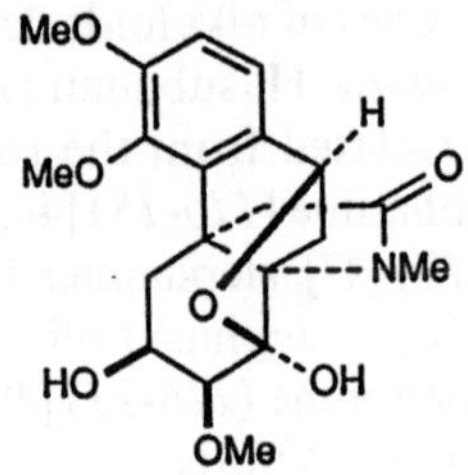
Oxostephasunoline (*116-25*)

Stephadiamine (*116-26*), the first example of a compound with a novel pentacyclic norhasubanan skeleton, was also isolated from *S. japonica*. The structure of stephadiamine was elucidated by spectroscopic methods and X-ray diffraction analysis [52]. In addition, alkaloids of the aporphine type, oxostephanine (*116-27*) and lanuginosine (*116-28*) [53], and alkaloids of the protoberberine type, cyclanoline [54] and steponine (*116-29*) [55], were isolated from *S. japonica*. Protostephanine (*116-30*) from *S. japonica* is an alkaloid of the bibenz[*d,f*]azonine type [53–58].

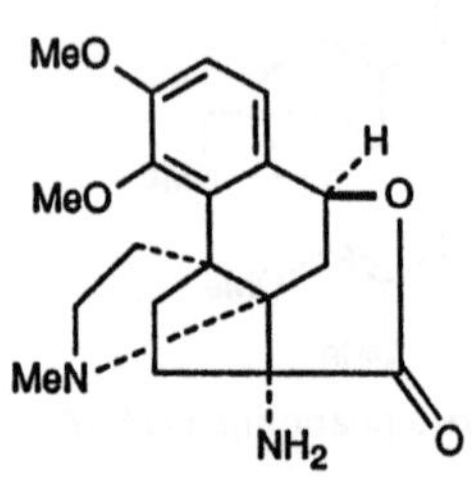
Stephadiamine (*116-26*)

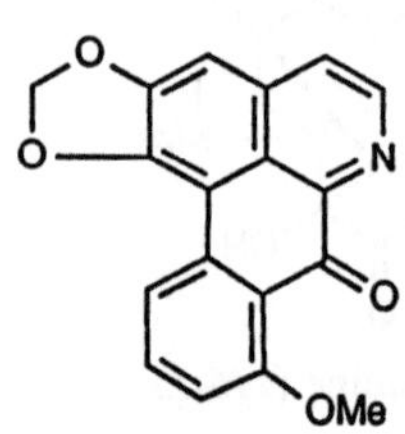
Oxostephanine (*116-27*)

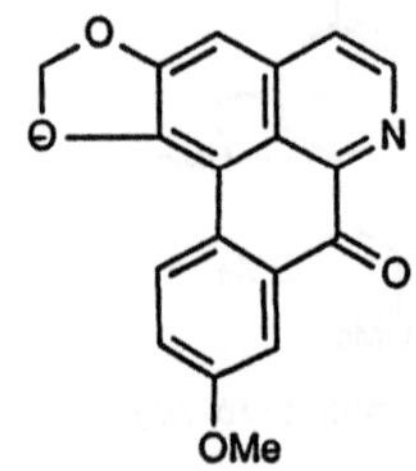
Lanuginosine (*116-28*)

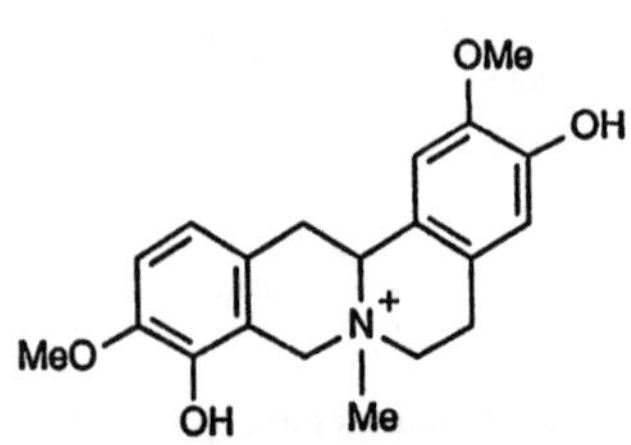
Steponine (*116-29*)

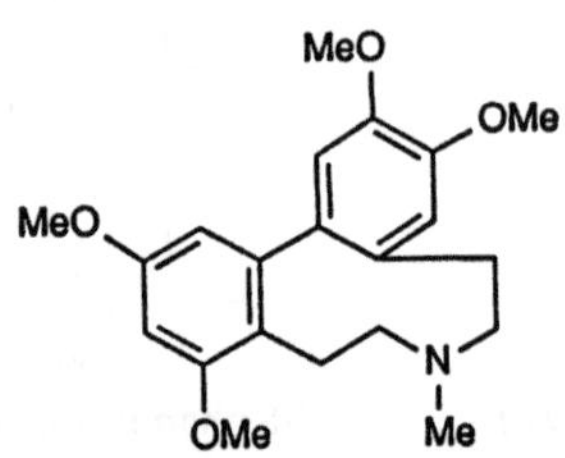
Protostephanine (*116-30*)

Whereas oxostephamiersine, 16-oxoprometaphanine, and stebisimine were isolated from the leaves of *S. japonica* [59], a hasubanan ester-ketal alkaloid named stephabenine (*116-31*) was isolated from the fruits [60]. The isolation of a new alkaloid of the hasubanan type named prostephanaberrine (*116-32*) from the fruits was further reported [61]. Treatment of prostephanaberrine with aqueous HCl gave stephanaberrine (*116-33*), which in its structure is closely related to that of metaphanine [61].

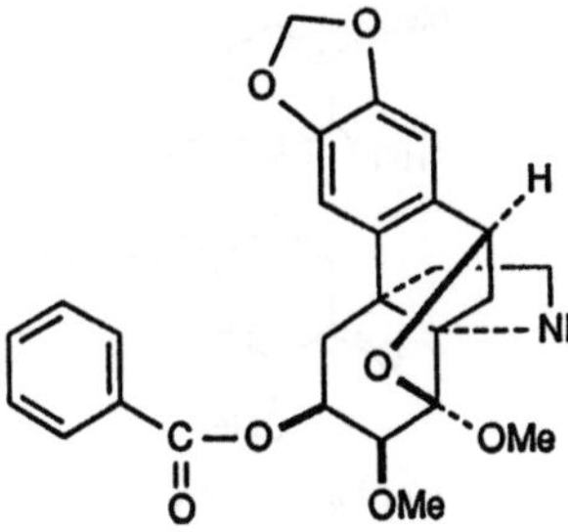

Stephabenine (*116-31*)

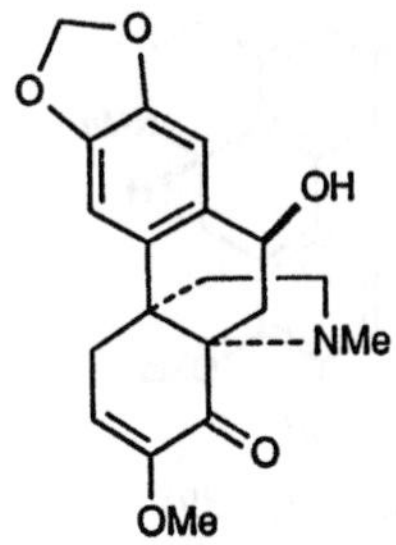

Prostephanaberrine (*116-32*)

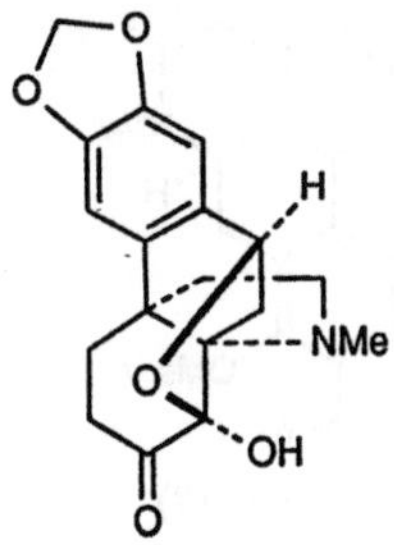

Stephanaberrine (*116-33*)

116.4.2 *Stephania cepharantha*

The tubers of *S. cepharantha* are another folk medicine in China that is used as an analgesic, diuretic, and tuberculostatic agent. A number of alkaloid constituents have been found.

116.4.2.1 Chemical Constituents

The main alkaloidal constituents in *S. cepharantha* are cepharanthine (*116-34*) and isotetrandrine (*116-35*) [62]. They are both alkaloids of the bisbenzylisoquinoline type.

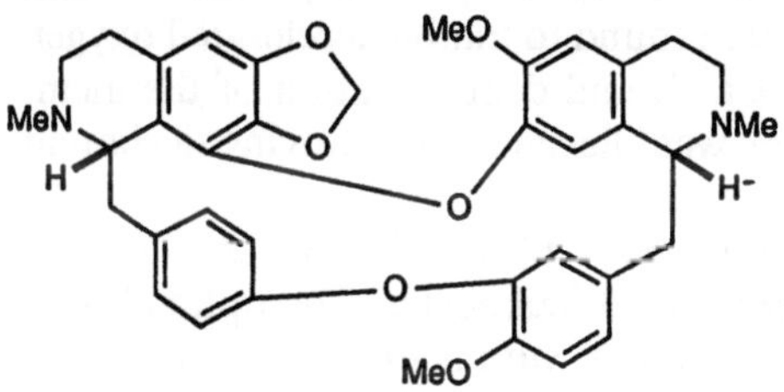

Cepharanthine (*116-34*)

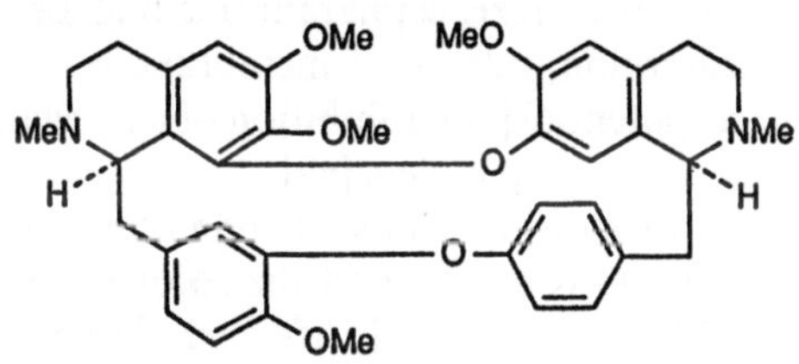

Isotetrandrine (*116-35*)

In addition to the main alkaloids, a number of minor alkaloids of different structural types were isolated. Berbamine [62], cepharanoline (*116-36*) [63], and cycleanine [62] are further bisbenzylisoquinoline type alkaloids, whereas stephanine (*116-37*), crebanine (*116-38*), *O*-nornuciferine, and stesakine (*116-39*) are aporphine derivatives isolated from *S. cepharantha* [62, 63]. Moreover, the protoberberine-type alkaloid palmatine [64] and the hasubanan-type alkaloid cepharamine (*116-40*) [65] were also isolated and identified.

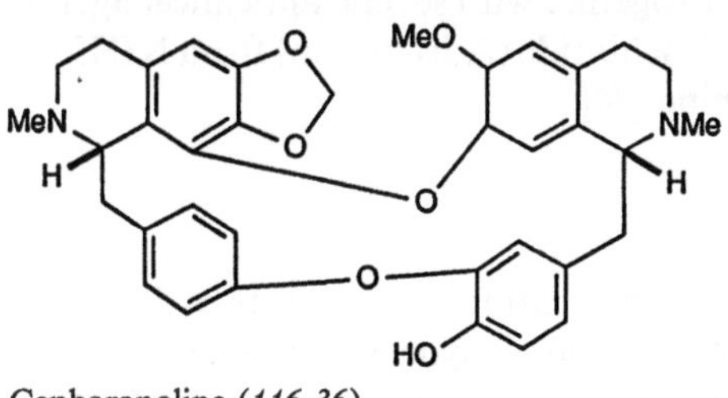

Cepharanoline (*116-36*)

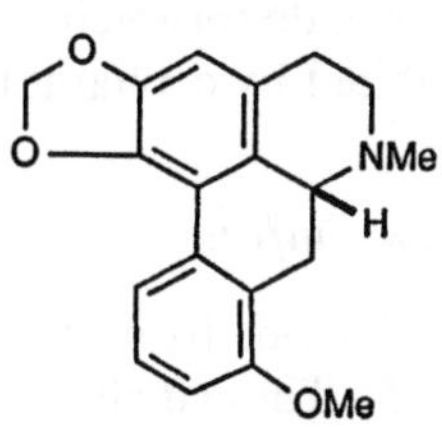

Stephanine (*116-37*)

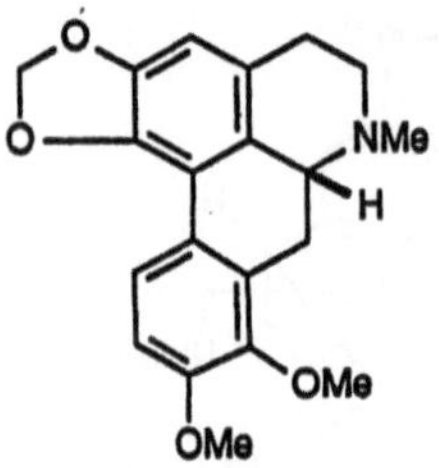

Crebanine (*116-38*)

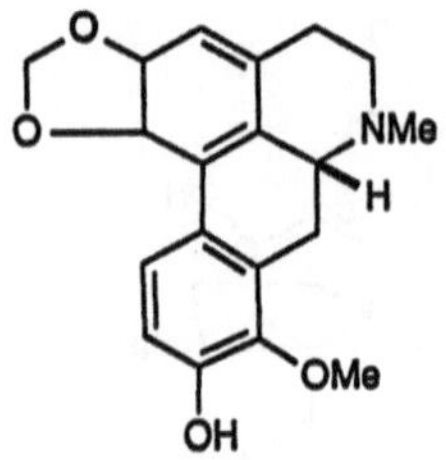

Stesakine (*116-39*)

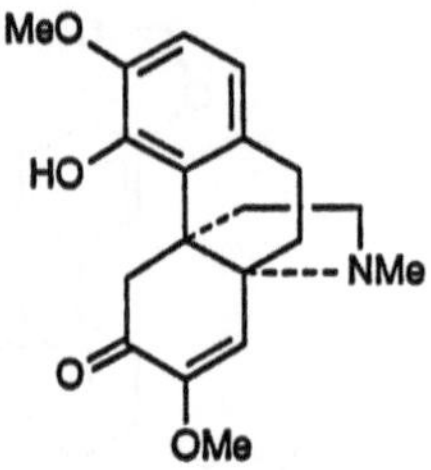

Cepharamine (*116-40*)

116.4.2.2 Pharmacology of Cepharanthine

Cepharanthine administered orally to volunteers at a dose of 30–60 mg inhibited collagen-induced blood platelet aggregation. Secondary aggregation induced by ADP was also inhibited 1–2 h after administration, but arachidonic acid induced aggregation was not affected. The inhibitory effect of cepharanthine disappeared within 2–4 h [66]. Platelet membrane morphology was not affected by 60 mg cepharanthine. Apparently, cepharanthine specifically inhibits the release of arachidonic acid from membrane phospholipids induced by collagen and ADP and consequently inhibits platelet aggregation [66]. The inhibition of thrombin-induced aggregation of platelets by cepharanthine was also correlated with the inhibition of arachidonate release from the phospholipids of the membrane, notably from phosphatidylcholine [67]. Furthermore, cepharanthine had also been found to inhibit accelerated oxygen consumption, release of membrane-bound Ca^{2+}, and depolarization of the membrane potential [68]. Inhibition of aggregation was alleviated by removing the bound drug from the platelets [69].

Intravenous injection of cepharanthine to mice for 14 days following X-irradiation at 300 R accelerated the recovery of peripheral leukocytes. The optimal dose was 0.1 mg/mouse [70]. After whole body irradiation with a single dose of 500 R, mice were given a daily dose of 0.1 mg cepharanthine i.p. for 20 days. The number of endogenous spleen colony-forming units increased markedly. The radiation-induced decrease in spleen weight recovered during 30 days [71]. Radiation-induced lipid peroxidation caused an increase in membrane permeability of phosphatidylcholine liposomes. Cepharanthine suppressed both lipid peroxidation and subsequent changes in permeability induced by radiation. The suppression of membrane permeability was mainly due to the inhibition of the radiation-induced lipid peroxidation. However, cepharanthine did not exhibit a radical trapping potential [72].

The simultaneous administration of myelosuppressive anticancer agents and ceparanthine reduced leukopenia. A significant improvement in leukocyte recovery was observed in mice receiving cepharanthine together with several anticancer agents [73]. Moreover, the efflux of doxorubicin hydrochloride (Adriamycin) from NIH3T3 cells was markedly inhibited by cepharanthine [74].

116.4.3 *Stephania dicentrinifera*

The isolation of seven alkaloids from the ethanolic extract of the roots of *S. dicentrinifera* was reported [75]. These include: stephanine, dehydrostephanine, sinoacutine (*116-41*), sinomenine (*116-42*), isocorydine (*116-43*), dicentrine (*116-44*), and dehydrodicentrine (*116-45*). Sinoacutine and sinomenine are two alkaloids with a morphinan skeleton.

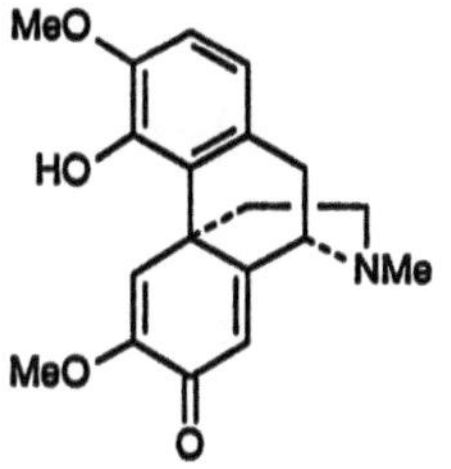

Sinoacutine (*116-41*)

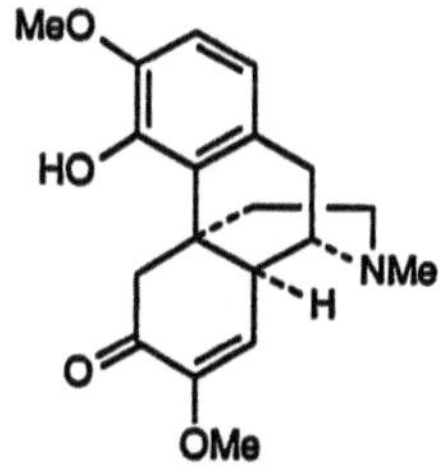

Sinomenine (*116-42*)

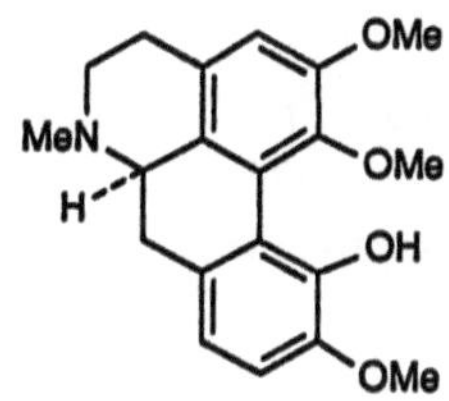

Isocorydine (*116-43*)

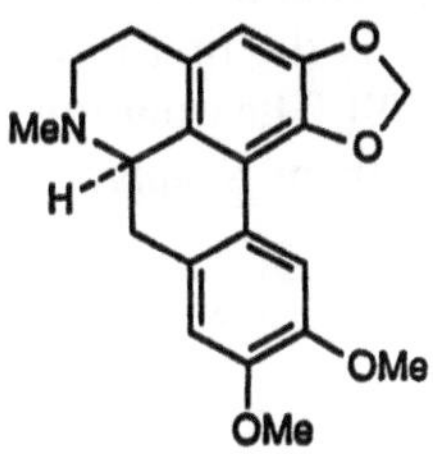

Dicentrine (*116-44*)

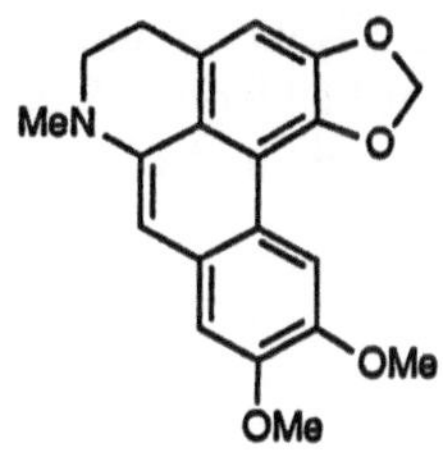

Dehydrodicentrine (*116-45*)

116.4.4 *Stephania dielsiana*

Six alkaloids were isolated from the roots of *S. dielsiana* and were identified as crebanine, stephanine, sinoacutine, dehydrostephanine, tetrahydropalmatine, and xylopinine (*116-46*) [76]. The crebanine content was the highest (0.3%), followed by sinoacutine (0.2%), stephanine (0.2%), and *l*-tetrahydropalmatine (0.1%) contents [77].

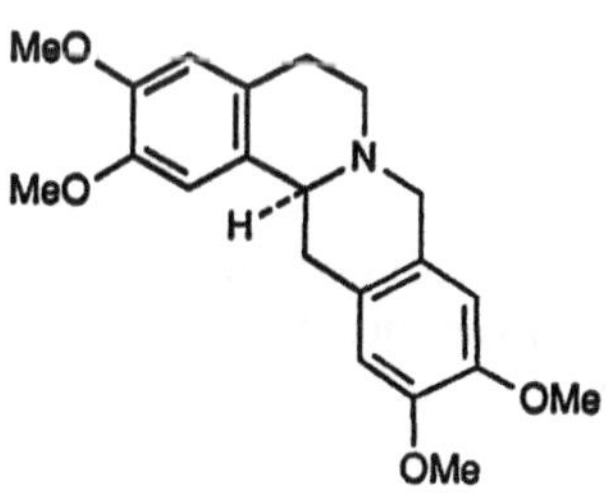

Xylopinine (*116-46*)

116.4.5 *Stephania mashanica*

Six alkaloids were isolated from the ethanolic extract of powdered tubers of *S. mashanica* and identified as sinoacutine, tetrahydrocolumbamine, tetrahydropalmatine, dicentrinone (*116-47*), tetrahydrojatrorrhizine, and dicentrine. Sinoacutine was the major constituent (0.5%) [78].

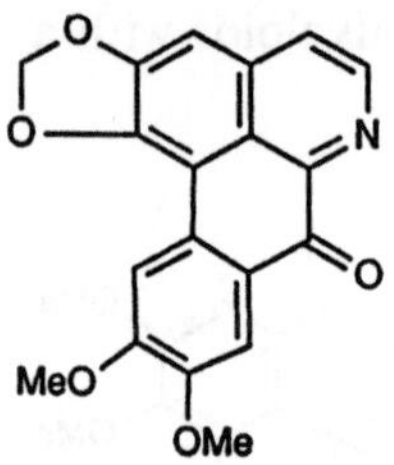

Dicentrinone (*116-47*)

116.4.6 *Stephania epigeae*

From the roots of *S. epigeae* the four known alkaloids cycleanine, cepharanthine, curine (*116-48*), and isocorydine were isolated and identified. This is the first time that curine has been isolated and identified from a *Stephania* plant [79]. The isolation of cepharanthine, dicentrine, sinomenine, and oliveroline (*116-49*) from *S. epigeae* was further reported [80].

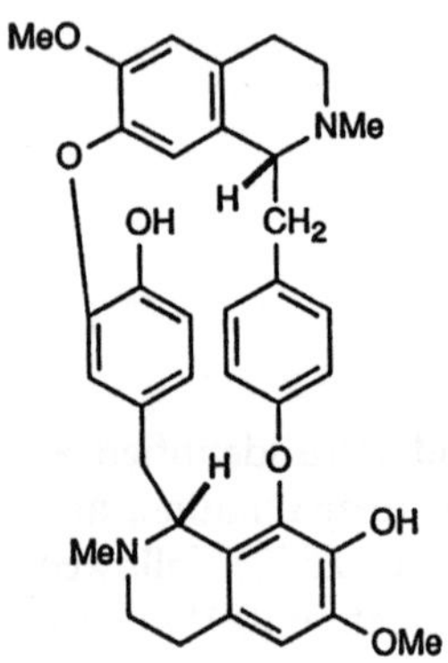

Curine (*116-48*)

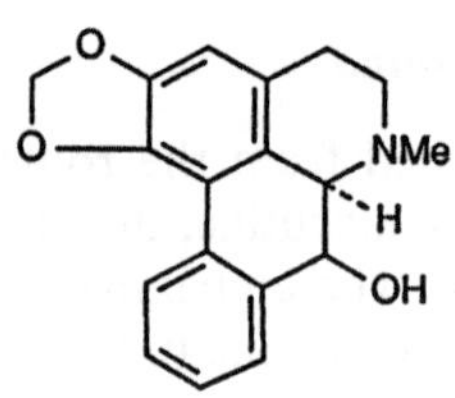

Oliveroline (*116-49*)

116.4.7 *Stephania viridiflavens*

From *S. viridiflavens* the following alkaloidal constituents were isolated and identified: xylopinine, palmatine, jatrorrhizine, and tetrahydropalmatine. The tetrahydropalmatine content in *S. viridiflavens* exceeded 4% [81].

116.4.8 *Stephania longa*

A new hasubanan alkaloid named longanone (*116-50*) was isolated from the roots and stems of *S. longa* and structurally elucidated by spectral analyses and chemical reactions [82]. In addition, four other alkaloids, longaninine (*116-51*), stephaboline (*116-52*), stephabyssine (*116-53*), and prostephabyssine (*116-54*), were also isolated and identified [83].

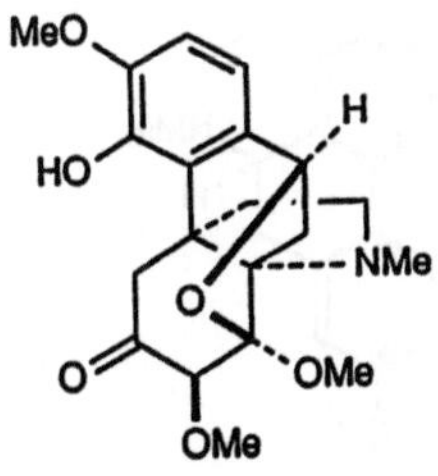

Longanone (*116-50*)

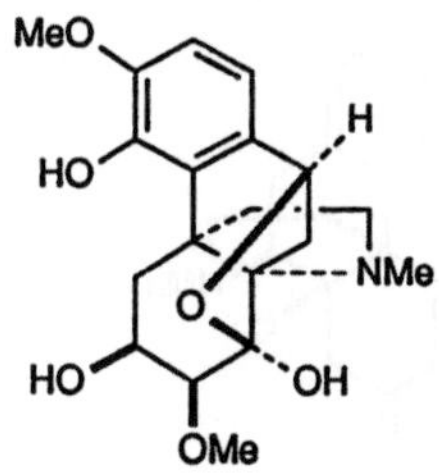

Longaninine (*116-51*)

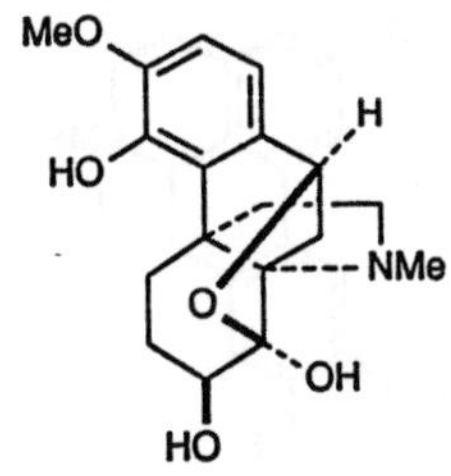

Stephaboline (*116-52*)

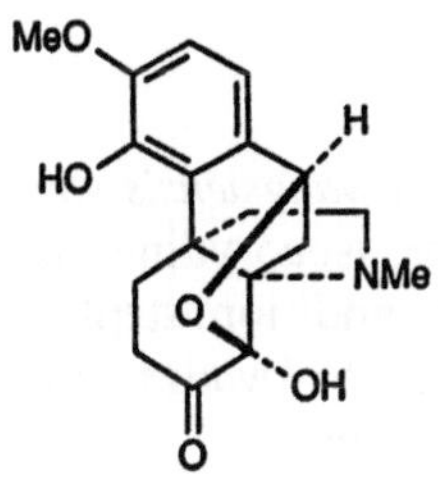

Stephabyssine (*116-53*)

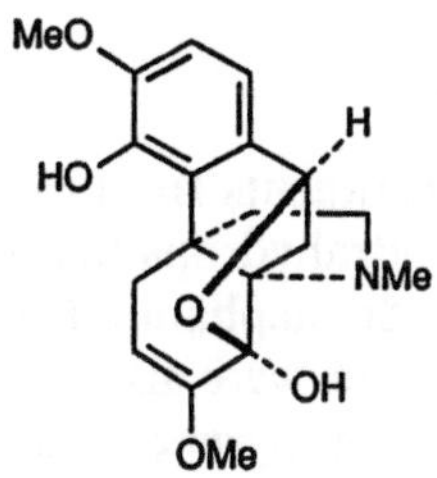

Prostephabyssine (*116-54*)

116.4.9 *Stephania micrantha*

Eleven alkaloids were isolated and identified from the tubers of *S. micrantha*. They are: dehydrostephanine, tetrahydropalmatine, capaurine (*116-55*), tetrahydrocolumbamine, corypalmine, xylopinine, dehydroremerine (*116-56*), stephanine, isocorydine, sinoacutine, and sinomenine [84].

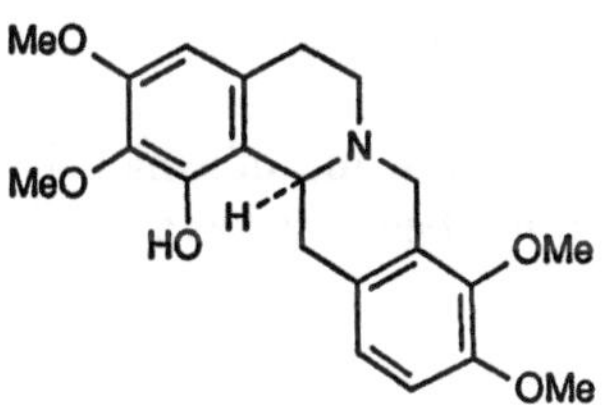

Capaurine (*116-55*)

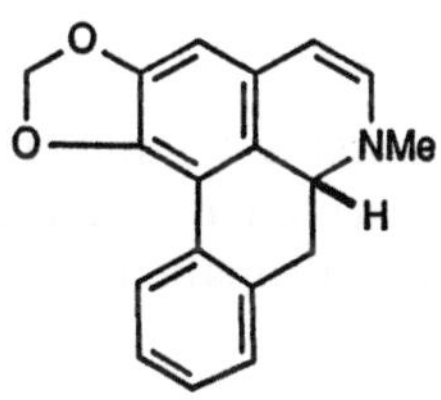

Dehydroremerine (*116-56*)

116.4.10 *Stephania brachyandra*

The following ten alkaloid constituents were isolated and identified from the tubers of *S. brachyandra* (yield): isocorydine (1.5%), tetrahydropalmatine (0.2%), dicentrine (0.3%), sinomenine (0.1%), corytuberine (0.04%), sinoacutine (0.006%), dehydrodicentrine (0.006%), isoboldine (*116-57*) (0.004%), dihydrosalutaridine (*116-58*) (0.001%), and *N*-methyllaurotetanine (*116-59*) (0.006%) [85].

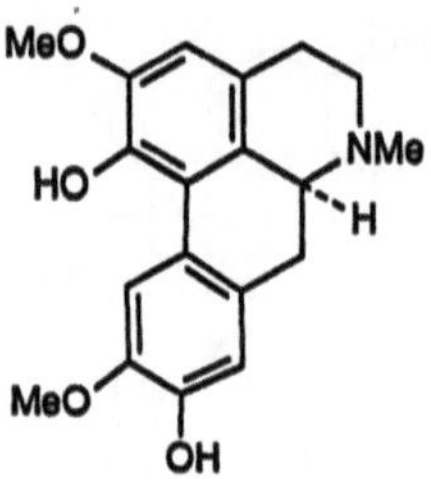

Isoboldine (*116-57*)

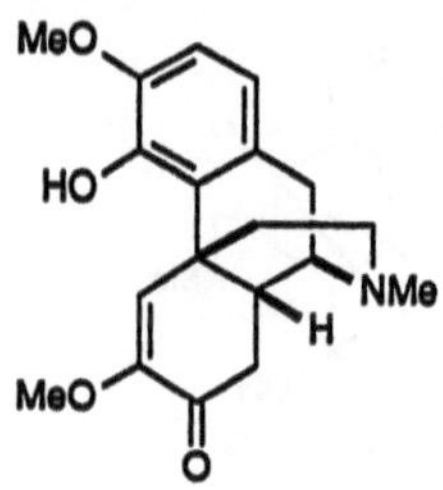

Dihydrosalutaridine (*116-58*)

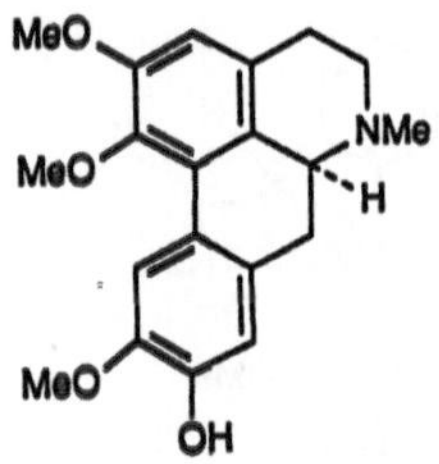

N-Methyllaurotetanine (*116-59*)

116.4.11 *Stephania kwangsiensis*

The isolation of six alkaloidal constituents from the tuber of *S. kwangsiensis* was reported. They were identified as tetrahydropalmatine, capaurine, isocorydine, remerine, dehydroremerine, and dehydrostephanine (*116-60*) [86]. In addition, stephanine, dihydropalmatine, and palmatine were also obtained [87, 88]. *S. kwangsiensis* is a medicinal plant with analgesic, antipyretic, and sedative activities.

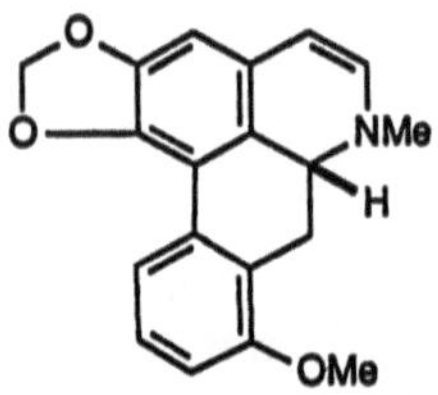

Dehydrostephanine (*116-60*)

116.4.12 *Stephania sinica*

The root of *S. sinica* contains a number of alkaloids, the main one of which is tetrahydropalmatine with a content of 1.2%–1.5%. Thus, it has been used as a plant resource to produce tetrahydropalmatine [89].

116.4.13 *Stephania longipes*

The alkaloid stepholidine (*116-61*) was isolated from *S. longipes* and showed sedative and antispastic effects in experimental animals. [90]. Intravenous injection of stepolidine lowered blood pressure in anesthetized dogs and rats [91].

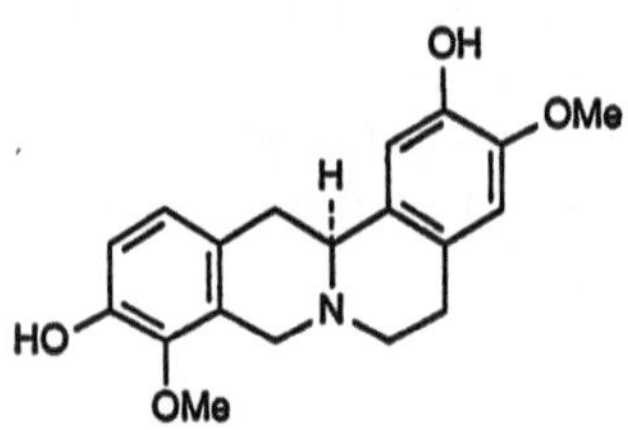

Stepholidine (*116-61*)

References

1. Chen KK, Chen AL (1935) The alkaloids of han-fang-chi. J Biol Chem 109:681–685
2. Chuang CK, Hsing CY, Kao YS, Chang KJ (1939) The alkaloids of han-fang-chi. Fangchinoline, a demethyltetrandrine. Chem Ber 72B:519–525
3. Hsing CY, Chang CH (1958) Constitution of fangchinoline. Sci Sin (Peking) 7:59–63
4. Feng YX, Chen H (1985) Pharmacognostic and chemical identification of Fang Ji (*Stephania tetrandra*). Chin J Pharm Anal 5:28–31
5. Yang YF (1985) Thin layer chromatography-UV-spectrophotometry for tetrandrine in *Stephania tetrandra*. Chin Trad Herb Drugs 16:281
6. Tomita M, Kozuka M, Lu ST (1967) Studies on the alkaloids of menispermaceous plants. CCXXIX. Alkaloids of *Stephania tetrandra*. Yakugaku Zasshi 87:316–318
7. Hu TM, Zhao SX (1986) The structure of oxofangchirine and stephenanthrine isolated from *Stephania tetrandra* S. Moore. Acta Pharm Sin 21:29–34
8. Cha ZL, Fang DC, Xia GJ, Jiang MX (1983) Antagonistic effects of tetrandrine on ouabain induced contraction of coronary arterial strips. Acta Pharmacol Sin 4:177–179
9. Fang DC, Jiang MS (1983) Study on the anticalcium effect of tetrandrine. Natl Med J China 63:772–774
10. Yao WX, Xia GJ, Fang DC, Jiang MX (1984) Effect of tetrandrine and verapamil on contractility and oxygen consumption in heart muscle. Acta Pharmacol Sin 5:97–100
11. Yao WX, Fang DC, Xia GJ, Qu L, Jang MX (1981) Effects of tetrandrine on guinea pig atrium. Acta Acad Med Wuhan 10:81–84
12. Jia JF, Gao LL, Xia GJ, Luo QF, Fang DC, Jiang MX (1984) Effects of tetrandine on contractility of isolated pig coronary artery strips. Acta Pharmacol Sin 5:32–35
13. Yao WX, Xia GJ, Fang DC, Jiang MX (1983) Effects of tetrandrine, verapamil and propranolol on contractility and cAMP level of isolated rabbit left atria. Acta Pharmacol Sin 4:29–32
14. Yao WX, Fang DC, Zhao JL, Jiang MX (1983) Effects of tetrandrine on contractility of isolated rat uterus. Acta Pharmacol Sin 4:130–134
15. Yu SL, Cao LS, Feng YB, Guo QG, Zhou BC, Cuo HP (1983) Effects of tetrandrine on acute experimental myocardial infarction. Chin J Cardiol 11:147–149
16. Yang XM, Fang DC, Jiang MX (1984) Protective effect of tetrandrine against the myocardial hypoxia and necrosis induced by isoprenaline in rats. Acta Acad Med Wuhan 13:201–204
17. Zhang GQ, Lin TF (1983) Electrophysiologic studies on the antiarrhythmic effect of tetrandrine. Chin J Cardiol 11:224–226
18. Zong XG, Jin MW, Xia GJ, Fang DC, Jiang MX (1983) Effects of tetrandrine on action potential and contraction of isolated guinea pig papillary muscle. Acta Pharmacol Sin 4:258–261
19. Ke J, Weng SA, Zhang GQ, Yang YH, Wang JK, Fu RF (1981) Effects of tetrandrine on experimental arrhythmias. Acta Pharmacol Sin 2:235–237
20. Cha L, Qian JQ, Lu FH (1981) Antiarrhythmic effect of tetrandrine and the total alkaloids of *Sophora flavescens*. Acta Pharmacol Sin 2:26–28
21. Qian YM, Huang YH (1989) Effects of tetrandrine on rabbit platelet aggregation, thromboxane A_2 generation and calmodulin activity. Acta Pharmacol Sin 10:61–65
22. Xie QM, Zheng QJ, Bian RL (1989) Inhibition of platelet aggregation by tetrandrine. J Zhejiang Med Univ 18:7–10
23. Yu XF, Zou CQ, Lin MB (1983) Observation of the effect of tetrandrine on experimental silicosis of rats. Ecotoxicol Environ Safety 7:306–312
24. Jiang HX, Hu TX, Peng BX, Shi DZ (1983) Preliminary study on the mechanism of therapeutic effect of tetrandrine on silicosis. Chin J Tuberc Resp Dis 6:92–94
25. Liu BC, Zou CQ, Li YR (1983) Studies on the contents of glycosaminoglycans from lungs of silicotic rats and tetrandine-treated silicotic rats. Exotoxicol Environ Safety 7:323–329
26. Vichkanova SA, Makarova LV, Gordeikina NI (1972) Tuberculostatic activity of preparations from plants. In: Aizenman BE (ed) Fitontsidy, Mater. Soveshch., 6th 1969, Kiev, pp 90–94 (CA 78:66905r)
27. Bian RL, Zhou HL, Xie QM, Tong FD, Yang W, Wang Y (1984) Observation on antiallergic effect of tetrandrine. Chin Trad Herb Drugs 15:262–264
28. Lin MB, Zhang W, Zhao XW, Lu JX, Wang M, Chen LG (1982) Biotransformation of tetrandrine in rats and in men. Acta Pharm Sin 17:728–735

29. He SX, Huang ZJ (1983) Biotransformation of tetrandrine. Tianjin Yiyao 11:231–233
30. Zhu ZY, Feng YX, Ho LY, Wang YC (1983) Study on the utilization of medicinal plant resources of the genus *Stephania* of the family Menispermaceae in China. Acta Pharm Sin 18:460–467
31. Kondo H, Tanaka K (1943) Alkaloids of Menispermaceae. LXIII. Alkaloids of *Stephania japonica*. 7. Structure of epistephanine 7. J Pharm Soc Jpn 63:267–273
32. Kondo H, Tanaka K (1943) Alkaloids of Menispermaceae. LXIV. Alkaloids of *Stephania japonica*. 8. Structure of epistephanine 8. J Pharm Soc Jpn 63:273–280
33. Kondo H, Nozoye T (1943) Alkaloids of Menispermaceae. LXV. Alkaloids of *Stephania japonica*. 9. Structure of hypoepistephanine (γ-epistephanine). J Pharm Soc Jpn 63:333–334
34. Tomita M, Ibuka T (1963) Alkaloids of menispermaceous plants. CCIV. Alkaloids of formosan *Stephania japonica*. 2. Structure of a new tertiary phenolic biscoclaurine type alkaloid, stepholine, Yakugaku Zasshi 83:940–944
35. Tomita M, Ibuka T (1965) Alkaloids of Menispermaceae plants. CCXI. Alkaloids of formosan *Stephania japonica*. 4. Synthesis of DL-*N*-methyl-*O,O*-diethylisococlaurine and the structure of stepholine. Yakugaku Zasshi 85:557–560
36. Barton DHR, Kirby GW, Wiechers A (1966) Phenol oxidation and biosynthesis. XI. The structure of stebisimine and the biosynthesis of epistephanine. J Chem Soc [C] 2313–2319
37. Satomi M (1955) Alkaloids of menispermaceous plants. CXXVIII. Alkaloids of *Stephania japonica*. 20. Alkaloids with a curare-like action; isolation of insularine. Annu Rept ITSUU Lab (Japan) 6:31–33 (CA 50:10112f)
38. Ibuka T, Konoshima T, Inubushi Y (1972) Stepinonine, a dimeric benzylisoquinoline-2-phenylhomotetrahydroisoquinoline alkaloid. Tetrahedron Lett 4001–4004
39. Ibuka T, Konoshima T, Inubushi Y (1975) Structure of stepinonine. New dimeric benzylisoquinoline-2-phenylhomotetrahydroisoquinoline alkaloid. Chem Pharm Bull (Tokyo) 23: 114–124
40. Satomi M (1952) Alkaloids of menispermaceous plants. XCVII. Alkaloids of *Stephania japonica*. 15. On Hasubanonine 2. J Pharm Soc Jpn 72:834–840
41. Watanabe Y, Matsumura H (1963) Alkaloids of menispermaceous plants. CCII. Alkaloids of *Stephania japonica*. 8. Structure of hasubanonine. 1. Hasubanol. Yakugaku Zasshi 83:991–996
42. Ibuka T, Kitano M (1967) Studies on the alkaloids of menispermaceous plants. CCXXXVIII. Alkaloids of *Stephania japonica*. 18. Structure of hasubanonine. 1. Chem Pharm Bull (Tokyo) 15:1809–1810
43. Tomita M, Ibuka T, Inubushi Y, Watanabe Y, Matsui M (1965) Alkaloids of menispermaceous plants. CCX. Alkaloids of *Stephania japonica*. 9. Structure of hasubanonine and homostephanoline. Chem Pharm Bull (Tokyo) 13:538–545
44. Tomita M, Ibuka T, Inubushi Y, Takeda K (1965) The alkaloids of menispermaceous plants. CCXIV. Alkaloids of *Stephania japonica*. 12. Structure of metaphanine. Chem Pharm Bull (Tokyo) 13:695–704
45. Tomita M, Ibuka T, Inubushi Y, Takeda K (1965) The alkaloids of menispermaceous plants. CCXV. Alkaloids of *Stephania japonica* Miers. 13. Structure of metaphanine. Chem Pharm Bull (Tokyo) 13:704–712
46. Ibuka T, Kitano M (1967) Alkaloids of menispermaceous plants. CCXXXVII. Alkaloids of *Stephania japonica*. 3. Structure of homostephanoline. Chem Pharm Bull (Tokyo) 15:1939–1943
47. Tomita M, Inubushi Y, Ibuka T (1967) Studies on the alkaloids of menispermaceous plants. CCXXX. Alkaloids of Formosan *Stephania japonica*. 5. Structure of prometaphanine. Yakugagu Zasshi 87:381–386
48. Matsui M, Watanabe Y, Ibuka T, Tanaka K (1973) Constitution of four new hasubanan alkaloids from *Stephania japonica*. Tetrahadron Lett 4263-4266
49. Watanabe Y, Matsui M, Uchida M (1975) Studies on the alkaloids of menispermaceous plants. 265. Hasubanan alkaloids of *Stephania japonica*. Phytochemistry 14:2695–2698
50. Matsui M, Yamamura Y, Takebayashi T, Iwaki K, Takami Y, Kunitake K, Koga F, Urasaki S, Watanabe Y (1984) Studies on alkaloids of menispermaceous plants. CCLXXXII. Oxoepistephamiersine, a new hasubalactam alkaloid from *Stephania japonica*. J Nat Prod 47:858–861
51. Matsui M, Watanabe Y (1984) Stuidies on alkaloids of menispermaceous plants. CCLXXX. Structure of oxostephasunoline, a new hasubanalactam alkaloid from *Stephania japonica*. J Nat Prod 47:465–469

52. Taga T, Akimoto N, Ibuka T (1984) Stephadiamine, a new skeletal alkaloid from *Stephania japonica*: the first example of a C-norhasubanan alkaloid. Chem Pharm Bull (Tokyo) 32:4223–4225
53. Watanabe Y, Matsui M, Iibuchi M, Hiroe S (1975) Alkaloids of *Stephania japonica*. Phytochemistry 14:2522–2523
54. Tomita M, Ibuka T, Tsuyama K (1964) The alkaloids of menispermaceous plants. CCVI. Alkaloids of formosan *Stephania japonica*. 3. The isolation of water soluble quaternary base, cyclanoline. Yakugaku Zasshi 84:776–778
55. Watanabe Y (1957) Alkaloids of menispermaceous plants. CLI. Alkaloids of *Stephania japonica*. 5. Constitution of steponine. 2. Yakugaku Zasshi 77:278–281
56. Kondo H, Takeda K (1953) Alkaloids of menispermaceous plants. CVIII. Alkaloids of *Stephania japonica*. 18. On protostephanine. 8. Annu Rep ITSUU Lab 4:51–58 (CA 49:1076b)
57. Kondo H, Takeda K (1955) Alkaloids of menispermaceous plants. CXXIX. Alkaloids of *Stephania japonica*. 21. Protostephanine. 10. Annu Rep ITSUU Lab 6:34–40 (CA 50:10112h)
58. Takeda K (1963) Alkaloids of menispermaceous plants. CCV. Alkaloids of *Stephania japonica*. Protostephanine 14. Itsuu Kenkyusho Nempo 13:45–49 (CA 60:5570f)
59. Matsui M, Kabashima T, Ishida K, Takebayashi T, Watanabe Y (1982) Studies on the alkaloids of menispermaceous plants. CCLXXIV. Alkaloids of the leaves of *Stephania japonica*. J Nat Prod 45:497–500
60. Kondo S, Matsui M, Watanabe Y (1983) Studies on the alkaloids of menispermaceous plants. CCLXXVIII. Alkaloids from the fruits of *Stephania japonica* Miers. 1. Structure of stephabenine: a new hasubanan esterketal alkaloid. Chem Pharm Bull (Tokyo) 31:2574–2577
61. Matsui M, Yamamura Y (1986) Alkaloids from the fruits of *Stephania japonica*. III. Structures of prostephanaberrine and stephanaberrine, two new hasubanan alkaloids. J Nat Prod 49:588–592
62. Tomita M, Kishikita H (1944) Alkaloids of Menispermaceae. LXIX. Alkaloids of *Stephania cepharantha*. 3. J Pharm Soc Jp 64:27–35
63. Tomita M, Sawada T, Kozuka M, Takeuchi M, Akasu M (1969) Alkaloids of menispermaceous plants. CCLI. Alkaloids of *Stephania cepharantha*. 8. Alkaloids of *Stephania cepharantha* cultivated in Japan. Yakugaku Zasshi 89:1678–1681
64. Xia GC, Qin YQ, Yang QZ, Zhang YS, Liu XM (1985) Palmitine from 16 Menispermaceae species. Chin Trad Herb Drugs 16:266–267
65. Tomita M, Kozuka M (1967) Alkaloids of menispermaceous plants. CCXLV. Alkaloids of *Stephania cepharantha*. 7. Structure of cepharamine. Yakugaku Zasshi 87:1203–1208
66. Kanaho Y, Sato T, Jujii T, Kanzaki M, Yasunaga K (1983) In vivo effect of cepharanthine on platelet aggregation. Naika Hokan 30:15–19 (CA 99:319e)
67. Kanaho Y, Sato T, Fujii T (1982) Mechanism of the inhibitory effect of cepharanthine on the aggregation of platelets. Cell Struct Funct 7:39–48
68. Watanabe S (1984) Inhibition of platelet aggregation by cepharanthine is accomplished during the early, membrane related activation process. Acta Med Okayama 38:101–115 (CA 101:17048y)
69. Kometani M, Kanaho Y, Sato T, Fujii T (1985) Inhibitory effect of cepharanthine on collagen induced activation in rabbit platelets. Eur J Pharmacol 111:97–105
70. Iida S (1984) Studies on biscoclaurine alkaloids in relation to radiation damage of hemopoietic tissue. I. Effect of cepharanthine on the recovery of peripheral blood cells after irradiation of the whole body. Okayama Igakkai Zasshi 96:883–890 (CA 103:19139u)
71. Iida S (1984) Studies on biscoclaurine alkaloids in relation to radiation damage of hemopoietic tissue. II. Effect of cepharanthine on the recovery from radiation damage in mouse hemopoietic tissue. Okayama Igakkai Zasshi 96:891–899 (CA 103:19138t)
72. Nagatsuka S, Nakazawa T (1982) Effects of membrane stabilizing agents, cholesterol and cepharanthine, on radiation induced lipid peroxidation and permeability in liposomes. Biochim Biophys Acta 691:171–177
73. Kasajima T, Yamakawa M, Maeda K, Matsuda M, Dobashi M, Imai Y (1983) An experimental study on the effect of cepharanthine on leukopenia induced by administration of anticancer drugs. Gan to Kagaku Ryoho 10:1188–1196 (CA 99:32825h)
74. Nagaoka S, Kawasaki S, Sasaki K, Nakanishi T (1985) Modification by an alkaloid (cepharanthine) of intracellular uptake and efflux of adriamycin in NIH 3T3 cells. Igaku no Ayumi 133:260–261 (CA 103:134381)

75. Min ZD, Zheng XZ (1984) Chemical constituents of *Stephania dicentrinifera* H.S. Lo et M. Yang. Chin Trad Herb Drugs 15:8
76. Xie PS, Hsu TY, Yang TH (1980) Alkaloids of gianguiqianjinteng (*Stephania dielsiana*). Chin Trad Herb Drugs 11:433–434
77. Min ZD (1983) Study on alkaloids of Qian Gui Qian Jin Teng (*Stephania dielsiana* C.Y. Wu). Chin Trad Herb Drugs 14:57–58
78. Wang XF, Wei RF (1983) Study on alkaloids on Ma Shan De Bu Rong (*Stephania mashanica*). Chin Trad Herb Drugs 14:249–251
79. Huang JX, Chen Y (1979) Studies of the alkaloids of *Stephania* species. I. Isolation and identification of alkaloids from *Stephania epigeae*. Acta Pharm Sin 14:612–616
80. Chen Y, Pan YP, Fang SD (1987) The alkaloids of *Stephania*. V. Nonphenolic alkaloids. Chin Trad Herb Drugs 18:438–440
81. Fang SD, Wang HN, Chen Y, Hu CP (1981) Studies on the alkaloids in genus *Stephania*. II. Alkaloids in *Stephania viridiflavens*. Chin Trad Herb Drugs 12:1–3
82. Lao AN, Tang ZJ, Xu RS (1981) Studies on the chemical constituents of *Stephania longa* L. (Menispermaceae). II. Acta Pharm Sin 16:940–942
83. Lao AN, Gao YL, Tang ZJ, Wang YH, Zhang XX, Wang CG, Xu RS (1982) Studies on the alkaloids of *Stephania longa* L. Acta Chim Sin 40:1038–1043
84. Min ZD, Lin XS, Sun WJ (1981) Studies on the alkaloids from *Stephania micrantha* H.S. Lo et M Yang. Acta Pharm Sin 16:557–559
85. Chan Y, Kang QS, Song GQ, Hu ZP, Huang JX (1982) Studies on the alkaloids of *Stephania* species. III. Isolation and identification of alkaloids from *Stephania brachyandra*. Chin Trad Herb Drugs 13:1–5
86. Min ZD, Zhong SM (1980) Studies on the alkaloids of *Stephania kwangsiensis* H.S. Lo. Acta Pharm Sin 15:532–537
87. Cheng KJ, Wang KC, Wang YH (1981) Studies on alkaloids in *Stephania kwangsiensis*. Nonphenolic basic fraction. Chin Pharm Bull 16:49–50
88. Cheng GR, Wang KC, Wen YH (1981) Studies on alkaloids of *Stephania kwangsiensis*. Chin Trad Herb Drugs 12:6–8
89. Lin CS (1977) Chemistry of the Constituents of Chinese Medicinal Herbs. Science, Beijing, p 746
90. Chen LF, Gao JZ, Wang FC, Yang CR (1985) Analgesic, sedative and antispastic effects of *l*-stepholidine. Acta Pharmacol Sin 6:156–158
91. Xiong ZL, Sun Z, Jin GZ, Chen Y (1987) Influence of *l*-stepholidine on blood pressure and its relation to α-adrenoceptors. Acta Pharmacol Sin 8:497–501

Swertia mileensis T. N. Ho et W. L. Shih 117

117.1 Introduction

Qingyedan, Herba Swertiae mileensis, is the dry whole herbs of *Swertia mileensis* T.N. Ho et W.L. Shih (Gentianaceae), which is collected in the fall during its flowering and fruiting season. It is officially listed in the Chinese Pharmacopoeia and is used in traditional Chinese medicine as a choleretic and diuretic agent.

A galenic preparation, Qingyedan Pian, Tabelle Swertiae mileensis, produced from the extract of the herb, is also included in the Chinese Pharmacopoeia and is used for the same indications as the herb.

117.2 Chemical Constituents

117.2.1 Chemical Constituents of *S. mileensis*

The bitter principles sweroside (*117-1*) [1], swertiamarin (*117-2*), 2′-*O*-acetylswertiamarin (*117-3*) [2], and a number of xanthones [3, 4] were isolated and identified from *S. mileensis*. Sweroside is a secoiridoid glycoside first isolated from *S. japonica* [5]. Xanthones isolated from the herb of *S. mileensis* were 1,8-dihydroxy-3,5-dimethoxyxanthone, named swerchirin (*117-4*), 1,8-dihydroxy-3,7-dimethoxyxanthone, 1-hydroxy-3,7,8-trimethoxyxanthone [3], and 1-hydroxy-2,3,4,5-tetramethoxyxanthone [4]. The secoiridoid glycoside and xanthone derivatives are characteristic constituents of gentianaceous plants of the genera *Gentiana* and *Swertia*.

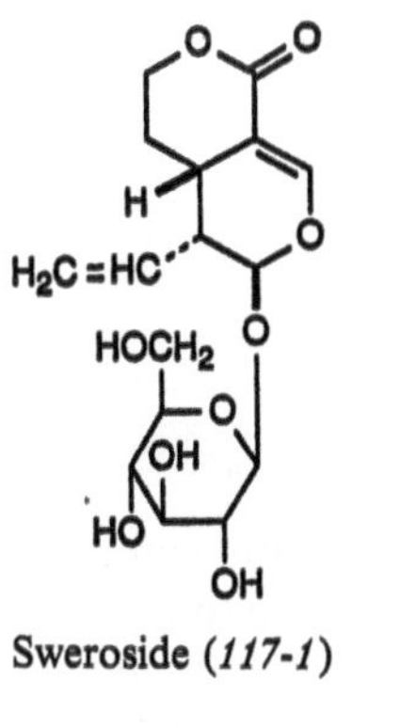

Sweroside (*117-1*)

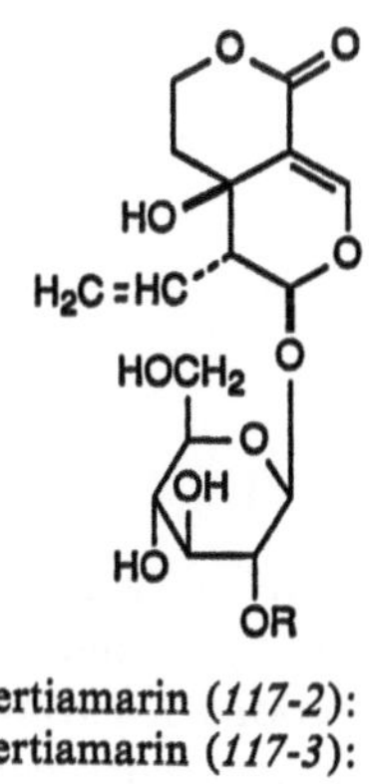

Swertiamarin (*117-2*): R=H
2′-*O*-Acetylswertiamarin (*117-3*): R=$COCH_3$

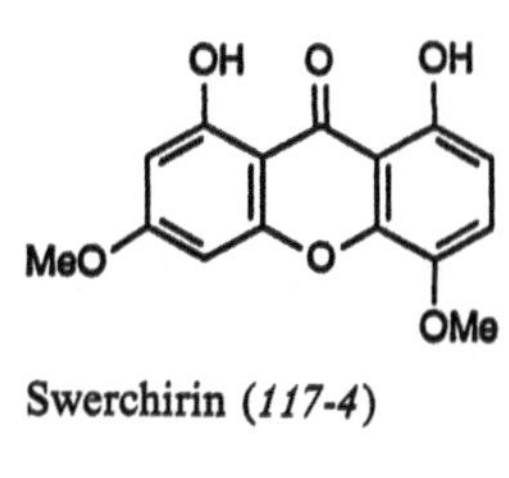

Swerchirin (*117-4*)

In addition, erythrocentaurin (*117-5*) and a new lactone named swermirin (*117-6*) were isolated from *S. mileensis* [6].

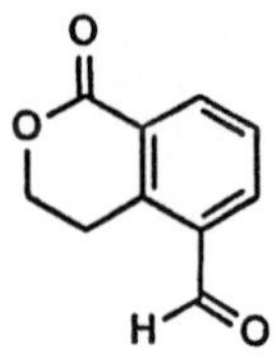

Erythrocentaurin (*117-5*)

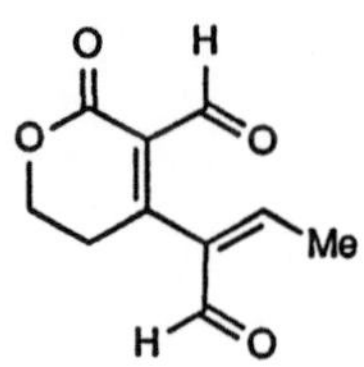

Swermirin (*117-6*)

117.2.2 Chemical Constituents of Other *Swertia* Species with Medicinal Use Occurring in China

Swertia cincta is a medicinal herb used in the treatment of infectious hepatitis. The main components of the ethanolic extract of *S. cincta* were found to be a new triterpene saponin, named swericinctoside (*117-7*), and oleanolic acid. The structure of swericinctoside was elucidated to be maslinic acid 28β-D-glucopyranosyl-(1→6)-β-D-glucopyranosyl-(1→2)-β-D-glucopyranosyl ester [7]. In addition, maslinic acid, 1,3,7,8-tetrahydroxyxanthone, and β-sitosterol were found. Swericinctoside exhibited antiinflammatory activity in pharmacological experiments [7].

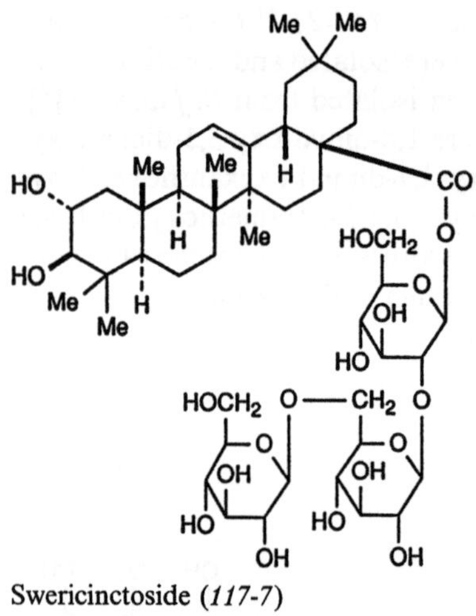

Swericinctoside (*117-7*)

Swertia devidi is a medicinal herb used for the treatment of hepatitis and enteritis. The isolation of gentianine [8], ursolic acid, and 1,5,8-trihydroxy-3-methoxyxanthone [9] from the whole plant was reported. From the medicinal plant *S. patens* the bitter principles swertiamarin [10, 11] and oleanolic acid [11], 1,8-dihydroxy-3,5-dimethoxyxanthone, 1,8-dihydroxy-3,7-dimethoxyxanthone, 1-hydroxy-3,5-dimethoxyxanthone, and 1-hydroxy-3,7,8-trimethoxyanthone [12] were isolated and identified. In addition, the isolation of ursolic acid, oleanolic acid, 1,5,8-trihydroxy-

3-methoxyxanthone (bellidifolin), and norswertianolin (*117-8*) from *S. randaiensis* [13] and oleanolic acid from *S. mussotii* [14] was reported.

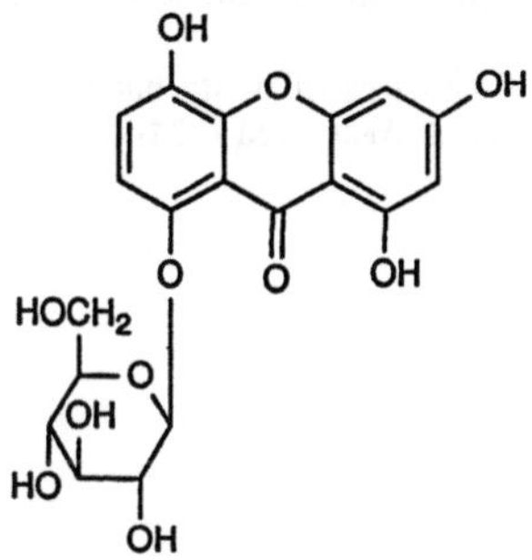

Norswertianolin (*117-8*)

117.3 Pharmacology

Swertiamarin is a spasmolytic agent that has been reported rapidly to alleviate acetylcholine-induced intestinal spasm in rabbits at an oral dosage of 60–100 mg/kg [10, 11]. Results from a pharmacological evaluation of norswertianolin in rats showed that it has depressant activity on the CNS [15]. Clinical effectivity of the extract of *S. devidi* against acute bacillary dysentery was also reported [16].

References

1. He RY, Nie RL (1980) Studies on bitter principles from *Swertia mileensis* T.N. Ho et W.L. Shih. Acta Bot Yunnan 2:480–482
2. Kikuzaki H, Kitamura S, Nakatani N (1988) Structure of iridoids from *Swertia mileensis* He et Shi. Chem Express 3:751–754
3. He RY, Feng SJ, Nie RL (1982) Isolation and identification of xanthones from *Swertia mileensis*. Acta Bot Yunnan 4:68
4. Liu JS, Huang MF (1982) Isolation and identification of xanthones from Qing Ye Dan (*Swertia mileensis*). Chin Trad Herb Drugs 13:433–434
5. Inouye H, Uedo S, Nakamura Y (1966) Struktur des Swerosides, eines neuen Glycosides aus *Swertia japonica* Makino. Tetrahedron Lett 5229–5234
6. Nie RL, He RY (1984) Structures of erythrocentaurin and swermirin from *Swertia mileensis*. Acta Bot Yunnan 6:325–328
7. Zhang JW, Mao Q (1984) Studies on the chemical constituents of *Swertia cincta* Burkill. Acta Pharm Sin 19:819–824
8. Guo XF, Chen CP (1980) Isolation and identification of gentianine from *Swertia devidi* Franch. Chin Trad Herb Drugs 11:200
9. Yu RS (1984) Studies on the constituents of *Swertia devidi* Franch. Acta Bot Sin 26:675–676
10. Liang JZ, Han DJ, Li H, Yuan XP (1982) Isolation and identification of swertiamarin, the active principle in *Swertia patens* Burk. Chin Pharm Bull 17:242–243
11. Liang JZ, Han DJ, Li H (1984) Studies on the active constituents of *Swertia patens*. Bull Chin Mater Med 9:226–228
12. He RY, Feng SJ, Nie RL (1984) Isolation and identification of xanthones from *Swertia patens*. Acta Bot Yunnan 6:341–343
13. Zong YY, Xu CY, Hu YJ (1986) Determination of oleanolic acid in tables of *Swertia mussotii* by TLC-densitometry. Chin J Pharm Anal 6:19–21

14. Liang JZ, Han DJ, Li H, Yuan XB (1982) Isolation and identification of swertiamarin from Jin Sha Qing Ye Dan (*Swertia patens* Burk.). Chin Trad Herbal Drugs 13:7–8
15. Chung MI, Gau KH, Lin CN, Chen IJ (1986) Studies on the constituents of formosan gentianaceous plants. VII. Constituents of *Swertia randaiensis* Hayata and pharmacological activity of norswertianolin. Kaoshiong I Hsueh Ko Hsueh Tsa Chih 2:131–135
16. Tian HY, Zhang XM, Zhou JQ, Wang JW (1986) Treatment of 75 cases of acute bacillary dysentery with *Swertia devidi* and trimethoprim. Chin J Integrat Trad West Med 6:34–35

Trichosanthes kirilowii Maxim. 118

118.1 Introduction

Trichosanthes kirilowii Maxim. (Cucurbitaceae) and several other *Trichosanthes* species are used in traditional Chinese medicine. The following items concerning *Trichosanthes* species are officially listed in the Chinese Pharmacopoeia:

- Tianhuafen, Radix Trichosanthis, is the dry roots of *T. kirilowii* or *T. japonica* Regel collected in the fall and winter. It is used in traditional Chinese medicine as an antiphlogistic. It was also found to be effective as an abortifacient.
- Gualou, Fructus Trichosanthis, is the dry ripe fruits of *T. kirilowii* or *T. rosthornii* Harms. The fruits are harvested in the fall when they have become ripe and are to be used as an antipyretic and expectorant, and in the treatment of constipation.
- Gualouzi, Semen Trichosanthis, is the dry ripe seeds of *T. kirilowii* or *T. rosthornii* collected in the fall. They are used as an expectorant and in the treatment of constipation.
- Gualoupi, Pericarpium Trichosanthis, is the dry pericarp of the ripe fruits of *T. kirilowii* or *T. rosthornii*, used as an expectorant.

118.2 Chemical Constituents

Several years ago, it was found that the active principle of the root of *T. kirilowii* which induced abortion and terminated early pregnancy is a basic protein named trichosanthin [1, 2]. The crude protein was obtained from the freshly pressed juice of the root tubers by successive precipitation with acetone at low temperature. Further purification allowed crystallization. Crystalline trichosanthin obtained by recrystallization from barbiturate buffer was established as homogeneous. It was biologically active, including being able to induce abortion in experimental animals [3]. The molecular weight of trichosanthin was estimated to be about 25 000. Chemical analysis indicated that crystallized trichosanthin did not contain carbohydrate residue or phosphorus. Thus, trichosanthin is a relatively simple protein with an isoelectric point at pH 9.4 [4].

Determination of the amino acid sequence gave the primary structure of trichosanthin [1, 5] (Fig. 118.1). It is a linear polypeptide with a C-terminus of Asn-Asn-Met and Asn-Asn-Met-Ala in a ratio of 7:3 [6, 7].

The secondary structure of trichosanthin was studied by circular dichroism spectral analysis at 185–330 nm [8], and was estimated by statistical analysis based on the confirmational preferences of amino acids [9]. X-ray diffraction analysis and laser Raman spectroscopy of trichosanthin revealed 8 segments of α-helix with about 85 amino acids (43.5%), 13 strands of β-sheet structure with about 70 amino acids

```
                                        10
H-Asp-Val-Ser-Phe-Arg-Leu-Ser-Gly-Ala-Thr-Ser-Ser-Ser-Tyr-Gly-Val-Phe-

           20                                          30
 Ile-Ser-Asn-Leu-Arg-Lys-Ala-Leu-Pro-Asn-Glu-Arg-Lys-Leu-Tyr-Asp-Leu-

                       40                                          50
 Pro-Leu-Ile-Arg-Ser-Ser-Leu-Pro-Gly-Ser-Gln-Arg-Tyr-Ala-Ile-Ile-His-

                                   60
 Leu-Thr-Asn-Tyr-Ala-Asp-Glu-Val-Ala-Leu-Asp-Val-Thr-Asn-Val-Asp-Ala-

       70                                      80
 Gly-Leu-Pro-Arg-Asn-Ala-Val-Leu-Tyr-Ile-Met-Gly-Tyr-Arg-Ala-Gly-Asp-

                   90                                          100
 Thr-Ser-Tyr-Phe-Phe-Asn-Glu-Ala-Ser-ala-Thr-Glu-Ala-Ala-Lys-Tyr-Val-

                               110
 Phe-Lys-Asp-Ala-Met-Arg-Lys-Val-Thr-Leu-Pro-Tyr-Ser-Gly-Asn-Tyr-Glu-

 120                                       130
 Arg-Leu-Gln-Thr-Ala-Ala-Gly-Gly-Leu-Arg-Glu-Asn-Ile-Pro-Leu-Gly-Leu-

           140                                         150
 Pro-Ala-Leu-Asp-Ser-Ala-Ile-Thr-Thr-Leu-Phe-Tyr-Tyr-Asn-Ala-Asn-Ser-

                       160                                         170
 Ala-Ala-Ser-Ala-Leu-Met-Val-Leu-Ile-Gln-Ser-Thr-Ser-Glu-Ala-Ala-Arg-

                                   180
 Tyr-Lys-Phe-Ile-Glu-Gln-Gln-Ile-Gly-Ser-Arg-Val-Asp-Lys-Thr-Phe-Leu-

       190                                     200
 Pro-Ser-Leu-Ala-Ile-Ile-Ser-Leu-Glu-Asn-Ser-Leu-Trp-Leu-Ala-Leu-Ser-

                   210                                         220
 Lys-Gln-Ile-Gln-Ile-Ala-Ser-Thr-Asn-Asn-Gly-Thr-Phe-Glu-Ser-Pro-Val-

                               230
 Val-Leu-Ile-Asn-Ala-Gln-Asn-Gln-Arg-Asn-Asn-Met-(Ala-)OH
```

Fig. 118.1. Primary structure of trichosanthin

(31.3%), as well as some extended chains (25.2%) [10–12]. The α-helices are in the center of the molecule and are surrounded by β-sheets (Fig. 118.2) [13].

Alignment of the complete sequences of trichosanthin with the ricin D subunit A revealed a striking homology [14]. This homology and the presence of identical pentapeptides in both proteins suggest that the sequence homologies are of biological significance. Both proteins are ribosome inactivators and may have a common evolution. Another crystal form of trichosanthin was obtained on crystallization at pH 5.4 with 14% KCl [15].

Besides trichosanthin, another peptide showing trypsin inhibition was isolated from the roots of *T. kirilowii*. This peptide contains 41 amino acids with 3 pairs of disulfide bonds. Two active domains were located at two disulfide loops, composed of eight (positions 17–24) and nine (positions 29–37) amino acid residues, respectively. The protein was found to inhibit two molecules of trypsin and may be regarded as the smallest double-headed trypsin inhibitor so far known, with a molecular

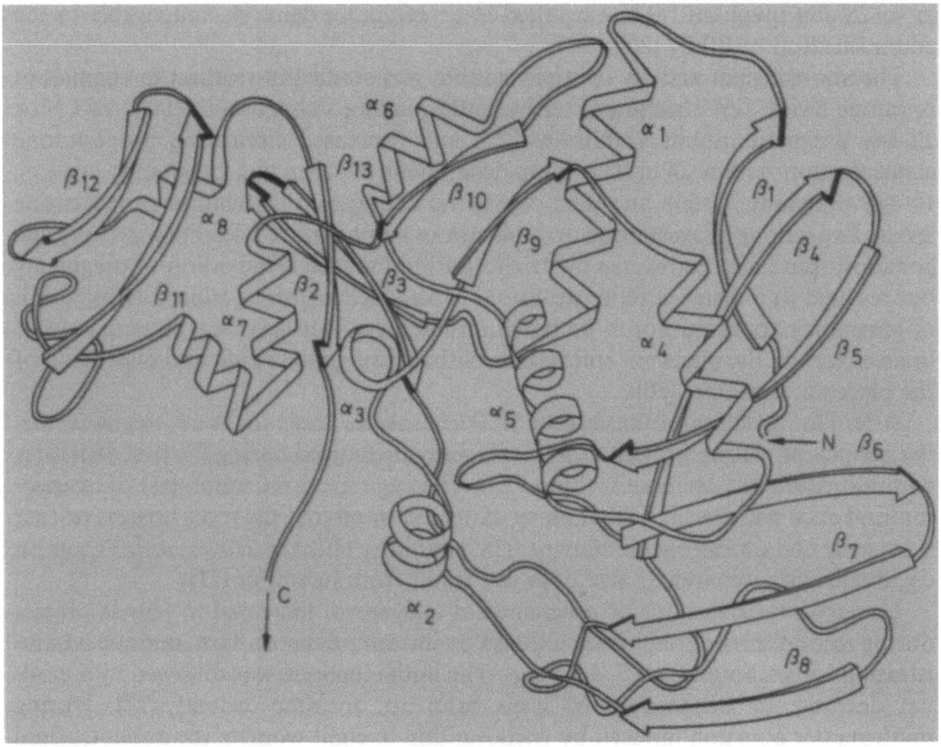

Fig. 118.2. The course of trichosanthin polypeptide chain

weight of 4575. Modification of the inhibitor with cyclohexanedione and citraconic anhydride showed that Arg^{20}-Gly^{21} and Lys^{30}-Leu^{31} corresponded to the two reactive sites [16].

In addition to the protein constituents, a polysaccharide with a molecular weight of 5200 was also isolated from the fresh juice of the tuber of *T. kirilowii.* Arabinose and galactose were identified as component sugars in a molar ratio of 1:0.06 [17].

118.3 Pharmacology

In experimental studies on mice, rats, hamsters, and rabbits, a single i.p. dose of trichosanthin induced mid-term abortion in mice and rabbits. The i.p. dose effective for induction of abortion in 10- or 11-day pregnant mouse was 50 µg. In rabbits, a response dependent on dose and state of pregnancy was observed. A dose of 0.5 mg trichosanthin was effective in 22-day pregnant rabbits, but 2.0 mg was needed in 17-day pregnant rabbits [18]. Attempts to induce mid-term abortion in rats and hamsters and in the termination of early pregnancy in studies of all four animal species including rats, hamsters, mice, and rabbits, were not successful, however, even at larger doses [18]. In contrast to the ineffectivity, it was reported that the extract of roots of *T. kirilowii* was active in inducing abortion after s.c. injection in 4-day pregnant mice [19]. Administration of trichosanthin to 19-day pregnant mice or to 28-day pregnant rabbits resulted in a premature delivery comparable to the administration of PGF_2 [20].

The abortifacient activity of trichosanthin was studied in relation to changes in hormone levels. I.p. injection of trichosanthin induced abortion in 100% of 17- or 22-day pregnant rabbits within 48–72 h and decreased circulating progesterone concentration within 24 h. The same dose failed to terminate pregnancy in 6- or 10-day pregnant rabbits and also caused no changes in circulation progesterone levels. Exogenous progesterone or prolactin in combination with human chorionic gonadotropin failed to reverse the trichosanthin-induced termination of pregnancy but resulted in a delay of fetal expulsion. Thus, trichosanthin-induced termination of pregnancy appeared not to be based on luteolysis but may be a consequence of toxic effects on the placenta, embryo, or both, causing fetal death and dislodging of the placenta in rabbits [20].

After i.m. injection of the extract of *Trichosanthes* roots to 7-day pregnant rats, the peripheral plasma progesterone levels were unchanged during the first 48 h after administration but decreased within 72 h. The vaginal cornification test of ovariectomized mice and uterine weight assay of mice showed that the roots possess neither estrogenic nor antiestrogenic activity. The inhibitory effect of *Trichosanthes* roots on decidualization apparently also does not result from luteolysis [21].

Urinary concentrations of pregnanediol and estriol increased in female objects during second trimester abortion induced by intramuscular or intraamniotic administrations of trichosanthin (5–12.5 mg). The initial increase was followed by a gradual decrease to extremely low levels prior to or after delivery [22]. During midtrimester abortion induced by trichosanthin in eight women, the amniotic fluid levels of prostaglandins PGE and $PGF_{2\alpha}$ increased from 175 to 2239 pg/ml and from 220 to 5220 pg/ml, respectively, whereas maternal plasma levels were not altered. This suggested that prostaglandins may be involved or may even mediate trichosanthin-induced abortion in humans [23].

Trichosanthin given i.m. to pregnant rats decreased plasma progesterone concentrations as well as the level of cytosolic progesterone receptor in uteri. Concomitantly, uterine $PGF_{2\alpha}$ concentration was enhanced. The response of uterine smooth muscle of pregnant rats to 15-methyl-$PGF_{2\alpha}$ and oxytocin was enhanced [24].

It was confirmed that i.p. administration of 1 or 2 mg trichosanthin to 10-day pregnant rabbits did not alter serum progesterone levels and did not interrupt pregnancy. However, in animals treated on day 10 of pregnancy with a combination of a noneffective dose of trichosanthin (1 mg) and a subeffective dose of $PGF_{2\alpha}$ (0.25 mg) pregnancy was terminated [25].

Trichosanthin has been utilized to induce passive lung anaphylaxis in rats. This has been proposed as a useful and simple experimental model for testing antiasthmatic substances. Aminophylline prevented pulmonary secretion and edema in rats immunized with trichosanthin [26].

In clinical studies trichosanthin was used to terminate mid-term pregnancy and also for treatment of dead fetus in utero, hydatidiform mole, and ectopic pregnancy. Success rates of usually greater than 90% have been reported [26]. Before treatment a preliminary intradermal hypersensitivity test of trichosanthin was carried out.

In the case of a negative skin test, 0.05 mg trichosanthin was given i.m. 20 min later. When there were no significant changes in blood pressure, pulse rate, and other signs or symptoms within 2 h of observation, a therapeutic dose of trichosanthin was administered by intraamniotic injection to terminate mid-term pregnancy [27]. Administration of trichosanthin induced an extensive coagulative necrosis of the trophoblastic tissues of the placental villi [27].

The use of trichosanthin either intraamniotically or i.m. was reported to be very effective in the induction of expulsion of trophoblastic tissues without the necessity for operative intervention and with considerably less blood loss especially in the case of hydatidiform mole; usually repeated curettage caused extensive uterine bleeding [27]. Anaphylactic reactions were reportedly avoided by the use of dexamethasone [27]. However, no significant differences in serum IgE levels were observed between women after trichosanthin treatment compared with untreated controls [28]. In another study on the termination of early pregnancy, trichosanthin was used together with testosterone propionate and reserpine, which resulted in abortion rates of more than 90% with an average induction time of 5–7 days. No changes in hematological and urinary parameters were apparent [29]. Production of mouse IgE monoclonal antibodies against trichosanthin and their use in the study of antigenic determinants has been reported [30–32].

References

1. Wang Y (1985) Chemistry of trichosanthin, a new biologically active plant protein. In: Chang HW, Yeung HW, Tso WW, Koo A (eds) Advances in Chinese Medicinal Research. World Science, Singapore, pp 289–293
2. Wang YH, Ling JF, Zhu LX (1976) Preliminary studies on an abortifacient plant protein, trichosanthin. Acta Zool Sin 22:137–143
3. Shanghai Institute of Organic Chemistry (1980) Chemistry of trichosanthin. In: Shen ZW (ed) Nucleic Acid Proteins. Science, Beijing, pp 318–323
4. Jin SW, Sun XX, Wang SF, Tian GY, Gu ZW, Qian WW, Liu YZ, She WY, Qian RQ, Wang Y (1981) Chemistry of trichosanthin. I. Physical and chemical properties of crystalline trichosanthin. Acta Chim Sin 39:917–925
5. Wang Y, Qian RQ, Gu ZW, Jin SW, Zhang LQ, Xia ZX, Tian GY, Ni CZ (1986) Scientific evaluation of Tian Hua Fen (THF) – history, chemistry and application. Pure Appl Chem 58:789–798
6. Gu ZW, Zhang XL, Zhu AQ, Zhang WQ, Fu YH, Weng QX, Wu YW, Liu YF, et al. (1981) Chemistry of trichosanthin II. Determination of N-terminal partial amino acid sequence of trichosanthin. Acta Chim Sin 39:927–931
7. Gu ZW, Qian RQ, Jin SW, Qian WW, Xu SZ, Zhang LQ, Zhang XL, Yao YZ, Liu YF, et al. (1984) Chemistry of trichosanthin. IV. The principal primary structure of trichosanthin. Acta Chim Sin 42:943–945
8. Zhuang PQ, Gao XC, Zhang YM (1983) Secondary structure studies of trichosanthin by circular dichroism. Jiegou Huaxue 2:47–52
9. Zhuang PQ, Gao XC, Zhang YM (1983) Secondary structure estimation of trichosanthin by statistical analysis. Jiegou Huaxue 2:53–56
10. Wu TW, Pang KC, Wu CC, Wu HT, Chang YM, Ni CC, Chu SC (1978) Growth of single crystals and determination of unit-cell parameters for trichosanthin. Kexue Tongbao 23:176–178
11. Pan KZ, Zhang YM, Lin YZ, Wu CW, Zheng A, Chen YZ, Dong YC, Chen SZ, Wu S, Ma XQ, Wang YP, Zhang MA, Xia ZX, Tian GY, Fan ZC, Ni CZ, Ma YL, Sun XX (1985) The secondary structure of trichosanthin. In: Chang HM, Yeung HW, Tso WW, Koo A (eds) Advances in Chinese Medicinal Material Research. World Science, Singapore, pp 297–303
12. Fang YX, Zhu ZY, He DJ, Tian GY (1985) Chemistry of trichosanthin. VI. Determination of the secondary structure of trichosanthin by laser Raman spectroscopy. Acta Chim Sin 43:965–969
13. Pan KZ, Zhang YM, Lin YJ, Wu ZW, Zheng A, Chen YZ, Dong YC, Ma XQ, Wang YP, Wu S, Zhang MA, Chen SZ, Xia ZX, Tian GY, Ni CZ, Fan ZC, Ma YL, Sun XX (1985) The

backbone structure of trichosanthin. In: Chang HM, Yeung HW, Tsa WW, Koo A (eds) Advances in Chinese Medicinal Material Research, World Science, Singapore, pp 305–309
14. Zhang XJ, Wang JH (1986) Homology of trichosanthin and ricin A chain. Nature 321:477–478
15. Wang JH, Wang YP, Tian GY (1985) A new crystal form of trichosanthin. Kexue Tongbao [Foreign Lang] 30:1396–1398
16. Tan FL, Zhang GD, Mu JF, Lin NQ, Chi CW (1984) Purification, characterization and sequence determination of a double-headed trypsin inhibitor peptide from *Trichosanthes kirilowii* (a Chinese medical herb). Hoppe Seylers Z Physiol Chem 365:1211–1217
17. Tian GY, Li ST, Tang TB, Wang DC (1985) *Trichosanthes* polysaccharide. I. Isolation and physical and chemical properties. Acta Biochim Biophys Sin 17:582–586
18. Chang MC, Saksena SK, Lau IF, Wang YH (1979) Induction of mid-term abortion by trichosanthin in laboratory animals. Contraception 19:175–184
19. Lau IF, Sakana SK, Chang MC (1980) Further studies on the trichosanthin-induced termination of pregnancy. Contraception 21:77–86
20. Zhou MH, Li Q, Shu HD, Bao YM, Chu YH (1982) Pharmacological study of the effect of Radix Trichosanthis on terminating early pregnancy. Acta Pharm Sin 17:176–181
21. Saksena SK, Chang MC, Lau IF (1979) Termination of pregnancy in rabbit and mouse by trichosanthin. Contraception 20:367–376
22. Jiang TJ, Gao KX, Zhou GZ, Zhu YP, Cai HP (1977) Changes in urinary pregnanediol and estriol excretions during second trimester abortion induced by trichosanthin – a protein from Radix Trichosanthis. Acta Zool Sin 23:243–254
23. Wang YF, Wen D, Liu JX, Fi C, Zhu WX, Chen YZ, Yu Z, Yan LM, Shen GS (1981) Prostaglandin E and $F_{2\alpha}$ levels in plasma and amniotic fluid during mid-trimester abortion induced by trichosanthin. Prostaglandins 22:289–294
24. Chu YH, Zhao ZF (1985) Effect of trichosanthin on progesterone, progesterone receptor and prostaglandin $F_{2\alpha}$ level in pregnant rat uteri. Acta Pharm Sin 20:262–266
25. Saksena SK, Lau IF (1980) Effects of prostaglandin $F_{2\alpha}$ and a plant protein "trichosanthin", on 10-day pregnant rabbits. Prostaglandins Med 5:383–390
26. Wu RS, Zhang SM, Ma JW (1985) Rat model of passive lung anaphylaxis induced by trichosanthin. Acta Pharmacol Sin 6:68–71
27. Jin YC (1985) Clinical study of trichosanthin. In: Chang HM, Yeung MW, Tso WW, Koo A (eds) Advances in Chinese Medicinal Material Research. World Science, Singapore, pp 319–326
28. Liu FY, Chen SY, Li YJ, Liu GW, Zhou ZR, Lu CL, Zhao ML (1986) Allergic reaction to trichosanthin assays of serum IgE and specific anti-trichosanthin IgE levels. Chin J Obst Gynecol 21:165–167
29. Liu GW, Liu FY, Li YJ, Yu SH (1985) A summary of 402 cases of termination of early pregnancy with crystalline preparations of trichosanthins. In: Chang HM, Yeung HW, Tso WW, Koo A (eds) Advances in Chinese Medicinal Material Research. World Science, Singapore, pp 327–333
30. Gu H, Ye M, Yao Z (1986) Preparation, isolation and characterization of mouse IgE monoclonal antibodies against trichosanthin protein. Acta Biol Exper Sin 19:109–119
31. Gu H, Ye M, Yao Z (1986) Investigation of antigenic determinants on trichosanthin by antibody competitive binding assay. Acta Biol Exper Sin 19:121–129
32. Ji YY, Jiang ZQ, Ye M (1986) The production of monoclonal antiidiotypic antibodies against trichosanthin-specific IgE by rat-mouse hybridomas. Acta Biol Exper Sin 19:91–98

Tripterygium wilfordii Hook 119

119.1 Introduction

Tripterygium wilfordii Hook (Celastraceae) is a traditional Chinese medicine that has been used in the past. It is no longer officially listed in the Chinese Pharmacopoeia and is only occasionally used in folk medicine as an insecticide because of its high toxicity. The drug has attracted some attention again after isolated constituents were found to have antileukemic activity.

119.2 Chemical Constituents

119.2.1 Terpene Lactones

The antileukemic activity of *Maytenus ovatus* and its active principle, maytansine, prompted investigations on other plants of the family Celastraceae. Three antileukemic triepoxidic diterpene lactones from the roots of *T. wilfordii* have been reported, triptolide (*119-1*), tripdiolide (*119-2*), and triptonide (*119-3*). Structures were elucidated by spectral analyses and by X-ray analysis [1]. They appeared to be the first diterpene triepoxides recognized. Roots collected in the fall showed a higher triptolide content (up to 0.02%) [2].

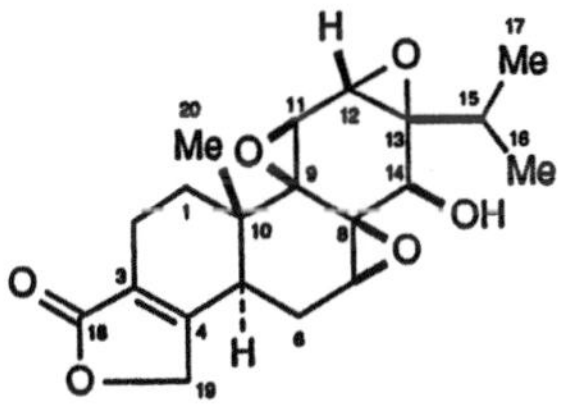

Triptolide (*119-1*)

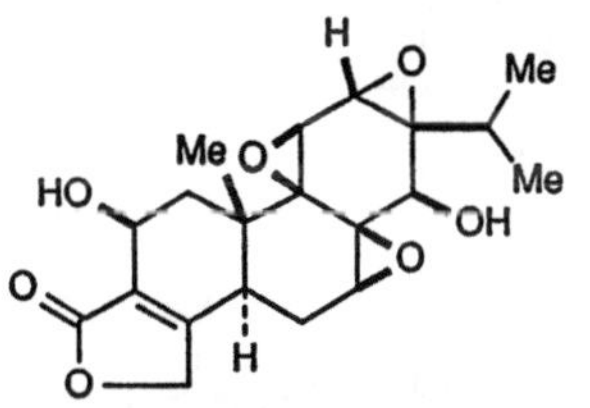

Tripdiolide (*119-2*)

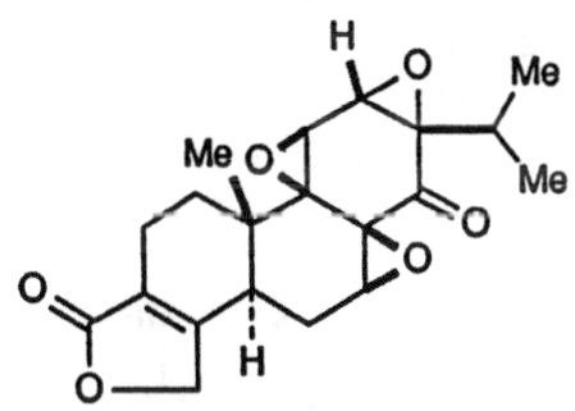

Triptonide (*119-3*)

During further studies on the chemical constituents of *T. wilfordii*, several new diterpene lactones were isolated and structurally determined. Thus, the isolation of triptonolide (*119-4*) [3], triptophenolide (*119-5*) and its methyl ether [4], neotriptophenolide (*119-6*) [5], triptolidenol (*119-7*) [6], and isoneotriptophenolide (*119-8*) [7] was reported.

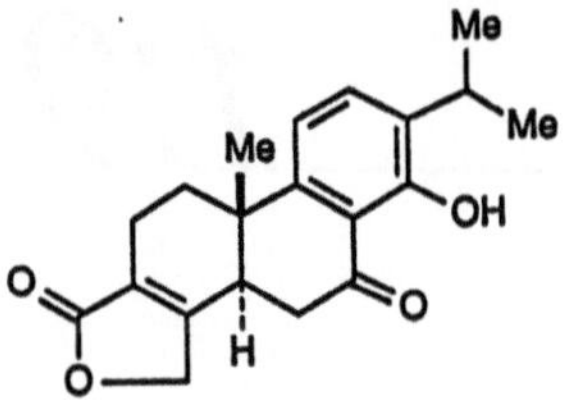

Triptonolide (*119-4*)

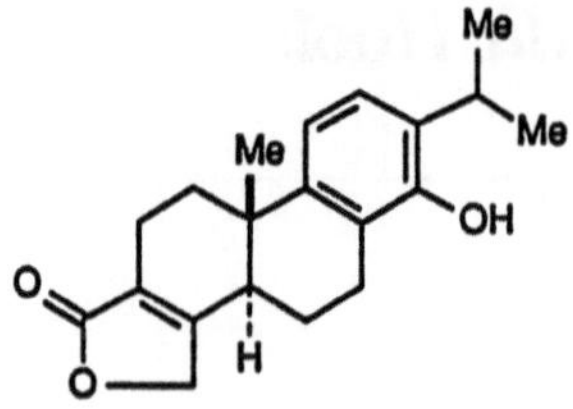

Triptophenolide (*119-5*)

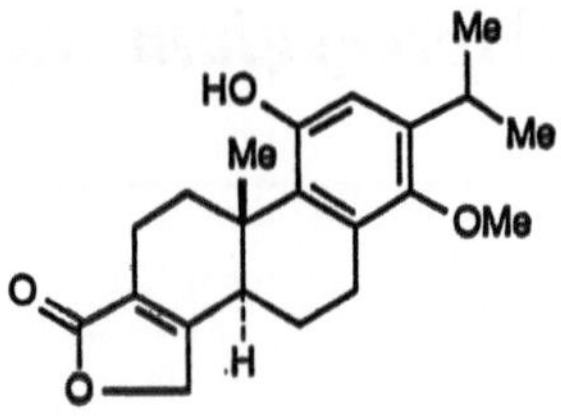

Neotriptophenolide (*119-6*)

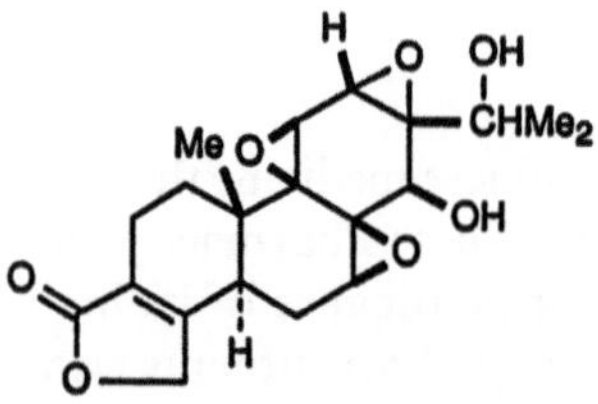

Triptolidenol (*119-7*)

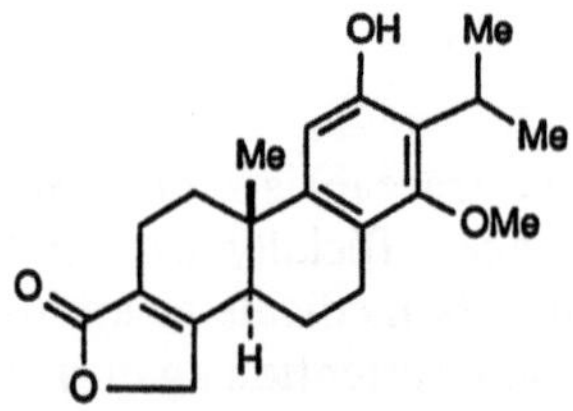

Isoneotriptophenolide (*119-8*)

Except for triptolidenol, these diterpene lactones (*119-4*–*119-6*, *119-8*) do not possess epoxy groups. Triptonoterpene (*119-9*) and its methyl ether [6], triptonoterpenol (*119-10*) [8], neotriptonoterpene (*119-11*), and triptonodiol (*119-12*) [9] are diterpenes without the lactone moiety isolated from *T. wilfordii*. Neotriptonolide (*119-13*) is a C_{14} lactone [9].

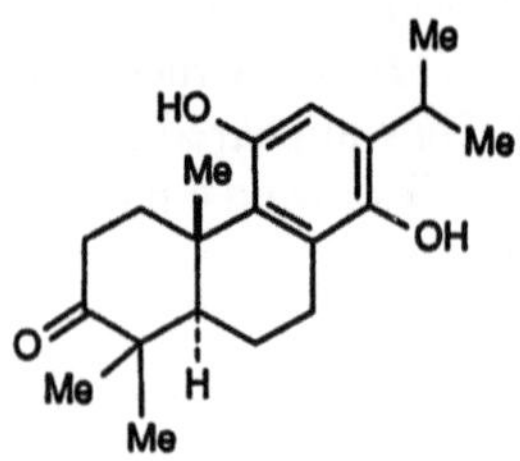

Triptonoterpene (*119-9*)

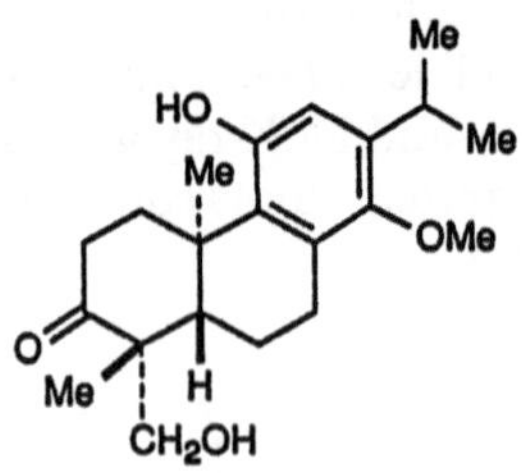

Triptonoterpenol (*119-10*)

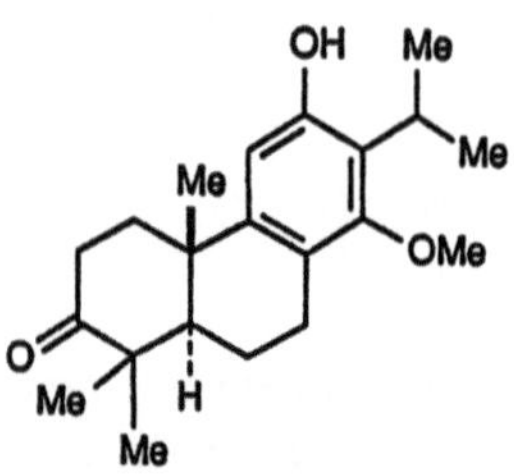

Neotriptonoterpene (*119-11*)

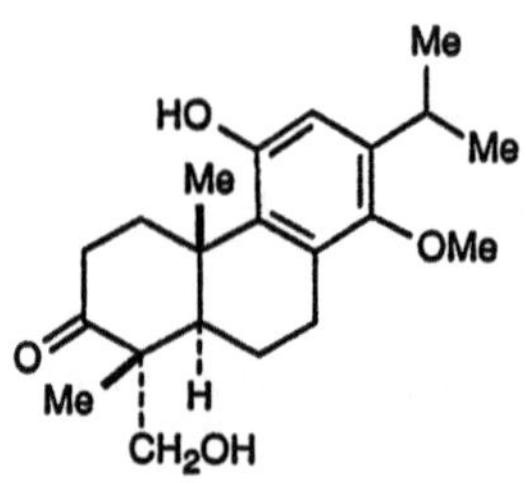

Triptonodiol (*119-12*)

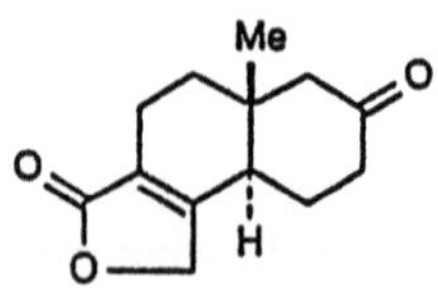

Neotriptonolide (*119-13*)

The isolation of triterpene lactones with an oleanane skeleton from *T. wilfordii*, wilforlide A (*119-14*), and wilforlide B (*119-15*) was further reported [10]. Triptotriterpenic acid A (*119-16*) [11, 12], triptodihydroxy acid methyl ester (*119-17*), and polpunonic acid (*119-18*) [13] are additional triterpenes isolated from *T. wilfordii.*

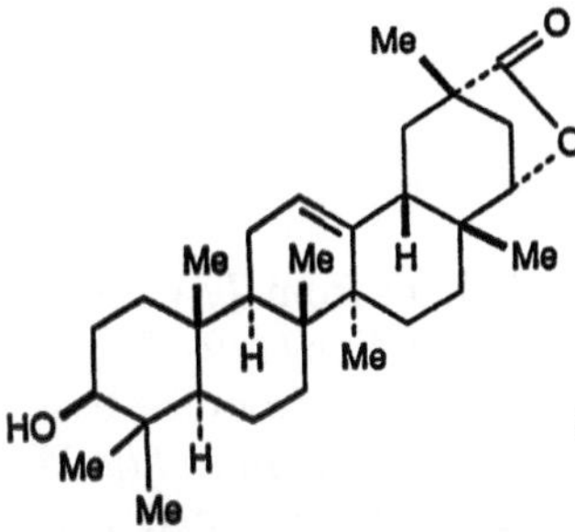

Wilforlide A (*119-14*)

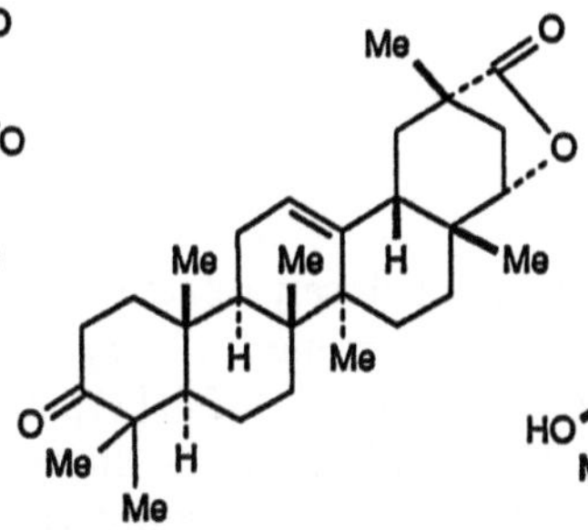

Wilforlide B (*119-15*)

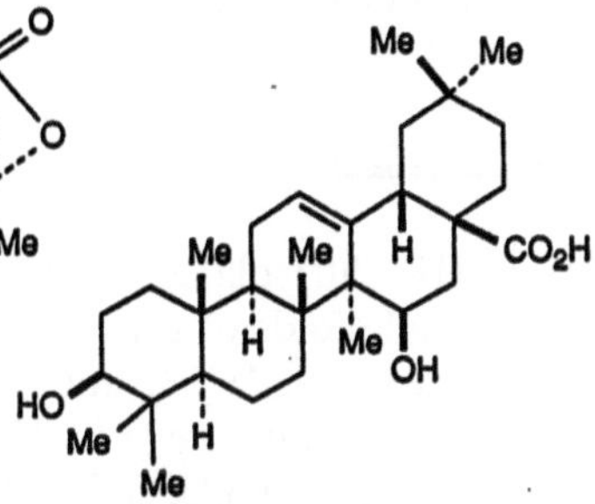

Triptotriterpenic acid A (*119-16*)

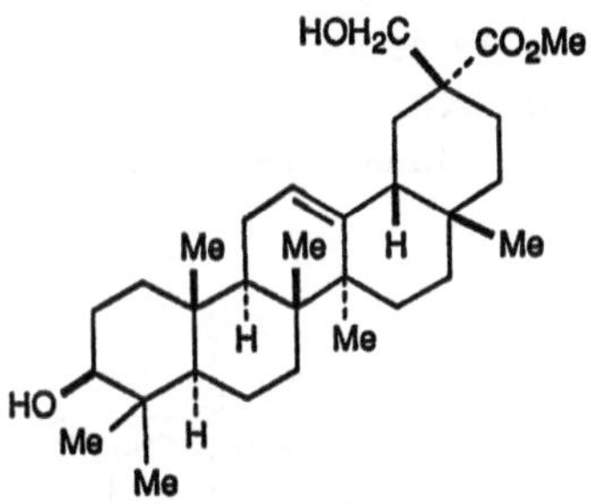

Triptodihydroxy acid methyl ester (*119-17*)

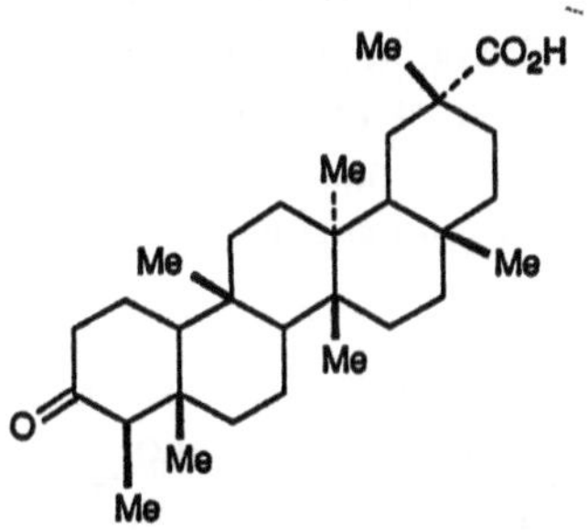

Polpunonic acid (*119-18*)

Cytotoxic diterpene lactones have also been isolated from tissue culture. Triptolide and tripdiolide were obtained from tissue culture in yields 3 and 16 times greater, respectively, than from the plant itself [14]. Other diterpenes and triterpenes isolated from tissue cultures of *T. wilfordii* were dehydroabietic acid (*119-19*), 1,4a-dimethyl-7-(1-hydroxy-1-methyl-ethyl)-hexahydrophenanthrene-carboxylic acid methyl ester, celastrol (*119-20*), and oleanolic acid. The isolation of dehydroabietic acid and of the phenanthrene-carboxylic acid derivative led to the suggestion that an important step in the biosynthetic pathway of triptolide and tripdiolide involves the rearrangement of dehydroabietic acid to the modified abietane skeleton of 1,4a-dimethyl-7-(1-hydroxy-1-methylethyl)-hexahydrophenanthrene-2-carboxylic acid. Subsequent enzymatic oxidations would lead ultimately to triptolide and tripdiolide.

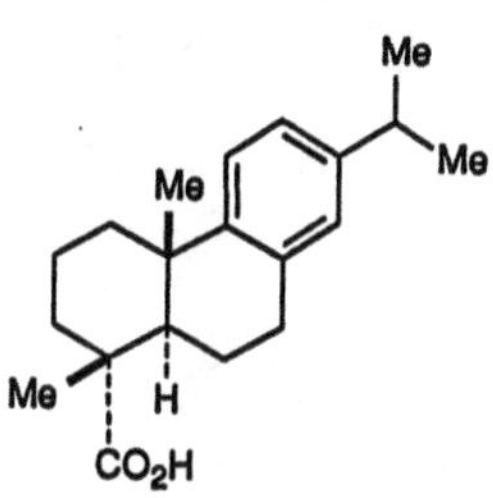

Dehydroabietic acid (*119-19*)

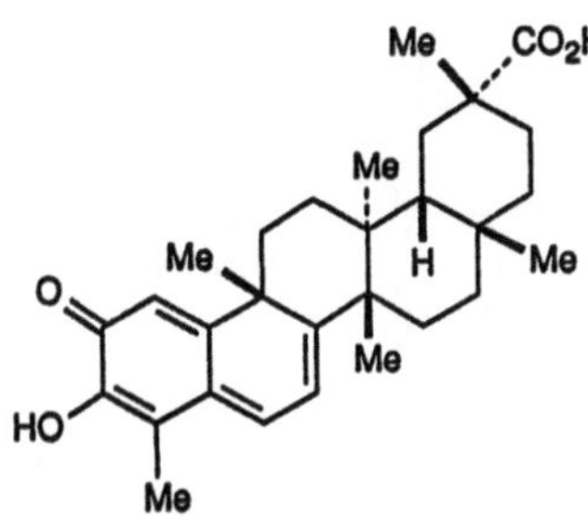

Celastrol (*119-20*)

The isolation of further abietane derivatives from the plant was recently reported [15]. A total synthesis of (±)-triptolide and (±)-triptonide was reported by van Tamelen and Leiden [16, 17].

119.2.2 Alkaloids

Beroza [18–20] isolated a number of closely related ester alkaloids from *T. wilfordii*. The new alkaloids of this series, wilfordine (*119-21*) [21], wilforidine (*119-22*) [22], wilfortrine (*119-23*) [13], and wilfornine (*119-24*) [21], are all derived from the basic structure of evonimine (*119-25*). In addition, the known alkaloid euonine (*119-26*) was also isolated [23, 24]. Wilfordine and wilfortrine were also isolated from *T. hypoglaucum* together with euonymine (*119-27*) [25], an alkaloid related to evonine (*119-28*). From the ester alkaloids, two acids, named wilfordic acid (*119-29*) and hydroxywilfordic acid (*119-30*), were obtained [26]. In addition to the ester alkaloids, a number of macrocyclic spermidine alkaloids were also isolated from *T. wilfordii*. They were named celacinnine (*119-31*), celallocinnine (*119-32*), celafurine (*119-33*), and celabenzine (*119-34*) [27, 28].

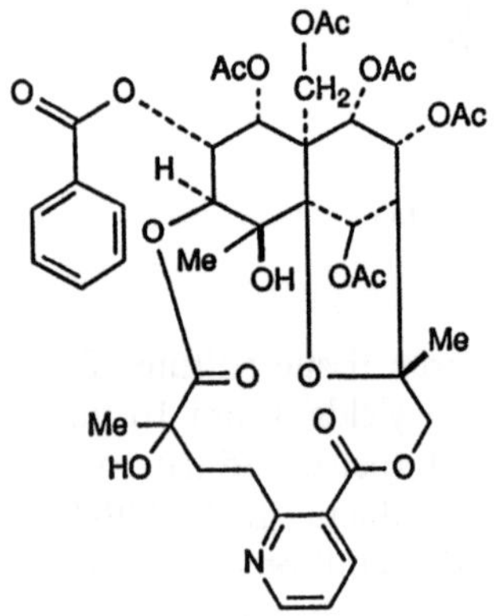

Wilfordine (*119-21*)

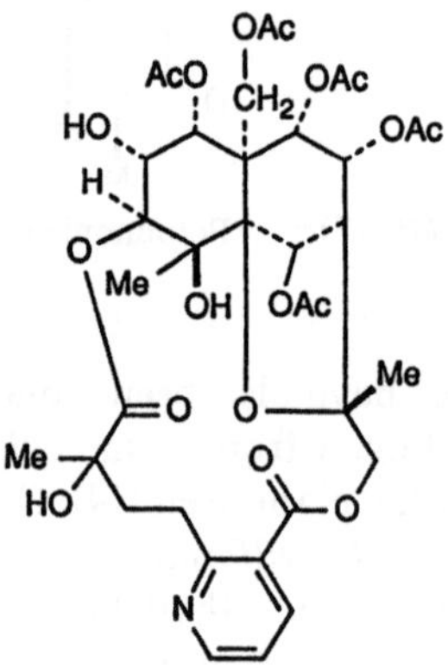

Wilforidine (*119-22*)

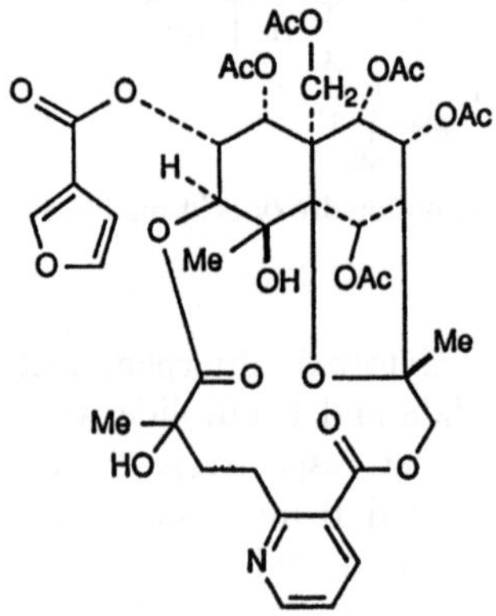

Wilfortrine (*119-23*)

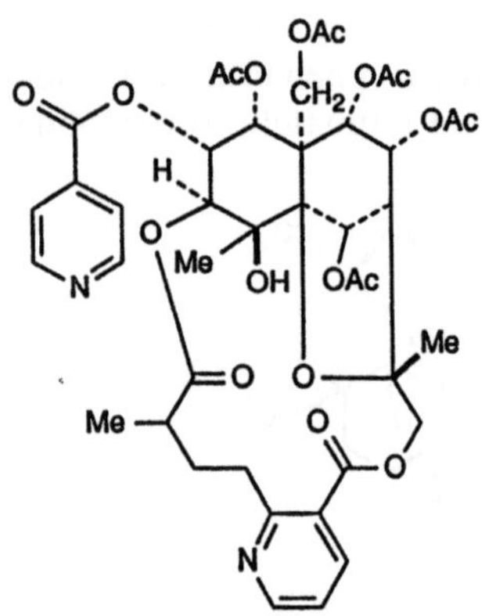

Wilfornine (*119-24*)

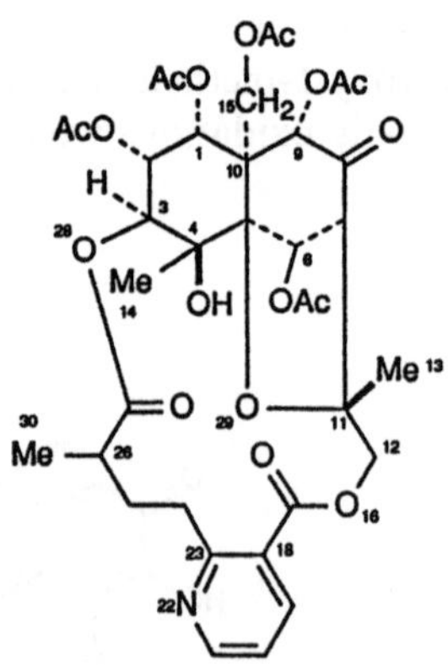

Evonimine (*119-25*)

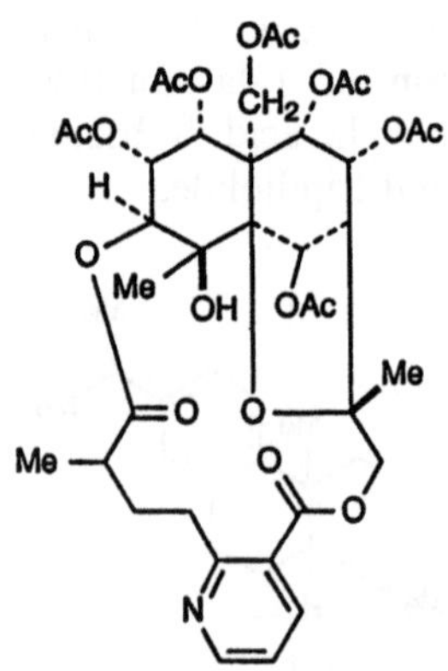

Euonine (*119-26*)

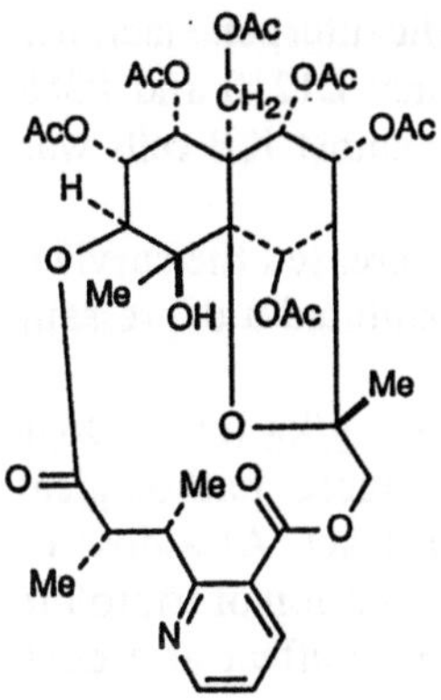

Euonymine (*119-27*)

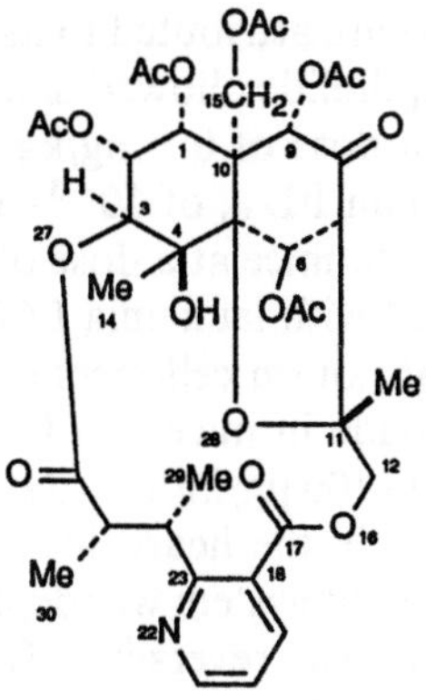

Evonine (*119-28*)

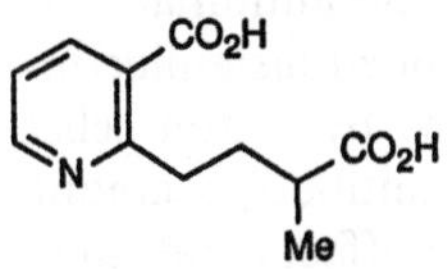

Wilfordic acid (*119-29*)

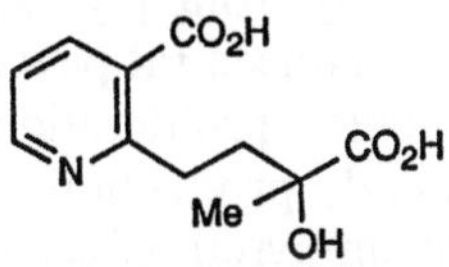

Hydroxywilfordic acid (*119-30*)

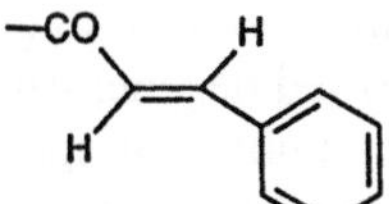

Celacinnine (*119-31*): R =

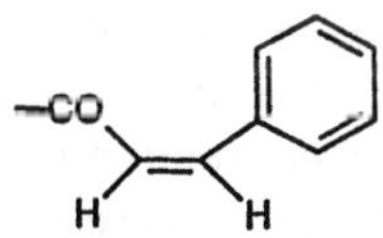

Celallocinnine (*119-32*): R =

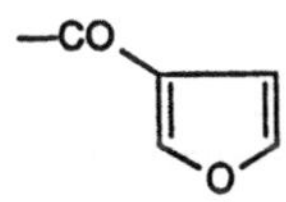

Celafurine (*119-33*): R =

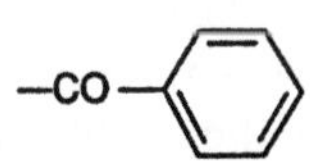

Celabenzine (*119-34*): R =

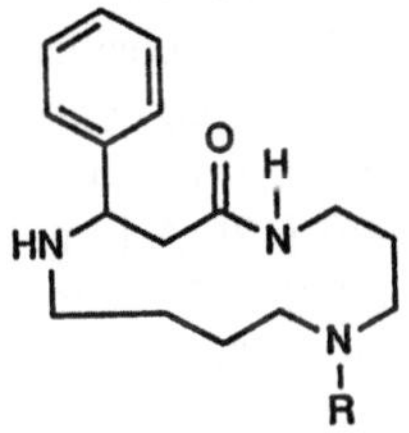

119.2.3 Chemical Constituents of Other Types

Besides the terpene lactones and alkaloids, the lignan syringaresinol was isolated from *T. wilfordii* [29]. It was also reported that *T. wilfordii* contains glycosides. Total glycosides were used for pharmacological studies [30–32].

119.3 Pharmacology

The alcoholic extract of *T. wilfordii* was found to show significant activity in vivo against leukemias L1210 and P388 in mice. In vitro activity against KB cells was also observed. These effects are attributed to its content of triepoxidic diterpene lactones [1]. Triptolide and tripdiolide showed significant activity against L1210 and P388 leukemias in mice at a dose of 0.1 mg/kg. Cytotoxic activity against KB cells was observed in vitro with an ED_{50} of 10^{-3}–10^{-4} μg/ml [1].

Triptolide given i.p. to mice at a dose of 0.2 or 0.25 mg/kg increased the survival time of mice inoculated with leukemia L615. Triptolide also exhibited a depressant effect on humoral, but not on cell-mediated, immunity [33].

The LD_{50} of triptolide in mice was 0.8 mg/kg (i.v.) and 0.9 mg/kg (i.p.). Dogs given triptolide at 20–100 μg/kg/per day i.v. for 7 days produced pathological and functional changes in the heart, liver, and gastrointestinal tract. At a dose of 60 μg/kg the hematopoietic system was depressed. After discontinuation of triptolide administration, toxicity was reversible. Triptolide had no adverse effect at a dose below 20 μg/kg per day [34].

Biological activities of the total glycosides isolated from *T. wilfordii* have also been investigated extensively. Total glycosides exhibited a significant antiinflammatory effect on acute agar-induced rat paw edema. They also inhibited histamine-induced capillary permeability, proliferation of granuloma induced by cotton pellet implantation, and antibody production in antigen-bound and antibody-secreting cells in rats [30]. A slight antipyretic effect and a mild bacteriostatic effect in rats were also noticed. The results suggested that the total glycosides represent the main antiinflammatory constituents of *T. wilfordii* [30].

After oral administration of total glycosides at a dose of 30 mg/kg in the diet for 80 days, male rats showed degenerative changes in the seminiferous tubule and the sperm, with lowered numbers of spermatocytes. The seminiferous tubule became atrophic, and a decreased plasma testosterone level was observed. The glycosides have been found to exhibit antispermatogenic activity similar to that of gossypol [31]. Loss of appetite, loss of body weight, leukopenia, and thrombocytopenia in experimental animals were related to dosage and course of treatment [32]. A large dose of total glycosides also caused thymus atrophy in young mice [30]. The combined effects of gossypol and *T. wilfordii* total glycosides on the fertility of male rats were also reported [35].

Following oral administration of total glycosides at a daily dose of 30 mg/kg for 35–80 days to female rats, the menstrual cycle became irregular in 80%–90% of the animals. The uterus weights were 28%–43% less than those of untreated controls [36]. Compared with the toxic effects of the glycosides on male organs, female reproductive organs are less affected. Fertility in male rats was restored to normal level 5 weeks after cessation of administration. Apparently, the antifertility activity of total glycosides is reversible [37]. Reproductive toxicity of the glycosides was higher in rats than in mice [38].

Studies on the antiinflammatory, immunosuppressive, and antifertility activities of the glycosides suggest that antifertility activity is closely connected with immunosuppressant activities. In contrast, antiinflammatory activity appears to be separable from immunosuppressant activities [39, 40].

Mice gavaged with 1 ml decoction, corresponding to 1 g of the root, died within 8 days. Autopsy revealed narrowing and occlusion of the glomerular capillaries, epithelial granulation, and partial necrosis of the distal tubules [41].

In the clinic, 4 out of 44 patients with psoriasis treated once daily with a decoction prepared from 12–15 g of the root of *T. wilfordii* experienced toxic effects such as dizziness and vomiting from the 3rd to 5th day of medication. In two cases, acute kidney failure, leukopenia, and thrombocytopenia developed [41].

References

1. Kupchan SM, Court WA, Dailey RG Jr, Gilmore CJ, Bryan RF (1972) Tumor inhibitors. LXXIV. Triptolide and tripdiolide, novel antileukemic diterpenoid triepoxides from *Tripterygium wilfordii*. J Am Chem Soc 94:7194–7195
2. Cheng ZZ, Fang H (1986) Quantitative determination of triptolide in common threewingnut (*Tripterygium wilfordii*) in different seasons and parts. Chin Trad Herb Drugs 17:12–13
3. Deng FX, Huang SQ, Wang ZD, Ma GG, Song GQ, Chen ZX (1981) Studies on the chemical constituents of *Tripterygium wilfordii* Hook. II. Structural determination of triptonolide, a new diterpenoid lactone. Acta Pharm Sin 16:155–157
4. Wu ZW, Chen YZ, Deng FX (1986) The crystal structure of triptophenolide methyl ether. Jiegou Huaxue 5:28–32
5. Deng FX, Zhou BN, Son GQ, Hu CQ (1982) Studies on the chemical contituents of *Tripterygium wilfordii* Hook. F. III. The isolation and structure of two new diterpenoid lactones triptophenolide methyl ether and neotriptophenolide. Acta Pharm Sin 17:146–150
6. Deng FX, Huang SQ, Cao JH, Xia ZL, Lin S, Zhu DY, Jiang SH, Zhu YL (1985) The isolation and structure of three new diterpenes from *Tripterygium wilfordii* Hook. F. Acta Bot Sin 27:516–519
7. Chen KC, Yang RZ, Wang CB (1986) The lactones from threewingnut (*Tripterygium wilfordii* Hook. f.) in Hubei. Chin Trad Herb Drugs 17:242–245
8. Deng FX, Cao JH, Xia ZL, Lin S, Zhu DY, Jiang SH (1987) The isolation and structure of triptonoterpenol. Acta Pharm Sin 22:377–379
9. Zhou BN, Zhu DY, Deng FX, Huang CG, Kutney JP, Roberts M (1988) Studies on new components and stereochemistry of diterpenoids from *Tripterygium wilfordii*. Planta Med 54:330–332
10. Qin GW, Yang XM, Ku WH, Wang BD, Chen ZX (1982) The structures of two new triterpene lactones, wilforlides A and B, from *Tripterygium wilfordii*. In: Wang Y (ed) Chemistry of Natural Products: Sino-American Symposium 1980. Science Press, Beijing, pp 177–179
11. Zhang CP, Zhang YG, Lu XY, Chen Y, Ma PC, Li SH, Zhang ZX (1984) A new triterpenoid, triptotriterpenic acid A, isolated from *Tripterygium wilfordii*. J Nanjing Coll Pharm 69
12. Zhang CP, Zhang YG, Lu XY, Chen Y, Ma PC, Yin YP, Xu LF (1986) A pentacyclic triterpene acid from *Tripterygium wilfordii*. Acta Acad Med Sin 8:204–206
13. Deng FX, Cao JH, Xia ZL, Lin L, Wang XY (1987) The structure of triptodihydroxy acid methyl ester and wilfortriene. Acta Bot Sin 29:73–76
14. Kutney JP, Hewitt GM, Kurihara T, Salisbury PJ, Sindelar RD, Stuart KL, Townsley PM, Chalmers WT, Jacoli GG (1981) Cytotoxic diterpenes triptolide, tripdiolide and cytotoxic triterpenes from tissue cultures of *Tripterygium wilfordii*. Can J Chem 59:2677–2683
15. Zhang WJ, Zhang RN, Pan DJ, Zhang LX, Xu GY (1986) Diterpenoids from *Tripterygium wilfordii*. Acta Acad Med Shanghai 13:267–272
16. Van Tamelen EE, Leiden TM (1982) Biogenetic type total synthesis of (±)-triptonide and (±)-triptolide. J Am Chem Soc 104:1785–1786
17. Van Tamelen EE, Demers JP, Taylor EG, Koller K (1980) Total synthesis of l-triptonide and l-triptolide. J Am Chem Soc 102:5424–5425
18. Beroza M (1951) Alkaloids from *Tripterygium wilfordii* Hook. – wilforine and wilfordine. J Am Chem Soc 73:3656–3659
19. Beroza M (1952) Alkaloids from *Tripterygium wilfordii* Hook: wilforgine and wilfortrine. J Am Chem Soc 74:1585–1588
20. Beroza M (1953) Alkaloids from *Tripterygium wilfordii* Hook. Isolation and structure of wilforzine. J Am Chem Soc 75:2136–2139
21. Shizuri Y, Yamada K, Hirata Y (1973) Structures of wilfordine and a new alkaloid alatamine from *Evonymus alatus* f. *striatus*. Tetrahedron Lett 741–744
22. He ZS, Hong SH, Li Y, Sha H, Yu XG (1985) Structure of the new alkaloid wilforidine from *Tripterygium wilfordii*. Acta Chim Sin 43:593–596

23. Deng FX, Cao JH, Xia ZL, Lin S, Wang XY (1987) Studies on the sesquiterpene alkaloids of *Tripterygium wilfordii* Hook. f. Acta Bot Sin 29:523–526
24. He ZS, Li Y, Fang SD, Hong SH (1987) Structures of wilforgine, and wilformine from *Tripterygium wilfordii*. Acta Chim Sin 45:510–513
25. Wu DG (1986) Application of COSY spectra to the identification and complete assignment of protons of the alkaloids from *Tripterygium hypoglaucum*. Acta Bot Yunnan 8:343–354
26. Beroza M (1963) Alkaloids from *Tripterygium wilfordii*. Chemical structure of wilfordic and hydroxywilfordic acids. J Org Chem 28:3562–3564
27. Kupchan SM, Hintz HPJ, Smith RM, Karim A, Cass MW, Court WA, Yatagai M (1974) Celacinnine, a novel macrocyclic spermidine alkaloid prototype. J Chem Soc Commun 329–330
28. Kupchan SM, Hintz HPJ, Smith RM, Karim A, Cass MW, Court WA, Yatagai M (1977) Macrocyclic spermidine alkaloids from *Maytenus serrata* and *Tripterygium wilfordii*. J Org Chem 42:3660–3664
29. Bryan RF, Fallon L (1976) Crystal structure of syringaresinol. J Chem Soc [Perkin Trans] 2:341–345
30. Zheng JR, Xu LF, Ma L, Wang DH, Gao JW (1983) Studies on pharmacological effects of total glycosides of *Tripterygium wilfordii*. Acta Acad Med Sin 5:1–8
31. Zheng JR, Fang JL, Xu LF, Gao JW, Guo HZ, Li ZR, Sun HZ (1985) Effects of total glycosides of *Tripterygium wilfordii* on animal reproductive organs. I. Experiments with male rats. Acta Acad Med Sin 7:1–5
32. Zheng JR, Liu JH, Hsu LF, Gao JW, Jiang BL (1983) Studies on toxicity of total glycosides in *Triterygium wilfordii*. Acta Acad Med Sin 5:73–78
33. Zhang TM, Chen ZY, Lin C (1981) Antineoplastic effect of triptolide and its effect on the immunologic function in mice. Acta Pharmacol Sin 2:128–131
34. Cheng YL, Ye JR, Lin DJ, Lin LJ, Zhu JN (1981) Some toxicities of triptolide in mice and dogs. Acta Pharmacol Sin 2:70–72
35. Xu Y, Tong JX, Qi AP, Zhong CQ, Qian SZ (1987) The effect of combined use of gossypol and *Tripterygium wilfordii* on the fertility of male rats. Acta Pharm Sin 22:818–821
36. Zheng JR, Fang JL, Xu LF, Gao JW, Guo HZ, Li ZR, Sun HZ (1985) Effects of total glycosides of *Tripterygium wilfordii* on the reproductive organs of experimental animals. II. Experiments in female rats. Acta Acad Med Sin 7:256–259
37. Xu Y, Wang SM, Zhong CQ, Qian SZ (1988) Study on the reversibility of infertility caused by polyglycosides of *Tripterygium wilfordii* root. Chin Pharm Bull 23:22–24
38. Zhang JR, Fang JL, Gao JW, Guo HZ, Xu LF, Yeng YP, Yu YH, Sun HZ (1986) Effects of the total glycosides of *Tripterygium wilfordii* on the reproductive organs of experimental animals. III. Dynamic observation on the reproductive organs and fertility in mice. Acta Acad Med Sin 8:19–23
39. Zheng JR, Fang JL, Gu KX, Xu LF, Gao JW, Guo HZ, Yu YH, Sun HZ (1987) Screening of active anti-inflammatory-immunosuppressive and antifertility compounds from *Tripterygium wilfordii*. I. Screening of 8 components from total glycosides of *Tripterygium wilfordii* (T_{II}). Acta Acad Med Sin 9:317–322
40. Zheng JR, Fang JL, Gu KX, Yin YP, Xu LF, Gao JW, Guo HZ, Yu YH, Sun HZ (1987) Screening of active anti-inflammatory-immunosuppressive and antifertility compounds from *Tripterygium wilfordii*. II. Screening of 5 monomers from the total glycosides of *Tripterygium wilfordii* (T_{II}). Acta Acad Med Sin 9:323–328
41. Zhang JZ, Ni RZ (1985) Four cases of nephrotoxicity by *Tripterygium wilfordii*. Chin J Dermatol 18:231–232

Uncaria rhynchophylla (Miq.) Jacks. 120

120.1 Introduction

Gouteng, Ramulus Uncariae cum Uncis, is the dry branches bearing hooks of *Uncaria rhynchophylla* (Miq.) Jacks., *U. macrophylla* Wall., *U. hirsuta* Havil., *U. sinensis* (Oliv.) Havil., or *U. sessilifructus* Roxb. (Rubiaceae), which are collected in the fall and winter. It is officially listed in the Chinese Pharmacopoeia and is used as an antipyretic and anticonvulsant for the treatment of headache, vertigo, and epilepsy.

120.2 Chemical Constituents

The main chemical constituents in the stems and hooks of *U. rhynchophylla* are alkaloids. The first isolated alkaloid was rhynchophylline (*120-1*) [1–3], which has a tetracyclic system composed of an indole moiety and an indolizidine moiety and is related to corynoxan (*120-2*). Later, isorhynchophylline (*120-3*), the 7-epimer of rhynchophylline, was isolated [4].

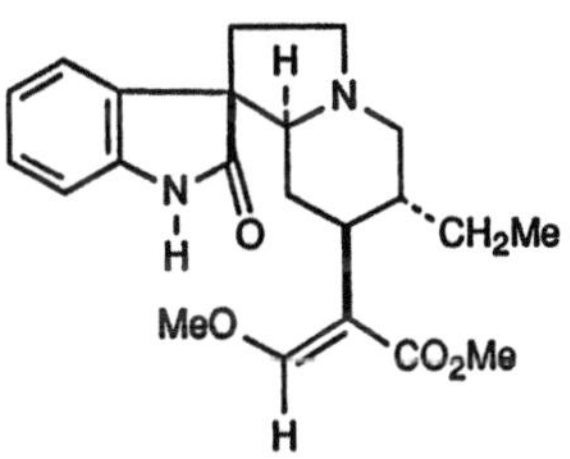

Rhynchophylline (*120-1*)

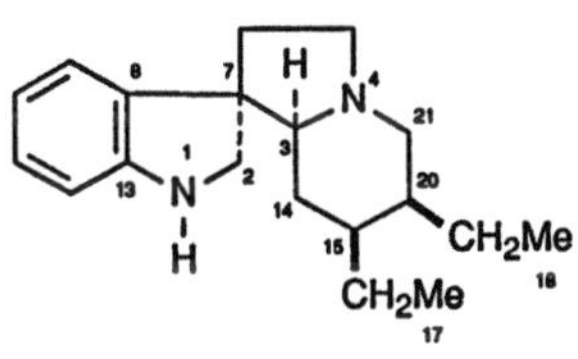

Corynoxan (*120-2*)

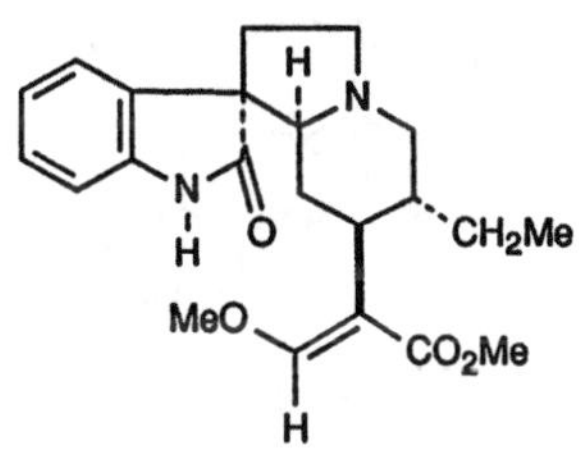

Isorhynchophylline (*120-3*)

Two further related alkaloids were the isomers corynoxeine (*120-4*) and isocorynoxeine (*120-5*) [5]. It was reported that approximately 97% of the total alkaloids detected in the hook, small stem, and leaf represented the oxindole alkaloids rhynchophylline, isorhynchophylline, corynoxeine, and isocorynoxeine. The bark of the underground part contains mainly the indole alkaloids hirsuteine (*120-6*) and hirsutine (*120-7*). Further more, the alkaloids corynantheine (*120-8*) and dihydrocorynantheine (*120-9*) were isolated from the wood of *U. rhynchophylla* [5]. The latter four alkaloids are derived from corynan (*120-10*).

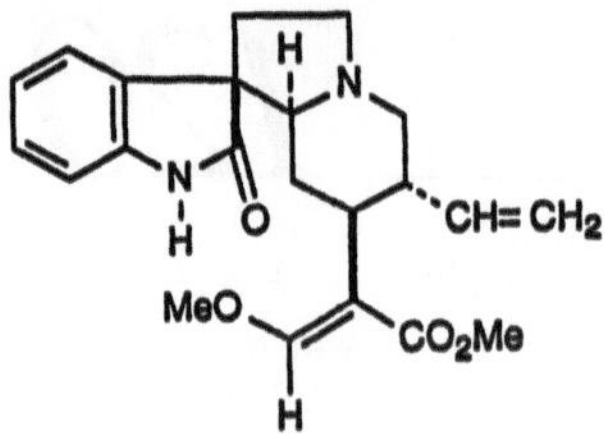

Corynoxeine (*120-4*)

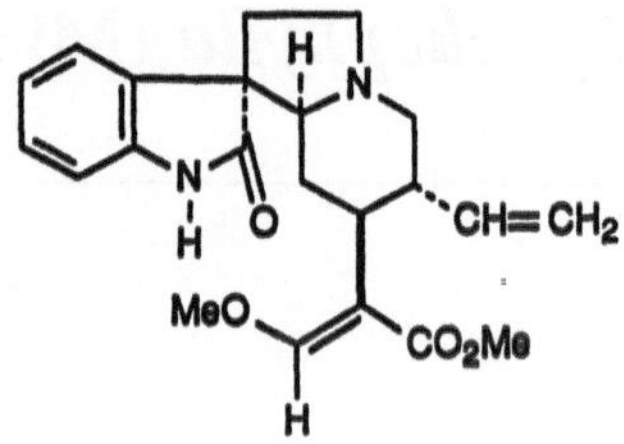

Isocorynoxeine (*120-5*)

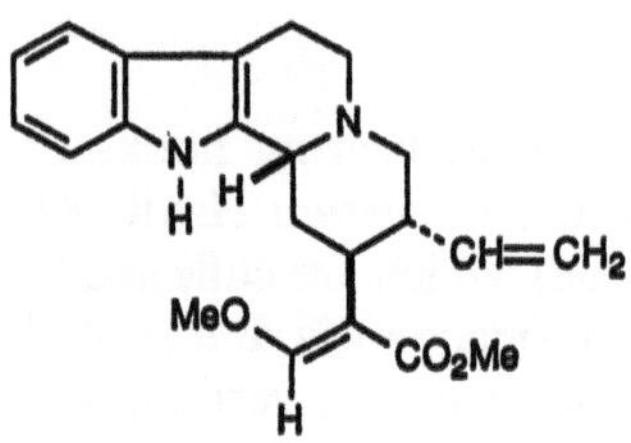

Hirsuteine (*120-6*)

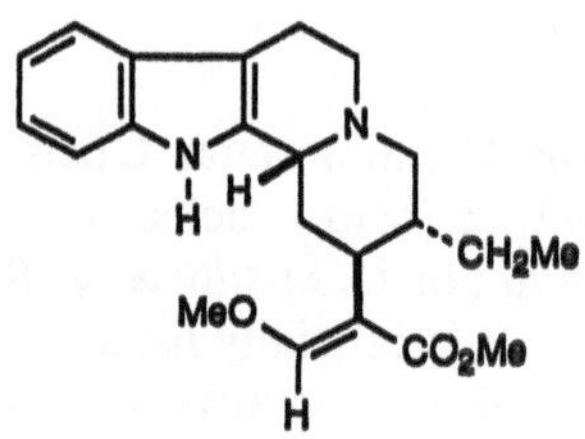

Hirsutine (*120-7*)

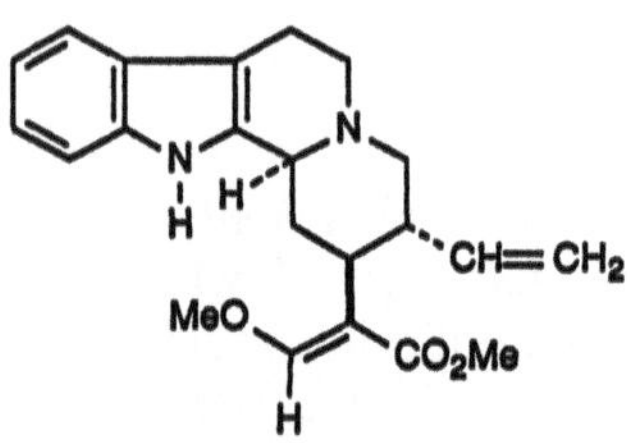

Corynantheine (*120-8*)

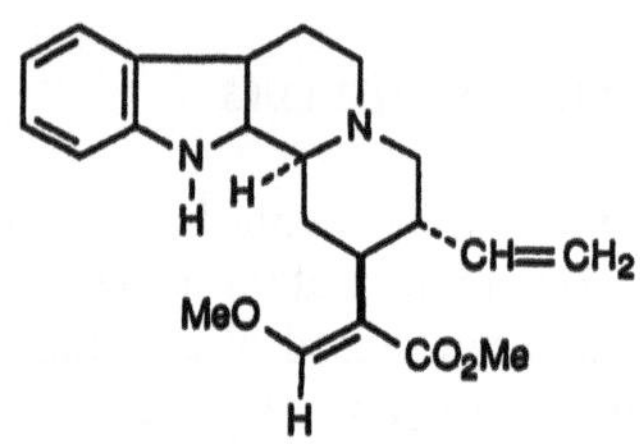

Dihydrocorynantheine (*120-9*)

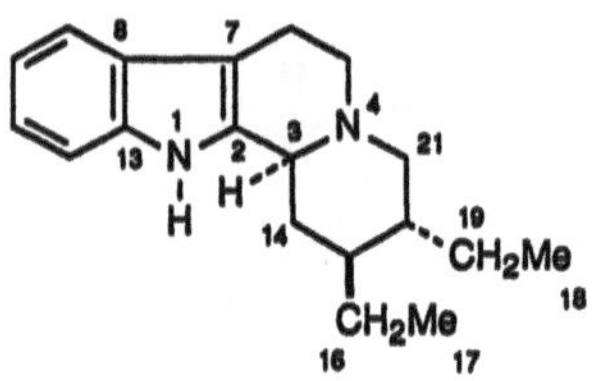

Corynan (*120-10*)

Also reported was the isolation from the stems and leaves of *U. rhynchophylla* of another corynan-type alkaloids, geissochizine (*120-11*) and vallesiachotamine (*120-12*), the oxayohimban-type alkaloid akummigine (*120-13*) [6], the oxayohimban alkaloid glycosides strictosamide (*120-14*), and its 3-epimer vincoside lactam together with the 6′-feruloylvincoside lactam named rhynchophine (*120-15*) [7].

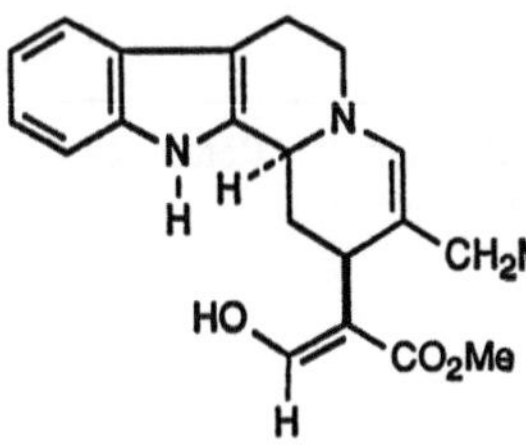

Geissoschizine (*120-11*)

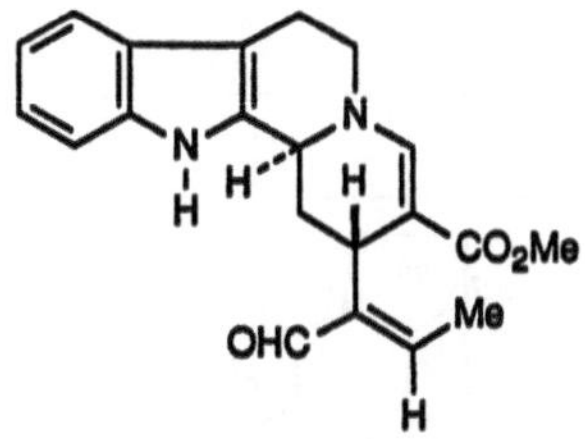

Vallesiachotamine (*120-12*)

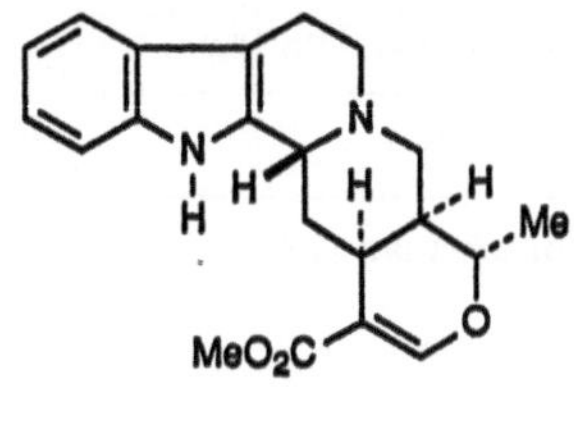

Akuammigine (*120-13*)

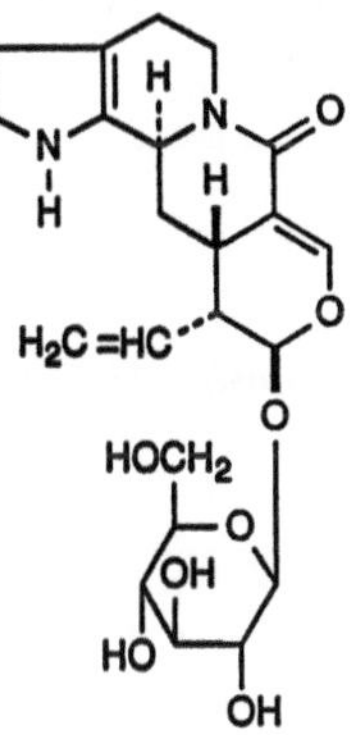

Strictosamide (*120-14*)

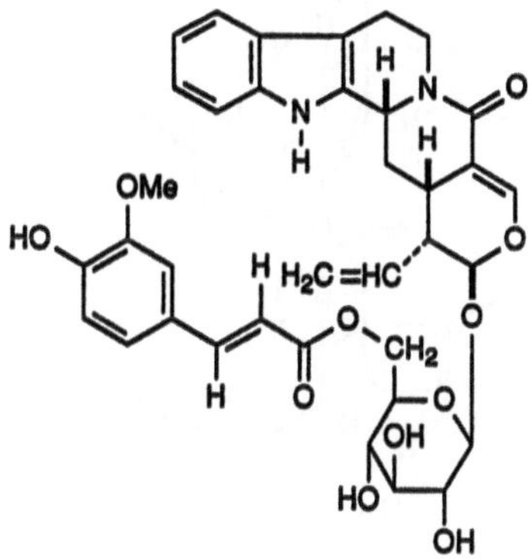

Rhynchophine (*120-15*)

Further oxindole-type alkaloids isolated from *Uncaria* species are derived from the basic skeleton formosanan (*120-16*), a pentacyclic system related to corynoxan. They are designated as uncarines A–F and are stereoisomers. The structures, synonyms, and plant origins of the uncarines are listed in Table 120.1.

Table 120.1. Uncarines

Compound	Structure	Plant origin	Reference
Formosanan (*120-16*)	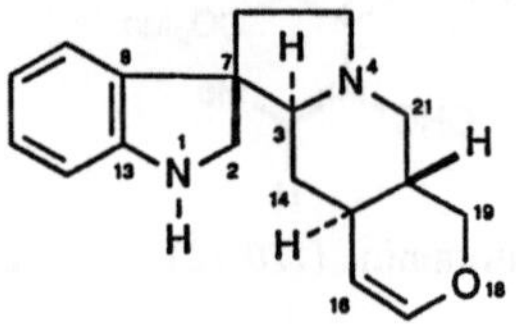		
Uncarine A (*120-17*) (Isoformosanine)	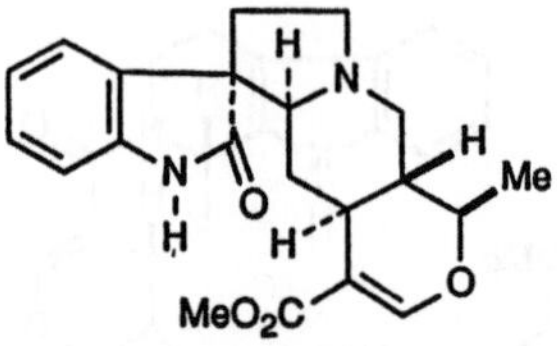	*Uncaria kawakami*	[8–11]
Uncarine B (*120-18*) (Formosanine)	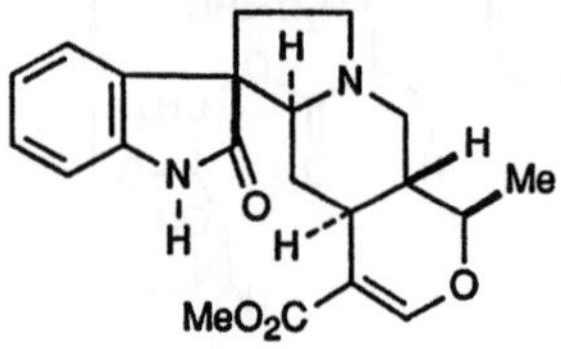	*U. formosana*	[8, 9, 12]
Uncarine C (*120-19*) (Pteropodine)	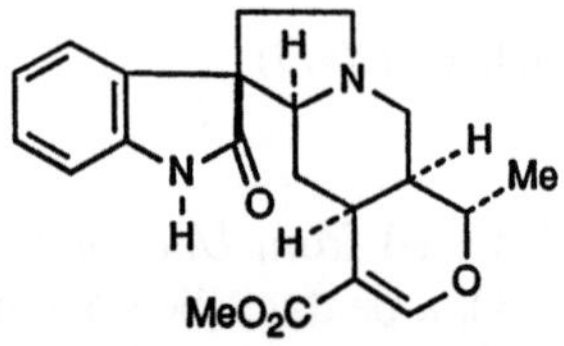	*U. bernaysii* *U. ferrea* *U. pteropoda*	[8, 9, 13–15]
Uncarine D (*120-20*)	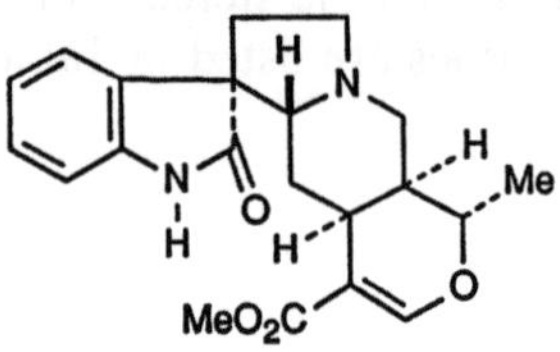	*U. bernaysii* *U. ferrea*	[8, 9, 13]
Uncarine E (*120-21*) (Isopteropodine)	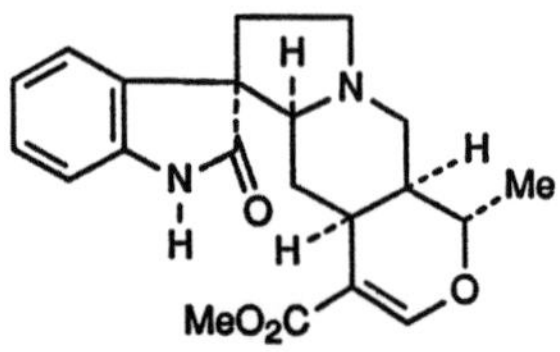	*U. bernaysii* *U. ferrea* *U. pteropoda*	[8, 9, 13–15]
Uncarine F (*120-22*)	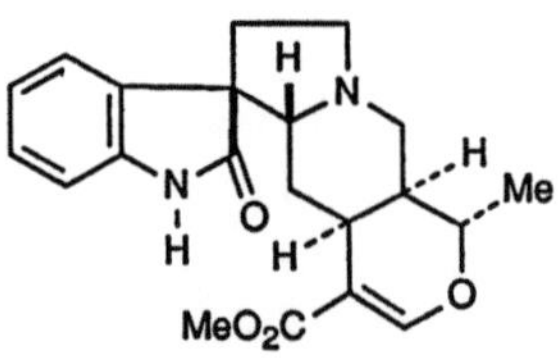	*U. bernaysii* *U. ferrea*	[8, 9, 13]

120.3 Pharmacology

The crude total alkaloids from *U. rhynchophylla* and the main alkaloid rhynchophyllin were found to have a hypotensive effect in cats after i.v. doses of 20 mg/kg. Marked hypotensive effects of total alkaloids and rhynchophyllin given at daily doses of 50 mg/kg for 20 or 15 days, respectively, were also observed in rats [16]. Intravenous infusion of total alkaloids of *U. macrophylla* into anesthetized dogs caused moderate hypotension, marked bradycardia, a rise in stroke volume, and a decrease in total peripheral resistance during the initial hypotension. Cardiac output increased briefly and then decreased. Hypotensive activity of total alkaloids from *U. macrophylla* was suggested to arise from decreases in cardiac output and peripheral resistance [17].

A methanolic extract of the hooks of an *Uncaria* species elicited a strong and long-lasting hypotensive effect in rats and the activity differed from the effects of rhynchophyllin and its analogs. Chemical and pharmacological studies on the extract resulted in the isolation of three indole alkaloid glycosides, cadambine (*120-23*), dihydrocadambine (*120-24*), and isodihydrocadambine (*120-25*). Cadambine and dihydrocadambine are derived from D-homooxayohimban, whereas isodihydrocadambine is derived from oxayohimban. Cadambine was found to be inactive but the other two alkaloids exhibited strong and long-lasting hypotensive action [18].

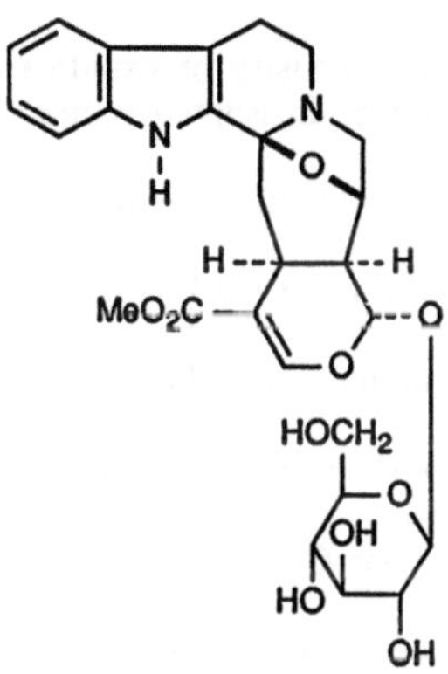

Cadambine (*120-23*)

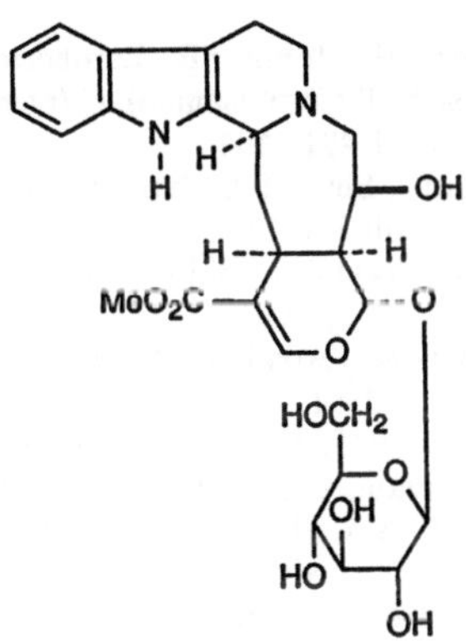

Dihydrocadambine (*120-24*)

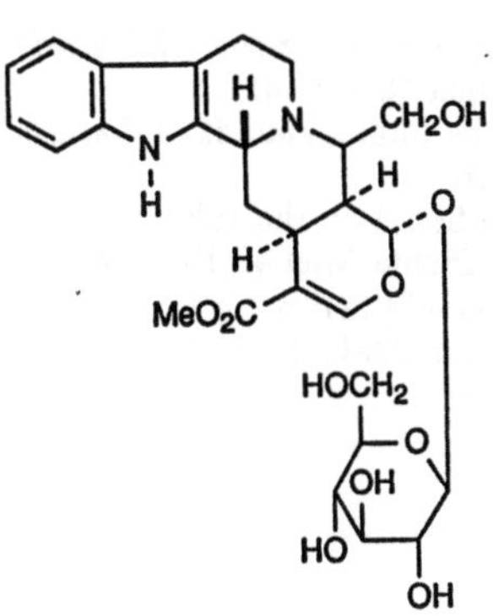

Isodihydrocadambine (*120-25*)

It was further reported that the oxindole alkaloids rhynchophylline, corynoxeine, and their isomers from *U. rhynchophylla* had Ca^{2+}-antagonistic activity, inhibiting Ca^{2+}-induced contraction of rat and rabbit thoracic aorta preparations [19].

References

1. Kondo H, Fukuda T, Tomita M (1928) Alkaloids of *Ouronparia rhynchophylla* Matsum. J Pharm Soc Jpn 48:321–337 (CA 22:3166)
2. Kondo H, Ikeda T (1937) Alkaloid of *Auronparia rhynchophylla* Matsum. III. Rhynchophylline. 2. J Pharm Soc Jpn 57:881–886
3. Nozoye T (1957) *Uncaria* alkaloids. XVII. Structure of rhynchophylline. 4. Structure of the side chain in rhynchophylline. Annu Rep Itsuu Lab 8:10–11 (CA 51:16504b)
4. Nozoye T (1958) *Uncaria* alkaloids. XX. Structure of rhynchophylline. 5. Structure of rhynchophylline and isorhynchophylline. Chem Pharm Bull (Tokyo) 6:309–312
5. Yamanaka E, Kimizuka Y, Aimi N, Sakai S, Haginiwa J (1983) Studies of plants containing indole alkaloids. IX. Determination of tertiary alkaloids in various parts of *Uncaria rhynchophylla*. Yakugaku Zasshi 103:1028–1033
6. Aimi N, Yamanaka E, Shinma N, Fujiu M, Kurita J, Sakai S, Haginiwa J (1977) Studies on plants containing indole alkaloids. VI. Minor bases of *Uncaria rhynchophylla* Miq. Chem Pharm Bull (Tokyo) 25:2067–2071
7. Aimi N, Shito T, Fukushima K, Itai Y, Aoyama C, Kunisawa K, Sakai S, Haginiwa J, Yamasaki K (1982) Studies on plants containing indole alkaloids. VIII. Indole alkaloid glycosides and other constituents of the leaves of *Uncaria rhynchophylla* Miq. Chem Pharm Bull (Tokyo) 30:4046–4051
8. Beecham AF, Hart NK, Johns SR, Lamberton JA (1968) The stereochemistry of oxindole alkaloids: uncarines A, B (formosanine), C (pteropodine), D (speciophylline), E (isopteropodine) and F. Aust J Chem 21:491–504
9. Pousset JL, Poisson J, Shine RJ, Shamma M (1967) Determination of the stereochemistry of oxindole alkaloids. Bull Soc Chim Fr 2766–2279
10. Kondo H, Ikeda T (1941) Alkaloid of *Uncaria* species. IV. Structure of uncarine. J Pharm Soc Jpn 61:416–429 (CA 45:2960d)
11. Kondo H, Ikeda T (1942) Alkaloid of *Uncaria* species. VII. Structure of uncarine. J Pharm Soc Jpn 62:15–22 (CA 45:2961a)
12. Raymont-Hamet M (1936) A new alkaloid, formosanine, extracted from *Ouronparia formosana* Matsumura and Hayata. C R Hebd Seances Acad Sci 203:1383–1384
13. Beecham AF (1967) A study of the C3/C7 stereochemistry of uncarines C, D, E and F by circular dichroism. Tetrahedron Lett 991–993
14. Chan KC (1969) Stereochemistry of pteropodine and isopteropodine. Phytochemistry 8:219–222
15. Yeoh GB, Chan KC, Morsingh F (1966) Pteropodine and isopteropodine, the alkaloids from *Uncaria pteropoda*. Tetrahedron Lett 931–938
16. Chang TH, Li HT, Li Y, Wang YF, Wu L, Li TH (1978) Hypotensive effect of *Uncaria rhynchophylla* total alkaloids and rhyncophyllin. Natl Med J China 58:408–411
17. Liu GX, Huang XN, Peng Y (1983) Hemodynamic effects of the total alkaloids of *Uncaria macrophylla* in anesthetized dogs. Acta Pharmacol Sin 4:114–116
18. Endo K, Oshima Y, Kikuchi H, Koshihara Y, Hikino H (1983) Series on the validity of the oriental medicines. L. Hypotensive principles of *Uncaria* hooks. Planta Med 49:188–190
19. Yamahara J, Miki S, Matsuda H, Kobayashi G, Fujimura H (1987) Screening for calcium antagonists in natural products. The active principles of uncariae ramulus et uncus. Yakurigaku Zasshi 90:133–140

Verbena officinalis L. 121

121.1 Introduction

Mabiancao, Herba Verbenae, is the dry aerial part of *Verbena officinalis* L. (Verbenaceae), which is collected in June–August when the plant blooms. It is officially listed in the Chinese Pharmacopoeia and is used as a hemostatic, antimalarial, and diuretic agent.

121.2 Chemical Constituents

The major components of *V. officinalis* are the iridoid glycosides. Thus, cornin and hastatoside were isolated from the aerial part of *V. officinalis*. Cornin (*121-1*), designated as verbenalin for a long time [1], was structurally determined by Büchi and Manning [2]. Hastatoside (*121-2*) was isolated and identified later [3,4] as was the iridoid glucoside aucubin [5].

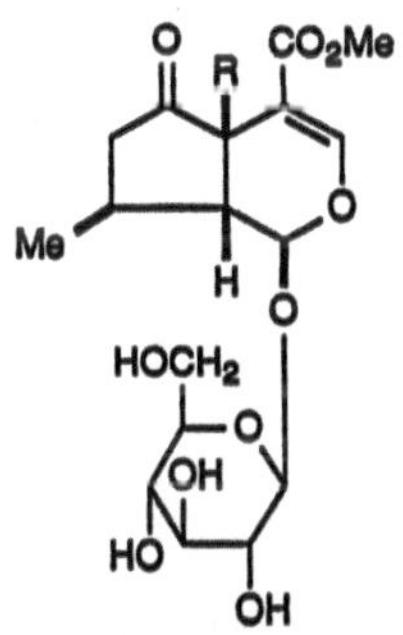

Cornin (Verbenalin, *121-1*): R=H
Hastatoside (*121-2*): R=OH

In addition to the flavone derivative artemetin (*121-3*), the triterpenes lupeol, ursolic acid, *β*-sitosterol [5, 6], and the phenylpropane glycosides verbascoside (*121-4*) and eukovoside (*121-5*) [7] were isolated and identified. Due to its relatively large content (0.8%), verbascoside may be used for the identification of this herbal medicine [8].

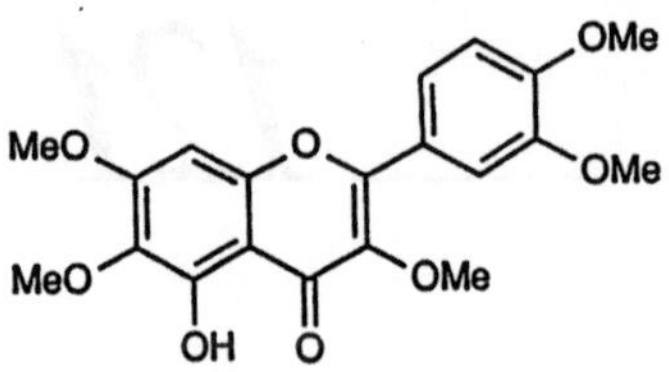

Artemetin (*121-3*)

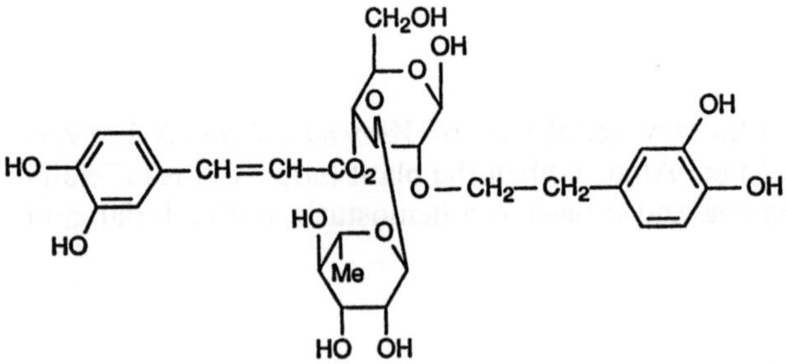

Verbascoside (*121-4*)

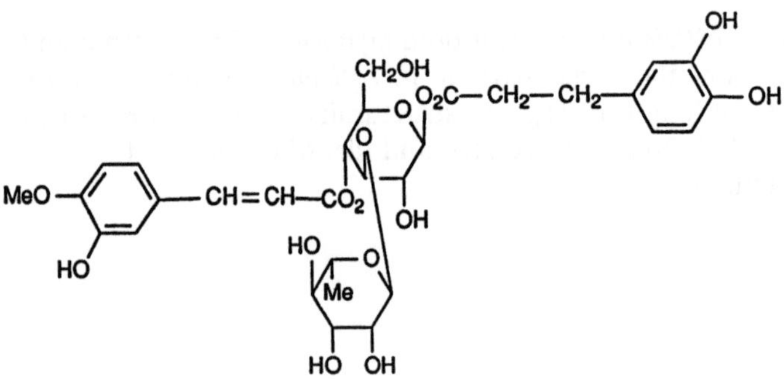

Eukovoside (*121-5*)

121.3 Pharmacology

The ethanolic extract of *V. officinalis* was reported to have antiphlogistic activity on conjunctivitis caused by croton oil in rabbits and also to exhibit analgesic activity [9]. The antitussive action of *V. officinalis* was found to be related to the content of cornin and β-sitosterol [6]. In a study on the interaction between *Verbena* constituents and prostaglandins, synergistic effects were reported on the contraction of isolated rat uterus preparations with PGE_2 and additive effects with $PGF_{2\alpha}$ [10].

References

1. Jensen SR, Kjaer A, Nielsen BJ (1973) Dihydrocornin, a novel natural iridoid glucoside. Acta Chem Scand 27:2581–2585
2. Büchi G, Manning RE (1960) Structure of verbenalin. Tetrahedron Lett 5–12
3. Rimpler H, Schäfer B (1973) Hastatosid, ein neues Iridoid aus *Verbena officinalis* und *Verbena hastata* (Verbenaceae). Tetrahedron Lett 1463–1464

4. Rimpler H, Schaefer B (1979) Hastatoside, a new iridoid from *Verbena hastata* L. and *Verbena officinalis* L. Z Naturforsch [C] 34:311–318
5. Makboul AM (1986) Chemical constituents of *Verbena officinalis*. Fitoterapia 57:50–51
6. Kui CH, Tang RJ (1985) Studies on the antitussive constituents of *Verbena officinalis*. Bull Chin Mater Med 10:467
7. Bianco A, Guiso M, Passacantilli P (1984) Iridoid and phenylpropanoid glycosides from new sources. J Nat Prod 47:901–902
8. Haensel R, Kallmann S (1986) Verbascoside, a main constituent of *Verbena officinalis*. Arch Pharm (Weinheim) 319:227–230
9. Sakai S (1963) Pharmacological actions of *Verbena officinalis* extracts. Gifu Ika Daigaku Kiyo 11:6–17 (CA 60:16384g)
10. Research Group on Reproductive Physiology (1974) Effect of *Verbena* herb (*Verbena officinalis*) on the uterus. II. Interaction between *Verbena* herb and prostaglandins. Acta Zool Sin 20:340–345

Vitex negundo L. var. *cannabifolia* (Sieb. et Zucc.) Hand.-Mazz. 122

122.1 Introduction

Mujingyou, Oleum Viticis negundo, is the essential oil of *Vitex negundo* L. var. *cannabifolia* (Sieb. et Zucc.) Hand.-Mazz. (Verbenaceae), obtained by steam distillation of fresh leaves. The essential oil is officially listed in the Chinese Pharmacopoeia and is used as an antitussive and expectorant agent in the treatment of chronic asthma. Mujinyou Jiaowan, Capsulae Olei Viticis negundo, is also listed in the Chinese Pharmacopoeia for the same medical indications.

122.2 Chemical Constituents

Essential oil from leaves of *V. negundo* var. *cannabifolia* was found to contain α-phellandrene, α-pinene, β-pinene, sabinene, 1,8-cineole, *p*-cymene, γ-terpinene, β-elemene, β-caryophyllene, and caryophyllene oxide, β-caryophyllene representing the main component [1]. The composition of the essential oils isolated from the leaves of *V. negundo* or *V. negundo* var. *heterophylla* was found to be similar to that from *V. negundo* var. *cannabifolia* [1]. The iridoid constituents of *V. negundo* were studied in more detail. The iridoid glycosides aucubin, agnuside, 2′-*p*-hydroxybenzoyl mussaenosidic acid (*122-1*) [2], 6′-*p*-hydroxybenzoyl mussaenosidic acid (*122-2*) [3], and nishindaside (*122-3*) [4] were isolated from the leaves.

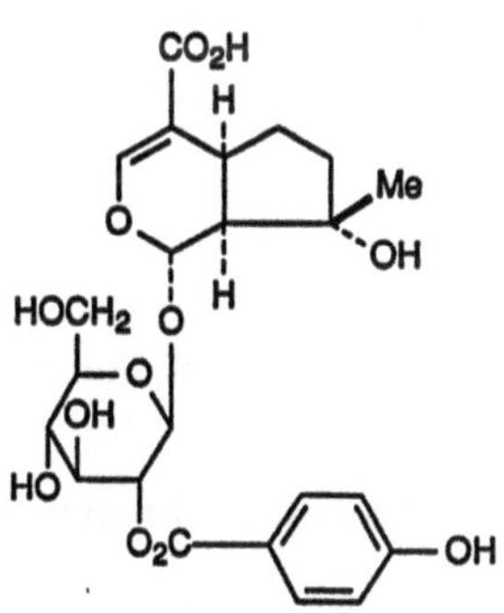

2′-*p*-Hydroxybenzoyl-mussaenosidic acid (*122-1*)

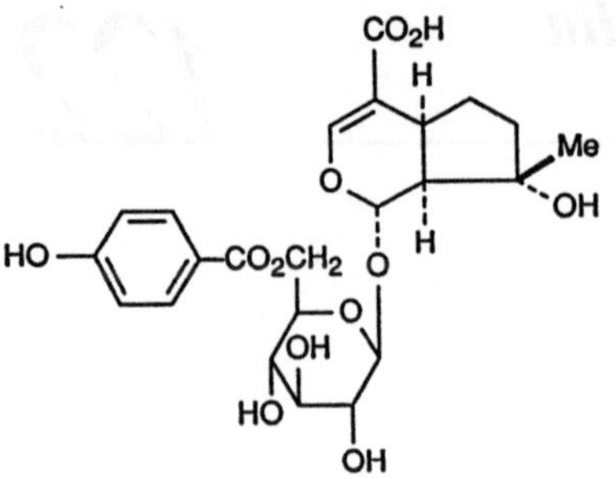

6′-*p*-Hydroxybenzoyl-mussaenosidic acid (*122-2*)

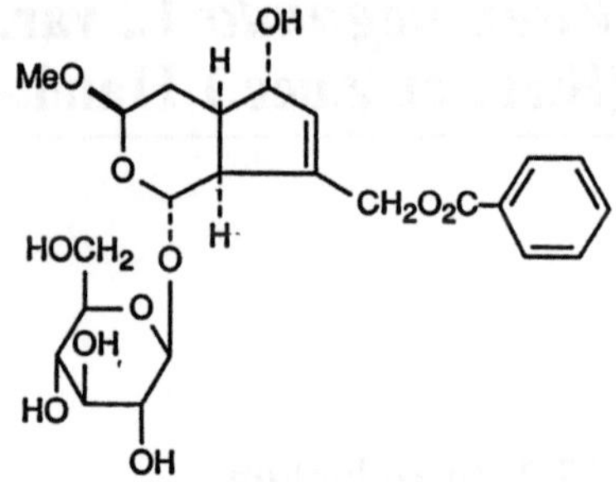

Nishindaside (*122-3*)

122.3 Pharmacology

The essential oil of *V. negundo* var *cannabifolia* was found to have therapeutic activity on bronchitis and asthma similar to that found in *Rhododendron racemosum* [5]. A high degree of similarity in chemical composition of the essential oils from *R. racemosum* and from some *Vitex* species was found [5].

122.4 *Vitex trifolia* L.

Another item in the *Chinese Pharmacopeia* for *Vitex* species is Manjingzi, Fructus Viticis. It is the dry ripe fruits of *V. trifolia* L. or *V. trifolia* var. *simplicifolia* Cham. which is collected in the fall when the fruits are ripe. It is used for the treatment of colds, headaches, and vertigo. The fruits of *V. trifolia* were found to contain aliphatic hydrocarbons, dulcitol, and vanillic acid [6]. The oil from the seeds contains myristic (0.4%), palmitic (35%), stearic (15%), palmitoleic (1.8%), oleic (28%), and linoleic (20%) acids [7]. The composition of the essential oil isolated from leaves of *V. trifolia* was similar to that of *V. negundo* var. *cannabifolia*, with α-pinene representing the main component [1]. Agnuside was also isolated from the leaves of *V. trifolia* [8].

References

1. Fan JF, Pan JQ, Sun YF, Ciu MS, He HJ, Duan SM, Wang XZ, Liu SM, Sun YR (1981) Studies on the chemical constituents of Chinese *Vitex*. I. Chemical constituents of essential oils of certain *Vitex* species. Chin Trad Herb Drug 12:393–396
2. Sehgal CK, Taneja SC, Dhar KL, Atal CK (1982) 2′-*p*-Hydroxybenzoyl mussaenosidic acid, a new iridoid glucoside from *Vitex negundo*. Phytochemistry 21:363–366
3. Sehgal CK, Taneja SC, Dhar KL, Atal CK (1983) 6′-*p*-Hydroxybenzoylmussaenosidic acid, an iridoid glycoside from *Vitex negundo*. Phytochemistry 22:1036–1038
4. Dutta PK, Chowdhury US, Chakravarty AK, Achari B, Pakrashi SC (1983) Studies on Indian medicinal plants. LXXV. Nishindaside, a novel iridoid glycoside from *Vitex negundo*. Tetrahedron 39:3067–3072
5. Fang HJ, Chen LS, Zhou TH (1980) Studies on the components of the essential oils. III. Studies of chemical constituents of the essential oil from *Rhododendron racemosum* Franch. Comparison of the constituents of *Vitex negundo* L. var. *cannabifolia* (Sieb. et Zucc.) Hand.-Mazz. and *V. negundo* L. var. *heterophylla* (Franch.) Rehd. Acta Pharm Sin 15:284–287

6. Higa M, Yogi S, Hokama K (1983) Studies on the constituents of *Vitex trifolia*. L. Bull Coll Sci Univ Rynkyns 35:61–66 (CA 99:3060z)
7. Prasad YR, Nigam SS (1982) Detailed chemical investigation of the seed oil of *Vitex trifolia* Linn. Proc Natl Acad Sci India [A] 52:336–339
8. Sirait M, Liemtjwanhoo F (1966) Isolation of agnuside from the leaves of *Vitex trifolia*. Suara Pharm 9:47–51 (CA 65:18993g)

Zingiber officinale (Willd.) Rosc. 123

123.1 Introduction

Ganjiang, Rhizoma Zingiberis, is the dry rootstock of *Zingiber officinale* (Willd.) Rosc. (Zingiberaceae), which is collected in winter. It is officially listed in the Chinese Pharmacopoeia and used in traditional Chinese medicine as stomachic, antiemetic, antidiarrheal, expectorant, antiasthmatic, hemostatic, and cardiotonic for the treatment of several gastrointestinal and respiratory diseases.

Shengjiang, Rhizoma Zingiberis recens, is the fresh rootstock of *Z. officinale*, which is collected in the fall and winter. It is the second entry for *Z. officinale* in the Chinese Pharmacopoeia and is used for the treatment of gastrointestinal and respiratory disorders.

Two galenic preparations of ginger rhizome are also listed:

- Jiang Liujingao, Extractum Zingiberis liquidum, an ethanol extract of ginger powder, is used as a stomachic.
- Jiang Ding, Tinctura Zingiberis, prepared from the extract, is also used as a stomachic.

123.2 Chemical Constituents

The chemical constituents from the ginger rhizome can be divided into two classes, the pungent and the flavoring principles.

123.2.1 Pungent Substances of *Zingiber officinale*

Gingerol, the major pungent principle of the ginger rhizome, was described more than 100 years ago, but separation and structural determination of different gingerols was first reported in 1969 [1].

Gingerols are 1-(4-hydroxy-3-methoxyphenyl)-5-hydroxyalkan-3-ones with an *S*(+)-configuration, having side chains of different length. Gingerols with side chains of varying length have been identified and designated as [3]-, [4]-, [5]-, [6]-, [8]-, and [10]-gingerol (*123-1–123-6*) [2].

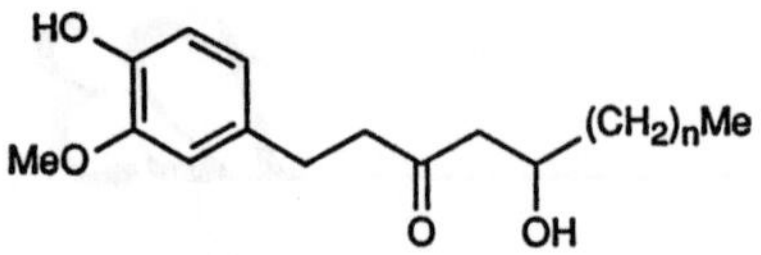

[3]-Gingerol (*123-1*): n=1
[4]-Gingerol (*123-2*): n=2
[5]-Gingerol (*123-3*): n=3
[6]-Gingerol (*123-4*): n=4
[8]-Gingerol (*123-5*): n=6
[10]-Gingerol (*123-6*): n=8

A biosynthetic study using ^{14}C- or ^{3}H-labeled precursors suggested a scheme of biosynthesis for *S*-(+)-[6]-gingerol in *Z. officinale* in which dihydroferulate arises from phenylalanine. In a "biological Claisen reaction" with malonate and hexanoate, a *β*-diketone is generated, which is reduced to [6]-gingerol [3].

Gingerols could be synthesized from zingerone by conversion into the corresponding *O*-trimethylsilyl ether, deprotonation with lithium bis(trimethylsilyl)amide [4] or lithium di-isopropylamide [5], and regioselective aldol condensation. A stereoselective synthesis of (*S*)-(+)-gingerols was carried out by using an optically active 3-(tolylsulfinylmethyl)-Δ^2-isoxazoline intermediate [6].

Zingerone (*123-7*), 4-(4-hydroxy-3-methoxy-phenyl)-butanone [7, 8], was the first constituent of ginger to be structurally determined. It has been shown, however, that it may result as an artifact from the isolation procedure [1]. It was not detectable by thin-layer chromatography after extraction under mild conditions [9].

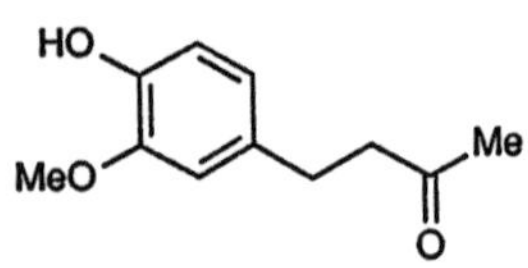

Zingerone (*123-7*)

Shogaols are gingerol analogs with a 4,5-double bond, resulting from elimination of the 5-hydroxy group. [6]-Shogaol (*123-8*), 1-(4-hydroxy-3-methoxy-phenyl)-4-decen-3-one, was the first compound reported in this series [10], and was synthesized by condensation of zingerone with hexanal [11]. Shogaols also appear mainly to be artifacts. If at all, they are at best very minor constituents of ginger rhizome [1]. They were derived by thermal dehydration from gingerols as two homologous series representing *cis*- and *trans*-isomers [12].

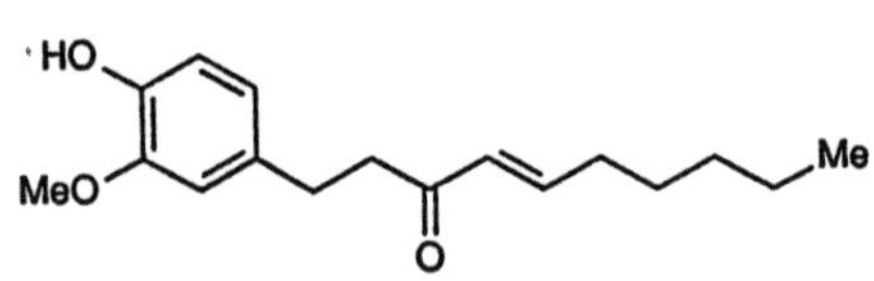

[6]-Shogaol (*123-8*)

Minor constituents related to gingerol are gingediols (*123-9*–*123-12*), methylgingediols, and their diacetates [13], hexahydrocurcumin and its demethyl analog [14], gingerdiones such as [6]-gingerdione (*123-13*) [15, 16] and dehydrogingerdiones [16]. Gingerdiones might be intermediates in gingerol biosynthesis.

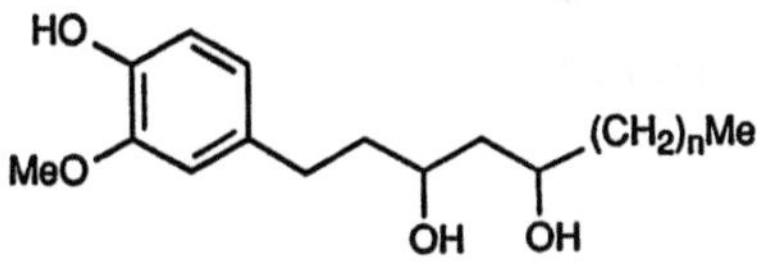

[4]-Gingediol (*123-9*): n=2
[6]-Gingediol (*123-10*): n=4
[8]-Gingediol (*123-11*): n=6
[10]-Gingediol (*123-12*): n=8

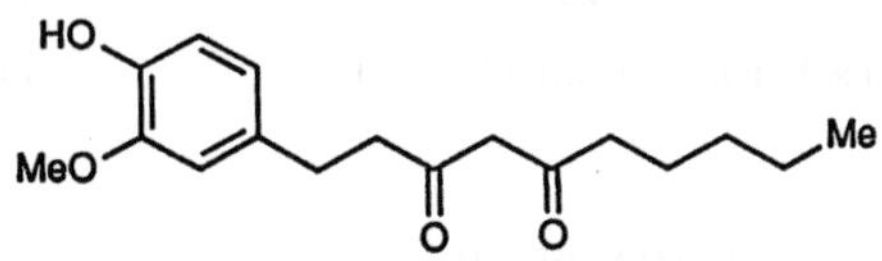

[6]-Gingerdione (*123-13*)

Furthermore, capsaicin (*123-14*), the pungent principle of *Capsicum* species, was isolated from ginger rhizome [17]. Purification and stabilization of ginger protease has also been described [18, 19].

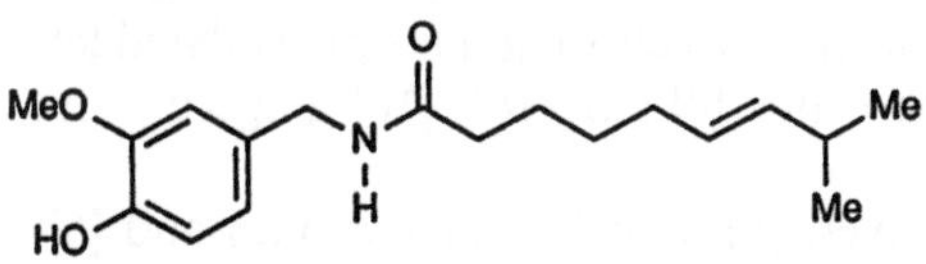

Capsaicin (*123-14*)

123.2.2 Flavoring Substances of *Zingiber officinalis*.

A number of components were isolated and identified from the essential oil of *Z. officinale*. They comprise α-pinene, β-pinene, cumene, camphene, myrcene, limonene, cineole, β-phellandrene, *p*-cymene, citral, linalool, borneol, bornyl acetate, geraniol, γ-selinene, β-elemene, zingiberene (*123-15*), β-bisabolene, ar-curcumene, farnesene, and sabinene [20, 21]. Zingiberol, a sesquiterpene isolated from ginger [22], was proven to be a stereoisomeric mixture of β-eudesmol with *trans*- and *cis*-ring junction [23]. Several new sesquiterpenes were also isolated and structurally investigated. Thus, the isolation of *cis*- and *trans*-sesquiphellandrol (*123-16*) [24], sesquithujene (*123-17*), sesquisabinene hydrate (*123-18*), and zingiberenol (*123-19*) [25] was reported.

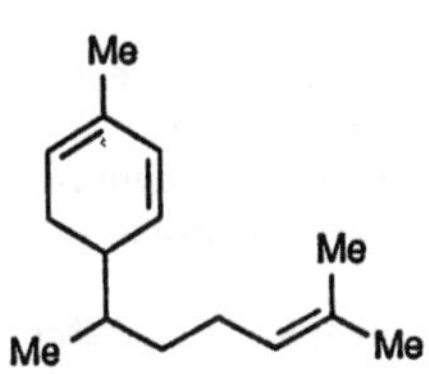

Zingiberene (*123-15*)

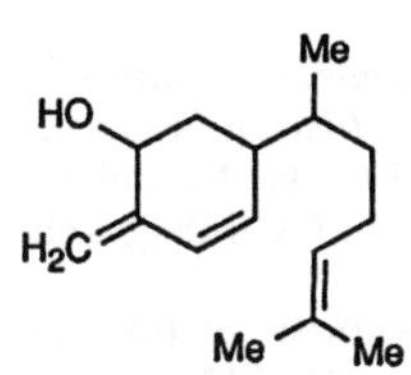

Sesquiphellandrol (*123-16*)

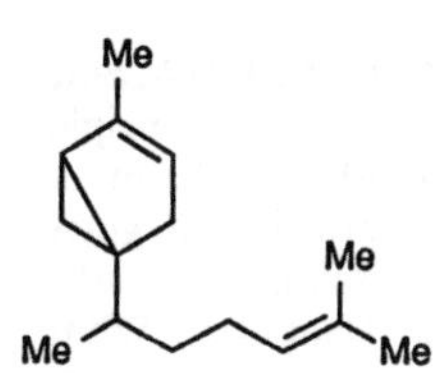

Sesquithujene (*123-17*)

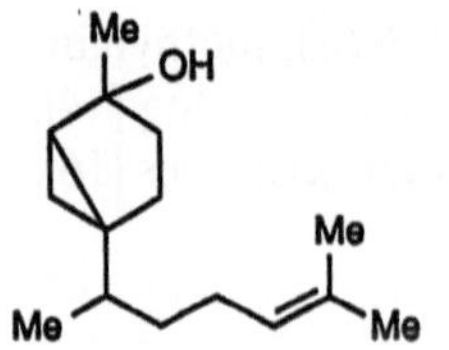

Sesquisabinene hydrate (*123-18*)

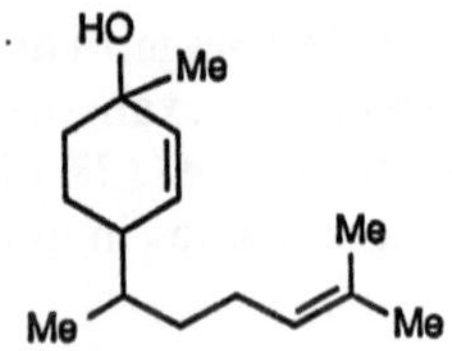

Zingiberenol (*123-19*)

123.3 Pharmacology

Oral administration of [6]-gingerol or [6]-shogaol at doses of 70–140 mg/kg and i.v. administration of both compounds at doses of 1.75–3.7 mg/kg to rats produced an inhibition of spontaneous motor activity, showed antipyretic and analgesic effects, and prolonged hexobarbital-induced sleeping time. [6]-Shogaol was more effective than [6]-gingerol. [6]-Shogaol was found to display an intense antitussive effect in comparison with dihydrocodeine phosphate. Both [6]-shogaol and [6]-gingerol suppressed gastric contraction in situ [26].

The methanolic extract of ginger rhizome or its active ingredients, the gingerols, has been found to possess potent positive inotropic effects on guinea pig isolated left atria. Cardiotonic activity decreased in the order [8]-gingerol, [10]-gingerol, and [6]-gingerol [27].

[6]-Gingerol and the gingerdione derivatives, [6]- and [8]-gingerdiones and [6]- and [8]-dehydrogingerdiones, are potent inhibitors of prostaglandin biosynthesis [16].

Gingerols and shogaols exhibited significant antihepatotoxic actions against CCl_4- and galactosamine-induced cytotoxicity in primary cultured rat hepatocytes. The antihepatotoxic activity of gingerols and shogaols was dependent on the length of the side chain, with the [7]- and [8]-homologs eliciting the strongest activity [28]. Intragastric or oral administration of ginger oleoresin to cholesterol-fed rats significantly lowered serum and hepatic cholesterol levels and increased fecal cholesterol excretion [29].

With 3% bovine serum albumin as substrate, the semipurified powdered protease of ginger showed a relatively high proteolytic activity at pH 4.5–6.0, with an optimum pH of 5.0. The ginger protease was protected by dithiothreitol during extraction and reaction, indicating the involvement of SH groups at the active site [30].

References

1. Connell DW, Sutherland MD (1969) A re-examination of gingerol, shogaol, and zingerone, the pungent principles of ginger (*Zingiber officinale* Roscoe). Aust J Chem 22:1033–1043
2. Masada Y, Inoue T, Hashimoto K, Fujioka M, Shiraki K (1973) Studies on the pungent principles of ginger (*Zingiber officinale*) by GC-MS (gas chromatography – mass spectrometry). Yakugaku Zasshi 93:318–321
3. Denniff P, Whiting DA (1976) Biosynthesis of (6)-gingerol, pungent principle of *Zingiber officinale*. J Chem Soc Chem Commun 711–712
4. Denniff P, Whiting DA (1976) Synthesis of (±)-(6)-gingerols (pungent principle of ginger) and relatives via directed aldol reactions. J Chem Soc Chem Commun 712–713
5. Denniff P, Macleod I, Whiting DA (1981) Synthesis of the (±)-[*n*]-gingerols (pungent principles

of ginger) and related compounds through regioselective aldol condensation: relative pungancy assays. J Chem Soc [Perkin Trans] 1:82–87
6. Cinquini M, Cozzi F, Gilardi A (1984) Synthesis of enantiomerically pure Δ^2-isoxazolines via sulphinyl derivatives. J Chem Soc Chem Commun 551–552
7. Nomura H (1917) Pungent principles of ginger. I. A new ketone, zingerone (4-hydroxy-3-methoxy-phenylethyl methyl ketone) occurring in ginger. J Chem Soc 769–776
8. Lapworth A, Wykes FH (1917) The pungent principle of ginger. II: Synthetic preparations of zingerone, methylzingerone and some related acids. J Chem Soc 790–798
9. Govindarajan B, Govindarajan VS (1979) Evaluation of spices and oleoresins. VIII. Improved separation and estimation of pungent and related components of ginger by thin-layer chromatography. J Food Qual 2:205–217
10. Nomura H, Tsurumi S (1926) The pungent principles of ginger. III. The constitution of shogaol. Ber Gesamte Physiol Exp Pharmakol 38:651
11. Nomura H, Tsurumi S (1927) Pungent principle of ginger. IV. Synthesis of shogaol. Proc Imp Acad (Japan) 3:159–160 (CA 21:2258)
12. Chen CC, Rosen RT, Ho CT (1986) Chromatographic analyses of isomeric shogaol compounds derived from isolated gingerol compounds of ginger (*Zingiber officinale* Roscoe). J Chromatogr 360:175–184
13. Masada Y, Inoue T, Hashimoto K, Fujika M, Uchino C (1974) Studies on the constituents of ginger (*Zingiber officinale*) by GC-MS (gas chromatography – mass spectrometry). Yakugaku Zasshi 94:735–738
14. Mutara T, Shinohara M, Miyamoto M (1972) Isolation of hexahydrocurcumin, dihydrogingerol, and two additional pungent principles from ginger. Chem Pharm Bull (Tokyo) 20:2291–2292
15. Harvey DJ (1981) Gas chromatographic and mass spectrometric studies of ginger constituents. Identification of gingerdiones and new hexahydrocurcumin analogs. J Chromatogr 212:75–84
16. Kiuchi F, Shibuya M, Sankawa U (1982) Inhibitors of prostaglandin biosynthesis from ginger. Chem Pharm Bull (Tokyo) 30:754–757
17. Bhagya (1977) Detection of capsaicin in adulterated ginger oleoresin. J Food Sci Technol 14:176–177
18. Ichikawa Y, Sasa H, Michi K (1973) Purification of ginger protease. Eiyo To Shokuryo 26:377–383 (CA 80:92529g)
19. Ohtsuki K, Kawabata M, Taguchi K (1978) Purification and stabilization of ginger protease. Kyoto-furitsu Daigaku Gakujutsu Hokoku, Rigaku, Seikatsu Kagaku 33–40 (CA 90:163874k)
20. Kami T, Nakayama M, Hayashi S (1972) Volatile constituents of *Zingiber officinale*. Phytochemistry 11:3377–3381
21. Nigam MC, Nigam IC, Levi L (1964) Essential oils and their constituents. XXII. Detection of new trace components in oil of ginger. Can J Chem 42:2610–2615
22. Brooks BT (1916) Zingiberol – a new sesquiterpene alcohol occurring in the essential oil of ginger. J Am Chem Soc 38:430–432
23. Varma KR, Jain TC, Bhattacharyya SC (1962) Terpenoids. XXXIV. Structure and stereochemistry of zingiberol and juniper camphor. Tetrahedron 18:979–984
24. Bednarzyk AA, Galetto WG, Kramer A (1975) *cis*- and *trans*-β-Sesquiphellandrol, two new sesquiterpene alcohols from oil of ginger (*Zingiber officinale*, Roscoe). J Agric Food Chem 23:499–501
25. Terhune SJ, Hogg JW, Bromstein AC, Lowrence BM (1975) Four new sesquiterpene analogs of common monoterpenes. Can J Chem 53:3285–3293
26. Suekawa M, Ishige A, Yuasa K, Sudo K, Aburada M, Hosoya E (1984) Pharmacological studies on ginger. I. Pharmacological actions of pungent constituents, [6]-gingerol and [6]-shogaol. J Pharmacobiodyn 7:836–848
27. Shoji N, Iwasa A, Takemoto T, Ishida Y, Ohizumi Y (1982) Cardiotonic principles of ginger (*Zingiber officinale* Roscoe). J Pharm Sci 71:1174–1175
28. Hikino H, Kiso Y, Kato N, Hamada Y, Shioiri T, Aiyama R, Itokawa H, Kiuchi F, Sankawa U (1985) Antihepatotoxic actions of gingerols and diarylheptanoids. J Ethnopharmacol 14:31–39
29. Gujral S, Bhumra H, Swaroop M (1978) Effect of ginger (*Zingiber officinale* Roscoe) oleoresin on serum and hepatic cholesterol levels in cholesterol fed rats. Nutr Rep Int 17:183–189
30. Thompson EH, Wolf ID, Allen CE (1974) Ginger rhizome. New source of proteolytic enzyme. J Food Sci 38:652–655

Ziziphus jujuba Mill. and *Z. spinosa* Hu 124

124.1 Introduction

Dazao, Fructus Jujubae, is the dry ripe fruits of *Ziziphus jujuba* Mill. (Rhamnaceae), which are collected in the fall. This officially listed herbal medicine in the Chinese Pharmacopoeia is mainly used as a tonic and sedative.

Suanzaoren, Semen Ziziphi spinosae, is the dry ripe seeds of *Z. spinosa* Hu, which are collected in the late fall and early winter when the fruits have ripened. It is officially listed in the Chinese Pharmacopoeia and is also used as a sedative and tonic.

124.2 Chemical Constituents

Ziziphus jujuba is a plant containing a great variety of chemical compounds. Saponins, alkaloids, flavone *C*-glycosides, vitamins, sugars, cyclic nucleotides, and other compounds were isolated and identified [1]. A crude saponin was obtained at about 0.2% yield from the seeds of *Z. jujuba.* Two saponins, named jujubosides A and B, containing the sapogenin jujubogenin (*124-1*), were separated. The structure of jujubogenin was elucidated by chemical reactions, spectral analyses and X-ray crystallography [2]. By acid hydrolysis of the saponin, the aglycon jujubogenin was converted almost quantitatively into ebelin lactone (*124-2*) (Fig. 124.1) [2, 3].

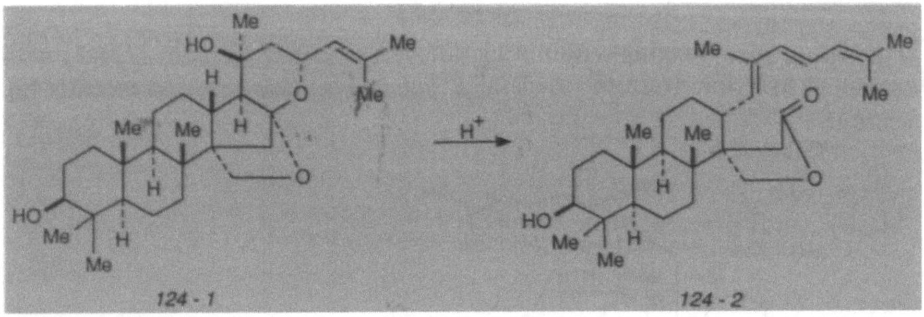

Fig. 124.1. Conversion of jujubogenin into ebelin lactone

The sequence and configuration of the sugar linkage in jujubosides and prosapogenins were determined from NMR data [4]. Jujuboside A (*124-3*) could be converted into jujuboside B (*124-4*) by enzymatic deglycosylation using a snail enzyme or naringinase [4].

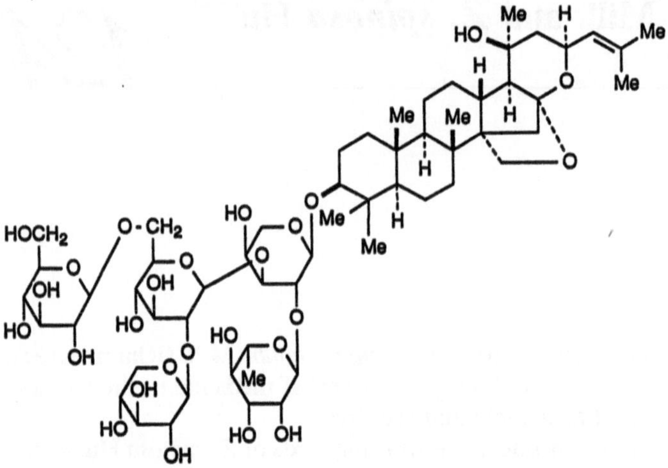

Jujuboside A (*124-3*)

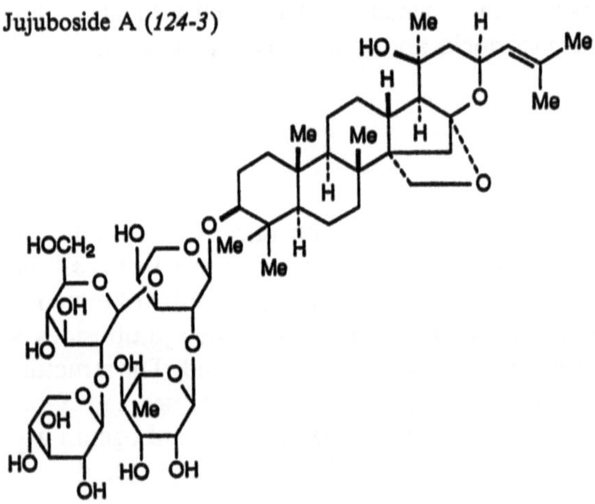

Jujuboside B (*124-4*)

The isolation of a sweetness-inhibiting substance named ziziphin (*124-5*) with jujubogenin as aglycone from the seeds of *Z. jujuba* was also reported recently [5].

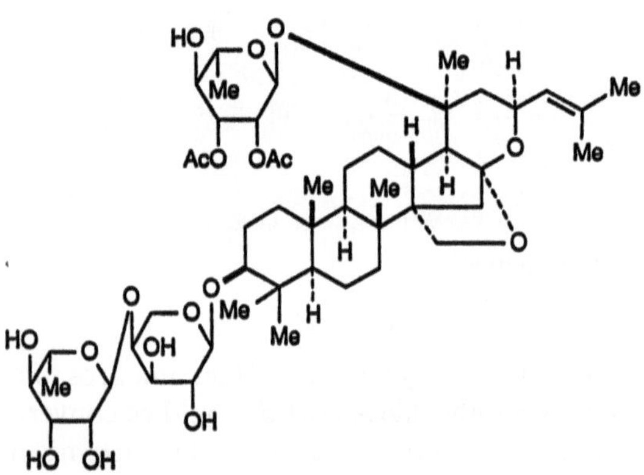

Ziziphin (*124-5*)

The triterpenes betulinic, alphitolic (*124-6*) [6], betulonic (*124-7*), oleanonic (*124-8*), maslinic (*124-9*) [7], oleanolic, and ursolic acid (*124-10*) [8] were isolated from ziziphus fruits with a number of isomeric *p*-coumaric acid esters of alphitolic [6] and maslinic acid [7].

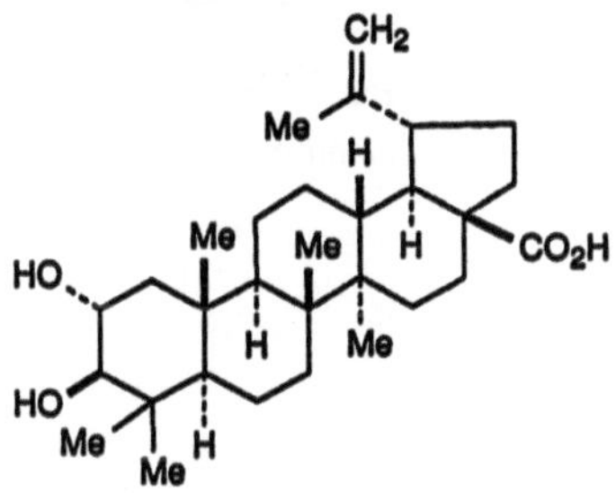

Alphitolic acid (*124-6*)

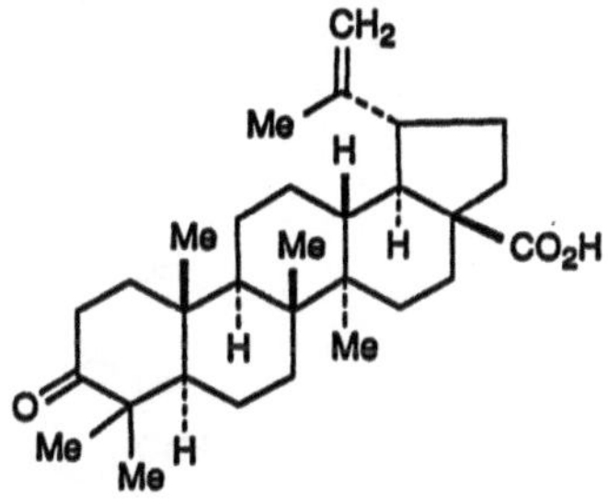

Betulonic acid (*124-7*)

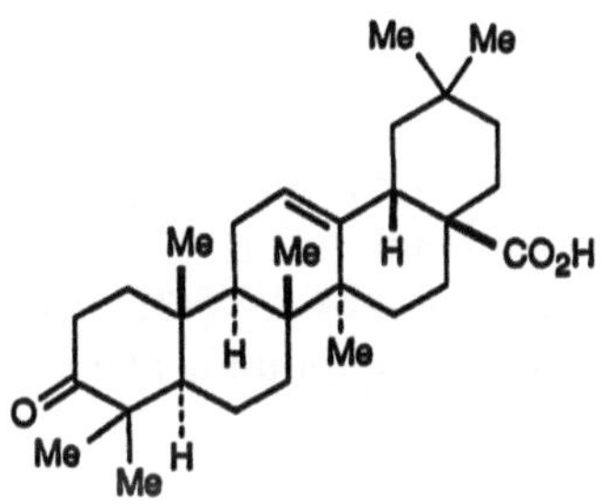

Oleanonic acid (*124-8*)

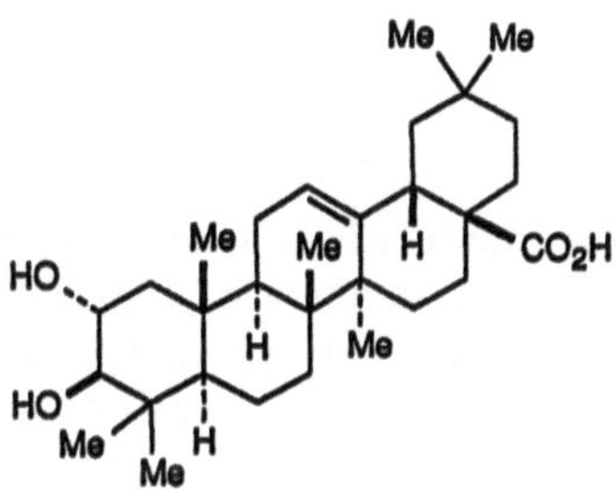

Maslinic acid (*124-9*)

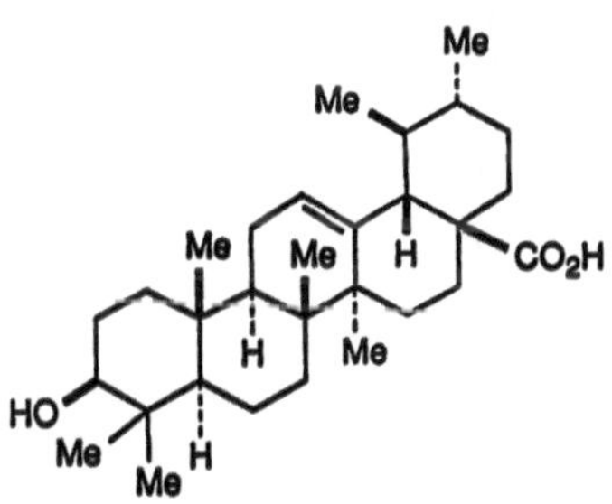

Ursolic acid (*124-10*)

Besides the saponins and triterpenes, a number of benzyl isoquinoline alkaloids were isolated from the leaves of *Z. jujuba*. Thus, the isolation and identification of coclaurine [9, 10], isoboldine (*116-57*), norisoboldine (*124-11*), asimilobine (*124-12*), yuziphine (*124-13*), and yuzirine (*124-14*) [10] were reported. The structure-related compound zizyphusine (*124-15*) was isolated from the fruits of *Z. jujuba* [11].

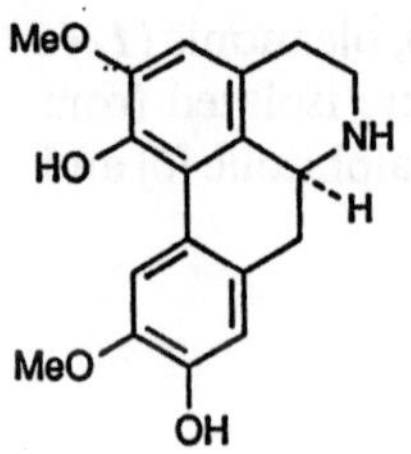

Norisoboldine (*124-11*)

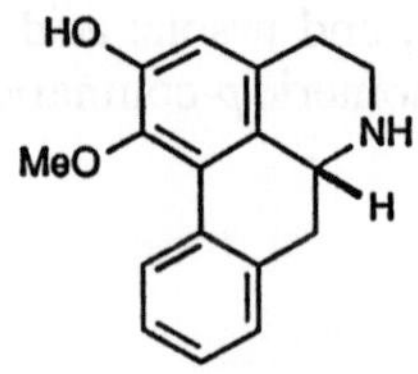

Asimilobine (*124-12*)

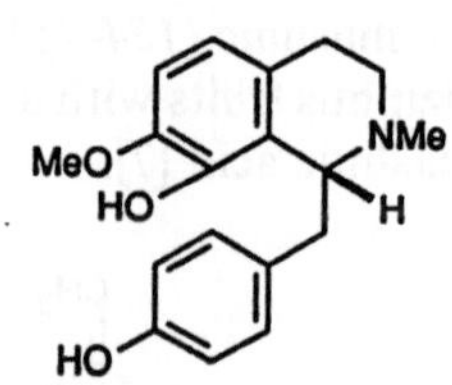

Yuziphine (*124-13*)

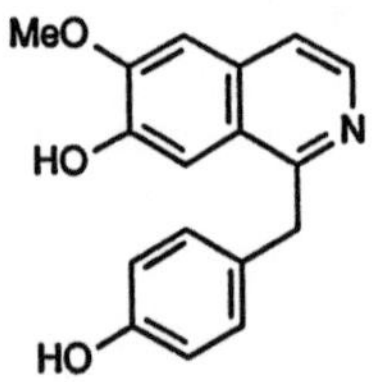

Yuzirine (*124-14*)

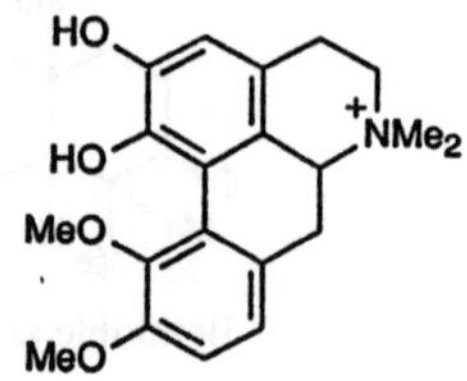

Zizyphusine (*124-15*)

Furthermore, the cyclopeptide alkaloids mauritine A (*124-16*), mucronine D (*124-17*), amphibine H (*124-18*), nummularine A (*124-19*), nummularine B (*124-20*), jubanine A (*124-21*), jubanine B (*124-22*) [12], and frangufoline (*124-23*) [13] were isolated from the stem bark of *Z. jujuba.* A structure-related compound, named daechucyclopeptide-1 (*124-24*), was isolated from the fruits [11].

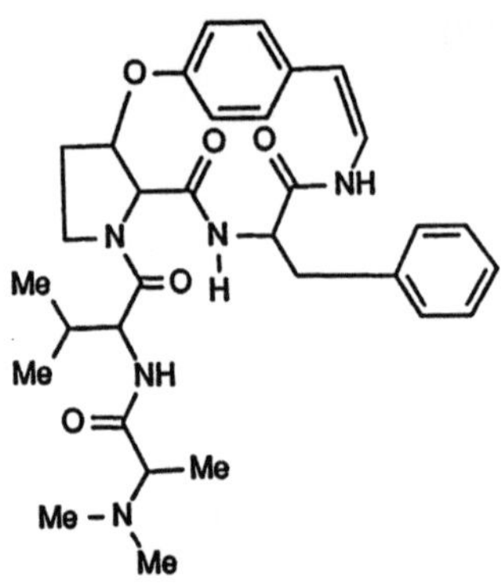

Mauritine A (*124-16*)

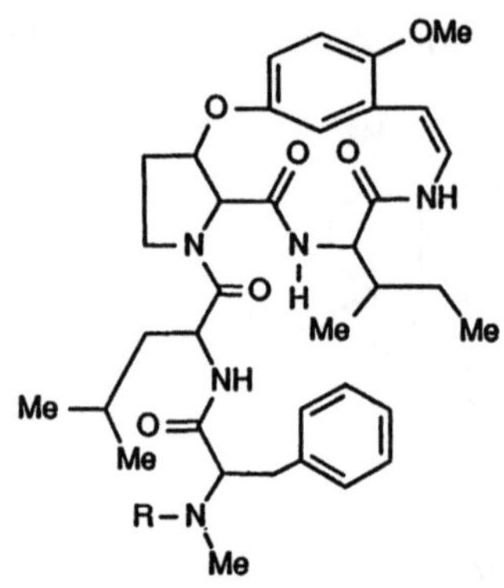

Mucronine D (*124-17*): R=CH_3
Nummularine A (*124-19*): R=H

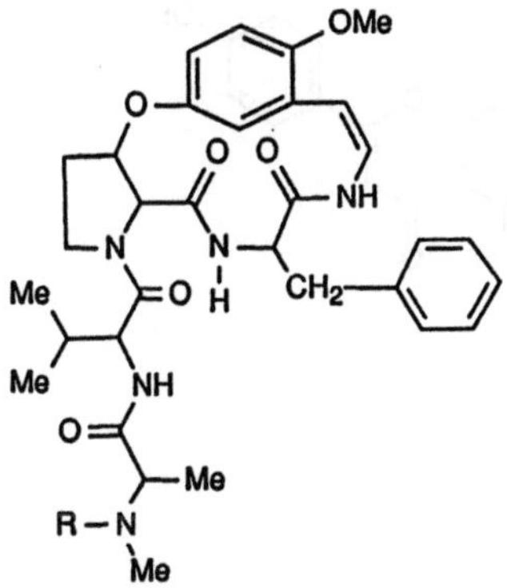

Amphibine H (*124-18*): R=CH_3
Nummularine B (*124-20*): R=H

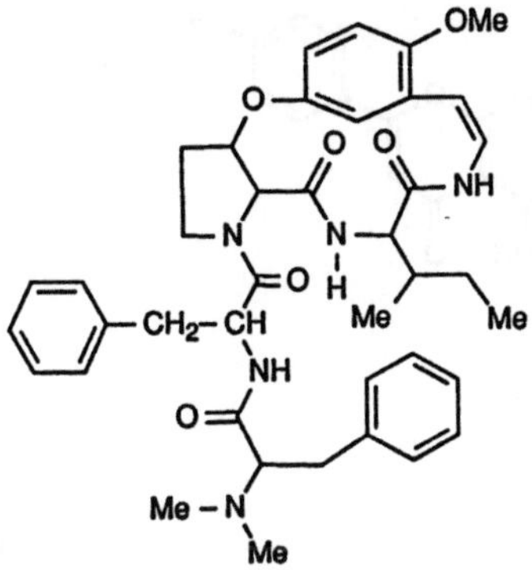

Jubanine A (*124-21*)

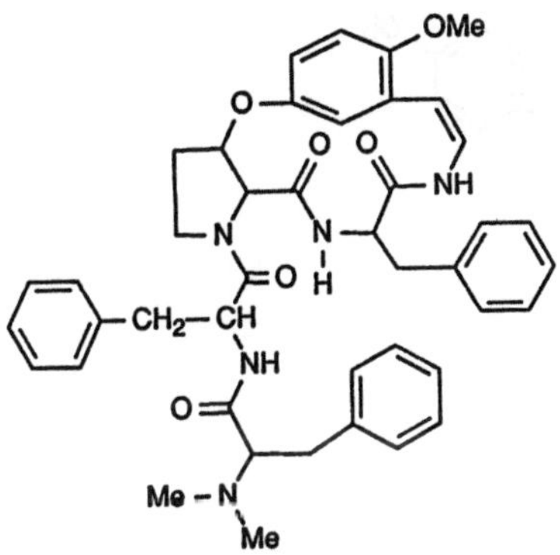

Jubanine B (*124-22*)

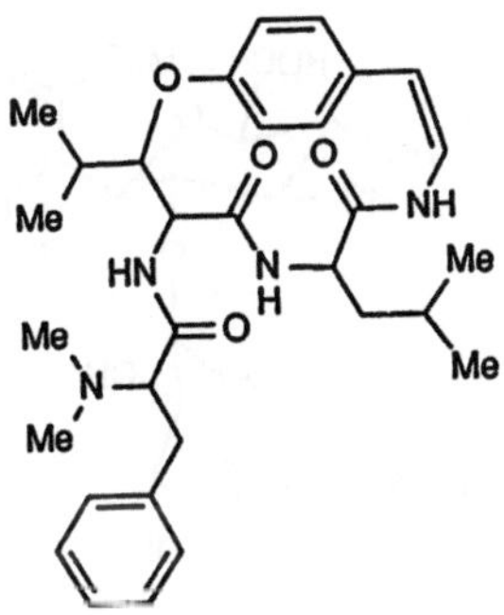

Frangufoline (*124-23*)

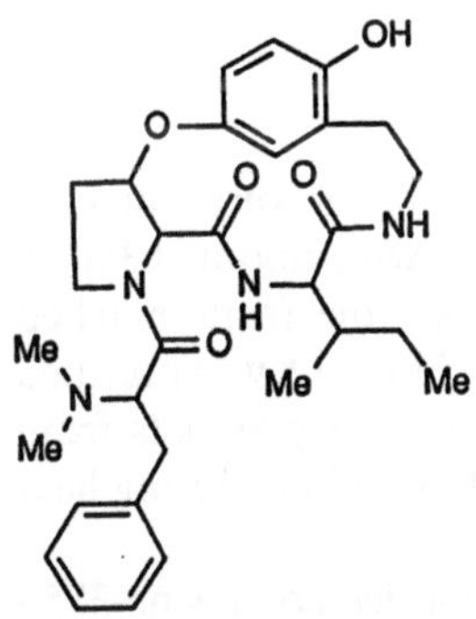

Daechucyclopeptide-1 (*124-24*)

Furthermore, the flavone *C*-glycosides swertisin (*124-25*) [14] and spinosin (*124-26*) as well as the acylated derivatives of spinosin sinapoyl- (*124-27*), feruloyl- (*124-28*), and coumaroylspinosin (*124-29*) [15] were isolated from the seeds of *Z. jujuba*.

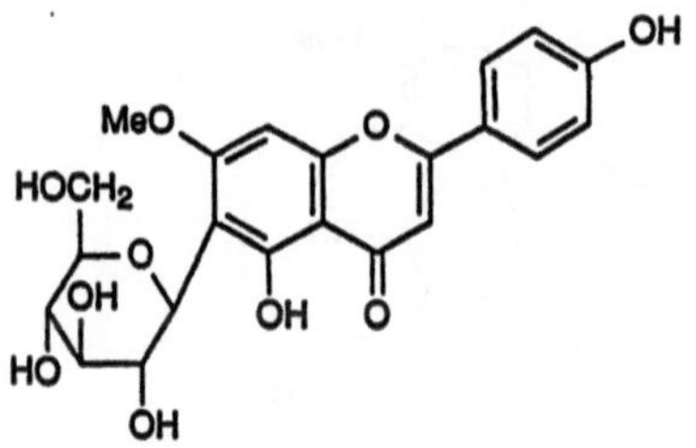

Swertisin (*124-25*)

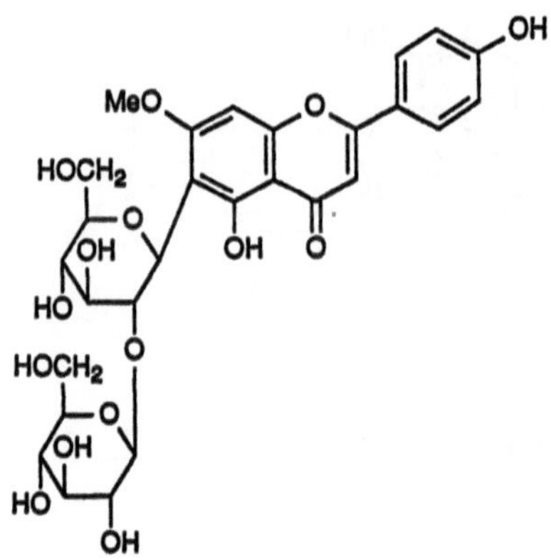

Spinosin (*124-26*)

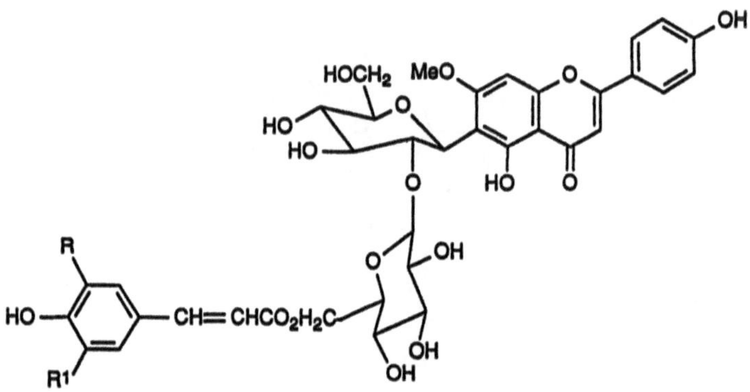

Sinapoylspinosin (*124-27*): $R = R^1 = CH_3$
Feruloylspinosin (*124-28*): $R = H$, $R^1 = CH_3$
p-Coumaroylspinosin (*124-29*): $R = R^1 = H$

The fruits of *Z. jujuba* are rich in ascorbic acid. Immature fruits were found to contain as much as 0.6%–0.8% of free vitamin C, together with about 0.3% in bound form [16]. The free form of vitamin C decreased when the fruits reached complete ripeness, and decreased further during storage [17]. This loss was ascribed to oxidation by ascorbic acid oxidase. Juices, compotes, and other products from jujube were high in vitamin C. Formation of a bound form of vitamin C during heat processing has been described [16].

Interestingly, relatively high amounts of the cyclic nucleotides adenosine-3′,5′-monophosphate (cAMP) and guanosine-3′,5′-monophosphate (cGMP) were found in the fruits of *Z. jujuba* [18–20]. cAMP and cGMP contents in dry fruits of *Z. jujuba* were reported to be 100–500 nmol/g and 30–50 nmol/g, respectively. In another study, however, only 3 nmol/g cAMP and cGMP, respectively were found by immunoassay in dry fruits of *Z. jujuba* [21].

The seeds of *Z. spinosa* were reported to contain betulin, betulinic acid, ceanothic acid (*124-30*), alphitolic acid, daucosterol [22], jujubosides A and B, ferulic acid, spinosin, and zivulgarin (*124-31*) [23].

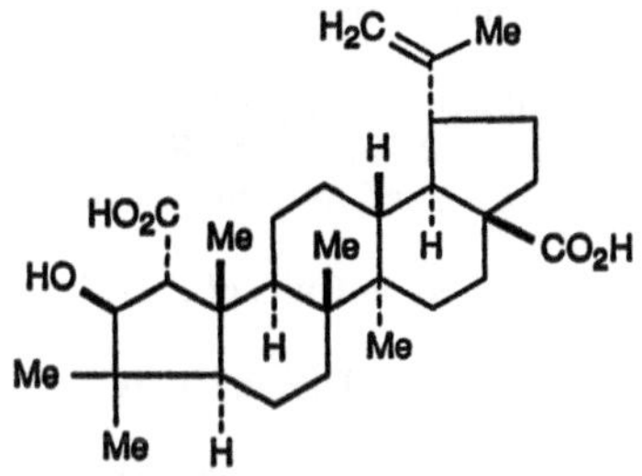

Ceanothic acid (*124-30*)

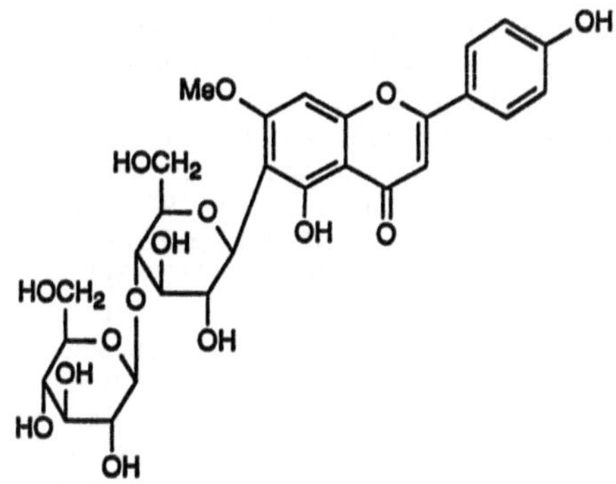

Zivulgarin (*124-31*)

124.3 Pharmacology

The flavone *C*-glycosides swertisin, spinosin, and the acylspinosins isolated from *Zizyphus* seeds were found to exhibited mild sedative activity in animal experiments. Of the flavones, swertisin showed the highest sedative activity [24].

References

1. Li SZ, Zhang B (1983) Pharmacological and chemical studies on Da Zao (*Ziziphus jujuba*). Chin Trad Herb Drugs 14:471–475
2. Kawai KI, Akiyama T, Ogihara Y, Shibata S (1974) A new sapogenin in the saponins of *Zizyphus jujuba*, *Hovenia dulcis*, and *Bacopa monniera*. Phytochemistry 13:2829–2832
3. Shibata S, Nagai Y, Tamaka O, Doi O (1970) Sapogenin of seeds of *Zizyphus jujuba* var. *spinosus*. Phytochemistry 9:677
4. Otsuka H, Akiyama T, Kawai KI, Shibata S, Inoue O, Ogihara Y (1978) The structure of jujubosides A and B, the saponins isolated from the seeds of *Zizyphus jujuba*. Phytochemistry 17:1349–1352
5. Kurihara Y, Ookubo K, Tasaki H, Kodama H, Akiyama Y, Yagi A, Halpern B (1988) Studies on the taste modifiers. I. Purification and structure determination of sweetness inhibiting substance in leaves of *Ziziphus jujuba*. Tetrahedron 44:61–66
6. Yagi A, Okamura N, Haraguchi Y, Noda K, Nishioka I (1978) Studies on the constituents of Zizyphi fructus. I. Structure of three new *p*-coumarates of alphitolic acid. Chem Pharm Bull (Tokyo) 26:1798–1802
7. Yagi A, Okamura N, Haraguchi Y, Noda K, Nishioka I (1978) Studies on the constituents of Zizyphi fructus. II. Structure of new *p*-coumaroylates of maslinic acid. Chem Pharm Bull (Tokyo) 26:3075–3079
8. Kozai K (1985) Isolation and mode of action of anti-plaque agents derived from Zizyphi fructus. Hiroshima Daigaku Shigaku Zasshi 17:1–20 (CA 104:115941r)
9. Otsuka H, Ogihara Y, Shibata S (1974) Isolation of coclaurine from *Zizyphus jujuba* by droplet counter-current chromatography. Phytochemistry 13:2016
10. Ziyaev R, Irgashev SY (1977) Alkaloids of *Ziziphus jujuba*. Structure of yuziphine and yuzirine. Khim Prir Soedin 239–243 (CA 87:114612c)
11. Han BH, Park MH (1987) Sedative activity and the active components of zizyphi fructus. Arch Pharmacal Res 10:208–211
12. Tschesche R, Kokhar I, Wilhelm H, Eckhardt G (1976) Jubanin A und Jubanin B, neue Cyclopeptidalkaloide aus *Ziziphus jujuba*. Phytochemistry 15:541–542
13. Devi S, Pandey VB, Singh JP, Shah AH (1987) Peptide alkaloids from *Zizyphus* species. Phytochemistry 26:3374–3375

14. Inoue O, Yokoi M, Ogihara Y (1974) Application of droplet countercurrent chromatography. *C*-glycosylflavone from the seeds of *Zizyphus jujuba*. Nagoya Shiritsu Daigaku Yakugakubu Kenkyu Nempo 22:36–37 (CA 83:55678w)
15. Woo WS, Kang SS, Wagner H, Seligmann O, Chari VM (1980) Acylated flavone-*C*-glycosides from seeds of *Zizyphus jujuba*. Phytochemistry 19:2791–2793
16. Korobkina ZV, Kruglyakov GN (1972) Vitamin value of foods produced from jujube fruits. Tr. Vses Semin Biol Aktiv (Lech) Veschestvam Plodov Yagod, 4th, 481–485 (CA 81:90040n)
17. Yao GS, Li YJ, Chang XQ, Lu JH (1983) Vitamin C content in vegetables and fruits in Shenyang (China) market during four seasons. Acta Nutr Sin 5:373–379
18. Cyong JC, Takahashi M (1982) Guanosine 3′,5′-monophosphate activity in fruits of *Zizyphus jujuba*. Chem Pharm Bull (Tokyo) 30:1081–1083
19. Cyong JC, Takahashi M (1982) Identification of guanosine 3′,5′-monophosphate in the fruit of *Zizyphus jujuba*. Phytochemistry 21:1871–1874
20. Cyong JC, Hanabusa K (1980) Cyclic adenosine monophosphate in fruits of *Zizyphus jujuba*. Phytochemistry 19:2747–2748
21. Liu JR, Liu ST (1983) Determination of cAMP- and cGMP-like substances in traditional Chinese medicines and their preparations. Shanxi Med J 12:131–132
22. Zeng L, Zhang RY, Wang X (1986) Constituents of the Chinese traditional drug – Suanzaoren. Acta Bot Sin 28:517–522
23. Zeng L, Zhang RY, Wang X (1987) Constituents of *Ziziphus spinosus*. Acta Pharm Sin 22:114–120
24. Woo WS, Shin KH, Kang SS (1980) Chemistry and pharmacology of flavone-*C*-glycoside from *Zizyphus* seeds. In: Woo WS, Han BH (eds) Recent Advances in Natural Products Research. Seoul National University Press, Seoul, pp 33–40 (CA 95:35560w)

Appendix 1. Herbal Medicines Contained in the *Chinese Pharmacopoeia* of 1985, Vol. I

Plant species	Item
Abutilon theophrastii Medic.	Semen Abutili
Acacia catechu (L.) Willd.	Catechu
Acanthopanax gracilistylus W. W. Smith	Cortex Acanthopanacis
Acanthopanax senticosus (Rupr. et Maxim.) Harms	Radix Acanthopanacis senticosi
Achyranthes bidentata Bl.	Radix Achyranthis bidentatae
Aconitum carmichaeli Debx.	Radix Aconiti, Radix Aconiti lateralis preparata
Aconitum Kusnezoffii Reichb.	Folium Aconiti kusnezoffii, Radix Aconiti kusnezoffii
Acorus gramineus Soland.	Rhizoma Acori graminei
Adenophora stricta Miq.	Radix Adenophorae
Adenophora tetraphylla (Thunb.) Fisch.	Radix Adenophorae
Aesculus chinensis Bge.	Semen Aesculi
Aesculus chinensis Bge. var. *chekingensis* (Hu et Fang) Fang	Semen Aesculi
Aesculus wilsonii Rehd.	Semen Aesculi
Agrimonia pilosa Ledeb.	Herba Agrimoniae
Ailanthus altissima (Mill.) Swingle	Cortex Ailanthi
Akebia quinata (Thunb.) Decne.	Fructus Akebiae
Akebia trifoliata (Thunb.) Koidz.	Fructus Akebiae
Akebia trifoliata (Thunb.) Koidz. var. *australis* (Diels) Rehd.	Fructus Akebiae
Albizia julibrissin Durazz.	Cortex Albiziae, Flos Albiziae
Alisma orientalis (Sam.) Juzep.	Rhizoma Alismatis
Allium macrostemon Bge.	Bulbus Allii macrostemi
Allium tuberosum Rottl.	Semen Allii tuberosi
Alpinia galanga Willd.	Fructus Galangae
Alpinia katsumadai Hayata	Semen Alpiniae katsumadai
Alpinia officinarum Hance	Rhizoma Alpiniae officinarum
Alpinia oxyphylla Miq.	Fructus Alpiniae oxyphyllae
Amomum compactum Soland ex Maton	Fructus Amomi rotundus
Amomum kravanh Pirre ex Gagnep.	Fructus Amomi rotundus
Amomum longiligulare T. L. Wu	Fructus Amomi
Amomum tsao-ko Crevost et Lemaire	Fructus Tsaoko
Amomum villosum Lour.	Fructus Amomi
Ampelopsis japonica (Thunb.) Makino	Radix Ampelopsis
Andrographis paniculata (Burm. f.) Nees	Herba Andrographitis
Anemarrhena asphodeloides Bge.	Rhizoma Anemarrhenae

Appendix 1. (continued)

Plant species	Item
Angelica dahurica (Fisch. ex Hoffm.) Benth. et Hook. f.	Radix Angelicae dahuricae
Angelica dahurica (Fisch. ex Hoffm.) Benth. et Hook. f. var. *formosana* (Boiss.) Shan et Yuan	Radix Angelicae dahuricae
Angelica pubescens Maxim. f. *bisserata* Shan et Yuan	Radix Angelicae pubescentis
Angelica sinensis (Oliv.) Diels	Radix Angelicae sinensis
Apocynum venetum L.	Folium Apocyni veneti
Aquilaria sinensis (Lour.) Gilg	Lignum Aquilariae resinatum
Arctium lappa L.	Fructus Arctii
Areca catechu L.	Pericarpium Arecae, Semen Arecae
Arisaema amurense Maxim.	Rhizoma Arisaematis
Arisaema erubescens (Wall.) Schott	Rhizoma Arisaematis
Arisaema heterophyllum Bl.	Rhizoma Arisaematis
Aristolochia contorta Bge.	Fructus Aristolochiae, Herba Aristolochiae
Aristolochia debilis Sieb. et Zucc.	Fructus Aristolochiae, Herba Aristolochiae, Radix Aristolochiae
Aristolochia fangchi Y. C. Wu et L. D. Chow et S. M. Hwang	Radix Aristolochiae fangchi
Aristolochia manshuriensis Kom.	Caulis Aristolochiae manshuriensis
Arnebia euchroma (Royle) Johnst.	Radix Arnebiae
Artemisia annua L.	Herba Artemisiae annuae
Artemisia argyi Levl. et Vant.	Folium Artemisiae argyi
Artemisia capillaris Thunb.	Herba Artemisiae scopariae
Artemisia scoparia Waldst. et Kit.	Herba Artemisiae scopariae
Asarum heterotropoides Fr. Schmidt var. *mandshuricum* (Maxim.) Kitag.	Herba Asari
Asarum sieboldii Miq.	Herba Asari
Asarum sieboldii Miq. var. *seoulense* Nakai	Herba Asari
Asparagus cochinchinensis (Lour.) Merr.	Radix Asparagi
Aster tataricus L. f.	Radix Asteris
Astragalus complanatus R. Br.	Semen Astragali complanati
Astragalus membranaceus (Fisch.) Bge.	Radix Astragali
Astragalus membranaceus Bge. var. *mongholicus* (Bge.) Hsiao	Radix Astragali
Atractylodes chinensis (DC.) Koidz.	Rhizoma Atractylodis
Atractylodes lancea (Thunb.) DC.	Rhizoma Atractylodis
Atractylodes macrocephala Koidz.	Rhizoma Atractylodis macrocephalae
Atropa belladonna L.	Herba Belladonnae
Aucklandia lappa Decne.	Radix Aucklandiae

Appendix 1. (continued)

Plant species	Item
Bambusa textilis McClure	Concretio Silicea bambusae
Bambusa tuldoides Munro	Caulis Bambusae in taeniam
Baphicacanthus cusia (Nees) Bremek	Indigo naturalis
Belamcanda chinensis (L.) DC.	Rhizoma Belamcandae
Benincasa hispida (Thunb.) Cogn.	Exocarpium Benincasae
Biota orientalis (L.) Endl.	Cacumen Biotae, Semem Biotae
Bletilla striata (Thunb.) Reichb. f.	Rhizoma Bletillae
Brassica juncea (L.) Czern. et Coss.	Semen Sinapis
Broussonetia papyrifera (L.) Vent.	Frcutus Broussonetiae
Brucea javanica (L.) Merr.	Fructus Bruceae
Buddleja officinalis Maxim.	Flos Buddlejae
Bupleurum chinensis DC.	Radix Bupleuri
Bupleurum scorzonerifolium Willd.	Radix Bupleuri
Caesalpinia sappan L.	Lignum Sappan
Calvatia gigantea (Batsch ex Pers.) Lloyd	Lasiosphaera seu Calvitia
Calvatia lilacina (Mont. et Berk.) Lloyd	Lasiosphaera seu Calvitia
Camellia meiocarpa Hu, ms.	Oleum Camelliae
Camellia oleifera Abel	Oleum Camelliae
Campsis grandiflora (Thunb.) K. Schum.	Flos Campsis
Campsis radicans (L.) Seem.	Flos Campsis
Canarium album Raeusch.	Fructus Canarii
Canavalia gladiata (Jacq.) DC.	Semen Canavaliae
Cannabis sativa L.	Fructus Cannabis
Carpesium abrotanoides L.	Fructus Carpesii
Carthamus tinctorius L.	Flos Carthami
Cassia acutifolia Delile	Folium Sennae
Cassia angustifolia Vahl	Folium Sennae
Cassia obtusifolia L.	Semen Cassiae
Cassia tora L.	Semen Cassiae
Celosia argentea L.	Semen Celosiae
Celosia cristata L.	Flos Celosiae cristatae
Centella asiatica (L.) Urb.	Herba Centellae
Centipeda minima (L.) A. Braun et Aschers.	Herba Centipedae
Cephalanoplos segetum (Bge.) Kitam.	Herba Cephalanoploris
Cephalanoplos setosum (Willd.) Kitam.	Herba Cephalanoploris
Chaenomeles speciosa (Sweet) Nakai	Fructus Chaenomelis
Changium smyrnioides Wolff	Radix Changii
Choerospondias axillaris (Roxb.) Burtt et Hill	Fructus Choerospondiatis
Chrysanthemum indicum L.	Flos Chrysanthemi indici
Chrysanthemum morifolium Ramat.	Flos Chrysanthemi

Appendix 1. (continued)

Plant species	Item
Cibotium barometz (L.) J. Sm.	Rhizoma Cibotii
Cichorium glandulosum Boiss. et Hout	Herba Cichorii
Cichorium intybus L.	Herba Cichorii
Cimicifuga dahurica (Turcz.) Maxim.	Rhizoma Cimicifugae
Cimicifuga foetida L.	Rhizoma Cimicifugae
Cimicifuga heracleifolia Kom.	Rhizoma Cimicifugae
Cinnamomum camphora (L.) Sieb.	Oleum Eucalypti
Cinnamomum cassia Presl	Cortex Cinnamomi, Ramulus Cinnamomi, Oleum Cinnamomi
Cirsium japonicum DC.	Herba Cirsii japonici, Radix Cirsii japonici
Cissampelos pareira L. var. *hirsuta* (Buch. ex DC.) Formen	Herba Cissampelotis
Cistanche deserticola Y. C. Ma	Herba Cistanches
Citrus aurantium L.	Fructus Aurantii, Fructus Aurantii Immaturus
Citrus grandis (L.) Osbeck	Exocarpium Citri grandis
Citrus medica L.	Fructus Citri
Citrus medica L. var. *sarcodactylis* Swingle	Fructus Citri sarcodactylis
Citrus reticulata Blanco	Exocarpium Citri rubrum, Pericarpium Citri reticulatae, Pericarpium Citri reticulatae viride, Semen Citri reticulatae
Citrus sinensis Osbeck	Fructus Aurantii immaturus
Citrus wilsonii Tanaka	Fructus Citri
Clematis armandii Franch.	Caulis Clematidis armandii
Clematis chinensis Osbeck	Radix Clematidis
Clematis montana Buch.-Ham.	Caulis Clematidis armandii
Clematis hexapetala Pall.	Radix Clematidis
Clematis manshurica Rupr.	Radix Clematidis
Cnidium monnieri (L.) Cuss.	Fructus Cnidii
Codonopsis pilosula (Franch.) Nannf.	Radix Codonopsis pilosulae
Coix lacryma-jobi L. var. *ma-yuen* (Roman.) Stapf	Semen Coicis
Commelina communis L.	Herba Commelinae
Coptis chinensis Franch	Rhizoma Coptidis
Coptis deltoidea C. Y. Cheng et Hsiao	Rhizoma Coptidis
Coptis teetoides C. Y. Cheng	Rhizoma Coptidis
Cordyceps sinensis (Berk.) Sacc.	Cordyceps
Cornus officinalis Sieb. et Zucc.	Fructus Corni
Corydalis turtschaninovii Bess. f. *yanhusuo* Y. H. Chou et C. C. Hsü	Rhizoma Corydalis
Crataegus cuneata Sieb. et Zucc.	Fructus Crataegi
Crataegus pinnatifida Bge.	Fructus Crataegi
Crataegus pinnatifida Bge. var. *major* N. E. Br.	Fructus Crataegi
Crocus sativus L.	Stigma Croci
Croton tiglium L.	Fructus Crotonis

Appendix 1. (continued)

Plant species	Item
Curculigo orchioides Gaertn.	Rhizoma Curculiginis
Curcuma aromatica Salisb.	Radix Curcumae, Rhizoma Zedoariae
Curcuma kwangsiensis S. Lee et C. F. Liang	Radix Curcumae, Rhizoma Zedoariae
Curcuma longa L.	Radix Curcumae, Rhizoma Curcumae longae
Curcuma zedoaria Rosc.	Radix Curcumae, Rhizoma Zedoariae
Cuscuta chinensis Lam.	Semen Cuscutae
Cyathula officinalis Kuan	Radix Cyathulae
Cymbopogon distans (Nees) A. Camus	Oleum Cymbopogonis
Cynanchum atratum Bge.	Radix Cynanchi atrati
Cynanchum glaucescens (Decne.) Hand.-Mazz.	Rhizoma Cynanchi stauntonii
Cynanchum stauntonii (Decne.) Schltr. ex Levl.	Rhizoma Cynanchi stauntonii
Cynanchum versicolor Bge.	Radix Cynanchi atrati
Cynomorium songaricum Rupr.	Herba Cynomorii
Cyperus rotundus L.	Rhizoma Cyperi
Dalbergia odorifera T. Chen	Lignum Dalbergiae odoriferae
Daphne genkwa Sieb et. Zucc.	Flos Genkwa
Datura metel L.	Flos Daturae
Daucus carota L.	Fructus Carotae
Dendrobium candidum Wall. ex Lindl.	Herba Dendrobii
Dendrobium chrysanthum Wall.	Herba Dendrobii
Dendrobium fimbriatum Hook. var. *oculatum* Hook.	Herba Dendrobii
Dendrobium loddigesii Rolfe.	Herba Dendrobii
Dendrobium nobile Lindl.	Herba Dendrobii
Descurainia sophia (L.) Webb ex Prantl	Semen Descurainiae
Desmodium styracifolium (Osb.) Merr.	Herba Desmodii styracifolii
Dianthus chinensis L.	Herba Dianthi
Dianthus superbus L.	Herba Dianthi
Dichroa febrifuga Lour.	Radix Dichroae
Dictamnus dasycarpus Turcz.	Cortex Dictamni
Digitalis purpurea L.	Folium Digitalis
Dioscorea futschauensis Uline	Rhizoma Dioscoreae septemlobae
Dioscorea hypoglaucea Palibin	Rhizoma Dioscoreae hypoglaucae
Dioscorea opposita Thunb.	Rhizoma Dioscoreae
Dioscorea septemloba Thunb.	Rhizoma Dioscoreae septemlobae
Diospyros kaki L. f.	Calyx Kaki
Dipsacus asper Wall.	Radix Dipsaci
Dilichos lablab L.	Semen Dolichoris album
Drynaria baronii (Christ) Diels	Rhizoma Drynariae
Drynaria fortunei (Kunze) J. Sm.	Rhizoma Drynariae

Appendix 1. (continued)

Plant species	Item
Echinops latifolius Tausch.	Radix Echinopsis
Ecklonia kurome Okam.	Thallus Eckloniae
Eclipta prostrata L.	Herba Ecliptae
Elsholtzia splendens Nakai ex F. Mackawa	Herba Elsholtziae
Ephedra equisetina Bge.	Herba Ephedrae
Ephedra intermedia Schrenk et C. C. Mey.	Herba Ephedrae, Radix Ephedrae
Ephedra sinica Stapf	Herba Ephedrae, Radix Ephedrae
Epimedium brevicornum Maxim.	Herba Epimedii
Epimedium koreanum Nakai	Herba Epimedii
Epimedium pubescens Maxim.	Herba Epimedii
Epimedium sagittatum (Sieb. et Zucc.) Maxim.	Herba Epimedii
Equisetum hiemale L.	Herba Equiseti hiemalis
Eriobotrya japonica (Thunb.) Lindl.	Folium Eriobotryae
Eriocaulon buergerianum Koern.	Flos Eriocauli
Erodium stephanianum Willd.	Herba Erodii seu Geranii
Eucalyptus globulus Labill.	Oleum Eucalypti
Eucommia ulmoides Oliv.	Cortex Eucommiae
Eugenia caryophyllata Thunb.	Flos Caryophylli
Eupatorium fortunei Turcz.	Herba Eupatorii
Euphorbia kansui T. N. Liou ex T. P. Wang	Radix Kansui
Euphorbia pekinensis Rupr.	Radix Euphorbiae pekinensis
Euphoria longan (Lour.) Steud.	Arillus Longan
Euryale ferox Salisb.	Semen Euryales
Evodia rutaecarpa (Juss.) Benth.	Fructus Evodiae
Evodia rutaecarpa (Juss.) Benth. var. *bodonieri* (Dode) Huang	Fructus Evodiae
Evodia rutaecarpa (Juss.) Benth. var. *officinalis* (Dode) Huang	Fructus Evodiae
Ferula fukanensis K. M. Shen	Resina Ferulae
Ferula sinkiangensis K. M. Shen	Resina Ferulae
Foeniculum vulgare Mill.	Fructus Foeniculi
Forsythia suspensa (Thunb.) Vahl.	Fructus Forsythiae
Fraxinus chinensis Roxb.	Cortex Fraxini
Fraxinus chinensis Roxb. var. *acuminata* Lingelsh.	Cortex Fraxini
Fraxinus rhynchophylla Hance	Cortex Fraxini
Fraxinus stylosa Lingelsh.	Cortex Fraxini
Fritillaria cirrhosa D. Don	Bulbus Fritillariae cirrhosae
Fritillaria delavayi Franch.	Bulbus Fritillariae cirrhosae
Fritillaria pallidiflora Schrenk	Bulbus Fritillariae pallidiflorae
Fritillaria przewalskii Maxim.	Bulbus Fritillariae cirrhosae
Fritillaria thunbergii Miq.	Bulbus Fritillariae thunbergii

Appendix 1. (continued)

Plant species	Item
Fritillaria unibracteata Hsiao et K. C. Hsia	Bulbus Fritillariae cirrhosae
Fritillaria walujewii Regel	Bulbus Fritillariae pallidiflorae
Gardenia jasminoides Ellis	Fructus Gardeniae
Gastrodia elata Bl.	Rhizoma Gastrodiae
Gentiana crassicaulis Duthie ex Burk.	Radix Gentianae macrophyllae
Gentiana dahurica Fisch.	Radix Gentianae macrophyllae
Gentiana macrophylla Pall.	Radix Gentianae macrophyllae
Gentiana manshurica Kitag.	Radix Gentianae
Gentiana rigescens Franch.	Radix Gentianae
Gentiana scabra Bge.	Radix Gentianae
Gentiana straminea Maxim.	Radix Gentianae macrophyllae
Gentiana triflora Pall.	Radix Gentianae
Geranium wilfordii Maxim.	Herba Erodii seu Geranii
Ginkgo biloba L.	Semen Ginkgo
Glechoma longituba (Nakai) Kupr.	Herba Glechomae
Gleditsia sinensis Lam.	Fructus Gleditsiae abnormalis, Spina Gleditsiae
Glehnia littoralis Fr. Schmidt ex Miq.	Radix Glehniae
Glycine max (L.) Merr.	Semen Sojae preparatum
Glycyrrhiza glabra L.	Radix Glycyrrhizae
Glycyrrhiza inflata Bat.	Radix Glycyrrhizae
Glycyrrhiza uralensis Fisch.	Radix Glycyrrhizae
Hedysarum polybotrys Hand.-Mazz.	Radix Hedysari
Hippophae rhamnoides L.	Fructus Hippophae
Hordeum vulgare L.	Fructus Hordei germinatus
Houttuynia cordata Thunb.	Herba Houttuyniae
Illicium verum Hook f.	Fructus Anisi stellati, Oleum Anisi stellati
Impatiens balsamina L.	Semen Impatientis
Imperata cylindrica Beauv. var. *major* (Nees) C. E. Hubb.	Rhizoma Imperatae
Indigofera suffruticosa Mill.	Indigo naturalis
Inula britannica L.	Flos Inulae
Inula helenium L.	Radix Inulae
Inula japonica Thunb.	Herba Inulae, Flos Inulae
Inula linariifolia Turcz.	Herba Inulae
Inula racemosa Hook. f.	Radix Inulae
Isatis indigodica Fort.	Folium Isatidis, Indigo naturalis, Radix Isatidis
Juncus effusus L.	Medulla Junci
Kaempferia galanga L.	Rhizoma Kaempferiae
Knoxia valerianoides Thorel et Pitard	Radix Knoxiae
Kochia scoparia (L.) Schrad	Fructus Kochiae
Laminaria japonica Aresch.	Thallus Laminariae
Lasiosphaera fenzlii Reich.	Lasiosphaera seu Calvatia
Ledebouriella divaricata (Turez.) Hiroe	Radix Ledebouriellae

Appendix 1. (continued)

Plant species	Item
Leonurus heterophyllus Sweet	Fructus Leonuri, Herba Leonuri
Lepidium apetalum Willd.	Semen Lepidii
Ligusticum chuanxiong Hort.	Rhizoma Chuanxiong
Ligusticum jeholense Nakai et Kitag.	Rhizoma Ligustici
Ligusticum sinensis Oliv.	Rhizoma Ligustici
Ligustrum lucidum Ait.	Fructus Ligustri lucidi
Lilium brownii F. E. Brown var. *viridulum* Baker	Bulbus Lilii
Lilium lancifolium Thunb.	Bulbus Lilii
Lilium pumilum DC.	Bulbus Lilii
Lindera communis Hemsl.	Oleum Linderae
Lindera strychnifolia (Sieb. et Zucc.) Vill.	Radix Linderae
Liquidambar formosana Hance	Fructus Liquidambaris, Resina Liquidambaris
Litchi chinensis Sonn.	Semen Litchi
Lithospermum erythrorhizon Sieb. et Zucc.	Radix Lithospermi
Litsea cubeba (Lour.) Pers.	Fructus Litseae
Lobelia chinensis Lour.	Herba Lobeliae chinensis
Lonicera confusa DC.	Flos Lonicerae
Lonicera dasystyla Rehd.	Flos Lonicerae
Lonicera hypoglauca Miq.	Flos Lonicerae
Lonicera japonica Thunb.	Caulis Lonicerae, Flos Lonicerae
Lophatherum gracile Brongn.	Herba Lophatheri
Luffa cylindrica (L.) Roem.	Retinervus Luffae fructus
Lycium barbarum L.	Cortex Lycii, Fructus Lycii
Lycium chinensis Mill.	Cortex Lycii
Lycopus lucidus Turcz. var. *hirtus* Regel	Herba Lycopi
Lygodium japonicum (Thunb.) Sw.	Spora Lygodii
Lysimachia christinae Hance	Herba Lysimachiae
Magnolia biondii Pamp.	Flos Magnoliae
Magnolia denudata Desr.	Flos Magnoliae
Magnolia officinalis Rehd. et Wils	Cortex Magnoliae officinalis, Flos Magnoliae officinalis
Magnolia officinalis Rehd. et Wils. var. *biloba* Rehd. et Wils	Cortex Magnoliae officinalis, Flos Magnoliae officinalis
Magnolia sprengeri Pamp.	Flos Magnoliae
Malva verticillata L.	Fructus Malvae
Melia azedarach L.	Cortex Meliae
Melia toosendan Sieb. et Zucc.	Cortex Meliae, Fructus Toosendan
Menispermum dauricum DC.	Rhizoma Menispermi
Mentha haplocalyx Briq.	Herba Menthae, Oleum Menthae
Momordica cochinchinensis (Lour.) Spreng.	Semen Momordicae

Appendix 1. (continued)

Plant species	Item
Momordica grosvenori Swingle	Fructus Momordicae
Morinda officinalis How	Radix Morindae officinalis
Morus alba L.	Cortex Mori, Folium Mori, Fructus Mori, Ramulus Mori
Nelumbo nucifera Gaertn.	Folium Nelumbinis, Nodus Nelumbinis rhizomatis, Plumula Nelumbinis, Receptaculum Nelumbinis, Stamen Nelumbinis, Semen Nelumbinis
Nigella glandulifera Freyn	Semen Nigellae
Notopterygium forbesii Boiss.	Rhizoma seu Radix Notopterygii
Notopterygium incisum Ting ex H. T. Chang	Rhizoma seu Radix Notopterygii
Ocimum gratissimum L.	Oleum Ocimi gratissimi
Omphalia lapidescens Schroet.	Omphalia
Ophiopogon japonicus (L. f.) Ker-Gawl.	Radix Ophiopogonis
Oroxylum indicum (L.) Vent.	Semen Oroxyli
Oryza sativa L.	Fructus Oryzae germinatus
Paeonia lactiflora Pall.	Radix Paeoniae alba, Radix Paeoniae rubra
Paeonia suffruticosa Andr.	Cortex Moutan
Paeonia veitchii Lynch	Radix Paeoniae rubra
Panax ginseng C. A. Mey.	Radix Ginseng
Panax japonicus C. A. Mey.	Rhizoma Panacis japonici
Panax japonicus C. A. Mey. var. *bipinnatifidus* (Seem.) C. Y. Wu et K. M. Feng	Rhizoma Panacis majoris
Panax japonicus C. A. Mey. var. *major* (Burk.) C. Y. Wu et K. M. Feng	Rhizoma Panacis majoris
Panax notoginseng (Burk.) F. H. Chen	Radix Notoginseng
Paris chinensis Franch.	Rhizoma Paridis
Paris yunnanensis Franch.	Rhizoma Paridis
Perilla frutescens (L.) Britt.	Caulis Perillae, Folium Perillae, Fructus Perillae
Periploca sepium Bge.	Cortex Periplocae
Peucedanum decursivum Maxim.	Radix Peucedani
Peucedanum praeruptorum Dunn	Radix Peucedani
Pharbitis nil (L.) Choisy	Semen Pharbitidis
Pharbitis purpurea (L.) Choisy	Semen Pharbitidis
Phaseolus angularis Wight	Semen Phaseoli
Phaseolus calcaratus Roxb.	Semen Phaseoli
Phellodendron amurense Rupr.	Cortex Phellodendri
Phellodendron chinense Schneid.	Cortex Phellodendri
Phragmites communis (L.) Trin.	Rhizoma Phragmitis
Phyllanthus emblica L.	Fructus Phyllanthi
Phyllostachys nigra var. *henonis* Stapf	Caulis Bambusae in taeniam
Physalis alkekengi L. var. *franchetii* (Mast.) Makino	Calyx seu Fructus Physalis
Physochlaina infundibularis Kuang	Radix Physochlainae

Appendix 1. (continued)

Plant species	Item
Phytolacca acinosa Roxb.	Radix Phytolaccae
Phytolacca americana L.	Radix Phytolaccae
Picrorhiza scrophulariiflora Pennell	Rhizoma Picrorhizae
Pinellia ternata (Thunb.) Breit.	Rhizoma Pinelliae
Pinus massoniana Lamb.	Pollen Pini
Pinus tabulaeformis Carr.	Pollen Pini
Piper futokadsura Sieb. et Zucc.	Caulis Piperis futokadsurae
Piper longum L.	Fructus Piperis longi
Piper nigrum L.	Fructus Piperis nigri
Plantago asiatica L.	Herba Plantaginis, Semen Plantaginis
Plantago depressa Willd.	Herba Plantaginis, Semen Plantaginis
Platycodon grandiflorum (Jacq.) A. DC.	Radix Platycodi
Podophyllum emodi Wall.	Fructus Podophylli
Pogostemon cablin (Blanco) Benth.	Herba Pogostemonis
Polygala sibirica L.	Radix Polygalae
Polygala tenuifolia Willd.	Radix Polygalae
Polygonatum cyrtonema Hua	Rhizoma Polygonati
Polygonatum kingianum Coll. et Hemsl.	Rhizoma Polygonati
Polygonatum odoratum (Mill.) Druce	Rhizoma Polygonati odorati
Polygonatum sibiricum Red.	Rhizoma Polygonati
Polygonum aviculare L.	Herba Polygoni avicularis
Polygonum bistorta L.	Rhizoma Bistortae
Polygonum cuspidatum Sieb. et Zucc.	Rhizoma Polygoni cuspidati
Polygonum multiflorum Thunb.	Caulis Polygoni multiflori, Radix Polygoni multiflori, Radix Polygoni multiflori preparata
Polygonum orientale L.	Fructus Polygoni orientialis
Polygonum tinctorium Ait.	Folium Polygoni tinctorii, Indigo naturalis
Polyporus umbellatus (Pers.) Fries	Polyporus
Poria cocos (Schw.) Wolf	Poria
Portulaca oleracea L.	Herba Portulacae
Potentilla chinensis Ser.	Herba Potentillae chinensis
Prinsepia uniflora Batal.	Nux Prinsepiae
Prunella vulgaris L.	Spica Prunellae
Prunus armeniaca L.	Semen Armeniacae amarum
Prunus armeniaca L. var. *ansu* Maxim.	Semen Armeniacae amarum
Prunus davidiana (Carr.) Franch.	Semen Persicae
Prunus humilis Bge.	Semen Pruni
Prunus japonica Thunb.	Semen Pruni
Prunus mandshurica (Maxim.) Koehne	Semen Armeniacae amarum
Prunus mume (Sieb.) Sieb. et Zucc.	Flos Mume, Fructus Mume
Prunus persica (L.) Batsch	Semen Persicae
Prunus sibirica L.	Semen Armeniacae amarum
Prunus tomentosa Thunb.	Semen Pruni

Appendix 1. (continued)

Plant species	Item
Pseudolarix kaempferi Gord.	Cortex Pseudolaricis
Pseudostellaria heterophylla (Miq.) Pax ex Pax et Hoffm.	Radix Pseudostellariae
Psoralea corylifolia L.	Fructus Psoraleae
Pueraria lobata (Willd.) Ohwi	Radix Puerariae
Pueraria thomsonii Benth.	Radix Puerariae
Pulsatilla chinensis (Bge.) Regel	Radix Pulsatillae
Punica granatum L.	Pericarpium Granati
Pyrola calliantha H. Andres	Herba Pyrolae
Pyrola decorata H. Andres	Herba Pyrolae
Pyrrosia lingue (Thunb.) Farwell	Folium Pyrrosiae
Pyrrosia petiolosa (Christ) Ching	Folium Pyrrosiae
Pyrrosia sheareri (Bak.) Ching	Folium Pyrrosiae
Quisqualis indica L.	Fructus Quisqualis
Raphanus sativus L.	Semen Raphani
Rehmannia glutinosa Libosch.	Radix Rehmanniae
Rhaponticum uniflorum (L.) DC.	Radix Rhapontici
Rheum officinale Baill.	Radix et Rhizoma Rhei
Rheum palmatum L.	Radix et Rhizoma Rhei
Rheum tanguticum Maxim. ex Balf.	Radix et Rhizoma Rhei
Rhododendron dauricum L.	Oleum Rhododendri daurici
Rhus chinensis Mill.	Galla Chinensis
Rhus potaninii Maxim.	Galla Chinensis
Rhus punjabensis Stew. var. *sinica* (Diels) Rehd. et Wils	Galla Chinensis
Ricinus communis L.	Semen Ricini, Oleum Ricini
Rosa chinensis Jacq.	Flos Rosae chinensis
Rosa laevigata Michx.	Fructus Rosae laevigatae
Rosa rugosa Thunb.	Flos Rosae fugosae
Rubia cordifolia L.	Radix Rubiae
Rubus chingii Hu	Fructus Rubi
Salvia miltiorrhiza Bge.	Radix Salviae miltiorrhizae
Sanguisorba officinalis L.	Radix Sanguisorbae
Sanguisorba officinalis L. var. *longifolia* (Bert.) Yü et Li	Radix Sanguisorbae
Sargassum fusiforme (Harv.) Setch.	Sargassum
Sargassum pallidum (Thurn.) C. Ag.	Sargassum
Sargentodoxa cuneata (Oliv.) Rehd. et Wils.	Caulis Sargentodoxae
Schisandra chinensis (Turcz.) Baill.	Fructus Schisandrae
Schisandra sphenanthera Rehd. et Wils.	Fructus Schisandrae
Schizonepeta tenuifolia Briq.	Herba Schizonepetae
Schizostachyum chinense Rendle	Concretio Silicea bambusae

Appendix 1. (continued)

Plant species	Item
Scrophularia ningpoensis Hemsl.	Radix Scrophulariae
Scutellaria baicalensis Georgi	Radix Scutellariae
Scutellaria barbata D. Don	Herba Scutellariae barbatae
Selaginella tamariscina (Beauv.) Spr.	Herba Selaginellae
Semiaquilegia adoxoides (DC.) Makino	Radix Semiaquilegiae
Sesamum indicum L.	Semen Sesami nigrum, Oleum Sesami
Setaria italica (L.) Beauv.	Fructus Setariae germinatus
Siegesbeckia glabrescens Makino	Herba Siegesbeckiae
Siegesbeckia orientalis L.	Herba Siegesbeckiae
Siegesbeckia pubescens Makino	Herba Siegesbechiae
Sinapis alba L.	Semen Sinapis
Sinocalamus beecheyanus (Munro) McClure var. *pubescens* P. F. Li	Caulis Bambusae in taeniam
Sinomenium acutum (Thunb.) Rehd. et Wils.	Caulis Sinomenii
Sinomenium acutum (Thunb.) Rehd. et Wils var. *cinereum* Rehd. et Wils.	Caulis Sinomenii
Smilax glabra Roxb.	Rhizoma Smilacis glabrae
Sophora flavescens Ait.	Radix Sophorae flavescentis
Sophora japonica L.	Flos Sophorae, Flos Sophorae immaturus, Fructus Sophorae
Sophora tonkinensis Gapnep.	Radix Sophorae tonkinensis
Sparganium stoloniferum Buch.-Ham.	Rhizoma Sparganii
Spatholobus suberectus Dunn	Caulis Spatholobi
Spirodela polyrrhiza (L.) Schleid.	Herba Spirodelae
Stellaria dichotoma L. var. *lanceolata* Bge.	Radix Stellariae
Stemona japonica (Bl.) Miq.	Radix Stemonae
Stemona sessilifolia (Miq.) Miq.	Radix Stemonae
Stemona tuberosa Lour.	Radix Stemonae
Stephania tetrandra S. Moore	Radix Stephaniae tetrandrae
Strychnos nux-vomica L.	Semen Strychni
Strychnos pierriana A.W. Hill	Semen Strychni
Styrax tonkinensis (Pierre) Craib ex Hart.	Benzoinum
Swertia mileensis T.N. Ho et W.L. Shih	Herba Swertiae mileensis
Tamarix chinensis Lour.	Cacumen Tamaricis
Taraxacum mongolicum Hand.-Mazz.	Herba Taraxaci
Taraxacum sinicum Kitag.	Herba Taraxaci
Taxillus chinensis (DC.) Danser	Herba Taxilli
Terminalia billerica (Gaertn.) Roxb.	Fructus Terminaliae billericae
Terminalia chebula Retz.	Fructus Chebulae
Terminalia chebula Retz. var. *tomentella* Kurt	Fructus Chebulae
Tetrapanax papyriferus (Hook.) K. Koch	Medulla Tetrapanacis
Tinospora capillipes Gagnep.	Radix Tinosporae
Tinospora sagittata (Oliv.) Gagnep.	Radix Tinosporae

Appendix 1. (continued)

Plant species	Item
Torreya grandis Fort.	Semen Torreyae
Trachelospermum jasminoides (Lindl.) Lem.	Caulis Trachelospermi
Trachycarpus fortunei H. Wendl.	Petiolus Trachycarpi
Tribulus terrestris L.	Fructus Tribuli
Trichosanthes japonica Regel	Radix Trichosanthis
Trichosanthes kirilowii Maxim.	Fructus Trichosanthis, Pericarpium Trichosanthis, Radix Trichosanthis, Semen Trichosanthis
Trichosanthes rosthornii Harms	Fructus Trichosanthis, Pericarpium Trichosanthis, Semen Trichosanthis
Trigonella foenum-graecum L.	Semen Trigonellae
Tussilago farfara L.	Flos Farfarae
Typha angustifolia L.	Pollen Typhae
Typha orientalis Presl	Pollen Typhae
Typhonium giganteum Engl.	Rhizoma Typhonii
Uncaria hirsuta Havil.	Ramulus Uncariae cum Uncis
Uncaria macrophylla Wall.	Ramulus Uncariae cum Uncis
Uncaria rhynchophylla (Miq.) Jacks.	Ramulus Uncariae cum Uncis
Uncaria sessilifructus Roxb.	Ramulus Uncariae cum Uncis
Uncaria sinensis (Oliv.) Havil.	Ramulus Uncariae cum Uncis
Vaccaria segetalis (Neck.) Garcke	Semen Vaccariae
Verbena officinalis L.	Herba Verbenae
Viola yedoensis Makino	Herba Violae
Viscum coloratum (Komar.) Nakai	Herba Visci
Vitex negundo L. var. *cannabifolia* (Sieb. et Zucc.) Hand.-Mazz.	Oleum Viticis negundo
Vitex trifolia L.	Fructus Viticis
Vitex trifolia L. var. *simplicifolia* Cham.	Fructus Viticis
Xanthium sibiricum Patr.	Fructus Xanthii
Zanthoxylum bungeanum Maxim.	Pericarpium Zanthoxyli
Zanthoxylum schinifolium Sieb. et Zucc.	Pericarpium Zanthoxyli
Zingiber officinale (Willd.) Rosc.	Rhizoma Zingiberis, Rhizoma Zingiberis recens
Ziziphus jujuba Mill.	Fructus Jujubae
Ziziphus spinosa Hu	Semen Ziziphi spinosae

Appendix 2. Herbal Medicines Contained in the Appendix of the *Chinese Pharmacopoeia* of 1985, Vol. I

Aconitum balfourii Stapf
Aconitum kusnezoffii Reichb.
Aconitum naviculare Stapf
Acontium szechenyianum Gay.
Aconitum tanguticum (Maxim.) Stapf
Acorus calamus L.
Allium sativum L.
Anemone raddeana Regel
Arctium lappa L.
Asparagus pseudofilicinus Wang et Tang
Benincasa hispida (Thunb.) Cogn.
Berberis soulieana Schneid.
Bupleurum marginatum Wall. ex DC.
Cinnamomum camphora (L.) Sieb.
Citrullus vulgaris Schrad.
Corydalis bungeana Turcz.
Corydalis stricta Steph.
Cremastra appendiculata (D. Don) Makino
Croton caudatus Geisel. var. *tomentosa* Hook.
Cucumis melo L.
Cynanchum thesioides (Freyn) K. Schum.
Dioscorea nipponica Makino
Dracocephalum tanguticum Maxim.
Entada phaseolides (L.) Merr.
Erycibe obtusifolia Benth.
Eugenia caryophyllata Thunb.
Glycine max (L.) Merr.
Gueldenstaedtia verna (Georgi) A. Bor.
Hedychium venustum Wight
Herpetospermum caudigerum Wall.
Hordeum vulgare L.
Hydnocarpus anthelmintica Pierre.
Hypecoum leptocarpum Hook. f. et. Thoms.
Impatiens balsamina L.
Inula cappa DC.
Ixeris chinensis (Thunb.) Nakai
Juglans regia L.
Lagotis brevituba Maxim.
Lagotis glauca Gaertn.
Liquidambar orientalis Mill.
Monochasma savatieri Franch. ex Maxim.
Myristica fragrans Houtt
Nigella sativa
Oldenlandia diffusa (Willd.) Roxb.
Oxytropis chiliophylla Royle
Oxytropis falcata Bge.
Oxytropis myriophylla (Pall.) DC.
Pegaeophyton scapiflora (Hook. f. et Thoms.) Marq. et Airy-Shaw
Phaseolus radiatus L.
Phlomis kawaguchii Murata
Pinus massoniana Lamb.
Pinus tabulaeformis Carr.
Polygonum aubertii Henry
Pterocarpus santalinus L.
Pterocephalus hookeri (C. B. Clarke) Höeck
Punica granatum L.
Rhodiola kirilowii (Regel) Regel
Rhododendron molle G. Don
Rhus verniciflua Stokes
Rubus idseus L.
Rubus sp.
Santalum album L.
Saxifraga tangutica Engl.
Sophora japonica L.
Streptocaulon griffithii Hook. f.
Syringa pinnatifolia Hemsl.
Tacca esquirolii (Levl.) Rehd.
Trapa bispinosa Roxb.
Trapa maximowiczii Korsch.
Vitex trifolia L.
Vitex trifolia L. var. *simplicifolia* Cham.
Vitis vinifera L.

Subject Index

D

E

H

L

M

S

T

U

V

W

X

Y

Z